Functional Groups Preparation—Master Table

Starting from	Acid anhydride	Acid chloride	Alcohol	Aldehyde	Alkane	Alkene	Alkyl-arene	Alkyl halide	Alkyne	Amide	Amine	Arene	Aryl halide	Carboxylic acid	Epoxide	Ester	Ether	Glycol	Ketone	Nitrile	Nitro-arene	Phenol
	Target Functional Group																					
Acid anhydride										17.7B				17.5B		17.6B						
Acid chloride	17.8*									17.7A				17.5A		17.6A			17.9C 20.1C			
Alcohol				9.9		9.7	20.1D	9.6A 9.6B						9.9 24.4D		16.8A 17.6A 17.6B	11.4 11.7		9.9			
Aldehyde			15.6 15.14A 15.14B		15.14C	15.7					15.10			15.13A					15.9			
Alkane								7.4														
Alkene			6.3B 6.3E 6.4	6.5C	6.6		20.1D	6.3A	10.6						11.9B 11.9C			6.5B	6.5C			
Alkylarene														19.6A								
Alkyl halide			8.1			15.7	20.1C		10.6		21.8A 21.8B			15.6A			11.4			8.1		
Alkyne				10.8	10.7A	10.7A 10.7B		10.9B											10.8 10.9C			
Amide											17.11B 17.12			17.5D								
Amine						21.10 21.11				17.7A 17.7B 17.7C			21.9E							21.9E		21.9E
Arene							20.1C 20.1D						20.1A						20.1C		20.1B	
Aryl halide											20.3B											20.3A
Carboxylic acid		16.9	16.7A													16.8A 16.8C						
Epoxide			11.10A 11.10B								11.10B							11.10				
Ester			17.9A 17.11A	17.11A						17.7C				17.5C								
Ether								11.6A														
Glycol				9.10															9.8 9.10			
Ketone			15.6 15.14		15.14C	15.7					15.10			15.12C		15.13B						
Nitrile											17.11C			17.5E								
Nitroarene											20.1B											
Phenol																	19.5D					

* Section numbers

Here Is What Your Colleagues Are Saying About Brown and Foote, *Organic Chemistry*, Second Edition

Pre-Publication Reviews

"One of the strengths based on the table of contents is that students are able to begin their study of chemical reactions very quickly and stereochemistry is introduced early enough so that it can be used when discussing all subsequent reactions. The separate chapter on acid-base concepts represents an important strength that not every text has. This topic is at the root of many of the problems students have understanding reactions."

John A. Landgrebe, University of Kansas

"The manuscript's greatest strengths are: (1) it is well written, (2) the Chemistry in Action series are really interesting and relevant, and (3) the end of chapter problems are very good. ...I found that many of the problems were based on 'every-day' and otherwise interesting molecules. This helps students realize that what they are learning is indeed relevant to areas of science that they themselves might find interesting."

Steven Pedersen, University of California, Berkeley

"The changes in the organization are welcome. I'm especially pleased with the earlier treatment of carbonyl chemistry. I feel, with many others, that this is the heart of organic chemistry and it needs to be treated early and often. ...The emphasis on bio-organic chemistry is also a strong point."

Kirk McMichael, Washington State University

"The strengths of these chapters lie mainly in the problems given for study purposes; they are not just 'predict the reagents and products' problems that are the bane of most students in organic chemistry. They are thought-provoking and instructional."

John W. Benbow, Lehigh University

"The apparent focus on the bioorganic aspects of the subject is appropriate for the authors' targeted audience, the health profession majors, and introducing a chapter on metabolism is an important and valuable change from the first edition of the textbook, one likely to appeal to faculty involved in this course. I also concur in introducing stereochemistry earlier, as this topic is of comparable significance to an understanding of bonding and electronic properties of organic molecules. The topical essays are timely, relevant, and augment the 'user-friendliness' of the textbook—a definite plus!"

John C. Gilbert, University of Texas

"Generally, the writing style of the text is excellent and the students should have no trouble understanding the material. ...The significance of organic chemistry in both the biological and industrial fields is very well presented."

Morris L. Fishman, New York University

"My overall impression about this text is that it has the potential to become one of the best textbooks on the market."

Lyle W. Castle, Idaho State University

"The in-text examples/problems are a real strength of the book. For most of the book they occur at just the right frequency. ...The distribution of problems throughout the text is done extremely well. The problems are well designed to illustrate ideas, often pointing the student forward in their thinking or covering an extra topic."

Daniel Singleton, Texas A&M University

"I find the presence of a real chapter on polymers very reassuring and useful."

Bruce E. Norcross,
State University of New York, Binghamton

"The greatest strength of the book is the large number of problems at the end of the chapters. These range from very basic questions to a substantial number of quite challenging problems, which require students to think more deeply about the material. The solutions are clearly and accurately presented. Many of these problems also help to illustrate interesting real-world world examples of organic chemistry."

Dale Drueckhammer, Stanford University

"I applaud the authors for including a brief description of time domain NMR. To my knowledge this is the only text that does this."

Carmelo J. Rizzo, Vanderbilt University

"The treatment of acids and bases in Chapter 3 is an excellent start. I'm impressed by the way in which this is carried through in detail in Chapter 18. This is an excellent feature of this text."

Kirk McMichael, Washington State University

ORGANIC CHEMISTRY

SECOND EDITION

WILLIAM H. BROWN
Beloit College

with CHRISTOPHER S. FOOTE
University of California, Los Angeles

SAUNDERS COLLEGE PUBLISHING
Harcourt Brace College Publishers

FORT WORTH PHILADELPHIA SAN DIEGO NEW YORK ORLANDO AUSTIN
SAN ANTONIO TORONTO MONTREAL LONDON SYDNEY TOKYO

To Carolyn,

with whom

life is a joy.

—Bill Brown

Publisher: Emily Barrosse
Vice President/Publisher: John Vondeling
Product Manager: Angus McDonald
Developmental Editor: Sandra Kiselica
Project Editor: Elizabeth Ahrens
Production Manager: Charlene Catlett Squibb
Art Director: Caroline McGowan
Text and Cover Designer: Ruth A. Hoover

Cover Credit: © Lester Lefkowitz/Tony Stone Images

Printed in the United States of America

ORGANIC CHEMISTRY, 2/E
0-03-020458-5

Library of Congress Catalog Card Number: 97-65257

890123456 039 10 9876543

Biographies

William H. Brown

William H. Brown is Professor of Chemistry at Beloit College, where he has twice been named Teacher of the Year. He is also the author of the college textbook *Introduction to Organic Chemistry,* published in 1996. His regular teaching responsibilities include organic chemistry, advanced organic chemistry, and, more recently, special topics in pharmacology and drug synthesis. He received his Ph.D. from Columbia University under the direction of Gilbert Stork and did postdoctoral work at California Institute of Technology and the University of Arizona.

Bill Brown and his wife Carolyn enjoy hiking in the Southwest and the study of petroglyphs and pictographs. Twice he has been the Director of Beloit College's World Outlook Seminar, a program coordinated with the University of Glasgow in Scotland.

Bill Brown crossing the finish line in a 10K race in Milwaukee.

Christopher S. Foote

Christopher S. Foote is Professor of Chemistry at the University of California, Los Angeles. He received his B.S. degree from Yale University and his Ph.D. in Organic Chemistry from Harvard University. In 1995, he received the Tolman Award of the ACS Southern California Section for his contributions to chemistry. Foote's research has focused on the chemistry of oxygen in organic and biological systems and on the chemistry of fullerenes. Other awards he has received include the Yale Science and Engineering Award for the Advancement of Basic and Applied Science and the ACS North Jersey Sections' Leo Hendrick Baekeland Award. He was also an ACS Cope Scholar in 1994 and is the author of more than 200 research papers.

Christopher S. Foote.

Preface

Time always seems so short between completion of one edition of a text and the need to consider a revision. But that is the nature of textbooks in the current market. In considering comments and suggestions from users of the first edition of this text, I (WHB) found those of Christopher Foote to be particularly perceptive and valuable. So much so that Saunders and I enlisted him to become involved in every stage of preparation of this second edition. His participation started with a reconsideration of the table of contents and has continued through to pages of the final manuscript. I am deeply indebted to Chris for the sense of pedagogy and depth of knowledge he has shared so freely.

The Audience

This book provides an introduction to organic chemistry for students majoring in chemistry and in related disciplines, most notably the health and biological sciences. Fundamental to the approach taken in this book is the fact that organic chemistry has an underlying rationale, namely the *mechanistic themes that unify the discipline.* As students learn organic reactions and their mechanisms, we hope that organic chemistry becomes for them an exciting area of scientific investigation. Organic chemistry is a dynamic and ever-expanding area of science waiting openly for those who are prepared, both by training and inquisitive nature, to ask questions and to explore.

Organization: An Overview

Chapters 1–23 lay the foundation for studying organic chemistry by covering the structures and typical reactions of the important classes of organic compounds: alkanes, alkenes, alkyl halides, alcohols, alkynes, ethers and epoxides, aldehydes and ketones, carboxylic acids and their derivatives, benzene and its derivatives, and finally amines. Chapters 12–14 within this group introduce mass spectrometry, ^{1}H-NMR and ^{13}C-NMR spectroscopy, and IR and ultraviolet spectroscopy. Chapter 23 presents a systematic introduction to organic polymer chemistry. Chapters 24–28 introduce students to the organic chemistry of the four major classes of biomolecules, namely carbohydrates, lipids, amino acids and proteins, and nucleic acids. In Chapter 26 the organic chemistry of two key metabolic pathways, namely glycolysis and β-oxidation of fatty acids, is discussed.

Chapter-by-Chapter

Chapter 1 begins with a review of the electronic structure of atoms and molecules, the Lewis model of bonding, and use of the VSEPR model to predict shapes of molecules and ions. Within this discussion, we introduce the structure of the hydroxyl, carbonyl, and carboxyl groups, the functional groups encountered most frequently in Chapters 1–16. The theory of resonance is introduced midway through Chapter 1, and with it, the use of curved arrows and electron pushing. The knowledge of resonance theory combined with a facility for moving electrons gives students two powerful tools for writing reaction mechanisms and understanding chemical reactivity. Chapter 1 concludes with an introduction to quantum mechanics and a description of covalent bonding in terms of the molecular orbital model.

Chapter 2 opens with a description of the structure, nomenclature, and conformational analysis of alkanes and cycloalkanes. Beginning here and continuing throughout the text, a clear distinction is made between IUPAC and common names. Where names are introduced, the IUPAC names are given first and the common names, where appropriate, follow in parentheses. The IUPAC system is introduced in Section 2.3 through the naming of alkanes, and in Section 2.5 it is presented as a general system of nomenclature. The concept of stereoisomerism is introduced in this chapter with a discussion of *cis-trans* isomerism in cycloalkanes.

Chapter 3 contains a general introduction to acid-base chemistry with emphasis on both qualitative and quantitative determination of the position of equilibrium in acid-base reactions. Introduced here are three concepts for relating structure and acidity, namely the hybridization of the atom bearing the negative charge in A^-, resonance effects in the delocalization of the negative charge in A^-, and the electron-withdrawing inductive effect of a nearby atom of high electronegativity in weakening the H—A bond. With these tools, students can then deal with questions such as "Why is a carboxylic acid a stronger acid than an alcohol?" and "Why is acetylene a stronger acid than ethane?"

Chapter 4 begins with a review of isomerism covered in earlier chapters (constitutional, conformational, *cis-trans,* and *E,Z* isomerism) and then introduces the concepts of chirality, enantiomerism, and diastereomerism.

Chapters 5 and 6 cover the chemistry of alkenes. Their structure and physical properties are presented in Chapter 5. This chapter concludes with the structure of terpenes and an introduction to one theme in the molecular logic of biomolecules. The focus of Chapter 6 is on the reactions of alkenes, which are organized in the order: electrophilic additions, hydroboration, oxidation, and reduction. The twin concepts of regioselectivity and stereoselectivity are introduced in the context of electrophilic additions to alkenes.

Chapter 7 has as its central theme the radical halogenation of alkanes. It provides an introduction to the mechanistic concepts of chain initiation, chain propagation, and chain termination steps. Regioselectivity of radical bromination compared with radical chlorination is interpreted in terms of Hammond's postulate. Also included in this chapter is a first encounter with organometallic compounds, namely organomagnesium, organolithium, and lithium diorganocopper compounds (Gilman reagents). Treatment of alkyl and alkenyl halides with Gilman reagents presents the first of what will later become many ways of forming new carbon-carbon bonds. Radical chain autoxidation, arguably the most important radical reaction in biological systems, is introduced.

Chapter 8 presents what, in our experience, is one of the most formidable and anxiety-producing aspects of introductory organic chemistry, namely S_N1, S_N2, E1, and E2 mechanisms and the attendant concepts of stereochemistry, kinetics, and relationships between structure and chemical reactivity. The difficulty does not lie in any single part of this material; no part of it is any more difficult than material already covered. The difficulty, rather, is in the number of concepts to be assimilated at one time. Now students must be able to "sing, dance, chew gum, snap their fingers, and whistle all at the same time." It is for these reasons that we present nucleophilic substitution after the chemistry of alkenes and alkyl halides. By this stage in the course, students have a good grounding in the structure of organic molecules, the theory of resonance, electron pushing, and reaction mechanisms. Nucleophilic substitution and β-elimination then become a vehicle for integration of previously covered chemistry into a larger pattern.

Chapter 9 continues the theme of the relationships between structure and reactivity by considering the chemistry of alcohols. A significant body of the chemistry of alcohols can be understood using the concepts of nucleophilic substitution, β-elimination, and the relative stability of carbocations, all of which have been developed in previous chapters.

Chapter 10 opens with alkylation of acetylide anions, and continues with hydroboration/oxidation and electrophilic additions. It concludes with an introduction to the strategy of organic synthesis, namely retrosynthetic analysis. The most important use of acetylene and substituted acetylenes in organic chemistry is as building blocks for the synthesis of larger molecules.

Chapter 11 is a logical extension of nucleophilic substitutions as applied to the synthesis and reactions of ethers and epoxides. The value of epoxides in organic synthesis is stressed, including their regioselective and stereoselective reactions with a variety of nucleophiles, as well as their regioselective reaction with Gilman reagents to form new carbon-carbon bonds.

Chapters 12–14 examine several instrumental methods for determining molecular structure and relate these methods to functional groups that have been studied to this point. First is mass spectrometry (Chapter 12), the instrumental technique by which precise molecular weights and molecular formulas can be determined. Given the placing of this chapter, students are prepared to deal with the mass spectrometry of alkanes, alkenes, alkynes, alkyl halides, alcohols, and ethers. The mass spectrometry of other classes of organic compounds is presented in subsequent chapters. Chapter 13 presents the fundamentals of both ^{1}H-NMR and ^{13}C-NMR spectroscopy, and IR and UV-visible spectroscopy are covered in Chapter 14. While this material is presented as a cluster of chapters midway through the text, the chapters are free-standing and can be used in other orders as appropriate to a particular course.

Chapters 15–18 concentrate on the chemistry of carbonyl-containing compounds. First is the chemistry of aldehydes and ketones in Chapter 15 followed by the chemistry of carboxylic acids and their functional derivatives in Chapters 16 and 17. Collected in Chapter 18 are various carbonyl condensation reactions, including the aldol reaction, and the Claisen and Dieckmann condensations. Also included are alkylation and acylation reactions of acetoacetic esters, malonic esters, and enamines.

Chapters 19 and 20 present the chemistry of aromatic compounds. The first of these chapters concentrates on structure and nomenclature of aromatic compounds, the concept of aromaticity, and the structure and acid/base properties of phenols. The second of these chapters is devoted to aromatic substitution reactions.

Chapter 21 presents the chemistry of aliphatic and aromatic amines.

Chapter 22 completes the introduction to organic functional groups with the particular chemistry of conjugated dienes, including both 1,2- and 1,4-addition and the Diels-Alder reaction. It has been our experience that a discussion of the Diels-Alder reaction in the first term, at a time when students have been introduced only to the chemistry of alkenes, places an undue burden on them. We find it preferable to postpone the chemistry of the Diels-Alder reaction until students have a firm grounding in the chemistry of all major functional groups to be introduced in the course.

Chapter 23, new to this edition, is a systematic introduction to organic polymer chemistry. Given the importance of organic polymers in the world around us, this chapter is more extensive than in most other organic textbooks.

Chapters 24 and 25 present an introduction to the chemistry of carbohydrates and lipids. Emphasis in these chapters is on the structure of these biomolecules.

Chapter 26 presents a discussion of two key metabolic pathways, namely glycolysis and β-oxidation of fatty acids. The purpose of this chapter is to show that the reactions of these pathways are biochemical equivalents of organic functional group reactions we have already studied in detail.

Chapters 27 and 28 present the organic chemistry of amino acids and proteins, and nucleic acids. Chapter 27 gives considerable attention to the acid-base properties of amino acids and then continues with an introduction to the primary, secondary, and tertiary structure of polypeptides. The concentration in Chapter 28 is on the primary, secondary, and tertiary structure of DNA with only a brief discussion of the structure of RNAs. Following the discussion of the genetic code, there is an introduction to the concepts of DNA sequencing, including DNA fingerprinting.

Special Features

- Full-Color Art Program. One of the most distinctive features of this text is its visual impact. The text's extensive full-color art program includes over 250 pieces of art created exclusively for this text by professional artists John and Bette Woolsey.
- Photo Art. Photos conceived and developed for this text show organic chemistry as it occurs in the laboratory and in everyday life, and depict the natural sources of many organic compounds.
- Stereoviews. A collection of 44 stereoviews has been prepared for this text, each chosen to reinforce the concept of organic chemistry as a three-dimensional science. Each copy of the text is equipped with a pair of stereoglasses for easy viewing. Stereoart is indicated by the following icon: STEREO
- Chemistry in Action Boxes. These boxes illustrate applications of organic chemistry to everyday settings. Topics range from "Radical Autoxidation," to "Drugs That Lower Plasma Levels of Cholesterol," to "Carbamate Insecticides."

- In-Chapter Examples. There is an abundance of in-chapter examples, each with a detailed solution. Following each in-chapter example is a comparable in-chapter problem designed to give students the opportunity to solve a related problem.
- End-of-Chapter Summaries and Key Reactions. End-of-chapter summaries highlight all important new terms found in a chapter. In addition, each reaction is annotated and keyed to the section where it is discussed.
- End-of-Chapter Problems. There are plentiful end-of-chapter problems. The majority of problems are categorized by topic. A tetrahedral icon (◈) indicates an applied problem, and a problem number set in blue indicates a more challenging problem.
- Glossary of Key Terms. Throughout the book in the margins, new terms are defined. In addition, at the end of the book is a section that gives the definitions of 311 important terms used in the text. Each glossary listing is keyed to the section of the text where the term is introduced.
- Functional Groups Preparation—Master Table is a unique overview of the methods of functional group interconversion and can be found in the front inside of this book.
- Color. Color is used to highlight parts of molecules and to follow the course of reactions. The graphic on the next page shows some of the colors used consistently in the artwork in this book.
- Interviews. Four interviews with prominent scientists describe how these people became interested in chemistry as a college student, then as an educator and/or research professional. Their enthusiasm for their work is evident, and they invite students to pursue similar interests in the sciences.
- Bio-Organic Chemistry. Bio-organic chemistry is emphasized throughout the text, in Chemistry in Action boxes, and in problems. Merck Index references are from *The Merck Index,* Susan Budavari, Editor, 12th Edition, Merck Research Laboratories, 1996. An invaluable reference for health-related organic chemistry is Goodman and Gilman's *The Pharmacological Basis of Therapeutics,* 8th Edition, A. Goodman Gilman, T. W. Rall, A. S. Nies, and P. Taylor, Editors, Pergamon Press, New York, 1990.

New to This Edition

There are several features new to the second edition, including

- Mechanisms, key elements in the organization of information in the study of organic chemistry, are now set apart and highlighted by a special design feature. All steps in a particular mechanism are presented together in its box. A list of the 124 mechanisms presented in the text can be found following the table of contents.
- Stereochemistry has been moved forward to Chapter 4, immediately following the introduction to the structure of organic compounds and the treatment of acid-base chemistry, and immediately before the chemistry of alkenes. With the earlier introduction to stereochemistry, the regioselectivity and stereoselectivity of alkene and alcohol reactions can now be discussed in greater depth.
- Given the importance of carbonyl chemistry for organic synthesis and to students engaged in the health and biological sciences, these chapters were moved forward to become Chapters 15 to 18.
- Chapter 23 is new to this edition and in it is presented a systematic introduction to organic polymer chemistry.

COLOR KEY

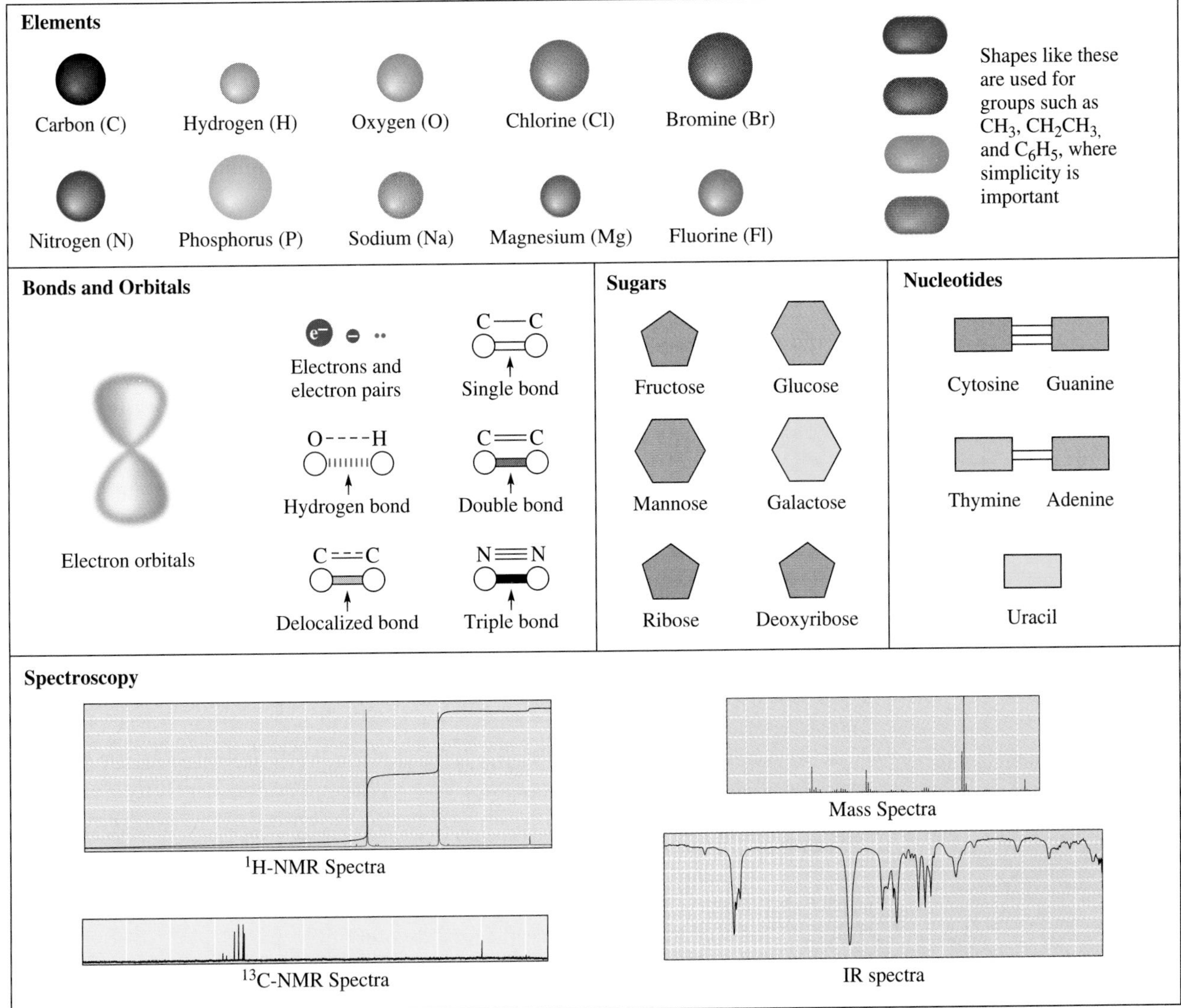

- Chapter 26 is also new to this text; it is the organic chemistry of two key metabolic pathways, namely glycolysis and β-oxidation of fatty acids. The intent of this chapter is to show that the reactions of these metabolic pathways have close analogies in laboratory reactions discussed in previous chapters.
- The strategy of retrosynthetic analysis is presented in Chapter 9 following the chemistry of alkynes. At this point, students can take the first steps to assemble carbon-carbon skeletons by treatment of alkyl and alkenyl halides with Gilman reagents (Chapter 7), and by alkylation of alkyne anions (Chapter 9).
- The number of worked out in-chapter examples and the number of end-of-chapter problems have been increased by approximately 40%. The text now contains approximately 250 solved in-chapter examples and 1200 in-text and end-of-chapter additional problems.

- A Master Table of Functional Group Preparation on the inside front cover provides students with section references to the transformation(s) discussed in the text for converting one functional group to another. Students can see from this table that alcohols, for example, can be prepared from aldehydes, alkenes, alkyl halides, carboxylic acids, epoxides, esters, and ketones.
- Appendix 7, Reagents and Their Uses, provides students with a list of all reagents described in the text and their uses in bringing about chemical transformations. Each entry is annotated with a section reference.

Support Package

- *PowerPoint™ Presentation.* A pre-built set of PowerPoint™ lecture notes corresponding to every chapter in the textbook. These slides can be used in conjunction with the freely distributable PowerPoint™ Viewer, or even edited and customized with the PowerPoint™ application program.
- *Student Study Guide and Problems Book* by Brent and Sheila Iverson of the University of Texas, Austin. Contains section-by-section overviews of each chapter and detailed solutions to all text problems. Each functional group chapter includes a "reactions grid" which organizes in a unique way the transformations of that functional group and the reagent(s) required to bring about each transformation.
- *Pushing Electrons: A Guide for Students of Organic Chemistry,* third edition, by Daniel P. Weeks, Northwestern University. A paperback workbook designed to help students learn techniques of electron pushing. Its programmed approach emphasizes repetition and active participation. Its organization is coordinated with that of *Organic Chemistry,* second edition.
- *1001 Ways to Pass Organic Chemistry: A Student's Resource Guide for Exams* by Shelton and Janet Bank, State University of New York, Albany. Contains 1001 problems with answers and tips for problem solving. The organization of this text is coordinated with that of *Organic Chemistry,* second edition.
- *Test Bank* by Robert Higgins of Fayetteville State University. Contains 1200 multiple-choice questions for instructors to use for tests, quizzes, or homework assignments. Available also in computerized form for IBM-compatible and Macintosh computers.
- *Saunders College Publishing Organic Chemistry Web Site* contains tutorials, animations, molecular modeling, and practice problems all keyed by book chapter. Available for Fall of '97 by accessing www.saunderscollege.com.
- *Introduction to Spectroscopy,* second edition, by Pavia, Lampman, and Kriz of Western Washington University is a paperback book designed to teach spectroscopic methods of structure determination.
- *Molecular Models.* A set of hand-held models, ball-and-stick or ball-and-stick and space-filling, is available at a reasonable price.
- *Organic Polymer Chemistry: A Primer* by Bruce M. Novak, Polymer Science and Engineering Department, University of Massachusetts. A paperback supplement containing an introduction to polymer chemistry with examples, problems, and essays dealing with step-growth, chain growth, and specialty polymers.
- *Overhead Transparency Acetates.* A selection of 125 colored figures from the text.
- *Saunders Multimedia Package.* This MediaActive CD-ROM includes 225 still images from this text, as well as hundreds from other Saunders chemistry texts.
- *CSC ChemOffice™ 2.1.* Includes ChemDraw and Chem3D, a drawing and modeling program, and is available at a very reasonable price to students. Also available is a

manual to use CSC ChemOffice™ with this text. Written by L. Kraig Steffen of Fairfield University, this manual enables students to work through the program on their own.

Saunders College Publishing may provide complimentary instructional aids and supplements or supplemental packages to those adopters qualified under our adoption policy. Please contact your sales representative for more information. If as an adopter or potential user you receive supplements you do not need, please return them to your sales representative or send them to

Attn: Returns Department
Troy Warehouse
465 South Lincoln Drive
Troy, MO 63379

Acknowledgments

While one or a few persons are listed as "author" of any textbook, the book is in fact the product of collaboration of many individuals, some obvious, others not so obvious. It is with gratitude that we acknowledge the contributions of the many. It is only fitting to begin with John Vondeling, Vice President and Publisher of Saunders College Publishing. John contributed the support systems necessary to bring this book from rough manuscript to bound book form, assembling the elements of the supplemental materials, and finally bringing to bear his keen sense of the marketplace.

Sandi Kiselica has been a rock of support as Developmental Editor. We so appreciate her ability to set challenging but manageable schedules and her constant encouragement as we worked to meet those deadlines. She was also an invaluable resource person with whom we could discuss everything from pedagogy to details of artwork.

Beth Ahrens as Project Editor shouldered with ease the daunting task of coordinating the transformation from manuscript to galleys to pages, including incorporation of the completed art program. Beth was always ready to answer questions or provide information, and always fully knowledgeable about every phase of the project.

We also want to acknowledge others at Saunders who contributed to this project, in particular, Caroline McGowan, Art Director; Angus McDonald, Executive Product Manager; and Charlene Squibb, Senior Production Manager.

E. Paul Papadopoulos of the University of New Mexico, Albuquerque, has been invaluable first as a very involved reviewer and then as a proofer/reviewer without peer. Paul has been deeply involved in all aspects of the final phases of this project and we are deeply indebted to him for his thoroughness and attention to detail.

We also want to acknowledge other colleagues who contributed their comments and suggestions to this book.

We enjoyed writing this text, and hope that instructors and students alike find in it a measure of the excitement we feel for organic chemistry.

William H. Brown
Beloit College
Beloit, WI

Christopher S. Foote
University of California, Los Angeles
Los Angeles, CA

Reviewers for the Second Edition of *Organic Chemistry:*

William Bailey, *University of Connecticut*
John Benbow, *Lehigh University*
Thomas Bryson, *University of South Carolina*
Mary Campbell, *Mount Holyoke College*
Lyle Castle, *Idaho State University*
Claire Castro, *University of San Francisco*
Clair Cheer, *San Jose State University*
Barry Coddens, *Northwestern University*
Mark DeCamp, *University of Michigan, Dearborn*
Dale Drueckhammer, *Stanford University*
William Epstein, *University of Utah*
Morris Fishman, *New York University*
Jack Gilbert, *University of Texas, Austin*
Stanley I. Goldberg, *University of New Orleans*
Scott Gronert, *San Francisco State University*
Dan Harvey, *University of California, San Diego*
Gene Hiegel, *California State University, Fullerton*
John Landgrebe, *University of Kansas*
Norman Lebel, *Wayne State University*
Richard Luibrand, *California State University, Hayward*
Eugene Mash, *University of Arizona*
Dominic McGrath, *University of Connecticut*
Kirk McMichael, *Washington State University*
Richard Morrison, *West Virginia University*
James Mulvaney, *University of Arizona*
Kathy Nabona, *Austin Community College*
Gary Newton, *University of Georgia*
Bruce Norcross, *State University of New York, Binghamton*
Steven Pedersen, *University of California, Berkeley*
Michael Rathke, *Michigan State University*
Carmelo Rizzo, *Vanderbilt University*
Alan Rosan, *Drew University*
Daniel Singleton, *Texas A&M University*
Robert Stern, *Oakland University*
J. William Suggs, *Brown University*
Michelle Sulikowski, *Texas A&M University*

Reviewers for the First Edition of *Organic Chemistry:*

Neil T. Allison, *University of Arkansas*
Rodney Badger, *Southern Oregon State College*
Shelton Bank, *State University of New York, Albany*
Nancy Barta, graduate student, *Michigan State University*
John L. Belletiri, *University of Cincinnati*
Edwin Bryant, graduate student, *Michigan State University*
Edward M. Burgess, *Georgia Institute of Technology*
Robert G. Carlson, *University of Kansas*
Dana S. Chatellier, *University of Delaware*
William D. Closson, *State University of New York, Albany*
David Crich, *University of Illinois, Chicago*
Dennis D. Davis, *New Mexico State University*
James A. Deyrup, *University of Florida*
Thomas A. Dix, *University of California, Irvine*
Paul Dowd, *University of Pittsburgh*
Michael B. East, *Florida Institute of Technology*
Raymond C. Fort, Jr., *University of Maine*
Warren Giering, *Boston University*
Leland Harris, *University of Arizona*
John Helling, *University of Florida*
John L. Hogg, *Texas A&M University*
John W. Huffman, *Clemson University*
Brent Iverson, *University of Texas, Austin*
Ronald Kluger, *University of Toronto*
Joseph B. Lambert, *Northwestern University*
Allan K. Lazarus, *Trenton State University*
Jerry March, *Adelphi University*
Kenneth L. Marsi, *California State University, Long Beach*
David M. McKinnon, *University of Manitoba*
James Mulvaney, *University of Arizona*
Walter Ott, *Emory University*
E. Paul Papadopoulos, *University of New Mexico, Albuquerque*
Russell C. Petter, *Sandoz Research Institute*
Joseph M. Prokipcak, *University of Guelph*
William A. Pryor, *Louisiana State University*
Michael Rathke, *Michigan State University*
Charles B. Rose, *University of Nevada, Reno*
James Schreck, *University of Northern Colorado*
Jonathan Sessler, *University of Texas*
Martin Sobczak, graduate student, *Michigan State University*
Steve Steffke, graduate student, *Michigan State University*
John Stille, *Michigan State University*
J. William Suggs, *Brown University*
Peter Trumper, *Bowdoin College*
Ken Turnbull, *Wright State University*
George Wahl, *North Carolina State University*
Michael Waldo, graduate student, *Michigan State University*
Daniel Weeks, *Northwestern University*
David F. Weimer, *University of Iowa*
Desmond Wheeler, *University of Nebraska*
Darrell J. Woodman, *University of Washington*
Ali Zand, graduate student, *Michigan State University*

Contents Overview

CONTENTS

Charles D. Winters

Stefan Eberhard/Fran Heyl Associates

Dr. E. R. Degginger

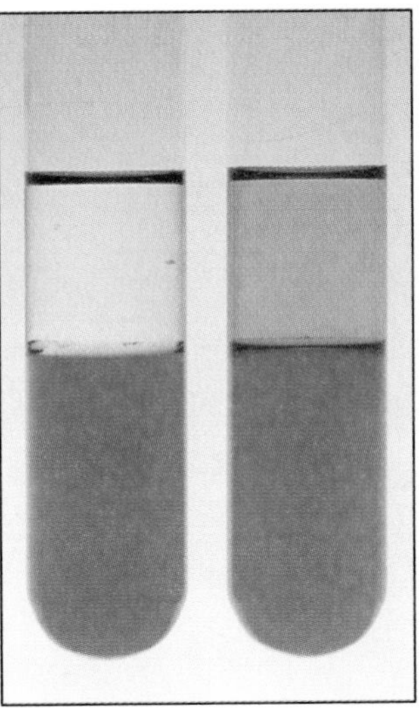

Charles D. Winters

Charles D. Winters

Courtesy of Ashland Oil, Inc.

Herb Charles Ohlmeyer/Fran Heyl Associates

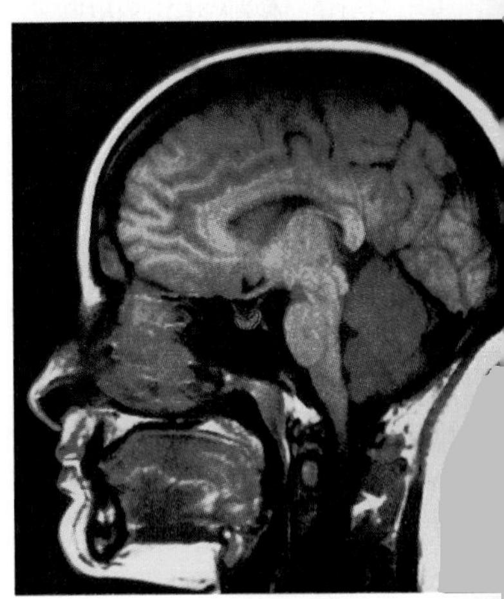

Scott Camazine/Photo Researchers

Dan McCoy/Rainbow

Ulof Björg Christianson/Fran Heyl Associates

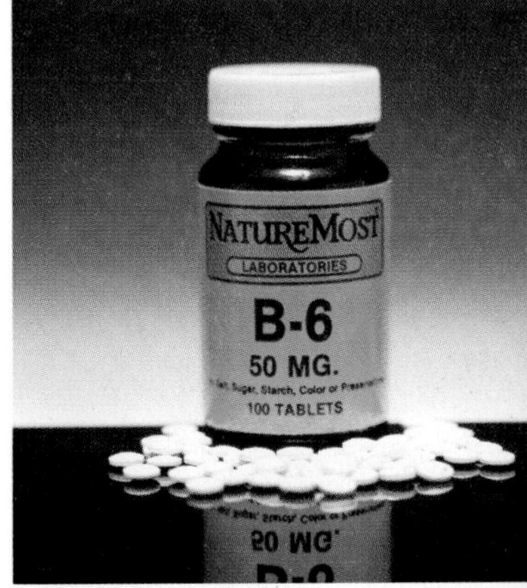

Charles D. Winters

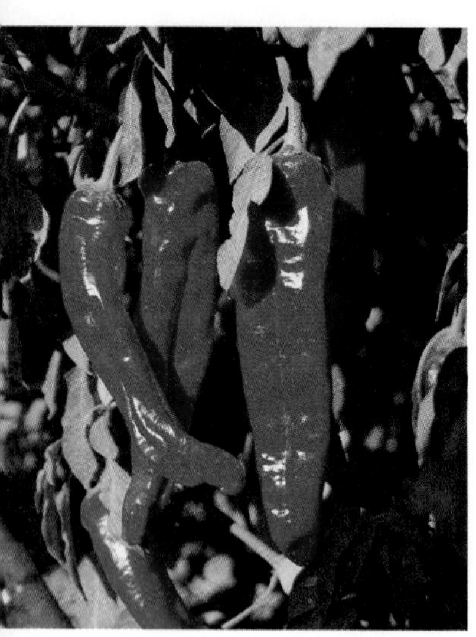

Grant Heilman/Grant Heilman Photography

© Juan M. Renjifo/Animals, Animals

The Stock Market

Charles D. Winters

25 LIPIDS 998

Charles D. Winters

26 THE ORGANIC CHEMISTRY OF METABOLISM 1032

27 AMINO ACIDS AND PROTEINS 1057

Stuart Westmoreland/Tony Stone Images

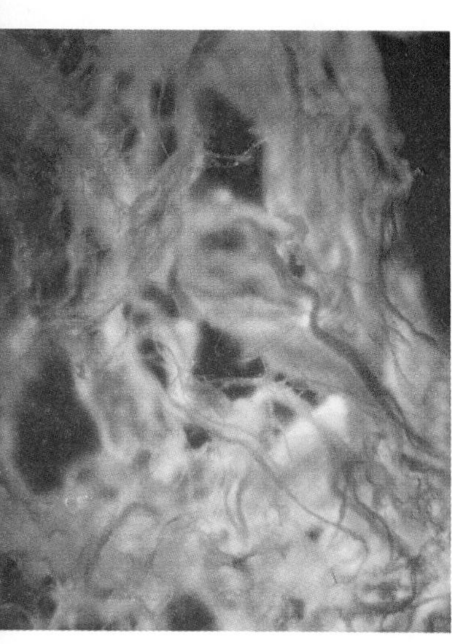

Phillip A. Harrington/Fran Heyl Associates

List of Mechanisms

Bulleted section numbers indicate mechanisms that are set off in a Mechanism box. Other mechanisms are discussed in the text, but not boxed.

List of Mechanisms (continued)

Chapter 22 Conjugated Dienes

- 22.1A 1,2- and 1,4-addition to a conjugated diene

22.2E The Diels-Alder reaction

- 22.3A The Claisen rearrangement

22.3B The Cope rearrangement

Chapter 23 Organic Polymer Chemistry

- 23.6A Radical polymerization of a substituted ethylene
- 23.6B Ziegler-Natta catalysis of alkene polymerization
- 23.6B Ziegler-Natta homogeneous coordination polymerization of alkenes
- 23.6D Initiation of anionic polymerization of alkenes
- 23.6D Initiation of anionic polymerization of conjugated dienes
- 23.6D Initiation of cationic polymerization of alkenes by a proton acid
- 23.6D Initiation of cationic polymerization of alkenes by a Lewis acid

Chapter 24 Carbohydrates

- 24.4A Acid-catalyzed formation of methyl α- and β-D-glucopyranosides

Chapter 25 Lipids

- 25.4B Formation of the dimethylallyl cation
- 25.4B Dimethylallyl cation to geranyl pyrophosphate
- 25.6D Activation of vitamin K by O_2

Chapter 26 The Organic Chemistry of Metabolism

- 26.1B Oxidation of an alcohol by NAD^+
- 26.1C Oxidation of a fatty acid $—CH_2—CH_2—$ to $—CH{=}CH—$ by FAD
- 26.3B A reverse Claisen condensation in β-oxidation of fatty acids
- 26.6 A reverse aldol reaction in Step 4 of glycolysis
- 26.6 Oxidation of glyceraldehyde 3-phosphate to 1,3-bisphosphoglycerate

Chapter 27 Amino Acids and Proteins

- 27.4B Cleavage of a peptide bond at methionine by cyanogen bromide
- 27.4B Edman degradation
- 27.5C Acid-catalyzed removal of a benzyloxycarbonyl protecting group
- 27.5E Dicyclohexylcarbodiimide (DCC) and formation of a peptide bond

List of Stereoviews

COVALENT BONDS AND SHAPES OF MOLECULES

1

■ A model of the structure of diamond. Diamond is one form of pure carbon. Each carbon atom in diamond is surrounded by four other C atoms at the corners of a tetrahedron. *(Charles D. Winters)*

According to the simplest definition, **organic chemistry** is the study of the compounds of carbon. Perhaps the most remarkable feature of organic chemistry is that it is the chemistry of carbon and only a few other elements—chiefly, hydrogen, oxygen, and nitrogen. Chemists have discovered or made well over ten million compounds composed of carbon and these three other elements. Organic compounds are everywhere around us—in our foods, flavors, and fragrances; in our medicines, toiletries, and cosmetics; in our plastics, films, fibers, and resins; in our paints and varnishes; in our glues and adhesives; and, of course, in our bodies and those of all living things. Let us begin our study of organic chemistry with a review of how the elements of C, H, O, and N combine by sharing electron pairs to form molecules.

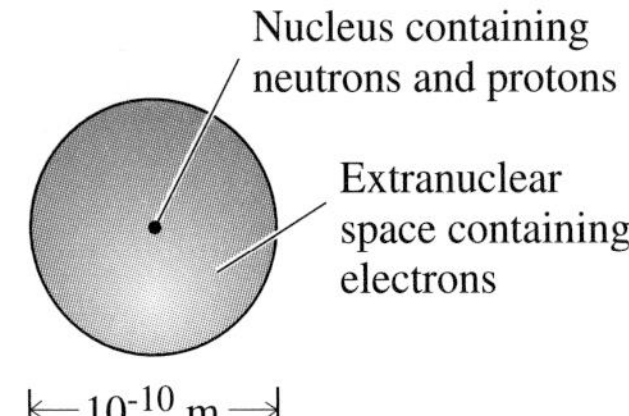

FIGURE 1.1
A schematic view of an atom. Most of the mass of an atom is concentrated in its small, dense nucleus.

1.1 Electronic Structure of Atoms

You are already familiar with the fundamentals of the electronic structure of atoms from a previous study of chemistry. Briefly, an atom contains a small, dense nucleus made of neutrons and positively charged protons. Most of the mass of an atom is contained in its nucleus. The nucleus is surrounded by a much larger extranuclear space containing negatively charged electrons. The nucleus of an atom has a diameter of 10^{-14} to 10^{-15} meter (m), which is 10^{-4} to 10^{-5} angstrom (Å). The extranuclear space where its electrons are found is a much larger area with a diameter of approximately 10^{-10} m, which is 1 Å (Figure 1.1).

Shell A region of space around a nucleus where electrons are found.

Electrons do not move freely in the space around the nucleus but are confined to regions of space called **principal energy levels** or, more simply, **shells.** Electron shells are identified by the principal quantum numbers 1, 2, 3, and so on. Each shell can contain up to **$2n^2$ electrons,** where n is the number of the shell. Thus, the first shell can contain 2 electrons, the second 8 electrons, the third 18 electrons, the fourth 32 electrons, and so on (Table 1.1). Electrons in the first shell are nearest to the positively charged nucleus and are held most strongly by it; these electrons are said to be lowest in energy. Electrons in higher numbered shells are farther from the positively charged nucleus and are held less strongly by it; these electrons are said to be higher in energy.

Orbital A region of space where an electron or pair of electrons spends 90%–95% of its time.

Shells are further subdivided into subshells designated by the letters *s*, *p*, *d*, and *f*, and, within these subshells, electrons are grouped in **orbitals** (Table 1.2). An **orbital** is a region of space that can hold electrons. Each orbital can hold two electrons. The first shell contains a single orbital called a 1*s* orbital. The second shell contains one *s* orbital and three *p* orbitals; these orbitals are designated 2*s*, $2p_x$, $2p_y$, and $2p_z$. The third shell contains one 3*s* orbital, three 3*p* orbitals, and five 3*d* orbitals.

One way to visualize the electron density associated with a particular orbital is to draw a boundary surface around the region of space that encompasses some arbitrary percent of the negative charge associated with that orbital. Most commonly, we draw the boundary surface at 95%. In this course, we concentrate on *s* and *p* orbitals, boundary surface shapes for which are shown in Section 1.7B.

TABLE 1.1 Distribution of Electrons in Shells

Shell	Number of Electrons Shell Can Hold	Relative Energies of Electrons in These Shells
4	32	higher
3	18	↑
2	8	
1	2	lower

TABLE 1.2 Distribution of Orbitals Within Shells

Shell	Orbitals Contained in That Shell
3	$3s$, $3p_x$, $3p_y$, $3p_z$, plus five $3d$ orbitals
2	$2s$, $2p_x$, $2p_y$, $2p_z$
1	$1s$

A. Electron Configuration of Atoms

The electron configuration of an atom is a description of the orbitals its electrons occupy. Every atom has an infinite number of possible electron configurations. At this stage, we are concerned primarily with the **ground-state electron configuration**—the electron configuration of lowest energy. We determine the ground-state electron configuration of an atom by using the following three rules.

Ground-state electron configuration The electron configuration of lowest energy for an atom, molecule, or ion.

Rule 1: The Aufbau Principle. Orbitals fill in order of increasing energy from lowest to highest.
Example: In this course, we are concerned primarily with the elements of the first, second, and third periods of the periodic table. Orbitals of these elements fill in the order $1s$, $2s$, $2p$, $3s$, $3p$, and so on.

Aufbau principle Orbitals fill in order of increasing energy, from lowest to highest.

Rule 2: The Pauli Exclusion Principle. No more than two electrons may be present in an orbital, one with **spin quantum number** $+\frac{1}{2}$, the other with spin quantum number $-\frac{1}{2}$.
Example: With four electrons, the $1s$ and $2s$ orbitals are filled and are written $1s^2 2s^2$. With an additional six electrons, a set of three $2p$ orbitals is filled and is written $2p_x^2\ 2p_y^2\ 2p_z^2$. Alternatively, a filled set of three $2p$ orbitals may be written $2p^6$.

Pauli exclusion principle No more than two electrons may be present in an orbital, one with spin $+\frac{1}{2}$, the other with spin $-\frac{1}{2}$.

Rule 3: Hund's Rule. When orbitals of equivalent energy are available but there are not enough electrons to fill all of them completely, then one electron is added to each orbital before a second electron is added to any one of them.
Example: After the $1s$ and $2s$ orbitals are filled with four electrons, a fifth electron is added to the $2p_x$ orbital, a sixth electron to the $2p_y$ orbital, and a seventh electron to the $2p_z$ orbital. Only after each $2p$ orbital contains one electron is a second electron added to the $2p_x$ orbital. For ten electrons, the ground-state electron configuration is $1s^2\ 2s^2\ 2p_x^2\ 2p_y^2\ 2p_z^2$, or alternatively $1s^2\ 2s^2\ 2p^6$. Table 1.3 shows ground-state electron configurations of the first 18 elements of the periodic table.

Hund's rule When orbitals of equivalent energy are available but there are not enough electrons to fill all of them completely, then one electron is added to each equivalent orbital before a second electron is added to any one of them.

EXAMPLE 1.1

Write ground-state electron configurations for these elements.

(a) Lithium **(b)** Oxygen **(c)** Chlorine

Solution

(a) Lithium (atomic number 3): $1s^2\ 2s^1$

(b) Oxygen (atomic number 8): $1s^2\ 2s^2\ 2p_x^2\ 2p_y^1\ 2p_z^1$. Alternatively, the four electrons of the $2p$ orbitals can be grouped and the electron configuration written $1s^2\ 2s^2\ 2p^4$.

(c) Chlorine (atomic number 17): $1s^2\ 2s^2\ 2p^6\ 3s^2\ 3p^5$

PROBLEM 1.1

Write and compare the ground-state electron configurations for the following:

(a) Carbon and silicon **(b)** Oxygen and sulfur **(c)** Nitrogen and phosphorus

B. Lewis Structures

Valence electrons Electrons in the valence (outermost) shell of an atom.

When discussing the physical and chemical properties of an element, chemists often focus on the outermost shell of its atoms because electrons in this shell are the ones involved in the formation of chemical bonds and in chemical reactions. Carbon, for example, with the ground-state electron configuration $1s^2\ 2s^2\ 2p^2$, has 4 outer shell electrons. Outer shell electrons are called **valence electrons,** and the energy level in

TABLE 1.3 Ground-State Electron Configurations for Elements 1–18

Element	Atomic Number	Orbital								
		$1s$	$2s$	$2p_x$	$2p_y$	$2p_z$	$3s$	$3p_x$	$3p_y$	$3p_z$
H	1	1								
He	2	2								
Li	3	2	1							
Be	4	2	2							
B	5	2	2	1						
C	6	2	2	1	1					
N	7	2	2	1	1	1				
O	8	2	2	2	1	1				
F	9	2	2	2	2	1				
Ne	10	2	2	2	2	2				
Na	11	2	2	2	2	2	1			
Mg	12	2	2	2	2	2	2			
Al	13	2	2	2	2	2	2	1		
Si	14	2	2	2	2	2	2	1	1	
P	15	2	2	2	2	2	2	1	1	1
S	16	2	2	2	2	2	2	2	1	1
Cl	17	2	2	2	2	2	2	2	2	1
Ar	18	2	2	2	2	2	2	2	2	2

which they are found is called the **valence shell.** To show the outermost electrons of an atom, we commonly use a representation called a **Lewis structure,** after the American chemist G. N. Lewis (1875–1946) who devised this notation. A Lewis structure shows the symbol of the element surrounded by a number of dots equal to the number of electrons in the outer shell of that element. In Lewis structures, the atomic symbol represents the "core," that is, the nucleus and all inner shells. Table 1.4 shows Lewis structures for the first 18 elements of the Periodic Table.

Valence shell The outermost electron shell of an atom.

Lewis structure of an atom The symbol of an element surrounded by a number of dots equal to the number of electrons in the valence shell of the atom.

The noble gases helium and neon have filled valence shells. The valence shell of helium is filled with two electrons; that of neon is filled with eight electrons. Neon, argon, and krypton have in common an electron configuration in which the *s* and *p* orbitals of their valence shells are filled with eight electrons. The valence shells of all other elements shown in Table 1.4 contain fewer than 8 electrons.

Compare the Lewis structures given in Table 1.4 with the ground-state electron configurations given in Table 1.3. The Lewis structure of boron (B), for example, is shown in Table 1.4 with three valence electrons; these are the paired 2*s* electrons and the single $2p_x$ electron shown in Table 1.3. The Lewis structure of carbon (C) is shown in Table 1.4 with four valence electrons; these are the two paired 2*s* electrons and the single $2p_x$ and $2p_y$ electrons shown in Table 1.3.

Notice also from Table 1.4 that for C, N, O, and F in period 2 of the Periodic Table, the valence electrons belong to the second shell. With eight electrons, this shell is completely filled. For Si, P, S, and Cl in period 3 of the Periodic Table, the valence electrons belong to the third shell. This shell is only partially filled with eight electrons; the 3*s* and 3*p* orbitals are fully occupied, but the five 3*d* orbitals can accommodate an additional ten valence electrons. Because of the differences in number and kind of valence shell orbitals available to elements of the second and third periods, significant differences exist in the covalent bonding of oxygen and sulfur, and of nitrogen and phosphorus (Section 1.2F). For example, although oxygen and nitrogen can accommodate no more than 8 electrons in their valence shells, many phosphorus-containing compounds have 10 electrons in the valence shell of phosphorus, and many sulfur-containing compounds have 10 and even 12 electrons in the valence shell of sulfur.

Gilbert N. Lewis introduced the theory of the shared electron pair bond and revolutionized chemistry. It is in his honor that we often refer to "electron dot" structures as Lewis structures. *(UPI/Bettmann)*

TABLE 1.4 Lewis Structures for Elements 1–18 of the Periodic Table

IA	IIA	IIIA	IVA	VA	VIA	VIIA	VIIIA
H·							He:
Li·	Be:	B:	·C:	·N:	:O:	:F:	:Ne:
Na·	Mg:	Al:	·Si:	·P:	:S:	:Cl:	:Ar:

1.2 The Lewis Model of Bonding

A. Formation of Ions

In 1916, Lewis devised a beautifully simple model that unified many of the observations about chemical bonding and reactions of the elements. He pointed out that the chemical inertness of the noble gases indicates a high degree of stability of the electron configurations of these elements: helium with a valence shell of two electrons ($1s^2$), neon with a valence shell of eight electrons ($2s^2\ 2p^6$), and argon with a valence shell of eight electrons ($3s^2\ 3p^6$). The tendency of atoms to react in ways that achieve an outer shell of eight valence electrons is particularly common among elements of Groups IA–VIIA (the main-group elements) and is given the special name **octet rule.**

Octet rule The tendency among atoms of Group IA–VIIA elements to react in ways that achieve an outer shell of eight valence electrons.

EXAMPLE 1.2

Show how sodium follows the octet rule in forming Na^+.

Solution

Following are ground-state electron configurations for Na and Na^+:

$$\text{Na (11 electrons): } 1s^2\ 2s^2\ 2p^6\ 3s^1$$
$$\text{Na}^+\text{ (10 electrons): } 1s^2\ 2s^2\ 2p^6$$

Thus, Na^+ has a complete octet of electrons in its outermost (valence) shell and has the same electron configuration as neon.

PROBLEM 1.2

Show that the following obey the octet rule:

(a) Sulfur forms S^{2-}. **(b)** Magnesium forms Mg^{2+}.

B. Formation of Chemical Bonds

According to Lewis' model, atoms bond together in such a way that each atom participating in a chemical bond acquires a completed outer-shell electron configuration resembling that of the noble gas nearest it in the Periodic Table. Atoms acquire completed outer shells in two ways.

Anion An atom or group of atoms bearing a negative charge.

Cation An atom or group of atoms bearing a positive charge.

1. An atom may lose or gain enough electrons to acquire a completely filled outer shell. An atom that gains electrons becomes an **anion** (a negatively charged ion), and an atom that loses electrons becomes a **cation** (a positively charged ion). A chemical bond between a positively charged ion and a negatively charged ion is called an **ionic bond.**
2. An atom may share electrons with one or more other atoms to complete its outer shell. A chemical bond formed by sharing electrons is called a **covalent bond.**

C. Electronegativity and Chemical Bonds

Electronegativity A measure of the force of an atom's attraction for electrons it shares in a chemical bond with another atom.

How do we estimate the degree of ionic or covalent character in a chemical bond? One way is to compare the electronegativities of the atoms involved. **Electronegativity** is a measure of the force of an atom's attraction for electrons that it shares in a chemical bond with another atom. The most widely used scale of electronegativities (Table 1.5) was devised by Linus Pauling in the 1930s and is based on bond energies.

TABLE 1.5 Electronegativity Values for Some Atoms (Pauling Scale)

IA	IIA	IIIB	IVB	VB	VIB	VIIB	VIIIB			IB	IIB	IIIA	IVA	VA	VIA	VIIA
								H 2.1								
Li 1.0	Be 1.5											B 2.0	C 2.5	N 3.0	O 3.5	F 4.0
Na 0.9	Mg 1.2											Al 1.5	Si 1.8	P 2.1	S 2.5	Cl 3.0
K 0.8	Ca 1.0	Sc 1.3	Ti 1.5	V 1.6	Cr 1.6	Mn 1.5	Fe 1.8	Co 1.8	Ni 1.8	Cu 1.9	Zn 1.6	Ga 1.6	Ge 1.8	As 2.0	Se 2.4	Br 2.8
Rb 0.8	Sr 1.0	Y 1.2	Zr 1.4	Nb 1.6	Mo 1.8	Tc 1.9	Ru 2.2	Rh 2.2	Pd 2.2	Ag 1.9	Cd 1.7	In 1.7	Sn 1.8	Sb 1.9	Te 2.1	I 2.5
Cs 0.7	Ba 0.9	La 1.1	Hf 1.3	Ta 1.5	W 1.7	Re 1.9	Os 2.2	Ir 2.2	Pt 2.2	Au 2.4	Hg 1.9	Tl 1.8	Pb 1.8	Bi 1.9	Po 2.0	At 2.2

- <1.0
- 1.0 – 1.4
- 1.5 – 1.9
- 2.0 – 2.4
- 2.5 – 2.9
- 3.0 – 4.0

On the Pauling scale, fluorine, the most electronegative element, is assigned an electronegativity of 4.0, and all other elements are assigned values in relation to fluorine. As you study the electronegativity values in this table, note that they increase from left to right within a period of the Periodic Table and decrease from top to bottom within a group. Values increase from left to right because of the increasing positive charge on the nucleus. They decrease from top to bottom because of the increasing distance of the valence electrons from the nucleus, which leads to weaker attraction.

EXAMPLE 1.3

Judging from their relative positions in the Periodic Table, which element in each set has the larger electronegativity?

(a) Lithium or carbon **(b)** Nitrogen or oxygen **(c)** Carbon or oxygen

Solution

The elements in these sets are all in the second period of the Periodic Table. Electronegativity in this period increases from left to right.

(a) C > Li **(b)** O > N **(c)** O > C

PROBLEM 1.3

Judging from their relative positions in the Periodic Table, which element in each set has the larger electronegativity?

(a) Lithium or potassium **(b)** Nitrogen or phosphorus **(c)** Carbon or silicon

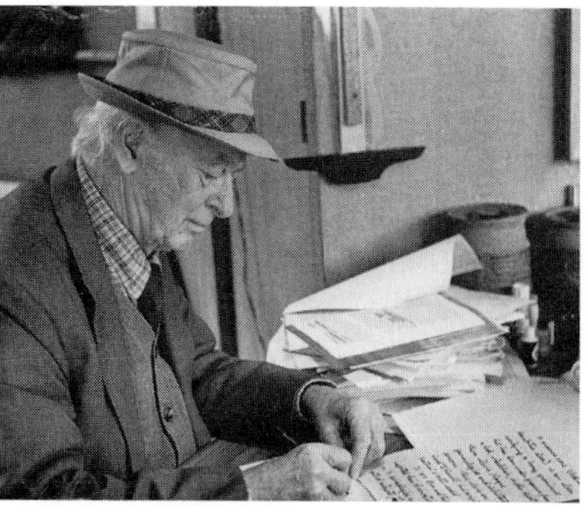

Linus Pauling (1901–1994) was the first person to receive two unshared Nobel prizes. He received the Nobel Prize for chemistry in 1954 for his contributions to our understanding of the nature of the chemical bond. He received the Nobel Prize for peace in 1962 in recognition of his efforts on behalf of international control of nuclear weapons and against nuclear testing. *(UPI/Bettmann)*

Ionic Bonds

Ionic bond A chemical bond resulting from the electrostatic attraction of an anion and a cation.

An **ionic bond** is a chemical bond formed by the attractive force between positive and negative ions. An ionic bond is formed by the transfer of electrons from the valence shell of an atom of lower electronegativity to the valence shell of an atom of higher electronegativity. The more electronegative atom gains one or more valence electrons and becomes an anion; the less electronegative atom loses one or more valence electrons and becomes a cation. We say that a chemical bond is ionic if the difference in electronegativity between the bonded atoms is 1.9 or greater. An example of an ionic bond is that formed between sodium (electronegativity 0.9) and fluorine (electronegativity 4.0). In the following equation, we use a single-headed (barbed) arrow to show the transfer of one electron from sodium to fluorine.

$$\mathrm{Na}\cdot + \cdot\ddot{\underset{..}{\mathrm{F}}}: \longrightarrow \mathrm{Na}^{+} :\ddot{\underset{..}{\mathrm{F}}}:^{-}$$

In forming $Na^{+}F^{-}$, the single 3*s* valence electron of sodium is transferred to the partially filled valence shell of fluorine:

$$\mathrm{Na}(1s^2\,2s^2\,2p^6\,\mathbf{3s^1}) + \mathrm{F}(1s^2\,2s^2\,\mathbf{2p^5}) \longrightarrow \mathrm{Na}^{+}(1s^2\,2s^2\,2p^6) + \mathrm{F}^{-}(1s^2\,2s^2\,\mathbf{2p^6})$$

As a result of this transfer of one electron, both sodium and fluorine form ions that have the same electron configuration as neon, the noble gas nearest each in the Periodic Table.

Covalent Bonds

Covalent bond A chemical bond formed by sharing one or more pairs of electrons.

A **covalent bond** is a chemical bond formed by sharing electron pairs between two atoms whose difference in electronegativity is less than 1.9. The simplest example of a covalent bond is that found in the hydrogen molecule. When two hydrogen atoms bond, the single electrons from each combine to form an electron pair. This shared pair completes the valence shell of each hydrogen. According to the Lewis model, a pair of electrons in a covalent bond functions in two ways simultaneously; it is shared by two atoms and at the same time fills the outer (valence) shell of each. We use a line between the two hydrogens to symbolize the covalent bond formed by the sharing of a pair of electrons.

$$\mathrm{H}\cdot + \cdot\mathrm{H} \longrightarrow \mathrm{H{-}H} \qquad \Delta H^0 = -104\ \mathrm{kcal/mol}\ (-435\ \mathrm{kJ/mol})$$

The Lewis model accounts for the stability of two covalently bonded atoms in the following way. In forming a covalent bond, an electron pair occupies the region be-

TABLE 1.6 Classification of Chemical Bonds

Difference in Electronegativity Between Bonded Atoms	Type of Bond
less than 0.5	nonpolar covalent
0.5 to 1.9	polar covalent
greater than 1.9	ionic

tween two nuclei and serves to shield one positively charged nucleus from the repulsive force of the other positively charged nucleus. At the same time, the electron pair attracts both nuclei. In other words, an electron pair in the space between two nuclei bonds them together and fixes the internuclear distance to within very narrow limits. The distance between nuclei participating in a chemical bond is called the **bond length.** Every covalent bond has a characteristic bond length. In H—H, it is 0.74 Å, where 1 Å = 10^{-10} meter.

Bond length The distance between atoms in a covalent bond.

Polar Covalent Bonds

Although all covalent bonds involve the sharing of electrons, they differ widely in the degree of sharing. We divide covalent bonds arbitrarily into two categories, polar and nonpolar, depending on the difference in electronegativity between bonded atoms (Table 1.6).

A **nonpolar covalent bond** is one in which the difference in electronegativity between bonded atoms is less than 0.5. A covalent bond between carbon and hydrogen, for example, is classified as nonpolar covalent because the difference in electronegativity between these two atoms is 2.5 − 2.1 = 0.4 unit on the Pauling scale.

Nonpolar covalent bond A covalent bond between atoms whose difference in electronegativity is less than 0.5.

A **polar covalent bond** is one in which the difference in electronegativity between bonded atoms is between 0.5 and 1.9. An example of a polar covalent bond is that of H—Cl. The difference in electronegativity between chlorine and hydrogen is 3.0 − 2.1 = 0.9 unit.

Polar covalent bond A covalent bond between atoms whose difference in electronegativity is between 0.5 and 1.9.

An important consequence of the unequal sharing of electrons in a polar covalent bond is that the more electronegative atom gains a greater fraction of the shared electrons and acquires a partial negative charge, indicated by the symbol $\delta-$. The less electronegative atom has a lesser fraction of the shared electrons and acquires a partial positive charge, indicated by the symbol $\delta+$.

EXAMPLE 1.4

Classify these bonds as nonpolar covalent, polar covalent, or ionic.

(a) O—H **(b)** N—H **(c)** Na—F **(d)** C—Mg

Solution

Based on differences in electronegativity between the bonded atoms, three of these bonds are polar covalent and one is ionic.

Bond	Difference in Electronegativity	Type of Bond
(a) O—H	3.5 − 2.1 = 1.4	polar covalent
(b) N—H	3.0 − 2.1 = 0.9	polar covalent
(c) Na—F	4.0 − 0.9 = 3.1	ionic
(d) C—Mg	2.5 − 1.2 = 1.3	polar covalent

PROBLEM 1.4

Classify these bonds as nonpolar covalent, polar covalent, or ionic.

(a) S—H **(b)** P—H **(c)** C—F **(d)** C—Cl

We must point out that electronegativity varies somewhat depending on the chemical environment and oxidation state of an atom, and, therefore, these rules are just guidelines and must be used with caution. Lithium iodide, for example, has a high melting point (449°C) and boiling point (1180°C) characteristic of ionic compounds. Yet, based on the difference in electronegativity between these two elements of 2.5 − 1.0 = 1.5 units, we would classify LiI as a polar covalent compound.

EXAMPLE 1.5

Using the symbols $\delta-$ and $\delta+$, indicate the direction of polarity in these polar covalent bonds.

(a) C—O **(b)** N—H **(c)** C—Mg

Solution

Electronegativity is given beneath each atom. The atom with the greater electronegativity has the partial negative charge, the atom with the lesser electronegativity has the partial positive charge.

$\overset{\delta+}{\text{C}}\text{—}\overset{\delta-}{\text{O}}$ **(a)** C—O with 2.5 3.5

(a) $\delta+$ C—O $\delta-$ (2.5, 3.5) **(b)** $\delta-$ N—H $\delta+$ (3.0, 2.1) **(c)** $\delta-$ C—Mg $\delta+$ (2.5, 1.2)

PROBLEM 1.5

Indicate the direction of polarity in these polar covalent bonds using the symbols $\delta-$ and $\delta+$.

(a) C—N **(b)** N—O **(c)** C—Cl

The polarity of a covalent bond is measured by a quantity called a **bond dipole** and is given the symbol μ (Greek mu). Bond dipole is defined as the product of one of the charges, *e*, (either the $\delta+$ or $\delta-$ because each is the same in absolute magnitude) of a dipole unit times the distance, *d*, separating the two dipolar charges. The SI units for charge and distance are coulombs (C) and meters (m) respectively. The charge on an electron is 1.60×10^{-19} C and the bond length of a typical single covalent bond is approximately 1×10^{-10} m. Hence, the bond dipole of a typical single covalent bond is on the order of 10^{-29} C·m. To simplify reporting, such moments are more commonly given in debye units, D, after the Dutch chemist Peter Debye (1884–1966), who was awarded the Nobel Prize for chemistry in 1936 for his contributions to our understanding of the polarity of molecules, and molecular structure. By definition, $1\text{D} = 3.34 \times 10^{-30}$ C·m. To show the direction of a bond dipole, we use an arrow with the head pointing toward the negative end of the bond dipole and the plus sign in the tail of the arrow at the positive end of the bond dipole. Table 1.7 lists bond dipoles for the types of covalent bonds we deal with most frequently in this course.

Bond dipole (μ) A measure of the polarity of a covalent bond. The product of the charge on each atom of a polar bond times the distance between the atoms.

D. Drawing Lewis Structures for Molecules and Polyatomic Ions

The ability to write Lewis structures for molecules and polyatomic ions is a fundamental skill for the study of organic chemistry. The following guidelines will help you to do this. As you study these guidelines, look at the examples in Table 1.8.

1. Determine the number of valence electrons in the molecule or ion. To do this, add the number of valence electrons contributed by each atom. For ions, add one electron for each negative charge on the ion, and subtract one electron for each positive charge on the ion. For example, the Lewis structure for the water molecule, H_2O, must show eight valence electrons; one from each hydrogen and six from oxygen. The Lewis structure for the hydroxide ion, OH^-, must also show eight valence electrons; one from hydrogen, six from oxygen, plus one for the negative charge on the ion.
2. Determine the arrangement of atoms in the molecule or ion. Except for the simplest molecules and ions, this arrangement must be determined experimentally. For some molecules and ions given as examples in the text, you are asked to propose an arrangement of atoms. For most, however, you are given the experimentally determined arrangement of atoms.

TABLE 1.7 Average Bond Dipoles of Selected Covalent Bonds

Bond	Bond Dipole (D)	Bond	Bond Dipole (D)	Bond	Bond Dipole (D)
⟼		⟼		⟼	
H—C	0.3	C—F	1.4	C—O	0.7
H—N	1.3	C—Cl	1.5	C=O	2.3
H—O	1.5	C—Br	1.4	C—N	0.2
H—S	0.7	C—I	1.2	C≡N	3.5

TABLE 1.8 Lewis Structures for Several Compounds

H—Ö—H	H—N̈—H (with H below N)	H—C—H (with H above and below C)	H—C̈l:
H_2O (8) Water	NH_3 (8) Ammonia	CH_4 (8) Methane	HCl (8) Hydrogen chloride
$H_2C{=}CH_2$	H—C≡C—H	$H_2C{=}\ddot{O}$:	$H{-}\ddot{O}{-}C({=}\ddot{O}){-}\ddot{O}{-}H$
C_2H_4 (12) Ethylene	C_2H_2 (10) Acetylene	CH_2O (12) Methanal	H_2CO_3 (24) Carbonic acid

3. Connect the atoms with single bonds. Then arrange the remaining electrons in pairs so that each atom in the molecule or ion has a complete outer shell. Each hydrogen atom must be surrounded by two electrons. Each atom of carbon, oxygen, nitrogen, and halogen must be surrounded by eight electrons (per the octet rule).
4. A pair of electrons involved in a covalent bond **(bonding electrons)** is shown as a single bond; an unshared **(nonbonding)** pair of electrons is shown as a pair of dots.
5. In a **single bond,** two atoms share one pair of electrons. In a **double bond** they share two pairs of electrons, and in a **triple bond** they share three pairs of electrons.

Bonding electrons Valence electrons involved in forming a covalent bond, i.e., shared electrons.

Nonbonding electrons Valence electrons not involved in forming covalent bonds, i.e., unshared electrons.

Table 1.8 shows Lewis structures, molecular formulas, and names for several compounds. The number of valence electrons each molecule contains is shown in parentheses. Notice that, in these molecules, each hydrogen is surrounded by two valence electrons, and each carbon, nitrogen, oxygen, and chlorine is surrounded by eight valence electrons. Furthermore, each carbon has four bonds, nitrogen has three bonds and one unshared pair of electrons, oxygen has two bonds and two unshared pairs of electrons, and chlorine (and other halogens as well) has one bond and three unshared pairs of electrons.

EXAMPLE 1.6

Draw Lewis structures, showing all valence electrons, for these molecules.

(a) CO_2 **(b)** CH_4O **(c)** CH_3Cl

Solution

Under the Lewis structure of each molecule is the number of valence electrons it contains.

(a) $\ddot{O}=C=\ddot{O}$ Carbon dioxide (16 valence electrons)

(b) $H-\overset{H}{\underset{H}{C}}-\ddot{\underset{\cdot\cdot}{O}}-H$ Methanol (14 valence electrons)

(c) $H-\overset{H}{\underset{H}{C}}-\ddot{\underset{\cdot\cdot}{Cl}}:$ Chloromethane (14 valence electrons)

PROBLEM 1.6

Draw Lewis structures, showing all valence electrons, for these molecules.

(a) C_2H_6 **(b)** CS_2 **(c)** HCN

E. Formal Charge

Throughout this course we deal not only with molecules but also with polyatomic cations and polyatomic anions. Examples of polyatomic cations are the hydronium ion, H_3O^+, and the ammonium ion, NH_4^+. An example of a polyatomic anion is the bicarbonate ion, HCO_3^-. It is important that you be able to determine which atom or atoms in a molecule or polyatomic ion bear the positive or negative charge. The charge on an atom in a molecule or polyatomic ion is called its **formal charge.** To derive a formal charge:

Formal charge The charge on an atom in an ion or molecule.

1. Write a correct Lewis structure for the molecule or ion.
2. Assign to each atom all its unshared (nonbonding) electrons and one-half its shared (bonding) electrons.
3. Compare this number with the number of valence electrons in the neutral, unbonded atom. If the number of electrons assigned to a bonded atom is less than that assigned to the unbonded atom, then more positive charges are in the nucleus than counterbalancing negative charges, and the atom has a formal positive charge. Conversely, if the number of electrons assigned to a bonded atom is greater than that assigned to the unbonded atom, then the atom has a formal negative charge.

Formal charge = number of valence electrons in the neutral unbonded atom − (all unshared electrons + one half of all shared electrons)

EXAMPLE 1.7

Draw Lewis structures for these ions, and show which atom in each bears the formal charge.

(a) H_3O^+ **(b)** HCO_3^-

Solution

(a) The Lewis structure for the hydronium ion must show 8 valence electrons; 3 from the three hydrogens, 6 from oxygen, minus 1 for the single positive charge. An oxygen atom has 6 valence electrons. The oxygen atom in H_3O^+ is assigned 2 unshared electrons and 1 from each shared pair of electrons, giving it a formal charge of 6 − (2 + 3) = +1, meaning that oxygen has 1 less electron than its normal 6 valence electrons.

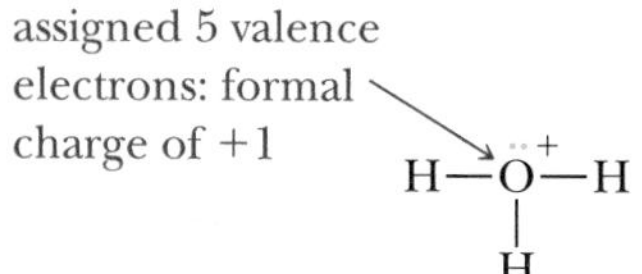

(b) The Lewis structure for the bicarbonate ion must show 24 valence electrons; 4 from carbon, 18 from the three oxygens, 1 from hydrogen, plus 1 for the single negative charge. Loss of a hydrogen ion from carbonic acid (Table 1.8) gives the bicarbonate ion. Carbon is assigned one electron from each shared pair and has no formal charge (4 − 4 = 0). Two oxygens are assigned six valence electrons each and have no formal charges (6 − 6 = 0). The third oxygen is assigned seven valence electrons and has a formal charge of 6 − (6 + 1) = −1.

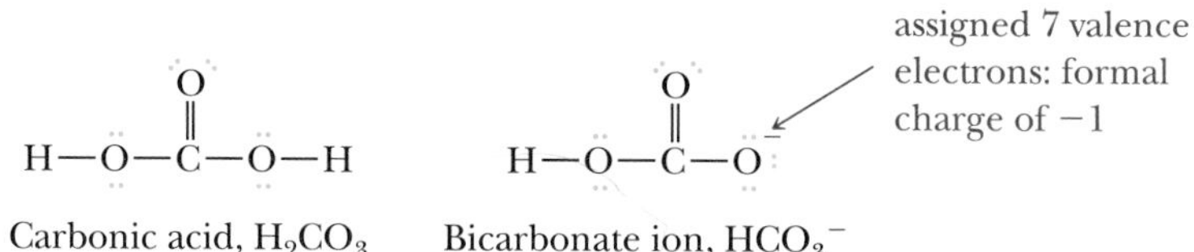

Carbonic acid, H_2CO_3 Bicarbonate ion, HCO_3^-

PROBLEM 1.7

Draw Lewis structures for these ions, and show which atom in each bears the formal charge.

(a) $CH_3NH_3^+$ **(b)** CO_3^{2-} **(c)** OH^-

In writing Lewis structures for molecules and ions, you must remember that elements of the second period, including carbon, nitrogen, and oxygen, can accommodate no more than eight electrons in the four orbitals ($2s$, $2p_x$, $2p_y$, and $2p_z$) of their valence shells. Following are two Lewis structures for nitric acid, HNO_3, each with the correct number of valence electrons, namely, 24:

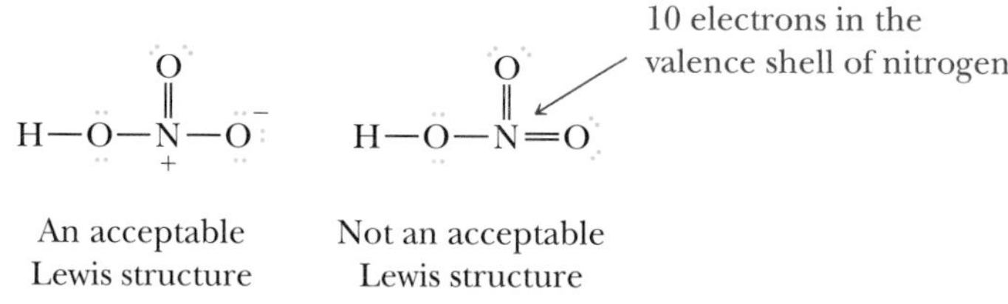

An acceptable Lewis structure Not an acceptable Lewis structure

The structure on the left is an acceptable Lewis structure. It shows the required 24 valence electrons, and each oxygen and nitrogen has a completed valence shell of eight electrons. Further, it shows a positive formal charge on nitrogen, and a negative formal charge on one of the oxygens. Note that the sum of the formal charges on the acceptable Lewis structure for HNO_3 is zero. The structure on the right is not an acceptable Lewis structure. Although it shows the correct number of valence electrons, it places 10 electrons in the valence shell of nitrogen.

F. Exceptions to the Octet Rule

The Lewis model of covalent bonding focuses on valence electrons and the necessity for each atom (other than H) in a covalent bond to have a completed valence shell of 8 electrons. Although most molecules formed by main-group elements (Groups IA–VIIA) have structures that satisfy the octet rule, there are two important exceptions to this rule.

The first group of exceptions are molecules containing atoms of Group IIIA elements. Following is a Lewis structure for BF_3. In this uncharged covalent compound, boron is surrounded by only six valence electrons. Aluminum chloride is an example of a compound in which aluminum, the element immediately below boron in group IIIA, has an incomplete valence shell. Because their valence shell is only partially filled, trivalent compounds of boron and aluminum are highly reactive.

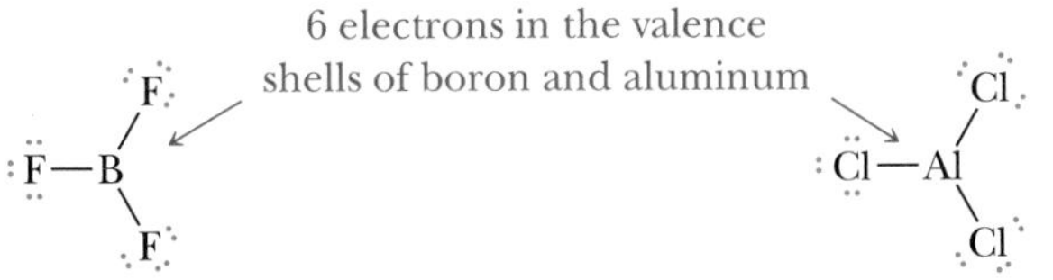

6 electrons in the valence shells of boron and aluminum

Boron trifluoride

Aluminum chloride

A second group of exceptions to the octet rule is made up of molecules and ions that contain an atom with more than eight electrons in its valence shell. Atoms of second-period elements use 2*s* and 2*p* orbitals for bonding, and these orbitals can contain only eight valence electrons, hence the octet rule. Atoms of third-period elements have 3*d* orbitals and can accommodate more than eight electrons in their valence shells. Following are Lewis structures for trimethylphosphine, phosphorus pentachloride, and phosphoric acid. The first compound has eight electrons in the valence shell of phosphorus; the second and third compounds have ten electrons in the valence shell of phosphorus.

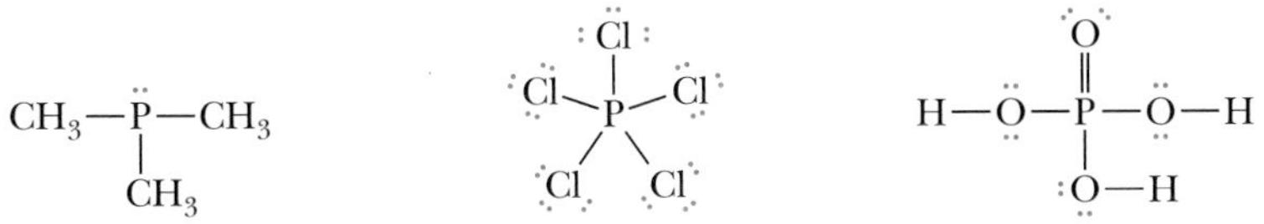

$CH_3—\ddot{P}—CH_3$ (with CH_3 below P)

Trimethylphosphine

Phosphorus pentachloride

Phosphoric acid

Sulfur, another third-period element, forms compounds in which its valence shell contains 8, 10, or 12 electrons.

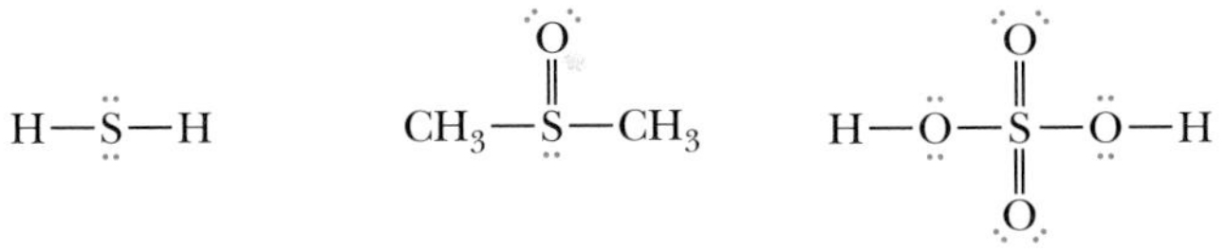

$H—\ddot{S}—H$

Hydrogen sulfide

$CH_3—\ddot{S}—CH_3$ (with =O above S)

Dimethyl sulfoxide

Sulfuric acid

CHEMISTRY IN ACTION

The Octet Rule

The octet rule of G. N. Lewis gives us a powerful and simple model for understanding bonding in organic compounds. Experiments designed to prepare molecules with ten electrons in the valence shell of carbon or nitrogen atoms might seem pointless or absurd. However, because chemistry is an experimental science, the truth of concepts such as the octet rule can be established only by experiment. No matter how many molecules obey the octet rule, a single exception would result in major modifications to the rule, or even its replacement.

In 1949, the German chemist Georg Wittig attempted to prepare pentamethylnitrogen, a compound with ten electrons in nitrogen's valence shell, by the following reaction:

$$(CH_3)_4N^+ \; Br^- \; + \; Li^+ :CH_3^- \xrightarrow{?} N(CH_3)_5 \; + \; Li^+ Br^-$$

Tetramethylammonium bromide; Methyllithium; Pentamethylnitrogen

Instead, an acid-base reaction took place with one of the C—H bonds in the tetramethylammonium ion to give an unstable compound that has a positive charge on nitrogen and a negative charge on carbon. This novel type of molecule had not been made before. Wittig gave this class of molecules the name "ylide." Nitrogen ylides cannot be isolated as stable compounds but they can be used as intermediates in other reactions.

$$(CH_3)_4N^+ \; Br^- \; + \; Li^+ :CH_3^- \longrightarrow (CH_3)_3N^+ — \ddot{C}H_2^- \; + \; Li^+ Br^- \; + \; CH_4$$

Tetramethylammonium bromide; Methyllithium; A nitrogen ylide

Reasoning that phosphorus (just below nitrogen in the Periodic Table) is capable of expanding its octet and might form stable ylides, Wittig carried out an analogous reaction with phosphorus. Once again, an acid-base reaction took place, this time to form a phosphorus ylide. Phosphorus ylides, Wittig discovered, can be isolated as stable compounds.

$$(CH_3)_4P^+ \; Br^- \; + \; Li^+ :CH_3^- \longrightarrow (CH_3)_3P^+ — \ddot{C}H_2^- \; + \; Li^+ Br^- \; + \; CH_4$$

Tetramethylphosphonium bromide; Methyllithium; A phosphorus ylide

Wittig soon abandoned his attempts to make compounds with five bonds to nitrogen and instead studied the chemistry of the newly discovered ylides (See Chapter 15). He found that phosphorus ylides are extraordinarily useful reagents for preparing complex organic molecules, including such important compounds as vitamin A. In 1979, Professor Wittig shared the Nobel Prize for chemistry for his discovery and work with phosphorus ylides.

See G. Wittig, *Science*, **210**, 600 (1980).

1.3 Functional Groups

Carbon combines with other atoms (e.g., C, H, N, O, S, halogens) to form structural units called **functional groups.** Functional groups are important for three reasons. First, they are the units by which we divide organic compounds into classes. Second, they are sites of characteristic chemical reactions; a particular functional group, in whatever compound it is found, undergoes the same types of chemical reactions. Third, functional groups serve as a basis for naming organic compounds.

Functional group An atom or group of atoms within a molecule that shows a characteristic set of physical and chemical properties.

Introduced here are several of the functional groups we encounter early in this course. At this point our concern is nothing more than pattern recognition. We have more to say about the structure and the physical and chemical properties of these functional groups in following chapters. A complete list of the major functional groups we will study in this course is presented at the front of the text.

A. Alcohols and Ethers

The functional group of an **alcohol** is an **—OH (hydroxyl)** group bonded to a carbon atom. The functional group of an **ether** is an atom of oxygen bonded to two carbon atoms.

Hydroxyl group An —OH group.

CH_3CH_2OH CH_3OCH_3

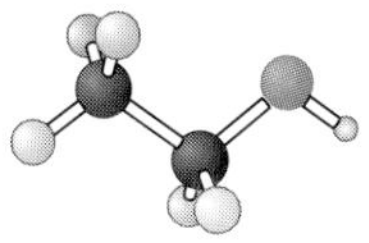

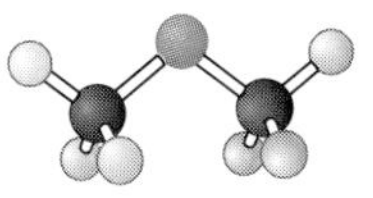

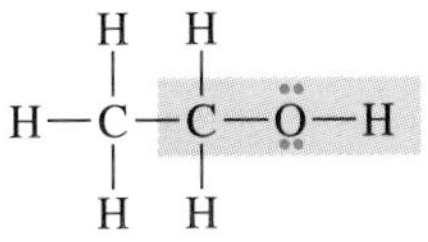

An alcohol (Ethanol)

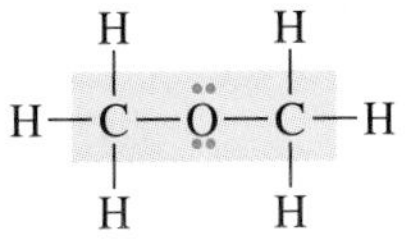

An ether (Dimethyl ether)

Functional group:

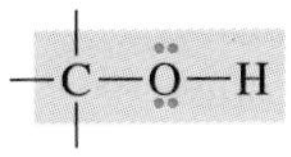

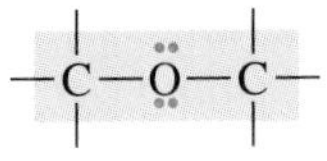

You may notice that ethanol and dimethyl ether have the same molecular formula, C_2H_6O, but differ in their connectivity. Compounds that have the same molecular formula but differ in their connectivity are called constitutional isomers. We discuss constitutional isomers in detail in Chapter 2.

We can write formulas for this alcohol and ether in an abbreviated form using what are called **condensed structural formulas.** In a condensed structural formula, CH_3 indicates a carbon with three attached hydrogens, CH_2 indicates a carbon with two attached hydrogens, and CH indicates a carbon with one attached hydrogen. Unshared pairs of electrons are generally not shown in condensed structural formulas. Following are condensed structural formulas for the alcohol and ether of molecular formula C_2H_6O. It is also common to write these formulas in an even more condensed manner, by omitting all single bonds in linear drawings.

$$CH_3{-}CH_2{-}OH \text{ or } CH_3CH_2OH \qquad CH_3{-}O{-}CH_3 \text{ or } CH_3OCH_3$$

EXAMPLE 1.8

Draw condensed structural formulas for the two alcohols of molecular formula C_3H_8O.

Solution

Bond the three carbon atoms in a chain with the —OH (hydroxyl) group attached to the end carbon of the chain or attached to the middle carbon of the chain. Then, to complete each structural formula, add seven hydrogens so that each carbon has four bonds to it.

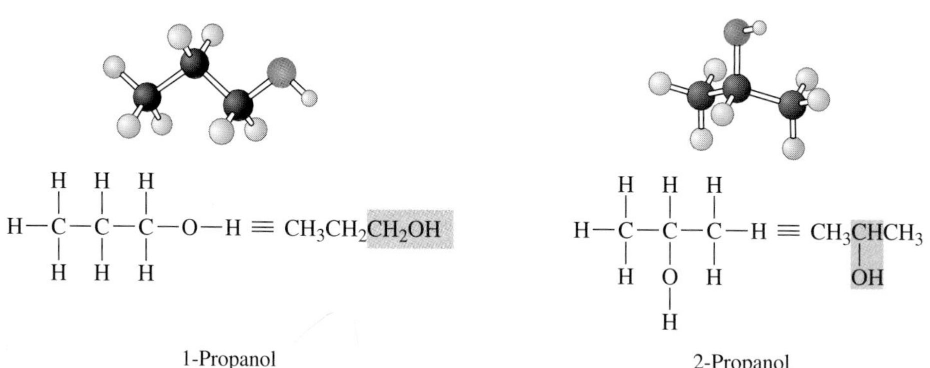

1-Propanol

2-Propanol

PROBLEM 1.8

Draw a condensed structural formula for the one ether of molecular formula C_3H_8O.

B. Aldehydes and Ketones

Carbonyl group A C=O group.

Both aldehydes and ketones contain a **C=O (carbonyl)** group. The functional group of an **aldehyde** is a carbonyl group bonded through its carbon to two hydrogens in

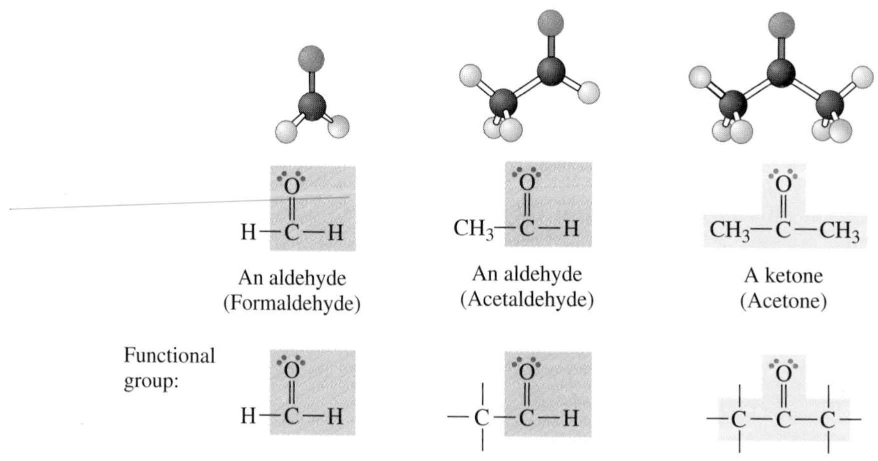

An aldehyde (Formaldehyde)

An aldehyde (Acetaldehyde)

A ketone (Acetone)

the case of methanal (formaldehyde), CH_2O, the simplest aldehyde, and to another carbon and a hydrogen in all other aldehydes. The functional group of a **ketone** is a carbonyl group bonded to two carbon atoms.

EXAMPLE 1.9

Draw condensed molecular formulas for the two aldehydes of molecular formula C_4H_8O.

Solution

First draw the functional group of an aldehyde and then add the remaining carbons. These may be attached in two different ways. Then, add seven hydrogens to complete the tetravalence of carbon and give the correct molecular formula. The aldehyde group may be written showing the carbon-oxygen double bond as C=O, or, alternatively, it may be written —CHO.

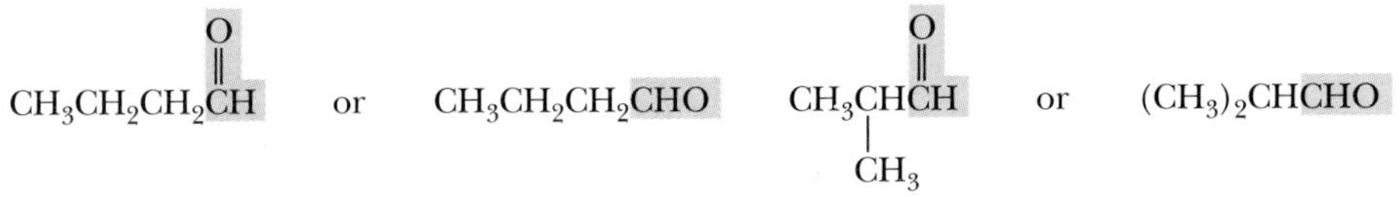

$$CH_3CH_2CH_2\overset{\overset{O}{\|}}{C}H \quad \text{or} \quad CH_3CH_2CH_2CHO \qquad CH_3\underset{\underset{CH_3}{|}}{C}H\overset{\overset{O}{\|}}{C}H \quad \text{or} \quad (CH_3)_2CHCHO$$

PROBLEM 1.9

Draw condensed structural formulas for the three ketones of molecular formula $C_5H_{10}O$.

C. Carboxylic Acids

The functional group of a carboxylic acid is a **$-CO_2H$ (carboxyl:** *carb*onyl + hydr*oxyl*) group. The carboxyl group may be written in any of the following ways, all of which are equivalent:

Carboxyl group A $-CO_2H$ group.

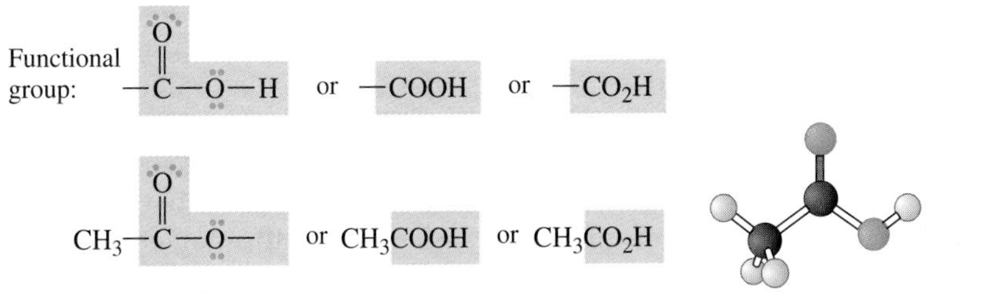

$$\text{Functional group:} \quad -\overset{\overset{\ddot{O}}{\|}}{C}-\ddot{O}-H \quad \text{or} \quad -COOH \quad \text{or} \quad -CO_2H$$

$$CH_3-\overset{\overset{\ddot{O}}{\|}}{C}-\ddot{O}- \quad \text{or} \quad CH_3COOH \quad \text{or} \quad CH_3CO_2H$$

A carboxylic acid
(Acetic acid)

EXAMPLE 1.10

Draw a condensed structural formula for the single carboxylic acid of molecular formula $C_3H_6O_2$.

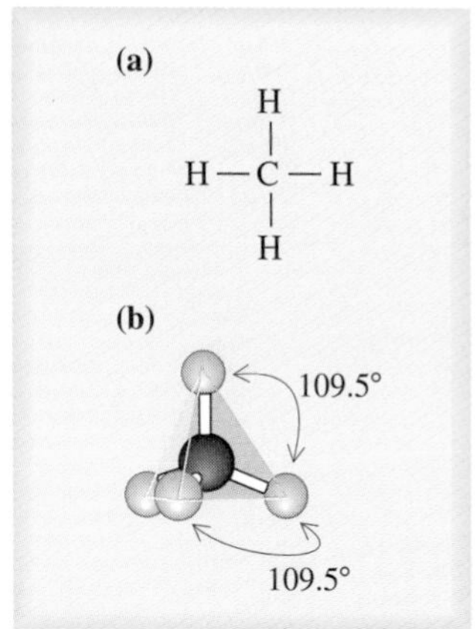

FIGURE 1.2
The shape of a methane molecule, CH_4. (a) Lewis structure and (b) the three-dimensional shape. The hydrogens occupy the four corners of a tetrahedron and all H—C—H bond angles are 109.5°.

Solution

$$CH_3—CH_2—\overset{\displaystyle O}{\overset{\|}{C}}—O—H \quad \text{or} \quad CH_3CH_2COOH \quad \text{or} \quad CH_3CH_2CO_2H$$

PROBLEM 1.10

Draw condensed structural formulas for the two carboxylic acids of molecular formula $C_4H_8O_2$.

1.4 Bond Angles and Shapes of Molecules

In Section 1.2, we used a shared pair of electrons as the fundamental unit of a covalent bond and drew Lewis structures for several molecules and ions containing various combinations of single, double, and triple bonds. We can predict bond angles in these and other molecules and ions in a very straightforward way using the **valence-shell electron-pair repulsion (VSEPR) model.** According to the VSEPR model, an atom is surrounded by an outer shell of valence electrons. These valence electrons may be involved in the formation of single, double, or triple bonds, or they may be unshared. Each of these combinations creates a negatively charged region of space, and because like charges repel each other, the various regions of electron density around an atom will spread out so that each is as far away from the others as possible.

We use the VSEPR model in the following way to predict the shape of methane, CH_4. The Lewis structure for CH_4 shows a carbon atom surrounded by four regions of electron density, each of which contains a pair of electrons forming a bond to a hydrogen atom. According to the VSEPR model, the four regions radiate from carbon so that they are as far away from each other as possible. This occurs when the angle between any two pairs of electrons is 109.5°. Therefore, we predict all H—C—H bond angles to be 109.5°, and the shape of the molecule to be **tetrahedral** (Figure 1.2). The H—C—H bond angles in methane have been measured experimentally and found to be 109.5°. Thus, the bond angles and shape of methane predicted by the VSEPR model are identical to those observed.

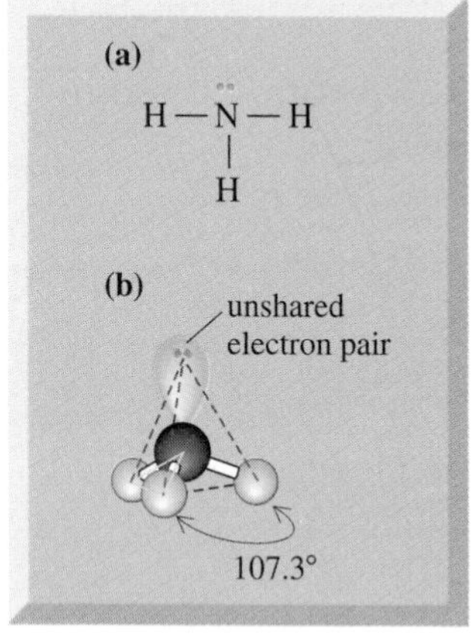

FIGURE 1.3
The shape of an ammonia molecule, NH_3. (a) Lewis structure and (b) the three-dimensional shape. The H—N—H bond angle is 107.3°, slightly smaller than the H—C—H bond angle of methane.

We predict the shape of the ammonia molecule, NH_3, in exactly the same manner. The Lewis structure of NH_3 shows nitrogen surrounded by four regions of electron density. Three regions contain single pairs of electrons forming covalent bonds with hydrogen atoms. The fourth region contains an unshared pair of electrons (Figure 1.3). Using the VSEPR model, we predict that the four regions of electron density around nitrogen are arranged in a tetrahedral manner and that all H—N—H bond angles are 109.5°. The observed bond angles are 107.3°. This small difference between the predicted and observed angles can be explained by proposing that the unshared pair of electrons on nitrogen repels adjacent electron pairs more strongly than do bonding pairs of electrons.

Figure 1.4 shows a Lewis structure and a ball-and-stick model of a water molecule. In H_2O, oxygen is surrounded by four regions of electron density. Two of these regions contain pairs of electrons used to form single covalent bonds to the two hydrogens; the remaining two contain unshared electron pairs. Using the VSEPR model, we predict that the four regions of electron density around oxygen are arranged in a tetrahedral manner and the H—O—H bond angle is 109.5°. Experimental measurements show the actual bond angle to be 104.5°, a value smaller than that predicted. This difference between the predicted and observed bond angles can be explained by proposing, as we did for NH_3, that unshared pairs of electrons repel adjacent pairs more strongly than do bonding pairs. Note that the distortion from 109.5° is greater in H_2O, which has two unshared pairs of electrons, than it is in NH_3, which has only one unshared pair.

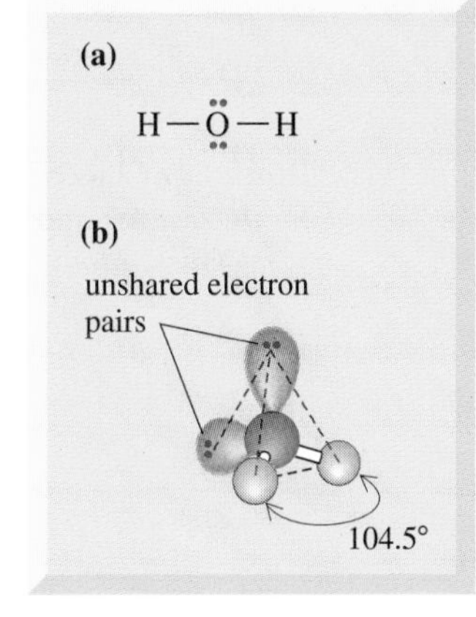

FIGURE 1.4
The shape of a water molecule, H_2O. (a) Lewis structure and (b) shape. The unshared pairs of electrons repel adjacent pairs of electrons giving the H—O—H bond angle of 104.5°.

A general prediction emerges from this discussion of the shapes of CH_4, NH_3, and H_2O molecules. If a Lewis structure shows four regions of electron density around a central atom, the VSEPR model predicts a tetrahedral distribution of electron density and bond angles of approximately 109.5°.

In many of the molecules we encounter, an atom is surrounded by three regions of electron density. Shown in Figure 1.5 are Lewis structures for methanal, CH_2O, and ethylene, C_2H_4.

According to the VSEPR model, a double bond is treated as a single region of electron density. In methanal, carbon is surrounded by three regions of electron density: two regions contain single pairs of electrons forming single bonds to hydrogen atoms; the third region of electron density contains two pairs of electrons forming a double bond to oxygen. In ethylene, each carbon atom is also surrounded by three regions of electron density; two contain single pairs of electrons, and the other contains two pairs of electrons.

Three regions of electron density about an atom are farthest apart when they are coplanar and make angles of 120° with each other. Thus, the predicted H—C—H and H—C—O bond angles in methanal and the predicted H—C—H and H—C—C bond angles in ethylene are all 120°. Further, all bonds in each compound lie in a plane. Both methanal and ethylene are **planar** molecules.

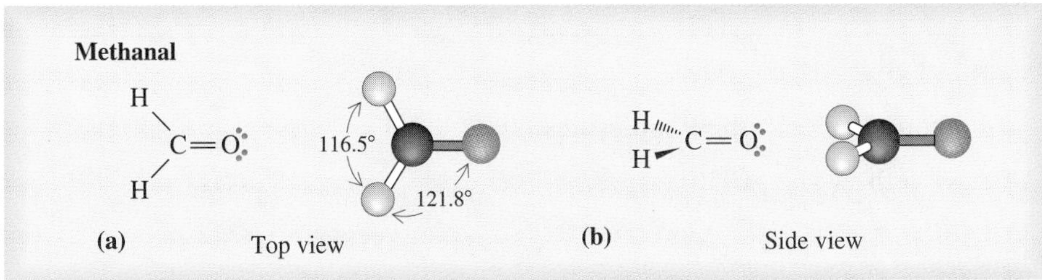

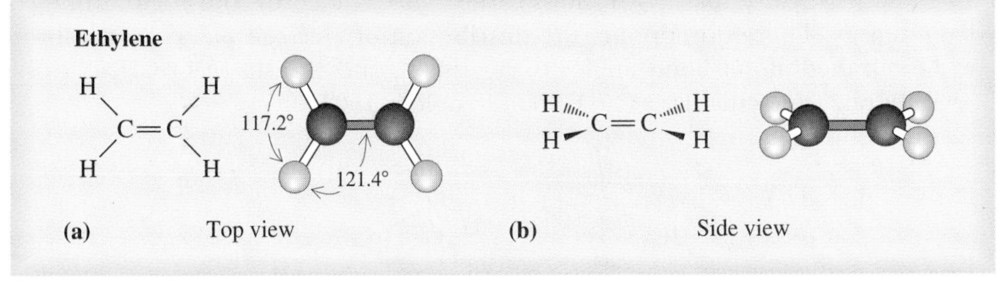

FIGURE 1.5
Shapes of methanal, CH_2O, and ethylene, C_2H_4, shown from (a) top view and (b) side view.

TABLE 1.9 Predicted Molecular Shapes (VSEPR Model)

Regions of Electron Density Around Central Atom	Predicted Distribution of Electron Density	Predicted Bond Angles	Examples		
4	tetrahedral	109.5°	H—O: with H below O	N with H, H, H	C with H, H, H, H
3	trigonal planar	120°	$H_2C{=}CH_2$	$H_2C{=}NH$	$H_2C{=}O$
2	linear	180°	H—C≡C—H		O=C=O

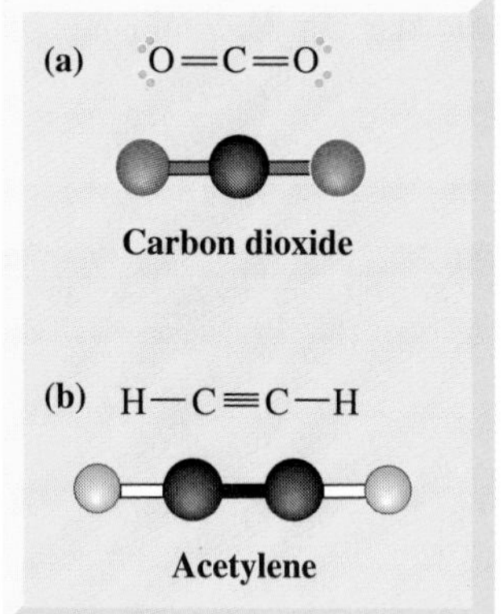

FIGURE 1.6 Shapes of (a) carbon dioxide, CO_2, and (b) acetylene, C_2H_2, molecules.

In still other types of molecules, a central atom is surrounded by only two regions of electron density. Shown in Figure 1.6 are Lewis structures and ball-and-stick models of carbon dioxide, CO_2, and acetylene, C_2H_2.

In carbon dioxide, carbon is surrounded by two regions of electron density: each contains two pairs of electrons and forms a double bond to an oxygen atom. In acetylene, each carbon is also surrounded by two regions of electron density: one contains a single pair of electrons and forms a single bond to a hydrogen atom, and the other contains three pairs of electrons and forms a triple bond to a carbon atom. In each case, the two regions of electron density are farthest apart if they form a straight line through the central atom and create an angle of 180°. Both carbon dioxide and acetylene are **linear** molecules.

Predictions of the VSEPR model are summarized in Table 1.9.

EXAMPLE 1.11

Predict all bond angles in these molecules.

(a) CH_3Cl **(b)** $CH_2{=}CHCl$

Solution

(a) The Lewis structure for CH_3Cl shows carbon surrounded by four regions of electron density. Therefore, predict the distribution of electron pairs about carbon to be tetrahedral, all bond angles to be 109.5°, and the shape of CH_3Cl to be tetrahedral. The actual H—C—Cl bond angle is 110°.

H—C—Cl: (with H above and H below C)   110° (H—C—Cl angle in the three-dimensional structure)

(b) The Lewis structure for $CH_2{=}CHCl$ shows each carbon surrounded by three regions of electron density. Therefore, predict all bond angles to be 120°.

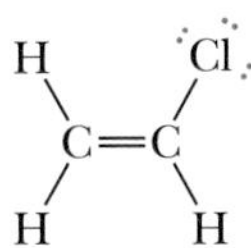

PROBLEM 1.11

Predict all bond angles for these molecules.

(a) CH_3OH **(b)** PF_3 **(c)** H_2CO_3

CHEMISTRY IN ACTION

Buckyball: A New Form of Carbon

A favorite chemistry question is: What are the elemental forms of carbon? The usual answer is that pure carbon is found in two forms: graphite and diamond. These forms have been known for centuries, and it was generally believed that they are the only forms of carbon having extended networks of C atoms in well-defined structures.

But that is not so! The scientific world was startled in 1985 when Richard Smalley of Rice University and Harry W. Kroto of the University of Sussex, UK, and their coworkers announced that they had detected a new form of carbon with a molecular formula C_{60}. They suggested that the molecule has a structure that resembles a soccer ball; it has 12 five-membered rings

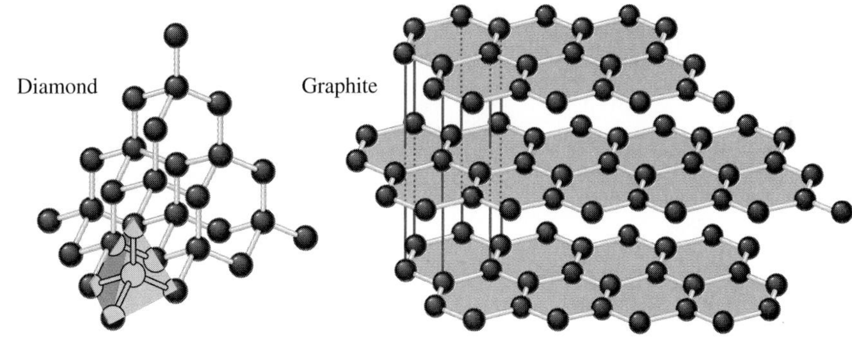

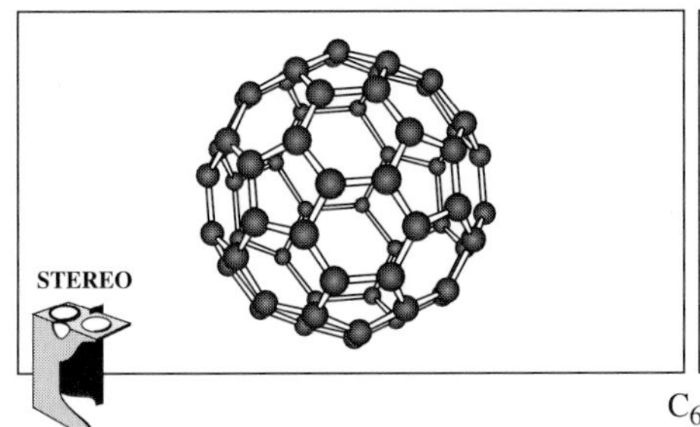

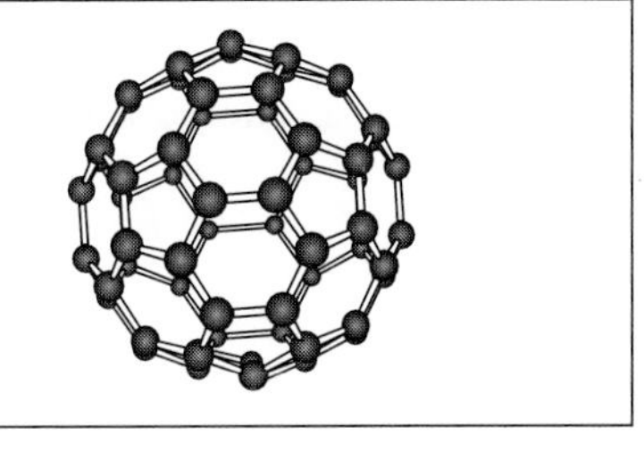

C_{60}

(continued)

and 20 six-membered rings arranged such that each five-membered ring is completely surrounded by six-membered rings. This structure reminded its discoverers of a geodesic dome, a structure invented by the innovative American engineer and philosopher R. Buckminster Fuller. Therefore, the official name of this allotrope of carbon has become buckminsterfullerene or more simply, "fullerene," and many chemists refer to C_{60} simply as "buckyball." Kroto, Smalley, and Robert F. Curl were awarded the Nobel Prize for chemistry in 1996 for this work. Many higher fullerenes, such as C_{70} and C_{84}, have also been isolated and studied. They have a rich chemistry.

Fullerenes behave like electron-deficient alkenes (Chapter 6) and are excellent dienophiles (Chapter 22). C_{60} and C_{70} are also excellent photosensitizers, and produce singlet oxygen with high efficiency (see the Chemistry in Action box "Singlet Oxygen"). Many different fullerene adducts with a variety of structures have been prepared. Cationic derivatives of these adducts, for example, bind tightly to DNA and can be used to visualize it by electron microscopy.

1.5 Polar and Nonpolar Molecules

Dipole moment The vector sum of individual bond moments in a molecule; given in Debye units, D.

We can now combine our understanding of bond dipoles (Section 1.2C) and molecular geometry (Section 1.4) to predict the polarity of polyatomic molecules. To predict whether a molecule is polar, we need to determine (1) if the molecule has polar bonds and (2) the arrangements of these bonds in space. The **dipole moment (μ)** of a molecule is the vector sum of its individual bond dipoles. In carbon dioxide, for example, each C—O bond is polar with oxygen, the more electronegative atom, bearing a partial negative charge and carbon bearing a partial positive charge. Because carbon dioxide is a linear molecule, the vector sum of its two bond dipoles is zero and, hence, the dipole moment of the CO_2 molecule is zero. Boron trifluoride is planar with bond angles of 120°. Although each B—F bond is polar, the vector sum of their bond dipoles is zero and BF_3 has no dipole moment. Carbon tetrachloride is tetrahedral with bond angles of 109.5°. Although this molecule has four polar C—Cl bonds, the vector sum of their bond dipoles is zero and CCl_4 has no dipole moment.

Carbon dioxide $\mu = 0$ D   Boron trifluoride $\mu = 0$ D   Carbon tetrachloride $\mu = 0$ D

Other molecules, such as water and ammonia, have polar bonds and dipole moments greater than zero; that is, they are polar molecules. Each O—H bond in a water molecule and each N—H bond in ammonia is polar with oxygen and nitrogen, the more electronegative atoms, bearing a partial negative charge and each hydrogen bearing a partial positive charge. The dipole moment of a water is 1.85 D and that of ammonia 1.47 D.

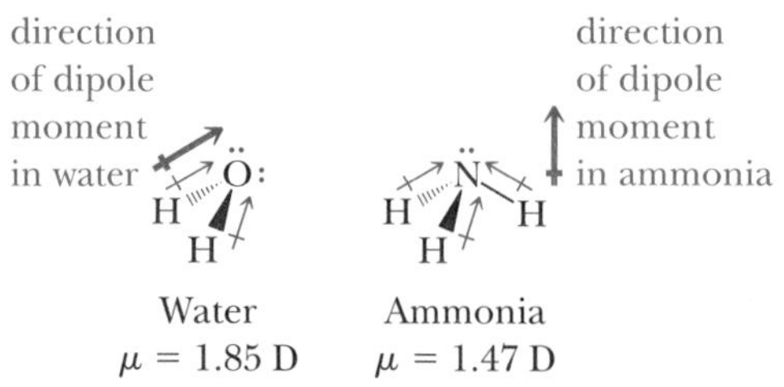

Water $\mu = 1.85$ D   Ammonia $\mu = 1.47$ D

EXAMPLE 1.12

Which of these molecules are polar? For each molecule that is, specify the direction of its dipole moment.

(a) CH_3Cl **(b)** CH_2O **(c)** C_2H_2

Solution

Both chloromethane, CH_3Cl, and formaldehyde, CH_2O, have polar bonds and, because of their geometry, are polar molecules. Because of its linear geometry, acetylene, C_2H_2, has no dipole moment.

(a) Chloromethane $\mu = 1.87$ D **(b)** Formaldehyde $\mu = 2.33$ D **(c)** H—C≡C—H Acetylene $\mu = 0$ D

PROBLEM 1.12

Which molecules are polar? For each molecule that is, specify the direction of its dipole moment.

(a) CH_2Cl_2 **(b)** HCN **(c)** H_2O_2

1.6 Resonance

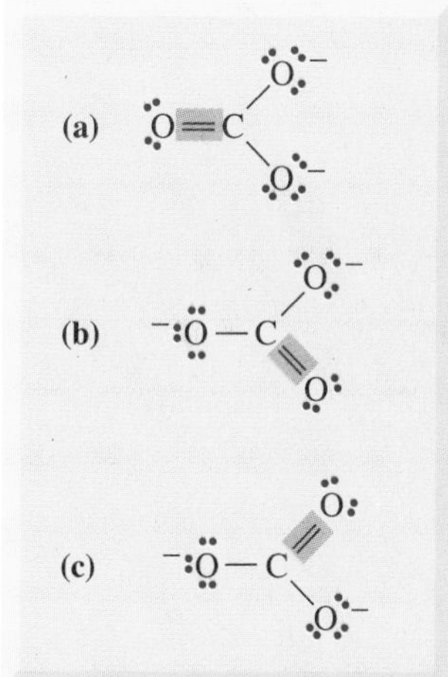

FIGURE 1.7
Three Lewis structures for the carbonate ion.

As chemists developed more understanding of covalent bonding in organic compounds, it became obvious that, for a great many molecules and ions, no single Lewis structure provides a truly accurate representation. For example, Figure 1.7 shows three Lewis structures for the carbonate ion, CO_3^{2-}, each of which shows carbon bonded to three oxygen atoms by a combination of one double bond and two single bonds. Each Lewis structure implies that one carbon-oxygen bond is different from the other two. However, this is not the case. All three carbon-oxygen bonds are identical.

To describe molecules and ions for which no single Lewis structure is adequate, we turn to the theory of resonance.

A. The Theory of Resonance

The **theory of resonance** was developed primarily by Linus Pauling in the 1930s. According to this theory, many molecules and ions are best described by writing two or more Lewis structures and considering the real molecule or ion to be a composite of these structures. Individual Lewis structures are called **contributing structures.** We show that the real molecule or ion is a **resonance hybrid** of the various contributing structures by interconnecting them with **double-headed arrows.** Do not confuse the double-headed arrow with the double arrow used to show chemical equilibrium. As we will explain shortly, resonance structures are not in equilibrium with each other.

Contributing structures Representations of a molecule or ion that differ only in the distribution of valence electrons.

Resonance hybrid A molecule or ion described as a composite of a number of contributing structures.

Double-headed arrow A symbol used to connect contributing structures.

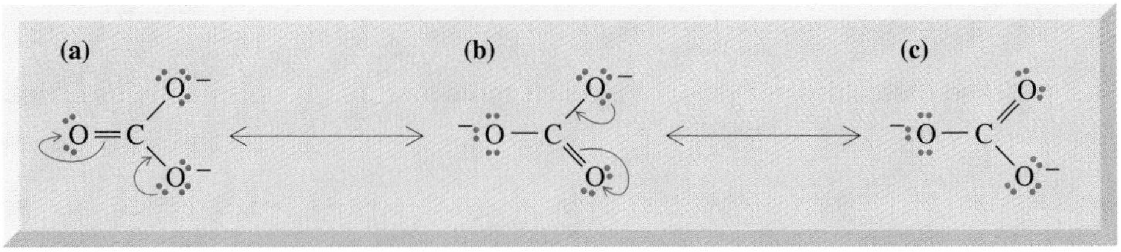

FIGURE 1.8
The carbonate ion represented as a resonance hybrid of three equivalent contributing structures. Curved arrows show the redistribution of valence electrons between one contributing structure and the next.

Three contributing structures for the carbonate ion are shown in Figure 1.8. These three contributing structures are said to be equivalent. **Equivalent contributing structures** have identical patterns of covalent bonding and are of equal energy.

The use of the term "resonance" for this theory of covalent bonding might suggest to you that bonds and electron pairs are constantly changing back and forth from one position to another over time. This notion is not at all correct. The carbonate ion, for example, has one and only one real structure. The problem is ours—how do we draw that one real structure? The resonance method is a way to describe the real structure and at the same time retain Lewis structures with electron-pair bonds. Thus, although we realize that the carbonate ion is not accurately represented by any one contributing structure shown in Figure 1.8, we continue to represent it by one of these for convenience. We understand, of course, that what is intended is the resonance hybrid.

B. Curved Arrows and Electron Pushing

Notice in Figure 1.8 that the only change from contributing structure (a) to (b) and then from (b) to (c) is a redistribution of valence electrons. To show how this redistribution of valence electrons occurs, chemists use a symbol called a curved arrow. A **curved arrow** shows the repositioning of an electron pair from its origin (the tail of the arrow) to its destination (the head of the arrow). This repositioning may be from an atom to an adjacent bond, or from a bond to an adjacent atom.

Curved arrow A symbol used to show the redistribution of valence electrons.

A curved arrow is nothing more than a bookkeeping symbol for keeping track of electron pairs, or, as some call it, **electron pushing.** Do not be misled by its simplicity. Electron pushing will help you see the relationship between contributing structures. Furthermore, it will help you follow bond-breaking and bond-forming steps in organic reactions. Stated even more directly, electron pushing is a survival skill in organic chemistry.

Following are contributing structures for the nitrite and ethanoate (acetate) ions. Curved arrows are used to show how the contributing structure on the left is converted to the one on the right. For each ion, the contributing structures are equivalent.

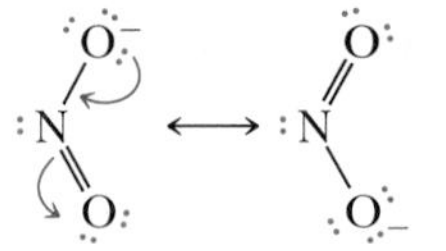

Nitrite ion
(equivalent contributing structures)

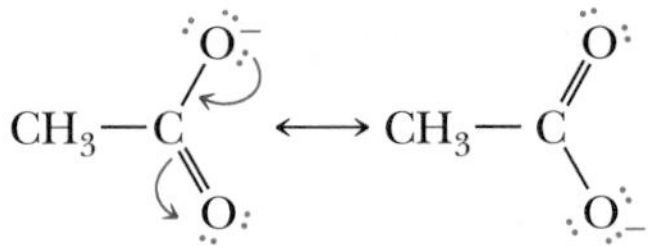

Ethanoate ion
(Acetate ion)
(equivalent contributing structures)

Following are contributing structures for the resonance hybrid of propanone (acetone). These contributing structures are nonequivalent; they have different patterns of covalent bonding and different energies.

$$H_3C-C(=\ddot{O}:)-CH_3 \longleftrightarrow H_3C-C^{+}(-\ddot{O}:^{-})-CH_3$$

Propanone
(Acetone)
(nonequivalent contributing structures)

EXAMPLE 1.13

Draw the contributing structure indicated by the curved arrows. Be certain to show all valence electrons and all formal charges.

(a) $CH_3-C(=O)-H \longleftrightarrow$ (b) $H-\overset{-}{C}H-C(=O)-H \longleftrightarrow$ (c) $CH_3-\ddot{O}-\overset{+}{C}H-H \longleftrightarrow$

Solution

(a) $CH_3-\overset{+}{C}(-\ddot{O}:^{-})-H$ (b) $H-CH=C(-\ddot{O}:^{-})-H$ (c) $CH_3-\overset{+}{\ddot{O}}=CH-H$

PROBLEM 1.13

Draw the contributing structure indicated by the curved arrows. Be certain to show all valence electrons and all formal charges.

(a) $H-C(=O)-\ddot{O}:^{-} \longleftrightarrow$ (b) $H-C(=O)-\ddot{O}:^{-} \longleftrightarrow$ (c) $CH_3-C(=O)-\ddot{O}-CH_3 \longleftrightarrow$

C. Rules for Writing Acceptable Contributing Structures

Certain rules must be followed in writing acceptable contributing structures:

1. All contributing structures must have the same number of valence electrons.
2. All contributing structures must obey the rules of covalent bonding; no contributing structure may have more than two electrons in the valence shell of hydrogen or more than eight electrons in the valence shell of a second-period element. Third-period elements, such as phosphorus and sulfur, may have up to 12 electrons in their valence shells.
3. The positions of all nuclei must be the same; that is, contributing structures differ only in the distribution of valence electrons.

4. All contributing structures must show the same number of paired and unpaired electrons.

EXAMPLE 1.14

Which sets are pairs of contributing structures?

(a) $CH_3-C(=\ddot{O}:)-CH_3 \longleftrightarrow CH_3-C^{+}(-:\ddot{O}:^{-})-CH_3$ (b) $CH_3-C(=\ddot{O}:)-CH_3 \longleftrightarrow CH_2=C(-:\ddot{O}-H)-CH_3$

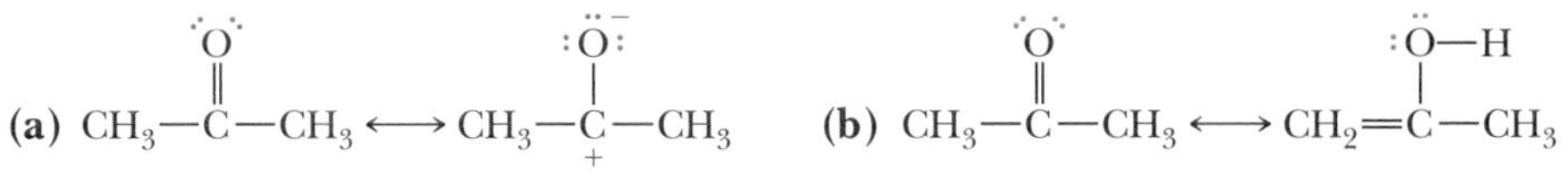

Solution

(a) Contributing structures. They differ only in the distribution of valence electrons.
(b) Not a set of contributing structures. They differ in the arrangement of their atoms.

PROBLEM 1.14

Which sets are pairs of contributing structures?

(a) $CH_3-C(=\ddot{O}:)(-:\ddot{O}:^{-}) \longleftrightarrow CH_3-C^{+}(-:\ddot{O}:^{-})(-:\ddot{O}:^{-})$ (b) $CH_3-C(=\ddot{O}:)(-:\ddot{O}:^{-}) \longleftrightarrow CH_3-\overset{-}{C}(=\ddot{O}:)(=\ddot{O}:)$

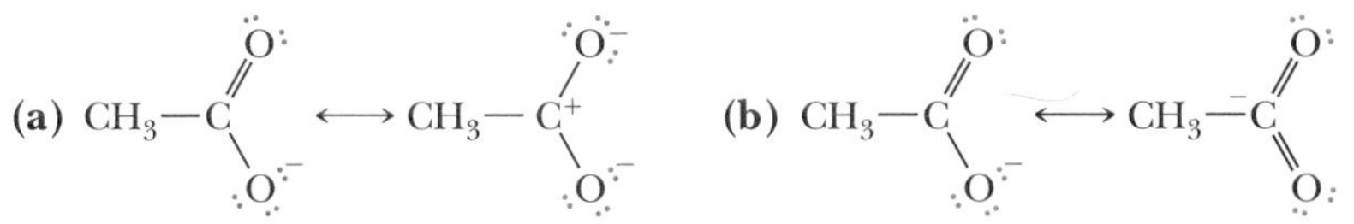

D. Estimating the Relative Importance of Contributing Structures

Not all structures contribute equally to a hybrid. The following preferences will help you to estimate the relative importance of various contributing structures. In fact, structures can be ranked by the number of preferences they follow. Those that follow the most preferences contribute most to the hybrid. Any structure that violates all four of these preferences can be ignored and never written.

Preference 1: Filled Valence Shells

Structures in which all atoms have filled valence shells (completed octets) contribute more than those in which one or more valence shells are unfilled. For example, the following are the contributing structures for Example 1.13(c) and its solution.

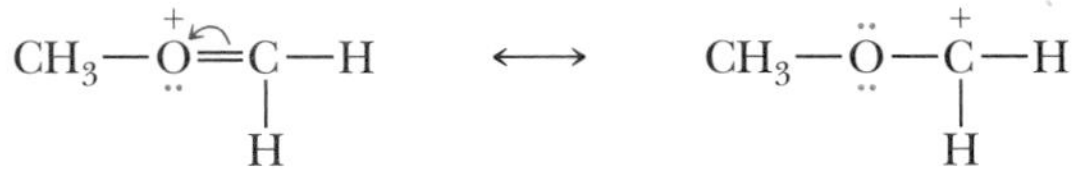

$$CH_3-\overset{+}{\ddot{O}}=CH-H \longleftrightarrow CH_3-\ddot{O}-\overset{+}{C}H-H$$

Greater contribution: both carbon and oxygen have complete valence shells

Lesser contribution: carbon has only 6 electrons in its valence shell

Preference 2: Maximum Number of Covalent Bonds

Structures with a greater number of covalent bonds contribute more than those with fewer covalent bonds. In the illustration for rule 1, the structure on the left has five covalent bonds in the $-OCH_2$ part of the molecule and makes the greater contribution to the hybrid. The structure on the right has only four covalent bonds in this part of the molecule.

Preference 3: Least Separation of Unlike Charges

Structures involving separation of unlike charges contribute less than those that do not involve charge separation.

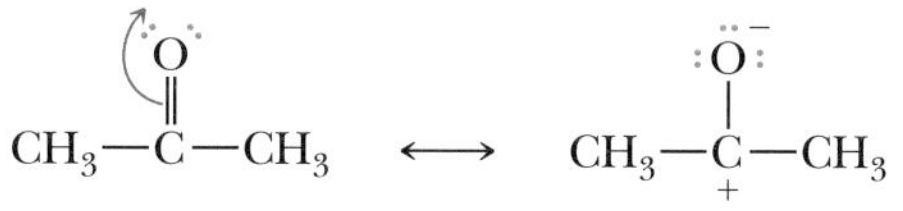

$$CH_3{-}\overset{\overset{\large O}{\|}}{C}{-}CH_3 \longleftrightarrow CH_3{-}\underset{+}{\overset{\overset{\large :\ddot{O}:^-}{|}}{C}}{-}CH_3$$

Greater contribution: no separation of unlike charge

Lesser contribution: separation of unlike charges

Preference 4: Negative Charge on a More Electronegative Atom

Structures that carry a negative charge on a more electronegative atom contribute more than those with a negative charge on a less electronegative atom. Conversely, structures that carry a positive charge on a less electronegative atom contribute more than those that carry a positive charge on a more electronegative atom. Following are three contributing structures for propanone:

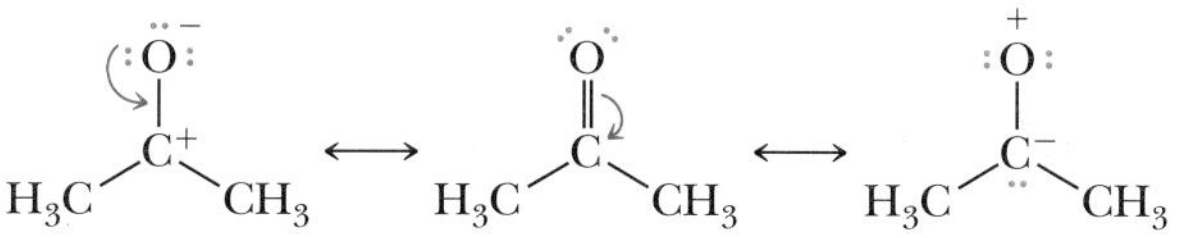

$$\underset{\text{(a) Lesser contribution}}{H_3C{-}\overset{\overset{:\ddot{O}:^-}{|}}{C^+}{-}CH_3} \longleftrightarrow \underset{\text{(b) Greater contribution}}{H_3C{-}\overset{\overset{\ddot{O}}{\|}}{C}{-}CH_3} \longleftrightarrow \underset{\text{(c) Can be ignored}}{H_3C{-}\overset{\overset{:O:^+}{|}}{\ddot{C}^-}{-}CH_3}$$

Structure (b) makes the largest contribution to the hybrid. Structure (a) contributes less because it involves separation of unlike charges. Structure (c) violates all four preference rules and can be ignored.

EXAMPLE 1.15

Estimate the relative contribution of the members in each set of contributing structures.

$$\textbf{(a)}\ H{-}\underset{\underset{H}{|}}{\ddot{C}^-}{-}\overset{\overset{\ddot{O}}{\|}}{C}{-}H \longleftrightarrow H{-}\underset{\underset{H}{|}}{C}{=}\overset{\overset{:\ddot{O}:^-}{|}}{C}{-}H \qquad \textbf{(b)}\ H{-}\overset{\overset{\ddot{O}}{\|}}{C}{-}\ddot{O}:^- \longleftrightarrow H{-}\overset{\overset{:\ddot{O}:^-}{|}}{C}{=}\ddot{O}$$

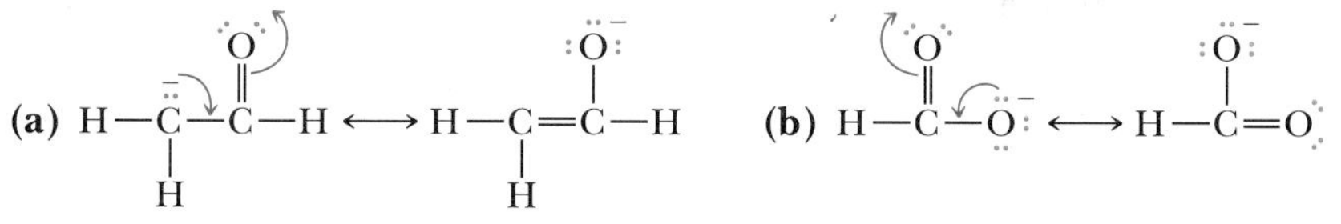

Solution

(a) The structure on the right makes a greater contribution to the hybrid because it places the negative charge on oxygen, the more electronegative atom.

(b) The structures are equivalent and make equal contributions to the hybrid.

PROBLEM 1.15

Estimate the relative contribution of the members in each set.

(a) $H_2C{=}CH_2 \longleftrightarrow H_2\overset{+}{C}{-}\overset{-}{C}H_2$

(b) $H{-}\underset{}{C}(=O){-}O{-}H \longleftrightarrow H{-}C(-O^{-}){=}\overset{+}{O}{-}H$

1.7 Quantum or Wave Mechanics

Thus far in this chapter, we have concentrated on the Lewis model of bonding and on the VSEPR model. The Lewis model deals primarily with coordination numbers of atoms (the number of bonds a given atom can form), and the VSEPR model deals primarily with bond angles and molecular geometries. Although both models are useful, each in its own way, neither gives us any means of accounting in a quantitative or even semiquantitative way for why atoms combine in the first place to form covalent bonds with the liberation of energy. At this point we need to study an entirely new approach to the theory of covalent bonding, one that provides a means of understanding not only the coordination numbers of atoms and molecular geometries but also the energetics of chemical bonding.

A. Moving Particles Exhibit the Properties of a Wave

Einstein received a Ph.D. in 1905, and in that same year, he published a paper in which he used his photon concept to explain the photoelectric effect. For this work he was awarded the Nobel Prize for physics in 1921.

The beginning of this new approach to the theory of chemical bonding was provided by Albert Einstein (1879–1955), a German-born American physicist. In 1905, Einstein postulated that light consists of photons of electromagnetic radiation. The energy, E, of a photon is proportional to the frequency, ν (Greek: nu), of the light. The proportionality constant in this equation is Planck's constant, h.

$$E = h\nu$$

de Broglie received a doctoral degree in physics from the University of Paris in 1924, and, for his research on the wave-particle theory of electromagnetic radiation, received the Nobel Prize for physics in 1929.

In 1923, a French physicist Louis de Broglie followed Einstein's lead and advanced the revolutionary idea that if light exhibits properties of a particle in motion, then a particle in motion should exhibit the properties of a wave. He proposed that a particle of mass m and speed v has an associated wavelength, λ (Greek: lambda), given by the equation

$$\lambda = \frac{h}{mv} \quad \text{(the de Broglie relationship)}$$

Node Any point in space where the value of a wave function is zero.

Illustrated in Figure 1.9 is a wave such as might result from plucking a guitar string. The mathematical equation that describes this wave is called a **wave equation.** The numerical value of the solution of a wave equation may be positive (corresponding to a wave crest), negative (corresponding to a wave trough), or zero. A **node** is any point where the value of the solution of a wave equation is zero. A **nodal plane** is any plane perpendicular to the direction of propagation that runs through a node. Shown in Figure 1.9 are three nodal planes.

Erwin Schrödinger built on the idea of de Broglie and in 1926 proposed an equation that could be used to describe the wave properties associated with an electron in

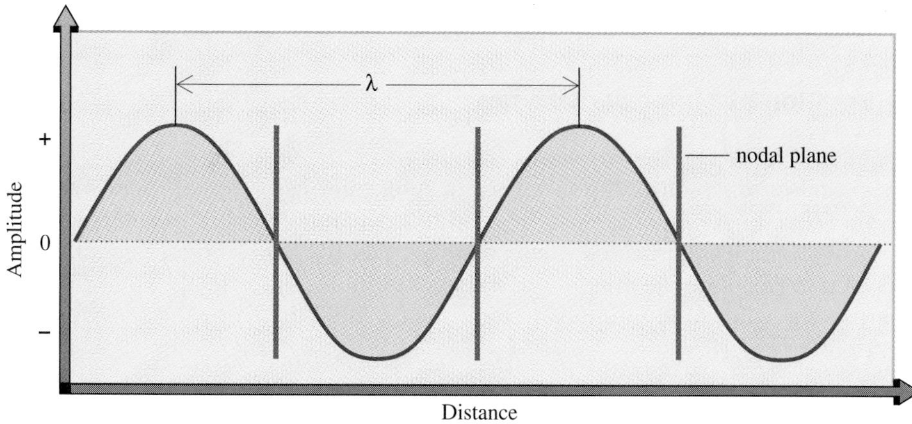

FIGURE 1.9
Characteristics of a wave associated with a moving particle. Wavelength is given the symbol λ.

an atom or a molecule. **Quantum mechanics (wave mechanics)** is the branch of science that studies particles and their associated waves.

Quantum mechanics A branch of science that studies particles and their associated waves.

Solving the Schrödinger equation gives a set of solutions called **wave functions.** Each wave function ψ (Greek: psi) is associated with a unique set of quantum numbers and with a particular atomic orbital. The value of ψ^2 is proportional to the probability of finding an electron at a given point in space; or looked at in another way, the value of ψ^2 at any point in space is proportional to the electron density at that point. Electron density theoretically reaches to infinity, but actually is vanishingly low at long distances from the nucleus.

In this course, we concentrate on wave functions and shapes associated with *s* and *p* atomic orbitals because they are the orbitals most often involved in covalent bonding in organic compounds.

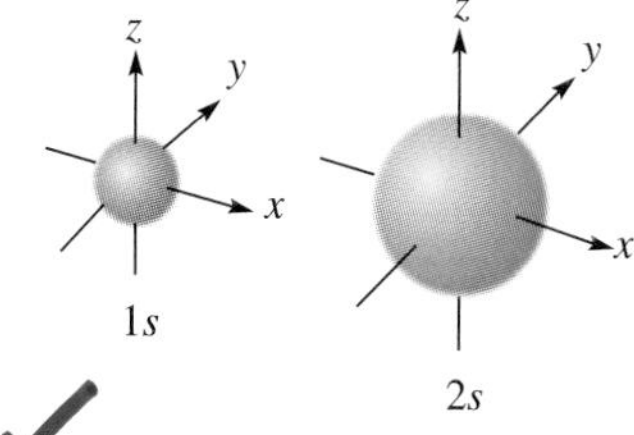

FIGURE 1.10
Probability distribution for 1*s* and 2*s* orbitals.

B. Shapes of Atomic *s* and *p* Orbitals

All *s* orbitals have the shape of a sphere, with the center of the sphere at the nucleus. Shown in Figure 1.10 are probability distributions for 1*s* and 2*s* orbitals.

Shown in Figure 1.11 are three-dimensional shapes of the three 2*p* orbitals, combined in one diagram to illustrate their relative orientations in space. Each 2*p* orbital consists of two lobes arranged in a straight line with the nucleus in the middle. The three 2*p* orbitals are mutually perpendicular and are designated $2p_x$, $2p_y$, and $2p_z$. The sign of the wave function of a 2*p* orbital is positive in one lobe, zero at the nucleus, and negative in the other lobe. Because the value of ψ^2 is always positive, the probability of finding an electron in the (+) lobe of a 2*p* orbital is the same as that of finding it in the (−) lobe.

Besides providing a way to determine the shapes of atomic orbitals, the Schrödinger equation also provides a way, at least in principle, to quantify the energetics of covalent bond formation. These approximations have taken two forms: (1) the valence bond model, and (2) the molecular orbital model. Both models for chemical bonding use the methods of quantum mechanics but each makes slightly different simplifying assumptions. At sufficiently high levels of theory, both models converge. We concentrate our discussion on the molecular orbital model.

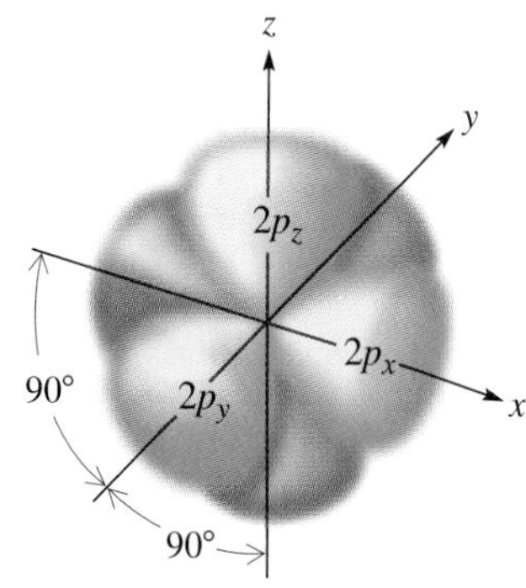

FIGURE 1.11
The 2*p* orbitals. Three-dimensional shapes of $2p_x$, $2p_y$, and $2p_z$ atomic orbitals and their orientation in space relative to one another. The three 2*p* orbitals are mutually perpendicular.

1.8 Molecular Orbital Theory of Covalent Bonding

A. Formation of Molecular Orbitals

Molecular orbital (MO) theory begins with the fact that electrons in atoms exist in **atomic orbitals** and assumes that electrons in molecules exist in **molecular orbitals.** Just as the Schrödinger equation can be used to calculate the energies and shapes of atomic orbitals, molecular orbital theory assumes that the Schrödinger equation can also be used to calculate the energies and shapes of molecular orbitals. Following is a summary of the rules used in applying molecular orbital theory to the formation of covalent bonds.

1. Combination of n atomic orbitals forms a set of n molecular orbitals. The number of molecular orbitals formed is equal to the number of atomic orbitals combined.
2. Just like atomic orbitals, molecular orbitals are arranged in order of increasing energy. It is possible to calculate the relative energies of a set of molecular orbitals with reasonable accuracy. Experimental measurements such as those derived from molecular spectroscopy can also be used to provide very detailed information about relative energies of molecular orbitals.
3. Filling of molecular orbitals is governed by the same principles as the filling of atomic orbitals. Molecular orbitals are filled beginning with the lowest unoccupied molecular orbital (the Aufbau principle). A molecular orbital can accommodate no more than two electrons, and the electrons must have opposite spins (the Pauli exclusion principle). When two or more equivalent molecular orbitals are available, one electron is added to each before any equivalent orbital is filled with two electrons (Hund's rule).

To illustrate the formation of molecular orbitals, consider the shapes and relative energies of the molecular orbitals arising from combination of two 1*s* atomic orbitals. Combination by addition of their wave functions gives the molecular orbital shown in Figure 1.12(a). When electrons occupy this MO, electron density is concentrated in the region between the two positively charged nuclei and serves to offset the repulsive interaction between them. The molecular orbital we have just described is called a sigma bonding molecular orbital and is given the symbol σ_{1s} (pronounced sigma one 'ess'). A **bonding molecular orbital** is one in which electrons have a lower energy than they would in the isolated atomic orbitals. A **sigma bonding molecular orbital** is one in which electron density lies between the two nuclei, along the axis joining them, and is cylindrically symmetric about the axis.

Bonding molecular orbital A molecular orbital in which electrons have a lower energy than they would in isolated atomic orbitals.

Sigma molecular orbital A molecular orbital in which electron density is concentrated between two nuclei along the axis joining them.

Antibonding molecular orbital A molecular orbital in which electrons have a higher energy than they would in isolated atomic orbitals.

Combination of two 1*s* atomic orbitals by subtraction of their wave functions gives the molecular orbital shown in Figure 1.12(b). If electrons occupy this MO, electron density is concentrated outside the region between the two nuclei, and consequently there is little or no electron density between nuclei to offset nuclear repulsion. This molecular orbital is called a sigma antibonding molecular orbital and is given the symbol σ^*_{1s} (pronounced sigma star one 'ess'). An **antibonding molecular orbital** is one in which electrons have a higher energy than they would in the isolated atomic orbitals. An asterisk shows that a molecular orbital is antibonding.

In representations of orbitals, we use blue to indicate the lobe of a particular orbital in which the sign of the wave function is negative, and red where it is positive.

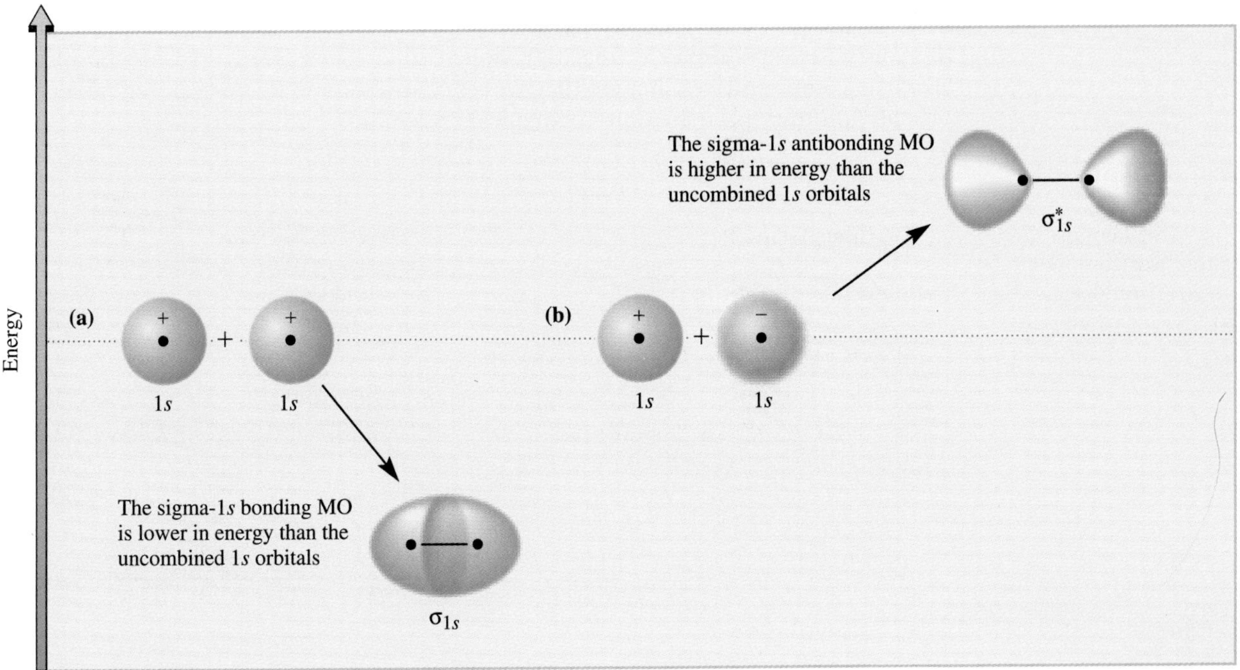

FIGURE 1.12
Molecular orbitals derived from combination of two 1*s* atomic orbitals: (a) combination by addition and (b) combination by subtraction. Electrons in the bonding MO spend most of their time in the region between the two nuclei to bond the atoms together. Electrons in the antibonding MO do not contribute to bonding.

The **ground state** of an atom or molecule is its state of lowest energy. In the ground state of the hydrogen molecule, the two electrons occupy the σ_{1s} MO with paired spins. An **excited state** is any electronic state other than the ground state. In the lowest excited state of the hydrogen molecule, one electron occupies the σ_{1s} MO and the other occupies the σ^*_{1s} MO. There is no net bonding in this excited state and dissociation will result from the electrostatic repulsion of the two hydrogen nuclei. Energy-level diagrams for the ground state and the lowest excited state of the hydrogen molecule are shown in Figure 1.13.

Combination of two 2*s* atomic orbitals produces two sigma molecular orbitals, designated σ_{2s} and σ^*_{2s}, that are similar in shape and relative energies to the sigma 1*s* molecular orbitals illustrated in Figure 1.13.

Next let us consider the molecular orbitals formed by the combination of 2*p* orbitals. According to MO theory, two 2*p* atomic orbitals can overlap end-on to form a sigma bonding and a sigma antibonding molecular orbital. We will not encounter these kinds of MOs in this course. What we encounter is a combination of parallel 2*p* atomic orbitals by addition of their wave functions to give the pi bonding (π_{2p}) molecular orbital shown in Figure 1.14(a). Combination of their wave functions by subtraction gives the pi antibonding (π^*_{2p}) molecular orbital shown in Figure 1.14(b). In a **pi MO,** electron density is concentrated not along the axis joining the bonded nuclei, but rather in two regions, one above and the other below the bond axis. We will have little need to refer to antibonding molecular orbitals throughout the remainder of the course, except when we treat ultraviolet-visible spectroscopy in Chapter 14 and aromatic compounds in Chapter 19. Our concentration will be on bonding molecular orbitals and their participation in chemical reactions.

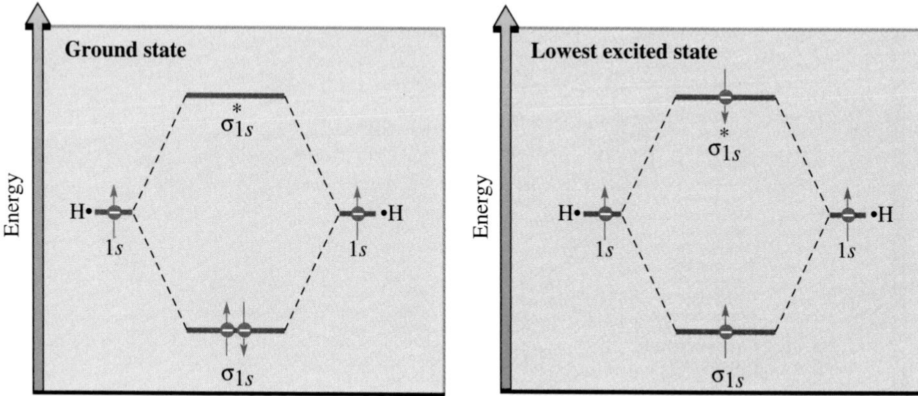

FIGURE 1.13
A molecular orbital energy diagram for the hydrogen molecule, H_2. (*left*) Ground state and (*right*) lowest excited state for H_2.

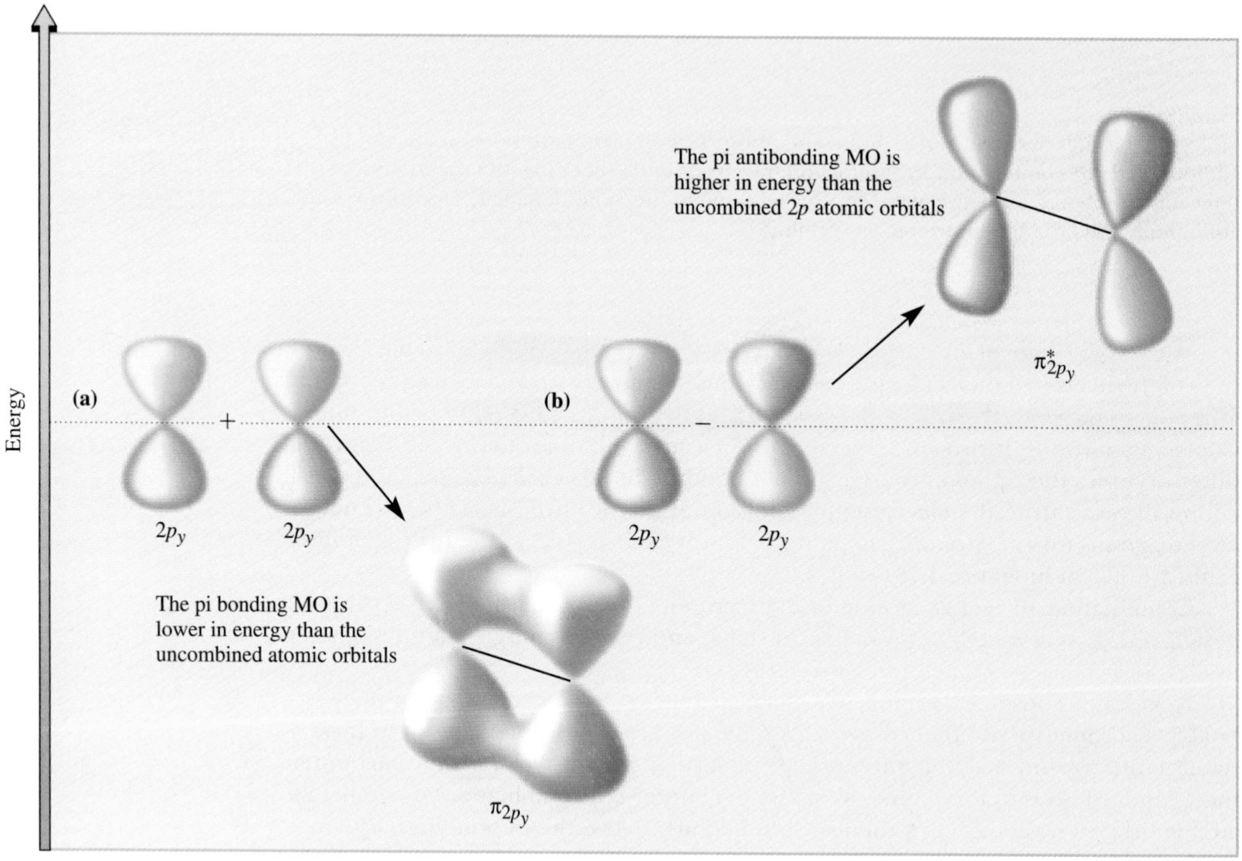

FIGURE 1.14
Molecular orbitals formed by combination of parallel 2*p* orbitals: (a) combination by addition and (b) combination by subtraction.

B. Hybridization of Atomic Orbitals

Erwin Schrödinger (1887–1961) wrote papers that gave the foundations for quantum mechanics. He shared the Nobel Prize for physics in 1933. *(The Bettmann Archive)*

Formation of a covalent bond between two hydrogen atoms is straightforward. Formation of covalent bonds between atoms of second-row elements, however, presents a new problem. In forming covalent bonds, atoms of carbon, nitrogen, and oxygen (all second-row elements) use 2*s* and 2*p* atomic orbitals. The three 2*p* atomic orbitals are at angles of 90° to each other (Figure 1.11), and, if atoms of second-row elements used these orbitals to form covalent bonds, bond angles around each would be approximately 90°. However, bond angles of 90° are only rarely observed in organic molecules. What we find instead are bond angles of approximately 109.5° in molecules with only single bonds, 120° in molecules with double bonds, and 180° in molecules with triple bonds.

109.5°   120° (approx.)   180°

CH_4 $H_2C{=}CH_2$ H—C≡C—H

To account for these observed bond angles, Pauling proposed that atomic orbitals combine to form new orbitals, called **hybrid orbitals,** which then interact to form bonds with the angles that we do observe. Hybrid orbitals are formed by combinations of atomic orbitals; the number of hybrid orbitals formed is equal to the number of atomic orbitals combined. Elements of the second row form three types of hybrid orbitals, designated sp^3, sp^2, and *sp*, each of which can contain up to two electrons. Let us see how hybrid orbitals are formed, how they are oriented in space, and how they account for the bond angles observed in organic molecules.

Hybrid orbital An orbital formed by the combination of two or more atomic orbitals.

C. sp^3 Hybrid Orbitals: Bond Angles of Approximately 109.5°

The combination of one 2*s* atomic orbital and three 2*p* atomic orbitals forms four equivalent **sp^3 hybrid orbitals.** Because they are derived from four atomic orbitals, sp^3 hybrid orbitals always occur in sets of four. Each sp^3 hybrid orbital consists of a larger lobe pointing in one direction and a smaller lobe of opposite sign pointing in the opposite direction. The axes of the four sp^3 hybrid orbitals are directed toward the corners of a regular tetrahedron, and sp^3 hybridization results in bond angles of approximately 109.5° (Figure 1.15).

sp^3 Hybrid orbital A hybrid atomic orbital formed by the combination of one *s* atomic orbital and three *p* atomic orbitals.

You must remember that superscripts in the designation of hybrid orbitals tell you how many atomic orbitals have been combined to form the hybrid orbitals. You know the designation sp^3 represents a hybrid orbital because it shows a combination of *s* and *p* orbitals. The superscripts in this case tell you *one s* atomic orbital and *three p* atomic orbitals are combined in forming the hybrid orbital. Do not confuse this use of superscripts with that used in writing a ground-state electron configuration, as for example $1s^2\ 2s^2\ 2p^5$ for fluorine. In the case of a ground-state electron configuration, superscripts tell you the number of electrons in each orbital or set of orbitals.

In Section 1.2, we described the covalent bonding in CH_4, NH_3, and H_2O in terms of the Lewis model, and in Section 1.4, we used the VSEPR model to predict bond angles of approximately 109.5° in each molecule. Now let us consider the bonding in these molecules in terms of the overlap of atomic orbitals. To bond with four other

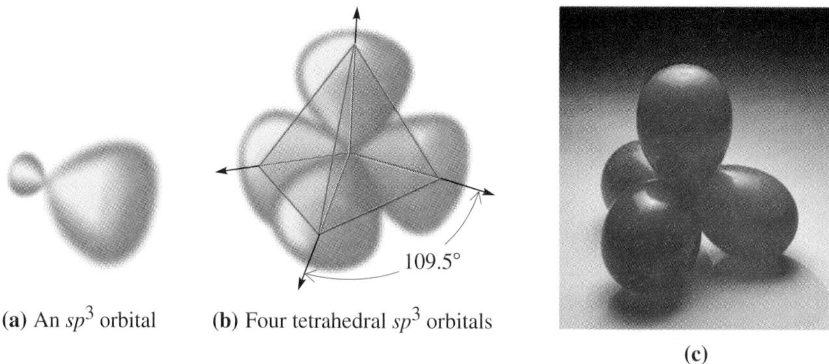

FIGURE 1.15
sp^3 Hybrid orbitals. (a) Representation of a single sp^3 hybrid orbital showing two lobes of unequal size. The sign of the wave function is positive in one lobe and negative in the other. (b) Three-dimensional representation of four sp^3 hybrid orbitals directed toward the corners of a regular tetrahedron. The smaller lobe of each sp^3 hybrid orbital is hidden behind the larger lobe. (c) If four balloons of similar size and shape are tied together, they will naturally assume a tetrahedral geometry.

atoms with bond angles of 109.5°, carbon uses sp^3 hybrid orbitals. Carbon has four valence electrons, and one electron is placed in each sp^3 hybrid orbital. Each partially filled sp^3 hybrid orbital then overlaps with a partially filled $1s$ atomic orbital of hydrogen to form a sigma (σ) bond, and hydrogen atoms occupy the corners of a regular tetrahedron (Figure 1.16).

In bonding with three other atoms, the five valence electrons of nitrogen are distributed so that one sp^3 hybrid orbital is filled with a pair of electrons and the other three sp^3 hybrid orbitals have one electron each. Overlapping of these partially filled sp^3 hybrid orbitals with $1s$ atomic orbitals of hydrogens produces the NH_3 molecule (Figure 1.16).

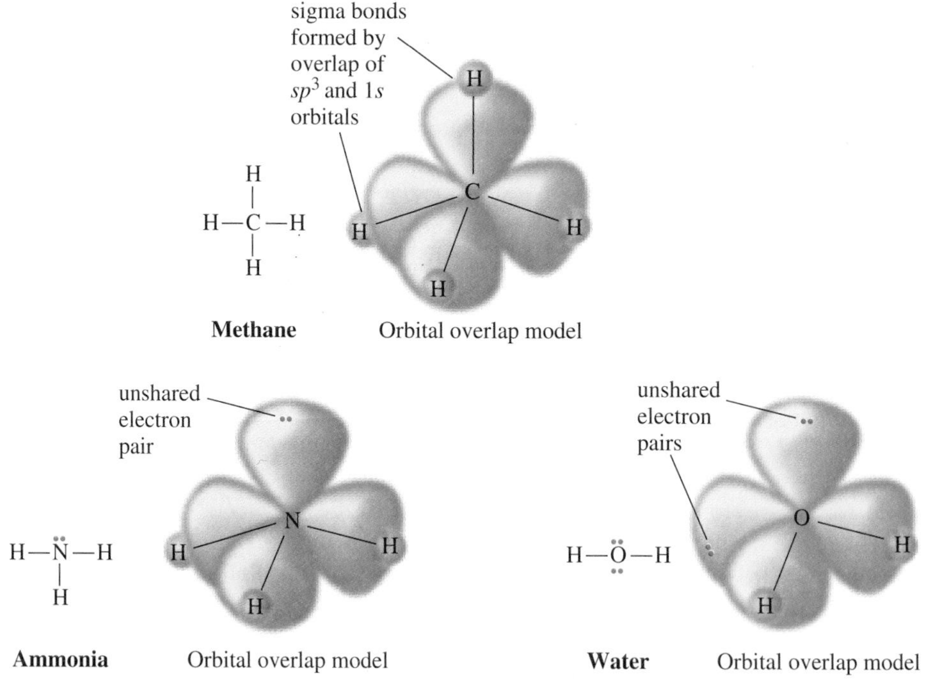

FIGURE 1.16
Orbital overlap pictures of methane, ammonia, and water.

In bonding with two other atoms, the six valence electrons of oxygen are distributed so that two sp^3 hybrid orbitals are filled and the remaining two have one electron each. Each partially filled sp^3 hybrid orbital overlaps with a 1*s* atomic orbital of hydrogen, and hydrogen atoms occupy two corners of a regular tetrahedron. The remaining two corners of the tetrahedron are occupied by unshared pairs of electrons (Figure 1.16).

D. sp^2 Hybrid Orbitals: Bond Angles of Approximately 120°

The combination of one 2*s* atomic orbital and two 2*p* atomic orbitals forms three equivalent **sp^2 hybrid orbitals.** Because they are derived from three atomic orbitals, sp^2 hybrid orbitals always occur in sets of three. Each sp^2 hybrid orbital consists of two lobes, one larger than the other. The axes of the three sp^2 hybrid orbitals lie in a plane and are directed toward the corners of an equilateral triangle; the angle between sp^2 hybrid orbitals is 120°. The third 2*p* atomic orbital (remember $2p_x$, $2p_y$, $2p_z$) is not involved in hybridization and consists of two lobes lying perpendicular to the plane of the sp^2 hybrid orbitals. Figure 1.17 shows three equivalent sp^2 orbitals along with the remaining unhybridized 2*p* atomic orbital.

sp^2 Hybrid orbital A hybrid atomic orbital formed by the combination of one *s* atomic orbital and two *p* atomic orbitals.

Second-row elements use sp^2 hybrid orbitals to form double bonds. Consider ethylene, C_2H_4, a Lewis structure for which is shown in Figure 1.18(a). A sigma bond between the carbons in ethylene is formed by overlap of sp^2 hybrid orbitals along a common axis as seen in Figure 1.18(b). Each carbon also forms sigma bonds to two hydrogens. The remaining 2*p* orbitals on the adjacent carbon atoms lie parallel to each other and overlap to form a pi bond [Figure 1.18(c)]. A **pi (π) bond** is a covalent bond formed by overlap of parallel *p* orbitals. Because of the lesser degree of overlap of orbitals forming pi bonds compared with those forming sigma bonds, pi bonds are generally weaker than are sigma bonds.

Pi (π) bond A covalent bond formed by the overlap of parallel *p* orbitals.

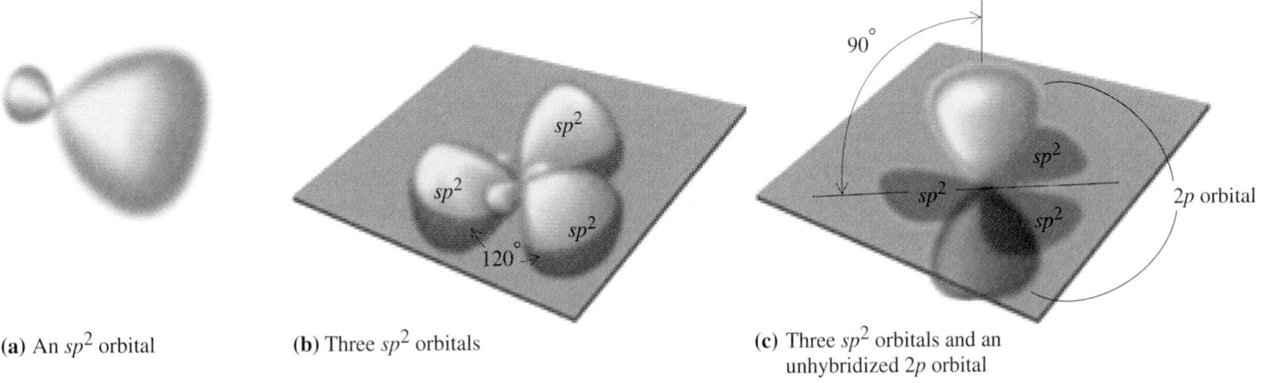

FIGURE 1.17
sp^2 Hybrid orbitals. (a) A single sp^2 hybrid orbital showing two lobes of unequal size. (b) The three sp^2 hybrid orbitals with their axes in a plane at angles of 120°. (c) The unhybridized 2*p* atomic orbital perpendicular to the plane created by the three sp^2 hybrid orbitals.

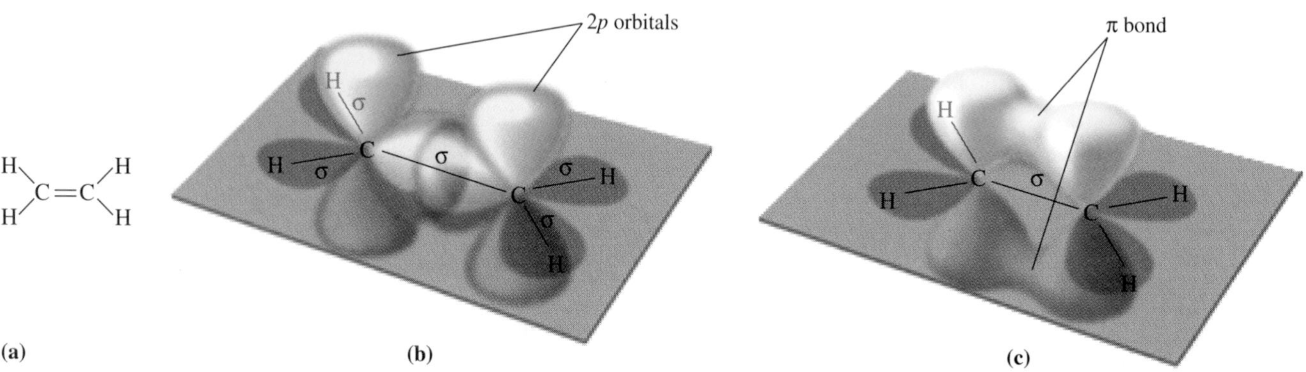

FIGURE 1.18
Covalent bond formation in ethylene. (a) Lewis structure, (b) overlap of sp^2 hybrid orbitals forms a sigma (σ) bond between the carbon atoms, and (c) overlap of parallel $2p$ atomic orbitals forms a pi (π) bond.

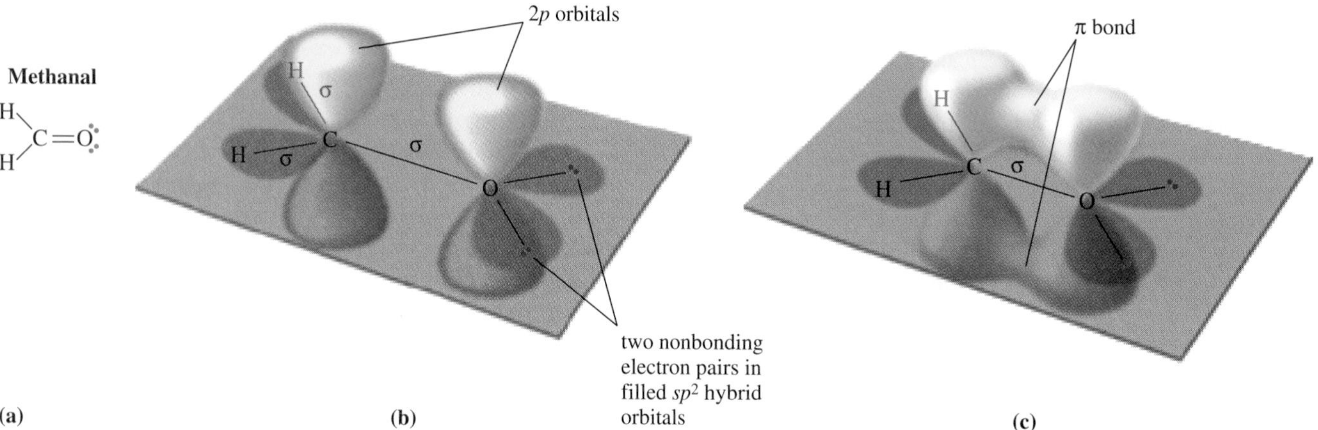

FIGURE 1.19
A carbon-oxygen double bond. (a) Lewis structure of $CH_2{=}O$, (b) the sigma (σ) bond framework and nonoverlapping parallel $2p$ atomic orbitals, and (c) overlap of parallel $2p$ atomic orbitals to form a pi (π) bond.

Molecular orbital theory describes all double bonds in the same manner we have already used to describe carbon-carbon double bonds. In formaldehyde, $CH_2{=}O$, the simplest organic molecule containing a carbon-oxygen double bond, carbon forms sigma bonds to two hydrogens by overlap of sp^2 orbitals of carbon and $1s$ atomic orbitals of hydrogens. Carbon and oxygen are joined by a sigma bond formed by overlap of sp^2 hybrid orbitals and a pi bond formed by overlap of unhybridized $2p$ atomic orbitals (Figure 1.19).

E. *sp* Hybrid Orbitals: Bond Angles of Approximately 180°

***sp* Hybrid orbital** A hybrid atomic orbital formed by the combination of one *s* atomic orbital and one *p* atomic orbital.

Combination of one $2s$ atomic orbital and one $2p$ atomic orbital forms two equivalent ***sp* hybrid orbitals.** Because they are derived from two atomic orbitals, *sp* hybrid orbitals always occur in sets of two. The two *sp* hybrid orbitals lie at an angle of 180°. The axes

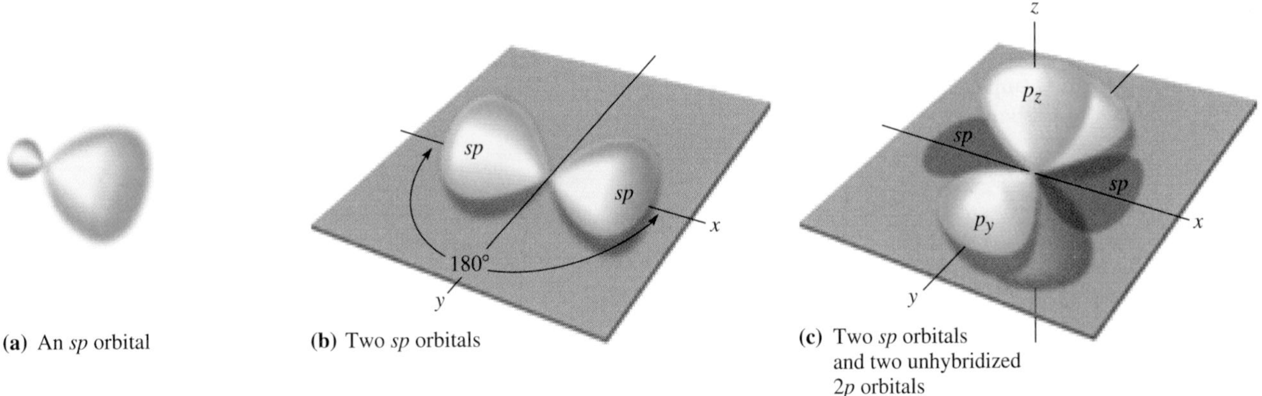

FIGURE 1.20
sp Hybrid orbitals. (a) A single *sp* hybrid orbital consisting of two lobes of unequal size. (b) Two *sp* hybrid orbitals in a linear arrangement. (c) Unhybridized $2p$ atomic orbitals are perpendicular to the line created by the axes of the two *sp* hybrid orbitals.

of the unhybridized $2p$ atomic orbitals are perpendicular to each other and to the axis of the two *sp* hybrid orbitals. In Figure 1.20, *sp* hybrid orbitals are shown on the *x* axis and unhybridized $2p$ orbitals on the *y* axis and *z* axis.

Figure 1.21 shows a Lewis structure and an orbital overlap diagram for acetylene, C_2H_2. A carbon-carbon triple bond consists of one sigma bond formed by overlap of *sp* hybrid orbitals and two pi bonds. One pi bond is formed by the overlap of a pair of parallel $2p$ atomic orbitals. The second pi bond is formed by overlap of another pair of parallel $2p$ atomic orbitals. The relationship among the number of groups bonded to carbon, orbital hybridization, and types of bonds involved is summarized in Table 1.10.

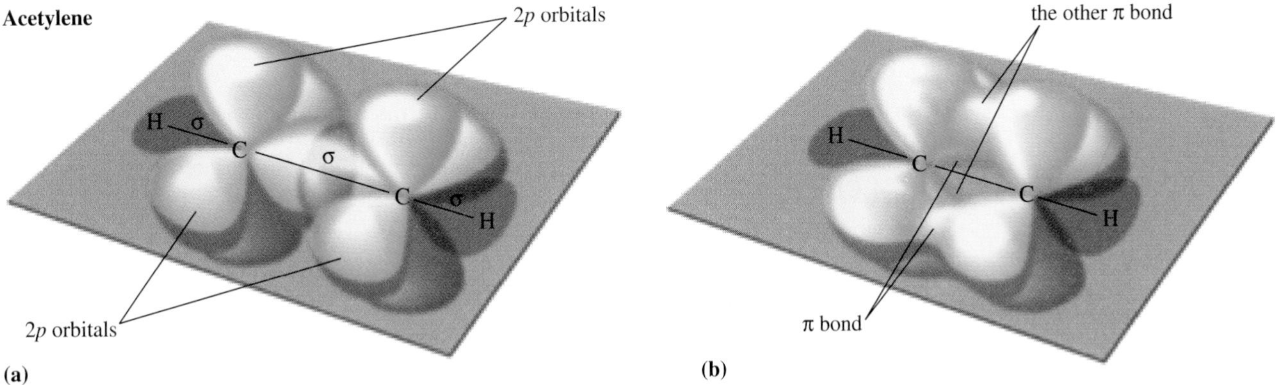

FIGURE 1.21
Covalent bonding in acetylene, C_2H_2. (a) The sigma bond framework shown along with non-overlapping $2p$ atomic orbitals. (b) Formation of two pi bonds by overlap of two sets of parallel $2p$ atomic orbitals.

TABLE 1.10 Covalent Bonding of Carbon

Number of Groups Bonded to Carbon	Orbital Hybridization	Types of Bonds to Carbon	Example	Name
4	sp^3	four sigma bonds	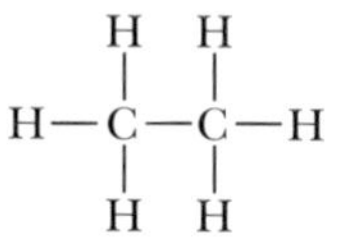 H_3C-CH_3	Ethane
3	sp^2	three sigma bonds and one pi bond	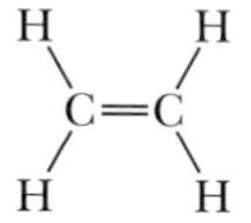 $H_2C=CH_2$	Ethylene
2	sp	two sigma bonds and two pi bonds	$H-C\equiv C-H$	Acetylene

EXAMPLE 1.16

Describe the bonding in acetic acid, CH_3CO_2H, in terms of the atomic orbitals involved, and predict all bond angles.

Solution

Following are three identical Lewis structures. Labels on the first Lewis structure point to atoms and show the hybridization of each atom. Labels on the second Lewis structure point to bonds and show the type of bond, either sigma or pi. Labels on the third point to atoms and show predicted bond angles about each atom.

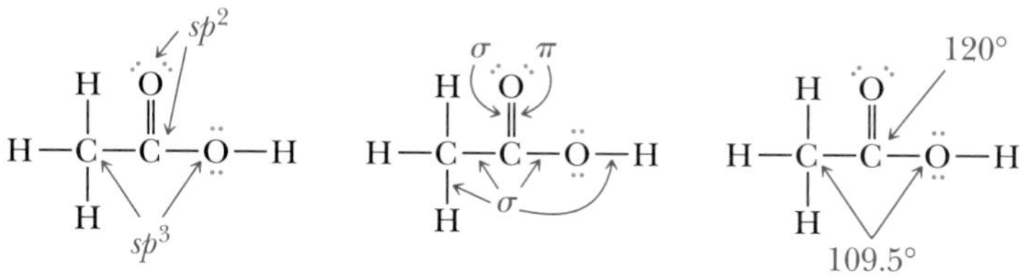

sp^2 sp^3 σ π σ 120° 109.5°

PROBLEM 1.16

Describe the bonding in these molecules in terms of atomic orbitals involved, and predict all bond angles.

(a) $CH_3CH{=}CH_2$ **(b)** CH_3NH_2

SUMMARY

Atoms consist of a nucleus and electrons distributed about the nucleus in regions of space called **principal energy levels** or **shells** (Section 1.1A). Each shell can contain as many as $\mathbf{2n^2}$ electrons where n is the number of the shell. Each principal energy level is subdivided into regions of space called **orbitals.**

The **Lewis structure** (Section 1.1B) of an element shows the symbol of the element surrounded by a number of dots equal to the number of electrons in the outermost, or **valence shell,** of the atom. According to the **Lewis model of covalent bonding** (Section 1.2), atoms bond together in such a way that each atom participating in a chemical bond acquires a completed valence-shell electron configuration resembling that of the noble gas nearest it in the Periodic Table. An **ionic bond** is a chemical bond formed by the attractive force between an anion and a cation. A **covalent bond** is a chemical bond formed by the sharing of electron pairs between atoms. The tendency of main-group elements (those of Groups IA–VIIA) to achieve an outer shell of eight valence electrons is called the **octet rule.**

Electronegativity (Section 1.2C) is a measure of the force of attraction by an atom for electrons it shares in a chemical bond with another atom. A **nonpolar covalent bond** (Section 1.2C) is a covalent bond in which the difference in electronegativity of the bonded atoms is less than 0.5. A **polar covalent bond** is a covalent bond in which the difference in electronegativity of the bonded atoms is between 0.5 and 1.9. In a polar covalent bond, the more electronegative atom bears a partial negative charge **($\delta-$)** and the less electronegative atom bears a partial positive charge **($\delta+$)**. A polar bond has a bond dipole equal to the product of the absolute value of the partial charge times the distance between the dipolar charges (the bond length). The dipole moment of a polyatomic molecule is the vector sum of its bond moments (Section 1.5).

An acceptable **Lewis structure** (Section 1.2D) for a molecule or an ion must show (1) the correct order of attachment of atoms, (2) the correct number of valence electrons, (3) no more than two electrons in the outer shell of hydrogen and no more than eight electrons in the outer shell of any second-row element, and (4) all **formal charges.**

Functional groups (Section 1.3) are characteristic structural units by which we divide organic compounds into classes and which serve as a basis for nomenclature. They are also sites of chemical reactivity; a particular functional group, in whatever compound it is found, undergoes the same types of chemical reactions.

Bond angles of molecules and polyatomic ions can be predicted using Lewis structures and the **valence-shell electron-pair repulsion (VSEPR) model** (Section 1.4). For atoms surrounded by four regions of electron density, predict bond angles of 109.5°; by three regions of electron density, predict bond angles of 120°; and for two regions of electron density, predict bond angles of 180°.

The **theory of resonance** (Section 1.6A) allows us to account for the structure of molecules and ions for which no single Lewis structure is adequate. These molecules and ions are best described by writing two or more Lewis structures, called **contributing structures,** and considering the real molecule or ion to be a **hybrid** of the various contributing structures. Contributing structures to the hybrid are interconnected by **double-headed arrows.** The manner in which valence electrons are redistributed from one contributing structure to the next is shown by **curved arrows** (Section 1.6B). Use of curved arrows in this way is commonly referred to as **electron pushing.** The most important contributing structures have (1) filled valence shells, (2) a maximum number of covalent bonds, (3) the least separation of unlike charge, and (4) carry any negative charge on a more electronegative atom and/or any positive charge on a less electronegative atom.

According to **molecular orbital theory** (Section 1.8), combination of n atomic orbitals gives n molecular orbitals. Molecular orbitals are divided into sigma and pi bonding and antibonding molecular orbitals. These orbitals are arranged in order of increasing energy, and their order of filling is governed by the same rules as for filling atomic orbitals.

The combination of atomic orbitals is called **hybridization** (Section 1.8B), and the resulting atomic orbitals are called **hybrid orbitals.** Combination of one $2s$ atomic orbital and three $2p$ atomic orbitals produces four equivalent **sp^3 hybrid orbitals,** each directed toward a corner of a regular tetrahedron at angles of 109.5° (Section 1.8C).

The combination of one $2s$ atomic orbital and two $2p$ atomic orbitals produces three equivalent **sp^2 hybrid orbitals,** the axes of which lie in a plane at angles of 120° (Section 1.8D). All C=C, C=O, C=N, N=N, and N=O double bonds are a combination of one sigma (σ) bond formed by the overlap of sp^2 hybrid orbitals and one pi (π) bond formed by overlap of parallel $2p$ orbitals.

The combination of one $2s$ atomic orbital and one $2p$ atomic orbital produces two equivalent **sp hybrid orbitals,** the axes of which lie at an angle of 180° (Section 1.8E). All C≡C and C≡N triple bonds are a combination of one sigma bond formed by the overlap of sp hybrid orbitals and two pi bonds formed by the overlap of two sets of parallel $2p$ orbitals.

Additional Problems

Electronic Structure of Atoms

1.17 Write the ground-state electron configuration for each atom. After each atom is given its atomic number.

(a) Sodium (11) **(b)** Magnesium (12) **(c)** Oxygen (8) **(d)** Nitrogen (7)

1.18 Which atom has the ground-state electron configuration of

(a) $1s^2\ 2s^2\ 2p^6\ 3s^2\ 3p^4$ **(b)** $1s^2\ 2s^2\ 2p^4$

1.19 Define valence shell and valence electron.

1.20 How many electrons are in the valence shell of each atom?

(a) Carbon **(b)** Nitrogen **(c)** Chlorine **(d)** Aluminum

Lewis Structures

1.21 Judging from their relative positions in the Periodic Table, which atom in each set is more electronegative?

(a) Carbon or nitrogen **(b)** Chlorine or bromine **(c)** Oxygen or sulfur

1.22 Which compounds have nonpolar covalent bonds, which have polar covalent bonds, and which have ionic bonds?

(a) LiF **(b)** CH_3F **(c)** $MgCl_2$ **(d)** HCl

1.23 Using the symbols $\delta-$ and $\delta+$, indicate the direction of polarity, if any, in each covalent bond.

(a) C—Cl **(b)** S—H **(c)** C—S **(d)** P—H

1.24 Write Lewis structures for these molecules. Show all valence electrons. None of them contains a ring of atoms.

(a) H_2O_2 Hydrogen peroxide
(b) N_2H_4 Hydrazine
(c) CH_3OH Methanol
(d) CH_3SH Methanethiol
(e) CH_3NH_2 Methylamine
(f) CH_2Cl_2 Dichloromethane
(g) CH_3OCH_3 Dimethyl ether
(h) H_2CO_3 Carbonic acid
(i) CH_2O Formaldehyde
(j) CH_3CO_2H Ethanoic acid
(k) CH_3COCH_3 Propanone
(l) HCN Hydrogen cyanide
(m) HNO_3 Nitric acid
(n) HNO_2 Nitrous acid
(o) HCO_2H Formic acid

1.25 Why are the following molecular formulas impossible?

(a) CH_5 **(b)** C_2H_7

1.26 Write Lewis structures for these ions. Show all valence electrons and all formal charges.

(a) NH_2^- Amide ion
(b) HCO_3^- Bicarbonate ion
(c) CO_3^{2-} Carbonate ion
(d) NO_3^- Nitrate ion
(e) HCO_2^- Methanoate ion
(f) $CH_3CO_2^-$ Ethanoate ion

1.27 Following the rule that each atom of carbon, oxygen, and nitrogen reacts to achieve a complete outer shell of eight valence electrons, add unshared pairs of electrons as necessary to complete the valence shell of each atom in these ions. Then assign formal charges as appropriate.

(a) H—C(H)(H)—C(H)(H)—O

(b) 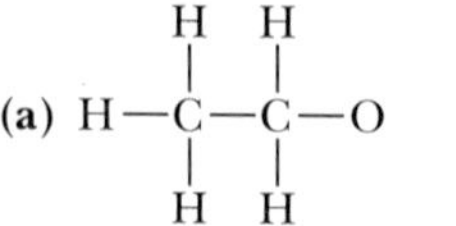

(c) 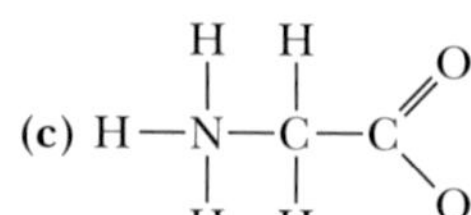

1.28 Following are several Lewis structures showing all valence electrons. Assign formal charges in each structure as appropriate.

```
         H  :O:                    :O:                           :O:
         |  ||                      |                             ||
(a) H—C—C—C—H       (b) H—N—C=C—H         (c) H—C—C—H
         |     |                   |     |                      |
         H     H                   H     H                      H

         H  :O:                H     H                          H
         |   |                 |     |                          |
(d) H—C—C=C—H       (e) H—C—C—C—H        (f) H—C—Ö—H
         |     |               |  |  |                          |  |
         H     H               H  H  H                          H  H
```

1.29 Each compound contains both ionic and covalent bonds. Draw the Lewis structure for each compound and show by dashes which are covalent bonds and, by indication of charges, which are ionic bonds.

(a) CH_3ONa
Sodium methoxide

(b) NH_4Cl
Ammonium chloride

(c) $NaHCO_3$
Sodium bicarbonate

(d) $NaBH_4$
Sodium borohydride

(e) $LiAlH_4$
Lithium aluminum hydride

Polarity of Covalent Bonds

1.30 Which statements are true about electronegativity?

(a) Electronegativity increases from left to right in a period of the Periodic Table.
(b) Electronegativity increases from top to bottom in a column of the Periodic Table.
(c) Hydrogen, the element with the lowest atomic number, has the smallest electronegativity.
(d) The higher the atomic number of an element, the greater its electronegativity.

1.31 Why does fluorine, the element in the upper right corner of the Periodic Table, have the largest electronegativity of any element?

1.32 Arrange the single covalent bonds within each set in order of increasing polarity.

(a) C—H, O—H, N—H
(b) C—H, B—H, O—H
(c) C—H, C—Cl, C—I
(d) C—S, C—O, C—N
(e) C—Li, C—B, C—Mg

1.33 Using the values of electronegativity given in Table 1.5, predict which indicated bond in each set is the more polar and, using the symbols $\delta+$ and $\delta-$, show the direction of its polarity.

(a) CH_3—OH or CH_3O—H
(b) H—NH_2 or CH_3—NH_2
(c) CH_3—SH or CH_3S—H
(d) CH_3—F or H—F

1.34 Identify the most polar bond in each molecule.

(a) $HSCH_2CH_2OH$
(b) $CHCl_2F$
(c) $HOCH_2CH_2NH_2$

1.35 Predict whether the carbon-metal bond in these organometallic compounds is nonpolar covalent, polar covalent, or ionic. For polar covalent bonds, show the direction of polarity by the symbols $\delta+$ and $\delta-$.

```
                  CH2CH3
                   |
(a) CH3CH2—Pb—CH2CH3        (b) CH3—Mg—Cl        (c) CH3—Hg—CH3
                   |
                  CH2CH3
```

(a) Tetraethyllead
(b) Methylmagnesium chloride
(c) Dimethylmercury

Bond Angles and Shapes of Molecules

1.36 Use the VSEPR model to predict bond angles about each highlighted atom.

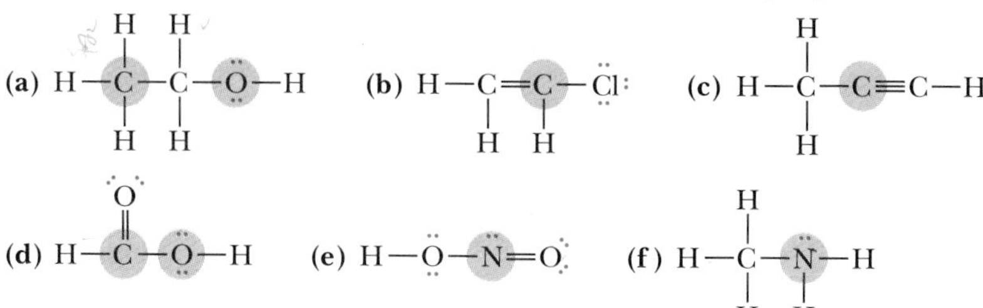

1.37 Use the VSEPR model to predict bond angles about each atom of carbon, nitrogen, and oxygen in these molecules. *Hint:* First add unshared pairs of electrons as necessary to complete the valence shell of each atom and then make your predictions.

(a) $CH_3-CH=CH_2$ (b) $CH_3-N(CH_3)-CH_3$ (c) $CH_3-CH_2-C(=O)-OH$

(d) $CH_2=C=CH_2$ (e) $CH_2=C=O$ (f) $CH_3-CH=N-OH$

1.38 Use the VSEPR model to predict the geometry of these ions:

(a) NH_2^- (b) NO_2^- (c) NO_2^+ (d) NO_3^-

(e) $CH_3CO_2^-$ (f) CH_3^- (g) $AlCl_4^-$

1.39 Silicon is immediately under carbon in the Periodic Table. Predict the geometry of silane, SiH_4.

1.40 Phosphorus is immediately under nitrogen in the Periodic Table. Predict the molecular formula for phosphine, the compound formed by phosphorus and hydrogen. Predict the H—P—H bond angle in phosphine.

Functional Groups

1.41 Draw Lewis structures for the following functional groups. Be certain to show all valence electrons on each.

(a) Carbonyl group (b) Carboxyl group (c) Hydroxyl group

1.42 Draw condensed structural formulas for all compounds of molecular formula C_4H_8O that contain

(a) A carbonyl group (there are two aldehydes and one ketone).
(b) A carbon-carbon double bond and a hydroxyl group (there are eight).
(c) A carbon-carbon double bond and an ether group (there are four).

1.43 Draw the structural formulas for

(a) The eight alcohols of molecular formula $C_5H_{12}O$.

(b) The six ethers of molecular formula $C_5H_{12}O$.

(c) The eight aldehydes of molecular formula $C_6H_{12}O$.

(d) The six ketones of molecular formula $C_6H_{12}O$.

(e) The eight carboxylic acids of molecular formula $C_6H_{12}O_2$.

Polar and Nonpolar Molecules

1.44 Draw a three-dimensional representation for each molecule. Indicate which have a dipole moment and in what direction it is pointing.

(a) CH_3F (b) CH_2Cl_2 (c) CH_2ClBr

(d) $CHCl_3$ (e) CCl_4 (f) $CH_2{=}CCl_2$

(g) $CH_2{=}CHCl$ (h) $HC{\equiv}C{-}C{\equiv}CH$ (i) $CH_3C{\equiv}N$

(j) $(CH_3)_2C{=}O$ (k) $BrCH{=}CHBr$ (two answers)

1.45 Account for the fact that the dipole moment of chloromethane, CH_3Cl (1.87 D), is larger than that of fluoromethane, CH_3F (1.85 D), even though the electronegativity of fluorine (4.0) is larger than that of chlorine (3.5).

1.46 Tetrafluoroethylene, C_2F_4, is the starting material for the synthesis of the polymer poly(tetrafluoroethylene), better known as Teflon. Tetrafluoroethylene has a zero dipole moment. Propose a structural formula for this molecule.

1.47 The dipole moment of chloromethane, CH_3Cl, is 1.87 D. Assume that the contribution of the three C—H bonds to its dipole moment is negligible and that the measured value of 1.87 D is due entirely to the polarity of the C—Cl bond. Given the fact that the charge on an electron is 1.60×10^{-19} coulomb (C) and the length of the C—Cl bond in CH_3Cl is 1.78 Å, calculate the partial negative charge of chlorine and the partial positive charge on carbon in this molecule.

Resonance and Contributing Structures

1.48 Which statements are true about resonance contributing structures?

(a) All contributing structures must have the same number of valence electrons.
(b) All contributing structures must have the same arrangement of atoms.
(c) All atoms in a contributing structure must have complete valence shells.
(d) All bond angles in sets of contributing structures must be the same.

1.49 Draw the contributing structure indicated by the curved arrow(s). Assign formal charges as appropriate.

(a) 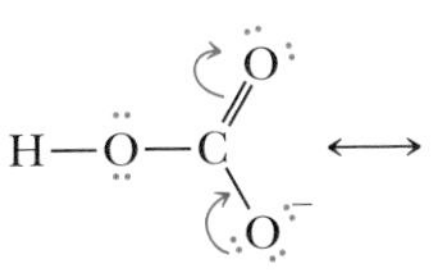 $H{-}\ddot{O}{-}C({=}\ddot{O}{:})({-}\ddot{O}{:}^-)$ ⟷

(b) $H{-}\ddot{O}{-}C({=}\ddot{O}{:})({-}\ddot{O}{:}^-)$ ⟷

(c) $CH_3{-}\ddot{O}{-}C({=}\ddot{O}{:})({-}\ddot{O}{:}^-)$ ⟷

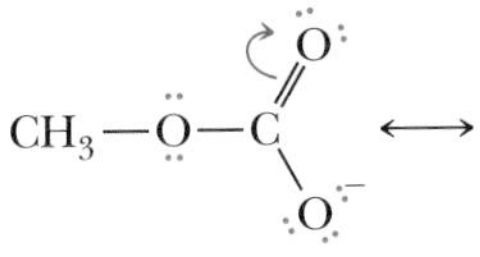

(d) 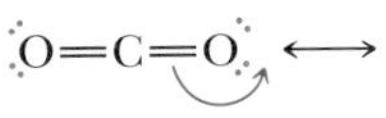 $\ddot{O}{=}C{=}\ddot{O}$ ⟷

(e) 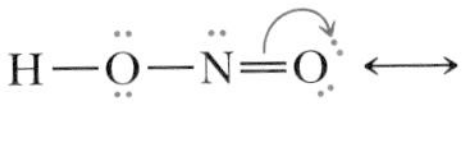 $H{-}\ddot{O}{-}\ddot{N}{=}\ddot{O}$ ⟷

(f) 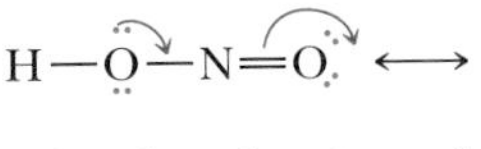 $H{-}\ddot{O}{-}N{=}\ddot{O}$ ⟷

1.50 Using the VSEPR model, predict the bond angles about the carbon atom in each pair of contributing structures in Problem 1.49. In what way do the bond angles change from one contributing structure to the other?

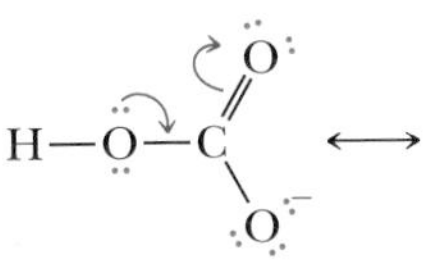

1.51 In Problem 1.49, you were given one contributing structure and asked to draw another. Label pairs of contributing structures that are equivalent. For those sets in which the contributing structures are not equivalent label the more important contributing structure.

1.52 Are the structures in each set valid contributing structures?

(a) 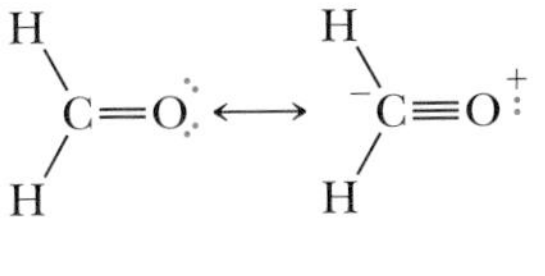 $H_2C{=}\ddot{O}{:}$ ⟷ $H_2\overset{-}{C}{\equiv}\overset{+}{O}{:}$ (b) 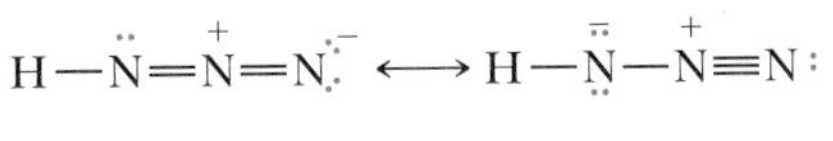 $H{-}\ddot{N}{=}\overset{+}{N}{=}\ddot{N}{:}^-$ ⟷ $H{-}\overset{=}{\ddot{N}}{-}\overset{+}{N}{\equiv}N{:}$

(c) $H-CH_2-C(=O)-H \longleftrightarrow H-CH_2-C(-O^-)-H$ (carbanion form)   (d) $H_2C=C(-OH)H \longleftrightarrow H-CH_2-C(=O)-H$

1.53 Following are three contributing structures for diazomethane, CH_2N_2.

$$H_2\overset{-}{C}-\overset{+}{N}\equiv N: \longleftrightarrow H_2C=\overset{+}{N}=\overset{-}{N} \longleftrightarrow H_2C=N-N$$

(a) Using curved arrows, show how each contributing structure is converted to the one on its right.

(b) Which contributing structure makes the largest contribution to the hybrid?

1.54 Draw a Lewis structure for the azide ion, N_3^-. (The order of atom attachment in this ion is N—N—N.) How does the resonance model account for the fact that the lengths of the N—N bonds in this ion are identical?

1.55 Draw a Lewis structure for the ozone molecule, O_3. (The order of atom attachment is O—O—O.) How does the resonance model account for the fact that the length of each O—O bond in ozone (1.28 Å) is shorter than the O—O single bond in hydrogen peroxide (1.47 Å), but longer than the O—O double bond in the oxygen molecule (1.23 Å)?

1.56 Cyanic acid, HOCN, and isocyanic acid, HNCO, dissolve in water to yield the same anion on loss of H^+.

(a) Write a Lewis structure for cyanic acid.

(b) Write a Lewis structure for isocyanic acid.

(c) Account for the fact that each acid gives the same anion on loss of H^+.

Molecular Orbital Theory

1.57 State the orbital hybridization of each circled atom.

(a) $H-CH_2-CH_2-H$ (b) $H_2C=CH_2$ (c) $H-C\equiv C-H$

(d) $H_2C=O$ (e) $H-C(=O)-O-H$ (f) $H-CH_2-O-H$

(g) $H-CH_2-NH-H$ (h) $H-O-N=O$ (i) $CH_2=C=CH_2$

1.58 Describe each circled bond in terms of the overlap of atomic orbitals.

(a) $H_2C=CH_2$ (b) $H-C\equiv C-H$ (c) $CH_2=C=CH_2$

(d) $H_2C{=}O$ (e) $H{-}C({=}O){-}O{-}H$ (f) $H{-}CH_2{-}O{-}H$

(g) $H{-}CH_2{-}NH{-}H$ (h) $H{-}CH_2{-}O{-}C({=}O){-}H$ (i) $H{-}O{-}N{=}O$

1.59 Following is the structural formula of benzene, C_6H_6.

(Structure: six-membered ring of C atoms with alternating single and double bonds, each C bearing one H)

(a) Predict each H—C—C bond angle in benzene and each C—C—C bond angle.
(b) State the hybridization of each carbon in benzene.
(c) Predict the shape of the benzene molecule.

1.60 Following is the structural formula of the prescription drug famotidine, manufactured by Merck Sharpe & Dohme under the name Pepcid. The primary clinical use of Pepcid is for the treatment of active duodenal ulcers and benign gastric ulcers. Pepcid is a competitive inhibitor of histamine H_2 receptors and reduces both gastric acid concentration and volume of gastric secretions.

(Structure: $(H_2N)_2C{=}N$–(thiazole ring C=N, C=C, S)–$CH_2{-}S{-}CH_2{-}CH_2{-}C(NH_2){=}N{-}S({=}O)_2{-}NH_2$)

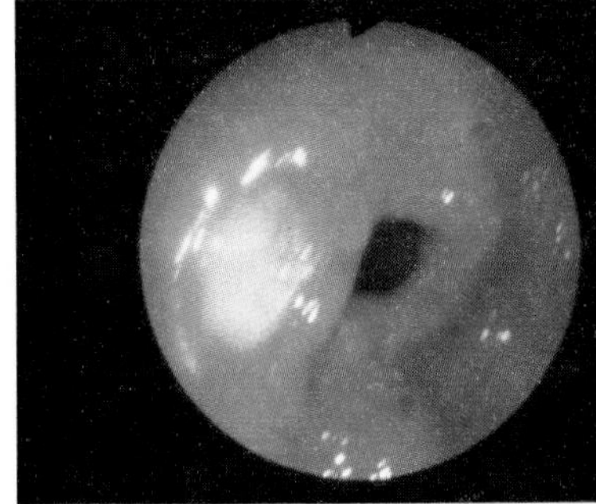

Endoscopic image of a superficial stomach (gastric) ulcer. *(Dr. Beer-Gabel/CNRL/Science Photo Library/Photo Researchers, Inc.)*

(a) Complete the Lewis structure of famotidine showing all valence electrons and any formal positive or negative charges.
(b) Describe each circled bond in terms of the overlap of atomic orbitals.

1.61 In Chapter 6, we study a group of organic cations called carbocations. Following is the structure of one such carbocation, the *tert*-butyl cation.

$(H_3C)_2C^{+}{-}CH_3$

tert-Butyl cation

(a) How many electrons are in the valence shell of the carbon bearing the positive charge?
(b) Predict the bond angles about this carbon.
(c) Given the bond angle you predicted in (b), what hybridization do you predict for this carbon?

1.62 Many reactions involve a change in hybridization of one or more atoms in the starting material. In each of the following, identify the atoms in the organic starting material that change hybridization and indicate what the change is. We will examine these reactions in more detail later in the course.

(a) $H_2C{=}CH_2 + Cl_2 \longrightarrow H{-}CH(Cl){-}CH(Cl){-}H$

(b) $H{-}C{\equiv}C{-}H + Cl_2 \longrightarrow HClC{=}CClH$

(c) $H{-}C{\equiv}C{-}H + H_2O \longrightarrow H{-}CH_2{-}C({=}O){-}H$

(d) $H_2C{=}O + H_2 \longrightarrow H{-}CH_2{-}O{-}H$

(e) $H{-}CH_2{-}\overset{+}{C}H{-}CH_2{-}H + H_2O \longrightarrow H{-}CH_2{-}CH(OH){-}CH_2{-}H$

ALKANES AND CYCLOALKANES

2

■ A petroleum refinery. *(Courtesy of Ashland Oil Company)*

In this chapter, we begin our study of organic compounds with the physical and chemical properties of alkanes and cycloalkanes, both saturated hydrocarbons and the simplest types of organic compounds.

A **hydrocarbon** is a compound that is composed of only carbon and hydrogen. A **saturated hydrocarbon** contains only single bonds. Saturated in this context means that the hydrocarbon has the maximum number of

Hydrocarbon A compound that is composed of only carbon and hydrogen atoms.

Saturated hydrocarbon A hydrocarbon containing only carbon-carbon single bonds.

Alkane A saturated hydrocarbon whose carbon atoms are arranged in an open chain.

Aliphatic hydrocarbon An alternative word to describe an alkane.

hydrogens for the number of carbons it contains. An **alkane** is a saturated hydrocarbon whose carbon atoms are arranged in an open chain. Alkanes are commonly referred to as **aliphatic hydrocarbons,** because the physical properties of the higher members of this class resemble those of the long carbon-chain molecules we find in animal fats and plant oils (Greek: *aleiphar,* fat or oil).

2.1 Structure of Alkanes

Methane, CH_4, and ethane, C_2H_6, are the first two members of the alkane family. Shown in Figure 2.1 are Lewis structures and stereorepresentations for these molecules. The shape of methane is tetrahedral and all H—C—H bond angles are 109.5°. Each of the carbon atoms in ethane is also tetrahedral, and all bond angles are approximately 109.5°.

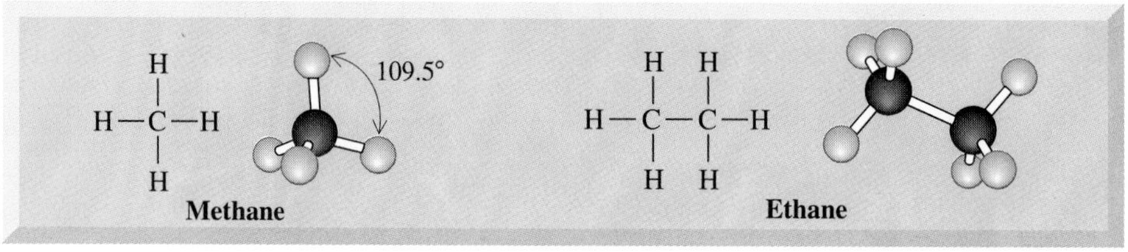

FIGURE 2.1
Methane and ethane: Lewis structures and ball-and-stick models.

Although the three-dimensional shapes of larger alkanes are more complex than those of methane and ethane, each carbon atom is still tetrahedral, and all bond angles are approximately 109.5°.

TABLE 2.1 Names, Molecular Formulas, and Condensed Structural Formulas for the First 20 Alkanes with Unbranched Chains

Name	Molecular Formula	Condensed Structural Formula	Name	Molecular Formula	Condensed Structural Formula
methane	CH_4	CH_4	undecane	$C_{11}H_{24}$	$CH_3(CH_2)_9CH_3$
ethane	C_2H_6	CH_3CH_3	dodecane	$C_{12}H_{26}$	$CH_3(CH_2)_{10}CH_3$
propane	C_3H_8	$CH_3CH_2CH_3$	tridecane	$C_{13}H_{28}$	$CH_3(CH_2)_{11}CH_3$
butane	C_4H_{10}	$CH_3(CH_2)_2CH_3$	tetradecane	$C_{14}H_{30}$	$CH_3(CH_2)_{12}CH_3$
pentane	C_5H_{12}	$CH_3(CH_2)_3CH_3$	pentadecane	$C_{15}H_{32}$	$CH_3(CH_2)_{13}CH_3$
hexane	C_6H_{14}	$CH_3(CH_2)_4CH_3$	hexadecane	$C_{16}H_{34}$	$CH_3(CH_2)_{14}CH_3$
heptane	C_7H_{16}	$CH_3(CH_2)_5CH_3$	heptadecane	$C_{17}H_{36}$	$CH_3(CH_2)_{15}CH_3$
octane	C_8H_{18}	$CH_3(CH_2)_6CH_3$	octadecane	$C_{18}H_{38}$	$CH_3(CH_2)_{16}CH_3$
nonane	C_9H_{20}	$CH_3(CH_2)_7CH_3$	nonadecane	$C_{19}H_{40}$	$CH_3(CH_2)_{17}CH_3$
decane	$C_{10}H_{22}$	$CH_3(CH_2)_8CH_3$	eicosane	$C_{20}H_{42}$	$CH_3(CH_2)_{18}CH_3$

The next members of the alkane family are propane, butane, and pentane.

$CH_3CH_2CH_3$	$CH_3CH_2CH_2CH_3$	$CH_3CH_2CH_2CH_2CH_3$
Propane	Butane	Pentane

Condensed structural formulas for alkanes can also be written in an abbreviated form. For example, the structural formula of pentane contains three $—CH_2—$ **(methylene)** groups in the middle of the chain. They can be grouped together and the structural formula can be written $CH_3(CH_2)_3CH_3$. Names, molecular formulas, and condensed structural formulas of the first 20 alkanes are given in Table 2.1. We have more to say about naming alkanes in Section 2.3. Alkanes have the general formula $\mathbf{C_nH_{2n+2}}$. Thus, given the number of carbon atoms in an alkane, it is easy to determine the number of hydrogens in the molecule and also its molecular formula. For example, decane with 10 carbon atoms must have $(2 \times 10) + 2 = 22$ hydrogens and a molecular formula of $C_{10}H_{22}$.

Some camping stoves use butane as a fuel. *(Charles D. Winters)*

2.2 Constitutional Isomerism in Alkanes

Compounds that have the same molecular formula but different structural formulas (different orders of attachment of atoms) are called **constitutional isomers.** In Section 1.3, we encountered several examples of constitutional isomers. We saw in Example 1.8 that there are two alcohols of molecular formula C_3H_8O; they are constitutional isomers. In this section, we study constitutional isomers in more detail, including how to recognize them and how to draw structures for them.

Constitutional isomers Compounds with the same molecular formula but a different order of attachment of their atoms.

For the molecular formulas CH_4, C_2H_6, and C_3H_8, only one order of attachment of atoms is possible. For the molecular formula C_4H_{10}, there are two possible orders of attachment of atoms. In one of these, the four carbons are attached in a chain; in the other, three carbons are attached in a chain with the fourth carbon as a branch on the chain. These two constitutional isomers are named butane and 2-methylpropane. We discuss how to name alkanes in the following section. Butane and 2-methylpropane are different compounds and have different physical and chemical properties.

$CH_3CH_2CH_2CH_3$

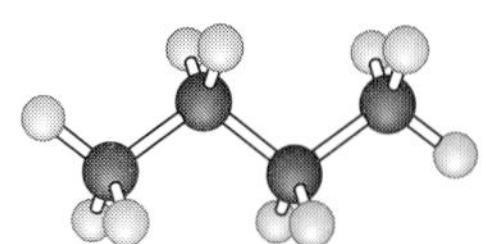

Butane (bp −0.5°C)

$$\begin{array}{c} CH_3 \\ | \\ CH_3CHCH_3 \end{array}$$

2-Methylpropane (bp −11.6°C)

To determine whether two or more structural formulas represent constitutional isomers, write the molecular formula of each and then compare them. All compounds that have the same molecular formula but different structural formulas are constitutional isomers.

EXAMPLE 2.1

Do the structural formulas in each set represent identical compounds or constitutional isomers?

(a) $CH_3CH_2CH_2CH_2CH_2CH_3$ and $\begin{array}{l} CH_3CH_2CH_2 \\ \quad\quad\quad\;\; | \\ \quad\quad\quad CH_2CH_2CH_3 \end{array}$ (each is C_6H_{14})

(b) $\begin{array}{l} \quad\quad\quad\quad\quad\; CH_3 \\ \quad\quad\quad\quad\quad\;\; | \\ CH_3CHCH_2CH \\ \quad\quad\; | \quad\quad\;\; | \\ \quad\;\; CH_3 \quad CH_3 \end{array}$ and $\begin{array}{l} \quad\quad\quad\quad\; CH_3 \\ \quad\quad\quad\quad\;\; | \\ CH_3CH_2CHCHCH_3 \\ \quad\quad\quad\quad\quad\;\; | \\ \quad\quad\quad\quad\quad CH_3 \end{array}$ (each is C_7H_{16})

Solution

(a) Each structural formula has an unbranched chain of six carbons; they are identical and represent the same compound.

$$\overset{1}{CH_3}\overset{2}{CH_2}\overset{3}{CH_2}\overset{4}{CH_2}\overset{5}{CH_2}\overset{6}{CH_3} \quad \text{and} \quad \begin{array}{l} \overset{1}{CH_3}\overset{2}{CH_2}\overset{3}{CH_2} \\ \qquad\quad |\,4 \quad 5 \quad 6 \\ \qquad\quad CH_2CH_2CH_3 \end{array}$$

(b) Each structural formula has a chain of five carbons with two CH_3— branches. Although the branches are identical, they are at different locations on the chains. Therefore, these structural formulas represent constitutional isomers.

$$\begin{array}{c} \qquad\quad \overset{5}{CH_3} \\ 1 \;\; 2 \;\; 3 \;\; 4| \\ CH_3CHCH_2CH \\ \quad | \qquad\;\; | \\ \;\; CH_3 \quad CH_3 \end{array} \quad \text{and} \quad \begin{array}{c} \qquad\; CH_3 \\ 5 \;\; 4 \;\; 3| \;\; 2 \;\; 1 \\ CH_3CH_2CHCHCH_3 \\ \qquad\quad | \\ \qquad\quad CH_3 \end{array}$$

PROBLEM 2.1

Do the structural formulas in each set represent identical compounds or constitutional isomers?

$$\text{(a)}\; \begin{array}{l} \quad CH_2CH_3 \\ \quad | \\ CH_3CHCHCH_3 \\ \qquad | \\ \qquad CH_2CH_3 \end{array} \quad \text{and} \quad \begin{array}{l} \qquad\quad CH_3 \quad CH_3 \\ \qquad\quad | \qquad\quad | \\ CH_3CH_2CHCH_2CHCH_3 \end{array}$$

$$\text{(b)}\; \begin{array}{l} \quad CH_3 \\ \quad | \\ CH_3CHCHCH_3 \\ \qquad | \\ \qquad CH_2CH_3 \end{array} \quad \text{and} \quad \begin{array}{l} \qquad CH_3 \\ \qquad | \\ CH_3CHCHCH_2CH_3 \\ \quad | \\ \quad CH_3 \end{array}$$

EXAMPLE 2.2

Draw structural formulas for the five constitutional isomers of molecular formula C_6H_{14}.

Solution

In doing problems of this type, you should devise a strategy and then follow it. Here is one such strategy. First, draw a structural formula for the constitutional isomer with all six carbons in an unbranched chain. Then, draw structural formulas for all constitutional isomers with five carbons in a chain and one carbon as a branch on the chain. Finally, draw structural formulas for all constitutional isomers with four carbons in a chain and two carbons as branches.

Six carbons in an unbranched chain:

$$\overset{1}{CH_3}\overset{2}{CH_2}\overset{3}{CH_2}\overset{4}{CH_2}\overset{5}{CH_2}\overset{6}{CH_3}$$

Five carbons in a chain; one carbon as a branch:

$$\begin{array}{l} \quad CH_3 \\ 1 \;\; 2| \;\; 3 \;\; 4 \;\; 5 \\ CH_3CHCH_2CH_2CH_3 \end{array} \qquad \begin{array}{l} \qquad\quad CH_3 \\ 1 \;\; 2 \;\; 3| \;\; 4 \;\; 5 \\ CH_3CH_2CHCH_2CH_3 \end{array}$$

Four carbons in a chain; two carbons as branches:

```
        CH3              CH3
   1  2| 3   4       1  2|  3  4
   CH3CCH2CH3        CH3CHCHCH3
       |                   |
       CH3                 CH3
```

No constitutional isomers with only three carbons in the longest chain are possible for C_6H_{14}.

PROBLEM 2.2

Draw structural formulas for the three constitutional isomers of molecular formula C_5H_{12}.

The ability of carbon atoms to form strong, stable bonds with other carbon atoms results in a staggering number of constitutional isomers. As shown in the following table, there are 3 constitutional isomers of molecular formula C_5H_{12}. For molecular formula $C_{10}H_{22}$, there are 75 constitutional isomers, and for $C_{25}H_{52}$, there are almost 37 million.

Carbon Atoms	Constitutional Isomers
1	1
5	3
10	75
15	4,347
25	36,797,588
30	4,111,846,763

CHEMISTRY IN ACTION

Counting the Number of Constitutional Isomers

We have seen in this section that, for the hydrocarbon $C_{30}H_{62}$, there are over four billion constitutional isomers. The number of constitutional isomers becomes even larger for organic compounds with a functional group. For example, although there are 75 alkanes of molecular formula $C_{10}H_{22}$, there are 507 alcohols of molecular formula $C_{10}H_{22}O$.

Obviously, these numbers cannot be obtained by drawing and counting all the possible constitutional isomers. Instead, they are counted using a type of mathematics called **graph theory.** The English mathematician, Arthur Cayley, was the first person to use these ideas for organic molecules, and his paper of 1874, titled "On the Mathematical Theory of Isomers," predates many fundamental organic concepts, including the idea that alkane carbons are tetrahedral.

To read further about the concept of graph theory for counting the number of constitutional isomers, see P. J. Hansen and P. C. Jurs, *Journal of Chemical Education,* **65,** 661, 1988.

Thus, for even a small number of carbon and hydrogen atoms, a very large number of constitutional isomers is possible. In fact, the potential for unique combinations of structural and functional groups among organic molecules made from just the basic building blocks of carbon, hydrogen, nitrogen, and oxygen is practically limitless.

2.3 Nomenclature of Alkanes

A. The IUPAC System

Ideally, every organic compound should have a name from which a structural formula can be drawn. For this purpose, chemists have adopted a set of rules established by an organization called the **International Union of Pure and Applied Chemistry (IUPAC).** The IUPAC name of an alkane with an unbranched chain of carbon atoms consists of two parts: (1) a prefix that indicates the number of carbon atoms in the chain and (2) the suffix **-ane** to show that the compound is a saturated hydrocarbon. Prefixes used to show the presence of 1 to 20 carbon atoms are given in Table 2.2.

The first four prefixes listed in Table 2.2 were chosen by the IUPAC because they were well established in the language of organic chemistry. In fact, they were well established even before there were hints of the structural theory underlying the discipline. For example, the prefix *but-* appears in the name butyric acid, a compound of four carbon atoms formed by air oxidation of butter (Latin: *butyrum,* butter). Prefixes to show five or more carbons are derived from Greek or Latin roots. Names, molecular formulas, and condensed structural formulas for the first 20 alkanes with unbranched chains were given in Table 2.1.

IUPAC names of alkanes with branched chains consist of a parent name, which indicates the longest chain of carbon atoms in the compound, and substituent names, which indicate the groups attached to the parent chain.

CH_3 ← substituent

$CH_3CH_2CH_2CHCH_2CH_2CH_2CH_3$ ← parent chain of eight carbon atoms

4-Methyloctane

TABLE 2.2 Prefixes Used in the IUPAC System to Show the Presence of 1 to 20 Carbon Atoms in a Chain

Prefix	Number of Carbon Atoms	Prefix	Number of Carbon Atoms
meth-	1	undec-	11
eth-	2	dodec-	12
prop-	3	tridec-	13
but-	4	tetradec-	14
pent-	5	pentadec-	15
hex-	6	hexadec-	16
hept-	7	heptadec-	17
oct-	8	octadec-	18
non-	9	nonadec-	19
dec-	10	eicos-	20

TABLE 2.3 Names of Common Alkyl Groups

Name	Condensed Structural Formula	Name	Condensed Structural Formula
methyl	$-CH_3$		
ethyl	$-CH_2CH_3$	*tert*-butyl	CH_3 | $-CCH_3$ | CH_3
propyl	$-CH_2CH_2CH_3$		
isopropyl	$-CHCH_3$ | CH_3	pentyl	$-CH_2CH_2CH_2CH_2CH_3$
butyl	$-CH_2CH_2CH_2CH_3$	isopentyl	$-CH_2CH_2CHCH_3$ | CH_3
isobutyl	$-CH_2CHCH_3$ | CH_3	neopentyl	CH_3 | $-CH_2CCH_3$ | CH_3
sec-butyl	$-CHCH_2CH_3$ | CH_3		

A substituent group derived from an alkane by removal of a hydrogen atom is called an **alkyl group.** The symbol **R—** is commonly used to represent an alkyl group. Alkyl groups are named by dropping the -ane from the name of the parent alkane and adding the suffix -yl. For example, the alkyl group CH_3CH_2- is named ethyl. Names and structural formulas for 11 of the most common alkyl groups are given in Table 2.3.

Alkyl group A group derived by removing a hydrogen from an alkane.

R— A symbol used to represent an alkyl group.

The rules of the IUPAC system for naming alkanes are as follows:

1. The general name of an open-chain saturated hydrocarbon is alkane.
2. For branched-chain hydrocarbons, the alkane corresponding to the longest chain of carbon atoms is taken as the parent chain and its name as the root name.
3. Groups attached to the parent chain are called substituents. Each substituent is given a name and a number. The number shows the carbon atom of the parent chain to which the substituent is attached.

$$\overset{1}{CH_3}\overset{2}{C}H(CH_3)\overset{3}{CH_3}$$

2-Methylpropane

   Note that the name of this hydrocarbon indicates the number of its carbon atoms. 2-Methylpropane contains 1 (methyl) + 3 (propane) = 4 carbon atoms.
4. If there is one substituent, number the parent chain from the end that gives it the lower number. The following alkane must be numbered as shown and named 2-methylpentane. Numbering from the other end of the chain gives the incorrect name 4-methylpentane.

$$\overset{5}{CH_3}\overset{4}{CH_2}\overset{3}{CH_2}\overset{2}{C}H(CH_3)\overset{1}{CH_3}$$

2-Methylpentane

5. If the same substituent occurs more than once, the number of each carbon of the parent chain on which the substituent occurs is given. In addition, the number of

times the substituent group occurs is indicated by a prefix di-, tri-, tetra-, penta-, hexa-, and so on.

$$\begin{array}{c} \quad CH_3 \quad\; CH_3 \\ 1\;\; 2|\;\; 3\;\; 4|\;\; 5 \\ CH_3CHCH_2CHCH_3 \end{array}$$

2,4-Dimethylpentane

The name of this compound shows that it contains 2 (dimethyl) + 5 (pentane) = 7 carbon atoms.

6. If there are two or more identical substituents, number the parent chain from the end that gives the lower number to the substituent encountered first.

$$\begin{array}{c} \quad\quad CH_3 \quad CH_3 \\ 6\;\; 5\;\; 4|\;\; 3\;\; 2|\;\; 1 \\ CH_3CH_2CHCH_2CHCH_3 \end{array}$$

2,4-Dimethylhexane

7. If there are two or more different substituents, list them in alphabetical order and number the chain from the end that gives the lower number to the substituent encountered first. If there are different substituents in equivalent positions on the parent chain, the substituent of lower alphabetical order is given the lower number.

$$\begin{array}{c} \quad\quad\quad\quad CH_3 \\ 1\;\; 2\;\; 3\;\; 4\;\; 5|\;\; 6\;\; 7 \\ CH_3CH_2CHCH_2CHCH_2CH_3 \\ | \\ CH_2CH_3 \end{array}$$

3-Ethyl-5-methylheptane

The name of this compound shows that it contains 2 (ethyl) + 1 (methyl) + 7 (heptane) = 10 carbon atoms.

8. The prefixes di-, tri-, tetra-, and so on are not included in alphabetizing. The names of substituents are alphabetized first, and then these multiplying prefixes are inserted. In this example, the alphabetized parts are *ethyl* and *methyl,* not ethyl and dimethyl.

$$\begin{array}{c} \quad CH_3\;\; CH_2CH_3 \\ 1\;\; 2|\; 3\;\; 4|\;\; 5\;\; 6 \\ CH_3CCH_2CHCH_2CH_3 \\ | \\ CH_3 \end{array}$$

4-Ethyl-2,2-dimethylhexane

The name of this compound shows that it contains 2 (ethyl) + 2 (dimethyl) + 6 (hexane) = 10 carbon atoms.

9. Hyphenated prefixes, such as *sec*- and *tert*-, are not considered when alphabetizing. The prefixes iso and neo are not hyphenated prefixes and are included when alphabetizing.

EXAMPLE 2.3

Give IUPAC names for these alkanes. Show that each name indicates the total number of carbons in the molecule.

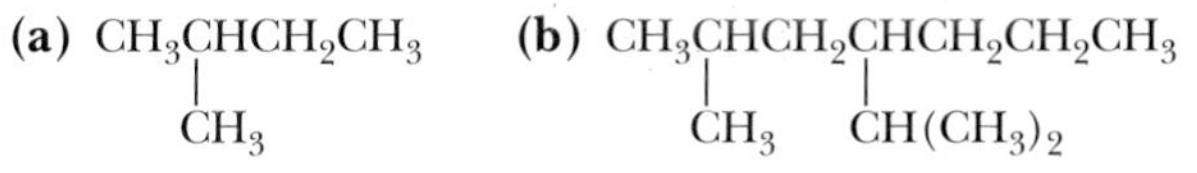

(a) $CH_3CHCH_2CH_3$ with CH_3 on C2

(b) $CH_3CHCH_2CHCH_2CH_2CH_3$ with CH_3 on C2 and $CH(CH_3)_2$ on C4

Solution

The longest chain in each is numbered from the end of the chain toward the substituent encountered first (rule 4). The substituents in (b) are listed in alphabetical order (rule 7).

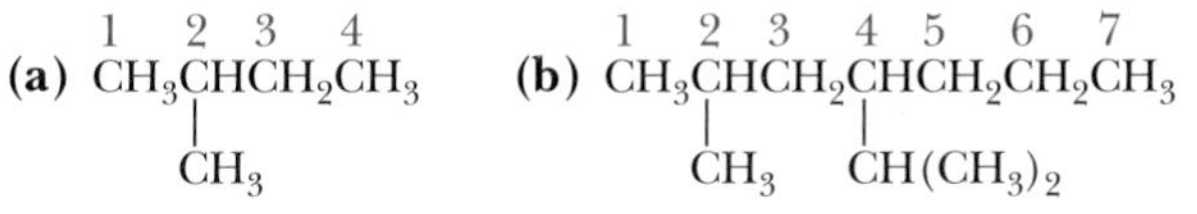

(a) $\overset{1}{C}H_3\overset{2}{C}H(CH_3)\overset{3}{C}H_2\overset{4}{C}H_3$

2-Methylbutane

(b) $\overset{1}{C}H_3\overset{2}{C}H(CH_3)\overset{3}{C}H_2\overset{4}{C}H(CH(CH_3)_2)\overset{5}{C}H_2\overset{6}{C}H_2\overset{7}{C}H_3$

4-Isopropyl-2-methylheptane

The name of compound (a) shows that it contains 1 (methyl) + 4 (butane) = 5 carbon atoms. The name of compound (b) shows that it contains 3 (isopropyl) + 1 (methyl) + 7 (heptane) = 11 carbon atoms.

PROBLEM 2.3

Write IUPAC names for these alkanes. Show that each name indicates the total number of carbons in the molecule.

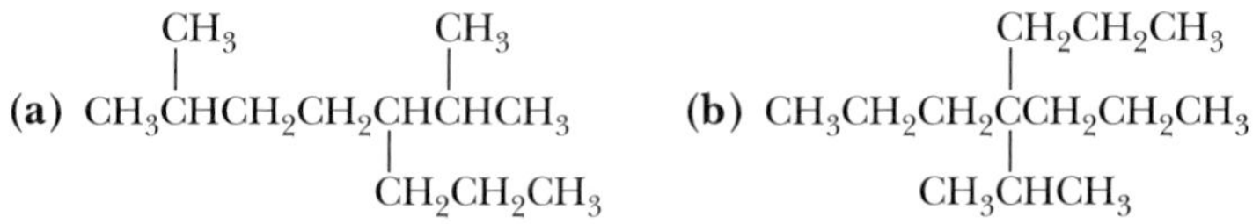

(a) $CH_3CHCH_2CH_2CHCHCH_3$ with CH_3 on C2, CH_3 on C6, and $CH_2CH_2CH_3$ on C5

(b) $CH_3CH_2CH_2CCH_2CH_2CH_3$ with $CH_2CH_2CH_3$ and CH_3CHCH_3 on C4

B. Common Names

In spite of the precision of the IUPAC system, routine communication in organic chemistry still relies on a combination of trivial, semisystematic, and systematic names. The reasons for this are rooted in both convenience and historical development.

In the older system of **common nomenclature,** the total number of carbon atoms in an alkane, regardless of their arrangement, determines the name. The first three alkanes are methane, ethane, and propane. All alkanes of formula C_4H_{10} are called butanes, all alkanes of formula C_5H_{12} are called pentanes, and those of formula C_6H_{14} are called hexanes. For alkanes beyond propane, **iso** is used to indicate that one end of an otherwise unbranched chain terminates in a $(CH_3)_2CH—$ group and **neo** is used to indicate that one end of an otherwise unbranched chain terminates in $(CH_3)_3C—$. Following are examples of common names.

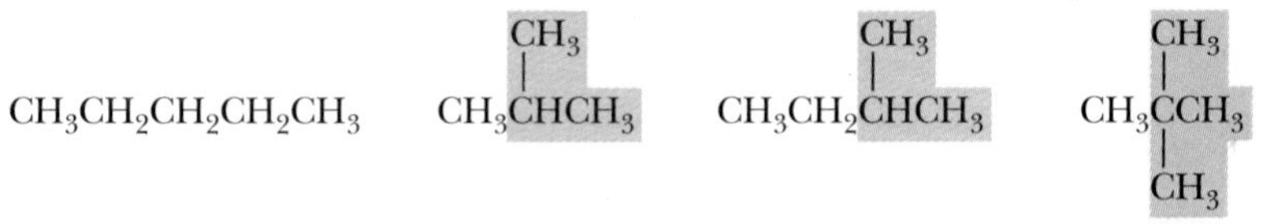

$CH_3CH_2CH_2CH_2CH_3$	$CH_3CH(CH_3)CH_3$	$CH_3CH_2CH(CH_3)CH_3$	$CH_3C(CH_3)_2CH_3$
Pentane	Isobutane	Isopentane	Neopentane

The common names amyl- and isoamyl- are sometimes used in place of pentyl- and isopentyl-.

	$-CH_2CH_2CH_2CH_2CH_3$	$-CH_2CH_2CH(CH_3)CH_3$
IUPAC name:	Pentyl	Isopentyl
Common name:	Amyl	Isoamyl

This system of common names has no good way of handling other branching patterns, and for more complex alkanes it is necessary to use the more flexible IUPAC system of nomenclature.

In this text we concentrate on IUPAC names. However, we also use common names, especially when the common name is used almost exclusively in the everyday discussions of chemists. When both IUPAC and common names are given in the text, we always give the IUPAC name first followed by the common name in parentheses. In this way, you should have no doubt about which name is which.

C. Classification of Carbon and Hydrogen Atoms

Primary (1°) carbon A carbon bonded to one other carbon atom.

Secondary (2°) carbon A carbon bonded to two other carbon atoms.

Tertiary (3°) carbon A carbon bonded to three other carbon atoms.

Quaternary (4°) carbon A carbon bonded to four other carbon atoms.

A carbon atom is classified as primary **(1°)**, secondary **(2°)**, tertiary **(3°)**, or quaternary **(4°)**, depending on the number of carbon atoms bonded to it. A carbon bonded to one carbon atom is a primary carbon, one bonded to two carbon atoms is a secondary carbon, and so forth. For example, propane contains two primary carbons and one secondary carbon. 2-Methylpropane contains three primary carbons and one tertiary carbon. 2,2,4-Trimethylpentane contains five primary carbons, one secondary carbon, one tertiary carbon, and one quaternary carbon.

two 1° carbons; a 2° carbon
$CH_3-CH_2-CH_3$
Propane

a 3° carbon
$CH_3-CH(CH_3)-CH_3$
2-Methylpropane

a 4° carbon
$CH_3-C(CH_3)_2-CH_2-CH(CH_3)-CH_3$
2,2,4-Trimethylpentane

Similarly, hydrogens are also classified as primary, secondary, or tertiary, depending on the type of carbon to which each is bonded. Those attached to primary carbons are classified as primary hydrogens, those on secondary carbons are secondary hydrogens, and those on tertiary carbons are tertiary hydrogens.

Equivalent hydrogens Hydrogens that have the same chemical environment.

Hydrogen atoms in a compound can be divided into equivalent sets. **Equivalent hydrogens** have the same chemical environment. A direct way to determine which hydrogens in a molecule are equivalent is to replace each in turn by a "test atom," as, for example, chlorine. If replacement of each of two hydrogens by chlorine gives the same compound, these hydrogens are equivalent. If such replacement gives different compounds, the hydrogens are nonequivalent. Using this test, we can show that propane contains two sets of equivalent hydrogens: a set of six equivalent primary hydrogens, and a set of two equivalent secondary hydrogens. Note that chloroalkanes are named by listing chlorine as a chloro- substituent.

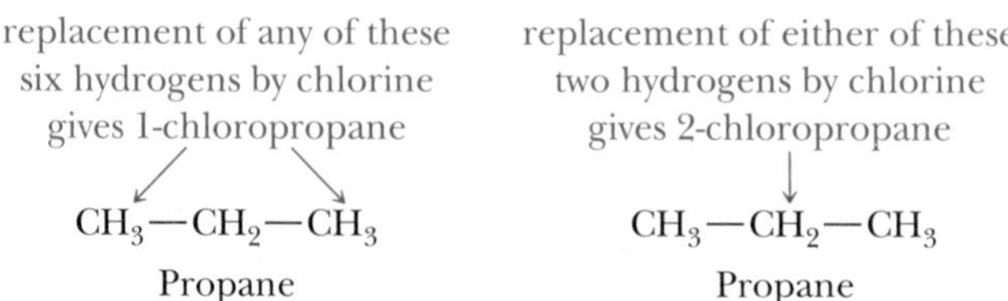
replacement of any of these six hydrogens by chlorine gives 1-chloropropane
$CH_3-CH_2-CH_3$
Propane
replacement of either of these two hydrogens by chlorine gives 2-chloropropane
$CH_3-CH_2-CH_3$
Propane

EXAMPLE 2.4

State the number of sets of equivalent hydrogens in each compound and the number of hydrogens in each set.

(a) $CH_3-\underset{}{\overset{CH_3}{\overset{|}{C}}}H-CH_3$ (b) $CH_3-CH_2-\overset{CH_3}{\overset{|}{C}}H-CH_3$

Solution

(a) 2-Methylpropane contains two sets of equivalent hydrogens.
(b) 2-Methylbutane contains four sets of equivalent hydrogens.

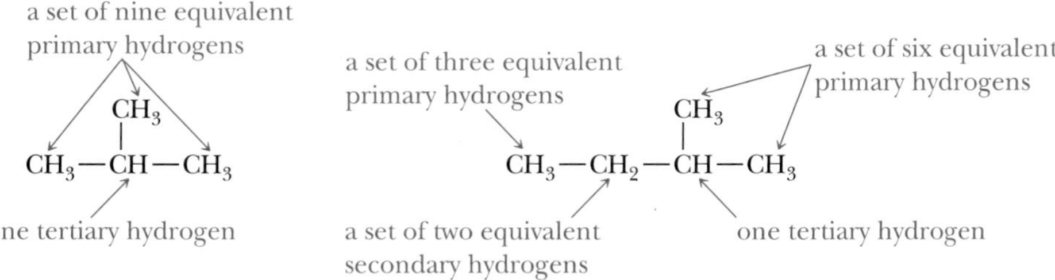

PROBLEM 2.4

State the number of sets of equivalent hydrogens in each compound and the number of hydrogens in each set.

(a) $CH_3-CH_2-\overset{CH_3}{\overset{|}{C}}H-CH_2-CH_3$ (b) $CH_3-\overset{CH_3}{\overset{|}{C}}H-CH_2-\underset{CH_3}{\underset{|}{\overset{CH_3}{\overset{|}{C}}}}-CH_3$

2.4 Cycloalkanes

So far we have considered only chains (branched and unbranched) of carbon atoms. The ends of these chains can also be joined to form a ring of carbon atoms. A hydrocarbon that contains carbon atoms joined to form a ring is called a **cyclic hydrocarbon.** When all carbons of the ring are saturated, the hydrocarbon is called a **cycloalkane.**

Cycloalkane A saturated hydrocarbon that contains carbon atoms joined to form a ring.

A. Structure and Nomenclature

Cycloalkanes of ring sizes ranging from 3 to over 30 are found in nature, and, in principle, there is no limit to ring size. Five-membered rings (cyclopentanes) and six-membered rings (cyclohexanes) are especially abundant in nature and, therefore, have received special attention.

Figure 2.2 shows structural formulas of cyclopropane, cyclobutane, cyclopentane, and cyclohexane. As a matter of convenience, organic chemists often do not show all carbons and hydrogens when writing structural formulas for cycloalkanes. Rather, the rings are represented by regular polygons having the same number of sides as there

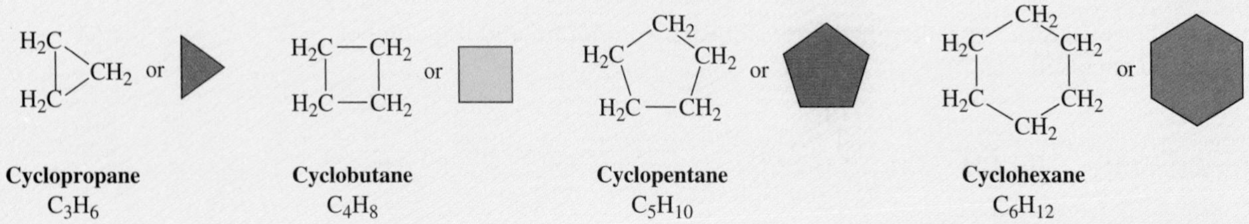

FIGURE 2.2
Examples of cycloalkanes.

are carbon atoms in the ring. For example, cyclopropane is represented by a triangle and cyclohexane by a hexagon.

Line-angle drawing An abbreviated way to draw structural formulas in which an angle represents a carbon atom and a line represents a bond.

The abbreviated structural formulas shown in Figure 2.2 are called line-angle drawings. In a **line-angle drawing** each angle and line terminus represents a carbon atom. Each single line represents a C—C bond, each double line represents a C=C bond, and each triple line represents a C≡C bond. Thus, only the carbon framework of the molecule is shown and you are left to fill in hydrogen atoms as necessary to complete the tetravalence of carbon.

Cycloalkanes contain two fewer hydrogen atoms than alkanes of the same number of carbon atoms. For example, compare the molecular formulas of cyclopropane (C_3H_6) and propane (C_3H_8) or those of cyclohexane (C_6H_{12}) and hexane (C_6H_{14}). The general formula of a cycloalkane is $\mathbf{C_nH_{2n}}$.

To name cycloalkanes, prefix the name of the corresponding open-chain alkane with cyclo-, and name each substituent on the ring. If there is only one substituent on the cycloalkane ring, there is no need to give it a number. If there are two or more substituents, number the ring beginning with the substituent of lowest alphabetical order. Where there is a choice, number to give the additional substituents the lowest possible number or set of numbers.

EXAMPLE 2.5

Write the molecular formula and IUPAC name for each cycloalkane.

(a) 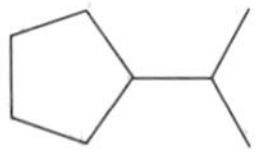 (b) 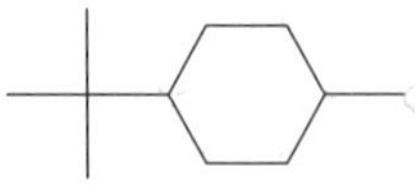 (c)

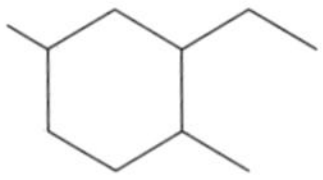

Solution

(a) First replace each angle and line terminus by a carbon and then add hydrogens as necessary to give each carbon four bonds. The molecular formula of this compound is C_8H_{16}. Because there is only an isopropyl group on the ring, there is no need to number the atoms of the ring. Its IUPAC name is isopropylcyclopentane.

$\Rightarrow C_8H_{16}$

(b) Number the atoms of the cyclohexane ring beginning with *tert*-butyl, the substituent of lower alphabetical order. Its name is 1-*tert*-butyl-4-methylcyclohexane and its molecular formula is $C_{11}H_{22}$.

(c) Number the ring beginning with ethyl. Its name is 1-ethyl-2,5-dimethylcyclohexane and its molecular formula is $C_{10}H_{20}$.

PROBLEM 2.5

Write the molecular formula and IUPAC name for each cycloalkane.

(a) **(b)** **(c)**

B. Bicycloalkanes and Spiroalkanes

An alkane that contains two rings that share two carbon atoms is classified as a **bicycloalkane.** The shared carbon atoms are called **bridgehead atoms,** and the carbon chains connecting them are referred to as **bridges.** The general formula of a bicycloalkane is C_nH_{2n-2}. In Figure 2.3, there are three examples of bicycloalkanes along with the IUPAC and common name of each.

Bicycloalkane An alkane containing two rings that share two carbon atoms.

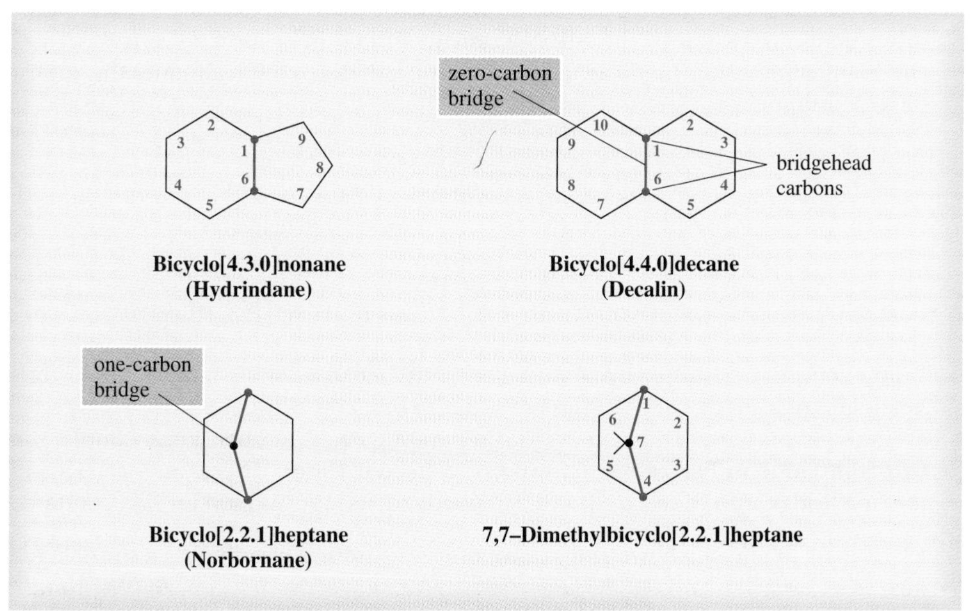

FIGURE 2.3
Examples of bicycloalkanes. Bridgehead atoms are shown in red.

EXAMPLE 2.6

Write the general formula for an alkane, a cycloalkane, and a bicycloalkane. How do these general formulas differ?

Solution

General formulas are alkane, C_nH_{2n+2}; cycloalkane, C_nH_{2n}; and bicycloalkane, C_nH_{2n-2}. Each general formula in this series has two fewer hydrogens than the previous member of the series.

PROBLEM 2.6

Write molecular formulas for each bicycloalkane, given its number of carbon atoms.

(a) Hydrindane (9 carbons) **(b)** Decalin (10 carbons) **(c)** Norbornane (7 carbons)

IUPAC names of bicycloalkanes are derived in the following way:

1. The parent name of a bicycloalkane is that of the unbranched alkane of the same number of carbon atoms as are in the bicyclic ring system. For example, the first compound in Figure 2.3 contains 9 carbons and is, therefore, a bicyclononane; the second compound contains 10 carbons and is a bicyclodecane.
2. Numbering begins at one bridgehead carbon atom and proceeds along the longest bridge to the second bridgehead carbon, then along the next longest bridge back to the original bridgehead carbon, and so on until all ring carbon atoms are numbered. If there are two bridges of the same length, proceed along the one that gives the lower number to the first encountered substituent. The name and location of substituents are shown by the rules already described in Section 2.3A. If there is a choice of numbering patterns, choose the one that gives substituents the lowest possible numbers.

2-Methylbicyclo[4.4.0]decane

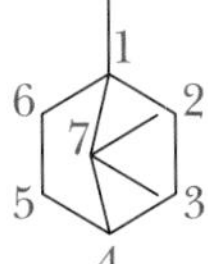

1,7,7-Trimethylbicyclo[2.2.1]heptane

3. Bridge lengths are shown by counting the number of carbons linking the bridgeheads and placing them in decreasing order in brackets between the prefix bicyclo- and the parent name and with periods separating each number. For example, the first compound in Figure 2.3 has two bridgehead carbons. There are 4 carbons in the first bridge, 3 in the second, and 0 in the third; its name is bicyclo[4.3.0]nonane.

Spiroalkane A cycloalkane in which two rings share only one carbon atom.

A cycloalkane in which two rings share only one atom is classified as a **spiroalkane** (Latin, *spira*, twisted), and the single carbon atom shared by the two rings is called a **spiro carbon.** Numbering a spiroalkane begins at the carbon on the shorter bridge next to the spirocarbon, around the shorter bridge, through the spirocarbon, and around the longer bridge. Following is a structural formula and stereoview for spiro[4.4]nonane. Note that the four bonds connected to a spiro carbon create two planes at right angles to each other; consequently, the two rings thus intersecting lie at right angles to each other.

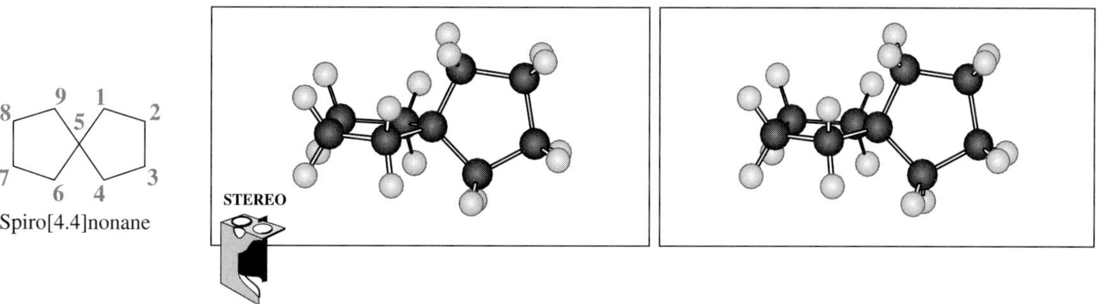

Spiro[4.4]nonane

EXAMPLE 2.7

Following are line-angle formulas, ball-and-stick models, and common names for three bicyclic compounds. Write the molecular formula of each compound and name the bicycloalkane from which it is derived.

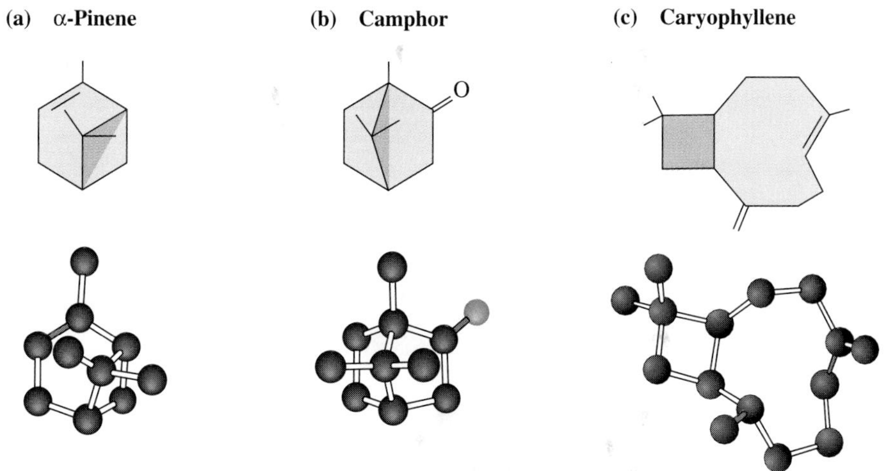

Solution

(a) The molecular formula of α-pinene is $C_{10}H_{16}$ and the bicycloalkane from which it is derived is bicyclo[3.1.1]heptane. α-Pinene is a major component, often as high as 65% by volume, of pine oil and turpentine.

(b) The molecular formula of camphor is $C_{10}H_{16}O$, and the bicycloalkane from which it is derived is bicyclo[2.2.1]heptane.

(c) The molecular formula of caryophyllene is $C_{15}H_{24}$ and the bicycloalkane from which it is derived is bicyclo[7.2.0]undecane. Caryophyllene is one of the fragrant components of oil of cloves.

PROBLEM 2.7

Draw structural formulas for the following bicycloalkanes and spiroalkanes.

(a) bicyclo[3.1.0]hexane **(b)** bicyclo[2.2.2]octane
(c) bicyclo[4.2.0]octane **(d)** 2,6,6-trimethylbicyclo[3.1.1]heptane
(e) spiro[2.4]heptane **(f)** spiro[2.5]octane

2.5 The IUPAC System: A General System of Nomenclature

The naming of alkanes and cycloalkanes in Sections 2.3 and 2.4 illustrated the application of the IUPAC system of nomenclature to two specific classes of organic compounds. Now, let us describe the general approach of the IUPAC system. The name assigned to any compound with a chain of carbon atoms consists of three parts: a **prefix,** an **infix** (a modifying element inserted into a word), and a **suffix.** Each part provides specific information about the structure of the compound.

1. The prefix indicates the number of carbon atoms in the parent chain. Prefixes to show the presence of 1 to 20 carbon atoms in a chain are given in Table 2.2.
2. The infix indicates the nature of the carbon-carbon bonds in the parent chain.

Infix	Nature of Carbon–Carbon Bonds in the Parent Chain
-an-	all single bonds
-en-	one or more double bonds
-yn-	one or more triple bonds

3. The suffix indicates the class of compound to which the substance belongs.

Suffix	Class of Compound
-e	hydrocarbon
-ol	alcohol
-al	aldehyde
-one	ketone
-oic acid	carboxylic acid

EXAMPLE 2.8

Following are IUPAC names and structural formulas for four compounds. Divide each name into a prefix, an infix, and a suffix, and specify the information about the structural formula that is contained in each part of the name. Also show that each name indicates the number of nonhydrogen atoms in a molecule of the compound.

(a) $CH_2{=}CHCH_3$ Propene **(b)** CH_3CH_2OH Ethanol **(c)** $CH_3CH_2CH_2CH_2\overset{\overset{\large O}{\|}}{C}OH$ Pentanoic acid **(d)** $HC{\equiv}CH$ Ethyne

Solution

(a) prop-en-e ← a hydrocarbon (prop: three carbon atoms; en: one carbon-carbon double bond)

(b) eth-an-ol ← an alcohol (eth: two carbon atoms; an: only carbon-carbon single bonds)

(c) pent-an-oic acid ← a carboxylic acid (pent: five carbon atoms; an: only carbon-carbon single bonds)

(d) eth-yn-e ← a hydrocarbon (eth: two carbon atoms; yn: one carbon-carbon triple bond)

Compound (a) contains 3 (prop) carbon atoms. Compound (b) contains 2 (eth) carbon atoms and 1 (ol) oxygen atom. Compound (c) contains 5 (pent) carbon atoms and 2 (oic acid) oxygen atoms. Compound (d) contains 2 (eth) carbon atoms.

PROBLEM 2.8

Combine the proper prefix, infix, and suffix, and write the IUPAC name for each compound.

(a) 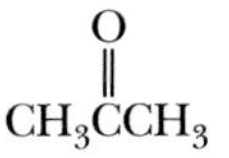 $CH_3C(=O)CH_3$ (b) 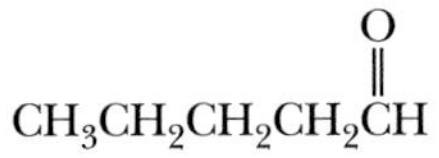 $CH_3CH_2CH_2CH_2CH(=O)$ (c) 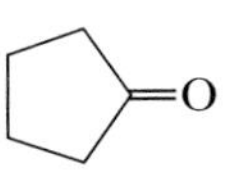 cyclopentanone (ring C=O) (d) 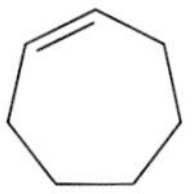 cycloheptene ring

2.6 Conformations of Alkanes and Cycloalkanes

Structural formulas are useful to show the order of attachment of atoms in a molecule. However, they usually do not show three-dimensional shapes. As chemists try to understand more and more about the relationships between structure and the chemical and physical properties of compounds, it becomes increasingly important to understand more about the three-dimensional shapes of molecules. In this section, you are asked to visualize molecules as three-dimensional objects, and to visualize not only bond angles within molecules but also distances within molecules between various atoms and groups not bonded to each other. We also describe intramolecular strain, which we divide into three types: torsional strain, nonbonded interaction strain, and angle strain. We strongly urge you to build models (either physically or by computer using software such as ChemDraw/Chem3D) of the molecules discussed in this section so that you become comfortable in dealing with them as three-dimensional objects and understand fully the origins of intramolecular strain.

Conformation Any three-dimensional arrangement of atoms in a molecule that results by rotation about a single bond.

A. Alkanes

Alkanes of two or more carbons can be twisted into a number of different three-dimensional arrangements of their atoms by rotation about a carbon-carbon bond or bonds. Any three-dimensional arrangement of atoms that results from rotation about single bonds is called a **conformation.** Figure 2.4(a) shows a ball-and-stick model of a **staggered conformation** of ethane. In this conformation, the three C—H bonds on one carbon are as far apart as possible from the three C—H bonds on the adjacent carbon. Figure 2.4(b) is a shorthand way to represent this conformation of ethane. It is called a **Newman projection.** In a Newman projection, a molecule is viewed along

Staggered conformation A conformation about a carbon-carbon single bond where the atoms or groups of atoms on one carbon are as far apart as possible from atoms or groups of atoms on an adjacent carbon.

Newman projection A way to view a molecule by looking along a carbon-carbon bond.

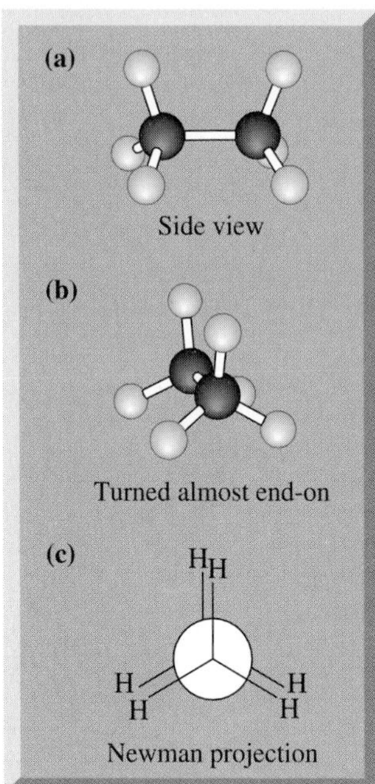

FIGURE 2.5
An eclipsed conformation of ethane. (a, b) Ball-and-stick models and (c) Newman projection.

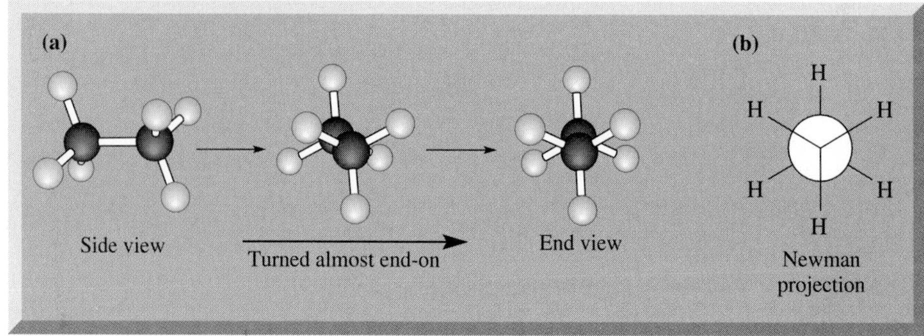

FIGURE 2.4
A staggered conformation of ethane. (a) Ball-and-stick models and (b) Newman projection.

the axis of a C—C bond. The three atoms or groups of atoms on the carbon nearer your eye are shown on lines extending from the center of the circle at angles of 120°. The three atoms or groups of atoms on the carbon farther from your eye are shown on lines extending from the circumference of the circle, also at angles of 120°. Remember that bond angles about each carbon in these Newman projections of ethane are approximately 109.5° and not 120° as this Newman projection might suggest.

Figure 2.5 shows a ball-and-stick model and a Newman projection for an **eclipsed conformation** of ethane. In this conformation, the three C—H bonds on one carbon are as close as possible to the three C—H bonds on the adjacent carbon. In other words, hydrogen atoms on the back carbon are eclipsed by the hydrogen atoms on the front carbon.

Eclipsed conformation A conformation about a carbon-carbon single bond where the atoms or groups of atoms on one carbon are as close as possible to the atoms or groups of atoms on an adjacent carbon.

To discuss potential energy relationships between conformations, it is convenient to define the term dihedral angle. A **dihedral angle, θ** (Greek letter theta), is the angle created by two intersecting planes. In the Newman projection of the eclipsed conformation of ethane in Figure 2.6(a), two H—C—C planes are shown. The angle at which these planes intersect (the dihedral angle) is 0°. Illustrated in Figure 2.6(b) is a staggered conformation in which the dihedral angle of the two H—C—C planes is 60°.

Dihedral angle The angle created by two intersecting planes.

In principle, there are an infinite number of conformations of ethane that differ only in the degree of rotation about the carbon-carbon single bond. At standard conditions of temperature and pressure (STP), ethane is a gas and, under these conditions, ethane molecules undergo collisions with sufficient energy so that rotation about the carbon-carbon single bond from one conformation to another occurs rapidly. There is a potential energy difference between conformations, however, and rotation is not completely free. The lowest-energy, most stable conformation of ethane is the staggered conformation. The highest-energy, least stable conformation is the eclipsed conformation.

The difference in energy between the eclipsed and staggered conformations of ethane is approximately 2.9 kcal/mol (12.1 kJ/mol) and is referred to as torsional strain. **Torsional strain** is the force that opposes the rotation of one part of a molecule about a bond while the other part is held fast. The relationship between potential energy and dihedral angle for the conformations of ethane is shown in Figure 2.7.

Torsional strain The force that opposes the rotation of one part of a molecule about a bond while the other part of the molecule is held fixed.

The torsional strain in eclipsed ethane is due to the slight repulsion of electron pairs of adjacent C—H bonds as they are rotated past each other in converting from one staggered conformation to another. If you make molecular models of staggered and eclipsed ethane, you will see that in an eclipsed conformation, electron pairs of adjacent C—H bonds are closer to one another than they are in a staggered conformation.

Nonbonded interaction strain The strain that arises when atoms not bonded to each other are forced abnormally close to one another.

Next let us look at the conformations of butane viewed along the bond between carbons 2 and 3. There are two types of staggered conformations for butane and two types of eclipsed conformations. The staggered conformation in which the methyl groups are the maximum distance apart ($\theta = 180°$) is called the **anti conformation;** that in which they are closer together ($\theta = 60°$) is called the **gauche conformation.** In one eclipsed conformation ($\theta = 0°$), methyl is eclipsed by methyl. In the other ($\theta = 120°$), methyl is eclipsed by hydrogen. The energy relationships for rotation from 0° to 360° are illustrated in Figure 2.8.

Note that both gauche and anti conformations of butane are staggered conformations, yet the gauche conformations are approximately 0.9 kcal/mol (3.8 kJ/mol) higher in energy than the anti conformation. The difference in energy between these conformations is due to nonbonded interaction strain. **Nonbonded interaction strain** arises when atoms or groups of atoms not bonded to each other are forced into close proximity. Nonbonded interaction strain in the case of gauche butane arises because the two methyl groups are closer to each other than they are in the anti conformation.

Given the difference in potential energy between the anti and gauche conformations of butane, we can calculate the ratio of these two conformations present at equilibrium using the following equation, which relates the change in Gibbs free energy, ΔG^0, for an equilibrium, and the equilibrium constant, K_{eq}

$$\Delta G^0 = -RT \ln K_{eq}$$

where ΔG^0 is the difference in free energy in kcal per mol between the gauche and anti conformations of butane, R is the gas constant ($1.987\ \text{cal} \cdot K^{-1} \cdot \text{mol}^{-1}$), and T is the temperature in degrees Kelvin. Converting from the natural logarithm to $\log_{10}$

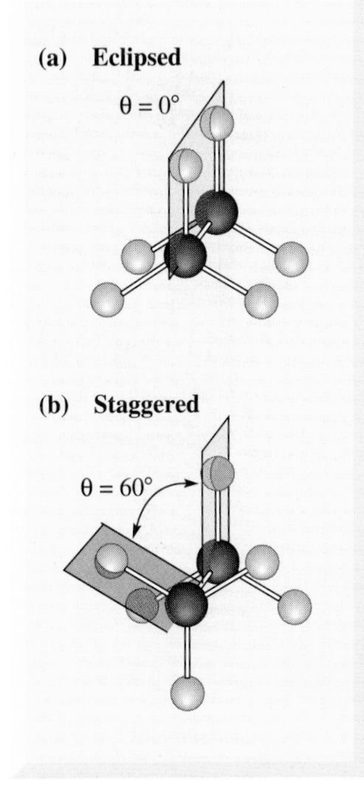

FIGURE 2.6
Dihedral angles in eclipsed and staggered conformations of ethane.

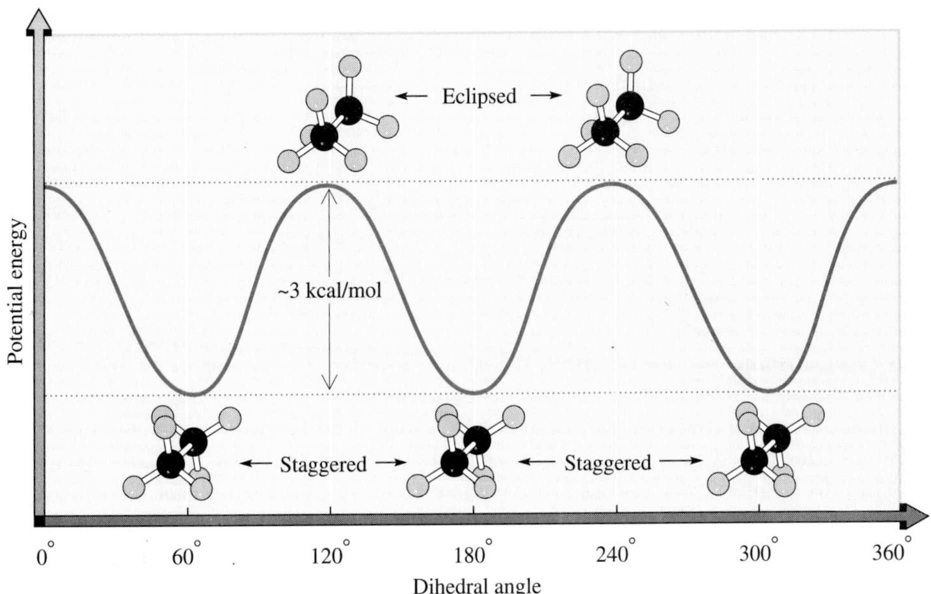

FIGURE 2.7
Potential energy of ethane as a function of dihedral angle. The eclipsed conformations are approximately 3 kcal/mol (12.6 kJ/mol) higher in energy than the staggered conformations.

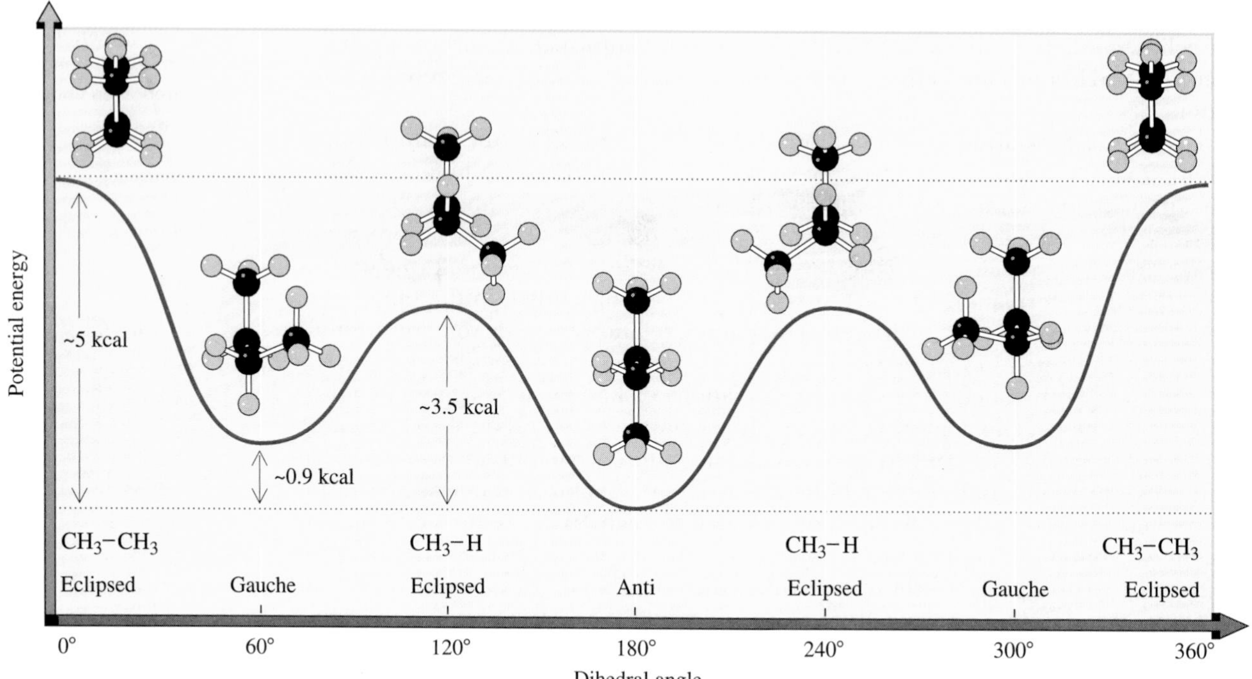

FIGURE 2.8
Potential energy of butane as a function of the dihedral angle about the bond between carbons 2 and 3. The lowest energy conformation occurs when the two methyl groups are the maximum distance apart (dihedral angle 180°). The energy maximum occurs when the two methyl groups are eclipsed (dihedral angle 0°).

(which requires the factor of 2.303) and solving this equation for log K_{eq} gives

$$\log K_{eq} = \frac{-\Delta G^0}{2.303\ RT}$$

Substituting the value of −900 cal/mol (gauche butane → anti butane) for ΔG^0 and solving the equation gives a value of 4.57 for the equilibrium constant at room temperature (25°C = 298K).

$$\log K_{eq} = \frac{-(-900\ \text{cal}\cdot\text{mol}^{-1})}{2.303\ (1.987\ \text{cal}\cdot\text{K}^{-1}\cdot\text{mol}^{-1})\ 298\text{K}} = 0.660$$

$$K_{eq} = 10^{0.660} = \frac{4.57\ \text{anti conformation}}{1\ \text{gauche conformation}}$$

At any given instant, therefore, there is a larger number of butane molecules in the anti conformation than in the gauche conformation.

You should note that, even though the two staggered conformations with methyl groups gauche (dihedral angles 60° and 300°) have equal energies, they are not identical. They are related by reflection; one is the reflection of the other just as your right hand is the reflection of your left hand. Notice that the conformations with eclipsed $—CH_3$ and —H groups (dihedral angles of 120° and 240°) are also related by reflection. We will have more to say about objects and their mirror reflections in Chapter 4.

EXAMPLE 2.9

Following is the structural formula of 1,2-dichloroethane.

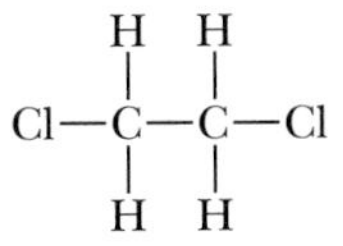

```
     H   H
     |   |
Cl—C—C—Cl
     |   |
     H   H
```

1,2-Dichloroethane

(a) Draw Newman projections for all staggered conformations formed by rotation from 0° to 360° about the carbon-carbon single bond.

(b) Which staggered conformation(s) has the lowest energy; which has the highest energy?

(c) Which, if any, of these conformations are related by reflection?

Solution

(a) If we take as a reference point the dihedral angle when the chlorines are eclipsed, staggered conformations occur at dihedral angles 60°, 180°, and 300°.

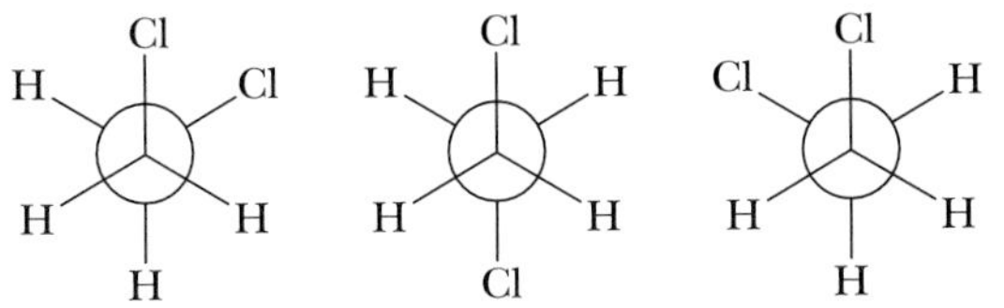

Gauche $\theta = 60°$ Anti $\theta = 180°$ Gauche $\theta = 300°$

(b) We predict that the anti conformation ($\theta = 180°$) has the lowest energy. The two gauche conformations ($\theta = 60°$ and $300°$) are of higher but equal energy. We are not given data in the problem to calculate the actual energy differences.

(c) The two gauche conformations are related by reflection.

PROBLEM 2.9

From studies of dipole moments, it has been estimated that, in the gas phase at room temperature, the ratio of molecules of 1,2-dichloroethane in the anti to gauche conformations is approximately 7.6 to 1. Calculate the difference in energy between these two conformations.

B. Cycloalkanes

Cyclopropane

The observed C—C—C bond angles in cyclopropane are 60° (Figure 2.9), a value considerably smaller than that of the 109.5° predicted for sp^3-hybridized carbon atoms. Furthermore, hydrogen atoms on adjacent carbons are forced into an eclipsed relationship.

Cyclopropane is a strained molecule due to both angle strain and torsional strain. **Angle strain** arises from the creation of abnormal bond angles, in this case compres-

Angle strain The strain that arises when a bond angle is either compressed or expanded compared to its normal value.

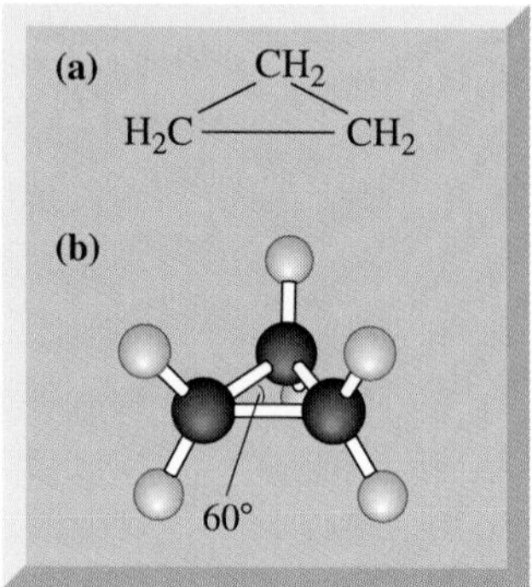

FIGURE 2.9
Cyclopropane. (a) Structural formula and (b) ball-and-stick model.

sion of C—C—C bond angles from 109.5° to 60°. **Torsional strain** in cyclopropane arises because of the eclipsed C—H bonds around the ring. The total strain energy in cyclopropane is approximately 28 kcal/mol (see Problem 2.53). It is because of its extreme degree of intramolecular strain that cyclopropane and its derivatives undergo several ring-opening reactions not shown by larger cycloalkanes.

Cyclobutane

In all cycloalkanes larger than cyclopropane, nonplanar or **puckered conformations** are favored. Figure 2.10 shows stereorepresentations for one planar and one puckered conformation of cyclobutane. If cyclobutane were planar, all C—C—C bond angles would be 90°, and there would be eight pairs of eclipsed hydrogen interactions. Puckering of the ring alters its potential energy in two ways: (1) it decreases the torsional strain associated with eclipsed hydrogen interactions, but (2) it increases further the angle strain due to compression of C—C—C bond angles. Since the decrease in torsional strain is greater than the increase in angle strain, puckered cyclobutane is more stable than planar cyclobutane. In the conformation of lowest energy, the measured bond angle is 88°. The strain energy in cyclobutane is approximately 26 kcal/mol (see Problem 2.53).

Cyclopentane

If cyclopentane were to adopt a planar conformation, all C—C—C bond angles would be 108°. This angle differs only slightly from the tetrahedral angle of 109.5°, and, consequently, there would be little angle strain in this conformation. There would be, however, ten fully eclipsed C—H bonds creating a torsional strain of approximately 10 kcal/mol (42 kJ/mol). To relieve at least a part of this torsional strain, the ring twists into an **"envelope" conformation,** shown in Figure 2.11. In this conformation, four carbon atoms are in a plane and the fifth is bent out of the plane, rather like an envelope with its flap bent upward. Cyclopentane is a dynamic equilibrium of five equivalent envelope conformations. In the envelope conformation, C—C—C bond angles are reduced (increasing angle strain), but the number of eclipsed hydrogen interactions is reduced (decreasing torsional strain). Overall, the molecule is more stable in the envelope conformation than in the planar conformation. The average C—C—C bond angle in cyclopentane is 105°, indicating that in its conformation of lowest energy, cyclopentane is slightly puckered. The strain energy in cyclopentane is approximately 6.5 kcal/mol (see Problem 2.53).

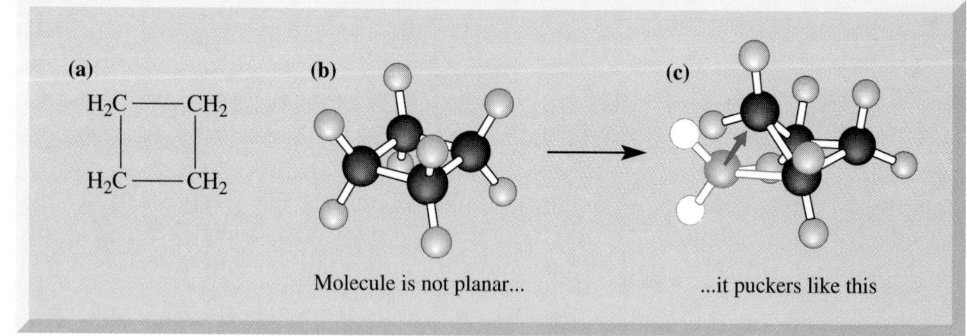

FIGURE 2.10
Cyclobutane. (a) Planar conformation and (b) a puckered or "butterfly" conformation in which C—C—C bond angles are approximately 88°. The potential energy of cyclobutane is a minimum in this puckered conformation.

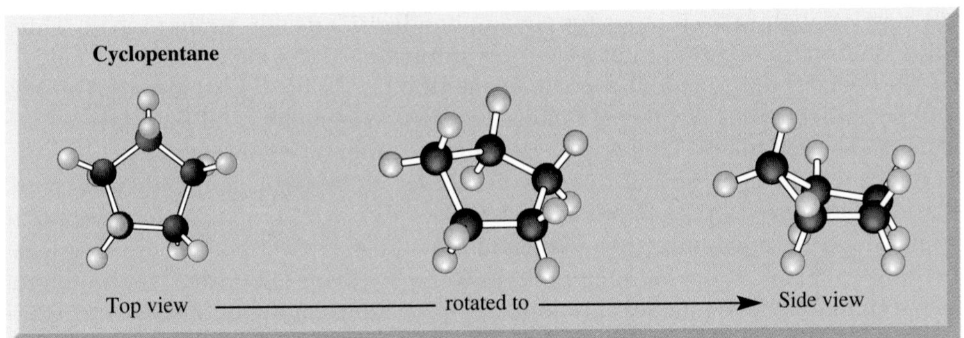

FIGURE 2.11 Cyclopentane. The most stable conformation is a puckered "envelope" conformation.

Cyclohexane

Cyclohexane adopts a number of puckered conformations, the most stable of which is the **chair conformation.** In a chair conformation (Figure 2.12), all C—C—C bond angles are 109.5° (precluding angle strain), and hydrogens on adjacent carbons are staggered with respect to one another (precluding torsional strain). Also, no two atoms

Chair conformation The most stable puckered conformation of a cyclohexane ring; all bond angles are approximately 109.5°, and all bonds on adjacent carbons are staggered.

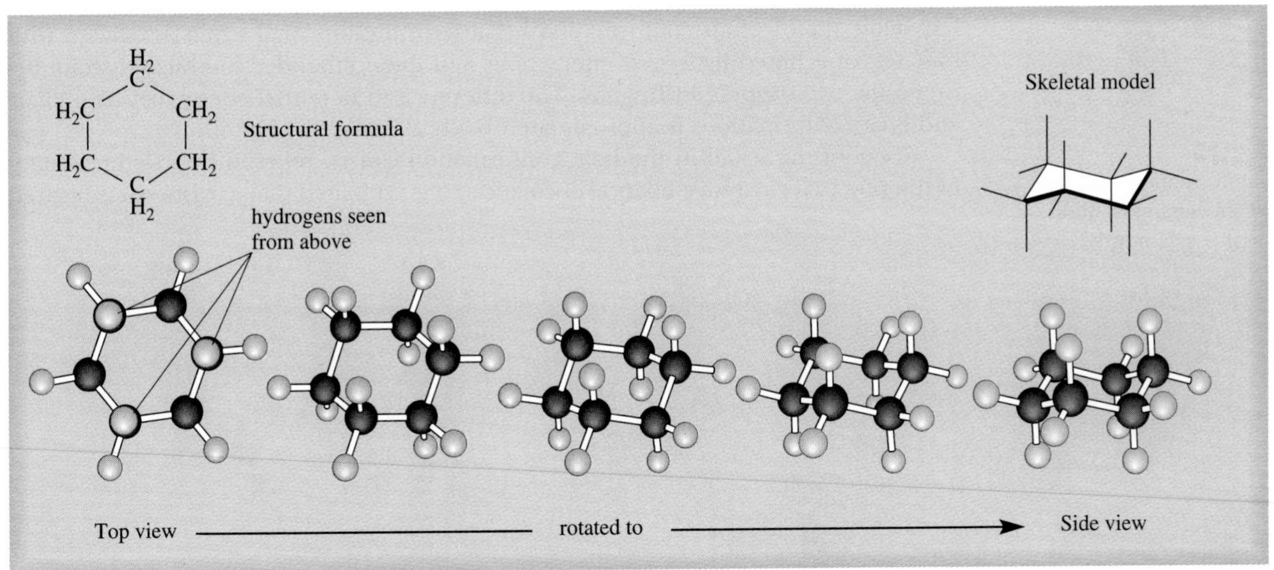

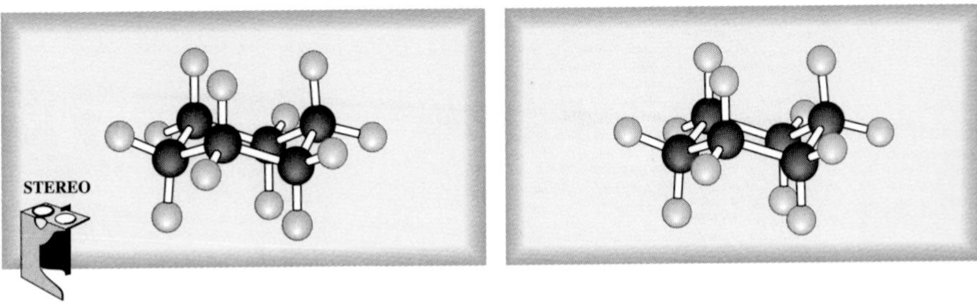

FIGURE 2.12 Cyclohexane.

Axial position A position on a chair conformation of a cyclohexane ring that extends from the ring parallel to the imaginary axis of the ring.

Equatorial position A position on a chair conformation of a cyclohexane ring that extends from the ring roughly perpendicular to the imaginary axis of the ring.

Boat conformation A puckered conformation of a cyclohexane ring in which carbons 1 and 4 of the ring are bent toward each other.

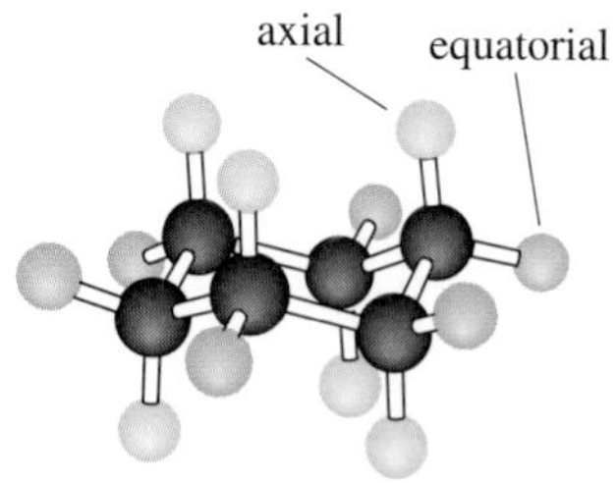

FIGURE 2.13
Chair conformation of cyclohexane, showing axial and equatorial C—H bonds.

are close enough to each other for any nonbonded interaction strain to exist. Thus, there is no strain of any kind in a chair conformation of cyclohexane.

In a chair conformation of cyclohexane, the C—H bonds are arranged in two different orientations. Six C—H bonds are called **axial bonds,** and the other six are called **equatorial bonds.** One way to visualize the difference between these two types of bonds is to imagine an axis through the center of the chair, perpendicular to the floor (Figure 2.13). Axial bonds are parallel to this axis. Three axial bonds point straight up; the other three axial bonds point straight down. Notice also that axial bonds alternate, first up and then down as you move from one carbon of the ring to the next. Equatorial bonds are approximately perpendicular to our imaginary axis and parallel to two ring carbon-carbon bonds. Equatorial bonds also alternate first slightly up and then slightly down as you move from one carbon of the ring to the next. Notice further that if the axial bond on a carbon points upward, then the equatorial bond on that carbon points slightly downward. Conversely, if the axial bond on a particular carbon points downward, then the equatorial bond on that carbon points slightly upward.

There are many other nonplanar conformations of cyclohexane, two of which, the **boat conformation** and **twist-boat conformation,** are shown in Figure 2.14. You can visualize interconversion of chair and boat conformations by twisting about a carbon-carbon bond as illustrated in Figure 2.15. A boat conformation is considerably less stable than a chair conformation because of the torsional strain associated with four sets of eclipsed hydrogen interactions and the nonbonded interaction strain between the two **flagpole hydrogens.** The difference in potential energy between chair and boat conformations is approximately 6.5 kcal/mol (27 kJ/mol).

Some of the strain in the boat conformation can be relieved by a slight twisting of the ring to form a twist-boat conformation. It is estimated that a twist-boat is favored

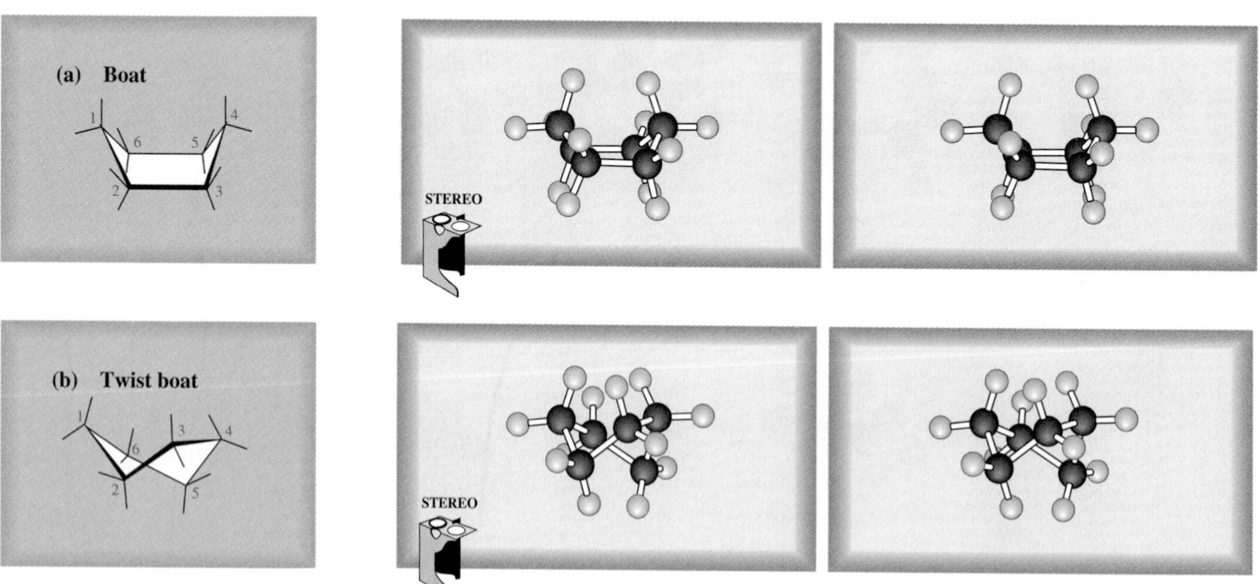

FIGURE 2.14
Boat and twist-boat conformations of cyclohexane.

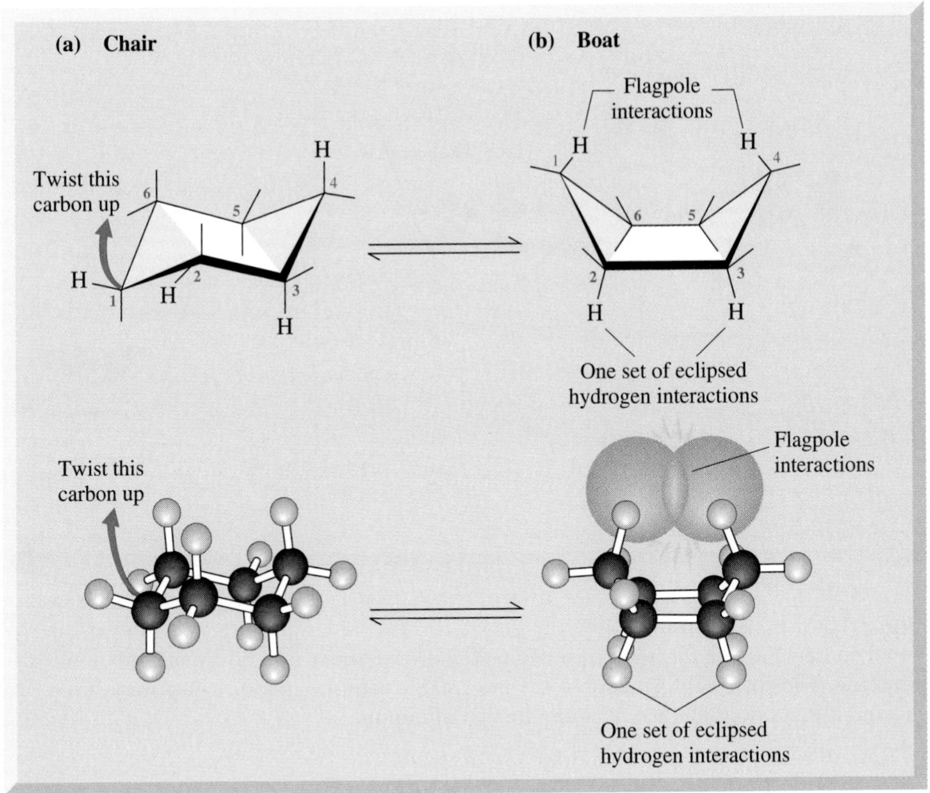

FIGURE 2.15
Interconversion of (a) a chair conformation to (b) a boat conformation produces one set of nonbonded "flagpole" interactions and four sets of eclipsed hydrogen interactions.

over a boat conformation by approximately 1.5 kcal/mol (6.3 kJ/mol). A potential energy diagram for interconversion between chair, twist-boat, and boat conformations is shown in Figure 2.16. The large difference in potential energy between chair and boat or twist-boat conformations means that at room temperature, molecules in chair conformations make up more than 99.99% of the equilibrium mixture.

For cyclohexane, the two equivalent chair conformations can be interconverted by twisting one chair first to a boat and then to the other chair.

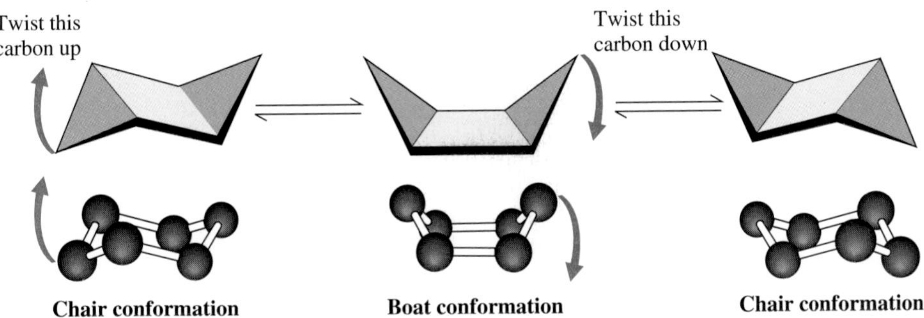

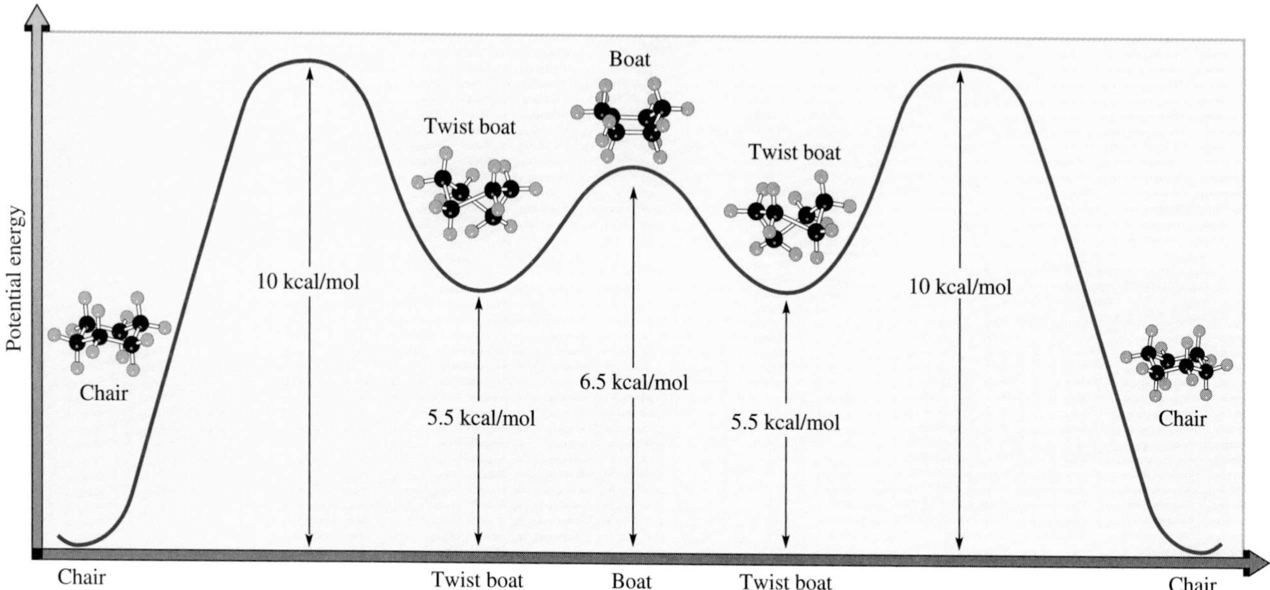

FIGURE 2.16
Potential energy diagram for interconversion of chair, twist-boat, and boat conformations of cyclohexane. The chair conformation is the most stable conformation because angle, torsional, and nonbonded interaction strain are at a minimum.

When one chair is converted to the other, a change occurs in the relative orientations in space of the hydrogen atoms attached to each carbon. All hydrogen atoms axial in one chair become equatorial in the other and vice versa (Figure 2.17). The conversion of one chair conformation of cyclohexane to the other occurs rapidly at room temperature.

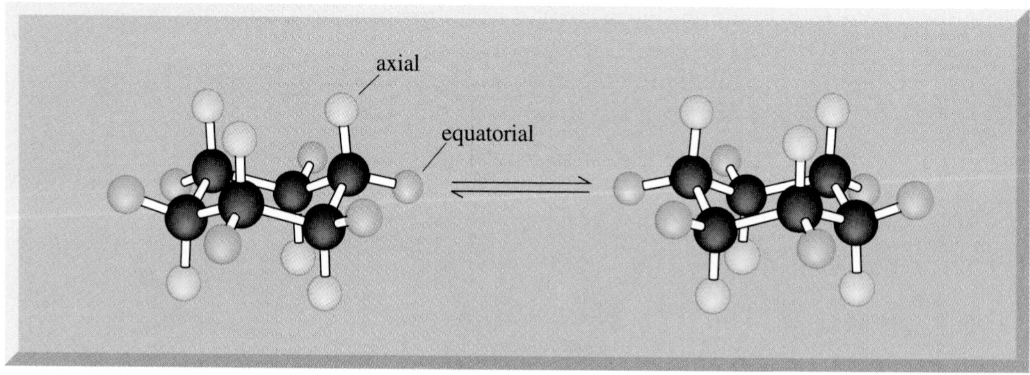

FIGURE 2.17
Interconversion of chair cyclohexanes. All C—H bonds equatorial in one chair are axial in the alternative chair, and vice versa.

EXAMPLE 2.10

Following is a chair conformation of cyclohexane showing two of its 12 hydrogens.

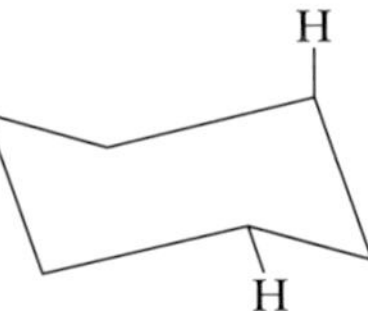

(a) Indicate by a label whether each hydrogen is equatorial or axial.
(b) Draw the other chair conformation and again label each hydrogen equatorial or axial.

Solution

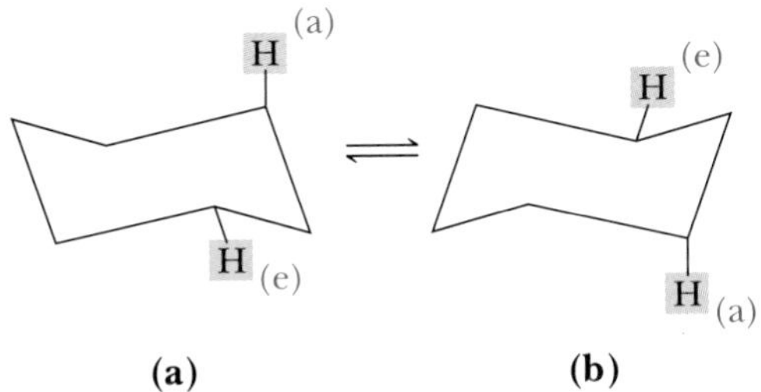

PROBLEM 2.10

Following is a chair conformation of cyclohexane with carbon atoms numbered 1 through 6.

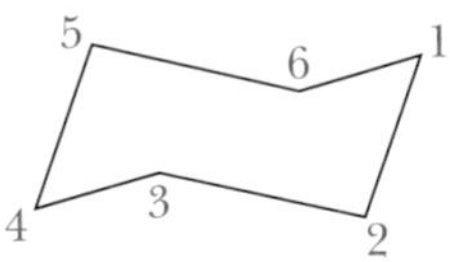

(a) Draw hydrogen atoms that are above the plane of the ring on carbons 1 and 2 and below the plane of the ring on carbon 4.
(b) Which of these hydrogens are equatorial; which are axial?
(c) Draw the other chair conformation. Now, which hydrogens are equatorial; which are axial?

If a hydrogen atom of cyclohexane is replaced by a methyl group or other alkyl group, the group occupies an equatorial position in one chair and an axial position in the other chair. This means that the two chairs are no longer equivalent and no longer of equal stability.

A convenient way to describe the relative stabilities of chair conformations with equatorial or axial substituents is in terms of a type of nonbonded interaction strain called **diaxial interaction.** Diaxial interaction refers to the repulsion between an axial substituent and an axial hydrogen (or other group) on the same side of the ring. Consider methylcyclohexane (Figure 2.18). When the $—CH_3$ is axial, it is parallel to

Diaxial interactions Interactions between groups in axial positions on the same side of a chair conformation of a cyclohexane ring.

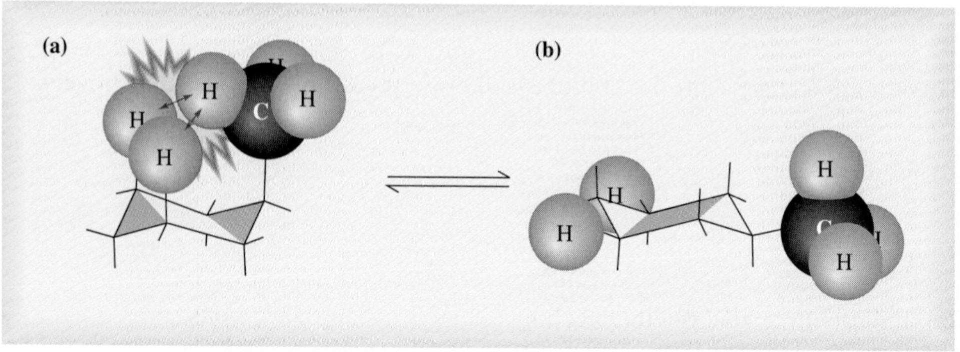

FIGURE 2.18
Two chair conformations of methylcyclohexane. The two diaxial interactions make conformation (a) less stable than conformation (b) by approximately 1.74 kcal/mol (7.28 kJ/mol).

the axial C—H bonds on carbons 3 and 5. Thus, for axial methylcyclohexane, there are two unfavorable methyl-hydrogen diaxial interactions. No such unfavorable interactions exist when the methyl group is in an equatorial position. For methylcyclohexane, the equatorial methyl conformation is favored over the axial methyl conformation by about 1.74 kcal/mol (7.28 kJ/mol). At equilibrium, approximately 95% of all methylcyclohexane molecules have the methyl group equatorial, and 5% have the methyl group axial.

To understand why axial methylcyclohexane is less stable than equatorial methylcyclohexane, recall our discussion of the conformations of butane. There we saw that gauche butane is less stable than anti butane by approximately 0.9 kcal/mol. Figure 2.19 is a stereoview of axial methylcyclohexane turned so that you sight along bonds C_1—C_2 and C_4—C_5 of the ring. As illustrated here, the conformation of the methyl group and an axial hydrogen is identical to a gauche butane conformation. There are actually two gauche butane-like interactions, one with the axial hydrogen on carbon 3, the other with the axial hydrogen on carbon 5. These two gauche butane-like interactions [$2 \times 0.9 = 1.8$ kcal/mol (7.5 kJ/mol)] in axial methylcyclohexane account for the intramolecular strain in this conformation [1.74 kcal/mol (7.3 kJ/mol)] compared to equatorial methylcyclohexane.

As the size of the alkyl substituent increases, preference for conformations with the group equatorial increases. Table 2.4 shows the difference in free energy between axial and equatorial substituents for monosubstituted cyclohexanes. With a group as large as *tert*-butyl, the equatorial conformation is approximately 4000 times more abun-

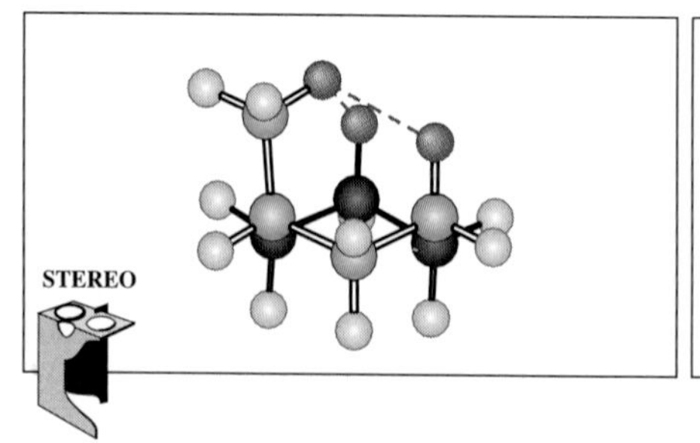

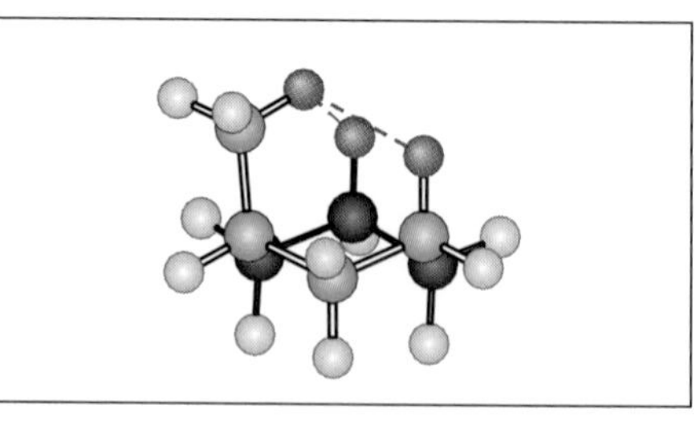

FIGURE 2.19
Intramolecular strain in axial methylcyclohexane. The ring is oriented to show the two diaxial interactions. The gauche butane skeleton is shown in color.

TABLE 2.4 ΔG^0 Between Equatorial and Axial Substituents for Monosubstituted Cyclohexanes at 25°C

axial $\rightleftharpoons$ equatorial

Group	$-\Delta G^0$ (kcal/mol)	Group	$-\Delta G^0$ (kcal/mol)
C≡N	0.20	NH_2	1.4
F	0.25	CO_2H	1.41
C≡CH	0.41	$CH{=}CH_2$	1.7
I	0.46	CH_3	1.74
Cl	0.52	CH_2CH_3	1.75
Br	0.55	$CH(CH_3)_2$	2.15
OH	0.95	$C(CH_3)_3$	4.9

dant at room temperature than the axial, and, in effect, a cyclohexane ring is "locked" into the chair conformation that has the *tert*-butyl group equatorial. In fact, a chair with an axial *tert*-butyl group is so unstable that, if a *tert*-butyl group is forced into an axial position, the ring adopts a twist-boat conformation.

EXAMPLE 2.11

Label all diaxial interactions in this chair conformation.

Solution

There are four diaxial interactions in this example; each axial methyl group has two sets of diaxial interactions with parallel hydrogen atoms on the same side of the ring. The equatorial methyl group has no diaxial interactions.

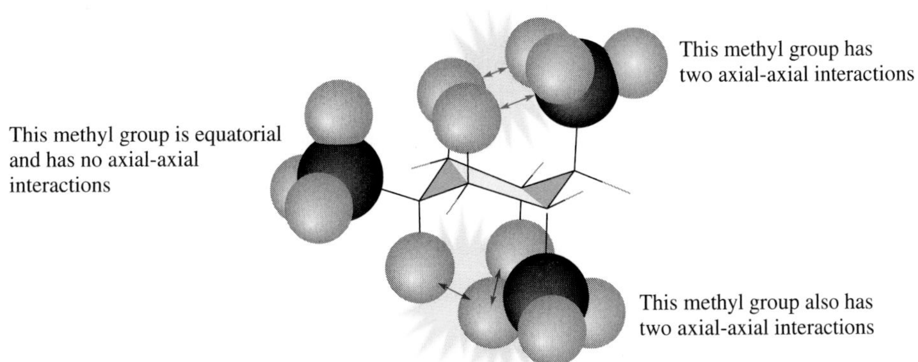
This methyl group has two axial-axial interactions
This methyl group is equatorial and has no axial-axial interactions
This methyl group also has two axial-axial interactions

CHEMISTRY IN ACTION

The Poisonous Puffer Fish

Nature is by no means limited to carbon in six-membered rings. Tetrodotoxin, one of the most potent toxins known, is composed of a set of interconnected six-membered rings, each in a chair conformation. All but one of these rings have atoms other than carbon in them. Tetrodotoxin is produced in the liver and ovaries of many species of *Tetraodontidae*, especially the puffer fish, so called because it inflates itself to an almost spherical spiny ball when alarmed. It is evidently a species highly preoccupied with defense, but the Japanese are not put off. They regard the puffer, called "fugu" in Japanese, as a delicacy. To serve it in a public restaurant, a chef must be registered as sufficiently skilled in removing the toxic organs so as to make the flesh safe to eat.

The puffer fish is considered a delicacy in Japan. (Gérard Lacz/Animals, Animals)

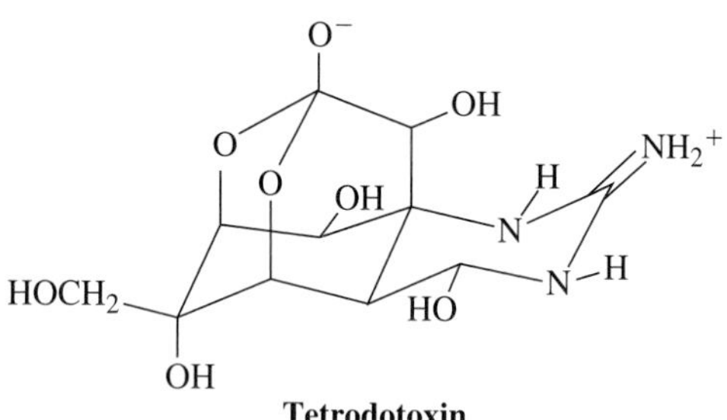

Tetrodotoxin

Symptoms of tetrodotoxin poisoning begin with attacks of severe weakness, progressing to complete paralysis and eventual death. Tetrodotoxin exerts its severe poisoning effect by blocking Na^+ ion channels in excitable membranes. The $=NH_2^+$ end of tetrodotoxin lodges in the mouth of a Na^+ ion channel, thus blocking further transport of Na^+ ions through the channel.

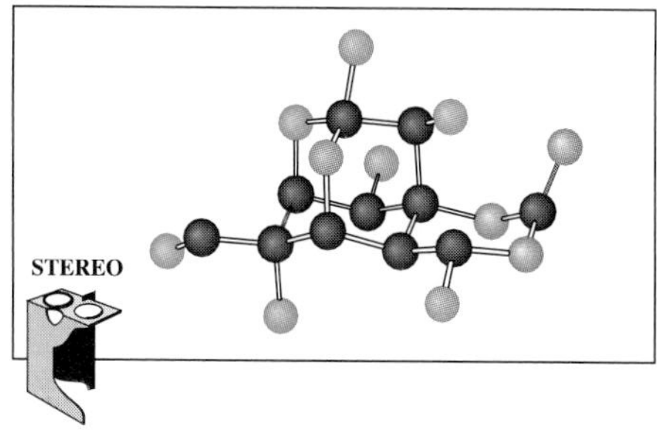

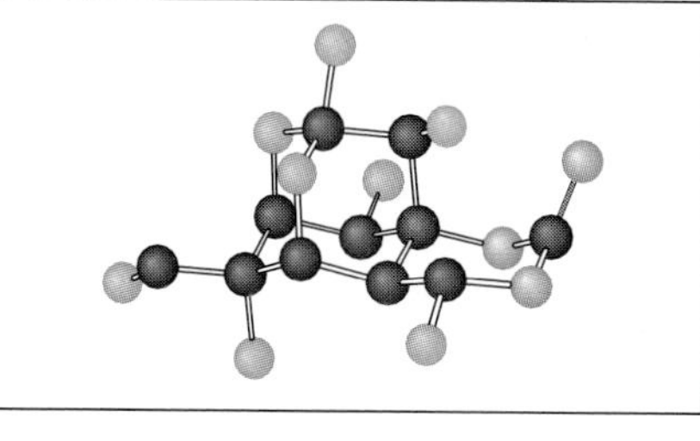

PROBLEM 2.11

Estimate the difference in energy between the alternative chair conformations of the trisubstituted cyclohexane given in Example 2.11.

EXAMPLE 2.12

Calculate the ratio of equatorial methylcyclohexane to axial methylcyclohexane at 25°C.

Solution

For these alternative chair conformations, ΔG^0 (axial $CH_3 \rightarrow$ equatorial CH_3) = -1.74 kcal/mol. Substituting this value in the equation $\Delta G^0 = -2.303\ RT \log K_{eq}$ gives a ratio 19 : 1.

$$\log K_{eq} = \frac{-(-1740 \text{ cal}\cdot\text{mol}^{-1})}{2.303\ (1.987 \text{ cal}\cdot\text{K}^{-1}\cdot\text{mol}^{-1})\ 298\text{K}} = 1.28$$

$$K_{eq} = 10^{1.28} = \frac{19}{1}$$

PROBLEM 2.12

For 1,4-dimethylcyclohexane:

(a) Draw the chair conformation in which both methyl groups are equatorial.
(b) Draw the chair conformation in which both methyl groups are axial.
(c) Estimate the difference in energy between these two conformations.
(d) Calculate the ratio of the diequatorial to the diaxial conformation for this isomer of 1,4-dimethylcyclohexane at 25°C.

Highly Strained Small-Ring Compounds

We have seen that small-ring compounds such as cyclopropane and cyclobutane have a high degree of both angle strain and torsional strain. It has been a particular challenge to chemists to attempt to synthesize even more highly strained rings, both for the challenge of devising new reaction sequences to make such molecules and to better understand relationships between molecular strain and chemical reactivity. Among the fascinating molecules synthesized in recent years are bicyclo[1.1.0]butane and a variety of propellanes. A **propellane** is a molecule in which two atoms joined by a single bond are also joined by three other bridges. The [1.1.1]propellane shown here is the smallest member of this class of compounds and, surprisingly, is more stable than the larger [2.1.1]propellane and [2.2.1]propellane.

Bicyclo[1.1.0]butane
(Bicyclobutane)

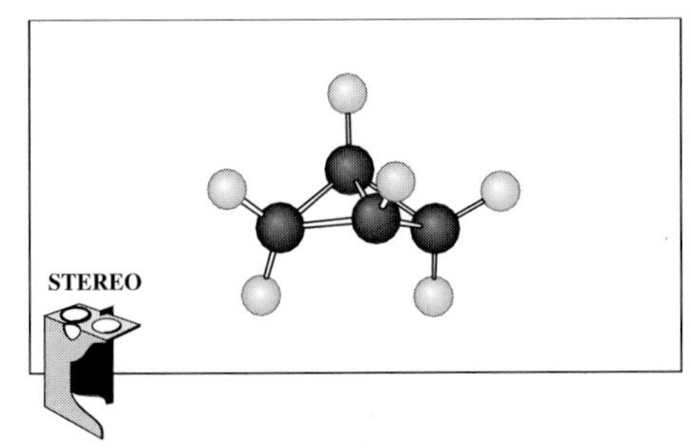

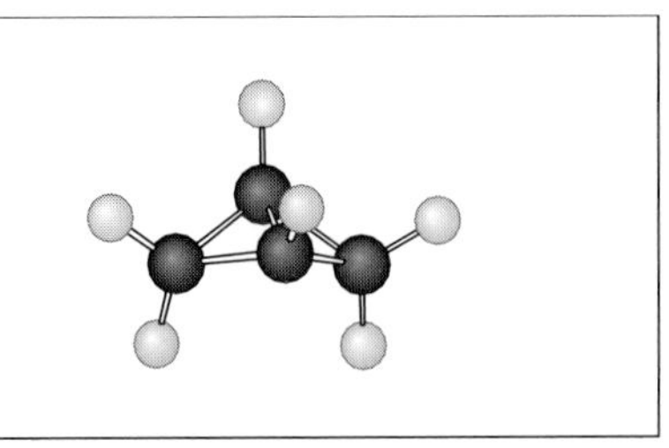

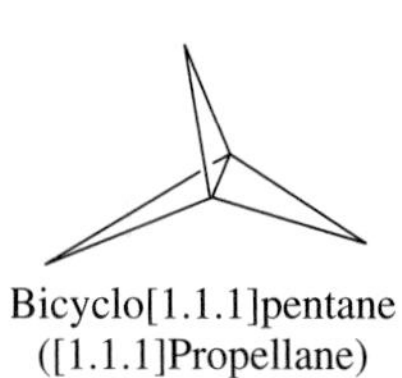

Bicyclo[1.1.1]pentane
([1.1.1]Propellane)

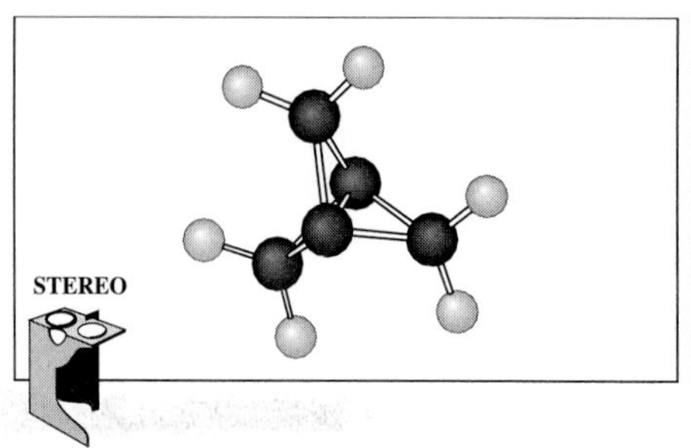

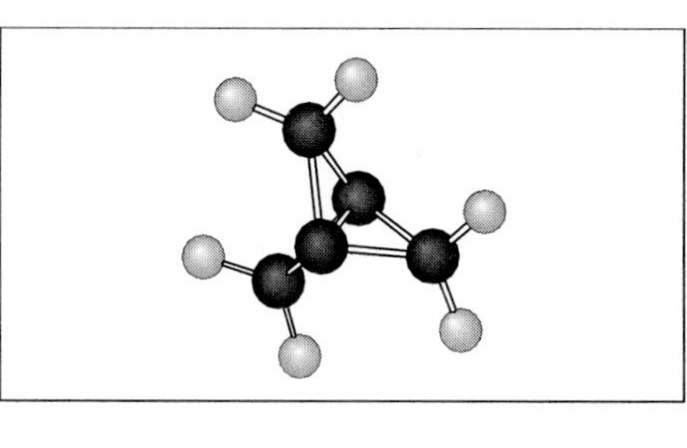

2.7 *Cis-Trans* Isomerism in Cycloalkanes and Bicycloalkanes

A. Cycloalkanes

Cis-trans isomers Isomers that have the same order of attachment of their atoms but a different arrangement of their atoms in space due to the presence of either a ring or a carbon-carbon double bond.

All cycloalkanes with substituents on two or more carbons of the ring show a type of isomerism called ***cis-trans* isomerism.** *Cis-trans* isomers have (1) the same molecular formula, (2) the same order of attachment of atoms, and (3) arrangements of atoms that cannot be interconverted by rotation about single bonds under ordinary conditions. By way of comparison, the potential energy difference between conformations is so small that they can be interconverted easily at or near room temperature by rotation about single bonds.

Cis A prefix meaning on the same side.

Trans A prefix meaning across from.

Cis-trans isomerism in cyclic structures can be illustrated by models of 1,2-dimethylcyclopentane. In the following drawings, the cyclopentane ring is shown as a planar pentagon viewed through the plane of the ring. Carbon-carbon bonds of the ring projecting forward are shown as heavy lines. When viewed from this perspective, substituents attached to the ring project above and below the plane of the ring. In one isomer of 1,2-dimethylcyclopentane, the methyl groups are on the same side of the ring; in the other, they are on opposite sides of the ring. The prefix ***cis*** (Latin: on the same side) is used to indicate that the substituents are on the same side of the ring; the prefix ***trans*** (Latin: across) is used to indicate that they are on opposite sides of the ring. In each isomer, the configuration of the methyl groups is fixed, and, because of the restricted rotation about a ring C—C bond, the *cis* isomer cannot be converted to the *trans* isomer or vice versa.

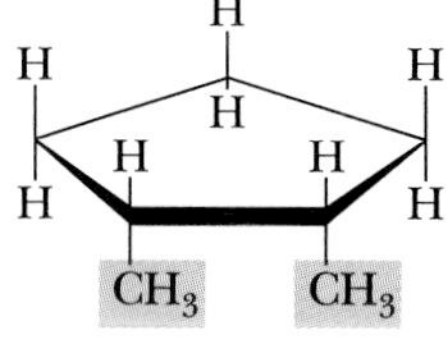

cis-1,2-Dimethylcyclopentane

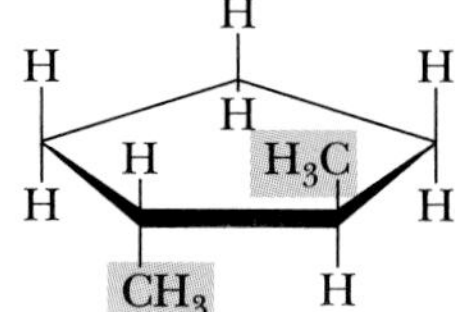

trans-1,2-Dimethylcyclopentane

EXAMPLE 2.13

Which cycloalkanes show *cis-trans* isomerism? For each that does, draw the *cis* and *trans* isomers.

(a) CH_3 (b) CH_3 CH_3 (c) H_3C CH_3

Solution

(a) Methylcyclopentane does not show *cis-trans* isomerism. It has only one substituent on the ring.

(b) 1,1-Dimethylcyclobutane does not show *cis-trans* isomerism. Only one arrangement is possible for the two methyl groups on the ring; they must be *trans* from each other.

(c) 1,3-Dimethylcyclobutane shows *cis-trans* isomerism. Note that in these structures, we show only the hydrogen atoms on carbons bearing the methyl groups.

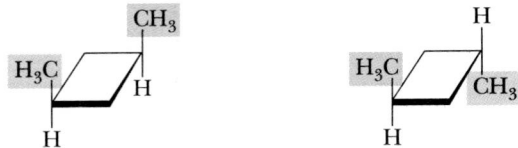

cis-1,3-Dimethylcyclobutane *trans*-1,3-Dimethylcyclobutane

PROBLEM 2.13

Which cycloalkanes show *cis-trans* isomerism? For each that does, draw the *cis* and *trans* isomers.

(a) H_3C CH_3 (b) CH_2CH_3 (c) CH_2CH_3 CH_3

For the purposes of determining the number of *cis-trans* isomers in substituted cycloalkanes, it is adequate to draw the cycloalkane ring as a planar polygon as is done in the following disubstituted cyclohexane. Two *cis-trans* isomers are possible for 1,4-dimethylcyclohexane. In these structural formulas, only the hydrogens on methyl-substituted carbons are shown.

trans-1,4-Dimethylcyclohexane *cis*-1,4-Dimethylcyclohexane

The *cis* and *trans* isomers of 1,4-dimethylcyclohexane can also be drawn as nonplanar chair conformations. When working with alternative chair conformations, it is helpful to remember that all axial groups in one chair become equatorial in the alternative chair, and vice versa.

In one chair conformation of *trans*-1,4-dimethylcyclohexane, the two methyl groups are axial; in the alternative chair conformation they are equatorial. Of these chair conformations, the one with both methyls equatorial is considerably more stable.

(less stable) (more stable)

trans-1,4-Dimethylcyclohexane

The alternative chair conformations of *cis*-1,4-dimethylcyclohexane are of equal energy. In one chair conformation of *cis*-1,4-dimethylcyclohexane, one methyl group occupies an equatorial position, and the other occupies an axial position. In the other chair, the orientations in space of the $-CH_3$ groups are reversed.

cis-1,4-Dimethylcyclohexane
(these conformations are of equal stability)

EXAMPLE 2.14

Following is a chair conformation of 1,3-dimethylcyclohexane.

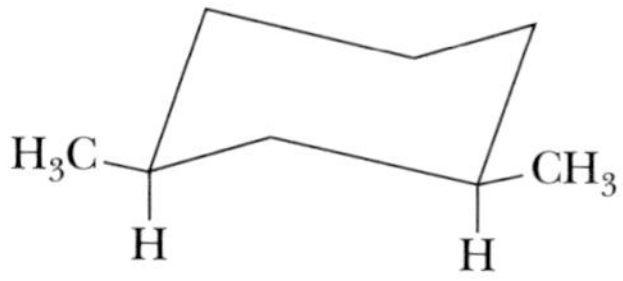

(a) Is this a chair conformation of *cis*-1,3-dimethylcyclohexane or of *trans*-1,3-dimethylcyclohexane?

(b) Draw the alternative chair conformation of this isomer. Of the two chair conformations, which is the more stable?

(c) Draw a planar hexagon representation for the isomer shown in this example.

Solution

(a) The isomer shown is *cis*-1,3-dimethylcyclohexane; the two methyl groups are on the same side of the ring.

(b) **(c)**

Diequatorial conformation (more stable)

Diaxial conformation (less stable)

PROBLEM 2.14

Following is a planar hexagon representation for one isomer of 1,2,4-trimethylcyclohexane. Draw alternative chair conformations of this isomer and state which chair conformation is the more stable.

B. Bicycloalkanes

Bicycloalkanes, particularly those with a carbon skeleton like that of bicyclo[4.4.0]decane (decalin), are abundant in the biological world. Figure 2.20 shows structural formulas for *trans*-decalin and *cis*-decalin. In the *trans* isomer, the two hydrogen atoms on the bridgehead carbons are on opposite sides of the molecule, and in the *cis* isomer, they are on the same side of the molecule. *Trans*-decalin and *cis*-decalin are different compounds and have different physical and chemical properties. The *cis* isomer, for example, has a boiling point of 195°C; the *trans* isomer has a boiling point of 185.5°C.

EXAMPLE 2.15

Compare the relative stabilities of *cis*-decalin and *trans*-decalin. Base your answer on the number of diaxial interactions in the most stable conformation of each isomer.

Solution

There are no diaxial interactions on the all-chair conformation of *trans*-decalin shown in Figure 2.20. In the *cis* isomer, there are three gauche butane interactions as shown in the stereoview. *trans*-Decalin is more stable than *cis*-decalin by approximately

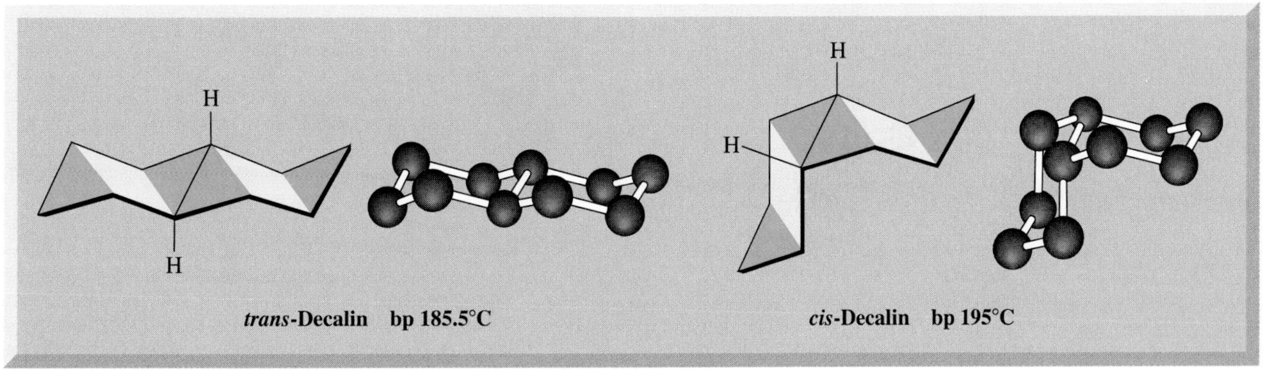

FIGURE 2.20
Cis-trans isomers of decalin.

2.9 kcal/mol (12.1 kJ/mol), which is consistent with three gauche butane interactions in the *cis* isomer.

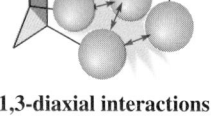

1,3-diaxial interactions

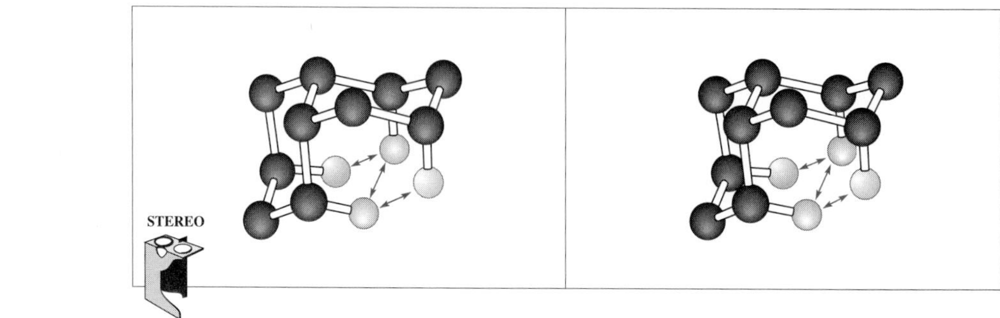

PROBLEM 2.15

Which stereoisomer is more stable?

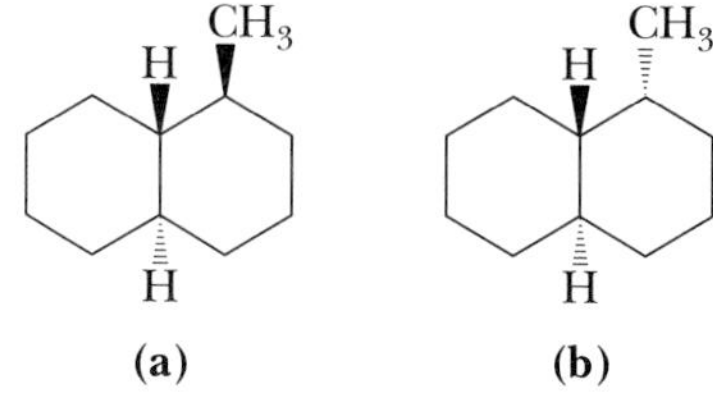

Let us also look at the possibilities for *cis-trans* isomerism in bicyclo[2.2.1]heptane (norbornane). This molecule is viewed in Figure 2.21 in three different ways. Figure

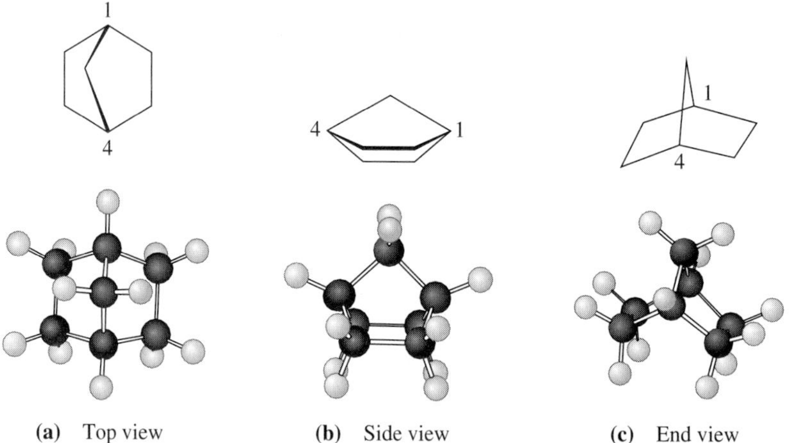

FIGURE 2.21
Bicyclo[2.2.1]heptane (norbornane) from three perspectives. *Cis-trans* isomerism is not possible in this bicycloalkane. It is seen in (b) that the one-carbon methylene bridge is connected by *cis*-1,4 bonds to the larger six-membered ring, which is held rigidly in a boat conformation. The four sets of eclipsed hydrogens are seen most clearly in (c).

2.21(a) is a view from above the six-membered ring with the one-carbon bridge bonded to carbons 1 and 4. Figure 2.21(b) shows that the six-membered ring is held rigidly in a boat conformation by the one-carbon methylene bridge between carbons 1 and 4. Thus, this bicycloalkane does not show *cis-trans* isomerism because the one-carbon methylene bridge can only be connected by *cis* bonds to carbons 1 and 4 of the boat conformation. Figure 2.21(c) shows the eclipsed hydrogens of the boat most clearly.

2.8 Physical Properties of Alkanes and Cycloalkanes

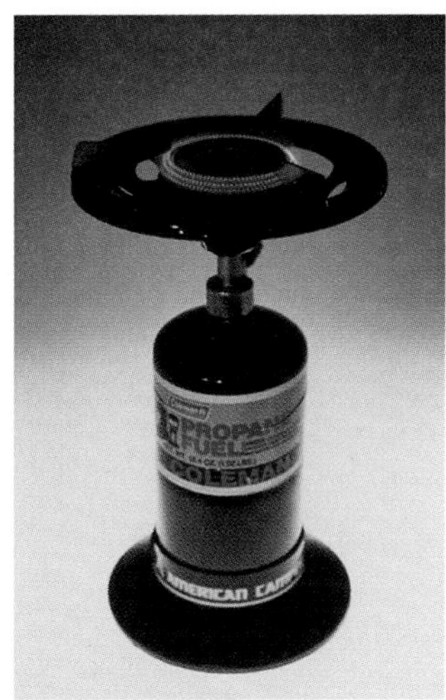

A tank of propane fuel.
(Charles D. Winters)

You are already familiar with the physical properties of some alkanes and cycloalkanes from your everyday experiences. The low-molecular-weight alkanes, such as methane, ethane, propane, and butane, are gases at room temperature and atmospheric pressure. Higher-molecular-weight alkanes, such as those in gasoline and kerosene, are liquids. Very high-molecular-weight alkanes, such as those found in paraffin wax, are solids. Melting points, boiling points, and densities of the first ten alkanes are listed in Table 2.5.

Methane is a gas at room temperature and atmospheric pressure. It can be converted to a liquid if cooled to − 164°C and to a solid if further cooled to − 182°C. The fact that methane (or any other compound, for that matter) can exist as a liquid or solid depends on the existence of **intermolecular forces of attraction** between particles of each pure compound. Although the forces of attraction between particles are all electrostatic in nature, they vary widely in their relative strengths. The strongest attractive forces are between ions, for example between Na^+ and Cl^- in NaCl (188 kcal/mol, 787 kJ/mol). Dipole-dipole interactions and hydrogen bonding (2–10 kcal/mol, 8–42 kJ/mol) are weaker attractive forces. We will have more to say about these intermolecular attractive forces in Chapter 9 when we discuss the physical properties of alcohols, compounds containing polar O—H groups.

Dispersion forces (0.02–2 kcal/mol, 0.08–8 kJ/mol) are the weakest intermolecular attractive forces. It is the existence of dispersion forces that accounts for the fact

TABLE 2.5 Physical Properties of Some Unbranched Alkanes

Name	Condensed Structural Formula	mp (°C)	bp (°C)	Density of Liquid (g/mL at 0°C)
methane	CH_4	−182	−164	(a gas)
ethane	CH_3CH_3	−183	−88	(a gas)
propane	$CH_3CH_2CH_3$	−190	−42	(a gas)
butane	$CH_3(CH_2)_2CH_3$	−138	0	(a gas)
pentane	$CH_3(CH_2)_3CH_3$	−130	36	0.626
hexane	$CH_3(CH_2)_4CH_3$	−95	69	0.659
heptane	$CH_3(CH_2)_5CH_3$	−90	98	0.684
octane	$CH_3(CH_2)_6CH_3$	−57	126	0.703
nonane	$CH_3(CH_2)_7CH_3$	−51	151	0.718
decane	$CH_3(CH_2)_8CH_3$	−30	174	0.730

that low-molecular-weight, nonpolar substances, such as hydrogen, neon, and methane, can be liquefied. To visualize the origin of dispersion forces, it is necessary to think in terms of instantaneous distributions of electron density rather than average distributions. Consider neon, for example. Neon is a gas at room temperature and 1.00 atm. It can be liquefied when cooled to −246°C. From the heat of vaporization, it can be calculated that the neon-neon attractive interaction in the liquid state is approximately 0.07 kcal/mol. We account for this intermolecular attractive interaction in the following way. Over time, the distribution of electron density in a neon atom is symmetrical and there is no dipole moment. However, at any instant, there is a probability that electron density is polarized (shifted) more toward one part of the atom than toward another. This temporary polarization creates a temporary dipole moment, which in turn induces temporary dipole moments in adjacent atoms. **Dispersion forces** are weak electrostatic attractive forces that occur between temporary induced dipoles of adjacent atoms or molecules. Creation of temporary dipole moments and the origin of dispersion forces are illustrated schematically in Figure 2.22.

Dispersion forces Very weak intermolecular coulombic forces of attraction.

The strength of dispersion forces depends on how easily an electron cloud can be polarized. Electrons in small atoms and molecules are held closer to their nuclei and, therefore, are not easily polarized. Electrons in larger atoms and molecules are more easily polarized. For this reason, the strength of dispersion forces tends to increase with increasing molecular mass and size. Intermolecular interactions between Cl_2 molecules and between Br_2 molecules are estimated to be 0.7 kcal/mol and 1.0 kcal/mol respectively.

Dispersion forces are inversely proportional to the sixth power of the distance between interacting particles, and for them to be important, the interacting particles must be in virtual contact with one another.

Because interactions between alkane molecules consist only of very weak dispersion forces, boiling points of alkanes are lower than those of almost any other type of compound of the same molecular weight. As the number of atoms and the molecular weight of an alkane increases, dispersion forces between its molecules become stronger, and boiling point increases.

Melting points of alkanes also increase with increasing molecular weight. The increase, however, is not as regular as that observed for boiling points because the packing of molecules into ordered patterns of solids changes as molecular size and shape change.

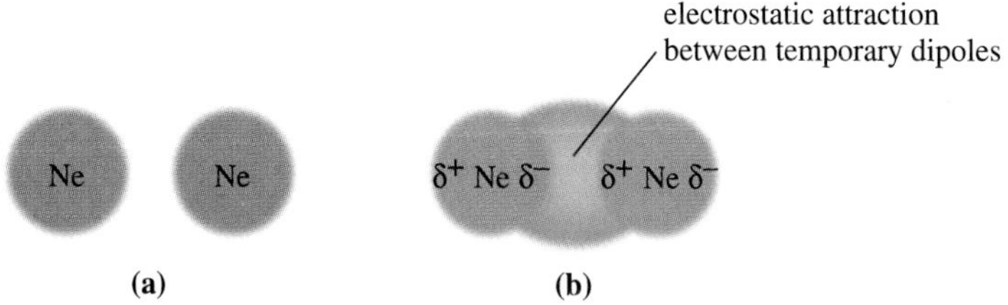

FIGURE 2.22
Dispersion forces. (a) The average distribution of electron density in a neon atom is symmetrical, and there is no polarity. (b) Temporary polarization of one atom induces temporary polarization in adjacent atoms. Electrostatic attractions between temporary dipoles are called dispersion forces.

TABLE 2.6 Physical Properties of the Isomeric Alkanes of Molecular Formula C_6H_{14}

Name	bp (°C)	mp (°C)	Density (g/mL)
hexane	68.7	−95	0.659
2-methylpentane	60.3	−154	0.653
3-methylpentane	63.3	−118	0.664
2,3-dimethylbutane	58.0	−129	0.661
2,2-dimethylbutane	49.7	−98	0.649

The average density of the alkanes listed in Table 2.5 is about 0.7 g/mL; that of higher molecular weight alkanes is about 0.8 g/mL. All liquid and solid alkanes are less dense than water (1.0 g/mL) and, therefore, float on water.

Alkanes that are constitutional isomers are different compounds and have different physical and chemical properties. Listed in Table 2.6 are boiling points, melting points, and densities of the five constitutional isomers of molecular formula C_6H_{14}. The boiling point of each of the branched-chain isomers of C_6H_{14} is lower than that of hexane itself, and the more branching there is, the lower the boiling point. These differences in boiling point are related to molecular shape in the following way. The only forces of attraction between alkane molecules are dispersion forces. As branching increases, the shape of an alkane molecule becomes more compact and its surface area decreases. As surface area decreases, contact among adjacent molecules decreases, the strength of dispersion forces decreases, and boiling points also decrease. Thus, for any group of alkane constitutional isomers, it is usually observed that the least branched isomer has the highest boiling point and the most branched isomer has the lowest boiling point.

EXAMPLE 2.16

Arrange the alkanes in each set in order of increasing boiling point.

(a) Butane, decane, and hexane
(b) 2-Methylheptane, octane, and 2,2,4-trimethylpentane

Solution

(a) All structures are unbranched alkanes. As the number of carbon atoms in the chain increases, dispersion forces between molecules increase and so do boiling points. Decane has the highest boiling point and butane has the lowest.

$CH_3CH_2CH_2CH_3$	$CH_3CH_2CH_2CH_2CH_2CH_3$	$CH_3(CH_2)_8CH_3$
Butane	Hexane	Decane
bp −0.5°C	bp 69°C	bp 174°C

(b) These three alkanes are constitutional isomers of molecular formula C_8H_{18}. Their relative boiling points depend on the degree of branching. 2,2,4-Trimethylpentane, the most highly branched isomer, has the smallest surface area and the lowest boiling point. Octane, the unbranched isomer, has the largest surface area and the highest boiling point.

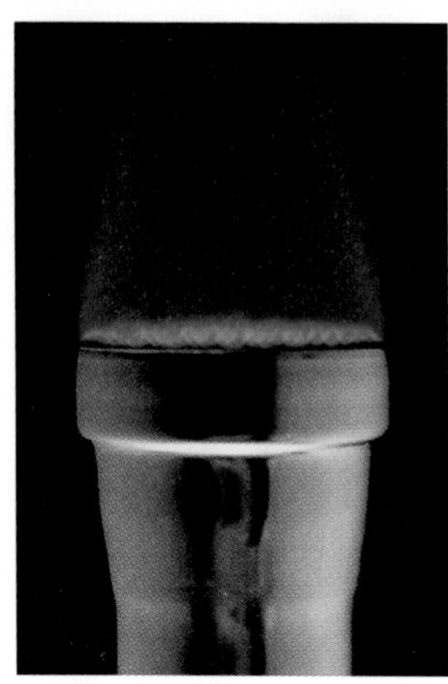

Burning natural gas, which is primarily methane with small amounts of ethane, propane, and butane. *(Charles D. Winters)*

2,2,4-Trimethylpentane: $CH_3C(CH_3)_2CH_2CH(CH_3)CH_3$	2-Methylheptane: $CH_3CH(CH_3)CH_2CH_2CH_2CH_2CH_3$	Octane: $CH_3(CH_2)_6CH_3$
bp 99°C	bp 118°C	bp 126°C

PROBLEM 2.16

Arrange the alkanes in each set in order of increasing boiling point.

(a) 2-Methylbutane, 2,2-dimethylpropane, and pentane
(b) 3,3-Dimethylheptane, 2,2,4-trimethylpentane, and nonane

2.9 Sources of Alkanes

The three major sources of alkanes throughout the world are the fossil fuels, namely natural gas, petroleum, and coal. These fossil fuels account for approximately 90% of the total energy consumed in the United States. Nuclear electric power and hydroelectric power make up most of the remaining 10%. In addition, these fossil fuels provide the bulk of the raw materials for the organic chemicals consumed worldwide.

A. Natural Gas

Natural gas consists of approximately 90% to 95% methane, 5% to 10% ethane, and a mixture of other relatively low boiling alkanes; chiefly propane, butane, and 2-methylpropane. The current widespread use of ethylene as the organic chemical industry's most important building block is due largely to the ease with which ethane can be separated from natural gas and cracked into ethylene. **Cracking** is a process whereby a saturated hydrocarbon is converted into an unsaturated hydrocarbon plus H_2. Ethane is cracked by heating it in a furnace at 800° to 900°C for a fraction of a second.

$$\underset{\text{Ethane}}{CH_3CH_3} \xrightarrow[\text{(thermal cracking)}]{800°\text{–}900°C} \underset{\text{Ethylene}}{CH_2{=}CH_2} + H_2$$

B. Petroleum

A petroleum refinery. *(Courtesy of Ashland Oil Company)*

Petroleum is a thick, viscous liquid mixture of literally thousands of compounds, most of them hydrocarbons, formed from the decomposition of ancient marine plants and animals. Petroleum and petroleum-derived products fuel automobiles, aircraft, and trains. They provide most of the greases and lubricants required for the machinery of our highly industrialized society. Furthermore, petroleum, along with natural gas, provides close to 90% of the organic raw materials for the synthesis and manufacture of synthetic fibers, plastics, detergents, drugs, dyes, and a multitude of other products.

It is the task of a petroleum refinery to produce usable products, with a minimum of waste, from the thousands of different hydrocarbons in this liquid mixture. The various physical and chemical processes for this purpose fall into two broad categories:

separation processes, which separate the complex mixture into various fractions, and **reforming processes**, which alter the molecular structure of the hydrocarbon components themselves.

The fundamental separation process in refining petroleum is fractional distillation (Figure 2.23). Practically all crude oil that enters a refinery goes to distillation units where it is heated gradually to temperatures as high as 370° to 425°C and separated into fractions. Each fraction contains a mixture of hydrocarbons that boils within a particular range. Following are the common names associated with several of these fractions along with the major uses of each.

1. Gases boiling below 20°C are taken off at the top of the distillation column. This fraction is a mixture of low-molecular-weight hydrocarbons, predominantly propane, butane, and 2-methylpropane, substances that can be liquefied under pressure at room temperature. The liquefied mixture, known as liquefied petroleum gas (LPG), can be stored and shipped in metal tanks and is a convenient gaseous fuel for home heating and cooking.
2. Naphthas, bp 20° to 200°C, are a mixture of C_5 to C_{12} alkanes and cycloalkanes. The naphthas also contain small amounts of benzene, toluene, xylene, and other aromatic hydrocarbons (Chapter 19). The light naphtha fraction, bp 20° to 150°C, is the source of straight-run gasoline and averages approximately 25% of crude petroleum. In a sense, naphthas are the most valuable distillation fractions because they are useful not only as fuel but also as sources of raw materials for the organic chemical industry.
3. Kerosene, bp 175° to 275°C, is a mixture of C_9 to C_{15} hydrocarbons.

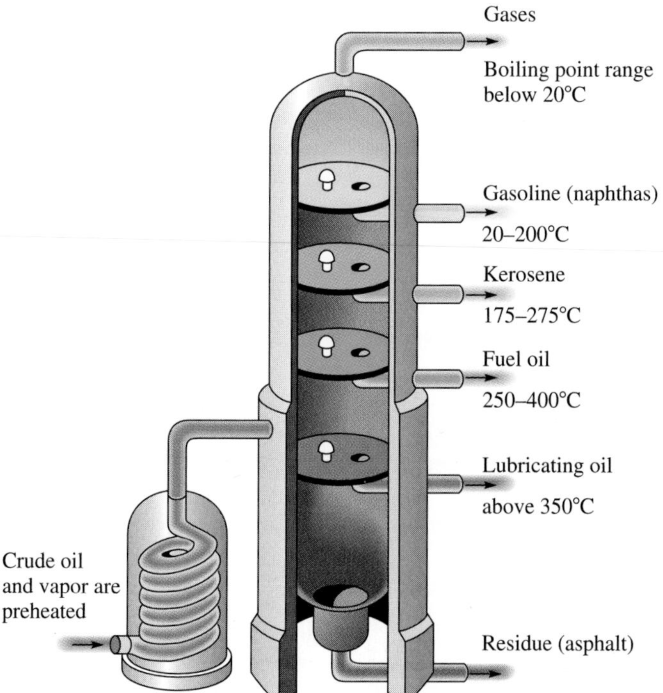

FIGURE 2.23
Fractional distillation of petroleum. The lighter, more volatile fractions are removed from higher up the column and the heavier, less volatile fractions, from lower down.

4. Fuel oil, bp 250° to 400°C, is a mixture of C_{15} to C_{18} hydrocarbons. It is from this fraction that diesel fuel is obtained.
5. Lubricating oil and heavy fuel oil distill from the column at temperatures above 350°C.
6. Asphalt is the name given to the black, tarry residue remaining after removal of the other volatile fractions.

The two most common reforming processes are cracking, as illustrated by thermal conversion of ethane to ethylene (Section 2.9A), and catalytic reforming. **Catalytic reforming** is illustrated by the conversion of hexane first to cyclohexane and then to benzene.

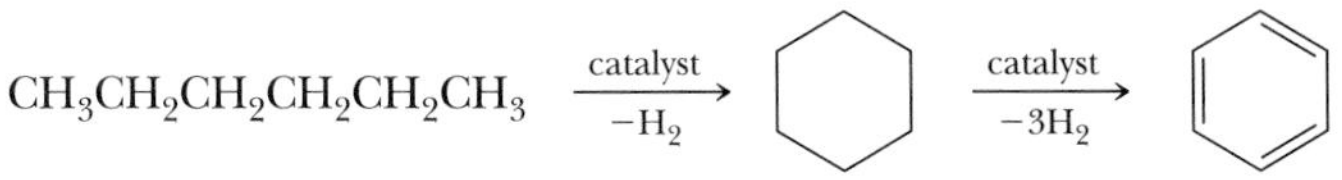

$$CH_3CH_2CH_2CH_2CH_2CH_3 \xrightarrow[-H_2]{\text{catalyst}} \text{Cyclohexane} \xrightarrow[-3H_2]{\text{catalyst}} \text{Benzene}$$

Hexane Cyclohexane Benzene

Gasoline is a complex mixture of C_6 to C_{12} hydrocarbons. The quality of gasoline as a fuel for internal combustion engines is expressed by its octane rating. Engine knocking occurs when a portion of the air-fuel mixture explodes prematurely (usually as a result of heat developed during compression) and independently of ignition by the spark plug. Two compounds were selected as reference fuels. One of these, 2,2,4-trimethylpentane ("isooctane"), has very good antiknock properties (the fuel/air mixture burns smoothly in the combustion chamber) and was assigned an octane rating of 100. (The name "isooctane" as used here is a trivial name.) Heptane, the other reference compound, has poor antiknock properties and was assigned an octane rating of 0.

$$CH_3(CH_2)_5CH_3 \qquad CH_3C(CH_3)_2CH_2CH(CH_3)CH_3$$

Heptane (octane rating 0)

2,2,4-Trimethylpentane (octane rating 100)

A gas pump showing octane rating as (R + M)/2 which is research octane number plus motor octane number divided by 2. *(David R. Frazier Photolibrary)*

The **octane rating** of a particular gasoline is the percent of isooctane in a mixture of isooctane and heptane that has equivalent knock properties. For example, the knock properties of 2-methylhexane are the same as those of a mixture of 42% isooctane and 58% heptane; therefore, the octane rating of 2-methylhexane is 42. Octane itself has an octane rating of −20, which means that it produces even more engine knocking than heptane.

Octane rating The percentage of isooctane in a test mixture of isooctane and heptane that has equivalent knock properties to a gasoline being rated.

C. Coal

To understand how coal can be used as a raw material for the production of organic compounds, it is necessary to discuss synthesis gas. **Synthesis gas** is a mixture of carbon monoxide and hydrogen in varying proportions depending on the means by which it is manufactured. Synthesis gas is prepared by passing steam over hot coal. It is also prepared by partial oxidation of methane with oxygen.

$$\underset{\text{Coal}}{C} + H_2O \xrightarrow{\text{heat}} CO + H_2$$

$$CH_4 + \tfrac{1}{2}O_2 \xrightarrow{\text{catalyst}} CO + 2H_2$$

Two important organic compounds produced today almost exclusively from carbon monoxide and hydrogen are methanol and acetic acid. In the production of methanol, the ratio of hydrogen to carbon monoxide is adjusted to 2 : 1 and the mixture is passed over a catalyst at elevated temperature and pressure.

$$CO + 2H_2 \xrightarrow{\text{catalyst}} \underset{\text{Methanol}}{CH_3OH}$$

Treatment of methanol, in turn, with carbon monoxide over a different catalyst gives acetic acid.

$$\underset{\text{Methanol}}{CH_3OH} + CO \xrightarrow{\text{catalyst}} \underset{\text{Acetic acid}}{CH_3\overset{\overset{\displaystyle O}{\|}}{C}OH}$$

Because the processes for making methanol and acetic acid directly from carbon monoxide are commercially proven, it is likely that the decades ahead will see the development of routes to other organic chemicals from coal via methanol.

2.10 Reactions of Alkanes

Alkanes and cycloalkanes are quite unreactive toward most reagents, a behavior consistent with the facts that they are nonpolar compounds and contain only strong sigma bonds. They do react, however, under certain conditions with oxygen and with the halogens Cl_2 and Br_2. At this point, we present only their reaction with oxygen. We discuss their reaction with halogens in Chapter 7.

Oxidation (combustion) of alkanes by O_2 to form carbon dioxide and water is by far their most economically important reaction. Oxidation of saturated hydrocarbons is the basis for their use as energy sources for heat (natural gas, liquefied petroleum gas [LPG], and fuel oil) and power (gasoline, diesel fuel, and aviation fuel). Following are balanced equations for the complete oxidation of methane, the major component of natural gas, and propane, the major component of LPG.

$$\underset{\text{Methane}}{CH_4} + 2O_2 \longrightarrow CO_2 + 2H_2O \qquad \Delta H^0 = -212 \text{ kcal/mol } (-886 \text{ kJ/mol})$$

$$\underset{\text{Propane}}{CH_3CH_2CH_3} + 5O_2 \longrightarrow 3CO_2 + 4H_2O \qquad \Delta H^0 = -530 \text{ kcal/mol } (-2220 \text{ kJ/mol})$$

SUMMARY

A **hydrocarbon** is a compound that is composed of only carbon and hydrogen. **Saturated hydrocarbons (alkanes** and **cycloalkanes)** contain only single bonds. Alkanes have the general formula $\mathbf{C_nH_{2n+2}}$. An **unsaturated hydrocarbon** contains one or more carbon-carbon double or triple bonds. Constitutional isomers (Section 2.2) have the same molecular formula but a different connectivity (a different order of attachment of their atoms).

Alkanes are named according to a set of rules developed by the **International Union of Pure and Applied Chemistry (IUPAC)** (Section 2.3A). The IUPAC system is a general system of nomenclature (Section 2.5). The IUPAC name of a compound consists of three parts: (1) a **prefix** that tells the number of carbon atoms in the parent chain, (2) an **infix** that tells the nature of the carbon-carbon bonds in the parent chain, and (3) a **suffix** that tells the class to which the compound belongs. Substituents derived from alkanes are known as **alkyl groups.** The name of an alkyl group is formed by dropping the suffix -ane from the name of the parent alkane and adding -yl in its place.

A carbon atom is classified as **primary (1°), secondary (2°), tertiary (3°),** or **quaternary (4°)** depending on the number of carbon atoms bonded to it (Section 2.3C). A hydrogen atom is classified as primary (1°), secondary (2°), or tertiary (3°), depending on the type of carbon atom to which it is bonded. Hydrogen atoms in a molecule or ion that have identical chemical environments are called **equivalent hydrogens** (Section 2.3C).

A saturated hydrocarbon that contains carbon atoms bonded to form a ring is called a **cycloalkane** (Section 2.4A). To name a cycloalkane, name and locate each substituent on the ring and prefix the name of the open-chain alkane by cyclo-. Five-membered rings (cyclopentanes) and six-membered rings (cyclohexanes) are especially abundant in the biological world. A **bicycloalkane** (Section 2.4B) is a hydrocarbon that contains two rings that share two carbon atoms. The shared carbons are called **bridgehead carbons.** A **spiroalkane** is an alkane in which a single carbon atom is shared by two rings (Section 2.4B).

A **conformation** is any three-dimensional arrangement of the atoms of a molecule that results by rotation about one or more single bonds (Section 2.6). One convention for showing conformations is the **Newman projection.** A **dihedral angle** is the angle created by two intersecting planes. For ethane, staggered conformations occur at dihedral angles of 60°, 180°, and 300°. Eclipsed conformations occur at dihedral angles of 0°, 120°, and 240°. For butane viewed along the C_2—C_3 bond, the staggered conformation of dihedral angle 180° is called an anti conformation; the staggered conformations of dihedral angle 60° and 300° are called gauche conformations. The anti conformation of butane is lower in energy than the gauche conformations by approximately 0.9 kcal/mol.

Intramolecular strain (Section 2.6A) is of three types: (1) **torsional strain,** which is the force that opposes the rotation of one part of a molecule about a bond while the other part is held fast; (2) **angle strain,** which arises from creation of either abnormally large or abnormally small bond angles; and (3) **nonbonded interaction strain,** which arises when atoms not bonded to each other are forced abnormally close to one another. The relationship between the change in Gibbs free energy and an equilibrium constant is given by the equation $\Delta G^0 = -2.303\, RT \log K_{eq}$.

In all cycloalkanes larger than cyclopropane, puckered conformations are favored. The lowest-energy conformation of cyclopentane is an envelope conformation (Section 2.6B). The lowest-energy conformations of cyclohexane are two interconvertible **chair conformations** (Section 2.6B). In a chair conformation, six bonds are **axial** and six bonds are **equatorial.** Bonds axial in one chair are equatorial in the alternative chair. **Boat** and **twist-boat conformations** are higher in energy than chair conformations. The more stable conformation of a substituted cyclohexane is the one that minimizes diaxial interactions.

Cis-trans* isomers** (Section 2.7A) have the same molecular formula, the same order of attachment of atoms, but an arrangement of atoms in space that cannot be interconverted by rotation about sigma bonds. ***Cis means that substituents are on the same side of the ring; ***trans*** means that substituents are on opposite sides of the ring. All cycloalkanes with substituents on two or more carbons show *cis-trans* isomerism.

Low-molecular-weight alkanes, such as methane, ethane, and propane, are gases at room temperature and atmospheric pressure. Higher-molecular-weight alkanes, such as those in gasoline and kerosene, are liquids. Very high-molecular-weight alkanes, such as those in paraffin wax, are solids.

Alkanes are nonpolar compounds and the only forces of attraction between their molecules are **dispersion forces** (Section 2.8), weak electrostatic interactions between temporary induced dipoles of adjacent atoms or molecules. Among a set of alkane constitutional isomers, the least branched isomer generally has the highest boiling point; the most branched isomer generally has the lowest boiling point.

Natural gas (Section 2.9A) consists of 90% to 95% methane with lesser amounts of ethane and other low-molecular-weight hydrocarbons. **Petroleum** (Section 2.9B) is a liquid mixture of literally millions of different hydrocarbons. The most important processes in petroleum refining are fractional distillation, catalytic cracking, and catalytic reforming. **Synthesis gas** (Section 2.9C), a mixture of carbon monoxide and hydrogen, can be derived from natural gas, coal, or petroleum. Chemicals from coal by way of synthesis gas may become significant as a means of reducing our dependence on petroleum as a raw material for the synthesis of organics required for industry and commerce.

KEY REACTIONS

1. Oxidation of Alkanes (Section 2.10)

Oxidation of alkanes to carbon dioxide and water is the basis for their use as energy sources for heat and power.

$$CH_3CH_2CH_3 + 5O_2 \longrightarrow 3CO_2 + 4H_2O + \text{energy}$$

ADDITIONAL PROBLEMS

Constitutional Isomerism

2.17 Which statements are true about constitutional isomers?

(a) They have the same molecular formula.
(b) They have the same molecular weight.
(c) They have the same order of attachment of atoms.
(d) They have the same physical properties.

2.18 Which structural formulas represent identical compounds and which represent constitutional isomers?

(a) $CH_3CH_2CH(Cl)CH_3$ (b) chlorocyclobutane (ring–Cl) (c) $ClCH_2$–cyclopropane (d) $CH_3CH(CH_2Cl)CH_3$

(e) $ClCH_2CH(CH_3)CH_3$ (f) $CH_3CH_2CH_2CH_2Cl$ (g) $CH_3CH(CH_2CH_3)Cl$ (h) $CH_3C(CH_3)(Cl)CH_3$

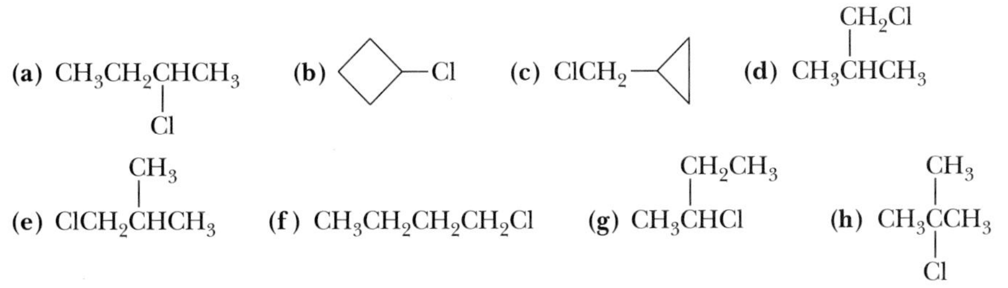

2.19 Tell whether the compounds in each set are constitutional isomers.

(a) CH_3CH_2OH and CH_3OCH_3 (b) $CH_3\overset{O}{\overset{\|}{C}}CH_3$ and $CH_3CH_2\overset{O}{\overset{\|}{C}}H$

(c) $CH_3\overset{O}{\overset{\|}{C}}OCH_3$ and $CH_3CH_2\overset{O}{\overset{\|}{C}}OH$ (d) $CH_3\overset{OH}{\overset{|}{C}}HCH_2CH_3$ and $CH_3\overset{O}{\overset{\|}{C}}CH_2CH_3$

(e) 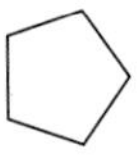 [cyclopentane] and $CH_3CH_2CH_2CH_2CH_3$ (f) 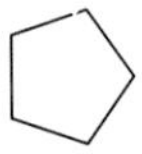 [cyclopentane] and $CH_2{=}CHCH_2CH_2CH_3$

2.20 Name and draw structural formulas for the nine constitutional isomers of molecular formula C_7H_{16}.

2.21 Draw structural formulas for all of the following:

- **(a)** Alcohols of molecular formula $C_4H_{10}O$
- **(b)** Aldehydes of molecular formula C_4H_8O
- **(c)** Ketones of molecular formula $C_5H_{10}O$
- **(d)** Carboxylic acids of molecular formula $C_5H_{10}O_2$

Nomenclature of Alkanes and Cycloalkanes

2.22 Write IUPAC names for these alkanes and cycloalkanes.

(a) $CH_3\underset{CH_3}{\underset{|}{C}}HCH_2CH_2CH_3$ (b) $CH_3\underset{CH_3}{\underset{|}{C}}HCH_2CH_2\underset{CH_3}{\underset{|}{C}}HCH_3$ (c) $CH_3(CH_2)_4\underset{CH_2CH_3}{\underset{|}{C}}HCH_2CH_3$

(d) $(CH_3)_2CHCH_2CH_2C(CH_3)_3$ (e) 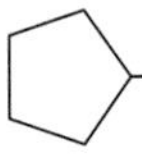 [cyclopentane]$-CH_2CH(CH_3)_2$

(f) [cyclohexane bearing $C(CH_3)_3$, CH_3, and H_3C substituents]

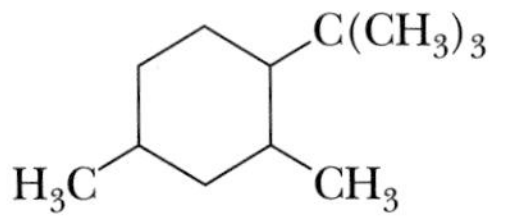

2.23 Write structural formulas for these alkanes and cycloalkanes.

- **(a)** 2,2,4-Trimethylhexane
- **(b)** 2,2-Dimethylpropane
- **(c)** 3-Ethyl-2,4,5-trimethyloctane
- **(d)** 5-Butyl-2,2-dimethylnonane
- **(e)** 4-Isopropyloctane
- **(f)** 3,3-Dimethylpentane
- **(g)** *trans*-1,3-Dimethylcyclopentane
- **(h)** *cis*-1,2-Diethylcyclobutane

2.24 Explain why each is an incorrect IUPAC name. Write a correct IUPAC name for the intended compound.

- **(a)** 1,3-Dimethylbutane
- **(b)** 4-Methylpentane
- **(c)** 2,2-Diethylbutane
- **(d)** 2-Ethyl-3-methylpentane
- **(e)** 2-Propylpentane
- **(f)** 2,2-Diethylheptane
- **(g)** 2,2-Dimethylcyclopropane
- **(h)** 1-Ethyl-5-methylcyclohexane

The IUPAC System of Nomenclature

2.25 For each of the following IUPAC names, draw the corresponding structural formula.

- **(a)** Butanone
- **(b)** Butanal
- **(c)** Butanoic acid
- **(d)** Ethanoic acid
- **(e)** Hexanoic acid
- **(f)** Propanoic acid
- **(g)** Propanal
- **(h)** Cyclopentene
- **(i)** Cyclopentanol
- **(j)** Cyclopentanone
- **(k)** Cyclopropanol
- **(l)** Propanone

2.26 Write IUPAC names for these compounds.

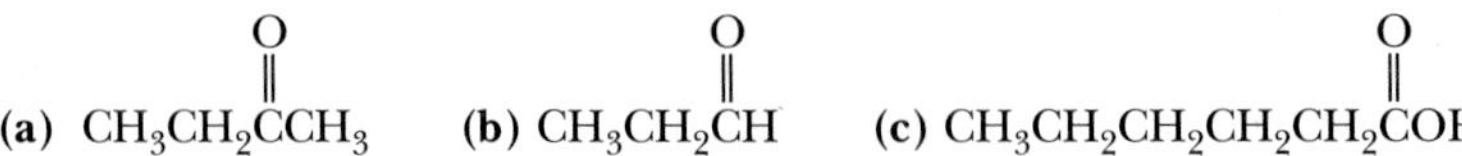

(d) $CH_3CH(OH)CH_3$ (e) cyclohexanone (=O) (f) cyclopropyl—OH

(g) $CH_3CH{=}CH_2$ (h) cyclohexene

Conformations of Alkanes and Cycloalkanes

2.27 Torsional strain resulting from eclipsed C—H bonds is approximately 1.0 kcal/mol (4.2 kJ/mol) and that for eclipsed C—H and C—CH_3 bonds is approximately 1.5 kcal/mol (6.3 kJ/mol). Given this information, sketch a graph of potential energy versus dihedral angle for propane.

2.28 How many different staggered conformations are there for 2-methylpropane? How many different eclipsed conformations are there?

2.29 Consider 1-bromopropane, $CH_3CH_2CH_2Br$.

(a) Draw a Newman projection for the conformation in which —CH_3 and —Br are anti (dihedral angle 180°).
(b) Draw Newman projections for the conformations in which —CH_3 and —Br are gauche (dihedral angles 60° and 300°).
(c) Which of these is the lowest energy conformation?
(d) Which of these conformations, if any, are related by reflection?

2.30 Consider 1-bromo-2-methylpropane and draw the following:

(a) The staggered conformation(s) of lowest energy
(b) The staggered conformation(s) of highest energy

2.31 In cyclohexane, an equatorial substituent is equidistant from the axial group and the equatorial group on an adjacent carbon. Build a molecular model, measure these distances, and convince yourself they are identical.

2.32 *trans*-1,4-Di-*tert*-butylcyclohexane exists in a normal chair conformation. *cis*-1,4-Di-*tert*-butylcyclohexane, however, adopts a twist boat conformation. Draw both isomers and explain why the *cis* isomer is more stable in the twist boat conformation.

Cis-Trans Isomerism in Cycloalkanes and Bicycloalkanes

2.33 Name and draw structural formulas for the *cis* and *trans* isomers of 1,2-dimethylcyclopropane.

2.34 Name and draw structural formulas for all cycloalkanes of molecular formula C_5H_{10}. Be certain to include *cis-trans* isomers as well as constitutional isomers.

2.35 Using a planar pentagon representation for the cyclopentane ring, draw structural formulas for the *cis* and *trans* isomers of:

(a) 1,2-Dimethylcyclopentane **(b)** 1,3-Dimethylcyclopentane

2.36 Energy differences between axial-substituted and equatorial-substituted chair conformations of cyclohexane were given in Table 2.4.

(a) Calculate the ratio of equatorial to axial *tert*-butylcyclohexane at 25°C.
(b) Explain, by using molecular models, why the conformational equilibria for methyl, ethyl, and isopropyl substituents are comparable but the conformational equilibrium for *tert*-butylcyclohexane lies considerably farther toward the equatorial conformation.

2.37 When cyclohexane is substituted by an ethynyl group, —C≡CH, the energy difference between axial and equatorial conformations is only 0.41 kcal/mol (1.75 kJ/mol). Compare the conformational equilibrium for methylcyclohexane with that for ethynylcyclohexane and account for the difference between the two.

2.38 Draw the alternative chair conformations for the *cis* and *trans* isomers of 1,2-dimethylcyclohexane, 1,3-dimethylcyclohexane, and 1,4-dimethylcyclohexane.

(a) Indicate by a label whether each methyl group is axial or equatorial.
(b) For which isomer(s) are the alternative chair conformations of equal stability?
(c) For which isomer(s) is one chair conformation more stable than the other?

2.39 Use your answers from Problem 2.38 to complete the table showing correlations between *cis, trans* and axial, equatorial for disubstituted derivatives of cyclohexane.

Position of Substitution	*cis*	*trans*
1,4-	a,e or e,a	e,e or a,a
1,3-	______ or ______	______ or ______
1,2-	______ or ______	______ or ______

2.40 Calculate the difference in energy in kcal/mol between the alternative chair conformations of:

(a) *trans*-1-Chloro-4-methylcyclohexane **(b)** *cis*-4-Methylcyclohexanol

Menthol is a component of mint oil, obtained from the plant, *Mentha piperita*. *(Wally Eberhart: PHOTO/NATS)*

2.41 There are four *cis, trans* isomers of 2-isopropyl-5-methylcyclohexanol.

(a) Using a planar hexagon representation for the cyclohexane ring, draw structural formulas for the four *cis, trans* isomers.
(b) Draw the more stable chair conformation for each of your answers in part (a).
(c) Of the four *cis, trans* isomers, which is the most stable? (If you answered this part correctly, you picked the isomer found in nature and given the name menthol.)

2.42 Draw alternative chair conformations for each substituted cyclohexane and state which chair is more stable.

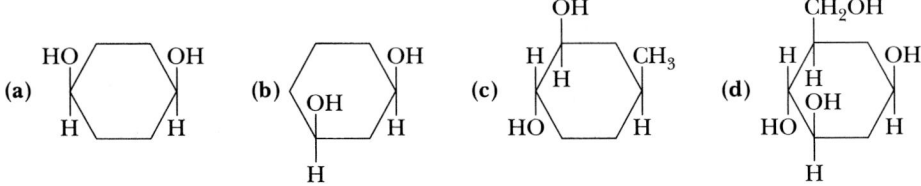

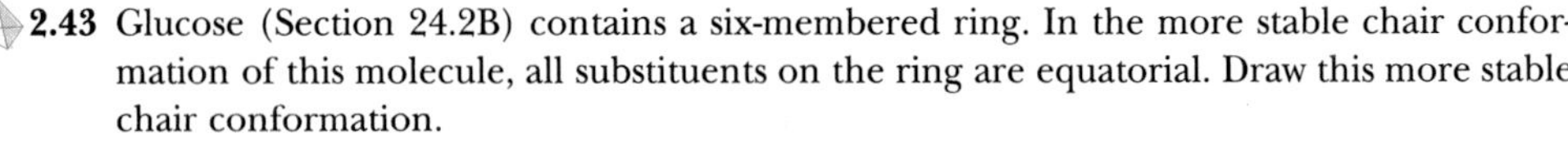

2.43 Glucose (Section 24.2B) contains a six-membered ring. In the more stable chair conformation of this molecule, all substituents on the ring are equatorial. Draw this more stable chair conformation.

Glucose

2.44 1,2,3,4,5,6-Hexachlorocyclohexane shows *cis-trans* isomerism. At one time a crude mixture of these isomers was sold as the insecticide benzene hexachloride (BHC) under the trade names Kwell and Gammexane. The insecticidal properties of the mixture arise from one isomer known as the γ-isomer (gamma-isomer) which is *cis*-1,2,4,5-*trans*-3,6-hexachlorocyclohexane.

(a) Draw a structural formula for 1,2,3,4,5,6-hexachlorocyclohexane disregarding, for the moment, the existence of *cis-trans* isomerism. What is the molecular formula of this compound?

(b) Using a planar hexagon representation for the cyclohexane ring, draw a structural formula for the γ-isomer.

(c) Draw a chair conformation for the γ-isomer and label which chlorine atoms are axial and which are equatorial.

(d) Draw the alternative chair conformation of the γ-isomer and again label which chlorine atoms are axial and which are equatorial.

(e) Which of the alternative chair conformations of the γ-isomer is more stable? Explain.

2.45 What kinds of conformations do the six-membered rings exhibit in adamantane and twistane? You may also find it helpful to build molecular models, particularly of twistane.

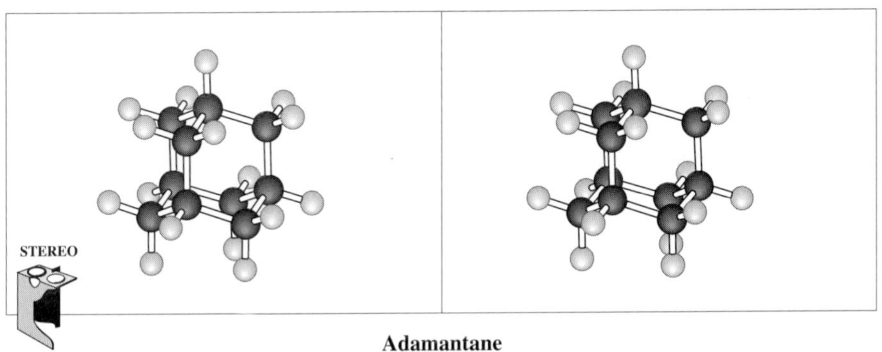

Adamantane

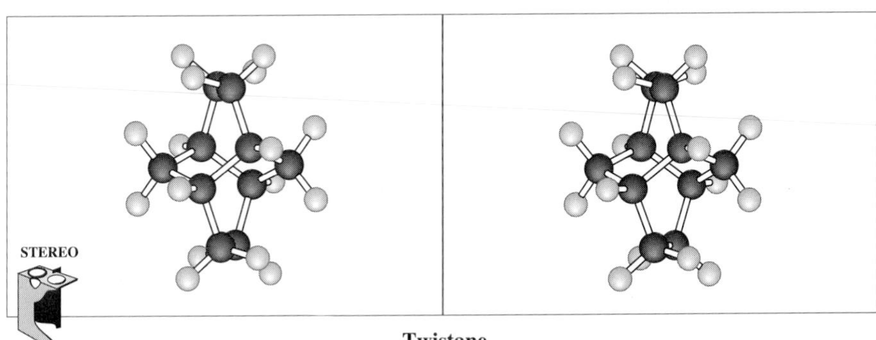

Twistane

2.46 Which of the following bicycloalkanes do you expect to show *cis-trans* isomerism? Explain. For each that does, draw suitable stereorepresentations of both *cis* and *trans* isomers.

(a) Bicyclo[2.2.2]octane
(b) Bicyclo[4.3.0]nonane
(c) 2-Methylbicylo[2.2.1]heptane
(d) 1-Chlorobicyclo[2.2.1]heptane
(e) 7-Chlorobicyclo[2.2.1]heptane

Physical Properties

2.47 In Problem 2.20, you drew structural formulas for all isomeric alkanes of molecular formula C_7H_{16}. Predict which isomer has the lowest boiling point and which has the highest boiling point.

2.48 What generalization can you make about the densities of alkanes relative to that of water?

2.49 What unbranched alkane has about the same boiling point as water? (Refer to Table 2.5 on the physical properties of alkanes.) Calculate the molecular weight of this alkane and compare it with that of water.

Reactions of Alkanes

2.50 Complete and balance the following combustion reactions. Assume that each hydrocarbon is converted completely to carbon dioxide and water.

(a) Propane + $O_2 \longrightarrow$ **(b)** Octane + $O_2 \longrightarrow$
(c) Cyclohexane + $O_2 \longrightarrow$ **(d)** 2-Methylpentane + $O_2 \longrightarrow$

2.51 Following are heats of combustion per mole for methane, propane, and 2,2,4-trimethylpentane. Each is a major source of energy. On a gram-for-gram basis, which of these hydrocarbons is the best source of heat energy?

Hydrocarbon	A Major Component of	ΔH^0 [kcal/mol (kJ/mol)]
CH_4	natural gas	−213 (−891 kJ/mol)
$CH_3CH_2CH_3$	LPG	−531 (−2221 kJ/mol)
$CH_3C(CH_3)_2CH_2CH(CH_3)CH_3$	gasoline	−1304 (−5452 kJ/mol)

2.52 Following are heats of combustion for ethane, propane, and pentane at 25°C.

Hydrocarbon	ΔH^0(kcal/mol)
ethane	−372.8
propane	−530.6
pentane	−845.2

(a) From these data, calculate the average heat of combustion of a methylene ($-CH_2-$) group in a gaseous hydrocarbon.
(b) Using the value of the heat of combustion of a methylene group calculated in part (a), estimate the heat of combustion of gaseous cyclopropane.
(c) Compare your estimated value and the experimentally determined value of −499.9 kcal/mol. How might you account for the difference between the two values?

2.53 Using the value of -157.4 kcal/mol as the average heat of combustion of a methylene group:

(a) Calculate the heat of combustion for the following cycloalkanes.
(b) Calculate the total strain energy for each cycloalkane.
(c) Calculate the strain energy per methylene group.
(d) Rank these cycloalkanes in order of most stable to least stable, based on strain energy per methylene group.

Cycloalkane	Calculated Heat of Combustion (kcal/mol)	Observed Heat of Combustion (kcal/mol)	Calculated Strain Energy (kcal/mol)
cyclopropane	______	−499.9	______
cyclobutane	______	−655.9	______
cyclopentane	______	−793.5	______
cyclohexane	______	−944.5	______
cycloheptane	______	−1108.2	______
cyclooctane	______	−1269.0	______
cyclononane	______	−1429.5	______
cyclodecane	______	−1586.0	______
cycloundecane	______	−1742.4	______
cyclododecane	______	−1891.2	______
cyclotetradecane	______	−2203.6	______

Molecular Modeling

2.54 Build a structural formula of ethane in ChemDraw, import it into Chem3D, and minimize its energy. Then click on one carbon and rotate it about the carbon-carbon single bond to create both the staggered and eclipsed conformations. Visually measure the distance between hydrogens on adjacent carbons in the staggered conformation and in the eclipsed conformation and estimate the ratio of eclipsed/staggered distance.

2.55 Draw a line-angle structure for cyclopentane in ChemDraw, import it into Chem3D, and minimize its energy. Then measure all C—C—C bond angles and compare them with the value of 105° given in the text for the "envelope" conformation of cyclopentane. How do you account for the difference in these values? (*Hint:* Consider the torsional strain of eclipsed hydrogen interactions. You recognize them, but does Chem3D?)

2.56 Build a line-angle structure of a chair cyclohexane in ChemDraw, import it into Chem3D, and minimize its energy. Now rotate the model so you view the chair from above, from the side, from the foot piece to the head piece, and so on. As you do these rotations, convince yourself that the six axial C—H bonds are parallel and that they alternate up, down, and so on. Also convince yourself that opposite (on positions 1 and 4) equatorial C—H bonds are parallel, and that one of them is above and the other below the plane of the ring.

2.57 Build a molecular model of axial methylcyclohexane and use a ruler to measure the distance between the methyl group and the ring hydrogens on carbons 2, 3, 4, 5, and 6. You should find that the axial methyl group is closer to axial hydrogens on carbons 3 and 5 than to any other ring hydrogens. Note that when you do this minimization, you find a "local minimum" because the energy required for ring flipping to the equatorial methyl conformation is too high.

2.58 Build a line-angle drawing of spiro[5.5]undecane in ChemDraw, import it into Chem3D, and minimize its energy. Show that both six-membered rings can assume chair conformations. It may take a bit of rotation of the stereoview but if you get it right, you will see both rings as strain-free chair conformations.

2.59 A large number of ChemDraw figures are given in this chapter, all of which are good exercises for you to practice using ChemDraw and Chem3D and for the creation of stereoviews. You might try reproducing and studying them as three-dimensional objects.

ACIDS AND BASES

3

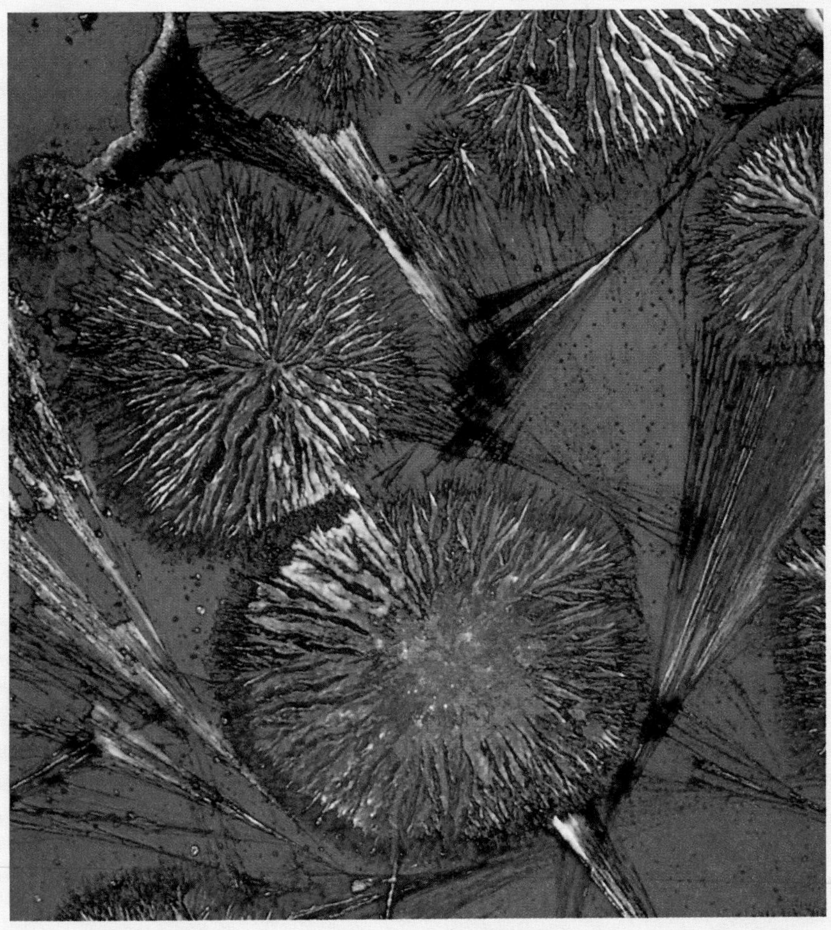

A great many organic reactions are either acid-base reactions or involve catalysis by an acid or base at some stage. Of the reactions involving acid catalysis, some use proton-donating acids, such as H_3O^+ and $CH_3CH_2OH_2^+$. Others use Lewis acids, such as BF_3 and $AlCl_3$. It is essential, therefore, that you have a good grasp of the fundamentals of acid-base chemistry. In this and following chapters, we study the acid-base properties of the major classes of organic compounds.

■ Folic acid crystals viewed under polarizing light. Folic acid is a biological carboxylic acid. *(© Stefan Eberhard, Fran Heyl Associates)*

3.1 Brønsted-Lowry Acids and Bases

Brønsted-Lowry acid A proton donor.

Brønsted-Lowry base A proton acceptor.

In 1923, the Danish chemist Johannes Brønsted and the English chemist Thomas Lowry independently proposed the following definitions: an **acid** is a **proton donor,** and a **base** is a **proton acceptor.** In the neutralization reaction, for example, between aqueous H_3O^+ and OH^-, a proton is transferred from H_3O^+, a Brønsted-Lowry acid, to OH^-, a Brønsted-Lowry base. We use curved arrows to show the flow of electrons in acid-base reactions. The curved arrow on the right in the following equation shows an unshared pair of electrons on oxygen forming a new bond with hydrogen; in donating electrons, this oxygen becomes neutral. The curved arrow on the left shows an electron pair from an O—H bond moving onto oxygen; this oxygen gains electrons and becomes neutral.

$$H_3O^+ + OH^- \longrightarrow H_2O + H_2O$$

Proton donor   Proton acceptor

The acid-base reaction between $H_3O^+(aq)$ and $NH_3(aq)$ involves transfer of a proton from H_3O^+, a proton donor, to NH_3, a proton acceptor. In this reaction, oxygen becomes neutral and nitrogen becomes positively charged.

$$H_3O^+ + NH_3 \longrightarrow H_2O + NH_4^+$$

Proton donor   Proton acceptor

The neutralization reaction between solutions of HCl and NH_3 dissolved in ethanol can be written as the sum of two proton-transfer reactions. HCl is a strong acid, and, when dissolved in ethanol, transfers a proton to the oxygen atom of ethanol to give an ethyloxonium ion and chloride ion. When this solution is mixed with a solution of ammonia in ethanol, a proton is transferred from the ethyloxonium ion to ammonia. The sum of these two **proton-transfer reactions** gives ammonium chloride, which precipitates as white crystals.

$$CH_3CH_2{-}O{-}H + H{-}Cl \longrightarrow CH_3CH_2{-}OH_2^+ + Cl^-$$

Ethanol (proton acceptor)   Hydrogen chloride (proton donor)

$$CH_3CH_2{-}OH_2^+ + NH_3 \longrightarrow CH_3CH_2{-}O{-}H + NH_4^+$$

Ethyloxonium ion (proton donor)   Ammonia (proton acceptor)

Net equation: $HCl + NH_3 \xrightarrow{\text{Ethanol}} NH_4Cl(s)$

The role of the solvent in proton-transfer reactions cannot be overemphasized. Although we often write H^+ as the acid in solution, H^+ is always associated with a solvent molecule. An acid-base reaction in solution is thus transfer of a proton from a protonated molecule of solvent to a base. In reactions in aqueous solutions, it is transfer of a proton from H_3O^+ to a molecule of base; in ethanol solutions, it is transfer of a proton from $CH_3CH_2OH_2^+$ to a molecule of base.

The reciprocal relationships in proton-transfer reactions are described by the following terminology. When an acid transfers a proton to a base, the acid is converted to its **conjugate base;** when a base accepts a proton, the base is converted to its **conjugate acid.** For example, in the reaction between hydronium ion and ammonia, H_3O^+ is converted into its conjugate base, H_2O, and NH_3 is converted into its conjugate acid, NH_4^+.

Conjugate base The species formed from an acid when it donates a proton to a base.

Conjugate acid The species formed from a base when it accepts a proton from an acid.

$$H_3O^+ + NH_3 \longrightarrow H_2O + NH_4^+$$

Acid | Base | Conjugate base | Conjugate acid

EXAMPLE 3.1

Write these reactions as proton-transfer reactions. Label which reactant is the acid and which the base; which product is the conjugate base of the original acid and which the conjugate acid of the original base. Use curved arrows to show the flow of electrons in each reaction.

(a) $H_2O + NH_4^+ \longrightarrow H_3O^+ + NH_3$ **(b)** $CH_3COH + NH_3 \longrightarrow CH_3CO^- + NH_4^+$ (with C=O in each acetyl group)

Solution

(a) First complete the Lewis structure of each reactant by showing all valence electrons. Water is the base (proton acceptor), and ammonium ion is the acid (proton donor).

$$H{-}\ddot{O}{:}(H) + H{-}N^+H_3 \longrightarrow H{-}\ddot{O}^+{-}H(H) + {:}NH_3$$

Base | Acid | Conjugate acid of H_2O | Conjugate base of NH_4^+

(b) Acetic acid is the acid (proton donor), and ammonia is the base (proton acceptor).

$$CH_3{-}C(=\ddot{O}){-}\ddot{O}{-}H + {:}NH_3 \longrightarrow CH_3{-}C(=\ddot{O}){-}\ddot{O}{:}^- + H{-}N^+H_3$$

Acid | Base | Conjugate base of CH_3CO_2H | Conjugate acid of NH_3

PROBLEM 3.1

Write these reactions as proton-transfer reactions. Label which reactant is the acid and which the base; which product is the conjugate base of the original acid and which the conjugate acid of the original base. Use curved arrows to show the flow of electrons in each reaction.

(a) $CH_3SH + OH^- \longrightarrow CH_3S^- + H_2O$
(b) $CH_3OH + NH_2^- \longrightarrow CH_3O^- + NH_3$

3.2 Quantitative Measure of Acid and Base Strength

A **strong acid** or a **strong base** is one that is completely ionized in aqueous solution. When HCl dissolves in water, there is complete transfer of a proton from HCl to H_2O to form H_3O^+ and Cl^-. There is no tendency for the reverse reaction to occur, namely for transfer of a proton from H_3O^+ to Cl^- to form HCl and H_2O. Therefore, when we compare the relative acidities of HCl and H_3O^+, we conclude that HCl is the stronger acid and H_3O^+ is the weaker acid. Similarly, H_2O is the stronger base, and Cl^- is the weaker base.

HCl	+	H_2O	$\longrightarrow$	Cl^-	+	H_3O^+
Acid (stronger acid)		Base (stronger base)		Conjugate base of HCl (weaker base)		Conjugate acid of H_2O (weaker acid)

Any acid stronger than H_3O^+ will protonate water to form H_3O^+. Therefore, the strongest acid that can exist in water is H_3O^+. Likewise, any base stronger than OH^- will be protonated by water to form OH^-. Therefore, the strongest base that can exist in water is OH^-. Examples of strong acids in aqueous solution are HCl, HBr, HI, HNO_3, $HClO_4$, and H_2SO_4. **Strong bases** in aqueous solution are LiOH, NaOH, KOH, $Ca(OH)_2$, and $Ba(OH)_2$.

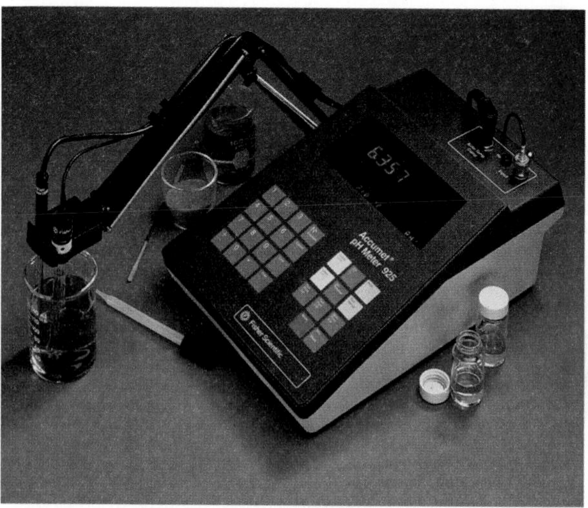

A pH meter used to measure the hydronium ion concentration of a solution. *(Courtesy of Fisher Scientific)*

A **weak acid** or **weak base** is one that is incompletely ionized in aqueous solution. Most organic acids and bases are weak. Among the most common organic acids we deal with are the carboxylic acids, which contain the functional group $—CO_2H$.

$$\underset{\substack{\text{Acid}\\\text{(weaker acid)}}}{CH_3\overset{\overset{\displaystyle O}{\|}}{C}OH} + \underset{\substack{\text{Base}\\\text{(weaker base)}}}{H_2O} \rightleftharpoons \underset{\substack{\text{Conjugate base}\\\text{of } CH_3CO_2H\\\text{(stronger base)}}}{CH_3\overset{\overset{\displaystyle O}{\|}}{C}O^-} + \underset{\substack{\text{Conjugate acid}\\\text{of } H_2O\\\text{(stronger acid)}}}{H_3O^+}$$

The equation for the ionization of a weak acid, HA, in water and the acid ionization constant, K_a, for this equilibrium are

$$HA + H_2O \rightleftharpoons A^- + H_3O^+$$

$$K_a = K_{eq}[H_2O] = \frac{[H_3O^+][A^-]}{[HA]}$$

Because acid ionization constants for weak acids are numbers with negative exponents, acid strengths are often expressed as pK_a where pK_a = $-\log_{10} K_a$. Table 3.1 gives names, molecular formulas, and values of pK_a for some organic and inorganic acids. Note that the larger the value of pK_a, the weaker the acid. Note also the inverse relationship between the strengths of the conjugate acid-conjugate base pairs; the stronger an acid, the weaker its conjugate base, and vice versa.

TABLE 3.1 pK_a Values for Some Organic and Inorganic Acids

	Acid	Formula	pK_a	Conjugate Base	
Weaker acid	ethane	CH_3CH_3	51	$CH_3CH_2^-$	Stronger base
↓	ethylene	$CH_2{=}CH_2$	44	$CH_2{=}CH^-$	↑
	ammonia	NH_3	38	NH_2^-	
	hydrogen	H_2	35	H^-	
	acetylene	$HC{\equiv}CH$	25	$HC{\equiv}C^-$	
	ethanol	CH_3CH_2OH	15.9	$CH_3CH_2O^-$	
	water	H_2O	15.7	HO^-	
	bicarbonate ion	HCO_3^-	10.33	CO_3^{2-}	
	ammonium ion	NH_4^+	9.24	NH_3	
	carbonic acid	H_2CO_3	6.36	HCO_3^-	
	acetic acid	CH_3CO_2H	4.76	$CH_3CO_2^-$	
	benzoic acid	$C_6H_5CO_2H$	4.19	$C_6H_5CO_2^-$	
	phosphoric acid	H_3PO_4	2.1	$H_2PO_4^-$	
	hydronium ion	H_3O^+	−1.74	H_2O	
	sulfuric acid	H_2SO_4	−5.2	HSO_4^-	
	hydrogen chloride	HCl	−7	Cl^-	
	hydrogen bromide	HBr	−8	Br^-	
Stronger acid	hydrogen iodide	HI	−9	I^-	Weaker base

EXAMPLE 3.2

For each value of pK_a, calculate the corresponding value of K_a. Which compound is the stronger acid?

(a) Ethanol, $pK_a = 15.9$ **(b)** Carbonic acid, $pK_a = 6.36$

Solution

(a) For ethanol, $K_a = 1.3 \times 10^{-16}$ **(b)** For carbonic acid, $K_a = 4.4 \times 10^{-7}$

Because the value of pK_a for carbonic acid is smaller than that for ethanol, carbonic acid is the stronger acid, and ethanol is the weaker acid.

PROBLEM 3.2

For each value of K_a, calculate the corresponding value of pK_a. Which compound is the stronger acid?

(a) Acetic acid, $K_a = 1.74 \times 10^{-5}$ **(b)** Water, $K_a = 1.82 \times 10^{-16}$

Values of pK_a in aqueous solution in the range 2–12 can be measured quite precisely. Values of pK_a smaller than 2 are less precise because very strong acids, such as HCl, HBr, and HI, are completely ionized in water and the only acid present in solutions of these acids is H_3O^+. For acids too strong to be measured accurately in water, less basic solvents such as acetic acid or mixtures of water and sulfuric acid are used. Although none of the halogen acids, for example, is completely ionized in acetic acid, HI shows a greater degree of ionization than either HBr or HCl and, therefore, is the strongest acid of the three. Values of pK_a greater than 12 are also less precise. For bases too strong to be measured in aqueous solution, more basic solvents, such as liquid ammonia and dimethyl sulfoxide, are used. Thus, because of the necessity of using different solvent systems to measure relative strengths at either end of the acidity scale, pK_a values smaller than 2 and greater than 12 should be used only in a qualitative way when comparing them with values in the middle of the scale.

3.3 Molecular Structure and Acidity

Let us now examine in some detail the relationships between molecular structure and acidity. As we will see, these relationships can be understood by considering (A) periodicity within a column or period of the Periodic Table, (B) hybridization of the atom bonded to H, (C) resonance, and (D) the inductive effect.

A. Chemical Periodicity: Acidity of HA Within a Period or Column of the Periodic Table

The relative acidity of the hydrogen acids of elements within a period of the Periodic Table is determined by the stability of A^-, the anion formed when a proton is transferred from HA to a base. The greater the electronegativity of A, the greater the stability of the anion, A^- and the stronger the acid HA is.

	$H_3C—H$	$H_2N—H$	HO—H	F—H
pK_a	51	38	15.7	3.5
Electronegativity of A in A—H	2.5	3.0	3.5	4.0

Increasing acid strength →

The relative acidities of hydrogen acids derived from elements within a column of the Periodic Table are determined by the strength of the H—A bond. Note that this trend is opposite that predicted by electronegativity. Among the hydrogen acids of column VIIA elements, for example, H—F, which has the strongest covalent bond, is the weakest acid; H—I, which has the weakest covalent bond, is the strongest acid.

	H—F	H—Cl	H—Br	H—I
pK_a	3.5	−7	−8	−9
Bond Dissociation Energy [kcal/mol (kJ/mol)]	136 (569)	103 (431)	88 (368)	71 (297)

Increasing acid strength →

We see the same relationship between acidity and bond strength among the hydrogen acids of column VIA elements. Compare, for example, the acidities of water (pK_a = 15.7) and hydrogen sulfide (pK_a = 7.0). The HO—H bond dissociation energy of water, the weaker acid, is 119 kcal/mol (498 kJ/mol). The HS—H bond dissociation energy of hydrogen sulfide, the stronger acid, is 91 kcal/mol (381 kJ/mol).

B. Hybridization: Acidity of Hydrocarbon C—H Bonds

Of the three classes of hydrocarbons shown in the table, alkynes are by far the most acidic.

Name (type)	Acid	Anion	Hybridization of Carbon	*s* Character of Hybrid Orbital (%)	pK_a
Acetylene (an alkyne)	HC≡CH	HC≡C:⁻	*sp*	50	25
Ethylene (an alkene)	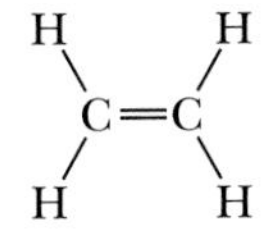 $H_2C{=}CH_2$	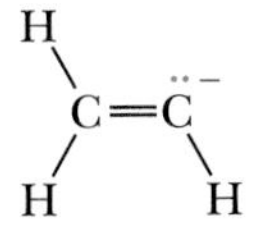 $H_2C{=}\ddot{C}H^-$	sp^2	33	44
Ethane (an alkane)	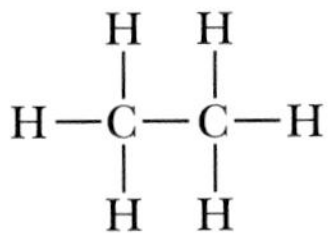 $H_3C—CH_3$	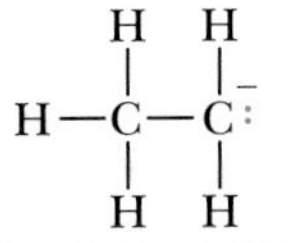 $H_3C—\ddot{C}H_2^-$	sp^3	25	51

Acid strength ↑

The greater acidity of alkyne hydrogens compared with alkene and alkane hydrogens is determined by the stability of the hydrocarbon anion, which is, in turn, determined by the hybridization of the atomic orbital containing the unshared pair of electrons of the anion. For any principal energy level, an *s* orbital lies at a lower energy than do *p* orbitals and, therefore, electrons in *s* orbitals are lower in energy (more stable) than those in *p* orbitals. It follows, then, that the greater the percent *s* character of a hybrid orbital, the lower in energy (more stable) are electrons in that orbital. The unshared pair of electrons in the *sp* hybrid orbital (50% *s* character) of the alkyne anion is lower in energy (more stable) than the unshared pair of electrons in the sp^2 hybrid orbital (33% *s* character) of the alkene anion, which is, in turn, lower in energy and more stable than the electron pair in the sp^3 hybrid orbital (25% *s* character) of the alkane anion. Because the anion derived from an alkyne is more stable than the anions derived from an alkene or alkane, an alkyne is the most acidic of the three types of hydrocarbons.

C. Resonance Effect: Delocalization of Charge in A^-

Carboxylic acids are weak acids. Values of pK_a for most unsubstituted carboxylic acids fall within the range 4 to 5. The value of pK_a for acetic acid, for example, is 4.76.

$$CH_3CO_2H + H_2O \rightleftharpoons CH_3CO_2^- + H_3O^+ \qquad pK_a = 4.76$$

Values of pK_a for most alcohols, compounds that also contain an —OH group, fall within the range 15 to 18; the value of pK_a for ethanol, for example, is 15.9. Thus, alcohols are slightly weaker acids than water ($pK_a = 15.7$) but much weaker acids than carboxylic acids.

We account for the greater acidity of carboxylic acids compared with alcohols using the resonance model and looking at the relative stabilities of the alkoxide ion and the carboxylate ion. Our guideline is this: The more stable the anion, the further the position of equilibrium is shifted toward the right and the more acidic the compound.

Here we take the acid ionization of an alcohol as a reference equilibrium.

$$\underset{\text{An alcohol}}{CH_3CH_2{-}\ddot{\underset{..}{O}}{-}H} + H_2O \rightleftharpoons \underset{\text{An alkoxide ion}}{CH_3CH_2{-}\ddot{\underset{..}{O}}{:}^-} + H_3O^+ \qquad pK_a = 15.9$$

There is no resonance stabilization in the alkoxide anion. Ionization of a carboxylic acid gives an anion for which we can write two equivalent contributing structures that result in delocalization of the negative charge of the anion. Because of this delocalization of negative charge, the carboxylate anion is significantly more stable than is the alkoxide anion. Therefore, the equilibrium for ionization of a carboxylic acid is shifted to the right relative to the ionization of an alcohol. That is, a carboxylic acid is a stronger acid than is an alcohol.

$$CH_3{-}C({=}\ddot{O}){-}\ddot{O}{-}H + H_2O \rightleftharpoons CH_3{-}C({=}\ddot{O}){-}\ddot{O}{:}^- \longleftrightarrow CH_3{-}C(\ddot{O}{:}^-){=}\ddot{O} + H_3O^+$$

These contributing structures are equivalent; the carboxylate anion is stabilized by delocalization of the negative charge.

D. The Inductive Effect: Withdrawal of Electron Density from the H—A Bond

The **inductive effect** is the polarization of electron density transmitted through covalent bonds by a nearby atom or atoms of higher electronegativity. We see the operation of the inductive effect in the carboxyl group in the following way. The carbonyl oxygen of the carboxyl group is more electronegative than the carbonyl carbon and polarizes the electrons of the C=O bond creating a partial negative charge on the carbonyl oxygen and a partial positive charge on the carbonyl carbon. The partial positive charge on the carbonyl carbon, in turn, withdraws electron density from the C—O and O—H bonds and promotes ionization of the carboxyl proton.

Inductive effect The polarization of the electron density of a covalent bond due to a nearby atom of high electronegativity.

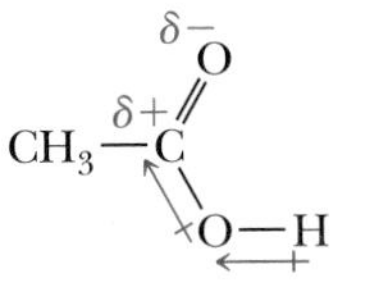

$$CH_3-\overset{\delta+}{C}(=\overset{\delta-}{O})-O-H$$

Thus, a carboxylic acid is more acidic than an alcohol because (1) the inductive effect of the adjacent carbonyl group promotes ionization of the carboxyl proton and (2) the resonance effect stabilizes the carboxylate anion by delocalization of its negative charge.

3.4 Position of Equilibrium in Acid-Base Reactions

Let us consider an acid-base reaction represented by the following general equation.

$$HA + B^- \rightleftharpoons A^- + HB$$

In acid-base reactions, the position of equilibrium favors reaction of the stronger acid with the stronger base to give the weaker acid and the weaker base. To determine the position of equilibrium for an acid-base reaction, we need to know which is the stronger acid or, conversely, which is the stronger base.

Let us consider an equilibrium in which HA is the stronger acid and HB is the weaker acid. If HA is the stronger acid, then A^- must be the weaker base and, conversely, if HB is the weaker acid, then B^- must be the stronger base. We can now label the components in this acid-base equilibrium as follows:

HA	+	B^-	$\rightleftharpoons$	A^-	+	HB
Stronger acid		Stronger base		Weaker base		Weaker acid

The position of equilibrium in this acid-base reaction favors reaction of HA, the stronger acid, with B^-, the stronger base, to give A^-, the weaker base, and HB, the weaker acid. The position of equilibrium lies toward the weaker base and weaker acid.

We can predict the position of equilibrium in acid-base reactions of this type using the data in Table 3.1. For example, consider the reaction between aqueous solutions of acetic acid and ammonia. We must consider the following equilibrium:

$$CH_3CO_2H + NH_3 \rightleftharpoons CH_3CO_2^- + NH_4^+$$

CH_3CO_2H	NH_4^+
Acetic acid	Ammonium ion
pK_a 4.76	pK_a 9.24
(stronger acid)	(weaker acid)

Acetic acid (pK_a 4.76) is a stronger acid than ammonium ion (pK_a 9.24); the position of equilibrium for this reaction, therefore, lies to the right. We can calculate the value of the equilibrium constant, K_{eq}, for this equilibrium in the following way. Write the equilibrium constant expression in the normal fashion, and then multiply it by $[H_3O^+]/[H_3O^+]$. Multiplying by unity does not change the value of K_{eq}. After rearranging terms, the expression for K_{eq} becomes the product of two terms, one of which is K_a for acetic acid, the other of which is the inverse of K_a for the ammonium ion. Substituting the values of K_a for each acid, we calculate that the equilibrium constant, K_{eq}, for this reaction is 3.02×10^4.

$$K_{eq} = \frac{[CH_3CO_2^-][NH_4^+]}{[CH_3CO_2H][NH_3]} \times \frac{[H_3O^+]}{[H_3O^+]} = \frac{[CH_3CO_2^-][H_3O^+]}{[CH_3CO_2H]} \times \frac{[NH_4^+]}{[NH_3][H_3O^+]}$$

$$= \frac{K_a(CH_3CO_2H)}{K_a(NH_4^+)} = \frac{1.74 \times 10^{-5}}{5.75 \times 10^{-10}} = 3.02 \times 10^4$$

Alternatively, we can calculate pK_{eq} for this reaction in the following way.

$$pK_{eq} = pK_a(CH_3CO_2H) - pK_a(NH_4^+) = 4.76 - 9.24 = -4.48$$

$$K_{eq} = 10^{4.48} = 3.02 \times 10^4$$

Consider also the reaction between aqueous solutions of acetic acid and sodium bicarbonate to give sodium acetate and carbonic acid. In the equation for this equilibrium, we omit the sodium ion, Na^+, because it does not undergo a chemical change in this reaction. Instead, we write the equilibrium as a net ionic equation, which shows only the species undergoing chemical change.

$$\underset{\substack{\text{Acetic acid} \\ pK_a\ 4.76}}{CH_3CO_2H} + \underset{\substack{\text{Bicarbonate} \\ \text{ion}}}{HCO_3^-} \rightleftharpoons \underset{\substack{\text{Acetate} \\ \text{ion}}}{CH_3CO_2^-} + \underset{\substack{\text{Carbonic acid} \\ pK_a\ 6.36}}{H_2CO_3} \qquad \begin{aligned} pK_{eq} &= 4.76 - 6.36 = -1.60 \\ K_{eq} &= 10^{1.60} = 40 \end{aligned}$$

Acetic acid is the stronger acid and, therefore, the position of this equilibrium lies to the right. Carbonic acid is formed, which then decomposes to carbon dioxide and water.

Vinegar, which contains acetic acid, and baking soda (sodium bicarbonate) react to produce sodium acetate, carbon dioxide, and water. The carbon dioxide fills the balloon, which expands. *(Charles D. Winters)*

EXAMPLE 3.3

Predict the position of equilibrium and calculate the equilibrium constant, K_{eq}, for these acid-base reactions. The pK_a of phenol is 9.95.

(a) $\underset{\text{Phenol}}{C_6H_5OH} + \underset{\text{Bicarbonate ion}}{HCO_3^-} \rightleftharpoons \underset{\text{Phenoxide ion}}{C_6H_5O^-} + \underset{\text{Carbonic acid}}{H_2CO_3}$

(b) $\underset{\text{Acetylene}}{HC{\equiv}CH} + \underset{\text{Amide ion}}{NH_2^-} \rightleftharpoons \underset{\text{Acetylide ion}}{HC{\equiv}C^-} + \underset{\text{Ammonia}}{NH_3}$

Solution

(a) Carbonic acid is the stronger acid; the position of this equilibrium lies to the left. Phenol does not transfer a proton to bicarbonate ion to form carbonic acid.

$$\underset{pK_a\ 9.95}{C_6H_5OH} + HCO_3^- \rightleftharpoons C_6H_5O^- + \underset{pK_a\ 6.36}{H_2CO_3} \qquad \begin{aligned} pK_{eq} &= 9.95 - 6.36 = 3.59 \\ K_{eq} &= 10^{-3.59} = 2.57 \times 10^{-4} \end{aligned}$$

(b) Acetylene is the stronger acid; the position of this equilibrium lies to the right.

$$\underset{pK_a\ 25}{HC{\equiv}CH} + NH_2^- \rightleftharpoons HC{\equiv}C^- + \underset{pK_a\ 38}{NH_3} \qquad pK_{eq} = 25 - 38 = -13 \quad K_{eq} = 10^{13}$$

PROBLEM 3.3

Predict the position of equilibrium and calculate the equilibrium constant, K_{eq}, for these acid-base reactions. The pK_a of methylammonium ion is 10.64.

(a) $\underset{\text{Methylamine}}{CH_3NH_2} + \underset{\text{Acetic acid}}{CH_3CO_2H} \rightleftharpoons \underset{\text{Methylammonium ion}}{CH_3NH_3^+} + \underset{\text{Acetate ion}}{CH_3CO_2^-}$

(b) $\underset{\text{Ethoxide ion}}{CH_3CH_2O^-} + \underset{\text{Ammonia}}{NH_3} \rightleftharpoons \underset{\text{Ethanol}}{CH_3CH_2OH} + \underset{\text{Amide ion}}{NH_2^-}$

3.5 Lewis Acids and Bases

Gilbert N. Lewis, who proposed that covalent bonds are formed by sharing one or more pairs of electrons (Section 1.2B), further expanded the theory of acids and bases to include a group of substances not included in the Brønsted-Lowry concept. According to the Lewis definition, a **Lewis acid** is a species that can form a new covalent bond by accepting a pair of electrons; a **Lewis base** is a species that can form a new covalent bond by donating a pair of electrons. In this general equation, the Lewis acid, A, accepts a pair of electrons in forming the new covalent bond and acquires a negative formal charge. The Lewis base, :B, donates the pair of electrons in forming the new covalent bond and acquires a positive formal charge.

Lewis acid Any molecule or ion that can form a new covalent bond by accepting a pair of electrons.

Lewis base Any molecule or ion that can form a new covalent bond by donating a pair of electrons.

$$\underset{\text{Lewis acid}}{A} + \underset{\text{Lewis base}}{:B} \rightleftharpoons \overset{-}{A}{-}\overset{+}{B}$$

new covalent bond formed in this Lewis acid-base reaction

Note that, although we speak of a Lewis base as "donating" a pair of electrons, the term is not fully accurate. Donating in this case does not mean that the electron pair under consideration is removed completely from the valence shell of the base. Rather, donating means that the electron pair becomes shared with another atom to form a covalent bond.

Consider the neutralization of HCl by NaOH in aqueous solution, which consists of a transfer of a proton from H_3O^+ to OH^-.

$$\underset{\text{Lewis acid (accepts an electron pair)}}{H{-}\overset{+}{O}(H){-}H} + \underset{\text{Lewis base (donates an electron pair)}}{:\overset{-}{\ddot{O}}{-}H} \longrightarrow H{-}\ddot{O}(H): + H{-}\ddot{O}{-}H$$

HCl + NaOH = NaCl H₂O

In creating the new covalent bond to form a water molecule, the proton accepts a pair of electrons; it is the Lewis acid. In forming the new covalent bond, hydroxide ion donates the pair of electrons; it is the Lewis base.

The Lewis concept of acids and bases includes proton-transfer reactions; all Brønsted-Lowry bases (proton acceptors) are also Lewis bases, and all Brønsted-Lowry acids (proton donors) are also Lewis acids. The Lewis model, however, is more general in that it is not restricted to proton-transfer reactions.

Consider the reaction that occurs when boron trifluoride gas is dissolved in diethyl ether.

$$(CH_3CH_2)_2\ddot{O}: + BF_3 \longrightarrow (CH_3CH_2)_2\overset{+}{\ddot{O}}-\overset{-}{B}F_3$$

Diethyl ether (a Lewis base) | Boron trifluoride (a Lewis acid) | A BF_3-ether complex

Diethyl ether is a Lewis base. Boron, a Group IIIA element, has three electrons in its valence shell, and, after forming single bonds with three fluorine atoms to give BF_3, boron still has only six electrons in its valence shell. Because it has an empty orbital in its valence shell and can accept two electrons into it, boron trifluoride is electron deficient and, therefore, a Lewis acid. In forming the O—B bond, the oxygen atom of diethyl ether donates an electron pair, and boron accepts the electron pair. The reaction between diethyl ether and boron trifluoride is classified as an acid-base reaction according to the Lewis model, but because there is no proton transfer involved, it is not classified as an acid-base reaction by the Brønsted-Lowry model.

EXAMPLE 3.4

Write an equation for the reaction between each Lewis acid-base pair, showing electron flow by means of curved arrows.

(a) $BF_3 + NH_3 \longrightarrow$ **(b)** $CH_3CH_2^+ + Cl^- \longrightarrow$

Solution

(a) BF_3 has an empty orbital in the valence shell of boron and is the Lewis acid. NH_3 has an unshared pair of electrons in the valence shell of nitrogen and is the Lewis base. In this example, each of these atoms takes on a formal charge; the resulting structure, however, has no net charge.

$$F_3B + :NH_3 \longrightarrow F_3\overset{-}{B}-\overset{+}{N}H_3$$

Lewis acid | Lewis base

(b) The trivalent carbon atom in the ethyl cation has an empty orbital in its valence shell, and, therefore, is the Lewis acid. Chloride ion is the Lewis base.

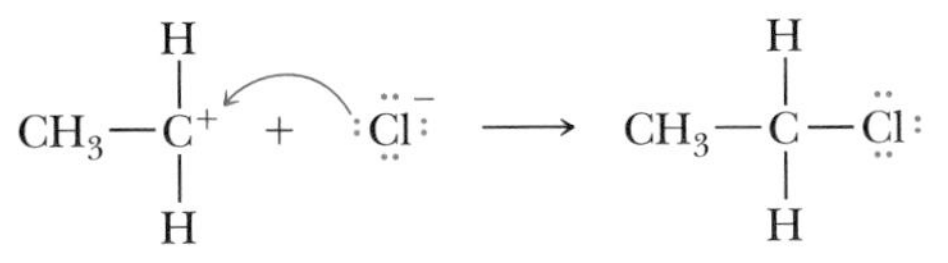

$$CH_3-\overset{H}{\underset{H}{C^+}} + :\ddot{Cl}:^- \longrightarrow CH_3-\overset{H}{\underset{H}{C}}-\ddot{Cl}:$$

Lewis acid Lewis base

PROBLEM 3.4

Write an equation for the reaction between each Lewis acid-base pair, showing electron flow by means of curved arrows.

(a) $(CH_3CH_2)_3B + OH^- \longrightarrow$ **(b)** $CH_3Cl + AlCl_3 \longrightarrow$

CHEMISTRY IN ACTION

The Strongest Acid?

What is the strongest acid? In recent years, organic chemists have prepared mixtures of protic and Lewis acids that have remarkable proton-donating power. Two of the most reactive of these mixtures, termed superacids, are HF with SbF_5 and FSO_3H (fluorosulfonic acid) with SbF_5. Either SO_2 or SO_2F_2 are used as solvents for superacids. The role of SbF_5 is to combine with the fluorine of each acid to form SbF_6^- and thus pull electrons away from hydrogen.

Both theoretical and experimental evidence exists that in superacids, the electron pairs that make up the carbon-carbon and carbon-hydrogen bonds of hydrocarbons act as Lewis bases and become protonated. The resulting cations have unusual structures. In the case of ethane, for example, the ion $C_2H_7^+$ is produced.

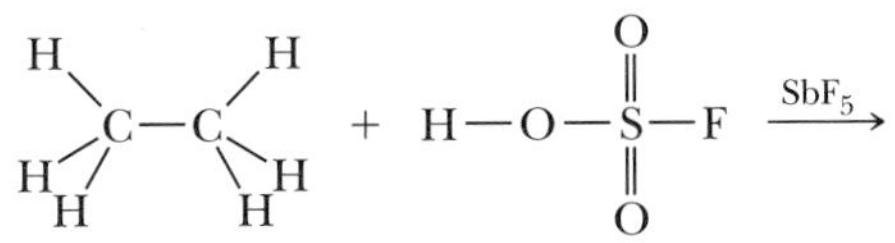

$$H_3C-CH_3 + H-O-\overset{O}{\underset{O}{\overset{\|}{\underset{\|}{S}}}}-F \xrightarrow{SbF_5}$$

Fluorosulfonic acid

$$\left[H_3C\overset{H}{\overset{+}{\cdots}}CH_3\right] + {}^-O-\overset{O}{\underset{O}{\overset{\|}{\underset{\|}{S}}}}-F$$

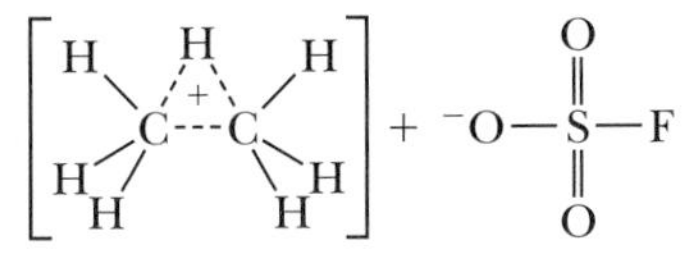

(unstable)

The dashed lines in the structure shown for $C_2H_7^+$ indicate the formation of a three-center, two-electron bond. Notice that because the incoming proton has no electrons of its own, the octet rule is not broken.

Even at low temperatures (− 78°C), the $C_2H_7^+$ ion is not very stable. One of its decomposition products is methane, CH_4. The observation of methane as a product supports protonation of the carbon-carbon bond electron pair over one of the six carbon-hydrogen electron pairs.

Among those studying the reactions of superacids with alkanes was George Olah, at the University of Southern California, who received the Nobel Prize for chemistry in 1994. Olah's discoveries completely transformed our understanding of the chemistry of carbocations.

Although the chemistry of alkanes in superacids may seem esoteric, in fact, these reactions provide a model for one of the most important reactions in industrial organic chemistry, namely the catalytic cracking of petroleum (Section 2.9B). Highly acidic sites on the solid catalysts used in petroleum refining promote protonation of C—C and C—H bonds. Protonation of C—C bonds leads to fragmentation and isomerization of larger hydrocarbons; protonation of C—H bonds leads to the production of hydrogen and alkenes.

SUMMARY

A **Brønsted-Lowry acid** is a proton donor, and a **Brønsted-Lowry base** is a proton acceptor (Section 3.1). Neutralization of an acid by a base is a **proton-transfer reaction.** In a proton transfer reaction, an acid is transformed into its **conjugate base,** and the base is transformed into its **conjugate acid.**

A **strong acid** or **strong base** is one that is completely ionized in water. A weak acid or weak base is one that is only partially ionized in water (Section 3.2). Among the most common weak organic acids are carboxylic acids, compounds that contain the —CO_2H (carboxyl) group. The value of K_a (the acid ionization constant) for acetic acid, a representative carboxylic acid, is 1.74×10^{-5}; the value of pK_a (the negative logarithm of K_a) for acetic acid is 4.76. The relative acidities of the organic acids we deal with most often are determined by (1) electron-withdrawing **inductive effects,** which weaken the H—A bond and promote its ionization, and (2) **resonance** and (3) **hybridization** effects, which stabilize the anion, A^-.

A **Lewis acid** (Section 3.5) is a species that can form a new covalent bond by accepting a pair of electrons; a **Lewis base** is a species that can form a new covalent bond by donating a pair of electrons.

KEY REACTIONS

1. Proton-Transfer Reaction (Section 3.1)

Involves transfer of a proton from a proton donor (a Brønsted-Lowry acid) to a proton acceptor (a Brønsted-Lowry base).

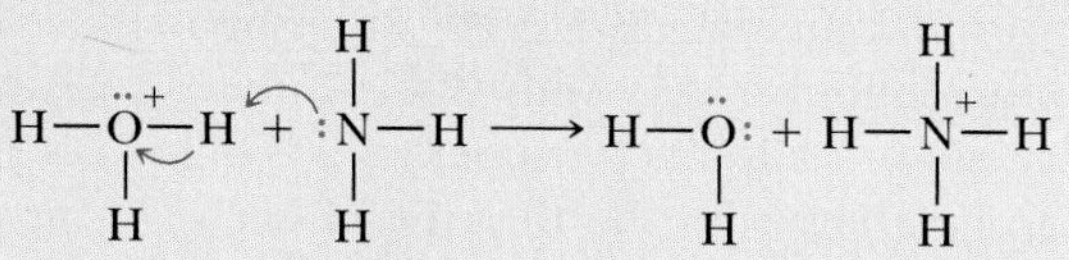

$$H_3O^+ + :NH_3 \longrightarrow H_2O + NH_4^+$$

Proton donor — Proton acceptor

2. Position of Equilibrium in an Acid-Base Reaction (Section 3.4)

The stronger acid reacts with the stronger base to give a weaker acid and a weaker base. K_{eq} for this equilibrium can be calculated from pK_a values for the two acids.

$$\underset{pK_a\ 4.76}{CH_3CO_2H} + CN^- \rightleftharpoons CH_3CO_2^- + \underset{pK_a\ 9.31}{HCN}$$

$$K_{eq} = \frac{K_a(CH_3CO_2H)}{K_a(HCN)} = \frac{1.74 \times 10^{-5}}{4.90 \times 10^{-10}} = 3.55 \times 10^4$$

3. Lewis Acid-Base Reaction (Section 3.5)

A Lewis acid-base reaction involves sharing an electron pair between an electron pair donor (a Lewis base) and an electron pair acceptor (a Lewis acid).

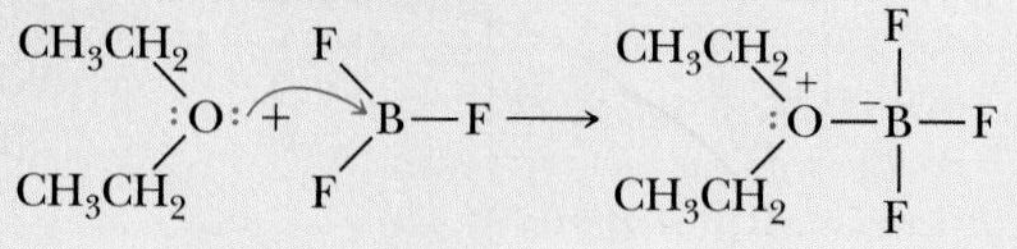

$$(CH_3CH_2)_2O: + BF_3 \longrightarrow (CH_3CH_2)_2\overset{+}{O}-\overset{-}{B}F_3$$

ADDITIONAL PROBLEMS

3.5 Complete a net ionic equation for each proton-transfer reaction using curved arrows to show the flow of electron pairs in each reaction. In addition, write Lewis structures for all starting materials and products. Label the original acid and its conjugate base; the original base and its conjugate acid. If you are uncertain about which substance in each equation is the proton donor, refer to Table 3.1 for the relative strengths of proton acids.

(a) $NH_3 + HCl \longrightarrow$ **(b)** $CH_3CH_2O^- + HCl \longrightarrow$
(c) $HCO_3^- + OH^- \longrightarrow$ **(d)** $CH_3CO_2^- + NH_4^+ \longrightarrow$

3.6 Complete a net ionic equation for each proton-transfer reaction using curved arrows to show the flow of electron pairs in each reaction. Label the original acid and its conjugate base; the original base and its conjugate acid.

(a) $NH_4^+ + OH^- \longrightarrow$ **(b)** $CH_3CO_2^- + CH_3NH_3^+ \longrightarrow$
(c) $CH_3CH_2O^- + NH_4^+ \longrightarrow$ **(d)** $CH_3NH_3^+ + OH^- \longrightarrow$

3.7 Each of the following molecules and ions can function as a base. Write the structural formula of the conjugate acid formed by reaction of each with H^+.

(a) CH_3CH_2OH **(b)** $\mathrm{H\overset{\overset{O}{\|}}{C}H}$ **(c)** $(CH_3)_2NH$ **(d)** HCO_3^-

3.8 In acetic acid, the —OH hydrogen is more acidic than the CH_3— hydrogens. Show how the concept of electronegativity can be used to account for this difference in acidity.

3.9 As we shall see in Chapter 15, hydrogens on a carbon adjacent to a carbonyl group are far more acidic than those not adjacent to a carbonyl group; that is, the anion derived from propanone is more stable than the anion derived from ethane. Account for the greater stability of the anion from propanone in terms (a) the inductive effect and (b) the resonance effect.

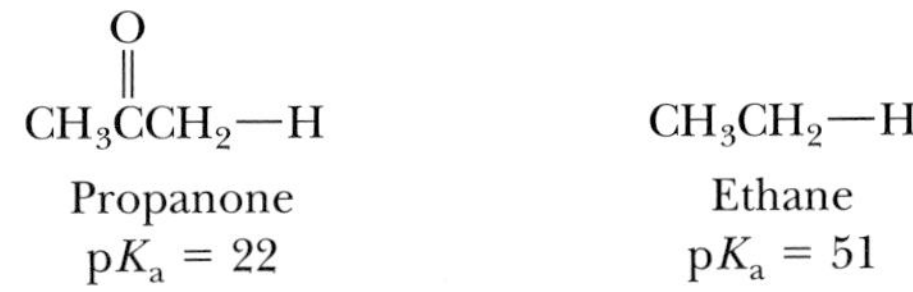

3.10 Offer an explanation for the following observations.

(a) H_3O^+ is a stronger acid than NH_4^+.
(b) Nitric acid, HNO_3, is a stronger acid than nitrous acid, HNO_2.
(c) Ethanol and water have approximately the same acidity.
(d) Trifluoroacetic acid, CF_3CO_2H, is a stronger acid than trichloroacetic acid, CCl_3CO_2H.

Quantitative Measure of Acid and Base Strength

3.11 Which has the larger numerical value:

(a) The pK_a of a strong acid or the pK_a of a weak acid?
(b) The K_a of a strong acid or the K_a of a weak acid?

3.12 In each pair, select the stronger acid:

(a) Pyruvic acid (pK_a 2.49) or lactic acid (pK_a 3.85).
(b) Citric acid (pK_{a1} 3.08) or phosphoric acid (pK_{a1} 2.10).
(c) Nicotinic acid (niacin, K_a 1.4×10^{-5}) or acetylsalicylic acid (aspirin, K_a 3.3×10^{-4}).
(d) Phenol (K_a 1.12×10^{-10}) or boric acid (K_a 7.24×10^{-10}).

3.13 Arrange the compounds in each set in order of increasing acid strength. Consult Table 3.1 for pK_a values of each acid.

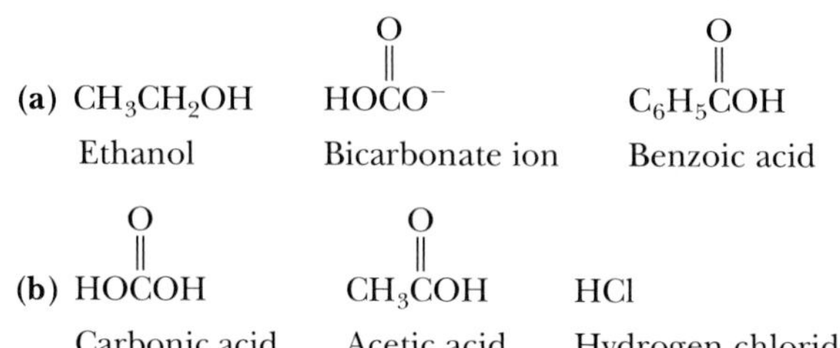

3.14 Arrange the compounds in each set in order of increasing base strength. Consult Table 3.1 for pK_a values of the conjugate acid of each base.

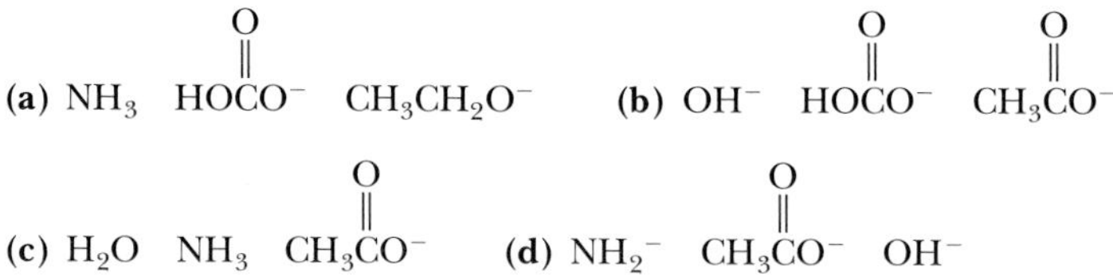

Position of Equilibrium in Acid-Base Reactions

3.15 Unless under pressure, carbonic acid in aqueous solution breaks down into carbon dioxide and water, and carbon dioxide is evolved as bubbles of gas. Write an equation for the conversion of carbonic acid to carbon dioxide and water.

3.16 Will carbon dioxide be evolved when sodium bicarbonate is added to an aqueous solution of these compounds?

(a) Sulfuric acid **(b)** Ethanol **(c)** Ammonium chloride

3.17 Acetic acid, CH_3CO_2H, is a weak organic acid, pK_a 4.76. Write equations for the equilibrium reactions of acetic acid with each base. Which equilibria lie considerably toward the left? Which lie considerably toward the right?

(a) $NaHCO_3$ **(b)** NH_3 **(c)** H_2O **(d)** NaOH

3.18 Alcohols are weak organic acids, pK_a 15–18. The pK_a of ethanol, CH_3CH_2OH, is 15.9. Write equations for the equilibrium reactions of ethanol with each base. Which equilibria lie considerably toward the left? Which lie considerably toward the right?

(a) HCO_3^- **(b)** OH^- **(c)** NH_2^- **(d)** NH_3

3.19 Benzoic acid, $C_6H_5CO_2H$, is only slightly soluble in water, but its sodium salt, $C_6H_5CO_2^-Na^+$, is quite soluble in water. Will benzoic acid dissolve in:

(a) Aqueous sodium hydroxide **(b)** Aqueous sodium bicarbonate
(c) Aqueous sodium carbonate

3.20 Phenol, C_6H_5OH (pK_a 9.95), is only slightly soluble in water, but its sodium salt, $C_6H_5O^-Na^+$, is quite soluble in water. Will phenol dissolve in:

(a) Aqueous NaOH **(b)** Aqueous $NaHCO_3$ **(c)** Aqueous Na_2CO_3

3.21 For an acid-base reaction, one way to determine the predominant species at equilibrium is to say that the reaction arrow points to the acid with the higher value of pK_a. For example:

$$\underset{\text{p}K_a\ 9.24}{NH_4^+} + H_2O \longleftarrow NH_3 + \underset{\text{p}K_a\ -1.74}{H_3O^+}$$

$$NH_4^+ + OH^- \longrightarrow NH_3 + H_2O$$
$$pK_a\ 9.24 \qquad\qquad pK_a\ 15.7$$

Explain why this rule works.

3.22 Will acetylene react with sodium amide according to the following equation to form a salt and ammonia? Calculate K_{eq} for this equilibrium.

$$\underset{\text{Acetylene}}{HC{\equiv}CH} + \underset{\text{Sodium amide}}{Na^+NH_2^-} \rightleftharpoons HC{\equiv}C^-Na^+ + NH_3$$

3.23 Will ethylene react with sodium amide according to the following equation to form a salt and ammonia? Calculate K_{eq} for this equilibrium.

$$\underset{\text{Ethylene}}{CH_2{=}CH_2} + \underset{\text{Sodium amide}}{Na^+NH_2^-} \rightleftharpoons CH_2{=}CH^-Na^+ + NH_3$$

3.24 Using pK_a values given in Table 3.1, predict the position of equilibrium in this acid-base reaction and calculate its K_{eq}.

$$H_3PO_4 + CH_3CH_2OH \rightleftharpoons H_2PO_4^- + CH_3CH_2OH_2^+$$

3.25 2,4-Pentanedione is a considerably stronger acid than is propanone (acetone). Write a structural formula for the conjugate base of each acid and account for the greater stability of the conjugate base from 2,4-pentanedione.

$$\underset{\substack{\text{Propanone}\\ pK_a = 22}}{CH_3\overset{O}{\overset{\|}{C}}CH_2{-}H} \qquad \underset{\substack{\text{2,4-Pentanedione}\\ pK_a = 9}}{CH_3\overset{O}{\overset{\|}{C}}\underset{H}{\underset{|}{C}}H\overset{O}{\overset{\|}{C}}CH_3}$$

3.26 Write an equation for the acid-base reaction between 2,4-pentanedione and sodium ethoxide and calculate its equilibrium constant, K_{eq}. Label the stronger acid, stronger base, and so on. The pK_a of 2,4-pentanedione is 9; that of ethanol is 15.9.

$$\underset{\text{2,4-Pentanedione}}{CH_3\overset{O}{\overset{\|}{C}}\underset{H}{\underset{|}{C}}H\overset{O}{\overset{\|}{C}}CH_3} + \underset{\text{Sodium ethoxide}}{CH_3CH_2O^-Na^+} \rightleftharpoons$$

Lewis Acids and Bases

3.27 For each equation, label the Lewis acid and the Lewis base. In addition, use curved arrows to show the flow of electrons in each reaction.

(a) $:\ddot{F}:^- + BF_3 \longrightarrow BF_4^-$ (:F:—B(—F:)(—F:) shown with lone pairs on each F)

(b) $H_2C{=}\ddot{O}: + H{-}\ddot{Cl}: \longrightarrow H_2C{=}\overset{+}{\ddot{O}}{-}H + :\ddot{Cl}:^-$

3.28 Complete the equation for the reaction between each Lewis acid-Lewis base pair. In each equation, label which starting material is the Lewis acid and which is the Lewis base; then use curved arrows to show the flow of electrons in each reaction. In doing this problem, it is essential that you show valence electrons for all atoms participating in each reaction.

(a) $CH_3-\underset{CH_3}{\overset{CH_3}{\underset{|}{\overset{|}{C}}}}-Cl + \underset{Cl}{\overset{Cl}{\underset{|}{\overset{|}{Al}}}}-Cl \longrightarrow$ (b) $CH_3-\underset{CH_3}{\overset{CH_3}{\underset{|}{\overset{|}{C^+}}}} + H-O-H \longrightarrow$

(c) $CH_3-\overset{+}{C}H-CH_3 + Br^- \longrightarrow$ (d) $CH_3-\overset{+}{C}H-CH_3 + CH_3-O-H \longrightarrow$

3.29 Each of these reactions can be written as a Lewis acid-Lewis base reaction. Label the Lewis acid, the Lewis base, and use curved arrows to show the flow of electrons in each reaction. In doing this problem, it is essential that you show valence electrons for all atoms participating in each reaction.

(a) $CH_3-CH=CH_2 + H-Cl \longrightarrow CH_3-\overset{+}{C}H-\overset{H}{\overset{|}{C}}H_2 + Cl^-$

(b) $CH_3-\underset{CH_3}{\underset{|}{C}}=CH_2 + Br-Br \longrightarrow CH_3-\underset{CH_3}{\underset{|}{\overset{+}{C}}}-CH_2-Br + Br^-$

STEREOCHEMISTRY

4

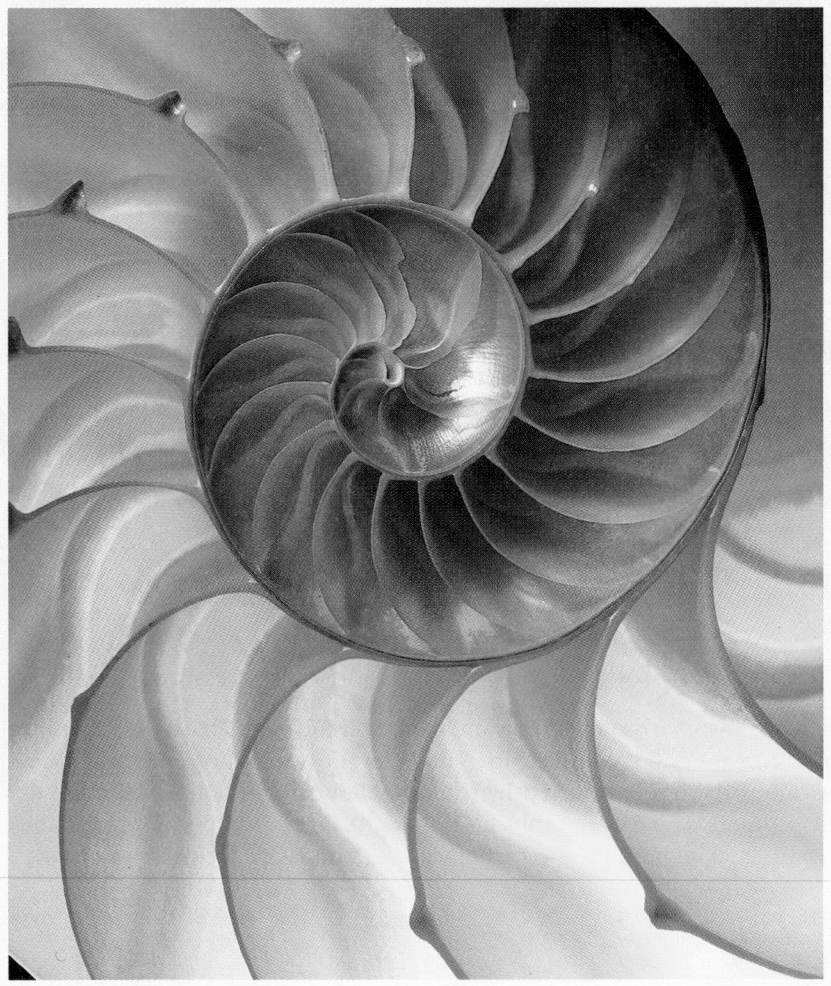

■ Median cross section through a shell of a chambered nautilus, *nautilus pompilius*, a cephalopod that lives in the southwest Pacific to depths of 1800 ft. This shell shows handedness; this cross section is a left-handed spiral. *(Lester Lefkowitz/Tony Stone Worldwide)*

Our goal in this chapter is to expand further our awareness of molecules as three-dimensional objects. In particular, we explore the relationships between three-dimensional objects and their mirror images. When you look in a mirror, you see a reflection, or mirror image, of yourself. Now suppose your mirror image became a three-dimensional object. We could then ask, "What is the relationship between you and your mirror image?" By relationship we mean, "Can your reflection be superposed on the original 'you' in such a way that every detail of the reflection corresponds exactly to the original?" The answer is that you and your mirror image are not superposable. If you have a ring on the little finger of your right hand, for example, your mirror image has the ring on the little finger of the left hand. If you part your hair on your right side, it will be parted on the left side in your reflection. Simply stated, you and your reflection are different objects. You cannot superpose one on the other.

An understanding of spatial relationships of this type is fundamental to an understanding of organic chemistry and biochemistry. In fact, the ability to deal with molecules as three-dimensional objects is a survival skill in organic chemistry and biochemistry.

4.1 Isomerism

Isomers Different compounds with the same molecular formula.

Constitutional isomers Isomers with a different connectivity, that is, a different order of attachment of their atoms.

Isomers are different compounds with the same molecular formula. Thus far, we have encountered two types of isomers. **Constitutional isomers** (Section 2.2) have the same molecular formula but a different order of attachment of atoms in their molecules. Examples of pairs of constitutional isomers are pentane and 2-methylbutane, and 1-pentene and cyclopentane.

Constitutional isomers:

$CH_3CH_2CH_2CH_2CH_3$ and $CH_3CH(CH_3)CH_2CH_3$ $CH_2{=}CHCH_2CH_2CH_3$ and [cyclopentane]

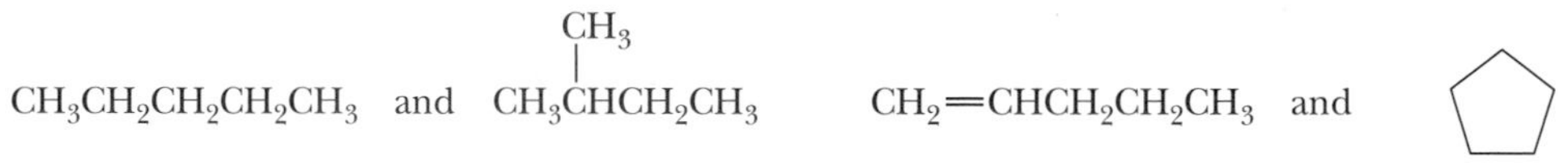

Pentane (C_5H_{12}) 2-Methylbutane (C_5H_{12}) 1-Pentene (C_5H_{10}) Cyclopentane (C_5H_{10})

Stereoisomers Isomers that have the same molecular formula and the same connectivity but different orientations of their atoms in space that cannot be interconverted by rotation about a single bond.

A second type of isomerism is stereoisomerism. **Stereoisomers** have the same molecular formula and the same connectivity, but different orientations of their atoms in space that cannot be interconverted by rotation about single bonds. The one example of stereoisomers we have seen thus far is that of *cis-trans* isomers in cycloalkanes (Section 2.7), which arise because the carbon-carbon bonds are locked into place by the ring.

Stereoisomers:

CH_3 CH_3 H H and CH_3 H H CH_3

cis-1,4-Dimethylcyclohexane *trans*-1,4-Dimethylcyclohexane

Mirror image The reflection of an object in a mirror.

Stereoisomers are divided into two groups: stereoisomers that are mirror images of each other and stereoisomers that are not mirror images of each other. A **mirror image** is the reflection of an object in a mirror. Stereoisomers that are mirror images

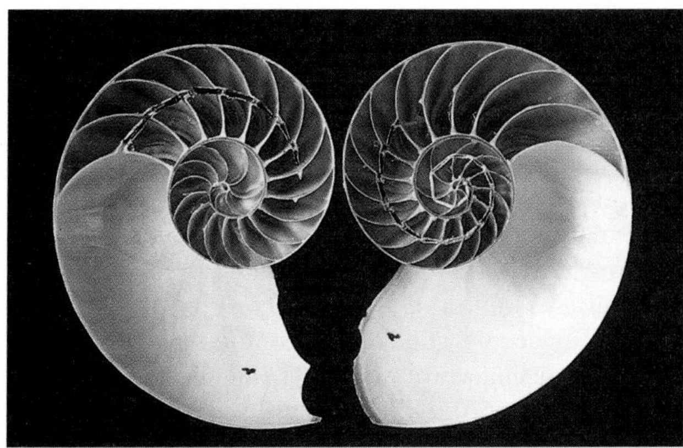

One half of this bisected nautilus shell has a left-handed spiral; the other half has a right-handed spiral. *(J. Kirk Cochran)*

of each other are called **enantiomers** (Greek: *enantios* + *meros,* opposite + part). Stereoisomers that are not mirror images are called **diastereomers.** Figure 4.1 shows these relationships. In the following sections, we deal with enantiomers and diastereomers and explain the similarities and differences among them.

Enantiomers Stereoisomers that are nonsuperposable mirror images; refers to a relationship between pairs of objects.

Diastereomers Stereoisomers that are not mirror images of each other; refers to relationships among two or more objects.

Isomers
Different compounds with the same molecular formula

Constitutional isomers
Isomers with a different order of attachment of atoms in their molecules

Stereoisomers
Isomers with the same order of attachment of atoms in their molecules, but a different orientation of their atoms or groups of atoms in space

Enantiomers
Stereoisomers whose molecules are mirror images of each other

Diastereomers
Stereoisomers whose molecules are not mirror images of each other

FIGURE 4.1
Relationships among isomers.

4.2 Chirality

Chiral From the Greek *cheir,* meaning hand; an object that is not superposable on its mirror image.

Molecules that are not superposable on their mirror images are said to be **chiral** (pronounced ki-ral, to rhyme with spiral; from the Greek: *cheir,* hand). That is, they show handedness. Chirality is encountered in three-dimensional objects of all sorts. Your left hand is chiral and so is your right hand. A spiral binding on a notebook is chiral. A machine screw with a right-handed twist is chiral. A ship's propeller is chiral. As you examine the objects in the world around you, you will undoubtedly conclude that the vast majority of them are chiral as well.

The contrasting situation to chirality occurs when an object and its mirror image are superposable. An object and its mirror image are superposable if one of them can be oriented in space so that all its features (corners, edges, points, designs, etc.) correspond exactly to those in the other member of the pair. If this can be done, the object and its mirror image are identical; the original object is achiral. An **achiral** object is one that lacks chirality. Examples of objects lacking chirality are an undecorated cup, a shell such as a sand dollar, a regular tetrahedron, a cube, and a perfect sphere.

Achiral An object that lacks chirality; an object that has no handedness.

An object is achiral if it possesses at least one plane of symmetry or one center of symmetry. A **plane of symmetry** (also called a **mirror plane**) is an imaginary plane passing through an object dividing it such that one half is the reflection of the other half. The beaker shown in Figure 4.2 has a single plane of symmetry, while the cube has several planes of symmetry. A **center of symmetry** is a point so situated that iden-

Plane of symmetry An imaginary plane passing through an object dividing it such that one half is the mirror image of the other half.

Center of symmetry A point so situated that identical components of the object are located on opposite sides and equidistant from the point along any axis passing through that point.

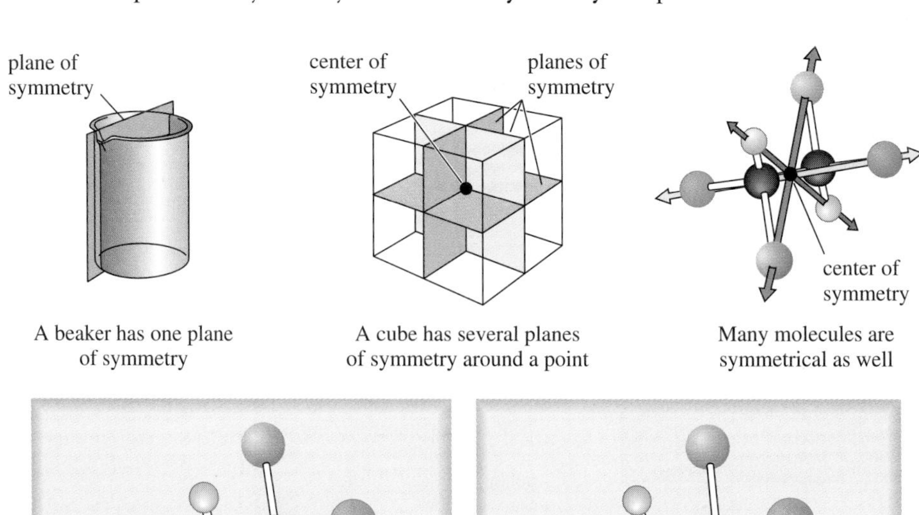
plane of symmetry
center of symmetry
planes of symmetry
center of symmetry
A beaker has one plane of symmetry
A cube has several planes of symmetry around a point
Many molecules are symmetrical as well

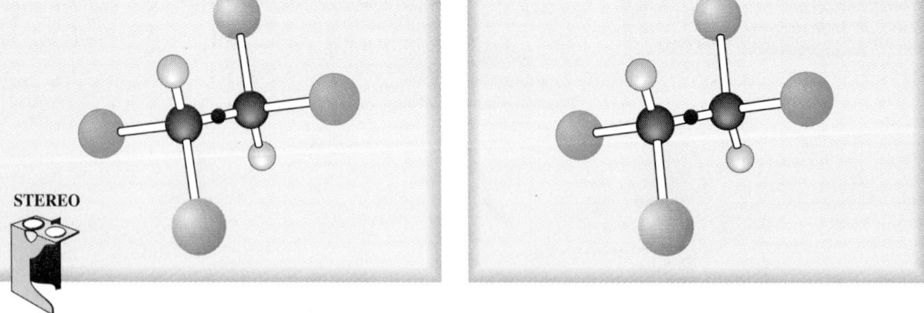
STEREO

FIGURE 4.2

Planes of symmetry (mirror planes) in a beaker and a cube. The cube and the staggered conformation of this disubstituted ethane each possess both one or more planes and a center of symmetry.

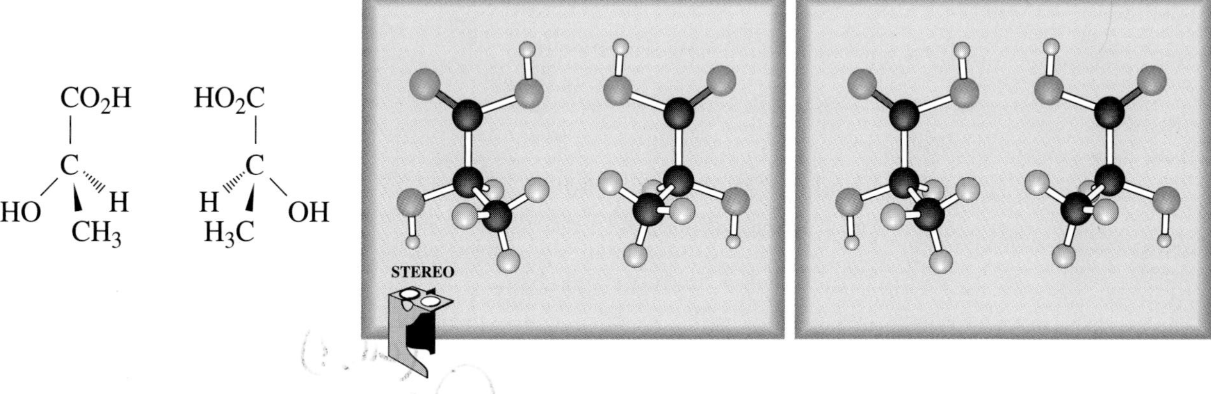

FIGURE 4.3
Stereorepresentations of lactic acid and its mirror image.

tical components of the object are located on opposite sides and equidistant from the point along any axis passing through that point. Both the cube and the staggered conformation of the disubstituted ethane shown in Figure 4.2 have a center of symmetry. No other conformation of this substituted ethane has such a center.

We can illustrate the chirality of an organic molecule by considering 2-hydroxypropanoic acid, more commonly named lactic acid. Figure 4.3 shows three-dimensional representations for lactic acid and its mirror image. In these representations, all bond angles about the central carbon atom are approximately 109.5°, and the four bonds from this carbon are directed toward the corners of a regular tetrahedron.

A model of lactic acid can be turned and rotated in any direction in space, but as long as bonds are not broken and rearranged, only two of the four groups attached to the central carbon can be made to coincide with those of its mirror image. Because lactic acid and its mirror image are nonsuperposable, they are classified as enantiomers. Enantiomers are nonsuperposable mirror images. Note that the terms chiral and achiral refer to objects; the term enantiomers refers to the relationship between objects.

The most common (but not the only) cause of chirality in organic molecules is a carbon atom bonded to four different groups. A carbon atom with four different atoms or groups of atoms attached to it is called a **stereocenter,** or alternatively, a **stereogenic center.** The carbon atom of lactic acid bearing the —OH, —H, $—CH_3$, and $—CO_2H$ groups is an example of a stereocenter (Figure 4.3). In more general terms, a stereocenter is an atom in a molecule at which interchange of two atoms or groups of atoms bonded to it gives a stereoisomer.

Stereocenter An atom that has four different atoms or groups of atoms attached to it; also called a stereogenic center.

EXAMPLE 4.1

Each molecule has one stereocenter. Draw stereorepresentations for the enantiomers of each.

(a) $CH_3CH(Cl)CH_2CH_3$ **(b)** $CH_3CH_2CH(OH)CH_2OH$

(c) [cyclohexene ring with Cl] (d) $CH_3\overset{OH}{C}HCH{=}CH_2$

Solution

You will find it helpful to build models of each pair of enantiomers and to view them from different perspectives as is done in these representations. As you work with these pairs of enantiomers, notice that each has a carbon atom bonded to four different groups, which makes the molecule chiral.

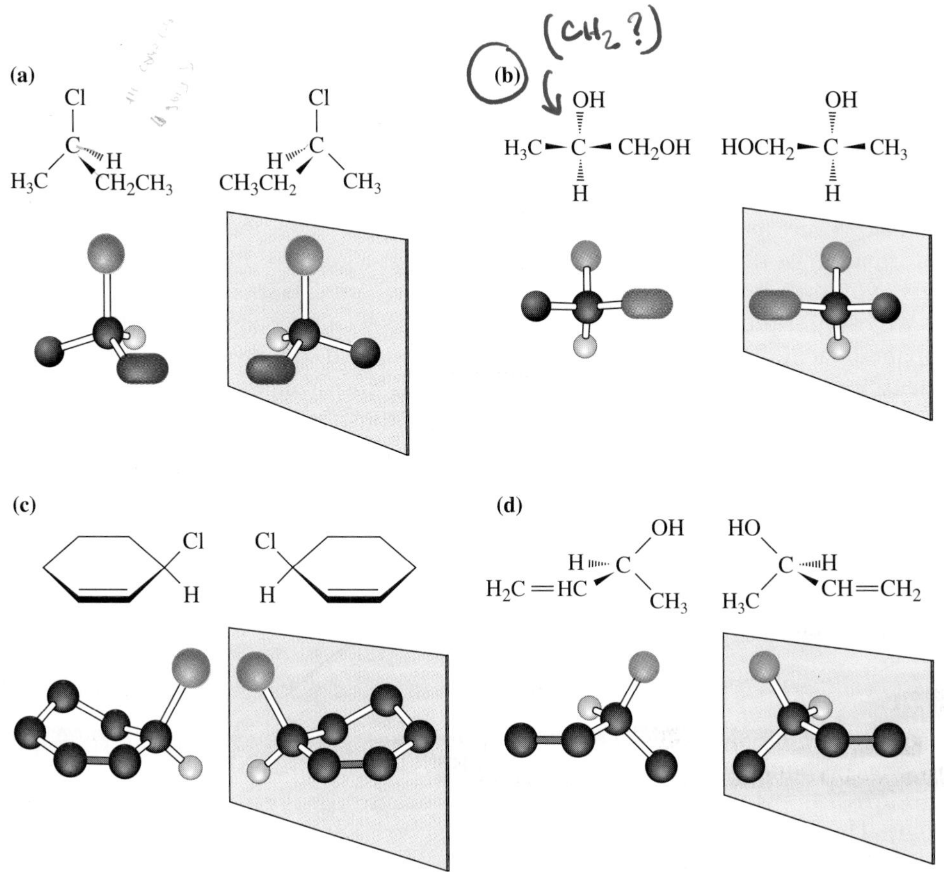

PROBLEM 4.1

Each molecule has one stereocenter. Draw stereorepresentations for the enantiomers of each.

(a) [cyclopentyl]$-\overset{OH}{C}HCH_3$ (b) $CH_3\overset{OH}{C}H\underset{CH_3}{C}HCH_3$

In all of the molecules studied thus far, chirality arises because of the presence of a carbon stereocenter. Stereocenters are not limited to carbon. Following are stereorepresentations of a chiral cation in which the stereocenter is nitrogen. We discuss the chirality of nitrogen stereocenters in more detail in Chapter 21.

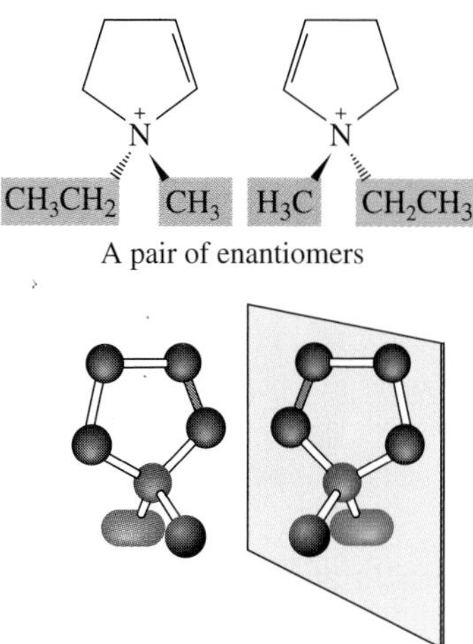

A pair of enantiomers

Enantiomers of tetrahedral silicon, phosphorus, and germanium compounds have also been isolated.

4.3 Naming Enantiomers: The (*R*,*S*) System

A system for designating the configuration of a stereocenter was devised in the late 1950s by R. S. Cahn and C. K. Ingold in England along with V. Prelog of Switzerland. The system, named the Cahn-Ingold-Prelog convention or, alternatively, the ***R,S* convention,** has been incorporated into the IUPAC rules of nomenclature. The orientation of groups about a stereocenter is specified using a set of priority rules.

***R,S* convention** A set of rules for specifying configuration about a stereocenter; also called the Cahn-Ingold-Prelog convention.

Priority Rules

1. Each atom bonded to the stereocenter is assigned a priority. Priority is based on atomic number—the higher the atomic number, the higher the priority. Following are several substituents arranged in order of increasing priority. The atomic number of the atom determining priority is shown in parentheses.

$$\begin{matrix} (1) & (6) & (7) & (8) & (16) & (17) & (35) & (53) \\ -H, & -CH_3, & -NH_2, & -OH, & -SH, & -Cl, & -Br, & -I \end{matrix}$$

Increasing priority →

2. If priority cannot be assigned on the basis of the atoms bonded to the stereocenter, look at the next set of atoms and continue until a priority can be assigned. Priority is assigned at the first point of difference. Following are a series of groups, arranged in order of increasing priority. Shown is the atomic number of the atom on which the assignment of priority is based.

(1)	(6)	(7)	(8)	(17)
$-CH_2-H$	$-CH_2-CH_3$	$-CH_2-NH_2$	$-CH_2-OH$	$-CH_2-Cl$

Increasing priority →

3. Atoms participating in a double or triple bond are considered to be bonded to an equivalent number of similar atoms by single bonds, that is, atoms of a double bond are replicated and atoms of a triple bond are triplicated.

$$-CH{=}CH_2 \xrightarrow{\text{is treated as}} -\underset{}{\overset{\displaystyle C}{\overset{|}{C}}}H-\overset{\displaystyle C}{\overset{|}{C}}H_2 \qquad -\overset{\displaystyle O}{\overset{\|}{C}}H \xrightarrow{\text{is treated as}} -\underset{\underset{\displaystyle H}{|}}{\overset{\overset{\displaystyle O-C}{|}}{C}}-O$$

$$-C{\equiv}CH \xrightarrow{\text{is treated as}} -\underset{\underset{\displaystyle C}{|}}{\overset{\overset{\displaystyle C}{|}}{C}}-\underset{\underset{\displaystyle C}{|}}{\overset{\overset{\displaystyle C}{|}}{C}}-H \qquad -C{\equiv}N \xrightarrow{\text{is treated as}} -\underset{\underset{\displaystyle N}{|}}{\overset{\overset{\displaystyle N}{|}}{C}}-\underset{\underset{\displaystyle C}{|}}{\overset{\overset{\displaystyle C}{|}}{N}}$$

EXAMPLE 4.2

Assign priorities to the groups in each set.

(a) $-\overset{O}{\overset{\|}{C}}OH$ and $-\overset{O}{\overset{\|}{C}}H$ **(b)** $-CH{=}CH_2$ and $-CH(CH_3)_2$

Solution

(a) The first point of difference is O in the $-OH$ of the carboxyl group compared to $-H$ in the aldehyde group. The carboxyl group is higher in priority.

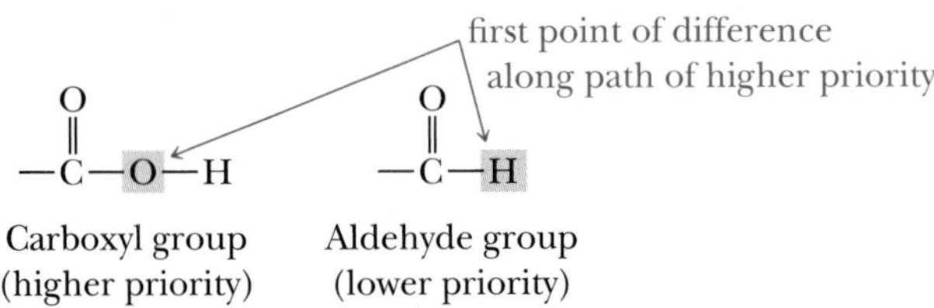

Carboxyl group (higher priority) Aldehyde group (lower priority)

(b) Carbon 1 in each group has the same pattern of atoms; namely C(C,C,H). Carbon 2 is the first point of difference. For the vinyl group, bonding at carbon 2 is C(C,H,H). For the isopropyl group, the bonding at carbon 2 is C(H,H,H). The vinyl group is higher in priority than the isopropyl group.

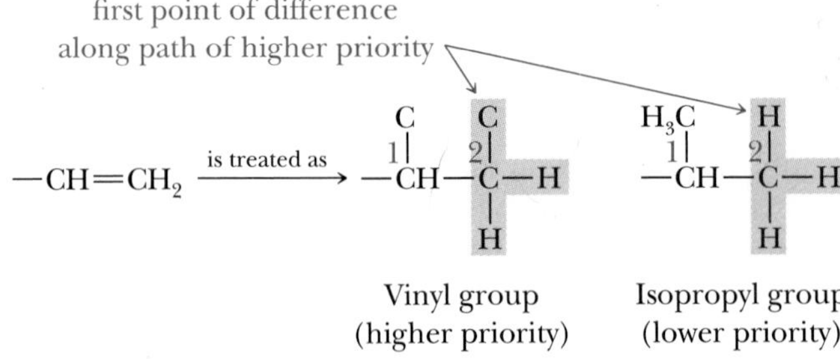

Vinyl group (higher priority) Isopropyl group (lower priority)

PROBLEM 4.2

Assign priorities to the groups in each set.

(a) $-CH_2OH$ and $-CH_2CH_2OH$ **(b)** $-CH_2OH$ and $-CH{=}CH_2$

To assign *R* or *S* configuration to a stereocenter:

1. Locate the stereocenter and identify its four substituents.
2. Assign a priority from 1 (highest) to 4 (lowest) to each substituent.
3. Orient the molecule in space such that the group of lowest priority (4) is directed away from you as would be, for instance, the steering column of a car. The three groups of higher priority (1–3) then project toward you as would the steering wheel.
4. Read the three groups projecting toward you in order from highest priority (1) to lowest priority (3).

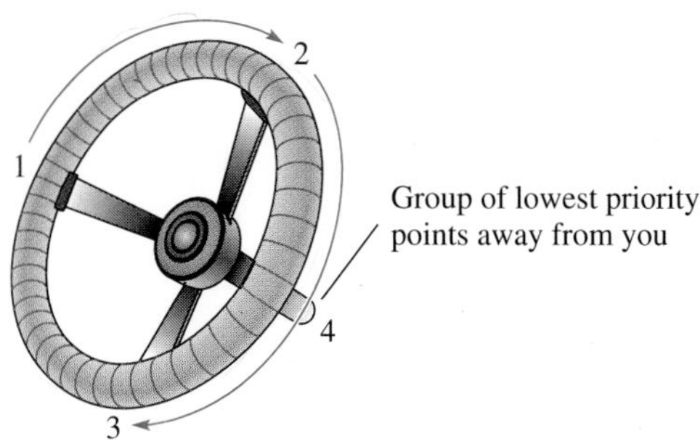

1
2
3
4
Group of lowest priority points away from you

5. If reading the groups proceeds in a clockwise direction, the configuration is designated as ***R*** (Latin: *rectus,* right); if reading proceeds in a counterclockwise direction, the configuration is ***S*** (Latin: *sinister,* left). You can also visualize this as follows: turning the steering wheel to the right equals *R*, and turning the steering wheel to the left equals *S*.

R From the Latin *rectus,* meaning right; used in the *R*,*S* convention to show that the order of priority of groups on a stereocenter is clockwise.

S From the Latin *sinister,* meaning left; used in the *R*,*S* convention to show that the order of priority of groups on a stereocenter is counterclockwise.

EXAMPLE 4.3

Assign an *R* or *S* configuration to each stereocenter.

(a) Cl, C, H, CH_3, CH_3CH_2 (b) Cl, H (c) H_3C, H, Br

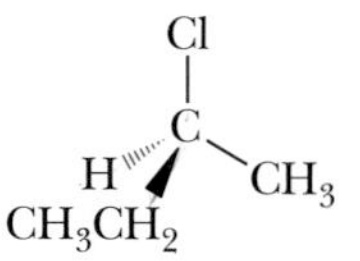

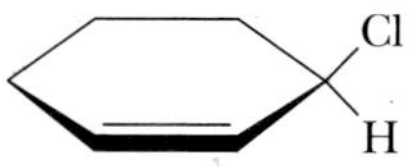

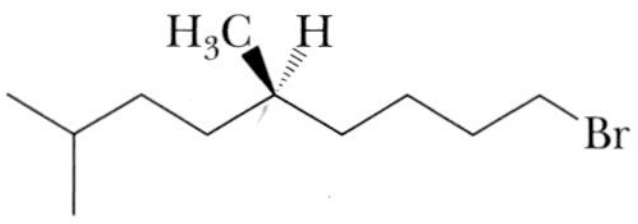

Solution

View each molecule through the stereocenter and along the bond from the stereocenter toward the group of lowest priority.

(a) The order of priority is $—Cl > —CH_2CH_3 > —CH_3 > —H$. The group of lowest priority, H, is pointed away from you. Reading the groups in the order 1, 2, 3 occurs in the counterclockwise direction, so the configuration is *S*.

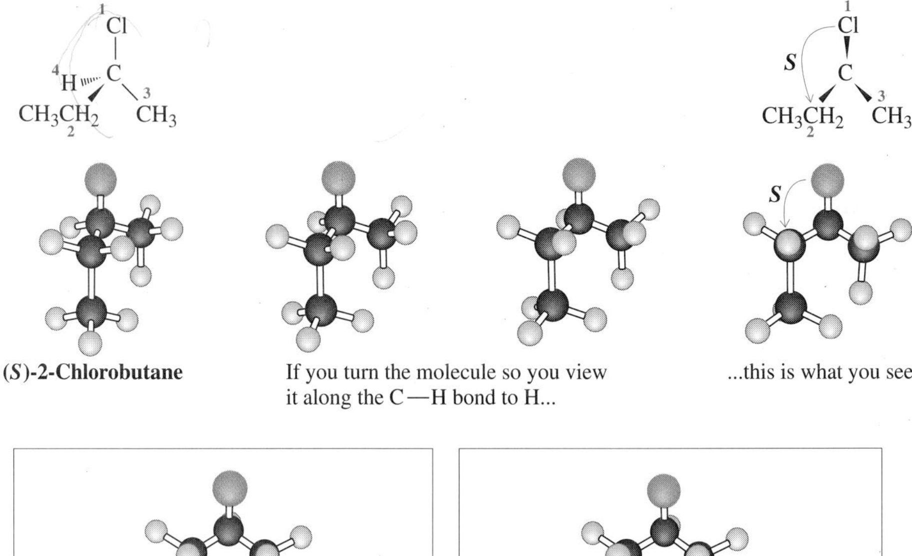

Cl
H
C
CH3CH2
CH3
S
(S)-2-Chlorobutane
If you turn the molecule so you view it along the C—H bond to H...
...this is what you see.
STEREO

(b) The order of priority is $—Cl > —CH{=}CH > —CH_2 > —H$. With hydrogen, the group of lowest priority pointing away from you, reading the groups in the order 1, 2, 3 occurs in the clockwise direction, so the configuration is *R*.

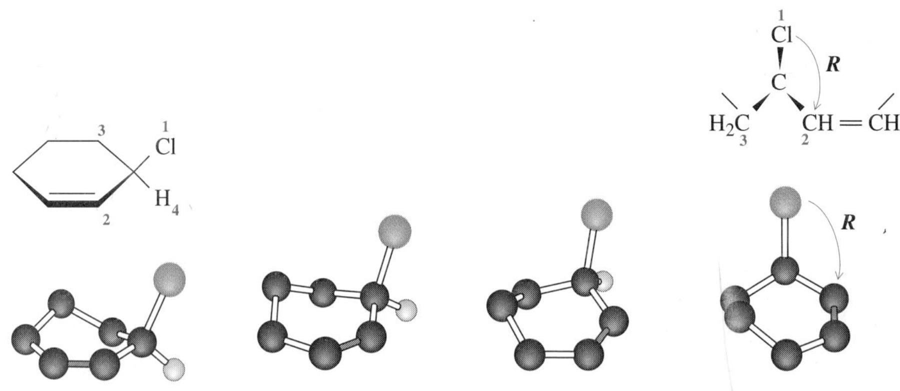

Cl
C
R
H2C
CH=CH
H
(R)-3-Chlorocyclohexene
If you turn the molecule so you view it along the C—H bond to H...
...this is what you see.

(c) The first point of difference and the priority of each group bonded to the stereocenter are shown. The configuration is *S*.

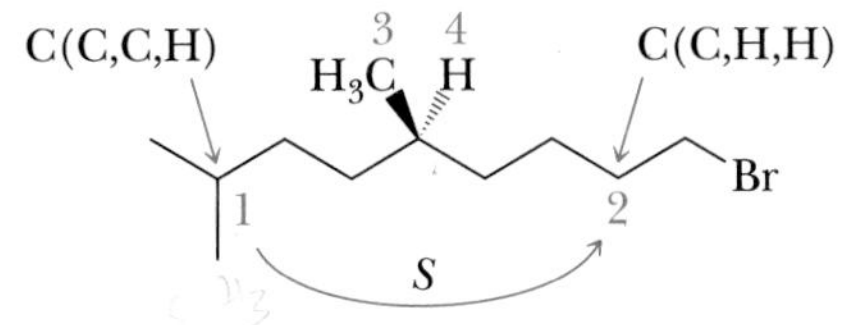

C(C,C,H) 3 4 H_3C H C(C,H,H) Br 1 2 *S*

PROBLEM 4.3

Assign an *R* or *S* configuration to each stereocenter.

(a) 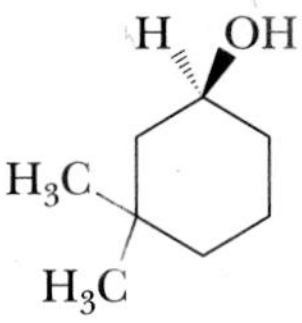 H_3C, H_3C, H, OH (b) 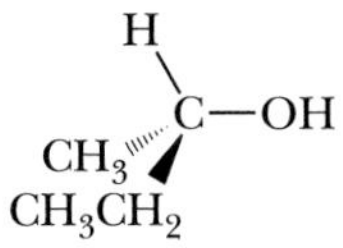 H, C—OH, CH_3, CH_3CH_2 (c) H—C—OH, CHO, CH_2OH

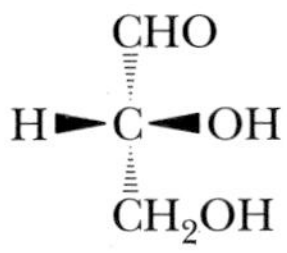

4.4 Fischer Projection Formulas

In writing structures for chiral molecules, we have thus far used lines, solid wedges, and broken wedges to indicate configuration about a stereocenter. The use of these drawing conventions accurately portrays stereochemistry and teaches you to treat molecules as three-dimensional objects.

Nonetheless, chemists sometimes use two-dimensional representations called **Fischer projections** to show the configuration of chiral molecules. This convention is especially useful in portraying the stereochemistry of compounds with several stereocenters, as we shall see in Chapter 24 when we deal with the configuration of carbohydrates.

Fischer projection A two-dimensional representation showing the configuration of a chiral molecule; horizontal lines represent bonds projecting forward and vertical lines represent bonds projecting to the rear. The only atom in the plane of the paper is the stereocenter.

To write a Fischer projection, orient the stereocenter of a chiral molecule so that the vertical bonds to the stereocenter are directed away from you and the horizontal bonds from the stereocenter are directed toward you. To do this for the stereorepresentation of (*S*)-2-butanol shown here, imagine that you hold the —OH group in your right hand and the —H in your left hand. Now turn the molecule so that these groups are both toward you, just as if you were turning the handlebars of a bicycle so that it moves straight ahead. This gives you the stereorepresentation shown here. Then write the molecule as a two-dimensional figure with the stereocenter indicated by the point at which the bonds cross. You now have a Fischer projection for (*S*)-2-butanol.

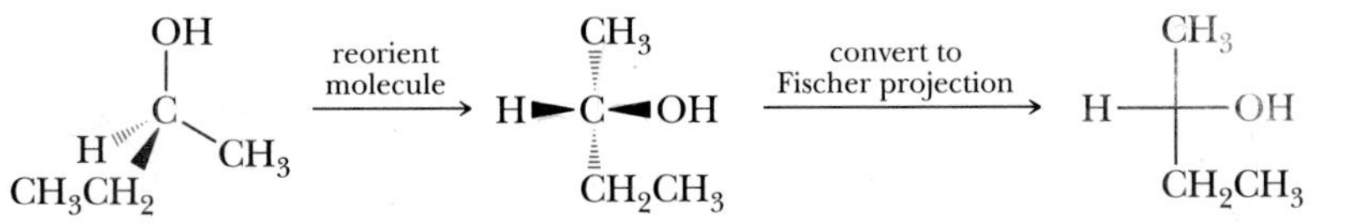

OH, C, H, CH_3, CH_3CH_2 —reorient molecule→ CH_3, H—C—OH, CH_2CH_3 —convert to Fischer projection→ CH_3, H—+—OH, CH_2CH_3

(*S*)-2-Butanol (three-dimensional formula)

(*S*)-2-Butanol (Fischer projection)

The two horizontal segments of a Fischer projection represent bonds directed toward you, and the two vertical segments represent bonds directed away from you. The only atom in the plane of the paper is the stereocenter.

Just as Fischer projections are derived in a very precise way, they can be manipulated only in very precise ways. When using Fischer projections to test the superposability of two structures, the only allowed manipulation that gives the same stereoisomer is rotation in the plane of the paper by 180°. Rotation by 90° in the plane of the paper, or rotation by 180° out of the plane of the paper, gives the enantiomer.

180°; CH_3 / H—OH / CH_2CH_3 — rotation by 180° in the plane of the paper → CH_2CH_3 / HO—H / CH_3

(*S*)-2-Butanol (*S*)-2-Butanol

Example 4.4

We said that rotation of a Fischer projection by 180° out of the plane of the paper gives a different molecule. Show that this manipulation of (*S*)-2-butanol gives (*R*)-2-butanol.

Solution

Rotation by 180° out of the plane of the paper converts the Fischer projection of (*S*)-2-butanol into a Fischer projection of (*R*)-2-butanol.

CH_3 / H—OH / CH_2CH_3 — rotate by 180° out of the plane of the paper → CH_3 / HO—H / CH_2CH_3 — convert to a three-dimensional formula → CH_3 / HO—C—H / CH_2CH_3

(*S*)-2-Butanol (Fischer projection) (*R*)-2-Butanol

Problem 4.4

We said that rotation of a Fischer projection by 90° in the plane of the paper gives a different molecule. Show that this manipulation of (*S*)-2-butanol gives (*R*)-butanol.

When dealing with Fischer projections, the interchange of any two groups at a stereocenter gives the enantiomer of the original molecule. The interchange of any three groups gives an alternative Fischer projection of the original molecule.

Example 4.5

Following is a Fischer projection of (*S*)-2-butanol. To its right are three additional Fischer projections of 2-butanol. Determine the *R,S* configuration of these three projections and the minimum number of group interchanges required to give each. Can you verify that one interchange of groups gives the enantiomer and two interchanges gives the original?

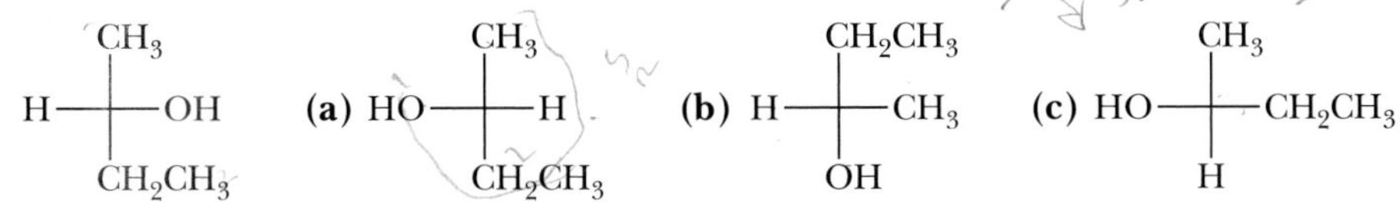

CH_3 — H—|—OH — CH_2CH_3
(*S*)-2-Butanol

(a) CH_3 / HO—|—H / CH_2CH_3 (b) CH_2CH_3 / H—|—CH_3 / OH (c) CH_3 / HO—|—CH_2CH_3 / H

Solution

Each projection is converted to a three-dimensional representation to help you see the stereochemistry at the stereocenter. Projection (a) is (*R*)-2-butanol; it is formed by the interchange of two groups (—H and —OH). To form projection (b), which is (*S*)-2-butanol, requires the interchange of three groups. One way to do this is to interchange —OH for —CH_2CH_3 and then —CH_2CH_3 for —CH_3. Projection (c) is also (*S*)-2-butanol.

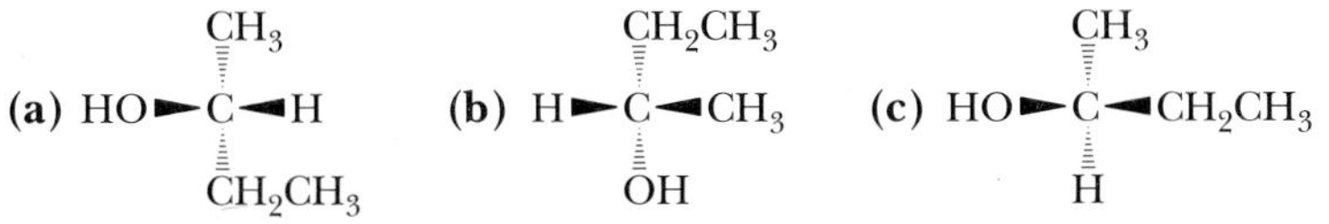

(a) CH_3 / HO►C◄H / CH_2CH_3 (b) CH_2CH_3 / H►C◄CH_3 / OH (c) CH_3 / HO►C◄CH_2CH_3 / H

(*R*)-2-Butanol (*S*)-2-Butanol (*S*)-2-Butanol

PROBLEM 4.5

Convert each three-dimensional representation to a Fischer projection. In so doing, orient the carbon chain vertically. Assign *R,S* configurations to each stereocenter.

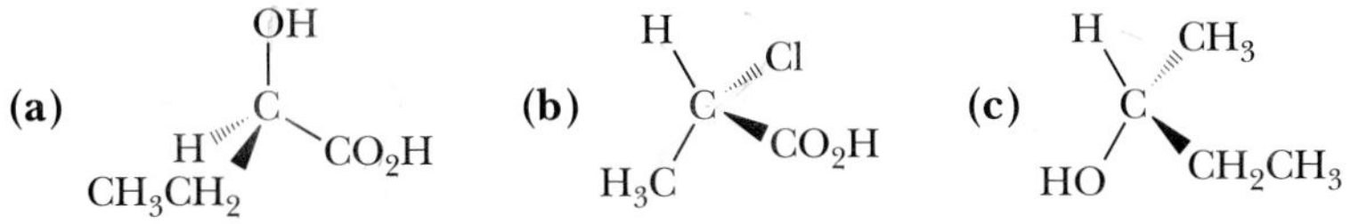

(a) OH, H, C, CO_2H, CH_3CH_2 (b) H, Cl, C, CO_2H, H_3C (c) H, CH_3, C, HO, CH_2CH_3

4.5 Acyclic Molecules with Two or More Stereocenters

For a molecule with ***n*** stereocenters, the maximum number of stereoisomers possible is $\mathbf{2^n}$. For molecules with one stereocenter, $2^1 = 2$; therefore, two stereoisomers are possible. We have now seen several examples of molecules with one stereocenter and verified that, for each molecule, two stereoisomers (one pair of enantiomers) are possible. If we now consider molecules with two or more stereocenters, we must be more precise in our wording, because now all we can specify by counting the number of stereocenters is the maximum number of stereoisomers possible. The maximum number of stereoisomers for a molecule with two stereocenters is $2^2 = 4$. For a molecule with three stereocenters, the maximum number of stereoisomers possible is $2^3 = 8$.

A. Enantiomers and Diastereomers

Let us begin our study of molecules with multiple stereocenters by considering 2,3,4-trihydroxybutanal, a molecule with two stereocenters, shown here in color.

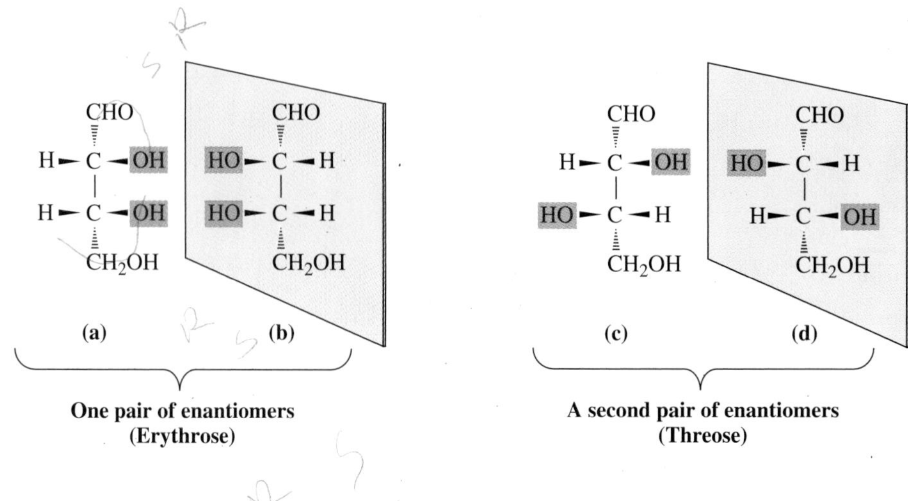

FIGURE 4.4 The four stereoisomers of 2,3,4-trihydroxybutanal, a compound with two stereocenters.

$$HOCH_2{-}\underset{\displaystyle OH}{\underset{|}{CH}}{-}\underset{\displaystyle OH}{\underset{|}{CH}}{-}CHO$$

2,3,4-Trihydroxybutanal

The maximum number of stereoisomers possible for this molecule is $2^2 = 4$, each of which is drawn in Figure 4.4. In these stereorepresentations, the carbon chain is drawn vertically with attached groups projecting forward toward you.

Stereoisomers (a) and (b) are nonsuperposable mirror images of each other and are, therefore, a pair of enantiomers. Stereoisomers (c) and (d) are also nonsuperposable mirror images of each other and are a second pair of enantiomers. One way to describe the four stereoisomers of 2,3,4-trihydroxybutanal is to say that they consist of two pairs of enantiomers. Enantiomers (a) and (b) are given the names (*R,R*)-erythrose and (*S,S*)-erythrose; enantiomers (c) and (d) are given the names (*R,S*)-threose and (*S,R*)-threose. Erythrose and threose belong to the class of compounds called carbohydrates, which we discuss in Chapter 24. Erythrose is found in erythrocytes (red blood cells), hence the derivation of its name.

We have specified the relationship between (a) and (b) and between (c) and (d). What is the relationship between (a) and (c), between (a) and (d), between (b) and (c), and between (b) and (d)? The answer is that they are diastereomers. **Diastereomers** are stereoisomers that are not enantiomers, that is, they are stereoisomers that are not mirror images of each other.

EXAMPLE 4.6

Following are Fischer projection formulas for the four stereoisomers of 1,2,3-butanetriol. *R,S* configurations are given for the stereocenters in (1) and (4).

(1)	(2)	(3)	(4)
CH_2OH	CH_2OH	CH_2OH	CH_2OH
S H—C—OH	H—C—OH	HO—C—H	HO—C—H *R*
S HO—C—H	H—C—OH	HO—C—H	H—C—OH *R*
CH_3	CH_3	CH_3	CH_3

(a) Write IUPAC names for each Fischer projection showing the *R,S* configuration of each stereocenter.
(b) Which compounds are enantiomers?
(c) Which compounds are diastereomers?

Solution

(a) (1) (2*S*,3*S*)-1,2,3-Butanetriol (2) (2*S*,3*R*)-1,2,3-Butanetriol
(3) (2*R*,3*S*)-1,2,3-Butanetriol (4) (2*R*,3*R*)-1,2,3-Butanetriol

(b) Enantiomers are stereoisomers that are nonsuperposable mirror images of each other. As you see from their configurations, compounds (1) and (4) are one pair of enantiomers, and compounds (2) and (3) are a second pair of enantiomers.

(c) Diastereomers are stereoisomers that are not mirror images of each other. Compounds (1) and (2), (1) and (3), (2) and (4), and (3) and (4) are diastereomers.

PROBLEM 4.6

Following are Fischer projection formulas for the four stereoisomers of 3-chloro-2-butanol.

(1)	(2)	(3)	(4)
CH_3	CH_3	CH_3	CH_3
H—+—OH	H—+—OH	HO—+—H	HO—+—H
Cl—+—H	H—+—Cl	H—+—Cl	Cl—+—H
CH_3	CH_3	CH_3	CH_3

(a) Show the *R,S* configuration of each stereocenter.
(b) Which compounds are enantiomers?
(c) Which compounds are diastereomers?

B. Meso Compounds

Certain molecules containing two or more stereocenters have special symmetry properties that reduce the number of stereoisomers to fewer than what is predicted by the 2^n rule. One such molecule is 2,3-dihydroxybutanedioic acid, more commonly named tartaric acid.

```
 O               O
 ||              ||
HOC—CH—CH—COH
     |    |
     OH   OH
```

2,3-Dihydroxybutanedioic acid
(Tartaric acid)

Tartaric acid is a colorless, crystalline compound occurring largely in the vegetable kingdom, especially in grapes. During fermentation of grape juice, potassium bitartrate (one $—CO_2H$ group is present as a potassium salt, $—CO_2^-\ K^+$) deposits as a crust on the sides of wine casks. When collected and purified, it is sold commercially as cream of tartar.

In tartaric acid, carbons 2 and 3 are stereocenters, and, using the 2^n rule, the maximum number of stereoisomers possible is $2^2 = 4$, stereorepresentations for which

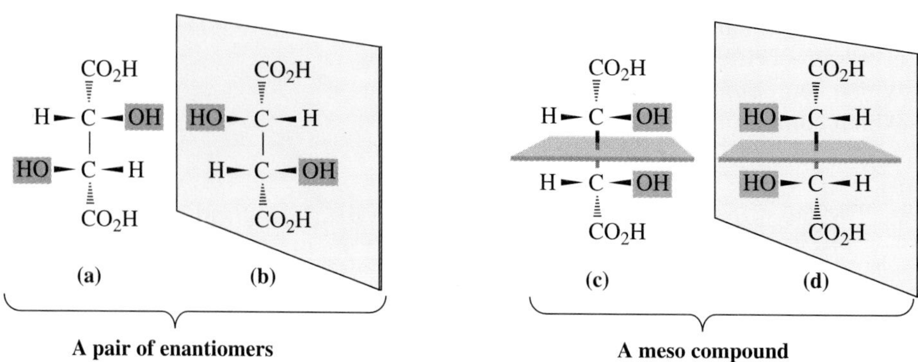

FIGURE 4.5
Stereoisomers of tartaric acid. One pair of enantiomers and one meso compound.

are drawn in Figure 4.5. Structures (a) and (b) are nonsuperposable mirror images and are, therefore, a pair of enantiomers. Structures (c) and (d) are also mirror images, but they are superposable. To see this, imagine that you first rotate (d) by 180° in the plane of the paper, then lift it out of the plane of the paper, and finally place it on top of (c). If you do this mental manipulation correctly, you find that (d) is superposable on (c). Therefore, (c) and (d) are not different molecules; they are the same molecule, just oriented differently. Because (c) and its mirror image are superposable, (c) is achiral.

Another way to determine that (c) is achiral is to see that it has a plane of symmetry that bisects the molecule in such a way that the top half is the reflection of the bottom half. Thus, even though it has two stereocenters, the molecule is achiral (Section 4.2).

Meso compound An achiral compound possessing two or more stereocenters.

The stereoisomer of tartaric acid represented by (c) or (d) is called a meso compound. A **meso compound** is an achiral compound that contains two or more stereocenters. We can now return to the original question: "How many stereoisomers are there of tartaric acid?" The answer is three: one meso compound and one pair of enantiomers. Note that the meso compound and either of the enantiomers are diastereomers; they are stereoisomers that are not mirror images of each other.

EXAMPLE 4.7

Following are Fischer projection formulas for the three stereoisomers of 2,3-butanediol.

(1)	(2)	(3)
CH_3	CH_3	CH_3
H—C—OH	H—C—OH	HO—C—H
HO—C—H	H—C—OH	H—C—OH
CH_3	CH_3	CH_3

(a) Write IUPAC names for each Fischer projection and specify the *R,S* configuration of each stereocenter.

(b) Which are enantiomers?

(c) Which is the meso compound?
(d) Which are diastereomers?

Solution

(a) (1) (2*S*,3*S*)-2,3-Butanediol (2) (2*S*,3*R*)-2,3-Butanediol
(3) (2*R*,3*R*)-Butanediol
(b) Compounds (1) and (3) are enantiomers.
(c) Compound (2) is a meso compound.
(d) (1) and (2) are diastereomers; (2) and (3) are also diastereomers.

PROBLEM 4.7

Following are four Newman projection formulas for tartaric acid.

(1): front: CO_2H, H, OH; back: H, OH, CO_2H
(2): front: CO_2H, HO, H; back: H, OH, CO_2H
(3): front: CO_2H, H, CO_2H; back: H, OH, OH
(4): front: CO_2H, HO, CO_2H; back: H, OH, H

(1) (2) (3) (4)

(a) Which represent the same compound?
(b) Which are enantiomers?
(c) Which represent a meso compound?
(d) Which are diastereomers?

4.6 Cyclic Molecules with Two or More Stereocenters

In this section, we concentrate on derivatives of cyclopentane and cyclohexane containing two stereocenters. We can analyze stereoisomerism in cyclic compounds in the same way we analyzed it in acyclic compounds.

A. Disubstituted Derivatives of Cyclopentane

Let us start with 2-methylcyclopentanol, a compound with two stereocenters. Using the 2^n rule, we predict a maximum of $2^2 = 4$ stereoisomers. Both the *cis* isomer and the *trans* isomer are chiral: the *cis* isomer exists as one pair of enantiomers, and the *trans* isomer exists as a second pair of enantiomers. The *cis* and *trans* isomers are stereoisomers that are not mirror images of each other, or more simply, the *cis* and *trans* isomers are diastereomers.

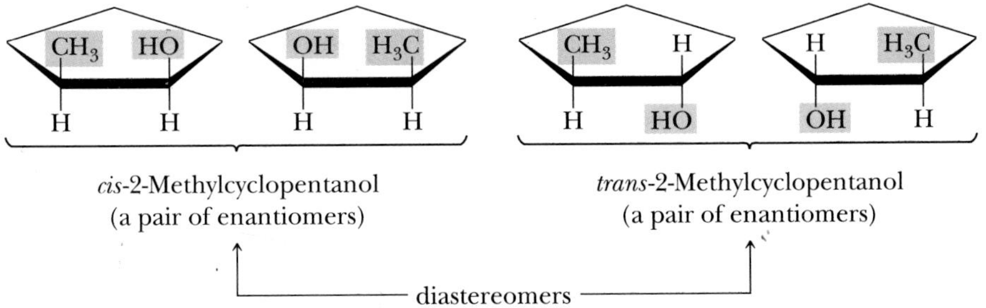

cis-2-Methylcyclopentanol
(a pair of enantiomers)

trans-2-Methylcyclopentanol
(a pair of enantiomers)

1,2-Cyclopentanediol also has two stereocenters and, therefore, the 2^n rule predicts a maximum of $2^2 = 4$ stereoisomers. As seen in the following stereodrawings, only three stereoisomers exist for this compound. The *cis* isomer is achiral (meso) because it and its mirror image are superposable. The *trans* isomer is chiral and exists as a pair of enantiomers.

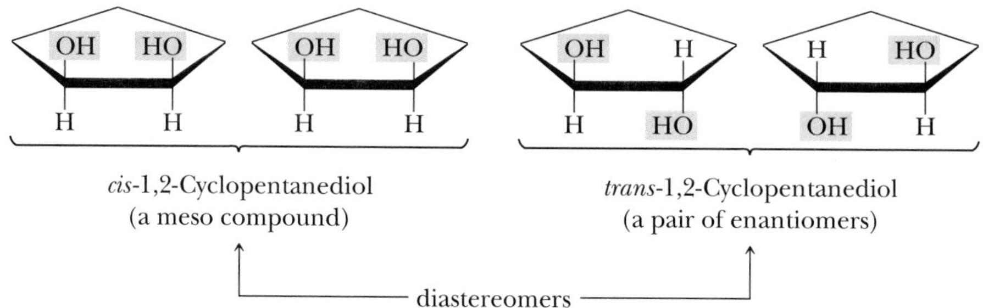

Alternatively, the *cis* isomer is achiral because it possesses a plane of symmetry that bisects the molecule into two mirror halves.

EXAMPLE 4.8

How many stereoisomers exist for 3-methylcyclopentanol?

Solution

There are four stereoisomers of 3-methylcyclopentanol. The *cis* isomer exists as one pair of enantiomers: the *trans* isomer, as a second pair of enantiomers.

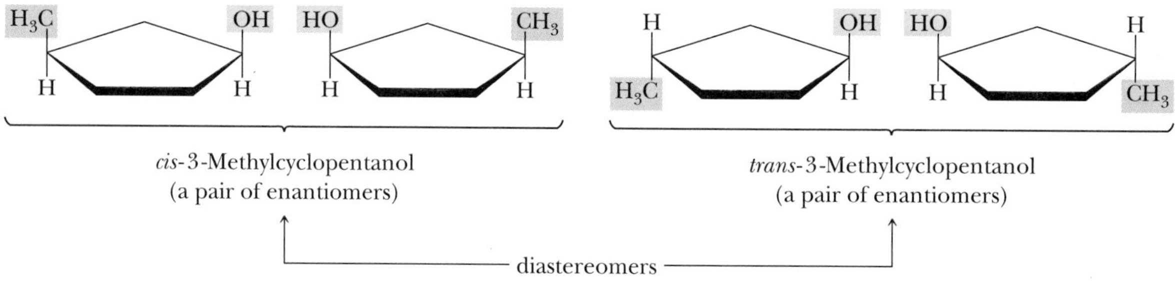

PROBLEM 4.8

How many stereoisomers exist for 1,3-cyclopentanediol?

B. Disubstituted Derivatives of Cyclohexane

As an example of a disubstituted cyclohexane, let us consider the methylcyclohexanols. 4-Methylcyclohexanol can exist as two stereoisomers: a pair of *cis-trans* isomers. Neither the *cis* nor the *trans* isomer has a stereocenter, and, therefore, each is achiral. A plane of symmetry runs through the CH_3—, the HO—, and the two attached carbons.

3-Methylcyclohexanol has two stereocenters and exists as $2^2 = 4$ stereoisomers. The *cis* isomer exists as one pair of enantiomers. The *trans* isomer exists as a second pair of enantiomers.

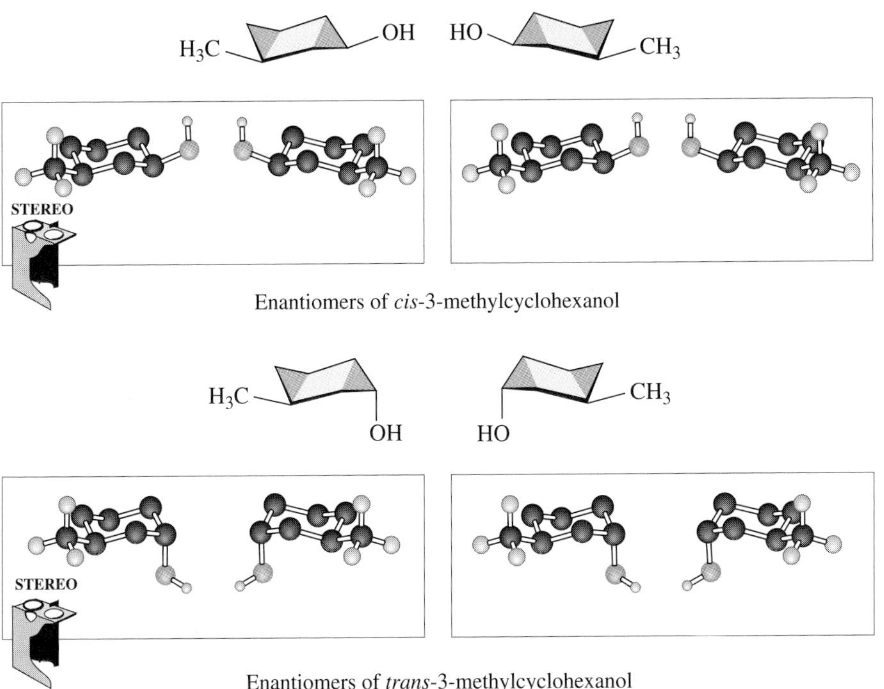

Enantiomers of *cis*-3-methylcyclohexanol

Enantiomers of *trans*-3-methylcyclohexanol

Similarly, 2-methylcyclohexanol has two stereocenters and exists as $2^2 = 4$ stereoisomers. The *cis* isomer exists as one pair of enantiomers; the *trans* isomer, as a second pair of enantiomers.

EXAMPLE 4.9

How many stereoisomers exist for 1,3-cyclohexanediol?

Solution

1,3-Cyclohexanediol has two stereocenters, and according to the 2^n rule, has a maximum of $2^2 = 4$ possible stereoisomers. The *trans* isomer of this compound exists as a pair of enantiomers. The *cis* isomer has a plane of symmetry and is a meso compound. Therefore, although the 2^n rule predicts a maximum of four stereoisomers for 1,3-cyclohexanediol, only three exist: one meso compound and one pair of enantiomers.

OH HO
OH HO OH HO
OH HO

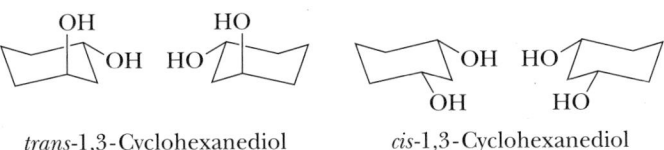

trans-1,3-Cyclohexanediol
(a pair of enantiomers)

cis-1,3-Cyclohexanediol
(meso)

PROBLEM 4.9

How many stereoisomers exist for 1,4-cyclohexanediol?

1,2-Cyclohexanediol has two stereocenters and, according to the 2^n rule, can exist as a maximum of four stereoisomers. The *cis* isomer exists as one pair of enantiomers and the *trans* isomer exists as a second pair of enantiomers.

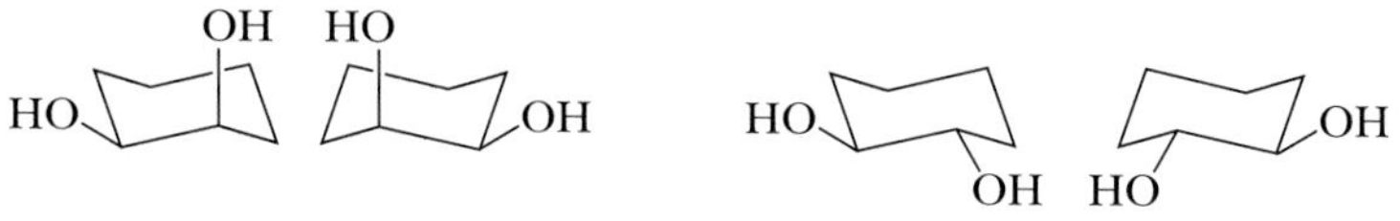

(I) (II)

cis-1,2-Cyclohexanediol
(a pair of enantiomers)

(III) (IV)

trans-1,2-Cyclohexanediol
(a pair of enantiomers)

The enantiomers of the *cis* isomer cannot be separated, however, because each is converted to the other by a rapid chair-to-alternative-chair conversion. As shown in the following structural formulas, the alternative chair of I is, in fact, the mirror image of I. Thus, each enantiomer interconverts to its mirror image and, therefore, the enantiomers cannot be separated.

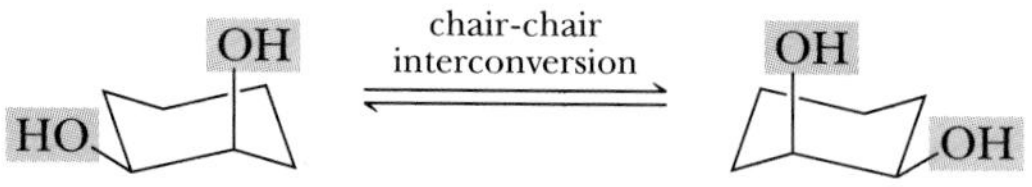

chair-chair interconversion

(I)

Alternative chair of I;
also the enantiomer of I

4.7 Properties of Stereoisomers

Enantiomers have identical physical and chemical properties in achiral environments. The enantiomers of tartaric acid (Table 4.1), for example, have the same melting point, the same boiling point, the same solubility in water and other common solvents,

TABLE 4.1 Some Physical Properties of the Stereoisomers of Tartaric Acid

	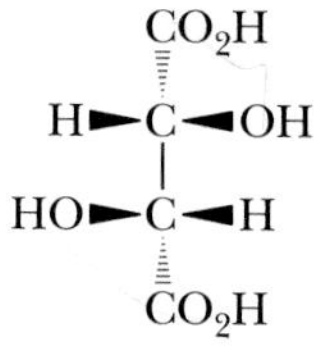 (*R*,*R*)-Tartaric acid	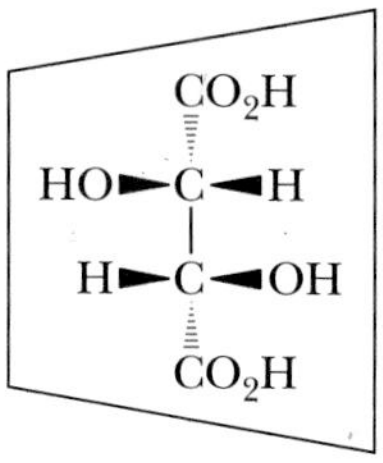 (*S*,*S*)-Tartaric acid	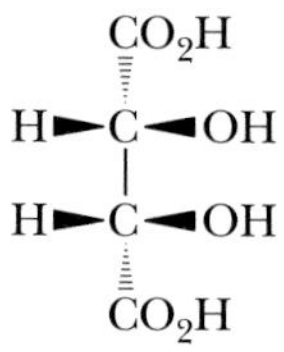 Meso tartaric acid
Specific rotation*	+12.7	−12.7	0
Melting point (°C)	171–174	171–174	146–148
Density at 20°C (g/cm^3)	1.7598	1.7598	1.660
Solubility in water at 20°C (g/100 mL)	139	139	125
pK_1 (25°C)	2.98	2.98	3.23
pK_2 (25°C)	4.34	4.34	4.82

* Specific rotation is discussed in the following section.

the same value of pK_a, and undergo the same acid-base reactions. The enantiomers of tartaric acid do, however, differ in optical activity (the ability to rotate the plane of plane polarized light), which we discuss in the following section. Diastereomers have different physical and chemical properties, even in achiral environments. Meso-tartaric acid has different physical properties from those of the enantiomers.

4.8 Optical Activity: How Chirality Is Detected in the Laboratory

As we have already established, enantiomers are different compounds, and we must expect, therefore, that they differ in some properties. One property that differs between enantiomers is their effect on the plane of polarized light. Each member of a pair of enantiomers rotates the plane of polarized light, and for this reason, enantiomers are said to be **optically active.** While each enantiomer is optically active, the two rotate the plane of polarized light in opposite directions.

Optically active Showing that a compound rotates the plane of polarized light.

The phenomenon of optical activity was discovered by the French physicist Jean Baptiste Biot in 1815. To understand how optical activity is detected in the laboratory, we must first understand something about plane-polarized light and a polarimeter, the device used to detect optical activity.

A. Plane-Polarized Light

Ordinary light consists of waves vibrating in all planes perpendicular to its direction of propagation (Figure 4.6). Certain materials such as calcite or Polaroid sheet (a plastic film containing properly oriented crystals of an organic substance embedded in it) selectively transmit light waves vibrating in one specific plane. Electromagnetic radiation vibrating in only one plane is said to be **plane polarized.**

Plane-polarized light Light vibrating in only one plane.

Plane-polarized light is actually an equal mixture of left and right **circularly polarized** light that propagates through space as left-handed and right-handed helices,

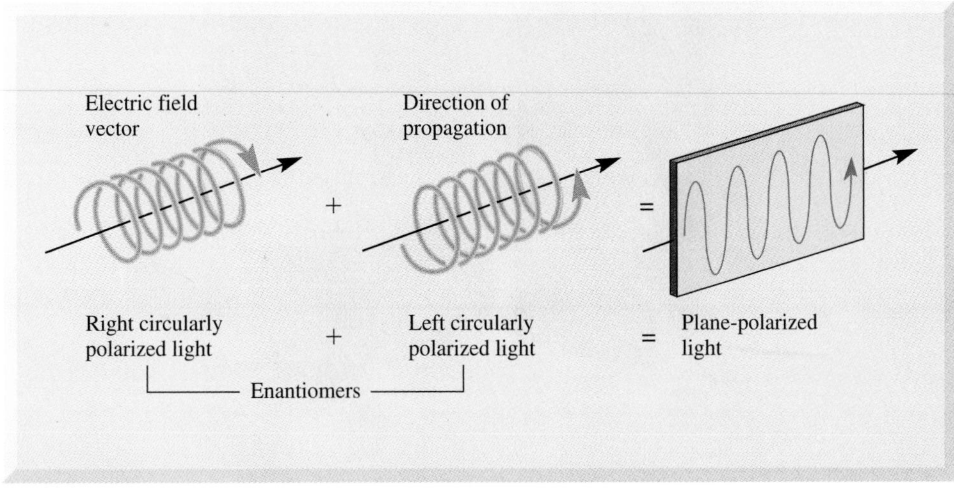

FIGURE 4.6
Plane-polarized light is a mixture of left and right circularly polarized light.

respectively (Figure 4.6). These two forms of light are nonsuperposable mirror images of each other and are, therefore, enantiomers. Because of the handedness of the circular components, plane-polarized light interacts one way with a stereocenter of *R*-handedness and differently with its enantiomer. The net effect is that the plane of polarization is rotated.

B. Polarimeters

Polarimeter A device for measuring the ability of a compound to rotate the plane of polarized light.

A **polarimeter** consists of a light source, a polarizing filter and an analyzing filter (each made of calcite or Polaroid film), and a sample tube (Figure 4.7). If the sample tube is empty, the intensity of light reaching the detector (your eye) is at its maximum when the polarizing axes of the two filters are parallel. If the analyzing filter is turned either clockwise or counterclockwise, less light is transmitted. When the axis of the analyzing filter is at right angles to the axis of the polarizing filter, the field of view is dark. This position of the analyzing filter is taken as 0° on the optical scale.

The ability of molecules to rotate the plane of polarized light can be observed using a polarimeter in the following way. First, a sample tube filled with solvent is placed in the polarimeter, and the analyzing filter is adjusted so that no light passes through to the observer; that is, it is set to 0°. When a solution of an optically active compound is placed in the sample tube, a certain amount of light now passes through the analyzing filter; the optically active compound has rotated the plane of polarized light from the polarizing filter so that it is now no longer at an angle of 90° to the analyzing filter. The analyzing filter is then rotated to restore darkness in the field of view. The number of degrees, α, through which the analyzing filter must be rotated to restore darkness to the field of view is called the **observed rotation.** If the analyzing filter must be turned to the right (clockwise) to restore darkness, that is, if the plane of polarized light has been rotated to the right, we say that the com-

Observed rotation The number of degrees through which a compound rotates the plane of polarized light.

A polarimeter is used to measure the rotation of plane-polarized light as it passes through a sample. *(Richard Megna, 1992, Fundamental Photographs)*

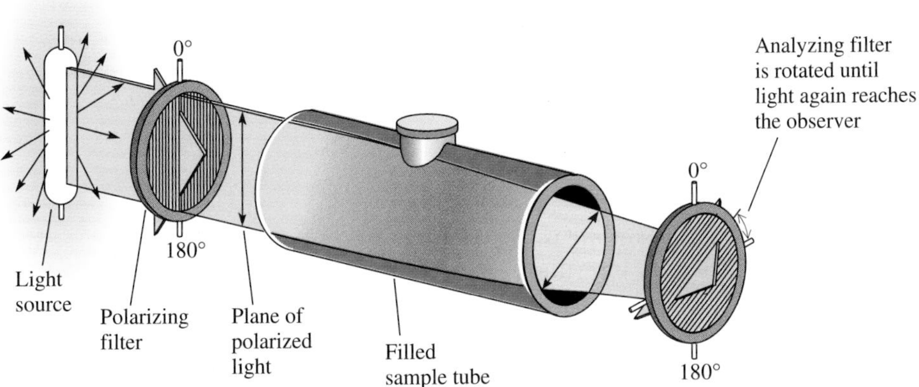

FIGURE 4.7
Schematic diagram of a polarimeter with its sample tube containing a solution of an optically active compound. The analyzing filter has been turned clockwise by α degrees to restore the light field.

pound is **dextrorotatory** (Latin: *dexter,* on the right side). If the analyzing filter must be turned to the left (counterclockwise), we say that the compound is **levorotatory** (Latin: *laevus,* on the left side).

Dextrorotatory Rotation of the plane of polarized light in a polarimeter to the right.

Levorotatory Rotation of the plane of polarized light in a polarimeter to the left.

Specific rotation Observed rotation of the plane of polarized light when a sample is placed in a tube 1.0 dm in length and at a concentration of 1 g/mL.

The magnitude of the observed rotation for a particular compound depends on its concentration, the length of the sample tube, the temperature, the solvent, and the wavelength of the light used. To standardize optical rotation data, chemists use the term "specific rotation." **Specific rotation, [α],** is defined as the observed rotation at a specific cell length and sample concentration.

$$\text{Specific rotation} = [\alpha]_{\lambda}^{T} = \frac{\text{observed rotation (degrees)}}{\text{length (dm)} \times \text{concentration (g/mL)}}$$

The standard cell length is 1 decimeter (1 dm or 10 cm). For a pure liquid sample, the concentration is expressed in grams per milliliter (g/mL; density). The concentration of a sample dissolved in a solvent is also usually expressed as grams per milliliter of solution. The temperature (T, in degrees centigrade) and wave length (λ) of light are designated, respectively, as superscript and subscript. The light source most commonly used in polarimetry is the sodium D line (λ = 589 nm), the line responsible for the yellow color of sodium-vapor lamps.

In reporting either observed or specific rotation, it is common to indicate a dextrorotatory compound with a plus sign in parentheses, (+), and a levorotatory compound with a minus sign in parentheses, (−). For any pair of enantiomers, one enantiomer is dextrorotatory, and the other is levorotatory. For each member, the value of the specific rotation is exactly the same, but the sign is opposite. Following are the specific rotations of the enantiomers of 2-butanol at 25°C using the D line of sodium.

(*S*)-(+)-2-Butanol	(*R*)-(−)-2-Butanol
$[\alpha]_{D}^{25}$ +13.52	$[\alpha]_{D}^{25}$ −13.52

EXAMPLE 4.10

A solution is prepared by dissolving 400 mg of testosterone, a male sex hormone, in 10.0 mL of ethanol and placing it in a sample tube 10.0 cm in length. The observed rotation of this sample at 25°C using the D line of sodium is +4.36. Calculate the specific rotation of testosterone.

Solution

The concentration of testosterone is 400 mg/10.0 mL = 0.0400 g/mL. The length of the sample tube is 1.00 dm. Inserting these values in the formula for calculating specific rotation gives

$$\text{Specific rotation} = \frac{\text{observed rotation (degrees)}}{\text{length (dm)} \times \text{concentration (g/mL)}} = \frac{+4.36}{1.00 \times 0.0400} = +109$$

PROBLEM 4.10

The specific rotation of progesterone, a female sex hormone, is +172, measured at 20°C. Calculate the observed rotation for a solution prepared by dissolving 300 mg of progesterone in 15.0 mL of dioxane and placing it in a sample tube 10.0 cm long.

C. Racemic Mixtures

Racemic mixture A mixture of equal amounts of two enantiomers.

An equimolar mixture of two enantiomers is called a **racemic mixture,** a term derived from the name "racemic acid" (Latin: *racemus,* a cluster of grapes). Racemic acid is the name originally given to an equimolar mixture of the enantiomers of tartaric acid (Table 4.1). Because a racemic mixture contains equal numbers of dextrorotatory and levorotatory molecules, its specific rotation is zero. Alternatively, we say that a racemic mixture is optically inactive. A racemic mixture is indicated by adding the prefix (±) or (*R,S*) to the name of the compound.

D. Optical Purity and Enantiomeric Excess

Optical purity The specific rotation of a mixture of enantiomers divided by the specific rotation of the enantiomerically pure substance.

When dealing with a pair of enantiomers, it is essential to have a means of describing the composition of that mixture and the degree to which one enantiomer is in excess relative to its mirror image. The most common way of describing the composition of a mixture of enantiomers is by its percent **optical purity,** a property that can be observed directly. Optical purity is the specific rotation of a mixture of enantiomers divided by the specific rotation of the enantiomerically pure substance.

$$\text{Percent optical purity} = \frac{[\alpha]_{\text{sample}}}{[\alpha]_{\text{pure enantiomer}}} \times 100$$

Enantiomeric excess (ee) The difference in number of moles of each enantiomer in a mixture compared with the total number of moles of both.

An alternative way to describe the composition of a mixture of enantiomers is by its **enantiomeric excess (ee),** which is the difference in the number of moles of each enantiomer in a mixture compared with the total number of moles of both.

$$\text{Percent optical purity} = \text{enantiomeric excess (ee)} = \frac{[R] - [S]}{[R] + [S]} \times 100 = \%R - \%S$$

EXAMPLE 4.11

Figure 4.8 presents a scheme for separation of the enantiomers of mandelic acid. The specific rotation of optically pure (*S*)-(−)-mandelic acid is −158. Suppose that instead of isolating pure (*S*)-(−)-mandelic acid from this scheme, the sample is a mixture of enantiomers with a specific rotation of −134. For this sample, calculate the following:

(a) The enantiomeric excess of (*S*)-(−)-mandelic acid.
(b) The percent of (*S*)-(−)-mandelic acid and of (*R*)-(+)-mandelic acid in the sample.

Solution

(a) The enantiomeric excess of (*S*)-(−)-mandelic acid is 84.8%.

$$\text{Enantiomeric excess} = \frac{-134}{-158} \times 100 = 84.8\%$$

(b) This sample is 84.8% (*S*)-(−)-mandelic acid and 15.2% (*R,S*)-mandelic acid. The (*R,S*)-mandelic acid is 7.6% (*S*)-enantiomer and 7.6% (*R*)-enantiomer. The sample, therefore, contains 92.4% of the (*S*)-enantiomer and 7.6% of the (*R*)-enantiomer. We can check these values by calculating the observed rotation of a mixture containing 92.4% (*S*)-(−)-mandelic acid and 7.6% (*R*)-(+)-mandelic acid as follows:

$$\text{Specific rotation} = 0.924 \times (-158) + 0.076 \times (+158) = -146 + 12 = -134$$

which agrees with the experimental specific rotation.

PROBLEM 4.11

One commercial synthesis of naproxen (the active ingredient in Aleve and a score of other over-the-counter nonsteroidal antiinflammatory drug preparations) gives the enantiomer shown in 97% enantiomeric excess.

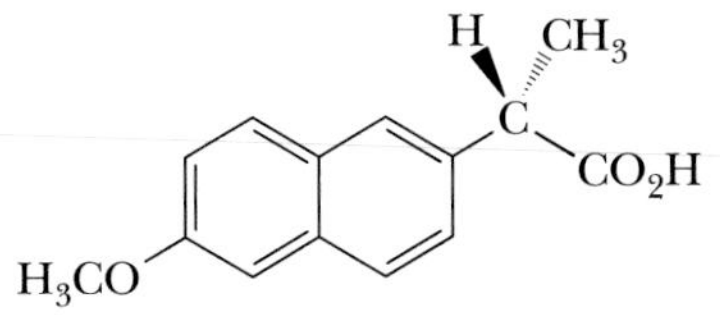

H CH$_3$ C CO$_2$H H$_3$CO

Naproxen
(a nonsteroidal antiinflammatory drug)

(a) Assign an *R,S* configuration to this enantiomer of naproxen.
(b) What are the percentages of *R* and *S* enantiomers in the mixture?

4.9 Separation of Enantiomers: Resolution

The separation of a racemic mixture into its two enantiomers is called optical resolution or, more simply, **resolution.**

Resolution Separation of a racemic mixture into its two enantiomers.

A. Resolution by Means of Diastereomeric Salts

One general scheme for separating enantiomers requires chemical conversion of a pair of enantiomers into two diastereomers with the aid of an enantiomerically pure chiral resolving agent. This chemical resolution is successful because the diastereomers thus formed have different physical properties and can often be separated by physical means (most commonly fractional crystallization or column chromatography) and purified. The final step in this scheme for resolution is chemical conversion of the separated diastereomers back to the individual enantiomers, and recovery of the chiral resolving agent.

A reaction that lends itself to chemical resolution is salt formation, because it is readily reversible.

$$\underset{\text{Carboxylic acid}}{RCO_2H} + \underset{\text{Base}}{:B} \rightleftharpoons \underset{\text{Salt}}{RCO_2^- \ HB^+}$$

Several enantiomerically pure bases available from plants have been used as chiral-resolving agents for racemic acids. Examples are cinchonine and quinine.

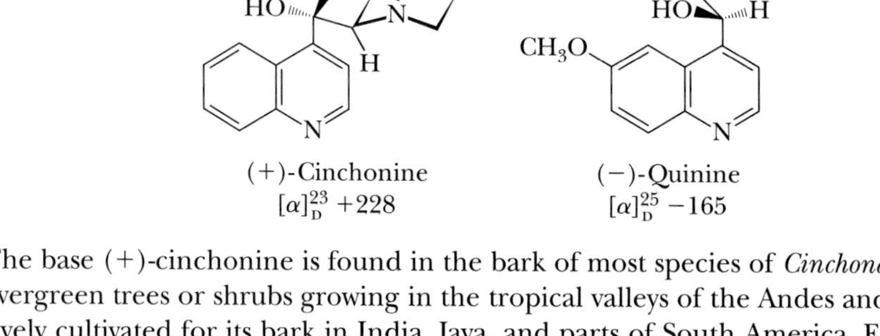

Cinchona bark, the source of quinine. *(© Walter H. Hodge/Peter Arnold, Inc.)*

The base (+)-cinchonine is found in the bark of most species of *Cinchona,* a genus of evergreen trees or shrubs growing in the tropical valleys of the Andes and now extensively cultivated for its bark in India, Java, and parts of South America. Extracts of its bark have been used for centuries as a tonic and to cure the fevers associated with malaria. The genus was named after the Countess of Cinchon, wife of the viceroy of Peru, who was cured of fever by cinchona bark and later brought a supply of it back to Spain. Also found in cinchona bark is (−)-quinine, an even more potent antimalarial drug than cinchonine.

(*R,S*)-Mandelic acid has been resolved into its enantiomers by way of its diastereomeric salts with cinchonine as illustrated in Figure 4.8. Racemic mandelic acid and optically pure (+)-cinchonine (Cin) are dissolved in boiling water, and the solution is allowed to cool, whereupon the less soluble diastereomeric salt crystallizes. This salt is collected and purified by further recrystallization. The filtrates, richer in the more soluble diastereomeric salt, are concentrated to give this salt, which is also purified by further recrystallization. The purified diastereomeric salts are treated with aqueous HCl to precipitate the nearly pure enantiomers of mandelic acid. Cinchonine remains in the aqueous solution as the water-soluble salt.

Optical rotations and melting points of racemic mandelic acid, cinchonine, the purified diastereomeric salts, and the pure enantiomers of mandelic acid are given in Figure 4.8. Note the following three points: (1) The melting point of racemic (*R,S*)-

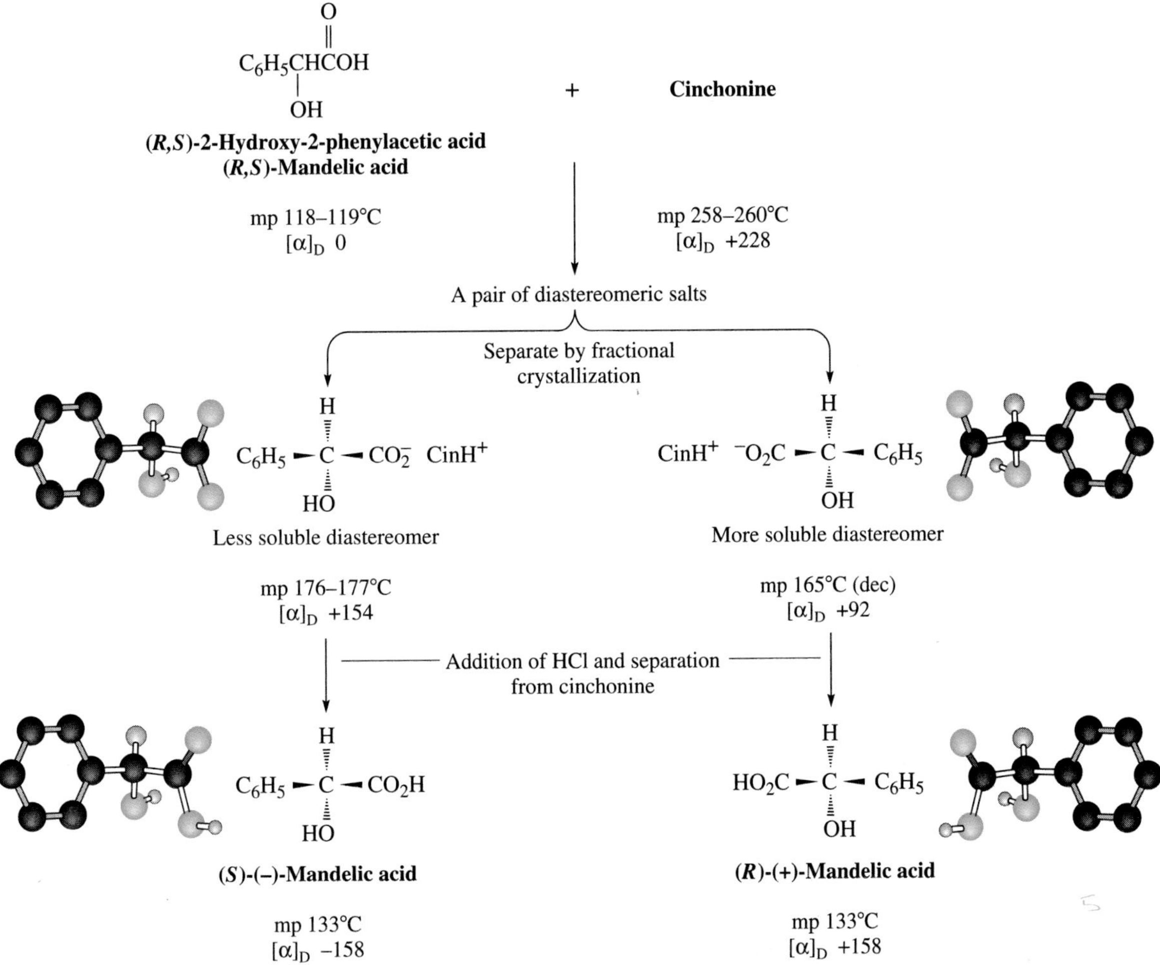

FIGURE 4.8
Resolution of mandelic acid.

mandelic acid is different from the melting point of the pure enantiomers. (2) The diastereomeric salts have different specific rotations and different melting points. (3) The enantiomers of mandelic acid have identical melting points and have specific rotations that are identical in magnitude but opposite in sign.

Resolution of a racemic base with a chiral acid is carried out in a similar way. Acids that are commonly used as chiral resolving agents are (+)-tartaric acid, (−)-malic acid, and (+)-camphoric acid (Figure 4.9). These and other naturally occurring chiral resolving agents are produced in plant and animal systems as single enantiomers.

B. Enzymes as Resolving Agents

In their quest for enantiomerically pure compounds, organic chemists have developed several new techniques for chiral synthesis. One approach is to use enzymes as chiral catalysts for the large-scale synthesis of enantiomerically pure substances. A class of

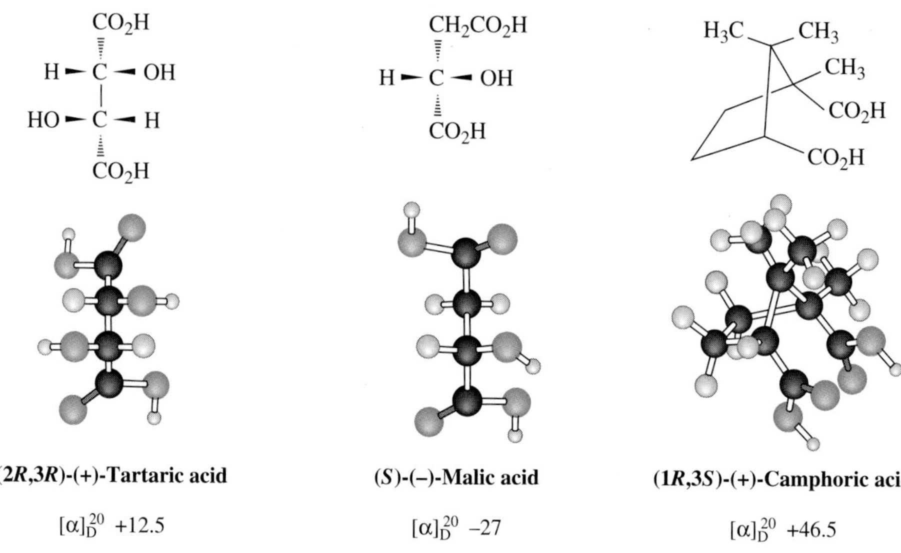

FIGURE 4.9
Some carboxylic acids used as chiral resolving agents.

enzymes under study are the lipases, which catalyze the formation of esters from alcohols and carboxylic acid anhydrides. In one resolution, shown on page 147, racemic (2*R*,2*S*)-5-norbornen-2-ol is treated with acetic anhydride in a reaction catalyzed by an (*R*)-lipase to give an ester. The enzyme is chiral and reacts with different enantiomers at different rates. The (*R*)-lipase-catalyzed reaction of the *R* enantiomer of 5-norbornen-2-ol with acetic anhydride is considerably faster than that of the *S* enantiomer, which leaves unreacted (2*S*)-5-norbornen-2-ol in greater than 98% enantiomeric excess. The unreacted *S* enantiomer is then separated from the product ester. The ester is a mixture of 89% (2*R*)-ester and 11% (2*S*)-ester. Because all catalysts catalyze the reverse reaction as well as the forward reaction, the ester product is treated with water and the lipase enzyme. The ester of the (2*R*)-5-norbornen-2-ol is hydrolyzed preferentially, and (2*R*)-5-norbornen-2-ol is recovered in greater than 98% enantiomeric excess.

Thus, by this skillful blend of enzymology and creative organic chemistry, both the (*R*)-alcohol and (*S*)-alcohol can be recovered in 98% enantiomeric excess and each used as a chiral precursor for other enantiomerically pure compounds.

4.10 The Significance of Chirality in the Biological World

Except for inorganic salts and a relatively few low-molecular-weight organic substances, the molecules in living systems, both plant and animal, are chiral. Although these molecules can exist as a number of stereoisomers, almost invariably only one stereoisomer is found in nature. Of course, instances do occur in which more than one stereoisomer is found, but these rarely exist together in the same biological system.

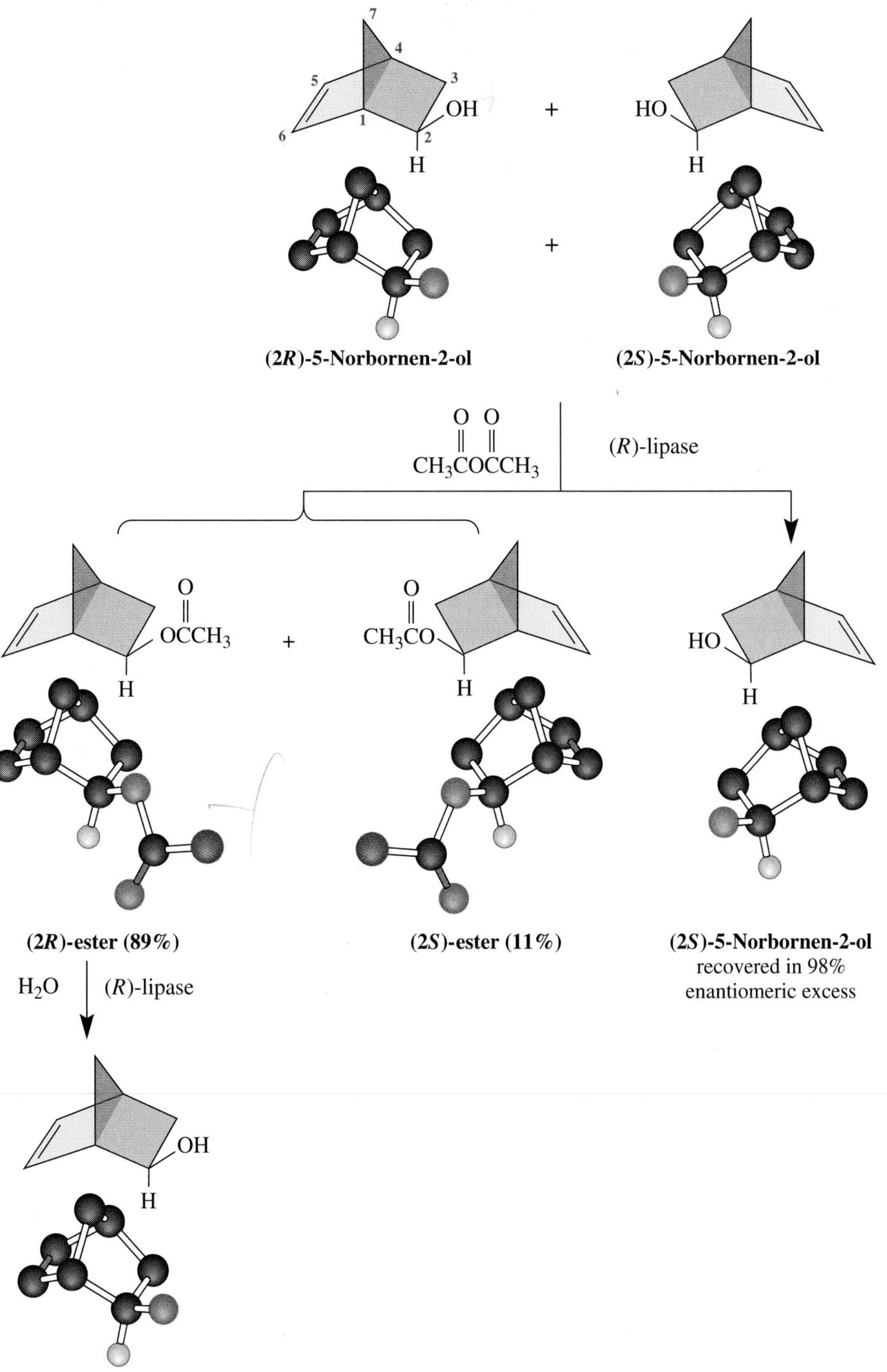

7
4
5
3
OH
+
HO
6
1
2
H
H
+
(2R)-5-Norbornen-2-ol
(2S)-5-Norbornen-2-ol
O O
CH3COCCH3
(R)-lipase
O
OCCH3
+
O
CH3CO
HO
H
H
H
(2R)-ester (89%)
(2S)-ester (11%)
(2S)-5-Norbornen-2-ol
recovered in 98%
enantiomeric excess
H2O
(R)-lipase
OH
H
(2R)-5-Norbornen-2-ol
recovered after hydrolysis
in 98% enantiomeric excess

A. Chirality in Biomolecules

Perhaps the most conspicuous examples of chirality among biological molecules are the enzymes, all of which have multiple stereocenters. An illustration is chymotrypsin, an enzyme that functions very efficiently in the intestines of animals between pH 7 and 8 in catalyzing the digestion of proteins. Chymotrypsin has 251 stereocenters. The maximum number of stereoisomers possible is 2^{251}, a staggeringly large number, almost beyond comprehension. Fortunately, nature does not squander its precious energy and resources unnecessarily; only one of these stereoisomers is produced and used by any given organism. Because enzymes are chiral substances, most either produce or react only with substances that match their own chirality.

B. How an Enzyme Distinguishes Between a Molecule and Its Enantiomer

Enzymes are chiral catalysts. Some are completely specific for the catalysis of reaction of only one particular compound, while others are less specific and catalyze similar reactions of the members of a family of compounds. An enzyme catalyzes a biological reaction of molecules by first positioning them at a **binding site** on its surface. These molecules may be held at the binding site by a combination of hydrogen bonds, electrostatic attractions, dispersion forces, or even covalent bonds.

An enzyme with specific binding sites for three of the four groups on a stereocenter can distinguish between a molecule and its enantiomer or one of its diastereomers. Assume, for example, that an enzyme involved in catalyzing a reaction of glyceraldehyde has three binding sites, one specific for —H, another specific for —OH, and the third specific for —CHO. Assume further that the three sites are arranged on the enzyme surface as shown in Figure 4.10. The enzyme can distinguish (R)-(+)-glyceraldehyde (the natural or biologically active form) from its enantiomer (S)-(−)-glyceraldehyde because the natural enantiomer can be adsorbed with three groups interacting with their appropriate binding sites; the other enantiomer can, at best, bind to only two of these sites.

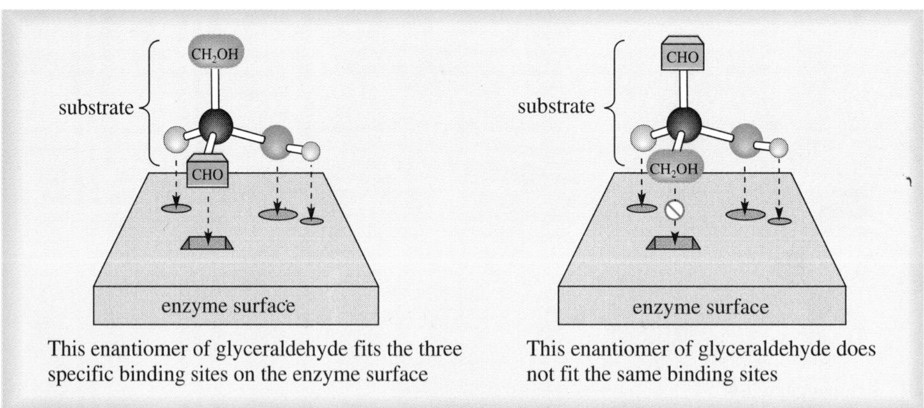
substrate
CH_2OH
CHO
enzyme surface
This enantiomer of glyceraldehyde fits the three specific binding sites on the enzyme surface
substrate
CHO
CH_2OH
enzyme surface
This enantiomer of glyceraldehyde does not fit the same binding sites

FIGURE 4.10
A schematic diagram of an enzyme surface capable of interacting with (R)-(+)-glyceraldehyde at three binding sites, but with (S)-(−)-glyceraldehyde at only two of these sites.

CHEMISTRY IN ACTION

Chiral Drugs

Some of the common drugs used in human medicine, for example aspirin (Section 17.6B), are achiral. Others are chiral and sold as single enantiomers. The penicillin and erythromycin classes of antibiotics and the drug captopril are all chiral drugs. Captopril, which is very effective for the treatment of high blood pressure and congestive heart failure, was developed in a research program designed to discover effective inhibitors for angiotensin-converting enzyme (ACE). It is manufactured and sold as the (*S,S*)-stereoisomer. A large number of chiral drugs, however, are sold as racemic mixtures. The popular analgesic ibuprofen (the active ingredient in Motrin, Advil, and many other nonaspirin analgesics) is an example.

Captopril

(*S*)-Ibuprofen

For racemic drugs, most often only one enantiomer exerts the beneficial effect, whereas the other enantiomer either has no effect or exerts a detrimental effect. Thus, enantiomerically pure drugs should, more often than not, be more effective than their racemic counterparts. A case in point is the drug dihydroxyphenylalanine used in the treatment of Parkinson's disease. The active drug is dopamine but, unfortunately, this compound does not cross the blood-brain barrier to the required site of action in the brain. What is administered, instead, is the prodrug 3,4-dihydroxyphenylalanine, which crosses the blood-brain barrier and then undergoes decarboxylation catalyzed by the enzyme dopamine decarboxylase. This enzyme is specific for the *S* enantiomer, which is commonly known as L-DOPA. It is essential, therefore, to administer the enantiomerically pure prodrug. Were the prodrug to be administered in a racemic form, there could be a dangerous buildup of the *R* enantiomer, which cannot be metabolized by the enzymes present in the brain.

enzyme-catalyzed decarboxylation →

(S)-(−)-3,4-Dihydroxyphenylalanine
(L-DOPA)
$[\alpha]_D^{13}$ −13.1°

+ CO_2

Dopamine

Recently, the U.S. Food and Drug Administration established new guidelines for the testing and marketing of chiral drugs. After reviewing these guidelines, many drug companies have decided to develop only single enantiomers of new chiral drugs. In addition to regulatory pressure, there are patent considerations. If a company has patents on a racemic drug, a new patent can often be taken out on one of its enantiomers. Only the *S* enantiomer of the pain reliever ibuprofen is biologically active. In the case of ibuprofen, however, the body converts the inactive *R* enantiomer to the active *S* enantiomer. It is not yet known if a chiral formulation for this particular compound is faster acting or superior in some other respect over the racemic mixture.

Because interactions between molecules in living systems take place in a chiral environment, it should be no surprise that a molecule and its enantiomer or diastereomers have different physiological properties. The tricarboxylic acid (TCA) cycle, for example, produces and then metabolizes only (*S*)-(+)-malic acid.

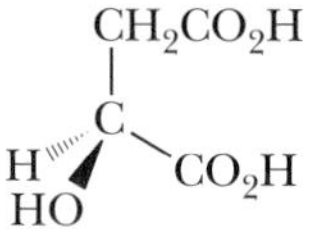

CH_2CO_2H
|
C
H ◀ CO_2H
HO

(*S*)-(+)-Malic acid

That interactions between molecules in the biological world are very specific in stereochemistry is not surprising, but just how these interactions are accomplished at the molecular level with such precision and efficiency is one of the great puzzles that modern science has only recently begun to unravel.

SUMMARY

Stereoisomers (Section 4.1) have the same order of attachment of atoms in their molecules but a different three-dimensional orientation of their atoms in space. Stereoisomers can be divided into enantiomers and diastereomers. **Enantiomers** are stereoisomers with molecules that are mirror images of each other. **Diastereomers** are stereoisomers that are not mirror images.

A **mirror image** (Section 4.2) is the reflection of an object in a mirror. Molecules that are not superposable on their mirror images are said to be **chiral.** Chirality is a property of an object as a whole, not of a particular atom. An **achiral** object is one that lacks chirality; that is, it is an object that has a superposable mirror image. Almost all achiral objects possess at least one plane or center of symmetry. A **plane of symmetry** is an imaginary plane passing through an object dividing it such that one half is the reflection of the other half. A **center of symmetry** is a point so situated that identical components of the object are located on opposite sides and equidistant from the point along any axis passing through that point.

A **stereocenter** (Section 4.2) is an atom in a molecule at which interchange of two atoms or groups of atoms bonded to it produces a different stereoisomer. A carbon atom with four different groups bonded to it is a **tetrahedral stereocenter.** Tetrahedral stereocenters are not limited to carbon. Enantiomers of tetrahedral nitrogen, silicon, phosphorus, and germanium compounds have also been prepared.

The **configuration** at any stereocenter can be specified by the **Cahn-Ingold-Prelog convention,** known alternatively as the ***R,S*** **convention** (Section 4.3). To apply this convention, each atom or group of atoms bonded to the stereocenter is (1) assigned a priority and (2) numbered from highest priority to lowest priority. (3) The molecule is oriented in space so that the group of lowest priority is directed away from the observer, and (4) the remaining three groups are read in order from highest priority to lowest priority. If reading of groups is clockwise, the configuration is ***R*** (Latin: *rectus,* right hand). If reading of groups is counterclockwise, the configuration is ***S*** (Latin: *sinister,* left hand).

A **Fischer projection** (Section 4.4) is a two-dimensional representation showing the configuration of a chiral molecule. Horizontal lines represent bonds projecting forward and vertical lines represent bonds projecting to the rear. The only atom in the plane of the drawing surface is the stereocenter.

For a molecule with ***n*** stereocenters, the maximum number of stereoisomers possible is 2^n (Sections 4.5 and 4.6). Certain molecules have special symmetry properties that reduce the number of stereoisomers to fewer than that predicted by the 2^n rule. A compound is **meso** (Section 4.5B) if it contains two or more stereocenters assembled in such a way that its molecules are achiral.

Light that vibrates in only one plane is said to be **plane polarized** (Section 4.8A). Plane-polarized light contains equal components of left and right circularly polarized light. A **polarimeter** (Section 4.8B) is an instrument used to detect and measure the magnitude of optical activity. **Observed rotation** is the number of degrees the plane of polarized light is rotated. **Specific rotation** is the observed rotation measured in a cell 1 dm long and at a concentration of 1 g/mL. If the analyzing prism must be turned clockwise to restore the zero point, the compound is **dextrorotatory.** If the analyzing prism must be turned counterclockwise to restore the zero point, the compound is **levorotatory.** A compound is said to be **optically active** if it rotates the plane of polarized light. Each member of a pair of enantiomers rotates the plane of polarized light an equal number of degrees but opposite in direction (Section 4.8B). A **racemic mixture** (Section 4.8C) is a mixture of equal amounts of two enantiomers and has a specific rotation of zero. Percent **optical purity** is defined as the specific rotation of a mixture of enantiomers divided by the specific rotation of the pure enantiomer times 100 (Section 4.8D).

Resolution (Section 4.9) is the experimental process of separating a mixture of enantiomers into the two pure enantiomers. A common chemical means of resolving organic compounds is to treat the racemic mixture with a chiral resolving agent that converts the mixture of enantiomers into a pair of diastereomers. Diastereomers have different physical properties, and can be separated based on these differences. Once the diastereomers are separated, each diastereomer is then converted back to a pure enantiomer. Enzymes are also used as resolving agents because of their ability to catalyze a reaction of one enantiomer but not that of its mirror image.

Enzymes catalyze biological reactions by first positioning the molecule or molecules at binding sites and holding them there by a combination of hydrogen bonds, electrostatic attractions, dispersion forces, and covalent bonds. An enzyme with specific binding sites for three of the four groups on a stereocenter can distinguish between a molecule and its enantiomer (Section 4.10B).

ADDITIONAL PROBLEMS

Chirality

4.12 Think about the helical coil of a telephone cord or a spiral binding and suppose that you view the spiral from one end and find that it is a left-handed twist. If you view the same spiral from the other end, is it a right-handed twist, or a left-handed twist from that end as well?

4.13 Next time you have the opportunity to view a collection of whelks, augers, or other sea shells that have a helical twist, study the chirality of their twists. Do you find an equal number of left-handed and right-handed whelks, for example, or are they all or mostly all of one chirality? What about the chirality of whelks compared with augers and other spiral shells?

4.14 One reason we can be sure that sp^3-hybridized carbon atoms are tetrahedral is the number of stereoisomers that can exist for different organic compounds.

(a) How many stereoisomers are possible for $CHCl_3$, CH_2Cl_2, and CHClBrF if the four bonds to carbon have a tetrahedral arrangement?

(b) How many stereoisomers are possible for each of these compounds if the four bonds to the carbon have a square planar geometry?

This Atlantic auger shell has a helical twist. *(© Carolina Biological Supply/Phototake, NYC)*

Enantiomers

4.15 Which compounds contain stereocenters?

(a) 2-Chloropentane **(b)** 3-Chloropentane
(c) 3-Chloro-1-pentene **(d)** 1,2-dichloropropane

4.16 Using only C, H, and O, write structural formulas for the lowest-molecular-weight chiral

(a) Alkane **(b)** Alcohol **(c)** Aldehyde **(d)** Ketone

4.17 Draw mirror images for these molecules:

(a) C with OH (up), H_3C (hashed), H (wedge), CO_2H

(b) CHO above; H—C—OH; CH_2OH below

(c) CO_2H above; H_2N—C—H; CH_3 below

(d) five-membered ring containing O, with OH and H on the carbon adjacent to O

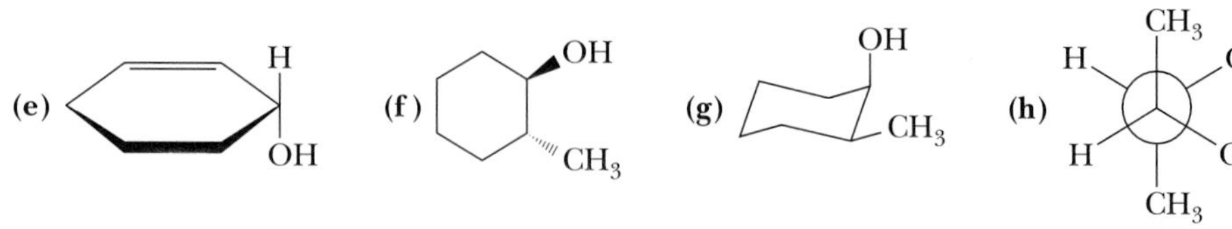

(e) H, OH (f) OH, CH_3 (g) OH, CH_3 (h) CH_3, H, OH, H, OH, CH_3

4.18 Following are several stereorepresentations for lactic acid. Take (a) as a reference structure. Which of the stereorepresentations are identical with (a) and which are mirror images of (a)?

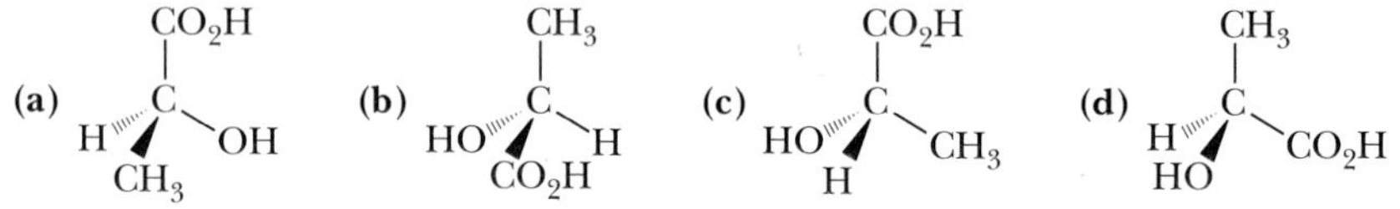

(a) CO_2H, H, C, OH, CH_3 (b) CH_3, HO, C, H, CO_2H (c) CO_2H, HO, C, CH_3, H (d) CH_3, H, C, CO_2H, HO

4.19 Mark each stereocenter in the following molecules with an asterisk. How many stereoisomers are possible for each molecule?

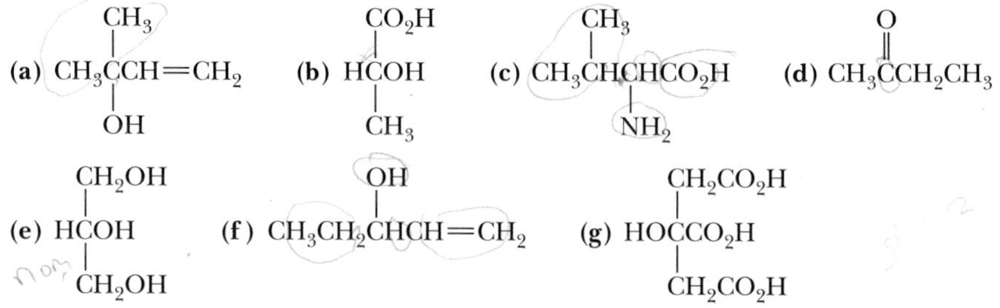

(a) $CH_3C(CH_3)(OH)CH{=}CH_2$ (b) $HC(CO_2H)(CH_3)OH$ (c) $CH_3CH(CH_3)CH(NH_2)CO_2H$ (d) $CH_3C(=O)CH_2CH_3$

(e) $HC(CH_2OH)(CH_2OH)OH$ (f) $CH_3CH_2CH(OH)CH{=}CH_2$ (g) $HOC(CH_2CO_2H)(CH_2CO_2H)CO_2H$

4.20 Show that butane in a gauche conformation is chiral. Do you expect that resolution of butane at room temperature is possible?

Designation of Configuration: The *R,S* Convention

4.21 Assign priorities to the groups in each set.

(a) $-H$ $-CH_3$ $-OH$ $-CH_2OH$

(b) $-CH_2CH{=}CH_2$ $-CH{=}CH_2$ $-CH_3$ $-CH_2CO_2H$

(c) $-CH_3$ $-H$ $-CO_2^-$ $-NH_3^+$

(d) $-CH_3$ $-CH_2SH$ $-NH_3^+$ $-CO_2^-$

4.22 Following are structural formulas for the enantiomers of carvone. Each has a distinctive odor characteristic of the source from which it can be isolated. Assign *R,S* configurations to each enantiomer.

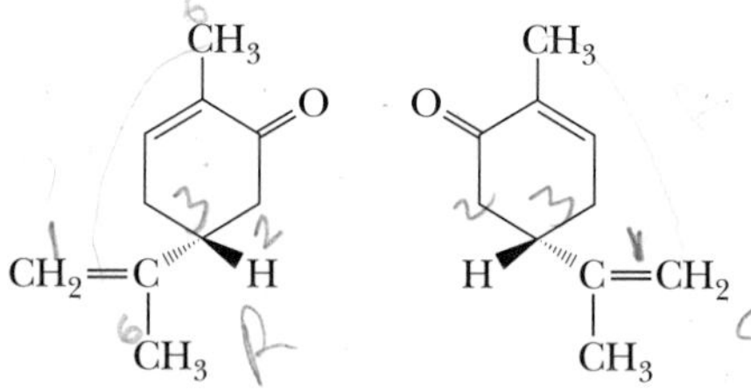

CH_3, O, $CH_2{=}C$, H, CH_3 — CH_3, O, H, $C{=}CH_2$, CH_3

(−)-Carvone
$[\alpha]_D^{20}$ −62.5
Spearmint oil

(+)-Carvone
$[\alpha]_D^{20}$ +62.5
Caraway oil

4.23 Following is a staggered conformation for one of the enantiomers of 2-butanol.

H_3C, H, H, H, OH, CH_3

(a) Is this (*R*)-2-butanol or (*S*)-2-butanol?
(b) Draw a Newman projection for this enantiomer, viewed along the bond between carbons 2 and 3.
(c) Draw a Newman projection for two more staggered conformations of this molecule. Which of your conformations is the most stable? Assume that —OH and $—CH_3$ are comparable in size.

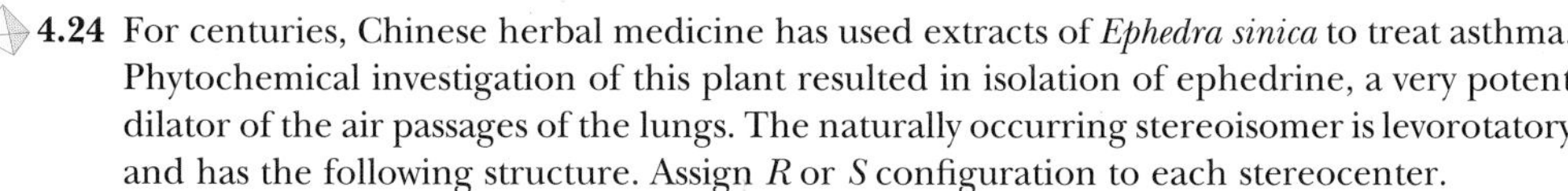

4.24 For centuries, Chinese herbal medicine has used extracts of *Ephedra sinica* to treat asthma. Phytochemical investigation of this plant resulted in isolation of ephedrine, a very potent dilator of the air passages of the lungs. The naturally occurring stereoisomer is levorotatory and has the following structure. Assign *R* or *S* configuration to each stereocenter.

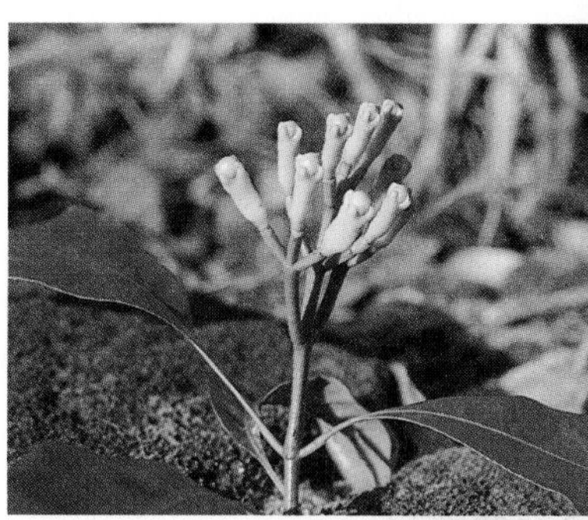

Ephedra sinica, the source of the drug ephedrine. *(© Paolo Koch/Photo Researchers, Inc.)*

C_6H_5, H, $NHCH_3$, C—C, HO, H, CH_3

Ephedrine
$[\alpha]_D^{21}$ −41

4.25 When oxaloacetic acid and acetyl-coenzyme A (acetyl-CoA) labeled with radioactive carbon-14 in position 2 are incubated with citrate synthase, an enzyme of the TCA cycle, only the following enantiomer of [2-^{14}C]-citric acid is formed. Note that citric acid containing only ^{12}C is achiral. Assign an *R* or *S* configuration to this enantiomer of [2-^{14}C] citric acid? *Note:* Carbon-14 has a higher priority than carbon-12.

$O{=}CCO_2H$ / CH_2CO_2H + $^{14}CH_3C(=O)SCoA$ $\xrightarrow{\text{citrate synthase}}$ HO_2C—C—OH with $^{14}CH_2CO_2H$ and CH_2CO_2H

Oxaloacetic acid — Acetyl-CoA — [2-^{14}C]Citric acid

Molecules with Two or More Stereocenters

4.26 Draw Newman projections for the three stereoisomers of 2,3-butanediol, showing the methyl groups anti (dihedral angle 180°).

4.27 Draw Fischer projections for the four stereoisomers of 3-chloro-2-butanol, showing the carbon chain vertical and —H, —OH, and —Cl horizontal.

4.28 Draw stereorepresentations for all stereoisomers of this compound. Label those that are meso compounds; those that are pairs of enantiomers.

H_3C CH_3, HO_2C, CO_2H

4.29 Mark each stereocenter in the following molecules with an asterisk. How many stereoisomers are possible for each molecule?

(a) 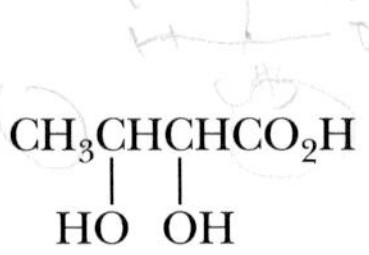 $CH_3CHCHCO_2H$ (with HO and OH on the two CH carbons)

(b) 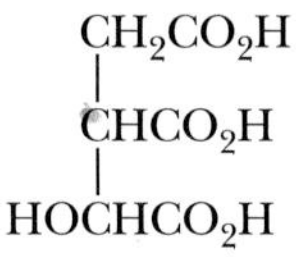 CH_2CO_2H / $CHCO_2H$ / $HOCHCO_2H$

(c) 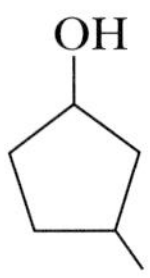 OH (d)  (e)

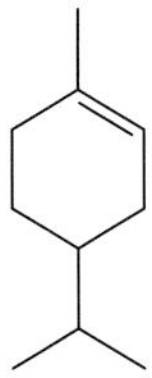

(f) 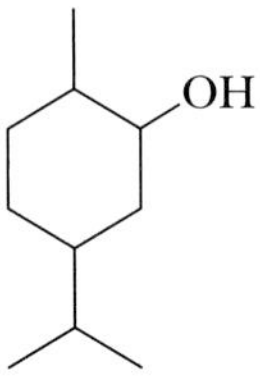 OH (g) 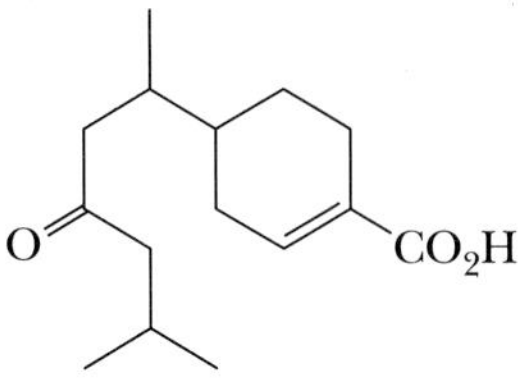 O, CO_2H (h) 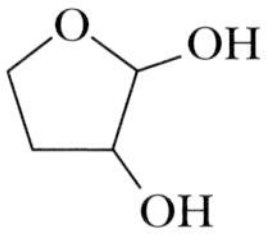 O, OH, OH (i) 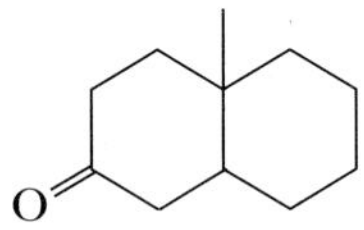 O

4.30 How many stereoisomers are possible for this compound, which is an aggregating pheromone for the Norway spruce beetle?

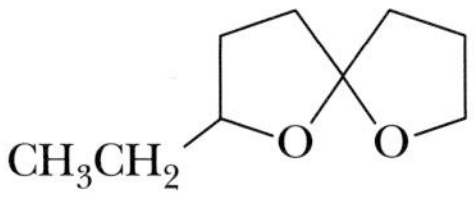

CH_3CH_2 O O

4.31 Mark all stereocenters and state the maximum number of stereoisomers possible for each molecule.

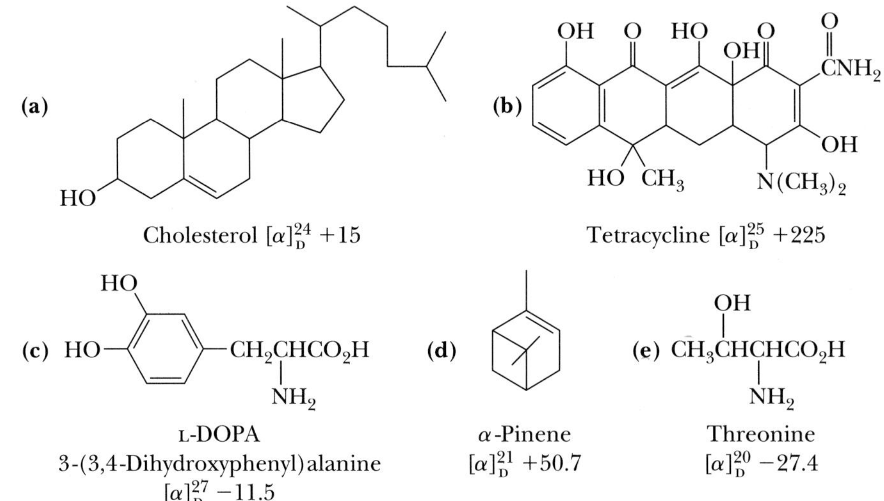

Cholesterol $[\alpha]_D^{24}$ +15

Tetracycline $[\alpha]_D^{25}$ +225

L-DOPA
3-(3,4-Dihydroxyphenyl)alanine
$[\alpha]_D^{27}$ −11.5

α-Pinene
$[\alpha]_D^{21}$ +50.7

Threonine
$[\alpha]_D^{20}$ −27.4

4.32 If the optical rotation of a new compound is measured and found to have a specific rotation of +40, how can you tell if the actual rotation is not really +40 plus some multiple of +360 (that is, the rotation is not actually $+40 + n(+360)$, where n has only integer values). In other words, how can you tell if the rotation is not actually a value such as +400 or +760?

4.33 Are the formulas within each set identical, enantiomers, or diastereomers?

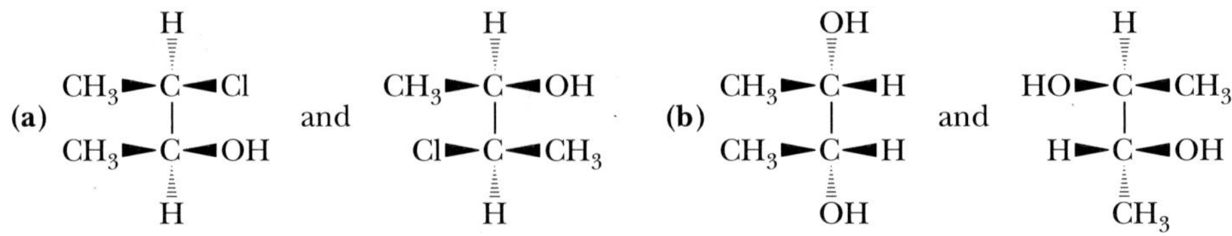

(c) 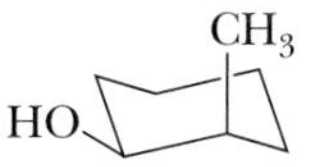 and 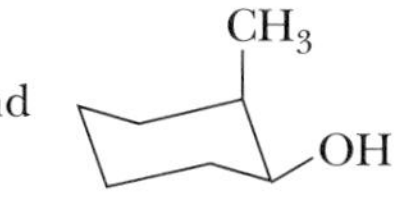 (d)  and

4.34 Which are meso compounds?

(a) 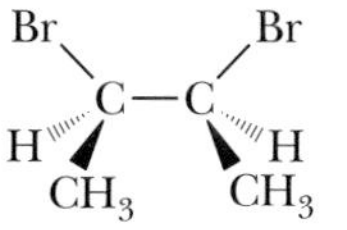 (b) 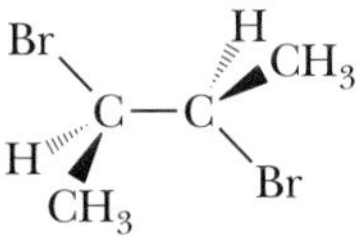 (c)

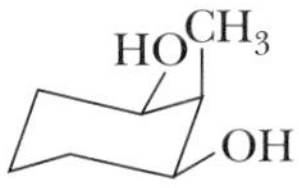

(d) 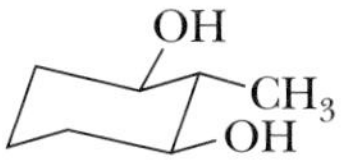 (e) 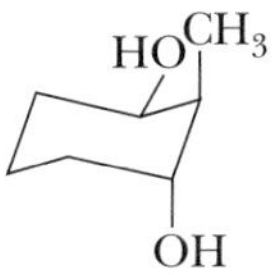 (f)

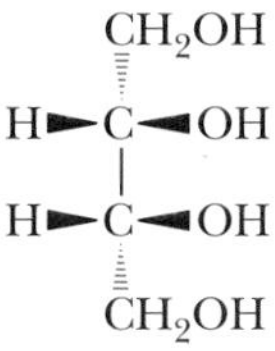

(g) 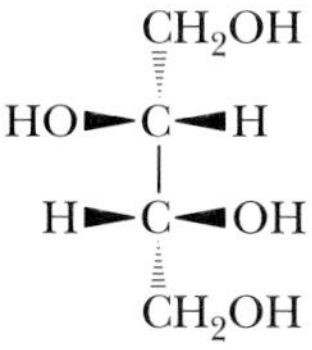 (h) 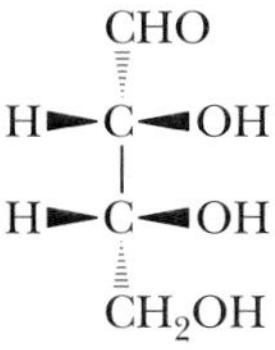 (i)

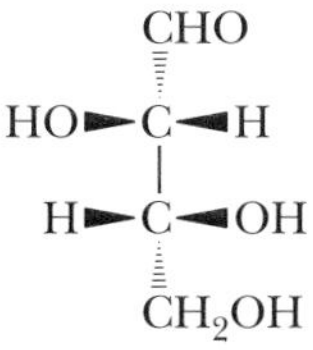

4.35 Vigorous oxidation of the following bicycloalkene gives 2,2-dimethylcyclopentane-1,3-dicarboxylic acid. Assume that the conditions of oxidation have no effect on the configuration of either the starting bicycloalkene or the resulting dicarboxylic acid. Is the dicarboxylic acid produced from this oxidation one enantiomer, a racemic mixture, or a meso compound?

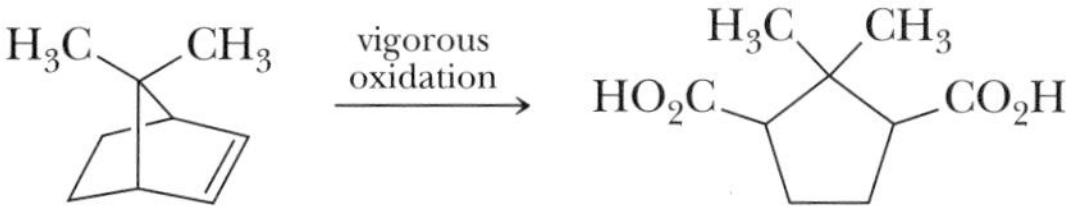

vigorous oxidation

7,7-Dimethylbicyclo-[2.2.1]hept-2-ene

2,2-Dimethylcyclopentane-1,3-dicarboxylic acid

4.36 A long polymer chain, such as polyethylene $-(CH_2CH_2)_n-$, can potentially exist in solution as a chiral object. Give two examples of chiral structures that a polyethylene chain could adopt.

Molecular Modeling

4.37 ChemDraw provides a very easy way to make mirror images. First, create a stereocenter in ChemDraw, make a copy, and place it adjacent to your original. Second, select the copy and, third, from the "Object" menu, select "Flip Horizontal." As shown here, this procedure converts an enantiomer to its mirror image.

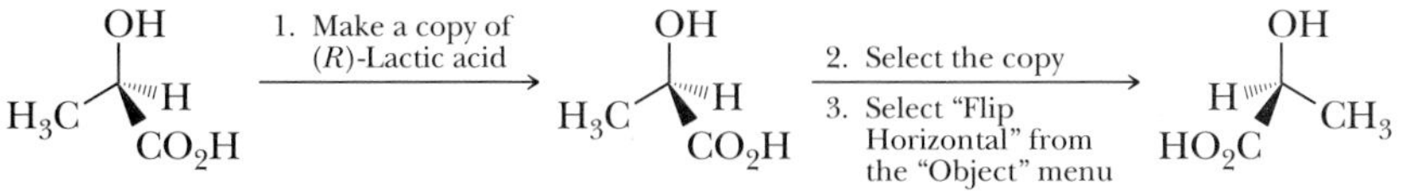

1. Make a copy of (*R*)-Lactic acid

2. Select the copy

3. Select "Flip Horizontal" from the "Object" menu

(*R*)-Lactic acid

Copy of (*R*)-Lactic acid

(*S*)-Lactic acid

Now try this procedure with these molecules chosen from the text.

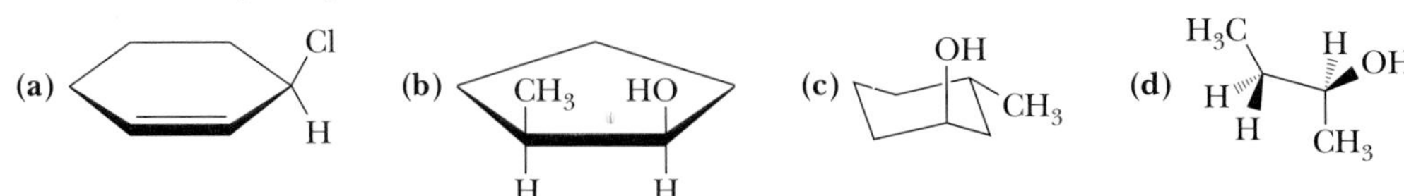

4.38 Chem3D provides a particularly effective way to create and view stereoisomers: (1) Create a stereoisomer, for example, (*R*)-lactic acid, in ChemDraw and then import it into Chem3D. (2) Under "Analyze," select "Minimize" to minimize the energy of the stereoisomer you have drawn. (3) Under "Build," select "Reflect in the Y-Z plane." You will now have the enantiomer, in this case (*S*)-lactic acid, of your original stereoisomer.

4.39 The following molecule is an attractant pheromone for the olive fly.

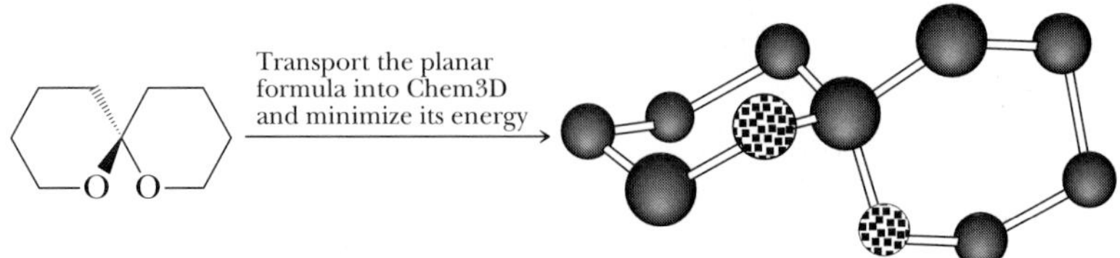

(a) Build a line-angle structure of this molecule in ChemDraw (as shown on the left). Using the directions given in Problem 4.37, create its mirror image. Are they superposable?

(b) Select your ChemDraw structure, paste it into Chem3D, and minimize its energy to give the model on the right. Rotate the model in Chem3D to convince yourself that each six-membered ring has a strain-free chair conformation.

(c) This molecule has no stereocenter and yet it is chiral. From examination of the three-dimensional model in Chem3D, convince yourself that it has no plane or center of symmetry and that it is, in fact, chiral.

(d) "The presence of a stereocenter in an organic molecule is a sufficient condition for chirality, but it is not a necessary condition." Explain.

4.40 The following molecule belongs to the class of compounds called allenes. The functional group of an allene is two adjacent carbon-carbon double bonds. Disubstituted allenes of this type are chiral. The specific rotation of the enantiomer shown is −314.

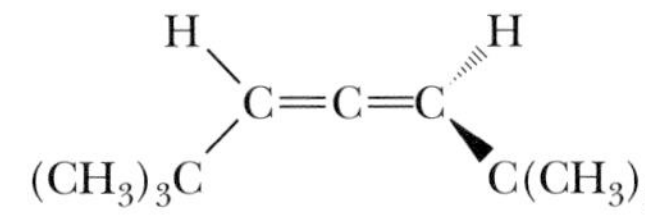

H H
C=C=C
$(CH_3)_3C$ $C(CH_3)_3$

$[\alpha]^{25}_{D}$ −314

(a) Build a line-angle structure of this allene in ChemDraw, import it into Chem3D, and minimize its energy. To make the ChemDraw and Chem3D models easier to see and manipulate, replace the *tert*-butyl groups by methyl groups.

(b) Make the mirror image of this stereoisomer (as you did in Problem 4.38) and convince yourself that the two are not superposable.

ALKENES I

5

■ Lily-of-the-valley, from which farnesol, a terpene (Figure 5.6) is isolated. *(Barry L. Runk/Grant Heilman Photography, Inc.)*

An **unsaturated hydrocarbon** contains one or more carbon-carbon double or triple bonds. The term unsaturation indicates that there are fewer hydrogens attached to carbon than in an alkane, C_nH_{2n+2}. There are three classes of unsaturated hydrocarbons: alkenes, alkynes, and arenes. **Alkenes** contain a carbon-carbon double bond and have the general formula C_nH_{2n}. **Alkynes** contain a carbon-carbon triple

Alkene An unsaturated hydrocarbon that contains a carbon-carbon double bond.

Alkyne An unsaturated hydrocarbon that contains a carbon-carbon triple bond.

bond and have the general formula C_nH_{2n-2}. The simplest alkene is ethene (ethylene) and the simplest alkyne is ethyne (acetylene).

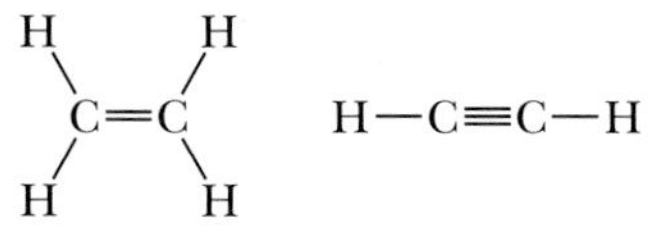

$$H_2C{=}CH_2 \qquad H{-}C{\equiv}C{-}H$$

Ethylene (an alkene) Acetylene (an alkyne)

In this chapter, we study the structure, nomenclature, and physical properties of alkenes. Alkynes are discussed separately in Chapter 10.

Arene A term used to classify benzene and its derivatives.

The third class of unsaturated hydrocarbons are the **arenes.** The Lewis structure of benzene, the simplest arene, is

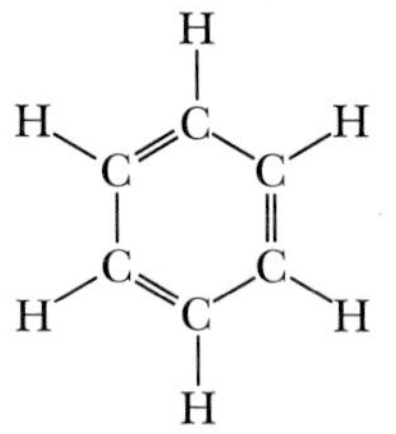

Benzene

Aryl group (Ar—) A group derived from an arene by removal of an H.

Just as a group derived by removal of an H from alkane is called an alkyl group and given the symbol R— (Section 2.3A), a group derived by removal of an H from an arene is called an **aryl group** and is given the symbol **Ar—.**

When a benzene ring occurs as a substituent on a parent chain, it is named a **phenyl group.** You might think that when present as a substituent, benzene would become benzyl, just as ethane becomes ethyl. This is not so! "Phene" is a now-obsolete name for benzene and, although this name is no longer used, a derivative has persisted in the name **phenyl.** Following is a structural formula for the phenyl group and two alternative representations for it.

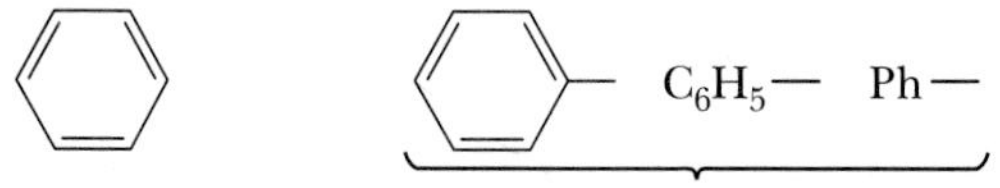

C_6H_5— Ph—

Benzene Alternative representations for the phenyl group

The chemistry of benzene and its derivatives is quite different from that of alkenes and alkynes, and, even though we do not study the chemistry of arenes until Chapters 19 and 20, we will show structural formulas of compounds containing aryl groups before that time. What you need to remember at this point is that an aryl group is not chemically reactive under any of the conditions we describe in Chapters 6–18.

5.1 Structure of Alkenes

A. Shapes of Alkenes

Double bond A covalent bond consisting of one sigma (σ) bond and one pi (π) bond.

Using the valence-shell electron-pair repulsion model (Section 1.4) of a **carbon-carbon double bond,** we predict a value of 120° for the bond angles about each carbon in a double bond. The observed H—C—C bond angle in ethylene is 121.7°, a value close to that predicted. In other alkenes, deviations from the predicted angle of 120° may be somewhat larger because of strain introduced by nonbonded interactions between groups attached to the carbons of the double bond. The C—C—C bond angle in propene, for example, is 124.7°.

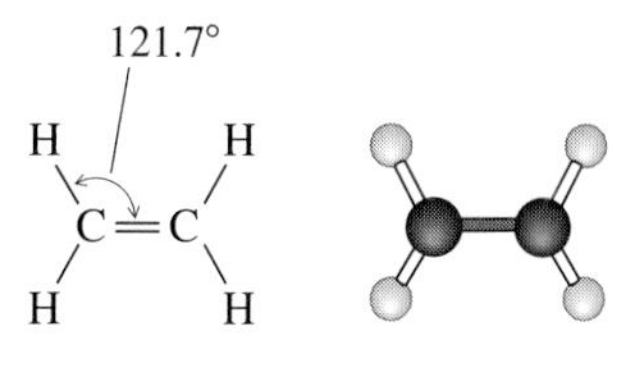

Ethene

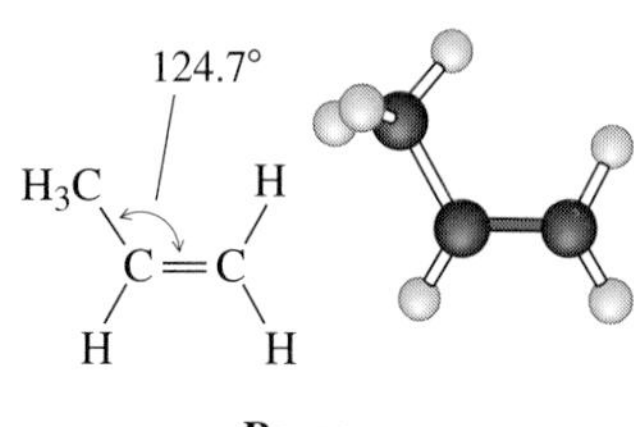

Propene

B. Molecular Orbital Model of a Carbon-Carbon Double Bond

In Section 1.8D, we described the formation of a carbon-carbon double bond in terms of the overlap of atomic orbitals. A carbon-carbon double bond consists of one sigma bond and one pi bond (Figure 5.1). Each carbon of the double bond uses its three sp^2 hybrid orbitals to form sigma bonds to three atoms. The unhybridized $2p$ atomic orbitals, which lie perpendicular to the plane created by the axes of the three sp^2 hybrid orbitals, combine to form two pi molecular orbitals: one bonding, the other antibonding. We are concerned here with the pi bonding MO only. For the unhybridized $2p$ orbitals to be parallel, thus giving maximum overlap, the two carbon atoms of the double bond and all four attached atoms must all lie in a plane.

C. Bond Lengths and Strengths of Alkanes, Alkenes, and Alkynes

Values of bond lengths and bond strengths (bond dissociation energies) for ethane, ethylene, and acetylene are given in Table 5.1. As you study Table 5.1, note the following points.

1. Carbon-carbon triple bonds are shorter than C—C double bonds, which in turn are shorter than C—C single bonds. This order of bond lengths is due to the fact that there are three bonds versus two bonds versus one bond holding the carbon atoms together.
2. The C—H bond in acetylene is shorter than that in ethylene, which in turn is shorter than that in ethane. The relative lengths of these C—H bonds are determined by the percent *s*-character in the hybrid orbital of carbon forming the sigma bond with hydrogen. The greater the percent *s*-character of a hybrid orbital, the closer to the nucleus electrons in it are held (Section 3.3B) and the shorter

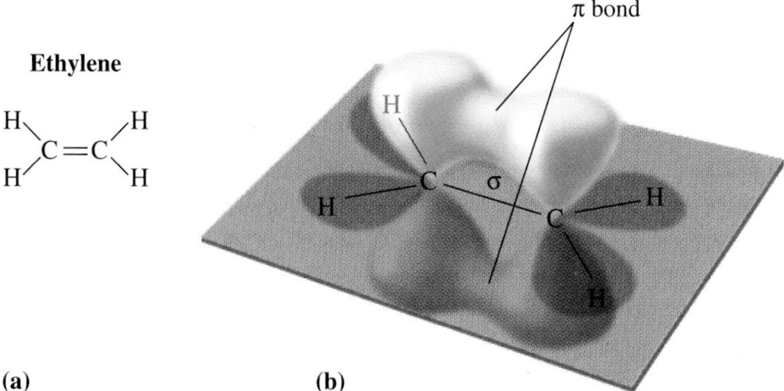

FIGURE 5.1
Covalent bonding in ethylene. (a) Lewis structure and (b) orbital overlap model showing the sigma and pi bonds.

TABLE 5.1 Bond Lengths and Bond Dissociation Energies for Ethane, Ethylene, and Acetylene

Molecule	Bond	Bond Orbital Overlap	Bond Length (Å)	Bond Strength [kcal/mol (kJ/mol)]
Ethane, $H_3C{-}CH_3$	C—C	sp^3-sp^3	1.54	90 (377)
	C—H	sp^3-1s	1.11	98 (410)
Ethylene, $H_2C{=}CH_2$	C—C	sp^2-sp^2, 2p-2p	1.34	172 (720)
	C—H	sp^2-1s	1.10	104 (435)
H—C≡C—H	C—C	sp-sp, two 2p-2p	1.21	230 (962)
	C—H	sp-1s	1.08	125 (523)

the bond. The relative lengths of C—H single bonds correlate with the fact that the percent s-character in an sp orbital is 50%, that in an sp^2 orbital is 33.3%, and that in an sp^3 orbital is 25%.

3. There is a correlation between bond length and bond strength; the shorter the bond, the stronger the bond. A carbon-carbon triple bond is the shortest C—C bond; it is also the strongest. The carbon-hydrogen bond in acetylene is the shortest; it is also the strongest.

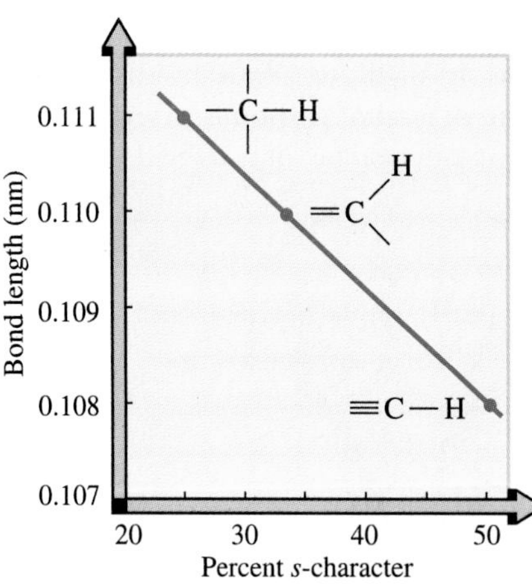

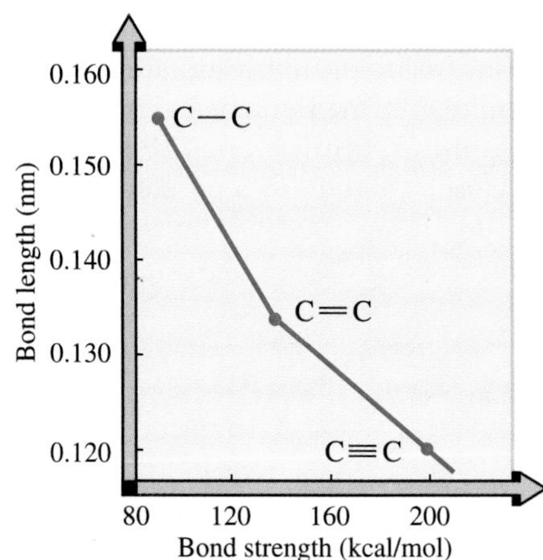

4. Although a C—C double bond is stronger than a C—C single bond, it is not twice as strong. It takes approximately 63 kcal/mol (264 kJ/mol) to break the pi bond in ethylene, that is, to rotate one carbon by 90° with respect to the other where zero overlap occurs between 2p orbitals on adjacent carbons (Figure 5.2). This energy is considerably greater than the thermal energy available at room temperature and, as a consequence, rotation about a carbon-carbon double bond is se-

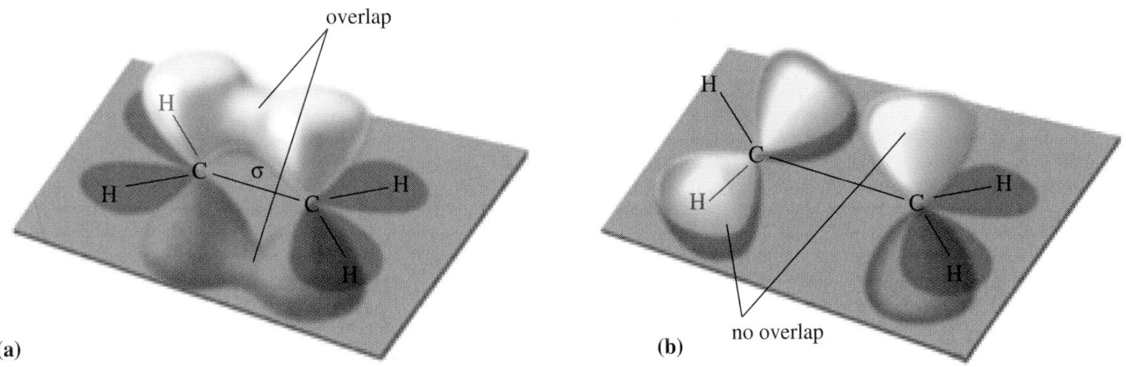

FIGURE 5.2
Restricted rotation about a carbon-carbon double bond. (a) Orbital overlap model showing the pi bond. (b) The pi bond is broken by rotating the plane of one H—C—H group by 90° with respect to the plane of the other H—C—H group.

verely restricted. You might compare rotation about a carbon-carbon double bond, such as in ethylene, with that about a carbon-carbon single bond, such as in ethane (Section 2.6A). Whereas rotation about the carbon-carbon single bond in ethane is relatively free (energy barrier ~ 3 kcal/mol), rotation about the carbon-carbon double bond in ethylene is restricted (energy barrier ~ 63 kcal/mol).

D. *Cis-Trans* Isomerism in Alkenes

Cis-trans **isomerism** Isomers that have the same order of attachment of their atoms, but a different arrangement of their atoms in space due to the presence of either a ring (Chapter 2) or a carbon-carbon double bond (Chapter 5).

Because of restricted rotation about the carbon-carbon double bond, any alkene in which each carbon of the double bond has two different groups attached to it shows *cis-trans* stereoisomerism. For example, 2-butene has two stereoisomers. In *cis*-2-butene, the two methyl groups are on one side of the double bond, and the two hydrogens are on the other side. In *trans*-2-butene, the two methyl groups are on opposite sides of the double bond. These two compounds cannot be converted into one another at room temperature because of the restricted rotation about the double bond; they are different compounds, with different physical and chemical properties.

H H
C=C
H_3C CH_3

***cis*-2-Butene**
mp −139°C
bp 4°C

H CH_3
C=C
H_3C H

***trans*-2-Butene**
mp −106°C
bp 1°C

Cis-alkenes are less stable than their *trans* isomers because of nonbonded interaction strain between alkyl substituents on the same side of the double bond, as can be seen in space-filling models of the *cis* and *trans* isomers of 2-butene (Figure 5.3).

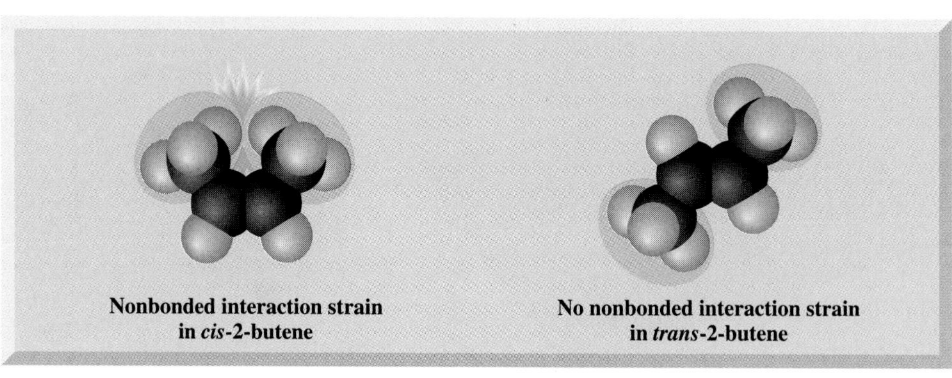

FIGURE 5.3
Space-filling models of *cis*-2-butene and *trans*-2-butene.

This is the same type of strain that results in the preference for equatorial methylcyclohexane over axial methylcyclohexane (Section 2.6B).

5.2 Nomenclature of Alkenes

Alkenes are named using the IUPAC system, but, as we shall see, some alkenes are still referred to by their common names.

A. IUPAC Names

IUPAC names of alkenes are formed by changing the **-an-** infix of the parent alkane to **-en-** (Section 2.5). Hence, $CH_2{=}CH_2$ is named ethene and $CH_3CH{=}CH_2$ is named propene. In higher alkenes, where isomers exist that differ in location of the double bond, a numbering system must be used. According to the IUPAC system, the longest carbon chain that contains the double bond is numbered in the direction that gives the carbon atoms of the double bond the lowest possible numbers. The location of the double bond is indicated by the number of its first carbon. Branched or substituted alkenes are named in a manner similar to alkanes. Carbon atoms are numbered, substituent groups are located and named, the double bond is located, and the main chain is named.

$$\overset{6}{C}H_3\overset{5}{C}H_2\overset{4}{C}H_2\overset{3}{C}H_2\overset{2}{C}H{=}\overset{1}{C}H_2$$

1-Hexene

$$\overset{6}{C}H_3\overset{5}{C}H_2\overset{4}{C}H(CH_3)\overset{3}{C}H_2\overset{2}{C}H{=}\overset{1}{C}H_2$$

4-Methyl-1-hexene

$$\overset{5}{C}H_3\overset{4}{C}H_2\overset{3}{C}H(CH_3)\overset{2}{C}(CH_2CH_3){=}\overset{1}{C}H_2$$

2-Ethyl-3-methyl-1-pentene

Note that there is a chain of six carbon atoms in 2-ethyl-3-methyl-1-pentene. However, because the longest chain that contains the double bond has only five carbons, the parent hydrocarbon is pentane, and the molecule is named as a disubstituted 1-pentene.

EXAMPLE 5.1

Write the IUPAC name of each alkene.

(a) $CH_2{=}CH(CH_2)_5CH_3$ (b) 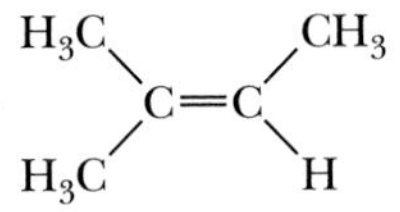 $(H_3C)_2C{=}C(CH_3)H$

Solution

(a) 1-Octene (b) 2-Methyl-2-butene

PROBLEM 5.1

Write the IUPAC name of each alkene.

(a) $CH_2{=}CHCH(CH_3)_2$ (b) $(CH_3)_2C{=}C(CH_3)_2$

B. Common Names

Despite the precision and universal acceptance of IUPAC nomenclature, some alkenes, particularly those of low molecular weight, are known almost exclusively by their common names, as illustrated by the common names of these alkenes.

	$CH_2{=}CH_2$	$CH_3CH{=}CH_2$	$CH_3C(CH_3){=}CH_2$
IUPAC name:	Ethene	Propene	2-Methylpropene
Common name:	Ethylene	Propylene	Isobutylene

Furthermore, the common names **methylene** (a CH_2 group), **vinyl,** and **allyl** are often used to show the presence of the following alkenyl groups:

Alkenyl Group	Common Name	Example
$CH_2{=}$	methylene	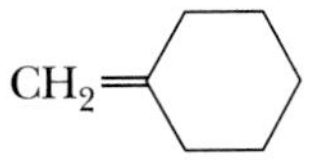 $CH_2{=}$cyclohexane methylenecyclohexane
$CH_2{=}CH{-}$	vinyl	$CH_2{=}CHCl$ vinyl chloride
$CH_2{=}CHCH_2{-}$	allyl	$CH_2{=}CHCH_2Cl$ allyl chloride

C. Methods for Designating Configuration in Alkenes

To designate the configuration of alkenes, we can use the *cis-trans* system or the *E,Z* system.

The *Cis-Trans* System

The most common method for specifying configuration in alkenes uses the prefixes *cis* and *trans*. There is no doubt whatsoever which isomers are intended by the names *cis*-2-butene and *trans*-3-hexene.

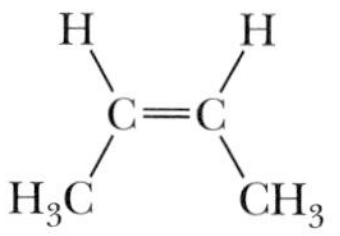

$$\underset{cis\text{-2-Butene}}{H_3C(H)C{=}C(H)CH_3} \qquad \underset{trans\text{-3-Hexene}}{CH_3CH_2(H)C{=}C(H)CH_2CH_3}$$

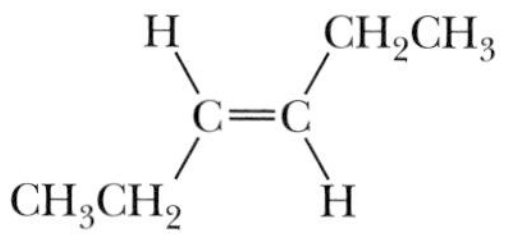

For more complex alkenes, the orientation of the atoms of the parent chain determines whether the alkene is *cis* or *trans*. Following is a structural formula for the *cis* isomer of 3,4-dimethyl-2-pentene. In this example, carbon atoms of the main chain (carbons 1 and 4) are on the same side of the double bond, and, therefore, this alkene is *cis*.

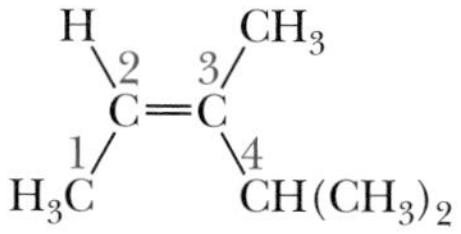

$$\underset{cis\text{-3,4-Dimethyl-2-pentene}}{\overset{1}{H_3C}(H)\overset{2}{C}{=}\overset{3}{C}(CH_3)\overset{4}{C}H(CH_3)_2}$$

EXAMPLE 5.2

Name each alkene and show the configuration about each double bond using the *cis-trans* system.

(a) $CH_3CH_2CH_2(H)C{=}C(H)CH_2CH_3$ (CH₃CH₂CH₂ upper left, H lower left; H upper right, CH₂CH₃ lower right)

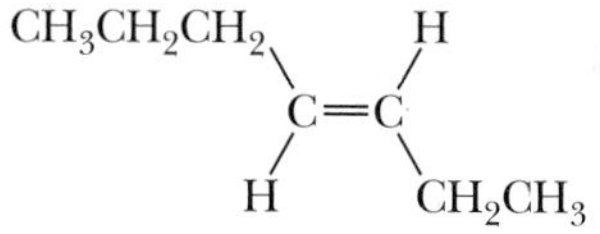

(b) $CH_3CH_2CH_2(CH_3)C{=}C(H)CH_2CH_3$ (CH₃CH₂CH₂ upper left, CH₃ lower left; CH₂CH₃ upper right, H lower right)

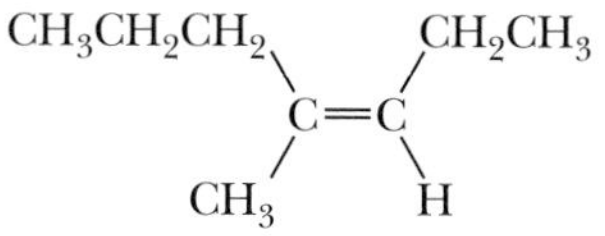

Solution

(a) The chain contains seven carbon atoms and is numbered from the end that gives the lower number to the first carbon of the double bond. Its name is *trans*-3-heptene.

(b) The longest chain contains seven carbon atoms and is numbered from the right so that the first carbon of the double bond is carbon 3 of the chain. Its name is *cis*-4-methyl-3-heptene.

PROBLEM 5.2

Which alkenes show *cis-trans* isomerism? For each alkene that does, draw the *trans* isomer.

(a) 2-Pentene **(b)** 2-Methyl-2-pentene **(c)** 3-Methyl-2-pentene

***E,Z* system** A system to specify the configuration of groups about a carbon-carbon double bond.

Z From the German, *zusammen,* together. Specifies that groups of higher priority on the carbons of a double bond are on the same side.

The *E,Z* System

The *E,Z* system uses the priority rules of the *R,S* system (Section 4.3) to assign priority to the substituents on each carbon of the double bond. Using these rules, we decide which of the two groups on each carbon of the double bond is of higher priority. If the groups of higher priority are on the same side of the double bond, the alkene is designated **Z** (German, *zusammen,* together). If the groups of higher priority are on opposite sides of the double bond, the alkene is designated ***E*** (German, *entgegen*, opposite).

E From the German, *entgegen*, opposite. Specifies that groups of higher priority on the carbons of a double bond are on opposite sides.

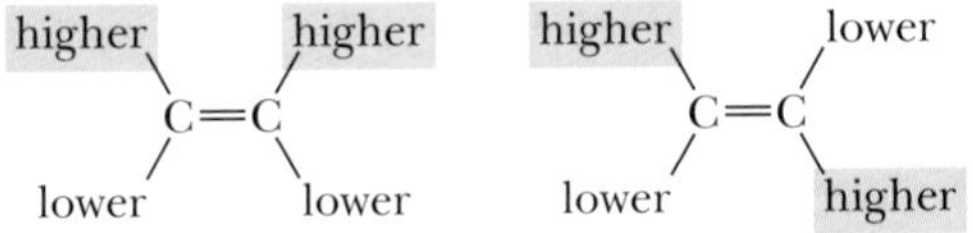

higher, higher / C=C / lower, lower — *Z* (zusammen)

higher, lower / C=C / lower, higher — *E* (entgegen)

Throughout this text, we use the *cis-trans* system for alkenes in which it is clear which is the main carbon chain. We use the *E,Z* system in all other cases.

EXAMPLE 5.3

Name each alkene and specify its configuration by the *E,Z* system.

(a) H, CH_3 / C=C / H_3C, $CH(CH_3)_2$

(b) Cl, H / C=C / H_3C, CH_2CH_3

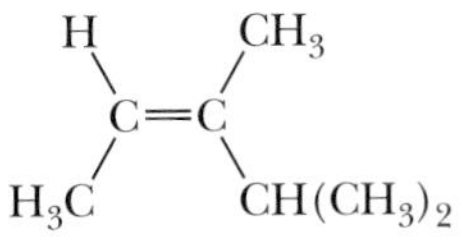

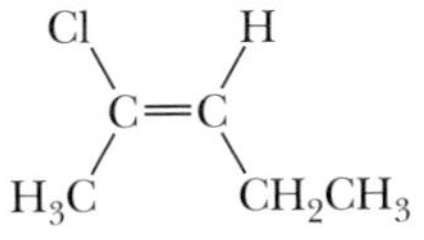

(c) [skeletal structure with Br]

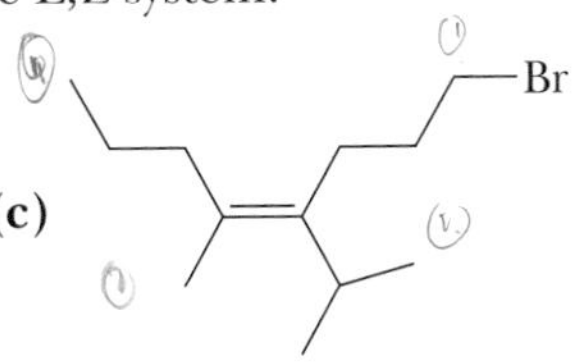

Solution

(a) The group of higher priority on carbon 2 is methyl; that of higher priority on carbon 3 is isopropyl. Because the groups of higher priority are on the same side of the carbon-carbon double bond, the alkene has the *Z* configuration. Its name is (*Z*)-3,4-dimethyl-2-pentene. Using the *cis-trans* system, it is named *cis*-3,4-dimethyl-2-pentene.

(b) Groups of higher priority on carbons 2 and 3 are —Cl and —CH_2CH_3. Because these groups are on opposite sides of the double bond, the configuration of this alkene is *E* and its name is (*E*)-2-chloro-2-pentene. Using the *cis-trans* system, it is *cis*-2-chloro-2-pentene.

(c) The groups of higher priority are on opposite sides of the double bond; the configuration is *E*. The name of this haloalkene is (*E*)-1-bromo-4-isopropyl-5-methyl-4-octene. Using the *cis-trans* system, it is *cis*-1-bromo-4-isopropyl-5-methyl-4-octene.

[numbered skeletal structure: 8, 7, 6, 5, 4, 3, 2, 1, Br]

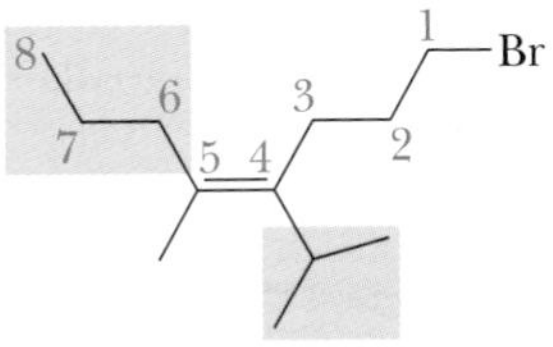

PROBLEM 5.3

Name each alkene and specify its configuration by the *E,Z* system.

(a) $ClCH_2$, CH_3 / C=C / H_3C, CH_2CH_3

(b) Cl, H / C=C / Br, CH_3

(c) $CH_3CH_2CH_2$, CH_3 / C=C / H_3C, $CH(CH_3)_2$

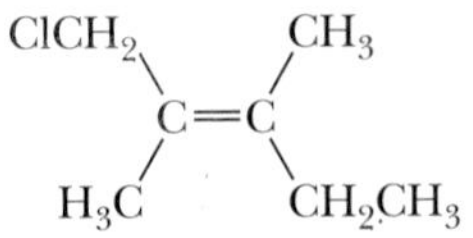

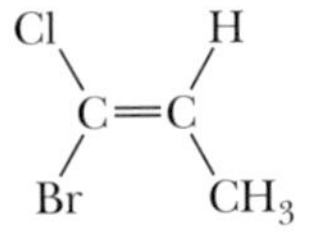

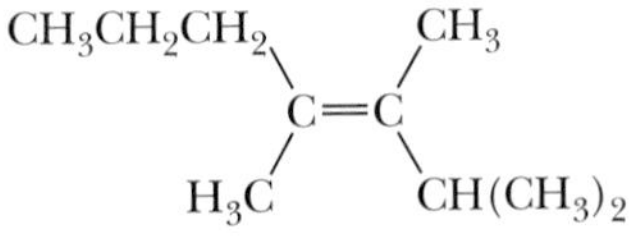

D. Cycloalkenes

In naming **cycloalkenes,** the carbon atoms of the ring double bond are numbered 1 and 2 in the direction that gives the substituent encountered first the smaller number.

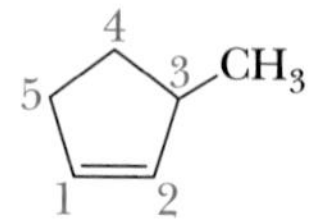

3-Methylcyclopentene 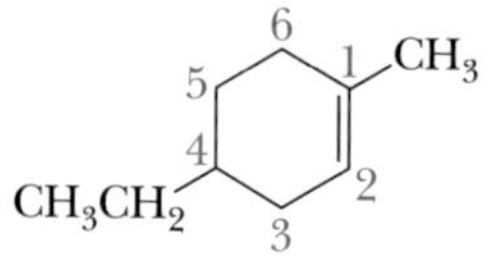 4-Ethyl-1-methylcyclohexene 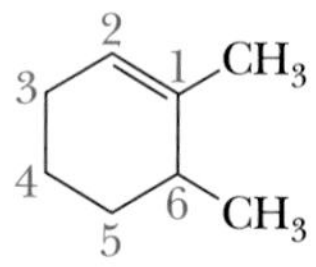 1,6-Dimethylcyclohexene

EXAMPLE 5.4

Write the IUPAC name of each cycloalkene.

(a) 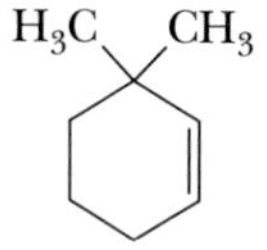 (b) 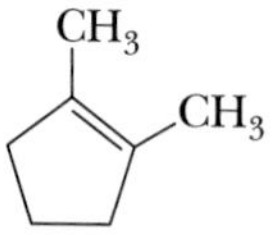 (c) 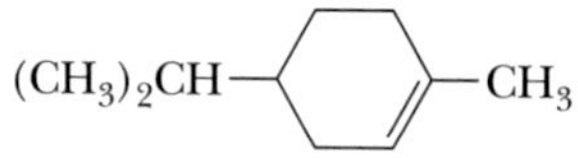 $(CH_3)_2CH$—⬡—CH_3

Solution

(a) 3,3-Dimethylcyclohexene **(b)** 1,2-Dimethylcyclopentene
(c) 4-Isopropyl-1-methylcyclohexene

PROBLEM 5.4

Write the IUPAC name of each cycloalkene.

(a) 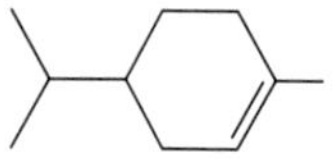 (b) 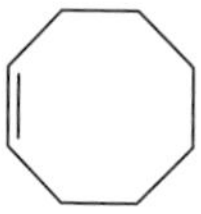 (c) 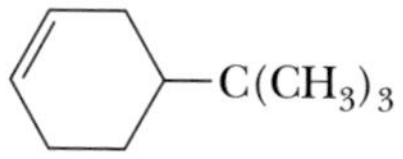 —$C(CH_3)_3$

E. *Cis-Trans* Isomerism in Cycloalkenes

Following are structural formulas for four cycloalkenes:

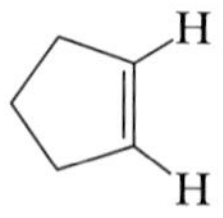

Cyclopentene 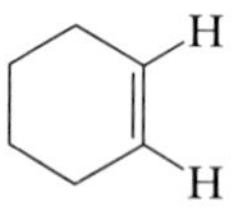 Cyclohexene 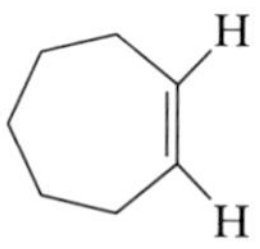 Cycloheptene 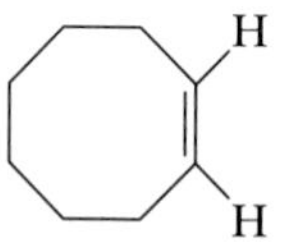 Cyclooctene

In these representations, the configuration about each double bond is *cis.* Is it possible to have a *trans* configuration in these and larger cycloalkenes? To date, *trans*-cyclooctene is the smallest *trans*-cycloalkene that has been prepared in pure form and is stable at room temperature. Yet, even in this *trans*-cycloalkene, there is considerable angle strain; the double bond's *p* orbitals make an angle of 44° to each other. *Cis*-cyclooctene is more stable than its *trans* isomer by 9.1 kcal/mol (38 kJ/mol).

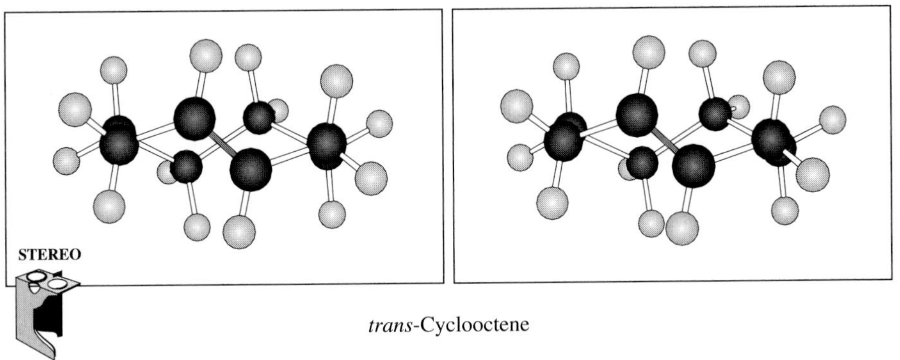

trans-Cyclooctene

F. Dienes, Trienes, and Polyenes

Alkenes that contain more than one double bond are named as alkadienes, alkatrienes, and so on. Those that contain several double bonds are also referred to more generally as polyenes (Greek, *poly*, many). Following are examples of three dienes.

$CH_2{=}CHCH_2CH{=}CH_2$
1,4-Pentadiene

$CH_2{=}C(CH_3)CH{=}CH_2$
2-Methyl-1,3-butadiene
(Isoprene)

1,3-Cyclopentadiene

G. *Cis-Trans* Isomerism in Dienes, Trienes, and Polyenes

Thus far we have considered *cis-trans* isomerism in alkenes containing only one carbon-carbon double bond. For an alkene with one carbon-carbon double bond that can show *cis-trans* isomerism, two stereoisomers are possible. For an alkene with ***n*** carbon-carbon double bonds, each of which can show *cis-trans* isomerism, $\mathbf{2^n}$ stereoisomers are possible.

EXAMPLE 5.5

How many stereoisomers are possible for 2,4-heptadiene?

Solution

This molecule has two carbon-carbon double bonds, each of which shows *cis-trans* isomerism. As shown in this table, there are $2^2 = 4$ stereoisomers. Two of these are drawn on the right.

Double Bond	
$C_2{-}C_3$	$C_4{-}C_5$
trans	*trans*
trans	*cis*
cis	*trans*
cis	*cis*

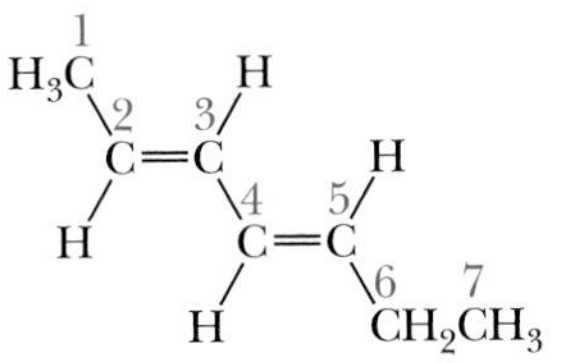

trans-trans-2,4-Heptadiene

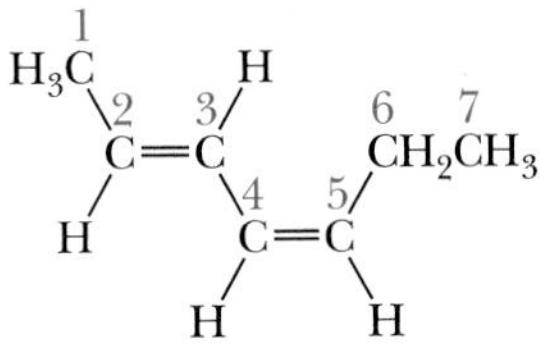

trans-cis-2,4-Heptadiene

CHEMISTRY IN ACTION

The Case of the Iowa and New York Strains of the European Corn Borer

Although humans communicate largely by sight and sound, chemicals are the primary means of communication for the vast majority of other species in the animal world. Often, communication within a species is specific for one of two configurational isomers. For example, a member of a given species responds to a *cis* isomer of a chemical but not to the *trans* isomer. Or, alternatively, it might respond to a quite precise blend of *cis* and *trans* isomers but not to other blends of these same isomers.

Several groups of scientists have studied the components of the sex pheromones of both the Iowa and the New York strains of the European corn borer. Females of these closely related species secrete the sex attractant 11-tetradecenyl acetate. Males of the Iowa strain show maximum response to a mixture containing 96% of the *cis* isomer and 4% of the *trans* isomer. When the pure *cis* isomer is used alone, males are only weakly attracted. Males of the New York strain show an entirely different response pattern. They respond maximally to a mixture containing 3% of the *cis* isomer and 97% of the *trans* isomer.

$$\underset{\text{H}}{\overset{\text{CH}_3\text{CH}_2}{}}\!\!\!\!\text{C}=\text{C}\!\!\!\!\underset{\text{H}}{\overset{(\text{CH}_2)_9\text{CH}_2\text{OCCH}_3 \;(\text{C}{=}\text{O})}{}}$$

cis-11-Tetradecenyl acetate

$$\underset{\text{CH}_3\text{CH}_2}{\overset{\text{H}}{}}\!\!\!\!\text{C}=\text{C}\!\!\!\!\underset{\text{H}}{\overset{(\text{CH}_2)_9\text{CH}_2\text{OCCH}_3 \;(\text{C}{=}\text{O})}{}}$$

trans-11-Tetradecenyl acetate

There is evidence that optimum response to a narrow range of stereoisomers as we see here is widespread in nature and that most insects maintain species isolation for mating and reproduction by the stereochemistry of their pheromones.

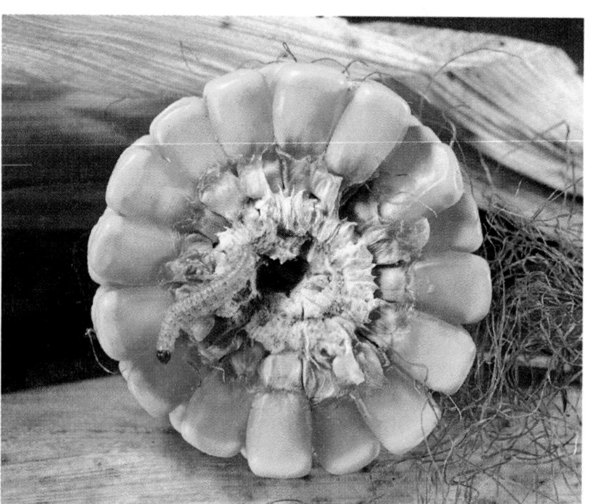

The European corn borer, Pyrausta nubilalis. *(Runk/Schoenberger/Grant Heilman Photography, Inc.)*

See Klun, J. A., *et al., Science,* **181:**661 (1973).

PROBLEM 5.5

Draw structural formulas for the other two stereoisomers of 2,4-heptadiene.

EXAMPLE 5.6

How many stereoisomers are possible for 10,12-hexadecadien-1-ol?

$$\text{CH}_3(\text{CH}_2)_2\text{CH}{=}\text{CHCH}{=}\text{CH}(\text{CH}_2)_8\text{CH}_2\text{OH}$$

10,12-Hexadecadien-1-ol

Solution

Cis-trans isomerism is possible about both double bonds. Four stereoisomers are possible.

PROBLEM 5.6

The sex pheromone of the silkworm is (10*E*,12*Z*)-10,12-hexadecadien-1-ol. Draw a structural formula for this compound.

An example of a biologically important compound for which a number of *cis-trans* isomers is possible is vitamin A. There are four carbon-carbon double bonds in the chain of atoms attached to the substituted cyclohexene ring, and each has the potential for *cis-trans* isomerism. There are 2^4, or 16, stereoisomers possible for this structural formula. Vitamin A is the all-*trans* isomer.

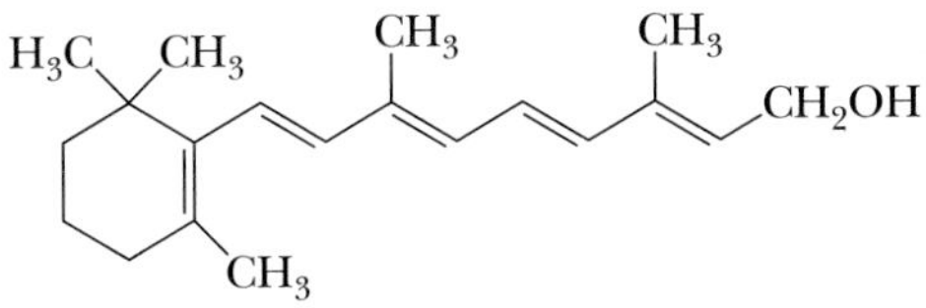

H_3C CH_3 CH_3 CH_3 CH_2OH CH_3

Vitamin A (retinol)

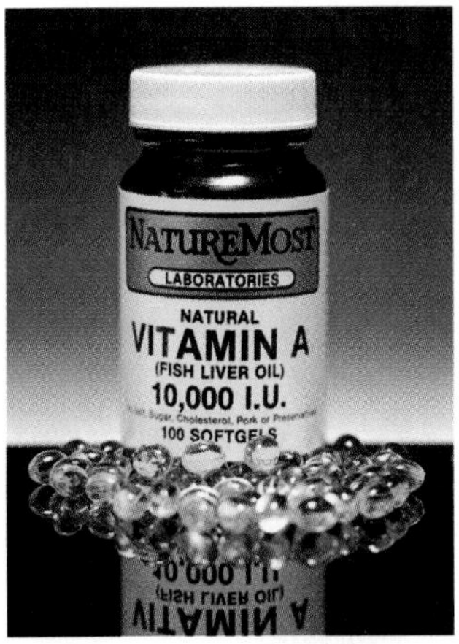

One International Unit (IU) of vitamin A is 0.30 μg, which is equivalent to 3.3×10^6 IU/g. The Recommended Daily Allowance (RDA) is 5000 IU. *(Charles D. Winters)*

5.3 Physical Properties of Alkenes

Alkenes are nonpolar compounds, and the only attractive forces between their molecules are dispersion forces (Section 2.8). Therefore, the physical properties of alkenes are similar to those of alkanes. Alkenes of two, three, and four carbon atoms are gases at room temperature. Those of five or more carbons are colorless liquids less dense than water. Alkenes are insoluble in water but soluble in one another, in other nonpolar organic liquids, and in ethanol. Table 5.2 lists physical properties of some alkenes.

5.4 Naturally Occurring Alkenes: Terpene Hydrocarbons

Terpenes are among the most widely distributed compounds in the biological world. The number of terpenes found in bacteria, plants, and animals is staggering. A **terpene** is a compound whose carbon skeleton can be divided into two or more units that are identical with the carbon skeleton of isoprene. Carbon 1 of an **isoprene unit** is called the tail; carbon 4 is called the head. Terpenes are formed by stringing together the tail of one isoprene unit to the head of another.

Terpene A compound whose carbon skeleton can be divided into two or more units identical with the carbon skeleton of isoprene.

$$CH_2{=}\underset{}{\overset{CH_3}{\overset{|}{C}}}{-}CH{=}CH_2$$

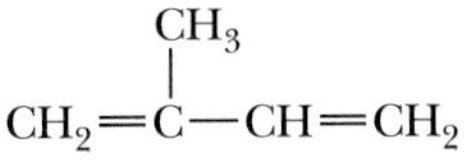

2-Methyl-1,3-butadiene
(Isoprene)

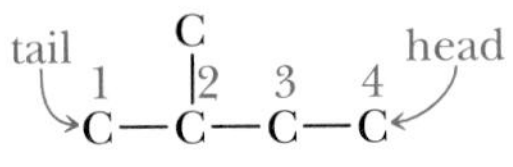

$$\text{tail} \rightarrow \overset{1}{C}-\overset{\overset{C}{|}}{\underset{2}{C}}-\overset{3}{C}-\overset{4}{C} \leftarrow \text{head}$$

Isoprene unit

TABLE 5.2 Physical Properties of Some Alkenes

Name	Structural Formula	mp (°C)	bp (°C)
ethylene	$CH_2{=}CH_2$	−169	−104
propene	$CH_3CH{=}CH_2$	−185	−47
1-butene	$CH_3CH_2CH{=}CH_2$	−185	−6
1-pentene	$CH_3CH_2CH_2CH{=}CH_2$	−138	30
cis-2-pentene	$(H_3C)(H)C{=}C(H)(CH_2CH_3)$ (H₃C and CH₂CH₃ on same side)	−151	37
trans-2-pentene	$(H_3C)(H)C{=}C(CH_2CH_3)(H)$ (H₃C and CH₂CH₃ on opposite sides)	−156	36
2-methyl-2-butene	$CH_3C(CH_3){=}CHCH_3$	−134	39

A study of terpenes provides a glimpse of the wondrous diversity that nature can generate from a simple carbon skeleton. Terpenes also illustrate an important principle of the molecular logic of living systems, namely, that in building large molecules, small subunits are strung together enzymatically by an iterative process and then chemically modified by precise enzyme-catalyzed reactions. Chemists use the same principles in the laboratory, but their methods do not have the precision and selectivity of the enzyme-catalyzed reactions of cellular systems.

Probably the terpenes most familiar to you, at least by odor, are components of the so-called "essential oils" obtained by steam distillation or ether extraction of various

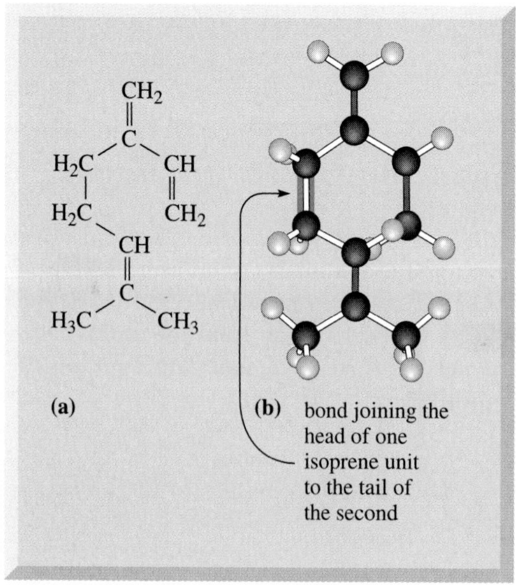

FIGURE 5.4
Myrcene. (a) The structural formula of myrcene and (b) a ball-and-stick model divided to show two isoprene units joined by a carbon-carbon bond between the head of one unit and the tail of the other.

Chemistry in Action

Why Plants Emit Isoprene

Names like Virginia's Blue Ridge, Jamaica's Blue Mountain Peak, and Australia's Blue Mountains remind us of the bluish haze that hangs over wooded hills in the summer time. It was discovered in the 1950s that this haze is rich in isoprene, which means that isoprene is far more abundant in the atmosphere than anyone thought. The haze itself is caused by light-scattering from an aerosol produced by photooxidation of isoprene and other hydrocarbons. Scientists now estimate that global emission of isoprene by plants is 3×10^{14} g/yr (3.3×10^{8} tons/year), which represents approximately 2% of all carbon fixed by photosynthesis. A recent study of hydrocarbon emissions in the Atlanta area showed that plants were by far the largest emitters of hydrocarbons, with plant-derived isoprene accounting for almost 60% of the total.

Haze over the Blue Ridge Mountains. (Photo/NATS)

Why do plants emit so much isoprene into the atmosphere rather than use it for the synthesis of terpenes and other natural products? Tom Sharkey, a University of Wisconsin plant physiologist, found that emission of isoprene is extremely sensitive to temperature. Plants grown at 20°C do not emit isoprene, but begin to emit it when leaf temperature is increased to 30°C. In certain plants, isoprene emission can increase as much as tenfold for each 10°C increase in leaf temperature. Sharkey studied the relationship between temperature-induced leaf damage and isoprene concentration in leaves of the kudzu [*Pueraria lobata* (Wild.) Ohwi.] plant. He discovered that leaf damage, as measured by chlorophyll destruction, begins to occur at 37.5°C in the absence of isoprene, but not until 45°C in the presence of isoprene. Sharkey speculates that isoprene dissolves in leaf membranes and in some way increases their tolerance to heat stress. Because isoprene is made rapidly and also lost rapidly, its concentration correlates with temperature throughout the day.

See *Why Plants Emit Isoprene* by T. D. Sharkey and E. L. Singsass, *Nature*, **374**:27 (April 1995), p. 769.

parts of plants. Essential oils contain the relatively low-molecular-weight substances that are in large part responsible for characteristic plant fragrances (essences). Many essential oils, particularly those from flowers, are used in perfumes.

One example of a terpene obtained from an essential oil is myrcene, $C_{10}H_{16}$, a component of bayberry wax and oils of bay and verbena. Myrcene is a triene with a parent chain of eight carbon atoms and two one-carbon branches [Figure 5.4(a)].

Head-to-tail linkages of isoprene units are vastly more common in nature than are the alternative head-to-head or tail-to-tail patterns. Figure 5.5 shows structural formulas of six more terpenes, all derived from two isoprene units. Geraniol and the aggregating pheromone of the bark beetle have the same carbon skeleton as myrcene. In addition, each has an —OH (hydroxyl) group. In the last four terpenes of Figure 5.5, the carbon

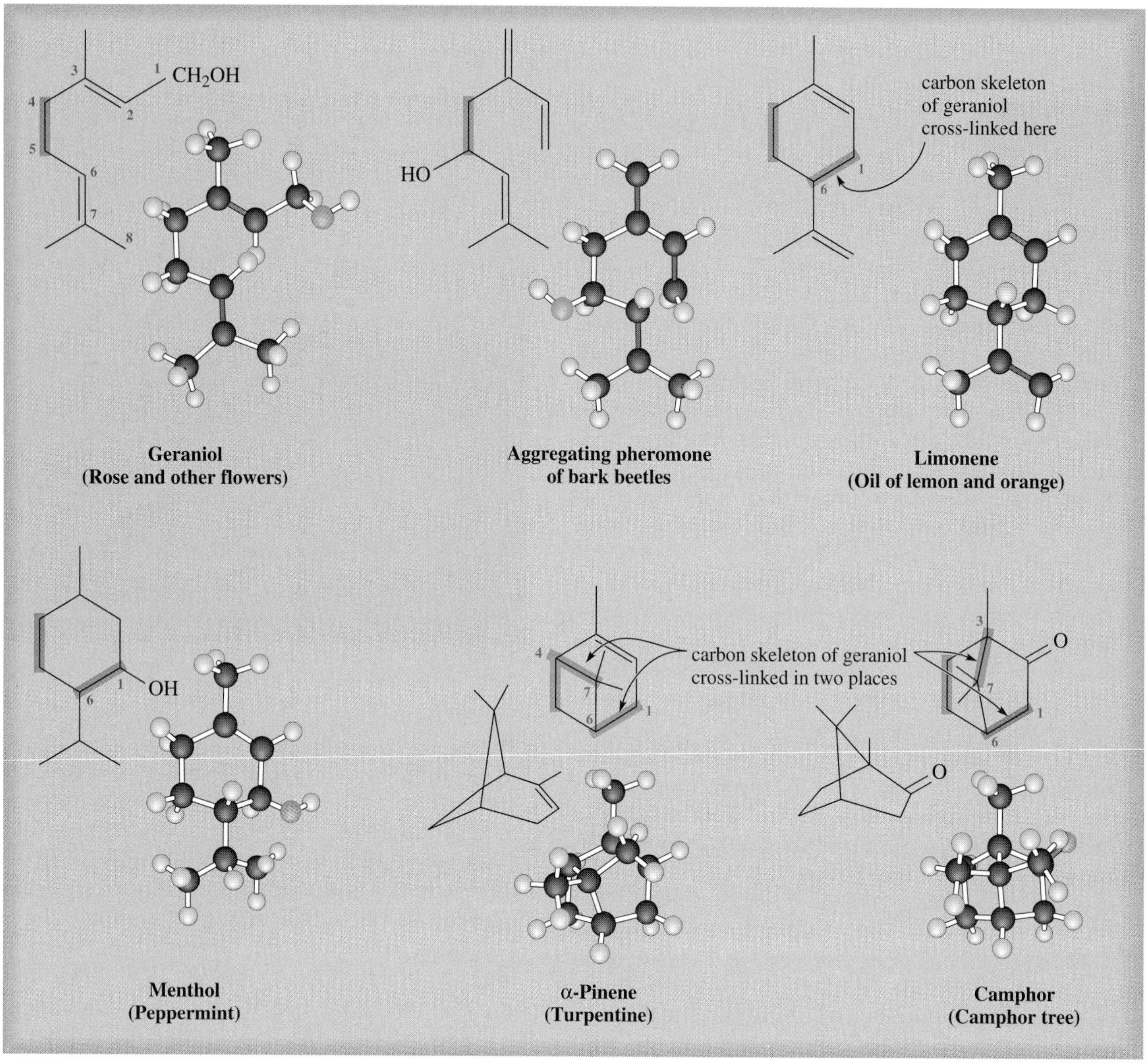

CH2OH

HO

carbon skeleton of geraniol cross-linked here

Geraniol (Rose and other flowers)

Aggregating pheromone of bark beetles

Limonene (Oil of lemon and orange)

OH

carbon skeleton of geraniol cross-linked in two places

O

Menthol (Peppermint)

α-Pinene (Turpentine)

Camphor (Camphor tree)

FIGURE 5.5 Six terpenes, each divisible into two isoprene units.

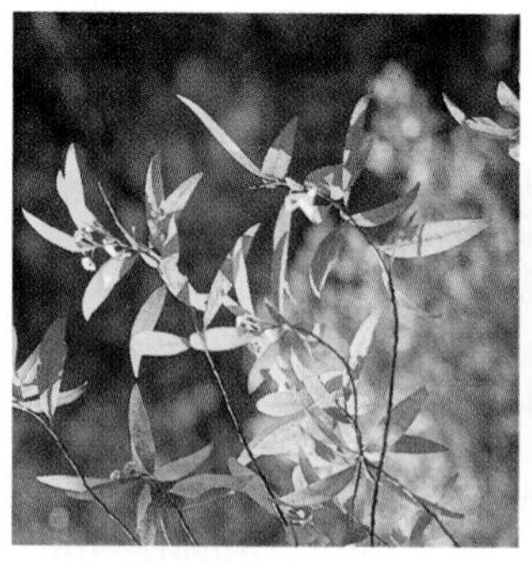

California laurel, *Umbellularia californica,* a source of myrcene. *(© Dan Suzio)*

Lemongrass, a source of geraniol. *(G. Büttner/Naturbild/Okapia/Photo Researchers, Inc.)*

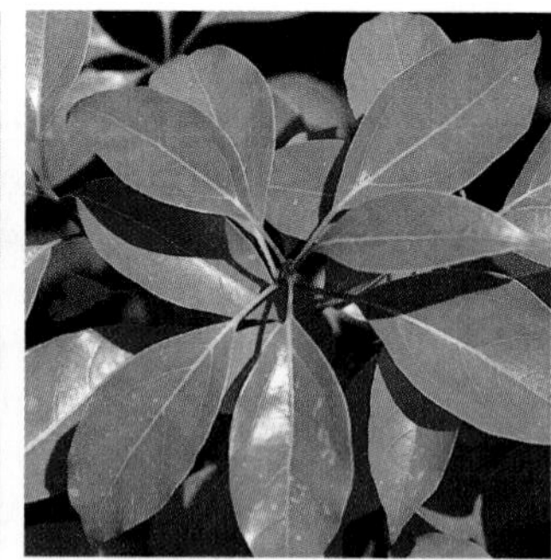

Leaves of the camphor tree, *Cinnamomum camphora. (Peter G. Aitken/Photo Researchers, Inc.)*

CHEMISTRY IN ACTION

Terpenes of the Cotton Plant

The floral fragrance of the cotton plant *(Gossypium)* is due in large part to a group of volatile, low-molecular-weight terpenes produced by leaf glands. At least a dozen scent components have been isolated and identified, including myrcene, limonene, and α-pinene. Among the terpenes divisible into three isoprene units are found spathulenol, gossonorol, and β-bisabolol.

HO OH HO

Spathulenol Gossonorol β-Bisabolol

What is notable about these three compounds is their diversity in structure. What is puzzling about them is why the cotton plant makes them. Are they, or were they at an earlier stage of evolution, important in the chemical defense of the cotton plant against insect predators or fungal infections? Are they important in these roles now? We do not know the answers to these questions.

A field of cotton. Fragrant terpenes, including myrcene, limonene, and α-pinene, can be isolated from the cotton plant. (©John Elk III)

For the role of two other terpenes, hemigossypol and gossypol, in the cotton plant's defense against fungal infections, see the box "Phytoalexins: Natural Plant Antibiotics" in Chapter 19.

atoms present in myrcene, geraniol, and the bark beetle pheromone are cross-linked to give cyclic structures. To help you identify the points of cross linkage and ring formation, the carbon atoms of the geraniol skeleton are numbered 1 through 8. This numbering pattern is used in the remaining terpenes to show points of cross-linking.

Shown in Figure 5.6 are structural formulas of several terpenes divisible into three isoprene units. For reference, the carbon atoms of the parent chain of farnesol are numbered 1 through 12. A bond between carbon atoms 1 and 6 of this skeleton gives the carbon skeleton of zingiberene. Try to discover for yourself what patterns of cross-linking give the carbon skeleton of caryophyllene.

Vitamin A (Section 5.2G), a terpene of molecular formula $C_{20}H_{30}O$, consists of four isoprene units linked head-to-tail and cross-linked at one point to form a six-membered ring.

The synthesis of substances in living systems is a fascinating area of research and one of the links between organic and biochemistry. However tempting it might be to propose that nature synthesizes terpenes by joining together molecules of isoprene, this is not quite the way it is done. We will discuss the synthesis of terpenes in Chapters 18 and 25.

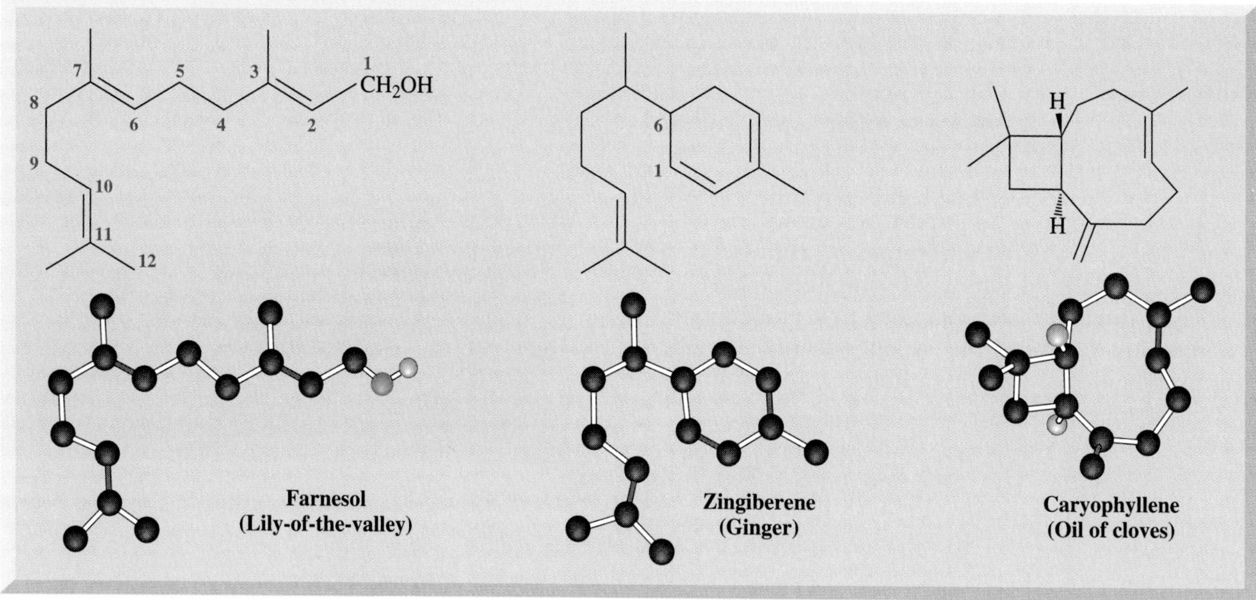

FIGURE 5.6
Terpenes containing three isoprene units.

SUMMARY

An **alkene** is an unsaturated hydrocarbon that contains a carbon-carbon double bond. The general formula of an alkene is C_nH_{2n}. A carbon-carbon double bond consists of one sigma bond formed by the overlap of sp^2 hybrid orbitals and one pi bond formed by the overlap of parallel $2p$ orbitals (Section 5.1B). A carbon-carbon double bond is both shorter and stronger than a carbon-carbon single bond (Section 5.1C). The relative strengths of the sigma and pi bonds in ethylene are approximately 85 kcal/mol and 63 kcal/mol, respectively. The structural feature that makes ***cis-trans* isomerism** possible in alkenes (Section 5.1D) is lack of rotation about the two carbons of the double bond.

According to the IUPAC system (Section 5.2A), the presence of the carbon-carbon double bond is shown by changing the infix of the parent hydrocarbon from -an- to **-en-**. For compounds containing two or more double bonds, the infix is changed to -adien-, -atrien-, and so on. The names **methylene, vinyl,** and **allyl** are commonly used to show the presence of $=CH_2$, $-CH=CH_2$, and $-CH_2CH=CH_2$ groups, respectively.

Whether an alkene is *cis* or *trans* is determined by the orientation of the main carbon chain about the double bond (Section 5.2C). The configuration of a carbon-carbon double bond is specified more precisely by the ***E,Z* system** using the same set of priority rules used for the *R,S* system (Section 4.3). If the two groups of higher priority are on the same side of the double bond, the alkene is designated ***Z*** (German, *zusammen,* together); if they are on opposite sides, the alkene is designated ***E*** (German, *entgegen,* opposite). To date, *trans*-cyclooctene is the smallest *trans*-cycloalkene that has been prepared in pure form and is stable at room temperature.

Because alkenes are essentially nonpolar compounds and the only attractive forces between their molecules are **dispersion forces,** their physical properties are similar to those of alkanes (Section 5.3).

The characteristic structural feature of a **terpene** (Section 5.4) is a carbon skeleton that can be divided into two or more **isoprene units.** Terpenes illustrate an important principle of the molecular logic of living systems, namely that in building large molecules, small subunits are strung together by an iterative process and then chemically modified by precise enzyme-catalyzed reactions.

ADDITIONAL PROBLEMS

Structure of Alkenes

5.7 Predict all bond angles about each highlighted carbon atom. To make these predictions, use the valence-shell electron-pair repulsion (VSEPR) model (Section 1.4).

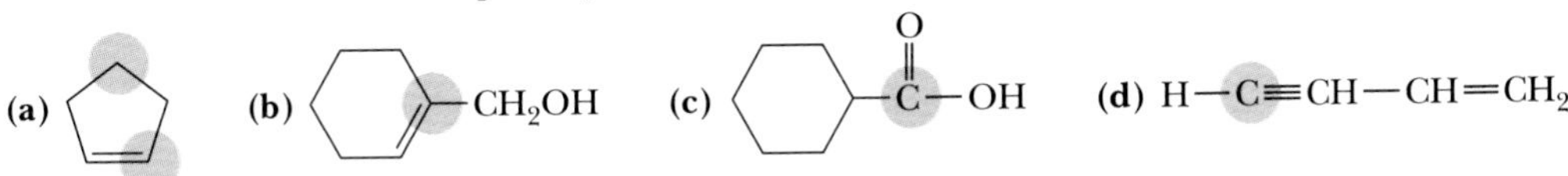

5.8 For each highlighted carbon atom in Problem 5.7, identify which atomic orbitals are used to form each sigma bond and which are used to form each pi bond.

5.9 Following is the structural formula of propadiene (allene).

$$CH_2{=}C{=}CH_2$$

Propadiene
(Allene)

(a) State the orbital hybridization of each carbon atom.
(b) Describe each carbon-carbon double bond in terms of the overlap of atomic orbitals.
(c) Predict all bond angles in allene.
(d) Draw a stereorepresentation showing the shape of this molecule.

5.10 Following are lengths for a series of C—C single bonds. Propose an explanation for the differences in bond lengths.

Structure	Length of C—C Single Bond (Å)
$CH_3{-}CH_3$	1.537
$CH_2{=}CH{-}CH_3$	1.510
$CH_2{=}CH{-}CH{=}CH_2$	1.465
$HC{\equiv}C{-}CH_3$	1.459

5.11 The best overlap between two adjacent p orbitals takes place when their axes are parallel. The overlap, and thus the strength of the pi component of a double bond, decreases approximately as $\cos^2\theta$, where θ is the angle that the axes of two p orbitals make with each other. How does the overlap decrease for p orbitals twisted 10°, 20°, 30°, 45°, 60°, and 90°?

5.12 Prepare a plot of potential energy versus angle of rotation about the carbon-carbon double bond in ethylene. How does the energy scale for this plot compare with the energy scale for a plot of potential energy versus angle of rotation about the carbon-carbon single bond in ethane?

Nomenclature of Alkenes

5.13 Draw structural formulas for these alkenes.

(a) *trans*-2-Methyl-3-hexene
(b) 2-Methyl-2-hexene
(c) 2-Methyl-1-butene
(d) 3-Ethyl-3-methyl-1-pentene
(e) 2,3-Dimethyl-2-butene
(f) *cis*-2-Pentene
(g) (*Z*)-1-Chloropropene
(h) 3-Methylcyclohexene
(i) 1-Isopropyl-4-methylcyclohexene
(j) (6*E*)-2,6-Dimethyl-2,6-octadiene

(k) Allylcyclopropane
(l) Vinylcyclopropane
(m) 2-Chloropropene
(n) Tetrachloroethylene
(o) 1-Chlorocyclohexene
(p) Bicyclo[2.2.1]-2-heptene
(q) Bicyclo[4.4.0]-1-decene

5.14 Name these alkenes and cycloalkenes.

(a) 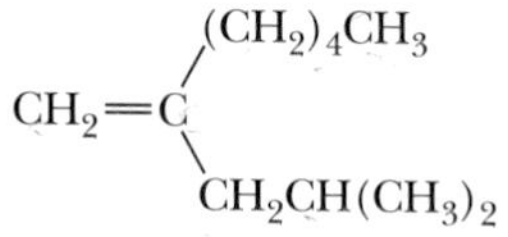 $CH_2{=}C(CH_2)_4CH_3$ / $CH_2CH(CH_3)_2$ (b) Cl, CH_3, CH_3 (c) $CH{=}CH_2$, $CH{=}CH_2$

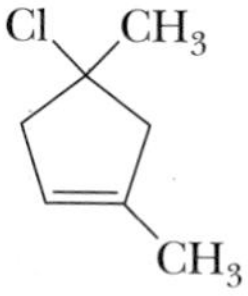

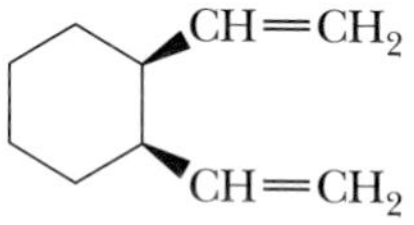

(d) $(CH_3)_2CHCH{=}C(CH_3)_2$ (e) 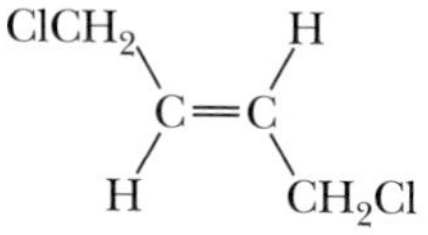 $ClCH_2$, H, C=C, H, CH_2Cl (f) F, F, C=C, F, F

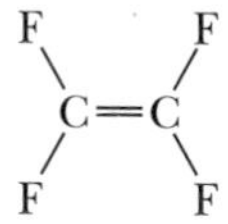

(g) 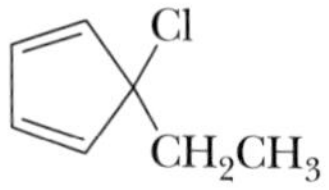 Cl, CH_2CH_3 (h) 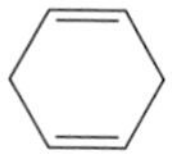 (i) 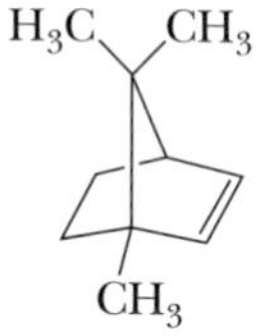 H_3C, CH_3, CH_3

5.15 Arrange the following groups in order of increasing priority.

(a) $-CH_3$ $-H$ $-Br$ $-CH_2CH_3$

(b) $-OCH_3$ $-CH(CH_3)_2$ $-B(CH_2CH_3)_2$ $-H$

(c) $-CH_3$ $-CH_2OH$ $-CH_2NH_2$ $-CH_2Br$

5.16 Assign an *E,Z* configuration and a *cis-trans* configuration to these dicarboxylic acids, each of which is an intermediate in the tricarboxylic acid cycle. Following each is given its common name.

(a) 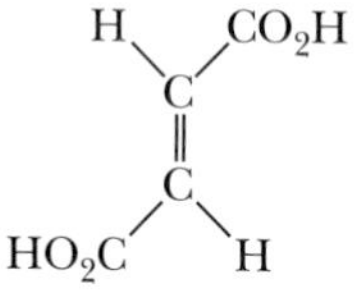 H, CO_2H, C=C, HO_2C, H
Fumaric acid

(b) 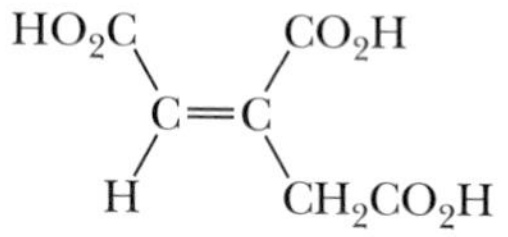 HO_2C, CO_2H, C=C, H, CH_2CO_2H
Aconitic acid

5.17 Name and draw structural formulas for all alkenes of molecular formula C_5H_{10}. As you draw these alkenes, remember that *cis* and *trans* isomers are different compounds and must be counted separately in drawing all alkenes possible for this molecular formula.

5.18 For each molecule that shows *cis-trans* isomerism, draw the *cis* isomer.

(a) 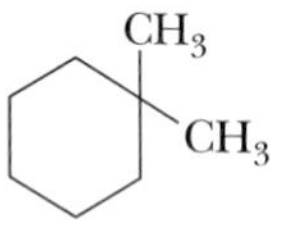 CH_3, CH_3 (b) 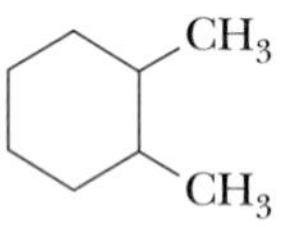 CH_3, CH_3 (c) 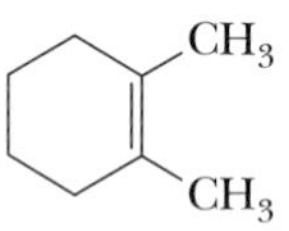 CH_3, CH_3 (d) 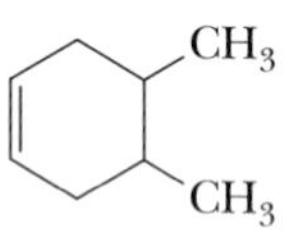 CH_3, CH_3

5.19 Draw structural formulas for all compounds of molecular formula C_5H_{10} that are:

(a) Alkenes that do not show *cis-trans* isomerism.
(b) Alkenes that do show *cis-trans* isomerism.
(c) Cycloalkanes that do not show *cis-trans* isomerism.
(d) Cycloalkanes that do show *cis-trans* isomerism.

5.20 β-Ocimene, a triene found in the fragrance of cotton blossoms and several essential oils, has the IUPAC name (3*Z*)-3,7-dimethyl-1,3,6-octatriene. Draw a structural formula for β-ocimene.

5.21 Draw the structural formula for at least one bromoalkene of molecular formula C_5H_9Br that shows:

(a) Neither *E,Z* isomerism nor chirality.
(b) *E,Z* isomerism but not chirality.
(c) Chirality but not *E,Z* isomerism.
(d) Both chirality and *E,Z* isomerism.

5.22 Following are structural formulas and common names for four molecules that contain both a carbon-carbon double bond and another functional group. Give each an IUPAC name.

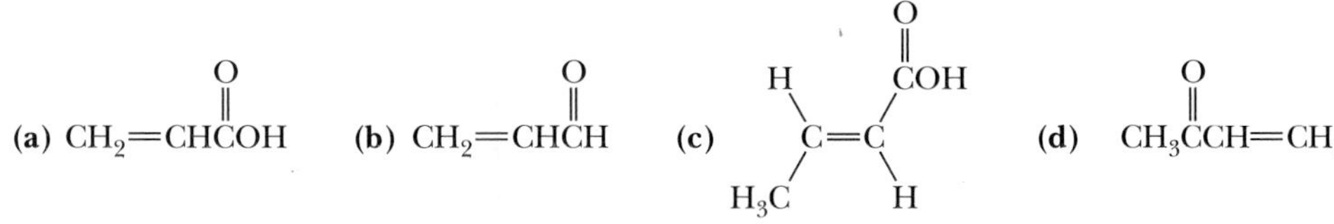

Acrylic acid | Acrolein | Crotonic acid | Methyl vinyl ketone

5.23 *Trans*-cyclooctene has been resolved, and its enantiomers are stable at room temperature. *Trans*-cyclononene has also been resolved, but it racemizes with a half-life of 4 min at 0°C. How can racemization of this cycloalkene take place without breaking any bonds? Why does *trans*-cyclononene racemize under these conditions but not *trans*-cyclooctene? You will find it especially helpful to build molecular models of these cycloalkenes.

5.24 How many stereoisomers are possible for the following natural products?

(a) Geraniol (Figure 5.5) **(b)** Limonene (Figure 5.5) **(c)** α-Pinene (Figure 5.5)
(d) Farnesol (Figure 5.6) **(e)** Zingiberene (Figure 5.6)

5.25 Which alkenes exist as pairs of *cis-trans* isomers? For each alkene that does, draw the *trans* isomer.

(a) $CH_2{=}CHBr$ **(b)** $CH_3CH{=}CHBr$ **(c)** $BrCH{=}CHBr$

(d) $(CH_3)_2C{=}CHCH_3$ **(e)** $(CH_3)_2CHCH{=}CHCH_3$

5.26 Four stereoisomers exist for 3-penten-2-ol.

$$CH_3{-}CH{=}CH{-}\underset{\displaystyle}{\overset{\displaystyle OH}{\overset{|}{C}}}H{-}CH_3$$

3-Penten-2-ol

(a) Explain how these four stereoisomers arise.
(b) Draw the stereoisomer having the *E* configuration about the carbon-carbon double bond and the *R* configuration at the stereocenter.

Terpenes

5.27 Show how the carbon skeleton of farnesol can be coiled and then cross-linked to give the carbon skeleton of zingiberene (Figure 5.6).

5.28 Show that the structural formula of vitamin A (Section 5.2G) can be divided into four isoprene units joined by head-to-tail linkages and cross-linked at one point to form the six-membered ring.

5.29 Following is the structural formula of lycopene, a deep-red compound that is partially responsible for the red color of ripe fruits, especially tomatoes. Approximately 20 mg of lycopene can be isolated from 1 kg of ripe tomatoes. (See *The Merck Index,* 12th ed., #5650.)

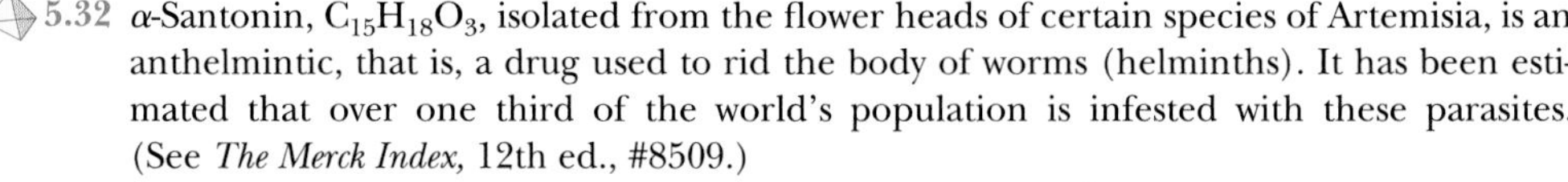

Lycopene

(a) Show that lycopene is a terpene, that is, its carbon skeleton can be divided into two sets of four isoprene units with the units in each set joined head-to-tail.

(b) How many of the carbon-carbon double bonds in lycopene have the possibility for *cis-trans* isomerism? Lycopene is the all-*trans* isomer.

5.30 The structural formula of β-carotene, precursor to vitamin A, is given in Section 25.6A. As you might suspect, it was first isolated from carrots. Dilute solutions of β-carotene are yellow, hence its use as a food coloring. Compare the carbon skeletons of β-carotene and lycopene. What are the similarities? What are the differences?

5.31 Following is the structural formula of warburganal, a crystalline solid isolated from the plant *Warburgia ugandensis, Canellaceae.* An important use of warburganal is its antifeeding activity against the African army worm. In addition, it acts as a plant growth regulator and has cytotoxic, antimicrobial, and molluscicidal properties. (See *The Merck Index,* 12th ed., #10173.)

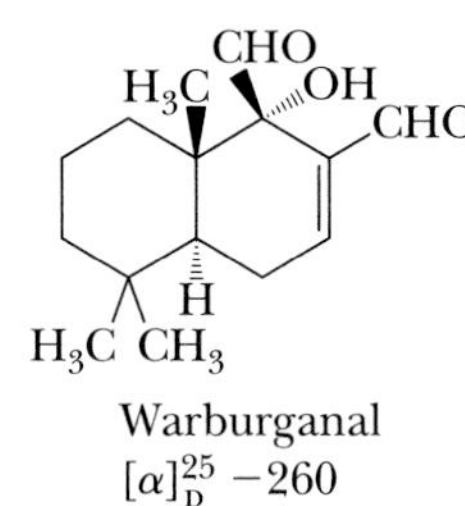

Warburganal
$[\alpha]_D^{25}$ −260

(a) Show that warburganal is a terpene.

(b) Label each stereocenter and specify the number of stereoisomers possible for a molecule of this structure.

Santonin can be isolated from the flower heads of wormwood, *Artemisia absinthium.* Principal habitats are Chinese Turkistan and the Southern Urals. (© *Kenneth J. Stein, Phototake, NYC)*

5.32 α-Santonin, $C_{15}H_{18}O_3$, isolated from the flower heads of certain species of Artemisia, is an anthelmintic, that is, a drug used to rid the body of worms (helminths). It has been estimated that over one third of the world's population is infested with these parasites. (See *The Merck Index,* 12th ed., #8509.)

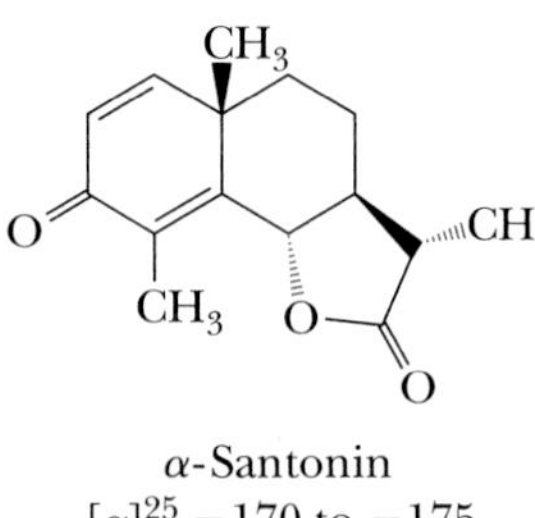

α-Santonin
$[\alpha]_D^{25}$ −170 to −175

(a) Locate the three isoprene units in santonin and show how the carbon skeleton of farnesol might be coiled and then cross-linked to give santonin. Two different coiling patterns of the carbon skeleton of farnesol can lead to santonin. Try to find them both.

(b) Label all stereocenters in santonin. How many stereoisomers are possible for this molecule?

5.33 In many parts of South America, extracts of the leaves and twigs of *Montanoa tomentosa* are brewed with water to make a "tea" used to stimulate menstruation, to facilitate labor, and as an abortifacient. Phytochemical investigations of this plant have resulted in isolation of a very potent fertility-regulating compound called zoapatanol. (See *The Merck Index,* 12th ed., #10318.)

Zoapatanol

(a) Show that the carbon skeleton of zoapatanol can be divided into four isoprene units bonded head-to-tail and then cross-linked in one point along the chain.

(b) Specify the configuration about the carbon-carbon double bond to the seven-membered ring according to the *E,Z* system.

(c) How many stereoisomers are possible for this molecule? In answering this problem, you must consider both *E,Z* isomerism and *R,S* isomerism.

5.34 Pyrethrin II and pyrethrosin are two natural products isolated from plants of the chrysanthemum family. Pyrethrin II is a natural insecticide and is marketed as such. (a) Label all stereocenters in each molecule and all carbon-carbon double bonds about which there is the possibility for *cis-trans* isomerism. (See *The Merck Index,* 12th ed., #8148 and #8149.)

(a) State the number of stereoisomers possible for each molecule.

(b) Show that the bicyclic ring system of pyrethrosin is composed of three isoprene units.

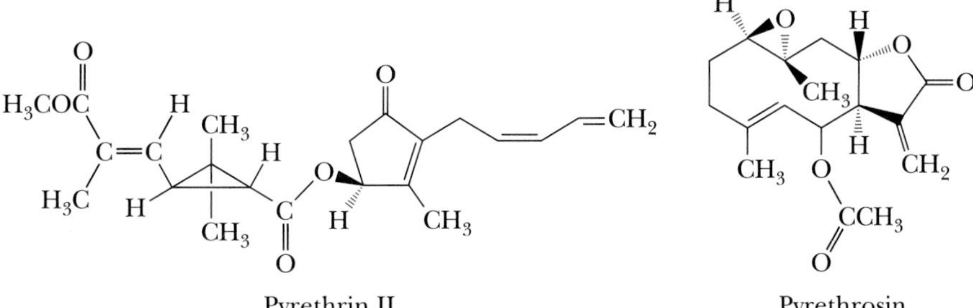

Pyrethrin II Pyrethrosin

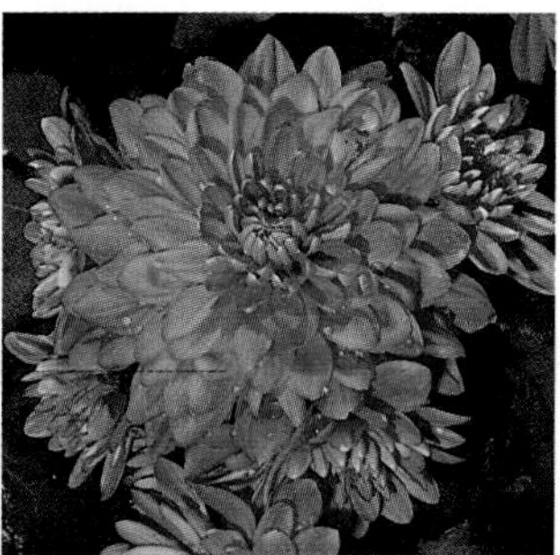

Pyrethrosin and pyrethrin II are natural products that can be isolated from chrysanthemum blossoms.

5.35 Show that the carbon skeletons of the three terpenes drawn in the Chemistry in Action box "Terpenes of the Cotton Plant" can be divided into three isoprene units bonded head-to-tail and then cross-linked at appropriate carbons.

Molecular Modeling

5.36 Construct line-angle drawings of *cis*- and *trans*-2-butene in ChemDraw, import each into Chem3D, and minimize its energy. Measure the CH_3,CH_3 distance in the *cis* isomer and the CH_3,H distance in the *trans* isomer. In which isomer is the nonbonded interaction strain greater?

5.37 Construct line-angle drawings of the *cis*- and *trans* isomers of 2,2,5,5-tetramethyl-3-hexene in ChemDraw, import each into Chem3D, and minimize its energy. Compare the degree of nonbonded interaction strain in these two isomers. Also compare the degree of nonbonded interaction strain in *cis*-2-butene and *cis*-2,2,5,5-tetramethyl-3-hexene.

5.38 Build a line-angle drawing of cyclohexene in ChemDraw, import it into Chem3D, and minimize its energy. Compare the calculated Chem3D bond angles with those predicted by the VSEPR model. Explain any differences.

5.39 Build line-angle drawings of *cis*- and *trans*-cyclooctene in ChemDraw, import each into Chem3D, and minimize its energy. Compare the calculated Chem3D bond angles with those predicted by the VSEPR model. In which isomer are deviations from the VSEPR model predictions greater?

5.40 Build a line-angle drawing of caryophyllene in ChemDraw. Be certain to show the correct stereochemistry, namely the *trans* fusion of the four- and nine-membered rings and the *trans* configuration of the carbon-carbon double bond in the nine-membered ring. As a guide, your structural formula should look like the formula here. Now import your line-angle drawing into Chem3D, minimize its energy, and construct a stereoview. One such stereoview is shown below. Try showing the stereoview with and without hydrogen atoms.

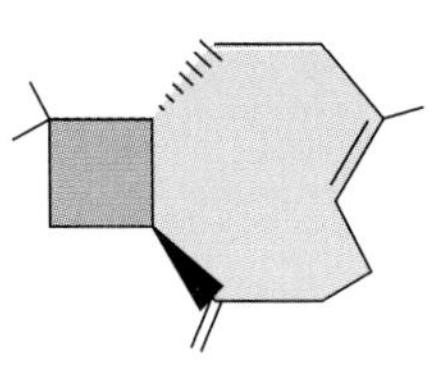

Caryophyllene (structural formula)

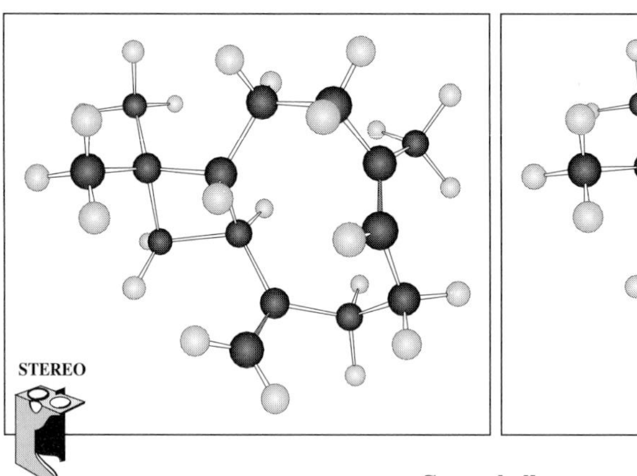

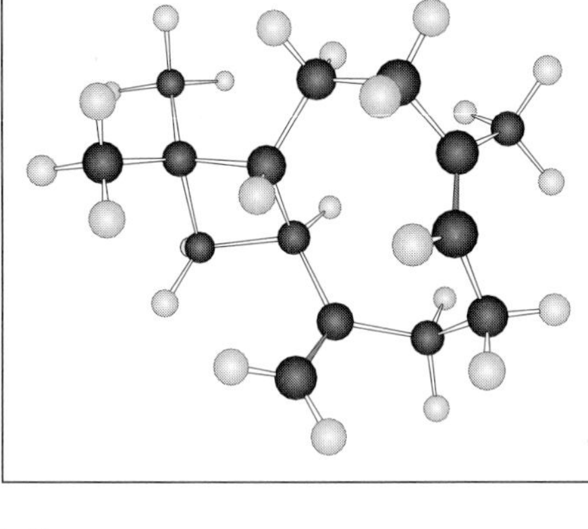

Caryophyllene (stereoview)

A CONVERSATION WITH . . .

CARL DJERASSI

Carl Djerassi, Professor of Chemistry at Stanford University, is a man of many achievements, in science and in letters. Born in Vienna in 1923, Professor Djerassi was forced to leave Europe in 1939 in the face of the Nazi occupation. Although he had not completed high school, he was allowed to enroll at Newark Junior College in New Jersey. He also went to Tarkio College in Missouri and then to Kenyon College where he graduated *summa cum laude* in 1942.

Still short of his nineteenth birthday, he joined the CIBA Corporation, where he and Charles Huttrer synthesized pyribenzamine, one of the first two antihistamines—drugs that act to control allergies. He left CIBA for the University of Wisconsin, where he received his Ph.D. in 1945, at the age of 22. For the next seven years he worked first at CIBA and then at Syntex, S. A., a fledgling pharmaceutical corporation then located in Mexico City. Here he was part of the research team that produced the first synthesis of cortisone, an important hormone, and later synthesized norethindrone, the first oral contraceptive and one that is still widely used all over the world.

In 1952, he moved to Wayne University (now Wayne State) in Detroit, as Associate Professor, and in 1959 ac-

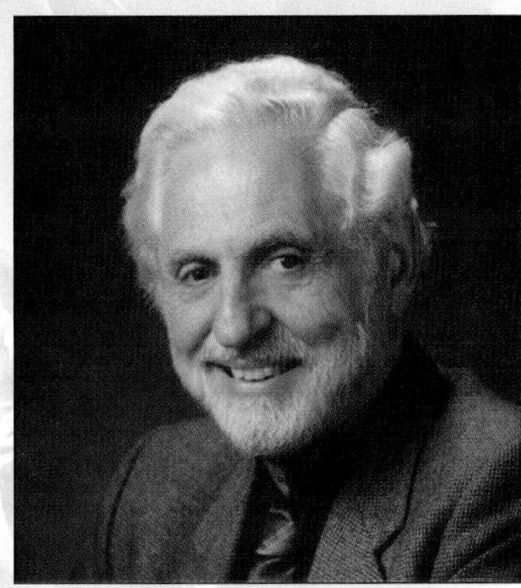

cepted a Professorship at Stanford University, where he has remained ever since. Dr. Djerassi has always been involved in steroid chemistry,* but his early interest in synthesis turned to the equally important area of the structural determination of naturally occurring steroids, of which there are many thousands. When he began his career, determining the structure of a single naturally occurring steroid was a difficult process that would often take years.

In the 1950s, instrumental methods of structure determination slowly began to replace the laborious, older chemical methods. Dr. Djerassi was one of the pioneers in developing and using techniques such as optical rotatory dispersion, circular dichroism, and mass spectrometry to determine the structures of steroids and other organic compounds. Because of the work of Dr. Djerassi and other pioneers, most unknown structures now can be determined in a few days or less.

In recent years Dr. Djerassi, without abandoning his interest in chemistry, has turned to literature. He has published a collection of short stories; two novels: *Cantor's Dilemma* and *The Bourlaxi Gambit;* a collection of poems; and two books of autobiography, one entitled *Steroids Made It Possible*, and the other, intended for a more general audience, called *The Pill, Pygmy Chimps, and Degas's Horse*. Under the auspices of the Djerassi Foundation, he has established an artist's colony near Woodside, California, that provides residences and studio space for visual, literary, choreographic, and musical artists.

CHOOSING A CAREER IN CHEMISTRY

"I really didn't choose to be an organic chemist; it sort of happened to me. I was the child of two physicians and it was always assumed that

*Steroids are discussed in Chapter 25.

I would go to medical school and become a practicing physician. When I arrived in this country at the age of 16 I had taken no chemistry and had not even graduated from high school. Fortunately, I got straight into a Junior College in New Jersey and began a pre-med curriculum that included first-year chemistry. I had a first-rate teacher there, named Nathan Washton, who got me interested in chemistry. After one semester at Tarkio College in Missouri (the alma mater of Wallace Carothers, the inventor of nylon), I went to Kenyon College, a small (at that time) college in Ohio, that had only two faculty members in chemistry. Chemistry classes were very small, but I got a first-class education, and that is where I decided to become a chemist, rather than go to medical school."

Synthesizing Antihistamines

"When I graduated from Kenyon I needed to earn some money, so I got a job as a junior chemist at CIBA Pharmaceutical Corporation, in New Jersey. At that time the company became interested in antihistamines, and I was one of only two chemists working on this project. So despite my youth I was involved in the synthesis of pyribenzamine, one of the first two antihistamines produced in this country. This compound, which was synthesized during the first year after I graduated from college, turned out to make a significant contribution in the treatment of allergies."

From Antihistamines to Steroids

"While working at CIBA, I began taking classes at NYU and Brooklyn Polytechnic Institute at night, so as to get an advanced degree, but commuting from New Jersey after a day of work was murderous. After a year of night school, I decided I would go to graduate school full time. CIBA was involved in steroid projects, and even though I was working on antihistamines, I started reading books on steroids, especially Louis Fieser's *Natural Products Related to Phenanthrene,* a superb book, which turned me on to steroid chemistry. When I went full time to the University of Wisconsin, I was prepared to work in this field.

> Because of the work of Dr. Djerassi and other pioneers, most unknown structures now can be determined in a few days or less.

Fortunately, Wisconsin had two young Assistant Professors in this area, and I did my Ph.D. with one of them, A. L. Wilds, on the conversion of androgens to estrogens, which at that time was a tough problem."

Developing Oral Contraceptives

"We did not set out with the objective of synthesizing oral contraceptives. Our goal was to develop an orally-active progestational hormone, in other words a compound that would mimic the biological properties of progesterone. At that time progesterone was clinically used for menstrual disorders and infertility, but there were ideas about using it as a contraceptive, because it is progesterone that naturally stops further ovulation after an ovum is fertilized. However, progesterone itself is not active by mouth, and daily injections would be needed. By this time I was Associate Director of Chemical Research at Syntex Corporation, located in Mexico City. By combining ideas discovered by previous investigators, we set out to synthesize a steroid that would not only be active by mouth, but would also have enhanced progestational activity. This compound was 19-nor-17 α-ethynyltestosterone (norethindrone), whose synthesis we completed on October 15, 1951. It was first tested for menstrual disorders and fertility problems and then as an oral contraceptive. Forty years after its synthesis it is still the active ingredient of about a third of all the oral contraceptives used throughout the world."

The Social Impact of Working on Contraceptives

"If I could do it over again, there is no question that I would work in the area of developing and synthesizing oral contraceptives. By now about 13–14 million women in the U.S. and 50–60 million in the world use the pill, making it the most widely used method of reversible birth control. Population growth is probably the biggest problem facing us in the world, assuming that the possibility of nuclear warfare is now greatly diminished, and the widespread use of oral contraceptives helps in controlling that. I think the development of these contraceptives is one of the most important contributions that chemistry has made to society. New drugs cure diseases of individuals during their lifetimes, but the use of contraceptives has implications for generations, because if you do not control the production of offspring, you do not control future generations. Furthermore, these compounds have had an enormous impact on women, empowering them to be in control of their own fertility."

From Oral Contraceptives to Insect Control

"Conceptually there was a relationship between our work on oral contraceptives and insect control. In a way, you could say that steroid oral contraceptives were true biorational methods of human birth control, since progesterone—our conceptional lead compound—is really nature's contraceptive. That was a model on which insect control could be based. At this time (the late 1960s) I was in charge of research at Syntex in addition to being Professor at Stanford University. Governments and the public realized that conventional methods of insect control—largely spraying with chlorinated hydrocarbons such as DDT—were damaging the environment. DDT and similar compounds were being banned and a new approach was needed. We formed a new company, called Zoecon, to try to synthesize insect controlling steroids. In the 1960s, a juvenile hormone, based on a sesquiterpene skeleton, had been discovered and we decided to focus on it. Insects pass through a juvenile stage controlled by the juvenile hormone, whose production is later shut off by another hormone so that the insect can then mature. Our biorational approach was to synthesize an artificial juvenile hormone which would continue to be applied to immature insects, so that the insect would never reach the stage at which it could reproduce. This turned out to be a new biorational approach to controlling mosquitoes, fleas, cockroaches, and other insects that do their damage as adults, and was approved by the Environmental Protection Agency for public use."

Passing the Torch

"I've trained somewhere between three hundred and four hundred graduate students and post-doctoral fellows in universities and many others in industry as well. Training these students is probably the biggest professional contribution that one makes. These are very intimate relationships, comparable to those between parents and children. The mentor becomes a role model for the student."

The Rewards of Scientific Research

"In my own work, material rewards never played any initial role, because they always came much later. To a large extent it was initially pure intellectual curiosity, but very soon I started to look for the potential benefits to society. My interest has always been in the biological areas and I was never interested in war-related research. I never had the slightest question about the societal appropriateness of the work I was doing."

The Interplay Between Academia and Industry in Developing New Inventions, Including Chemical Inventions

"I have always been connected simultaneously with a university and with chemical companies. This made me a much better Professor, because I became aware of the many steps needed to take a laboratory discovery up to practical realization. Conversely, I was a much better industrial research director because of what I learned in the university. For example, should scientists be concerned with societal implications of their work? Obviously yes, but it is not easy for a scientist who stays only in the ivory tower. You ought to have a responsibility for taking your work a step further. There should however be guidelines to prevent actual or potential conflicts of interest."

6 ALKENES II

■ Polyethylene wash bottles. *(Charles D. Winters)*

In this chapter, we begin our systematic study of one of the most important unifying concepts in organic chemistry, namely the study of how reactions of organic molecules occur or, to use the terminology of chemists, the study of **reaction mechanisms.** We use the reactions of alkenes as the vehicle by which to introduce this concept.

6.1 Reactions of Alkenes: An Overview

The most characteristic reaction of alkenes is **addition to the carbon-carbon double bond** in such a way that the pi bond is broken and in its place sigma bonds are formed to two new atoms or groups of atoms. Several examples of reactions at the carbon-carbon double bond are shown in Table 6.1 along with the descriptive name(s) associated with each. In the following sections, we study these alkene reactions in considerable detail, with particular attention to the mechanisms by which they occur.

A second characteristic reaction of alkenes is the formation of **step-growth polymers** (Greek, *poly*, many, and *meros*, part). In the presence of certain catalysts called initiators, many alkenes form polymers made by the addition of **monomers** (Greek, *mono*, one, and *meros*, part) to a growing polymer chain as illustrated by the formation of polyethylene from ethylene. In alkene polymers of industrial and commercial importance, n is a large number, typically several thousand.

$$n\mathrm{CH_2{=}CH_2} \xrightarrow{\text{initiator}} \mathrm{-\!\!(CH_2CH_2)_{\overline{n}}}$$

We discuss this alkene reaction in Chapter 23.

6.2 Reaction Mechanisms

A **reaction mechanism** describes in detail how a reaction occurs. It describes which bonds are broken and which new ones are formed as well as the order and relative rates of the various bond-breaking and bond-forming steps. If the reaction takes place

Reaction mechanism A step-by-step description of how a chemical reaction occurs.

TABLE 6.1 Characteristic Alkene Addition Reactions

Reaction	Descriptive Name(s)
$\mathrm{C{=}C} + \mathrm{HCl}\ (\mathrm{HX}) \longrightarrow -\overset{\vert}{\underset{\mathrm{H}}{\mathrm{C}}}-\overset{\vert}{\underset{\mathrm{Cl\ (X)}}{\mathrm{C}}}-$	hydrochlorination (hydrohalogenation)
$\mathrm{C{=}C} + \mathrm{H_2O} \longrightarrow -\overset{\vert}{\underset{\mathrm{H}}{\mathrm{C}}}-\overset{\vert}{\underset{\mathrm{OH}}{\mathrm{C}}}-$	hydration
$\mathrm{C{=}C} + \mathrm{Br_2}\ (\mathrm{X_2}) \longrightarrow -\overset{\vert}{\underset{\mathrm{(X)\ Br}}{\mathrm{C}}}-\overset{\vert}{\underset{\mathrm{Br\ (X)}}{\mathrm{C}}}-$	bromination (halogenation)
$\mathrm{C{=}C} + \mathrm{BH_3} \longrightarrow -\overset{\vert}{\underset{\mathrm{H}}{\mathrm{C}}}-\overset{\vert}{\underset{\mathrm{BH_2}}{\mathrm{C}}}-$	hydroboration
$\mathrm{C{=}C} + \mathrm{OsO_4} \longrightarrow -\overset{\vert}{\underset{\mathrm{HO}}{\mathrm{C}}}-\overset{\vert}{\underset{\mathrm{OH}}{\mathrm{C}}}-$	hydroxylation (oxidation)
$\mathrm{C{=}C} + \mathrm{H_2} \longrightarrow -\overset{\vert}{\underset{\mathrm{H}}{\mathrm{C}}}-\overset{\vert}{\underset{\mathrm{H}}{\mathrm{C}}}-$	hydrogenation (reduction)

in solution, it describes the role of the solvent. If the reaction involves a catalyst, it describes the role of the catalyst. A complete reaction mechanism describes the positions of all atoms during the reaction and the energy of the entire system during the course of the reaction. This ideal can rarely be approached in practice, however.

A. Potential Energy Diagrams and Transition States

Let us consider a general reaction in which A and B react to give C and D. The **total energy** *E* of the reacting molecules is the sum of their **kinetic energy** (KE) and **potential energy** (PE).

$$A + B \longrightarrow C + D \qquad E_{total} = KE + PE$$

Potential energy (PE) diagram A graph showing the changes in energy that occur during a chemical reaction; potential energy is plotted on the vertical axis and reaction progress is plotted on the horizontal axis.

Reaction coordinate A measure of the change in positions of atoms during a reaction, plotted on the horizontal axis in a potential energy diagram.

Reactants A and B have a certain kinetic energy. As they collide, a part of the kinetic energy is absorbed and converted to potential energy in the form of vibrational motions of bonds. To understand the relationship between chemical reaction and potential energy, think of a chemical bond as a spring. As a spring is stretched from its resting position, its potential energy is increased. As it returns to its resting position, its potential energy is decreased. Similarly, during a chemical reaction, bond breaking corresponds to an increase in potential energy and bond forming corresponds to a decrease in potential energy. We use a **potential energy diagram** to show the changes in energy that occur in going from reactants to products. Potential energy is measured on the vertical axis and the change in position of the atoms during the reaction is represented on the horizontal axis, called the **reaction coordinate.** Figure 6.1 shows a potential energy diagram for the proton transfer reaction that occurs when hydrogen chloride is dissolved in water. This reaction occurs in one step, meaning that bond breaking in starting materials and bond forming to give the observed products occur simultaneously.

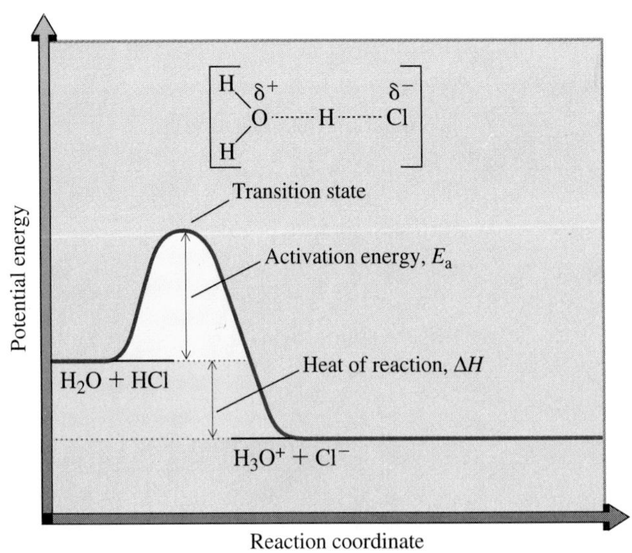

FIGURE 6.1
A potential energy diagram for the proton transfer reaction between HCl and H_2O. The dashed lines in the transition state indicate that the new O—H bond is partially formed and the H—Cl bond is partially broken. The energy of the reactants is higher than that of products, and heat is released in this reaction; the reaction is exothermic.

The difference in potential energy between the reactants and products is called the **heat of reaction.** If the energy of products is lower than that of reactants, heat is released; the reaction is **exothermic.** If the energy of products is higher than that of reactants, heat is absorbed; the reaction is **endothermic.** The one-step proton transfer reaction shown in Figure 6.1 is exothermic; the energy of the H_3O^+ and Cl^- is lower than that of H_2O and HCl.

Exothermic reaction A reaction in which the energy of the products is lower than the energy of the reactants; a reaction in which heat is liberated.

Endothermic reaction A reaction in which the energy of the products is higher than the energy of the reactants; a reaction in which heat is absorbed.

A **transition state** is the point on the reaction coordinate at which the potential energy is a maximum. At the transition state, sufficient potential energy has become concentrated in the proper bonds so that bonds in reactants break and new bonds form, giving products. Once the transition state is reached, the reaction proceeds to give products with the release of energy. A transition state has a definite geometry, a definite arrangement of bonding and nonbonding electrons, and a definite distribution of electron density and charge. Because a transition state is at an energy maximum on a potential energy diagram, it cannot be isolated and its structure cannot be determined experimentally. However, as we shall see soon, even though we cannot observe a transition state directly by any experimental means, we can often infer a great deal about its probable structure from other experimental observations.

Transition state An unstable species of maximum potential energy formed during the course of a reaction; an energy maximum on a potential energy diagram.

For the proton-transfer reaction illustrated in Figure 6.1, we use dashed lines to show the partial bonding in the transition state. As an unshared pair of electrons on oxygen begins to form a new covalent bond with hydrogen (shown by the dashed line), oxygen develops a partial positive charge. Conversely, as the bonding pair of electrons in H—Cl becomes transferred to chlorine (shown by another dashed line), chlorine develops a partial negative charge.

The difference in potential energy between reactants and the transition state is called the **activation energy,** E_a. E_a determines the rate of a reaction, that is, how fast the reaction occurs. If E_a is large, only a very few molecular collisions occur with sufficient energy to reach the transition state, and the reaction is slow. If E_a is small, many collisions generate sufficient energy to reach the transition state and the reaction is fast.

Activation energy The difference in potential energy between reactants and the transition state.

In a reaction that occurs in two or more steps, each step has its own transition state and activation energy. Shown in Figure 6.2 is a potential energy diagram for

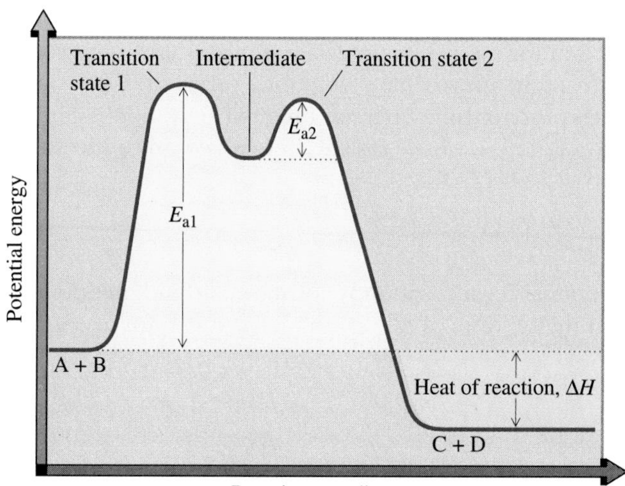

FIGURE 6.2
Potential energy diagram for a two-step reaction involving formation of an intermediate. The energy of the reactants is higher than that of the products, and heat is released in the conversion of A + B to C + D. The reaction is exothermic.

Intermediate A species, formed during a two-step reaction, that lies in a potential energy minimum between two transition states.

Rate-limiting step The step in a multistep reaction sequence that crosses the highest potential energy barrier.

conversion of reactants to products in two steps. An **intermediate** corresponds to a potential energy minimum between two transition states, in this case between transition state 1 and transition state 2. Note that because the energies of the intermediates we describe in this chapter are higher than that of either reactants or products, they are highly reactive, and rarely if ever can such an intermediate be isolated.

The slowest step in a multistep reaction, called the **rate-limiting step,** is the step that crosses the highest potential energy barrier. In the two-step sequence shown in Figure 6.2, step 1 crosses the higher energy barrier and is, therefore, the rate-limiting step.

B. Activation Energy and Rate

The relationship between **rate constant, *k*,** and activation energy for most chemical reactions is given by the following equation, known as the Arrhenius equation.

frequency of collisions with the correct geometry required for reaction

$$k = \text{reaction rate constant} = Ae^{-E_a/RT}$$

the fraction of molecules with the minimum energy required for reaction

where ***A*** is a frequency factor in units of $\sec^{-1}$, $\boldsymbol{E_a}$ is the activation energy in kcal/mol, ***R*** is the gas constant with a value of 1.987×10^{-3} kcal·mol^{-1}·deg^{-1}, and ***T*** is the temperature in degrees Kelvin.

A great many organic reactions we deal with in this course have activation energies in the range of 10–35 kcal/mol (42–146 kJ/mol). Those with activation energies of 10–20 kcal/mol (42–84 kJ/mol) proceed rapidly at room temperature. Those with higher activation energies require heating to provide more molecules with sufficient energy to overcome the activation energy barrier.

EXAMPLE 6.1

An often stated generalization is that for many reactions taking place at or near room temperature (25°C), the rate of reaction approximately doubles for every 10°C rise in temperature. This generalization is valid for only a narrow range of activation energies. What is the activation energy for a reaction whose rate at 35°C is two times its rate at 25°C?

Solution

If we assume that the activation energy for a reaction is constant and independent of temperature, then we can determine the ratio of rate constants k_2 and k_1 for a reaction run at temperatures T_2 and T_1.

$$\frac{k_2}{k_1} = \frac{Ae^{-E_a/RT_2}}{Ae^{-E_a/RT_1}}$$

Taking the logarithm of each side of the equation, converting to base 10, and rearranging terms gives the following equation.

$$\log \frac{k_2}{k_1} = \frac{E_a}{2.303R}\left(\frac{1}{T_1} - \frac{1}{T_2}\right)$$

The reaction is carried out at 25°C (298K) and 35°C (308K), respectively. Substituting values of the relative rate constants, T_2, T_1, and R in the Arrhenius equation and solving for the activation energy gives:

$$\log \frac{2}{1} = \frac{E_a(\text{kcal/mol})}{2.303 \times 1.987 \times 10^{-3}\ \text{kcal}\cdot\text{mol}^{-1}\cdot\text{K}^{-1}}\left(\frac{1}{298\text{K}} - \frac{1}{308\text{K}}\right)$$

$$E_a = 12.6\ \text{kcal/mol}\ (52.7\ \text{kJ/mol})$$

PROBLEM 6.1

Suppose that the activation energy for a particular chemical reaction is 25.2 kcal/mol. By what factor is the rate of reaction increased when the reaction takes place at 35°C compared with the rate at 25°C?

EXAMPLE 6.2

If the rate constants for two reactions, each taking place at 25°C and each with the same value of A, differ by a factor of 10, what is the difference in their energies of activation in kcal/mol?

Solution

The following relationship, derived from the Arrhenius equation, relates rate constants k_2 and k_1 of two reactions to their relative energies of activation.

$$\Delta E_a = -2.303\ RT \log \frac{k_2}{k_1}$$

Substituting the ratio of 10/1 in the derived Arrhenius equation, we find that their energies of activation differ by approximately 1.4 kcal/mol.

$$\Delta E_a = -2.303 \times \frac{1.987 \times 10^{-3}\ \text{kcal}}{\text{mol} \times \text{K}} \times 298\text{K} \times \log \frac{10}{1} = -1.36\ \text{kcal/mol}$$

PROBLEM 6.2

Complete the first three entries in this table for reactions taking place at 25°C. Given the pattern of these first three entries, estimate the approximate values for the remaining two entries. How many kilocalories per mole in activation energy corresponds to a power of 10 in relative rates?

ΔE_a (kcal/mol)	$\frac{k_2}{k_1}$
______	1
______	10
______	100
______	1000
______	10000

C. Developing a Reaction Mechanism

To develop a reaction mechanism, chemists begin by designing experiments that will reveal details of a particular chemical reaction. Next, through a combination of experience and intuition, they propose several sets of steps or mechanisms, each of which might account for the overall chemical transformation. Finally, each proposed mechanism is tested against the experimental observations to exclude those mechanisms that are not consistent with the facts. A mechanism becomes generally established by excluding reasonable alternatives and by showing that it is consistent with every test that can be devised. This, of course, does not mean that a generally accepted mechanism is a completely accurate description of the chemical events, but only that it is the best chemists have been able to devise. It is important to keep in mind that, as new experimental evidence is obtained, it may be necessary to modify a generally accepted mechanism or possibly even discard it and start all over again.

Before we go on to consider reactions and reaction mechanisms, we might ask why it is worth the trouble to establish them and worth your time to learn about them. One reason is very practical: Mechanisms provide a framework within which to organize a great deal of descriptive chemistry. For example, with insight into how reagents add to particular alkenes, it is possible to make generalizations and then to predict how the same reagents might add to other alkenes. A second reason lies in the intellectual satisfaction derived from constructing models that accurately reflect the behavior of chemical systems. Finally, to a creative scientist, a mechanism is a tool to be used in the search for new information and new understanding. A mechanism consistent with all that is known about a reaction can be used to make predictions about chemical interactions as yet unexplored and experiments can be designed to test these predictions. Thus, reaction mechanisms provide a way not only to organize knowledge but also to extend it.

6.3 Electrophilic Additions

We begin our introduction to the chemistry of alkenes with an examination of five addition reactions, namely, addition of hydrogen halides (HCl, HBr, and HI), water (H_2O), halogens (Cl_2 and Br_2), mercuric acetate [$Hg(OAc)_2$] in the presence of water, and finally Cl_2 and Br_2 in the presence of water. We first study some of the experimental observations about each addition reaction and then its mechanism. Through the study of these particular reactions, we will develop a general understanding of how alkenes undergo addition reactions.

A. Addition of Hydrogen Halides

The hydrogen halides HCl, HBr, and HI add to alkenes to give haloalkanes (alkyl halides). These additions may be carried out either with the pure reagents (neat) or in the presence of a polar solvent such as acetic acid. Addition of HCl to ethylene gives chloroethane (ethyl chloride).

$$\underset{\text{Ethylene}}{CH_2{=}CH_2} + HCl \longrightarrow \underset{\text{Chloroethane}}{\overset{H}{\overset{|}{C}}H_2-\overset{Cl}{\overset{|}{C}}H_2}$$

Addition of HCl to propene gives 2-chloropropane (isopropyl chloride); hydrogen adds to carbon 1 of propene and chlorine adds to carbon 2. If the orientation of addition were reversed, 1-chloropropane (propyl chloride) would be formed. The observed result is that 2-chloropropane is formed to the virtual exclusion of 1-chloropropane. We say that addition of HCl to propene is highly regioselective. A **regioselective reaction** is a reaction in which one direction of bond forming or breaking occurs preferentially to all other directions.

Regioselective reaction A reaction in which one direction of bond forming or bond breaking occurs in preference to all other directions.

$$CH_3CH{=}CH_2 + HCl \longrightarrow CH_3\underset{}{\overset{Cl}{\overset{|}{C}}}H{-}\overset{H}{\overset{|}{C}}H_2 + CH_3\overset{H}{\overset{|}{C}}H{-}\overset{Cl}{\overset{|}{C}}H_2$$

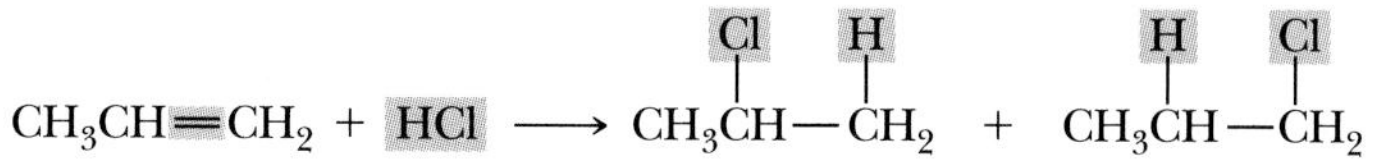

Propene — 2-Chloropropane — 1-Chloropropane (not observed)

This regioselectivity was noted by Vladimir Markovnikov who made the generalization known as **Markovnikov's rule;** in addition of H—X to alkenes, hydrogen adds to the double-bonded carbon that has the greater number of hydrogens already attached to it. Although Markovnikov's rule provides a way to predict the product of many alkene addition reactions, it does not explain why one product predominates over other possible products.

Markovnikov's rule In the addition of HX or H_2O to an alkene, hydrogen adds to the carbon of the double bond having the greater number of hydrogens.

EXAMPLE 6.3

Name and draw the structural formula for the product of each alkene addition reaction.

(a) 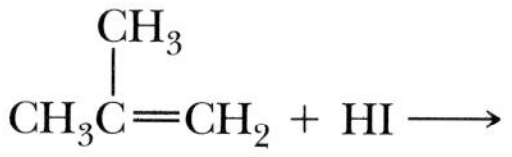 $CH_3\overset{CH_3}{\overset{|}{C}}{=}CH_2 + HI \longrightarrow$ **(b)** 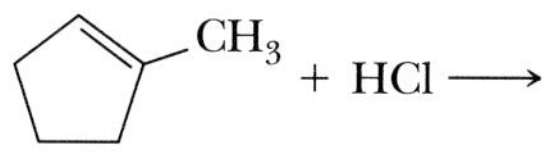 1-methylcyclopentene (ring with CH_3 on the double-bonded carbon) $+ HCl \longrightarrow$

Solution

Using Markovnikov's rule, predict that 2-iodo-2-methylpropane is the product in (a) and 1-chloro-1-methylcyclopentane is the product in (b).

(a) 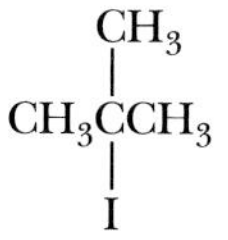 $CH_3\underset{I}{\underset{|}{\overset{CH_3}{\overset{|}{C}}}}CH_3$ **(b)** 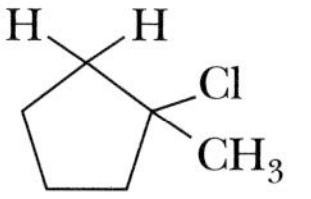 cyclopentane ring bearing Cl and CH_3 on one carbon (H, H shown on adjacent carbon)

2-Iodo-2-methylpropane — 1-Chloro-1-methylcyclopentane

PROBLEM 6.3

Name and draw the structural formula for the product of each alkene addition reaction.

(a) $CH_3CH{=}CH_2 + HI \longrightarrow$ **(b)** 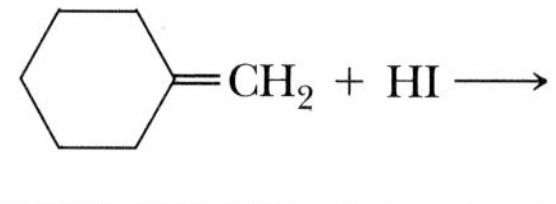 cyclohexane ring$=CH_2 + HI \longrightarrow$

Chemists account for the addition of HX to an alkene by a two-step mechanism, which we illustrate by the reaction of 2-butene with hydrogen chloride to give 2-chlorobutane. Let us first look at this two-step mechanism in overview and then go back and study each step in detail. In overview, addition begins with transfer of a proton from HCl to 2-butene, as shown by the two curved arrows on the left side of Step 1. The first curved arrow shows that the pi bond of the alkene is broken and its electron pair used to form a new covalent bond with the hydrogen atom of HCl. The second curved arrow shows that the polar covalent bond in HCl is broken and its electron pair given entirely to chlorine, forming chloride ion. The first step in this mechanism results in formation of an organic cation and chloride ion. The second step is reaction of the organic cation with chloride ion to form a 2-chlorobutane.

MECHANISM Electrophilic Addition of HCl to 2-Butene

Step 1: Formation of the *sec*-butyl cation, a 2° carbocation intermediate

$$CH_3CH{=}CHCH_3 + \overset{\delta+}{H}{-}\overset{\delta-}{\ddot{C}l}: \xrightarrow[\text{limiting step}]{\text{slow, rate-}} CH_3\overset{+}{C}H{-}\underset{}{\overset{H}{C}}HCH_3 + :\ddot{C}l:^-$$

sec-Butyl cation
(a 2° carbocation
intermediate)

Step 2: Reaction of the *sec*-butyl cation with chloride ion

$$:\ddot{C}l:^- + CH_3\overset{+}{C}HCH_2CH_3 \xrightarrow{\text{fast}} CH_3\overset{:\ddot{C}l:}{C}HCH_2CH_3$$

Chloride ion (a Lewis base) — *sec*-Butyl cation (a Lewis acid)

Now that you have looked at this two-step mechanism in overview, let us go back and look at the individual steps in more detail. There is a great deal of important organic chemistry embedded in these two steps and it is important that you understand it now.

Carbocation A species containing a carbon atom with only six electrons in its valence shell and bearing a positive charge.

Electrophile Any species that can accept a pair of electrons to form a new covalent bond; a Lewis acid.

Step 1 results in formation of an organic cation. One carbon atom in this cation has only six electrons in its valence shell and carries a charge of +1. A species containing a positively charged carbon atom is called a **carbocation** (*carbon* + *cation*). Such carbon-containing cations have also been called carbonium ions and carbenium ions. While these terms are little used now, you will see them in the older literature. Carbocations are classified as **primary (1°), secondary (2°)**, or **tertiary (3°)** depending on the number of carbon atoms bonded to the carbon bearing the positive charge. All carbocations are Lewis acids (Section 3.5). They are also electrophiles. The term **electrophile** quite literally means "electron lover." An electrophile is any species that can accept a pair of electrons to form a new covalent bond.

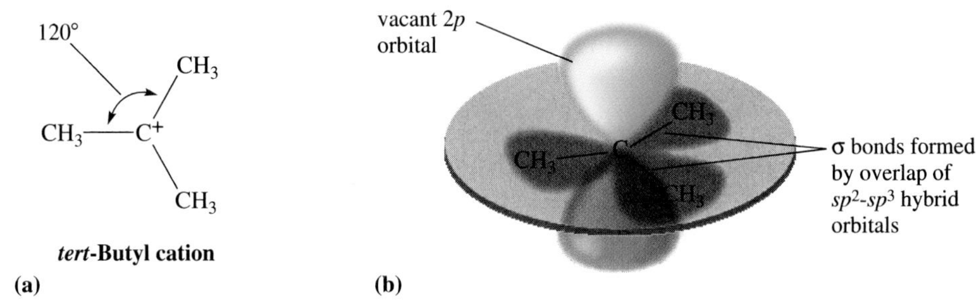

FIGURE 6.3
The structure of the *tert*-butyl cation. (a) Lewis structure and (b) an orbital picture.

In a carbocation, the carbon bearing the positive charge is bonded to three other atoms, and as predicted by the valence-shell electron-pair repulsion (VSEPR) model, the three bonds are coplanar and form angles of 120°. According to the molecular orbital model, the electron-deficient carbon of a carbocation uses sp^2 hybrid orbitals to form sigma bonds to the three attached groups. The unhybridized $2p$ orbital lies perpendicular to the sigma bond framework and contains no electrons. A Lewis structure and orbital overlap diagram for the *tert*-butyl cation are shown in Figure 6.3.

Figure 6.4 shows a potential energy diagram for the two-step reaction of 2-butene with HCl. The slower, rate-limiting step (the one that crosses the higher potential energy barrier) is the first one, which leads to formation of the 2° carbocation intermediate. This carbocation intermediate lies in an energy minimum between the transition states for Steps 1 and 2. As soon as the carbocation intermediate (a Lewis acid) is formed, it reacts with chloride ion (a Lewis base) in a Lewis acid-base reaction to give 2-chlorobutane. Note that the energy level for 2-chlorobutane (the product) is lower than the energy level for 2-butene and HCl (the reactants). Thus, in this alkene addition reaction, heat is released; the reaction is exothermic.

Relative Stabilities of Carbocations: Regioselectivity and Markovnikov's Rule

Reaction of HX and an alkene can, at least in principle, give two different carbocation intermediates depending on which of the doubly bonded carbons forms a bond with H^+, as illustrated by the reaction of HCl with propene.

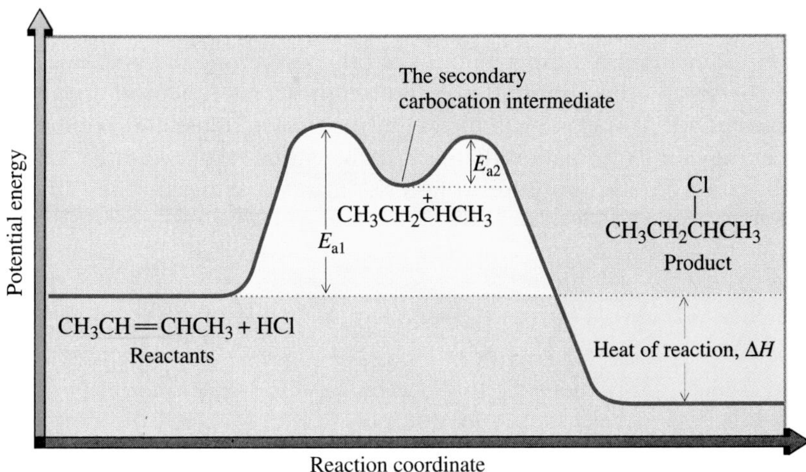

FIGURE 6.4
An energy diagram for the two-step addition of HCl to 2-butene. The reaction is exothermic.

$$CH_3CH{=}CH_2 + H{-}\ddot{Cl}: \longrightarrow CH_3CH_2\overset{+}{C}H_2 \xrightarrow{:\ddot{Cl}:^-} CH_3CH_2CH_2\ddot{Cl}:$$

Propene — Propyl cation (a 1° carbocation) — 1-Chloropropane (not formed)

$$CH_3CH{=}CH_2 + H{-}\ddot{Cl}: \longrightarrow CH_3\overset{+}{C}HCH_3 \xrightarrow{:\ddot{Cl}:^-} CH_3CH(Cl)CH_3$$

Propene — Isopropyl cation (a 2° carbocation) — 2-Chloropropane (product formed)

The propyl cation is a 1° carbocation and the isopropyl cation is a 2° carbocation. The propyl cation reacts with chloride ion to give 1-chloropropane and the isopropyl cation reacts with chloride ion to give 2-chloropropane. The observed product is 2-chloropropane, indicating that the 2° carbocation is formed in preference to the 1° carbocation.

Similarly, in the reaction of HCl with 2-methylpropene, addition of H^+ to the carbon-carbon double bond might form either an isobutyl cation (a 1° carbocation) or a *tert*-butyl cation (a 3° carbocation).

$$CH_3C(CH_3){=}CH_2 + H{-}\ddot{Cl}: \longrightarrow CH_3CH(CH_3)\overset{+}{C}H_2 \xrightarrow{:\ddot{Cl}:^-} CH_3CH(CH_3)CH_2\ddot{Cl}:$$

2-Methylpropene — Isobutyl cation (a 1° carbocation) — 1-Chloro-2-methylpropane (not formed)

$$CH_3C(CH_3){=}CH_2 + H{-}\ddot{Cl}: \longrightarrow CH_3\overset{+}{C}(CH_3)CH_3 \xrightarrow{:\ddot{Cl}:^-} CH_3C(CH_3)(Cl)CH_3$$

2-Methylpropene — *tert*-Butyl cation (a 3° carbocation) — 2-Chloro-2-methylpropane (product formed)

Reaction of the isobutyl cation with chloride ion gives 1-chloro-2-methylpropane (isobutyl chloride); reaction of the *tert*-butyl cation with chloride ion gives 2-chloro-2-methylpropane (*tert*-butyl chloride). The observed product is 2-chloro-2-methylpropane, indicating that the 3° carbocation is formed in preference to the 1° carbocation.

From such experiments and a great amount of other experimental evidence, we know that a 3° carbocation is both more stable and requires a lower activation energy for its formation than a 2° carbocation, which is in turn more stable and requires a lower activation energy for its formation than a 1° carbocation. It follows, then, that a more stable carbocation intermediate forms faster than a less stable carbocation intermediate. Following is the order of stability of four types of alkyl carbocations.

$H{-}\overset{+}{C}H_2$	$CH_3{-}\overset{+}{C}H_2$	$CH_3{-}\overset{+}{C}H(CH_3)$	$CH_3{-}\overset{+}{C}(CH_3)_2$
Methyl cation (methyl)	Ethyl cation (1°)	Isopropyl cation (2°)	*tert*-Butyl cation (3°)

Order of increasing carbocation stability →

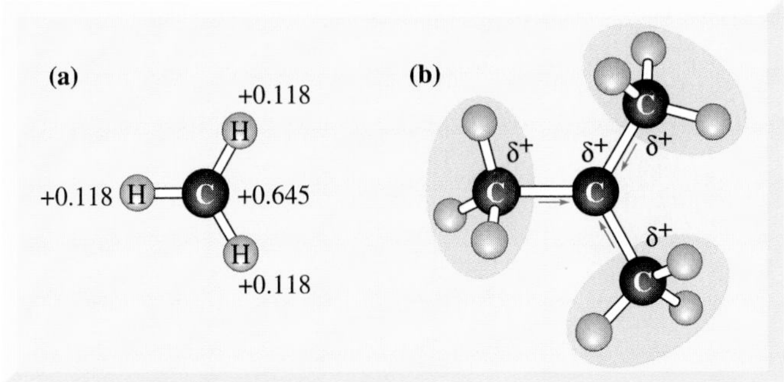
(a)
+0.118
H
+0.118 H
C +0.645
H
+0.118
(b)
δ+
δ+
δ+
δ+
C
C
C
C

FIGURE 6.5
Delocalization of charge by the electron-withdrawing inductive effect of the trivalent, positively charged carbon. (a) The distribution of positive charge in the methyl cation according to molecular orbital calculations, and (b) the *tert*-butyl cation.

Now that we know the order of stability of carbocations, how do we account for that order? The principles of physics teach us that a system bearing a charge (either positive or negative) is more stable if the charge is delocalized. Using this principle, we can explain the order of stability of carbocations if we make the assumption that alkyl groups attached to positively charged carbon are electron releasing and thereby help delocalize the charge on the cation. The electron-releasing ability of alkyl groups bonded to a cationic carbon is accounted for by two effects: the inductive effect and hyperconjugation.

The inductive effect operates in the following way. The electron deficiency of the carbon atom bearing a positive charge exerts an electron-withdrawing **inductive effect** that polarizes electrons from adjacent sigma bonds towards it. Thus, the positive charge of the cation is not localized on the trivalent carbon, but rather delocalized over nearby atoms. The larger the volume over which the positive charge is delocalized, the greater the stability of the cation. Thus, as the number of alkyl groups bonded to the cationic carbon increases, the stability of the cation increases. The electron-withdrawing inductive effect of the positively charged carbon and the resulting delocalization of charge are illustrated in Figure 6.5. According to quantum mechanical calculations, the charge on carbon in the methyl cation is approximately +0.645, and the charge on each of the hydrogen atoms is +0.118. Thus, even in the methyl cation, the positive charge is not localized on carbon. Rather, it is delocalized over the volume of space occupied by the entire ion. Polarization of electron density and delocalization of charge is even more extensive in the *tert*-butyl cation.

The second effect operating to stabilize carbocations is hyperconjugation. **Hyperconjugation** involves partial overlap of the sigma bonding orbital of an adjacent C—H or C—C bond with the vacant 2*p* orbital of the cationic carbon (Figure 6.6).

Hyperconjugation Interaction of electrons in a sigma bonding orbital with the vacant 2*p* orbital of an adjacent positively charged carbon.

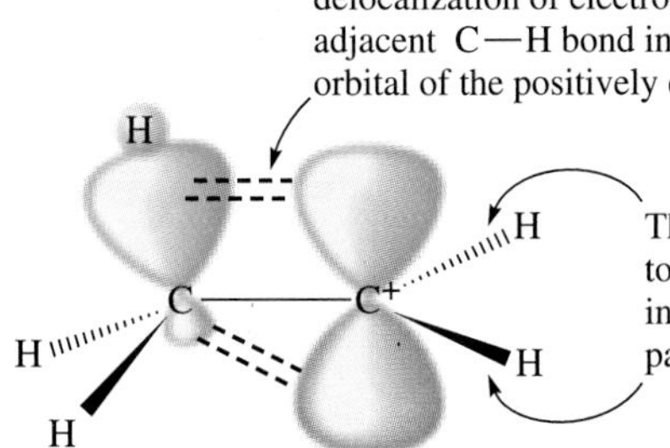
delocalization of electrons from an adjacent C—H bond into the empty 2p orbital of the positively charged carbon
H
H
H
C
C+
H
H
The orbitals of these C—H bonds are perpendicular to the vacant 2p orbital of the cationic carbon. Electrons in them cannot flow into the vacant 2p orbital and cannot participate in hyperconjugation

FIGURE 6.6
Hyperconjugation.

In this way, electrons of the adjacent sigma bond are partially delocalized, which also partially delocalizes the positive charge of the cation. Replacing any C—H bond to the cationic carbon by an alkyl group increases the possibility for electron delocalization by hyperconjugation and increases the stability of the carbocation. Note that, because of the geometry of the cationic carbon and the C—H bonds on an adjacent carbon, only electrons in the adjacent C—H bond can participate in hyperconjugation.

EXAMPLE 6.4

Arrange these carbocations in order of increasing stability.

(a) $CH_3\overset{+}{C}HC(CH_3)_2CH_3$ **(b)** $CH_3\overset{+}{C}(CH_3)CHCH_3$ with CH_3 **(c)** $CH_3C(CH_3)_2CH_2CH_2^+$

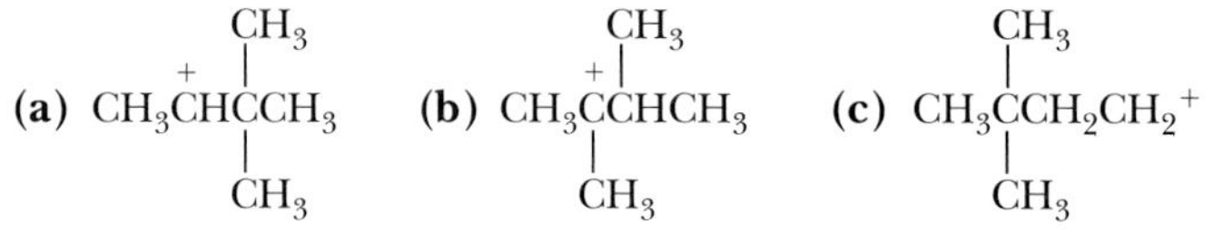

Solution

Carbocation (a) is secondary, (b) is tertiary, and (c) is primary. In order of increasing stability they are *c* < *a* < *b*.

PROBLEM 6.4

Arrange these carbocations in order of increasing stability.

(a) 1-methylcyclohexyl cation ($\overset{+}{C}$—CH_3) **(b)** 2-methylcyclohexyl cation (ring $\overset{+}{C}$ adjacent to C—CH_3) **(c)** cyclohexyl—$\overset{+}{C}H_2$

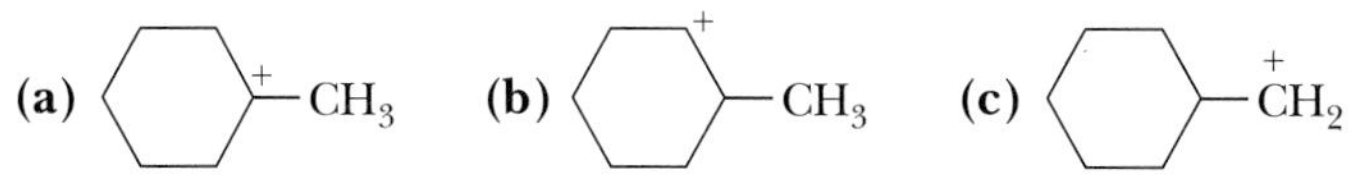

EXAMPLE 6.5

Propose a mechanism for the addition of HI to methylenecyclohexane to give 1-iodo-1-methylcyclohexane. Which step in your mechanism is rate limiting?

Solution

Propose a two-step mechanism. Step 1 is a rate-limiting proton transfer from HI to the carbon-carbon double bond to form a 3° carbocation intermediate. Reaction of this intermediate (a Lewis acid) with iodide ion (a Lewis base) in Step 2 gives the product.

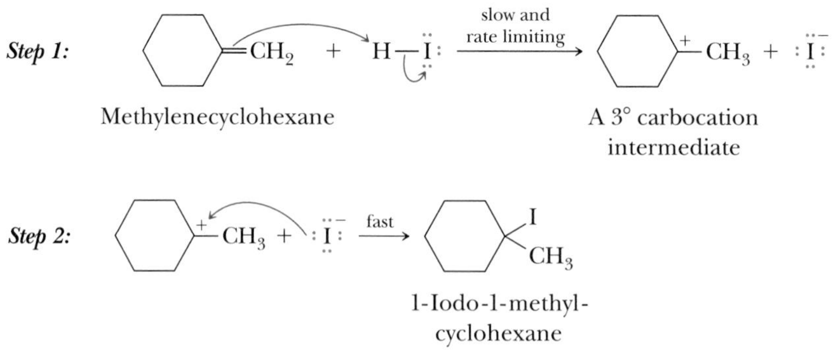

Step 1: Methylenecyclohexane ($=CH_2$) + H—I: $\xrightarrow{\text{slow and rate limiting}}$ A 3° carbocation intermediate ($\overset{+}{C}$—CH_3) + :I:$^-$

Step 2: ($\overset{+}{C}$—CH_3) + :I:$^-$ $\xrightarrow{\text{fast}}$ 1-Iodo-1-methyl-cyclohexane

PROBLEM 6.5

Propose a mechanism for addition of HI to 1-methylcyclohexene to give 1-iodo-1-methylcyclohexane. Which step in your mechanism is rate limiting?

B. Addition of Water: Acid-Catalyzed Hydration

In the presence of an acid catalyst—most commonly sulfuric acid—water adds to an alkene to give an alcohol. Addition of water is called **hydration.** In the case of simple alkenes, —H adds to the carbon of the double bond with the greater number of hydrogens and —OH adds to the carbon with the fewer hydrogens. Thus, H—OH adds to alkenes in accordance with Markovnikov's rule.

Hydration Addition of water.

$$CH_3CH{=}CH_2 + H_2O \xrightarrow{H_2SO_4} CH_3\underset{}{\overset{OH}{\overset{|}{C}}}H{-}\overset{H}{\overset{|}{C}}H_2$$

Propene → 2-Propanol

$$CH_3\overset{CH_3}{\overset{|}{C}}{=}CH_2 + H_2O \xrightarrow{H_2SO_4} CH_3\overset{CH_3}{\overset{|}{\underset{HO}{\underset{|}{C}}}}{-}\underset{H}{\underset{|}{C}}H_2$$

2-Methylpropene → 2-Methyl-2-propanol

EXAMPLE 6.6

Draw a structural formula for the product of acid-catalyzed hydration of 1-methylcyclohexene.

Solution

1-Methylcyclohexene (CH$_3$) + H_2O $\xrightarrow{H_2SO_4}$ 1-Methylcyclohexanol (CH$_3$, OH)

PROBLEM 6.6

Draw the structural formula for the product of each alkene hydration reaction.

(a) $CH_3\overset{CH_3}{\overset{|}{C}}{=}CHCH_3 + H_2O \xrightarrow{H_2SO_4}$ (b) $CH_2{=}\overset{CH_3}{\overset{|}{C}}CH_2CH_3 + H_2O \xrightarrow{H_2SO_4}$

The mechanism for acid-catalyzed hydration of alkenes is quite similar to what we already proposed for addition of HCl, HBr, and HI to alkenes and is illustrated by conversion of propene to 2-propanol. In Step 1, proton transfer from H_3O^+ to

Oxonium ion An ion in which oxygen is bonded to three other atoms and bears a positive charge.

the alkene forms a 2° carbocation intermediate (a Lewis acid). This intermediate then completes its valence shell in Step 2 by forming a new covalent bond with an unshared pair of electrons of the oxygen atom of H_2O (a Lewis base) to give an oxonium ion. An **oxonium ion** is any ion in which oxygen is bonded to three other atoms or groups and bears a positive charge. Finally, proton transfer to H_2O in Step 3 gives the alcohol and regenerates the acid catalyst. Formation of the carbocation intermediate in Step 1 is the rate-limiting step. This mechanism is consistent with the fact that acid is a catalyst. An H_3O^+ is consumed in Step 1 but another is generated in Step 3.

MECHANISM Acid-Catalyzed Hydration of Propene

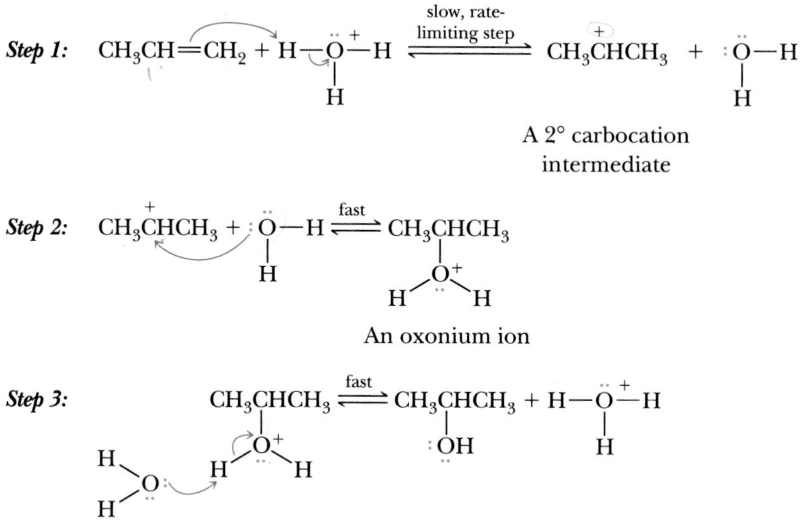

Step 1: $CH_3CH{=}CH_2 + H{-}\overset{+}{O}{-}H$ (with H on O) $\xrightarrow{\text{slow, rate-limiting step}}$ $CH_3\overset{+}{C}HCH_3 + H{-}\ddot{O}{-}H$

A 2° carbocation intermediate

Step 2: $CH_3\overset{+}{C}HCH_3 + H{-}\ddot{O}{-}H \xrightarrow{\text{fast}} CH_3CH(\overset{+}{O}H_2)CH_3$

An oxonium ion

Step 3: $CH_3CH(\overset{+}{O}H_2)CH_3 + H_2\ddot{O}: \xrightarrow{\text{fast}} CH_3CH(\ddot{O}H)CH_3 + H{-}\overset{+}{O}{-}H$ (with H on O)

EXAMPLE 6.7

Propose a mechanism for the acid-catalyzed hydration of methylenecyclohexane to give 1-methylcyclohexanol. Which step in your mechanism is the rate-limiting step?

Solution

Propose a three-step mechanism beginning with proton transfer in Step 1 from the acid catalyst to the alkene to form a 3° carbocation intermediate. Reaction of this intermediate (a Lewis acid) with water (a Lewis base) in Step 2 completes the valence shell of carbon and gives an oxonium ion. Proton transfer from the oxonium ion to water in Step 3 gives the product. Formation of the 3° carbocation intermediate in Step 1 is rate limiting.

Step 1: Methylenecyclohexane $+ H_3O^+$ $\xrightleftharpoons{\text{slow, rate-limiting step}}$ 1-methylcyclohexyl cation $+ H_2O$

(A 3° carbocation intermediate)

Step 2: 1-methylcyclohexyl cation $+ H_2O$ $\xrightleftharpoons{\text{fast}}$ oxonium ion

(An oxonium ion)

Step 3: oxonium ion $+ H_2O$ $\xrightleftharpoons{\text{fast}}$ 1-methylcyclohexanol $+ H_3O^+$

PROBLEM 6.7

Propose a mechanism for the acid-catalyzed hydration of 1-methylcyclohexene to give 1-methylcyclohexanol. Which step in your mechanism is the rate-limiting step?

C. Carbocation Rearrangements

As we have seen in the preceding discussions, the expected product of electrophilic addition to a carbon-carbon double bond involves rupture of the pi bond and formation of two new sigma bonds in its place. In addition of HCl to 3-methyl-1-butene, however, only 40% of the expected product is formed. The major product is 2-chloro-2-methylbutane, a compound with a different connectivity (a different order of attachment of its atoms) compared with the connectivity in the starting alkene. The connectivity shown at carbons 2 and 3 in the starting alkene is CH—CH, and it remains CH—CH in the expected product. The connectivity about these carbons in the second product has changed and is now C—CH_2. We say that formation of 2-chloro-2-methylbutane has involved a rearrangement. A **rearrangement** is any change in connectivity of atoms in a product compared with their connectivity in the starting material. Typically, either a hydrogen or an alkyl group migrates with its bonding electrons to an electron-deficient atom. In the rearrangements we examine in this chapter, migration is to an electron-deficient atom bearing a positive charge.

Rearrangement A change in connectivity of the atoms in a product compared with the connectivity of the same atoms in the starting material.

$$CH_3CH(CH_3)CH{=}CH_2 + HCl \longrightarrow CH_3CH(CH_3)CHClCH_3 + CH_3CCl(CH_3)CH_2CH_3$$

3-Methyl-1-butene → 2-Chloro-3-methylbutane (the expected product; 40%) + 2-Chloro-2-methylbutane (a product of rearrangement; 60%)

Formation of the rearranged product in this reaction can be accounted for in the following way. Reaction of 3-methyl-1-butene with HCl in Step 1 forms a 2° carbocation. In Step 2, migration of a hydrogen atom with its bonding electron pair from an adjacent carbon atom gives a more stable 3° carbocation. This type of rearrangement is called a 1,2-shift. In a **1,2-shift,** an atom or group of atoms with its bonding electrons moves from one atom to an adjacent electron-deficient atom. In the migration shown in Step 2, the migrating group is a hydride ion (a hydrogen nucleus with two valence electrons). Finally, in Step 3, the more stable carbocation reacts with a chloride ion to give the rearrangement product.

MECHANISM Carbocation Rearrangement in the Addition of HCl to an Alkene

Step 1: Proton transfer to the alkene to form a 2° carbocation intermediate

$$CH_3\underset{H}{\overset{CH_3}{C}}H{=}CH_2 + H{-}\ddot{\underset{..}{Cl}}: \xrightarrow[\text{limiting}]{\text{slow, rate}} CH_3\underset{H}{\overset{CH_3}{C}}{-}\overset{+}{C}HCH_3 + :\ddot{\underset{..}{Cl}}:^-$$

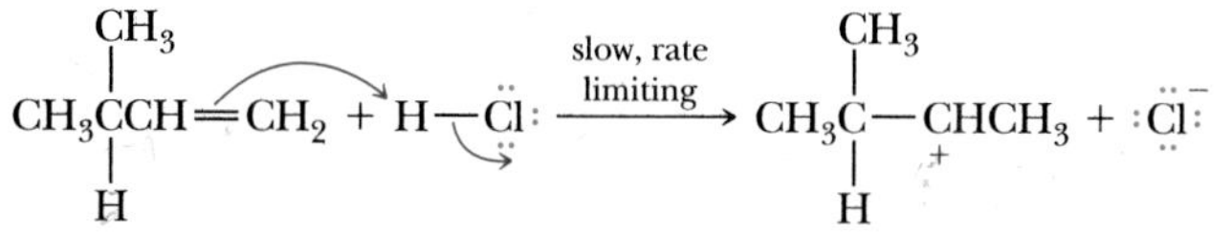

3-Methyl-1-butene — A 2° carbocation intermediate

Step 2: Rearrangement of a less stable 2° carbocation intermediate to a more stable 3° carbocation intermediate by migration of a hydrogen with its bonding electrons

$$CH_3\underset{H}{\overset{CH_3}{C}}{-}\overset{+}{C}HCH_3 \xrightarrow{\text{fast}} CH_3\overset{CH_3}{\underset{+}{C}}{-}\underset{H}{C}HCH_3$$

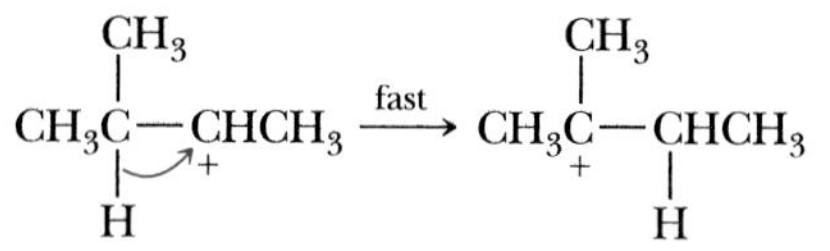

A 3° carbocation intermediate

Step 3: Reaction of the 3° carbocation (a Lewis acid) with chloride ion (a Lewis base) to give the product

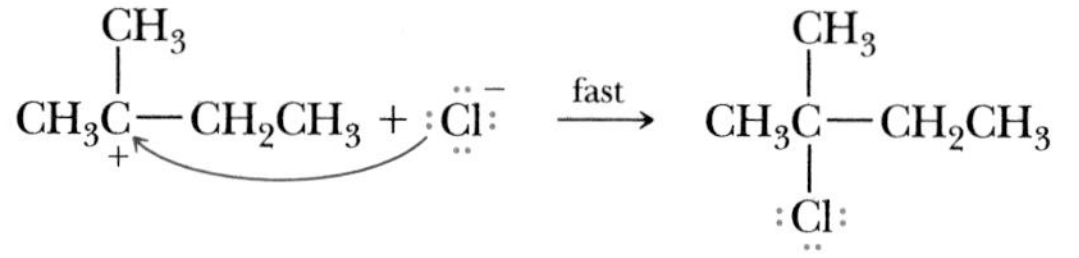

$$CH_3\overset{CH_3}{\underset{+}{C}}{-}CH_2CH_3 + :\ddot{\underset{..}{Cl}}:^- \xrightarrow{\text{fast}} CH_3\overset{CH_3}{\underset{:\ddot{Cl}:}{C}}{-}CH_2CH_3$$

2-Chloro-2-methylbutane

The driving force for this rearrangement is the fact that the less stable 2° carbocation originally formed is converted to a more stable 3° carbocation. From the study of this and other carbocation rearrangements, we find that 1° carbocations rearrange to more stable 2° or 3° carbocations, and 2° carbocations rearrange to more stable 3° carbocations. They rarely rearrange in the opposite direction.

Rearrangements also occur in acid-catalyzed hydration of alkenes, especially when the carbocation formed in the first step can rearrange to a more stable carbocation. For example, acid-catalyzed hydration of 3-methyl-1-butene gives 2-methyl-2-butanol.

$$\underset{\text{3-Methyl-1-butene}}{\mathrm{CH_3\overset{\overset{\Large CH_3}{|}}{C}HCH{=}CH_2}} + \mathrm{H_2O} \xrightarrow{\mathrm{H^+}} \underset{\text{2-Methyl-2-butanol}}{\mathrm{CH_3\overset{\overset{\Large CH_3}{|}}{\underset{\underset{\Large OH}{|}}{C}}CH_2CH_3}}$$

EXAMPLE 6.8

In addition of HCl to 3,3-dimethyl-1-butene, only 17% of the expected product is formed. Formation of 2-chloro-2,3-dimethylbutane, the major product, has involved a rearrangement. Propose a mechanism for the formation of this rearranged product.

$$\underset{\text{3,3-Dimethyl-1-butene}}{\mathrm{CH_3\overset{\overset{\Large CH_3}{|}}{\underset{\underset{\Large CH_3}{|}}{C}}CH{=}CH_2}} + \mathrm{HCl} \longrightarrow \underset{\substack{\text{3-Chloro-2,2-dimethylbutane}\\(17\%)}}{\mathrm{CH_3\overset{\overset{\Large CH_3}{|}}{\underset{\underset{\Large H_3C}{|}}{C}}{-}\underset{\underset{\Large Cl}{|}}{C}HCH_3}} + \underset{\substack{\text{2-Chloro-2,3-dimethylbutane}\\(83\%)}}{\mathrm{CH_3\overset{\overset{\Large CH_3}{|}}{\underset{\underset{\Large Cl}{|}}{C}}{-}\underset{\underset{\Large CH_3}{|}}{C}HCH_3}}$$

Solution

Propose a three-step mechanism involving rearrangement of a $-CH_3$ group with its bonding electron pair from an adjacent carbon to the positively charged carbon.

Step 1: Proton transfer from HCl to the alkene to form a 2° carbocation intermediate

$$\mathrm{CH_3\overset{\overset{\Large CH_3}{|}}{\underset{\underset{\Large CH_3}{|}}{C}}{-}CH{=}CH_2} + \mathrm{H{-}\ddot{\underset{..}{Cl}}:} \xrightarrow[\text{limiting}]{\text{slow, rate}} \underset{\substack{\text{A 2° carbocation}\\\text{intermediate}}}{\mathrm{CH_3\overset{\overset{\Large CH_3}{|}}{\underset{\underset{\Large H_3C}{|}}{C}}{-}\overset{+}{C}H{-}CH_3}} + \mathrm{:\ddot{\underset{..}{Cl}}:^-}$$

Step 2: Rearrangement of the less stable 2° carbocation intermediate to a more stable 3° carbocation intermediate

$$\mathrm{CH_3\overset{\overset{\Large CH_3}{|}}{\underset{\underset{\Large H_3C}{|}}{C}}{-}\overset{+}{C}H{-}CH_3} \xrightarrow{\text{fast}} \mathrm{CH_3\overset{\overset{\Large CH_3}{|}}{\overset{+}{C}}{-}\underset{\underset{\Large CH_3}{|}}{C}H{-}CH_3}$$

Step 3: Reaction of the 3° carbocation (a Lewis acid) with chloride ion (a Lewis base) to give the product

$$\mathrm{:\ddot{\underset{..}{Cl}}:^-} + \mathrm{CH_3\overset{\overset{\Large CH_3}{|}}{\overset{+}{C}}{-}\underset{\underset{\Large CH_3}{|}}{C}H{-}CH_3} \xrightarrow{\text{fast}} \mathrm{CH_3\overset{\overset{\Large CH_3}{|}}{\underset{\underset{\Large :\ddot{Cl}:}{|}}{C}}{-}\underset{\underset{\Large CH_3}{|}}{C}H{-}CH_3}$$

PROBLEM 6.8

Acid-catalyzed hydration of 3-methyl-1-butene gives 2-methyl-2-butanol as the major product. Propose a mechanism for formation of this rearranged alcohol.

D. Addition of Bromine and Chlorine

Chlorine, Cl_2, and bromine, Br_2, react with alkenes at room temperature by addition of halogen atoms to the two carbon atoms of the double bond with formation of two new carbon-halogen bonds. Fluorine, F_2, also adds to alkenes, but because its reactions are very fast and difficult to control, this reaction is not a useful laboratory procedure. Iodine, I_2, is so unreactive that it does not add to alkenes.

Halogenation with bromine or chlorine is generally carried out either with the pure reagents or by mixing them in an inert solvent such as CCl_4 or CH_2Cl_2.

$$CH_3CH{=}CHCH_3 + Br_2 \xrightarrow{CCl_4} CH_3\underset{|}{\overset{Br}{C}}H{-}\overset{Br}{\underset{|}{C}}HCH_3$$

2-Butene → 2,3-Dibromobutane

Addition of bromine or chlorine to a cycloalkene gives a *trans* dihalide. Addition of bromine to cyclohexene, for example, gives *trans*-1,2-dibromocyclohexane; the *cis* isomer is not formed. Thus, addition of a halogen to an alkene is stereoselective. A **stereoselective reaction** is one in which one stereoisomer is formed or destroyed in preference to all others that may be formed or destroyed.

Stereoselective reaction A reaction in which one stereoisomer is formed or destroyed in preference to all others that may be formed or destroyed.

Cyclohexene + Br_2 $\xrightarrow{CCl_4}$ *trans*-1,2-Dibromocyclohexane

A solution of bromine in carbon tetrachloride is red. Add a few drops of an alkene and the color disappears. *(Charles D. Winters)*

Reaction of bromine with an alkene is a particularly useful qualitative test for the presence of a carbon-carbon double bond. If we dissolve bromine in carbon tetrachloride, the solution is red. Both alkenes and dibromoalkanes are colorless. If we now mix a few drops of the bromine solution with an alkene, a dibromoalkane is formed, and the red solution becomes colorless.

EXAMPLE 6.9

Complete these reactions, showing the stereochemistry of the product.

(a) cyclopentene + Br_2 $\xrightarrow{CH_2Cl_2}$ **(b)** 1-methylcyclohexene (CH_3) + Cl_2 $\xrightarrow{CH_2Cl_2}$

Solution

Addition of both Br_2 and Cl_2 is stereoselective. The halogen atoms are *trans* to each other in each product.

(a) 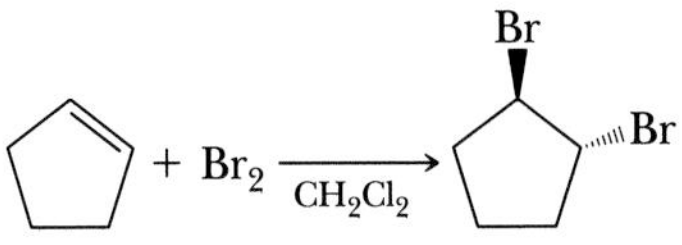 cyclopentene + Br_2 $\xrightarrow{CH_2Cl_2}$ trans-1,2-dibromocyclopentane (Br, Br)

(b) 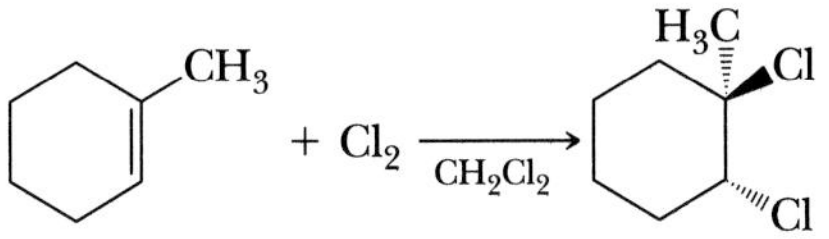 1-methylcyclohexene (CH_3) + Cl_2 $\xrightarrow{CH_2Cl_2}$ product (H_3C, Cl, Cl)

PROBLEM 6.9

Complete these reactions.

(a) $CH_3C(CH_3)_2CH{=}CH_2 + Br_2 \xrightarrow{CH_2Cl_2}$ (b) 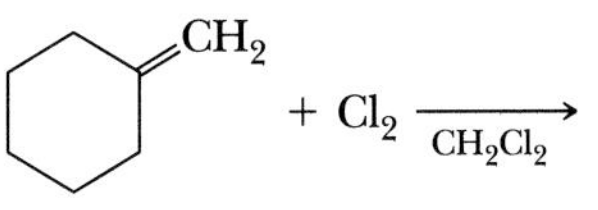 methylenecyclohexane (CH_2) $+ Cl_2 \xrightarrow{CH_2Cl_2}$

Stereoselectivity and Bridged Halonium Ion Intermediates

We explain the addition of bromine and chlorine to alkenes and their stereoselectivity by the following two-step mechanism for the reaction of bromine with an alkene. Reaction is initiated in Step 1 by interaction of the pi electrons of the alkene with bromine to form an intermediate in which bromine bears a positive charge. A bromine atom bearing a positive charge is called a **bromonium ion** and the cyclic structure of which it is a part is called a bridged bromonium ion. This intermediate is shown as a hybrid of three contributing structures of which the bridged bromonium ion is the most important. Then, in Step 2, a bromide ion reacts with this bridged intermediate from the side opposite that which is occupied by the bromine atom to give the dibromoalkane. Thus, bromine atoms are added from opposite faces of the carbon-carbon double bond. We say that this addition occurs with **anti stereoselectivity.**

Anti stereoselectivity The addition of atoms or groups of atoms to opposite faces of a carbon-carbon double bond.

MECHANISM Addition of Bromine with Anti Stereoselectivity

Step 1: Reaction of the pi electrons of the carbon-carbon double bond with bromine forms a bridged bromonium ion intermediate

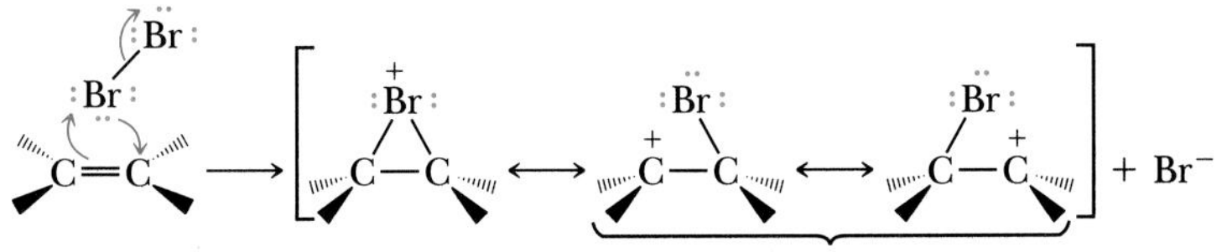

:Br: / :Br: , C=C ⟶ [bridged :Br:⁺ C—C ⟷ :Br: C—C⁺ ⟷ :Br: ⁺C—C] + Br^-

The bridged bromonium ion is the most important contributing structure

These carbocations are only minor contributing structures

Step 2: Attack of bromide ion on carbon from the side opposite the bridged bromonium ion and opening of the three-membered ring

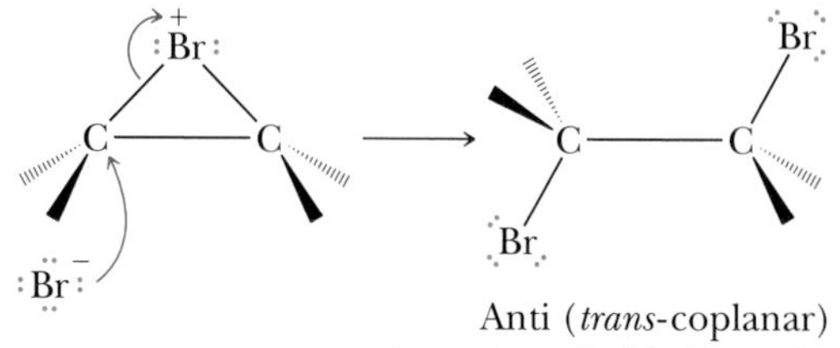

Anti (*trans*-coplanar) orientation of added bromine atoms

Addition of chlorine or bromine to cyclohexene and its derivatives gives a *trans*-diaxial product because only axial positions on adjacent atoms of a cyclohexane ring are anti and coplanar. The initial *trans*-diaxial conformation of the product is in equilibrium with the *trans*-diequatorial conformation, and, in simple derivatives of cyclohexane, the latter is more stable and predominates.

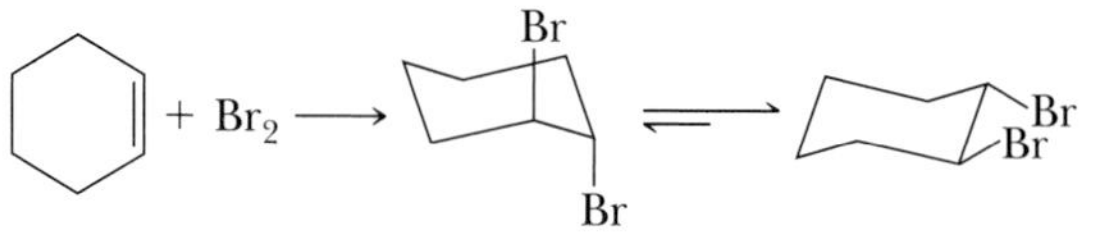

$$+ Br_2 \longrightarrow$$

trans-Diaxial

trans-Diequatorial (more stable)

In derivatives of cyclohexane in which interconversion between one chair conformation and the other is not possible or is severely restricted, the *trans*-diaxial product is isolated. If a cyclohexane ring, for example, contains a bulky alkyl group such as *tert*-butyl (Section 2.6), then the molecule exists overwhelmingly in a conformation in which the *tert*-butyl group is equatorial. Bromination of 4-*tert*-butylcyclohexene gives 1,2-dibromo-4-*tert*-butylcyclohexane. In the favored chair conformation of this product, *tert*-butyl is equatorial and the bromine atoms remain axial.

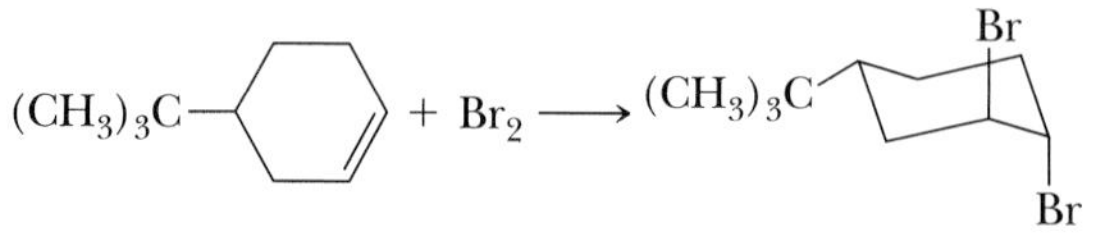

$$(CH_3)_3C\text{–} + Br_2 \longrightarrow (CH_3)_3C\text{–}$$

4-*tert*-Butylcyclohexene

1,2-Dibromo-4-*tert*-butylcyclohexane

E. Addition of HOCl and HOBr

Treatment of an alkene with Br_2 or Cl_2 in the presence of water results in addition of —OH and —Br, or —OH and —Cl to the carbon-carbon double bond to give a halohydrin. A **halohydrin** is a compound containing a halogen atom and a hydroxyl group on adjacent carbon atoms; those containing —Br and —OH are called **bromohydrins,** those containing —Cl and —OH are called **chlorohydrins.**

$$CH_3CH{=}CH_2 + Cl_2 + H_2O \longrightarrow CH_3\overset{\displaystyle HO}{\overset{|}{C}H}-\overset{\displaystyle Cl}{\overset{|}{C}H_2} + HCl$$

Propene

1-Chloro-2-propanol (a chlorohydrin)

Addition of HOCl and HOBr is regioselective (halogen adds to the less substituted carbon atom) and anti stereoselective. Both the regioselectivity and anti stereoselectivity are illustrated by the addition of HOBr to 1-methylcyclopentene. Bromine and the hydroxyl group add anti to each other with Br bonding to the less substituted carbon and OH bonding to the more substituted carbon.

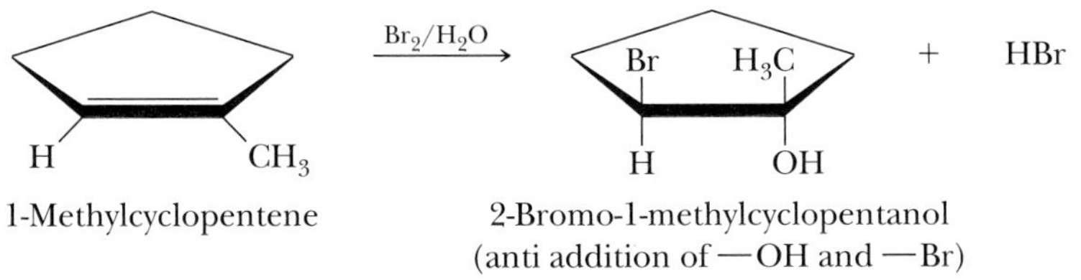

1-Methylcyclopentene $\xrightarrow{Br_2/H_2O}$ 2-Bromo-1-methylcyclopentanol (anti addition of —OH and —Br) + HBr

To account for the regioselectivity and anti stereoselectivity of these reactions, chemists propose a three-step mechanism. Step 1 involves reaction of halogen with the pi bond of the alkene to form the same bridged halonium ion intermediate as in the halogenation of an alkene. This intermediate has some of the character of a carbocation (to account for the regioselectivity) and some of the character of a halonium ion (to account for the stereoselectivity). Reaction of this halonium ion intermediate with H_2O in Step 2 followed by loss of a proton in Step 3 completes the reaction.

MECHANISM Halohydrin Formation and Its Anti Stereoselectivity

Step 1: Reaction of the pi electrons of the carbon-carbon double bond with bromine to form a bridged bromonium ion intermediate

+ Br^-

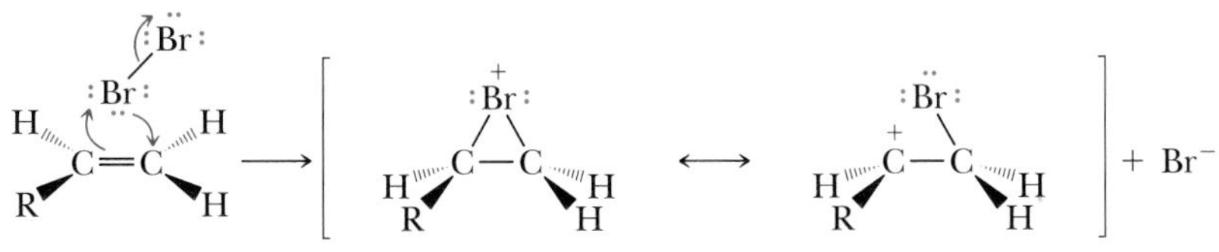

The bridged bromonium ion is the most important contributing structure

This carbocation is only a minor contributing structure

Step 2: Attack of H_2O on the more substituted carbon and opening of the three-membered ring

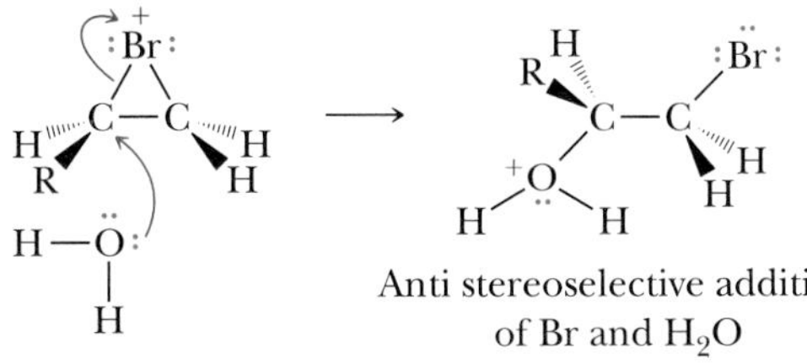

Anti stereoselective addition of Br and H_2O

Step 3: Proton transfer to H_2O to complete the reaction

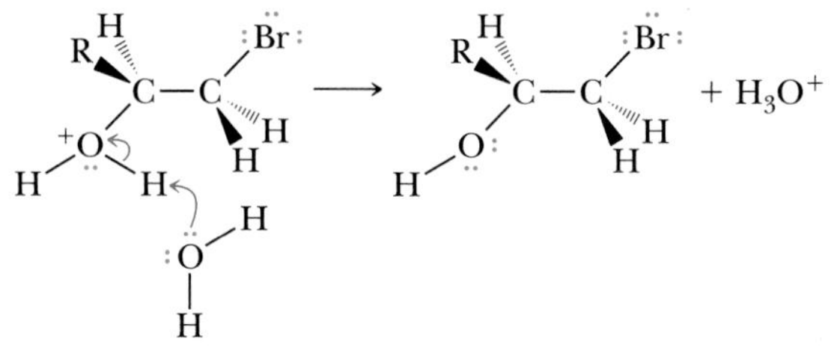

$$\text{R(H)C(}\overset{+}{\text{O}}\text{H}_2\text{)}-\text{C(Br)H}_2 + :\ddot{\text{O}}\text{H}_2 \longrightarrow \text{R(H)C(}\ddot{\text{O}}\text{H)}-\text{C(Br)H}_2 + H_3O^+$$

EXAMPLE 6.10

Draw the structure of the bromohydrin formed when 2-methylpropene is treated with Br_2/H_2O.

Solution

Addition is regioselective, with —OH adding to the more substituted carbon and —Br adding to the less substituted carbon.

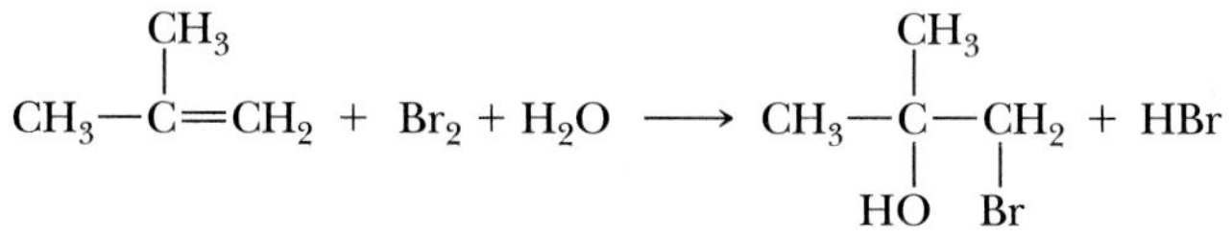

$$CH_3-\underset{}{\overset{CH_3}{\overset{|}{C}}}=CH_2 + Br_2 + H_2O \longrightarrow CH_3-\underset{HO}{\overset{CH_3}{\overset{|}{\underset{|}{C}}}}-\underset{Br}{\underset{|}{CH_2}} + HBr$$

2-Methylpropene → 1-Bromo-2-Methyl-2-propanol

PROBLEM 6.10

Draw the structure of the chlorohydrin formed when 1-methylcyclohexene is treated with Cl_2/H_2O.

F. Oxymercuration/Reduction

Hydration of alkenes can also be accomplished by treatment of an alkene with mercury(II) acetate (mercuric acetate) in water followed by reduction of the resulting product with sodium borohydride, $NaBH_4$. In the following structural formulas for mercury(II) acetate, the acetate group is written in full in the first formula. In the second formula, it is abbreviated as AcO—.

$$CH_3\overset{O}{\overset{\|}{C}}O-Hg-O\overset{O}{\overset{\|}{C}}CH_3 \qquad AcO-Hg-OAc$$

Mercury(II) acetate
(Mercuric acetate)

Oxymercuration, the addition of mercury(II) to one carbon of the double bond and oxygen to the other, is illustrated by the first step in the two-step conversion of 2-butene to 2-butanol.

Oxymercuration:

$$\mathrm{CH_3CH{=}CHCH_3} + \mathrm{Hg(OAc)_2} + \mathrm{H_2O} \longrightarrow \mathrm{CH_3\underset{\displaystyle|}{\overset{\displaystyle OH}{\overset{|}{C}}}H\underset{\underset{\displaystyle HgOAc}{|}}{C}HCH_3} + \mathrm{CH_3\overset{\displaystyle O}{\overset{\|}{C}}OH}$$

2-Butene · Mercury(II) acetate · An organomercury compound · Acetic acid

The initial organomercury compound is reduced by sodium borohydride, $NaBH_4$, to replace Hg by H.

Reduction:

$$\mathrm{CH_3\overset{\displaystyle OH}{\overset{|}{C}}H\underset{\underset{\displaystyle HgOAc}{|}}{C}HCH_3} \xrightarrow{\mathrm{NaBH_4}} \mathrm{CH_3\overset{\displaystyle OH}{\overset{|}{C}}H\underset{\underset{\displaystyle H}{|}}{C}HCH_3} + \mathrm{CH_3\overset{\displaystyle O}{\overset{\|}{C}}OH} + \mathrm{Hg}$$

An organomercury compound · 2-Butanol · Acetic acid

Oxymercuration is regioselective: HgOAc becomes bonded to the less substituted carbon of the alkene and —OH of water becomes bonded to the more substituted carbon. The result of oxymercuration followed by sodium borohydride reduction is Markovnikov addition of H—OH to an alkene.

$$\mathrm{CH_3CH_2CH_2CH_2CH{=}CH_2} \xrightarrow[\text{2. NaBH}_4]{\text{1. Hg(OAc)}_2\text{, H}_2\text{O}} \mathrm{CH_3CH_2CH_2CH_2\overset{\displaystyle OH}{\overset{|}{C}}HCH_3}$$

1-Hexene · 2-Hexanol

$$\mathrm{CH_3\overset{\displaystyle CH_3}{\overset{|}{C}}HCH{=}CH_2} \xrightarrow[\text{2. NaBH}_4]{\text{1. Hg(OAc)}_2\text{, H}_2\text{O}} \mathrm{CH_3\overset{\displaystyle CH_3}{\overset{|}{C}}H\underset{\underset{\displaystyle OH}{|}}{C}HCH_3}$$

3-Methyl-1-butene · 3-Methyl-2-butanol

Oxymercuration of 3-methyl-1-butene followed by $NaBH_4$ reduction gives 3-methyl-2-butanol exclusively and illustrates a very important feature of this reaction sequence; it occurs without rearrangement. You might compare the product of oxymercuration-reduction of 3-methyl-1-butene with the product formed by acid-catalyzed hydration of the same alkene. In the former, no rearrangement occurs. In the latter, the major product is 2-methyl-2-butanol, a compound formed by rearrangement. The fact that no rearrangement occurs during oxymercuration-reduction indicates that no free carbocation intermediate is formed.

The stereoselectivity of oxymercuration is illustrated by the reaction of mercury(II) acetate in the presence of water with cyclopentene.

Cyclopentene (H, H) $\xrightarrow[\mathrm{H_2O}]{\mathrm{Hg(OAc)_2}}$ (OH, H / H, HgOAc) $\xrightarrow{\mathrm{NaBH_4}}$ Cyclopentanol (OH, H / H, H)

Cyclopentene · (Anti addition of —OH and —HgOAc) · Cyclopentanol

The fact that oxymercuration is both regioselective and stereoselective has led chemists to propose the following mechanism for this reaction, which is closely analogous to that for the addition of Br_2 and Cl_2 to an alkene (Section 6.3C). Dissociation of mercury(II) acetate in Step 1 gives acetate ion (a Lewis base) and $AcOHg^+$ (a Lewis acid). Reaction then proceeds by interaction of $AcOHg^+$ with the electron pair of the pi bond to give a **bridged mercurinium ion** intermediate. This bridged ion closely resembles a bridged bromonium ion intermediate (Section 6.3C) with the difference that, unlike bromine, mercury has no electron pair to donate to form fully covalent bonds in the intermediate. Rather, in the bridged mercurinium ion intermediate, the two pi electrons of the carbon-carbon double bond now form a ring containing three atoms bonded by two electrons. The open cation structure, which places the positive charge on the carbon giving the more stable carbocation, is a minor contributing structure to the mercurinium ion intermediate.

MECHANISM Oxymercuration

Step 1: Dissociation of mercury(II) acetate to give $AcOHg^+$, a Lewis acid, and acetate ion

$$\text{AcO—Hg—OAc} \longrightarrow \text{AcO—Hg}^+ + \text{AcO}^-$$

Step 2: Attack of $AcO—Hg^+$ on the carbon-carbon double bond to form a bridged mercurinium ion intermediate. In this bridged intermediate, the two electrons of the pi bond now form a three-center bond, here indicated by dashed lines, in which each participating atom bears a fraction of the positive charge

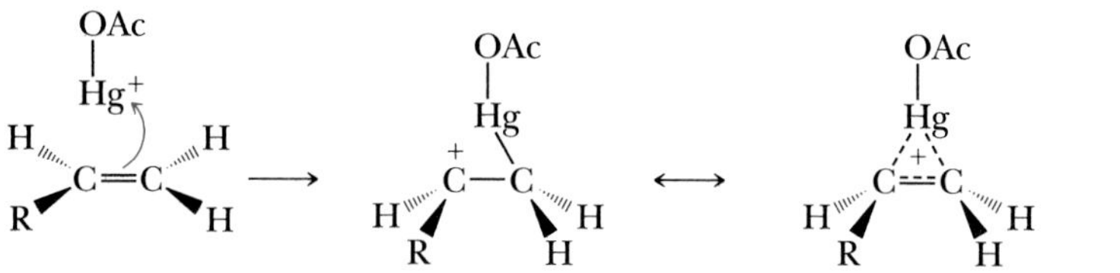

OAc
Hg+
H H
C=C
R H
⟶
OAc
Hg
+
C—C
H H
R H
⟷
OAc
Hg
+
C=C
H H
R H

An open carbocation intermediate (minor contributor)

A bridged mercurinium ion intermediate (major contributor)

Step 3: Anti stereoselective attack of water on the bridged mercurinium ion intermediate at the more substituted carbon and opening of the three-membered ring. Proton transfer from this product to water completes oxymercuration of the alkene

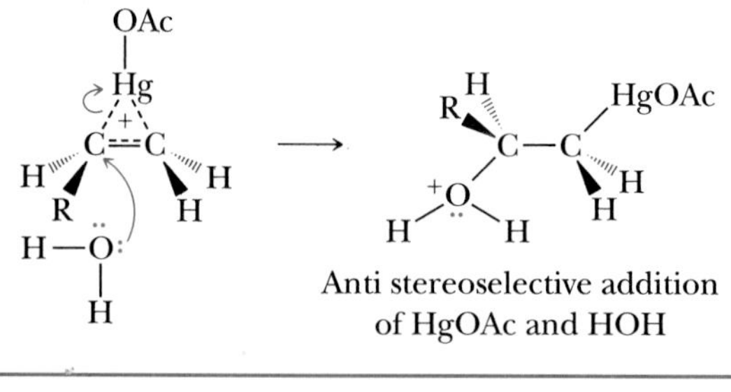

OAc
Hg
+
C=C
H H
R H
H—Ö:
H
⟶
R H HgOAc
C—C
+O H
H H H

Anti stereoselective addition of HgOAc and HOH

The fact that oxymercuration occurs without rearrangement indicates that the intermediate is not a true carbocation but is a resonance hybrid with largely the character of a bridged intermediate. Furthermore, the bridged structure allows us to account for the fact that the stereochemistry of this electrophilic addition is predominantly anti; the nucleophile attacks the bridged intermediate from the face opposite that occupied by mercury, as shown in Step 3.

Yet the fact that the electrophile ($AcOHg^+$) adds to the less substituted carbon and the nucleophile (HOH) adds to the more substituted carbon indicates that the intermediate must have some carbocation character. It is probable that, in the actual intermediate, mercury is bonded partially to each carbon, thereby preventing rearrangement. Of the two carbon atoms of the mercurinium ion, the more substituted one has a greater degree of partial positive charge and is the one attacked by the nucleophile, H_2O. This is well accounted for by resonance theory; only the more stable carbocation participates appreciably in the resonance-stabilized mercurinium ion intermediate.

6.4 Hydroboration/Oxidation

Hydroboration/oxidation of an alkene is an extremely valuable laboratory method for the stereoselective (syn addition of both —H and —OH) and regioselective hydration of an alkene. Furthermore, this sequence of reactions occurs without rearrangement.

Hydroboration is the addition of borane, BH_3, to an alkene to form a trialkylborane. The overall reaction occurs in three successive steps. BH_3 reacts with one molecule of alkene, then a second, and finally a third until all three hydrogens of borane have been replaced by alkyl groups.

$$\text{H—BH}_2 \xrightarrow{CH_2=CH_2} \underset{\text{An alkylborane}}{CH_3CH_2\text{—BH}_2} \xrightarrow{2CH_2=CH_2} \underset{\text{A trialkylborane}}{CH_3CH_2\text{—B}(CH_2CH_3)_2}$$

Borane cannot be prepared as a pure compound because it dimerizes to diborane, B_2H_6, a toxic gas that ignites spontaneously in air.

$$\underset{\text{Borane}}{2BH_3} \rightleftharpoons \underset{\text{Diborane}}{B_2H_6}$$

However, BH_3 exists as a stable complex with an ether such as tetrahydrofuran (THF) by formation of a Lewis acid-base complex. Borane is most often used as a commercially available solution of BH_3 in THF.

$$2\ \underset{\substack{\text{Tetrahydrofuran}\\ \text{(THF)}}}{C_4H_8\ddot{O}\!:} + B_2H_6 \rightleftharpoons 2\ \underset{BH_3 \cdot \text{THF}}{C_4H_8\overset{+}{O}\text{—}\overset{-}{B}H_3}$$

Boron, atomic number 5, has three electrons in its valence shell. To bond with three other atoms, boron uses sp^2 hybrid orbitals. The unoccupied $2p$ orbital of boron is perpendicular to the plane created by boron and the three other atoms to which it is bonded. An example of a stable compound in which boron is bonded to three other atoms is boron trifluoride, BF_3, a planar molecule with F—B—F bond angles of 120°

(Section 1.2F). Because of the vacant $2p$ orbital in the valence shell of boron, BH_3, BF_3, and all other tricovalent compounds of boron are Lewis acids and act as electron-pair acceptors. These compounds of boron closely resemble carbocations, except that they are electrically neutral.

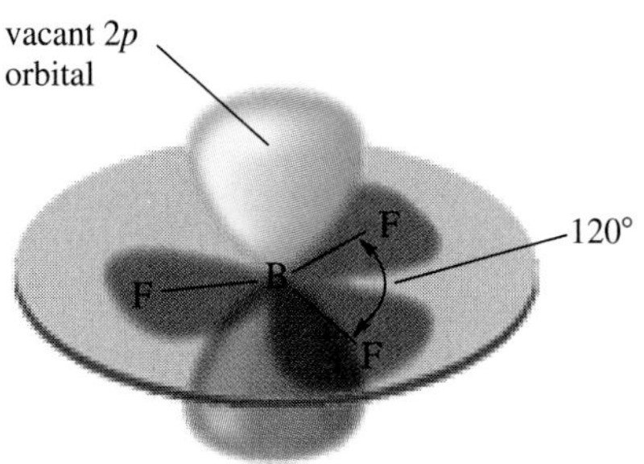

Addition of borane to alkenes is regioselective. In addition of borane to an unsymmetrical alkene, boron becomes bonded to the less substituted carbon of the double bond, as illustrated by the hydroboration of 1-hexene.

$$3CH_3CH_2CH_2CH_2CH{=}CH_2 + BH_3 \longrightarrow (CH_3CH_2CH_2CH_2CH_2CH_2)_3B$$

1-Hexene Trihexylborane

Syn stereoselectivity The addition of atoms or groups of atoms to the same face of a carbon-carbon double bond.

Hydroboration is also stereoselective. The major product is that in which hydrogen and boron add from the same side of the double bond, that is, the reaction is a **syn stereoselective** addition as illustrated by hydroboration of 1-methylcyclopentene.

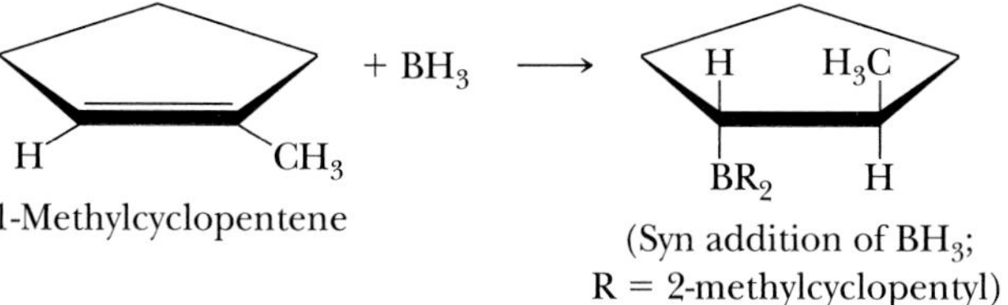

1-Methylcyclopentene

(Syn addition of BH_3; R = 2-methylcyclopentyl)

Addition of borane to an alkene is initiated by coordination of the vacant $2p$ orbital of boron (a Lewis acid) with the electron pair of the pi bond (a Lewis base). We account for the stereoselectivity of hydroboration by proposing formation of a cyclic, four-center transition state. Boron and hydrogen are added simultaneously and from the same side of the double bond.

MECHANISM OF HYDROBORATION A Concerted Regioselective and Stereoselective Addition

$$CH_3CH_2CH_2CH{=}CH_2 + \overset{\delta-}{H}{-}\overset{\delta+}{B} \longrightarrow CH_3CH_2CH_2\underset{|}{CH}(H){-}CH_2(B)$$

Bond breaking and bond forming occur simultaneously

Boron adds to the less substituted carbon of the double bond

An acceptable mechanism for hydroboration must account for both the regioselectivity and the stereoselectivity. We account for the regioselectivity of hydroboration by a combination of steric and electronic factors. In terms of steric effects, boron, the larger part of the reagent, adds to the less hindered carbon of the double bond, and hydrogen, the smaller part of the reagent, adds to the more hindered carbon.

Alternatively, we account for the regioselectivity in terms of electronic effects. The electronegativity of hydrogen (2.1) is slightly greater than that of boron (2.0) and, hence, there is a small degree of polarity (approximately 5%) to each B—H bond with boron bearing a partial positive charge and hydrogen a partial negative charge. It is proposed that there is some degree of carbocation character in the transition state and that the partial positive charge is on the carbon better able to accommodate it. It is believed that the observed regioselectivity is largely due to steric effects.

$$\overset{\delta-\quad\delta+}{H-B}\quad CH_3CH_2CH_2CH{=}CH_2 \longrightarrow \overset{\delta-}{H-B}\quad CH_3CH_2CH_2\overset{\delta+}{C}H{=\!=}CH_2$$

Starting reagents

Transition state (some carbocation character exists in the transition state)

Trialkylboranes are rarely isolated. Rather, they are converted directly to other products formed by substitution of another atom (H, O, N, C, or halogen) for boron. One of the most important reactions of trialkylboranes is with hydrogen peroxide in aqueous sodium hydroxide. Hydrogen peroxide is an oxidizing agent and under these conditions reacts with trialkylboranes to form an alcohol and sodium borate, Na_3BO_3. Reaction begins with donation of an electron pair from hydroperoxide anion (a Lewis base) to the boron atom of the trialkylborane (a Lewis acid). The intermediate thus formed has eight electrons in the valence of boron, which gives boron a negative formal charge. This intermediate then undergoes a rearrangement in which an R group with its pair of bonding electrons migrates from boron to oxygen and ejects hydroxide ion.

$$R_3B + {}^{-}\!:\ddot{O}-\ddot{O}-H \longrightarrow R_3\overset{-}{B}-\ddot{O}-\ddot{O}-H \longrightarrow R_2B-\ddot{O}R + {}^{-}\!:\ddot{O}-H$$

Two more reactions with hydroperoxide ion followed by rearrangements give a trialkyl borate, which then reacts with sodium hydroxide to form the alcohol and sodium borate.

$$(RO)_3B + 3NaOH \longrightarrow 3ROH + Na_3BO_3$$

A trialkyl borate

The net reaction from hydroboration and subsequent oxidation is hydration of a carbon-carbon double bond. Because hydrogen is added to the more substituted carbon of the original alkene, we refer to the results of hydroboration and subsequent oxidation as anti-Markovnikov hydration.

$$(CH_3CH_2CH_2CH_2CH_2CH_2)_3B + 3H_2O_2 + 3NaOH \longrightarrow 3CH_3CH_2CH_2CH_2CH_2CH_2OH + Na_3BO_3 + 3H_2O$$

Trihexylborane — 1-Hexanol — Sodium borate

Hydrogen peroxide oxidation of a trialkylborane is stereoselective in that configuration is retained; whatever the position of boron in relation to other groups in the trialkylborane, the —OH group by which it is replaced occupies the same position. Thus, both hydroboration of the alkene and subsequent peroxide oxidation of the resulting trialkylborane are syn stereoselective.

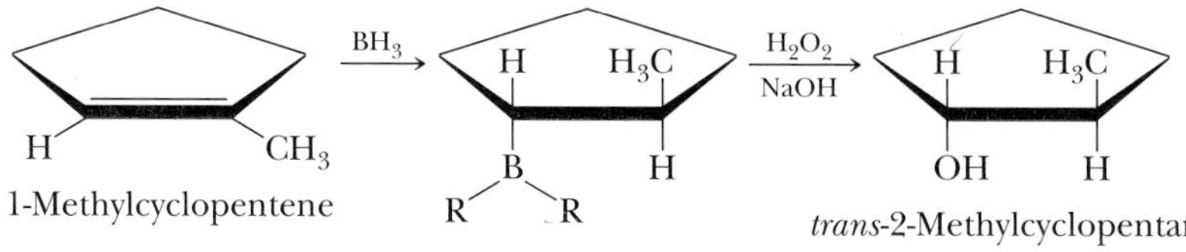

1-Methylcyclopentene $\xrightarrow{BH_3}$ A trialkylborane (R = 2-methylcyclopentyl) $\xrightarrow[NaOH]{H_2O_2}$ *trans*-2-Methylcyclopentanol

EXAMPLE 6.11

Draw structural formulas for the trialkylborane and alcohol formed in the following reaction sequences.

(a) $CH_3C(CH_3){=}CHCH_3 \xrightarrow{BH_3}$ a trialkylborane $\xrightarrow[NaOH]{H_2O_2}$ an alcohol

(b) 1-methylcyclohexene $\xrightarrow{BH_3}$ a trialkylborane $\xrightarrow[NaOH]{H_2O_2}$ an alcohol

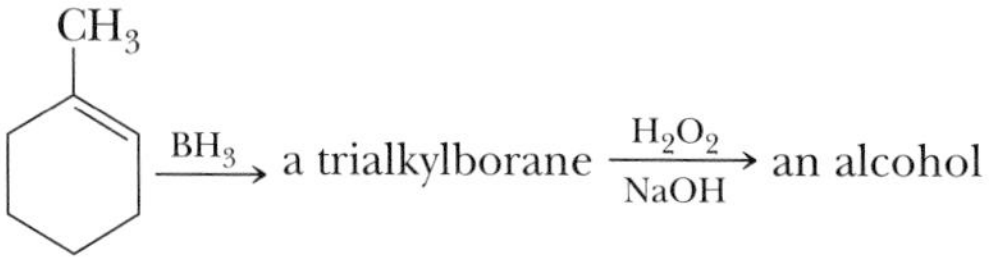

Solution

(a) $CH_3C(CH_3){=}CHCH_3 \xrightarrow{BH_3} (CH_3CH(CH_3)CH(CH_3){-})_3B \xrightarrow[NaOH]{H_2O_2} CH_3CH(CH_3)CH(OH)CH_3$

(b) 1-methylcyclohexene $\xrightarrow{BH_3}$ 2-methylcyclohexyl-BR_2 (R = 2-Methylcyclohexyl) $\xrightarrow[NaOH]{H_2O_2}$ *trans*-2-Methylcyclohexanol

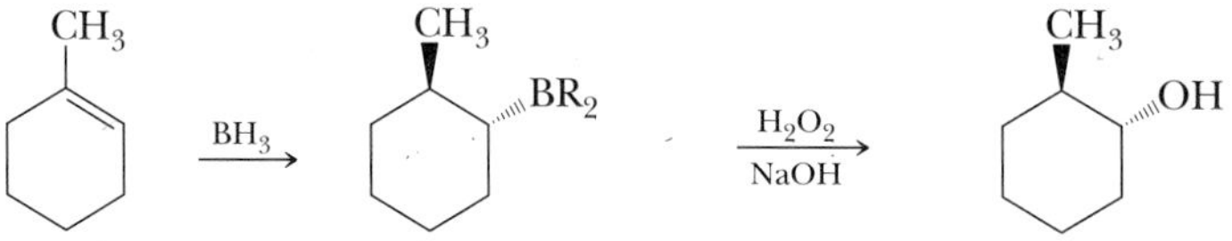

PROBLEM 6.11

Draw structural formulas for the trialkylborane and alkene that give the following alcohols under the reaction conditions shown.

(a) an alkene $\xrightarrow{BH_3}$ a trialkylborane $\xrightarrow[NaOH]{H_2O_2} CH_3CH(CH_3)CH_2CH_2OH$

(b) an alkene $\xrightarrow{BH_3}$ a trialkylborane $\xrightarrow[NaOH]{H_2O_2}$ cyclopentyl-CH_2OH

6.5 Oxidation of Alkenes

In this and the following section, we study oxidation and reduction of alkenes. Oxidation and reduction reactions are among the most important methods for transforming one functional group into another. It is, therefore, essential that you be able to recognize reactions that involve oxidation, those that involve reduction, and those that involve neither oxidation nor reduction. We begin with a general method by which you can recognize oxidation-reduction reactions and then consider special oxidation-reduction reactions of alkenes.

A. How to Recognize Oxidation-Reduction

In the following reactions, propene is transformed into three different compounds by reactions we study in this chapter. The first reaction involves reduction, the third involves oxidation, and the second involves neither oxidation nor reduction. These equations are not complete because they do not specify any reactant other than propene and they do not specify what reagents are necessary to bring about the particular transformation. Each does specify, however, that the carbon atoms of the products are derived from those of propene.

$$CH_3CH{=}CH_2 \xrightarrow{\text{reduction}} CH_3\underset{\text{H}}{\overset{|}{C}}H{-}\underset{\text{H}}{\overset{|}{C}}H_2 \quad \text{Propane}$$

$$CH_3CH{=}CH_2 \longrightarrow CH_3\underset{\text{OH}}{\overset{|}{C}}H{-}\underset{\text{H}}{\overset{|}{C}}H_2 \quad \text{2-Propanol}$$

$$CH_3CH{=}CH_2 \xrightarrow{\text{oxidation}} CH_3\underset{\text{OH}}{\overset{|}{C}}H{-}\underset{\text{OH}}{\overset{|}{C}}H_2 \quad \text{1,2-Propanediol}$$

(In the printed structures, the H, OH substituents are drawn above the carbons.)

It is possible to decide if transformations such as these involve oxidation, reduction, or neither by the use of **balanced half-reactions.** To write a balanced half-reaction:

1. Write a half-reaction showing the organic reactant(s) and product(s).
2. Complete a material balance, that is, balance the number of atoms on each side of the half-reaction. To balance the number of oxygens and hydrogens for a reaction taking place in acid solution, use H_2O for oxygens and then H^+ for hydrogens. For a reaction taking place in basic solution, use OH^- and H_2O.
3. Complete a charge balance, that is, balance the charge on both sides of the half-reaction. To balance charge, add electrons, e^-, to one side or the other. The equation completed in this step is a balanced half-reaction.

Oxidation is the loss of electrons. If electrons appear on the right side of a balanced half-reaction, the reactant has given up electrons and has been oxidized. **Reduction** is the gain of electrons. If electrons appear on the left side of a balanced half-reaction, the reactant has gained electrons and has been reduced. If no electrons appear in the balanced half-reaction, then the transformation involves neither oxidation nor reduction. Let us apply these steps to the transformation of propene to propane.

Oxidation The loss of electrons.

Reduction The gain of electrons.

Step 1: Half-reaction $CH_3CH{=}CH_2 \longrightarrow CH_3CH_2CH_3$

Step 2: Material balance $CH_3CH{=}CH_2 + 2H^+ \longrightarrow CH_3CH_2CH_3$

Step 3: Balanced half-reaction $CH_3CH{=}CH_2 + 2H^+ + 2e^- \longrightarrow CH_3CH_2CH_3$

Because two electrons appear on the left side of the balanced half-reaction (Step 3), conversion of propene to propane is a two-electron reduction. To bring it about requires use of a reducing agent.

Following is a balanced half-reaction for the transformation of propene to 2-propanol:

$$\text{Balanced half-reaction: } \underset{\text{Propene}}{CH_3CH{=}CH_2} + H_2O \longrightarrow \underset{\text{2-Propanol}}{CH_3\overset{\overset{\displaystyle OH}{|}}{C}HCH_3}$$

Because no electrons are required to achieve an electrical balance, conversion of propene to 2-propanol is neither oxidation nor reduction.

A balanced half-reaction for the transformation of propene to 1,2-propanediol requires two electrons on the right side of the equation for a charge balance; this transformation is a two-electron oxidation.

$$\text{Balanced half-reaction: } \underset{\text{Propene}}{CH_3CH{=}CH_2} + 2H_2O \longrightarrow \underset{\text{1,2-Propanediol}}{CH_3\overset{\overset{\displaystyle HO}{|}}{C}H\overset{\overset{\displaystyle OH}{|}}{C}H_2} + 2H^+ + 2e^-$$

It is important to realize that this strategy for recognizing oxidation and reduction is only that, a strategy. In no way does it give any indication of how a particular oxidation or reduction might be carried out in the laboratory. For example, the balanced half-reaction for the transformation of propene to propane requires $2H^+$ and $2e^-$. Yet by far the most common laboratory procedure for reducing propene to propane does not involve H^+ at all; rather it involves molecular hydrogen, H_2, and a transition metal catalyst.

We will use this method of balanced half-reactions throughout the text to recognize transformations that involve oxidation, those which involve reduction, and those which do not involve either oxidation or reduction.

EXAMPLE 6.12

Use a balanced half-reaction to show that each transformation involves an oxidation.

$$\textbf{(a)}\ CH_3CH_2CH_2OH \longrightarrow CH_3CH_2\overset{\overset{\displaystyle O}{\|}}{C}H \qquad \textbf{(b)}\ CH_3CH{=}CH_2 \longrightarrow CH_3\overset{\overset{\displaystyle O}{\|}}{C}H + H\overset{\overset{\displaystyle O}{\|}}{C}H$$

Solution

First complete a material balance and then a charge balance. The first transformation is a two-electron oxidation, the second is a four-electron oxidation. To bring each about requires an oxidizing agent.

$$\textbf{(a)}\ CH_3CH_2CH_2OH \longrightarrow CH_3CH_2\overset{\overset{\displaystyle O}{\|}}{C}H + 2H^+ + 2e^-$$

$$\textbf{(b)}\ CH_3CH{=}CH_2 + 2H_2O \longrightarrow CH_3\overset{\overset{\displaystyle O}{\|}}{C}H + H\overset{\overset{\displaystyle O}{\|}}{C}H + 4H^+ + 4e^-$$

PROBLEM 6.12

Use a balanced half-reaction to show that each transformation involves a reduction.

(a) 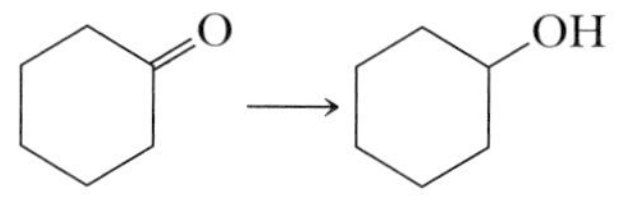 cyclohexanone ⟶ cyclohexanol (OH)

(b) CH_3CH_2COH (C=O) ⟶ $CH_3CH_2CH_2OH$

B. OsO_4: Oxidation of an Alkene to a Glycol

Certain transition metal oxides, in particular Os(VIII) oxide, are effective oxidizing agents for the conversion of an alkene to a **glycol,** a compound with two hydroxyl groups on adjacent carbons. Oxidation of an alkene by osmium tetroxide, OsO_4, is stereoselective in that it involves syn addition of —OH groups to the carbons of the double bond. For example, oxidation of cyclopentene by OsO_4 gives *cis*-1,2-cyclopentanediol, a ***cis* glycol.** Note that both *cis* and *trans* isomers are possible for this glycol. Because only the *cis*-glycol is formed, this oxidation is said to be syn stereoselective.

Glycol A compound with two hydroxyl (—OH) groups on adjacent carbons.

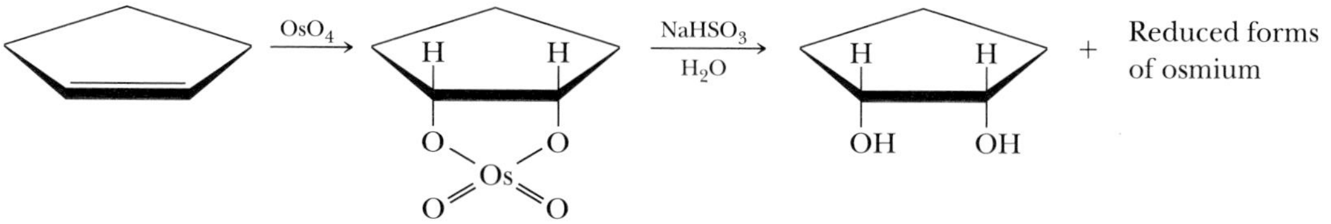

Cyclopentene $\xrightarrow{OsO_4}$ (A cyclic osmate ester) $\xrightarrow[H_2O]{NaHSO_3}$ *cis*-1,2-Cyclopentanediol (a *cis* glycol) + Reduced forms of osmium

The stereoselectivity of osmium tetroxide oxidation of alkenes is accounted for by the formation of a cyclic osmate ester in which oxygen atoms of OsO_4 form new covalent bonds with each carbon of the double bond in such a way that the five-membered osmium-containing ring is fused in a *cis* configuration to the original alkene. Osmate esters can be isolated and characterized. Usually, however, the cyclic osmate ester is treated directly with a reducing agent, such as $NaHSO_3$, which cleaves osmium-oxygen bonds to give the *cis* glycol and reduced forms of osmium.

The drawbacks of OsO_4 are that it is both expensive and highly toxic. One strategy to circumvent the high cost of OsO_4 is to use it in catalytic amounts along with stoichiometric amounts of another oxidizing agent which reoxidizes the reduced forms of osmium and thus recycles Os(VIII). Oxidizing agents commonly used for this purpose are hydrogen peroxide and *tert*-butyl hydroperoxide. When this procedure is used, there is no need for a reducing step using $NaHSO_3$.

HOOH — Hydrogen peroxide

$CH_3C(CH_3)_2OOH$ — *tert*-Butyl hydroperoxide (*t*-BuOOH)

C. Ozone: Cleavage of a Carbon-Carbon Double Bond (Ozonolysis)

Treatment of an alkene with ozone, O_3, followed by a suitable workup cleaves the carbon-carbon double bond and forms two carbonyl (C=O) groups in its place. The alkene is dissolved in an inert solvent, such as CH_2Cl_2, and a stream of ozone is bubbled through the solution. The products isolated from ozonolysis depend on the reaction conditions. Hydrolysis of the reaction mixture with water yields hydrogen peroxide, an oxidizing agent that can bring about further oxidations. To prevent side reactions from reactive intermediate peroxides, a weak reducing agent is added during the work-up to reduce peroxides to water. The reducing agent most commonly used for this purpose is dimethyl sulfide, $(CH_3)_2S$, as illustrated in the following example.

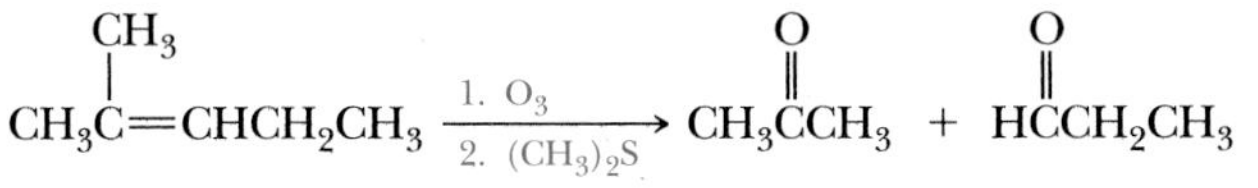

$$\underset{\text{2-Methyl-2-pentene}}{CH_3\overset{\overset{CH_3}{|}}{C}{=}CHCH_2CH_3} \xrightarrow[2.\ (CH_3)_2S]{1.\ O_3} \underset{\substack{\text{Propanone}\\ \text{(a ketone)}}}{CH_3\overset{\overset{O}{\|}}{C}CH_3} + \underset{\substack{\text{Propanal}\\ \text{(an aldehyde)}}}{H\overset{\overset{O}{\|}}{C}CH_2CH_3}$$

EXAMPLE 6.13

Draw structural formulas for the products of the following ozonolysis reactions. Name the new functional groups formed in each oxidation.

(a) 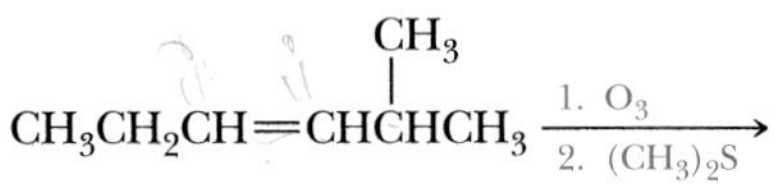 $CH_3CH_2CH{=}CH\overset{\overset{CH_3}{|}}{C}HCH_3 \xrightarrow[2.\ (CH_3)_2S]{1.\ O_3}$ (b) 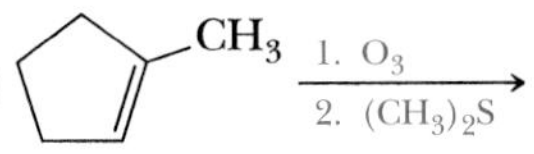 1-methylcyclopentene $\xrightarrow[2.\ (CH_3)_2S]{1.\ O_3}$

Solution

(a) 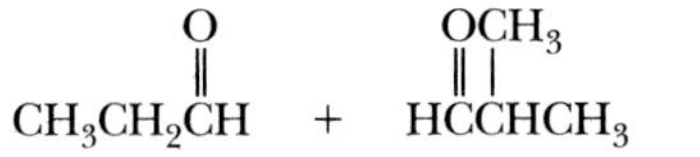 $\underset{\substack{\text{Propanal}\\ \text{(an aldehyde)}}}{CH_3CH_2\overset{\overset{O}{\|}}{C}H} + \underset{\substack{\text{2-Methylpropanal}\\ \text{(an aldehyde)}}}{H\overset{\overset{O}{\|}}{C}\overset{\overset{CH_3}{|}}{C}HCH_3}$

(b) 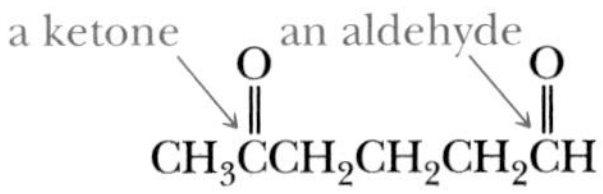 $\underset{\substack{\text{5-Oxohexanal}\\ \text{(a ketoaldehyde)}}}{CH_3\overset{\overset{O}{\|}}{C}CH_2CH_2CH_2\overset{\overset{O}{\|}}{C}H}$ (a ketone; an aldehyde)

PROBLEM 6.13

What alkene of molecular formula C_6H_{12}, when treated with ozone and then dimethyl sulfide, gives the following product(s)?

(a) 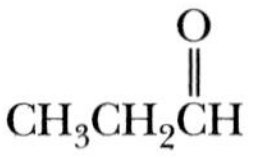 $CH_3CH_2\overset{\overset{O}{\|}}{C}H$ (only product)

(b) 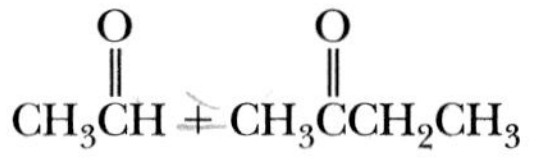 $CH_3\overset{\overset{O}{\|}}{C}H + CH_3\overset{\overset{O}{\|}}{C}CH_2CH_3$ (equal moles of each)

(c) 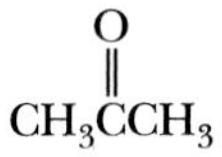 $CH_3\overset{\overset{O}{\|}}{C}CH_3$ (only product)

6.6 Reduction of Alkenes

Most alkenes react quantitatively with molecular hydrogen, H_2, in the presence of a transition metal catalyst to give alkanes. Commonly used transition metals include platinum, palladium, ruthenium, and nickel. Yields are usually quantitative or nearly so. Because conversion of an alkene to an alkane involves reduction by hydrogen in the presence of a catalyst, the process is called **catalytic reduction** or, alternatively, **catalytic hydrogenation.**

$$CH_2{=}CH_2 + H_2 \xrightarrow[25°C,\ 3\ atm]{Pd} CH_3{-}CH_3$$

Ethylene Ethane

$$\text{Cyclohexene} + H_2 \xrightarrow[25°C,\ 3\ atm]{Pd} \text{Cyclohexane}$$

Cyclohexene Cyclohexane

Monosubstituted and disubstituted carbon-carbon double bonds react readily at room temperature under a few atmospheres pressure of hydrogen. Trisubstituted carbon-carbon double bonds require slightly elevated temperatures and pressures of up to 100 atmospheres. Tetrasubstituted carbon-carbon double bonds are difficult to reduce and may require temperatures up to 275°C and pressures of 1000 atmospheres of hydrogen.

Although addition of hydrogen to an alkene is exothermic (the heat of hydrogenation of ethylene is −32.8 kcal/mol and that of cyclohexene is −28.6 kcal/mol), reduction is immeasurably slow in the absence of a catalyst. The metal catalyst is used as a finely powdered solid or may be supported on some inert material, such as finely powdered charcoal or alumina. The reaction is usually carried out by dissolving the alkene in ethanol or another nonreacting organic solvent, adding the solid catalyst, and then shaking the mixture under hydrogen gas. Alternatively, the metal may be complexed with certain organic molecules and used in the form of a soluble complex.

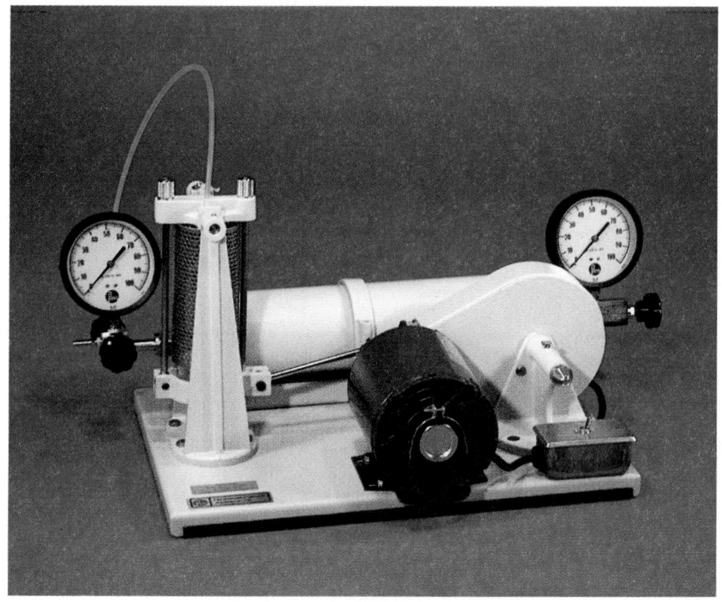

Parr shaker-type hydrogenation apparatus. *(Parr Instrument Co., Moline, IL)*

The most common pattern in catalytic reduction of an alkene is **syn addition** of hydrogens to the carbon-carbon double bond. Catalytic reduction of 1,2-dimethylcyclohexene, for example, yields predominantly *cis*-1,2-dimethylcyclohexane. Along with the *cis*-isomer are formed lesser amounts of *trans*-1,2-dimethylcyclohexane.

$$\text{1,2-Dimethylcyclohexene} + H_2 \xrightarrow{Pt} \text{70\% to 85\% } cis\text{-1,2-Dimethylcyclohexane} + \text{30\% to 15\% } trans\text{-1,2-Dimethylcyclohexane}$$

A. Mechanism of Catalytic Reduction

Certain transition metals are able to adsorb large quantities of hydrogen onto their surfaces, probably by forming metal-hydrogen sigma bonds. Similarly, alkenes are also adsorbed on metal surfaces with formation of carbon-metal bonds [Figure 6.7(a)]. Addition of hydrogen atoms to the alkene most probably occurs in two steps. First, one new C—H bond is formed to give an intermediate in which the alkene remains partially adsorbed to the metal surface [Figure 6.7(b)]. The second hydrogen is then added from the same side as the first hydrogen.

If addition of hydrogens is syn stereoselective, how then is the formation of a *trans* product accounted for? It is proposed that before a second hydrogen can be delivered from the metal surface to complete the reduction, there is transfer of a hydrogen from a carbon atom adjacent to the original double bond to the metal surface. This hydrogen transfer in effect reverses the first step and forms a new alkene that is isomeric with the original alkene. As shown in the following equation, 1,2-dimethylcyclohexene undergoes isomerization on the metal surface to form 1,6-dimethylcyclohexene. This alkene then leaves the metal surface. When it is later readsorbed and reduced, hydrogens are still added to it with syn stereoselectivity, but not necessarily from the same side as the original hydrogen.

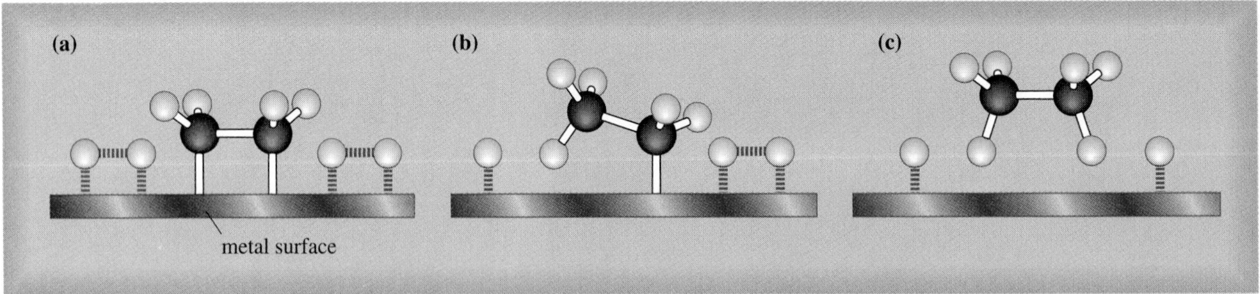

FIGURE 6.7
Addition of hydrogen to an alkene involving a transition metal catalyst. (a) Hydrogen and the alkene are adsorbed on the metal surface and (b) one hydrogen atom is transferred to the alkene forming a new C—H bond. The other carbon remains absorbed on the metal surface. (c) A second C—H bond is formed and the alkane is desorbed.

1,2-Dimethylcyclohexene $\xrightarrow{H_2/Pt}$ [Pt-bound intermediate] $\longrightarrow$ 1,6-Dimethylcyclohexene

B. Heats of Hydrogenation and the Relative Stabilities of Alkenes

Heat of hydrogenation of an alkene is defined as its heat of reaction, ΔH^0, with hydrogen to form an alkane. Table 6.2 lists heats of hydrogenation for several alkenes. Three important points are derived from information given in this table.

1. Reduction of an alkene to an alkane is exothermic. This observation is consistent with the fact that during hydrogenation, there is net conversion of a weaker pi bond to a stronger sigma bond, that is, one sigma bond (H—H) and one pi bond (C=C) are broken, and two new sigma bonds (C—H) are formed. For a comparison of the relative strengths of sigma and pi bonds, refer to Section 5.1D.
2. Heats of hydrogenation depend on the degree of substitution of the double bond; the greater the substitution, the lower the heat of hydrogenation. Compare, for example, heats of hydrogenation of ethylene (no substituents), propene (one substituent), 1-butene (one substituent), and the *cis-trans* isomers of 2-butene (two substituents).

TABLE 6.2 Heats of Hydrogenation of Several Alkenes

Name	Structural Formula	ΔH^0 [kcal/mol (kJ/mol)]
ethylene	$CH_2{=}CH_2$	−32.8 (−137.2)
propene	$CH_3CH{=}CH_2$	−30.1 (−125.9)
1-butene	$CH_3CH_2CH{=}CH_2$	−30.3 (−126.8)
cis-2-butene	$(H_3C)(H)C{=}C(CH_3)(H)$ (both CH_3 on same side)	−28.6 (−119.7)
trans-2-butene	$(H_3C)(H)C{=}C(H)(CH_3)$ (CH_3 on opposite sides)	−27.6 (−115.5)
2-methyl-2-butene	$(H_3C)_2C{=}C(CH_3)(H)$	−26.9 (−113)
2,3-dimethyl-2-butene	$(H_3C)_2C{=}C(CH_3)_2$	−26.6 (−111)

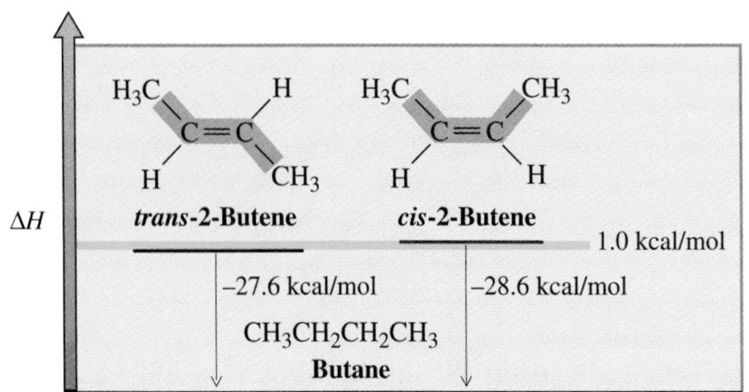

FIGURE 6.8
Heats of hydrogenation of *cis*-2-butene and *trans*-2-butene. *Trans*-2-butene is more stable than the *cis* isomer by 1.0 kcal/mol (4.2 kJ/mol).

3. The heat of hydrogenation of a *trans* alkene is lower than that of the isomeric *cis* alkene. Compare, for example, the heats of hydrogenation of *cis*-2-butene and *trans*-2-butene. Because reduction of each alkene gives butane, any difference in heats of hydrogenation must be due to a difference in relative energy between the two alkenes (Figure 6.8). The alkene with the lower (less negative) value of ΔH^0 is the more stable alkene.

We explain the greater stability of *trans* alkenes relative to *cis* alkenes in terms of nonbonded interaction strain, which can be visualized using space-filling models (Figure 6.9). In *cis*-2-butene, the two $—CH_3$ groups are sufficiently close to each other that there is repulsion between the electron clouds of each. This repulsion is reflected in the larger heat of hydrogenation (decreased stability) of *cis*-2-butene compared with *trans*-2-butene (approximately 1.0 kcal/mol).

C. Stability of Conjugated Dienes

Dienes are compounds that contain two carbon-carbon double bonds. Dienes can be divided into three groups: unconjugated, conjugated, and cumulated. An **unconjugated diene** is one in which the double bonds are separated by two or more single bonds. A **conjugated diene** is a diene in which the double bonds are separated by one single bond. A **cumulated diene** is one in which two double bonds share a carbon. We discuss the reactions of conjugated dienes in Chapter 22.

Conjugated diene A diene in which the double bonds are separated by one single bond.

$CH_2{=}CHCH_2CH{=}CH_2$	$CH_2{=}CHCH{=}CHCH_3$	$CH_2{=}C{=}CHCH_2CH_3$
1,4-Pentadiene (an unconjugated diene)	1,3-Pentadiene (a conjugated diene)	1,2-Pentadiene (a cumulated diene)

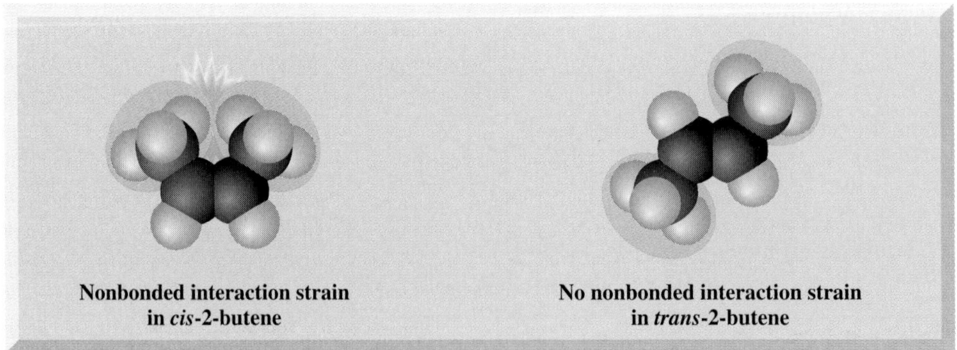

FIGURE 6.9
Space-filling models of *cis*-2-butene and *trans*-2-butene.

EXAMPLE 6.14

Which of these molecules contain conjugated double bonds?

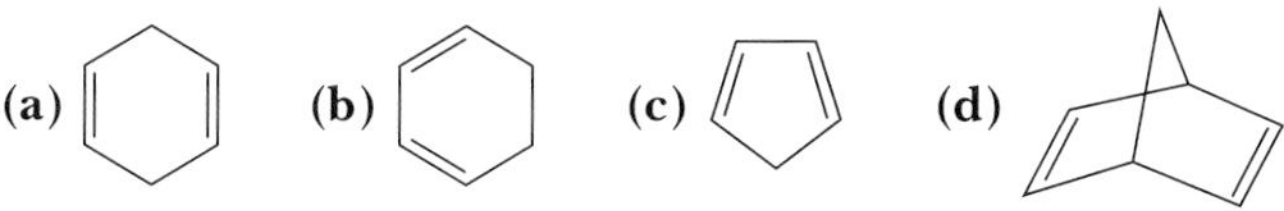

Solution

Compounds (b) and (c) contain conjugated double bonds. The double bonds in compounds (a) and (d) are unconjugated.

PROBLEM 6.14

Which of these terpenes (Figure 5.5) contains conjugated double bonds?

(a) CH_2OH Geraniol

(b) Limonene

(c) HO An aggregating pheromone of bark beetles

Given in Table 6.3 are heats of hydrogenation for several alkenes and conjugated dienes. By using these data, we can compare the relative stabilities of conjugated and unconjugated dienes.

The simplest conjugated diene is 1,3-butadiene, but because this molecule has only four carbon atoms, it has no unconjugated constitutional isomer. Nonetheless, we can estimate the effect of conjugation of two double bonds in this molecule in the following way. The heat of hydrogenation of 1-butene is −30.3 kcal/mol (127 kJ/mol). A molecule of 1,3-butadiene has two terminal double bonds, each with the same degree of substitution as the one double bond in 1-butene and, thus, we can estimate that the heat of hydrogenation of 1,3-butadiene should be 2(−30.3 kcal/mol)

TABLE 6.3 Heats of Hydrogenation of Several Alkenes and Conjugated Dienes

Name	Structural Formula	ΔH^0 kcal/mol (kJ/mol)
1-butene	$CH_2{=}CHCH_2CH_3$	−30.3 (127)
1-pentene	$CH_2{=}CHCH_2CH_2CH_3$	−30.1 (126)
cis-2-butene	$CH_3CH{=}CHCH_3$	−28.6 (120)
trans-2-butene	$CH_3CH{=}CHCH_3$	−27.6 (115)
1,3-butadiene	$CH_2{=}CHCH{=}CH_2$	−56.5 (236)
trans-1,3-pentadiene	$CH_2{=}CHCH{=}CHCH_3$	−54.1 (226)
1,4-pentadiene	$CH_2{=}CHCH_2CH{=}CH_2$	−60.8 (254)

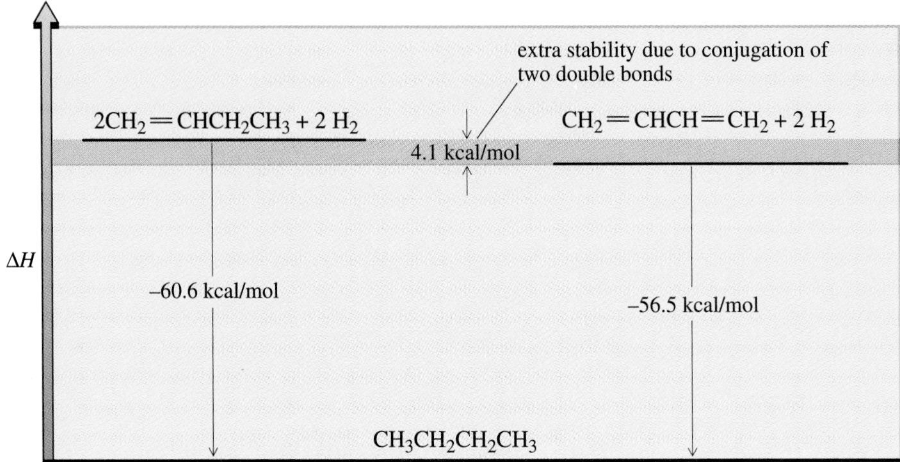

FIGURE 6.10 Conjugation of double bonds in butadiene gives the molecule an additional stability of approximately 4.1 kcal/mol (17 kJ/mol).

or -60.6 kcal/mol (254 kJ/mol). The observed heat of hydrogenation of 1,3-butadiene is -56.5 kcal/mol (236 kJ/mol), a value 4.1 kcal/mol (17 kJ/mol) less than estimated.

$$2CH_2{=}CHCH_2CH_3 + 2H_2 \xrightarrow[\text{}]{\text{transition metal catalyst}} 2CH_3CH_2CH_2CH_3 \qquad \Delta H^0 = 2(-30.3\ \text{kcal/mol}) = -60.6\ \text{kcal/mol}$$

$$CH_2{=}CHCH{=}CH_2 + 2H_2 \xrightarrow[\text{}]{\text{transition metal catalyst}} CH_3CH_2CH_2CH_3 \qquad \Delta H^0 = -56.5\ \text{kcal/mol}$$

The conclusion from this calculation is that conjugation of two double bonds in 1,3-butadiene gives an extra stability to the molecule of approximately 4.1 kcal/mol. These energy relationships are displayed graphically in Figure 6.10.

Calculations of this type for other conjugated and unconjugated dienes give similar results: Conjugated dienes are more stable than isomeric unconjugated dienes by approximately 3.5 to 4.0 kcal/mol. The effects of conjugation on stability are even more general. Compounds containing conjugated double bonds, not just those in dienes, are more stable than isomeric compounds containing unconjugated double bonds. For example, 2-cyclohexenone is more stable than its isomer 3-cyclohexenone.

2-Cyclohexenone (more stable) | 3-Cyclohexenone (less stable)

The additional stability of conjugated dienes relative to unconjugated dienes arises from delocalization of electron density in the conjugated diene. In two unconjugated double bonds, each pair of pi electrons is localized between two carbons. In a conjugated diene, however, the four pi electrons are delocalized over the entire set of four parallel $2p$ orbitals (Figure 6.11).

According to the molecular orbital model, the conjugated system of a diene is described as a set of four pi molecular orbitals arising from combination of four $2p$

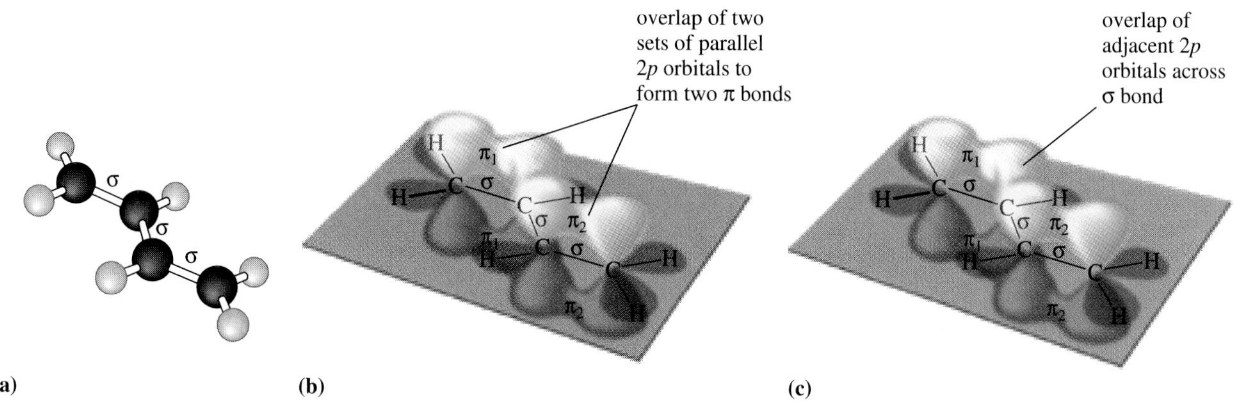

FIGURE 6.11
Structure of 1,3-butadiene: orbital overlap model. (a) The skeletal framework; all bond angles are approximately 120° and all atoms lie in the same plane or nearly so. (b) Overlap of pairs of parallel $2p$ orbitals on C_1 and C_2 and on C_3 and C_4 forms two pi bonds. (c) Overlap of the $2p$ orbitals on C_2 and C_3 allows for even greater delocalization of electron density.

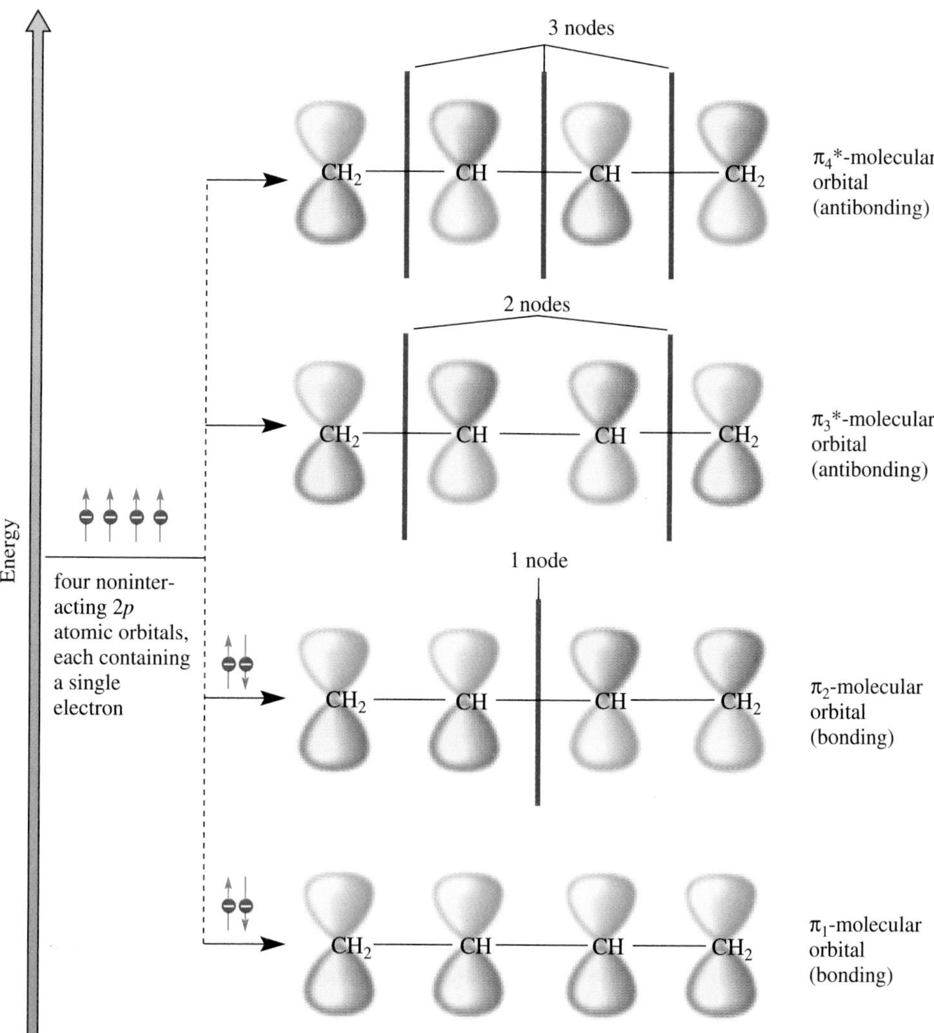

FIGURE 6.12
Structure of 1,3-butadiene: molecular orbital model. Combination of four parallel $2p$ atomic orbitals gives two pi bonding MOs and two pi antibonding MOs. In the ground state, each pi bonding MO is filled with two spin-paired electrons. The pi antibonding MOs are unoccupied.

atomic orbitals. These MOs have zero, one, two, and three nodes, respectively, as illustrated in Figure 6.12. In the ground state, all four pi electrons lie in pi bonding MOs.

EXAMPLE 6.15

Using data from Table 6.3, estimate the extra stability due to the conjugation of double bonds in *trans*-1,3-pentadiene. Display the results of your calculations in graphical form.

Solution

Compare the sum of heats of hydrogenation of 1-pentene and *trans*-2-butene with the heat of hydrogenation of *trans*-1,3-pentadiene. Conjugation of double bonds in *trans*-1,3-pentadiene imparts an added stability of approximately 3.6 kcal/mol.

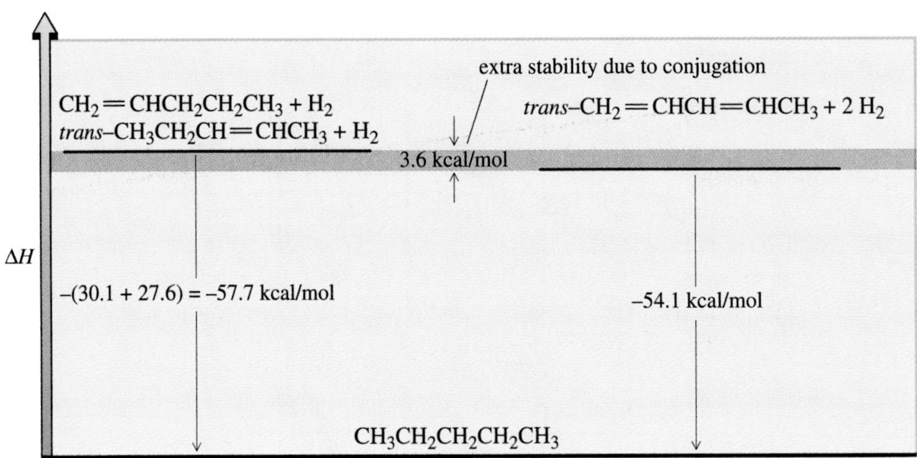

PROBLEM 6.15

Estimate the stabilization gained due to conjugation when 1,4-pentadiene is converted to *trans*-1,3-pentadiene. Note that the answer is not as simple as comparing the heats of hydrogenation of 1,4-pentadiene and *trans*-1,3-pentadiene because, although the double bonds are moved from unconjugated to conjugated, the degree of substitution of one of the double bonds is also changed, in this case from a monosubstituted double bond to a *trans* disubstituted double bond. To answer this question you must separate the effect due to conjugation from that due to change in degree of substitution.

6.7 Molecules Containing Stereocenters as Reactants or Products

As the structure of an organic compound is altered in the course of a reaction, one or more stereocenters, usually at carbon, may be created, inverted, or destroyed. Let us consider two examples from this chapter to illustrate these possibilities.

Addition of bromine to 2-butene (Section 6.3D) gives 2,3-dibromobutane, a molecule with two stereocenters. Three stereoisomers are possible for this compound: a meso compound and a pair of enantiomers (Section 4.5). We now ask, "Is the product one enantiomer, a pair of enantiomers, the meso compound, or a mixture of all three stereoisomers?" A partial answer is that the product formed depends on the configuration of the alkene. Let us first examine addition of bromine to *cis*-2-butene.

Attack of bromine on *cis*-2-butene from either face of the planar part of the molecule gives the same bridged bromonium ion intermediate (Figure 6.13). Although this intermediate has two stereocenters, it has a plane of symmetry and is, therefore,

Step 1: Formation of a meso bromonium ion intermediate

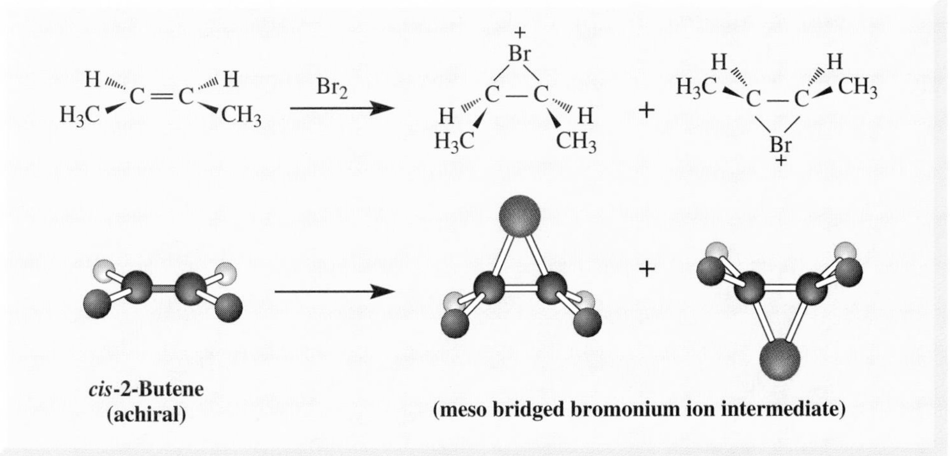

Step 2: Attack of bromide ion at carbon 2 gives one enantiomer; attack at carbon 3 gives the other enantiomer. Attack occurs with equal probability and rate at carbons 2 and 3 and gives a racemic mixture

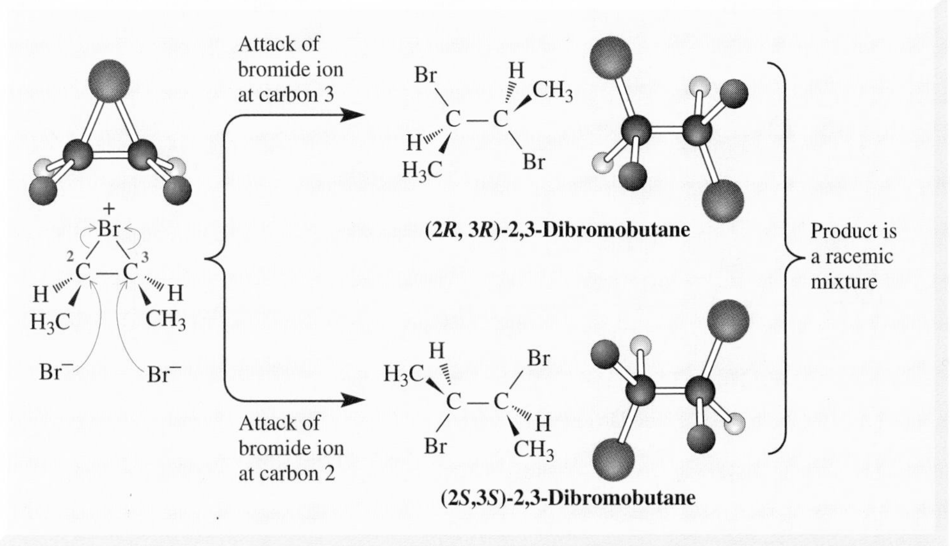

FIGURE 6.13
Anti stereoselective addition of bromine to *cis*-2-butene gives racemic 2,3-dibromobutane.

meso. Attack of Br^- on this intermediate from the side opposite that of the bromonium ion bridge gives a pair of enantiomers. Attack of bromide ion on carbon 2 of this meso intermediate gives the (2*S*,3*S*) enantiomer. Attack of bromide ion on carbon 3 gives the (2*R*,3*R*) enantiomer. Attack of bromide ion occurs equally at each carbon and, therefore, the enantiomers are formed in equal amounts and 2,3-dibromobutane is obtained as a racemic mixture (Figure 6.13). We have shown attack of Br^+ from one side of the carbon-carbon double bond. Attack of Br^+ from the opposite side followed by opening of the resulting bromonium ion intermediate also produces these same two stereoisomers.

Addition of Br_2 to *trans*-2-butene leads to two enantiomeric bridged bromonium ion intermediates. Attack by Br^- at either carbon atom of either bromonium ion intermediate gives the meso product, which is optically inactive (Figure 6.14).

In Section 6.5B we studied oxidation of alkenes by osmium tetroxide in the presence of hydrogen peroxide. This oxidation brings about syn stereoselective hydroxylation of the alkene to form a glycol. In the case of cycloalkenes, the product is a *cis* glycol. The first step in each oxidation involves formation of a cyclic osmate ester and

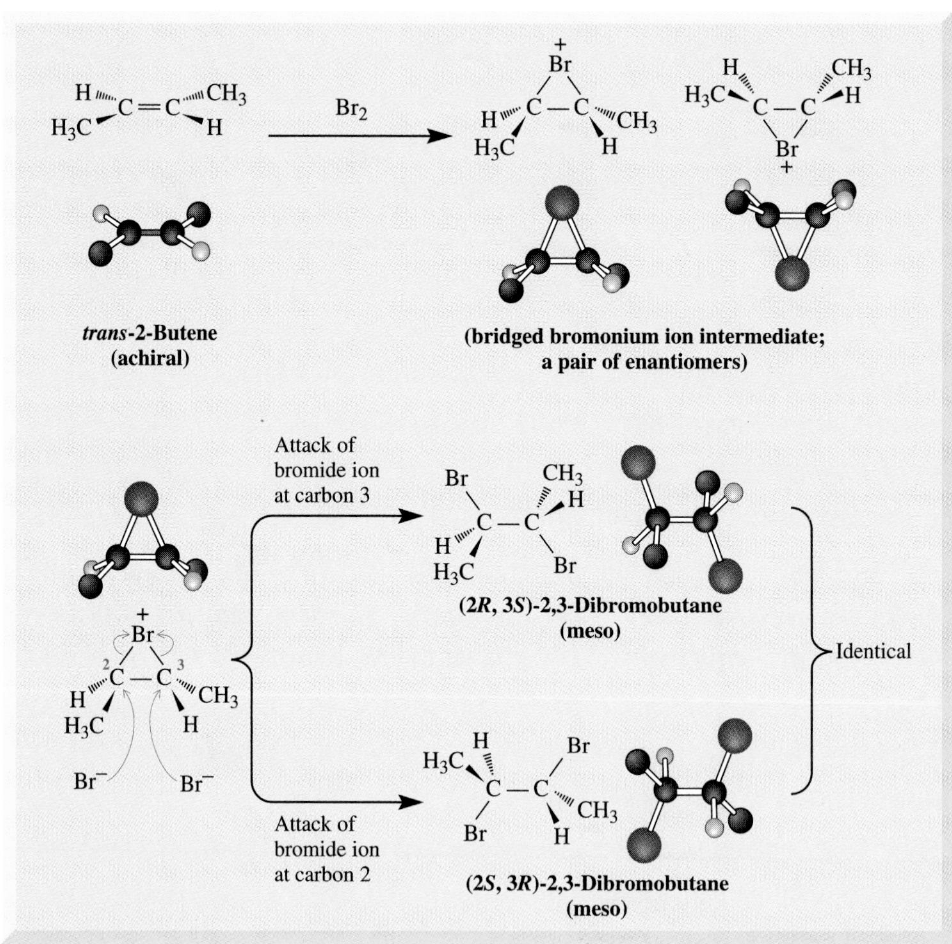

FIGURE 6.14
Anti stereoselective addition of bromine to *trans*-2-butene gives meso-2,3-dibromobutane.

is followed intermediately by reaction with water to give a glycol. As shown in the following sequences, syn hydroxylation of *cis*-2-butene gives meso-2,3-butanediol. Syn hydroxylation of *trans*-2-butene gives racemic 2,3-butanediol.

cis-2-Butene (achiral) $\xrightarrow[\text{ROOH}]{\text{OsO}_4}$ (2*S*,3*R*)-2,3-Butanediol + (2*R*,3*S*)-2,3-Butanediol — Identical; a meso compound

trans-2-Butene (achiral) $\xrightarrow[\text{ROOH}]{\text{OsO}_4}$ (2*S*,3*S*)-2,3-Butanediol + (2*R*,3*R*)-2,3-Butanediol — A pair of enantiomers; a racemic mixture

Note that the stereochemistry of the product of osmium tetroxide oxidation of *trans*-2-butene is opposite to that formed on addition of bromine to *trans*-2-butene. Osmium tetroxide oxidation gives the glycol as a pair of enantiomers forming a racemic mixture. Addition of bromine to *trans*-2-butene gives the dibromide as a meso compound. A similar difference is observed between the stereochemical outcomes of these reactions with *cis*-2-butene.

In this section, we have seen several examples of reactions in which achiral starting materials give chiral products. In each case, the chiral product is formed as a racemic mixture. This illustrates a very important point about the creation of chiral molecules, namely that an enantiomerically pure compound can never be produced from achiral starting materials and reagents. Although the molecules of the product may be chiral, their enantiomers are produced in equal amounts as a racemic mixture. Conversely, an enantiomerically pure compound can be generated in a reaction if at least one of the reactants is enantiomerically pure, or if the reaction is carried out in the presence of a catalyst that is itself chiral and enantiomerically pure. Recall the example given in Section 4.9B in which the enzyme (*R*)-lipase was used to catalyze the formation of an ester with a high degree of enantiomeric purity.

We will encounter many reactions throughout the remainder of this course where achiral starting materials are converted into chiral products. For convenience, we often draw just one of the enantiomeric products, but we must always keep in mind that both are formed in equal amounts.

SUMMARY

A **reaction mechanism** (Section 6.2) is a detailed description of how and why a chemical reaction occurs as it does, which bonds are broken and which new ones are formed, the order in which the various bond-breaking and bond-forming steps take place and their relative rates, the role of the solvent if the reaction takes place in solution, and the role of the catalyst if the reaction involves a catalyst. **Transition state theory** (Section 6.2A) provides a model for understanding the relationships between molecular structure, reaction rates, and energetics. A potential energy diagram shows the changes in energy that occur in going from reactants to products. Potential energy is measured on the vertical axis and the change in position of the atoms during reaction is measured on the horizontal axis (the reaction coordinate). A **transition state** is a point on the reaction coordinate at which the potential energy is a maximum. The difference in potential energy between reactants and a transition state is called the **activation energy, E_a**. An **intermediate** corresponds to a potential energy minimum between two transition states. In a multistep reaction, the step that crosses the highest potential energy barrier is called the **rate-limiting step.** The relationship between activation energy and rate constant for a reaction is given by the Arrhenius equation, $k = Ae^{-E_a/RT}$ (Section 6.2B).

A characteristic reaction of alkenes is **addition,** during which a pi bond is broken and sigma bonds to two new atoms are formed. A **regioselective reaction** is a reaction in which one direction of bond-forming or bond-breaking occurs in preference to all other directions (Section 6.3A). According to **Markovnikov's rule** (Section 6.3A), in addition of HX or H_2O to an alkene, hydrogen adds to the carbon of the double bond having the greater number of hydrogens. An **electrophile** (Section 6.3A) is any atom, molecule, or ion that can accept a pair of electrons to form a new covalent bond. By definition, an electrophile is also a Lewis acid (Section 3.5). The rate-limiting step in **electrophilic addition** to an alkene is formation of a **carbocation** intermediate. A carbocation is a positively charged ion that contains a carbon atom with only six electrons in its valence shell. Carbocations are planar with bond angles of approximately 120° about the positive carbon. The order of stability of carbocations is 3° > 2° > 1° > methyl. Carbocations are stabilized by the electron-releasing **inductive effect** of alkyl groups bonded to the cationic carbon and by **hyperconjugation** (Section 6.3A).

A **stereoselective reaction** is a reaction in which one stereoisomer is formed or destroyed in preference to all others that might be formed or destroyed (Section 6.3D). Addition of new atoms or groups of atoms from opposite sides or faces of a double bond is called **anti addition.** In cyclic systems, anti addition is equivalent to *trans*-coplanar addition. **Syn addition** is the addition of atoms or groups of atoms to the same side or face of a double bond.

The driving force for **carbocation rearrangement** (Section 6.3C) is conversion of a primary or secondary carbocation to a more stable secondary or tertiary carbocation. Rearrangement is by a 1,2-shift in which an atom or group of atoms moves with its bonding electrons from one atom to an adjacent atom within the same molecule.

To determine if a transformation is an oxidation, a reduction, or neither, use a **balanced half-reaction** (Section 6.5A). **Oxidation** is the loss of electrons; **reduction** is the gain of electrons.

From **heats of hydrogenation** of a series of alkenes (Section 6.6B), we conclude that in general (1) the greater the degree of substitution of the carbon-carbon double bond, the more stable the alkene and (2) a *trans* alkene is more stable than a *cis* alkene. Compounds containing **conjugated double bonds** are more stable than isomeric compounds containing unconjugated double bonds (Section 6.6C). The extra stability of two conjugated double bonds arises because the overlap of four parallel $2p$ orbitals results in delocalization of electron density over the entire pi framework.

KEY REACTIONS

1. Addition of HX (Section 6.3A)

Addition is regioselective and follows Markovnikov's rule. Reaction involves a carbocation intermediate (a Lewis acid), which may rearrange before it combines with a halide ion (a Lewis base) to complete the reaction.

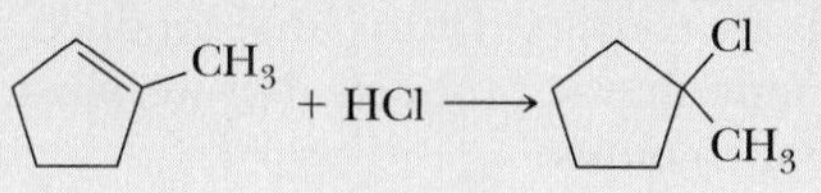

CH_3 + HCl ⟶ Cl, CH_3

2. Acid-Catalyzed Hydration (Section 6.3B)

Addition is regioselective and follows Markovnikov's rule. Reaction involves a carbocation intermediate that may rearrange before it combines with water (a Lewis base) to complete the reaction.

$$CH_3\overset{\overset{\displaystyle CH_3}{|}}{C}{=}CH_2 + H_2O \xrightarrow{H_2SO_4} CH_3\underset{\underset{\displaystyle OH}{|}}{\overset{\overset{\displaystyle CH_3}{|}}{C}}CH_3$$

3. Addition of Bromine and Chlorine (Section 6.3D)

Addition is anti stereoselective; it involves anti addition of halogen atoms by way of a bridged halonium ion intermediate with no rearrangement.

$$\text{cyclohexene} + Br_2 \xrightarrow[CCl_4]{} \text{trans-1,2-dibromocyclohexane (Br wedge, Br hash)}$$

4. Addition of HOCl and HOBr (Section 6.3E)

Addition is regioselective (—X adds to the less substituted carbon via a bridged halonium ion intermediate and —OH adds to the more substituted carbon), anti stereoselective, and occurs without rearrangement.

$$\text{1-methylcyclopentene (H, } CH_3) \xrightarrow{Br_2/H_2O} \text{product: Br, } H_3C \text{ (up); H, OH (down)}$$

5. Oxymercuration-Reduction (Section 6.3F)

Oxymercuration is regioselective (HgOAc adds to the less substituted carbon and OH adds to the more substituted carbon) and anti stereoselective. Anti addition occurs via a bridged mercurinium ion intermediate with no rearrangement. The result of oxymercuration-reduction is Markovnikov hydration of an alkene.

$$CH_3\overset{\overset{\displaystyle CH_3}{|}}{C}HCH{=}CH_2 \xrightarrow[2.\ NaBH_4]{1.\ Hg(OAc)_2,\ H_2O} CH_3\overset{\overset{\displaystyle CH_3}{|}}{C}H\underset{\underset{\displaystyle OH}{|}}{C}HCH_3$$

6. Hydroboration-Oxidation (Section 6.4)

Addition of BH_3 is syn stereoselective and regioselective (boron adds to the less substituted carbon and hydrogen to the more substituted carbon). Hydroboration-oxidation results in anti-Markovnikov hydration of the alkene without rearrangement.

$$\text{1-methylcyclohexene} \xrightarrow[2.\ H_2O_2,\ NaOH]{1.\ BH_3} \text{trans-2-methylcyclohexanol } (CH_3 \text{ wedge, OH hash})$$

7. Oxidation to a Glycol by OsO_4 (Section 6.5B)

Oxidation gives a glycol resulting from syn addition of —OH groups to the double bond via a cyclic osmate ester with no rearrangement.

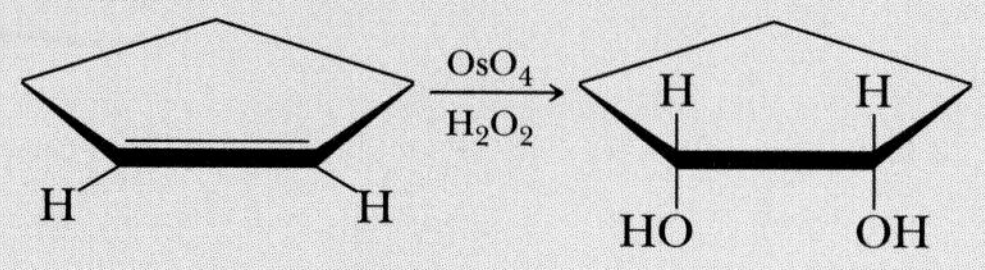

8. Oxidation by Ozone (Section 6.5C)

Treatment with ozone followed by dimethyl sulfide cleaves a carbon-carbon double bond and gives two carbonyl groups in its place.

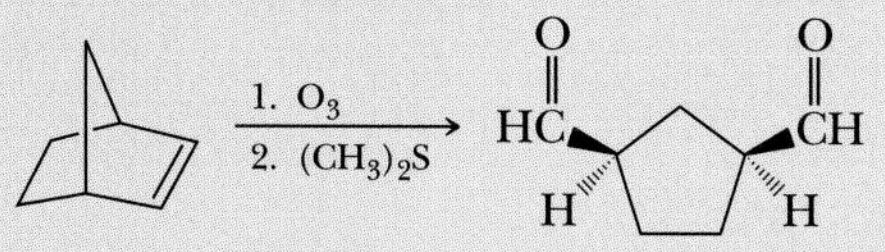

9. Addition of H_2; Catalytic Reduction (Section 6.6)

Catalytic reduction involves predominantly syn addition of hydrogens.

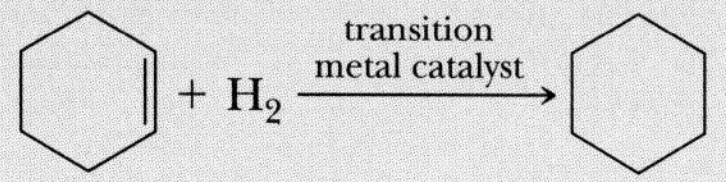

ADDITIONAL PROBLEMS

Energetics of Chemical Reactions

6.16 Most chemical reactions can occur as written if they are exothermic, that is, if the bonds that are formed in the products are stronger than the ones broken in the starting materials. To determine if a reaction is exothermic as written, add the bond dissociation energies of all bonds broken in the starting materials (it costs energy to break bonds). Subtract from this the total of bond dissociation energies of all bonds formed in the products (formation of bonds liberates energy). If the sum of these numbers is negative, the reaction is exothermic (energy is liberated) and the reaction proceeds to the right as written. If the sum of these numbers is positive, the reaction is endothermic (it requires energy) and it does not proceed to the right as written. Using the table of bond dissociation energies at 25°C, determine which of the following reactions are energetically favorable at room temperature, that is, if a suitable catalyst could be found, which would proceed to the right as written?

Bond	Bond Dissociation Energy [kcal/mol (kJ/mol)]	Bond	Bond Dissociation Energy [kcal/mol (kJ/mol)]
H—H	104 (535)	C—Si	72 (201)
O—H	110.6 (462.8)	C=C	146 (611)
C—H	98.7 (413)	C=O (aldehyde)	174 (728)
N—H	93.4 (391)	C=O (CO_2)	192 (803)
Si—H	76 (318)	C≡O (CO)	257 (1075)
C—C	82.6 (346)	N≡N	227 (950)
C—N	73 (305)	C≡C	230 (962)
C—O	85.5 (358)	O=O	119 (498)
C—I	51 (213)		

(a) $CH_2{=}CH_2 + 2H_2 + N_2 \longrightarrow H_2NCH_2CH_2NH_2$

(b) $CH_2{=}CH_2 + CH_4 \longrightarrow CH_3CH_2CH_3$

(c) $CH_2{=}CH_2 + (CH_3)_3SiH \longrightarrow CH_3CH_2Si(CH_3)_3$

(d) $CH_2{=}CH_2 + CHI_3 \longrightarrow CH_3CH_2CI_3$

(e) $CH_2{=}CH_2 + CO + H_2 \longrightarrow CH_3CH_2CH{=}O$

(f) 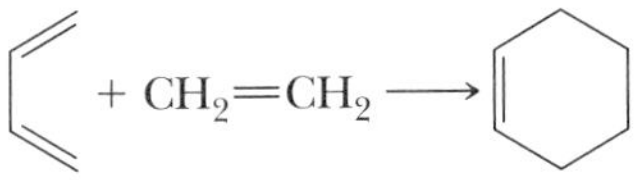 1,3-butadiene $+ CH_2{=}CH_2 \longrightarrow$ cyclohexene

(g) 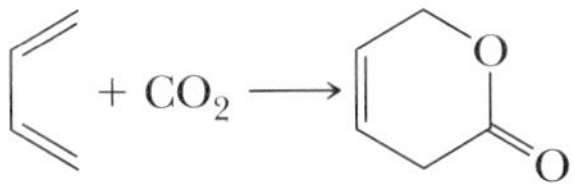 1,3-butadiene $+ CO_2 \longrightarrow$ (cyclic lactone, O in ring, C=O)

(h) $HC{\equiv}CH + O_2 \longrightarrow HC(=O){-}CH(=O)$

(i) $2CH_4 + O_2 \longrightarrow 2CH_3OH$

Electrophilic Additions

6.17 Draw structural formulas for the isomeric carbocations formed by the addition of H^+ to each alkene. Label each carbocation primary, secondary, or tertiary, and state which of the isomeric carbocations is formed more readily.

(a) $CH_3CH_2C(CH_3){=}CHCH_3$ (b) $CH_3CH_2CH{=}CHCH_3$ (c) 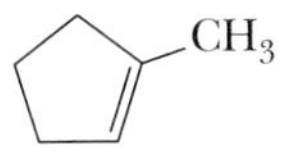 1-methylcyclopentene (CH_3) (d) 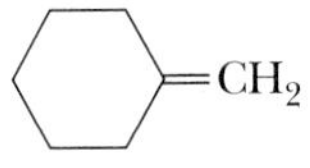 methylenecyclohexane ($=CH_2$)

6.18 Arrange the alkenes in each set in order of increasing rate of reaction with HI. Draw the structural formula of the major product formed in each case, and explain the basis for your ranking.

(a) $CH_3CH{=}CHCH_3$ and $CH_3C(CH_3){=}CHCH_3$ (b) 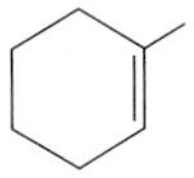 1-methylcyclohexene and cyclohexene

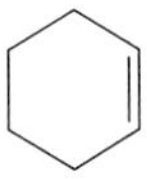

6.43 Draw the structural formula of the alkene that reacts with ozone followed by dimethyl sulfide to give each product or set of products.

(a) $C_7H_{12} \xrightarrow[\text{2. } (CH_3)_2S]{\text{1. } O_3} CH_3C(=O)CH_2CH_2CH_2C(=O)CH_3$

(b) $C_{10}H_{18} \xrightarrow[\text{2. } (CH_3)_2S]{\text{1. } O_3} CH_3C(=O)CH_3 + CH_3C(=O)CH_2CH_3 + HC(=O)CH_2C(=O)H$

(c) $C_{10}H_{18} \xrightarrow[\text{2. } (CH_3)_2S]{\text{1. } O_3} CH_3CH(CH_3)CH_2C(=O)CH_2CH_2CH_2CH_2C(=O)H$

6.44 Bicyclo[2.2.1]-2-heptene (norbornene) is oxidized by ozone/dimethyl sulfide to cyclopentane-1,3-dicarbaldehyde.

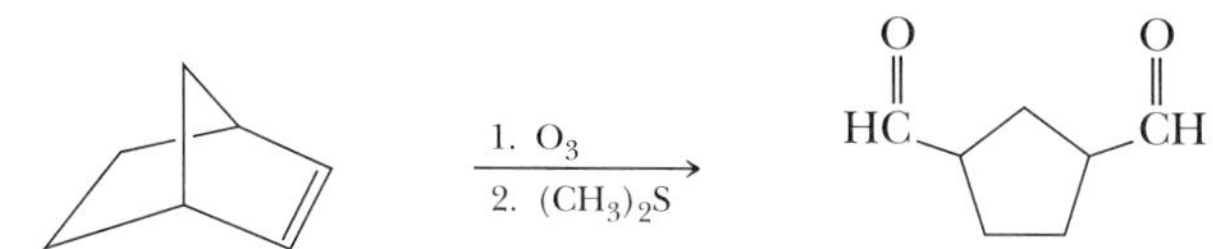

Bicyclo[2.2.1]-2-heptene (Norbornene)   Cyclopentane-1,3-dicarbaldehyde

(a) How many stereoisomers are possible for this dicarbaldehyde?
(b) Which of the possible stereoisomers is/are formed by ozonolysis of norbornene?

6.45 **(a)** Draw a structural formula for the bicycloalkene of molecular formula C_8H_{12} that, on treatment with ozone followed by dimethyl sulfide, gives cyclohexane-1,4-dicarbaldehyde.
(b) Do you predict the product to be the *cis* isomer, the *trans* isomer, or a mixture of *cis* and *trans* isomers? Explain.
(c) Draw a suitable stereorepresentation for the more stable chair conformation of the dicarbaldehyde formed in this oxidation.

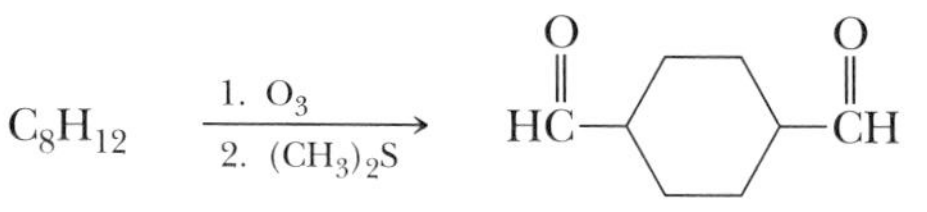

$C_8H_{12} \xrightarrow[\text{2. } (CH_3)_2S]{\text{1. } O_3}$ HC(=O)–C₆H₁₀–C(=O)H

Cyclohexane-1,4-dicarbaldehyde

Reduction

6.46 Predict the major organic product(s) of the following reactions. Show stereochemistry where appropriate.

(a) Geraniol (CH_2OH) $+ 2H_2 \xrightarrow{Pt}$ **(b)** α-Pinene $+ H_2 \xrightarrow{Pt}$

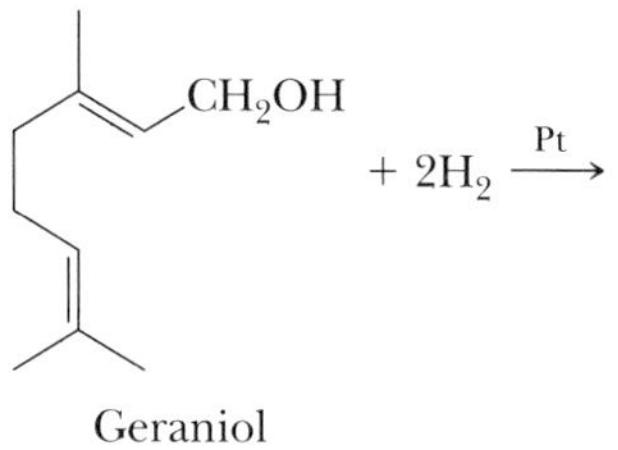

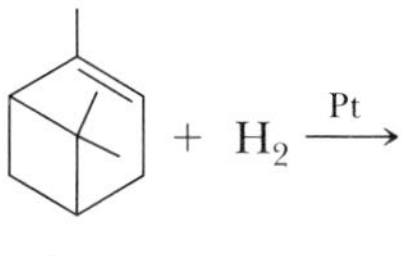

6.47 The heat of hydrogenation of allene (1,2-propadiene) to propene is −35.3 kcal/mol (177 kJ/mol). Compare this value with the heat of hydrogenation of 1,3-butadiene to 1-butene. Does allene have the characteristics of a conjugated or a nonconjugated diene?

6.48 The heat of hydrogenation of *cis*-di-*tert*-butylethylene is −36.7 kcal/mol (154 kJ/mol) while that of the *trans* isomer is only −26.9 kcal/mol (113 kJ/mol).

(a) Why is the heat of hydrogenation of the *cis* isomer so much larger than that of the *trans* isomer?

(b) If a catalyst could be found that allowed equilibration of the *cis* and *trans* isomers at room temperature (such catalysts do exist), what would be the ratio of *trans* to *cis* isomers?

Synthesis

6.49 Show how to convert ethylene to these compounds.

(a) Ethane **(b)** Ethanol **(c)** Bromoethane
(d) 2-Chloroethanol **(e)** 1,2-Dibromoethane **(f)** 1,2-Ethanediol
(g) Chloroethane

6.50 Show how to convert cyclopentene into these compounds.

(a) *trans*-1,2-Dibromocyclopentane **(b)** *cis*-1,2-Cyclopentanediol
(c) Cyclopentanol **(d)** Iodocyclopentane
(e) Cyclopentane **(f)** Pentanedial

Reactions that Produce Chiral Compounds

6.51 State the number and kind of stereoisomers formed when (*R*)-3-methyl-1-pentene is treated with these reagents.

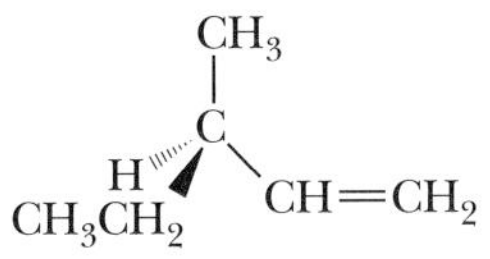

CH_3
C
H, CH_3CH_2, $CH{=}CH_2$

(*R*)-3-Methyl-1-pentene

(a) $Hg(OAc)_2$, H_2O followed by $NaBH_4$ **(b)** H_2/Pt

(c) BH_3 followed by H_2O_2 in NaOH **(d)** Br_2 in CCl_4

6.52 Describe the stereochemistry of the bromohydrin formed in each reaction.

(a) *cis*-3-Hexene + Br_2/H_2O **(b)** *trans*-3-Hexene + Br_2/H_2O

6.53 In each of these reactions, the organic starting material is achiral. The structural formula of the product is given. For each reaction state:

(1) How many stereoisomers are possible for the product.
(2) Which of the possible stereoisomers is/are formed in the reaction shown.
(3) Whether the product is optically active or optically inactive.

(a) cyclopentyl–$CH{=}CH_2$ $\xrightarrow[\text{2. NaBH}_4]{\text{1. Hg(OAc)}_2\text{, H}_2\text{O}}$ cyclopentyl–$\underset{}{CHCH_3}$ (with OH on CH)

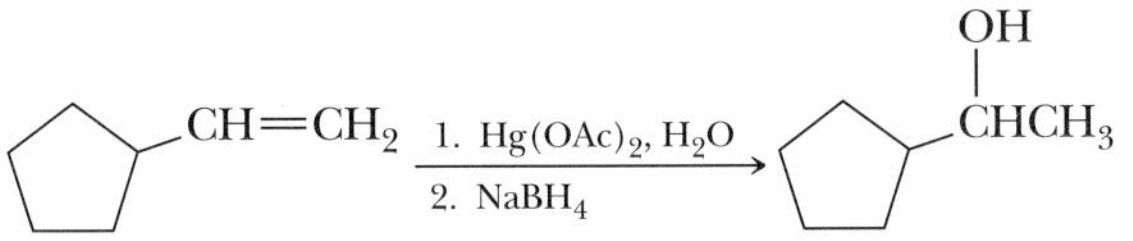

(b) H_3C, H / C=C / CH_3, H + Br_2 $\xrightarrow{CCl_4}$ $CH_3CH{-}CHCH_3$ (Br on each CH)

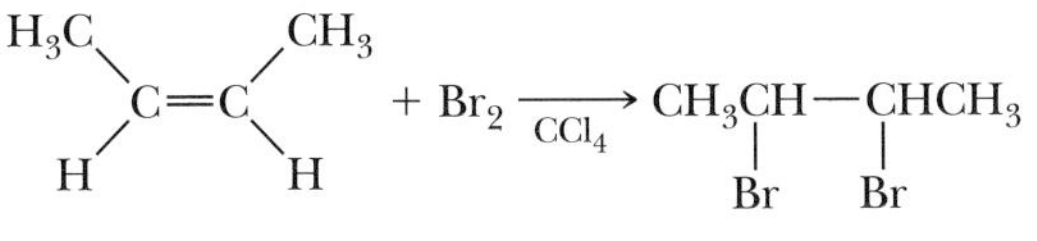

(c) $H_3C(H)C{=}C(H)CH_2CH_3 + Br_2 \xrightarrow{CCl_4} CH_3\underset{Br}{CH}{-}\underset{Br}{CH}CH_2CH_3$

(d) $CH_3CH_2CH{=}C(CH_3)_2 + HCl \longrightarrow CH_3CH_2CH_2\overset{CH_3}{\underset{Cl}{C}}CH_3$

(e) cyclohexene + Cl_2 in H_2O ⟶ 2-chlorocyclohexanol (OH, Cl)

(f) $\xrightarrow[ROOH]{OsO_4}$ (OH, OH)

(g) 1-methylcyclohexene (CH_3) $\xrightarrow[2.\ H_2O_2,\ NaOH]{1.\ BH_3}$ (OH, CH_3)

(h) 1-methylcyclopentene (CH_3) + HBr ⟶ (CH_3, Br)

7 ALKYL HALIDES AND RADICAL REACTIONS

■ Compact disks are made of polyvinyl chloride. *(Charles D. Winters)*

Compounds containing a halogen atom covalently bonded to an sp^3 hybridized carbon atom are named haloalkanes or, in the common system of nomenclature, alkyl halides. Our purpose in this chapter is to use the preparation of alkyl halides as a means to introduce an important type of reaction mechanism, namely a radical chain mechanism.

7.1 Structure

Alkyl halide A compound containing a halogen atom covalently bonded to an sp^3 hybridized carbon atom. Given the symbol R—X.

Vinylic halide A compound containing a halogen bonded to one of the carbons of a carbon-carbon double bond in an alkene.

Aryl halide A compound containing a halogen atom bonded to a benzene ring. Given the symbol Ar—X.

The general symbol for an **alkyl halide** is R—X, where —X may be —F, —Cl, —Br, or —I. If the halogen is bonded to a doubly bonded carbon, the compound belongs to a class called **vinylic halides.** If it is bonded to a benzene ring, the compound belongs to a class called **aryl halides.**

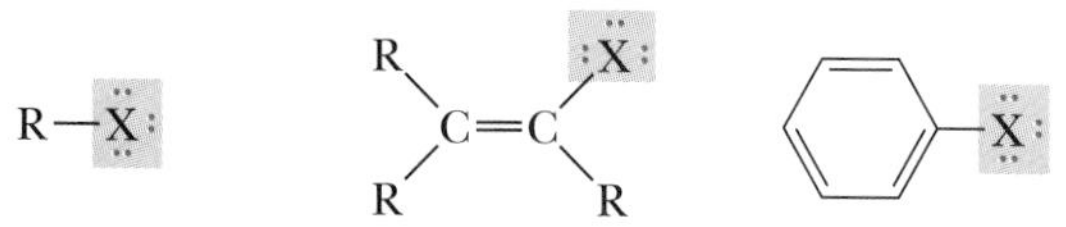

A haloalkane (an alkyl halide) | A haloalkene (a vinylic halide) | A haloarene (an aryl halide)

In this chapter, we are concerned only with the synthesis and physical properties of alkyl halides. Neither vinylic nor aryl halides can be prepared by the reactions we describe in this chapter.

7.2 Nomenclature

A. IUPAC System

IUPAC names for haloalkanes are derived by naming the parent alkane according to the rules given in Section 2.3A. The parent chain is numbered from the direction that gives the substituent encountered first the lower number, be it halogen or an alkyl group. Halogen substituents are indicated by the prefixes fluoro-, chloro-, bromo-, and iodo- and are listed in alphabetical order with other substituents. The location of each halogen atom on the parent chain is given by a number preceding the name of the halogen.

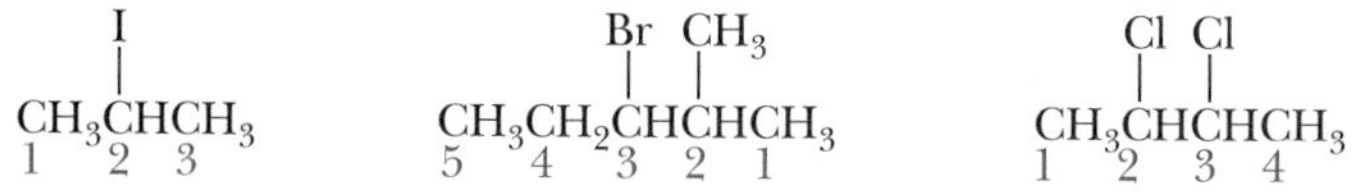

$CH_3CH(I)CH_3$ — 2-Iodopropane

$CH_3CH_2CH(Br)CH(CH_3)CH_3$ — 3-Bromo-2-methylpentane

$CH_3CH(Cl)CH(Cl)CH_3$ — 2,3-Dichlorobutane

In haloalkenes, numbering the parent hydrocarbon is determined by the location of the carbon-carbon double bond. Numbering is done in the direction that gives the carbon atoms of the double bond the lowest set of numbers.

$CH_3C(CH_3)(Cl)CH{=}CH_2$ — 3-Chloro-3-methyl-1-butene

4-Bromocyclohexene

B. Common Names

Common names of haloalkanes and haloalkenes consist of the common name of the alkyl group followed by the name of the halide as a separate word. Hence, the name

alkyl halide is a common name for this class of compounds. In the following examples, the IUPAC name of the compound is given first and then, in parentheses, its common name.

CH_3CH_2Cl	$CH_2{=}CHCl$	$CH_2{=}CHCH_2Cl$
Chloroethane (Ethyl chloride)	Chloroethene (Vinyl chloride)	3-Chloropropene (Allyl chloride)
$CH_3CH_2CH_2Cl$	$CH_3CH(Br)CH_3$	$CH_3CH(I)CH_2CH_3$
1-Chloropropane (Propyl chloride)	2-Bromopropane (Isopropyl bromide)	2-Iodobutane (*sec*-Butyl iodide)

Several polyhaloalkanes are common solvents and are generally referred to by their common, or trivial, names. Dichloromethane (methylene chloride) is the most widely used haloalkane solvent. Compounds of the type CHX_3 are called **haloforms.** The common name for $CHCl_3$, for example, is chloroform. It is from the name "chloroform" that the common name "methyl chloroform" is derived for the compound CH_3CCl_3. Methyl chloroform and trichloroethylene are common solvents for commercial dry cleaning.

Haloform A compound of the type CHX_3 where X is a halogen.

CH_2Cl_2	$CHCl_3$	CH_3CCl_3	$CCl_2{=}CHCl$
Dichloromethane (Methylene chloride)	Trichloromethane (Chloroform)	1,1,1-Trichloroethane (Methyl chloroform)	Trichloroethylene (Trichlor)

Hydrocarbons in which all hydrogens are replaced by halogens are commonly named as perhaloalkanes or perhaloalkenes.

$Cl_3C{-}CCl_3$	$F_3C{-}CF_2{-}CF_3$	$Cl_2C{=}CCl_2$
Perchloroethane	Perfluoropropane	Perchloroethylene

EXAMPLE 7.1

Write the IUPAC name and, where possible, the common name of each compound.

(a) $CH_3C(CH_3)_2CH_2Br$ (b) $(H_3C)(H)C{=}C(CH_3)(CHBrCH_3)$ (c) $CH_3(CH_2)_6$–C(H)(CH_3)–Br

Solution

(a) 1-Bromo-2,2-dimethylpropane. Its common name is neopentyl bromide.
(b) (*E*)-4-bromo-3-methyl-2-pentene or *trans*-4-bromo-3-methyl-2-pentene
(c) (*S*)-2-Bromononane

PROBLEM 7.1

Write the IUPAC name, and where possible, the common name of each compound.

(a) $CH_3CH(CH_3)CH_2Cl$ (b) $(H_3C)(H)C{=}C(Cl)(CH_3)$ (c) Cl, $C(CH_3)_3$ (cyclohexane chair) (d) $CH_2{=}C(Cl)CH{=}CH_2$

7.3 Physical Properties

A. Polarity

Fluorine, chlorine, and bromine are all more electronegative than carbon (Table 1.5) and hence C—X bonds with these atoms are polarized with a partial negative charge on halogen and a partial positive charge on carbon. Table 7.1 shows that each of the halomethanes has a substantial dipole moment. The magnitude of a dipole moment depends on the size of the partial charges, the distance between them, and the polarizability of the three pairs of unshared electrons on each halogen. For the halomethanes, dipole moment increases as the electronegativity of halogen and the bond length increase. These two trends run counter to each other, with the result that chloromethane has the largest dipole moment of the series.

B. Boiling Point

Van der Waals forces A group of intermolecular attractive forces including dipole-dipole, dipole-induced dipole, and induced dipole-induced dipole (dispersion) forces.

In the liquid state, alkyl halides are held together by a combination of dipole-dipole, dipole-induced dipole, and induced dipole-induced dipole (dispersion) forces. These forces are grouped together under the term **van der Waals forces,** in honor of J. D. van der Waals, the 19th century Dutch physicist who pioneered our modern understanding of the effects of these forces on physical properties of compounds. Van der Waals attractive forces pull molecules together. As atoms or molecules are brought closer and closer, van der Waals attractive forces are overcome by repulsive forces between electron clouds of adjacent atoms. The energy minimum is where the net attractive forces are the strongest. Nonbonded interatomic and intermolecular distances at these minima can be measured by x-ray crystallography of the solid compounds, and each atom and group of atoms can be assigned an atomic or molecular radius called a **van der Waals radius.** Thus, a useful way to picture an atom or group

TABLE 7.1 Dipole Moments (Gas Phase) of Halomethanes

Halomethane	Electronegativity of Halogen	Carbon-Halogen Bond Length (Å)	Dipole Moment (debyes: D)
CH_3F	4.0	1.39	1.85
CH_3Cl	3.0	1.78	1.87
CH_3Br	2.8	1.93	1.81
CH_3I	2.5	2.14	1.62

TABLE 7.2 Van der Waals Radii (Å) for Selected Atoms and Groups of Atoms

H	F	Cl	Br	CH_2	CH_3	I
1.2	1.35	1.80	1.95	2.0	2.0	2.15

Increasing van der Waals radius →

of atoms in a molecule is as a set of spheres in contact with each other. Van der Waals radii for selected atoms and groups of atoms are given in Table 7.2. Notice from Table 7.2 that the van der Waals radius of fluorine is only slightly greater than that of hydrogen and that, among the halogens, only iodine has a larger van der Waals radius than methyl.

Boiling points of several low-molecular-weight alkyl halides and the alkanes from which they are derived are given in Table 7.3. There are several trends to be noticed from these data. First, constitutional isomers with branched chains have lower boiling points than unbranched-chain isomers (Section 2.8). Compare, for example, the boiling points of straight-chain butyl bromide (bp 100°C) with the more spherical and compact *tert*-butyl bromide (bp 72°C). Branched-chain isomers have lower boiling points because they have a more spherical shape and, therefore, decreased area of contact and magnitude of van der Waals forces between their molecules.

Second, for an alkane and alkyl halide of comparable size and shape, the alkyl halide has a higher boiling point. Compare, for example, the boiling points of ethane (bp −89°C) and methyl bromide (bp 4°C). Although both molecules are roughly the same size (the van der Waals radii of $—CH_3$ and —Br are almost identical) and have roughly the same effective contact area, the boiling point of methyl bromide is considerably higher than that of ethane. This difference is due almost entirely to the greater **polarizability** of the three unshared pairs of electrons on the halogen compared with the shared electron pairs in the hydrocarbon of comparable size and shape. Recall from Section 2.8 that the strength of dispersion forces, the weakest of all inter-

TABLE 7.3 Boiling Points of Some Low-Molecular-Weight Alkanes and Alkyl Halides

		Boiling Point (°C)				
Alkyl Group	**Name**	**H**	**F**	**Cl**	**Br**	**I**
$CH_3—$	methyl	−161	−78	−24	4	43
$CH_3CH_2—$	ethyl	−89	−37	13	38	72
$CH_3(CH_2)_2—$	propyl	−45	3	46	71	102
$(CH_3)_2CH—$	isopropyl	−45	−11	35	60	89
$CH_3(CH_2)_3—$	butyl	0	32	77	100	130
$CH_3CH_2(CH_3)CH—$	*sec*-butyl	0	25	67	90	119
$(CH_3)_2CHCH_2—$	isobutyl	−1	16	68	91	120
$(CH_3)_3C—$	*tert*-butyl	−1	12	51	72	98
$CH_3(CH_2)_4—$	pentyl	36	63	108	129	157
$CH_3(CH_2)_5—$	hexyl	69	92	134	155	181

TABLE 7.4 Densities of Some Low-Molecular-Weight Alkyl Halides

		Density of Liquid (g/mL) at 25°C		
Alkyl Group	**Name**	**Cl**	**Br**	**I**
CH_3—	methyl	—	—	2.279
CH_3CH_2—	ethyl	—	1.460	1.936
$CH_3(CH_2)_2$—	propyl	0.891	1.354	1.749
$(CH_3)_2CH$—	isopropyl	0.862	1.314	1.703
$CH_3(CH_2)_3$—	butyl	0.886	1.276	1.615
$(CH_3)_3C$—	*tert*-butyl	0.842	1.221	1.545
$CH_3(CH_2)_5$—	hexyl	0.879	1.174	1.440

molecular forces, depends on the polarizability of electrons, which in turn depends on how tightly they are held to the nucleus. The further electrons are from the nucleus, the greater their polarizability. In addition, unshared electron pairs have a higher polarizability than electrons shared in a covalent bond.

Third, the boiling points of alkyl fluorides are even lower than those of hydrocarbons of comparable molecular weight. Compare, for example, the boiling points of hexane (MW 86.2, bp 69°C) and 1-fluoropentane (MW 90.1, bp 63°C), and the boiling points of 2-methylpropane (MW 58.1, bp − 1°C) and 2-fluoropropane (MW 62.1, bp − 11°C). This difference is due to the small size of fluorine, the tightness with which its electrons are held, and their particularly low polarizability. The distinctive properties of fluorocarbons, for example, the nonstick properties of polytetrafluoroethylene (Teflon), are also a consequence of the uniquely low polarizability of the three electron pairs of fluorine.

C. Density

The densities of liquid alkyl halides are greater than those of hydrocarbons of comparable molecular weight because of the halogens' large mass per volume ratio. A bromine atom and a methyl group have almost identical van der Waals radii and yet bromine has a mass of 79.9 atomic mass units (amu) compared with 15 amu for methyl. Table 7.4 gives densities for some low-molecular-weight alkyl halides that are liquid at 25°C. The densities of all liquid alkyl bromides and alkyl iodides are greater than that of water. Although the densities of liquid alkyl chlorides are less than that of water, further substitution of chlorine for hydrogen increases the density to the point where di- and polychloroalkanes have a greater density than water (Table 7.5).

TABLE 7.5 Density of Polyhalomethanes

	Density of Liquid (g/mL) at 25°C		
Haloalkane	**X = Cl**	**Br**	**I**
CH_2X_2	1.327	2.497	3.325
CHX_3	1.483	2.890	4.008
CX_4	1.594	3.273	4.23

D. Bond Lengths and Bond Strengths

With the exception of C—F bonds, C—X bonds are weaker than C—H bonds (Table 7.6). As the size of the halogen atom increases, the C—X bond length increases and its strength decreases. These relationships between bond strength and bond length help us to understand the difference in the ease with which alkyl halides undergo reactions that involve carbon-halogen bond breaking. Alkyl fluorides, for example, with the strongest and shortest C—X bonds, are highly resistant to bond breaking under most conditions. It is this characteristic inertness that makes polyfluoroalkanes, such as Teflon, such useful materials.

Of all the fluoroalkanes, **chlorofluorocarbons (CFCs)** manufactured under the trade name **Freons** are the most widely known. CFCs are nontoxic, nonflammable, odorless, and noncorrosive and were ideal replacements for the hazardous compounds used as heat transfer media in refrigeration systems. Among the CFCs most widely used for this purpose were trichlorofluoromethane, CCl_3F, marketed under the trade name Freon-11, and dichlorodifluoromethane, CCl_2F_2, marketed as Freon-12.

The CFCs, along with 1,1,1-trichloroethane (methyl chloroform) and 1,1,2,2-tetrachloroethene (perchloroethylene), also found wide use as industrial cleaning solvents to prepare surfaces for coatings, to remove cutting oils and waxes from millings, and to remove protective coatings. CFCs were also used as propellants for aerosol sprays.

Concern about the environmental impact of CFCs arose in the 1970s when it was shown that more than 1 billion pounds per year of CFCs were being emitted into the atmosphere. Then, in 1974 Drs. Sherwood Rowland of the University of California, Irvine and Mario Molina of the Massachusetts Institute of Technology announced their theory of ozone destruction by these compounds. When released into the air, CFCs escape to the lower atmosphere, but because of their inertness, they do not decompose there. Slowly, they find their way to the stratosphere where they absorb ultraviolet radiation from the sun and then decompose. As they decompose, they set up a chemical reaction that may lead to destruction of the stratospheric ozone layer, which acts as a shield for the earth against excess ultraviolet radiation from the sun. An increase in ultraviolet radiation reaching the earth, it is theorized, may lead to destruction of certain crops and agricultural species, and even increased incidence of skin cancer in sensitive individuals.

The results of this concern were two conventions, one in Vienna in 1985 and the other in Montreal in 1987, held by the United Nations Environmental Program. The 1987 meeting produced the "Montreal Protocol," which set limits on the production

TABLE 7.6 Average Bond Dissociation Energies for C—H and C—X Bonds

Bond	Bond Length (Å)	Bond Dissociation Energy [kcal/mol (kJ/mol)]
C—H	1.09	90–100 (377–418)
C—F	1.42	105 (439)
C—Cl	1.78	80 (345)
C—Br	1.93	65 (272)
C—I	2.14	50 (209)

and use of ozone-depleting CFCs and urged complete phaseout of their production by the year 1996. The fact that an international agreement on the environment that set limits on the production of any substance could be reached is indeed amazing and bodes well for the health of the planet.

Rowland, Molina, and Paul Crutzen, a Dutch chemist at the Max Planck Institute for Chemistry, Germany, were awarded the Nobel Prize for chemistry in 1995. The Royal Swedish Academy of Science said in its citation, "By explaining the chemical mechanisms that affect the thickness of the ozone layer, these three researchers have contributed to our salvation from a global environmental problem that could have catastrophic consequences."

The chemical industry is responding by developing ozone nondepleting alternatives to CFCs, among which are the hydrofluorocarbons (HFCs) and hydrochlorofluorocarbons (HCFCs)

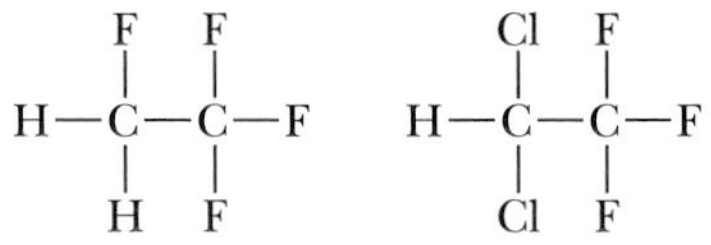

```
     F  F           Cl F
     |  |           |  |
  H—C—C—F       H—C—C—F
     |  |           |  |
     H  F           Cl F
```

HFC-134a HCFC-123

We must not assume, however, that haloalkanes are introduced into the environment only by human action. It is estimated, for example, that annual production of bromomethane from natural sources is 300,000 tons, largely from marine algae, giant kelp, and volcanoes. Furthermore, global emission of chloromethane is estimated to be 5 million tons per year, most of it from terrestrial and marine biomass. These haloalkanes, however, have only short atmospheric lifetimes and only a tiny fraction reach the stratosphere. It is the CFCs that are the problem; they have longer atmospheric lifetimes, reach the stratosphere, and do their damage there.

7.4 Halogenation of Alkanes

As we saw in Sections 6.3A and 6.3C, alkyl halides can be prepared by addition of HX and X_2 to alkenes. They are also prepared by halogenation of alkanes and by replacement of the —OH group of alcohols by halogen. We study preparation of alkyl halides from alcohols in Chapter 9. In this chapter, we concentrate on preparation of alkyl halides by halogenation of alkanes, illustrated here by the treatment of 2-methylpropane with bromine at an elevated temperature.

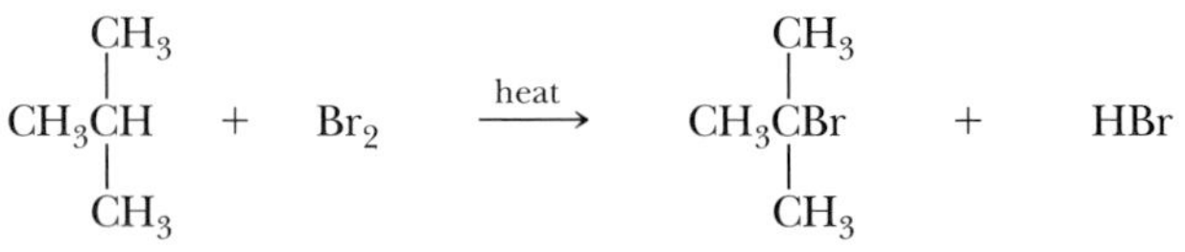

```
    CH3                          CH3
    |                            |
  CH3CH   +   Br2   --heat-->  CH3CBr   +   HBr
    |                            |
    CH3                          CH3
```

2-Methylpropane (Isobutane) → 2-Bromo-2-methylpropane (*tert*-Butyl bromide)

Halogenation of alkanes is common with Br_2 and Cl_2. Fluorine, F_2, is seldom used because its reactions with alkanes are highly exothermic and difficult to control, which can cause explosions. Iodine, I_2, is seldom used because the reaction is endothermic and the position of equilibrium favors alkane and I_2 rather than iodoalkane and HI.

Chemistry in Action

Artificial Blood

Today, the best known use of fluorocarbons is in nonstick cookware. In the future, this same class of compounds may serve a life-saving role as a temporary substitute for blood.

Blood picks up oxygen using hemoglobin (an iron-containing protein) in red blood cells. Whole blood can transport approximately 20 mL of oxygen per 100 mL, or 20 vol % oxygen. Fluorinated hydrocarbons, such as perfluorodecalin and perfluorotripropylamine, are able to dissolve up to 50 vol % oxygen.

trans-Perfluorodecalin

$(CF_3CF_2CF_2)_2N$–$CF_2CF_2CF_3$

Perfluorotripropylamine

Fluorocarbons cannot be used directly in the bloodstream because they are nonpolar and do not mix with water. They can be used as an oxygen-carrying blood substitute, however, as a fluorocarbon-water emulsion. One such preparation is Fluosol DA, a 20% emulsion of perfluorodecalin and perfluorotripropylamine in water. It now appears that mammals can live with a majority of their hemoglobin replaced by fluorocarbons such as these. It has been found, for example, that dogs treated with a 25% perfluorotributylamine-water emulsion replacing 70% of their blood can still live a normal life span.

In 1989, a new drug application was approved in the United States for perflubron (LiquiVent), a water emulsion of 1-bromoperfluorooctane, as a blood substitute.

$$CF_3(CF_2)_6CF_2Br$$

Bromoperfluorooctane

In 1996, clinical trials on this blood substitute were undertaken for patients suffering from a variety of life-threatening lung illnesses and injuries, including infections, near drowning, and smoke inhalation. If results continue to be promising, we can expect that organic chemists will be able to synthesize new perfluorocarbon compounds optimized for use as blood substitutes.

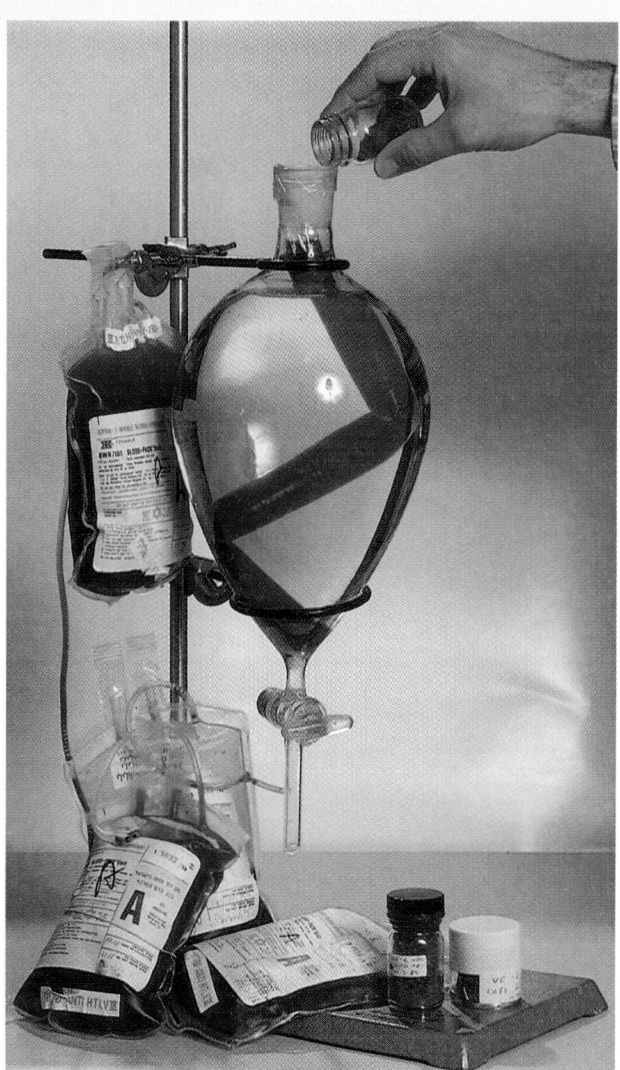

Artificial blood research. (© Richard Nowitz/Phototake NYC)

See *The Merck Index*, 12th ed., #4221 and #7299.

If a mixture of methane and chlorine gas is kept in the dark at room temperature, no detectable change occurs. If, however, the mixture is heated or exposed to visible or ultraviolet light, a reaction begins almost at once with the evolution of heat. The products are chloromethane and hydrogen chloride. What occurs is a substitution reaction, in this case substitution of a chlorine atom for a hydrogen atom in methane and the production of an equivalent amount of hydrogen chloride.

$$\underset{\text{Methane}}{CH_4} + Cl_2 \xrightarrow{\text{heat}} \underset{\substack{\text{Chloromethane}\\ \text{(Methyl chloride)}}}{CH_3Cl} + HCl \qquad \Delta H^0 = -25 \text{ kcal/mol } (-105 \text{ kJ/mol})$$

Substitution A reaction in which an atom or group of atoms in a compound is replaced by another atom or group of atoms.

Substitution is a reaction in which an atom or group of atoms in a compound is replaced by another atom or group of atoms.

If chloromethane is allowed to react with more chlorine, further chlorination produces a mixture of dichloromethane (methylene chloride), trichloromethane (chloroform), and tetrachloromethane (carbon tetrachloride).

$$\underset{\substack{\text{Chloromethane}\\ \text{(Methyl chloride)}}}{CH_3Cl} + Cl_2 \xrightarrow{\text{heat}} \underset{\substack{\text{Dichloromethane}\\ \text{(Methylene chloride)}}}{CH_2Cl_2} + HCl$$

$$\underset{\substack{\text{Dichloromethane}\\ \text{(Methylene chloride)}}}{CH_2Cl_2} \xrightarrow[\text{heat}]{Cl_2} \underset{\substack{\text{Trichloromethane}\\ \text{(Chloroform)}}}{CHCl_3} \xrightarrow[\text{heat}]{Cl_2} \underset{\substack{\text{Tetrachloromethane}\\ \text{(Carbon tetrachloride)}}}{CCl_4}$$

Notice that in the last equation, the reagent Cl_2 is placed over the reaction arrow and the equivalent amount of HCl formed is not shown. Placing reagents over reaction arrows and omitting byproducts is a common practice among organic chemists. Treatment of ethane with chlorine gives chloroethane (ethyl chloride); treatment with bromine gives bromoethane (ethyl bromide).

$$\underset{\text{Ethane}}{CH_3CH_3} + Br_2 \xrightarrow{\text{heat}} \underset{\substack{\text{Bromoethane}\\ \text{(Ethyl bromide)}}}{CH_3CH_2Br} + HBr$$

A. Regioselectivity

Treatment of propane with bromine gives a mixture consisting of approximately 8% of 1-bromopropane and 92% of 2-bromopropane.

$$\underset{\text{Propane}}{CH_3CH_2CH_3} + Br_2 \xrightarrow[\text{or light}]{\text{heat}} \underset{\substack{\text{2-Bromopropane}\\ (92\%)}}{CH_3\overset{\overset{\large Br}{|}}{C}HCH_3} + \underset{\substack{\text{1-Bromopropane}\\ (8\%)}}{CH_3CH_2CH_2Br} + HBr$$

Propane contains eight hydrogens: one set of six equivalent primary hydrogens and one set of two equivalent secondary hydrogens (Section 2.3C). Substitution of bromine for a primary hydrogen gives 1-bromopropane; substitution of bromine for a secondary hydrogen gives 2-bromopropane. On the basis of random substitution of any one of the eight hydrogens in propane, we predict the isomeric bromopropanes to be formed in the ratio of 6:2 or 75% 1-bromopropane and 25% 2-bromopropane. In fact, in the bromination of propane, substitution of a secondary hydrogen is favored over a primary hydrogen and 2-bromopropane is the major product.

Product Distribution	$CH_3CH_2CH_2Br$	$CH_3CH(Br)CH_3$
prediction based on ratio of 6 primary hydrogens to 2 secondary hydrogens	75%	25%
experimental observation	8%	92%

Other experiments have shown that substitution at a tertiary hydrogen is favored over both secondary and primary hydrogens. For example, monobromination of 2-methylpentane gives almost exclusively 2-bromo-2-methylpentane.

$$CH_3CH(CH_3)CH_2CH_2CH_3 + Br_2 \xrightarrow[\text{or light}]{\text{heat}} CH_3C(CH_3)(Br)CH_2CH_2CH_3 + HBr$$

2-Methylpentane 2-Bromo-2-methylpentane

The reaction of bromine with an alkane occurs regioselectively in the order 3° hydrogen > 2° hydrogen > 1° hydrogen. Chlorination of alkanes is also regioselective, but to a much lesser degree than bromination. For example, treatment of propane with chlorine gives a mixture consisting of approximately 57% 2-chloropropane and 43% 1-chloropropane.

$$CH_3CH_2CH_3 + Cl_2 \xrightarrow[\text{or light}]{\text{heat}} CH_3CH(Cl)CH_3 + CH_3CH_2CH_2Cl + HCl$$

Propane 2-Chloropropane (57%) 1-Chloropropane (43%)

Thus, we can conclude that, although both bromine and chlorine are regioselective in hydrogen replacement in the order 3° > 2° > 1°, regioselectivity is far greater for bromination than for chlorination. From data on product distribution, it has been determined that regioselectivity for bromination is approximately 1600:80:1, while for chlorination, it is approximately 5:4:1. We will discuss reasons for this regioselectivity in the following section.

EXAMPLE 7.2

Name and draw structural formulas for all monobromination products formed by treatment of 2-methylpropane with Br_2. Predict the major product based on the regioselectivity of the reaction of Br_2 with alkanes.

$$CH_3CH(CH_3)CH_3 + Br_2 \xrightarrow[\text{heat}]{\text{light or}} \text{monobromoalkanes} + HBr$$

Solution

In 2-methylpropane, there are nine equivalent primary hydrogens and one tertiary hydrogen. Substitution of bromine for a primary hydrogen gives 1-bromo-2-methylpropane; substitution for the tertiary hydrogen gives 2-bromo-2-methylpropane. Given the regioselectivity of bromination for 3° > 2° > 1° hydrogens, predict that 2-bromo-2-methylpropane is the major product.

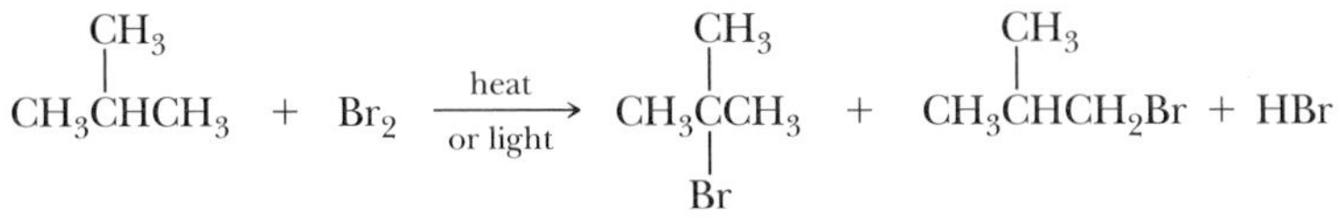

$$CH_3\overset{\overset{\displaystyle CH_3}{|}}{C}HCH_3 + Br_2 \xrightarrow[\text{or light}]{\text{heat}} CH_3\underset{\underset{\displaystyle Br}{|}}{\overset{\overset{\displaystyle CH_3}{|}}{C}}CH_3 + CH_3\overset{\overset{\displaystyle CH_3}{|}}{C}HCH_2Br + HBr$$

2-Methylpropane; 2-Bromo-2-methylpropane (major product); 1-Bromo-2-methylpropane

$$\text{Predicted \% 2-bromo-2-methylpropane} = \frac{1 \times 1600}{1 \times 1600 + 9 \times 1} \times 100 = 99.4\%$$

PROBLEM 7.2

Name and draw structural formulas for all monochlorination products formed by treatment of butane with Cl_2. Predict the major product based on the regioselectivity of the reaction of chlorine with alkanes.

$$CH_3CH_2CH_2CH_3 + Cl_2 \xrightarrow[\text{or light}]{\text{heat}} \text{monochlorobutanes} + HCl$$

B. Energetics

Using data from the table of Bond Dissociation Energies (Appendix 3), we can calculate the heat of reaction for the halogenation of an alkane. In this calculation for the chlorination of methane, energy is required to break CH_3—H and Cl—Cl bonds (105 and 59 kcal/mol, respectively). Energy is released in making the CH_3—Cl and H—Cl bonds (−85 and −103 kcal/mol, respectively). Summing these energies, we calculate that chlorination of methane to form chloromethane and hydrogen chloride liberates 24 kcal/mol (100 kJ/mol).

	CH_4	+ Cl_2	⟶ CH_3Cl	+ HCl	$\Delta H^0 = -24$ kcal/mol
BDE (kcal/mol)	+105	+59	−85	−103	(−100 kJ/mol)

EXAMPLE 7.3

Using the table of bond dissociation energies, calculate ΔH^0 for bromination of propane to give 2-bromopropane and hydrogen bromide.

Solution

Under each molecule is given the energy for breaking or forming each corresponding bond. The calculated heat of reaction is −14 kcal/mol (−59 kJ/mol).

$$\mathrm{CH_3\overset{\overset{\displaystyle H}{|}}{C}HCH_3 + Br{-}Br \longrightarrow CH_3\overset{\overset{\displaystyle Br}{|}}{C}HCH_3 + H{-}Br} \qquad \Delta H^0 = -14\ \mathrm{kcal/mol}$$

BDE (kcal/mol)	+96	+46	−68	−88	(−59 kJ/mol)

PROBLEM 7.3

Using the table of bond dissociation energies, calculate ΔH^0 for bromination of propane to give 1-bromopropane and hydrogen bromide.

7.5 Mechanism of Halogenation of Alkanes

From detailed studies of the conditions and products for halogenation of alkanes, chemists have concluded that these reactions occur by a type of mechanism called a radical chain mechanism. A **radical** is any chemical species that contains one or more unpaired electrons.

Radical Any chemical species that contains one or more unpaired electrons.

A. Formation of Radicals

Radicals are produced from a molecule by cleavage of a bond in such a way that each atom or fragment participating in the bond retains one electron. In the following equations, **fishhook arrows** are used to show the change in position of single electrons.

Fishhook arrow A barbed curved arrow used to show the change in position of a single electron.

$$\underset{\text{Diethyl peroxide}}{\mathrm{CH_3CH_2O{-}OCH_2CH_3}} \xrightarrow{80^\circ} \underset{\text{Ethoxy radicals}}{\mathrm{CH_3CH_2O\cdot + \cdot OCH_2CH_3}} \qquad \Delta H^0 = +36\ \mathrm{kcal/mol}\ (+150\ \mathrm{kJ/mol})$$

$$\underset{\text{Chlorine}}{\mathrm{Cl{-}Cl}} \xrightarrow{\text{light}} \underset{\text{Chlorine atoms}}{\mathrm{Cl\cdot + \cdot Cl}} \qquad \Delta H^0 = +59\ \mathrm{kcal/mol}\ (+247\ \mathrm{kJ/mol})$$

$$\underset{\text{Ethane}}{\mathrm{CH_3{-}CH_3}} \xrightarrow{\text{heat}} \underset{\text{Ethyl radicals}}{\mathrm{CH_3\cdot + \cdot CH_3}} \qquad \Delta H^0 = +90\ \mathrm{kcal/mol}\ (+377\ \mathrm{kJ/mol})$$

Energy to cause bond cleavage and generation of radicals can be supplied either by light or heat. The energy of visible and ultraviolet radiation (wavelength from 200–800 nm) falls in the range of 140 to 36 kcal/mol (580 to 151 kJ/mol) and is of the same order of magnitude as the bond dissociation energies of single covalent bonds. The bond dissociation energy of Br_2 is 46 kcal/mol (192 kJ/mol); that for Cl_2 is 59 kcal/mol (247 kJ/mol). Dissociation of these halogens can also be brought about by heating to temperatures above 350°C.

Oxygen-oxygen single bonds in peroxides (ROOR) and hydroperoxides (ROOH) have dissociation energies in the range of 30 to 40 kcal/mol (126 to 167 kJ/mol), and compounds containing these bonds are cleaved to radicals at considerably lower temperatures than those required for rupture of carbon-carbon bonds. Diethyl peroxide, for example, begins to dissociate to ethoxy radicals at 80°C.

B. A Radical Chain Mechanism

To account for the products formed from halogenation of alkanes, chemists propose a radical chain mechanism involving three types of steps: chain initiation, chain propagation, and chain termination. We illustrate radical halogenation of alkanes by the reaction of ethane with chlorine.

MECHANISM Radical Halogenation of an Alkane

Chain initiation

1. $Cl_2 \xrightarrow[\text{or light}]{\text{heat}} Cl\cdot + \cdot Cl$

Chain propagation

2. $CH_3CH_3 + \cdot Cl \longrightarrow CH_3CH_2\cdot + HCl$
3. $CH_3CH_2\cdot + Cl_2 \longrightarrow CH_3CH_2Cl + \cdot Cl$

Chain termination

4. $2CH_3CH_2\cdot \longrightarrow CH_3CH_2CH_2CH_3$
5. $CH_3CH_2\cdot + \cdot Cl \longrightarrow CH_3CH_2Cl$
6. $Cl\cdot + \cdot Cl \longrightarrow Cl_2$

Chain initiation A step in a chain reaction characterized by the formation of radicals from nonradical compounds.

The characteristic feature of a **chain initiation** step is formation of radicals from nonradical compounds. In the case of chlorination of ethane, chain initiation is by thermal or light induced cleavage of the Cl—Cl bond to give two chlorine atoms (radicals).

Chain initiation: 1. $Cl_2 \xrightarrow[\text{or light}]{\text{heat}} Cl\cdot + \cdot Cl$

Chain propagation A step in a chain reaction characterized by the reaction of a radical and a molecule to give a new radical.

The characteristic feature of a **chain propagation** step is reaction of a radical and a molecule to give a new radical. A chlorine atom is consumed in Step 2 but an ethyl radical is produced. Similarly, an ethyl radical is consumed in Step 3 but a chlorine atom is produced. Steps 2 and 3 can repeat thousands of times as long as neither radical is removed by a different reaction.

Chain propagation:
$$\begin{cases} 2.\ CH_3CH_3 + \cdot Cl \longrightarrow CH_3CH_2\cdot + HCl \\ 3.\ CH_3CH_2\cdot + Cl_2 \longrightarrow CH_3CH_2Cl + \cdot Cl \end{cases}$$

A second characteristic feature of chain propagation steps is that, when added together, they give the observed stoichiometry of the reaction. Adding Steps 2 and 3 and canceling structures that appear on both sides of the equation gives the balanced equation for the radical chlorination of ethane.

Chain length The number of times the cycle of chain propagation steps repeats in a chain reaction.

Steps 2 + 3: $CH_3CH_3 + Cl_2 \longrightarrow CH_3CH_2Cl + HCl$

The number of times a cycle of chain propagation steps repeats is called **chain length.**

Chain propagation steps continue until two radicals react with each other to terminate the process. Among the characteristic features of a **chain termination** step is destruction of radicals. The most important chain termination reactions during halogenation of alkanes are radical couplings.

Chain termination A step in a radical chain mechanism that involves destruction of radicals.

Chain termination:

4. $2CH_3CH_2\cdot \longrightarrow CH_3CH_2CH_2CH_3$
5. $CH_3CH_2\cdot + \cdot Cl \longrightarrow CH_3CH_2Cl$
6. $Cl\cdot + \cdot Cl \longrightarrow Cl_2$

The structures, geometries, and relative stabilities of simple alkyl radicals are similar to those of alkyl carbocations. They are planar or almost so, with bond angles of 120° about the carbon with the unpaired electron. This geometry indicates that carbon is sp^2 hybridized and that the unpaired electron occupies its unhybridized $2p$ orbital. The order of stability of alkyl radicals, like that of alkyl carbocations, is $3° > 2° > 1° >$ methyl.

C. Energetics of Chain Propagation Steps

Once the radical chain is initiated, the heat of reaction is derived entirely from the heat of reaction of the individual chain propagation steps. In Step 2 of radical chlorination of ethane, for example, energy is required for breaking the $CH_3CH_2{-}H$ bond (BDE = 100 kcal/mol), but energy is released on formation of the H—Cl bond (BDE = 103 kcal/mol). Similarly, energy is required in Step 3 for breaking the Cl—Cl bond (BDE = 59 kcal/mol), but energy is released on formation of the $CH_3CH_2{-}Cl$ bond (BDE = 80 kcal/mol). We see that, just as the sum of the chain propagation steps for radical halogenation gives the observed stoichiometry, the sum of the heats of reaction for each propagation step gives the observed heat of reaction.

Reaction Step	**ΔH^0 [kcal/mol (kJ/mol)]**
Step 2: $CH_3CH_3 + \cdot Cl \longrightarrow CH_3CH_2\cdot + HCl$	−3 (−13)
$+100$ $\qquad -103$	
Step 3: $CH_3CH_2\cdot + Cl_2 \longrightarrow CH_3CH_2Cl + \cdot Cl$	−21 (−88)
$+59$ $\qquad -80$	
Sum: $CH_3CH_3 + Cl_2 \longrightarrow CH_3CH_2Cl + HCl$	−24 (−100)

Example 7.4

Using the table of bond dissociation energies in Appendix 3, calculate ΔH^0 for each propagation step in the radical bromination of propane to give 2-bromopropane and HBr.

Solution

Here are the two chain propagation steps along with bond dissociation energies for the bonds broken and the bonds formed. The first chain propagation step is endothermic, the second is exothermic, and the overall reaction is exothermic ($\Delta H^0 = +8 - 22 = -14$ kcal/mol).

$$\underset{\text{+96}}{\mathrm{CH_3\overset{\overset{\displaystyle H}{|}}{C}HCH_3}} + \mathrm{Br}\cdot \longrightarrow \mathrm{CH_3\dot{C}HCH_3} + \underset{-88}{\mathrm{H{-}Br}} \qquad \Delta H^0 = +8 \text{ kcal/mol}$$

BDE (kcal/mol): +96, −88

$$\mathrm{CH_3\dot{C}HCH_3} + \underset{+46}{\mathrm{Br{-}Br}} \longrightarrow \underset{-68}{\mathrm{CH_3\overset{\overset{\displaystyle Br}{|}}{C}HCH_3}} + \mathrm{Br}\cdot \qquad \Delta H^0 = -22 \text{ kcal/mol}$$

BDE (kcal/mol): +46, −68

$$\Delta H^0 = -14 \text{ kcal/mol}$$

PROBLEM 7.4

Write a pair of chain propagation steps for the radical bromination of propane to give 1-bromopropane and calculate ΔH^0 for each propagation step, and ΔH^0 for the overall reaction.

D. Regioselectivity of Bromination Versus Chlorination: Hammond's Postulate

Hammond's postulate The structure of the transition state for an exothermic step looks more like the reactants of that step. Conversely, the structure of the transition state for an endothermic step looks more like the products of that step.

The regioselectivity in halogenation of alkanes can be accounted for in terms of the relative stabilities of radicals (3° > 2° > 1° > methyl). But how do we account for the greater regioselectivity in bromination of alkanes compared with chlorination of alkanes? To do so, we need to consider a refinement of transition state theory proposed in 1955 by George Hammond, then at Iowa State University.

According to what has become known as **Hammond's postulate,** the structure of the transition state for an exothermic reaction looks more like the reactants of that reaction. Conversely, the structure of the transition state for an endothermic reaction looks more like the products of that reaction. Shown in Figure 7.1 are potential energy diagrams for a highly exothermic reaction and a highly endothermic reaction, each

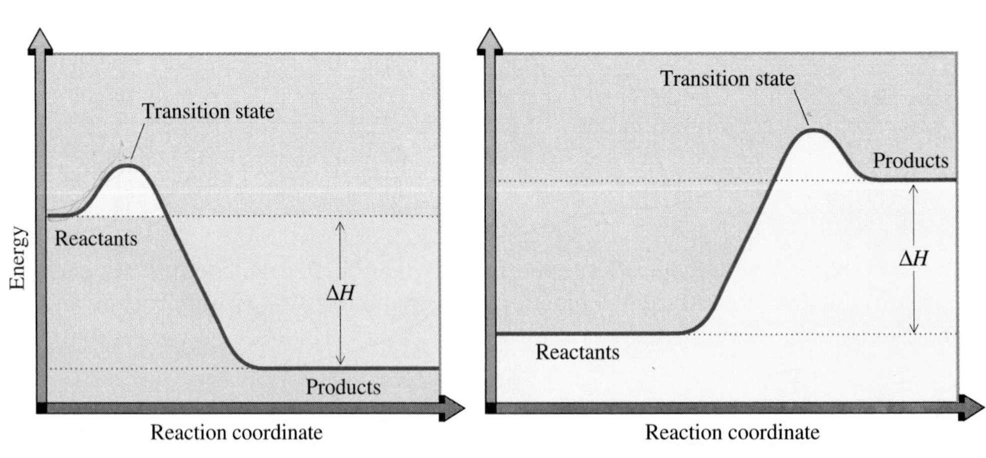

FIGURE 7.1
Hammond's postulate. Potential energy diagrams for two one-step reactions. (a) In an exothermic reaction, the structure of the transition state resembles that of the reactants of that reaction. (b) In an endothermic reaction, the structure of the transition state resembles that of the products of that reaction.

occurring in one step. It is important to realize that we cannot observe a transition state directly; we can only infer its existence, structure, and stability. What Hammond's postulate does is give us a reasonable way of deducing something about the structure of the transition state by examining things we can observe; the structure of reactants and products and heats of reaction. Hammond's postulate applies equally well to a multistep reaction. The transition state of any exothermic step in a multistep sequence looks like the starting material(s) of that step; the transition state of any endothermic step in the sequence looks like the product(s) of that step.

Now let us apply Hammond's postulate to explain the relative regioselectivities of chlorination versus bromination of alkanes. In applying Hammond's postulate, we deal with the rate-limiting step which, in radical halogenation of alkanes, is abstraction of a hydrogen atom by a halogen radical. Given in the following table are heats of reaction for the hydrogen abstraction steps in chlorination and bromination of ethane. Also given under the formulas of ethane, HCl, and HBr are bond dissociation energies for the bonds broken and formed in each step.

Reaction Step				**ΔH^0 [kcal/mol (kJ/mol)]**
$CH_3CH_3 + \cdot Cl \longrightarrow$	$CH_3CH_2\cdot$	+	HCl	−5.0 (21 kJ/mol)
+98			−103	
$CH_3CH_3 + \cdot Br \longrightarrow$	$CH_3CH_2\cdot$	+	HBr	+10.0 (41.8 kJ/mol)
+98			−88	

Abstraction of hydrogen by chlorine is exothermic and, according to Hammond's postulate, the transition state is reached sooner in the course of the reaction [Figure 7.2(a)] and its structure resembles the reactants, namely the alkane and a chlorine

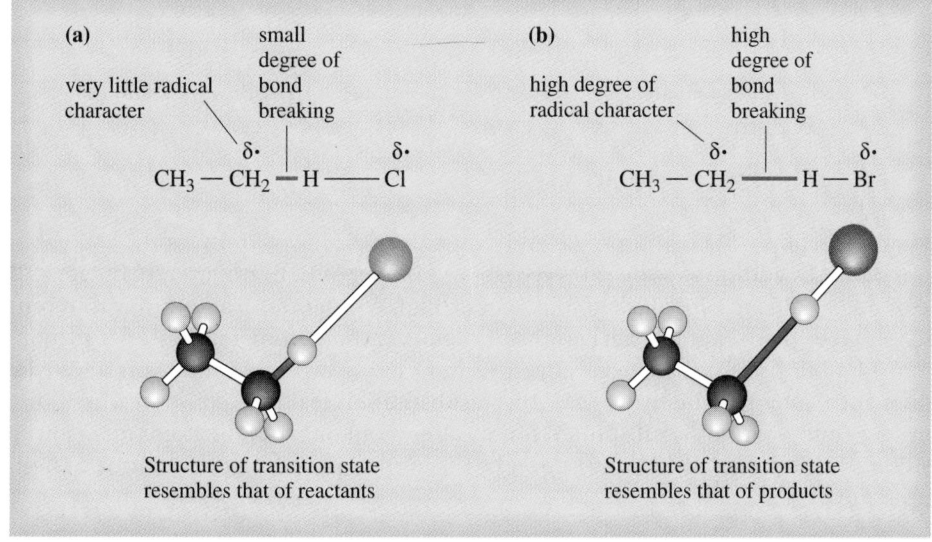

FIGURE 7.2
Transition states for hydrogen abstraction in the radical halogenation of ethane. (a) Chlorination and (b) bromination.

atom. Because there is relatively little radical character on carbon in the transition state, regioselectivity in radical chlorination is only slightly influenced by the relative stabilities of possible radical intermediates. Products are determined more by whether a chlorine atom collides with a primary, secondary, or tertiary hydrogen.

Conversely, abstraction of hydrogen by bromine is endothermic. The transition state is reached later in the course of the reaction [Figure 5.7(b)] and its structure resembles that of the products, namely, an alkyl radical and HBr. Since the transition states resemble the radicals, and the radical stabilities are in the order 3° > 2° > 1°, the transition state stabilities and, thus, the regioselectivities are in the same order.

E. Stereochemistry of Radical Halogenation

When radical halogenation produces a stereocenter or takes place at a hydrogen on an existing stereocenter, the product is an equal mixture of *R* and *S* enantiomers. Consider, for example, radical bromination of butane, which produces 2-bromobutane. In this example, both of the starting materials are achiral and, as is true for any reaction of achiral starting materials that gives a chiral product (Section 6.7), the product is a racemic mixture.

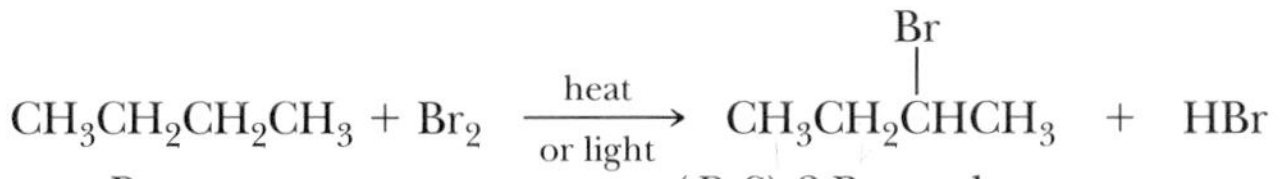

$$\underset{\text{Butane}}{CH_3CH_2CH_2CH_3} + Br_2 \xrightarrow[\text{or light}]{\text{heat}} \underset{(R,S)\text{-2-Bromobutane}}{CH_3CH_2\overset{\overset{\large Br}{|}}{C}HCH_3} + HBr$$

In the case of the *sec*-butyl radical, the carbon bearing the unpaired electron is sp^2 hybridized and the unpaired electron lies in the unhybridized $2p$ orbital. Reaction of the alkyl radical intermediate with halogen in the second chain propagation step occurs with equal probability from either face to give an equal mixture of the *R* and *S* configurations at the newly created stereocenter.

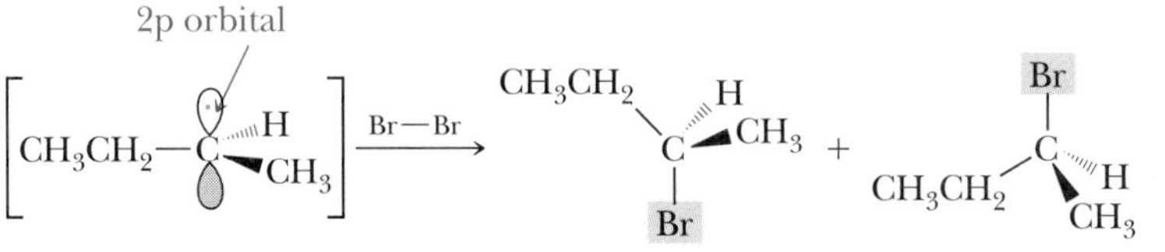

the unhybridized 2p orbital

$$[CH_3CH_2\text{—}\dot{C}(H)(CH_3)] \xrightarrow{Br\text{—}Br} (R)\text{-}CH_3CH_2C(H)(CH_3)Br + (S)\text{-}CH_3CH_2C(H)(CH_3)Br$$

Radical intermediate (*R*)-2-Bromobutane (*S*)-2-Bromobutane

7.6 Allylic Halogenation

We saw in Section 6.3D that propene and other alkenes react with Br_2 and Cl_2 at room temperature by addition to the carbon-carbon double bond. If, however, propene and one of these halogens are allowed to react at a high temperature, an entirely different reaction takes place, namely allylic substitution. A carbon atom adjacent to a carbon-carbon double bond is called an **allylic carbon** and a hydrogen atom on an allylic carbon is called an **allylic hydrogen. Allylic substitution** is any reaction in which one atom or group of atoms is substituted for another atom or group of atoms at an allylic position. We illustrate allylic substitution by reaction of propene with chlorine at high temperature.

Allylic substitution Any reaction in which one atom or group of atoms is substituted for another atom or group of atoms at an allylic position.

$$\underset{\text{Propene}}{CH_2{=}CHCH_3} + Cl_2 \xrightarrow{350°C} \underset{\substack{\text{3-Chloropropene}\\ \text{(Allyl chloride)}}}{CH_2{=}CHCH_2Cl} + HCl$$

A comparable reaction takes place when propene is treated with bromine at an elevated temperature.

To predict which of the various C—H bonds in propene is most likely to break when a mixture of propene and bromine or chlorine is heated, we need to look at bond dissociation energies. We find that the bond dissociation energy of an allylic C—H in propene is approximately 20 kcal/mole (84 kJ/mol) less than that of a vinylic C—H. The reason that the allylic C—H bond is weaker is discussed in Section 7.6B.

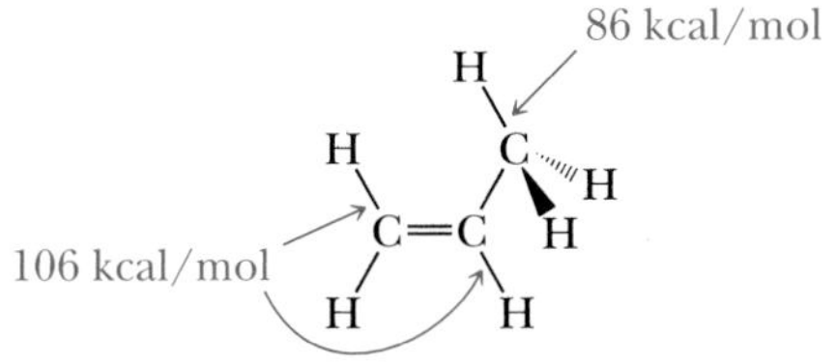

86 kcal/mol

106 kcal/mol

Treatment of propene with bromine or chlorine at elevated temperature illustrates a very important point about organic reactions: It is often possible to change the product(s) of a reaction by changing the reaction conditions and thereby changing the mechanism of the reaction.

$$\underset{\text{Propene}}{CH_2{=}CHCH_3} + Br_2 \xrightarrow{\text{high temp.}} CH_2{=}CHCH_2Br + HBr \quad \text{(allylic substitution)}$$

$$CH_2{=}CHCH_3 + Br_2 \xrightarrow[CCl_4]{\text{room temp.}} \underset{\text{Br}\quad\text{Br}}{CH_2CHCH_3} \quad \text{(electrophilic addition)}$$

A very useful way to carry out allylic bromination in the laboratory at or slightly above room temperature is to use the reagent ***N*-bromosuccinimide (NBS)** in carbon tetrachloride. Reaction between an alkene and NBS is most commonly initiated by light or a catalytic amount of a peroxide, such as benzoyl peroxide.

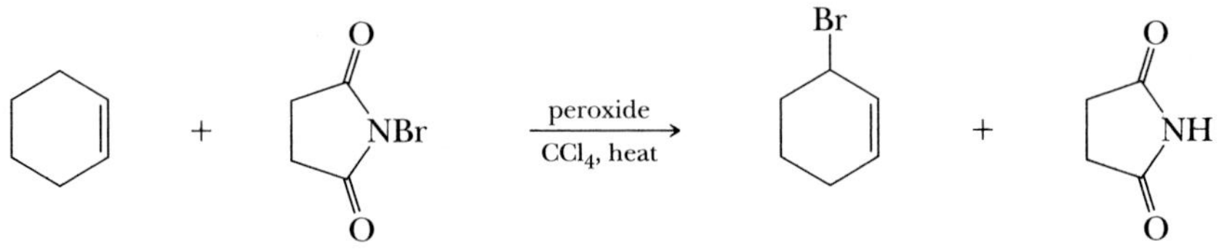

Cyclohexene + *N*-Bromosuccinimide (NBS) $\xrightarrow[CCl_4,\ \text{heat}]{\text{peroxide}}$ 3-Bromocyclohexene + Succinimide

In the NBS reaction, a double substitution occurs, bromine for a hydrogen in the alkene and hydrogen for bromine in NBS.

A. Mechanism of Allylic Halogenation

Allylic bromination and chlorination proceed by a radical chain reaction involving the same type of chain initiation, chain propagation, and chain termination steps involved in the radical halogenation of alkanes.

Following is a set of chain propagation steps for allylic bromination of propene. In the first propagation step, a bromine atom abstracts an allylic hydrogen (the weakest C—H bond in propene) to produce an allyl radical. The allyl radical in turn reacts with a bromine molecule to form allyl bromide and a new bromine atom. Note that, as must be the case, this combination of chain propagation steps adds up to the observed stoichiometry. This reaction is exactly like halogenation of alkanes, but is strongly regioselective for an allylic hydrogen because of its weaker bond. Propagation of the chain reaction continues until termination steps produce nonradical products and thus stop further reaction.

Chain propagation:

$$CH_2{=}CHCH_3 + \cdot Br \longrightarrow CH_2{=}CHCH_2\cdot + HBr$$

$$CH_2{=}CHCH_2\cdot + Br_2 \longrightarrow CH_2{=}CHCH_2Br + \cdot Br$$

Sum of chain propagation steps:

$$CH_2{=}CHCH_3 + Br_2 \longrightarrow CH_2{=}CHCH_2Br + HBr$$

Possible chain termination steps:

$$Br\cdot + \cdot Br \longrightarrow Br_2$$

$$CH_2{=}CHCH_2\cdot + \cdot Br \longrightarrow CH_2{=}CHCH_2Br$$

$$2CH_2{=}CHCH_2\cdot \longrightarrow CH_2{=}CHCH_2{-}CH_2CH{=}CH_2$$

Allylic bromination using *N*-bromosuccinimide in CCl_4 also proceeds by way of a radical chain mechanism. The Br_2 necessary for allylic bromination is formed by reaction of HBr with NBS.

$$\text{NBr (N-bromosuccinimide)} + HBr \longrightarrow \text{NH (succinimide)} + Br_2$$

Bromine formed in this step then reacts with an allyl radical to continue the chain propagation reactions. Thus, in effect, NBS reacts with the HBr formed in the first chain propagation step to yield Br_2, which then continues the chain reaction.

The mechanism we described for allylic bromination by NBS poses the following problem. NBS is the indirect source of Br_2, which then takes part in chain propagation. But if Br_2 is present in the reaction mixture, why does it not react instead with the carbon-carbon double bond by addition to the double bond? In other words, why is the observed reaction allylic substitution rather than electrophilic addition? The answer is that when radicals are present, the rates of the chain propagation steps are much faster than the rate of electrophilic addition of bromine to the alkene. Furthermore, the concentration of Br_2 is very low throughout the reaction, which also slows the rate of electrophilic addition.

B. Structure of the Allyl Radical

The allyl radical can be represented as a hybrid of two contributing structures. Here fishhook arrows show the movement of single electrons and how the contributing structure on the left is converted to the one on the right.

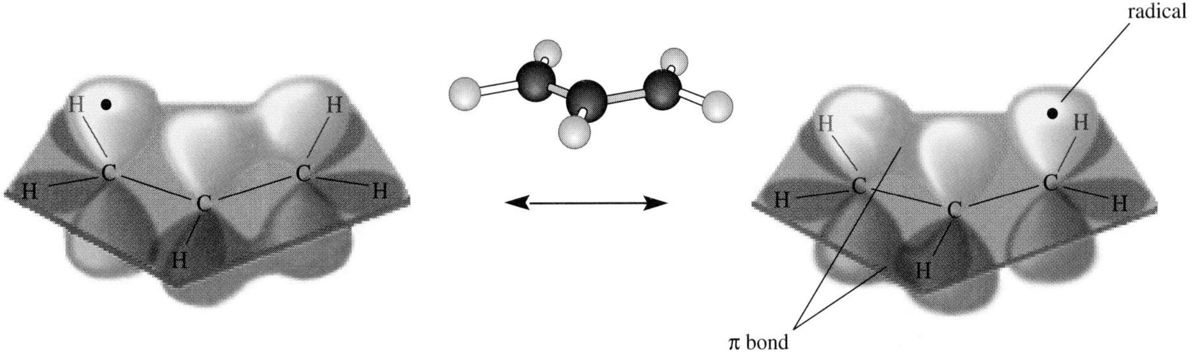

FIGURE 7.3
Covalent bonding in the allyl radical.

$$\dot{C}H_2{=}CH{-}\dot{C}H_2 \longleftrightarrow \dot{C}H_2{-}CH{=}CH_2$$

(Equivalent contributing structures)

The eight atoms of the allyl radical lie in a plane, and all bond angles are approximately 120°. Each carbon atom is sp^2 hybridized, and the three $2p$ orbitals participating in resonance stabilization of the radical are parallel to one another as shown in Figure 7.3. Because of delocalization of electrons in the allyl radical (the two electrons of the pi bond and the single electron of the radical), the allyl radical is considerably more stable than might be expected by looking at any one of its contributing structures.

Based on bond dissociation energies, we conclude that an allyl radical is even more stable than a 3° alkyl radical.

	Bond Dissociation Energy [kcal/mol (kJ/mol)]	**Type of Radical Formed**	
$CH_2{=}CHCH_2{-}H$	86 (360)	allylic	↑ Increasing stability
$(CH_3)_3C{-}H$	93 (389)	3°	
$(CH_3)_2CH{-}H$	96 (402)	2°	
$CH_3CH_2{-}H$	100 (418)	1°	
$CH_3{-}H$	105 (439)	methyl	

According to the molecular orbital description, the conjugated system of the allyl radical involves formation of three molecular orbitals by overlap of three $2p$ atomic orbitals (Figure 7.4). The lowest energy MO has zero nodes, the next MO has one node, and the highest energy MO has two nodes. The molecular orbital of intermediate energy in this case leads to neither net stabilization or destabilization, and is therefore called a nonbonding MO. According to MO theory, in the lowest energy (ground) state of the allyl radical, two electrons of the pi system lie in the pi bonding

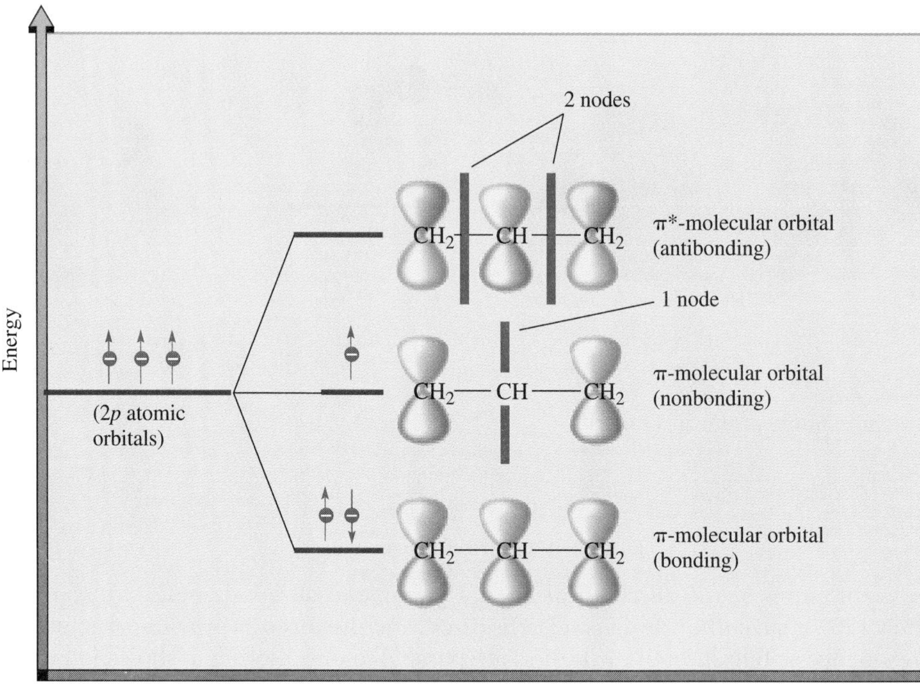

FIGURE 7.4
Molecular orbital model of covalent bonding in the allyl radical. Combination of three $2p$ atomic orbitals gives three pi molecular MOs. The lowest, a pi bonding MO, has no node. The next in energy, a pi nonbonding MO, has one node and the highest in energy, a pi antibonding MO, has two nodes.

MO, and the third lies in the pi nonbonding MO; the pi antibonding MO is unoccupied. The lone electron of the allyl radical is associated with the pi nonbonding MO which places electron density on carbons 1 and 3 only. Thus both the resonance model and molecular orbital theory are equivalent in predicting radical character on carbons 1 and 3 of the allyl radical but no radical character on carbon 2.

Since the lowest pi MO is at a lower energy than the isolated $2p$ atomic orbitals, putting two electrons in this MO releases considerable energy, which accounts for the stability of the allyl radical.

EXAMPLE 7.5

Account for the fact that allylic bromination of 1-octene by NBS gives these products.

$$\underset{\text{1-Octene}}{CH_3(CH_2)_4CH_2CH{=}CH_2} \xrightarrow[CCl_4]{NBS} \underset{\substack{\text{3-Bromo-1-octene}\\(17\%)}}{CH_3(CH_2)_4\overset{\overset{\displaystyle Br}{|}}{C}HCH{=}CH_2} + \underset{\substack{\text{1-Bromo-2-octene}\\(83\%)}}{CH_3(CH_2)_4CH{=}CHCH_2Br}$$

Solution

The rate-limiting step in this radical chain mechanism is hydrogen abstraction from the allylic position on 1-octene to give a 2° allylic radical. This radical is stabilized by delocalization of the two pi electrons and the unpaired electron as shown in these contributing structures. Reaction of the radical at carbon 1, the less hindered carbon, gives the major product. Reaction of the radical at carbon 3, the more hindered carbon, gives the minor product.

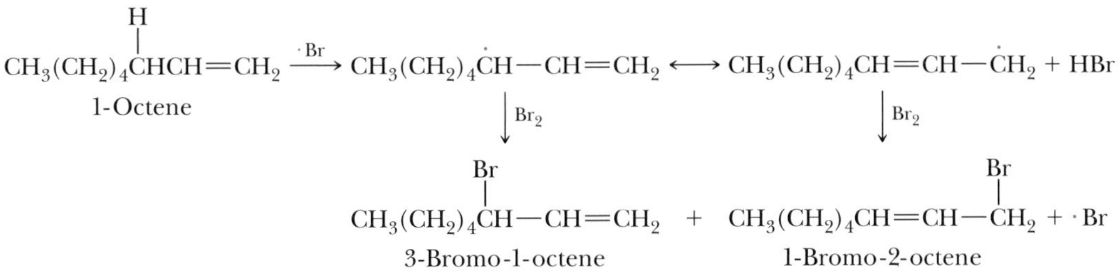

$$\underset{\text{1-Octene}}{CH_3(CH_2)_4\overset{H}{C}HCH{=}CH_2} \xrightarrow{\cdot Br} CH_3(CH_2)_4\dot{C}H{-}CH{=}CH_2 \longleftrightarrow CH_3(CH_2)_4CH{=}CH{-}\dot{C}H_2 + HBr$$

$$\downarrow Br_2 \qquad\qquad \downarrow Br_2$$

$$\underset{\text{3-Bromo-1-octene}}{CH_3(CH_2)_4\overset{Br}{C}H{-}CH{=}CH_2} + \underset{\text{1-Bromo-2-octene}}{CH_3(CH_2)_4CH{=}CH{-}\overset{Br}{C}H_2} + \cdot Br$$

PROBLEM 7.5

Given the solution to Example 7.5, predict the structure of the product(s) formed when 3-hexene is treated with NBS?

CHEMISTRY IN ACTION

Radical Autoxidation

One of the most important reactions for materials, foods, and also for living systems is called **autoxidation,** that is, oxidation requiring oxygen and no other reactant. This reaction takes place by a radical chain mechanism very similar to that for allylic bromination. If you open a bottle of cooking oil that has stood for a long time, you will notice the hiss of air entering the bottle. This is because there is a negative pressure caused by the consumption of oxygen by autoxidation of the oil.

Cooking oil contains **polyunsaturated fatty acid esters** (Section 25.1). The most common of these compounds have 16 or 18 carbon chains containing 1,4-diene functional groups, as shown in the following structure. (Both double bonds are *cis*; the nature of R_1 and R_2 need not concern us at this stage.) The hydrogens on the CH_2 group between the double bonds are doubly allylic, that is, allylic with respect to two different double bonds. As you might expect, the radical formed by abstraction of one of these hydrogens is even more stabilized by delocalization of the unpaired electron than an allylic radical. Just as an allylic C—H bond is much weaker than a corresponding alkane C—H, the doubly allylic C—H is even weaker.

Autoxidation begins when a radical initiator, X·, formed either by light activation of an impurity in the oil or by thermal decomposition of peroxide impurities, abstracts the doubly allylic hydrogen to form a radical.

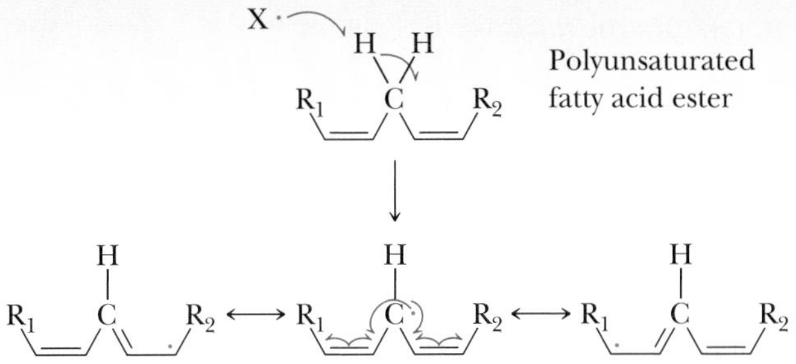

Polyunsaturated fatty acid ester

This radical reacts with oxygen, itself a diradical, to form a peroxy radical, which then reacts with another 1,4-diene fatty acid ester (H—R) to give a new radical (R·) and a hydroperoxide. This causes a radical chain reaction in which hundreds of molecules of fatty acid ester are oxidized for each initiator radical formed.

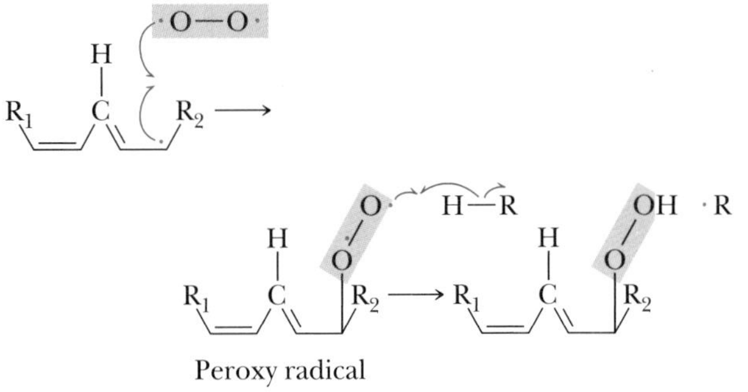

Peroxy radical

(continued)

The ultimate fate of the peroxide, and some of the peroxy radical as well, is complex. Some autoxidation products degrade to short-chain aldehydes and carboxylic acids with unpleasant "rancid" smells familiar to anyone who has smelled old cooking oil or spoiled foods that contain polyunsaturated oils. It has been suggested that some products of autoxidation of oils are toxic and/or carcinogenic. Oils lacking the 1,4-diene structure are much less easily oxidized. Similar oxidative degradation of materials in low density lipoproteins (LDL, Section 25.4) deposited on the walls of arteries leads to cardiovascular disease in humans. Many effects of aging in humans and damage to materials such as rubber and plastic occur by similar mechanisms.

Many plants contain polyunsaturated fatty acid esters in their leaves or seeds. In the natural state, they are protected against autoxidation by a variety of agents. One of the most important of these agents is α-tocopherol (vitamin E, Section 25.6 C). This compound is a phenol (Section 19.5). The characteristic of phenols that makes them effective as protective agents against autoxidation is their O—H bond, which is even weaker than the doubly allylic C—H bond. Vitamin E reacts preferentially with the initial peroxy radical to give a resonance-stabilized phenoxy radical, which is very unreactive and survives to scavenge another peroxy radical, $RO_2\cdot$. The resulting peroxide is stable under the conditions. The antioxidant thus removes two molecules of peroxy radical and stops the radical chain oxidation dead in its tracks.

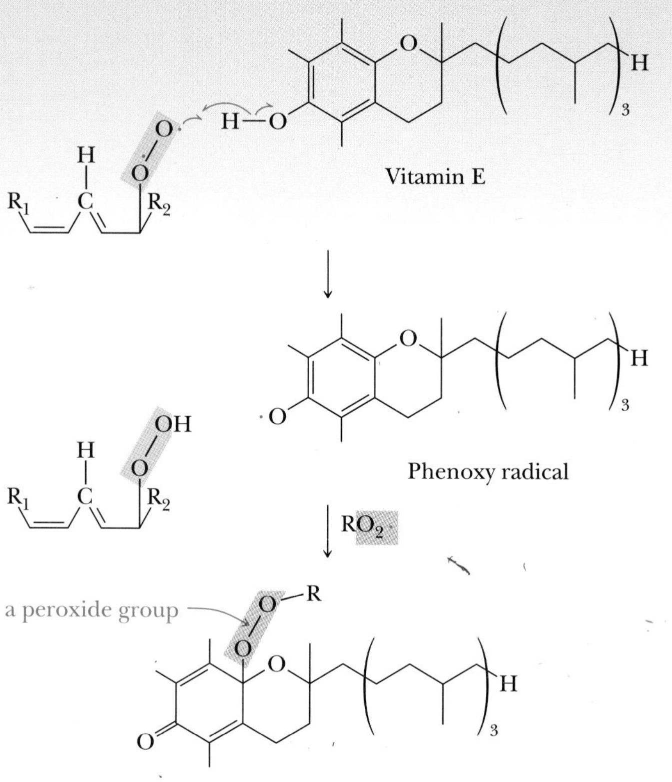

Vitamin E

Phenoxy radical

$RO_2\cdot$

a peroxide group

A peroxide derived from vitamin E

Because vitamin E is removed in the processing of many food products, similar phenols such as BHA and BHT (Problem 20.23) are often added to foods to prevent spoilage by autoxidation. These compounds are all called **radical inhibitors** because of their ability to terminate radical chains. Similar compounds are also added to many materials, such as plastics and rubber, to protect them against autoxidation.

Walling, C., "Autoxidation," in *Active Oxygen in Chemistry;* Foote, C. S., Valentine, J. S., Greenberg, A., Liebman, J. F., Eds.; Chapman and Hall, London, 1995; pp. 24–65.

7.7 Formation of Organometallic Compounds

Organometallic compound A compound that contains a metal bonded to a carbon atom.

In this section, we undertake our first discussion of a broad class of organic compounds called organometallic compounds. An **organometallic compound** is one that contains a metal bonded to a carbon atom. Here we introduce organomagnesium, organolithium, and organocopper compounds and concentrate on their formation and basicity. We continue the discussion on organometallics in considerably more detail in Chapters 15 and 17.

RMgX	RLi	R_2CuLi
An organomagnesium compound (a Grignard reagent)	An organolithium compound	A lithium diorganocopper compound (a Gilman reagent)

A. Formation of Organomagnesium and Organolithium Compounds

Alkyl and aryl halides react with Group I, Group II, and certain other metals to form organometallic compounds. Organomagnesium compounds are called **Grignard reagents** after Victor Grignard, who was awarded a Nobel Prize for chemistry in 1912 for their discovery and their application to organic synthesis. Butylmagnesium chloride, for example, is prepared by treating 1-chlorobutane dissolved in diethyl ether with magnesium metal.

Grignard reagent An organomagnesium compound.

$$CH_3CH_2CH_2CH_2Cl + Mg \xrightarrow{\text{ether}} CH_3CH_2CH_2CH_2MgCl$$

1-Chlorobutane — Butylmagnesium chloride

Grignard reagents form on the surface of the metal and dissolve as coordination complexes consisting of one molecule of Grignard reagent and two molecules of ether. In this ether-soluble complex, magnesium acts as a Lewis acid and the ether acts as a Lewis base.

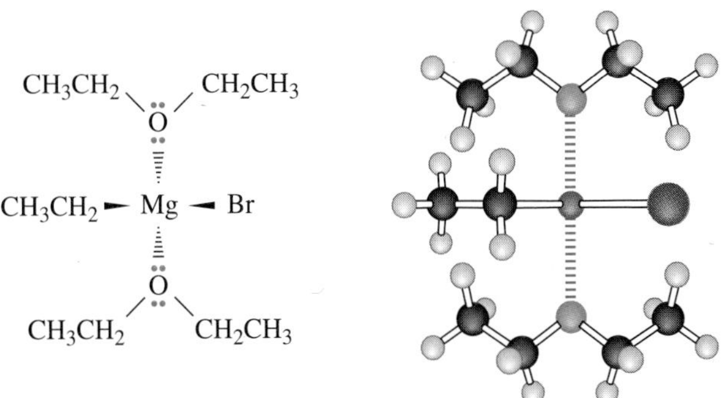

Ethylmagnesium bromide dietherate

Organolithium reagents are formed by the reaction of an alkyl or aryl halide with two equivalents of lithium metal as illustrated by the preparation of butyllithium. In this reaction, a solution of 1-chlorobutane in pentane is added to lithium wire at −10°C. Many organolithium reagents are now commercially available.

$$CH_3CH_2CH_2CH_2Cl + 2Li \xrightarrow{\text{pentane}} CH_3CH_2CH_2CH_2Li + LiCl$$

1-Chlorobutane — Butyllithium

Carbon-metal bonds in organometallics are best described as polar covalent, with carbon bearing a partial negative charge and the metal bearing a partial positive charge. Shown in Table 7.7 are electronegativity differences (Pauling scale, Table 1.5) between carbon and various metals. From this difference, we can estimate the percent ionic character of each carbon-metal bond. Organolithium and organomagnesium bonds have the highest partial ionic character, while organocopper and organomer-

TABLE 7.7 Percent Ionic Character of Some C—M Bonds

C—M Bond	Difference in Electronegativity	Percent Ionic Character*
C—Li	2.5 − 1.0 = 1.5	60
C—Mg	2.5 − 1.2 = 1.3	52
C—Al	2.5 − 1.5 = 1.0	40
C—Zn	2.5 − 1.6 = 0.9	36
C—Sn	2.5 − 1.8 = 0.7	28
C—Cu	2.5 − 1.9 = 0.6	24
C—Hg	2.5 − 1.9 = 0.6	24

* Percent ionic character $= \frac{E_C - E_M}{E_C} \times 100$

cury compounds have lower partial ionic character. In spite of the polarity of the carbon-metal bond, these compounds do not behave as salts. Organolithium compounds, for example, which have the highest partial ionic character, dissolve in hydrocarbon solvents such as pentane.

B. Reaction with Proton Acids

Both organomagnesium and organolithium compounds are strong bases and react rapidly with any acid (proton donor) stronger than the alkane from which they are derived. Ethylmagnesium bromide, for example, reacts instantly with water to give ethane and magnesium salts. This reaction is an example of a stronger acid and a stronger base reacting to give a weaker acid and a weaker base (Section 3.4).

$$\overset{\delta-}{CH_3CH_2}\text{—}\overset{\delta+}{MgBr} + H\text{—}OH \longrightarrow CH_3CH_2\text{—}H + Mg^{2+} + OH^- + Cl^- \qquad pK_{eq} = -35,\; K_{eq} = 10^{35}$$

Stronger base	pK_a 15.7 Stronger acid	pK_a 51 Weaker acid	Weaker base

Following are several classes of proton donors that react readily with Grignard and organolithium reagents.

R_2NH	$RC{\equiv}CH$	ROH	HOH	ArOH	RSH	RCO_2H
pK_a 38–40	pK_a 25	pK_a 16–18	pK_a 15.7	pK_a 9–10	pK_a 8–9	pK_a 4–5
Amines	Alkynes	Alcohols	Water	Phenols	Thiols	Carboxylic acids

EXAMPLE 7.6

Write an equation for the acid-base reaction between ethylmagnesium iodide and an alcohol. Use curved arrows to show the flow of electrons in this reaction. In addition, show that this reaction is an example of a stronger acid and stronger base reacting to form a weaker acid and weaker base.

Solution

The alcohol is the stronger acid and the partially negatively charged ethyl group is the stronger base.

$$CH_3CH_2{-}MgI + H{-}OR \longrightarrow CH_3CH_2{-}H + RO^-MgI^+$$

	An alcohol	An alkane	A magnesium alkoxide
	pK_a 16–18	pK_a 51	
(Stronger base)	(Stronger acid)	(Weaker acid)	(Weaker base)

PROBLEM 7.6

Explain how these Grignard reagents will react with molecules of their own kind to "self-destruct."

(a) $HOCH_2CH_2CH_2MgBr$ **(b)** $HC{\equiv}C(CH_2)_4CH_2MgBr$

C. Lithium Diorganocopper Reagents and the Synthesis of Alkanes

One of the most important uses of organolithium reagents (Section 7.6A) is in the preparation of diorganocopper reagents, often called Gilman reagents after Henry Gilman of Iowa State University. Formation of a Gilman reagent is illustrated by the preparation of lithium dibutylcopper from butyllithium and copper(I) iodide.

$$2CH_3CH_2CH_2CH_2Li + CuI \xrightarrow[\text{or THF}]{\text{diethyl ether}} (CH_3CH_2CH_2CH_2)_2Cu^-\ Li^+ + LiI$$

Butyllithium; Copper(I) iodide; Lithium dibutylcopper (a Gilman reagent)

Lithium diorganocopper reagents are especially valuable for the formation of new carbon-carbon bonds. They react with alkyl chlorides, bromides, and iodides (but not fluorides) to give cross-coupling products, as illustrated by the reaction of lithium dimethylcopper and 1-iododecane to give undecane. Notice that only one of the Gilman-reagent alkyl groups is transferred in the cross-coupling.

$$(CH_3)_2CuLi + CH_3(CH_2)_8CH_2I \xrightarrow[\text{or THF}]{\text{diethyl ether}} CH_3(CH_2)_8CH_2CH_3 + CH_3Cu + LiI$$

Lithium dimethylcopper; 1-Iododecane; Undecane

Good yields of cross-coupling products are obtained with methyl, primary alkyl, secondary cycloalkyl, allylic, vinylic, and aryl halides. Yields are lower with secondary and tertiary alkyl halides. Gilman reagents giving the best yields of cross-coupling products are those prepared from methyl, primary alkyl, allylic, vinylic, and aryl halides.

In cross-coupling with a vinylic halide, the configuration of the carbon-carbon double bond is preserved, as illustrated by the synthesis of *trans*-5-tridecene.

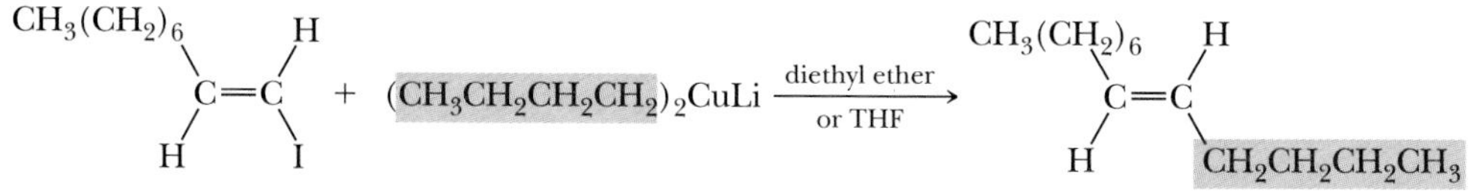

$$\underset{trans\text{-1-Iodo-1-nonene}}{CH_3(CH_2)_6(H)C{=}C(H)I} + (CH_3CH_2CH_2CH_2)_2CuLi \xrightarrow[\text{or THF}]{\text{diethyl ether}} \underset{trans\text{-5-Tridecene}}{CH_3(CH_2)_6(H)C{=}C(H)CH_2CH_2CH_2CH_3}$$

The mechanism of these coupling reactions is not fully understood, and is the subject of active investigation.

EXAMPLE 7.7

Show how to bring about these conversions using a lithium diorganocopper reagent.

(a) 1-Bromocyclohexene to 1-methylcyclohexene
(b) 1-Bromo-2-methylpropane to 2,5-dimethylhexane

Solution

(a) 1-Bromocyclohexene is a vinylic bromide. Treat it with lithium dimethylcopper.

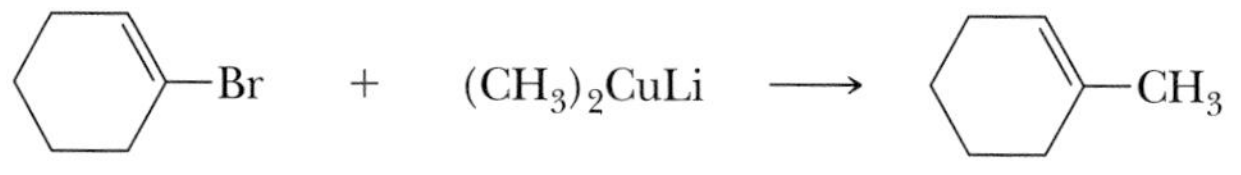

$$\underset{\text{1-Bromocyclohexene}}{C_6H_9\text{–}Br} + \underset{\text{Lithium dimethylcopper}}{(CH_3)_2CuLi} \longrightarrow \underset{\text{1-Methylcyclohexene}}{C_6H_9\text{–}CH_3}$$

(b) The combination of Gilman reagent and alkyl halide shown here will give 2,5-dimethylhexane in good yield.

$$\underset{\substack{\text{1-Bromo-2-methylpropane}\\\text{(Isobutyl bromide)}}}{(CH_3)_2CHCH_2Br} + \underset{\substack{\text{Lithium}\\\text{diisobutylcopper}}}{[(CH_3)_2CHCH_2]_2CuLi} \longrightarrow \underset{\text{2,5-Dimethylhexane}}{CH_3CH(CH_3)CH_2{-}CH_2CH(CH_3)CH_3}$$

PROBLEM 7.7

Show how to bring about these conversions using a lithium diorganocopper reagent.

(a) $C_6H_5I \longrightarrow C_6H_5CH_2CH_2CH_2CH_3$ **(b)** Bromocyclohexane $\longrightarrow$ (3-methylbutyl)cyclohexane [skeletal structures; Br on cyclohexane]

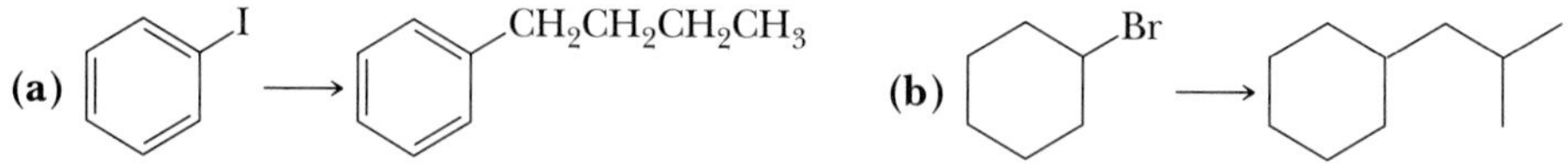

SUMMARY

Haloalkanes contain a halogen covalently bonded to an sp^3 hybridized carbon (Section 7.1). In the IUPAC system, halogen atoms are named as fluoro-, chloro-, bromo-, and iodo- substituents and are listed in alphabetical order with other substituents (Section 7.2). The van der Waals radius of fluorine is only slightly greater than that of hydrogen and, among the other halogens, only iodine has a larger van der Waals radius than methyl.

Among alkanes and chloro-, bromo-, and iodoalkanes of comparable size and shape, the haloalkanes have the higher boiling points because of the greater polarizability of the unshared electrons of the halogen atom (Section 7.3B). Boiling points of fluoroalkanes are generally lower than those of alkanes of comparable size and shape because of the uniquely low polarizability of the valence electrons of fluorine. The density of liquid haloalkanes is greater than that of hydrocarbons of comparable molecular weight because of the halogen's larger mass to volume ratio (Section 7.3C).

A **radical chain reaction** consists of three types of steps: chain initiation, chain propagation, and chain termination (Section 7.5). In **chain initiation,** radicals are formed from nonradical compounds. In a **chain propagation** step, a radical and a molecule react to give a new radical. When summed, chain propagation steps give the observed stoichiometry of the reaction. **Chain length** is the number of times a cycle of chain propagation steps repeats. In a **chain termination** step, radicals are destroyed. Simple alkyl radicals are planar or almost so with bond angles of 120° about the carbon with the unpaired electron. Heats of reaction for a radical reaction and for individual chain initiation, propagation, and termination steps can be calculated from bond dissociation energies.

According to **Hammond's postulate** (Section 7.5D), the structure of the transition state of an exothermic reaction looks more like the reactants than like the products. Conversely, the structure of the transition state of an endothermic reaction looks more like the products than it does the reactants. Hammond's postulate accounts for the fact that bromination of an alkane is more regioselective than chlorination. For both bromination and chlorination of alkanes, the rate-limiting step is hydrogen abstraction to form an alkyl radical. Hydrogen abstraction is endothermic for bromination and exothermic for chlorination.

Allylic substitution is any reaction in which an atom or group of atoms is substituted for another atom or group of atoms at a carbon adjacent to a carbon-carbon double bond (Section 7.6). **Allylic halogenation** proceeds by a radical chain mechanism. Because of delocalization of electrons, the allyl radical is more stable than the *tert*-butyl radical.

An **organometallic compound** is one that contains a carbon-metal bond (Section 7.7). Organomagnesium compounds are named Grignard reagents after their discoverer, Victor Grignard. Lithium diorganocopper reagents (Gilman reagents) undergo cross-coupling reactions with methyl, primary alkyl, secondary cycloalkyl, vinylic, allylic, and aryl halides to yield hydrocarbons.

KEY REACTIONS

1. Chlorination and Bromination of Alkanes (Section 7.4)

Chlorination and bromination of alkanes are regioselective in the order 3°H > 2°H > 1°H. Bromination has a higher regioselectivity than chlorination.

$$CH_3CH_2CH_3 + Br_2 \xrightarrow[\text{heat}]{\text{light or}} \underset{\substack{| \\ Br \\ 92\%}}{CH_3CHCH_3} + \underset{8\%}{CH_3CH_2CH_2Br} + HBr$$

2. Allylic Bromination and Chlorination (Section 7.6)

This is a radical chain reaction and occurs at high temperatures (heat is the radical initiator) using the halogens themselves, or at room temperature using *N*-bromosuccinimide (NBS) and a peroxide as the radical initiator.

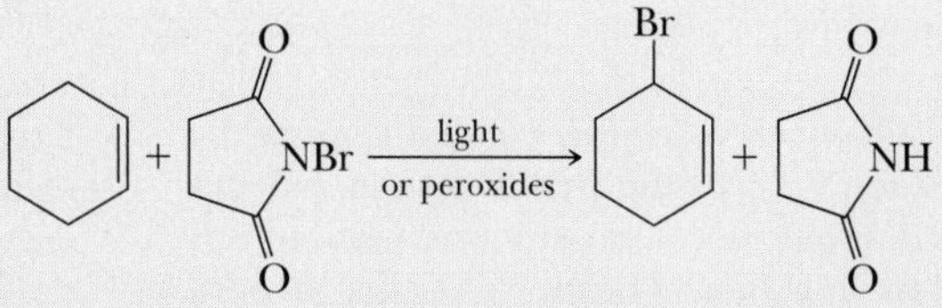

Cyclohexene + NBr (N-bromosuccinimide) $\xrightarrow[\text{or peroxides}]{\text{light}}$ 3-bromocyclohexene + NH (succinimide)

3. Formation of Organomagnesium (Grignard) and Organolithium Reagents (Section 7.7)

Both organomagnesium and organolithium compounds are strong bases.

$$CH_3CH_2CH_2CH_2Cl + Mg \xrightarrow{\text{ether}} CH_3CH_2CH_2CH_2MgCl$$

$$CH_3CH_2CH_2CH_2Cl + 2Li \xrightarrow{\text{pentane}} CH_3CH_2CH_2CH_2Li + LiCl$$

4. Formation of Gilman Reagents (Section 7.7 C)

Lithium diorganocopper (Gilman) reagents are prepared by treating an organolithium compound with copper(I) iodide. Gilman reagents undergo cross-coupling with alkyl, alkenyl, and aryl halides.

$$(CH_3)_2CuLi + CH_3(CH_2)_8CH_2I \xrightarrow[\text{or THF}]{\text{diethyl ether}} CH_3(CH_2)_8CH_2CH_3 + CH_3Cu + LiI$$

ADDITIONAL PROBLEMS

Nomenclature

7.8 Give IUPAC names for the following compounds. Where stereochemistry is shown, include a designation of configuration in your answer.

(a) 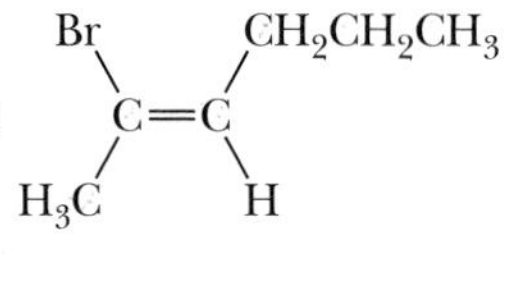 Br(H$_3$C)C=C(H)CH$_2$CH$_2$CH$_3$

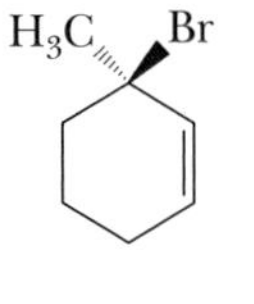

(b) cyclohexene ring bearing H_3C and Br on the same carbon

(c) H_3C–cyclohexane (chair)–Br

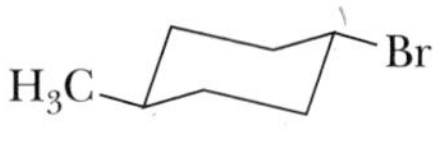

(d) $ClCH_2CH_2CH_2CH_2Cl$

(e) 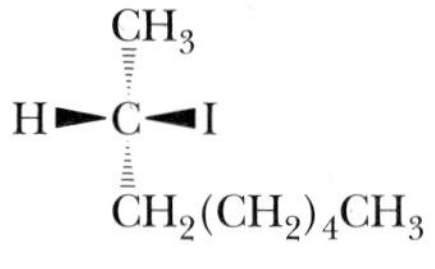 H–C(CH$_3$)(I)–$CH_2(CH_2)_4CH_3$

(f) CH_3–C(H)(Br)–$(CH_2)_2CH_3$

(g) cycloheptene ring bearing F

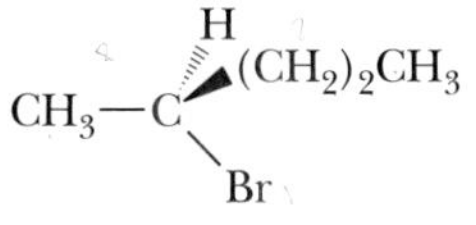

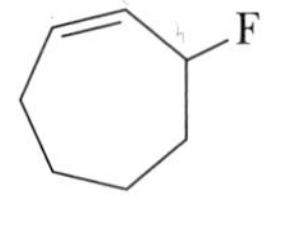

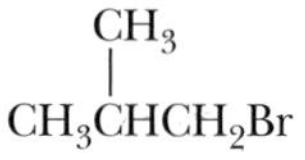

(h) $CH_3\overset{CH_3}{\overset{|}{C}}HCH_2Br$

(i) bicyclic ring bearing Cl

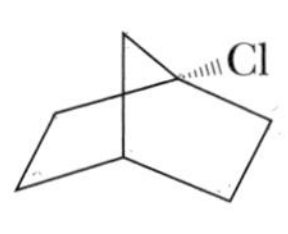

7.9 Draw structural formulas for the following compounds.

(a) Allyl iodide

(b) (*R*)-2-Chlorobutane

(c) meso-2,3-Dibromobutane

(d) *trans*-1-Bromo-3-isopropylcyclohexane

(e) Neopentyl iodide

(f) Cyclobutyl bromide

Physical Properties

7.10 Water and methylene chloride are insoluble in each other. When each is added to a test tube, two layers form. Which layer is water and which layer is methylene chloride?

7.11 The boiling point of methylcyclohexane (C_7H_{14}, MW 98.2) is 101°C. The boiling point of perfluoromethylcyclohexane (C_7F_{14}, MW 350) is 76°C. Account for the fact that although the molecular weight of perfluoromethylcyclohexane is over 3 times that of methylcyclohexane, its boiling point is lower than that of methylcyclohexane.

7.12 Account for the fact that among the chlorinated derivatives of methane, chloromethane has the largest dipole moment and tetrachloromethane has the smallest dipole moment.

Name	Molecular Formula	Dipole Moment (debyes: D)
chloromethane	CH_3Cl	1.87
dichloromethane	CH_2Cl_2	1.60
trichloromethane	$CHCl_3$	1.01
tetrachloromethane	CCl_4	0

Halogenation of Alkanes

7.13 Name and draw structural formulas for all possible monohalogenation products that might be formed in the following reactions.

(a) 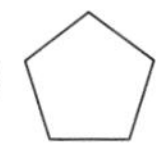 (cyclopentane) $+ Cl_2 \xrightarrow{\text{light}}$

(b) $CH_3CH(CH_3)CH_2CH_2CH_3 + Cl_2 \xrightarrow{\text{light}}$

(c) $CH_3C(CH_3)HCH(CH_3)CH_3 + Br_2 \xrightarrow{\text{light}}$

(d) (cyclopropane) $+ Br_2 \xrightarrow{\text{light}}$

7.14 Which compounds can be prepared in high yield by regioselective halogenation of an alkane?

(a) 2-Chloropentane
(b) Chlorocyclopentane
(c) 2-Bromo-2-methylheptane
(d) 2-Bromo-3-methylbutane
(e) 2-Bromo-2,4,4-trimethylpentane
(f) Iodoethane

7.15 There are three constitutional isomers of molecular formula C_5H_{12}. When treated with chlorine gas at 300°C, isomer A gives a mixture of four monochlorination products. Under the same conditions, isomer B gives a mixture of three monochlorination products and isomer C gives only one monochlorination product. From this information, assign structural formulas to isomers A, B, and C.

7.16 Following is a balanced equation for bromination of propane.

$$CH_3CH_2CH_3 + Br_2 \longrightarrow CH_3CH(Br)CH_3 + HBr$$

(a) Using the values for bond dissociation energies given in Appendix 3, calculate ΔH^0 for this reaction.

(b) Propose a pair of chain propagation steps and show that they add up to the observed reaction.

(c) Calculate ΔH^0 for each chain propagation step.

(d) Which propagation step is rate limiting, that is, which crosses the higher potential energy barrier?

7.17 Write a balanced equation and calculate ΔH^0 for reaction of CH_4 and I_2 to give CH_3I and HI. Explain why this reaction cannot be used as a method of preparation of iodomethane.

7.18 Following are balanced equations for fluorination of propane to produce a mixture of 1-fluoropropane and 2-fluoropropane.

$$\underset{\text{Propane}}{CH_3CH_2CH_3} + F_2 \longrightarrow \underset{\text{1-Fluoropropane}}{CH_3CH_2CH_2F} + HF$$

$$\underset{\text{Propane}}{CH_3CH_2CH_3} + F_2 \longrightarrow \underset{\text{2-Fluoropropane}}{CH_3\overset{F}{\overset{|}{C}}HCH_3} + HF$$

Assume that each product is formed by a radical chain mechanism.

(a) Calculate ΔH^0 for each reaction.

(b) Propose a pair of chain propagation steps for each reaction, and calculate ΔH^0 for each step.

(c) Reasoning from the Hammond postulate, predict the regioselectivity of radical fluorination relative to that of radical chlorination and bromination.

7.19 As you demonstrated in Problem 7.18, radical fluorination of alkanes is highly exothermic. As per Hammond's postulate, assume that the transition state for radical fluorination is almost identical to the starting material. With this assumption, estimate the fraction of each monofluoro product formed in the fluorination of 2-methylbutane.

7.20 Cyclobutane reacts with bromine to give bromocyclobutane, but bicyclo[1.1.0]butane reacts with bromine to give 1,3-dibromocyclobutane. Account for the difference between the reactions of these two compounds.

Cyclobutane + Br_2 $\xrightarrow{\text{heat}}$ Bromocyclobutane (—Br) + HBr

Bicyclo[1.1.0]butane + Br_2 $\xrightarrow{\text{heat}}$ 1,3-Dibromocyclobutane (Br— —Br)

7.21 The first chain propagation step of all radical halogenation reactions we considered in Sections 7.5B and 7.6 is abstraction of hydrogen by the halogen atom to give an alkyl radical and HX, as for example

$$CH_3CH_3 + \cdot Br \longrightarrow CH_3CH_2\cdot + HBr$$

Suppose, instead, that radical halogenation occurs by an alternative pair of chain propagation steps, beginning with this step:

$$CH_3CH_3 + \cdot Br \longrightarrow CH_3CH_2Br + H\cdot$$

(a) Propose a second chain propagation step. Remember that a characteristic of chain propagation steps is that they add to the observed reaction.

(b) Calculate the heat of reaction, ΔH^0, for each of the two steps.

(c) Compare the energetics and relative rates of the set of chain propagation steps in Section 7.5B with the set proposed here.

Allylic Halogenation

7.22 Following is a balanced equation for the allylic bromination of propene.

$$CH_2{=}CHCH_3 + Br_2 \longrightarrow CH_2{=}CHCH_2Br + HBr$$

(a) Calculate the heat of reaction, ΔH^0, for this conversion.
(b) Propose a pair of chain propagation steps and show that they add up to the observed stoichiometry.
(c) Calculate the ΔH^0 for each chain propagation step and show that they add up to the observed ΔH^0 for the overall reaction.

7.23 Using Appendix 3, estimate the bond dissociation energy of each indicated bond in cyclohexene.

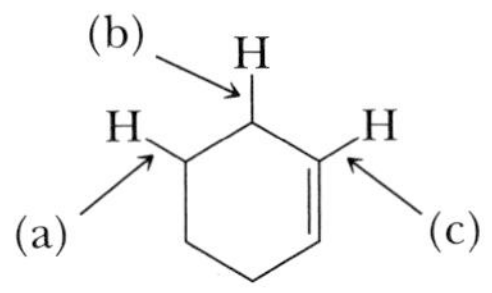

(b) H, (a) H, (c) H

7.24 Propose a series of chain initiation, propagation, and termination steps for this reaction and estimate its heat of reaction.

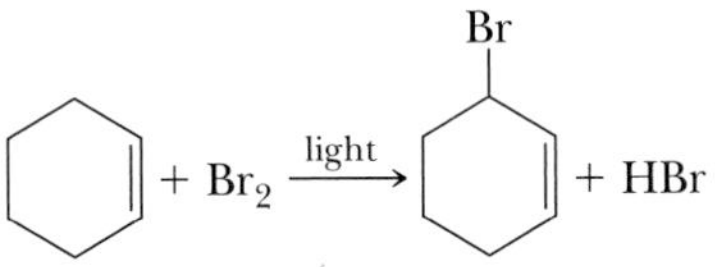

$$\text{cyclohexene} + Br_2 \xrightarrow{\text{light}} \text{3-bromocyclohexene} + HBr$$

7.25 The major product formed when methylenecyclohexane is treated with NBS in carbon tetrachloride is 1-bromomethylcyclohexene. Account for the formation of this product.

7.26 Draw the structural formula of the products formed when each alkene is treated with one equivalent of NBS in CCl_4 in the presence of benzoyl peroxide. (Two products are possible from each alkene.)

(a) $CH_3CH{=}CHCH_2CH_3$ **(b)**  [4,4-dimethylcyclohexene] **(c)** 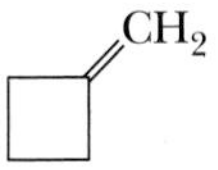 [methylenecyclobutane, CH_2]

7.27 The activation energy for hydrogen abstraction from ethane by a chlorine atom is 1.0 kcal/mol; that for hydrogen abstraction by a bromine atom is 13.2 kcal/mol (Section 7.5D). Calculate the ratio of rate constants, k_{Cl}/k_{Br}, for these two reactions. *Hint:* Review the Arrhenius equation, Section 6.2.

Synthesis

7.28 Show reagents and conditions to bring about these conversions.

(a) 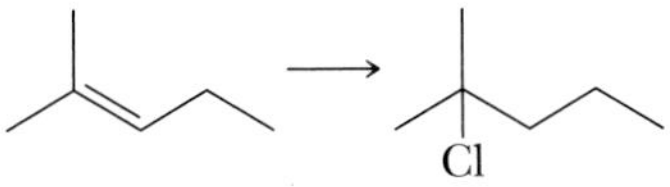 [2-methyl-2-pentene] $\longrightarrow$ [2-chloro-2-methylpentane, Cl]

(b) $CH_3CH{=}CHCH_3 \longrightarrow CH_3CH{=}CHCH_2Br$

(c) $CH_3CH{=}CHCH_3 \longrightarrow CH_3\underset{Br}{CH}{-}\underset{Br}{CH}CH_3$

(d) 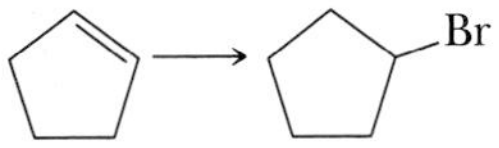 [cyclopentene] $\longrightarrow$ [bromocyclopentane, Br]

(e) 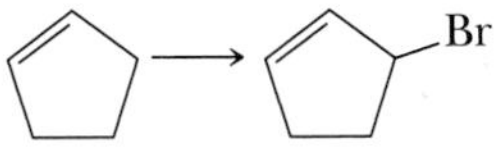 [cyclopentene] $\longrightarrow$ [3-bromocyclopentene, Br]

7.29 Complete these reactions involving lithium diorganocopper (Gilman) reagents.

(a) Cyclohexyl–Br + $(CH_2{=}C(CH_3))_2CuLi \xrightarrow{\text{ether}}$

(b) 1-Bromocyclohexene + $(CH_3CH_2CH_2CH_2)_2CuLi \xrightarrow{\text{ether}}$

(c) $CH_3CH_2CH_2CH_2I + [(CH_3)_3C)]_2CuLi \xrightarrow{\text{ether}}$

(d) Cyclohexylidene=C(H)(Cl) + $\left(\overset{H_3C}{\underset{H}{}}C{=}C\overset{H}{\underset{CH_2}{}}\right)_2CuLi \xrightarrow{\text{ether}}$

7.30 Show how to convert 1-bromopentane to each of these compounds using a lithium diorganocopper (Gilman) reagent. Write an equation, showing structural formulas, for each synthesis.

(a) Nonane **(b)** 3-Methyloctane **(c)** 2,2-Dimethylheptane
(d) 1-Heptene **(e)** 1-Octene

7.31 In Problem 7.30, you used a series of lithium diorganocopper (Gilman) reagents. Show how to prepare each Gilman reagent from an appropriate alkyl or vinylic halide.

7.32 Show how to prepare each compound from the given starting compound through the use of a lithium diorganocopper (Gilman) reagent.

(a) 4-Methylcyclopentene from 4-bromocyclopentene
(b) (Z)-2-Undecene from (Z)-1-bromopropene
(c) 1-Butylcyclohexene from 1-iodocyclohexene
(d) 1-Decene from 1-iodooctane
(e) 1,8-Nonadiene from 1,5-dibromopentane

NUCLEOPHILIC SUBSTITUTION AND β-ELIMINATION

8

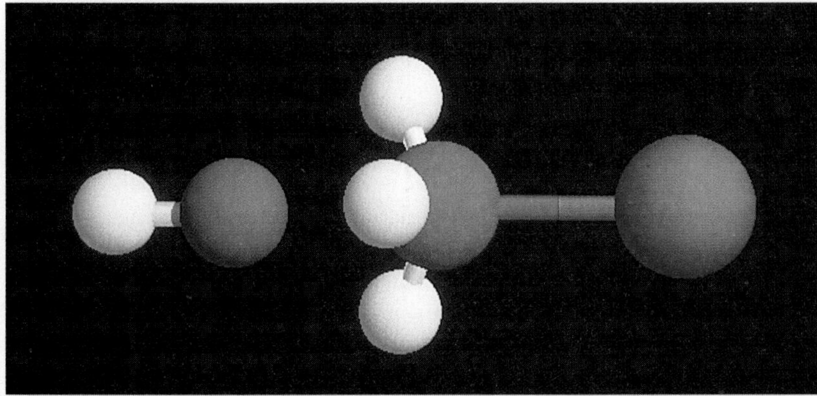

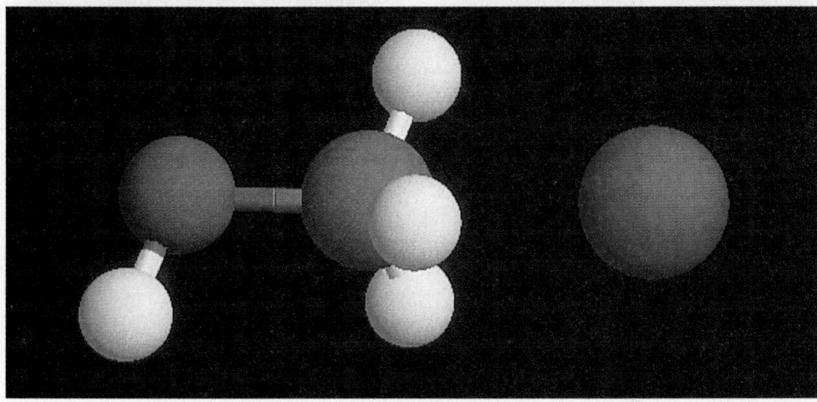

■ Hydroxide ion reacts with bromomethane in an S_N2 reaction (Section 8.3). *(Charles D. Winters)*

Nucleophilic substitution is any reaction in which one **nucleophile** is substituted for another. In the following general equation, $Nu:^-$ is the nucleophile and X is the leaving group. By far the most carefully studied behavior of alkyl halides is their reaction with nucleophilic reagents, such as HO^-, $CH_3CH_2O^-$, and NH_3.

$$\underset{\text{Nucleophile}}{Nu:^-} + -\overset{|}{\underset{|}{C}}-\underset{\substack{\text{leaving}\\\text{group}}}{X} \xrightarrow[\text{substitution}]{\text{nucleophilic}} -\overset{|}{\underset{|}{C}}-Nu + :X^-$$

Nucleophilic substitution Any reaction in which one nucleophile is substituted for another.

Nucleophile A molecule or ion that donates a pair of electrons to another atom or ion to form a new covalent bond; a Lewis base.

Because all nucleophiles are also bases, nucleophilic substitution and base-promoted ***β*-elimination** are competing reactions. The ethoxide ion, for example, is both a nucleophile and a base. With bromocyclohexane, it reacts as a nucleophile to give ethoxycyclohexane (cyclohexyl ethyl ether), and as a base to give cyclohexene and ethanol.

$$\text{C}_6\text{H}_{11}\text{Br} + CH_3CH_2O^-Na^+ \xrightarrow[\text{ethanol}]{\text{nucleophilic substitution}} \text{C}_6\text{H}_{11}OCH_2CH_3 + Na^+Br^-$$

$$\text{C}_6\text{H}_{11}\text{Br} + CH_3CH_2O^-Na^+ \xrightarrow[\text{ethanol}]{\beta\text{-elimination}} \text{cyclohexene} + CH_3CH_2OH + Na^+Br^-$$

In this chapter, we study these two types of organic reactions. Using them, alkyl halides can be converted to compounds with other functional groups including alcohols, ethers, thiols, sulfides, amines, nitriles, alkenes, and alkynes. Thus, an understanding of nucleophilic substitution and β-elimination opens entirely new areas of organic chemistry for you.

8.1 Nucleophilic Aliphatic Substitution

Nucleophilic substitution is one of the most important reactions of alkyl halides and can lead to a wide variety of new functional groups, many of which are illustrated in Table 8.1. As you study the entries in this table, note these points.

TABLE 8.1 Some Nucleophilic Substitution Reactions

Reaction: $Nu: + CH_3Br \longrightarrow CH_3Nu + Br^-$

Nucleophile		Product	Class of Compound Formed
$HO:^-$	$\longrightarrow$	CH_3OH	an alcohol
$RO:^-$	$\longrightarrow$	CH_3OR	an ether
$HS:^-$	$\longrightarrow$	CH_3SH	a thiol (a mercaptan)
$RS:^-$	$\longrightarrow$	CH_3SR	a sulfide (a thioether)
$HC{\equiv}C:^-$	$\longrightarrow$	$CH_3C{\equiv}CH$	an alkyne
$^-:C{\equiv}N:$	$\longrightarrow$	$CH_3C{\equiv}N:$	a nitrile
$:I:^-$	$\longrightarrow$	$CH_3I:$	an alkyl iodide
$^-:N{=}\overset{+}{N}{=}N:^-$	$\longrightarrow$	$CH_3{-}N{=}\overset{+}{N}{=}N:^-$	an alkyl azide
$:NH_3$	$\longrightarrow$	$CH_3NH_3^+$	an alkylammonium ion
HOH	$\longrightarrow$	$CH_3\overset{+}{O}(H){-}H$	an alcohol (after proton transfer)
CH_3OH	$\longrightarrow$	$CH_3\overset{+}{O}(H){-}CH_3$	an ether (after proton transfer)

1. If the nucleophile is negatively charged, as for example OH^- and $HC{\equiv}C^-$, then the atom donating the pair of electrons in a substitution reaction becomes neutral in the product.
2. If the nucleophile is uncharged, as for example NH_3 and CH_3OH, then the atom donating the pair of electrons in the substitution reaction becomes positively charged in the product.

EXAMPLE 8.1

Complete these nucleophilic substitution reactions.

(a) $CH_3CH_2CH_2CH_2Br + CH_3O^-Na^+ \xrightarrow[\text{methanol}]{}$

(b) $CH_3CH_2CH_2CH_2Cl + NH_3 \xrightarrow[\text{ethanol}]{}$

Solution

(a) Methoxide ion is the nucleophile and bromine is the leaving group.

$$\underset{\text{1-Bromobutane}}{CH_3CH_2CH_2CH_2Br} + \underset{\substack{\text{Sodium}\\\text{methoxide}}}{CH_3O^-Na^+} \xrightarrow[\text{methanol}]{} \underset{\substack{\text{1-Methoxybutane}\\\text{(Butyl methyl ether)}}}{CH_3(CH_2)_3OCH_3} + \underset{\substack{\text{Sodium}\\\text{bromide}}}{Na^+Br^-}$$

(b) Ammonia is the nucleophile and chlorine is the leaving group.

$$\underset{\text{1-Chlorobutane}}{CH_3CH_2CH_2CH_2Cl} + \underset{\text{Ammonia}}{NH_3} \longrightarrow \underset{\text{Butylammonium chloride}}{CH_3CH_2CH_2CH_2NH_3^+\ Cl^-}$$

PROBLEM 8.1

Complete the following nucleophilic substitution reactions.

(a) Chlorocyclohexane $+ CH_3\overset{O}{\overset{\|}{C}}O^-Na^+ \xrightarrow{\text{ethanol}}$

(b) $CH_3\underset{}{\overset{I}{\overset{|}{C}}}HCH_2CH_3 + CH_3CH_2S^-Na^+ \xrightarrow{\text{acetone}}$

(c) $CH_3\overset{CH_3}{\overset{|}{C}}HCH_2CH_2Br + CH_3C{\equiv}C^-Na^+ \xrightarrow{\text{dimethyl sulfoxide}}$

8.2 Solvents for Nucleophilic Substitution Reactions

Solvents provide a medium in which the reactants are dissolved and in which nucleophilic substitution reactions take place. Common solvents for these reactions are divided into two groups: protic solvents and aprotic solvents. **Protic solvents** are hydrogen bond donors; the most common of these contain —OH groups. **Aprotic solvents** cannot serve as hydrogen bond donors; nowhere in the molecule is a hydrogen bonded to an atom of high electronegativity.

Protic solvent A solvent that is a hydrogen bond donor; the most common protic solvents contain —OH groups.

Aprotic solvent A solvent that cannot serve as a hydrogen bond donor; nowhere in the molecule is a hydrogen bonded to an atom of high electronegativity.

TABLE 8.2 Common Polar Protic Solvents

Solvent	Structure	Dielectric constant (25°C)
water	H_2O	79
formic acid	HCO_2H	59
methanol	CH_3OH	33
ethanol	CH_3CH_2OH	24

Furthermore, solvents are classified as polar and nonpolar. Dielectric constant, the most commonly used measure of solvent polarity, is defined by reference to the electrostatic force of attraction between two charges, Q and Q', of opposite sign, separated by a distance r in a vacuum. This force of attraction is reduced if a substance that provides a degree of insulation is placed between the two charges.

$$\text{Force of attraction} = \frac{Q \times Q'}{\text{dielectric constant} \times r^2}$$

Dielectric constant A measure of a solvent's ability to insulate opposite charges from one another.

Thus, **dielectric constant** is a measure of the ability of a solvent to insulate opposite charges from one another. The greater the value of the dielectric constant of a solvent, the smaller the interaction between ions of opposite charge dissolved in that solvent. As an arbitrary guideline, we say that a solvent is polar if it has a dielectric constant of 15 or greater. A solvent is nonpolar if it has a dielectric constant of less than 15.

The common protic solvents for nucleophilic substitution reactions are water, low-molecular-weight alcohols, and low-molecular-weight carboxylic acids (Table 8.2).

TABLE 8.3 Common Aprotic Solvents

Solvent	Structure	Dielectric Constant
Polar		
dimethyl sulfoxide (DMSO)	$CH_3S(=O)CH_3$	48.9
acetonitrile	$CH_3C{\equiv}N$	37.5
N,N-dimethylformamide (DMF)	$HC(=O)N(CH_3)_2$	36.7
acetone	$CH_3C(=O)CH_3$	20.7
Nonpolar		
dichloromethane	CH_2Cl_2	9.1
diethyl ether	$CH_3CH_2OCH_2CH_3$	4.3
toluene	$C_6H_5{-}CH_3$	2.3
hexane	$CH_3(CH_2)_4CH_3$	1.9

Increasing solvent polarity

Each is able to solvate ionic substances by electrostatic interaction between its partially negatively charged oxygen(s) and cations, and between its partially positively charged hydrogen and anions. By our arbitrary guideline, water, formic acid, methanol, and ethanol are classified as **polar protic solvents.**

The aprotic solvents most commonly used for nucleophilic substitution reactions are given in Table 8.3. Of these, dimethyl sulfoxide (DMSO), acetonitrile, *N, N*-dimethylformamide (DMF), and acetone are classified as **polar aprotic solvents.** Dichloromethane, diethyl ether, toluene, and hexane are classified as **nonpolar aprotic solvents.**

8.3 Mechanisms of Nucleophilic Aliphatic Substitution

On the basis of a wealth of experimental observations developed over a 50-year period, two limiting mechanisms for nucleophilic substitutions have been proposed. A fundamental difference between them is the timing of bond breaking between carbon and the leaving group and bond forming between carbon and the nucleophile. At one extreme, the two processes are concerted, meaning that bond breaking and bond forming occur simultaneously. Thus, departure of the leaving group is assisted by the incoming nucleophile. This mechanism is designated $\mathbf{S_N2}$. Here S stands for *S*ubstitution, N stands for *N*ucleophilic, and 2 stands for *bi*molecular. A **bimolecular reaction** is one in which two reactants (for an S_N2 reaction of an alkyl halide, both the alkyl halide and the nucleophile) are involved in the rate-limiting step.

Bimolecular reaction A reaction in which two species are involved in the rate-limiting step.

In an S_N2 reaction, the rate of disappearance of RX depends on the concentration of both RX and the nucleophile; that is, the reaction follows second order kinetics—it is first order in RX and first order in Nu. In the following equation, the term *k* is a proportionality constant called the rate constant.

$$\text{Rate of reaction of RX} = -\frac{d[\text{RX}]}{dt} = k[\text{RX}][\text{Nu}]$$

Following is an S_N2 mechanism for the reaction of hydroxide ion and bromomethane to form methanol and bromide ion. Note that the nucleophile is shown attacking the reactive center from the side of the molecule opposite from the leaving group, that is, the reaction involves **backside attack** by the nucleophile. We will present evidence in the following section to show that backside attack is a necessary characteristic of all S_N2 reactions.

MECHANISM An S_N2 Reaction

$$HO^- + CH_3\text{—}Br \longrightarrow \left[\overset{\delta-}{HO}\text{---}C(H_3)\text{---}\overset{\delta-}{Br} \right] \longrightarrow HO\text{—}CH_3 + Br^-$$

Transition state with simultaneous bond breaking and bond forming

Figure 8.1 shows a potential energy diagram for an S_N2 reaction. There is a single transition state and no reactive intermediate.

In the other limiting mechanism, termed S_N1, bond breaking between carbon and the leaving group is entirely completed before bond forming with the nucleophile begins. In the designation $\mathbf{S_N1}$, S stands for *S*ubstitution, N stands for *N*ucleophilic, and 1 stands for *uni*molecular. A **unimolecular reaction** is one in which only one reactant (in this case only the alkyl halide) is involved in the rate-limiting step.

Unimolecular reaction A reaction in which only one species is involved in the rate-limiting step.

The rate of disappearance of RX in an S_N1 reaction depends only on the concentration of RX; it is independent of the concentration of the nucleophile. This reaction follows first order kinetics. It is first order in RX and zero order in Nu.

$$\text{Rate of reaction of RX} = -\frac{d[\text{RX}]}{dt} = k[\text{RX}]$$

An S_N1 mechanism is illustrated by the reaction between 2-bromo-2-methylpropane (*tert*-butyl bromide) and methanol to form *tert*-butyl methyl ether and HBr. In this example, the nucleophile is also the solvent and the reaction is called a solvolysis. **Solvolysis** is any reaction in which the solvent is also the nucleophile. Ionization of the C—X bond in Step 1 forms a 3° carbocation intermediate, which then undergoes reaction with methanol, the nucleophile, in Step 2 to give an oxonium ion. Attack of the nucleophile occurs with equal probability from either face of the planar carbocation intermediate. Loss of H^+ by proton transfer from the oxonium ion in Step 3 then gives *tert*-butyl methyl ether. This last step is an acid-base reaction following the S_N1 reaction.

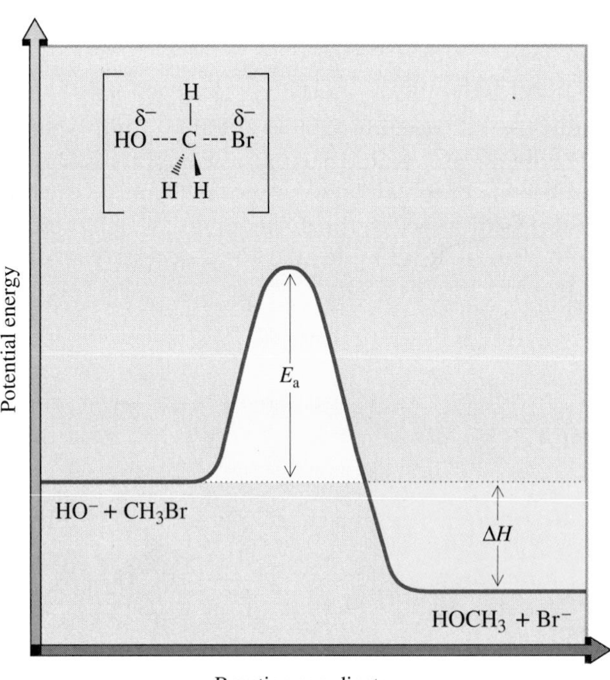

FIGURE 8.1
A potential energy diagram for an S_N2 reaction. There is one transition state and no reactive intermediate.

MECHANISM An S_N1 Reaction

Step 1: Ionization of C—X bond to form a carbocation intermediate.

$$(CH_3)_3C{-}Br \xrightarrow{\text{slow, rate-limiting step}} (CH_3)_3C^+ + Br^-$$

A carbocation intermediate, its shape is trigonal planar

Step 2: Reaction of the carbocation intermediate with methanol, the nucleophile, from either side.

$$(CH_3)_3C^+ + CH_3\ddot{O}H \longrightarrow CH_3\overset{+}{O}(H){-}C(CH_3)_3 + (CH_3)_3C{-}\overset{+}{O}(H)CH_3$$

Step 3: Proton transfer to methanol to give *tert*-butyl methyl ether.

$$(CH_3)_3C{-}\overset{+}{O}(H)CH_3 + H\ddot{O}CH_3 \underset{}{\overset{\text{fast}}{\rightleftharpoons}} (CH_3)_3C{-}\ddot{O}{-}CH_3 + H{-}\overset{+}{O}(H){-}CH_3$$

Figure 8.2 shows a potential energy diagram for the S_N1 reaction of 2-bromo-2-methylpropane with methanol. There is one transition state leading to formation of the carbocation intermediate in Step 1 and a second transition state for the reaction of the carbocation intermediate with methanol in Step 2 to give the oxonium ion. The reaction leading to formation of the carbocation intermediate crosses the higher potential energy barrier and, therefore, is the rate-limiting step.

8.4 Experimental Evidence for S_N1 and S_N2 Mechanisms

Let us now examine the experimental evidence on which these two contrasting mechanisms are based. As we do, we will consider the following questions.

1. What effect does the structure of the nucleophile have on the rate of reaction?
2. What is the stereochemical outcome of nucleophilic substitution when the leaving group is displaced from a stereocenter?
3. What effect does the structure of the alkyl halide have on the rate of reaction?
4. What effect does the structure of the leaving group have on the rate of reaction?
5. What is the role of the solvent?
6. Under what conditions are skeletal rearrangements observed?

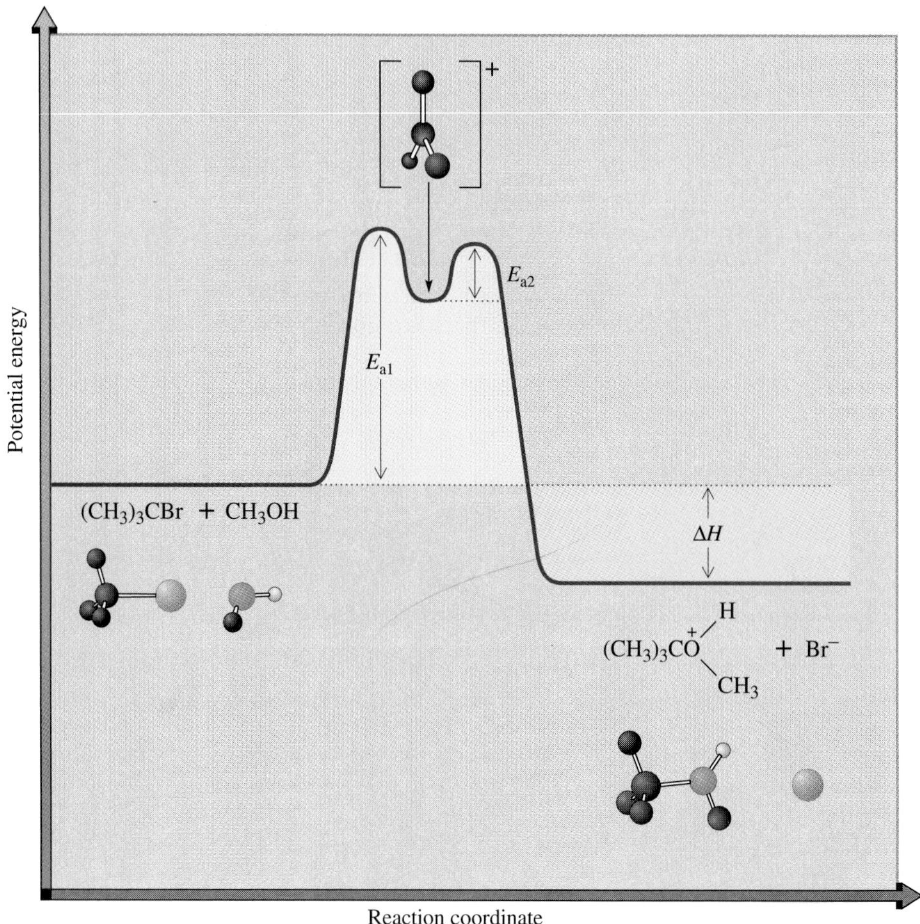

FIGURE 8.2
A potential energy diagram for an S_N1 reaction. There are two transition states and one reactive intermediate. Step 1 crosses the higher potential energy barrier and, hence, is the rate-limiting step.

A. **Kinetics**

The kinetic order of nucleophilic substitutions can be studied by measuring the effect on rate of varying concentrations of alkyl halide and nucleophile. Those reactions whose rate is dependent only on the concentration of alkyl halide are classified as S_N1; those whose rate is dependent on the concentration of both alkyl halide and nucleophile are classified as S_N2.

Because the transition state for formation of the carbocation intermediate in an S_N1 mechanism involves only the alkyl halide and not the nucleophile, it is a unimolecular process; the rate of the slow step depends only on the concentration of alkyl halide. The result is a first-order reaction; that is, the overall rate of an S_N1 reaction is proportional to [RX] and independent of [Nu]. In the following rate equation, the term k is called a rate constant. In this instance, the rate of reaction is expressed as the rate of disappearance of the starting material, 2-bromo-2-methylpropane.

$$\begin{array}{c} CH_3 \\ | \\ CH_3CBr \\ | \\ CH_3 \end{array} \quad + \quad CH_3OH \longrightarrow \begin{array}{c} CH_3 \\ | \\ CH_3COCH_3 \\ | \\ CH_3 \end{array} \quad + \quad HBr$$

2-Bromo-2-methylpropane (*tert*-butyl bromide) | Methanol | 2-Methoxy-2-methylpropane (*tert*-butyl methyl ether)

$$\text{Rate} = -\frac{d[(CH_3)_3CBr]}{dt} = k[(CH_3)_3CBr]$$

By contrast, there is only one step in the S_N2 mechanism, and both OH^- and CH_3Br are present in the transition state; that is, the reaction is bimolecular. The reaction between CH_3Br and NaOH to give CH_3OH and NaBr is second order: first order in CH_3Br and first order in OH^-.

$$\underset{\text{Bromomethane}}{CH_3Br} + Na^+OH^- \longrightarrow \underset{\text{Methanol}}{CH_3OH} + Na^+Br^-$$

$$\text{Rate} = -\frac{d[CH_3Br]}{dt} = k[CH_3Br][OH^-]$$

B. Structure of the Nucleophile

Nucleophilicity is a kinetic property measured by the rate at which a nucleophile attacks a reference compound to cause nucleophilic substitution under a standard set of experimental conditions. Relative nucleophilicities for a series of nucleophiles are established by measuring the rate at which each displaces bromide ion from ethyl bromide in ethanol at 25°C.

Nucleophilicity A kinetic property measured by the rate at which a nucleophile attacks a reference compound to cause nucleophilic substitution under a standard set of experimental conditions.

$$CH_3CH_2Br + NH_3 \longrightarrow CH_3CH_2\overset{+}{N}H_3 + Br^-$$

From these studies, we can then make correlations between the structure of a nucleophile and its relative nucleophilicity. Listed in Table 8.4 are the types of nucleophiles we deal most commonly with in this text.

Basicity is an equilibrium property measured by the position of equilibrium in an acid-base reaction, as, for example, the acid-base reaction between ammonia and water.

Basicity An equilibrium property measured by the position of equilibrium in an acid-base reaction, as, for example, the acid-base reaction between ammonia and water.

$$NH_3 + H_2O \rightleftharpoons NH_4^+ + OH^- \qquad K_b = \frac{[NH_4^+][OH^-]}{[NH_3]} = 1.8 \times 10^{-5}$$

Because all nucleophiles are bases as well, we also study correlations between nucleophilicity and basicity. Basicity also has a strong correlation with leaving group ability, which we shall see in Section 8.4F.

The solvent in which nucleophilic substitution is carried out has a marked effect on relative nucleophilicities. For a fuller understanding of the role of the solvent, let us consider nucleophilic substitution reactions carried out in polar aprotic solvents and in polar protic solvents. An organizing principle for substitution reactions in solution is the following: All other factors being equal, the freer the nucleophile, the stronger its nucleophilicity. Conversely, the greater the stabilization of the nucleophile by solvation, the weaker its nucleophilicity.

TABLE 8.4 Common Nucleophiles and Their Relative Nucleophilicities

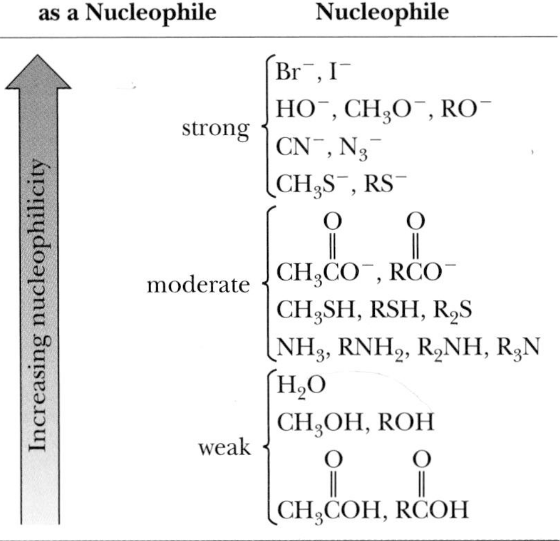

Effectiveness as a Nucleophile	Nucleophile
strong	Br^-, I^-
	HO^-, CH_3O^-, RO^-
	CN^-, N_3^-
	CH_3S^-, RS^-
moderate	$CH_3\overset{\overset{O}{\|}}{C}O^-$, $R\overset{\overset{O}{\|}}{C}O^-$
	CH_3SH, RSH, R_2S
	NH_3, RNH_2, R_2NH, R_3N
weak	H_2O
	CH_3OH, ROH
	$CH_3\overset{\overset{O}{\|}}{C}OH$, $R\overset{\overset{O}{\|}}{C}OH$

Nucleophilicity in Polar Aprotic Solvents

The most commonly used polar aprotic solvents (DMSO, acetone, acetonitrile, and DMF) are very effective in solvating cations but are not nearly so effective in solvating anions. Consider, for example, DMSO. Because the negative end of its dipole can come close to the center of positive charge in a cation, it is effective in solvating cations. The positive end of its dipole, however, is shielded by surrounding groups (two methyls and one oxygen) and is less effective in solvating anions. The sodium ion of sodium iodide, for example, is effectively solvated by DMSO and acetone, but the iodide ion is only poorly solvated. Because anions are only poorly solvated in polar aprotic solvents, they participate readily in nucleophilic substitution reactions and their relative nucleophilicities parallel their relative basicities. The relative nucleophilicities of halide ions in polar aprotic solvents, for example, are $F^- > Cl^- > Br^- > I^-$.

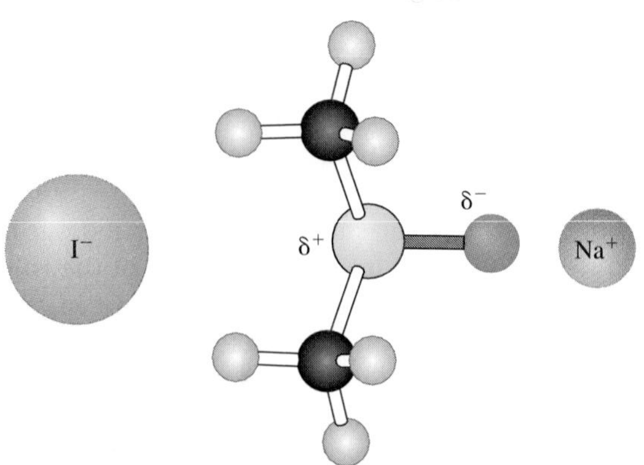

TABLE 8.5 Relative Nucleophilicities of Halide Ions in Aprotic and Protic Solvents

Solvent	Increasing nucleophilicity →
polar aprotic	$I^- < Br^- < Cl^- < F^-$
polar protic	$F^- < Cl^- < Br^- < I^-$

Nucleophilicity in Polar Protic Solvents

The relative nucleophilicities of halide ions in polar protic solvents are quite different from those in aprotic solvents (Table 8.5). In polar protic solvents, iodide ion, the least basic of the halide ions, has the greatest nucleophilicity. Conversely, fluoride ion, the most basic of the halide ions, has the smallest nucleophilicity. The reason for this reversal of correlation between nucleophilicity and basicity lies in the degree of solvation of anions in protic solvents compared with aprotic solvents. In polar aprotic solvents, anions are only weakly solvated and, therefore, relatively free to participate in nucleophilic substitution reactions. In polar protic solvents, however, anions are highly solvated by hydrogen bonding with solvent molecules. The negative charge on the fluoride ion, the smallest of the halide ions, is concentrated in a small volume, and the very tightly held solvent shell constitutes a barrier between fluoride ion and substrate. The fluoride ion must be at least partially removed from its tightly held solvation shell before it can participate in nucleophilic substitution. The negative charge on the iodide ion, the largest of the halide ions, is far less concentrated, the solvent shell is less tightly held, and iodide is considerably freer to participate in nucleophilic substitution reactions.

We can make the following additional generalizations about nucleophilicity:

1. Within a period of the Periodic Table, nucleophilicity increases from right to left (Table 8.6), that is, it increases with increasing basicity.
2. In a series of reagents with the same nucleophilic atom, anionic reagents are stronger nucleophiles than neutral reagents (Table 8.7). This trend also parallels the base strength of the nucleophile.
3. When comparing a group of reagents in which the nucleophilic atom is the same, the stronger the base, the greater the nucleophilicity. The oxygen nucleophiles in

TABLE 8.6 Relative Nucleophilicities of Atoms Within a Period

Period	← Increasing nucleophilicity
period 2	$CH_3^- > NH_2^- > OH^- > F^-$
period 3	$PH_2^- > SH^- > Cl^-$

TABLE 8.7 The Effect of Charge on Nucleophilicity

Increasing nucleophilicity →
$H_2O < OH^-$
$ROH < RO^-$
$NH_3 < NH_2^-$
$RSH < RS^-$

TABLE 8.8 Correlation of Nucleophilicity and Basicity for Reagents with the Same Nucleophilic Atom

nucleophile	RCO_2^- carboxylate ion	HO^- hydroxide ion	RO^- alkoxide ion
	Increasing nucleophilicity →		
conjugate acid	RCO_2H	HOH	ROH
pK_a	4–5	15.7	16–18
	Decreasing acidity →		

Table 8.8 are listed in order of increasing nucleophilicity. Below each, for comparison, is given the formula and pK_a of its conjugate acid. In this series, the carboxylic acid is the strongest acid, and, consequently, its anion is the weakest base and the poorest nucleophile.

C. Stereochemistry

S_N1 Reactions

Experiments in which nucleophilic substitution takes place at a stereocenter provide us with information about the stereochemical outcome of the reaction. One of the compounds studied to determine the stereochemistry of an S_N1 reaction was the alkyl chloride shown in the accompanying figure. When either enantiomer of this molecule undergoes nucleophilic substitution by an S_N1 pathway, the product is almost completely racemized. On ionization, this secondary halide forms an achiral carbocation. Attack of the nucleophile from the right gives the *R* enantiomer; attack from the left gives the *S* enantiomer. The *R* and *S* enantiomers are formed in equal amounts, and the product is a racemic mixture.

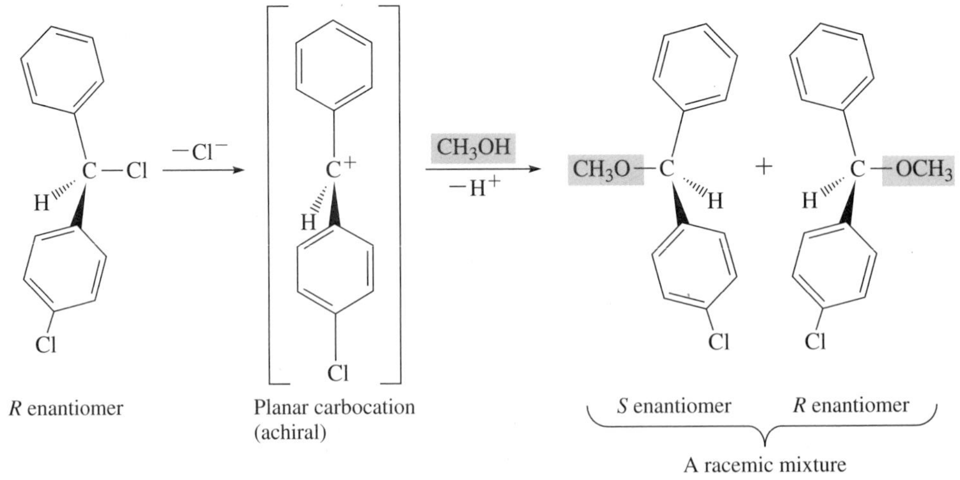

R enantiomer

Planar carbocation (achiral)

S enantiomer *R* enantiomer

A racemic mixture

The S_N1 mechanism as initially described requires complete racemization of any product in which the carbon at which substitution takes place is a stereocenter. Although examples of complete racemization have been observed, it is more common to find only partial racemization, with the more prevalent product being the one with inversion of configuration at the stereocenter undergoing reaction. This observation is accounted for by proposing that although bond breaking between carbon and the leaving group is complete, the leaving group (chloride ion in this example) remains associated for a short time with the carbocation as an ion pair.

Approach of the nucleophile from this side is less hindered → R_1–C^+(H)(R_2) $:\ddot{Cl}:^-$ ← Approach of the nucleophile from this side is partially blocked by chloride ion, which remains associated with the carbocation as an ion pair

To the extent that the leaving group remains associated with the carbocation as an ion pair, it hinders approach of the nucleophile (methanol in this example) from that side of the carbocation. The result is that slightly more than 50% of the product is formed by attack of the nucleophile from the side of the carbocation opposite that of the leaving group.

S_N2 Reactions

That every S_N2 reaction proceeds with inversion of configuration was shown in an ingenious experiment designed by the English chemists E. D. Hughes and C. K. Ingold. They studied the exchange reaction between optically active 2-iodooctane and iodine-131, a radioactive isotope of iodine. Iodine-127, the naturally occurring isotope of iodine, is stable; it does not undergo radioactive decay.

$$CH_3(CH_2)_5CH(I)CH_3 + {}^{131}I^- \xrightarrow[\text{acetone}]{S_N2} CH_3(CH_2)_5CH({}^{131}I)CH_3 + I^-$$

Hughes and Ingold first demonstrated that the reaction is second order: first order in 2-iodooctane and first order in iodide ion. Therefore, the reaction proceeds by an S_N2 mechanism. They observed further that the rate of racemization of optically pure 2-iodooctane is exactly 2 times the rate of incorporation of iodine-131. This observation must mean, they reasoned, that each displacement of iodine-127 by iodine-131 proceeds with inversion of configuration as illustrated in the following equation.

$$^{131}I^- + (H_3C)(H)(CH_3(CH_2)_5)C{-}I \xrightarrow[\text{acetone}]{S_N2} {}^{131}I{-}C(CH_3)(H)((CH_2)_5CH_3) + I^-$$

(*R*)-2-Iodooctane (*S*)-2-Iodooctane

Substitution with inversion of configuration in one molecule cancels the rotation of one molecule that has not reacted, so that for each molecule undergoing inversion, one racemic pair is formed. Inversion of configuration in 50% of the molecules results in 100% racemization.

For the majority of second-order substitution reactions studied in these early investigations, the incoming group was a negative ion and the leaving group departed as a negative ion. The transition state of minimum energy, it was argued, would have these two negatively charged species separated as far as possible, that is, at an angle of 180°. Thus, backside attack provides the transition state of minimum energy. But what would happen, it was asked, if the incoming nucleophile and leaving group were of opposite charge types? Would nucleophilic substitution then proceed with retention of configuration at the site of reaction? As a test case, Hughes and Ingold designed an experiment in which the incoming nucleophile was negatively charged and the leaving group was positively charged and departed as a neutral molecule.

Question: Might there be frontside attack if incoming and leaving groups are of opposite charge types?

They began with optically pure 1-chloro-1-phenylethane (Figure 8.3). In one set of reactions, the (*S*)-chloroalkane was treated in Step 1 with azide ion, an S_N2 reaction known to involve inversion of configuration. The resulting (*R*)-azide was then treated in Step 2 with hydrogen in the presence of a platinum catalyst to give an (*R*)-amine.

In another set of reactions, the (*S*)-chloroalkane was treated in Step 3 with hydrosulfide ion in an S_N2 reaction also known to involve inversion of configuration. The (*R*)-thiol from Step 3 was then treated in Step 4 with iodomethane to give an (*R*)-sulfonium ion. Reaction of this sulfonium ion with azide ion in Step 5 was the key

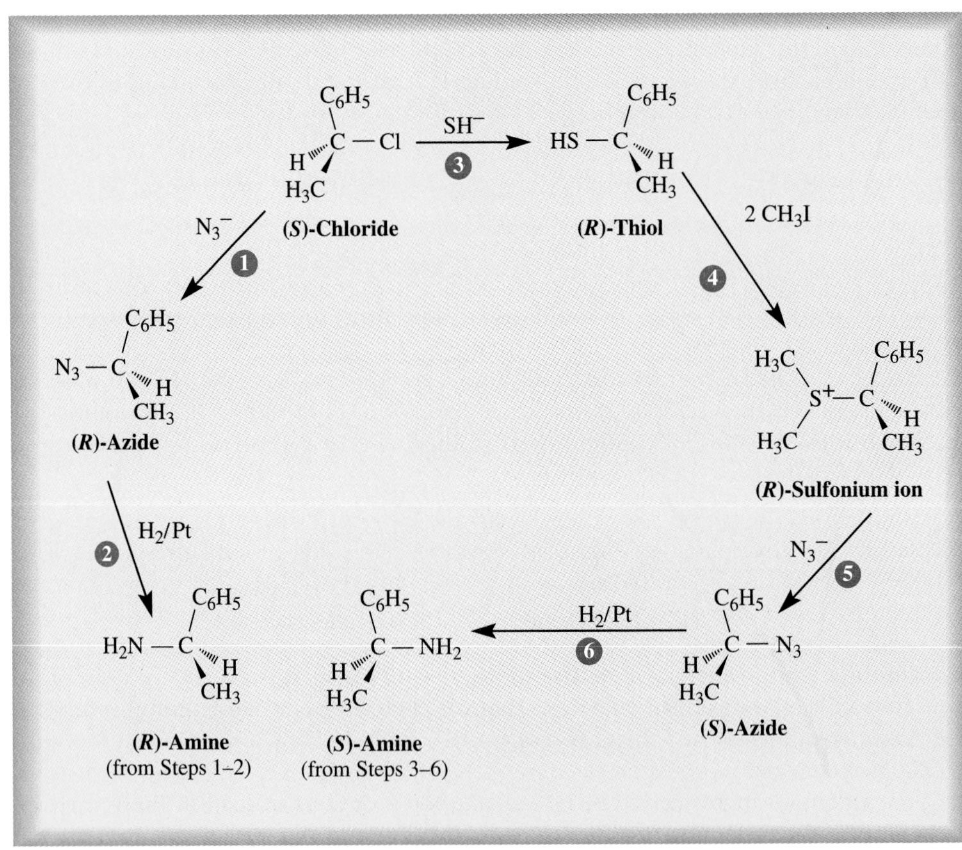

FIGURE 8.3
Inversion of configuration in S_N2 reactions, even when the nucleophile and leaving group are of opposite charge types.

reaction. If this nucleophilic substitution involved inversion of configuration, then the azide formed would have the (*S*) configuration, and reduction in Step 6 would give the (*S*)-amine. If, however, Step 5 involved front-side displacement, then the product would be the (*R*)-azide, which in turn would give the (*R*)-amine. The observed products of Steps 5 and 6 were the (*S*)-azide and the (*S*)-amine. The conclusion, then, was that even this nucleophilic substitution, involving incoming and leaving groups of opposite charge types, proceeds with inversion of configuration.

EXAMPLE 8.2

Complete these nucleophilic substitution reactions, showing the configuration of the product.

(a) 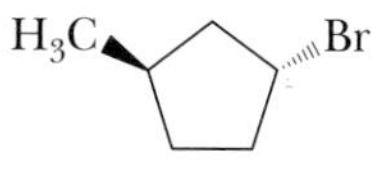 H_3C–cyclopentane–Br (wedge H_3C, hashed Br) $+ CH_3CO_2^-Na^+ \longrightarrow$

(b) $CH_3CH(Br)(CH_2)_4CH_3 + CH_3NH_2 \longrightarrow$
(the *R* enantiomer)

Solution

Each is an S_N2 reaction and occurs with inversion of configuration at the stereocenter.

(a) 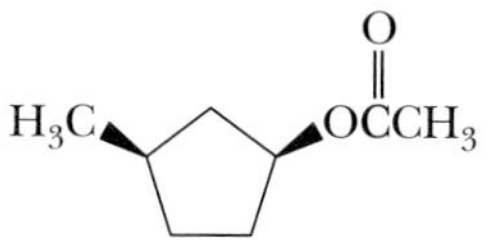 H_3C–cyclopentane–$OCCH_3$ (C=O; both wedges)
(the *cis* isomer)

(b) $CH_3CH(\overset{+}{N}H_2CH_3\ Br^-)CH_2CH_2CH_3$
(the *S* enantiomer)

PROBLEM 8.2

Complete these S_N2 reactions, showing the configuration of the product.

(a) 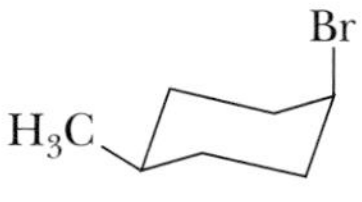 H_3C–cyclohexane–Br (chair) $+ Na^+N_3^- \longrightarrow$

(b) $C_6H_5CH(Br)CH_2CH_3 + CH_3S^-Na^+ \longrightarrow$
(the *S* enantiomer)

D. Structure of the Alkyl Halide

S_N1 reactions are governed mainly by **electronic factors,** namely the relative stabilities of carbocation intermediates. S_N2 reactions, on the other hand, are governed mainly by **steric factors** and their transition states are particularly sensitive to nonbonded interactions (steric hindrance) about the site of reaction. The competition between steric factors and electronic factors and their effect on relative rates of nucleophilic substitution reactions for alkyl halides are summarized in Figure 8.4.

Tertiary halides, for example, react by an S_N1 mechanism because 3° carbocation intermediates are particularly stable; they never react by an S_N2 mechanism because of the extreme crowding at the tertiary carbon. Methyl halides and primary halides have little crowding around the reaction site and react readily by an S_N2 mechanism; they never react by an S_N1 mechanism because methyl and primary carbocations are

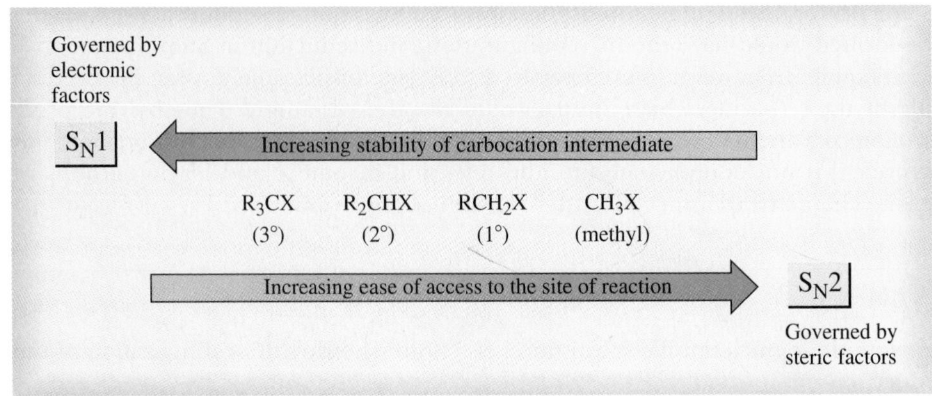

FIGURE 8.4
Effect of steric factors and electronic factors in competition between S_N1 and S_N2 reactions of alkyl halides.

so unstable. Secondary halides may be made to react by either S_N1 or S_N2 pathways, depending on the choice of nucleophile and solvent.

We see a similar effect of steric hindrance on S_N2 reactions in molecules with branching at the β-carbon. Table 8.9 shows the relative rates of S_N2 reactions on a series of primary alkyl bromides. In these data, the rate of nucleophilic substitution of ethyl bromide is taken as a reference and is given the value 1.0. As CH_3 branches are added to the β-carbon, the relative rate of reaction decreases. Compare the relative rates of ethyl bromide (no β-branch) with that of 1-bromo-2,2-dimethylpropane, a compound with three β-branches. The rate of S_N2 substitution of this compound is only 10^{-5} that of ethyl bromide.

As shown in Figure 8.5, the carbon of the C—Br bond in ethyl bromide is unhindered and open to attack by a nucleophile in an S_N2 reaction. The corresponding carbon in 1-bromo-2,2-dimethylpropane, however, is screened by the three β-methyl groups. Thus, although the carbon bearing the leaving group is primary, approach of

TABLE 8.9 Effect of β-Branches on the Rate of S_N2 Reactions

Alkyl Bromide	Number of β-Branches	Relative Rate
$\overset{\beta}{C}H_3CH_2Br$	0	1.0
$CH_3\overset{\beta}{C}H_2CH_2Br$	1	4.1×10^{-1}
$CH_3\overset{\beta}{C}H(CH_3)CH_2Br$	2	1.2×10^{-3}
$CH_3\overset{\beta}{C}(CH_3)_2CH_2Br$	3	1.2×10^{-5}

CH_3CH_2Br

Bromoethane
(Ethyl bromide)

$CH_3C(CH_3)_2CH_2Br$

2-Bromo-2,2-dimethylpropane
(Neopentyl bromide)

FIGURE 8.5
The effect of β-branching in S_N2 reactions on a primary alkyl halide. With bromoethane, attack of the nucleophile is unhindered. With 1-bromo-2,2-dimethylpropane, the three β-branches block approach of the nucleophile to the backside of the C—Br bond, thus drastically reducing the rate of S_N2 reaction of this compound.

the nucleophile to it is so hindered that the rate of S_N2 reaction of this compound is greatly reduced compared to the rate of S_N2 reaction of bromoethane.

E. Allylic Halides

At this point, we need to introduce a new type of carbocation, namely the allyl cation and allylic carbocations. An **allylic carbocation** is any carbocation in which the positive charge is on an allylic carbon (Section 7.6). The **allyl cation** is the simplest allylic carbocation. Because the allyl cation has only one substituent on the carbon bearing the positive charge, it is said to be a primary allylic carbocation.

Allylic carbocation Any carbocation in which the positive charge is on an allylic carbon.

Just as was the case with allylic radicals, allylic carbocations are considerably more stable than comparably substituted alkyl carbocations because of resonance interaction between the positively charged carbon and the vinyl group. The allyl cation, for example, can be represented as a hybrid of two equivalent contributing structures.

$$CH_2{=}CH{-}\overset{+}{C}H_2 \longleftrightarrow \overset{+}{C}H_2{-}CH{=}CH_2$$

Allyl cation
(hybrid of two equivalent contributing structures)

Because of this delocalization of both the pi electrons of the double bond and the positive charge in the hybrid, the allyl cation is considerably more stable than a primary carbocation. It has been determined experimentally that the presence of one vinyl group provides approximately as much stabilization as two alkyl groups. Thus, the allyl cation and isopropyl cation are of comparable stability.

carbocation with one vinyl substituent — carbocation with two alkyl substituents

$$CH_2{=}CH{-}\overset{+}{C}H_2 \qquad CH_3{-}\overset{+}{C}H{-}CH_3$$

1° Allylic cation — 2° Alkyl cation

These cations are of comparable stability

The classification of allylic cations as 1°, 2°, and 3° is determined by the location of the positive charge in the more important contributing structure. Following are examples of a 2° and a 3° allylic carbocation.

$$CH_2{=}CH{-}\overset{+}{C}H{-}CH_3 \qquad CH_2{=}CH{-}\overset{+}{C}(CH_3){-}CH_3$$

(A 2° allylic carbocation) (A 3° allylic carbocation)

Benzylic carbocations show approximately the same stability as allylic carbocations. Both are stabilized by resonance delocalization of the positive charge. We discuss formation and reactions of benzylic carbocations in Section 19.6.

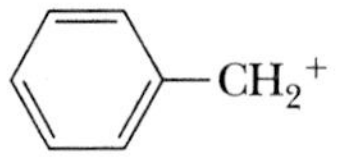

$C_6H_5-CH_2^+$

Benzyl cation (a benzylic carbocation)

The benzyl cation is also written in this abbreviated form

In Section 5.3D, we presented the order of stability of methyl, primary, secondary, and tertiary carbocations. We can now expand this order to include primary, secondary, and tertiary allylic carbocations, and benzylic carbocations as well.

methyl < 1° alkyl < {2° alkyl, 1° allylic, 1° benzylic} < {3° alkyl, 2° allylic, 2° benzylic} < {3° allylic, 3° benzylic}

Increasing stability of carbocations

EXAMPLE 8.3

Write an additional contributing structure for each carbocation and state which of the two makes the greater contribution to the resonance hybrid. Classify each as a 1°, 2°, or 3° allylic cation.

(a) 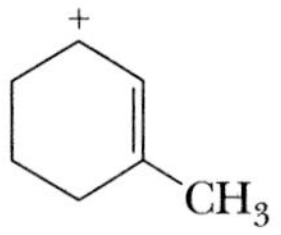 (b) 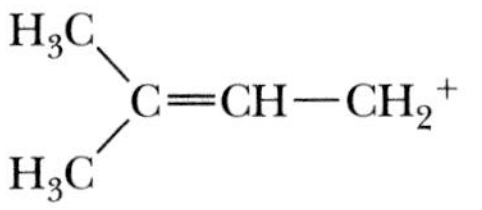 $(H_3C)_2C=CH-CH_2^+$

Solution

The additional structure in each case is a 3° allylic cation. The contributing structure having the greater degree of substitution on the positively charged carbon makes the greater contribution to the hybrid.

(a) 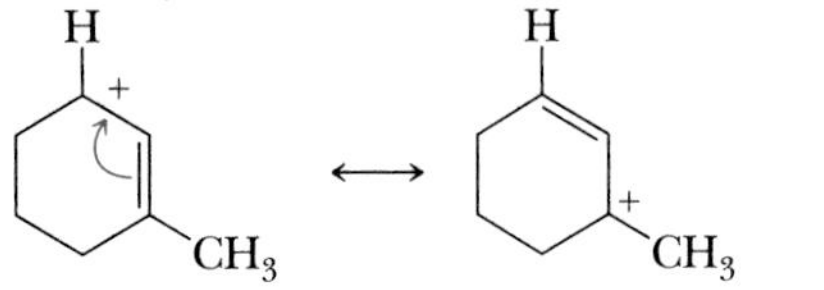 ⟷ Greater contribution

(b) 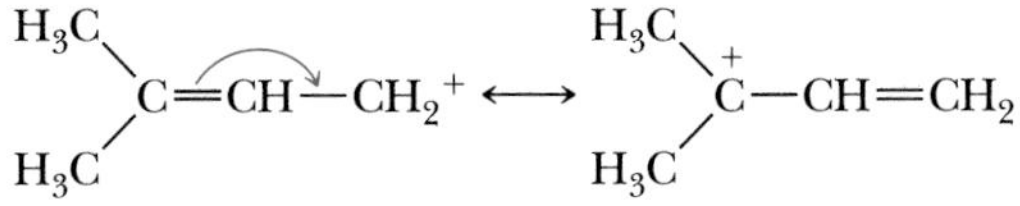 $(H_3C)_2C=CH-CH_2^+ \longleftrightarrow (H_3C)_2\overset{+}{C}-CH=CH_2$

Greater contribution

PROBLEM 8.3

Write an additional contributing structure for each carbocation and state which of the two makes the greater contribution to the resonance hybrid.

(a) 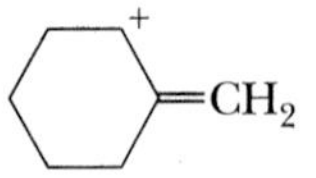 (b)

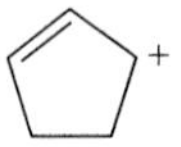

Primary allylic and benzylic halides can be made to react by either S_N1 or S_N2 mechanisms depending on the solvent and the nucleophile; they are primary (the steric factor favoring S_N2), and, at the same time, they can lose a halide ion to form a stable allylic or benzylic carbocation (the electronic factor favoring S_N1). In polar protic solvents, they undergo solvolysis by an S_N1 mechanism. They can be made to undergo S_N2 reactions in aprotic solvents by treatment with good nucleophiles. Secondary and tertiary allylic and benzylic halides react almost exclusively by an S_N1 mechanism.

F. The Leaving Group

In the transition state for nucleophilic substitution, the leaving group develops a partial negative charge in both S_N1 and S_N2 mechanisms and, therefore, the group's ability to function as a leaving group is related to how stable it is as an anion. The most stable anions and the best leaving groups are the conjugate bases of strong acids. The relative rates of displacement of halogens from an alkyl halide are shown as follows.

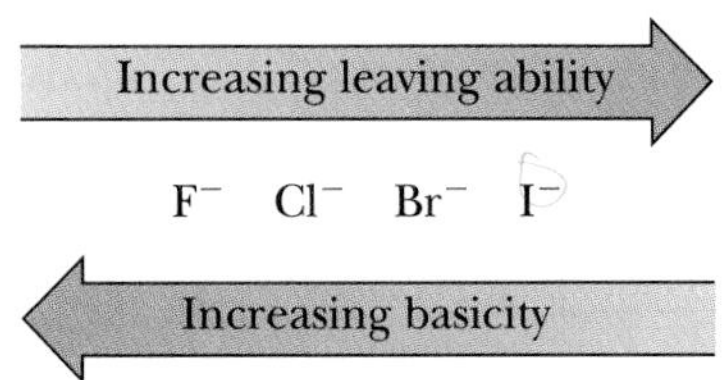

Iodide ion, the most stable anion and the weakest base, is the best leaving group. Fluoride ion, the least stable anion and the strongest base, is the poorest leaving group. The correlation between anion stability, basicity, and leaving group ability holds for other classes of compounds as well as alkyl halides. Hydroxide ion, for example, is such a poor leaving group that it is never displaced in nucleophilic substitution reactions.

G. The Solvent

To appreciate the important role of the solvent in nucleophilic substitution reactions we need to be specific about whether the substitution is S_N2 or S_N1. Let us first take up the effect of solvent on S_N2 reactions.

The Effect of Solvent on S_N2 Reactions

The most common type of S_N2 reaction involves a negative nucleophile and a negative leaving group.

negatively charged nucleophile

negative charge dispersed in the transition state

negatively charged leaving group

$$\mathrm{Nu}^{-} + \mathrm{C{-}X} \longrightarrow \left[\overset{\delta-}{\mathrm{Nu}}\cdots\mathrm{C}\cdots\overset{\delta-}{\mathrm{X}}\right] \longrightarrow \mathrm{Nu{-}C} + \mathrm{X}^{-}$$

Transition state

TABLE 8.10 Rates of an S_N2 Reaction as a Function of Solvent

$$CH_3CH_2CH_2CH_2Br + N_3^- \xrightarrow[\text{solvent}]{S_N2} CH_3CH_2CH_2CH_2N_3 + Br^-$$

Solvent Type	Solvent	$\frac{k_{(solvent)}}{k_{(methanol)}}$
polar aprotic	$CH_3C{\equiv}N$	5000
	$(CH_3)_2NCHO$	2800
	$(CH_3)_2S{=}O$	1300
polar protic	H_2O	7
	CH_3OH	1

The stronger the solvation of the nucleophile, the greater the energy required to remove the nucleophile from its solvation shell to reach the transition state, and hence the lower the rate of the S_N2 reaction.

Because of their good ability to solvate cations but only poor ability to solvate anions (nucleophiles), polar aprotic solvents have a particularly dramatic effect on the rate of S_N2 reactions. Reactions of the same substrate and nucleophile are accelerated, often by several orders of magnitude, when carried out in aprotic solvents compared with the rate obtained in protic solvents. Shown in Table 8.10 are ratios of specific rate constants for the S_N2 reaction of 1-bromobutane with sodium azide as a function of solvent. The rate of reaction in methanol is taken as a reference.

The Effect of Solvent on S_N1 Reactions

Because nucleophilic substitution by an S_N1 pathway involves creation and separation of opposite charges in the transition state of the rate-limiting step, the rate of S_N1 reactions depends on the ability of the solvent to keep opposite charges separated as well as its ability to stabilize both positive and negative sites by solvation. The solvents that meet these requirements best are polar protic solvents such as H_2O, ROH, and to a lesser degree RCO_2H. As seen in Table 8.11, the rate of solvolysis of 2-chloro-2-methylpropane (*tert*-butyl chloride) increases by a factor of 10^5 when the solvent is changed from ethanol to water. **Solvolysis** is any nucleophilic substitution in which the solvent is also the nucleophile.

Solvolysis Any nucleophilic substitution in which the solvent is also the nucleophile.

H. Skeletal Rearrangement

As we saw in Section 6.3D, skeletal rearrangement is typical of reactions involving a carbocation intermediate which can rearrange to a more stable one. Because there is little or no carbocation character at the substitution center, S_N2 reactions are free of rearrangement. In contrast, S_N1 reactions often proceed with rearrangement. An example of an S_N1 reaction involving rearrangement is solvolysis of 2-chloro-3-phenylbutane in methanol, a polar protic solvent and a weak nucleophile. The mechanism of this reaction is shown directly beneath it. Reaction is initiated by ionization of the carbon-chlorine bond to form a 2° carbocation, which rearranges to a considerably more stable 3° carbocation by shift of a hydrogen with its pair of electrons (a hydride ion) from the adjacent benzylic carbon. The major substitution product is the ether

TABLE 8.11 Rates of an S_N1 Reaction as a Function of Solvent

$$(CH_3)_3CCl + ROH \xrightarrow{\text{solvolysis}} (CH_3)_3COR + H{-}Cl$$

Solvent	$\frac{k_{(solvent)}}{k_{(ethanol)}}$
water	100,000
80% water: 20% ethanol	14,000
40% water: 60% ethanol	100
ethanol	1

Increasing polarity of solvent (arrow pointing up) — Increasing rate of solvolysis (arrow pointing up)

with rearranged structure. Note that not only is the rearranged carbocation tertiary, it is also benzylic, which adds resonance-stabilization.

$$C_6H_5CH(CH_3)CHClCH_3 + CH_3OH \longrightarrow C_6H_5C(CH_3)(OCH_3)CH_2CH_3 + CH_3OH_2^+ + Cl^-$$

2-Chloro-3-phenylbutane → 2-Methoxy-2-phenylbutane

2-Chloro-3-phenylbutane —($-Cl^-$, loss of chloride)→ $C_6H_5{-}CH(CH_3){-}\overset{+}{C}H{-}CH_3$ —(hydride ion migration)→ $C_6H_5{-}\overset{+}{C}(CH_3){-}CH_2{-}CH_3$ + $H{-}\ddot{O}{-}CH_3$ —(reaction with nucleophile)→ $C_6H_5{-}C(CH_3)(\overset{+}{O}(H)CH_3){-}CH_2{-}CH_3$ + $:\ddot{O}(H)CH_3$ —(proton transfer to solvent)→ 2-Methoxy-2-phenylbutane

In general, migration of a hydrogen atom or an alkyl group with its bonding electrons occurs when a more stable carbocation can be formed.

The factors favoring S_N1 or S_N2 reactions are summarized in Table 8.12. Also shown is the change in configuration when substitution takes place at a stereocenter.

TABLE 8.12 Summary of S_N1 Versus S_N2 Reactions of Alkyl Halides

Type of Alkyl Halide	S_N2	S_N1
methyl (CH_3X)	S_N2 favored.	S_N1 does not occur; the methyl cation is so unstable, it is never observed in solution.
primary (RCH_2X)	S_N2 favored.	S_N1 rarely occurs; primary cations are so unstable, they are rarely observed in solution.
secondary (R_2CHX)	S_N2 favored in aprotic solvents with good nucleophiles.	S_N1 favored in protic solvents with poor nucleophiles; carbocation rearrangements may occur.
tertiary (R_3CX)	S_N2 does not occur because of steric hindrance around the reaction center.	S_N1 favored because of the ease of formation of tertiary carbocations.
substitution at a stereocenter	Inversion of configuration: The nucleophile attacks the stereocenter from the side opposite the leaving group.	Racemization favored: The carbocation intermediate is planar, and attack of the nucleophile occurs with equal probability from either side; there is often some net inversion of configuration.

8.5 Neighboring Group Participation

Thus far we have considered two limiting mechanisms for nucleophilic substitutions that focus on the degree of covalent bonding between the nucleophile and the substitution center during departure of the leaving group. In an S_N2 mechanism, the leaving group is assisted in its departure by the nucleophile. In an S_N1 mechanism, the leaving group is not assisted in its departure by the nucleophile. An essential criterion for distinguishing between these two pathways is the order of reaction. Nucleophile-assisted substitutions are second-order, first-order in RX and first-order in nucleophile. Nucleophile-unassisted substitutions are first-order; first-order in RX and zero-order in nucleophile.

Chemists recognize that certain nucleophilic substitutions, which have the kinetic characteristics of first-order (S_N1) substitution, in fact involve two successive S_N2 reactions. A characteristic feature of a great many of these nucleophilic substitutions is the presence of an internal nucleophile, most commonly sulfur, nitrogen, or oxygen, on the carbon atom beta to the leaving group. This neighboring nucleophile participates in the departure of the leaving group to give an intermediate, which then reacts with an external nucleophile to complete the reaction.

The **mustard gases** are one group of compounds that react by participation of a neighboring group. The characteristic structural feature of a mustard gas is a two-carbon chain, with a halogen on one carbon and a divalent sulfur or trivalent nitrogen on the other carbon (S—C—C—X or N—C—C—X). An example of a mustard gas is bis(2-chloroethyl)sulfide, a poison gas used extensively in World War I. This compound is a deadly vesicant (blistering agent) and quickly causes conjunctivitis and blindness.

$$ClCH_2CH_2SCH_2CH_2Cl \qquad CH_3N(CH_2CH_2Cl)_2$$

Bis(2-chloroethyl)sulfide (a sulfur mustard gas) — Bis(2-chloroethyl)methylamine (a nitrogen mustard gas)

Bis(2-chloroethyl)sulfide and bis(2-chloroethyl)methylamine are not gases at all. They are oily liquids. They do, however, have a high vapor pressure, hence the designation "gas." Nitrogen and sulfur mustards react very rapidly with moisture in the air and in the mucous membranes of the eye, nose, and throat to produce HCl, which then burns and blisters these sensitive tissues. What is unusual about the reactivity of the mustard gases is that they react very rapidly with water, a very poor nucleophile.

$$ClCH_2CH_2SCH_2CH_2Cl + 2H_2O \longrightarrow HOCH_2CH_2SCH_2CH_2OH + 2HCl$$

Mustard gases also react rapidly with other nucleophiles, such as those in biological molecules, which makes them particularly dangerous chemicals. For insight into how mustard gases were recognized as potential starting points for the synthesis of effective drugs for the treatment of certain kinds of cancer, see the Chemistry in Action box "Mustard Gases and the Treatment of Neoplastic Diseases" in Chapter 28.

The reason for the extremely rapid hydrolysis of the sulfur mustards is neighboring group participation by sulfur in the ionization of the carbon-chlorine bond to form a cyclic sulfonium ion.

$$ClCH_2CH_2-\ddot{S}(CH_2-CH_2-Cl) \xrightarrow[\text{(an internal } S_N2 \text{ reaction)}]{\text{reaction with an internal nucleophile}} ClCH_2CH_2-\overset{+}{S}<(CH_2CH_2) + Cl^-$$

A cyclic sulfonium ion

At this point, you should review halogenation and oxymercuration of alkenes (Sections 6.3C and 6.3F) and compare the cyclic halonium ion formed there with the cyclic sulfonium ion formed here.

The cyclic sulfonium ion contains a highly strained three-membered ring and reacts rapidly with an external nucleophile to open the ring followed by proton transfer to H_2O to give H_3O^+. The net effect of these reactions is nucleophilic substitution of —OH for —Cl.

$$ClCH_2CH_2-\overset{+}{S}<(CH_2CH_2) + H-\ddot{O}-H \xrightarrow[\text{(a second } S_N2 \text{ reaction)}]{\text{reaction with an external nucleophile}} ClCH_2CH_2-\ddot{S}-CH_2CH_2\ddot{O}H + H_3O^+$$

Thus, although this reaction has the kinetic characteristics of an S_N1 reaction, it actually involves two successive S_N2 reactions. Of the two steps, the first one is the slower and rate-limiting step of the overall reaction. As a result, the rate of reaction is proportional to the concentration of the mustard but independent of the concentration of external nucleophile.

We continue to use the terms S_N2 and S_N1 to describe nucleophilic substitution reactions. We realize, however, that these mechanisms do not adequately describe all nucleophilic substitution reactions. More general terms for these reactions are nucleophile-assisted and unassisted reactions.

EXAMPLE 8.4

Write a mechanism for the hydrolysis of the nitrogen mustard bis(2-chloroethyl)methylamine.

Solution

Following is a three-step mechanism. Reaction begins in Step 1 with participation of the neighboring nitrogen in the ionization of the carbon-chlorine bond to form a three-membered ring ammonium ion intermediate. This is followed in Step 2 by reaction of this intermediate with water to open the strained three-membered ring. Proton transfer to solvent in Step 3 completes the reaction and gives an alcohol.

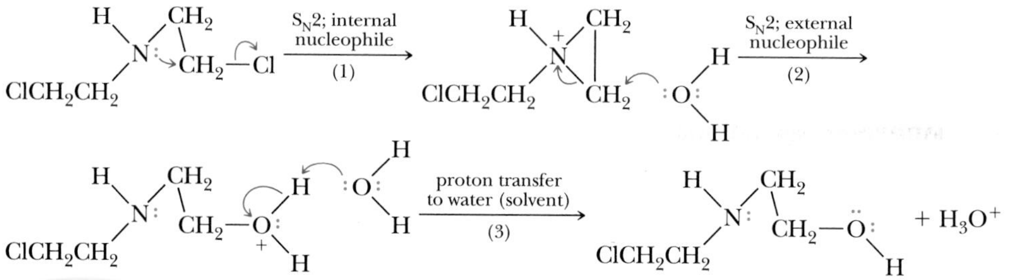

S_N2; internal nucleophile (1)

S_N2; external nucleophile (2)

proton transfer to water (solvent) (3)

$$+ H_3O^+$$

PROBLEM 8.4

Knowing what you do about the regioselectivity of S_N2 reactions, predict the product of hydrolysis of this compound.

$$(CH_3)_2NCH_2\underset{\displaystyle Cl}{CH}CH_2CH_3 + H_2O \longrightarrow$$

8.6 An Analysis of Several Nucleophilic Substitution Reactions

Predictions about the mechanism for a particular nucleophilic substitution reaction must be based on considerations of the structure of the alkyl halide, the nucleophile, the leaving group, and the solvent. Following are five nucleophilic substitution reactions and an analysis of the factors that contribute to favoring an S_N1 or S_N2 mechanism for each.

Nucleophilic Substitution 1

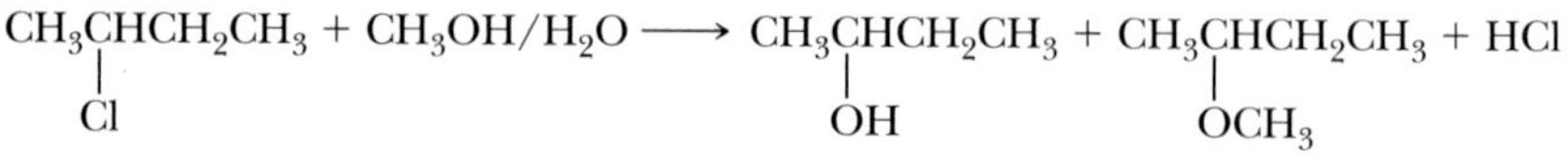

$$CH_3\underset{\displaystyle Cl}{CH}CH_2CH_3 + CH_3OH/H_2O \longrightarrow CH_3\underset{\displaystyle OH}{CH}CH_2CH_3 + CH_3\underset{\displaystyle OCH_3}{CH}CH_2CH_3 + HCl$$

R enantiomer

The mixture of methanol and water is a polar protic solvent and a good ionizing solvent in which to form carbocations. 2-Chlorobutane ionizes in this solvent to form a fairly stable 2° carbocation intermediate. Both water and methanol are weak nucleophiles. From this analysis, we predict that reaction is by an S_N1 mechanism, a nucleophile-unassisted ionization of the secondary alkyl chloride to give a carbocation intermediate that then reacts with either water or methanol as the nucleophile to give the observed products. Each product is formed as an approximately 50 : 50 mixture of *R* and *S* enantiomers.

Nucleophilic Substitution 2

$$CH_3\underset{\large CH_3}{\overset{|}{C}}HCH_2Br + Na^+CN^- \xrightarrow[\text{DMSO}]{} CH_3\underset{\large CH_3}{C}HCH_2CN + NaBr$$

$$\overset{\displaystyle CH_3}{\underset{\displaystyle CH_3CHCH_2Br}{|}} + Na^+CN^- \xrightarrow[\text{DMSO}]{} \overset{\displaystyle CH_3}{\underset{\displaystyle CH_3CHCH_2CN}{|}} + NaBr$$

This is a primary alkyl bromide in the presence of cyanide ion, a strong nucleophile. Dimethyl sulfoxide (DMSO), a polar aprotic solvent, is a particularly good solvent in which to carry out nucleophile-assisted substitution reactions because its ability to solvate cations (in this case, Na^+) is good, but its ability to solvate anions (in this case, CN^-) is poor. We predict an S_N2 mechanism.

Nucleophilic Substitution 3

$$\overset{\displaystyle Br}{\underset{\displaystyle CH_3CHCH_2CH_3}{|}} + CH_3S^-Na^+ \xrightarrow[\text{acetone}]{} \overset{\displaystyle SCH_3}{\underset{\displaystyle CH_3CHCH_2CH_3}{|}} + NaBr$$

Bromine is a good leaving group and it is on a secondary carbon. The sulfide ion is a strong nucleophile. Acetone, a polar aprotic solvent, is a good medium in which to carry out S_N2 reactions but a poor medium in which to carry out S_N1 reactions. We predict reaction by an S_N2 mechanism.

Nucleophilic Substitution 4

$$\text{(3-bromocyclohexene)}\!-\!Br + CH_3\overset{\displaystyle O}{\overset{\|}{C}}OH \xrightarrow[\text{acetic acid}]{} \text{(cyclohexenyl)}\!-\!O\overset{\displaystyle O}{\overset{\|}{C}}CH_3 + HBr$$

Ionization of the carbon-bromine bond forms a resonance-stabilized secondary, allylic carbocation. Acetic acid is a weak nucleophile, which reduces the likelihood of an S_N2 reaction. Further, acetic acid is a protic (hydroxylic) solvent that favors S_N1 reaction. We predict reaction by an S_N1 mechanism.

Nucleophilic Substitution 5

$$CH_3(CH_2)_5CH_2Br + (CH_3)_3P \xrightarrow[\text{toluene}]{} CH_3(CH_2)_5CH_2{-}\overset{+}{P}(CH_3)_3\ Br^-$$

The alkyl bromide is primary and bromine is a good leaving group. Trivalent compounds of phosphorus, a third row element, are strong nucleophiles. Toluene is a nonpolar aprotic solvent. Given the combination of a primary halide, a good leaving group, a good nucleophile, and a nonpolar aprotic solvent, we predict reaction by an S_N2 pathway.

EXAMPLE 8.5

Write the expected substitution product(s) for each reaction and predict the mechanism by which each product is formed. Note that each alkyl halide is chiral and one pure enantiomer.

(a) 3-Chlorocyclopentene + $CH_3OH \xrightarrow{\text{methanol}}$ (b) $CH_3(CH_2)_5CH(I)CH_3 + CH_3C(=O)O^-Na^+ \xrightarrow{\text{DMSO}}$

R enantiomer — *R* enantiomer

Solution

(a) This 2° allylic chloride is treated with methanol, a weak nucleophile and a polar protic solvent. Ionization of the carbon-chlorine bond forms a resonance-stabilized secondary allylic cation. Therefore, we predict reaction by an S_N1 mechanism.

$$\text{3-chlorocyclopentene} + CH_3OH \xrightarrow[\text{methanol}]{S_N1} \text{3-methoxycyclopentene } (OCH_3) + HCl$$

R enantiomer — *R*,*S* mixture

(b) Iodide is a good leaving group on a moderately accessible secondary carbon. Acetate ion dissolved in a polar aprotic solvent is a moderate nucleophile. We predict substitution by an S_N2 pathway with inversion of configuration at the stereocenter.

$$CH_3(CH_2)_5CH(I)CH_3 + CH_3C(=O)O^-Na^+ \xrightarrow[\text{DMSO}]{S_N2} CH_3(CH_2)_5CH(OC(=O)CH_3)CH_3 + NaI$$

R enantiomer — *S* enantiomer

PROBLEM 8.5

Write the expected substitution product(s) for each reaction and predict the mechanism by which each product is formed.

(a) $(CH_3)_3C$–cyclohexyl–Br (chair) $+ Na^+SH^- \xrightarrow{\text{acetone}}$ (b) $CH_3CH(Cl)CH_2CH_3 + HC(=O)OH \longrightarrow$

R enantiomer

8.7 Phase-Transfer Catalysis

Very often, nucleophilic substitution involves reaction between a covalent organic compound and an ionic compound, as, for example, between 1-chlorooctane and sodium cyanide.

$$CH_3(CH_2)_6CH_2Cl + Na^+CN^- \longrightarrow CH_3(CH_2)_6CH_2CN + Na^+Cl^-$$

1-Chlorooctane — Nonanenitrile

This reaction is simple to write, but, for it to occur, both compounds must be brought together so that they can react. The solubility characteristics of these reactants are quite different. Sodium cyanide is an ionic solid, soluble in water and a few other polar solvents, but insoluble in nonpolar organic solvents such as methylene chloride. 1-Chlorooctane, on the other hand, is insoluble in water but quite soluble in methylene chloride. One way to bring these reactants together is to dissolve them in DMSO, DMF, or other polar aprotic solvents. The advantages of DMSO and DMF are that each dissolves both organic and ionic compounds. When 1-chlorooctane and sodium cyanide are dissolved in DMSO, reaction between them occurs very readily.

Although DMSO and DMF are excellent solvents in which to carry out organic reactions, they have certain disadvantages. Both are several times more expensive than solvents such as methylene chloride and ethanol; and, on an industrial scale, solvent cost can be an important consideration. Furthermore, because DMSO and DMF are so soluble in water, it is often difficult to recover them from mixtures with water. Finally, because they have higher boiling points than other common solvents (189°C for DMSO and 153°C for DMF compared with 78°C for ethanol and 40°C for methylene chloride), it is often difficult to remove them entirely from organic reaction products by distillation.

Phase-transfer catalyst A substance that transfers ions from an aqueous phase into an organic phase and vice versa.

Hydrophilic From the Greek, meaning water-loving

Hydrophobic From the Greek, meaning water-fearing.

Another way to bring about reaction between 1-chlorooctane and cyanide ion is by a technique called phase-transfer catalysis. A **phase-transfer catalyst** is a substance that transfers ions from an aqueous phase into an organic phase and vice versa. The characteristics necessary for an effective phase-transfer catalyst for anions are a balanced combination of (1) hydrophilic character to dissolve in water and form an ion pair with the anion to be transported and (2) hydrophobic character to dissolve in the organic phase and transport the anion into it. Two effective phase-transfer catalysts are tetrabutylammonium chloride and benzyltrimethylammonium chloride, both water-soluble salts.

$(CH_3CH_2CH_2CH_2)_4N^+\ Cl^-$	$C_6H_5CH_2\overset{+}{N}(CH_3)_3\ Cl^-$
Tetrabutylammonium chloride (TBACl)	Benzyltrimethylammonium chloride

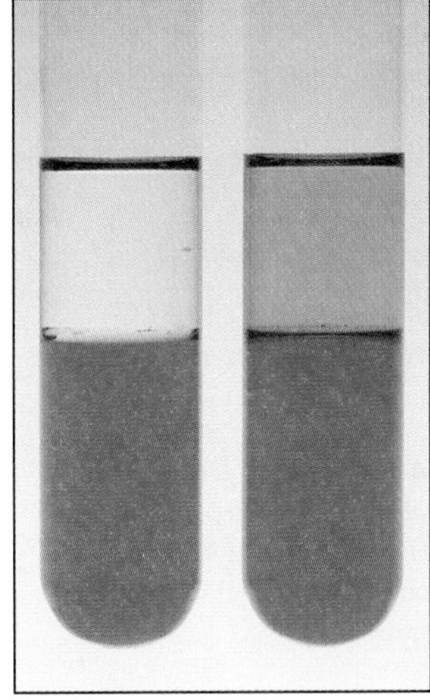

When aqueous potassium permanganate and hexane are mixed in the test tube on the left, the purple color of permanganate ion is present only in the lower aqueous layer. When a phase-transfer catalyst has been added (test tube on the right), permanganate ion is transported into the upper hexane layer. *(Charles D. Winters)*

The tetrabutylammonium (TBA^+) and benzyltrimethylammonium ions each have hydrophilic and hydrophobic regions. **Hydrophilic** means water-loving; **hydrophobic** means water-fearing. The positively charged nitrogen atom of the tetrabutylammonium ion is a hydrophilic site that interacts with water, with other polar molecules, and with anions. The four butyl groups attached to nitrogen in TBA^+ are hydrophobic sites and do not interact with water. Because of its hydrophilic positively charged nitrogen atom, $(Bu)_4N^+$ is soluble in water and, because of its four hydrophobic butyl groups, it is also soluble in nonpolar organic solvents.

A phase-transfer catalyst works in the following way, illustrated here using tetrabutylammonium chloride. Suppose sodium cyanide is dissolved in water, 1-chlorooctane is dissolved in methylene chloride, and the solutions are mixed. Because water and methylene chloride are immiscible, a two-phase system results [Figure 8.6(a)]. No reaction takes place between 1-chlorooctane and cyanide ion because they are in different phases.

When added to this two-phase system, tetrabutylammonium chloride dissolves in the aqueous phase [Figure 8.6(a)]. TBA^+ and CN^- then form an ion pair that is transferred into the organic phase. CN^- displaces the chlorine atom, and Cl^- is trans-

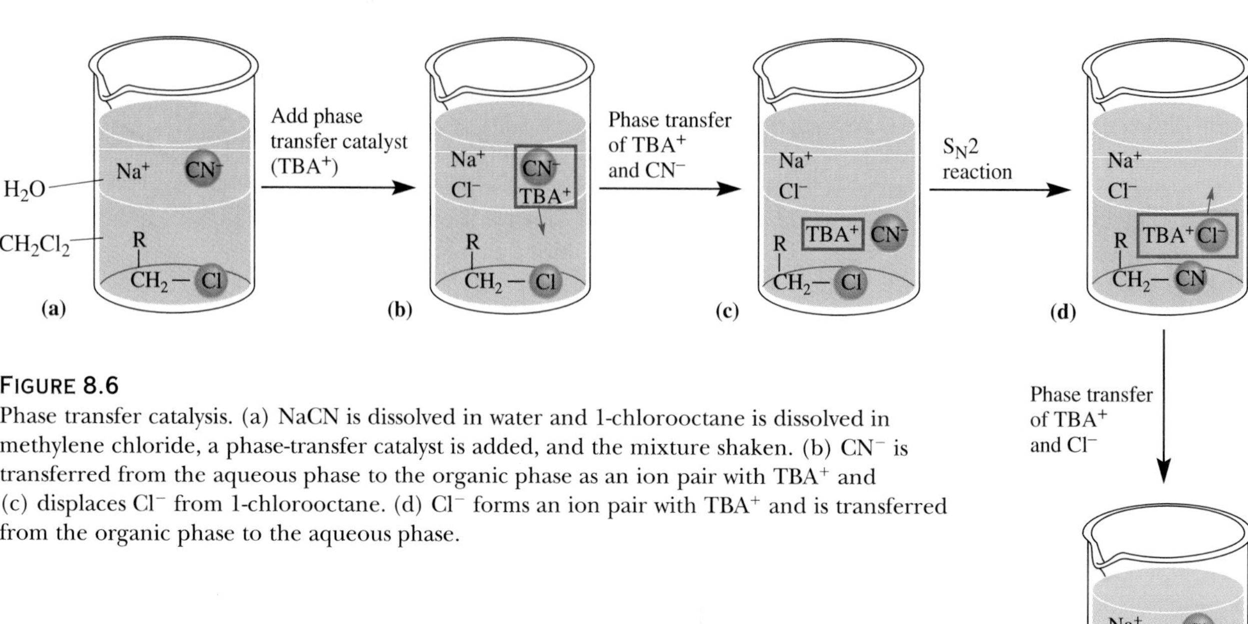

FIGURE 8.6
Phase transfer catalysis. (a) NaCN is dissolved in water and 1-chlorooctane is dissolved in methylene chloride, a phase-transfer catalyst is added, and the mixture shaken. (b) CN^- is transferred from the aqueous phase to the organic phase as an ion pair with TBA^+ and (c) displaces Cl^- from 1-chlorooctane. (d) Cl^- forms an ion pair with TBA^+ and is transferred from the organic phase to the aqueous phase.

ferred as an ion pair with TBA^+ to the aqueous phase. This process is repeated until all cyanide ions have been transferred to the organic phase and have reacted with 1-chlorooctane.

8.8 β-Elimination

β-Elimination A reaction in which a small molecule, such as HCl, HBr, HI, or HOH, is split out or eliminated from adjacent carbons.

Dehydrohalogenation Removal of —H and —X from adjacent carbons; a type of β-elimination.

In ***β*-elimination,** a small molecule, such as HCl, HBr, HI, or HOH, is split out or eliminated from adjacent carbons of a larger molecule. Here, we study a type of β-elimination called **dehydrohalogenation.** In the presence of base, halogen can be removed from one carbon of an alkyl halide and hydrogen from an adjacent carbon to form an alkene. The carbon bearing the halogen in an alkyl halide is called the α-carbon, and the next carbon is called a β-carbon.

$$\underset{\text{An alkyl halide}}{-\overset{|}{\underset{\underset{H}{|}}{C}}^{\beta}-\overset{|}{\underset{\underset{X}{|}}{C}}^{\alpha}-} + \underset{\text{Base}}{CH_3CH_2O^-Na^+} \xrightarrow[CH_3CH_2OH]{} \underset{\text{An alkene}}{>C=C<} + CH_3CH_2OH + Na^+X^-$$

Strong bases that serve effectively in β-eliminations of alkyl halides are OH^-, OR^-, and NH_2^-. Following are three examples of base-promoted β-elimination reactions. In the first example, the base is shown as a reactant. In the second and third examples, the base is a reactant but is shown over the reaction arrow.

$$\overset{\beta}{CH_3(CH_2)_7CH_2}\overset{\alpha}{CH_2}Br + (CH_3)_3CO^-K^+ \longrightarrow CH_3(CH_2)_7CH{=}CH_2 + (CH_3)_3COH + K^+Br^-$$

1-Bromodecane; Potassium *tert*-butoxide; 1-Decene; 2-Methyl-2-propanol (*tert*-butyl alcohol)

$$CH_3CH_2C(Br)(CH_3)CH_3 \xrightarrow[CH_3CH_2OH]{CH_3CH_2O^-Na^+} CH_3CH{=}C(CH_3)CH_3 + CH_3CH_2C(CH_3){=}CH_2$$

2-Bromo-2-methylbutane; 2-Methyl-2-butene (major product); 2-Methyl-1-butene

$$\text{1-Bromo-1-methylcyclopentane} \xrightarrow[CH_3OH]{CH_3O^-Na^+} \text{1-Methylcyclopentene} + \text{Methylenecyclopentane}$$

1-Bromo-1-methylcyclopentane; 1-Methylcyclopentene (major product); Methylenecyclopentane

In the second and third illustrations, there are two nonequivalent β-carbons, each bearing a hydrogen; therefore, two alkenes are possible. Each reaction is said to follow **Zaitsev's rule:** the major product of a β-elimination reaction is the more stable alkene, that is, the more highly substituted alkene (Section 6.6B).

Zaitsev's rule The major product of a β-elimination reaction is the more stable alkene, that is, the one having the greater number of substituents on the carbon-carbon double bond.

EXAMPLE 8.6

Predict the β-elimination product(s) formed when each bromoalkane is treated with sodium ethoxide in ethanol. If two or more products might be formed, predict which is the major product.

(a) $CH_3CH(CH_3)CH(Br)CH_3$ **(b)** $CH_3CH(CH_3)CH_2CH_2Br$

Solution

(a) There are two nonequivalent β-carbons in this bromoalkane, and two alkenes are possible. 2-Methyl-2-butene, the more substituted alkene, is the major product.

$$CH_3\overset{\beta}{C}H(CH_3)CH(Br)\overset{\beta}{C}H_3 \xrightarrow[CH_3CH_2OH]{CH_3CH_2O^-Na^+} CH_3C(CH_3){=}CHCH_3 + CH_3CH(CH_3)CH{=}CH_2$$

(major product)

(b) There is only one β-carbon in this bromoalkane and only one alkene is possible.

$$CH_3\overset{\beta}{C}H(CH_3)\overset{\alpha}{C}H_2CH_2Br \xrightarrow[CH_3CH_2OH]{CH_3CH_2O^-Na^+} CH_3CH(CH_3)CH{=}CH_2$$

PROBLEM 8.6

Predict the β-elimination product(s) formed when each chloroalkane is treated with sodium ethoxide in ethanol. If two or more products might be formed, predict which is the major product.

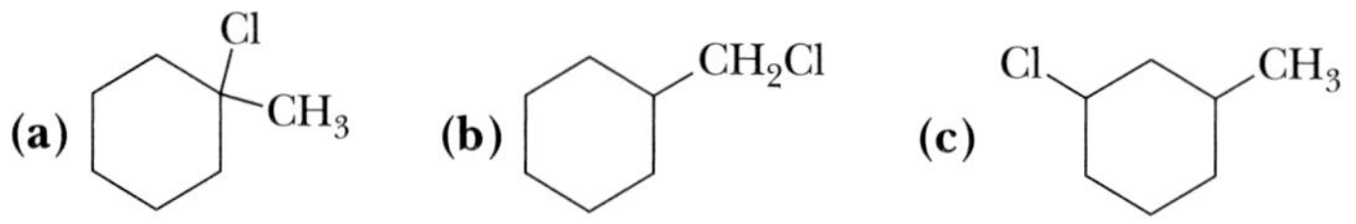

(a) 1-chloro-1-methylcyclohexane (Cl, CH_3 on same ring carbon) (b) cyclohexyl–CH_2Cl (c) 1-chloro-3-methylcyclohexane (Cl, CH_3)

8.9 Mechanisms of β-Elimination

There are two limiting mechanisms for β-eliminations. A fundamental difference between them is the timing of the bond-breaking and bond-forming steps. Recall that we made the same statement about the two limiting mechanisms for nucleophilic substitution reactions (Section 8.3).

A. E1 Mechanism

At one extreme, breaking of the C—X bond is complete before any reaction occurs with base to lose a hydrogen and form the carbon-carbon double bond. This mechanism is designated **E1** where E stands for *E*limination and 1 stands for *uni*molecular; one species, in this case the alkyl halide, is involved in the rate-limiting step. It is proposed that an E1 mechanism involves two steps. Step 1 is a slow, rate-limiting ionization of the C—X bond to form a carbocation intermediate. In Step 2, the carbocation intermediate reacts with a base to lose a hydrogen and to form the alkene. The reaction of 2-bromo-2-methylpropane with methanol to form 2-methylpropene is an example of an E1 reaction.

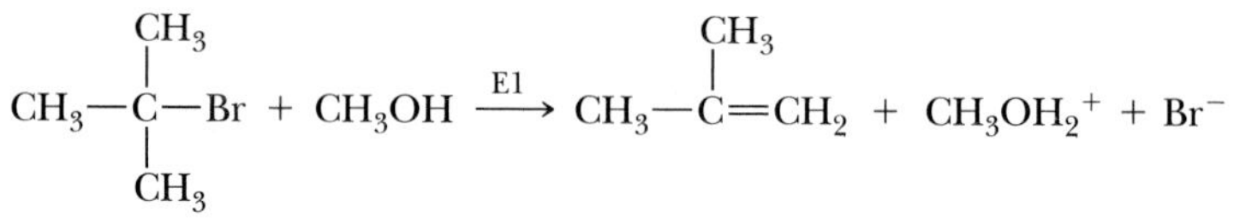

$$(CH_3)_3C{-}Br + CH_3OH \xrightarrow{E1} CH_3{-}\underset{CH_3}{\overset{|}{C}}{=}CH_2 + CH_3OH_2^{+} + Br^{-}$$

2-Bromo-2-methylpropane → 2-Methylpropene

MECHANISM E1 Reaction of 2-Bromo-2-Methylpropane

Step 1: Rate-limiting ionization of the C—Br bond to form a carbocation intermediate

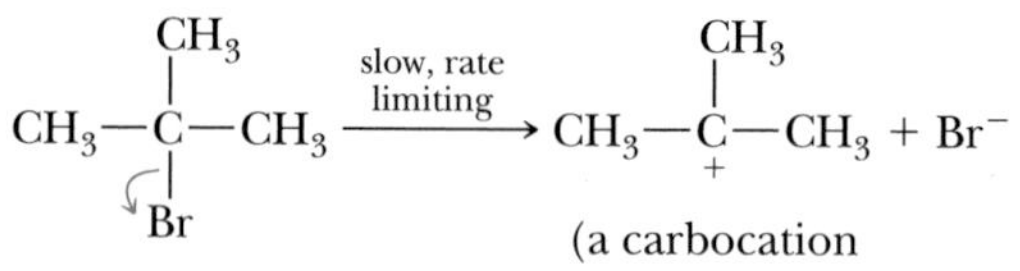

$$(CH_3)_3C{-}Br \xrightarrow{\text{slow, rate limiting}} (CH_3)_3C^{+} + Br^{-}$$

(a carbocation intermediate)

Step 2: Proton transfer from the carbocation intermediate to solvent (in this example methanol) to give the alkene

$$H_3C(H)\ddot{O}: \; H{-}CH_2{-}\overset{+}{C}(CH_3){-}CH_3 \xrightarrow{\text{fast}} H_3C(H)\ddot{O}^{+}{-}H + CH_2{=}C(CH_3){-}CH_3$$

In an E1 mechanism, one transition state exists for the formation of the carbocation in Step 1 and a second exists for the loss of a hydrogen in Step 2 (Figure 8.7). Formation of the carbocation intermediate in Step 1 crosses the higher potential energy barrier and is the rate-limiting step.

B. E2 Mechanism

At the other extreme is a concerted process. In an E2 reaction, here illustrated by the reaction of 1-bromopropane with sodium ethoxide, proton transfer to the base, formation of the carbon-carbon double bond, and ejection of bromide ion occur simultaneously; all bond-breaking and bond-forming steps are concerted (Figure 8.8). Because base removes a β-hydrogen at the same time the C—Br bond is broken to form a halide ion, the transition state has considerable double bond character.

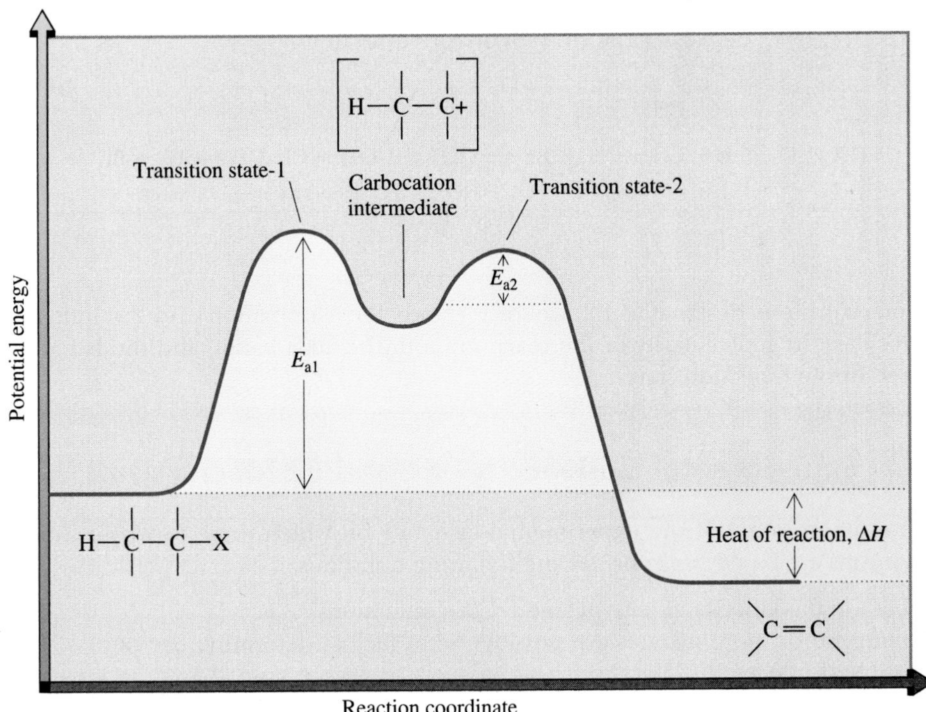

FIGURE 8.7
A potential energy diagram for an E1 reaction showing two transition states and one carbocation intermediate.

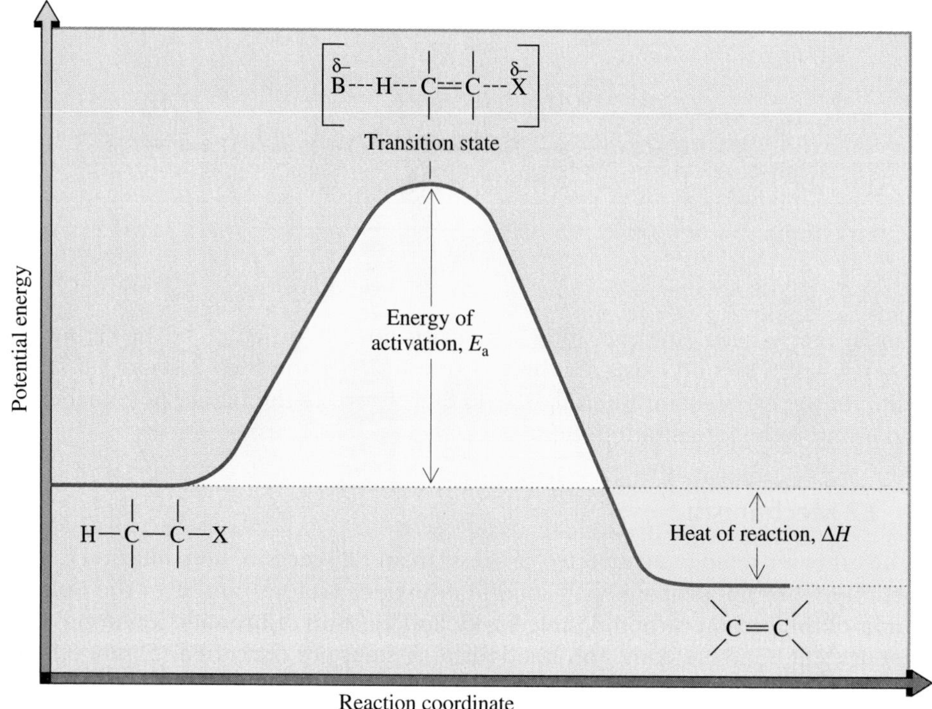

FIGURE 8.8
A potential energy diagram for an E2 reaction. There is considerable double bond character in the transition state.

MECHANISM E2 Reaction of 1-Bromopropane

$$CH_3CH_2O^- + H{-}\underset{}{\overset{CH_3}{\overset{|}{CH}}}{-}CH_2{-}Br \longrightarrow CH_3CH_2OH + CH_3CH{=}CH_2 + Br^-$$

This mechanism is designated **E2**, where E stands for *E*limination and 2 stands for *bi*molecular, the molecularity of the reaction; both the alkyl halide and the base are involved in the transition state.

8.10 Experimental Evidence for E1 and E2 Mechanisms

As we examine some of the experimental evidence on which these two contrasting mechanisms are based, we consider the following questions:

1. What are the kinetics of base-promoted β-eliminations?
2. Where two or more alkenes are possible, what factors determine the ratio of the possible products?
3. What is the stereochemistry of β-elimination?

A. Kinetics

E1 Reactions

The rate-limiting step in an E1 reaction is ionization of the halide to form a carbocation. Because this step involves only the alkyl halide and not the base, the reaction is said to be unimolecular.

$$\text{Rate} = -\frac{d[\text{RX}]}{dt} = k[\text{RX}]$$

Recall that the first step in an S_N1 reaction is also formation of a carbocation. Thus, for both S_N1 and E1 reactions, formation of the carbocation is the rate-limiting step.

E2 Reactions

Only one step occurs in an E2 mechanism, and the transition state is bimolecular. The reaction is second-order: first-order in RX and first-order in base.

$$\text{Rate} = -\frac{d[\text{RX}]}{dt} = k[\text{RX}][\text{Base}]$$

B. Regioselectivity

E1 Reactions

The major product in E1 reactions is the more stable alkene, that is, the alkene with the more highly substituted carbon-carbon double bond. Once the carbocation is formed in the rate-limiting step of an E1 reaction, it may lose a hydrogen to complete β-elimination or it may rearrange and then lose a hydrogen.

E2 Reactions

For E2 reactions using strong bases and in which the leaving group is a halide ion, the major product is that formed following Zaitsev's rule. Double bond character is so highly developed in the transition state that the relative stability of possible alkenes determines which alkene is the major product. Thus, the transition state of lowest energy is that leading to the most stable, most highly substituted alkene.

$$CH_3(CH_2)_3\underset{\text{Br}}{\overset{|}{C}}HCH_3 + CH_3O^-Na^+ \xrightarrow[CH_3OH]{E2} CH_3CH_2CH_2CH{=}CHCH_3 + CH_3CH_2CH_2CH_2CH{=}CH_2 + CH_3OH + NaBr$$

2-Bromohexane; 2-Hexene (74%); 1-Hexene (26%)

C. Stereoselectivity

The E2 reaction is most favorable when X and H are oriented anti and coplanar (at a dihedral angle of 180°) to one another.

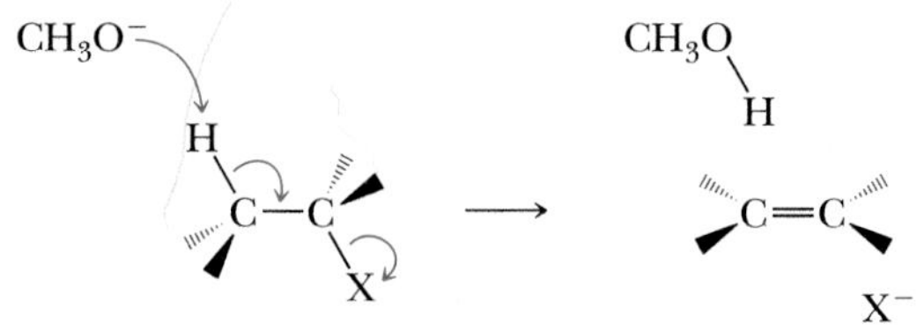

CH_3O^- H C—C X → CH_3O—H C=C X^-

—H and —X are anti and coplanar
(dihedral angle 180°)

This requirement can be demonstrated clearly in chlorocyclohexanes. In these molecules, anti and coplanar corresponds to *trans*, diaxial. Consider base-promoted E2 reaction of the *cis* and *trans* isomers of 1-chloro-2-isopropylcyclohexane. From the *cis* isomer, the major product is 1-isopropylcyclohexene, the more substituted cycloalkene. From the *trans* isomer, only 3-isopropylcyclohexene, the less substituted alkene, is formed.

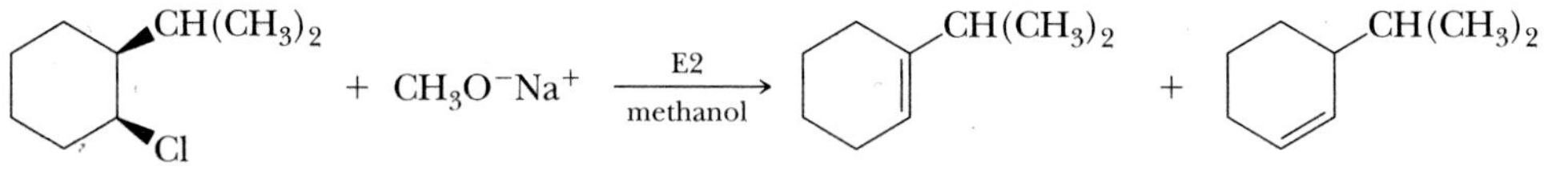

$CH(CH_3)_2$, Cl + $CH_3O^-Na^+$ $\xrightarrow[\text{methanol}]{\text{E2}}$ $CH(CH_3)_2$ + $CH(CH_3)_2$

cis-1-Chloro-2-isopropylcyclohexane — 1-Isopropylcyclohexene (major product) — 3-Isopropylcyclohexene

$CH(CH_3)_2$, Cl + $CH_3O^-Na^+$ $\xrightarrow[\text{methanol}]{\text{E2}}$ $CH(CH_3)_2$

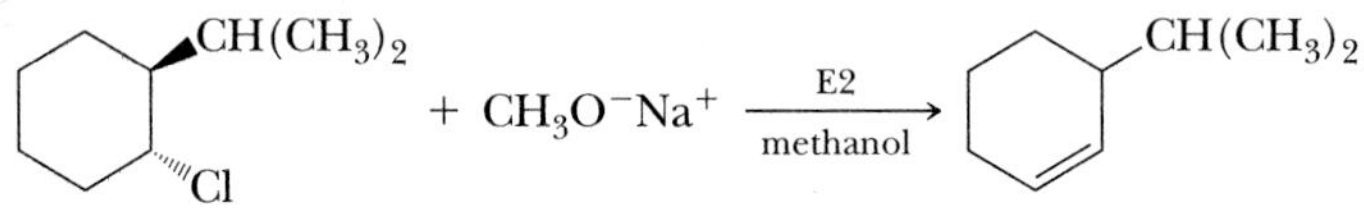

trans-1-Chloro-2-isopropylcyclohexane — 3-Isopropylcyclohexene

In the more stable chair conformation of the *cis* isomer, the considerably larger isopropyl group is equatorial and the smaller chlorine is axial. In this chair conformation, —H on carbon 2 and —Cl on carbon 1 are anti and coplanar. Concerted E2 elimination gives 1-isopropylcyclohexene, a trisubstituted alkene, as the major product. Note that —H on carbon 6 and —Cl are also anti and coplanar. Dehydrohalogenation of this combination of —H and —Cl gives 3-isopropylcyclohexene, a disubstituted and, therefore, less stable alkene.

E2 Reaction of *cis*-1-chloro-2-isopropylcyclohexane

CH_3O^- H H 2 6 $CH(CH_3)_2$ H H 1 Cl $\xrightarrow{\text{E2}}$ $CH(CH_3)_2$

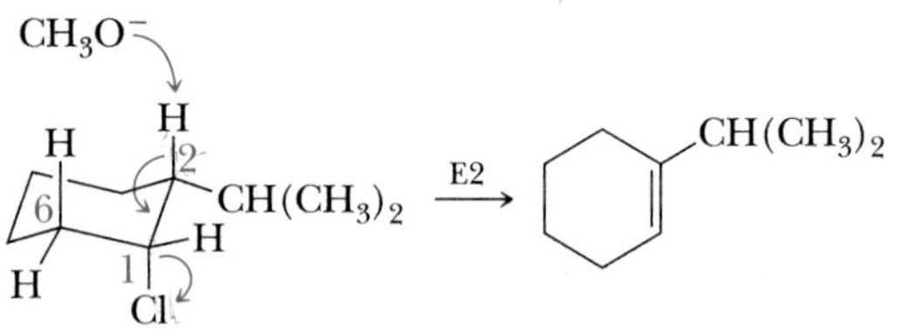

More stable chair
(—H and —Cl are *trans* and coplanar)

1-Isopropylcyclohexene

In the more stable chair conformation of the *trans* isomer, both isopropyl and chlorine are equatorial. In this conformation, the hydrogen atom on carbon 2 is *cis* to the chlorine atom. One of the hydrogen atoms on carbon 6 is *trans* to —Cl, but it is not anti and coplanar. In the alternative, less stable chair conformation of the *trans* isomer, both isopropyl and chlorine are axial. In this conformation, the axial hydrogen in carbon 6 is anti and coplanar to chlorine and E2 β-elimination can occur to give 3-isopropylcyclohexene. E2 Reaction of *trans*-1-chloro-2-isopropylcyclohexane.

E2 Reaction of *trans*-1-chloro-2-isopropylcyclohexane

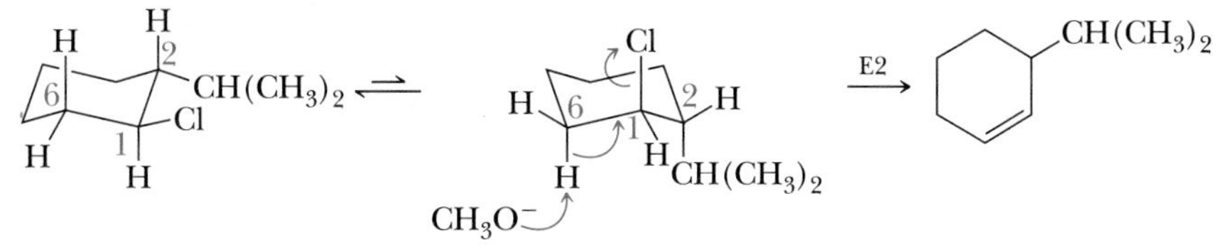

More stable chair (no —H is *trans* and coplanar to —Cl) ⇌ Less stable chair (—H on carbon-6 is *trans* and coplanar to —Cl) $\xrightarrow{E2}$ 3-Isopropylcyclohexene

The rate at which the *cis* isomer undergoes E2 reaction is considerably greater than the rate for the *trans* isomer. We can account for this observation in the following manner. The more stable chair conformation of the *cis* isomer has —H and —Cl anti and coplanar, and the activation energy for the reaction is that required to reach the E2 transition state. The more stable chair conformation of the *trans* isomer, however, cannot undergo anti elimination. To react, it must first be converted to the less stable chair. Thus, the activation energy for the *trans* isomer includes (1) the energy necessary to convert the more stable chair to the less stable chair, and (2) the energy to reach the E2 transition state from this conformation.

EXAMPLE 8.7

Treatment of 1,2-dibromo-1,2-diphenylethane with sodium methoxide in methanol gives 1-bromo-1,2-diphenylethylene. The meso isomer of the starting material gives (*E*)-1-bromo-1,2-diphenylethylene whereas the racemic mixture produces (*Z*)-1-bromo-1,2-diphenylethylene. How do you account for the stereoselectivity of these β-eliminations?

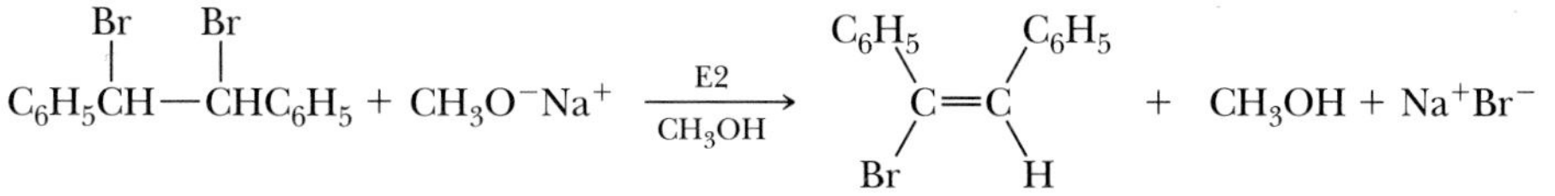

$$C_6H_5CH(Br)-CH(Br)C_6H_5 + CH_3O^-Na^+ \xrightarrow[CH_3OH]{E2} (C_6H_5)(Br)C{=}C(C_6H_5)(H) + CH_3OH + Na^+Br^-$$

meso-1,2-Dibromo-1,2-diphenylethane → (*E*)-1-Bromo-1,2-diphenylethylene

$$C_6H_5CH(Br)-CH(Br)C_6H_5 + CH_3O^-Na^+ \xrightarrow[CH_3OH]{E2} (C_6H_5)(Br)C{=}C(H)(C_6H_5) + CH_3OH + Na^+Br^-$$

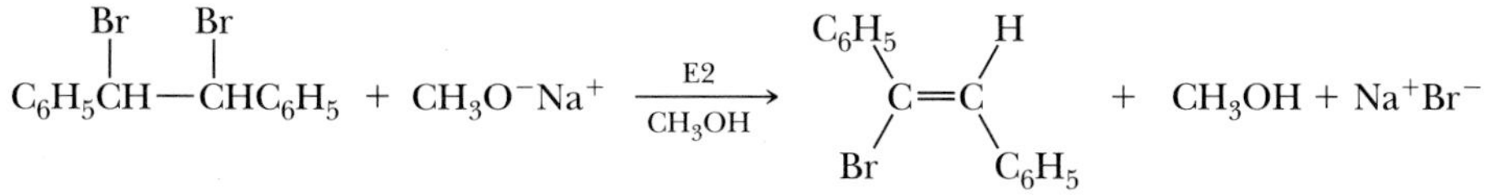

racemic-1,2-Dibromo-1,2-diphenylethane → (*Z*)-1-Bromo-1,2-diphenylethylene

Solution

It is a requirement for an E2 reaction that —H and —X be anti and coplanar. Following is a stereorepresentation of the meso isomer, drawn to show the plane of symmetry. Rotation of the left carbon by 60° brings —H and —Br into an anti and coplanar relationship. E2 reaction of this conformation gives the (*E*)-alkene. E2 reaction of the proper conformation of either enantiomer of the racemic mixture gives the (*Z*)-alkene.

plane of symmetry

Meso isomer $\xrightarrow{\text{rotate left carbon counterclockwise by 60°}}$ $\xrightarrow{E2}$ (*E*)-1-Bromo-1,2-diphenylethylene

One enantiomer of the racemic mixture $\xrightarrow{\text{rotate left carbon clockwise by 60°}}$ $\xrightarrow{E2}$ (*Z*)-1-Bromo-1,2-diphenylethylene

PROBLEM 8.7

1-chloro-4-isopropylcyclohexane exists as two stereoisomers: one *cis* and one *trans.* Treatment of either isomer with sodium ethoxide in ethanol gives 4-isopropylcyclohexene by an E2 reaction.

$$Cl{-}C_6H_{10}{-}CH(CH_3)_2 \xrightarrow[CH_3CH_2OH]{CH_3CH_2O^-Na^+} C_6H_9{-}CH(CH_3)_2$$

1-Chloro-4-isopropylcyclohexane — 4-Isopropylcyclohexene

The *cis* isomer undergoes E2 reaction several orders of magnitude faster than the *trans* isomer. How do you account for this experimental observation?

8.11 Substitution Versus Elimination

Thus far, we have considered two types of reactions of alkyl halides, namely nucleophilic substitution and β-elimination. Many of the nucleophiles we have considered, for example hydroxide ion and alkoxide ions, are also strong bases. Thus, nucleophilic substitution and β-elimination often compete with each other, and the ratio of products formed by these reactions depends on the relative rates of the two reactions.

$$H{-}C{-}C{-}X + {:}Nu^- \xrightarrow{\text{nucleophilic substitution}} H{-}C{-}C{-}Nu + X^-$$

$$H{-}C{-}C{-}X + {:}Nu^- \xrightarrow{\beta\text{-elimination}} C{=}C + H{-}Nu + X^-$$

A. S_N1 Versus E1

Reactions of secondary and tertiary alkyl halides in polar protic solvents give mixtures of substitution and elimination products. In both reactions, the first step is the formation of a carbocation intermediate that is then followed by characteristic carbocation reactions: (1) loss of a hydrogen (E1) to give an alkene, (2) reaction with solvent (S_N1) to give a substitution product, or (3) rearrangement followed by reaction (1) or (2). In polar protic solvents, the products formed depend only on the structure of the particular carbocation. For example, *tert*-butyl chloride and *tert*-butyl iodide in 80% aqueous ethanol both react with solvent giving the same mixture of substitution and elimination products. Because iodide ion is a better leaving group than chloride ion, *tert*-butyl iodide reacts over 100 times faster than *tert*-butyl chloride. Yet the ratio of products is the same.

$$(CH_3)_3C{-}I \xrightarrow{-I^-} (CH_3)_3C^+ \xleftarrow{-Cl^-} (CH_3)_3C{-}Cl$$

$$(CH_3)_3C^+ \xrightarrow{E1} CH_2{=}C(CH_3)_2 + H^+$$

$$(CH_3)_3C^+ \xrightarrow[H_2O]{S_N1} (CH_3)_3C{-}OH + H^+$$

$$(CH_3)_3C^+ \xrightarrow[CH_3CH_2OH]{S_N1} (CH_3)_3C{-}OCH_2CH_3 + H^+$$

It is difficult to predict the ratio of substitution to elimination products for first-order reactions of alkyl halides. For the majority of cases, however, S_N1 predominates over E1.

B. S_N2 Versus E2

It is considerably easier to predict the ratio of substitution to elimination products for second-order reactions of alkyl halides with reagents that act both as nucleophiles and bases. The guiding principles are:

1. Branching at the α-carbon or β-carbon(s) increases steric hindrance about the α-carbon and significantly retards S_N2 reactions. Conversely, branching at the α-carbon or β-carbon(s) increases the rate of E2 reaction because of the increased stability of the alkene product.
2. The greater the nucleophilicity of the attacking reagent, the greater the S_N2 to E2 ratio. Conversely, the greater the basicity of the attacking reagent, the greater the E2 to S_N2 ratio.

Primary halides react with bases/nucleophiles to give predominantly substitution products. With strong bases, such as hydroxide ion and ethoxide ion, a percentage of the product is formed by an E2 reaction, but it is generally small compared with that formed by S_N2 reaction. The E2 product becomes the major product when a bulky base, such as *tert*-butoxide, is used. Tertiary halides react with all strong bases/nucleophiles to give only elimination products.

Secondary halides are borderline, and substitution or elimination may be favored depending on the particular base/nucleophile, solvent, and temperature at which the reaction is carried out. With bases/nucleophiles for which the pK_a of the conjugate acid is 11 or above (as, for example, hydroxide ion and ethoxide ion), elimination is favored. With bases/nucleophiles for which the pK_a of the conjugate acid is 11 or below (as, for example, acetate ion), substitution is favored. The reason for this change in the ratio of S_N2/E2 is that as the pK_a of the conjugate acid increases, basicity increases faster than nucleophilicity. These generalizations about substitution versus elimination reactions of methyl, primary, secondary, and tertiary alkyl halides are summarized in Table 8.13.

TABLE 8.13 Summary of Substitution Versus Elimination Reactions of Alkyl Halides

Halide	Reaction	Comments
methyl (CH_3—X)	S_N2	S_N1 reactions of methyl halides are never observed. The methyl cation is so unstable that it is never formed in solution.
primary (RCH_2—X)	S_N2	The main reaction with good nucleophiles/weak bases such as I^- and $CH_3CO_2^-$.
	E2	The main reaction with strong, bulky bases such as potassium *tert*-butoxide.
		Primary cations are rarely formed in solution, and, therefore, S_N1 and E1 reactions of primary halides are rarely observed.
secondary (R_2CH—X)	S_N2	The main reaction with bases/nucleophiles where pK_a of the conjugate acid is 11 or less; as, for example, I^- and $CH_3CO_2^-$.
	E2	The main reaction with bases/nucleophiles where the pK_a of the conjugate acid is 11 or greater; as, for example, OH^- and $CH_3CH_2O^-$.
	S_N1/E1	Common in reactions with weak nucleophiles in polar protic solvents, such as water, methanol, and ethanol.
tertiary (R_3C—X)	E2	Main reaction with strong bases such as HO^- and RO^-.
	S_N1/E1	Main reactions with poor nucleophiles.
		S_N2 reactions of tertiary halides are rarely observed because of the extreme crowding around the 3° carbon.

EXAMPLE 8.8

Predict whether each reaction proceeds predominantly by substitution (S_N1 or S_N2) or elimination (E1 or E2) or whether the two compete. Write structural formulas for the major organic product(s).

(a) $CH_3C(CH_3)(Cl)CH_2CH_3 + NaOH \xrightarrow[H_2O]{80°C}$ (b) $CH_3CH(CH_3)CH_2CH_2Br + (C_2H_5)_3N \xrightarrow[CH_2Cl_2]{30°C}$

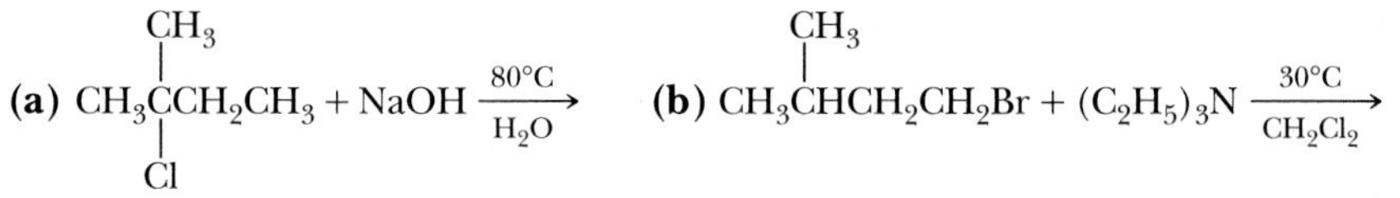

Solution

(a) A 3° halide is heated with a strong base/strong nucleophile. Elimination by an E2 reaction predominates to give 2-methyl-2-butene as the major product.

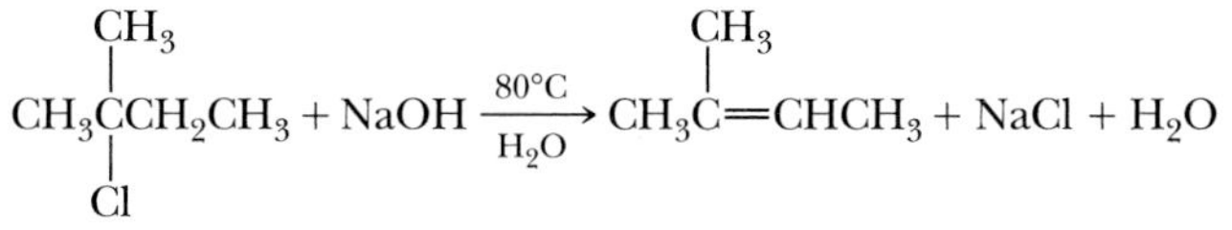

$$CH_3C(CH_3)(Cl)CH_2CH_3 + NaOH \xrightarrow[H_2O]{80°C} CH_3C(CH_3){=}CHCH_3 + NaCl + H_2O$$

(b) Reaction of a 1° halide and a moderate nucleophile/weak base gives substitution by an S_N2 reaction.

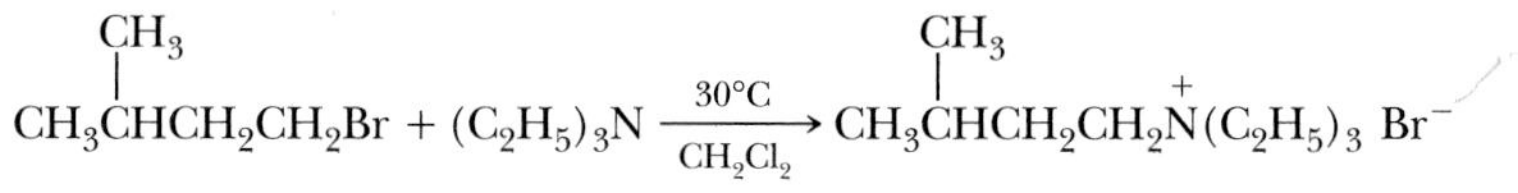

$$CH_3CH(CH_3)CH_2CH_2Br + (C_2H_5)_3N \xrightarrow[CH_2Cl_2]{30°C} CH_3CH(CH_3)CH_2CH_2\overset{+}{N}(C_2H_5)_3\ Br^-$$

PROBLEM 8.8

Predict whether each reaction proceeds predominantly by substitution (S_N1 or S_N2) or elimination (E1 or E2) or whether the two compete. Write structural formulas for the major organic product(s).

(a) $CH_3CH_2CH(I)CH_2CH_3 + CH_3O^-Na^+ \xrightarrow[\text{methanol}]{}$

(b) cyclohexane with Cl (hashed wedge) and H_3C (solid wedge) in a 1,4 relationship $+ Na^+I^- \xrightarrow[\text{acetone}]{}$

(c) $C_6H_5CH_2CH_2Br + Na^+CN^- \xrightarrow[\text{methanol}]{}$

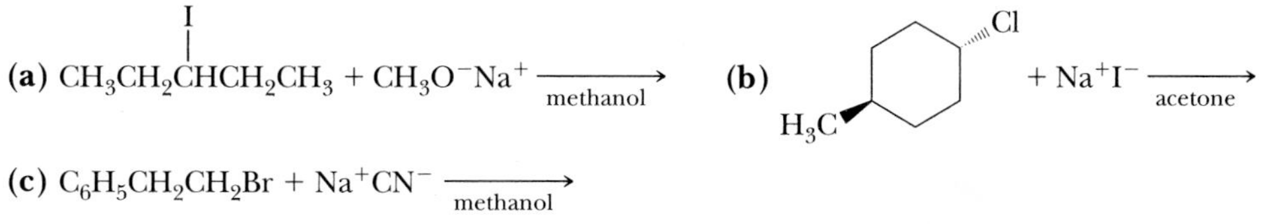

SUMMARY

A **nucleophile** is a molecule or ion that donates a pair of electrons to another atom or ion to form a new covalent bond; a Lewis base. Nucleophilic substitution is any reaction in which one nucleophile is substituted for another.

Protic solvents are hydrogen bond donors (Section 8.2). The most common protic solvents are those containing —OH groups. **Aprotic solvents** cannot serve as hydrogen bond donors. Common aprotic solvents are acetone, diethyl ether, dimethyl sulfoxide, and dimethylformamide. **Polar solvents** interact strongly with ions and polar molecules. **Nonpolar solvents** do not interact strongly with ions and polar molecules. The **dielectric constant** is the most commonly used measure of solvent polarity.

An **S_N2 reaction** (Section 8.3) occurs in one step. Departure of the leaving group is assisted by the incoming nucleophile and both nucleophile and leaving group are involved in the transition state of the rate-limiting step. An **S_N1 reaction** occurs in two steps. Step 1 is a slow, rate-limiting ionization of the C—X bond to form a carbocation followed in Step 2 by rapid reaction of the carbocation with a nucleophile to complete the substitution.

The **nucleophilicity** of a reagent is measured by the rate of its reaction in a reference nucleophilic substitution (Section 8.4B). Relative nucleophilicities depend on whether a reaction is carried out in a polar protic solvent or a polar aprotic solvent. A general principle is that the freer a nucleophile is from a surrounding solvation shell, the greater its nucleophilicity.

For S_N1 reactions taking place at a tetrahedral stereocenter, the major reaction occurs with racemization (Section 8.4C). To the extent that the leaving group remains in association with the carbocation as an ion pair and hinders approach of the nucleophile from that face of the carbocation, the result is racemization with some net inversion of configuration. An S_N2 reaction proceeds with complete inversion of configuration.

S_N1 reactions are governed by **electronic factors,** namely, the relative stabilities of carbocation intermediates (Section 8.4D). S_N2 reactions are governed by **steric factors,** namely the degree of crowding around the site of substitution. Both allylic and benzylic cations are stabilized by delocalization of the positive charge of the cation (Section 8.4E).

The ability of a group to function as a leaving group is related to its stability as an anion (Section 8.4F). The most stable anions and the best leaving groups are the conjugate bases of strong acids.

S_N2 reactions in polar aprotic solvents are often several orders of magnitude faster than the same reactions in polar protic solvents (Section 8.4G).

Because S_N1 reactions involve formation of a carbocation intermediate, the greater the ability of the solvent to solvate the carbocation intermediate, the lower the activation energy for carbocation formation and, hence, the greater the rate of reaction. Skeletal rearrangements are characteristic of S_N1 reactions in which a carbocation initially formed can rearrange to a more stable carbocation (Section 8.4H).

Certain nucleophilic displacements that have the kinetic characteristic of S_N1 reactions (first order in alkyl halide and zero order in nucleophile) involve two successive S_N2 reactions. Many such reactions involve participation of a neighboring nucleophile. The **mustard gases** are one group of compounds whose nucleophilic substitution reactions involve neighboring group participation (Section 8.5).

A **phase-transfer catalyst** is a substance that transports ions from an aqueous phase into an organic phase, and vice versa (Section 8.7). Effective phase-transfer catalysts for anions have a balanced combination of (1) hydrophilic character to dissolve in water and form an ion pair with the anion to be transferred and (2) hydrophobic character to dissolve in the organic phase and transport the anion into it.

A ***β*-elimination reaction** (Section 8.8) involves removal of atoms or groups of atoms from adjacent carbon atoms. *β*-Elimination to give the more highly substituted alkene is called **Zaitsev elimination.** An **E1 reaction** occurs in two steps: breaking the C—X bond to form a carbocation intermediate followed by proton transfer to form an alkene. An E1 reaction is first order in alkyl halide and zero order in base. An E2 reaction occurs in one step: simultaneous reaction with base to remove a hydrogen, formation of the alkene, and departure of the leaving group. E2 reactions are stereoselective for **anti and coplanar elimination** of H and X.

KEY REACTIONS

1. Nucleophilic Aliphatic Substitution: S_N2 (Section 8.3)

A second-order reaction involving inversion of configuration at the reaction center. The nucleophile may be negatively charged as in the first example, or neutral as in the second example. S_N2 reactions are accelerated in polar aprotic solvents compared with polar protic solvents.

$$:\ddot{I}:^- + (CH_3CH_2)(H)(CH_3)C{-}\ddot{Cl}: \longrightarrow :\ddot{I}{-}C(CH_2CH_3)(H)(CH_3) + :\ddot{Cl}:^-$$

$$(CH_3)_3N: + (CH_3CH_2)(H)(CH_3)C{-}\ddot{Cl}: \longrightarrow (CH_3)_3\overset{+}{N}{-}C(CH_2CH_3)(H)(CH_3) + :\ddot{Cl}:^-$$

2. Nucleophilic Aliphatic Substitution: S_N1 (Section 8.3)

A first-order reaction involving a carbocation intermediate. Reaction at a stereocenter gives largely racemization often accompanied with some slight excess of inversion of configuration. Reactions often involve carbocation rearrangements and are accelerated by polar protic solvents. In this example, the rearranged cation is both tertiary and benzylic.

$$C_6H_5CH(CH_3)CH(Cl)CH_3 + CH_3OH \longrightarrow C_6H_5C(CH_3)(OCH_3)CH_2CH_3 + HCl$$

3. Neighboring Group Participation (Section 8.5)

First-order kinetics but with participation of an internal nucleophile in the departure of the leaving group, as in hydrolysis of a mustard gas. The mechanism for the hydrolysis of each halide involves two successive S_N2 reactions.

$$ClCH_2CH_2SCH_2CH_2Cl + 2H_2O \longrightarrow HOCH_2CH_2SCH_2CH_2OH + 2HCl$$

4. β-Elimination: E1 (Section 8.9)

Elimination of atoms or groups of atoms from adjacent carbons by a reaction that is zero order in the base and first order in the molecule undergoing elimination. The reaction involves a carbocation intermediate and carbocation rearrangements are common.

$$CH_3CH(CH_3)CH(Cl)CH_3 \xrightarrow[CH_3CO_2H]{E1} (CH_3)_2C{=}C(H)CH_3 + HCl$$

5. β-Elimination: E2 (Section 8.9)

Elimination of atoms or groups of atoms from adjacent carbon atoms by a reaction that is first order in the base and first order in the compound undergoing elimination. Elimination is concerted and stereoselective requiring an anti and coplanar arrangement of groups being eliminated.

$$\text{1-chloro-2-methylcyclohexane} \; (CH_3, Cl) \xrightarrow[CH_3CH_2OH]{CH_3CH_2O^-Na^+} \text{3-methylcyclohexene} \; (CH_3)$$

Additional Problems

Nucleophilic Aliphatic Substitution

8.9 Draw a structural formula for the most stable carbocation of each molecular formula and indicate how each might be formed.

(a) $C_4H_9^+$ **(b)** $C_3H_7^+$ **(c)** $C_8H_{15}^+$ **(d)** $C_3H_7O^+$

8.10 Reaction of 1-bromopropane and sodium hydroxide in ethanol follows an S_N2 mechanism. What happens to the rate of this reaction if the:

(a) Concentration of NaOH is doubled?
(b) Concentrations of both NaOH and 1-bromopropane are doubled?
(c) Volume of the solution in which the reaction is carried out is doubled?

8.11 From each pair, select the stronger nucleophile.

(a) H_2O or OH^- **(b)** $CH_3CO_2^-$ or OH^- **(c)** CH_3SH or CH_3S^-

(d) Cl^- or I^- in DMSO **(e)** Cl^- or I^- in methanol **(f)** CH_3OCH_3 or CH_3SCH_3

8.12 Draw the structural formula for the product of each S_N2 reaction. Where configuration of the starting material is given, show the configuration of the product.

(a) $CH_3CH_2CH_2Cl + CH_3CH_2ONa \xrightarrow{\text{ethanol}}$ **(b)** $(CH_3)_3N: + CH_3I \xrightarrow{\text{acetone}}$

(c) (cyclohexenyl)$-CH_2Br + NaCN \xrightarrow{\text{acetone}}$ **(d)** H_3C–(cyclohexane)–$Cl + CH_3SNa \xrightarrow{\text{ethanol}}$

(e) $CH_3CH_2CH_2Cl + CH_3C{\equiv}C:^-Na^+ \longrightarrow$ **(f)** (cyclopentyl)$-CH_2Cl + :NH_3 \xrightarrow{\text{ethanol}}$

(g) O(ring):NH $+ CH_3(CH_2)_6CH_2Cl \xrightarrow{\text{ethanol}}$ **(h)** $CH_3CH_2CH_2Br + NaCN \xrightarrow{\text{acetone}}$

8.13 You were told that each reaction in the previous problem proceeds by an S_N2 mechanism. Suppose you were not told the mechanism. Describe how you could conclude from the structure of the alkyl halide, the nucleophile, and the solvent that each reaction is in fact an S_N2 reaction.

8.14 Treatment of 1,3-dichloropropane with potassium cyanide results in formation of 1,3-dicyanopropane. The rate of this reaction is about 1000 times greater in DMSO than it is in ethanol. Account for this difference in rate.

$$Cl{-}CH_2CH_2CH_2{-}Cl + 2K^+CN^- \longrightarrow NC{-}CH_2CH_2CH_2{-}CN + 2K^+Cl^-$$

1,3-Dichloropropane 1,3-Dicyanopropane

8.15 Treatment of 1-aminoadamantane, $C_{10}H_{17}N$, with methyl 2,4-dibromobutanoate involves two successive S_N2 reactions and gives compound A, an intermediate in the synthesis of carmantidine. Propose a structural formula for this intermediate. Carmantidine has been used in treating the spasms associated with Parkinson's disease.

(adamantyl)$-NH_2$ + $BrCH_2CH_2CH(Br)CO_2CH_3 \xrightarrow{R_3N} C_{15}H_{23}NO_2$

1-Aminoadamantane Methyl 2,4-dibromobutanoate A

8.16 Select the member of each pair that shows the faster rate of S_N2 reaction with KI in acetone.

(a) $CH_3CH_2CH_2CH_2Cl$ or $CH_3CH(CH_3)CH_2Cl$ **(b)** $CH_3CH_2CH_2CH_2Cl$ or $CH_3CH_2CH_2CH_2Br$

(c) $CH_3CH(CH_3)CH_2CH_2Cl$ or $CH_3C(CH_3)_2CH_2Cl$ (d) $CH_3CH_2CH_2CH(Br)CH_3$ or $CH_3CH(Br)CH(CH_3)CH_3$

8.17 Select the member of each pair that shows the faster rate of S_N2 reaction with KN_3 in acetone.

(a) bromocyclohexane or 1-bromo-1-methylcyclohexane (b) 1-bromo-4,4-dimethylcyclohexane (H_3C, H_3C) or 1-bromo-2,2-dimethylcyclohexane (Br, CH_3, CH_3)

8.18 What hybridization best describes the reacting carbon in the S_N2 transition state? Would electron-withdrawing groups or electron-donating groups stabilize the transition state better?

8.19 Each carbocation is capable of rearranging to a more stable carbocation. Limiting yourself to a single 1,2-shift, suggest a structure for the rearranged carbocation.

(a) $(CH_3)_2CH\overset{+}{C}HCH_3$ (b) $(CH_3)_3C\overset{+}{C}HCH_3$ (c) $CH_2{=}CHCH_2\overset{+}{C}HCH_2CH_3$

(d) $CH_3OCH_2\overset{+}{C}HC(CH_3)_3$ (e) $C_6H_5CH_2\overset{+}{C}HCH_3$ (f) cyclohexyl cation bearing a 1,1-dimethyl-substituted adjacent carbon (CH_3, CH_3)

8.20 Attempts to prepare optically active iodides by nucleophilic displacement on optically active compounds with I^- normally produce racemic alkyl iodides. Why are the product alkyl iodides racemic?

8.21 Draw a structural formula for the product of each S_N1 reaction. Where configuration of the starting material is given, show the configuration of the product.

(a) (S)-$CH_3CH(Cl)CH_2CH_3 + CH_3CH_2OH \xrightarrow{\text{ethanol}}$ (b) 1-chloro-1-methylcyclopentane (CH_3, Cl) $+ CH_3OH \xrightarrow{\text{methanol}}$

(c) $CH_3C(CH_3)_2Cl + CH_3COOH \xrightarrow{\text{acetic acid}}$ (d) 3-bromocyclohexene (—Br) $+ CH_3OH \xrightarrow{\text{methanol}}$

8.22 You were told that each reaction in the previous problem proceeds by an S_N1 mechanism. Suppose you were not told the mechanism. Describe how you could conclude from the structure of the alkyl halide, the nucleophile, and the solvent that each reaction is in fact an S_N1 reaction.

8.23 Vinylic halides, such as vinyl bromide, $CH_2{=}CHBr$, undergo neither S_N1 nor S_N2 reactions. What factors account for this lack of reactivity of vinylic halides?

8.24 Select the member of each pair that undergoes S_N1 solvolysis in aqueous ethanol more rapidly.

(a) $CH_3CH_2CH_2CH_2Cl$ or $CH_3C(CH_3)_2Cl$ (b) $CH_3C(CH_3)_2Cl$ or $CH_3C(CH_3)_2Br$

(c) $CH_2{=}CHCH_2Cl$ or $CH_3CH_2CH_2Cl$

(d) 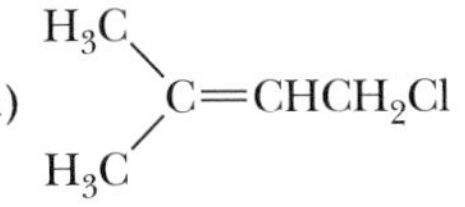 $(H_3C)_2C{=}CHCH_2Cl$ or $CH_2{=}CHCH_2Cl$

(e) $CH_3(CH_2)_3CH_2Cl$ or $CH_3(CH_2)_2CHClCH_3$ (f) 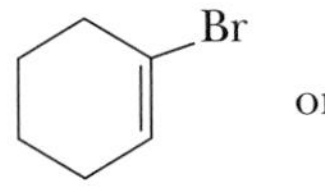 3-bromocyclohexene or 4-bromocyclohexene (structures; Br)

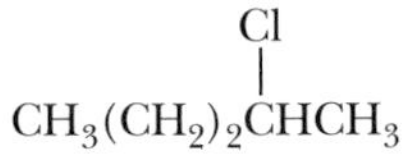

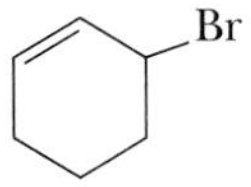

8.25 Account for the following relative rates of solvolysis under experimental conditions favoring S_N1 reaction.

	$CH_3\ddot{O}CH_2CH_2Cl$	$CH_3CH_2CH_2CH_2Cl$	$CH_3CH_2\ddot{O}CH_2Cl$
Relative rate of solvolysis (S_N1)	0.2	1	10^9

8.26 Not all tertiary halides undergo S_N1 reactions readily. For example, 1-iodobicyclo[2.2.2]octane is very unreactive under S_N1 conditions. What feature of this molecule is responsible for such lack of reactivity?

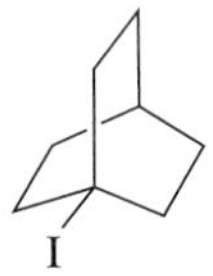

I

1-Iodobicyclo[2.2.2]octane

8.27 Show how you might synthesize the following compounds from an alkyl halide and a nucleophile:

(a) 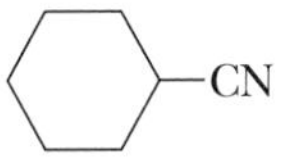 cyclohexyl—CN (b) 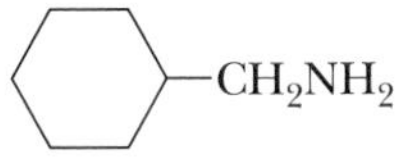 cyclohexyl—CH_2NH_2 (c) 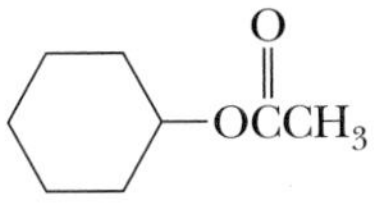 cyclohexyl—$OCCH_3$ (with C=O)

(d) $CH_3(CH_2)_3CH_2SH$ (e) $CH_3(CH_2)_5C{\equiv}CH$ (f) $CH_3CH_2OCH_2CH_3$

(g) 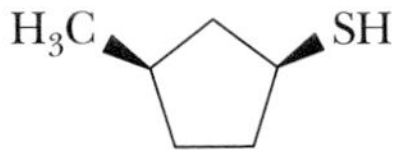 cyclopentane with H_3C and SH substituents (wedged)

8.28 3-Chloro-1-butene reacts with sodium ethoxide in ethanol to produce 3-ethoxy-1-butene. The rate of this reaction is second order; first order in 3-chloro-1-butene and first order in sodium ethoxide. In the absence of sodium ethoxide, 3-chloro-1-butene reacts with ethanol to produce both 3-ethoxy-1-butene and 1-ethoxy-2-butene. Explain these results.

8.29 1-Chloro-2-butene undergoes hydrolysis in warm water to give a mixture of these allylic alcohols. Propose a mechanism for their formation.

$$\underset{\text{1-Chloro-2-butene}}{CH_3CH{=}CHCH_2Cl} \xrightarrow{H_2O} \underset{\text{2-Buten-1-ol}}{CH_3CH{=}CHCH_2OH} + \underset{\text{3-Buten-2-ol}}{CH_3CH(OH)CH{=}CH_2}$$

8.30 In the following reaction, nucleophilic substitution occurs with rearrangement. Suggest a mechanism for formation of the observed product. If the starting material is optically active, the product is also optically active. (*Hint:* An intermediate, $C_8H_{18}NCl$, can, with care, be isolated from the reaction mixture. This intermediate is water soluble.)

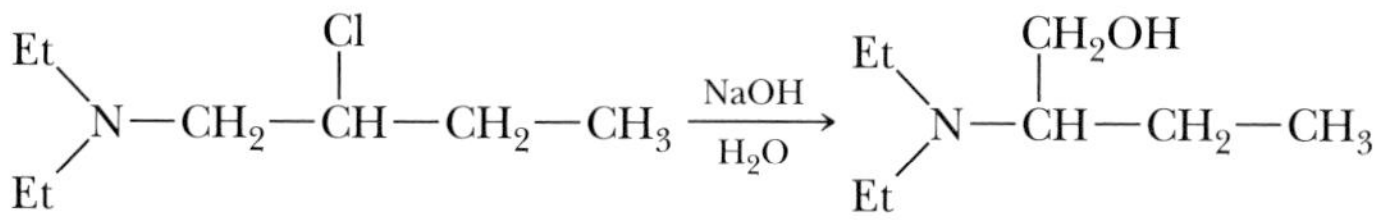

$$Et_2N{-}CH_2{-}CH(Cl){-}CH_2{-}CH_3 \xrightarrow[H_2O]{NaOH} Et_2N{-}CH(CH_2OH){-}CH_2{-}CH_3$$

8.31 Propose a mechanism for the formation of these products in the solvolysis of this alkyl bromide.

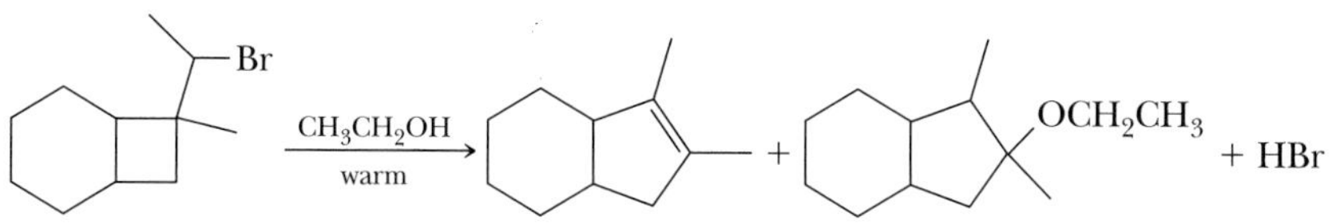

Br; CH_3CH_2OH, warm; + OCH_2CH_3 + HBr

8.32 Solvolysis of the following bicyclic compound in acetic acid gives a mixture of products, two of which are shown. The leaving group is the anion of a sulfonic acid, Ar—SO_3H. A sulfonic acid is a strong acid and its anion, $ArSO_3^-$, is a weak base and a good leaving group. Propose a mechanism for this reaction. (*Hint:* The connectivity of the carbons in the products is different from that in the starting bicyclic compound.)

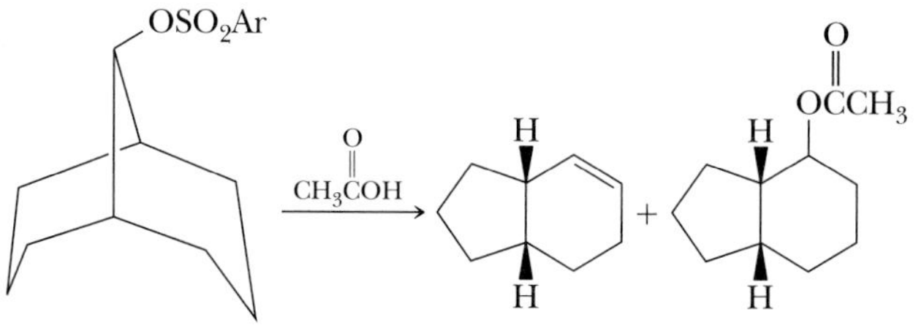

OSO_2Ar; $CH_3\overset{O}{\overset{\|}{C}}OH$; H, H; + H, $\overset{O}{\overset{\|}{OCCH_3}}$, H

8.33 Which compound in each set undergoes more rapid solvolysis when refluxed in ethanol? Show the major product formed from the more reactive compound.

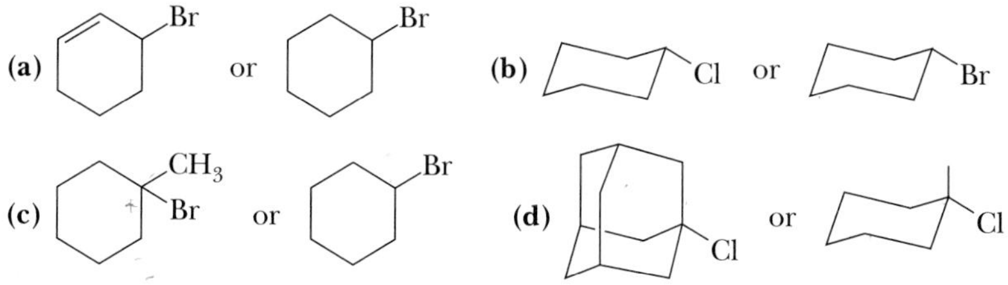

(a) Br or Br (b) Cl or Br

(c) CH_3, Br or Br (d) Cl or Cl

8.34 Account for the relative rates of solvolysis of these compounds in aqueous acetic acid.

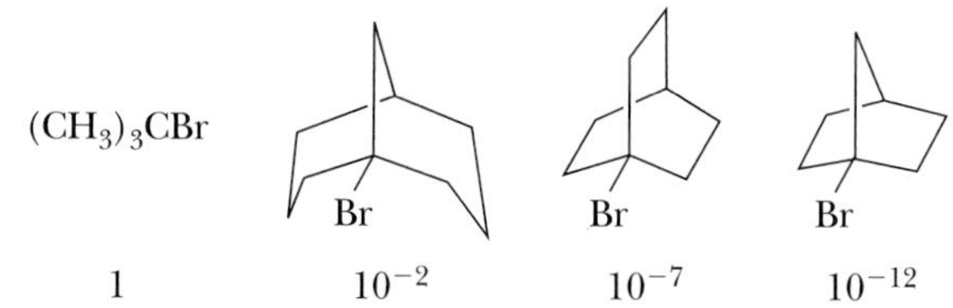

$(CH_3)_3CBr$	Br	Br	Br
1	10^{-2}	10^{-7}	10^{-12}

8.35 A comparison of the rates of S_N1 solvolysis of the bicyclic compounds (1) and (2) in acetic acid shows that compound (1) reacts 10^{11} times faster than compound (2). Furthermore, solvolysis of (1) occurs with complete retention of configuration: The nucleophile occupies the same position on the one-carbon bridge as did the leaving OSO_2Ar group.

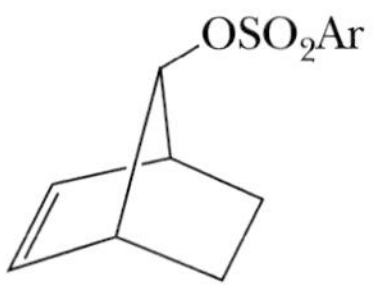

OSO_2Ar (1) OSO_2Ar (2)

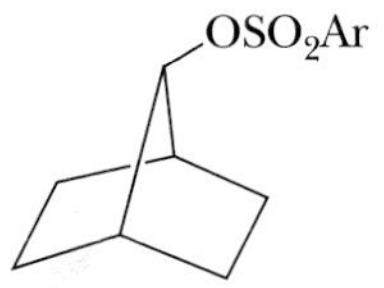

(a) Draw structural formulas for the products of solvolysis of each compound.
(b) Account for difference in rate of solvolysis between (1) and (2).
(c) Account for complete retention of configuration in the solvolysis of (1).

β-Eliminations

8.36 Draw structural formulas for the alkene(s) formed by treatment of each alkyl halide with sodium ethoxide in ethanol. Assume that elimination occurs by an E2 mechanism.

(a) $CH_3CH(Br)C(CH_3)_2CH_3$ **(b)** 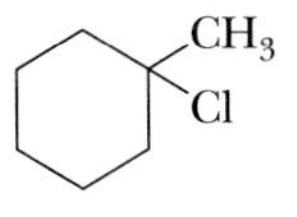 CH_3, Cl **(c)** CH_3, Cl **(d)** CH_3, Cl

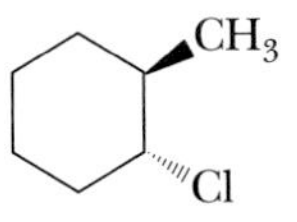

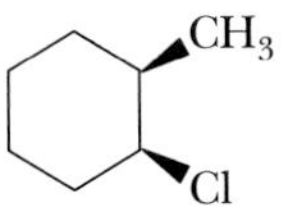

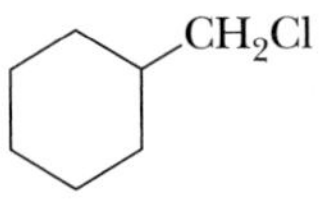

(e) CH_2Cl **(f)** $CH_2{=}CHCH_2C(CH_3)_2Br$

8.37 Draw the structural formulas of all chloroalkanes that undergo dehydrohalogenation when treated with KOH to give each alkene as the major product. For some parts, only one chloroalkane gives the desired alkene as the major product. For other parts, two chloroalkanes may give the desired alkene as the major product.

(a) 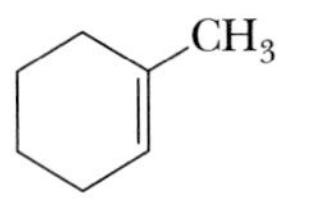 CH_3 **(b)** CH_2 **(c)** $CH_2{=}C(CH_3)CH(CH_3)CH_2CH_3$

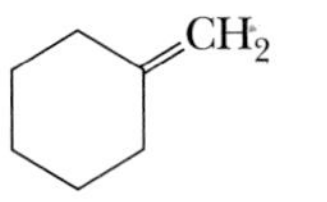

(d) $CH_3C(CH_3){=}C(CH_3)CH_2CH_3$ **(e)** $CH_3CH(CH_3)C(CH_3){=}CHCH_3$

8.38 Following are diastereomers (A) and (B) of 3-bromo-3,4-dimethylhexane. On treatment with sodium ethoxide in ethanol, each gives 3,4-dimethyl-3-hexene as the major product. One of these diastereomers gives the (*E*)-alkene, and the other gives the (*Z*)-alkene. Which diastereomer gives which alkene? Account for the stereoselectivity of each β-elimination.

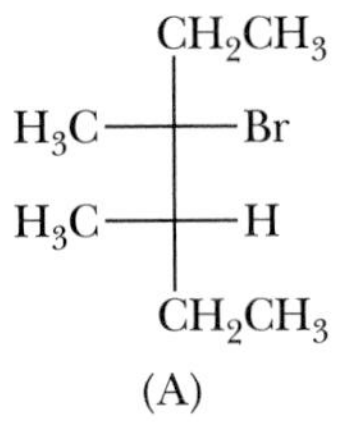

(A): CH_2CH_3; H_3C—Br; H_3C—H; CH_2CH_3

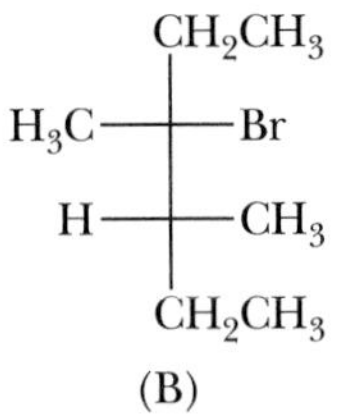

(B): CH_2CH_3; H_3C—Br; H—CH_3; CH_2CH_3

8.39 Treatment of the following stereoisomer (Fischer projection) of 1-bromo-2,3-diphenylpropane with sodium ethoxide in ethanol gives a single stereoisomer of 1,2-diphenylpropene. Knowing what you do about the stereoselectivity of E2 reactions, predict whether the product has the *E* configuration or the *Z* configuration.

$$\begin{array}{c} \text{Br} \\ | \\ C_6H_5 - \overset{|}{\underset{|}{C}} - H \\ C_6H_5 - \overset{|}{\underset{|}{C}} - H \\ | \\ CH_3 \end{array} \xrightarrow[CH_3CH_2OH]{CH_3CH_2O^-Na^+} C_6H_5CH{=}\overset{\overset{\displaystyle CH_3}{|}}{C}C_6H_5$$

1-Bromo-2,3-diphenylpropane → 1,2-Diphenylpropene

8.40 Elimination of HBr from 2-bromonorbornane gives only 2-norbornene and no 1-norbornene. How do you account for the regioselectivity of this dehydrohalogenation? In answering this question, you may find it helpful to make molecular models of both 1-norbornene and 2-norbornene and analyze the angle strain in each.

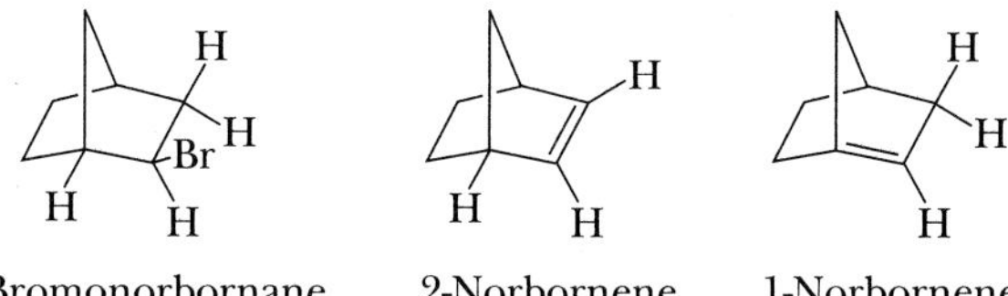

2-Bromonorbornane 2-Norbornene 1-Norbornene

8.41 Which compound reacts faster when refluxed with potassium *tert*-butoxide in *tert*-butyl alcohol, *cis*-1-bromo-3-isopropylcyclohexane or *trans*-1-bromo-3-isopropylcyclohexane? Draw the structure of the expected product from the faster reacting compound.

Substitution Versus Elimination

8.42 Consider the following statements in reference to S_N1, S_N2, E1, and E2 reactions of alkyl halides. To which mechanism(s), if any, does each statement apply?

(a) Involves a carbocation intermediate
(b) Is first-order in alkyl halide and first-order in nucleophile
(c) Involves inversion of configuration at the site of substitution
(d) Involves retention of configuration at the site of substitution
(e) Substitution at a stereocenter gives predominantly a racemic product
(f) Is first-order in alkyl halide and zero-order in base
(g) Is first-order in alkyl halide and first-order in base
(h) Is greatly accelerated in protic solvents of increasing polarity
(i) Rearrangements are common
(j) Order of reactivity is 3° > 2° > 1° > methyl
(k) Order of reactivity is methyl > 1° > 2° > 3°

8.43 Arrange these alkyl halides in order of increasing ratio of E2 to S_N2 products observed on reaction of each with sodium ethoxide in ethanol.

(a) CH_3CH_2Br **(b)** $CH_3\overset{\overset{\displaystyle CH_3}{|}}{C}HCH_2Br$ **(c)** $CH_3\underset{\underset{\displaystyle Cl}{|}}{\overset{\overset{\displaystyle CH_3}{|}}{C}}CH_2CH_3$ **(d)** $CH_3\overset{\overset{\displaystyle CH_3}{|}}{C}HCH_2CH_2Br$

8.44 Draw a structural formula for the major organic product of each reaction and specify the most likely mechanism for formation of the product you have drawn.

(a) (cyclohexene ring)$-Br + CH_3OH \xrightarrow[\text{methanol}]{}$ **(b)** $CH_3\underset{\underset{\displaystyle Cl}{|}}{\overset{\overset{\displaystyle CH_3}{|}}{C}}CH_2CH_3 + NaOH \xrightarrow[H_2O]{80°}$

(c) $(R)\text{-}CH_3\overset{Cl}{\overset{|}{C}}HCH_2CH_3 + CH_3\overset{O}{\overset{\|}{C}}O^-Na^+ \xrightarrow{DMSO}$

(d) 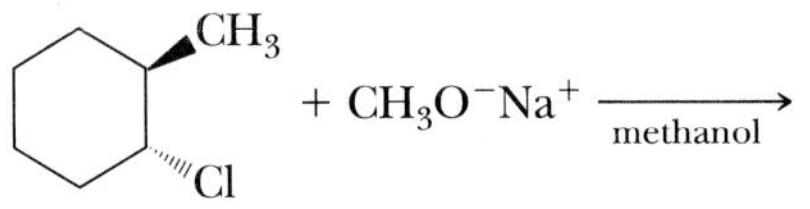 cyclohexane with CH_3 (wedge) and Cl (hash) $+ CH_3O^-Na^+ \xrightarrow{methanol}$ (e) 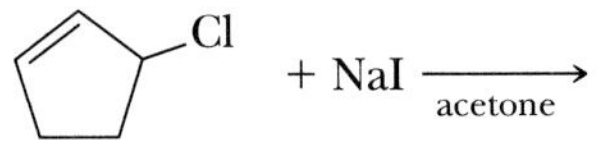 3-chlorocyclopentene $+ NaI \xrightarrow{acetone}$

R enantiomer

(f) $CH_3\overset{Cl}{\overset{|}{C}}HCH_2CH_3 + H\overset{O}{\overset{\|}{C}}OH \xrightarrow{formic\ acid}$

R enantiomer

(g) $CH_3CH_2ONa + CH_2{=}CHCH_2Cl \xrightarrow{ethanol}$

8.45 When *cis*-4-chlorocyclohexanol is treated with sodium hydroxide in ethanol, it gives only the substitution product *trans*-1,4-cyclohexanediol (1). Under the same reaction conditions, *trans*-4-chlorocyclohexanol gives 3-cyclohexenol (2) and the bicyclic ether (3).

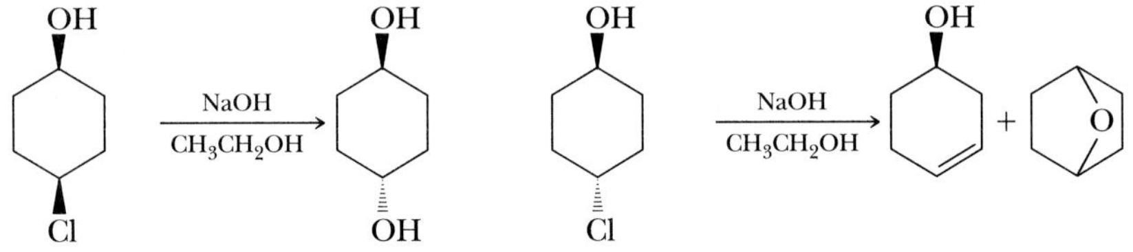

cis-4-Chlorocyclohexanol $\xrightarrow[CH_3CH_2OH]{NaOH}$ (1) (OH, OH)

trans-4-Chlorocyclohexanol $\xrightarrow[CH_3CH_2OH]{NaOH}$ (2) (OH) + (3) (O)

(a) Propose a mechanism for formation of product (1), and account for its configuration.
(b) Propose a mechanism for formation of product (2).
(c) Account for the fact that the bicyclic ether (3) is formed from the *trans* isomer but not from the *cis* isomer.

Synthesis

8.46 Show how to convert the given starting material into the desired product. Note that some syntheses require only one step whereas others require two or more steps.

(a) $CH_3\overset{CH_3}{\overset{|}{C}}HCH_2Cl \longrightarrow CH_3\overset{CH_3}{\overset{|}{C}}{=}CH_2$ (b) $CH_3\overset{CH_3}{\overset{|}{C}}{=}CH_2 \longrightarrow CH_3\overset{CH_3}{\overset{|}{C}}CH_3$ (with Br)

(c) $CH_3\overset{CH_3}{\overset{|}{C}}HCH_2Cl \longrightarrow CH_3\overset{CH_3}{\overset{|}{C}}CH_3$ (with OH) (d) 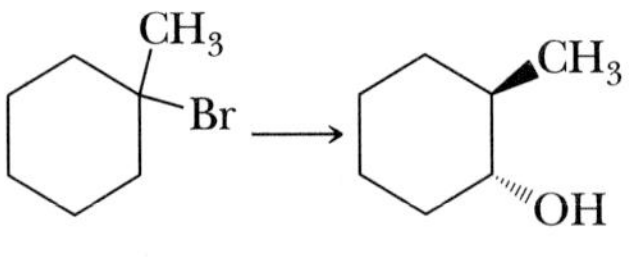 1-bromo-1-methylcyclohexane (CH_3, Br) $\longrightarrow$ cyclohexane with CH_3 (wedge) and OH (hash)

(e) 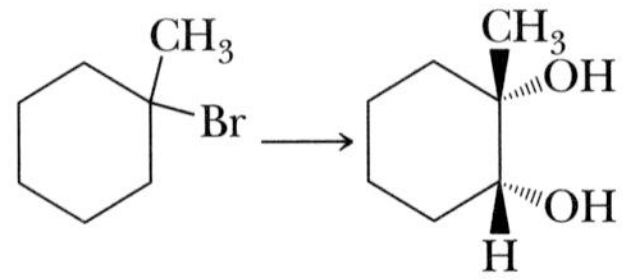 1-bromo-1-methylcyclohexane (CH_3, Br) $\longrightarrow$ cyclohexane with CH_3, OH, OH, H (f) 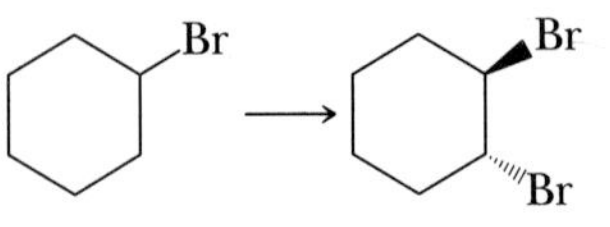 bromocyclohexane (Br) $\longrightarrow$ cyclohexane with Br (wedge) and Br (hash)

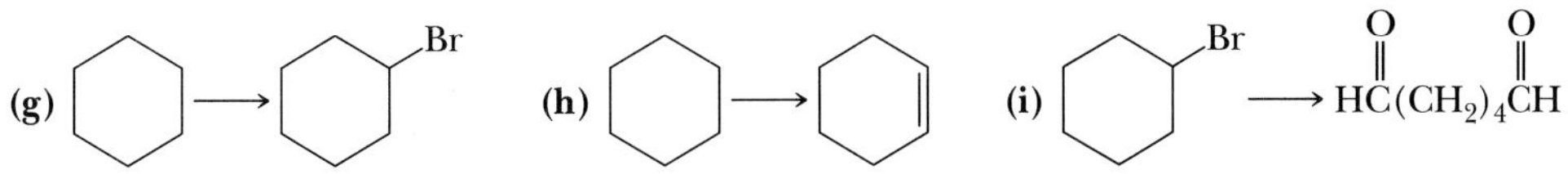

(g) [cyclohexane] ⟶ [bromocyclohexane, Br] (h) [cyclohexane] ⟶ [cyclohexene] (i) [bromocyclohexane, Br] ⟶ $HC(O)(CH_2)_4CH(O)$

8.47 The Williamson ether synthesis involves treatment of an alkyl halide with a metal alkoxide. Following are two reactions intended to give *tert*-butyl ethyl ether. One reaction gives the ether in good yield, the other reaction does not. Which reaction gives the ether? What is the product of the other reaction, and how do you account for its formation?

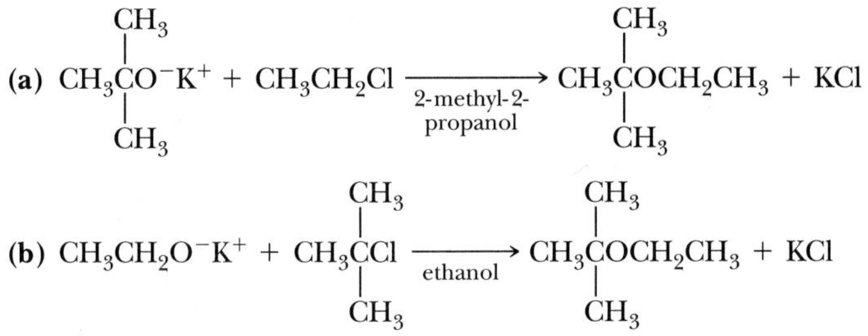

(a) $(CH_3)_3CO^-K^+ + CH_3CH_2Cl \xrightarrow{\text{2-methyl-2-propanol}} (CH_3)_3COCH_2CH_3 + KCl$

(b) $CH_3CH_2O^-K^+ + (CH_3)_3CCl \xrightarrow{\text{ethanol}} (CH_3)_3COCH_2CH_3 + KCl$

8.48 The following ethers can, in principle, be synthesized by two different combinations of alkyl halide and metal alkoxide. Show one combination of alkyl halide and alkoxide that forms ether bond (1) and another that forms ether bond (2). Which combination gives the higher yield of ether?

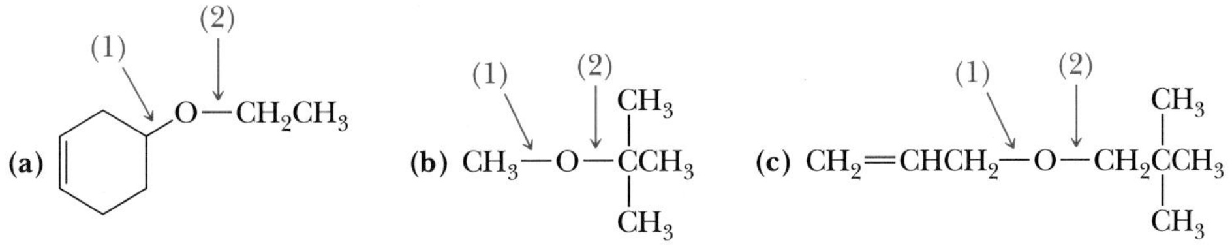

(a) [cyclohexenyl]–(1)–O–(2)–CH_2CH_3 (b) CH_3–(1)–O–(2)–$C(CH_3)_3$ (c) $CH_2{=}CHCH_2$–(1)–O–(2)–$CH_2C(CH_3)_3$

8.49 Propose a mechanism for this reaction.

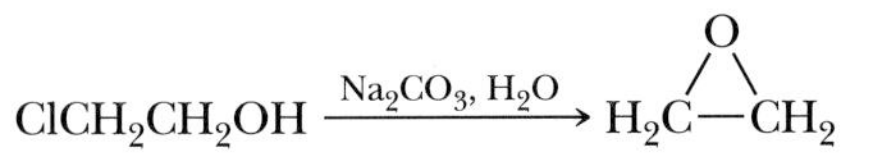

$$ClCH_2CH_2OH \xrightarrow{Na_2CO_3,\ H_2O} H_2C{-}CH_2 \text{ (epoxide, bridged by O)}$$

8.50 Each of these compounds can be synthesized by an S_N2 reaction. Suggest a combination of an alkyl halide and a nucleophile that will give each product.

(a) CH_3OCH_3 (b) CH_3SH (c) $CH_3CH_2CH_2PH_2$ (d) CH_3CH_2CN

(e) $CH_3SCH_2C(CH_3)_3$ (f) $(CH_3)_3NH^+\ Cl^-$ (g)  $C_6H_5C(=O)OCH_2C_6H_5$

(h)  (*R*)-$CH_3CH(N_3)CH_2CH_2CH_3$ (i) $CH_2{=}CHCH_2OCH(CH_3)_2$

(j) $CH_2{=}CHCH_2OCH_2CH{=}CH_2$ (k) 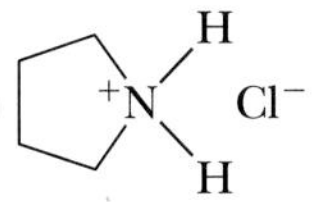 [pyrrolidinium, $^+NH_2$] Cl^- (l) 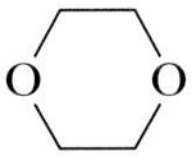 [1,4-dioxane, O ring O]

9

ALCOHOLS AND THIOLS

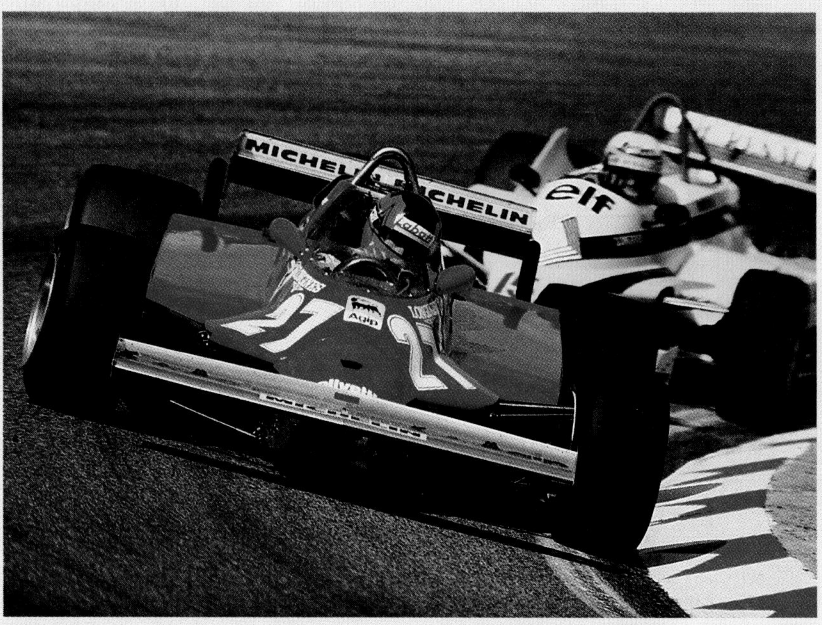

■ Methanol, CH_3OH, is used as a fuel in cars of the type that race in the Indianapolis 500. *(Thomas Zimmerman, © Tony Stone Images)*

In this chapter, we study the physical and chemical properties of alcohols, a class of oxygen-containing compounds. We also study thiols, a class of sulfur-containing compounds. A thiol is like an alcohol in structure, except that it contains an —SH group rather than an —OH group.

CH_3CH_2OH	CH_3CH_2SH
Ethanol	Ethanethiol
(an alcohol)	(a thiol)

Ethanol is the fuel additive in gasohol, the alcohol in alcoholic beverages, and an important industrial solvent. Ethanethiol, like all other low-molecular-weight thiols, has a stench; such smells from skunks, rotten eggs, and sewage are caused by thiols.

Alcohols are particularly important in the laboratory and biochemical transformations of organic compounds. They can be converted into many other types of compounds, such as alkenes, alkyl halides, aldehydes, ketones, carboxylic acids, and esters. Not only can alcohols be converted to

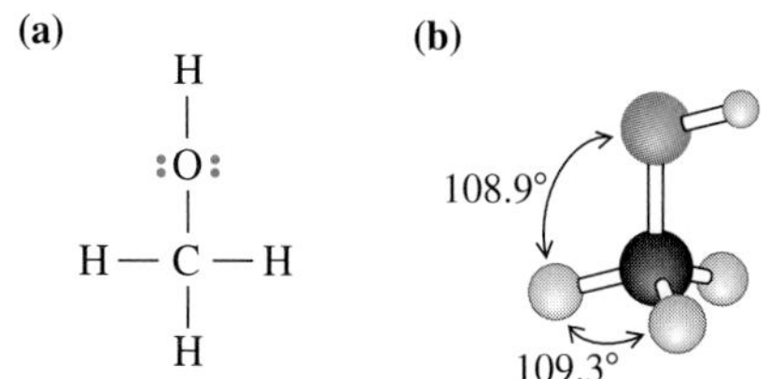

FIGURE 9.1
Methanol, CH_3OH. (a) Lewis structure and (b) ball-and-stick model.

these compounds, but alcohols also can be prepared from them. Thus, alcohols play a central role in the interconversion of organic functional groups.

Because sulfur and oxygen are both Group VI elements, thiols and alcohols undergo many of the same types of reactions. Sulfur, a third row element, however, can expand its valence shell to include more than eight electrons and, therefore, thiols undergo some reactions that are not possible for alcohols. In addition, sulfur's electronegativity and basicity are less than oxygen.

Alcohol A compound containing an —OH (hydroxyl) group bonded to an sp^3 hybridized carbon.

Thiol A compound containing an —SH (sulfhydryl) group bonded to an sp^3 hybridized carbon.

9.1 Structure of Alcohols and Thiols

A. Alcohols

The functional group of an alcohol is an **—OH (hydroxyl) group** bonded to an sp^3 hybridized carbon (Section 1.3A). The oxygen atom of an alcohol is also sp^3 hybridized. Two sp^3 hybrid orbitals of oxygen form sigma bonds to atoms of carbon and hydrogen, and the remaining two sp^3 hybrid orbitals each contain an unshared pair of electrons. Figure 9.1 shows a Lewis structure and a ball-and-stick model of methanol, CH_3OH, the simplest alcohol. The measured H—C—O bond angle in methanol is 108.9°, very close to the tetrahedral angle of 109.5°.

B. Thiols

The functional group of a thiol is an **—SH (sulfhydryl) group** bonded to an sp^3 hybridized carbon. Figure 9.2 shows a Lewis structure and a ball-and-stick model of methanethiol, CH_3SH, the simplest thiol. The C—S—H bond angle in methanethiol is 100.3°. By way of comparison, the H—S—H bond angle in H_2S is 93.3°. If a sulfur atom were bonded to two other atoms by sp^3 hybrid orbitals, bond angles about sulfur would be approximately 109.5°; if, instead, a sulfur atom were bonded to two other atoms by $3p$ orbitals, bond angles would be approximately 90°. The fact that the C—S—H bond angle in methanethiol is 100.3° and the H—S—H bond angle in H_2S is 93.3° indicates that there is considerably more p character to the bonding orbitals of divalent sulfur than there is to those of divalent oxygen.

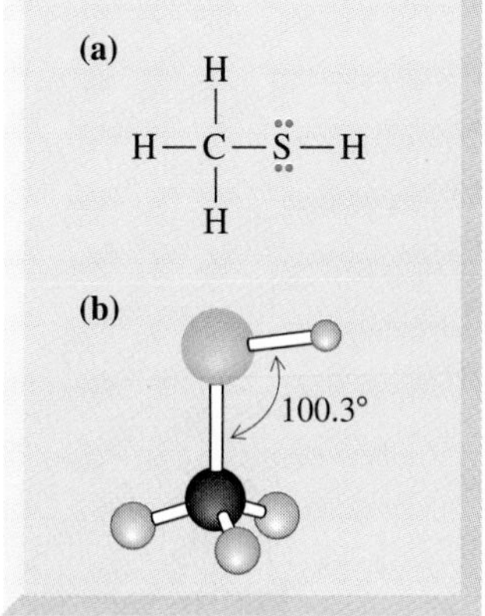

FIGURE 9.2
Methanethiol, CH_3SH.
(a) Lewis structure and (b) ball-and-stick model.

9.2 Nomenclature

A. Alcohols

In the IUPAC system, the longest chain of carbon atoms containing the —OH group is selected as the parent alkane and numbered from the end closer to —OH. To show that the compound is an alcohol, the suffix **-e** of the parent alkane is changed to **-ol**

(Section 2.5), and a number is added to show the location of the —OH group. The location of the —OH group takes precedence over alkyl groups and halogens in numbering the parent chain. For cyclic alcohols, numbering begins with the carbon bearing the —OH group unless there is some other numbering system that takes precedence, as, for example, the numbering of bicyclic rings.

Common names for alcohols are derived by naming the alkyl group attached to —OH and then adding the word "alcohol." Here are IUPAC names and, in parentheses, common names for several low-molecular-weight alcohols.

$CH_3CH_2CH_2OH$
1-Propanol
(Propyl alcohol)

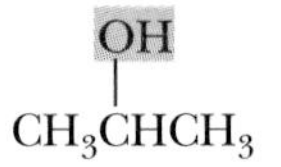

$CH_3CH(OH)CH_3$
2-Propanol
(Isopropyl alcohol)

$CH_3CH_2CH_2CH_2OH$
1-Butanol
(Butyl alcohol)

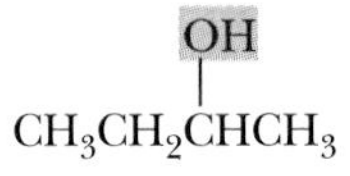

$CH_3CH_2CH(OH)CH_3$
2-Butanol
(*sec*-Butyl alcohol)

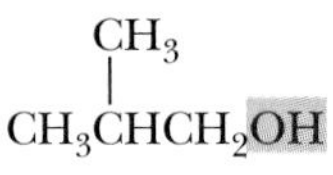

$CH_3CH(CH_3)CH_2OH$
2-Methyl-1-propanol
(Isobutyl alcohol)

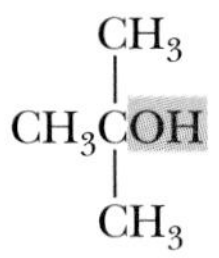

$CH_3C(CH_3)_2OH$
2-Methyl-2-propanol
(*tert*-Butyl alcohol)

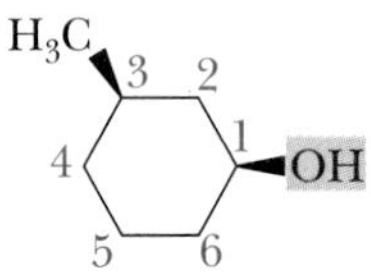

cis-3-Methylcyclohexanol

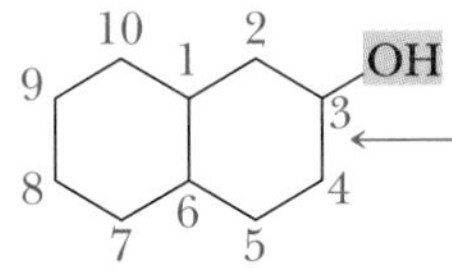

Bicyclo[4.4.0]-3-decanol

Numbering of the bicyclic ring takes precedence over the location of —OH

EXAMPLE 9.1

Write IUPAC names of these alcohols:

(a) 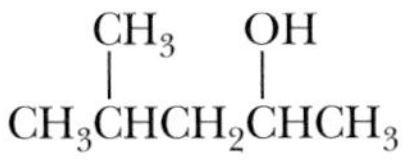 $CH_3CH(CH_3)CH_2CH(OH)CH_3$ (b) 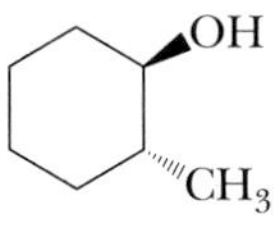 (c)

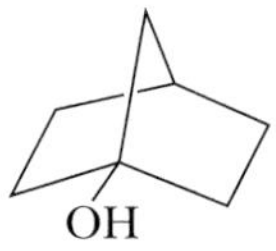

Solution

(a) 4-Methyl-2-pentanol (b) *trans*-2-Methylcyclohexanol
(c) Bicyclo[2.2.1]-1-heptanol

PROBLEM 9.1

Write IUPAC names of these alcohols:

(a) 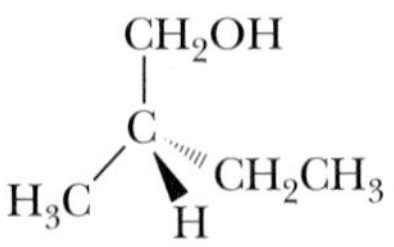 (b) 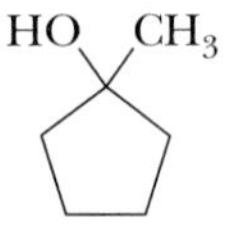 (c)

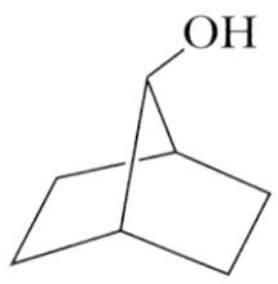

We classify alcohols as **primary (1°), secondary (2°),** or **tertiary (3°)** depending on whether the —OH group is on a primary carbon, a secondary carbon, or a tertiary carbon (Section 2.3C). Here are general formulas for 1°, 2°, and 3° alcohols.

$$\underset{\text{Primary (1°)}}{R{-}\overset{H}{\underset{H}{C}}{-}OH} \qquad \underset{\text{Secondary (2°)}}{R{-}\overset{R'}{\underset{H}{C}}{-}OH} \qquad \underset{\text{Tertiary (3°)}}{R{-}\overset{R'}{\underset{R''}{C}}{-}OH}$$

EXAMPLE 9.2

Classify each alcohol as primary, secondary, or tertiary.

(a) cyclohexyl–C(OH)(H)CH_3 (b) $CH_3\overset{CH_3}{\underset{CH_3}{C}}OH$ (c) cyclopentyl–CH_2OH

Solution

(a) Secondary (2°) **(b)** Tertiary (3°) **(c)** Primary (1°)

PROBLEM 9.2

Classify each alcohol as primary, secondary, or tertiary.

(a) $CH_3\overset{CH_3}{\underset{CH_3}{C}}CH_2OH$ **(b)** cyclopropyl–OH **(c)** $CH_2{=}CHCH_2OH$ **(d)** 1-methylcyclopentanol (ring with CH_3 and OH)

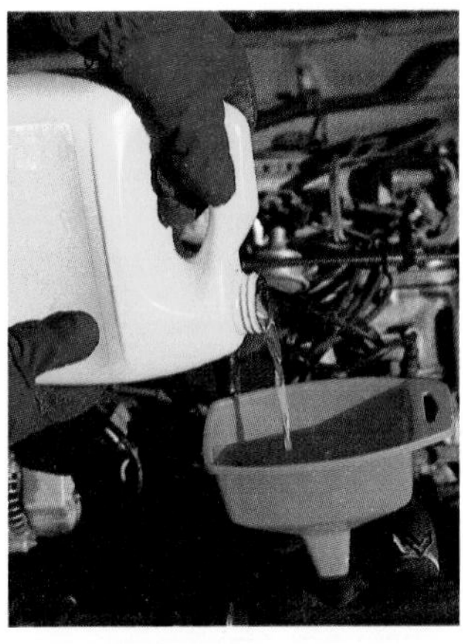

Ethylene glycol is the major component of commercial antifreeze. *(Dan McCoy, Rainbow)*

In the IUPAC system, a compound containing two hydroxyl groups is named as a **diol,** one containing three hydroxyl groups is named as a **triol,** and so on. In IUPAC names for diols, triols, and so forth, the final -e (the suffix) of the parent alkane name is retained, as, for example, in the name 1,2-ethanediol. As with many organic compounds, common names for certain diols and triols have persisted. Compounds containing two hydroxyl groups on adjacent carbons are often referred to as **glycols** (Section 6.5B). Ethylene glycol and propylene glycol are synthesized from ethylene and propylene, respectively, hence their common names.

Diol A compound containing two hydroxyl groups.

Glycol A compound containing two hydroxyl groups on adjacent carbons.

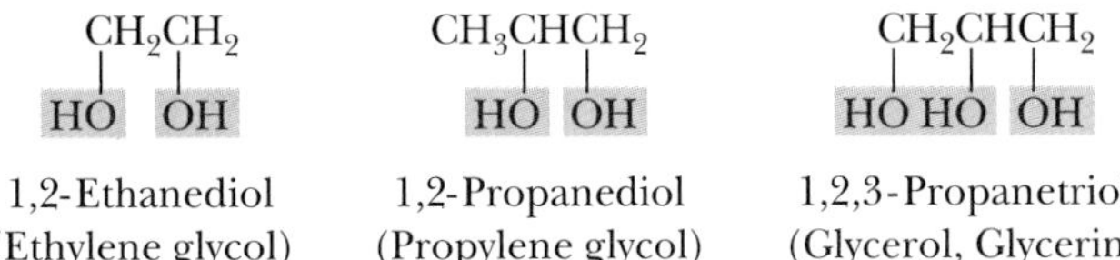

1,2-Ethanediol (Ethylene glycol) 1,2-Propanediol (Propylene glycol) 1,2,3-Propanetriol (Glycerol, Glycerin)

Compounds containing —OH and C=C groups are often referred to as **unsaturated alcohols,** because of the presence of the carbon-carbon double bond. In the IUPAC system, the double bond is shown by changing the infix of the parent alkane from -an- to -en- (Section 2.5), and the hydroxyl group is shown by changing the suffix of the parent alkane from -e to -ol. Numbers must be used to show the location of both the carbon-carbon double bond and the hydroxyl group. The parent alkane is numbered to give the —OH group the lowest possible number, that is, the group shown by the suffix (in this case, -ol) takes precedence over the group shown by an infix (in this case, -en-).

$$\overset{1}{HOCH_2}\text{ and H on }\overset{2}{C}=\overset{3}{C}\text{ (trans)}; \overset{2}{C}\text{–H}, \overset{3}{C}\text{–}\overset{4}{C}H_2\overset{5}{C}H_2\overset{6}{C}H_3$$

(*E*)-2-Hexene-1-ol
(*trans*-2-Hexen-1-ol)

EXAMPLE 9.3

Write IUPAC names for these unsaturated alcohols.

(a) $CH_2{=}CHCH_2OH$ **(b)** $CH_3CH_2(H)C{=}C(H)CH(OH)CH_3$ (cis)

Solution

(a) 2-Propen-1-ol. Its common name is allyl alcohol.
(b) (*Z*)-3-Hexen-2-ol (*cis*-3-Hexen-2-ol)

PROBLEM 9.3

Write IUPAC names for these unsaturated alcohols.

(a) $CH_2{=}CHCH_2CH_2OH$ **(b)** $CH_3CH(OH)CH{=}CH_2$

B. Thiols

Mercaptan A common name for a compound containing an —SH group.

The sulfur analog of an alcohol is called a **thiol** (thi- from the Greek, *theion*, sulfur) or, in the older literature, a **mercaptan,** which literally means "mercury capturing." Thiols react with Hg^{2+} in aqueous solution to give sulfide salts as insoluble precipitates. Thiophenol, C_6H_5SH, for example, gives $(C_6H_5S)_2Hg$.

According to the IUPAC system, thiols are named by selecting as the parent alkane the longest chain of carbon atoms that contains the —SH group. To show that the compound is a thiol, the final -e in the name of the parent alkane is retained and the suffix **-thiol** is added. The location of the —SH group takes precedence over alkyl groups and halogens in numbering the parent chain.

Common names for simple thiols are derived by naming the alkyl group attached to —SH and adding the word "mercaptan." Here are IUPAC names and, in parentheses, common names for several low-molecular-weight thiols.

$CH_3CH_2CH_2CH_2SH$	$CH_3CH_2CH(SH)CH_3$	$HSCH_2CH_2SH$
1-Butanethiol (Butyl mercaptan)	2-Butanethiol (*sec*-Butyl mercaptan)	1,2-Ethanedithiol (Ethylene dimercaptan)

In compounds containing other functional groups of higher precedence, the presence of an —SH group is indicated by the prefix **mercapto-**. According to the IUPAC system, —OH takes precedence over —SH in both numbering and naming.

$HSCH_2CH_2OH$

2-Mercaptoethanol

EXAMPLE 9.4

Write names for these thiols.

(a) $CH_3(CH_2)_3CH_2SH$ **(b)** $H_3C(H)C{=}C(H)CH_2SH$ (with H_3C and CH_2SH trans)

Solution

(a) 1-Pentanethiol (pentyl mercaptan) **(b)** (*E*)-2-Butene-1-thiol

PROBLEM 9.4

Write IUPAC names for these thiols.

(a) $(CH_3)_2CHCH_2CH_2SH$ **(b)** $H_3C(H)C{=}C(H)CH_2CH(SH)CH_3$ (with H_3C and $CH_2CH(SH)CH_3$ cis)

9.3 Physical Properties

A. Alcohols

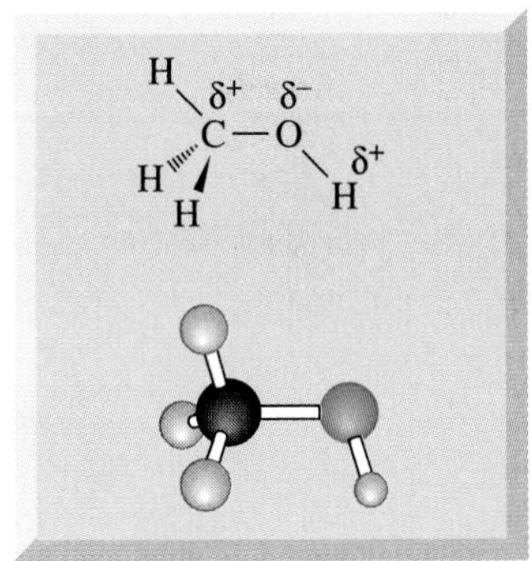

FIGURE 9.3 Polarity of the C—O—H bond in an alcohol.

Because of the presence of the polar —OH group, alcohols are polar compounds, with partial positive charges on carbon and hydrogen and a partial negative charge on oxygen (Figure 9.3).

The attraction between the positive end of one dipole and the negative end of another is called **dipole-dipole interaction.** When the positive end of one of the dipoles is a hydrogen atom bonded to F, O, or N, atoms of high electronegativity, and the negative end of the other dipole is an F, O, or N atom, the attractive interaction between dipoles is particularly strong and is given the special name of **hydrogen bonding.** The length of a hydrogen bond in water is 1.77 Å, about 80% longer than an O—H covalent bond. The strength of a hydrogen bond in water is approximately 5 kcal/mol (21 kJ/mol). For comparison, the strength of the O—H covalent bond in water is approximately 119 kcal/mol (498 kJ/mol). As can be seen by comparing these

Dipole-dipole interaction The attraction between the positive end of one dipole and the negative end of another.

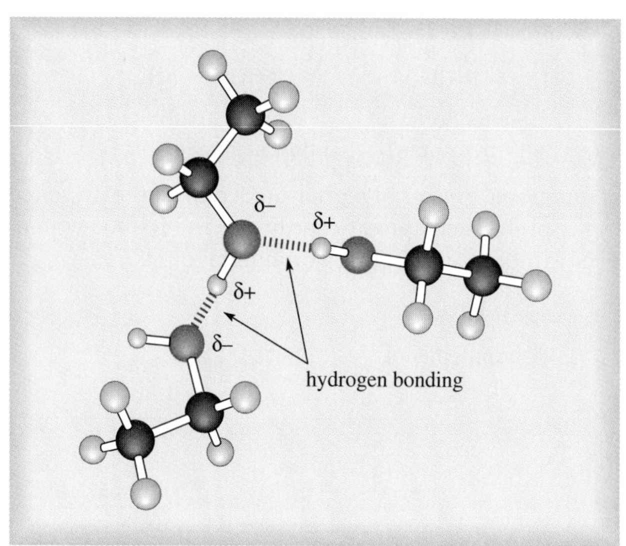

FIGURE 9.4
The association of ethanol molecules in the liquid state. Each O—H can participate in up to three hydrogen bonds (one through hydrogen and two through oxygen). Only two of these three possible hydrogen bonds per molecule are shown here. Hydrogen bonding gives alcohols an added attractive force between their molecules.

Hydrogen bonding The attractive interaction between dipoles when the positive end of one of the dipoles is a hydrogen atom bonded to F, O, or N and the negative end of the other dipole is an F, O, or N atom.

numbers, an O--------H hydrogen bond is considerably weaker than an O—H covalent bond. The presence of a large number of hydrogen bonds in liquid water, however, has an important effect on the physical properties of water. Because of hydrogen bonding, extra energy is required to separate each hydrogen-bonded water molecule from its neighbors, hence the relatively high boiling point of water. Similarly, there is extensive hydrogen bonding between alcohol molecules in the pure liquid. Figure 9.4 shows the association of ethanol molecules by hydrogen bonding between the partially negative oxygen atom of one ethanol molecule and the partially positive hydrogen atom of another ethanol molecule.

Table 9.1 lists the boiling points and solubilities in water for several groups of alcohols and hydrocarbons of similar molecular weight. Of the compounds compared in each group, the alcohols have the higher boiling points because more energy is needed to overcome the attractive forces of hydrogen bonding between their polar —OH groups. The presence of additional hydroxyl groups in a molecule further increases the extent of hydrogen bonding, as can be seen by comparing the boiling points of hexane (bp 69°C), 1-pentanol (bp 138°C), and 1,4-butanediol (bp 230°C), all of which have approximately the same molecular weight. Because of increased dispersion forces between larger molecules, boiling points of all types of compounds, including alcohols, increase with increasing molecular weight. To see this, compare the boiling points of ethanol, 1-propanol, 1-butanol, and 1-pentanol.

The effect of hydrogen bonding in alcohols is illustrated dramatically by comparing the boiling points of ethanol (bp 78°C) and its constitutional isomer dimethyl ether (bp −24°C). The difference in boiling point between these two compounds is due to the presence of a polar O—H group in the alcohol, which is capable of forming intermolecular hydrogen bonds. This hydrogen bonding increases the attractive forces between molecules of ethanol, and, thus, ethanol has a higher boiling point than dimethyl ether.

CH_3CH_2OH	CH_3OCH_3
Ethanol	Dimethyl ether
bp 78°C	bp −24°C

TABLE 9.1 Boiling Points and Solubilities in Water of Five Groups of Alcohols and Hydrocarbons of Similar Molecular Weight

Structural Formula	Name	Molecular Weight	bp (°C)	Solubility in Water
CH_3OH	methanol	32	65	infinite
CH_3CH_3	ethane	30	−89	insoluble
CH_3CH_2OH	ethanol	46	78	infinite
$CH_3CH_2CH_3$	propane	44	−42	insoluble
$CH_3CH_2CH_2OH$	1-propanol	60	97	infinite
$CH_3CH_2CH_2CH_3$	butane	58	0	insoluble
$CH_3CH_2CH_2CH_2OH$	1-butanol	74	117	8 g/100 g
$CH_3CH_2CH_2CH_2CH_3$	pentane	72	36	insoluble
$HOCH_2CH_2CH_2CH_2OH$	1,4-butanediol	90	230	infinite
$CH_3CH_2CH_2CH_2CH_2OH$	1-pentanol	88	138	2.3 g/100 g
$CH_3CH_2CH_2CH_2CH_2CH_3$	hexane	86	69	insoluble

Because alcohols can interact by hydrogen bonding with water, they are more soluble in water than are alkanes and alkenes of comparable molecular weight. Methanol, ethanol, and 1-propanol are soluble in water in all proportions. As molecular weight increases, the physical properties of alcohols become more like those of hydrocarbons of comparable molecular weight. Alcohols of higher molecular weight are much less soluble in water because of the increase in size of the hydrocarbon portion of the molecule.

EXAMPLE 9.5

Following are three alcohols of molecular formula $C_4H_{10}O$. Their boiling points, listed from lowest to highest, are 82.3°C, 99.5°C, and 117°C. Which alcohol has which boiling point?

$$CH_3CH_2CH_2CH_2OH \qquad CH_3CH_2\underset{}{\overset{\overset{\Large OH}{|}}{C}}HCH_3 \qquad CH_3\overset{\overset{\Large CH_3}{|}}{\underset{\underset{\Large CH_3}{|}}{C}}OH$$

1-Butanol — 2-Butanol — 2-Methyl-2-propanol

Solution

Boiling points of these constitutional isomers depend on the strength of intermolecular hydrogen bonding. The primary —OH group of 1-butanol is the most accessible for intermolecular hydrogen bonding; this alcohol has the highest boiling point, 117°C. The tertiary —OH group of 2-methyl-2-propanol is the least accessible for intermolecular hydrogen bonding; this alcohol has the lowest boiling point, 82.3°C.

Ethylene glycol is a polar compound and dissolves readily in water. *(Charles D. Winters)*

PROBLEM 9.5

Arrange these compounds in order of increasing boiling point.

$CH_3CH_2CH_2CH_2CH_2CH_3$ $HOCH_2CH_2CH_2CH_2OH$ $CH_3CH_2CH_2CH_2CH_2OH$

EXAMPLE 9.6

Arrange these compounds in order of increasing solubility in water.

$CH_3CH_2CH_2CH_2CH_2CH_3$ $HOCH_2CH_2CH_2CH_2OH$ $CH_3CH_2CH_2CH_2CH_2OH$

Solution

Hexane, C_6H_{14}, a nonpolar hydrocarbon, has the lowest solubility in water. Both 1-pentanol and 1,4-butanediol are polar compounds due to the presence of —OH groups, and each interacts with water molecules by hydrogen bonding. Because 1,4-butanediol has more sites within its molecule for hydrogen bonding than 1-pentanol, it is more soluble in water than 1-pentanol. The water solubilities of these compounds are given in Table 9.1.

$CH_3CH_2CH_2CH_2CH_2CH_3$	$CH_3CH_2CH_2CH_2CH_2OH$	$HOCH_2CH_2CH_2CH_2OH$
Insoluble	2.3 g/100 g water	Infinitely soluble

PROBLEM 9.6

Arrange these compounds in order of increasing solubility in water.

$ClCH_2CH_2Cl$ $CH_3CH_2CH_2OH$ $CH_3CH_2CH_2CH_2OH$

The scent of the spotted skunk is a mixture of two thiols, 3-methyl-1-butanethiol and 2-butene-1-thiol. *(Animals, Animals © E. R. Degginger)*

B. Thiols

The most outstanding physical characteristic of low-molecular-weight thiols is their stench. The scent of skunks is due primarily to two thiols:

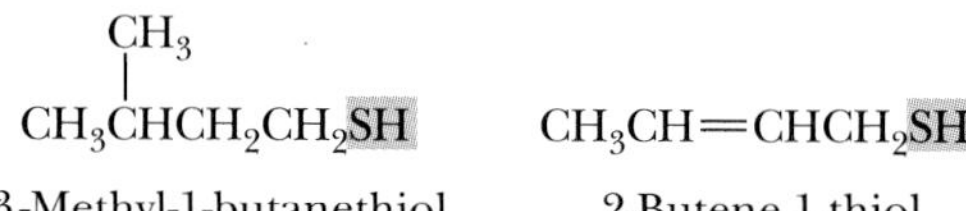

3-Methyl-1-butanethiol 2-Butene-1-thiol

Because of the very low polarity of the S—H bond, thiols show little association by hydrogen bonding. Consequently, they have lower boiling points and are less soluble in water and other polar solvents than alcohols of comparable molecular weights. Table 9.2 gives names and boiling points for three low-molecular-weight thiols. Shown

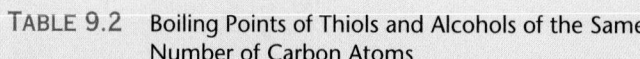

TABLE 9.2 Boiling Points of Thiols and Alcohols of the Same Number of Carbon Atoms

Thiol	bp (°C)	Alcohol	bp (°C)
methanethiol	6	methanol	65
ethanethiol	35	ethanol	78
1-butanethiol	98	1-butanol	117

for comparison are boiling points for alcohols of the same number of carbon atoms. Earlier we illustrated the importance of hydrogen bonding in alcohols by comparing the boiling points of ethanol (bp 78°C) and its constitutional isomer dimethyl ether (bp −24°C). By comparison, the boiling point of ethanethiol is 35°C, and that of its constitutional isomer dimethyl sulfide is 37°C. The fact that the boiling points of these constitutional isomers are almost identical indicates that little or no association by hydrogen bonding occurs between thiol molecules.

CH_3CH_2SH	CH_3SCH_3
Ethanethiol	Dimethyl sulfide
bp 35°C	bp 37°C

9.4 Acidity and Basicity of Alcohols

In dilute aqueous solution, alcohols are very weakly acidic as shown by the ionization of methanol.

$$CH_3\ddot{O}{-}H + :\ddot{O}{-}H \rightleftharpoons CH_3\ddot{O}:^- + H{-}\ddot{O}^+{-}H \qquad K_a = \frac{[CH_3O^-][H_3O^+]}{[CH_3OH]} = 15.5$$

Shown in Table 9.3 are acid ionization constants for several low-molecular-weight alcohols. Methanol and ethanol are about as acidic as water. Higher-molecular-weight, water-soluble alcohols are slightly weaker acids than water. Thus, while alcohols have some acidity, they are not strong enough acids to react with weak bases such as sodium bicarbonate or sodium carbonate. (At this point, it would be wise to review Section 3.4 and the discussion of the position of equilibrium in acid-base reactions.)

For simple alcohols, such as methanol and ethanol, acidity depends primarily on the degree of solvation and stabilization of the alkoxide ion by water molecules. The negatively charged oxygen atoms of the methoxide and ethoxide ions are almost equally accessible for solvation as is the hydroxide ion and, hence, these alcohols are about as acidic as water. As the bulk of the alkyl group attached to oxygen increases, the ability of water molecules to solvate the alkoxide ion decreases. *tert*-Butyl alcohol is a weaker acid than either methanol or ethanol, primarily because of the bulk of the

TABLE 9.3 pK_a Values for Selected Alcohols in Dilute Aqueous Solution*

Compound	Structural Formula	pK_a	
hydrogen chloride	HCl	−7	Stronger acid
acetic acid	CH_3CO_2H	4.8	↑
methanol	CH_3OH	15.5	
water	H_2O	**15.7**	
ethanol	CH_3CH_2OH	15.9	
2-propanol	$(CH_3)_2CHOH$	17	Weaker
2-methyl-2-propanol	$(CH_3)_3COH$	18	acid

*Also given for comparison are pK_a values for water, acetic acid, and hydrogen chloride.

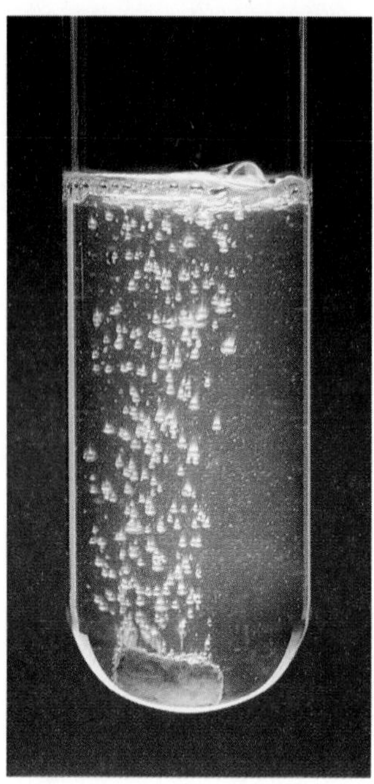

Sodium metal reacts with methanol with evolution of hydrogen gas. *(Charles D. Winters)*

tert-butyl group, which reduces solvation of the *tert*-butoxide anion by surrounding water molecules.

In the presence of strong acids, the oxygen atom of an alcohol behaves as a base and reacts with an acid by proton transfer to form an oxonium ion.

$$CH_3CH_2-\ddot{O}-H + H-\overset{+}{\ddot{O}}(H)-H \xrightleftharpoons{H_2SO_4} CH_3CH_2-\overset{+}{\ddot{O}}(H)-H + :\ddot{O}(H)-H$$

Ethanol; Hydronium ion (pK_a −1.7); Ethyloxonium ion (pK_a −2.4)

Thus, we see that alcohols can function as both weak acids and weak bases.

9.5 Reaction with Active Metals

Like water, alcohols react with Li, Na, K, and other active metals to liberate hydrogen and form alkoxide salts. In this oxidation/reduction reaction, Na is oxidized to Na^+ and methanol is reduced to methoxide ion and H_2.

$$2CH_3OH + 2Na \longrightarrow 2CH_3O^-Na^+ + H_2$$

Sodium methoxide

To name salts of alcohols, name the cation first, followed by the name of the anion. The name of the anion is derived from the prefix showing the number of carbon atoms and their arrangement (meth-, eth-, isoprop-, *tert*-but-, and so on) followed by the suffix -oxide.

Alkoxide ions are somewhat stronger bases than the hydroxide ion. In addition to sodium methoxide, the following metal salts of alcohols are commonly used in organic reactions requiring a strong base in a nonaqueous solvent, as, for example sodium ethoxide in ethanol and potassium *tert*-butoxide in 2-methyl-2-propanol (*tert*-butyl alcohol).

$$CH_3CH_2O^-Na^+ \qquad CH_3\underset{CH_3}{\overset{CH_3}{C}}O^-K^+$$

Sodium ethoxide; Potassium *tert*-butoxide

Alcohols can also be converted to salts by reaction with bases stronger than alkoxide ions. One such base is sodium hydride, NaH. Hydride ion, $H:^-$, the conjugate base of H_2, is an extremely strong base.

$$CH_3CH_2OH + Na^+H^- \longrightarrow CH_3CH_2O^-Na^+ + H_2$$

Ethanol; Sodium hydride; Sodium ethoxide

Reactions of sodium hydride with compounds containing acidic hydrogens are irreversible and driven to completion by the formation of H_2, which is given off as a gas.

EXAMPLE 9.7

Write a balanced equation for the reaction of cyclohexanol with sodium metal.

Solution

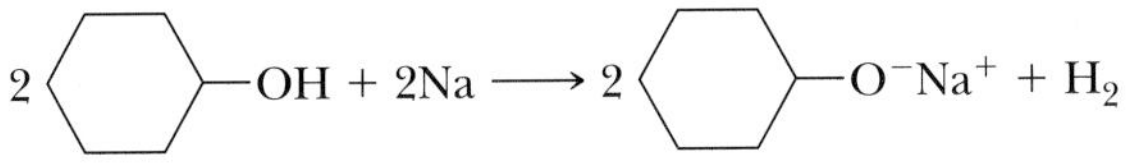

$$2\ C_6H_{11}\text{—OH} + 2Na \longrightarrow 2\ C_6H_{11}\text{—O}^-Na^+ + H_2$$

PROBLEM 9.7

Predict the position of equilibrium for this acid-base reaction. (*Hint:* Review Section 3.4.)

$$CH_3CH_2O^-Na^+ + CH_3\overset{O}{\overset{\|}{C}}OH \rightleftharpoons CH_3CH_2OH + CH_3\overset{O}{\overset{\|}{C}}O^-Na^+$$

9.6 Conversion to Alkyl Halides

Conversion of an alcohol to an alkyl halide involves substitution of halogen for —OH at a saturated carbon. The most common reagents for this conversion are the halogen acids (HCl, HBr, and HI) and the inorganic halides PBr_3 and $SOCl_2$.

A. Reaction with HCl, HBr, and HI

Tertiary alcohols react very rapidly with HCl, HBr, and HI. Mixing a tertiary alcohol, for example, 2-methyl-2-propanol, with concentrated HCl for a few minutes at room temperature results in conversion of the alcohol to 2-chloro-2-methylpropane. Reaction is evident by formation of a water-insoluble chloroalkane that separates from the aqueous layer. Low-molecular-weight, water-soluble primary and secondary alcohols are unreactive under these conditions.

$$(CH_3)_3COH + HCl \xrightarrow{25°C} (CH_3)_3CCl + H_2O$$

2-Methyl-2-propanol → 2-Chloro-2-methylpropane

Water-insoluble tertiary alcohols are converted to tertiary halides by bubbling gaseous HX through a solution of the alcohol dissolved in diethyl ether or tetrahydrofuran (THF).

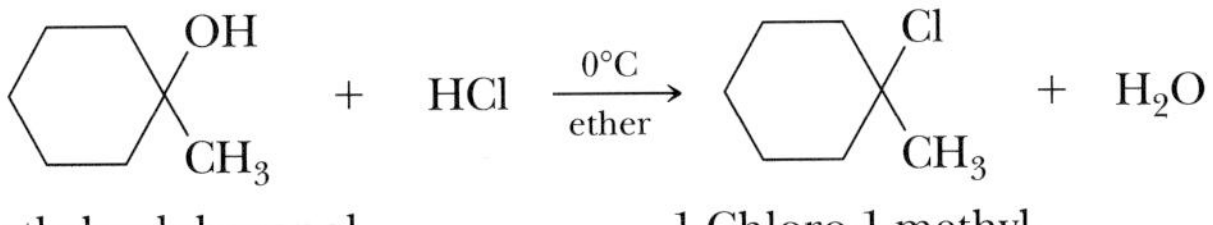

$$\text{1-Methylcyclohexanol} + HCl \xrightarrow[\text{ether}]{0°C} \text{1-Chloro-1-methylcyclohexane} + H_2O$$

1-Methylcyclohexanol → 1-Chloro-1-methylcyclohexane

Primary and secondary alcohols react only slowly with HCl.

Primary and secondary alcohols are converted to bromoalkanes and iodoalkanes by treatment with HBr and HI. For example, when heated to reflux with concentrated HBr, 1-butanol is converted smoothly to 1-bromobutane.

$$CH_3CH_2CH_2CH_2OH + HBr \xrightarrow[\text{reflux}]{H_2O} CH_3CH_2CH_2CH_2Br + H_2O$$

1-Butanol → 1-Bromobutane

Secondary alcohols generally give at least some rearranged product, evidence for the formation of carbocation intermediates during their reaction. For example, reaction of 3-pentanol with HBr gives 3-bromopentane as the major product, along with some 2-bromopentane.

$$\underset{\text{3-Pentanol}}{CH_3CH_2\underset{|}{\overset{}{C}}\!\!\underset{OH}{H}CH_2CH_3} + HBr \xrightarrow[\text{heat}]{} \underset{\text{3-Bromopentane (major product)}}{CH_3CH_2\underset{Br}{CH}CH_2CH_3} + \underset{\text{2-Bromopentane}}{CH_3CH_2CH_2\underset{Br}{CH}CH_3} + H_2O$$

(a product of rearrangement: 2-Bromopentane)

Primary alcohols with extensive β-branching give large amounts of a product derived from rearrangement. For example, treatment of 2,2-dimethyl-1-propanol (neopentyl alcohol) with HBr gives a rearranged product almost exclusively.

$$\underset{\text{2,2-Dimethyl-1-propanol}}{CH_3\overset{CH_3}{\underset{CH_3}{\overset{\beta}{C}}}\overset{\alpha}{C}H_2OH} + HBr \longrightarrow \underset{\text{2-Bromo-2-methylbutane (a product of rearrangement)}}{CH_3\overset{CH_3}{\underset{Br}{C}}CH_2CH_3} + H_2O$$

Based on observations of the relative ease of reaction of alcohols with HX ($3° > 2° > 1°$) and the occurrence of rearrangements, chemists have concluded that conversion of tertiary and secondary alcohols to haloalkanes by concentrated HX occurs by an S_N1 mechanism and involves formation of a carbocation intermediate. Step 1 is a rapid, reversible proton transfer from HX (or H_3O^+ when aqueous acid is used) to the hydroxyl group of the alcohol to form an oxonium ion. This is followed in Step 2, the rate-limiting step, by loss of a molecule of water to give a carbocation intermediate. The carbocation intermediate then reacts with halide ion in Step 3 to give the product. Note that the acid in this reaction converts OH^-, a poor leaving group, into H_2O, a better leaving group.

MECHANISM Conversion of a Tertiary Alcohol to a Tertiary Alkyl Chloride: An S_N1 Reaction

Step 1:

$$\underset{\substack{\text{2-Methyl-2-propanol}\\(\textit{tert}\text{-Butyl alcohol})}}{CH_3-\overset{CH_3}{\underset{CH_3}{C}}-\ddot{\underset{..}{O}}-H} + H-\overset{+}{\ddot{O}}-H \;(\text{with } H \text{ below O}) \xrightleftharpoons[\text{reversible}]{\text{rapid and}} \underset{\text{An oxonium ion}}{CH_3-\overset{CH_3}{\underset{CH_3}{C}}-\overset{+}{O}{<}^{H}_{H}} + :\ddot{O}-H \;(\text{with } H \text{ below O})$$

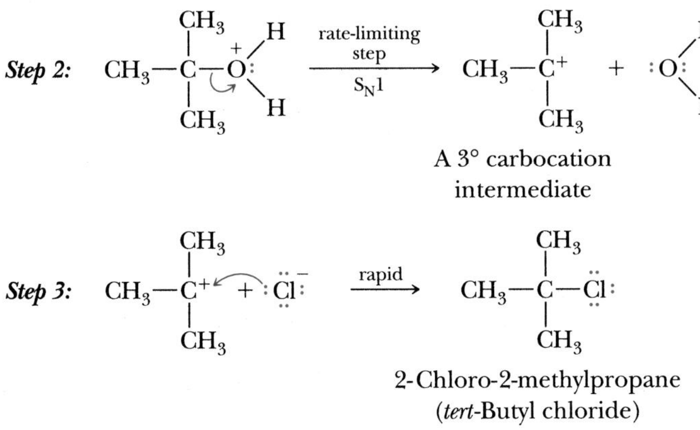

Step 2: $CH_3-\overset{CH_3}{\underset{CH_3}{C}}-\overset{+}{O}H_2 \xrightarrow[S_N1]{\text{rate-limiting step}} CH_3-\overset{CH_3}{\underset{CH_3}{C^+}} + :\ddot{O}H_2$

A 3° carbocation intermediate

Step 3: $CH_3-\overset{CH_3}{\underset{CH_3}{C^+}} + :\ddot{Cl}:^- \xrightarrow{\text{rapid}} CH_3-\overset{CH_3}{\underset{CH_3}{C}}-\ddot{Cl}:$

2-Chloro-2-methylpropane (*tert*-Butyl chloride)

Primary alcohols react with HX by an S_N2 mechanism. Step 1 is a rapid, reversible reaction of the hydroxyl group with H_3O^+ to form an oxonium ion. In Step 2, the rate-limiting step, halide ion reacts at the carbon bearing the oxonium ion to displace H_2O and form the C—X bond.

MECHANISM Conversion of a Primary Alcohol to a Primary Alkyl Halide: An S_N2 Reaction

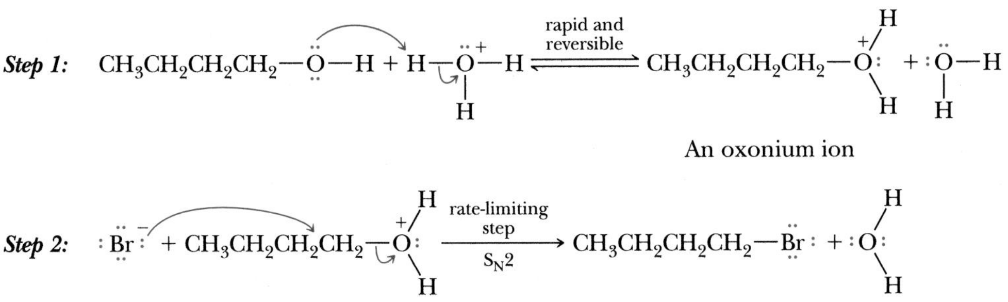

Step 1: $CH_3CH_2CH_2CH_2-\ddot{O}-H + H-\overset{+}{\ddot{O}}(H)-H \underset{}{\overset{\text{rapid and reversible}}{\rightleftharpoons}} CH_3CH_2CH_2CH_2-\overset{+}{O}H_2 + :\ddot{O}H_2$

An oxonium ion

Step 2: $:\ddot{Br}:^- + CH_3CH_2CH_2CH_2-\overset{+}{O}H_2 \xrightarrow[S_N2]{\text{rate-limiting step}} CH_3CH_2CH_2CH_2-\ddot{Br}: + :\ddot{O}H_2$

For primary alcohols with extensive β-branching, such as 2,2-dimethyl-1-propanol, direct displacement of OH_2 from the primary carbon is difficult, if not impossible. Furthermore, formation of a primary carbocation is also difficult, if not impossible. It is believed that primary alcohols with extensive β-branching react by a mechanism involving formation of an intermediate 3° carbocation by simultaneous (concerted) loss of H_2O and migration of an alkyl group. As illustrated with 2,2-dimethyl-1-propanol (neopentyl alcohol), Step 1 involves formation of an oxonium ion. Two changes take place simultaneously in Step 2. First, the C—O bond is cleaved and OH_2 is eliminated. Second, a methyl group and its pair of electrons migrates to the site occupied by the departing OH_2 group. The result of these molecular changes is elimination of a molecule of water and formation of a tertiary carbocation, which then reacts with chloride ion to form the product. Because the slow, rate-limiting step of this nucleophilic substitution involves only one reactant, namely the protonated alcohol, it is classified as an S_N1 reaction.

MECHANISM Rearrangement on Treatment of Neopentyl Alcohol with HX

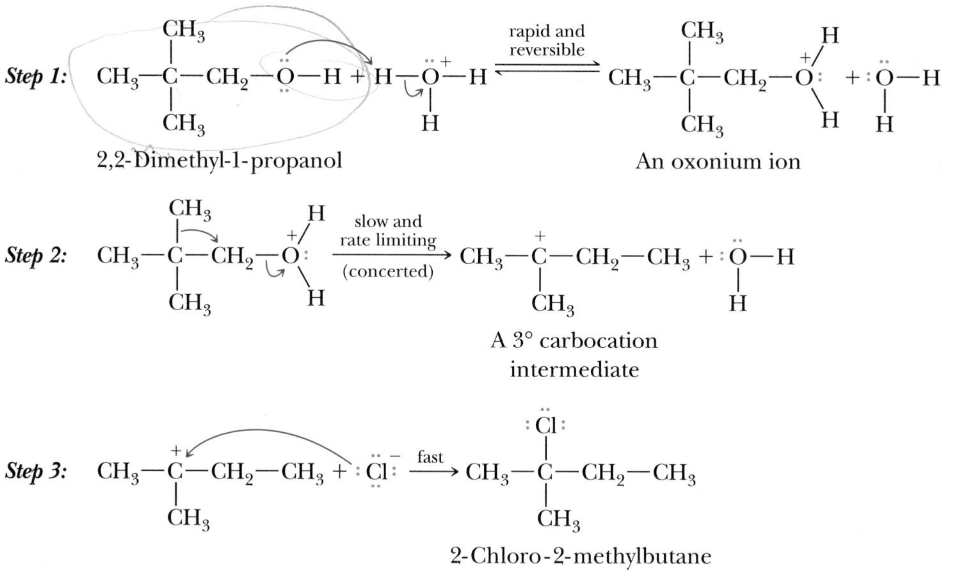

Step 1: $(CH_3)_3C{-}CH_2{-}\ddot{O}{-}H + H{-}\overset{+}{O}H_2 \underset{\text{reversible}}{\overset{\text{rapid and}}{\rightleftharpoons}} (CH_3)_3C{-}CH_2{-}\overset{+}{O}H_2 + :\ddot{O}H_2$

2,2-Dimethyl-1-propanol — An oxonium ion

Step 2: $(CH_3)_3C{-}CH_2{-}\overset{+}{O}H_2 \xrightarrow[\text{(concerted)}]{\text{slow and rate limiting}} CH_3{-}\overset{+}{C}(CH_3){-}CH_2{-}CH_3 + :\ddot{O}H_2$

A 3° carbocation intermediate

Step 3: $CH_3{-}\overset{+}{C}(CH_3){-}CH_2{-}CH_3 + :\ddot{Cl}:^- \xrightarrow{\text{fast}} CH_3{-}C(Cl)(CH_3){-}CH_2{-}CH_3$

2-Chloro-2-methylbutane

In summary, preparation of haloalkanes by treatment of ROH with HX is most useful for primary and tertiary alcohols. Because of the possibility of rearrangement, this process is less useful for secondary alcohols (except for simple cycloalkanols) and for primary alcohols with branching on the β-carbon.

B. Reaction with Phosphorus Tribromide

An alternative method for the synthesis of bromoalkanes from 1° and 2° alcohols is through the use of phosphorus tribromide, PBr_3. This method of preparation of bromoalkanes takes place under milder conditions than treatment with HBr. Although rearrangement sometimes occurs with PBr_3, the extent is considerably less than that with HBr, especially when the reaction mixture is kept at or below 0°C.

$$3(CH_3)_2CHCH_2OH + PBr_3 \xrightarrow{0^\circ C} 3(CH_3)_2CHCH_2Br + H_3PO_3$$

2-Methyl-1-propanol (Isobutyl alcohol) + Phosphorus tribromide → 1-Bromo-2-methylpropane (Isobutyl bromide) + Phosphorous acid

When treated with PBr_3, an alcohol is converted in Step 1 to a protonated dibromophosphite, $ROPBr_2$. By this reaction, OH^-, a poor leaving group, is transformed into a derivative of phosphorous acid, a good leaving group. S_N2 displacement by Br^- in Step 2 gives the RBr. The other two bromine atoms on phosphorus are replaced similarly, giving three moles of RBr and one of phosphorous acid, H_3PO_3.

MECHANISM Reaction of a Primary Alcohol with PBr_3

Step 1: A Lewis acid-base reaction to form a protonated dibromophosphite group

$$R-CH_2-\ddot{O}-H + Br-\ddot{P}(Br)-Br \longrightarrow R-CH_2-\overset{+}{\ddot{O}}(H)-PBr_2 + Br^-$$

—OH, a poor leaving group, is converted into this good leaving group

Step 2: Nucleophilic displacement of the protonated dibromophosphite group

$$Br^- + R-CH_2-\overset{+}{\ddot{O}}(H)-\ddot{P}Br_2 \longrightarrow R-CH_2-Br + H\ddot{O}-\ddot{P}Br_2$$

C. Reaction with Thionyl Chloride

Probably the most widely used reagent for the conversion of primary and secondary alcohols to alkyl chlorides is thionyl chloride, $SOCl_2$. The byproducts of the reaction are HCl and SO_2. Thionyl chloride converts primary and secondary alcohols to alkyl chlorides in good yield, usually with no rearrangement.

$$\underset{\text{1-Heptanol}}{CH_3(CH_2)_5CH_2OH} + \underset{\text{Thionyl chloride}}{SOCl_2} \xrightarrow{\text{pyridine}} \underset{\text{1-Chloroheptane}}{CH_3(CH_2)_5CH_2Cl} + SO_2 + HCl$$

Reactions are most commonly carried out in the presence of pyridine (Section 21.6C) or a tertiary amine such as triethylamine $(CH_3CH_2)_3N$ (Section 21.2). The function of the amine (a weak base) is to neutralize the HCl generated during the reaction and in this way prevent unwanted side reactions.

$$\underset{\text{Pyridine}}{C_5H_5\ddot{N}} + H-Cl \longrightarrow \underset{\text{Pyridinium chloride}}{C_5H_5\overset{+}{N}-H\ Cl^-}$$

A particular value of the reaction of alcohols with thionyl chloride is that it is stereoselective; it proceeds with inversion of configuration. Reaction of thionyl chloride with (S)-2-octanol, for example, in the presence of a tertiary amine proceeds with inversion of configuration and gives (R)-2-chlorooctane.

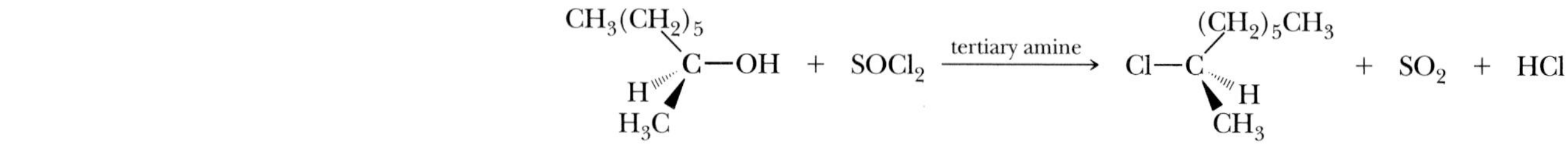

(S)-2-Octanol + Thionyl chloride $\xrightarrow{\text{tertiary amine}}$ (R)-2-Chlorooctane + SO_2 + HCl

In Steps 1 and 2, the alcohol reacts with thionyl chloride to form an **alkyl chlorosulfite.** This reaction is analogous to the first step of the reaction of an alcohol with PBr_3. If the reaction between the alcohol and thionyl chloride is carried out at 0°C or below, alkyl chlorosulfites can be isolated. In Step 3, chloride ion (a moderately good nucleophile) displaces the chlorosulfite group as SO_2 and Cl^- (both good leaving groups) in an S_N2 reaction with inversion of configuration.

MECHANISM Conversion of an Alcohol to an Alkyl Chloride by Thionyl Chloride

Step 1: Nucleophilic displacement of chlorine by oxygen

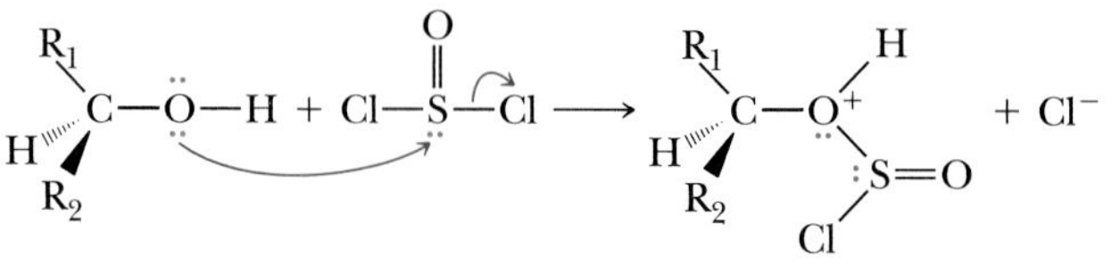

Thionyl chloride

Step 2: Proton transfer to the tertiary amine to form an alkyl chlorosulfite

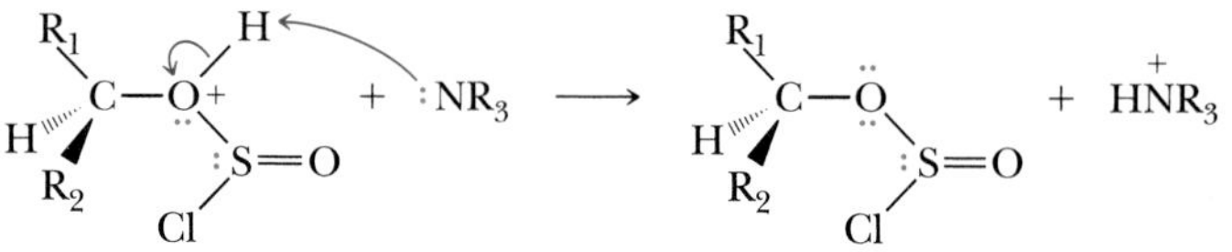

A tertiary amine

An alkyl chlorosulfite

Step 3: Backside attack of chloride ion and decomposition of the chlorosulfite

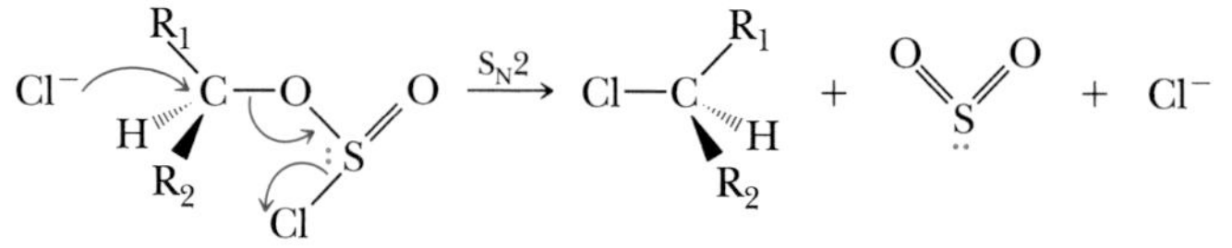

D. Formation of Alkyl Sulfonates

As we have just seen, alcohols react with thionyl chloride with displacement of chloride ion to form alkyl chlorosulfites. They also react with compounds called sulfonyl chlorides to form alkyl sulfonates. Sulfonyl chlorides are derived from sulfonic acids, compounds that are very strong acids, comparable in strength to sulfuric acid. What is important for us at this point is that a sulfonate anion is a very weak base and stable anion and, therefore, a very good leaving group in nucleophilic substitution reactions.

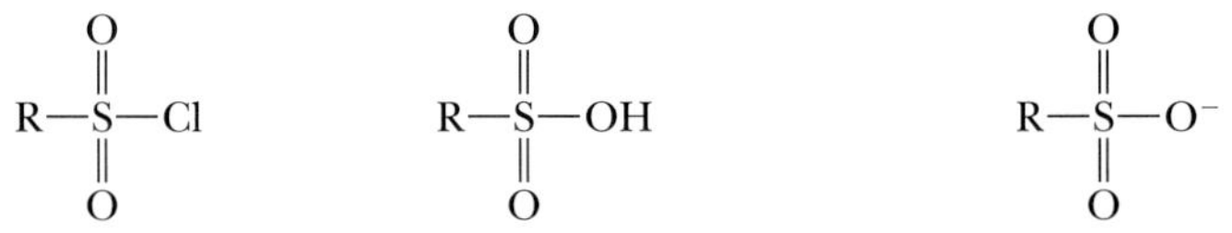

$$\mathrm{R{-}S(=O)_2{-}Cl} \qquad \mathrm{R{-}S(=O)_2{-}OH} \qquad \mathrm{R{-}S(=O)_2{-}O^-}$$

A sulfonyl chloride

A sulfonic acid (a very strong acid)

A sulfonate anion (a very weak base and stable anion; a very good leaving group)

Two of the most commonly used sulfonyl chlorides are *p*-toluenesulfonyl chloride (abbreviated tosyl chloride, Ts-Cl) and methanesulfonyl chloride (abbreviated mesyl chloride, Ms-Cl). Treatment of ethanol with *p*-toluenesulfonyl chloride in the presence of pyridine gives ethyl *p*-toluenesulfonate (tosylate). Pyridine is added to neutralize HCl formed as a byproduct. Cyclohexanol is converted to cyclohexyl methanesulfonate (mesylate) by a similar reaction of cyclohexanol with methanesulfonyl chloride. In each case, reaction involves breaking the O—H bond of the alcohol; it does not affect the C—O bond in any way. If the carbon bearing the —OH group is a stereocenter, sulfonate ester formation takes place with retention of configuration.

$$\mathrm{CH_3CH_2OH + Cl{-}S(=O)_2{-}C_6H_4{-}CH_3 \xrightarrow{pyridine} CH_3CH_2O{-}S(=O)_2{-}C_6H_4{-}CH_3 + HCl}$$

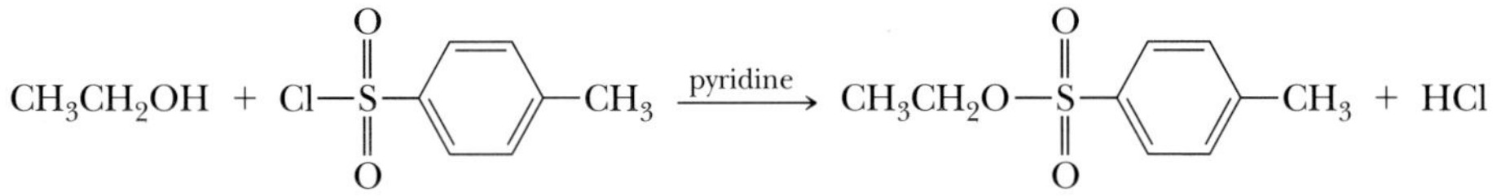

Ethanol — *p*-Toluenesulfonyl chloride — Ethyl *p*-toluenesulfonate (Ethyl tosylate)

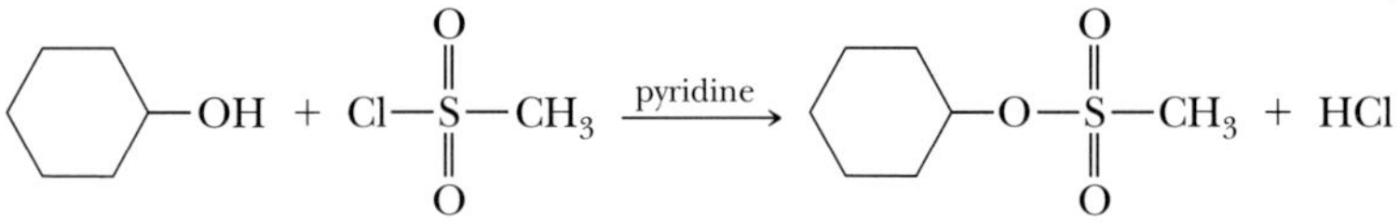

$$\mathrm{C_6H_{11}{-}OH + Cl{-}S(=O)_2{-}CH_3 \xrightarrow{pyridine} C_6H_{11}{-}O{-}S(=O)_2{-}CH_3 + HCl}$$

Cyclohexanol — Methanesulfonyl chloride — Cyclohexyl methanesulfonate (Cyclohexyl mesylate)

A particular advantage of sulfonate esters is that, through their use, a hydroxyl group, a very poor leaving group, can be converted to a tosylate or mesylate group, often shown as OTs and OMs, respectively, both very good leaving groups that are readily displaced by nucleophilic substitution.

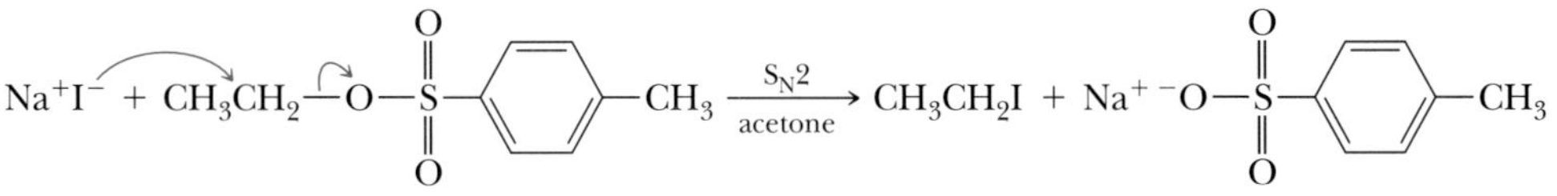

$$\mathrm{Na^+I^- + CH_3CH_2{-}O{-}S(=O)_2{-}C_6H_4{-}CH_3 \xrightarrow[acetone]{S_N2} CH_3CH_2I + Na^{+\,-}O{-}S(=O)_2{-}C_6H_4{-}CH_3}$$

Ethyl *p*-toluenesulfonate — *p*-Toluenesulfonate anion (a very good leaving group)

Following is a two-step sequence for conversion of (*S*)-2-octanol to (*R*)-2-octyl acetate via a tosylate. The first step involves cleavage of the O—H bond and proceeds with retention of configuration at the stereocenter. The second step involves nucleophilic displacement of tosylate by acetate ion, and proceeds with inversion of configuration at the stereocenter.

$$CH_3(CH_2)_5(H)(CH_3)C{-}OH + Cl{-}Ts \xrightarrow{\text{pyridine}} CH_3(CH_2)_5(H)(CH_3)C{-}OTs + HCl$$

(*S*)-2-Octanol (*S*)-2-Octyl tosylate

$$CH_3\overset{O}{\overset{\|}{C}}O^-Na^+ + CH_3(CH_2)_5(H)(CH_3)C{-}OTs \xrightarrow[\text{ethanol}]{S_N2} CH_3\overset{O}{\overset{\|}{C}}O{-}C(CH_2)_5CH_3(H)(CH_3) + Na^+OTs^-$$

(*S*)-2-Octyl tosylate (*R*)-2-Octyl acetate

EXAMPLE 9.8

Show how to convert *trans*-4-methylcyclohexanol to *cis*-1-iodo-4-methylcyclohexane via a tosylate.

Solution

Treat the alcohol with *p*-toluenesulfonyl chloride in pyridine to form a tosylate with retention of configuration. Then treat the tosylate with sodium iodide in acetone. The S_N2 reaction with inversion of configuration gives the product.

$$CH_3{-}C_6H_{10}{-}OH \xrightarrow[\text{pyridine}]{Cl{-}Ts} CH_3{-}C_6H_{10}{-}OTs \xrightarrow[\text{acetone}]{NaI} CH_3{-}C_6H_{10}{-}I$$

PROBLEM 9.8

Show how you might convert (*R*)-2-butanol to (*S*)-2-butanethiol via a tosylate.

$$CH_3CH_2\overset{OH}{\overset{|}{C}}HCH_3 \xrightarrow{?} CH_3CH_2\overset{SH}{\overset{|}{C}}HCH_3$$

(*R*)-2-Butanol (*S*)-2-Butanethiol

9.7 Acid-Catalyzed Dehydration of Alcohols

Dehydration Elimination of water.

An alcohol can be converted to an alkene by elimination of a molecule of water from adjacent carbon atoms. Elimination of water is called **dehydration.** In the laboratory, dehydration of an alcohol is most often brought about by heating it with either 85% phosphoric acid or concentrated sulfuric acid. Primary alcohols are the most difficult to dehydrate and generally require heating in concentrated sulfuric acid at temperatures as high as 180°C. Secondary alcohols undergo acid-catalyzed dehydration at somewhat lower temperatures. Acid-catalyzed dehydration of tertiary alcohols often requires temperatures only slightly above room temperature.

$$CH_3CH_2OH \xrightarrow[180°C]{H_2SO_4} CH_2{=}CH_2 + H_2O$$

$$\text{Cyclohexanol} \xrightarrow[140°C]{H_2SO_4} \text{Cyclohexene} + H_2O$$

$$(CH_3)_3COH \xrightarrow[50°C]{H_2SO_4} CH_3C(CH_3){=}CH_2 + H_2O$$

2-Methyl-2-propanol (*tert*-butyl alcohol) → 2-Methylpropene (Isobutylene)

Thus, the ease of acid-catalyzed dehydration of alcohols is in this order:

1° alcohol 2° alcohol 3° alcohol

Ease of dehydration of alcohols →

When isomeric alkenes are obtained in acid-catalyzed dehydration of an alcohol, the alkene having the greater number of substituents on the double bond (the more stable alkene) generally predominates (Zaitsev's rule, Section 8.8).

$$CH_3CH_2CH(OH)CH_3 \xrightarrow[heat]{85\%\ H_3PO_4} CH_3CH{=}CHCH_3 + CH_3CH_2CH{=}CH_2$$

2-Butanol → 2-Butene (80%) + 1-Butene (20%)

EXAMPLE 9.9

Draw structural formulas for the alkenes formed on acid-catalyzed dehydration of these alcohols. Where isomeric alkenes are possible, predict which alkene is the major product.

(a) $CH_3CH(CH_3)CH(OH)CH_3 \xrightarrow{\text{acid-catalyzed dehydration}}$

(b) 2-methylcyclopentanol $\xrightarrow{\text{acid-catalyzed dehydration}}$

Solution

(a) Elimination of H_2O from carbons 2–3 gives 2-methyl-2-butene; elimination of H_2O from carbons 1–2 gives 3-methyl-1-butene. 2-Methyl-2-butene, with three alkyl groups (three methyl groups) on the double bond, is the major product. 3-Methyl-1-butene, with only one alkyl group (an isopropyl group) on the double bond, is the minor product.

$$\overset{4}{C}H_3\overset{3}{C}H(CH_3)\overset{2}{C}H(OH)\overset{1}{C}H_3 \xrightarrow{\text{acid-catalyzed dehydration}} CH_3C(CH_3){=}CHCH_3 + CH_3CH(CH_3)CH{=}CH_2 + H_2O$$

3-Methyl-2-butanol → 2-Methyl-2-butene (major product) + 3-Methyl-1-butene

migration of an atom or group, with its pair of electrons, from the β-carbon to the carbon bearing the positive charge. The driving force for rearrangements of this type is conversion of a less stable carbocation to a more stable carbocation. Proton transfer to H_2O gives the alkenes.

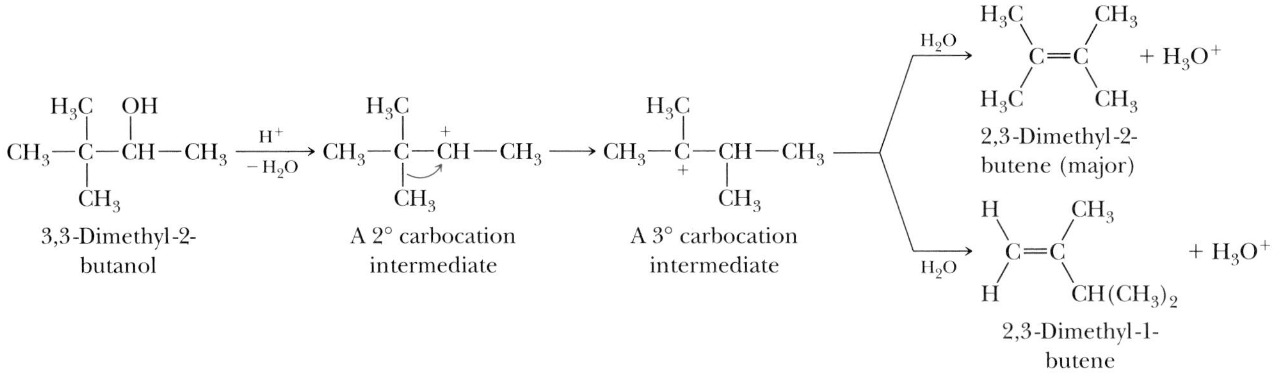

3,3-Dimethyl-2-butanol $\xrightarrow[-H_2O]{H^+}$ A 2° carbocation intermediate $\longrightarrow$ A 3° carbocation intermediate $\xrightarrow{H_2O}$ 2,3-Dimethyl-2-butene (major) + H_3O^+; $\xrightarrow{H_2O}$ 2,3-Dimethyl-1-butene + H_3O^+

EXAMPLE 9.10

Propose a mechanism to account for this reaction.

$\xrightarrow[heat]{H_2SO_4}$ + H_2O

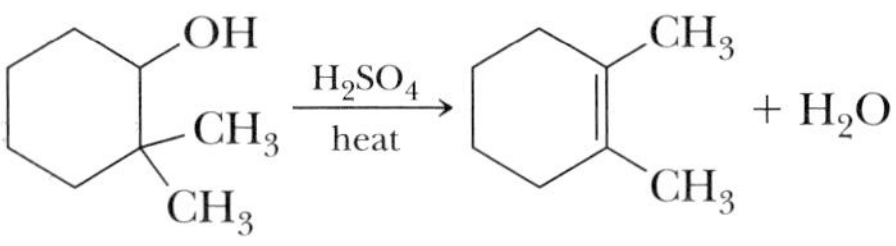

Solution

Proton transfer to the —OH group to form an oxonium ion followed by loss of OH_2 gives a 2° carbocation intermediate. Migration of a methyl group with its pair of electrons from the adjacent carbon to the positively charged carbon gives a more stable 3° carbocation. Loss of H^+ from this intermediate gives the observed product.

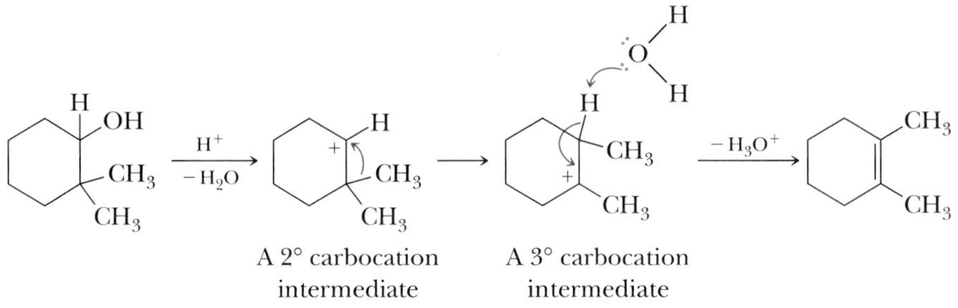

$\xrightarrow[-H_2O]{H^+}$ A 2° carbocation intermediate $\longrightarrow$ A 3° carbocation intermediate $\xrightarrow{-H_3O^+}$

PROBLEM 9.10

Propose a mechanism to account for this reaction.

$\xrightarrow[warm]{H_2SO_4}$ + H_2O

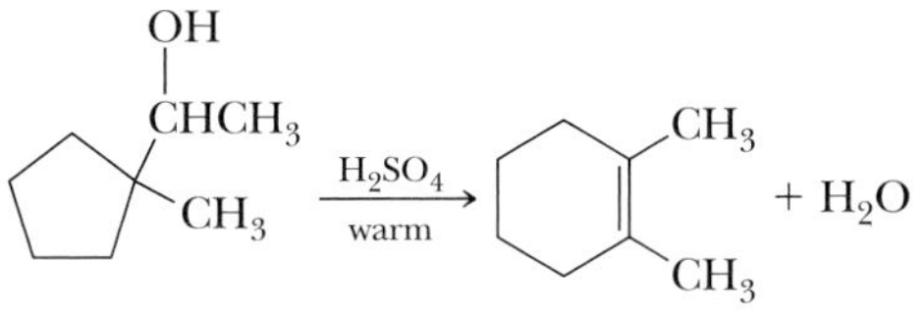

In Section 6.3B we discussed a mechanism for acid-catalyzed hydration of alkenes to give alcohols. In the present section, we discussed a mechanism for acid-catalyzed dehydration of alcohols to give alkenes. In fact, hydration-dehydration reactions are reversible. Alkene hydration and alcohol dehydration are competing processes, and the following equilibrium exists.

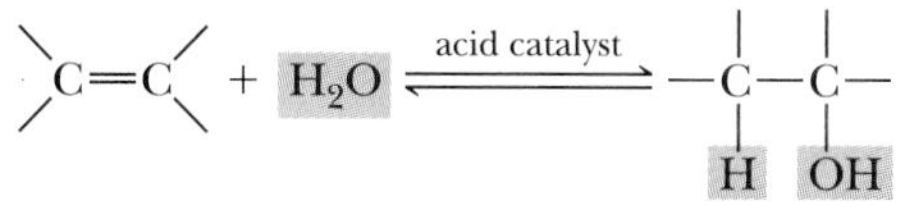

$$\mathrm{>C{=}C<} + \mathrm{H_2O} \underset{}{\overset{\text{acid catalyst}}{\rightleftharpoons}} \mathrm{-\underset{H}{\underset{|}{\overset{|}{C}}}-\underset{OH}{\underset{|}{\overset{|}{C}}}-}$$

An alkene An alcohol

Large amounts of water (use of dilute aqueous acid) favor alcohol formation, whereas scarcity of water (use of concentrated acid) or experimental conditions where water is removed (heating the reaction mixture above 100°C) favor alkene formation. Thus, depending on experimental conditions, it is possible to use the hydration-dehydration equilibrium to prepare either alcohols or alkenes, each in high yields.

This hydration-dehydration equilibrium illustrates a very important principle in the study of reaction mechanisms—the **principle of microscopic reversibility.** According to the principle of microscopic reversibility, the sequence of transition states and reactive intermediates (that is, the mechanism) for any reversible reaction must be the same, but in reverse order, for the backward reaction as for the forward reaction.

Principle of microscopic reversibility The sequence of transition states and reactive intermediates in the mechanism of any reversible reaction must be the same but in reverse order for the backward reaction as for the forward reaction.

To apply the principle of microscopic reversibility to acid-catalyzed hydration-dehydration equilibria, the mechanism we presented in this section for the acid-catalyzed dehydration of 2-butanol to give 2-butene is exactly the reverse of that presented in Section 6.3B for the acid-catalyzed hydration of 2-propene to give 2-propanol.

9.8 The Pinacol Rearrangement

Compounds containing two hydroxyl groups on adjacent carbon atoms are called **1,2-diols,** or alternatively, **glycols.** Such compounds can be synthesized by a variety of methods, including oxidation of alkenes by OsO_4 (Section 6.5B). Following is the structure of 2,3-dimethyl-2,3-butanediol, commonly called pinacol.

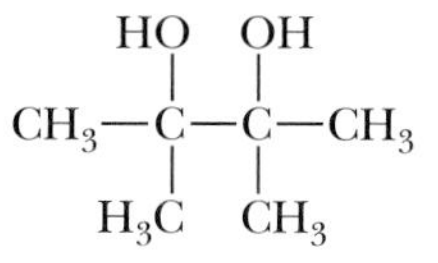

$$\mathrm{CH_3-\underset{H_3C}{\underset{|}{\overset{HO}{\overset{|}{C}}}}-\underset{CH_3}{\underset{|}{\overset{OH}{\overset{|}{C}}}}-CH_3}$$

2,3-Dimethyl-2,3-butanediol
(Pinacol)

The products of acid-catalyzed dehydration of glycols are quite different from those of acid-catalyzed dehydration of alcohols. For example, treatment of pinacol with concentrated sulfuric acid gives 3,3-dimethyl-2-butanone, commonly called pinacolone.

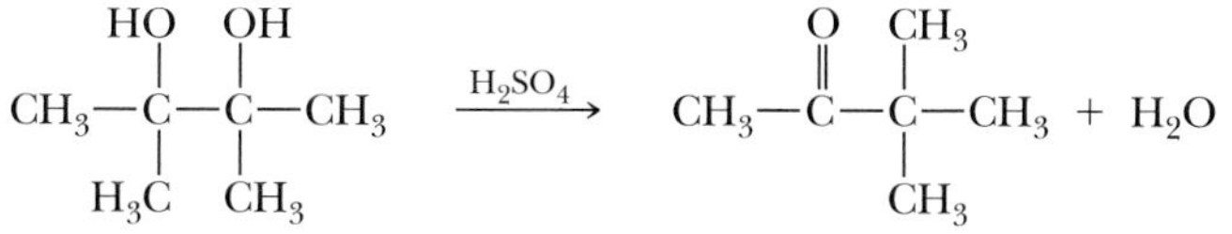

$$\mathrm{CH_3-\underset{H_3C}{\underset{|}{\overset{HO}{\overset{|}{C}}}}-\underset{CH_3}{\underset{|}{\overset{OH}{\overset{|}{C}}}}-CH_3} \xrightarrow{\mathrm{H_2SO_4}} \mathrm{CH_3-\overset{O}{\overset{\|}{C}}-\underset{CH_3}{\underset{|}{\overset{CH_3}{\overset{|}{C}}}}-CH_3} + \mathrm{H_2O}$$

2,3-Dimethyl-2,3-butanediol (Pinacol) 3,3-Dimethyl-2-butanone (Pinacolone)

Note two features of this reaction. It involves (1) dehydration of a glycol to form a ketone, and (2) migration of a methyl group from one carbon to an adjacent carbon. Acid-catalyzed conversion of pinacol to pinacolone is an example of a type of reaction called the **pinacol rearrangement.**

We account for the conversion of pinacol to pinacolone in the following way. Proton transfer to one of the hydroxyl groups of pinacol in Step 1 generates an oxonium ion, which then loses a molecule of water in Step 2 to form a tertiary carbocation intermediate. Migration of a methyl group with its pair of electrons in Step 3 gives a resonance-stabilized cation. Of the two contributing structures drawn for it, the one on the right makes the greater contribution because, in it, both carbon and oxygen have complete octets of valence electrons (Section 1.6D). The resonance-stabilized cation then transfers a proton to solvent in Step 4 to form pinacolone.

MECHANISM The Pinacol Rearrangement of 2,3-Dimethyl-2,3-Butanediol (Pinacol)

Step 1: Proton transfer from the acid to form an oxonium ion

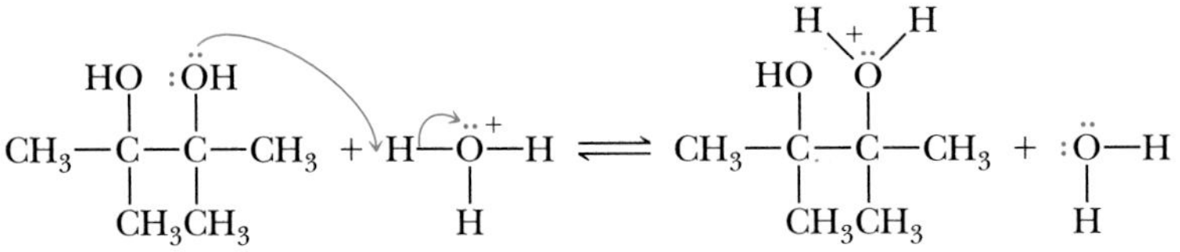

$$CH_3-C(OH)(CH_3)-C(\ddot{O}H)(CH_3)-CH_3 + H-\ddot{O}^+(H)-H \rightleftharpoons CH_3-C(OH)(CH_3)-C(\ddot{O}^+H_2)(CH_3)-CH_3 + :\ddot{O}(H)-H$$

An oxonium ion

Step 2: Loss of H_2O to form a 3° carbocation intermediate

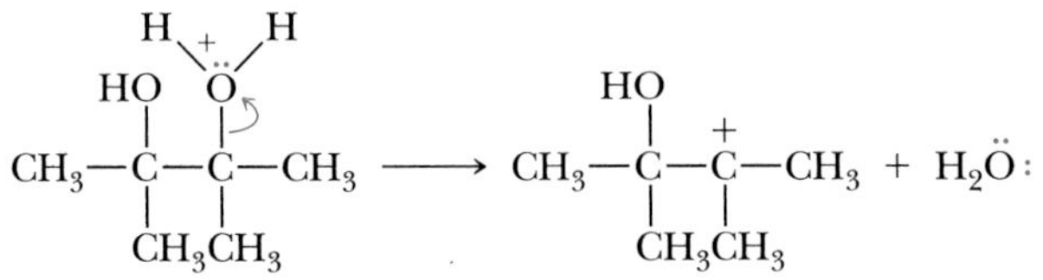

$$CH_3-C(OH)(CH_3)-C(\ddot{O}^+H_2)(CH_3)-CH_3 \longrightarrow CH_3-C(OH)(CH_3)-\overset{+}{C}(CH_3)-CH_3 + H_2\ddot{O}:$$

A 3° carbocation intermediate

Step 3: Migration of a methyl group from the adjacent carbon to form a new, more stable cation intermediate

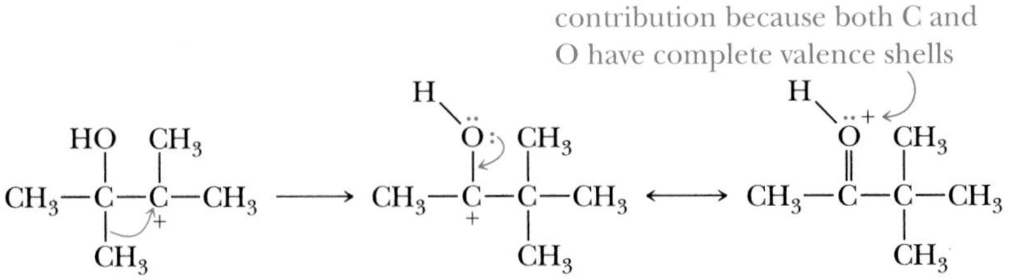

$$CH_3-C(OH)(CH_3)-\overset{+}{C}(CH_3)-CH_3 \longrightarrow CH_3-\overset{+}{C}(\ddot{O}H)-C(CH_3)_2-CH_3 \longleftrightarrow CH_3-C(=\ddot{O}^+H)-C(CH_3)_2-CH_3$$

this structure makes the greater contribution because both C and O have complete valence shells

A resonance-stabilized cation intermediate

Step 4: Proton transfer to give pinacolone

$$H_2\ddot{O}: + CH_3-\underset{}{C}(=\overset{+}{O}H)-C(CH_3)_2-CH_3 \longrightarrow CH_3-C(=\ddot{O}:)-C(CH_3)_2-CH_3 + H_3O^+$$

The pinacol rearrangement is general for all 1,2-diols. In the rearrangement of pinacol, a symmetrical diol, equivalent carbocations are formed no matter which —OH is protonated and leaves. Studies of unsymmetrical 1,2-diols have revealed that the —OH group that is protonated and leaves is the one that gives rise to the more stable carbocation. For example, treatment of 2-methyl-1,2-propanediol with cold concentrated sulfuric acid gives a tertiary carbocation. Subsequent migration of hydride ion (H:$^-$) followed by loss of a proton from the new cation gives 2-methylpropanal.

$$CH_3-C(OH)(CH_3)-CH_2OH \xrightarrow[-H_2O]{H_2SO_4} CH_3-\overset{+}{C}(CH_3)-CH_2OH \longrightarrow CH_3-CH(CH_3)-CH=O$$

2-Methyl-1,2-propanediol — A 3° carbocation intermediate — 2-Methylpropanal

EXAMPLE 9.11

Predict the product of treatment of each glycol with H_2SO_4.

(a) $CH_3CH_2C(OH)(CH_3)CH_2OH \xrightarrow[-H_2O]{H_2SO_4}$ (b) 1,1'-bi(cyclopentane)-1,1'-diol (OH, HO) $\xrightarrow[-H_2O]{H_2SO_4}$

Solution

(a) Protonation of the tertiary hydroxyl group followed by loss of a molecule of water gives a tertiary carbocation. Migration of a hydride ion from the adjacent carbon and loss of a proton gives 2-methylbutanal.

$$\text{Compound (a)} \xrightarrow[-H_2O]{H_2SO_4} CH_3CH_2\overset{+}{C}(CH_3)CH_2OH \longrightarrow CH_3CH_2CH(CH_3)CH{=}O$$

A 3° carbocation — 2-Methylbutanal

(b) Protonation of either hydroxyl group followed by loss of water gives a tertiary carbocation. The group that then migrates is a CH_2 group of the five-membered ring, and the product is a ketone. This compound belongs to the class of compounds called spiroketones, ketones in which two rings share only one carbon atom (Section 2.4B).

Compound (b) $\xrightarrow[-H_2O]{H_2SO_4}$ A 3° carbocation $\longrightarrow$ $\equiv$ Spiro[4.5]decan-6-one

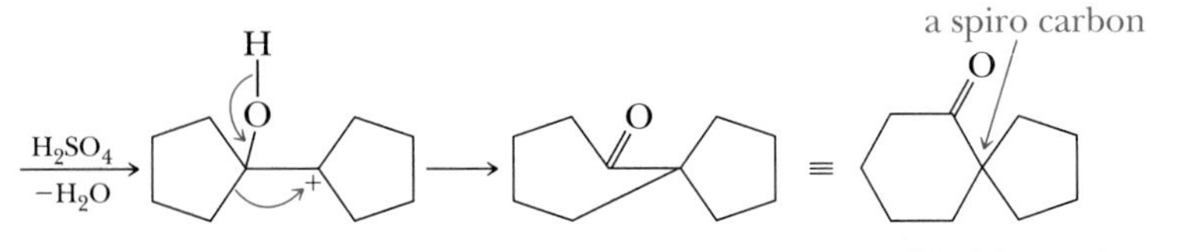

a spiro carbon

PROBLEM 9.11

Propose a mechanism to account for the following transformation:

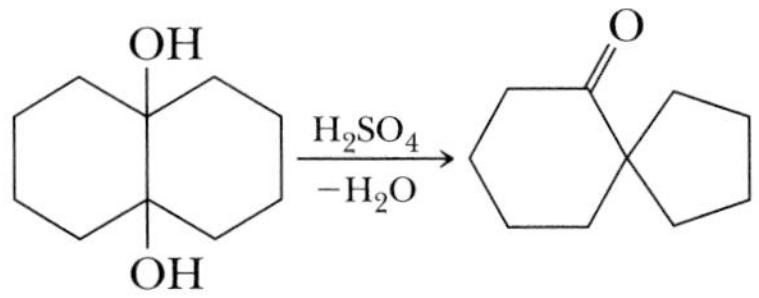

$\xrightarrow[-H_2O]{H_2SO_4}$

9.9 Oxidation of Primary and Secondary Alcohols

A primary alcohol is oxidized to an aldehyde or a carboxylic acid, depending on experimental conditions. Secondary alcohols are oxidized to ketones. Tertiary alcohols are not oxidized. Following is a series of transformations in which a primary alcohol is oxidized first to an aldehyde and then to a carboxylic acid. The fact that each transformation involves oxidation is indicated by the symbol O in brackets over the reaction arrow.

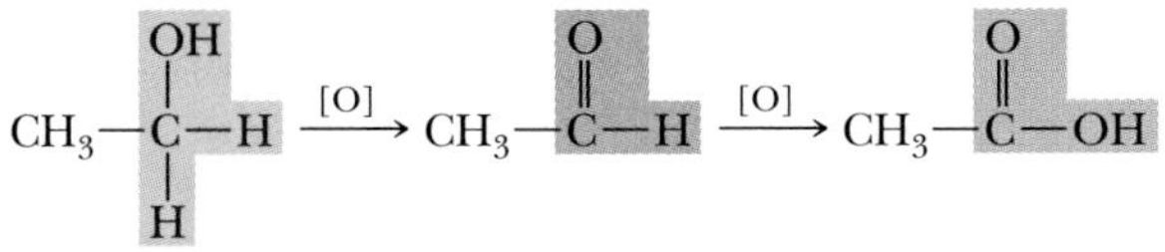

$$CH_3-CH_2-OH \xrightarrow{[O]} CH_3-\overset{O}{\overset{\|}{C}}-H \xrightarrow{[O]} CH_3-\overset{O}{\overset{\|}{C}}-OH$$

A primary alcohol — An aldehyde — A carboxylic acid

Inspection of balanced half-reactions (Section 6.5A) shows that each transformation in this series is a two-electron oxidation.

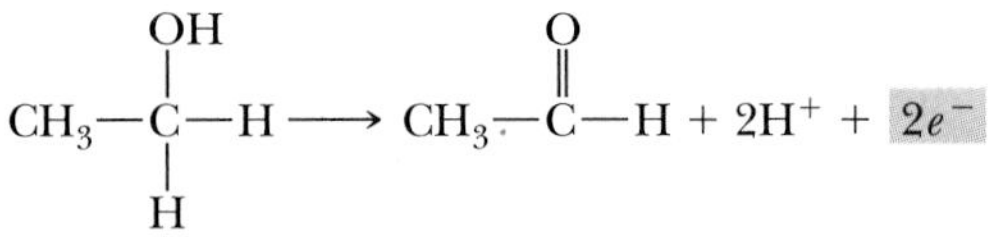

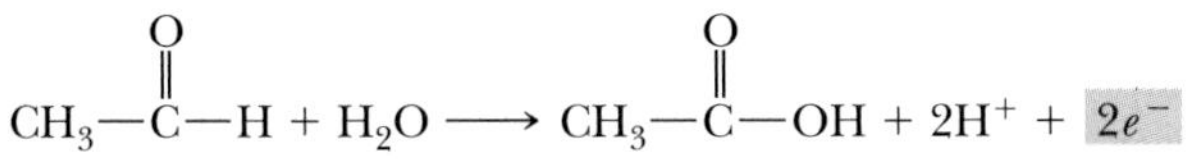

$$CH_3-\underset{H}{\underset{|}{\overset{OH}{\overset{|}{C}}}}-H \longrightarrow CH_3-\overset{O}{\overset{\|}{C}}-H + 2H^+ + 2e^-$$

$$CH_3-\overset{O}{\overset{\|}{C}}-H + H_2O \longrightarrow CH_3-\overset{O}{\overset{\|}{C}}-OH + 2H^+ + 2e^-$$

The oxidizing agent most commonly used in the laboratory for the conversion of a primary alcohol to a carboxylic acid is chromic acid, H_2CrO_4. Chromic acid is prepared by dissolving either chromium(VI) oxide or potassium dichromate in aqueous sulfuric acid.

$$CrO_3 + H_2O \xrightarrow{H_2SO_4} H_2CrO_4$$

Chromium(VI) oxide — Chromic acid

$$K_2Cr_2O_7 \xrightarrow{H_2SO_4} H_2Cr_2O_7 \xrightarrow{H_2O} 2H_2CrO_4$$

Potassium dichromate — Chromic acid

Oxidation of 1-octanol using chromic acid in aqueous sulfuric acid gives octanoic acid in high yield. These experimental conditions are more than sufficient to oxidize the intermediate aldehyde to a carboxylic acid.

$$CH_3(CH_2)_6CH_2OH \xrightarrow[H_2SO_4,\ H_2O]{CrO_3} \left[CH_3(CH_2)_6\overset{O}{\overset{\|}{C}}H\right] \longrightarrow CH_3(CH_2)_6\overset{O}{\overset{\|}{C}}OH$$

1-Octanol — Octanal (not isolated) — Octanoic acid

The form of Cr(VI) commonly used for oxidation of a primary alcohol to an aldehyde is prepared by dissolving CrO_3 in aqueous HCl and adding pyridine to precipitate **pyridinium chlorochromate (PCC)** as a solid.

CrO_3 + HCl + Pyridine ⟶ Pyridinium chlorochromate (PCC), CrO_3Cl^-

(pyridinium ion; chlorochromate ion)

Pyridine — Pyridinium chlorochromate (PCC)

This reagent is not only very selective for the oxidation of primary alcohols to aldehydes, but it also has little effect on carbon-carbon double bonds or other easily oxidized functional groups. In the following example, geraniol, a primary terpene alcohol, is oxidized to geranial without affecting either carbon-carbon double bond.

Geraniol (CH_2OH) $\xrightarrow{PCC}$ Geranial ($\overset{O}{\overset{\|}{C}}H$)

Geraniol — Geranial

Secondary alcohols are oxidized to ketones by both chromic acid and PCC.

$$\text{Menthol} + H_2CrO_4 \xrightarrow{\text{acetone}} \text{Menthone} + Cr^{3+}$$

(substituents: $CH(CH_3)_2$, OH / =O, CH_3)

2-Isopropyl-5-methylcyclohexanol (Menthol) — 2-Isopropyl-5-methylcyclohexanone (Menthone)

Tertiary alcohols are resistant to oxidation because the carbon that bears the —OH is already bonded to three other carbon atoms and, therefore, cannot form an additional carbon-oxygen bond.

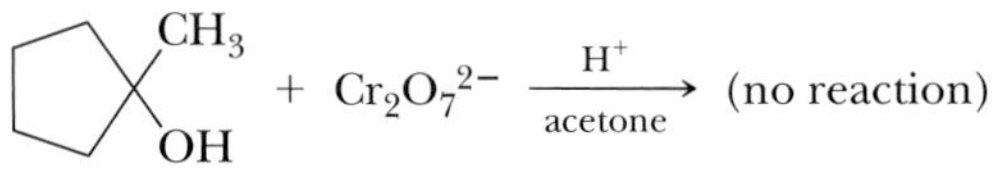

$$\text{1-Methylcyclopentanol} + Cr_2O_7^{2-} \xrightarrow[\text{acetone}]{H^+} \text{(no reaction)}$$

1-Methylcyclopentanol

Step 1 in chromic acid oxidation of an alcohol is fast and reversible formation of an alkyl chromate. This is followed in Step 2 by a slow, rate-limiting reaction in which the alkyl chromate reacts with a base (here shown as a water molecule). In this concerted process, which is analogous to an E2 reaction, a C—H bond is broken, the carbonyl group is formed, and chromium(VI) is reduced to chromium(IV). Chromium(IV) then participates in further oxidation by a similar mechanism and eventually is transformed to Cr^{3+}. Note that the lack of an alpha-hydrogen in tertiary alcohols prevents such a reaction from taking place.

MECHANISM Chromic Acid Oxidation of an Alcohol

Step 1: Formation of a chromate ester

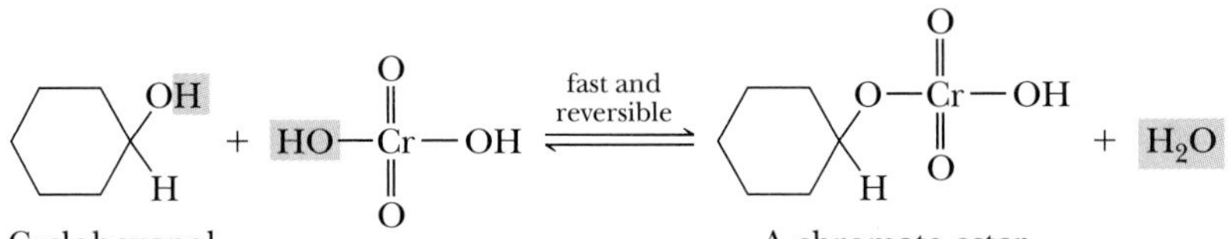

$$\text{Cyclohexanol} + HO{-}CrO_2{-}OH \underset{}{\overset{\text{fast and reversible}}{\rightleftharpoons}} \text{A chromate ester } (C_6H_{11}{-}O{-}CrO_2{-}OH) + H_2O$$

Cyclohexanol

A chromate ester

Step 2: Proton transfer to solvent and decomposition of the chromate ester

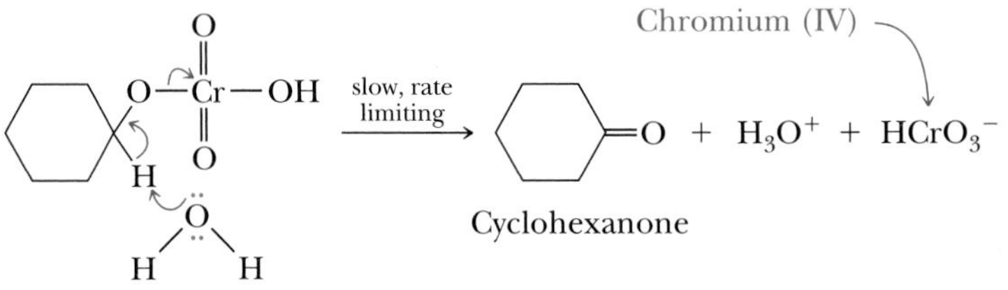

$$C_6H_{11}{-}O{-}CrO_2{-}OH + H_2O \xrightarrow{\text{slow, rate limiting}} \text{Cyclohexanone} + H_3O^+ + HCrO_3^-$$

Chromium (IV)

Cyclohexanone

We have shown that in aqueous chromic acid, a primary alcohol is oxidized first to an aldehyde and then to a carboxylic acid. In the second oxidation, it is not the aldehyde that is oxidized, but rather the aldehyde hydrate formed by reaction of the aldehyde carbonyl group with a molecule of water. We will study the hydration of aldehydes and ketones in more detail in Chapter 15. It is an —OH of the aldehyde hydrate that now forms an ester with chromic acid to complete the oxidation of the aldehyde to a carboxylic acid.

CHEMISTRY IN ACTION

Blood Alcohol Screening

Potassium dichromate is a strong oxidizing agent, and under appropriate experimental conditions, it oxidizes primary alcohols to carboxylic acids. Potassium dichromate oxidation of ethanol to acetic acid is the basis for the original breath alcohol screening test used by law enforcement agencies to determine a person's blood alcohol content (BAC). The test is based on the difference in color between the dichromate anion (reddish orange) in the reagent and the chromium(III) ion (green) in the product. Thus, color change can be used as a measure of the quantity of ethanol present in a breath sample.

$$\underset{\text{Ethanol}}{CH_3CH_2OH} + \underset{\substack{\text{Dichromate ion}\\\text{(reddish orange)}}}{Cr_2O_7^{2-}} \xrightarrow[H_2O]{H_2SO_4} \underset{\text{Acetic acid}}{CH_3\overset{O}{\overset{\|}{C}}OH} + \underset{\substack{\text{Chromium(III) ion}\\\text{(green)}}}{Cr^{3+}}$$

In its simplest form, a breath alcohol screening test consists of a sealed glass tube containing a potassium dichromate-sulfuric acid reagent impregnated on silica gel. To administer the test, the ends of the tube are broken off, a mouthpiece is fitted to one end and the other end is inserted into the neck of a plastic bag. The person being tested then blows into the mouthpiece until the plastic bag is inflated.

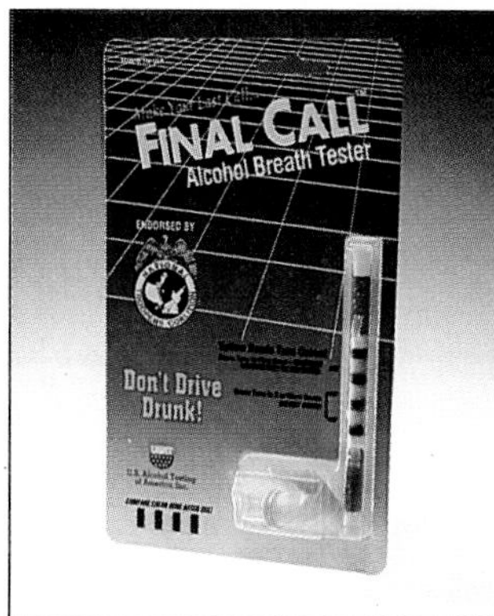
FINAL CALL
Alcohol Breath Tester
Don't Drive Drunk!

A device for testing the breath for the presence of ethanol. The test works because ethanol is oxidized to acetic acid by potassium dichromate, $K_2Cr_2O_7$. The reddish-orange color of dichromate ion turns to green as it is reduced to chromium(III) ion. (Charles D. Winters)

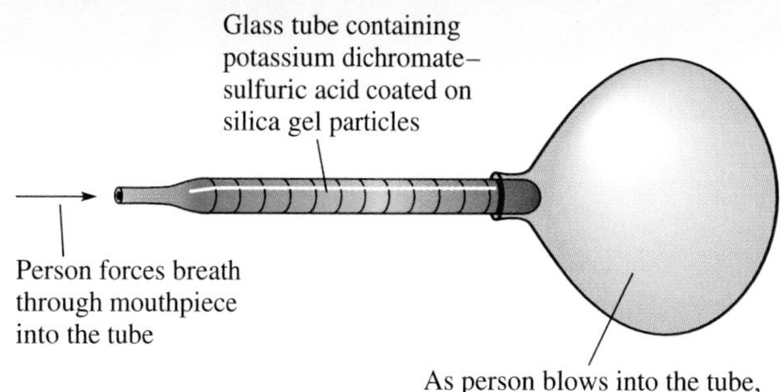
Glass tube containing potassium dichromate–sulfuric acid coated on silica gel particles
Person forces breath through mouthpiece into the tube
As person blows into the tube, the plastic bag becomes inflated

As breath containing ethanol vapor passes through the tube, reddish orange dichromate ion is reduced to green chromium(III) ion. The concentration of ethanol in the breath is then estimated by measuring how far the green chromium(III) ion color extends along the length of the tube. When the green color extends beyond the halfway point, the person is judged as having a sufficiently high blood alcohol content to warrant further, more precise testing.

The Breathalyzer, a more precise testing device, operates on the same principle as the simplified screening test. In a Breathalyzer test, a measured volume of breath is bubbled through a solution of potassium dichromate in aqueous sulfuric acid, and the color change is measured spectrophotometrically.

These tests measure alcohol in the breath. The legal definition of being under the influence of alcohol, however, is based on blood alcohol content, not breath alcohol content. The chemical correlation between these two measurements is that air deep within the lungs is in equilibrium with blood passing through the pulmonary arteries, and an equilibrium is established between blood alcohol and breath alcohol. It has been determined by tests in a person drinking alcohol that 2100 mL of breath contains the same amount of ethanol as 1.00 mL of blood.

See W. C. Timmer, *J. Chem. Ed.*, **63:**897 (1986).

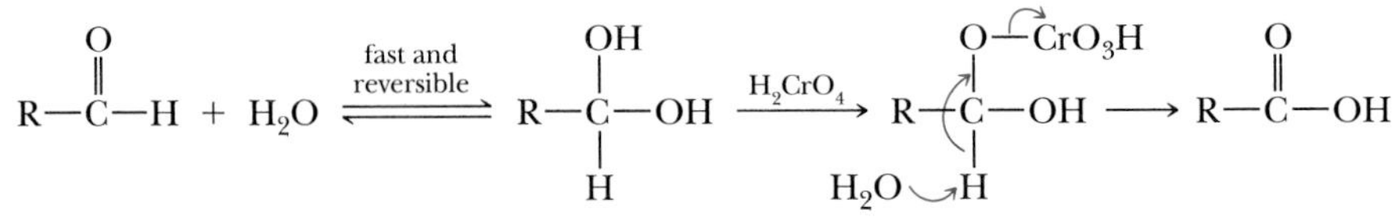

$$\underset{\text{An aldehyde}}{R-\overset{\overset{\displaystyle O}{\|}}{C}-H} + H_2O \xrightleftharpoons[\text{reversible}]{\text{fast and}} \underset{\text{An aldehyde hydrate}}{R-\overset{OH}{\underset{H}{C}}-OH} \xrightarrow{H_2CrO_4} \underset{\text{A chromate ester}}{R-\overset{O-CrO_3H}{\underset{H}{C}}-OH} \longrightarrow \underset{\text{A carboxylic acid}}{R-\overset{\overset{\displaystyle O}{\|}}{C}-OH}$$

Herein lies the reason why PCC is specific for the oxidation of primary alcohols to aldehydes. It does not oxidize aldehydes further because there is no water present in the PCC reagent.

EXAMPLE 9.12

Draw the product of treatment of each alcohol with PCC.

(a) 1-Hexanol **(b)** 2-Hexanol **(c)** Cyclohexanol

Solution

1-Hexanol is a primary alcohol, and is oxidized to hexanal. 2-Hexanol, a secondary alcohol, is oxidized to 2-hexanone. Cyclohexanol, a secondary alcohol, is oxidized to cyclohexanone.

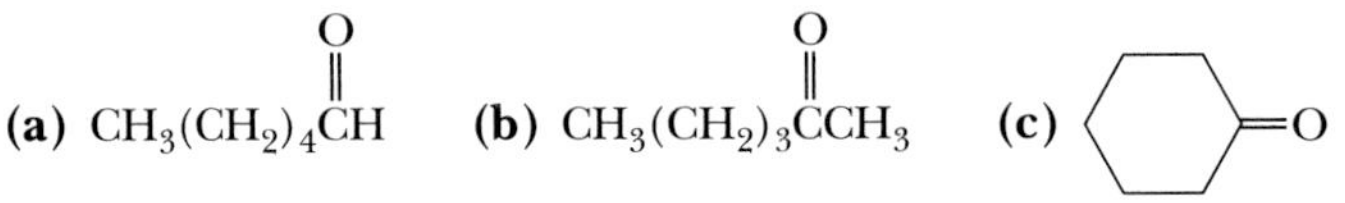

(a) $CH_3(CH_2)_4\overset{\overset{\displaystyle O}{\|}}{C}H$ Hexanal **(b)** $CH_3(CH_2)_3\overset{\overset{\displaystyle O}{\|}}{C}CH_3$ 2-Hexanone **(c)** (cyclohexane ring)=O Cyclohexanone

PROBLEM 9.12

Draw the product of treatment of each alcohol in Example 9.12 with chromic acid.

9.10 Oxidation of Glycols by Periodic Acid

Periodic acid, H_5IO_6 (or alternatively $HIO_4 \cdot 2H_2O$), is a white crystalline solid, mp 122°C. Its major use in organic chemistry is cleavage of glycols to carbonyl-containing compounds (a two-electron oxidation). In the process, periodic acid is reduced to iodic acid (a two-electron reduction).

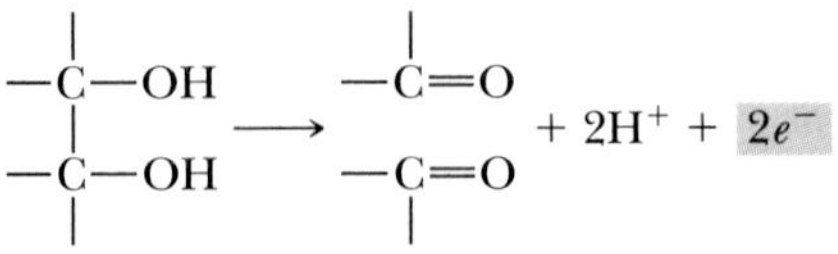

$$\begin{matrix} -\overset{|}{C}-OH \\ -\underset{|}{C}-OH \end{matrix} \longrightarrow \begin{matrix} -\overset{|}{C}=O \\ -\underset{|}{C}=O \end{matrix} + 2H^+ + 2e^- \quad \text{(a two-electron oxidation)}$$

$$\underset{\text{Periodic acid}}{HIO_4} + 2H^+ + 2e^- \longrightarrow \underset{\text{Iodic acid}}{HIO_3} + H_2O \quad \text{(a two-electron reduction)}$$

For example, when treated with periodic acid, 1,2-cyclohexanediol is oxidized to hexanedial.

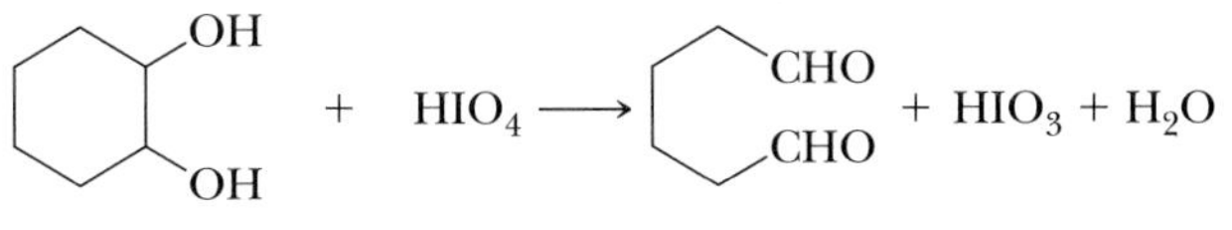

$$\text{1,2-Cyclohexanediol} + HIO_4 \longrightarrow \text{Hexanedial} + HIO_3 + H_2O$$

1,2-Cyclohexanediol   Hexanedial

Step 1 of the mechanism of periodic acid oxidation of a glycol involves formation of a five-membered cyclic diester between the diol and HIO_4. Redistribution of electrons within this cyclic ester in Step 2 generates the two carbonyl groups and HIO_3.

MECHANISM Oxidation of a Glycol by Periodic Acid

Step 1: Formation of a five-membered cyclic periodic ester

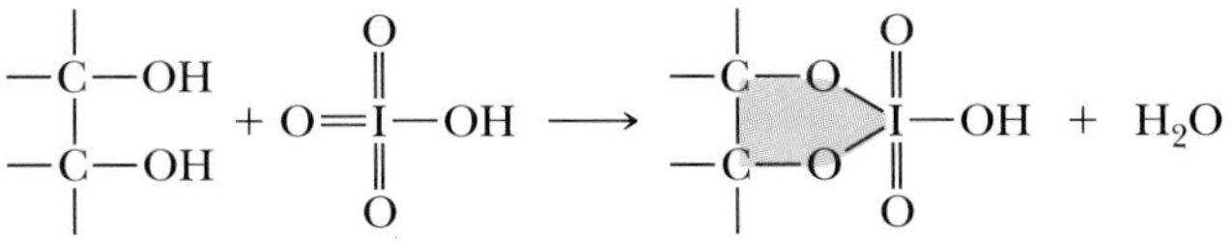

A cyclic periodic ester

Step 2: Redistribution of electrons within the cyclic periodic ester to give HIO_3 and two carbonyl groups

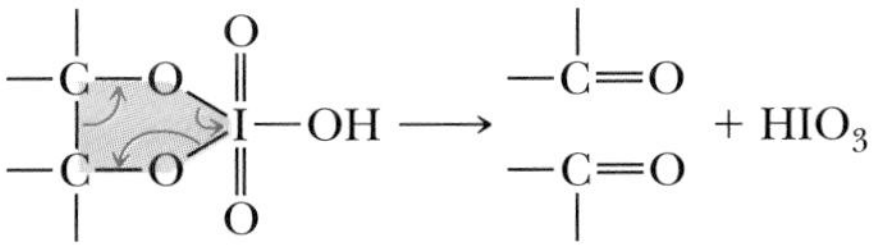

$$+ HIO_3$$

This mechanism is consistent with the fact that HIO_4 oxidations are restricted to glycols that can form a five-membered cyclic periodic ester. Any glycol that cannot form such a cyclic ester is not cleaved by periodic acid. Following are structural formulas for two isomeric *cis* and *trans* decalindiols. Only the *cis* glycol can form a cyclic ester with periodic acid and only it is cleaved by this reagent.

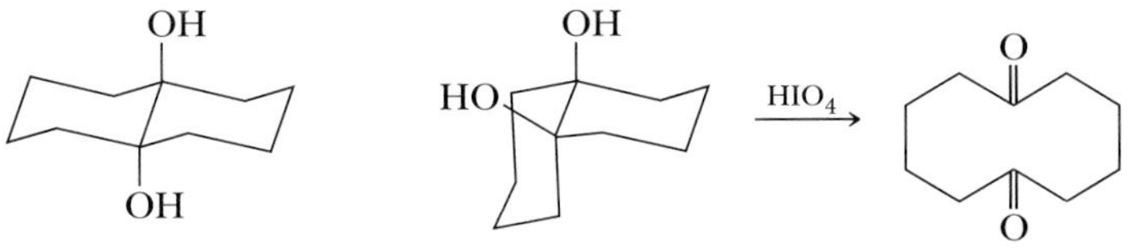

The *trans* isomer is unreactive toward periodic acid

The *cis* isomer forms a cyclic ester with periodic acid and is cleaved

EXAMPLE 9.13

What products are formed when each glycol is treated with HIO_4?

(a) 1-(hydroxymethyl)cyclopentan-1-ol: cyclopentane ring bearing OH and CH_2OH on the same carbon **(b)** $(CH_3)_2\overset{\text{HO}}{\overset{|}{C}}-\overset{\text{OH}}{\overset{|}{C}}HCH_3$

Solution

The bond between the carbons bearing the —OH groups is cleaved and each —OH group is converted to a carbonyl group.

(a) cyclopentanone $=O$ $+ H-\overset{O}{\overset{\|}{C}}-H$ **(b)** $(H_3C)_2C{=}O + H-\overset{O}{\overset{\|}{C}}-CH_3$

PROBLEM 9.13

α-Hydroxyketones and α-hydroxyaldehydes are also cleaved by treatment with periodic acid. It is not the α-hydroxyketone or aldehyde, however, that undergoes reaction with periodic acid, but rather the compound formed by addition of a molecule of water to the carbonyl group of the α-hydroxyketone or aldehyde. Draw the structural formula for the product formed by addition of a molecule of H_2O to the carbonyl group of the following compound and write a mechanism for the reaction of this product with HIO_4. *Hint:* Notice the similarity of this oxidation to the oxidation of an aldehyde by chromic acid (Section 9.9).

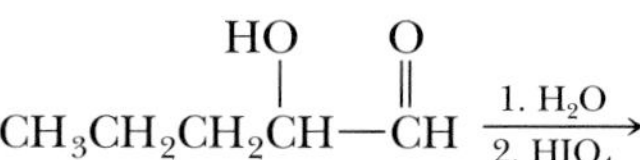

$$CH_3CH_2CH_2\overset{\text{HO}}{\overset{|}{C}}H-\overset{O}{\overset{\|}{C}}H \xrightarrow[2.\ HIO_4]{1.\ H_2O}$$

9.11 Thiols

A. Preparation

The most common preparation of thiols, RSH, depends on the very high nucleophilicity of hydrosulfide ion, HS^- (Section 8.4B). Sodium hydrosulfide is prepared by bubbling H_2S through a solution of NaOH in water or aqueous ethanol. Reaction of HS^- with an alkyl halide gives a thiol.

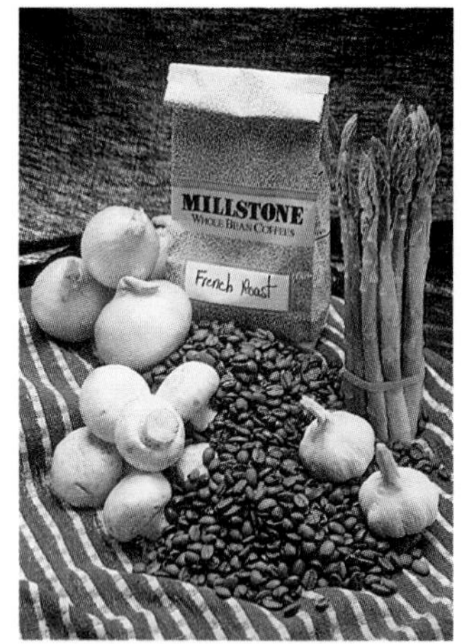

Mushrooms, onions, garlic, and coffee all contain sulfur compounds. One of the compounds responsible for the aroma of coffee is 2-furanylmethanethiol (furan ring bearing $-CH_2SH$). *(Charles D. Winters)*

$$\underset{\text{1-Iododecane}}{CH_3(CH_2)_8CH_2I} + \underset{\text{Sodium hydrosulfide}}{Na^+\,SH^-} \xrightarrow{S_N2} \underset{\text{1-Decanethiol}}{CH_3(CH_2)_8CH_2SH} + Na^+\,I^-$$

In practice, the scope and limitations of this reaction are governed by competition between substitution and β-elimination (Section 8.11). The reaction is most useful for preparation of thiols from primary halides. Yields are low from secondary halides because of the competing β-elimination reaction. With tertiary halides, β-elimination predominates, and the alkene formed by dehydrohalogenation is the major product.

In a commercial application of thiol formation by nucleophilic substitution, thioglycolic acid is prepared by the reaction of sodium hydrosulfide and sodium iodoacetate.

$$\underset{\text{Sodium hydrosulfide}}{Na^+ SH^-} + \underset{\text{Sodium iodoacetate}}{ICH_2\overset{\overset{O}{\|}}{C}O^- Na^+} \xrightarrow{S_N2} \underset{\substack{\text{Sodium mercaptoacetate} \\ \text{(sodium thioglycolate)}}}{HSCH_2\overset{\overset{O}{\|}}{C}O^- Na^+} + Na^+ I^-$$

The sodium and ammonium salts of thioglycolic acid are used in cold waving of hair; the calcium salt is used as a depilatory, that is, to remove body hair.

$$(HSCH_2CO_2^-)_2Ca^{2+}$$

Calcium mercaptoacetate
(Calcium thioglycolate)

B. Acidity

Hydrogen sulfide is a stronger acid than water.

$$H_2O + H_2O \rightleftharpoons HO^- + H_3O^+ \qquad pK_a = 15.7$$

$$H_2S + H_2O \rightleftharpoons HS^- + H_3O^+ \qquad pK_a = 7.0$$

Similarly, thiols are stronger acids than alcohols. Compare, for example, the values of pK_a of ethanol and ethanethiol in dilute aqueous solution.

$$CH_3CH_2OH + H_2O \rightleftharpoons CH_3CH_2O^- + H_3O^+ \qquad pK_a = 15.9$$

$$CH_3CH_2SH + H_2O \rightleftharpoons CH_3CH_2S^- + H_3O^+ \qquad pK_a = 8.5$$

Thiols are sufficiently strong acids that, when dissolved in aqueous sodium hydroxide, they are converted completely to alkylsulfide salts.

$$\underset{\substack{pK_a = 8.5 \\ \text{Stronger acid}}}{CH_3CH_2SH} + \underset{\text{Stronger base}}{Na^+OH^-} \longrightarrow \underset{\text{Weaker base}}{CH_3CH_2S^-Na^+} + \underset{\substack{pK_a = 15.7 \\ \text{Weaker acid}}}{H_2O}$$

To name salts of thiols, give the name of the cation first, followed by the name of the alkyl group to which is attached the suffix -sulfide. For example, the sodium salt derived from ethanethiol is named sodium ethylsulfide.

C. Oxidation

Many of the chemical properties of thiols stem from the fact that the sulfur atom of a thiol is oxidized easily to several higher oxidation states, the most important of which are summarized in Table 9.4. To bring about each requires an oxidizing agent. The relationships between the relative oxidation states of a thiol, a disulfide, a sulfinic acid, and a sulfonic acid are shown in the form of a sequence of balanced half-reactions. Note that the valence shell of sulfur contains 8 electrons in a thiol and a disulfide, 10 electrons in a sulfinic acid, and 12 electrons in a sulfonic acid (Section 1.2F).

The most common oxidation-reduction reaction of sulfur compounds in biological systems is interconversion between a thiol and a disulfide. The functional group of a disulfide is an —S—S— group.

$$\underset{\text{A thiol}}{2RSH} + I_2 \longrightarrow \underset{\text{A disulfide}}{RSSR} + 2HI$$

TABLE 9.4 Functional Groups Formed by Oxidation of a Thiol

A thiol can be converted to the following higher oxidation states:

$$2\text{R—S—H} \longrightarrow \text{R—S—S—R} + 2\text{H}^+ + 2e^-$$

A thiol → A disulfide

$$\text{R—S—H} + 2\text{H}_2\text{O} \longrightarrow \text{R—S(=O)—O—H} + 4\text{H}^+ + 4e^-$$

A thiol → A sulfinic acid

$$\text{R—S—H} + 3\text{H}_2\text{O} \longrightarrow \text{R—S(=O)_2—O—H} + 6\text{H}^+ + 6e^-$$

A thiol → A sulfonic acid

Thiols are also oxidized to disulfides by molecular oxygen. In fact, thiols are so susceptible to oxidation that they must be protected from contact with air during storage.

$$2\text{RSH} + \tfrac{1}{2}\text{O}_2 \longrightarrow \text{RSSR} + \text{H}_2\text{O}$$

A thiol → A disulfide

The disulfide bond is an important structural feature of many biomolecules, including proteins (Chapter 27).

SUMMARY

The functional group of an alcohol (Section 9.1A) is an **—OH (hydroxyl)** group bonded to an sp^3 hybridized carbon. A **thiol** (Section 9.1B) is a sulfur analog of an alcohol; it contains an **—SH (sulfhydryl)** group in place of an —OH group. IUPAC names of alcohols (Section 9.2A) are derived by changing the suffix of the parent alkane from -e to -ol. The chain is numbered from the direction that gives the carbon bearing —OH the lower number. In compounds containing other functional groups of higher precedence, the presence of —OH is indicated by the prefix **hydroxy-**. Common names for alcohols are derived by naming the alkyl group bonded to —OH and adding the word "alcohol." Alcohols are classified as **1°, 2°,** or **3°** (Section 9.2A) depending on whether the —OH group is bonded to a primary, secondary, or tertiary carbon. Thiols (Section 9.2B) are named in the same manner as alcohols, but the suffix -e of the parent alkane is retained and **-thiol** added. Common names for thiols are derived by naming the alkyl group bonded to —SH and adding the word "mercaptan." In compounds containing other functional groups of higher precedence, the presence of —SH is indicated by the prefix **mercapto-**.

Alcohols are polar compounds (Section 9.3A) with oxygen bearing a partial negative charge and both the α-carbon and hydrogen bearing partial positive charges. Because of intermolecular association caused by **hydrogen bonding,** the boiling points of alcohols are higher than those of hydrocarbons of comparable molecular weight. Because of increased dispersion forces, the boiling points of alcohols increase with increasing molecular weight. Alcohols interact with water by hydrogen bonding and, therefore, are more soluble in water than hydrocarbons of comparable molecular weight. Because the S—H bond is almost nonpolar, the physical properties of thiols are more like those of hydrocarbons of comparable molecular weight.

According to the **principle of microscopic reversibility** (Section 9.7), the sequence of transition states and reactive intermediates for any reversible reaction must be the same, but in reverse order, for the backward reaction as for the forward reaction. As an application of this principle, the mechanism of acid-catalyzed dehydration of an alcohol to an alkene is the same, but in reverse, as that for acid-catalyzed hydration of an alkene to an alcohol.

Key Reactions

1. Acidity of Alcohols (Section 9.4)

In dilute aqueous solution, methanol and ethanol are comparable in acidity to water. Secondary and tertiary alcohols are weaker acids.

$$CH_3OH + H_2O \rightleftharpoons CH_3O^- + H_3O^+ \qquad pK_a = 15.5$$

2. Reaction with Active Metals (Section 9.5)

Alcohols react with Li, Na, K, and other active metals to form metal alkoxides, which are somewhat stronger bases than NaOH and KOH.

$$CH_3CH_2OH + Na \longrightarrow CH_3CH_2O^-Na^+ + H_2$$

3. Reaction with HCl, HBr, and HI to Form Haloalkanes (Section 9.6A)

Primary alcohols react by an S_N2 mechanism.

$$CH_3CH_2CH_2CH_2OH + HBr \xrightarrow[\text{reflux}]{} CH_3CH_2CH_2CH_2Br + H_2O$$

Tertiary alcohols react by an S_N1 mechanism with formation of a carbocation intermediate which may undergo rearrangement.

$$\underset{\underset{\displaystyle CH_3}{|}}{\overset{\overset{\displaystyle CH_3}{|}}{CH_3COH}} + HCl \xrightarrow{25°C} \underset{\underset{\displaystyle CH_3}{|}}{\overset{\overset{\displaystyle CH_3}{|}}{CH_3CCl}} + H_2O$$

Secondary alcohols may react by an S_N2 or an S_N1 mechanism depending on the alcohol and experimental conditions. Primary alcohols with extensive β-branching react by an S_N1 mechanism involving formation of a rearranged carbocation.

$$\underset{\underset{\displaystyle CH_3}{|}}{\overset{\overset{\displaystyle CH_3}{|}}{CH_3CCH_2OH}} + HBr \longrightarrow \underset{\underset{\displaystyle Br}{|}}{\overset{\overset{\displaystyle CH_3}{|}}{CH_3CCH_2CH_3}} + H_2O$$

4. Reaction with PBr_3 (Section 9.6B)

Although some rearrangement may occur with this reagent, it is less likely than in the reaction of an alcohol with HBr.

$$3CH_3\overset{\overset{\displaystyle OH}{|}}{CH}CH_3 + PBr_3 \longrightarrow 3CH_3\overset{\overset{\displaystyle Br}{|}}{CH}CH_3 + H_3PO_3$$

5. Reaction with $SOCl_2$ (Section 9.6C)

This is often the method of choice for converting a primary or secondary alcohol to an alkyl chloride.

$$CH_3(CH_2)_5CH_2OH + SOCl_2 \xrightarrow{\text{pyridine}} CH_3(CH_2)_5CH_2Cl + SO_2 + HCl$$

6. Acid-Catalyzed Dehydration (Section 9.7)

When isomeric alkenes are possible, the major product is generally the more substituted alkene (Zaitsev's rule). Rearrangements are common with secondary alcohols and also with primary alcohols with extensive β-branching.

Reactions of Alcohols

9.33 Write equations for the reaction of 1-butanol with each reagent. Where you predict no reaction, write NR.

(a) Na metal	**(b)** HBr, heat	**(c)** HI, heat
(d) PBr_3	**(e)** $SOCl_2$, pyridine	**(f)** $K_2Cr_2O_7$, H_2SO_4, H_2O, heat
(g) HIO_4	**(h)** PCC	**(i)** CH_3SO_2Cl, pyridine

9.34 Write equations for the reaction of 2-butanol with each reagent listed in Problem 9.33. Where you predict no reaction, write NR.

9.35 Complete these equations. Show structural formulas for the major products, but do not balance.

(a) $CH_3CH_2CH_2OH + H_2CrO_4 \longrightarrow$

(b) $CH_3\underset{}{\overset{CH_3}{\overset{|}{C}}}HCH_2CH_2OH + SOCl_2 \longrightarrow$

(c) 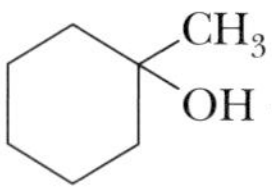 1-methylcyclohexanol (ring C bearing CH_3 and OH) + HCl $\longrightarrow$

(d) $HOCH_2CH_2CH_2CH_2OH + HBr \longrightarrow$

(e) 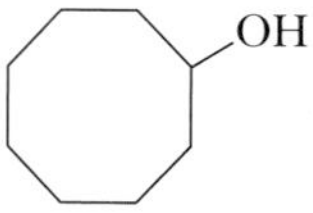 cyclooctanol (ring–OH) $+ H_2CrO_4 \longrightarrow$

(f) 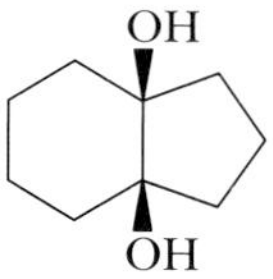 bicyclic diol (OH at each ring-fusion carbon) $+ HIO_4 \longrightarrow$

(g) 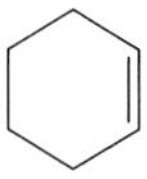 cyclohexene $\xrightarrow[2.\ HIO_4]{1.\ OsO_4,\ H_2O_2}$

(h) 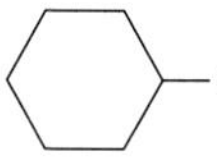 cyclohexane–OH $+ SOCl_2 \longrightarrow$

9.36 When (*R*)-2-butanol is left standing in aqueous acid, it slowly loses its optical activity. Account for this observation.

9.37 Two diastereomeric sets of enantiomers, A/B and C/D, exist for 3-bromo-2-butanol. When enantiomer A or B is treated with HBr, only racemic 2,3-dibromobutane is formed; no meso isomer is formed. When enantiomer C or D is treated with HBr, only meso 2,3-dibromobutane is formed; no racemic 2,3-dibromobutane is formed. Account for these observations. (*Hint:* Consider neighboring group participation (Section 8.5) and the type of intermediate that could produce this stereoselectivity.)

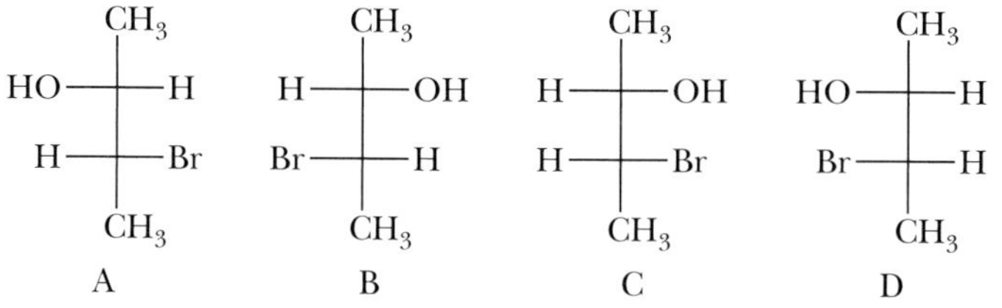

	A	B	C	D
	CH_3	CH_3	CH_3	CH_3
C2	HO—C—H	H—C—OH	H—C—OH	HO—C—H
C3	H—C—Br	Br—C—H	H—C—Br	Br—C—H
	CH_3	CH_3	CH_3	CH_3

9.38 Acid-catalyzed dehydration of 3-methyl-2-butanol gives three alkenes: 2-methyl-2-butene, 3-methyl-1-butene, and 2-methyl-1-butene. Propose a mechanism to account for the formation of each product.

9.39 Show how you might bring about the following conversions. For any conversion involving more than one step, show each intermediate compound formed.

(a) $CH_3\overset{CH_3}{\overset{|}{C}}HCH_2OH \longrightarrow CH_3\overset{CH_3}{\overset{|}{C}}{=}CH_2$

(b) $CH_3\overset{CH_3}{\overset{|}{C}}HCH_2OH \longrightarrow CH_3\underset{OH}{\underset{|}{\overset{CH_3}{\overset{|}{C}}}}CH_3$

(c) 1-methylcyclohexanol $\longrightarrow$ trans-2-methylcyclohexanol (CH_3, OH)

(d) 1-methylcyclohexanol (CH_3, OH) $\longrightarrow$ $CH_3\overset{O}{\overset{\|}{C}}CH_2CH_2CH_2CH_2\overset{O}{\overset{\|}{C}}H$

(e) methylenecyclohexane ($=CH_2$) $\longrightarrow$ cyclohexyl$-CH_2Cl$

(f) ethylidenecyclohexane ($=CHCH_3$) $\longrightarrow$ cyclohexyl$-\overset{O}{\overset{\|}{C}}CH_3$

(g) $CH_3(CH_2)_6CH_2OH \longrightarrow CH_3(CH_2)_6\overset{O}{\overset{\|}{C}}H$

(h) cyclooctanol (OH) $\longrightarrow$ cis-1,2-cyclooctanediol (OH, OH)

(i) methylenecyclohexane ($=CH_2$) $\longrightarrow$ cyclohexyl$-\overset{O}{\overset{\|}{C}}OH$

Pinacol Rearrangement

9.40 Propose a mechanism for the following pinacol rearrangement catalyzed by boron trifluoride etherate:

HO OH (1,1'-bicyclohexyl-1,1'-diol) $\xrightarrow{BF_3 \cdot (CH_3CH_2)_2O}$ Spiro[5.6]dodecan-7-one (O) $+ H_2O$

Spiro[5.6]dodecan-7-one

Synthesis

9.41 Give reactions for the synthesis of each alcohol from a suitable alkene.

(a) 2-Pentanol **(b)** 1-Pentanol **(c)** 2-Methyl-2-pentanol
(d) 2-Methyl-2-butanol **(e)** 3-Pentanol **(f)** 3-Ethyl-3-pentanol

9.42 Dihydropyran is synthesized by treating tetrahydrofurfuryl alcohol with acid. Propose a mechanism for this conversion.

Tetrahydrofurfuryl alcohol (H, O, CH_2OH) $\xrightarrow{ArSO_3H}$ Dihydropyran (O) $+ H_2O$

Tetrahydrofurfuryl alcohol Dihydropyran

9.43 Show how to convert propene to each of these compounds. Use any inorganic reagents as necessary.

(a) Propane **(b)** 1,2-Propanediol **(c)** 1-Propanol
(d) 2-Propanol **(e)** Propanal **(f)** Propanone
(g) Propanoic acid **(h)** 1-Bromo-2-propanol **(i)** 3-Chloropropene
(j) 1,2,3-trichloropropane **(k)** 1-Chloropropane **(l)** 2-Chloropropane
(m) 2-Propen-1-ol **(n)** Propenal

9.44 Show how to bring about this conversion in good yield.

OH (decalin-4a-ol) $\longrightarrow$ cyclodecane-1,6-dione (O, O)

9.45 The tosylate of a primary alcohol normally undergoes an S_N2 reaction with hydroxide ion to give a primary alcohol. Reaction of this tosylate, however, gives a compound of molecular formula $C_7H_{12}O$. Propose a structural formula for this compound and a mechanism for its formation.

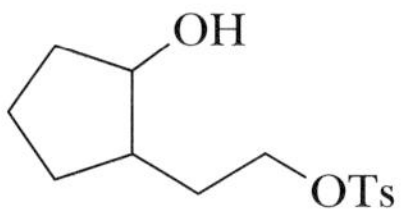

OH, OTs; NaOH → $C_7H_{12}O$

9.46 Show how to convert cyclohexene to each compound in good yield.

(a) 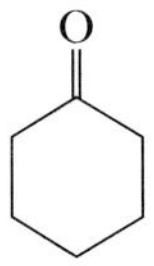 (b) 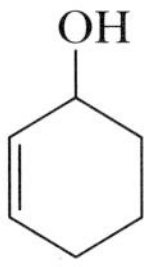 (c) 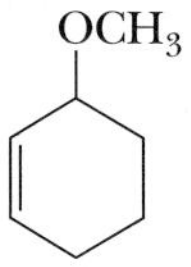 (d) 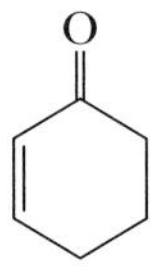 (e) $HC(=O)(CH_2)_4CH(=O)$

Molecular Modeling

9.47 Oxymercuration of an alkene followed by reduction with sodium borohydride is regioselective. Oxymercuration of this bicycloalkene followed by reduction gives a single alcohol in better than 95% yield. Propose a structural formula for the alcohol formed. *Hint:* Build a line angle structure in ChemDraw, import it into Chem3D, minimize its energy, and then see if you can determine which face of the double bond is more accessible to the oxymercuration reagent.

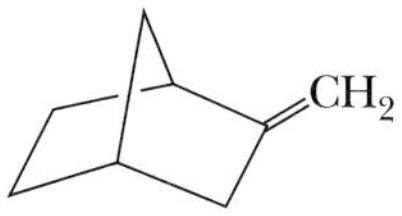

CH_2

ALKYNES

10

In this chapter, we continue our discussion of the chemistry of carbon-carbon pi bonds as we consider the chemistry of alkynes. Just as alkenes and alkynes are similar in that the multiple bond in each is a combination of sigma and pi bonds, they also show similarities in the types of chemical reactions they undergo. Alkynes undergo electrophilic additions of X_2, HX, and H_2O. They undergo hydroboration/oxidation and the carbon-carbon triple bond can be reduced first to a double bond and then to a single bond. Another important reaction of alkynes is conversion of terminal alkynes to their alkali metal salts, which are good nucleophiles and, therefore, valuable building blocks for the construction of larger molecules.

■ Cutting with an oxyacetylene torch. *(© T. J. Florian, Rainbow)*

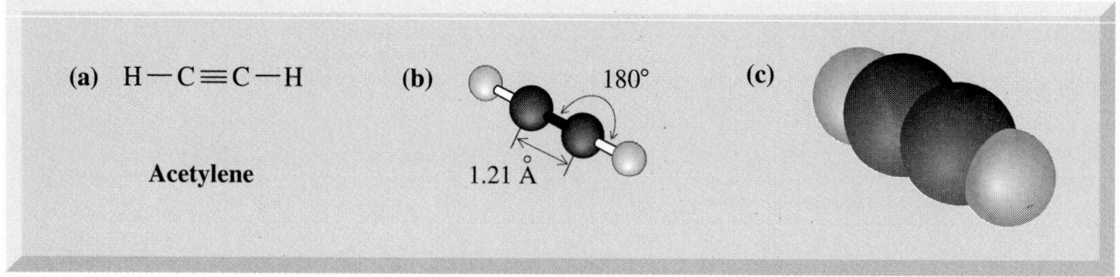

FIGURE 10.1
Acetylene. (a) Lewis structure, (b) ball-and-stick model, and (c) space-filling model.

10.1 Structure

Alkyne An unsaturated compound that contains one or more carbon-carbon triple bonds.

The functional group of an **alkyne** is a carbon-carbon triple bond. The simplest alkyne is ethyne, C_2H_2, more commonly named acetylene (Figure 10.1). Acetylene is a linear molecule; all bond angles are 180°. The carbon-carbon bond length in acetylene is 1.21 Å (Table 5.1). By comparison, the length of the carbon-carbon double bond in ethylene is 1.34 Å and that of the carbon-carbon single bond in ethane is 1.54 Å. Thus, triple bonds are shorter than double bonds, which, in turn, are shorter than single bonds. The bond dissociation energy of the carbon-carbon triple bond in acetylene [230 kcal/mol (962 kJ/mol)] is also considerably larger than that for the carbon-carbon double bond in ethylene [163 kcal/mol (682 kJ/mol)] and the carbon-carbon single bond in ethane [90 kcal/mol (377 kJ/mol)].

A triple bond is described in terms of overlap of *sp* hybrid orbitals of adjacent carbon atoms to form a sigma bond, overlap of parallel $2p_y$ orbitals to form one pi bond, and overlap of parallel $2p_z$ orbitals to form a second pi bond (Figure 1.21). In acetylene, each carbon-hydrogen bond is formed by overlap of a 1*s* orbital of hydrogen with an *sp* orbital of carbon.

10.2 Nomenclature

A. IUPAC Names

According to rules of the IUPAC system, the infix **-yn-** is used to show the presence of a carbon-carbon triple bond (Section 2.5). Thus, HC≡CH is named ethyne and CH_3C≡CH is named propyne. The IUPAC system retains the name acetylene and, therefore, there are two acceptable names for HC≡CH: ethyne and acetylene. Of these two names, acetylene is used much more frequently. For larger molecules, the longest carbon chain that contains the triple bond is numbered from the end that gives the triply bonded carbons the smaller numbers. The location of the triple bond is indicated by the number of the first carbon of the triple bond. If a hydrocarbon chain contains more than one triple bond, the infixes **-adiyn-**, **-atriyn-**, and so forth, are used.

4 3 2 1
$CH_3CHC{\equiv}CH$ (with CH_3 on C3)
3-Methyl-1-butyne

1 2 3 45 6 7
$CH_3CH_2C{\equiv}CCH_2CCH_3$ (with two CH_3 on C6)
6,6-Dimethyl-3-heptyne

1 23 4 5 6 7
$HC{\equiv}CCH_2CH_2CH_2C{\equiv}CH$
1,6-Heptadiyne

EXAMPLE 10.1

Write the IUPAC name for each compound.

(a) $CH_3CH_2C{\equiv}CCH_3$ **(b)** $ClCH_2C{\equiv}CH$ **(c)** $HC{\equiv}C(CH_2)_3CH_2OH$

Solution

(a) 2-Pentyne **(b)** 3-Chloro-1-propyne
(c) 5-Hexyn-1-ol. The hydroxyl group is indicated by the suffix -ol, and its location determines the numbering of the carbon chain.

PROBLEM 10.1

Write the IUPAC name for each compound.

(a) $CH_3(CH_2)_5C{\equiv}CH$ **(b)** $CH_3C(OH)(CH_3)C{\equiv}CH$ **(c)** $CH_3C(CH_3)(CH_3)C{\equiv}CH$

B. Common Names

Common names for alkynes are derived by prefixing the names of substituents on the carbon-carbon triple bond to the name acetylene as shown in the following examples. Note in the third example that when a carbon-carbon double bond (indicated by -en-) and a carbon-carbon triple bond (indicated by -yn-) are both present in the same molecule, the IUPAC rules specify that the location of the double bond takes precedence in numbering the compound.

	$CH_3C{\equiv}CH$	$CH_3C{\equiv}CCH_3$	$CH_2{=}CHC{\equiv}CH$
IUPAC name:	Propyne	2-Butyne	1-Buten-3-yne
Common name:	Methylacetylene	Dimethylacetylene	Vinylacetylene

EXAMPLE 10.2

Write the common name for each compound.

(a) $CH_3CH(CH_3)CH_2C{\equiv}CCH_3$ **(b)** $CH_3CH_2CH(CH_3)C{\equiv}CH$ **(c)** $HC{\equiv}CC(CH_3)_2CH_3$

Solution

(a) Isobutylmethylacetylene **(b)** *sec*-Butylacetylene **(c)** *tert*-Butylacetylene

PROBLEM 10.2

Write the common name for each compound.

(a) $\underset{}{CH_3\overset{\displaystyle CH_3}{\overset{|}{C}}HC{\equiv}C\overset{\displaystyle CH_3}{\overset{|}{C}}HCH_3}$ (b) cyclohexyl$-C{\equiv}CH$ (c) $HC{\equiv}CCH_2CH_2CH_2CH_3$

The smallest cycloalkyne that has been isolated is cyclooctyne. This molecule is quite unstable, and polymerizes rapidly at room temperature. The C—C≡C bond angle in cyclooctyne is approximately 155°, indicating a high degree of angle strain. Cyclononyne has also been prepared and is stable at room temperature. The C—C≡C bond angles in this cycloalkyne are approximately 160°, which still represents a considerable distortion from the optimal C—C≡C bond angle of 180°. You can see the distortion of the C—C≡C bond angles in the accompanying stereoview. You can also see the degree to which C—C and C—H bonds on adjacent carbons are staggered, thus minimizing torsional strain.

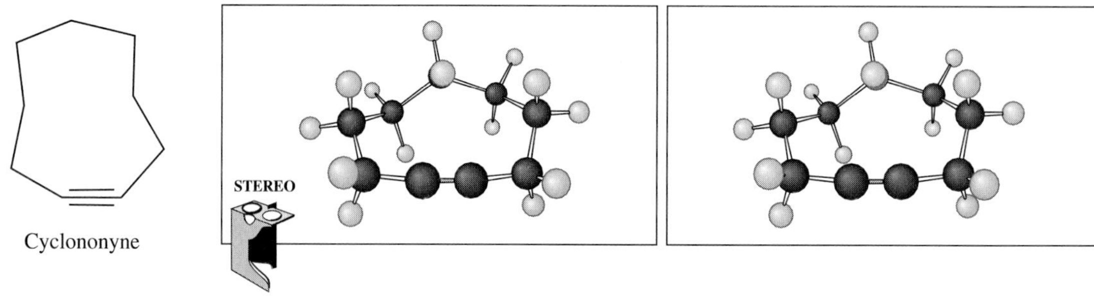

Cyclononyne

10.3 Physical Properties

The physical properties of alkynes are quite similar to those of alkanes and alkenes with similar carbon skeletons. The lower-molecular-weight alkynes are gases at room temperature. Those that are liquids at room temperature all have densities less than 1.0 g/mL (they are less dense than water). Listed in Table 10.1 are melting points,

TABLE 10.1 Physical Properties of Some Low-Molecular-Weight Alkynes

Name	Formula	Melting Point (°C)	Boiling Point (°C)	Density at 20°C (g/mL)
ethyne	$HC{\equiv}CH$	−81	−84	(a gas)
propyne	$CH_3C{\equiv}CH$	−102	−23	(a gas)
1-butyne	$CH_3CH_2C{\equiv}CH$	−126	8	(a gas)
2-butyne	$CH_3C{\equiv}CCH_3$	−32	27	0.691
1-pentyne	$CH_3(CH_2)_2C{\equiv}CH$	−90	40	0.690
1-hexyne	$CH_3(CH_2)_3C{\equiv}CH$	−132	71	0.716
1-octyne	$CH_3(CH_2)_5C{\equiv}CH$	−79	125	0.746
1-decyne	$CH_3(CH_2)_7C{\equiv}CH$	−36	174	0.766

boiling points, and densities of several low-molecular-weight alkynes. Because alkynes are nonpolar compounds, they are insoluble in water and other polar solvents. They are soluble in each other and in other nonpolar organic solvents.

10.4 Acidity

One of the major differences between the chemistry of alkynes and alkenes is that a hydrogen attached to a triply bonded carbon atom is sufficiently acidic that it can be removed by a strong base, such as sodium amide or sodium hydride. Shown in Table 10.2 are pK_a values for acetylene, ethylene, and ethane. Also shown for comparison is the value for water.

We discussed the acidity of terminal alkynes compared with terminal alkenes and alkanes in Section 3.3C. The lone pair of electrons of the anion from an alkyne lies in an *sp* hybrid orbital that has 50% *s* character; an alkyne anion is the most stable and, therefore, the least basic. Conversely, the lone pair of electrons of an anion derived from an alkane lies in an sp^3 orbital that has only 25% *s* character; the anion derived from an alkane is the least stable and, therefore, the most basic.

Because acetylene is a stronger acid than ammonia, it reacts with the amide ion to form the acetylide anion and ammonia. The position of the equilibrium for this reaction lies very much toward the right.

$$HC{\equiv}CH + {}^{-}{:}\ddot{N}H_2 \rightleftharpoons HC{\equiv}C{:}^{-} + {:}NH_3 \qquad K_{eq} = 10^{13}$$

pK_a 25			pK_a 38
Stronger acid	Stronger base	Weaker base	Weaker acid

Other strong bases commonly used to form acetylide anions are sodium hydride and lithium diisopropylamide (LDA).

$Na^{+}\ {}^{-}{:}H$ — Sodium hydride

$[(CH_3)_2CH]_2\ddot{N}{:}^{-}\ Li^{+}$ — Lithium diisopropylamide (LDA)

Water is a stronger acid than acetylene and, therefore, the hydroxide ion is not a strong enough base to convert acetylene to acetylide anion.

$$HC{\equiv}CH + {}^{-}{:}\ddot{O}H \rightleftharpoons HC{\equiv}C{:}^{-} + H_2\ddot{O}{:} \qquad K_{eq} = 10^{-9.3}$$

pK_a 25			pK_a 15.7
Weaker acid	Weaker base	Stronger base	Stronger acid

TABLE 10.2 Acidity of Alkanes, Alkenes, and Alkynes

Weak Acid	Example	Conjugate Base	pK_a
water	$H\ddot{O}{-}H$	$H\ddot{O}{:}^{-}$	15.7
alkyne	$HC{\equiv}C{-}H$	$HC{\equiv}C{:}^{-}$	25
alkene	$CH_2{=}CH{-}H$	$CH_2{=}CH{:}^{-}$	44
alkane	$CH_3CH_2{-}H$	$CH_3CH_2{:}^{-}$	51

Increasing acidity (upward arrow: alkane → water)

The pK_a values for alkene and alkane hydrogens are so large (they are so weakly acidic) that neither the commonly used alkali metal hydroxides nor sodium hydride, sodium amide, or lithium diisopropylamide are strong enough bases to remove a proton from alkanes or alkenes.

10.5 Alkylation of Acetylide Anions

Alkylation Any reaction in which a new carbon-carbon bond to an alkyl group is formed.

Acetylide anions are both strong bases and good nucleophiles. As nucleophiles, they undergo S_N2 reactions with alkyl halides, alkyl tosylates, and alkyl mesylates to form new carbon-carbon bonds to alkyl groups, that is, they undergo alkylation reactions. **Alkylation** is any reaction in which a new carbon-carbon bond to an alkyl group is formed. Treatment of sodium acetylide, for example, with 1-bromobutane gives 1-hexyne.

$$HC{\equiv}C{:}^-Na^+ + CH_3(CH_2)_2CH_2{-}Br \xrightarrow{S_N2} HC{\equiv}CCH_2(CH_2)_2CH_3 + Na^+Br^-$$

Sodium acetylide — 1-Bromobutane — 1-Hexyne

Because of the ready availability of acetylene and the ease with which it is converted to a good nucleophilic anion, alkylation of the acetylide anion is the most convenient laboratory method for the synthesis of terminal alkynes. The process of alkylation can be repeated and a terminal alkyne can, in turn, be converted to an internal alkyne.

$$CH_3CH_2C{\equiv}C{:}^-Na^+ + CH_3CH_2{-}Br \xrightarrow{S_N2} CH_3CH_2C{\equiv}CCH_2CH_3 + Na^+Br^-$$

Sodium butynide — Bromoethane — 3-Hexyne

Because acetylide anions are strong bases as well as good nucleophiles, alkylation of acetylide anions is practical only with methyl and primary halides. With secondary and tertiary halides, E2 elimination becomes the main reaction (Section 8.11B).

$$HC{\equiv}C{:}^-Na^+ + \text{Bromocyclohexane} \xrightarrow{E2} HC{\equiv}CH + \text{Cyclohexene} + Na^+Br^-$$

Sodium acetylide — Bromocyclohexane — Acetylene — Cyclohexene

10.6 Preparation

In addition to alkylation of salts of terminal alkynes, there are several other important types of reactions by which alkynes can be made. Starting materials for these methods include alkyl halides and alkenes.

Alkynes can be synthesized from alkenes by the combination of an addition reaction followed by two successive β-elimination reactions. In the addition reaction, the alkene is treated with bromine or chlorine to form a dibromoalkane or dichloroalkane (Section 6.3C).

$$\mathrm{RCH{=}CHR} + \mathrm{Br_2} \xrightarrow{\mathrm{CCl_4}} \underset{\mathrm{Br}}{\mathrm{RCH}}{-}\underset{\mathrm{Br}}{\mathrm{CHR}}$$

An alkene — A dibromoalkane

Treatment of such a dibromide with 2 mol of strong base, most commonly sodium amide, results in the elimination of 2 mol of HBr by successive E2 reactions and formation of an alkyne. The two halogens may be on adjacent carbons (a **vicinal dihalide,** from the Latin, *vicinalis*, neighbor) or they may be on the same carbon (a **geminal dihalide,** from the Latin, *geminatus*, twin).

Vicinal (vic-) dihalide Vicinal, from the Latin, *vicinalis*, neighbor. Indicates the presence of two halogen atoms, on adjacent carbon atoms.

Geminal (gem-) dihalide Geminal, from the Latin, *geminatus*, twin. Indicates the presence of two halogens, both on the same carbon atom.

$$\mathrm{R{-}\overset{H}{\underset{Br}{C}}{-}\overset{H}{\underset{Br}{C}}{-}R} + 2\mathrm{NaNH_2} \xrightarrow[-33°\mathrm{C}]{\mathrm{NH_3(l)}} \mathrm{R{-}C{\equiv}C{-}R} + 2\mathrm{NaBr} + 2\mathrm{NH_3}$$

A vicinal dibromide — Sodium amide — An alkyne

Double dehydrohalogenation to form an alkyne may also be carried out using potassium *tert*-butoxide in dimethyl sulfoxide (DMSO).

$$\mathrm{R{-}\overset{H}{\underset{H}{C}}{-}\overset{Br}{\underset{Br}{C}}{-}R} + 2(\mathrm{CH_3})_3\mathrm{CO^-K^+} \xrightarrow{\mathrm{DMSO}} \mathrm{R{-}C{\equiv}C{-}R} + 2(\mathrm{CH_3})_3\mathrm{COH} + 2\mathrm{K^+Br^-}$$

A geminal dibromide — Potassium *tert*-butoxide — An alkyne — 2-Methyl-2-propanol

With a strong base such as sodium amide, both dehydrohalogenations occur readily. However, with weaker bases, such as sodium hydroxide or potassium hydroxide in ethanol, it is often possible to stop the reaction after the first dehydrohalogenation and isolate the haloalkene. In practice, it is much more common to use a stronger base and go directly to the alkyne.

Given the ease of converting alkenes to dihaloalkanes followed by base-promoted double dehydrohalogenations, alkenes are valuable starting materials for the preparation of alkynes. The following equations show the conversion of 1-hexene to 1-hexyne. Note that three equivalents of sodium amide are used in this sequence. Two equivalents are required for the double dehydrohalogenation reaction, which gives 1-hexyne. As soon as any 1-hexyne (a weak acid, pK_a 25) is formed, it reacts with sodium amide (a strong base) to form an alkyne salt. Thus, a third mol of sodium amide is required to complete the dehydrohalogenation of the remaining bromoalkene. Addition of water or aqueous acid completes the sequence and gives 1-hexyne.

$$\mathrm{CH_3(CH_2)_3CH{=}CH_2} \xrightarrow{\mathrm{Br_2}} \mathrm{CH_3(CH_2)_3\underset{}{\overset{Br}{C}}H{-}\overset{Br}{C}H_2} \xrightarrow[-2\mathrm{HBr}]{2\mathrm{NaNH_2}} \mathrm{CH_3(CH_2)_3C{\equiv}CH} \xrightarrow{\mathrm{NaNH_2}}$$

1-Hexene — 1,2-Dibromohexane — 1-Hexyne

$$\mathrm{CH_3(CH_2)_3C{\equiv}C^-Na^+} \xrightarrow{\mathrm{H_2O}} \mathrm{CH_3(CH_2)_3C{\equiv}CH}$$

Sodium salt of 1-hexyne — 1-Hexyne

In dehydrohalogenation of a haloalkene with at least one hydrogen on each adjacent carbon, a side reaction occurs that must be considered, namely, formation of an allene.

$$\underset{\text{A haloalkene (a vinylic halide)}}{\mathrm{RC(H)(R){-}C(X){=}C(H)R}} \xrightarrow[-\mathrm{HBr}]{\mathrm{NaNH_2}} \underset{\text{An allene}}{\mathrm{R_2C{=}C{=}C(H)R}} \;+\; \underset{\text{An alkyne}}{\mathrm{RC(H)(R)C{\equiv}CR}}$$

Allene The compound $CH_2{=}C{=}CH_2$. Any compound that contains adjacent carbon-carbon double bonds, that is, $C{=}C{=}C$.

An **allene** has two adjacent carbon-carbon double bonds, that is, $C{=}C{=}C$. The simplest allene is 1,2-propadiene, commonly named allene. In it, each end carbon is sp^2 hybridized and the middle carbon is sp hybridized. Each carbon-carbon sigma bond is formed by overlap of sp and sp^2 hybrid orbitals. One pi bond is formed by overlap of parallel $2p_y$ orbitals, the other by overlap of parallel $2p_z$ orbitals. The two pi bonding molecular orbitals are in planes perpendicular to each other, as are the two CH_2 groups.

$CH_2{=}C{=}CH_2$

1,2-Propadiene
(Allene)

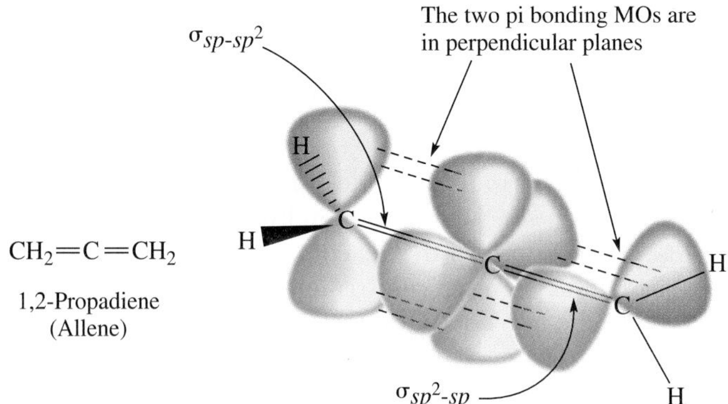

Most allenes are less stable than their isomeric alkynes. For example, allene itself is less stable by 1.6 kcal/mol than its isomer propyne, and 1,2-butadiene is less stable by 4.0 kcal/mol than its isomer 2-butyne.

$$CH_2{=}C{=}CH_2 \longrightarrow CH_3C{\equiv}CH \qquad \Delta H^0 = -1.6\ \text{kcal/mol}$$

$$CH_2{=}C{=}CHCH_3 \longrightarrow CH_3C{\equiv}CCH_3 \qquad \Delta H^0 = -4.0\ \text{kcal/mol}$$

Because of their lower stability relative to isomeric alkynes, allenes are generally only minor products of alkyne-forming dehydrohalogenation reactions.

EXAMPLE 10.3

Show how you might prepare 2-butyne from each starting material:

(a) $CH_3CH{=}CHCH_3$ **(b)** $CH_3CH_2CHCH_3$ (with Cl on the CH carbon: $CH_3CH_2CH(Cl)CH_3$)

Solution

(a) This synthesis can be done in two steps: treatment of 2-butene with 1 mol of bromine followed by dehydrohalogenation with 2 mol of sodium amide.

$$CH_3CH{=}CHCH_3 + Br_2 \longrightarrow CH_3\underset{\displaystyle Br}{\underset{|}{C}}H{-}\underset{\displaystyle Br}{\underset{|}{C}}HCH_3$$

$$CH_3\underset{\displaystyle Br}{\underset{|}{C}}H{-}\underset{\displaystyle Br}{\underset{|}{C}}HCH_3 + 2NaNH_2 \longrightarrow CH_3C{\equiv}CCH_3 + 2NaBr + 2NH_3$$

(b) This synthesis requires three steps. First, dehydrohalogenation of 2-chlorobutane with sodium hydroxide (Section 8.8), followed by addition of chlorine (or bromine), and then a double dehydrohalogenation as in part (a) of this problem.

$$CH_3CH_2\underset{\displaystyle Cl}{\underset{|}{C}}HCH_3 + NaOH \longrightarrow CH_3CH{=}CHCH_3 + NaCl + H_2O$$

$$CH_3CH{=}CHCH_3 \xrightarrow{Cl_2} CH_3\underset{\displaystyle Cl}{\underset{|}{C}}H{-}\underset{\displaystyle Cl}{\underset{|}{C}}HCH_3 \xrightarrow{2NaNH_2} CH_3C{\equiv}CCH_3$$

PROBLEM 10.3

Draw a structural formula for an alkene and dichloroalkane of the given molecular formula that yields the indicated alkyne by each reaction sequence.

(a) $C_6H_{12} \xrightarrow{Cl_2} C_6H_{12}Cl_2 \xrightarrow{2NaNH_2} CH_3CH_2C{\equiv}CCH_2CH_3$

(b) $C_7H_{14} \xrightarrow{Cl_2} C_7H_{14}Cl_2 \xrightarrow{2NaNH_2} CH_3C{\equiv}C\overset{\displaystyle CH_3}{\overset{|}{\underset{\displaystyle CH_3}{\underset{|}{C}}}}CH_3$

10.7 Reduction

Two types of reactions are used to convert alkynes to alkenes and to alkanes, namely, catalytic reduction and chemical reduction.

A. Catalytic Reduction

Treatment of an alkyne with hydrogen in the presence of a transition metal catalyst, most commonly palladium, platinum, or nickel, results in the addition of 2 mol of hydrogen to the alkyne and its conversion to an alkane. Catalytic reduction of an alkyne can be brought about at or slightly above room temperature and with moderate pressures of hydrogen gas.

$$\underset{\text{2-Butyne}}{CH_3C{\equiv}CCH_3} + 2H_2 \xrightarrow[\text{3 atm}]{\text{Pd, Pt, or Ni}} \underset{\text{Butane}}{CH_3CH_2CH_2CH_3}$$

Lindlar catalyst Finely powdered palladium metal deposited on solid calcium carbonate that has been specially modified with lead salts. Its particular use is as a catalyst for the reduction of an alkyne to a *cis* alkene.

Reduction of an alkyne occurs in two stages: first addition of 1 mol of hydrogen to form an alkene and then addition of the second mol of hydrogen to the alkene to form the alkane.

By careful choice of catalyst, it is possible to stop the reduction after the addition of 1 mol of hydrogen. The catalyst most commonly used for this purpose consists of finely powdered palladium metal deposited on solid calcium carbonate that has been specially modified with lead salts. This combination is known as the **Lindlar catalyst.** Reduction (hydrogenation) of alkynes over a Lindlar catalyst is stereoselective; **syn addition** of two hydrogen atoms to the carbon-carbon triple bond gives a *cis* alkene.

$$CH_3C{\equiv}CCH_3 + H_2 \xrightarrow{\text{Lindlar catalyst}} \underset{H\quad\quad H}{\overset{H_3C\quad\quad CH_3}{C{=}C}}$$

2-Butyne — *cis*-2-Butene

Because addition of hydrogen in the presence of the Lindlar catalyst is stereoselective for syn-addition, it has been proposed that reduction proceeds by simultaneous or nearly simultaneous transfer of two hydrogen atoms from the surface of the metal catalyst to the alkyne. We presented a similar mechanism in Section 6.6A for the catalytic reduction of an alkene to an alkane.

B. Chemical Reduction

Alkynes can also be reduced to alkenes by using either sodium or lithium metal in liquid ammonia or in low-molecular-weight primary or secondary amines. The alkali metal is the reducing agent and in the process is oxidized to M^+. Reduction of an alkyne to an alkene by lithium or sodium in liquid ammonia, NH_3 (l), is 100% stereoselective; it involves **anti addition** of 2 hydrogen atoms to the triple bond.

$$RC{\equiv}CR' + 2Na + 2NH_3 \xrightarrow[NH_3(l)]{} \underset{H\quad\quad R'}{\overset{R\quad\quad H}{C{=}C}} + 2NaNH_2$$

An alkyne — A *trans*-alkene

$$CH_3CH_2CH_2C{\equiv}CCH_2CH_2CH_3 \xrightarrow[NH_3(l)]{2Na} \underset{H\quad\quad CH_2CH_2CH_3}{\overset{CH_3CH_2CH_2\quad\quad H}{C{=}C}}$$

4-Octyne — *trans*-4-Octene

Catalytic and alkali metal reduction of alkynes are complementary reactions; by the proper choice of reagents and reaction conditions, it is possible to reduce an alkyne to either a *cis*-alkene or a *trans*-alkene.

The stereoselectivity of alkali metal reduction of alkynes to alkenes can be accounted for by the following mechanism. In Step 1, the alkyne undergoes a one-electron reduction to form an **alkenyl radical anion** (an ion containing an unpaired electron on one carbon and a negative charge on an adjacent carbon). In Step 2, the alkenyl radical anion (a very strong base) abstracts a proton from a molecule of ammonia (under these conditions, a weak acid) to form an **alkenyl radical.** There follows a second one-electron reduction in Step 3 to give an **alkenyl anion.** The *trans*-alkenyl anion is more stable than its *cis* isomer and it is at this step that the stereochemistry of the final product is determined. A second proton-transfer reaction in Step 4 completes the reduction and gives the *trans*-alkene. Adding the four steps and canceling species which appear on both sides of the equation gives the overall equation for the reaction.

MECHANISM Reduction of an Alkyne by Sodium in Liquid Ammonia

Step 1: A one-electron reduction of the alkyne to a radical anion

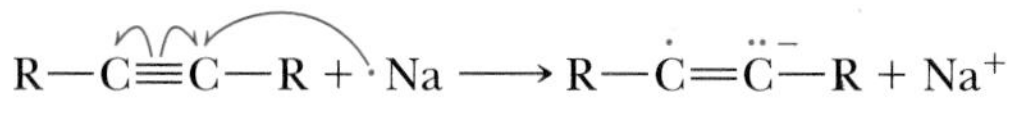

$$R—C{\equiv}C—R + \cdot Na \longrightarrow R—\dot{C}{=}\ddot{C}^{-}—R + Na^{+}$$

An alkenyl radical anion

Step 2: An acid-base reaction to form an alkenyl radical and amide ion

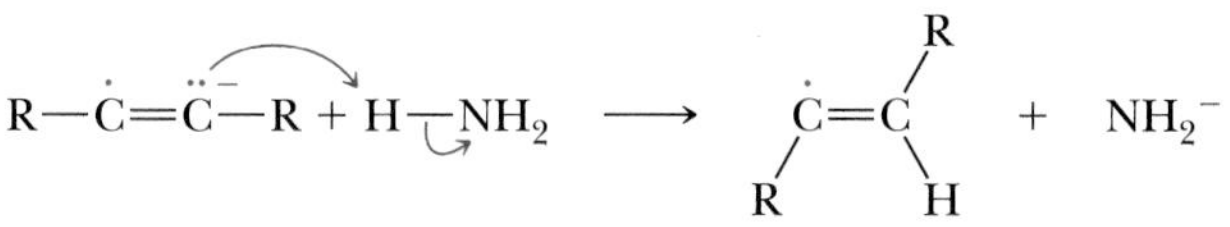

$$R—\dot{C}{=}\ddot{C}^{-}—R + H—NH_2 \longrightarrow \dot{C}(R){=}C(R)(H) + NH_2^{-}$$

An alkenyl radical

Step 3: A second one-electron reduction to form an alkenyl anion

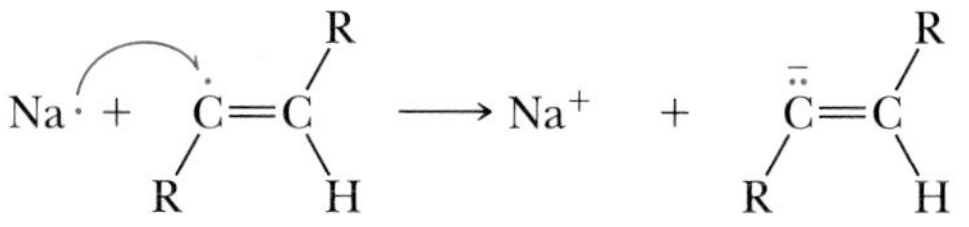

$$Na\cdot + \dot{C}(R){=}C(R)(H) \longrightarrow Na^{+} + \ddot{C}^{-}(R){=}C(R)(H)$$

An alkenyl anion

Step 4: A second acid-base reaction to give the *trans*-alkene

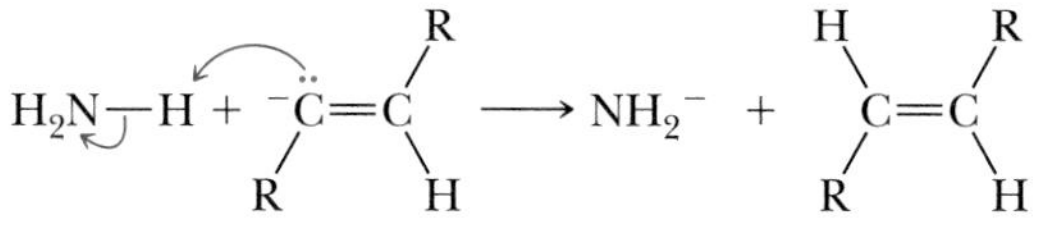

$$H_2N—H + {}^{-}\ddot{C}(R){=}C(R)(H) \longrightarrow NH_2^{-} + C(H)(R){=}C(R)(H)$$

A *trans*-alkene

10.8 Hydroboration

Borane adds readily to an internal alkyne as illustrated by its reaction with 3-hexyne. The product of **hydroboration** of an alkyne is a trialkenylborane (the infix -enyl- shows the presence of a carbon-carbon double bond on the carbon attached to boron).

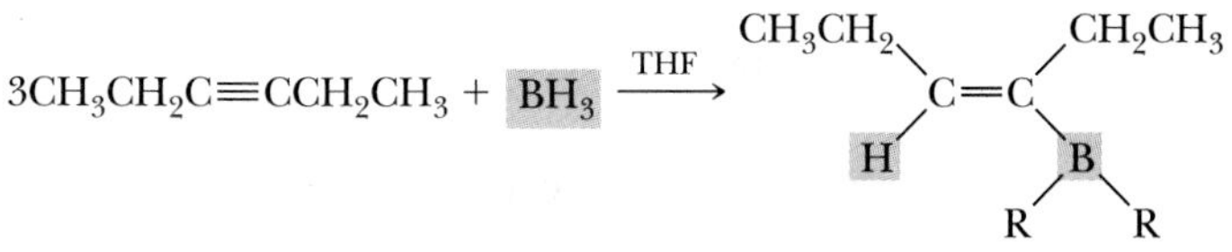

$$3CH_3CH_2C{\equiv}CCH_2CH_3 + BH_3 \xrightarrow{THF} (CH_3CH_2)(H)C{=}C(CH_2CH_3)(BR_2)$$

3-Hexyne

A trialkenylborane (R = *cis*-3-Hexenyl group)

As for alkenes, hydroboration of alkynes is also stereoselective; it involves **syn addition** of hydrogen and boron.

Treatment of a trialkenylborane with a carboxylic acid, such as acetic acid, results in stereoselective replacement of boron by hydrogen: a *cis*-alkenyl group attached to boron is converted to a *cis*-alkene.

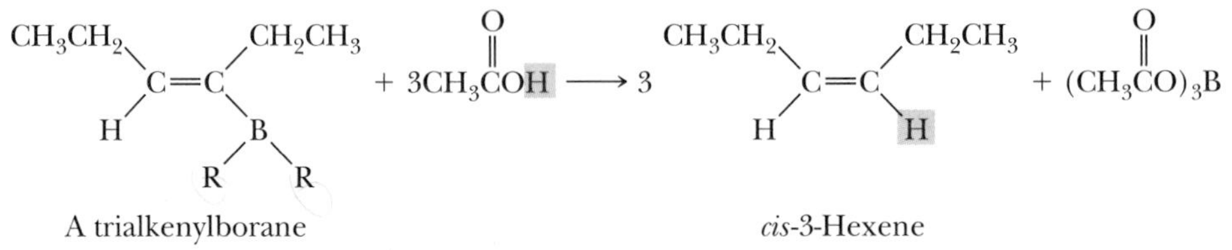

$$(CH_3CH_2)(H)C{=}C(CH_2CH_3)(BR_2) + 3CH_3\overset{O}{\overset{\|}{C}}OH \longrightarrow 3\ (CH_3CH_2)(H)C{=}C(CH_2CH_3)(H) + (CH_3\overset{O}{\overset{\|}{C}}O)_3B$$

A trialkenylborane   *cis*-3-Hexene

The net effect of hydroboration of an internal alkyne followed by treatment with acetic acid is reduction of the alkyne to a *cis*-alkene. Thus, hydroboration-protonolysis and catalytic reduction over a Lindlar catalyst provide alternative schemes for conversion of an alkyne to a *cis*-alkene.

Terminal alkynes react regioselectively with borane to form trialkenylboranes. In practice, however, the reaction is difficult to stop at this stage because the alkenyl groups react further with borane to undergo a second hydroboration.

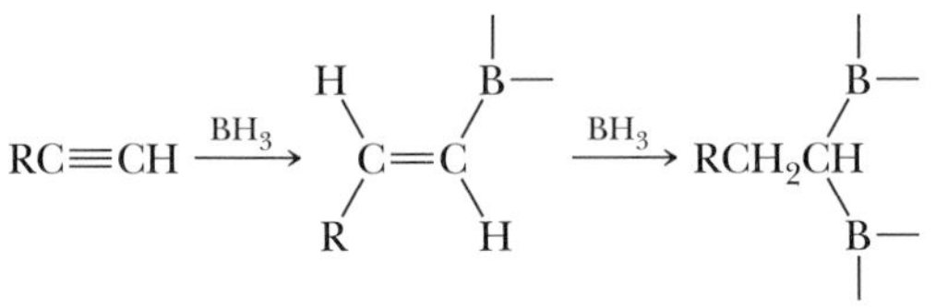

$$RC{\equiv}CH \xrightarrow{BH_3} (H)(R)C{=}C(B{<})(H) \xrightarrow{BH_3} RCH_2CH(B{<})_2$$

It is possible to prevent the second hydroboration step and, in effect, stop the reaction at the alkenylborane stage by using a sterically hindered disubstituted borane. One of the most widely used of these is disiamylborane $[(sia)_2BH]$, prepared by treatment of borane with 2 equivalents of 2-methyl-2-butene.

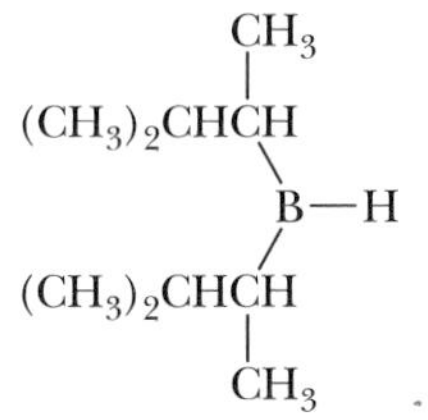

$$[(CH_3)_2CHCH(CH_3)]_2B{-}H$$

di-**s**ec-**i**so**a**mylborane
$[(\mathbf{sia})_2BH]$

Reaction of this hindered dialkylborane with a terminal alkyne results in a single addition and formation of an alkenylborane.

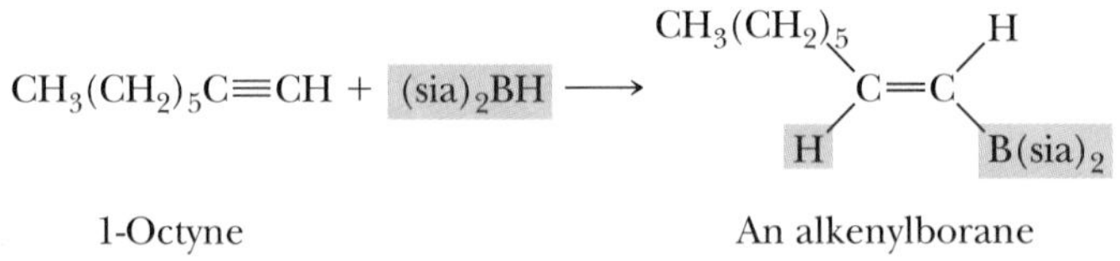

$$CH_3(CH_2)_5C{\equiv}CH + (sia)_2BH \longrightarrow (CH_3(CH_2)_5)(H)C{=}C(H)(B(sia)_2)$$

1-Octyne   An alkenylborane

Note that, as for unsymmetrical alkenes, the addition of $(sia)_2BH$ to a carbon-carbon triple bond of a terminal alkene is regioselective; boron adds to the less substituted carbon.

Treatment of an alkenylborane with hydrogen peroxide in aqueous sodium hydroxide gives a product that corresponds to hydration of an alkyne, that is, addition of H to one carbon of the triple bond and OH to the other as illustrated by the hydroboration-oxidation of 2-butyne.

$$\underset{\text{2-Butyne}}{CH_3C{\equiv}CCH_3} \xrightarrow[\text{2. } H_2O_2,\ NaOH]{\text{1. } BH_3} \underset{\substack{\text{2-Buten-2-ol}\\ \text{(an enol)}}}{CH_3CH{=}\overset{OH}{\overset{|}{C}}CH_3} \rightleftharpoons \underset{\substack{\text{2-Butanone}\\ \text{(a ketone)}}}{CH_3\overset{H}{\overset{|}{C}}H{-}\overset{O}{\overset{\|}{C}}CH_3} \qquad \underset{\substack{\text{(for keto-enol}\\ \text{tautomerism)}}}{K_{eq} = 6.7 \times 10^6}$$

The initial product of hydroboration-oxidation is an **enol,** a compound containing a hydroxyl group attached to a carbon-carbon double bond. The name enol is derived from the IUPAC designation of it as both an alkene (-en-) and an alcohol (-ol). Enols are in equilibrium with an isomer formed by migration of a hydrogen atom from oxygen to carbon and rearrangement of the carbon-carbon double bond to form a carbon-oxygen double bond.

Enol A compound containing a hydroxyl group attached to a doubly bonded carbon atom.

The keto and enol forms of 2-butanone are said to be tautomers. **Tautomers** are constitutional isomers in equilibrium with each other that differ in the location of a hydrogen atom and a double bond relative to a heteroatom, most commonly O, N, and S. This type of isomerism is called **tautomerism.** Because the type of tautomerism we are dealing with in this section involves keto (from ketone) and enol forms, it is commonly called **keto-enol tautomerism.** As can be seen from the value of K_{eq}, 2-butanone (the keto form) is much more stable than its enol. We discuss keto-enol tautomerism in more detail in Section 15.11.

Tautomers Constitutional isomers in equilibrium with each other that differ in location of a hydrogen atom and a double bond relative to a heteroatom, most commonly O, N, or S.

Keto-enol tautomerism A type of isomerism involving keto (from ketone) and enol tautomers.

Thus, the product isolated after hydroboration-oxidation of 2-butyne is 2-butanone and that from hydroboration-oxidation of 3-hexyne is 3-hexanone.

$$\underset{\text{3-Hexyne}}{CH_3CH_2C{\equiv}CCH_2CH_3} \xrightarrow[\text{2. } H_2O_2,\ NaOH]{\text{1. } BH_3} \underset{\text{3-Hexanone}}{CH_3CH_2\overset{O}{\overset{\|}{C}}CH_2CH_2CH_3}$$

Hydroboration of a terminal alkyne using disiamylborane followed by oxidation in alkaline hydrogen peroxide gives an enol that is in equilibrium with the more stable aldehyde. Thus, hydroboration-oxidation of a terminal alkyne under these conditions gives an aldehyde.

$$\underset{\text{1-Octyne}}{CH_3(CH_2)_5C{\equiv}CH} \xrightarrow[\text{2. } H_2O_2,\ NaOH]{\text{1. } (sia)_2BH} \underset{\text{An enol}}{(H)(CH_3(CH_2)_5)C{=}C(OH)(H)} \rightleftharpoons \underset{\text{Octanal}}{CH_3(CH_2)_5CH_2\overset{O}{\overset{\|}{C}}H}$$

EXAMPLE 10.4

Hydroboration-oxidation of 2-pentyne gives a mixture of two ketones, each of molecular formula $C_5H_{10}O$. Propose structural formulas for these two ketones and for the enol from which each is derived.

$$CH_3CH_2C{\equiv}CCH_3 \xrightarrow[\text{2. } H_2O_2,\ NaOH]{\text{1. } BH_3} C_5H_{10}O$$

Solution

Because each carbon of the triple bond in 2-pentyne has the same degree of substitution, very little regioselectivity occurs during hydroboration. Two enols are formed, and from them, the isomeric ketones are formed.

$$\underset{\text{2-Pentyne}}{CH_3CH_2C{\equiv}CCH_3} \xrightarrow[\text{2. } H_2O_2,\ NaOH]{\text{1. } BH_3} \underset{\substack{\text{2-Pentene-3-ol}\\ \text{(an enol)}}}{CH_3CH_2\overset{\overset{\displaystyle OH}{|}}{C}{=}CHCH_3} + \underset{\substack{\text{2-Pentene-2-ol}\\ \text{(an enol)}}}{CH_3CH_2CH{=}\overset{\overset{\displaystyle OH}{|}}{C}CH_3}$$

$$\downarrow \qquad\qquad\qquad \downarrow$$

$$\underset{\text{3-Pentanone}}{CH_3CH_2\overset{\overset{\displaystyle O}{\|}}{C}CH_2CH_3} + \underset{\text{2-Pentanone}}{CH_3CH_2CH_2\overset{\overset{\displaystyle O}{\|}}{C}CH_3}$$

PROBLEM 10.4

Draw the structural formula for a hydrocarbon of the given molecular formula that undergoes hydroboration-oxidation to give the indicated product.

(a) $C_7H_{10} \xrightarrow[\text{2. } H_2O_2,\ NaOH]{\text{1. } (sia)_2BH}$ cyclopentyl$-CH_2\overset{\overset{\displaystyle O}{\|}}{C}H$

(b) $C_7H_{12} \xrightarrow[\text{2. } H_2O_2,\ NaOH]{\text{1. } BH_3}$ cyclopentyl$-CH_2CH_2OH$

10.9 Electrophilic Additions

Alkynes undergo many of the same electrophilic additions as alkenes. In this section, we study addition of bromine and chlorine, the hydrogen halides, and water.

A. Addition of Bromine and Chlorine

Addition of 1 mol of Br_2 to an alkyne gives a dibromoalkene. As illustrated by the reaction of 2-butyne with 1 mol of Br_2, addition of bromine to a triple bond is stereoselective; the major product corresponds to **anti addition** of the two bromine atoms. The preference for anti addition can be increased significantly by carrying out bromination in acetic acid in which is dissolved a source of bromide ion, for example LiBr.

$$\underset{\text{2-Butyne}}{CH_3C{\equiv}CCH_3} + Br_2 \xrightarrow[\substack{\text{anti}\\ \text{addition}}]{CH_3CO_2H,\ LiBr} \underset{(E)\text{-2,3-Dibromo-2-butene}}{(H_3C)(Br)C{=}C(Br)(CH_3)}$$

Addition of bromine to alkynes follows much the same type of mechanism as it does for addition to alkenes (Section 6.3C), namely, formation of a bridged bromonium ion intermediate, which is then attacked by bromide ion from the face opposite that occupied by the positively charged bromine atom. Alkynes similarly undergo addition of Cl_2, although less stereoselectively than with Br_2.

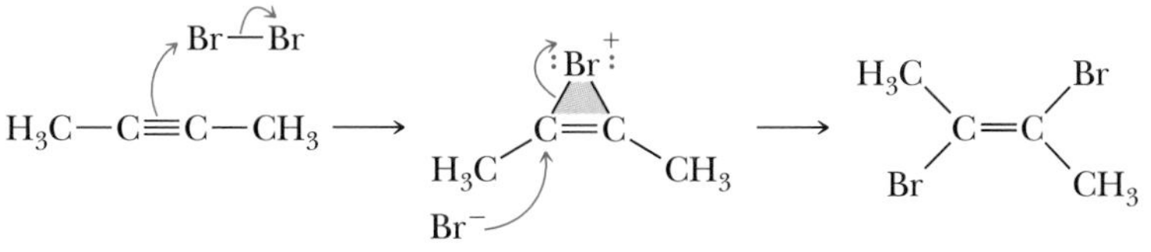

$$H_3C-C\equiv C-CH_3 \xrightarrow{Br-Br} \text{(bridged } Br^+ \text{ across } C=C\text{; } H_3C, CH_3\text{; attacked by } Br^-) \longrightarrow (H_3C)(Br)C=C(Br)(CH_3)$$

A bridged bromonium ion intermediate

Addition of a second mole of Br_2 gives a tetrabromoalkane.

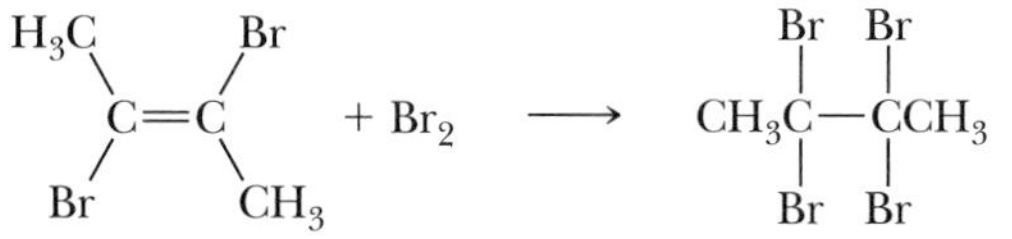

$$(H_3C)(Br)C=C(Br)(CH_3) + Br_2 \longrightarrow CH_3CBr_2-CBr_2CH_3$$

2,2,3,3-Tetrabromobutane

B. Addition of Hydrogen Halides

Alkynes add either 1 or two 2 mol of HBr and HCl depending on the ratios in which the alkyne and halogen acid are mixed.

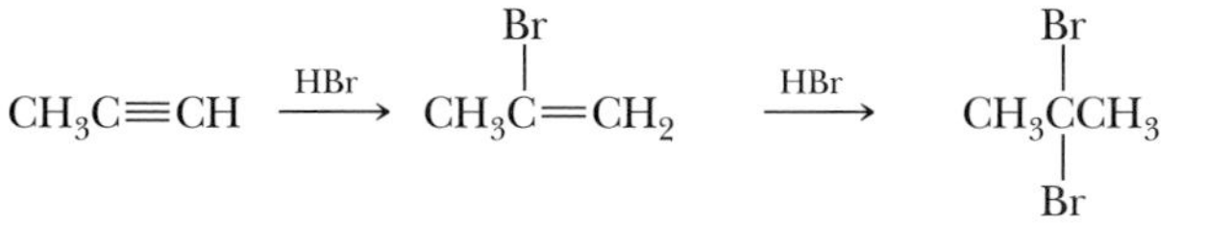

$$CH_3C\equiv CH \xrightarrow{HBr} CH_3C(Br)=CH_2 \xrightarrow{HBr} CH_3CBr_2CH_3$$

Propyne 2-Bromopropene 2,2-Dibromopropane

As shown in this equation, additions of both the first and second moles of HBr are regioselective. They follow **Markovnikov's rule** (Section 6.3A); hydrogen adds to the carbon that bears the greater number of hydrogens.

We can account for this regioselectivity of addition of HX by proposing a two-step mechanism for each. In Step 1, reaction of the pair of electrons of a pi bond of the alkyne with HBr forms a **vinylic carbocation.** The more stable 2° vinylic carbocation is formed in preference to a less stable 1° vinylic carbocation. The vinylic carbocation then reacts with bromide ion to give the observed product.

Vinylic carbocation A carbocation in which the positive charge of the cation is on one of the carbons of a carbon-carbon double bond.

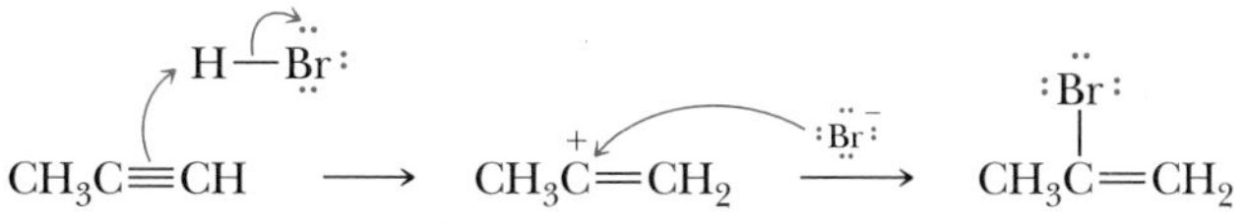

$$CH_3C\equiv CH \xrightarrow{H-\ddot{B}r:} CH_3\overset{+}{C}=CH_2 \xrightarrow{:\ddot{B}r:^-} CH_3C(:\ddot{B}r:)=CH_2$$

A 2° vinylic carbocation

Alkynes are considerably less reactive toward most electrophilic additions than are alkenes. The major reason for this difference is the instability of the vinylic carbocation intermediate formed from an alkyne compared with the alkyl carbocation formed from an alkene.

In the case of addition of the second mole of HX, Step 1 is again reaction of the electron pair of the remaining pi bond with HBr to form a carbocation. Of the two possible carbocations, the one with the positive charge on the carbon bearing the halogen is secondary and, therefore, favored over the primary carbocation. The sec-

Some products made of polyvinyl chloride (PVC). The symbol for recyclable PVC materials has a 3 in the center of the triangle of curved arrows with the letter V under the symbol. *(Charles D. Winters)*

ondary carbocation is also favored because of the possibility for resonance stabilization by the adjacent halogen atom.

$$CH_3C(Br){=}CH_2 + H{-}Br \xrightarrow{\text{slower}} CH_3CH(Br){-}\overset{+}{C}H_2$$

less stable primary carbocation

$$CH_3C(Br){=}CH_2 + H{-}Br \xrightarrow{\text{faster}} \left[CH_3\overset{+}{C}(Br){-}CH_3 \longleftrightarrow CH_3C(={}^{+}Br){-}CH_3 \right] \xrightarrow{:Br^-} CH_3CBr_2CH_3$$

More stable secondary carbocation (resonance stabilized)

Addition of 1 mol of HCl to acetylene gives chloroethene (vinyl chloride), a compound of considerable industrial importance.

$$\underset{\text{Ethyne (Acetylene)}}{HC{\equiv}CH} + HCl \longrightarrow \underset{\text{Chloroethene (Vinyl chloride)}}{CH_2{=}CHCl}$$

In 1995, the United States produced 15 billion pounds of vinyl chloride for use as a monomer in the production of the polymer poly(vinyl chloride). Poly(vinyl chloride) dominates much of the plumbing and construction market for plastics. Approximately 67% of all pipe, fittings, and conduit, along with 42% of all plastics used in construction at the present time, are fabricated from poly(vinyl chloride). We will describe the synthesis of PVC and its properties in Chapter 23. Our purpose here is to describe the synthesis of vinyl chloride.

$$\underset{\text{Vinyl chloride}}{nCH_2{=}CHCl} \xrightarrow{\text{catalyst}} \underset{\text{Poly(vinyl chloride) (PVC)}}{\left(CH_2CH(Cl) \right)_n}$$

At one time, hydrochlorination of acetylene was the major source of vinyl chloride. As the cost of production of acetylene increased, however, manufacturers of vinyl chloride sought other routes to this material. The starting material chosen was ethylene. What was sought was a way to convert ethylene to vinyl chloride.

$$\underset{\text{Ethylene}}{CH_2{=}CH_2} \xrightarrow{???} \underset{\text{Vinyl chloride}}{CH_2{=}CHCl}$$

One answer to this problem was to treat ethylene with chlorine gas to form 1,2-dichloroethane. When heated in the presence of charcoal or another catalyst, 1,2-dichloroethane loses a molecule of HCl to form vinyl chloride.

$$\underset{\text{Ethylene}}{CH_2{=}CH_2} \xrightarrow{Cl_2} \underset{\text{1,2-Dichloroethane}}{CH_2(Cl)CH_2(Cl)} \xrightarrow[\text{heat}]{} \underset{\text{Vinyl chloride}}{CH_2{=}CHCl} + HCl$$

A vinyl chloride plant in Texas. *(Courtesy of Occidental Petroleum Corporation)*

However, the byproduct HCl is corrosive and presents problems in handling and disposal. Although the problem of how to process this HCl was new, the solution had been discovered almost a century earlier. In 1868, Henry Deacon discovered that HCl can be oxidized to Cl_2 and H_2O by the oxygen in air when the gaseous mixture is passed over a copper(II) catalyst.

$$4HCl + O_2 \xrightarrow{CuCl_2} 2H_2O + 2Cl_2 \qquad \text{(Deacon process)}$$

This process has been improved by the use of new technology, and today vinyl chloride is made by passing ethylene, hydrogen chloride, and air over a copper(II) chloride-potassium chloride catalyst to give 1,2-dichloroethane, which is then thermally cracked to vinyl chloride and HCl. HCl is then recycled. The three reactions in this scheme and the net result are as follows.

$$\begin{array}{rl} 4HCl + O_2 \longrightarrow & 2H_2O + 2Cl_2 \\ 2CH_2{=}CH_2 + 2Cl_2 \longrightarrow & 2ClCH_2CH_2Cl \\ 2ClCH_2CH_2Cl \longrightarrow & 2CH_2{=}CHCl + 2HCl \\ \hline \end{array}$$

$$\text{Net reaction:} \quad 2CH_2{=}CH_2 + 2HCl + O_2 \longrightarrow 2CH_2{=}CHCl + 2H_2O$$

We described the production of vinyl chloride first from acetylene and then from ethylene to illustrate an important point about industrial organic chemistry. The aim is to produce a desired chemical from readily available and inexpensive starting materials by a direct scheme of reactions in which byproducts can be recycled in some useful way. The objective, increasingly sought after by all chemical companies, is to minimize both costs and production of materials that require disposal or can harm the environment.

C. Addition of Water: Hydration

In the presence of sulfuric acid and Hg(II) salts as catalysts, alkynes undergo the addition of water. The Hg(II) salts most often used for this purpose are mercury(II) sulfate or mercury(II) acetate. For terminal alkynes, addition of water follows Markovnikov's rule; hydrogen adds to the carbon atom of the triple bond bearing the hydrogen atom. The resulting enol is in equilibrium with a keto form (Section 10.8), and the product isolated is a ketone (an aldehyde in the case of acetylene itself).

$$\underset{\text{Propyne}}{CH_3C{\equiv}CH} + H_2O \xrightarrow[HgSO_4]{H_2SO_4} \underset{\substack{\text{1-Propen-2-ol}\\\text{(an enol)}}}{CH_3\overset{OH}{\overset{|}{C}}{=}CH_2} \rightleftharpoons \underset{\substack{\text{Propanone}\\\text{(Acetone)}}}{CH_3\overset{O}{\overset{\|}{C}}CH_3}$$

The mechanism for this reaction is illustrated by the hydration of propyne to give propanone (acetone). The first step in this Hg^{2+} catalyzed hydration is formation of a bridged mercurinium ion intermediate in a reaction similar to that between mercury(II) acetate and an alkene (Section 6.3E). Reaction of this mercurinium ion intermediate with water followed by loss of H^+ gives an organomercury enol. Note that this addition follows Markovnikov's rule: the electrophile, Hg^{2+}, adds to the less substituted carbon, and the nucleophile, H_2O, adds to the more substituted carbon.

MECHANISM $HgSO_4/H_2SO_4$ Catalyzed Hydration of an Alkyne

Step 1: Attack of Hg^{2+} on the carbon-carbon triple bond to form a bridged mercurinium ion intermediate, which contains a three-center, two-electron bond

$$H_3C{-}C{\equiv}C{-}H \;\; + Hg^{2+} \longrightarrow \text{bridged } Hg^+ \text{ over } C{\equiv}C \text{ (with } H_3C \text{ and } H\text{)}$$

A bridged mercurinium ion intermediate

Step 2: Attack of water on the bridged mercurinium ion intermediate from the side opposite the bridge

$$\text{mercurinium ion} + :\ddot{O}H_2 \longrightarrow (H_3C)(H{-}\overset{+}{O}H{-})C{=}C(Hg^+)(H)$$

Step 3: Proton transfer to solvent

$$H_2\ddot{O}: + (H_3C)(H{-}\overset{+}{O}H{-})C{=}C(Hg^+)(H) \longrightarrow (H_3C)(H\ddot{O}{:}{-})C{=}C(Hg^+)(H) + H_3O^+$$

Step 4: Cleavage of the carbon-mercury bond by water followed by keto-enol tautomerism

$$\underset{HO}{\overset{H_3C}{}}\!\!>\!C{=}C\!<\!\underset{H}{\overset{Hg^+}{}} \xrightarrow{H_2O} \underset{HO}{\overset{H_3C}{}}\!\!>\!C{=}C\!<\!\underset{H}{\overset{H}{}} \rightleftharpoons \underset{O}{\overset{H_3C}{}}\!\!>\!C{-}CH_3$$

EXAMPLE 10.5

Show reagents to bring about the following conversions:

(a) $C_6H_{11}{-}C{\equiv}CH \longrightarrow C_6H_{11}{-}\overset{O}{\overset{\|}{C}}CH_3$ (b) $C_6H_{11}{-}C{\equiv}CH \longrightarrow C_6H_{11}{-}CH_2\overset{O}{\overset{\|}{C}}H$

Solution

(a) Hydration of the monosubstituted alkyne gives an enol that is in equilibrium with the more stable keto form.

$$C_6H_{11}{-}C{\equiv}CH + H_2O \xrightarrow[HgSO_4]{H_2SO_4} C_6H_{11}{-}\overset{OH}{\overset{|}{C}}{=}CH_2 \longrightarrow C_6H_{11}{-}\overset{O}{\overset{\|}{C}}CH_3$$

(b) Hydroboration using disiamylborane followed by treatment with alkaline hydrogen peroxide gives an enol that is in equilibrium with the more stable aldehyde.

$$C_6H_{11}{-}C{\equiv}CH \xrightarrow[2.\ H_2O_2,\ NaOH]{1.\ (sia)_2BH} C_6H_{11}{-}CH{=}\overset{OH}{\overset{|}{C}}H \longrightarrow C_6H_{11}{-}CH_2\overset{O}{\overset{\|}{C}}H$$

PROBLEM 10.5

Acid-catalyzed hydration of 2-pentyne gives a mixture of two ketones, each of molecular formula $C_5H_{10}O$. Propose structural formulas for these two ketones and for the enol from which each is derived.

$$CH_3CH_2C{\equiv}CCH_3 + H_2O \xrightarrow[HgSO_4]{H_2SO_4} C_5H_{10}O$$

D. Addition of Acetic Acid: Formation of Vinyl Esters

In the presence of sulfuric acid and Hg(II) salts as catalysts, carboxylic acids add to alkynes to form vinylic esters. Perhaps the most important example of this reaction is that between acetylene and acetic acid to form the ester with the common name vinyl acetate.

$$\underset{\text{Acetylene}}{HC{\equiv}CH} + \underset{\text{Acetic acid}}{CH_3\overset{O}{\overset{\|}{C}}OH} \xrightarrow[HgSO_4]{H_2SO_4} \underset{\text{Vinyl acetate}}{CH_3\overset{O}{\overset{\|}{C}}OCH{=}CH_2}$$

Annual production of vinyl acetate, the monomer for the production of poly(vinyl acetate) in the United States in 1995 was 2.9 billion pounds. The major use of poly(vinyl acetate) is as an adhesive in the construction and packaging industries and in the paint and coatings industry.

Until 1967, the bulk of vinyl acetate manufactured in the United States was by addition of acetic acid to acetylene catalyzed by sulfuric acid-mercury(II) sulfate. As the price of acetylene rose, manufacturers turned to ethylene as a starting material for the production of vinyl acetate. Using a modification of technology originally developed by Wacker-Chemie in 1959 and known as the **Wacker process,** ethylene, acetic acid, and molecular oxygen react at elevated temperatures and in the presence of a palladium(II) chloride-copper(II) chloride catalyst to produce vinyl acetate.

$$\underset{\text{Ethylene}}{2CH_2{=}CH_2} + \underset{\text{Acetic acid}}{2CH_3\overset{\overset{\displaystyle O}{\|}}{C}OH} + O_2 \xrightarrow[\text{(Wacker process)}]{PdCl_2 \cdot CuCl_2} \underset{\text{Vinyl acetate}}{2CH_3\overset{\overset{\displaystyle O}{\|}}{C}OCH{=}CH_2} + 2H_2O$$

10.10 Organic Synthesis

We have now seen how to prepare both terminal and internal alkynes from acetylene and substituted acetylenes, and we have seen several common reactions of alkynes, including additions (HX, X_2, and H_2O), hydroboration/oxidation, and reduction. Now let us move a step farther to consider what might be called the art of organic synthesis. Synthesis is an important objective of organic chemists, be it the synthesis of compounds for use as pharmaceuticals, agrochemicals, plastics, elastomers, textile fibers, and so on. A successful synthesis must provide the desired product in maximum yield and with maximum control of stereochemistry at all stages of the synthesis. Furthermore, there is increasing desire to develop "green" syntheses, that is, syntheses that do not produce or release byproducts harmful to the environment.

Our goal in this section is to develop an ability to plan a successful synthesis. To this end, our strategy is to work backwards from the desired product. First, we analyze the desired product in the following way.

1. The carbon skeleton. This is often the most challenging problem. Here you need to consider what carbon-carbon bond-forming reactions are available to you. At this stage in the course, you have only two, namely alkylation of Gilman reagents (Section 7.7C) and alkylation of alkynide anions (Section 10.5).
2. The functional groups. What are they and how can they be changed to facilitate formation of the carbon skeleton? How can they then be changed to give the final set of functional groups in the desired product?

Let us use these steps to plan a synthesis for *cis*-3-hexene. As readily available starting materials, we use acetylene and alkyl halides.

Target molecule:

$$\begin{matrix} CH_3CH_2 & & CH_2CH_3 \\ & \diagdown\; C{=}C \;\diagup & \\ & \diagup \qquad \diagdown & \\ H & & H \end{matrix}$$

cis-3-Hexene

Analysis: The functional group in the product is a *cis* carbon-carbon double bond, which can be prepared by catalytic reduction of a carbon-carbon triple bond using the Lindlar catalyst. We call this transformation a **functional group interconversion (FGI).** We then disconnect the carbon skeleton into possible starting materials, which we can later reconnect by known reactions during the synthesis. In the example here, we disconnect at points (a), the two carbon-carbon bonds adjacent to the triple bond. These bonds can be formed during the synthesis by alkylation of the acetylide dianion using a haloalkane (Section 10.5). This type of scheme, in which we work from the desired product back to a set of starting materials, is called a **retrosynthesis.** We use an **open arrow** to symbolize a step in a retrosynthesis.

Retrosynthesis A process of reasoning backwards from a target molecule to a suitable set of starting materials.

$$\underset{\textit{cis}\text{-3-Hexene}}{(CH_3CH_2)HC{=}CH(CH_2CH_3)} \xRightarrow{\text{FGI}} \underset{\text{3-Hexyne}}{CH_3CH_2 \overset{a}{\wr} C{\equiv}C \overset{a}{\wr} CH_2CH_3} \xRightarrow{S_N2} \underset{\text{Acetylide dianion}}{{}^-{:}C{\equiv}C{:}^-} + \underset{\text{Bromoethane}}{CH_3CH_2Br}$$

Thus, our starting materials for this synthesis of *cis*-3-hexene are acetylene and bromoethane, both readily available compounds. The synthesis is carried out in five steps as follows.

$$\underset{\text{Acetylene}}{HC{\equiv}CH} \xrightarrow[\text{2. } CH_3CH_2Br]{\text{1. } NaNH_2} \underset{\text{1-Butyne}}{CH_3CH_2{-}C{\equiv}CH} \xrightarrow[\text{4. } CH_3CH_2Br]{\text{3. } NaNH_2} \underset{\text{3-Hexyne}}{CH_3CH_2{-}C{\equiv}C{-}CH_2CH_3} \xrightarrow[\text{Lindlar catalyst}]{\text{5. } H_2} \underset{\textit{cis}\text{-3-Hexene}}{(CH_3CH_2)HC{=}CH(CH_2CH_3)}$$

Let us use the same approach to devise a synthesis for 2-heptanone, a compound with a penetrating fruity odor which is responsible for the "peppery" odor of cheeses of the Roquefort type. As readily available starting materials, we again use acetylene and alkyl halides.

$$\text{Target molecule:} \quad \underset{\text{2-Heptanone}}{CH_3\overset{O}{\overset{\|}{C}}CH_2CH_2CH_2CH_2CH_3}$$

Analysis: The functional group in the target molecule is a ketone, which can be prepared by hydration of a carbon-carbon triple bond. Hydration of 1-heptyne gives only 2-heptanone. Hydration of 2-heptyne gives a mixture of 2-heptanone and 3-heptanone. Therefore, we choose a functional group interconversion via 1-heptyne.

$$\underset{\text{2-Heptanone}}{CH_3\overset{O}{\overset{\|}{C}}CH_2CH_2CH_2CH_2CH_3} \xRightarrow{\text{FGI}} \underset{\text{2-Heptyne}}{CH_3C{\equiv}CCH_2CH_2CH_2CH_3}$$

An acid-catalyzed hydration gives a mixture of 2-heptanone and 3-heptanone

$$\underset{\text{2-Heptanone}}{CH_3\overset{O}{\overset{\|}{C}}CH_2CH_2CH_2CH_2CH_3} \xRightarrow{\text{FGI}} \underset{\text{1-Heptyne}}{HC{\equiv}C \overset{a}{\wr} CH_2CH_2CH_2CH_2CH_3} \Longrightarrow \underset{\text{Acetylide anion}}{HC{\equiv}C{:}^-} + \underset{\text{1-Bromopentane}}{CH_3(CH_2)_3CH_2Br}$$

Our starting materials are acetylene and 1-bromopentane and the synthesis is carried out in three steps as follows.

$$HC{\equiv}CH \xrightarrow[\text{2. } CH_3(CH_2)_3CH_2Br]{\text{1. } NaNH_2} \underset{\text{1-Heptyne}}{HC{\equiv}CCH_2(CH_2)_3CH_3} \xrightarrow[H_2SO_4,\ HgSO_4]{\text{3. } H_2O} \underset{\text{2-Heptanone}}{CH_3\overset{O}{\overset{\|}{C}}CH_2(CH_2)_3CH_3}$$

EXAMPLE 10.6

How might the scheme for the synthesis of 2-heptanone be modified so that the product is heptanal?

Solution

Steps (1) and (2) are the same and give 1-heptyne. Instead of acid-catalyzed hydration of 1-heptyne, treat the alkyne with $(sia)_2BH$ followed by alkaline hydrogen peroxide.

$$HC{\equiv}CH \xrightarrow[2.\ CH_3(CH_2)_3CH_2Br]{1.\ NaNH_2} \underset{\text{1-Heptyne}}{HC{\equiv}C(CH_2)_4CH_3} \xrightarrow[4.\ H_2O_2,\ NaOH]{3.\ (sia)_2BH} \underset{\text{Heptanal}}{H\overset{O}{\overset{\|}{C}}(CH_2)_5CH_3}$$

PROBLEM 10.6

Show how the synthetic scheme in Example 10.6 might be modified to give

(a) 1-Heptanol **(b)** 2-Heptanol

SUMMARY

Alkynes contain one or more carbon-carbon triple bonds. The triple bond is a combination of one sigma bond formed by overlap of *sp* hybrid orbitals and two pi bonds formed by overlap of two sets of parallel 2*p* orbitals (Section 10.1). In the IUPAC system (Section 10.2A), the infix **-yn-** is used to show the presence of the carbon-carbon triple bond. Common names of alkynes are derived by prefixing the names of substituents on the carbon-carbon triple bond to the name -acetylene (Section 10.2B). The physical properties of alkynes (Section 10.3) are similar to those of alkanes and alkenes of comparable carbon skeleton.

The **pK_a** values for terminal alkynes are approximately 25 (Section 10.4); they are less acidic than water but more acidic than alkanes, alkenes, or ammonia. Hydrogen attached to a carbon-carbon triple bond is sufficiently acidic that it can be removed by a strong base, most commonly with $NaNH_2$, NaH, or LDA. The greater acidity of an alkyne hydrogen compared to those of alkenes and alkanes lies in the fact that an anion derived from an alkyne is more stable than one derived from an alkene or alkane. The lone pair of electrons of the alkyne anion lies in an *sp* hybrid orbital that has 50% *s* character; the alkyne anion is the most stable and, therefore, the least basic. Conversely, the lone pair of an alkane anion lies in an sp^3 hybrid orbital (25% *s* character); the alkane anion is the least stable and, therefore, the most basic.

Tautomers (Section 10.8) are constitutional isomers in equilibrium with each other that differ in the location of a hydrogen and a double bond relative to a heteroatom, most commonly O, N, or S. **Keto-enol tautomerism** is the most common type of tautomerism we encounter in this course. The functional group of an **enol** is an —OH group attached to a doubly bonded carbon atom. The enol form is in equilibrium with the keto form, and equilibrium most commonly lies far on the side of the keto form.

KEY REACTIONS

1. Acidity of Alkynes (Section 10.4)

Terminal alkynes react with strong bases, most commonly $NaNH_2$ or NaH, to give salts.

$$CH_3C{\equiv}CH + NaNH_2 \longrightarrow CH_3C{\equiv}C^-Na^+ + NH_3$$

2. Alkylation of Acetylide Anions (Section 10.5)

S_N2 reactions using acetylide anions are effective using only methyl and 1° alkyl halides. With 2° and 3° alkyl halides, E2 is the major reaction.

$$HC{\equiv}C{:}^{-}Na^{+} + CH_3(CH_2)_2CH_2{-}Br \xrightarrow{S_N2} HC{\equiv}CCH_2(CH_2)_2CH_3 + Na^{+}Br^{-}$$

3. Synthesis of an Alkyne from an Alkene (Section 10.6)

Addition of Br_2 or Cl_2 to an alkene followed by a double dehydrohalogenation using $NaNH_2$ or other strong base gives an alkyne.

$$RCH{=}CHR \xrightarrow[CCl_4]{Br_2} R\overset{Br}{\overset{|}{C}}H\overset{Br}{\overset{|}{C}}HR \xrightarrow[NH_3(l)]{2NaNH_2} RC{\equiv}CR$$

4. Catalytic Reduction (Section 10.7A)

Reaction of an alkyne with 2 mol of H_2 under moderate pressure, in the presence of a transition metal catalyst, at room temperature, gives an alkane.

$$CH_3C{\equiv}CCH_3 + 2H_2 \xrightarrow{\text{Pd, Pt, or Ni}} CH_3CH_2CH_2CH_3$$

Reaction in the presence of the Lindlar catalyst is stereoselective; syn addition of 1 mol of H_2 to an internal alkyne gives a *cis*-alkene.

$$CH_3C{\equiv}CCH_3 + H_2 \xrightarrow{\text{Lindlar catalyst}} \begin{matrix} H_3C & & CH_3 \\ & C{=}C & \\ H & & H \end{matrix}$$

5. Reduction Using Na or Li Metal in NH_3(l) (Section 10.7B)

Alkali metal reduction of an internal alkyne is stereoselective: anti addition of hydrogens gives a *trans*-alkene.

$$CH_3CH_2CH_2C{\equiv}CCH_2CH_2CH_3 \xrightarrow[NH_3(l)]{2Na} \begin{matrix} CH_3CH_2CH_2 & & H \\ & C{=}C & \\ H & & CH_2CH_2CH_3 \end{matrix}$$

6. Hydroboration-Oxidation (Section 10.8)

Hydroboration of an internal alkyne is stereoselective; syn addition of BH_3. Oxidation using H_2O_2/NaOH gives a vinyl alcohol that is in equilibrium by keto-enol tautomerism with a ketone.

$$CH_3CH_2C{\equiv}CCH_2CH_3 \xrightarrow[2.\ H_2O_2,\ NaOH]{1.\ BH_3} CH_3CH_2\overset{O}{\overset{\|}{C}}CH_2CH_2CH_3$$

Hydroboration of a terminal alkyne using a hindered dialkylborane followed by oxidation using H_2O_2/NaOH and then keto-enol tautomerism gives an aldehyde.

$$CH_3(CH_2)_5C{\equiv}CH \xrightarrow[2.\ H_2O_2,\ NaOH]{1.\ (sia)_2BH} CH_3(CH_2)_5CH_2\overset{O}{\overset{\|}{C}}H$$

7. Keto-Enol Tautomerism (Section 10.8)

In an equilibrium between a keto form and an enol form, the keto form generally predominates.

$$\underset{\text{Enol form}}{CH_3CH{=}\overset{\overset{\displaystyle OH}{|}}{C}CH_3} \rightleftharpoons \underset{\text{Keto form}}{CH_3CH_2\overset{\overset{\displaystyle O}{\|}}{C}CH_3}$$

8. Addition of Br_2 and Cl_2 (Section 10.9A)

Addition of 1 mol of Br_2 or Cl_2 is stereoselective; anti addition gives an (*E*)-dihaloalkene. Addition of a second mol of halogen gives a tetrahaloalkane.

$$CH_3C{\equiv}CCH_3 \xrightarrow{Br_2} \underset{Br}{\overset{CH_3}{}}C{=}C\underset{CH_3}{\overset{Br}{}} \xrightarrow{Br_2} CH_3\overset{Br}{\underset{Br}{C}}{-}\overset{Br}{\underset{Br}{C}}CH_3$$

9. Addition of HX (Section 10.9B)

Addition of HX is regioselective. Reaction by way of a vinylic carbocation intermediate follows Markovnikov's rule. Addition of 2HX gives a geminal dihaloalkane.

$$CH_3C{\equiv}CH \xrightarrow{HBr} CH_3\overset{\overset{\displaystyle Br}{|}}{C}{=}CH_2 \xrightarrow{HBr} CH_3\overset{Br}{\underset{Br}{C}}CH_3$$

10. Acid-Catalyzed Hydration (Section 10.9C)

Acid-catalyzed addition of water in the presence of Hg(II) salts is regioselective. Keto-enol tautomerism of the resulting enol gives a ketone.

$$CH_3C{\equiv}CH + H_2O \xrightarrow[HgSO_4]{H_2SO_4} CH_3\overset{\overset{\displaystyle OH}{|}}{C}{=}CH_2 \rightleftharpoons CH_3\overset{\overset{\displaystyle O}{\|}}{C}CH_3$$

ADDITIONAL PROBLEMS

Structure and Nomenclature

10.7 Write IUPAC names for the following compounds:

(a) $CH_3C{\equiv}C\overset{CH_3}{\underset{CH_3}{C}}CH_3$ (b) $HC{\equiv}CCH_2Br$ (c) cyclopentyl$-C{\equiv}CH$

(d) $HC{\equiv}CCH_2CH_2CH_2C{\equiv}CH$ (e) $CH_3(CH_2)_5C{\equiv}CCH_2OH$

(f) $CH_3(CH_2)_6C{\equiv}CH$ (g) $CH_3C{\equiv}CCH_2OH$ (h) $CH_3(CH_2)_7C{\equiv}C(CH_2)_7CO_2H$

10.8 Draw structural formulas for the following compounds:

(a) 3-Hexyne
(b) Vinylacetylene
(c) 3-Chloro-1-butyne
(d) 5-Isopropyl-3-octyne
(e) 3-Pentyn-2-ol
(f) 2-Butyne-1,4-diol
(g) Diisopropylacetylene
(h) *tert*-Butylmethylacetylene
(i) Cyclodecyne

10.9 Predict all bond angles about each circled atom.

(a) $CH_3C{\equiv}CCH_3$ (b) $CH_2{=}CHC{\equiv}CH$

(c) $CH_2{=}C{=}CHCH_3$ (d) $CH_2{=}CHCH{=}CH_2$

10.10 State the orbital hybridization of each circled atom.

(a) $CH_3C{\equiv}CCH_3$ (b) $CH_2{=}CHC{\equiv}CH$

(c) $CH_2{=}C{=}CHCH_3$ (d) $O{=}C{=}O$

10.11 Describe each circled carbon-carbon bond in terms of the overlap of atomic orbitals.

(a) $CH_3C{\equiv}CCH_3$ (b) $CH_2{=}CH{-}C{\equiv}CH$

(c) $CH_2{=}C{=}CHCH_3$ (d) $CH_2{=}CH{-}CH{=}CH_2$

10.12 Enanthotoxin is an extremely poisonous organic compound found in hemlock water dropwart, which is reputed to be the most poisonous plant in England. It is believed that no British plant has been responsible for more fatal accidents. The most poisonous part of the plant are the roots, which resemble small white carrots, giving the plant the name "five finger death." Also poisonous are its leaves, which look like parsley. Enanthotoxin is thought to interfere with the Na^+ current in nerve cells, which leads to convulsions and death. (See *The Merck Index,* 12th ed., #3608.)

How many stereoisomers are possible for enanthotoxin?

Preparation of Alkynes

10.13 Show how to prepare each alkyne from the given starting material.

(a) $CH_3CH_2CH_2CH{=}CH_2 \longrightarrow CH_3CH_2CH_2C{\equiv}CH$

(b) $CH_3(CH_2)_5CH(Cl)CH_3 \longrightarrow CH_3(CH_2)_4C{\equiv}CCH_3$

(c) $CH_3CH_2CH_2C{\equiv}CH \longrightarrow CH_3CH_2CH_2C{\equiv}CD$

10.14 If a catalyst could be found that would establish an equilibrium between 1,2-butadiene and 2-butyne, what would be the ratio of the more stable isomer to the less stable isomer at 25°C?

$$CH_2{=}C{=}CHCH_3 \rightleftharpoons CH_3C{\equiv}CCH_3 \qquad \Delta G^0 = -4.0 \text{ kcal/mol}$$

Reactions of Alkynes

10.15 Complete these acid-base reactions and predict whether the position of equilibrium lies toward the left or toward the right.

(a) $CH_3C{\equiv}CH + (CH_3)_3CO^-K^+ \overset{(CH_3)_3COH}{\rightleftharpoons}$ (b) $CH_2{=}CH_2 + Na^+NH_2^- \overset{NH_3(l)}{\rightleftharpoons}$

(c) $CH_3C{\equiv}CCH_2OH + Na^+NH_2^- \overset{NH_3(l)}{\rightleftharpoons}$

10.16 Draw structural formulas for the major product(s) formed by reaction of 3-hexyne with each of these reagents. Where you predict no reaction, write NR.

(a) H_2(excess)/Pt **(b)** H_2/Lindlar catalyst

(c) Na in NH_3(l) **(d)** BH_3 followed by H_2O_2/NaOH

(e) BH_3 followed by CH_3CO_2H **(f)** BH_3 followed by CH_3CO_2D

(g) Cl_2 (1 mol) **(h)** $NaNH_2$ in NH_3(l)

(i) HBr (1 mol) **(j)** HBr (2 mol)

(k) H_2O in $H_2SO_4/HgSO_4$ **(l)** CH_3CO_2H in $H_2SO_4/HgSO_4$

10.17 Draw the structural formula of the enol formed in each alkyne hydration reaction and then draw the structural formula of the carbonyl compound with which each enol is in equilibrium.

(a) $CH_3(CH_2)_5C{\equiv}CH + H_2O \xrightarrow[H_2SO_4]{HgSO_4}$ (an enol) $\longrightarrow$

(b) $CH_3(CH_2)_5C{\equiv}CH \xrightarrow[2.\ NaOH/H_2O_2]{1.\ (sia)_2BH}$ (an enol) $\longrightarrow$

10.18 Propose a mechanism for this reaction.

$$HC{\equiv}CH + CH_3\overset{O}{\overset{\|}{C}}OH \xrightarrow[HgSO_4]{H_2SO_4} CH_3\overset{O}{\overset{\|}{C}}OCH{=}CH_2$$

Acetylene Acetic acid Vinyl acetate

Syntheses

10.19 Show how to convert 9-octadecynoic acid to the following:

$$CH_3(CH_2)_7C{\equiv}C(CH_2)_7CO_2H$$

9-Octadecynoic acid

(a) (*E*)-9-Octadecenoic acid (eliadic acid) **(b)** (*Z*)-9-Octadecenoic acid (oleic acid)
(c) 9,10-Dihydroxyoctadecanoic acid **(d)** Octadecanoic acid (stearic acid)

10.20 For small-scale and consumer welding applications, many hardware stores sell cylinders of MAAP gas, which is a mixture of propyne (methylacetylene) and 1,2-propadiene (allene) gases, with other hydrocarbons. How would you prepare the mixture of methyl acetylene and allene in the laboratory?

10.21 Show reagents and experimental conditions you might use to convert propyne into each product. Some of these syntheses can be done in one step. Others require two or more steps.

(a) $CH_3CBr_2{-}CHBr_2$ (b) $CH_3CBr_2CH_3$ (c) $CH_3\overset{O}{\overset{\|}{C}}CH_3$ (d) $CH_3CH_2\overset{O}{\overset{\|}{C}}H$

10.22 Show reagents and experimental conditions you might use to convert each starting material into the desired product. Some of these syntheses can be done in one step. Others require two or more steps.

(a) $CH_3CH_2CH_2C\equiv CCH_3 \longrightarrow$ $(CH_3CH_2CH_2)(H)C=C(H)(CH_3)$ [trans]

(b) $CH_3CH_2CH_2C\equiv CCH_3 \longrightarrow$ $(CH_3CH_2CH_2)(H)C=C(CH_3)(H)$ [cis]

(c) $CH_3(CH_2)_4C\equiv CH \longrightarrow CH_3(CH_2)_4C\equiv C{:}^- Na^+$

(d) $CH_3CH_2C\equiv CH \longrightarrow CH_3CH_2C\equiv CD$

(e) $CH_3CH_2C\equiv CH \longrightarrow$ $(CH_3CH_2)(H)C=C(H)(D)$

(f) $CH_3CH_2C\equiv CH \longrightarrow$ $(CH_3CH_2)(D)C=C(H)(H)$

(g) $(HOCH_2)(H)C=C(H)(CH_2OH) \longrightarrow HOCH_2CH_2CH_2CH_2OH$

(h) 1-ethynylcyclohexanol (HO, $C\equiv CH$ on cyclohexane) $\longrightarrow$ 1-acetylcyclohexanol (HO, $\overset{O}{\|}C-CH_3$ on cyclohexane)

10.23 Rimantadine is effective in preventing naturally occurring infections caused by the influenza A virus and in treating established illnesses. It is thought to exert its antiviral effect by blocking a late stage in the assembly of the virus. Rimantadine is synthesized from adamantane by the following sequence. We will discuss the chemistry of Step 5 in Chapter 15. (See *The Merck Index,* 12th ed., #8390.)

Adamantane $\xrightarrow[(1)]{Br_2}$ 1-Bromoadamantane (Br) $\xrightarrow[(2)]{CH_2=CHBr,\ AlBr_3}$ adamantyl–CH_2CHBr_2

$\xrightarrow{(3)}$ adamantyl–$C\equiv CH$ $\xrightarrow{(4)}$ adamantyl–$\overset{O}{\|}{C}CH_3$ $\xrightarrow{(5)}$ adamantyl–$\overset{NH_2}{|}{C}HCH_3$

Rimantadine
(an antiviral agent)

(a) Describe experimental conditions to bring about Step 1. By what type of mechanism does this reaction occur? Account for the regioselectivity of bromination in Step 1.

(b) Propose a mechanism for Step 2. *Hint*: As we shall see in Chapter 20, reaction of a bromoalkane such as 1-bromoadamantane with aluminum bromide (a Lewis acid, Section 3.4) results in formation of a carbocation and $AlBr_4^-$. Assume that adamantyl cation is formed in Step 2 and proceed from there to describe a mechanism.

(c) Account for the regioselectivity of carbon-carbon bond formation in Step 2.

(d) Describe experimental conditions to bring about Step 3.

(e) Describe experimental conditions to bring about Step 4.

10.24 Show reagents and experimental conditions required to bring about the following transformations.

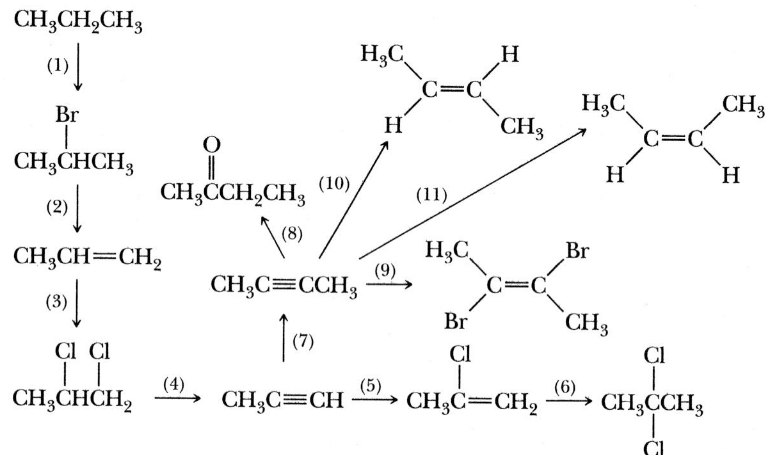

10.25 Show reagents to bring about each conversion.

$$CH_3(CH_2)_5C{\equiv}CH \xrightarrow{(a)} CH_3(CH_2)_5C{\equiv}C^-Na^+ \xrightarrow{(b)} CH_3(CH_2)_5C{\equiv}CCH_2CH_3 \xrightarrow{(c)} \text{(Z)-}CH_3(CH_2)_5CH{=}CHCH_2CH_3$$

$$CH_3(CH_2)_5C{\equiv}CH \xrightarrow{(d)} CH_3(CH_2)_5CH_2\overset{O}{\overset{\|}{C}}H$$

$$CH_3(CH_2)_5C{\equiv}CH \xrightarrow{(e)} CH_3(CH_2)_5\overset{O}{\overset{\|}{C}}CH_3$$

10.26 Which of these alkynes can be prepared in good yield by monoalkylation or dialkylation of acetylene? For each that cannot, explain why not.

(a) 3-Methyl-1-butyne **(b)** 4,4-Dimethyl-1-pentyne **(c)** 2-Octyne

10.27 Propose a synthesis for (*Z*)-9-tricosene (muscalure), the sex pheromone for the common house fly *(Musca domestica)* starting with acetylene and alkyl halides as sources of carbon atoms. (See *The Merck Index*, 12th ed., #6388.)

$$\text{(Z)-}CH_3(CH_2)_7CH{=}CH(CH_2)_{12}CH_3$$

Muscalure

10.28 Propose an efficient synthesis of each compound starting from acetylene and any necessary organic and inorganic reagents.

(a) 4-Octyne (b) 4-Octanone (c) *cis*-4-Octene
(d) *trans*-4-Octene (e) 4-Octanol (f) meso-4,5-Octanediol

10.29 Show how to prepare each compound from ethylene.

(a) 1,2-Dichloroethane (b) Chloroethene (vinyl chloride)
(c) 1,1-Dichloroethane

10.30 Show how to bring about this conversion.

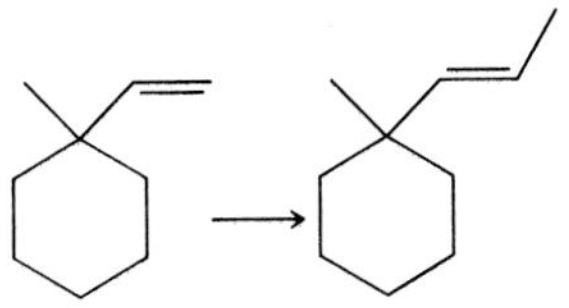

A CONVERSATION WITH . . .
ROALD HOFFMANN

Roald Hoffmann is a remarkable individual. When he was only 44 years old, he shared the 1981 Nobel Prize for Chemistry with Kenichi Fukui of Japan for work in applied theoretical chemistry. In addition, he has received awards from the American Chemical Society in both organic chemistry and inorganic chemistry, the only person to have achieved this honor. And, in 1990, he was awarded the Priestley Medal, the highest award given by the American Chemical Society.

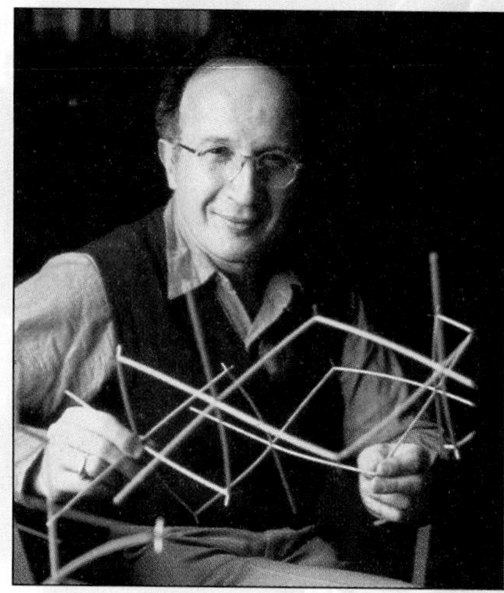

The numerous honors celebrating his achievements in chemistry tell only part of the story of his life. He was born to a Polish Jewish family in Zloczow, Poland in 1937, and was named Roald after the famous Norwegian explorer Roald Amundsen. Shortly after World War II began in 1939, the Nazis first forced him and his parents into a ghetto and then into a labor camp. However, his father smuggled Hoffmann and his mother out of the camp, and they were hidden for more than a year in the attic of a schoolhouse in the Ukraine. His father was later killed by the Nazis after trying to organize an attempt to break out of the labor camp. After the war, Hoffmann, his mother, and his stepfather made their way west to Czechoslovakia, Austria, and then Germany. They finally emigrated to the United States, arriving in New York in 1949. That Hoffmann and his mother survived these years is our good fortune. Of the 12,000 Jews living in Zloczow in 1941 when the Nazis took over, only 80 people, three of them children, survived the Holocaust. One of those three children was Roald Hoffmann.

On arriving in New York, Hoffmann learned his sixth language, English. He went to public schools in New York City and then to Stuyvesant High School, one of the city's select science schools. From there he went to Columbia University and then on to Harvard University, where he earned his Ph.D. in 1962. Shortly thereafter, he began the work with Professor R. B. Woodward that eventually led to the Nobel Prize. Since 1965 he has been a professor at Cornell University.

In addition to his work in chemistry, Professor Hoffmann also writes popular articles on science for the *American Scientist* and other magazines, and he has published two volumes of his poetry. Finally, he appeared in a series of 26 half-hour television programs for a chemistry course called "World of Chemistry," which airs on public television and cable channels.

FROM MEDICINE TO CEMENT TO THEORETICAL CHEMISTRY

Professor Hoffmann's office at Cornell University is full of mineral samples, molecular models, and Japanese art. When asked what brought him into chemistry he said, "I came rather late to chemistry, I was not interested in it from childhood." However, he clearly feels that one can come late to chemistry, and that it can be a very positive thing. "I am always worried about fields in which people exhibit precocity, like music and mathematics. Precocity is some sort of evidence that you have to have talent. I don't like that. I like the idea that human beings can do anything they want to. They need to be trained sometimes. They need a teacher to awaken the intelligence within them. But to be a chemist requires no special talent, I'm glad to say. Anyone can do it, with hard work."

He took a standard chemistry course in high school. He recalls that it was a fine course, but apparently, he found biology more enjoyable because, in his high school yearbook, "under the picture of me with a crew cut, it says 'medical research' under my name." Indeed, he says that "medical research was a compromise between my interest in science and the typical Jewish middle class family pressures to become a medical doctor. The same kind of pressures seem to apply to Asian-Americans today."

When he went to Columbia University, Hoffmann enrolled as a pre-med student, but says that there were several factors that shifted him away from a career in medicine. One of these was his work at the National Bureau of Standards in Washington, D. C., for two summers and then at Brookhaven National Laboratory for a third summer. He says that these experiences gave him a feeling for the excitement of chemical research. Nonetheless, during his first summer at the Bureau of Standards, he "did some not very exciting work on the thermochemistry of cement." During his second summer there, he went over to the National Institutes of Health to find out what medical research was about. "To my amazement," he says, "most of the people had Ph.D.'s and not M.D.'s. I just didn't know. Young people do not often know what is required for a given profession. Once I found that out, and found that I did well in chemistry, it made me feel that I didn't really have to do medicine, that I could do some research in chemistry or biology. Later, what influenced me to decide on theoretical chemistry was an excellent instructor.

"At the very same time I was being exposed to the humanities, in part because of Columbia's core curriculum—which I think is a great idea—that had so-called contemporary civilization and humanities courses. I took advantage of the liberal arts education to the hilt, and that has remained with me all my life. The humanities teachers have remained permanently fixed in my mind and have changed my ways of thinking. These were the people who really had the intellectual impact on me and helped to shape my life.

"To trace the path, I was a latecomer to chemistry and was inspired by research. I think *research* is the way in. It just gives you a different perception."

A Love for Complexity

Having discussed what brought him into chemistry, we were interested in his view of the qualities that a student should possess to pursue a career in the field. He said, "One thing one needs to be a chemist is a love for complexity and richness. To some extent that is true of biology and natural history, too. I think one of the things that is beautiful about chemistry is that there are 10,000,000 compounds, each with different properties. What's beautiful when you make a molecule is that you can make derivatives in which you can vary substituents, the pieces of a molecule, and we know that those substituents give a molecule function, give it complexity and richness. That's why a protein or nucleic acid with all its variety is essential for life. That's why to me, intellectually, isomerism and stereochemistry in organic chemistry are at the heart of chemistry. I think we should teach that much earlier. It requires no mathematics, only a little model-building; you can do this without theory. I think it is no accident that organic chemistry drew to it the intellects of its time."

Experiment and Theory

Professor Hoffmann has spent his career immersed in the theories of chemistry. However, he believes that fundamentally "chemistry is an experimental science, in spite of some of my colleagues saying otherwise. However, the educational process certainly favors theory. It's in the nature of things for teacher and student both to want to understand and then give primacy to the soluble and the understood at the expense of other things. We also have this reductionist philosophy of science, the idea that the social sciences derive from biology, that biology follows from chemistry, chemistry from physics, and so on. This notion gives an inordinate amount of importance to theoretical thinking, the more mathematical the better. Of course this is not true in reality, but it's an ideology; it is a religion of science."

There is of course a role for theory. "You can't report just the facts and nothing but the facts; by themselves they are dull. They have to be woven into a framework so that there is understanding. That's accomplished usually by a theory. It may not be mathematical, but a qualitative network of relationships." Indeed, Hoffmann believes that the incorporation of theory into chemistry "is what made American science better than that in many other countries. The emphasis in chemistry on theory and theoretical understanding is very important, but not nearly as important as the syntheses and reactions of molecules.

"Although I think chemists need to like to do experiments, that doesn't mean there is no role for people like me. It turns out that I am really an experimental chemist hiding as a theoretician. I think that is the key to my success. That is, I think I can empathize with what bothers the experimentalists. In another day I could have become an experimentalist."

Major Issues in Chemistry and Science Today

Professor Hoffmann has worked, and is presently working, at the forefront of several major areas of chemistry. He is presently quite intrigued by surface science. "For instance, there is the Fischer-Tropsch process, a pretty incredible thing in detail. Carbon monoxide and hydrogen gas come onto a metal catalyst, a surface of some sort, and off come long chain hydrocarbons and alcohols. The richness of all these things happening is intriguing, and we are on the verge of understanding. We now have structural information on surfaces that's reliable, and we are just beginning to get kinetic information. Surface science is at a crossing of chemistry, physics, and engineering. The field is in some danger of being spun off on its own, but I would like to keep it in chemistry.

"Bioinorganic chemistry is another such field. In my research group we are doing some work trying to understand the mechanism of oxygen production in photosynthesis, the last steps. What is known is very little. There is an enzyme in photosystem II that involves 3 to 4 manganese atoms, and they are at oxidation state 3 to 4. And they somehow take oxide or hydroxide to peroxide and eventually to molecular oxygen. That's all we know. Experimentally, not theoretically, I think bioinorganic chemistry is a very interesting field."

> "That's why to me, intellectually, isomerism and stereochemistry in organic chemistry are at the heart of chemistry."

Finally, Hoffmann remarked, "There are going to be finer and finer ways of controlling the synthesis of molecules, the most essential activity of chemists. If I were to point to a single thing that chemists do, it would be that they make molecules. Chemistry is the science of molecules and their transformations. The transformations are the essential part. I think there are exciting possibilities for chemical intervention into biological systems with an ever finer degree of control. We need not be afraid of nature. We can mimic it, and even surpass its synthetic capabilities. And find a way to cooperate with it."

Scientific Literacy and Democracy

Roald Hoffmann is very concerned not only about science in general and chemistry in particular, but also about our society. One of his concerns is scientific literacy because "some degree of scientific literacy is absolutely necessary today for the population at large as part of a democratic system of government. People have to make intelligent decisions about all kinds of technological issues." He recently offered his comments on this important issue in *The New York Times.* He wrote, "What concerns me about scientific, or humanistic, illiteracy is the barrier it poses to rational democratic governance. Democracy occasionally gives in to *technocracy*—a reliance on experts on matters such as genetic engineering, nuclear waste disposal, or the cost of medical care. That is fine, but the people must be able to vote intelligently on these issues. The less we know as a nation, the more we must rely on experts, and the more likely we are to be misled by demagogues. We must know more."

The Responsibility of Scientists

"Scientists have a great obligation to speak to the public," Hoffmann says. "We have an obligation as educators to train the next generation of people. We should pay as much attention to those people who are *not* going to be chemists, and sometimes need to make compromises about what is to be taught and what is the nature of our courses. I think scientists have an obligation to speak to the public broadly, and here I think they have been negligent. I think society is paying scientists money to do research, and can demand an accounting in plain language. That's why I put in a lot of time on that television show [*The World of Chemistry*]."

A Teacher of Chemistry—and Proud of It

In the Nobel Yearbook, Professor Hoffmann wrote that the technical description of his work "does not communicate what I think is my major contribution. I am a teacher, and I am proud of it. At Cornell University I have taught primarily undergraduates. . . . I have also taught chemistry courses to nonscientists and graduate courses in bonding theory and quantum mechanics. To the chemistry community at large, and to my fellow scientists, I have tried to teach 'applied theoretical chemistry': a special blend of computations stimulated by experiment and coupled to the construction of general models—frameworks for understanding." His success in this area is unquestioned.

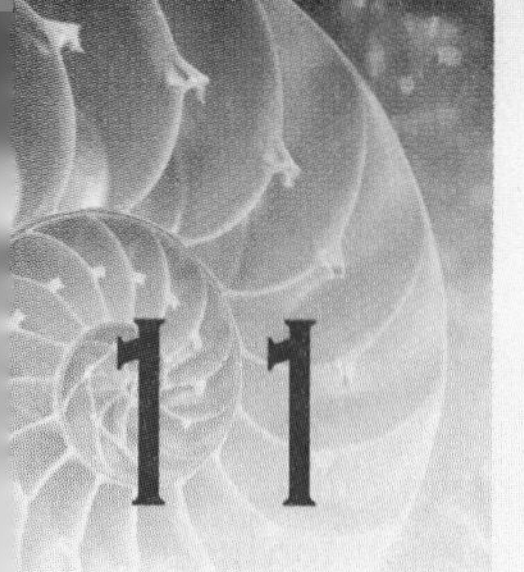
11

Ethers, Sulfides, and Epoxides

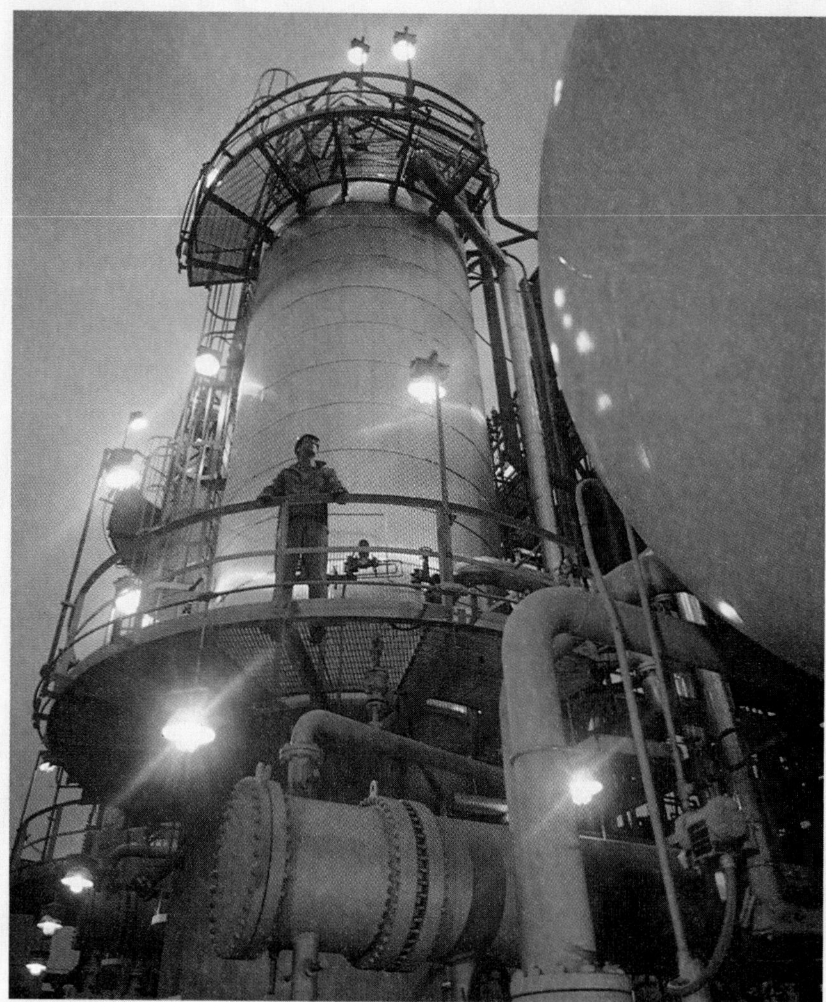

■ A *tert*-butyl methyl ether manufacturing plant. *(Courtesy of Ashland Petroleum)*

In this chapter, we first discuss the structure, nomenclature, and physical properties of ethers and then compare their physical properties with those of isomeric alcohols. Next, we study the preparation and reactions of ethers. As we shall see, their most important reactions involve nucleophilic substitution. In a sense then, this chapter is an extension of the discussion of S_N1 and S_N2 reaction mechanisms begun in Chapter 8 and continued in Chapters 9 and 10.

(a)

```
      H       H
      |  ..   |
  H—C—O—C—H
      |  ..   |
      H       H
```

(b)

110.3°

FIGURE 11.1
The structure of dimethyl ether, CH_3OCH_3. (a) Lewis structure and (b) ball-and-stick model.

11.1 Structure of Ethers

The functional group of an **ether** is an atom of oxygen bonded to two carbon atoms. Figure 11.1 shows a Lewis structure and a ball-and-stick model of dimethyl ether, CH_3OCH_3, the simplest ether. In dimethyl ether, two sp^3 hybrid orbitals of oxygen form sigma bonds to the two carbon atoms. The other two sp^3 hybrid orbitals each contain an unshared pair of electrons. The C—O—C bond angle in dimethyl ether is 110.3°, a value close to the tetrahedral angle of 109.5°.

Ether A molecule containing an oxygen atom bonded to two carbon atoms.

11.2 Nomenclature of Ethers

In the IUPAC system, ethers are named by selecting the longest carbon chain as the parent alkane and naming the —OR group attached to it as an **alkoxy** substituent. Common names are derived by listing the alkyl groups attached to oxygen in alphabetical order and adding the word "ether." Following are the IUPAC names and, in parentheses, the common names for several low-molecular-weight ethers.

Alkoxy group An —OR group, where R is an alkyl group.

$CH_3CH_2OCH_2CH_3$	$CH_3OC(CH_3)_2CH_3$	(cyclohexane ring with OH and OCH_2CH_3)
Ethoxyethane (Diethyl ether)	2-Methoxy-2-methylpropane (*tert*-Butyl methyl ether)	*trans*-2-Ethoxycyclohexanol

Chemists almost invariably use common names for low-molecular-weight ethers. For example, although ethoxyethane is the IUPAC name for $CH_3CH_2OCH_2CH_3$, it is rarely called that; rather, it is called diethyl ether, ethyl ether, or even more commonly, simply ether. The abbreviation for *tert*-butyl methyl ether, an ether that is becoming increasingly important as an octane-improving additive to gasolines, is MTBE, after the common name of methyl *tert*-butyl ether.

Three other ethers deserve special mention. 2-Methoxyethanol and 2-ethoxyethanol, more commonly known as Methyl Cellosolve and Cellosolve, are good polar protic solvents in which to carry out organic reactions and are also used commercially in some paint strippers. *Di*ethylene *gly*col di*me*thyl *e*ther, more commonly known by its acronym, diglyme, is a common solvent for hydroboration and $NaBH_4$ reductions.

$CH_3OCH_2CH_2OH$	$CH_3CH_2OCH_2CH_2OH$	$CH_3OCH_2CH_2OCH_2CH_2OCH_3$
2-Methoxyethanol (Methyl Cellosolve)	2-Ethoxyethanol (Cellosolve)	Diethylene glycol dimethyl ether (Diglyme)

Cyclic ethers, that is, heterocyclic compounds in which the ether oxygen is one of the atoms in a ring, are given special names. The presence of oxygen in a saturated

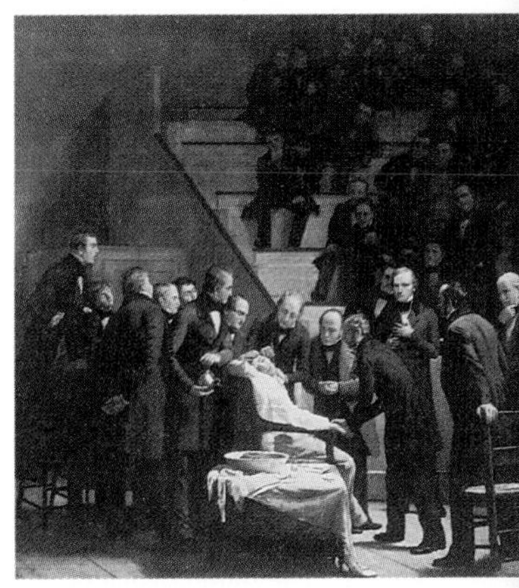

This painting by Robert Hinckley shows the first use of ether as an anesthetic in 1846. The patient, Gilbert Abbott, was having a tumor removed from his neck by Dr. Robert John Collins. The ether was administered by the dentist W. T. G. Morton, who discovered the anesthetic properties of ether. *(Boston Medical Library in the Francis A. Countway Library of Medicine)*

ring is indicated by the prefix ox- and ring sizes from three to six are indicated by the endings -irane, -etane, -olane, and -ane, respectively. Several of these smaller-ring cyclic ethers are more often referred to by their common names, here shown in parentheses. Numbering of the atoms of the ring begins with the oxygen atom.

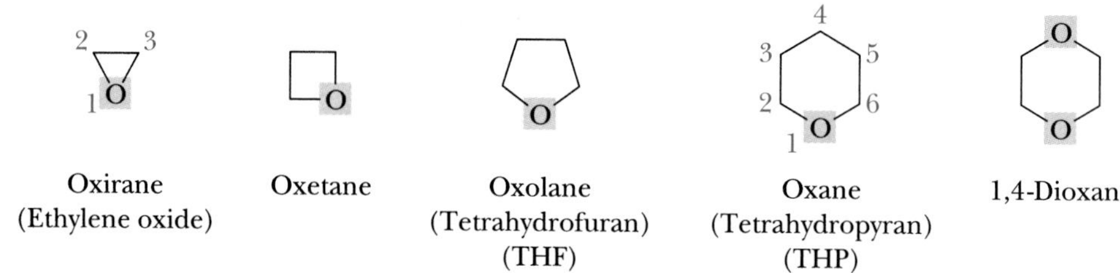

Oxirane (Ethylene oxide)	Oxetane	Oxolane (Tetrahydrofuran) (THF)	Oxane (Tetrahydropyran) (THP)	1,4-Dioxane

EXAMPLE 11.1

Write IUPAC and common names for these ethers.

(a) $CH_3C(CH_3)_2OCH_2CH_3$ **(b)** (cyclohexyl)–O–(cyclohexyl) **(c)** $CH_2{=}CHOCH_3$

Solution

(a) 2-Ethoxy-2-methylpropane. Its common name is *tert*-butyl ethyl ether.
(b) Cyclohexyloxycyclohexane. Its common name is dicyclohexyl ether.
(c) Methoxyethene. Its common name is methyl vinyl ether.

PROBLEM 11.1

Write IUPAC and common names for these ethers.

(a) $CH_3CH(CH_3)CH_2OCH_2CH_3$ **(b)** $CH_3OCH_2CH_2OCH_3$ **(c)** cyclohexane with OCH_2CH_3 and OCH_2CH_3 on adjacent carbons

Sulfide The sulfur analog of an ether; a molecule containing a sulfur atom bonded to two carbon atoms.

Sulfur analogs of ethers are named by using the word **sulfide** to show the presence of the —S— group. Following are common names of two sulfides:

CH_3SCH_3 — Dimethyl sulfide

$CH_3CH_2SCH(CH_3)CH_3$ — Ethyl isopropyl sulfide

Disulfide A molecule containing an —S—S— group.

The functional group of a **disulfide** is an —S—S— group. Common names of disulfides are derived by listing the names of the groups attached to sulfur and adding the word "disulfide."

$CH_3{-}S{-}S{-}CH_3$

Dimethyl disulfide

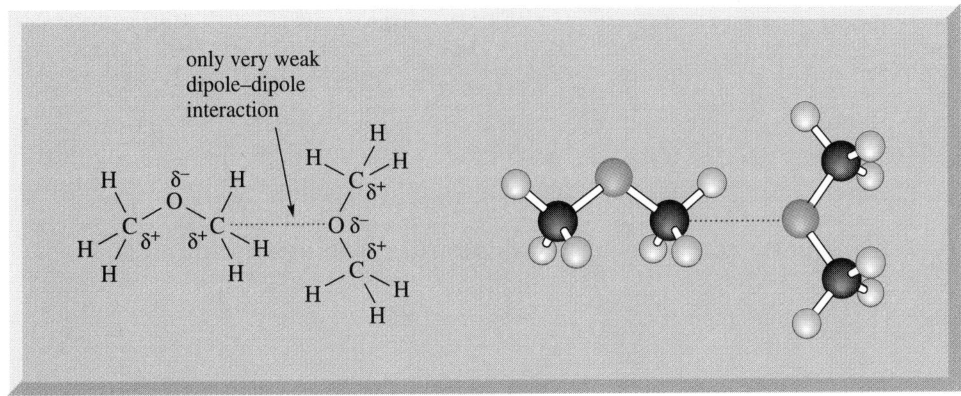

FIGURE 11.2
Ethers are polar compounds, but because of steric hindrance, there are only weak dipole-dipole interactions between their molecules in the pure liquid.

11.3 Physical Properties of Ethers

Ethers are polar molecules in which oxygen bears a partial negative charge and each attached carbon bears a partial positive charge (Figure 11.2). However, only weak dipole-dipole interactions exist between their molecules in the pure state (Figure 11.2). Consequently, boiling points of ethers are much lower than those of alcohols of comparable molecular weight (Table 11.1) and are close to those of hydrocarbons of comparable molecular weight (compare Tables 2.5 and 11.1). Because ethers cannot donate hydrogen bonds, they are much less soluble in water than alcohols (Table 11.1).

Because the oxygen atom of an ether carries a partial negative charge, ethers form hydrogen bonds with water (Figure 11.3); therefore, they are more soluble in water

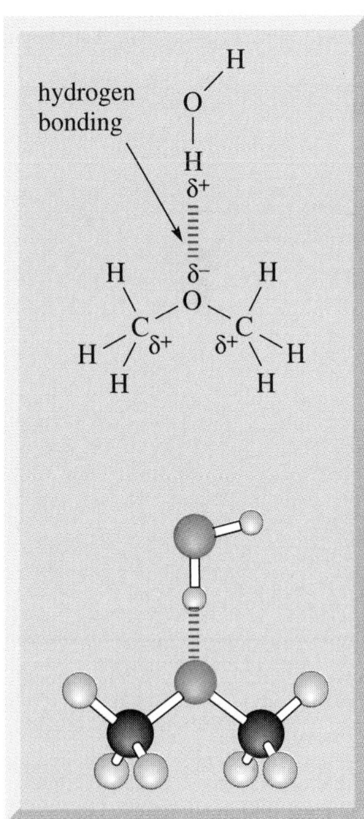

FIGURE 11.3
Ethers are hydrogen bond acceptors only. They are not hydrogen bond donors.

TABLE 11.1 Boiling Points and Solubilities in Water of Some Ethers and Alcohols of Comparable Molecular Weight

Structural Formula	Name	Molecular Weight	bp (°C)	Solubility in Water
CH_3CH_2OH	ethanol	46	78	infinite
CH_3OCH_3	dimethyl ether	46	−24	7.8 g/100 g
$CH_3CH_2CH_2CH_2OH$	1-butanol	74	117	7.4 g/100 g
$CH_3CH_2OCH_2CH_3$	diethyl ether	74	35	6.1 g/100 g
$HOCH_2CH_2CH_2CH_2OH$	1,4-butanediol	90	230	infinite
$CH_3CH_2CH_2CH_2CH_2OH$	1-pentanol	88	138	2.3 g/100 g
$CH_3OCH_2CH_2OCH_3$	ethylene glycol dimethyl ether	90	84	infinite
$CH_3CH_2CH_2CH_2OCH_3$	butyl methyl ether	88	71	slight

than hydrocarbons of comparable molecular weight and shape (compare data in Tables 2.5 and 11.1).

EXAMPLE 11.2

Arrange these compounds in order of increasing solubility in water.

$CH_3OCH_2CH_2OCH_3$	$CH_3CH_2OCH_2CH_3$	$CH_3CH_2CH_2CH_2CH_2CH_3$
Ethylene glycol dimethyl ether	Diethyl ether	Hexane

Solution

Water is a polar solvent. Hexane, a nonpolar hydrocarbon, has the lowest solubility in water. Both diethyl ether and ethylene glycol dimethyl ether are polar compounds due to the presence of the polar C—O—C bond, and each interacts with water as a hydrogen bond acceptor. Because ethylene glycol dimethyl ether has more sites within its molecules for hydrogen bonding than diethyl ether, it is more soluble in water than diethyl ether.

$CH_3CH_2CH_2CH_2CH_2CH_3$	$CH_3CH_2OCH_2CH_3$	$CH_3OCH_2CH_2OCH_3$
Insoluble	8 g/100 g water	Soluble in all proportions

PROBLEM 11.2

Arrange these compounds in order of increasing boiling point:

$CH_3OCH_2CH_2OCH_3$ $\quad HOCH_2CH_2OH$ $\quad CH_3OCH_2CH_2OH$

11.4 Preparation of Ethers

A. Williamson Ether Synthesis

The most common general method for the synthesis of ethers, the **Williamson ether synthesis,** involves second-order nucleophilic displacement of a halide ion or other good leaving group by an alkoxide ion.

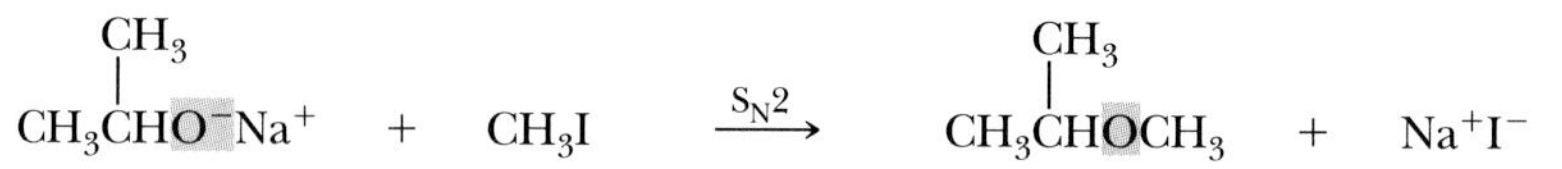

$$\underset{\text{Sodium isopropoxide}}{CH_3\overset{CH_3}{\overset{|}{C}}HO^-Na^+} + \underset{\substack{\text{Iodomethane}\\\text{(Methyl iodide)}}}{CH_3I} \xrightarrow{S_N2} \underset{\substack{\text{2-Methoxypropane}\\\text{(Isopropyl methyl ether)}}}{CH_3\overset{CH_3}{\overset{|}{C}}HOCH_3} + Na^+I^-$$

In planning a Williamson ether synthesis, it is essential to use a combination of reactants that maximizes nucleophilic substitution and minimizes the competing β-elimination reaction (Section 8.11B). Yields of ether are highest when the halide is displaced from methyl or a primary carbon. Yields are lower in the displacement from secondary halides (because of competing β-elimination), and the Williamson ether synthesis fails altogether with tertiary halides, in which β-elimination by an E2 mechanism is the exclusive reaction. For example, *tert*-butyl ethyl ether can be prepared by the reaction of potassium *tert*-butoxide and ethyl bromide. With the alternative combination of sodium ethoxide and *tert*-butyl bromide, no ether is formed. Rather, 2-methylpropene is formed by dehydrohalogenation.

$$\underset{\substack{\text{Potassium}\\ \textit{tert}\text{-butoxide}}}{CH_3\overset{\overset{CH_3}{|}}{\underset{\underset{CH_3}{|}}{C}}O^-K^+} + \underset{\text{Ethyl bromide}}{CH_3CH_2Br} \xrightarrow{S_N2} \underset{\textit{tert}\text{-Butyl ethyl ether}}{CH_3\overset{\overset{CH_3}{|}}{\underset{\underset{CH_3}{|}}{C}}OCH_2CH_3} + K^+Br^-$$

$$\underset{\substack{\textit{tert}\text{-Butyl}\\ \text{bromide}}}{CH_3\overset{\overset{CH_3}{|}}{\underset{\underset{CH_3}{|}}{C}}Br} + \underset{\substack{\text{Sodium}\\ \text{ethoxide}}}{CH_3CH_2O^-Na^+} \xrightarrow{E2} \underset{\text{2-Methylpropene}}{CH_3\overset{\overset{CH_3}{|}}{C}{=}CH_2} + CH_3CH_2OH + Na^+Br^-$$

EXAMPLE 11.3

Show the combination of alcohol and alkyl halide that can best be used to prepare these ethers by the Williamson ether synthesis:

(a) $CH_3(CH_2)_3OCH(CH_3)_2$ **(b)** $(R)\text{-}CH_3CH_2\overset{\overset{OCH_2CH_3}{|}}{C}HCH_3$

Solution

(a) Treat 2-propanol with sodium metal to form sodium isopropoxide. Then treat this metal alkoxide with 1-bromobutane. The alternative combination of sodium butoxide and 2-bromopropane would give considerably more elimination product.

$$\underset{\text{1-Bromobutane}}{CH_3(CH_2)_3Br} + \underset{\substack{\text{Sodium}\\ \text{isopropoxide}}}{(CH_3)_2CHO^-Na^+} \longrightarrow CH_3(CH_2)_3OCH(CH_3)_2 + Na^+Br^-$$

(b) Treat (R)-2-butanol with sodium metal to form the metal alkoxide. This reaction involves only the O—H bond and does not affect the stereocenter. Then, treat this metal alkoxide with an ethyl halide to give the desired product.

$$(R)\text{-}CH_3CH_2\overset{\overset{OH}{|}}{C}HCH_3 \xrightarrow[\text{2. } CH_3CH_2I]{\text{1. Na}} (R)\text{-}CH_3CH_2\overset{\overset{OCH_2CH_3}{|}}{C}HCH_3$$

An alternative synthesis is to convert the (S)-2-butanol to its tosylate followed by treatment with sodium ethoxide. This synthesis, however, gives only a low yield of the desired product. Recall from Section 8.11B that when a 2° halide or tosylate is treated with a strong base/good nucleophile, E2 is the major reaction.

$$(S)\text{-}CH_3CH_2\underset{\underset{OH}{|}}{C}HCH_3 \xrightarrow[\text{2. } CH_3CH_2O^-Na^+]{\text{1. TsCl, pyridine}} \underset{\substack{\text{Product of E2}\\ \text{(major product)}}}{CH_3CH{=}CHCH_3} + \underset{\substack{\text{Product of } S_N2\\ \text{(minor product)}}}{(R)\text{-}CH_3CH_2\overset{\overset{OCH_2CH_3}{|}}{C}HCH_3}$$

PROBLEM 11.3

Show how you might use the Williamson ether synthesis to prepare these ethers:

(a) 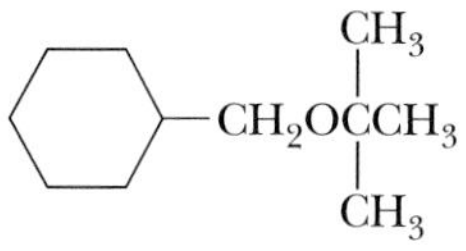 $C_6H_{11}{-}CH_2OC(CH_3)_2CH_3$ (cyclohexyl–CH_2OCCH_3 with two CH_3 groups on the quaternary carbon) **(b)** $(CH_3CH_2CH_2CH_2)_2O$

B. Acid-Catalyzed Dehydration of Alcohols

Diethyl ether and several other commercially available ethers are synthesized on an industrial scale by acid-catalyzed dehydration of primary alcohols.

$$2CH_3CH_2OH \xrightarrow[140°C]{H_2SO_4} CH_3CH_2OCH_2CH_3 + H_2O$$

Ethanol Diethyl ether

Acid-catalyzed intermolecular dehydration of alcohols is a specific example of an S_N2 reaction in which a poor leaving group, OH^-, is transformed into a better leaving group, OH_2, in the presence of acid.

MECHANISM Acid-Catalyzed Intermolecular Dehydration of a Primary Alcohol

Step 1: Proton transfer from the acid to form an oxonium ion

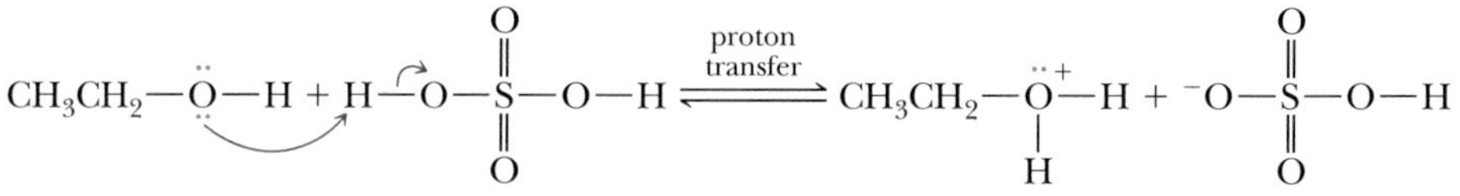

$$CH_3CH_2{-}\ddot{O}{-}H + H{-}O{-}SO_2{-}O{-}H \overset{\text{proton transfer}}{\rightleftharpoons} CH_3CH_2{-}\overset{+}{O}H{-}H + {}^{-}O{-}SO_2{-}O{-}H$$

An oxonium ion

Step 2: Nucleophilic displacement of OH_2 by the OH group of the alcohol to give a new oxonium ion

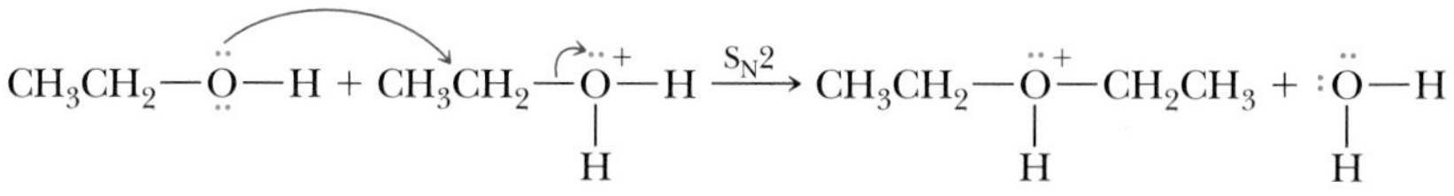

$$CH_3CH_2{-}\ddot{O}{-}H + CH_3CH_2{-}\overset{+}{O}H{-}H \xrightarrow{S_N2} CH_3CH_2{-}\overset{+}{O}H{-}CH_2CH_3 + :\ddot{O}H{-}H$$

An new oxonium ion

Step 3: Proton transfer to solvent to complete the reaction

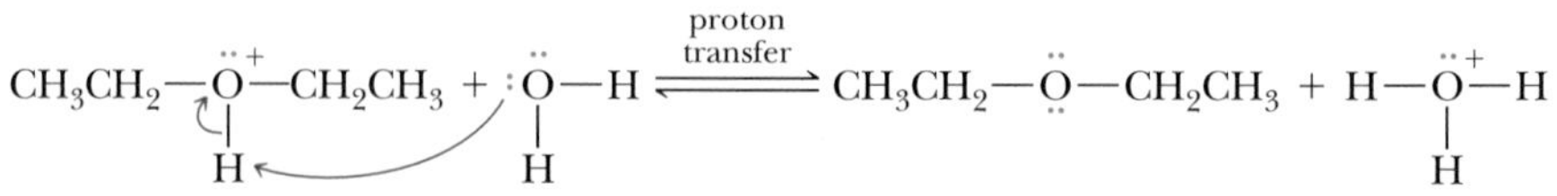

$$CH_3CH_2{-}\overset{+}{O}H{-}CH_2CH_3 + :\ddot{O}H{-}H \overset{\text{proton transfer}}{\rightleftharpoons} CH_3CH_2{-}\ddot{O}{-}CH_2CH_3 + H{-}\overset{+}{O}H{-}H$$

Intramolecular dehydration of ethanol to ethylene is a competing reaction, but it requires a higher temperature. In practice, experimental conditions can be adjusted so as to favor formation of either ethylene or diethyl ether.

Yields of ethers from acid-catalyzed intermolecular dehydration of alcohols are highest for symmetrical ethers formed from primary unbranched alcohols. Examples of symmetrical ethers formed in good yield by this method are dimethyl ether, diethyl ether, and dibutyl ether. From secondary alcohols, yields of ether are lower because of competition from acid-catalyzed dehydration (Section 9.7). In the case of tertiary alcohols, dehydration is the major reaction.

EXAMPLE 11.4

Explain why this reaction does not give a good yield of ethyl hexyl ether.

$$CH_3(CH_2)_4CH_2OH + CH_3CH_2OH \xrightarrow[\text{heat}]{H_2SO_4} CH_3(CH_2)_4CH_2OCH_2CH_3 + H_2O$$

Solution

From this reaction we expect a mixture of three ethers: diethyl ether, ethyl hexyl ether, and dihexyl ether.

PROBLEM 11.4

Show how ethyl hexyl ether might be prepared by a Williamson ether synthesis.

C. Acid-Catalyzed Addition of Alcohols to Alkenes

Under suitable conditions, alcohols can be added to the carbon-carbon double bond of an alkene to give an ether. The usefulness of this method of ether synthesis is limited to the interaction of alkenes that can form stable carbocations and primary alcohols. An example is the commercial synthesis of *tert*-butyl methyl ether, an antiknock, octane-improving gasoline additive. 2-Methylpropene and methanol are passed over an acid catalyst to give the ether.

$$CH_3\overset{\overset{\large CH_3}{|}}{C}{=}CH_2 + CH_3OH \xrightarrow{\text{acid catalyst}} CH_3\underset{\underset{\large CH_3}{|}}{\overset{\overset{\large CH_3}{|}}{C}}OCH_3$$

2-Methoxy-2-methylpropane
(*tert*-Butyl methyl ether)

The mechanism for this ether synthesis involves electrophilic attack of H^+ on the carbon-carbon double bond to generate a carbocation followed by addition of the nucleophile and loss of H^+ to give the final product.

MECHANISM Acid-Catalyzed Addition of an Alcohol to an Alkene

Step 1: Proton transfer from the acid to the alkene to form a carbocation

$$CH_3\overset{CH_3}{\overset{|}{C}}{=}CH_2 + H{-}\overset{+}{\underset{H}{\ddot{O}}}{-}CH_3 \longrightarrow CH_3\overset{CH_3}{\overset{|}{\underset{+}{C}}}CH_3 + :\underset{H}{\ddot{O}}{-}CH_3$$

Step 2: A Lewis acid-base reaction between the carbocation (a Lewis acid) and the alcohol (a Lewis base) to give an oxonium ion

$$CH_3\overset{CH_3}{\overset{|}{\underset{+}{C}}}CH_3 + H\ddot{O}CH_3 \longrightarrow CH_3\overset{CH_3}{\overset{|}{\underset{\overset{|+}{\ddot{O}}(H)(CH_3)}{C}}}CH_3$$

Step 3: Proton transfer to solvent (in this case the alcohol) to complete the reaction

$$CH_3{-}\ddot{O}{-}H + CH_3\overset{CH_3}{\overset{|}{\underset{\overset{|+}{\ddot{O}}(H)(CH_3)}{C}}}CH_3 \longrightarrow CH_3{-}\overset{+}{\underset{H}{\ddot{O}}}{-}H + CH_3\overset{CH_3}{\overset{|}{\underset{\overset{|}{:\ddot{O}}{-}CH_3}{C}}}CH_3$$

As an octane-improving additive, MTBE is superior to ethanol (the additive in gasohol). A blend of 15% MTBE with gasoline improves octane rating by approximately 5 units. Production of MTBE was 13.6 billion pounds in 1994 and 17.6 billion pounds in 1995. It is expected that demand for MTBE will continue to grow steadily in the coming years, and the factor limiting its expansion will be the availability of methanol.

11.5 Preparation of Sulfides

Symmetrical sulfides, RSR (also called symmetrical thioethers), are prepared by the treatment of 1 mol of Na_2S (where S^{2-} is the nucleophile) with 2 mol of alkyl halide.

$$2RX + Na_2S \longrightarrow \underset{\text{A sulfide}}{RSR} + 2NaX$$

This same reaction can also be used to prepare five- and six-membered cyclic sulfides. Treatment of a 1,4-dihalide with Na_2S gives a five-membered cyclic sulfide; treatment of a 1,5-dihalide with Na_2S gives a six-membered cyclic sulfide.

$$\underset{\text{1,4-Dichlorobutane}}{ClCH_2CH_2CH_2CH_2Cl} + Na_2S \xrightarrow{S_N2} \underset{\text{Thiolane (Tetrahydrothiophene)}}{\text{(five-membered ring with S)}} + 2Na^+Cl^-$$

$$\text{ClCH}_2\text{CH}_2\text{CH}_2\text{CH}_2\text{CH}_2\text{Cl} + \text{Na}_2\text{S} \xrightarrow{S_N2} \text{(thiane ring)} + 2\text{Na}^+\text{Cl}^-$$

1,5-Dichloropentane — Thiane (Tetrahydrothiopyran)

Unsymmetrical sulfides, RSR′, are prepared by converting a thiol to a sodium salt with either sodium hydroxide or sodium ethoxide and then allowing the salt to react with an alkyl halide. This method of thioether formation is the sulfur analog of the Williamson ether synthesis (Section 11.4A).

$$\text{CH}_3(\text{CH}_2)_8\text{CH}_2\text{S}^-\text{Na}^+ + \text{CH}_3\text{I} \xrightarrow{S_N2} \text{CH}_3(\text{CH}_2)_8\text{CH}_2\text{SCH}_3 + \text{Na}^+\text{I}^-$$

Sodium 1-decanethiolate — 1-(Methylthio)decane (Decyl methyl sulfide)

Note that all of these reactions leading to sulfides (thioethers) are direct applications of nucleophilic substitution reactions (Chapter 8).

11.6 Reactions of Ethers

Ethers resemble hydrocarbons in their resistance to chemical reaction. They do not react with oxidizing agents, such as potassium dichromate or potassium permanganate. They are stable toward even very strong bases, and except for tertiary alkyl ethers, they are not affected by most weak acids at moderate temperatures. Because of their good solvent properties and general inertness to chemical reaction, ethers are excellent solvents in which to carry out many organic reactions.

A. Acid-Catalyzed Cleavage by Concentrated HX

Hydrolysis of ethers requires both a strong acid and a good nucleophile, hence the use of concentrated aqueous HI (57%) or HBr (48%). Dibutyl ether, for example, reacts with concentrated HBr to give two molecules of 1-bromobutane.

$$(\text{CH}_3\text{CH}_2\text{CH}_2\text{CH}_2)_2\text{O} + 2\text{HBr} \xrightarrow[\text{heat}]{} 2\text{CH}_3\text{CH}_2\text{CH}_2\text{CH}_2\text{Br} + \text{H}_2\text{O}$$

Dibutyl ether — 1-Bromobutane

Concentrated HCl (38%) is far less effective in cleaving ethers, primarily because Cl^- is a weaker nucleophile in water than either I^- or Br^-. Cleavage of primary and secondary ethers proceeds by an S_N2 pathway. Cleavage of tertiary ethers proceeds by an S_N1 pathway.

The mechanism of acid-catalyzed hydrolysis of dialkyl ethers begins in Step 1 with protonation of the ether oxygen to form an oxonium ion. Step 2 depends on the nature of the group attached to oxygen. In the mechanism shown, both carbons bonded to oxygen are primary, and cleavage involves an S_N2 reaction in which a halide ion is the nucleophile. In this example, the leaving group is CH_3CH_2OH, a weak base and a weak nucleophile.

MECHANISM Acid-Catalyzed Hydrolysis of an Ether

Step 1: Proton transfer to the oxygen atom of the ether to form an oxonium ion

$$CH_3CH_2-\ddot{O}-CH_2CH_3 + H-\overset{..+}{O}(H)-H \xrightleftharpoons{\text{proton transfer}} CH_3CH_2-\overset{..+}{O}(H)-CH_2CH_3 + :\ddot{O}(H)-H$$

An oxonium ion

Step 2: Nucleophilic displacement on the primary carbon

$$Br^- + CH_3CH_2-\overset{..+}{O}(H)-CH_2CH_3 \xrightarrow{S_N2} CH_3CH_2Br + :\ddot{O}(H)-CH_2CH_3$$

This cleavage produces one molecule of alkyl bromide and one molecule of alcohol. In the presence of excess concentrated HBr, the alcohol is converted to a second molecule of alkyl bromide by another S_N2 process (Section 8.3).

Tertiary, allylic, and benzylic ethers are particularly susceptible to cleavage by acid, often under quite mild conditions. Tertiary-butyl ethers, for example, are cleaved by aqueous HCl at room temperature.

$$C_6H_{11}-O-C(CH_3)_3 + HCl \xrightarrow{S_N1} C_6H_{11}-OH + Cl-C(CH_3)_2-CH_3$$

EXAMPLE 11.5

Account for the fact that reaction of most methyl ethers with concentrated HI gives CH_3I and ROH as the initial major products rather than CH_3OH and RI. For example:

$$CH_3CH(CH_3)CH_2CH_2OCH_3 + HI \longrightarrow CH_3CH(CH_3)CH_2CH_2OH + CH_3I$$

Solution

The first step is protonation of the ether oxygen to give an oxonium ion. Cleavage is by an S_N2 pathway on the less hindered methyl carbon.

$$CH_3CH(CH_3)CH_2CH_2-\overset{..+}{O}(H)-CH_3$$

site of attack by nucleophile (the CH_3 on oxygen)

PROBLEM 11.5

Account for the fact that treatment of *tert*-butyl methyl ether with a limited amount of concentrated HI gives methanol and *tert*-butyl iodide rather than methyl iodide and *tert*-butyl alcohol.

EXAMPLE 11.6

Draw structural formulas for the major products of the following reactions:

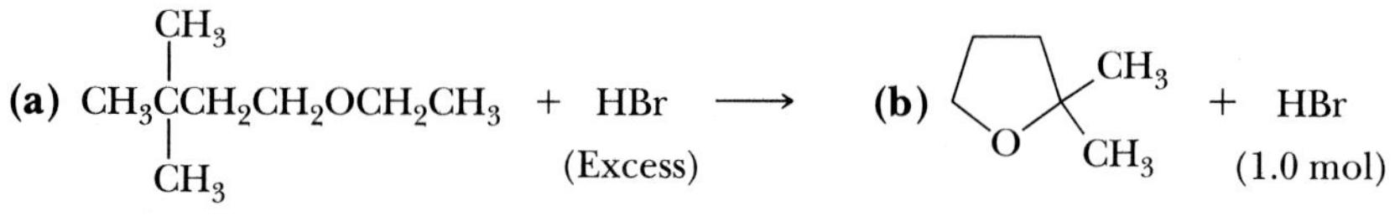

(a) $CH_3C(CH_3)_2CH_2CH_2OCH_2CH_3$ + HBr (Excess) ⟶ (b) 2,2-dimethyltetrahydrofuran + HBr (1.0 mol)

Solution

(a) Cleavage on either side of the ether oxygen by an S_N2 mechanism gives an alcohol and an alkyl bromide. Reaction of the alcohol then gives a second molecule of alkyl bromide.

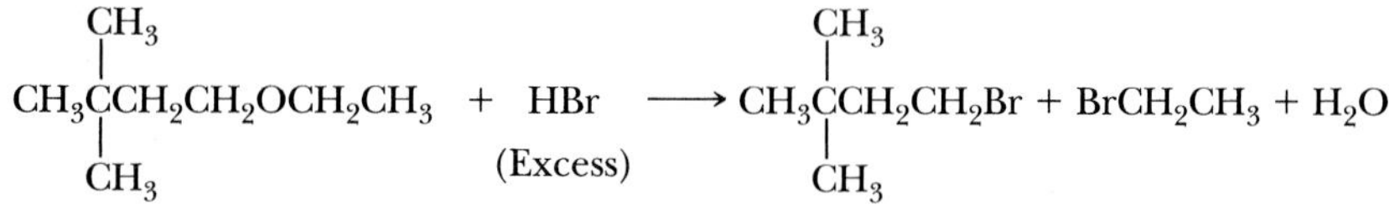

$$CH_3C(CH_3)_2CH_2CH_2OCH_2CH_3 + \underset{\text{(Excess)}}{HBr} \longrightarrow CH_3C(CH_3)_2CH_2CH_2Br + BrCH_2CH_3 + H_2O$$

(b) Protonation of the ether oxygen followed by cleavage gives a tertiary carbocation, which may then (1) react with bromide ion to give a bromoalcohol or (2) lose a proton to give an unsaturated alcohol.

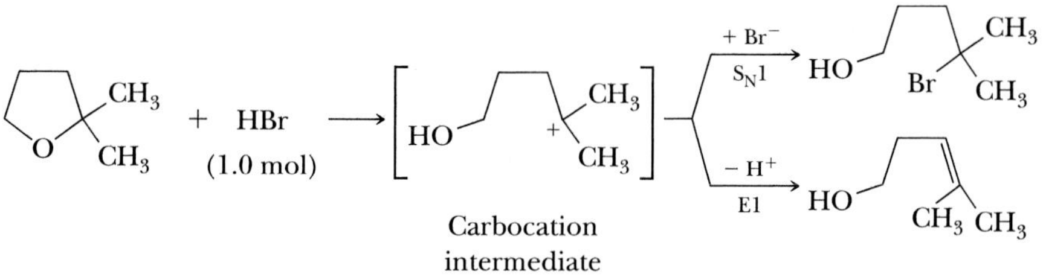

2,2-dimethyltetrahydrofuran + HBr (1.0 mol) ⟶ [HO–CH₂CH₂CH₂–C⁺(CH₃)₂] (Carbocation intermediate)

⟶ (+ Br⁻, S_N1) HO–CH₂CH₂CH₂–CBr(CH₃)₂

⟶ (− H⁺, E1) HO–CH₂CH₂CH=C(CH₃)₂

PROBLEM 11.6

Draw structural formulas for the major products of these reactions:

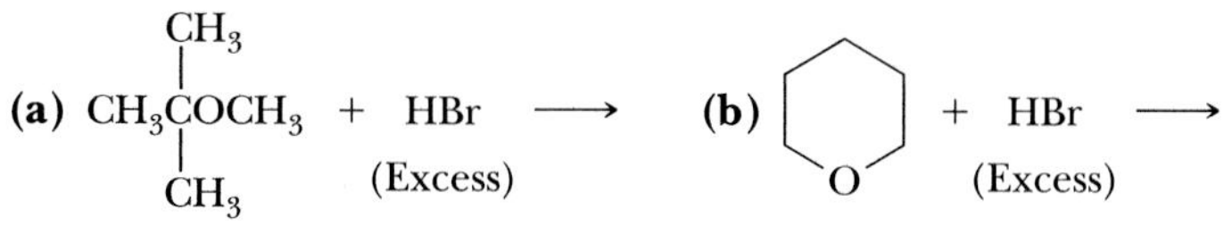

(a) $CH_3C(CH_3)_2OCH_3$ + HBr (Excess) ⟶ (b) tetrahydropyran + HBr (Excess) ⟶

B. Oxidation of Ethers: Formation of Hydroperoxides

Two hazards must be avoided when working with diethyl ether and other low-molecular-weight ethers. First, they are highly flammable. Consequently, open flames and electric appliances with sparking contacts must be avoided where ethers are being used. Because diethyl ether is so volatile (it has a low boiling point), it should be used in a fume hood to prevent the build-up of vapors and possible explosion. Second, anhydrous ethers react with molecular oxygen at a C—H bond adjacent to the ether oxygen to form explosive hydroperoxides. The functional group of a **hydroperoxide** is an OOH group.

Hydroperoxide A molecule containing an —OOH group.

$$\underset{\text{Diethyl ether}}{CH_3CH_2OCH_2CH_3} + O_2 \longrightarrow \underset{\text{A hydroperoxide}}{CH_3CH_2O\overset{\overset{\displaystyle OOH}{|}}{C}HCH_3}$$

Hydroperoxidation proceeds by a radical chain mechanism (See the Chemistry in Action box "Radical Autoxidation," pp. 261–262). Rates of hydroperoxide formation increase dramatically if the C—H bond adjacent to oxygen is tertiary, as for example in diisopropyl ether, because of favored generation of a relatively stable 3° radical intermediate. This hydroperoxide precipitates from solution as a waxy solid.

$$\underset{\text{Diisopropyl ether}}{CH_3\underset{\underset{\displaystyle CH_3}{|}}{C}HO\underset{\underset{\displaystyle CH_3}{|}}{C}HCH_3} + O_2 \longrightarrow \underset{\text{A hydroperoxide}}{CH_3\underset{\underset{\displaystyle CH_3}{|}}{C}HO\overset{\overset{\displaystyle OOH}{|}}{\underset{\underset{\displaystyle CH_3}{|}}{C}}CH_3}$$

Hydroperoxides are dangerous because they are explosive. Furthermore, they react with some metals to form hydroperoxide salts, which are also explosive and especially sensitive to shock. Peroxides can be detected by shaking an ether with an acidified 10% aqueous solution of potassium iodide, KI, or with starch iodine paper with a drop of acetic acid. Peroxides oxidize iodide ion to iodine, I_2, which gives a yellow color to the solution. This is converted to a deep blue-purple color by the addition of starch, due to the formation of a starch-iodine complex. Hydroperoxides can be removed by treatment with a reducing agent. One effective procedure is to shake the hydroperoxide-contaminated ether with a solution of iron(II) sulfate in dilute aqueous sulfuric acid. Bottles containing diisopropyl ether with precipitated hydroperoxide should be removed and disposed of by a bomb squad!

C. Oxidation of Sulfides

Many of the properties of sulfides stem from the fact that divalent sulfur is a reducing agent; it is easily oxidized to two higher oxidation states. Treatment of a sulfide with 1 mol of 30% aqueous hydrogen peroxide at room temperature gives a sulfoxide, as illustrated by oxidation of methyl phenyl sulfide to methyl phenyl sulfoxide. Several other oxidizing agents, including sodium periodate, $NaIO_4$, also bring about the same conversion. Treatment of a sulfoxide with HIO_4 brings about its oxidation to a sulfone.

$$\underset{\substack{\text{Methyl phenyl sulfide}\\\text{(a sulfide)}}}{C_6H_5-\ddot{S}-CH_3} \xrightarrow[25°C]{H_2O_2} \underset{\substack{\text{Methyl phenyl sulfoxide}\\\text{(a sulfoxide)}}}{C_6H_5-\overset{\overset{\displaystyle O}{\|}}{\underset{\cdot\cdot}{S}}-CH_3} \xrightarrow[25°C]{HIO_4} \underset{\substack{\text{Methyl phenyl sulfone}\\\text{(a sulfone)}}}{C_6H_5-\overset{\overset{\displaystyle O}{\|}}{\underset{\underset{\displaystyle O}{\|}}{S}}-CH_3}$$

Dimethyl sulfoxide (DMSO) is manufactured on an industrial scale by air oxidation of dimethyl sulfide in the presence of oxides of nitrogen.

$$\underset{\text{Dimethyl sulfide}}{2CH_3-\ddot{S}-CH_3} + O_2 \xrightarrow{\text{oxides of nitrogen}} \underset{\text{Dimethyl sulfoxide}}{2CH_3-\overset{\overset{\displaystyle :O:}{\|}}{\underset{\cdot\cdot}{S}}-CH_3}$$

11.7 Ethers as Protecting Groups

When dealing with organic compounds containing two or more functional groups, it is often necessary to protect one functional group (to prevent its reaction) while carrying out a reaction at another functional group. Suppose, for example, that you wish to convert 4-pentyn-1-ol to 4-heptyn-1-ol.

$$\underset{\text{4-Pentyn-1-ol}}{HC{\equiv}CCH_2CH_2CH_2OH} \xrightarrow{?} \underset{\text{4-Heptyn-1-ol}}{CH_3CH_2C{\equiv}CCH_2CH_2CH_2OH}$$

The new carbon-carbon bond in the product can be formed by alkylation of the acetylide anion from 4-pentyn-1-ol (Section 10.5) with ethyl bromide. 4-Pentyn-1-ol, however, contains two acidic hydrogens, one on the hydroxyl group (pK_a 16–18) and the other on the carbon-carbon triple bond (pK_a 25). Treatment of this compound with one equivalent of $NaNH_2$ forms the alkoxide anion (the —OH group is the stronger acid) rather than the acetylide anion.

pK_a 25 (on the $\equiv$C–H); pK_a 16–18 (on the OH)

$$\underset{\text{4-Pentyn-1-ol}}{HC{\equiv}CCH_2CH_2CH_2OH} + Na^+NH_2^- \longrightarrow HC{\equiv}CCH_2CH_2CH_2O^-Na^+ + NH_3$$

It is necessary, therefore, to protect the —OH group to prevent its reaction with sodium amide. A good protecting group is one that (1) is easy to add and to remove and (2) is resistant to the reagents used to transform the unprotected functional group. Chemists have devised protecting groups for many functional groups, and we will encounter several of them in this text. In this section, we concentrate on three protecting groups for hydroxyl groups.

The first of these is the *tert*-butyl group, formed by treatment of an alcohol with 2-methylpropene (isobutylene) in the presence of an acid catalyst. This treatment of 4-pentyn-1-ol with 2-methylpropene in the presence of sulfuric acid as a catalyst gives the *tert*-butyl ether. Treatment of this ether with sodium amide followed by ethyl bromide forms the new carbon-carbon bond. The protecting group is then removed by treatment with aqueous acid. Thus, the *tert*-butyl protecting group is added in the presence of an acid catalyst, is stable under neutral and basic conditions, and is removed by treatment with aqueous acid.

$$\underset{\text{4-Pentyn-1-ol}}{HC{\equiv}CCH_2CH_2CH_2OH} \xrightarrow[H_2SO_4]{1.\ CH_2{=}C(CH_3)_2} HC{\equiv}CCH_2CH_2CH_2OC(CH_3)_2CH_3 \xrightarrow[3.\ CH_3CH_2Br]{2.\ Na^+NH_2^-}$$

$$CH_3CH_2C{\equiv}CCH_2CH_2CH_2OC(CH_3)_2CH_3 \xrightarrow{4.\ H_3O^+/H_2O}$$

$$\underset{\text{4-Heptyn-1-ol}}{CH_3CH_2C{\equiv}CCH_2CH_2CH_2OH} + H_2C{=}C(CH_3)_2$$

EXAMPLE 11.7

When the *tert*-butyl protecting group is removed by treatment with aqueous acid, its products are *tert*-butyl alcohol and 2-methylpropene (isobutylene). Propose a mechanism for the acid-catalyzed removal of the *tert*-butyl protecting group and the formation of these two products.

Solution

Proton transfer from the acid catalyst to the oxygen of the ether gives an oxonium ion. Cleavage on the side of the *tert*-butyl group gives the more stable *tert*-butyl cation, which then either transfers a proton to water to give isobutylene (E1) or adds water to give *tert*-butyl alcohol (S_N1).

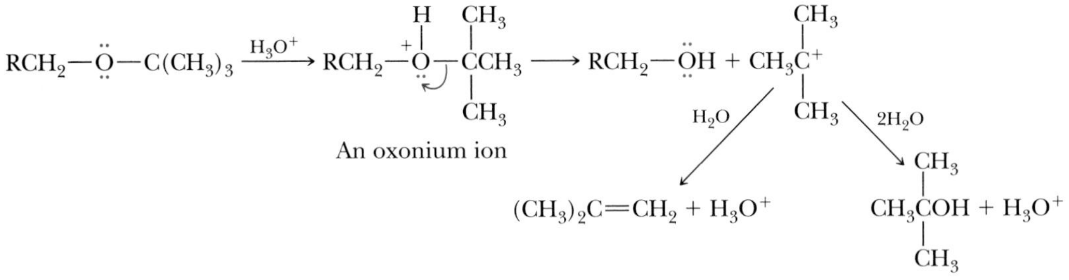

$$RCH_2{-}\ddot{O}{-}C(CH_3)_3 \xrightarrow{H_3O^+} RCH_2{-}\overset{+}{O}(H){-}C(CH_3)_3 \longrightarrow RCH_2{-}\ddot{O}H + (CH_3)_3C^+$$

An oxonium ion

$$(CH_3)_3C^+ \xrightarrow{H_2O} (CH_3)_2C{=}CH_2 + H_3O^+$$

$$(CH_3)_3C^+ \xrightarrow{2H_2O} (CH_3)_3COH + H_3O^+$$

PROBLEM 11.7

Why is the use of the *tert*-butyl protecting group limited to protection of primary alcohols?

An —OH group can also be protected by converting it to a trimethylsilyl ether, $-OSi(CH_3)_3$, by treating the alcohol with chlorotrimethylsilane in the presence of a tertiary amine, such as triethylamine or pyridine. The function of the tertiary amine is to neutralize the HCl formed by reaction of the alcohol with chlorotrimethylsilane.

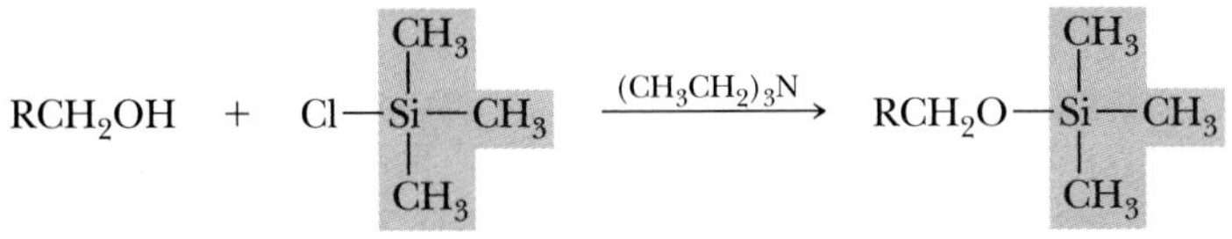

$$RCH_2OH + Cl{-}Si(CH_3)_3 \xrightarrow{(CH_3CH_2)_3N} RCH_2O{-}Si(CH_3)_3$$

Chlorotrimethylsilane — A trimethylsilyl ether

Like the *tert*-butyl ether protecting group, the trimethylsilyl ether group is removed by treatment with aqueous acid, or alternatively using fluoride ion, F^-, in the form of tetrabutylammonium fluoride, $(CH_3CH_2CH_2CH_2)_4N^+F^-$.

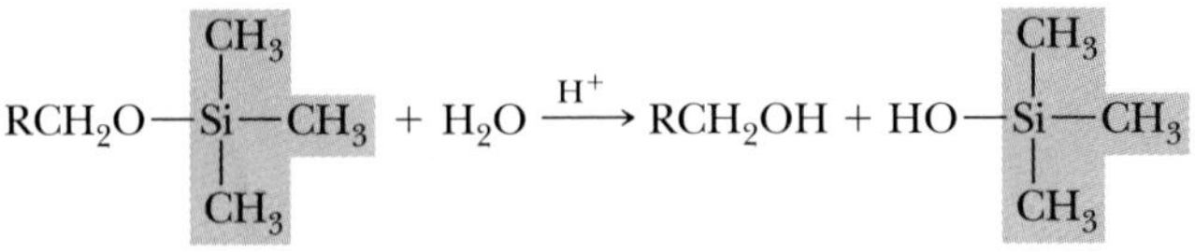

$$RCH_2O{-}Si(CH_3)_3 + H_2O \xrightarrow{H^+} RCH_2OH + HO{-}Si(CH_3)_3$$

A trimethylsilyl ether

A third method of protecting an —OH group of primary and secondary alcohols is treatment of the alcohol with dihydropyran in the presence of an acid catalyst, commonly anhydrous HCl, or a sulfonic acid, RSO_3H. The product is a tetrahydropyranyl ether.

$$RCH_2OH + \text{Dihydropyran} \xrightarrow{H^+} \text{A tetrahydropyranyl ether } (RCH_2O\text{–THP})$$

Dihydropyran A tetrahydropyranyl ether

The tetrahydropyranyl (THP) group is stable in neutral and basic solutions and to most oxidizing and reducing agents. It is removed easily by treatment with dilute aqueous acid to regenerate the original primary or secondary alcohol. We will discuss the chemistry of formation and removal of the THP group in Chapter 15.

11.8 Epoxides: Structure and Nomenclature

An **epoxide** is a cyclic ether in which oxygen is one atom of a three-membered ring. Although epoxides are technically classed as ethers, we discuss them separately because of their exceptional chemical reactivity compared with other ethers. Simple epoxides are named as derivatives of oxirane, the parent epoxide. Where the epoxide is a part of another ring system, it is named using the prefix epoxy-.

Epoxide A cyclic ether in which oxygen is one atom of a three-membered ring.

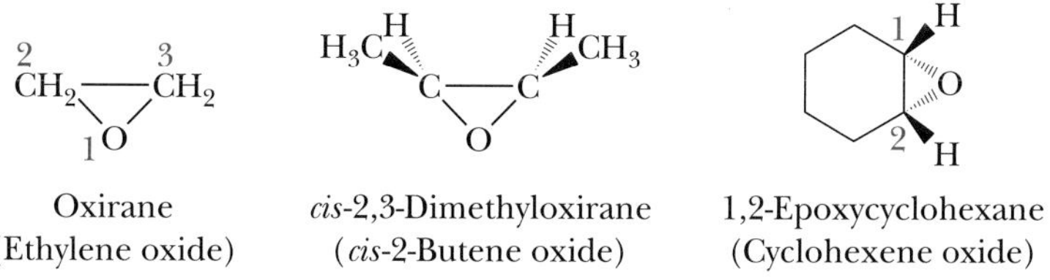

Oxirane (Ethylene oxide) *cis*-2,3-Dimethyloxirane (*cis*-2-Butene oxide) 1,2-Epoxycyclohexane (Cyclohexene oxide)

Common names of epoxides are derived by giving the name of the alkene from which the epoxide is formally derived followed by the word "oxide"; an example is *cis*-2-butene oxide.

As can be seen in the stereoview of 1,2-epoxycyclohexane, the presence of the three-membered epoxide ring fused by *cis* bonds to the larger cyclohexane ring results in considerable distortion of the cyclohexane ring.

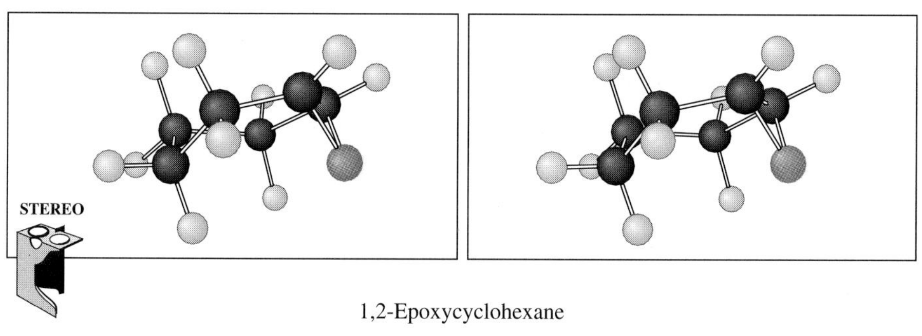

1,2-Epoxycyclohexane

11.9 Synthesis of Epoxides

Ethylene oxide is prepared on an industrial scale by catalyzed air oxidation of ethylene. The most common methods for preparation of other epoxides are (1) oxidation of an alkene with a peroxycarboxylic acid and (2) treatment of a halohydrin with base.

A. Ethylene Oxide

Ethylene oxide, one of the few epoxides manufactured on an industrial scale, is prepared by passing a mixture of ethylene and air (or oxygen) over a silver catalyst. In the United States, the 1995 production of ethylene oxide by this method was 7.62 billion pounds.

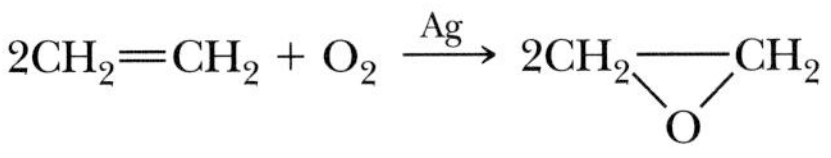

$$2CH_2{=}CH_2 + O_2 \xrightarrow{Ag} 2CH_2\text{—}CH_2 \text{ (bridged by O)}$$

Oxirane
(Ethylene oxide)

This method fails when applied to other low-molecular-weight alkenes.

B. Oxidation of Alkenes with Peroxycarboxylic Acids

The most common laboratory method for the synthesis of epoxides from alkenes is oxidation with a peroxycarboxylic acid (a peracid). Two of the most widely used peroxy acids are the magnesium salt of monoperoxyphthalic acid (MMPP) and peroxyacetic acid.

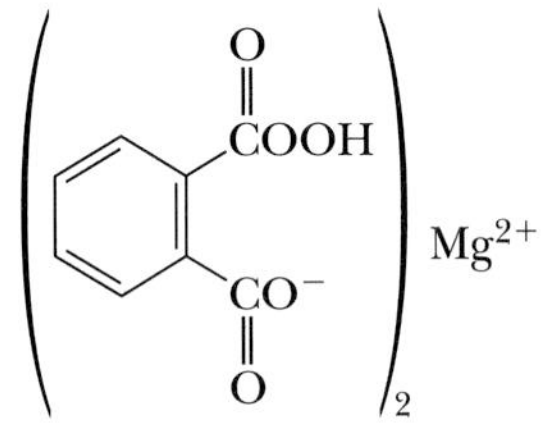

$$\left(C_6H_4(\overset{O}{\overset{\|}{C}}OOH)(\overset{O}{\overset{\|}{C}}O^-)\right)_2 Mg^{2+} \qquad CH_3\overset{O}{\overset{\|}{C}}OOH$$

Magnesium monoperoxyphthalate (MMPP) — Peroxyacetic acid (Peracetic acid)

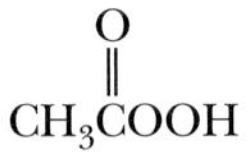

Following is a balanced equation for the epoxidation of cyclohexene by a peroxycarboxylic acid, RCO_3H. In the process, the peroxycarboxylic acid is reduced to a carboxylic acid.

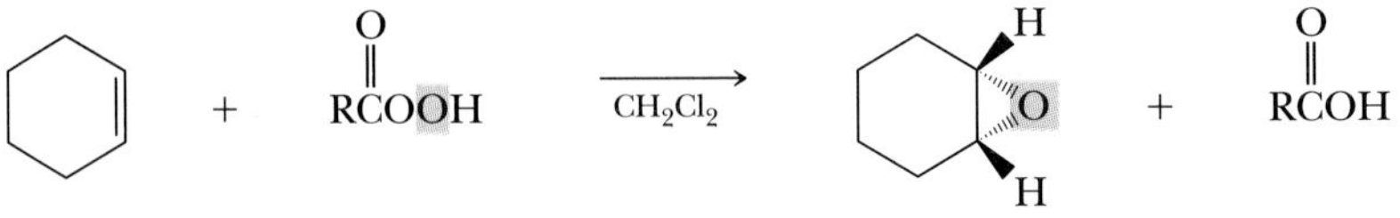

$$\text{Cyclohexene} + R\overset{O}{\overset{\|}{C}}OOH \xrightarrow{CH_2Cl_2} \text{1,2-Epoxycyclohexane} + R\overset{O}{\overset{\|}{C}}OH$$

Cyclohexene — A peroxycarboxylic acid — 1,2-Epoxycyclohexane (Cyclohexene oxide) — A carboxylic acid

Epoxidation of an alkene is stereoselective. Epoxidation of *cis*-2-butene, for example, yields only *cis*-2,3-dimethyloxirane and epoxidation of *trans*-2-butene yields only *trans*-2,3-dimethyloxirane.

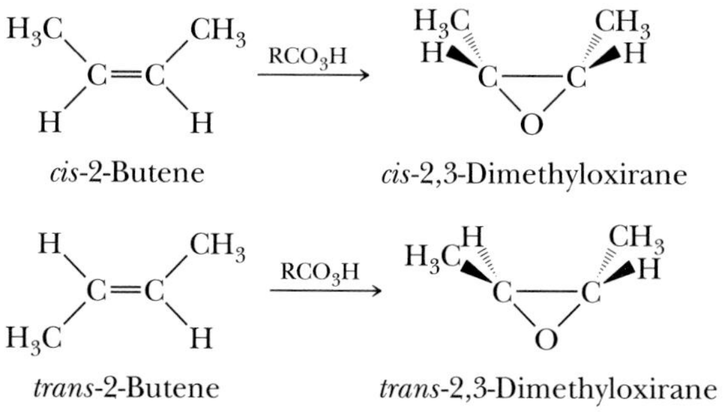

cis-2-Butene $\xrightarrow{RCO_3H}$ cis-2,3-Dimethyloxirane

trans-2-Butene $\xrightarrow{RCO_3H}$ trans-2,3-Dimethyloxirane

A mechanism for epoxidation by a peroxyacid must take into account the following facts. (1) The reaction takes place in nonpolar solvents, which means that the reaction cannot involve the formation of ions or any species with significant separation of unlike charges. (2) The reaction is stereoselective, with retention of the alkene configuration, which means that even though the pi bond of the carbon-carbon double bond is broken, at no time is there free rotation about the remaining sigma bond. Following is a mechanism consistent with these observations.

MECHANISM Epoxidation of an Alkene by RCO_3H

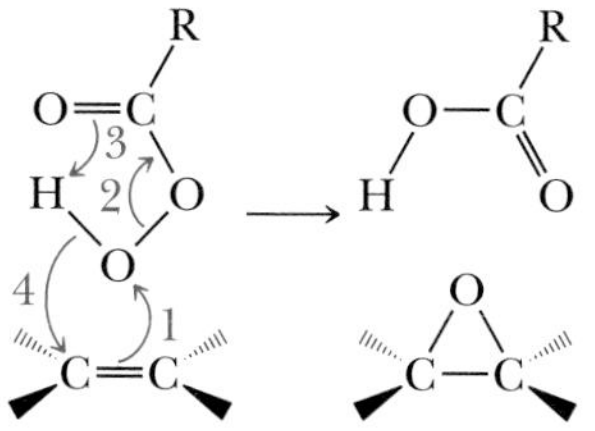

To help you see a pattern to the flow of electron pairs, curved arrows are numbered 1 through 4. Arrow 1 shows interaction of the pi electrons of the carbon-carbon double bond with the end oxygen atom of the peroxyacid and formation of a new C—O bond. Arrows 2 and 3 show shifts of electron pairs within the peroxyacid, and arrow 4 shows formation of the second carbon-oxygen bond. The numbering of these arrows does not imply an order in which covalent bonds are broken and made. Rather, they are meant as a guide to help you understand the mechanism. It is thought that the entire combination of bond-making and bond-breaking steps is concerted, or nearly so.

C. Internal Nucleophilic Substitution in Halohydrins

A second general method for the preparation of epoxides from alkenes involves (1) treatment of the alkene with chlorine or bromine in water to form a chlorohydrin or bromohydrin, followed by (2) treatment of the halohydrin with base and intramolecular displacement of X^-. By these steps, propene is first converted to 1-chloro-2-propanol and then to methyloxirane (propylene oxide).

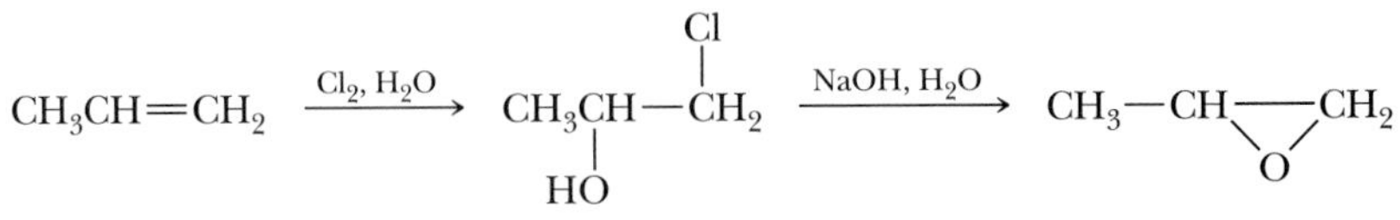

$$CH_3CH{=}CH_2 \xrightarrow{Cl_2,\ H_2O} CH_3CH(OH){-}CH_2Cl \xrightarrow{NaOH,\ H_2O} CH_3{-}CH{-}CH_2 \text{ (bridged by O)}$$

Propene — 1-Chloro-2-propanol (a chlorohydrin) — Methyloxirane (Propylene oxide)

We studied the reaction of alkenes with chlorine or bromine in water to form halohydrins (Section 6.3F) and saw that it is both regioselective and stereoselective. Conversion of a halohydrin to an epoxide with base is stereoselective as well and can be viewed as an internal S_N2 reaction. Hydroxide ion or other base abstracts a proton from the halohydrin hydroxyl group to form an alkoxide ion, a nucleophile, which then displaces halogen on the adjacent carbon. As with all S_N2 reactions, attack of the nucleophile is from the backside of the C—X bond and causes inversion of configuration at the site of substitution.

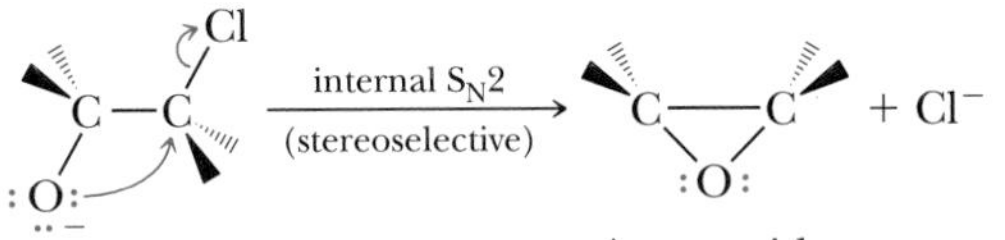

internal S_N2 (stereoselective) → + Cl^-

An epoxide

Note that this displacement of halide by the alkoxide anion can also be viewed as an intramolecular variation of the Williamson ether synthesis we studied in Section 11.4A. In this case, the displacing alkoxide and leaving halide ions are on adjacent carbon atoms.

EXAMPLE 11.8

Conversion of an alkene to a halohydrin and internal displacement of a halide ion by alkoxide ion are both stereoselective. Use this information to demonstrate that the configuration of the alkene is preserved in the epoxide. As an illustration, show that reaction of *cis*-2-butene by this two-step sequence gives *cis*-2,3-dimethyloxirane (*cis*-2-butene oxide).

Solution

Addition of HOCl to an alkene occurs by anti addition of —OH and —Cl to the double bond (Section 6.3F). The conformation of this product is also the conformation necessary for backside displacement of the halide ion by the alkoxide ion. Thus, a *cis*-alkene gives a *cis*-disubstituted oxirane and a *trans*-alkene gives a *trans*-disubstituted oxirane.

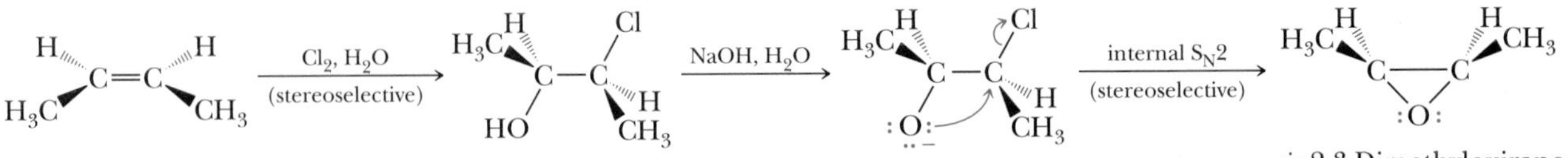

cis-2-Butene $\xrightarrow[\text{(stereoselective)}]{Cl_2,\ H_2O}$ A chlorohydrin $\xrightarrow{NaOH,\ H_2O}$ $\xrightarrow[\text{(stereoselective)}]{\text{internal } S_N2}$ *cis*-2,3-Dimethyloxirane

PROBLEM 11.8

Consider the possibilities for stereoisomerism in the halohydrin and epoxide formed in Example 11.8.

(a) How many stereoisomers are possible for the chlorohydrin? Which of the possible chlorohydrins are formed by reaction of *trans*-2-butene with Cl_2/H_2O?

(b) How many stereoisomers are possible for the epoxide? Which of the possible stereoisomers is/are formed in this reaction sequence?

11.10 Reactions of Epoxides

Because of the strain associated with the three-membered ring, epoxides undergo a variety of ring-opening reactions, the characteristic feature of which is nucleophilic substitution at one of the carbons of the epoxide ring with the oxygen atom as the leaving group.

Characteristic reaction of epoxides:

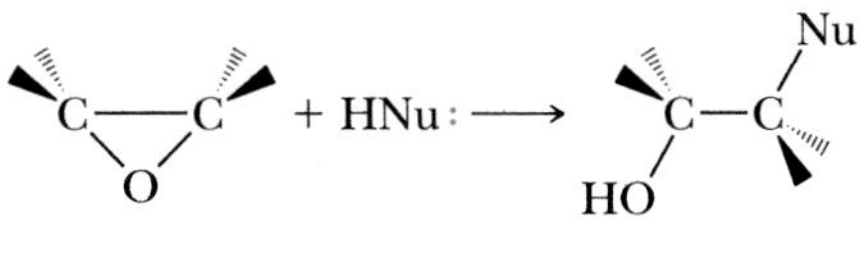

$$\text{C—C (epoxide, O bridge)} + \text{HNu:} \longrightarrow \text{HO—C—C—Nu}$$

A. Acid-Catalyzed Ring Opening

In the presence of an acid catalyst, such as sulfuric acid, epoxides are hydrolyzed to 1,2-diols. As an example, acid-catalyzed hydrolysis of oxirane gives 1,2-ethanediol (ethylene glycol). Production of ethylene glycol in the United States in 1995 was 5.2 billion pounds.

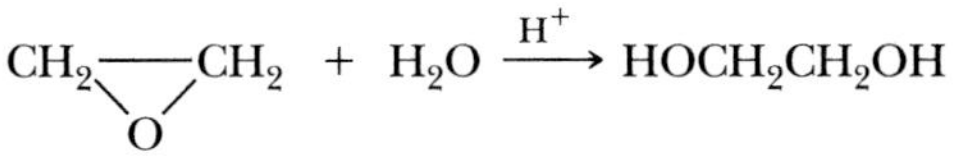

$$CH_2\text{—}CH_2 \text{ (O bridge)} + H_2O \xrightarrow{H^+} HOCH_2CH_2OH$$

Oxirane (Ethylene oxide) 1,2-Ethanediol (Ethylene glycol)

Under acidic conditions, the oxygen atom of the epoxide is protonated in Step 1 to form a bridged oxonium ion intermediate. The oxonium ion intermediate is then attacked by water in Step 2 from the side opposite the oxonium ion bridge, which results in the ring opening. Proton transfer in Step 3 completes the reaction.

MECHANISM Acid-Catalyzed Hydrolysis of an Epoxide

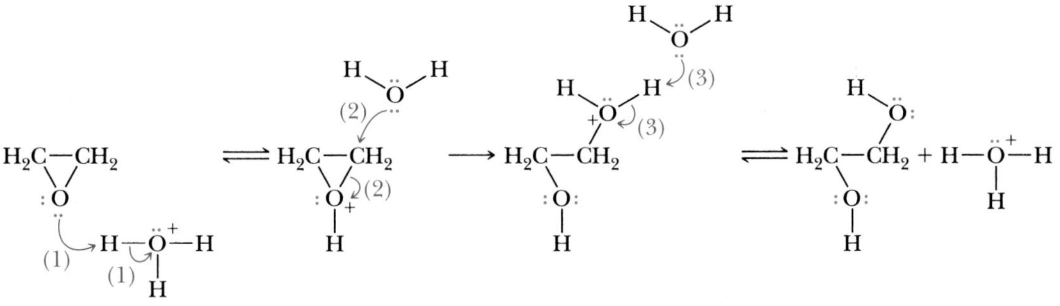

$$H_2C\text{—}CH_2 \text{ (:O: bridge)} + H\text{—}\ddot{O}^+\text{—}H\,(H) \;\overset{(1)}{\rightleftharpoons}\; H_2C\text{—}CH_2 \text{ (:O}^+\text{—H bridge)} \xrightarrow{(2)\; H_2\ddot{O}} H_2C(\text{:O—H})\text{—}CH_2(\text{—}O^+H_2) \;\overset{(3)\; H_2\ddot{O}}{\rightleftharpoons}\; H_2C(\text{:O—H})\text{—}CH_2(\text{—}\ddot{O}\text{—H}) + H\text{—}\ddot{O}^+\text{—}H\,(H)$$

Attack of a nucleophile on a protonated epoxide shows a stereoselectivity typical of S_N2 reactions; the nucleophile attacks anti to the leaving hydroxyl group and the

—OH groups in the glycol thus formed are anti. As a result, hydrolysis of an epoxycycloalkane yields a *trans*-1,2-cycloalkanediol.

1,2-Epoxycyclopentane (Cyclopentene oxide) + H_2O $\xrightarrow{H^+}$ *trans*-1,2-Cyclopentanediol

Note the similarity in ring opening of this bridged oxonium ion intermediate, the bridged halonium ion intermediate in electrophilic addition of halogens to an alkene (Section 6.3C), and the bridged mercurinium ion intermediate in oxymercuration (Section 6.3E). In each case, the intermediate is a three-membered ring with a heteroatom bearing a positive charge, and attack of the nucleophile is anti to the leaving group.

Because there is some carbocation character developed in the transition state for acid-catalyzed epoxide ring opening, attack of the nucleophile on unsymmetrical epoxides occurs preferentially at the carbon better able to bear a partial positive charge.

1-Methyl-1,2-epoxycyclohexane + CH_3OH $\xrightarrow{H_2SO_4}$ 2-Methoxy-2-methylcyclohexanol

Thus, the stereochemistry of acid-catalyzed ring openings is S_N2-like in that attack of the nucleophile is from the side opposite the oxonium ion bridged intermediate. The regiochemistry, however, is S_N1-like. Because of the partial carbocation character of the transition state, attack of the nucleophile on the oxonium ion intermediate occurs preferentially on the more substituted carbon, that is at the one better able to bear the partial positive charge that develops on carbon in the transition state.

At this point, let us compare the stereochemistry of the glycol formed by acid-catalyzed hydrolysis of an epoxide with that formed by oxidation of an alkene with osmium tetroxide (Section 6.5B). Each reaction sequence is stereoselective, but gives a different stereoisomer. Acid-catalyzed hydrolysis of cyclopentene oxide gives *trans*-1,2-cyclopentanediol; osmium tetroxide oxidation of cyclopentene gives *cis*-1,2-cyclopentanediol. Thus, a cycloalkene can be converted to either a *cis*-glycol or a *trans*-glycol by the proper choice of reagents.

Cyclopentene $\xrightarrow{RCO_3H}$ epoxide $\xrightarrow[H_2O]{H^+}$ *trans*-1,2-Cyclopentanediol

Cyclopentene $\xrightarrow{OsO_4,\ t\text{-}BuOOH}$ *cis*-1,2-Cyclopentanediol

B. Nucleophilic Ring Opening

Ethers are not normally susceptible to reaction with nucleophiles. Because of the strain associated with a three-membered ring, however, epoxides undergo ring-opening reactions with a variety of nucleophiles. Good nucleophiles attack an epoxide ring by an S_N2 mechanism and show a regioselectivity typical of such reactions, namely, attack of the nucleophile at the less hindered carbon. Following is an equation for the reaction of methyloxirane (propylene oxide) with sodium methoxide in methanol.

$$CH_3CH\overset{O}{\frown}CH_2 + CH_3O^-Na^+ \xrightarrow{S_N2} CH_3\underset{O^-Na^+}{CH}-CH_2-OCH_3$$

$$CH_3\underset{O^-Na^+}{CH}-CH_2-OCH_3 + CH_3OH \longrightarrow CH_3\underset{OH}{CH}-CH_2-OCH_3 + CH_3O^-Na^+$$

Net reaction:

$$CH_3CH\overset{O}{\frown}CH_2 + CH_3OH \xrightarrow{CH_3O^-Na^+} CH_3\underset{OH}{CH}CH_2OCH_3$$

Methyloxirane (Propylene oxide) → 1-Methoxy-2-propanol

In the reaction of propylene oxide with methanol in the presence of sodium methoxide, methoxide ion, the nucleophile, attacks the primary carbon in preference to the secondary carbon and displaces $-O^-$. Nucleophilic ring opening of epoxides is also stereoselective; as expected of an S_N2 reaction, attack of the nucleophile is anti to the leaving group as illustrated by the reaction of cyclohexene oxide with methanol in the presence of sodium methoxide to give *trans*-2-methoxycyclohexanol.

$$\text{Cyclohexene oxide} + CH_3OH \xrightarrow{CH_3O^-Na^+} \textit{trans}\text{-2-Methoxycyclohexanol}$$

Cyclohexene oxide — *trans*-2-Methoxycyclohexanol

The reaction of epoxides with organolithium and lithium diorganocopper (Gilman) reagents is an important method for the formation of new carbon-carbon bonds. These reagents bring about regioselective ring opening of substituted epoxides at the less substituted carbon to give alcohols. Treatment of styrene oxide with lithium divinylcopper followed by workup in aqueous acid, for example, gives 1-phenyl-3-buten-1-ol.

$$C_6H_5-CH\overset{O}{\frown}CH_2 \xrightarrow[2.\ H_2O,\ HCl]{1.\ [(CH_2{=}CH)_2Cu]Li} C_6H_5-\underset{}{\overset{OH}{CH}}-CH_2-CH{=}CH_2$$

Styrene oxide — 1-Phenyl-3-buten-1-ol

EXAMPLE 11.9

Show a combination of epoxide and Gilman reagent that can be used to prepare this alcohol.

$$C_6H_5-CH_2CH_2\overset{OH}{\overset{|}{C}}HCH_2CH_2CH_3$$

1-Phenyl-3-hexanol

Solution

Two combinations of epoxide and Gilman reagent give this alcohol. In disconnection of this compound, realize that the carbon bearing the hydroxyl group was one of the carbon atoms of the epoxide. The second carbon of the epoxide was either the one to the right of the carbon now bearing the —OH, or the one to the left of it. In these solutions, the phenyl group is written C_6H_5—.

$$C_6H_5-CH_2CH_2-\overset{OH}{\overset{|}{C}}H-CH_2CH_2CH_3$$

if the epoxide was between these carbons (left: CH_2—CH) — if the epoxide was between these carbons (right: CH—CH_2)

$[(C_6H_5CH_2)_2Cu]Li$ + $\overset{O}{CH_2-CH}CH_2CH_2CH_3$ (epoxide)

$C_6H_5-CH_2CH_2\overset{O}{CH-CH_2}$ (epoxide) + $[(CH_3CH_2)_2Cu]Li$

PROBLEM 11.9

Show how to prepare each Gilman reagent in Example 11.9 from an appropriate alkyl halide.

The value of epoxides lies in the number of nucleophiles that bring about ring opening and the combinations of functional groups that can be prepared from them. The most important of these ring-opening reactions are summarized in the following chart.

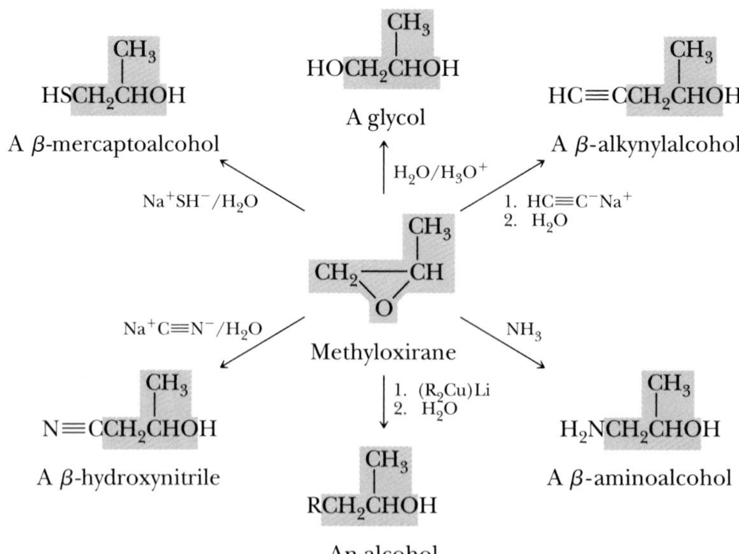

Chemistry in Action

Ethylene Oxide: A Chemical Sterilant

Because ethylene oxide is such a strained molecule, it reacts with the types of nucleophilic groups present in biological materials, including amino groups and thiol groups. At sufficiently high concentrations, ethylene oxide reacts with enough molecules in cells to cause the death of microorganisms. This toxic property is the basis for using ethylene oxide as a chemical sterilant. In hospitals, surgical instruments and other items that cannot be made disposable or heat treated are now sterilized by exposure to ethylene oxide.

In addition to reacting with biomolecules, ethylene oxide also reacts with chloride ion to give 2-chloroethanol, which in turn can be oxidized to chloroacetaldehyde. Treatment of deoxyribonucleic acids (DNAs) with either ethylene oxide or chloroacetaldehyde can produce mutations and chromosomal abnormalities.

$$\underset{\text{Ethylene oxide}}{CH_2{-}CH_2\ (\text{-O-})} \xrightarrow{Cl^-,\ H_2O} \underset{\text{2-Chloroethanol}}{ClCH_2CH_2OH} \xrightarrow{\text{oxidation}} \underset{\text{Chloroacetaldehyde}}{ClCH_2\overset{O}{\overset{\|}{C}}H}$$

The nucleophilic —SH group of glutathione, the most prevalent sulfur-containing molecule in many cells, is able to trap reactive electrophiles before they damage DNAs. The glutathione adducts are then broken down into smaller, harmless fragments, which are excreted in the urine.

$$\underset{\text{Glutathione}}{H_3\overset{+}{N}\underset{CO_2^-}{CH}CH_2CH_2\overset{O}{\overset{\|}{C}}NH\underset{CH_2SH}{CH}\overset{O}{\overset{\|}{C}}NHCH_2CO_2H} + \underset{\text{2-Chloroethanol}}{ClCH_2CH_2OH} \xrightarrow{S_N2} \underset{\text{A 2-Hydroxyethylglutathione}}{H_3\overset{+}{N}\underset{CO_2^-}{CH}CH_2CH_2\overset{O}{\overset{\|}{C}}NH\underset{CH_2SCH_2CH_2OH}{CH}\overset{O}{\overset{\|}{C}}NHCH_2CO_2H}$$

Finally, treatment with $LiAlH_4$ reduces an epoxide to an alcohol. Lithium aluminum hydride is similar to sodium borohydride, $NaBH_4$, in that it is a donor of hydride ion, $H{:}^-$, which is both a strong base and a good nucleophile. In the reduction of a substituted epoxide by $LiAlH_4$, preferential attack of the hydride reducing agent occurs at the less hindered carbon of the epoxide, an observation consistent with S_N2 reactivity.

$$\underset{\substack{\text{Phenyloxirane}\\ \text{(Styrene oxide)}}}{C_6H_5{-}CH{-}CH_2\ (\text{-O-})} \xrightarrow[2.\ H_2O]{1.\ LiAlH_4} \underset{\text{1-Phenylethanol}}{C_6H_5{-}\underset{OH}{CH}CH_2{-}H}$$

11.11 Crown Ethers

In the early 1960s, Charles Pedersen of Du Pont discovered a family of cyclic polyethers derived from ethylene glycol and substituted ethylene glycols. Compounds of this structure were given the name **crown ethers** because one of their most stable conformations resembles the shape of a crown. Although crown ethers do have IUPAC names, they are more commonly referred to by the shorthand notation devised by Pedersen. The parent name is "crown." It is preceded by a number describing the size of the ring and followed by a number describing the number of oxygen atoms in the

Crown ether A family of cyclic polyethers derived from ethylene glycol and substituted ethylene glycols.

ring. The figure gives names and structural formulas for 12-crown-4 and 18-crown-6. Pedersen shared the 1987 Nobel Prize for chemistry for this work along with Donald J. Cram of the United States and Jean-Marie Lehn of France.

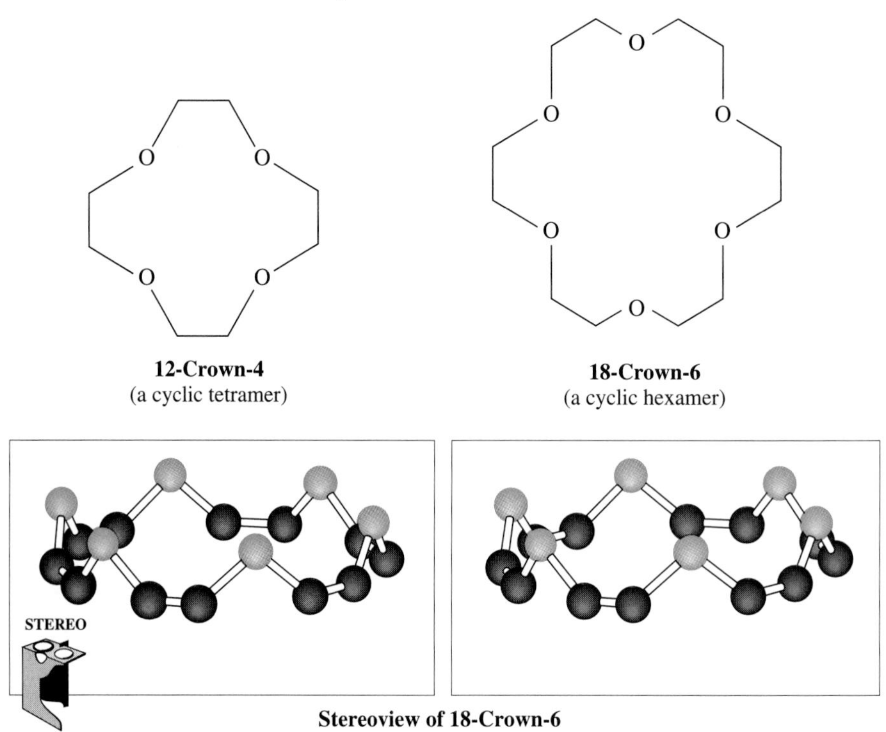

12-Crown-4
(a cyclic tetramer)

18-Crown-6
(a cyclic hexamer)

Stereoview of 18-Crown-6

A remarkable structural feature of crown ethers is that the diameter of the cavity created by the repeating oxygen atoms of the ring is comparable to the diameter of alkali metal ions. The diameter of the cavity in 18-crown-6, for example, is approximately the diameter of a potassium ion. When a potassium ion is inserted into the cavity of 18-crown-6, the unshared electron pairs on the six oxygens of the crown ether are close enough to the potassium ion to provide very effective solvation for K^+. 18-Crown-6 forms somewhat weaker complexes with rubidium ion (a somewhat larger ion) and with sodium ion (a somewhat smaller ion). It does not coordinate to any appreciable degree with lithium ion (a considerably smaller ion). 12-Crown-4, however, with its smaller cavity, does form a strong complex with lithium ion.

Ion	Diameter (nm)
Li^+	0.136
Na^+	0.194
K^+	0.266
Rb^+	0.294
Mg^{2+}	0.164
Ca^{2+}	0.286

diameter of K^+ = 0.266 nm

K^+

diameter of cavity created by 6 oxygens = 0.260 – 0.320 nm

A complex of K^+ and 18-crown-6

The cavity of a crown ether is a polar region, and the unshared pairs of electrons on the oxygen atoms lining the cavity provide effective solvation for alkali metal ions. The outer surface of the crown is nonpolar and hydrocarbon-like, and, thus, crown ethers and their alkali metal ion complexes dissolve readily in nonpolar organic solvents.

Crown ethers have proven to be particularly valuable for the same reasons as phase-transfer catalysts (Section 8.7), namely, for their ability to cause inorganic salts to dissolve in nonpolar aprotic organic solvents such as methylene chloride, hexane, and benzene. Potassium permanganate, for example, does not dissolve in benzene. If 18-crown-6 is added to benzene, the solution takes on the purple color characteristic of permanganate ion. In this instance, the crown-potassium ion complex is soluble in benzene and brings permanganate ion into solution with it. The resulting "purple benzene" is a valuable reagent for the oxidation of water-insoluble organic compounds.

Crown ethers have also proven valuable in nucleophilic displacement reactions. The cations of potassium salts, such as KF, KCN, or KN_3, are very tightly bound within the solvation cavity of 18-crown-6 molecules. The anions, however, are only weakly solvated, and because of the geometry of cation binding within the cavity of the crown, only loose ion-pairing occurs between the anion and cation. Thus, in nonpolar aprotic solvents these anions are without any appreciable solvent shell and are, therefore, highly reactive as nucleophiles. The nucleophilicity of F^-, CN^-, N_3^-, and other anions in nonpolar aprotic solvents containing an 18-crown-6 equals and often exceeds that in polar aprotic solvents such as DMSO and acetonitrile.

SUMMARY

An **ether** (Section 11.1) contains an atom of oxygen bonded to two carbon atoms. In the IUPAC name, the parent chain is named and the —OR group is named as an **alkoxy** substituent (Section 11.2). Common names are derived by naming the two groups attached to oxygen followed by the word "ether." **Heterocyclic ethers** have an oxygen atom as one of the members of a ring. For **thioethers** name the two groups attached to sulfur followed by the word "sulfide." Ethers are weakly polar compounds (Section 11.3) and associate by weak dipole-dipole interactions and dispersion forces. Boiling points of ethers are close to those of hydrocarbons of comparable molecular weight but much lower than those of alcohols of comparable molecular weight. Because ethers are hydrogen-bond acceptors, they are more soluble in water than are hydrocarbons of comparable molecular weight.

Crown ethers (Section 11.10) are cyclic polyethers having 12 or more atoms in a ring. The cavity of a crown ether is a polar region, and the unshared pairs of electrons on the ether oxygens can solvate alkali metal ions. The cavity of 18-crown-6, for example, has approximately the same diameter as potassium ion. The outer surface of a crown ether is nonpolar and hydrocarbon-like. Crown ethers are valuable for their ability to cause ionic compounds to dissolve in nonpolar organic solvents.

KEY REACTIONS

1. Williamson Ether Synthesis (Section 11.4A)

S_N2 reaction of a metal alkoxide with an alkyl halide. Yields are highest with primary halides and lower with secondary halides because of competition from E2 elimination. The reaction fails altogether with tertiary halides.

$$\mathrm{CH_3\overset{\overset{\textstyle CH_3}{|}}{\underset{\underset{\textstyle CH_3}{|}}{C}}O^-K^+ + CH_3CH_2Br \xrightarrow{S_N2} CH_3\overset{\overset{\textstyle CH_3}{|}}{\underset{\underset{\textstyle CH_3}{|}}{C}}OCH_2CH_3 + K^+Br^-}$$

2. Acid-Catalyzed Dehydration of Alcohols (Section 11.4B)

Yields are highest for symmetrical ethers formed from unbranched primary alcohols.

$$\mathrm{2CH_3CH_2OH \xrightarrow[140°C]{H_2SO_4} CH_3CH_2OCH_2CH_3 + H_2O}$$

3. Acid-Catalyzed Addition of Alcohols to Alkenes (Section 11.4C)

Proton transfer to the alkene generates a carbocation. Nucleophilic addition of an alcohol to the carbocation followed by proton transfer to the solvent gives the ether.

$$\mathrm{CH_3\overset{\overset{\textstyle CH_3}{|}}{C}{=}CH_2 + CH_3OH \xrightarrow[\text{catalyst}]{\text{acid}} CH_3\overset{\overset{\textstyle CH_3}{|}}{\underset{\underset{\textstyle CH_3}{|}}{C}}OCH_3}$$

4. Reaction of Alcohols with Chlorotrimethylsilane (Section 11.7)

The trimethylsilyl group is used as a protecting group for primary and secondary alcohols. It is removed by treatment with fluoride ion to regenerate the original alcohol.

$$\mathrm{RCH_2OH + \underset{\text{Chlorotrimethylsilane}}{ClSi(CH_3)_3} \xrightarrow{(CH_3CH_2)_3N} \underset{\text{A trimethylsilyl ether}}{RCH_2OSi(CH_3)_3}}$$

5. Reaction of Alcohols with Dihydropyran (Section 11.7)

The tetrahydropyranyl (THP) group is a protecting group for primary and secondary alcohols. It is removed by treatment with dilute aqueous acid to regenerate the original alcohol.

$$\mathrm{RCH_2OH + \text{dihydropyran} \xrightarrow{H^+} \text{2-}(OCH_2R)\text{-tetrahydropyran}}$$

6. Acid-Catalyzed Cleavage of Dialkyl Ethers (Section 11.6A)

Cleavage of ethers requires both a strong acid and a good nucleophile, hence the use of concentrated HBr and HI. Cleavage of primary and secondary alkyl ethers is by an S_N2 pathway. Cleavage of tertiary alkyl ethers is by an S_N1 pathway.

$$\mathrm{C_6H_{11}{-}OCH_3 + HI\ (excess) \longrightarrow C_6H_{11}{-}I + CH_3I + H_2O}$$

7. Oxidation of Sulfides (Section 11.6C)

Oxidation of a sulfide gives either a sulfoxide or a sulfone, depending on the oxidizing agent and experimental conditions. Air oxidation of dimethyl sulfide is a common route to dimethyl sulfoxide, a polar aprotic solvent.

$$CH_3{-}S{-}CH_3 + O_2 \xrightarrow[\text{nitrogen}]{\text{oxides of}} CH_3{-}\overset{\overset{\displaystyle O}{\|}}{S}{-}CH_3$$

8. Oxidation of Alkenes by Peroxycarboxylic Acids (Section 11.9B)

Two commonly used peroxycarboxylic acid oxidizing agents are the magnesium salt of monoperoxyphthalic acid (MMPP) and peroxyacetic acid.

$$\text{2-cyclohexen-1-ol} + RCO_3H \longrightarrow \text{epoxide (OH, H, O, H)} + RCO_2H$$

9. Synthesis of Epoxides from Halohydrins (Section 11.9C)

Both formation of the halohydrin and the following intramolecular S_N2 reaction are stereoselective.

$$\text{(H, H}_3\text{C)C=C(H, CH}_3) \xrightarrow{Cl_2,\ H_2O} H_3C(H)(HO)C{-}C(Cl)(H)(CH_3) \xrightarrow{NaOH,\ H_2O} H_3C(H)C{-}C(H)CH_3 \text{ (epoxide, O)}$$

10. Acid-Catalyzed Hydrolysis of Epoxides (Section 11.10A)

Hydrolysis of an epoxide derived from a cycloalkene gives a *trans*-glycol.

$$\text{cyclopentene oxide (H, O, H)} \xrightarrow[H_2O]{H^+} \text{trans-1,2-cyclopentanediol (OH, OH)}$$

11. Nucleophilic Opening of Epoxides (Section 10.10B)

Attack on the epoxide is regioselective with the nucleophile attacking the less substituted carbon of the epoxide.

$$\text{1-methylcyclohexene oxide (CH}_3\text{, O, H)} + CH_3OH \xrightarrow{CH_3O^-Na^+} \text{cyclohexane (CH}_3\text{, OH; H, OCH}_3)$$

12. Treatment of an Epoxide with a Gilman Reagent (Section 11.10B)

Treatment of an epoxide with a Gilman reagent followed by hydrolysis gives an alcohol.

$$C_6H_5{-}\overset{O}{CH{-}CH_2} \xrightarrow[\text{2. } HCl,\ H_2O]{\text{1. } (CH_3)_2CuLi} C_6H_5\overset{\overset{\displaystyle OH}{|}}{C}HCH_2CH_3$$

13. Reduction of an Epoxide to an Alcohol (Section 11.10B)

Regioselective hydride ion transfer from lithium aluminum hydride to the less hindered carbon of the epoxide gives an alcohol.

$$C_6H_5{-}\overset{O}{CH{-}CH_2} \xrightarrow[\text{2. } H_2O]{\text{1. } LiAlH_4} C_6H_5{-}\underset{\underset{\displaystyle OH}{|}}{C}HCH_3$$

ADDITIONAL PROBLEMS

Structure and Nomenclature

11.10 Write names for these compounds. Where possible, write both IUPAC names and common names.

(a) 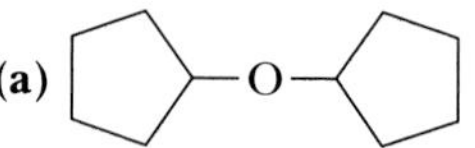

(b) $-OCH_3$

(c) $CH_3CH_2OCH_2CH_2OH$

(d) $CH_3CH_2OCH_2CH_2OCH_2CH_3$

(e) 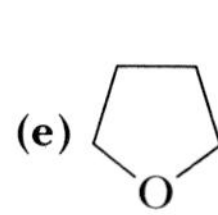

(f) O O

(g) $CH_3CH(CH_2)_5CH_3$ with SCH_2CH_3 on C-2

(h) $[CH_3(CH_2)_4]_2O$

11.11 Draw structural formulas for these compounds.

(a) Diisopropyl ether
(b) *trans*-2,3-Diethyloxirane
(c) *trans*-2-Ethoxycyclopentanol
(d) Divinyl ether
(e) Cyclohexene oxide
(f) Allyl cyclopropyl ether
(g) (*R*)-2-Methyloxirane
(h) 1,1-Dimethoxycyclohexane

Physical Properties

11.12 Each compound given in this problem is a common organic solvent. From each pair of compounds, select the solvent with the greater solubility in water.

(a) CH_2Cl_2 and CH_3CH_2OH

(b) $CH_3CH_2OCH_2CH_3$ and CH_3CH_2OH

(c)  CH_3CCH_3 (C=O) and $CH_3CH_2OCH_2CH_3$

(d) $CH_3CH_2OCH_2CH_3$ and $CH_3(CH_2)_3CH_3$

11.13 Following are structural formulas, boiling points, and solubilities in water for diethyl ether and tetrahydrofuran (THF). Account for the fact that tetrahydrofuran is so much more soluble in water than diethyl ether.

$CH_3CH_2OCH_2CH_3$	H_2C-CH_2, H_2C, CH_2, O (ring)
Diethyl ether	Tetrahydrofuran
bp 35°C	bp 67°C
8 g/100 mL water	very soluble in water

11.14 Because of the Lewis base properties of ether oxygen atoms, crown ethers are excellent complexing agents for Na^+, K^+, and NH_4^+. What kind of molecule might serve as a complexing agent for Cl^- or Br^-?

Preparation of Ethers

11.15 Write equations to show a combination of reactants to prepare each ether. Which ethers can be prepared in good yield by a Williamson ether synthesis? If there are any that cannot be prepared by the Williamson method, explain why not.

(a) 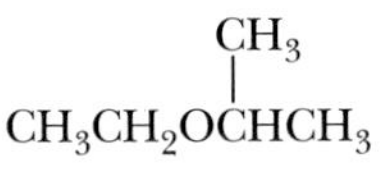 $CH_3CH_2OCHCH_3$ (with CH_3 on the CH)　(b) 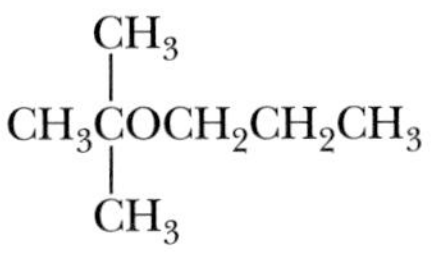 $CH_3C(CH_3)_2OCH_2CH_2CH_3$　(c) 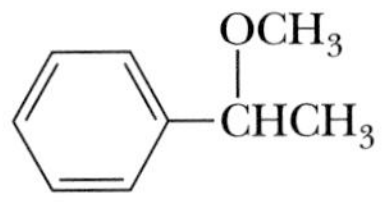 $C_6H_5CH(OCH_3)CH_3$

(d) 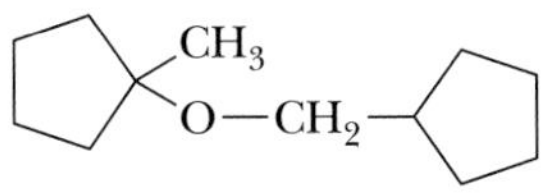 cyclopentyl with CH_3 and $O{-}CH_2{-}$cyclopentyl　(e) 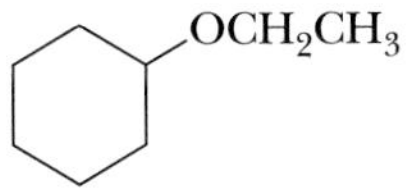 cyclohexyl$-OCH_2CH_3$　(f) 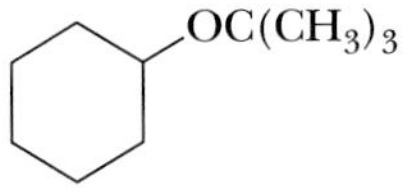 cyclohexyl$-OC(CH_3)_3$

11.16 Propose a mechanism for this reaction.

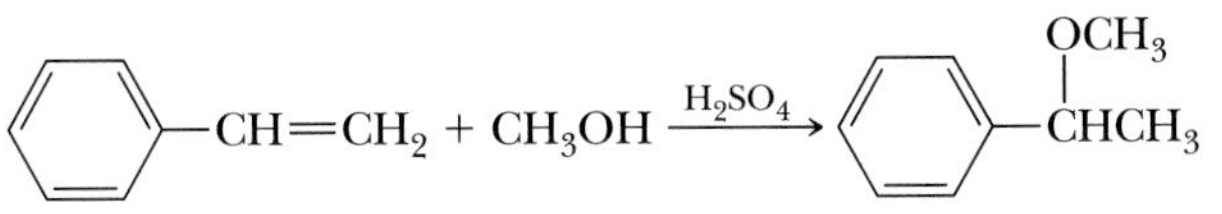

$$C_6H_5{-}CH{=}CH_2 + CH_3OH \xrightarrow{H_2SO_4} C_6H_5{-}CHCH_3(OCH_3)$$

Reactions of Ethers

11.17 Draw structural formulas for the products formed when each compound is refluxed in concentrated HI.

(a) $CH_3CH_2OCH_2CH_2CH_3$　(b) 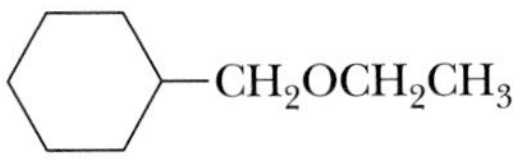 cyclohexyl$-CH_2OCH_2CH_3$

(c) 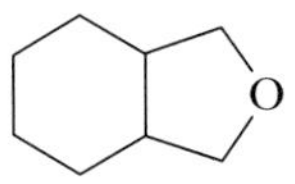 [bicyclic: cyclohexane fused to tetrahydrofuran ring, O]　(d) 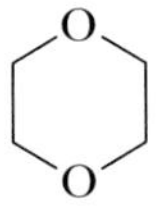 [1,4-dioxane ring, O, O]

11.18 Following is an equation for the reaction of diisopropyl ether and oxygen to form a hydroperoxide.

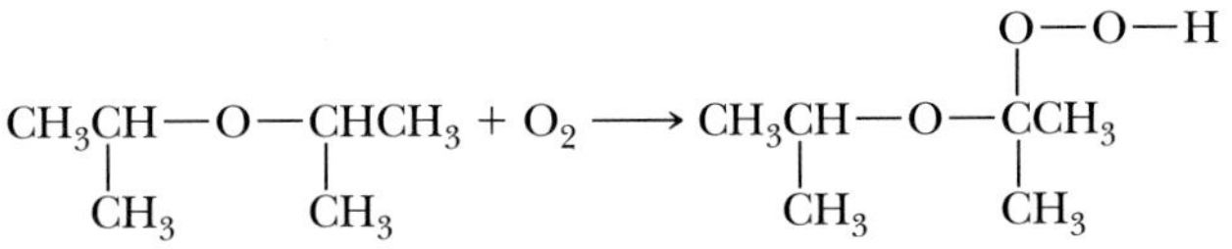

$$\underset{\text{Diisopropyl ether}}{CH_3CH(CH_3){-}O{-}CH(CH_3)CH_3} + O_2 \longrightarrow \underset{\text{A hydroperoxide}}{CH_3CH(CH_3){-}O{-}C(O{-}O{-}H)(CH_3)CH_3}$$

Formation of an ether hydroperoxide can be written as a radical chain reaction.

(a) Write a pair of chain propagation steps that accounts for the formation of this ether hydroperoxide. Assume that initiation is by radical, R·.

(b) Account for the fact that hydroperoxidation of ethers is regioselective, that is, it occurs preferentially at a carbon adjacent to the ether oxygen.

Synthesis and Reactions of Epoxides

11.19 Triethanolamine (TEA) is a widely used biological buffer, with maximum buffering capacity at pH 7.8. Propose a synthesis of this compound from ethylene oxide and ammonia. The structural formula of triethanolamine is $(HOCH_2CH_2)_3N$.

11.20 Ethylene oxide is the starting material for the synthesis of both methyl cellosolve and cellosolve, two important industrial solvents. Propose a mechanism for these reactions.

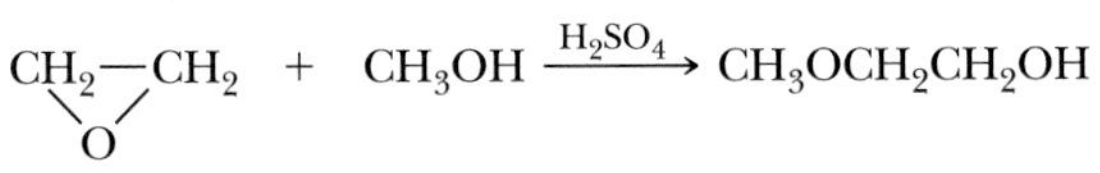

$$\underset{\substack{\text{Oxirane}\\ \text{(Ethylene oxide)}}}{CH_2{-}CH_2\ (\text{bridged by O})} + CH_3OH \xrightarrow{H_2SO_4} \underset{\substack{\text{2-Methoxyethanol}\\ \text{(Methyl cellosolve)}}}{CH_3OCH_2CH_2OH}$$

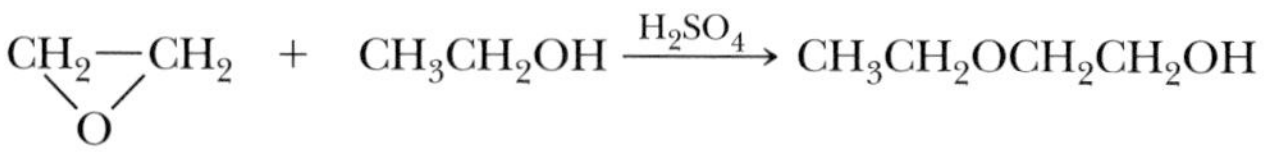

$$\underset{\text{(Ethylene oxide)}}{\underset{\text{Oxirane}}{CH_2\text{—}CH_2 \text{ (—O—)}}} + CH_3CH_2OH \xrightarrow{H_2SO_4} \underset{\text{(Cellosolve)}}{\underset{\text{2-Ethoxyethanol}}{CH_3CH_2OCH_2CH_2OH}}$$

11.21 Ethylene oxide is the starting material for the synthesis of 1,4-dioxane. Propose a mechanism for each step in this synthesis.

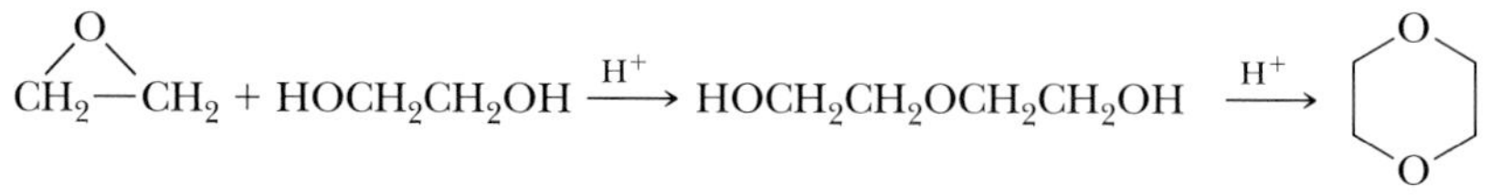

$$CH_2\text{—}CH_2 \text{ (—O—)} + HOCH_2CH_2OH \xrightarrow{H^+} HOCH_2CH_2OCH_2CH_2OH \xrightarrow{H^+} \text{1,4-Dioxane}$$

1,4-Dioxane

11.22 Propose a synthesis for each ether starting with ethylene oxide and any readily available alcohols.

(a) $CH_3OCH_2CH_2OCH_3$ **(b)** $CH_3OCH_2CH_2OCH_2CH_2OCH_3$

11.23 Propose a synthesis for 18-crown-6. If a base is used in your synthesis, does it make a difference if it is a lithium salt or a potassium salt?

11.24 Propose a mechanism for the reaction of a primary alcohol, RCH_2OH, with dihydropyran to give a tetrahydropyranyl ether.

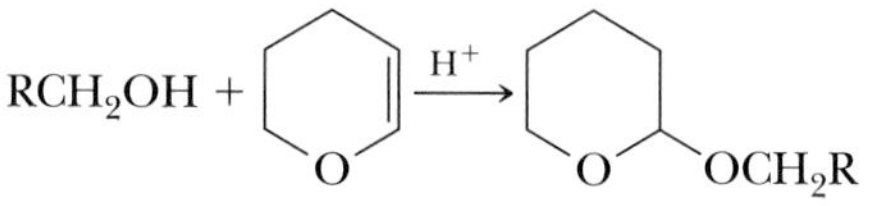

$$RCH_2OH + \text{(dihydropyran)} \xrightarrow{H^+} \text{(tetrahydropyran)}\text{—}OCH_2R$$

11.25 Predict the structural formula of the major product of the reaction of 2,2,3-trimethyloxirane with each set of reagents.

(a) $CH_3OH/CH_3O^-Na^+$ **(b)** CH_3OH/H^+

11.26 The following equation shows the reaction of *trans*-2,3-diphenyloxirane with hydrogen chloride in benzene to form 2-chloro-1,2-diphenylethanol.

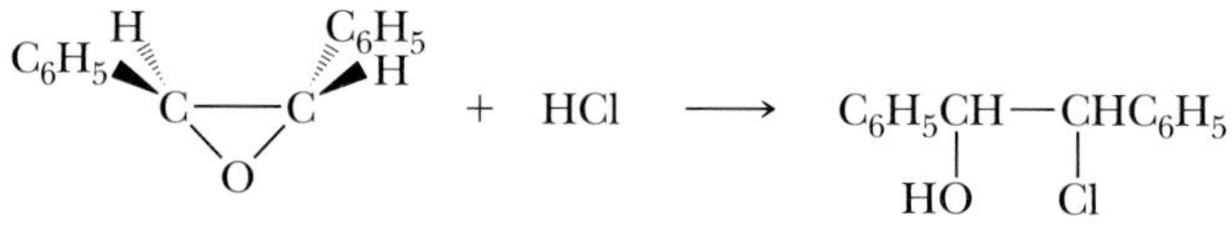

$$\underset{\textit{trans}\text{-2,3-Diphenyloxirane}}{C_6H_5(H)C\text{—}C(H)C_6H_5 \text{ (—O—)}} + HCl \longrightarrow \underset{\text{2-Chloro-1,2-diphenylethanol}}{C_6H_5CH(OH)\text{—}CH(Cl)C_6H_5}$$

(a) How many stereoisomers are possible for 2-chloro-1,2-diphenylethanol?

(b) Given that opening of the epoxide ring in this reaction is stereoselective, predict which of the possible stereoisomers of 2-chloro-1,2-diphenylethanol is/are formed in the reaction.

11.27 Propose a mechanism to account for this rearrangement.

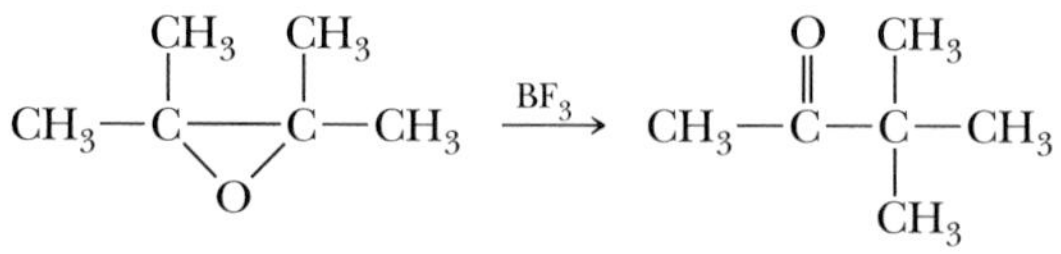

$$\underset{\text{Tetramethyloxirane}}{CH_3\text{—}C(CH_3)\text{—}C(CH_3)\text{—}CH_3 \text{ (—O—)}} \xrightarrow{BF_3} \underset{\text{3,3-Dimethyl-2-butanone}}{CH_3\text{—}C(=O)\text{—}C(CH_3)_2\text{—}CH_3}$$

11.28 Following is the structural formula for an epoxide derived from 9-methyldecalin. Acid-catalyzed hydrolysis of this epoxide gives a *trans*-diol. Of the two possible *trans*-diols, only one is formed. How do you account for this stereoselectivity? *Hint:* Begin by drawing *trans*-decalin with each six-membered ring in the more stable chair conformation (Section 2.7B), and then determine whether each substituent in the isomeric glycols is axial or equatorial.

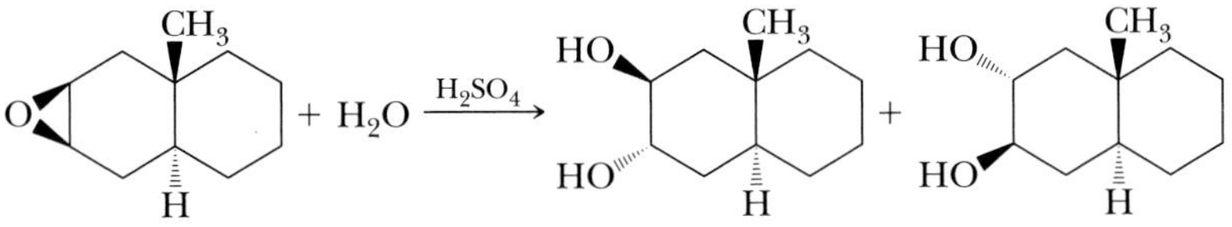

$$+ H_2O \xrightarrow{H_2SO_4} \quad + $$

Only this glycol is formed

This glycol is not formed

11.29 Following are two reaction sequences for converting 1,2-diphenylethylene into 2,3-diphenyloxirane.

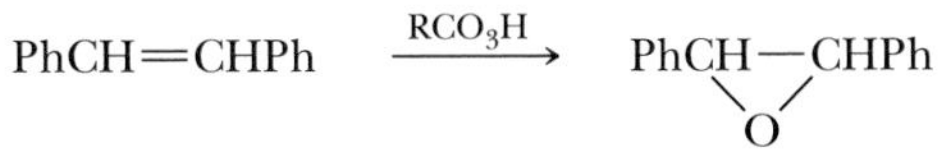

$$PhCH{=}CHPh \xrightarrow{RCO_3H} PhCH{-}CHPh \text{ (O bridging)}$$

1,2-Diphenylethylene — 2,3-Diphenyloxirane

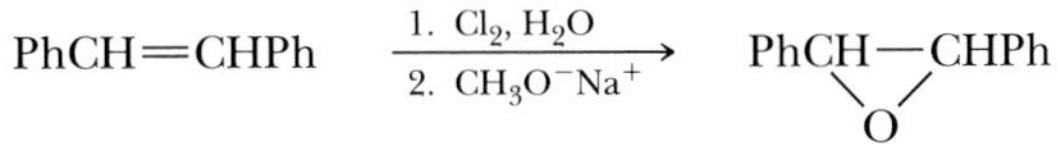

$$PhCH{=}CHPh \xrightarrow[\text{2. } CH_3O^-Na^+]{\text{1. } Cl_2,\ H_2O} PhCH{-}CHPh \text{ (O bridging)}$$

1,2-Diphenylethylene — 2,3-Diphenyloxirane

Suppose that the starting alkene is *trans*-1,2-diphenylethylene.

(a) What is the configuration of the oxirane formed in each sequence?

(b) Does the oxirane formed in either sequence rotate the plane of polarized light? Explain.

11.30 Complete these reactions. *Hint:* Each reaction shows a stereoselectivity typical of nucleophilic opening of the epoxide ring.

(a) 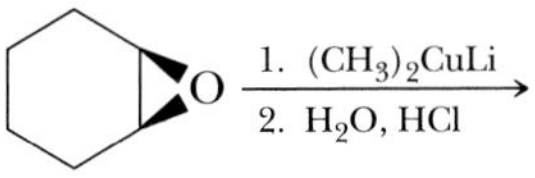 $\xrightarrow[\text{2. } H_2O,\ HCl]{\text{1. } (CH_3)_2CuLi}$ **(b)** 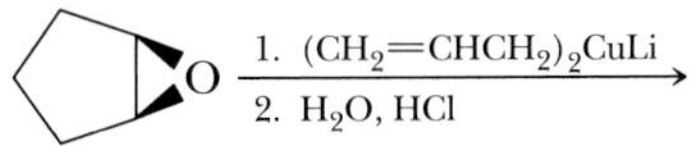 $\xrightarrow[\text{2. } H_2O,\ HCl]{\text{1. } (CH_2{=}CHCH_2)_2CuLi}$

11.31 One of the most useful organic reactions discovered in the last 15 years is the titanium-catalyzed asymmetric epoxidation of allylic alcohols developed by Professor Barry Sharpless and coworkers [see K. B. Sharpless et al., *Pure and Appl. Chem.*, **55,** 589 (1983)]. The reagent combination consists of $Ti[(OCH(CH_3)_2]_4$, a hydroperoxide, and a chiral molecule such as (+)-diethyl tartrate.

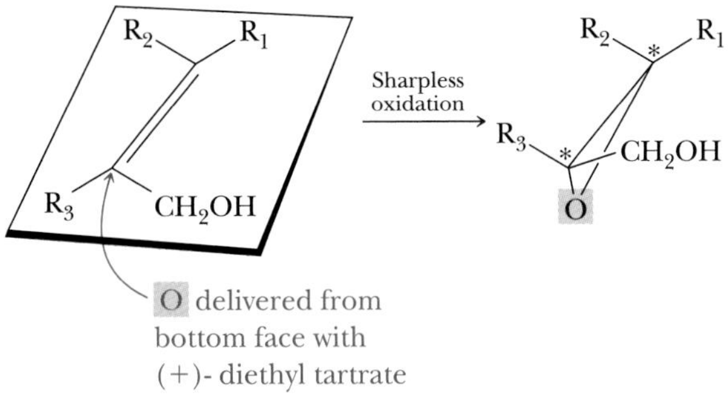

Sharpless oxidation

O delivered from bottom face with (+)- diethyl tartrate

Two new stereocenters are created in the product. If (+)-diethyl tartrate is used in the reaction, the product arises by delivery of O from the hydroperoxide to the bottom face of the molecule, and the product epoxide is the stereoisomer shown. If (−)-diethyl tartrate is used, the product is the enantiomer of the stereoisomer shown. Draw the expected products of Sharpless epoxidation of the following allylic alcohols using (+)-diethyl tartrate.

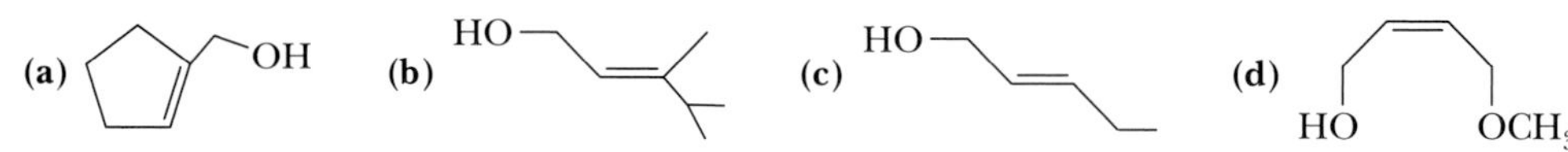

11.32 The following chiral epoxide is an intermediate in the synthesis of the insect pheromone frontalin. How can this epoxide be prepared from an allylic alcohol precursor, using the Sharpless epoxidation reaction described in Problem 11.31?

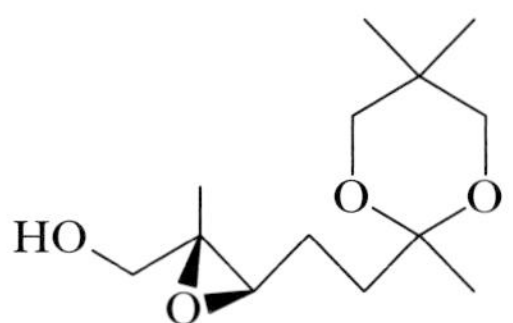

11.33 Human white cells produce an enzyme called myeloperoxidase. This enzyme catalyzes the reaction between hydrogen peroxide and chloride ion to produce hypochlorous acid, HOCl, which reacts as if it is Cl^+OH^-. When attacked by white cells, cholesterol reacts to give the following chlorohydrin as the major product.

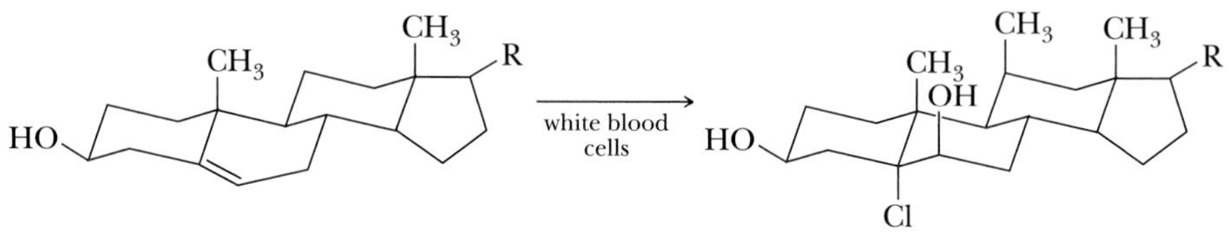

white blood cells

Cholesterol

(a) Propose a mechanism for this reaction. Account for both the regioselectivity and the stereoselectivity.

(b) On standing or (much more rapidly) on treatment with base, the chlorohydrin is converted to an epoxide. Show the structure of the epoxide and a mechanism for its formation.

11.34 Propose a mechanism for the following acid-catalyzed rearrangement.

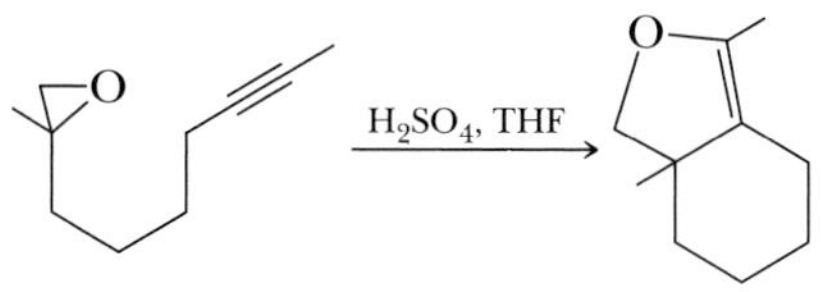

H_2SO_4, THF

Synthesis

11.35 Show reagents and experimental conditions to synthesize the following compounds from 1-propanol. Any derivative of 1-propanol prepared in an earlier part of this problem may then be used for a later synthesis.

(a) Propanal
(b) Propanoic acid
(c) Propene
(d) 2-Propanol
(e) 2-Bromopropane
(f) 1-Chloropropane
(g) 1,2-Dibromopropane
(h) Propyne
(i) 2-Propanone
(j) 1-Chloro-2-propanol
(k) Methyloxirane
(l) Dipropyl ether
(m) Isopropyl propyl ether
(n) 1-Mercapto-2-propanol
(o) 1-Amino-2-propanol
(p) 1,2-Propanediol

11.36 Starting with *cis*-3-hexene, show how to prepare the following:

(a) Meso 3,4-hexanediol (b) Racemic 3,4-hexanediol

11.37 Show reagents and experimental conditions to convert cycloheptene to the following. Any compound made in an earlier part of this problem may be used as an intermediate in any following conversion.

(a) 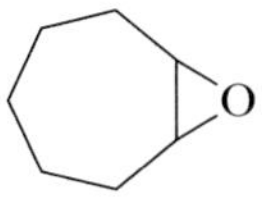 O (b) 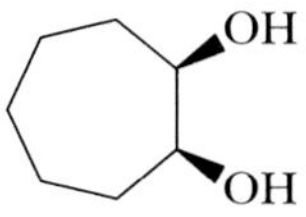 OH OH (c) 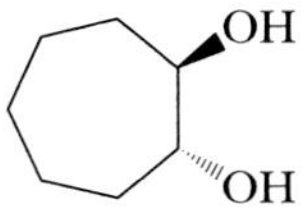 OH OH

(d) 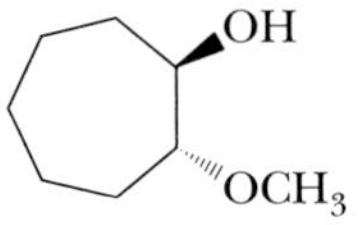 OH OCH_3 (e) 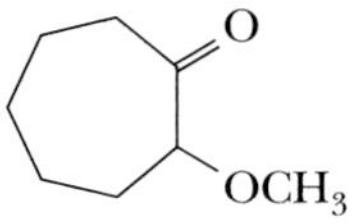 O OCH_3 (f) 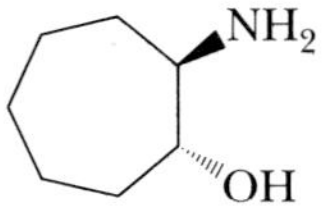 NH_2 OH

(g) 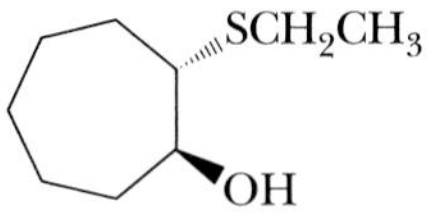 SCH_2CH_3 OH (h) 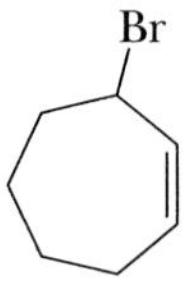 Br (i)

(j) 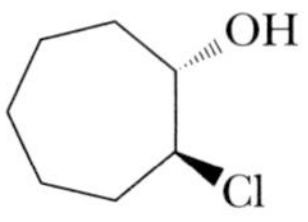 OH Cl (k) 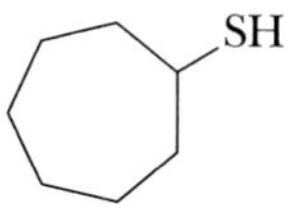 SH (l) 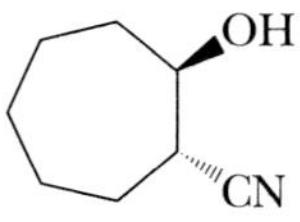 OH CN

(m) 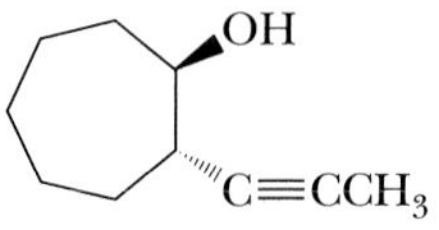 OH $C{\equiv}CCH_3$ (n) $\overset{O}{\overset{\|}{HC}}(CH_2)_5\overset{O}{\overset{\|}{CH}}$

11.38 Show reagents to bring about each reaction:

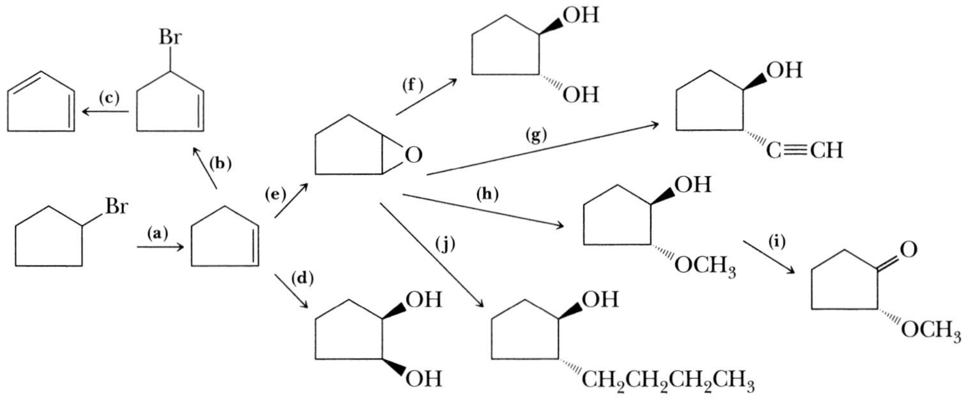

Br (c) Br (b) (a) (e) O (f) OH OH (g) OH $C{\equiv}CH$ (h) OH OCH_3 (i) O OCH_3 (d) OH OH (j) OH $CH_2CH_2CH_2CH_3$

11.39 Propose a synthesis of the following alcohol from styrene and 1-chloro-3-methyl-2-butene.

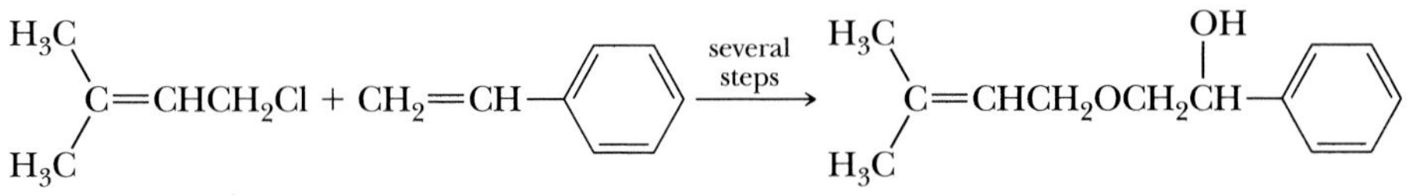

$(H_3C)_2C{=}CHCH_2Cl + CH_2{=}CH{-}C_6H_5 \xrightarrow{\text{several steps}} (H_3C)_2C{=}CHCH_2OCH_2\overset{OH}{\overset{|}{C}H}{-}C_6H_5$

Styrene

11.40 Starting with acetylene and ethylene oxide as the only sources of carbon atoms, show how to prepare these compounds.

(a) 3-Butyn-1-ol **(b)** 3-Hexyn-1,6-diol **(c)** 1,6-Hexanediol
(d) (*Z*)-3-hexen-1,6-diol **(e)** (*E*)-3-Hexen-1,6-diol **(f)** Hexanedial

11.41 Following are the steps in the industrial synthesis of glycerin. Provide structures for all intermediates and describe the type of mechanism by which each is formed.

$$\underset{\text{Propene}}{CH_2{=}CHCH_3} \xrightarrow{Cl_2,\ heat} A\ (C_3H_5Cl) \xrightarrow{NaOH,\ H_2O} B\ (C_3H_6O) \xrightarrow{Cl_2,\ H_2O}$$

$$C\ (C_3H_7ClO_2) \xrightarrow[heat]{Ca(OH)_2} D\ (C_3H_6O_2) \xrightarrow{H_2O,\ NaOH} HOCH_2CH(OH)CH_2OH$$

1,2,3-Propanetriol
(glycerol, glycerin)

11.42 The following is a retrosynthetic scheme for the preparation of *trans*-2-allylcyclohexanol. Show reagents to bring about the synthesis of this compound from cyclohexane.

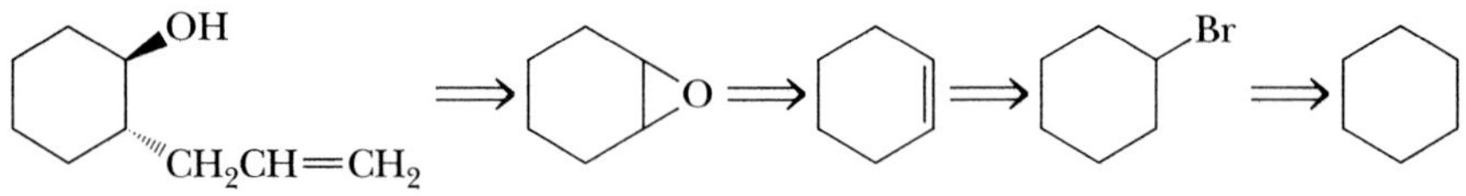

[trans-2-allylcyclohexanol: OH, $CH_2CH{=}CH_2$] ⟹ [cyclohexene oxide, O] ⟹ [cyclohexene] ⟹ [bromocyclohexane, Br] ⟹ [cyclohexane]

11.43 Show reagents and experimental conditions to bring about this conversion in good yield.

$$CH_3(CH_2)_4CH_2OH \longrightarrow CH_3(CH_2)_4CH_2CH_2COH \;(\text{C}{=}\text{O})$$

A pink bollworm. *(Fran Heyl Associates)*

11.44 Gossyplure, the sex pheromone of the pink bollworm, is the acetic ester of 7,11-hexadecadien-1-ol. The active pheromone has the *Z* configuration at the C7-C8 double bond and is a mixture of *E,Z* isomers at the C11-C12 double bond. Shown here is the *Z,E* isomer.

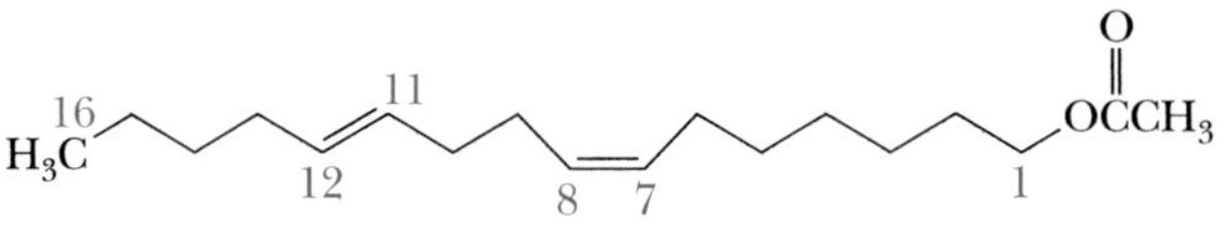

[Structure: H_3C (16) … 12, 11 … 8, 7 … 1 $OCCH_3$ (O)]

(7*Z*,11*E*)-7,11-hexadecadienyl acetate

Following is a retrosynthetic analysis for (7*Z*,11*E*)-7,11-hexadecadien-1-ol, which then led to a successful synthesis of gossyplure.

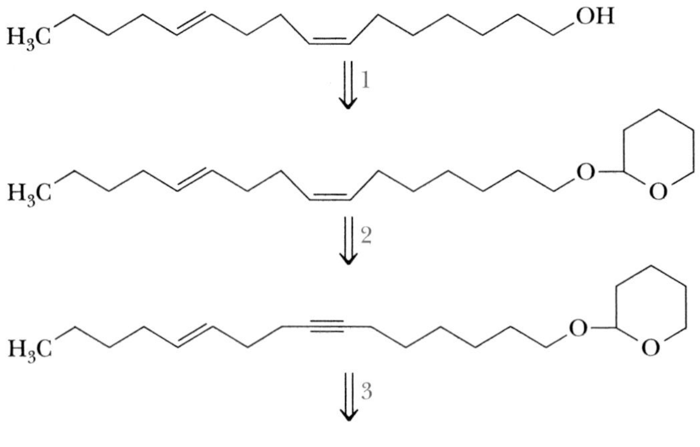

[Structure: H_3C … OH]

⇓ 1

[Structure: H_3C … O–tetrahydropyranyl (O)]

⇓ 2

[Structure: H_3C … alkyne … O–tetrahydropyranyl (O)]

⇓ 3

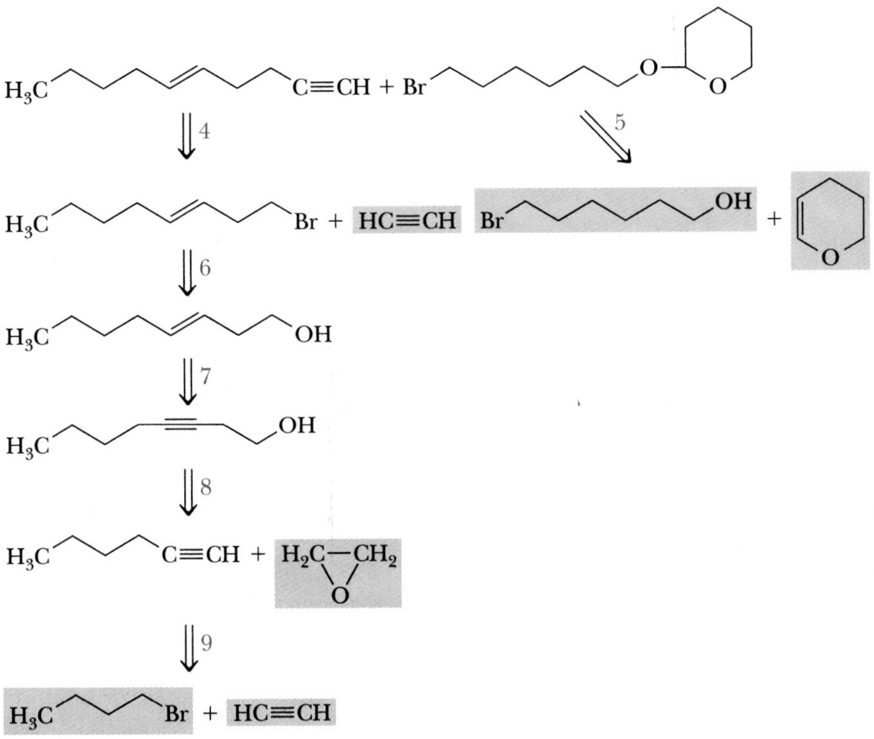

H_3C–(chain)–C≡CH + Br–(chain)–O–(tetrahydropyranyl)

⇓ 4 ⇘ 5

H_3C–(chain)–Br + HC≡CH Br–(chain)–OH + (dihydropyran)

⇓ 6

H_3C–(chain)–OH

⇓ 7

H_3C–(chain)–≡–(chain)–OH

⇓ 8

H_3C–(chain)–C≡CH + H_2C—CH_2 (O)

⇓ 9

H_3C–(chain)–Br + HC≡CH

(a) Suggest reagents and experimental conditions for each step in this synthesis.
(b) Why is it necessary to protect the —OH group of 6-bromohexanol?
(c) How might you modify this synthesis to prepare the 7*Z*,11*Z* isomer of gossyplure?

12

Mass Spectrometry

■ Crystals of dopamine viewed under a polarizing light. The mass spectrum of dopamine is shown on page 437 (Figure 12.2). *(© Herb Charles Ohlmeyer/Fran Heyl Associates)*

Determination of molecular structure is one of the central themes of organic chemistry. For this purpose, chemists today rely almost exclusively on instrumental methods, four of which we discuss in this text. We begin this chapter with the study of mass spectrometry (MS). Then, in the following two chapters, we first study nuclear magnetic resonance (NMR) spectroscopy, and then infrared (IR) spectroscopy and ultraviolet-visible (UV-Vis) spectroscopy.

Mass spectrometry is an analytical technique for measuring the mass-to-charge ratio (m/z) of ions, most commonly positive ions. The principles of mass spectrometry were first recognized in 1898. In 1911,

J. J. Thomson recorded the first mass spectrum, that of neon, and discovered that this element can be separated into a more abundant isotope, ^{20}Ne, and a less abundant isotope, ^{22}Ne. Using improved instrumentation, F. W. Aston showed that most of the naturally occurring elements are mixtures of isotopes. It was found, for example, that approximately 75% of chlorine atoms in nature are ^{35}Cl, and 25% are ^{37}Cl. Mass spectrometry did not come into common use, however, until the 1950s at which time commercial instruments became available that offered high resolution, reliability, and relatively inexpensive maintenance. Today, mass spectrometry is our most valuable analytical tool for the determination of precise molecular weights. Furthermore, extensive information about the molecular formula and structure of a compound can be obtained from analysis of its mass spectrum.

Mass spectrometry An analytical technique for measuring the mass-to-charge ratio (m/z) of an ion.

12.1 A Mass Spectrometer

A mass spectrometer (Figure 12.1) is designed to do three things:

1. Convert neutral atoms or molecules into a beam of positive or negative ions.
2. Separate the ions on the basis of their mass-to-charge (m/z) ratio.
3. Measure the relative abundance of each ion.

From this information, we can determine both the molecular weight and the molecular formula of an unknown compound. In addition, we can also obtain valuable clues about the molecular structure of the compound.

Samples of gases and volatile liquids can be introduced directly into the **ionization chamber.** For less volatile liquids and solids, the sample may be placed on the tip of a heated probe that is then inserted directly into the ionization chamber. Because the interior of a mass spectrometer is kept at a high vacuum, liquid and even solid samples vaporize almost instantly. Another extremely useful method for introducing a sample

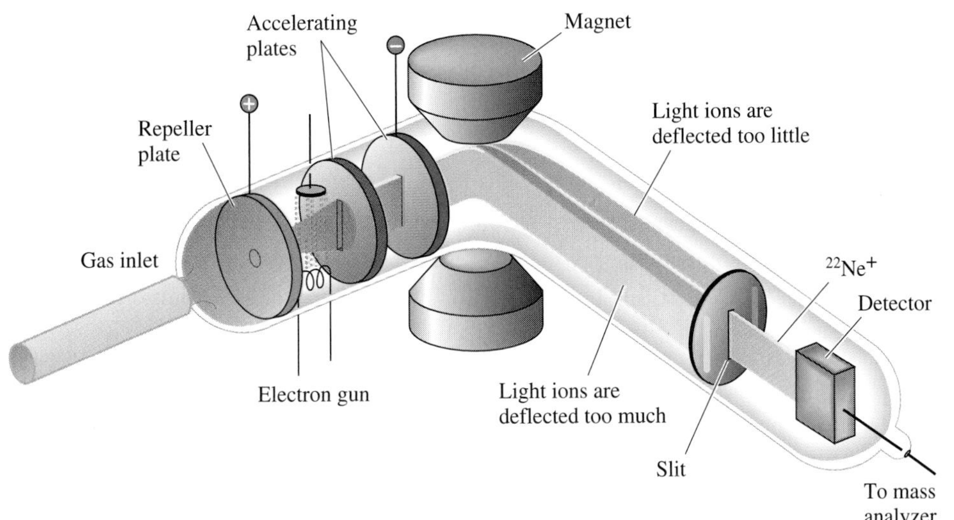

FIGURE 12.1
A simplified schematic diagram of an electron ionization mass spectrometer (EI MS). A vaporized sample in the ionization chamber is bombarded with high-energy electrons that cause electrons to be stripped from molecules of the sample. The resulting positive ions are accelerated by a series of negatively charged accelerator plates into an analyzing chamber about which is placed a magnetic field, perpendicular to the direction of the ion beam. The magnetic field causes the ion beam to curve. The radius of curvature of each ion depends on the charge on the ion (z), its mass (m), the accelerating voltage, and the strength of the magnetic field. A mass spectrum is a plot of relative ion abundance versus m/z ratio.

into the ionization chamber is to link a gas chromatograph (GC) directly to the mass spectrometer. Each fraction eluted from the GC is monitored and passed directly into the ionization chamber of the mass spectrometer.

Once in the ionization chamber, molecules of the sample are bombarded with a stream of high-energy electrons that are emitted from a hot filament and then accelerated by an electric field to energies of approximately 70 eV [1 eV = 23.05 kcal/mol (96.4 kJ/mol)]. Collisions between molecules of the sample and these high-energy electrons result in loss of electrons from sample molecules to form positive ions. A **molecular ion, M^+,** or alternatively, **M,** is the species formed by removal of a single electron from a molecule. A molecular ion belongs to a class of ions called radical cations. A **radical cation** is a species formed when a neutral molecule loses one electron; it contains both an odd number of electrons and a positive charge. When methane, for example, is bombarded with high-energy electrons, an electron is dislodged from the molecule to give a molecular ion of m/z 16.

Molecular ion (M^+, or alternatively, M) The species formed by removal of a single electron from a molecule.

$$CH_4 + e^- \longrightarrow [CH_4]^{\cdot +} + 2e^-$$

Molecular ion
(a radical cation)

Which electron is lost in forming the molecular ion is determined by the ionization potential of the atom or molecule. The **ionization potential (IP)** is the minimum energy required to remove an electron from an atom or molecule to a distance where there is no electrostatic interaction between ion and electron. Ionization potentials for most organic molecules are between 8 and 15 eV. They are at the lower end of this range for nonbonding electrons of oxygen and nitrogen, and for pi electrons of unsaturated compounds such as alkenes, alkynes, and aromatic hydrocarbons. Ionization potentials for sigma electrons, such as those of C—C, C—H, and C—O sigma bonds, are at the higher end of the range.

Ionization potential (IP) The minimum energy required to remove an electron from an atom or molecule to a distance where there is no electrostatic interaction between ion and electron.

For our purposes, it doesn't matter which electron is lost, because, in general, the radical cation character is delocalized throughout the molecule. Therefore, we write the molecular formula of the parent molecule in brackets with a plus sign to show that it is a cation and a dot to show that it has an odd number of electrons. See, for example, the molecular ion for ethyl isopropyl ether, shown here on the left. At times, however, we will find it useful to depict the radical cation in a certain position in order to better understand its reactions.

$$[CH_3CH_2OCH(CH_3)_2]^{\cdot +} \qquad CH_3CH_2\overset{\cdot +}{\underset{\cdot\cdot}{O}}CH(CH_3)_2$$

As we shall see in Section 12.2D, a molecular ion can undergo fragmentation to form a variety of smaller cations (which themselves may undergo further fragmentation), as well as radicals and smaller molecules. Of these smaller fragments, a mass spectrometer detects only cations.

Once molecular ions and their fragmentation ions have been formed, a positively charged **repeller plate** directs them toward a series of negatively charged **accelerator plates,** a process which produces a rapidly traveling ion beam. The ion beam is then focused by one or more focusing slits and passed into a **mass analyzer.** Ions with larger values of m/z are deflected less than those with smaller m/z values. The mass analyzer of a mass spectrometer is constructed with a fixed radius of curvature. Thus, by varying

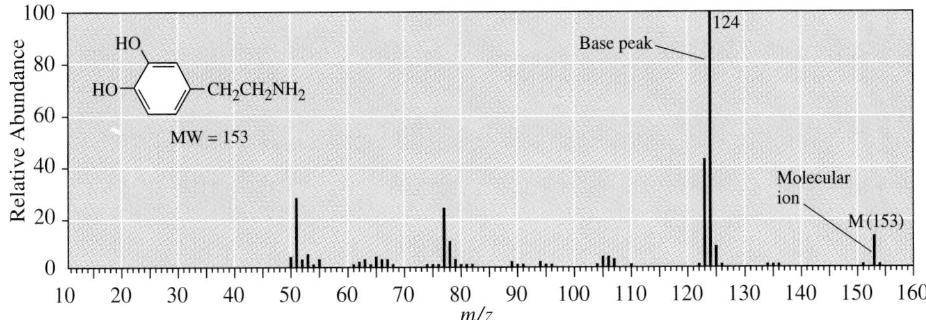

FIGURE 12.2
A partial mass spectrum of dopamine showing all peaks with intensity equal to or greater than 0.5% of the base peak.

either the accelerating voltage or the strength of the magnetic field, ions of the same m/z ratio can be focused on a detector, where the ion current is recorded. Modern ion current detectors are capable of detecting single ions and of scanning a desired mass-to-charge region in a few tenths of a second or less.

A **mass spectrum** is a plot of the relative abundance of each ion versus mass-to-charge ratio. The peak due to the most abundant ion is called the **base peak** and is assigned an arbitrary intensity of 100. The relative abundances of all other ions in a mass spectrum are reported as percentages of abundance of the base peak. Shown in Figure 12.2 is a partial mass spectrum of dopamine, a neurotransmitter in the brain's caudate nucleus, a center involved with coordination and integration of fine muscle movement. A deficiency of dopamine receptors is the underlying biochemical defect in Parkinson's disease. As can be seen in the table accompanying Figure 12.2, the number of peaks recorded depends on the sensitivity of the detector. If we record all peaks with intensity equal to or greater than 0.5% of the base peak, as in Figure 12.2, we find 45 peaks for dopamine. If we record all peaks with intensity equal to or greater than 0.05% of the base peak, we find 120 peaks.

Mass spectrum A plot of the relative abundance of ions versus their mass-to-charge ratio.

Base peak The peak due to the most abundant ion in a mass spectrum. It is assigned an arbitrary intensity of 100.

Number of Peaks Recorded in the Mass Spectrum of Dopamine as a Function of Detector Sensitivity

Peak Intensity Relative to Base Peak	Number of Peaks Recorded
>5%	8
>1%	31
>0.5%	45
>0.05%	120

The technique we have described is called **electron ionization mass spectrometry** (EI MS). This technique was the first developed and for a time the one most widely used. It is limited, however, to relatively low molecular weight compounds that are vaporized easily in the evacuated ionization chamber. In recent years, a revolution in ionization techniques has extended the use of mass spectrometry to very high molecular weight compounds and others that cannot be vaporized directly. Among the new techniques is fast-atom bombardment (FAB), which uses high-energy particles, such as xenon atoms, to bombard a dispersion of a compound in a nonvolatile matrix,

CHEMISTRY IN ACTION

Martian Mass Spectrometry

In the summer of 1976, two spacecraft, Viking 1 and 2, landed on the surface of Mars. On board were a variety of instruments including mass spectrometers, which were used to analyze the gases making up the Martian atmosphere as Viking descended to the surface. In a separate experiment, a gas chromatograph mass spectrometer (GCMS) was employed for surveillance of the Martian soil, in search of organic molecules. This latter type of mass spectrometer, developed by Professor Klaus Biemann and coworkers at the Massachusetts Institute of Technology, is capable of analyzing mixtures of organic compounds by first separating them on a gas chromatograph and then detecting the volatile materials by mass spectrometry. Identification of organic compounds in the soil was considered to be fundamental to any conclusion about the existence of past or present life on Mars. However, the GCMS experiments detected no organic material down to the parts per billion range. This information, together with data from a host of other experiments, temporarily quenched the world's long-standing fantasy about life on Mars.

Twenty years later, in the summer of 1996, researchers from America and Canada stunned the world with the announcement that they had evidence suggestive of a past Martian biota. The evidence came from careful examination of a Martian meteorite, ALH84001, that had landed in Antarctica 13,000 years ago. The age of this piece of Martian terrain is placed at 4.5 billion years and it is estimated that it spent 16 million years in space before falling through our atmosphere.

Several lines of research allowed the scientists to make such a bold announcement. One of the most crucial analyses involved a sophisticated technique known as microprobe two-step laser mass spectrometry (μL^2MS) developed at Stanford University by Professor Richard Zaire and coworkers. This technique was used to probe interior fracture surfaces of the meteorite for the presence of organic compounds. Several different polycyclic aromatic hydrocarbons (PAH), molecules well known to be associated with fossilized remains on earth, were identified using this methodology. The fact that the quantity of PAHs increased as a function of depth inside the fractures allowed the researchers to conclude that these compounds are not earthly contaminants.

Is all of the information presented by these researchers enough to conclude that there was past life on Mars? Only time will help answer that question. NASA is currently working on several spacecraft intended to study the Martian surface in greater detail. The space agency's ultimate goal is to bring back samples of Martian terrain to Earth. There is no doubt that mass spectrometry will be a cornerstone method for analyzing such extraterrestrial samples.

See Horowitz, Norman H, *Scientific American,* November 1977, p. 52; and McKay, David S., et. al., *Science,* Vol. 273, p. 924, 1996.

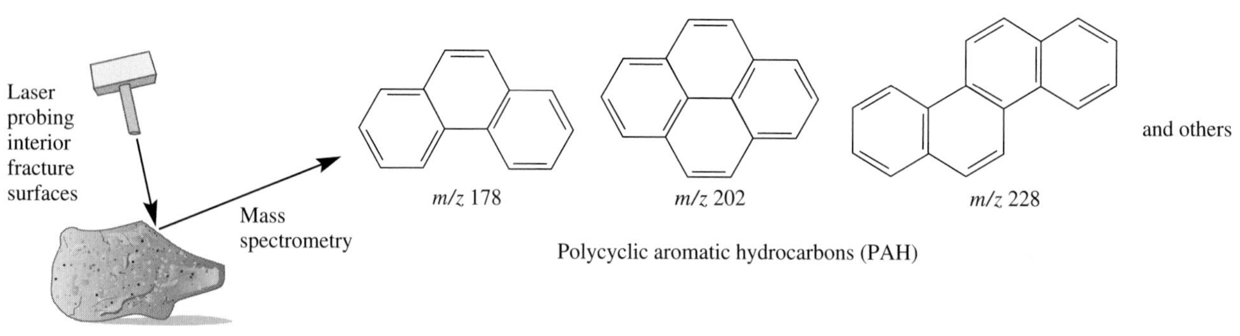

producing ions of the compound and expelling them into the gas phase. A second technique is matrix-assisted laser desorption ionization (MALDI), which uses photons from an energetic laser for the same purpose. A third technique is chemical ionization (CI), which uses gas-phase acid-base reactions to produce ions, and is particularly useful for identifying the molecular mass of a base (Brønsted-Lowry or Lewis) as its conjugate acid, MH^+.

12.2 Features of a Mass Spectrum

To understand the complexity of a mass spectrum, we need to understand some of the relationships between mass spectra and resolution, the presence of isotopes, and fragmentation of molecules and molecular ions in both the ionization chamber and the analyzing chamber.

A. Resolution

An important operating characteristic of a mass spectrometer is its **resolution,** that is, how well it separates ions of different mass. **Low-resolution mass spectrometry** refers to instruments capable of distinguishing among ions of different **nominal mass,** that is, ions that differ by one or more mass units. **High-resolution mass spectrometry** refers to instruments capable of distinguishing among ions that differ in **precise mass** by 0.0001 amu or less.

To illustrate, compounds of molecular formulas C_3H_6O and C_3H_8O have nominal masses of 58 and 60 and can be resolved by low-resolution mass spectrometry. The compounds C_3H_8O and $C_2H_4O_2$, however, have the same nominal mass and cannot be distinguished by low-resolution mass spectrometry. If we calculate the precise mass of each compound using the data in Table 12.1 (Section 12.2B), we see that they differ by 0.03642 amu and can be distinguished by high-resolution mass spectrometry. Observation of a molecular ion with a mass of 60.058 or 60.021 would establish the molecular formula of the unknown compound.

Resolution In mass spectrometry, a measure of how well a mass spectrometer separates ions of different mass.

Low-resolution mass spectrometry Data obtained from a mass spectrometer that is capable of distinguishing only between ions that differ by at least one or more mass units.

High-resolution mass spectrometry Data from a mass spectrometer that is capable of distinguishing between ions that differ in mass by as little as 0.0001 mass unit.

Molecular Formula	Nominal Mass	Precise Mass
C_3H_8O	60	60.05754
$C_2H_4O_2$	60	60.02112

B. The Presence of Isotopes

In the mass spectrum of dopamine (Figure 12.2), the molecular ion appears at *m/z* 153. If you look more closely at this mass spectrum, you will see that there is a small peak at *m/z* 154, one amu heavier than the molecular ion of dopamine. This peak is actually the sum of four separate peaks, each of amu 154 and each corresponding to the presence in the ion of a single heavier isotope of H, C, N, or O in dopamine. Because this peak corresponds to an ion one amu heavier than the molecular ion, it

TABLE 12.1 Precise Masses and Natural Abundances of Isotopes Relative to 100 Atoms of the Most Abundant Isotope

Element	Atomic Weight	Isotope	Precise Mass (amu)	Relative Abundance
hydrogen	1.0079	^{1}H	1.00783	100
		^{2}H	2.01410	0.016
carbon	12.011	^{12}C	12.0000	100
		^{13}C	13.0034	1.11
nitrogen	14.007	^{14}N	14.0031	100
		^{15}N	15.0001	0.38
oxygen	15.999	^{16}O	15.9949	100
		^{17}O	16.9991	0.04
		^{18}O	17.9992	0.20
sulfur	32.066	^{32}S	31.9721	100
		^{33}S	32.9715	0.78
		^{34}S	33.9679	4.40
chlorine	35.453	^{35}Cl	34.9689	100
		^{37}Cl	36.9659	32.5
bromine	79.904	^{79}Br	78.9183	100
		^{81}Br	80.9163	98.0

is called an M + 1 peak. We are concerned in this section primarily with M + 1 and M + 2 peaks.

Virtually all of the elements common to organic compounds, including H, C, N, O, S, Cl, and Br, are mixtures of isotopes. Exceptions are fluorine, phosphorus, and iodine, which occur in nature exclusively as ^{19}F, ^{31}P, and ^{127}I. Table 12.1 shows average atomic weights for the elements most common to organic compounds along with the masses and relative abundances in nature of the stable isotopes of each. In this table, the relative abundances are tabulated according to the number of atoms of heavier isotope per 100 atoms of the most abundant isotope. Carbon in nature, for example, is 98.90% ^{12}C and 1.10% ^{13}C. Thus, there are 1.11 atoms of carbon-13 in nature for every 100 atoms of carbon-12.

$$1.10 \times \frac{100}{98.90} = 1.11 \text{ atoms } ^{13}\text{C per 100 atoms } ^{12}\text{C}$$

EXAMPLE 12.1

Calculate the precise mass of each ion to five significant figures. Unless otherwise indicated, use the mass of the most abundant isotope of each element.

(a) $[CH_2Cl_2]^{\dot{+}}$ **(b)** $[^{13}CH_2Cl_2]^{\dot{+}}$ **(c)** $[CH_2Cl^{37}Cl]^{\dot{+}}$

Solution

(a) $12.0000 + 2(1.00783) + 2(34.9689) = 83.954$ **(b)** 84.957 **(c)** 85.951

PROBLEM 12.1

Calculate the nominal mass of each ion. Unless otherwise indicated, use the mass of the most abundant isotope of each element.

(a) $[CH_3Br]^{\cdot+}$ **(b)** $[CH_3{}^{81}Br]^{\cdot+}$ **(c)** $[{}^{13}CH_3Br]^{\cdot+}$

C. Relative Abundance of M, M + 2, and M + 1 Peaks

The most common elements giving rise to significant M + 2 peaks are chlorine and bromine. Chlorine in nature is 75.77% ^{35}Cl and 24.23% ^{37}Cl. Thus, a ratio of M to M + 2 peaks of approximately 3 : 1 indicates the presence of a single chlorine atom in the compound. Similarly, bromine in nature is 50.7% ^{79}Br and 49.3% ^{81}Br and a ratio of M to M + 2 of approximately 1 : 1 indicates the presence of a single bromine atom in the compound. Sulfur is the only other element common to organic compounds that gives a significant M + 2 peak.

Let us use pentane, C_5H_{12}, to illustrate the relationship between M and M + 1 peaks. Pentane has a nominal mass of 72 and its molecular ion appears at m/z 72. In any sample of pentane, there is a probability that there will be a molecule in which one of the atoms of carbon is ^{13}C, the heavier isotope of carbon present in nature. This molecule has a nominal mass of 73 and its molecular ion will appear at m/z 73. Similarly, there is a probability that there will be a molecule in which one of the atoms of hydrogen is the heavier isotope of hydrogen, namely deuterium, ^2H. The probability of each of these isotope substitutions occurring is related to the natural abundance of each isotope in the following way:

$$\%(M + 1) = \Sigma \text{ (abundance of heavier isotope} \times \text{number of atoms in the formula)}$$

Using this formula, we calculate that the relative intensity of the M + 1 peak for pentane is:

$$\begin{aligned}(M + 1) &= \Sigma\ (1.11 \times 5C + 0.016 \times 12H)\\ &= 5.55 + 0.19\\ &= 5.74\% \text{ of molecular ion peak}\end{aligned}$$

Notice that the M + 1 peak for pentane is due almost entirely to ^{13}C. The same is true for other common compounds containing only C and H. Because M + 1 peaks are relatively low in intensity compared to the molecular ion and often difficult to measure with any precision, they are generally not useful for accurate determinations of molecular formulas.

D. Fragmentation of Molecular Ions

To attain high efficiency of molecular ion formation and to give reproducible mass spectra, it is common to use electrons with energies of 70 eV [approximately 1600 kcal/mol (6700 kJ/mol)]. This energy is sufficient not only to dislodge one or more electrons from a molecule but also to cause extensive fragmentation. These fragments may be unstable as well and, in turn, break apart into even smaller fragments.

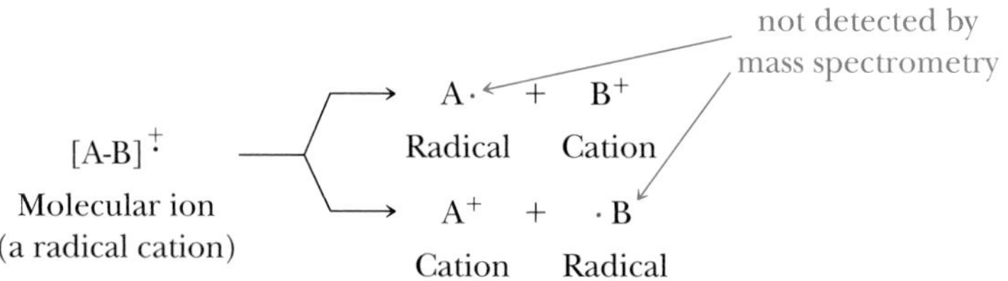

The molecular ions of some compounds have a sufficiently long lifetime in the analyzing chamber that they are observed in the mass spectrum, sometimes as the base (most intense) peak. Molecular ions of other compounds have a shorter lifetime and are observed either in only low abundance or not at all. Furthermore, many fragment ions undergo further fragmentation. As a result, the mass spectrum of a compound typically (but not always) consists of a peak for the molecular ion and series of peaks for fragment ions. The fragmentation pattern and relative abundances of peaks is unique for each compound and characteristic of that compound. Fragmentation patterns give us valuable information about molecular structure.

A great deal of the chemistry of ion fragmentation can be understood in terms of the formation and relative stabilities of carbocations in solution. Where fragmentation occurs, forming new carbocations, the mode of fragmentation that gives the most stable carbocation is favored. Thus, the probability of fragmentation to form a new carbocation increases in the order

$CH_3^+ < 1° < 2° \cong 1°$ allylic/benzylic $< 3° \cong 2°$ allylic/benzylic $< 3°$ allylic/benzylic

Increasing carbocation stability

Molecular rearrangements are also characteristic of certain types of functional groups. We concentrate in this chapter on the fragmentation patterns common to the classes of compounds we have already encountered, and describe fragmentation patterns of new classes of compounds as we encounter them in later chapters.

12.3 Interpreting Mass Spectra

Chemists often use mass spectra primarily for the determination of molecular weight and molecular formula. Very rarely do they attempt a full interpretation of a mass spectrum, which can be very time consuming and difficult. The mass spectrum of dopamine (Figure 12.2), for example, contains at least 45 peaks with intensity equal to or greater than 0.5% of the intensity of the base peak. We have neither the need nor the time to attempt to interpret this level of complexity. Rather, we concentrate in this section on the fragmentation mechanisms giving rise to major peaks.

As we now look at typical mass spectra of the classes of organic compounds we have seen so far, keep the following two points in mind. They provide you with valuable information about the molecular composition of an unknown compound.

Nitrogen rule If a compound has an odd number of nitrogen atoms, its molecular ion will appear as an odd *m/z* value.

1. The only elements giving rise to significant M + 2 peaks are ^{34}S (4.40%), ^{37}Cl (32.5%), and ^{81}Br (98%). If no large M + 2 peak is present, then these elements are absent.
2. Is the mass of the molecular ion odd or even? According to the **nitrogen rule,** if a compound has zero or an even number of nitrogen atoms, its molecular ion will

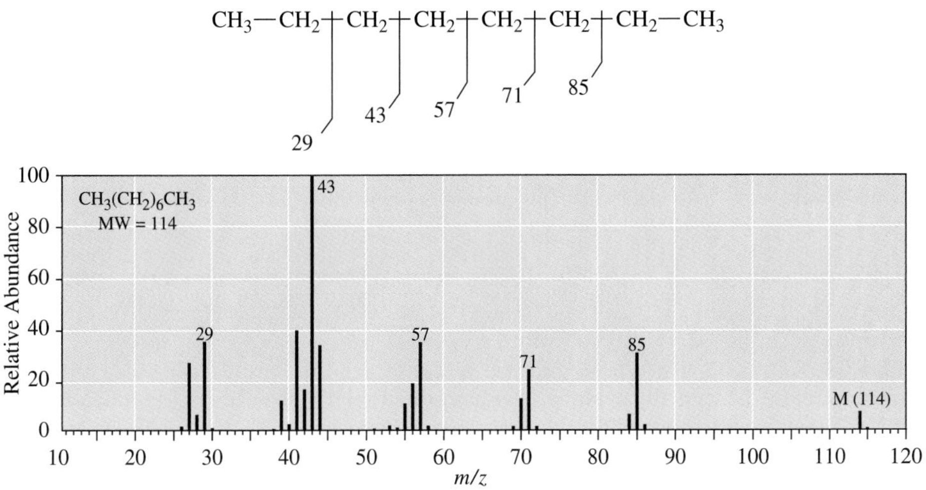

FIGURE 12.3
Mass spectrum of octane.

appear at an even *m/z* value. Conversely, if a compound has an odd number of nitrogen atoms, its molecular ion will appear at an odd *m/z* value. This rule is most helpful when there is an odd number of nitrogens. Where there is an even number of nitrogens, you may need additional experimental information to establish their presence.

A. Alkanes

Two rules will help you interpret the mass spectra of alkanes. (1) Fragmentation tends to occur in the middle of unbranched chains rather than at the ends. (2) The differences in energy among allylic, benzylic, tertiary, secondary, primary, and methyl carbocations in the gas phase are much greater than the differences among comparable radicals. Therefore, where alternative modes of fragmentation are possible, the more stable carbocation tends to form in preference to the more stable radical.

Unbranched alkanes fragment to form a series of cations differing by 14 amu (a CH_2 group). The mass spectrum of octane (Figure 12.3), for example, shows a peak for the molecular ion (*m/z* 114), as well as peaks for $C_6H_{13}^+$ (*m/z* 85), $C_5H_{11}^+$ (*m/z* 71), $C_4H_9^+$ (*m/z* 57), $C_3H_7^+$ (*m/z* 43), and $C_2H_5^+$ (*m/z* 29). Fragmentation of the $CH_2—CH_3$ bond is not observed; there is no peak corresponding to a methyl cation (*m/z* 15), nor is there one corresponding to a heptyl cation (*m/z* 99).

Fragmentation of branched-chain alkanes leads to preferential formation of secondary and tertiary carbocations, and because these cations are more easily formed than methyl and primary carbocations, extensive fragmentation is likely. For this reason, the molecular ion of branched-chain hydrocarbons is often very weak or absent entirely from the spectrum. The molecular ion corresponding to *m/z* 114 is not observed, for example, in the highly branched 2,2,4-trimethylpentane (Figure 12.4). The base peak for this hydrocarbon is at *m/z* 57, due to the *tert*-butyl cation ($C_4H_9^+$). Other prominent peaks are at *m/z* 43, due to the isopropyl cation, and *m/z* 41, due to the allyl cation ($CH_2=CHCH_2^+$).

Sometimes peaks occur in a mass spectrum the origin of which seems to defy any of the rules of chemical logic we have encountered thus far. For example, the prominent peak at *m/z* 29 in the mass spectrum of 2,2,4-trimethylpentane (Figure 12.4) is

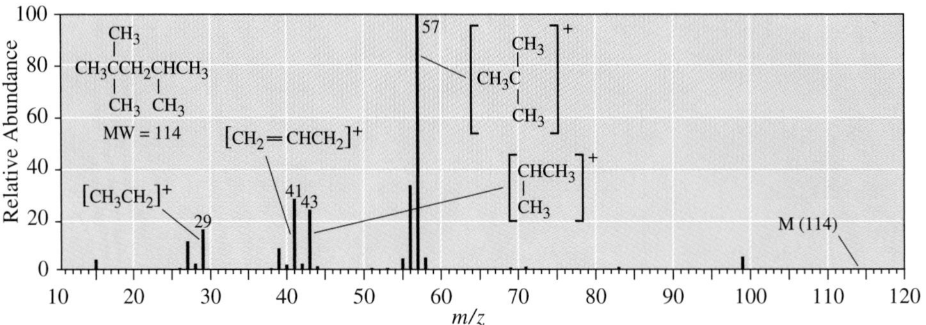

FIGURE 12.4
Mass spectrum of 2,2,4-trimethylpentane. The peak due to the molecular ion is of such low abundance that it does not appear in this spectrum.

due to the ethyl cation, $CH_3CH_2^+$. There is, however, no ethyl group in the parent molecule! We can only conclude that this cation must be formed by some combination of fragmentation and rearrangement quite beyond anything that we have seen up to this point.

The most common fragmentation patterns of cycloalkanes are loss of side chains and loss of ethylene, $CH_2{=}CH_2$. The peak at m/z 69 in the mass spectrum of methylcyclopentane (Figure 12.5) is due to loss of the one-carbon side chain to give the cyclopentyl cation, $C_5H_9^+$. The base peak at m/z 56 is due to loss of ethylene and corresponds to a cation of molecular formula $C_4H_8^+$.

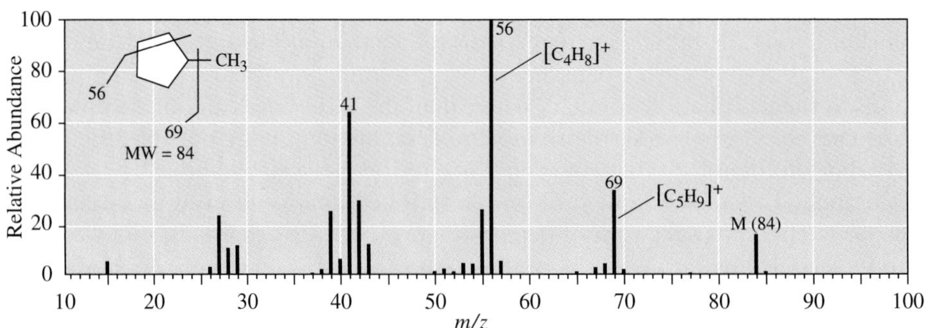

FIGURE 12.5
Mass spectrum of methylcyclopentane.

EXAMPLE 12.2

The base peak at m/z 56 in the mass spectrum of methylcyclopentane corresponds to loss of ethylene to give a radical cation of molecular formula C_4H_8. Propose a structural formula for this radical cation and show how it might be formed.

Solution

Following is a structural formula for the radical cation that might be formed in the ionizing chamber. In it, a single electron is dislodged from a carbon-carbon single bond. Rearrangement of bonding electrons in this radical cation gives ethylene and a new radical cation.

$$\dot{C}H_2\text{—}CH_2\text{—}CH_2\text{—}\overset{+}{C}H\text{—}CH_3 \longrightarrow CH_2{=}CH_2 + \cdot CH_2\text{—}CH_2\text{—}\overset{+}{C}H\text{—}CH_3$$

Molecular ion (a radical cation, *m/z* 84) — Ethylene — A new radical cation (2° carbocation, *m/z* 56)

PROBLEM 12.2

Propose a structural formula for the cation of *m/z* 41 observed in the mass spectrum of methylcyclopentane.

B. Alkenes

Alkenes characteristically show a strong molecular ion peak, most probably formed by removal of one pi electron from the double bond. Furthermore, they cleave readily to form resonance-stabilized allylic cations, such as the allyl cation seen at *m/z* 41 in the mass spectrum of 1-butene (Figure 12.6).

Cyclohexenes undergo fragmentation to give a 1,3-diene and an alkene in a process that is the reverse of a Diels-Alder reaction (Chapter 22). The terpene, limonene, a disubstituted cyclohexene, for example, fragments by a reverse Diels-Alder reaction to give two molecules of 2-methyl-1,3-butadiene (isoprene): one formed as a neutral diene and the other formed as a diene radical cation. Even mass fragments are diagnostic of two bond cleavages.

$$[\text{Limonene}]^{+\cdot} \longrightarrow CH_2{=}CH\text{—}C(CH_3){=}CH_2 + [CH_2{=}C(CH_3)\text{—}CH{=}CH_2]^{+\cdot}$$

Limonene (*m/z* 136) — A neutral diene (*m/z* 68) — A radical cation (*m/z* 68)

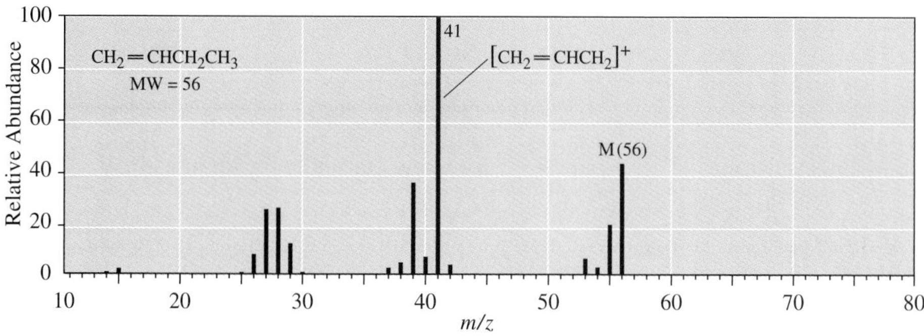

FIGURE 12.6
Mass spectrum of 1-butene.

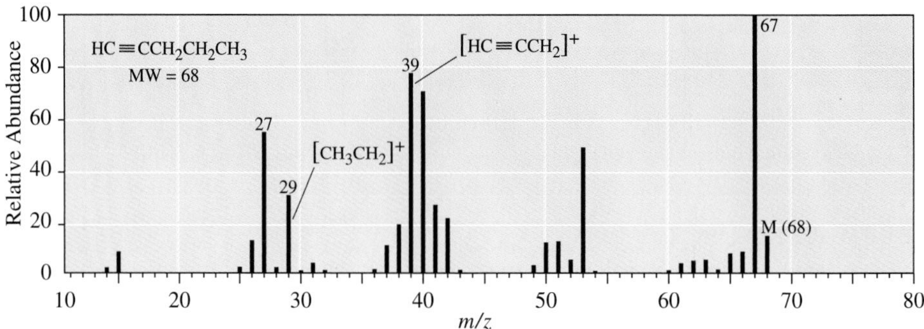

FIGURE 12.7
Mass spectrum of 1-pentyne.

C. Alkynes

As with alkenes, alkynes show a strong peak due to the molecular ion. Their fragmentation patterns are also similar to those of alkenes. One of the most prominent peaks in the mass spectrum of most alkynes is due to the resonance-stabilized propargyl cation (m/z 39) or a substituted propargyl cation.

$$HC{\equiv}C{-}CH_2^{+} \longleftrightarrow H\overset{+}{C}{=}C{=}CH_2$$

Propargyl cation

Both the molecular ion, m/z 68, and the propargyl cation, m/z 39, are seen in the mass spectrum of 1-pentyne (Figure 12.7). Also seen is the ethyl cation, m/z 29.

D. Alcohols

The intensity of the molecular ion from primary and secondary alcohols is normally quite low, and there usually is no molecular ion detectable for tertiary alcohols. One of the most common fragmentation patterns for alcohols is loss of a molecule of water to give a peak corresponding to the molecular ion minus 18 (M − 18). Another common pattern is loss of an alkyl group from the carbon bearing the —OH group to form a resonance-stabilized oxonium ion and an alkyl radical. The oxonium ion is particularly stable because of delocalization of charge.

$$R'{-}\underset{R''}{\overset{R}{C}}{-}\ddot{O}^{\cdot +}{-}H \longrightarrow R\cdot + R'{-}\underset{R''}{C}{=}\overset{+}{O}{-}H \longleftrightarrow R'{-}\underset{R''}{\overset{+}{C}}{-}\ddot{O}{-}H$$

Molecular ion (a radical cation) — A radical — A resonance-stabilized oxonium ion

Each of these patterns is found in the mass spectrum of 1-butanol (Figure 12.8). The molecular ion appears at m/z 74. The prominent peak at m/z 56 corresponds to loss of a molecule of water from the molecular ion (M − 18). The base peak at m/z 31 corresponds to cleavage of a propyl group (M − 43) from the carbon bearing the —OH group. The propyl cation is visible at m/z 43.

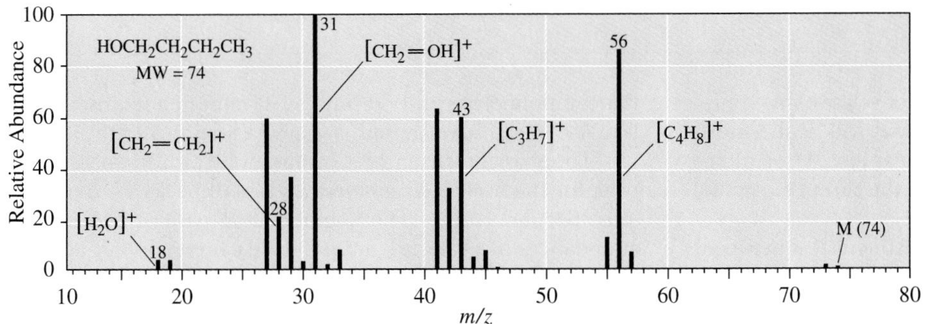

FIGURE 12.8
Mass spectrum of 1-butanol.

EXAMPLE 12.3

A low-resolution mass spectrum of 2-methyl-2-butanol (MW 88) shows 16 peaks. The molecular ion is absent. Account for the formation of peaks at *m/z* 73, 70, 59, and 55, and propose a structural formula for each cation.

Solution

The peak at *m/z* 73 (M − 15) corresponds to loss of a methyl radical from the molecular ion. The peak at *m/z* 59 (M − 29) corresponds to loss of an ethyl radical. Loss of water as a neutral molecule from the molecular ion gives an alkene of *m/z* 70 (M − 18) as a radical cation. Loss of methyl from this radical cation gives an allylic carbocation of *m/z* 55.

$$\left[CH_3-\underset{OH}{\overset{CH_3}{C}}-CH_2-CH_3\right]^{+\cdot} \longrightarrow CH_3\cdot + \underset{{}^{+}OH}{\overset{CH_3}{C}}-CH_2-CH_3 \quad m/z\ 73$$

$$\longrightarrow CH_3-\underset{{}^{+}OH}{\overset{CH_3}{C}} + \cdot CH_2-CH_3 \quad m/z\ 59$$

$$\longrightarrow H_2O + \left[CH_2=\overset{CH_3}{C}-CH_2-CH_3\right]^{+\cdot} \longrightarrow CH_2=\overset{CH_3}{C}-CH_2{}^{+} + CH_3\cdot$$

Molecular ion (a radical cation) — *m/z* 70 — *m/z* 55

PROBLEM 12.3

The low-resolution mass spectrum of 2-pentanol shows 15 peaks. Account for the formation of the peaks at *m/z* 73, 70, 55, 45, 43, and 41. (*Hint:* Consider (1) the loss of water to form an alkene and then fragmentations that the resulting alkene might undergo and (2) the fragmentation of bonds to the carbon bearing the —OH group.

SUMMARY

A **mass spectrum** (Section 12.1) is a plot of relative ion abundance versus mass-to-charge ratio. The **base peak** is the most intense peak in a mass spectrum. A **molecular ion,** M^+ or, alternatively, **M**, is a radical cation derived from the parent molecule by loss of one electron.

Low-resolution mass spectrometry distinguishes between ions that differ in **nominal mass,** that is, ions that differ by one amu (Section 12.2A). **High-resolution mass spectrometry** distinguishes between ions that differ by 0.0001 amu or less.

$(M + 1)$ and higher peaks in a mass spectrum are due to the presence of heavier isotopes (Section 12.2B). The abundance of these higher mass-to-charge peaks relative to the molecular ion provides information about the elemental composition of the molecular ion (Section 12.2C). The presence of a single chlorine atom, for example, is indicated by M to $(M + 2)$ peaks in a ratio of 3:1. According to the **nitrogen rule,** if a compound has an odd number of nitrogen atoms, its molecular ion will have an odd m/z value (Section 12.3).

The mass spectrum of a compound typically consists of a peak for the molecular ion and a series of peaks for fragment ions (Section 12.2D). The fragmentation pattern and relative abundances of peaks are unique for each compound and are characteristic of that compound.

A great many of the observed fragmentation patterns can be understood in terms of the relative stability of carbocations (Section 12.2D). Where alternative modes of fragmentation are possible, the more stable carbocation tends to be formed in preference to the more stable radical.

KEY REACTIONS

1. Formation of a Molecular Ion (Section 12.1)

A molecular ion (a radical cation) is formed by loss of a single electron from a molecule.

$$CH_4 + e^- \longrightarrow [CH_4]^{\dot{+}} + 2e^-$$

Molecular ion
(a radical cation)

2. Mass Spectrometry of Unbranched Alkanes (Section 12.3A)

Unbranched alkanes commonly undergo fragmentation by breaking carbon-carbon bonds to form a series of cations differing in mass by 14 units (a methylene group, CH_2).

$$[CH_3(CH_2)_6CH_3]^{\dot{+}} \longrightarrow CH_3CH_2CH_2^+ + \cdot CH_2(CH_2)_3CH_3$$

3. Mass Spectrometry of Branched-Chain Alkanes (Section 12.3A)

Branched alkanes commonly undergo fragmentation at a branch point to give secondary and tertiary carbocations. Because of the ease of fragmentation of the molecular ion of branched-chain alkanes, the molecular ion itself is often absent from the spectrum.

$$\left[CH_3\overset{\overset{CH_3}{|}}{\underset{\underset{CH_3}{|}}{C}}CH_2\overset{\overset{CH_3}{|}}{C}HCH_3\right]^{\dot{+}} \longrightarrow CH_3\overset{\overset{CH_3}{|}}{\underset{\underset{CH_3}{|}}{C^+}} + \cdot CH_2\overset{\overset{CH_3}{|}}{C}HCH_3$$

4. Mass Spectrometry of Cycloalkanes (Section 12.3A)

Common fragmentation patterns are loss of side chains and cleavage of the ring with loss of ethylene or a substituted ethylene.

$$\text{cyclo-}[\dot{C}H_2\text{—}{}^{+}CH\text{—}CH_3] \longrightarrow CH_2{=}CH_2 + \cdot CH_2\text{—}CH_2\text{—}\overset{+}{C}H\text{—}CH_3$$

5. Mass Spectrometry of Alkenes (Section 12.3B)

A common fragmentation pattern is cleavage of the molecular ion to give a resonance-stabilized allylic cation.

$$[CH_2{=}CHCH_2CH_2CH_3]^{\dot{+}} \longrightarrow \underset{\text{Allyl cation}}{CH_2{=}CHCH_2^{\ +}} + \cdot CH_2CH_3$$

6. Mass Spectrometry of Cyclohexenes (Section 12.3B)

A common fragmentation pattern of cyclohexene and substituted cyclohexenes is to give a 1,3-diene and an alkene by a reverse Diels-Alder reaction.

$$[\text{cyclohexene}]^{\dot{+}} \xrightarrow{\text{reverse Diels-Alder}} [\text{1,3-butadiene}]^{\dot{+}} + CH_2{=}CH_2$$

7. Mass Spectrometry of Alkynes (Section 12.3C)

Fragmentation patterns of alkynes are similar to those of alkenes. A prominent peak in the mass spectrum of most alkynes is the resonance-stabilized propargyl cation or a substituted propargyl cation.

$$[HC{\equiv}CCH_2CH_2CH_3]^{\dot{+}} \longrightarrow \underset{\text{Propargyl cation}}{HC{\equiv}CCH_2^{\ +}} + \cdot CH_2CH_3$$

8. Mass Spectrometry of Alcohols (Section 12.3D)

A common fragmentation of alcohols is loss of water to give a peak at (M − 18). A second common fragmentation pattern of alcohols is loss of an alkyl radical from the carbon bearing the —OH group to give a resonance-stabilized oxonium ion. Where alternative fragmentations are possible, loss of the larger alkyl radical is favored.

$$\left[CH_3\underset{OH}{\overset{CH_3}{C}}CH_2CH_3\right]^{\dot{+}} \longrightarrow CH_3\underset{\overset{\|}{{}^{+}OH}}{\overset{CH_3}{C}} + \cdot CH_2CH_3$$

ADDITIONAL PROBLEMS

12.4 Draw acceptable Lewis structures for the molecular ions (radical cations) formed from the following molecules when each is bombarded by high-energy electrons in a mass spectrometer.

(a) 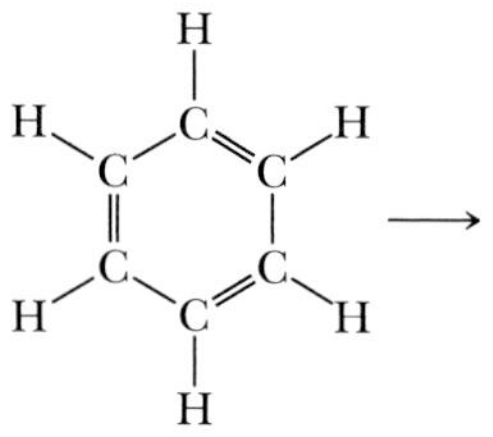 $\longrightarrow$ (b) 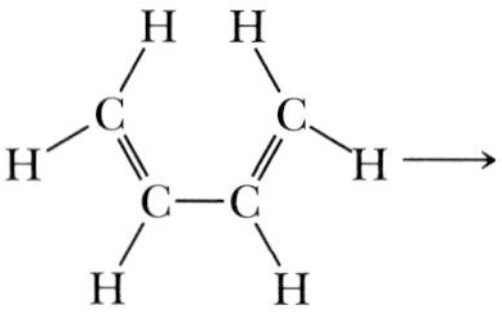 $\longrightarrow$

(c) 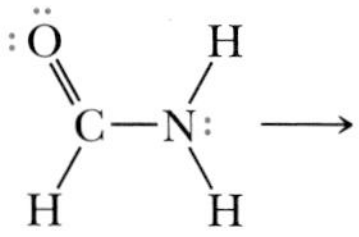 $\longrightarrow$ (d) $H-C\equiv C-H \longrightarrow$

12.5 Some organic molecules can add a single electron to form unstable species called radical anions. A radical anion possesses both a negative charge and an unpaired electron. For example:

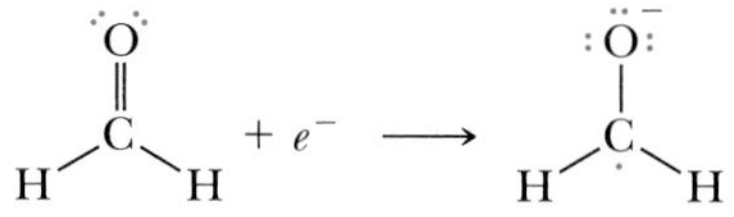

$$H_2C{=}O + e^- \longrightarrow H_2C{-}O^-$$

A radical anion

Draw an acceptable Lewis structure for a radical anion formed from these molecules:

(a) 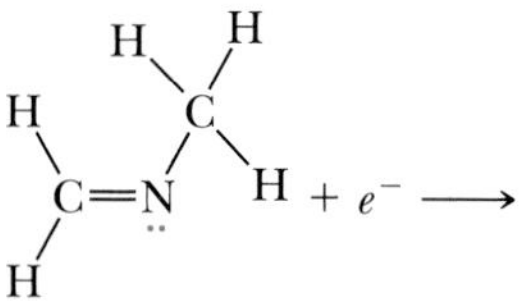 $+ e^- \longrightarrow$ (b) 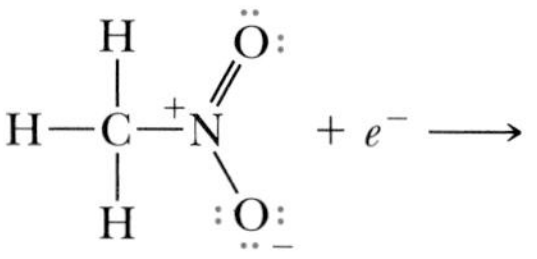 $+ e^- \longrightarrow$

12.6 The molecular ion for compounds containing only C, H, and O is always at an even mass-to-charge value. Why is this so? What can you say about mass-to-charge ratios of ions that arise from fragmentation of one bond in the molecular ion? From fragmentation of two bonds in the molecular ion?

12.7 For which compounds containing a heteroatom (an atom other than carbon or hydrogen) does a molecular ion have an even-numbered mass and for which does it have an odd-numbered mass?

(a) A chloroalkane of molecular formula $C_nH_{2n+1}Cl$
(b) A bromoalkane of molecular formula $C_nH_{2n+1}Br$
(c) An alcohol of molecular formula $C_nH_{2n+1}OH$
(d) A primary amine of molecular formula $C_nH_{2n+1}NH_2$
(e) A thiol of molecular formula $C_nH_{2n+1}SH$

12.8 The so-called nitrogen rule states that if a compound has an odd number of nitrogen atoms, the value of m/z for its molecular ion will be an odd number. Why is this so?

12.9 Both $C_6H_{10}O$ and C_7H_{14} have the same nominal mass, namely 98. Show how these compounds can be distinguished by the m/z ratio of their molecular ions in high-resolution mass spectrometry.

12.10 Show how the compounds of molecular formulas C_6H_9N and C_5H_5NO can be distinguished by the m/z ratio of their molecular ions in high-resolution mass spectrometry.

12.11 What rule would you expect for the m/z values of fragment ions resulting from the cleavage of one bond in a compound with an odd number of nitrogen atoms?

12.12 In a natural sample of ethane, what is the probability that

(a) One carbon in an ethane molecule is ^{13}C?
(b) Both carbons in an ethane molecule are ^{13}C?
(c) Two hydrogens in an ethane molecule are replaced by deuterium atoms?

12.13 The molecular ions of both $C_5H_{10}S$ and $C_6H_{14}O$ appear at m/z 102 in low-resolution mass spectrometry. Show how determination of the correct molecular formula can be made from the appearance and relative intensity of the (M + 2) peak of each compound.

12.14 In Section 12.3, we saw several examples of fragmentation of molecular ions to give resonance-stabilized cations. Make a list of these resonance-stabilized cations and write important contributing structures of each. Estimate the relative importance of the contributing structures in each set.

12.15 Carboxylic acids often give strong fragment ions at m/z (M − 17). What is the likely structure of these cations, and how might they be formed? Show by drawing contributing structures that these cations are stabilized by resonance.

12.16 For primary amines with no branching on the carbon bearing the nitrogen, the base peak occurs at m/z 30. What cation does this peak represent and how is it formed? Show by drawing contributing structures that this cation is stabilized by resonance.

12.17 The base peak in the mass spectrum of propanone (acetone) occurs at m/z 43. What cation does this peak represent?

12.18 A characteristic peak in the mass spectrum of most aldehydes occurs at m/z 29. What cation does this peak represent? (No, it is not an ethyl cation, $CH_3CH_2^+$.)

Interpretation of Mass Spectra

12.19 Predict the relative intensities of the M and M + 2 peaks for

(a) CH_3CH_2Cl **(b)** CH_3CH_2Br **(c)** $BrCH_2CH_2Br$ **(d)** CH_3CH_2SH

12.20 The mass spectrum of Compound B shows the molecular ion at m/z 85, an M + 1 peak at 86 of approximately 6% abundance relative to M, and an M + 2 peak at 87 of less than 0.1% abundance relative to M.

(a) Propose a molecular formula for Compound B. (*Hint:* Remember the nitrogen rule.)
(b) Draw at least 10 possible structural formulas for this molecular formula.

12.21 The mass spectrum of Compound E, a colorless liquid, shows these peaks in its mass spectrum. The base peak is at m/z 43. Other peaks are given in relative abundance to the base peak. Determine the molecular formula of Compound E and propose a structural formula for it. *Hint:* Calculate the ratio of the M to M + 2 peaks.

m/z	Relative Abundance
43	100 (base)
78	23.6 (M)
79	1.00
80	7.55
81	0.25

12.22 Write molecular formulas for the five possible molecular ions of m/z 88 containing the elements C, H, N, and O. (*Note:* Because the value of the mass of this set of molecular ions is an even number, members of the set must have either no nitrogen atoms or an even number of nitrogen atoms.)

12.23 Write molecular formulas for the six possible molecular ions of m/z 100 containing only the elements C, H, N, and O.

12.24 The molecular ion in the mass spectrum of 2-methyl-1-pentene appears at m/z 84. Propose structural formulas for the prominent peaks at m/z 69, 55, 41, and 29.

12.25 Following is the mass spectrum of 1,2-dichloroethane.

(a) Account for the appearance of an (M + 2) peak with approximately two-thirds the intensity of the molecular ion peak.

(b) Predict the intensity of the (M + 4) peak.

(c) Propose structural formulas for the cations of m/z 64, 63, 62, 51, 49, 27, and 26.

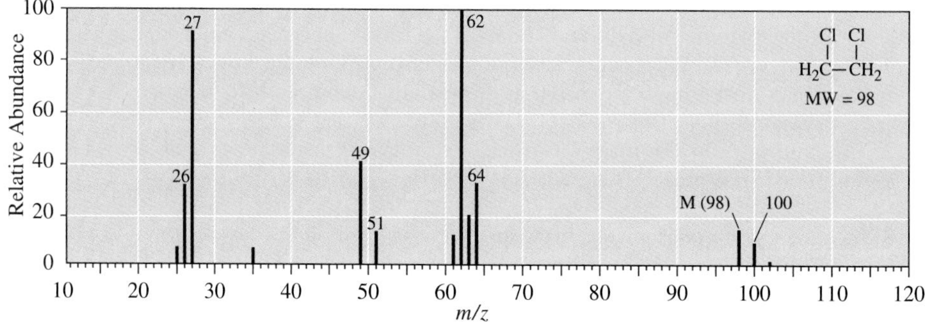

12.26 Following is the mass spectrum of 1-bromobutane.

(a) Account for the appearance of the (M + 2) peak of approximately 95% the intensity of the molecular ion peak.

(b) Propose structural formulas for the cations of m/z 57, 41, and 29.

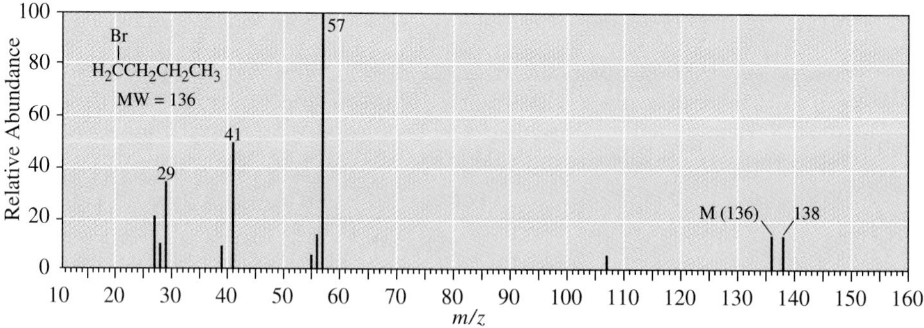

12.27 Following is the mass spectrum of bromocyclopentane. The molecular ion m/z 148 is of such low intensity that it does not appear in this spectrum. Assign structural formulas for the cations of m/z 69 and 41.

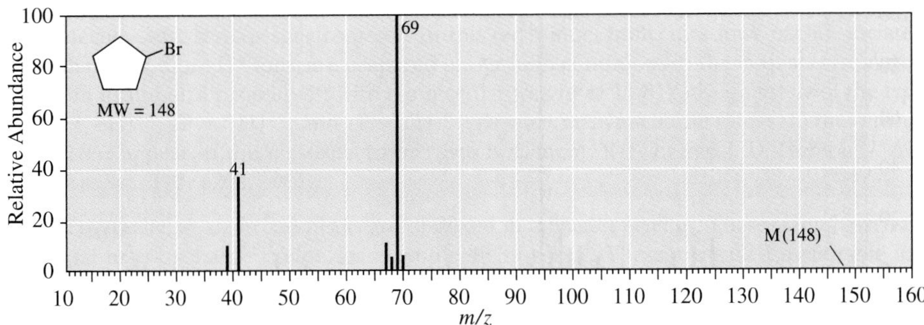

12.28 Following is the mass spectrum of 3-methyl-2-butanol. The molecular ion *m/z* 88 does not appear in this spectrum. Propose structural formulas for the cations of *m/z* 45, 43, and 41.

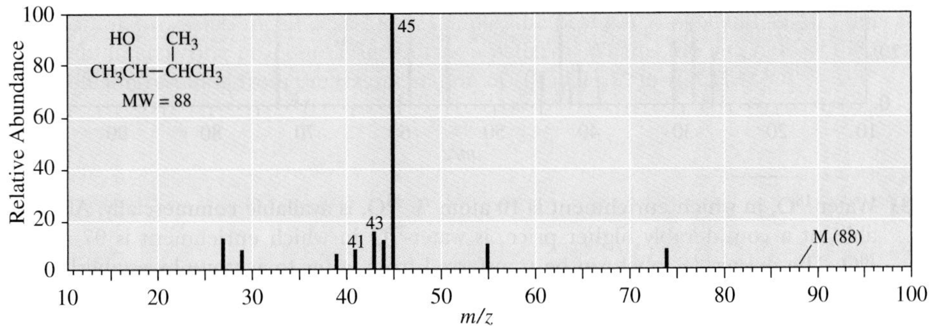

12.29 The following is the mass spectrum of compound A, C_3H_8O. Compound A is infinitely soluble in water, undergoes reaction with sodium metal with the evolution of a gas, and undergoes reaction with thionyl chloride to give a water-insoluble chloroalkane. Propose a structural formula for compound A, and write equations for each of its reactions.

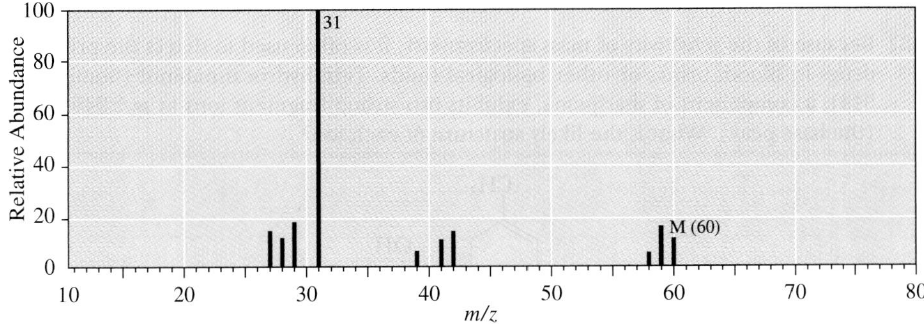

12.30 Following are mass spectra for the constitutional isomers 2-pentanol and 2-methyl-2-butanol. Assign each isomer its correct spectrum.

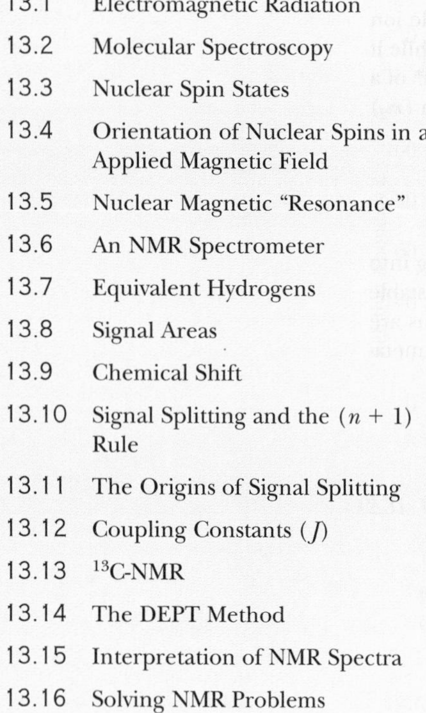

13 NUCLEAR MAGNETIC RESONANCE SPECTROSCOPY

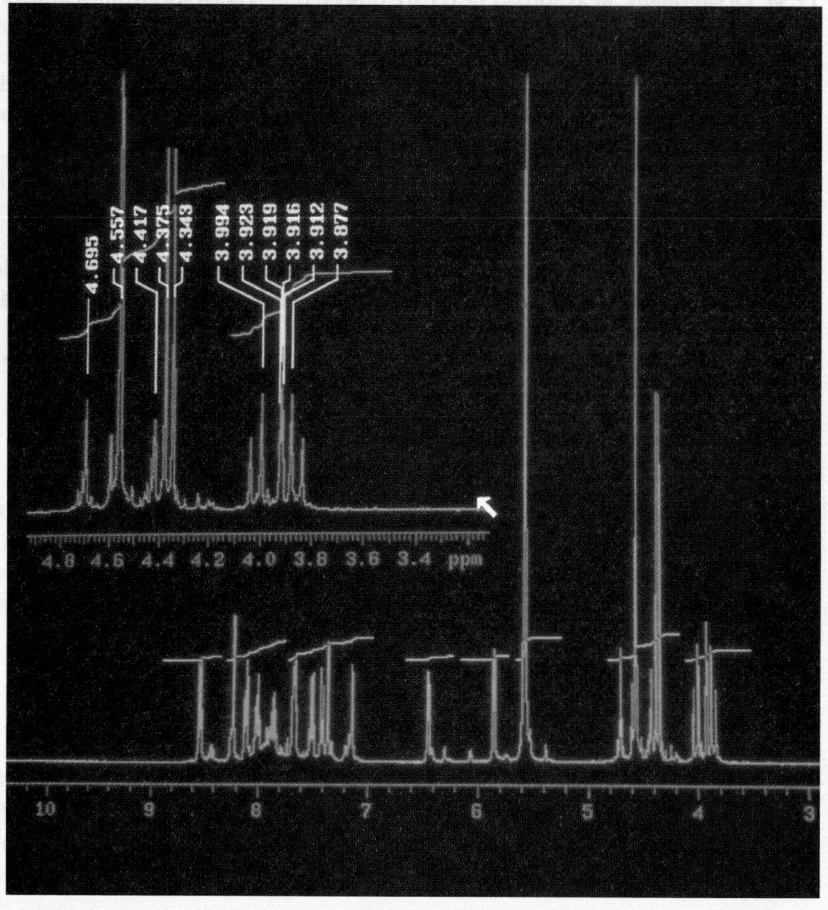

■ A screen display of a ^{1}H-NMR spectrum. *(Hank Morgan/Photo Researchers, Inc.)*

Nuclear magnetic resonance spectroscopy was developed in the late 1950s, and, within a decade, it became the single most important technique available to chemists for the determination of molecular structure. In this chapter, we first develop a basic understanding of the theory behind this type of spectroscopy, and then we concentrate on the interpretation of spectra and the information they can provide us about details of molecular structure.

13.1 Electromagnetic Radiation

Gamma rays, x-rays, ultraviolet light, visible light, infrared radiation, microwaves, and radio waves are all parts of the electromagnetic spectrum. Because **electromagnetic radiation** behaves as a wave traveling at the speed of light, it can be described in terms of its wavelength and its frequency. Wavelengths, frequencies, and energies of various regions of the electromagnetic spectrum are summarized in Table 13.1.

Electromagnetic radiation Light and other forms of radiant energy.

Wavelength is the distance between any two consecutive identical points on a wave. Wavelength is given the symbol $\boldsymbol{\lambda}$ (Greek lambda) and is usually expressed in the SI base unit of meters. Other derived units commonly used to express wavelength are given in Table 13.2.

Wavelength ($\boldsymbol{\lambda}$) The distance between consecutive peaks on a wave.

Frequency, the number of full cycles of a wave that pass a given point in a second, is given the symbol ν (Greek nu) and is reported in **hertz (Hz).** One MHz $= 10^6$ Hz. Wavelength and frequency are inversely proportional, and one can be calculated from the other using the following relationship:

Frequency ($\boldsymbol{\nu}$) The number of full cycles of a wave that pass a point in a second.

Hertz (Hz) The unit in which radiation frequency is reported; s^{-1} (read "per second").

$$\lambda\nu = c$$

where ν is frequency in hertz; c is the velocity of light, 3.00×10^8 m/s; and λ is the wavelength in meters. For example, consider infrared radiation, or heat radiation as it is also called, of wavelength 1.5×10^{-5} m (15 μm). The frequency of this radiation is 2.0×10^{13} Hz.

$$\nu = \frac{3.0 \times 10^8 \text{ m/s}}{1.5 \times 10^{-5} \text{ m}} = 2.0 \times 10^{13} \text{ Hz}$$

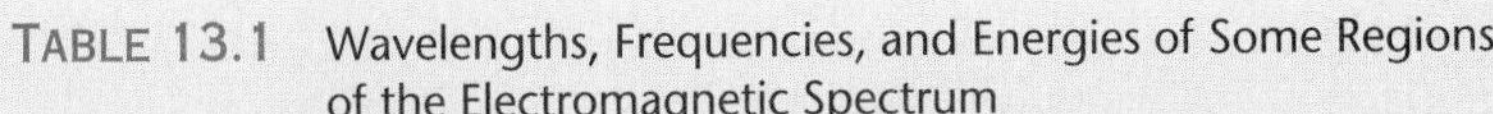

TABLE 13.1 Wavelengths, Frequencies, and Energies of Some Regions of the Electromagnetic Spectrum

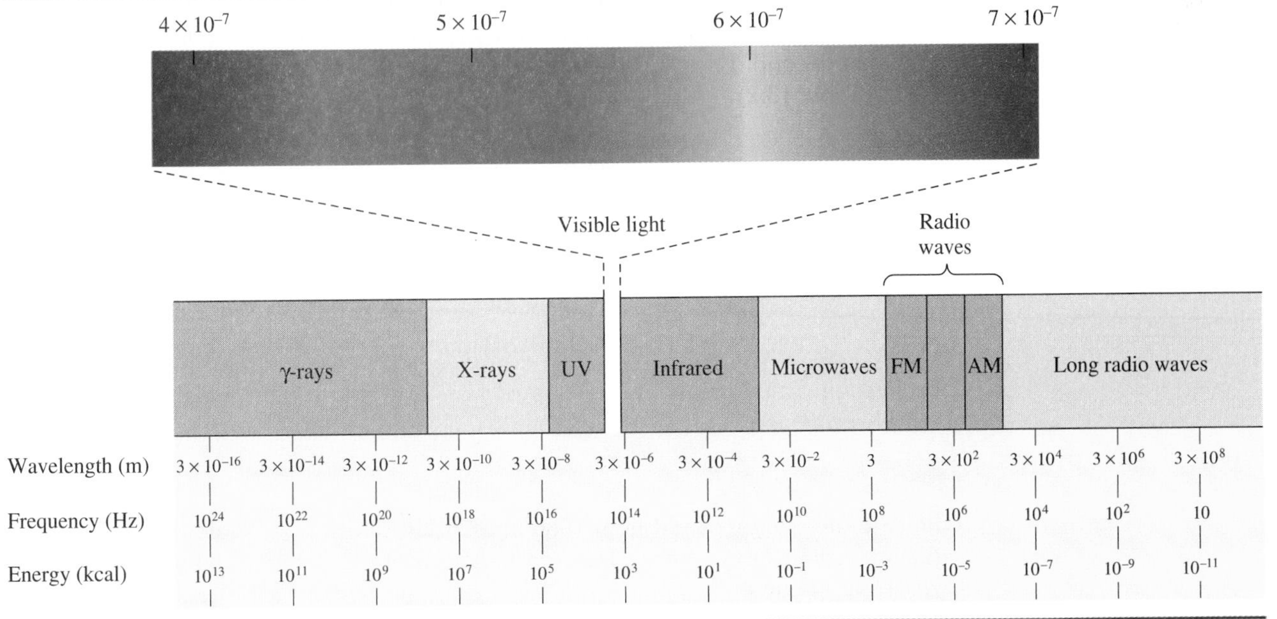

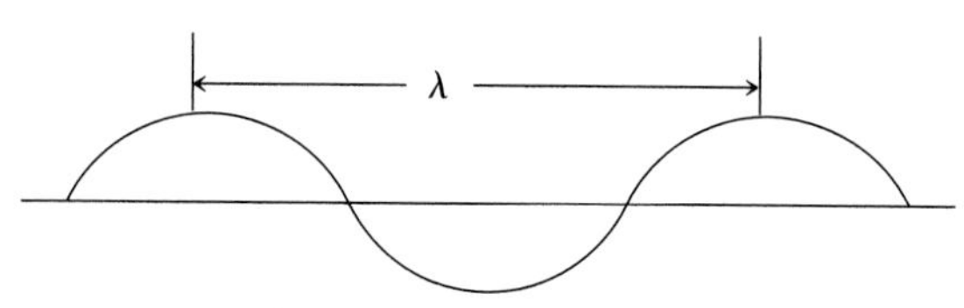

TABLE 13.2 Common Units Used to Express Wavelength (λ)

Unit	Relation to Meter
meter (m)	—
millimeter (mm)	$1\ \text{mm} = 10^{-3}\ \text{m}$
micrometer (μm)	$1\ \mu\text{m} = 10^{-6}\ \text{m}$
nanometer (nm)	$1\ \text{nm} = 10^{-9}\ \text{m}$
Angstrom (Å)	$1\ \text{Å} = 10^{-10}\ \text{m}$

An alternative way to describe electromagnetic radiation is in terms of its properties as a stream of particles, called **photons.** The energy in a mole of photons is related to the frequency of the radiation by the equation

$$E = h\nu = h\frac{c}{\lambda}$$

where E is the energy in kcal/mol (kJ/mol) and h is Planck's constant, $9.537 \times 10^{-14}\ \text{kcal} \cdot \text{s} \cdot \text{mol}^{-1}$ ($3.99 \times 10^{-13}\ \text{kJ} \cdot \text{s} \cdot \text{mol}^{-1}$).

EXAMPLE 13.1

Calculate the energy in kilocalories per mole of radiation of wavelength 2.50 μm. What type of radiant energy is this? (Refer to Table 13.1.)

Solution

Use the relationship $E = hc/\lambda$. Make certain that dimensions for distance are consistent; if the dimension of wavelength is meters, then express the velocity of light in meters per second. First convert 2.50 μm to meters, using the relationship $1\ \mu\text{m} = 10^{-6}\ \text{m}$ (Table 13.2).

$$2.50\ \mu\text{m} \times \frac{10^{-6}\ \text{m}}{1\ \mu\text{m}} = 2.50 \times 10^{-6}\ \text{m}$$

Now substitute this value in the equation $E = hc/\lambda$.

$$E = \frac{hc}{\lambda} = 9.54 \times 10^{-14}\ \frac{\text{kcal} \cdot \text{s}}{\text{mol}} \times 3.00 \times 10^{8}\ \frac{\text{m}}{\text{s}} \times \frac{1}{2.50 \times 10^{-6}\ \text{m}}$$
$$= 11.4\ \text{kcal/mol}\ (47.7\ \text{kJ/mol})$$

Electromagnetic radiation with energy of 11.4 kcal/mol corresponds to radiation in the infrared region.

PROBLEM 13.1

Calculate the energy of red light (680 nm) in kilocalories per mole. Which form of radiation carries more energy, infrared radiation of wavelength 2.50 μm or red light of wavelength 680 nm?

13.2 Molecular Spectroscopy

Organic molecules are flexible structures. As we discussed in Chapter 2, atoms and groups of atoms can rotate about single covalent bonds. In addition, covalent bonds can stretch and bend just as if their atoms are joined by flexible springs. Furthermore, electrons within molecules can move from one electronic energy level to another as, for example, promotion of an electron from a pi bonding molecular orbital to a pi* antibonding molecular orbital. Finally, certain nuclei behave as if they are spinning charged particles and can change from one nuclear spin energy level to another.

An atom or molecule can be made to undergo a transition from energy state E_1 to a higher energy state E_2 by irradiating it with electromagnetic radiation corresponding to the energy difference between states E_1 and E_2 as illustrated schematically in Figure 13.1. When the atom or molecule returns to the ground state E_1, an equivalent amount of energy is emitted.

Molecular spectroscopy is the experimental process of measuring which frequencies of radiation are absorbed or emitted by a particular substance and then attempting to correlate patterns of energy absorption or emission with details of molecular structure. The regions of the electromagnetic spectrum of most interest to us and the relationships of each to changes in atomic and molecular energy levels are summarized in Table 13.3.

Molecular spectroscopy The study of which frequencies of electromagnetic radiation are absorbed or emitted by substances and the correlation between these frequencies and specific types of molecular structure.

In this chapter, we concentrate on absorption of radio-frequency radiation, which causes transitions between nuclear spin energy levels; that is, we concentrate on nuclear magnetic resonance spectroscopy. The phenomenon of nuclear magnetic resonance was first detected in 1946 by Felix Bloch and Edward Purcell, both of the United States, who shared the Nobel Prize for physics in 1952. The particular value of **nuclear magnetic resonance (NMR) spectroscopy** is that it gives us information about the number and types of atoms in a molecule: for example, about the number and types of hydrogens using **^{1}H-NMR spectroscopy,** and about the number and types of carbons using **^{13}C-NMR spectroscopy.**

Nuclear magnetic resonance (NMR) spectroscopy A spectroscopic technique that gives us information about the number and types of atoms in a molecule (for example, about the number and types of hydrogens using ^{1}H-NMR spectroscopy, and about the number and types of carbons using ^{13}C-NMR spectroscopy).

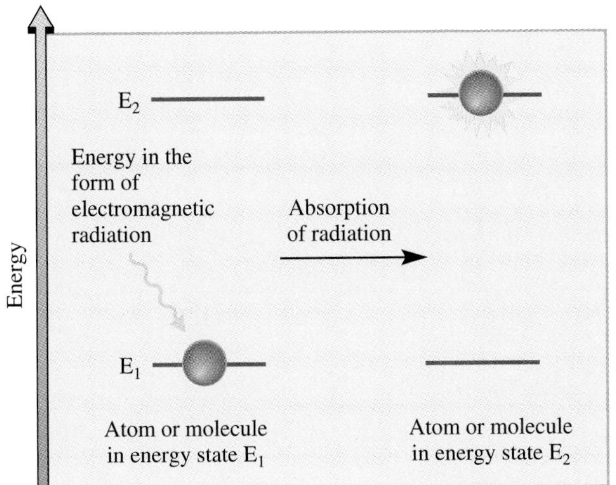

FIGURE 13.1
Absorption of energy in the form of electromagnetic radiation causes an atom or molecule in energy state E_1 to change to a higher energy state E_2.

TABLE 13.3 Types of Energy Transitions Resulting from Absorption of Energy from Three Regions of the Electromagnetic Spectrum

Region of the Electromagnetic Spectrum	Absorption of Electromagnetic Radiation Results in Transition Between
radio frequency	nuclear spin energy levels
infrared	vibrational energy levels
ultraviolet-visible	electronic energy levels

13.3 Nuclear Spin States

You are already familiar from general chemistry with the concept that an electron has a spin quantum number of $\frac{1}{2}$, with allowed values of $+\frac{1}{2}$ and $-\frac{1}{2}$, and that a spinning charge creates an associated magnetic field. In effect, an electron behaves as if it is a tiny bar magnet and has what is called a **magnetic moment.** According to the Pauli exclusion principle (Section 1.1), two electrons can occupy the same atomic or molecular orbital only if their spins are paired (opposite in sign).

Any atomic nucleus that has an odd mass number, an odd atomic number, or both also has a spin and a resulting nuclear magnetic moment. The allowed nuclear spin states are determined by the spin quantum number, I, of the nucleus. A nucleus with spin quantum number I has $2I + 1$ spin states. Our focus in this chapter is on nuclei of ^{1}H and ^{13}C, isotopes of the two elements most common to organic compounds. Each has a nuclear spin quantum number of $\frac{1}{2}$ and each, therefore, has $2(\frac{1}{2}) + 1 = 2$ allowed spin states. Spin quantum numbers and allowed **nuclear spin states** for these nuclei along with the nuclei of other elements common to organic compounds are shown in Table 13.4. Note that ^{12}C and ^{16}O each have a spin quantum number of zero and only one allowed nuclear spin state.

TABLE 13.4 Spin Quantum Numbers and Allowed Nuclear Spin States for Selected Isotopes of Elements Common to Organic Compounds

Element	^{1}H	^{2}H	^{12}C	^{13}C	^{14}N	^{16}O	^{31}P	^{32}S
nuclear spin quantum number (I)	$\frac{1}{2}$	1	0	$\frac{1}{2}$	1	0	$\frac{1}{2}$	0
number of spin states	2	3	1	2	3	1	2	1

13.4 Orientation of Nuclear Spins in an Applied Magnetic Field

Within a collection of ^1H and ^{13}C atoms, the spins of their tiny nuclear bar magnets are completely random in orientation. When placed between the poles of a powerful magnet of **field strength B_0**, however, interactions between their nuclear spins and the applied magnetic field are quantized, with the result that only certain orientations of nuclear magnetic moments are allowed. For ^1H for ^{13}C nuclei, only two orientations are allowed as illustrated in Figure 13.2. By convention, nuclei with spin $+\frac{1}{2}$ are assumed to be aligned with the applied field and in the lower energy state; nuclei with spin $-\frac{1}{2}$ are assumed to be aligned against the applied field and in the higher energy state.

The SI unit for magnetic field strength is the tesla (T). A unit still in common use, however, is the gauss (G). Values of T and G are related by the equation 1 T = 10^4 G.

The difference in energy between nuclear spin states increases with the strength of the applied field (Figure 13.3). At an applied field strength of 7.05 T, which is

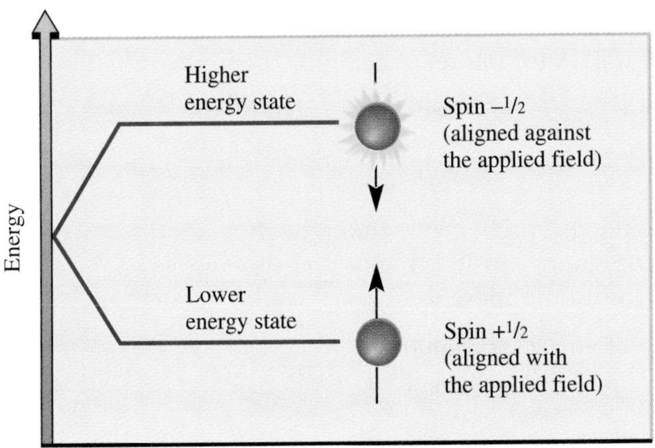

FIGURE 13.2
^1H and ^{13}C nuclei with spin $+\frac{1}{2}$ are aligned with the applied magnetic field, B_0, and are in the lower spin energy state; those with spin $-\frac{1}{2}$ are aligned against the applied magnetic field and are in the higher spin energy state.

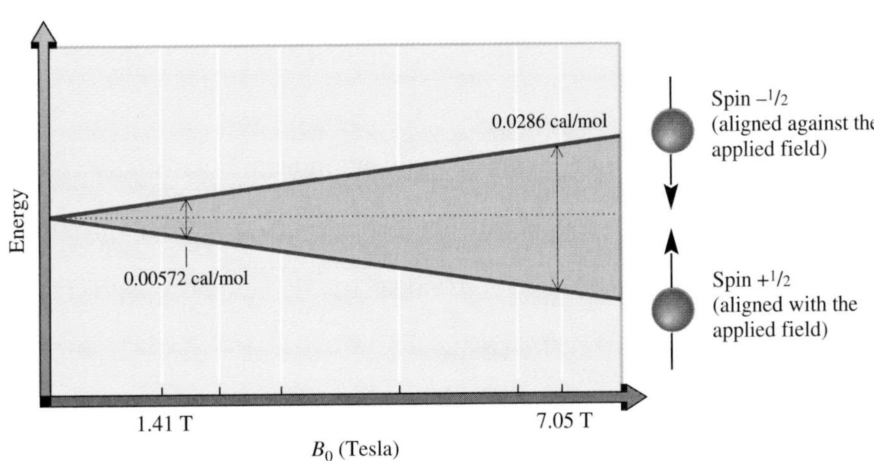

FIGURE 13.3
The energy difference between the allowed nuclear spin states increases linearly with applied field strength. Values shown here are for ^1H nuclei.

readily available with present-day superconducting electromagnets, the difference in energy between nuclear spin states for ^{1}H is approximately 0.0286 cal/mol (0.120 J/mol), which corresponds to electromagnetic radiation of approximately 300 MHz (300,000,000 Hz). At 7.05 T, the difference in energy between nuclear spin states for ^{13}C nuclei is approximately 0.00715 cal/mol, which corresponds to radiation of approximately 75 MHz.

To put these values for nuclear spin energy levels in perspective, energies for transitions between vibrational energy levels observed in infrared spectroscopy are 2 to 15 kcal/mol (8 to 63 kJ/mol) and energies for transitions between electronic energy levels observed in ultraviolet-visible spectroscopy are 30 to 140 kcal/mol (126 to 586 kJ/mol), except as noted.

EXAMPLE 13.2

Calculate the ratio of nuclei in the higher spin state to those in the lower spin state, N_h/N_l, for ^{1}H at 25°C in an applied field strength of 7.05 T.

Solution

Use the equation given in Section 2.6A for the relationship between the difference in energy states and equilibrium constant. In this problem, this relationship has the form

$$\Delta E = -2.303\ RT \log \frac{N_h}{N_l}$$

The difference in energy between the higher and lower nuclear spin states in an applied field of 7.05 T is approximately 0.0286 cal/mol and the temperature in degrees K is 25 + 273 = 298 K. Substituting these values in the above equation gives

$$\log \frac{N_h}{N_l} = \frac{-\Delta E}{2.303\ RT} = \frac{-0.0286\ \text{cal}\cdot\text{mol}^{-1}}{2.303 \times 1.987\ \text{cal}\cdot\text{deg}^{-1}\cdot\text{mol}^{-1} \times 298\ \text{deg}} = -2.097 \times 10^{-5}$$

$$\frac{N_h}{N_l} = 0.9999517 = \frac{1.000000}{1.000048}$$

From this calculation, we determine that, for every 1,000,000 hydrogen atoms in the higher energy state in this applied field, there are 1,000,048 in the lower energy state. The excess population of the lower energy state under these conditions is only 48 per million!

PROBLEM 13.2

Calculate the ratio of nuclei in the higher spin state to those in the lower spin state, N_h/N_l, for ^{13}C at 25°C in an applied field strength of 7.05 T. The difference in energy between nuclear spin states at this field strength is approximately 0.00715 cal/mol.

13.5 Nuclear Magnetic "Resonance"

When nuclei with spin quantum number $\frac{1}{2}$ are placed in an applied magnetic field, a small majority of nuclear spins are aligned with the applied field in the lower energy state. When nuclei in the lower energy spin state are irradiated with a radio frequency of the appropriate energy, they absorb energy and their nuclear spins flip from the lower energy state to the higher energy state.

To understand the mechanism by which a spinning nucleus absorbs energy and the meaning of resonance in this context, imagine a spinning nucleus. When an applied field of strength B_0 is turned on, the nucleus becomes aligned with the applied field in an allowed spin energy state. The nucleus then begins to **precess** as shown in Figure 13.4(a) and traces out a cone-shaped surface in much the same manner as a spinning top or gyroscope traces out a cone-shaped surface as it precesses in the earth's gravitational field. We can express the **rate of precession** as a frequency in hertz.

If the precessing nucleus is irradiated with electromagnetic radiation of the same frequency as the precession frequency, then the two frequencies couple, energy is absorbed, and the nuclear spin is "flipped" from spin state $+\frac{1}{2}$ (with the applied field) to spin state $-\frac{1}{2}$ (against the applied field) as illustrated in Figure 13.4(b). For ^1H in an applied magnetic field of 7.05 T, the frequency of precession is approximately 300 MHz. For ^{13}C in the same field, it is approximately 75 MHz. Herein lies the reason for the use of the term "resonance." **Resonance** in this context is the absorption of electromagnetic radiation by a precessing nucleus and the resulting flip of its nuclear spin from a lower energy state to a higher energy state. The instrument used to detect this coupling of precession frequency and electromagnetic radiation records it as a **signal.**

Resonance, in NMR spectroscopy The absorption of electromagnetic radiation by a precessing nucleus and the resulting "flip" of its nuclear spin from a lower energy state to a higher energy state.

Signal A recording in an NMR spectrum of a nuclear magnetic resonance.

If we were dealing with ^1H nuclei isolated from all other atoms and electrons, any combination of applied field and electromagnetic radiation that produces a signal for one hydrogen nucleus would produce a signal for all hydrogen nuclei. In other words, hydrogens would be indistinguishable one from another. Hydrogens in an organic molecule, however, are not isolated; they are surrounded by electrons, which are caused to circulate by the presence of an applied magnetic field. You may think of

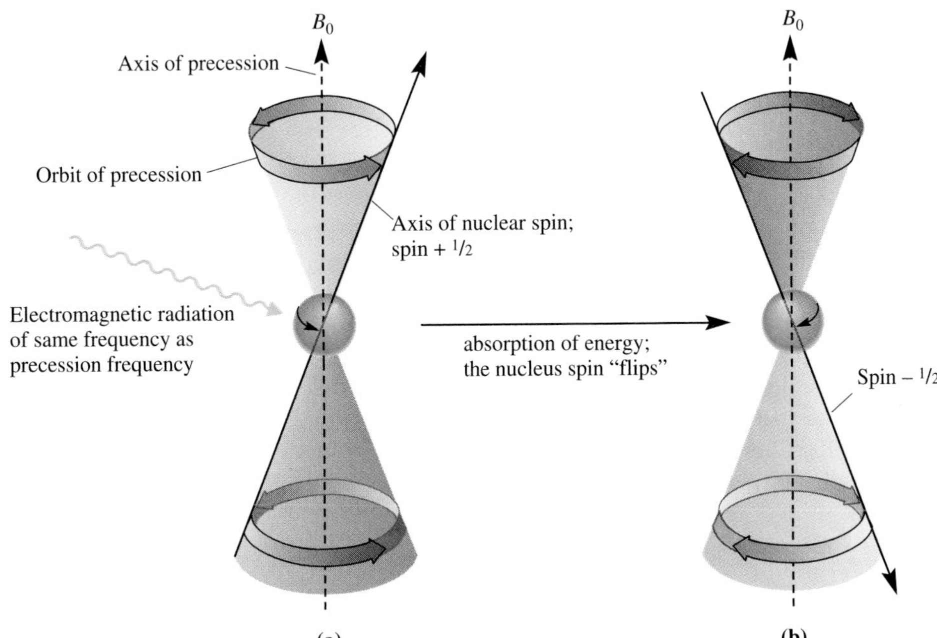

FIGURE 13.4
The origin of nuclear magnetic "resonance." (a) Precession of a spinning nucleus in an applied magnetic field. (b) Absorption of electromagnetic radiation occurs when the frequency of radiation is equal to the frequency of precession.

this electron circulation as you would the flow of electrons in a wire. We know from the laws of physics that the flow of electric current through a wire generates a magnetic field. In the case of electrons about a hydrogen nucleus, their circulation generates a magnetic field opposed to the applied field and thereby acts to shield the hydrogen nucleus from the applied field. The circulation of electrons around a nucleus in an applied field is called a **diamagnetic current** and the nuclear shielding resulting from it is called **diamagnetic shielding.** Although the diamagnetic shielding created by circulating electrons is orders of magnitude weaker than the applied fields used in NMR spectroscopy, it is nonetheless significant at the molecular level.

Diamagnetic current The circulation of electrons about a nucleus in an applied field. The nuclear shielding resulting from it is called diamagnetic shielding.

Shielding, in NMR spectroscopy An effect produced when electron density is increased about a nucleus, causing it to absorb toward the right (upfield) on the chart paper.

Deshielding, in NMR spectroscopy An effect produced when electron density is decreased around a nucleus, causing it to absorb toward the left (downfield) on the chart paper.

The degree of **shielding** depends on several factors, which we will take up soon. For the moment, however, it is sufficient to realize that the greater the shielding of a particular nucleus by local magnetic fields, the greater the strength of the applied field required to bring it into resonance. Conversely, the less the shielding of a nucleus, or as it is more commonly expressed, the greater its **deshielding,** the lower the strength of the applied field required to bring it into resonance.

The differences in resonance frequencies among the various hydrogen nuclei within a molecule due to shielding/deshielding are generally very small. The difference between the resonance frequencies of hydrogens in chloromethane compared with those in fluoromethane, for example, is only 360 Hz under an applied field of 7.05 T. Considering that the radio-frequency radiation used at this applied field is approximately 300 MHz, the difference in resonance frequencies between these two sets of hydrogens is only slightly greater than 1 part per million (1 ppm) compared with the irradiating frequency.

$$\frac{360 \text{ Hz}}{300 \times 10^6 \text{ Hz}} = \frac{1.2}{10^6} = 1.2 \text{ ppm}$$

It is customary to measure the resonance frequencies of individual nuclei relative to the resonance frequency of nuclei in a reference compound. The reference compound now universally accepted for ^{1}H-NMR and ^{13}C-NMR spectroscopy is **tetramethylsilane (TMS).**

$$\begin{array}{c} CH_3 \\ | \\ H_3C-Si-CH_3 \\ | \\ CH_3 \end{array}$$

Tetramethylsilane (TMS)

When the ^{1}H-NMR spectrum of a compound is determined, the resonance signals of its hydrogens are reported by how far they are shifted from the resonance signal of the 12 equivalent hydrogens in TMS. When a ^{13}C-NMR spectrum is determined, the resonance frequencies of its carbons are reported by how far they are shifted from the resonance signal of the four equivalent carbons in TMS.

To standardize reporting of NMR data, workers have adopted the parameter called chemical shift. **Chemical shift (δ),** in parts per million (ppm), is the frequency shift from TMS divided by the operating frequency of the spectrometer.

Chemical shift, δ The shift in parts per million (ppm) of an NMR signal from the signal of TMS.

$$\delta = \frac{\text{shift in frequency from TMS (Hz)}}{\text{frequency of spectrometer (Hz)}}$$

Thus, by definition, chemical shift is independent of the operating frequency of the spectrometer. On chart paper used to record NMR spectra, chemical shift (δ) values are shown in increasing order to the left of the TMS signal.

13.6 An NMR Spectrometer

The essential elements of an NMR spectrometer are a powerful magnet, a radio-frequency generator, a radio-frequency detector, and a sample tube (Figure 13.5).

The sample is dissolved in a solvent, most commonly carbon tetrachloride (CCl_4), deuteriochloroform ($CDCl_3$), or deuterium oxide (D_2O), which have little or no NMR activity in the region being analyzed. The sample cell is a small glass tube suspended in the magnetic field and set spinning on its long axis to ensure that all parts of the sample experience a homogeneous applied field.

Modern Fourier transform NMR (FT-NMR) spectrometers operate in the following way. The magnetic field is held constant, and the sample is irradiated with a short pulse (approximately 10^{-5} s) of radio-frequency energy that flips the spins of all susceptible nuclei simultaneously. The process by which each nucleus returns to its equilibrium state produces a sine wave emission at the frequency of its resonance. The intensity of the sine wave decays with time and falls to zero as nuclei resonating at that frequency return to their equilibrium state. A computer records this intensity-versus-time information and then uses a mathematical process called Fourier transformation (FT) to convert it to intensity-versus-frequency information. An FT-NMR spectrum can be recorded in about 2 seconds. A particular advantage of FT-NMR spectroscopy is that a large number of spectra (as many as several thousand per sample) can be recorded and digitally summed to give a time-averaged spectrum. Instrumental electronic noise is random and partially cancels out when spectra are time-averaged, but sample signals accumulate and become much stronger relative to the noise than those from a single spectrum.

All NMR spectra shown in this text were recorded and displayed using FT techniques. All ^{1}H-NMR spectra were recorded at an applied magnetic field strength of 7.05 T using a radio frequency of 300 MHz. All ^{13}C-NMR spectra were recorded at 7.05 T and 75 MHz, except as noted.

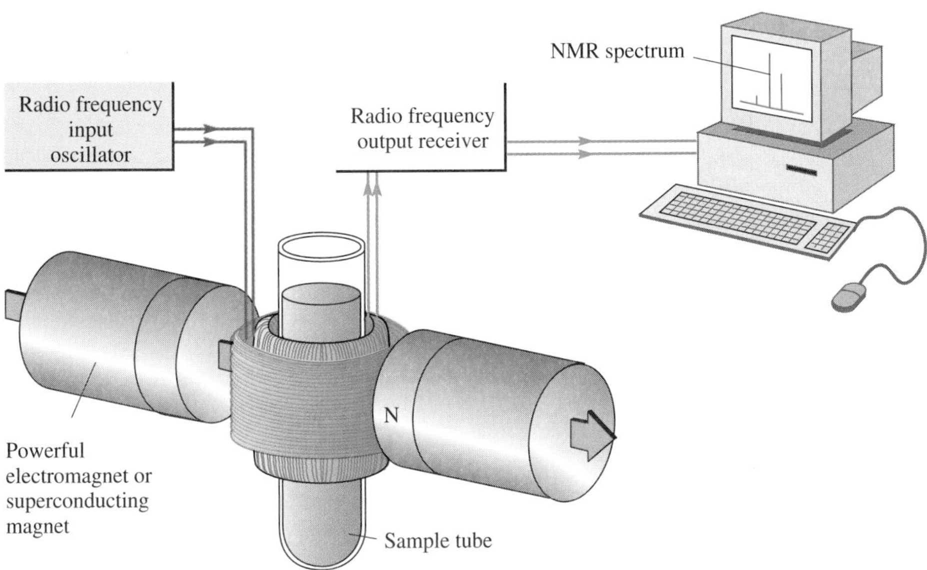

FIGURE 13.5
Schematic diagram of a nuclear magnetic resonance spectrometer.

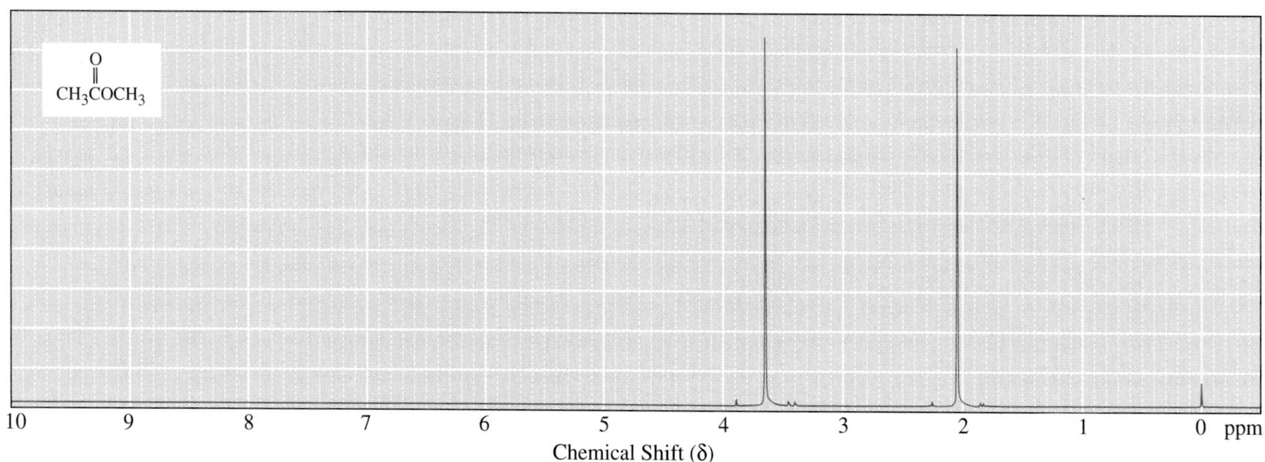

FIGURE 13.6
^{1}H-NMR spectrum of methyl acetate.

Downfield The shift of an NMR signal to the left on the chart paper. A downfield shift results when a nucleus is deshielded and its signal is produced by a weaker applied field.

Upfield The shift of an NMR signal to the right in the chart paper. An upfield shift results when a nucleus is shielded and a stronger applied field is required to produce its signal.

Figure 13.6 shows a 300 MHz ^{1}H-NMR spectrum of methyl acetate. The lower axis is calibrated in units of the delta scale, parts per million (ppm). The small signal at δ 0 is due to the hydrogens of the reference compound, TMS. The remainder of the spectrum consists of two signals: one for the three hydrogens on the methyl adjacent to oxygen and one for the three hydrogens on the methyl adjacent to the carbonyl group. It is not our purpose at the moment to determine which hydrogens give rise to which signal, but only to recognize the form in which an NMR spectrum is recorded and the origin of the calibration marks.

A note on terminology. If a signal is shifted toward the left on the chart paper, we say that it is shifted **downfield,** meaning that nuclei giving rise to that signal are more deshielded and come into resonance at a weaker applied field. Conversely, if a signal is shifted toward the right on the chart paper, we say that it is shifted **upfield,** meaning that nuclei giving rise to that signal are more shielded and come into resonance at a stronger applied field.

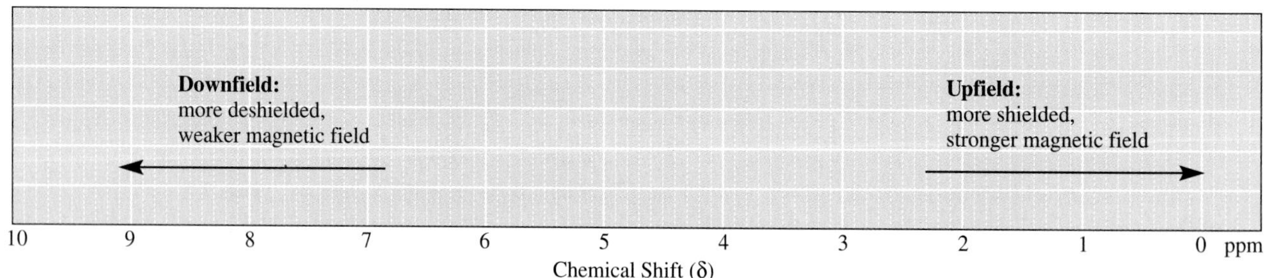

13.7 Equivalent Hydrogens

Equivalent hydrogens Hydrogens that have the same chemical environment.

Given the structural formula of a compound, how do we know how many signals to expect in its ^{1}H-NMR spectrum? The answer is that all equivalent hydrogens give the same ^{1}H-NMR signal; each set of nonequivalent hydrogens gives a different ^{1}H-NMR signal. **Equivalent hydrogens** have the same chemical environment (Section 2.3C).

Example 13.3

State the number of sets of equivalent hydrogens in each compound and the number of hydrogens in each set.

(a) 2-Methylpropane **(b)** 2-Methylbutane

Solution

(a) 2-Methylpropane contains two sets of equivalent hydrogens: a set of nine equivalent primary hydrogens and one tertiary hydrogen.

(b) 2-Methylbutane contains four sets of equivalent hydrogens. Nine primary hydrogens are in this molecule: three in one set and six in a second set. Replacement by chlorine of any hydrogen in the set of three gives 1-chloro-3-methylbutane. Replacement of any hydrogen in the set of six gives 1-chloro-2-methylbutane. In addition, the molecule contains a set of two equivalent secondary hydrogens and one tertiary hydrogen.

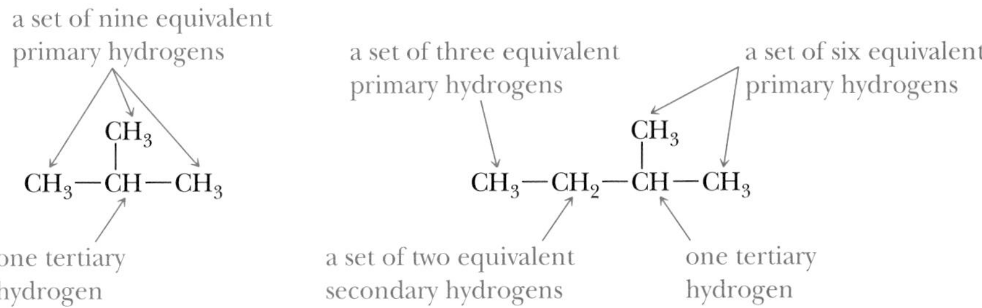

Problem 13.3

State the number of sets of equivalent hydrogens in each compound and the number of hydrogens in each set.

(a) $CH_3-CH_2-CH(CH_3)-CH_2-CH_3$ **(b)** $CH_3-CH(CH_3)-CH_2-C(CH_3)_2-CH_3$

Here are four organic compounds, each of which has one set of equivalent hydrogens and gives one signal in its ^{1}H-NMR spectrum.

$CH_3C(=O)CH_3$	$ClCH_2CH_2Cl$	(cyclopentane ring)	$(H_3C)_2C{=}C(CH_3)_2$
Propanone (Acetone)	1,2-Dichloroethane	Cyclopentane	2,3-Dimethyl-2-butene

Molecules with two or more sets of equivalent hydrogens give rise to a different resonance signal for each set. Chloroethane, for example, has a set of three equivalent hydrogens and a set of two equivalent hydrogens; two resonance signals exist in its ^{1}H-NMR spectrum.

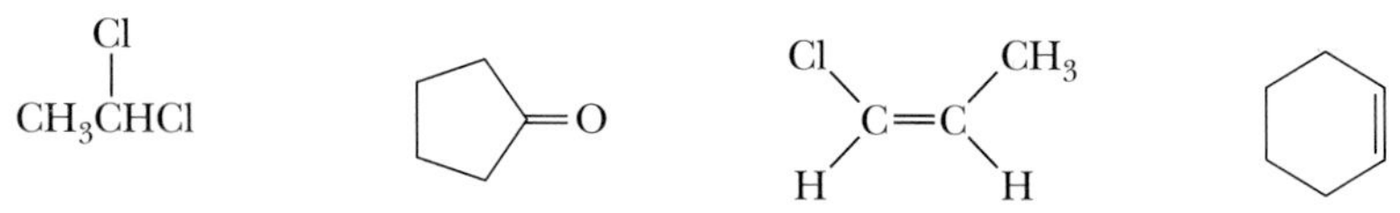

1,1-Dichloroethane (2 signals) — CH_3CHCl_2 (Cl above C); Cyclopentanone (2 signals); (Z)-1-Chloropropene (3 signals); Cyclohexene (3 signals)

You should see immediately that valuable information about molecular structure can be obtained simply by counting the number of signals in the ^{1}H-NMR spectrum of a compound. Consider, for example, the two constitutional isomers of molecular formula $C_2H_4Cl_2$. The compound 1,2-dichloroethane has one set of equivalent hydrogens and one signal in its ^{1}H-NMR spectrum. Its constitutional isomer 1,1-dichloroethane has two sets of equivalent hydrogens and two signals in its ^{1}H-NMR spectrum. Thus, simply counting signals allows you to distinguish between these two compounds.

EXAMPLE 13.4

Each compound gives only one signal in its ^{1}H-NMR spectrum. Propose a structural formula for each compound.

(a) C_2H_6O **(b)** $C_3H_6Cl_2$ **(c)** C_6H_{12} **(d)** C_4H_6

Solution

(a) CH_3OCH_3 **(b)** 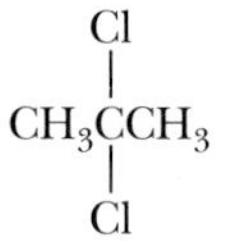 $CH_3CCl_2CH_3$ **(c)** 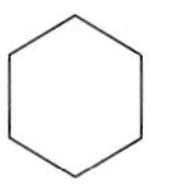 cyclohexane or $(CH_3)_2C{=}C(CH_3)_2$ **(d)** $CH_3C{\equiv}CCH_3$

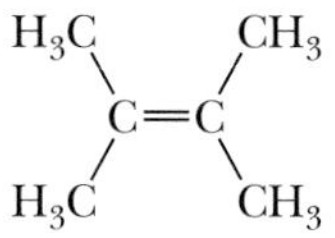

PROBLEM 13.4

Each compound gives only one signal in its ^{1}H-NMR spectrum. Propose a structural formula for each compound.

(a) C_3H_6O **(b)** C_5H_{10} **(c)** C_5H_{12} **(d)** $C_4H_6Cl_4$

13.8 Signal Areas

We have just seen that the number of signals in an ^{1}H-NMR spectrum gives us information about the number of sets of equivalent hydrogens. The relative areas of these signals provide additional information in the following way. As a spectrum is being run, the instrument's computer numerically adds the area under each signal. In the spectra shown in this text, this information is displayed in the form of a **line of integration** superposed on the original spectrum. The vertical rise of the line of integration over each signal is proportional to the area under that signal, which, in turn, is proportional to the number of hydrogens giving rise to that signal. Figure 13.7 shows an integrated ^{1}H-NMR spectrum of *tert*-butyl acetate, $C_6H_{12}O_2$. The spectrum shows signals at δ 1.44 and 1.95. The integrated signal heights are 23 plus 67, or 90, chart divisions, which correspond to 12 hydrogens. From these numbers we calculate that $23/90 \times 12$ or 3 hydrogens are in one set, and $67/90 \times 12$ or 9 hydrogens are in the second set.

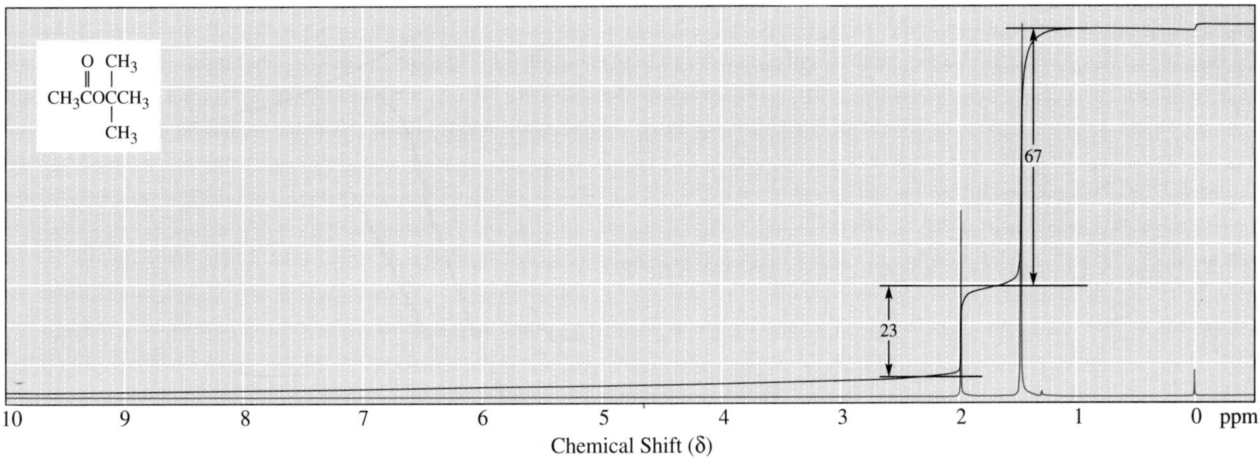

FIGURE 13.7
^{1}H-NMR spectrum of *tert*-butyl acetate showing a line of integration. The total vertical rise of 90 chart divisions corresponds to 12 hydrogens, 9 in one set and 3 in the other.

EXAMPLE 13.5

Following is an ^{1}H-NMR spectrum for a compound of molecular formula $C_9H_{10}O_2$. From an analysis of the integration line, calculate the number of hydrogens giving rise to each signal.

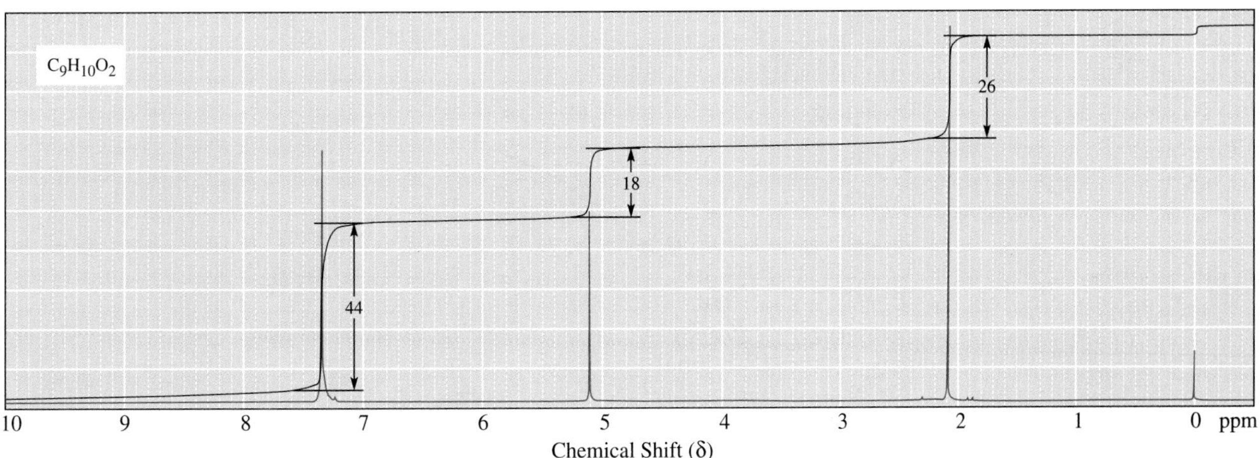

Solution

The total vertical rise in the line of integration is 88 chart divisions and corresponds to 10 hydrogens. From these numbers, we calculate that $44/88 \times 10$, or 5, of the hydrogens give rise to the signal at δ 7.34. By similar calculations, the signals at δ 5.08 and 2.06 correspond to two hydrogens and three hydrogens, respectively.

PROBLEM 13.5

The line of integration of the two signals in the ^{1}H-NMR spectrum of a ketone of molecular formula $C_7H_{14}O$ rises 62 and 10 chart divisions, respectively. Calculate the number of hydrogens giving rise to each signal, and propose a structural formula for this ketone.

13.9 Chemical Shift

The chemical shift for a signal in an ^{1}H-NMR spectrum can give valuable information about the type of hydrogen giving rise to that absorption. Hydrogens on methyl groups bonded to sp^3 hybridized carbons, for example, give signals near δ 1.0 (compare Figure 13.7). Hydrogens on methyl groups bonded to a carbonyl carbon give signals near δ 2.1–2.3 (compare Figures 13.6 and 13.7), and hydrogens on a methyl group bonded to oxygen give signals near δ 3.4–3.8 (Figure 13.6). Tabulated in Table 13.5 are average chemical shifts for most of the types of hydrogens we deal with in this course. Notice that most of these values fall within a rather narrow range of 0–8 δ units (ppm).

TABLE 13.5 Average Values of Chemical Shifts of Representative Types of Hydrogens

Type of Hydrogen (R = alkyl, Ar = aryl)	Chemical Shift (δ)*	Type of Hydrogen (R = alkyl, Ar = aryl)	Chemical Shift (δ)*
$(CH_3)_4Si$	0 (by definition)	$RC(=O)OCH_3$	3.7–3.9
RCH_3	0.9		
RCH_2R	1.2–1.4	$RC(=O)OCH_2R$	4.1–4.7
R_3CH	1.4–1.7		
$R_2C{=}CRCHR_2$	1.6–2.6	RCH_2I	3.1–3.3
$RC{\equiv}CH$	2.0–3.0	RCH_2Br	3.4–3.6
$ArCH_3$	2.2–2.5	RCH_2Cl	3.6–3.8
$ArCH_2R$	2.3–2.8	RCH_2F	4.4–4.5
ROH	0.5–6.0		
RCH_2OH	3.4–4.0	$R_2C{=}CH_2$	4.6–5.0
RCH_2OR	3.3–4.0	$R_2C{=}CHR$	5.0–5.7
R_2NH	0.5–5.0	ArH	6.5–8.5
$RC(=O)CH_3$	2.1–2.3	$RC(=O)H$	9.5–10.1
$RC(=O)CH_2R$	2.2–2.6	$RC(=O)OH$	10–13

* Values are approximate. Other atoms within the molecule may cause the signal to appear outside these ranges.

EXAMPLE 13.6

Following are structural formulas for two constitutional isomers of molecular formula $C_6H_{12}O_2$.

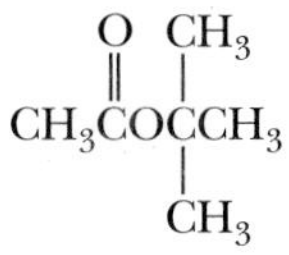

$$\underset{(1)}{CH_3\overset{O}{\overset{\|}{C}}O\underset{CH_3}{\underset{|}{\overset{CH_3}{\overset{|}{C}}}}CH_3} \qquad \underset{(2)}{CH_3O\overset{O}{\overset{\|}{C}}-\underset{CH_3}{\underset{|}{\overset{CH_3}{\overset{|}{C}}}}CH_3}$$

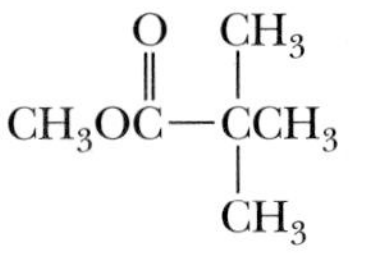

(a) Predict the number of signals in the ^{1}H-NMR spectrum of each isomer.
(b) Predict the ratio of areas of the signals in each spectrum.
(c) Show how you can distinguish between these isomers on the basis of chemical shift.

Solution

The ^{1}H-NMR spectrum of each consists of two signals in the ratio 9:3, or 3:1. Distinguish between these constitutional isomers by the chemical shift of the single $-CH_3$ group. The hydrogens of CH_3O are deshielded (appear farther downfield) than the hydrogens of $CH_3C{=}O$. See Table 13.5 for approximate values for each chemical shift. Experimentally determined values are:

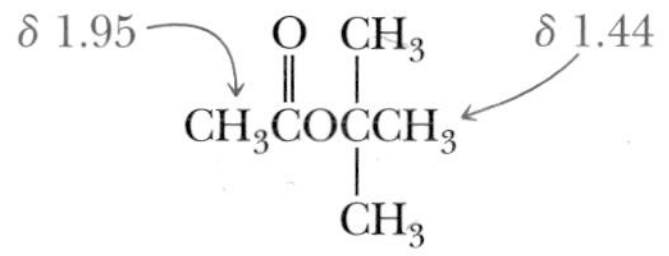

$$\underset{(1)}{\overset{\delta\ 1.95}{CH_3}\overset{O}{\overset{\|}{C}}O\underset{CH_3}{\underset{|}{\overset{CH_3}{\overset{|}{C}}}}\overset{\delta\ 1.44}{CH_3}} \qquad \underset{(2)}{\overset{\delta\ 3.67}{CH_3}O\overset{O}{\overset{\|}{C}}-\underset{CH_3}{\underset{|}{\overset{CH_3}{\overset{|}{C}}}}\overset{\delta\ 1.20}{CH_3}}$$

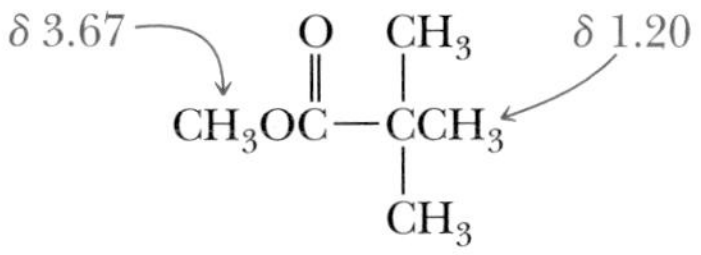

PROBLEM 13.6

Following are two constitutional isomers of molecular formula C_4H_8O.

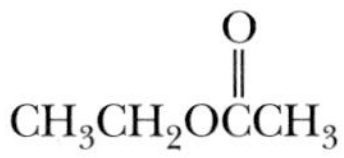

$$\underset{(1)}{CH_3CH_2O\overset{O}{\overset{\|}{C}}CH_3} \qquad \underset{(2)}{CH_3CH_2\overset{O}{\overset{\|}{C}}OCH_3}$$

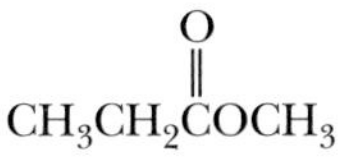

(a) Predict the number of signals in the ^{1}H-NMR spectrum of each isomer.
(b) Predict the ratio of areas of the signals in each spectrum.
(c) Show how you can distinguish between these isomers on the basis of chemical shift.

The chemical shift of a particular type of hydrogen depends primarily on three factors: (1) the electronegativity of nearby atoms, (2) the hybridization of the adjacent atoms, and (3) magnetic induction within an adjacent pi bond. Let us consider these one at a time.

TABLE 13.6 Dependence of Chemical Shift of CH_3X on the Electronegativity of X

CH_3—X	Electronegativity of X	Chemical Shift (δ) of Methyl Hydrogens
CH_3F	4.0	4.26
CH_3OH	3.5	3.47
CH_3Cl	3.1	3.05
CH_3Br	2.8	2.68
CH_3I	2.5	2.16
$(CH_3)_4C$	2.1	0.86
$(CH_3)_4Si$	1.8	0.00 (by definition)

A. Electronegativity of Nearby Atoms

As illustrated in Table 13.6 for the chemical shift of methyl hydrogens in the series CH_3—X, the greater the electronegativity of X, the greater the chemical shift. The effect of an electronegative substituent falls off quickly with distance. The effect of an electronegative substituent two atoms away is only about 10% of that when it is on the adjacent atom. The effect of an electronegative substituent three atoms away is almost negligible.

How are electronegativity and chemical shift related? The answer starts with the fact that an applied magnetic field also affects the electrons around each hydrogen and creates local **induced magnetic fields** (Section 13.5) which oppose the applied field and thereby shield nearby hydrogens. The presence of an electronegative atom reduces electron density and its shielding effect on nearby hydrogens. In effect, the electronegative atom deshields nearby hydrogens and causes them to resonate farther downfield, with a larger chemical shift.

B. Hybridization of Adjacent Atoms

Hydrogens attached to an sp^3 hybridized carbon typically absorb at δ 0.9 to 1.7. Vinylic hydrogens (those on a carbon of a carbon-carbon double bond) are considerably deshielded and resonate at δ 4.6 to 5.7 (Table 13.7). Part of the explanation for the greater deshielding of vinylic hydrogens compared with alkyl hydrogens lies in the state of hybridization of carbon. Because a sigma bonding orbital of an sp^2 hybridized carbon has more s character than a sigma bonding orbital of an sp^3 hybridized carbon (33% compared with 25%), an sp^2 hybridized carbon atom is more electronegative. Vinylic hydrogens are deshielded by this electronegativity effect and resonate farther downfield (at a weaker applied field and with a larger chemical shift) relative to alkyl hydrogens. Similarly, acetylene and aldehyde hydrogens also appear farther downfield compared with alkyl hydrogens.

Differences in chemical shifts of vinylic and acetylenic hydrogens cannot be accounted for on the basis of the hybridization of carbon alone. If the chemical shift of

TABLE 13.7 The Effect of Hybridization on Chemical Shift

Type of Hydrogen (R = alkyl)	Name of Hydrogen	Chemical Shift (δ)
RCH_3, R_2CH_2, R_3CH	alkyl	0.8–1.7
$R_2C{=}C(R)CHR_2$	allylic	1.6–2.6
$RC{\equiv}CH$	acetylenic	2.0–3.0
$R_2C{=}CHR$, $R_2C{=}CH_2$	vinylic	4.6–5.7
RCHO	aldehydic	9.5–10.1

vinylic hydrogens (δ 4.6–5.7) were due entirely to the hybridization of carbon, then the chemical shift of acetylenic hydrogens should be even greater than that of vinylic hydrogens. Yet the chemical shift of acetylenic hydrogens is only δ 2.0 to 3.0. It seems that either the chemical shift of acetylenic hydrogens is abnormally small or the chemical shift of vinylic hydrogens is abnormally large. In either case, another factor must be contributing to the magnitude of the chemical shift. Theoretical and experimental evidence (neither of which we take the time to develop here) suggest that the chemical shifts of hydrogens attached to pi-bonded carbons are influenced not only by the relative electronegativities of the sp^2 and sp-hybridized carbon atoms but also by magnetic induction in pi bonds.

C. Magnetic Induction in Pi Bonds

To understand the influence of pi bonds on the chemical shift of an acetylenic hydrogen, imagine that the carbon-carbon triple bond is oriented as shown in Figure 13.8 with respect to the applied field. In accord with the laws of magnetic induction, the applied field induces a circulation of the pi electrons, which in turn produces an induced magnetic field. Given the geometry of an alkyne and the cylindrical nature of its pi cloud, the induced magnetic field is shielding in the vicinity of the acetylenic

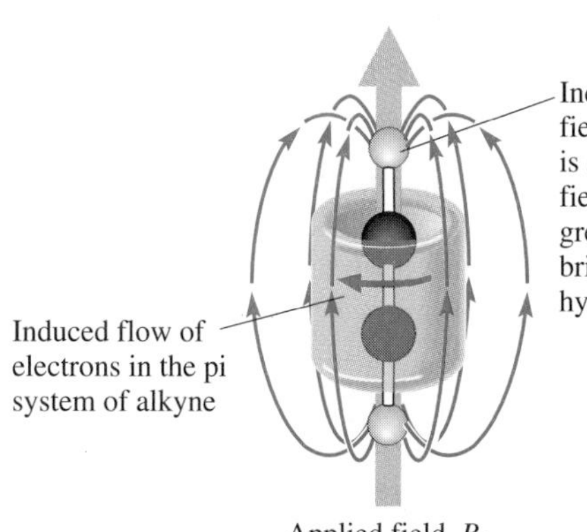
Induced local magnetic field of the pi electrons is against the applied field; it requires a greater applied field to bring an acetylenic hydrogen into resonance.
Induced flow of electrons in the pi system of alkyne
Applied field, B_0

FIGURE 13.8
A magnetic field induced in the pi bonds of a carbon-carbon triple bond shields an acetylenic hydrogen and shifts its signal upfield.

FIGURE 13.9
A magnetic field induced in the pi bond of a carbon-carbon double bond deshields vinylic hydrogens and shifts their signal downfield to a larger δ value.

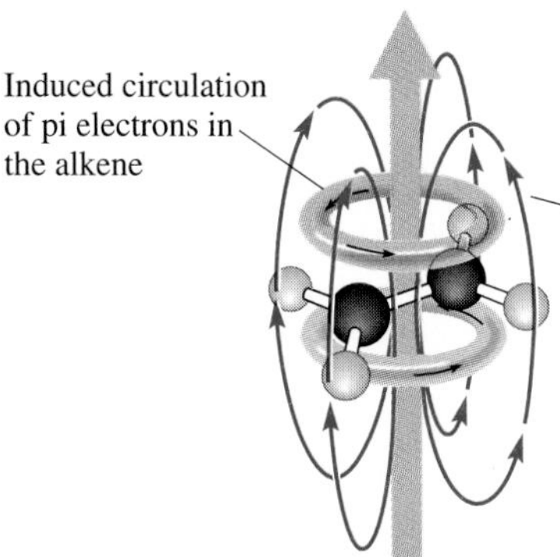
Induced circulation of pi electrons in the alkene
Induced local magnetic field of the pi electrons reinforces the applied field and provides part of the field necessary to bring a vinyl hydrogen into resonance.
Applied field, B_0

hydrogen, and, therefore, a stronger applied field is required to make an acetylenic hydrogen resonate; the local magnetic field induced in the pi bonds shifts the signal of an acetylenic hydrogen upfield to a smaller δ value.

The effect of the induced circulation of pi electrons on a vinylic hydrogen (Figure 13.9) is opposite to that on an acetylenic hydrogen. The direction of the induced magnetic field in the pi bond of a carbon-carbon double bond is parallel to the applied field in the region of the vinylic hydrogens. The induced magnetic field deshields vinylic hydrogens and, thus, shifts their signal downfield to a larger δ value. The presence of the pi electrons in the carbonyl group (Section 15.4B) has a similar effect on the chemical shift of the hydrogen of an aldehyde group.

13.10 Signal Splitting and the $(n + 1)$ Rule

We have now seen three kinds of information that can be derived from examination of an ^{1}H-NMR spectrum:

1. From the number of signals, we can determine the number of sets of equivalent hydrogens.
2. From integration of signal areas, we can determine the relative numbers of hydrogens giving rise to each signal.
3. From the chemical shift of each signal, we derive information about the types of hydrogens in each set.

A fourth kind of information can be derived from the splitting pattern of each signal. Consider, for example, the ^{1}H-NMR spectrum of 1,1-dichloroethane shown in Figure 13.10. This molecule contains two sets of hydrogens and, according to what we have learned so far, we predict two signals with relative areas 3:1 corresponding to the three hydrogens of the $—CH_3$ group and the one hydrogen of the $—CHCl_2$ group. You see from the spectrum, however, that there are in fact six **peaks.** A peak is a unit into which an NMR signal is split: two peaks in a doublet, three peaks in a triplet, and so on. The grouping of two peaks at δ 2.1 is the signal for the three hydrogens of the $—CH_3$ group, and the grouping of four peaks at δ 5.9 is the signal for the single

Peaks The units into which an NMR signal is split; doublet, triplet, quartet, etc.

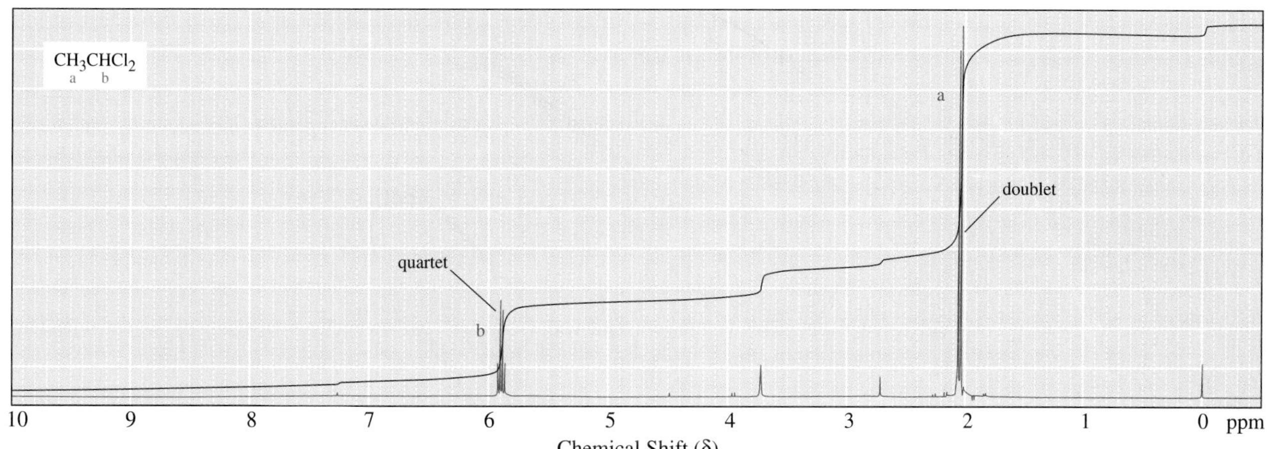

FIGURE 13.10
^{1}H-NMR spectrum of 1,1-dichloroethane.

hydrogen of the $—CHCl_2$ group. We say that the CH_3 resonance at δ 2.1 is split into a doublet and that the CH resonance at δ 5.9 is split into a quartet.

The degree of **signal splitting** can be predicted on the basis of the **$(n + 1)$ rule.** According to this rule, if a hydrogen has ***n*** hydrogens nonequivalent to it but equivalent among themselves on the same or adjacent atom(s), its ^{1}H-NMR signal is split into $(n + 1)$ peaks.

Signal splitting Splitting of an NMR signal into a set of peaks by the influence of neighboring nonequivalent nuclei.

Let us apply the $(n + 1)$ rule to the analysis of the spectrum of 1,1-dichloroethane. The three hydrogens of the $—CH_3$ group have one nonequivalent neighbor hydrogen $(n = 1)$, and, therefore, their signal is split into a doublet. The single hydrogen of the $—CHCl_2$ group has a set of three nonequivalent neighbor hydrogens $(n = 3)$, and its signal is split into a quartet.

$(n + 1)$ Rule The ^{1}H-NMR signal of a hydrogen or set of equivalent hydrogens is split into $(n + 1)$ peaks by a nonequivalent set of n equivalent neighboring hydrogens.

For these hydrogens, $n = 1$; their signal is split into $(1 + 1)$ or 2 peaks (a **doublet**)

$$CH_3—CH(Cl)—Cl$$

For this hydrogen, $n = 3$; its signal is split into $(3 + 1)$ or 4 peaks (a **quartet**)

EXAMPLE 13.7

Predict the number of signals and the splitting pattern of each signal in the ^{1}H-NMR spectrum of each molecule.

(a) $CH_3\overset{O}{\overset{\|}{C}}CH_2CH_3$ **(b)** $CH_3CH_2\overset{O}{\overset{\|}{C}}CH_2CH_3$ **(c)** $CH_3\overset{O}{\overset{\|}{C}}CH(CH_3)_2$

Solution

The sets of equivalent hydrogens in each molecule are labeled a, b, and c. Molecule (a) has three sets of equivalent hydrogens; its ^{1}H-NMR spectrum shows a singlet, a quartet, and a triplet in the ratio 3:2:3. Molecule (b) has two sets of equivalent hydrogens; its ^{1}H-NMR spectrum shows a triplet and a quartet in the ratio 3:2. Molecule (c) has three sets of equivalent hydrogens; its ^{1}H-NMR spectrum shows a singlet, a septet, and a doublet in the ratio 3:1:6.

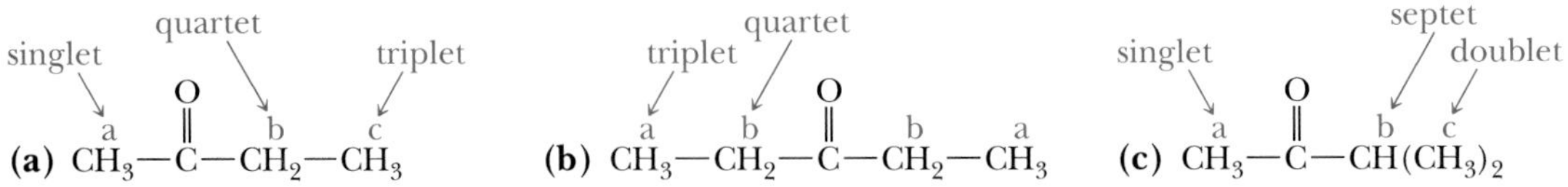

(a) $\overset{a}{CH_3}-\overset{O}{\overset{\|}{C}}-\overset{b}{CH_2}-\overset{c}{CH_3}$ (a: singlet; b: quartet; c: triplet)

(b) $\overset{a}{CH_3}-\overset{b}{CH_2}-\overset{O}{\overset{\|}{C}}-\overset{b}{CH_2}-\overset{a}{CH_3}$ (a: triplet; b: quartet)

(c) $\overset{a}{CH_3}-\overset{O}{\overset{\|}{C}}-\overset{b}{CH}(\overset{c}{CH_3})_2$ (a: singlet; b: septet; c: doublet)

PROBLEM 13.7

Following are pairs of constitutional isomers. Predict the number of signals in the ^{1}H-NMR spectrum of each isomer and the splitting pattern of each signal.

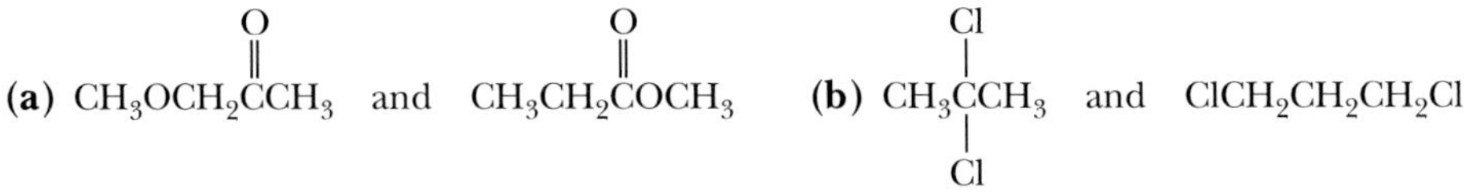

(a) $CH_3OCH_2\overset{O}{\overset{\|}{C}}CH_3$ and $CH_3CH_2\overset{O}{\overset{\|}{C}}OCH_3$ (b) $CH_3CCl_2CH_3$ and $ClCH_2CH_2CH_2Cl$

13.11 The Origins of Signal Splitting

Coupling The magnetic interaction of the nuclear spins of nearby atoms.

When the chemical shift of one nucleus is influenced by the spin of another, the two are said to be **coupled.** Consider, for example, the situation in which nonequivalent hydrogens, H_a and H_b, exist on adjacent carbons.

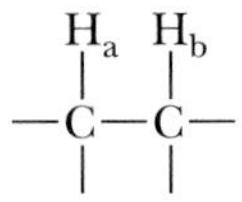

$$-\overset{H_a}{\overset{|}{\underset{|}{C}}}-\overset{H_b}{\overset{|}{\underset{|}{C}}}-$$

The chemical shift of H_a is influenced by whether the spin of H_b is aligned with the applied field or aligned against it. If the field caused by H_b adds to the applied field, then H_a absorbs at a lower applied field. If, on the other hand, the field caused by H_b subtracts from it, then H_a absorbs at a higher applied field.

As a result of spin-spin coupling between nonequivalent hydrogens H_a and H_b, the signal of hydrogen H_a is split into two peaks. Because there is an almost equal probability of H_a experiencing the $+\frac{1}{2}$ and $-\frac{1}{2}$ spin states of H_b, each peak of the doublet is of equal area and each is half the area of what the H_a signal would be if hydrogen H_b were not present (Figure 13.11). Note that spin-spin coupling and the resulting signal splitting are reciprocal; if H_a is split by H_b, then H_b is equally split by H_a. If neither of these hydrogens has any other neighbors, then their ^{1}H-NMR signals are two doublets of equal area.

Signal splitting patterns for a hydrogen with zero, one, two, and three equivalent neighbors are summarized in Table 13.8. The ratios of the areas of the peaks in these and any other splitting patterns can be derived from an analysis of spin combinations for adjacent hydrogens. With three adjacent hydrogens, for example, areas within the resulting quartet are 1:3:3:1. Alternatively, the ratio of peak areas in any multiplet can be derived from a mathematical mnemonic device called **Pascal's triangle** (Figure 13.12). A note of caution in counting the number of peaks in a multiplet: If the signal

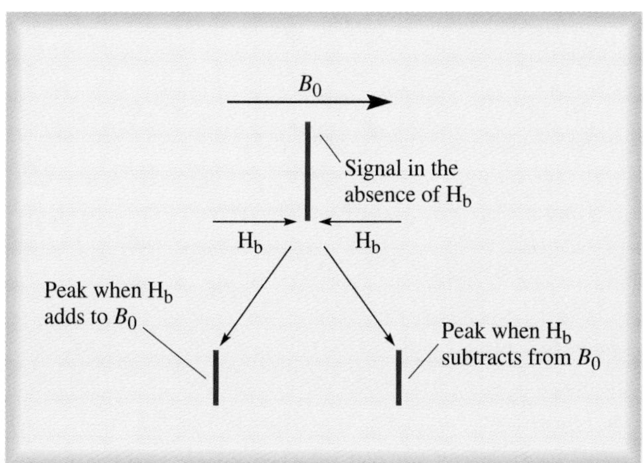

FIGURE 13.11
The signal of H_a is split into two peaks of equal area (a doublet) by its nonequivalent neighbor H_b.

of a particular hydrogen is of low intensity compared with others in the spectrum, it may not be possible to distinguish some of the smaller side peaks because of electronic noise in the baseline. It should be noted here that the nuclei of all adjacent hydrogens couple. It is only when coupling is between nonequivalent hydrogens that it results in signal splitting; coupling between equivalent hydrogens, whether they are on the same or adjacent atoms, does not result in signal splitting.

TABLE 13.8 Observed Signal Splitting Patterns for a Hydrogen with Zero, One, Two, and Three Equivalent Neighboring Hydrogens

Structure	Spin States of H_b	Signal of H_a*
H_a on C—C (no H on adjacent C)		single peak
H_a on C, H_b on adjacent C: $-CH_a-CH_b-$	↑ ↓	1 1
$-CH_a-CH_b-H_b$	↑↑ ↑↓ ↓↑ ↓↓	1 2 1
$-CH_a-CH_b(H_b)-H_b$	↑↑↑ ↑↑↓ ↑↓↑ ↓↑↑ ↑↓↓ ↓↑↓ ↓↓↑ ↓↓↓	1 3 3 1

spaced as per Pascal: triangle 2+1=3

equal spacing among the four sets in the quartet

δ H_a

* The area of integration is the same for all four signals.

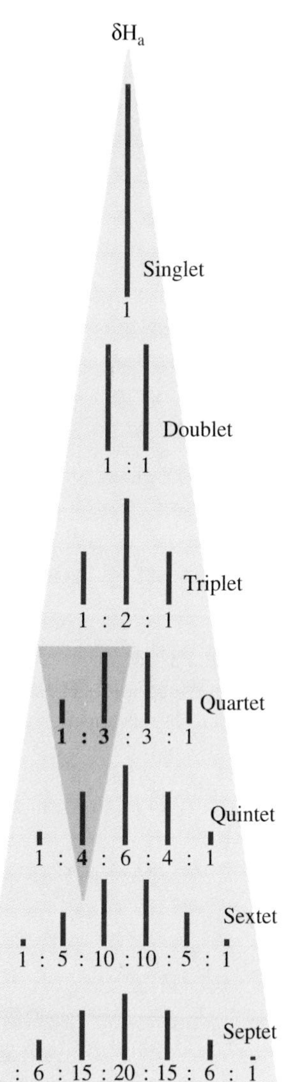

FIGURE 13.12
Pascal's triangle. As illustrated by the highlighted entries, each entry is the sum of the values immediately above it to the left and the right.

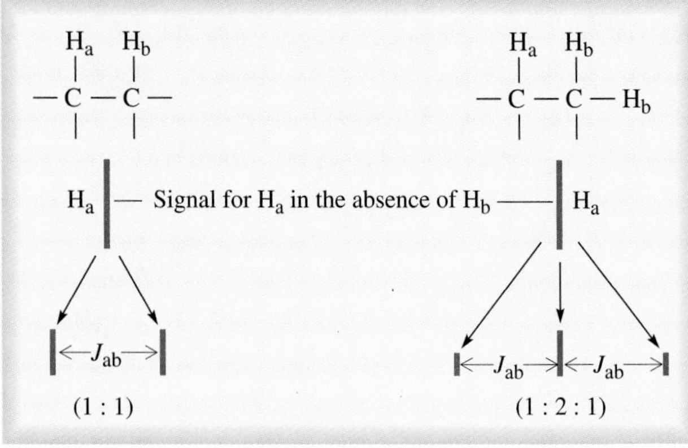

FIGURE 13.13
Measurement of the coupling constant, J_{ab}, for a doublet (ratio 1:1) and a triplet (ratio 1:2:1) derived from splitting of the H_a signal by nonequivalent hydrogen(s) H_b.

13.12 Coupling Constants (J)

Coupling constant (J) The distance between peaks in an NMR split signal expressed in hertz. The value of J is a quantitative measure of the magnetic interaction of nuclei whose spins are coupled.

A **coupling constant** (J) is the separation between adjacent peaks in a multiplet and is a quantitative measure of the shielding/deshielding influence of the magnetic moments of adjacent hydrogens. The magnitude of a coupling constant is expressed in hertz and, in ^{1}H-NMR spectroscopy, falls generally in the range 0–18 Hz. The value of J depends only on internal fields within a molecule and is independent of the external field strength and the radio-frequency radiation.

Measurements of coupling constants for hydrogens with one and two equivalent neighboring hydrogens are illustrated in Figure 13.13. The coupling constant for two hydrogens on adjacent sp^3 hybridized carbon atoms is approximately 7 Hz. In the case of older NMR spectrometers operating at 60 MHz, this value of J corresponds to approximately 0.117 ppm. For a modern NMR spectrometer operating at 300 MHz, a coupling constant of 10 Hz corresponds to only 0.023 ppm. Because peaks with this and comparable values of J are so narrowly spaced, splitting patterns from 300 MHz spectra are often very difficult to determine. In such cases, we show expansion of appropriate signals. Shown in Figure 13.14 is the ^{1}H-NMR spectrum of 3-pentanone

FIGURE 13.14
The quartet-triplet ^{1}H-NMR signals of 3-pentanone showing the original trace and a scale expansion to show the signal splitting pattern more clearly.

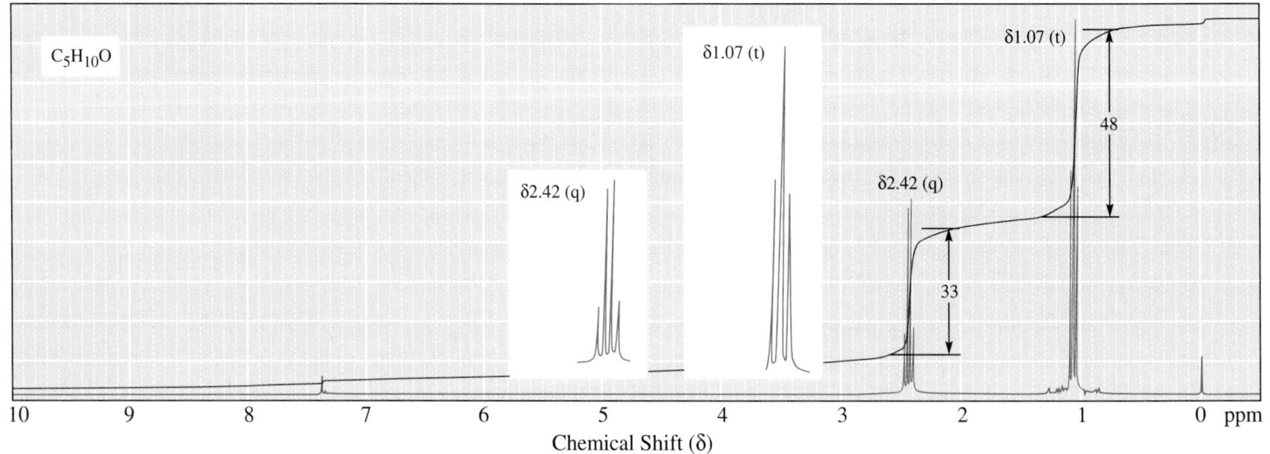

TABLE 13.9 Approximate Values of *J* for Compounds Containing Alkyl and Alkenyl Groups

H_a–C–C–H_b	cyclohexane, H_a axial / H_b axial	cyclohexane, H_a axial / H_b equatorial	cyclohexane, H_a equatorial / H_b equatorial
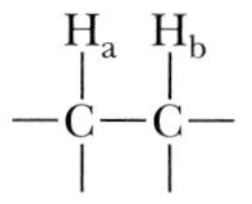 6–8 Hz	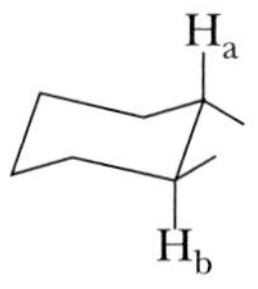 8–14 Hz	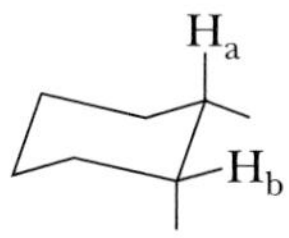 0–5 Hz	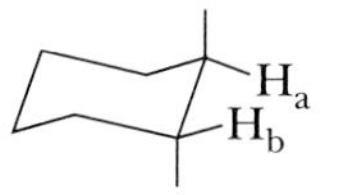 0–5 Hz

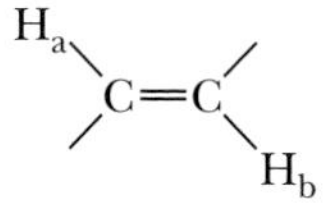 H_a–C=C–H_b (trans)	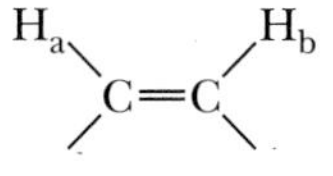 H_a–C=C–H_b (cis)	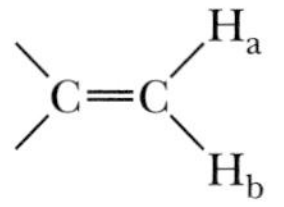 C=C(H_a)(H_b)	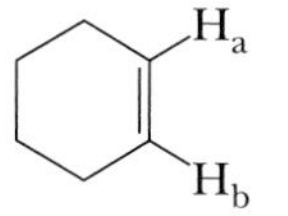 cyclohexene H_a, H_b
11–18 Hz	5–10 Hz	0–5 Hz	8–11 Hz

showing the quartet-triplet pattern of the ethyl group. Shown above the original trace of each signal is a scale expansion of each, in which you can see the triplet-quartet pattern more easily. You should note also that the types of splitting patterns we have described hold only when the separation between coupled signals is much greater than the coupling constant. When this is not the case, spectra can become much more complex.

Given in Table 13.9 are approximate values for coupling constants for hydrogens on singly bonded and doubly bonded carbons. Notice that these values are relatively small compared with the chemical shifts observed in ^{1}H-NMR spectroscopy. A coupling constant of 10 Hz at a radio frequency of 300 MHz, for example, corresponds to a chemical shift of only δ 0.033 ppm.

13.13 ^{13}C-NMR

The development of **^{13}C-NMR spectroscopy** lagged behind ^{1}H-NMR spectroscopy primarily because of two problems. One is the particularly low natural abundance of ^{13}C (only 1.1%) and the resulting weak signal. The second problem is that the magnetic moment of ^{13}C is considerably smaller than that of ^{1}H, which causes the populations of the higher and lower energy states to differ much less than for ^{1}H. Taken in combination, these two factors mean that ^{13}C-NMR signals in natural samples (those not artificially enriched with carbon-13) are only about 10^{-4} times the strength of ^{1}H-NMR signals. Whereas ^{1}H-NMR spectroscopy became a routine analytical tool in the mid-1960s, it was not until 20 years later with the development of FT-NMR techniques that ^{13}C-NMR spectroscopy became widely available as a routine analytical tool.

As with ^{1}H-NMR spectra, splitting patterns in ^{13}C-NMR spectra are also explained according to the $(n + 1)$ rule. Since in natural abundance, only 1.1% of carbon atoms are ^{13}C, almost all ^{13}C atoms in a molecule have only magnetically inactive ^{12}C next to them and, therefore, ^{13}C-^{13}C signal splitting is not normally observed. The signal from a ^{13}C nucleus is split, however, by ^{1}H bonded to it. Whereas signals of hydrogens are split by nonequivalent hydrogens on adjacent atoms, signals of carbon-13 are split

CHEMISTRY IN ACTION

Magnetic Resonance Imaging

The NMR phenomenon was discovered and explained by physicists in the 1950s; by the 1960s, it had become an invaluable analytical tool for chemists. It was realized by the early 1970s that imaging of parts of the body using NMR could be a valuable addition to diagnostic medicine. Because the term "nuclear magnetic resonance" sounds to many people as if the technique might involve radioactive material, health care personnel call the technique "magnetic resonance imaging (MRI)."

The body contains several nuclei that, in principle, could be used for MRI. Of these, hydrogens, most of which come from water, triglycerides (Section 25.2), and membrane phospholipids (Section 25.5), give the most useful signals. Phosphorus MRI is also used in diagnostic medicine.

Recall that in NMR spectroscopy, energy in the form of radio-frequency radiation is absorbed by nuclei in the sample. The relaxation time is the characteristic time at which excited nuclei give up this energy and relax to their ground state.

In 1971, it was discovered that relaxation of water in certain cancerous tumors takes much longer than the relaxation of water in normal cells. Thus, if a relaxation image of the body could be obtained, it might be possible to identify tumors at an early stage. Subsequent work demonstrated that many tumors can be identified in this way. Another important application of MRI is in the examination of the brain and spinal cord. White and gray matter are easily distinguished

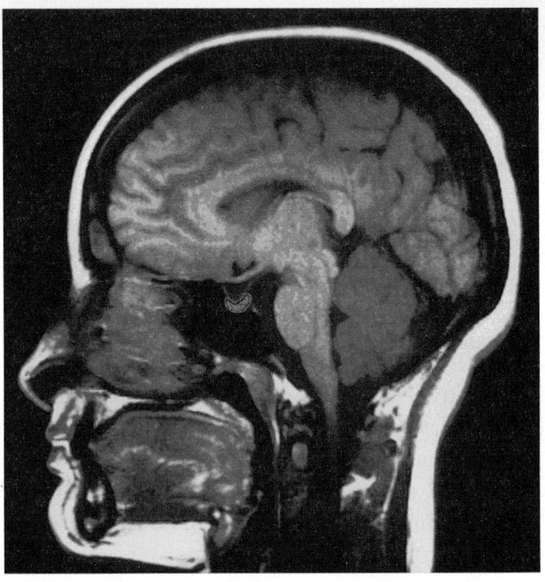

Computer-enhanced MRI scan of a normal human brain with pituitary gland highlighted. (Scott Camazine/Photo Researchers)

by the hydrogens attached directly to them. The signal for an atom of carbon-13 with three attached hydrogens is split to a quartet, the signal for an atom of carbon-13 with two attached hydrogens is split to a triplet, and so on. The value of ^{13}C-H signal splitting is that it provides important information about the number of hydrogen atoms bonded to carbon. The disadvantage of ^{13}C-H signal splitting is that coupling constants of 100–250 Hz are common. Coupling constants of this magnitude correspond to 1.33–3.33 ppm at 75 MHz, which means that there can be significant overlap between signals and splitting patterns are very often difficult to determine. For this reason, the most common mode of operation of a ^{13}C-NMR spectrometer is a hydrogen-decoupled mode. (See Problem 13.30 for an interesting problem on the use of coupling constants to determine orbital hybridization.)

In the **hydrogen-decoupled mode,** the sample is irradiated with two different radio frequencies. The first radio frequency is used to excite ^{13}C nuclei. The second radio frequency is a broad spectrum of frequencies that causes all hydrogens in the

Magnetic resonance imaging is a useful medical diagnostic tool. *(Paul Shambroom/Science Source/Photo Researchers, Inc.)*

by MRI, which is useful in the study of such diseases as multiple sclerosis. Magnetic resonance imaging and x-ray imaging are in many cases complementary. The hard, outer layer of bone is essentially invisible to MRI, but shows up extremely well in x-ray images, whereas soft tissue is nearly transparent to x-rays but shows up in MRI.

The key to any medical imaging technique is to know which part of the body gives rise to which signal. In MRI, this type of spatial information is encoded using magnetic field gradients. Recall that a linear relationship exists between the frequency at which a nucleus resonates and the intensity of the magnetic field. In ^{1}H-NMR spectroscopy, we use a homogeneous magnetic field, in which all equivalent hydrogens absorb at the same radio frequency and have the same chemical shift. In MRI, the patient is placed in a magnetic field gradient that can be varied from place to place. Nuclei in a weaker magnetic field absorb at a lower frequency. Nuclei elsewhere in a stronger magnetic field absorb at a higher frequency. In a magnetic field gradient, a correlation exists between the absorption frequency of a nucleus and its position in space. A gradient along a single axis images a plane. Two mutually perpendicular gradients image a line segment, and three mutually perpendicular gradients image a point. In practice, more complicated procedures are used to obtain magnetic resonance images, but they are all based on the idea of magnetic field gradients.

molecule to undergo rapid transitions between their nuclear spin states. On the time scale of a ^{13}C-NMR spectrum, each hydrogen is in an average or effectively constant nuclear spin state with the result that ^{1}H-^{13}C spin-spin interactions are not observed; they are said to be decoupled. In a hydrogen-decoupled spectrum, all ^{13}C signals appear as singlets. The hydrogen-decoupled ^{13}C-NMR spectrum of 1-bromobutane (Figure 13.15) consists of four singlets.

Table 13.10 shows approximate chemical shifts for carbon-13 spectroscopy. Notice how much wider the range of chemical shifts is for ^{13}C-NMR spectroscopy than for ^{1}H-NMR spectroscopy. Whereas most chemical shifts for ^{1}H-NMR spectroscopy fall within a rather narrow range of 0–10 ppm, those for ^{13}C-NMR spectroscopy cover 0–220 ppm. Because of this expanded scale, it is very unusual to find any two nonequivalent carbons in the same molecule with identical chemical shifts. Most commonly, each different type of carbon within a molecule has a distinct signal clearly resolved from all other signals.

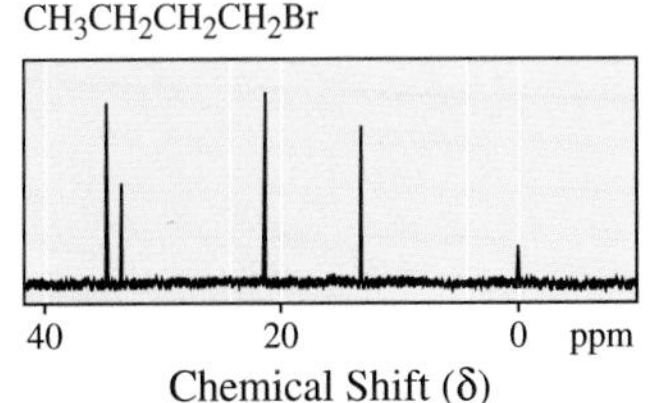

FIGURE 13.15 Hydrogen-decoupled ^{13}C-NMR spectrum of 1-bromobutane.

TABLE 13.10 ^{13}C-NMR Chemical Shifts

Type of Carbon	Chemical Shift (δ)	Type of Carbon	Chemical Shift (δ)
RCH_3	0–40	(benzene ring)C—R	110–160
RCH_2R	15–55		
R_3CH	20–60		
RCH_2I	0–40	RC(=O)OR	160–180
RCH_2Br	25–65		
RCH_2Cl	35–80	$RC(=O)NR_2$	165–180
R_3COH	40–80		
R_3COR	40–80	RC(=O)OH	175–185
RC≡CR	65–85		
$R_2C{=}CR_2$	100–150	RC(=O)H, RC(=O)R	180–210

Notice further that the chemical shift of carbonyl carbons is quite distinct from those of sp^3 hybridized carbons and of other types of sp^2 hybridized carbons. The presence or absence of a carbonyl carbon is quite easy to recognize in a ^{13}C-NMR spectrum. Note also that signals from sp^2 hybridized carbons fall in a distinctive range of 100–160 ppm.

A great advantage of ^{13}C-NMR spectroscopy is that it generally makes it possible to count the number of types of carbon in a molecule. One caution here, however. Because of certain complications, including the long relaxation times of spin-flipped ^{13}C nuclei, it is generally not possible to determine the number of carbons of each type by integration of signal areas.

EXAMPLE 13.8

Predict the number of signals in the proton-decoupled ^{13}C-NMR spectrum of each compound.

(a) $CH_3C(=O)OCH_3$ **(b)** $CH_3CH_2CH_2C(=O)CH_3$ **(c)** $CH_3CH_2C(=O)CH_2CH_3$

Solution

Following are the number of signals in the proton-decoupled spectrum of each compound, along with the chemical shifts of each signal. The chemical shifts of the carbonyl carbons are quite distinctive (Table 13.10) and in these examples occur at δ 171.37, 208.85, and 211.97.

(a) Methyl acetate: three signals (δ 171.37, 51.53, and 20.63)
(b) 2-Pentanone: five signals (δ 208.85, 45.68, 29.79, 17.35, and 13.68)
(c) 3-Pentanone: three signals (δ 211.97, 35.45, and 7.92)

PROBLEM 13.8

Explain how to distinguish between the members of each pair of constitutional isomers based on the number of signals in the proton-decoupled ^{13}C-NMR spectrum of each member.

(a) 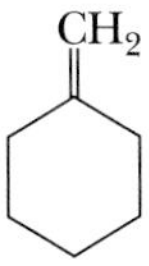 [methylenecyclohexane, CH_2] and [1-methylcyclohexene, CH_3] **(b)** 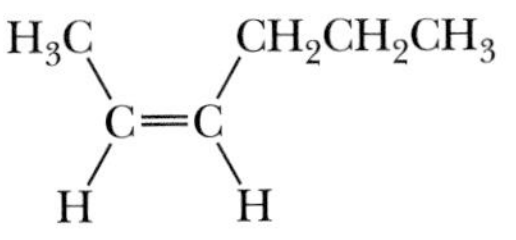 $H_3C(H)C{=}C(H)CH_2CH_2CH_3$ and $CH_3CH_2(H)C{=}C(H)CH_2CH_3$

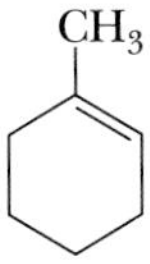

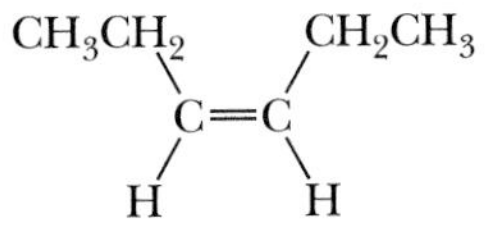

13.14 The DEPT Method

We saw in Section 13.13 that there is spin-spin coupling between a ^{13}C atom and its attached hydrogens but, because coupling constants are large and overlap of peaks is considerable, proton-coupled ^{13}C-NMR spectra are often very difficult to interpret. For these reasons, ^{13}C-NMR spectra are commonly run in the proton-decoupled mode, in which case information on C/H ratios is lost. DEPT-NMR spectroscopy, one of the more advanced techniques in NMR spectroscopy, provides a way to acquire this information. **Distortionless enhancement by polarization transfer,** or **DEPT** as it is more commonly known, uses complex sequences of pulses in both the ^{1}H and ^{13}C resonance ranges with the result that the ^{13}C signals for CH_3, CH_2, and CH exhibit different phases. Signals for CH_3 and CH carbons are recorded as positive signals and those for CH_2 carbons are recorded as negative signals. Using a slightly different pulse sequence, CH_3 and CH signals can be distinguished. In the DEPT technique, a carbon with no attached hydrogens, as, for example, a carbonyl carbon or a quaternary carbon, gives no signal.

DEPT-NMR Distortionless Enhancement by Polarization Transfer. A spectroscopic technique for distinguishing among ^{13}C signals for CH_3, CH_2, CH, and quaternary carbons.

DEPT spectra may be run in several ways. In one variation, a first trace records only CH_3 signals (as positive signals), a second trace records only CH_2 signals (as negative signals), and a third trace records only CH signals (as positive signals). In another variation, CH_3, CH_2, and CH signals are recorded on one spectrum. The first trace shows CH_3 and CH as positive signals and the second trace shows CH_2 as negative signals. Shown in Figure 13.16(a) (page 486) is a proton-decoupled ^{13}C-NMR spectrum of isopentyl acetate showing six signals. Figure 13.16(b) is a DEPT spectrum showing color-coded signals corresponding to CH_3, CH_2, and CH groups. Note that the carbonyl carbon appears in the proton-decoupled spectrum but does not appear in the DEPT spectrum.

CHEMISTRY IN ACTION

^{31}P-NMR Spectroscopy as a pH Meter

We have seen that NMR spectra can be obtained from ^{1}H and ^{13}C nuclei. ^{31}P (Table 13.4) is another nucleus that provides useful spectroscopic information. The chemical shift for ^{31}P in phosphate ion depends on the pH of its environment. At or near neutral pH, phosphate ion is an equilibrium mixture of $H_2PO_4^-$ and HPO_4^{2-}. $H_2PO_4^-$ is a weak acid and ionizes according to the following equilibrium equation. The acid ionization constant K_a for this equilibrium is 6.2×10^{-8}. The chemical shift of ^{31}P in HPO_4^{2-} is 21.05, whereas its chemical shift in $H_2PO_4^-$ is 24.90.

$$\underset{\delta\ 24.90}{H_2PO_4^-} + H_2O \rightleftharpoons \underset{\delta\ 21.05}{HPO_4^{2-}} + H_3O^+$$

$$K_a = \frac{[HPO_4^{2-}][H_3O^+]}{[H_2PO_4^-]} = 6.2 \times 10^{-8}$$

Solving the K_a expression for hydrogen ion concentration and then taking the negative logarithm of both sides of this equation gives an equation that relates pH directly to the ratio of the molar concentrations of $H_2PO_4^-$ and HPO_4^{2-}.

$$pH = 7.21 - \log\frac{[H_2PO_4^-]}{[HPO_4^{2-}]}$$

The equilibrium between $H_2PO_4^-$ and HPO_4^{2-} is very fast. Because a ^{31}P-NMR spectrum is like a camera with a slow shutter speed, only one signal, the chemical shift of which is a mole-weighted average of the chemical shifts of $H_2PO_4^-$ and HPO_4^{2-}, is observed for phosphate. The value of the ^{31}P chemical shift indicates what the $[H_2PO_4^-]/[HPO_4^{2-}]$ is, which allows calculation of the pH by the above equation. In other words, measuring the chemical shift of ^{31}P is equivalent to measuring the pH of the phosphate's environment.

Thus, ^{31}P-NMR provides a noninvasive way to measure the pH of regions that are difficult to measure, such as the inside of cells. In an early application of ^{31}P-NMR to the measurement of cellular pH, a group from Bell Laboratories* recorded ^{31}P-NMR spectra of *Escherichia coli* cells, which were supplied with glucose and either deprived of oxygen or supplied with oxygen. The type of spectra they recorded are shown in the figure.

Signal B is intracellular phosphate ion, and signal C is from phosphate ion in the surrounding fluid. Initially, the cells were under nitrogen and their interior was slightly acidic (pH 6.93). However, as oxygen is supplied, the cells are able to begin active respiration, which causes the interior to become more basic (pH 7.55 for this bacterium). An advantage of the NMR method is that, under conditions of higher sensitivity, the concentrations of other phosphorus-containing biological molecules, such as ATP, can be followed along with changes in pH.

* See G. Navon, S. Ogawa, R. G. Shulman, and T. Yamane, *Proc. Natl. Acad. Sci.*, USA, **74**, 888, 1977.

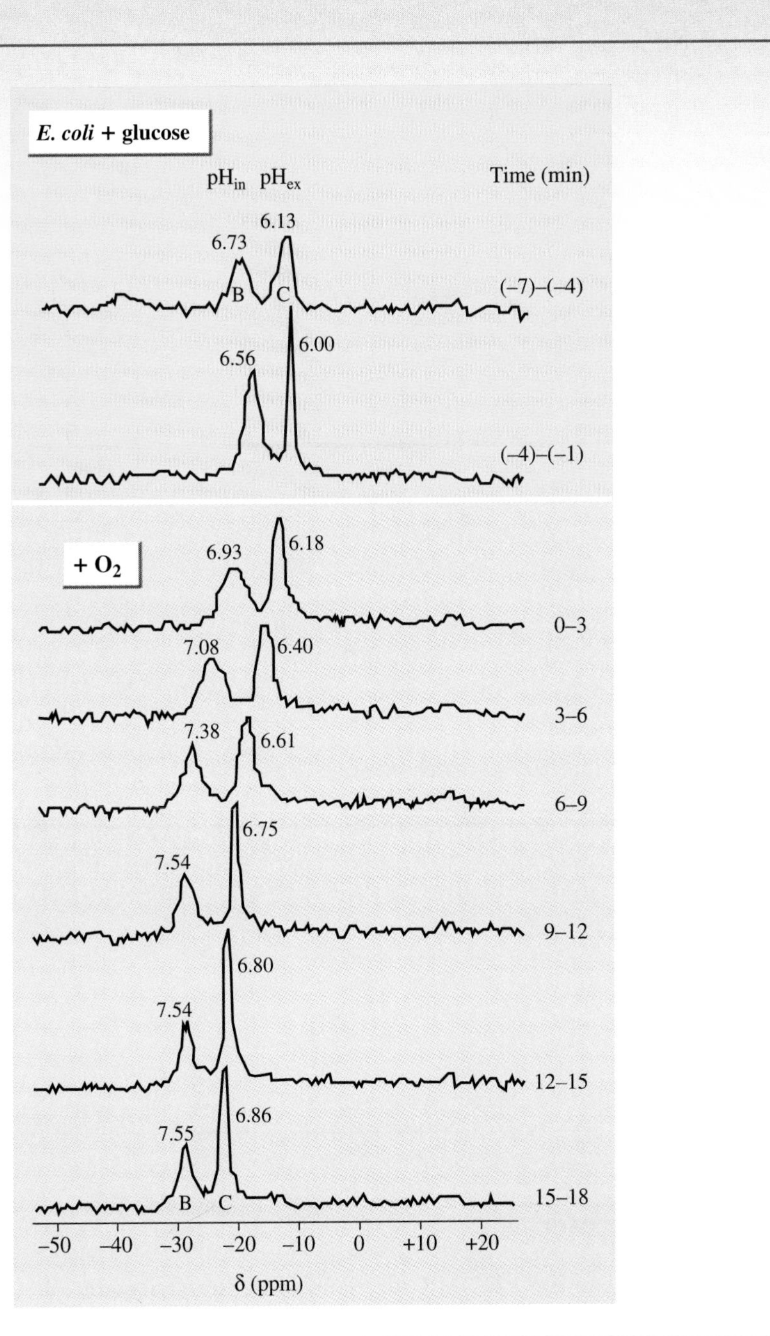
E. coli + glucose
pHin pHex
Time (min)
6.73
6.13
B
C
(−7)–(−4)
6.56
6.00
(−4)–(−1)
+ O2
6.93
6.18
0–3
7.08
6.40
3–6
7.38
6.61
6–9
6.75
7.54
9–12
6.80
7.54
12–15
6.86
7.55
B
C
15–18
−50 −40 −30 −20 −10 0 +10 +20
δ (ppm)

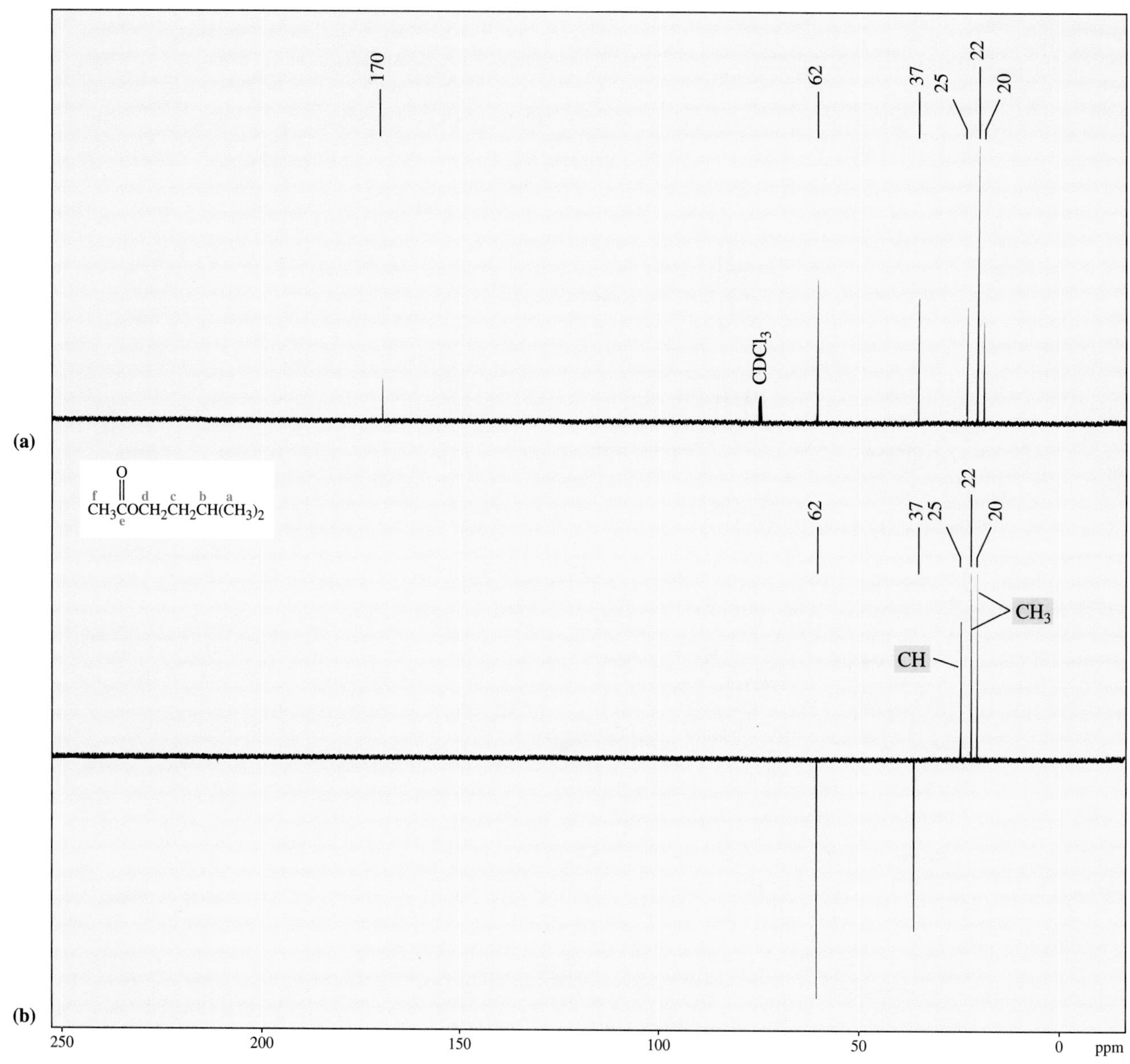

FIGURE 13.16
^{13}C-NMR spectra (400 MHz) of isopentyl acetate. (a) Proton-decoupled spectrum and (b) DEPT spectrum. (*Courtesy Robert Stackow, UCLA*)

EXAMPLE 13.9

Assign all signals in the ^{13}C-NMR spectrum of isopentyl acetate.

Solution

The positive DEPT signals at 20, 22, and 24 represent CH_3 and CH groups. The taller methyl signal at 22 represents the two equivalent methyl groups (a), and the shorter methyl signal at 20 represents the single methyl group (f). The signal at 24, the only other positive DEPT signal, represents the CH group (b). The signal at 62 represents (d), the CH_2 group nearer to and more deshielded by the adjacent oxygen atom. The signal at 37 represents (c), the other CH_2 group. The signal at 170, which is not present in the DEPT spectrum, represents the carbon (e) of the carbonyl group.

PROBLEM 13.9

Assign all signals in the ^{13}C-NMR spectrum of 4-methyl-2-pentanone.

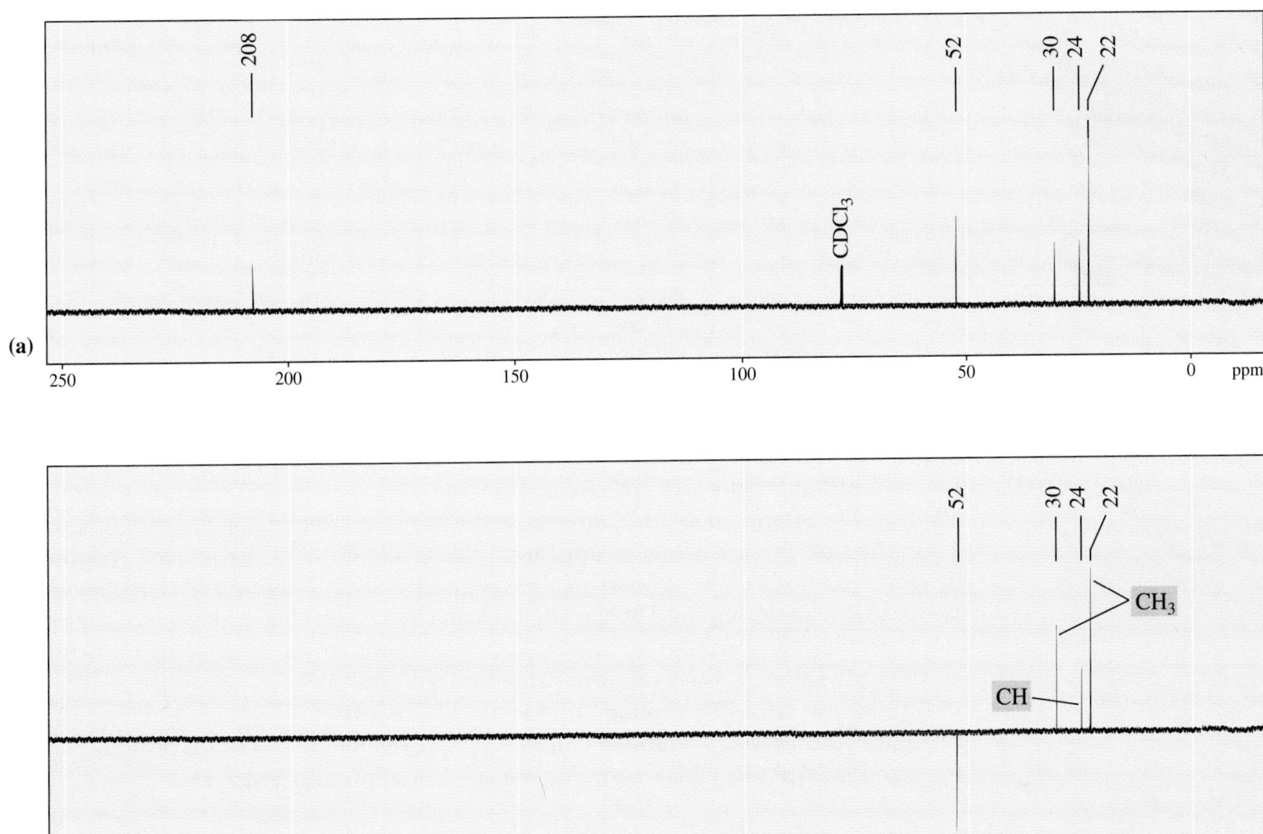

13.15 Interpretation of NMR Spectra

A. Alkanes

All hydrogens in alkanes are in very similar chemical environments, and, therefore, ^{1}H-NMR chemical shifts of alkane hydrogens fall within a narrow range of δ 0.8–1.7. Chemical shifts for alkane carbons in ^{13}C-NMR spectroscopy fall within the considerably wider range of δ 0–60.

B. Alkenes

The chemical shifts of vinylic hydrogens are larger than those of alkane hydrogens and typically fall in the range δ 4.6–5.7. Vinylic hydrogens are deshielded by the sp^2 hybridized carbons of the double bond (Section 13.9B) and the local magnetic field

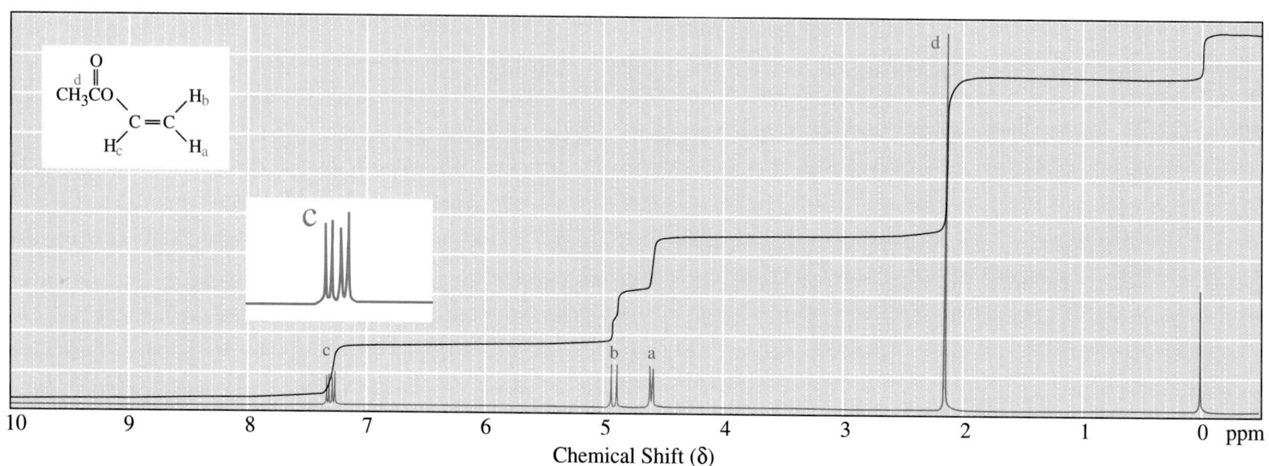

FIGURE 13.17
^{1}H-NMR spectrum of vinyl acetate.

induced in the pi bond of alkenes (Section 13.9C). The splitting pattern observed in the ^{1}H-NMR spectrum of vinyl acetate (Figure 13.17) is typical of monosubstituted alkenes. The singlet at δ 2.12 represents the three hydrogens of the methyl group. The terminal vinylic hydrogens appear at δ 4.58 and δ 4.90. The internal vinylic hydrogen, which normally appears in the range δ 5.0–5.7, is shifted farther downfield to δ 7.30 due to deshielding by the adjacent electronegative oxygen atom of the ester.

Coupling constants are generally larger for *trans* vinylic hydrogens than for *cis* vinylic hydrogens, and it is often possible to distinguish between *cis* and *trans* alkenes by an analysis of their coupling constants. Protons *trans* to each other couple most strongly, typically with a coupling constant of approximately 15 Hz. *Cis* protons couple less strongly, typically with a coupling constant of approximately 8 Hz.

H_b, H_a on C=C	H_b, H_c on C=C	H_c, H_a on C=C
Terminal	*cis*	*trans*
J_{ab} = 0–5 Hz	J_{bc} = 5–10 Hz	J_{ac} = 11–18 Hz

Shown in Figure 13.18 is a graphical analysis of the spectrum of vinyl acetate. The signal of each vinylic hydrogen is a doublet of doublets. The signal for H_c, for example, is split to a doublet by coupling with H_a and further split to a doublet of doublets by coupling with H_b.

The *sp*^{2} hybridized carbons of alkenes give ^{13}C-NMR signals in the range δ 100–150 (Table 13.10), which is considerably downfield from resonances of *sp*^{3} hybridized carbons.

C. Alcohols

The chemical shift of a hydroxyl hydrogen in an ^{1}H-NMR spectrum is variable and depends on the purity of the sample, the concentration, the solvent, and the temperature. It normally appears in the range δ 3.0–4.5, but depending on experimental conditions, it may appear as high as δ 0.5. Hydrogens on the carbon bearing the

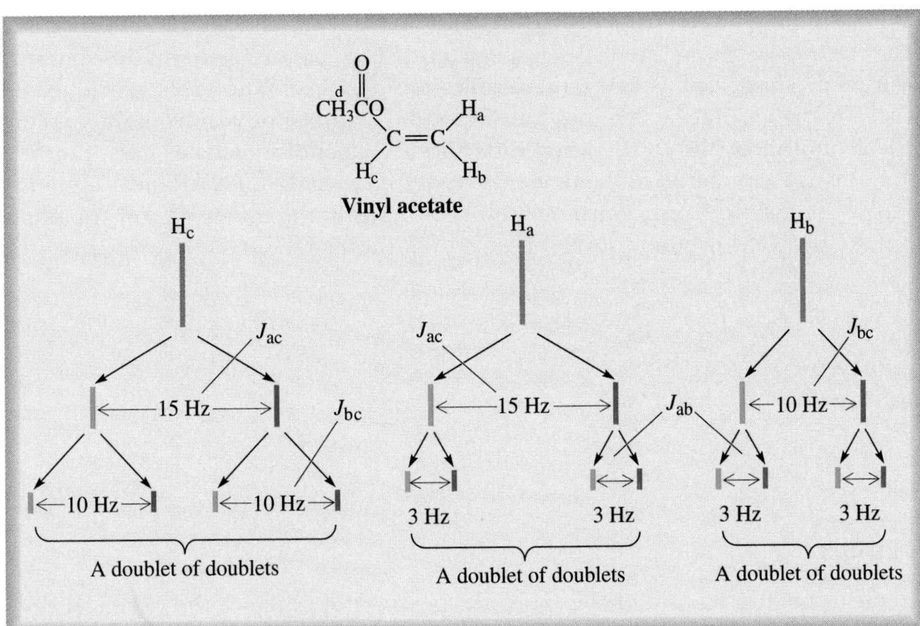

FIGURE 13.18
Graphical analysis of the signal splitting patterns of the three vinylic hydrogens in vinyl acetate. The largest coupling constant, J_{ac} (15 Hz), is that for the two *trans* vinylic hydrogens. The smallest coupling constant, J_{ab} (3 Hz), is that for the two terminal vinylic hydrogens.

—OH group are deshielded by the electron-withdrawing inductive effect of the oxygen atom and their absorptions also typically appear in the range δ 3.4–4.3. Shown in Figure 13.19 is the ^{1}H-NMR spectrum of 1-propanol. This spectrum consists of four signals. The hydroxyl hydrogen appears at δ 3.18 as a narrowly spaced triplet. The signal of hydrogens on the oxygen-bearing carbon in 1-propanol appears as a narrowly spaced multiplet at δ 3.56.

FIGURE 13.19
^{1}H-NMR spectrum of 1-propanol.

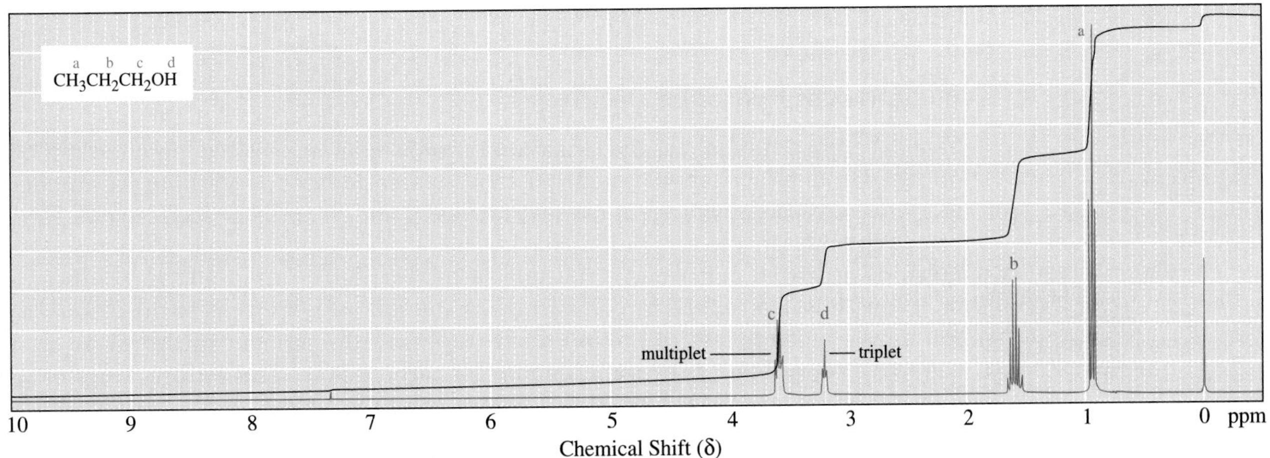

Signal splitting between the hydrogen on O—H and its neighbors on the adjacent —CH_2— group is seen in the ^{1}H-NMR spectrum of 1-propanol. However, this splitting is rarely seen. The reason is that most samples of alcohol contain traces of acid, base, or other impurities that catalyze the transfer of the hydroxyl proton from the oxygen of one alcohol molecule to the oxygen of another alcohol molecule. This transfer, which is very fast compared to the time scale required to make an NMR measurement, decouples the hydroxyl proton from all other protons in the molecule. For the same reason, the hydroxyl proton does not usually split the signal of any α-hydrogens.

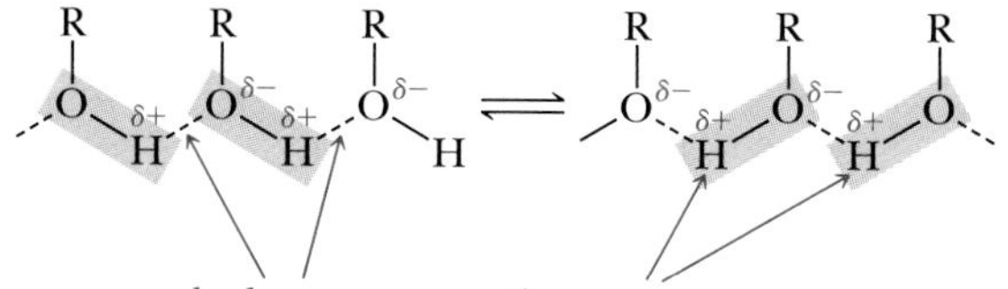

hydrogen bonding

these protons have been exchanged

D. Ethers

The most distinctive feature of the ^{1}H-NMR spectrum of ethers is the chemical shift of hydrogens on carbon attached to the ether oxygen. Resonance signals for this type of hydrogen fall in the range δ 3.3–4.0, which corresponds to a downfield shift of approximately 2.4 units compared with their normal position in alkanes. The chemical shifts of **H**—C—O— hydrogens in ethers are similar to those seen for comparable **H**—C—OH hydrogens of alcohols.

13.16 Solving NMR Problems

One of the first steps in determining molecular structure is establishing the molecular formula. In the past, this was most commonly done by elemental analysis, combustion analysis to determine percent composition, molecular weight determination, and so forth. More commonly today, molecular weight and molecular formula are often determined by mass spectrometry (Chapter 12). In the examples that follow, we assume that the molecular formula of any unknown compound has already been determined, and we proceed from that point, using spectral analysis to determine a structural formula.

A. Index of Hydrogen Deficiency

Valuable information about the structural formula of an unknown compound can be obtained from inspection of its molecular formula. In addition to learning the number of atoms of carbon, hydrogen, oxygen, nitrogen, and so forth in a molecule of the compound, we can also determine what is called its index of hydrogen deficiency. The **index of hydrogen deficiency** is the sum of the number of rings and pi bonds in a molecule. It is determined by comparing the number of hydrogens in the molecular formula of a compound of unknown structure with the number of hydrogens in a **reference compound** of the same number of carbon atoms and with no rings or pi bonds. The molecular formula of a reference hydrocarbon is C_nH_{2n+2} (Section 2.1).

$$\text{Index of hydrogen deficiency} = \frac{(\text{H}_{\text{reference}} - \text{H}_{\text{molecule}})}{2}$$

EXAMPLE 13.10

Calculate the index of hydrogen deficiency for 1-hexene, C_6H_{12}, and account for this deficiency by reference to its structural formula.

Solution

The molecular formula of the reference hydrocarbon of six carbon atoms is C_6H_{14}. The index of hydrogen deficiency of 1-hexene $(14 - 12)/2 = 1$ and is accounted for by the one pi bond in 1-hexene.

PROBLEM 13.10

Calculate the index of hydrogen deficiency of cyclohexene, C_6H_{10}, and account for this deficiency by reference to its structural formula.

To determine the molecular formula for a reference compound containing elements besides carbon and hydrogen, write the formula of the reference hydrocarbon, add to it the atoms of other elements contained in the unknown compound, and make the following adjustments to the number of hydrogen atoms.

1. For each atom of a Group VII element (F, Cl, Br, I) added to the reference hydrocarbon, subtract one hydrogen; halogen substitutes for hydrogen and reduces the number of hydrogens by one per halogen. The general formula of an acyclic monochloroalkane, for example, is $C_nH_{2n+1}Cl$.
2. No correction is necessary for the addition of atoms of Group VI elements (O, S, Se) to the reference hydrocarbon. Insertion of a divalent Group VI element into a reference hydrocarbon does not change the number of hydrogens.
3. For each atom of a Group V element (N, P, As) added to the formula of the reference hydrocarbon, add one hydrogen. Insertion of a trivalent Group V element adds one hydrogen to the molecular formula of the reference compound. The general molecular formula for an acyclic alkylamine, for example, is $C_nH_{2n+3}N$.

EXAMPLE 13.11

Isopentyl acetate, a compound with a banana-like odor, is a component of the alarm pheromone of honey bees. The molecular formula of isopentyl acetate is $C_7H_{14}O_2$. Calculate the index of hydrogen deficiency of this compound.

Solution

The molecular formula of the reference hydrocarbon is C_7H_{16}. Adding oxygens to this formula does not require any correction in the number of hydrogens. The molecular formula of the reference compound is $C_7H_{16}O_2$, and the index of hydrogen deficiency is $(16 - 14)/2 = 1$, indicating either one ring or one pi bond. Following is the structural formula of isopentyl acetate. It contains one pi bond, in this case, a carbon-oxygen pi bond.

$$CH_3\overset{\overset{\displaystyle O}{\|}}{C}OCH_2CH_2CH(CH_3)_2$$

Isopentyl acetate

PROBLEM 13.11

The index of hydrogen deficiency of niacin is 5. Account for this index of hydrogen deficiency by reference to the structural formula of niacin.

Nicotinamide
(Niacin)

B. From an ^{1}H-NMR Spectrum to a Structural Formula

The following steps may prove helpful as a systematic approach to solving spectral problems.

Step 1: **Molecular formula and index of hydrogen deficiency.** Examine the molecular formula, calculate the index of hydrogen deficiency, and deduce what information you can about the presence or absence of rings or pi bonds.

Step 2: **Number of signals.** Count the number of signals to determine the number of sets of equivalent hydrogens present in the compound.

Step 3: **Integration.** Use the integration and the molecular formula to determine the numbers of hydrogens present in each set.

Step 4: **Pattern of chemical shifts.** Examine the NMR spectrum for signals characteristic of the following types of equivalent hydrogens. Keep in mind that these are broad ranges and that hydrogens of each type may be shifted either farther upfield or farther downfield, depending on details of molecular structure.

Types of Hydrogens	Descriptive Name	Typically Absorb in the Range (δ)
RCH_3, RCH_2R, R_3CH	alkyl hydrogens	0.8–1.7
$R_2C{=}CRCHR_2$	allylic hydrogens	1.6–2.6
RCH_2OH, RCH_2OR	hydrogens on a carbon adjacent to an sp^3 hybridized oxygen	3.3–4.0
$R_2C{=}CH_2$, $R_2C{=}CHR$	vinylic hydrogens	4.6–5.7
$RCH{=}O$ (RCH with C=O)	aldehyde hydrogens	9.5–10.1
$RC({=}O)OH$ (RCOH with C=O)	carboxyl hydrogens	10–13

Step 5: **Signal splitting patterns.** Examine splitting patterns for information about the number of nearest nonequivalent hydrogen neighbors.

Step 6: **Structural formula.** Write a structural formula consistent with the previous information.

Spectral Problem 1: Molecular formula $C_5H_{10}O$.

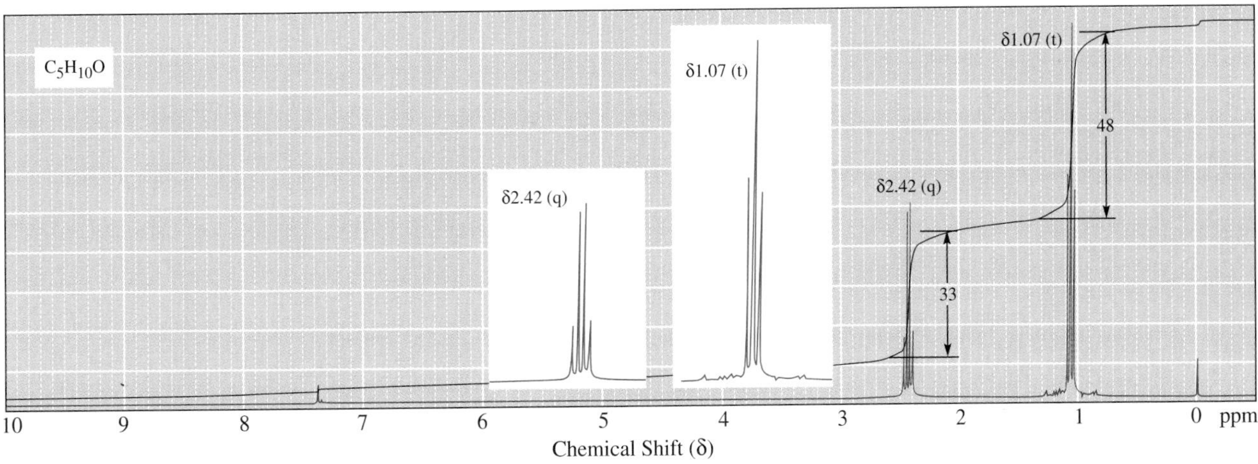

Analysis of Spectral Problem 1

Step 1: **Molecular formula and index of hydrogen deficiency.** The reference compound is $C_5H_{12}O$, and, therefore, the index of hydrogen deficiency is 1; the molecule contains either one ring or one pi bond.

Step 2: **Number of signals.** There are two signals (a triplet and a quartet), and, therefore, two sets of equivalent hydrogens.

Step 3: **Integration.** From the integration, calculate that the number of hydrogens in each set are in the ratio 3:2. Because there are 10 hydrogens, conclude that 6H give rise to the signal at δ 1.07, and 4H give rise to the signal at δ 2.42.

Step 4: **Pattern of chemical shifts.** The signal at δ 1.07 is in the alkyl region and, based on its chemical shift, most probably represents a methyl group. No signal occurs between δ 4.6–5.9; there are no vinylic hydrogens. If a carbon-carbon double bond is in the molecule, no hydrogens are on it (that is, it is tetrasubstituted). The chemical shift of the four protons at δ 2.42 is consistent with two CH_2 groups next to a carbonyl group.

Step 5: **Signal splitting pattern.** The methyl signal at δ 1.07 is split into a triplet (t); it must have two neighbors, indicating $—CH_2CH_3$. The signal at δ 2.42 is split into a quartet (q); it must have three neighbors. An ethyl group accounts for these two signals. No other signals occur in the spectrum, and, therefore, there are no other types of hydrogens in the molecule.

Step 6: **Structural formula.** Put this information together to arrive at the following structural formula. The chemical shift of the methylene group ($—CH_2—$) at δ 2.42 is consistent with an alkyl group adjacent to a carbonyl group.

$$CH_3—CH_2—\overset{\overset{\displaystyle O}{\|}}{C}—CH_2—CH_3$$

δ 2.42 (q) → CH_2; δ 1.07 (t) → CH_3

3-Pentanone

Spectral Problem 2: Molecular formula $C_7H_{14}O$.

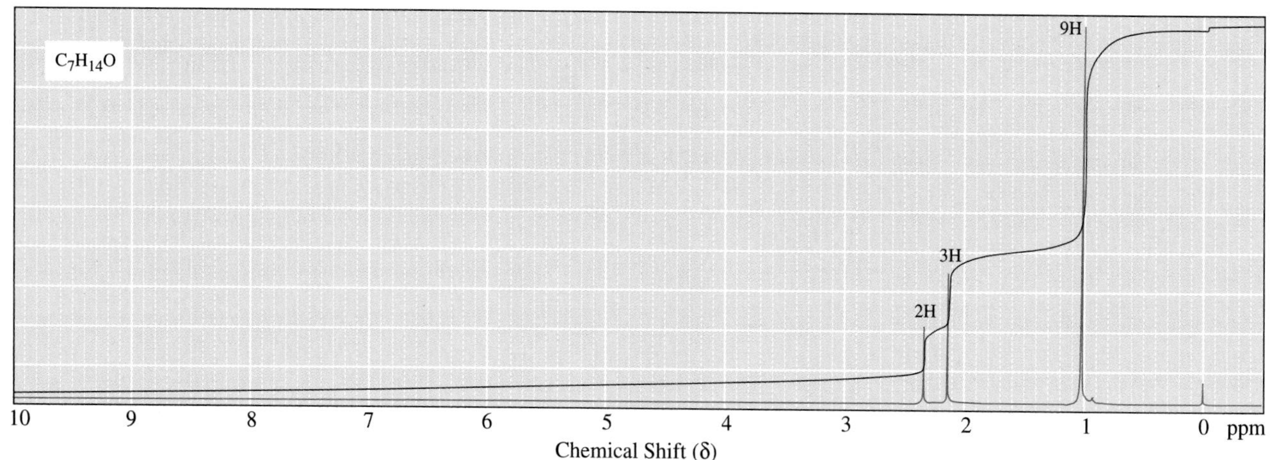

Analysis of Spectral Problem 2

Step 1: **Molecular formula and index of hydrogen deficiency.** The index of hydrogen deficiency is 1; the compound contains one ring or one pi bond.

Step 2: **Number of signals.** There are three signals and, therefore, three sets of equivalent hydrogens.

Step 3: **Line of integration.** Reading from right to left, the number of hydrogens giving rise to these signals are 9, 3, and 2.

Step 4: **Pattern of chemical shifts.** The signal at δ 1.01 is characteristic of a methyl group adjacent to an sp^3 hybridized carbon. The signals at δ 2.11 and 2.32 are characteristic of alkyl groups adjacent to a carbonyl group.

Step 5: **Signal splitting pattern.** All signals are singlets (s). Therefore, none of the groups has hydrogens on neighboring carbons.

Step 6: **Structural formula.** The compound is 4,4-dimethyl-2-pentanone.

δ 1.01 (s) δ 2.32 (s) δ 2.11 (s)

$$CH_3-\overset{CH_3}{\underset{CH_3}{C}}-CH_2-\overset{O}{\overset{\|}{C}}-CH_3$$

4,4-Dimethyl-2-pentanone

SUMMARY

Electromagnetic radiation (Section 13.1) can be described in terms of its **wavelength (λ)** and its **frequency (ν).** Frequency is reported in **hertz (Hz)**. An alternative way to describe electromagnetic radiation is in terms of its energy where $E = h\nu$ (Section 13.1).

Molecular spectroscopy (Section 13.2) is the experimental process of measuring which frequencies of radiation are absorbed or emitted by a substance and then attempting to correlate patterns of energy absorption or emission with details of molecular structure. Interaction of molecules with **radio-**

frequency radiation gives us information about nuclear spin energy levels; interaction with **infrared radiation** gives us information about vibrational energy levels, and interaction with **ultraviolet-visible radiation** gives us information about electronic energy levels of pi and nonbonding electrons.

Nuclei of ^1H and ^{13}C have a **nuclear spin quantum number** of $\frac{1}{2}$ and allowed **nuclear spin states** of $+\frac{1}{2}$ and $-\frac{1}{2}$ (Section 13.3). In the presence of an **applied magnetic field, B_0,** nuclei with spin $+\frac{1}{2}$ are aligned with the applied field and are in a lower energy state; nuclei with spin $-\frac{1}{2}$ are aligned against the applied field and are in a higher energy state (Section 13.4).

When placed in a powerful magnetic field (Section 13.5), ^1H and ^{13}C nuclei become aligned in an allowed spin state and begin to **precess** about an axis parallel to the direction of the applied field. **Resonance** is the absorption of electromagnetic radiation by a precessing nucleus and the resulting "flip" of its nuclear spin from the lower energy spin state to the higher energy spin state. An NMR spectrometer records such resonance as a **signal.**

The experimental conditions required to cause nuclei to resonate are affected by the local chemical and magnetic environment. Electrons around a hydrogen or carbon create local magnetic fields that **shield** the nuclei of these atoms from the applied field (Section 13.5). Any factor that increases the exposure of nuclei to an applied field is said to **deshield** them and shifts their signal downfield to a larger δ value. Conversely, any factor that decreases the exposure of nuclei to an applied field is said to **shield** them and shifts their signal upfield to a smaller δ value.

Equivalent hydrogens within a molecule have identical chemical shifts (Section 13.7). The chemical shift of a particular set of equivalent hydrogens depends primarily on three factors: (1) nearby electronegative atoms have a deshielding effect; (2) the greater the percent of s character in a hybrid orbital, the greater the deshielding effect of the atom to which the orbital belongs; and (3) induced local magnetic fields in pi bonds either add to or subtract from the applied field. The area of an ^1H-NMR signal is proportional to the number of equivalent hydrogens giving rise to that signal (Section 13.8).

The resonance signals in ^1H-NMR spectra are reported by how far they are shifted from the resonance signal of the 12 equivalent hydrogens in **tetramethylsilane (TMS).** The resonance signals in ^{13}C-NMR spectra are reported by how far they are shifted from the resonance signal of the four equivalent carbons in TMS. **Chemical shift, δ,** (Section 13.9) is defined as the frequency shift from TMS divided by the operating frequency of the spectrometer.

According to the **$(n + 1)$ rule,** if a hydrogen has n hydrogens nonequivalent to it but equivalent among themselves on the same or adjacent atom(s), its ^1H-NMR signal is split into $(n + 1)$ peaks (Section 13.10). Splitting patterns are commonly referred to as singlets (s), doublets (d), triplets (t), quartets (q), quintets, and multiplets (m). The relative intensities of peaks in a multiplet can be predicted from an analysis of spin combinations for adjacent hydrogens (Table 13.8) or from the mnemonic device called **Pascal's triangle** (Figure 13.12).

A **coupling constant** (J) is the distance between adjacent peaks in a multiplet and is reported in hertz (Section 13.12). The value of J depends only on internal forces within a molecule and is independent of machine operating parameters.

^{13}C-NMR spectra (Section 13.13) are commonly recorded in a hydrogen-decoupled instrumental mode. In this mode, all ^{13}C signals appear as singlets. The DEPT method can be used to identify CH_3, CH_2, and CH signals separately.

ADDITIONAL PROBLEMS

Index of Hydrogen Deficiency

13.12 Complete the following table:

Class of Compound	General Molecular Formula	Index of Hydrogen Deficiency	Reason for Hydrogen Deficiency
alkane	C_nH_{2n+2}	0	(reference hydrocarbon)
alkene	C_nH_{2n}	1	one pi bond
alkyne	______	—	______
alkadiene	______	—	______
cycloalkane	______	—	______
cycloalkene	______	—	______
bicycloalkane	______	—	______

13.13 Calculate the index of hydrogen deficiency of these compounds:

(a) Aspirin, $C_9H_8O_4$

(b) Ascorbic acid (vitamin C), $C_6H_8O_6$

(c) Pyridine, C_5H_5N

(d) Urea, CH_4N_2O

(e) Cholesterol, $C_{27}H_{46}O$

(f) Dopamine, $C_8H_{11}NO_2$

Interpretation of ^{1}H-NMR and ^{13}C-NMR Spectra

13.14 Complete the following table. Which nucleus requires the least energy to flip its spin at this applied field? Which nucleus requires the most energy?

Nucleus	Applied Field (tesla, T)	Radio Frequency (MHz)	Energy (cal/mol)
^{1}H	7.05	300	______
^{13}C	7.05	75.5	______
^{19}F	7.05	282	______

13.15 The natural abundance of ^{13}C is only 1.1%. Furthermore, its sensitivity in NMR spectroscopy (a measure of the energy difference between a spin aligned with or against an external magnetic field) is only 1.6% that of ^{1}H. What are the relative signal intensities expected for the ^{1}H-NMR and ^{13}C-NMR spectra of the same sample of $Si(CH_3)_4$?

13.16 Following are structural formulas for three constitutional isomers of molecular formula $C_7H_{16}O$ and three sets of ^{13}C-NMR spectral data. Assign each constitutional isomer its correct spectral data.

(a) $CH_3CH_2CH_2CH_2CH_2CH_2CH_2OH$

(b) $CH_3C(OH)(CH_3)CH_2CH_2CH_2CH_3$

(c) $CH_3CH_2C(OH)(CH_2CH_3)CH_2CH_3$

Spectrum 1	Spectrum 2	Spectrum 3
74.66	70.97	62.93
30.54	43.74	32.79
7.73	29.21	31.86
	26.60	29.14
	23.27	25.75
	14.09	22.63
		14.08

13.17 Following are structural formulas for the *cis* isomers of 1,2-, 1,3-, and 1,4-dimethylcyclohexane and three sets of ^{13}C-NMR spectral data. Assign each constitutional isomer its correct spectral data.

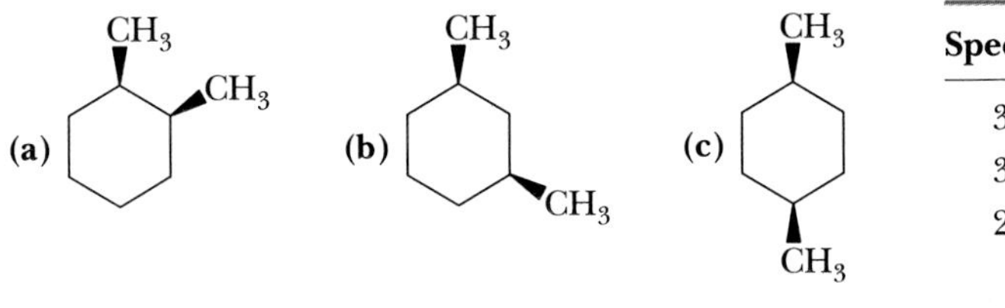

Spectrum 1	Spectrum 2	Spectrum 3
31.35	34.20	44.60
30.67	31.30	35.14
20.85	23.56	32.88
	15.97	26.54
		23.01

13.18 Following are structural formulas, dipole moments, and ^1H-NMR chemical shifts for acetonitrile, fluoromethane, and chloromethane.

$CH_3C{\equiv}N$	CH_3F	CH_3Cl
Acetonitrile	Fluoromethane	Chloromethane
3.92 D	1.85 D	1.87 D
δ 1.97	δ 4.26	δ 3.05

(a) How do you account for the fact that the dipole moments of fluoromethane and chloromethane are almost identical even though fluorine is considerably more electronegative than chlorine?

(b) How do you account for the fact that the dipole moment of acetonitrile is considerably greater than that of either fluoromethane or chloromethane?

(c) How do you account for the fact that the chemical shift of the methyl hydrogens in acetonitrile is considerably less than that for either fluoromethane or chloromethane? (*Hint:* Consider the magnetic induction in the pi bonds of acetonitrile.)

13.19 Following are three compounds of molecular formula $C_4H_8O_2$, and three ^1H-NMR spectra. Assign each compound its correct spectrum and assign all signals to their corresponding hydrogens.

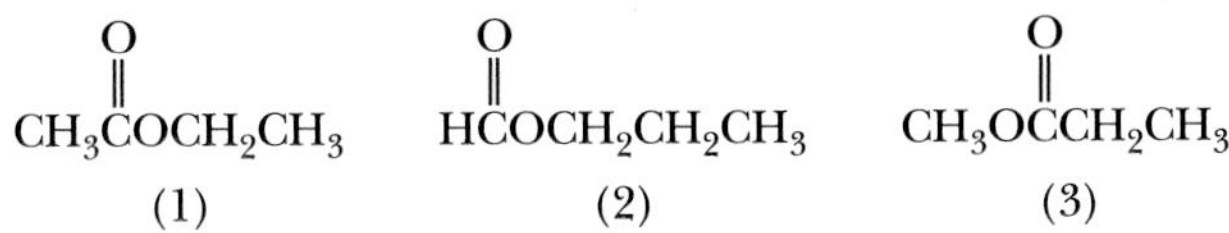

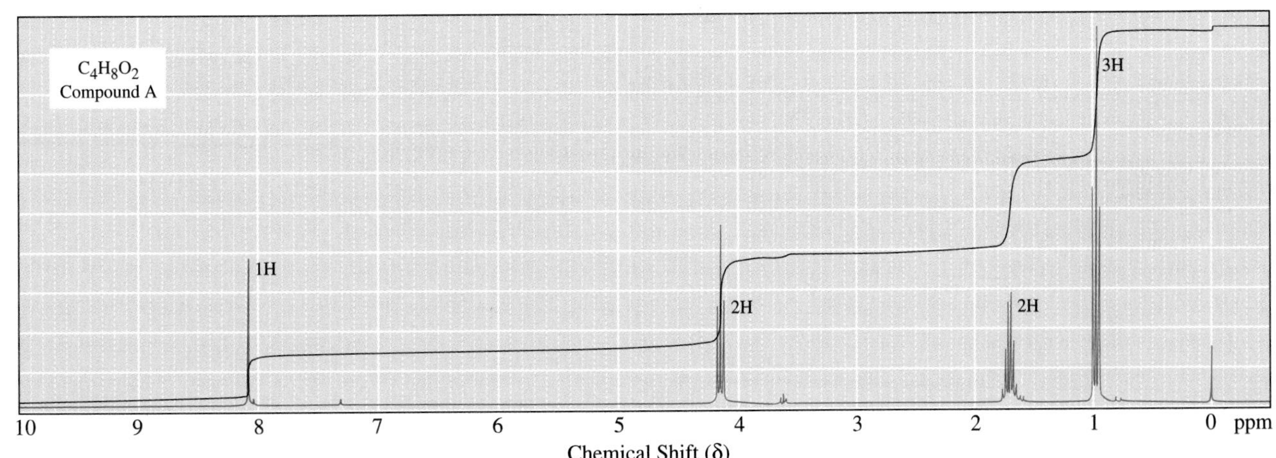

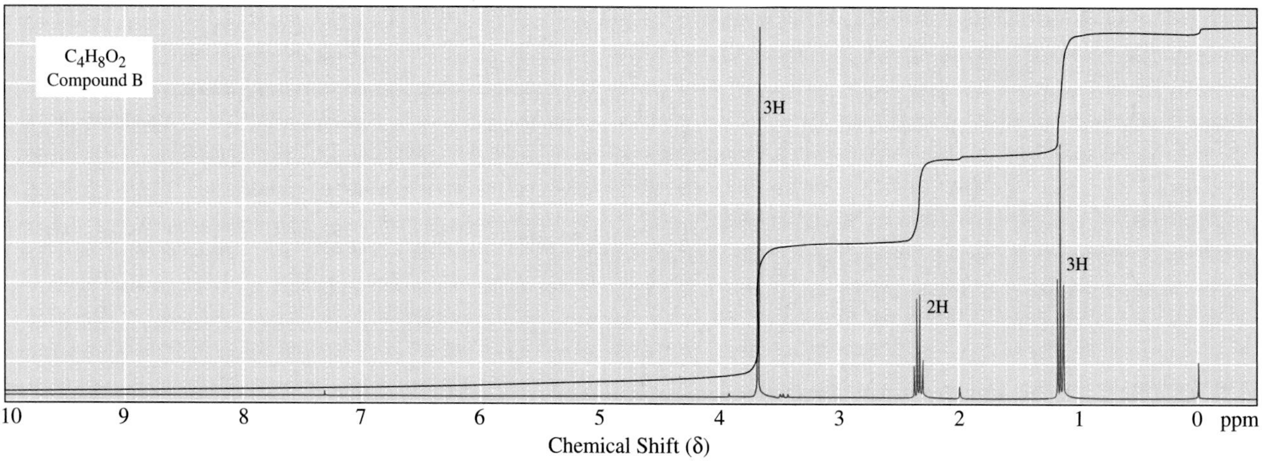

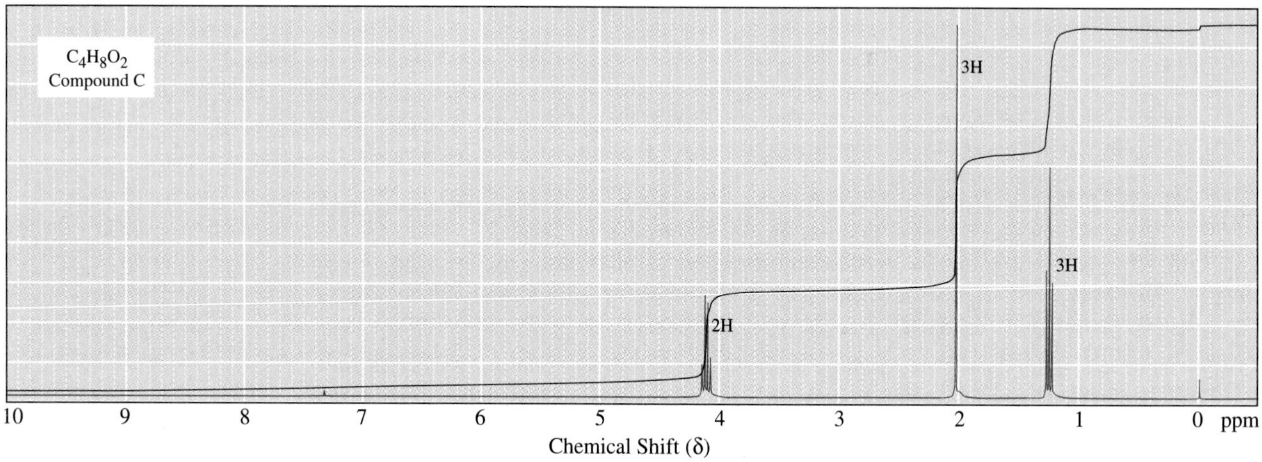

13.20 Following are ^{1}H-NMR spectra for compounds A, B, and C, each of molecular formula C_6H_{12}. Each readily decolorizes a solution of Br_2 in CCl_4. Propose structural formulas for compounds A, B, and C, and account for the observed patterns of signal splitting.

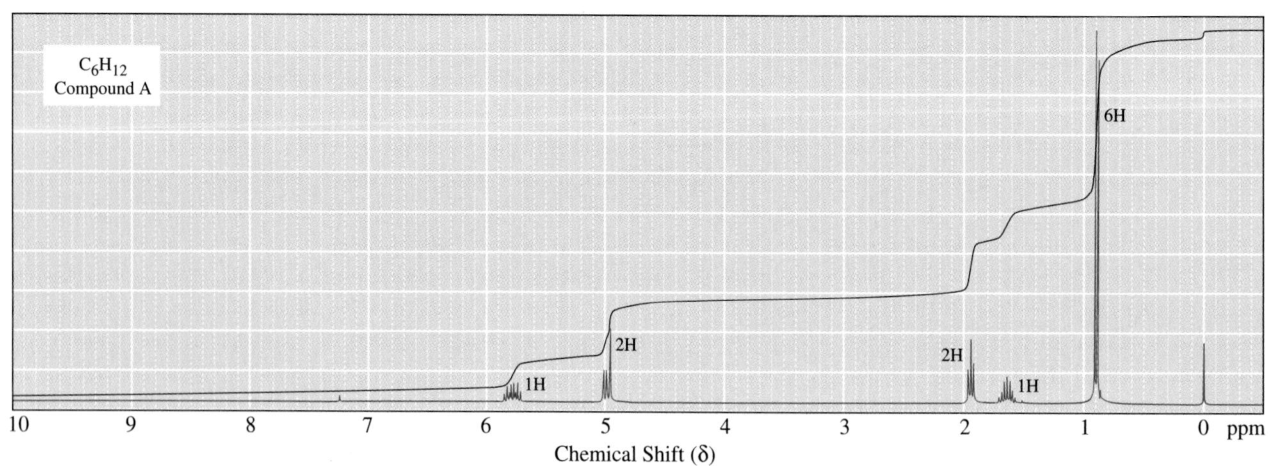

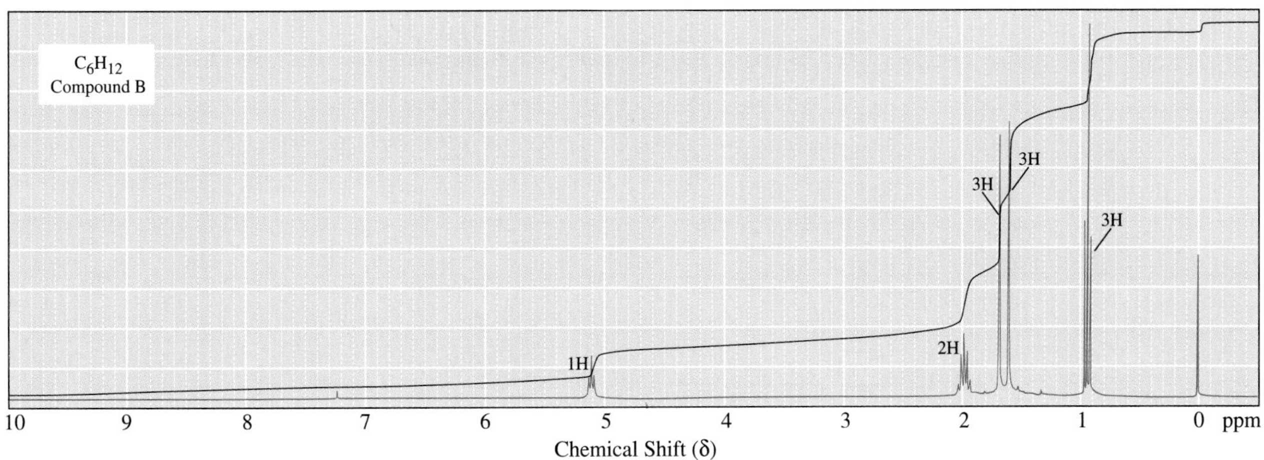

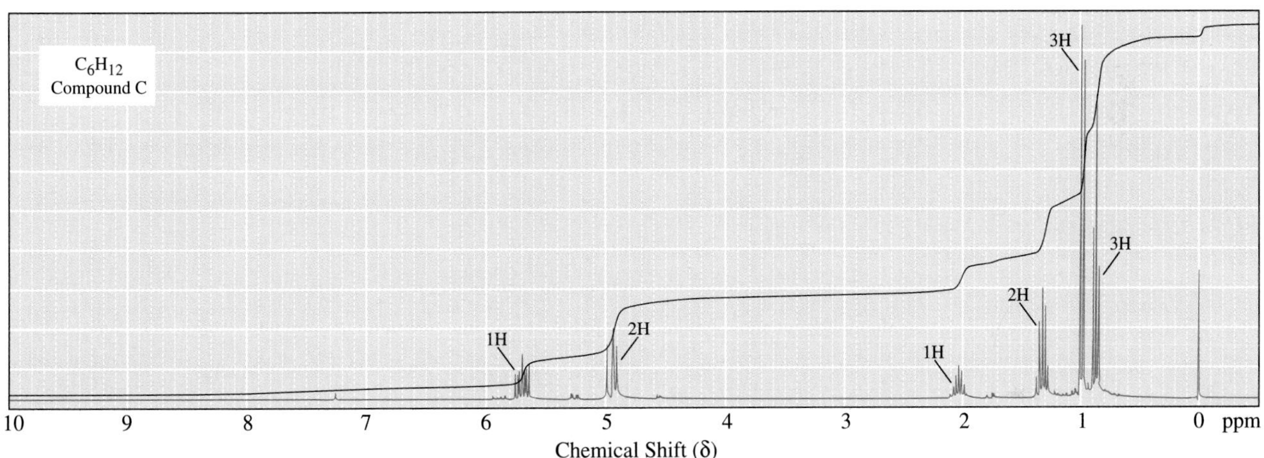

13.21 Following are ^{1}H-NMR spectra for compounds D, E, and F, each of molecular formula $C_5H_{12}O$. Each is a liquid at room temperature, slightly soluble in water, and reacts with sodium metal with the evolution of a gas. Propose structural formulas of compounds D, E, and F. (*Hints:* For compound D, the signal at δ 0.9 results from two overlapping doublets. For compound E, the signal at δ 0.9 results from the overlapping of a doublet and a triplet.)

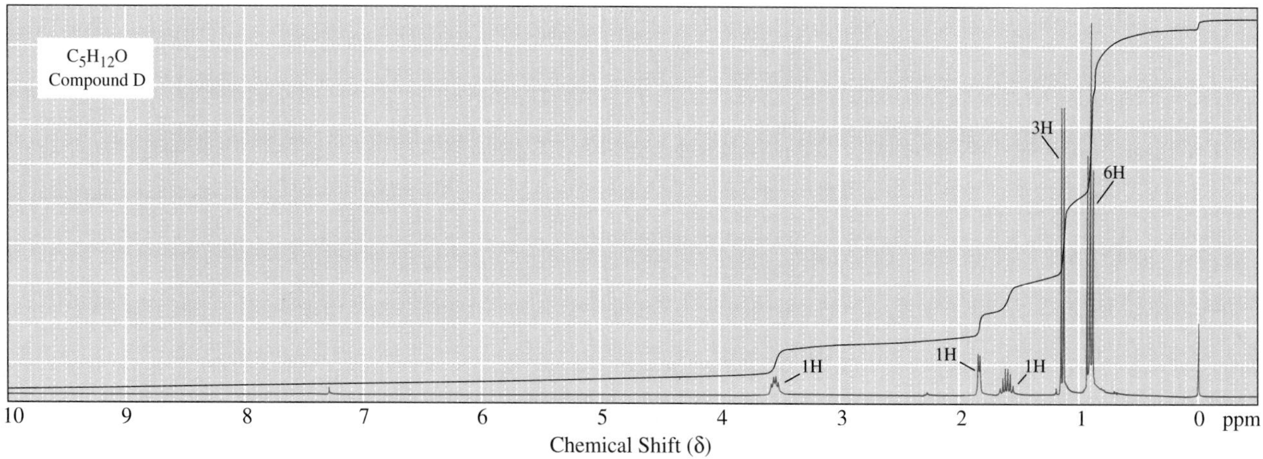
$C_5H_{12}O$
Compound D
3H
6H
1H
1H
1H
10 9 8 7 6 5 4 3 2 1 0 ppm
Chemical Shift (δ)

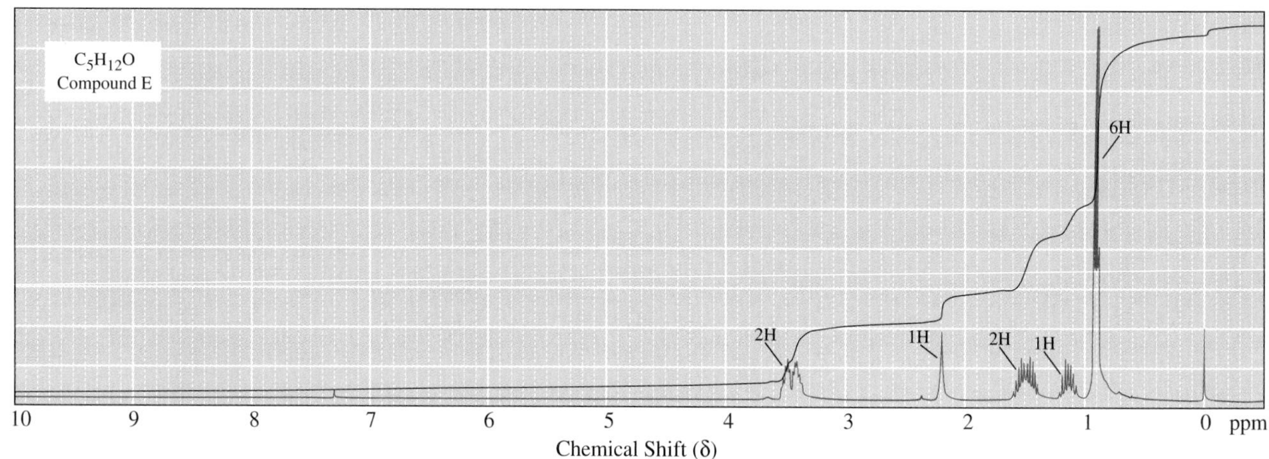
$C_5H_{12}O$
Compound E
6H
2H
1H
2H
1H
10 9 8 7 6 5 4 3 2 1 0 ppm
Chemical Shift (δ)

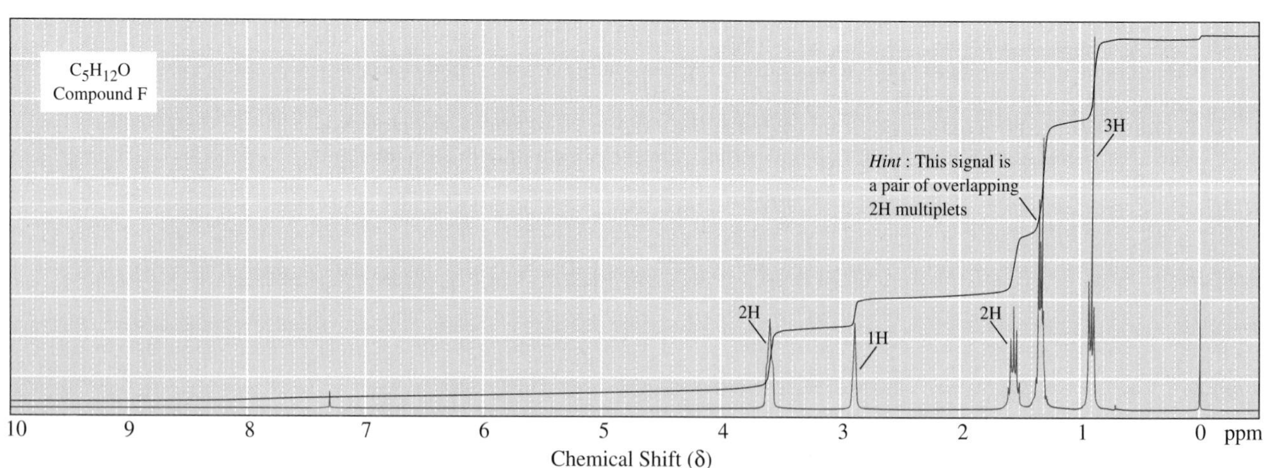
$C_5H_{12}O$
Compound F
3H
Hint : This signal is
a pair of overlapping
2H multiplets
2H
1H
2H
10 9 8 7 6 5 4 3 2 1 0 ppm
Chemical Shift (δ)

13.22 Propose a structural formula for compound G, molecular formula C_3H_6O, consistent with the following ^{1}H-NMR spectrum.

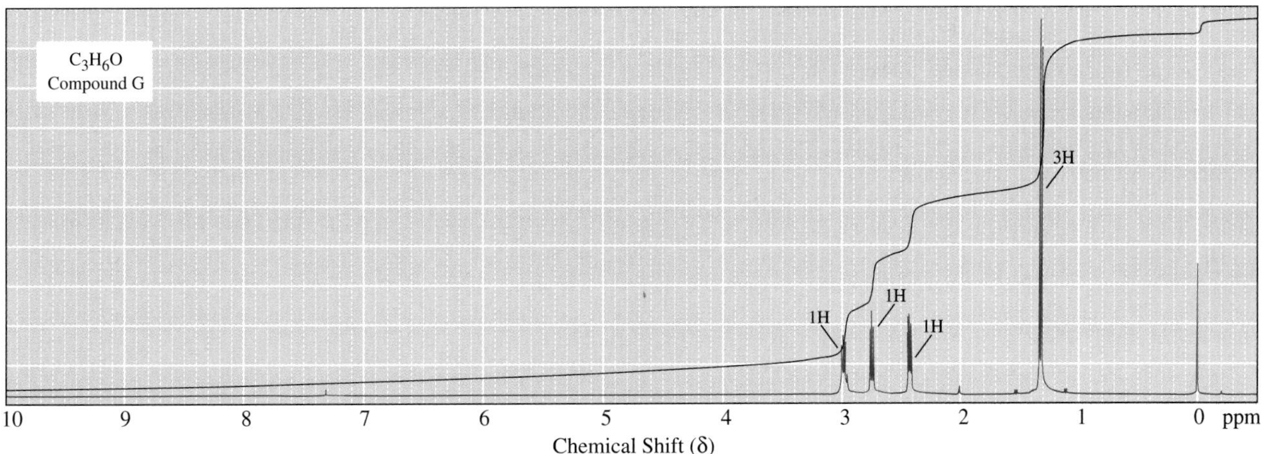

13.23 Compound H, molecular formula $C_6H_{14}O$, readily undergoes acid-catalyzed dehydration when warmed with phosphoric acid to give compound I, molecular formula C_6H_{12}, as the major organic product. The ^{1}H-NMR spectrum of compound H shows signals at δ 0.90 (t, 6H), 1.12 (s, 3H), 1.38 (s, 1H), and 1.48 (q, 4H). The ^{13}C-NMR spectrum of compound H shows signals at 72.98, 33.72, 25.85, and 8.16. Deduce the structural formulas of compounds H and I.

13.24 Compound J, molecular formula $C_5H_{10}O$, readily decolorizes Br_2 in CCl_4, and is converted by H_2/Ni into compound K, molecular formula $C_5H_{12}O$. Following is the ^{1}H-NMR spectrum of compound J. The ^{13}C-NMR spectrum of compound J shows signals at 146.12, 110.75, 71.05, and 29.38. Deduce the structural formulas of compounds J and K.

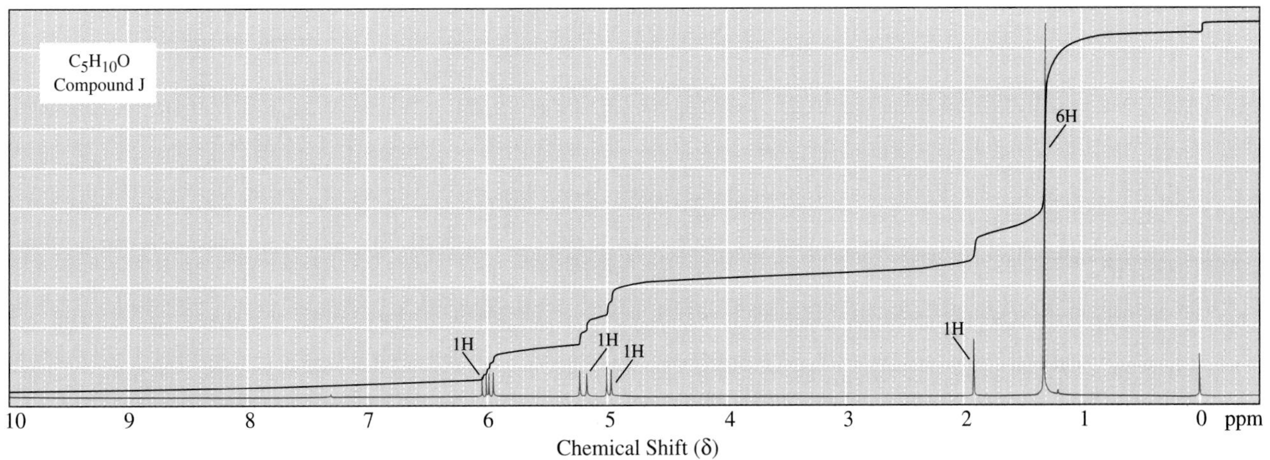

13.25 Following is the ^{1}H-NMR spectrum of compound L, molecular formula C_7H_{12}. Compound L reacts with bromine in carbon tetrachloride to give a compound of molecular formula $C_7H_{12}Br_2$. The ^{13}C-NMR spectrum of compound L shows signals at 150.12, 106.43, 35.44, 28.36, and 26.36. Deduce the structural formula of compound L.

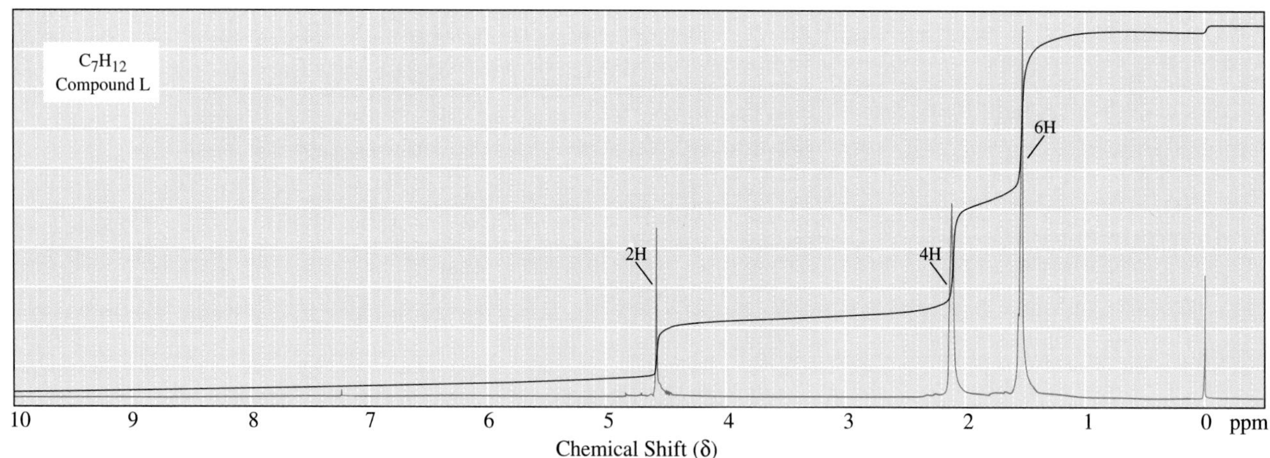

13.26 Treatment of compound M with BH_3 followed by H_2O_2/NaOH gives compound N. Following are ^{1}H-NMR spectra for compounds M and N, along with ^{13}C-NMR spectral data. From this information, deduce structural formulas for compounds M and N.

$$\underset{(M)}{C_7H_{12}} \xrightarrow[\text{2. } H_2O_2,\ NaOH]{\text{1. } BH_3} \underset{(N)}{C_7H_{14}O}$$

^{13}C-NMR	
(M)	**(N)**
132.38	72.71
32.12	37.59
29.14	28.13
27.45	22.68

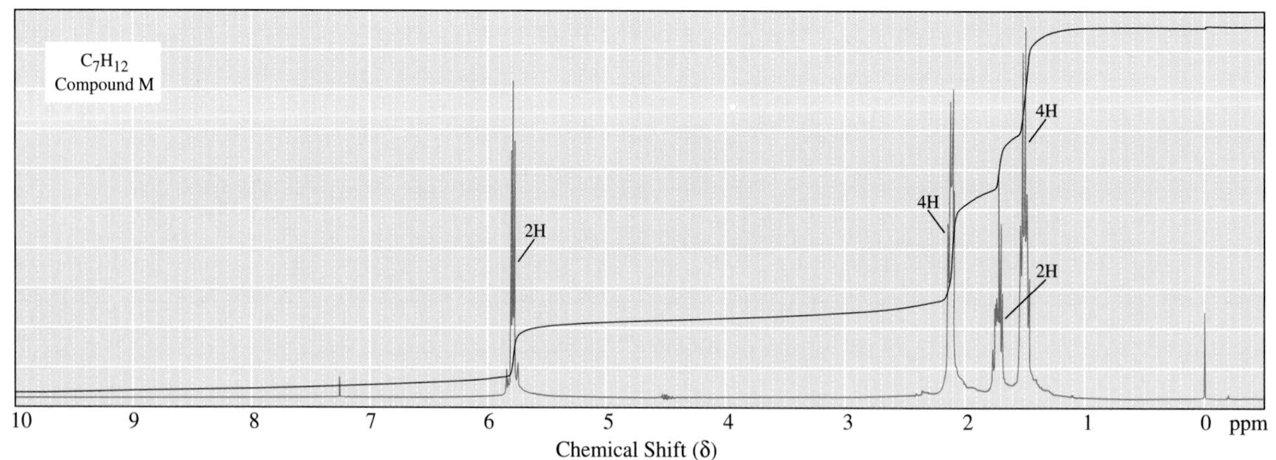

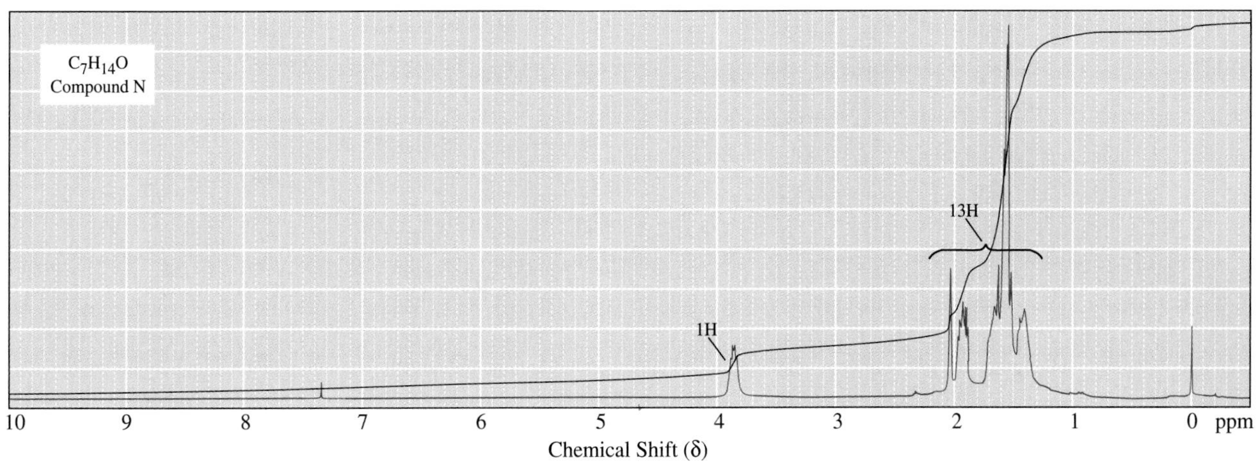

13.27 Compound O is known to contain only C, H, and O. Its mass spectrum shows a weak molecular ion peak at *m/z* 102 and prominent peaks at *m/z* 87, 45, and 43. Its ^{1}H-NMR spectrum consists of two signals: δ 1.1 (doublet) and δ 3.6 (septet) in the ratio 6:1. Propose a structural formula for compound O consistent with this information.

13.28 Following are eight structural formulas along with their ^{13}C-NMR and DEPT spectral data. Given this data, assign each carbon in each compound its correct ^{13}C chemical shift.

(a) $CH_3CH_2CH_2CH(Br)CH_3$

^{13}C	DEPT
51.55	CH
43.22	CH_2
26.46	CH_2
21.00	CH_3
13.40	CH_3

(b) $CH_3CH_2C(CH_3){=}CH_2$

^{13}C	DEPT
147.70	—
108.33	CH_2
30.56	CH_2
22.47	CH_3
12.23	CH_3

(c) $CH_2{=}CHCH_2CH(CH_3)CH_3$

^{13}C	DEPT
137.81	CH
115.26	CH_2
43.35	CH_2
28.12	CH
22.26	CH_3

(d) $CH_3C(CH_3)_2CH_2Br$

^{13}C	DEPT
49.02	CH_2
33.15	—
28.72	CH_3

(e) $CH_3CH_2C({=}O)CH_2CH_3$

^{13}C	DEPT
207.8	—
35.1	CH_2
7.5	CH_3

(f) $CH_3CH_2C({=}O)CH_3$

^{13}C	DEPT
208.7	—
37.6	CH_2
30.1	CH_3
9.2	CH_3

(g) $CH_3CH(CH_3)COCH_3$ (with C=O: $CH_3CHCOCH_3$, O double-bonded to the C of CO, CH_3 branch on CH)

^{13}C	DEPT
177.48	—
51.50	CH_3
33.94	CH
19.01	CH_3

(h) $CH_3COCH_2CH_2CH(CH_3)CH_3$ (O double-bonded to the C of CO, CH_3 branch on CH)

^{13}C	DEPT
171.17	—
63.12	CH_2
37.21	CH_2
25.05	CH
24.45	CH_3
21.02	CH_3

13.29 Write structural formulas for the following compounds:

(a) $C_2H_4Br_2$: δ 2.5 (d, 3H) and 5.9 (q, 1H)

(b) $C_4H_8Cl_2$: δ 1.60 (d, 3H), 2.15 (m, 2H), 3.72 (t, 2H), and 4.27 (m, 1H)

(c) $C_5H_8Br_4$: δ 3.6 (s, 8H)

(d) C_4H_8O: δ 1.0 (t, 3H), 2.1 (s, 3H), and 2.4 (quartet, 2H)

(e) $C_4H_8O_2$: δ 1.2 (t, 3H), 2.1 (s, 3H), and 4.1 (quartet, 2H); contains an ester group

(f) $C_4H_8O_2$: δ 1.2 (t, 3H), 2.3 (quartet, 2H), and 3.6 (s, 3H); contains an ester group

(g) C_4H_9Br: δ 1.1 (d, 6H), 1.9 (m, 1H), and 3.4 (d, 2H)

(h) $C_6H_{12}O_2$: δ 1.5 (s, 9H) and 2.0 (s, 3H)

(i) $C_7H_{14}O$: δ 0.9 (t, 6H), 1.6 (sextet, 4H), and 2.4 (t, 4H)

(j) $C_5H_{10}O_2$: δ 1.2 (d, 6H), 2.0 (s, 3H), and 5.0 (septet, 1H)

(k) $C_5H_{11}Br$: δ 1.1 (s, 9H) and 3.2 (s 2H)

(l) $C_7H_{15}Cl$: δ 1.1 (s, 9H) and 1.6 (s, 6H)

13.30 The percent s character of carbon participating in a C—H bond can be established by measuring the ^{13}C-^{1}H coupling constant and using the relationship

$$\text{Percent s character} = 0.2\, J(^{13}\text{C-}^{1}\text{H})$$

The ^{13}C-^{1}H coupling constant observed for methane, for example, is 125 Hz, which gives 25% s character, the value expected for an sp^3 hybridized carbon atom.

(a) Calculate the expected ^{13}C-^{1}H coupling constant in ethylene and acetylene.

(b) In cyclopropane, the ^{13}C-^{1}H coupling constant is 160 Hz. What is the hybridization of carbon in cyclopropane?

INFRARED AND ULTRAVIOLET-VISIBLE SPECTROSCOPY

14

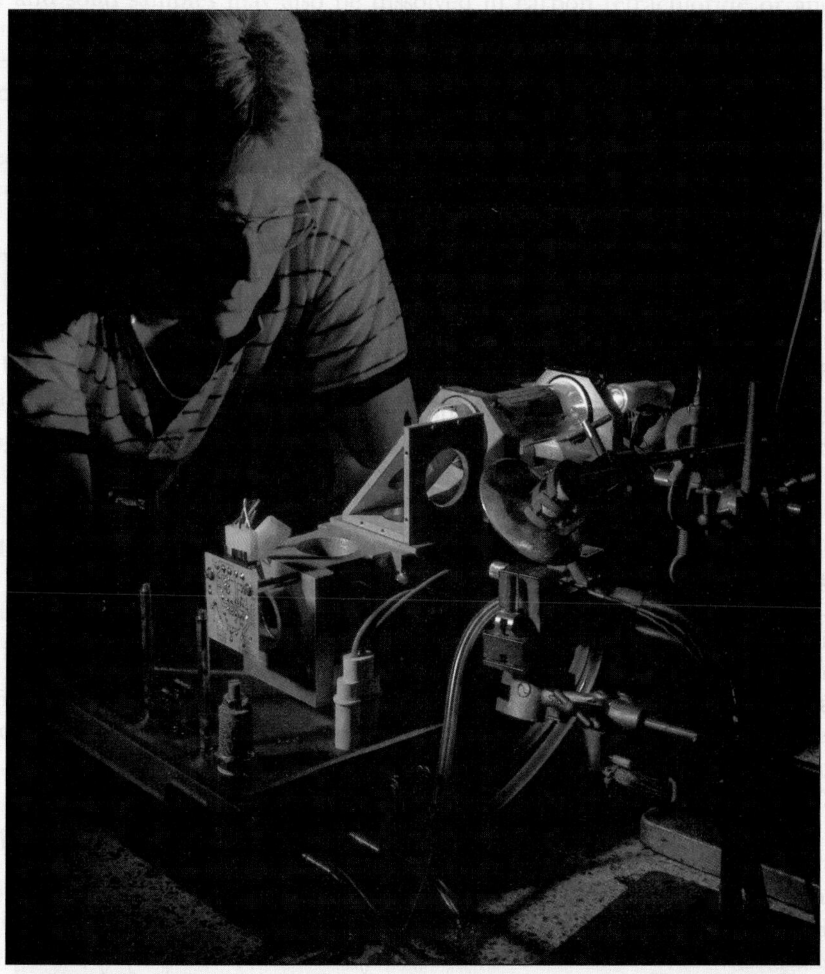

■ A scientist working with a Fourier-Transform infrared spectroscopy instrument. *(Chris Taylor/CSIRO/Science Photo Library/Photo Researchers, Inc.)*

In **infrared (IR) spectroscopy,** a compound is irradiated with infrared radiation, absorption of which causes covalent bonds to change from a lower vibrational energy level to a higher one. Thus, in infrared spectroscopy, we detect functional groups by the vibrations of the atoms constituting the functional group. Using **ultraviolet-visible (UV-vis) spectroscopy,** we detect electronic transitions of pi and nonbonding electrons.

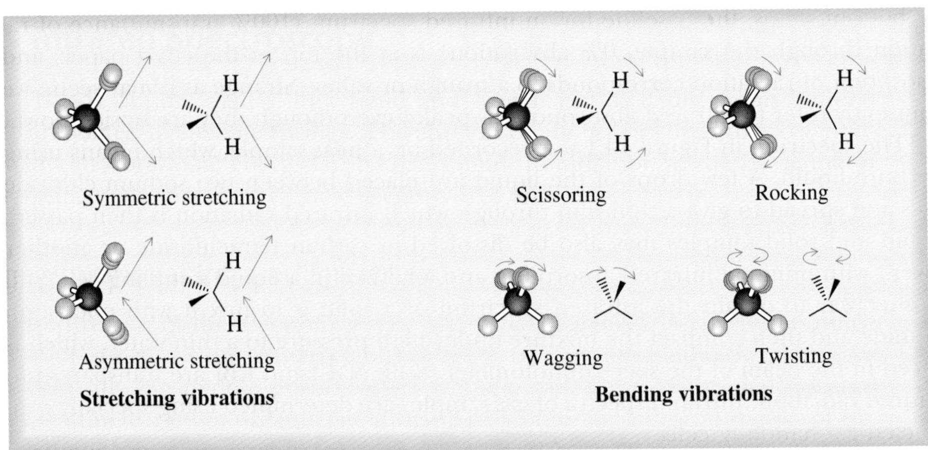

FIGURE 14.2
Fundamental modes of vibration for a methylene group.

$$\underset{\text{2,3-Dimethyl-2-butene}}{(H_3C)_2C{=}C(CH_3)_2} \qquad \underset{\text{2-Butyne}}{H_3C{-}C{\equiv}C{-}CH_3}$$

For a nonlinear molecule containing ***n*** atoms, $\mathbf{3n - 6}$ allowed fundamental vibrations exist. For a molecule as simple as ethanol, C_2H_6O, there are 21 fundamental vibrations, and for hexanoic acid, $C_6H_{12}O_2$, there are 54. Thus, for even relatively simple molecules, a large number of vibrational energy levels exist, and the patterns of energy absorption for these and larger molecules are very complex.

The simplest vibrational motions in molecules giving rise to absorption of infrared radiation are **stretching** and **bending** motions. Illustrated in Figure 14.2 are the fundamental stretching and bending vibrations for a methylene group.

C. Characteristic Absorption Patterns

Analysis of the modes of vibration for a molecule is very complex because all of the atoms contribute to the vibrational modes. It is possible, however, to make a series of useful generalizations about where absorptions due to particular vibration modes will appear in the infrared spectrum by considering each individual bond and ignoring other bonds in the molecule. As a simplifying assumption, let us consider two covalently bonded atoms as two vibrating masses connected by a spring. As the bond vibrates, its energy continually changes from kinetic to potential and vice versa. The total energy (the sum of potential and kinetic energies) is proportional to the frequency of vibration. For a simple harmonic oscillator, the frequency of a stretching vibration is given by this equation, which is derived from **Hooke's law** for a vibrating spring.

$$\bar{\nu} = \frac{1}{2\pi c}\sqrt{\frac{NK}{\mu}}$$

where $\bar{\nu}$ is the wavenumber of the vibration in reciprocal centimeters (cm^{-1}); c is the velocity of light, 2.998×10^{10} cm/s; N is Avogadro's number, 6.022×10^{23} atoms/mol; K is the force constant of the bond, which is a measure of the bond's strength, in dynes per centimeter (g/s^2); and μ is the reduced mass in grams per atom, $(m_1 m_2)/(m_1 + m_2)$, where m is the mass of an atom in grams. In calculating reduced mass, the mass of each atom must be expressed in units of grams per atom. The mass of one atom of carbon-12, for example, is calculated as follows:

$$\text{For carbon-12:} \quad \text{Mass per atom} = \frac{12 \text{ g}}{6.022 \times 10^{23} \text{ atoms}} = 1.99 \times 10^{-23} \text{ g/atom}$$

Collecting and substituting values for all constants (Avogadro's number, pi, and the velocity of light) gives

$$\bar{\nu} = \frac{\sqrt{6.022 \times 10^{23}}}{2 \times 3.142 \times 2.998 \times 10^{10}} \sqrt{\frac{K}{\mu}} = 4.120 \sqrt{\frac{K}{\mu}}$$

Force constants for single, double, and triple bonds are approximately 5×10^5, 10×10^5, and 15×10^5 dynes/cm, respectively. Using the value for the force constant for a single bond, we calculate the wavenumber for the stretching vibration of a single bond between carbon-12 and hydrogen-1 as follows:

$$\text{For } ^{12}\text{C-}^{1}\text{H stretching:} \quad \text{Reduced mass } (\mu) = \frac{(12 \times 1) \text{ g}^2/\text{atom}^2}{(12 + 1) \text{ g/atom}} = 0.9231 \text{ g/atom}$$

$$\bar{\nu} = 4.120 \sqrt{\frac{5 \times 10^5}{0.9231}} = 3032 \text{ cm}^{-1}$$

The experimentally determined value for the wavenumber of an alkyl C—H stretching vibration is approximately 2900 cm^{-1}. Given the simplifying assumptions made in this calculation and the fact that the value of the force constant for a single bond is an average value and expressed only to one significant figure, the agreement between the calculated value and the experimental value is remarkably good.

From a practical standpoint, Hooke's law indicates that the *position* of absorption of a stretching vibration in an IR spectrum depends both on the strength of the vibrating bond and on the masses of the atoms connected by the bond. The stronger the bond and the lighter the atoms, the higher the wavenumber of the stretching vibration. As we saw earlier, the *intensity* of the absorption depends primarily on the polarity of the vibrating bond.

EXAMPLE 14.2

Calculate the stretching frequency in cm^{-1} for a carbon-carbon double bond. Assume each carbon is the most abundant isotope, namely carbon-12.

Solution

Assume a force constant for 10×10^5 dynes/cm for C=C. The calculated wavenumber is 1682 cm^{-1}, a value close to the experimental value of 1630 cm^{-1}.

$$\text{For C=C stretching:} \quad \bar{\nu} = 4.120 \sqrt{\frac{10 \times 10^5}{(12 \times 12)/(12 + 12)}} = 1682 \text{ cm}^{-1}$$

$$\text{Experimental value} = 1630 \text{ cm}^{-1}$$

PROBLEM 14.2

Without doing the calculation, which member of each pair do you expect to occur at the higher wavenumber?

(a) C=O or C=C stretching? **(b)** C=O or C—O stretching?
(c) C≡C or C=O stretching? **(d)** C—H or C—Cl stretching?

Detailed interpretation of most infrared spectra is difficult because of the complexity of vibrational modes. In addition to the fundamental vibrational modes we have described, other types of absorptions occur, resulting in so-called overtone and coupling peaks that are usually quite weak. **Overtone peaks** are higher frequency harmonics of fundamental vibrations and occur at or near integral multiples of the wavenumber of the fundamental vibration. For example, an infrared absorption at 600 cm^{-1} may well have weaker overtone peaks near 1200 cm^{-1}, 1800 cm^{-1}, 2400 cm^{-1}, and so on. **Coupling peaks** result from the coupling of two vibrations by addition ($\bar{\nu}_1 + \bar{\nu}_2$) and by subtraction ($\bar{\nu}_1 - \bar{\nu}_2$). Only certain combinations of these coupling vibrations are allowed, meaning that only certain combinations are possible within the constraints of quantum mechanics.

To one skilled in the interpretation of infrared spectra, the absorption patterns can yield an enormous amount of information about chemical structure. We, however, have neither the time nor the need to develop this level of competence. The value of infrared spectra for us is that they can be used to determine the presence or absence of characteristic functional groups. A carbonyl group, for example, typically shows strong absorption at approximately 1630–1810 cm^{-1}. The position of absorption for a particular carbonyl group depends on whether it is that of an aldehyde, a ketone, a carboxylic acid, an ester, or, if in a ring, on the size of the ring. In Chapters 15, 16, and 17 we discuss how structural variations, such as ring size or conjugation, affect this value.

D. Correlation Tables

Data on absorption patterns of functional groups are collected in tables called **correlation tables.** The correlation table given in Table 14.1 is a listing of characteristic infrared absorptions for the types of bonds and functional groups we deal with most often. With each new functional group introduced in the following chapters, we also present a correlation table for that functional group. A cumulative correlation chart

TABLE 14.1 Characteristic Infrared Absorptions of Selected Functional Groups

Bond	Frequency, cm^{-1}	Intensity
O—H	3200–3650	strong and broad
N—H	3100–3500	medium
C—H	2850–3100	medium to strong
C=O	1630–1810	strong
C=C	1600–1680	weak
C—O	1000–1250	strong

can be found in Appendix 6. In these tables, the intensity of a particular absorption is often referred to as **strong (s), medium (m),** or **weak (w).**

In general, we pay most attention to the region from 3500 cm^{-1} to 1000 cm^{-1}, because the characteristic stretching and bending vibrations for most functional groups are found in this region. Vibrations in the region 1000 cm^{-1} to 400 cm^{-1} are much more complex and far more difficult to analyze. Because even slight variations in molecular structure and absorption patterns are most obvious in this region, it is often called the **fingerprint region.** If two compounds have even slightly different structures, the differences in their infrared spectra are most clearly discernible in this region.

Fingerprint region The portion of the vibrational infrared region that extends from 1000 cm^{-1} to 400 cm^{-1}.

EXAMPLE 14.3

What functional group is most likely present if a compound shows IR absorption at

(a) 1705 cm^{-1} **(b)** 2950 cm^{-1}

Solution

(a) A C=O group **(b)** Aliphatic C—H groups

PROBLEM 14.3

A compound shows strong, very broad IR absorption in the region 3300–3600 cm^{-1} and strong, sharp absorption at 1715 cm^{-1}. What functional groups account for these absorptions?

EXAMPLE 14.4

Propanone and 2-propen-1-ol are constitutional isomers. Show how to distinguish between them by IR spectroscopy.

$$\underset{\text{Propanone (Acetone)}}{CH_3-\overset{\overset{\displaystyle O}{\|}}{C}-CH_3} \qquad \underset{\text{2-Propen-1-ol (Allyl alcohol)}}{CH_2{=}CH-CH_2-OH}$$

Solution

Only propanone shows strong absorption in the C=O stretching region, 1630–1815 cm^{-1}. Alternatively, only 2-propen-1-ol shows strong absorption in the O—H stretching region, 3200–3650 cm^{-1}.

PROBLEM 14.4

Propanoic acid and methyl ethanoate are constitutional isomers. Show how to distinguish between them by IR spectroscopy.

$$\underset{\text{Propanoic acid}}{CH_3CH_2\overset{\overset{\displaystyle O}{\|}}{C}OH} \qquad \underset{\text{Methyl ethanoate (Methyl acetate)}}{CH_3\overset{\overset{\displaystyle O}{\|}}{C}OCH_3}$$

TABLE 14.2 Characteristic Infrared Absorptions of Alkanes, Alkenes, and Alkynes

Hydrocarbon	Vibration	Frequency (cm^{-1})	Intensity
Alkane			
C—H	stretching	2850–3000	strong
CH_2	bending	1450	medium
CH_3	bending	1375 and 1450	weak to medium
C—C	(not useful for interpretation—too many bands)		
Alkene			
C—H	stretching	3000–3100	weak to medium
C=C	stretching	1600–1680	weak to medium
		conjugation moves this peak to the left (to lower frequency) and increases its intensity	
Alkyne			
C—H	stretching	3300	medium to strong
C≡C	stretching	2100–2250	weak

14.2 Interpreting Infrared Spectra

A. Alkanes

Infrared spectra of alkanes are usually simple, with few peaks, the most common of which are given in Table 14.2.

Shown in Figure 14.3 is an infrared spectrum of decane. The strong peak with multiple splittings between 2850 and 3000 cm^{-1} is characteristic of C—H stretching. The other prominent peaks in this spectrum are methylene bending absorption at 1465 cm^{-1} and methyl bending absorption at 1380 cm^{-1}.

FIGURE 14.3 Infrared spectrum of decane (neat, salt plates).

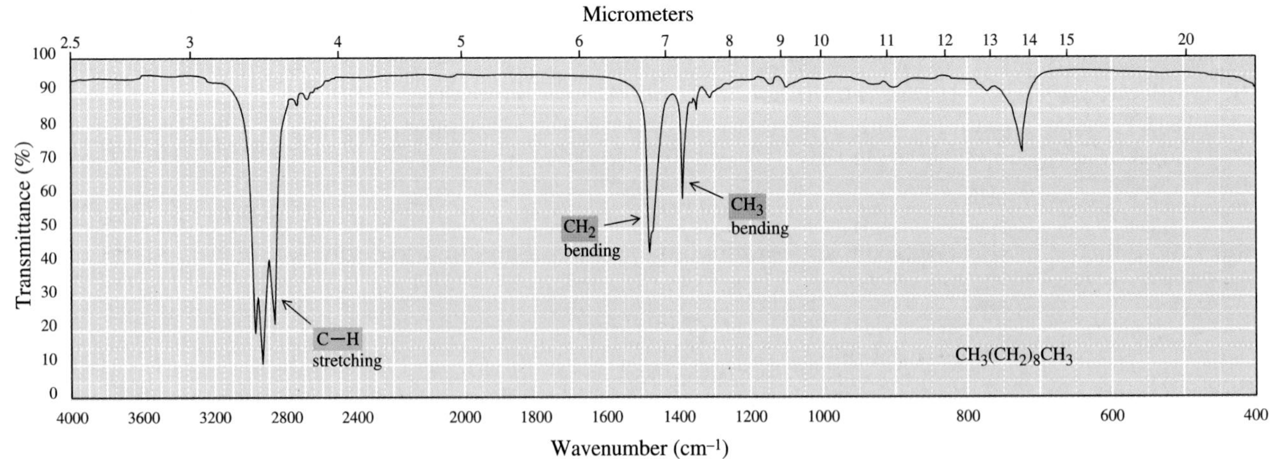

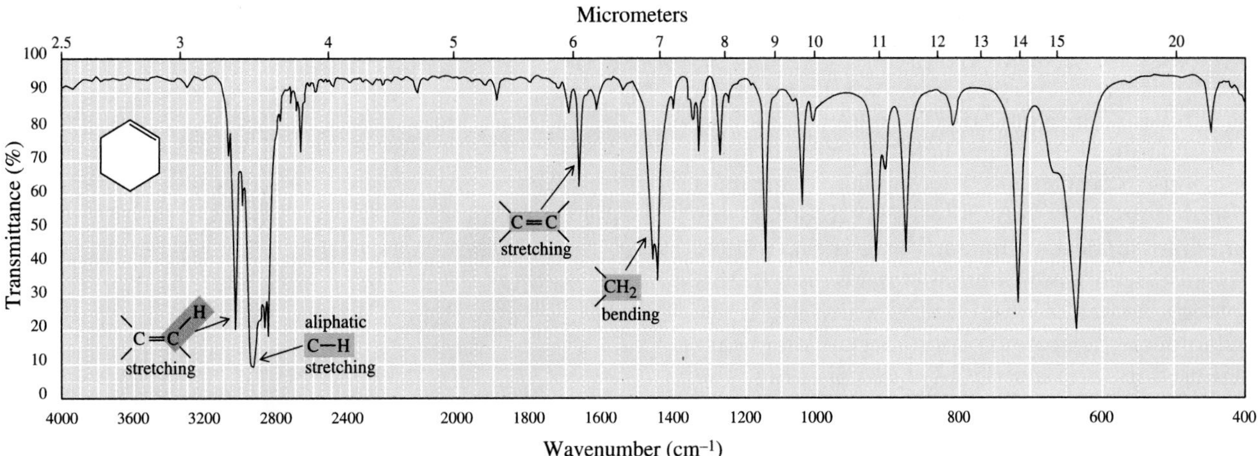

FIGURE 14.4
Infrared spectrum of cyclohexene (neat, salt plates).

B. Alkenes

An easily recognized alkene absorption is the vinylic C—H stretching slightly to the left (greater wavenumber) of 3000 cm^{-1}. Also characteristic of alkenes is C=C stretching at 1600–1680 cm^{-1}. This vibration, however, is often weak and difficult to observe. Conjugation moves the C=C stretch to the left and increases its intensity. Both vinylic C—H stretching and C=C stretching can be seen in the infrared spectrum of cyclohexene (Figure 14.4). Also visible are the aliphatic C—H stretching near 2900 cm^{-1} and methylene bending near 1440 cm^{-1}.

C. Alkynes

Alkyne ≡C—H stretching occurs near 3300 cm^{-1}, a frequency higher than that for either alkyl C—H or vinylic C—H stretching. This peak is usually sharp and strong. Also observed in terminal alkynes is absorption near 2150 cm^{-1} due to C≡C stretching. Both of these peaks can be seen in the infrared spectrum of 1-octyne (Figure 14.5).

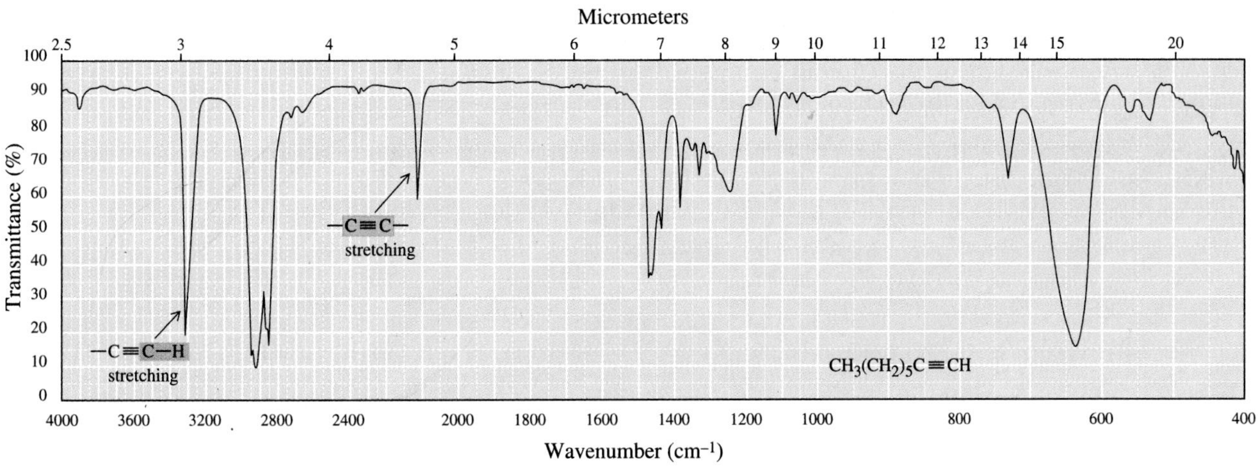

FIGURE 14.5
Infrared spectrum of 1-octyne.

Chemistry in Action

Dyes That Absorb in the Infrared

With the discovery of solid state lasers based on gallium aluminum arsenide and indium phosphide that emit light in the near infrared region ($<14{,}000\ cm^{-1}$) computer companies have started research to find dyes that absorb in this region. One application for such dyes is the WORM (*w*rite *o*nce, *r*ead *m*any) optical storage disc. In this technology, a layer of near-infrared active dye is dispersed under a transparent plastic sheet. A laser beam is then used to form a series of small pits where the dye absorbs energy. Then, a second, lower power laser detects the pits due to differences in reflectivity. An important class of near-infrared active dyes are the phthalocyanines (1). These compounds can have many different groups attached to the aromatic rings. Some also have a metal atom (such as lead) at the center of the molecule.

Optical storage disks use light sensitive dyes. (©*Dan McCoy, Rainbow*)

(1)

A great deal of current research is underway to perfect optical discs that can be erased and then rewritten, similar to the magnetic discs found in most computers. What is needed for this technology to work is a compound that can be switched from colorless to colored with a beam of light. The class of compounds called spiropyrans are under investigation for this application. The spiro compound (2), for example, is colorless. However, when it is irradiated with a beam of ultraviolet light, the molecule rearranges to give the colored compound (3). With an infrared laser, (3) can be switched back into the colorless spiro compound (2), erasing the information stored as a colored bit. Unfortunately, these particular compounds are not suitable as components for erasable storage discs because they degrade after only a few cycles of color changes. Related molecules are under study, and it is hoped that through them will come a major advance in the technology of information storage on optical discs.

(2) Colorless ⇌ (UV / IR) (3) Colored

J. Fabian, H. Nakazumi, and M. Masuoka, *Chem Rev.*, **92**, 1197, 1992.

TABLE 14.3 Characteristic Stretching Vibrations of Alcohols

Bond	Frequency, cm^{-1}	Intensity
O—H (free)	3600–3650	weak
O—H (hydrogen bonded)	3200–3500	medium, broad
C—O	1050–1250	medium

For internal alkynes, the C≡C stretching absorption is often very weak or completely absent.

D. Alcohols

Both the position of the O—H stretching absorption and its intensity depend on the extent of hydrogen bonding. Under normal conditions, where there is extensive hydrogen bonding between alcohol molecules, O—H stretching absorption occurs as a broad peak at 3200–3500 cm^{-1}. The C—O stretching absorption appears in the range 1050–1250 cm^{-1} (Table 14.3).

Shown in Figure 14.6 is an infrared spectrum of 1-hexanol. The hydrogen-bonded O—H stretching appears as a broad band of medium intensity centered at 3340 cm^{-1}. The C—O stretching appears near 1050 cm^{-1}, a value characteristic of primary alcohols.

E. Ethers

The C—O stretching absorptions of ethers are similar to those observed in alcohols. Dialkyl ethers typically show a single absorption in the region between 1000 and 1150 cm^{-1} as can be seen in the infrared spectrum of dibutyl ether (Figure 14.7).

The presence or absence of O—H stretching absorption at 3200 to 3500 cm^{-1} for hydrogen-bonded O—H can be used to distinguish between an ether and an

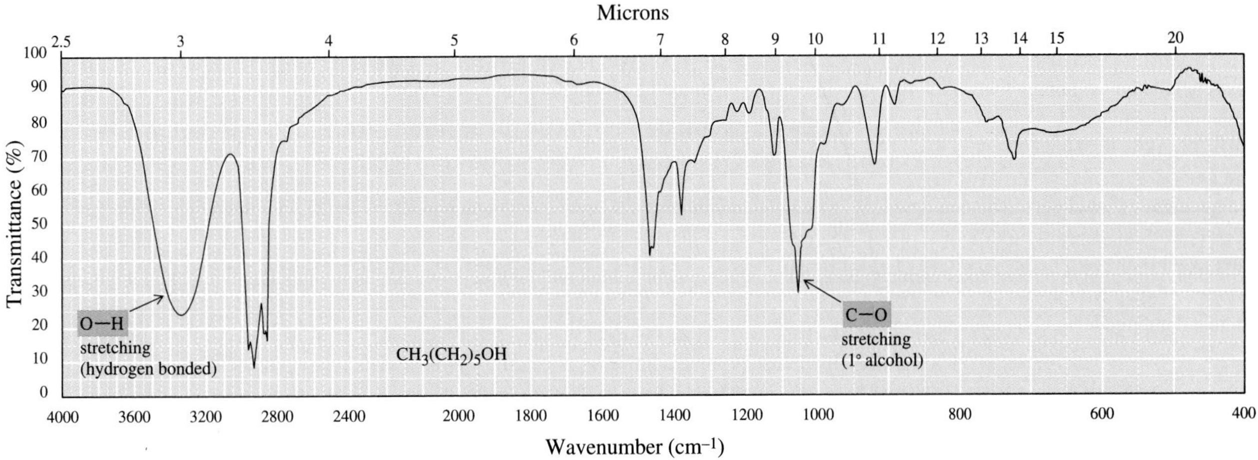

FIGURE 14.6
Infrared spectrum of 1-hexanol (neat, salt plates).

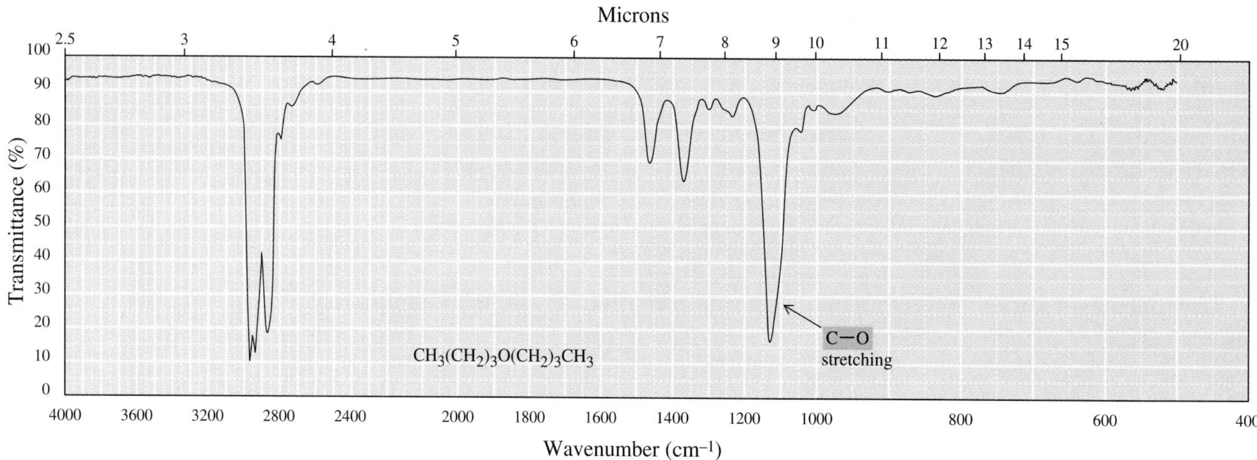

FIGURE 14.7
Infrared spectrum of dibutyl ether.

isomeric alcohol. As we shall see in Chapter 17, a C—O stretching vibration is also present in esters. In this case, the presence or absence of C═O stretching can be used to distinguish between an ether and an ester.

14.3 Ultraviolet-Visible Spectroscopy

At the molecular level, absorption of ultraviolet-visible radiation (Table 13.2) causes transitions of pi and nonbonding electrons between electronic energy levels. In this section, we study correlations between absorption of ultraviolet-visible radiation and the information this absorption gives us about the presence and substitution patterns of carbon-carbon double bonds and of conjugation of carbon-carbon and carbon-oxygen double bonds.

A. Introduction

The region of the electromagnetic spectrum covered by most ultraviolet spectrophotometers runs from 200 to 400 nm, a region commonly referred to as the **near-ultraviolet.** Because oxygen of the atmosphere absorbs ultraviolet (UV) radiation at wavelengths below 200 nm, spectra obtained in this region must be run in an atmosphere of pure nitrogen or in a vacuum, hence, the name **vacuum ultraviolet.** Because of the special instrumentation required for vacuum ultraviolet, this region is little used for routine analysis. The region covered by most visible spectrophotometers runs from 400 nm (violet light) to 700 nm (red light), often with extensions into the IR to 800 or 1000 nm.

Ultraviolet spectroscopy A spectroscopic technique in which a compound is irradiated with ultraviolet radiation, absorption of which causes electrons to change from a lower electronic energy level to a higher one. Ultraviolet spectroscopy is particularly valuable for determining the extent of conjugation in organic molecules containing pi bonds.

EXAMPLE 14.5

Calculate the energy of radiation at either end of the near-ultraviolet spectrum, that is, at 200 nm and 400 nm (review Section 13.1).

Solution

Use the relationship $E = hc/\lambda$. Be certain to express the dimension of length in consistent units.

$$E = \frac{hc}{\lambda}$$

$$= 9.54 \times 10^{-14} \frac{\text{kcal} \cdot \text{sec}}{\text{mol}} \times 3.00 \times 10^{8} \frac{\text{m}}{\text{sec}} \times \frac{1}{200 \times 10^{-9}\ \text{m}} = 143\ \text{kcal/mol}$$

By a similar calculation, the energy of radiation of wavelength 400 nm is found to be 71.5 kcal/mol.

PROBLEM 14.5

Wavelengths in ultraviolet-visible spectroscopy are commonly expressed in nanometers; wavelengths in infrared spectroscopy are commonly expressed in micrometers. Carry out the following conversions:

(a) 2.5 μm to nanometers **(b)** 200 nm to micrometers

Wavelengths and corresponding energies for near-ultraviolet and visible radiation are summarized in Table 14.4.

Ultraviolet and visible spectral data are recorded on chart paper as plots of **absorbance (*A*)** on the vertical axis versus wavelength on the horizontal axis.

Absorbance (A) A quantitative measure of the extent to which a compound absorbs ultraviolet-visible radiation of a particular wavelength.

$$\text{Absorbance } (A) = \log \frac{I_0}{I}$$

where I_0 is the intensity of radiation incident on the sample, and I is the intensity of the radiation transmitted through the sample.

Typically, UV-visible spectra consist of a small number of peaks, sometimes just one peak. Figure 14.8 is an ultraviolet absorption spectrum of 2,5-dimethyl-2,4-hexadiene. Absorption of ultraviolet radiation by this conjugated diene begins at wavelengths below 200 nm and continues to almost 270 nm with maximum absorption at 242 nm. This spectrum is reported as a single absorption peak using the notation λ_{max} 242 nm.

The extent of absorption of ultraviolet-visible radiation is proportional to the number of molecules capable of undergoing the observed electronic transition, and, therefore, ultraviolet-visible spectroscopy can be used for quantitative analysis of samples. The relationship between absorbance, concentration, and length of the sample cell (cuvette) is known as the **Beer-Lambert law.** The proportionality constant in this equation is given the name **molar absorptivity (ϵ)** or extinction coefficient.

$$\text{Beer-Lambert law:} \quad A = \epsilon c l$$

TABLE 14.4 Wavelengths and Energies for Near-Ultraviolet and Visible Radiation

Region of Spectrum	Wavelength (nm)	Energy (kcal/mol)
ultraviolet	200–400	71.5–143
visible	400–700	40.9–71.5

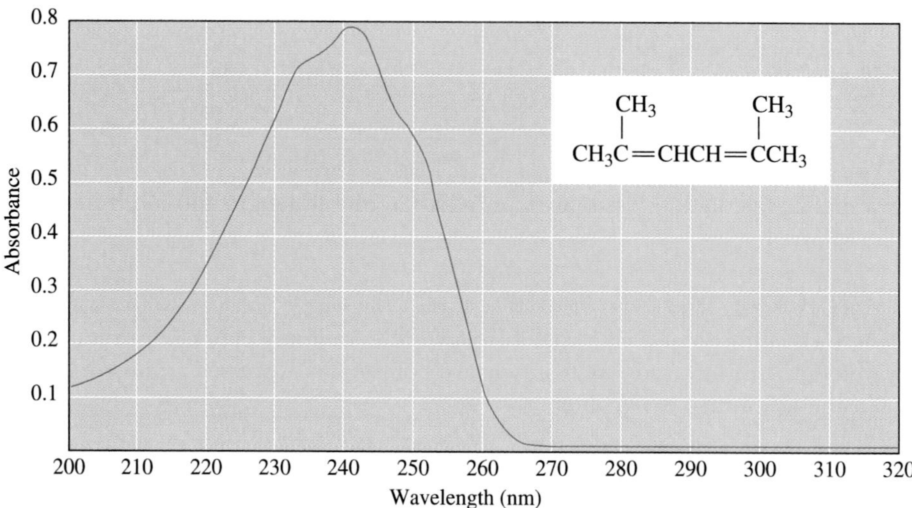

FIGURE 14.8
Ultraviolet spectrum of 2,5-dimethyl-2,4-hexadiene (in methanol).

where A is the absorbance, ϵ is the molar absorptivity (in liters per mol per centimeter, $M^{-1}cm^{-1}$), c is the concentration of solute (in moles per liter, M), and l is the length of the sample cell (in centimeters, cm).

The molar absorptivity is a characteristic property of a compound and is not affected by its concentration or the length of the light path. Values range from zero to 10^6 $M^{-1}cm^{-1}$. Values above 10^4 $M^{-1}cm^{-1}$ correspond to high-intensity absorptions; values below 10^4 $M^{-1}cm^{-1}$ to low-intensity absorptions. The molar absorptivity of 2,5-dimethyl-2,4-hexadiene, for example, is 13,100 $M^{-1}cm^{-1}$, a high-intensity absorption.

EXAMPLE 14.6

The molar absorptivity of 2,5-dimethyl-2,4-hexadiene in methanol is 13,100 $M^{-1}cm^{-1}$. What concentration of this diene in methanol is required to give an absorbance of 1.6? Assume a light path of 1.00 cm. Calculate concentration in units of:

(a) moles per liter **(b)** milligrams per milliliter

Solution

Solve the Beer-Lambert equation for concentration and substitute appropriate values for length, absorbance, and molar absorptivity.

(a) $$c = \frac{A}{l \times \epsilon} = \frac{1.6}{1.00 \text{ cm} \times 13{,}100 \text{ L} \cdot \text{mol}^{-1} \cdot \text{cm}^{-1}} = 1.22 \times 10^{-4} \text{ mol/L}$$

(b) The molecular weight of 2,5-dimethyl-2,4-hexadiene is 110 g/mol. The concentration of the sample in milligrams per milliliter is

$$1.22 \times 10^{-4} \frac{\text{mol}}{\text{L}} \times \frac{110 \text{ g}}{\text{mol}} \times \frac{1\text{L}}{1000 \text{ mL}} \times \frac{1000 \text{ mg}}{\text{g}} = 1.34 \times 10^{-2} \text{ mg/mL}$$

PROBLEM 14.6

The visible spectrum of the tetraterpene β-carotene ($C_{40}H_{56}$, MW 536.88, the orange pigment of carrots) dissolved in hexane shows intense absorption maxima at 463 nm and 494 nm, both in the blue-green region. Because light of these wavelengths is absorbed by β-carotene, we perceive the color of this compound as that of the complement to blue-green, namely, red-orange.

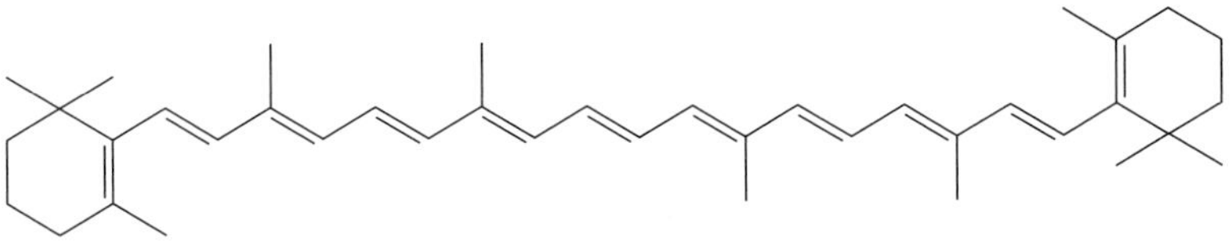

β-carotene

λ_{max} 463 (log ε 5.10); 494 (log ε 4.77)

Calculate the concentration in milligrams per milliliter of β-carotene that gives an absorbance of 1.8 at λ_{max} 463.

B. The Origin of Transitions Between Electronic Energy Levels

Absorption of radiation in the ultraviolet-visible spectrum results in a transition of electrons from a lower energy occupied MO to a higher energy unoccupied MO. The energy of this radiation is generally insufficient to affect electrons in sigma bonding molecular orbitals but sufficient to affect $\pi \rightarrow \pi^*$ transitions of electrons in pi MOs, most particularly pi electrons of conjugated systems. Following are three examples of conjugated systems.

$CH_2{=}CH{-}CH{=}CH_2$ — 1,3-Butadiene

$CH_2{=}CH{-}\overset{O}{\overset{\|}{C}}{-}CH_3$ — 3-Buten-2-one

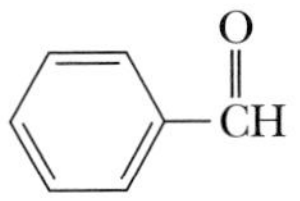

$C_6H_5{-}\overset{O}{\overset{\|}{C}}H$ — Benzaldehyde

As we shall see in Section 15.4, ultraviolet-visible radiation is also sufficient to cause $n \rightarrow \pi^*$ transitions of nonbonding electrons (i.e., unshared) on oxygen in carbonyl groups.

As an example of a $\pi \rightarrow \pi^*$ transition, consider ethylene. The double bond in ethylene consists of one sigma bond formed by combination of sp^2 orbitals and one pi bond formed by combination of $2p$ orbitals. The relative energies of the pi bonding and antibonding molecular orbitals are shown schematically in Figure 14.9. The $\pi \rightarrow \pi^*$ transitions for simple, unconjugated alkenes occur below 200 nm (at 165 nm for ethylene). Because these transitions occur in the vacuum ultraviolet, they are not observed in conventional ultraviolet spectroscopy and, therefore, are not useful to us for determination of molecular structure.

For 1,3-butadiene, the difference in energy between the highest occupied pi molecular orbital and the lowest unoccupied pi antibonding molecular orbital is less than it is for ethylene with the result that a $\pi \rightarrow \pi^*$ transition for 1,3-butadiene (Figure 14.10) takes less energy (occurs at longer wavelength) than that for ethylene. This transition for 1,3-butadiene occurs at 217 nm.

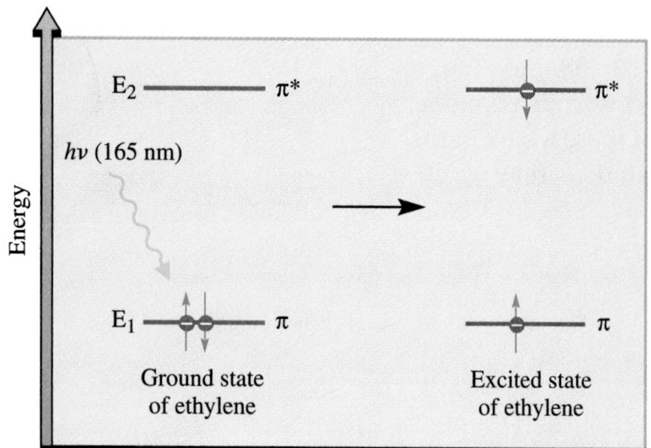

FIGURE 14.9
Electronic excitation of ethylene, a $\pi \rightarrow \pi^*$ transition. Absorption of ultraviolet radiation causes a transition of an electron from a pi bonding MO in the ground state to a pi* antibonding MO in the excited state. There is no change in electron spin.

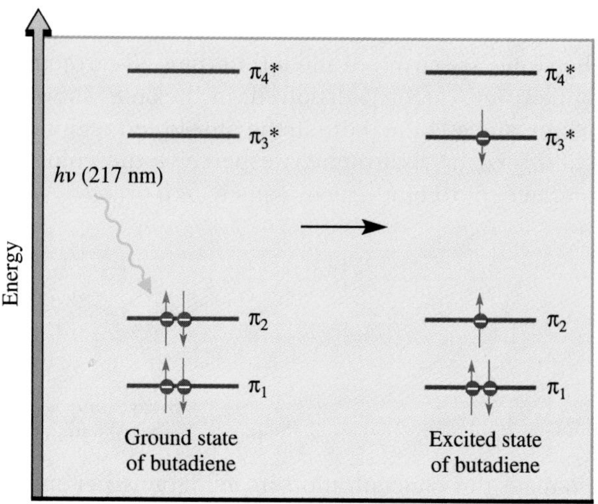

FIGURE 14.10
Electronic excitation of 1,3-butadiene; a $\pi \rightarrow \pi^*$ transition.

Electronic excitations are routinely accompanied by changes in vibrational or rotational energy levels. The energy levels for these excitations are quite closely spaced and their differences are considerably smaller than the energy differences between electronic excitations (Figure 14.11). Transitions between vibration and rotation energy levels are superposed on the electronic excitations, which results in a large number of absorption peaks so closely spaced that the spectrophotometer cannot resolve

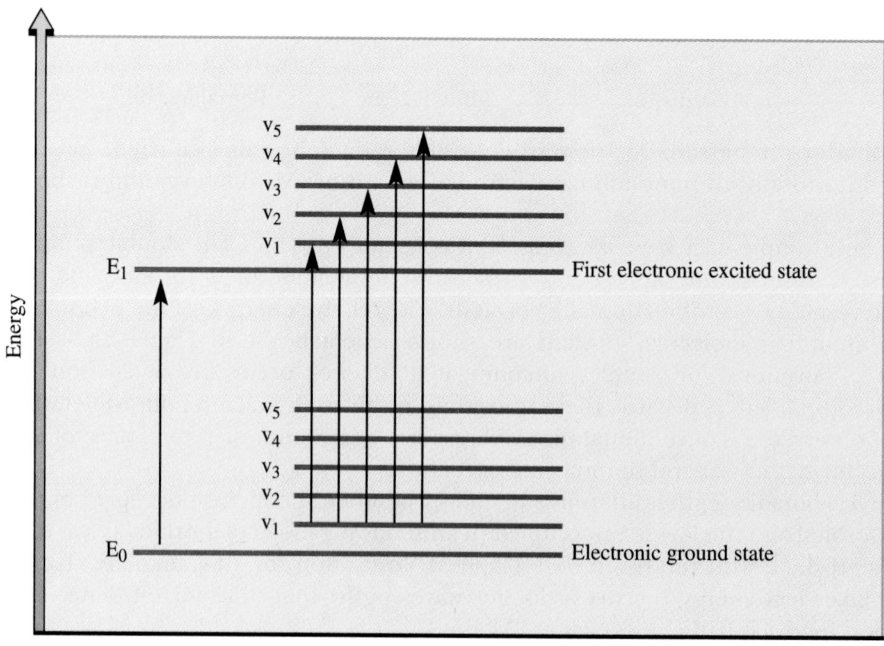

FIGURE 14.11
Electronic energy levels E_0 and E_1 with associated vibrational energy levels. Not shown are associated rotational energy levels.

CHEMISTRY IN ACTION

Chemiluminescence

Spectroscopy deals with the interaction of electromagnetic radiation, in this case light, with matter. Some compounds can emit light as well as absorb it. Chemiluminescence occurs when a chemical reaction produces light as one of its products. For light to be emitted, the reaction must be highly exothermic.

One class of compounds that exhibit chemiluminescence are the 1,2-dioxetanes. Their molecules possess a strained, four-membered ring and a weak oxygen-oxygen bond. At or above room temperature, 1,2-dioxetanes fragment into two carbonyl compounds. The relief of ring strain and the generation of two strong carbon-oxygen double bonds liberate so much energy that one of the carbonyl compounds is formed in an electronically excited state. The electronically excited carbonyl compound then gives off light as it relaxes to its ground state.

Chemical reactions often release energy. The reaction in this tube and beaker releases energy in the form of light in a process called chemiluminescence. (Charles D. Winters)

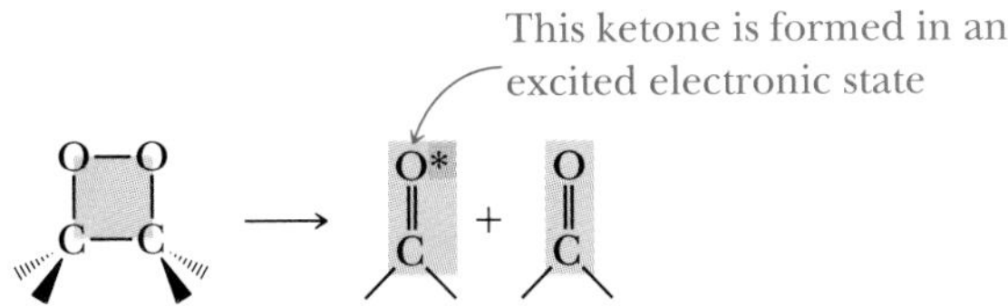
This ketone is formed in an excited electronic state
O—O
C—C
O*
C
+
O
C

A 1,2-dioxetane

Recently, new 1,2-dioxetanes have been synthesized, which make possible new and ultrasensitive detection procedures. For example, the 1,2-dioxetane (1) is stable at room temperature and neutral pH. When (1) is treated with water in the presence of alkaline phosphatase, an enzyme which catalyzes hydrolysis of esters of phosphoric acid, the unstable dioxetane anion (2) is produced. Compound (2) undergoes rapid fragmentation, and light from the electronically excited ester is readily detected.

(continued)

(1) OCH_3, O—O, OPO_3^{2-} $+ H_2O$ $\xrightarrow{\text{alkaline phosphatase}}$ (2) OCH_3, O—O, O^- $+ H_2PO_4^-$ $\xrightarrow{25°C}$ O $+$ OCH_3, C, *O, O^-

Research in clinical chemistry has now demonstrated that chemiluminescence has the potential to become a very useful noninvasive diagnostic technique. It can be used, for example, to detect antigens in the following way. First, the alkaline phosphatase enzyme is attached to a specific antibody and then mixed with serum or another clinical sample. If the corresponding antigen is present in the sample, an antibody-antigen complex is formed and then separated from unreacted antibody. To detect the complex, the 1,2-dioxetane (1) is added. Alkaline phosphatase catalyzes the hydrolysis of (1) to (2), the unstable dioxetane anion decomposes, and light is given off as the electronically excited ester returns to its ground state. The amount of light generated is proportional to the amount of antigen in the sample. Because the enzyme can catalyze the destruction of many 1,2-dioxetane molecules, and because light is so easily detected, these enzyme-linked chemiluminescence reactions can detect as little as an attomole (10^{-18} mole) of antigen. Because these sensitivity limits equal or exceed the sensitivity of analytical techniques based on radioactivity, chemiluminescent assays may, in time, replace many assays that now rely on radioactivity for signal detection. This will lessen radioactive waste disposal problems and enhance safety for clinical workers.

A. P. Schaap, H. Akhavan, and L. J. Romano, *Clin. Chem.*, **35**, 1863, 1989.

them. For this reason, UV-visible absorption peaks usually are much broader than IR absorption peaks. Thus, the ultraviolet-visible spectrum of a compound consists of a band or bands of absorption with maxima corresponding to the major electronic transitions.

It has been found from analysis of spectral data that the greater the number of double bonds in conjugation, the longer the wavelength of ultraviolet radiation absorbed (the smaller the energy required for the $\pi \rightarrow \pi^*$ transition). As a result, UV-visible spectroscopy is mainly useful for the study of compounds containing conjugated double bonds. Shown in Table 14.5 are wavelengths and energies required for $\pi \rightarrow \pi^*$ transitions in several conjugated alkenes.

TABLE 14.5 Wavelengths and Energies Required for $\pi \rightarrow \pi^*$ Transitions of Ethylene and Three Conjugated Polyenes

Name	Structural Formula	λ_{max}	Energy (kcal/mol)
ethylene	$CH_2{=}CH_2$	165	173
1,3-butadiene	$CH_2{=}CHCH{=}CH_2$	217	132
(3*E*)-1,3,5-hexatriene	$CH_2{=}CHCH{=}CHCH{=}CH_2$	268	107
(3*E*,5*E*)-1,3,5,7-octatetraene	$CH_2{=}CH(CH{=}CH)_2CH{=}CH_2$	290	92

SUMMARY

The **vibrational infrared** spectrum (Section 14.1A) extends from 4000 to 400 cm^{-1}. To be **infrared-active** (Section 14.1B), a bond must be polar and its vibration must change the dipole moment of the molecule. There are $\mathbf{3n - 6}$ allowed fundamental vibrations for a nonlinear molecule containing n atoms. The simplest vibrations that give rise to absorption of infrared radiation are **stretching** and **bending** vibrations. Stretching may be symmetrical or asymmetrical. Scissoring, rocking, wagging, and twisting are names given to types of bending vibrations.

The frequency of vibration for an infrared-active bond can be derived from Hooke's law for the vibration of a simple harmonic oscillator (Section 14.1C). From this equation, we can make the following correlations. The frequency of vibration increases when (1) bond strength increases and (2) the reduced mass of the vibrating system decreases. In addition to fundamental vibration peaks, infrared spectra also contain overtone and coupling peaks that are usually much weaker.

A **correlation chart** is a list of the absorption patterns of characteristic functional groups. The intensity of a peak is referred to as strong (s), medium (m), or weak (w). Characteristic bending and stretching vibrations for most functional groups appear in the region 3600 to 1000 cm^{-1}. The region 1000 to 400 cm^{-1} is called the **fingerprint region** (Section 14.1D).

The region of the electromagnetic spectrum covered by most ultraviolet spectrophotometers runs from 200 to 400 nm, a region commonly referred to as the **near-ultraviolet** (Section 14.3A). Wavelengths and corresponding energies for near-ultraviolet and visible radiation are summarized in Table 14.4. Ultraviolet and visible spectral data are recorded as plots of **absorbance (A)** versus wavelength. The relationship between absorbance, concentration, and length of the sample cell is known as the **Beer-Lambert law.**

Beer-Lambert law: $A = \epsilon cl$

where A is the absorbance, ϵ is the molar absorptivity (in liters per mole per centimeter), c is the concentration of solute (in moles per liter), and l is the length of the sample cell (in centimeters).

Absorption of radiation in the near-ultraviolet-visible spectrum is generally sufficient to cause $\pi \rightarrow \pi^*$ transitions of electrons in double bonds, particularly pi electrons of conjugated systems. It is also sufficient to affect $n \rightarrow \pi^*$ transitions of nonbonding electrons associated with carbonyl groups. Absorption of ultraviolet-visible radiation usually occurs over a relatively wide range of wavelengths because transitions between vibrational and rotational energy levels are superposed on the electronic transitions. The result is a vast number of absorption peaks so closely spaced that the spectrophotometer cannot resolve them. Thus, the ultraviolet-visible spectrum of a compound consists of a band or small number of bands of absorption with maxima centered on the major electronic transitions.

ADDITIONAL PROBLEMS

Molecular Mechanics

14.7 In molecular mechanics calculations, the energy required to stretch or compress a bond is given by $E_b = k_b(r - r_0)^2$ where k_b is a constant for a given type of bond, r_0 is the equilibrium bond length, and r is the length of the bond in the stretched or compressed state. Values of these parameters for some common types of bonds are shown in the table:

Bond Type	k_b **(kcal/mol · nm²)**	r_0 **(nm)**
C=O	58.0	0.123
$C(sp^3)$—$C(sp^3)$	20.0	0.153
$C(sp^3)$—H	30.0	0.108
O—H	45.0	0.096

How much energy is required to stretch each type of bond by 5% of its length? By 10% of its length?

14.8 In molecular mechanics calculations, the energy required to bend a bond is given by $E_q = k_q(q - q_0)^2$ where k_q is a constant for a given type of bond, q_0 is the equilibrium bond angle, and q is the angle of the bond in its bent state. Values of these parameters for some common types of bonds are shown in the table.

Bond Type	k_q **(kcal/mol-radians²)**	q_0 **(degrees)**
$C(sp^3)$—C=O	86	121.6
$C(sp^3)$—$C(sp^3)$—$C(sp^3)$	70	109.5
H—$C(sp^3)$—H	40	109.5
$C(sp^3)$—$C(sp^3)$—O	50	109.5

How much energy is required to bend each type of bond by 5%? By 10%?

14.9 Given your answers to the two previous problems, is it easier to bend bonds or to stretch bonds?

Infrared Spectra

14.10 Following are infrared spectra of methylenecyclopentane and 2,3-dimethyl-2-butene. Assign each compound its correct spectrum.

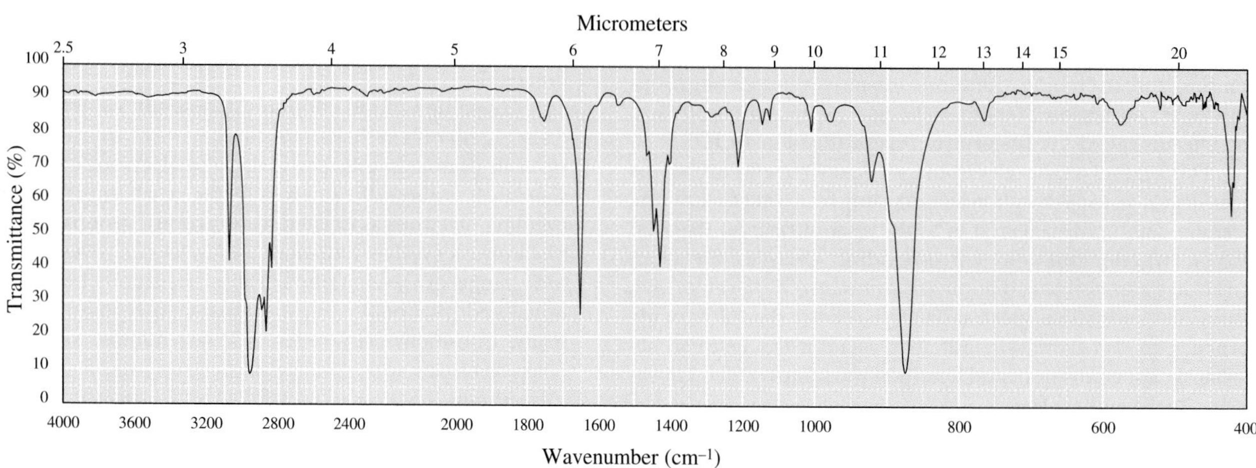

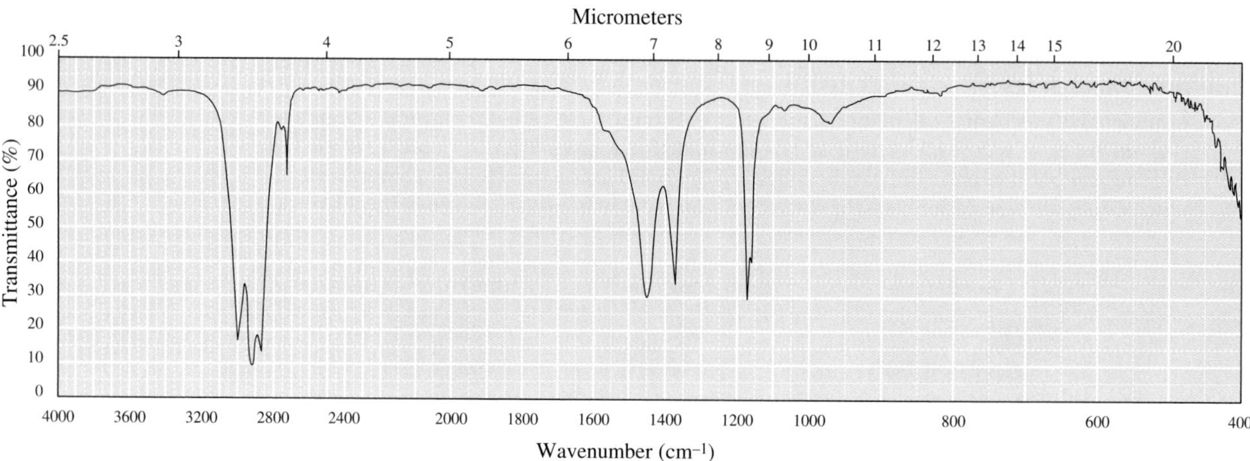

14.11 Following are infrared spectra of nonane and 1-hexanol. Assign each compound its correct spectrum.

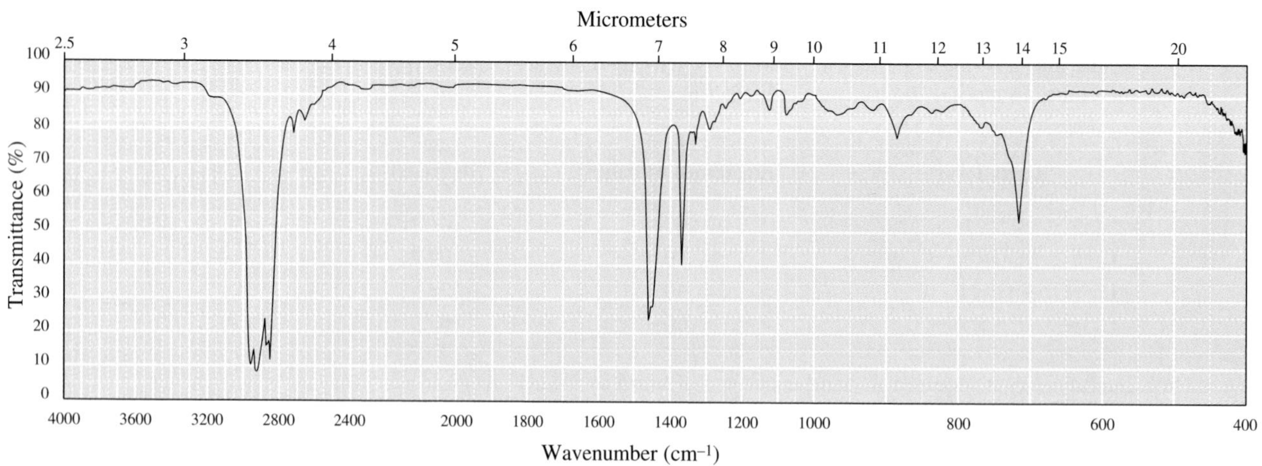

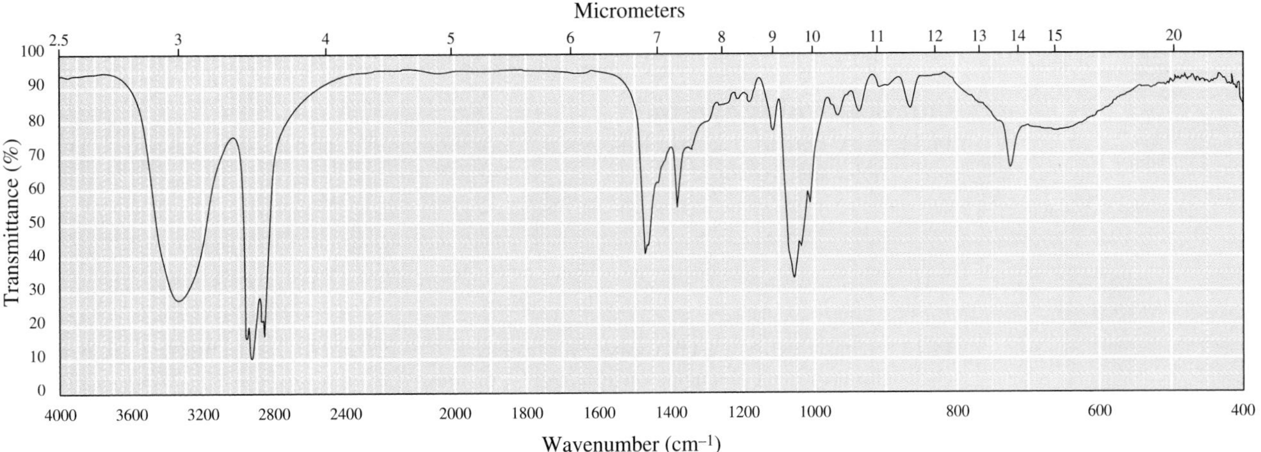

14.12 Following are infrared spectra of 2-methyl-1-butanol and *tert*-butyl methyl ether. Assign each compound its correct spectrum.

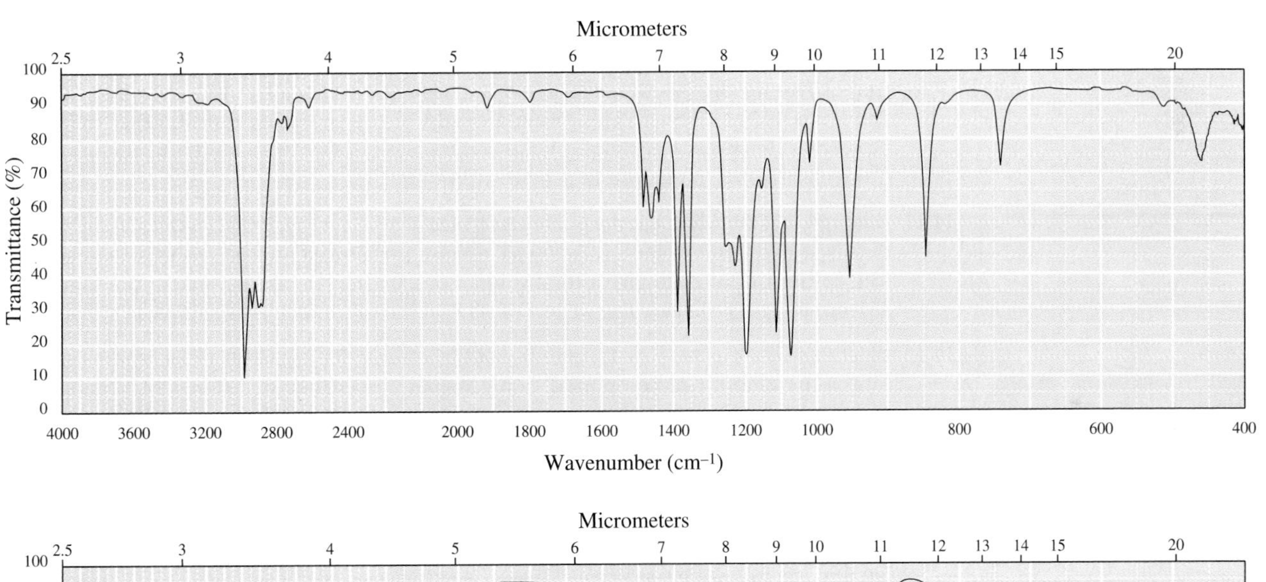

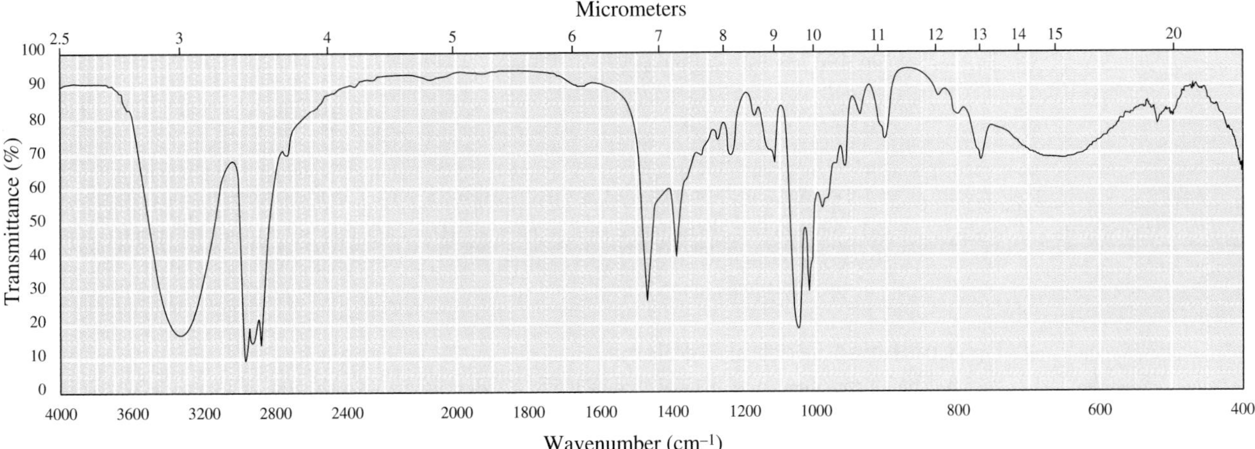

14.13 The IR C≡C stretch absorptions in symmetrical alkynes are usually absent. Why is this so?

Ultraviolet-Visible Spectra

14.14 Show how to distinguish between 1,3-cyclohexadiene and 1,4-cyclohexadiene by ultraviolet spectroscopy.

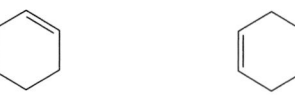

1,3-Cyclohexadiene 1,4-Cyclohexadiene

14.15 Pyridine exhibits a UV transition of the type $n \rightarrow \pi^*$ at 270 nm. In this transition, one of the unshared electrons on nitrogen is promoted from a nonbonding MO to a pi* antibonding MO. What is the effect on this UV peak if pyridine is protonated?

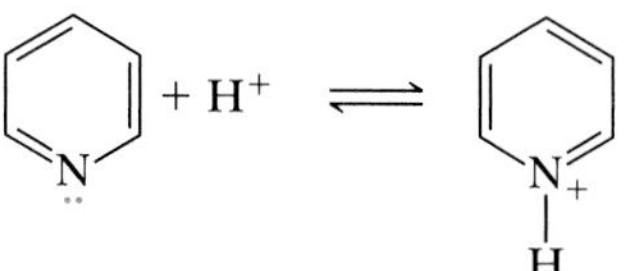

Pyridine Pyridinium ion

14.16 The weight of proteins or nucleic acids in solution is commonly determined by UV spectroscopy using the Beer-Lambert law. For example, the ϵ of double-stranded DNA at 260 nm is 6670 $M^{-1}cm^{-1}$. The formula weight of the repeating unit in DNA (650) can be used as the molecular weight. What is the weight of DNA in 2.0 mL of aqueous buffer if the absorbance, measured in a 1-cm cuvette, is 0.75?

14.17 A sample of adenosine triphosphate (ATP) (MW 507, ϵ = 14,700 $M^{-1}cm^{-1}$ at 257 nm) is dissolved in 5.0 mL of buffer. A 250-μL aliquot is removed and placed in a 1-cm cuvette with sufficient buffer to give a total volume of 2.0 mL. The absorbance of the sample at 257 nm is 1.15. Calculate the weight of ATP in the original 5.0-mL sample.

14.18 Biochemical molecules are frequently sold by optical density (OD) units, where one OD unit is the amount of compound that gives an absorbance of 1.0 at its UV maximum in 1.0 mL of solvent in a 1-cm cuvette. If the cost of 10.0 OD units of a DNA polymer, ϵ = 6600 $M^{-1}cm^{-1}$ at 262 nm, is $51, what is the cost per gram of this biochemical?

14.19 The Beer-Lambert law applies to IR spectroscopy as well as to UV. Whereas ultraviolet spectra are a plot of absorbance (*A*) versus wavelength, IR spectra are a plot of percent transmittance (% T) versus wavenumber. Absorbance and percent transmittance are related in the following way:

$$\text{Absorbance } (A) = \log \frac{T_0}{T}$$

where T_0 is the baseline (100%) transmittance and T is the transmittance at the peak.

(a) What is the absorbance of a peak with 10% transmittance in an IR spectrum?

(b) If the concentration of this sample is halved, how does the absorbance and percent transmittance change?

Combined Spectral Problems

14.20 Compound A, a hydrocarbon, bp 81°C, decolorizes a solution of bromine in carbon tetrachloride. Following are its mass spectrum, ^{1}H-NMR spectrum, and infrared spectrum. (*Note:* Compound A is transparent to ultraviolet-visible radiation.)

(a) What is the molecular formula of compound A?

(b) From its molecular formula, calculate the index of hydrogen deficiency of compound A. How many rings are possible for this compound? How many pi bonds?

(c) Propose a structural formula for compound A consistent with the spectral information.

(d) Account for the presence of peaks in the mass spectrum of compound A at *m/z* 67, 53, 41, 29, and 15.

(e) The ^{13}C-NMR spectrum of compound A shows signals at δ 12.4, 14.4, and 80.9. Assign each carbon in compound A its appropriate ^{13}C chemical shift.

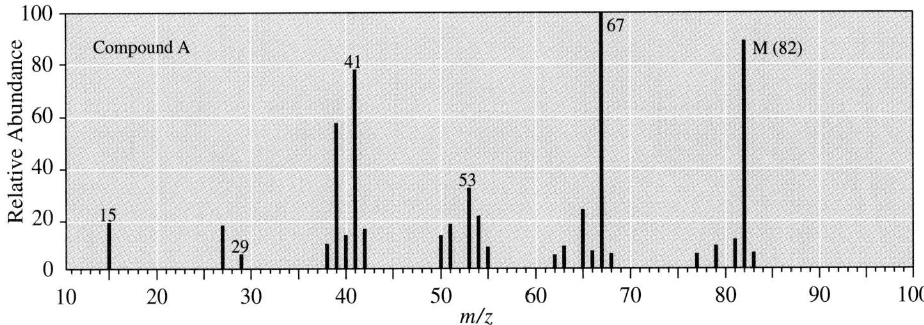

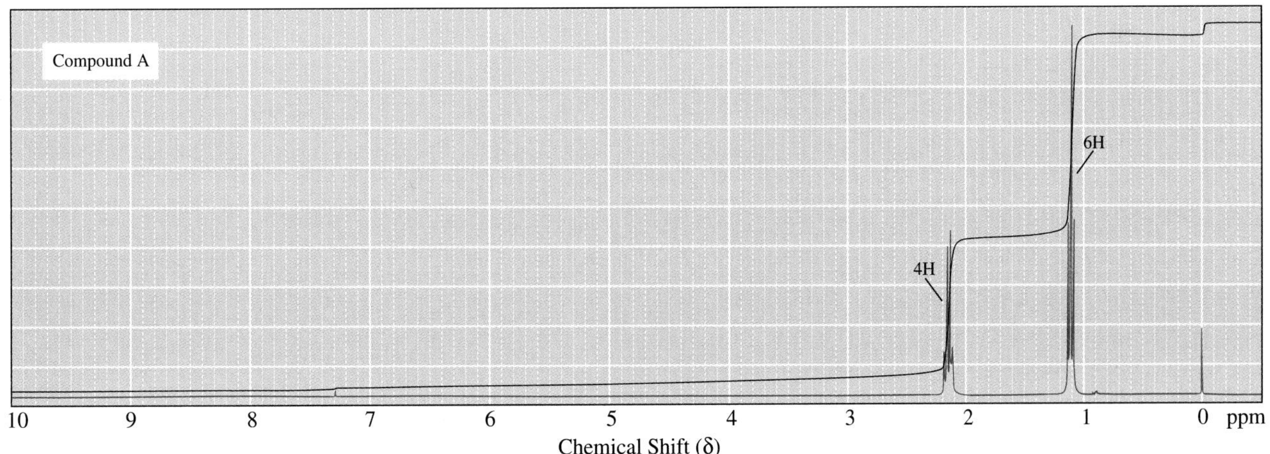

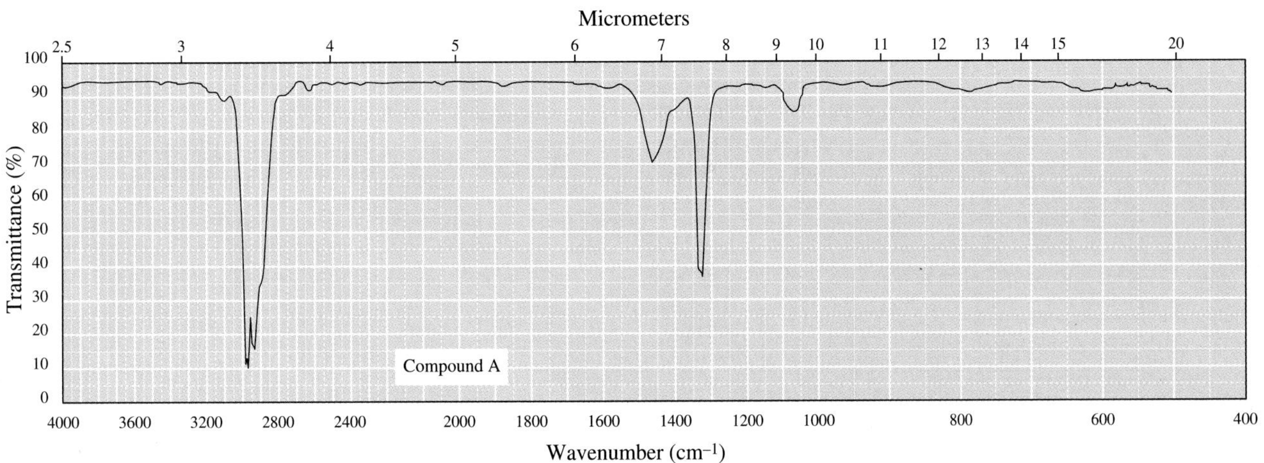

14.21 Compound B is a liquid, bp 122°C. Following are its ^{1}H-NMR and infrared spectra. Compound B shows a molecular ion peak at m/z 136 and an M + 2 peak of approximately equal intensity at m/z 138.

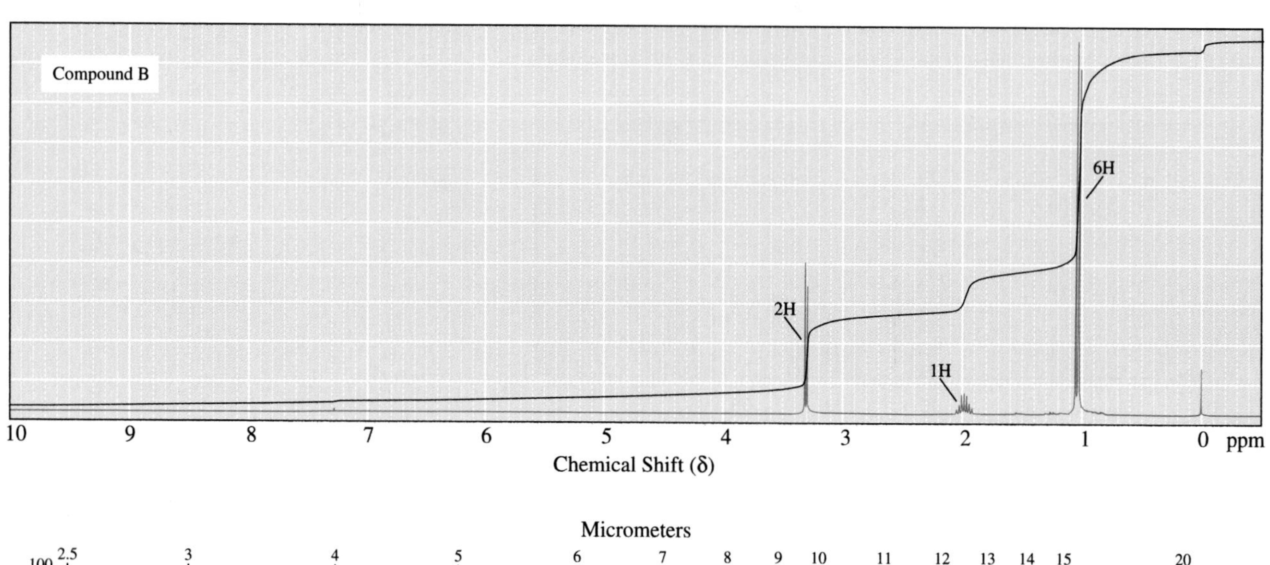

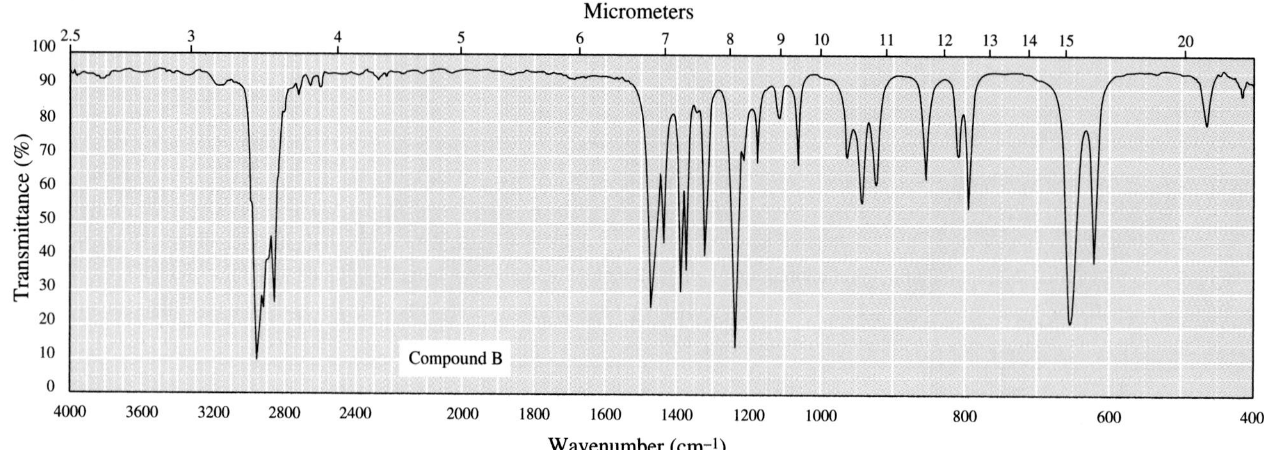

(a) From this information, deduce the structural formula of compound B.
(b) Account for the presence of peaks in its mass spectrum at m/z 123, 121, 43, and 41.
(c) The DEPT ^{13}C-NMR spectrum of compound B shows signals at δ 21.0 (CH_3), 30.47 (CH), and 42.45 (CH_2). Assign each carbon in compound B its appropriate ^{13}C chemical shift.

14.22 Compound C, C_4H_8O, is a liquid, bp 97°C. Following are its ^{1}H-NMR spectrum and IR spectrum.

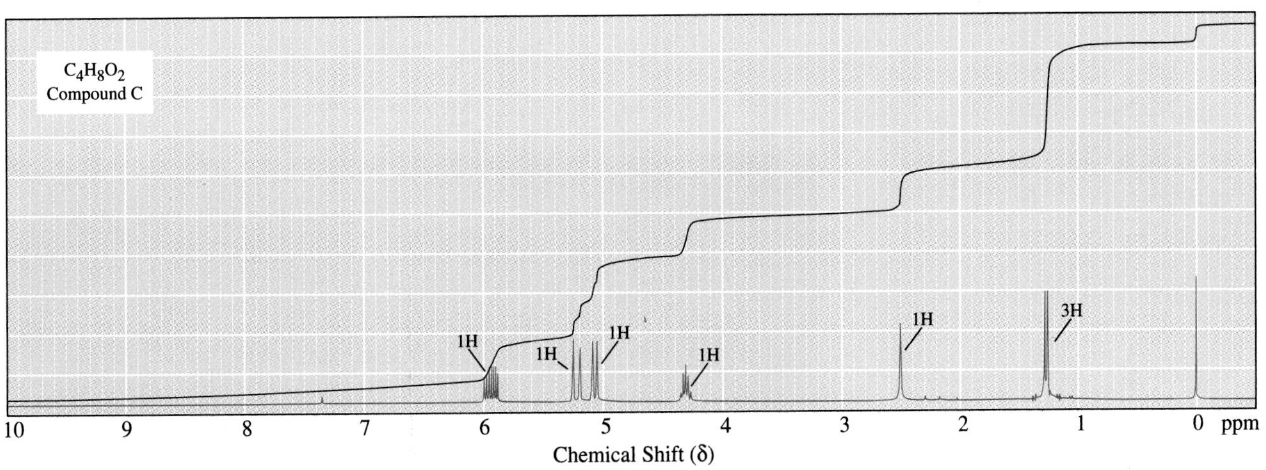

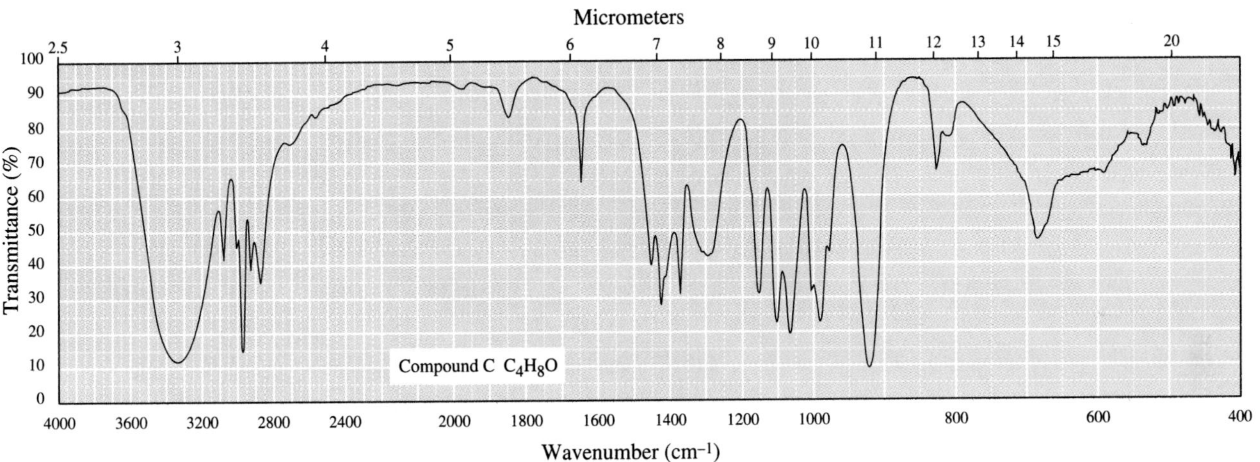

(a) What is the index of hydrogen deficiency of compound C? How many pi bonds can it contain?

(b) What information can you get from the infrared spectrum about the oxygen-containing functional group?

(c) What information can you get from the ^{1}H-NMR spectrum about the presence or absence of vinylic hydrogens?

(d) Propose a structural formula for compound C consistent with the spectral information.

(e) Account for the presence of peaks in the mass spectrum of compound C at m/z 72 (M), 71, 57, 45, 27, and 15.

(f) Account for the splitting patterns of single hydrogens at δ 5.1, 5.3, and 5.8. [*Hint:* Review the ^{1}H-NMR spectrum of vinyl acetate (Figure 13.19).]

(g) The ^{13}C-NMR spectrum of compound C shows peaks at δ 23.1, 68.92, 113.5, and 143.3. Assign these peaks to the appropriate carbons in compound C.

15 Aldehydes and Ketones

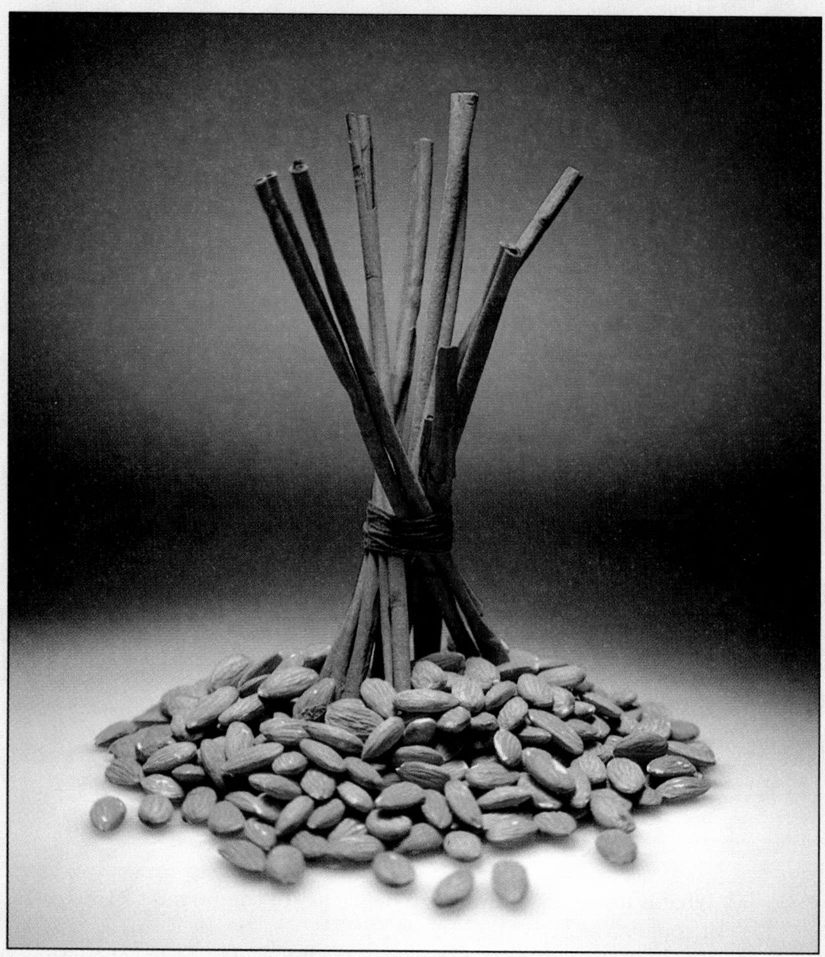

■ Benzaldehyde is found in the kernels of bitter almonds. Cinnamaldehyde is found in Ceylon and Chinese cinnamon oils. *(Charles D. Winters)*

In this and several following chapters, we study the physical and chemical properties of compounds containing the **carbonyl group, C=O.** Because the carbonyl group is the functional group of aldehydes, ketones, and carboxylic acids and their derivatives, it is one of the most important functional groups in organic chemistry. The chemical properties of this functional group are straightforward, and an understanding of its few characteristic reaction themes leads very quickly to an understanding of a wide variety of organic reactions.

15.1 Structure and Bonding

The functional group of an **aldehyde** is a carbonyl group bonded to a hydrogen atom and a carbon atom (Section 1.3B). In methanal, the simplest aldehyde, the carbonyl group is bonded to two hydrogen atoms. In other aldehydes, it is bonded to one hydrogen atom and one carbon atom. Following are Lewis structures for methanal and ethanal. The functional group of a **ketone** is a carbonyl group bonded to two carbon atoms (Section 1.3B). Following is a Lewis structure for propanone, the simplest ketone.

Aldehyde A compound containing a carbonyl group bonded to a hydrogen (a CHO group).

Ketone A compound containing a carbonyl group bonded to two carbons.

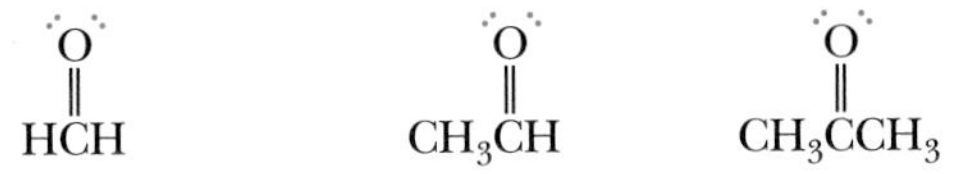

$\overset{\ddot{O}}{\Vert}$ HCH	$\overset{\ddot{O}}{\Vert}$ CH_3CH	$\overset{\ddot{O}}{\Vert}$ CH_3CCH_3
Methanal (Formaldehyde)	Ethanal (Acetaldehyde)	Propanone (Acetone)

The carbon-oxygen double bond consists of one sigma bond formed by overlap of sp^2 hybrid orbitals of carbon and oxygen and one pi bond formed by the overlap of parallel $2p$ orbitals. The two nonbonding pairs of electrons on oxygen lie in the remaining sp^2 hybrid orbitals (Figure 1.19).

According to the molecular orbital model, combination of two atomic orbitals from carbon and four atomic orbitals from oxygen gives six molecular orbitals: a set of sigma bonding and sigma antibonding MOs, a set of pi bonding and pi antibonding MOs, and a set of two nonbonding MOs. The sigma bonding and antibonding MOs are formed by combination of sp^2 atomic orbitals lying along a common axis; the pi bonding and antibonding MOs are formed by combination of parallel $2p$ atomic orbitals. The relative energies of the uncombined atomic orbitals and these three sets of MOs are shown in Figure 15.1. In the ground state, the eight electrons of the uncombined atomic orbitals fill the sigma and pi bonding MOs and the two nonbonding MOs.

The highest occupied MOs are the two nonbonding MOs, and the lowest unoccupied MO is the pi antibonding MO. When a carbonyl group enters into a chemical

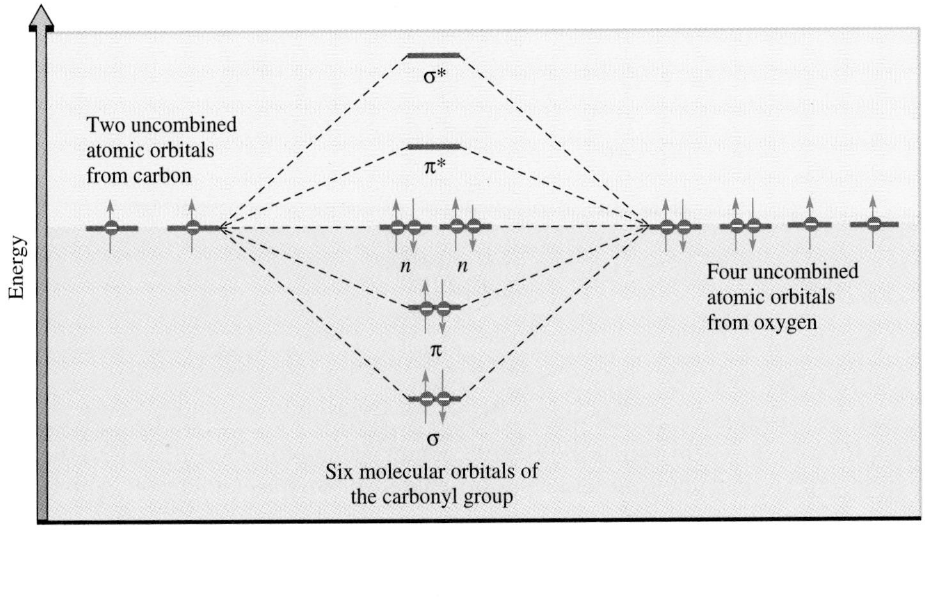

FIGURE 15.1
Molecular orbital description of the covalent bonding in a carbonyl group.

reaction as an electron-pair donor (a nucleophile), it donates a pair of electrons from a nonbonding MO. When a carbonyl group enters into a chemical reaction as an electron-pair acceptor (an electrophile), the pi antibonding MO is the electron-pair acceptor. As we will see in Section 15.4D, the strongest electronic transition seen in the ultraviolet spectrum of a carbonyl-containing compound is a π to π^* transition.

15.2 Nomenclature

A. IUPAC Nomenclature

The IUPAC system of nomenclature for aldehydes and ketones follows the familiar pattern of selecting as the parent alkane the longest chain of carbon atoms that contains the functional group. The aldehyde group is shown by changing the suffix **-e** of the parent alkane to **-al** (Section 2.5). Because the carbonyl group of an aldehyde can only appear at the end of a parent chain and because numbering must start with it as carbon-1, its position is unambiguous, and, therefore, there is no need to use a number to locate it.

For **unsaturated aldehydes,** the presence of a carbon-carbon double or triple bond is indicated by the infix **-en-** or **-yn-.** As with other molecules with both an infix and a suffix, the location of the group corresponding to the suffix determines the numbering pattern.

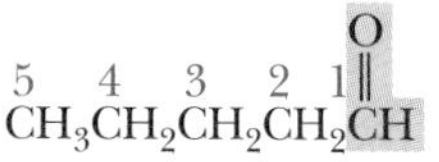

$CH_3CH_2CH_2CH_2CH{=}O$ (carbons numbered 5 4 3 2 1)
Pentanal

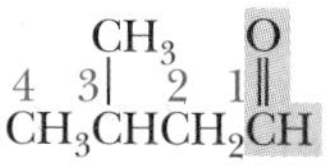

$CH_3CH(CH_3)CH_2CH{=}O$ (carbons numbered 4 3 2 1)
3-Methylbutanal

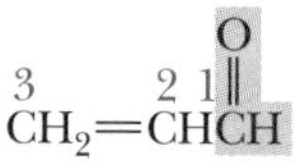

$CH_2{=}CHCH{=}O$ (carbons numbered 3 2 1)
2-Propenal
(Acrolein)

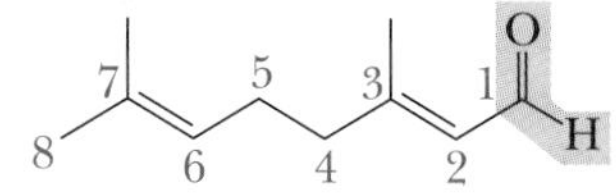

(2*E*)-3,7-Dimethyl-2,6-octadienal
(Geranial)

For cyclic molecules in which —CHO is attached directly to the ring, the molecule is named by adding the suffix **-carbaldehyde** to the name of the ring. The atom of the ring to which the aldehyde group is attached is numbered 1 unless the ring (as, for example, a bicyclic ring) has some other fixed numbering pattern, in which case the —CHO group is given a number as low as possible.

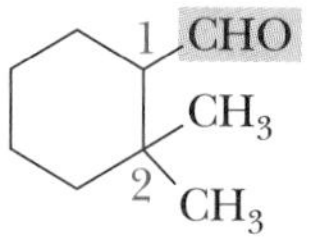

2,2-Dimethylcyclohexanecarbaldehyde

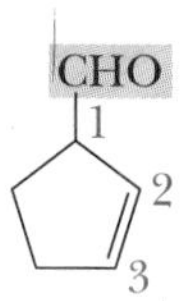

2-Cyclopentenecarbaldehyde

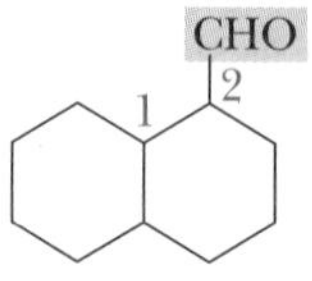

Bicyclo[4.4.0]decane-2-carbaldehyde

Among the aldehydes for which the IUPAC system retains common names are benzaldehyde and cinnamaldehyde, as well as formaldehyde and acetaldehyde. Note here the alternative ways of writing the phenyl group. In benzaldehyde, it is written as a line-angle drawing; in cinnamaldehyde, it is written C_6H_5—.

Benzaldehyde — *trans*-3-Phenyl-2-propenal (Cinnamaldehyde)

Benzaldehyde is responsible for the strong odor of almonds. *(Charles D. Winters)*

In the IUPAC system, ketones are named by selecting as the parent alkane the longest chain that contains the carbonyl group and then indicating the presence of the carbonyl group by changing the suffix from **-e** to **-one** (Section 2.5). The parent chain is numbered from the direction that gives the carbonyl carbon the smaller number. The IUPAC system retains the common names acetone, acetophenone, and benzophenone.

$CH_3\overset{O}{\overset{\|}{C}}CH_3$ — Propanone (Acetone)

$CH_3CH_2\overset{O}{\overset{\|}{C}}CH_2\overset{CH_3}{\overset{|}{C}H}CH_3$ — 5-Methyl-3-hexanone

Bicyclo[2.2.1]-2-heptanone

Acetophenone — Benzophenone — 1-Phenyl-1-pentanone

EXAMPLE 15.1

Give IUPAC names for these compounds. Specify configuration of all stereocenters in (c).

(a) $CH_3CH_2\overset{CH_3}{\overset{|}{C}H}\underset{CH_2CH_3}{\underset{|}{C}H}CHO$ **(b)** **(c)**

Solution

(a) The longest chain has six carbons, but the longest chain that contains the aldehyde group has five carbons. Therefore, the parent chain is pentane. The name is 2-ethyl-3-methylpentanal.

(b) Number the six-membered ring beginning with the carbonyl carbon. The IUPAC name is 3-methyl-2-cyclohexenone.

(c) The name (2*S*,5*R*)-2-isopropyl-5-methylcyclohexanone provides a complete description of both the configuration of each stereocenter and also the *trans* relationship between the isopropyl and methyl groups. The common name of this compound is menthone.

PROBLEM 15.1

Write IUPAC names for these compounds. Specify configuration in (c).

(a) 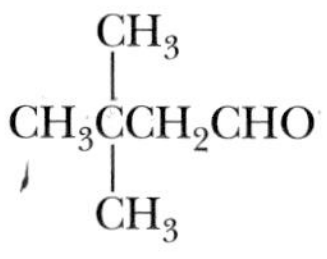 $CH_3C(CH_3)_2CH_2CHO$ (b) 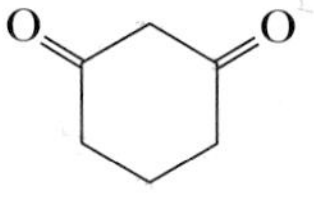 cyclohexane-1,3-dione (c) 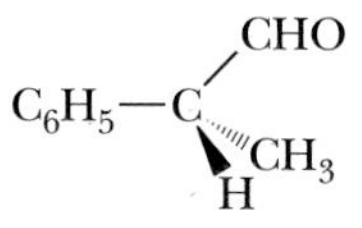 $C_6H_5—C(CHO)(CH_3)H$

EXAMPLE 15.2

Write structural formulas for all ketones of molecular formula $C_6H_{12}O$ and give each its IUPAC name. Which of these ketones are chiral?

Solution

Following are line-angle drawings and IUPAC names for the five ketones of this molecular formula. Only 3-methyl-2-pentanone has a stereocenter and is chiral.

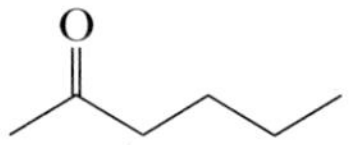

Only this ketone has a stereocenter

2-Hexanone 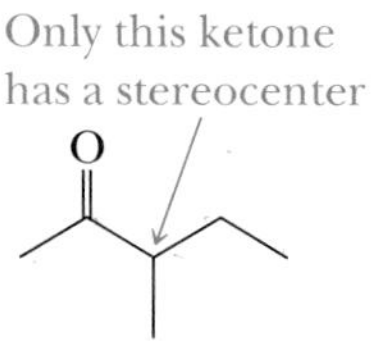 3-Methyl-2-pentanone 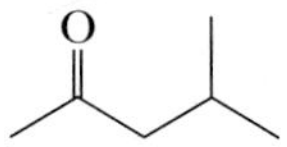 4-Methyl-2-pentanone

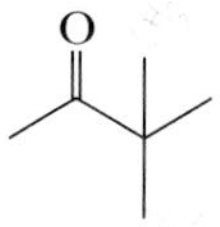

3,3-Dimethyl-2-butanone 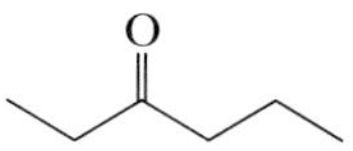 3-Hexanone

PROBLEM 15.2

Write structural formulas for all aldehydes of molecular formula $C_6H_{12}O$ and give each its IUPAC name. Which of these aldehydes are chiral?

B. IUPAC Names for More Complex Aldehydes and Ketones

Order of Precedence A system for ranking functional groups in order of priority for the purposes of IUPAC nomenclature.

In naming compounds that contain more than one functional group that might be indicated by a suffix, the IUPAC system has established an **order of precedence of functions.** The order of precedence for the functional groups we have studied so far is given in Table 15.1.

EXAMPLE 15.3

Write IUPAC names for these compounds. Show configuration for (b) and (c).

(a) 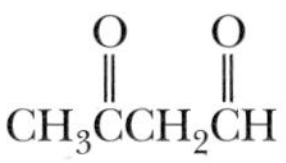 $CH_3\overset{O}{\overset{\|}{C}}CH_2\overset{O}{\overset{\|}{C}}H$ (b) H–C(CHO)(OH)–CH_2OH (c) H_3C, H, OH–C–$CH_2CH_2\overset{O}{\overset{\|}{C}}CH_3$

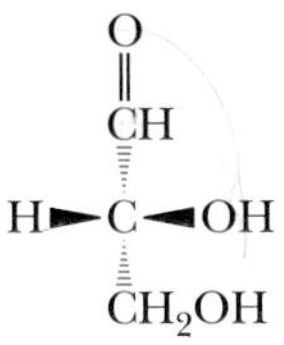

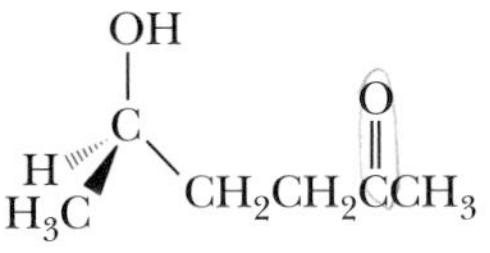

Solution

(a) An aldehyde has higher precedence than a ketone. The presence of the carbonyl group of the ketone is indicated by the prefix oxo- (Table 15.1). The IUPAC name of this compound is 3-oxobutanal.

(b) The C=O group has higher precedence than the hydroxyl groups. The —OH groups are indicated by the prefix hydroxy-. The order of priority among the four groups at the stereocenter is —OH > —CHO > CH_2OH > H. The IUPAC name for this compound is (*R*)-2,3-dihydroxypropanal. Its common name is glyceraldehyde. Glyceraldehyde is the simplest carbohydrate (Chapter 24).

(c) The C=O group has higher precedence than the —OH group. The IUPAC name of this compound is (*R*)-5-hydroxy-2-hexanone.

PROBLEM 15.3

Write IUPAC names for these compounds.

(a) 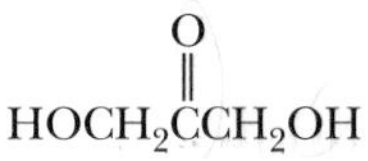 $HOCH_2\overset{O}{\overset{\|}{C}}CH_2OH$ (b) cyclohexane-1,2-dione (c) $H_2NCH_2CH_2CH_2\overset{O}{\overset{\|}{C}}H$

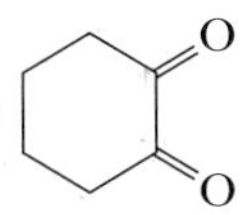

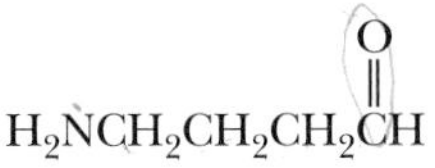

C. Common Names

The common name for an aldehyde is derived from the common name of the corresponding carboxylic acid by dropping the word "acid" and changing the suffix -ic or -oic to -aldehyde. Because we have not yet studied common names for carboxylic

TABLE 15.1 Increasing Order of Precedence of Six Functional Groups

Functional Group	Suffix if Higher in Precedence	Prefix if Lower in Precedence
$—CO_2H$	-oic acid	—
—CHO	-al	oxo-
>C=O	-one	oxo-
—OH	-ol	hydroxy-
$—NH_2$	-amine	amino-
—SH	-thiol	mercapto-

(Arrow: Increasing precedence, from —SH up to $—CO_2H$)

acids, we are not in a position to discuss common names for aldehydes. However, we can illustrate how they are derived by reference to a few common names with which you are familiar. The name formaldehyde is derived from formic acid; the name acetaldehyde is derived from acetic acid.

HCH (C=O)	$HCOH$ (C=O)	CH_3CH (C=O)	CH_3COH (C=O)
Formaldehyde	Formic acid	Acetaldehyde	Acetic acid

Common names for ketones are derived by naming the two alkyl or aryl groups attached to the carbonyl group as separate words, followed by the word "ketone."

$CH_3CH(CH_3)C(=O)CH_2CH_3$	$CH_3CH_2C(=O)CH_2CH_3$	$C_6H_{11}C(=O)C_6H_{11}$
Ethyl isopropyl ketone	Diethyl ketone	Dicyclohexyl ketone

15.3 Physical Properties

Oxygen is more electronegative than carbon (3.5 compared with 2.5) and, therefore, a carbon-oxygen double bond is polar, with oxygen bearing a partial negative charge and carbon bearing a partial positive charge. The bond moment of a carbonyl group is 2.3D (Table 1.7).

$$\overset{\delta+}{C}=\overset{\delta-}{\ddot{O}}:$$

Polarity of a carbonyl group

Because of the polarity of the carbonyl group, aldehydes and ketones are polar compounds and interact in the pure state by dipole-dipole interactions; they have higher boiling points than nonpolar compounds of comparable molecular weight. Table 15.2 lists boiling points of six compounds of comparable molecular weight.

Pentane and diethyl ether have the lowest boiling points. Diethyl ether is a polar molecule, but because of steric hindrance, only weak dipole-dipole interactions exist

TABLE 15.2 Boiling Points of Six Compounds of Comparable Molecular Weight

Name	Structural Formula	Molecular Weight	bp (°C)
diethyl ether	$CH_3CH_2OCH_2CH_3$	74	34
pentane	$CH_3CH_2CH_2CH_2CH_3$	72	36
butanal	$CH_3CH_2CH_2CHO$	72	76
2-butanone	$CH_3CH_2COCH_2CH_3$	72	80
1-butanol	$CH_3CH_2CH_2CH_2OH$	74	117
propanoic acid	$CH_3CH_2CO_2H$	74	141

TABLE 15.3 Physical Properties of Selected Aldehydes and Ketones

IUPAC Name	Common Name	Structural Formula	bp (°C)	Solubility (g/100 g water)
methanal	formaldehyde	HCHO	−21	infinite
ethanal	acetaldehyde	CH_3CHO	20	infinite
propanal	propionaldehyde	CH_3CH_2CHO	49	16
butanal	butyraldehyde	$CH_3CH_2CH_2CHO$	76	7
hexanal	caproaldehyde	$CH_3(CH_2)_4CHO$	129	slight
propanone	acetone	CH_3COCH_3	56	infinite
2-butanone	ethyl methyl ketone	$CH_3COCH_2CH_3$	80	26
3-pentanone	diethyl ketone	$CH_3CH_2COCH_2CH_3$	101	5

between its molecules (Section 11.3). Both butanal and 2-butanone are polar compounds and, because of the intermolecular attraction between their carbonyl groups, their boiling points are higher than those of pentane and diethyl ether. Alcohols (Section 9.3A) and carboxylic acids (Section 16.3) are polar compounds, and their molecules associate by hydrogen bonding; the boiling points of 1-butanol and propanoic acid are higher than those of butanal and 2-butanone, compounds whose molecules cannot associate by hydrogen bonding.

The carbonyl groups of aldehydes and ketones interact with water molecules as hydrogen bond acceptors and, therefore, low-molecular-weight aldehydes and ketones are more soluble in water than are nonpolar compounds of comparable molecular weight. Listed in Table 15.3 are boiling points and solubilities in water for several low-molecular-weight aldehydes and ketones.

15.4 Spectroscopic Properties

A. Mass Spectrometry

A characteristic fragmentation pattern of aliphatic aldehydes and ketones is cleavage of one of the bonds to the carbonyl group (α-cleavage). α-Cleavage of 2-octanone, for example, gives carbonyl-containing ions of m/z 43 and 113. α-Cleavage of the aldehyde proton gives an M − 1 peak, which is often quite distinct and provides a useful way to distinguish between an aldehyde and a ketone.

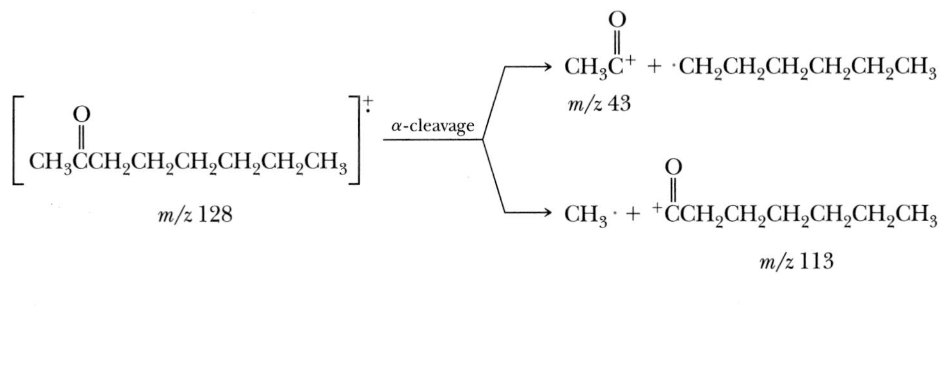

$$\left[CH_3\overset{O}{\overset{\|}{C}}CH_2CH_2CH_2CH_2CH_2CH_3\right]^{+\cdot} \xrightarrow{\alpha\text{-cleavage}} CH_3\overset{O}{\overset{\|}{C}}{}^+ + \cdot CH_2CH_2CH_2CH_2CH_2CH_3$$

m/z 128 → m/z 43

$$\longrightarrow CH_3\cdot + {}^+\overset{O}{\overset{\|}{C}}CH_2CH_2CH_2CH_2CH_2CH_3$$

m/z 113

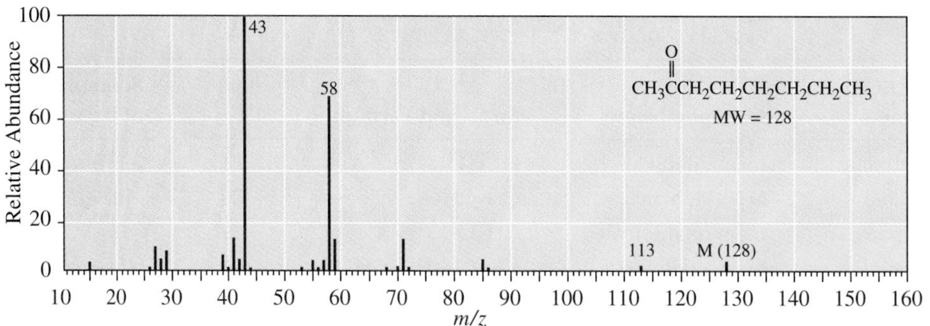

FIGURE 15.2
Mass spectrum of 2-octanone. Ions of m/z 43 and 113 result from α-cleavage. The ion of m/z 58 results from McLafferty rearrangement.

Aldehydes and ketones with a sufficiently long carbon chain show a fragmentation pattern called a McLafferty rearrangement. In a **McLafferty rearrangement** of an aldehyde or ketone, the carbonyl oxygen abstracts a hydrogen five atoms away to form an alkene and a new radical cation. Reaction occurs through a six-membered ring transition state. McLafferty rearrangement of 2-octanone, for example, gives 1-butene and a radical cation of m/z 58, which is the enol of acetone.

$$\left[CH_3CH_2CH(H)-CH_2-CH_2-C(=O)-CH_3 \right]^{\cdot+} \xrightarrow{\text{McLafferty rearrangement}} CH_3CH_2CH{=}CH_2 + \left[H_2C{=}C(OH)CH_3 \right]^{\cdot+}$$

Molecular ion $\qquad\qquad$ m/z 58

The results of both α-cleavage and McLafferty rearrangement can be seen in the mass spectrum of 2-octanone (Figure 15.2).

B. Nuclear Magnetic Resonance Spectroscopy

^{1}H-NMR spectroscopy is an important means for identifying aldehydes and for distinguishing between aldehydes and other carbonyl-containing compounds. Just as the pi system of a carbon-carbon double bond causes a downfield shift in the signal of a vinylic hydrogen (Figure 13.9), the pi system of a carbon-oxygen double bond also causes a large downfield shift in the signal of an aldehyde hydrogen, typically to δ 9.5–10.1. Coupling constants between this hydrogen and those on the adjacent α-carbon are small (approximately 1 to 3 Hz) and, consequently, the aldehyde hydrogen signal often appears as a singlet rather than a doublet or triplet as the case may be. The ^{1}H-NMR spectrum of butanal, for example, shows a singlet at δ 9.78 for the aldehyde hydrogen at the resolution shown (Figure 15.3).

Hydrogens on an α-carbon of an aldehyde or ketone typically appear around δ 2.2–2.6. Because coupling constants between α-hydrogens and an aldehyde hydrogen are small (approximately 1–3 Hz), the ^{1}H-NMR signals of α-hydrogens may appear unsplit by aldehyde hydrogen. The carbonyl carbons of aldehydes and ketones are readily identifiable in ^{13}C-NMR spectroscopy by the position of their signal between δ 180 and 210.

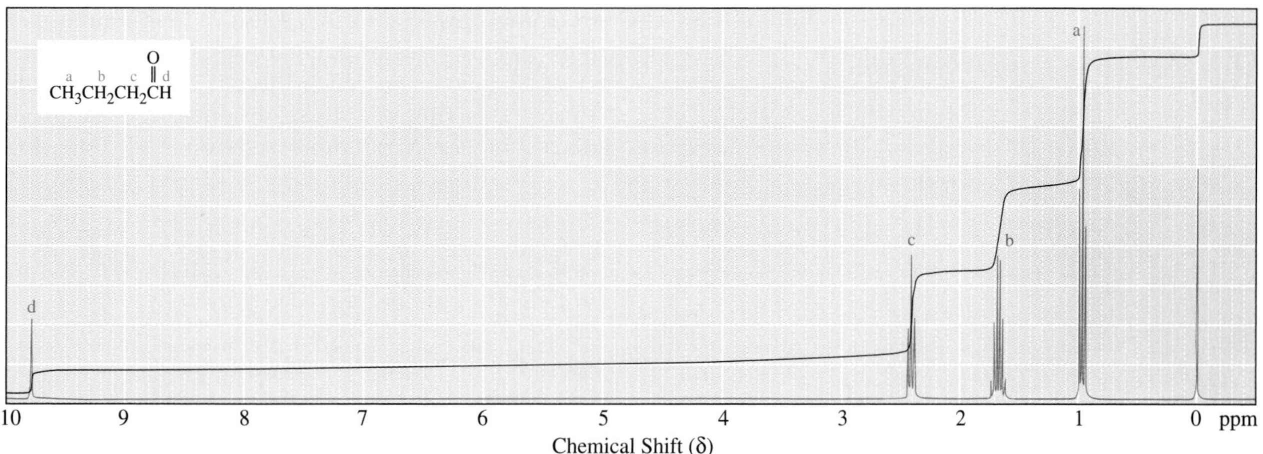

FIGURE 15.3
^{1}H-NMR spectrum of butanal.

C. Infrared Spectroscopy

Aldehydes and ketones show characteristic strong infrared absorption between 1630 and 1810 cm^{-1} associated with the stretching vibration of the carbon-oxygen double bond. The stretching vibration for the carbonyl group of menthone occurs at 1705 cm^{-1} (Figure 15.4).

Because few other bond vibrations absorb energy between 1630 and 1810 cm^{-1}, absorption in this region of the spectrum is a reliable means for confirming the presence of a carbonyl group. Because several different functional groups contain a carbonyl group, it is often not possible to tell from absorption in this region alone whether the carbonyl-containing compound is an aldehyde, a ketone, or, as we shall see in Chapters 16 and 17, a carboxylic acid, an ester, an amide, or an anhydride.

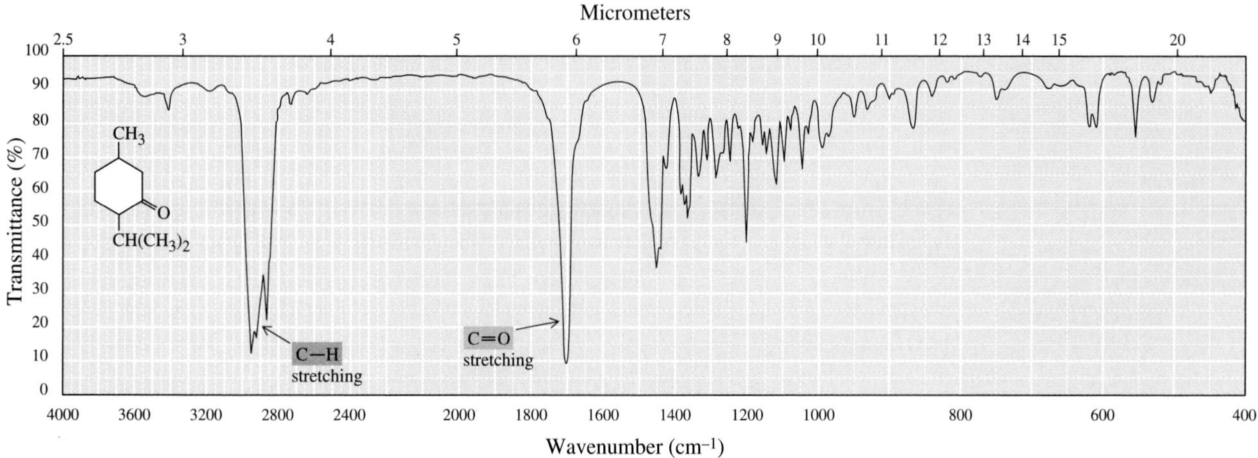

FIGURE 15.4
Infrared spectrum of menthone.

Menthone can be isolated from the oil of the peppermint plant *(Mentha piperita)*. *(Hans Reinhard/OKAPIA/Photo Researchers, Inc.)*

The position of the C=O stretching vibration is quite sensitive to the molecular environment of the carbonyl group, as illustrated by comparison of these cycloalkanones. Cyclohexanone and larger cyclic ketones in which there is little or no angle strain show absorption near 1715 cm^{-1}. As ring size decreases and angle strain increases, the C=O absorption shifts to a higher wavenumber as shown in the following series.

1715 cm^{-1} 1745 cm^{-1} 1780 cm^{-1} 1850 cm^{-1}

The presence of a carbon-carbon double bond or benzene ring in conjugation with the carbonyl group results in a shift of the C=O absorption to a lower wavenumber as seen by comparing the carbonyl stretching frequencies of the following molecules.

$(H_3C)_2CH-CH_2-\overset{O}{\overset{\|}{C}}-CH_3$ 1717 cm^{-1}

$(H_3C)_2C=CH-\overset{O}{\overset{\|}{C}}-CH_3$ 1690 cm^{-1}

$C_6H_5-\overset{O}{\overset{\|}{C}}H$ 1700 cm^{-1}

D. Ultraviolet-Visible Spectroscopy

Simple aldehydes and ketones show only weak absorption in the ultraviolet region of the spectrum due to an n to π^* electronic transition of the carbonyl group. If, however, the carbonyl group is conjugated with one or more carbon-carbon double bonds, intense absorption ($\epsilon = 8{,}000 - 20{,}000\ M^{-1}cm^{-1}$) occurs due to a π to π^* transition; the position of absorption is shifted to longer wavelengths, and the molar absorptivity, ϵ, of the absorption maximum increases sharply. For the α,β-unsaturated ketone 3-pentene-2-one, for example, λ_{max} is 224 nm (log ϵ 4.10).

$CH_3CH_2CH_2\overset{O}{\overset{\|}{C}}CH_3$ — 3-Pentanone, λ_{max} 180 nm (ϵ 900)

$CH_3CH=CH\overset{O}{\overset{\|}{C}}CH_3$ — 3-Penten-2-one, λ_{max} 224 nm (ϵ 12,590)

$C_6H_5\overset{O}{\overset{\|}{C}}CH_3$ — Acetophenone, λ_{max} 246 nm (ϵ 9,800)

The greater the extent of conjugation of unsaturated systems with the carbonyl group, the more the absorption maximum is shifted toward the visible region of the spectrum.

15.5 Reactions

One of the most common reaction themes of a carbonyl group is addition of a nucleophile to form a **tetrahedral carbonyl addition compound.** In the following general reaction, the nucleophilic reagent is written as $Nu:^-$ to emphasize the presence of the unshared pair of electrons on the nucleophile.

$$\text{Nu}:^- + \text{R}_2\text{C}{=}\ddot{\text{O}}: \longrightarrow \text{Nu}{-}\text{CR}_2{-}\ddot{\text{O}}:^-$$

Tetrahedral carbonyl addition compound

A second common reaction theme of a carbonyl group is reaction with a proton or Lewis acid to form a resonance-stabilized cation. Protonation increases the electron deficiency of the carbonyl carbon and makes it more reactive toward nucleophiles. This cation then reacts with a nucleophile to give a tetrahedral carbonyl addition compound.

$$\text{R}_2\text{C}{=}\ddot{\text{O}}: + \text{H}{-}\text{B} \xrightleftharpoons{\text{fast}} \text{R}_2\text{C}{=}\ddot{\text{O}}^+{-}\text{H} + \text{B}^-$$

$$\text{H}{-}\text{Nu}: + \text{R}_2\text{C}{=}\ddot{\text{O}}^+{-}\text{H} \xrightarrow{\text{slow}} \text{Nu}{-}\text{CR}_2{-}\ddot{\text{O}}{-}\text{H} + \text{H}{-}\text{B}$$

Tetrahedral carbonyl addition compound

15.6 Addition of Carbon Nucleophiles

In this section, we examine reactions of aldehydes and ketones with the following types of carbon nucleophiles:

RMgX	RLi	$RC{\equiv}C:^-$	$^-:C{\equiv}N:$
A Grignard reagent	An organolithium reagent	An anion of a terminal alkyne	Cyanide ion

From the perspective of the organic chemist, addition of a carbon nucleophile is one of the most important types of nucleophilic additions to a carbonyl group because a new carbon-carbon bond is formed in the process.

A. Addition of Grignard Reagents

The special value of Grignard reagents (Section 7.7) is that they provide excellent ways to form new carbon-carbon bonds. Given the difference in electronegativity between carbon and magnesium ($2.5 - 1.2 = 1.3$), the carbon-magnesium bond of a Grignard reagent is polar covalent with carbon bearing a partial negative charge and magnesium bearing a partial positive charge. In its reactions, a Grignard reagent behaves as a **carbanion.** A carbanion is a good nucleophile and adds to the carbonyl group of an aldehyde or ketone to form a tetrahedral carbonyl addition compound. The driving force for these reactions is the attraction of the partial negative charge on the carbon of the organometallic compound to the partial positive charge of the carbonyl carbon. In the following examples, the magnesium oxygen bond is written $—O^-[MgBr]^+$ to emphasize its ionic character. The alkoxide ions formed in these

Carbanion An anion in which carbon has an unshared pair of electrons and bears a negative charge.

reactions are strong bases (Section 9.5) and, when treated with an aqueous acid such as HCl during workup, form alcohols.

Addition to Formaldehyde Gives a Primary Alcohol

Treatment of a Grignard reagent with formaldehyde followed by hydrolysis in aqueous acid gives a primary alcohol.

$$\overset{\delta-}{CH_3CH_2}-\overset{\delta+}{MgBr} + H-\overset{\overset{\delta-}{O}}{\overset{\|}{\underset{\delta+}{C}}}-H \xrightarrow{\text{ether}} CH_3CH_2-\overset{O^-[MgBr]^+}{\overset{|}{CH_2}} \xrightarrow[H_2O]{HCl} CH_3CH_2-\overset{OH}{\overset{|}{CH_2}} + Mg^{2+}$$

Formaldehyde; A magnesium alkoxide; 1-Propanol (a primary alcohol)

Addition to an Aldehyde (Except Formaldehyde) Gives a Secondary Alcohol

Treatment of a Grignard reagent with any other aldehyde followed by hydrolysis in aqueous acid gives a secondary alcohol.

$$C_6H_{11}\overset{\delta-}{-}\overset{\delta+}{MgBr} + CH_3-\overset{\overset{\delta-}{O}}{\overset{\|}{\underset{\delta+}{C}}}-H \xrightarrow{\text{ether}} C_6H_{11}-\overset{O^-[MgBr]^+}{\overset{|}{CHCH_3}} \xrightarrow[H_2O]{HCl} C_6H_{11}-\overset{OH}{\overset{|}{CHCH_3}} + Mg^{2+}$$

Acetaldehyde (an aldehyde); A magnesium alkoxide; 1-Cyclohexylethanol (a secondary alcohol)

Addition to a Ketone Gives a Tertiary Alcohol

Treatment of a Grignard reagent with a ketone followed by hydrolysis in aqueous acid gives a tertiary alcohol.

$$C_6H_5\overset{\delta-}{-}\overset{\delta+}{MgBr} + CH_3-\overset{\overset{\delta-}{O}}{\overset{\|}{\underset{\delta+}{C}}}-CH_3 \xrightarrow{\text{ether}} C_6H_5-\underset{CH_3}{\underset{|}{\overset{O^-[MgBr]^+}{\overset{|}{CCH_3}}}} \xrightarrow[H_2O]{HCl} C_6H_5-\underset{CH_3}{\underset{|}{\overset{OH}{\overset{|}{CCH_3}}}} + Mg^{2+}$$

Acetone (a ketone); A magnesium alkoxide; 2-Phenyl-2-propanol (a tertiary alcohol)

Addition to Carbon Dioxide Gives a Carboxylic Acid

Treatment of a Grignard reagent with carbon dioxide gives the magnesium salt of a carboxylic acid, which, on treatment with aqueous acid, gives a carboxylic acid. Thus, carbonation of a Grignard reagent is a convenient way to convert an alkyl or aryl halide to a carboxylic acid, as illustrated by the conversion of bromocyclopentane via a Grignard reagent to cyclopentanecarboxylic acid.

$$C_5H_9\overset{\delta-}{-}\overset{\delta+}{MgBr} + \underset{O}{\underset{\|}{\overset{\overset{\delta-}{O}}{\overset{\|}{C^{\delta+}}}}} \longrightarrow C_5H_9-\overset{O}{\overset{\|}{C}}O^-[MgBr]^+ \xrightarrow[H_2O]{HCl} C_5H_9-\overset{O}{\overset{\|}{C}}OH + Mg^{2+}$$

Carbon dioxide; Cyclopentanecarboxylic acid

EXAMPLE 15.4

2-Phenyl-2-butanol can be synthesized by three different combinations of a Grignard reagent and a ketone. Show each combination.

Solution

In each solution, curved arrows show formation of the new carbon-carbon bond and the alkoxide ion. The new carbon-carbon bond formed by each set of reagents is labeled in the final product.

(a) $C_6H_5MgBr + CH_3C(=O)CH_2CH_3$

(b) $C_6H_5C(=O)CH_3 + CH_3CH_2MgBr$

(c) $C_6H_5C(=O)CH_2CH_3 + CH_3MgBr$

$$\longrightarrow C_6H_5C(CH_3)(CH_2CH_3)O^-[MgBr]^+ \xrightarrow[H_2O]{HCl} C_6H_5C(CH_3)(CH_2CH_3)OH$$

PROBLEM 15.4

Show how these three products can be synthesized from the same Grignard reagent; that is, treating the Grignard reagent with one compound gives (a), with another compound gives (b), and so on. (*Hint:* Grignard reagents react with ethylene oxide in the same manner as Gilman reagents, Section 11.10B.)

(a) $C_6H_{11}CO_2H$ (cyclohexyl) (b) $C_6H_{11}CH_2OH$

(c) $C_6H_{11}CH(OH)CH_3$ (d) $C_6H_{11}CH_2CH_2OH$

B. Addition of Organolithium Compounds

Organolithium compounds are generally more reactive in carbonyl addition reactions than organomagnesium compounds and typically give higher yields of products. They are more troublesome to use, however, because they must be prepared and used under an atmosphere of nitrogen or other inert gas. Following is a synthesis illustrating the use of an organolithium compound to form a sterically hindered tertiary alcohol.

$$C_6H_5Li + CH_3C(=O)C(CH_3)_3 \longrightarrow C_6H_5C(CH_3)(O^-Li^+)C(CH_3)_3 \xrightarrow[H_2O]{HCl} C_6H_5C(CH_3)(OH)C(CH_3)_3$$

3,3-Dimethyl-2-butanone → 3,3-Dimethyl-2-phenyl-2-butanol

C. Addition of Salts of Terminal Alkynes

The anion of a terminal alkyne is a nucleophile (Section 10.5) and adds to the carbonyl group of an aldehyde or ketone to form a tetrahedral carbonyl addition compound. These addition compounds contain both a hydroxyl group and a carbon-carbon triple bond, each of which can be further modified. In the following example, addition of sodium acetylide to cyclohexanone followed by hydrolysis in aqueous acid gives 1-ethynylcyclohexanol.

$$HC{\equiv}C\!:^-\ Na^+ + \text{cyclohexanone} \longrightarrow \text{(HC}{\equiv}\text{C, O}^-\ Na^+\text{)} \xrightarrow[H_2O]{HCl} \text{(HC}{\equiv}\text{C, OH)}$$

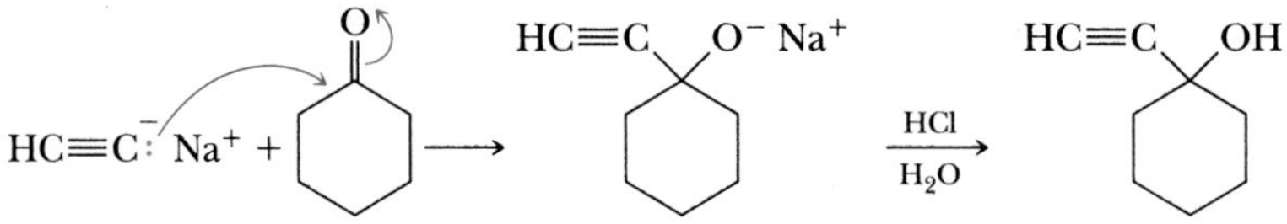

A sodium alkoxide 1-Ethynylcyclohexanol

Acid-catalyzed hydration of this hydroxyalkyne (Section 10.9C) gives an α-hydroxyketone. Alternatively, hydroboration followed by oxidation with alkaline hydrogen peroxide (Section 10.8) gives a β-hydroxyaldehyde.

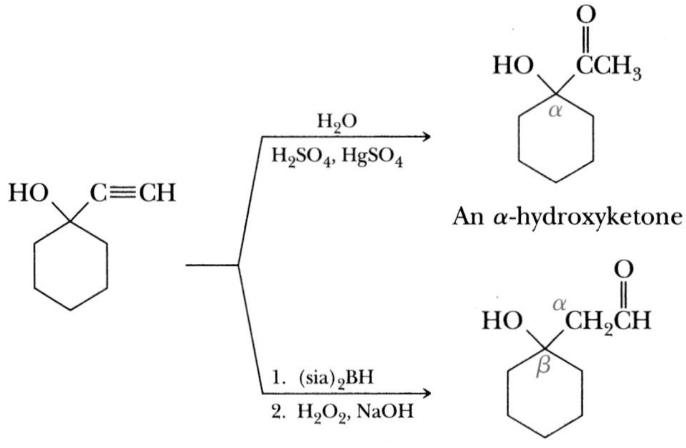

HO, C≡CH (cyclohexane) $\xrightarrow[H_2SO_4,\ HgSO_4]{H_2O}$ HO, $\overset{O}{\overset{\|}{C}}CH_3$ (cyclohexane)

An α-hydroxyketone

HO, C≡CH (cyclohexane) $\xrightarrow[2.\ H_2O_2,\ NaOH]{1.\ (sia)_2BH}$ HO, $CH_2\overset{O}{\overset{\|}{C}}H$ (cyclohexane)

A β-hydroxyaldehyde

This example illustrates two of the most valuable reactions of alkynes: (1) addition of the anion of a terminal alkyne to the carbonyl group of an aldehyde or ketone to form an alkynyl alcohol and (2) hydration of an alkyne to give either an aldehyde or ketone depending on the substitution pattern of the alkyne and the method of hydration.

D. Addition of Hydrogen Cyanide

Cyanohydrin A molecule containing an —OH group and a —CN group bonded to the same carbon.

Hydrogen cyanide, HCN, adds to the carbonyl group of an aldehyde or ketone to form a tetrahedral carbonyl addition compound called a cyanohydrin. A **cyanohydrin** is a molecule containing a carbon atom bonded to an —OH group and a —CN group. For example, HCN adds to acetaldehyde to form acetaldehyde cyanohydrin in 75% yield. We study the naming of compounds containing the nitrile group in Section 17.1E.

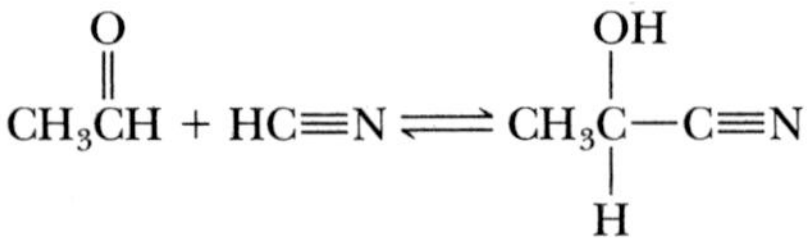

$$CH_3\overset{O}{\overset{\|}{C}}H + HC{\equiv}N \rightleftharpoons CH_3\underset{H}{\underset{|}{\overset{OH}{\overset{|}{C}}}}{-}C{\equiv}N$$

2-Hydroxypropanenitrile
(Acetaldehyde cyanohydrin)

Addition of hydrogen cyanide is catalyzed by cyanide ion. Because HCN is a weak acid, pK_a 9.31, the concentration of cyanide ion in aqueous HCN is too low for cyanohydrin formation to proceed at a reasonable rate. For this reason, cyanohydrin formation is generally carried out by dissolving NaCN or KCN in water and adding acid to adjust the pH of the solution to approximately 10.0, giving a solution in which HCN and CN^- are present in approximately equal concentrations. Reaction is initiated by nucleophilic addition of cyanide ion to the carbonyl group to form an alkoxide ion that in turn reacts with HCN to form the cyanohydrin and CN^-.

MECHANISM Formation of a Cyanohydrin

$(CH_3)_2C{=}O + {}^-:C{\equiv}N: \rightleftharpoons (CH_3)_2C(O^-)C{\equiv}N \xrightleftharpoons{H-C\equiv N} (CH_3)_2C(OH)C{\equiv}N + {}^-:C{\equiv}N:$

Propanone (Acetone)

2-Hydroxy-2-methylpropanenitrile (Acetone cyanohydrin)

For aldehydes and most aliphatic ketones, the position of equilibrium favors cyanohydrin formation. For many aryl ketones and sterically hindered aliphatic ketones, however, the position of equilibrium favors starting materials, and for these types of compounds, cyanohydrin formation is not a useful reaction. The following synthesis of ibuprofen, for example, failed because the cyanohydrin was formed only in low yield.

4-Isobutylacetophenone $\xrightleftharpoons[HCN]{KCN}$ A cyanohydrin $\xrightarrow{\text{several steps}}$ Ibuprofen

Benzaldehyde cyanohydrin (mandelonitrile) provides an interesting example of a chemical defense mechanism in the biological world. This substance is synthesized by millipedes *(Apheloria corrigata)* and stored in special glands. When a millipede is threatened, the cyanohydrin is released from the storage gland, undergoes enzyme-catalyzed dissociation to produce HCN, and is then released to ward off predators. The quantity of HCN emitted by a single millipede is sufficient to kill a small mouse. Mandelonitrile is also found in bitter almond and peach pits. Its function there is unknown, as is how millipedes survive exposure to hydrogen cyanide.

When the millipede *Apheloria corrigata* is attacked, benzaldehyde cyanohydrin is released from its storage glands. It then undergoes enzyme-catalyzed hydrolysis to yield benzaldehyde and HCN, which is emitted as a spray to ward off predators. *(© Michael J. Doolittle, Rainbow)*

$C_6H_5CH(OH)C{\equiv}N \xrightarrow{\text{(enzyme catalysis)}} C_6H_5CHO + HC{\equiv}N$

Benzaldehyde cyanohydrin — Benzaldehyde

The main value of cyanohydrins lies in the new functional groups into which they can be converted. First, the secondary or tertiary alcohol group of the cyanohydrin may undergo acid-catalyzed dehydration to form an unsaturated nitrile. For example, acid-catalyzed dehydration of acetaldehyde cyanohydrin gives acrylonitrile, the monomer from which polyacrylonitrile (Orlon, Table 23.1) is made.

$$\underset{\text{(Acetaldehyde cyanohydrin)}}{\underset{\text{2-Hydroxypropanenitrile}}{CH_3\overset{\overset{OH}{|}}{C}HC{\equiv}N}} \xrightarrow[\text{catalyst}]{\text{acid}} \underset{\text{(Acrylonitrile)}}{\underset{\text{Propenenitrile}}{CH_2{=}CHC{\equiv}N}} + H_2O$$

Second, acid-catalyzed hydrolysis of the cyano group (Section 17.5E) gives a carboxylic acid. For example, hydrolysis of benzaldehyde cyanohydrin (mandelonitrile) gives 2-hydroxy-2-phenylethanoic acid (mandelic acid).

$$\underset{\text{(Mandelonitrile)}}{\underset{\text{Benzaldehyde cyanohydrin}}{C_6H_5\overset{\overset{OH}{|}}{C}HC{\equiv}N}} + H_2O \xrightarrow[\text{catalyst}]{\text{acid}} \underset{\text{(Mandelic acid)}}{\underset{\text{2-Hydroxy-2-phenylethanoic acid}}{C_6H_5\overset{\overset{HO}{|}}{C}H\overset{\overset{O}{\|}}{C}OH}}$$

Third, a nitrile is reduced to a primary amine by hydrogen in the presence of nickel or other transition metal catalyst (Section 17.11C). Catalytic reduction of benzaldehyde cyanohydrin, for example, gives 2-amino-1-phenylethanol.

$$\underset{\text{Benzaldehyde cyanohydrin}}{C_6H_5\overset{\overset{OH}{|}}{C}HC{\equiv}N} + 2H_2 \xrightarrow{\text{Ni}} \underset{\text{2-Amino-1-phenylethanol}}{C_6H_5\overset{\overset{OH}{|}}{C}HCH_2NH_2}$$

15.7 The Wittig Reaction

Ylide A molecule which, when written in a Lewis structure showing all atoms with complete valence shells, has positive and negative charges on adjacent atoms.

In 1954, Georg Wittig reported a method for the synthesis of alkenes from aldehydes and ketones using compounds called phosphonium ylides (Chemistry in Action: "The Octet Rule," Chapter 1). An **ylide** is a molecule that, when written in a Lewis structure showing all atoms with complete valence shells, has positive and negative charges on adjacent atoms. For his pioneering study and development of this reaction into a major synthetic tool, Professor Wittig shared the Nobel Prize for chemistry in 1979. (The other recipient was Herbert C. Brown for his studies of hydroboration and the chemistry of organoboron compounds.) A Wittig synthesis is illustrated by the conversion of cyclohexanone to methylenecyclohexane. In this reaction a C=O double bond is converted to a C=C double bond.

$$C_6H_{10}{=}O + \underset{\text{A phosphonium ylide}}{Ph_3\overset{+}{P}{-}\overset{-}{C}H_2} \longrightarrow \underset{\text{Methylenecyclohexane}}{C_6H_{10}{=}CH_2} + \underset{\text{Triphenylphosphine oxide}}{Ph_3\overset{+}{P}{-}O^-}$$

We study the Wittig reaction in two stages: first, the formation and structure of phosphonium ylides themselves and, second, the reaction of a phosphonium ylide with the carbonyl group of an aldehyde or ketone to give an alkene.

Phosphorus is the second element in Group VA of the Periodic Table and, like nitrogen, has five electrons in its valence shell. Examples of trivalent phosphorus compounds are phosphine and triphenylphosphine. Phosphine is a highly toxic, flammable gas. Triphenylphosphine is a colorless, odorless solid.

$$\mathrm{H{-}\ddot{P}H_2} \qquad \mathrm{Ph{-}\ddot{P}Ph_2}$$

Phosphine (bp −87°C) Triphenylphosphine (mp 81°C)

Because phosphorus is below nitrogen in the Periodic Table, phosphines are weaker bases than amines and also better nucleophiles (Section 8.4B). Treatment of a phosphine with a primary or secondary alkyl halide gives a phosphonium salt by an S_N2 pathway. Because phosphines are also weak bases, treatment of a tertiary halide with a phosphine gives largely an alkene by an E2 pathway.

$$\mathrm{Ph_3P\!:} + \mathrm{CH_3{-}I} \xrightarrow{S_N2} \mathrm{Ph_3\overset{+}{P}{-}CH_3\ I^-}$$

Triphenylphosphine; Methyltriphenylphosphonium iodide (an alkyltriphenylphosphonium salt)

α-Hydrogen atoms on the alkyl group of an alkyltriphenylphosphonium ion are weakly acidic and can be removed by reaction with a very strong base, typically butyllithium, sodium hydride (NaH), or sodium amide ($NaNH_2$).

$$\mathrm{CH_3CH_2CH_2\ddot{C}H_2^-\,Li^+} + \mathrm{H{-}CH_2{-}\overset{+}{P}Ph_3\ I^-} \longrightarrow \mathrm{\ddot{C}H_2^-{-}\overset{+}{P}Ph_3} + \mathrm{CH_3CH_2CH_2CH_3} + \mathrm{Li^+I^-}$$

Butyllithium; An alkyltriphenylphosphonium iodide; A phosphonium ylide; Butane

The product of removal of hydrogen from an alkyltriphenylphosphonium ion is a phosphonium ylide.

Phosphonium ylides react with the carbonyl groups of aldehydes and ketones by a cycloaddition reaction to form a four-membered ring called an **oxaphosphetane.** The name for this ring system is derived by combination of the following units: oxa- to show that it contains oxygen, -phosph- to show that it contains phosphorus, -et- to show that it is a four-membered ring, and -ane to show only carbon-carbon single bonds in the ring. Oxaphosphetanes can be isolated at low temperature. On warming to room temperature, they fragment to give triphenylphosphine oxide and an alkene. The driving force for this reaction is formation of the very strong phosphorus-oxygen bond in triphenylphosphine oxide.

MECHANISM The Wittig Reaction

$$\mathrm{\ddot{O}{=}CR_2} + \mathrm{Ph_3\overset{+}{P}{-}\ddot{C}H_2^-} \longrightarrow \text{(ring: } \mathrm{O{-}CR_2 / Ph_3P{-}CH_2}) \longrightarrow \mathrm{Ph_3P^+{-}\ddot{O}^-} + \mathrm{CR_2{=}CH_2}$$

An oxaphosphetane; Triphenylphosphine oxide; An alkene

The Wittig reaction is effective with a wide variety of aldehydes and ketones and with ylides derived from a wide variety of primary, secondary, and allylic halides as shown by the following examples:

$$CH_3\overset{O}{\overset{\|}{C}}CH_3 + Ph_3\overset{+}{P}-\overset{-}{C}H(CH_2)_3CH_3 \longrightarrow CH_3\overset{CH_3}{\overset{|}{C}}=CH(CH_2)_3CH_3 + Ph_3\overset{+}{P}-O^-$$

2-Methyl-2-heptene

$$PhCH_2\overset{O}{\overset{\|}{C}}H + Ph_3\overset{+}{P}-\overset{-}{C}HCH_3 \longrightarrow PhCH_2CH=CHCH_3 + Ph_3\overset{+}{P}-O^-$$

1-Phenyl-2-butene
(87% *Z* isomer; 13% *E* isomer)

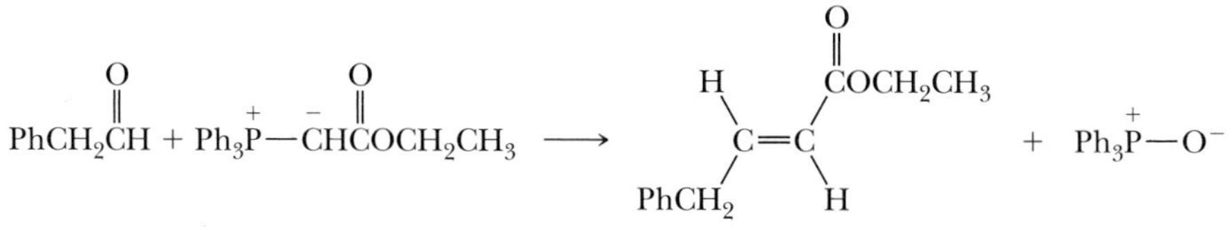

$$PhCH_2\overset{O}{\overset{\|}{C}}H + Ph_3\overset{+}{P}-\overset{-}{C}H\overset{O}{\overset{\|}{C}}OCH_2CH_3 \longrightarrow \underset{PhCH_2}{\overset{H}{}}C=C\underset{H}{\overset{\overset{O}{\|}}{COCH_2CH_3}} + Ph_3\overset{+}{P}-O^-$$

Ethyl (*E*)-4-phenyl-2-butenoate
(only the *E* isomer is formed)

EXAMPLE 15.5

Show how this alkene can be synthesized by a Wittig reaction.

(cyclopentylidene)=CHCH$_3$

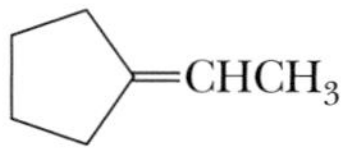

Solution

Starting materials are either cyclopentanone and the triphenylphosphonium ylide derived from chloroethane, or acetaldehyde and the triphenylphosphonium ylide derived from chlorocyclopentane.

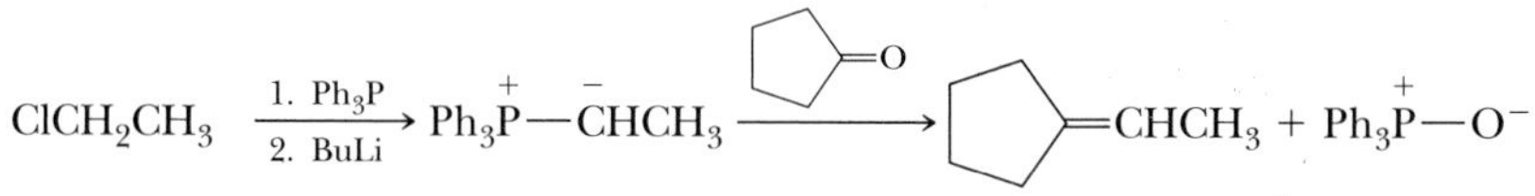

$$ClCH_2CH_3 \xrightarrow[2.\ BuLi]{1.\ Ph_3P} Ph_3\overset{+}{P}-\overset{-}{C}HCH_3 \xrightarrow{\text{cyclopentanone}} \text{(cyclopentylidene)}=CHCH_3 + Ph_3\overset{+}{P}-O^-$$

Chloroethane

$$\text{(cyclopentyl)}-Cl \xrightarrow[2.\ NaH]{1.\ Ph_3P} \text{(cyclopentyl)}\overset{-}{}-\overset{+}{P}Ph_3 \xrightarrow{CH_3\overset{O}{\overset{\|}{C}}H} \text{(cyclopentylidene)}=CHCH_3 + Ph_3\overset{+}{P}-O^-$$

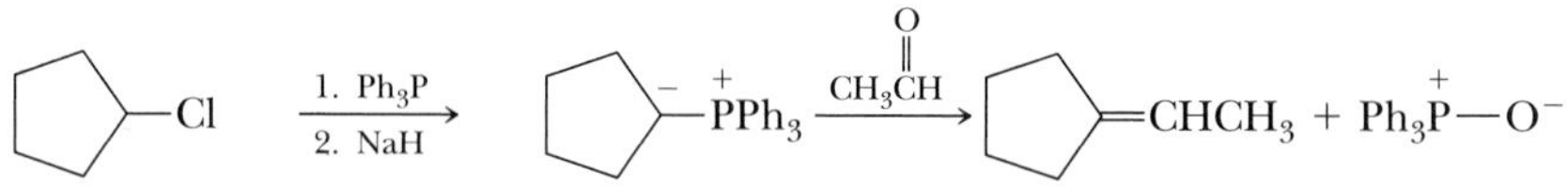

Chlorocyclopentane

PROBLEM 15.5

Show how each alkene can be synthesized by a Wittig reaction. There are two routes to each.

(a) $(CH_3)_2C=CHCH_2CH_3$ **(b)** 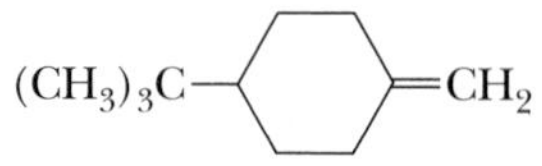 $(CH_3)_3C$–(cyclohexylidene)=CH$_2$

15.8 Addition of Oxygen Nucleophiles

A. Addition of Water

Addition of water (hydration) to a carbonyl group of an aldehyde or ketone forms a 1,1-diol (commonly referred to as a gem-diol). Note that here, the numbers 1,1 do not refer to an IUPAC numbering system, but rather to the fact that the two hydroxyl groups are on the same carbon. A 1,1-diol is commonly referred to as the hydrate of the corresponding aldehyde or ketone. These compounds are unstable and are rarely isolated.

$$\underset{\text{An aldehyde}}{R-\overset{O}{\overset{\|}{C}}-H} + H_2O \rightleftharpoons \underset{\substack{\text{A 1,1-diol}\\\text{(hydrate of}\\\text{an aldehyde)}}}{R-\overset{OH}{\overset{|}{\underset{H}{\underset{|}{C}}}}-OH} \qquad \underset{\text{A ketone}}{R-\overset{O}{\overset{\|}{C}}-R} + H_2O \rightleftharpoons \underset{\substack{\text{A 1,1-diol}\\\text{(hydrate of}\\\text{a ketone)}}}{R-\overset{OH}{\overset{|}{\underset{R}{\underset{|}{C}}}}-OH}$$

These reactions are readily reversible and the diol can eliminate water to regenerate the aldehyde or ketone. Although equilibrium strongly favors the carbonyl group, for a few simple aldehydes the 1,1-diol is favored. For example, when formaldehyde is dissolved in water at 20°C, the position of equilibrium is such that it is more than 99% hydrated. A 37% solution of formaldehyde in water is called formalin. Formalin is commonly used for preservation of biological specimens.

$$\underset{\text{Formaldehyde}}{H\overset{O}{\overset{\|}{C}}H} + H_2O \rightleftharpoons \underset{\substack{\text{Formaldehyde}\\\text{hydrate}\\(>99\%)}}{H\overset{OH}{\overset{|}{\underset{H}{\underset{|}{C}}}}OH}$$

In contrast, an aqueous solution of the ketone, acetone, consists of less than 0.1% of the diol and 99.9% of acetone at equilibrium.

$$\underset{\substack{\text{Acetone}\\(99.9\%)}}{(H_3C)_2C=O} + H_2O \rightleftharpoons \underset{\substack{\text{2,2-Propanediol}\\(0.1\%)}}{(H_3C)_2C(OH)_2}$$

B. Addition of Alcohols: Formation of Acetals

Alcohols add to aldehydes and ketones in the same manner as described for water. Addition of one molecule of alcohol to the carbonyl group of an aldehyde or ketone forms a **hemiacetal** (a half-acetal).

Hemiacetal A molecule containing an —OH and an —OR or —OAr group bonded to the same carbon.

$$CH_3\overset{O}{\overset{\|}{C}}CH_3 + \overset{H}{\overset{|}{O}}CH_2CH_3 \rightleftharpoons \underset{\text{A hemiacetal}}{CH_3\overset{OH}{\overset{|}{\underset{CH_3}{\underset{|}{C}}}}OCH_2CH_3}$$

The functional group of a hemiacetal is a carbon bonded to an —OH group and an —OR or —OAr group.

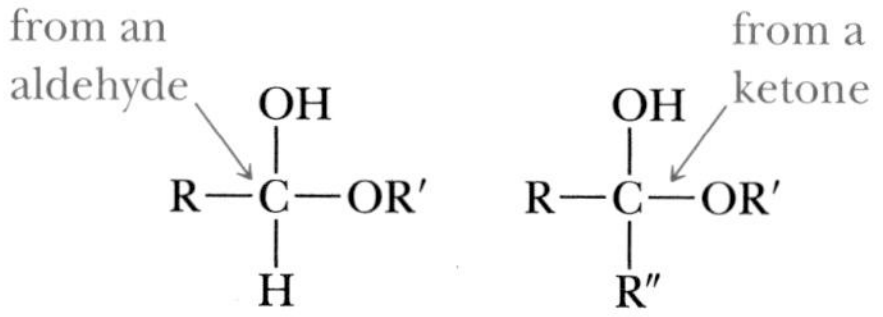

from an aldehyde — R—C(OH)(H)—OR′ ; R—C(OH)(R″)—OR′ — from a ketone

Hemiacetals

Hemiacetals are generally unstable and are only minor components of an equilibrium mixture except in one very important type of compound. When a hydroxyl group is part of the same molecule that contains the carbonyl group, and a five- or six-membered ring can form, the compound exists almost entirely in the cyclic hemiacetal form. We have much more to say about cyclic hemiacetals when we consider the chemistry of carbohydrates in Chapter 24.

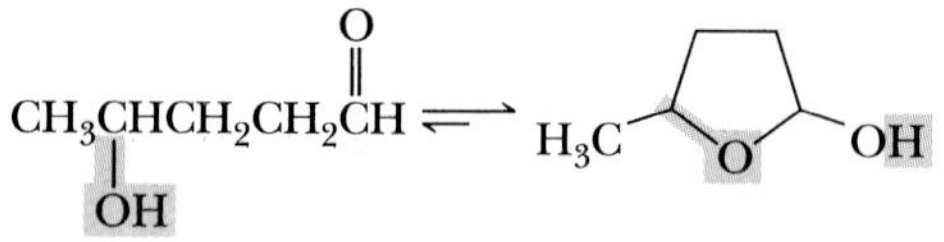

$CH_3CH(OH)CH_2CH_2CHO \rightleftharpoons$ H_3C–(cyclic five-membered ring with O)–OH

4-Hydroxypentanal

A cyclic hemiacetal (major form present at equilibrium)

Formation of hemiacetals is catalyzed by acid, most commonly sulfuric acid, *p*-toluenesulfonic acid, or hydrogen chloride. The function of the acid catalyst, here represented by H—A, is to protonate the carbonyl oxygen, thus rendering the carbonyl carbon more electrophilic and susceptible to attack by the nucleophilic oxygen atom of the alcohol.

MECHANISM Acid-Catalyzed Formation of a Hemiacetal

Step 1: Proton transfer from acid H—A to the carbonyl carbon

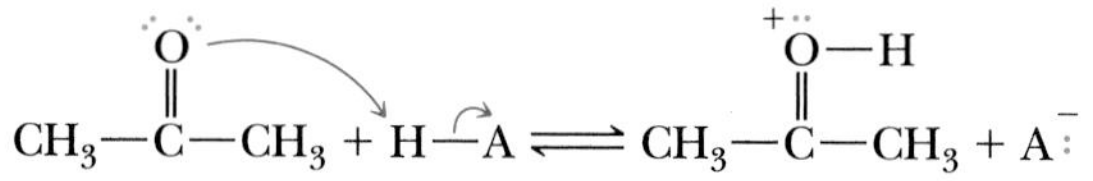

$$CH_3-\overset{O}{\overset{\|}{C}}-CH_3 + H-A \rightleftharpoons CH_3-\overset{\overset{+}{O}-H}{\overset{\|}{C}}-CH_3 + A^-$$

Step 2: Attack of ROH on the carbonyl carbon followed by proton transfer to A^- to give the hemiacetal and regenerate the acid catalyst

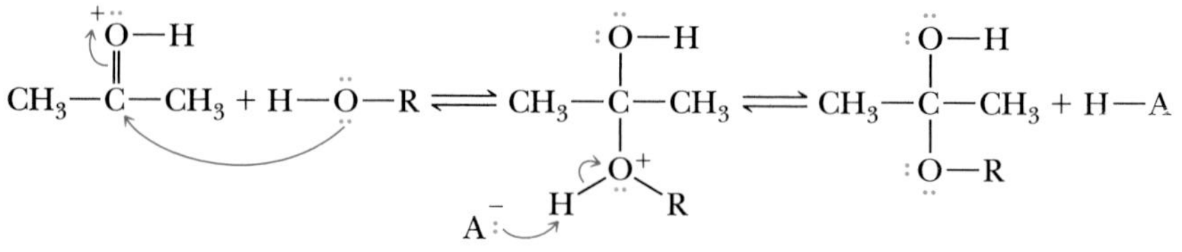

$$CH_3-\overset{\overset{+}{O}-H}{\overset{\|}{C}}-CH_3 + H-O-R \rightleftharpoons CH_3-\overset{O-H}{\overset{|}{\underset{\underset{H\quad R}{\overset{|}{O^+}}}{C}}}-CH_3 \rightleftharpoons CH_3-\overset{O-H}{\overset{|}{\underset{\underset{}{\overset{|}{O-R}}}{C}}}-CH_3 + H-A$$

A^-

Hemiacetals react further with alcohols to form **acetals** plus a molecule of water. This reaction also is acid-catalyzed.

Acetal A molecule containing two —OR or —OAr groups bonded to the same carbon.

$$\underset{\text{A hemiacetal}}{CH_3\overset{OH}{\underset{CH_3}{C}}OCH_2CH_3} + CH_3CH_2OH \overset{H^+}{\rightleftharpoons} \underset{\text{A diethyl acetal}}{CH_3\overset{OCH_2CH_3}{\underset{CH_3}{C}}OCH_2CH_3} + H_2O$$

The functional group of an acetal is a carbon bonded to two —OR or —OAr groups.

$$\underset{\text{from an aldehyde}}{R-\overset{OR'}{\underset{H}{C}}-OR'} \qquad \underset{\text{from a ketone}}{R-\overset{OR'}{\underset{R''}{C}}-OR'}$$

Acetals

The mechanism for the acid-catalyzed conversion of a hemiacetal to an acetal is as follows. Proton transfer in Step 1 from the acid, H—A, to the hemiacetal —OH forms an intermediate that then loses a molecule of water in Step 2 to form a resonance-stabilized oxonium ion. The oxonium ion is an electrophile and in Step 3 reacts with a nucleophile (in this case, a second molecule of alcohol) to form a protonated acetal. Proton transfer to A^- in Step 4 completes the reaction. Note that acid H—A is a true catalyst in this reaction. It is used in Step 1 but another HA is generated in Step 4.

MECHANISM Acid-Catalyzed Formation of an Acetal

$$R-\overset{HO}{\underset{H}{C}}-OCH_3 + H-A \overset{(1)}{\rightleftharpoons} R-\overset{H_2O^+}{\underset{H}{C}}-OCH_3 \overset{(2)}{\rightleftharpoons} \underset{\text{An oxonium ion}}{R-\underset{H}{C}=\overset{+}{O}CH_3} + H_2O + A^-$$

$$CH_3-\overset{H}{O} + R-\underset{H}{C}=\overset{+}{O}CH_3 \overset{(3)}{\rightleftharpoons} \underset{\text{A protonated acetal}}{R-\overset{H\overset{+}{O}CH_3}{\underset{H}{C}}-OCH_3} \overset{(4)}{\rightleftharpoons} \underset{\text{An acetal}}{R-\overset{O-CH_3}{\underset{H}{C}}-OCH_3} + H-A$$

Formation of acetals is often carried out using the alcohol as the solvent and dissolving either dry HCl (hydrogen chloride gas) or *p*-toluenesulfonic acid in the

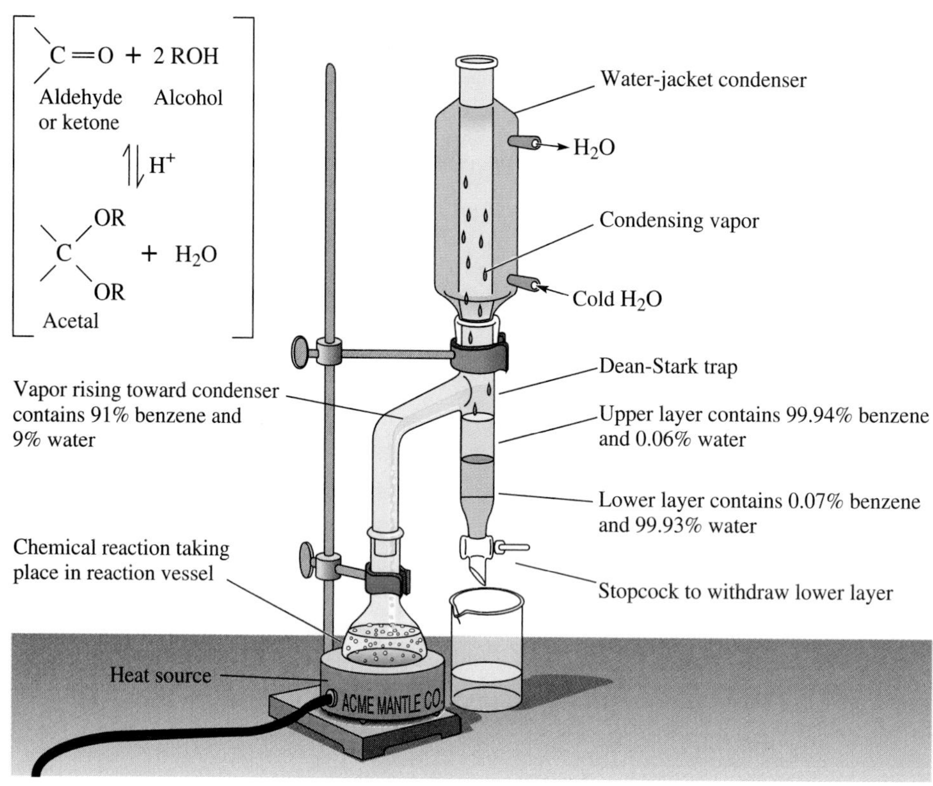

FIGURE 15.5
A Dean-Stark trap for removing water by azeotropic distillation.

alcohol. Because the alcohol is both a reactant and solvent, it is present in large molar excess, which forces the equilibrium to the right and favors acetal formation.

Azeotrope A liquid mixture with constant composition and a boiling point that is different from that of any of its components.

In another experimental technique to force the equilibrium to the right, water is removed from the reaction vessel by azeotropic distillation using a **Dean-Stark trap** (Figure 15.5). An **azeotrope** is a liquid mixture of constant composition with a boiling point that is different from that of any of its components. An azeotropic mixture boils at a constant temperature without change in composition.

In this method for preparing an acetal, the aldehyde or ketone, alcohol, acid catalyst, and benzene are brought to reflux. The component in this mixture with the lowest boiling point is an azeotrope, bp 69°C, consisting of 91% benzene and 9% water. This vapor is condensed and collected in a side trap where it separates into two layers. At room temperature, the composition of the upper, less dense layer is 99.94% benzene and 0.06% water. The composition of the lower, more dense layer is almost the reverse, 0.07% benzene and 99.93% water. As reflux continues, benzene from the top layer is returned to the refluxing mixture, and water is drawn off at the bottom through a stopcock. A Dean-Stark trap "pumps" water out of the reaction mixture, thus forcing the equilibrium to the right. This same apparatus is used in many other reactions where water needs to be removed, as for example, in formation of enamines (Section 15.10A).

EXAMPLE 15.6

Show the reaction of the carbonyl group of each aldehyde or ketone with one molecule of alcohol to form a hemiacetal and then with a second molecule of alcohol to form an acetal. Note that in part (b), ethylene glycol is a diol, and one molecule provides both —OH groups.

(a) $CH_3CH_2C(=O)CH_3 + 2CH_3CH_2OH \overset{H^+}{\rightleftharpoons}$ (b) cyclopentanone $+ HOCH_2CH_2OH \overset{H^+}{\rightleftharpoons}$

Solution

Given are structural formulas of the hemiacetal and then the acetal:

(a) $CH_3CH_2C(OH)(CH_3)OCH_2CH_3 \longrightarrow CH_3CH_2C(OCH_2CH_3)(CH_3)OCH_2CH_3$

A hemiacetal — An acetal

(b) 1-hydroxy-1-(OCH_2CH_2OH)cyclopentane $\longrightarrow$ cyclopentane spiro-fused with $O—CH_2—CH_2—O$ ring

A hemiacetal — A cyclic acetal

PROBLEM 15.6

Hydrolysis of an acetal in aqueous acid forms an aldehyde or ketone and two molecules of alcohol. Following are structural formulas for three acetals. Draw the structural formulas for the products of hydrolysis of each in aqueous acid.

(a) $CH_3O—C_6H_4—CH(OCH_3)OCH_3$ (b) $(H_3C)_2C$ in ring with $O—CH_2—CH_2—O$ (c) H_3C–tetrahydrofuran ring (O)–OCH_3

Like ethers (Section 11.6), acetals are also unreactive to bases, hydride reducing agents such as $LiAlH_4$ and $NaBH_4$, Grignard and other organometallic reagents, oxidizing agents (except, of course, for those involving use of aqueous acid), and conditions for catalytic reduction. Because of their lack of reactivity toward these reagents, acetals are often used to protect the carbonyl groups of aldehydes and ketones while reactions are carried out on functional groups in other parts of the molecule.

C. Acetals as Carbonyl-Protecting Groups

Acetals are valuable in synthetic organic chemistry as carbonyl protecting groups, as illustrated by the synthesis of 4-hydroxy-4-phenylbutanal from benzaldehyde and 3-bromopropanal.

$$C_6H_5CH(=O) + BrCH_2CH_2CH(=O) \xrightarrow{??} C_6H_5CH(OH)CH_2CH_2CH(=O)$$

Benzaldehyde — 3-Bromopropanal — 4-Hydroxy-4-phenylbutanal

One obvious way to form a new carbon-carbon bond between these two molecules is to treat benzaldehyde with the Grignard reagent from 3-bromopropanal. A Grignard reagent formed from 3-bromopropanal, however, reacts immediately with the carbonyl group of another molecule of 3-bromopropanal, causing it to self-destruct during preparation. A way to avoid this problem is to protect the carbonyl group of 3-bromopropanal by conversion to an acetal. Cyclic acetals are often used because they are particularly easy to prepare.

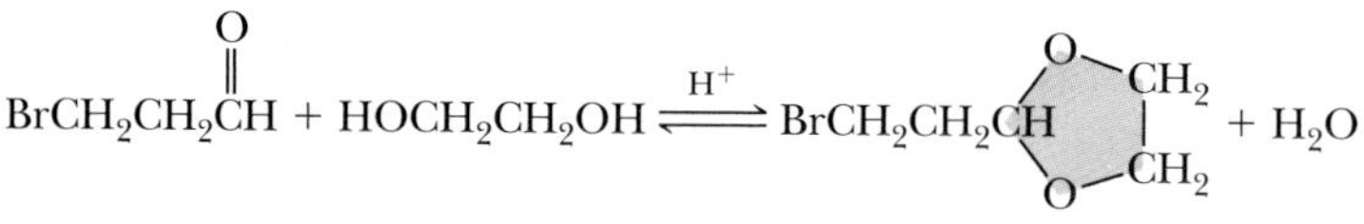

$$BrCH_2CH_2CHO + HOCH_2CH_2OH \xrightleftharpoons{H^+} BrCH_2CH_2CH\begin{matrix} O-CH_2 \\ | \\ O-CH_2 \end{matrix} + H_2O$$

A cyclic acetal

Treatment of the protected bromoaldehyde with magnesium in diethyl ether followed by addition of benzaldehyde gives a magnesium alkoxide.

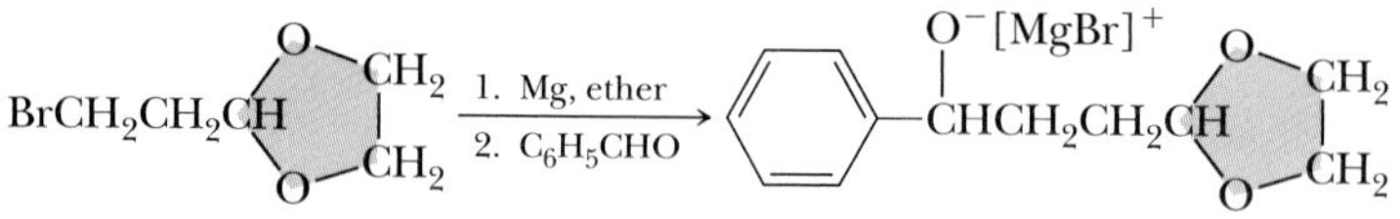

$$BrCH_2CH_2CH\begin{matrix} O-CH_2 \\ | \\ O-CH_2 \end{matrix} \xrightarrow[2.\ C_6H_5CHO]{1.\ Mg,\ ether} C_6H_5-\underset{}{\overset{O^-[MgBr]^+}{CH}}CH_2CH_2CH\begin{matrix} O-CH_2 \\ | \\ O-CH_2 \end{matrix}$$

Treatment of the magnesium alkoxide with aqueous acid accomplishes two things. First, protonation of the alkoxide anion gives the desired hydroxyl group, and second, hydrolysis of the cyclic acetal regenerates the aldehyde group.

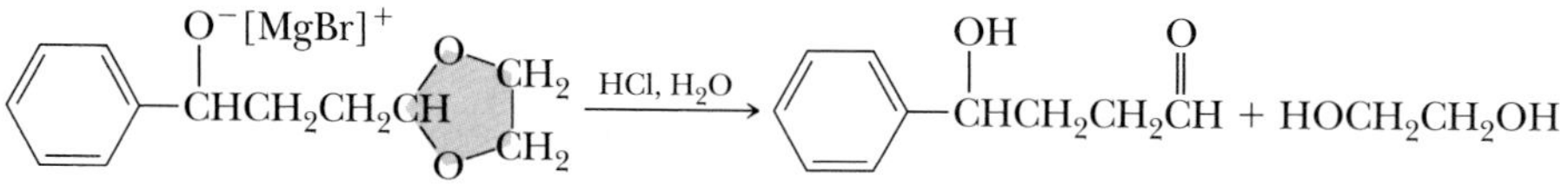

$$C_6H_5-\overset{O^-[MgBr]^+}{CH}CH_2CH_2CH\begin{matrix} O-CH_2 \\ | \\ O-CH_2 \end{matrix} \xrightarrow{HCl,\ H_2O} C_6H_5-\overset{OH}{CH}CH_2CH_2\overset{O}{\overset{\|}{C}}H + HOCH_2CH_2OH$$

15.9 Addition of Sulfur Nucleophiles

The sulfur atom of a thiol is a better nucleophile than the oxygen atom of an alcohol (Section 8.4B). Thiols, like alcohols, add to the carbonyl group of aldehydes and ketones to form tetrahedral carbonyl addition compounds. A common sulfur nucleophile used for this purpose is 1,3-propanedithiol. The carbonyl groups of both aldehydes and ketones react with this compound in the presence of an acid catalyst to form cyclic thioacetals. Products of this reaction are also called **1,3-dithianes;** "1,3-dithi-" indicates that atoms of sulfur occur at positions 1 and 3 of the ring, and "-ane" indicates that the ring is six-membered (Section 11.2). Following is an equation for formation of a 1,3-dithiane from an aldehyde.

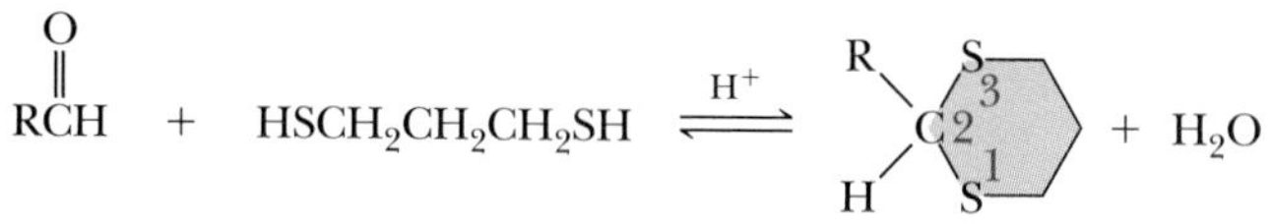

$$R\overset{O}{\overset{\|}{C}}H + HSCH_2CH_2CH_2SH \xrightleftharpoons{H^+} \text{(R)(H)C2 in ring with S1, S3} + H_2O$$

An aldehyde

A 1,3-dithiane (a cyclic thioacetal)

The special value of 1,3-dithianes derived from an aldehyde is that the hydrogen atom on carbon 2 of the ring is weakly acidic; it has a pK_a of approximately 31. In the presence of a very strong base, such as butyllithium, a 1,3-dithiane derived from an aldehyde is converted into a lithium salt.

$$\text{(1,3-dithiane, C–H, C–R)} + CH_3CH_2CH_2CH_2Li \longrightarrow \text{(1,3-dithiane, } \overset{-}{C}\text{: Li}^+\text{, C–R)} + CH_3CH_2CH_2CH_3$$

Stronger acid pK_a 31 — Stronger base — Weaker base — Weaker acid pK_a 50

A 1,3-dithiane anion is an excellent nucleophile and reacts by an S_N2 pathway with primary alkyl halides to give a disubstituted dithiane. Treatment of this product with mercuric chloride, $HgCl_2$, in aqueous acetonitrile brings about hydrolysis of the dithiane to give a ketone and a mercury salt of the dithiol.

$$\text{(dithiane } \overset{-}{C}\text{: Li}^+\text{, R)} + R'CH_2\text{—Br} \xrightarrow{S_N2} \text{(dithiane C(CH}_2\text{R')(R))} \xrightarrow[CH_3CN]{H_2O,\ HgCl_2} R\text{—}\overset{O}{\overset{\|}{C}}\text{—}CH_2R'$$

Lithium salt of a 1,3-dithiane

The result of this series of reactions is conversion of an aldehyde to a ketone.

$$R\text{—}\overset{O}{\overset{\|}{C}}\text{—}H \xrightarrow[\text{3. nucleophilic substitution on } R'CH_2\text{—X; 4. hydrolysis of the dithiane}]{\text{1. formation of a 1,3-dithiane; 2. treatment with BuLi}} R\text{—}\overset{O}{\overset{\|}{C}}\text{—}CH_2R'$$

Anions derived from 1,3-dithianes also add to the carbonyl groups of aldehydes and ketones to form tetrahedral carbonyl addition compounds. Hydrolysis of the lithium salt in aqueous acetonitrile containing mercury(II) chloride gives an α-hydroxyketone.

$$\text{(dithiane } \overset{-}{C}\text{: Li}^+\text{, R)} + H\text{—}\overset{O}{\overset{\|}{C}}\text{—}R' \longrightarrow \text{(dithiane C(R)—}\overset{O^-Li^+}{\overset{|}{C}H}\text{—}R'\text{)} \xrightarrow[CH_3CN]{H_2O,\ HgCl_2} R\text{—}\overset{O}{\overset{\|}{C}}\text{—}\overset{OH}{\overset{|}{C}H}\text{—}R'$$

Lithium salt of a 1,3-dithiane

The result of this series of reactions is replacement of the hydrogen atom of the original aldehyde by either a secondary alcohol or, if carbonyl addition takes place on a ketone, a tertiary alcohol.

These two reactions of 1,3-dithianes provide examples of what is called by the German name **Umpolung** (note the capitalization of the noun), which means pole reversal. Under normal circumstances, the carbon atom of a carbonyl group bears a partial positive charge and is, therefore, an electrophile. When converted to a 1,3-dithiane and then treated with butyllithium, the same carbon becomes an anion and, therefore, a nucleophile.

EXAMPLE 15.7

Give structural formulas for the lettered compounds in each reaction sequence.

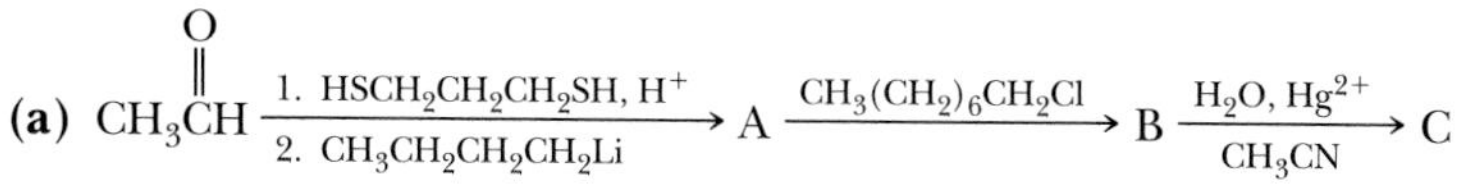

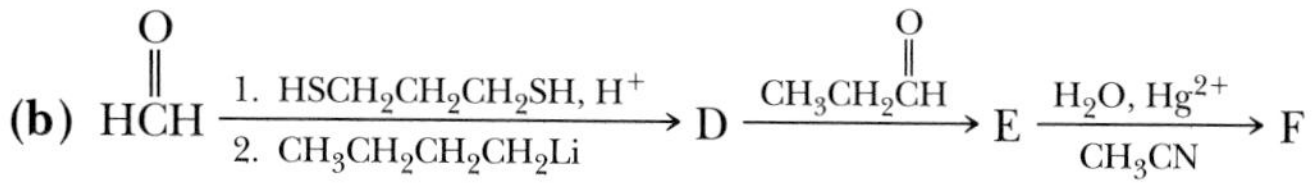

(a) $CH_3\overset{O}{\overset{\|}{C}}H \xrightarrow[\text{2. } CH_3CH_2CH_2CH_2Li]{\text{1. } HSCH_2CH_2CH_2SH,\ H^+} A \xrightarrow{CH_3(CH_2)_6CH_2Cl} B \xrightarrow[CH_3CN]{H_2O,\ Hg^{2+}} C$

(b) $H\overset{O}{\overset{\|}{C}}H \xrightarrow[\text{2. } CH_3CH_2CH_2CH_2Li]{\text{1. } HSCH_2CH_2CH_2SH,\ H^+} D \xrightarrow{CH_3CH_2\overset{O}{\overset{\|}{C}}H} E \xrightarrow[CH_3CN]{H_2O,\ Hg^{2+}} F$

Solution

(a) A (2-methyl-1,3-dithiane anion, $C:^-$ bearing CH_3) → B (1,3-dithiane C bearing $CH_2(CH_2)_6CH_3$ and CH_3) → C: $CH_3\overset{O}{\overset{\|}{C}}CH_2(CH_2)_6CH_3$

(b) D (1,3-dithiane anion, $C:^-$ bearing H) → E (1,3-dithiane C bearing $\overset{OH}{\overset{|}{C}}HCH_2CH_3$ and H) → F: $H\overset{O}{\overset{\|}{C}}—\overset{OH}{\overset{|}{C}}HCH_2CH_3$

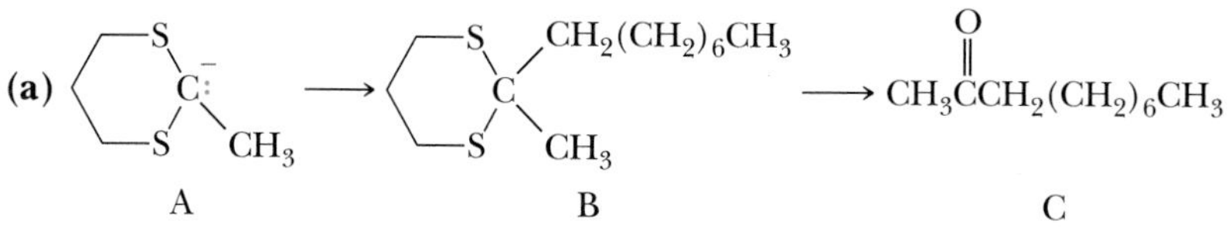

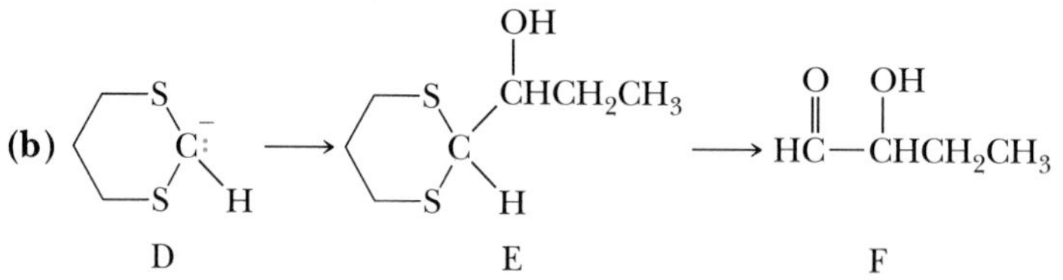

PROBLEM 15.7

Show how to convert the given starting material into the indicated product using a 1,3-dithiane as an intermediate.

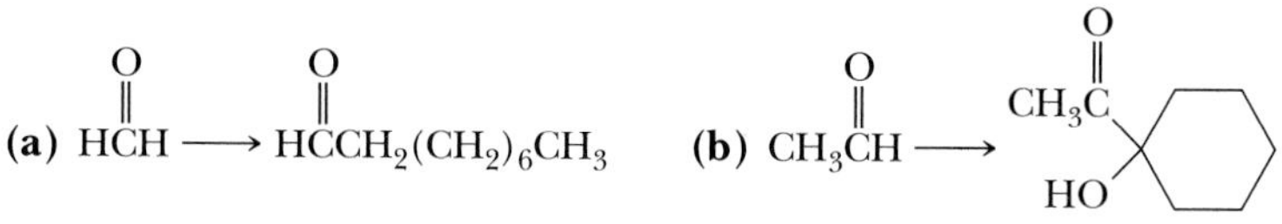

(a) $H\overset{O}{\overset{\|}{C}}H \longrightarrow H\overset{O}{\overset{\|}{C}}CH_2(CH_2)_6CH_3$ (b) $CH_3\overset{O}{\overset{\|}{C}}H \longrightarrow$ 1-acetylcyclohexanol ($CH_3\overset{O}{\overset{\|}{C}}$ and HO on the same cyclohexane carbon)

15.10 Addition of Nitrogen Nucleophiles

A. Ammonia and Its Derivatives

Imine A compound containing a carbon-nitrogen double bond; also called a Schiff base.

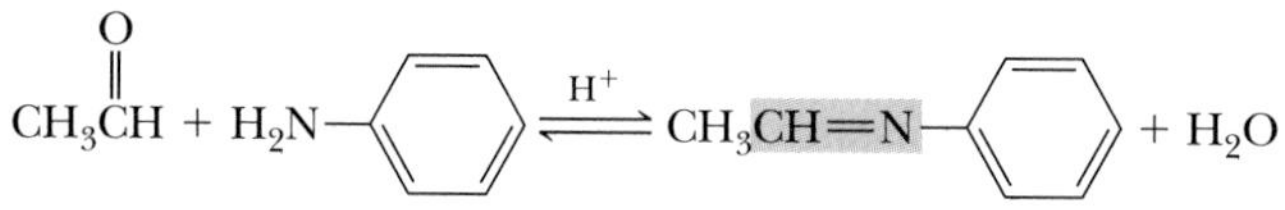

Ammonia, primary aliphatic amines (RNH_2), and primary aromatic amines ($ArNH_2$) react with the carbonyl group of aldehydes and ketones in the presence of an acid catalyst to give a product that contains a carbon-nitrogen double bond. A molecule containing a carbon-nitrogen double bond is called an **imine** or, alternatively, a **Schiff base.**

$$CH_3\overset{O}{\overset{\|}{C}}H + H_2N—C_6H_5 \underset{}{\overset{H^+}{\rightleftharpoons}} CH_3CH{=}N—C_6H_5 + H_2O$$

Ethanal Aniline An imine (a Schiff base)

$$\text{Cyclopentanone}{=}O + H_2NCH_3 \overset{H^+}{\rightleftharpoons} \text{Cyclopentane}{=}NCH_3 + H_2O$$

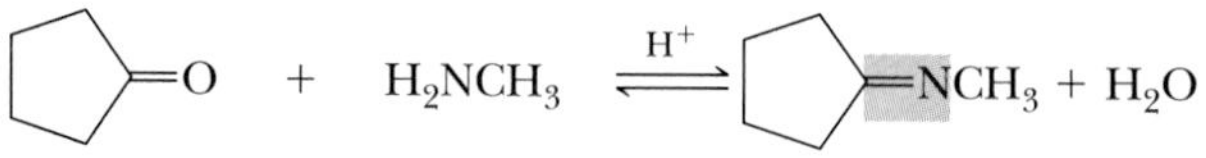

Cyclopentanone Methylamine An imine (a Schiff base)

The mechanism of imine formation can be divided into two steps. In Step 1, the nitrogen atom of ammonia or the primary amine, both good nucleophiles, adds to the carbonyl carbon to form a tetrahedral carbonyl addition compound. This reaction is exactly the same as with water and alcohols. Acid-catalyzed dehydration of this addition compound in Step 2 is the slow, rate-limiting step.

MECHANISM Formation of an Imine From an Aldehyde or Ketone

Step 1: Addition of the nucleophile to the carbonyl carbon followed by proton transfer to form a tetrahedral carbonyl addition intermediate

$$\text{C=O} + H_2\ddot{N}\text{—R} \rightleftharpoons \text{—C(—}\ddot{O}\text{:}^-\text{)—}\overset{+}{N}H_2\text{—R} \rightleftharpoons \text{—C(—}\ddot{O}H\text{)—}\ddot{N}H\text{—R}$$

A tetrahedral carbonyl addition compound

Step 2: Protonation of the —OH group followed by loss of H_2O and proton transfer to solvent gives the imine

$$H_3O^+ + \text{—C(—}\ddot{O}H\text{)—}\ddot{N}H\text{—R} \rightleftharpoons \text{—C(—}\overset{+}{O}H_2\text{)—}\ddot{N}H\text{—R} \rightleftharpoons \text{C=}\ddot{N}\text{—R} + H_2O + H_3O^+$$

An imine

One of the chief values of imines is that the carbon-nitrogen double bond can be reduced by hydrogen in the presence of a nickel or other transition metal catalyst to a carbon-nitrogen single bond. Thus, a primary amine is converted to a secondary amine by way of an imine as illustrated by the conversion of cyclohexylamine to dicyclohexylamine.

$$C_6H_{10}\text{=O} + H_2N\text{—}C_6H_{11} \xrightarrow[-H_2O]{H^+} C_6H_{10}\text{=N—}C_6H_{11} \xrightarrow{H_2/Ni} C_6H_{11}\text{—NH—}C_6H_{11}$$

Cyclohexanone Cyclohexylamine (An imine) Dicyclohexylamine

To give but one example of the importance of imines in biological systems, the active form of **vitamin A aldehyde** (retinal) is bound to the protein **opsin** in the human retina in the form of an imine. The primary amino group for this reaction is provided by the side chain of the amino acid lysine (Section 27.1). The imine is called **rhodopsin** or **visual purple.**

A ball-and-stick model of 11-*cis*-retinal. *(Charles D. Winters)*

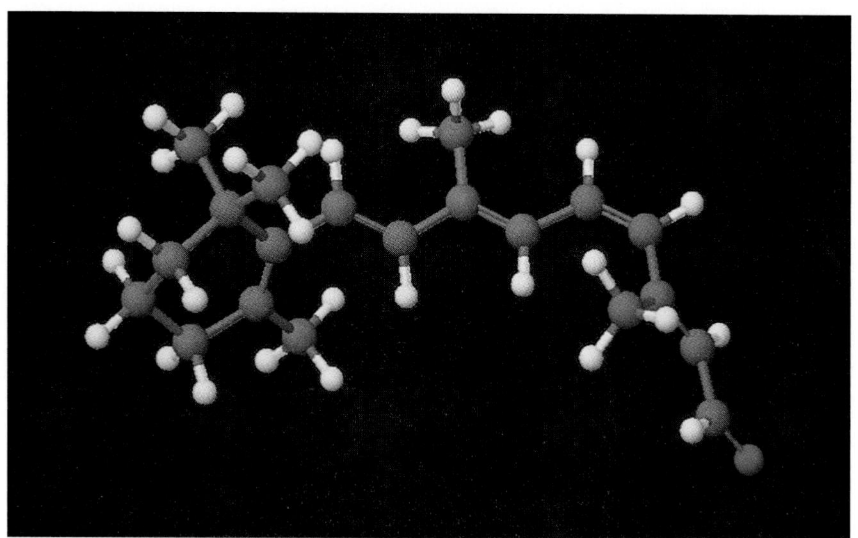

11-*cis*-Retinal + H_2N—opsin ⟶ Rhodopsin (visual purple)

EXAMPLE 15.8

Write structural formulas for the imines formed in these reactions:

(a) cyclopentanone + $CH_3CH(NH_2)CH_2CH_3$ $\xrightarrow[-H_2O]{H^+}$ (b) $CH_3\overset{O}{\overset{\|}{C}}CH_3$ + $CH_3O-C_6H_4-NH_2$ $\xrightarrow[-H_2O]{H^+}$

Solution

Given is a structural formula for each imine:

(a) cyclopentylidene=$NCH(CH_3)CH_2CH_3$ (b) $(H_3C)_2C{=}N-C_6H_4-OCH_3$

PROBLEM 15.8

Acid-catalyzed hydrolysis of an imine gives an amine and an aldehyde or ketone. When one equivalent of acid is used, the amine is converted to an ammonium salt. Write structural formulas for the products of hydrolysis of the following imines using one equivalent of HCl:

(a) 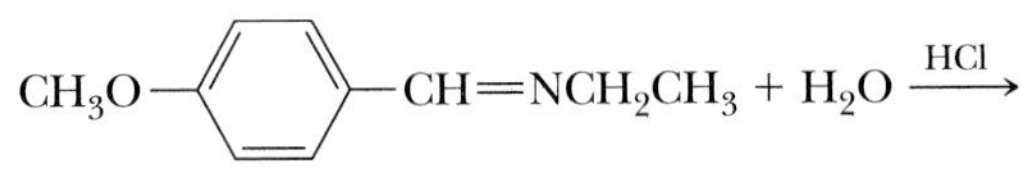 CH_3O—(benzene ring)—$CH{=}NCH_2CH_3 + H_2O \xrightarrow{HCl}$ **(b)** 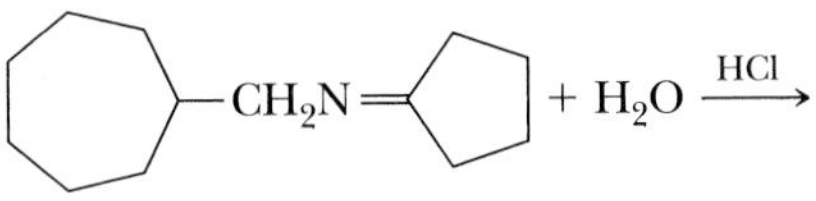 (cycloheptane ring)—$CH_2N{=}$(cyclopentane ring) $+ H_2O \xrightarrow{HCl}$

Secondary amines react with aldehydes and ketones to form enamines. The name **enamine** is derived from **en-** to indicate the presence of a carbon-carbon double bond and **-amine** to indicate the presence of an amino group. An example is enamine formation between cyclohexanone and piperidine, a cyclic secondary amine.

Enamine An unsaturated compound derived by the reaction of an aldehyde or ketone and a secondary amine followed by loss of H_2O; $CR_2{=}CR{-}NR_2$.

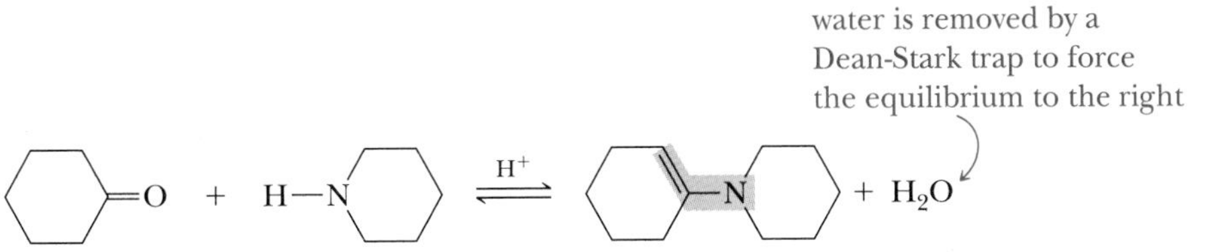

water is removed by a Dean-Stark trap to force the equilibrium to the right

Cyclohexanone $+$ H—N(piperidine) $\overset{H^+}{\rightleftharpoons}$ enamine $+ H_2O$

Piperidine (a secondary amine) An enamine

The mechanism for formation of an enamine is very similar to that for formation of an imine. In Step 1, nucleophilic addition of the secondary amine to the carbonyl carbon of the aldehyde or ketone followed by proton transfer from nitrogen to oxygen gives a tetrahedral carbonyl addition compound. Acid-catalyzed dehydration in Step 2 gives the enamine. It is at this stage that enamine formation differs from imine formation. The nitrogen has no proton to be lost. Instead, a proton is lost from the α-carbon of the ketone or aldehyde portion of the molecule in an elimination reaction.

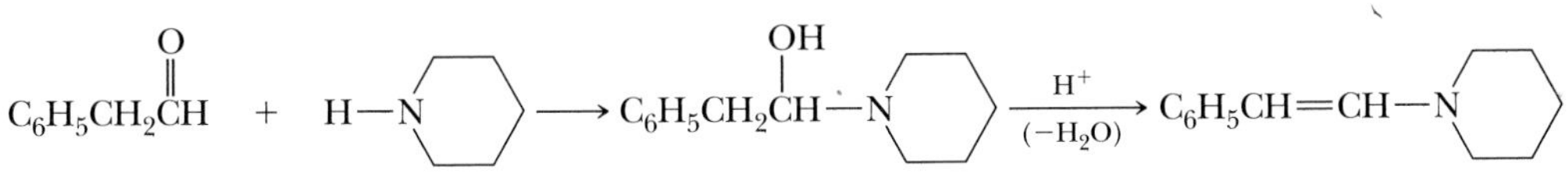

$C_6H_5CH_2\overset{O}{\overset{\|}{C}}H$ $+$ H—N(piperidine) $\longrightarrow$ $C_6H_5CH_2\overset{OH}{\overset{|}{C}}H$—N(piperidine) $\xrightarrow[(-H_2O)]{H^+}$ $C_6H_5CH{=}CH$—N(piperidine)

Phenylacetaldehyde Piperidine A tetrahedral carbonyl addition compound An enamine

We will return to the chemistry of enamines and their use in synthesis in Chapter 18.

B. Hydrazine and Related Compounds

Aldehydes and ketones react with hydrazine to form compounds called hydrazones as illustrated by the treatment of cyclopentanone with hydrazine in a reaction similar to imine formation. A common use of hydrazones is as intermediates in the Wolff-Kishner reduction of carbonyl groups to methylene groups (Section 15.14C).

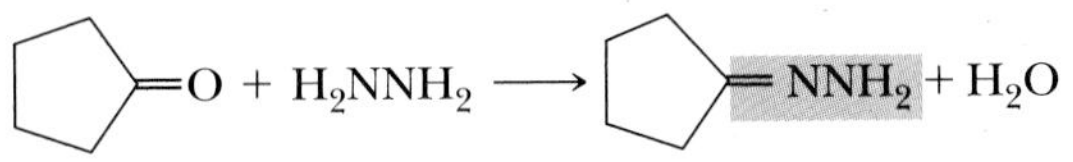

Cyclopentanone$=O + H_2NNH_2 \longrightarrow$ cyclopentane$=NNH_2 + H_2O$

Hydrazine A hydrazone

Other derivatives of ammonia and hydrazine that react with aldehydes and ketones to give imines are listed in Table 15.4. The chief value of the nitrogen nucleophiles listed in this table is that most aldehydes and ketones react with them to give crystalline solids with sharp melting points. Historically, these derivatives often provided a con-

TABLE 15.4 Derivatives of Ammonia and Hydrazine Used for Forming Imines

Reagent, $H_2N—R$	Name of Reagent	Name of Derivative Formed
$H_2N—OH$	hydroxylamine	an oxime
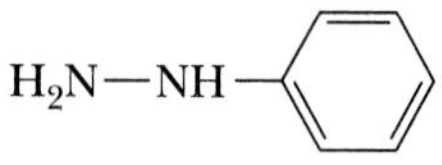 $H_2N—NH—C_6H_5$	phenylhydrazine	a phenylhydrazone
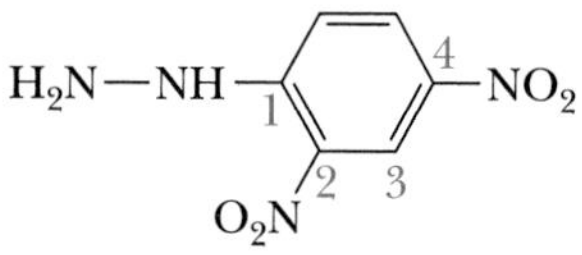 $H_2N—NH—C_6H_3(NO_2)_2$ (2,4-dinitrophenyl)	2,4-dinitrophenylhydrazine	a 2,4-dinitrophenylhydrazone
$H_2N—NHCNH_2$ (with C=O)	semicarbazide	a semicarbazone

venient way to identify liquid aldehydes or ketones. Now, these compounds are more readily identified by IR and NMR spectroscopy.

15.11 Keto-Enol Tautomerism

A. Acidity of α-Hydrogens

α-Carbon A carbon atom adjacent to a carbonyl group.

α-Hydrogen A hydrogen on a carbon alpha to a carbonyl group.

A carbon atom adjacent to a carbonyl group is called an ***α*-carbon,** and hydrogen atoms bonded to it are called ***α*-hydrogens.**

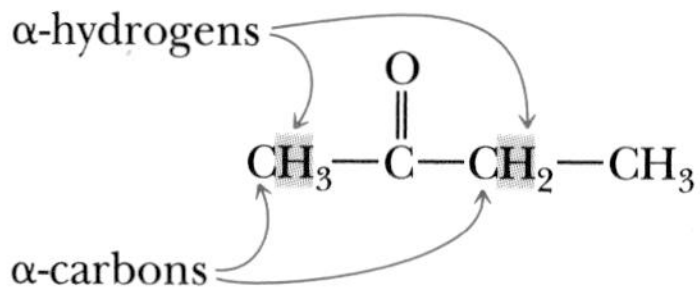

$$CH_3—\overset{O}{\overset{\|}{C}}—CH_2—CH_3$$

α-hydrogens

α-carbons

Because carbon and hydrogen have comparable electronegativities, a C—H bond normally has little polarity and a hydrogen atom bonded to carbon shows very low acidity. The situation is different, however, for hydrogens alpha to a carbonyl group. α-Hydrogens are more acidic than acetylenic, vinylic, and alkane hydrogens but less acidic than —OH hydrogens of alcohols.

Type of Bond	pK_a
$CH_3CH_2O—H$	16
$CH_3\overset{O}{\overset{\|}{C}}CH_2—H$	20
$CH_3C{\equiv}C—H$	25
$CH_2{=}CH—H$	44
$CH_3CH_2—H$	51

Hydrogens alpha to the carbonyl group of an aldehyde or ketone are more acidic than hydrogens of alkanes, alkenes, and alkynes because of two factors. First, the electron-withdrawing inductive effect of the adjacent carbonyl group weakens the bond to the alpha hydrogen. Second, the negative charge on the resulting **enolate anion** is delocalized by resonance, thus stabilizing it relative to the anion from an alkane, alkene, or alkyne. Note that we used these same two factors in Sections 3.3C and D to explain the greater acidity of carboxylic acids compared with alcohols.

Enolate anion An anion derived by loss of a hydrogen from a carbon alpha to a carbonyl group.

$$CH_3-\overset{\overset{\large O}{\|}}{C}-CH_2-H + :A^- \rightleftharpoons \left[CH_3-\overset{\overset{\large O}{\|}}{C}-CH_2^- \longleftrightarrow CH_3-\overset{\overset{\large O^-}{|}}{C}=CH_2 \right] + H-A$$

Resonance-stabilized enolate anion

EXAMPLE 15.9

Predict the position of the following equilibrium.

$$C_6H_5-\overset{\overset{\large O}{\|}}{C}-CH_3 + CH_3CH_2O^- \xrightleftharpoons{\text{ethanol}} C_6H_5-\overset{\overset{\large O^-}{|}}{C}=CH_2 + CH_3CH_2OH$$

Acetophenone

Solution

The pK_a of ethanol is 16 (Table 3.1, rounded to two significant figures). Assume that the pK_a of acetophenone is approximately equal to that of acetone, namely approximately 20. Ethanol is the stronger acid and, therefore, the equilibrium lies to the left.

$$\underset{pK_a\ 20}{C_6H_5-\overset{\overset{\large O}{\|}}{C}-CH_3} + CH_3CH_2O^- \rightleftharpoons C_6H_5-\overset{\overset{\large O^-}{|}}{C}=CH_2 + \underset{pK_a\ 16}{CH_3CH_2OH} \qquad pK_{eq} = 4 \quad K_{eq} = 10^{-4}$$

PROBLEM 15.9

Predict the position of the following equilibrium.

$$C_6H_5-\overset{\overset{\large O}{\|}}{C}-CH_3 + NH_2^- \rightleftharpoons C_6H_5-\overset{\overset{\large O^-}{|}}{C}=CH_2 + NH_3$$

Acetophenone

When an enolate anion reacts with a proton donor, it may do so either on oxygen or on the α-carbon. Protonation of the enolate anion on the α-carbon gives the original molecule in what is called the **keto form.** Protonation on oxygen gives an **enol form.** Because the anion can be derived by loss of a proton from the enol form, it is called an enolate anion.

$$\underset{\text{Keto form}}{A^- + CH_3-\overset{\overset{\large O}{\|}}{C}-CH_3} \xleftarrow{H-A} \underset{\text{Enolate anion}}{\left[CH_3-\overset{\overset{\large O}{\|}}{C}-CH_2^- \longleftrightarrow CH_3-\overset{\overset{\large O^-}{|}}{C}=CH_2 \right]} \xrightarrow{H-A} \underset{\text{Enol form}}{CH_3-\overset{\overset{\large OH}{|}}{C}=CH_2 + A^-}$$

Enol formation can also be catalyzed by acid. Step 1 is rapid and reversible proton transfer from the acid, H—A, to the carbonyl oxygen. This is then followed by a second proton transfer from the α-carbon to A^- to give the enol. The only difference between the base- and acid-catalyzed reactions is the order of proton addition and elimination. In acid-catalyzed reactions, the proton is added first; in base-catalyzed reactions, the proton is removed first.

MECHANISM Acid-Catalyzed Equilibration of Keto and Enol Tautomers

Step 1: Proton transfer to the carbonyl oxygen

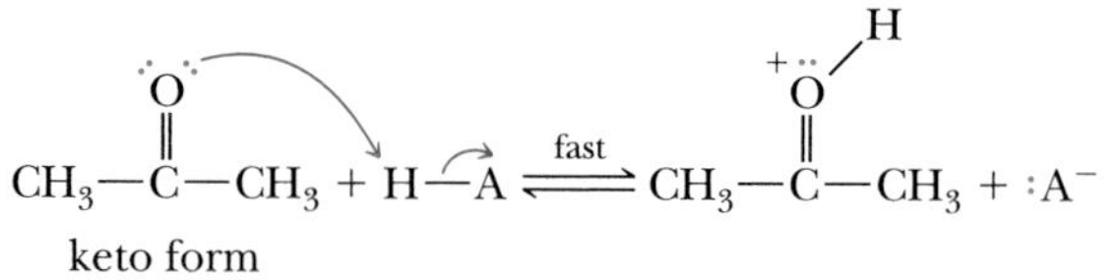

$$CH_3-\overset{\overset{\Large O}{\|}}{C}-CH_3 + H-A \underset{}{\overset{\text{fast}}{\rightleftharpoons}} CH_3-\overset{\overset{\Large \overset{+}{O}-H}{\|}}{C}-CH_3 + :A^-$$

keto form

Step 2: Proton transfer to A^- to give the enol

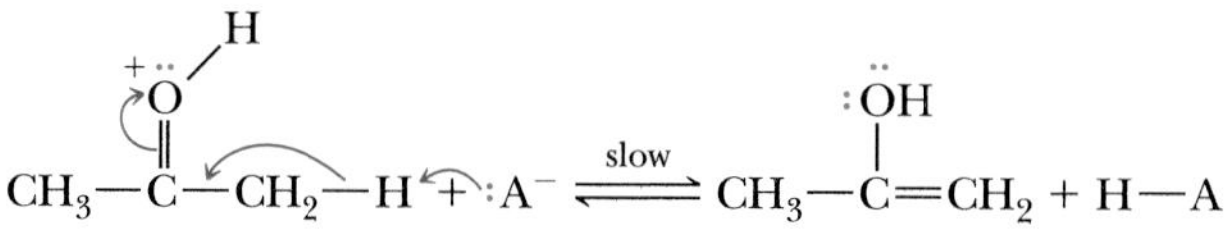

$$CH_3-\overset{\overset{\Large \overset{+}{O}-H}{\|}}{C}-CH_2-H + :A^- \underset{}{\overset{\text{slow}}{\rightleftharpoons}} CH_3-\overset{\overset{\Large :\ddot{O}H}{|}}{C}=CH_2 + H-A$$

enol form

B. Keto-Enol Tautomerism

Aldehydes and ketones with at least one α-hydrogen are in equilibrium with their enol forms. We first encountered this type of equilibrium in Section 10.8 in our study of the hydroboration/oxidation and acid-catalyzed hydration of alkynes. As seen in Table 15.5, the position of keto-enol equilibrium for simple aldehydes and ketones lies far on the side of the keto form because a carbon-oxygen double bond is stronger than a carbon-carbon double bond.

For certain types of molecules, the enol form may be the major form and, in some cases, the only form present at equilibrium. For β-diketones, such as 1,3-cyclohexanedione and 2,4-pentanedione, where an α-carbon is between two carbonyl groups, the position of equilibrium shifts in favor of the enol form.

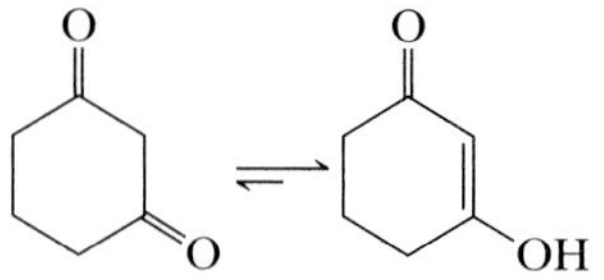

1,3-Cyclohexanedione (keto form ⇌ enol form with OH)

$$CH_3-\overset{\overset{\Large O}{\|}}{C}-CH_2-\overset{\overset{\Large O}{\|}}{C}-CH_3 \rightleftharpoons H_3C-\overset{\overset{\Large O^{\delta-}\cdots H^{\delta+}}{\|}}{C}-\underset{\underset{\Large H}{|}}{C}=\overset{\overset{\Large O}{|}}{C}-CH_3$$

hydrogen bonding

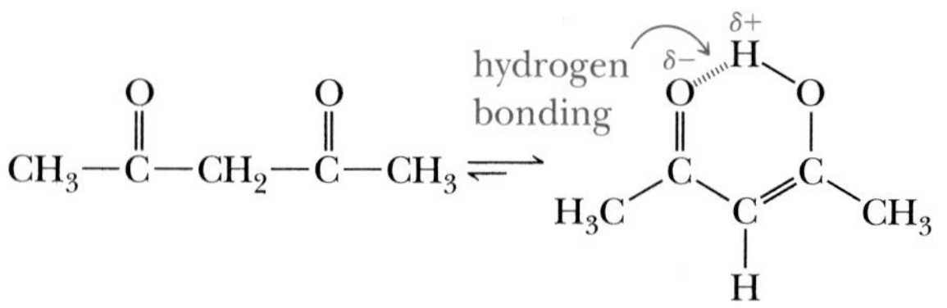

20% 80%

2,4-Pentanedione (Acetylacetone)

TABLE 15.5 The Position of Keto-Enol Equilibrium for Some Simple Aldehydes and Ketones*

Keto Form		Enol Form	% Enol Present at Equilibrium
CH_3CHO	$\rightleftharpoons$	$CH_2{=}CH(OH)$	6×10^{-5}
$CH_3CH_2CH_2CHO$	$\rightleftharpoons$	$CH_3CH_2CH{=}CH(OH)$	5.5×10^{-4}
CH_3COCH_3	$\rightleftharpoons$	$CH_3C(OH){=}CH_2$	6×10^{-7}
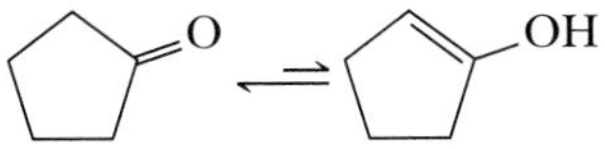 cyclopentanone	$\rightleftharpoons$	1-cyclopentenol	1×10^{-6}
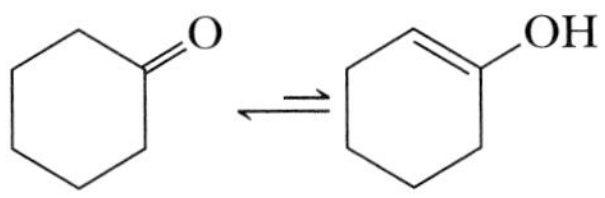 cyclohexanone	$\rightleftharpoons$	1-cyclohexenol	4×10^{-5}

*Data from J. March, *Advanced Organic Chemistry*, 4th ed., (New York: Wiley Interscience), 1992, p. 70.

These enols are stabilized by conjugation of the pi systems of the carbon-carbon double bond and the carbonyl group. The enol of 2,4-pentanedione, an open-chain β-diketone, is further stabilized by intramolecular hydrogen bonding.

EXAMPLE 15.10

Write two enol structures for each compound. Which enol of each predominates at equilibrium?

(a) 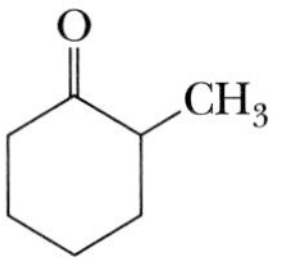 2-methylcyclohexanone (cyclohexanone with O and CH_3)

(b) $CH_3CCH_2(CH_2)_2CH_3$ (C=O at C-2)

Solution

(a) 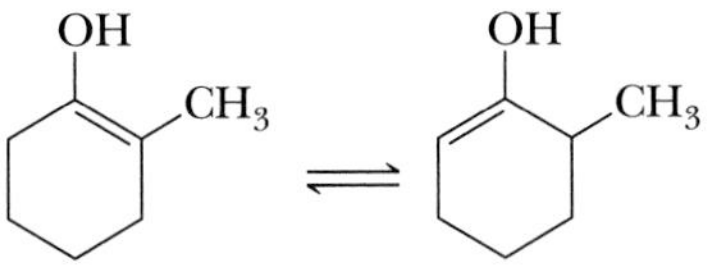 2-methyl-1-cyclohexenol (OH, CH_3) $\rightleftharpoons$ 6-methyl-1-cyclohexenol (OH, CH_3)

Major enol form (more substituted double bond)

(b) $CH_2{=}C(OH)CH_2(CH_2)_2CH_3 \rightleftharpoons CH_3C(OH){=}CH(CH_2)_2CH_3$

Major enol form (more substituted double bond)

PROBLEM 15.10

Draw the structural formula for the keto form of each enol.

(a) 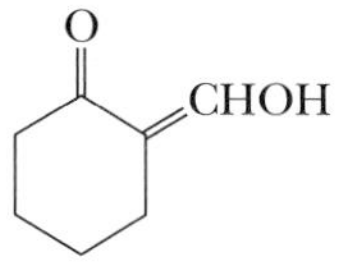 2-(hydroxymethylene)cyclohexanone: O, CHOH (b) 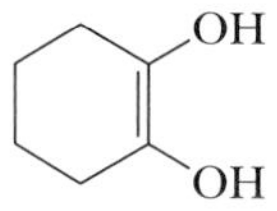 OH, OH (c) 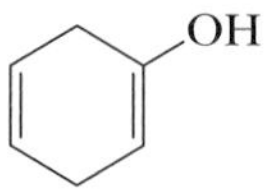 OH

15.12 Reactions at the α-Carbon

A. Racemization

When enantiomerically pure (either *R* or *S*) 3-phenyl-2-butanone is dissolved in ethanol, no change occurs in the optical activity of the solution over time. If, however, a trace of either acid (for example, aqueous or gaseous HCl) or base (for example, sodium ethoxide) is added, the optical activity of the solution begins to decrease and gradually drops to zero. When 3-phenyl-2-butanone is isolated from this solution, it is found to be a racemic mixture (Section 4.8C). Furthermore, the rate of racemization is proportional to the concentration of acid or base. These observations can be explained by a rate-limiting acid- or base-catalyzed formation of an achiral enol intermediate. Tautomerism of the achiral enol to the chiral keto form generates the *R* and *S* enantiomers with equal probability.

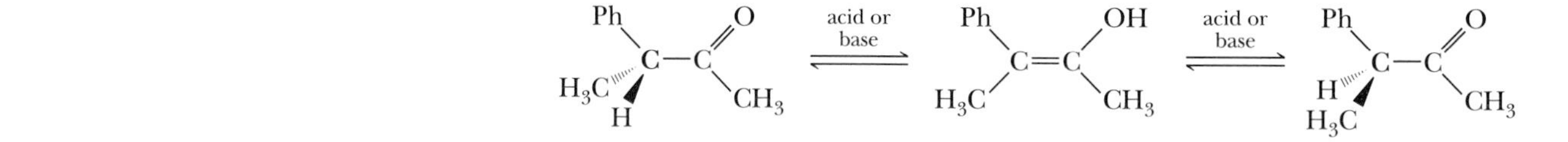

Ph, O, H_3C, H, CH_3 $\xrightleftharpoons{\text{acid or base}}$ Ph, OH, H_3C, CH_3 $\xrightleftharpoons{\text{acid or base}}$ Ph, O, H, H_3C, CH_3

(*R*)-3-Phenyl-2-butanone An achiral enol (*S*)-3-Phenyl-2-butanone

Racemization by this mechanism occurs only at α-carbon stereocenters with at least one α-hydrogen.

B. Deuterium Exchange

When an aldehyde or ketone with one or more α-hydrogens is dissolved in an aqueous solution enriched with D_2O and also containing catalytic amounts of either D^+ or OD^-, exchange of α-hydrogens occurs at a rate which is proportional to the concentration of the acid or base catalyst. We account for incorporation of deuterium by proposing a rate-limiting acid- or base-catalyzed enolization followed by incorporation of deuterium as the enol form converts to the keto form.

$$CH_3\overset{O}{\overset{\|}{C}}CH_3 + 6D_2O \xrightleftharpoons[\text{or } OD^-]{D^+} CD_3\overset{O}{\overset{\|}{C}}CD_3 + 6HOD$$

Acetone Acetone-d_6

In naming compounds, the presence of deuterium is shown by the symbol "d" and the number of deuterium atoms is shown by a subscript following the symbol "d."

Deuterium exchange has two values. First, by observing changes in hydrogen ratios before and after deuterium exchange, it is possible to determine the number of exchangeable α-hydrogens in a molecule. Second, exchange of α-hydrogens is a convenient way to introduce an isotopic label into molecules. At present, more than 225 deuterium-labeled compounds are available commercially in isotopic enrichments of up to 99.8 atom % D. Among these are

$CDCl_3$	$CD_3C(=O)OD$	$CH_3C(=O)OD$	$NaBD_4$	CH_3CH_2OD
Chloroform-d	Acetic-d_3 acid-d	Acetic acid-d	Sodium borodeuteride	Ethanol-d

Deuterated solvents, such as $CDCl_3$, acetone-d_6, and benzene-d_6, are used in ^{1}H-NMR spectroscopy because they lack protons that might otherwise obscure the spectrum of a compound.

C. α-Halogenation

Aldehydes and ketones with at least one α-hydrogen react at an α-carbon with bromine and chlorine to form α-haloaldehydes and α-haloketones as illustrated by bromination of acetophenone.

$$C_6H_5C(=O)CH_3 + Br_2 \xrightarrow{CH_3CO_2H} C_6H_5C(=O)CH_2Br + HBr$$

Acetophenone

Bromination or chlorination at an α-carbon is catalyzed by both acid and base. For acid-catalyzed halogenation, acid generated by the reaction catalyzes further reaction. The slow step of acid-catalyzed halogenation is formation of an enol. This is followed by rapid reaction of the double bond with halogen to give the α-haloketone.

MECHANISM Acid-Catalyzed α-Halogenation of a Ketone

Step 1: Acid-catalyzed enolization (Section 15.11A)

$$R'-C(=O)-CH(R)-R \underset{}{\overset{\text{slow}}{\rightleftharpoons}} (H-O)(R')C=CR_2$$

Step 2: Nucleophilic attack of the enol on halogen

$$(H-\ddot{O}:)(R')C=CR_2 + Br-Br \xrightarrow{\text{fast}} (:\ddot{O}=)(R')C-C(Br)(R)-R + H-Br$$

α-Halogenation gives HX as a byproduct and, in order to keep the solution basic, it is necessary to add slightly more than one mole of base per mole of aldehyde or ketone. Because base is a reactant required in equimolar amounts, we say that this reaction is base promoted. The slow step in base-promoted α-halogenation is removal of an α-hydrogen to form an **enolate anion,** which then reacts with halogen by nucleophilic displacement to form the final product.

MECHANISM Base-Promoted α-Halogenation of a Ketone

Step 1: Formation of a resonance-stabilized enolate anion

$$R'-\overset{\overset{O}{\|}}{C}-\underset{\underset{R}{|}}{\overset{\overset{H}{|}}{C}}-R + :\ddot{O}H^- \underset{}{\overset{\text{slow}}{\rightleftharpoons}} R'(C{=}O)-\ddot{C}^-R_2 \longleftrightarrow R'(^-O)C{=}CR_2 + H_2O$$

Resonance-stabilized enolate anion

Step 2: Nucleophilic attack of the enolate anion on halogen

$$R'(^-O)C{=}CR_2 + Br{-}Br \xrightarrow{\text{fast}} R'-\overset{\overset{O}{\|}}{C}-\underset{\underset{R}{|}}{\overset{\overset{Br}{|}}{C}}-R + Br^-$$

A major difference exists between acid-catalyzed and base-promoted halogenation. In principle, both can lead to polyhalogenation. In practice, the rate of acid-catalyzed introduction of a second halogen is considerably less than the rate of the first halogenation because introduction of the α-halogen makes the carbonyl less basic for protonation. Thus, it is generally possible to stop acid-catalyzed halogenation at a single substitution. For base-promoted halogenation, each successive halogenation is more rapid than the previous one because introduction of an electronegative halogen atom on an α-carbon further increases the acidity of remaining α-hydrogens and, thus, each successive α-hydrogen is removed more rapidly than the previous one. For this reason, base-promoted halogenation is generally not a useful synthetic reaction.

However, one case in which base-promoted halogenation is useful is the halogenation of methyl ketones. In the presence of base, a methyl ketone reacts with three equivalents of halogen to form a 1,1,1-trihaloketone, which then reacts with an additional mol of hydroxide ion to form a carboxylic salt and a trihalomethane. Reaction of the carboxylic salt with aqueous HCl or other strong acid gives the carboxylic acid.

$$R\overset{\overset{O}{\|}}{C}CH_3 + 3Br_2 + 3NaOH \longrightarrow R\overset{\overset{O}{\|}}{C}CBr_3 + 3NaBr + 3H_2O$$

$$R\overset{\overset{O}{\|}}{C}CBr_3 + NaOH \longrightarrow R\overset{\overset{O}{\|}}{C}O^-Na^+ + \underset{\text{Tribromomethane (Bromoform)}}{CHBr_3}$$

Common names for the trihalomethanes are chloroform, bromoform, and iodoform. For this reason, reaction of a methyl ketone with a halogen in base is called the **haloform reaction.**

The final step of the haloform reaction is cleavage of a carbon-carbon bond to give a carboxylate anion and a haloform. In Step 1, hydroxide ion adds to the carbonyl carbon to form a tetrahedral carbonyl addition intermediate, which collapses to give a haloform anion and a carboxyl group. A CX_3^- carbanion is stabilized by three electron-withdrawing halogen atoms, making it a good leaving group (Section 8.4F). The haloform reaction is one of the rare cases where a carbanion is a leaving group. The haloform anion and carboxyl group react in Step 2 to give the haloform and carboxylate anion.

MECHANISM The Final Step of the Haloform Reaction

Step 1: Addition of hydroxide ion to the carbonyl carbon to form a tetrahedral carbonyl addition intermediate followed by its collapse

$$RC(=O){-}CBr_3 + {}^-{:}\ddot{O}H \rightleftharpoons RC(O^-)(OH){-}CBr_3 \longrightarrow RC(=O){-}OH + {:}CBr_3^-$$

Step 2: Proton transfer from the carboxyl group to the haloform anion

$$RC(=O){-}\ddot{O}{-}H + {}^-{:}CBr_3 \longrightarrow RC(=O){-}\ddot{O}{:}^- + H{-}CBr_3$$

Bromoform

The haloform reaction is an indirect way to oxidize a methyl ketone to a carboxylic acid as illustrated by the oxidation of the following unsaturated methyl ketone to an unsaturated carboxylic acid.

$$CH_3CH(CH_3)CH{=}CHC(=O)CH_3 \xrightarrow[\text{2. } HCl/H_2O]{\text{1. } Cl_2/NaOH} CH_3CH(CH_3)CH{=}CHC(=O)OH + CHCl_3$$

5-Methyl-3-hexen-2-one; 4-Methyl-2-pentenoic acid; Trichloromethane (Chloroform)

15.13 Oxidation

A. Oxidation of Aldehydes

Aldehydes are oxidized to carboxylic acids by a variety of common oxidizing agents, including chromic acid and molecular oxygen. In fact, aldehydes have one of the most easily oxidized of all functional groups. Oxidation by chromic acid is illustrated by the conversion of hexanal to hexanoic acid. For the mechanism of this oxidation, review Section 9.9.

By adding Tollens' reagent to an aldehyde, a silver mirror has been deposited in the inside of this flask. *(Charles D. Winters)*

$$CH_3(CH_2)_4\overset{O}{\overset{\|}{C}}H \xrightarrow{H_2CrO_4} CH_3(CH_2)_4\overset{O}{\overset{\|}{C}}OH$$

Hexanal — Hexanoic acid

Aldehydes are also oxidized to carboxylic acids by Ag(I) ion. One laboratory procedure is to shake a solution of the aldehyde in aqueous ethanol or aqueous tetrahydrofuran (THF) with a slurry of Ag_2O.

$$\text{Vanillin} + Ag_2O \xrightarrow[\text{NaOH}]{\text{THF, } H_2O} \xrightarrow[H_2O]{HCl} \text{Vanillic acid} + Ag$$

Vanillin — Vanillic acid

Tollens' reagent, another form of Ag(I), is prepared by dissolving silver nitrate in water, adding sodium hydroxide to precipitate silver(I) as Ag_2O, and then adding aqueous ammonia to redissolve silver(I) as the silver-ammonia complex ion.

$$Ag^+NO_3^- + 2NH_3 \xrightleftharpoons{NH_3,\ H_2O} Ag(NH_3)_2{}^+NO_3^-$$

When Tollens' reagent is added to an aldehyde, the aldehyde is oxidized to a carboxylate anion and silver(I) is reduced to metallic silver. If this reaction is carried out properly, silver precipitates as a smooth, mirror-like deposit, hence the name **silver-mirror test.** Silver(I) is rarely used at the present time for the oxidation of aldehydes because of the cost of silver and because other, more convenient methods exist for this oxidation. This reaction, however, is still used for silvering mirrors. In this process, formaldehyde or glucose is generally used as the aldehyde to reduce Ag(I).

Aldehydes are also oxidized to carboxylic acids by molecular oxygen and by hydrogen peroxide. Reaction with oxygen is a radical chain reaction (see the Chemistry in Action box "Radical Autoxidation" in Chapter 7) made possible because the aldehyde C—H bond is unusually weak.

$$2\,C_6H_5\overset{O}{\overset{\|}{C}}H + O_2 \longrightarrow 2\,C_6H_5\overset{O}{\overset{\|}{C}}OH$$

Benzaldehyde — Benzoic acid

Molecular oxygen is the least expensive and most readily available oxidizing agent, and on an industrial scale, air oxidation of organic compounds, including aldehydes, is very common. Air oxidation of aldehydes can also be a problem. Aldehydes that are liquid at room temperature are so sensitive to oxidation by molecular oxygen that they must be protected from contact with air during storage. Often this is done by sealing the aldehyde in a container under an atmosphere of nitrogen.

EXAMPLE 15.11

Draw a structural formula for the product formed by treating each compound with Tollens' reagent followed by acidification with aqueous HCl.

(a) Pentanal **(b)** Cyclopentanecarbaldehyde

Solution

The aldehyde group in each compound is oxidized to a carboxyl group.

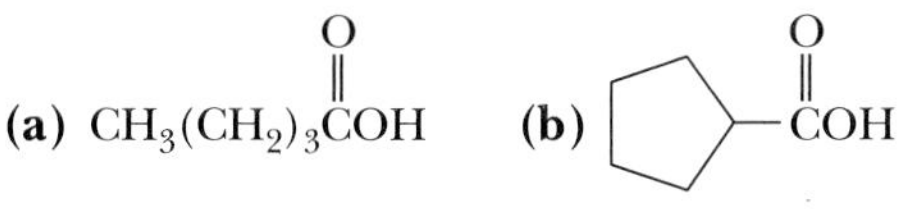

(a) $CH_3(CH_2)_3\overset{O}{\overset{\|}{C}}OH$ (b) cyclopentyl–$\overset{O}{\overset{\|}{C}}OH$

Pentanoic acid Cyclopentanecarboxylic acid

PROBLEM 15.11

Complete these oxidations.

(a) Hexanal + $H_2O_2 \longrightarrow$ **(b)** 3-Phenylpropanal + Tollens' reagent $\longrightarrow$

B. Oxidation of Ketones

Ketones, in contrast to aldehydes, are oxidized only under rather special conditions. For example, they are not normally oxidized by potassium permanganate or chromic acid. In fact, these reagents are used routinely to oxidize secondary alcohols to ketones in good yield.

Ketones are oxidized to esters by peroxyacids. The net effect of this oxidation is to insert an oxygen atom between the carbonyl carbon and an α-carbon of the ketone as illustrated by the following reactions. Reagents most commonly used for this purpose are peroxyacetic acid and peroxytrifluoroacetic acid. Oxidation of ketones by peroxyacids is often called **Baeyer-Villiger oxidation** after the chemists who developed this synthetically useful reaction.

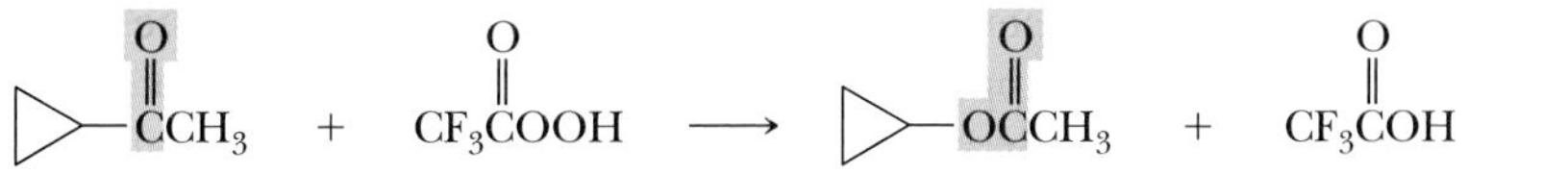

cyclopropyl–$\overset{O}{\overset{\|}{C}}CH_3$ + $CF_3\overset{O}{\overset{\|}{C}}OOH \longrightarrow$ cyclopropyl–$O\overset{O}{\overset{\|}{C}}CH_3$ + $CF_3\overset{O}{\overset{\|}{C}}OH$

Peroxytrifluoroacetic acid Trifluoroacetic acid

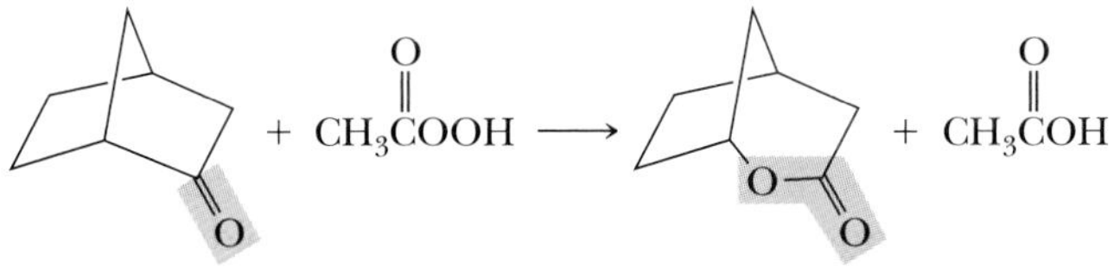

bicyclic ketone + $CH_3\overset{O}{\overset{\|}{C}}OOH \longrightarrow$ bicyclic lactone + $CH_3\overset{O}{\overset{\|}{C}}OH$

Peroxyacetic acid Acetic acid

The mechanism of the Baeyer-Villiger oxidation involves addition of the peroxyacid to the ketone carbonyl to form a tetrahedral carbonyl addition compound. Note that this addition is closely analogous to the addition of an alcohol to the carbonyl group of an aldehyde or ketone to form a hemiacetal (Section 15.8B). Decomposition by rearrangement involving migration of a group attached to the carbonyl carbon completes the reaction. This sequence of steps is illustrated by reaction of an unsymmetrical ketone with peroxyacetic acid. The rate-limiting step in this mechanism is the simultaneous migration of the alkyl or aryl group and departure of $^-OCOR''$. This reaction resembles the rearrangement of the peroxyborane intermediate in the sec-

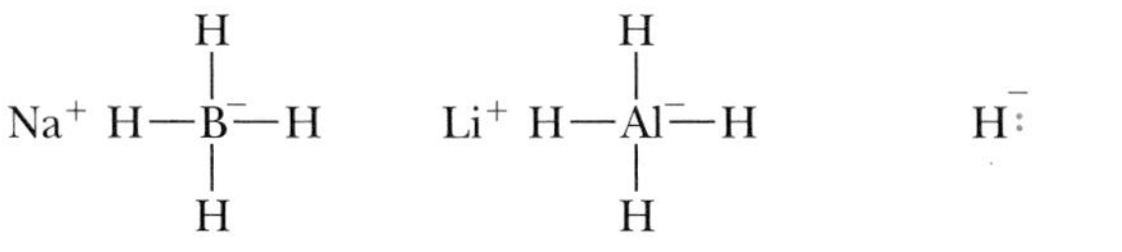

$$Na^+\ H\text{—}\overset{\overset{\displaystyle H}{|}}{\underset{\underset{\displaystyle H}{|}}{B^-}}\text{—}H \qquad Li^+\ H\text{—}\overset{\overset{\displaystyle H}{|}}{\underset{\underset{\displaystyle H}{|}}{Al^-}}\text{—}H \qquad H{:}^-$$

Sodium borohydride — Lithium aluminum hydride (LAH) — Hydride ion

Lithium aluminum hydride is a very powerful reducing agent; it reduces not only the carbonyl groups of aldehydes and ketones rapidly but also those of carboxylic acids (Section 16.7A) and their functional derivatives (Section 17.11). Sodium borohydride is a much more selective reagent, reducing only aldehydes and ketones rapidly.

Reductions using sodium borohydride are most commonly carried out in aqueous methanol, in pure methanol, or in ethanol. The initial product of reduction is a tetraalkyl borate, which, on warming with water, is converted to an alcohol and sodium borate salts. One mol of sodium borohydride reduces 4 mol of aldehyde or ketone.

$$4R\overset{\overset{\displaystyle O}{\|}}{C}H + NaBH_4 \xrightarrow{\text{methanol}} (RCH_2O)_4B^-\ Na^+ \xrightarrow{H_2O} 4RCH_2OH + \text{borate salts}$$

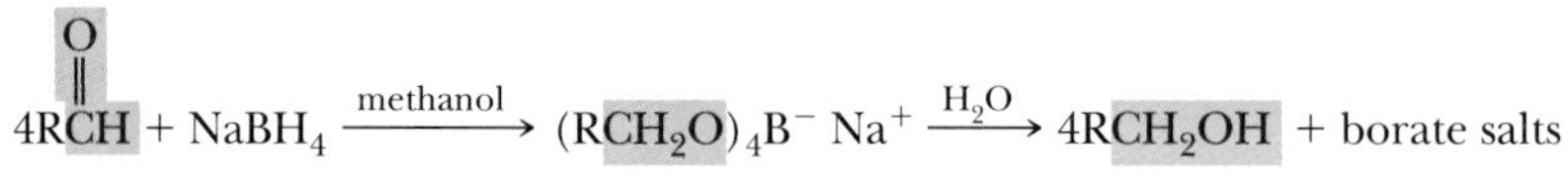

A tetraalkyl borate

The key step in the metal hydride reduction of an aldehyde or ketone is transfer of a hydride ion from the reducing agent to the carbonyl carbon to form a tetrahedral carbonyl addition compound. In the reduction of an aldehyde or ketone to an alcohol, only the hydrogen atom attached to carbon comes from the hydride reducing agent; the hydrogen atom attached to oxygen comes from water during hydrolysis of the metal alkoxide salt.

MECHANISM Sodium Borohydride Reduction of an Aldehyde or Ketone

$$Na^+H\text{—}\overset{\overset{\displaystyle H}{|}}{\underset{\underset{\displaystyle H}{|}}{B^-}}\text{—}H + R\text{—}\overset{\overset{\displaystyle O}{\|}}{C}\text{—}R' \longrightarrow R\text{—}\overset{\overset{\displaystyle O\text{—}\bar{B}H_3\ Na^+}{|}}{\underset{\underset{\displaystyle H}{|}}{C}}\text{—}R' \xrightarrow{H_2O} R\text{—}\overset{\overset{\displaystyle OH}{|}}{\underset{\underset{\displaystyle H}{|}}{C}}\text{—}R'$$

This H comes from water during hydrolysis

This H comes from the hydride reducing agent

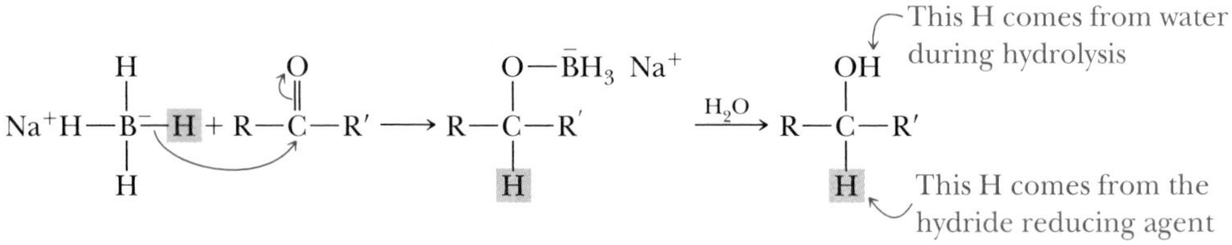

Unlike sodium borohydride, LAH reacts violently with water, methanol, and other protic solvents to liberate hydrogen gas and form metal hydroxides. Therefore, reductions of aldehydes and ketones using this reagent must be carried out in aprotic solvents, most commonly diethyl ether or tetrahydrofuran. The stoichiometry for LAH reductions is the same as that for sodium borohydride reductions: 1 mol of LAH per 4 mol of aldehyde or ketone. Because of the formation of gelatinous aluminum salts, aqueous acid or base work-up is usually used to solubilize these salts.

$$4RCR + LiAlH_4 \xrightarrow{\text{ether}} (R_2CHO)_4Al^- \, Li^+ \xrightarrow{H_2O} 4RCHR + \text{aluminum salts}$$

(4RC(=O)R; product 4RCH(OH)R)

A tetraalkyl aluminate

The following equations illustrate selective reduction of a carbonyl group in the presence of a carbon-carbon double bond and, alternatively, selective reduction of a carbon-carbon double bond in the presence of a carbonyl group using rhodium on powdered charcoal as a catalyst.

Selective reduction of a carbonyl group:

$$RCH{=}CHCR' \xrightarrow[2.\ H_2O]{1.\ NaBH_4} RCH{=}CHCHR'$$

(reactant: RCH=CHC(=O)R'; product: RCH=CHCH(OH)R')

Selective reduction of a carbon-carbon double bond:

$$RCH{=}CHCR' + H_2 \xrightarrow{Rh} RCH_2CH_2CR'$$

(reactant: RCH=CHC(=O)R'; product: $RCH_2CH_2C(=O)R'$)

EXAMPLE 15.12

Complete these reductions.

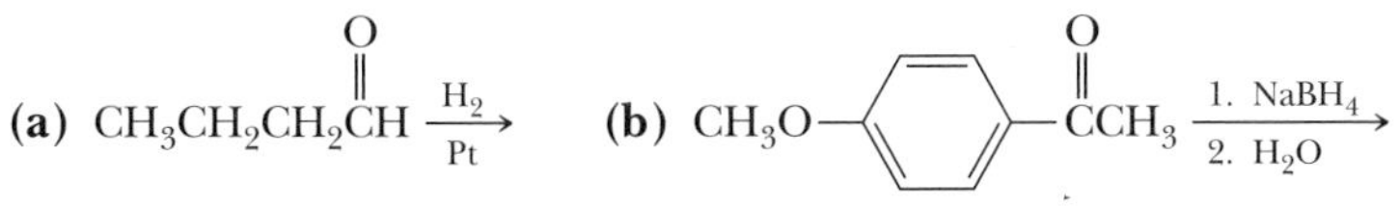

(a) $CH_3CH_2CH_2CH \xrightarrow[Pt]{H_2}$ (with C=O on the terminal CH)

(b) $CH_3O{-}C_6H_4{-}CCH_3 \xrightarrow[2.\ H_2O]{1.\ NaBH_4}$ (para-substituted benzene ring; C=O on the CCH₃ carbon)

Solution

The carbonyl group of the aldehyde in (a) is reduced to a primary alcohol group, and the carbonyl group of the ketone in (b) is reduced to a secondary alcohol group.

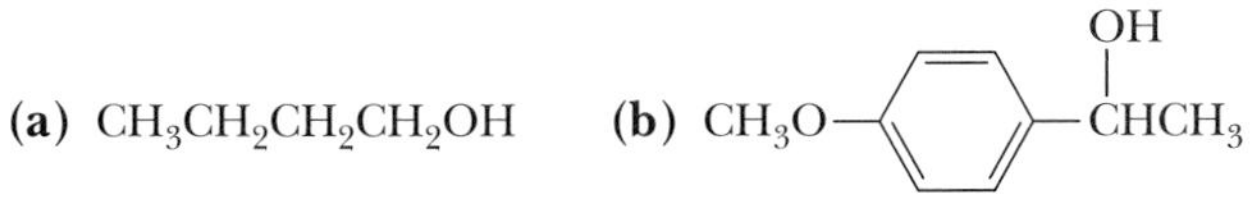

(a) $CH_3CH_2CH_2CH_2OH$ **(b)** $CH_3O{-}C_6H_4{-}CHCH_3$ (with OH on the CH carbon)

PROBLEM 15.12

What aldehyde or ketone gives these alcohols on reduction with $NaBH_4$?

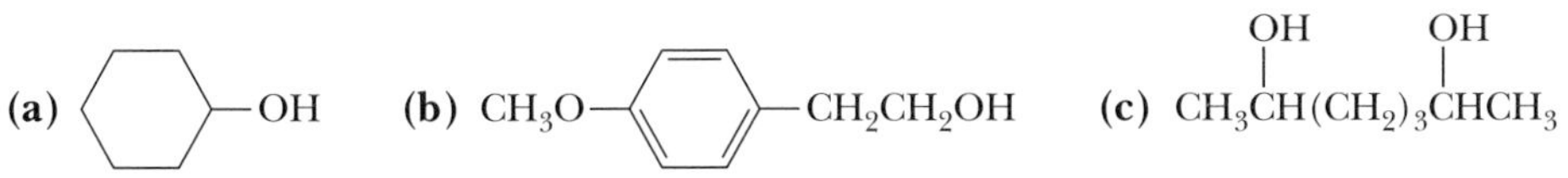

(a) cyclohexane ring–OH **(b)** $CH_3O{-}C_6H_4{-}CH_2CH_2OH$ (para-substituted benzene ring) **(c)** $CH_3CH(CH_2)_3CHCH_3$ (with OH on both CH carbons)

C. Reduction of a Carbonyl Group to a Methylene Group

Several methods are available for reducing the carbonyl group of an aldehyde or ketone to a methylene group ($-CH_2-$). One of the first discovered involves refluxing the aldehyde or ketone with amalgamated zinc (zinc with a surface layer of mercury) in concentrated HCl. This reaction is known as the **Clemmensen reduction** after the German chemist, E. Clemmensen, who developed it in 1912.

$$\text{2-HOC}_6\text{H}_4\text{C(=O)(CH}_2)_5\text{CH}_3 \xrightarrow{\text{Zn(Hg), HCl}} \text{2-HOC}_6\text{H}_4\text{CH}_2(\text{CH}_2)_5\text{CH}_3$$

Because the Clemmensen reduction requires the use of concentrated HCl, it cannot be used to reduce a carbonyl group in a molecule that also contains acid-sensitive groups, as, for example, a tertiary alcohol that might undergo dehydration or an acetal that is hydrolyzed and the resulting carbonyl group also reduced. The mechanism of Clemmensen reduction is not well understood.

The **Wolff-Kishner reduction,** discovered independently by N. Kishner in 1911 and L. Wolff in 1912 and reported within months of Clemmensen's discovery, is an alternative method for reduction of a carbonyl group to a methylene group. In this reduction, a mixture of the aldehyde or ketone, hydrazine, and concentrated potassium hydroxide is refluxed in a high-boiling solvent such as diethylene glycol (bp 245°C).

$$\text{C}_6\text{H}_5\text{C(=O)CH}_3 + \underset{\text{Hydrazine}}{H_2NNH_2} \xrightarrow[\text{Diethylene glycol (reflux)}]{\text{KOH}} \text{C}_6\text{H}_5\text{CH}_2\text{CH}_3 + N_2 + H_2O$$

More recently it has been found possible to bring about the same reaction in dimethyl sulfoxide (DMSO) with potassium *tert*-butoxide and hydrazine at room temperature.

MECHANISM Wolff-Kishner Reduction

Step 1: Reaction of the aldehyde or ketone with hydrazine gives a hydrazone (Section 15.10B)

$$R_2C{=}O + \underset{\text{Hydrazine}}{H_2N{-}NH_2} \longrightarrow \underset{\text{A hydrazone}}{R_2C{=}N{-}NH_2} + H_2O$$

Step 2: Base catalyzed tautomerism; compare keto-enol tautomerism (Section 15.11B)

$$R_2C{=}N{-}NH{-}H + {:}\ddot{O}{-}H^- \rightleftharpoons R_2C{=}N{-}\ddot{N}H^- \longleftrightarrow R_2\bar{C}{-}N{=}N{-}H + H{-}\ddot{O}{-}H$$

$$\rightleftharpoons R_2CH{-}N{=}N{-}H + {:}\ddot{O}{-}H^-$$

Step 3: Proton transfer to hydroxide ion followed by loss of N_2 gives a carbanion; proton transfer from water to the carbanion gives the hydrocarbon and hydroxide ion

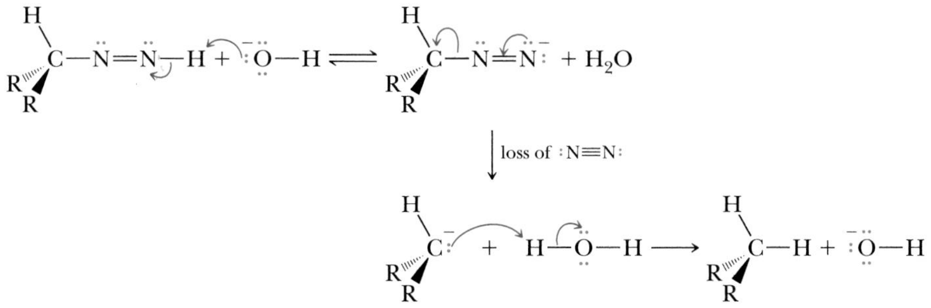

$$R_2C(H)-\ddot{N}=\ddot{N}-H + {}^{-}:\ddot{O}-H \rightleftharpoons R_2C(H)-\ddot{N}=\ddot{N}:^{-} + H_2O$$

$$\downarrow \text{loss of } :N\equiv N:$$

$$R_2C(H):^{-} + H-\ddot{O}-H \longrightarrow R_2C(H)-H + {}^{-}:\ddot{O}-H$$

A carbanion

Each of the reductions has its special conditions, advantages, and disadvantages. The Clemmensen reduction cannot be used in the presence of groups sensitive to concentrated acid, and the Wolff-Kishner reduction cannot be used in the presence of groups sensitive to concentrated base. However, the carbonyl group of almost any aldehyde or ketone can be reduced to a methylene group by one of these methods.

EXAMPLE 15.13

Complete the following reactions:

(a) 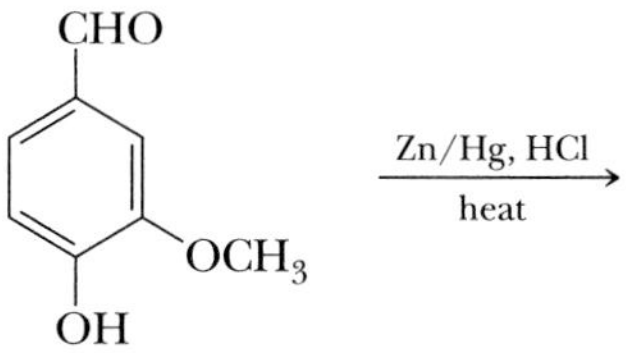 Vanillin (from vanilla beans) (benzene ring with CHO, OCH_3, OH) $\xrightarrow[\text{heat}]{\text{Zn/Hg, HCl}}$

(b) 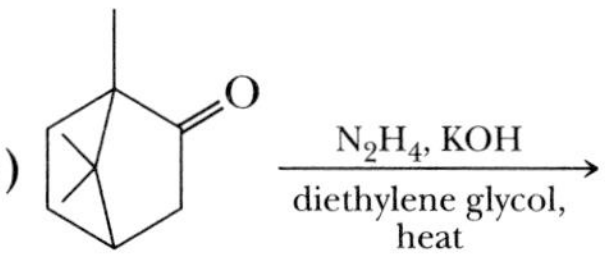 Camphor $\xrightarrow[\text{diethylene glycol, heat}]{N_2H_4\text{, KOH}}$

Solution

The reaction for (a) is a Clemmensen reduction; for (b) it is a Wolff-Kishner reduction:

(a) 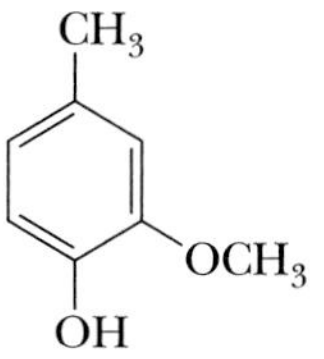 benzene ring with CH_3, OCH_3, OH

(b) 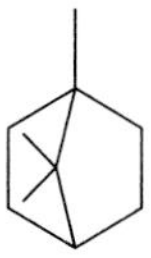 camphor skeleton without the C=O group

PROBLEM 15.13

Complete the following reactions:

(a) 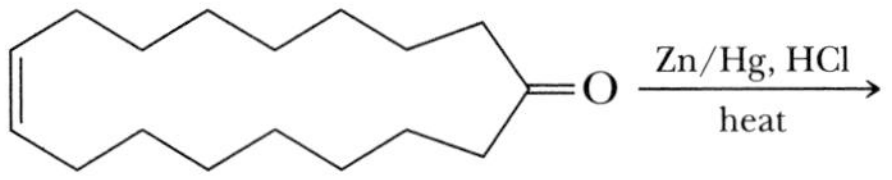 [Civetone] =O $\xrightarrow[\text{heat}]{\text{Zn/Hg, HCl}}$

Civetone
(from the civet cat; used in perfumery)

(b) 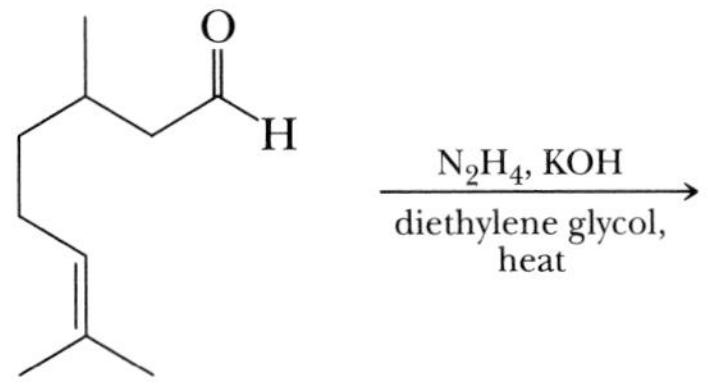 [Citronellal] $\xrightarrow[\text{diethylene glycol, heat}]{N_2H_4,\ KOH}$

Citronellal
(from citronella and lemongrass oils)

SUMMARY

An **aldehyde** (Section 15.1) contains a carbonyl group bonded to a hydrogen atom and a carbon atom. A **ketone** contains a carbonyl group bonded to two carbons. An aldehyde is named by changing -e of the parent alkane to **-al** (Section 15.2). A ketone is named by changing -e of the parent alkane to **-one** and using a number to locate the carbonyl group. In naming compounds that contain more than one functional group, the IUPAC system has established an **order of precedence of functions** (Section 15.2B). If the carbonyl group of an aldehyde or ketone is lower in precedence than other functional groups in the molecule, it is indicated by the infix **-oxo-.**

Aldehydes and ketones are polar compounds (Section 15.3) and interact in the pure state by dipole-dipole interactions; they have higher boiling points and are more soluble in water than nonpolar compounds of comparable molecular weight.

One of the most common reaction themes of aldehydes and ketones is addition of a nucleophile to the carbonyl carbon to form a tetrahedral carbonyl addition compound (Section 15.5). Among the nucleophiles adding to the carbonyl carbon are carbon nucleophiles from organomagnesium and organolithium reagents, terminal alkynes, and cyanide ion (Section 15.6); carbon nucleophiles of Wittig reagents (Section 15.7); oxygen nucleophiles such as water and alcohols (Section 15.8); sulfur nucleophiles such as thiols (Section 15.9); and nitrogen nucleophiles such as amines and hydrazines (Section 15.10).

The carbon atom adjacent to a carbonyl group is called an ***α*-carbon** (Section 15.11A), and a hydrogen attached to it is called an ***α*-hydrogen.** The pK_a of an α-hydrogen of an aldehyde or ketone is approximately 20, which makes it less acidic than alcohols but more acidic than terminal alkynes.

Reactions at the α-carbon of an aldehyde or ketone carbonyl include **racemization** (Section 15.12A), **deuterium exchange** (Section 15.12B), and **halogenation** (Section 15.12C). The rate-limiting step is formation of an enol for acid-catalyzed halogenation and formation of an enolate anion for base-promoted halogenation.

KEY REACTIONS

1. McLafferty Rearrangement of Aldehydes and Ketones (Section 15.4D)

Transfer of a hydrogen five atoms away to the carbonyl oxygen by way of a six-membered ring transition state gives an alkene and a new radical cation. The new radical cation is the enol of an aldehyde or ketone.

$$\left[CH_3CH(H)CH_2CH_2C(=O)CH_3\right]^{+\cdot} \xrightarrow{\text{McLafferty rearrangement}} CH_3CH{=}CH_2 + \left[H_2C{=}C(OH)CH_3\right]^{+\cdot}$$

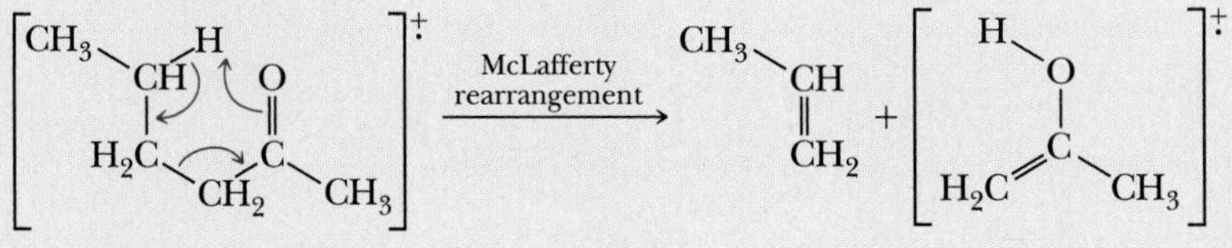

Molecular ion $\qquad$ m/z 58

2. Reaction with Grignard Reagents (Section 15.6A)

Treatment of formaldehyde with a Grignard reagent followed by hydrolysis gives a primary alcohol. Similar treatment of any other aldehyde gives a secondary alcohol. Treatment of a ketone gives a tertiary alcohol.

$$CH_3\overset{\overset{\displaystyle O}{\|}}{C}CH_3 \xrightarrow[\text{2. HCl, } H_2O]{\text{1. } C_6H_5MgBr} C_6H_5\overset{\overset{\displaystyle OH}{|}}{C}(CH_3)_2$$

3. Reaction with Organolithium Reagents (Section 15.6B)

Reactions of aldehydes and ketones with organolithium reagents are similar to those with Grignard reagents.

$$CH_3\overset{\overset{\displaystyle O}{\|}}{C}C(CH_3)_3 \xrightarrow[\text{2. HCl, } H_2O]{\text{1. } C_6H_5Li} C_6H_5\underset{\underset{\displaystyle CH_3}{|}}{\overset{\overset{\displaystyle OH}{|}}{C}}C(CH_3)_3$$

4. Reaction with Anions of Terminal Alkynes (Section 15.6C)

Treatment of an aldehyde or ketone with the alkali metal salt of a terminal alkyne followed by hydrolysis gives an α-alkynylalcohol.

$$\text{cyclohexanone} \xrightarrow[\text{2. HCl, } H_2O]{\text{1. } HC\equiv C{:}^-Na^+} \text{1-(OH)-1-(C}\equiv\text{CH)cyclohexane}$$

5. Reaction with HCN to Form Cyanohydrins (Section 15.6D)

For aldehydes and most sterically unhindered aliphatic ketones, equilibrium favors formation of the cyanohydrin. For aryl ketones, equilibrium favors starting materials, and little cyanohydrin is obtained.

$$C_6H_5\overset{\overset{\displaystyle O}{\|}}{C}H + HC\equiv N \xrightarrow{NaCN} C_6H_5\overset{\overset{\displaystyle OH}{|}}{C}HC\equiv N$$

6. The Wittig Reaction (Section 15.7)

Treatment of an aldehyde or ketone with a triphenylphosphonium ylide gives an oxaphosphetane intermediate, which fragments to give triphenylphosphine oxide and an alkene.

$$\text{cyclohexane}{=}O + Ph_3\overset{+}{P}{-}\overset{-}{C}H_2 \longrightarrow \text{cyclohexane}{=}CH_2 + Ph_3\overset{+}{P}{-}O^-$$

7. Hydration (Section 15.8A)

The degree of hydration is greater for aldehydes than for ketones.

$$H\overset{\overset{\displaystyle O}{\|}}{C}H + H_2O \rightleftharpoons H\underset{\underset{\displaystyle H}{|}}{\overset{\overset{\displaystyle OH}{|}}{C}}OH$$

(>99°)

8. Addition of Alcohols to Form Hemiacetals (Section 15.8B)

Hemiacetals are only minor components of an equilibrium mixture of aldehyde or ketone and alcohol, except where the —OH and the C=O are parts of the same molecule and a five- or six-membered ring can form.

$$CH_3CH(OH)CH_2CH_2CHO \rightleftharpoons \text{(2-methyl-5-hydroxytetrahydrofuran)}$$

4-Hydroxypentanal — A cyclic hemiacetal (major form present at equilibrium)

9. Addition of Alcohols to Form Acetals (Section 15.8B)

Formation of acetals is catalyzed by acid. Acetals are stable to water and aqueous base but are hydrolyzed in aqueous acid. Acetals are valuable as carbonyl-protecting groups.

$$\text{cyclopentanone} + HOCH_2CH_2OH \overset{H^+}{\rightleftharpoons} \text{(cyclopentane spiro-}O\text{—}CH_2\text{—}CH_2\text{—}O\text{ acetal)} + H_2O$$

10. Addition of Sulfur Nucleophiles: Formation of 1,3-Dithianes (Section 15.9)

The most commonly used thiol for preparation of thioacetals is 1,3-propanedithiol. The product is called a 1,3-dithiane.

$$CH_3\text{—}\overset{O}{\overset{\|}{C}}\text{—}H + HSCH_2CH_2CH_2SH \overset{H^+}{\rightleftharpoons} \text{2-methyl-1,3-dithiane} + H_2O$$

11. Alkylation of Anions Derived from Aldehyde 1,3-Dithianes (Section 15.9)

Treatment of an aldehyde 1,3-dithiane (pK_a 31) with butyllithium gives an anion. This anion can enter into substitution reactions with primary alkyl, allylic, and benzylic halides and addition reactions with the carbonyl group of aldehydes and ketones.

$$\text{2-lithio-2-methyl-1,3-dithiane } (\ddot{C}^- Li^+,\ CH_3) \xrightarrow[\text{2. } H_2O,\ HgCl_2]{\text{1. } C_6H_5CH_2Cl} CH_3\overset{O}{\overset{\|}{C}}CH_2C_6H_5$$

12. Addition of Ammonia and Its Derivatives: Formation of Imines (Section 15.10A)

Addition of ammonia or a primary amine to the carbonyl group of an aldehyde or ketone forms a tetrahedral carbonyl addition compound. Loss of water from this intermediate gives an imine.

$$\text{cyclopentanone} + H_2NCH_3 \overset{H^+}{\rightleftharpoons} \text{cyclopentylidene}{=}NCH_3 + H_2O$$

13. Addition of Secondary Amines: Formation of Enamines (Section 15.10A)

Addition of a secondary amine to the carbonyl group of an aldehyde or ketone forms a tetrahedral carbonyl addition intermediate. Acid-catalyzed dehydration of this intermediate gives an enamine.

$$\text{(cyclohexanone)}=O + H{-}N\text{(piperidine)} \overset{H^+}{\rightleftharpoons} \text{(1-cyclohexenyl)}{-}N\text{(piperidine)} + H_2O$$

14. Addition of Hydrazine and Its Derivatives (Section 15.10B)

Treatment of an aldehyde or ketone with hydrazine gives a hydrazone. Derivatives of hydrazine react similarly.

$$\text{(cyclopentanone)}=O + H_2NNH_2 \longrightarrow \text{(cyclopentylidene)}=NNH_2 + H_2O$$

15. Keto-Enol Tautomerism (Section 15.11B)

The keto form predominates at equilibrium, except for those aldehydes and ketones in which the enol is stabilized by resonance or hydrogen bonding.

$$\underset{\substack{\text{Keto form} \\ (\sim 99.9\%)}}{CH_3\overset{O}{\overset{\|}{C}}CH_3} \rightleftharpoons \underset{\text{Enol form}}{CH_3\overset{OH}{\overset{|}{C}}{=}CH_2}$$

16. Deuterium Exchange at the α-Carbon (Section 15.12B)

Acid- or base-catalyzed deuterium exchange at an α-carbon involves formation of an enol or enolate anion intermediate.

$$CH_3\overset{O}{\overset{\|}{C}}CH_3 + 6D_2O \overset{DCl}{\rightleftharpoons} CD_3\overset{O}{\overset{\|}{C}}CD_3 + 6HOD$$

17. Halogenation at the α-Carbon (Section 15.12C)

The rate-limiting step in acid-catalyzed α-halogenation is formation of an enol. In base-promoted α-halogenation, it is formation of an enolate anion.

$$Br{-}C_6H_4{-}\overset{O}{\overset{\|}{C}}CH_3 + Br_2 \xrightarrow[CH_3CO_2H]{} Br{-}C_6H_4{-}\overset{O}{\overset{\|}{C}}CH_2Br + HBr$$

18. The Haloform Reaction (Section 15.12B)

The haloform reaction oxidizes a methyl ketone to a carboxylic acid.

$$CH_3\overset{CH_3}{\overset{|}{C}}HCH{=}CH\overset{O}{\overset{\|}{C}}CH_3 \xrightarrow[2.\ HCl/H_2O]{1.\ Cl_2/NaOH} CH_3\overset{CH_3}{\overset{|}{C}}HCH{=}CH\overset{O}{\overset{\|}{C}}OH + CHCl_3$$

19. Oxidation of an Aldehyde to a Carboxylic Acid (Section 15.13A)

The aldehyde group is among the most easily oxidized functional groups. Oxidizing agents include $KMnO_4$, $K_2Cr_2O_7$, Tollens' reagent, H_2O_2, and O_2.

$$\text{2-HOC}_6\text{H}_4\text{CHO} + Ag_2O \xrightarrow[\text{2. } H_2O,\ HCl]{\text{1. THF, } H_2O,\ NaOH} \text{2-HOC}_6\text{H}_4\text{COOH} + Ag$$

20. Oxidation of a Ketone to an Ester: The Baeyer-Villiger Rearrangement (Section 15.13B)

Oxidation of a ketone by a peroxyacid involves nucleophilic addition to the carbonyl group of the ketone to form a tetrahedral carbonyl addition intermediate followed by molecular rearrangement to give an ester.

$$\text{c-C}_3\text{H}_5\text{-CCH}_3(=O) + CF_3COOH(=O) \longrightarrow \text{c-C}_3\text{H}_5\text{-OCCH}_3(=O) + CF_3COH(=O)$$

21. Catalytic Reduction (Section 15.14A)

Catalytic reduction of the carbonyl group of an aldehyde or ketone to an alcohol group is simple to carry out and yields of alcohol are high. A disadvantage of this method is that some other functional groups, including carbon-carbon double and triple bonds, may also be reduced.

$$\text{cyclohexanone} + H_2 \xrightarrow[25°C,\ 2\ atm]{Pt} \text{cyclohexanol (-OH)}$$

22. Metal Hydride Reduction (Section 15.14B)

Both $LiAlH_4$ and $NaBH_4$ are selective in that neither reduces isolated carbon-carbon double or triple bonds.

$$\text{cyclohex-3-enone (=O)} \xrightarrow[\text{2. } H_2O]{\text{1. } NaBH_4} \text{cyclohex-3-enol (-OH)}$$

23. Clemmensen Reduction of an Aldehyde or Ketone (Section 15.14C)

Reduction of the carbonyl group of an aldehyde or ketone using amalgamated zinc in the presence of concentrated hydrochloric acid gives a methylene group.

$$\text{2-HOC}_6\text{H}_4\text{C(=O)(CH}_2)_5\text{CH}_3 \xrightarrow{Zn(Hg),\ HCl} \text{2-HOC}_6\text{H}_4\text{CH}_2(\text{CH}_2)_5\text{CH}_3$$

24. Wolff-Kishner Reduction of an Aldehyde or Ketone (Section 15.14C)

Formation of a hydrazone followed by treatment with base, commonly KOH in diethylene glycol or potassium *tert*-butoxide in dimethyl sulfoxide, reduces the carbonyl group of an aldehyde or ketone to a methylene group.

$$\text{C}_6\text{H}_5\text{CCH}_3(=O) \xrightarrow[\text{2. KOH, heat}]{\text{1. } H_2NNH_2} \text{C}_6\text{H}_5\text{CH}_2\text{CH}_3 + N_2 + H_2O$$

Additional Problems

Structure and Nomenclature

15.14 Name these compounds.

(a) $(CH_3CH_2CH_2)_2C{=}O$

(b) O, CH_3, H

(c) H_3C, CHO, C=C, H, CH_3

(d) CHO, H, C, CH_2OH, CH_3

(e) $CH_3O{-}C_6H_4{-}\overset{O}{\overset{\|}{C}}CH_2CH_3$

(f) $CH_3\overset{OH}{\overset{|}{C}}HCH_2\overset{O}{\overset{\|}{C}}CH_2CH_2{-}C_6H_5$

(g) O, $CH_2CH_2CH_3$, O

(h) $H\overset{O}{\overset{\|}{C}}CH_2CH_2CH_2CH_2\overset{O}{\overset{\|}{C}}H$

(i) $CH_3CH_2\overset{O}{\overset{\|}{C}}\underset{Br}{\underset{|}{C}}HCH_3$

15.15 Draw structural formulas for these compounds.

(a) 1-Chloro-2-propanone
(b) 3-Hydroxybutanal
(c) 4-Hydroxy-4-methyl-2-pentanone
(d) 3-Methyl-3-phenylbutanal
(e) 1,3-Cyclohexanedione
(f) 3-Methyl-3-buten-2-one
(g) 5-Oxohexanal
(h) 2,2-Dimethylcyclohexanecarbaldehyde
(i) 3-Oxobutanoic acid

Spectroscopy

15.16 2-Methylpentanal and 4-methyl-2-pentanone are constitutional isomers of molecular formula $C_6H_{12}O$. Each shows a molecular ion peak in its mass spectrum at m/z 100. Spectrum A shows significant peaks at m/z 85, 58, 57, 43, and 42. Spectrum B shows significant peaks at m/z 71, 58, 57, 43, and 29. Assign each compound its correct spectrum.

15.17 The infrared spectrum of Compound A, $C_6H_{12}O$, shows a strong, sharp peak at 1724 cm^{-1}. From this information and its ^{1}H-NMR spectrum, deduce the structure of compound A.

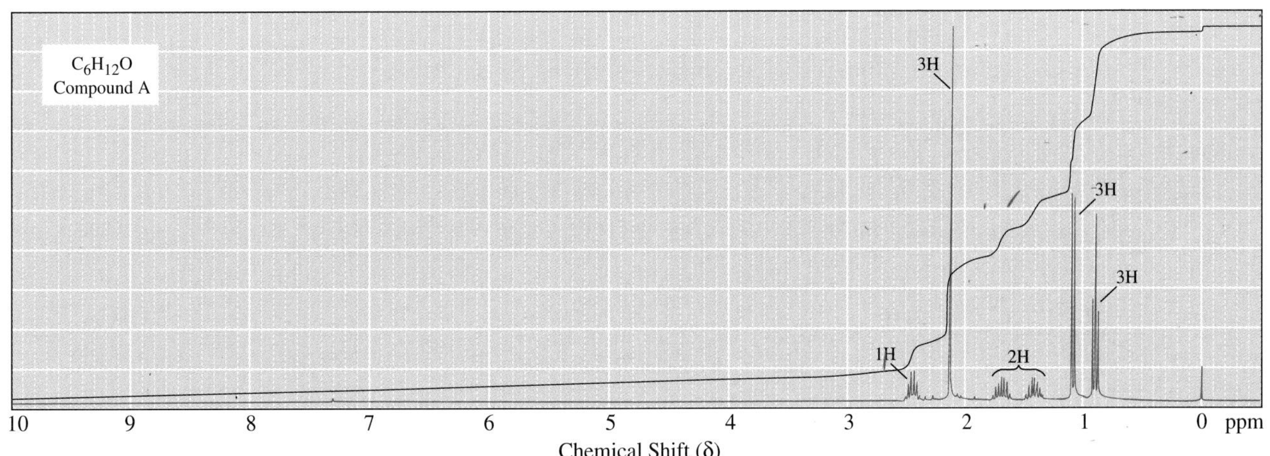

15.18 Following are ^{1}H-NMR spectra for compounds B, $C_6H_{12}O_2$, and C, $C_6H_{10}O$. On warming in dilute acid, compound B is converted to compound C. Deduce the structural formulas for compounds B and C.

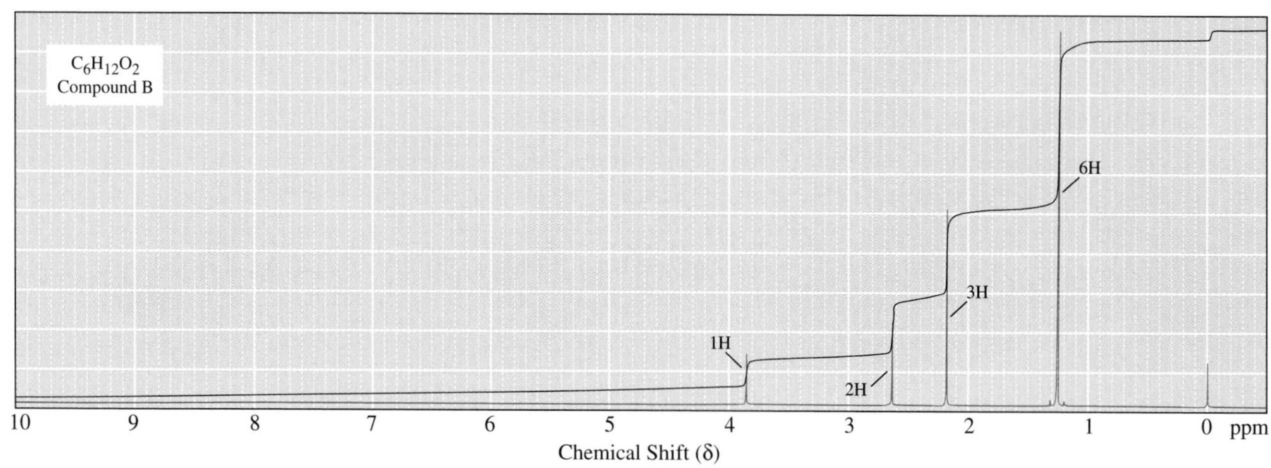

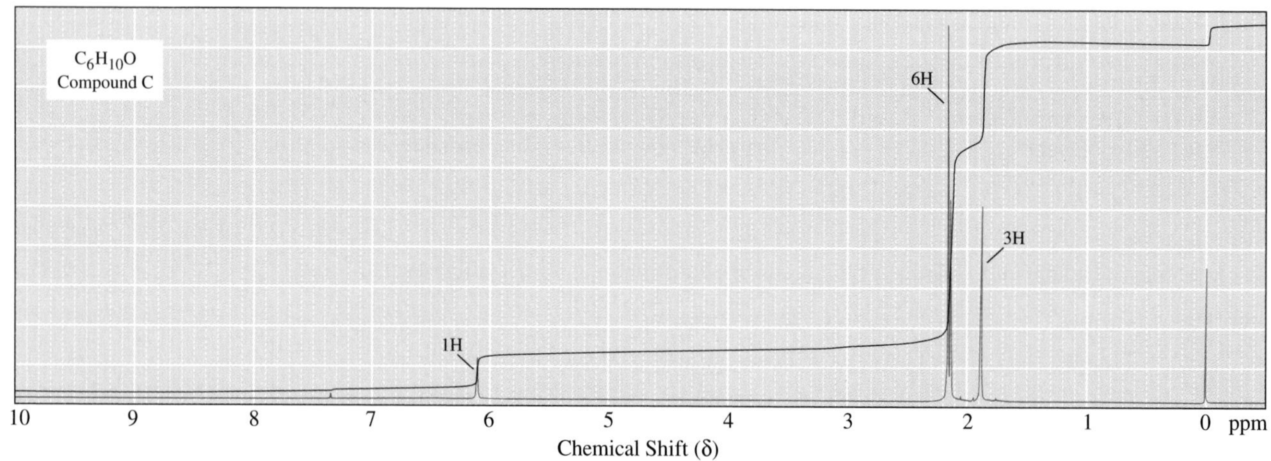

Addition of Carbon Nucleophiles

15.19 Draw structural formulas for the product formed by treating each compound with propylmagnesium bromide followed by aqueous HCl.

(a) CH_2O **(b)** $CH_2{-}CH_2$ (bridged by O, epoxide) **(c)** $CH_3CH_2\overset{O}{\overset{\|}{C}}CH_2CH_3$ **(d)** cyclopentanone (ring =O) **(e)** CO_2

15.20 Suggest a synthesis for the following alcohols starting from an aldehyde or ketone and an appropriate Grignard reagent. Below each target molecule is shown the number of combinations of Grignard reagent and aldehyde or ketone that might be used.

(a) $CH_3\overset{OH}{\overset{|}{C}}CH_2CH_2CH_3$ with CH_2CH_3 on the central carbon
3 Combinations

(b) $CH_3CH_2\overset{OH}{\overset{|}{C}}HCH{=}CHCH_3$
2 Combinations

(c) $CH_3O{-}C_6H_4{-}\overset{OH}{\overset{|}{C}}H{-}C_6H_5$
2 Combinations

15.21 Show how to synthesize the following alcohol using 1-bromopropane, propanal, and ethylene oxide as the only sources of carbon atoms. It can be done using each compound only once. (*Hint:* Do one Grignard reaction to form an alcohol, convert the alcohol to an alkyl halide, and then do a second Grignard reaction.)

$$CH_3CH_2CH_2Br + \underset{}{\overset{CH_2CH_3}{\overset{|}{C}}HO} + \underset{\diagdown O \diagup}{CH_2{-}CH_2} \xrightarrow[\text{steps}]{\text{several}} CH_3CH_2CH_2\overset{CH_2CH_3}{\overset{|}{C}}HCH_2CH_2OH$$

15.22 1-Phenyl-2-butanol is used in perfumery. Show how to synthesize this alcohol from bromobenzene, 1-butene, and any necessary inorganic reagents.

$$C_6H_5{-}Br + CH_3CH_2CH{=}CH_2 \xrightarrow[\text{steps}]{\text{several}} C_6H_5{-}CH_2\overset{OH}{\overset{|}{C}}HCH_2CH_3$$

Bromobenzene 1-Butene 1-Phenyl-2-butanol

15.23 With organolithium and organomagnesium compounds, approach to the carbonyl carbon from the less hindered direction is generally preferred. Assuming this is the case, predict the structure of the major product formed by reaction of methylmagnesium bromide with 4-*tert*-butylcyclohexanone.

Wittig Reaction

15.24 Draw structural formulas for (1) the triphenylphosphonium salt formed by treatment of each haloalkane with triphenylphosphine, (2) the phosphonium ylide formed by treatment of each phosphonium salt with butyllithium, and (3) the alkene formed by treatment of each phosphonium ylide with acetone.

(a) $(CH_3)_2CHBr$ **(b)** $CH_2{=}CHCH_2Br$ **(c)** cyclopentyl$-CH_2Cl$

(d) $ClCH_2\overset{O}{\overset{\|}{C}}OCH_2CH_3$ **(e)** $CH_3CH_2CH_2CH_2Br$ **(f)** $C_6H_5{-}CH{=}CHCH_2Cl$

15.25 Show how to bring about the following conversions using a Wittig reaction.

(a) $CH_3\overset{O}{\overset{\|}{C}}CH_3 \longrightarrow CH_3\overset{CH_3}{\overset{|}{C}}{=}CH(CH_2)_3CH_3$ **(b)** $C_6H_5{-}\overset{O}{\overset{\|}{C}}CH_3 \longrightarrow C_6H_5{-}\overset{CH_2}{\overset{\|}{C}}CH_3$

(c) cyclopentanone ($C_5H_8{=}O$) $\longrightarrow$ cyclopentylidene$={CH}-C_6H_3(OCH_3)_2$ (3,4-dimethoxyphenyl: $-OCH_3$, OCH_3)

15.26 The Wittig reaction can be used for the synthesis of conjugated dienes, as, for example, 1-phenyl-1,3-pentadiene. Propose two sets of reagents that might be combined in a Wittig reaction to give this conjugated diene.

$$C_6H_5{-}CH{=}CHCH{=}CHCH_3$$

1-Phenyl-1,3-pentadiene

15.27 Show how to convert heptanal into the following:

(a) $CH_3(CH_2)_5CH{=}CH_2$

(b) $CH_3(CH_2)_5\overset{O}{CH{-}CH_2}$ (epoxide)

(c) $CH_3(CH_2)_5CH{=}CHCH{=}CH_2$

(d) $CH_3(CH_2)_5CH(OH)CH_2OH$

(e) $CH_3(CH_2)_5CH_2C(=O)CH_3$

(f) $CH_3(CH_2)_5CH_2CH_2CH_2OH$

15.28 Wittig reactions with the following α-chloroethers can be used for the synthesis of aldehydes and ketones:

$ClCH_2OCH_3$ (A) $\quad$ $ClCH(CH_3)OCH_3$ (B)

(a) Draw the structure of the triphenylphosphonium salt and Wittig reagent formed from each chloroether.
(b) Draw the structural formula of the product formed by treatment of each Wittig reagent with cyclopentanone. Note that the functional group is an enol ether, or, alternatively, a vinyl ether.
(c) Draw the structural formula of the product formed on acid-catalyzed hydrolysis of each enol ether from part (b).

15.29 It is possible to generate sulfur ylides in a manner similar to that used to produce phosphonium ylides. For example, treating a sulfonium salt with a strong base gives the sulfur ylide.

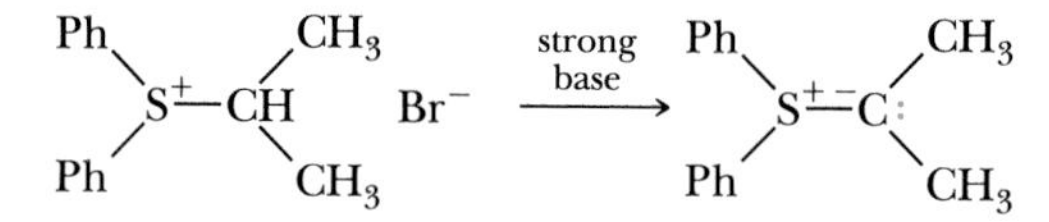

$$Ph_2\overset{+}{S}{-}CH(CH_3)_2\ Br^- \xrightarrow{\text{strong base}} Ph_2\overset{+}{S}{-}\overset{-}{C}(CH_3)_2$$

A sulfonium bromide salt

Sulfur ylides react with ketones to give epoxides. Suggest a mechanism for this reaction.

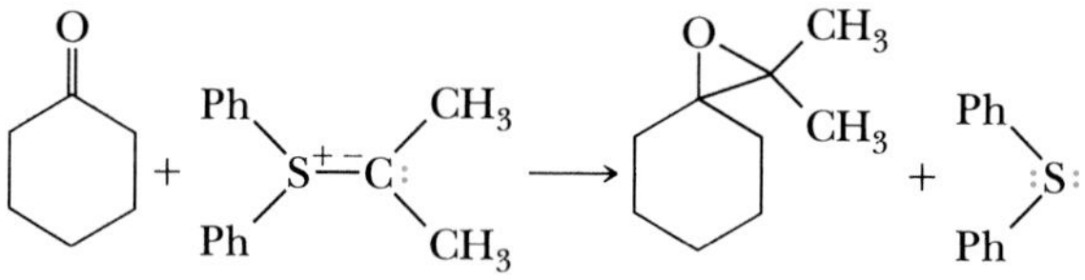

$$\text{cyclohexanone} + Ph_2\overset{+}{S}{-}\overset{-}{C}(CH_3)_2 \longrightarrow \text{epoxide (spiro, with } C(CH_3)_2) + Ph_2S$$

A sulfur ylide

15.30 Propose a structural formula for Compound A and the final product, $C_9H_{14}O$ formed in this reaction sequence.

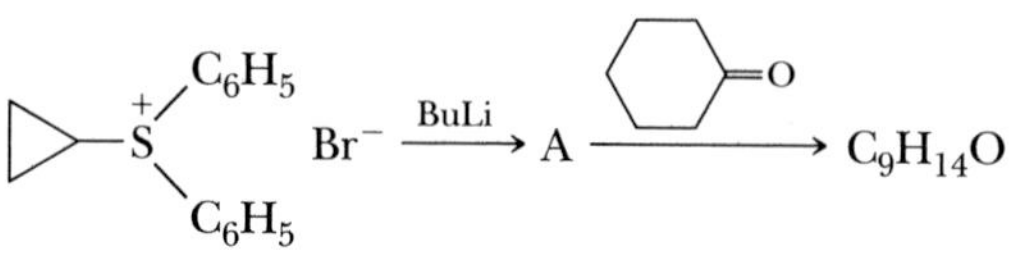

$$\text{cyclopropyl}{-}\overset{+}{S}(C_6H_5)_2\ Br^- \xrightarrow{\text{BuLi}} A \xrightarrow{\text{cyclohexanone}} C_9H_{14}O$$

Addition of Oxygen Nucleophiles

15.31 5-Hydroxyhexanal forms a six-membered cyclic hemiacetal, which predominates at equilibrium in aqueous solution.

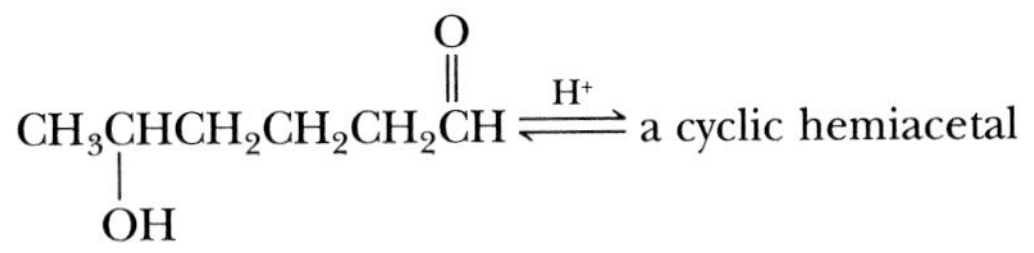

$$\underset{\text{OH}}{\underset{|}{CH_3CH}}CH_2CH_2CH_2\overset{\text{O}}{\overset{\|}{C}}H \underset{}{\overset{H^+}{\rightleftharpoons}} \text{a cyclic hemiacetal}$$

5-Hydroxyhexanal

(a) Draw a structural formula for this cyclic hemiacetal.
(b) How many stereoisomers are possible for 5-hydroxyhexanal?
(c) How many stereoisomers are possible for this cyclic hemiacetal?
(d) Draw alternative chair conformations for each stereoisomer of the cyclic hemiacetal and label groups axial or equatorial. Also predict which of the alternative chair conformations for each stereoisomer is the more stable.

15.32 Draw structural formulas for the hemiacetal and then the acetal formed from each pair of reactants in the presence of an acid catalyst.

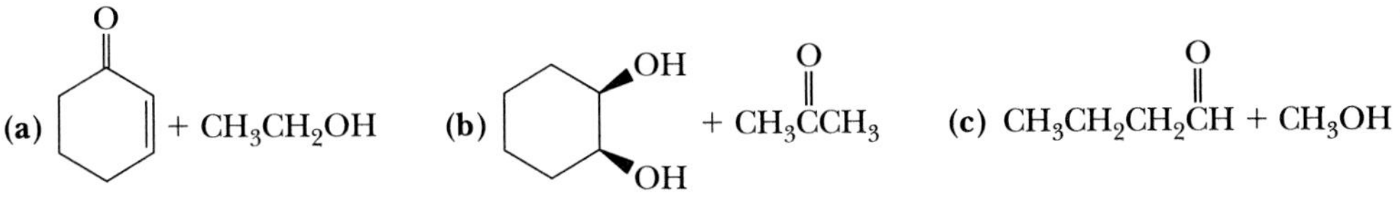

(a) (cyclohexenone, O) $+ CH_3CH_2OH$ **(b)** (cyclohexane-1,2-diol, OH, OH) $+ CH_3\overset{\text{O}}{\overset{\|}{C}}CH_3$ **(c)** $CH_3CH_2CH_2\overset{\text{O}}{\overset{\|}{C}}H + CH_3OH$

15.33 Draw structural formulas for the products of hydrolysis of the following acetals.

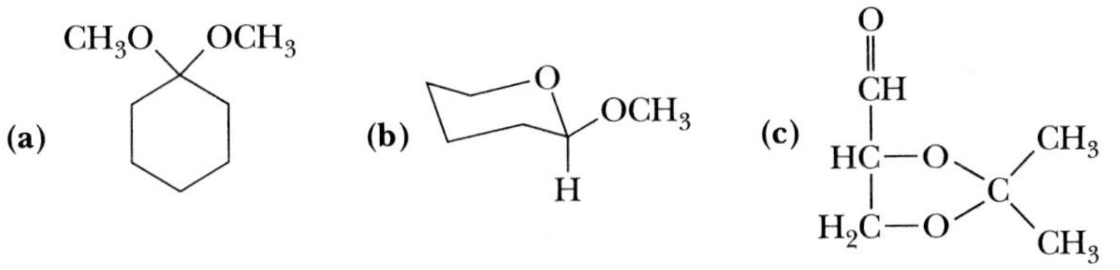

(a) CH_3O OCH_3 (cyclohexane) **(b)** (tetrahydropyran, O, OCH_3, H) **(c)** $\overset{\text{O}}{\overset{\|}{C}}H$, $HC{-}O$, $H_2C{-}O$, C, CH_3, CH_3

15.34 Propose a mechanism to account for the formation of a cyclic acetal from 4-hydroxypentanal and one equivalent of methanol. If the carbonyl oxygen of 4-hydroxypentanal is enriched with oxygen-18, do you predict that the oxygen label appears in the cyclic acetal or in the water?

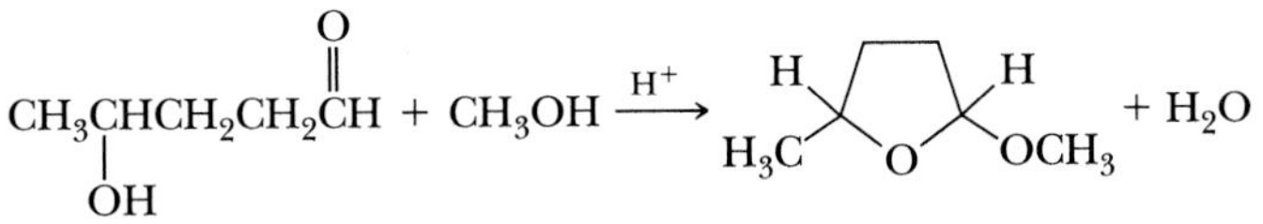

$$\underset{\text{OH}}{\underset{|}{CH_3CH}}CH_2CH_2\overset{\text{O}}{\overset{\|}{C}}H + CH_3OH \xrightarrow{H^+} \text{(H, } H_3C\text{, O, H, } OCH_3\text{)} + H_2O$$

4-Hydroxypentanal A cyclic acetal

15.35 Propose a mechanism for this acid-catalyzed hydrolysis.

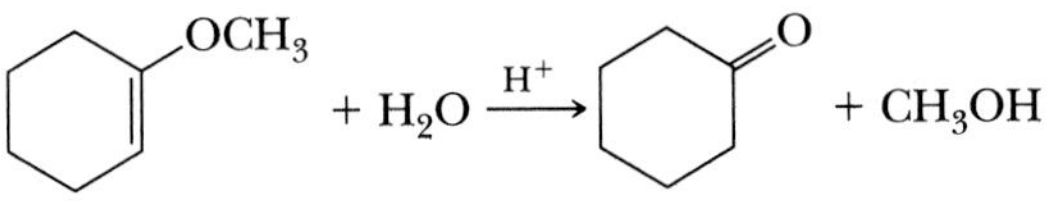

$$\text{(cyclohexene, } OCH_3\text{)} + H_2O \xrightarrow{H^+} \text{(cyclohexanone, O)} + CH_3OH$$

15.36 All rearrangements we have discussed so far involve generation of a positively charged electron-deficient carbon atom (a carbocation) followed by a 1,2-shift of an atom or group of atoms from an adjacent carbon to the carbocation. A mechanism that can be written for the following reaction also involves generation of an electron-deficient atom (in this case an oxygen) followed by a 1,2-shift from an adjacent carbon to the electron-deficient atom.

$$C_6H_5C(OOH)(CH_3)CH_3 \xrightarrow{H_2SO_4} C_6H_5OH + CH_3\overset{O}{\overset{\|}{C}}CH_3$$

Cumene hydroperoxide — Phenol — Acetone

Propose a mechanism for the acid-catalyzed rearrangement of cumene hydroperoxide to phenol and acetone based on the previous rationale. In completing a mechanism, you will want to review the characteristic structural features of a hemiacetal (Section 15.8B) and its equilibration with a ketone by loss of an alcohol.

15.37 In Section 11.6A we saw that ethers, such as diethyl ether and tetrahydrofuran, are quite resistant to the action of dilute acids and require hot concentrated HI or HBr for cleavage. Acetals, however, in which two ether groups are linked to the same carbon, undergo hydrolysis readily, even in dilute aqueous acid. How do you account for this marked difference in chemical reactivity toward dilute aqueous acid between ethers and acetals?

15.38 Draw a structural formula for the magnesium alkoxide formed in the following Grignard reaction and then the product formed on hydrolysis of the magnesium alkoxide with aqueous acid.

$$CH_3O\text{-}C_6H_4\text{-}\overset{O}{\overset{\|}{C}}H + BrMgCH_2CH_2CH\langle\begin{smallmatrix}O-CH_2\\ \quad\ |\\ O-CH_2\end{smallmatrix}\rangle$$

15.39 Show how to bring about the following conversion:

$$CH_2{=}CH\overset{O}{\overset{\|}{C}}H \xrightarrow{\text{several steps}} \underset{HO}{CH_2}\underset{OH}{CH}\overset{O}{\overset{\|}{C}}H$$

15.40 Multistriatin is a component of the aggregating pheromone of the European elm bark beetle, the insect vector of Dutch elm disease. In one laboratory synthesis of this molecule, (*Z*)-2-butene-1,4-diol was used as a starting material. Because of the requirements of

European elm bark beetle, the vector of Dutch elm disease. *(© Stephen Dalton, Photo Researchers, Inc.)*

subsequent steps, it was necessary to protect the two —OH groups of this molecule, which was done by making a cyclic acetal with acetone. The cyclic acetal was then treated with 3-chloroperoxybenzoic acid followed by methylmagnesium iodide and acid hydrolysis to give a compound of molecular formula $C_5H_{12}O_3$.

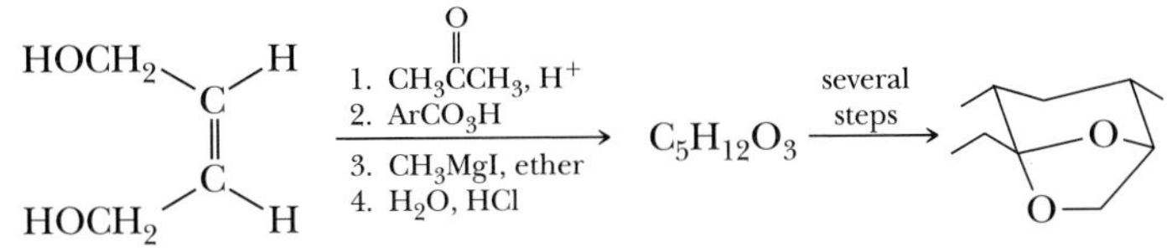

(Z)-2-Butene-1,4-diol: $HOCH_2(H)C{=}C(H)CH_2OH$ → 1. $CH_3\overset{O}{\overset{\|}{C}}CH_3$, H^+; 2. $ArCO_3H$; 3. CH_3MgI, ether; 4. H_2O, HCl → $C_5H_{12}O_3$ → (several steps) → Multistriatin

(Z)-2-Butene-1,4-diol Multistriatin

(a) Draw structural formulas for the products of Steps 1, 2, 3, and 4.
(b) Explain why it was necessary to protect the —OH groups in the starting diol.

15.41 Which of these molecules will cyclize to give the insect pheromone frontalin?

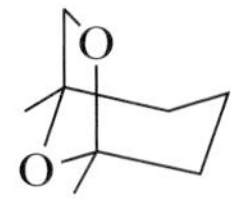

Frontalin 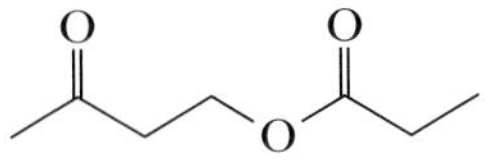 (A) 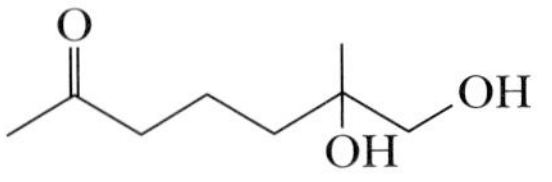 (B) 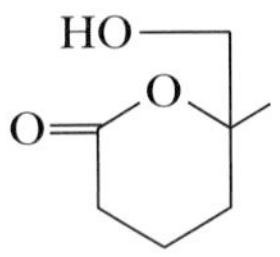 (C)

Addition of Sulfur Nucleophiles

15.42 Draw a structural formula for the product of reaction of benzaldehyde with the following dithiols in the presence of an acid catalyst.

(a) 1,2-ethanedithiol **(b)** 1,3-propanedithiol

15.43 Draw a structural formula for the product formed by treating each of these compounds with (1) the lithium salt of the 1,3-dithiane derived from acetaldehyde and then (2) H_2O, $HgCl_2$.

(a) 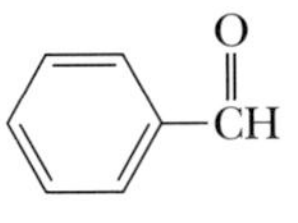 $C_6H_5\overset{O}{\overset{\|}{C}}H$ **(b)** 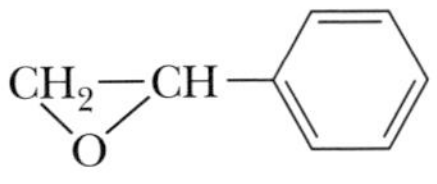 $\underset{\diagdown O \diagup}{CH_2{-}CH}{-}C_6H_5$ **(c)** $ClCH_2CH{=}CH_2$

15.44 Show how to bring about the following conversions using a 1,3-dithiane:

(a) 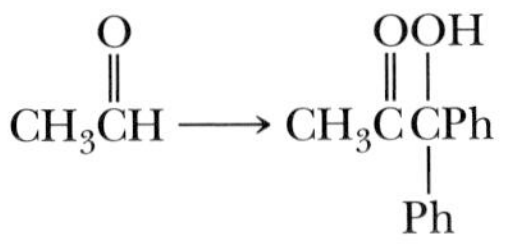 $CH_3\overset{O}{\overset{\|}{C}}H \longrightarrow CH_3\overset{OH}{\overset{|}{\underset{Ph}{\underset{|}{C}}}}\overset{O}{\overset{\|}{C}}Ph$ **(b)** 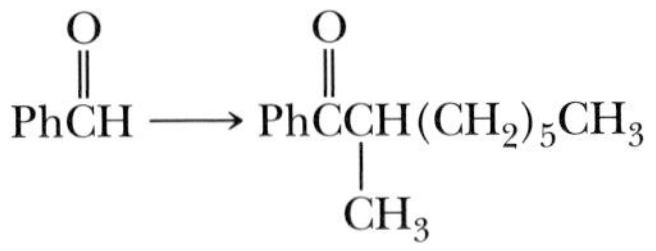 $Ph\overset{O}{\overset{\|}{C}}H \longrightarrow Ph\overset{O}{\overset{\|}{C}}\underset{CH_3}{\underset{|}{C}H}(CH_2)_5CH_3$

(c) $C_5H_9\overset{O}{\overset{\|}{C}}H \longrightarrow C_5H_9\overset{O}{\overset{\|}{C}}CH_2CH_2OH$

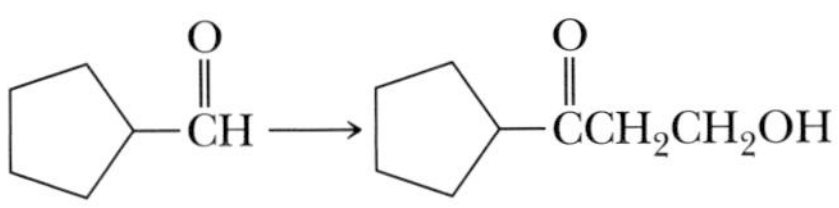

Addition of Nitrogen Nucleophiles

15.45 Draw structural formulas for the products of the following acid-catalyzed reactions:

(a) Phenylacetaldehyde + hydrazine ⟶
(b) Cyclopentanone + semicarbazide ⟶
(c) Acetophenone + 2,4-dinitrophenylhydrazine ⟶
(d) Benzaldehyde + hydroxylamine ⟶

15.46 The following ketone reacts with hydroxylamine to form a pair of isomeric oximes related in the same manner that *cis* and *trans* alkenes are related. Draw structural formulas for these isomeric oximes, and specify the configuration of each using the *E,Z* convention.

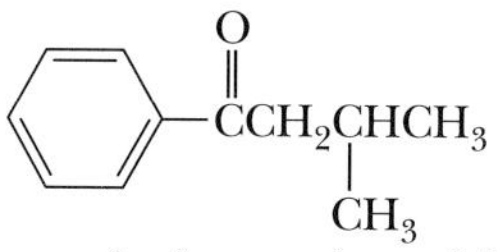

$$C_6H_5-\overset{O}{\overset{\|}{C}}CH_2\underset{CH_3}{\underset{|}{C}}HCH_3$$

15.47 Methenamine (hexamethylenetetramine), a product of the reaction of formaldehyde and ammonia, is an example of a prodrug, a compound that is inactive itself but is converted to an active drug by a biochemical transformation. The strategy behind use of methenamine as a prodrug is that nearly all bacteria are sensitive to formaldehyde at concentrations of 20 mg/mL or higher. Formaldehyde cannot be used directly in human medicine, however, because an effective concentration in plasma cannot be achieved with safe doses. Methenamine is stable at pH 7.4 (the pH of blood plasma) but undergoes acid-catalyzed hydrolysis to formaldehyde and ammonium ion under the acidic conditions of renal tubules and the urinary tract. Thus, methenamine can be used as a site-specific drug to treat urinary infections.

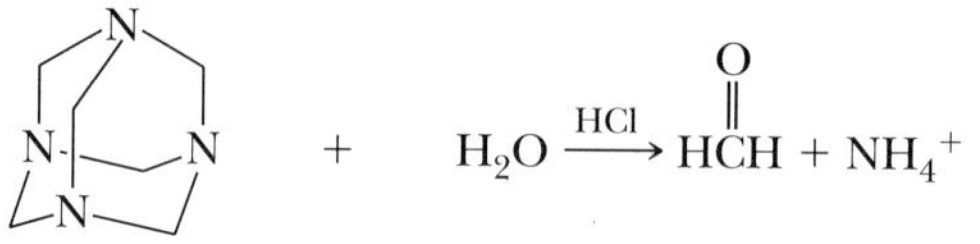

$$\text{Methenamine} + H_2O \xrightarrow{HCl} H\overset{O}{\overset{\|}{C}}H + NH_4^+$$

Methenamine
(Hexamethylenetetramine)

(a) Write a balanced equation for the hydrolysis of methenamine to formaldehyde and ammonium ion.

(b) Does the pH of an aqueous solution of methenamine increase, remain the same, or decrease as a result of hydrolysis? Explain.

(c) Explain the meaning of the following statement: The functional group in methenamine is the nitrogen analog of an acetal.

(d) Account for the observation that methenamine is stable in blood plasma but undergoes hydrolysis in the urinary tract.

(e) Propose a mechanism for the acid-catalyzed hydrolysis of methenamine to formaldehyde and ammonium ion.

Keto-Enol Tautomerism

15.48 The following molecule belongs to a class of compounds called enediols; each carbon of the double bond carries an —OH group. Draw structural formulas for the α-hydroxyketone and the α-hydroxyaldehyde with which this enediol is in equilibrium.

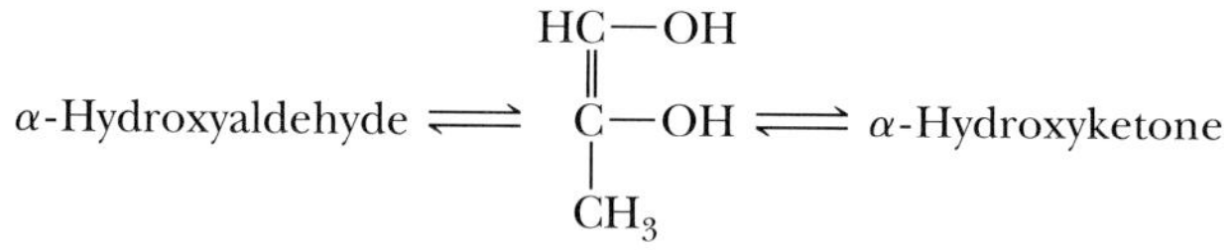

$$\alpha\text{-Hydroxyaldehyde} \rightleftharpoons \begin{array}{c} HC{-}OH \\ \| \\ C{-}OH \\ | \\ CH_3 \end{array} \rightleftharpoons \alpha\text{-Hydroxyketone}$$

An enediol

15.49 In dilute aqueous base, (*R*)-glyceraldehyde is converted into an equilibrium mixture of (*R,S*)-glyceraldehyde and dihydroxyacetone. Propose a mechanism for this isomerization.

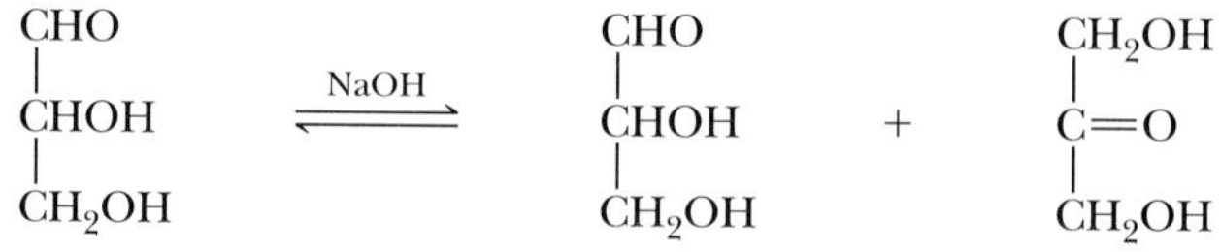

$$\begin{array}{c} CHO \\ | \\ CHOH \\ | \\ CH_2OH \end{array} \overset{NaOH}{\rightleftharpoons} \begin{array}{c} CHO \\ | \\ CHOH \\ | \\ CH_2OH \end{array} + \begin{array}{c} CH_2OH \\ | \\ C{=}O \\ | \\ CH_2OH \end{array}$$

(*R*)-Glyceraldehyde (*R,S*)-Glyceraldehyde Dihydroxyacetone

15.50 When *cis*-2-decalone is dissolved in ether containing a trace of HCl, an equilibrium is established with *trans*-2-decalone. The latter ketone predominates in the equilibrium mixture. Propose a mechanism for this isomerization and account for the fact that the *trans* isomer predominates at equilibrium.

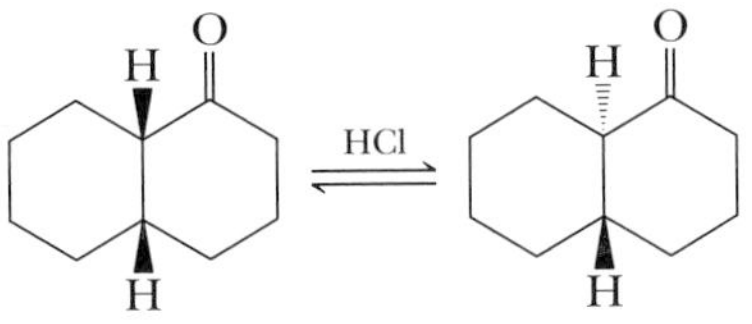

cis-2-Decalone $\xrightarrow{HCl}$ *trans*-2-Decalone

Reactions at the α-Carbon

15.51 The following bicyclic ketone has two α-carbons and three α-hydrogens. When this molecule is treated with D_2O in the presence of an acid catalyst, only two α-hydrogens exchange with deuterium. The α-hydrogen at the bridgehead does not exchange. How do you account for the fact that two α-hydrogens do exchange but the third does not? You will find it helpful to build a model of this molecule and of the enols by which exchange of α-hydrogens occurs.

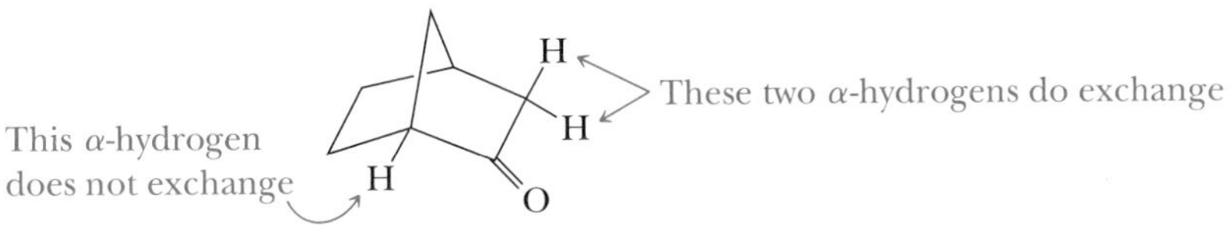

This α-hydrogen does not exchange

These two α-hydrogens do exchange

15.52 Propose a mechanism for this reaction.

$$\text{cyclohexanone} + Cl_2 \xrightarrow{CH_3CO_2H} \text{2-chlorocyclohexanone} + HCl$$

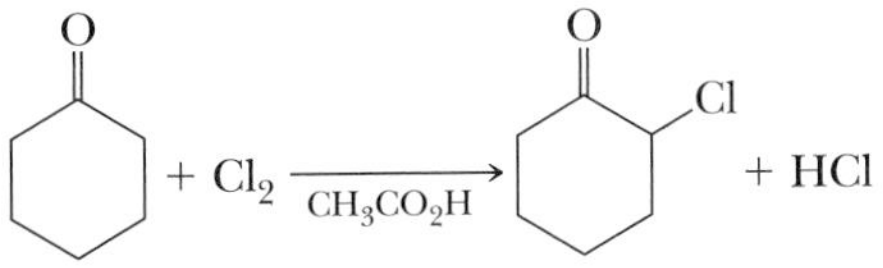

15.53 The base-promoted rearrangement of an α-haloketone to a carboxylic acid, known as the Favorskii rearrangement, is illustrated by the conversion of 2-chlorocyclohexanone to cyclopentanecarboxylic acid. It is proposed that NaOH first converts the α-haloketone to the substituted cyclopropanone shown in brackets, and then to the sodium salt of cyclopentanecarboxylic acid.

$$\xrightarrow[THF]{NaOH} [\ \ ] \xrightarrow[THF]{NaOH} \text{COO}^-Na^+ \xrightarrow[H_2O]{HCl} \text{COH}$$

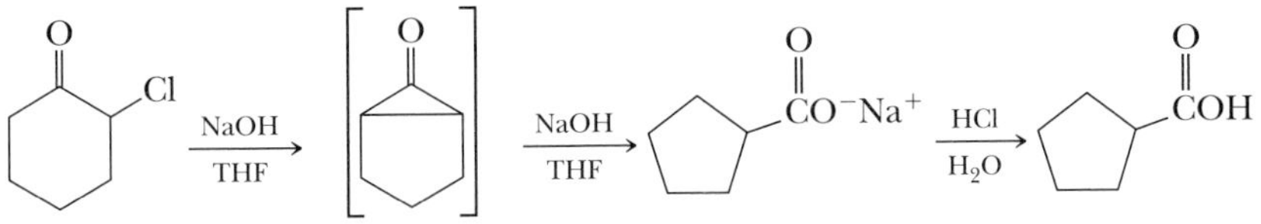

A proposed intermediate

(a) Propose a mechanism for base-promoted conversion of 2-chlorocyclohexanone to the bracketed intermediate.

(b) Propose a mechanism for base-promoted conversion of the bracketed intermediate to sodium cyclopentanecarboxylate. (*Hint:* Begin by adding hydroxide ion to the carbonyl carbon to form a tetrahedral carbonyl addition intermediate.)

15.54 If the Favorskii rearrangement of 2-chlorocyclohexanone is carried out using sodium ethoxide in ethanol, the product is ethyl cyclopentanecarboxylate. Propose a mechanism for this reaction (see next page).

$$\text{2-chlorocyclohexanone} \xrightarrow[CH_3CH_2OH]{CH_3CH_2O^-Na^+} \text{cyclopentyl}-\overset{O}{\overset{\|}{C}}OCH_2CH_3$$

Pennyroyal *(Mentha pulegium)*, from which pulegone is isolated. *(Kenneth W. Fink/Photo Researchers, Inc.)*

15.55 (*R*)-(+)-Pulegone, readily available from the pennyroyal oils (*Merck Index*, 12th Ed., #8124), is an important enantiopure building block for organic syntheses. Propose a mechanism for each step in this transformation of pulegone. (*Hint:* The first stages of the mechanism for the second reaction are essentially identical to those of the Favorskii rearrangement.)

$$(R)\text{-}(+)\text{-Pulegone} \xrightarrow{Br_2} \text{dibromide (Br, Br)} \xrightarrow[CH_3OH]{CH_3O^-Na^+} \text{cyclopentyl}-\overset{O}{\overset{\|}{C}}OCH_2CH_3$$

(*R*)-(+)-Pulegone

15.56 (*R*)-(+)-Pulegone is converted to (*R*)-citronellic acid by addition of HCl followed by treatment with NaOH. Propose a mechanism for each step in this transformation, including the regioselectivity of HCl addition.

$$(R)\text{-}(+)\text{-Pulegone} \xrightarrow{HCl} \text{chloroketone (Cl)} \xrightarrow[2.\ HCl]{1.\ NaOH} \overset{O}{\overset{\|}{C}}OH$$

(*R*)-(+)-Pulegone

(*R*)-3,7-Dimethyl-6-octenoic acid
(*R*)-Citronellic acid

Oxidation/Reduction of Aldehydes and Ketones

15.57 Draw structural formulas for the products formed by treatment of butanal with the following reagents.

(a) $LiAlH_4$ followed by H_2O
(b) $NaBH_4$ in CH_3OH/H_2O
(c) H_2/Pt
(d) $Ag(NH_3)_2^+$ in NH_3/H_2O
(e) H_2CrO_4, heat
(f) $HOCH_2CH_2OH$, HCl
(g) Zn(Hg)/HCl
(h) N_2H_4, KOH at 250°C
(i) $C_6H_5NH_2$
(j) $C_6H_5NHNH_2$

15.58 Draw structural formulas for the products of the reaction of acetophenone with the reagents given in Problem 15.57.

Synthesis

15.59 Starting with cyclohexanone, show how to prepare these compounds. In addition to the given starting material, use any other organic or inorganic reagents as necessary.

(a) Cyclohexanol
(b) Cyclohexene
(c) *cis*-1,2-Cyclohexanediol
(d) 1-Methylcyclohexanol
(e) 1-Methylcyclohexene
(f) 1-Phenylcyclohexanol
(g) 1-Phenylcyclohexene
(h) Cyclohexene oxide
(i) *trans*-1,2-Cyclohexanediol

15.60 Show how to convert cyclopentanone to these compounds. In addition to cyclopentanone, use other organic or inorganic reagents as necessary.

(a) cyclopentanol —OH (b) chlorocyclopentane —Cl (c) =CH—CH=CH_2 (d) OH

15.61 Disparlure is a sex attractant of the gypsy moth *(Porthetria dispar)*. It has been synthesized in the laboratory from the following (*Z*)-alkene.

(*Z*)-2-Methyl-7-octadecene → Disparlure

(a) Propose two sets of reagents that might be combined in a Wittig reaction to give the indicated (*Z*)-alkene. Note that at least for simple phosphonium ylides, the product of a Wittig reaction is predominantly the (*Z*)-alkene.

(b) How might the (*Z*)-alkene be converted to disparlure?

(c) How many stereoisomers are possible for disparlure? How many are formed in the sequence you chose?

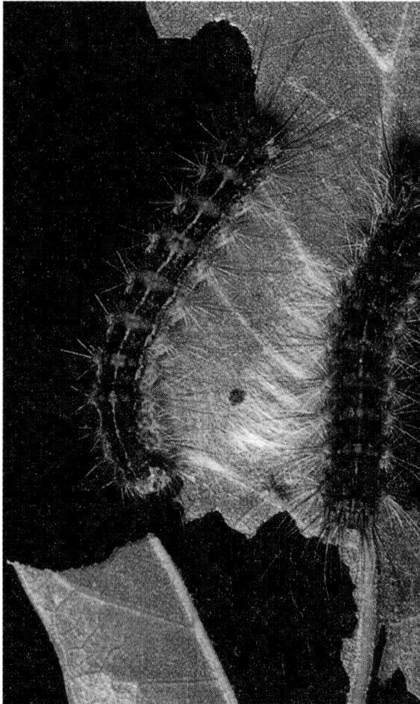

Disparlure is the sex attractant of the gypsy moth, *Porthetria dispar.* See Problem 15.61. *(Animals, Animals © William D. Griffin)*

15.62 Starting with the given two compounds and any other necessary organic or inorganic reagents, show how to make the compound shown at the right.

$$CH_3-C_6H_4-\overset{O}{\overset{\|}{C}}CH_3 + BrCH_2CH_2\overset{O}{\overset{\|}{C}}H \xrightarrow{\text{several steps}} CH_3-C_6H_4-\underset{CH_3}{\underset{|}{\overset{OH}{\overset{|}{C}}}}CH_2CH_2\overset{O}{\overset{\|}{C}}H$$

15.63 Propose structural formulas for compounds A, B, and C in the following conversion. Show also how to prepare compound C by a Wittig reaction.

$$\text{cyclopentanone} \xrightarrow[\text{2. } H_2O]{\text{1. } HC\equiv CH,\ NaNH_2} \underset{(A)}{C_7H_{10}O} \xrightarrow[\text{Lindlar catalyst}]{H_2} \underset{(B)}{C_7H_{12}O} \xrightarrow[\text{heat}]{KHSO_4} \underset{(C)}{C_7H_{10}}$$

15.64 Following is a retrosynthetic scheme for the synthesis of *cis*-3-penten-2-ol. Write a synthesis for this compound from the indicated starting materials.

$$\text{—OH} \Longrightarrow \text{—OH} \Longrightarrow \text{—}\equiv + \overset{O}{\overset{\|}{HC}}CH_3$$

$$CH_3I + HC\equiv CH$$

15.65 Following is the structural formula of the tranquilizer Oblivon (meparfynol). Propose a synthesis for this molecule starting with acetylene and a ketone.

$$CH_3CH_2\underset{CH_3}{\underset{|}{\overset{OH}{\overset{|}{C}}}}C\equiv CH$$

Oblivon

15.66 Following is the structural formula of surfynol, a defoaming surfactant. Describe the synthesis of this molecule from acetylene and a ketone. You may wish to look up the word "surfactant" in a science reference book and find out what surfactants are, how they work, and what they are used for.

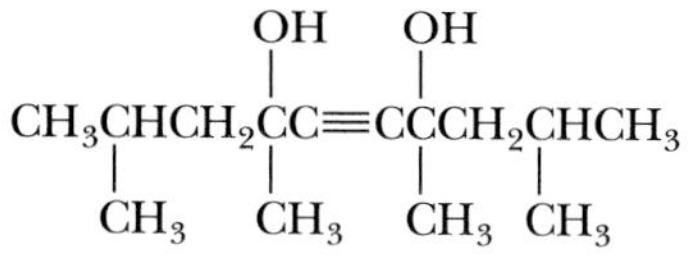

$$\underset{\displaystyle CH_3}{\overset{}{CH_3\underset{|}{C}HCH_2}}\overset{OH}{\underset{CH_3}{C}}C\equiv C\overset{OH}{\underset{CH_3}{C}}CH_2\underset{CH_3}{CH}CH_3$$

Surfynol

15.67 Propose a mechanism for this isomerization.

$$\text{1-ethynylcyclohexanol (OH, C}\equiv\text{CH)} \xrightarrow{H^+} \text{cyclohexylidene}=C(H)CHO$$

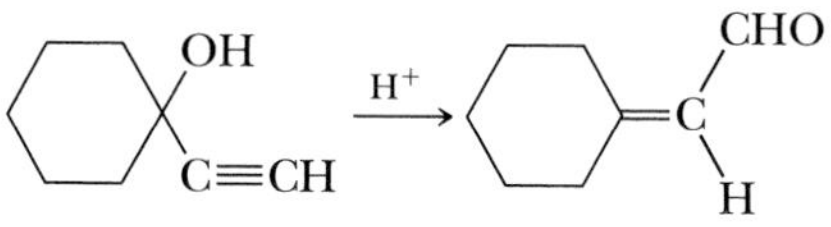

15.68 Propose a mechanism for this isomerization.

$$\text{spiro epoxide (O)} \xrightarrow{ArSO_3H} \text{spiro cyclobutanone (O)}$$

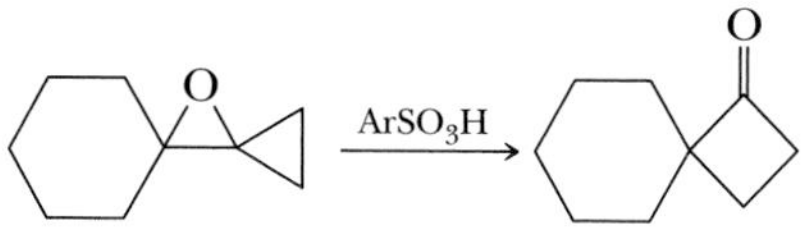

15.69 Starting with acetylene and 1-bromobutane as the only sources of carbon atoms, show how to synthesize the following:

(a) Meso 5,6-decanediol (b) Racemic 5,6-decanediol
(c) 5-Decanone (d) 5,6-Epoxydecane
(e) 5-Decanol (f) Decane
(g) 6-Methyl-5-decanol (h) 6-Methyl-5-decanone

15.70 Following are the final steps in one industrial synthesis of vitamin A acetate:

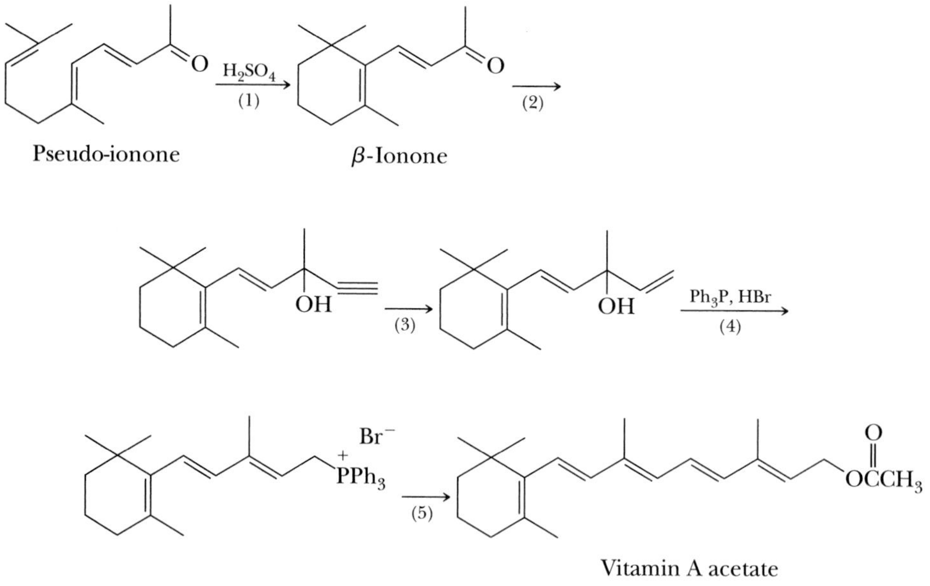

$$\text{Pseudo-ionone} \xrightarrow[(1)]{H_2SO_4} \beta\text{-Ionone} \xrightarrow{(2)}$$

Pseudo-ionone β-Ionone

$$\text{(OH, C}\equiv\text{CH)} \xrightarrow{(3)} \text{(OH, CH=CH}_2) \xrightarrow[(4)]{Ph_3P,\ HBr}$$

$$\text{(}\overset{+}{P}Ph_3\ Br^-\text{)} \xrightarrow{(5)} \text{Vitamin A acetate (}O\overset{O}{\overset{\|}{C}}CH_3\text{)}$$

Vitamin A acetate

(a) Propose a mechanism for the acid-catalyzed cyclization in Step 1.
(b) Propose reagents to bring about Step 2.
(c) Propose reagents to bring about Step 3.
(d) Propose a mechanism for formation of the phosphonium salt in Step 4.
(e) Show how Step 5 can be completed by a Wittig reaction.

A CONVERSATION WITH . . .
JACQUELYN GERVAY

The life of a young faculty member is composed of a myriad of new responsibilities. Tasks include teaching several courses for the first time, establishing an independent research program, writing grant proposals, and working with graduate and undergraduate students in the lab.

Jacquelyn Gervay (jur-vay), an organic chemist and, since 1992, an assistant professor of chemistry at the University of Arizona in Tucson, finds that teaching and research consume most of her days. However, Gervay still finds time for some of her other interests, such as running, inline skating, and even a game of basketball or tennis on occasion.

Gervay characterizes her research approach as one of "bridging" chemistry, biology, and immunology in an effort to understand biological processes at the molecular level. Her current projects include the synthesis of carbohydrates and novel helical materials; the development of targeted drug delivery systems for treatment of spinal meningitis, cancer, and HIV infections; and the development of new nuclear magnetic resonance (NMR) techniques for understanding reaction mechanisms. Gervay is also interested in chemical education. With a grant from the National Science Foundation, she is developing an approach that brings research into the classroom.

Gervay received B.S. and Ph.D. degrees in chemistry from the University of California at Los Angeles and held a postdoctoral position in the Yale University laboratory of synthetic organic chemist Samuel J. Danishefsky.

A LATE START

"My interest in chemistry got started late. I never took chemistry in high school; in fact, when I entered UCLA, I planned to major in psychobiology. The degree required chemistry, so I enrolled in the introductory courses my first year at UCLA. I did fine, although I considered it nothing more than work I had to complete to meet other goals. Then in my second year, I took organic chemistry. Even though I worked very hard, I just couldn't get it. So I understand students who have difficulty with organic chemistry in spite of their efforts. I waited until my senior year to finish the organic sequence. The instructor, Mike [Michael E.] Jung was a terrific teacher. Suddenly organic chemistry made sense and I became captivated by the subject matter. What turned me on was hearing about and discussing the research Mike was conducting in the laboratory. Research made chemistry a living science for me. It was during that course that I decided to become a chemistry professor. But since it was so late in my undergraduate career, I had to take another year and a half of chemistry courses to graduate as a chemistry major. That was a great time for me because I knew exactly what I wanted to do."

A TASTE OF GRADUATE SCHOOL

"After I decided to become a chemistry major, I wanted to get some hands-on lab experience. I approached Professor Christopher Foote and asked if I could work in his lab over the summer. He had an opening and let me join his group.

There, I worked with singlet oxygen. I looked at the singlet oxygen "ene" reaction and other kinds of photochemical experiments; I found the research fun and challenging. Following that experience, I joined Mike Jung's research group to get experience in the synthesis of organic compounds. When it came time to apply to graduate school, I applied only to UCLA. First of all, because I wanted to continue my work in Mike Jung's group, and second, because my husband had a business in Los Angeles, and it would have been difficult for me to go elsewhere for graduate school."

Organic Synthesis

"In graduate school, I spent a good two years trying to develop a route to the synthesis of reserpine [a natural product with tranquilizing properties] via an intramolecular Diels-Alder reaction (see Section 22.5). Unfortunately, we could not overcome the unreactive nature of the Diels-Alder dienophile. Then Mike [Jung] came back from a conference with an interesting idea about doing an intramolecular Diels-Alder cyclization and we decided to look at the effects of substituents on the rate of the reaction. I synthesized a number of test compounds with different alkyl groups near the reaction site, and then studied their cyclization. We did all of the classic experiments to prove the reactions were essentially irreversible, and then we published a paper reporting this somewhat surprising result. Later I started doing the reactions in a different solvent and saw some amazing and unexpected differences in reaction rates. We were really intrigued, and in the end, we discovered that during the reaction some unfavorable dipole interactions took place, interactions accentuated in polar solvents that accelerated the reaction."

A New Interest Sends Gervay East

"For postdoctoral study, I went to Yale University to work with Samuel J. Danishefsky, a noted synthetic organic chemist. There I wanted to work on a molecule called sialic acid, which is a 3-deoxy sugar with a hydroxyl and a carboxyl group at the anomeric carbon.

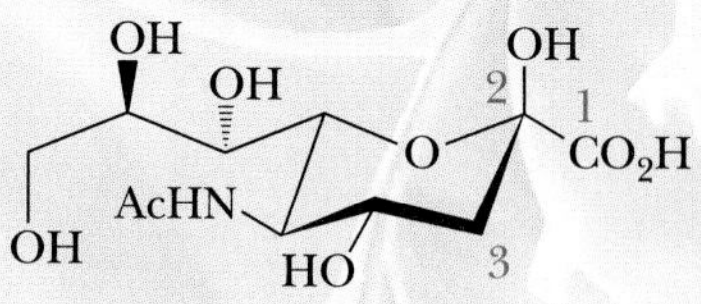

OH OH OH 2 1 O CO_2H AcHN OH HO 3

Sialic acid

This compound is found in cell glycolipids, and I was captivated by sialic acid, both for its biological role and as a synthetic project.

> "Suddenly organic chemistry made sense and I became captivated by the subject matter."

"My years as a postdoctoral student were very formative. Almost everything I now study in my lab is based on the work I did at Yale. There I learned carbohydrate chemistry and helped to complete the total synthesis of Sialyl-Lewis X glycal [a tetrasaccharide that can be converted to compounds known to be important in tumor metastasis and in attracting white blood cells to damaged tissue]. Also exciting for me was interacting with graduate students and other postdocs, who had come there from all over the world."

On Being a Professor

"In the fall on 1992, I joined the faculty of the University of Arizona. The most rewarding part of being a professor is working with students. I always learn something; they always learn something. Using grant money from the National Science Foundation, one of my main goals in my undergraduate classes is to show the students that, given a basic understanding of chemistry, they can come up with original ideas for research, and they can learn to think critically about important areas of organic chemistry. I give them original research articles and ask them to propose some ideas for further research. Many of the ideas they come up with have been done and are in the literature, and that is exciting to them. They can't believe that they—just sophomores in organic chemistry—can think of ideas that scientists studied and wrote about in journal articles. Then I ask them to take their thinking a step further, and a step further, until they come up with questions that have not been explored. At that point, I ask them to design some experiments that might help answer their questions. Based on their ideas, the students are placed into research teams and they present their combined ideas in a poster session. This endeavor constitutes the final exam of the first semester. In the second semester, the research teams conduct their proposed experiments and report their findings in an electronic conference on the Worldwide Web."

On Being a Woman Scientist

"I know that some women face barriers in science, but that hasn't been my personal experience. Some of my women students have told me it makes a difference to them that I'm a woman chemist. They say that they are much more comfortable talking to me than to a male professor, especially when they don't understand something. I have the sense that they don't feel intimidated when they approach me. If talking to women students helps them reach their potential, then I am happy to help."

Advice on Science for Students

"I don't want to make everyone a chemist. I don't think that's prudent. My goal is to make everybody in my classes appreciate chemistry. At the same time, I tell my sophomore students that chemistry, and certainly organic chemistry, is not strictly intuitive. If you have difficulty understanding the concepts, it's not because you're not smart; organic chemistry is challenging, much like a foreign language. So I think a large part of being successful is just sticking to it. For students who are contemplating a career in chemistry, one of the most important things to do is to get laboratory experience in a research setting, not just lab classes. That is how you'll know if a career in chemistry is for you."

16 Carboxylic Acids

■ Crystals of ibuprofen viewed under polarizing light. Its structure is

$$CH_3CH(CH_3)CH_2-C_6H_4-CH(CH_3)COOH$$

(© *Ulof Björg Christianson/Fran Heyl Associates*)

The most important chemical property of carboxylic acids, another class of organic compounds containing the carbonyl group, is their acidity. Furthermore, carboxylic acids form numerous important derivatives, including esters, amides, anhydrides, and acid halides. In this chapter, we study carboxylic acids themselves; in Chapter 17, we study derivatives of carboxylic acids.

16.1 Structure

The functional group of a carboxylic acid is the **carboxyl group,** so named because it is made up of a **carb**onyl group and a hydr**oxyl** group. Following is a Lewis structure of the carboxyl group as well as two alternative representations for it:

Carboxyl group A $-CO_2H$ group.

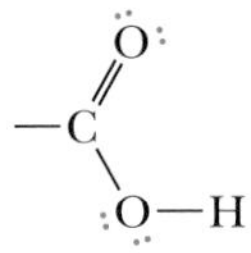

$$-C(=\ddot{O}:)-\ddot{O}-H \qquad -COOH \qquad -CO_2H$$

Alternative representations of a carboxyl group

The general formula for an aliphatic carboxylic acid is RCO_2H; the general formula for an aromatic carboxylic acid is $ArCO_2H$.

RCO_2H	$ArCO_2H$
An aliphatic carboxylic acid	An aromatic carboxylic acid

16.2 Nomenclature

A. IUPAC System

The IUPAC name of a carboxylic acid is derived from the name of the longest carbon chain that contains the carboxyl group by dropping the final **-e** from the name of the parent alkane and adding the suffix **-oic** followed by the word **acid** (Section 2.5). The chain is numbered, beginning with the carbon of the carboxyl group. The carboxyl carbon is understood to be carbon 1. The IUPAC system retains the common names formic acid and acetic acid.

HCO_2H	CH_3CO_2H	$CH_3CH(CH_3)CH_2CO_2H$
Methanoic acid (Formic acid)	Ethanoic acid (Acetic acid)	3-Methylbutanoic acid

If the carboxylic acid contains a carbon-carbon double or triple bond, change the infix from **-an-** to **-en-** or **-yn-** to indicate the presence of the multiple bond, and show the location of the multiple bond by a number.

$CH_2{=}CHCO_2H$	$(C_6H_5)(H)C{=}C(H)(CO_2H)$	$(H_3C)(H)C{=}C(H)(CO_2H)$
Propenoic acid (Acrylic acid)	*trans*-3-Phenylpropenoic acid (Cinnamic acid)	*trans*-2-Butenoic acid (Crotonic acid)

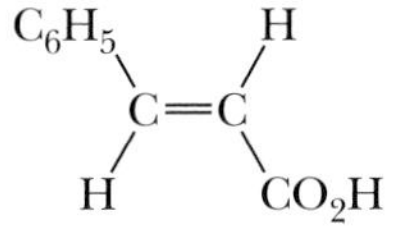

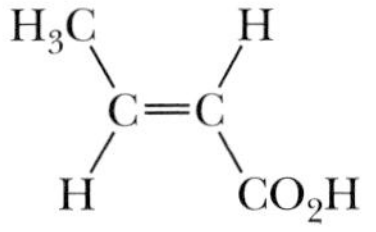

In the IUPAC system, a carboxyl group takes precedence over most other functional groups (Table 15.1), including hydroxyl groups and amino groups, as well as the carbonyl groups of aldehydes and ketones. As illustrated in the following examples, an $-OH$ group is indicated by the prefix hydroxy-, an $-NH_2$ group is indicated by the prefix amino-, and the $C{=}O$ group of an aldehyde or ketone is indicated by the prefix oxo-.

Leaves of the rhubarb plant contain the poison oxalic acid. *(© Ann Reilly: PHOTO/NATS)*

$CH_3CH(OH)CH_2CH_2CH_2CO_2H$ — 5-Hydroxyhexanoic acid

$H_2NCH_2CH_2CH_2CO_2H$ — 4-Aminobutanoic acid

$CH_3C(=O)CH_2CH_2CH_2CO_2H$ — 5-Oxohexanoic acid

Dicarboxylic acids are named by adding the suffix **-dioic acid** to the name of the carbon chain that contains both carboxyl groups. The numbers of the carboxyl carbons are not indicated because they can be only at the ends of the parent chain. Following are IUPAC and common names for several important aliphatic dicarboxylic acids.

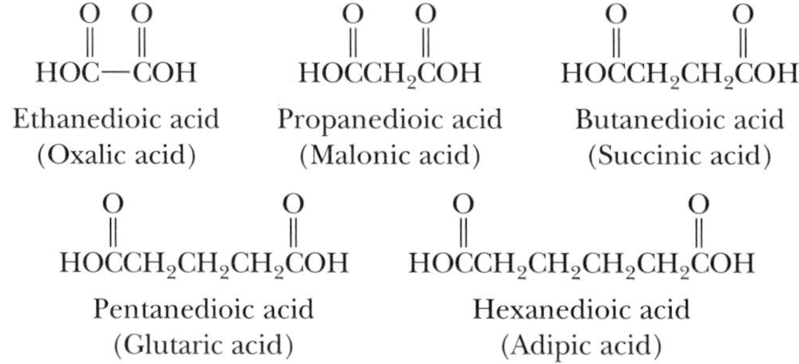

The name "oxalic acid" is derived from one of its sources in the biological world, namely, plants of the genus *Oxalis,* one of which is rhubarb. Adipic acid is one of the two monomers required for the synthesis of the polymer nylon 66. In 1995, the U.S. chemical industry produced 1.8 billion pounds of adipic acid, solely for the synthesis of nylon 66.

Tri- and higher carboxylic acids are named by using the suffixes **-tricarboxylic acid, -tetracarboxylic acid,** and so on. An example of a tricarboxylic acid is 2-hydroxy-1,2,3-propanetricarboxylic acid, whose common name, "citric acid," is retained by the IUPAC system. Citric acid can be extracted from citrus juices; lemon juice, for example, contains 5% to 8% citric acid. Citric acid is important in a metabolic pathway known alternatively as the tricarboxylic acid (TCA) cycle or Krebs cycle.

$HO_2C-C(OH)(CH_2CO_2H)_2$

2-Hydroxy-1,2,3-propanetricarboxylic acid
(Citric acid)

A carboxylic acid containing a carboxyl group attached to a cycloalkane ring is named by giving the name of the ring and adding the suffix **-carboxylic acid.** The atoms of the ring are numbered beginning with the carbon bearing the $-CO_2H$ group.

2-Cyclohexenecarboxylic acid (CO_2H at C-1; double bond C-2–C-3)

trans-1,3-Cyclopentanedicarboxylic acid (HO_2C, CO_2H)

The simplest aromatic carboxylic acid is benzoic acid. Derivatives are named by using numbers to show the location of substituents relative to the carboxyl group. Certain aromatic carboxylic acids have common names by which they are more usually known. For example, 2-hydroxybenzoic acid is more often called salicylic acid, a name derived from the fact that this aromatic carboxylic acid was first isolated from the bark of the willow, a tree of the genus *Salix*. 4-Aminobenzoic acid is a growth factor required by microorganisms for the synthesis of folic acid.

Benzoic acid ($C_6H_5CO_2H$) | 2-Hydroxybenzoic acid (Salicylic acid) | 4-Aminobenzoic acid (H_2N–C_6H_4–CO_2H)

Aromatic dicarboxylic acids are named by adding the words "dicarboxylic acid" to "benzene." Following are structural formulas for 1,2-benzenedicarboxylic acid and 1,4-benzenedicarboxylic acid. Each is more usually known by its common name: phthalic acid and terephthalic acid, respectively. Terephthalic acid is one of the two organic components required for the synthesis of the textile fiber known as Dacron polyester, or Dacron (Section 23.5B).

1,2-Benzenedicarboxylic acid (Phthalic acid) | 1,4-Benzenedicarboxylic acid (Terephthalic acid) (HO_2C–C_6H_4–CO_2H)

B. Common Names

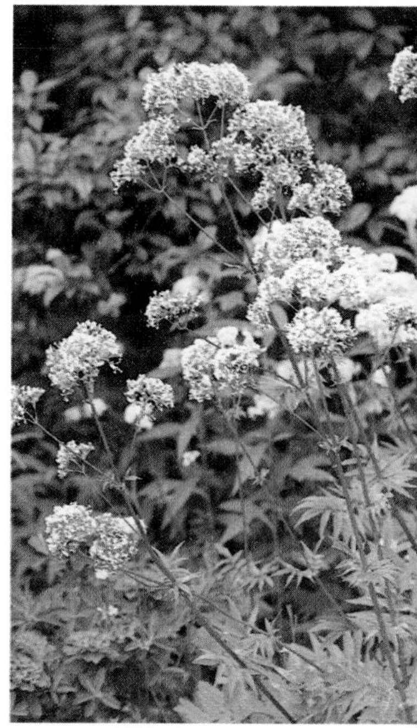

Garden heliotrope, a plant of the genus *Valeriana,* is a source of valeric acid. *(© Ann Reilly: PHOTO/NATS)*

Aliphatic carboxylic acids, many of which were known long before the development of structural theory and IUPAC nomenclature, are named according to their source or for some characteristic property. For example, formic acid was so named because it was first isolated from ants; acetic acid is the acidic component of vinegar; propionic acid was the first acid to be classified as a fatty acid; butyric acid was first isolated from butter; and valeric acid was first isolated from garden heliotrope, a plant of the genus *Valeriana* and native to Europe and Asia. Table 16.1 lists several of the unbranched aliphatic carboxylic acids found in the biological world along with the common name and the Latin or Greek derivation of each. Those of 16, 18, and 20 carbon atoms are particularly abundant in fats and oils (Section 25.1), and in the phospholipid components of biological membranes (Section 25.5).

When common names are used, the Greek letters α, β, γ, and δ are often used as prefixes to locate substituents. The α-position in a carboxylic acid is the one next to the carboxyl group; an α-substituent in a common name is equivalent to a 2-substituent in an IUPAC name. Following are two examples of the use of Greek letters in common names to show the position of substituents:

$$\overset{\delta}{\underset{5}{C}}-\overset{\gamma}{\underset{4}{C}}-\overset{\beta}{\underset{3}{C}}-\overset{\alpha}{\underset{2}{C}}-\underset{1}{C}(=O)-OH$$

$HOCH_2CH_2CH_2COOH$ — 4-Hydroxybutanoic acid (γ-Hydroxybutyric acid)

$CH_3CH(NH_2)COOH$ — 2-Aminopropanoic acid (α-Aminopropionic acid; Alanine)

TABLE 16.1 Several Aliphatic Carboxylic Acids and Their Common Names and Derivations

Structure	IUPAC Name	Common Name	Derivation
HCO_2H	methanoic acid	formic acid	Latin, *formica,* ant
CH_3CO_2H	ethanoic acid	acetic acid	Latin, *acetum,* vinegar
$CH_3CH_2CO_2H$	propanoic acid	propionic acid	Greek, *propion,* first fat
$CH_3(CH_2)_2CO_2H$	butanoic acid	butyric acid	Latin, *butyrum,* butter
$CH_3(CH_2)_3CO_2H$	pentanoic acid	valeric acid	Latin, *valeriana,* a flowering plant
$CH_3(CH_2)_4CO_2H$	hexanoic acid	caproic acid	Latin, *caper,* goat
$CH_3(CH_2)_6CO_2H$	octanoic acid	caprylic acid	Latin, *caper,* goat
$CH_3(CH_2)_8CO_2H$	decanoic acid	capric acid	Latin, *caper,* goat
$CH_3(CH_2)_{10}CO_2H$	dodecanoic acid	lauric acid	Latin, *laurus,* laurel
$CH_3(CH_2)_{12}CO_2H$	tetradecanoic acid	myristic acid	Greek, *myristikos,* fragrant
$CH_3(CH_2)_{14}CO_2H$	hexadecanoic acid	palmitic acid	Latin, *palma,* palm tree
$CH_3(CH_2)_{16}CO_2H$	octadecanoic acid	stearic acid	Greek, *stear,* solid fat
$CH_3(CH_2)_{18}CO_2H$	eicosanoic acid	arachidic acid	Greek, *arachis,* peanut

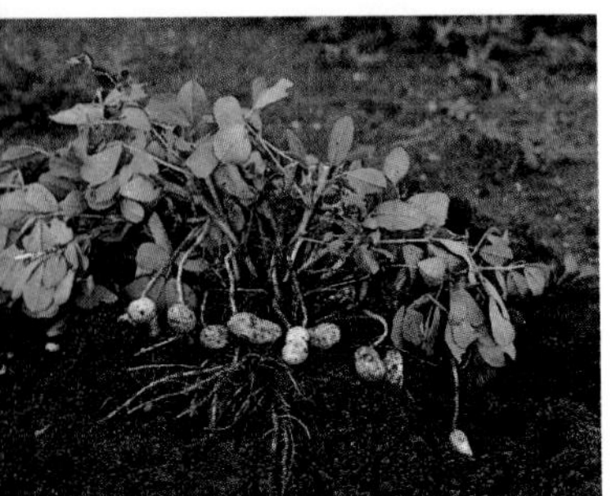

The peanut plant, from which arachidic acid was first isolated. *(Kenneth W. Fink/Photo Researchers, Inc.)*

In common names, the presence of a ketone carbonyl in a substituted carboxylic acid is indicated by the prefix keto-, illustrated by the common name β-ketobutyric acid.

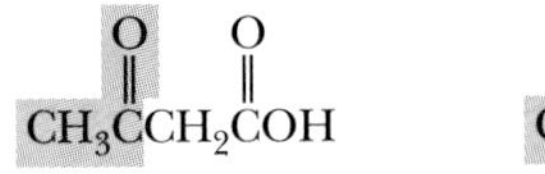

$CH_3C(=O)—$

3-Oxobutanoic acid (β-Ketobutyric acid; acetoacetic acid)

Acetyl group (Aceto group)

This substituted carboxylic acid is also named acetoacetic acid. In deriving this name, 3-oxobutanoic acid is regarded as a substituted acetic acid. In the common nomenclature, the substituent is named an **aceto group.**

Aceto group A $CH_3CO—$ group.

EXAMPLE 16.1

Write IUPAC names for the following carboxylic acids:

(a) $CH_3(CH_2)_7CH=CH(CH_2)_7CO_2H$ (cis; both H on the same side)

(b) 2-hydroxycyclohexane with CO_2H and OH (trans)

(c) $CH_3CH(OH)CO_2H$ (wedge structure with H, OH, CO_2H, CH_3)

(d) $ClCH_2CO_2H$

Solution

(a) (Z)-9-Octadecenoic acid **(b)** *trans*-2-Hydroxycyclohexanecarboxylic acid
(c) (R)-2-Hydroxypropanoic acid [(R)-lactic acid] **(d)** Chloroacetic acid

CHEMISTRY IN ACTION

Abscisic Acid: A Potent Plant Growth Regulator

Throughout their growth, plants are exposed to environmental stresses such as drought, waterlogged soil, pathogenic fungi, and feeding insects and animals. It is not surprising, therefore, that they have developed a chemistry to protect themselves against these stresses. Abscisic acid, one of these chemicals, is nearly universally distributed in higher plants and has a variety of actions.

OH CO_2H O

Abscisic acid

Abscisic acid promotes abscission (leaf fall), development of dormancy in buds, and formation of potato tubers. It has also been discovered that during times of water stress (lack of adequate water for growth), the concentration of abscisic acid in tomato plants increases by over 50-fold. Not only does the increased concentration of abscisic acid inhibit the hormones responsible for extension growth and cell division, but its presence in leaves also leads to closure of stoma and thus retards loss of water. Thus, plants protect themselves against times of drought by producing a chemical that conserves energy, reduces the rate of growth, and reduces loss of water.

See S. T. C. Wright and R. W. P. Hiron, *Nature* (London) **224,** 719, 1969.

PROBLEM 16.1

Each of these compounds has a well-recognized common name. A derivative of glyceric acid is an intermediate in glycolysis. Maleic acid is an intermediate in the tricarboxylic acid (TCA) cycle. Mevalonic acid is an intermediate in the biosynthesis of steroids. Write the IUPAC name for each compound. Be certain to specify configuration.

(a) CO_2H / H—C—OH / CH_2OH — Glyceric acid

(b) HO_2C, CO_2H / C=C / H, H — Maleic acid

(c) HO, CH_3 / C / $HOCH_2CH_2$, CH_2CO_2H — Mevalonic acid

16.3 Physical Properties

In the liquid and solid states, carboxylic acids are associated by hydrogen bonding into dimeric structures as shown for acetic acid in the liquid state.

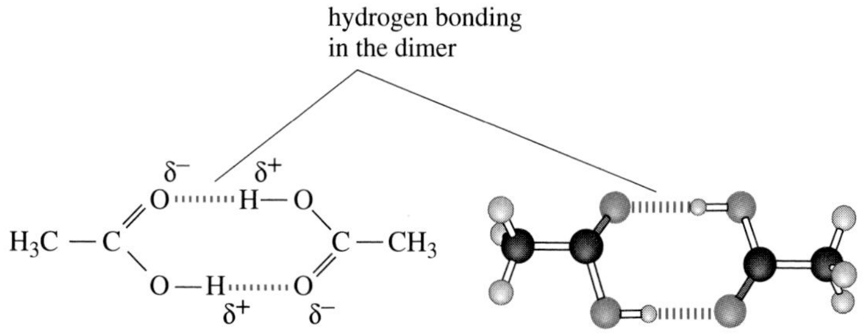

TABLE 16.2 Boiling Points of Selected Carboxylic Acids, Alcohols, and Aldehydes of Comparable Molecular Weight

Structure	Name	Molecular Weight	Boiling Point (°C)	Solubility (g/100 mL H_2O)
CH_3CO_2H	acetic acid	60.5	118	infinite
$CH_3CH_2CH_2OH$	1-propanol	60.1	97	infinite
CH_3CH_2CHO	propanal	58.1	48	16
$CH_3(CH_2)_2CO_2H$	butanoic acid	88.1	163	infinite
$CH_3(CH_2)_3CH_2OH$	1-pentanol	88.1	137	2.3
$CH_3(CH_2)_3CHO$	pentanal	86.1	103	slight
$CH_3(CH_2)_4CO_2H$	hexanoic acid	116.2	205	1.0
$CH_3(CH_2)_5CH_2OH$	1-heptanol	116.2	176	0.2
$CH_3(CH_2)_5CHO$	heptanal	114.1	153	0.1

Carboxylic acids have significantly higher boiling points than other types of organic compounds of comparable molecular weight, such as alcohols, aldehydes, and ketones. For example, butanoic acid (Table 16.2) has a higher boiling point than either 1-pentanol or pentanal. The higher boiling points of carboxylic acids result from their polarity and from the fact that they form very strong intermolecular hydrogen bonds.

Carboxylic acids also interact with water molecules by hydrogen bonding through both the carbonyl and hydroxyl groups. Because of greater hydrogen-bonding interactions, carboxylic acids are more soluble in water than alcohols, ethers, aldehydes, and ketones of comparable molecular weight. Like other compounds with polar functional groups, the solubility of a carboxylic acid in water decreases as its molecular weight increases. We account for this trend of solubility in water for carboxylic acids in the following way. A carboxylic acid consists of two regions of distinctly different polarity: a polar **hydrophilic** carboxyl group and, except for formic acid, a nonpolar **hydrophobic** hydrocarbon chain. The hydrophilic carboxyl group increases water solubility; the hydrophobic hydrocarbon chain decreases water solubility.

Hydrophilic From the Greek meaning "water-loving."

Hydrophobic From the Greek meaning "water-fearing."

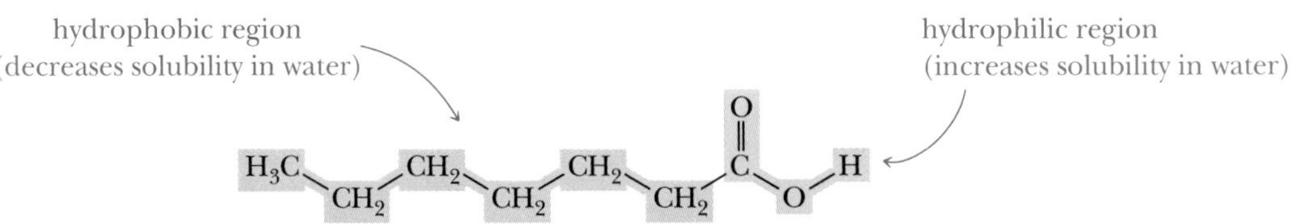

The first four aliphatic carboxylic acids (formic, acetic, propanoic, and butanoic acids) are infinitely soluble in water because the hydrophobic character of the hydrocarbon chain is more than counterbalanced by the hydrophilic character of the carboxyl group. As the size of the hydrocarbon chain increases relative to the size of the hydrophilic group, water solubility decreases. The solubility of hexanoic acid in water is 1.0 g/100 mL water. That of decanoic acid is only 0.2 g/100 mL water.

16.4 Acidity

A. Acid Ionization Constants

Carboxylic acids are weak acids. Values of K_a for most unsubstituted aliphatic and aromatic carboxylic acids fall within the range 10^{-4} to 10^{-5}. The value of K_a for acetic acid, for example, is 1.74×10^{-5}. Its pK_a is 4.76.

$$CH_3CO_2H + H_2O \rightleftharpoons CH_3CO_2^- + H_3O^+ \qquad K_a = \frac{[CH_3CO_2^-][H_3O^+]}{[CH_3CO_2H]} = 1.74 \times 10^{-5}$$

$$pK_a = 4.76$$

As we discussed in Section 3.3, two factors contribute to the greater acidity of carboxylic acids (pK_a 4–5) compared with alcohols (pK_a 15–18). First, the electron-withdrawing inductive effect of the carbonyl group weakens the adjacent O—H bond and facilitates its ionization. Second, resonance stabilizes the carboxylate anion by delocalizing its negative charge. Neither of these effects exists in alcohols.

We see another illustration of inductive effects in the fact that electron-withdrawing substituents near the carboxyl group increase the acidity of carboxylic acids, often by several orders of magnitude. Compare, for example, the acidities of acetic acid and the halogen-substituted acetic acids. As the electronegativity of the halogen increases, its inductive effect increases, and the strength of the halogen-substituted acid increases. Fluoroacetic acid is the strongest of the monohalogenated acetic acids, while iodoacetic acid is the weakest.

Formula:	CH_3CO_2H	ICH_2CO_2H	$BrCH_2CO_2H$	$ClCH_2CO_2H$	FCH_2CO_2H
Name:	Acetic acid	Iodoacetic acid	Bromoacetic acid	Chloroacetic acid	Fluoroacetic acid
pK_a:	4.76	3.18	2.90	2.86	2.59

Increasing acid strength →

To see the effects of multiple halogen substitution, compare the values of pK_a for acetic acid with its mono-, di-, and trichloro- derivatives. A single chlorine substituent increases acid strength by nearly 100. Trichloroacetic acid, the strongest of the three acids, is a stronger acid than H_3PO_4.

Formula:	CH_3CO_2H	$ClCH_2CO_2H$	Cl_2CHCO_2H	Cl_3CCO_2H
Name:	Acetic acid	Chloroacetic acid	Dichloroacetic acid	Trichloroacetic acid
pK_a:	4.76	2.86	1.48	0.70

Increasing acid strength →

The inductive effect of halogen substitution falls off rapidly as its distance from the carboxyl group increases. Although the acid ionization constant for 2-chlorobutanoic acid is 100 times that for butanoic acid, the acid ionization constant for 4-chlorobutanoic is only about twice that for butanoic acid.

Formula:	$CH_3CH_2CH_2CO_2H$	$ClCH_2CH_2CH_2CO_2H$	$CH_3CH(Cl)CH_2CO_2H$	$CH_3CH_2CH(Cl)CO_2H$
Name:	Butanoic acid	4-Chlorobutanoic acid	3-Chlorobutanoic acid	2-Chlorobutanoic acid
pK_a:	4.82	4.52	3.98	2.83

Increasing acid strength →

We also see an example of the importance of the inductive effect in a comparison of the relative acidities of benzoic acid and acetic acid. Because of the stronger electron-withdrawing inductive effect of the sp^2 hybridized carbon of the benzene ring compared with the sp^3 hybridized carbon of the methyl group of acetic acid, benzoic acid is a stronger acid than acetic acid; its K_a is approximately four times that of acetic acid.

$C_6H_5-C(=O)-OH$ Benzoic acid pK_a 4.19

$CH_3-C(=O)-OH$ Acetic acid pK_a 4.76

EXAMPLE 16.2

Which is the stronger acid in each pair?

(a) $HC{\equiv}CH$ (Acetylene) or $HC{\equiv}N$ (Hydrogen cyanide)

(b) $CH_3CH(OH)COOH$ (2-Hydroxypropanoic acid (Lactic acid)) or CH_3CH_2COOH (Propanoic acid)

Solution

(a) Hydrogen cyanide ($pK_a = 9.31$) is a considerably stronger acid than acetylene ($pK_a = 25$), because of the electron-withdrawing inductive effect of the adjacent nitrogen atom on the C—H bond.

(b) 2-Hydroxypropanoic acid ($pK_a = 3.08$) is a stronger acid than propanoic acid ($pK_a = 4.87$) because of the electron-withdrawing inductive effect of the hydroxyl oxygen.

CHEMISTRY IN ACTION

New Organic Acids

Carboxylic acids are the major type of organic acids. Delocalization of the negative charge in a carboxylate anion over two electronegative oxygen atoms provides the driving force for ionization. The same kind of charge delocalization, but on a different carbon framework, can produce exceptionally strong organic acids. The following organic acids all have the formula $(C_nO_n)H_2$.

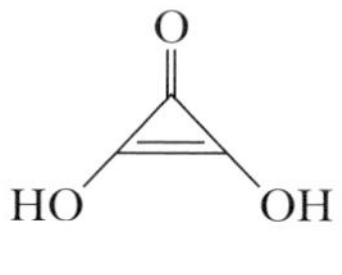

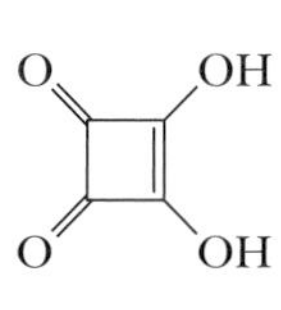

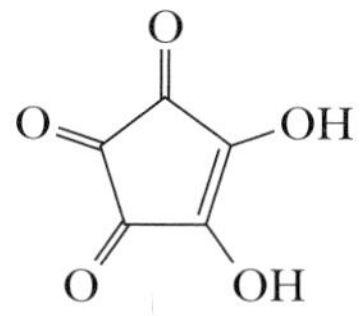

	Deltic acid	Squaric acid	Croconic acid
pK_{a1}	2.5	0.5	0.8
pK_{a2}	6.0	3.5	2.2

Although croconic acid has been known for almost 100 years, squaric acid and deltic acid were first made in 1959 and 1976, respectively. On loss of its two protons, each acid produces a dianion in which the negative charge is highly delocalized. The pK_{a1} values of squaric and croconic acids are comparable to those of phosphoric acid (H_3PO_4, pK_{a1} 2.1), trichloroacetic acid (CCl_3CO_2H, pK_a 0.7), and trifluoroacetic acid (CF_3CO_2H, pK_a 0.2).

Structural studies of squaric acid indicate that its dianion is highly symmetrical, with equal C—C bond lengths and equal C—O bond lengths. The dotted lines in the structure on the right represent the extensive delocalization of electrons.

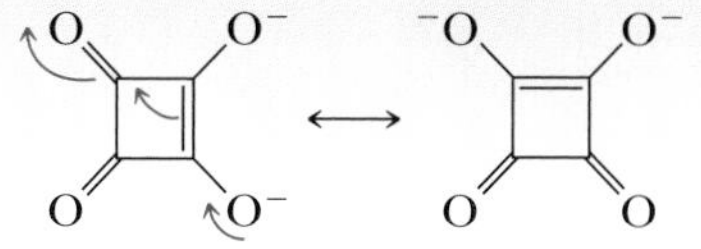

Two of four equivalent contributing structures

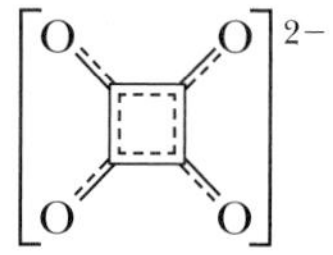

Resonance-stabilized dianion; equal C—C bond lengths and C—O bond lengths

Recently, these acids, especially deltic acid, have been used in a novel way by inorganic chemists. In the solid state, molecules of deltic acid exist as layered sheets, stacked one upon another. Many useful inorganic materials also exist in the solid state as layered sheets. In attempts to make new layered structures, salts of various metal cations and deltic acid anions have been prepared to determine if the flat deltic acid anion can promote the crystallization of new layered materials.

D. Eggerding and R. West, *J. Am. Chem. Soc.*, **98,** 3641, 1976.

PROBLEM 16.2

Which is the stronger acid in each pair?

(a) CH_3CO_2H (Acetic acid) or CH_3SO_3H (Methanesulfonic acid)

(b) CH_3CCO_2H (with C=O on the central carbon) (2-Oxopropanoic acid (Pyruvic acid)) or $CH_3CH_2CO_2H$ (Propanoic acid)

Sodium benzoate and calcium propanoate are used as preservatives in baked goods. *(Charles D. Winters)*

B. Reaction with Bases

All carboxylic acids, whether soluble or insoluble in water, react with NaOH, KOH, and other strong bases to form water-soluble salts.

$$C_6H_5CO_2H + NaOH \xrightarrow[H_2O]{} C_6H_5CO_2^-\,Na^+ + H_2O$$

Benzoic acid (slightly soluble in water) — Sodium benzoate (60 g/100 mL water)

Sodium benzoate is a fungal growth inhibitor and is often added to baked goods "to retard spoilage." Calcium propanoate is also used for the same purpose. Carboxylic acids also form water-soluble salts with ammonia and amines.

$$C_6H_5CO_2H + NH_3 \xrightarrow[H_2O]{} C_6H_5CO_2^-\,NH_4^+$$

Benzoic acid (slightly soluble in water) — Ammonium benzoate (20 g/100 mL water)

As described in Section 3.4, carboxylic acids react with sodium bicarbonate and sodium carbonate to form water-soluble sodium salts and carbonic acid (a weaker acid). Carbonic acid, in turn, decomposes to give water and carbon dioxide, which evolves as a gas.

$$CH_3CO_2H + NaHCO_3 \longrightarrow CH_3CO_2^-Na^+ + CO_2 + H_2O$$

Salts of carboxylic acids are named in the same manner as salts of inorganic acids; the cation is named first and then the anion. The name of the anion is derived from the name of the carboxylic acid by dropping the suffix **-ic acid** and adding the suffix **-ate.**

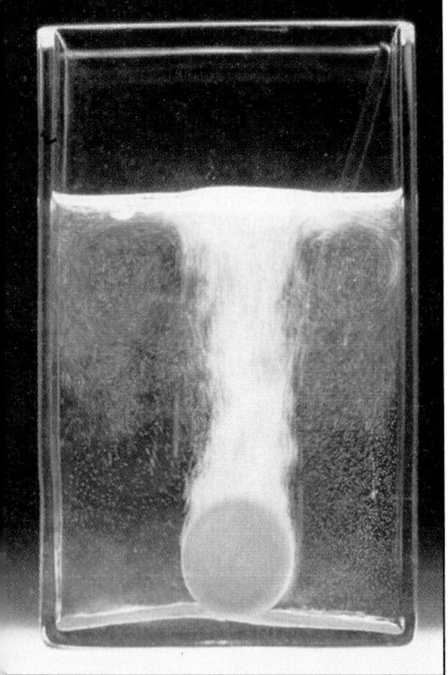

A commercial remedy for excess stomach acid. The bubbles are carbon dioxide, CO_2, from the reaction between citric acid, $C_6H_8O_7$, and sodium bicarbonate, $NaHCO_3$. *(Charles D. Winters)*

Example 16.3

Complete these acid-base reactions and name each carboxylic salt.

(a) $CH_3(CH_2)_2CO_2H + NaOH \longrightarrow$ **(b)** $CH_3CH(OH)CO_2H + NaHCO_3 \longrightarrow$

Solution

Each carboxylic acid is converted to its sodium salt. In (b), carbonic acid is formed, which decomposes to carbon dioxide and water.

(a) $CH_3(CH_2)_2CO_2H + NaOH \longrightarrow CH_3(CH_2)_2CO_2^-Na^+ + H_2O$

Butanoic acid — Sodium butanoate

(b) $CH_3CH(OH)CO_2H + NaHCO_3 \longrightarrow CH_3CH(OH)CO_2^-Na^+ + H_2O + CO_2$

2-Hydroxypropanoic acid (Lactic acid) — Sodium 2-hydroxypropanoate (Sodium lactate)

PROBLEM 16.3

Write equations for the reaction of each acid in Example 16.3 with ammonia and name the carboxylic salt formed.

A consequence of the water solubility of carboxylic acid salts is that water-insoluble carboxylic acids can be converted to water-soluble ammonium or alkali metal salts and extracted into aqueous solution. The salt, in turn, can be transformed back to the free carboxylic acid by addition of HCl, H_2SO_4, or other strong acid. These reactions allow an easy separation of carboxylic acids from water-insoluble neutral compounds.

Shown in Figure 16.1 is a flowchart for separation of benzoic acid, a water-insoluble carboxylic acid, from benzyl alcohol, a nonacidic compound. First, the mixture of benzoic acid and benzyl alcohol is dissolved in diethyl ether. When the ether solution is shaken with aqueous NaOH or other strong base, benzoic acid is converted to its water-soluble salt. Then the ether and aqueous phases are separated. The ether solution is distilled, yielding first diethyl ether (bp 35°C) and then benzyl alcohol

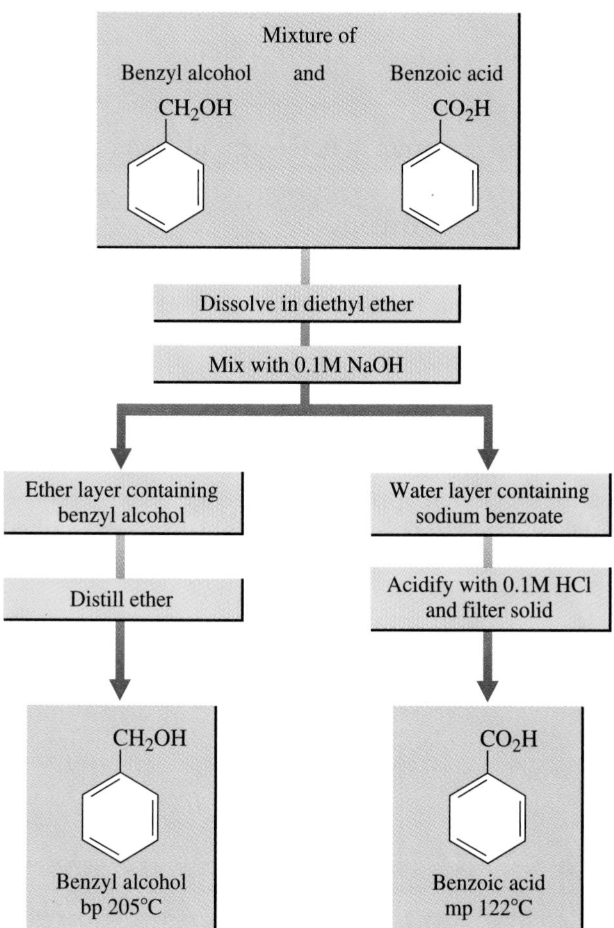

FIGURE 16.1
Flowchart for separation of benzoic acid from benzyl alcohol.

(bp 205°C). The aqueous solution is acidified with HCl, and benzoic acid precipitates as a crystalline solid (mp 122°C) and is recovered by filtration.

16.5 Spectroscopic Properties

A. Mass Spectrometry

The molecular ion peak from a carboxylic acid is generally observed, although it is often very weak. The most common fragmentation patterns are α-cleavage of the carboxyl group to give the ion $[CO_2H]^+$ of m/z 45 and McLafferty rearrangement. The base peak is very often that due to the McLafferty rearrangement product.

$$\left[CH_3CH_2CH_2-\overset{\overset{\displaystyle O}{\|}}{C}-OH\right]^{+\cdot} \xrightarrow{\alpha\text{-cleavage}} CH_3CH_2CH_2\cdot + :\ddot{O}=C=\overset{+}{\ddot{O}}-H$$

m/z 45

$$\left[\begin{matrix} H_2C(-H) & & O \\ | & & \| \\ H_2C & -CH_2- & C-OH \end{matrix}\right]^{+\cdot} \xrightarrow{\text{McLafferty rearrangement}} \begin{matrix} H_2C \\ \| \\ H_2C \end{matrix} + \left[\begin{matrix} H-O \\ | \\ CH_2=C-OH \end{matrix}\right]^{+\cdot}$$

m/z 60

Each of these patterns is seen in the mass spectrum of butanoic acid (Figure 16.2).

B. Nuclear Magnetic Resonance Spectroscopy

Hydrogens on the α-carbon to a carboxyl group give a signal in the ^{1}H-NMR spectrum in the range δ 2.0–2.5. The hydrogen of the carboxyl group appears in the range δ 10 to 13. The chemical shift of the carboxyl hydrogen is large, even larger than the chemical shift of an aldehyde hydrogen (δ 9.5–10.1), and serves to distinguish carboxyl hydrogens from most other types of hydrogens. The ^{1}H-NMR signal for the carboxyl hydrogen of 2-methylpropanoic acid appears at δ 12.2 and is shown at the left in Figure 16.3.

The ^{13}C absorption of the carboxyl carbon appears in the range δ 160–180 and is similar to that of the carboxyl derivatives (esters, amides, and anhydrides) described in Section 17.3B.

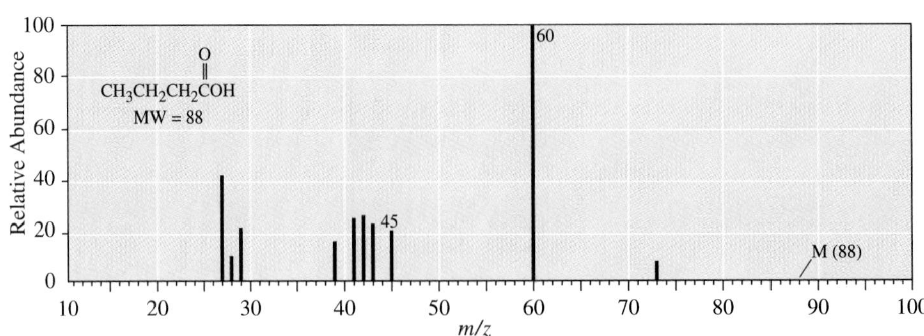

FIGURE 16.2 Mass spectrum of butanoic acid. Common fragmentation patterns of carboxylic acids are α-cleavage to give the ion $[CO_2H]^+$ of m/z 45 and McLafferty rearrangement.

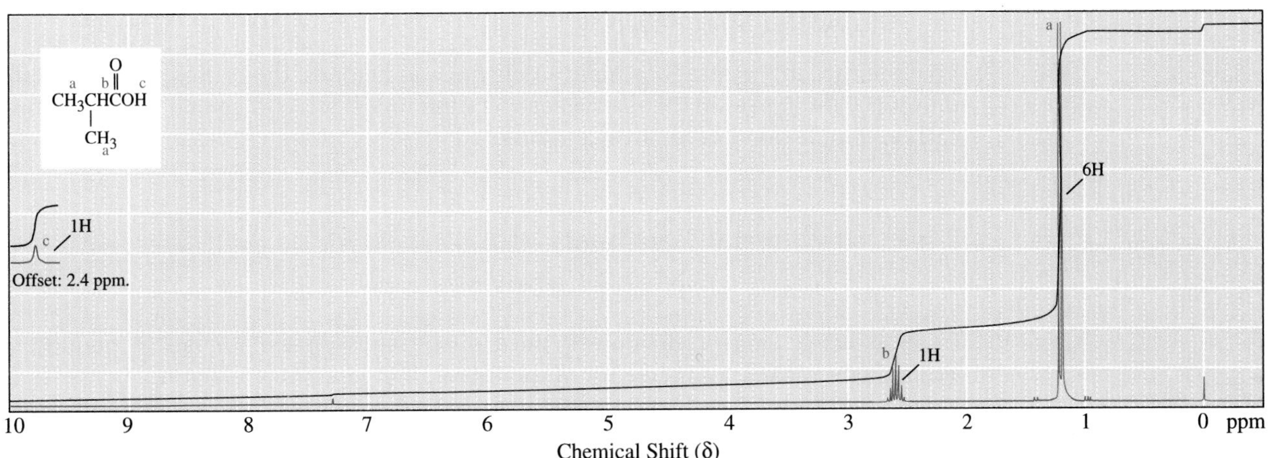

FIGURE 16.3
^{1}H-NMR spectrum of 2-methylpropanoic acid (isobutyric acid).

C. Infrared Spectroscopy

The carboxyl group gives rise to two characteristic absorptions in the infrared spectrum. One of these occurs in the region 1700–1725 cm^{-1} and is associated with the stretching vibration of the carbonyl group. This is essentially the same range of absorption as for the carbonyl group of aldehydes and ketones but the peak is usually broader in the case of the carboxyl carbonyl because of intermolecular hydrogen bonding. The other infrared absorption characteristic of the carboxyl group is a peak between 2400 and 3400 cm^{-1} due to the stretching vibration of the O—H group, which often overlaps the C—H stretching absorptions. This absorption is generally very broad due to hydrogen bonding between molecules of the carboxylic acid. Both C=O and O—H stretchings can be seen in the infrared spectrum of pentanoic acid in Figure 16.4.

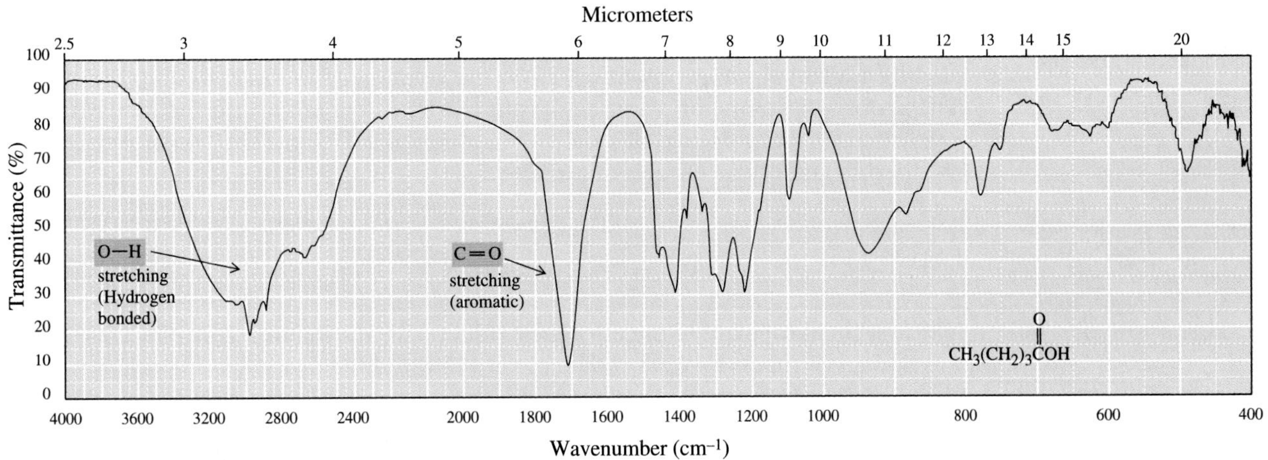

FIGURE 16.4
IR spectrum of pentanoic acid.

D. Ultraviolet Spectroscopy

Like the carbonyl groups of simple aldehydes and ketones, the carboxyl group shows only weak absorption in the ultraviolet spectrum unless it is conjugated with a carbon-carbon double bond or an aromatic ring.

16.6 Industrial Synthesis of Acetic Acid: Transition Metal Catalysis

In 1995, acetic acid production in the United States totaled 4.7 billion pounds, a volume that ranked it nineteenth in the list of all organics manufactured by the U.S. chemical industry. The first industrial synthesis of acetic acid was commercialized in 1916 in Canada and Germany, using acetylene as a feedstock. The process involved two stages: (1) hydration of acetylene to acetaldehyde, followed by (2) oxidation of acetaldehyde to acetic acid by molecular oxygen, catalyzed by cobalt(III) acetate.

$$\underset{\text{Acetylene}}{HC{\equiv}CH} + H_2O \xrightarrow[HgSO_4]{H_2SO_4} \underset{\text{Enol of acetaldehyde}}{[CH_2{=}CHOH]} \longrightarrow \underset{\text{Acetaldehyde}}{CH_3\overset{\overset{O}{\|}}{C}H} \xrightarrow[Co^{3+}]{O_2} \underset{\text{Acetic acid}}{CH_3\overset{\overset{O}{\|}}{C}OH}$$

The light of one kind of headlamp used by miners and cave explorers is given off by the combustion of acetylene, which is produced by the slow addition of water to calcium carbide. *(Charles D. Winters)*

The technology of producing acetic acid from acetylene is simple and yields are high, factors that made this the major route to acetic acid for over 50 years. Acetylene was prepared by the reaction of calcium carbide with water. Calcium carbide, in turn, was prepared by heating calcium oxide (from limestone, $CaCO_3$) with coke (from coal) to between 2000° and 2500°C in an electric furnace.

$$\underset{\text{Calcium oxide}}{CaO} + 2C \xrightarrow{2500°C} \underset{\text{Calcium carbide}}{CaC_2} \xrightarrow{2H_2O} \underset{\text{Acetylene}}{HC{\equiv}CH} + Ca(OH)_2$$

This procedure required enormous amounts of energy and, as the cost of energy rose, acetylene ceased to be an economical feedstock from which to manufacture acetic acid.

As an alternative feedstock, chemists and chemical engineers turned to ethylene, already available in huge quantities from the refining of natural gas and petroleum. The process of producing acetic acid from ethylene depends on the fact, known since 1894, that in the presence of catalytic amounts of Pd^{2+} and Cu^{2+} salts, ethylene is oxidized by molecular oxygen to acetaldehyde.

$$2CH_2{=}CH_2 + O_2 \xrightarrow[\text{(Wacker process)}]{Pd^{2+},\ Cu^{2+}} 2CH_3\overset{\overset{O}{\|}}{C}H$$

The first chemical plant to use ethylene oxidation for the manufacture of acetaldehyde was built in Germany by Wacker-Chemie in 1959, and the process itself became known as the **Wacker process.**

In 1973, the Monsanto Company in the United States developed an entirely new synthesis for acetic acid based on the reaction of methanol with carbon monoxide in the presence of small amounts of soluble rhodium(III) salts, HI and H_2O.

$$CH_3OH + CO \xrightarrow[\text{(Monsanto process)}]{Rh^{3+},\ HI,\ H_2O} CH_3\overset{\overset{\large O}{\|}}{C}OH$$

A plant for the production of acetic acid from methanol and carbon monoxide. *(Courtesy of B.P. America)*

The **Monsanto process** for low-pressure carbonylation is based on methanol, a compound readily available by catalytic reduction of carbon monoxide. Carbon monoxide and hydrogen, a mixture called **synthesis gas** (Section 2.9C), is in turn available from the reaction of water with methane, coal, and various petroleum products. Synthesis gas is now the major source of both methanol and acetic acid, and it is likely that it will be a major feedstock for the production of other organics in the decades ahead.

16.7 Reduction

The carboxyl group is one of the organic functional groups most resistant to reduction. It is not affected by catalytic hydrogenation under conditions that easily reduce aldehydes and ketones to alcohols and that reduce alkenes and alkynes to alkanes. The most common reagent for the reduction of carboxylic acids to primary alcohols is the very powerful reducing agent lithium aluminum hydride (Section 15.14B).

A. Lithium Aluminum Hydride

Lithium aluminum hydride, $LiAlH_4$, reduces a carboxylic acid to a primary alcohol in excellent yield. Reduction is carried out in diethyl ether, tetrahydrofuran (THF), or some other nonreactive, aprotic solvent. The initial product is an aluminum alkoxide, which is then treated with water to give the primary alcohol and lithium and aluminum hydroxides. These hydroxides are insoluble in diethyl ether or THF and are removed by filtration. Evaporation of the solvent then yields the primary alcohol.

$$\text{3-Cyclopentenecarboxylic acid} \xrightarrow[2.\ H_2O]{1.\ LiAlH_4,\ \text{ether}} \text{4-Hydroxymethylcyclopentene } (-CH_2OH) + LiOH + Al(OH)_3$$

3-Cyclopentene-carboxylic acid — 4-Hydroxymethyl-cyclopentene

Alkenes are generally not affected by metal hydride reducing reagents. These reagents function as hydride ion donors, that is, as nucleophiles, and alkenes are not attacked by nucleophiles. The less reactive sodium borohydride (Section 15.14B) does not reduce carboxylic acids. In the reduction of a carboxyl group by lithium aluminum hydride, the first hydride ion reacts with the carboxyl hydrogen to give H_2. The three remaining hydride ions are then used for carboxyl reduction.

Following are balanced equations for treatment of a carboxylic acid with LAH to form a tetraalkyl aluminate ion, followed by its hydrolysis in water.

$$4R\overset{\overset{\large O}{\|}}{C}OH + 3LiAlH_4 \longrightarrow [(RCH_2O)_4Al]Li + 2LiAlO_2 + 4H_2$$

$$[(RCH_2O)_4Al]Li + 4H_2O \longrightarrow 4RCH_2OH + LiOH + Al(OH)_3$$

PROBLEM 16.5

Complete the following equations:

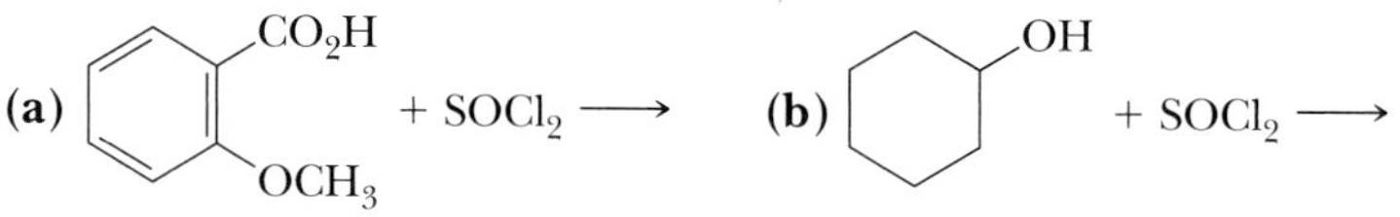

(a) 2-methoxybenzoic acid (CO_2H, OCH_3) $+ SOCl_2 \longrightarrow$ (b) cyclohexanol (OH) $+ SOCl_2 \longrightarrow$

16.10 Decarboxylation

A. β-Ketoacids

Decarboxylation Loss of CO_2 from a carboxyl group.

Decarboxylation is the loss of CO_2 from the carboxyl group of a molecule. Almost any carboxylic acid, heated to a very high temperature, undergoes thermal decarboxylation.

$$\mathrm{R\overset{O}{\overset{\|}{C}}OH} \xrightarrow{\text{decarboxylation}} \mathrm{RH} + \mathrm{CO_2}$$

Most carboxylic acids, however, are quite resistant to moderate heat and melt or even boil without decarboxylation. Exceptions are carboxylic acids that have a carbonyl group β to the carboxyl group. This type of carboxylic acid undergoes decarboxylation quite readily on mild heating. For example, when 3-oxobutanoic acid is heated moderately, it undergoes decarboxylation to give acetone and carbon dioxide.

$$\mathrm{CH_3{-}\overset{O}{\overset{\|}{C}}{-}CH_2{-}\overset{O}{\overset{\|}{C}}{-}OH} \xrightarrow{\text{warm}} \mathrm{CH_3{-}\overset{O}{\overset{\|}{C}}{-}CH_3} + \mathrm{CO_2}$$

3-Oxobutanoic acid (Acetoacetic acid) (β and α carbons labeled) → Acetone

Decarboxylation on mild heating is a unique property of 3-oxocarboxylic acids (β-ketoacids) and is not observed with other classes of ketoacids.

The mechanism of decarboxylation of a β-ketoacid involves rearrangement of six electrons in a cyclic six-membered transition state to give carbon dioxide and an enol. The enol is then converted to the keto form by keto-enol tautomerism (Section 15.11B).

MECHANISM Decarboxylation of a β-Ketocarboxylic Acid

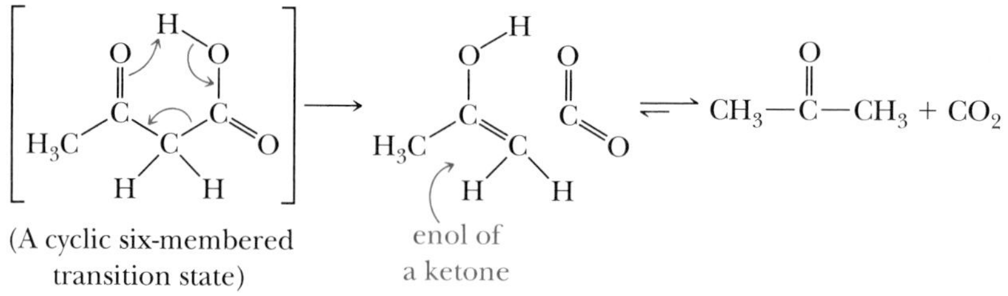

[Transition state structure] $\longrightarrow$ enol + $O{=}C{=}O$ $\rightleftharpoons \mathrm{CH_3{-}\overset{O}{\overset{\|}{C}}{-}CH_3} + \mathrm{CO_2}$

(A cyclic six-membered transition state)

enol of a ketone

(c) $CH_3CH_2\overset{\displaystyle OH}{\overset{|}{C}}{=}CHCH_3 \longrightarrow CH_3CH_2\overset{\displaystyle O}{\overset{\|}{C}}CH_2CH_3 + CO_2$

PROBLEM 16.6

Account for the observation that the following β-ketoacid does not u
boxylation. It can be heated for extended periods of time at tempera
melting point without noticeable decomposition.

2-Oxobicyclo[2.2.1]heptane-1-carboxylic acid

SUMMARY

The functional group of a **carboxylic acid** (Section 16.1) is the **carboxyl group, —CO_2H.** IUPAC names of carboxylic acids (Section 16.2A) are derived from the parent alkane by dropping the suffix -e and adding **-oic acid.** Dicarboxylic acids are named as **-dioic acids.**

Carboxylic acids are polar compounds (Section 16.3) and in the liquid and solid states are associated by hydrogen bonding into dimers. Carboxylic acids have higher boiling points and are more soluble in water than alcohols, aldehydes, ketones, and ethers of comparable molecular weight. A carboxylic acid consists of two regions of distinctly different polarity; a polar, **hydrophilic** carboxyl group, which increases solubility in water, and a nonpolar **hydrophobic** hydrocarbon chain, which decreases solubility in water. The first four aliphatic carboxylic

acids are inf
boxyl group
drocarbon c
ever, the hyc
in water dec

Values c
4.0 to 5.0 (S
ids compare
withdrawing
ens the O—
resonance st
alkoxide anic
boxyl group i

KEY REACTIONS

1. Acidity of Carboxylic Acids (Section 16.4A)

Values of pK_a for most unsubstituted aliphatic and aromatic carboxylic acids ar
range pK_a 4–5. Substitution by electron-withdrawing groups decreases pK_a (inc
ity).

$$CH_3\overset{O}{\overset{\|}{C}}OH + H_2O \rightleftharpoons CH_3\overset{O}{\overset{\|}{C}}O^- + H_3O^+ \qquad K_a = 1.74 \times 10^{-5}$$

CHEMISTRY IN ACTION

Ketone Bodies and Diabetes Mellitus

3-Oxobutanoic acid (acetoacetic acid) and its reduction product, 3-hydroxybutanoic acid are synthesized in the liver from acetyl-CoA, a product of the metabolism of fatty acids and certain amino acids. 3-Hydroxybutanoic acid and 3-oxobutanoic acid are known collectively as ketone bodies. The concentration of ketone bodies in the blood of healthy, well-fed humans is approximately 0.01 mM/L. However, in persons suffering from starvation or diabetes mellitus, the concentration of ketone bodies may increase to as much as 500 times normal. Under these conditions, the concentration of acetoacetic acid increases to the point where it undergoes spontaneous decarboxylation to form acetone and carbon dioxide. Acetone is not metabolized by humans and is excreted through the kidneys and the lungs. The odor of acetone is responsible for the characteristic "sweet smell" of the breath of severely diabetic patients.

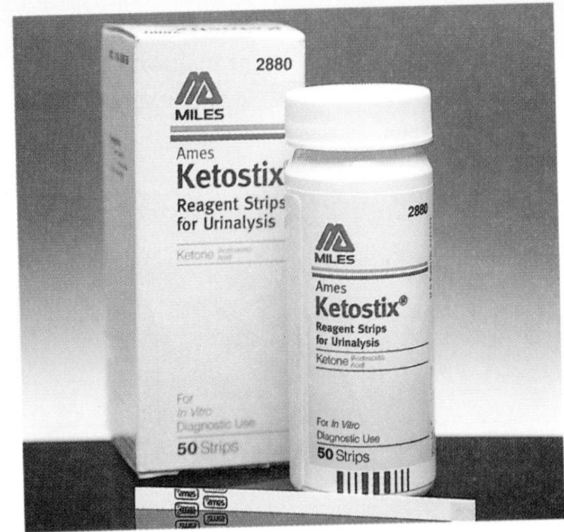

Pharmacy test kit for presence of ketone bodies in urine. (Charles D. Winters)

An important example of decarboxylation of a β-ketoacid in the biological world occurs during the oxidation of foodstuffs in the TCA cycle. One of the intermediates in this cycle is oxalosuccinic acid, which undergoes spontaneous decarboxylation to produce α-ketoglutaric acid. Only one of the three carboxyl groups of oxalosuccinic acid has a carbonyl group in the β-position to it, and it is this carboxyl group that is lost as CO_2.

Oxalosuccinic acid ($HO\overset{O}{\overset{\|}{C}}$—CH($CH_2CO_2H$)—$\overset{O}{\overset{\|}{C}}CO_2H$) ⟶ α-Ketoglutaric acid (CH_2CO_2H—CH_2—$\overset{O}{\overset{\|}{C}}CO_2H$) + CO_2

Only this carboxyl has a C=O beta to it

B. Decarboxylation of Malonic Acid and Substituted Malonic Acids

The presence of a ketone or aldehyde carbonyl group beta to the carboxyl group is sufficient to facilitate decarboxylation. In the more general reaction, decarboxylation is facilitated by the presence of any carbonyl group at the β position, including that of a carboxyl group or ester. Malonic acid and substituted malonic acids, for example, undergo decarboxylation on heating, as illustrated by the decarboxylation of malonic acid when it is heated slightly above its melting point of 135°–137°C.

$HO\overset{O}{\overset{\|}{C}}CH_2\overset{O}{\overset{\|}{C}}OH$ $\xrightarrow{140}$

Propanedioic acid
(Malonic acid)

The mechanism of decarboxylation of ı
just seen for the decarboxylation of β-ke
transition state involving rearrangement
a carboxylic acid, which is in turn isome

MECHANISM Decarboxylation o

A cyclic six-membered transition state

enol o
carboxylic

EXAMPLE 16.6

Each of these carboxylic acids undergoes t
formula for the enol intermediate and fin

(a) (b) CO_2H CO_2H (c

Solution

Following is a structural formula for the en
each decarboxylation:

(a) Enol intermediate $\longrightarrow$ + CO_2 (b) En

potassium sorbate is potassium (*E,E*)-2,4-hexadienoate. Draw the structural formula of potassium sorbate.

16.13 Zinc 10-undecenoate, the zinc salt of 10-undecenoic acid, is used to treat certain fungal infections, particularly *tinea pedis* (athlete's foot). Draw the structural formula of this zinc salt.

16.14 On a cyclohexane ring, an axial carboxyl group has a conformational energy of +1.4 kcal/mol (5.9 kJ/mol) relative to an equatorial carboxyl group. Consider the equilibrium for the alternative chair conformations of *trans*-1,4-cyclohexanedicarboxylic acid. Draw the less stable chair conformation on the left of the equilibrium arrows and the more stable chair on the right. Calculate ΔG^0 for the equilibrium as written and calculate the ratio of more stable chair to less stable chair at 25°C.

Physical Properties

16.15 Arrange the compounds in each set in order of increasing boiling point:

(a) $CH_3(CH_2)_5\overset{O}{\overset{\|}{C}}OH$ $\quad CH_3(CH_2)_6\overset{O}{\overset{\|}{C}}H$ $\quad CH_3(CH_2)_6CH_2OH$

(b) $CH_3CH_2\overset{O}{\overset{\|}{C}}OH$ $\quad CH_3CH_2CH_2CH_2OH$ $\quad CH_3CH_2OCH_2CH_3$

16.16 Acetic acid has a boiling point of 118°C, whereas its methyl ester has a boiling point of 57°C. Account for the fact that the boiling point of acetic acid is higher than that of its methyl ester, even though acetic acid has a lower molecular weight.

Spectroscopy

16.17 Account for the presence of peaks at *m/z* 135 and 107 in the mass spectrum of 4-methoxybenzoic acid (*p*-anisic acid).

16.18 Account for the presence of the following peaks in the mass spectrum of hexanoic acid.

(a) *m/z* 60
(b) A series of peaks differing by 14 mass units at *m/z* 45, 59, 73, and 87
(c) A series of peaks differing by 14 mass units at *m/z* 29, 43, 57, and 71

16.19 Given here are ^{1}H-NMR and ^{13}C-NMR spectral data for ten compounds. Each compound shows strong absorption between 1720 and 1700 cm^{-1}, and strong, broad absorption over the region 2500–3500 cm^{-1}. Propose a structural formula for each compound.

(a) $C_5H_{10}O_2$

^{1}H-NMR	^{13}C-NMR
0.94 (t, 3H)	180.71
1.39 (m, 2H)	33.89
1.62 (m, 2H)	26.76
2.35 (t, 2H)	22.21
12.0 (s, 1H)	13.69

(b) $C_6H_{12}O_2$

^{1}H-NMR	^{13}C-NMR
1.08 (s, 9H)	179.29
2.23 (s, 2H)	47.82
12.1 (s, 1H)	30.62
	29.57

(c) $C_5H_8O_4$

^{1}H-NMR	^{13}C-NMR
0.93 (t, 3H)	170.94
1.80 (m, 2H)	53.28
3.10 (t, 1H)	21.90
12.7 (s, 2H)	11.81

(d) $C_5H_8O_4$

^{1}H-NMR	^{13}C-NMR
1.29 (s, 6H)	174.01
12.8 (s, 2H)	48.77
	22.56

(e) $C_4H_6O_2$

^{1}H-NMR	^{13}C-NMR
1.91 (d, 3H)	172.26
5.86 (d, 1H)	147.53
7.10 (m, 1H)	122.24
12.4 (s, 1H)	18.11

(f) $C_3H_4Cl_2O_2$

^{1}H-NMR	^{13}C-NMR
2.34 (s, 3H)	171.82
11.3 (s, 1H)	79.36
	34.02

(g) $C_5H_8Cl_2O_2$

^{1}H-NMR	^{13}C-NMR
1.42 (s, 6H)	180.15
6.10 (s, 1H)	77.78
12.4 (s, 1H)	51.88
	20.71

(h) $C_5H_9BrO_2$

^{1}H-NMR	^{13}C-NMR
0.97 (t, 3H)	176.36
1.50 (m, 2H)	45.08
2.05 (m, 2H)	36.49
4.25 (t, 1H)	20.48
12.1 (s, 1H)	13.24

(i) $C_4H_8O_3$

^{1}H-NMR	^{13}C-NMR
2.62 (t, 2H)	177.33
3.38 (s, 3H)	67.55
3.68 (s, 2H)	58.72
11.5 (s, 1H)	34.75

Preparation of Carboxylic Acids

16.20 Complete these reactions:

(a) 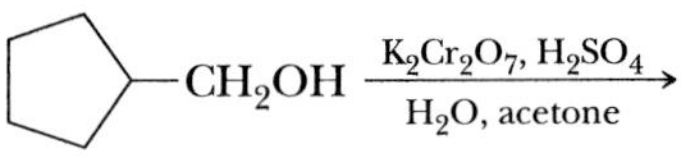 cyclopentyl–CH_2OH $\xrightarrow[H_2O,\ acetone]{K_2Cr_2O_7,\ H_2SO_4}$

(b) 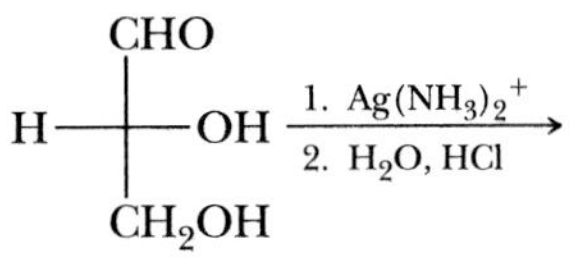 H–C(CHO)(CH$_2$OH)–OH $\xrightarrow[2.\ H_2O,\ HCl]{1.\ Ag(NH_3)_2^+}$

(c) $(CH_3)_2C{=}CHCCH_3$ (with C=O) $\xrightarrow[2.\ HCl,\ H_2O]{1.\ Cl_2,\ KOH\ in\ water/dioxane}$

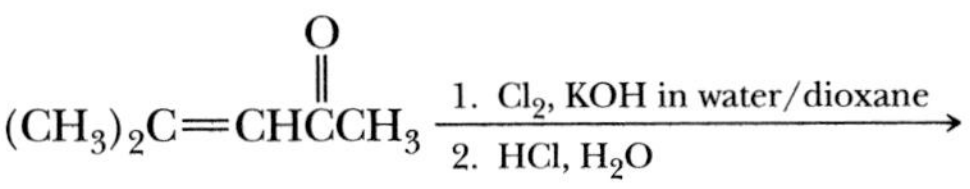

(d) 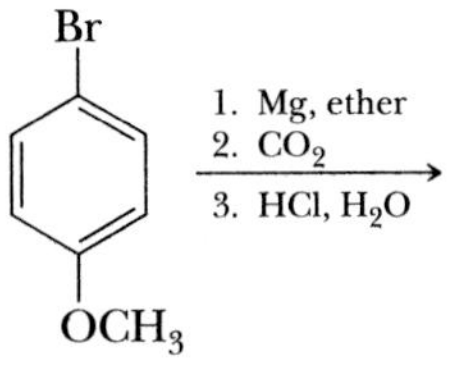 4-Br-C$_6$H$_4$-OCH$_3$ $\xrightarrow[3.\ HCl,\ H_2O]{1.\ Mg,\ ether;\ 2.\ CO_2}$

16.21 Show how to bring about each conversion in good yield.

(a) 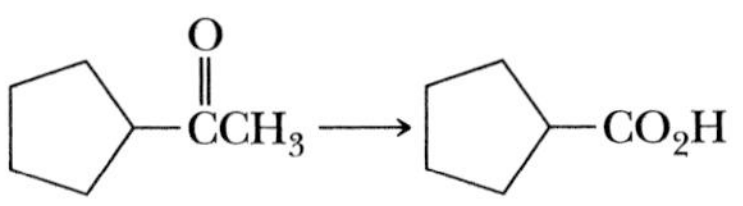 cyclopentyl–$\overset{O}{\parallel}{C}CH_3$ ⟶ cyclopentyl–CO_2H

(b) 3-chlorocyclohexene (Cl) ⟶ cyclohex-2-ene-1-carboxylic acid (CO_2H)

(c) 4-(hydroxymethyl)cyclohexanol (OH, CH_2OH) ⟶ 4-oxocyclohexanecarboxylic acid (O, CO_2H)

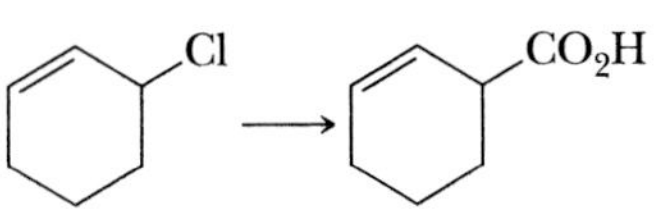

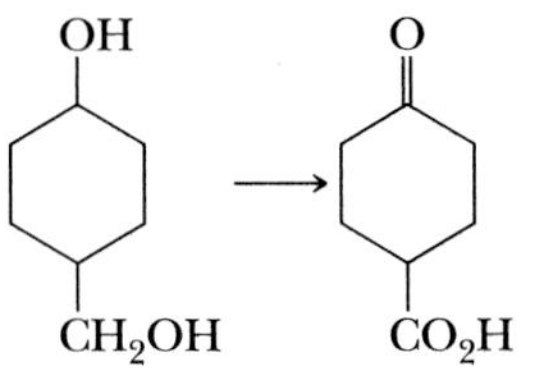

(d) $PhCH_2CH_2OH \longrightarrow PhCH_2CO_2H$

16.22 Show how to prepare pentanoic acid from these compounds:

(a) 1-Pentanol **(b)** Pentanal **(c)** 1-Pentene **(d)** 1-Butanol
(e) 1-Bromopropane **(f)** 2-Hexanone **(g)** 1-Hexene

16.23 Draw the structural formula of a compound of the given molecular formula that, on oxidation by potassium dichromate in aqueous H_2SO_4, gives the carboxylic acid or dicarboxylic acid shown.

(a) $C_6H_{14}O \xrightarrow{\text{oxidation}} CH_3(CH_2)_4\overset{O}{\overset{\|}{C}}OH$ **(b)** $C_6H_{12}O \xrightarrow{\text{oxidation}} CH_3(CH_2)_4\overset{O}{\overset{\|}{C}}OH$

(c) $C_6H_{14}O_2 \xrightarrow{\text{oxidation}} HO\overset{O}{\overset{\|}{C}}(CH_2)_4\overset{O}{\overset{\|}{C}}OH$

16.24 Show the reagents and experimental conditions necessary to bring about each conversion in good yield. Shown over each reaction arrow is the number of steps (not including workup in aqueous acid) required for each conversion.

(a) 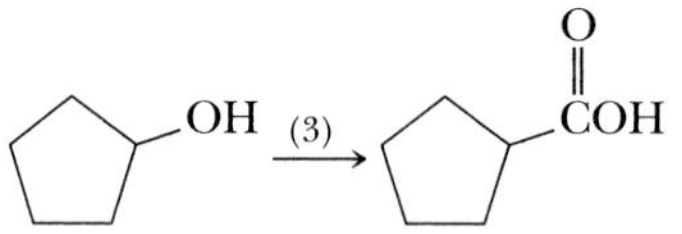 cyclopentyl–OH $\xrightarrow{(3)}$ cyclopentyl–$\overset{O}{\overset{\|}{C}}OH$

(b) $CH_3C(CH_3)_2OH \xrightarrow{(3)} CH_3C(CH_3)_2CO_2H$

(c) $CH_3C(CH_3)_2OH \xrightarrow{(3)} CH_3CH(CH_3)CO_2H$

(d) $CH_3C(CH_3)_2OH \xrightarrow{(5)} CH_3CH(CH_3)CH_2CO_2H$

(e) $CH_3CH{=}CHCH_3 \xrightarrow{(3)} CH_3CH{=}CHCH_2CO_2H$

16.25 Succinic acid can be synthesized by the following series of reactions from acetylene. Show the reagents and experimental conditions necessary to carry out this synthesis in good yield.

$$HC{\equiv}CH \longrightarrow HOCH_2C{\equiv}CCH_2OH \longrightarrow HOCH_2CH_2CH_2CH_2OH \longrightarrow HO\overset{O}{\overset{\|}{C}}CH_2CH_2\overset{O}{\overset{\|}{C}}OH$$

Acetylene; 2-Butyne-1,4-diol; 1,4-Butanediol; Butanedioic acid (Succinic acid)

16.26 The reaction of an α-diketone with concentrated sodium or potassium hydroxide to give the salt of an α-hydroxyacid is given the general name benzil-benzilic acid rearrangement. It is illustrated by the conversion of benzil to sodium benzilate and then to benzilic acid.

$$Ph{-}\overset{O}{\overset{\|}{C}}{-}\overset{O}{\overset{\|}{C}}{-}Ph + NaOH \xrightarrow{H_2O} Ph{-}\underset{Ph}{\overset{HO}{C}}{-}\overset{O}{\overset{\|}{C}}{-}O^-Na^+ \xrightarrow[H_2O]{HCl} Ph{-}\underset{Ph}{\overset{HO}{C}}{-}\overset{O}{\overset{\|}{C}}{-}OH$$

Benzil (an α-diketone); Sodium benzilate; Benzilic acid

Propose a mechanism for this rearrangement of benzil to sodium benzilate.

Acidity of Carboxylic Acids

16.27 Select the stronger acid in each set.

(a) Phenol (pK_a 9.95) and benzoic acid (pK_a 4.17)
(b) Lactic acid (K_a 8.4×10^{-4}) and ascorbic acid (K_a 7.9×10^{-5})

16.28 Assign the acid in each set its appropriate pK_a.

(a) $C_6H_5CO_2H$ and $C_6H_5SO_3H$ (pK_a 4.19 and 0.70)

(b) $CH_3CCH_2CO_2H$ (with C=O) and CH_3CCO_2H (with C=O) (pK_a 3.58 and 2.49)

(c) $CH_3CH_2CO_2H$ and $N{\equiv}CCH_2CO_2H$ (pK_a 4.78 and 2.45)

16.29 Low-molecular-weight dicarboxylic acids normally exhibit two different pK_a values. Ionization of the first carboxyl group is easier than the second. This effect diminishes with molecular size, and, for adipic acid and longer chain dicarboxylic acids, the two acid ionization constants differ by about one pK unit.

Dicarboxylic Acid	Structural Formula	pK_{a1}	pK_{a2}
oxalic	HO_2CCO_2H	1.23	4.19
malonic	$HO_2CCH_2CO_2H$	2.83	5.69
succinic	$HO_2C(CH_2)_2CO_2H$	4.16	5.61
glutaric	$HO_2C(CH_2)_3CO_2H$	4.31	5.41
adipic	$HO_2C(CH_2)_4CO_2H$	4.43	5.41

Why do the two pK_a values differ more for the shorter chain dicarboxylic acids than for the longer chain dicarboxylic acids?

16.30 Complete the following acid-base reactions:

(a) C_6H_5–$CH_2CO_2H + NaOH \longrightarrow$

(b) $CH_3CH{=}CHCH_2CO_2H + NaHCO_3 \longrightarrow$

(c) 2-methoxybenzoic acid (benzene ring with CO_2H and ortho OCH_3) $+ NaHCO_3 \longrightarrow$

(d) CH_3CHCO_2H (with OH on the CH) $+ H_2NCH_2CH_2OH \longrightarrow$

(e) $CH_3CH{=}CHCH_2CO_2{}^-Na^+ + HCl \longrightarrow$

(f) $CH_3CH_2CH_2CH_2Li + CH_3CO_2H \longrightarrow$

(g) $CH_3CH_2CH_2CH_2MgBr + CH_3CH_2OH \longrightarrow$

16.31 The normal pH range for blood plasma is 7.35–7.45. Under these conditions, would you expect the carboxyl group of lactic acid (pK_a 3.07) to exist primarily as a carboxyl group or as a carboxylate anion? Explain.

16.32 The pK_{a1} of ascorbic acid (Section 24.6) is 7.94×10^{-5}. Would you expect ascorbic acid dissolved in blood plasma to exist primarily as ascorbic acid or as ascorbate anion? Explain.

16.33 Excess ascorbic acid is excreted in the urine, the pH of which is normally in the range 4.8–8.4. What form of ascorbic acid would you expect to be present in urine of pH 8.4, free ascorbic acid or ascorbate anion? Explain.

Reactions of Carboxylic Acids

16.34 Give the expected organic products when $PhCH_2CO_2H$, phenylacetic acid, is treated with each of these reagents.

(a) $SOCl_2$ (b) $NaHCO_3$, H_2O

(c) NaOH, H_2O (d) CH_3MgBr (1 equivalent)

(e) $LiAlH_4$ followed by H_2O (f) CH_2N_2

(g) $CH_3OH + H_2SO_4$ (catalyst)

16.35 Show how to convert *trans*-3-phenyl-2-propenoic acid (cinnamic acid) to these compounds.

(a) 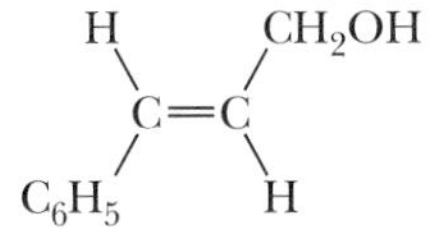 $C_6H_5(H)C{=}C(H)CH_2OH$ (trans) (b) $C_6H_5CH_2CH_2CO_2H$ (c) $C_6H_5CH_2CH_2CH_2OH$

16.36 Show how to convert 3-oxobutanoic acid (acetoacetic acid) to these compounds.

(a) $CH_3CH(OH)CH_2CO_2H$ (b) $CH_3CH(OH)CH_2CH_2OH$ (c) $CH_3CH{=}CHCO_2H$

16.37 Complete these examples of Fischer esterification. Assume alcohol is present in excess.

(a) $CH_3CO_2H + HOCH_2CH_2CH(CH_3)_2 \overset{H^+}{\rightleftharpoons}$ (b) 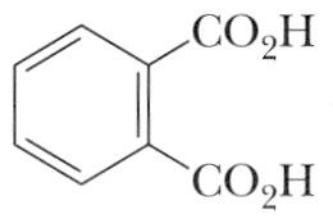 1,2-benzenedicarboxylic acid ($C_6H_4(CO_2H)_2$) $+ CH_3OH \overset{H^+}{\rightleftharpoons}$

(c) $HO_2C(CH_2)_2CO_2H + CH_3CH_2OH \overset{H^+}{\rightleftharpoons}$

16.38 Benzocaine, a topical anesthetic, is prepared by treatment of 4-aminobenzoic acid with ethanol in the presence of an acid catalyst followed by neutralization. Draw the structural formula of benzocaine.

16.39 From what carboxylic acid and what alcohol is each ester derived?

(a) 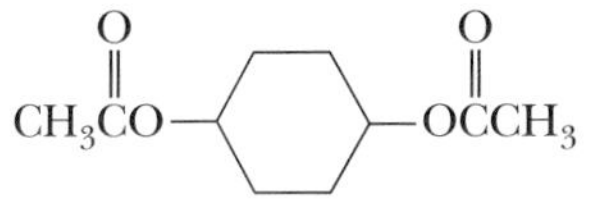 $CH_3C(=O)O$–(1,4-cyclohexylene)–$OC(=O)CH_3$ (b) $CH_3OC(=O)CH_2CH_2C(=O)OCH_3$

(c) 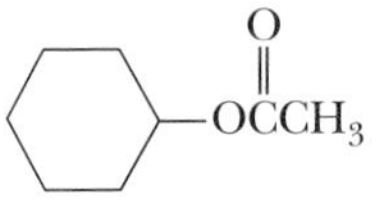 cyclohexyl–$OC(=O)CH_3$ (d) $CH_3CH_2CH{=}CHC(=O)OCH(CH_3)_2$

16.40 When 4-hydroxybutanoic acid is treated with an acid catalyst, it forms a lactone (a cyclic ester). Draw the structural formula of this lactone and propose a mechanism for its formation.

16.41 Fischer esterification cannot be used to prepare *tert*-butyl esters. Instead, carboxylic acids are treated with 2-methylpropene and an acidic catalyst to generate them.

(a) Why does the Fischer esterification fail for the synthesis of *tert*-butyl esters?
(b) Propose a mechanism for the 2-methylpropene method.

$$RC(=O)OH + CH_2{=}C(CH_3)CH_3 \xrightarrow{H^+} RC(=O)OC(CH_3)_2CH_3$$

2-Methylpropene (Isobutylene) A *tert*-butyl ester

16.42 Draw the product formed on thermal decarboxylation of these compounds.

(a) $C_6H_5\overset{\overset{O}{\|}}{C}CH_2CO_2H$ (b) $C_6H_5CH_2\overset{\overset{CO_2H}{|}}{C}HCO_2H$ (c) 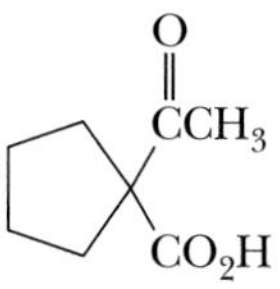 1-acetylcyclopentanecarboxylic acid (cyclopentane ring bearing $\overset{\overset{O}{\|}}{C}CH_3$ and CO_2H on the same carbon)

16.43 When heated, carboxylic salts, in which there is a good leaving group on the carbon beta to the carboxylate group, undergo decarboxylation/elimination to give an alkene. Propose a mechanism for this type of decarboxylation/elimination. Compare the mechanism of these decarboxylations with the mechanism for decarboxylation of β-ketoacids; in what way(s) are the mechanisms similar?

(a) 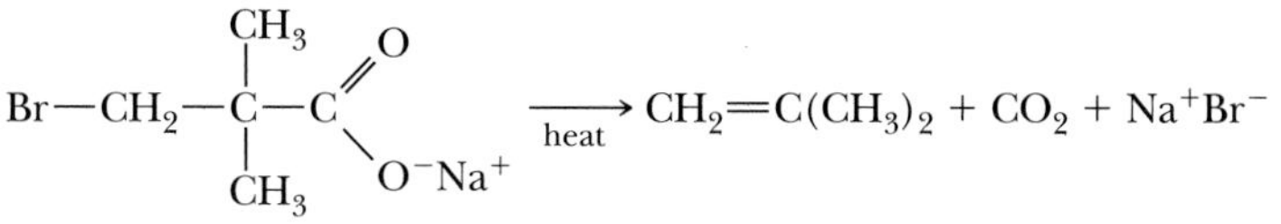 $Br{-}CH_2{-}C(CH_3)_2{-}CO_2^-Na^+ \xrightarrow{\text{heat}} CH_2{=}C(CH_3)_2 + CO_2 + Na^+Br^-$

(b) 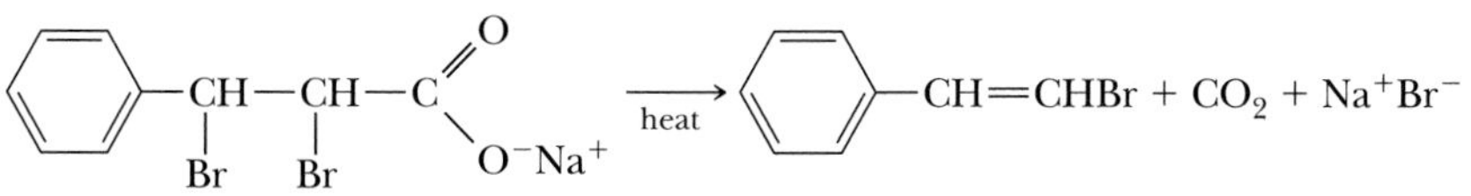 $C_6H_5{-}CHBr{-}CHBr{-}CO_2^-Na^+ \xrightarrow{\text{heat}} C_6H_5{-}CH{=}CHBr + CO_2 + Na^+Br^-$

17 FUNCTIONAL DERIVATIVES OF CARBOXYLIC ACIDS

■ Crystals of 4-acetamidophenol (acetaminophen, Tylenol) viewed under polarized light.

$HO-C_6H_4-NHC(=O)CH_3$

(Phillip A. Harrington/Fran Heyl Associates)

In previous chapters, we studied the structure, nomenclature, physical properties, and characteristic reactions of organic compounds one class at a time. In this chapter, we study five classes of organic compounds, each related to the carboxyl group: acid halides, acid anhydrides, esters, amides, and nitriles. Under the general formula of each functional group is a drawing to help you see how it is formally related to a carboxylic acid. Loss of —OH from a carboxyl group and H— from H—Cl, for example, gives an acid chloride. Similarly, loss of —OH from a carboxyl group and H— from ammonia gives an amide.

$RC(=O)Cl$	$RC(=O)OC(=O)R'$	$RC(=O)OR'$	$RC(=O)NH_2$	$RC{\equiv}N$
An acid chloride	An acid anhydride	An ester	An amide	A nitrile
$RC(=O)-OH\quad H-Cl$	$RC(=O)-OH\quad H-OC(=O)R'$	$RC(=O)-OH\quad H-OR'$	$RC(=O)-OH\quad H-NH_2$	$RC(OH)=NH$ The enol of an amide

The first four classes show many of the same types of reactions. In this chapter, we concentrate on one of them, namely addition to the carbonyl group to form a tetrahedral carbonyl addition intermediate, which then collapses to regenerate the carbonyl group. The result of this sequence is **nucleophilic acyl substitution.**

Nucleophilic acyl substitution A reaction in which a nucleophile bonded to a carbonyl carbon is replaced by another nucleophile.

MECHANISM Nucleophilic Substitution at an Acyl Carbon

$$R{-}\overset{\overset{\displaystyle O}{\|}}{C}{-}Y + :Nu^- \longrightarrow \left[R{-}\overset{\overset{\displaystyle :\ddot{O}:^-}{|}}{\underset{\underset{\displaystyle Y}{|}}{C}}{-}Nu \right] \longrightarrow R{-}\overset{\overset{\displaystyle O}{\|}}{C}{-}Nu + :Y^-$$

Tetrahedral carbonyl addition intermediate

We have already seen one important example of this reaction, namely Fischer esterification (Section 16.8).

17.1 Structure and Nomenclature

A. Acid Halides

The functional group of an **acid halide** (acyl halide) is an **acyl group (RCO—)** bonded to a halogen atom. Acid chlorides are the most common acid halides. Acid halides are named by changing the suffix **-ic acid** in the name of the parent carboxylic acid to **-yl halide.**

Acyl halide A derivative of a carboxylic acid in which the —OH of the carboxyl group is replaced by halogen, most commonly chlorine.

Acyl group An RCO— or ArCO— group.

$R\overset{O}{\overset{\|}{C}}{-}$	 $R\overset{O}{\overset{\|}{C}}Cl$	$CH_3\overset{O}{\overset{\|}{C}}Cl$	$C_6H_5\overset{O}{\overset{\|}{C}}Cl$	$Cl\overset{O}{\overset{\|}{C}}(CH_2)_4\overset{O}{\overset{\|}{C}}Cl$
An acyl group	An acyl chloride (An acid chloride)	Ethanoyl chloride (Acetyl chloride)	Benzoyl chloride	Hexanedioyl chloride (Adipoyl chloride)

Replacement of —OH in a sulfonic acid by chlorine gives a derivative called a **sulfonyl chloride.** Following are structural formulas for two sulfonic acids and the acid chloride derived from each:

$CH_3\overset{O}{\overset{\|}{\underset{\underset{O}{\|}}{S}}}OH$	$CH_3\overset{O}{\overset{\|}{\underset{\underset{O}{\|}}{S}}}Cl$	$H_3C{-}C_6H_4{-}\overset{O}{\overset{\|}{\underset{\underset{O}{\|}}{S}}}OH$	$H_3C{-}C_6H_4{-}\overset{O}{\overset{\|}{\underset{\underset{O}{\|}}{S}}}Cl$
Methanesulfonic acid	Methanesulfonyl chloride	*p*-Toluenesulfonic acid	*p*-Toluenesulfonyl chloride (Tosyl chloride, TsCl)

B. Acid Anhydrides

Carboxylic Acid Anhydrides

Carboxylic acid anhydride A functional group in which two acyl groups are bonded to an oxygen.

The functional group of a **carboxylic acid anhydride** is two acyl groups bonded to an oxygen atom. These compounds are called acid anhydrides because they are formally derived from two carboxylic acid molecules by loss of water (Chapter 17, Introduction). The acyl groups may be derived from either aliphatic or aromatic carboxylic acids. Furthermore, the anhydride may be symmetrical (two identical acyl groups), or it may be mixed (two different acyl groups). Anhydrides are named by replacing the word *acid* in the name of the parent carboxylic acid by the word *anhydride.*

CH_3COCCH_3 (with C=O on each carbonyl) — Acetic anhydride

$C_6H_5COCC_6H_5$ — Benzoic anhydride

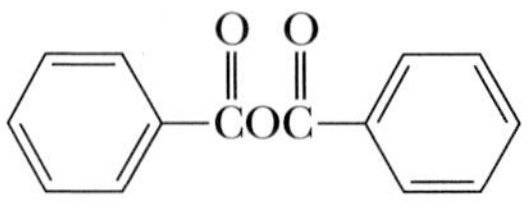

$CH_3COCC_6H_5$ — Acetic benzoic anhydride

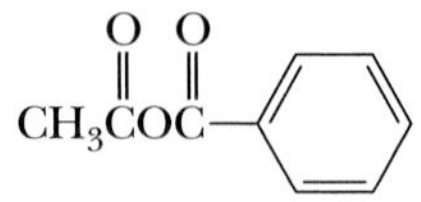

Cyclic anhydrides are named from the dicarboxylic acids from which they are derived. Here are the cyclic anhydrides derived from succinic acid, maleic acid, and phthalic acid.

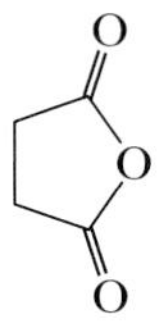

Succinic anhydride 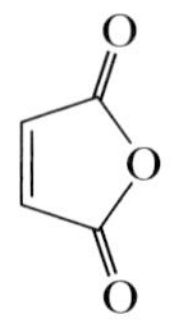 Maleic anhydride 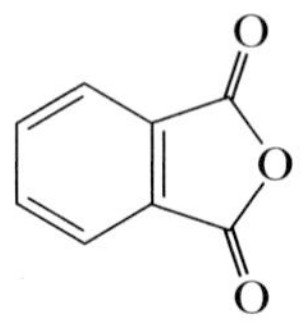 Phthalic anhydride

Phosphoric Acid Anhydrides

Phosphoric acid anhydride A functional group in which two phosphoryl groups are bonded to an oxygen atom.

Because of the special importance of anhydrides of phosphoric acid in biochemical systems (Chapters 25, 26, and 28), we include them here to show the similarity between them and the anhydrides of carboxylic acids. The functional group of a **phosphoric acid anhydride** is two phosphoryl groups bonded to an oxygen atom. Here are structural formulas for two anhydrides of phosphoric acid and the ions derived by ionization of each acidic hydrogen.

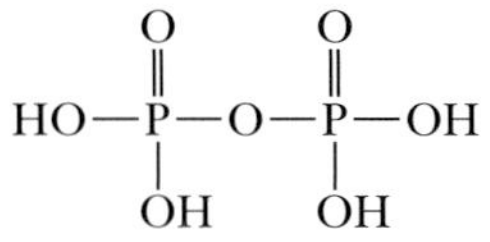

$HO-P(=O)(OH)-O-P(=O)(OH)-OH$ — Diphosphoric acid (Pyrophosphoric acid)

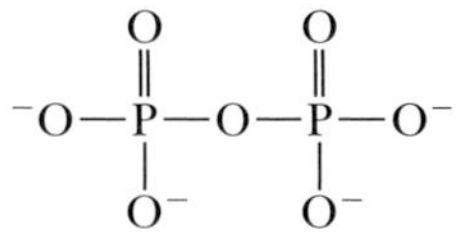

$^-O-P(=O)(O^-)-O-P(=O)(O^-)-O^-$ — Diphosphate ion (Pyrophosphate ion)

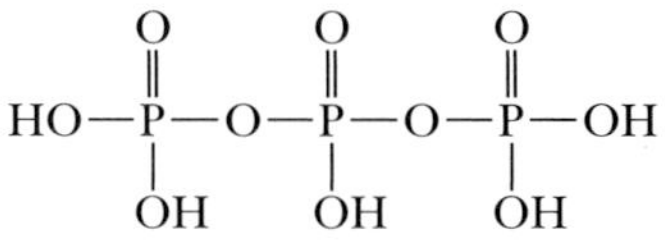

$HO-P(=O)(OH)-O-P(=O)(OH)-O-P(=O)(OH)-OH$ — Triphosphoric acid

$^-O-P(=O)(O^-)-O-P(=O)(O^-)-O-P(=O)(O^-)-O^-$ — Triphosphate ion

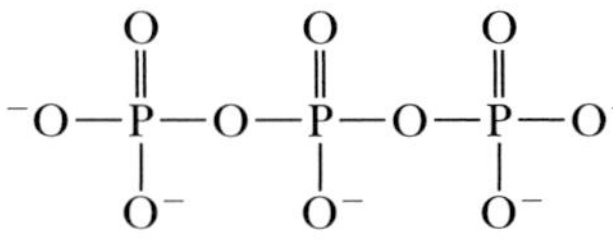

C. Esters

Esters of Carboxylic Acids

Ester A functional group in which an acyl group is bonded to —OR or to —OAr.

The functional group of a **carboxylic acid ester** is an acyl group bonded to —OR or —OAr. Both IUPAC and common names of esters are derived from the names of the parent carboxylic acids. The alkyl or aryl group bonded to oxygen is named first, followed by the name of the acid in which the suffix -ic acid is replaced by the suffix -ate.

$CH_3COCH_2CH_3$ (C=O)	C_6H_5—$COCH(CH_3)_2$ (C=O)	$CH_3CH_2OCCH_2CH_2COCH_2CH_3$ (two C=O)
Ethyl ethanoate (Ethyl acetate)	Isopropyl benzoate	Diethyl butanedioate (Diethyl succinate)

Lactones: Cyclic Esters

Cyclic esters are called **lactones.** The IUPAC system has developed a set of rules for naming these compounds. Nonetheless, the simplest lactones are still named by dropping the suffix -ic or -oic acid from the name of the parent carboxylic acid and adding the suffix **-olactone.** The location of the oxygen atom in the ring is indicated by a number if the IUPAC name of the acid is used, or by a Greek letter α, β, γ, δ, ϵ, and so forth, if the common name of the acid is used.

Lactone A cyclic ester.

3-Butanolactone (β-Butyrolactone)	4-Butanolactone (γ-Butyrolactone)	5-Hexanolactone (δ-Caprolactone)	6-Hexanolactone (ϵ-Caprolactone)

Esters of Phosphoric Acid

Phosphoric acid has three —OH groups and forms mono-, di-, and triesters, which are named by giving the name(s) of the alkyl or aryl group(s) attached to oxygen followed by the word "phosphate," as, for example, dimethyl phosphate. In more complex phosphoric esters, it is common to name the organic molecule and then indicate the presence of the phosphoric ester using either the word "phosphate" or the prefix phospho-. On the right are two phosphoric esters, each of special importance in the biological world. The first reaction in the metabolism of glucose is formation of a phosphoric ester to give D-glucose-6-phosphate (Section 26.6). Pyridoxal phosphate is one of the metabolically active forms of vitamin B_6. Each of these esters is shown as it is ionized at pH 7.4, the pH of blood plasma; the two hydroxyl groups of each phosphate group are ionized, giving each phosphate group a charge of -2.

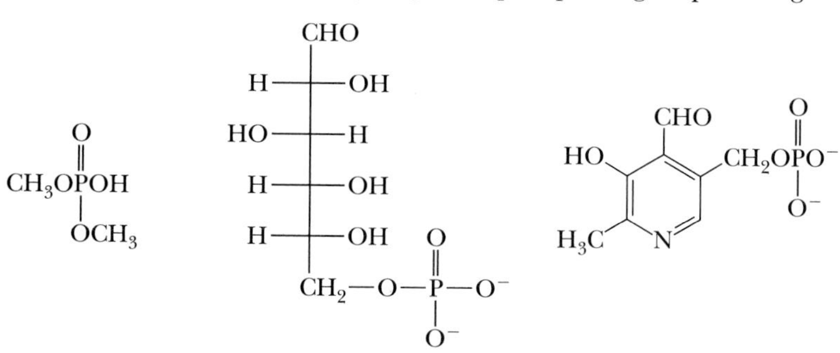

Dimethyl phosphate D-Glucose-6-phosphate Pyridoxal phosphate

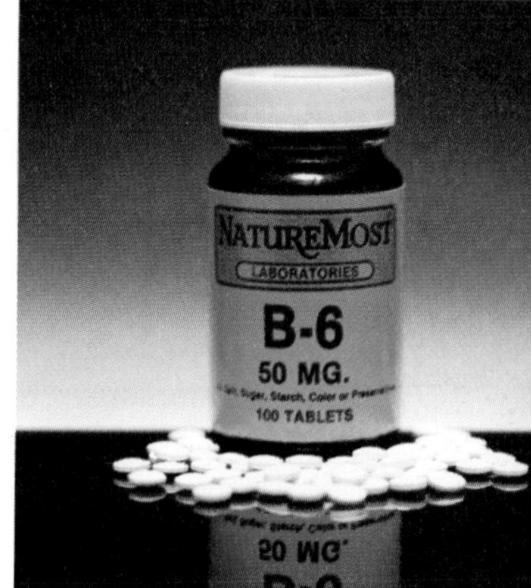

Vitamin B_6, pyridoxal. *(Charles D. Winters)*

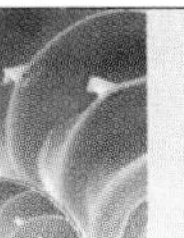

CHEMISTRY IN ACTION

From Moldy Clover to a Blood Thinner

In 1933, a disgruntled farmer delivered a bale of moldy clover, a pail of unclotted blood, and a dead cow to the laboratory of Dr. Carl Link at the University of Wisconsin. Six years and many bales of moldy clover later, Link and his collaborators isolated the anticoagulant dicoumarol, a substance that delays or prevents blood clotting. When cows are fed moldy clover, they ingest dicoumarol, their blood clotting is inhibited, and they bleed to death from minor cuts and scratches. Dicoumarol exerts its anticoagulation effect by interfering with vitamin K activity (Section 25.6D). Within a few years after its discovery, dicoumarol became widely used to treat victims of heart attack and others at risk for developing blood clots.

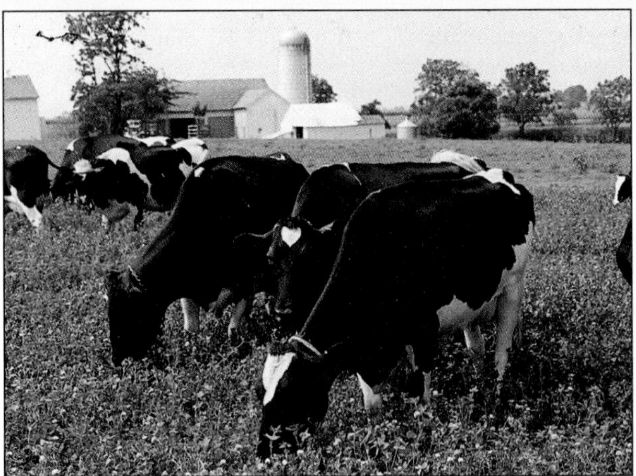

The powerful anticoagulant dicoumarol was first isolated from moldy clover. (Grant Heilman/Grant Heilman Photography, Inc.)

Dicoumarol is a derivative of coumarin, a lactone that gives sweet clover its pleasant smell. Coumarin, which does not interfere with blood clotting, is converted to dicoumarol as sweet clover becomes moldy.

O O

Coumarin
(from sweet clover)

as sweet clover becomes moldy →

OH HO

O O O O

Dicoumarol
(an anticoagulant)

In a search for even more potent anticoagulants, Link developed warfarin (named for the Wisconsin Alumni Research Foundation), now used primarily as a rat poison. When rats consume it, their blood fails to clot and they bleed to death. Warfarin is also used as a blood anticoagulant in humans. In 1989, physicians prescribed more than 2700 pounds of this "rat poison" for human medical use. The synthesis of warfarin is described in Problem 19.53. (See also *The Merck Index*, 12th ed., #10174.)

O

OH H CH_3

O O

Warfarin
(a synthetic anticoagulant)

The *S* enantiomer, shown here, is more active than the *R* enantiomer. The commercial product is sold as a racemic mixture.

D. Amides and Imides

Amide The functional group in which an acyl group is bonded to a trivalent nitrogen atom.

The functional group of an **amide** is an acyl group bonded to a nitrogen atom. Amides are named by dropping the suffix -oic acid from the IUPAC name of the parent acid, or -ic acid from its common name, and adding **-amide.** If the nitrogen atom of an

CHEMISTRY IN ACTION

Insecticides of Plant Origin: Polyunsaturated *N*-Isobutylamides

Plant-derived substances, largely abandoned during the era of synthetic insecticides such as DDT, are once again under study as safe and selective agents for insect control. It is now well recognized that almost every plant species has developed a unique group of chemicals that protects it from insect predation. In a search for natural means of small-scale control of mosquitoes in rural areas, scientists at the International Centre of Insect Physiology and Ecology (ICIPE) in Nairobi, Kenya investigated *Spilanthes mauritiana* (Compositae), a medicinal plant used to treat mouth infections, stomachaches, diarrhea, and toothaches. A methanolic extract of its leaves yields the *N*-isobutylamide of a tetraunsaturated fatty acid, which, at a concentration of 10^{-5} mg/mL, causes 100% mortality in larvae of *Aedes aegypti,* the insect vector of yellow fever.

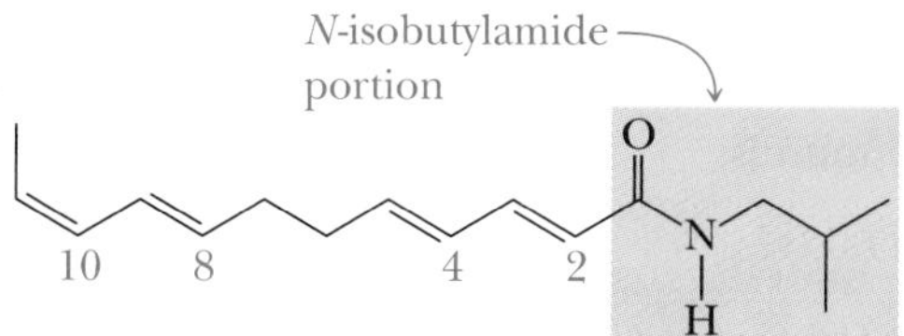

N-Isobutyl-(2E,4E,8E,10*Z*)-2,4,8,10-dodecatetraenamide

It is hoped that such plant extracts may be useful in rural communities where mosquitoes breed in small collections of water, such as in temporary rain puddles, containers, and drums. Periodic treatment of these breeding pools with native plant-derived insecticides could considerably reduce the multiplication of disease-bearing mosquitoes.

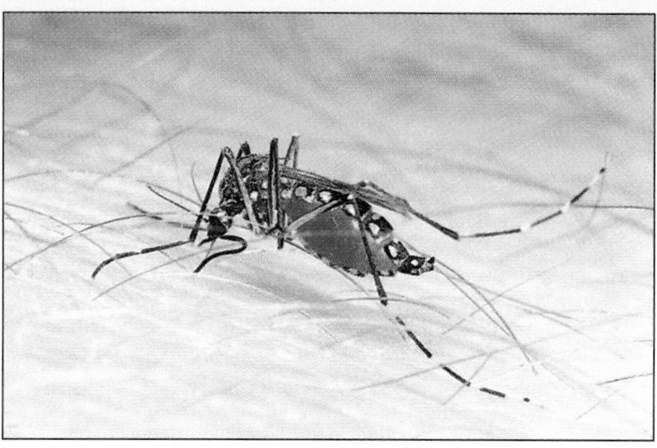

Aedes aegypti, the mosquito of yellow fever. (D. R. Specker/Animals, Animals)

See J. T. Arnason, B. J. R. Philogène, and P. Morand, eds., "Insecticides of Plant Origin," *ACS Symposium Series,* no. 387, American Chemical Society, Washington, D. C., 1989.

amide is bonded to an alkyl or aryl group, the group is named and its location on nitrogen is indicated by ***N*-**. Two alkyl or aryl groups on nitrogen are indicated by *N,N*-di. *N,N*-Dimethylformamide (DMF) is a widely used polar aprotic solvent (Section 8.2).

$CH_3C(=O)NH_2$	$C_6H_5C(=O)NH_2$	$CH_3C(=O)NHCH_3$	$HC(=O)N(CH_3)_2$
Acetamide (a 1° amide)	Benzamide (a 1° amide)	*N*-Methylacetamide (a 2° amide)	*N,N*-Dimethylformamide (DMF) (a 3° amide)

Lactam A cyclic amide.

Cyclic amides are given the special name **lactam.** Their names are derived in a manner similar to those of lactones, with the difference that the suffix -lactone is replaced by -lactam.

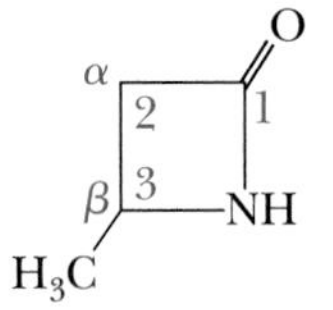

3-Butanolactam (β-Butyrolactam)

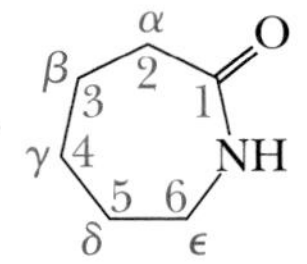

6-Hexanolactam (ϵ-Caprolactam)

Imide A functional group in which two acyl groups, RCO— or ArCO—, are bonded to a nitrogen atom.

The functional group of an **imide** is two acyl groups bonded to nitrogen. Both succinimide and phthalimide are cyclic imides.

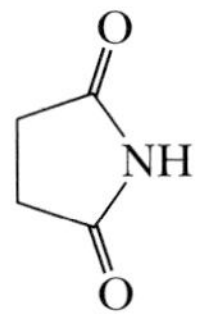

Succinimide

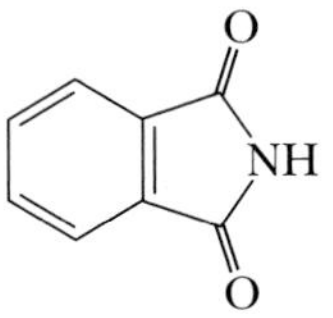

Phthalimide

EXAMPLE 17.1

Write IUPAC names for these compounds:

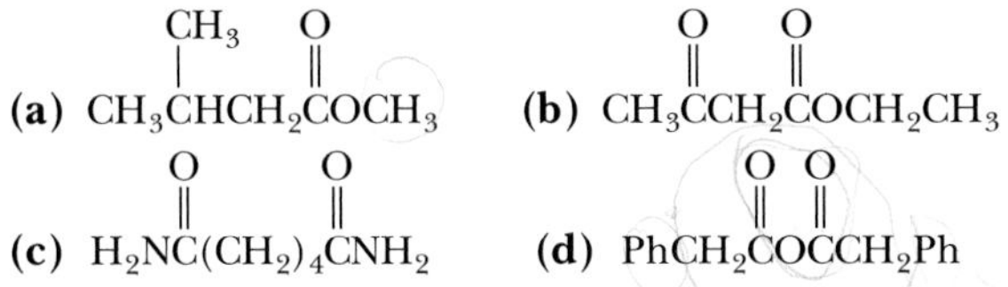

(a) $CH_3CH(CH_3)CH_2C(=O)OCH_3$

(b) $CH_3C(=O)CH_2C(=O)OCH_2CH_3$

(c) $H_2NC(=O)(CH_2)_4C(=O)NH_2$

(d) $PhCH_2C(=O)OC(=O)CH_2Ph$

Solution

Given first are IUPAC names and then, in parentheses, common names:

(a) Methyl 3-methylbutanoate (methyl isovalerate, from isovaleric acid).
(b) Ethyl 3-oxobutanoate (ethyl β-ketobutyrate, from β-ketobutyric acid). It is also known as ethyl acetoacetate (Section 18.6).
(c) Hexanediamide (adipamide, from adipic acid).
(d) Phenylethanoic anhydride (phenylacetic anhydride, from phenylacetic acid).

PROBLEM 17.1

Draw structural formulas for these compounds.

(a) *N*-Cyclohexylacetamide **(b)** *sec*-Butyl acetate
(c) Cyclobutyl butanoate **(d)** *N*-(2-Octyl)succinimide
(e) Diethyl adipate **(f)** 2-Aminopropanamide

E. Nitriles

The functional group of a **nitrile** is a cyano (C≡N) group bonded to carbon. IUPAC names follow the pattern alkanenitrile: for example, ethanenitrile. Common names are derived by dropping the suffix -ic or -oic acid from the name of the parent carboxylic acid and adding the suffix **-onitrile.**

Nitrile A compound containing a —C≡N (cyano) group bonded to a carbon atom.

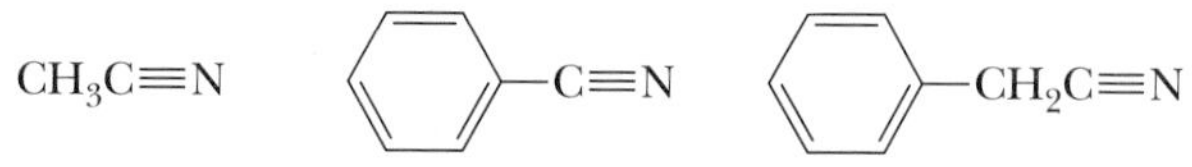

$CH_3C{\equiv}N$ — Ethanenitrile (Acetonitrile)

$C_6H_5{-}C{\equiv}N$ — Benzonitrile

$C_6H_5{-}CH_2C{\equiv}N$ — Phenylethanenitrile (Phenylacetonitrile)

17.2 Acidity of Amides, Imides, and Sulfonamides

Following are structural formulas of a primary amide, a sulfonamide, and two cyclic imides, along with pK_a values for each.

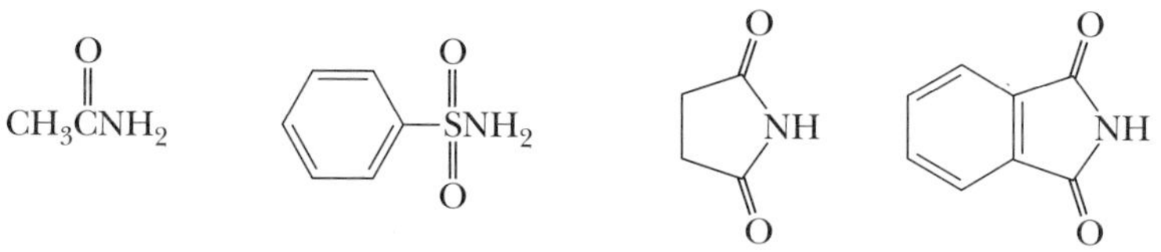

CH_3CONH_2	$C_6H_5SO_2NH_2$	(cyclic imide, NH)	(cyclic imide, NH)
Acetamide	Benzenesulfonamide	Succinimide	Phthalimide
pK_a 15–17	pK_a 10	pK_a 9.7	pK_a 8.3

Values of pK_a for amides of carboxylic acids are in the range of 15 to 17, which means they are comparable in acidity to alcohols. Amides show no evidence of acidity in aqueous solution, that is, water-insoluble amides do not react with aqueous solutions of NaOH or other alkali metal hydroxides to form water-soluble salts.

Imides are considerably more acidic than amides and readily dissolve in 5% aqueous NaOH by forming water-soluble salts. We account for the acidity of imides in the same manner as for the acidity of carboxylic acids (Section 16.4), namely (1) the electron-withdrawing inductive effect of the two adjacent carbonyl groups weakens the N—H bond, and (2) the imide anion is stabilized by delocalization of the negative charge. The more important contributing structures for the anion formed by ionization of an imide delocalize the negative charge on the two carbonyl oxygens.

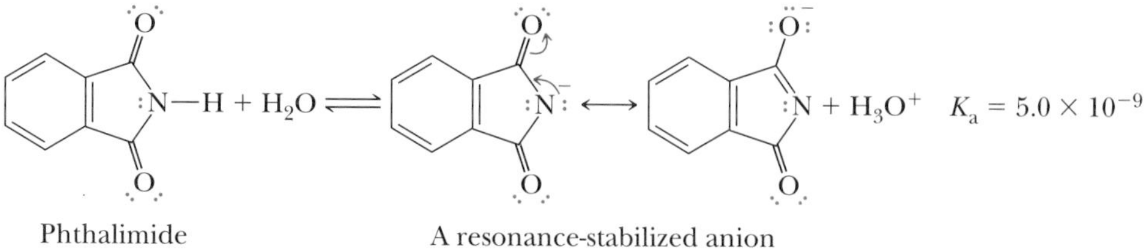

$$\text{Phthalimide (N—H)} + H_2O \rightleftharpoons [\text{anion} \longleftrightarrow \text{anion}] + H_3O^+ \quad K_a = 5.0 \times 10^{-9}$$

Phthalimide

A resonance-stabilized anion (the negative charge is delocalized on both oxygens)

Sulfonamides derived from ammonia and primary amines are also sufficiently acidic to dissolve in aqueous solutions of NaOH or other alkali metal hydroxides by forming water-soluble salts. The pK_a of benzenesulfonamide is approximately 10. We

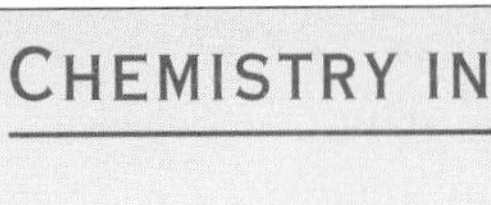

CHEMISTRY IN ACTION

The Penicillins and Cephalosporins: β-Lactam Antibiotics

The **penicillins** were discovered in 1928 by the Scottish bacteriologist Sir Alexander Fleming. As a result of the brilliant experimental work of Sir Howard Florey, an Australian pathologist, and Ernst Chain, a German chemist who fled Nazi Germany, penicillin G was introduced into the practice of medicine in 1943. For their pioneering work in developing one of the most effective antibiotics of all time, Fleming, Florey, and Chain were awarded the Nobel Prize for medicine and physiology in 1945. The mold from which Fleming discovered penicillin was *Penicillium notatum,* a strain that gives a relatively low yield of penicillin. It was replaced in commercial production of the antibiotic by *P. chrysogenum,* a strain cultured from a mold found growing on a grapefruit in a market in Peoria, Illinois.

The structural feature common to all penicillins is a ***β*-lactam** ring fused to a five-membered thiazolidine ring. The penicillins owe their antibacterial activity to a common mechanism that inhibits the biosynthesis of a vital part of bacterial cell walls.

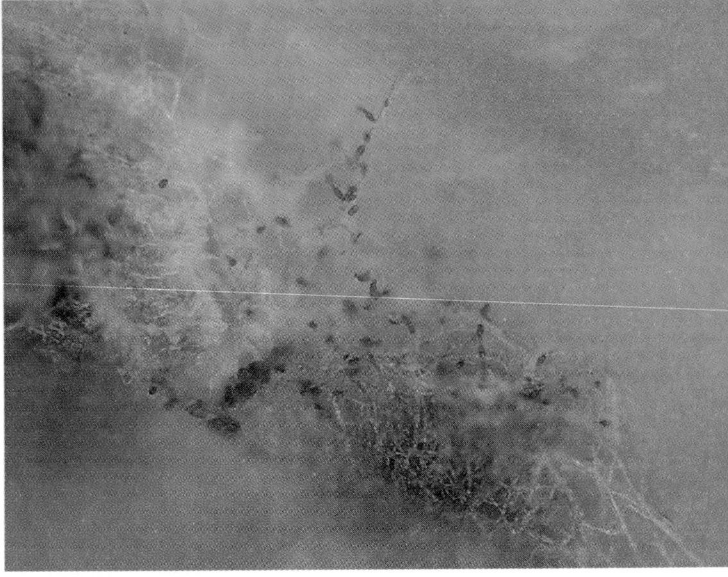

The blue-green mold Penicillium on the rind of an orange. (© Herb Charles Ohlmeyer/Fran Heyl Associates)

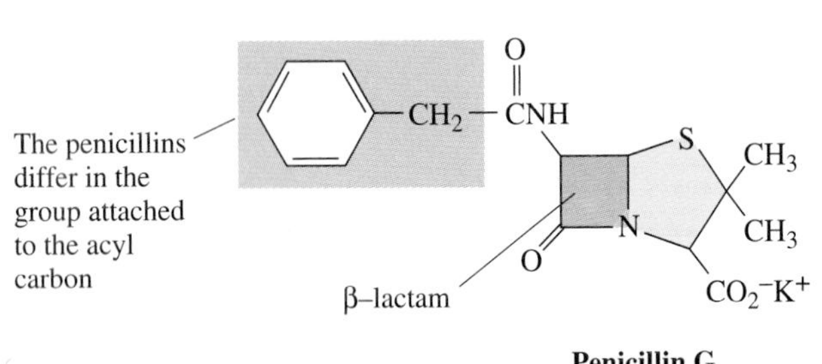

Penicillin G
(a β–lactam antibiotic)

Soon after the penicillins were introduced into medical practice, penicillin-resistant strains of bacteria began to appear and have since proliferated. One approach to combating resistant strains is to synthesize

account for the acidity of sulfonamides in the same way we did for imides, namely (1) the electron-withdrawing inductive effect of the adjacent $-SO_2-$ group, which weakens the N—H bond, and (2) the resonance stabilization of the resulting anion.

$$C_6H_5-SO_2-NH-H + H_2O \rightleftharpoons C_6H_5-SO_2-\ddot{N}H^- \longleftrightarrow C_6H_5-S(=O)(-O^-)=NH + H_3O^+ \quad K_a = 10^{-10}$$

Benzenesulfonamide   A resonance-stabilized anion

newer, more effective penicillins. Among those developed are ampicillin, methicillin, and amoxicillin. Another approach is to search for newer, more effective β-lactam antibiotics. At the present time, the most effective of these are the **cephalosporins.** The first cephalosporin was isolated from the fungus *Cephalosporium acremonium.* This class of β-lactam antibiotics has an even broader spectrum of antibacterial activity than the penicillins and is effective against many penicillin-resistant bacterial strains. The most common mechanism of resistance in bacteria involves their production of a specific enzyme that cleaves the β-lactam.

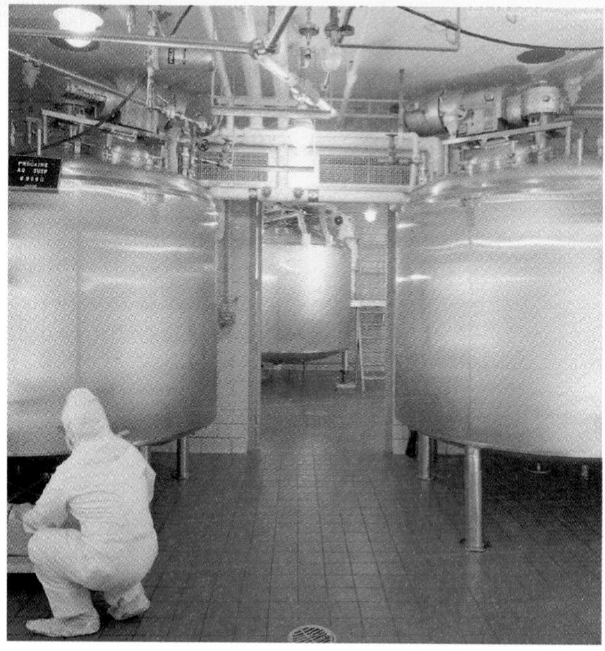

Fermentation tanks used in modern penicillin production. (Courtesy of Pfizer, Inc.)

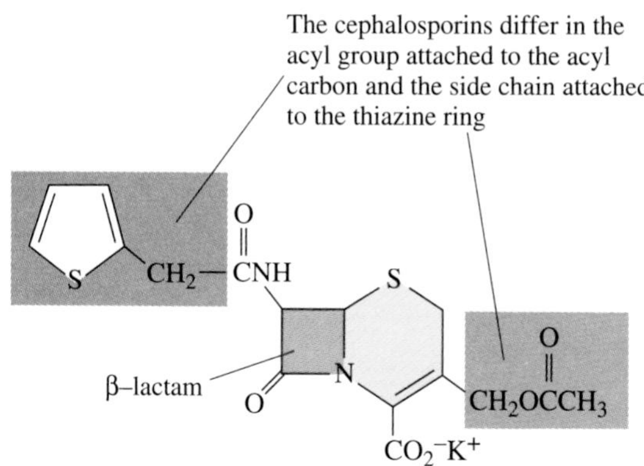

A cephalosporin, a newer generation β–lactam antibiotic

EXAMPLE 17.2

Phthalimide is insoluble in water. Will phthalimide dissolve in aqueous NaOH?

Solution

Phthalimide is the stronger acid and NaOH is the stronger base. The position of equilibrium, therefore, lies to the right. $K_{eq} = 2.5 \times 10^7$. Phthalimide dissolves in aqueous NaOH by forming a water-soluble sodium salt.

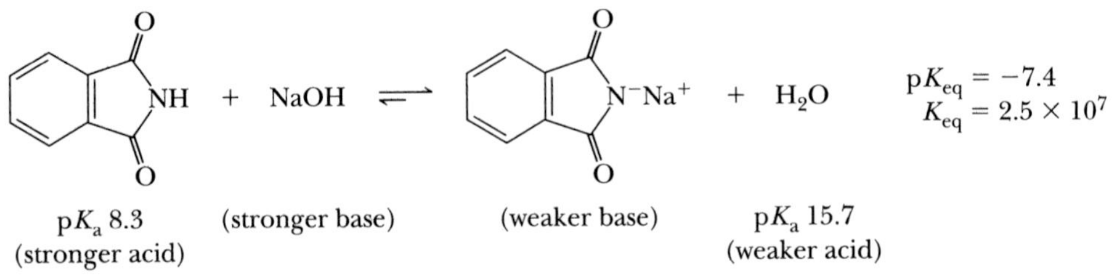

$$\text{Phthalimide (NH)} + NaOH \rightleftharpoons \text{(N}^-Na^+\text{)} + H_2O \qquad pK_{eq} = -7.4,\quad K_{eq} = 2.5 \times 10^7$$

pK_a 8.3 (stronger acid) — (stronger base) — (weaker base) — pK_a 15.7 (weaker acid)

PROBLEM 17.2

Will phthalimide dissolve in aqueous sodium bicarbonate?

The noncaloric artificial sweetener, saccharin, is an imide. The imide hydrogen of saccharin is sufficiently acidic that it reacts with sodium hydroxide and aqueous ammonia to form water-soluble salts. The ammonium salt is used to make liquid sweeteners. Saccharin is used in solid form as the Ca^{2+} salt. (See *The Merck Index,* 12th ed., #8463.)

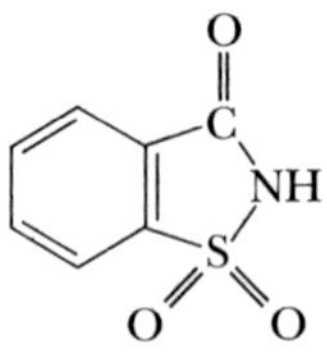

Saccharin

Saccharin is approximately 500 times sweeter than sugar, and at one time, was the most important noncaloric sweetener used in foods. At the present time, the most widely used noncaloric artificial sweetener is aspartame (Nutrasweet), which we will discuss further in Chapter 27.

17.3 Spectroscopic Properties

Esters and amides are the most common functional derivatives of carboxylic acids and also the most commonly analyzed. Therefore, we concentrate on the spectroscopic properties of these two functional derivatives.

A. Mass Spectrometry

Esters and amides generally show discernible molecular ion peaks. Their most characteristic fragmentation patterns are α-cleavage and McLafferty rearrangement, both of which can be seen in the mass spectrum of methyl butanoate (Figure 17.1). Peaks at m/z 71 and 59 are the result of α-cleavage.

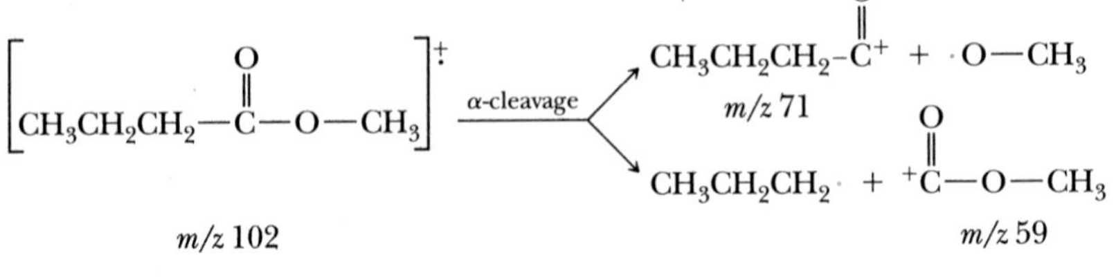

$$\left[CH_3CH_2CH_2-\overset{O}{\overset{\|}{C}}-O-CH_3\right]^{\dot{+}} \xrightarrow{\alpha\text{-cleavage}} \begin{cases} CH_3CH_2CH_2-\overset{O}{\overset{\|}{C}}{}^{+} + \cdot O-CH_3 & m/z\ 71 \\ CH_3CH_2CH_2\cdot + {}^{+}\overset{O}{\overset{\|}{C}}-O-CH_3 & m/z\ 59 \end{cases}$$

m/z 102

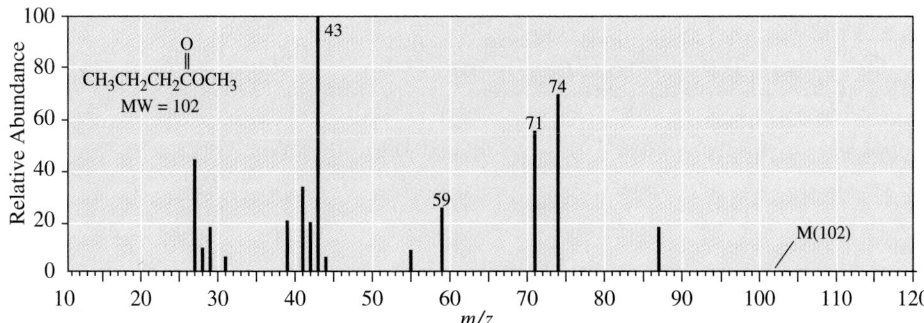

FIGURE 17.1
Mass spectrum of methyl butanoate. Characteristic fragmentation patterns of esters are α-cleavage and McLafferty rearrangement.

The peak at m/z 74 is the result of McLafferty rearrangement.

$$\left[\begin{array}{c} H_2C\text{—H} \quad O \\ | \qquad\quad \| \\ H_2C\text{—}CH_2\text{—}C\text{—}OCH_3 \end{array} \right]^{\dot{+}} \xrightarrow{\text{McLafferty rearrangement}} \begin{array}{c} H_2C \\ \| \\ H_2C \end{array} + \left[\begin{array}{c} H\text{—}O \\ | \\ CH_2{=}C\text{—}OCH_3 \end{array} \right]^{\dot{+}}$$

m/z 74

B. Nuclear Magnetic Resonance Spectroscopy

Hydrogens on the α-carbon to a carbonyl group are slightly deshielded and come into resonance at δ 2.0–2.6. Hydrogens on the carbon attached to the ester oxygen are more strongly deshielded and come into resonance at δ 3.6–4.1. It is possible to distinguish between ethyl acetate and its constitutional isomer, methyl propanoate, by the chemical shifts of either the $—CH_3$ singlet absorption (compare δ 2.04 and 3.68), or the $—CH_2—$ quartet absorption (compare δ 4.11 and 2.33). Carbonyl carbons of esters show characteristic ^{13}C-NMR absorption at δ 160–180.

δ 2.04(s) → $CH_3—C({=}O)—O—CH_2—CH_3$ ← δ 4.11(q)

Ethyl acetate

δ 2.33(q) → $CH_3—CH_2—C({=}O)—O—CH_3$ ← δ 3.68(s)

Methyl propanoate

C. Infrared Spectroscopy

The most important infrared absorption of carboxylic acids and their functional derivatives is due to the C═O stretching vibration. Infrared spectroscopic data for these compounds are summarized in Table 17.1.

The carbonyl stretching of amides occurs at 1630–1680 cm^{-1}, at a lower wavenumber than for other carbonyl compounds. Primary and secondary amides show N—H stretching in the region 3200–3400 cm^{-1}; primary amides ($RCONH_2$) usually show two N—H absorptions, whereas secondary amides (RCONHR) show only a single N—H absorption (Figure 17.2).

Esters display strong C═O stretching absorption in the region between 1735 and 1800 cm^{-1}. In addition, they also display strong C—O stretching absorptions in the region 1000 to 1100 cm^{-1} for the sp^3 C—O stretch and 1200–1250 cm^{-1} for the sp^2 C—O stretch (Figure 17.3).

TABLE 17.1 Infrared Absorptions for Carboxylic Acids and Their Functional Derivatives

Compound	Stretching Absorption (cm^{-1})	Additional Absorptions (cm^{-1})
RCCl (with C=O)	1790–1800	
RCOCR (two C=O)	1740–1760 and 1800–1850	C—O stretching at 900–1300
RCOR (with C=O)	1735–1800	C—O stretching at 1000–1100 and 1200–1250
RCOH (with C=O)	1700–1725	O—H stretching at 2400–3400 C—O stretching at 1210–1320
$RCNH_2$ (with C=O)	1630–1680	N—H stretching at 3200 and 3400 (1° amides have two N—H peaks) (2° amides have one N—H peak)
RC≡N	2200–2250	

Anhydrides have two carbonyl stretching absorptions, one near 1760 cm^{-1}, the other near 1810 cm^{-1}. In addition, anhydrides display strong C—O stretching absorption in the region 900 to 1300 cm^{-1}. Nitriles can be distinguished by strong C≡N stretching absorption at 2200 to 2250 cm^{-1}.

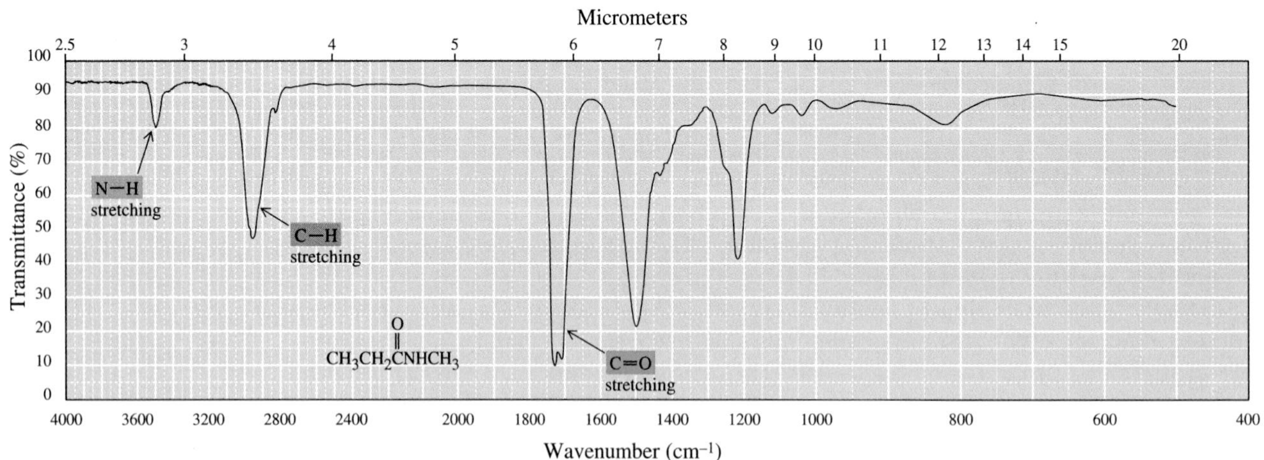

FIGURE 17.2
Infrared spectrum of *N*-methylpropanamide (a secondary amide).

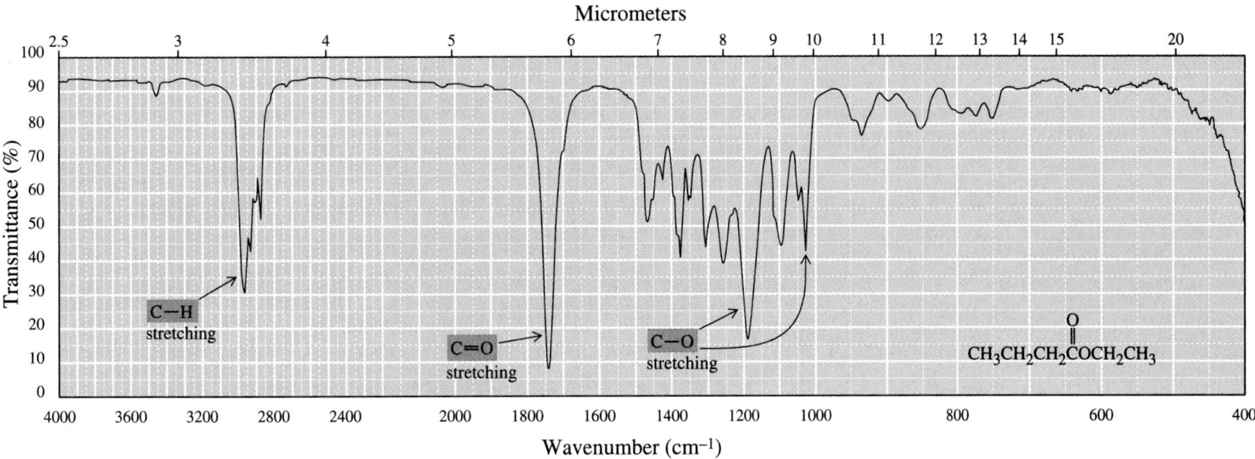

FIGURE 17.3
Infrared spectrum of ethyl butanoate.

17.4 Characteristic Reactions

The most common reaction theme of acid halides, anhydrides, esters, and amides is addition of a nucleophile to the carbonyl carbon to form a tetrahedral carbonyl addition intermediate. In this sense, the reaction of these functional groups is similar to nucleophilic addition to the carbonyl groups in aldehydes and ketones (Section 15.5). The tetrahedral carbonyl addition intermediate formed from an aldehyde or ketone then adds H^+ to give the product. The result of this reaction is **nucleophilic acyl addition.** This reaction can also be catalyzed by acid, in which case protonation of the carbonyl oxygen precedes the attack of the nucleophile.

Nucleophilic acyl addition: $R-\overset{O}{\overset{\|}{C}}-R + :Nu^- \longrightarrow \left[R_2C(-O^-)(Nu) \right] \xrightarrow{H^+} R_2C(-OH)(Nu)$

An aldehyde or ketone — Tetrahedral carbonyl addition intermediate — Addition product

For functional derivatives of carboxylic acids, the fate of the tetrahedral carbonyl addition intermediate is quite different; the intermediate collapses to expel the leaving group Y and regenerate the carbonyl group. The result of this addition-elimination sequence is **nucleophilic acyl substitution.**

Nucleophilic acyl substitution: $R-\overset{O}{\overset{\|}{C}}-Y + :Nu^- \longrightarrow \left[RC(-O^-)(Y)(Nu) \right] \longrightarrow R-\overset{O}{\overset{\|}{C}}-Nu + :Y^-$

Tetrahedral carbonyl addition intermediate — Substitution product

The major difference between these two types of carbonyl addition reactions is that aldehydes and ketones do not have a group that can leave as a relatively stable anion. Aldehydes and ketones undergo only nucleophilic acyl addition. The four carboxylic acid derivatives we study in this chapter do have a group, Y, that can leave as a relatively stable anion; they undergo nucleophilic acyl substitution.

In this general reaction, we show the nucleophile and the leaving group as anions. This need not be the case. Neutral molecules, such as water, alcohols, ammonia, and amines, may also serve as nucleophiles and leaving groups in the acid-catalyzed version of this reaction. We show the leaving group here as an anion, however, to illustrate an important point about leaving groups: the weaker the base, the better the leaving group (Section 8.4F).

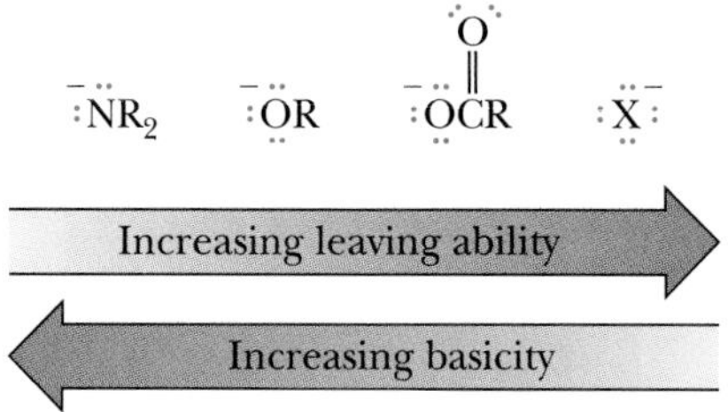

The weakest base in the series, and the best leaving group, is halide ion; acid halides are the most reactive toward nucleophilic acyl substitution. The strongest base, and the poorest leaving group, is amide ion; amides are the least reactive toward nucleophilic acyl substitution. Acid halides and acid anhydrides are so reactive that they are not found in nature. Esters and amides, however, are universally present.

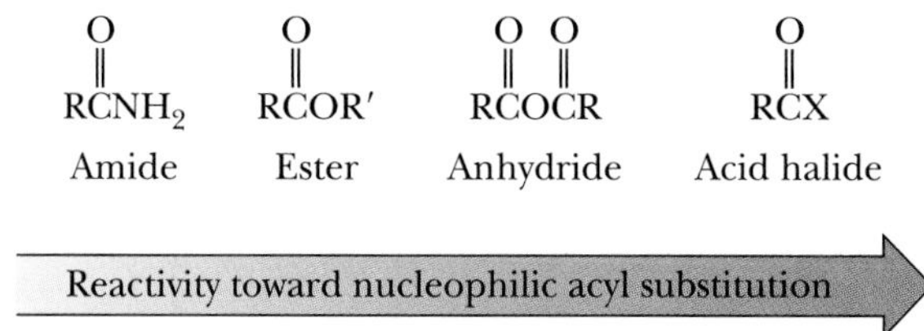

In addition, many reactions of carboxyl derivatives occur by acid catalysis. In these reactions, the carbonyl group is protonated in the first step, which increases its electronegativity. Then the leaving group is protonated in a later step to decrease its basicity and make it a better leaving group. We will see detailed mechanisms for many examples of both acid- and base-catalyzed mechanisms in this chapter.

17.5 Reaction with Water: Hydrolysis

A. Acid Chlorides

Low-molecular-weight acid chlorides react very rapidly with water to form carboxylic acids and HX. Higher-molecular-weight acid halides are less soluble and consequently react less rapidly with water.

$$CH_3\overset{O}{\overset{\|}{C}}Cl + H_2O \longrightarrow CH_3\overset{O}{\overset{\|}{C}}OH + HCl$$

B. Acid Anhydrides

Anhydrides are generally less reactive than acid chlorides. However, the lower-molecular-weight anhydrides also react readily with water to form two molecules of carboxylic acid.

$$CH_3\overset{O}{\overset{\|}{C}}O\overset{O}{\overset{\|}{C}}CH_3 + H_2O \longrightarrow CH_3\overset{O}{\overset{\|}{C}}OH + HO\overset{O}{\overset{\|}{C}}CH_3$$

The following mechanism is divided into two stages: first, formation of a tetrahedral carbonyl addition intermediate, and second, collapse of this intermediate by elimination of acetate ion, a moderate base, and a good leaving group.

MECHANISM Hydrolysis of an Acid Anhydride

Step 1: Formation of a tetrahedral carbonyl addition intermediate

$$CH_3-\overset{O}{\overset{\|}{C}}-O-\overset{O}{\overset{\|}{C}}-CH_3 + H-\ddot{O}-H \longrightarrow CH_3-\underset{H-\overset{+}{O}-H}{\overset{O^-}{\overset{|}{C}}}-O-\overset{O}{\overset{\|}{C}}-CH_3 \xrightarrow{:\ddot{O}H_2} CH_3-\underset{O-H}{\overset{O^-}{\overset{|}{C}}}-O-\overset{O}{\overset{\|}{C}}-CH_3 + H-\overset{+}{O}(H)-H$$

Tetrahedral carbonyl addition intermediate

Step 2: Collapse of the tetrahedral carbonyl addition intermediate

$$CH_3-\underset{O-H}{\overset{O^-}{\overset{|}{C}}}-O-\overset{O}{\overset{\|}{C}}-CH_3 \longrightarrow CH_3-\underset{O-H}{\overset{O}{\overset{\|}{C}}} + {}^-:\ddot{O}-\overset{O}{\overset{\|}{C}}-CH_3 \xrightarrow{H-\overset{+}{O}(H)-H} 2CH_3-\overset{O}{\overset{\|}{C}}-\ddot{O}-H + H_2O$$

Tetrahedral carbonyl addition intermediate

C. Esters

Esters are hydrolyzed only very slowly, even in boiling water. Hydrolysis becomes considerably more rapid, however, when they are refluxed in aqueous acid or base. We already discussed acid-catalyzed (Fischer) esterification in Section 16.8B and pointed out that it is an equilibrium reaction. Hydrolysis of esters in aqueous acid is also an equilibrium reaction and proceeds by the same mechanism as esterification, except in reverse. The role of the acid catalyst is to protonate the carbonyl oxygen. In doing so, it increases the electrophilic character of the carbonyl carbon toward attack by water to form a **tetrahedral carbonyl addition intermediate.** Collapse of this intermediate gives the carboxylic acid and an alcohol. In this reaction, acid is a catalyst; it is consumed in the first step but is regenerated at the end of the reaction.

$$RCOOCH_3 + H_2O \overset{H^+}{\rightleftharpoons} \left[R\text{—}C(OH)_2\text{—}OCH_3 \right] \overset{H^+}{\rightleftharpoons} RCOOH + CH_3OH$$

Tetrahedral carbonyl addition intermediate

Although formation of a tetrahedral carbonyl addition intermediate is the most common mechanism for the hydrolysis of esters in aqueous acid, alternative mechanistic pathways are followed in special cases. Such a case occurs when the alkyl group attached to oxygen can form an especially stable carbocation. Then protonation of the carbonyl oxygen is followed by cleavage of the O—C bond to give a carboxylic acid and a carbocation. Benzyl and *tert*-butyl esters readily undergo this type of ester hydrolysis.

MECHANISM Hydrolysis of a *tert*-Butyl Ester in Aqueous Acid

Proton transfer to the carbonyl oxygen followed by rearrangement of electron pairs gives the carboxylic acid and *tert*-butyl cation. This cation then either reacts with water to give *tert*-butyl alcohol (S_N1) or transfers a proton to water to give 2-methylpropene (E1).

$$R\text{—}C(=O)\text{—}O\text{—}C(CH_3)_3 \xrightarrow{H_3O^+} R\text{—}C(=O^+H)\text{—}O\text{—}C(CH_3)_3 \longrightarrow R\text{—}C(OH){=}O + {}^+C(CH_3)_3$$

tert-Butyl cation

Saponification Hydrolysis of an ester in aqueous NaOH or KOH to an alcohol and the sodium or potassium salt of a carboxylic acid.

Hydrolysis of esters may also be carried out using hot aqueous base, such as aqueous NaOH. Hydrolysis of esters in aqueous base is often called **saponification,** a reference to the use of this reaction in the manufacture of soaps (Section 25.2A). Although the carbonyl carbon of an ester is not strongly electrophilic, hydroxide ion is a good nucleophile and adds to the carbonyl carbon to form a tetrahedral carbonyl addition intermediate, which in turn collapses to give a carboxylic acid and an alkoxide ion. The carboxylic acid reacts with the alkoxide ion or other base present to form a carboxylic acid anion.

MECHANISM Hydrolysis of an Ester in Aqueous Base

$$R-\overset{\overset{\Large O}{\|}}{C}-\ddot{O}-CH_3 + {}^{-}\!:\ddot{O}H \rightleftharpoons R-\underset{\underset{:\ddot{O}H}{|}}{\overset{\overset{:\ddot{O}:^{-}}{|}}{C}}-\ddot{O}-CH_3 \rightleftharpoons R-\underset{\underset{:\ddot{O}-H}{|}}{\overset{\overset{\Large O}{\|}}{C}} + {}^{-}\!:\ddot{O}-CH_3 \longrightarrow R-\underset{\underset{:\ddot{O}:^{-}}{|}}{\overset{\overset{\Large O}{\|}}{C}} + :\underset{\underset{H}{|}}{\ddot{O}}-CH_3$$

Thus, each mol of ester hydrolyzed requires 1 mol of base as shown in the following balanced equation:

$$R\overset{\overset{O}{\|}}{C}OCH_3 + NaOH \xrightarrow{H_2O} R\overset{\overset{O}{\|}}{C}O^-Na^+ + CH_3OH$$

There are two major differences between hydrolysis of esters in aqueous acid and aqueous base.

1. For hydrolysis of an ester in aqueous acid, acid is required in only catalytic amounts. For hydrolysis in aqueous base, base in required in equimolar amounts because it is a reactant, not just a catalyst.
2. Hydrolysis of an ester in aqueous acid is reversible. Because a carboxylic acid anion (weakly electrophilic, if at all) is not attacked by ROH (a weak nucleophile), hydrolysis of an ester in aqueous base is irreversible.

EXAMPLE 17.3

Complete and balance equations for the hydrolysis of each ester in aqueous sodium hydroxide. Show all products as they are ionized under these conditions.

(a) $C_6H_5-\overset{\overset{O}{\|}}{C}OCH(CH_3)_2 + NaOH \xrightarrow{H_2O}$ **(b)** $CH_3\overset{\overset{O}{\|}}{C}OCH_2CH_2O\overset{\overset{O}{\|}}{C}CH_3 + NaOH \xrightarrow{H_2O}$

Solution

The products of hydrolysis of (a) are benzoic acid and 2-propanol. In aqueous NaOH, benzoic acid is converted to its sodium salt. Therefore, 1 mol of NaOH is required for hydrolysis of 1 mol of this ester. Compound (b) is a diester of ethylene glycol. Two mol of NaOH are required for complete hydrolysis.

(a) $C_6H_5-\overset{\overset{O}{\|}}{C}O^-Na^+ + HOCH(CH_3)_2$ **(b)** $2CH_3\overset{\overset{O}{\|}}{C}O^-Na^+ + HOCH_2CH_2OH$

Sodium benzoate — 2-Propanol (Isopropyl alcohol) — Sodium acetate — 1,2-Ethanediol (Ethylene glycol)

PROBLEM 17.3

Complete and balance equations for the hydrolysis of each ester in aqueous solution. Show each product as it is ionized under the indicated experimental conditions.

(a) 1,2-$C_6H_4(CO_2CH_3)_2$ $+ NaOH \xrightarrow{H_2O}$

(b) $CH_3\overset{O}{\overset{\|}{C}}CH_2CH_2CH_2\overset{O}{\overset{\|}{C}}OCH_2CH_3 + H_2O \xrightarrow{HCl}$

D. Amides

Amides require considerably more vigorous conditions for hydrolysis in both acid and base than esters. Amides undergo hydrolysis in hot aqueous acid to give a carboxylic acid and an ammonium ion. Hydrolysis is driven to completion by the acid-base reaction between ammonia or the amine and acid to form an ammonium ion. One mol of acid is required per mol of amide.

$$CH_3CH_2\underset{Ph}{\underset{|}{C}}H\overset{O}{\overset{\|}{C}}NH_2 + H_2O + HCl \xrightarrow[\text{heat}]{H_2O} CH_3CH_2\underset{Ph}{\underset{|}{C}}H\overset{O}{\overset{\|}{C}}OH + NH_4{}^+Cl^-$$

2-Phenylbutanamide → 2-Phenylbutanoic acid

In aqueous base, the products of amide hydrolysis are a carboxylic acid salt and ammonia or an amine. Hydrolysis in aqueous base is driven to completion by the acid-base reaction between the resulting carboxylic acid and base to form a salt. One mol of base is required per mol of amide.

$$CH_3\overset{O}{\overset{\|}{C}}NH\text{—}C_6H_5 + NaOH \xrightarrow[\text{heat}]{H_2O} CH_3\overset{O}{\overset{\|}{C}}O^-Na^+ + H_2N\text{—}C_6H_5$$

N-Phenylethanamide (*N*-Phenylacetamide, Acetanilide) → Sodium acetate + Aniline

The steps in the mechanism for the hydrolysis of amides in aqueous acid are similar to those for the hydrolysis of esters in aqueous acid. The role of hydrogen ion in Step 1 is to protonate the carbonyl oxygen to increase the electrophilic character of the carbonyl carbon. Following protonation, the polarized carbonyl group reacts with a molecule of water in Step 2 to give a tetrahedral carbonyl addition intermediate. Collapse of the tetrahedral carbonyl addition intermediate in Step 3 completes the reaction. Note that the leaving group in this case is a neutral amine (a weaker base), a far better leaving group than amide ion (a much stronger base).

MECHANISM Hydrolysis of an Amide in Aqueous Acid

Step 1: Protonation of the carbonyl oxygen

$$R-\overset{\overset{\displaystyle O}{\|}}{C}-NH_2 + H-\overset{+}{\underset{\underset{\displaystyle H}{|}}{O}}-H \longrightarrow \left[R-\overset{\overset{\displaystyle \overset{+}{O}H}{\|}}{C}-NH_2 \longleftrightarrow R-\overset{\overset{\displaystyle OH}{|}}{C}=\overset{+}{N}H_2\right] + H_2O$$

Resonance-stabilized cation

Step 2: Addition of water to the carbonyl carbon followed by proton transfer gives a tetrahedral carbonyl addition intermediate

$$R-\overset{\overset{\displaystyle OH}{|}}{C}=\overset{+}{N}H_2 + \underset{\underset{\displaystyle H}{|}}{O}-H \longrightarrow R-\overset{\overset{\displaystyle OH}{|}}{\underset{\underset{\displaystyle H\diagup \overset{+}{O} \diagdown H}{|}}{C}}-NH_2 \xrightarrow[\text{to nitrogen}]{\text{proton transfer from oxygen}} R-\overset{\overset{\displaystyle OH}{|}}{\underset{\underset{\displaystyle OH}{|}}{C}}-NH_3^+$$

Tetrahedral carbonyl addition intermediate

Step 3: Collapse of the tetrahedral carbonyl addition intermediate coupled with proton transfer gives a carboxylic acid and ammonium ion

$$R-\overset{\overset{\displaystyle OH}{|}}{\underset{\underset{\displaystyle OH}{|}}{C}}-NH_3^+ \longrightarrow R-\overset{\overset{\displaystyle \overset{+}{O}-H}{\|}}{\underset{\underset{\displaystyle OH}{|}}{C}} + NH_3 \longrightarrow R-\overset{\overset{\displaystyle O}{\|}}{C}-OH + NH_4^+$$

The mechanism for the hydrolysis of amides in aqueous base is similar to that for the hydrolysis of esters in aqueous base. Addition of hydroxide ion to the carbonyl carbon in Step 1 forms a tetrahedral carbonyl addition intermediate, which then collapses in Step 2 to regenerate the carbonyl group and eject the nitrogen-containing group. In the following mechanism, loss of nitrogen and proton transfer from water to nitrogen are concerted so that the leaving group is not amide ion, NH_2^-, a stronger base and poorer leaving group, but rather ammonia, NH_3, a weaker base and better leaving group.

MECHANISM Hydrolysis of an Amide in Aqueous Base

Step 1: Addition of hydroxide ion to the carbonyl carbon to give a tetrahedral carbonyl addition intermediate

$$R-\overset{\overset{\displaystyle O}{\|}}{C}-NH_2 + {}^-OH \longrightarrow R-\overset{\overset{\displaystyle O^-}{|}}{\underset{\underset{\displaystyle OH}{|}}{C}}-NH_2$$

Tetrahedral carbonyl addition intermediate

Step 2: Collapse of the tetrahedral carbonyl addition intermediate to form a carboxylic acid and ammonia

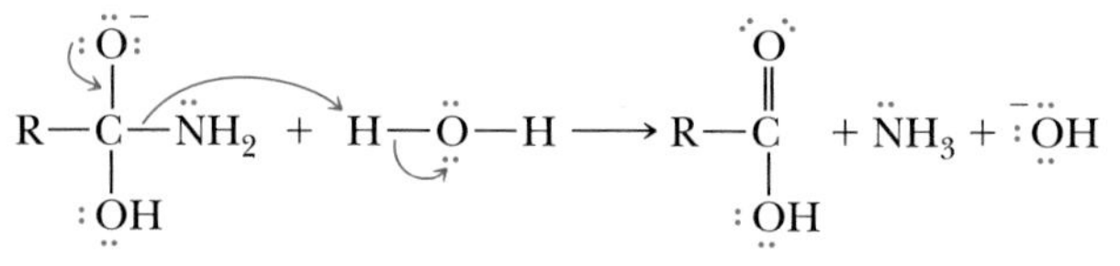

$$R-\underset{:\ddot{O}H}{\overset{:\ddot{O}:^-}{C}}-\ddot{N}H_2 + H-\ddot{O}-H \longrightarrow R-\underset{:\ddot{O}H}{\overset{\ddot{O}}{\overset{\|}{C}}} + \ddot{N}H_3 + :\ddot{O}H^-$$

Tetrahedral carbonyl addition intermediate

Step 3: Proton transfer to form the carboxylate anion and water: hydrolysis is driven to completion by this acid-base reaction

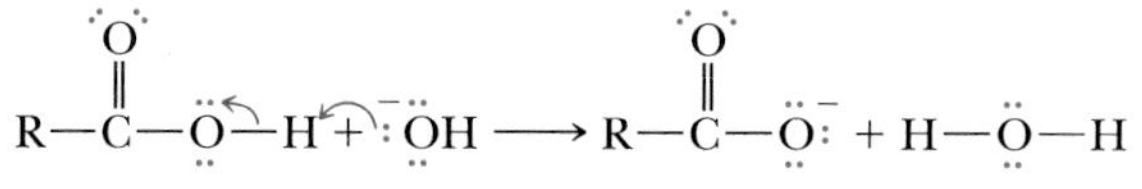

$$R-\overset{\ddot{O}}{\overset{\|}{C}}-\ddot{O}-H + :\ddot{O}H^- \longrightarrow R-\overset{\ddot{O}}{\overset{\|}{C}}-\ddot{O}:^- + H-\ddot{O}-H$$

Example 17.4

Write equations for hydrolysis of these amides in concentrated aqueous HCl. Show all products as they exist in aqueous HCl, and the number of mol of HCl required for hydrolysis of each amide.

(a) $CH_3\overset{O}{\overset{\|}{C}}N(CH_3)_2$ **(b)** 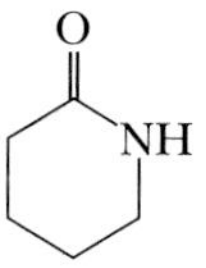 (six-membered ring lactam: C=O adjacent to NH)

Solution

(a) Hydrolysis of *N,N*-dimethylacetamide gives acetic acid and dimethylamine. Dimethylamine, a base, is protonated by HCl to form dimethylammonium ion and is shown in the balanced equation as dimethylammonium chloride. One mol of HCl is required per mol of amide.

$$CH_3\overset{O}{\overset{\|}{C}}N(CH_3)_2 + H_2O + HCl \xrightarrow[\text{heat}]{} CH_3\overset{O}{\overset{\|}{C}}OH + (CH_3)_2NH_2^+Cl^-$$

(b) Hydrolysis of this δ-lactam gives the protonated form of 5-aminopentanoic acid. One mol of HCl is required per mol of amide.

$$\text{(δ-lactam)} + H_2O + HCl \xrightarrow[\text{heat}]{} HO\overset{O}{\overset{\|}{C}}CH_2CH_2CH_2CH_2NH_3^+Cl^-$$

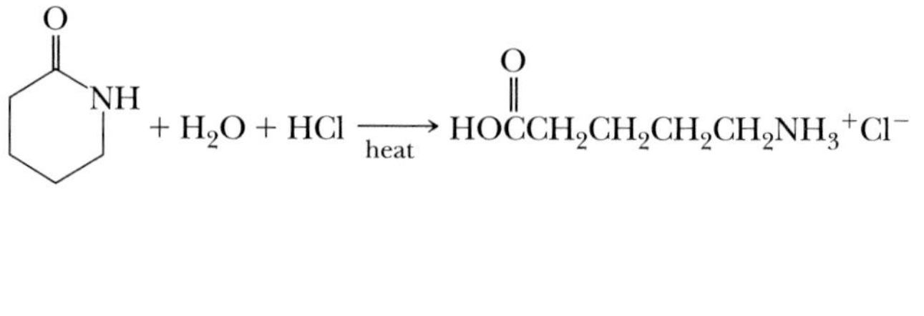

PROBLEM 17.4

Complete equations for the hydrolysis of the amides in Example 17.4 in concentrated aqueous NaOH. Show all products as they exist in aqueous NaOH, and the number of mol of NaOH required for hydrolysis of each amide.

E. Nitriles

The cyano group of a nitrile is hydrolyzed in aqueous acid to a carboxyl group and ammonium ion as shown in the following equation.

$$C_6H_5CH_2C{\equiv}N + 2H_2O + H_2SO_4 \xrightarrow[\text{heat}]{H_2O} C_6H_5CH_2\overset{\overset{\displaystyle O}{\|}}{C}OH + NH_4^+HSO_4^-$$

Phenylacetonitrile — Phenylacetic acid — Ammonium hydrogen sulfate

In hydrolysis of a cyano group in aqueous acid, protonation of the nitrogen atom gives a cation that reacts with water to give an imidic acid (the enol of an amide). Keto-enol tautomerism of the imidic acid gives an amide. The amide is then hydrolyzed, as already described, to a carboxylic acid and ammonium ion.

$$R{-}C{\equiv}N + H_2O \xrightarrow{H^+} R{-}\overset{\overset{\displaystyle OH}{|}}{C}{=}NH \rightleftharpoons R{-}\overset{\overset{\displaystyle O}{\|}}{C}{-}NH_2$$

An imidic acid (enol of an amide) — An amide

The reaction conditions required for acid-catalyzed hydrolysis of a cyano group are typically more vigorous than those required for hydrolysis of an amide, and in the presence of excess water, a cyano group is hydrolyzed first to an amide and then to a carboxylic acid. It is possible, however, to stop at the amide by using sulfuric acid as a catalyst and one mol of water per mol of nitrile.

Hydrolysis of a cyano group in aqueous base gives a carboxylic acid anion and ammonia. The reaction is driven to completion by the acid-base reaction between the carboxylic acid and base to form a carboxylic acid anion. Acidification of the reaction mixture during workup converts the carboxylate anion to the carboxylic acid.

$$CH_3(CH_2)_9C{\equiv}N + H_2O + NaOH \xrightarrow[\text{heat}]{H_2O} CH_3(CH_2)_9\overset{\overset{\displaystyle O}{\|}}{C}O^-Na^+ + NH_3$$

Undecanenitrile — Sodium undecanoate

$$CH_3(CH_2)_9\overset{\overset{\displaystyle O}{\|}}{C}O^-Na^+ + HCl \xrightarrow{H_2O} CH_3(CH_2)_9\overset{\overset{\displaystyle O}{\|}}{C}OH + NaCl$$

Sodium undecanoate — Undecanoic acid

Hydrolysis of a cyano group in aqueous base involves initial formation of the anion of an imidic acid, which, after proton transfer from water, undergoes keto-enol tautomerism to give an amide. The amide is then hydrolyzed by aqueous base, as we have seen earlier, to the carboxylic acid anion and ammonia.

MECHANISM Hydrolysis of a Cyano Group to an Amide in Aqueous Base

Step 1: Addition of hydroxide ion to carbon of the cyano group followed by proton transfer from water gives an imidic acid

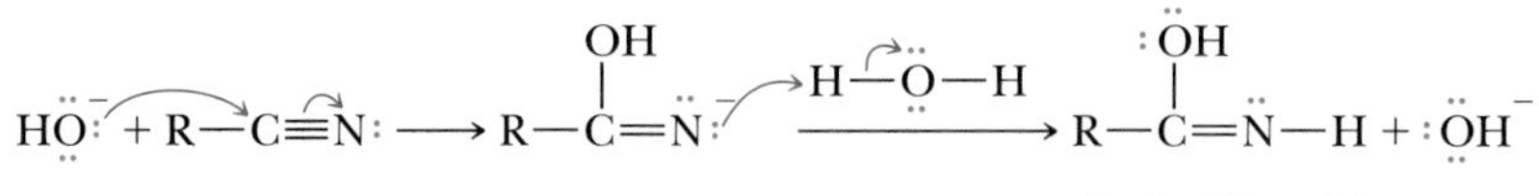

$$HO^{-} + R-C{\equiv}N: \longrightarrow R-\underset{}{\overset{OH}{\overset{|}{C}}}=N:^{-} \xrightarrow{H-O-H} R-\overset{:OH}{\overset{|}{C}}=N-H + :OH^{-}$$

An imidic acid

Step 2: Tautomerism of the imidic acid gives the amide

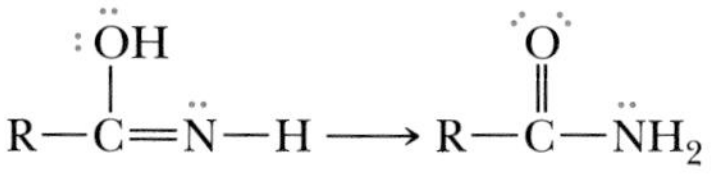

$$R-\overset{:OH}{\overset{|}{C}}=N-H \longrightarrow R-\overset{O}{\overset{\|}{C}}-NH_2$$

An imidic acid

Hydrolysis of nitriles is a valuable route to the synthesis of carboxylic acids from primary or secondary alkyl halides. In this route, one carbon in the form of a cyano group (Table 8.1) is added to a carbon chain and then converted to a carboxyl group.

$$CH_3(CH_2)_8CH_2Cl \xrightarrow[\text{ethanol, water}]{KCN} CH_3(CH_2)_9C{\equiv}N \xrightarrow[\text{heat}]{H_2SO_4,\ H_2O} CH_3(CH_2)_9\overset{O}{\overset{\|}{C}}OH$$

1-Chlorodecane | Undecanenitrile | Undecanoic acid

Hydrolysis of cyanohydrins, which are obtained by the addition of HCN to an aldehyde or ketone (Section 15.6D), provides a valuable route to α-hydroxycarboxylic acids, as illustrated by the synthesis of mandelic acid.

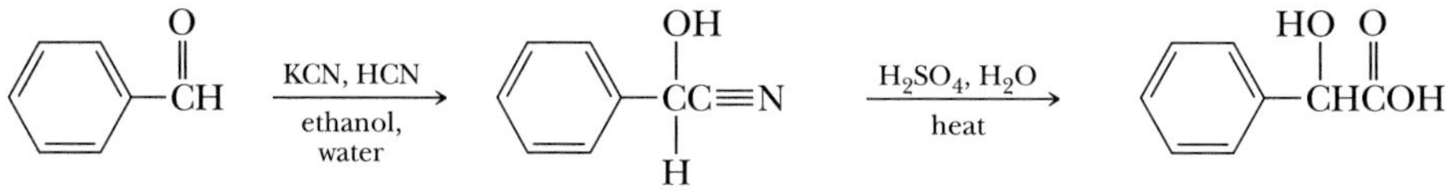

$$C_6H_5\overset{O}{\overset{\|}{C}}H \xrightarrow[\text{ethanol, water}]{KCN,\ HCN} C_6H_5\overset{OH}{\overset{|}{\underset{H}{\underset{|}{C}}}}C{\equiv}N \xrightarrow[\text{heat}]{H_2SO_4,\ H_2O} C_6H_5\overset{HO}{\overset{|}{C}}H\overset{O}{\overset{\|}{C}}OH$$

Benzaldehyde | Benzaldehyde cyanohydrin (Mandelonitrile) | 2-Hydroxyphenylacetic acid (Mandelic acid)

EXAMPLE 17.5

Show how to bring about the following conversions using as one step the hydrolysis of a cyano group:

(a) $(CH_3CH_2CH_2)_2CHCl \longrightarrow (CH_3CH_2CH_2)_2CHCO_2H$

4-Chloroheptane → 2-Propylpentanoic acid (Valproic acid)

(b) Cyclohexanone ($C_6H_{10}{=}O$) $\longrightarrow$ 1-hydroxycyclohexanecarboxylic acid (ring carbon bearing CO_2H and OH)

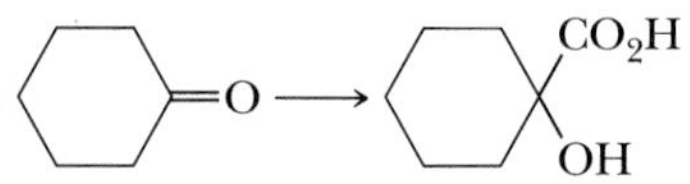

Solution

(a) Treatment of 4-chloroheptane with KCN in aqueous ethanol by an S_N2 pathway gives a nitrile. Hydrolysis of the cyano group in aqueous sulfuric acid gives the product.

$$(CH_3CH_2CH_2)_2CHCl \xrightarrow[\text{ethanol, water}]{\text{KCN}} (CH_3CH_2CH_2)_2CHCN \xrightarrow[\text{heat}]{H_2SO_4,\ H_2O} (CH_3CH_2CH_2)_2CHCO_2H$$

(b) Treatment of cyclohexanone with KCN/HCN in aqueous ethanol gives a cyanohydrin. Hydrolysis of the cyano group in concentrated sulfuric acid gives the carboxyl group of the product.

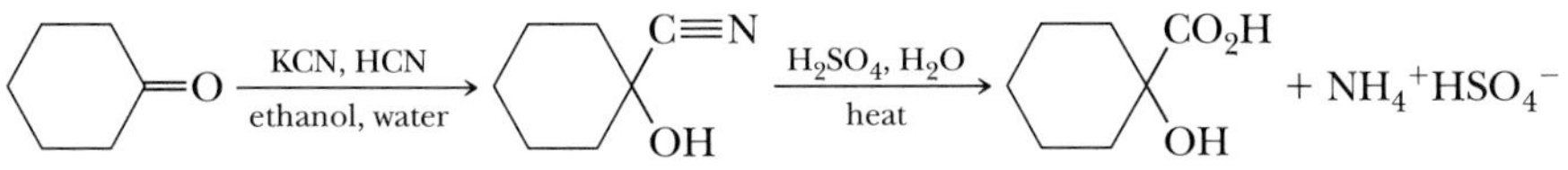

$$\text{cyclohexanone} \xrightarrow[\text{ethanol, water}]{\text{KCN, HCN}} \text{1-hydroxycyclohexane-1-}C{\equiv}N \xrightarrow[\text{heat}]{H_2SO_4,\ H_2O} \text{1-hydroxycyclohexane-1-}CO_2H + NH_4^+HSO_4^-$$

PROBLEM 17.5

Synthesis of nitriles by nucleophilic displacement of halide from an alkyl halide is practical only with primary and secondary alkyl halides. It fails with tertiary alkyl halides. Why? What is the major product of the following reaction?

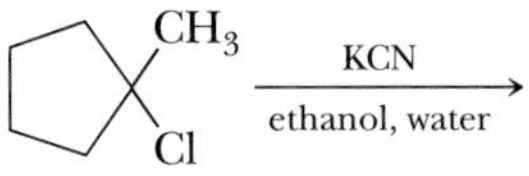

$$\text{1-chloro-1-methylcyclopentane } (CH_3, Cl) \xrightarrow[\text{ethanol, water}]{\text{KCN}}$$

17.6 Reaction with Alcohols

A. Acid Halides

An acid halide reacts with an alcohol to give an ester. Because acid halides are so reactive toward even weak nucleophiles, such as alcohols, no catalyst is necessary for these reactions.

$$R\overset{O}{\overset{\|}{C}}Cl + HOR' \longrightarrow R\overset{O}{\overset{\|}{C}}OR' + HCl$$

$$CH_3CH_2CH_2\overset{O}{\overset{\|}{C}}Cl + HO\text{—}C_6H_{11} \longrightarrow CH_3CH_2CH_2\overset{O}{\overset{\|}{C}}O\text{—}C_6H_{11} + HCl$$

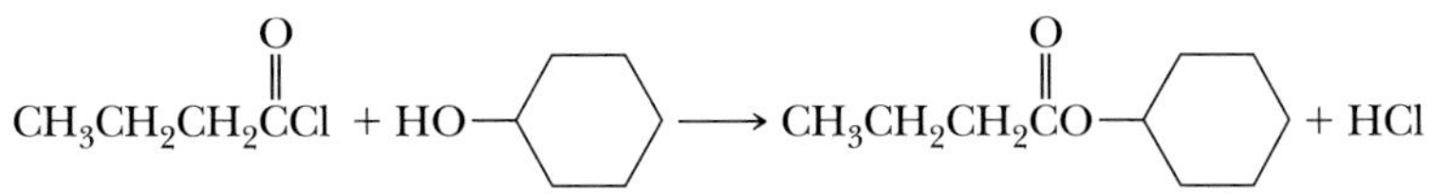

Butanoyl chloride Cyclohexanol Cyclohexyl butanoate

In cases in which the alcohol or resulting ester is sensitive to acid, the reaction is carried out in the presence of a tertiary amine to neutralize the HCl as it is formed. The amines most commonly used for this purpose are pyridine and triethylamine.

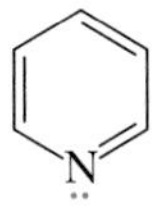

$$(CH_3CH_2)_3N\!:$$

Pyridine Triethylamine

When used for this purpose, each amine is converted to its hydrochloride salt. Pyridine, for example, is converted to pyridinium chloride, as illustrated by its use in the synthesis of isoamyl benzoate.

$$C_6H_5\overset{O}{\overset{\|}{C}}Cl + HOCH_2CH_2\overset{CH_3}{\overset{|}{C}}HCH_3 + C_5H_5N: \longrightarrow C_6H_5\overset{O}{\overset{\|}{C}}OCH_2CH_2\overset{CH_3}{\overset{|}{C}}HCH_3 + C_5H_5\overset{+}{N}H\ Cl^-$$

Benzoyl chloride | 3-Methyl-1-butanol (Isoamyl alcohol) | Pyridine | 3-Methylbutyl benzoate (Isoamyl benzoate) | Pyridinium chloride

Sulfonic acid esters are prepared by the reaction of an alkane- or arenesulfonyl chloride with an alcohol or phenol. Two of the most common sulfonyl chlorides are *p*-toluenesulfonyl chloride, abbreviated TsCl, and methanesulfonyl chloride, abbreviated MsCl (Section 17.1A).

$$CH_3{-}C_6H_4{-}SO_2{-}Cl + CH_3(CH_2)_5CH(CH_3)OH \xrightarrow{\text{pyridine}} CH_3(CH_2)_5CH(CH_3){-}O{-}SO_2{-}C_6H_4{-}CH_3$$

p-Toluenesulfonyl chloride (Tosyl chloride; TsCl) | (*R*)-2-Octanol | (*R*)-2-Octyl *p*-toluenesulfonate

As discussed in Section 9.6D, a special value of *p*-toluenesulfonic (tosylate) and methanesulfonic (mesylate) esters is that in forming them, an —OH is converted from a poor leaving group (hydroxide ion) in nucleophilic displacement to an excellent leaving group, the *p*-toluenesulfonate (tosylate) or methanesulfonate (mesylate) anions.

B. Acid Anhydrides

Acid anhydrides react with alcohols to give one mol of ester and one mol of a carboxylic acid. Thus, the reaction of an alcohol with an anhydride is a useful method for the synthesis of esters.

$$CH_3\overset{O}{\overset{\|}{C}}O\overset{O}{\overset{\|}{C}}CH_3 + HOCH_2CH_3 \longrightarrow CH_3\overset{O}{\overset{\|}{C}}OCH_2CH_3 + CH_3\overset{O}{\overset{\|}{C}}OH$$

Acetic anhydride | Ethanol | Ethyl acetate | Acetic acid

$$\text{Phthalic anhydride} + HO\overset{CH_3}{\overset{|}{C}}HCH_2CH_3 \longrightarrow C_6H_4(\overset{O}{\overset{\|}{C}}O\overset{CH_3}{\overset{|}{C}}HCH_2CH_3)(\overset{O}{\overset{\|}{C}}OH)$$

Phthalic anhydride | 2-Butanol (*sec*-Butyl alcohol) | *sec*-Butyl hydrogen phthalate

Aspirin is synthesized on an industrial scale by the reaction of acetic anhydride and salicylic acid.

$$\text{2-Hydroxybenzoic acid (Salicylic acid)} + \underset{\text{Acetic anhydride}}{CH_3\overset{O}{\overset{\|}{C}}O\overset{O}{\overset{\|}{C}}CH_3} \longrightarrow \text{Acetylsalicylic acid (Aspirin)} + \underset{\text{Acetic acid}}{CH_3\overset{O}{\overset{\|}{C}}OH}$$

2-Hydroxybenzoic acid (Salicylic acid) — Acetic anhydride — Acetylsalicylic acid (Aspirin) — Acetic acid

C. Esters

Esters react with alcohols in an acid-catalyzed reaction called **transesterification.** For example, it is possible to convert methyl acrylate to butyl acrylate by heating the methyl ester with 1-butanol in the presence of an acid catalyst. The acids most commonly used are HCl as a gas bubbled into the reaction medium and *p*-toluenesulfonic acid.

Transesterification Exchange of the —OR or —OAr group of an ester for another —OR or —OAr group.

$$\underset{\substack{\text{Methyl propenoate}\\\text{(Methyl acrylate)}\\\text{(bp 81°C)}}}{CH_2{=}CH\overset{O}{\overset{\|}{C}}OCH_3} + \underset{\substack{\text{1-Butanol}\\\text{(bp 117°C)}}}{HOCH_2CH_2CH_2CH_3} \overset{HCl}{\rightleftharpoons} \underset{\substack{\text{Butyl propenoate}\\\text{(Butyl acrylate)}\\\text{(bp 147°C)}}}{CH_2{=}CH\overset{O}{\overset{\|}{C}}OCH_2CH_2CH_2CH_3} + \underset{\substack{\text{Methanol}\\\text{(bp 65°C)}}}{CH_3OH}$$

Transesterification is an equilibrium reaction and can be driven in either direction by control of experimental conditions. For example, in the reaction of methyl acrylate with 1-butanol, transesterification is carried out at a temperature slightly above the boiling point of methanol (the lowest boiling component in the mixture). Methanol distills from the reaction mixture, thus shifting the position of equilibrium in favor of butyl acrylate. Conversely, reaction of butyl acrylate with a large excess of methanol shifts the equilibrium to favor formation of methyl acrylate.

EXAMPLE 17.6

Complete the following transesterification reactions. The stoichiometry of each is given in the problem.

(a) $CH_3\overset{O}{\overset{\|}{C}}OCH_2(CH_2)_8CH_3 + CH_3OH \xrightarrow{H^+}$ **(b)** $CH_3\overset{O}{\overset{\|}{C}}OCH_2CH_2O\overset{O}{\overset{\|}{C}}CH_3 + 2CH_3OH \xrightarrow{H^+}$

Solution

(a) $CH_3\overset{O}{\overset{\|}{C}}OCH_3 + HOCH_2(CH_2)_8CH_3$ **(b)** $2CH_3\overset{O}{\overset{\|}{C}}OCH_3 + HOCH_2CH_2OH$

PROBLEM 17.6

Complete the following transesterification reaction. The stoichiometry is given in the equation.

$$2\,C_6H_5\text{—}\overset{O}{\overset{\|}{C}}OCH_3 + HOCH_2CH_2OH \xrightarrow{H^+}$$

D. Amides

Amides, the least reactive of the functional derivatives of carboxylic acids, do not react with alcohols. Thus, the reaction of an amide with an alcohol cannot be used to prepare an ester.

17.7 Reactions with Ammonia and Amines

A. Acid Halides

Acid halides react readily with ammonia and primary or secondary amines to form amides. For complete conversion of an acid halide to an amide, two equivalents of ammonia or amine are used; one to form the amide and one to neutralize the hydrogen halide formed.

$$\underset{\text{Hexanoyl chloride}}{CH_3(CH_2)_4\overset{O}{\overset{\|}{C}}Cl} + \underset{\text{Ammonia}}{2NH_3} \longrightarrow \underset{\text{Hexanamide}}{CH_3(CH_2)_4\overset{O}{\overset{\|}{C}}NH_2} + \underset{\text{Ammonium chloride}}{NH_4^+Cl^-}$$

The mechanism for the reaction between an acid chloride and ammonia or a primary or secondary amine begins in Step 1 by addition of ammonia or the amine to the carbonyl carbon to form a tetrahedral carbonyl addition intermediate, which collapses in Step 2 to eject chloride, a good leaving group. Proton transfer in Step 3 completes formation of the amide and ammonium ion.

MECHANISM Reaction Between an Acid Chloride and Ammonia to Form an Amide

Addition to the carbonyl carbon in Step 1 forms a tetrahedral carbonyl addition intermediate, which collapses in Step 2 to regenerate the carbonyl group. Proton transfer to ammonia in Step 3 completes formation of the amide.

$$R-\overset{O}{\overset{\|}{C}}-Cl + :NH_3 \xrightarrow[(1)]{} R-\underset{H-\underset{H}{\overset{|}{N^+}}-H}{\overset{:\ddot{O}:^-}{\overset{|}{\underset{|}{C}}}}-Cl \xrightarrow[(2)]{-Cl^-} R-\underset{H-\underset{H}{\overset{|}{N^+}}-H}{\overset{O}{\overset{\|}{\underset{|}{C}}}} \xrightarrow[(3)]{:NH_3} R-\overset{O}{\overset{\|}{C}}-NH_2 + NH_4^+$$

B. Acid Anhydrides

Acid anhydrides react with ammonia, as well as with primary and secondary amines to form amides. As with acid halides, 2 mol of amine are required: 1 mol to form the amide and 1 mol to neutralize the carboxylic acid byproduct.

$$\underset{\text{Acetic anhydride}}{CH_3\overset{O}{\overset{\|}{C}}O\overset{O}{\overset{\|}{C}}CH_3} + \underset{\text{Ammonia}}{2NH_3} \longrightarrow \underset{\substack{\text{Ethanamide}\\\text{(Acetamide)}}}{CH_3\overset{O}{\overset{\|}{C}}NH_2} + \underset{\text{Ammonium acetate}}{CH_3\overset{O}{\overset{\|}{C}}O^- NH_4^+}$$

Alternatively, if the amine used to make the amide is expensive, a tertiary amine such as triethylamine may be used to neutralize the carboxylic acid.

C. Esters

Esters react with ammonia and primary or secondary amines to form amides. Because an alkoxide anion is a poorer leaving group than either a halide ion or a carboxylate ion, esters are less reactive toward ammonia, primary amines, and secondary amines than are acid halides or acid anhydrides.

$$\underset{\text{Ethyl phenylacetate}}{C_6H_5CH_2\overset{O}{\overset{\|}{C}}OCH_2CH_3} + NH_3 \longrightarrow \underset{\text{Phenylacetamide}}{C_6H_5CH_2\overset{O}{\overset{\|}{C}}NH_2} + CH_3CH_2OH$$

D. Amides

Amides do not react with ammonia, or primary or secondary amines.

EXAMPLE 17.7

Complete the following reactions. The stoichiometry of each reaction is given in the equation.

(a) $\underset{\text{Ethyl 2-butenoate}}{CH_3CH{=}CH\overset{O}{\overset{\|}{C}}OCH_2CH_3} + NH_3 \longrightarrow$ **(b)** $\underset{\text{Diethyl carbonate}}{CH_3CH_2O\overset{O}{\overset{\|}{C}}OCH_2CH_3} + 2NH_3 \longrightarrow$

Solution

(a) $\underset{\text{2-Butenamide}}{CH_3CH{=}CH\overset{O}{\overset{\|}{C}}NH_2} + CH_3CH_2OH$ **(b)** $\underset{\text{Urea}}{H_2N\overset{O}{\overset{\|}{C}}NH_2} + 2CH_3CH_2OH$

PROBLEM 17.7

Complete and balance equations for the following reactions. The stoichiometry of each reaction is given in the equation.

(a) $CH_3\overset{O}{\overset{\|}{C}}O-C_6H_4-O\overset{O}{\overset{\|}{C}}CH_3 + 2NH_3 \longrightarrow$ (para-disubstituted benzene ring) **(b)** (six-membered cyclic ester, δ-valerolactone) $+ NH_3 \longrightarrow$

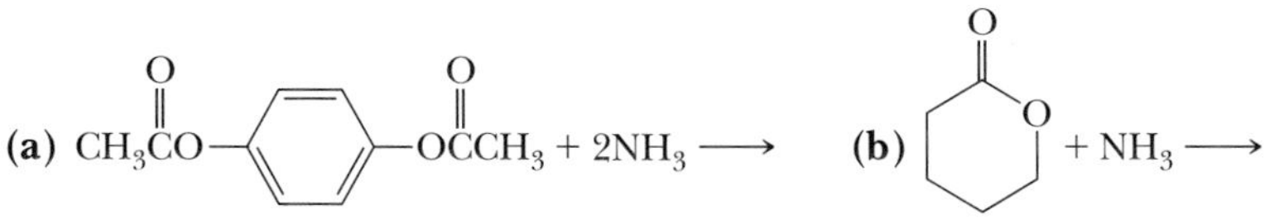

17.8 Reaction of Acid Chlorides with Salts of Carboxylic Acids

Acid chlorides react with salts of carboxylic acids to give anhydrides. Most commonly used are the sodium or potassium salts.

$$\underset{\text{Acetyl chloride}}{CH_3\overset{O}{\overset{\|}{C}}Cl} + \underset{\text{Sodium benzoate}}{Na^{+}\,{}^{-}O\overset{O}{\overset{\|}{C}}-C_6H_5} \longrightarrow \underset{\text{Acetic benzoic anhydride}}{CH_3\overset{O}{\overset{\|}{C}}O\overset{O}{\overset{\|}{C}}-C_6H_5} + Na^{+}Cl^{-}$$

Reaction of an acid halide with the anion of a carboxylic acid is a particularly useful method for synthesis of mixed anhydrides.

17.9 Reactions with Organometallic Compounds

A. Grignard Reagents

Treatment of a formic ester with two mol of a Grignard reagent followed by hydrolysis of the magnesium alkoxide salt in aqueous acid gives a secondary alcohol. Treatment of an ester other than a formate with a Grignard reagent gives a tertiary alcohol in which two of the groups bonded to the carbon bearing the —OH group are the same.

$$\underset{\substack{\text{An ester of}\\\text{formic acid}}}{H\overset{O}{\overset{\|}{C}}OCH_3} + 2RMgX \longrightarrow \begin{matrix}\text{magnesium}\\\text{alkoxide}\\\text{salt}\end{matrix} \xrightarrow{H_2O,\ HCl} \underset{\text{A secondary alcohol}}{H\overset{OH}{\overset{|}{\underset{R}{\underset{|}{C}}}}R} + CH_3OH$$

$$\underset{\substack{\text{An ester of any acid}\\\text{other than formic acid}}}{CH_3\overset{O}{\overset{\|}{C}}OCH_3} + 2RMgX \longrightarrow \begin{matrix}\text{magnesium}\\\text{alkoxide}\\\text{salt}\end{matrix} \xrightarrow{H_2O,\ HCl} \underset{\text{A tertiary alcohol}}{CH_3\overset{OH}{\overset{|}{\underset{R}{\underset{|}{C}}}}R} + CH_3OH$$

The reaction of an ester and a Grignard reagent begins with (1) addition of one mol of Grignard reagent to the carbonyl carbon to form a tetrahedral carbonyl addition intermediate. Because alkoxide ion is a moderately good leaving group, this addition intermediate (2) collapses to give a new carbonyl-containing compound and a magnesium alkoxide salt. This new carbonyl-containing compound then (3) reacts with a second mol of Grignard reagent to form a second tetrahedral carbonyl addition compound, which, after hydrolysis in aqueous acid, gives a tertiary alcohol (or a secondary alcohol if the starting ester was a formate). It is important to realize that it is not possible to use RMgX and an ester to prepare a ketone; the intermediate ketone is more reactive than the ester and reacts immediately with Grignard reagent to give a tertiary alcohol.

MECHANISM Reaction of an Ester with a Grignard Reagent

$$CH_3-\overset{\overset{\ddot{O}}{\|}}{C}-OCH_3 + R-MgX \longrightarrow CH_3-\underset{R}{\overset{:\ddot{O}:^-[MgX]^+}{|}}{C}-OCH_3 \longrightarrow CH_3-\underset{R}{\overset{\overset{\ddot{O}}{\|}}{C}} + CH_3O^-[MgX]^+$$

Magnesium salt — A ketone

$$CH_3-\underset{R}{\overset{\overset{\ddot{O}}{\|}}{C}} + R-MgX \longrightarrow CH_3-\underset{R}{\overset{:\ddot{O}:^-[MgX]^+}{C}}-R \xrightarrow{H_2O,\ HCl} CH_3-\underset{R}{\overset{:\ddot{O}H}{C}}-R$$

A ketone — Magnesium salt — A tertiary alcohol

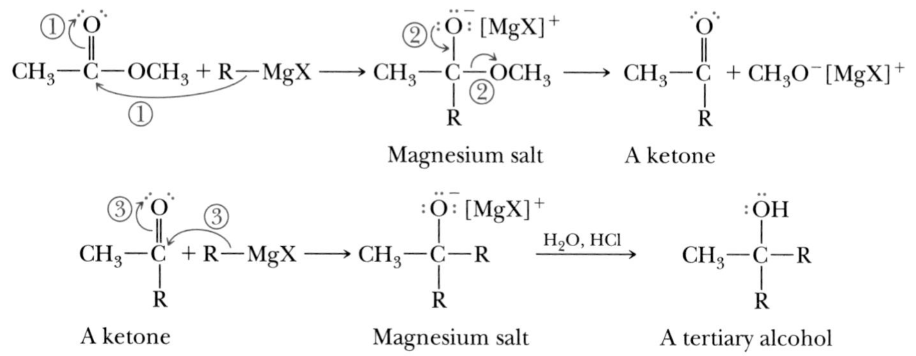

EXAMPLE 17.8

Complete these Grignard reactions:

(a) $H\overset{O}{\overset{\|}{C}}OCH_3 \xrightarrow[2.\ H_2O,\ HCl]{1.\ 2CH_3CH_2CH_2MgBr}$ (b) $CH_3CH_2CH_2\overset{O}{\overset{\|}{C}}OCH_3 \xrightarrow[2.\ H_2O,\ HCl]{1.\ 2PhMgBr}$

Solution

Sequence (a) gives a secondary alcohol, and sequence (b) gives a tertiary alcohol.

(a) $CH_3CH_2CH_2\overset{OH}{\overset{|}{C}}HCH_2CH_2CH_3$ (b) $CH_3CH_2CH_2\underset{Ph}{\overset{OH}{C}}Ph$

PROBLEM 17.8

Show how to prepare these alcohols by treatment of an ester with a Grignard reagent.

(a) 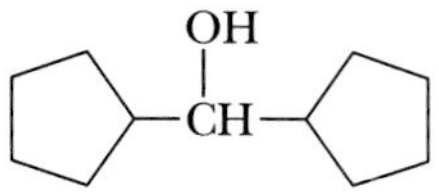 cyclopentyl–$\overset{OH}{\overset{|}{C}}H$–cyclopentyl (b) $CH_2=CHCH_2\underset{Ph}{\overset{OH}{C}}CH_2CH=CH_2$

B. Organolithium Compounds

Organolithium compounds are even more powerful nucleophiles than Grignard reagents and react with esters to give the same types of secondary and tertiary alcohols as shown for Grignard reagents, often in higher yields.

$$R\overset{O}{\overset{\|}{C}}OCH_3 \xrightarrow[2.\ H_2O,\ HCl]{1.\ 2R'Li} R-\underset{R'}{\overset{OH}{C}}-R'$$

C. Lithium Diorganocuprates

Acid chlorides react readily with lithium diorganocopper (Gilman) reagents to give ketones, as illustrated by the conversion of pentanoyl chloride to 2-hexanone. The reaction is carried out at − 78°C in either diethyl ether or tetrahydrofuran. Following hydrolysis in aqueous acid, the ketone is isolated in good yield.

$$\underset{\text{Pentanoyl chloride}}{CH_3(CH_2)_3\overset{\overset{\large O}{\|}}{C}Cl} \xrightarrow[2.\ H_2O]{1.\ (CH_3)_2CuLi,\ ether,\ -78°C} \underset{\text{2-Hexanone}}{CH_3(CH_2)_3\overset{\overset{\large O}{\|}}{C}CH_3}$$

Notice that, under these conditions, the ketone does not react further. This contrasts with the reaction of an ester with a Grignard reagent or organolithium compound, where the intermediate ketone reacts with a second mol of the organometallic compound to give an alcohol. The reason for this difference in reactivity is that the tetrahedral carbonyl addition intermediate is stable at −78°C; it survives until the workup causes it to decompose to the ketone. R_2CuLi reagents react readily only with acid chlorides; they do not react with esters, amides, acid anhydrides, or nitriles. The following molecule contains both an acid chloride and an ester group. When treated with lithium dimethylcopper, only the acid chloride reacts.

$$CH_3O\overset{\overset{\large O}{\|}}{C}CH_2CH_2\overset{\overset{\large O}{\|}}{C}Cl \xrightarrow[2.\ H_2O]{1.\ (CH_3)_2CuLi,\ ether,\ -78°C} CH_3O\overset{\overset{\large O}{\|}}{C}CH_2CH_2\overset{\overset{\large O}{\|}}{C}CH_3$$

EXAMPLE 17.9

Show how to bring about these conversions in good yield.

(a) 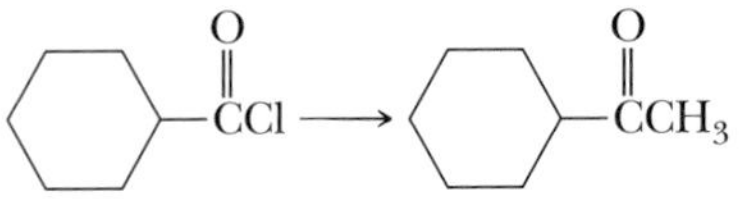 $C_6H_{11}\overset{\overset{\large O}{\|}}{C}Cl \longrightarrow C_6H_{11}\overset{\overset{\large O}{\|}}{C}CH_3$ (cyclohexyl)

(b) 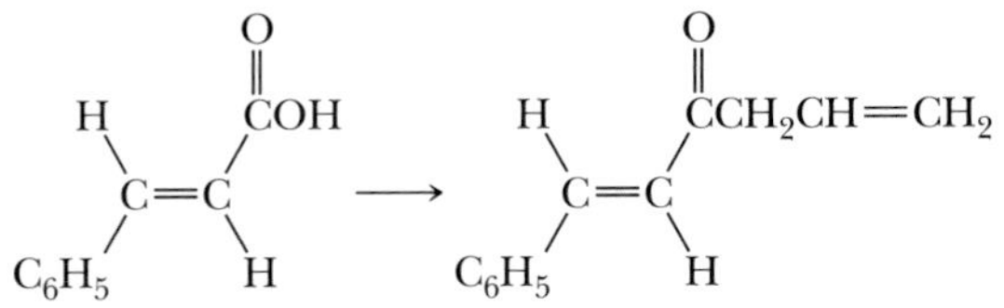 (E)-$C_6H_5CH{=}CH\overset{\overset{\large O}{\|}}{C}OH \longrightarrow$ (E)-$C_6H_5CH{=}CH\overset{\overset{\large O}{\|}}{C}CH_2CH{=}CH_2$ (H and H trans on the C=C)

Solution

(a) Treat the acid chloride with lithium dimethylcopper followed by H_2O.

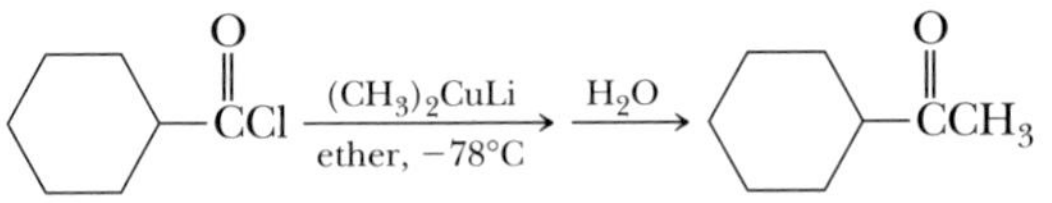

$$C_6H_{11}\overset{\overset{\large O}{\|}}{C}Cl \xrightarrow[ether,\ -78°C]{(CH_3)_2CuLi} \xrightarrow{H_2O} C_6H_{11}\overset{\overset{\large O}{\|}}{C}CH_3$$

(b) Treat the carboxylic acid with thionyl chloride to form the acid chloride, followed by treatment with lithium diallylcopper and then aqueous acid.

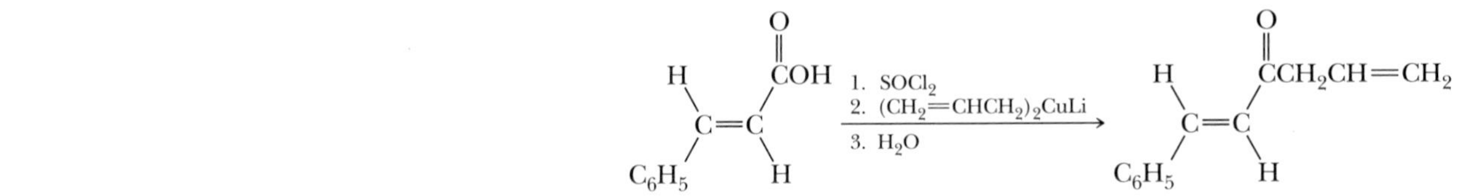

$$C_6H_5CH{=}CH\overset{\overset{\large O}{\|}}{C}OH \xrightarrow[3.\ H_2O]{1.\ SOCl_2\quad 2.\ (CH_2{=}CHCH_2)_2CuLi} C_6H_5CH{=}CH\overset{\overset{\large O}{\|}}{C}CH_2CH{=}CH_2$$

PROBLEM 17.9

Show how to bring about these conversions in good yield.

(a) $C_6H_5COH \longrightarrow C_6H_5C(CH_2)_5CH_3$ (each carbonyl carbon bearing =O) (b) $CH_2{=}CHCl \longrightarrow CH_2{=}CHC(CH_2)_4CH_3$ (C=O)

17.10 Interconversion of Functional Derivatives

We have seen throughout the past several sections that acid chlorides are the most reactive toward nucleophilic acyl substitution and that amides are the least reactive.

Amide < Ester < Acid anhydride < Acid halide

Increasing reactivity toward nucleophilic acyl substitution

Another useful way to think about the relative reactivities of these four functional derivatives of carboxylic acids is summarized in Figure 17.4. Any functional group lower in this figure can be prepared from any functional group above it by treatment with an appropriate oxygen or nitrogen nucleophile. An acid chloride, for example, can be converted to an acid anhydride, an ester, an amide, or a carboxylic acid. Acid anhydrides, esters, and amides, however, do not react with chloride ion to give acid chlorides.

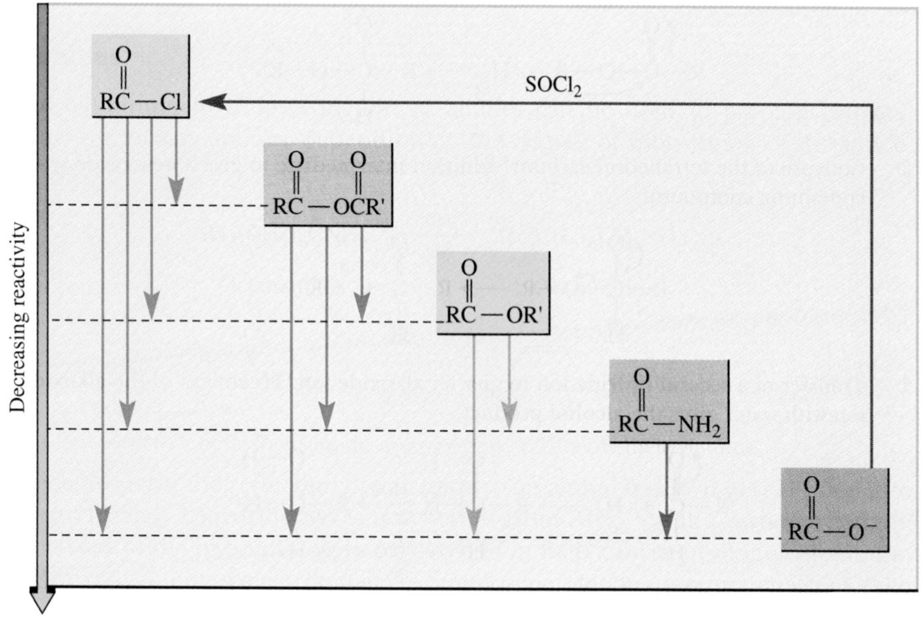

FIGURE 17.4 Relative reactivities of carboxyl derivatives toward nucleophilic acyl substitution. A more reactive derivative may be converted to a less reactive derivative by treatment with an appropriate reagent. Treatment of a carboxylic acid with thionyl chloride (the acid chloride of sulfurous acid) converts it to the more reactive acid chloride. Carboxylic acids are about as reactive as esters under acidic conditions, but are converted to the unreactive carboxylates under basic conditions.

adds a second hydride ion in Step 4 to complete the reduction. R_3AlO^- is a reasonably good leaving group; recall that $Al(OH)_3$ is somewhat acidic.

MECHANISM Reduction of an Amide by Lithium Aluminum Hydride

$$RC(=O)NH_2 + H{-}AlH_3^- \xrightarrow{(1)} R{-}CH(O^-{-}AlH_3)NH_2 \xrightarrow{(2)} R{-}CH(O{-}AlH_3^-)NH_2 \xrightarrow{(3)} R{-}CH{=}N^+H_2 \xrightarrow{(4)\ H^-} R{-}CH_2{-}NH_2$$

An iminium ion

EXAMPLE 17.10

Show how to bring about the following conversions in good yield.

(a) $C_6H_5COOH \longrightarrow C_6H_5CH_2{-}N(CH_2)_4$ (pyrrolidine) **(b)** $C_6H_{11}COOH \longrightarrow C_6H_{11}CH_2NHCH_3$ (cyclohexyl)

Solution

The key in each part is to convert the carboxylic acid to an amide and then reduce the amide with $LiAlH_4$ (Section 17.11B). The amide can be prepared by treatment of the carboxylic acid with $SOCl_2$ to form the acid chloride (Section 16.9) and then treatment of the acid chloride with an amine (Section 17.7A). Alternatively, the carboxylic acid can be converted to an ethyl ester by Fischer esterification (Section 16.8A) and the ester treated with an amine to give the amide. Solution (a) uses the acid chloride route and solution (b) uses the ester route.

(a) $C_6H_5COOH \xrightarrow{SOCl_2} C_6H_5COCl \xrightarrow{HN(CH_2)_4} C_6H_5C(=O){-}N(CH_2)_4 \xrightarrow[2.\ H_2O]{1.\ LiAlH_4} C_6H_5CH_2{-}N(CH_2)_4$

(b) $C_6H_{11}COOH \xrightarrow[2.\ CH_3NH_2]{1.\ CH_3CH_2OH,\ H^+} C_6H_{11}CNHCH_3\ (C{=}O) \xrightarrow[4.\ H_2O]{3.\ LiAlH_4} C_6H_{11}CH_2NHCH_3$

PROBLEM 17.10

Show how to convert hexanoic acid to each amine in good yield.

(a) $CH_3(CH_2)_5N(CH_3)_2$ **(b)** $CH_3(CH_2)_5NHCH(CH_3)_2$

EXAMPLE 17.11

Show how to convert phenylacetic acid to these compounds:

(a) $PhCH_2COCH_3$ (C=O) **(b)** $PhCH_2CNH_2$ (C=O) **(c)** $PhCH_2CH_2NH_2$ **(d)** $PhCH_2CH_2OH$

Solution

Prepare methyl ester (a) by Fischer esterification (Section 16.8A) of phenylacetic acid with methanol and then treat this ester with ammonia to prepare amide (b). Alternatively, treat phenylacetic acid with thionyl chloride (Section 16.9) to give an acid chloride. Then treat this acid chloride with ammonia to give amide (b). Reduction of amide (b) by $LiAlH_4$ gives the primary amine (c). Similar reduction of either phenylacetic acid or ester (a) gives alcohol (d).

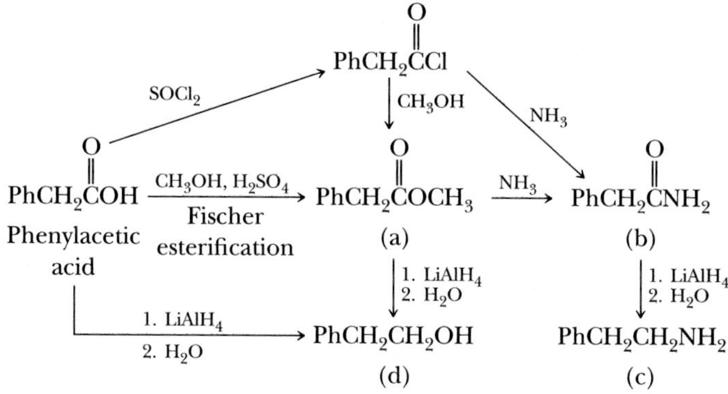

PROBLEM 17.11

Show how to convert (*R*)-2-phenylpropanoic acid to these compounds:

(a) (*R*)-$PhCHCH_2OH$ with CH_3 on the CH — (*R*)-2-Phenyl-1-propanol

(b) (*R*)-$PhCHCH_2NH_2$ with CH_3 on the CH — (*R*)-2-Phenyl-1-propanamine

C. Nitriles

A cyano group of a nitrile is reduced by lithium aluminum hydride to a primary amino group. Reduction of cyano groups is useful for the preparation of primary amines only.

$$\text{Br-}C_6H_3(\text{Br})\text{-C}\equiv\text{N} \xrightarrow[\text{2. } H_2O]{\text{1. } LiAlH_4\text{, ether}} \text{Br-}C_6H_3(\text{Br})\text{-}CH_2NH_2$$

17.12 The Hofmann Rearrangement

When a primary amide is treated with bromine or chlorine in aqueous sodium or potassium hydroxide, the following rearrangement takes place; the amide is converted to a primary amine with one fewer carbon atom than the starting amide and the carbonyl carbon of the amide is lost as carbonate ion. Furthermore, when the migrating group is chiral, it migrates with complete retention of configuration.

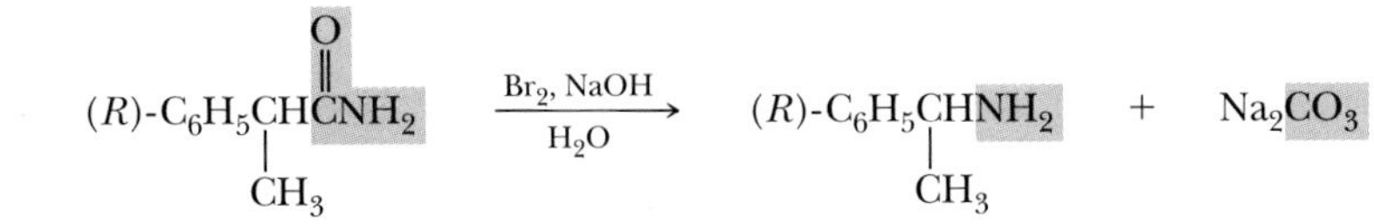

$$(R)\text{-}C_6H_5CH(CH_3)C(=O)NH_2 \xrightarrow[H_2O]{Br_2,\ NaOH} (R)\text{-}C_6H_5CH(CH_3)NH_2 + Na_2CO_3$$

(*R*)-2-Phenylpropanamide (*R*)-1-Phenylethanamine

This rearrangement of primary amides was discovered by the German chemist August Hofmann, for whom it has been named. A mechanism for the Hofmann rearrangement can be divided into four stages, each having some analogy to reactions we have already studied.

MECHANISM The Hofmann Rearrangement of Primary Amides

Stage 1: An acid-base reaction between the amide and hydroxide ion gives an amide anion, which is both a base and a nucleophile. Reaction of this anion with bromine by nucleophilic displacement gives an *N*-bromoamide.

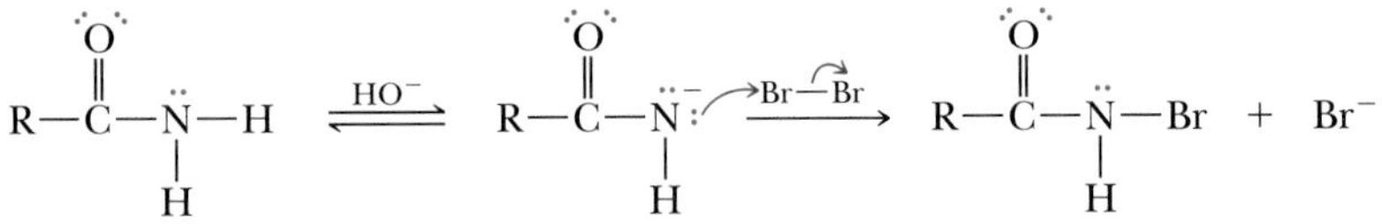

$$R{-}C(=O){-}NH{-}H \overset{HO^-}{\rightleftharpoons} R{-}C(=O){-}\ddot{N}^-{-}H \xrightarrow{Br-Br} R{-}C(=O){-}NH{-}Br + Br^-$$

An amide anion An *N*-bromoamide

Stage 2: The second amide hydrogen is abstracted by base followed by elimination of Br^- (an α-elimination) to give an acyl nitrene, an unstable species containing a neutral, electron-deficient nitrogen atom. Migration of the adjacent R group with its bonding electrons to the electron-deficient nitrogen gives an isocyanate, a molecule in which all atoms have complete valence shells.

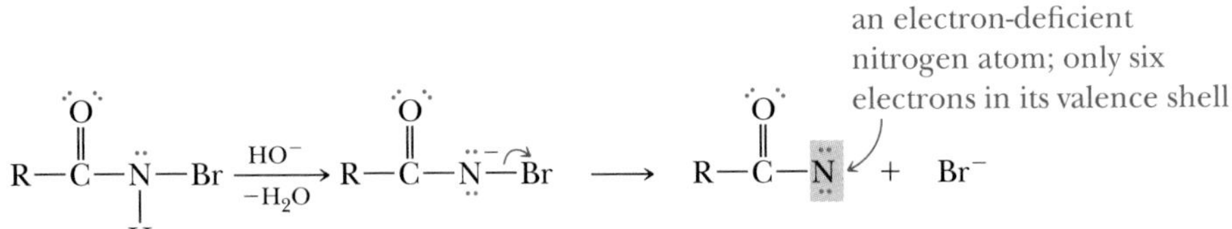

$$R{-}C(=O){-}NH{-}Br \xrightarrow[-H_2O]{HO^-} R{-}C(=O){-}\ddot{N}^-{-}Br \longrightarrow R{-}C(=O){-}\ddot{N}: + Br^-$$

An acyl nitrene

an electron-deficient nitrogen atom; only six electrons in its valence shell

$$R{-}C(=O){-}\ddot{N}: \longrightarrow R{-}\ddot{N}{=}C{=}\ddot{O}:$$

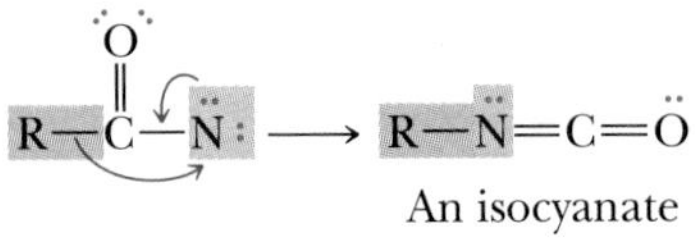

An isocyanate

Stage 3: Reaction of the isocyanate with water gives a **carbamic acid,** the functional group of which is a carboxyl group bonded to nitrogen.

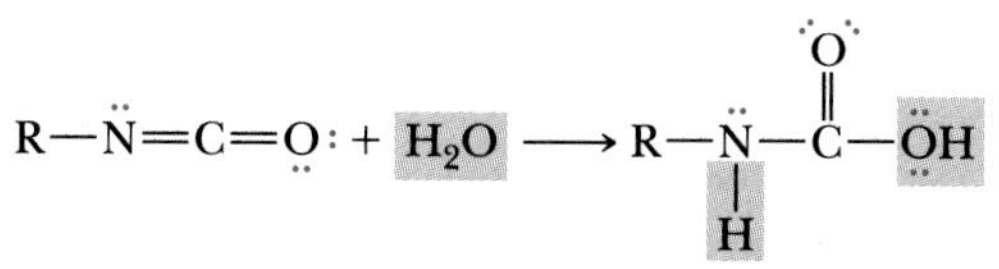

$$R{-}\ddot{N}{=}C{=}\ddot{O}: + H_2O \longrightarrow R{-}\underset{\displaystyle H}{\underset{|}{\ddot{N}}}{-}\overset{\displaystyle :\ddot{O}:}{\overset{\|}{C}}{-}\ddot{O}H$$

A carbamic acid

Stage 4: Carbamic acids are unstable species and undergo decarboxylation to give a primary amine and carbon dioxide.

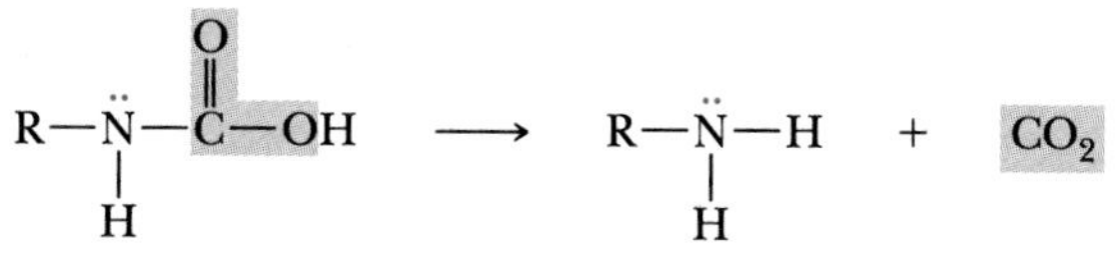

$$R{-}\underset{\displaystyle H}{\underset{|}{\ddot{N}}}{-}\overset{\displaystyle O}{\overset{\|}{C}}{-}OH \longrightarrow R{-}\underset{\displaystyle H}{\underset{|}{\ddot{N}}}{-}H + CO_2$$

A primary amine

EXAMPLE 17.12

Complete these equations:

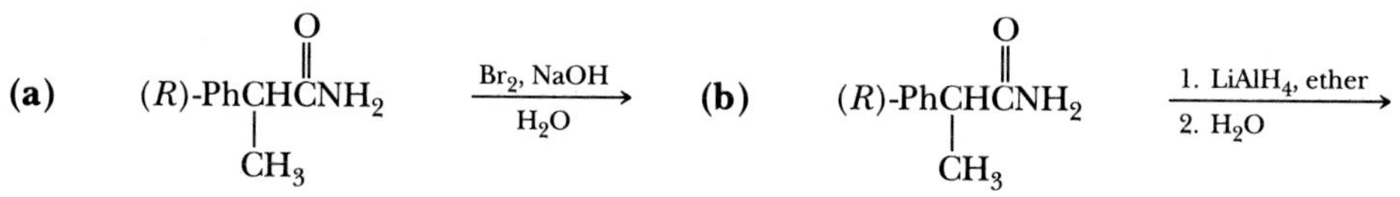

(a) $(R)\text{-PhCH}\overset{\displaystyle O}{\overset{\|}{C}}NH_2$ with CH_3 on the CH; $\xrightarrow[H_2O]{Br_2,\ NaOH}$

(*R*)-2-Phenylpropanamide

(b) $(R)\text{-PhCH}\overset{\displaystyle O}{\overset{\|}{C}}NH_2$ with CH_3 on the CH; $\xrightarrow[2.\ H_2O]{1.\ LiAlH_4,\ ether}$

(*R*)-2-Phenylpropanamide

Solution

Each reaction proceeds with complete retention of configuration at the stereocenter.

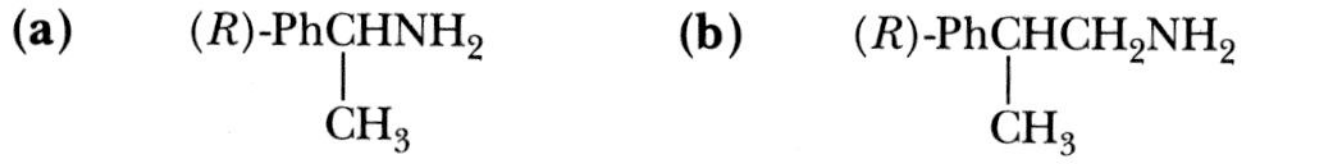

(a) $(R)\text{-PhCHNH}_2$ with CH_3 on the CH

(*R*)-1-Phenylethanamine

(b) $(R)\text{-PhCHCH}_2NH_2$ with CH_3 on the CH

(*R*)-2-Phenyl-1-propanamine

PROBLEM 17.12

Show how to convert phenylacetic acid into the following in good yield.

(a) $PhCH_2CH_2NH_2$ **(b)** $PhCH_2NH_2$

CHEMISTRY IN ACTION

Carbamate Insecticides

Acetylcholinesterase (AChE) is the enzyme that functions in synapses to catalyze the hydrolysis of the neurotransmitter acetylcholine into choline and acetate ion.

$$\underset{\text{Acetylcholine}}{CH_3\overset{O}{\overset{\|}{C}}OCH_2CH_2\overset{+}{N}(CH_3)_3} + 2H_2O \xrightarrow{\text{AChE}}$$

$$CH_3\overset{O}{\overset{\|}{C}}O^- + \underset{\text{Choline}}{HOCH_2CH_2\overset{+}{N}(CH_3)_3} + H_3O^+$$

If this enzyme is blocked, acetylcholine builds up, leading to continuous nerve stimulation, which can produce convulsions, paralysis, and eventually death.

Certain carbamates are effective blockers of acetylcholinesterase, especially the variety found in insects. This makes carbamates one of the most useful families of pesticides. Carbamates are esters of carbamic acid, H_2NCO_2H, which in turn is the monoamide of carbonic acid. Carbamic acid is unstable and undergoes decarboxylation to give carbon dioxide and ammonia. Esters of carbamic acid are stable compounds.

At the molecular level, AChE has a binding site for the trimethylammonium group of acetylcholine and another for the ester group. The binding site for the ester group is also the catalytic site for hydrolysis of the ester group of acetylcholine. This enzyme catalyzes the hydrolysis of the ester group by first transferring the acetyl group from the oxygen of choline to an oxygen in the binding-catalytic site of the enzyme. This transfer leads to the transient formation of an "acyl enzyme" intermediate. The acetyl group is then hydrolyzed from the enzyme's —OH, and the enzyme is regenerated for further catalysis.

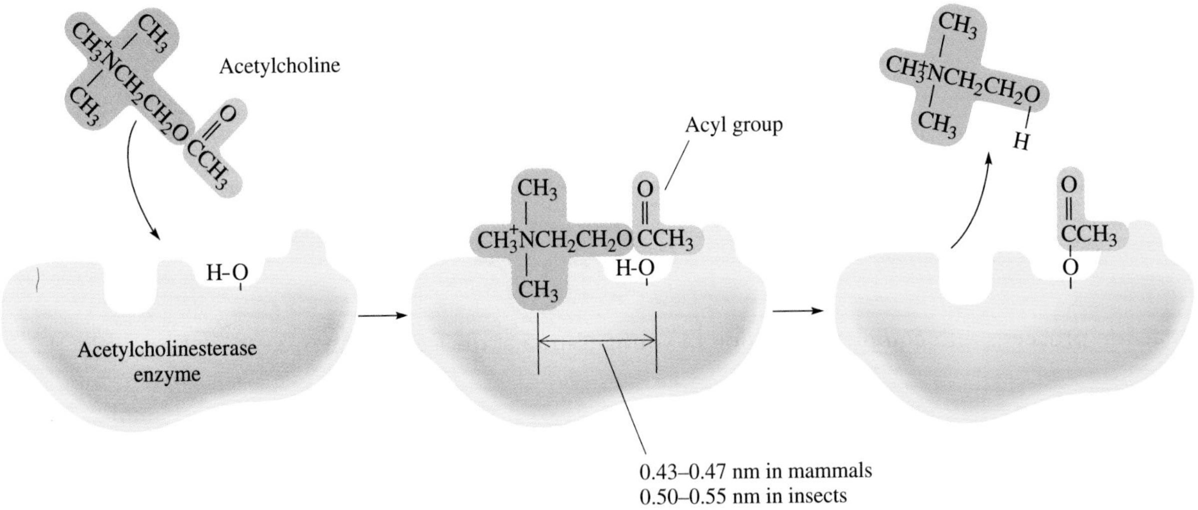

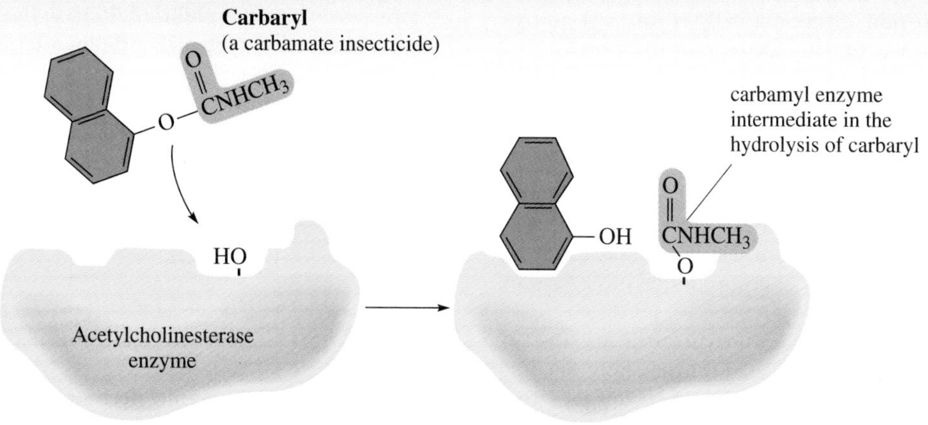

Carbaryl and other carbamate insecticides also bind to the active site of AChE and undergo hydrolysis to form a carbamyl-enzyme intermediate.

The carbamyl group is hydrolyzed much more slowly from the enzyme's OH group than an acetyl group. As a result, the catalytic activity of the enzyme is decreased, the concentration of acetylcholine in synapses increases, and nerve stimulation increases. Carbaryl and other carbamate insecticides are much more toxic to insects than to humans because humans and insects have slightly different shapes of their AChE binding sites. The distance between the two pockets in the enzyme is greater for insects than for humans, so that compounds that fit snugly in the binding site of insect AChE do not fit as well in the binding site of human AChE.

Carbamate insecticides are environmentally safe. Once they are released into the environment, they are broken down rapidly by microorganisms in the soil and thus exit the food chain quickly. In water, their hydrolysis is catalyzed by sunlight.

SUMMARY

The functional group of an **acyl halide** (Section 17.1A) is an acyl group bonded to a halogen. The most common and widely used of these are the acyl chlorides. The functional group of an **carboxylic acid anhydride** (Section 17.1B) is two acyl groups bonded to an oxygen. The functional group of an **ester** (Section 17.1C) is an acyl group bonded to —OR or to —OAr. A cyclic ester is given the name **lactone.** Phosphoric acid has three —OH groups and can form mono-, di-, and triesters. The functional group of an **amide** (Section 17.1D) is an acyl group bonded to a nitrogen. A cyclic amide is given the name **lactam.** The functional group of an **imide** is two acyl groups bonded to a trivalent nitrogen. The functional group of a **nitrile** (Section 17.1E) is a C≡N group bonded to a carbon atom.

Values of pK_a for amides of carboxylic acids are 15–17, which means they are comparable in acidity to alcohols (Section 9.4). Values of pK_a for imides are 8–10, which means that they dissolve in aqueous NaOH to form water-soluble salts. Sulfonamides derived from ammonia and primary amines have pK_a values of approximately 10 and dissolve in aqueous NaOH to form water-soluble salts.

A common reaction theme of functional derivatives of carboxylic acids is **nucleophilic addition** to the carbonyl carbon to form **a tetrahedral carbonyl addition intermediate,** which then collapses to regenerate the carbonyl group. The result is **nucleophilic acyl substitution** (Section 17.4). Listed in order of increasing reactivity toward nucleophilic acyl substitution, these functional derivatives are

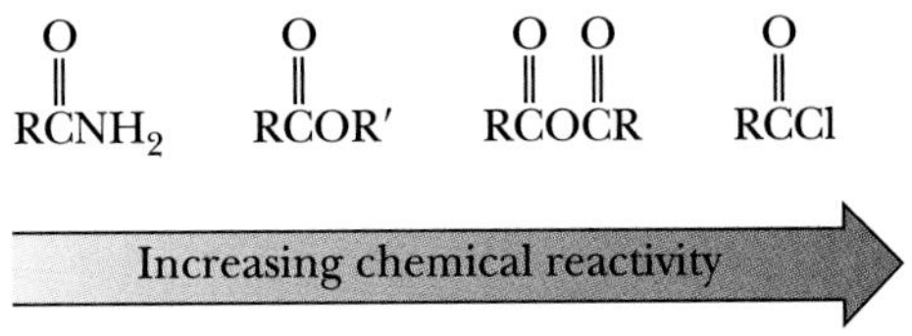

$$\underset{\text{RCNH}_2}{\overset{\text{O}}{\|}} \quad \underset{\text{RCOR}'}{\overset{\text{O}}{\|}} \quad \underset{\text{RCOCR}}{\overset{\text{O}\ \ \text{O}}{\|\ \ \|}} \quad \underset{\text{RCCl}}{\overset{\text{O}}{\|}}$$

Increasing chemical reactivity →

Any more reactive functional derivative can be converted to any less reactive functional derivative by reaction with an appropriate oxygen or nitrogen nucleophile (Section 17.10).

KEY REACTIONS

1. Acidity of Imides (Section 17.2)

Imides (pK_a 8.0–9.5) dissolve in aqueous NaOH by forming water-soluble salts. Imides are stronger acids than amides because (1) the electron-withdrawing inductive effect of the two adjacent carbonyl groups weakens the N—H bond more and (2) the imide anion is stabilized more by delocalization of the negative charge onto the two carbonyl oxygens.

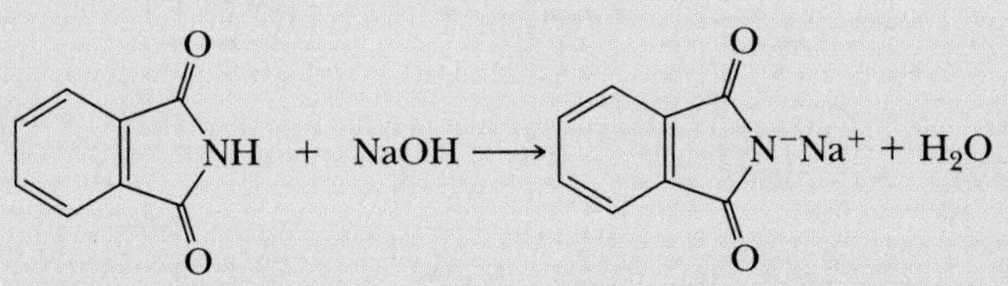

$$\text{(phthalimide)NH} + \text{NaOH} \longrightarrow \text{(phthalimide)N}^-\text{Na}^+ + H_2O$$

Insoluble in water — A water-soluble salt

2. Acidity of Sulfonamides (Section 17.2)

Sulfonamides (pK_a 9–10), like imides, also dissolve in aqueous NaOH by forming water-soluble salts. The stability of the sulfonamide anion is due to (1) the electron-withdrawing inductive effect of the adjacent $—SO_2—$ group and (2) delocalization of the negative charge in the anion.

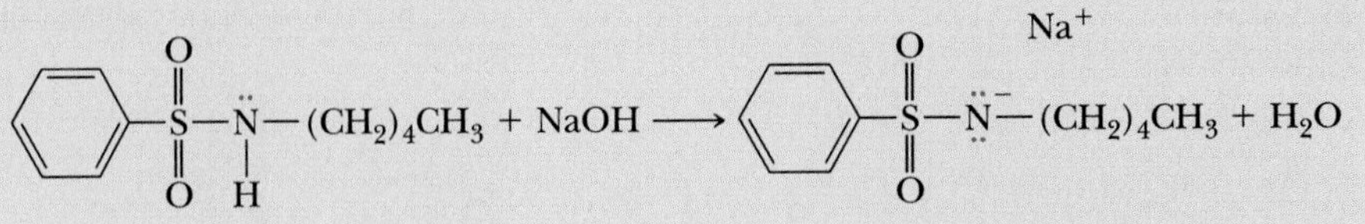

$$C_6H_5SO_2NH(CH_2)_4CH_3 + \text{NaOH} \longrightarrow C_6H_5SO_2\ddot{N}^-(CH_2)_4CH_3\ Na^+ + H_2O$$

Insoluble in water — A water-soluble salt

3. Hydrolysis of an Acid Chloride (Section 17.5A)

Low-molecular-weight acid chlorides react vigorously with water. Higher-molecular-weight acid chlorides react less rapidly.

$$CH_3\overset{O}{\overset{\|}{C}}Cl + H_2O \longrightarrow CH_3\overset{O}{\overset{\|}{C}}OH + HCl$$

4. Hydrolysis of an Acid Anhydride (Section 17.5B)

Low-molecular-weight acid anhydrides react readily with water. Higher-molecular-weight acid anhydrides react less rapidly.

$$CH_3\overset{O}{\overset{\|}{C}}O\overset{O}{\overset{\|}{C}}CH_3 + H_2O \longrightarrow CH_3\overset{O}{\overset{\|}{C}}OH + HO\overset{O}{\overset{\|}{C}}CH_3$$

5. Hydrolysis of an Ester (Section 17.5C)

Esters are hydrolyzed only in the presence of acid or base. Acid is a catalyst. Base is required in an equimolar amount.

$$CH_3\overset{O}{\overset{\|}{C}}O\text{-cyclohexyl} + NaOH \xrightarrow{H_2O} CH_3\overset{O}{\overset{\|}{C}}O^-Na^+ + HO\text{-cyclohexyl}$$

6. Hydrolysis of an Amide (Section 17.5D)

Either acid or base is required in an amount equivalent to that of the amide.

$$CH_3CH_2CH_2\overset{O}{\overset{\|}{C}}NH_2 + H_2O + HCl \xrightarrow[heat]{H_2O} CH_3CH_2CH_2\overset{O}{\overset{\|}{C}}OH + NH_4^+Cl^-$$

$$CH_3\overset{O}{\overset{\|}{C}}NH\text{-phenyl} + NaOH + \xrightarrow[heat]{H_2O} CH_3\overset{O}{\overset{\|}{C}}O^-Na^+ + H_2N\text{-phenyl}$$

7. Hydrolysis of a Nitrile (Section 17.5E)

Either acid or base is required in an amount equivalent to that of the nitrile.

$$\text{phenyl-}CH_2C{\equiv}N + 2H_2O + H_2SO_4 \xrightarrow[heat]{H_2O} \text{phenyl-}CH_2\overset{O}{\overset{\|}{C}}OH + NH_4^+HSO_4^-$$

$$CH_3(CH_2)_9C{\equiv}N + H_2O + NaOH \xrightarrow[heat]{H_2O} CH_3(CH_2)_9\overset{O}{\overset{\|}{C}}O^-Na^+ + NH_3$$

8. Reaction of an Acid Chloride with an Alcohol (Section 17.6A)

Treatment of an acid chloride with an alcohol gives an ester plus HCl. Preparation of an acid-sensitive ester is carried out using an equimolar amount of triethylamine or pyridine to neutralize the HCl.

$$CH_3CH_2CH_2\overset{O}{\overset{\|}{C}}Cl + HO\text{-cyclohexyl} \longrightarrow CH_3CH_2CH_2\overset{O}{\overset{\|}{C}}O\text{-cyclohexyl} + HCl$$

9. Reaction of an Acid Anhydride with an Alcohol (Section 17.6B)

Treatment of an acid anhydride with an alcohol gives one mol of ester and one mol of carboxylic acid.

$$CH_3\overset{O}{\overset{\|}{C}}O\overset{O}{\overset{\|}{C}}CH_3 + HOCH_2CH_3 \longrightarrow CH_3\overset{O}{\overset{\|}{C}}OCH_2CH_3 + CH_3\overset{O}{\overset{\|}{C}}OH$$

10. Reaction of an Ester with an Alcohol: Transesterification (Section 17.6C)

Transesterification requires an acid catalyst and an excess of alcohol to drive the reaction to completion.

$$CH_2{=}CH\overset{O}{\overset{\|}{C}}OCH_3 + HO(CH_2)_3CH_3 \xrightarrow{HCl} CH_2{=}CH\overset{O}{\overset{\|}{C}}O(CH_2)_3CH_3 + CH_3OH$$

11. Reaction of an Acid Chloride with Ammonia or an Amine (Section 17.7A)

Reaction requires two mol of ammonia or amine: one mol to form the amide and one mol to neutralize the HCl byproduct.

$$CH_3\overset{O}{\overset{\|}{C}}Cl + 2NH_3 \longrightarrow CH_3\overset{O}{\overset{\|}{C}}NH_2 + NH_4^+\ Cl^-$$

12. Reaction of an Acid Anhydride with Ammonia or an Amine (Section 17.7B)

The reaction requires two mol of ammonia or amine: one mol to form the amide and one mol to neutralize the carboxylic acid byproduct.

$$CH_3\overset{O}{\overset{\|}{C}}O\overset{O}{\overset{\|}{C}}CH_3 + 2NH_3 \longrightarrow CH_3\overset{O}{\overset{\|}{C}}NH_2 + CH_3\overset{O}{\overset{\|}{C}}O^-\ NH_4^+$$

13. Reaction of an Ester with Ammonia or an Amine (Section 17.7C)

Treatment of an ester with ammonia, a primary or secondary amine gives an amide.

$$C_6H_5{-}CH_2\overset{O}{\overset{\|}{C}}OCH_2CH_3 + NH_3 \longrightarrow C_6H_5{-}CH_2\overset{O}{\overset{\|}{C}}NH_2 + CH_3CH_2OH$$

14. Reaction of an Acid Chloride with a Carboxylic Acid Salt (Section 17.8)

Treatment of an acid chloride with the salt of a carboxylic acid is a valuable method for synthesizing mixed anhydrides.

$$CH_3\overset{O}{\overset{\|}{C}}Cl + Na^+\ {}^-O\overset{O}{\overset{\|}{C}}{-}C_6H_5 \longrightarrow CH_3\overset{O}{\overset{\|}{C}}O\overset{O}{\overset{\|}{C}}{-}C_6H_5 + Na^+\ Cl^-$$

15. Reaction of an Ester with a Grignard Reagent (Section 17.9A)

Treatment of a formic ester with a Grignard reagent followed by hydrolysis gives a secondary alcohol. Treatment of any other ester with a Grignard reagent gives a tertiary alcohol.

$$C_5H_9{-}\overset{O}{\overset{\|}{C}}OCH_3 \xrightarrow[\text{2. } H_2O,\ HCl]{\text{1. } 2CH_3CH_2MgBr} C_5H_9{-}\overset{OH}{\overset{|}{C}}(CH_2CH_3)_2$$

16. Reaction of an Acid Chloride with a Lithium Diorganocuprate (Section 17.9C)

Acid chlorides react readily with lithium diorganocuprates at $-78°C$ to give ketones.

$$CH_3(CH_2)_3\overset{O}{\overset{\|}{C}}Cl \xrightarrow[2.\ H_2O]{1.\ (CH_3)_2CuLi,\ ether,\ -78°C} CH_3(CH_2)_3\overset{O}{\overset{\|}{C}}CH_3$$

17. Reduction of an Ester (Section 17.11A)

Reduction of an ester by lithium aluminum hydride gives two alcohols. Reduction by diisobutylaluminum hydride (DIBALH) at low temperature gives an aldehyde and an alcohol.

$$C_6H_5\underset{CH_3}{\underset{|}{C}}H\overset{O}{\overset{\|}{C}}OCH_3 \xrightarrow[2.\ H_2O]{1.\ LiAlH_4,\ ether} C_6H_5\underset{CH_3}{\underset{|}{C}}HCH_2OH + CH_3OH$$

$$C_6H_5\underset{CH_3}{\underset{|}{C}}H\overset{O}{\overset{\|}{C}}OCH_3 \xrightarrow[2.\ H_2O,\ HCl]{1.\ DIBALH,\ -78°C} C_6H_5\underset{CH_3}{\underset{|}{C}}H\overset{O}{\overset{\|}{C}}H$$

18. Reduction of an Amide (Section 17.11B)

Reduction of an amide by lithium aluminum hydride gives an amine.

$$C_6H_5\overset{O}{\overset{\|}{C}}NH_2 \xrightarrow[2.\ H_2O]{1.\ LiAlH_4,\ THF} C_6H_5CH_2NH_2$$

19. Reduction of the Cyano Group of a Nitrile (Section 17.11C)

Reduction of a cyano group by lithium aluminum hydride gives a primary amino group.

$$Br\text{-}C_6H_3(Br)\text{-}C{\equiv}N \xrightarrow[2.\ H_2O]{1.\ LiAlH_4,\ ether} Br\text{-}C_6H_3(Br)\text{-}CH_2NH_2$$

20. The Hofmann Rearrangement of a Primary Amide (Section 17.12)

When treated with bromine and aqueous sodium hydroxide, a primary amide is converted to an amine with one fewer carbon. The carbon atom of the carbonyl group is lost as carbon dioxide.

$$(H)(H_3C)(PhCH_2)C\text{-}\overset{O}{\overset{\|}{C}}NH_2 \xrightarrow[H_2O]{Br_2,\ NaOH} (H)(H_3C)(PhCH_2)C\text{-}NH_2 + Na_2CO_3$$

ADDITIONAL PROBLEMS

Structure and Nomenclature

17.13 Draw structural formulas for these compounds:

(a) Dimethyl carbonate
(b) Benzonitrile
(c) Isopropyl 3-methylhexanoate
(d) Diethyl oxalate
(e) Ethyl (*Z*)-2-pentenoate
(f) Butanoic anhydride
(g) Dodecanamide
(h) Ethyl 3-hydroxybutanoate
(i) Octanoyl chloride
(j) Diethyl *cis*-1,2-cyclohexanedicarboxylate
(k) Methanesulfonyl chloride
(l) *p*-Toluenesulfonyl chloride

17.14 Write the IUPAC name for each compound.

(a) 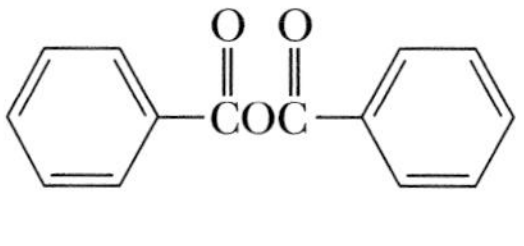 $C_6H_5\overset{\overset{O}{\|}}{C}O\overset{\overset{O}{\|}}{C}C_6H_5$

(b) $C_6H_5\underset{\underset{O}{\|}}{\overset{\overset{O}{\|}}{S}}NH_2$

(c) $CH_3(CH_2)_4\overset{\overset{O}{\|}}{C}NHCH_3$

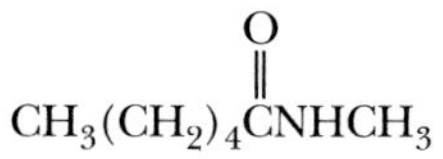

(d) 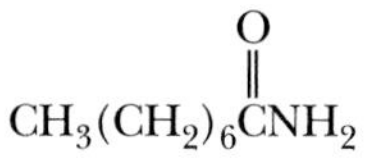 $CH_3(CH_2)_6\overset{\overset{O}{\|}}{C}NH_2$

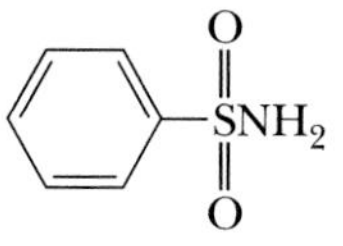

(e) $CH_2(CO_2CH_2CH_3)_2$

(f) 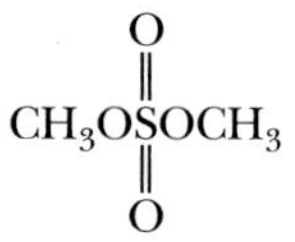 $CH_3O\underset{\underset{O}{\|}}{\overset{\overset{O}{\|}}{S}}OCH_3$

(g) 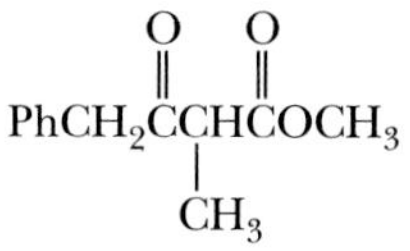 $PhCH_2\overset{\overset{O}{\|}}{C}\underset{\underset{CH_3}{|}}{C}H\overset{\overset{O}{\|}}{C}OCH_3$

(h) $Cl\overset{\overset{O}{\|}}{C}(CH_2)_4\overset{\overset{O}{\|}}{C}Cl$

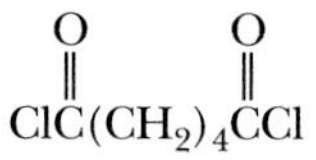

(i) $CH_3(CH_2)_5CN$

Physical Properties

17.15 Both the melting point and boiling point of acetamide are higher than those of its *N,N*-dimethyl derivative.

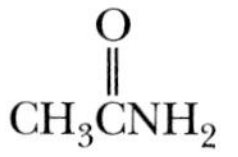

$CH_3\overset{\overset{O}{\|}}{C}NH_2$	$CH_3\overset{\overset{O}{\|}}{C}N(CH_3)_2$
Acetamide	*N,N*-Dimethylacetamide
mp 82.3°C, bp 221.2°C	mp −20°C, bp 165°C

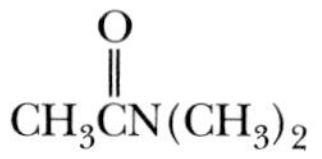

How do you account for these differences?

Spectroscopy

17.16 All methyl esters of long-chain aliphatic acids (for example, methyl tetradecanoate, $C_{13}H_{27}CO_2CH_3$) show significant fragment ions at m/z 74, 59, and 31. What are the structures of these ions and how are they formed?

17.17 The two hydrogens of primary amides typically have separate ^{1}H-NMR resonances, as illustrated by the separate signals for the two amide hydrogens of propanamide. Furthermore, each methyl group of the *N,N*-dimethylformamide has a separate resonance. How do you account for these observations?

Propanamide: $CH_3CH_2C(=O)N(H)H$ — δ 6.22 and 6.58

N,*N*-Dimethylformamide: $HC(=O)N(CH_3)CH_3$ — δ 3.88 and 3.98

Hint: Consider the covalent bonding in a resonance-stabilized amide group and the likely orientation in space of the six atoms of the amide group.

17.18 Propose a structural formula for compound A, $C_7H_{14}O_2$, consistent with its ^{1}H-NMR and infrared spectra.

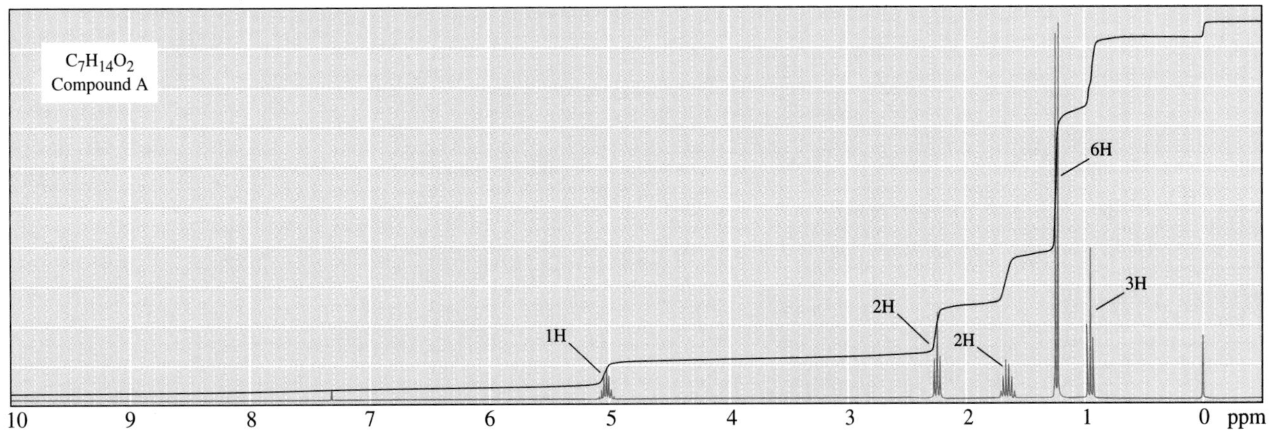

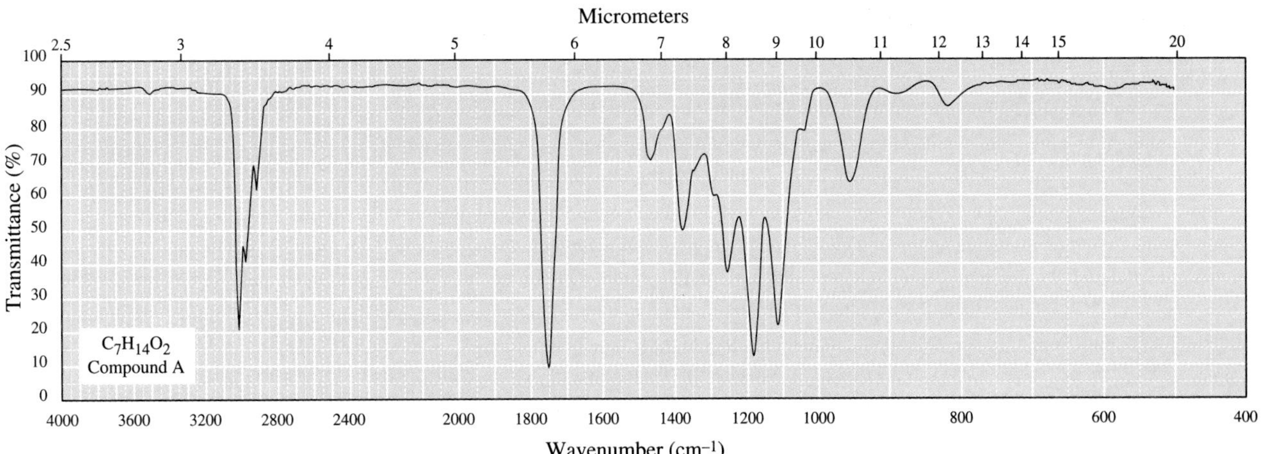

17.19 Propose a structural formula for compound B, $C_6H_{13}NO$, consistent with its ^{1}H-NMR and infrared spectra.

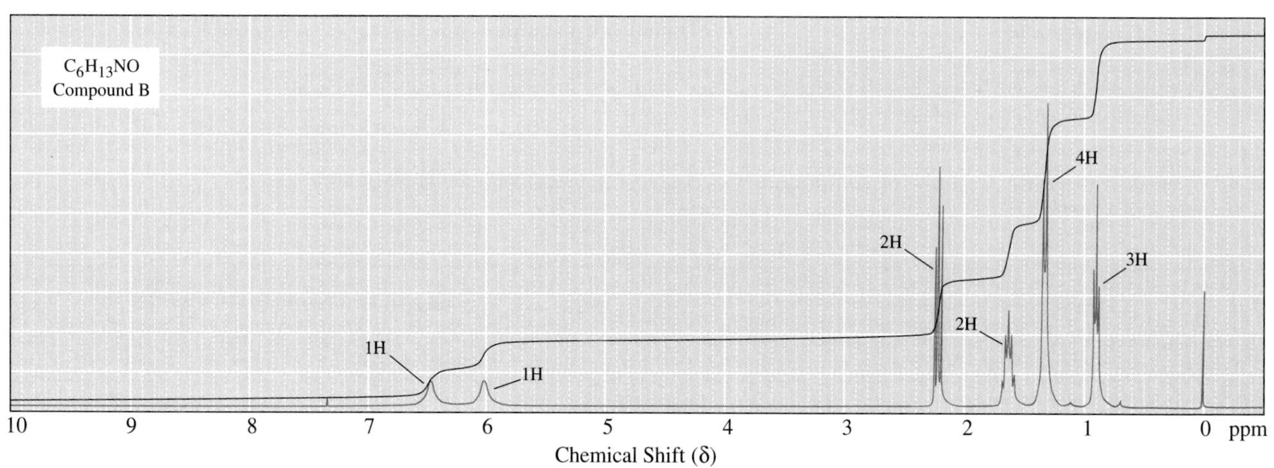

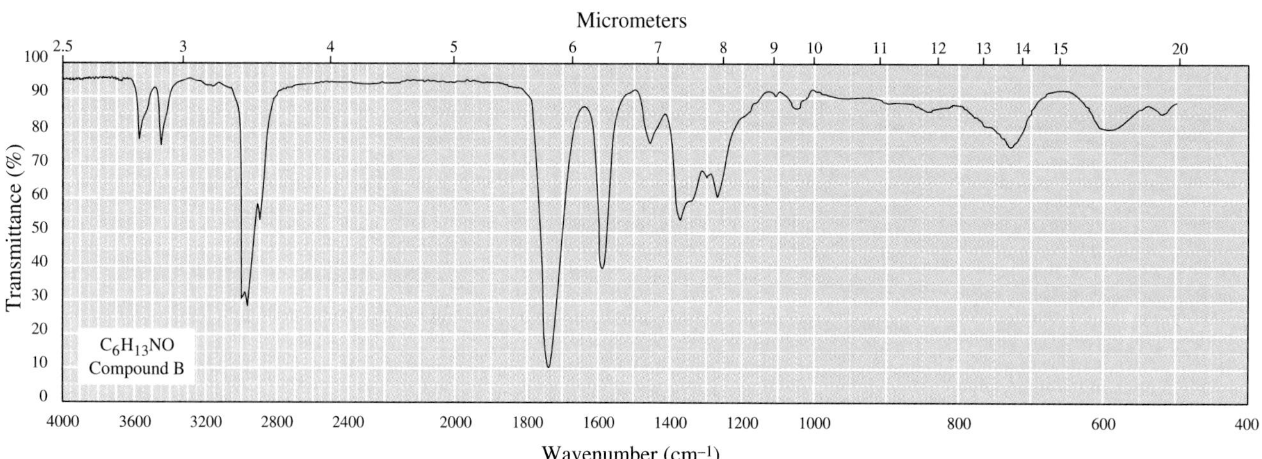

17.20 Propose a structural formula for each compound consistent with its ^{1}H-NMR and ^{13}C-NMR spectra.

(a) $C_5H_{10}O_2$

^{1}H-NMR	^{13}C-NMR
0.96 (d, 6H)	161.11
1.96 (m, 1H)	70.01
3.95 (d, 2H)	27.71
8.08 (s, 1H)	19.00

(b) $C_7H_{14}O_2$

^{1}H-NMR	^{13}C-NMR
0.92 (d, 6H)	171.15
1.52 (m, 2H)	63.12
1.70 (m, 1H)	37.31
2.09 (s, 3H)	25.05
4.10 (t, 2H)	22.45
	21.06

(c) $C_6H_{12}O_2$

^{1}H-NMR	^{13}C-NMR
1.18 (d, 6H)	177.16
1.26 (t, 3H)	60.17
2.51 (m, 1H)	34.04
4.13 (q, 2H)	19.01
	14.25

(d) $C_7H_{12}O_4$

^{1}H-NMR	^{13}C-NMR
1.28 (t, 6H)	166.52
3.36 (s, 2H)	61.43
4.21 (q, 4H)	41.69
	14.07

(e) $C_4H_7ClO_2$

^{1}H-NMR	^{13}C-NMR
1.68 (d, 3H)	170.51
3.80 (s, 3H)	52.92
4.42 (q, 1H)	52.32
	21.52

(f) $C_4H_6O_2$

^{1}H-NMR	^{13}C-NMR
2.29 (m, 2H)	177.81
2.50 (t, 2H)	68.58
4.36 (t, 2H)	27.79
	22.17

Reactions

17.21 Arrange these compounds in order of decreasing reactivity toward nucleophilic acyl substitution.

O OCH$_2$CF$_3$ (1) O Cl (2) O NEt$_2$ (3) O O O (4) O OEt (5)

17.22 Write the structural formula of the principal product formed when benzoyl chloride is treated with the following reagents:

(a) Cyclohexanol

(b) $CH_3CH_2CH_2CH_2OH$, pyridine

(c) $CH_3CH_2CH_2CH_2SH$, pyridine

(d) $CH_3CH_2CH_2CH_2NH_2$ (two equivalents)

(e) $CH_3CH_2CH_2CO_2^-\ Na^+$

(f) $(CH_3)_2CuLi$ (-78°C), then H_3O^+

(g) $CH_3O-C_6H_4-NH_2$, pyridine

(h) C_6H_5MgBr (two equivalents), then H_3O^+

17.23 Write the structural formula of the principal product formed when ethyl benzoate is treated with the following reagents:

(a) H_2O, NaOH, heat

(b) H_2O, H_2SO_4, heat

(c) $CH_3CH_2CH_2CH_2NH_2$

(d) DIBALH (-78°C), then H_2O

(e) $LiAlH_4$, then H_2O

(f) C_6H_5MgBr (two equivalents), then H_2O

17.24 The mechanism for hydrolysis of an ester in aqueous acid involves formation of a tetrahedral carbonyl addition intermediate. Evidence in support of this mechanism comes from an experiment designed by Myron Bender. He first prepared ethyl benzoate enriched with oxygen-18 in the carbonyl oxygen and then carried out acid-catalyzed hydrolysis of the ester in water containing no enrichment in oxygen-18. He discovered that if he stopped the experiment after only partial hydrolysis and isolated the remaining ester, the recovered ethyl benzoate had lost a portion of its enrichment in oxygen-18. In other words, some exchange had occurred between oxygen-18 of the ester and oxygen-16 of water. Show how this observation bears on the formation of a tetrahedral carbonyl addition intermediate during acid-catalyzed ester hydrolysis.

17.25 Predict the distribution of oxygen-18 in the products obtained from hydrolysis of ethyl benzoate labeled in the ethoxy oxygen under the following conditions:

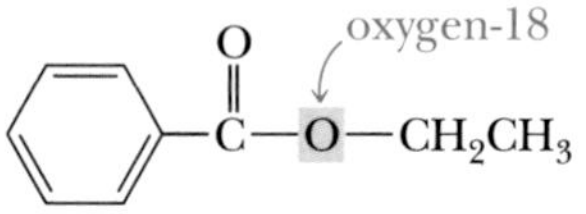

$$C_6H_5-\overset{O}{\overset{\|}{C}}-\underset{\text{oxygen-18}}{O}-CH_2CH_3$$

(a) In aqueous NaOH **(b)** In aqueous HCl

17.26 Write the structural formula of the principal product formed when benzamide is treated with the following reagents:

(a) H_2O, HCl, heat **(b)** NaOH, H_2O, heat

(c) $LiAlH_4$, then H_2O **(d)** Br_2, NaOH, heat

17.27 Write the structural formula of the principal product formed when benzonitrile is treated with the following reagents:

(a) H_2O (one equivalent), H_2SO_4, heat **(b)** H_2O (excess), H_2SO_4, heat

(c) NaOH, H_2O, heat **(d)** $LiAlH_4$, then H_2O

17.28 Show the product expected when the following unsaturated δ-ketoester is treated with each reagent.

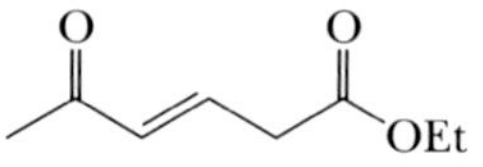

$$CH_3COCH{=}CHCH_2COOEt$$

(a) $\xrightarrow[\text{Pd, EtOH}]{H_2\text{ (1 mol)}}$ **(b)** $\xrightarrow[CH_3OH]{NaBH_4}$ **(c)** $\xrightarrow[\text{2. } H_2O]{\text{1. } LiAlH_4\text{, THF}}$ **(d)** $\xrightarrow[\text{2. } H_2O]{\text{1. DIBALH, } -78°}$

17.29 The reagent diisobutylaluminum hydride (DIBALH) reduces esters to aldehydes. When nitriles are treated with DIBALH, followed by mild acid hydrolysis, the product is also an aldehyde. Give a mechanism for this reaction.

17.30 Show the product of treatment of this anhydride with each reagent.

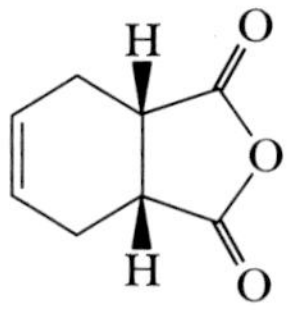

(a) $\xrightarrow[\text{heat}]{H_2O\text{, HCl}}$ **(b)** $\xrightarrow[\text{heat}]{H_2O\text{, NaOH}}$ **(c)** $\xrightarrow[\text{2. } H_2O]{\text{1. } LiAlH_4}$

(d) $\xrightarrow{CH_3OH}$ **(e)** $\xrightarrow{NH_3\text{ (2 mol)}}$

17.31 The analgesic acetaminophen is synthesized by treating 4-aminophenol with one equivalent of acetic anhydride. Write an equation for the formation of acetaminophen. (*Hint:* An $-NH_2$ group is a better nucleophile than an $-OH$ group.)

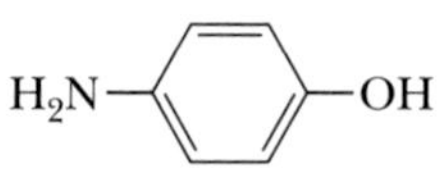

$$H_2N-C_6H_4-OH$$

4-Aminophenol

17.32 Treatment of choline with acetic anhydride gives acetylcholine, a neurotransmitter. Write an equation for the formation of acetylcholine.

$$(CH_3)_3\overset{+}{N}CH_2CH_2OH$$

Choline

17.33 Nicotinic acid, more commonly named niacin, is one of the B vitamins. Show how nicotinic acid can be converted to (a) ethyl nicotinate and then to (b) nicotinamide.

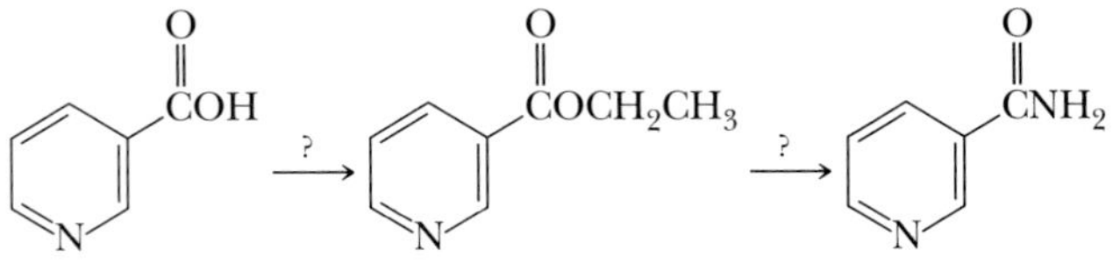

Nicotinic acid (Niacin) $\xrightarrow{?}$ Ethyl nicotinate $\xrightarrow{?}$ Nicotinamide

17.34 Complete these reactions.

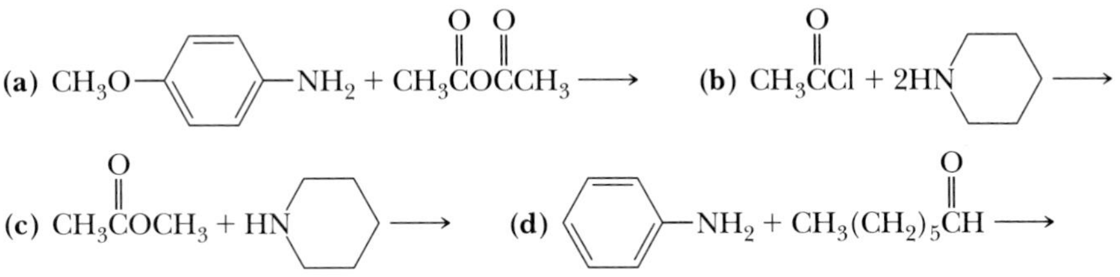

(a) $CH_3O-C_6H_4-NH_2 + CH_3COCCH_3$ (acetic anhydride) $\longrightarrow$

(b) CH_3CCl (acetyl chloride) + 2HN(piperidine) $\longrightarrow$

(c) CH_3COCH_3 (methyl acetate) + HN(piperidine) $\longrightarrow$

(d) $C_6H_5-NH_2 + CH_3(CH_2)_5CH$ (O) $\longrightarrow$

17.35 Show the product of treatment of the following γ-butyrolactone with each reagent.

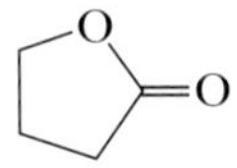

(a) $NH_3 \longrightarrow$

(b) $\xrightarrow[2.\ H_2O]{1.\ LiAlH_4}$

(c) $\xrightarrow[2.\ H_2O,\ HCl]{1.\ 2PhMgBr,\ ether}$

(d) $NaOH \xrightarrow[heat]{H_2O}$

(e) $\xrightarrow[2.\ H_2O,\ HCl]{1.\ 2CH_3Li,\ ether}$

(f) $\xrightarrow[2.\ H_2O,\ HCl]{1.\ DIBALH,\ ether,\ -78°C}$

17.36 Show the product of treatment of the following γ-lactam with each reagent.

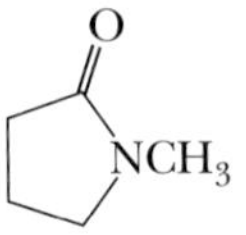

(N-methyl lactam, NCH_3)

(a) $\xrightarrow[heat]{H_2O,\ HCl}$

(b) $\xrightarrow[heat]{H_2O,\ NaOH}$

(c) $\xrightarrow[2.\ H_2O]{1.\ LiAlH_4}$

17.37 Draw structural formulas for the products of complete hydrolysis of meprobamate, phenobarbital, and pentobarbital in hot aqueous acid. Meprobamate is a tranquilizer prescribed under one of 58 different trade names, including Equanil and Miltown. Phenobarbital is a long-acting sedative, hypnotic, and anticonvulsant. Luminal is one of over a dozen names under which it is prescribed. Pentobarbital is a short-acting sedative, hypnotic, and anticonvulsant. Nembutal is one of several trade names under which it is prescribed. (*Hint:* Remember that when heated, β-dicarboxylic acids and β-ketoacids undergo decarboxylation.)

(a) 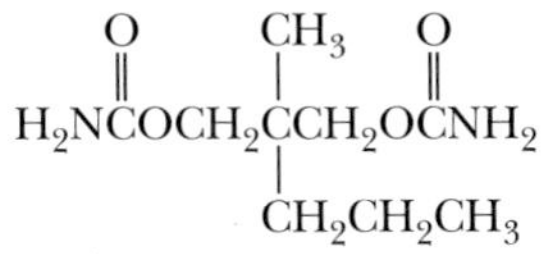 $H_2NCOCH_2C(CH_3)(CH_2CH_2CH_3)CH_2OCNH_2$

Meprobamate

(b) 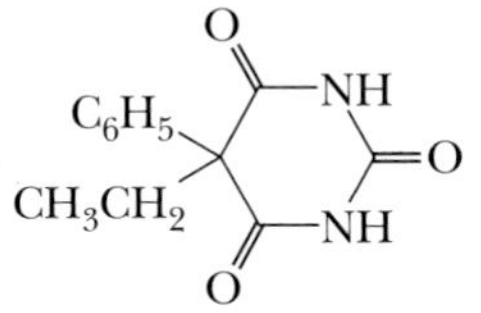 C_6H_5, CH_3CH_2 barbiturate ring (C=O, NH, C=O, NH, C=O)

Phenobarbital

(c) 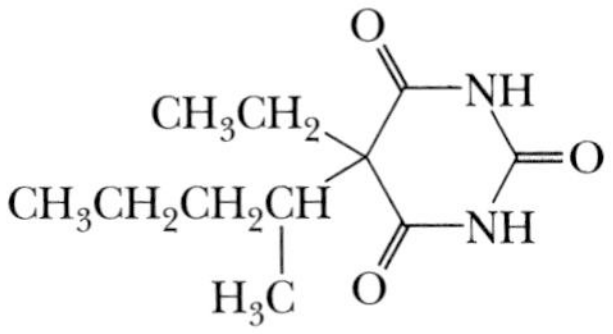 CH_3CH_2, $CH_3CH_2CH_2CH(CH_3)$ barbiturate ring (C=O, NH, C=O, NH, C=O)

Pentobarbital

The Hofmann Rearrangement

17.38 Following are steps in a synthesis of anthranilic acid from phthalic anhydride. Describe how you could bring about each step.

$$\text{Phthalic anhydride} \xrightarrow{(1)} C_6H_4(CO^-NH_4^+)(CONH_2) \xrightarrow{(2)} C_6H_4(COOH)(CONH_2) \xrightarrow{(3)} C_6H_4(COOH)(NH_2)$$

Phthalic anhydride Anthranilic acid

17.39 Hofmann rearrangements of lower-molecular-weight primary amides can be brought about using bromine in aqueous NaOH. Primary amides larger than about seven or eight carbon atoms are not sufficiently soluble in aqueous solution to react. Instead, they are dissolved in methanol or ethanol, and the corresponding sodium alkoxide is used as the base. Under these conditions, the isocyanate intermediate reacts with the alcohol to form a carbamate.

$$\underset{\text{A primary amide}}{RCONH_2} + Br_2 \xrightarrow[CH_3OH]{CH_3O^-Na^+} \underset{\text{An isocyanate}}{RN{=}C{=}O} \xrightarrow{CH_3OH} \underset{\text{A carbamate}}{RNHCOCH_3}$$

Propose a mechanism for the reaction of an isocyanate with methanol to form a methyl carbamate.

Synthesis

17.40 *N,N*-Diethyl-*m*-toluamide (DEET) is the active ingredient in several common insect repellents. DEET can be synthesized in two steps: (1) treatment of 3-methylbenzoic acid (*m*-toluic acid) with thionyl chloride to form an acid chloride followed by (2) treatment of the acid chloride with diethylamine. Write equations for each step in this synthesis of DEET.

$$H_3C\text{–}C_6H_4\text{–}CO_2H$$

3-Methylbenzoic acid

17.41 Following is the structural formula of isoniazid, a drug used to treat tuberculosis. It is estimated that one-third of the world's population is infected with tuberculosis, which results in approximately 3 million TB-related deaths per year. Show how to prepare isoniazid from pyridine-4-carboxylic acid. (See *The Merck Index,* 12th ed., #5203.)

$$\underset{\text{Pyridine-4-carboxylic acid}}{C_5H_4N\text{–}COH} \xrightarrow{?} \underset{\text{Isoniazid}}{C_5H_4N\text{–}C(=O)\text{–}NH\text{–}NH_2}$$

17.42 Show how to convert phenylacetylene to allyl phenylacetate.

$$\underset{\text{Phenylacetylene}}{C_6H_5\text{–}C{\equiv}CH} \longrightarrow \underset{\text{Allyl phenylacetate}}{C_6H_5\text{–}CH_2COCH_2CH{=}CH_2}$$

17.43 A step in a synthesis of PGE_1 (prostaglandin E_1, alprostadil) is reaction of a trisubstituted cyclohexene with bromine to form a bromolactone. Alprostadil is used as a temporary therapy for infants born with congenital heart defects that restrict pulmonary blood flow. It brings about dilation of the ductus arteriosus, which in turn increases blood flow in the lungs and blood oxygenation. Propose a mechanism for formation of this bromolactone and account for the observed stereochemistry of each substituent on the cyclohexane ring.

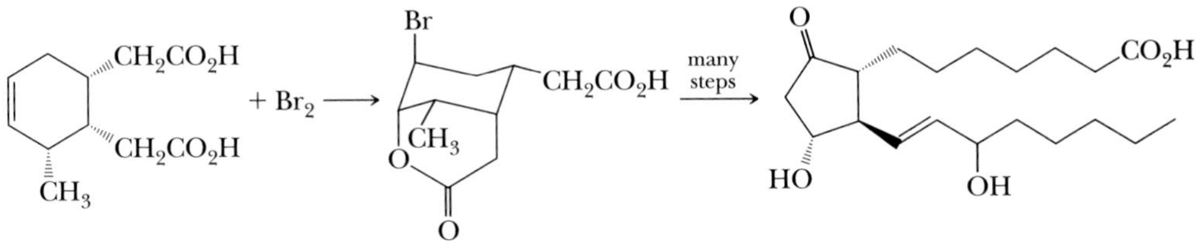

CH_2CO_2H, CH_2CO_2H, CH_3 + Br_2 ⟶ (Br, CH_2CO_2H, CH_3, O, O) —many steps→ (O, CO_2H, HO, OH)

A bromolactone PGE_1 (alprostadil)

17.44 Barbiturates are prepared by treatment of a derivative of diethyl malonate with urea in the presence of sodium ethoxide as a catalyst. Following is an equation for the preparation of barbital, a long-duration hypnotic and sedative, from diethyl diethylmalonate and urea. Barbital is prescribed under one of a dozen or more trade names.

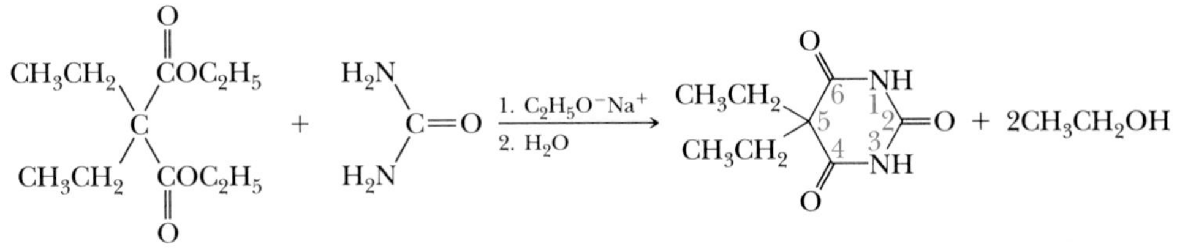

$(CH_3CH_2)_2C(COC_2H_5)_2$ (C=O) + $(H_2N)_2C{=}O$ —1. $C_2H_5O^-Na^+$; 2. H_2O⟶ 5,5-diethylbarbituric acid (ring atoms numbered 1–6; NH, NH, =O) + $2CH_3CH_2OH$

Diethyl diethylmalonate Urea 5,5-Diethylbarbituric acid (Barbital)

(a) Propose a mechanism for this reaction.

(b) The pK_a of barbital is 7.4. Which is the most acidic hydrogen in this molecule and how do you account for its acidity?

17.45 The following compound is one of a group of β-chloroamines, many of which have antitumor activity. Describe a synthesis of this compound from anthranilic acid and ethylene oxide. [*Hint:* To see how the seven-membered ring might be formed, review the chemistry of the nitrogen mustards (Section 8.5).]

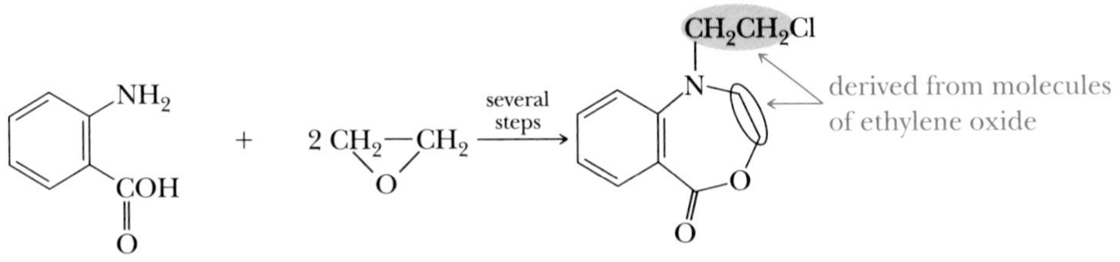

2-Aminobenzoic acid (NH_2, COH, O) + $2\ CH_2{-}CH_2$ (O) —several steps⟶ (CH_2CH_2Cl, N, O, O)

derived from molecules of ethylene oxide

2-Aminobenzoic acid
(Anthranilic acid)

17.46 Following is a retrosynthetic scheme for the synthesis of 5-nonanone from 1-bromobutane as the only organic starting material. Show reagents and experimental conditions to bring about this synthesis.

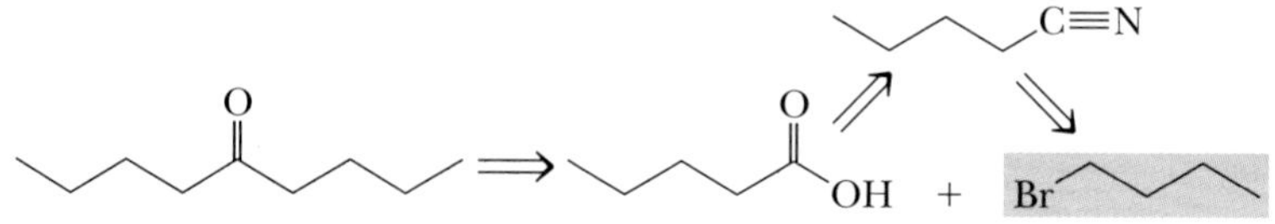

5-Nonanone $\Longrightarrow$ pentanoic acid (OH) + Br–butane; pentanenitrile (C≡N)

5-Nonanone 1-Bromobutane

17.47 Procaine (its hydrochloride is marketed as Novocaine) was one of the first local anesthetics for infiltration and regional anesthesia. According to this retrosynthetic scheme, procaine can be synthesized from 4-aminobenzoic acid, ethylene oxide, and diethylamine as sources of carbon atoms. Provide reagents and experimental conditions to carry out the synthesis of procaine from these three compounds.

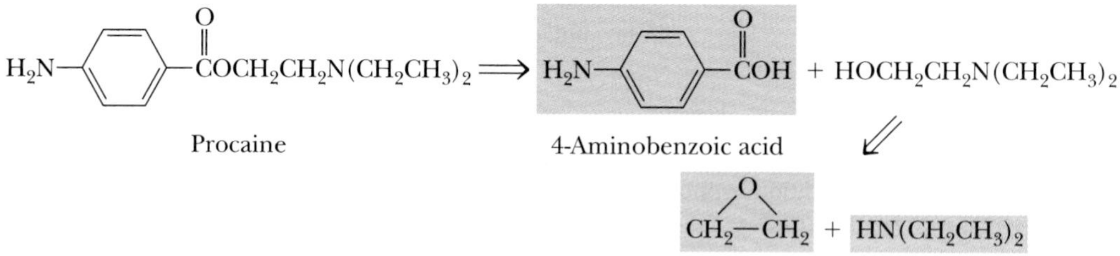

$$H_2N{-}C_6H_4{-}\overset{O}{\overset{\|}{C}}OCH_2CH_2N(CH_2CH_3)_2 \Longrightarrow H_2N{-}C_6H_4{-}\overset{O}{\overset{\|}{C}}OH + HOCH_2CH_2N(CH_2CH_3)_2$$

Procaine 4-Aminobenzoic acid

$$HOCH_2CH_2N(CH_2CH_3)_2 \Longrightarrow CH_2{-}CH_2\ (\text{epoxide, O}) + HN(CH_2CH_3)_2$$

Oxirane (Ethylene oxide) Diethylamine

17.48 The following sequence of steps converts (*R*)-2-octanol to (*S*)-2-octanol. Propose structural formulas for intermediates A and B, specify the configuration of each, and account for the inversion of configuration in this sequence.

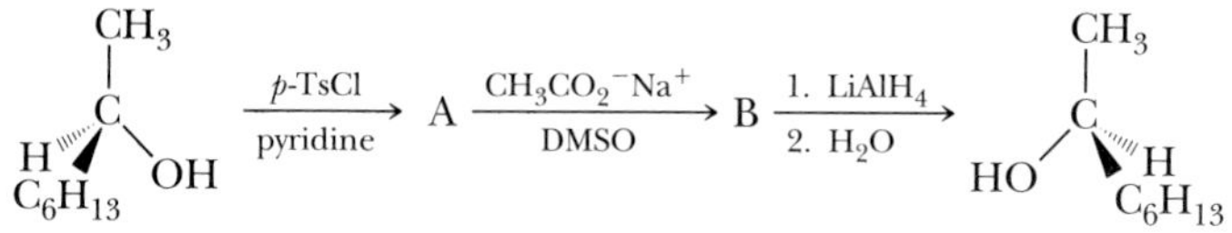

$$(R)\text{-2-Octanol} \xrightarrow[\text{pyridine}]{p\text{-TsCl}} A \xrightarrow[\text{DMSO}]{CH_3CO_2^-Na^+} B \xrightarrow[2.\ H_2O]{1.\ LiAlH_4} (S)\text{-2-Octanol}$$

(*R*)-2-Octanol (*S*)-2-Octanol

17.49 Reaction of a primary or secondary amine with diethyl carbonate under controlled conditions gives a carbamic ester. Propose a mechanism for this reaction.

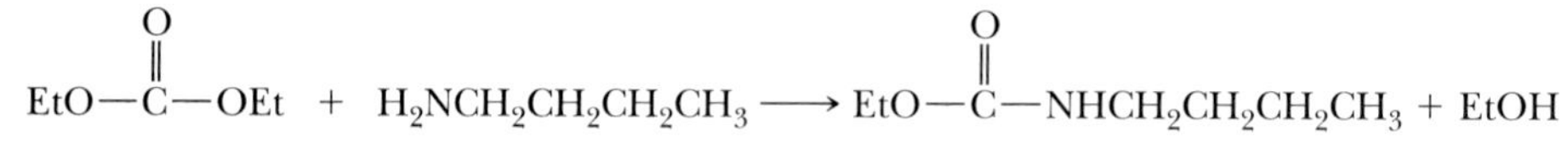

$$EtO{-}\overset{O}{\overset{\|}{C}}{-}OEt + H_2NCH_2CH_2CH_2CH_3 \longrightarrow EtO{-}\overset{O}{\overset{\|}{C}}{-}NHCH_2CH_2CH_2CH_3 + EtOH$$

Diethyl carbonate Butylamine Ethyl *N*-butylcarbamate

17.50 Several sulfonylureas, a class of compounds containing $RSO_2NHCONHR$, are useful drugs as orally active replacements for injected insulin in patients with adult-onset diabetes. It was discovered in 1942 that certain members of this class cause hypoglycemia in laboratory animals. Clinical trials of tolbutamide were begun in the early 1950s and since that time, more than 20 sulfonylureas have been introduced into clinical medicine. The sulfonylureas decrease blood glucose concentrations by stimulating β cells of the pancreas to release insulin and by increasing the sensitivity of insulin receptors on peripheral tissues to insulin stimulation.

Tolbutamide is synthesized by the reaction of the sodium salt of *p*-toluenesulfonamide and ethyl *N*-butylcarbamate (see the previous problem for the synthesis of this carbamic ester). Propose a mechanism for the following step in the synthesis of tolbutamide:

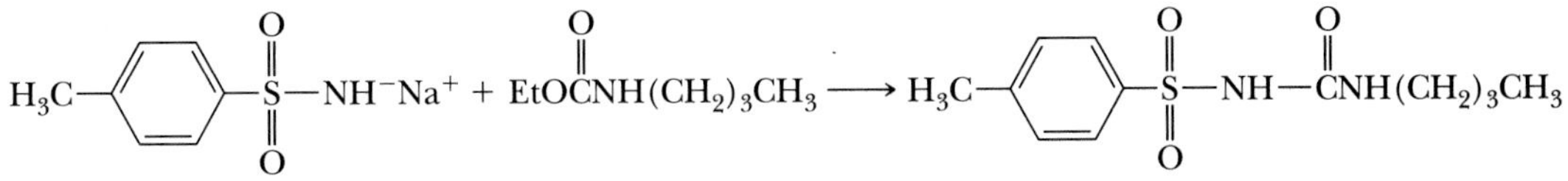

$$H_3C\text{—}C_6H_4\text{—}\overset{O}{\underset{O}{\overset{\|}{\underset{\|}{S}}}}\text{—}NH^-Na^+ + EtO\overset{O}{\overset{\|}{C}}NH(CH_2)_3CH_3 \longrightarrow H_3C\text{—}C_6H_4\text{—}\overset{O}{\underset{O}{\overset{\|}{\underset{\|}{S}}}}\text{—}NH\text{—}\overset{O}{\overset{\|}{C}}NH(CH_2)_3CH_3$$

Sodium salt of *p*-toluenesulfonamide

A carbamic ester

Tolbutamide (Oramide, Orinase)

17.51 Following are structural formulas for two more widely used sulfonylurea hypoglycemic agents. Show how each might be synthesized by converting an appropriate amine to a carbamic ester and then treating the carbamate with the sodium salt of a substituted benzenesulfonamide.

(a) 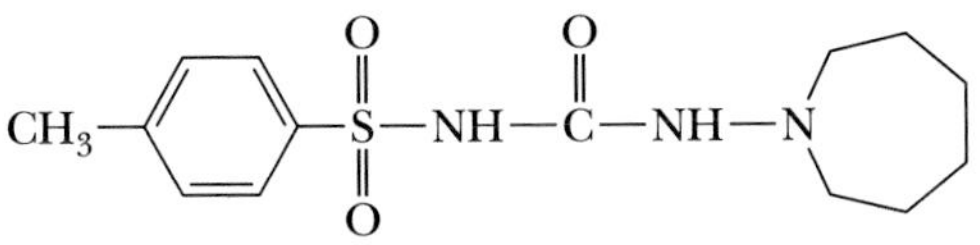 $CH_3\text{—}C_6H_4\text{—}\overset{O}{\underset{O}{\overset{\|}{\underset{\|}{S}}}}\text{—}NH\text{—}\overset{O}{\overset{\|}{C}}\text{—}NH\text{—}N$ (azepane ring)

Tolazamide (Tolamide, Tolinase)

(b) 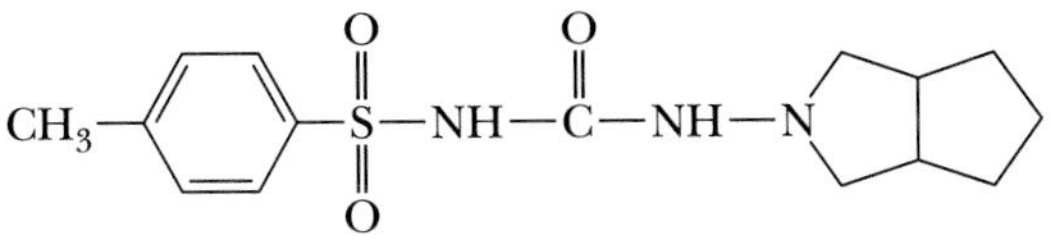 $CH_3\text{—}C_6H_4\text{—}\overset{O}{\underset{O}{\overset{\|}{\underset{\|}{S}}}}\text{—}NH\text{—}\overset{O}{\overset{\|}{C}}\text{—}NH\text{—}N$ (bicyclic ring)

Gliclazide (Diamicron)

17.52 Amantadine is one of the very few available antiviral agents and is effective in preventing infections caused by the influenza A virus and in treating established illnesses. It is thought to block a late stage in the assembly of the virus. Amantadine is synthesized as follows. Treatment of 1-bromoadamantane with acetonitrile in sulfuric acid gives *N*-adamantylacetamide, which is then converted to amantadine. (See the *The Merck Index,* 12th ed., #389.)

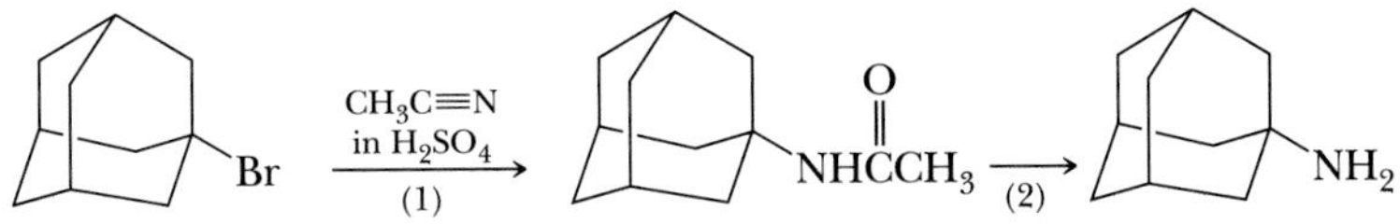

1-Bromoadamantane (Br) $\xrightarrow[(1)]{CH_3C{\equiv}N \text{ in } H_2SO_4}$ adamantyl–$NH\overset{O}{\overset{\|}{C}}CH_3$ $\xrightarrow[(2)]{}$ adamantyl–NH_2

1-Bromoadamantane

Amantadine

(a) Propose a mechanism for the transformation in Step 1. Consider the possibility of forming an adamantyl cation under these experimental conditions and then the manner in which this carbocation might undergo reaction with acetonitrile.

(b) Describe experimental conditions to bring about Step 2.

17.53 The following four-step sequence converts natural camphor, an enantiomerically pure compound, to the product shown on the right. The configuration of camphor is shown, but the configuration of the product is not shown.

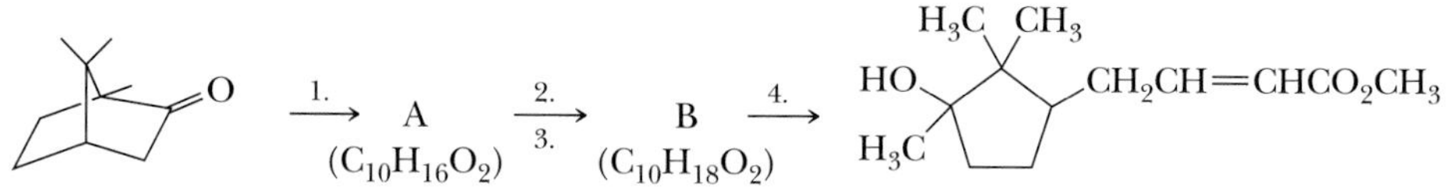

Camphor $\xrightarrow{1.}$ A ($C_{10}H_{16}O_2$) $\xrightarrow[3.]{2.}$ B ($C_{10}H_{18}O_2$) $\xrightarrow{4.}$ HO, H_3C, H_3C CH_3 cyclopentane with $CH_2CH{=}CHCO_2CH_3$

Camphor
($C_{10}H_{16}O$)

Steps/reagents: 1. $ArCO_3H$ 2. DIBALH (−78°C)
3. H_2O, HCl 4. $Ph_3P{=}CHCO_2CH_3$

(a) Propose structural formulas for intermediates A and B. Also specify the configuration at all stereocenters in A and B.

(b) Label all stereocenters in the product of this sequence and tell how many stereoisomers are possible for this compound?

(c) This synthesis is enantioselective, meaning that only one of the possible stereoisomers is formed. Which of the possible stereoisomers is formed and how do you account for its formation?

17.54 When natural camphor is treated first with peroxybenzoic acid and then with lithium aluminum hydride in diethyl ether, there is formed an optically active compound of molecular formula $C_{10}H_{20}O_2$. Propose a structural formula for this compound and show configuration at all of its stereocenters.

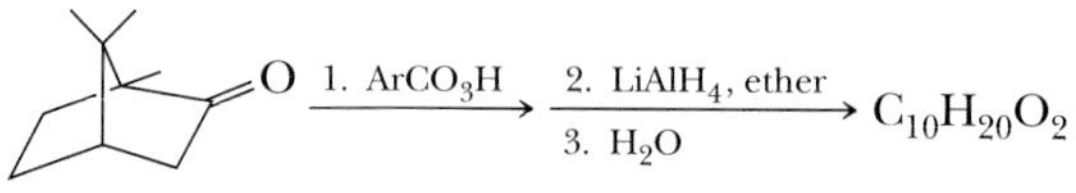

Camphor ($C_{10}H_{16}O$) $\xrightarrow{1.\ ArCO_3H}$ $\xrightarrow[3.\ H_2O]{2.\ LiAlH_4,\ ether}$ $C_{10}H_{20}O_2$

17.55 In a series of seven steps, (*S*)-malic acid is converted to the bromoepoxide shown on the right in 50% overall yield. (*S*)-Malic acid occurs in apples and many other fruits. It is also available in enantiopure form from microbiological fermentation. This synthesis is enantioselective: of the stereoisomers possible for the bromoepoxide, only one is formed. In thinking about the chemistry of these steps, you will want to review the use of dihydropyran as an —OH protecting group (Section 11.7) and the use of *p*-toluenesulfonyl chloride to convert the —OH, a poor leaving group, into a tosylate, a good leaving group (Section 9.6D).

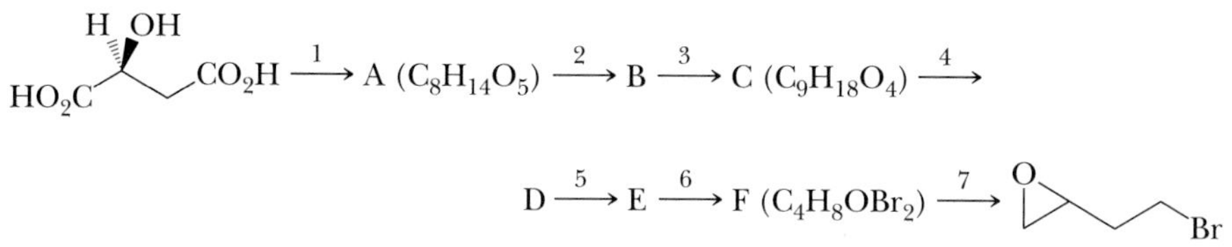

H OH, HO_2C, CO_2H $\xrightarrow{1}$ A ($C_8H_{14}O_5$) $\xrightarrow{2}$ B $\xrightarrow{3}$ C ($C_9H_{18}O_4$) $\xrightarrow{4}$

D $\xrightarrow{5}$ E $\xrightarrow{6}$ F ($C_4H_8OBr_2$) $\xrightarrow{7}$ O, Br

Steps/reagents: 1. CH_3CH_2OH, H^+ 2. 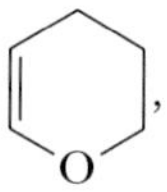 (dihydropyran, O), H^+ 3. $LiAlH_4$, then H_2O
4. TsCl, pyridine 5. NaBr, DMSO 6. HOAc, H_2O 7. KOH

(a) Propose structural formulas for intermediates A through F. Also specify configuration at each stereocenter.

(b) What is the configuration of the stereocenter in the bromoepoxide and how do you account for the stereoselectivity of this seven-step conversion?

17.56 Following is a retrosynthetic analysis for the synthesis of the herbicide (*S*)-Metolachlor. Ciba synthesizes approximately 10,000 tons of this agrochemical per year. According to this analysis, starting materials are 2-ethyl-6-methylaniline, chloroacetic acid, acetone, and methanol.

(*S*)-Metolachlor $\overset{1}{\Longrightarrow}$ $ClCH_2\overset{O}{\overset{\|}{C}}OH$ (Chloroacetic acid) + [secondary amine] $\overset{2}{\Longrightarrow}$ [imine] $\overset{3}{\Longrightarrow}$ NH_2 2-Ethyl-6-methylaniline + $CH_3\overset{O}{\overset{\|}{C}}CH_2OCH_3$ $\overset{4}{\Longrightarrow}$ $CH_3\overset{O}{\overset{\|}{C}}CH_2Cl$ + CH_3OH (Methanol)

$CH_3\overset{O}{\overset{\|}{C}}CH_2Cl$ $\overset{5}{\Longrightarrow}$ $CH_3\overset{O}{\overset{\|}{C}}CH_3$ (Acetone)

Show reagents and experimental conditions for the conversion of these three organic starting materials to Metolachlor. Note that your synthesis will most likely give a racemic mixture. The chiral catalyst used by Ciba for Step 2 gives 80% enantiomeric excess of the *S* enantiomer.

18 Enolate Anions and Enamines

■ Crystals of tamoxifen (see Problem 18.44) viewed under polarizing light. *(Herb Charles Ohymeyer/Fran Heyl Associates)*

This chapter is a continuation of the chemistry of carbonyl compounds. In Chapters 15–17, we concentrated on the carbonyl group itself and on nucleophilic additions to the carbonyl carbon to form tetrahedral carbonyl addition intermediates or products. In this chapter, we expand on the chemistry of carbonyl-containing compounds and consider the acidity of α-hydrogens and the formation of enolate anions. Following is a resonance-stabilized **enolate anion** formed from a ketone.

$$CH_3\overset{\overset{\displaystyle O}{\|}}{C}CH_3 + NaOH \rightleftharpoons \left[H-\underset{\underset{\displaystyle H}{|}}{\overset{-}{C}}-\overset{\overset{\displaystyle :\ddot{O}}{\|}}{C}CH_3 \longleftrightarrow H-\underset{\underset{\displaystyle H}{|}}{C}=\overset{\overset{\displaystyle :\ddot{O}:^-}{|}}{C}CH_3 \right] Na^+ + H_2O$$

An enolate anion

Enolate anions function as nucleophiles in S_N2 reactions as shown in this general reaction.

Enolate anion An anion formed by removal of an α-hydrogen from a carbonyl-containing compound.

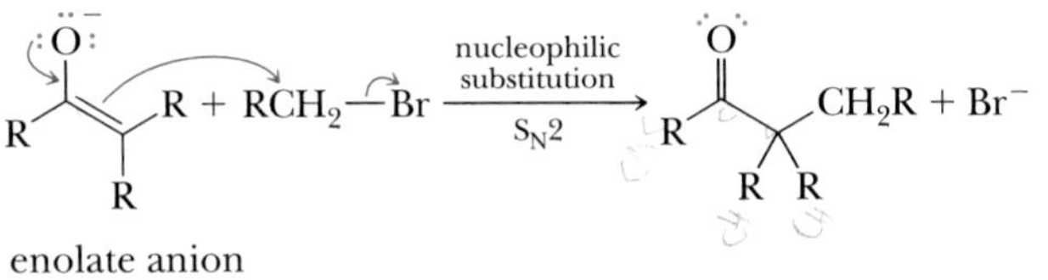

$$\text{An enolate anion} + RCH_2{-}Br \xrightarrow[S_N2]{\text{nucleophilic substitution}} R{-}C(=O){-}CR_2{-}CH_2R + Br^-$$

An enolate anion

They also function as nucleophiles in carbonyl addition reactions. Here, we show nucleophilic addition to the carbonyl carbon of a ketone. Enolate anions also add in this manner to the carbonyl groups of aldehydes and esters.

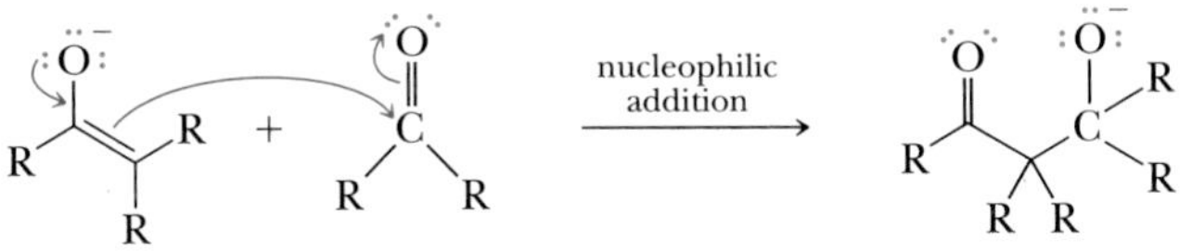

$$\text{An enolate anion} + R_2C{=}O \xrightarrow{\text{nucleophilic addition}} R{-}C(=O){-}CR_2{-}CR_2{-}O^-$$

An enolate anion | A ketone | A tetrahedral carbonyl addition intermediate

18.1 The Aldol Reaction

Unquestionably, the most important reaction of enolate anions is nucleophilic addition to the carbonyl group of another molecule of the same or different compound, as illustrated by the following reactions. Although such reactions may be catalyzed by either acid or base, base catalysis is more common.

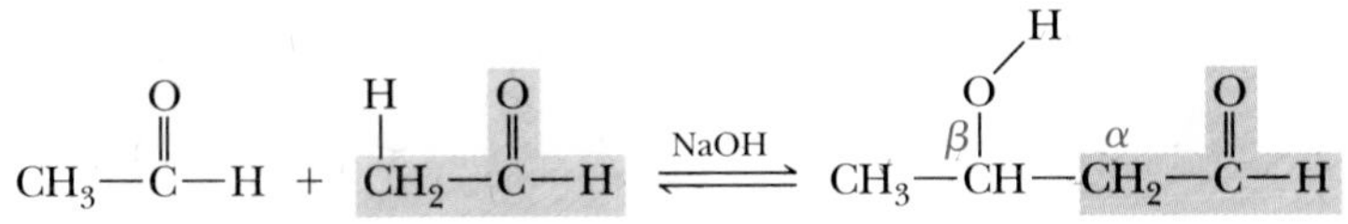

$$CH_3{-}\overset{O}{\overset{\|}{C}}{-}H + CH_2(H){-}\overset{O}{\overset{\|}{C}}{-}H \overset{NaOH}{\rightleftharpoons} CH_3{-}\overset{\beta}{CH}(OH){-}\overset{\alpha}{CH_2}{-}\overset{O}{\overset{\|}{C}}{-}H$$

Ethanal (Acetaldehyde) | Ethanal (Acetaldehyde) | 3-Hydroxybutanal (a β-hydroxyaldehyde)

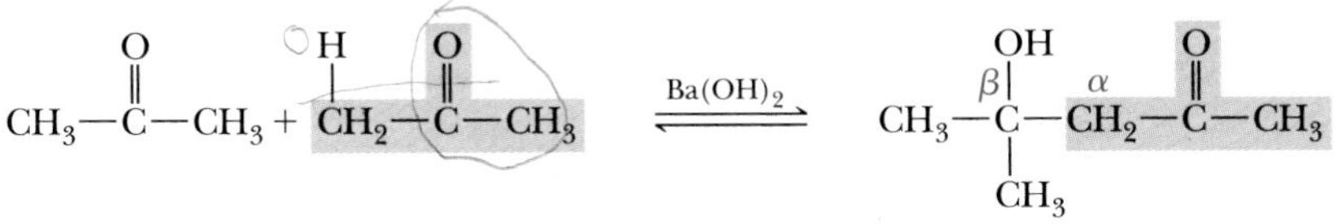

$$CH_3{-}\overset{O}{\overset{\|}{C}}{-}CH_3 + CH_2(H){-}\overset{O}{\overset{\|}{C}}{-}CH_3 \overset{Ba(OH)_2}{\rightleftharpoons} CH_3{-}\overset{\beta}{C}(OH)(CH_3){-}\overset{\alpha}{CH_2}{-}\overset{O}{\overset{\|}{C}}{-}CH_3$$

Propanone (Acetone) | Propanone (Acetone) | 4-Hydroxy-4-methyl-2-pentanone (a β-hydroxyketone)

The common name of the product derived from the reaction of acetaldehyde in base is aldol, so named because it is both an **ald**ehyde and an alcoh**ol**. Aldol is also the generic name given to any product formed in this type of reaction. The product of an **aldol reaction** is a β-hydroxyaldehyde or a β-hydroxyketone.

The key step in a base-catalyzed aldol reaction is nucleophilic addition of the enolate anion of one carbonyl-containing molecule to the carbonyl group of another to form a tetrahedral carbonyl addition intermediate. This mechanism is illustrated by the aldol reaction between two molecules of acetaldehyde. Notice in this three-step mechanism that OH^- is a catalyst; an OH^- is used in Step 1, but another OH^- is generated in Step 3.

MECHANISM Base-Catalyzed Aldol Reaction

Step 1: Formation of a resonance-stabilized enolate anion

$$HO^- + H{-}CH_2{-}\overset{\overset{\displaystyle O}{\|}}{C}{-}H \rightleftharpoons HO{-}H + \left[\ddot{:}\overset{-}{C}H_2{-}\overset{\overset{\displaystyle O}{\|}}{C}{-}H \longleftrightarrow CH_2{=}\overset{\overset{\displaystyle :O:^-}{|}}{C}{-}H \right]$$

Enolate anion

Step 2: Nucleophilic addition of the enolate anion to a carbonyl carbon to form a tetrahedral carbonyl addition intermediate

$$CH_3{-}\overset{\overset{\displaystyle O}{\|}}{C}{-}H + \overset{-}{:}CH_2{-}\overset{\overset{\displaystyle O}{\|}}{C}{-}H \rightleftharpoons CH_3{-}\overset{\overset{\displaystyle :O:^-}{|}}{CH}{-}CH_2{-}\overset{\overset{\displaystyle O}{\|}}{C}{-}H$$

A tetrahedral carbonyl addition intermediate

Step 3: Reaction of the tetrahedral carbonyl addition intermediate with a proton donor to form the aldol product

$$CH_3{-}\overset{\overset{\displaystyle :O:^-}{|}}{CH}{-}CH_2{-}\overset{\overset{\displaystyle O}{\|}}{C}{-}H + H{-}OH \rightleftharpoons CH_3{-}\overset{\overset{\displaystyle :\ddot{O}H}{|}}{CH}{-}CH_2{-}\overset{\overset{\displaystyle O}{\|}}{C}{-}H + OH^-$$

The mechanism of the acid-catalyzed aldol reaction involves two steps, the first of which is acid-catalyzed establishment of equilibrium between the keto and enol forms of the ketone or aldehyde (Section 15.11B). The second step is formation of the new carbon-carbon bond by attack of the enol on the carbon of the protonated carbonyl group.

MECHANISM Acid-Catalyzed Aldol Reaction

Step 1: Acid-catalyzed equilibration of keto and enol forms of one molecule

$$CH_3{-}\overset{\overset{\displaystyle O}{\|}}{C}{-}H \underset{}{\overset{HA}{\rightleftharpoons}} CH_2{=}\overset{\overset{\displaystyle OH}{|}}{C}{-}H$$

Step 2: Proton transfer from the acid, HA, to the carbonyl oxygen of another molecule

$$CH_3{-}\overset{\overset{\displaystyle \ddot{O}}{\|}}{C}{-}H + H{-}A \longrightarrow CH_3{-}\overset{\overset{\displaystyle \overset{+}{O}{-}H}{\|}}{C}{-}H + :A^-$$

Step 3: Attack of the enol of one molecule on the protonated carbonyl group of another molecule; this is followed by proton transfer to A^- to regenerate the acid catalyst

$$CH_3{-}\overset{\overset{\displaystyle \overset{+}{O}{-}H}{\|}}{C}{-}H + CH_2{=}\overset{\overset{\displaystyle :\ddot{O}{-}H}{|}}{C}{-}H \xrightarrow{A^-} CH_3{-}\overset{\overset{\displaystyle :\ddot{O}{-}H}{|}}{CH}{-}CH_2{-}\overset{\overset{\displaystyle \ddot{O}}{\|}}{C}{-}H + H{-}A$$

EXAMPLE 18.1

Draw the product of base-catalyzed aldol reaction of these compounds.

(a) Butanal **(b)** Cyclohexanone

Solution

The aldol product is formed by nucleophilic addition of the α-carbon of one molecule to the carbonyl carbon of another.

(a) $CH_3CH_2CH_2CH(OH)-CH(CH_2CH_3)CHO$ (the new C—C bond)

(b) 2-(1-hydroxycyclohexyl)cyclohexanone (the new C—C bond)

PROBLEM 18.1

Draw the product of the base-catalyzed aldol reaction of these compounds.

(a) Phenylacetaldehyde, $C_6H_5CH_2CHO$ **(b)** Cyclopentanone

β-Hydroxyaldehydes and β-hydroxyketones are very easily dehydrated, and often the conditions necessary to bring about an aldol reaction are sufficient to cause dehydration. This is particularly true in the case of acid-catalyzed aldol reactions. Alternatively, dehydration can be brought about by warming the aldol product in dilute acid. The major product from dehydration of an aldol is one in which the carbon-carbon double bond is conjugated with the carbonyl group, that is, the product is an α,β-unsaturated aldehyde or ketone.

$$CH_3CH(OH)CH_2CHO \xrightarrow{\text{warm in either acid or base}} \overset{\beta}{CH_3CH}=\overset{\alpha}{CH}CHO + H_2O$$

An α,β-unsaturated aldehyde

In base-catalyzed dehydration, an α-hydrogen is removed to form a new enolate anion, which then expels a hydroxide ion.

$$CH_3-CH(OH)-CH(H)-CH{=}O \xrightarrow{OH^-} \left[CH_3-CH(OH)-CH=CH-O^-\right] \xrightarrow{-OH^-} \overset{\beta}{CH_3CH}=\overset{\alpha}{CH}CHO$$

An enolate anion — An α,β-unsaturated aldehyde

In acid-catalyzed dehydration, proton transfer from H_3O^+ to the —OH group of the aldol product forms an oxonium ion. Loss of water from this protonated enol gives a protonated form of the product. Proton transfer from the protonated carbonyl group to H_2O completes the dehydration of the aldol product.

transfer of this proton to solvent completes the dehydration reaction

$$H_2\ddot{O}^+{-}H \;\; CH_3{-}\underset{H_3C}{\overset{:\ddot{O}H}{C}}{-}CH{=}\overset{:\ddot{O}{-}H}{C}{-}CH_3 \rightleftharpoons CH_3{-}\underset{H_3C}{\overset{H_2\ddot{O}^+}{C}}{-}CH{=}\overset{:\ddot{O}{-}H}{C}{-}CH_3 \longrightarrow CH_3\underset{CH_3}{\overset{\beta}{C}}{=}\overset{\alpha}{C}H\overset{\ddot{O}^+{-}H}{\overset{\|}{C}}CH_3 + H_2O$$

The enol of the ketone

Protonated α,β-unsaturated ketone

Example 18.2

Draw the product of dehydration of each aldol product in Example 18.1.

Solution

Loss of H_2O from aldol product (a) gives an α,β-unsaturated aldehyde; loss of H_2O from (b) gives an α,β-unsaturated ketone.

(a) $CH_3CH_2CH_2CH{=}\underset{CH_2CH_3}{C}\overset{O}{\overset{\|}{C}}H$ **(b)** [2-cyclohexylidenecyclohexanone]

Problem 18.2

Draw the product of dehydration of each aldol product in Problem 18.1.

Aldol reactions are readily reversible, especially when catalyzed by base, and, for aldol reactions of ketones especially, there is generally little aldol product present at equilibrium. Equilibrium constants for dehydration, however, are generally large so that, if reaction conditions are sufficiently vigorous to bring about dehydration, good yields of product can be obtained.

The reactants in the key step of an aldol reaction are an enolate anion and an enolate anion acceptor. In self-reactions, both roles are played by one kind of molecule. **Crossed aldol reactions** are also possible, as, for example, the crossed aldol reaction between acetone and formaldehyde. Formaldehyde cannot provide an enolate anion because it has no α-hydrogen, but it can function as a particularly good anion acceptor because its carbonyl group is unhindered. Acetone forms an enolate anion but its carbonyl group, which is bonded to two alkyl groups, is less reactive than that of formaldehyde. Consequently, the crossed aldol reaction between acetone and formaldehyde gives 4-hydroxy-2-butanone.

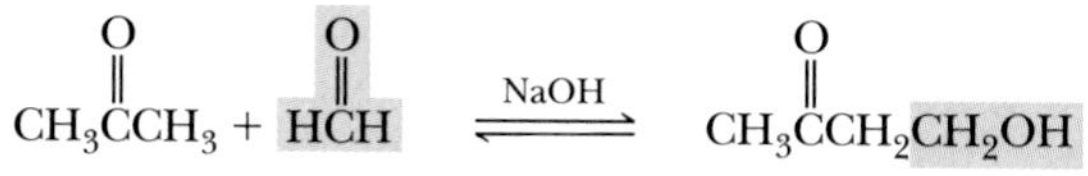

$$CH_3\overset{O}{\overset{\|}{C}}CH_3 + H\overset{O}{\overset{\|}{C}}H \xrightleftharpoons{NaOH} CH_3\overset{O}{\overset{\|}{C}}CH_2CH_2OH$$

4-Hydroxy-2-butanone

As this example illustrates, for a crossed aldol reaction to be successful, one of the two reactants should have no α-hydrogen so that an enolate anion does not form. It also helps if the compound with no α-hydrogen has the more reactive carbonyl, for example, an aldehyde. Following are examples of aldehydes that have no α-hydrogens and can be used in crossed aldol reactions. If these requirements are not met, a complex mixture of products results.

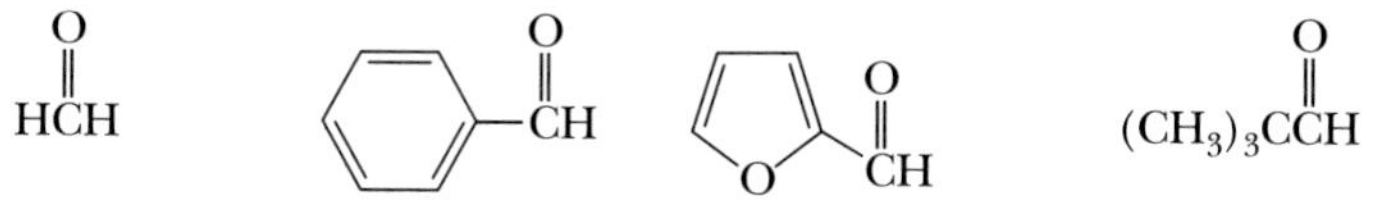

$H\overset{O}{\overset{\|}{C}}H$ Methanal (Formaldehyde)

Benzaldehyde

Furfural

$(CH_3)_3C\overset{O}{\overset{\|}{C}}H$ 2,2-Dimethylpropanal

EXAMPLE 18.3

Draw the product of the base-catalyzed crossed aldol reaction between furfural and cyclohexanone and the product formed by its dehydration.

Solution

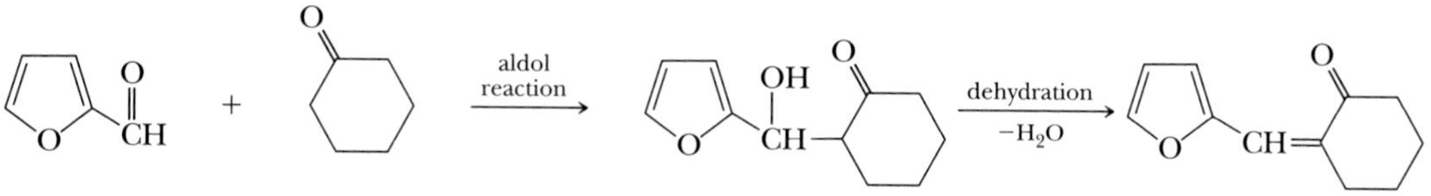

Furfural + Cyclohexanone —aldol reaction→ Aldol product —dehydration, $-H_2O$→

PROBLEM 18.3

Draw the product of the base-catalyzed crossed aldol reaction between benzaldehyde and 3-pentanone and the product formed by its dehydration.

When both the enolate anion and the carbonyl group to which it adds are in the same molecule, aldol reaction results in formation of a ring. This type of **intramolecular aldol reaction** is particularly useful for formation of five- and six-membered rings. Intramolecular aldol reaction of 2,7-octanedione via its enolate anion at α_3 gives a five-membered ring.

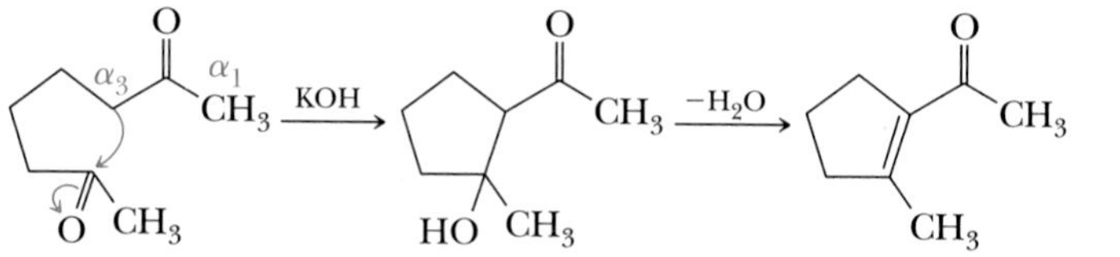

2,7-Octanedione —KOH→ —$-H_2O$→

Note that in 2,7-octanedione, two enolate anions are possible, one of which leads to the five-membered ring just shown. Aldol reaction via enolate anion at α_1 leads to a seven-membered ring. Formation of five- and six-membered rings is favored relative to formation of four- and seven-membered rings.

In general, smaller rings form faster than larger rings because the reacting groups are closer together. However, the formation of three- and four-membered rings is disfavored because of the strain in these rings.

Following is another example in which either a four-membered ring (via an enolate anion at α_3) or a six-membered ring (via an enolate anion at α_1) could be formed. Because of the greater stability of six-membered rings compared to four-membered rings, it is the six-membered ring that is formed in this intramolecular aldol reaction.

$CH_2CH_2CCH_3$ (α_3, α_1) $\xrightarrow{KOH}$ HO α_1 $\longrightarrow$ + H_2O

Nitro groups can be introduced into aliphatic compounds by way of an aldol reaction between the anion of a nitroalkane and an aldehyde or a ketone. The α-hydrogens of nitroalkanes are sufficiently acidic that they are removed by bases such as aqueous NaOH and KOH. The pK_a of nitromethane, for example, is 10.2. The acidity of the α-hydrogen of a nitroalkane is due to (1) the electron-withdrawing inductive effect of the nitro group, which weakens the H—C bond, and (2) the stabilization of the resulting anion by resonance delocalization of its negative charge.

$HO^- + H{-}CH_2{-}N^+(=O){-}O^- \rightleftharpoons HO{-}H + {}^-CH_2{-}N^+(=O){-}O^- \longleftrightarrow CH_2{=}N^+(O^-){-}O^-$

Nitromethane pK_a 10.2 (stronger acid) — Water pK_a 15.7 (weaker acid) — Resonance stabilized anion

Following is an aldol reaction between nitromethane and cyclohexanone. Reduction of the nitro group in the aldol product thus formed is a convenient synthetic route to β-aminoalcohols.

Cyclohexanone + CH_3NO_2 (Nitromethane) $\xrightarrow[\text{(aldol)}]{NaOH}$ 1-(Nitromethyl)-cyclohexanol (HO, CH_2NO_2) $\xrightarrow{H_2,\ Ni}$ 1-(Aminomethyl)-cyclohexanol (HO, CH_2NH_2)

18.2 Directed Aldol Reactions

A. Formation of Enolate Anions: Kinetic Versus Thermodynamic Control

Enolate anions are formed when a carbonyl compound containing an α-hydrogen is treated with base. When alkali metal hydroxides or alkoxides are used as bases, the position of equilibrium favors reactants rather than products.

$$\underset{\substack{\text{p}K_a\ 20\\\text{(weaker acid)}}}{CH_3\overset{\overset{\displaystyle O}{\|}}{C}CH_3} + \underset{\text{(weaker base)}}{NaOH} \rightleftharpoons \underset{\substack{\text{A sodium}\\\text{enolate}\\\text{(stronger base)}}}{CH_2{=}\overset{\overset{\displaystyle O^-Na^+}{|}}{C}CH_3} + \underset{\substack{\text{p}K_a\ 15.7\\\text{(stronger acid)}}}{H_2O} \qquad K_{eq} = 5 \times 10^{-5}$$

$$\underset{\substack{\text{p}K_a\ 23\\\text{(weaker acid)}}}{CH_3\overset{\overset{\displaystyle O}{\|}}{C}OC_2H_5} + \underset{\text{(weaker base)}}{C_2H_5O^-Na^+} \rightleftharpoons \underset{\substack{\text{A sodium}\\\text{enolate}\\\text{(stronger base)}}}{CH_2{=}\overset{\overset{\displaystyle O^-Na^+}{|}}{C}OC_2H_5} + \underset{\substack{\text{p}K_a\ 16\\\text{(stronger acid)}}}{C_2H_5OH} \qquad K_{eq} = 10^{-7}$$

With a stronger base, however, the formation of an enolate anion can be driven to the right. One of the most widely used bases for this purpose is **lithium diisopropylamide (LDA).**

$$[(CH_3)_2CH]_2\ddot{N}{:}^-\ Li^+$$

Lithium diisopropylamide
(LDA)

Lithium diisopropylamide is prepared by dissolving diisopropylamine in tetrahydrofuran and treating this solution with butyllithium.

$$\underset{\substack{\text{Diisopropylamine}\\\text{p}K_a\ 40\\\text{(stronger acid)}}}{[(CH_3)_2CH]_2NH} + \underset{\substack{\text{Butyllithium}\\\text{(stronger base)}}}{CH_3(CH_2)_3Li} \longrightarrow \underset{\substack{\text{Lithium diisopropyl-}\\\text{amide (LDA)}\\\text{(weaker base)}}}{[(CH_3)_2CH]_2N^-Li^+} + \underset{\substack{\text{Butane}\\\text{p}K_a\ 50\\\text{(weaker acid)}}}{CH_3(CH_2)_2CH_3} \qquad K_{eq} = 10^{10}$$

Although LDA is a very strong base, it is a very poor nucleophile because of steric crowding around the nitrogen and does not add to carbonyl groups. It is, therefore, ideal for generation of enolate anions from carbonyl-containing compounds, as illustrated by treatment of ethyl acetate with LDA. By using a molar equivalent of LDA, an aldehyde, ketone, or ester can be converted completely to its lithium enolate.

$$\underset{\substack{\text{p}K_a\ 23\\\text{(stronger acid)}}}{CH_3\overset{\overset{\displaystyle O}{\|}}{C}OC_2H_5} + \underset{\substack{\text{LDA}\\\text{(stronger base)}}}{[(CH_3)_2CH]_2N^-Li^+} \longrightarrow \underset{\substack{\text{A lithium}\\\text{enolate}\\\text{(weaker base)}}}{CH_2{=}\overset{\overset{\displaystyle O^-Li^+}{|}}{C}OC_2H_5} + \underset{\substack{\text{p}K_a\ 40\\\text{(weaker acid)}}}{[(CH_3)_2CH]_2NH} \qquad K_{eq} = 10^{17}$$

For a ketone with two sets of nonequivalent α-hydrogens, the following question arises: Is formation of an enolate anion regioselective, and if so, what are the factors that determine the degree of regioselectivity? It has been determined experimentally that a very high degree of regioselectivity often exists, and that it depends on experimental conditions. When 2-methylcyclohexanone, for example, is added to a slight excess of LDA, the ketone is converted to its lithium enolate, which consists almost entirely of the salt of the less substituted enolate anion.

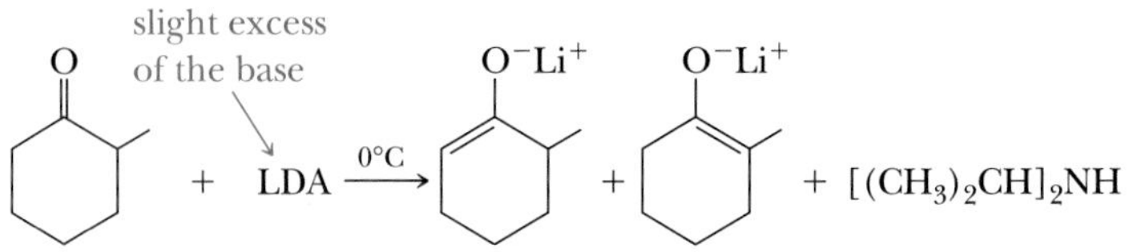

slight excess of the base

2-Methylcyclohexanone + LDA $\xrightarrow{0°C}$ O^-Li^+ (99%) + O^-Li^+ (1%) + $[(CH_3)_2CH]_2NH$

When 2-methylcyclohexanone is treated with LDA under conditions in which the ketone is in a slight excess, then the composition of the product is quite different; it is richer in the more substituted enolate anion.

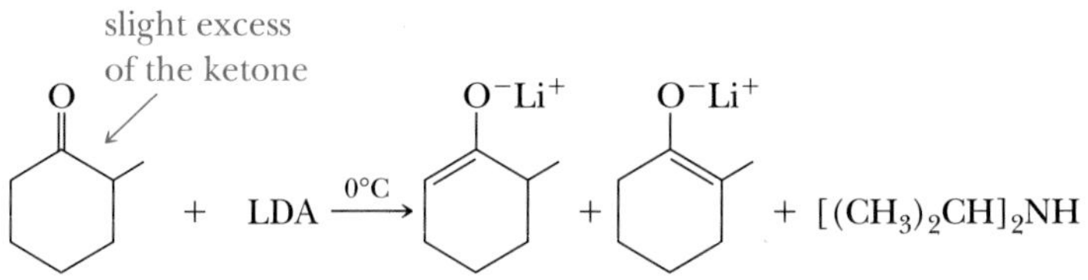

slight excess of the ketone

2-Methylcyclohexanone + LDA $\xrightarrow{0°C}$ O^-Li^+ (10%) + O^-Li^+ (90%) + $[(CH_3)_2CH]_2NH$

Thermodynamic control Experimental conditions that permit the establishment of equilibrium between two or more products of a reaction. The composition of the product mixture is determined by the relative stabilities of the products.

The most important factor determining the composition of an enolate anion mixture is whether the reaction forming it is under kinetic (rate) control or thermodynamic (equilibrium) control. In a reaction under **thermodynamic control:**

1. The reaction conditions permit the equilibration of alternative products, and
2. The composition of the product mixture is determined by the relative stabilities of the products.

Equilibrium among enolate anions is established when the ketone is in slight excess, a condition under which it is possible for proton-transfer reactions to occur. An enolate anion can undergo proton transfer from the α-carbon of an unreacted molecule of ketone to form the alternative enolate anion, and vice versa. Thus, an equilibrium is established between alternative enolate anions.

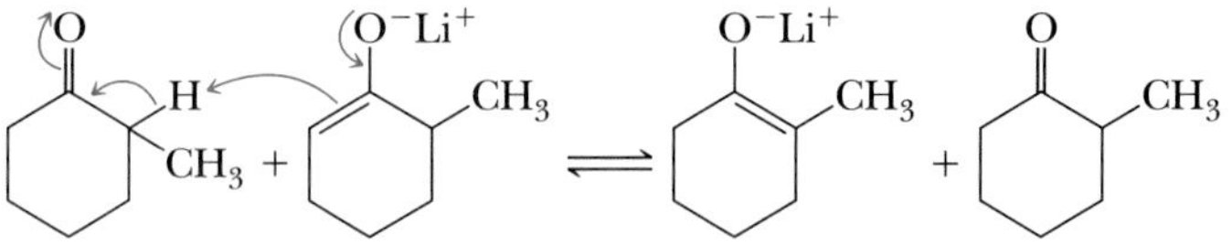

O, H, CH_3 + O^-Li^+, CH_3 ⇌ O^-Li^+, CH_3 + O, CH_3

Less stable enolate anion

More stable enolate anion

Under these conditions, it is the more stable enolate anion that predominates. The factors that determine the relative stabilities of enolate anions are the same as those that determine the relative stabilities of alkenes (Section 6.6B), namely, the more substituted the double bond of the enolate anion, the greater its stability. Thus, the composition of the enolate anion mixture formed under conditions of thermodynamic control reflects the relative stabilities of the individual enolate anions.

In a reaction under **kinetic control,** the composition of the product mixture is determined by the relative rates of formation of each product. In the case of formation of enolate anions, kinetic control refers to the relative rates of removal of the alternative α-hydrogens. The less hindered α-hydrogen is removed more rapidly, and thus the major product is the less substituted enolate anion. Because a slight excess of base is used, there is no ketone to serve as a proton donor, and the less stable enolate anion cannot equilibrate with the more stable enolate anion.

Kinetic control Experimental conditions under which the composition of the product mixture is determined by the relative rates of formation of each product.

B. Directed Aldol Reactions

In Section 18.1, we discussed aldol reactions and pointed out a problem inherent in carrying out **crossed aldol reactions.** Crossed aldol reactions between an aldehyde with no α-hydrogens and a ketone generally give good yields of a single product because (1) only the ketone can form an enolate anion, and (2) the carbonyl group of an aldehyde is a better enolate anion acceptor than the more crowded carbonyl group of a ketone. Aldol reaction of benzaldehyde and acetone followed by dehydration of the aldol intermediate, for example, gives a single product in good yield.

$$\underset{\text{Benzaldehyde}}{C_6H_5\overset{O}{\overset{\|}{C}}H} + \underset{\text{Acetone}}{CH_3\overset{O}{\overset{\|}{C}}CH_3} \xrightarrow[C_2H_5OH]{NaOH} \underset{\text{4-Phenyl-3-buten-2-one}}{C_6H_5CH{=}CH\overset{O}{\overset{\|}{C}}CH_3} + H_2O$$

As mentioned in the previous section, a problem arises in a crossed aldol reaction when both the aldehyde and ketone have α-hydrogens. Consider, for example, the problem of how to prepare the following aldol product from phenylacetaldehyde and acetone.

$$\underset{\text{Phenylacetaldehyde}}{C_6H_5CH_2\overset{O}{\overset{\|}{C}}H} + \underset{\text{Acetone}}{CH_3\overset{O}{\overset{\|}{C}}CH_3} \xrightarrow{?} \underset{\text{4-Hydroxy-5-phenyl-2-pentanone}}{C_6H_5CH_2\overset{OH}{\overset{|}{C}}HCH_2\overset{O}{\overset{\|}{C}}CH_3}$$

Four aldol products are possible and a mixture of the four is formed if these two compounds are mixed in the presence of $C_2H_5O^-Na^+$. The desired aldol reaction may be carried out successfully by treatment of acetone with LDA to convert it completely and irreversibly to its enolate anion. The preformed enolate anion is then treated with the aldehyde followed by work-up in water to give the crossed aldol reaction product.

$$CH_3\overset{O}{\overset{\|}{C}}CH_3 \xrightarrow[-78°C]{LDA} \underset{\text{A lithium enolate}}{CH_2{=}\overset{O^-Li^+}{\overset{|}{C}}CH_3} \xrightarrow[2.\ H_2O]{1.\ C_6H_5CH_2\overset{O}{\overset{\|}{C}}H} C_6H_5CH_2\overset{OH}{\overset{|}{C}}HCH_2\overset{O}{\overset{\|}{C}}CH_3$$

EXAMPLE 18.4

Show how to prepare this β-hydroxyketone by a directed aldol reaction.

$$\underset{\substack{|\\CH_3}}{CH_3\overset{OH}{\overset{|}{C}}H\,CH\overset{O}{\overset{\|}{C}}CH_2CH_3}$$

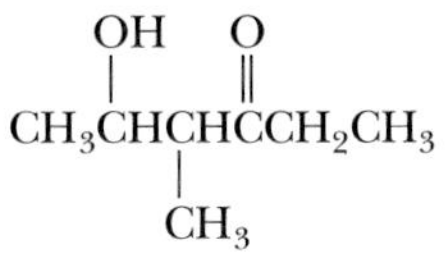

Solution

First recognize that the two carbonyl-containing compounds to be joined in the aldol reaction are 3-pentanone and acetaldehyde. Treat the symmetrical ketone with LDA to form its lithium enolate. Treatment of this enolate anion with acetaldehyde followed by aqueous work-up gives the desired aldol product.

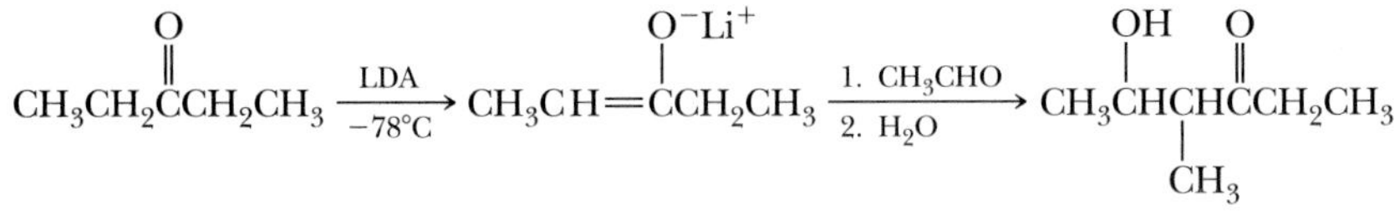

$$CH_3CH_2\overset{O}{\overset{\|}{C}}CH_2CH_3 \xrightarrow[-78°C]{LDA} CH_3CH{=}\overset{O^-Li^+}{\overset{|}{C}}CH_2CH_3 \xrightarrow[2.\ H_2O]{1.\ CH_3CHO} CH_3\overset{OH}{\overset{|}{C}}H\underset{CH_3}{\underset{|}{C}H}\overset{O}{\overset{\|}{C}}CH_2CH_3$$

PROBLEM 18.4

Show how you might prepare these compounds by directed aldol reactions.

(a) 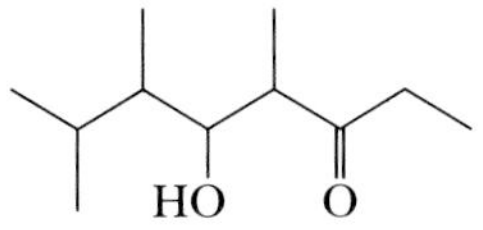 [skeletal structure; labels: HO, O]

(b) 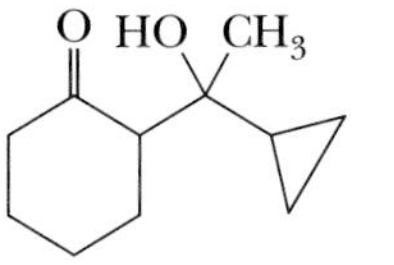 [structure; labels: O, HO, CH_3]

(c) 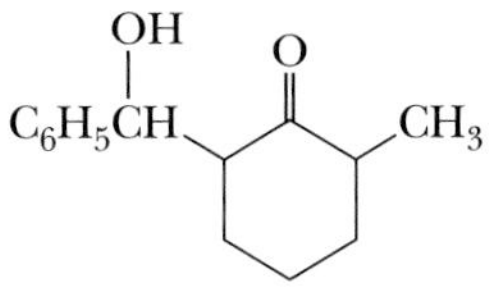 $C_6H_5\overset{OH}{\overset{|}{C}}H$ [cyclohexanone ring structure; labels: O, CH_3]

18.3 Claisen and Dieckmann Condensations

A. Claisen Condensation

In Chapter 17, we described reactions of esters, all of which take place at the carbonyl carbon and involve nucleophilic acyl substitution. In this section, we examine a second type of reaction characteristic of esters, namely one that involves both formation of an enolate anion at the α-carbon and nucleophilic acyl substitution. One of the first discovered of these reactions is the **Claisen condensation,** named after the German chemist Ludwig Claisen (1851–1930). A Claisen condensation is illustrated by the condensation of two molecules of ethyl acetate in the presence of sodium ethoxide followed by acidification to give ethyl acetoacetate. Note that in this equation and many of the following equations, the ethyl group, C_2H_5, is written as —Et.

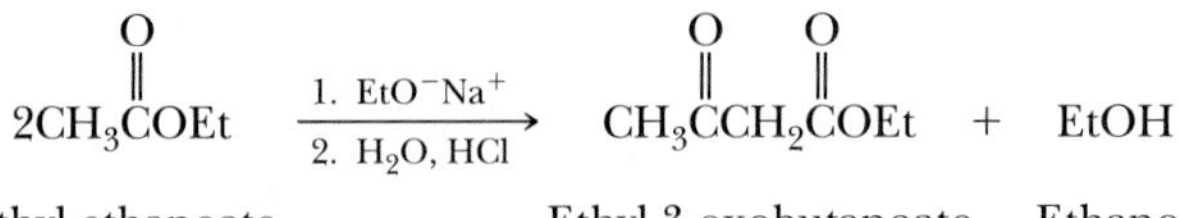

$$2CH_3\overset{O}{\overset{\|}{C}}OEt \xrightarrow[2.\ H_2O,\ HCl]{1.\ EtO^-Na^+} CH_3\overset{O}{\overset{\|}{C}}CH_2\overset{O}{\overset{\|}{C}}OEt + EtOH$$

Ethyl ethanoate (Ethyl acetate) | Ethyl 3-oxobutanoate (Ethyl acetoacetate) | Ethanol

The product of a Claisen condensation is a ***β*-ketoester.**

$$-\overset{|}{\underset{|}{C}}-\overset{O}{\overset{\|}{C}}\overset{\beta}{-}\overset{|}{\underset{|}{C}}-\overset{O}{\overset{\|}{C}}-OR$$

A β-ketoester

Claisen condensations, like the aldol reaction, require a base. Aqueous bases, such as NaOH, however, cannot be used in Claisen condensations because they would bring about the hydrolysis of the ester. Rather, the bases most commonly used in Claisen condensations are nonaqueous, such as sodium ethoxide in ethanol and sodium methoxide in methanol.

Claisen condensation of two molecules of ethyl propanoate gives the following β-ketoester.

$$CH_3CH_2\overset{O}{\overset{\|}{C}}OEt + \underset{CH_3}{\underset{|}{CH_2}}\overset{O}{\overset{\|}{C}}OEt \xrightarrow[2.\ H_2O,\ HCl]{1.\ EtO^-Na^+} CH_3CH_2\overset{O}{\overset{\|}{C}}\underset{CH_3}{\underset{|}{CH}}\overset{O}{\overset{\|}{C}}OEt + EtOH$$

Ethyl propanoate — Ethyl propanoate — Ethyl 2-methyl-3-oxopentanoate

The first steps of a Claisen condensation bear a close resemblance to the first steps of an aldol reaction (Section 18.1). In Step 1, base removes a proton from an α-carbon to form a **resonance-stabilized enolate anion.** Because the α-hydrogen of an ester is the weaker acid and ethoxide ion is the weaker base, the position of this equilibrium lies very much toward the left, and the concentration of enolate anion is very low compared with that of ethoxide ion and ester. In Step 2, the enolate anion attacks the carbonyl carbon of another ester molecule to form a tetrahedral carbonyl addition intermediate, which in turn collapses in Step 3 to give a β-ketoester. The position of equilibrium for Steps 1–3 lies far toward the starting ester. The overall reaction is driven to completion, however, because the β-ketoester formed in Step 3 is a stronger acid than ethanol. The β-ketoester (a stronger acid) reacts with ethoxide ion (a stronger base) in Step 4 to give ethanol (a weaker acid) and the anion of the β-ketoester (a weaker base).

MECHANISM Claisen Condensation

Step 1: Formation of an enolate anion

$$C_2H_5\ddot{O}{:}^- + H-CH_2-\overset{O}{\overset{\|}{C}}OEt \rightleftharpoons Et\ddot{O}-H + \left[{:}\overset{-}{C}H_2-\overset{O}{\overset{\|}{C}}OEt \longleftrightarrow CH_2=\overset{{:}\ddot{O}{:}^-}{\overset{|}{C}}OEt\right]$$

(weaker base) — $pK_a = 22$ (weaker acid) — $pK_a = 15.9$ (stronger acid) — Resonance-stabilized enolate anion (stronger base)

Step 2: Attack of the enolate anion on a carbonyl carbon to give a tetrahedral carbonyl addition intermediate

$$CH_3-\overset{O}{\overset{\|}{C}}OEt + {}^{-}:CH_2-\overset{O}{\overset{\|}{C}}OEt \rightleftharpoons \left[CH_3-\underset{:\ddot{O}Et}{\overset{:\ddot{O}:^{-}}{C}}-CH_2-\overset{O}{\overset{\|}{C}}OEt\right]$$

A tetrahedral carbonyl addition intermediate

Step 3: Collapse of the tetrahedral carbonyl addition intermediate to form a β-ketoester and ethoxide ion

$$CH_3-\underset{:\ddot{O}Et}{\overset{:\ddot{O}:^{-}}{C}}-CH_2-\overset{O}{\overset{\|}{C}}OEt \rightleftharpoons CH_3-\overset{O}{\overset{\|}{C}}-CH_2-\overset{O}{\overset{\|}{C}}OEt + Et\ddot{O}:^{-}$$

Step 4: Formation of the enolate anion of the β-ketoester, which drives the Claisen condensation to the right

$$\underset{\substack{pK_a\ 10.7 \\ \text{(stronger acid)}}}{CH_3-\overset{O}{\overset{\|}{C}}-\overset{H}{\overset{|}{C}H}-\overset{O}{\overset{\|}{C}}OEt} + \underset{\text{(stronger base)}}{Et\ddot{O}:^{-}} \rightleftharpoons \underset{\text{(weaker base)}}{CH_3-\overset{O}{\overset{\|}{C}}-\ddot{C}H^{-}-\overset{O}{\overset{\|}{C}}OEt} + \underset{\substack{pK_a\ 15.9 \\ \text{(weaker acid)}}}{Et\ddot{O}H}$$

Thus, the structural feature required for a successful Claisen condensation is an ester with two α-hydrogens: one to form the initial enolate anion and the second to form the enolate anion of the resulting β-ketoester. The β-ketoester is formed and isolated upon acidification with aqueous acid during work-up.

$$CH_3-\overset{O}{\overset{\|}{C}}-\ddot{C}H^{-}-\overset{O}{\overset{\|}{C}}-OEt + H^{+} \xrightarrow{HCl,\ H_2O} CH_3-\overset{O}{\overset{\|}{C}}-CH_2-\overset{O}{\overset{\|}{C}}-OEt$$

EXAMPLE 18.5

Show the product of Claisen condensation of ethyl butanoate in the presence of sodium ethoxide followed by acidification with aqueous HCl.

Solution

The new bond formed in a Claisen condensation is between the carbonyl group of one ester and the α-carbon of another.

the new C—C bond

$$CH_3CH_2CH_2\overset{O}{\overset{\|}{C}}-\underset{CH_2CH_3}{\underset{|}{C}H}\overset{O}{\overset{\|}{C}}OEt + EtOH$$

Ethyl 2-ethyl-3-oxohexanoate

PROBLEM 18.5

Show the product of Claisen condensation of ethyl 3-methylbutanoate in the presence of sodium ethoxide followed by acidification with aqueous HCl.

B. The Dieckmann Condensation

An intramolecular Claisen condensation of a dicarboxylic ester to give a five- or six-membered ring is given the special name of **Dieckmann condensation.** In the presence of one equivalent of sodium ethoxide, for example, diethyl hexanedioate (diethyl adipate) undergoes an intramolecular condensation to form a five-membered ring.

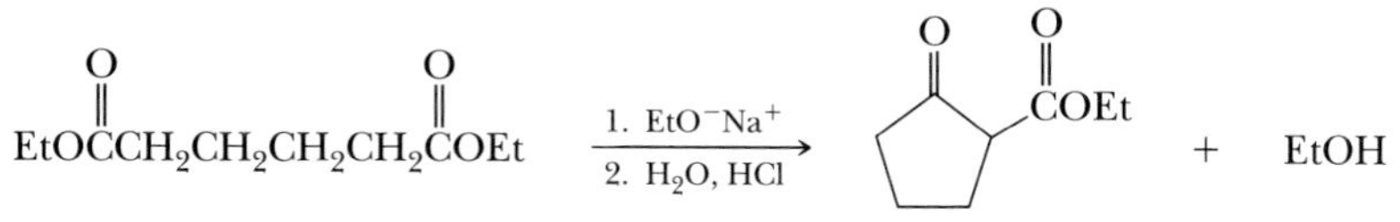

$$\mathrm{EtO\overset{O}{\overset{\|}{C}}CH_2CH_2CH_2CH_2\overset{O}{\overset{\|}{C}}OEt} \xrightarrow[\text{2. } H_2O,\ HCl]{\text{1. } EtO^-Na^+} \text{(cyclopentanone ring bearing } \overset{O}{\overset{\|}{C}}OEt\text{)} + EtOH$$

Diethyl hexanedioate (Diethyl adipate) — Ethyl 2-oxocyclopentanecarboxylate

The mechanism of a Dieckmann condensation is identical to the mechanism we have described for the Claisen condensation. An anion formed at the α-carbon of one ester group in Step 1 adds to the carbonyl of the other ester group in Step 2 to form a tetrahedral carbonyl addition intermediate. This intermediate ejects ethoxide ion in Step 3 to regenerate the carbonyl group. Cyclization is followed by formation of the conjugate base of the β-ketoester, just as in the Claisen condensation. The β-ketoester is isolated after acidification with aqueous acid.

MECHANISM Dieckmann Condensation

Step 1: Proton transfer to ethoxide ion giving an enolate anion

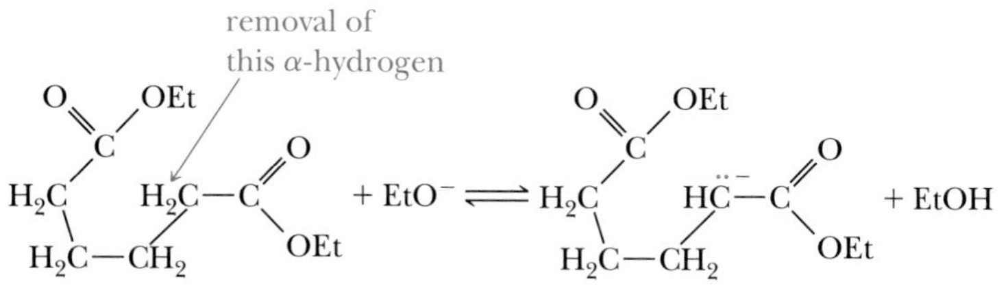

removal of this α-hydrogen

$$\text{EtO(O=)C–CH}_2\text{CH}_2\text{CH}_2\text{CH}_2\text{–C(=O)OEt} + EtO^- \rightleftharpoons \text{EtO(O=)C–CH}_2\text{CH}_2\text{CH}_2\ddot{C}H^-\text{–C(=O)OEt} + EtOH$$

An enolate anion

Step 2: Attack of the enolate anion on the other carbonyl carbon to give a tetrahedral carbonyl addition intermediate

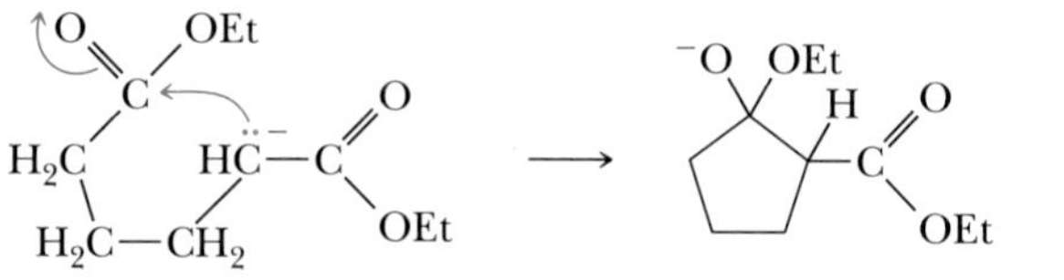

$$\text{EtO(O=)C–CH}_2\text{CH}_2\text{CH}_2\ddot{C}H^-\text{–C(=O)OEt} \longrightarrow \text{cyclopentane with } C(O^-)(OEt) \text{ and } CH\text{–C(=O)OEt}$$

A tetrahedral carbonyl addition intermediate

Step 3: Collapse of the tetrahedral carbonyl addition intermediate to give a β-ketoester and ethoxide ion

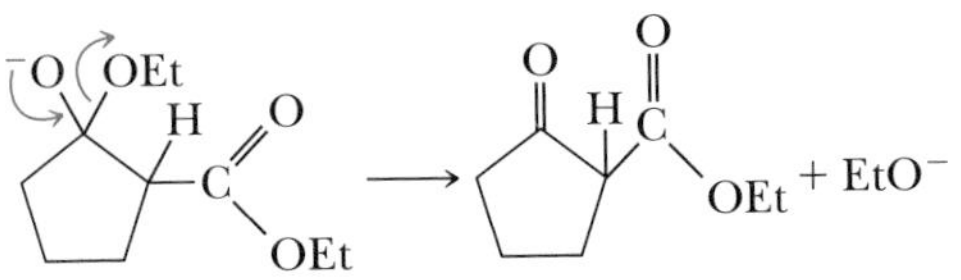

Step 4: Proton transfer from the β-ketoester to ethoxide to form the anion of the β-ketoester, which drives the Dieckmann condensation to completion; final work-up with aqueous acid gives the β-ketoester

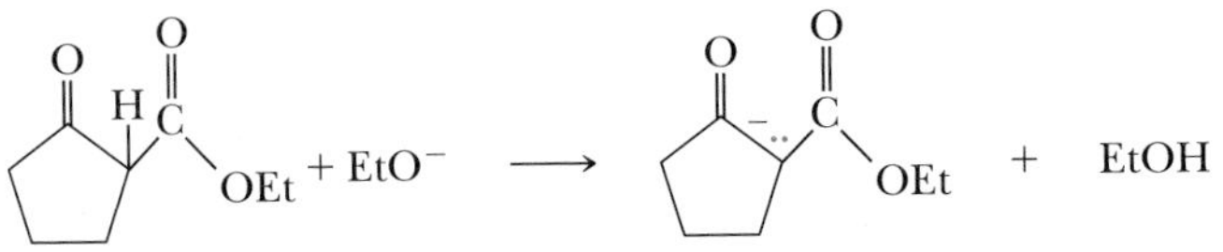

Resonance-stabilized enolate anion of the β-ketoester

C. Crossed Claisen Condensations

As in crossed aldol reactions, in **crossed Claisen condensations** between two different esters, each with two α-hydrogens, a mixture of four β-ketoesters is possible, and, therefore, crossed Claisen condensations of this type are not synthetically useful. Such condensations are useful, however, if appreciable differences in reactivity exist between the two esters, as, for example, when one of the esters has no α-hydrogens and can function only as an enolate anion acceptor. Following are four examples of esters without α-hydrogens:

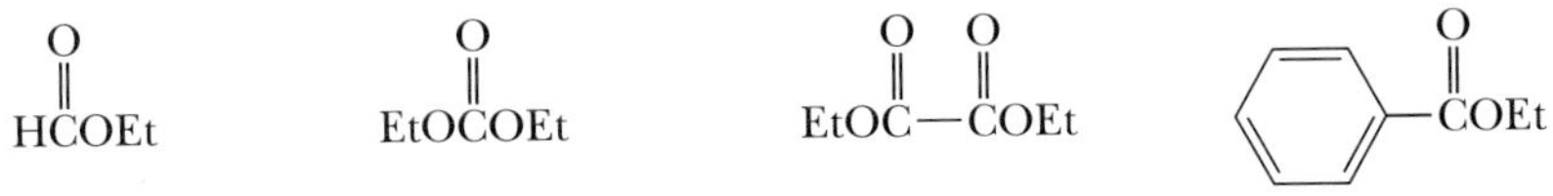

$HCOEt$ (C=O)	$EtOCOEt$ (C=O)	$EtOC{-}COEt$ (two C=O)	C_6H_5COEt (C=O)
Ethyl formate	Diethyl carbonate	Diethyl ethanedioate (Diethyl oxalate)	Ethyl benzoate

Crossed Claisen condensations of this type are usually carried out by using the ester with no α-hydrogens in excess. In the following illustration, methyl benzoate is used in excess.

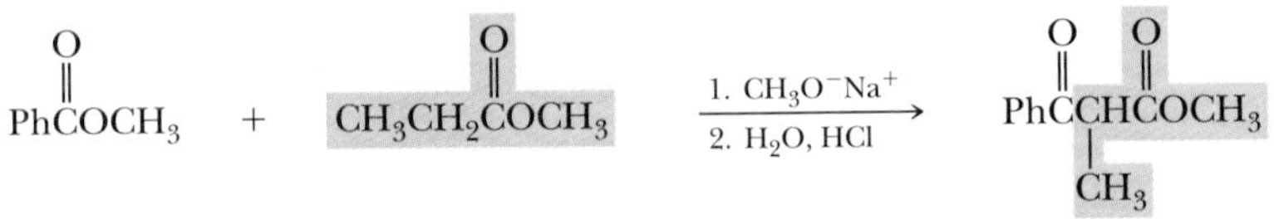

$$PhCOCH_3 + CH_3CH_2COCH_3 \xrightarrow[\text{2. } H_2O,\ HCl]{\text{1. } CH_3O^-Na^+} PhCCH(CH_3)COCH_3$$

Methyl benzoate — Methyl propanoate — Methyl 2-methyl-3-oxo-3-phenylpropanoate

EXAMPLE 18.6

Complete the equation for this crossed Claisen condensation:

$$CH_3CH_2\overset{O}{\overset{\|}{C}}OEt + H\overset{O}{\overset{\|}{C}}OEt \xrightarrow[2.\ H_2O,\ HCl]{1.\ EtO^-Na^+}$$

Solution

$$\underset{}{H\overset{O}{\overset{\|}{C}}}\underset{\underset{CH_3}{|}}{C}H\overset{O}{\overset{\|}{C}}OEt + EtOH$$

PROBLEM 18.6

Complete the equation for this crossed Claisen condensation:

$$C_6H_5\overset{O}{\overset{\|}{C}}OEt + C_6H_5CH_2\overset{O}{\overset{\|}{C}}OEt \xrightarrow[2.\ H_2O,\ HCl]{1.\ EtO^-Na^+}$$

D. Hydrolysis and Decarboxylation of β-Ketoesters

Recall from Section 17.5C that hydrolysis of an ester in aqueous sodium hydroxide (saponification) followed by acidification of the reaction mixture with aqueous HCl converts an ester to a carboxylic acid. Recall also from Section 16.10 that β-ketoacids and β-dicarboxylic acids (substituted malonic acids) readily undergo decarboxylation (lose CO_2) when heated. The following equations illustrate the results of a Claisen condensation followed by hydrolysis of the ester, acidification, and finally decarboxylation.

Claisen condensation:

$$2CH_3CH_2\overset{O}{\overset{\|}{C}}OEt \xrightarrow[2.\ H_2O,\ HCl]{1.\ EtO^-Na^+} CH_3CH_2\overset{O}{\overset{\|}{C}}\underset{\underset{CH_3}{|}}{C}H\overset{O}{\overset{\|}{C}}OEt$$

Saponification followed by acidification:

$$CH_3CH_2\overset{O}{\overset{\|}{C}}\underset{\underset{CH_3}{|}}{C}H\overset{O}{\overset{\|}{C}}OEt \xrightarrow[4.\ H_2O,\ HCl]{3.\ NaOH,\ H_2O,\ heat} CH_3CH_2\overset{O}{\overset{\|}{C}}\underset{\underset{CH_3}{|}}{C}H\overset{O}{\overset{\|}{C}}OH$$

Decarboxylation:

$$CH_3CH_2\overset{O}{\overset{\|}{C}}\underset{\underset{CH_3}{|}}{C}H\overset{O}{\overset{\|}{C}}OH \xrightarrow{5.\ heat} CH_3CH_2\overset{O}{\overset{\|}{C}}CH_2CH_3 + CO_2$$

The result of this five-step sequence is reaction between two molecules of ester (one furnishing a carbonyl group and the other furnishing an enolate anion) to give a ketone and carbon dioxide. In the general reaction, both ester molecules are the same, and, hence, the product is a symmetrical ketone.

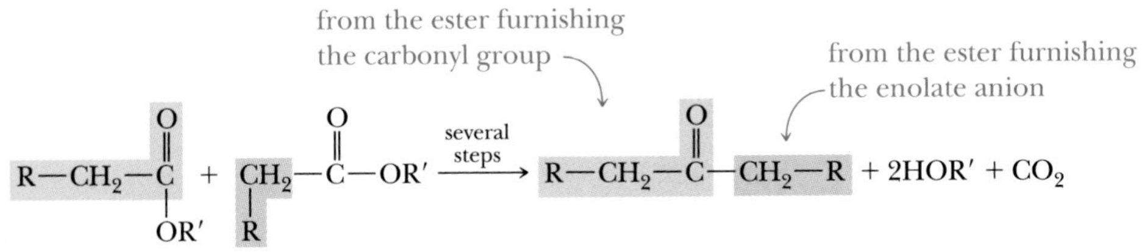

from the ester furnishing the carbonyl group

from the ester furnishing the enolate anion

$$R{-}CH_2{-}\overset{\overset{O}{\|}}{\underset{\underset{OR'}{|}}{C}} + \underset{\underset{R}{|}}{CH_2}{-}\overset{\overset{O}{\|}}{C}{-}OR' \xrightarrow[\text{steps}]{\text{several}} R{-}CH_2{-}\overset{\overset{O}{\|}}{C}{-}CH_2{-}R + 2HOR' + CO_2$$

The same sequence of reactions starting with a crossed Claisen condensation gives an unsymmetrical ketone.

EXAMPLE 18.7

Each set of compounds undergoes (1, 2) Claisen condensation, (3) saponification, followed by (4) acidification, and (5) thermal decarboxylation. Draw the structural formula of the product isolated after completion of this reaction sequence.

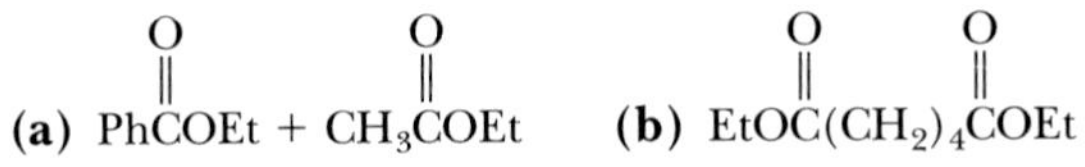

(a) $Ph\overset{\overset{O}{\|}}{C}OEt + CH_3\overset{\overset{O}{\|}}{C}OEt$ **(b)** $EtO\overset{\overset{O}{\|}}{C}(CH_2)_4\overset{\overset{O}{\|}}{C}OEt$

Solution

Steps 1 and 2 bring about a crossed Claisen or Dieckmann condensation to form a β-ketoester. Steps 3 and 4 bring about hydrolysis of the β-ketoester to give a β-ketoacid, and Step 5 brings about decarboxylation to give a ketone.

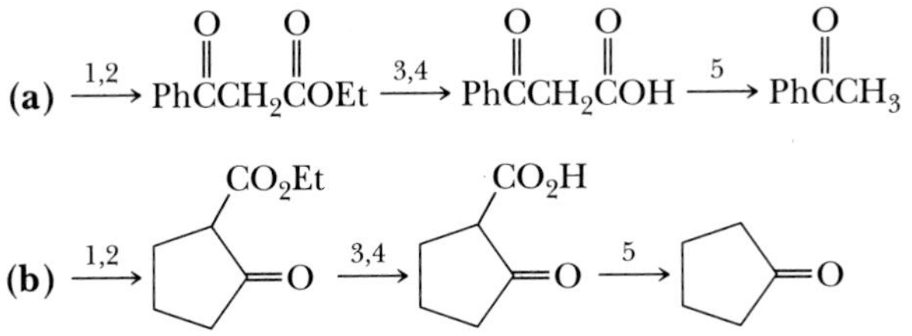

(a) $\xrightarrow{1,2} Ph\overset{\overset{O}{\|}}{C}CH_2\overset{\overset{O}{\|}}{C}OEt \xrightarrow{3,4} Ph\overset{\overset{O}{\|}}{C}CH_2\overset{\overset{O}{\|}}{C}OH \xrightarrow{5} Ph\overset{\overset{O}{\|}}{C}CH_3$

(b) $\xrightarrow{1,2}$ 2-(CO_2Et)cyclopentanone $\xrightarrow{3,4}$ 2-(CO_2H)cyclopentanone $\xrightarrow{5}$ cyclopentanone

PROBLEM 18.7

Show how to convert benzoic acid to 3-methyl-1-phenyl-1-butanone (isobutyl phenyl ketone) by the following synthetic strategies, each of which uses a different type of reaction to form the new carbon-carbon bond to the carbonyl group of benzoic acid.

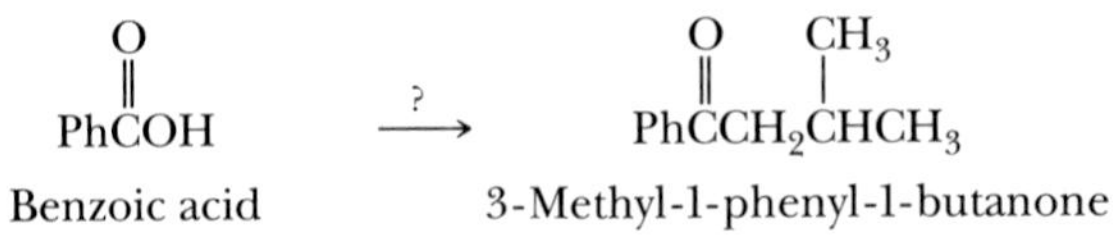

$$\underset{\text{Benzoic acid}}{Ph\overset{\overset{O}{\|}}{C}OH} \xrightarrow{?} \underset{\text{3-Methyl-1-phenyl-1-butanone}}{Ph\overset{\overset{O}{\|}}{C}CH_2\overset{\overset{CH_3}{|}}{C}HCH_3}$$

(a) A lithium diorganocopper (Gilman) reagent **(b)** A Claisen condensation

---

18.4 Claisen and Aldol Condensations in the Biological World

Carbonyl condensations are among the most widely used reactions in the biological world for the assembly of new carbon-carbon bonds in such important biomolecules as fatty acids, cholesterol, steroid hormones, and terpenes. One source of carbon atoms for the synthesis of these biomolecules is **acetyl-CoA,** a thioester of acetic acid and the thiol group of coenzyme A (Problem 24.37). Note that in the discussions that follow, we will not be concerned with the mechanism by which each of these enzyme-catalyzed reactions occurs. Rather, our concern is on recognizing the type of reaction that takes place.

In a Claisen condensation catalyzed by the enzyme thiolase, acetyl-CoA is converted to its enolate anion, which then attacks the carbonyl group of a second molecule of acetyl-CoA to form a tetrahedral carbonyl addition intermediate. Collapse of this intermediate by loss of coenzyme A anion ($CoAS^-$) gives acetoacetyl-CoA. Subsequent proton transfer to coenzyme A anion gives coenzyme A.

$$\underset{\text{Acetyl-CoA}}{CH_3\overset{O}{\overset{\|}{C}}SCoA} + \underset{\text{Acetyl-CoA}}{CH_3\overset{O}{\overset{\|}{C}}SCoA} \xrightarrow[\text{Claisen condensation}]{\text{thiolase}} \underset{\text{Acetoacetyl-CoA}}{CH_3\overset{O}{\overset{\|}{C}}CH_2\overset{O}{\overset{\|}{C}}SCoA} + \underset{\text{Coenzyme A}}{CoASH}$$

Enzyme-catalyzed aldol reaction with a third molecule of acetyl-CoA on the ketone carbonyl of acetoacetyl-CoA gives (*S*)-3-hydroxy-3-methylglutaryl-CoA. Note three features of this reaction. First, only the *S* enantiomer is formed. Condensation takes place in a chiral environment created by the enzyme, 3-hydroxy-3-methylglutaryl-CoA synthetase, which induces the formation of one enantiomer of the product to the exclusion of the other. Second, hydrolysis of the thioester group of acetyl-CoA is coupled with the aldol reaction. Third, the carboxyl group is shown as it is ionized at pH 7.4, the approximate pH of blood plasma and many cellular fluids.

The second carbonyl condensation takes place at this carbonyl

$$\underset{\text{Acetoacetyl-CoA}}{CH_3\overset{O}{\overset{\|}{C}}CH_2\overset{O}{\overset{\|}{C}}SCoA} + \underset{\text{Acetyl-CoA}}{CH_3\overset{O}{\overset{\|}{C}}SCoA} \xrightarrow[CoASH]{\text{3-hydroxy-3-methyl-glutaryl-CoA synthetase}} \underset{(S)\text{-3-Hydroxy-3-methylglutaryl-CoA}}{{}^-O\overset{O}{\overset{\|}{C}}CH_2\overset{H_3C\quad OH}{C}CH_2\overset{O}{\overset{\|}{C}}SCoA}$$

Enzyme-catalyzed reduction by NADH of the thioester group of 3-hydroxy-3-methylglutaryl-CoA to a primary alcohol gives mevalonic acid, here shown as its anion. Note that, in this reduction, a change occurs in the designation of configuration from *S* to *R*, not because of any change in configuration at the stereocenter, but rather because there is a change in priority among the four groups bonded to the stereocenter. The priorities of the groups on the stereocenter in the *S* starting material and the *R* product are shown on the following structural formulas.

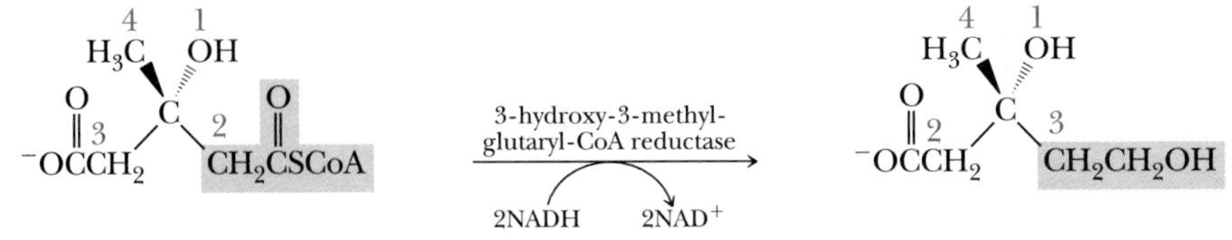

$^{-}OCCH_2$(=O)–C(H_3C)(OH)–CH_2CSCoA(=O) $\xrightarrow[2NADH \;\; 2NAD^+]{\text{3-hydroxy-3-methyl-glutaryl-CoA reductase}}$ $^{-}OCCH_2$(=O)–C(H_3C)(OH)–CH_2CH_2OH

(*S*)-3-Hydroxy-3-methylglutaryl-CoA (*R*)-Mevalonate

Enzyme-catalyzed transfer of a phosphate group from adenosine triphosphate (ATP) to the 3-hydroxyl group of mevalonate gives a phosphoric ester at carbon 3. Enzyme-catalyzed transfer of a pyrophosphate group from a second molecule of ATP gives a pyrophosphoric ester at carbon 5. Enzyme-catalyzed β-elimination from this molecule results in loss of CO_2 and PO_4^{3-}, both good leaving groups.

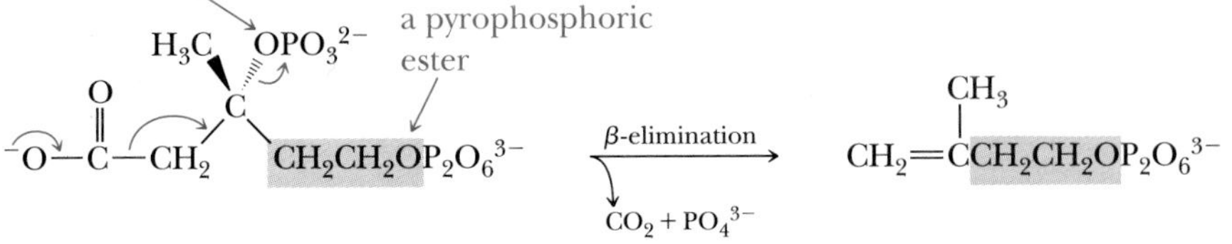

a phosphoric ester; a pyrophosphoric ester

$^{-}O-\overset{O}{\overset{\|}{C}}-CH_2$–C($H_3C$)($OPO_3^{2-}$)–$CH_2CH_2OP_2O_6^{3-}$ $\xrightarrow[CO_2 + PO_4^{3-}]{\beta\text{-elimination}}$ $CH_2{=}\overset{CH_3}{\overset{|}{C}}CH_2CH_2OP_2O_6^{3-}$

3-Phospho-5-pyrophospho-(*R*)-mevalonate Isopentenyl pyrophosphate

Isopentenyl pyrophosphate has the carbon skeleton of isoprene, the unit into which terpenes can be divided (Section 5.4). This compound is in fact a key intermediate in the synthesis of terpenes, as well as of cholesterol and steroid hormones. We shall return to the chemistry of isopentenyl pyrophosphate in Section 25.4B and discuss its conversion to cholesterol and terpenes.

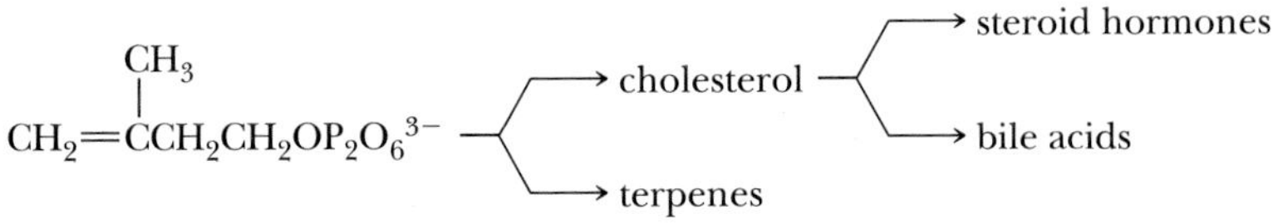

$CH_2{=}\overset{CH_3}{\overset{|}{C}}CH_2CH_2OP_2O_6^{3-}$ → cholesterol (→ steroid hormones; → bile acids); → terpenes

Isopentenyl pyrophosphate

18.5 Enamines

Enamine An unsaturated compound derived by the reaction of an aldehyde or ketone and a secondary amine followed by loss of H_2O; $CR_2{=}CR{-}NR_2$.

Enamines are formed by the reaction of a secondary amine with an aldehyde or ketone (Section 15.10A). The secondary amines most commonly used to prepare enamines are pyrrolidine and morpholine.

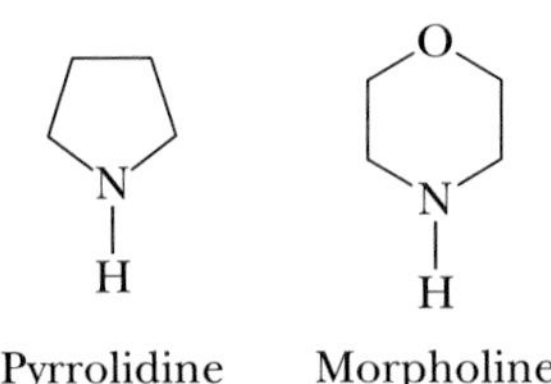

Pyrrolidine Morpholine

EXAMPLE 18.8

Draw structural formulas for the aminoalcohol and enamine formed in the following reactions:

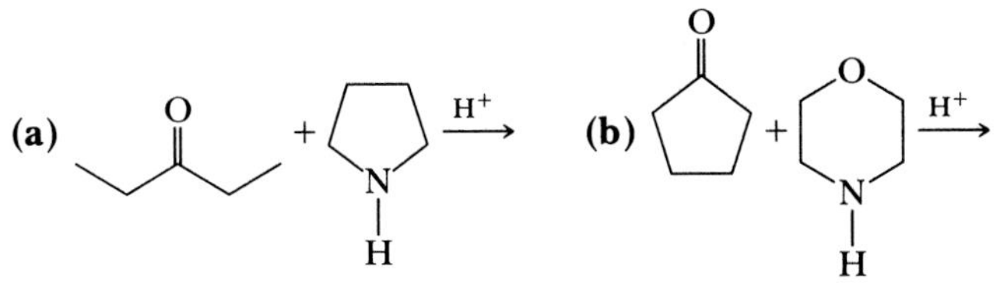

(a) O + N—H $\xrightarrow{H^+}$ (b) O + O N—H $\xrightarrow{H^+}$

Solution

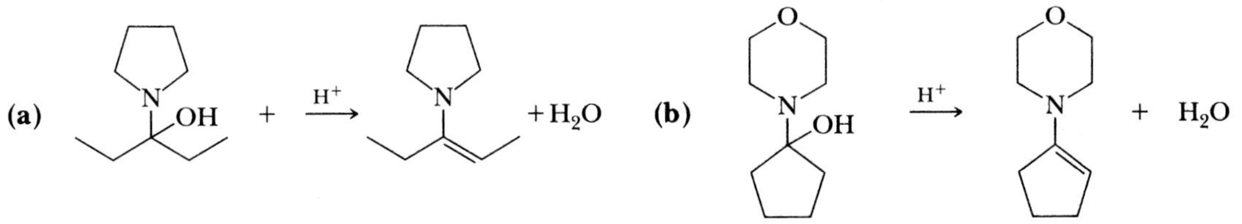

(a) N OH + $\xrightarrow{H^+}$ N + H_2O (b) O N OH $\xrightarrow{H^+}$ O N + H_2O

An aminoalcohol An enamine An aminoalcohol An enamine

PROBLEM 18.8

Following are structural formulas for two enamines. Draw structural formulas for the secondary amine and carbonyl compound from which each is derived.

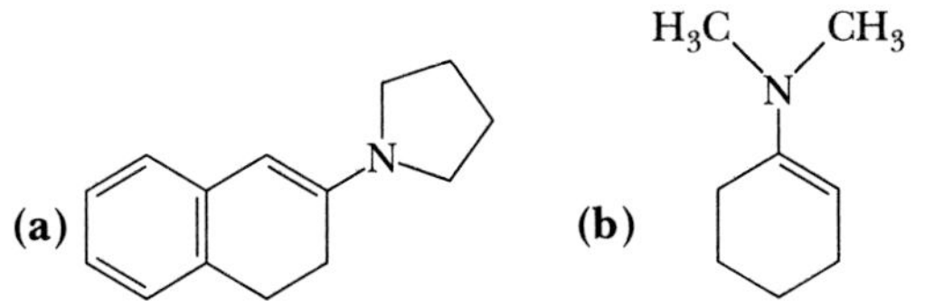

(a) N (b) H_3C CH_3 N

The particular value of enamines in synthetic organic chemistry is the fact that the β-carbon of an enamine is a nucleophile by virtue of the conjugation of the carbon-carbon double bond with the electron pair on nitrogen. Enamines resemble enols and enolate ions in their reactions.

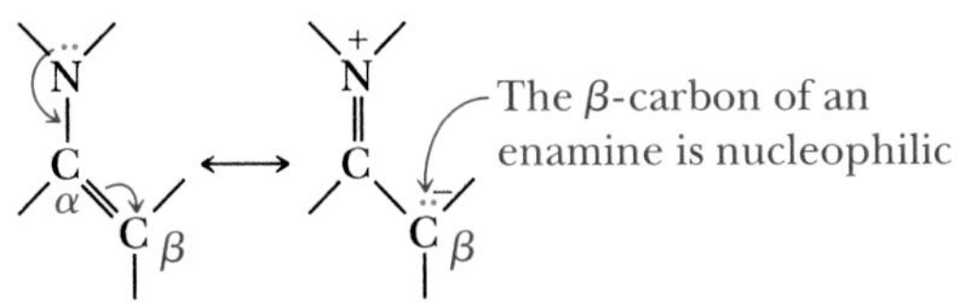

The β-carbon of an enamine is nucleophilic

An enamine as a resonance hybrid of two contributing structures

The use of enamines as synthetic intermediates for the alkylation and acylation at the α-carbon of aldehydes and ketones was pioneered by Gilbert Stork of Columbia University. This use of enamines is called the Stork enamine reaction.

A. Alkylation of Enamines

Enamines readily undergo S_N2 reactions with methyl and primary alkyl halides, α-haloketones, and α-haloesters. Alkylation is carried out in two steps. In Step 1, the enamine is treated with one equivalent of the alkylating agent to give an iminium halide. Hydrolysis of the iminium halide in Step 2 gives the alkylated aldehyde or ketone.

MECHANISM Alkylation of an Enamine

Step 1: Alkylation of the enamine

$CH_2{=}CHCH_2{-}Br \xrightarrow{S_N2}$ (iminium bromide with $CH_2CH{=}CH_2$, Br^-)

The morpholine enamine of cyclohexanone + Allyl bromide → An iminium bromide

Step 2: Hydrolysis of the iminium halide

(iminium bromide with $CH_2CH{=}CH_2$, Br^-) + HCl/H_2O → (2-allylcyclohexanone with $CH_2CH{=}CH_2$) + (morpholinium with Cl^-)

2-Allylcyclohexanone Morpholinium chloride

EXAMPLE 18.9

Show how to use an enamine to bring about this synthesis:

(2-tetralone) → (2-tetralone with CH_2CO_2Et at C-1)

Solution

Prepare an enamine by treating the ketone with either morpholine or pyrrolidine. The intermediate aminoalcohol can undergo dehydration in two directions. The direction shown here is favored because of the stabilization gained by conjugation of the carbon-carbon double bond of the enamine with the aromatic ring. Treatment of the enamine with ethyl 2-chloroacetate followed by hydrolysis of the iminium chloride in aqueous hydrochloric acid gives the product.

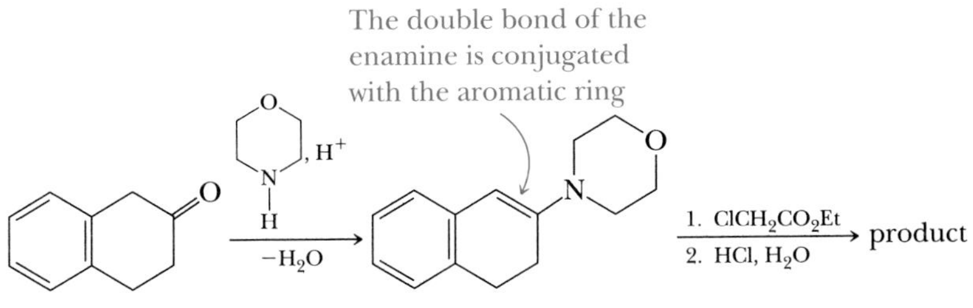

The double bond of the enamine is conjugated with the aromatic ring

morpholine (N–H), H^+; $-H_2O$ → enamine; 1. $ClCH_2CO_2Et$ 2. HCl, H_2O → product

PROBLEM 18.9

Write a mechanism for the hydrolysis of the following iminium chloride in aqueous HCl:

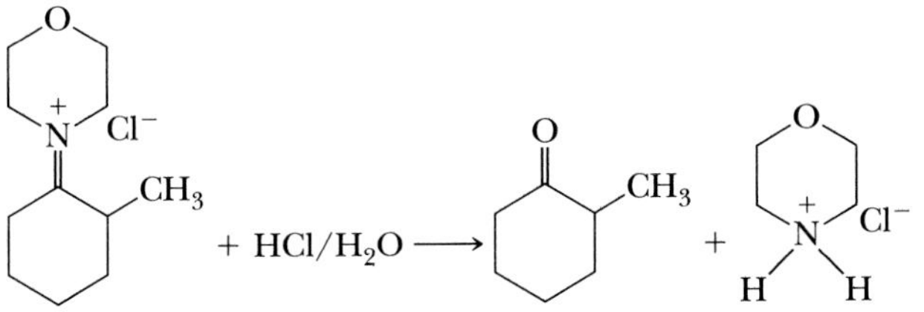

Cl^-, CH_3 + HCl/H_2O ⟶ CH_3 + Cl^- (H, H)

B. Acylation of Enamines

Enamines undergo acylation when treated with acid chlorides and acid anhydrides. The reaction is a nucleophilic acyl substitution as illustrated by the conversion of cyclohexanone, via its pyrrolidine enamine, to 2-acetylcyclohexanone. Thus, we can attach an acyl group to the α-carbon of an aldehyde or ketone using its enamine as an intermediate. The process of introducing an acyl group onto an organic molecule is called **acylation.**

Acylation The process of introducing an acyl group, RCO— or ArCO—, onto an organic molecule.

+ CH_3CCl (O) ⟶ Cl^- $\overset{+}{N}$ CCH_3 (O) $\xrightarrow{HCl/H_2O}$ CCH_3 (O) + Cl^- (H, H)

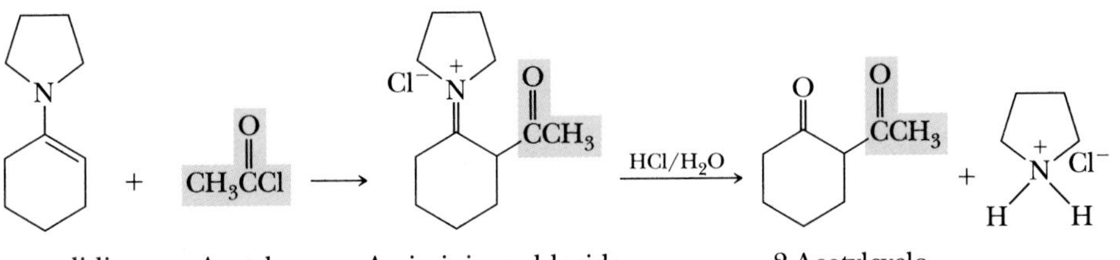

The pyrrolidine enamine of cyclohexanone | Acetyl chloride | An iminium chloride | 2-Acetylcyclohexanone

EXAMPLE 18.10

Show how to use an enamine to bring about this synthesis:

Cyclopentanone → 2-hexanoylcyclopentanone ($C(=O)(CH_2)_4CH_3$)

Solution

Treatment of cyclopentanone with pyrrolidine gives an enamine. Treatment of the enamine with hexanoyl chloride followed by hydrolysis in aqueous HCl gives the desired β-diketone.

Cyclopentanone —(1. pyrrolidine (N–H), H^+)→ enamine —(2. $CH_3(CH_2)_4\overset{O}{\overset{\|}{C}}Cl$; 3. H_2O, HCl)→ 2-hexanoylcyclopentanone ($\overset{O}{\overset{\|}{C}}(CH_2)_4CH_3$)

PROBLEM 18.10

Show how to use alkylation or acylation of an enamine to convert acetophenone to the following compounds.

(a) $C_6H_5\overset{O}{\overset{\|}{C}}CH_2\overset{O}{\overset{\|}{C}}CH_3$ (b) $C_6H_5\overset{O}{\overset{\|}{C}}CH_2CH_2\overset{O}{\overset{\|}{C}}CH_3$ (c) $C_6H_5\overset{O}{\overset{\|}{C}}CH_2CH_2\overset{O}{\overset{\|}{C}}OEt$

18.6 The Acetoacetic Ester Synthesis

What makes acetoacetic ester and other β-ketoesters such versatile starting materials for formation of new carbon-carbon bonds is (1) the acidity of α-hydrogens between the two carbonyl groups (pK_a 10–11), and (2) the nucleophilicity of the enolate anion resulting from loss of an α-hydrogen. The **acetoacetic ester synthesis** is useful for the preparation of monosubstituted and disubstituted acetones of the following types:

$CH_3\overset{O}{\overset{\|}{C}}CH_2\overset{O}{\overset{\|}{C}}OEt$ Ethyl acetoacetate (Acetoacetic ester)

→ $CH_3\overset{O}{\overset{\|}{C}}CH_2R$ A monosubstituted acetone

→ $CH_3\overset{O}{\overset{\|}{C}}CHR$ (with R′ on the CH) A disubstituted acetone

We have already seen the chemistry of the individual steps in this synthesis but have not put them together in this particular sequence. Let us illustrate the acetoacetic ester synthesis by choosing 5-hexen-2-one as a target molecule. The three carbons shown in color are provided by ethyl acetoacetate. The remaining three carbons represent the —R group of a substituted acetone.

These three carbons are from ethyl acetoacetate → The —R group of a monosubstituted acetone

$$CH_3-\overset{\overset{\displaystyle O}{\|}}{C}-CH_2-CH_2-CH{=}CH_2$$

5-Hexen-2-one

1. The methylene hydrogens of ethyl acetoacetate are more acidic (pK_a 10.7) than ethanol (pK_a 15.9), and, therefore, ethyl acetoacetate is converted completely to its anion by sodium ethoxide or other alkali metal alkoxides.

$$CH_3\overset{\overset{\displaystyle O}{\|}}{C}-\underset{\underset{\displaystyle H}{|}}{CH}-\overset{\overset{\displaystyle O}{\|}}{C}OEt + EtO^-Na^+ \longrightarrow CH_3\overset{\overset{\displaystyle O}{\|}}{C}-\overset{..-}{CH}-\overset{\overset{\displaystyle O}{\|}}{C}OEt\;Na^+ + EtOH$$

Ethyl acetoacetate pK_a 10.7 (stronger acid)	Sodium ethoxide (stronger base)	Sodium salt of ethyl acetoacetate (weaker base)	Ethanol pK_a 15.9 (weaker acid)

2. The enolate anion of ethyl acetoacetate is a nucleophile and reacts by an S_N2 pathway with methyl and primary alkyl halides, α-haloketones, and α-haloesters. Secondary halides give lower yields, and tertiary halides undergo E2 elimination. In the following example, the anion of ethyl acetoacetate is alkylated with allyl bromide:

$$CH_3\overset{\overset{\displaystyle O}{\|}}{C}-\overset{..-}{CH}-\overset{\overset{\displaystyle O}{\|}}{C}OEt\;Na^+ + CH_2{=}CHCH_2Br \xrightarrow{S_N2} CH_3\overset{\overset{\displaystyle O}{\|}}{C}-\underset{\underset{\displaystyle CH_2CH{=}CH_2}{|}}{CH}-\overset{\overset{\displaystyle O}{\|}}{C}OEt + Na^+Br^-$$

3,4. Hydrolysis of the alkylated acetoacetic ester in aqueous NaOH followed by acidification with aqueous HCl (Section 17.5C) gives a β-ketoacid.

$$CH_3\overset{\overset{\displaystyle O}{\|}}{C}-\underset{\underset{\displaystyle CH_2CH{=}CH_2}{|}}{CH}-\overset{\overset{\displaystyle O}{\|}}{C}OEt \xrightarrow[\text{4. HCl, }H_2O]{\text{3. NaOH, }H_2O} CH_3\overset{\overset{\displaystyle O}{\|}}{C}-\underset{\underset{\displaystyle CH_2CH{=}CH_2}{|}}{CH}-\overset{\overset{\displaystyle O}{\|}}{C}OH + EtOH$$

5. Heating the β-ketoacid brings about decarboxylation (Section 16.10A) to give 5-hexen-2-one.

$$CH_3\overset{\overset{\displaystyle O}{\|}}{C}-\underset{\underset{\displaystyle CH_2CH{=}CH_2}{|}}{CH}-\overset{\overset{\displaystyle O}{\|}}{C}-OH \xrightarrow{\text{heat}} CH_3\overset{\overset{\displaystyle O}{\|}}{C}-CH_2-CH_2CH{=}CH_2 + CO_2$$

5-Hexen-2-one (a monosubstituted acetone)

A disubstituted acetone can be prepared by interrupting this sequence after Step 2, treating the monosubstituted acetoacetic ester with a second equivalent of base, carrying out a second alkylation, and then proceeding with Steps 3 through 5.

1′. Treatment with a second equivalent of base to form a second enolate anion:

$$CH_3\overset{O}{\overset{\|}{C}}-\underset{CH_2CH=CH_2}{\underset{|}{CH}}-\overset{O}{\overset{\|}{C}}OEt + EtO^-Na^+ \longrightarrow CH_3\overset{O}{\overset{\|}{C}}-\underset{CH_2CH=CH_2}{\underset{|}{\ddot{C}^-}}-\overset{O}{\overset{\|}{C}}OEt \; (Na^+) + EtOH$$

2′. The second alkylation:

$$CH_3\overset{O}{\overset{\|}{C}}-\underset{CH_2CH=CH_2}{\underset{|}{\ddot{C}^-}}-\overset{O}{\overset{\|}{C}}OEt \; (Na^+) + CH_3I \xrightarrow{S_N2} CH_3\overset{O}{\overset{\|}{C}}-\underset{CH_2CH=CH_2}{\underset{|}{\overset{CH_3}{\overset{|}{C}}}}-\overset{O}{\overset{\|}{C}}OEt + Na^+I^-$$

3,4. Hydrolysis of the ester in aqueous base followed by acidification gives the β-ketoacid:

$$CH_3\overset{O}{\overset{\|}{C}}-\underset{CH_2CH=CH_2}{\underset{|}{\overset{CH_3}{\overset{|}{C}}}}-\overset{O}{\overset{\|}{C}}OEt \xrightarrow[\text{4. HCl, } H_2O]{\text{3. NaOH, } H_2O} CH_3\overset{O}{\overset{\|}{C}}-\underset{CH_2CH=CH_2}{\underset{|}{\overset{CH_3}{\overset{|}{C}}}}-\overset{O}{\overset{\|}{C}}OH + EtOH$$

5. Decarboxylation of the β-ketoacid gives the ketone and carbon dioxide:

$$CH_3\overset{O}{\overset{\|}{C}}-\underset{CH_2CH=CH_2}{\underset{|}{\overset{CH_3}{\overset{|}{C}}}}-\overset{O}{\overset{\|}{C}}OH \xrightarrow{\text{heat}} CH_3\overset{O}{\overset{\|}{C}}-\overset{CH_3}{\overset{|}{CH}}-CH_2CH=CH_2 + CO_2$$

3-Methyl-5-hexen-2-one
(a disubstituted acetone)

EXAMPLE 18.11

Show how the acetoacetic ester synthesis can be used to prepare this ketone.

$$CH_3\overset{O}{\overset{\|}{C}}CH_2CH_2-C_6H_5$$

Solution

First determine which three carbons of the product originate from ethyl acetoacetate, then the location on the carbon chain of the $-CO_2H$ lost in decarboxylation, and finally the bond formed in the alkylation step. By this analysis, determine that the starting materials are ethyl acetoacetate and a benzyl halide.

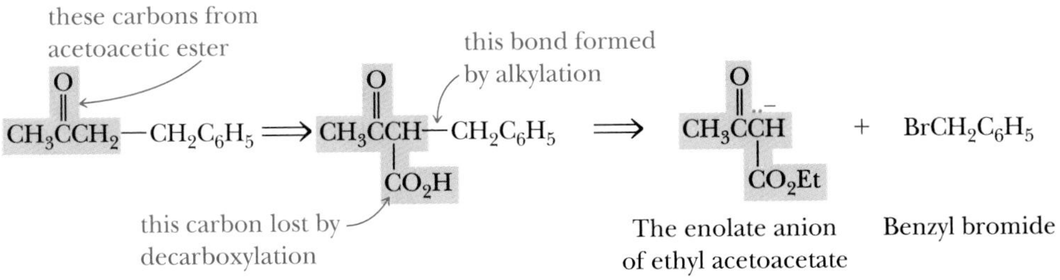

these carbons from acetoacetic ester

this bond formed by alkylation

$$\overset{\text{O}}{\overset{\|}{\text{CH}_3\text{C}}}\text{CH}_2\text{—CH}_2\text{C}_6\text{H}_5 \Longrightarrow \overset{\text{O}}{\overset{\|}{\text{CH}_3\text{C}}}\underset{\text{CO}_2\text{H}}{\underset{|}{\text{CH}}}\text{—CH}_2\text{C}_6\text{H}_5 \Longrightarrow \overset{\text{O}}{\overset{\|}{\text{CH}_3\text{C}}}\underset{\text{CO}_2\text{Et}}{\underset{|}{\overset{..-}{\text{CH}}}} + \text{BrCH}_2\text{C}_6\text{H}_5$$

this carbon lost by decarboxylation

The enolate anion of ethyl acetoacetate — Benzyl bromide

Now combine these reagents in the following way to prepare the desired ketone.

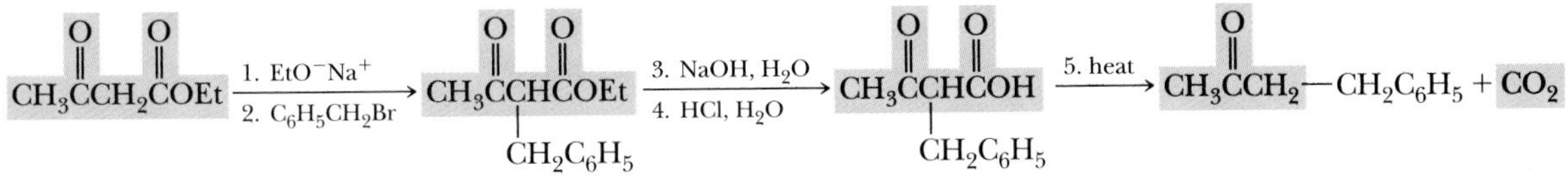

$$\overset{\text{O}}{\overset{\|}{\text{CH}_3\text{C}}}\text{CH}_2\overset{\text{O}}{\overset{\|}{\text{C}}}\text{OEt} \xrightarrow[\text{2. C}_6\text{H}_5\text{CH}_2\text{Br}]{\text{1. EtO}^-\text{Na}^+} \overset{\text{O}}{\overset{\|}{\text{CH}_3\text{C}}}\underset{\text{CH}_2\text{C}_6\text{H}_5}{\underset{|}{\text{CH}}}\overset{\text{O}}{\overset{\|}{\text{C}}}\text{OEt} \xrightarrow[\text{4. HCl, H}_2\text{O}]{\text{3. NaOH, H}_2\text{O}} \overset{\text{O}}{\overset{\|}{\text{CH}_3\text{C}}}\underset{\text{CH}_2\text{C}_6\text{H}_5}{\underset{|}{\text{CH}}}\overset{\text{O}}{\overset{\|}{\text{C}}}\text{OH} \xrightarrow{\text{5. heat}} \overset{\text{O}}{\overset{\|}{\text{CH}_3\text{C}}}\text{CH}_2\text{—CH}_2\text{C}_6\text{H}_5 + \text{CO}_2$$

PROBLEM 18.11

Show how the acetoacetic ester synthesis can be used to prepare these compounds:

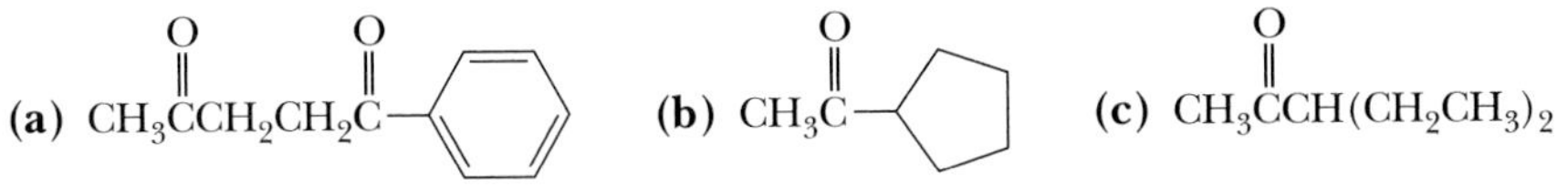

(a) $\text{CH}_3\overset{\text{O}}{\overset{\|}{\text{C}}}\text{CH}_2\text{CH}_2\overset{\text{O}}{\overset{\|}{\text{C}}}\text{—C}_6\text{H}_5$ (phenyl ring) **(b)** $\text{CH}_3\overset{\text{O}}{\overset{\|}{\text{C}}}\text{—}$(cyclopentyl) **(c)** $\text{CH}_3\overset{\text{O}}{\overset{\|}{\text{C}}}\text{CH(CH}_2\text{CH}_3)_2$

We have described what is commonly known as the acetoacetic ester synthesis and have illustrated the use of ethyl acetoacetate as the starting reagent. This same synthetic strategy is applicable to any β-ketoester, as, for example, those that are available by the Claisen condensation (Section 18.3A) and the Dieckmann condensation (Section 18.3B). Following are structural formulas for two β-ketoesters that can be made to undergo (1) formation of an enolate anion, (2) alkylation or acylation, (3) hydrolysis followed by (4) acidification, and finally (5) decarboxylation just as we have shown for ethyl acetoacetate.

(cyclopentanone ring)—$\overset{\text{O}}{\overset{\|}{\text{C}}}\text{OEt}$

Ethyl 2-oxocyclopentanecarboxylate

$$\text{CH}_3\text{CH}_2\overset{\text{O}}{\overset{\|}{\text{C}}}\underset{\text{CH}_3}{\underset{|}{\text{CH}}}\overset{\text{O}}{\overset{\|}{\text{C}}}\text{OEt}$$

Ethyl 2-methyl-3-oxopentanoate

EXAMPLE 18.12

Show how to convert ethyl 2-oxocyclopentanecarboxylate to 2-allylcyclopentanone.

Solution

Treat this β-ketoester with one equivalent of sodium ethoxide to form an anion followed by alkylation of the anion with one equivalent of an allyl halide. Subsequent hydrolysis of the ester in aqueous base followed by acidification and thermal decarboxylation gives the desired product.

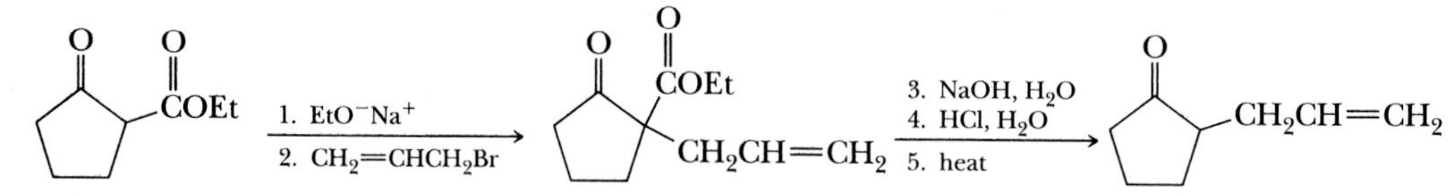

Ethyl 2-oxocyclopentanecarboxylate $\xrightarrow[\text{2. } CH_2{=}CHCH_2Br]{\text{1. } EtO^-Na^+}$ (2-allyl-2-carbethoxycyclopentanone, $COEt$, $CH_2CH{=}CH_2$) $\xrightarrow[\text{5. heat}]{\text{3. NaOH, } H_2O \text{; 4. HCl, } H_2O}$ 2-Allylcyclopentanone ($CH_2CH{=}CH_2$)

Ethyl 2-oxocyclopentanecarboxylate

2-Allylcyclopentanone

PROBLEM 18.12

Show how to convert ethyl 2-oxocyclopentanecarboxylate to this compound.

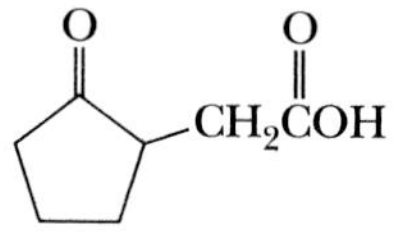

(2-oxocyclopentyl)–CH_2COH

18.7 The Malonic Ester Synthesis

The factors that make malonic esters and other β-diesters such versatile starting materials for formation of new carbon-carbon bonds are the same as those we have already seen for the acetoacetic ester synthesis, namely (1) the acidity of α-hydrogens between the two carbonyl groups and (2) the nucleophilicity of the enolate anion resulting from loss of such an α-hydrogen. The **malonic ester synthesis** is useful for the preparation of monosubstituted and disubstituted acetic acids of the following types.

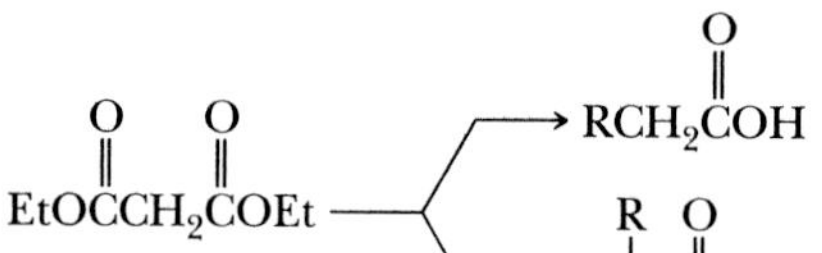

EtOCCH$_2$COEt (Diethyl malonate (Malonic ester)) → RCH_2COH A monosubstituted acetic acid; → $RCHCOH$ (with R on the α-carbon) A disubstituted acetic acid

As with the acetoacetic ester synthesis, we have already encountered all of the important chemistry of the malonic ester synthesis, although not in this particular pattern. Let us illustrate this synthesis by choosing 5-methoxypentanoic acid as a target molecule. The two carbons shown in color are provided by diethyl malonate. The remaining three carbons and the methoxy group represent the —R group of a monosubstituted acetic acid.

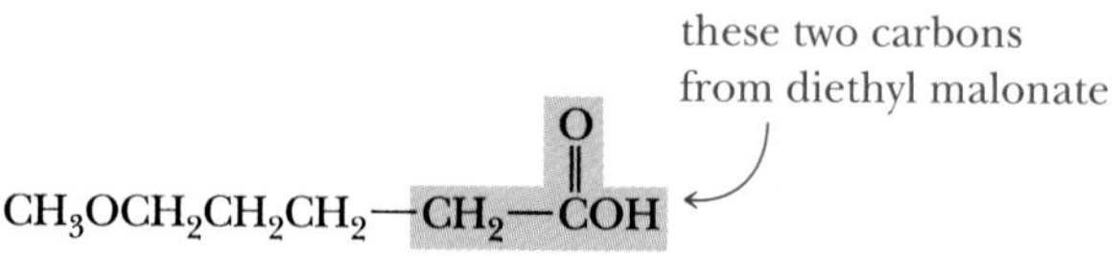

5-Methoxypentanoic acid

1. The α-hydrogens of diethyl malonate (pK_a 13.3) are more acidic than ethanol (pK_a 15.9), and, therefore, diethyl malonate is converted completely to its anion by sodium ethoxide or other alkali metal alkoxide.

$$\underset{\substack{\text{Diethyl malonate}\\ \text{p}K_a\ 13.3\\ \text{(stronger acid)}}}{\text{EtO}\overset{\text{O}}{\overset{\|}{\text{C}}}-\underset{\text{H}}{\underset{|}{\text{CH}}}-\overset{\text{O}}{\overset{\|}{\text{C}}}\text{OEt}} + \underset{\substack{\text{Sodium}\\ \text{ethoxide}\\ \text{(stronger base)}}}{\text{EtO}^-\text{Na}^+} \longrightarrow \underset{\substack{\text{Sodium salt of}\\ \text{diethyl malonate}\\ \text{(weaker base)}}}{\text{EtO}\overset{\text{O}}{\overset{\|}{\text{C}}}-\overset{..\,-}{\text{CH}}-\overset{\text{O}}{\overset{\|}{\text{C}}}\text{OEt}\ \ \text{Na}^+} + \underset{\substack{\text{Ethanol}\\ \text{p}K_a\ 15.9\\ \text{(weaker acid)}}}{\text{EtOH}}$$

2. The enolate anion of diethyl malonate is a nucleophile and reacts by an S_N2 pathway with methyl and primary alkyl halides, α-haloketones, and α-haloesters. In the following example, the anion of diethyl malonate is alkylated with 1-bromo-3-methoxypropane.

$$\text{EtO}\overset{\text{O}}{\overset{\|}{\text{C}}}-\overset{..\,-}{\text{CH}}-\overset{\text{O}}{\overset{\|}{\text{C}}}\text{OEt}\ \text{Na}^+ + \text{CH}_3\text{OCH}_2\text{CH}_2\text{CH}_2\text{Br} \xrightarrow{S_N2} \text{EtO}\overset{\text{O}}{\overset{\|}{\text{C}}}-\underset{\text{CH}_2\text{CH}_2\text{CH}_2\text{OCH}_3}{\underset{|}{\text{CH}}}-\overset{\text{O}}{\overset{\|}{\text{C}}}\text{OEt} + \text{Na}^+\text{Br}^-$$

3,4. Hydrolysis of the alkylated malonic ester in aqueous NaOH followed by acidification with aqueous HCl gives a β-dicarboxylic acid.

$$\text{EtO}\overset{\text{O}}{\overset{\|}{\text{C}}}-\underset{\text{CH}_2\text{CH}_2\text{CH}_2\text{OCH}_3}{\underset{|}{\text{CH}}}-\overset{\text{O}}{\overset{\|}{\text{C}}}\text{OEt} \xrightarrow[\text{4. HCl, H}_2\text{O}]{\text{3. NaOH, H}_2\text{O}} \text{HO}\overset{\text{O}}{\overset{\|}{\text{C}}}-\underset{\text{CH}_2\text{CH}_2\text{CH}_2\text{OCH}_3}{\underset{|}{\text{CH}}}-\overset{\text{O}}{\overset{\|}{\text{C}}}\text{OH} + 2\text{EtOH}$$

5. Heating the β-dicarboxylic acid slightly above its melting point brings about decarboxylation to give 5-methoxypentanoic acid.

$$\text{HO}\overset{\text{O}}{\overset{\|}{\text{C}}}-\underset{\text{CH}_2\text{CH}_2\text{CH}_2\text{OCH}_3}{\underset{|}{\text{CH}}}-\overset{\text{O}}{\overset{\|}{\text{C}}}\text{OH} \xrightarrow{\text{heat}} \underset{\text{5-methoxypentanoic acid}}{\text{CH}_3\text{OCH}_2\text{CH}_2\text{CH}_2-\text{CH}_2\overset{\text{O}}{\overset{\|}{\text{C}}}\text{OH}} + \text{CO}_2$$

A disubstituted acetic acid can be prepared by interrupting the previous sequence after Step 2, treating the monosubstituted diethyl malonate with a second equivalent of base, carrying out a second alkylation, and then proceeding with Steps 3 through 5.

EXAMPLE 18.13

Show how the malonic ester synthesis can be used to prepare 3-phenylpropanoic acid.

Solution

Determine first which two carbons of the product originated from diethyl malonate, then the location on the carbon chain of the $-CO_2H$ lost in decarboxylation, and finally the bond formed in the alkylation step. By this analysis, determine that the starting materials are diethyl malonate and a benzyl halide.

these two carbons from malonic ester — this bond formed by alkylation

$$C_6H_5CH_2{-}CH_2\overset{O}{\overset{\|}{C}}OH \Longrightarrow C_6H_5CH_2{-}\underset{CO_2H}{\underset{|}{C}}H\overset{O}{\overset{\|}{C}}OH \Longrightarrow C_6H_5CH_2Br \;+\; {}^{-}\underset{CO_2Et}{\underset{|}{\ddot{C}}}H\overset{O}{\overset{\|}{C}}OEt$$

this carbon lost by decarboxylation — Benzyl bromide — The enolate anion of diethyl malonate

Now combine these reagents in the following way to get the desired product:

$$EtO\overset{O}{\overset{\|}{C}}{-}CH_2{-}\overset{O}{\overset{\|}{C}}OEt \xrightarrow[2.\ C_6H_5CH_2Br]{1.\ EtO^-Na^+} EtO\overset{O}{\overset{\|}{C}}{-}\underset{CH_2C_6H_5}{\underset{|}{C}H}{-}\overset{O}{\overset{\|}{C}}OEt \xrightarrow[4.\ HCl,\ H_2O]{3.\ NaOH,\ H_2O}$$

$$HO\overset{O}{\overset{\|}{C}}{-}\underset{CH_2C_6H_5}{\underset{|}{C}H}{-}\overset{O}{\overset{\|}{C}}OH \xrightarrow{5.\ heat} \underset{\text{3-Phenylpropanoic acid}}{C_6H_5CH_2CH_2\overset{O}{\overset{\|}{C}}OH} + CO_2$$

PROBLEM 18.13

Show how the malonic ester synthesis can be used to prepare the following substituted acetic acids.

(a) $C_6H_5\overset{O}{\overset{\|}{C}}CH_2CH_2\overset{O}{\overset{\|}{C}}OH$ **(b)** $(CH_3CH_2)_2CHCO_2H$ **(c)** cyclopentyl$-CO_2H$

18.8 Conjugate Addition to α,β-Unsaturated Carbonyl Compounds

Thus far we have used a variety of carbon nucleophiles to form new carbon-carbon bonds:

1. Anions of terminal acetylenes (Section 10.5).
2. Organomagnesium compounds (Grignard reagents), organolithium compounds, and lithium diorganocopper (Gilman) reagents (Section 15.9).
3. Anions derived from 1,3-dithianes (Section 15.9).
4. Enolate anions derived from aldehydes and ketones (aldol reactions), esters (Claisen and Dieckmann condensations), β-diesters (malonic ester syntheses), and β-ketoesters (acetoacetic ester syntheses).
5. Enamines (which are synthetically equivalent to enolate anions).

These species have been used to form new carbon-carbon bonds by two synthetic strategies: (1) substitution by the carbon nucleophile in an S_N2 reaction and (2) addition of the carbon nucleophile to a carbonyl carbon. Conjugate addition, as it is also known, presents a different synthetic strategy: addition of a carbon nucleophile to an electrophilic carbon-carbon double or triple bond. In this section, we study two types of conjugate addition to electrophilic double bonds: addition of enolate anions (the Michael reaction) and addition of lithium diorganocopper (Gilman) reagents.

A. Michael Addition of Enolate Anions

Addition of enolate anions to α,β-unsaturated carbonyl compounds was first reported in 1887 by the American chemist, Arthur Michael. Following are two examples of **Michael reactions.** In the first example, the nucleophile adding to the conjugated system is the enolate anion of diethyl malonate. In the second example, the nucleophile is the enolate anion of ethyl acetoacetate.

$$(EtOC(=O))_2CH_2 + CH_2{=}CHCCH_3(=O) \xrightarrow[EtOH]{EtO^-Na^+} (EtOC(=O))_2CHCH_2CH_2CCH_3(=O)$$

Diethyl propanedioate (Diethyl malonate) — 3-Buten-2-one (Methyl vinyl ketone)

$$CH_3C(=O)CH_2COEt(=O) + \text{2-cyclohexenone} \xrightarrow[EtOH]{EtO^-Na^+} CH_3C(=O)CH(COEt(=O))\text{-(3-oxocyclohexyl)}$$

Ethyl 3-oxobutanoate (Ethyl acetoacetate) — 2-Cyclohexenone

The Michael reaction takes place with a wide variety of α,β-unsaturated carbonyl compounds as well as with α,β-unsaturated nitriles and nitro compounds (Table 18.1).

We can write the following general mechanism for a Michael reaction. In Step 1, treatment of H—Nu with base forms the nucleophile, Nu^-. The most commonly used types of nucleophiles in Michael reactions are summarized in Table 18.1. The most commonly used bases are metal alkoxides, pyridine, and piperidine. Nucleophilic addition of Nu^- to the β-carbon of the conjugated system in Step 2 gives a resonance-stabilized enolate anion. Proton transfer in Step 3 from H—B to the enolate anion gives an enol. The enol corresponds to 1,4-addition to the conjugated system of the α,β-unsaturated carbonyl compound, and it is because of formation of this intermediate that the Michael reaction is classified as a 1,4- or conjugate addition. Note that in Step 3, the base, B^-, is regenerated, in accord with the experimental observation that a Michael reaction requires only a catalytic amount of base rather than a molar equivalent. In Step 4, the less stable enol undergoes keto-enol tautomerism (Section 15.11B) to the more stable keto form.

TABLE 18.1 Combinations of Reagents for Effective Michael Reactions

These Types of α,β-Unsaturated Compounds Are Nucleophile Acceptors in Michael Reactions		These Types of Compounds Provide Effective Nucleophiles for Michael Reactions	
$CH_2{=}CHCH{=}O$ ($CH_2{=}CH\overset{O}{\overset{\|}{C}}H$)	aldehyde	$CH_3\overset{O}{\overset{\|}{C}}CH_2\overset{O}{\overset{\|}{C}}CH_3$	β-diketone
$CH_2{=}CH\overset{O}{\overset{\|}{C}}CH_3$	ketone	$CH_3\overset{O}{\overset{\|}{C}}CH_2\overset{O}{\overset{\|}{C}}OEt$	β-ketoester
$CH_2{=}CH\overset{O}{\overset{\|}{C}}OCH_2CH_3$	ester	$CH_3\overset{O}{\overset{\|}{C}}CH_2CN$	β-ketonitrile
$CH_2{=}CH\overset{O}{\overset{\|}{C}}NH_2$	amide	$EtO\overset{O}{\overset{\|}{C}}CH_2\overset{O}{\overset{\|}{C}}OEt$	β-diester
$CH_2{=}CHC{\equiv}N$	nitrile	(pyrrolidine N)–$CH_3C{=}CH_2$	enamine
$CH_2{=}CHNO_2$	nitro compound		

MECHANISM Michael Reaction: Conjugate Addition of Enolate Anions

Step 1: Proton transfer to the base $B:^-$ to form a nucleophile, $Nu:^-$

$$Nu{-}H + :B^- \rightleftharpoons Nu:^- + H{-}B$$

Base

Step 2: Addition of the nucleophile to the β-carbon of the α,β-unsaturated carbonyl compound

$$Nu:^- + {-}\overset{|}{\underset{|}{C}}{=}\underset{|}{C}{-}\overset{O}{\overset{\|}{C}}{-} \longrightarrow Nu{-}\overset{|}{\underset{|}{C}}{-}\underset{|}{C}{=}\overset{:\ddot{O}:^-}{\overset{|}{C}}{-} \longleftrightarrow Nu{-}\overset{|}{\underset{|}{C}}{-}\overset{-\,\cdot\cdot}{\underset{|}{C}}{-}\overset{O}{\overset{\|}{C}}{-}$$

Resonance-stabilized enolate anion

Step 3: Proton transfer from H—B to form an enol

$$Nu{-}\overset{|}{\underset{|}{C}}{-}\underset{|}{C}{=}\overset{:\ddot{O}:^-}{\overset{|}{C}}{-} + H{-}B \longrightarrow Nu{-}\overset{|4}{\underset{|}{C}}{-}\overset{3}{\underset{|}{C}}{=}\overset{\overset{1}{:\ddot{O}-H}}{\overset{2\,|}{C}}{-} + :B^-$$

An enol
(a product of 1,4-addition)

Step 4: Tautomerism of the less stable enol to the more stable keto form

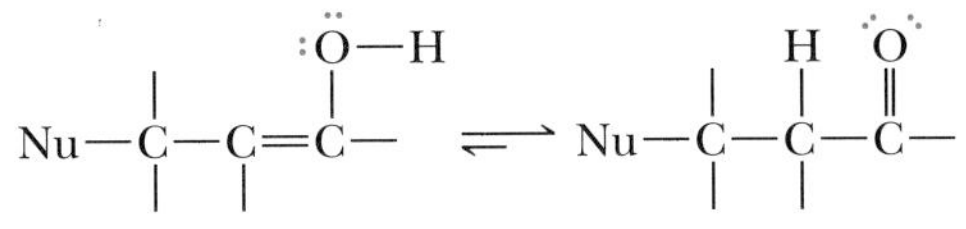

$$\mathrm{Nu{-}\overset{|}{\underset{|}{C}}{-}\underset{|}{C}{=}\overset{\ddot{O}-H}{\overset{|}{C}}{-}} \rightleftharpoons \mathrm{Nu{-}\overset{|}{\underset{|}{C}}{-}\overset{H}{\overset{|}{\underset{|}{C}}}{-}\overset{\ddot{O}}{\overset{\|}{C}}{-}}$$

Less stable enol form — More stable keto form

EXAMPLE 18.14

Draw a structural formula for the product formed when each set of reactants is treated with sodium ethoxide in ethanol under conditions of the Michael reaction:

(a) $CH_3\overset{O}{\overset{\|}{C}}CH_2\overset{O}{\overset{\|}{C}}OEt + CH_2{=}CH\overset{O}{\overset{\|}{C}}OEt$ **(b)** $EtO\overset{O}{\overset{\|}{C}}CH_2\overset{O}{\overset{\|}{C}}OEt$ + 2-cyclopentenone

Solution

(a) $(CH_3\overset{O}{\overset{\|}{C}})(EtO\overset{O}{\overset{\|}{C}})CHCH_2CH_2\overset{O}{\overset{\|}{C}}OEt$

(b) $(EtO\overset{O}{\overset{\|}{C}})_2CH$– attached to C-3 of cyclopentanone

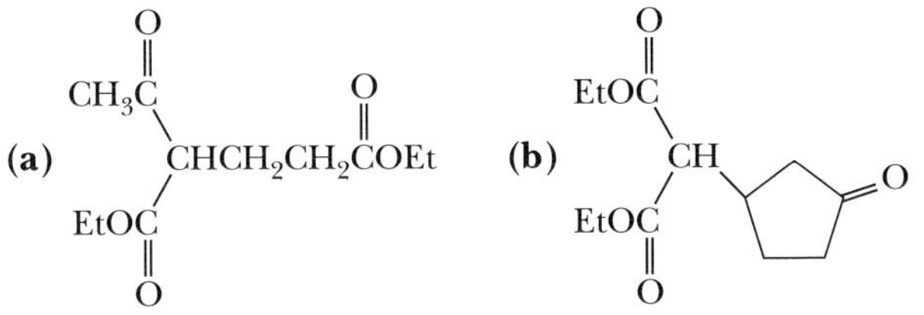

PROBLEM 18.14

Show the product formed from each Michael product in the solution to Example 18.14 after (1) hydrolysis in aqueous NaOH, (2) acidification, and (3) thermal decarboxylation of each β-ketoacid or β-dicarboxylic acid. These reactions illustrate the usefulness of the Michael reaction for the synthesis of 1,5-dicarbonyl compounds.

EXAMPLE 18.15

Show how the series of reactions in the Example 18.14 and Problem 18.14 (Michael reaction, hydrolysis, acidification, and thermal decarboxylation) can be used to prepare 2,6-heptanedione.

Solution

As shown in the following retrosynthetic analysis, this molecule can be constructed from the carbon skeletons of ethyl acetoacetate and methyl vinyl ketone.

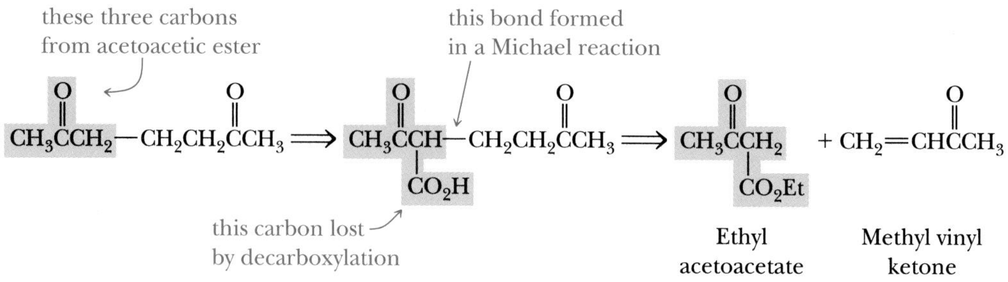

these three carbons from acetoacetic ester

this bond formed in a Michael reaction

$$CH_3\overset{O}{\overset{\|}{C}}CH_2{-}CH_2CH_2\overset{O}{\overset{\|}{C}}CH_3 \Longrightarrow CH_3\overset{O}{\overset{\|}{C}}\underset{CO_2H}{\underset{|}{C}H}{-}CH_2CH_2\overset{O}{\overset{\|}{C}}CH_3 \Longrightarrow CH_3\overset{O}{\overset{\|}{C}}\underset{CO_2Et}{\underset{|}{C}H_2} + CH_2{=}CH\overset{O}{\overset{\|}{C}}CH_3$$

this carbon lost by decarboxylation

Ethyl acetoacetate

Methyl vinyl ketone

Following are the steps in their conversion to 2,6-heptanedione.

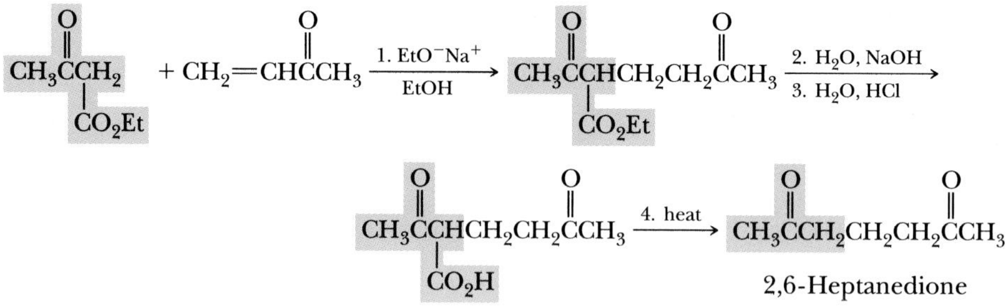

$$CH_3\overset{O}{\overset{\|}{C}}\underset{CO_2Et}{\underset{|}{C}H_2} + CH_2{=}CH\overset{O}{\overset{\|}{C}}CH_3 \xrightarrow[\text{EtOH}]{1.\ \text{EtO}^-\text{Na}^+} CH_3\overset{O}{\overset{\|}{C}}\underset{CO_2Et}{\underset{|}{C}H}CH_2CH_2\overset{O}{\overset{\|}{C}}CH_3 \xrightarrow[3.\ H_2O,\ HCl]{2.\ H_2O,\ NaOH}$$

$$CH_3\overset{O}{\overset{\|}{C}}\underset{CO_2H}{\underset{|}{C}H}CH_2CH_2\overset{O}{\overset{\|}{C}}CH_3 \xrightarrow{4.\ \text{heat}} CH_3\overset{O}{\overset{\|}{C}}CH_2CH_2CH_2\overset{O}{\overset{\|}{C}}CH_3$$

2,6-Heptanedione

PROBLEM 18.15

Show how the sequence of Michael reaction, hydrolysis, acidification, and thermal decarboxylation can be used to prepare pentanedioic acid (glutaric acid).

As noted in Table 18.1, enamines also participate in Michael reactions as illustrated by the addition of the enamine of cyclohexanone to acrylonitrile.

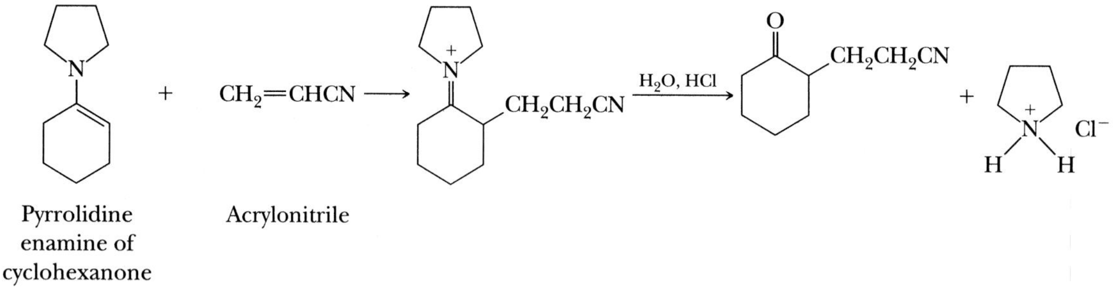

Pyrrolidine enamine of cyclohexanone + $CH_2{=}CHCN$ ⟶ (iminium ion with CH_2CH_2CN) $\xrightarrow{H_2O,\ HCl}$ 2-(CH_2CH_2CN)cyclohexanone + pyrrolidinium (Cl^-)

Pyrrolidine enamine of cyclohexanone

Acrylonitrile

A final word about addition of nucleophiles to α,β-unsaturated carbonyl compounds. The Michael reaction is an example of 1,4-addition (conjugate addition) to an α,β-unsaturated carbonyl compound. In general, resonance-stabilized enolate anions and enamines are weak bases, react slowly, and give 1,4-addition products. Organolithium and organomagnesium compounds, on the other hand, are strong bases, react rapidly, and give primarily 1,2-addition products, that is, products formed by addition to the carbonyl carbon

$$C_6H_5Li + (H_3C)_2C{=}CHCCH_3(=O) \longrightarrow (H_3C)_2C{=}CHC(O^-Li^+)(C_6H_5)CH_3 \xrightarrow[HCl]{H_2O} (H_3C)_2C{=}CHC(OH)(C_6H_5)CH_3$$

Phenyl-lithium 4-Methyl-3-penten-2-one 4-Methyl-2-phenyl-3-penten-2-ol

Why do the nucleophiles listed in Table 18.1 react with conjugated carbonyl compounds by 1,4-addition rather than 1,2-addition? The answer has to do with **kinetic control** versus **thermodynamic control** of product formation. It has been shown that 1,2-addition of nucleophiles to the carbonyl carbon of α,β-unsaturated carbonyl compounds is faster than conjugate addition. If formation of the 1,2-addition product is irreversible, then that is the product observed. If, however, formation of the 1,2-addition product is reversible, then an equilibrium is established between the more rapidly formed 1,2-addition product and the more slowly formed 1,4-addition product. A carbon-oxygen double bond is stronger than a carbon-carbon double bond. Recall, for example, the relative percentages of keto and enol forms present at equilibrium for simple aldehydes and ketones (Section 15.11B). Thus, under conditions of thermodynamic (equilibrium) control, the more stable 1,4-Michael addition product is formed.

$$Nu{:}^- + C{=}C{-}C({=}O){-} \underset{\text{fast}}{\rightleftharpoons} C{=}C{-}C(O^-)(Nu){-} \underset{\text{fast}}{\overset{ROH}{\rightleftharpoons}} C{=}C{-}C(OH)(Nu){-} + RO^-$$

1,2-Addition
(less stable product)

$$Nu{:}^- + C{=}C{-}C({=}O){-} \underset{\text{slow}}{\rightleftharpoons} Nu{-}C{-}C{=}C(O^-){-} \underset{\text{fast}}{\overset{ROH}{\rightleftharpoons}} Nu{-}C{-}C(H){-}C({=}O){-} + RO^-$$

1,4-Addition
(more stable product)

Michael reaction with an α,β-unsaturated ketone followed by an intramolecular aldol reaction has proven to be a valuable method for the synthesis of 2-cyclohexenones. An especially important example of a Michael-aldol sequence is the **Robinson annulation** in which treatment of a cyclic ketone, β-ketoester, or β-diketone with an α,β-unsaturated ketone in the presence of a base catalyst forms a cyclohexenone ring fused to the original ring. When the following β-ketoester, for example, is treated with methyl vinyl ketone in the presence of sodium ethoxide in ethanol, the Michael adduct is first formed and then, in the presence of sodium ethoxide, undergoes a base-catalyzed aldol reaction followed by dehydration to give a cyclohexenone.

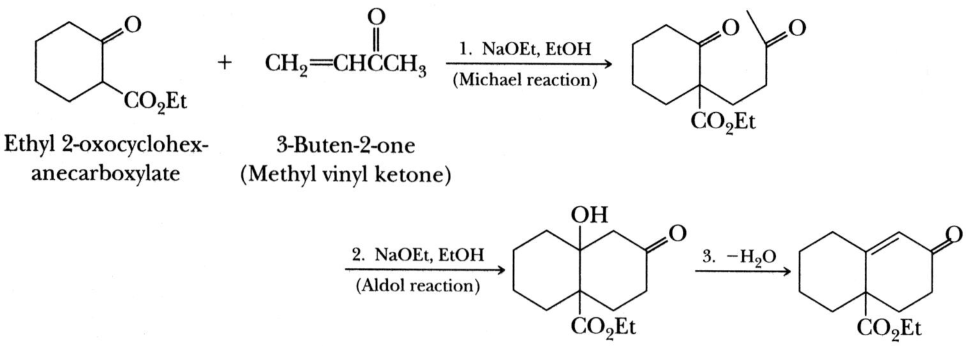

+ $CH_2{=}CH\overset{O}{\overset{\|}{C}}CH_3$ —1. NaOEt, EtOH (Michael reaction)→ ... —2. NaOEt, EtOH (Aldol reaction)→ ... —3. $-H_2O$→ ...

CO_2Et / Ethyl 2-oxocyclohexanecarboxylate; 3-Buten-2-one (Methyl vinyl ketone); OH; CO_2Et

EXAMPLE 18.16

Draw structural formulas for the lettered compounds in the following synthetic sequence.

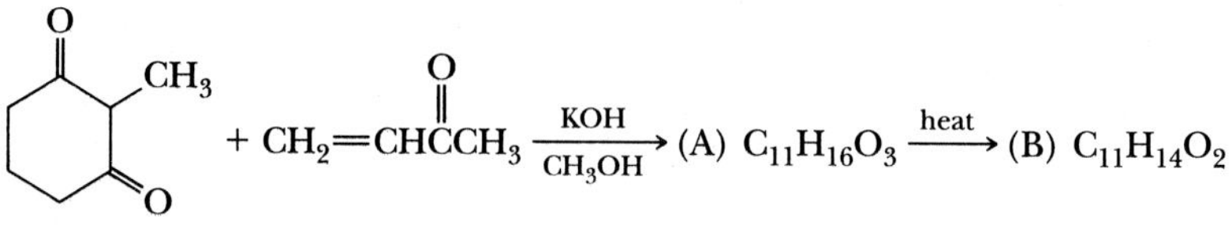

CH_3 + $CH_2{=}CH\overset{O}{\overset{\|}{C}}CH_3$ $\xrightarrow[CH_3OH]{KOH}$ (A) $C_{11}H_{16}O_3$ $\xrightarrow{heat}$ (B) $C_{11}H_{14}O_2$

Solution

The product is the result of Michael addition to an α,β-unsaturated ketone followed by base-catalyzed aldol reaction and dehydration.

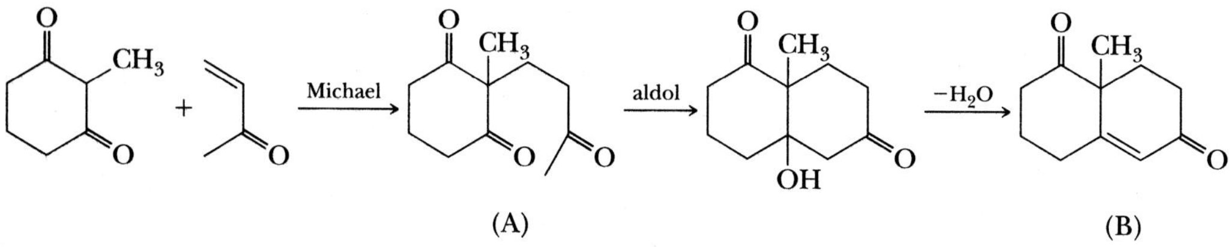

CH_3 + —Michael→ CH_3 (A) —aldol→ CH_3, OH —$-H_2O$→ CH_3 (B)

PROBLEM 18.16

Show how to bring about the following conversion.

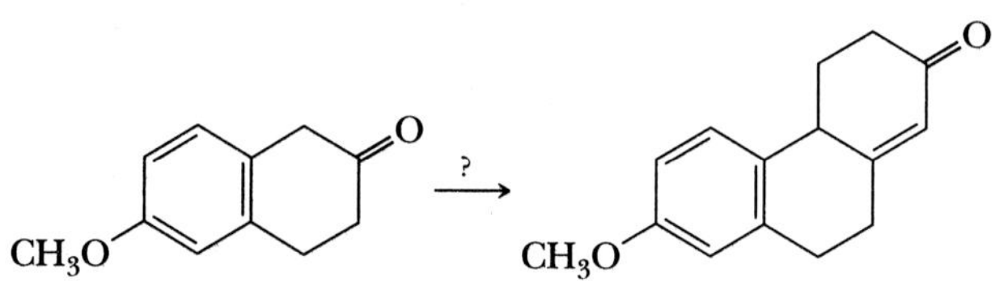

CH_3O —?→ CH_3O

B. Conjugate Addition of Lithium Diorganocopper Reagents

Lithium diorganocopper reagents undergo 1,4-addition to α,β-unsaturated aldehydes and ketones in a reaction that is closely related to the Michael reaction. Yields are highest with primary alkyl, vinyl, and aryl organocopper reagents.

CH_3

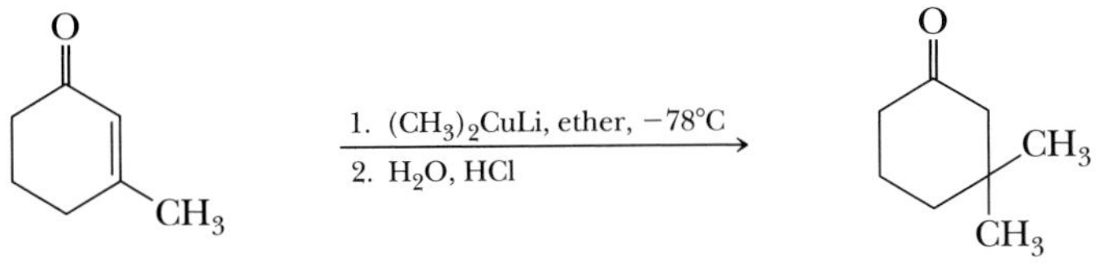

3-Methyl-2-cyclohexenone

1. $(CH_3)_2CuLi$, ether, $-78°C$
2. H_2O, HCl

CH_3 CH_3

3,3-Dimethylcyclohexanone

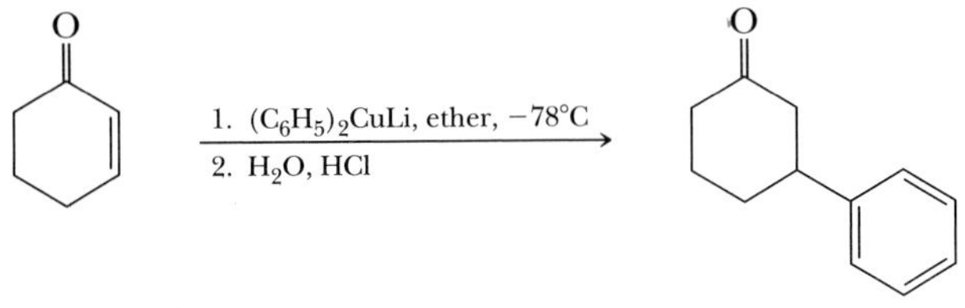

2-Cyclohexenone

1. $(C_6H_5)_2CuLi$, ether, $-78°C$
2. H_2O, HCl

3-Phenylcyclohexanone

Lithium diorganocopper reagents are unique among organometallic compounds in that they give almost exclusively 1,4-addition, which makes them very valuable reagents in synthetic organic chemistry. The mechanism of conjugate addition of lithium diorganocopper reagents is not fully understood.

EXAMPLE 18.17

Propose two syntheses of 4-octanone, each involving conjugate addition of a lithium diorganocopper reagent.

Solution

A lithium diorganocopper reagent adds to the beta carbon of an α,β-unsaturated aldehyde or ketone. Therefore, locate each beta carbon to the carbonyl group in this target molecule and disconnect at those points.

Synthesis 1:

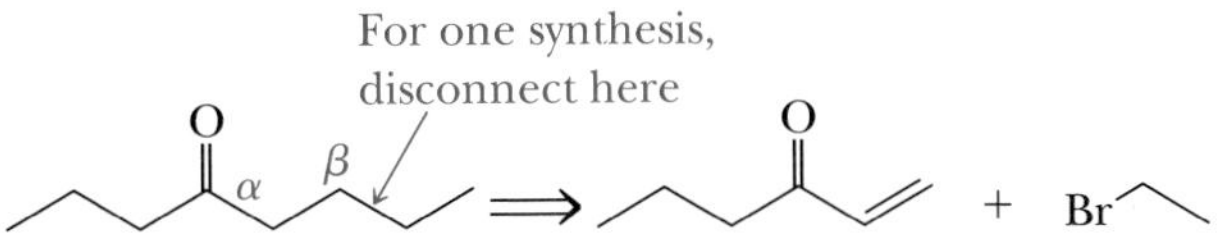

For one synthesis, disconnect here

α β

4-Octanone $\Longrightarrow$ 1-Hexen-3-one + Br Bromoethane

For this synthesis, add lithium diethylcopper to 1-hexen-3-one.

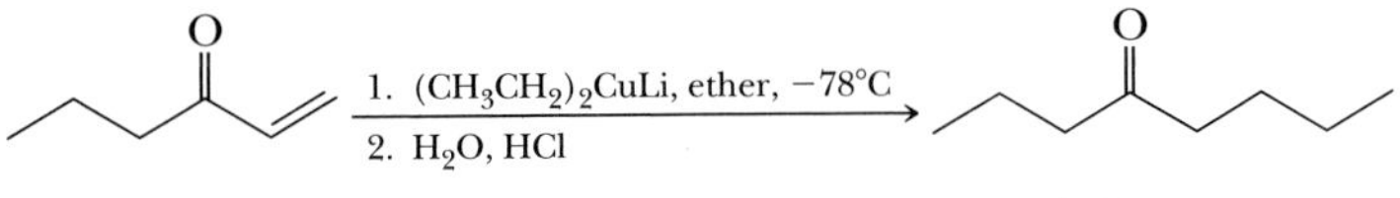

1-Hexen-3-one

1. $(CH_3CH_2)_2CuLi$, ether, $-78°C$
2. H_2O, HCl

4-Octanone

(Text continues on page 724.)

CHEMISTRY IN ACTION

The Bergman Reaction and Anticancer Drugs

In the early 1970s, Professor Robert Bergman, then at the California Institute of Technology, discovered an unusual cyclization reaction of "enediynes," compounds with two acetylene units linked by a double bond. When heated to 200° to 300°C, an enediyne cyclizes to a compound containing a benzene ring in which two of the carbon atoms have no hydrogens. This benzene diradical is extremely reactive and can abstract hydrogen atoms from many different molecules.

$\xrightarrow{2H\cdot}$

A (Z)-enediyne Benzene diradical Benzene

Almost twenty years later, chemists from Bristol-Myers and from American Cyanamid corporations discovered two naturally occurring and incredibly potent families of antitumor compounds, which they named esperamicins and calichemicins. Both have very similar structural formulas and show activity at doses of 1 $\mu g/kg$ of body weight, which is over 1000 times more active than many currently available antitumor drugs. Common to each compound is an enediyne unit, a most unusual functional group to find in a naturally occurring substance. An obvious question arises: Is the enediyne unit involved in the drug's antitumor activity? Research on the mechanism by which these drugs kill cancer cells has established that the enediyne unit is vital to their function.

In the body, enzyme-catalyzed reduction of the trisulfide CH_3—S—S—S— unit generates a thiol anion, RS^-. This nucleophile then participates in a Michael reaction with the double bond of the nearby α,β-unsaturated ketone to form a new five-membered, sulfur-containing ring. As a result of this addition, a marked change occurs in molecular geometry. The two carbon atoms that were sp^2 hybridized with trigonal planar geometry now are sp^3 hybridized with bond angles of 109.5°. This rehybridization and change of bond angles brings the ends of the enediyne unit closer together, something like a molecular nutcracker. The terminal carbons of each triple bond, initially 3.35 Å apart, are now squeezed together to 3.16 Å, a distance short enough for the pi clouds of the two triple bonds to interact and undergo a Bergman cyclization. Instead of taking place at 200° to 300°C, this cyclization occurs at body temperature. The resulting benzene diradical then abstracts hydrogen atoms from whatever molecule is nearby. It is believed that the sugar group in each drug helps to steer it toward DNA, so that often it is DNA that is attacked by the very reactive benzene diradical. The resulting DNA damage is difficult for the cell to repair, and so the cell dies. Cancer cells are rapidly dividing so that their DNA is more exposed than that of normal cells, which accounts for the relative selectivity towards killing tumor cells.

The enediyne story illustrates the remarkable way in which different areas of science can become connected by new discoveries. Professor Bergman certainly never imagined that his studies on the high-temperature rearrangements of enediynes would someday help explain the chemistry behind potent, naturally occurring antitumor drugs. Chemists can now use these ideas to synthesize completely new anticancer drugs.

See K. C. Nicolaou, W. M. Dai, S.-C. Tsay, V. A. Estevez, and W. Wrasidlo, *Science,* **256,** 1172, 1992.

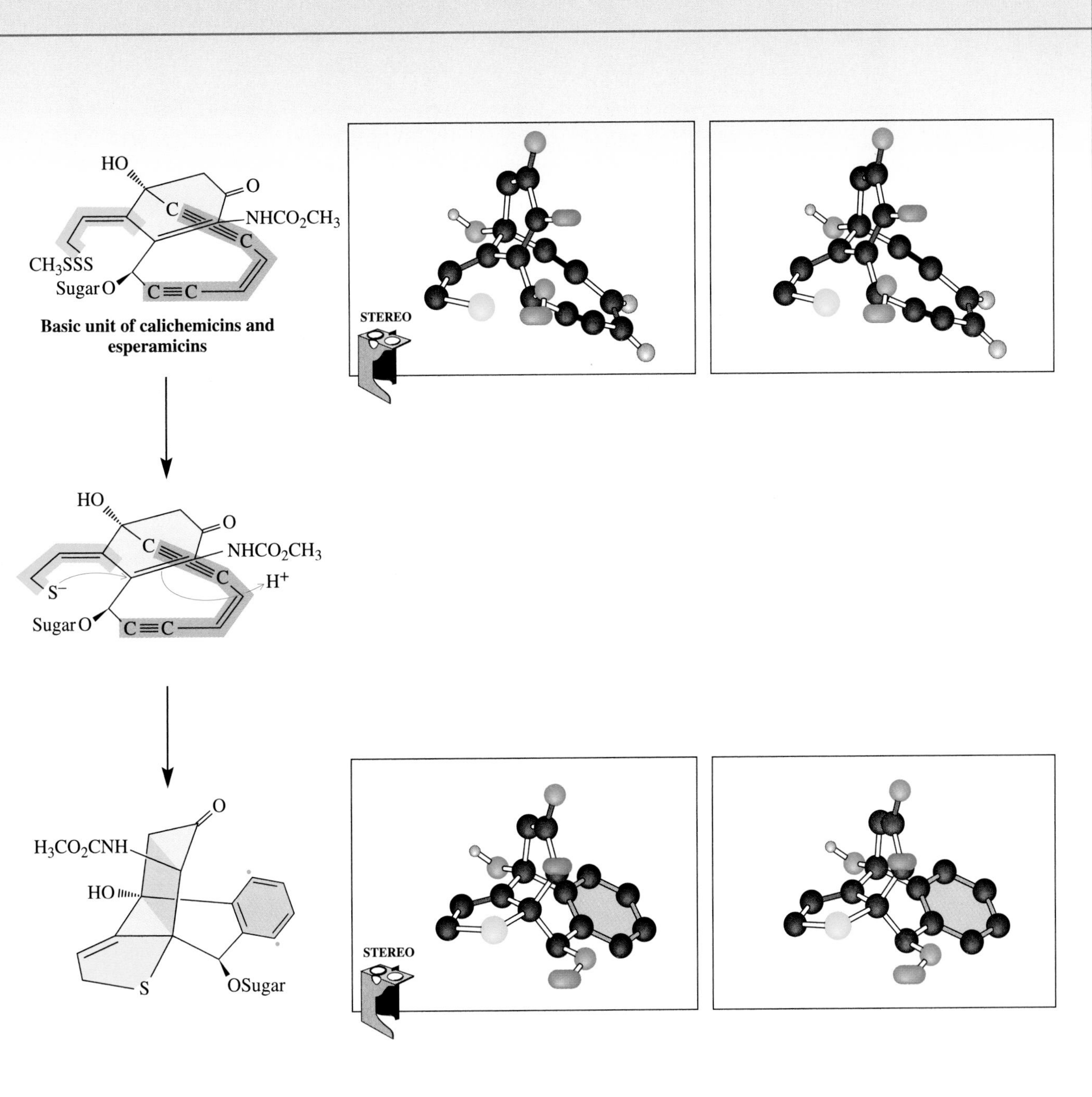
HO
O
C
NHCO2CH3
C
CH3SSS
Sugar O
C≡C
Basic unit of calichemicins and esperamicins
STEREO
HO
O
C
NHCO2CH3
C
S−
H+
Sugar O
C≡C
O
H3CO2CNH
HO
S
OSugar
STEREO

Synthesis 2:

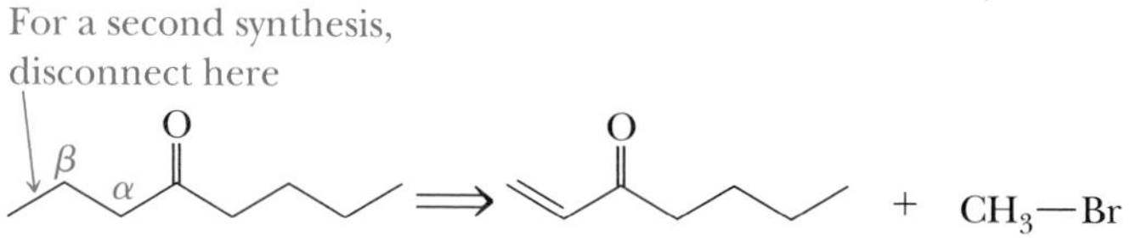

For a second synthesis, disconnect here

4-Octanone $\Rightarrow$ 1-Hepten-3-one + CH_3—Br Bromomethane

For this synthesis, add lithium dimethylcopper to 1-hepten-3-one.

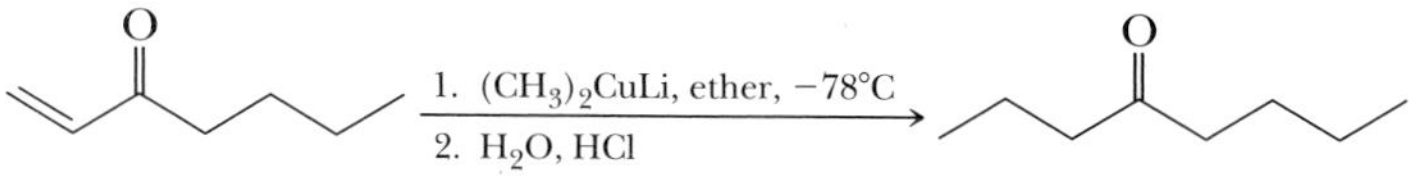

1-Hepten-3-one $\xrightarrow[\text{2. } H_2O,\ HCl]{\text{1. } (CH_3)_2CuLi,\ \text{ether},\ -78°C}$ 4-Octanone

PROBLEM 18.17

Propose two syntheses of 4-phenyl-2-pentanone, each involving conjugate addition of a lithium diorganocopper reagent.

SUMMARY

An **enolate anion** is an anion formed by removal of an α-hydrogen from a carbonyl-containing compound (Section 18.1). Aldehydes, ketones, and esters can be converted completely to their enolate anions by treatment with a strong base such as lithium diisopropylamide (LDA).

Acetyl-Co A (Section 18.4) is the source of the carbon atoms for the synthesis of terpenes, cholesterol, steroid hormones, and fatty acids. Key intermediates in the synthesis of these biomolecules are mevalonic acid and isopentenyl pyrophosphate.

An **enamine** is a compound in which a trivalent nitrogen is bonded to a carbon-carbon double bond (Section 18.5). The β-carbon of an enamine is nucleophilic and participates in both nucleophilic alkylation and acylation reactions.

KEY REACTIONS

1. The Aldol Reaction (Section 18.1)

The aldol reaction involves nucleophilic addition of the enolate anion of one aldehyde or ketone to the carbonyl group of another aldehyde or ketone. The product of an aldol reaction is a β-hydroxyaldehyde or a β-hydroxyketone.

$$2CH_3CH_2CH_2\overset{O}{\overset{\|}{C}}H \underset{}{\overset{NaOH}{\rightleftharpoons}} CH_3CH_2CH_2\overset{OH}{\overset{|}{C}}H\underset{CH_2CH_3}{\underset{|}{C}}H\overset{O}{\overset{\|}{C}}H$$

2. Dehydration of the Product of an Aldol Reaction (Section 18.1)

Dehydration of the β-hydroxyaldehyde or ketone from an aldol reaction occurs very readily under acidic or basic conditions and gives an α,β-unsaturated aldehyde or ketone.

$$\mathrm{CH_3\overset{\overset{OH}{|}}{C}HCH_2\overset{\overset{O}{\|}}{C}H \xrightarrow{H^+} CH_3CH{=}CH\overset{\overset{O}{\|}}{C}H + H_2O}$$

3. Formation of Enolate Anions (Section 18.2A)

Aldehydes, ketones, esters, and other compounds with acidic α-hydrogens are converted completely to their enolate anions by strong bases, such as lithium diisopropylamide (LDA).

$$\underset{\substack{pK_a\ 23\\ \text{(stronger acid)}}}{\mathrm{CH_3\overset{\overset{O}{\|}}{C}OEt}} + \underset{\text{LDA}}{\mathrm{[(CH_3)_2CH]_2N^-Li^+}} \longrightarrow \underset{\text{A lithium enolate}}{\mathrm{CH_2{=}\overset{\overset{O^-Li^+}{|}}{C}OEt}} + \underset{\substack{pK_a\ 40\\ \text{(weaker acid)}}}{\mathrm{[(CH_3)_2CH]_2NH}} \quad K_{eq} = 10^{17}$$

4. Formation of Enolate Anions Under Kinetic Control (Section 18.2A)

The composition of the enolate anion mixture is determined by the relative rates of removal of alternative α-hydrogens; the less substituted enolate anion predominates, often to the exclusion of alternative enolate anions.

2-Methylcyclohexanone + LDA (Slight excess of LDA) $\xrightarrow{0°C}$ lithium enolate with C=C away from CH_3 (99%) + lithium enolate with C=C toward CH_3 (1%) + $[(CH_3)_2CH]_2NH$

5. Formation of Enolate Anions Under Thermodynamic Control (Section 18.2A)

The composition of an enolate anion mixture is determined by the relative stabilities of the individual enolate anions; generally, the more substituted enolate anion predominates.

2-Methylcyclohexanone (Slight excess of the ketone) + LDA $\longrightarrow$ lithium enolate with C=C away from CH_3 (10%) + lithium enolate with C=C toward CH_3 (90%) + $[(CH_3)_3CH]_2NH$

6. Directed Aldol Reactions (Section 18.2B)

An enolate anion is preformed and then treated with a carbonyl compound acting as an enolate anion acceptor.

$$\mathrm{CH_3\overset{\overset{O}{\|}}{C}CH_3 \xrightarrow[-78°C]{1.\ LDA} \underset{\text{A lithium enolate}}{CH_2{=}\overset{\overset{O^-Li^+}{|}}{C}CH_3} \xrightarrow[3.\ H_2O]{2.\ C_6H_5CH_2\overset{\overset{O}{\|}}{C}H} C_6H_5CH_2\overset{\overset{OH}{|}}{C}HCH_2\overset{\overset{O}{\|}}{C}CH_3}$$

7. The Claisen Condensation (Section 18.3A)

The product of a Claisen condensation is a β-ketoester. Condensation occurs by nucleophilic acyl substitution in which the attacking nucleophile is the enolate anion of an ester.

$$2CH_3CH_2\overset{O}{\overset{\|}{C}}OEt \xrightarrow[2.\ H_2O,\ HCl]{1.\ EtO^-Na^+} CH_3CH_2\overset{O}{\overset{\|}{C}}\underset{CH_3}{\underset{|}{C}H}\overset{O}{\overset{\|}{C}}OEt + EtOH$$

8. The Dieckmann Condensation (Section 18.3B)

An intramolecular Claisen condensation is called a Dieckmann condensation.

$$EtO\overset{O}{\overset{\|}{C}}(CH_2)_4\overset{O}{\overset{\|}{C}}OEt \xrightarrow[2.\ H_2O,\ HCl]{1.\ EtO^-Na^+} \text{2-oxocyclopentane-}\overset{O}{\overset{\|}{C}}OEt + EtOH$$

9. Crossed Claisen Condensations (Section 18.3C)

Crossed Claisen condensations are useful only where there is an appreciable difference in the reactivity between the two esters. Such is the case when one of the two esters has no α-hydrogens and, therefore, can function only as an enolate anion acceptor.

$$Ph\overset{O}{\overset{\|}{C}}OCH_3 + CH_3CH_2\overset{O}{\overset{\|}{C}}OCH_3 \xrightarrow[2.\ H_2O,\ HCl]{1.\ CH_3O^-Na^+} Ph\overset{O}{\overset{\|}{C}}\underset{CH_3}{\underset{|}{C}H}\overset{O}{\overset{\|}{C}}OCH_3$$

10. Alkylation of an Enamine Followed by Hydrolysis (Section 18.5A)

Enamines are reactive nucleophiles with methyl and primary alkyl halides, α-haloketones, and α-haloesters.

$$\text{cyclohexanone} \xrightarrow[2.\ CH_2{=}CHCH_2Br]{1.\ \text{morpholine (N–H)},\ H^+} \text{2-}(CH_2CH{=}CH_2)\text{cyclohexylidene-morpholinium } (N^+)\ Br^- \xrightarrow{3.\ H_2O,\ HCl} \text{2-}(CH_2CH{=}CH_2)\text{cyclohexanone}$$

11. Acylation of an Enamine Followed by Hydrolysis (Section 18.5B)

$$\text{cyclohexanone} \xrightarrow[2.\ CH_3\overset{O}{\overset{\|}{C}}Cl]{1.\ \text{pyrrolidine (N–H)},\ H^+} \text{2-}(\overset{O}{\overset{\|}{C}}CH_3)\text{cyclohexylidene-pyrrolidinium } (N^+)\ Cl^- \xrightarrow{3.\ H_2O,\ HCl} \text{2-}(\overset{O}{\overset{\|}{C}}CH_3)\text{cyclohexanone}$$

12. Acetoacetic Ester Synthesis (Section 18.6)

This sequence is useful for the synthesis of monosubstituted and disubstituted acetones.

$$\underset{\text{Ethyl acetoacetate}}{CH_3\overset{O}{\overset{\|}{C}}CH_2\overset{O}{\overset{\|}{C}}OEt} \xrightarrow[\substack{4.\ HCl,\ H_2O \\ 5.\ heat}]{\substack{1.\ EtO^-Na^+ \\ 2.\ CH_2{=}CHCH_2Br \\ 3.\ NaOH,\ H_2O}} \underset{\text{A monosubstituted acetone}}{CH_3\overset{O}{\overset{\|}{C}}CH_2{-}CH_2CH{=}CH_2}$$

13. Malonic Ester Synthesis (Section 18.7)

This sequence is useful for the synthesis of monosubstituted and disubstituted acetic acids.

$$\underset{\text{Diethyl malonate}}{EtO\overset{O}{\overset{\|}{C}}CH_2\overset{O}{\overset{\|}{C}}OEt} \xrightarrow[\substack{4.\ HCl,\ H_2O \\ 5.\ heat}]{\substack{1.\ EtO^-Na^+ \\ 2.\ CH_2{=}CHCH_2Br \\ 3.\ NaOH,\ H_2O}} \underset{\text{A monosubstituted acetic acid}}{CH_2{=}CHCH_2{-}CH_2\overset{O}{\overset{\|}{C}}OH}$$

14. Michael Reaction (Section 18.8A)

Addition of a relatively weakly basic nucleophile to a carbon-carbon double bond made electrophilic by conjugation with the carbonyl group of an aldehyde, ketone, or ester or with a nitro group or cyano group.

$$C_6H_5CH{=}CHCO_2Et + CH_2(CO_2Et)_2 \xrightarrow[EtOH]{EtO^-Na^+} C_6H_5\underset{\displaystyle CH(CO_2Et)_2}{\underset{|}{C}H}CH_2CO_2Et$$

15. Robinson Annulation (Section 18.8A)

A Michael reaction followed by an intramolecular aldol reaction and dehydration to form a substituted 2-cyclohexenone.

$$\text{2-(ethoxycarbonyl)cyclohexanone} + CH_2{=}CH\overset{O}{\overset{\|}{C}}CH_3 \xrightarrow{1.\ Michael} \text{Michael adduct } (CO_2Et) \xrightarrow[3.\ -H_2O]{2.\ aldol} \text{bicyclic enone } (CO_2Et)$$

16. Conjugate Addition of Lithium Diorganocopper Reagents (Section 18.8B)

In a reaction closely related to the Michael reaction, lithium diorganocopper reagents undergo conjugate addition to the electrophilic double bond of α,β-unsaturated aldehydes and ketones.

$$\text{2-cyclohexenone} \xrightarrow[2.\ H_2O,\ HCl]{1.\ (CH_3)_2CuLi,\ ether,\ -78°C} \text{3-methylcyclohexanone } (CH_3)$$

ADDITIONAL PROBLEMS

The Aldol Reaction

18.18 Draw structural formulas for the product of the aldol reaction of each compound and for the α,β-unsaturated aldehyde or ketone formed from dehydration of each aldol product.

(a) CH_3CH_2CHO (b) $C_6H_5COCH_3$ (c) cyclohexanone

18.19 Draw structural formulas for the product of each crossed aldol reaction and for the compound formed by dehydration of each aldol product.

(a) $(CH_3)_3CCHO + CH_3COCH_3$ (b) $C_6H_5COCH_3 + C_6H_5CHO$

(c) cyclohexanone $+ HCHO$ (d) $C_6H_5CHO + CH_3(CH_2)_4CHO$

18.20 When a 1:1 mixture of acetone and 2-butanone is treated with base, six aldol products are possible. Draw structural formulas for these six aldol products.

18.21 Show how to prepare these α,β-unsaturated ketones by an aldol reaction followed by dehydration of the aldol product.

(a) $C_6H_5CH{=}CHCOCH_3$ (b) $CH_3C(CH_3){=}CHCOCH_3$

18.22 Show how to prepare these α,β-unsaturated aldehydes by an aldol reaction followed by dehydration of the aldol product.

(a) $C_6H_5CH{=}CHCHO$ (b) $C_7H_{15}CH{=}C(C_6H_{13})CHO$

18.23 When treated with base, the following compound undergoes an intramolecular aldol reaction to give a product containing a ring (yield 78%). Propose a structural formula for this product.

$$CH_3CH_2CH{=}CHCH_2CH_2CCH_2CH_2CH \xrightarrow[\text{aldol reaction}]{\text{base}} C_{10}H_{14}O + H_2O$$

(with $C{=}O$ at the ketone carbon and at the terminal aldehyde carbon)

18.24 Cyclohexene can be converted to 1-cyclopentenecarbaldehyde by the following series of reactions. Propose a structural formula for each intermediate compound.

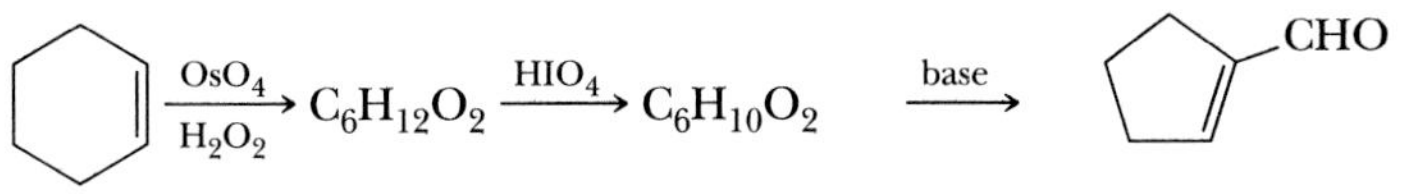

$\xrightarrow[H_2O_2]{OsO_4} C_6H_{12}O_2 \xrightarrow{HIO_4} C_6H_{10}O_2 \xrightarrow{\text{base}}$ CHO

1-Cyclopentenecarbaldehyde

18.25 Propose a structural formula for each lettered compound.

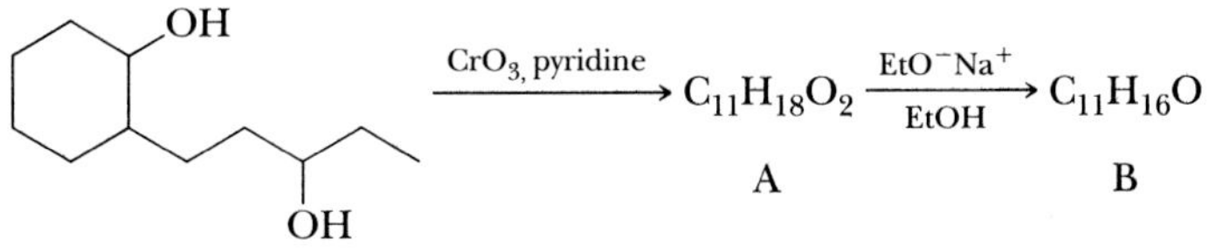

OH

$\xrightarrow{CrO_3, \text{ pyridine}} C_{11}H_{18}O_2 \xrightarrow[EtOH]{EtO^-Na^+} C_{11}H_{16}O$

A B

OH

18.26 How might you bring about the following conversion?

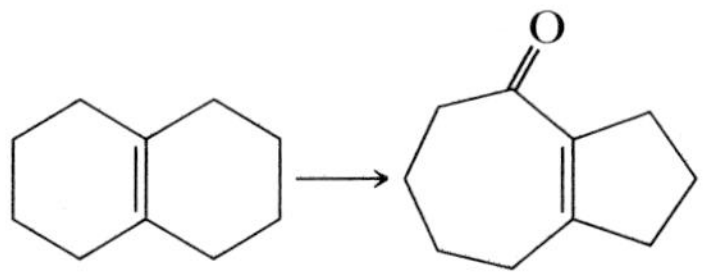

O

18.27 Pulegone, $C_{10}H_{16}O$, a compound from oil of pennyroyal, has a pleasant odor midway between peppermint and camphor. (See *The Merck Index,* 12th ed., #8124.) Treatment of pulegone with steam produces acetone and 3-methylcyclohexanone.

O $\xrightarrow{\text{steam}}$ O + O

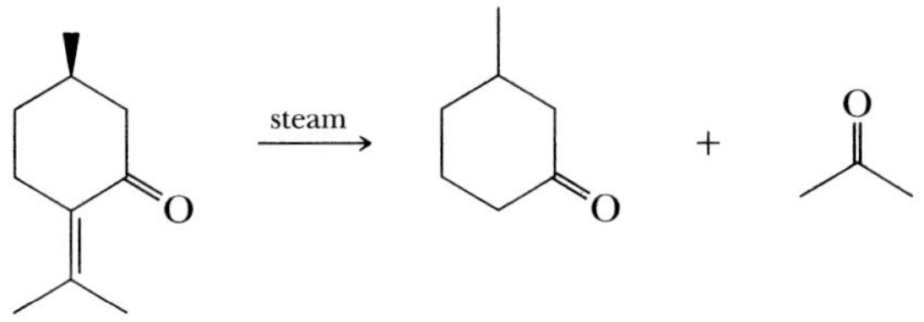

Pulegone 3-Methylcyclohexanone Acetone

(a) Natural pulegone has the configuration shown. Assign an *R,S* configuration to its stereocenter.

(b) Propose a mechanism for the steam hydrolysis of pulegone to the compounds shown.

(c) In what way does this steam hydrolysis affect the configuration of the stereocenter in pulegone? Assign an *R,S* configuration to the 3-methylcyclohexanone formed in this reaction.

18.28 Propose a mechanism for this acid-catalyzed aldol reaction and the dehydration of the resulting aldol product.

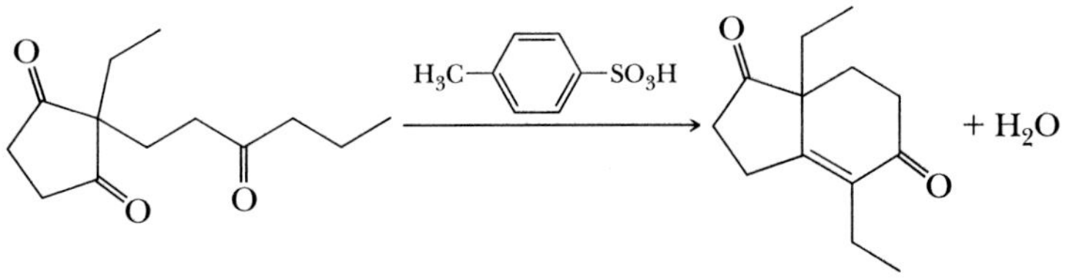

O O O $\xrightarrow{H_3C-C_6H_4-SO_3H}$ O O $+ H_2O$

Directed Aldol Reactions

18.29 In Section 18.2B, it was stated that four possible aldol products are formed when phenylacetaldehyde and acetone are mixed in the presence of base. Draw structural formulas for each of these aldol products.

18.30 In the synthesis of a lithium enolate from a ketone and LDA, is it preferable (a) to add a solution of LDA to a solution of the ketone, or (b) to add a solution of the ketone to a solution of LDA, or (c) to conclude that the order in which the solutions are mixed makes no difference? Explain your answer.

The Claisen Condensation

18.31 Show the product of Claisen condensation of these esters.

(a) Ethyl phenylacetate in the presence of sodium ethoxide.
(b) Methyl hexanoate in the presence of sodium methoxide.

18.32 When a 1:1 mixture of ethyl propanoate and ethyl butanoate is treated with sodium ethoxide, four Claisen condensation products are possible. Draw structural formulas for these four products.

18.33 Draw structural formulas for the β-ketoesters formed by Claisen condensation of ethyl propanoate with each ester:

(a) EtOC(=O)—C(=O)OEt **(b)** PhC(=O)OEt **(c)** HC(=O)OEt

18.34 Draw a structural formula for the product of saponification, acidification, and decarboxylation of each β-ketoester formed in Problem 18.33.

18.35 The Claisen condensation can be used as one step in the synthesis of ketones, as illustrated by this reaction sequence. Propose structural formulas for compounds A, B, and the ketone formed in this sequence.

$$2CH_3CH_2CH_2CH_2\overset{O}{\overset{\|}{C}}OEt \xrightarrow[2.\ HCl,\ H_2O]{1.\ EtO^-Na^+} A \xrightarrow[heat]{NaOH,\ H_2O} B \xrightarrow[heat]{HCl,\ H_2O} C_9H_{18}O$$

18.36 Propose a synthesis for each ketone, using as one step in the sequence a Claisen condensation and the reaction sequence illustrated in Problem 18.35.

(a) $PhCH_2CH_2\overset{O}{\overset{\|}{C}}CH_2CH_2Ph$ **(b)** $PhCH_2\overset{O}{\overset{\|}{C}}CH_2Ph$ **(c)** [bicyclic ketone: cyclohexene ring fused to cyclopentanone, with ring-fusion H atoms]

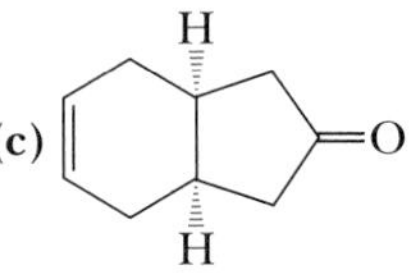

18.37 Propose a mechanism for the following conversion.

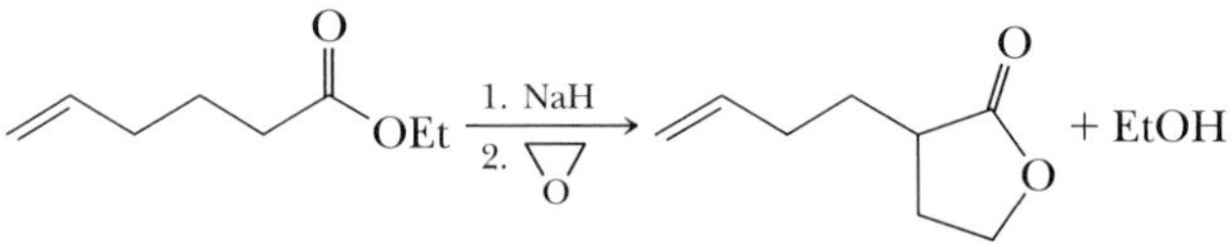

[ethyl hept-6-enoate] $\xrightarrow[2.\ \text{ethylene oxide}]{1.\ NaH}$ [α-(pent-4-enyl)-γ-butyrolactone] + EtOH

18.38 Claisen condensation between diethyl phthalate and ethyl acetate followed by saponification, acidification, and decarboxylation forms a diketone, $C_9H_6O_2$. Propose structural formulas for compounds A, B, and the diketone.

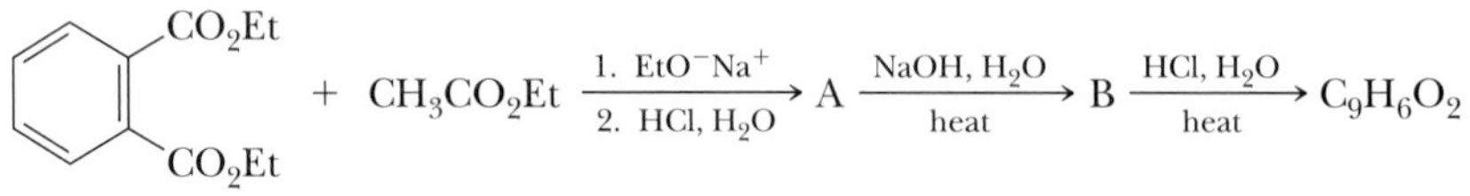

Diethyl phthalate ($C_6H_4(CO_2Et)_2$) + CH_3CO_2Et (Ethyl acetate) $\xrightarrow[2.\ HCl,\ H_2O]{1.\ EtO^-Na^+} A \xrightarrow[heat]{NaOH,\ H_2O} B \xrightarrow[heat]{HCl,\ H_2O} C_9H_6O_2$

18.39 In 1887, the Russian chemist Sergei Reformatsky at the University of Kiev discovered that treatment of an α-haloester with zinc metal in the presence of an aldehyde or ketone followed by hydrolysis in aqueous acid results in formation of a β-hydroxyester. This reaction is similar to a Grignard reaction in that a key intermediate is an organometallic compound, in this case a zinc salt of an ester enolate anion. Grignard reagents, however, are so reactive that they undergo self-condensation with the ester.

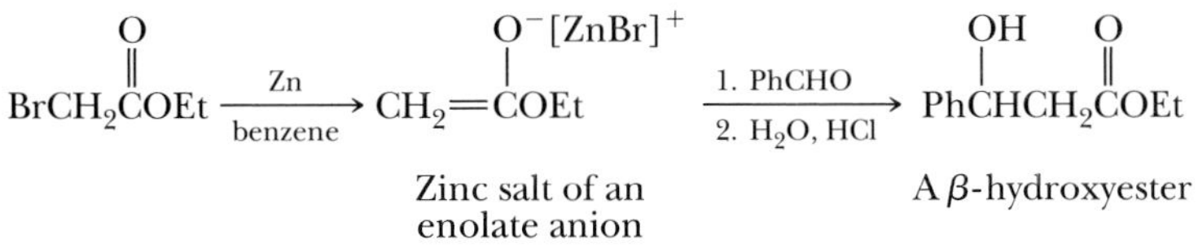

$$\mathrm{BrCH_2\overset{O}{\overset{\|}{C}}OEt} \xrightarrow[\text{benzene}]{\text{Zn}} \mathrm{CH_2{=}\overset{O^-[ZnBr]^+}{\overset{|}{C}}OEt} \xrightarrow[\text{2. } H_2O,\ HCl]{\text{1. PhCHO}} \mathrm{Ph\overset{OH}{\overset{|}{C}}HCH_2\overset{O}{\overset{\|}{C}}OEt}$$

Zinc salt of an enolate anion

A β-hydroxyester

Show how a Reformatsky reaction can be used to synthesize these compounds from an aldehyde or ketone and an α-haloester.

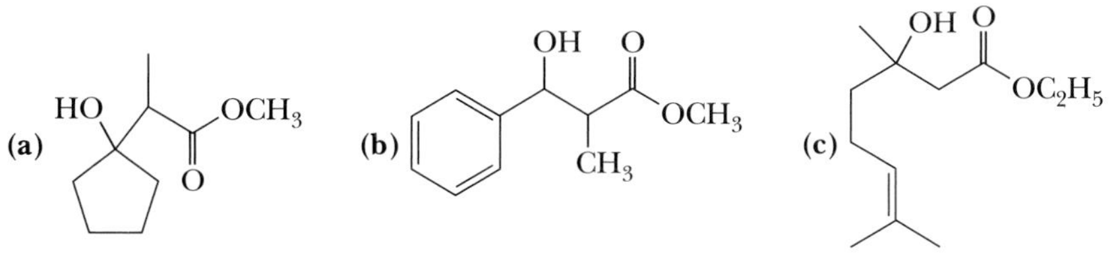

(a) HO, OCH_3, O

(b) OH, O, OCH_3, CH_3

(c) OH, O, OC_2H_5

18.40 Many types of carbonyl condensation reactions have acquired specialized names, after the 19th century organic chemists who first studied them. Propose mechanisms for the following named condensations.

(a) Perkin condensation: Condensation of an aromatic aldehyde with a carboxylic acid anhydride.

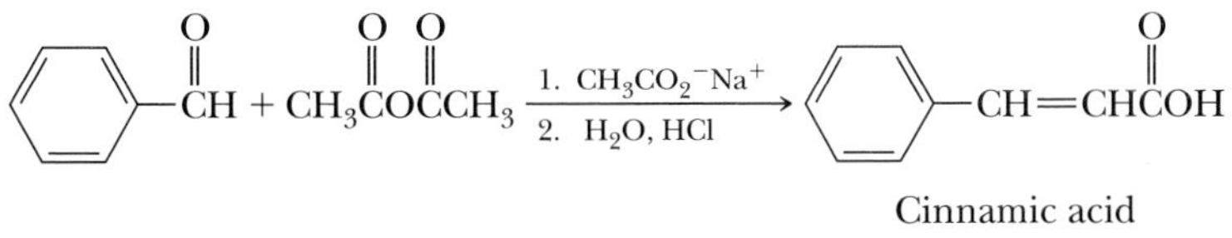

$$\mathrm{C_6H_5{-}\overset{O}{\overset{\|}{C}}H + CH_3\overset{O}{\overset{\|}{C}}O\overset{O}{\overset{\|}{C}}CH_3} \xrightarrow[\text{2. } H_2O,\ HCl]{\text{1. } CH_3CO_2^-Na^+} \mathrm{C_6H_5{-}CH{=}CH\overset{O}{\overset{\|}{C}}OH}$$

Cinnamic acid

(b) Darzens condensation: Condensation of an α-haloester with a ketone or an aromatic aldehyde.

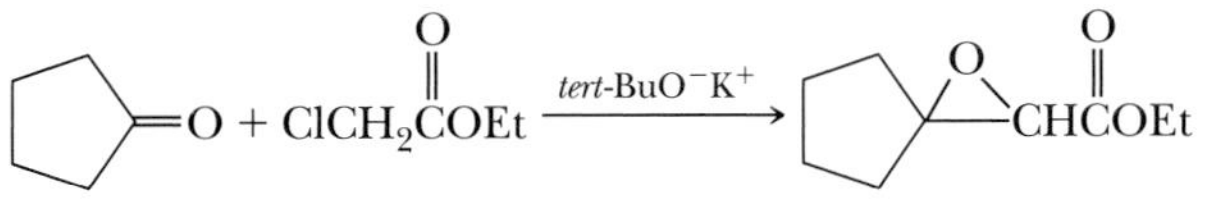

$$\text{(cyclopentanone)}{=}\mathrm{O} + \mathrm{ClCH_2\overset{O}{\overset{\|}{C}}OEt} \xrightarrow{tert\text{-}\mathrm{BuO^-K^+}} \text{(spiro-epoxide)}\mathrm{CH\overset{O}{\overset{\|}{C}}OEt}$$

Enamines

18.41 When 2-methylcyclohexanone is treated with pyrrolidine, two isomeric enamines are formed. Why is enamine A with the less substituted double bond the thermodynamically favored product? You will find it helpful to build models.

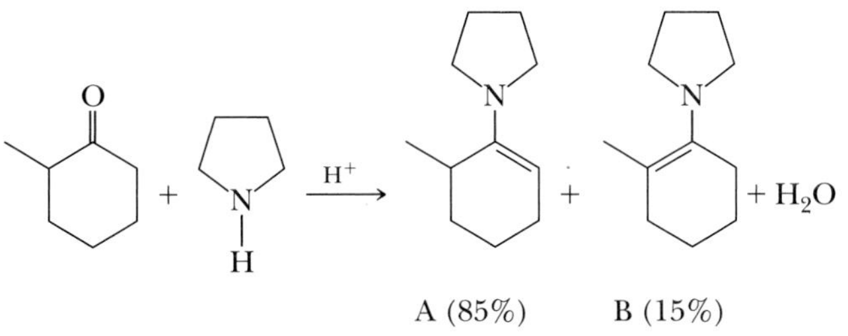

$$\text{2-methylcyclohexanone} + \text{pyrrolidine (N–H)} \xrightarrow{H^+} \text{A} + \text{B} + H_2O$$

A (85%) B (15%)

18.42 Enamines normally react with methyl iodide to give two products: one arising from alkylation at nitrogen, the second arising from alkylation at carbon. For example:

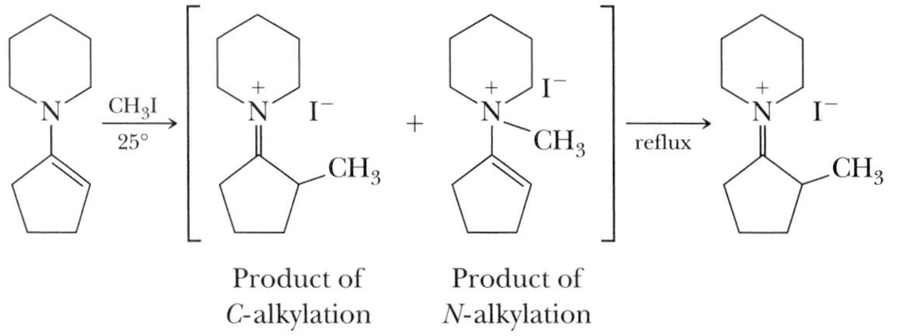

CH_3I, 25°; reflux

Product of *C*-alkylation

Product of *N*-alkylation

Heating the mixture of *C*-alkylation and *N*-alkylation products gives only the product from *C*-alkylation. Propose a mechanism for this isomerization.

18.43 Propose a mechanism for the following conversion.

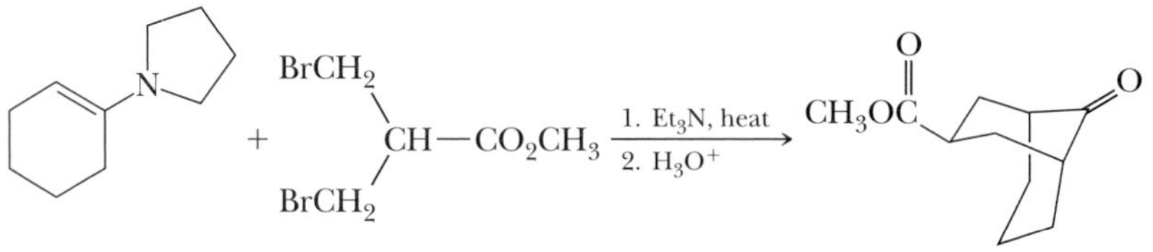

$BrCH_2$ / $CH{-}CO_2CH_3$ / $BrCH_2$ + enamine; 1. Et_3N, heat 2. H_3O^+ → $CH_3OC(=O)$ bicyclic ketone

18.44 Many tumors of the breast are estrogen-dependent. Drugs that interfere with estrogen binding have antitumor activity and may even help prevent tumor occurrence. A widely used antiestrogen drug is tamoxifen. (See *The Merck Index,* 12th ed., #9216.)

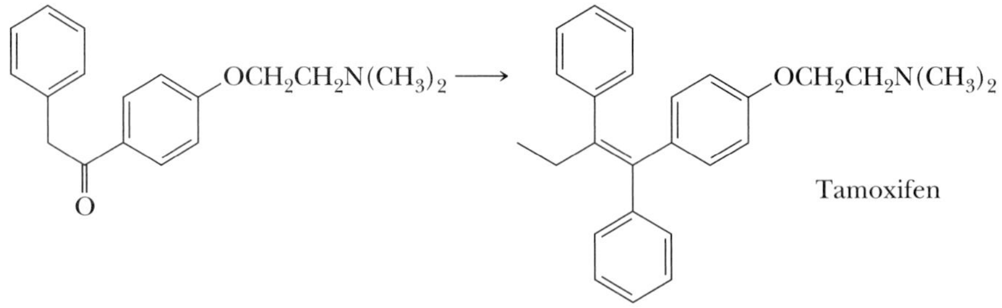

$OCH_2CH_2N(CH_3)_2$ ⟶ $OCH_2CH_2N(CH_3)_2$

Tamoxifen

(a) How many stereoisomers are possible for tamoxifen?

(b) Specify the configuration of the stereoisomer shown here.

(c) Show how tamoxifen can be synthesized from the given ketone using an enamine and a Grignard reaction.

18.45 Propose a mechanism for the following reaction.

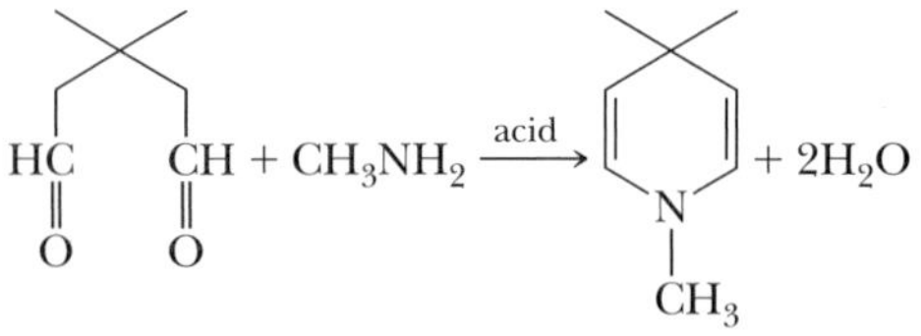

$$HC(=O)\ \ CH(=O) + CH_3NH_2 \xrightarrow{\text{acid}} \text{N-}CH_3 + 2H_2O$$

Acetoacetic Ester and Malonic Ester Syntheses

18.46 Propose syntheses of the following derivatives of diethyl malonate, each being a starting material for synthesis of a barbiturate currently available in the United States.

(a) $(CH_3CH_2)((CH_3)_2CHCH_2CH_2)C(COEt)_2$ [each COEt with C=O]

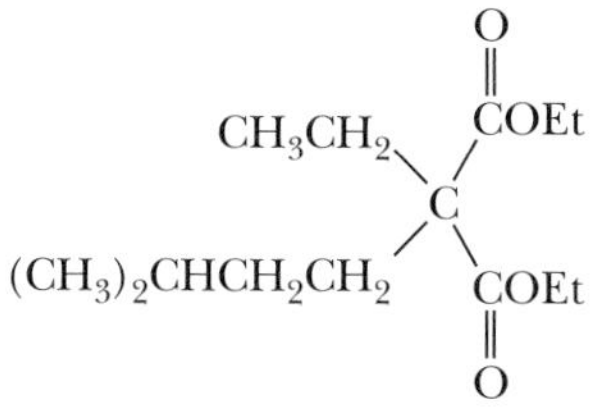

Needed for the synthesis of amobarbital

(b) $(CH_3CH_2CH_2CH(CH_3))(CH_2{=}CHCH_2)C(COEt)_2$ [each COEt with C=O]

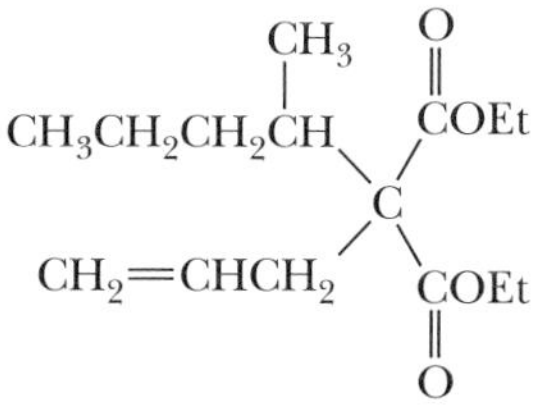

Needed for the synthesis of secobarbital

18.47 2-Propylpentanoic acid (valproic acid) is an effective drug for treatment of several types of epilepsy, particularly absence seizures, which are generalized epileptic seizures characterized by brief and abrupt loss of consciousness. (See *The Merck Index,* 12th ed., #10049.) Propose a synthesis of valproic acid starting with diethyl malonate.

18.48 Show how to synthesize the following compounds using either the malonic ester synthesis or the acetoacetic ester synthesis.

(a) 4-Phenyl-2-butanone
(b) 2-Methylhexanoic acid
(c) 3-Ethyl-2-pentanone
(d) 2-Propyl-1,3-propanediol
(e) 4-Oxopentanoic acid
(f) 3-Benzyl-5-hexene-2-one
(g) Cyclopropanecarboxylic acid
(h) Cyclobutyl methyl ketone

18.49 Propose a mechanism for formation of 2-carbethoxy-4-butanolactone and then 4-butanolactone (γ-butyrolactone) in the following sequence of reactions.

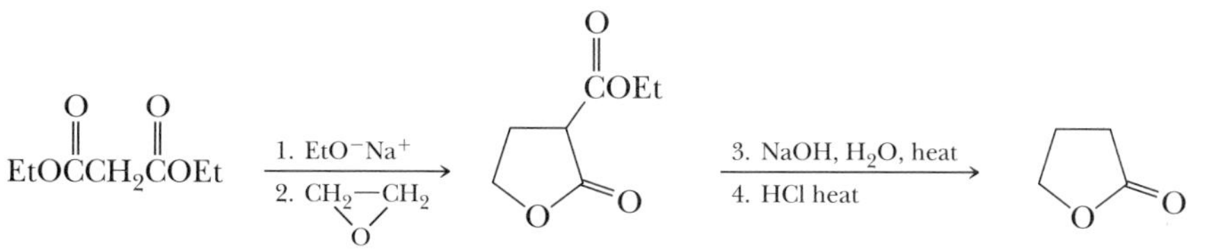

$EtOCCH_2COEt$ (both C=O) $\xrightarrow[\text{2. } CH_2-CH_2 \text{ (epoxide, O)}]{\text{1. } EtO^-Na^+}$ 2-Carbethoxy-4-Butanolactone (five-membered lactone ring bearing COEt) $\xrightarrow[\text{4. HCl heat}]{\text{3. NaOH, } H_2O\text{, heat}}$ 4-Butanolactone (γ-Butyrolactone)

18.50 Show how the scheme for formation of 4-butanolactone in Problem 18.49 can be used to synthesize lactones (a) and (b). Each has a peach odor and is used in perfumery. As sources of carbon atoms for these syntheses, use diethyl malonate, ethylene oxide, 1-bromoheptane, and 1-nonene.

(a) [γ-lactone with heptyl chain at the 4-position] (b) [γ-lactone with octyl chain at the 3-position]

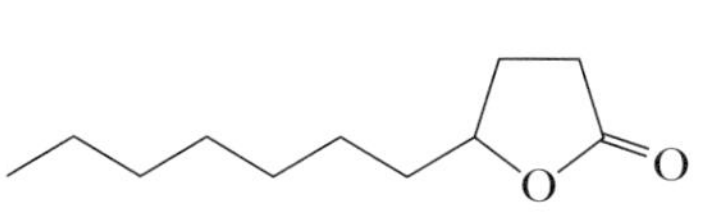

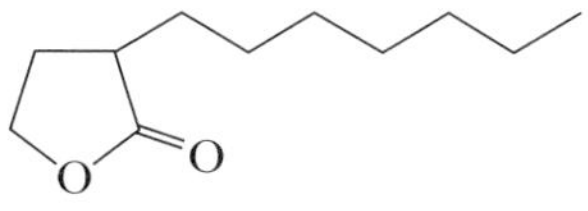

Michael Reactions

18.51 The following synthetic route is used to prepare an intermediate in the total synthesis of the anticholinergic drug benzilonium bromide. Write structural formulas for intermediates (A), (B), (C), and (D) in this synthesis.

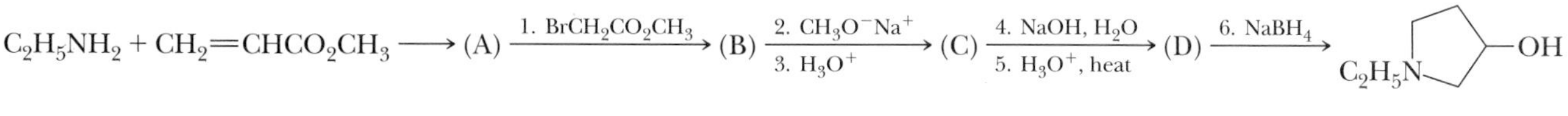

$$C_2H_5NH_2 + CH_2{=}CHCO_2CH_3 \longrightarrow (A) \xrightarrow{\text{1. } BrCH_2CO_2CH_3} (B) \xrightarrow[\text{3. } H_3O^+]{\text{2. } CH_3O^-Na^+} (C) \xrightarrow[\text{5. } H_3O^+\text{, heat}]{\text{4. NaOH, } H_2O} (D) \xrightarrow{\text{6. } NaBH_4} \text{1-ethyl-3-pyrrolidinol } (C_2H_5N\text{ ring, OH})$$

18.52 Propose a mechanism for formation of the bracketed intermediate, and for the bicyclic ketone formed in the following reaction sequence.

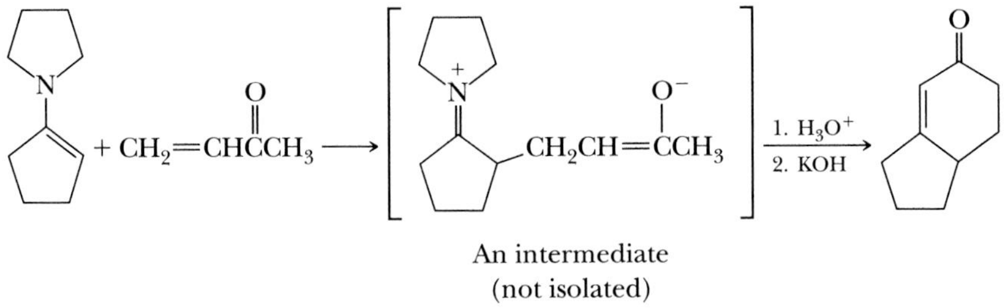

$$\text{(1-pyrrolidinyl)cyclopentene} + CH_2{=}CH\overset{O}{\overset{\|}{C}}CH_3 \longrightarrow \left[\text{iminium}{-}CH_2CH{=}\overset{O^-}{\overset{|}{C}}CH_3\right] \xrightarrow[2.\ KOH]{1.\ H_3O^+} \text{bicyclic enone}$$

An intermediate
(not isolated)

Synthesis

18.53 Show experimental conditions by which to carry out the following synthesis starting with benzaldehyde and methyl acetoacetate.

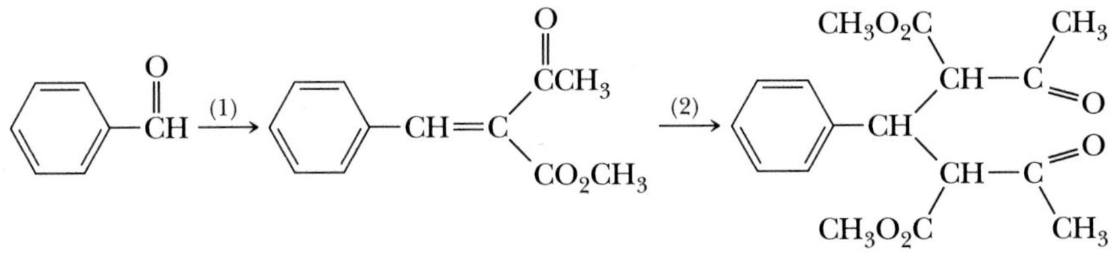

$$C_6H_5\overset{O}{\overset{\|}{C}}H \xrightarrow{(1)} C_6H_5CH{=}C(\overset{O}{\overset{\|}{C}}CH_3)(CO_2CH_3) \xrightarrow{(2)} C_6H_5CH[CH(CO_2CH_3)C(=O)CH_3]_2$$

18.54 Nifedipine (Procardia, Adalat) belongs to a class of drugs called calcium channel blockers and is effective in the treatment of various types of angina, including that induced by exercise. (See *The Merck Index,* 12th ed., #6617.) Show how nifedipine can be synthesized from 2-nitrobenzaldehyde, methyl acetoacetate, and ammonia. (*Hint:* Review the chemistry of your answers to Problems 18.45 and 18.53, and then combine that chemistry to solve this problem.)

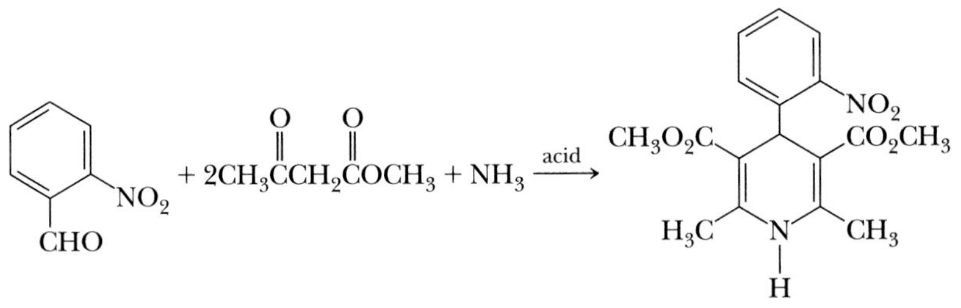

$$\text{2-nitrobenzaldehyde (CHO, } NO_2) + 2CH_3\overset{O}{\overset{\|}{C}}CH_2\overset{O}{\overset{\|}{C}}OCH_3 + NH_3 \xrightarrow{\text{acid}} \text{Nifedipine}$$

Nifedipine
(Procardia)

18.55 The compound 3,5,5-trimethyl-2-cyclohexenone can be synthesized using acetone and ethyl acetoacetate as sources of carbon atoms. New carbon-carbon bonds in this synthesis are formed by a combination of aldol reactions and Michael reactions. Show reagents and conditions by which this synthesis might be accomplished.

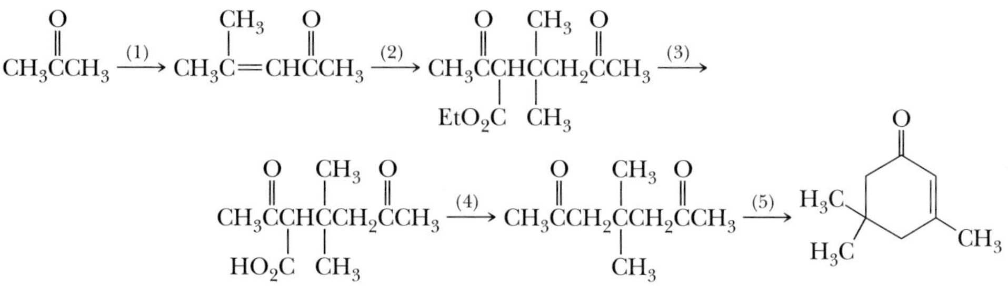

$$CH_3\overset{O}{\overset{\|}{C}}CH_3 \xrightarrow{(1)} CH_3\overset{CH_3}{\overset{|}{C}}{=}CH\overset{O}{\overset{\|}{C}}CH_3 \xrightarrow{(2)} CH_3\overset{O}{\overset{\|}{C}}\underset{EtO_2C}{\underset{|}{C}}H\overset{CH_3}{\overset{|}{\underset{CH_3}{\underset{|}{C}}}}CH_2\overset{O}{\overset{\|}{C}}CH_3 \xrightarrow{(3)}$$

$$CH_3\overset{O}{\overset{\|}{C}}\underset{HO_2C}{\underset{|}{C}}H\overset{CH_3}{\overset{|}{\underset{CH_3}{\underset{|}{C}}}}CH_2\overset{O}{\overset{\|}{C}}CH_3 \xrightarrow{(4)} CH_3\overset{O}{\overset{\|}{C}}CH_2\overset{CH_3}{\overset{|}{\underset{CH_3}{\underset{|}{C}}}}CH_2\overset{O}{\overset{\|}{C}}CH_3 \xrightarrow{(5)}$$

3,5,5-Trimethyl-2-cyclohexenone

18.56 The Weiss reaction, discovered in 1968 by Dr. Ulrich Weiss at the National Institutes of Health, is a route to fused five-membered rings. An example of a Weiss reaction is treating dimethyl 3-oxopentanedioate with ethanedial (glyoxal) in aqueous base under carefully controlled conditions. The bicyclo[3.3.0]octane derivative (A) is formed in 90% yield.

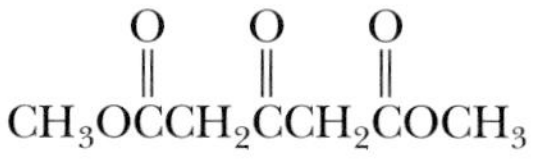

$$CH_3O\overset{O}{\overset{\|}{C}}CH_2\overset{O}{\overset{\|}{C}}CH_2\overset{O}{\overset{\|}{C}}OCH_3$$

Dimethyl 3-oxopentanedioate

The mechanism of the Weiss reaction has been investigated, and the overall steps, as presently understood, involve a combination of aldol, Michael, and dehydration reactions. The molecule shown in brackets is assumed to be an intermediate, but it is not isolated.

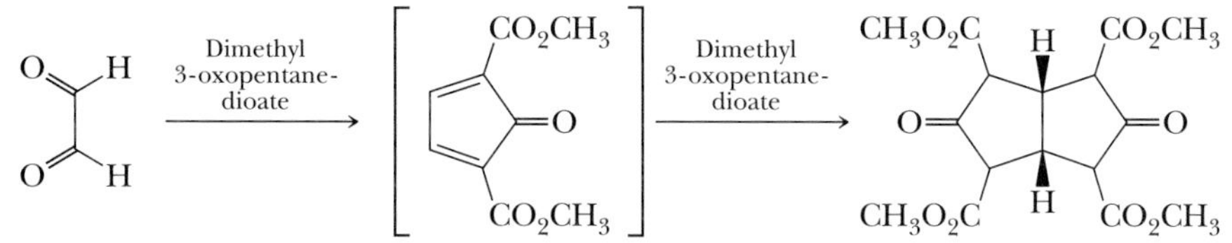

Ethanedial (Glyoxal) —Dimethyl 3-oxopentanedioate→ [An intermediate (not isolated)] —Dimethyl 3-oxopentanedioate→ (A)

Propose a mechanism for the formation of compound A.

18.57 The following β-diketone (A) can be synthesized from cyclopentanone and an acid chloride using an enamine reaction.

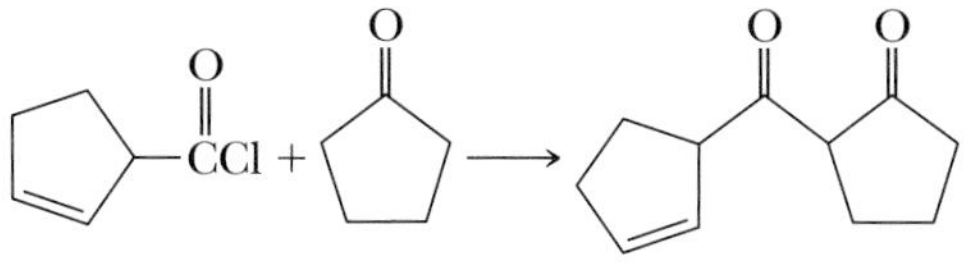

(A)

(a) Propose a synthesis of the starting acid chloride from cyclopentene.
(b) Show the steps in the synthesis of compound A using a morpholine enamine.

18.58 Cisplatin (see *The Merck Index,* 12th ed., #2378) was first prepared in 1844, but it was not until 1964 that its value as an anticancer drug was realized. In that year, Barnett Rosenberg and coworkers at Michigan State University observed that when platinum electrodes are inserted into a growing bacterial culture and an electric current passed through the culture, all cell division ceased within 1 to 2 hours. The result was surprising. Equally surprising was their finding that cell division was inhibited by *cis*-diamminedichloroplat-

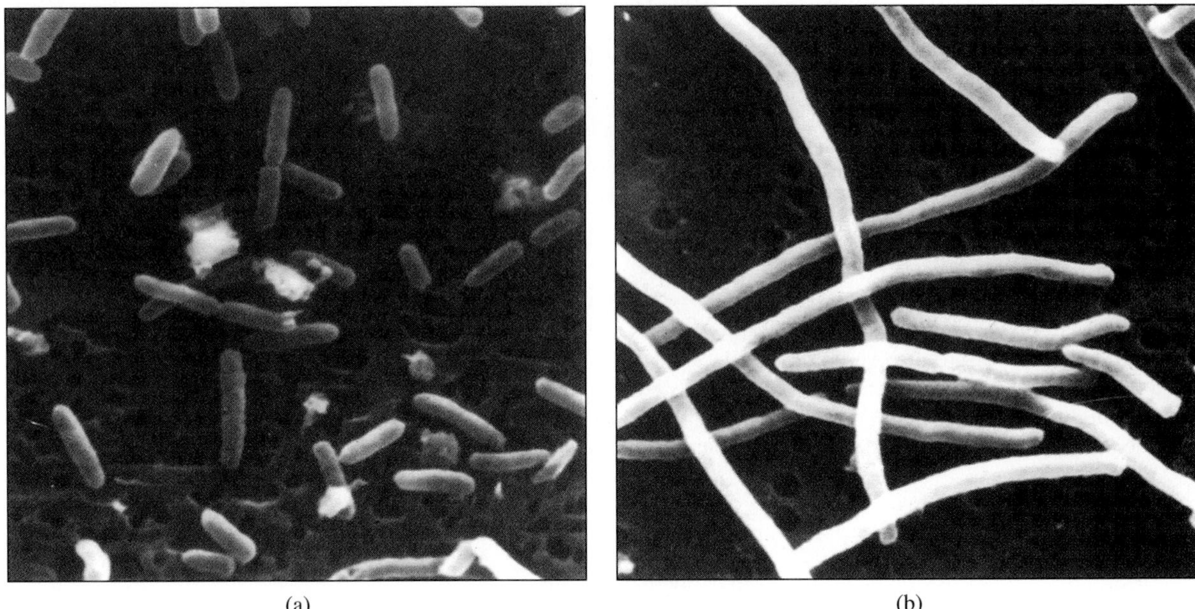

Scanning electromicrographs of (a) normal *E. coli* and (b) *E. coli* grown in a medium containing a few ppm of cisplatin. Cisplatin inhibits cell division, but not growth, leading to long filaments. *(© Doris J. Beck, Biological Sciences, Bowling Green State University)*

inum(II), more commonly named cisplatin, a platinum complex formed in the presence of ammonia and chloride ion. Cisplatin has a broad spectrum of anticancer activity and is particularly useful for treatment of epithelial malignancies. Evidence suggests that platinum(II) in the complex bonds to DNA and forms intrachain and interchain cross linkages. More than 1000 platinum complexes have since been prepared and tested in attempts to discover even more active cytotoxic drugs. In spiroplatin, the two NH_3 groups are replaced by primary amino groups. This drug showed excellent antileukemic activity in animal models, but was disappointing in human trials. In carboplatin, the two chloride ions are replaced by carboxylate groups (see *The Merck Index,* 12th ed., #1870). In 1989, carboplatin was approved by the FDA for treatment of ovarian cancers.

Cisplatin Spiroplatin Carboplatin

(a) Devise a synthesis for the diamine required in the synthesis of spiroplatin starting with diethyl malonate and 1,5-dibromopentane as the sources of carbon atoms.

(b) Devise a synthesis for the dicarboxylic acid required in the synthesis of carboplatin starting with diethyl malonate and 1,3-dibromopropane as sources of carbon atoms.

18.59 Oxanamide is a mild sedative belonging to a class of molecules called oxanamides (it contains an *oxirane* (epoxide) group and an *amide* group). (See *The Merck Index,* 12th ed., #7053.) As seen in this retrosynthetic scheme, the source of carbon atoms for the synthesis of oxanamide is butanal.

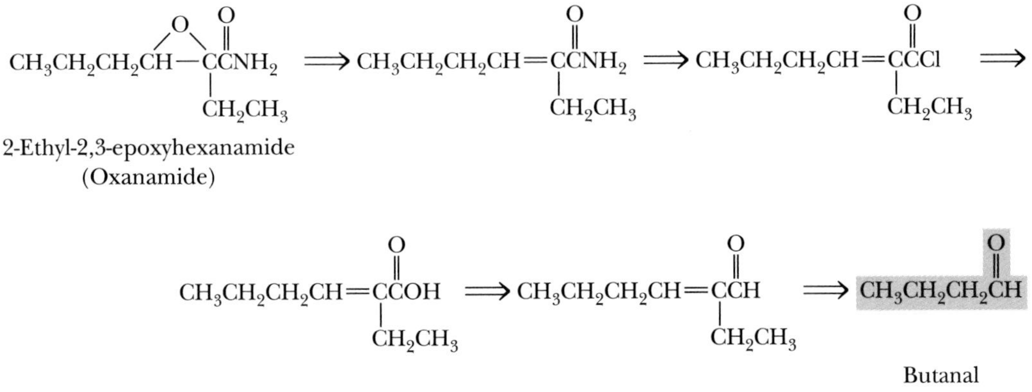

$$CH_3CH_2CH_2CH\overset{O}{—}C(CH_2CH_3)CNH_2(=O) \Longrightarrow CH_3CH_2CH_2CH{=}C(CH_2CH_3)\overset{O}{\overset{\|}{C}}NH_2 \Longrightarrow CH_3CH_2CH_2CH{=}C(CH_2CH_3)\overset{O}{\overset{\|}{C}}Cl \Longrightarrow$$

2-Ethyl-2,3-epoxyhexanamide
(Oxanamide)

$$CH_3CH_2CH_2CH{=}C(CH_2CH_3)\overset{O}{\overset{\|}{C}}OH \Longrightarrow CH_3CH_2CH_2CH{=}C(CH_2CH_3)\overset{O}{\overset{\|}{C}}H \Longrightarrow CH_3CH_2CH_2\overset{O}{\overset{\|}{C}}H$$

Butanal

(a) Show reagents and experimental conditions by which oxanamide can be synthesized from butanal.

(b) How many stereocenters are there in oxanamide? How many stereoisomers are possible for this compound?

18.60 The widely used anticoagulant warfarin (see the Chemistry in Action box "From Moldy Clover to a Blood Thinner" in Chapter 17) is synthesized from 4-hydroxycoumarin, benzaldehyde, and acetone as shown in this retrosynthesis. Show how warfarin is synthesized from these reagents.

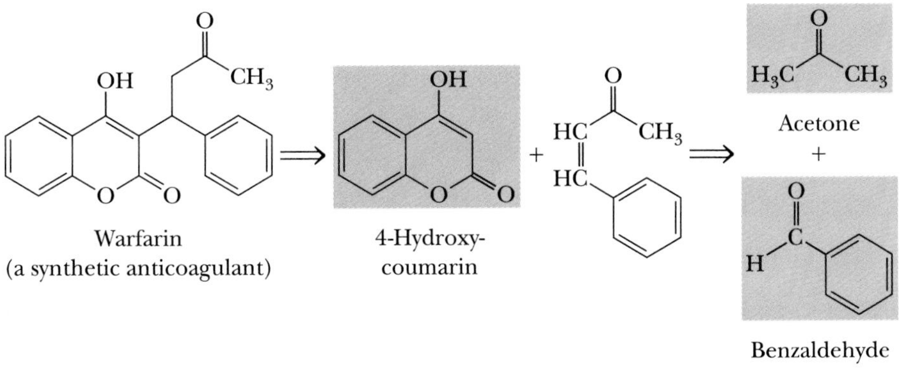

Warfarin
(a synthetic anticoagulant) $\Longrightarrow$ 4-Hydroxy-coumarin + $HC(=O)... $ (HC=CH–C(=O)CH$_3$ with phenyl) $\Longrightarrow$ Acetone ($H_3C\overset{O}{\overset{\|}{C}}CH_3$) + Benzaldehyde (H–C(=O)–phenyl)

19 Aromatics I: Benzene and Its Derivatives

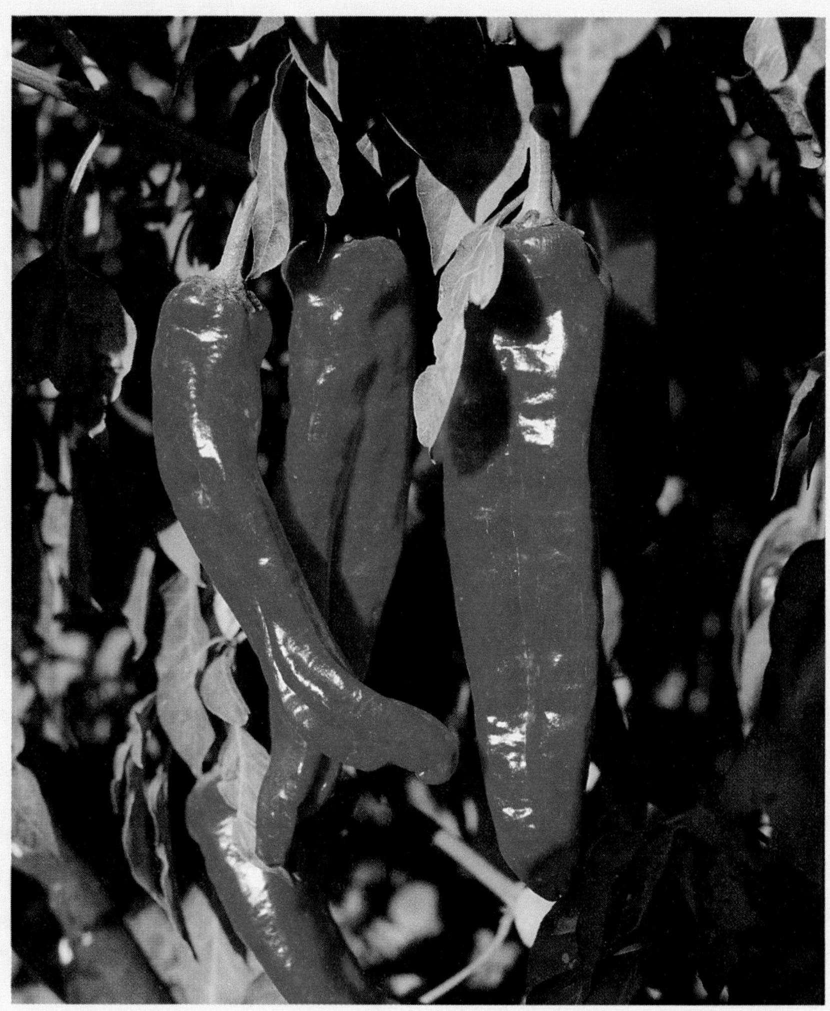

■ The candle pepper, *Capsicum annum conoides,* contains capsaicin. See the box on page 767. *(Grant Heilman/Grant Heilman Photography)*

Benzene is a colorless compound with a melting point of 6°C and a boiling point of 80°C. It was first isolated by Michael Faraday in 1825 from the oily residue that collected in the illuminating gas lines of London. Benzene's molecular formula, C_6H_6, suggests a high degree of unsaturation. Compared with a saturated alkane of molecular formula C_6H_{14}, its index of hydrogen deficiency is four, which can be met by an appropriate combination of rings, double bonds, and triple bonds. For example, a compound of molecular formula C_6H_6 might have four double bonds, or three double bonds and one ring, or two double bonds and two rings, or one triple bond and two rings, and so on. Considering benzene's high degree of unsaturation, it might be expected to show many of the reactions characteristic of alkenes and alkynes. Yet benzene is remarkably stable! It does not undergo addition, oxidation, and reduction reactions characteristic of alkenes and alkynes. For example, benzene does not react with bromine, hydrogen chloride, or other reagents that usually add to carbon-carbon double and triple bonds. It is not oxidized by potassium permanganate or chromic acid under conditions that readily oxidize alkenes and alkynes. When benzene reacts, it does so by substitution in which a hydrogen atom is replaced by another atom or group of atoms.

As noted in the introduction to Chapter 5, the term **aromatic** was originally used to classify benzene and its derivatives because many of them have distinctive odors. The term "aromatic," as it is now used, refers instead to the fact that these compounds are highly unsaturated and unexpectedly stable toward reagents that attack alkenes and alkynes. The term **arene** is used to describe aromatic hydrocarbons, by analogy with alkane, alkene, and alkyne. Benzene is the parent arene. Just as a group derived by removal of an H from an alkane is called an alkyl group and given the symbol R—, a group derived by removal of an H from an arene is called an **aryl group** and given the symbol **Ar**—.

Aromatic compound A term used initially to classify benzene and its derivatives. More accurately, it is used to classify any compound that meets the Hückel criteria for aromaticity (Section 19.2A).

Arene An aromatic hydrocarbon.

Aryl group A group derived from an aromatic compound (an arene) by removal of an H; given the symbol Ar—.

Ar— The symbol used for an aryl group, by analogy with R— for an alkyl group.

19.1 The Structure of Benzene

Let us put ourselves in the mid-19th century and examine the evidence on which chemists attempted to build a model for the structure of benzene. First, because the molecular formula of benzene is C_6H_6, it seemed clear that the molecule must be highly unsaturated. Yet, benzene does not show the chemical properties of alkenes, the only unsaturated hydrocarbons known at that time. Benzene does undergo chemical reactions, but its characteristic reaction is substitution rather than addition. Furthermore, when benzene is treated with bromine in the presence of ferric chloride as a catalyst, only one compound of molecular formula C_6H_5Br is formed.

$$\underset{\text{Benzene}}{C_6H_6} + Br_2 \xrightarrow{FeCl_3} \underset{\text{Bromobenzene}}{C_6H_5Br} + HBr$$

Chemists concluded, therefore, that all six hydrogens of benzene must be equivalent. When bromobenzene is treated with bromine in the presence of ferric chloride as a catalyst, three isomeric dibromobenzenes are formed.

$$\underset{\text{Bromobenzene}}{C_6H_5Br} + Br_2 \xrightarrow{FeCl_3} \underset{\substack{\text{Three isomeric}\\ \text{dibromobenzenes}}}{C_6H_4Br_2} + HBr$$

For chemists in the mid-19th century, the problem was to incorporate these observations, along with the accepted tetravalence of carbon, into a structural formula for benzene. Before we examine these proposals, we should note that the problem of the structure of benzene and other aromatic hydrocarbons has occupied the efforts of chemists for over a century. Only since the 1930s has a general understanding of this problem been realized.

A. Kekulé's Model of Benzene

The first structure for benzene was proposed by August Kekulé in 1865 and consisted of a six-membered ring with one hydrogen attached to each carbon. Although Kekulé's original structural formula provided for the equivalency of the C—H and C—C bonds, it was inadequate because all of the carbon atoms were trivalent. To maintain the tetravalence of carbon, Kekulé proposed in 1872 that the ring contains three double bonds that shift back and forth so rapidly that the two forms cannot be separated. Each structure became known as a **Kekulé structure.**

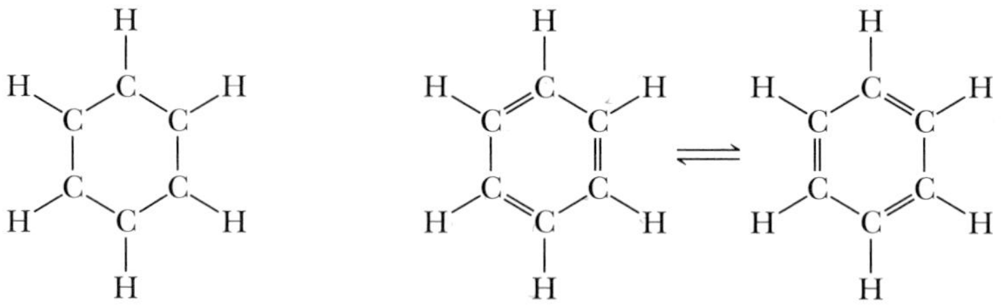

Structure for benzene, proposed by Kekulé in 1865

Kekulé structures for benzene, proposed in 1872

Now, more than 125 years after the time of Kekulé, we are apt to take for granted what scientists in his time knew and did not know. For example, it is a given to us that covalent bonds consist of one or more pairs of shared electrons. We must remember, however, that it was not until 1897 that J. J. Thomson, professor of physics at the Cavendish Laboratory of Cambridge University, discovered the electron. Thomson was awarded the Nobel prize for physics in 1906. That the electron played any role in chemical bonding did not become clear for another thirty years. Thus, at the time Kekulé made his proposal for the structure of benzene, the existence of electrons and their role in chemical bonding was completely unknown.

Kekulé's proposal accounted nicely for the fact that bromination of benzene gives only one bromobenzene, and bromination of bromobenzene gives three isomeric dibromobenzenes.

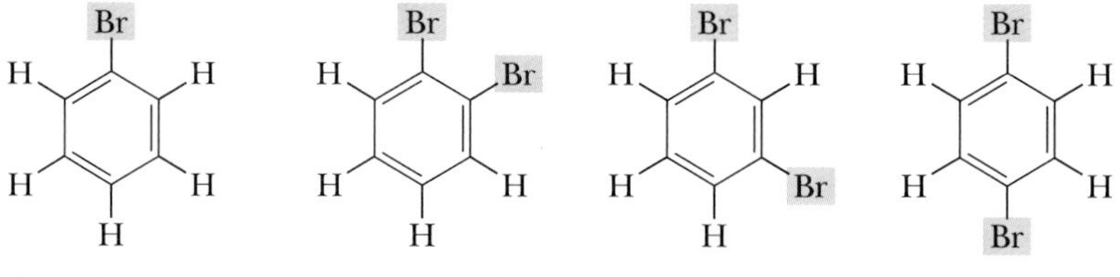

Bromobenzene

Three isomeric dibromobenzenes

Although his proposal was consistent with many experimental observations, it did not totally solve the problem and was contested for years. The major objection was that it did not account for the unusual chemical behavior of benzene. If benzene contains three double bonds, Kekulé's critics asked, why does it not show reactions typical of alkenes? Why, for example, does benzene not add three moles of bromine to form 1,2,3,4,5,6-hexabromocyclohexane? We now understand the surprising unreactivity of benzene on the basis of two complementary descriptions, the molecular orbital model and the resonance model.

B. The Molecular Orbital Model of Benzene

The concepts of **hybridization of atomic orbitals** and the **theory of resonance,** developed by Linus Pauling in the 1930s, provided the first adequate description of the structure of benzene. The carbon skeleton of benzene forms a regular hexagon with C—C—C and H—C—C bond angles of 120°. For this type of bonding, carbon uses sp^2 hybrid orbitals. Each carbon forms sigma bonds to two adjacent carbons by overlap of sp^2-sp^2 hybrid orbitals, and one sigma bond to hydrogen by overlap of sp^2-$1s$ orbitals. As determined experimentally, all carbon-carbon bonds are 1.39 Å in length, a value almost midway between the length of a single bond between sp^3 hybridized carbons (1.54 Å) and a double bond between sp^2 hybridized carbons (1.33 Å).

Each carbon also has a single unhybridized $2p$ orbital perpendicular to the plane of the ring and containing one electron. Combination of these six parallel $2p$ atomic orbitals gives a set of six pi MOs, three bonding pi MOs and three antibonding pi MOs. These six molecular orbitals and their relative energies are shown in Figure 19.1. Note that π_2 and π_3 MOs are degenerate (they have the same energy). Similarly, $\pi_4{}^*$ and $\pi_5{}^*$ are a degenerate pair of pi antibonding MOs. In the ground-state electron configuration of benzene, the six electrons of the pi system occupy the three bonding MOs (Figure 19.1). The great stability of benzene results from the fact that these three bonding MOs are of much lower energy compared with the six uncombined $2p$ atomic orbitals.

It is common to represent the pi system of benzene as one torus (a donut-shaped region) above the plane of the ring and a second torus below the plane of the ring, as shown in Figure 19.2. While this picture is useful in thinking about the electron density of the pi system, you must use it with caution because it represents only the lowest-lying pi bonding molecular orbital.

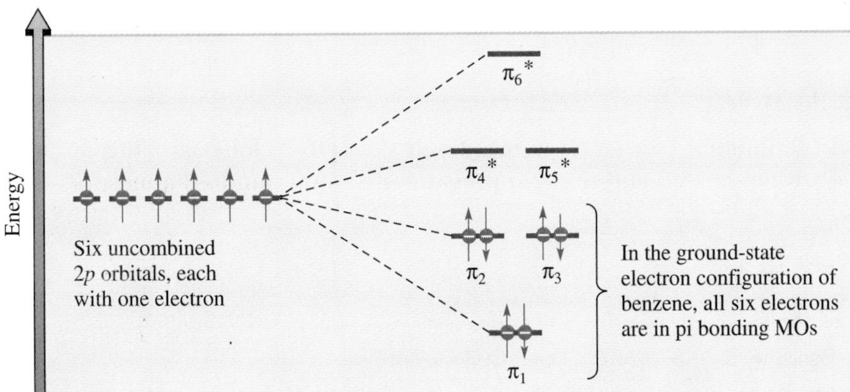

FIGURE 19.1
The molecular orbital representation of the pi bonding in benzene. Benzene has 6 pi electrons ($4n + 2$, where $n = 1$) in a closed loop and is, therefore, aromatic.

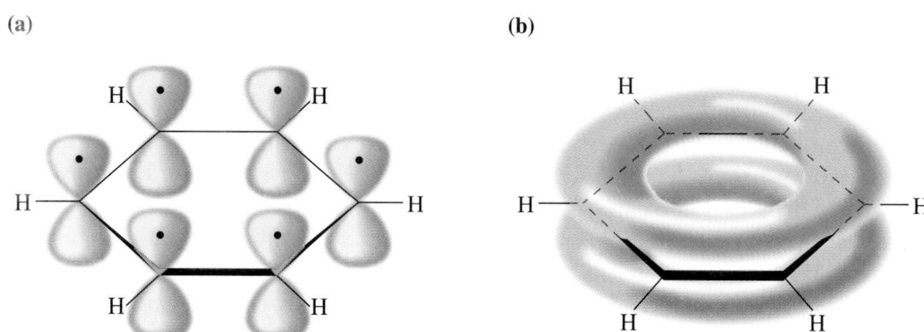

FIGURE 19.2
The pi system of benzene. (a) The carbon-hydrogen framework. The six $2p$ orbitals, each with one electron, are shown uncombined. (b) Overlap of parallel $2p$ orbitals forms a continuous pi cloud, shown by one torus above the plane of the ring and a second torus below the plane of the ring.

C. The Resonance Model of Benzene

One of the postulates of resonance theory is that when a molecule or ion can be represented by two or more contributing structures, then it is not adequately represented by any single contributing structure. We can represent benzene as a hybrid of two equivalent contributing structures, often referred to as Kekulé structures. Each Kekulé structure makes an equal contribution to the hybrid, and thus the C—C bonds are neither single nor double bonds but something intermediate.

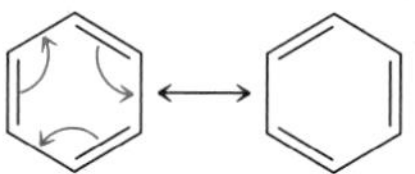

Benzene as a hybrid of two equivalent contributing structures

We recognize that neither of these contributing structures exists; they are merely alternative ways to pair $2p$ orbitals with no reason to prefer one or the other. Nevertheless, chemists continue to use a single contributing structure to represent this molecule because it is as close as we can come to an accurate structure within the limitations of classical valence bond structures and the tetravalence of carbon.

Resonance energy The difference in energy between a resonance hybrid and the most stable of its hypothetical contributing structures in which electrons are localized on particular atoms and in particular bonds.

Resonance energy is the difference in energy between a resonance hybrid and its most stable contributing structure. One way to estimate the resonance energy of benzene is to compare the heats of hydrogenation of cyclohexene and benzene. Cyclohexene is readily reduced to cyclohexane by hydrogen in the presence of a transition metal catalyst (Section 6.6A).

$$\text{Cyclohexene} + H_2 \xrightarrow[\text{1–2 atm}]{\text{Ni}} \text{Cyclohexane} \qquad \Delta H^0 = -28.6 \text{ kcal/mol } (-120 \text{ kJ/mol})$$

Under these conditions, benzene is reduced only slowly to cyclohexane. It is reduced more rapidly when heated and under a pressure of several hundred atmospheres of hydrogen.

$$\text{Benzene} + 3H_2 \xrightarrow[\text{200–300 atm}]{\text{Ni}} \text{Cyclohexane} \qquad \Delta H^0 = -49.8 \text{ kcal/mol } (-208 \text{ kJ/mol})$$

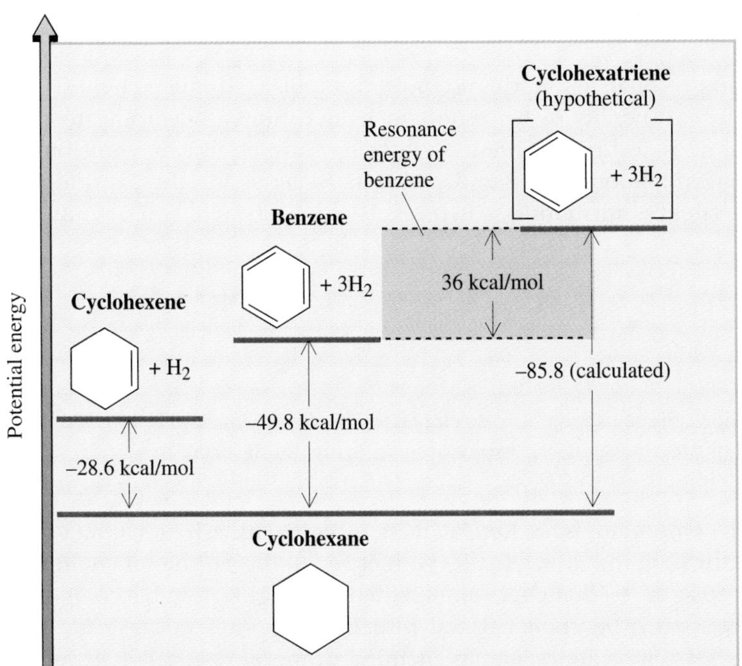

FIGURE 19.3
The resonance energy of benzene as determined by comparison of the heats of hydrogenation of cyclohexene, benzene, and the hypothetical compound 1,3,5-cyclohexatriene.

Catalytic hydrogenation of an alkene is an exothermic reaction. The heat of hydrogenation per double bond varies somewhat with the degree of substitution of the particular alkene; for cyclohexene, $\Delta H^0 = -28.6$ kcal/mol (-120 kJ/mol). If we consider benzene to be 1,3,5-cyclohexatriene, a hypothetical unsaturated compound with alternating single and double bonds, we calculate that $\Delta H^0 = 3(-28.6$ kcal/mol$) = -85.8$ kcal/mol (-359 kJ/mol). The ΔH^0 for reduction of benzene to cyclohexane is only -49.8 kcal/mol (-208 kJ/mol), considerably less than that calculated for 1,3,5-cyclohexatriene. The difference between these values, 36.0 kcal/mol (151 kJ/mol), is the **resonance energy of benzene.** These experimental results are shown graphically in Figure 19.3.

There have been several other experimental determinations of the resonance energy of benzene using different model compounds, and, although these determinations differ somewhat in their results, they all agree that the resonance stabilization of benzene is large. Following are resonance energies for several other aromatic hydrocarbons.

	Benzene	Naphthalene	Anthracene	Phenanthrene
Resonance energy in kcal/mol (kJ/mol):	36 (151)	61 (255)	83 (347)	91 (381)

19.2 The Concept of Aromaticity

Molecular orbital and resonance theories are powerful tools with which chemists can understand the unusual stability of benzene and its derivatives. According to resonance theory, benzene is best represented as a hybrid of two equivalent contributing structures. By analogy, cyclobutadiene and cyclooctatetraene can also be drawn as hybrids of two equivalent contributing structures. Is either of these compounds aromatic?

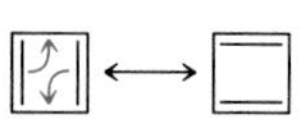

Cyclobutadiene as a hybrid of two equivalent contributing structures

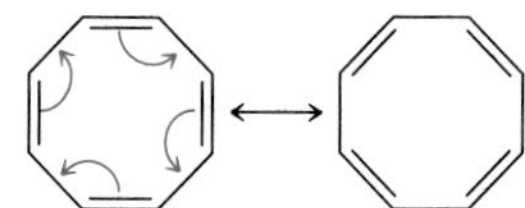

Cyclooctatetraene as a hybrid of two equivalent contributing structures

The answer for both compounds is no. Repeated attempts to synthesize cyclobutadiene failed. It was not until 1965 that it was finally synthesized and even then it could only be detected if trapped at 4 K (−269°C). Cyclobutadiene is a highly unstable compound and does not show any of the chemical and physical properties we associate with aromatic compounds. Cyclooctatetraene has chemical properties typical of alkenes. It reacts readily with halogens and halogen acids, as well as with mild oxidizing and reducing agents.

We are then faced with the broad question, "What are the fundamental principles underlying aromatic character?" What are the structural characteristics of unsaturated compounds that have a large resonance energy and do not undergo reactions typical of alkenes but rather undergo substitution reactions?

A. Hückel's Criteria for Aromaticity

The underlying criteria for aromaticity were recognized in the early 1930s by Erich Hückel, a German chemical physicist. He carried out MO energy calculations for monocyclic, planar molecules in which each atom of the ring has one $2p$ orbital available for forming sets of molecular orbitals. His calculations demonstrated that monocyclic, planar molecules with a closed loop of 2, 6, 10, 14, 18, and so on pi electrons in a fully conjugated system should be aromatic. These numbers are generalized in the **($4n + 2$) pi electron rule,** where n is a positive integer (0, 1, 2, 3, 4, . . .). Conversely, monocyclic, planar molecules with $4n$ pi electrons (4, 8, 12, 16, 20, . . .) are especially unstable and are said to be **antiaromatic.** We have more to say about antiaromaticity shortly. Hückel's **criteria for aromaticity** are summarized as follows. To be aromatic, a compound must

Antiaromatic compound A monocyclic compound that is planar or nearly so, has one p orbital on each atom of the ring, and has $4n$ pi electrons in the cyclic arrangement of p orbitals. Antiaromatic compounds are especially unstable.

Hückel criteria for aromaticity To be aromatic, a monocyclic compound must have one p orbital on each atom of the ring, be planar or nearly so, and have $4n + 2$ pi electrons in the cyclic arrangement of p orbitals.

1. Be cyclic.
2. Have one p orbital on each atom of the ring.
3. Be planar or nearly planar so that there is continuous or nearly continuous overlap of all p orbitals of the ring.
4. Have a closed loop of $4n + 2$ pi electrons in the cyclic arrangement of p orbitals.

To appreciate the reasons for aromaticity and antiaromaticity, we must examine MO energy diagrams for the molecules and ions we will consider in this and the

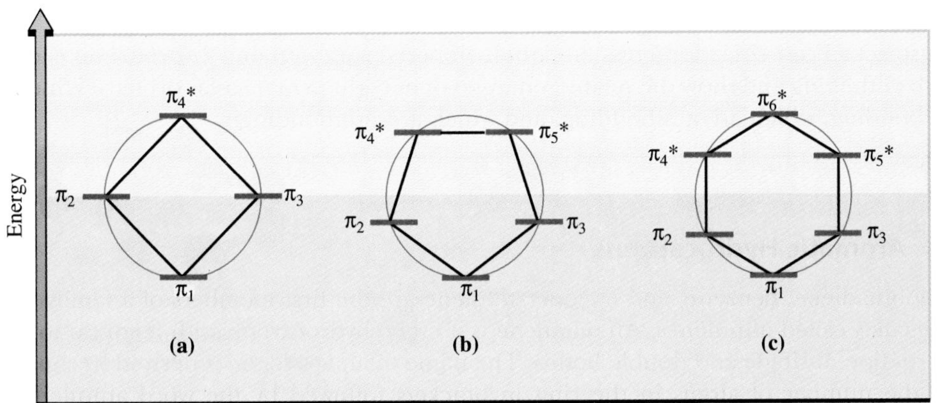

FIGURE 19.4
Frost circles showing the number and relative energies of the pi MOs for planar, fully conjugated four-, five-, and six-membered rings.

following section. The relative energies of the pi MOs for planar, monocyclic, fully conjugated systems can be constructed quite easily using the **Frost circle,** or **inscribed polygon method.** To construct such a diagram, draw a circle and then inscribe in it a polygon of the same number of sides as the ring in question. Inscribe the polygon so that one of its vertices is at the bottom of the circle. The relative energies of the MOs in the ring are then given by the points where the vertices touch the circle. Those MOs below the horizontal line through the center of the circle are bonding MOs. Those on the horizontal line are nonbonding MOs, and those above the line are antibonding MOs. Shown in Figure 19.4 are Frost circles describing the MOs of monocyclic, planar, and fully conjugated four-, five-, and six-membered rings.

Frost circle A graphic method for determining the relative energies of pi MOs for planar, fully conjugated, monocyclic compounds.

EXAMPLE 19.1

Construct a Frost circle for a planar seven-membered ring with one 2*p* orbital on each atom of the ring and show the relative energies of its seven pi molecular orbitals. Which are bonding, which are antibonding, and which are nonbonding?

Solution

Of the seven pi molecular orbitals, three are bonding and four are antibonding.

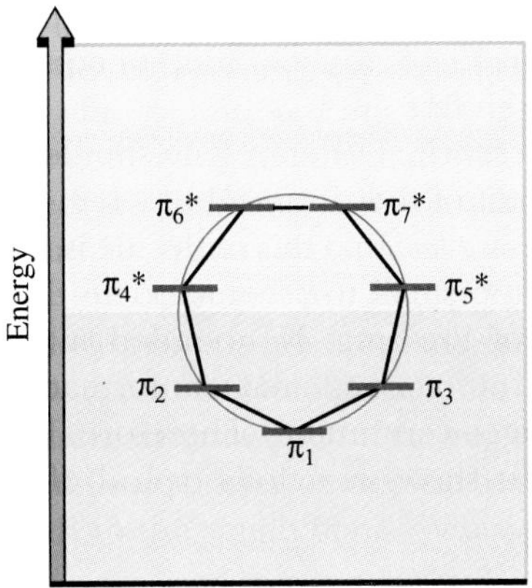

CHEMISTRY IN ACTION

Isolation of Cyclobutadiene

Cyclobutadiene, an antiaromatic compound, is one of the most reactive and unstable compounds in organic chemistry. By far the most unusual approach to stabilizing cyclobutadiene came from D. J. Cram (University of California, Los Angeles), a pioneer in "host-guest chemistry." In host-guest chemistry, the host is a relatively large molecule with an internal cavity that can be occupied by a smaller, guest molecule. Compound 1 is an example of a host molecule called a hemicarcerand (because it is capable of incarcerating guest molecules).

Compound 1

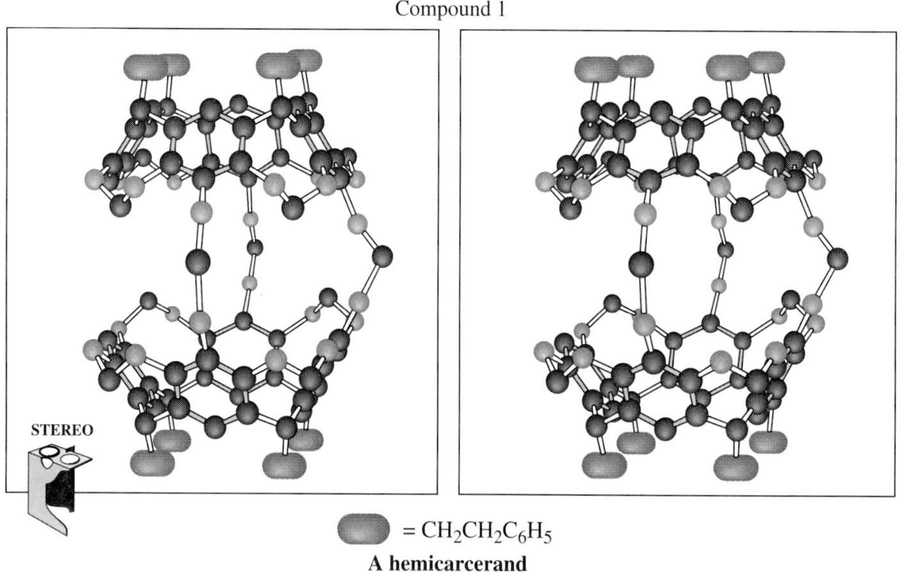

= $CH_2CH_2C_6H_5$

A hemicarcerand

Each molecule meets the Hückel criteria for aromaticity. Each is monocyclic and planar, has one $2p$ orbital on each atom of the ring, and has six electrons in the pi system. In pyridine, nitrogen is sp^2 hybridized, and its unshared pair of electrons occupies an sp^2 orbital perpendicular to the $2p$ orbitals of the pi system and, thus, is not a part of the pi system. In pyrimidine, neither unshared pair of electrons of nitrogen is part of the pi system. The resonance energy of pyridine is 32 kcal/mol (134 kJ/mol), slightly less than that of benzene. The resonance energy of pyrimidine is 26 kcal/mol (108 kJ/mol).

Notice that in Compound 1 there are two bands of four benzene rings, with three of the top benzene rings each linked to a different benzene ring along the bottom band. This molecule behaves like a clam shell. The three pairs of linked benzene rings act as hinges and the fourth benzene rings act as a mouth. At low temperatures, the mouth is closed, but as the temperature is raised to 100°C the molecule undergoes a conformational change, and the mouth opens. Molecules of the right size can enter into the cavity of the hemicarcerand, and, as the temperature is reduced to room temperature, the mouth closes and the guest is trapped inside.

To trap cyclobutadiene inside Compound 1, first a precursor molecule had to be introduced into the cavity. This was done by heating Compound 1 with a large excess of α-pyrone and then cooling the system. When the α-pyrone trapped in the cavity is irradiated with UV light, it loses CO_2 and forms cyclobutadiene.

α-Pyrone —UV (200–250 nm), 25° for 30 min→ Cyclobutadiene + CO_2

The formation of the host-guest complex Compound 1 – (α-pyrone) was established by ^{1}H-NMR spectroscopy and elemental analysis. When the host-guest complex was irradiated with UV light at 25°C, the ^{1}H-NMR signals of the guest α-pyrone disappeared, and a signal for cyclobutadiene appeared δ 2.35. Further evidence that cyclobutadiene was trapped inside the molecular cavity came when the product of irradiation was heated at 220°C. At this temperature, the mouth of the cavity opened, allowing the guest molecule to escape. Heating under these conditions gave Compound 1 and cyclooctatetraene. Cyclooctatetraene is a major dimerization product of cyclobutadiene.

2 Cyclobutadiene → [dimer] —Δ→ Cyclooctatetraene

Additional spectroscopic and chemical evidence clearly established that the highly reactive cyclobutadiene had been trapped inside Compound 1 and was stable at room temperature in this environment (a "molecular bottle"). Because unreactive C—C and C—O bonds lined its molecular cage, cyclobutadiene could undergo no chemical reactions with its surroundings. Like some vicious beast, it could now be studied without being a danger to itself or others.

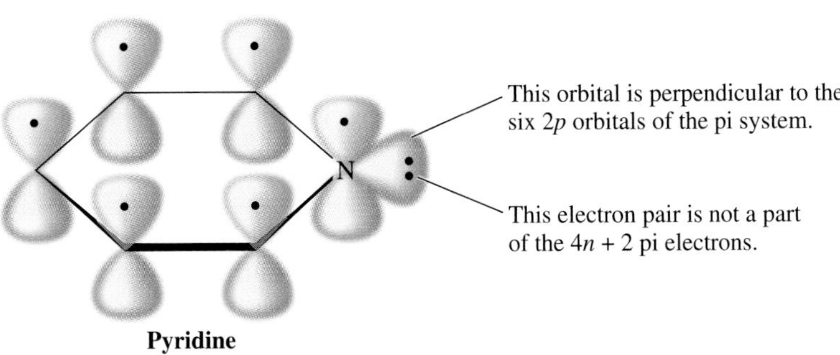
N
This orbital is perpendicular to the six 2p orbitals of the pi system.
This electron pair is not a part of the $4n + 2$ pi electrons.
Pyridine

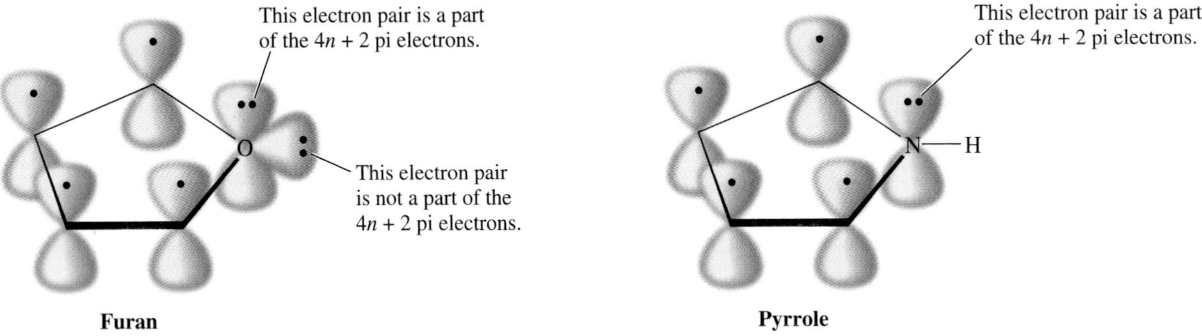
This electron pair is a part of the 4n + 2 pi electrons.
This electron pair is not a part of the 4n + 2 pi electrons.
Furan
This electron pair is a part of the 4n + 2 pi electrons.
Pyrrole

FIGURE 19.7
Origin of the $4n + 2$ pi electrons in furan and pyrrole. The resonance energy of furan is 16 kcal/mol (67 kJ/mol); the resonance energy of pyrrole is 21 kcal/mol (88 kJ/mol).

The five-membered ring heterocyclic compounds furan, thiophene, pyrrole, and imidazole are also aromatic.

Furan   Thiophene   Pyrrole   Imidazole

In these planar compounds, each heteroatom is sp^2 hybridized, and its unhybridized $2p$ orbital is part of a continuous cycle of five p orbitals. In furan and thiophene, one unshared pair of electrons of the heteroatom lies in the unhybridized $2p$ orbital and is a part of the pi system (Figure 19.7). The other unshared pair of electrons lies in an sp^2 hybrid orbital perpendicular to the $2p$ orbitals, and is not a part of the pi system. In pyrrole, the unshared pair of electrons on nitrogen is part of the aromatic sextet. In imidazole, the unshared pair on one nitrogen is part of the aromatic sextet, and the unshared pair on the other nitrogen is not.

Nature abounds with compounds having a heterocyclic ring fused to one or more other rings. Two such compounds especially important in the biological world are indole and purine.

Indole   Serotonin (a neurotransmitter)   Purine   Caffeine

Indole contains a pyrrole ring fused with a benzene ring. Compounds derived from indole include the essential amino acid L-tryptophan (Section 27.1C) and the neurotransmitter serotonin. Purine contains a six-membered pyrimidine ring fused with a five-membered imidazole ring. Caffeine is a trimethyl derivative of an oxidized purine. Compounds derived from purine and pyrimidine are building blocks of deoxyribonucleic acids (DNA) and ribonucleic acids (RNA).

Coffee beans, a source of caffeine. *(Dr. E. R. Degginger)*

E. Aromatic Hydrocarbon Ions

Any neutral monocyclic unsaturated hydrocarbon with an odd number of carbons in the ring must of necessity have at least one $—CH_2—$ group in the ring and, therefore, cannot be aromatic. Examples of such hydrocarbons are cyclopropene, cyclopentadiene, and cycloheptatriene.

Cyclopropene Cyclopentadiene Cycloheptatriene

Cyclopropene has the correct number of pi electrons to be aromatic, namely $4(0) + 2 = 2$, but it does not have a continuous closed loop of $2p$ orbitals. If, however, the $—CH_2—$ group becomes a $—CH^+—$ group in which the carbon atom is sp^2 hybridized and has a vacant $2p$ orbital, then the overlap of orbitals is continuous and, according to molecular orbital theory, the **cyclopropenyl cation** should be aromatic. The cyclopropenyl cation can be drawn as a resonance hybrid of three equivalent contributing structures. That we can draw three equivalent contributing structures is not the key to the aromaticity of this cation; the key is that it meets the Hückel criteria of aromaticity. Cyclopropenyl cation may also be represented by a triangle with an inscribed circle and plus sign.

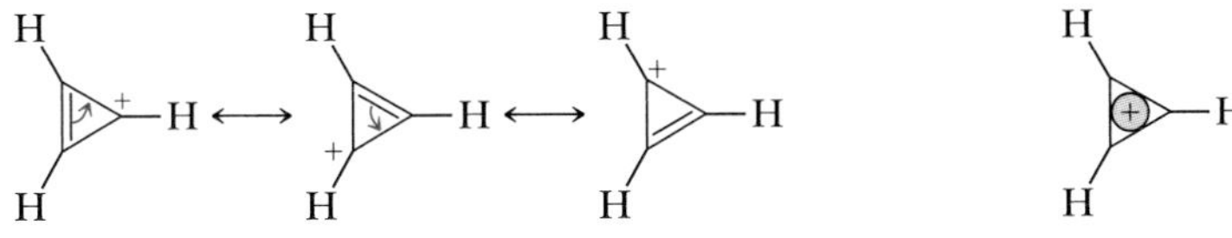

Cyclopropenyl cation represented as a hybrid of three equivalent contributing structures

Aromaticity of the cyclopropenyl cation shown by an inscribed circle and plus sign

As an example of the aromatic stabilization of this cation, 3-chlorocyclopropene reacts readily with antimony(V) chloride to form a stable salt.

$+ \; SbCl_5 \longrightarrow$ $H\ SbCl_6^-$

3-Chlorocyclopropene (Cyclopropenyl chloride) | Antimony(V) chloride (a Lewis acid) | Cyclopropenyl hexachloroantimonate

This chemical behavior is to be contrasted with that of 5-chloro-1,3-cyclopentadiene, which cannot be made to form a stable salt. In fact, a cyclic, planar, conjugated cyclopentadienyl cation has four pi electrons and, if it were to be synthesized, it would be antiaromatic.

$+ \; AgBF_4 \nrightarrow$ $H\ BF_4^- + AgCl$

5-Chloro-1,3-cyclopentadiene | Cyclopentadienyl tetrafluoroborate

Note that it is possible to draw five equivalent contributing structures for the cyclopentadienyl cation. Yet this cation is not aromatic because it has only $4n$ pi electrons rather than the required $4n + 2$ pi electrons.

To form an aromatic ion from cyclopentadiene, it is necessary to convert the CH_2 group to a CH^- group in which the carbon becomes sp^2 hybridized and has two electrons in its unhybridized $2p$ orbital. The resulting **cyclopentadienyl anion** is aromatic. Its aromatic character may also be represented by an inscribed circle with a minus sign.

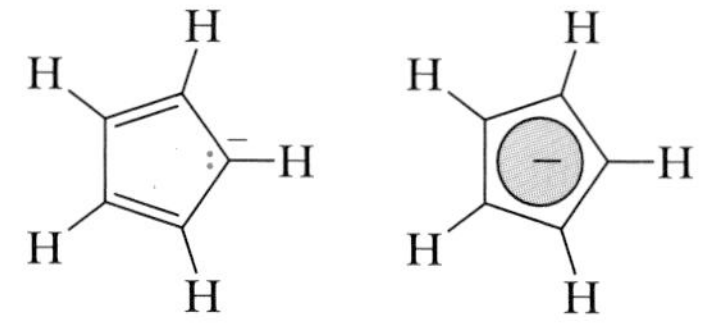

Cyclopentadienyl anion
(aromatic)

Evidence of the stability of this anion is the fact that cyclopentadiene has a pK_a of approximately 16.0 and is one of the most acidic hydrocarbons known. The acidity of cyclopentadiene is comparable to that of water ($pK_a = 15.7$). Consequently, when cyclopentadiene is treated with aqueous sodium hydroxide, an equilibrium is established in which some of the hydrocarbon is converted to its aromatic anion. K_{eq} for this equilibrium is approximately 0.5. Cyclopentadiene is converted completely to its anion by treatment with sodium amide.

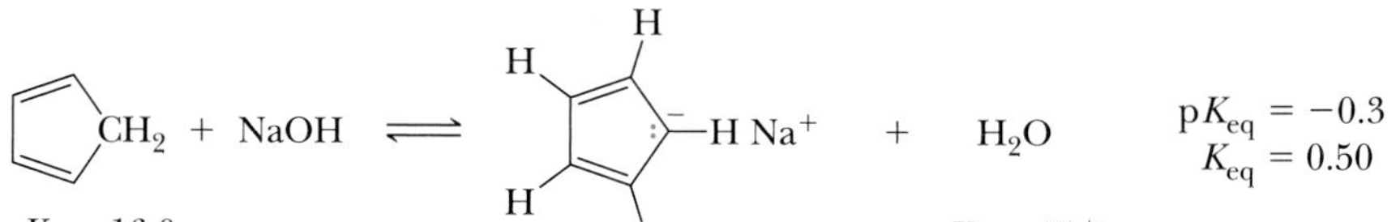

CH_2 + NaOH ⇌ H Na^+ + H_2O $pK_{eq} = -0.3$, $K_{eq} = 0.50$

$pK_a = 16.0$ $pK_a = 15.7$

CH_2 + $NaNH_2$ ⟶ H Na^+ + NH_3

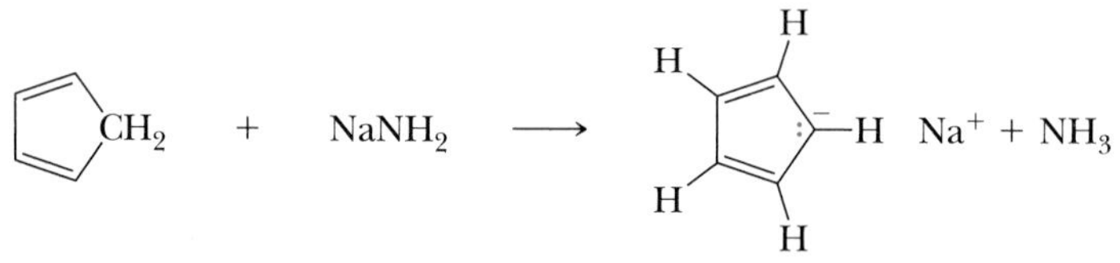

Cyclopentadiene Cyclopentadienyl sodium

EXAMPLE 19.2

Construct a MO energy diagram for the cyclopentadienyl anion and describe its ground-state electron configuration.

Solution

Refer to the Frost circle shown in Figure 19.1 for a planar, fully conjugated five-membered ring. The 6 pi electrons occupy the π_1, π_2, and π_3 molecular orbitals, all of which are bonding MOs.

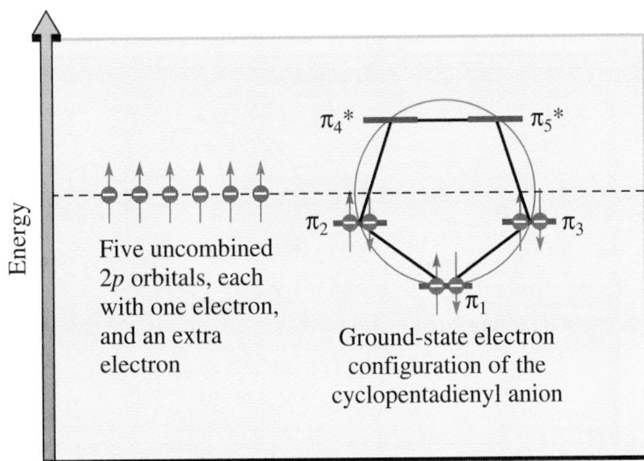

PROBLEM 19.2

Describe the ground-state electron configuration of the cyclopentadienyl cation and radical. Assuming each species is planar, would you expect it to be aromatic or antiaromatic?

Cycloheptatriene forms an aromatic cation by conversion of its CH_2 group to a CH^+ group with this sp^2 hybridized carbon having a vacant $2p$ atomic orbital. The **cycloheptatrienyl (tropylium) cation** is planar and has six pi electrons in seven $2p$ orbitals, one from each atom of the ring. It can be drawn as a resonance hybrid of seven equivalent contributing structures.

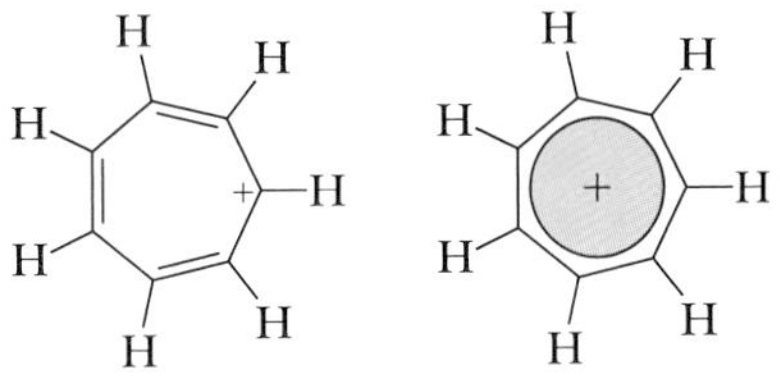

Cycloheptatrienyl cation
(tropylium ion)
(aromatic)

EXAMPLE 19.3

Construct a MO energy diagram for the cycloheptatrienyl cation and describe its ground-state electron configuration.

Solution

Refer to the Frost circle you constructed in answer to Example 19.1. In the ground-state electron configuration of the cycloheptatrienyl cation, the six pi electrons occupy the π_1, π_2, and π_3 molecular orbitals, all of which are bonding.

The mass spectra of toluene and most other alkylbenzenes show a fragment ion of m/z 91. Although it might seem that the most likely structure for this ion is that of the benzyl cation, experimental evidence suggests a molecular rearrangement to form the more stable tropylium ion (Section 19.2E). In the tropylium ion, an aromatic cation, the positive charge is delocalized equally over all seven carbon atoms of the cycloheptatrienyl ring.

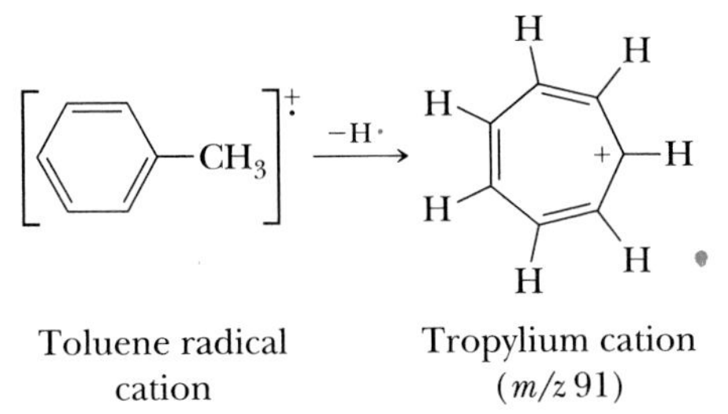

$$[C_6H_5-CH_3]^{\dot{+}} \xrightarrow{-H\cdot} C_7H_7^+$$

Toluene radical cation

Tropylium cation (m/z 91)

B. NMR Spectroscopy

All six hydrogens of benzene are equivalent and their signal appears in its ^{1}H-NMR spectrum as a sharp singlet at δ 7.27. Hydrogens attached to a substituted benzene ring appear in the region δ 6.5 to δ 8.5. Few other hydrogens absorb in this region and, thus, aryl hydrogens are quite easily identifiable by their distinctive chemical shifts.

Recall that vinylic hydrogens (hydrogens attached to a carbon of carbon-carbon double bond) give signals in the range δ 4.6–δ 5.7 (Section 13.15B). That aryl hydrogens absorb even farther downfield than vinylic hydrogens is accounted for by the **ring current,** a special property of aromatic rings (Figure 19.9). When the plane of an aromatic ring is oriented perpendicularly to an applied magnetic field, the applied field causes the pi electrons to circulate around the ring, which constitutes the so-called "ring current." This induced ring current has associated with it a magnetic field that opposes the applied field in the middle of the ring but reinforces the applied field on the outside of the ring. Thus, given the position of aromatic hydrogens relative to the induced ring current, they come into resonance at a lower applied field, that is, at a larger chemical shift.

Induced circulation of pi electrons in the aromatic ring
Induced local magnetic field of the circulating pi electrons reinforces the applied field and provides part of the field necessary to bring aromatic hydrogens into resonance
Applied field

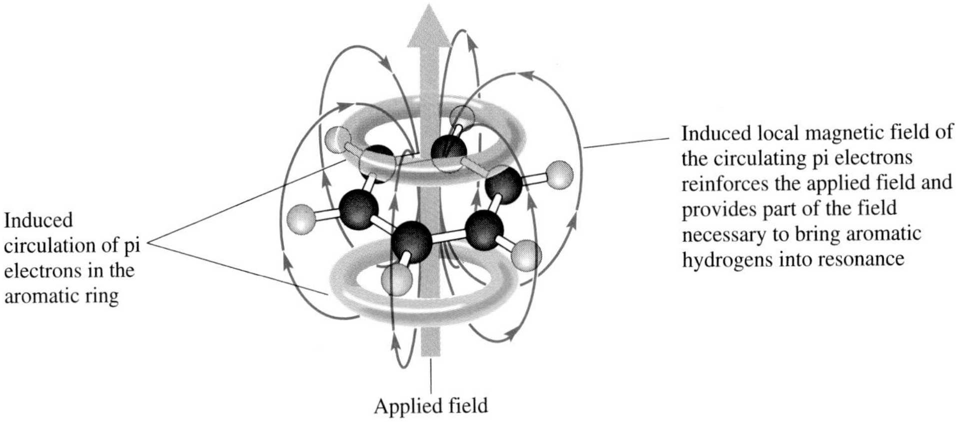

FIGURE 19.9 The magnetic field induced by circulation of pi electrons in the aromatic ring reinforces the applied field near aromatic hydrogens with the result that they come into resonance at a lower applied field (at a larger chemical shift relative to the TMS standard).

Note that the increase in chemical shift of aromatic hydrogens arises in much the same way as the increase in chemical shift of vinylic hydrogens (Figure 13.9). The increase is larger for aromatic hydrogens, however, by approximately δ 1.5 to 2.5. The effect of induced ring current is characteristic not only of benzene and its derivatives, but also of all compounds that meet the Hückel criteria for aromaticity (Section 19.2A). Note that this concept of a circulating ring current and of an induced magnetic field correctly predicts that hydrogen atoms on the outside of the ring should come into resonance with a downfield shift. It also predicts that a hydrogen atom in the inside of the ring should come into resonance farther upfield. Of course, no hydrogens are on the inside of the benzene ring, but with larger aromatic annulenes, as for example, [18]annulene, there are both "inside" hydrogens and "outside" hydrogens. The degree of the upfield chemical shift of the inside hydrogens of [18]annulene is remarkable. They come into resonance at $\delta - 3.00$, that is at 3.00 δ units to the right of the TMS standard.

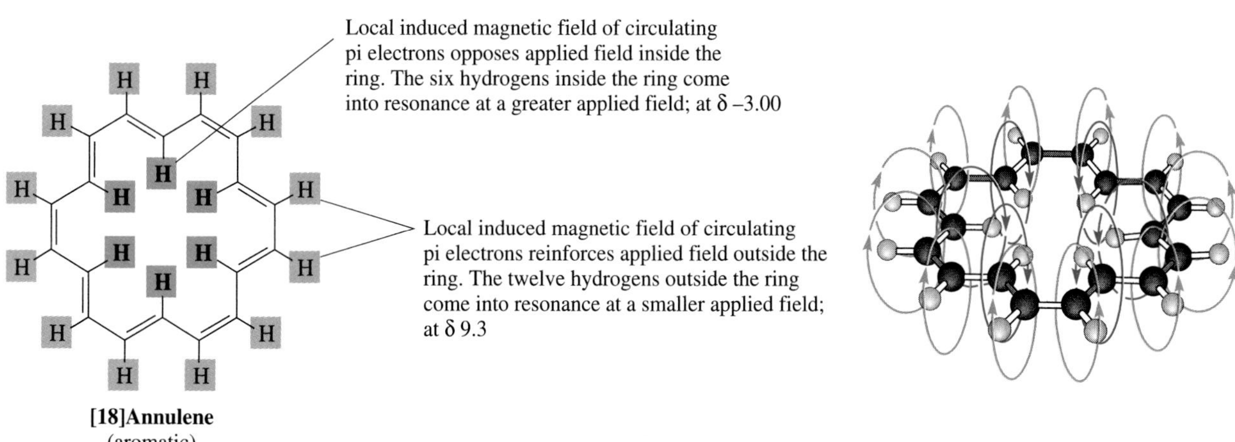

Local induced magnetic field of circulating pi electrons opposes applied field inside the ring. The six hydrogens inside the ring come into resonance at a greater applied field; at δ –3.00
Local induced magnetic field of circulating pi electrons reinforces applied field outside the ring. The twelve hydrogens outside the ring come into resonance at a smaller applied field; at δ 9.3
[18]Annulene
(aromatic)

EXAMPLE 19.5

Which hydrogens have a larger chemical shift, the six hydrogens of benzene or the eight hydrogens of cyclooctatetraene? Explain.

Solution

Benzene is an aromatic compound; its six equivalent hydrogens appear as a sharp singlet at δ 7.27. Cyclooctatetraene does not meet the Hückel criteria for aromaticity because it has $4n$ pi electrons and is nonplanar. Therefore, the eight equivalent hydrogens of the cyclooctatetraene ring appear as a singlet at δ 5.8 in the region of vinylic hydrogens (δ 4.6 to δ 5.7).

PROBLEM 19.5

Which compound gives a signal in the ^{1}H-NMR spectrum at a lower applied field (with a larger chemical shift): furan or cyclopentadiene? Explain.

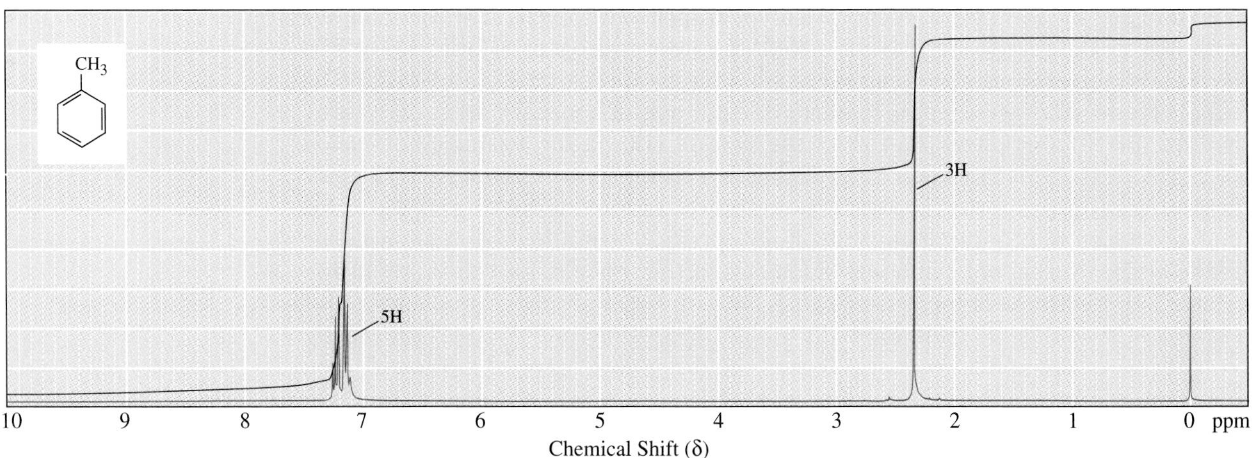

FIGURE 19.10
^{1}H-NMR spectrum of toluene.

In monosubstituted benzenes in which the ring substituent is neither strongly electron-withdrawing nor electron-releasing, as for example the alkylbenzenes, all ring hydrogens have very similar chemical shifts. The ^{1}H-NMR spectrum of toluene (Figure 19.10) shows a 3H singlet at δ 2.3 for the three hydrogens of the methyl group and a 5H closely spaced multiplet at δ 7.3 for the five hydrogens of the aromatic ring. The signal for the methyl hydrogens appears at δ 2.3, about 1.4 farther downfield than the methyl hydrogens of an alkane. This downfield shift, like that of aryl hydrogens, is due to the effect of the induced ring current and its associated magnetic field.

When a substituent is strongly electron-releasing or electron-withdrawing, the ortho, meta, and para hydrogens have different chemical shifts, and ^{1}H-NMR spectra become significantly more complex. The ^{1}H-NMR signal for the five aromatic hydrogens of anisole, for example, consists of two complex multiplets in the ratio of 3:2. The two ortho hydrogens and the one para hydrogen are more shielded and come into resonance at the higher applied field (smaller chemical shift); the two meta hydrogens are more deshielded and come into resonance at a lower applied field (larger chemical shift). We will not attempt to analyze these complex splitting patterns here.

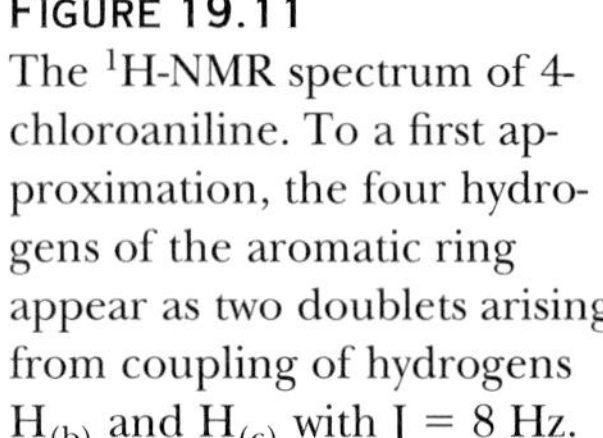

FIGURE 19.11
The ^{1}H-NMR spectrum of 4-chloroaniline. To a first approximation, the four hydrogens of the aromatic ring appear as two doublets arising from coupling of hydrogens $H_{(b)}$ and $H_{(c)}$ with J = 8 Hz.

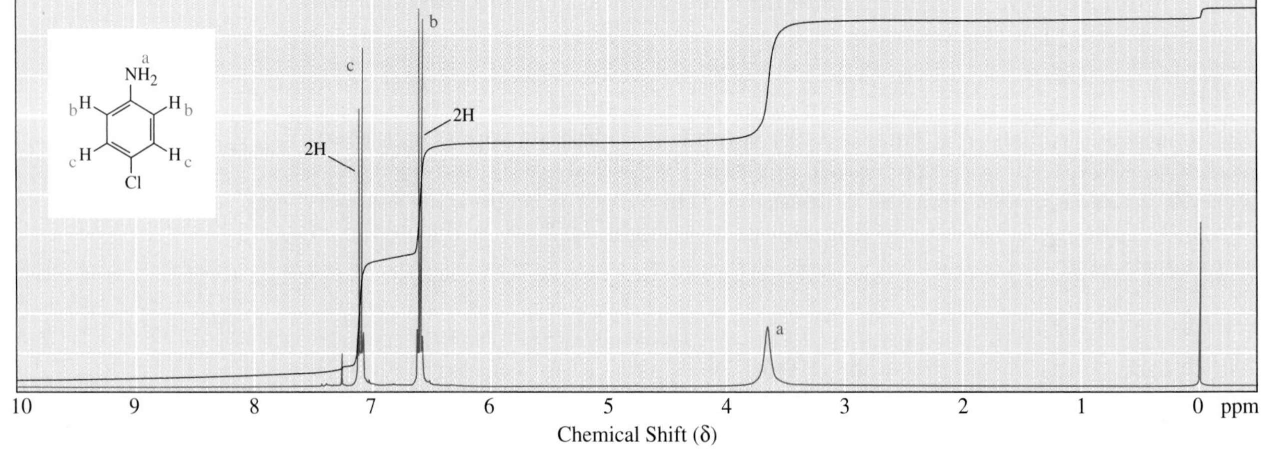

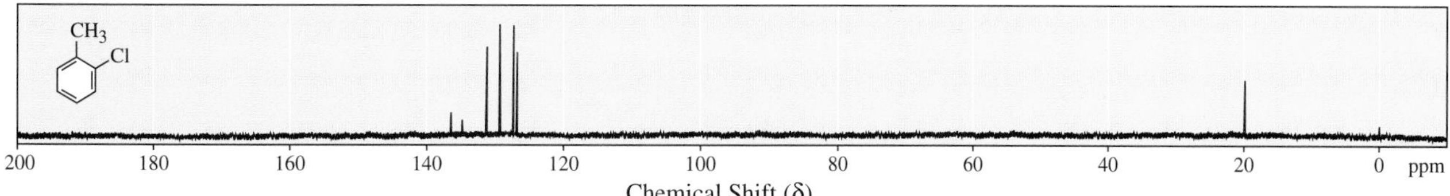

FIGURE 19.12
^{13}C-NMR spectrum of 2-chlorotoluene.

The splitting patterns of di-, tri-, and polysubstituted benzene rings are very complex. One splitting pattern, however, is quite easy to recognize, namely, that of a para-disubstituted benzene ring. If the two substituents are of sufficiently different electronic effects, then to a first approximation, the spectrum appears as a pair of doublets. Shown in Figure 19.11 is the ^{1}H-NMR spectrum of 4-chloroaniline.

singlet δ 3.65
NH_2 (a)
doublet δ 6.58
(b) H H (b)
(c) H H (c)
Cl
doublet δ 7.10

The higher field ^{1}H-NMR spectrum (200 MHz or higher) of 4-chloroaniline is considerably more complex than just a pair of doublets because of long-range coupling. It is not our purpose in this text to deal with this level of complexity.

In ^{13}C-NMR spectroscopy, carbon atoms of aromatic rings appear in the range δ 110–160. Benzene, for example, shows a single signal at δ 128. Because carbon-13 signals for alkene carbons also appear in the range δ 110–160, it is generally not possible to establish the presence of an aromatic ring by ^{13}C-NMR spectroscopy alone. ^{13}C-NMR spectroscopy is particularly useful, however, in establishing substitution patterns of aromatic rings. The ^{13}C-NMR spectrum of 2-chlorotoluene (Figure 19.12) shows six signals in the aromatic region; its more symmetric isomer 4-chlorotoluene (Figure 19.13) shows four signals in the aromatic region. Thus, all one needs to do is count signals to distinguish between these constitutional isomers.

C. Infrared Spectroscopy

Aromatic rings show a medium to weak peak in the C—H stretching region at approximately 3030 cm^{-1} characteristic of *sp*^{2} C—H bonds. In addition, aromatic rings show strong absorption in the region 690 to 900 cm^{-1} due to out-of-plane C—H bending. Finally, aromatic rings show several absorptions due to C=C stretching be-

FIGURE 19.13
^{13}C-NMR spectrum of 4-chlorotoluene.

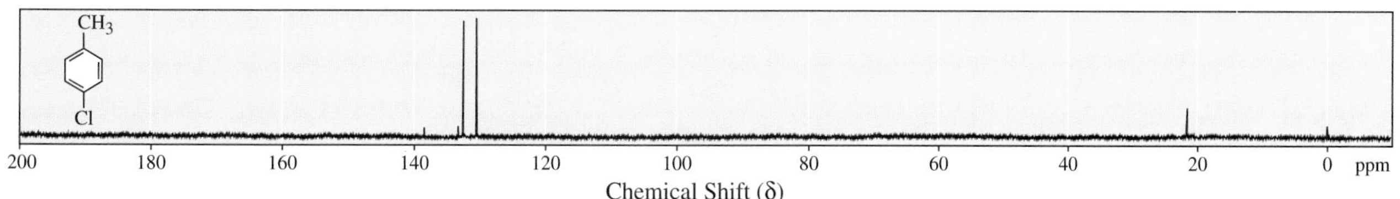

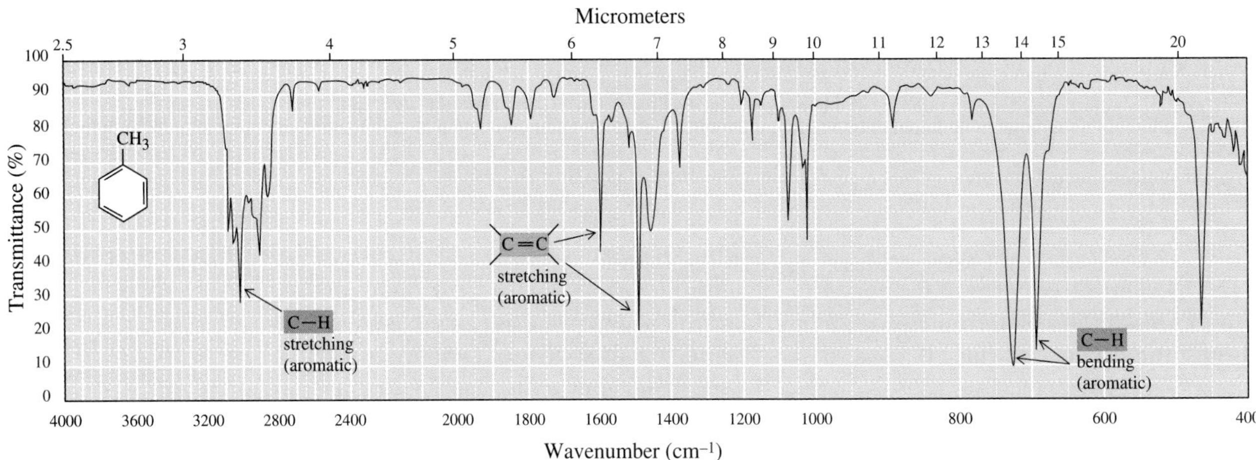

FIGURE 19.14
Infrared spectrum of toluene.

tween 1450 cm^{-1} and 1600 cm^{-1}. Each of these characteristic absorption patterns can be seen in the infrared spectrum of toluene (Figure 19.14).

Aromatic and vinyl ethers typically show two C—O stretching vibrations, one at either end of the range for C—O stretching. Anisole (Figure 19.15), for example, shows C—O stretching vibrations at 1050 cm^{-1} (sp^3 C—O) and 1250 cm^{-1} (sp^2 C—O).

D. Ultraviolet-Visible Spectroscopy

Aromatic rings absorb radiation in the ultraviolet region of the spectrum as a result of π to π^* transitions (Figure 19.1). Most commonly, two broad absorptions occur: the first of high intensity near 205 nm and a second, less intense absorption near 270 nm. The presence of these two absorptions in the ultraviolet spectrum is clear evidence for the presence of an aromatic ring. Shown in Table 19.1 are values of λ_{max} for benzene and several of its monosubstituted derivatives.

FIGURE 19.15
Infrared spectrum of anisole.

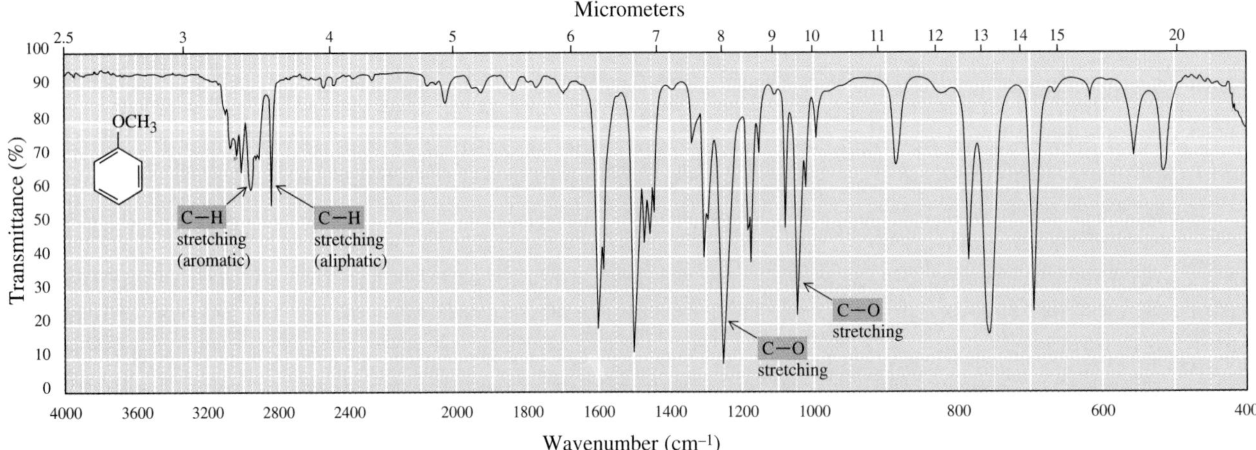

TABLE 19.1 Ultraviolet Maxima for Several Aromatic Compounds ($C_6H_5—R$)

R	λ_{max} (nm)	ϵ	λ_{max} (nm)	ϵ
—H	203	7,400	254	204
$—CH_3$	207	7,000	261	225
—OH	211	6,200	270	1,450
$—CO_2H$	230	11,600	273	970

19.5 Phenols

A. Structure and Nomenclature

The functional group of a **phenol** is a hydroxyl group bonded directly to a benzene ring. Following is the structural formula of phenol, the simplest member of this class of compounds. Other phenols are named either as derivatives of phenol, as benzenols, or by common names.

Phenol A compound that contains an —OH bonded to a benzene ring; a benzenol.

Phenol	3-Methylphenol (*m*-Cresol)	1,2-Benzenediol (Catechol)	1,3-Benzenediol (Resorcinol)	1,4-Benzenediol (Hydroquinone)

Phenols are widely distributed in nature. Phenol itself and the isomeric cresols (*o*-, *m*-, and *p*-cresol) are found in coal tar and petroleum. Thymol and vanillin are important constituents of thyme and vanilla beans.

2-Isopropyl-5-methylphenol (Thymol)

4-Hydroxy-3-methoxybenzaldehyde (Vanillin)

Thymol is a constituent of garden thyme, *Thymus vulgaris*. (© *Connie Toops*)

Phenol, or carbolic acid as it was once called, is a low-melting solid only slightly soluble in water. In sufficiently high concentrations, it is corrosive to all kinds of cells. In dilute solutions, it has some antiseptic properties, and was introduced into the practice of surgery by Joseph Lister who demonstrated his technique of aseptic surgery in the surgical theater of the University of Glasgow School of Medicine in 1865. Phenol's medical use is now limited. It has been replaced by antiseptics that are both more powerful and have fewer undesirable side effects. Among these is *n*-hexylresorcinol, which is widely used in nonprescription preparations as a mild antiseptic and disin-

The plant *Eugenia aromatica* from which eugenol is isolated. Eugenol is used in dentistry as an antiseptic agent. *(G. Büttner/Naturbid/OKAPIA/ Photo Researchers, Inc.)*

fectant. Eugenol, which can be isolated from the flower buds (cloves) of *Eugenia aromatica*, is used as a dental antiseptic and analgesic.

[Structures: *n*-Hexylresorcinol (benzene ring with OH, OH, and $CH_2(CH_2)_4CH_3$); Eugenol (benzene ring with OH, OCH_3, and $CH_2CH{=}CH_2$)]

n-Hexylresorcinol — Eugenol

B. Acidity of Phenols

Phenols and alcohols both contain a hydroxyl group, —OH. However, phenols are grouped as a separate class of compounds because their chemical properties are quite different from those of alcohols. One of the most important of these differences is that phenols are significantly more acidic than alcohols. The acid ionization constant of phenol is 10^6 times larger than that of ethanol.

$$C_6H_5{-}OH + H_2O \rightleftharpoons C_6H_5{-}O^- + H_3O^+ \qquad K_a = 1.1 \times 10^{-10} \qquad pK_a = 9.95$$

$$CH_3CH_2OH + H_2O \rightleftharpoons CH_3CH_2O^- + H_3O^+ \qquad K_a = 1.3 \times 10^{-16} \qquad pK_a = 15.9$$

Another way to compare the relative acid strengths of ethanol and phenol is to look at the hydrogen ion concentration and pH of a 0.1 *M* aqueous solution of each (Table 19.2). For comparison, the hydrogen ion concentration and pH of 0.1 *M* HCl are also included.

In aqueous solution, alcohols are neutral substances, and the hydrogen ion concentration of 0.1 *M* ethanol is the same as that of pure water. A 0.1 *M* solution of phenol is slightly acidic and has a pH of 5.4. By contrast, 0.1 *M* HCl, a strong acid (completely ionized in aqueous solution), has a pH of 1.0.

We account for the enhanced acidity of phenols compared with alcohols in the following way. First, the electron-withdrawing inductive effect of the sp^2 hybridized carbon weakens the O—H bond of the phenol compared with the effect of the sp^3 hybridized carbon of the alcohol. Second, delocalization of charge by resonance stabilizes the phenoxide ion relative to the alkoxide ion. Delocalization of charge in the phenoxide ion can be seen in the following contributing structures. In the two contributing structures on the left, the negative charge is placed on oxygen. The three contributing structures on the right place the negative charge on the ortho and para

TABLE 19.2 Relative Acidities of 0.1 *M* Solutions of Ethanol, Phenol, and HCl

Acid Ionization Equation	$[H^+]$	pH
$CH_3CH_2OH + H_2O \rightleftharpoons CH_3CH_2O^- + H_3O^+$	1×10^{-7}	7.0
$C_6H_5OH + H_2O \rightleftharpoons C_6H_5O^- + H_3O^+$	3.3×10^{-6}	5.4
$HCl + H_2O \rightleftharpoons Cl^- + H_3O^+$	0.1	1.0

Chemistry in Action

Capsaicin: For Those Who Like It Hot

Capsaicin, the pungent principle from the fruit of various species of peppers (Capsicum and Solanaceae), was isolated in 1876 and its structure was determined in 1919.

Capsaicin
(from various types of peppers)

The inflammatory properties of capsaicin are well known; as little as one drop in 5 L of water can be detected by the human tongue. We all know of the burning sensation in the mouth and sudden tearing in the eyes caused by a good dose of hot chili peppers. Capsaicin-containing extracts from these flaming foods are also used in sprays to ward off dogs or other animals that might nip at your heels while you are running or cycling.

Ironically, capsaicin is able to cause pain and relieve it as well. Currently, two capsaicin-containing creams, Mioton and Zostrix, are prescribed to treat the burning pain associated with postherpetic neuralgia, a complication of shingles. They are also prescribed for diabetics to relieve persistent foot and leg pain.

The mechanism by which capsaicin relieves these pains is not fully understood. It has been suggested, however, that, after application, the nerve endings in the area responsible for the transmission of pain remain temporarily numb. Capsaicin remains bound to specific receptor sites on these pain-transmitting neurons, blocking them from further action. Eventually, capsaicin is removed from these receptor sites, but in the meantime, its presence provides needed relief from pain.

Peppers of the capsicum family. (Dr. E. R. Degginger)

positions of the ring. These contributing structures spread the negative charge of the phenoxide ion so that it is delocalized over four atoms. It is this delocalization of the negative charge that stabilizes the phenoxide ion relative to an alkoxide ion, for which no delocalization is possible.

These two Kekulé structures are equivalent

These three contributing structures delocalize the negative charge onto carbon atoms of the ring

CHEMISTRY IN ACTION

Phytoalexins: Natural Plant Antibiotics

Insect pests and fungal infections cause enormous losses in food and fiber crops throughout the world. Even though these losses attract attention, they often obscure the fact that resistance rather than susceptibility is the rule in nature. Although all plants are exposed to a very wide range of potentially pathogenic fungi, they are completely resistant to most of them.

A sweet pea plant, Lathyrus latifolius. *(Runk/Schoenberger, from Grant Heilman)*

Scientists have discovered that one mechanism of resistance is a build-up of protective chemicals at the site of attempted fungal infection. These natural plant antibiotics are given the name phytoalexins, a word derived from the Greek *phyto* (a plant) and *alexein* (to ward off, protect). While systematic investigations of these natural plant defense mechanisms are only now being carried out, it is clear that phytoalexin synthesis is a common mechanism of disease resistance in higher plants.

Among the several hundred phytoalexins characterized to date, there are wide variations in structure. Pisatin, for example, a fungitoxin induced in the sweet pea, has five phenol ether groups within its structure. Wyerone, from the broad bean, contains a carbon-carbon triple bond.

CH_3O … O … OH … O … O … O

Pisatin
(from the sweet pea)

$H(H)C{=}C(CH_2CH_3)-C{\equiv}C-C(=O)-$(furan)$-CH{=}CH-COCH_3$ (C=O)

Wyerone
(from the broad bean)

following examples illustrate the Williamson synthesis of alkyl-aryl ethers. The synthesis of allyl phenyl ether involves nucleophilic displacement on a primary halide; the synthesis of anisole illustrates the use of dimethyl sulfate as a methylating agent.

$$C_6H_5{-}OH + CH_2{=}CHCH_2Cl \xrightarrow[Bu_4N^+Br^-]{NaOH,\ H_2O,\ CH_2Cl_2} C_6H_5{-}OCH_2CH{=}CH_2$$

Phenol — 3-Chloropropene (Allyl chloride) — Phenyl 2-propenyl ether (Allyl phenyl ether)

The most thoroughly studied phytoalexins, those of cotton *(Gossypium),* include the terpene hemigossypol, its oxidation product gossypol (the yellow pigment of cottonseed), and a host of related terpene compounds. In addition to its role as a phytoalexin, gossypol is toxic to a variety of herbivores and various insects. There have also been reports from the People's Republic of China that it acts as a male infertility agent in humans.

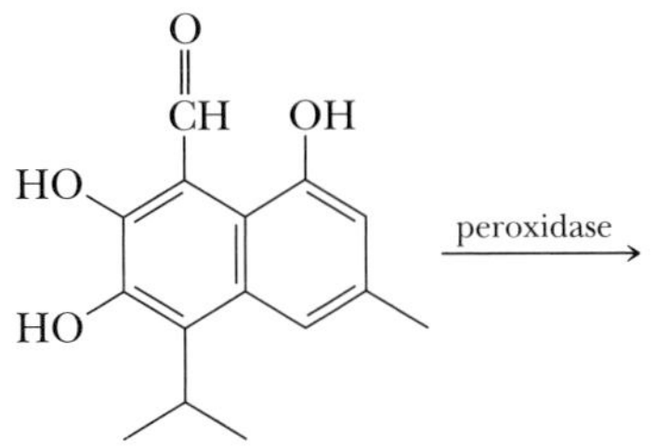

peroxidase

Hemigossypol
(from cotton)

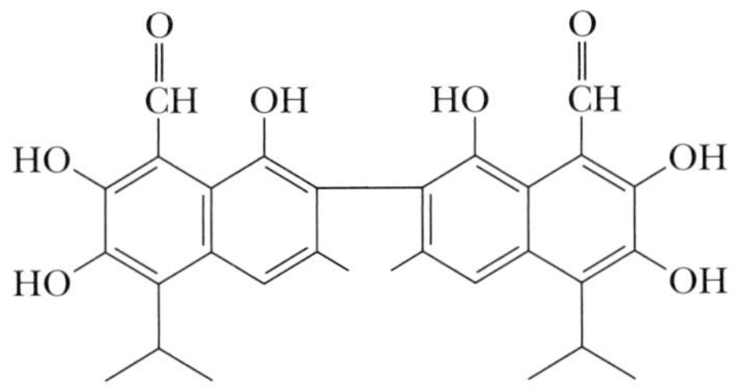

Gossypol
(from cotton)

The failure of these plant antifungal agents to halt the progress of certain pathogenic fungi is not necessarily an indication of their ineffectiveness. What it means, instead, is that fungal parasites have coevolved with the plant and developed a biochemical machinery to detoxify the plant phytoalexins as soon as they are produced.

When we breed new resistant varieties of plants, for example, perhaps we are selecting an ability to produce more effective phytoalexins. If we can understand the biochemistry of their production, perhaps we can select and then transfer this ability by genetic engineering from resistant to susceptible varieties. Alternatively, an understanding of the chemical structure of phytoalexins and their modes of action may provide us with clues to the development of new and safer synthetic fungicides.

See M. B. Green and P. A. Hedin, eds., "Natural Resistance of Plants to Pests; Roles of Allelochemicals," *ACS Symposium Series,* no. 387, American Chemical Society, Washington, DC, 1986.

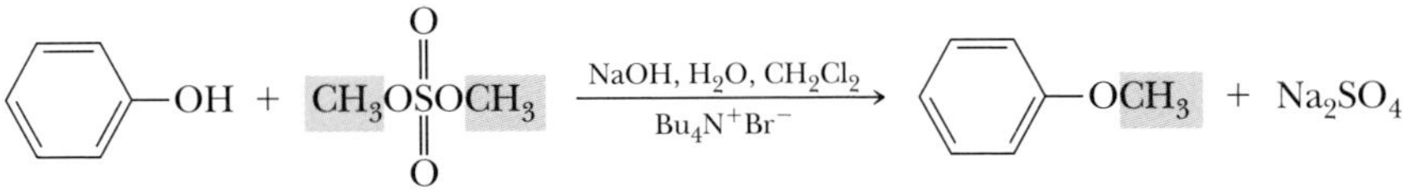

$$C_6H_5OH + CH_3OSO_2OCH_3 \xrightarrow[Bu_4N^+Br^-]{NaOH,\ H_2O,\ CH_2Cl_2} C_6H_5OCH_3 + Na_2SO_4$$

Phenol　Dimethyl sulfate　Methyl phenyl ether (Anisole)

An alkyl aryl ether, ArOR, is cleaved by hydrohalic acids, HX, to form an alkyl halide and a phenol. This illustrates the fact that nucleophilic substitution is not likely to occur at an aromatic carbon and that phenols, unlike alcohols, are not converted to aryl halides by treatment with concentrated HCl, HBr, or HI.

$$C_6H_5{-}OCH(CH_3)_2 + HI \longrightarrow C_6H_5{-}OH + (CH_3)_2CHI$$

2-Phenoxypropane (Isopropyl phenyl ether) — Phenol — 2-Iodopropane (Isopropyl iodide)

E. Kolbe Carboxylation; Synthesis of Salicylic Acid

Phenoxide ions react with carbon dioxide to give a carboxylic acid salt as shown by the industrial synthesis of salicylic acid, the starting material for the production of aspirin (Section 17.6B). Phenol is dissolved in aqueous NaOH, and this solution is then saturated with CO_2 under pressure to give sodium salicylate. This process is referred to as high-pressure carboxylation of sodium phenoxide. Upon acidification of the alkaline solution, salicylic acid is isolated as a solid, mp 157–159°C.

Phenol $\xrightarrow[H_2O]{NaOH}$ Sodium phenoxide ($C_6H_5O^-Na^+$) $\xrightarrow[H_2O]{CO_2}$ Sodium salicylate (o-$HOC_6H_4CO^-Na^+$, with C=O) $\xrightarrow[H_2O]{HCl}$ Salicylic acid (o-HOC_6H_4COH, with C=O)

The importance of salicylic acid in industrial organic chemistry is demonstrated by the fact that over 3 million pounds of aspirin are synthesized in the United States each year.

The reaction between sodium phenoxide and carbon dioxide begins by nucleophilic attack of the phenoxide anion on the carbonyl of carbon dioxide to give a cyclohexadienone intermediate which then undergoes keto-enol tautomerism (Section 15.11B) to give the salicylate anion. Note that the enol in this case, due to its aromatic character, is the more stable of the two tautomers.

MECHANISM Kolbe Carboxylation of Phenol

Sodium phenoxide + O=C=O ⟶ [A cyclohexadienone intermediate (C=O ring, C(H)—C(=O)—O⁻)] $\xrightarrow{\text{keto-enol tautomerism}}$ Salicylate anion (o-$HOC_6H_4CO^-$, with C=O)

Sodium phenoxide — A cyclohexadienone intermediate — Salicylate anion

F. Oxidation to Quinones

Because of the presence of the electron-donating —OH group on the ring, phenols are susceptible to oxidation by a variety of strong oxidizing agents. For example, oxidation of phenol itself by potassium dichromate gives 1,4-benzoquinone (*p*-quinone). By definition, a quinone is a cyclohexadienedione. Those with carbonyl groups ortho to each other are called *o*-quinones; those with carbonyl groups para to each other are called *p*-quinones.

Phenol $\xrightarrow[H_2SO_4]{K_2Cr_2O_7}$ 1,4-Benzoquinone (*p*-Quinone)

Quinones can also be obtained by oxidation of 1,2-benzenediol (catechol) or 1,4-benzenediol (hydroquinone).

1,2-Benzenediol (Catechol) $\xrightarrow[H_2SO_4]{K_2Cr_2O_7}$ 1,2-Benzoquinone (*o*-Quinone)

1,4-Benzenediol (Hydroquinone) $\xrightarrow[H_2SO_4]{K_2Cr_2O_7}$ 1,4-Benzoquinone (*p*-Quinone)

Perhaps the most important chemical property of quinones is that they are readily reduced to benzenediols. For example, *p*-quinone is readily reduced to hydroquinone by sodium dithionite in neutral or alkaline solution. There are other ways to carry out

The bombadier beetle generates *p*-quinone, an irritating chemical, by the enzyme-catalyzed oxidation of hydroquinone using hydrogen peroxide as the oxidizing agent. Heat generated in this oxidation produces superheated steam which is ejected, along with *p*-quinone, with explosive force. *(Thomas Eisner and David Aneshansley, Cornell University)*

this reduction. The point is that it can be done very easily, as can the corresponding oxidation of a hydroquinone.

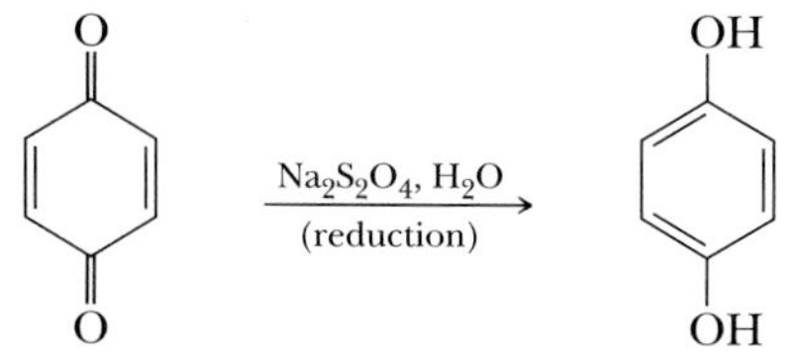

$Na_2S_2O_4$, H_2O (reduction)

1,4-Benzoquinone (p-Quinone) → 1,4-Benzenediol (Hydroquinone)

There are many examples in which the reversible oxidation/reduction of hydroquinones or quinones is important. One such example is coenzyme Q, alternatively known as ubiquinone. The name of this important biomolecule is derived from the Latin *ubique* (everywhere) plus quinone.

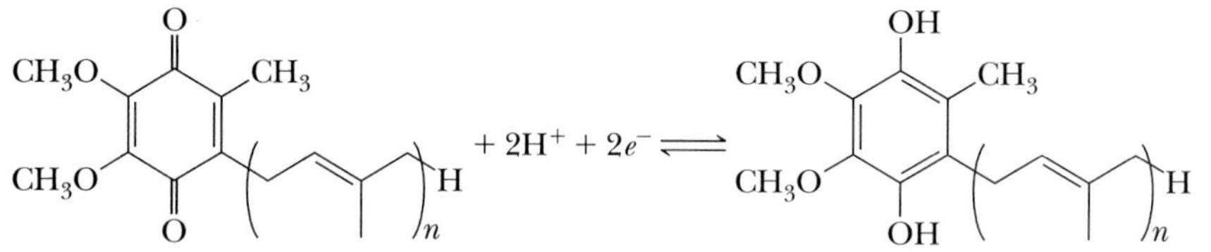

$$+ 2H^+ + 2e^- \rightleftharpoons$$

Coenzyme Q (oxidized form) ⇌ Coenzyme Q (reduced form)

Coenzyme Q, a carrier of electrons in the respiratory chain, contains a long hydrocarbon chain of between 6 and 10 isoprene units that serves to anchor it firmly in the nonpolar environment of the mitochondrial inner membrane. As can be seen from the balanced half-reaction, the oxidized form of coenzyme Q is a two-electron oxidizing agent. In subsequent steps of the respiratory chain, the reduced form of coenzyme Q transfers these two electrons to another link until they are eventually delivered to a molecule of oxygen, which is in turn reduced to water.

Another quinone important in biological systems is vitamin K_2. This compound was discovered in 1935 as a result of a study of newly hatched chicks with a fatal disease in which their blood was slow to clot. It was later discovered that the delayed clotting time of blood was caused by a deficiency of prothrombin, and it is now known that vitamin K_2 is essential to the synthesis of prothrombin in the liver. The natural form of vitamin K_2 has a chain of 5 to 8 isoprene units attached to a 1,4-naphthoquinone ring. The following structure shows 7 isoprene units in the side chain.

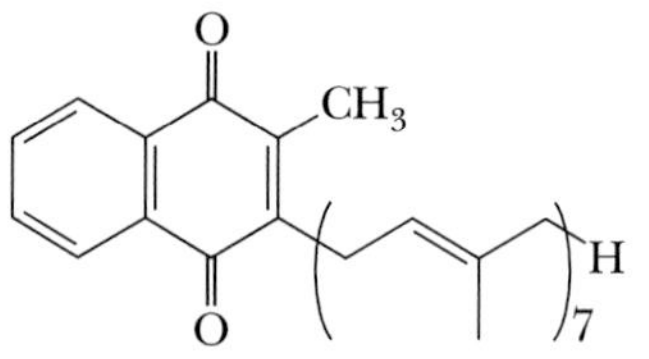

Vitamin K_2

The natural vitamins of the K family have for the most part been replaced by synthetic preparations. Menadione, one such synthetic material with vitamin K activity, has only hydrogen in place of the long alkyl side chain. Menadione is prepared by chromic acid oxidation of 2-methylnaphthalene under mild conditions.

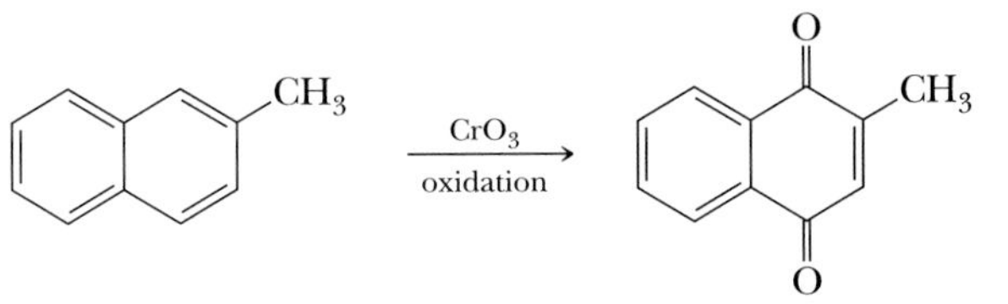

CH_3 — $\xrightarrow[\text{oxidation}]{CrO_3}$ — O, CH_3, O

2-Methylnaphthalene — 2-methyl-1,4-naphthoquinone (Menadione)

A commercial process that uses a quinone is black-and-white photography. Black-and-white film is coated with an emulsion containing silver bromide or silver iodide crystals, which become activated by exposure to light. The activated silver ions are reduced in the developing stage to metallic silver by hydroquinone, which at the same time is oxidized to quinone. Following is an equation showing the relationship between these species.

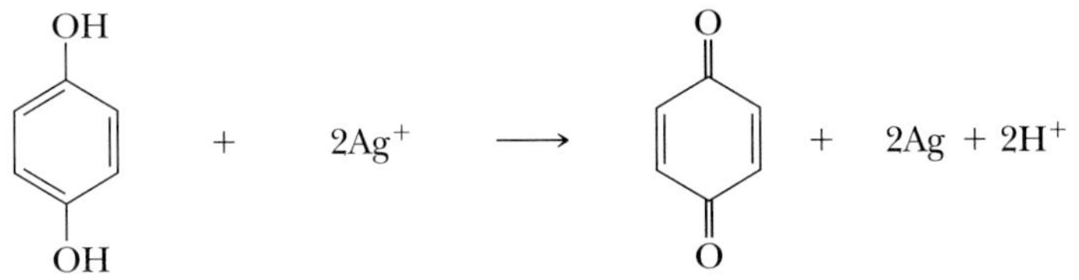

OH, OH + $2Ag^+$ ⟶ O, O + $2Ag + 2H^+$

1,4-Benzenediol (Hydroquinone) — 1,4-Benzoquinone (*p*-Quinone)

All silver halide not activated by light and then reduced by interaction with hydroquinone is removed in the fixing process, and the result is a black image (a negative) left by deposited metallic silver where the film has been struck by light. Other compounds are now used to reduce "light-activated" silver bromide, but the result is the same—a deposit of metallic silver in response to exposure of film to light.

19.6 Reactions at a Benzylic Position

In this section, we study two reactions of substituted aromatic hydrocarbons that occur preferentially at the **benzylic position.**

Benzylic position An sp^3 hybridized carbon attached to a benzene ring.

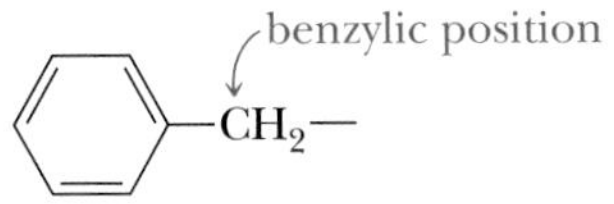

benzylic position

$-CH_2-$

Benzyl group

These reactions occur preferentially at the benzylic position for two reasons. First, the benzene ring is especially resistant to reaction with many of the reagents that normally attack alkanes. Second, benzylic cations and benzylic radicals are easily formed because of resonance stabilization of these intermediates. A benzylic cation or radical is a hybrid of five contributing structures: two Kekulé structures and three that delocalize the positive charge (or the lone electron) onto carbons of the aromatic ring. Following are contributing structures for a benzylic cation. Similar contributing structures can be written for a benzylic radical and a benzylic anion. Thus, benzylic contributing structures are closely analogous to allylic structures in stabilizing cations, radicals, and anions.

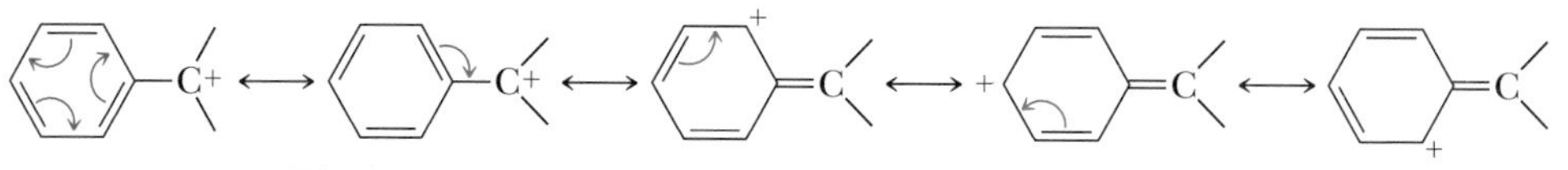

The benzyl cation as a hybrid of five contributing structures

A. Oxidation

Benzene is unaffected by strong oxidizing agents, such as H_2CrO_4 and $KMnO_4$. When toluene is treated with these oxidizing agents under quite vigorous conditions, the side chain methyl group is oxidized to a carboxyl group to give benzoic acid.

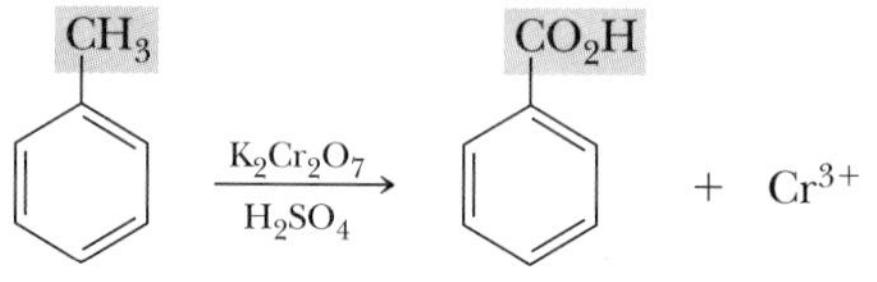

$$\text{Toluene} \xrightarrow[H_2SO_4]{K_2Cr_2O_7} \text{Benzoic acid} + Cr^{3+}$$

Toluene Benzoic acid

Halogen and nitro substituents on an aromatic ring are unaffected by these oxidations. 2-Chloro-4-nitrotoluene, for example, is oxidized to 2-chloro-4-nitrobenzoic acid.

$$\text{2-Chloro-4-nitrotoluene} \xrightarrow[H_2SO_4]{K_2Cr_2O_7} \text{2-Chloro-4-nitrobenzoic acid}$$

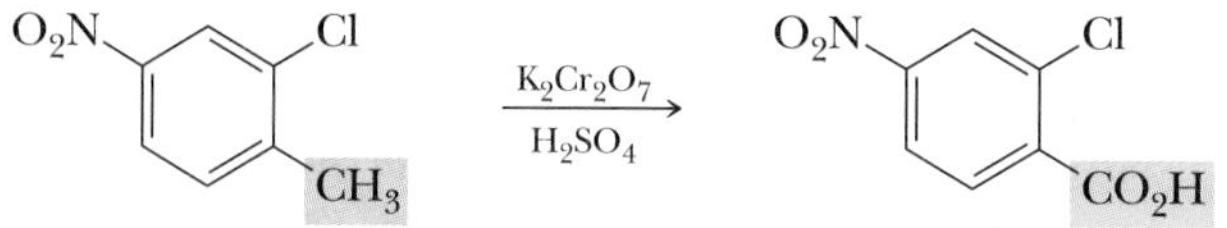

2-Chloro-4-nitrotoluene 2-Chloro-4-nitrobenzoic acid

Ethyl and isopropyl side chains are also oxidized to carboxyl groups. The side chain of *tert*-butylbenzene, however, is not oxidized.

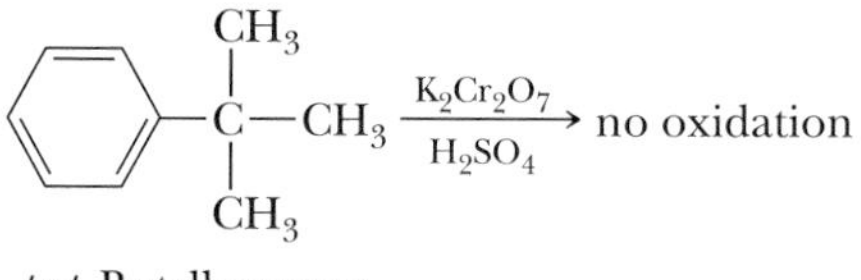

$$C_6H_5-C(CH_3)_2-CH_3 \xrightarrow[H_2SO_4]{K_2Cr_2O_7} \text{no oxidation}$$

tert-Butylbenzene

From these observations, we conclude that if a benzylic hydrogen exists, then the benzylic carbon is oxidized to a carboxyl group and all other carbons of the side chain are removed. If no benzylic hydrogen exists, as in the case of *tert*-butylbenzene, no oxidation of the side chain occurs.

If more than one alkyl side chain exists, each is oxidized to $-CO_2H$. Oxidation of *m*-xylene gives 1,3-benzenedicarboxylic acid, more commonly named isophthalic acid.

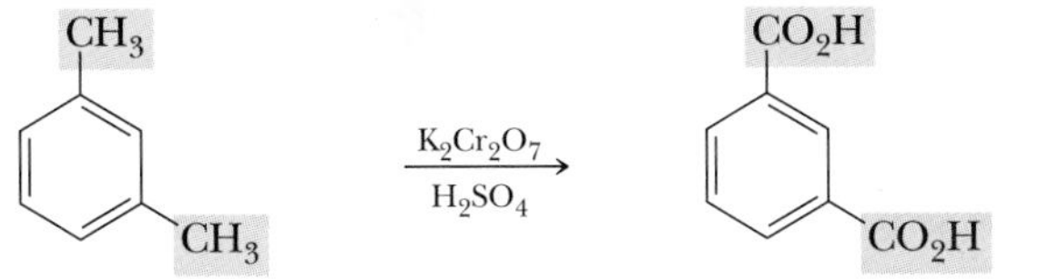

CH_3 / CH_3 $\xrightarrow[H_2SO_4]{K_2Cr_2O_7}$ CO_2H / CO_2H

m-Xylene — 1,3-Benzenedicarboxylic acid (Isophthalic acid)

EXAMPLE 19.7

Draw structural formulas for the product of vigorous oxidation 1,4-dimethylbenzene (*p*-xylene) by $K_2Cr_2O_7$ in aqueous H_2SO_4.

Solution

Both alkyl groups are oxidized to $-CO_2H$ groups. The product is terephthalic acid, one of two monomers required for the synthesis of Dacron polyester and Mylar (Section 23.5B).

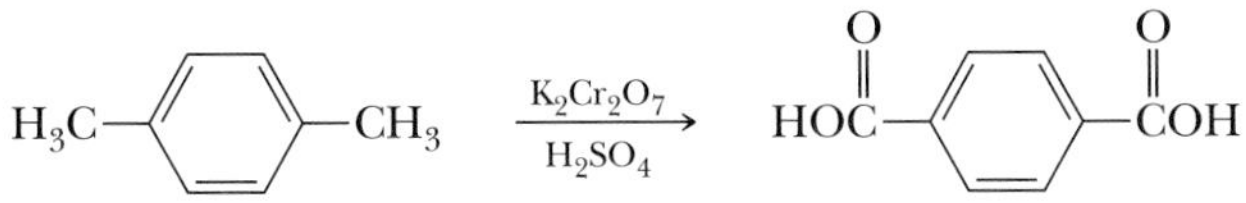

H_3C—C₆H₄—CH_3 $\xrightarrow[H_2SO_4]{K_2Cr_2O_7}$ HOC(=O)—C₆H₄—C(=O)OH

1,4-Dimethylbenzene (*p*-Xylene) — 1,4-Benzenedicarboxylic acid (Terephthalic acid)

PROBLEM 19.7

Predict the products resulting from vigorous oxidation of these compounds by $K_2Cr_2O_7$ in aqueous H_2SO_4.

(a) 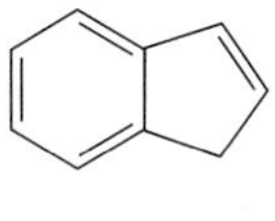 (b) 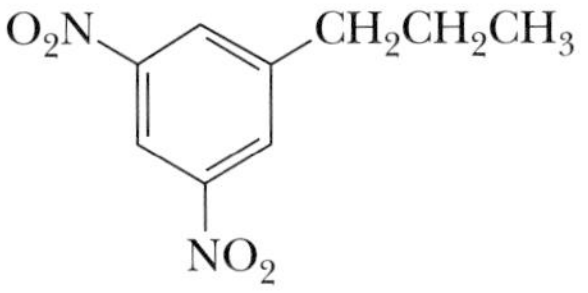 O_2N, $CH_2CH_2CH_3$, NO_2

It has been difficult to study these side chain oxidations and to formulate mechanisms for them. Available evidence, however, supports the formation of unstable intermediates that are either benzylic radicals or benzylic carbocations.

Naphthalene is oxidized to phthalic acid by molecular oxygen in the presence of a vanadium(V) oxide (vanadium pentoxide) catalyst. This conversion, which is the basis for an industrial synthesis of this aromatic dicarboxylic acid, illustrates the ease of oxidation of condensed benzene rings compared with benzene itself.

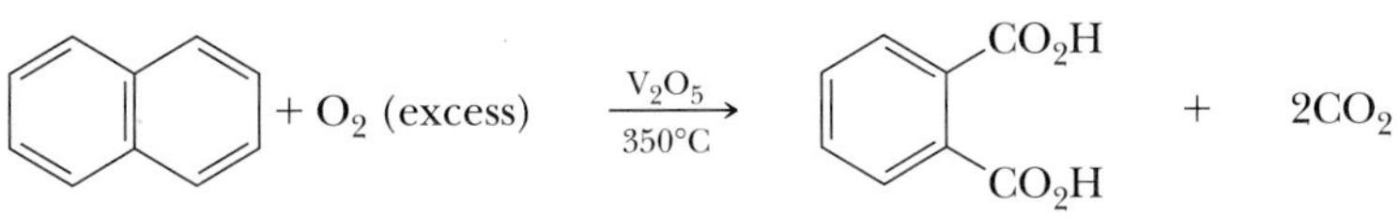

Naphthalene $+ O_2$ (excess) $\xrightarrow[350°C]{V_2O_5}$ CO_2H / CO_2H $+ 2CO_2$

Naphthalene — 1,2-Benzenedicarboxylic acid (Phthalic acid)

B. Halogenation

Reaction of toluene with chlorine in the presence of heat or light results in formation of chloromethylbenzene and HCl. Bromination is easily accomplished by using *N*-bromosuccinimide (NBS) in the presence of a peroxide catalyst.

$$C_6H_5-CH_3 + Cl_2 \xrightarrow[\text{or light}]{\text{heat}} C_6H_5-CH_2Cl + HCl$$

Toluene → Chloromethylbenzene (Benzyl chloride)

$$C_6H_5-CH_3 + \text{NBS} \xrightarrow[CCl_4]{(PhCO_2)_2} C_6H_5-CH_2Br + \text{Succinimide (N-H)}$$

Toluene, *N*-Bromosuccinimide (NBS) → Bromomethylbenzene (Benzyl bromide), Succinimide

Halogenation of a larger alkyl side chain is highly regioselective, as illustrated by the halogenation of ethylbenzene. When treated with NBS, the only monobromo organic product formed is 1-bromo-1-phenylethane. This regioselectivity is dictated by resonance effects, namely, the resonance stabilization of the benzylic radical intermediate. The mechanism of radical bromination at a benzylic position is identical to that for allylic bromination (Section 7.6).

$$C_6H_5-CH_2CH_3 \xrightarrow[(PhCO_2)_2,\ CCl_4]{\text{NBS}} C_6H_5-CHBrCH_3$$

Ethylbenzene → 1-Bromo-1-phenylethane

When ethylbenzene is treated with chlorine under radical reaction conditions, two products are formed in the ratio of 9:1.

$$C_6H_5-CH_2CH_3 + Cl_2 \xrightarrow[\text{or light}]{\text{heat}} C_6H_5-CHClCH_3 + C_6H_5-CH_2CH_2Cl + HCl$$

Ethylbenzene → 1-Chloro-1-phenylethane (90%), 1-Chloro-2-phenylethane (10%)

Thus, chlorination of alkyl side chains is also regioselective but not to the same high degree as bromination. Recall that we observed this same pattern in the regioselectivities of bromination and chlorination of alkanes (Section 7.4A).

Combining the information on product distribution for bromination and chlorination of hydrocarbons, we conclude that the order of stability of radicals is:

$$\text{methyl} < 1° < 2° < 3° < \text{allylic} = \text{benzylic}$$

Increasing radical stability →

This order reflects the C—H bond dissociation energies (BDE) for formation of these radicals (Table 19.3).

TABLE 19.3 C—H Bond Dissociation Energies for Formation of Selected Radicals

Hydrocarbon	Radical	Type of Radical	BDE (kcal/mol)
$CH_2{=}CHCH_2{-}H$	$CH_2{=}CHCH_2\cdot$	allylic	86
$C_6H_5CH_2{-}H$	$C_6H_5CH_2\cdot$	benzylic	88
$(CH_3)_3C{-}H$	$(CH_3)_3C\cdot$	tertiary	93
$(CH_3)_2CH{-}H$	$(CH_3)_2CH\cdot$	secondary	96
$CH_3CH_2{-}H$	$CH_3CH_2\cdot$	primary	100
$CH_3{-}H$	$CH_3\cdot$	methyl	105

C. Hydrogenolysis of Benzyl Ethers

Among ethers, benzylic ethers are unique in that they are cleaved under the conditions of catalytic hydrogenation as illustrated by the hydrogenolysis of benzyl hexyl ether. **Hydrogenolysis** is the cleavage of a single bond by H_2. In the hydrogenolysis of a benzylic ether, it is the single bond between the benzylic carbon and its attached oxygen that is cleaved and replaced by a carbon-hydrogen bond. In this illustration, the benzyl group is converted to toluene and the alkyl group is converted to an alcohol.

Hydrogenolysis Cleavage of a single bond by H_2, most commonly accomplished by treatment of a compound with H_2 in the presence of a transition metal catalyst.

$$C_6H_5{-}CH_2O(CH_2)_5CH_3 + H_2 \xrightarrow{Pd/C} C_6H_5{-}CH_3 + HO(CH_2)_5CH_3$$

Benzyl hexyl ether — Toluene — 1-Hexanol

Benyzl ethers are formed by treatment of an alcohol or phenol with benzyl chloride in the presence of a base such as triethylamine or pyridine (Section 19.5D). The function of the base is to convert the phenol (a poor nucleophile) into its phenoxide ion (a good nucleophile) and also to neutralize the HCl byproduct.

The particular value of benzylic ethers is that they can serve as protecting groups for the —OH groups of alcohols and phenols. Suppose, for example, we want to treat 2-allylphenol with diborane followed by hydrogen peroxide to bring about anti-Markovnikov hydration of the carbon-carbon double bond. This scheme will not give the desired result because the phenolic —OH group is sufficiently acidic to react with BH_3 and destroy it. The desired product can be prepared, however, by protection of the phenolic —OH group as the benzylic ether, hydroboration/oxidaton of the carbon-carbon double bond, and hydrogenolysis of the benzylic ether.

$$\text{2-}(CH_2CH{=}CH_2)C_6H_4{-}OH \xrightarrow[(CH_3CH_2)_3N]{1.\ ClCH_2C_6H_5} \text{2-}(CH_2CH{=}CH_2)C_6H_4{-}OCH_2C_6H_5 \xrightarrow[3.\ H_2O_2/NaOH]{2.\ BH_3\cdot THF}$$

2-(2-Propenyl)phenol (2-Allylphenol)

$$\text{2-}(CH_2CH_2CH_2OH)C_6H_4{-}OCH_2C_6H_5 \xrightarrow[Pd/C]{H_2} \text{2-}(CH_2CH_2CH_2OH)C_6H_4{-}OH$$

2-(3-Hydroxypropyl)phenol

SUMMARY

Benzene and its alkyl derivatives are classified as **aromatic hydrocarbons,** or **arenes.** The structure of benzene, proposed by August Kekulé in 1865, represented benzene as two rapidly interconverting Kekulé structures (Section 19.1A). The concepts of **hybridization of atomic orbitals** and the **theory of resonance** (Section 19.1B), developed by Linus Pauling in the 1930s, provided the first adequate description of the structure of benzene. According to the **molecular orbital model,** the six $2p$ atomic orbitals of the sp^2 hybridized ring carbon atoms combine to give three pi bonding MOs and three pi antibonding MOs. In the ground state, the six pi electrons of benzene lie in the three pi bonding MOs. The **resonance energy** of benzene (Section 19.1C), as calculated from experimental values for heats of hydrogenation of benzene and cyclohexene, is approximately 36 kcal/mol.

According to the **Hückel criteria for aromaticity** (Section 19.2A), a monocyclic compound is aromatic if it (1) has one p orbital on each atom of the ring, (2) is planar so that overlap of all p orbitals of the ring is continuous or nearly continuous, and (3) has $(4n + 2)$ pi electrons in the cyclic, overlapping arrangement of p orbitals. An **annulene** (Section 19.2B) is a cyclic hydrocarbon with an alternation of single and double bonds. Many have been synthesized to test the validity of Hückel's criteria for aromaticity. It has been found, for example, that [14]annulene and [18]annulene are aromatic as predicted. **Antiaromatic compounds** (Section 19.2C) have only $4n$ pi electrons in a monocyclic, planar system of continuously overlapping p orbitals.

A **heterocyclic aromatic compound** (Section 19.2D) contains one or more atoms other than carbon in an aromatic ring. Particularly abundant in the biological world are derivatives of the heterocyclic aromatic amines pyridine, pyrimidine, imidazole, and pyrrole.

The **cyclopropenyl cation,** the **cyclopentadienyl anion,** and the **cycloheptatrienyl cation** (Section 19.2E) each meet the Hückel criteria for aromaticity and are particularly stable hydrocarbon ions.

Aromatic compounds are named by the IUPAC system. The common names toluene, xylene, cumene, styrene, phenol, aniline, and benzoic acid (Section 19.3A) are retained. The $C_6H_5—$ group is named **phenyl** and the $C_6H_5CH_2—$ group is named **benzyl.** Two substituents on a benzene ring may be located by numbering the atoms of the ring or by using the locators **ortho (*o*), meta (*m*),** and **para (*p*). Polynuclear aromatic hydrocarbons** (Section 19.3C) contain two or more fused benzene rings. Particularly abundant are naphthalene, anthracene, phenanthrene, and their derivatives.

The functional group of a **phenol** (Section 19.5A) is an —OH group bonded to a benzene ring. Phenol and its derivatives are weak acids, pK_a approximately 10. The greater acidity of phenols substituted with electron-withdrawing groups, for example $—NO_2$, is accounted for by a combination of **inductive effects** and **resonance effects.**

Reactions of aromatic compounds containing alkyl side chains occur preferentially at the benzylic carbon (Section 19.6). Benzylic cations and radicals are especially stable because of delocalization of their positive charge or unpaired electron, respectively, onto the ortho and para positions of the aromatic ring.

KEY REACTIONS

1. Acidity of Phenols (Section 19.5B)

Phenols are weak acids, pK_a approximately 10. Ring substituents may increase or decrease acidity by a combination of resonance and inductive effects.

$$C_6H_5{-}OH + H_2O \rightleftharpoons C_6H_5{-}O^- + H_3O^+ \quad K_a = 1.1 \times 10^{-10} \quad pK_a = 9.95$$

2. Reaction of Phenols with Strong Bases (Section 19.5C)

Water-insoluble phenols react quantitatively with strong bases to form water-soluble salts.

$$C_6H_5{-}OH + NaOH \longrightarrow C_6H_5{-}O^-Na^+ + H_2O$$

Phenol	Sodium hydroxide	Sodium phenoxide	Water
$pK_a = 9.95$			$pK_a = 15.7$
(stronger acid)	(stronger base)	(weaker base)	(weaker acid)

3. Kolbe Synthesis: Carboxylation of Phenols (Section 19.5E)

Nucleophilic addition of a phenoxide ion to carbon dioxide gives a substituted cyclohexadienone which then undergoes keto-enol tautomerism to regenerate the aromatic ring.

$$\text{4-CH}_3\text{C}_6\text{H}_4\text{O}^-\text{Na}^+ \xrightarrow[H_2O]{CO_2} \text{cyclohexadienone (C=O, } CO_2^-Na^+\text{, H, } CH_3\text{)} \xrightarrow[H_2O]{HCl} \text{2-hydroxy-5-methylbenzoic acid (OH, } CO_2H\text{, } CH_3\text{)}$$

4. Oxidation of Phenols to Quinones (Section 19.5F)

Oxidation by $K_2Cr_2O_7$ gives 1,2-quinones (*o*-quinones) or 1,4-quinones (*p*-quinones), depending on the structure of the particular phenol.

$$\text{C}_6\text{H}_5\text{OH} \xrightarrow[H_2SO_4]{K_2Cr_2O_7} \text{1,4-benzoquinone}$$

5. Oxidation at a Benzylic Position (Section 19.6A)

A benzylic carbon bonded to at least one hydrogen is oxidized to a carboxyl group.

$$H_3C{-}C_6H_4{-}CH(CH_3)_2 \xrightarrow[H_2SO_4]{K_2Cr_2O_7} HO_2C{-}C_6H_4{-}CO_2H$$

6. Halogenation at a Benzylic Position (Section 19.6B)

Halogenation is regioselective for a benzylic position and occurs by a radical chain mechanism. Bromination shows a higher regioselectivity for a benzylic position than chlorination.

$$C_6H_5{-}CH_2CH_3 \xrightarrow[\text{peroxide}]{\text{NBS}} C_6H_5{-}CHBrCH_3$$

7. Hydrogenolysis of Benzylic Ethers (Section 19.6C)

Benzylic ethers are cleaved under the conditions of catalytic hydrogenation.

$$C_6H_5{-}CH_2O(CH_2)_5CH_3 \xrightarrow[Pd/C]{H_2} C_6H_5{-}CH_3 + HO(CH_2)_5CH_3$$

ADDITIONAL PROBLEMS

Nomenclature and Structural Formulas

19.8 Name the following molecules and ions.

(a)  NO_2, Cl (b) 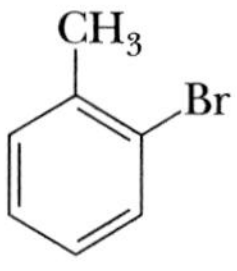 CH_3, Br (c) 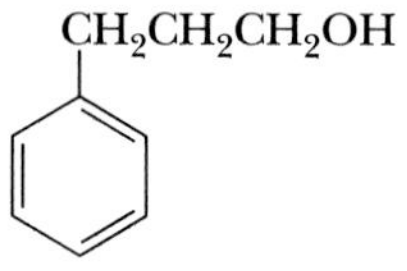 $CH_2CH_2CH_2OH$ (d) 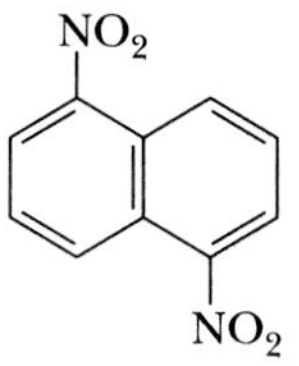 NO_2, NO_2

(e) 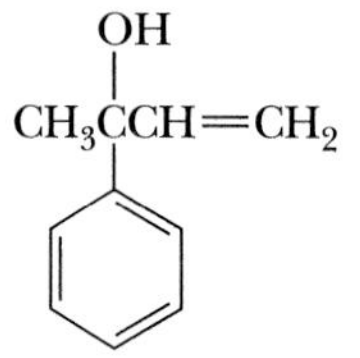 OH, $CH_3CCH{=}CH_2$ (f) 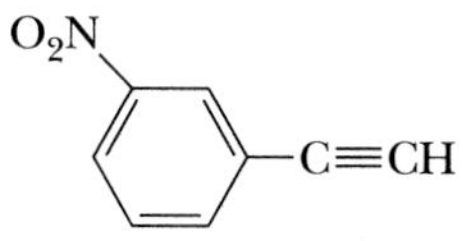 O_2N, $-C{\equiv}CH$ (g) 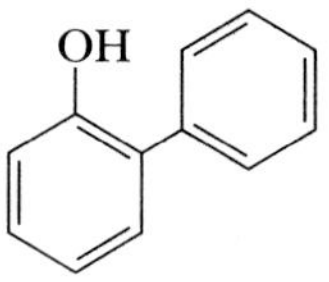 OH

(h) 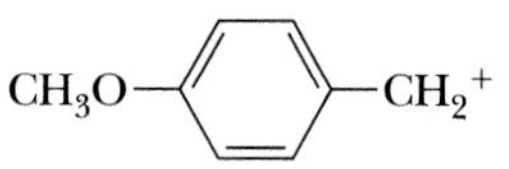 CH_3O- $-CH_2^+$ (i) 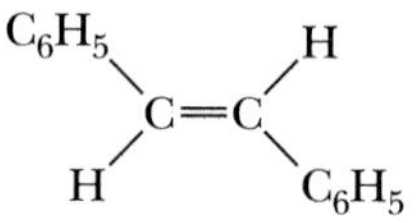 C_6H_5, H, C=C, H, C_6H_5 (j) 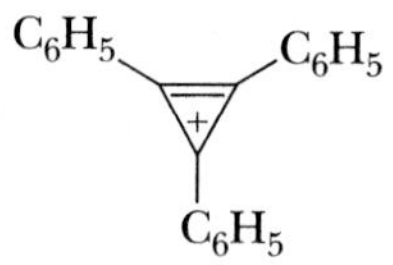 C_6H_5, C_6H_5, +, C_6H_5

19.9 Draw structural formulas for these molecules.

(a) 1-Bromo-2-chloro-4-ethylbenzene
(b) *m*-Nitrocumene
(c) 4-Chloro-1,2-dimethylbenzene
(d) 3,5-Dinitrotoluene
(e) 2,4,6-Trinitrotoluene
(f) 4-Phenyl-2-pentanol
(g) *p*-Cresol
(h) Pentachlorophenol
(i) 1-Phenylcyclopropanol
(j) Triphenylmethane
(k) Phenylethylene (styrene)
(l) Benzyl bromide
(m) 1-Phenyl-1-butyne
(n) 3-Phenyl-2-propen-1-ol

19.10 Draw structural formulas for these molecules.

(a) 1-Nitronaphthalene
(b) 1,6-Dichloronaphthalene
(c) 9-Bromoanthracene
(d) 2-Methylphenanthrene

19.11 Molecules of 6,6′-dinitrobiphenyl-2,2′-dicarboxylic acid have no tetrahedral stereocenter, and yet they can be resolved to a pair of enantiomers. Account for this chirality. (*Hint:* It will help to build a model and study the possibility of rotation about the single bond joining the two benzene rings.)

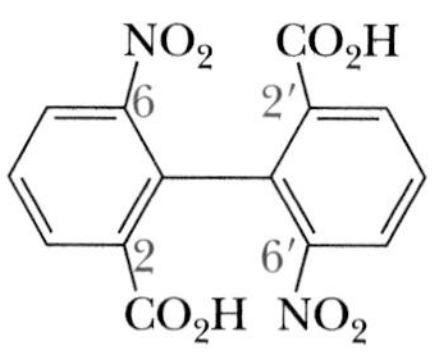

NO_2 CO_2H 6 2′ 2 6′ CO_2H NO_2

6,6′-Dinitrobiphenyl-2,2′-dicarboxylic acid

Resonance in Aromatic Compounds

19.12 Following each name is the number of Kekulé structures that can be drawn for it. Draw these Kekulé structures, and show, using curved arrows, how the first contributing structure for each molecule is converted to the second and so forth.

(a) Naphthalene (3) (b) Phenanthrene (5)

19.13 Each molecule can be drawn as a hybrid of five contributing structures: two Kekulé structures and three that involve creation and separation of unlike charges. For (a) and (b), the creation and separation of unlike charges places a positive charge on the substituent and a negative charge on the ring. For (c), a positive charge is placed on the ring and an additional negative charge is placed on the $—NO_2$ group. Draw these five contributing structures for each molecule.

(a) Chlorobenzene **(b)** Phenol **(c)** Nitrobenzene

19.14 Following are structural formulas for furan and pyridine.

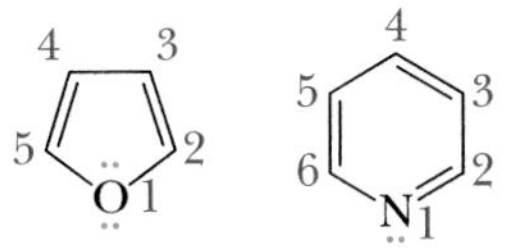

Furan Pyridine

(a) Write four contributing structures for the furan hybrid that place a positive charge on oxygen and a negative charge first on carbon 3 of the ring and then on each other carbon of the ring.

(b) Write three contributing structures for the pyridine hybrid that place a negative charge on nitrogen and a positive charge first on carbon 2, then on carbon 4, and finally on carbon 6.

The Concept of Aromaticity

19.15 State the number of *p* orbital electrons in each of the following.

(a) 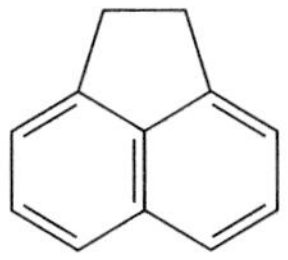 (b) 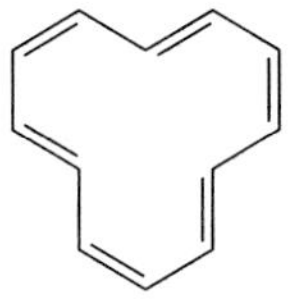 (c) 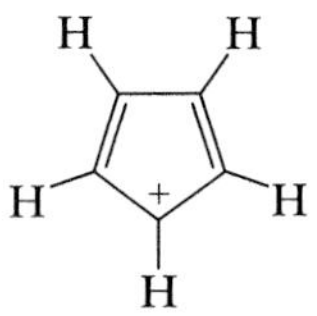 (d)

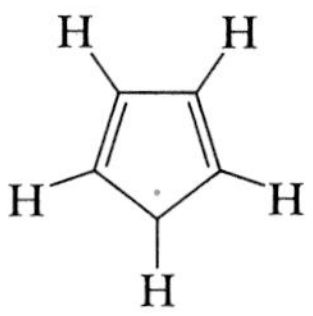

(e) 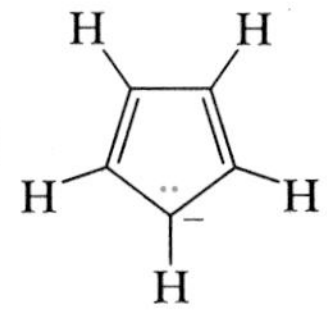 (f) 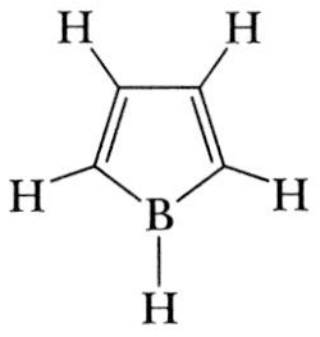 (g) 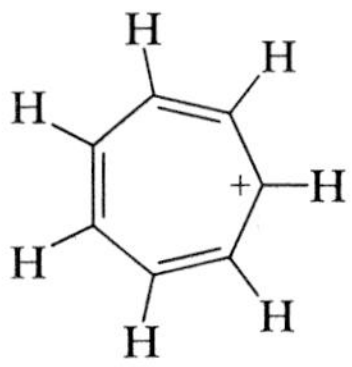 (h)

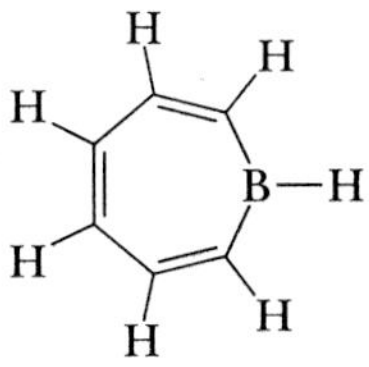

(i) 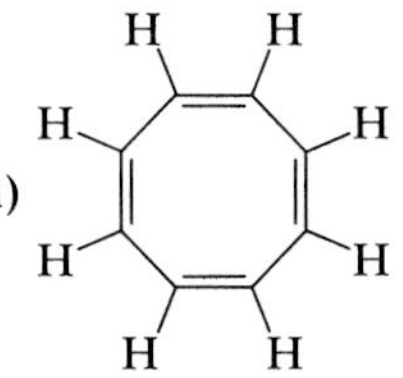 (j)

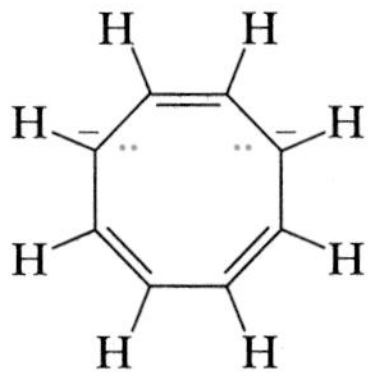

19.16 Which of the molecules and ions given in the previous problem are aromatic according to the Hückel criteria? Which, if planar, would be antiaromatic?

19.17 Construct MO energy diagrams for the cyclopropenyl cation, radical, and anion. Which of these species is aromatic according to the Hückel criteria?

19.18 Naphthalene and azulene are constitutional isomers of molecular $C_{10}H_8$. Naphthalene is a colorless solid with a dipole moment of zero. Azulene is a solid with an intense blue color and a dipole moment of 1.0 D. Account for the difference in dipole moments of these constitutional isomers.

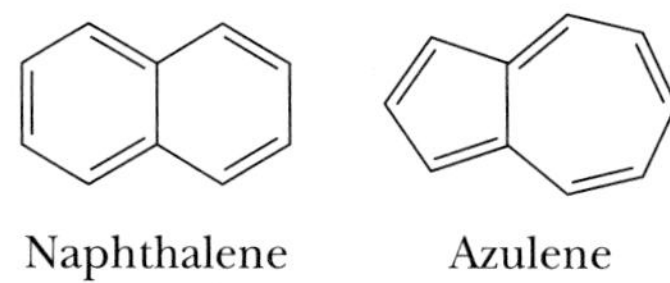

Spectroscopy

19.19 Compound A, molecular formula C_9H_{12}, shows prominent peaks in its mass spectrum at m/z 120 and 105. Compound B, also molecular formula C_9H_{12}, shows prominent peaks at m/z 120 and 91. On vigorous oxidation by potassium dichromate in sulfuric acid, both compounds give benzoic acid. From this information, deduce the structural formulas of compounds A and B.

19.20 Compound C shows a molecular ion at m/z 148, and other prominent peaks at m/z 105, and 77. Following are its ^{1}H-NMR and infrared spectra.

(a) Deduce the structural formula of compound C.

(b) Account for the appearance of peaks in its mass spectrum at m/z 105 and 77.

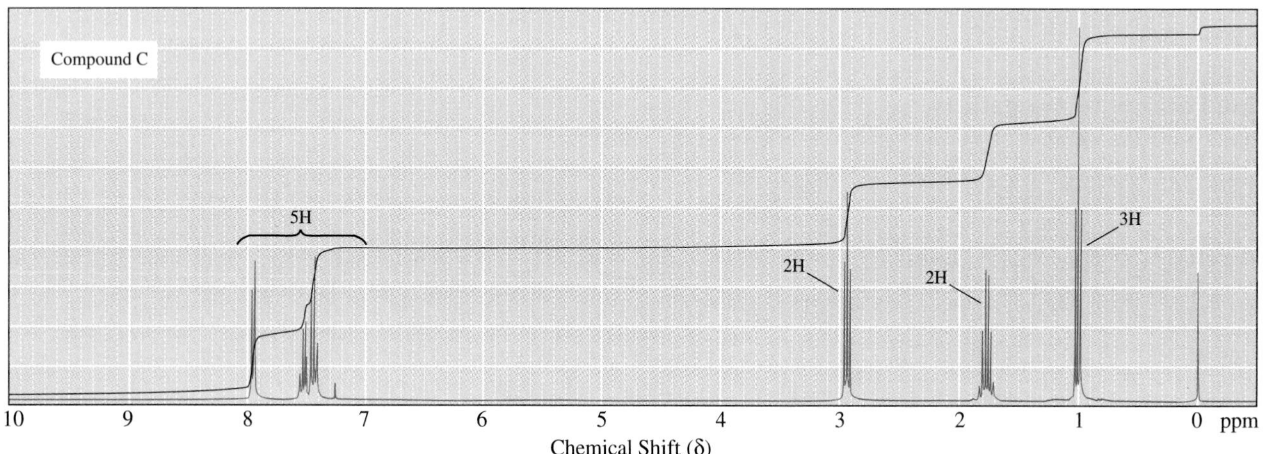

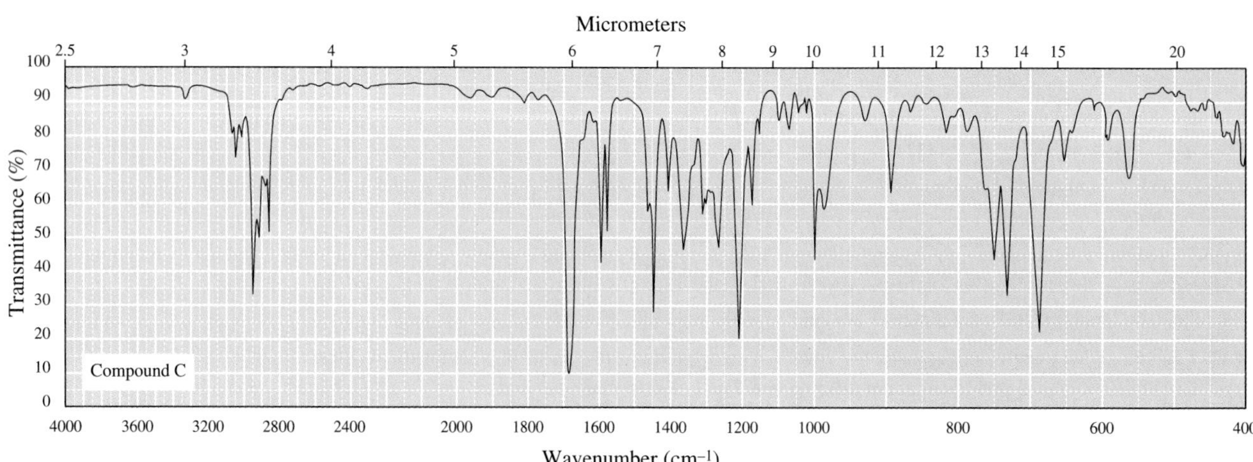

19.21 Following are IR and ^{1}H-NMR spectra of compound D. The mass spectrum of compound D shows a molecular ion peak at m/z 136, a base peak at m/z 107, and other prominent peaks at m/z 118 and 59.

(a) Propose a structural formula for compound D based on this information.
(b) Propose structural formulas for ions in the mass spectrum at m/z 118, 107, and 59.

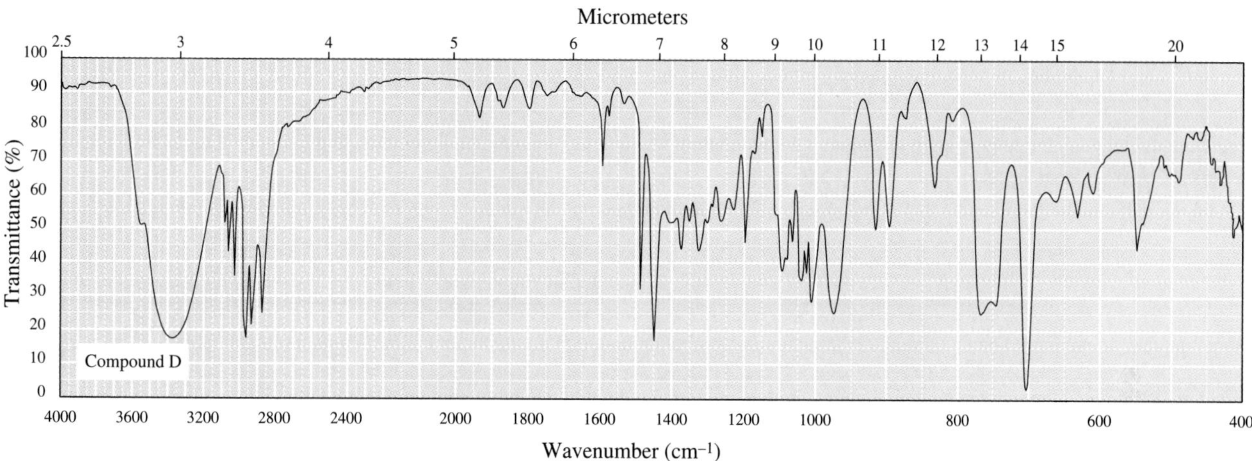

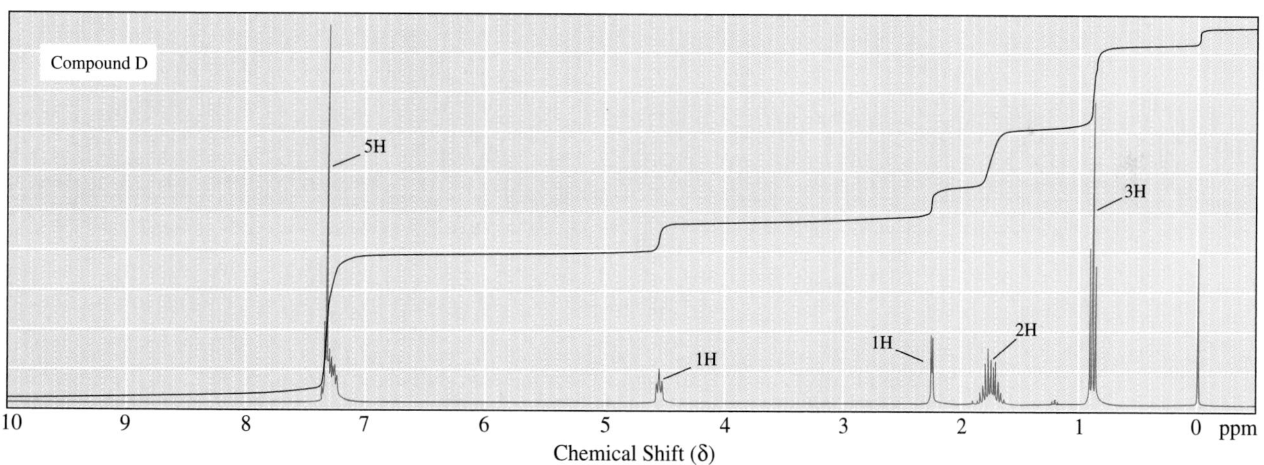

19.22 Compound E is a neutral solid of molecular formula $C_8H_{10}O_2$. Its mass spectrum shows a molecular ion a m/z 138 and prominent peaks at M-1 and M-17. Following are IR and ^{1}H-NMR spectra of compound E. Deduce the structure of compound E.

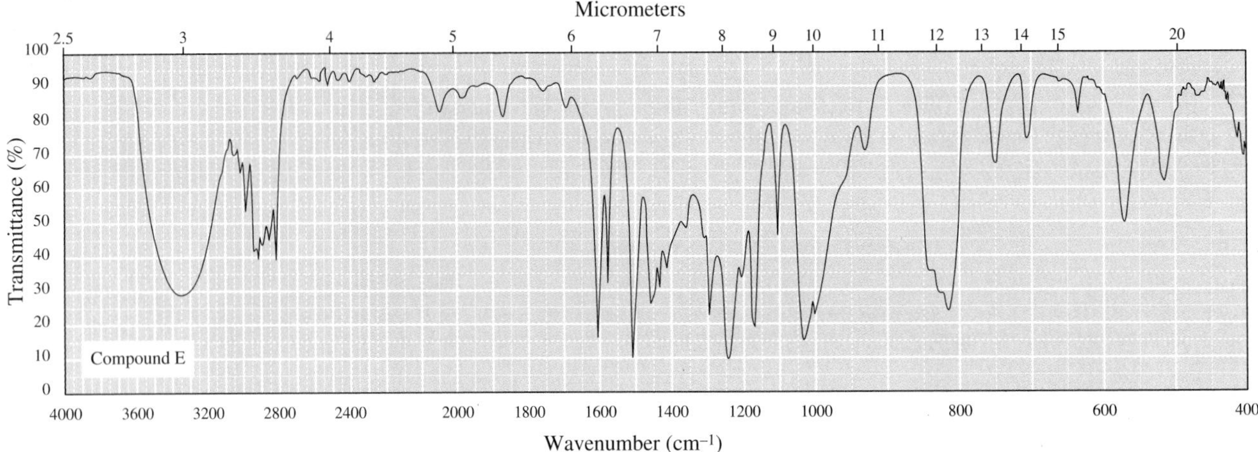

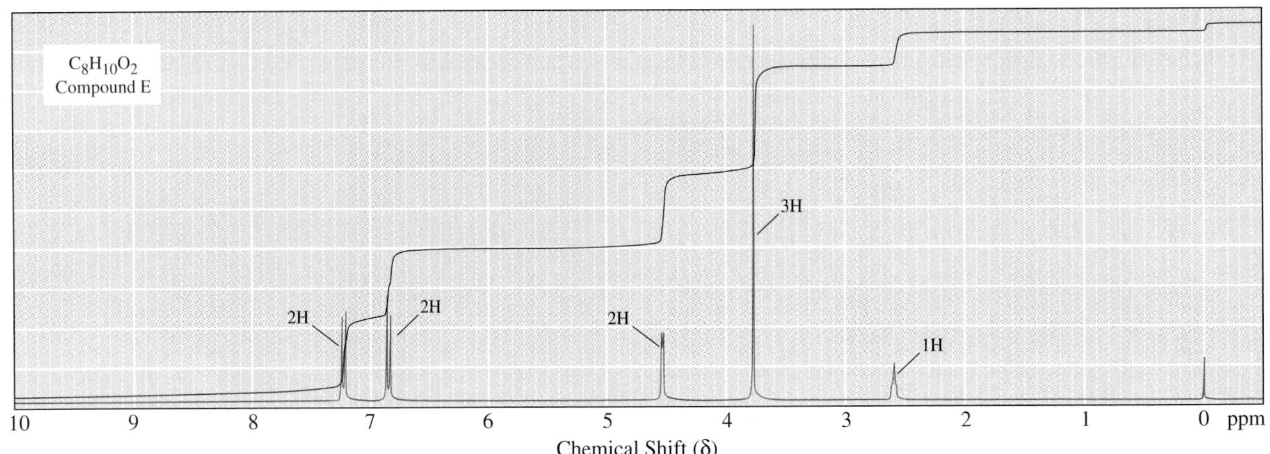

19.23 Following are ^{1}H-NMR and ^{13}C-NMR spectral data for compound F, $C_{12}H_{16}O$. From this information, deduce the structure of compound F.

^{1}H-NMR	^{13}C-NMR	
0.83 d (6H)	207.82	50.88
2.11 m (1H)	134.24	50.57
2.30 d (2H)	129.36	24.43
3.64 s (2H)	128.60	22.48
7.2–7.4 m (5H)	126.86	

19.24 Following are ^{1}H-NMR and ^{13}C-NMR spectral data for compound G, $C_{10}H_{10}O$. From this information, deduce the structure of compound G.

^{1}H-NMR	^{13}C-NMR	
2.50 t (2H)	210.19	126.82
3.05 t (2H)	136.64	126.75
3.58 s (2H)	133.25	45.02
7.1 to 7.3 m (4H)	128.14	38.11
	127.75	28.34

19.25 Compound H, $C_8H_6O_3$, gives a precipitate when treated with hydroxylamine in aqueous ethanol, and a silver mirror when treated with Tollens' solution. Following is its ^{1}H-NMR spectrum. Deduce the structure of compound H.

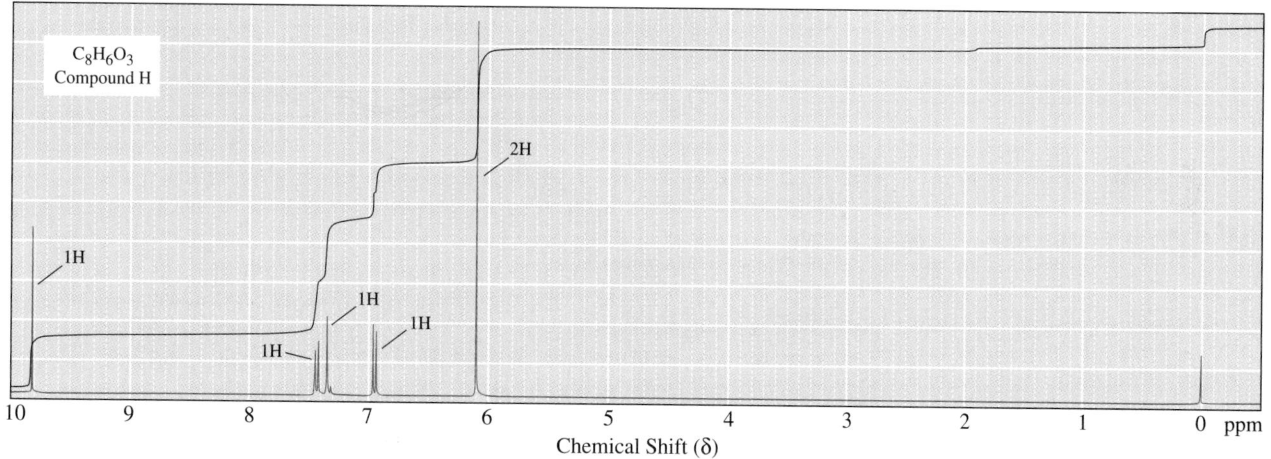

19.26 Compound I, $C_{11}H_{14}O_2$, is insoluble in water, aqueous acid, and aqueous $NaHCO_3$ but dissolves readily in 10% Na_2CO_3 and 10% NaOH. When these alkaline solutions are acidified with 10% HCl, compound I is recovered unchanged. Given this information and its ^{1}H-NMR spectrum, deduce the structure of compound I.

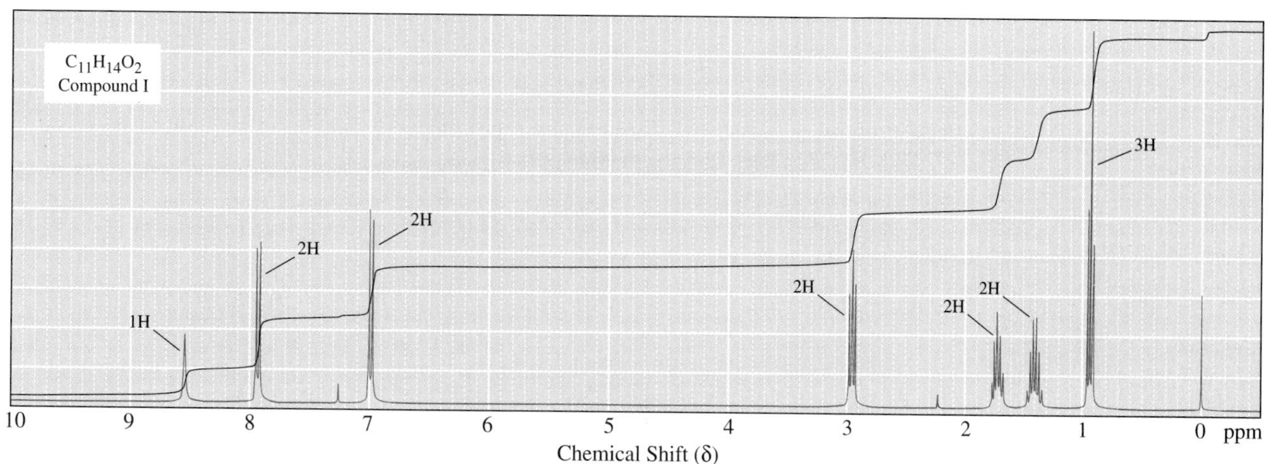

19.27 Propose a structural formula for compound J, $C_{11}H_{14}O_3$, consistent with its ^{1}H-NMR and infrared spectra.

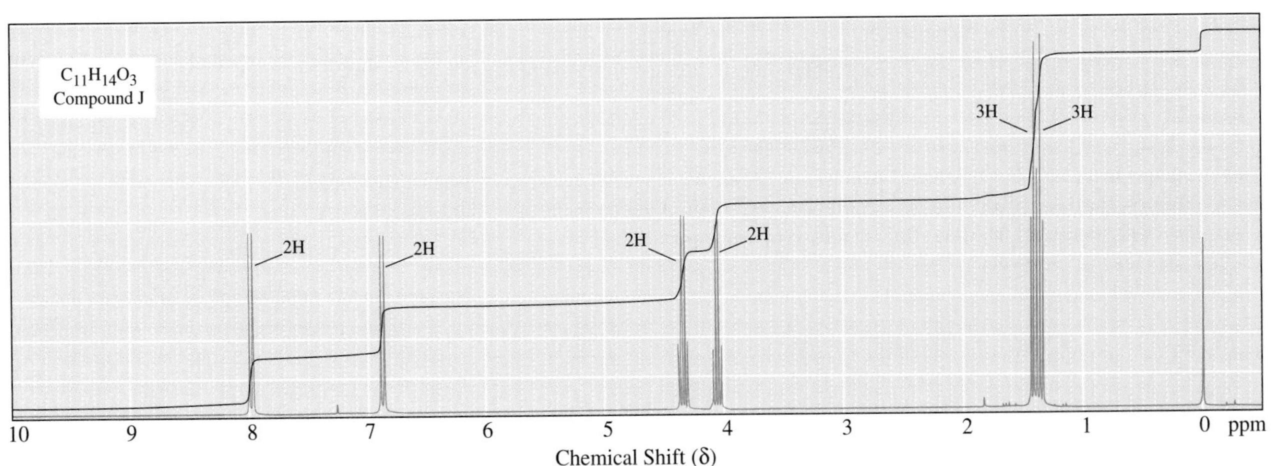

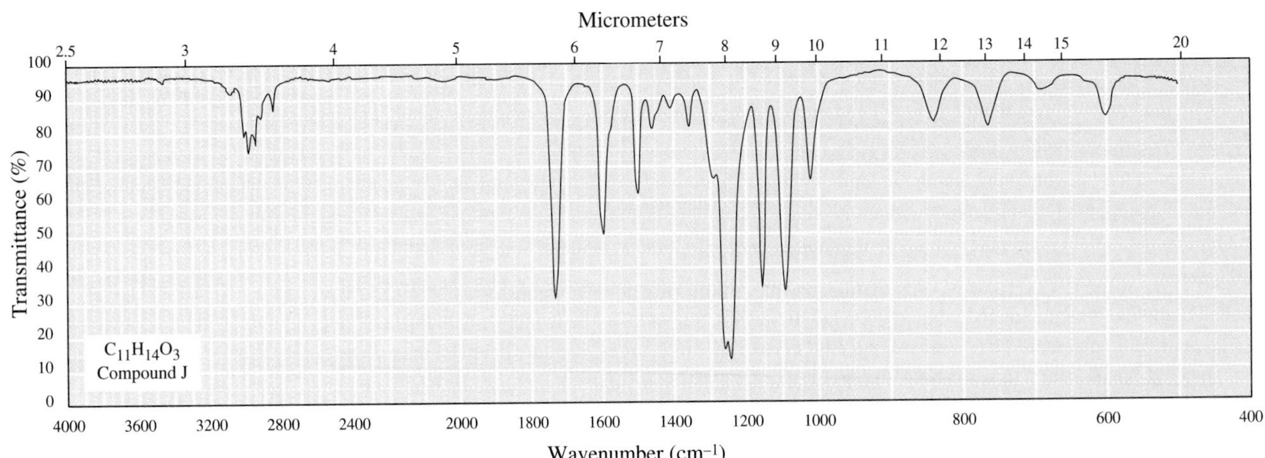

19.28 Propose a structural formula for the analgesic phenacetin, molecular formula $C_{10}H_{13}NO_2$, based on its ^{1}H-NMR spectrum.

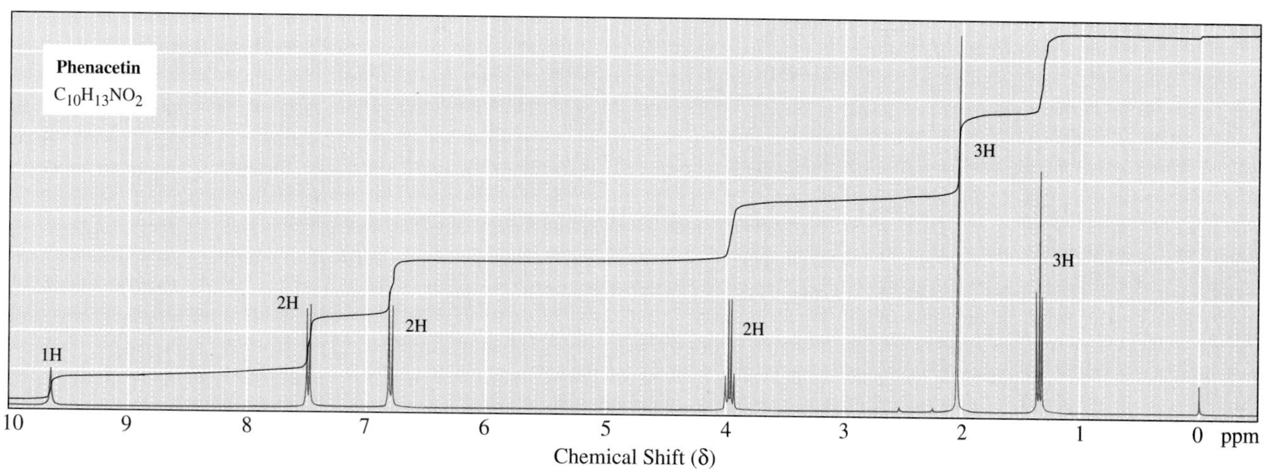

19.29 Compound K, $C_{10}H_{12}O_2$, is insoluble in water, 10% NaOH, and 10% HCl. Given this information and the following ^{1}H-NMR and ^{13}C-NMR spectral information, deduce the structural formula of compound K.

^{1}H-NMR	^{13}C-NMR	
2.10 s (3H)	206.51	114.17
3.61 s (2H)	158.67	55.21
3.77 s (3H)	130.33	50.07
6.86 d (2H)	126.31	29.03
7.12 d (2H)		

19.30 Propose a structural formula for each compound given these NMR data.

(a) $C_9H_9BrO_2$

^{1}H-NMR	^{13}C-NMR
1.39 (t, 3H)	165.73
4.38 (q, 2H)	131.56
7.57 (d, 2H)	131.01
7.90 (d, 2H)	129.84
	127.81
	61.18
	24.28
	21.13

(b) C_8H_9NO

^{1}H-NMR	^{13}C-NMR
2.06 (s, 3H)	168.14
7.01 (t, 1H)	139.24
7.30 (m, 2H)	128.51
7.59 (d, 2H)	122.83
9.90 (s, 1H)	118.90
	23.93

(c) $C_9H_9NO_3$

^{1}H-NMR	^{13}C-NMR
2.10 (s, 3H)	168.74
7.72 (d, 2H)	166.85
7.91 (d, 2H)	143.23
10.3 (s, 1H)	130.28
12.7 (s, 1H)	124.80
	118.09
	24.09

19.31 Given here are ^{1}H-NMR and ^{13}C-NMR spectral data for two compounds. Each shows strong, sharp absorption between 1700 and 1720 cm^{-1}, and strong, broad absorption over the region 2500–3000 cm^{-1}. Propose a structural formula for each compound.

(a) $C_{10}H_{12}O_3$

^{1}H-NMR	^{13}C-NMR
2.49 (t, 2H)	173.89
2.80 (t, 2H)	157.57
3.72 (s, 3H)	132.62
6.78 (d, 2H)	128.99
7.11 (d, 2H)	113.55
12.4 (s, 1H)	54.84
	35.75
	29.20

(b) $C_{10}H_{10}O_2$

^{1}H-NMR	^{13}C-NMR
2.34 (s, 3H)	167.82
6.38 (d, 1H)	143.82
7.18 (d, 1H)	139.96
7.44 (d, 2H)	131.45
7.56 (d, 2H)	129.37
12.0 (s, 1H)	127.83
	111.89
	21.13

Acidity of Phenols

19.32 Account for the fact that *p*-nitrophenol is a stronger acid than phenol. Consider both the resonance and inductive effects of the nitro group.

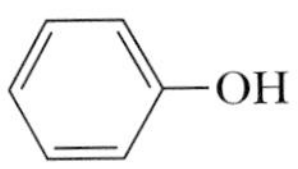

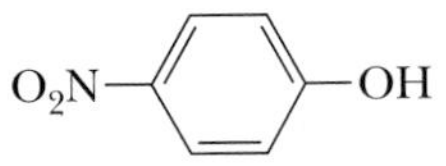

C_6H_5–OH $\quad K_a = 1.1 \times 10^{-10}$

O_2N–C_6H_4–OH $\quad K_a = 7.0 \times 10^{-8}$

19.33 Account for the fact that water-insoluble carboxylic acids (pK_a 4–5) dissolve in 10% aqueous sodium bicarbonate (pH 8.5) with the evolution of a gas but water-insoluble phenols (pK_a 9.5–10.5) do not dissolve in 10% sodium bicarbonate.

19.34 Match each compound with its appropriate pK_a value.

(a) 4-Nitrobenzoic acid, benzoic acid, 4-chlorobenzoic acid
pK_a = 4.19, 3.98, and 3.41

(b) Benzoic acid, cyclohexanol, phenol pK_a = 18.0, 9.95, and 4.19

(c) 4-Nitrobenzoic acid, 4-nitrophenol, 4-nitrophenylacetic acid
pK_a = 7.15, 3.85, and 3.41

19.35 Arrange the molecules and ions in each set in order of increasing acidity (from least acidic to most acidic).

(a) 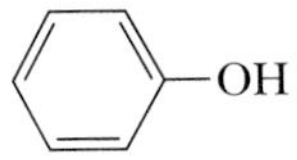 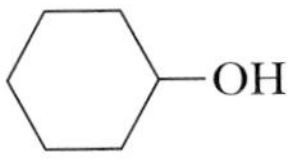 C_6H_5–OH $\quad C_6H_{11}$–OH $\quad CH_3CO_2H$

(b) 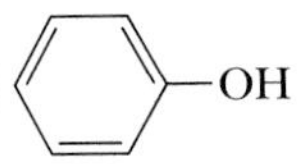 C_6H_5–OH $\quad HCO_3^-$ $\quad H_2O$

(c) 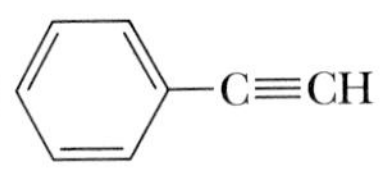 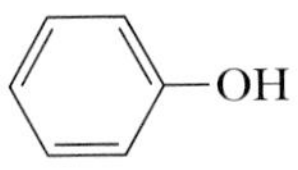 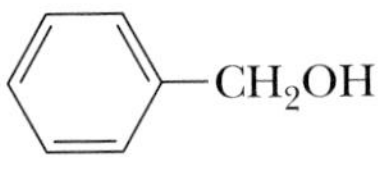 C_6H_5–C≡CH $\quad C_6H_5$–OH $\quad C_6H_5$–CH_2OH

19.36 Explain the trends in the acidity of phenol and the monofluoro derivatives of phenol:

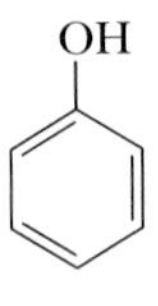

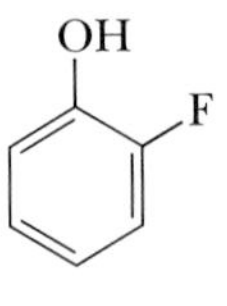

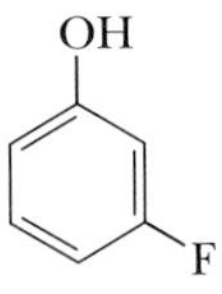

Phenol (OH) pK_a 10.0 $\quad$ 2-fluorophenol (OH, F) pK_a 8.81 $\quad$ 3-fluorophenol (OH, F) pK_a 9.28 $\quad$ 4-fluorophenol (OH, F) pK_a 9.81

19.37 You wish to determine the inductive effects of a series of functional groups, for example Cl, Br, CN, CO_2H, and C_6H_5. Is it best to use a series of ortho-, meta-, or para-substituted phenols? Explain your answer.

19.38 From each pair, select the stronger base.

(a) 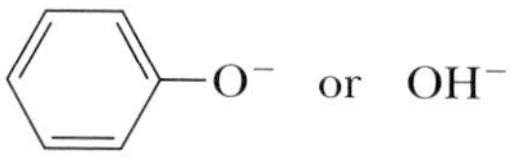 $C_6H_5O^-$ or OH^- **(b)** $C_6H_5O^-$ or $C_6H_{11}O^-$

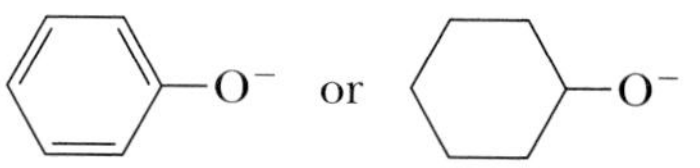

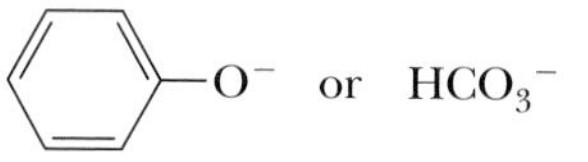

(c) $C_6H_5O^-$ or HCO_3^- **(d)** 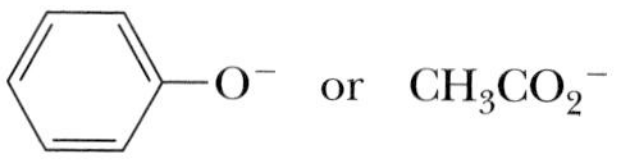 $C_6H_5O^-$ or $CH_3CO_2^-$

19.39 Describe a chemical procedure to separate a mixture of benzyl alcohol and *o*-cresol and recover each in pure form.

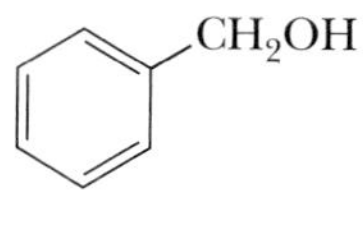

CH_2OH CH_3 OH

Benzyl alcohol 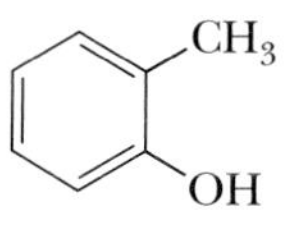 *o*-Cresol

19.40 The compound 2-hydroxypyridine, a derivative of pyridine, is in equilibrium with 2-pyridone. 2-Hydroxypyridine is aromatic. Does 2-pyridone have comparable aromatic character? Explain.

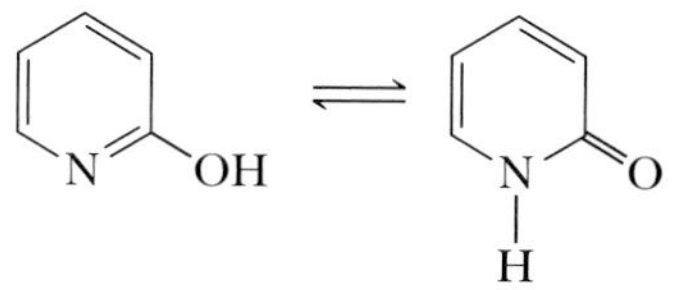

N OH ⇌ N O H

2-Hydroxypyridine 2-Pyridone

Reactions at the Benzylic Position

19.41 Write a balanced equation for the oxidation of *p*-xylene to 1,4-benzenedicarboxylic acid (terephthalic acid) using potassium dichromate in aqueous sulfuric acid. How many milligrams of $K_2Cr_2O_7$ is required to oxidize 250 mg of *p*-xylene to terephthalic acid?

19.42 Each of the following reactions occurs by a radical chain mechanism. (Consult Appendix 3 for bond dissociation energies.)

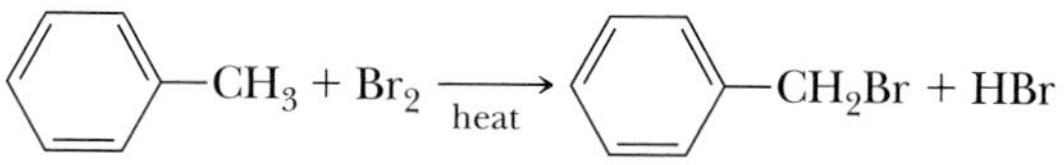

$$C_6H_5CH_3 + Br_2 \xrightarrow{\text{heat}} C_6H_5CH_2Br + HBr$$

Toluene Benzyl bromide

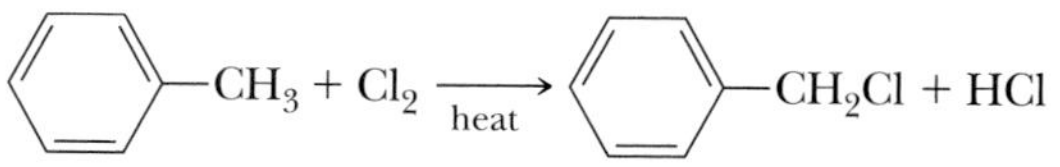

$$C_6H_5CH_3 + Cl_2 \xrightarrow{\text{heat}} C_6H_5CH_2Cl + HCl$$

Toluene Benzyl chloride

(a) Calculate the heat of reaction, ΔH^0, in kilocalories per mole for each reaction.

(b) Write a pair of chain propagation steps for each mechanism, and show that the net result of the chain propagation steps is the observed reaction.

(c) Calculate ΔH^0 for each chain propagation step, and show that the sum for each pair of chain propagation steps is identical with the ΔH^0 value calculated in part (a).

19.43 Following is an equation for iodination of toluene.

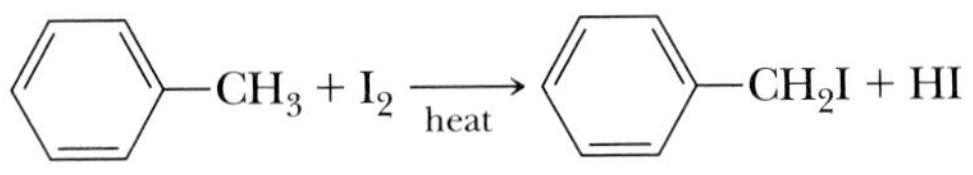

$$C_6H_5{-}CH_3 + I_2 \xrightarrow{\text{heat}} C_6H_5{-}CH_2I + HI$$

Toluene Benzyl iodide

This reaction does not take place. All that happens under experimental conditions for the formation of radicals is initiation to form iodine radicals, I·, followed by termination to reform I_2. How do you account for these observations?

19.44 Although most alkanes react with chlorine by a radical chain mechanism when reaction is initiated by light or heat, benzene fails to react under the same conditions. Benzene cannot be converted to chlorobenzene by treatment with chlorine in the presence of light or heat.

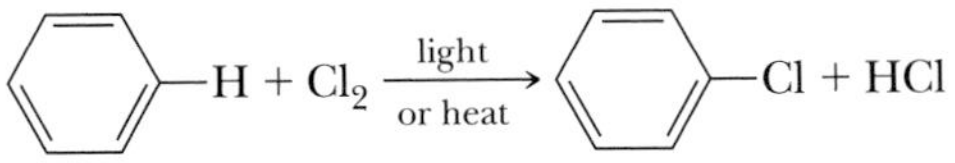

$$C_6H_5{-}H + Cl_2 \xrightarrow[\text{or heat}]{\text{light}} C_6H_5{-}Cl + HCl$$

(a) Explain why benzene fails to react under these conditions. Consult Appendix 3 for relevant bond dissociation energies. (*Hint:* Consider a possible radical chain mechanism, and the energetics of its rate-limiting step.)

(b) Explain why the bond dissociation energy of a C—H bond in benzene is significantly greater than that in alkanes. (*Hint:* Think of the orbitals forming the C—H bond in alkanes compared with those in benzene.)

19.45 Following is an equation for hydroperoxidation of cumene.

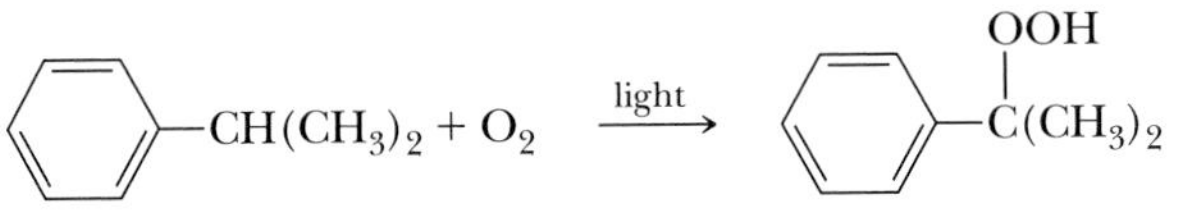

$$C_6H_5{-}CH(CH_3)_2 + O_2 \xrightarrow{\text{light}} C_6H_5{-}C(OOH)(CH_3)_2$$

Cumene Cumene hydroperoxide

Propose a radical chain mechanism for this reaction. Assume that initiation is by an unspecified radical, R·.

19.46 Para-substituted benzyl halides undergo reaction with methanol by an S_N1 mechanism to give a benzyl ether. Account for the following order of reactivity under these conditions.

$$R{-}C_6H_4{-}CH_2Br + CH_3OH \xrightarrow[\text{methanol}]{} R{-}C_6H_4{-}CH_2OCH_3 + HBr$$

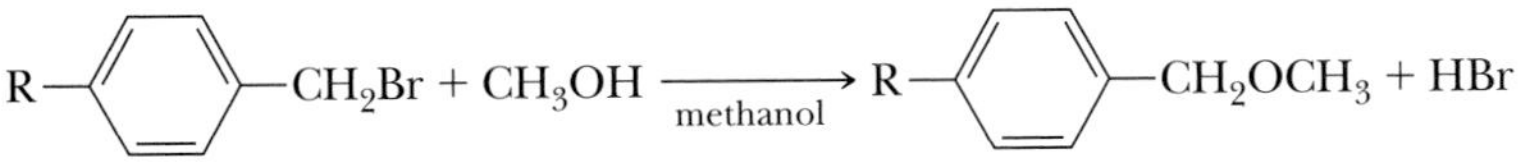

Rate of S_N1 reaction: R = $CH_3O{-} > CH_3{-} > H{-} > NO_2{-}$

19.47 When warmed in dilute sulfuric acid, 1-phenyl-1,2-propanediol undergoes dehydration and rearrangement to give 2-phenylpropanal.

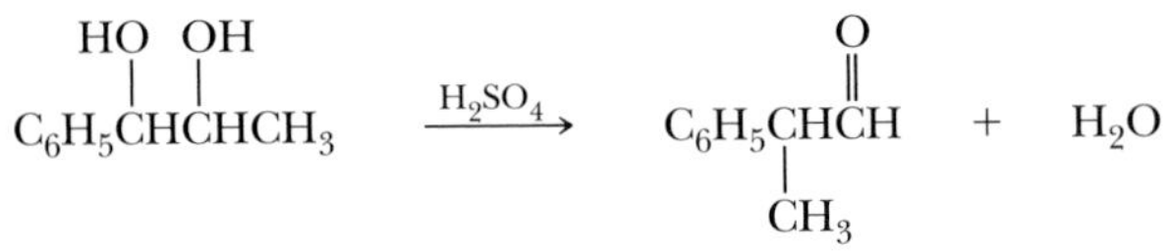

$$C_6H_5CH(OH)CH(OH)CH_3 \xrightarrow{H_2SO_4} C_6H_5CH(CH_3)CHO + H_2O$$

1-Phenyl-1,2-propanediol 2-Phenylpropanal

(a) Propose a mechanism for this example of a pinacol rearrangement (Section 9.8).

(b) Account for the fact that 2-phenylpropanal is formed rather than its constitutional isomer, 1-phenyl-1-propanone.

19.48 In the chemical synthesis of DNA and RNA, hydroxyl groups are normally converted to triphenylmethyl (trityl) ethers to protect the hydroxyl group from reaction with other reagents.

$$\underset{}{RCH_2OH} + \underset{\substack{\text{Triphenylmethyl chloride}\\\text{(trityl chloride)}}}{Ph_3CCl} \xrightarrow{\text{tertiary amine}} \underset{\substack{\text{A triphenylmethyl ether}\\\text{(a trityl ether)}}}{RCH_2OCPh_3} + \overset{\text{neutralized by the tertiary amine}}{HCl}$$

Triphenylmethyl ethers are stable to aqueous base but are rapidly cleaved in aqueous acid.

$$RCH_2OCPh_3 + H_2O \xrightarrow{H^+} RCH_2OH + Ph_3COH$$

(a) Why are triphenylmethyl ethers so readily hydrolyzed by aqueous acid?
(b) How might the structure of the triphenylmethyl group be modified in order to increase or decrease its acid sensitivity?

Synthesis

19.49 Using ethylbenzene as the only aromatic starting material, show how to synthesize the following compounds. In addition to ethylbenzene, use any other necessary organic or inorganic chemicals. Note that any compound already synthesized in one part of this problem may then be used to make any other compound in the problem.

(a) C_6H_5COOH **(b)** $C_6H_5CHBrCH_3$ **(c)** $C_6H_5CH{=}CH_2$ **(d)** $C_6H_5CH(OH)CH_3$

(e) $C_6H_5COCH_3$ **(f)** $C_6H_5CH_2CH_2OH$ **(g)** $C_6H_5CH_2CHO$ **(h)** $C_6H_5CH_2COOH$

(i) $C_6H_5CHBrCH_2Br$ **(j)** $C_6H_5C{\equiv}CH$ **(k)** $C_6H_5C{\equiv}CCH_2CH{=}CH_2$

(l) $C_6H_5C{\equiv}C(CH_2)_5CH_3$ **(m)** (E)-$C_6H_5CH{=}CH(CH_2)_5CH_3$ (C_6H_5 and H on one carbon, H and $(CH_2)_5CH_3$ on the other; C_6H_5 and $(CH_2)_5CH_3$ trans) **(n)** (Z)-$C_6H_5CH{=}CH(CH_2)_5CH_3$ (C_6H_5 and $(CH_2)_5CH_3$ cis)

19.50 Show how to convert 1-phenylpropane into the following compounds. In addition to this starting material, use any necessary inorganic reagents. Any compound synthesized in one part of this problem may then be used to make any other compound in the problem.

(a) $C_6H_5CHBrCH_2CH_3$ **(b)** $C_6H_5CH{=}CHCH_3$ **(c)** $C_6H_5CHClCHClCH_3$

(d) $C_6H_5C{\equiv}CCH_3$ **(e)** (E)-$C_6H_5CH{=}CHCH_3$ (C_6H_5 and CH_3 trans) **(f)** (Z)-$C_6H_5CH{=}CHCH_3$ (C_6H_5 and CH_3 cis)

(g) $C_6H_5CH(OH)CH(OH)CH_3$ **(h)** $C_6H_5CH(OH)CH_2CH_3$ **(i)** $C_6H_5COCH_2CH_3$

19.51 Propranolol is a β-adrenergic receptor antagonist. Members of this class have received enormous clinical attention because of their effectiveness in treating hypertension (high blood pressure), migraine headaches, glaucoma, ischemic heart disease, and certain cardiac arrhythmias. Starting materials for the synthesis of propranolol are propene, 1-naphthol, and isopropylamine. Show how to convert propene to epichlorohydrin in stage 1, and then complete the synthesis of propranolol in stage 2.

Stage 1: Synthesis of epichlorohydrin

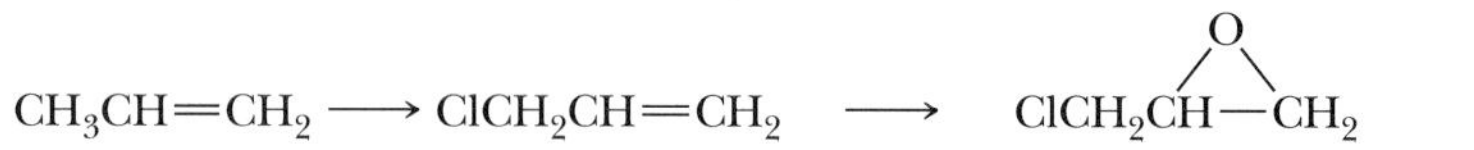

$$CH_3CH{=}CH_2 \longrightarrow ClCH_2CH{=}CH_2 \longrightarrow ClCH_2\underset{\backslash O /}{CH{-}CH_2}$$

Propene — 3-Chloropropene (Allyl chloride) — 3-Chloro-1,2-epoxypropane (Epichlorohydrin)

Stage 2: Synthesis of propranolol

1-Naphthol $\xrightarrow{ClCH_2CHCH_2 \text{ (epoxide)}}$ 1-naphthyl–OCH_2CHCH_2 (epoxide) $\xrightarrow{(CH_3)_2CHNH_2}$ 1-naphthyl–$OCH_2CH(OH)CH_2NHCH(CH_3)_2$

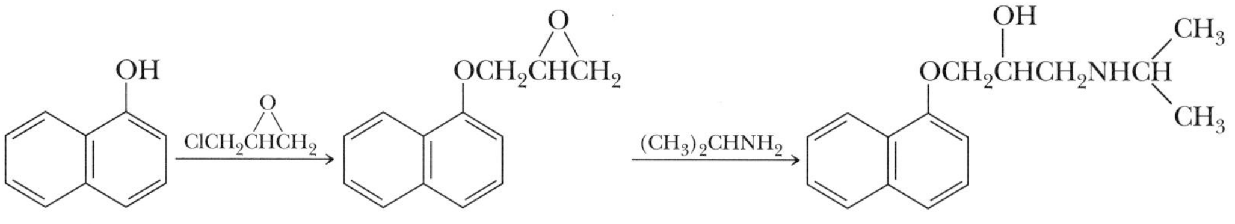

1-Naphthol — Propranolol

19.52 Side effects of propranolol (Problem 19.51) are disturbances of the central nervous system (CNS) such as fatigue, sleep disturbances (including insomnia and nightmares), and depression. Pharmaceutical companies wondered if this drug could be redesigned to eliminate or at least reduce these side effects. Propranolol itself is highly lipophilic (hydrophobic) and readily passes through the blood-brain barrier, a lipid-like protective membrane that surrounds the capillary system in the brain and prevents hydrophilic compounds from entering the brain by passive diffusion. Propranolol, it was reasoned, enters the CNS by passive diffusion because of the lipid-like character of its naphthalene ring. The challenge, then, was to design a more hydrophilic drug that does not cross the blood-brain barrier but still retains a β-adrenergic antagonist property. A product of this research is atenolol, a potent β-adrenergic blocker that is hydrophilic enough that it crosses the blood-brain barrier to only a very limited extent.

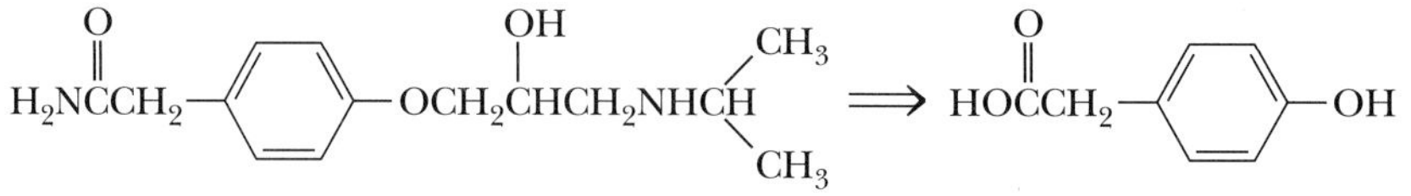

$$H_2N\overset{O}{\overset{\|}{C}}CH_2-C_6H_4-OCH_2\overset{OH}{\overset{|}{C}H}CH_2NHCH(CH_3)_2 \Longrightarrow HO\overset{O}{\overset{\|}{C}}CH_2-C_6H_4-OH$$

Atenolol (a β-adrenergic antagonist) — 4-Hydroxyphenylacetic acid

Propose a synthesis for atenolol starting with 4-hydroxyphenylacetic acid, epichlorohydrin (Problem 19.51), and isopropylamine.

19.53 Benzylic bromination followed by loss of HBr by heating at high temperatures can be used to generate reactive intermediates such as (1). How do you take advantage of this observation to synthesize hexaradialene. (See L. G. Harruff, M. Brown, and V. Boekelheide, *J. Am. Chem. Soc.*, **100,** 2893, 1978).

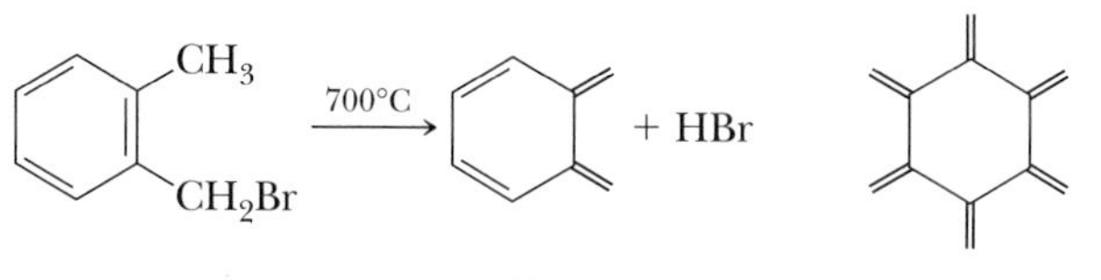

CH_3 / CH_2Br $\xrightarrow{700°C}$ (1) + HBr Hexaradialene

19.54 Carbinoxamine is a histamine antagonist, specifically an H_1-antagonist. The maleic acid salt of the levorotatory isomer is sold as the prescription drug Rotoxamine. Propose a synthesis of carbinoxamine from the three compounds shown on the left of the reaction arrow. (*Note:* Aryl bromides form Grignard reagents much more readily than aryl chlorides.)

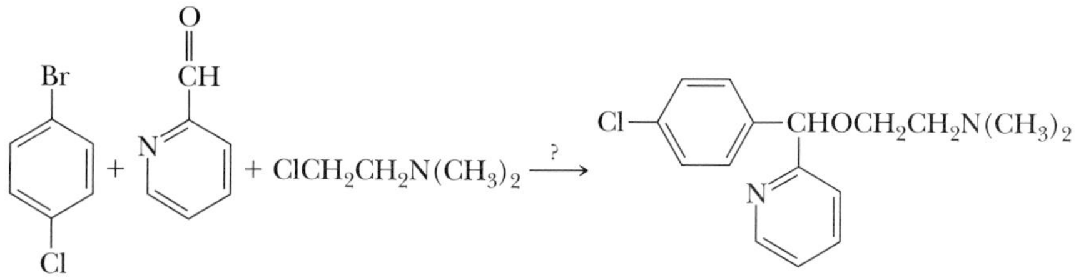

Br, Cl + (pyridine-2-CHO) + $ClCH_2CH_2N(CH_3)_2$ $\xrightarrow{?}$ $Cl-C_6H_4-CHOCH_2CH_2N(CH_3)_2$ (2-pyridyl)

Carbinoxamine

19.55 Cromolyn sodium, developed in the 1960s, is used to prevent allergic reactions primarily affecting the lungs, as for example exercise-induced emphysema. It is thought to block the release of histamine, which prevents the sequence of events leading to swelling, itching, and to constriction of bronchial tubes. Cromolyn sodium is synthesized in the following series of steps. Treatment of one mol of epichlorohydrin with two mol of 2,6-dihydroxyacetophenone in the presence of base gives I. Treatment of I with two mol of diethyl oxalate in the presence of sodium ethoxide gives a diester II. Saponification of the diester with aqueous NaOH gives cromolyn sodium.

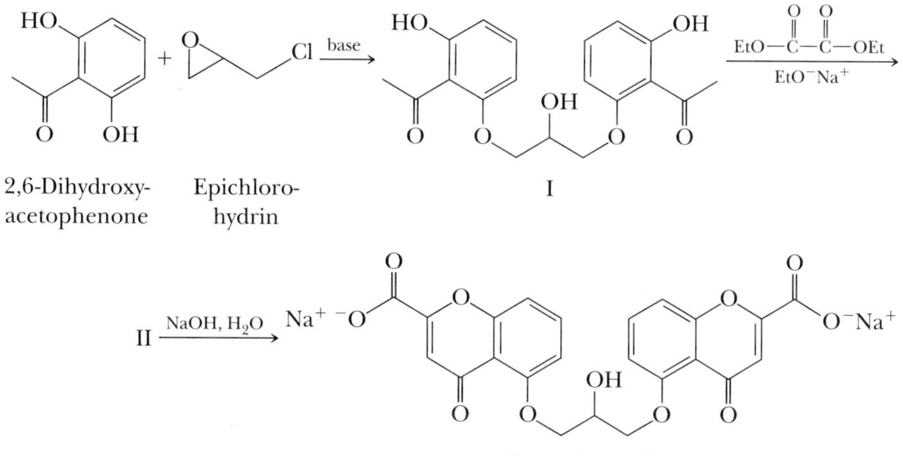

2,6-Dihydroxyacetophenone + Epichlorohydrin $\xrightarrow{\text{base}}$ I $\xrightarrow[EtO^-Na^+]{EtO-\overset{O}{\overset{\|}{C}}-\overset{O}{\overset{\|}{C}}-OEt}$

II $\xrightarrow{NaOH,\ H_2O}$ Cromolyn sodium ($Na^+\ ^-O$ … O^-Na^+)

(a) Propose a mechanism for the formation of compound I.

(b) Propose a structural formula for compound II and a mechanism for its formation.

20 AROMATICS II: REACTIONS OF BENZENE AND ITS DERIVATIVES

■ Crystals of bisphenol A (see Problem 20.22) viewed under polarizing light. *(Herb Charles Ohlmeyer/Fran Heyl Associates)*

By far the most characteristic reaction of aromatic compounds is substitution at a ring carbon. Some groups that can be introduced directly on the ring are the halogens, the nitro ($-NO_2$) group, the sulfonic acid ($-SO_3H$) group, alkyl ($-R$) groups, and acyl ($RCO-$) groups. Each of these substitution reactions is represented in the following equations.

Halogenation:

$$C_6H_5-H + Cl_2 \xrightarrow{FeCl_3} C_6H_5-Cl + HCl$$

Chlorobenzene

Nitration:

$$C_6H_5-H + HNO_3 \xrightarrow{H_2SO_4} C_6H_5-NO_2 + H_2O$$

Nitrobenzene

Sulfonation:

$$C_6H_5{-}H + SO_3 \xrightarrow{H_2SO_4} C_6H_5{-}SO_3H$$

Benzenesulfonic acid

Alkylation:

$$C_6H_5{-}H + RX \xrightarrow{AlX_3} C_6H_5{-}R + HX$$

An alkylbenzene

Acylation:

$$C_6H_5{-}H + RCX(=O) \xrightarrow{AlX_3} C_6H_5{-}CR(=O) + HX$$

An acylbenzene

We take these reactions one at a time and examine their common mechanistic theme.

20.1 Electrophilic Aromatic Substitution

An **electrophile** is an electron-deficient species that can accept a pair of electrons from a nucleophile to form a new covalent bond. A reaction in which a hydrogen atom of an aromatic ring is replaced by an electrophile is called **electrophilic aromatic substitution.**

Electrophilic aromatic substitution A reaction in which there is substitution of an electrophile for a hydrogen on an aromatic ring.

$$C_6H_5{-}H + E^+ \longrightarrow C_6H_5{-}E + H^+$$

In this and the following sections, we study several common types of electrophiles, how they are generated, and the mechanisms by which they replace hydrogen on an aromatic ring.

A. Chlorination and Bromination

Chlorine alone does not react with benzene, in contrast to the instantaneous addition of chlorine to cyclohexene. In the presence of a Lewis acid catalyst, such as ferric chloride or aluminum chloride (Section 3.5), a reaction does take place to give chlorobenzene and HCl. The first step in this reaction involves interaction of chlorine and the Lewis acid catalyst. A chlorine atom of Cl_2 donates a pair of electrons to the Lewis acid to form a molecular complex with a positive charge on chlorine and a negative charge on iron. Redistribution of electrons in this complex generates a **chloronium ion** as part of an ion pair. Reaction of the $Cl_2{-}FeCl_3$ complex with the pi electron cloud of the aromatic ring forms a resonance-stabilized cation intermediate, here represented as a hybrid of three contributing structures. Proton transfer from the cation intermediate to $FeCl_4^-$ forms HCl, regenerates the Lewis acid catalyst, and gives chlorobenzene.

MECHANISM Electrophilic Aromatic Substitution: Chlorination

Step 1: Reaction between chlorine and the Lewis acid to form a chloronium ion

$$:\ddot{Cl}-\ddot{Cl}: + FeCl_3 \rightleftharpoons :\ddot{Cl}-\overset{+}{\ddot{Cl}}-\overset{-}{Fe}Cl_3 \rightleftharpoons :\ddot{Cl}^+ \; FeCl_4^-$$

Chlorine (a Lewis base) — Ferric chloride (a Lewis acid) — A molecular complex with a positive charge on chlorine and a negative charge on iron — An ion pair containing the chloronium ion

Step 2: Attack of the chloronium ion on the aromatic ring to give a cation intermediate

$$C_6H_6 + Cl^+ \xrightarrow{\text{rate-limiting step}} [C_6H_6Cl]^+$$

Resonance-stabilized cation intermediate; the positive charge is delocalized onto three atoms of the ring

Step 3: Proton transfer to regenerate the aromatic character of the ring

$$[C_6H_6Cl]^+ + Cl-\overset{-}{Fe}Cl_3 \xrightarrow{\text{fast}} C_6H_5Cl + HCl + FeCl_3$$

Cation intermediate — Chlorobenzene

Treatment of benzene with bromine in the presence of ferric chloride or aluminum chloride gives bromobenzene and HBr. The mechanism for this reaction is the same as that for chlorination of benzene.

We can write the following two-step general mechanism for electrophilic aromatic substitution. The first and rate-limiting step is attack of the electrophile, E^+, on the aromatic ring to give a resonance-stabilized cation intermediate. In the second and faster step, loss of H^+ from the cation intermediate regenerates the aromatic ring and gives the product.

Step 1:

$$C_6H_5-H + E^+ \xrightarrow{\text{rate-limiting step}} [C_6H_5(H)E]^+$$

Electrophile — Resonance-stabilized cation intermediate

Step 2:

$$[C_6H_5(H)E]^+ \xrightarrow{\text{fast}} C_6H_5-E + H^+$$

The major difference between addition of halogen to an alkene and substitution by halogen on an aromatic ring centers on the fate of the cationic intermediate formed in the first step of each reaction. Recall from Section 6.3D that addition of chlorine

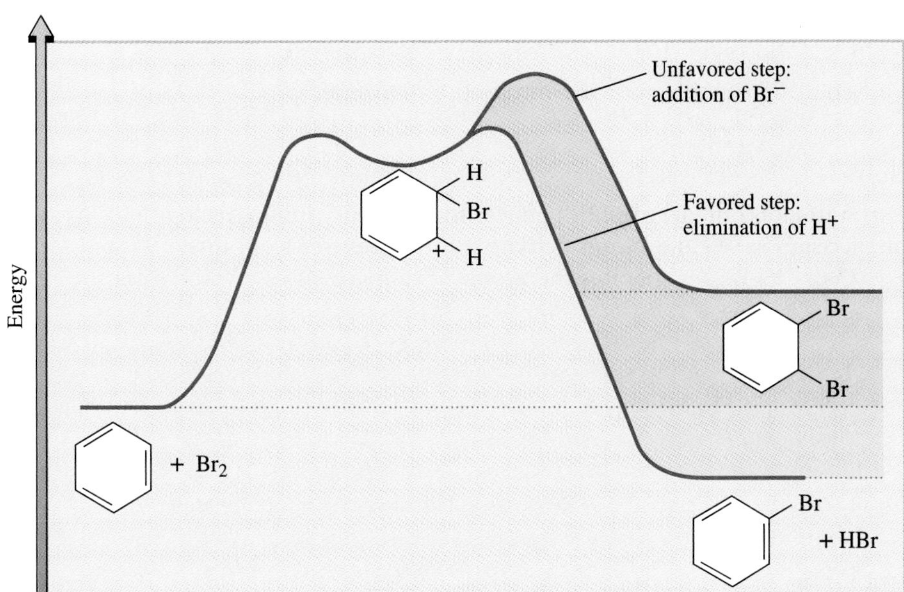

FIGURE 20.1
Potential energy diagram for the reaction of benzene with bromine. Reaction of the cation intermediate to form an addition product results in loss of the resonance stabilization of the aromatic ring. Reaction of the cation intermediate to form a substitution product regenerates the resonance-stabilized aromatic ring.

to an alkene is a two-step process, the first and slower step of which is formation of a bridged chloronium ion intermediate. This intermediate then reacts with chloride ion to complete the addition. With aromatic compounds, the cation intermediate loses H^+ to regenerate the aromatic ring and regain the large resonance stabilization. There is no such resonance stabilization to be regained in the case of an alkene. The energy diagram in Figure 20.1 shows both addition and substitution reactions of benzene.

B. Nitration and Sulfonation

The sequence of steps for nitration and sulfonation of benzene is similar to that for chlorination and bromination. For nitration, the electrophile is the **nitronium ion,** NO_2^+, generated by reaction of nitric acid and sulfuric acid.

MECHANISM Formation of the Nitronium Ion

$$\underset{\text{Nitric acid}}{H{-}\ddot{O}{-}NO_2} + H{-}O{-}SO_3H \rightleftharpoons H{-}\overset{\overset{H}{|}}{\overset{+}{O}}{-}NO_2 + HSO_4^-$$

$$H{-}\overset{\overset{H}{|}}{\overset{+}{O}}{-}NO_2 \rightleftharpoons H{-}\overset{\overset{H}{|}}{\ddot{O}}: + \underset{\text{Nitronium ion}}{:\ddot{O}{=}\overset{+}{N}{=}\ddot{O}:}$$

EXAMPLE 20.1

Write a stepwise mechanism for the nitration of benzene.

Solution

The nitronium ion (an electrophile) attacks the benzene ring (a nucleophile) to form a resonance-stabilized cation intermediate. Proton transfer from this intermediate to either H_2O or HSO_4^- regenerates the aromatic ring and gives nitrobenzene.

$$C_6H_6 + NO_2^+ \longrightarrow [\text{H, NO}_2 \text{ cation} \longleftrightarrow \text{H, NO}_2 \text{ cation} \longleftrightarrow \text{H, NO}_2 \text{ cation}]$$

Resonance-stabilized cation intermediate

$$H_2\ddot{O}: + \text{(H, NO}_2\text{ cation)} \longrightarrow C_6H_5NO_2 + H_3O^+$$

Nitrobenzene

A particular value of nitration is that the resulting nitro group, $-NO_2$, can be reduced to a primary amino group, $-NH_2$, by hydrogenation in the presence of a transition metal catalyst such as nickel, palladium, or platinum under fairly mild conditions. This method has the potential disadvantage that other susceptible groups such as carbon-carbon double bonds and aldehyde and ketone carbonyl groups are also reduced under these conditions. Note that neither the $-CO_2H$ nor the aromatic ring is reduced under these conditions.

$$O_2N-C_6H_4-CO_2H + 3H_2 \xrightarrow[\text{(3 atm)}]{\text{Ni}} H_2N-C_6H_4-CO_2H + 2H_2O$$

4-Nitrobenzoic acid → 4-Aminobenzoic acid

Alternatively, a nitro group can be reduced to a primary amino group by a metal in acid. The most commonly used metal reducing agents are iron, zinc, and tin in dilute HCl. When reduced with a metal and hydrochloric acid, the amine is obtained as a salt which is then treated with strong base to liberate the free amine.

$$\text{2,4-}(NO_2)_2C_6H_3CH_3 \xrightarrow[C_2H_5OH,\ H_2O]{\text{Fe, HCl}} \text{2,4-}(NH_3^+Cl^-)_2C_6H_3CH_3 \xrightarrow{\text{NaOH, } H_2O} \text{2,4-}(NH_2)_2C_6H_3CH_3$$

2,4-Dinitrotoluene → 4-Methyl-1,3-benzenediamine (2,4-Diaminotoluene)

Sulfonation of benzene is carried out using concentrated sulfuric acid containing dissolved sulfur trioxide (fuming sulfuric acid). In the following equation, the sulfonating agent is shown as sulfur trioxide. The electrophile is either SO_3 or HSO_3^+, depending on experimental conditions.

$$\text{Benzene} + SO_3 \xrightarrow{H_2SO_4} C_6H_5{-}SO_3H$$

Benzene — Benzenesulfonic acid

PROBLEM 20.1

Write the stepwise mechanism for sulfonation of benzene by hot, concentrated sulfuric acid. In this reaction, the electrophile is SO_3 formed as shown in the following equation. (*Hint:* In thinking about a mechanism for this reaction, consider formal charges on sulfur and oxygen in the Lewis structure of sulfur trioxide.)

$$H_2SO_4 \rightleftharpoons SO_3 + H_2O$$

C. Friedel-Crafts Alkylation and Acylation

Alkylation of aromatic hydrocarbons was discovered in 1877 by the French chemist Charles Friedel and a visiting American chemist, James Crafts. They discovered that mixing benzene, an alkyl halide, and $AlCl_3$ results in formation of an alkylbenzene and HX. **Friedel-Crafts alkylation** forms a new carbon-carbon bond between benzene and an alkyl group, as illustrated by the reaction of benzene with 2-chloropropane in the presence of aluminum chloride.

$$\text{Benzene} + CH_3CH(CH_3)Cl \xrightarrow{AlCl_3} C_6H_5{-}CH(CH_3)_2 + HCl$$

Benzene — 2-Chloropropane (Isopropyl chloride) — Cumene (Isopropylbenzene)

Friedel-Crafts alkylation is among the most important methods for forming new carbon-carbon bonds to aromatic rings. It begins with formation of a complex between the alkyl halide and aluminum chloride. In this complex, aluminum has a negative charge, and the halogen of the alkyl halide has a positive charge. The alkyl group can also be written as a carbocation, although it is unlikely that a free carbocation is actually formed, especially in the case of the relatively unstable primary and secondary carbocations. Nonetheless, we very often represent the reactive intermediate as a carbocation to simplify the mechanism. Reaction of an alkyl carbocation with an aromatic ring gives a resonance-stabilized cation intermediate, which then loses a hydrogen to give an alkylbenzene.

MECHANISM Friedel-Crafts Alkylation

Step 1: Formation of an alkyl carbocation as an ion pair

$$R-\ddot{Cl}: + AlCl_3 \rightleftharpoons R-\overset{+}{Cl}-\overset{-}{Al}Cl_3 \rightleftharpoons R^+ \; AlCl_4^-$$

A molecular complex with a positive charge on chlorine and a negative charge on aluminum

An ion pair containing a carbocation

Step 2: Attack of the alkyl cation on the aromatic ring to form a cation intermediate

The positive charge is delocalized onto three atoms of the ring

Step 3: Proton transfer to regenerate the aromatic character of the ring

$$Cl-Al^-Cl_3 \quad \longrightarrow \quad C_6H_5-R + AlCl_3 + HCl$$

Vinylic halides and aryl halides generally do not react under conditions of the Friedel-Crafts alkylation because of the high activation energy required to form vinylic and aryl carbocations.

There are two major limitations on Friedel-Crafts alkylation. First is the possibility for rearrangement of the alkyl group, which occurs in the following way. Friedel-Crafts alkylation involves the generation of a carbocation and, as we have already seen in Section 6.3D, carbocations may undergo any of three different reactions: (1) reaction with a nucleophile; (2) loss of H^+ to form an alkene; and (3) rearrangement to a more stable carbocation followed by reactions (1) or (2), or by both (1) and (2). **Carbocation rearrangements** are common in Friedel-Crafts alkylations. For example, reaction of benzene with 1-chloro-2-methylpropane (isobutyl chloride) gives only 2-methyl-2-phenylpropane (*tert*-butylbenzene).

$$C_6H_6 + CH_3\underset{}{\overset{CH_3}{CH}}CH_2Cl \xrightarrow{AlCl_3} C_6H_5-C(CH_3)_3 + HCl$$

Benzene | 1-Chloro-2-methylpropane (Isobutyl chloride) | 2-Methyl-2-phenylpropane (*tert*-Butylbenzene)

In this case, the isobutyl chloride/$AlCl_3$ complex rearranges directly to the *tert*-butyl cation/$AlCl_4^-$ ion pair, which then is the electrophile in this example of Friedel-Crafts alkylation.

$$CH_3CH(CH_3)CH_2{-}\ddot{Cl}: + AlCl_3 \longrightarrow CH_3C(CH_3)(H){-}CH_2{-}\overset{+}{Cl}{-}\overset{-}{Al}Cl_3 \longrightarrow CH_3\overset{+}{C}(CH_3)_2\ AlCl_4^-$$

Isobutyl chloride — Isobutyl chloride-aluminum chloride complex — *tert*-Butyl cation/$AlCl_4^-$ ion pair

In practice, alkylation with primary halides is not useful, and alkylbenzenes containing a primary alkyl group other than $-CH_2CH_3$ must be prepared by other means. Alkylation is useful for introducing isopropyl, *tert*-butyl, and other alkyl groups, the cations of which do not tend to rearrange.

A second limitation on Friedel-Crafts alkylation is that it fails altogether on benzene rings bearing one or more strongly electron-withdrawing groups. We shall see in the following section that substituents on a benzene ring have a dramatic effect on the ring's reactivity toward further electrophilic aromatic substitution.

$$C_6H_5Y + RX \xrightarrow{AlCl_3} \text{no reaction}$$

When Y Equals Any of These Groups, the Benzene Ring Does Not Undergo Friedel-Crafts Alkylation

$-\overset{O}{\overset{\|\|}{C}}H$	$-\overset{O}{\overset{\|\|}{C}}R$	$-\overset{O}{\overset{\|\|}{C}}OH$	$-\overset{O}{\overset{\|\|}{C}}OR$	$-\overset{O}{\overset{\|\|}{C}}NH_2$
$-SO_3H$	$-C{\equiv}N$	$-NO_2$	$-NR_3^+$	
$-CF_3$	$-CCl_3$			

Friedel and Crafts also discovered that treatment of an aromatic hydrocarbon with an acyl halide (Section 17.1A) in the presence of aluminum chloride gives a ketone. An RCO— group is known as an acyl group; hence, reaction of an aromatic hydrocarbon with an acyl halide is known as **Friedel-Crafts acylation,** as illustrated by the reaction of benzene and acetyl chloride in the presence of aluminum chloride to form acetophenone.

$$C_6H_6 + CH_3\overset{O}{\overset{\|\|}{C}}Cl \xrightarrow{AlCl_3} C_6H_5\overset{O}{\overset{\|\|}{C}}CH_3 + HCl$$

Benzene — Acetyl chloride (an acyl halide) — Acetophenone (a ketone)

The following example of electrophilic aromatic substitution involves intramolecular acylation to form a six-membered ring.

$$\text{4-Phenylbutanoyl chloride} \xrightarrow{AlCl_3} \alpha\text{-Tetralone} + HCl$$

4-Phenylbutanoyl chloride

α-Tetralone

Friedel-Crafts acylation begins with donation of a pair of electrons from the halogen of the acyl halide to aluminum chloride to form a molecular complex similar to what we drew for Friedel-Crafts alkylations. In this complex, halogen has a positive formal charge and aluminum has a negative formal charge. Redistribution of electrons of the carbon-chlorine bond gives an ion pair containing an **acylium ion.**

Acylium ion A resonance-stabilized cation with the structure $[RC{=}O]^+$ or $[ArC{=}O]^+$. The positive charge is delocalized over both the carbonyl carbon and the carbonyl oxygen.

MECHANISM Friedel-Crafts Acylation: Generation of an Acylium Ion

$$R{-}\overset{O}{\overset{\|}{C}}{-}Cl + AlCl_3 \rightleftharpoons R{-}\overset{O}{\overset{\|}{C}}{-}\overset{+}{Cl}{-}\overset{-}{Al}Cl_3 \rightleftharpoons R{-}\overset{O}{\overset{\|}{C}}{}^{+} \; AlCl_4^-$$

Acyl chloride (a Lewis base)

Aluminum chloride (a Lewis acid)

A molecular complex with a positive charge on chlorine and a negative charge on aluminum

An ion pair containing an acylium ion

Of the two major contributing structures that can be drawn for an acylium ion, the one with complete valence shells for both carbon and oxygen makes the greater contribution to the hybrid.

$$R{-}\overset{+}{C}{=}\ddot{O}: \longleftrightarrow R{-}C{\equiv}\overset{+}{O}:$$

Both atoms have complete valence shells

More important contributing structure

Friedel-Crafts acylation is free of a major limitation on Friedel-Crafts alkylations: acyl cations do not rearrange. Thus, the carbon skeleton of an acyl halide is transferred unchanged to the aromatic ring.

EXAMPLE 20.2

Write structural formulas for the products you expect from Friedel-Crafts alkylation or acylation of benzene with

(a) $C_6H_5CH_2Cl$
Benzyl chloride

(b) C_6H_5CCl (with C=O)
Benzoyl chloride

Solution

(a) Benzyl chloride in the presence of a Lewis acid catalyst gives the benzyl cation, which then attacks benzene followed by proton transfer to give diphenylmethane. In this example, the benzyl cation, although primary, cannot rearrange.

$$C_6H_5CH_2^+ + C_6H_6 \longrightarrow C_6H_5CH_2C_6H_5 + H^+$$

Benzyl cation — Diphenylmethane

(b) Treatment of benzoyl chloride with aluminum chloride gives an acylium ion, an electrophile, which then attacks benzene to give benzophenone, a ketone.

$$C_6H_5C^+{=}O + C_6H_6 \longrightarrow C_6H_5C(=O)C_6H_5 + H^+$$

Benzoyl cation — Benzophenone

PROBLEM 20.2

Write structural formulas for the products you expect from Friedel-Crafts alkylation or acylation of benzene with

(a) $(CH_3)_3CCCl$ (with C=O) **(b)** $CH_3CH_2CH_2Cl$ **(c)** $C_6H_5CHClCH_3$

A special value of Friedel-Crafts acylations in synthesis is the preparation of unrearranged alkylbenzenes, as illustrated by the preparation of isobutylbenzene. Treatment of benzene with 2-methylpropanoyl chloride in the presence of aluminum chloride gives 2-methyl-1-phenyl-1-propanone. Wolff-Kishner or Clemmensen reduction (Section 15.14C) of the carbonyl group to a methylene group gives isobutylbenzene.

$$C_6H_6 + Cl{-}C(=O){-}CH(CH_3)CH_3 \xrightarrow{AlCl_3} C_6H_5{-}C(=O){-}CH(CH_3)CH_3 \xrightarrow[\text{diethylene glycol}]{N_2H_4,\ KOH} C_6H_5{-}CH_2CH(CH_3)CH_3$$

2-Methylpropanoyl chloride — 2-Methyl-1-phenyl-1-propanone — Isobutylbenzene

D. Other Electrophilic Aromatic Alkylations

Once it was discovered that Friedel-Crafts alkylations and acylations involve cationic electrophiles, it was realized that the same reactions can be accomplished by other combinations of reagents and catalysts. We study two of these: generation of carbocations from alkenes and from alcohols.

As we saw in Section 6.3, treatment of an alkene with a strong acid, most commonly HX, H_2SO_4, H_3PO_4, or HF/BF_3, generates a carbocation. Cumene is synthesized industrially (over 5.6 billion pounds in 1995) by the reaction of benzene with propene in the presence of phosphoric acid as a catalyst.

$$\text{Benzene} + CH_3CH{=}CH_2 \xrightarrow{H_3PO_4} C_6H_5{-}CH(CH_3)_2$$

Benzene Propene (Propylene) Cumene

Alkylation with an alkene can also be carried out with a Lewis acid catalyst. Treatment of benzene with cyclohexene in the presence of aluminum chloride gives phenylcyclohexane.

$$\text{Benzene} + \text{Cyclohexene} \xrightarrow{AlCl_3} \text{Phenylcyclohexane}$$

Benzene Cyclohexene Phenylcyclohexane

Carbocations can also be generated by treatment of an alcohol with H_2SO_4, H_3PO_4, or HF (Section 9.6A).

$$\text{Benzene} + (CH_3)_3COH \xrightarrow{H_3PO_4} C_6H_5{-}C(CH_3)_3 + H_2O$$

Benzene 2-Methyl-2-propanol (*tert*-Butyl alcohol) *tert*-Butylbenzene

EXAMPLE 20.3

Write a mechanism for the formation of isopropylbenzene (cumene) from benzene and propene in the presence of phosphoric acid.

Solution

As a first step, propose a proton transfer from phosphoric acid to propene to form the isopropyl cation. Electrophilic attack of this cation on the benzene ring forms a resonance-stabilized carbocation intermediate. Proton transfer from this intermediate to dihydrogen phosphate ion gives cumene.

$$CH_3CH{=}CH_2 + H{-}O{-}P(OH)(=O){-}O{-}H \xrightleftharpoons[\text{reversible}]{\text{fast and}} CH_3\overset{+}{C}HCH_3 + {}^{-}O{-}P(OH)(=O){-}O{-}H$$

$$C_6H_6 + {}^{+}CH(CH_3)_2 \xrightleftharpoons[\text{limiting}]{\text{slow, rate-}} [C_6H_6{-}CH(CH_3)_2]^{+}$$

$$[C_6H_5(H){-}CH(CH_3)_2]^{+} + {}^{-}O{-}P(OH)(=O){-}O{-}H \xrightarrow{\text{fast}} C_6H_5{-}CH(CH_3)_2 + H{-}O{-}P(OH)(=O){-}O{-}H$$

Cumene

PROBLEM 20.3

Write a mechanism for the formation of *tert*-butylbenzene from benzene and *tert*-butyl alcohol in the presence of phosphoric acid.

20.2 Disubstitution and Polysubstitution

A. Effects of a Substituent Group on Further Substitution

In electrophilic aromatic substitution of a monosubstituted benzene, three products are possible: substitution ortho, meta, or para to the existing group on the ring. We can make the following generalizations about the manner in which existing groups influence further substitution reactions.

Ortho-para director Any substituent on a benzene ring that directs electrophilic aromatic substitution preferentially to ortho and para positions.

Meta director Any substituent on a benzene ring that directs electrophilic aromatic substitution preferentially to a meta position.

Activating group Any substituent on a benzene ring that causes the rate of electrophilic aromatic substitution to be greater than that for benzene.

Deactivating group Any substituent on a benzene ring that causes the rate of electrophilic aromatic substitution to be lower than that for benzene.

1. Substituents affect the orientation of new groups. Certain substituents direct an incoming group preferentially to the ortho and para positions; other substituents direct it preferentially to the meta position. In other words, substituents on a benzene ring can be classified as **ortho-para directing** or as **meta directing.**
2. Substituents affect the rate of further substitution. Certain substituents cause the rate of a second substitution to be greater than that for benzene itself, whereas other substituents cause the rate of a second substitution to be lower than that for benzene. In other words, groups on a benzene ring can be classified as **activating** or **deactivating** toward further substitution.

These directing and activating-deactivating effects can be seen by comparing the products and rates of bromination of anisole and nitration of nitrobenzene. Bromination of anisole proceeds at a rate considerably greater than that for benzene (the methoxy group is activating), and the product is a mixture of *o*-bromoanisole and *p*-bromoanisole (the methoxy group is ortho-para directing).

A CONVERSATION WITH...

JAMES A. CUSUMANO

James A. Cusumano obtained a B.A. in 1964 and a Ph.D. in chemistry in 1967 from Rutgers University. He is the co-founder and chairman of Catalytica Inc. of Mountain View, California, a company that develops new catalysts, substances that speed up and control chemical reactions.

By 1990, it is estimated that the United States was producing about eight pounds per person per day of hazardous wastes and air pollutants. Cusumano believes that the most cost-effective, long-term solution to pollution control is primary prevention—avoiding the formation of pollutants at the source. Therefore, Catalytica focuses on ways to replace hazardous raw materials with safer substances and to prevent pollution during the manufacturing process. The company's goals, and its method of operation, represent a new direction for the chemical industry.

A BASEMENT LAB

James Cusumano grew up in Elizabeth, New Jersey, the first of 10 children of a postal worker and his homemaker wife. As a young child he had two loves: chemistry and music. His father wanted James to be a medical doctor. However, his father bought him a Gilbert Chemistry Set when he was 9. James set up a laboratory in his basement and decided he wanted to be a chemist. "On every birthday and Christmas all I ever wanted were chemistry books or chemicals. I started what I called O & O Research Laboratories when I was about 12, and I started to make ink. There was a grocery store around the corner. I told them I could make ink, and we could sell it and make a small profit. I guess they wanted to humor me so they let me do it. I made my own labels and packaging, and kids started to buy the stuff. Then I had my first problem. You had to put gum arabic in ink, and I put in too much. It clogged up all the pens, so I had to go back and clean the pens, and I realized I had the wrong formulation. That was my first experience with technical service."

"I got the bug to use chemistry to make other products such as cosmetics and spot removers. Then, in order to get chemicals, I wrote all over the United States to chemical companies asking for samples. I actually got visits from salesmen trying to sell 10,000 pounds of chemicals."

POP MUSIC AND CHEMISTRY

"I was also very interested in music, mainly because I wanted to start a band to play at proms. My dad had me take lessons from a band leader because I said I did not want to learn classical music. I just wanted to learn to play popular music to play at dances. Then I started to write music. I went to New York City to sell it and started to make a hundred dollars at a time. Both chemistry and music were entrepreneurial, although I didn't think of it at that time. I learned to provide a service and got paid for it."

"When I was 16, one of the songs I sold was recorded ("Short

Shorts Twist"), and I started making money. When I was 17 years old I was playing in Las Vegas, and I started making a significant amount of money and was helping my family. All the time my family said music is nice, but you have to continue school. So, I went to Rutgers to study chemistry."

"In my senior year I did an honors research project. That was my next exposure to chemistry and entrepreneurship. My professor said if you can solve this project [the direct combination of benzene (C_6H_6) and ammonia (NH_3) in the presence of oxygen to make the useful compound aniline ($C_6H_5NH_2$)] you may become a millionaire. I thought that was an exciting thing to do and so I worked with him. But I quickly found out the problem was too difficult to solve in the time I had. Nonetheless, that was my first exposure to catalysis."

Cusumano stayed at Rutgers to complete his Ph.D. in chemistry. He continued to play music to support himself. Graduate fellowships in those days were only a few thousand dollars. However, he bought a brand new Cadillac from his earnings with music. "I was the only first year graduate student who had a Cadillac at Rutgers University," he said.

Starting a New Company

After his Ph.D. was completed, he went to work for Exxon. "I saw how industry solved problems from an economic point of view. I think industry is a good place to learn about catalysis because you have to worry about how to do science, how to do engineering, and how to make it work economically in the real world."

While at Exxon Cusumano met Ricardo Levy, who is now his partner at Catalytica, and Michel Boudart, a professor of chemical engineering at Stanford University. They came up with the idea of Catalytica Associates, a consulting company. Cusumano, Levy, and Boudart each put up $10,000, and started doing business in Levy's basement in 1974. "The first day we started work we got a contract with Merck, Sharpe, and Dohme in New Jersey. We agreed to help them with a problem on a compound that prevents worms in chickens."

> "We have chemists work together with chemical engineers and marketing people. They all have a vested interest in the technology and the business, and they work in parallel. If you put these people together you increase the probability of success."

Within a few weeks they had more contracts and decided to move to California. "In the period from 1974 to 1984 we built to about a 50-person business. Very comfortable and very profitable. We worked on about 200 projects with a hundred companies." The company continued to grow, and by the time they first offered stock for public sale in February 1993, they had raised many millions of dollars from venture capitalists. They also had begun to develop their own technology. They wanted to "use catalysis to create technology that is economically attractive and environmentally more friendly. We do that by designing a catalyst that takes the raw materials in a selective way directly to the product. There are not many byproducts, so you don't have to get rid of them, and you don't have any waste."

Solving Environmental Problems

Catalytica has picked several areas where there are environmental problems and where the problems could be solved economically. Cusumano said they first turned to the problem of reducing nitrogen oxide (NO_x) emissions in gas-fired turbines used to generate electricity. When natural gas is burned, the temperature of the gases formed by its combustion ($CO_2 + H_2O$) is about 1800°C. The problem is that this gas cannot be put into the turbine because it will destroy the turbine blades. It must be at 1300°C. In addition, at 1800°C the N_2 and O_2 in the air combine to give NO_x (a designation for $NO + NO_2$), pollutants that must be removed. Catalytica's solution has been to develop a catalyst that allows natural gas to burn flamelessly at 1300°C, where NO_x does not form. Furthermore, the cooler gas can be introduced into the turbine directly.

Catalytica's combustion system is being tested in conjunction with General Electric and is also being adapted to automobiles. In the near future, 2% of the cars in California and other states will have to be "zero emission vehicles." "Right now that means the electric car. The oil companies, Detroit, and others are not doing much with electric cars because battery technology is not such that it is possible to fuel a car that way. So, we have designed a completely different system that uses our technology. The consumer can use gasoline, and all that comes out the tail pipe is CO_2 and water. The electric car is zero pollution when it is on the highway, but not when you recharge it. It puts the pollution somewhere else."

As a result of recent environmental legislation, it is more difficult for petroleum refiners to produce high-octane gasoline economically without environmental risks. One way to make high-octane gasoline is to add "alkylates," alkanes with special characteristics that make them burn efficiently. The problem now is that liquid hydrogen fluoride (HF) or concentrated sulfuric acid (H_2SO_4) is required to make alkylates. Indeed, some communities have passed laws that ban the use of HF in local chemical plants. Catalytica has developed a new process that replaces these highly acidic compounds with a solid catalyst that is safe enough to put on the kitchen counter.

Teamwork in Industry

The way technology is developed in companies in the United States is changing, and Catalytica demonstrates these new directions. "It used to be that chemists worked by themselves and came up with some interesting reactions. Then they would go to a chemical engineer, whom they had never met and who had no vested interest in the process of chemistry. The engineer would say 'that is not economically attractive' or 'it can't be done.' If they got past that group they would go to the marketing group, which may not have an interest in the product. Catalytica can't afford those kinds of mistakes. We have chemists work together with chemical engineers and marketing people. They all have a vested interest in the technology and the business, and they work in parallel. If you put these people together you increase the probability of success. That is extremely important for commercialized technology."

The Next Step

What does the future hold for the chemical industry? Cusumano believes that we will be using alkanes to make chemicals for the next 40 to 50 years. After that, he believes, the future is in biochemistry. "We will take things that grow and are renewable and convert them into chemicals and advanced materials and products. There is a little bit of that being done right now, but it is probably the next step. It will be a natural progression."

AMINES

21

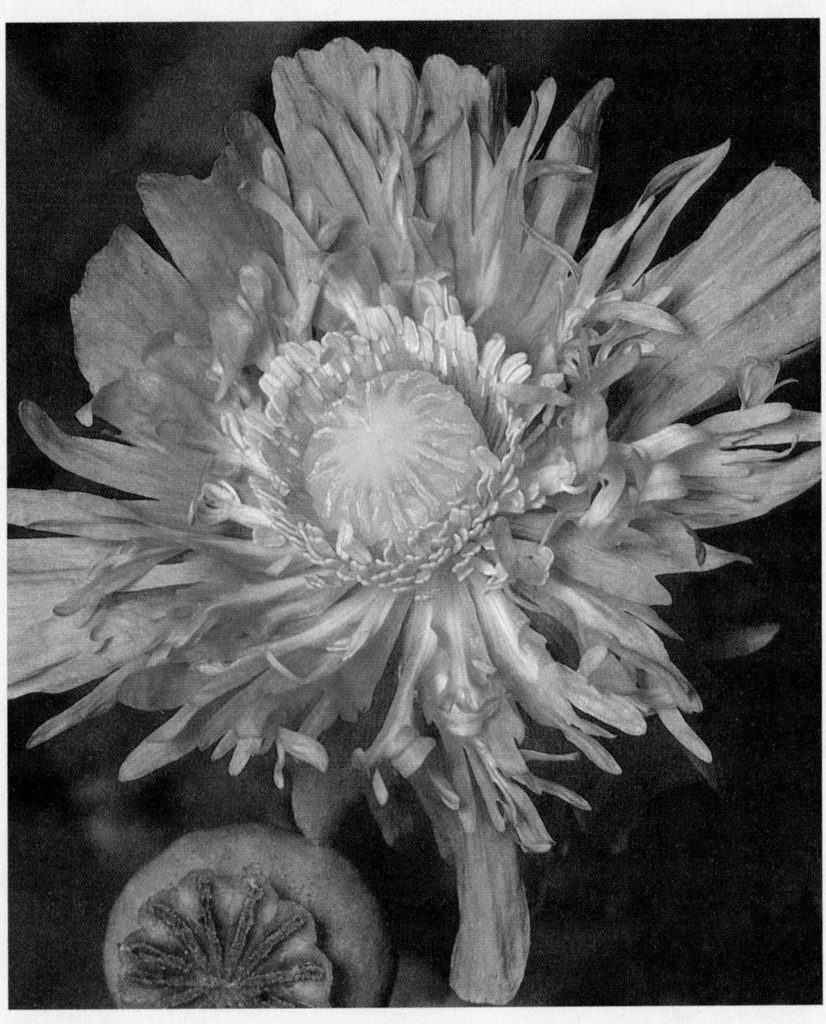

■ The opium poppy is the source of the amines morphine and codeine. See Problem 21.20. *(© Barbara Woike, PHOTO/NATS)*

Carbon, hydrogen, and oxygen are the three most common elements in organic compounds. Because of the wide distribution of amines in the biological world, nitrogen is the fourth most common element in organic compounds. The most important chemical properties of amines are their basicity and their nucleophilicity.

21.1 Structure and Classification

Primary (1°) amine An amine in which one hydrogen of ammonia has been replaced by an alkyl or aryl group.

Secondary (2°) amine An amine in which two hydrogens of ammonia have been replaced by alkyl or aryl groups.

Tertiary (3°) amine An amine in which all three hydrogens of ammonia have been replaced by alkyl or aryl groups.

Amines are derivatives of ammonia in which one or more hydrogens are replaced by alkyl or aryl groups. Amines are classified as primary, secondary, or tertiary, depending on the number of carbon atoms bonded directly to nitrogen. In a **primary amine,** one hydrogen of ammonia is replaced by an alkyl or aryl group. In a **secondary amine,** two hydrogens are replaced by alkyl or aryl groups, and in a **tertiary amine** all three hydrogens are replaced by alkyl or aryl groups.

$:NH_3$	$CH_3-\ddot{N}H_2$	$CH_3-\ddot{N}H-CH_3$	$(CH_3)_3N:$
Ammonia	Methylamine (a 1° amine)	Dimethylamine (a 2° amine)	Trimethylamine (a 3° amine)

Alcohols (Section 9.2A) are also classified as primary, secondary, or tertiary, but the basis for their classification is different from that for amines. For alcohols, classification depends on the number of carbon atoms attached to the carbon bearing the —OH group.

$(CH_3)_3C-OH$ — —OH bonded to a tertiary carbon

A 3° alcohol

$(CH_3)_3C-NH_2$ — only one alkyl group bonded to nitrogen

A 1° amine

Quaternary (4°) ammonium ion An ion in which nitrogen is bonded to four carbons and bears a positive charge.

An ion containing a nitrogen atom bonded to any combination of four alkyl or aryl groups is classified as a **quaternary ammonium ion.** Compounds containing such ions have properties characteristic of salts.

$(CH_3)_4N^+Cl^-$	$C_5H_5\overset{+}{N}CH_2(CH_2)_{12}CH_3\ Cl^-$	$C_6H_5CH_2\overset{+}{N}(CH_3)_3\ OH^-$
Tetramethylammonium chloride	Tetradecylpyridinium chloride (Cetylpyridinium chloride)	Benzyltrimethylammonium hydroxide

Aliphatic amine An amine in which nitrogen is bonded only to alkyl groups.

Aromatic amine An amine in which nitrogen is bonded to one or more aryl groups.

Amines are further divided into aliphatic and aromatic amines. In an **aliphatic amine,** all of the carbons bonded directly to nitrogen are derived from alkyl groups; in an **aromatic amine,** one or more of the groups bonded to nitrogen are aryl groups.

$C_6H_5-\ddot{N}H_2$	$C_6H_5-\ddot{N}(CH_3)-H$	$C_6H_5-CH_2-\ddot{N}(CH_3)-CH_3$
Aniline (a 1° aromatic amine)	*N*-Methylaniline (a 2° aromatic amine)	Benzyldimethylamine (a 3° aliphatic amine)

Heterocyclic amine An amine in which nitrogen is one of the atoms of a ring.

An amine in which the nitrogen atom is part of a ring is classified as a **heterocyclic amine.** When the nitrogen is part of an aromatic ring (Section 19.2D), the amine is classified as a **heterocyclic aromatic amine.** Following are structural formulas for two heterocyclic aliphatic amines and two heterocyclic aromatic amines.

Heterocyclic aromatic amine An amine in which nitrogen is one of the atoms of an aromatic ring.

Piperidine Pyrrolidine (heterocyclic aliphatic amines) Pyrrole Pyridine (heterocyclic aromatic amines)

EXAMPLE 21.1

Alkaloids are basic nitrogen-containing compounds of plant origin, many of which are physiologically active when administered to humans. Classify each amino group in these alkaloids according to type (i.e., primary, secondary, tertiary, aliphatic, heterocyclic, aromatic, heterocyclic aromatic). Coniine, isolated from water hemlock, is highly toxic. Ingestion can cause weakness, labored respiration, paralysis, and eventually death. It is the toxic substance in "poison hemlock" used in the death of Socrates. Nicotine occurs in the tobacco plant. In small doses, it is an addictive stimulant. In larger doses, it causes depression, nausea, and vomiting. In still larger doses, it is a deadly poison. Solutions of nicotine in water are used as insecticides. Cocaine is a central nervous system stimulant obtained from the leaves of the coca plant.

Alkaloids Basic nitrogen-containing compounds of plant origin, many of which are physiologically active when administered to humans.

(a) (*S*)-(+)-Coniine ($CH_2CH_2CH_3$) (b) (*S*)-(−)-Nicotine (CH_3) (c) Cocaine (CH_3, CO_2CH_3, OCC_6H_5)

The death of Socrates, as painted by the French artist Jacques David (1748–1825). Socrates was sentenced to death and carried out his own death sentence by drinking a cup of poison hemlock. *(Courtesy of the Metropolitan Museum of Art, New York)*

Solution

(a) A secondary heterocyclic aliphatic amine.
(b) A tertiary heterocyclic aliphatic amine and a heterocyclic aromatic amine.
(c) A tertiary heterocyclic aliphatic amine.

PROBLEM 21.1

Identify all carbon stereocenters in coniine, nicotine, and cocaine.

21.2 Nomenclature

A. Systematic Names

Systematic names for aliphatic amines are derived just as they are for alcohols. The suffix **-e** of the parent alkane is dropped and is replaced by **-amine.**

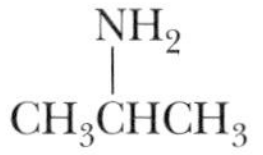

$CH_3CH(NH_2)CH_3$ — 2-Propanamine

$C_6H_5CH(NH_2)CH_3$ — (*S*)-1-Phenylethanamine

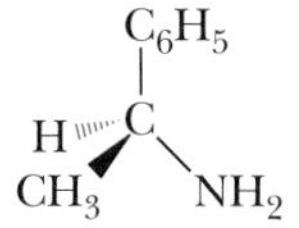

$H_2NCH_2CH_2CH_2CH_2CH_2CH_2NH_2$ — 1,6-Hexanediamine

EXAMPLE 21.2

Give systematic names to these amines.

(a) $CH_3(CH_2)_5NH_2$ **(b)** $H_2N(CH_2)_4NH_2$ **(c)** 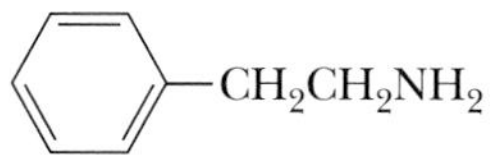 C_6H_5—$CH_2CH_2NH_2$

Solution

(a) 1-Hexanamine **(b)** 1,4-Butanediamine **(c)** 2-Phenylethanamine

PROBLEM 21.2

Write structural formulas for these amines.

(a) 2-Methyl-1-propanamine **(b)** Cyclohexanamine **(c)** (*R*)-2-Butanamine

IUPAC nomenclature retains the common name **aniline** for $C_6H_5NH_2$, the simplest aromatic amine. Its simple derivatives are named using the prefixes *o*-, *m*-, and *p*-, or numbers to locate substituents. Several derivatives of aniline have common names that are still widely used. Among these are **toluidine** for a methyl-substituted aniline and **anisidine** for a methoxyl-substituted aniline.

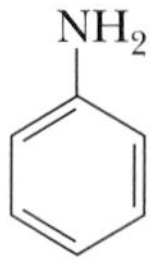

Aniline 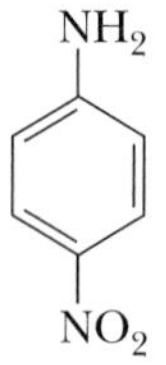   4-Nitroaniline    4-Methylaniline (*p*-Toluidine)   3-Methoxyaniline (*m*-Anisidine)

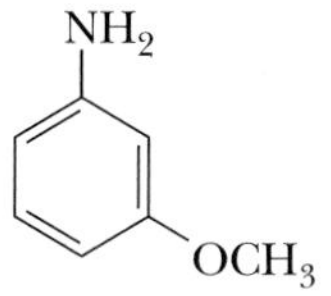

CHEMISTRY IN ACTION

Polyamines

A vigorous subdiscipline of bioorganic chemistry and biochemistry involves the study of polyamines, including putrescine, cadaverine, spermine, and spermidine.

$H_2N(CH_2)_4NH_2$ Putrescine	$H_2N(CH_2)_4NH(CH_2)_3NH_2$ Spermidine
$H_2N(CH_2)_5NH_2$ Cadaverine	$H_2N(CH_2)_3NH(CH_2)_4NH(CH_2)_3NH_2$ Spermine

Putrescine and cadaverine were first isolated from the cholera bacterium *Vibro cholerae* and were named for the foul odors that arise from decaying flesh. It was later discovered that putrescine has only a slight odor; it is other amines coexisting with it that are the source of the unpleasant odors. Spermine and spermidine were first isolated from human seminal fluid, where they are highly concentrated, hence their names. Only a relatively small number of scientists have worked with these compounds, especially with putrescine and cadaverine. Some have suggested that the compounds' names have discouraged more workers from entering this field.

At the body's pH, these polyamines are protonated and associated with negatively charged biological molecules, especially with RNA and DNA. This association is apparently necessary for life because there is a strong correlation between high polyamine concentrations and cells that engage in high rates of protein, RNA, and DNA synthesis.

The study of polyamines exploded in 1971 when it was discovered that human cancer patients have elevated concentrations of polyamines in their urine (D. H. Russell, *Nature,* **233,** 144, 1971). Might it be that measuring polyamine levels could lead to a simple blood test for the presence of cancer? Unfortunately, the answer is no. It has been found that any condition that leads to cell loss or pathological cell growth (such as inflammation) produces elevated levels of polyamines. This understanding, however, has produced hundreds of studies on polyamine levels as markers of cell regeneration, tumor regression, and even pregnancy. One wonders if even more studies might have been done if putrescine and cadaverine had been given more pleasant names.

Secondary and tertiary amines are commonly named as *N*-substituted primary amines. For unsymmetrical amines, the largest group is taken as the parent amine; the smaller group(s) attached to nitrogen are named and their location is indicated by the prefix *N* (indicating that they are attached to nitrogen).

C_6H_5–$NHCH_3$ C_5H_9–$N(CH_3)_2$

N-Methylaniline   *N,N*-Dimethylcyclopentanamine

Following are names and structural formulas for four heterocyclic aromatic amines, the common names of which have been retained in the IUPAC system.

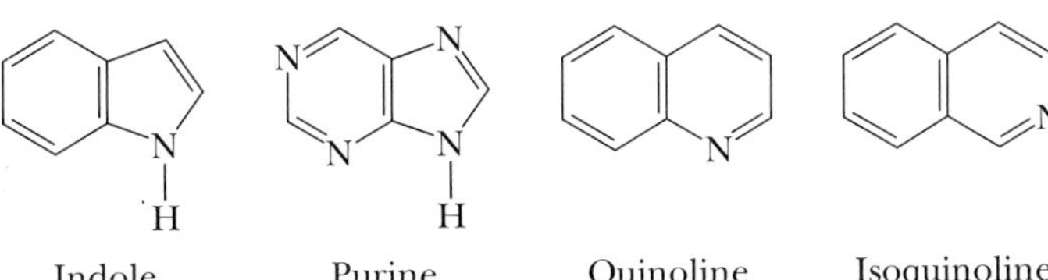

Indole   Purine   Quinoline   Isoquinoline

Among the various functional groups discussed in this text, the $—NH_2$ group is one of the lowest in precedence (Table 15.1). The following compounds each contain a functional group of higher precedence than the amino group, and accordingly, the amino group is indicated by the prefix **amino.**

$H_2NCH_2CH_2OH$	$(CH_3)_2CH$–C(H)(CH_2OH)–NH_2	H_2N–C_6H_4–CO_2H
2-Aminoethanol	(*S*)-2-Amino-3-methyl 1-butanol	4-Aminobenzoic acid

B. Common Names

Common names for most aliphatic amines are derived by listing the alkyl groups attached to nitrogen in alphabetical order in one word ending in the suffix -amine; that is, they are named as alkylamines.

CH_3NH_2	$CH_3C(CH_3)_2NH_2$	$(C_5H_9)_2NH$	$CH_3CH_2N(CH_2CH_3)CH_2CH_3$
Methylamine	*tert*-Butylamine	Dicyclopentylamine	Triethylamine

Example 21.3

Write structural formulas for these amines.

(a) Isopropylamine **(b)** Cyclohexylmethylamine **(c)** Benzylamine

Solution

(a) $(CH_3)_2CHNH_2$ **(b)** C_6H_{11}–$NHCH_3$ **(c)** C_6H_5–CH_2NH_2

Problem 21.3

Write structural formulas for these amines.

(a) Isobutylamine **(b)** Triphenylamine **(c)** Diisopropylamine

When four atoms or groups of atoms are bonded to a nitrogen atom, the compound is named as a salt of the corresponding amine. The ending -amine (or -aniline, or pyridine, and so on) is replaced by **ammonium** (or anilinium, or pyridinium, etc.) and the name of the anion (chloride, acetate, etc.) is added.

$(CH_3CH_2)_3NH^+\ Cl^-$	$C_5H_5NH^+\ CH_3CO_2^-$
Triethylammonium chloride	Pyridinium acetate

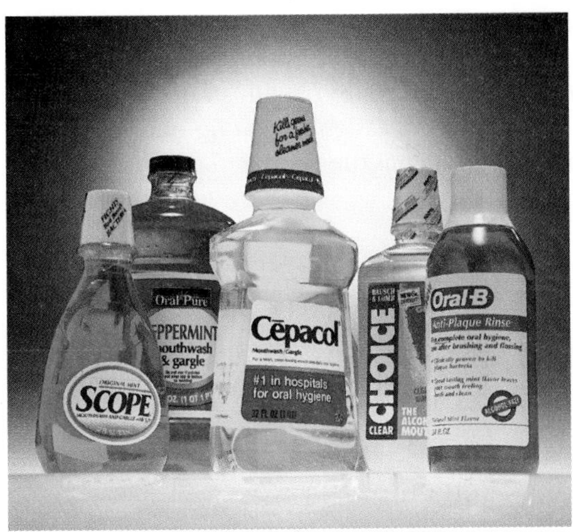

Several over-the-counter mouthwashes contain an *N*-alkylpyridinium chloride salt as an antibacterial agent. *(Charles D. Winters)*

EXAMPLE 21.4

Give each compound an acceptable name.

(a) $(C_6H_5)_2NH$ (b) 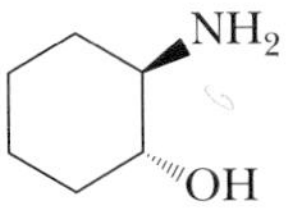 cyclohexane ring bearing NH_2 (wedge) and OH (hash) on adjacent carbons (c) 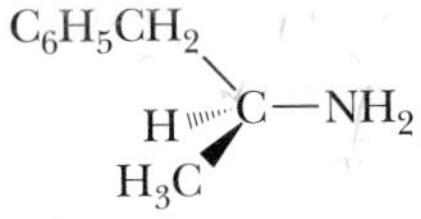 $C_6H_5CH_2$, H (hash), H_3C (wedge) on C—NH_2

Solution

(a) Diphenylamine **(b)** *trans*-2-Aminocyclohexanol

(c) Its systematic name is (*S*)-1-phenyl-2-propanamine. The common name of this compound is amphetamine. The dextrorotatory isomer of amphetamine (shown here) is a central nervous system stimulant and is manufactured and sold under several trade names. The salt with sulfuric acid is marketed as Dexedrine sulfate.

PROBLEM 21.4

Write IUPAC and, where possible, common names for these amines.

(a) 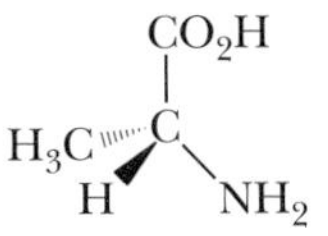 CO_2H above C; H_3C (hash), H (wedge), NH_2 on C (b) $H_2NCH_2CH_2CH_2CO_2H$ (c) $(CH_3)_3CCH_2NH_2$

21.3 Chirality of Amines and Quaternary Ammonium Ions

The geometry of a nitrogen atom bonded to three other atoms or groups of atoms is trigonal pyramidal (Section 1.4). The sp^3 hybridized nitrogen atom is at the apex of the pyramid, and the three groups bonded to it extend downward to form the triangular base of the pyramid. If we consider the unshared pair of electrons on nitrogen as a fourth group, then the arrangement of "groups" around nitrogen is approximately tetrahedral. Because of this geometry, an amine with three different groups bonded to nitrogen is chiral and can exist as a pair of enantiomers, as illustrated by the non-superposable mirror images of ethylmethylamine. In assigning configuration to these enantiomers, the group of lowest priority on nitrogen is the unshared pair of electrons.

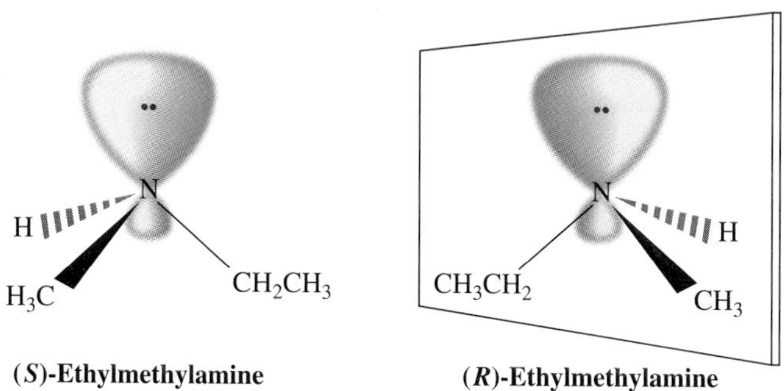

(*S*)-Ethylmethylamine **(*R*)-Ethylmethylamine**

In principle, a chiral amine can be resolved; that is, it can be separated into a pair of enantiomers. Except for special cases, however, chiral amines cannot be resolved because they undergo rapid interconversion by a process known as pyramidal inversion. **Pyramidal inversion** is the rapid oscillation of a nitrogen atom from one side of the plane of the three atoms bonded to it to the other side of that plane.

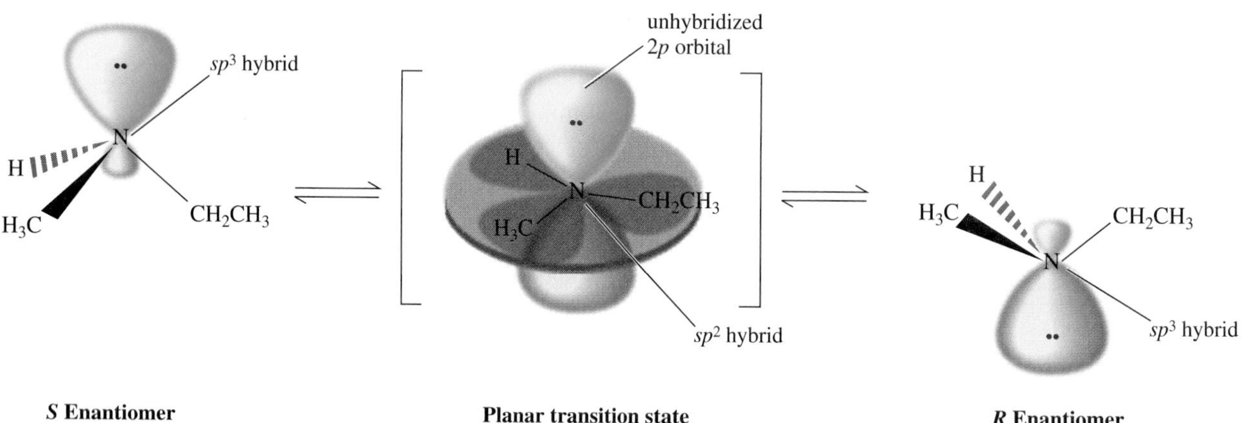

***S* Enantiomer** **Planar transition state** ***R* Enantiomer**

To visualize this process, imagine the sp^3 hybridized nitrogen atom lying above the plane of the three atoms to which it is bonded. In the transition state for pyramidal inversion, the nitrogen atom and the three groups to which it is bonded become coplanar and the molecule becomes achiral. In this planar transition state, nitrogen is sp^2 hybridized, and its lone pair of electrons lies in its unhybridized $2p$ orbital.

Nitrogen then completes the inversion, becomes sp^3 hybridized again, and now lies below the plane of the three atoms to which it is bonded. As a result of pyramidal inversion, a chiral amine quite literally turns itself inside out, like an umbrella in a stormy wind, and in the process becomes a racemic mixture. The activation energy for pyramidal inversion of simple amines is about 6 kcal/mol. For ammonia at room temperature, the rate of nitrogen inversion is approximately $2 \times 10^{11}\ s^{-1}$. For simple amines, the rate is less rapid but nonetheless sufficient to make it impossible to resolve them.

Pyramidal inversion is not possible for quaternary ammonium ions, and their salts can be resolved.

C_6H_5, Cl^-, N^+, H_3C, CH_3CH_2, $CH_2CH{=}CH_2$ — *R* enantiomer

C_6H_5, Cl^-, N^+, CH_3, $CH_2{=}CHCH_2$, CH_2CH_3 — *S* enantiomer

Phosphorus, in the same family as nitrogen, forms trivalent compounds called phosphines, which also have trigonal pyramidal geometry. The activation energy for pyramidal inversion of trivalent phosphorus compounds is considerably greater than it is for trivalent compounds of nitrogen, with the result that a number of chiral phosphines have been resolved.

21.4 Physical Properties

Amines are polar compounds, and both primary and secondary amines form intermolecular hydrogen bonds (Figure 21.1). An N—H---N hydrogen bond is not as strong as an O—H---O hydrogen bond because the difference in electronegativity between nitrogen and hydrogen (3.0 − 2.1 = 0.9) is not as great as that between oxygen and hydrogen (3.5 − 2.1 = 1.4). The effect of intermolecular hydrogen bonding can be illustrated by comparing the boiling points of methylamine and methanol. Both compounds have polar molecules that interact in the pure liquid by hydrogen bonding. Hydrogen bonding is stronger in methanol than in methylamine, and, therefore, methanol has the higher boiling point.

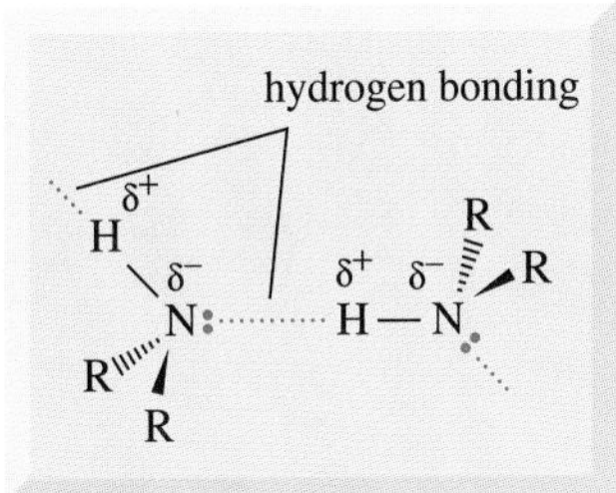

FIGURE 21.1
Intermolecular association by hydrogen bonding in primary and secondary amines. Nitrogen is approximately tetrahedral in shape with the axis of the hydrogen bond along the fourth position of the tetrahedron.

	CH_3NH_2	CH_3OH
MW (g/mol)	31.1	32.0
bp (°C)	−6.3	65.0

All classes of amines form hydrogen bonds with water and are more soluble in water than hydrocarbons of comparable molecular weight. Most low-molecular-weight amines are completely soluble in water (Table 21.1). Higher-molecular-weight amines are only moderately soluble or insoluble.

TABLE 21.1 Physical Properties of Selected Amines

Name	Structural Formula	mp (°C)	bp (°C)	Solubility in Water
Ammonia	NH_3	−78	−33	very soluble
Primary Amines				
methylamine	CH_3NH_2	−95	−6	very soluble
ethylamine	$CH_3CH_2NH_2$	−81	17	very soluble
propylamine	$CH_3CH_2CH_2NH_2$	−83	48	very soluble
isopropylamine	$(CH_3)_2CHNH_2$	−95	32	very soluble
butylamine	$CH_3(CH_2)_3NH_2$	−49	78	very soluble
benzylamine	$C_6H_5CH_2NH_2$	—	185	very soluble
cyclohexylamine	$C_6H_{11}NH_2$	−17	135	slightly soluble
Secondary Amines				
dimethylamine	$(CH_3)_2NH$	−93	7	very soluble
diethylamine	$(CH_3CH_2)_2NH$	−48	56	very soluble
Tertiary Amines				
trimethylamine	$(CH_3)_3N$	−117	3	very soluble
triethylamine	$(CH_3CH_2)_3N$	−114	89	slightly soluble
Aromatic Amines				
aniline	$C_6H_5NH_2$	−6	184	slightly soluble
Heterocyclic Aromatic Amines				
pyridine	C_5H_5N	−42	116	very soluble

21.5 Spectroscopic Properties

A. Mass Spectrometry

Of the compounds containing C, H, N, O, and the halogens, only those containing an odd number of nitrogen atoms have a molecular ion of odd m/z ratio (the Nitrogen Rule, Section 12.3). Thus, mass spectrometry can be a particularly valuable tool for identifying amines. The molecular ion for aliphatic amines, however, is often very weak. The most characteristic fragmentation of amines, and the one often giving rise to the base peak, is β-cleavage. Where alternative possibilities for β-cleavage exist, it is generally the largest R group that is lost. The most prominent peak by far in the mass spectrum of 3-methyl-1-butanamine (Figure 21.2) is due to $[CH_2{=}NH_2]^+$, m/z 30, resulting from β-cleavage. β-Cleavage is also characteristic of secondary and tertiary amines.

$$CH_3-\underset{\displaystyle CH_3}{\overset{}{CH}}-CH_2-CH_2-\overset{\cdot +}{N}H_2 \xrightarrow{\beta\text{-cleavage}} CH_3-\underset{\displaystyle CH_3}{CH}-CH_2\cdot + CH_2{=}\overset{+}{N}H_2$$

m/z 30

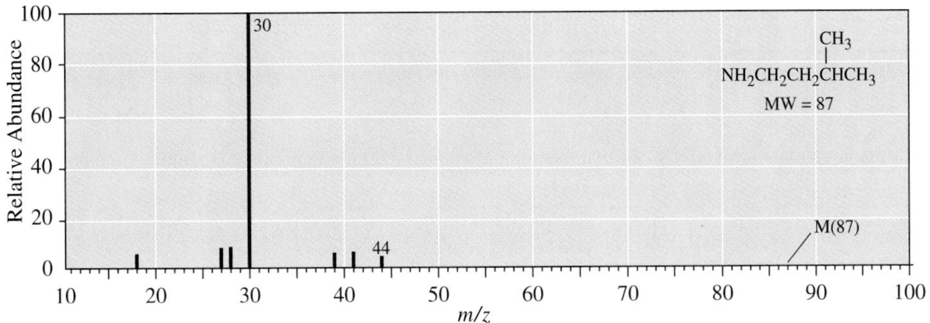

FIGURE 21.2
Mass spectrum of 3-methyl-1-butanamine (isopentylamine). The most characteristic fragmentation pattern of aliphatic amines is β-cleavage.

B. Nuclear Magnetic Resonance Spectroscopy

The chemical shifts of amine hydrogens, like those of hydroxyl hydrogens (Section 13.15C), are variable and may be found in the region δ 0.5 to δ 5.0, depending on the solvent, the concentration, and the temperature. Furthermore, the rate of intermolecular exchange of amine hydrogens is sufficiently rapid compared with the time scale of a ^{1}H-NMR measurement that spin-spin splitting between amine hydrogens and hydrogens on an adjacent α-carbon is prevented. Thus, amine hydrogens generally appear as singlets. The amine hydrogens of 1-butanamine, for example, appear as a broad singlet at δ 1.11 (Figure 21.3).

Carbons bonded to nitrogen are deshielded by approximately 20 ppm in the ^{13}C-NMR spectrum relative to their signal in an alkane of comparable structure. Compare, for example, the chemical shift of carbon 1 in 1-butanamine (42.0 ppm) with that of carbon 2 in butane (25.0 ppm). The chemical shift of carbons adjacent to oxygen is in turn approximately 20 ppm greater than for carbons adjacent to nitrogen. Compare, for example, the chemical shift of carbon 1 in 1-butanamine (42.0 ppm) with that of carbon 1 in 1-butanol (62.4 ppm).

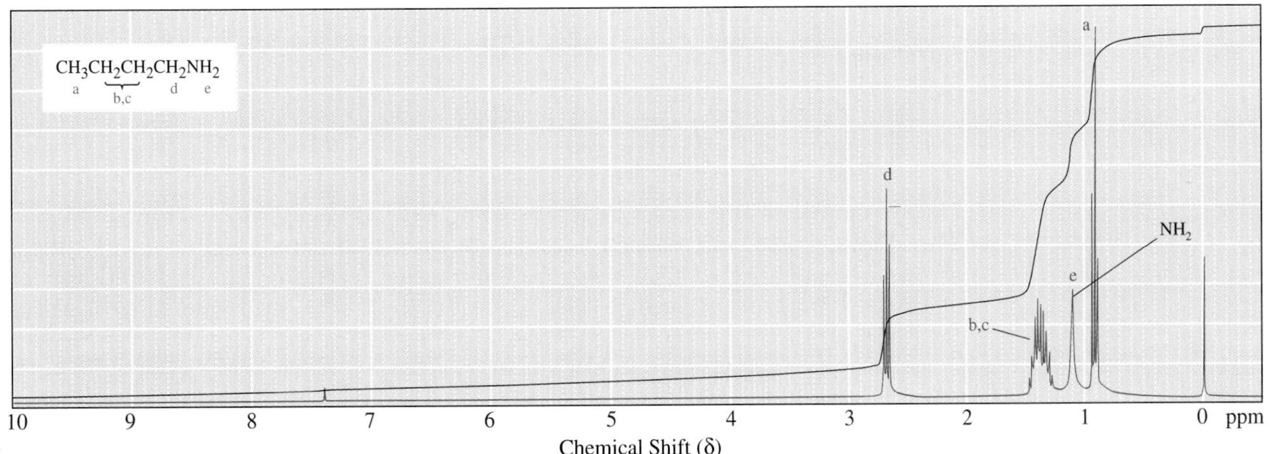

FIGURE 21.3
^{1}H-NMR spectrum of 1-butanamine. The amine hydrogens appear as a broad singlet at δ 1.11.

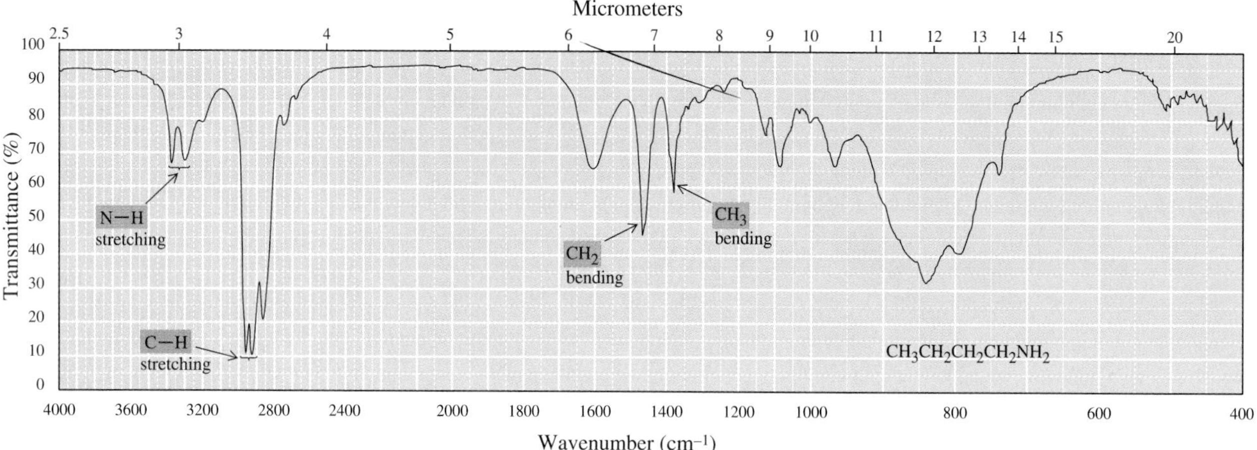

FIGURE 21.4
Infrared spectrum of 1-butanamine, a primary amine.

Formula	Carbon Atom (δ, ppm)			
	1	2	3	4
$CH_3CH_2CH_2CH_2OH$	62.4	34.9	19.0	13.9
$CH_3CH_2CH_2CH_2NH_2$	42.0	36.1	20.2	13.9
$CH_3CH_2CH_2CH_3$	13.2	25.0	25.0	13.2

C. Infrared Spectroscopy

The most important and readily observed infrared absorptions of primary and secondary amines are due to N—H stretching vibrations, and appear in the region 3300–3500 cm^{-1}. Primary amines have two bands in this region: one due to symmetric stretching, the other due to asymmetric stretching. The two N—H stretching absorptions characteristic of a primary amine can be seen in the IR spectrum of 1-butanamine (Figure 21.4). Secondary amines give only one absorption in this region. Tertiary amines have no N—H and, therefore, are transparent in this region of the infrared spectrum.

21.6 Basicity of Amines

Like ammonia, all amines are weak bases, and aqueous solutions of amines are basic. The following acid-base reaction between an amine and water is written using curved arrows to emphasize that, in these proton-transfer reactions, the unshared pair of electrons on nitrogen forms a new covalent bond with hydrogen and displaces hydroxide ion.

$$CH_3-NH_2: + H-\ddot{O}-H \rightleftharpoons CH_3-NH_3^{+} \quad :\ddot{O}-H^{-}$$

Methylamine | Methylammonium hydroxide

Chemistry in Action

The Poison Dart Frogs of South America

The Noanamá and Embrá peoples of the jungles of western Colombia have used poison blow darts for centuries, perhaps millennia. The poisons are obtained from the skin secretions of several brightly colored frogs of the genus *Phyllobates* (*neará* and *kokoi* in the language of the native peoples). A single frog contains enough poison for up to twenty darts. For the most poisonous species (*Phyllobates terribilis*), just rubbing a dart over the frog's back suffices to charge the dart with poison.

Scientists at the National Institutes of Health in the United States became interested in studying these poisons when it was discovered that they act on cellular ion channels, which would make them useful tools in basic research on mechanisms of ion transport. A field station was, therefore, established in western Colombia to collect the relatively common poison dart frogs. From 5000 frogs, 11 mg of two toxins, given the names batrachotoxin and batrachotoxinin A, was isolated. These names are derived from *batrachos,* the Greek word for frog. A combination of NMR spectroscopy, mass spectrometry, and single-crystal x-ray diffraction was used to determine the structures of these compounds.

The poison dart frog, Phyllobates terribilis. *(Animals, Animals)*

Batrachotoxin and batrachotoxinin A are among the most lethal poisons ever discovered. It is estimated that as little as 200 μg of batrachotoxin is sufficient to induce irreversible cardiac arrest in a human being. It has been determined that they act by causing voltage-gated Na^+ channels in nerve and muscle cells to be blocked in the open position, which leads to a huge influx of Na^+ ions into the affected cell.

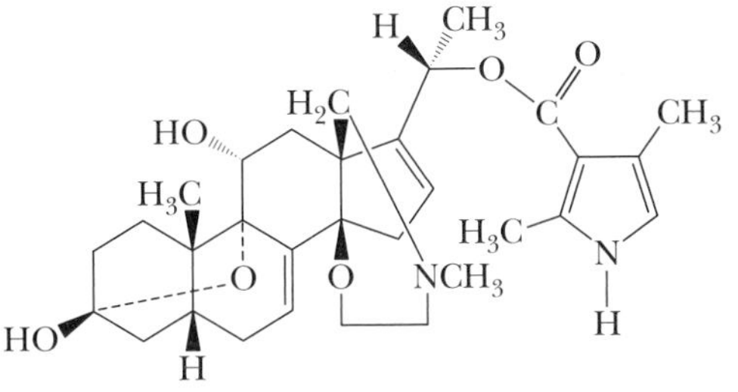

Batrachotoxin

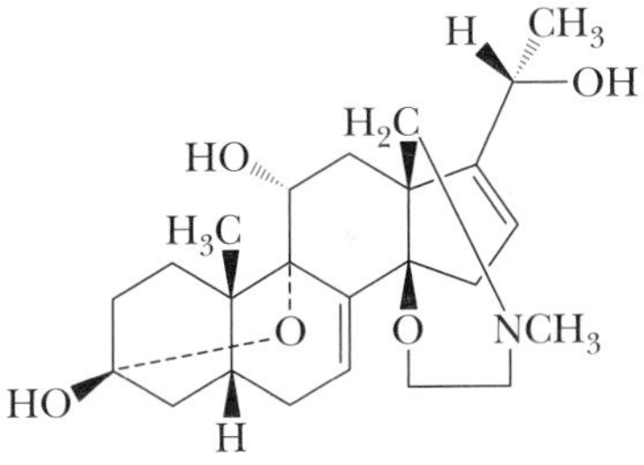

Batrachotoxinin A

The batrachotoxin story illustrates several common themes in drug discovery. First, information about the kinds of biologically active compounds and their sources is often obtained from the native peoples of a region. Second, tropical rain forests are a rich source of structurally complex, biologically active substances. Third, the entire ecosystem, not only the plants, are potential sources of fascinating organic molecules.

J. W. Daly, "Progress in the Chemistry of Organic Natural Products," **41**, edited by W. Herz, H. Grisebach, and G. W. Kirby, Springer-Verlag, Wien, p. 205, 1982.

The equilibrium constant for the reaction of an amine with water, K_{eq}, has the following form, illustrated for the reaction of methylamine with water to give methylammonium hydroxide:

$$K_{eq} = \frac{[CH_3NH_3^+][OH^-]}{[CH_3NH_2][H_2O]}$$

Because the concentration of water in dilute solutions of methylamine in water is essentially a constant ($[H_2O] = 55.5$ mol/L), it is combined with K_{eq} in a new constant called a base ionization constant and given the symbol K_b. The value of K_b for methylamine is 4.37×10^{-4} ($pK_b = 3.36$).

$$K_b = K_{eq}[H_2O] = \frac{[CH_3NH_3^+][OH^-]}{[CH_3NH_2]} = 4.37 \times 10^{-4}$$

It is more common to discuss the basicity of amines by reference to the acid ionization constant of the corresponding conjugate acid as illustrated for the ionization of the methylammonium ion.

$$CH_3NH_3^+ + H_2O \rightleftharpoons CH_3NH_2 + H_3O^+$$

$$K_a = \frac{[CH_3NH_2][H_3O^+]}{[CH_3NH_3^+]} = 2.29 \times 10^{-11} \qquad pK_a = 10.64$$

Values for pK_a and pK_b for any acid-conjugate base pair are related by the following equation:

$$pK_a + pK_b = 14.00$$

Values of pK_a and pK_b for selected aliphatic amines, aromatic amines, and heterocyclic aromatic amines are given in Table 21.2. Using values of pK_a, we can compare acidities of amine conjugate acids with other acids. It is this approach we developed in Chapter 3 to predict the position of equilibrium in acid-base reactions. Equilibrium favors formation of the weaker acid and weaker base, as illustrated by reactions of methylamine with water and with acetic acid.

CH_3NH_2	+ H_2O	$\rightleftharpoons$ $CH_3NH_3^+$	+ OH^-	$pK_{eq} = 5.1$
	pK_a 15.7	pK_a 10.64		$K_{eq} = 8.7 \times 10^{-6}$
(weaker base)	(weaker acid)	(stronger acid)	(stronger base)	

CH_3NH_2	+ CH_3CO_2H	$\rightleftharpoons$ $CH_3NH_3^+$	+ $CH_3CO_2^-$	$pK_{eq} = -5.88$
	pK_a 4.76	pK_a 10.64		$K_{eq} = 7.6 \times 10^5$
(stronger base)	(stronger acid)	(weaker acid)	(weaker base)	

A. Aliphatic Amines

All aliphatic amines have about the same base strength, pK_b 3.0–4.0, and are slightly stronger bases than ammonia. The increase in basicity compared with ammonia can be attributed to the greater stability of an alkylammonium ion, as for example, $RCH_2NH_3^+$, compared with the ammonium ion, NH_4^+. This greater stability arises from the electron-releasing effect of alkyl groups and the resulting partial delocalization of the positive charge from nitrogen onto carbon in the alkylammonium ion.

TABLE 21.2 Base Strengths, pK_b, of Selected Amines and Acid Strengths, pK_a, of Their Conjugate Acids*

Amine	Structure	pK_b	pK_a
Ammonia	NH_3	4.74	9.26
Primary Amines			
methylamine	CH_3NH_2	3.36	10.64
ethylamine	$CH_3CH_2NH_2$	3.19	10.81
cyclohexylamine	$C_6H_{11}NH_2$	3.34	10.66
Secondary Amines			
dimethylamine	$(CH_3)_2NH$	3.27	10.73
diethylamine	$(CH_3CH_2)_2NH$	3.02	10.98
Tertiary Amines			
trimethylamine	$(CH_3)_3N$	4.19	9.81
triethylamine	$(CH_3CH_2)_3N$	3.25	10.75
Aromatic Amines			
aniline	C_6H_5—NH_2	9.37	4.63
4-methylaniline	CH_3—C_6H_4—NH_2	8.92	5.08
4-chloroaniline	Cl—C_6H_4—NH_2	9.85	4.15
4-nitroaniline	O_2N—C_6H_4—NH_2	13.0	1.0
Heterocyclic Aromatic Amines			
pyridine	N (pyridine ring)	8.75	5.25
imidazole	N, N—H (imidazole ring)	7.05	6.95

*For each amine, pK_a + pK_b = 14.00.

Positive charge is partially delocalized onto the alkyl group

$$\mathrm{R{-}\overset{\delta+}{CH_2}\longrightarrow \overset{\delta+}{N}H_3}$$

Recall that we invoked a similar argument in Section 6.3A to account for the effect of alkyl groups in stabilizing carbocations.

B. Aromatic Amines

Aromatic amines are considerably weaker bases than aliphatic amines. Compare, for example, values of pK_b for aniline and cyclohexylamine. The base ionization constant for aniline is smaller (the larger the value of pK_b, the weaker the base) than that for cyclohexylamine by a factor of 10^6.

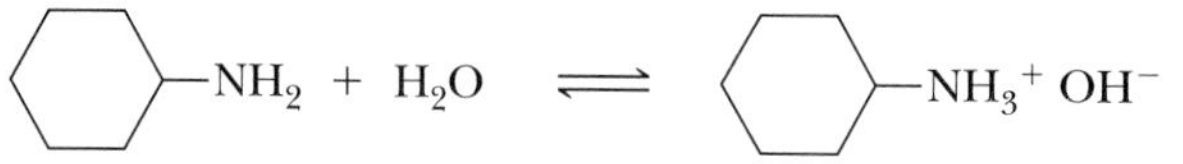

$$C_6H_{11}NH_2 + H_2O \rightleftharpoons C_6H_{11}NH_3^+\ OH^- \qquad pK_b = 3.34;\ K_b = 4.5 \times 10^{-4}$$

Cyclohexylamine — Cyclohexylammonium hydroxide

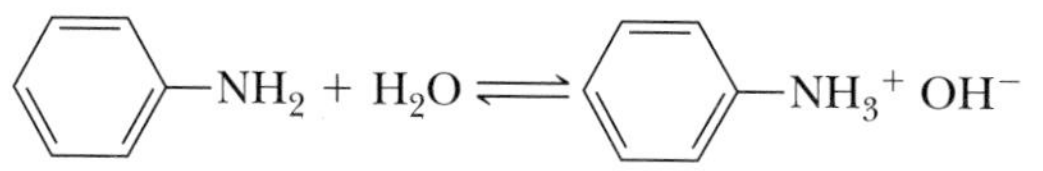

$$C_6H_5NH_2 + H_2O \rightleftharpoons C_6H_5NH_3^+\ OH^- \qquad pK_b = 9.37;\ K_b = 4.3 \times 10^{-10}$$

Aniline — Anilinium hydroxide

Aromatic amines are less basic than aliphatic amines because of a combination of two factors. First is the resonance stabilization of the free base form of aromatic amines. For aniline and other arylamines, this resonance stabilization is the result of interaction of the unshared pair on nitrogen with the pi system of the aromatic ring. The resonance energy of benzene is approximately 36 kcal/mol. For aniline, it is 39 kcal/mol. Because of this interaction, the electron pair on nitrogen is less available for reaction with acid. No such resonance stabilization is possible for alkylamines, and, therefore, the electron pair on the nitrogen of an alkylamine is more available for reaction with an acid; alkylamines are stronger bases than arylamines.

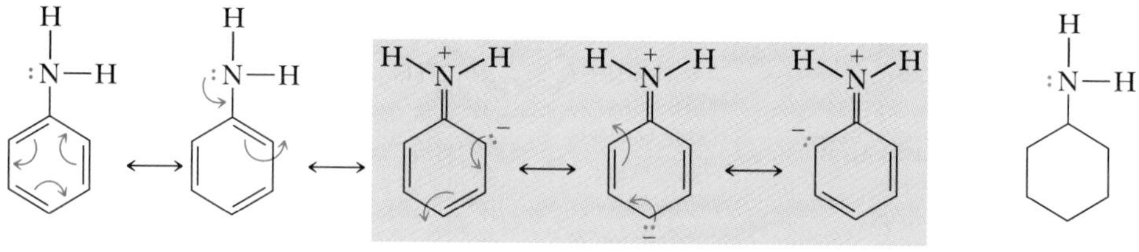

Interaction of the electron pair on nitrogen with the pi system of the aromatic ring

No resonance is possible with alkylamines

The second factor contributing to the decreased basicity of aromatic amines is the electron-withdrawing inductive effect of the sp^2-hybridized carbons of the aromatic ring compared with the sp^3-hybridized carbons of aliphatic amines. The unshared pair of electrons on nitrogen in an aromatic amine is pulled toward the ring and, therefore, less available for protonation to form the conjugate acid of the amine.

Electron-releasing groups (e.g., methyl, ethyl, and other alkyl groups) increase the basicity of aromatic amines, whereas electron-withdrawing groups (halogen, nitro, carbonyl) decrease their basicity. The decrease in basicity on halogen substitution is due to the electron-withdrawing inductive effect of the electronegative halogen. The decrease in basicity on nitro substitution is due to a combination of inductive and resonance effects, as can be seen by comparing the base ionization constants of 3-nitroaniline and 4-nitroaniline. Note that the conjugate acid of 4-nitroaniline (pK_a 1.0) is a stronger acid than phosphorous acid (pK_{a1} = 2.0).

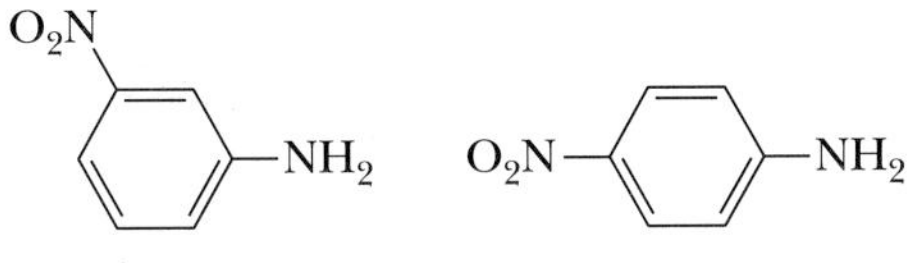

3-Nitroaniline
pK_b 11.53

4-Nitroaniline
pK_b 13.0

The base-decreasing effect of nitro substitution in the 3 position is due almost entirely to its inductive effect, whereas that of nitro substitution in the 4 position is due to both inductive and resonance effects. In the case of para substitution (and ortho substitution as well), delocalization of the lone pair on the amino nitrogen involves not only the carbons of the aromatic ring but also oxygen atoms of the nitro group.

delocalization of the nitrogen lone pair onto the oxygen atoms of the nitro group

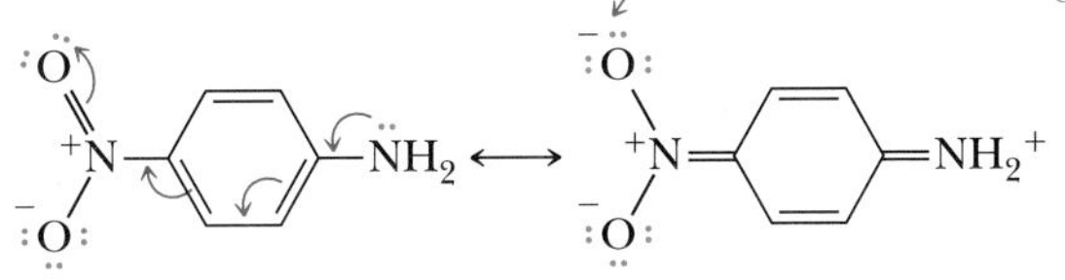

C. Heterocyclic Amines

Heterocyclic aromatic amines are weaker bases than heterocyclic aliphatic amines. Compare, for example, pK_b values for piperidine, pyridine, and imidazole. Or, alternatively, compare the pK_a values for the conjugate acid of each amine.

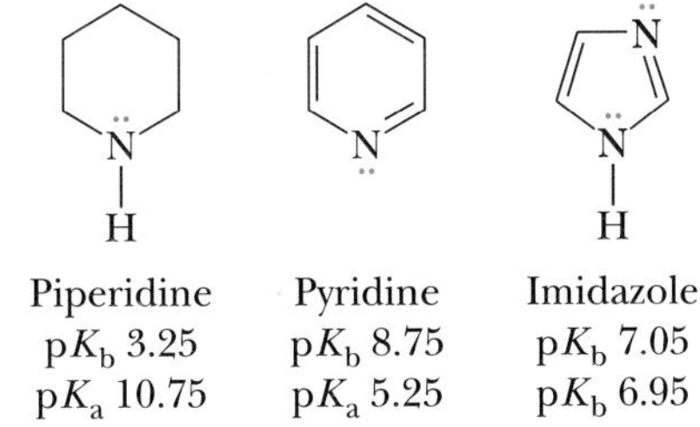

Piperidine	Pyridine	Imidazole
pK_b 3.25	pK_b 8.75	pK_b 7.05
pK_a 10.75	pK_a 5.25	pK_b 6.95

We discussed the structure and bonding in pyridine and imidazole in Section 19.2D. In accounting for the relative basicities of these and other heterocyclic aromatic amines, it is important to determine first if the unshared pair of electrons on nitrogen is or is not a part of the $(4n + 2)$ pi electrons giving rise to aromaticity. In the case of pyridine, the unshared pair of electrons is not a part of the aromatic sextet. Rather, it lies in an sp^2 hybrid orbital in the plane of the ring and perpendicular to the six $2p$ orbitals containing the aromatic sextet.

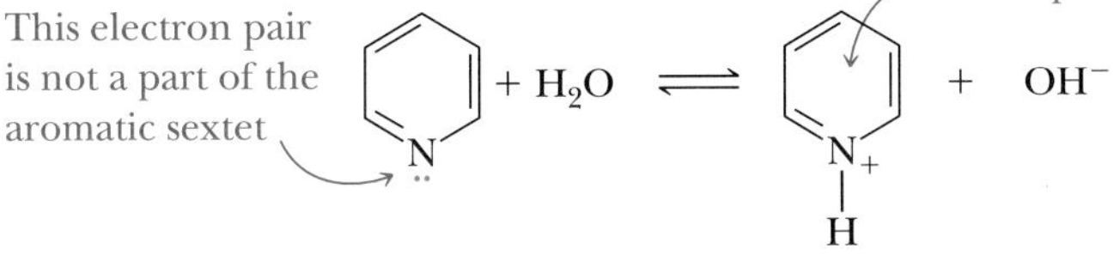

This electron pair is not a part of the aromatic sextet

Aromaticity is maintained, even when protonated

Pyridine $+ H_2O \rightleftharpoons$ Pyridinium ion $+ OH^-$

Proton transfer from water or other acid to pyridine does not involve the electrons of the aromatic sextet. Why, then, is pyridine a considerably weaker base than aliphatic amines? The answer is that the unshared pair of electrons of the pyridine nitrogen lies in an sp^2 hybrid orbital, whereas in aliphatic amines, the unshared pair lies in an sp^3 hybrid orbital. Electrons in an sp^2 hybrid orbital (33% *s* character) are held more tightly by the nucleus than electrons in an sp^3 hybrid orbital (25% *s* character). It is this effect that decreases markedly the basicity of the electron pair on an sp^2 hybridized nitrogen compared with that on an sp^3 hybridized nitrogen.

There are two nitrogen atoms in imidazole, each with an unshared pair of electrons. One unshared pair lies in a $2p$ orbital and is an integral part of the $(4n + 2)$ pi electrons of the aromatic system. The other unshared pair lies in an sp^2 hybrid orbital and is not a part of the aromatic sextet. It is the pair of electrons not part of the pi system that functions as the proton acceptor.

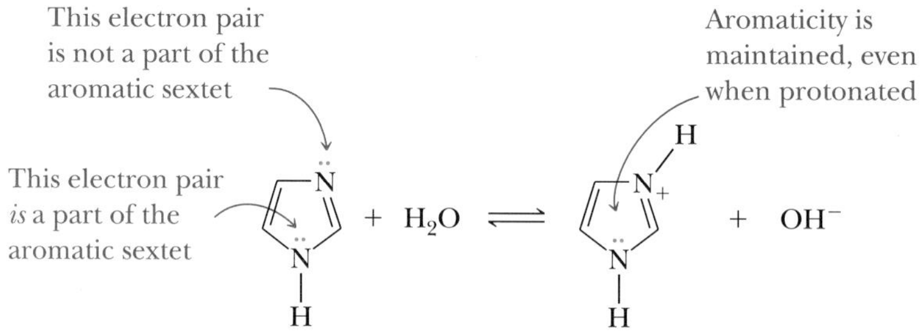

This electron pair is not a part of the aromatic sextet

Aromaticity is maintained, even when protonated

This electron pair *is* a part of the aromatic sextet

$$\text{Imidazole} + H_2O \rightleftharpoons \text{Imidazolium ion} + OH^-$$

Imidazole   Imidazolium ion

As is the case with pyridine, the unshared pair of electrons functioning as the proton acceptor in imidazole lies in an sp^2 hybrid orbital and has markedly decreased basicity compared with an unshared pair of electrons in an sp^3 hybrid orbital. The positive charge on the imidazolium ion is delocalized on both nitrogen atoms of the ring and, therefore, imidazole is a stronger base than pyridine.

EXAMPLE 21.5

Select the stronger base in each pair of amines.

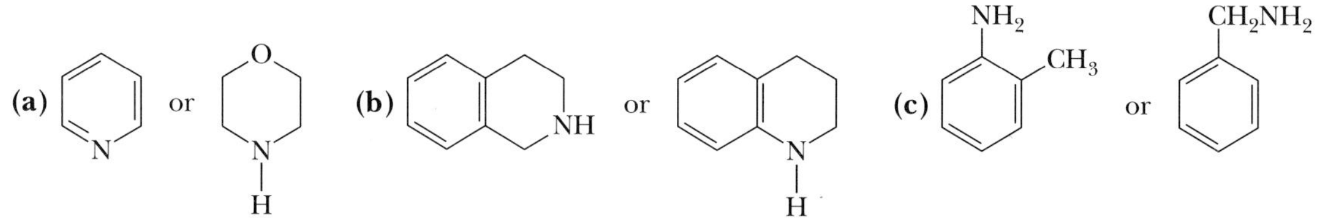

(a) (A) or (B)   **(b)** (C) or (D)   **(c)** (E) or (F)

(A) pyridine (N); (B) morpholine (O, N–H); (C) tetrahydroisoquinoline (NH); (D) tetrahydroquinoline (N–H); (E) NH_2, CH_3 substituted benzene; (F) CH_2NH_2 substituted benzene

Solution

(a) Morpholine (B) is the stronger base (pK_b 5.8, pK_a 8.2). It has a basicity comparable to that of secondary aliphatic amines. Pyridine (A), a heterocyclic aromatic amine (pK_b 8.75, pK_a 5.25), is considerably less basic than aliphatic amines.

(b) Tetrahydroisoquinoline (C) has a basicity comparable to that of secondary aliphatic amines ($pK_b \sim 3.2$, $pK_a \sim 10.8$) and is the stronger base. Tetrahydroquinoline (D) has a basicity comparable to an *N*-substituted aniline (pK_b ~ 9.6, pK_a ~ 4.4) and is the weaker base.

(c) Benzylamine (F) is the stronger base (pK_b 4.4, pK_a 9.6). Its basicity is comparable to that of other aliphatic amines. The basicity of *o*-toluidine (E), an aromatic amine, is comparable to that of aniline (pK_b 9.4, pK_a 4.6).

PROBLEM 21.5

Select the stronger acid from each pair of compounds.

(a) O_2N–C_6H_4–NH_3^+ (A) or CH_3–C_6H_4–NH_3^+ (B) **(b)** pyridinium $\overset{+}{N}H$ (C) or cyclohexyl–NH_3^+ (D)

D. Guanidine

Guano deposits on Galapagos Islands. Guanidine is the compound by which migratory birds excrete waste metabolic nitrogen. *(Animals, Animals, © B. G. Murray, Jr.)*

Guanidine, pK_b 0.4, is the strongest base among neutral compounds. Alternatively, its conjugate acid is a weaker acid (pK_a 13.6) than almost any other protonated amine.

$$H_2N{-}\overset{\overset{NH}{\|}}{C}{-}NH_2 + H_2O \rightleftharpoons H_2N{-}\overset{\overset{^+NH_2}{\|}}{C}{-}NH_2 + OH^- \quad pK_b = 0.4$$

Guanidine — Guanidinium ion

The remarkable basicity of guanidine is attributed to the fact that the positive charge on the guanidinium ion is delocalized equally over the three nitrogen atoms as shown by these three equivalent contributing structures.

$$H_2\ddot{N}{-}\overset{\overset{^+NH_2}{\|}}{C}{-}\ddot{N}H_2 \longleftrightarrow H_2\overset{+}{N}{=}\overset{\overset{\ddot{N}H_2}{|}}{C}{-}\ddot{N}H_2 \longleftrightarrow H_2\ddot{N}{-}\overset{\overset{\ddot{N}H_2}{|}}{C}{=}\overset{+}{N}H_2$$

Three equivalent contributing structures

21.7 Reaction with Acids

Amines, whether soluble or insoluble in water, react quantitatively with strong acids to form water-soluble salts as illustrated by the reaction of norepinephrine (noradrenaline) with aqueous HCl to form a hydrochloride salt. Norepinephrine, secreted by the medulla of the adrenal gland, is a neurotransmitter. It has been suggested that it acts in those areas of the brain that mediate emotional behavior.

(*R*)-(−)-Norepinephrine (only slightly soluble in water) + HCl $\xrightarrow{H_2O}$ (*R*)-(−)-Norepinephrine hydrochloride (a water-soluble salt) [NH_2 → NH_3^+ Cl^-]

EXAMPLE 21.6

Complete each acid-base reaction and name the salt formed.

(a) $(CH_3CH_2)_2NH + HCl \longrightarrow$ **(b)** $C_6H_5CH_2NH_2 + CH_3CO_2H \longrightarrow$

Solution

(a) $(CH_3CH_2)_2NH_2^+ \ Cl^-$ **(b)** $C_6H_5CH_2NH_3^+ \ CH_3CO_2^-$

Diethylammonium chloride Benzylammonium acetate

PROBLEM 21.6

Complete each acid-base reaction and name the salt formed.

(a) $(CH_3CH_2)_3N + HCl \longrightarrow$ **(b)** (piperidine ring) $NH + CH_3CO_2H \longrightarrow$

EXAMPLE 21.7

Following are two structural formulas for alanine (2-aminopropanoic acid), one of the building blocks of proteins (Chapter 27). Is alanine better represented by structural formula (A) or (B)? Explain.

$$\underset{\text{(A)}}{CH_3CH(NH_2)\overset{O}{\overset{\|}{C}}OH} \rightleftharpoons \underset{\text{(B)}}{CH_3CH(NH_3^+)\overset{O}{\overset{\|}{C}}O^-}$$

Solution

Structural formula (A) contains both an amino group (a base) and a carboxyl group (an acid). Proton transfer from the stronger acid ($-CO_2H$) to the stronger base ($-NH_2$) gives an internal salt and, therefore, (B) is the better representation for alanine. Within the field of amino acid chemistry, the internal salt represented by (B) is called a **zwitterion** (Section 27.2).

PROBLEM 21.7

Following are structural formulas for propanoic acid and the conjugate acids of isopropylamine and alanine, along with pK_a values for each functional group:

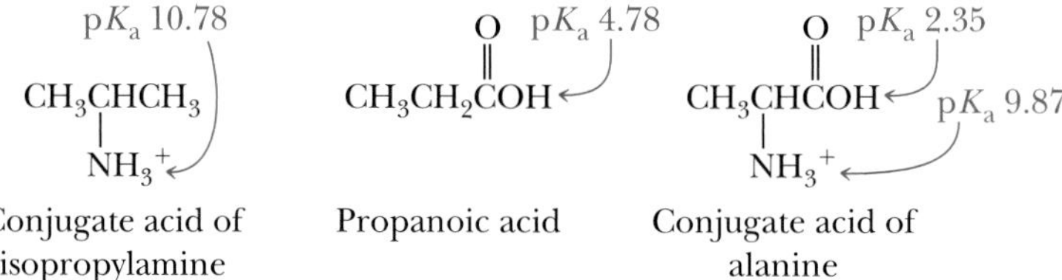

(a) How do you account for the fact that the $-NH_3^+$ group of the conjugate acid of alanine is a stronger acid than the $-NH_3^+$ group of the conjugate acid of isopropylamine?

(b) How do you account for the fact that the $-CO_2H$ group of the conjugate acid of alanine is a stronger acid than the $-CO_2H$ group of propanoic acid?

The basicity of amines and the solubility in water of amine salts can be used to separate water-insoluble amines from water-insoluble, nonbasic compounds. Shown in Figure 21.5 is a flowchart for the separation of aniline from acetanilide, a neutral compound.

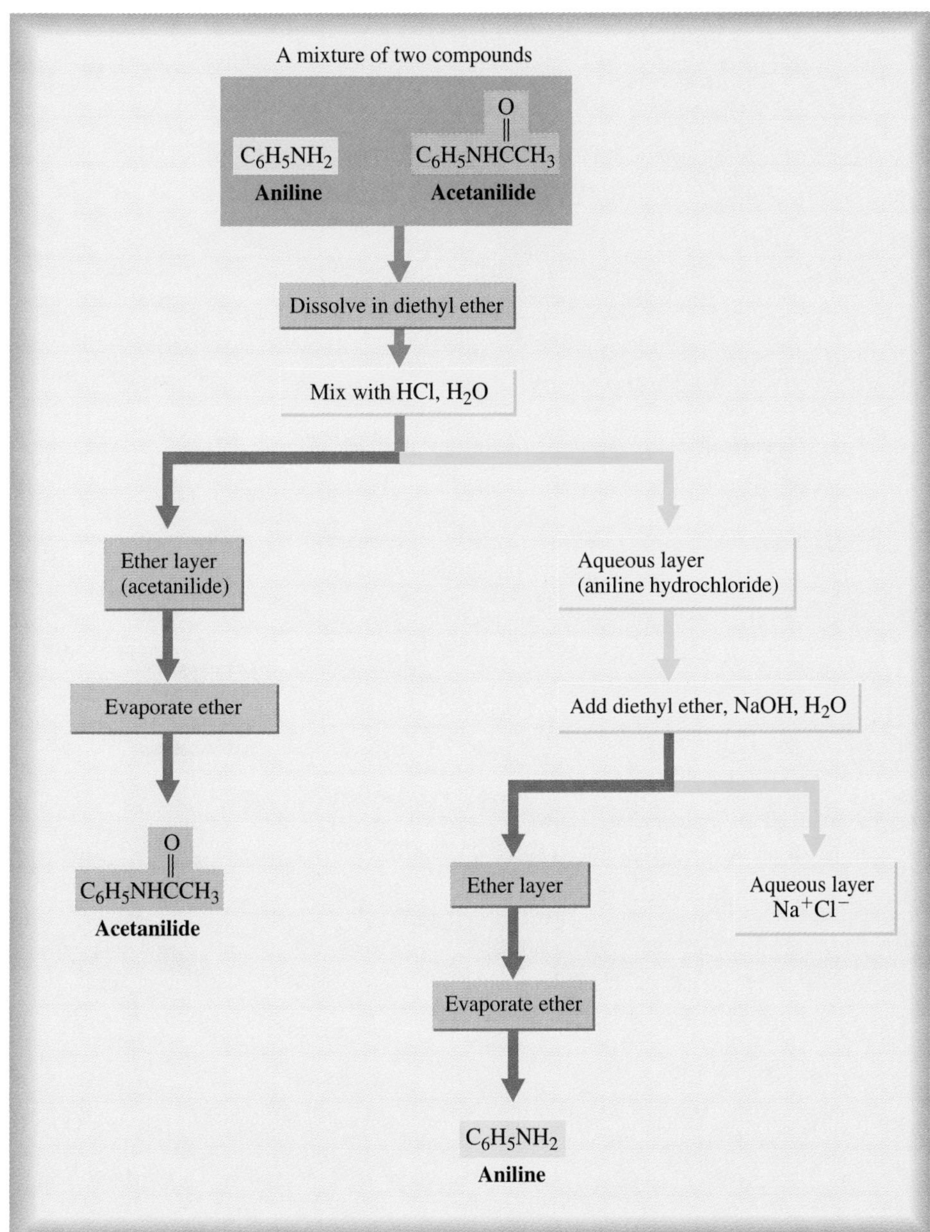

FIGURE 21.5
Separation and purification of an amine and a neutral compound.

EXAMPLE 21.8

Here is a flowchart for the separation of a mixture of a primary aliphatic amine (RNH_2, pK_a 10.8), a carboxylic acid (RCO_2H, pK_a 5), and a phenol (ArOH, pK_a 10). Assume that each is insoluble in water but soluble in diethyl ether. The mixture is separated into fractions A, B, and C. Which fraction contains the amine, which the carboxylic acid, and which the phenol?

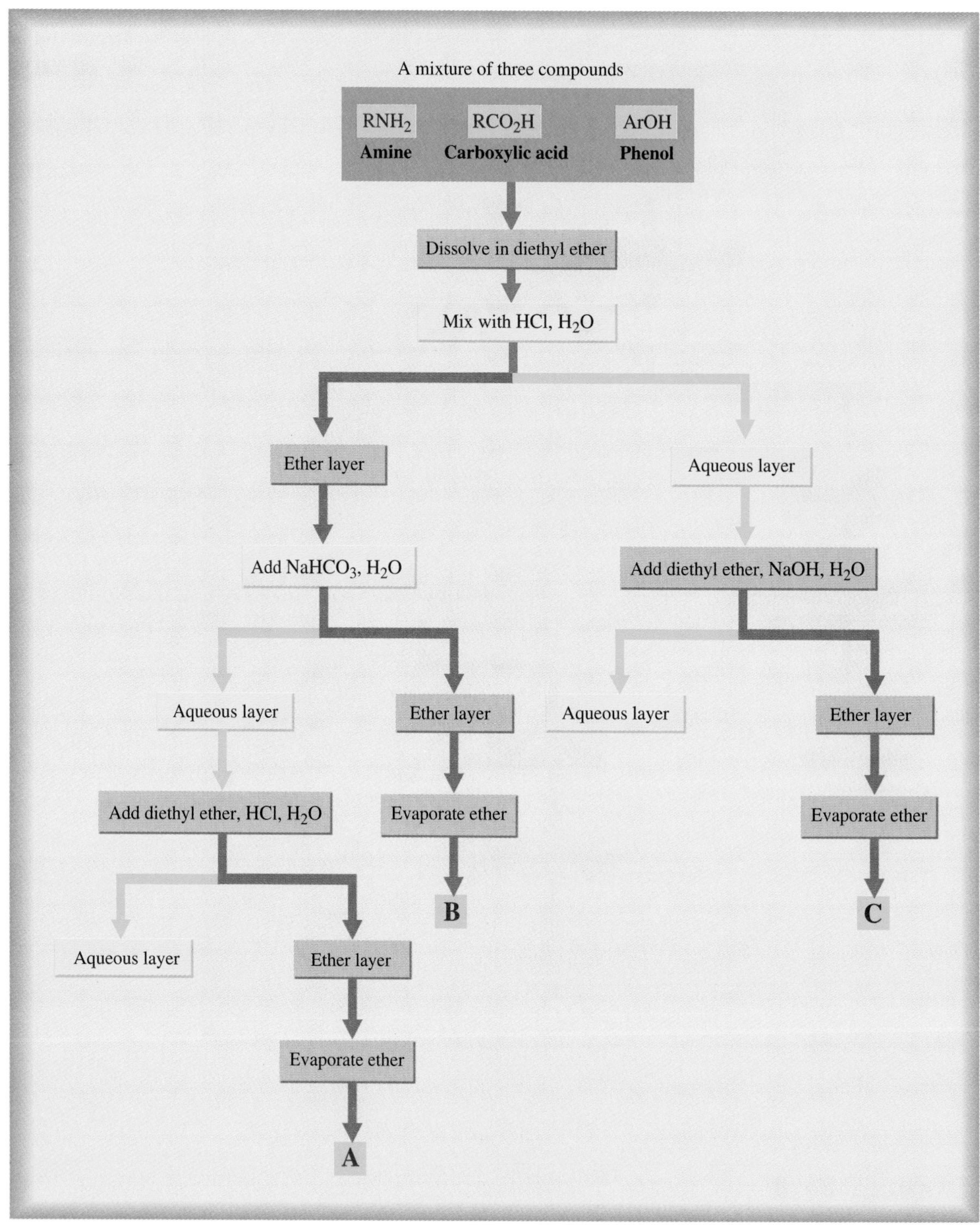

Separation of an amine, a carboxylic acid, and a phenol.

Solution

Fraction C contains RNH_2, fraction B contains ArOH, and fraction A contains RCO_2H.

PROBLEM 21.8

In what way(s) might the results of the separation and purification procedure outlined in Example 21.8 be different if

(a) Aqueous NaOH is used in place of aqueous $NaHCO_3$?
(b) The starting mixture contains an aromatic amine, $ArNH_2$, rather than an aliphatic amine, RNH_2?

21.8 Preparation of Amines

The synthesis of amines is primarily a problem of how to form a carbon-nitrogen bond and, if the newly formed nitrogen-containing compound is not already an amine, how to convert it into an amine. We have already seen the following methods for the preparation of amines.

1. Nucleophilic ring opening of epoxides by ammonia and amines (Section 11.10B).
2. Addition of nitrogen nucleophiles to the carbonyl group of aldehydes and ketones to form imines (Section 15.10).
3. Reduction of imines to amines (Section 15.10).
4. Reduction of amides by $LiAlH_4$ (Section 17.11B).
5. Reduction of a nitrile to a primary amine (Section 17.11C).
6. Hofmann rearrangement of amides (Section 17.12).
7. Nitration of an arene followed by reduction of the nitro group to a primary amine (Section 20.1B).

In this chapter, we present two additional methods for the preparation of amines.

A. Alkylation of Ammonia and Amines

Surely one of the most direct synthetic routes to an amine would seem to be treatment of an alkyl halide with ammonia or an amine. Reaction between these two compounds by a second-order nucleophilic substitution (S_N2) reaction gives an alkylammonium salt, as illustrated by treatment of bromomethane with ammonia to give methylammonium bromide.

$$CH_3Br + NH_3 \xrightarrow{S_N2} \underset{\text{Methylammonium bromide}}{CH_3NH_3^+ \; Br^-}$$

Unfortunately, reaction does not stop at this stage but continues to give a complex mixture of products as shown in the following equation.

$$CH_3Br + NH_3 \longrightarrow CH_3NH_3^+Br^- + (CH_3)_2NH_2^+Br^- + (CH_3)_3NH^+Br^- + (CH_3)_4N^+Br^-$$

This complex mixture is formed in the following way. Proton transfer between ammonia and methylammonium ion gives ammonium ion and methylamine, also a good nucleophile, which then undergoes reaction with bromomethane to give dimethylammonium bromide. A second proton transfer reaction converts the dimethylammonium ion to dimethylamine, yet another good nucleophile, which also participates in nucleophilic substitution, and so on.

$$CH_3NH_3^+\ Br^- + NH_3 \xrightleftharpoons{\text{proton transfer}} CH_3NH_2 + NH_4^+\ Br^-$$

Methylammonium bromide — Methylamine

$$CH_3Br + CH_3NH_2 \xrightarrow{S_N2} (CH_3)_2NH_2^+\ Br^-$$

Dimethylammonium bromide

$$(CH_3)_2NH_2^+\ Br^- + NH_3 \xrightleftharpoons{\text{proton transfer}} (CH_3)_2NH + NH_4^+\ Br^-$$

Dimethylamine

The final product from such a series of nucleophilic substitution and proton transfer reactions is a tetraalkylammonium halide. The relative proportions of the various alkylation products depend on the ratio of alkyl halide to ammonia in the reaction mixture. Whatever the starting mixture, however, the product is almost invariably a mixture of alkylated products and, for this reason, alkylation of ammonia or amines is not a generally useful laboratory method for the preparation of more complex amines. Primary amines are easily prepared, however, because ammonia is inexpensive and can be used in large excess. Other amines can also be prepared in this way if the nucleophilic amine is inexpensive enough to be used in large excess.

B. Alkylation of Azide Ion

As we have just seen in the previous subsection, alkylation of ammonia or amines is generally not a useful method for the preparation of amines. One strategy for eliminating the problem of overalkylation is to use a form of nitrogen that can function as nucleophile, but is no longer an effective nucleophile once it has formed a new carbon-nitrogen bond. One such nucleophilic form of nitrogen is the azide ion, N_3^-. Alkyl azides are easily prepared from sodium or potassium azide and a primary or secondary alkyl halide by an S_N2 reaction. Azides are in turn reduced to primary amines by a variety of reducing agents including lithium aluminum hydride.

$$N_3^- \quad :\ddot{N}^{-}{=}\overset{+}{N}{=}\ddot{N}^{-}: \qquad RN_3 \quad R{-}\ddot{N}{=}\overset{+}{N}{=}\ddot{N}^{-}:$$

Azide ion (a good nucleophile) — An alkyl azide

$$C_6H_5CH_2Cl \xrightarrow[\text{(phase-transfer catalyst)}]{K^+\ N_3^-} C_6H_5CH_2N_3 \xrightarrow[2.\ H_2O]{1.\ LiAlH_4} C_6H_5CH_2NH_2$$

Benzyl chloride — Benzyl azide — Benzylamine

The azide ion can also be used for stereoselective ring opening of epoxides. Reduction of the resulting β-azidoalcohol gives a β-aminoalcohol as illustrated by the conversion of cyclohexene to *trans*-2-aminocyclohexanol. Oxidation of cyclohexene by a peroxyacid (Section 11.9B) gives an epoxide. Stereoselective nucleophilic attack by azide ion anti to the leaving oxygen of the epoxide ring (Section 11.10B) followed by reduction of the azide with lithium aluminum hydride gives *trans*-2-aminocyclohexanol.

Cyclohexene $\xrightarrow{ArCO_3H}$ 1,2-Epoxycyclohexane $\xrightarrow[2.\ H_2O]{1.\ K^+N_3^-}$ *trans*-2-Azidocyclohexanol $\xrightarrow[2.\ H_2O]{1.\ LiAlH_4}$ *trans*-2-Amino-cyclohexanol

EXAMPLE 21.9

Show how to convert 4-methoxybenzyl chloride to each amine in good yield.

(a) $CH_3O-C_6H_4-CH_2NH_2$ **(b)** $CH_3O-C_6H_4-CH_2CH_2NH_2$

Solution

(a) Two methods might be used: (1) alkylation of NH_3 using a large molar excess of NH_3 to reduce the extent of overalkylation, or (2) nucleophilic displacement of chloride using azide ion (from NaN_3) followed by $LiAlH_4$ reduction of the azide. Of these methods, nucleophilic displacement by azide is the more convenient on a laboratory scale.

(b) Nucleophilic displacement of chloride by cyanide ion followed by reduction of the cyano group with lithium aluminum hydride.

$$CH_3O-C_6H_4-CH_2Cl \xrightarrow{1.\ Na^+N_3^-;\ 2.\ LiAlH_4;\ 3.\ H_2O} CH_3O-C_6H_4-CH_2NH_2 \quad \text{(a)}$$

$$CH_3O-C_6H_4-CH_2Cl \xrightarrow{1.\ Na^+CN^-;\ 2.\ LiAlH_4;\ 3.\ H_2O} CH_3O-C_6H_4-CH_2CH_2NH_2 \quad \text{(b)}$$

PROBLEM 21.9

Show how to bring about each conversion in good yield. In addition to the given starting material, use any other reagents as necessary.

(a) 3-methoxyaniline (NH_2, OCH_3 on benzene) $\longrightarrow$ $N(CH_2CH_2OH)_2$, OCH_3 on benzene

(b) $C_6H_5-CH-CH_2$ (epoxide, O bridging) $\longrightarrow$ $C_6H_5-CH(OH)CH_2N(CH_3)_2$

21.9 Reaction with Nitrous Acid

Nitrous acid, HNO_2, is an unstable compound that is prepared by adding aqueous sulfuric acid or hydrochloric acid to an aqueous solution of sodium nitrite, $NaNO_2$. Nitrous acid is a weak oxygen acid and ionizes according to the following equation:

$$HNO_2 + H_2O \rightleftharpoons H_3O^+ + NO_2^- \qquad K_a = 4.26 \times 10^{-4}$$
$$pK_a = 3.37$$

Nitrous acid undergoes reaction with amines in different ways, depending on whether the amine is primary, secondary, or tertiary and whether it is aliphatic or aromatic. These reactions are all related by the facts that nitrous acid (1) participates in proton-transfer reactions and (2) is a source of the **nitrosyl cation,** a weak electrophile. The nitrosyl cation is generated by protonation of the —OH of nitrous acid followed by loss of H_2O. It is here represented as a resonance hybrid of two contributing structures.

MECHANISM Formation of the Nitrosyl Cation

$$\mathrm{H{-}\ddot{O}{-}\ddot{N}{=}\ddot{O}: + H^+ \rightleftharpoons H{-}\overset{+}{\ddot{O}}(H){-}\ddot{N}{=}\ddot{O}: \rightleftharpoons H{-}\ddot{O}(H): + :\overset{+}{N}{=}\ddot{O}: \longleftrightarrow :N{\equiv}\overset{+}{O}:}$$

Nitrosyl cation as a hybrid of two contributing structures

A. Tertiary Aliphatic Amines

When treated with nitrous acid, tertiary aliphatic amines, whether water-soluble or water-insoluble, are protonated to form water-soluble salts. No further reaction occurs beyond salt formation. This reaction is of little practical use.

B. Tertiary Aromatic Amines

Tertiary aromatic amines are bases and can also form salts with nitrous acid. An alternative pathway, however, is open to tertiary aromatic amines, namely, electrophilic aromatic substitution involving the nitrosyl cation. The nitrosyl cation is a very weak electrophile and reacts only with aromatic rings containing strongly activating, ortho-para directing groups, such as the hydroxyl and dialkylamino groups. When treated with nitrous acid, these compounds undergo **nitrosation,** predominantly in the para position to give blue or green aromatic nitroso compounds.

$$(CH_3)_2N{-}C_6H_5 \xrightarrow[\text{2. NaOH, }H_2O]{\text{1. NaNO}_2\text{, HCl, 0°–5°C}} (CH_3)_2N{-}C_6H_4{-}N{=}O$$

N,N-Dimethylaniline — *N,N*-Dimethyl-4-nitrosoaniline

C. Secondary Aliphatic and Aromatic Amines

Secondary amines, whether aliphatic or aromatic, undergo reaction with nitrous acid to give ***N*-nitrosoamines,** more commonly called *N*-nitrosamines, as illustrated by the reaction of piperidine with nitrous acid.

$$C_5H_{10}N{-}H + HNO_2 \longrightarrow C_5H_{10}N{-}N{=}O + H_2O$$

Piperidine — *N*-Nitrosopiperidine

Formation of an *N*-nitrosoamine involves interaction of the unshared pair of electrons of the amine (a nucleophile) with the positively charged nitrosyl cation (an electrophile) to form a new carbon-nitrogen bond. Proton transfer from the ammonium ion to water gives the *N*-nitrosoamine.

MECHANISM Reaction of a 2° Amine with the Nitrosyl Cation to Give an *N*-Nitrosoamine

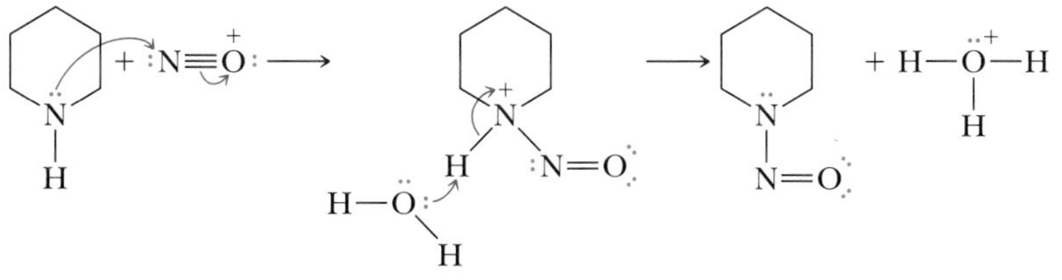

N-Nitrosamines are of little synthetic or commercial value. They have received considerable attention in recent years, however, because many of them are potent carcinogens. Following are structural formulas of two *N*-nitrosamines, each of which is a known carcinogen.

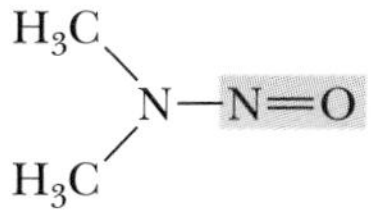

N-Nitrosodimethylamine (found in cigarette smoke and when bacon "preserved" with sodium nitrite is fried)

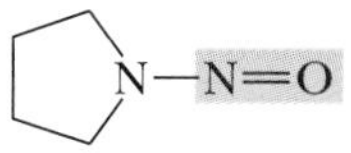

N-Nitrosopyrrolidine (formed when bacon "preserved" with sodium nitrite is fried)

It has been common practice within the food industry to add sodium nitrite to processed meats to "retard spoilage," that is, to inhibit the growth of *Clostridium botulinum*, the bacterium responsible for botulism poisoning. Although this practice was well grounded before the days of adequate refrigeration, it is of questionable value today. Sodium nitrite is also added to prevent red meats from turning brown. Controversy over the use of sodium nitrite has been generated by the demonstration that nitrite ion in the presence of acid converts secondary amines to *N*-nitrosamines and that many *N*-nitrosamines are powerful carcinogens. This demonstration led in turn to pressure by consumer groups to force the Food and Drug Administration to ban the use of nitrite additives in foods. The strength of the argument to ban nitrites was weakened with the finding that enzymes in our mouths and intestinal tracts have the ability to catalyze reduction of nitrate to nitrite. Nitrate ion is normally found in a wide variety of foods and in drinking water. To date, there is no evidence that nitrite as a food additive poses any risk not already present through our existing dietary habits. The FDA has established the current permissible level of sodium nitrite in processed meats as 50 to 125 ppm (that is, 50–125 μg nitrite per gram of cured meat).

D. Primary Aliphatic Amines

Treatment of a primary aliphatic amine with nitrous acid results in loss of nitrogen, N_2, and formation of substitution, elimination, and rearrangement products as illustrated by the treatment of butylamine with nitrous acid.

$$CH_3CH_2CH_2CH_2NH_2 \xrightarrow[0°\text{–}5°C]{NaNO_2,\ HCl}$$

- $CH_3CH_2CH_2CH_2OH$ (25%) — Substitution
- $CH_3CH_2CH_2CH_2Cl$ (5.2%) — Substitution
- $CH_3CH_2CH(OH)CH_3$ (13.2%) — Substitution involving rearrangement
- $CH_3CH_2CH{=}CH_2$ (25.9%) — Elimination
- $CH_3CH{=}CHCH_3$ (10.6%) — Elimination involving rearrangement

Diazonium ion An ArN_2^+ or RN_2^+ ion.

We can account for formation of this mixture of products in the following way. As a first step, treatment of a primary aliphatic amine with nitrous acid gives an *N*-nitrosoamine, which can then undergo keto-enol tautomerism (Section 15.11B) to give a diazotic acid, so named because it has two (di-) nitrogen (-azot-) atoms within its structure. Protonation of the —OH group of the diazotic acid followed by loss of water gives a **diazonium ion.** This conversion of a primary amine to a diazonium ion is called diazotization.

MECHANISM Reaction of a Primary Amine with Nitrous Acid to Give a Diazonium Ion

Step 1: Reaction of the primary amine with nitrous acid to form a diazotic acid

$$R{-}NH_2 + HNO_2 \longrightarrow R{-}N(H){-}N{=}O \xrightleftharpoons{\text{keto-enol tautomerism}} R{-}N{=}N{-}O{-}H$$

A 1° aliphatic amine — An *N*-nitrosamine — A diazotic acid

Step 2: Protonation of the diazotic acid followed by loss of H_2O and N_2 to give a carbocation

$$R{-}N{=}N{-}O{-}H \xrightleftharpoons{H^+} R{-}N{=}N{-}\overset{+}{O}(H){-}H \xrightarrow{-H_2O} R{-}\overset{+}{N}{\equiv}N{:} \longrightarrow R^+ + {:}N{\equiv}N{:}$$

A diazotic acid — A diazonium acid — A carbocation

Aliphatic diazonium ions are unstable, even at 0°C, and immediately lose nitrogen to give carbocations and nitrogen gas. The driving force for this reaction is the fact that N_2 is one of the best leaving groups because it is an extraordinarily weak base. It is removed from the reaction mixture as a gas as it is formed. The carbocation now has open to it the three reactions in the repertoire of aliphatic carbocations: (1) loss of a proton to form an alkene, (2) reaction with a nucleophile to give a substitution product, and (3) rearrangement to a more stable carbocation and then reaction further by (1) or (2).

Because treatment of a primary aliphatic amine with HNO_2 gives a mixture of products, it is generally not a useful reaction. An exception is the Tiffeneau-Demjanov reaction in which a cyclic ***β*-aminoalcohol** is treated with nitrous acid to form a ring-expanded ketone, with evolution of nitrogen.

primary amino group beta to the —OH group

OH α β CH_2NH_2 + HNO_2 ⟶ =O + H_2O + N_2

A β-aminoalcohol Cycloheptanone

We can account for this molecular rearrangement in the following way. Treatment of the primary amine with nitrous acid generates a diazonium ion. Concerted loss of nitrogen and rearrangement by a 1,2-shift gives a resonance-stabilized, ring-expanded cation. Loss of a proton from this cation gives cycloheptanone.

MECHANISM The Tiffeneau-Demjanov Reaction

:ÖH $CH_2\ddot{N}H_2$ —HNO_2⟶ :Ö—H CH_2—$\overset{+}{N}$≡N: —loss of N_2 and rearrangement by a 1,2-shift⟶

A diazonium ion

:Ö—H + CH_2 ⟷ +Ö—H CH_2 —proton transfer to H_2O⟶ Ö:

A resonance-stabilized cation Cycloheptanone

The driving force for this molecular rearrangement is precisely what we already saw for other cation rearrangements: transformation of a less stable cation into a more stable cation. This reaction is analogous to the pinacol rearrangement (Section 9.8) with the leaving group being N_2 rather than H_2O.

EXAMPLE 21.10

The following sequence of reactions gives cyclooctanone. Propose a structural formula for compound (A) and a mechanism for its conversion to cyclooctanone.

$$\text{1-hydroxycycloheptanecarbonitrile (HO, C}\equiv\text{N)} \xrightarrow{H_2,\ Pt} \text{A } (C_8H_{17}NO) \xrightarrow{HNO_2} \text{cyclooctanone}$$

Solution

Catalytic hydrogenation using hydrogen over a platinum catalyst reduces the carbon-nitrogen triple bond to a single bond (Section 17.11C) and gives a β-aminoalcohol. Treatment of the β-aminoalcohol with nitrous acid results in loss of N_2 and expansion of the seven-membered ring to an eight-membered cyclic ketone.

$$\text{(HO, C}\equiv\text{N cycloheptane)} \xrightarrow{H_2,\ Pt} \underset{\text{A}}{\text{(HO, CH}_2\text{NH}_2\text{ cycloheptane)}} \xrightarrow{HNO_2} \text{cyclooctanone}$$

PROBLEM 21.10

How might you bring about this conversion?

$$\text{cycloheptanone} \xrightarrow{?} \text{(HO, CH}_2\text{NO}_2\text{ cycloheptane)} \xrightarrow{?} \text{cyclooctanone}$$

E. Primary Aromatic Amines

Primary aromatic amines, like primary aliphatic amines, undergo reaction with nitrous acid to form arenediazonium salts. Arenediazonium salts, unlike their aliphatic counterparts, are stable at 0°C and can be kept in solution for short periods without decomposition. When an arenediazonium salt is treated with an appropriate reagent, nitrogen is lost and replaced by another atom or functional group. What makes reactions of primary aromatic amines with nitrous acid so valuable is the fact that the amino group can be replaced in a totally regioselective manner by the groups shown.

$$Ar{-}NH_2 \xrightarrow[0°\text{–}5°C]{NaNO_2,\ H_3O^+} Ar{-}N_2^+ \xrightarrow{(-N_2)}$$

- $\xrightarrow{H_2O}$ Ar—OH
- $\xrightarrow{HBF_4}$ Ar—F (Schiemann reaction)
- $\xrightarrow{HCl,\ CuCl}$ Ar—Cl (Sandmeyer reaction)
- $\xrightarrow{HBr,\ CuBr}$ Ar—Br (Sandmeyer reaction)
- $\xrightarrow{KCN,\ CuCN}$ Ar—CN (Sandmeyer reaction)
- $\xrightarrow{KI}$ Ar—I
- $\xrightarrow{H_3PO_2}$ Ar—H

Aromatic amines can be converted to phenols by first forming the arenediazonium salt in aqueous sulfuric acid and then heating the solution. In this manner, 2-bromo-4-methylaniline is converted to 2-bromo-4-methylphenol.

$$\text{2-Bromo-4-methylaniline} \xrightarrow[\text{2. } H_2O\text{, heat}]{\text{1. } NaNO_2,\ H_2SO_4,\ H_2O} \text{2-Bromo-4-methylphenol}$$

2-Bromo-4-methylaniline

2-Bromo-4-methylphenol

The intermediate in the decomposition of an arenediazonium ion in water is an aryl cation, which then undergoes reaction with water to form the phenol. Note that the aryl cation is so unstable that it is only with N_2 as the leaving group that it can be formed.

$$C_6H_5-\overset{+}{N}\equiv N: \xrightarrow[\text{(loss of } N_2)]{\text{heat}} [C_6H_5^+] \xrightarrow{2H_2O} C_6H_5-OH + H_3O^+$$

Benzenediazonium ion

An aryl cation

Phenol

This reaction of arenediazonium salts represents the main laboratory preparation of phenols.

EXAMPLE 21.11

What reagents and experimental conditions will bring about each step in the conversion of toluene to 4-hydroxybenzoic acid?

$$\text{Toluene } (CH_3) \xrightarrow{(1)} \text{4-}NO_2\text{-toluene} \xrightarrow{(2)} \text{4-}NO_2\text{-}C_6H_4CO_2H \xrightarrow{(3)} \text{4-}NH_2\text{-}C_6H_4CO_2H \xrightarrow{(4)} \text{4-}OH\text{-}C_6H_4CO_2H$$

Solution

Step 1: Nitration of the aromatic ring (Section 20.1B) using HNO_3 in H_2SO_4 followed by separation of the ortho and para isomers gives 4-nitrotoluene.

Step 2: Oxidation at the benzylic carbon (Section 19.6A) using $K_2Cr_2O_7$ in H_2SO_4 gives 4-nitrobenzoic acid.

Step 3: Reduction of the nitro group (Section 20.1B) to an amino group can be brought about using H_2 in the presence of Ni or other transition metal catalyst. Alternatively, it can be brought about using Zn, Sn, or Fe in aqueous HCl followed by aqueous NaOH.

Step 4: Reaction of the aromatic amine with $NaNO_2$ in aqueous H_2SO_4 followed by heating gives 4-hydroxybenzoic acid.

PROBLEM 21.11

Show how to convert toluene to 3-hydroxybenzoic acid using the same set of reactions as in Example 21.11, but changing the order in which some of the steps are carried out.

The **Schiemann reaction** is the most common method for introduction of fluorine onto an aromatic ring. It is carried out by treatment of a primary aromatic amine with sodium nitrite in aqueous HCl followed by addition of HBF_4 or $NaBF_4$. The diazonium fluoroborate salt precipitates and is collected and dried. Heating the dry salt brings about its decomposition to an aryl fluoride, nitrogen, and boron trifluoride. The Schiemann reaction is also thought to involve an aryl cation as an intermediate

$$C_6H_5NH_2 \xrightarrow[\text{2. } HBF_4]{\text{1. } NaNO_2, HCl, 0°C} C_6H_5N_2^+ \; BF_4^- \xrightarrow{\text{heat}} C_6H_5F + N_2 + BF_3$$

A diazonium fluoroborate — Fluorobenzene

Treatment of a primary aromatic amine with nitrous acid followed by heating with HCl/CuCl, HBr/CuBr, or KCN/CuCN results in replacement of the diazonium group by —Cl, —Br, or —CN, respectively, and is known as the **Sandmeyer reaction.** The Sandmeyer reaction fails when attempted with CuI or CuF.

2-Methylaniline (*o*-Toluidine) $\xrightarrow[0°–5°C]{NaNO_2, H_3O^+}$ 2-methylbenzenediazonium (N_2^+)

- $\xrightarrow[\text{heat}]{HCl/CuCl}$ 2-Chlorotoluene
- $\xrightarrow[\text{heat}]{HBr/CuBr}$ 2-Bromotoluene
- $\xrightarrow[\text{heat}]{KCN/CuCN}$ 2-Methylbenzonitrile ($C{\equiv}N$)

Treatment of an arenediazonium ion with iodide ion, generally from potassium iodide, is the best and most convenient method for introduction of iodine onto an aromatic ring.

2-Methylaniline (*o*-Toluidine) $\xrightarrow[0°–5°C]{NaNO_2, HCl}$ $N_2^+Cl^-$ $\xrightarrow[25°C]{KI}$ 2-Iodotoluene

Treatment of an arenediazonium ion with **hypophosphorous acid, H_3PO_2,** results in reduction of the diazonium group and its replacement by —H as illustrated by the conversion of aniline to 1,3,5-trichlorobenzene. Recall that $—NH_2$ is a powerful activating group (Section 20.2A). Treatment of aniline with chlorine requires no catalyst and gives 2,4,6-trichloroaniline. To complete the conversion, the $—NH_2$ group is removed by treatment with nitrous acid followed by hypophosphorous acid to give 1,3,5-trichlorobenzene.

Aniline $\xrightarrow{Cl_2}$ 2,4,6-Trichloroaniline $\xrightarrow[0°–5°C]{NaNO_2,\ HCl}$ 2,4,6-Trichlorobenzene-diazonium chloride $\xrightarrow{H_3PO_2}$ 1,3,5-Trichlorobenzene

EXAMPLE 21.12

Show reagents and conditions that will bring about the conversion of toluene to 3-bromo-4-methylphenol.

Toluene $\xrightarrow{(1)}$ 4-nitrotoluene $\xrightarrow{(2)}$ 2-bromo-4-nitrotoluene $\xrightarrow{(3)}$ 3-bromo-4-methylaniline $\xrightarrow{(4)}$ 3-bromo-4-methylphenol

Solution

Step 1: HNO_3 in H_2SO_4. Methyl is ortho-para directing and slightly activating.

Step 2: Treatment of 4-nitrotoluene with bromine in the presence of $FeCl_3$.

Step 3: Reduction of the nitro group using either H_2/Ni or using Sn, Zn, or Fe in aqueous HCl followed by aqueous NaOH.

Step 4: Diazotization of the amine with $NaNO_2$ in aqueous sulfuric acid followed by heating of the solution to replace $—N_2^+$ by —OH.

PROBLEM 21.12

Starting with 3-nitroaniline, show how to prepare the following compounds.

(a) 3-Nitrophenol
(b) 3-Bromoaniline
(c) 1,3-Dihydroxybenzene (resorcinol)
(d) 3-Fluoroaniline
(e) 3-Fluorophenol
(f) 3-Hydroxybenzonitrile

21.10 Hofmann Elimination

When a quaternary ammonium halide is treated with moist silver oxide (a slurry of Ag_2O in H_2O), silver halide precipitates, leaving a solution of a quaternary ammonium hydroxide.

$$\text{C}_6\text{H}_{11}\text{—CH}_2\text{—}\overset{+}{\text{N}}(\text{CH}_3)_3\ \text{I}^- + \text{Ag}_2\text{O} \xrightarrow{\text{H}_2\text{O}} \text{C}_6\text{H}_{11}\text{—CH}_2\text{—}\overset{+}{\text{N}}(\text{CH}_3)_3\ \text{OH}^- + \text{AgI}$$

(Cyclohexylmethyl)trimethylammonium iodide — Silver oxide — (Cyclohexylmethyl)trimethylammonium hydroxide

In the mid-19th century, Augustus Hofmann (for whom the Hofmann rearrangement is named) discovered that when a quaternary ammonium hydroxide is heated, it decomposes to an alkene, a tertiary amine, and water. Thermal decomposition of a quaternary ammonium hydroxide to an alkene is known as **Hofmann elimination.**

$$\text{C}_6\text{H}_{11}\text{—CH}_2\text{—}\overset{+}{\text{N}}(\text{CH}_3)_3\ \text{OH}^- \xrightarrow{160°} \text{C}_6\text{H}_{10}\text{=CH}_2 + (\text{CH}_3)_3\text{N} + \text{H}_2\text{O}$$

(Cyclohexylmethyl)trimethylammonium hydroxide — Methylenecyclohexane — Trimethylamine

The Hofmann elimination has most of the characteristics of an E2 elimination reaction (Section 8.9B). First, Hofmann eliminations are concerted, meaning that bond-breaking and bond-forming steps occur simultaneously, or nearly so. Second, Hofmann eliminations are stereoselective anti eliminations, meaning that —H and the leaving group must be anti to each other. The following mechanism illustrates the concerted nature of bond forming and bond breaking, and the anti arrangement of —H and the trialkylamino group.

MECHANISM The Hofmann Elimination

$$\text{HO}^- + \text{CH}_3\text{CH}_2\text{CH(H)—CH}_2\text{—}\overset{+}{\text{N}}(\text{CH}_3)_3 \xrightarrow[\text{(concerted elimination)}]{\text{E2 reaction}} \text{HOH} + \text{CH}_3\text{CH}_2\text{CH=CH}_2 + :\text{N}(\text{CH}_3)_3$$

When we studied E2 reactions of alkyl halides in Section 8.9B, we saw that a β-hydrogen must be anti to the leaving group. If only one β-hydrogen meets this requirement, then the double bond is formed in that direction. If, however, two β-hydrogens meet this requirement, then elimination follows Zaitsev's rule: elimination occurs preferentially to form the more substituted double bond.

$$\text{CH}_3\text{CH}_2\text{CH(Br)CH}_3 \xrightarrow[\text{E2}]{\text{CH}_3\text{CH}_2\text{O}^-\text{Na}^+} \underset{(75\%)}{\text{CH}_3\text{CH=CHCH}_3} + \underset{(25\%)}{\text{CH}_3\text{CH}_2\text{CH=CH}_2}$$

Thermal decomposition of quaternary ammonium hydroxides is different because elimination occurs preferentially to form the least substituted double bond. Thermal decomposition of *sec*-butyltrimethylammonium hydroxide, for example, gives 1-butene as the major product.

$$\underset{\substack{|\\ \overset{+}{N}(CH_3)_3 \\ HO^-}}{CH_3CH_2CHCH_3} \xrightarrow[\text{heat}]{\text{E2}} \underset{(5\%)}{CH_3CH{=}CHCH_3} + \underset{(95\%)}{CH_3CH_2CH{=}CH_2} + (CH_3)_3N + H_2O$$

Elimination reactions that give the less substituted alkene as the major product are said to follow Hofmann's rule. According to **Hofmann's rule,** elimination occurs preferentially to give the least substituted double bond. Thus, we say that thermal decomposition of quaternary ammonium hydroxides follows Hofmann's rule.

Hofmann's rule Any β-elimination that occurs preferentially to give the least substituted double bond as the major product is said to follow Hofmann's rule.

The following examples further illustrate Hofmann elimination.

$$\text{1-methylcyclohexyl-}\overset{+}{N}(CH_3)_3\ OH^- \xrightarrow{\text{heat}} \text{methylenecyclohexane } ({=}CH_2) + (CH_3)_3N + H_2O$$

EXAMPLE 21.13

Draw the structural formula of the major alkene formed in each β-elimination.

(a) $\underset{\substack{|\\NH_2}}{CH_3(CH_2)_5CHCH_3} \xrightarrow[\text{3. heat}]{\text{1. } CH_3I \text{ (excess)}, K_2CO_3 \quad \text{2. } Ag_2O, H_2O}$

(b) $\underset{\substack{|\\I}}{CH_3(CH_2)_5CHCH_3} \xrightarrow{CH_3O^-Na^+}$

Solution

Thermal decomposition of a quaternary ammonium hydroxide in (a) follows Hofmann's rule and gives 1-octene as the major product. E2 elimination from an alkyl iodide in (b) by sodium methoxide follows Zaitsev's rule and gives 2-octene as the major product.

(a) $\underset{\text{1-Octene}}{CH_3(CH_2)_5CH{=}CH_2} + (CH_3)_3N + H_2O$

(b) $\underset{\text{2-Octene}}{CH_3(CH_2)_4CH{=}CHCH_3} + CH_3OH$

PROBLEM 21.13

The procedure of methylation of amines and thermal decomposition of quaternary ammonium hydroxides was first reported by Hofmann in 1851, but its value as a means of structure determination was not appreciated until 1881 when he published a report of its use in determining the structure of piperidine. Following are the results obtained by Hofmann:

$$\underset{\text{Piperidine}}{C_5H_{11}N} \xrightarrow[\text{3. heat}]{\text{1. } CH_3I \text{ (excess)}, K_2CO_3 \quad \text{2. } Ag_2O, H_2O} \underset{\text{(A)}}{C_7H_{15}N} \xrightarrow[\text{6. heat}]{\text{4. } CH_3I \text{ (excess)}, K_2CO_3 \quad \text{5. } Ag_2O, H_2O} \underset{\text{1,4-Pentadiene}}{CH_2{=}CHCH_2CH{=}CH_2}$$

(a) Show that these results are consistent with the structure of piperidine (Section 21.1).

(b) Propose two additional structural formulas (excluding stereoisomers) for $C_5H_{11}N$ that are also consistent with the results obtained by Hofmann.

In summary, both Hofmann and Zaitsev eliminations are always anti. If only one β-hydrogen is anti to the leaving group, then that is the one removed. If more than one β-hydrogen is anti, then there is competition between Hofmann and Zaitsev elimination.

1. Eliminations involving a negatively charged leaving group, for example Cl^-, Br^-, I^-, and OTs^-, almost always follow Zaitsev's rule, unless a bulky base is used.
2. Elimination involving a neutral leaving group, for example $N(CH_3)_3$ and $S(CH_3)_2$, almost always follow Hofmann's rule.
3. The bulkier the base, as for example $(CH_3)_3CO^-K^+$ compared with $CH_3O^-Na^+$, the greater the percentage of Hofmann product.

One of the likeliest explanations for formation of the less stable double bond is that the Hofmann elimination is governed largely by steric factors, namely the bulk of the $—NR_3^+$ group. The hydroxide ion preferentially approaches and removes the least hindered β-hydrogen and gives the least substituted alkene as product. For the same reason, bulky bases, such as $(CH_3)_3CO^-K^+$, give Hofmann elimination also from alkyl halides.

21.11 Cope Elimination

Treatment of a tertiary amine with hydrogen peroxide results in a two-electron oxidation of the amine to an **amine oxide.**

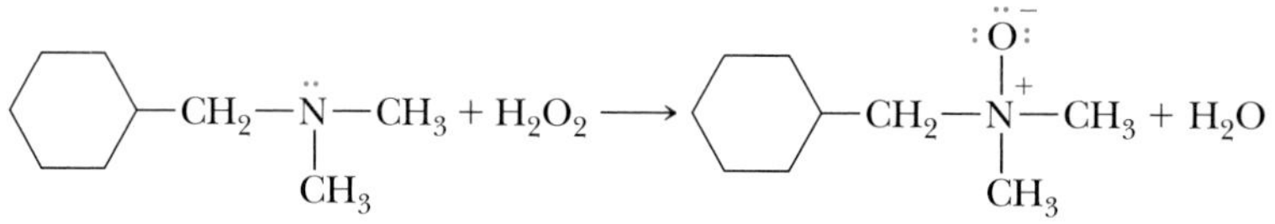

$$C_6H_{11}—CH_2—\ddot{N}(CH_3)—CH_3 + H_2O_2 \longrightarrow C_6H_{11}—CH_2—\overset{+}{N}(\ddot{O}:^-)(CH_3)—CH_3 + H_2O$$

An amine oxide

When an amine oxide with at least one β-hydrogen is heated, it undergoes thermal decomposition to form an alkene and an *N,N*-dialkylhydroxylamine. Thermal decomposition of an amine oxide to an alkene is known as a **Cope elimination** after its discoverer, Arthur C. Cope, of the Massachusetts Institute of Technology.

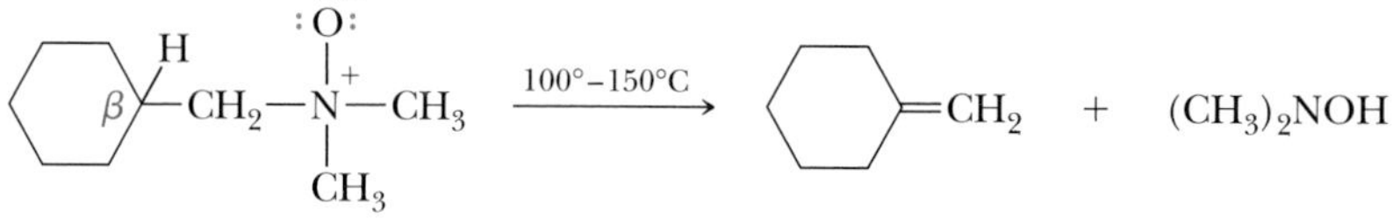

$$C_6H_{10}(\beta\text{-}H)—CH_2—\overset{+}{N}(\ddot{O}:^-)(CH_3)—CH_3 \xrightarrow{100°-150°C} C_6H_{10}{=}CH_2 + (CH_3)_2NOH$$

Methylenecyclohexane; *N,N*-Dimethylhydroxylamine

All experimental evidence indicates that the Cope elimination is syn stereoselective and concerted. As shown in the following mechanism, the transition state involves a planar or nearly planar arrangement of the five participating atoms and a cyclic flow of three pairs of electrons.

MECHANISM The Cope Elimination

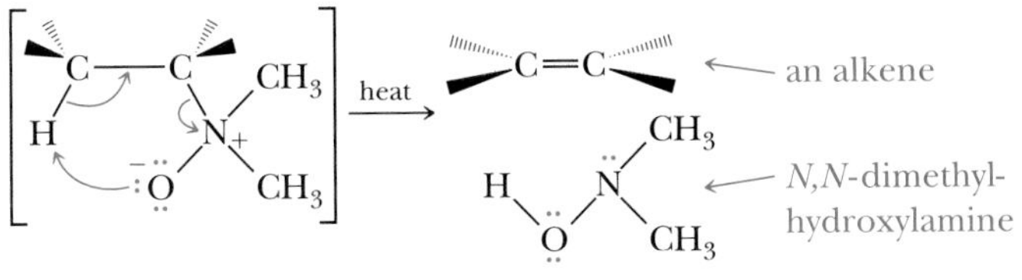

$$\left[\begin{array}{c} \text{C}-\text{C}-\text{CH}_3 \\ \text{H} \quad \text{N}^+ \\ ^-\text{:O:} \quad \text{CH}_3 \end{array}\right] \xrightarrow{\text{heat}} \text{C}=\text{C} + \text{H}-\ddot{\text{O}}-\ddot{\text{N}}(\text{CH}_3)_2$$

← an alkene

← *N,N*-dimethylhydroxylamine

Transition state (cyclic flow of three pairs of electrons)

If two or more syn β-hydrogens can be removed in a Cope elimination, there is little preference for one over another except when the double bond is conjugated with an aromatic ring. Therefore, as a method of preparation of alkenes, Cope eliminations are best used where only one alkene is possible.

EXAMPLE 21.14

Following is a formula for 2-dimethylamino-3-phenylbutane. When it is treated with hydrogen peroxide and then made to undergo a Cope elimination, the major alkene formed is 2-phenyl-2-butene.

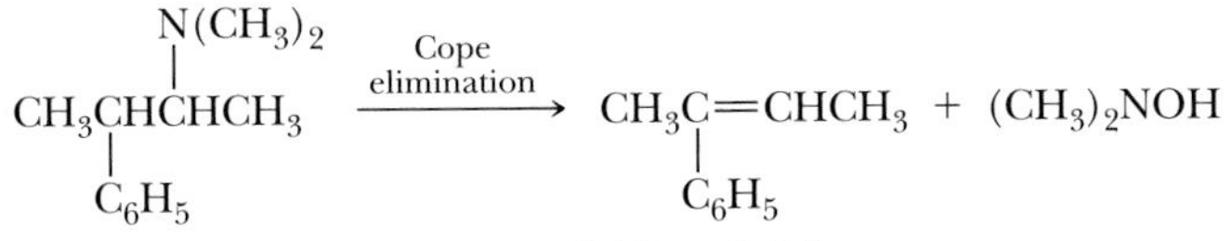

$$CH_3CH(C_6H_5)CH(N(CH_3)_2)CH_3 \xrightarrow{\text{Cope elimination}} CH_3C(C_6H_5)=CHCH_3 + (CH_3)_2NOH$$

2-Dimethylamino-3-phenylbutane → 2-Phenyl-2-butene

(a) How many stereoisomers are possible for 2-dimethylamino-3-phenylbutane?
(b) How many stereoisomers are possible for 2-phenyl-2-butene?
(c) Suppose the starting amine is the (2*R*,3*S*)-isomer. What is the configuration of the product?

Solution

(a) There are two stereocenters in the starting amine. Four stereoisomers are possible: two pair of enantiomers.
(b) There is one carbon-carbon double bond about which stereoisomerism is possible. Two stereoisomers are possible: one *E,Z* pair.

(c) Following is a stereodrawing of the (2*R*,3*S*) stereoisomer showing a syn conformation of the dimethylamino group and the β-hydrogen. Cope elimination on this stereoisomer gives (*E*)-2-phenyl-2-butene.

3*S*; C_6H_5, H_3C, H, C—C, H, CH_3, $N(CH_3)_2$, 2*R* $\xrightarrow[\text{2. heat}]{\text{1. } H_2O_2}$ C_6H_5, H_3C C=C H, CH_3 + $(CH_3)_2NOH$

(*E*)-2-Phenyl-2-butene

PROBLEM 21.14

In Example 21.14, you considered the product of Cope elimination from the (2*R*,3*S*) stereoisomer of 2-dimethylamino-3-phenylbutane. What is the product of Cope elimination from each of these stereoisomers? What is the product of Hofmann elimination from each of these stereoisomers?

(a) (2*S*,3*R*) stereoisomer? **(b)** (2*S*,3*S*) stereoisomer?

SUMMARY

Amines are classified as **primary, secondary,** or **tertiary,** depending on the number of carbon atoms bonded to nitrogen (Section 21.1). In an **aliphatic amine,** all carbon atoms attached directly to nitrogen are derived from alkyl groups. In an **aromatic amine,** one or more of the groups attached to nitrogen are aromatic rings. A **heterocyclic amine** is one in which the nitrogen atom is part of a ring. A **heterocyclic aromatic amine** is one in which the nitrogen atom is part of an aromatic ring.

In systematic nomenclature, aliphatic amines are named **alkanamines** (Section 21.2A). A cation in which a nitrogen bonded to four alkyl or aryl groups is named as a quaternary ammonium ion. In common nomenclature (Section 21.2B), aliphatic amines are named **alkylamines;** the alkyl groups are listed in alphabetical order in one word ending in the suffix **-amine.**

Amines in which nitrogen is bonded to three different groups are chiral (Section 21.3) and, in principle, can be resolved. In practice, however, they undergo rapid **pyramidal inversion** with the result that a chiral amine is converted into a racemic mixture. Pyramidal inversion is not possible for quaternary ammonium ions, and chiral quaternary ammonium ions can be resolved. Chiral phosphines invert slowly and have also been prepared and resolved.

Amines are polar compounds, and primary and secondary amines associate by intermolecular hydrogen bonding (Section 21.4). An N—H---N hydrogen bond is weaker than an O—H---O hydrogen bond and, therefore, amines have lower boiling points than alcohols of comparable molecular weight and structure. All classes of amines form hydrogen bonds with water and are more soluble in water than hydrocarbons of comparable molecular weight.

Amines are weak bases, and aqueous solutions of amines are basic (Section 21.6). The base dissociation constant of an amine in water is given the symbol K_b. It is also common to discuss the acid-base properties of amines by reference to the acid dissociation constant, K_a, for their corresponding conjugate acids. Acid and base dissociation constants for amines in water are related by the equation $pK_a + pK_b = 14.0$.

Most aliphatic amines have comparable basicity and all are slightly stronger bases than ammonia (Section 21.6A). For representative aliphatic amines, values of pK_b are in the range 3.0 to 5.0. Aromatic amines are considerably weaker bases than aliphatic amines. For representative aromatic amines, values of pK_b are in the range 9.0 to 11.0 (Section 21.6B). Unprotonated aromatic amines have resonance stabilization due to interaction of the unshared pair of electrons on nitrogen with the pi system of the aromatic ring. This interaction is lost on protonation.

Heterocyclic aromatic amines are considerably weaker bases than aliphatic amines because the unshared pair of electrons on nitrogen giving rise to basicity lies in an sp^2 hybrid orbital (Section 21.6C). The electron-withdrawing inductive effect of an sp^2 hybridized nitrogen compared with an sp^3 hybridized nitrogen is responsible for the decreased basicity of heterocyclic aromatic amines.

All amines, whether soluble or insoluble in water, react quantitatively with strong acids to form water-soluble salts (Section 21.7). The basicity of amines and the solubility of amine salts in water can be used to separate water-insoluble amines from water-insoluble nonbasic compounds.

KEY REACTIONS

1. Basicity of Aliphatic Amines (Section 21.6A)

Aliphatic amines are slightly stronger bases than ammonia due to the electron-releasing effect of alkyl groups and partial delocalization of positive charge in the alkylammonium ion.

$$CH_3NH_2 + H_2O \rightleftharpoons CH_3NH_3^+ + OH^- \qquad pK_b = 3.36$$

2. Basicity of Aromatic Amines (Section 21.6B)

Aromatic amines are considerably weaker bases than aliphatic amines. Resonance stabilization by interaction of the unshared electron pair on nitrogen with the pi system of the aromatic ring is lost on protonation.

$$C_6H_5\ddot{N}H_2 + H_2O \rightleftharpoons C_6H_5NH_3^+ + OH^- \qquad pK_b = 9.37$$

3. Basicity of Heterocyclic Aromatic Amines (Section 21.6C)

Heterocyclic aromatic amines are considerably weaker bases than aliphatic amines.

$$\text{imidazole} + H_2O \rightleftharpoons \text{imidazolium (N–H}^+\text{)} + OH^- \qquad pK_b = 7.05$$

4. Reaction of Amines with Strong Acids (Section 21.7)

All amines react quantitatively with strong acids to form water-soluble salts.

$$C_6H_5\ddot{N}(CH_3)_2 + HCl \longrightarrow C_6H_5\overset{+}{N}H(CH_3)_2\ Cl^-$$

Insoluble in water — A water-soluble salt

5. Alkylation of Ammonia and Amines (Section 21.8A)

This method is seldom used for preparation of pure amines because of overalkylation and the difficulty of separating products.

$$C_6H_{11}CH_2NH_2 + CH_3I \longrightarrow C_6H_{11}CH_2\overset{+}{N}H_2CH_3\ I^-$$

6. Alkylation of Azide Ion Followed by Reduction (Section 21.8B)

Azides are prepared by treatment of a primary or secondary alkyl halide, or an epoxide with NaN_3 and are reduced to primary amines by a variety of reducing agents including lithium aluminum hydride.

$$\text{cyclohexene oxide} \xrightarrow[\text{2. } H_2O]{\text{1. } K^+N_3^-} \text{trans-2-azidocyclohexanol} \xrightarrow[\text{2. } H_2O]{\text{1. } LiAlH_4} \text{trans-2-aminocyclohexanol}$$

7. Nitrosation of Tertiary Aromatic Amines (Section 21.9B)

The nitrosyl cation is a very weak electrophile and participates in electrophilic aromatic substitution only with highly activated aromatic rings.

$$(CH_3)_2N\text{—}C_6H_5 \xrightarrow[\text{2. NaOH, } H_2O]{\text{1. } NaNO_2\text{, HCl, 0°–5°C}} (CH_3)_2N\text{—}C_6H_4\text{—}N{=}O$$

8. Formation of *N*-Nitrosamines from Secondary Amines (Section 21.9C)

Reaction of the nitrosyl cation, a weak electrophile, with a 2° amine, a nucleophile, gives an *N*-nitrosamine.

$$\text{(piperidine)N—H} + HNO_2 \longrightarrow \text{(piperidine)N—N{=}O} + H_2O$$

9. Reaction of Primary Aliphatic Amines with Nitrous Acid (Section 21.9D)

Treatment of a primary aliphatic amine with nitrous acid gives an unstable diazonium ion that loses N_2 to give a carbocation. The carbocation may (1) lose a proton to give an alkene, (2) react with a nucleophile, or (3) rearrange, followed by (1) or (2).

$$CH_3CH_2CH_2CH_2NH_2 \xrightarrow[\text{0°–5°C}]{NaNO_2\text{, HCl}} [CH_3CH_2CH_2CH_2N_2^+] \longrightarrow CH_3CH_2CH_2CH_2^+ + N_2$$

10. Reaction of Cyclic *β*-Aminoalcohols with Nitrous Acid (Section 21.9D)

Treatment of a cyclic *β*-aminoalcohol with nitrous acid leads to rearrangement and a ring-expanded ketone.

$$\text{1-(aminomethyl)cyclohexanol (OH, } CH_2NH_2) + HNO_2 \longrightarrow \text{cycloheptanone (=O)} + H_2O + N_2$$

11. Formation of Arenediazonium Salts (Diazotization) (Section 21.9E)

Arenediazonium salts are stable in aqueous solution at 0°C for short periods.

$$C_6H_5\text{—}NH_2 + HNO_2 \xrightarrow[\text{0°C}]{HCl} C_6H_5\text{—}\overset{+}{N}{\equiv}N\text{:}\ Cl^- + H_2O$$

12. Conversion of a Primary Arylamine to a Phenol (Section 21.9E)

Formation of an arenediazonium salt followed by loss of nitrogen gives an aryl cation intermediate, which then reacts with water to give a phenol.

NH_2, Br, CH_3 (4-methyl-2-bromoaniline) $\xrightarrow[\text{2. } H_2O\text{, heat}]{\text{1. } NaNO_2\text{, } H_2SO_4\text{, } H_2O}$ OH, Br, CH_3

13. Schiemann Reaction (Section 21.9E)

Heating an arenediazonium fluoroborate is the most common synthetic method for introduction of fluorine onto an aromatic ring.

$$C_6H_5\text{—}NH_2 \xrightarrow[\text{2. } HBF_4]{\text{1. } NaNO_2\text{, HCl, 0°C}} C_6H_5\text{—}N_2^+\ BF_4^- \xrightarrow{\text{heat}} C_6H_5\text{—}F + N_2 + BF_3$$

An arenediazonium fluoroborate — Fluorobenzene

14. Sandmeyer Reaction (Section 21.9E)

Treatment of an arenediazonium salt with CuCl, CuBr, or CuCN results in replacement of the diazonium group by —Cl, —Br, or —CN, respectively.

CH_3, NH_2 $\xrightarrow[\text{0°–5°C}]{NaNO_2\text{, HCl}}$ CH_3, N_2^+ $\xrightarrow[\text{heat}]{\text{CuCN}}$ CH_3, C≡N

15. Reaction of an Arenediazonium Salt with KI (Section 21.9E)

Treatment of an arenediazonium salt with KI is the most convenient method for introduction of iodine onto an aromatic ring.

CH_3, NH_2 $\xrightarrow[\text{0°–5°C}]{NaNO_2\text{, HCl}}$ CH_3, N_2^+ $\xrightarrow[\text{25°C}]{\text{KI}}$ CH_3, I

16. Reduction of an Arenediazonium Salt with Hypophosphorous Acid (Section 21.9E)

An —NO_2 or —NH_2 group can be used to control orientation of further substitution and then removed once it has served its purpose.

NH_2, Cl, Cl $\xrightarrow[\text{0°–5°C}]{NaNO_2\text{, HCl}}$ N_2^+, Cl, Cl $\xrightarrow{H_3PO_2}$ Cl, Cl

17. Hofmann Elimination (Section 21.10)

Anti stereoselective elimination of quaternary ammonium hydroxides occurs preferentially to form the least substituted carbon-carbon double bond (Hofmann's rule).

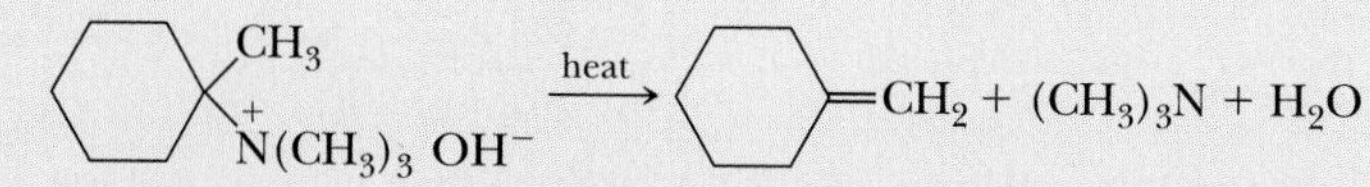

$$\text{cyclohexane-1-}CH_3\text{-1-}\overset{+}{N}(CH_3)_3\ OH^- \xrightarrow{\text{heat}} \text{cyclohexane}{=}CH_2 + (CH_3)_3N + H_2O$$

18. Cope Elimination: Pyrolysis of a Tertiary Amine Oxide (Section 21.11)

Elimination is syn stereoselective and involves a cyclic flow of six electrons in a planar transition state.

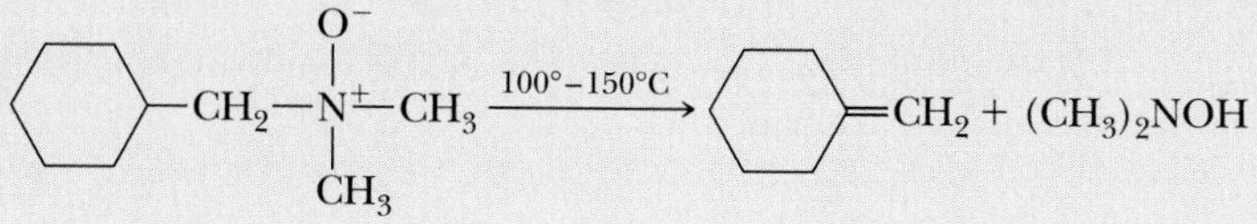

$$\text{C}_6\text{H}_{11}{-}CH_2{-}\overset{O^-}{\overset{|}{N^+}}(CH_3){-}CH_3 \xrightarrow{100°-150°C} \text{cyclohexane}{=}CH_2 + (CH_3)_2NOH$$

ADDITIONAL PROBLEMS

Structure and Nomenclature

21.15 Draw structural formulas for these amines and amine derivatives.

(a) *N,N*-Dimethylaniline
(b) Triethylamine
(c) *tert*-Butylamine
(d) 1,4-Benzenediamine
(e) 4-Aminobutanoic acid
(f) (*R*)-2-Butanamine
(g) Benzylamine
(h) *trans*-2-Aminocyclohexanol
(i) 1-Phenyl-2-propanamine (amphetamine)
(j) Lithium diisopropylamide (LDA)
(k) Benzyltrimethylammonium hydroxide (Triton B)

21.16 Give an acceptable name for these compounds.

(a) 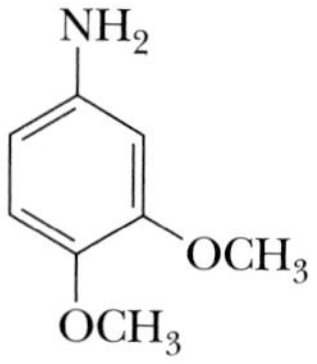 NH_2, OCH_3, OCH_3 (substituted benzene)
(b) 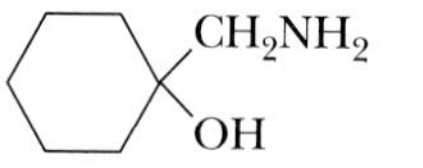 CH_2NH_2, OH (cyclohexane)
(c) 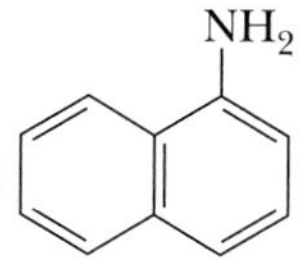 NH_2 (naphthalene)
(d) $CH_3NH(CH_2)_2CH_3$
(e) 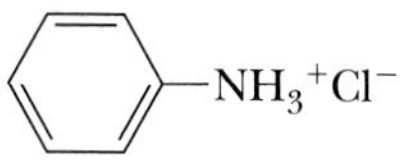 $C_6H_5{-}NH_3^+Cl^-$
(f) 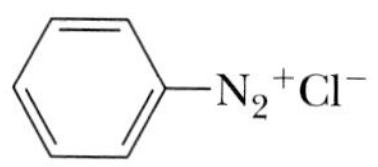 $C_6H_5{-}N_2^+Cl^-$
(g) 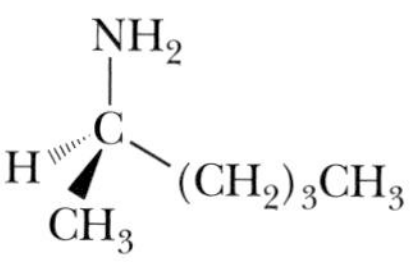 NH_2, H, CH_3, $(CH_2)_3CH_3$ on C
(h) CO_2H on pyridine (N)

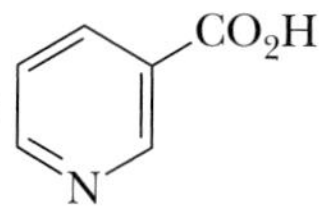

21.17 Classify each amine as primary, secondary, or tertiary; as aliphatic or aromatic.

(a) 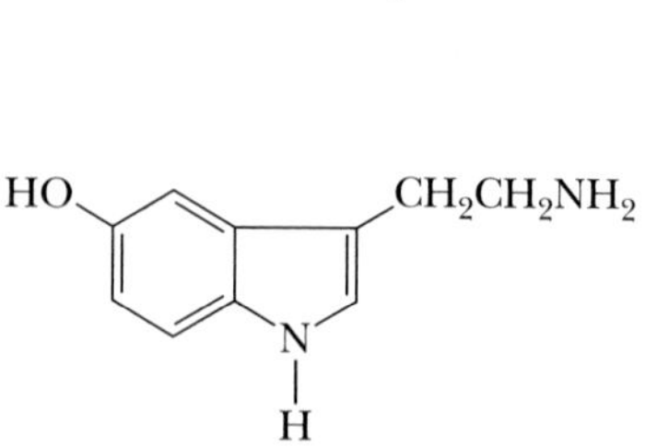 HO, $CH_2CH_2NH_2$, N–H (indole)

Serotonin
(a neurotransmitter)

(b) 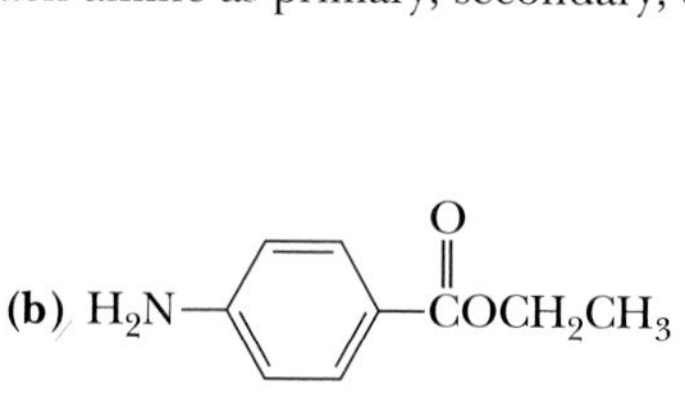 $H_2N{-}C_6H_4{-}\overset{O}{\overset{\|}{C}}OCH_2CH_3$

Benzocaine
(a topical anesthetic)

(c) 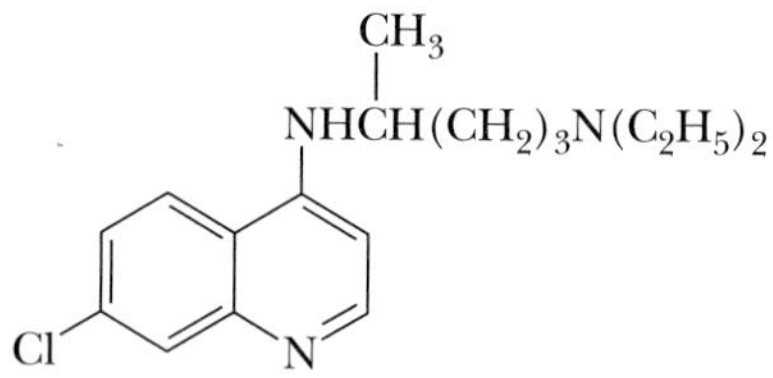 $\overset{CH_3}{\overset{|}{NHCH}}(CH_2)_3N(C_2H_5)_2$, Cl, N (quinoline)

Chloroquine
(a drug for the
treatment of malaria)

21.18 Epinephrine is a hormone secreted by the adrenal medulla. Among its actions, it is a bronchodilator. Salbutamol, sold as the *R* enantiomer, is one of the most effective and widely prescribed antiasthma drugs. The *R* enantiomer is 68 times more effective in the treatment of asthma than the *S* enantiomer.

Epinephrine (Adrenaline)

(*R*)-Salbutamol (Proventil, an antiasthma drug)

(a) Classify each as a primary, secondary, or tertiary amine.
(b) Compare the similarities and differences between their structural formulas.

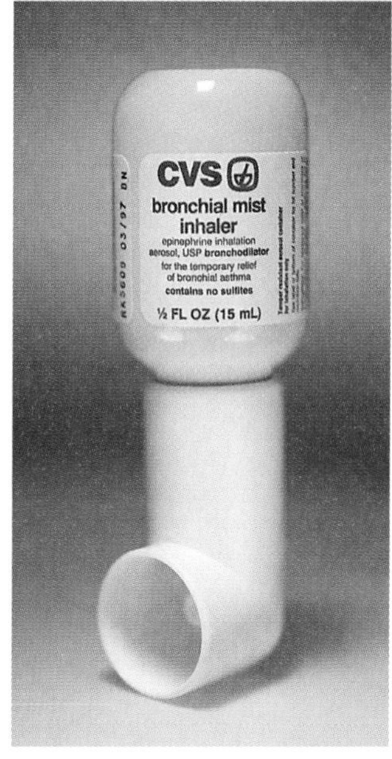

Epinephrine is used in the treatment of asthma. See Problem 21.18. *(Charles D. Winters)*

21.19 Draw the structural formula for a compound of the given molecular formula.

(a) A 2° arylamine, C_7H_9N
(b) A 3° arylamine, $C_8H_{11}N$
(c) A 1° aliphatic amine, C_7H_9N
(d) A chiral 1° amine, $C_4H_{11}N$
(e) A 3° heterocyclic amine, $C_6H_{11}N$
(f) A trisubstituted 1° arylamine, $C_9H_{13}N$
(g) A chiral quaternary ammonium salt, $C_6H_{16}NCl$

21.20 Morphine and its *O*-methylated derivative, codeine, are among the most effective pain killers known. However, they possess two serious drawbacks: they are addictive, and repeated use induces a tolerance to the drug. Increasingly large doses become necessary, which can lead to respiratory arrest. Many morphine analogs have been prepared in an effort to find drugs that are equally effective as pain killers but that have less risk of physical dependence and potential for abuse. Following are structural formulas for pentazocine (one-third the potency of codeine), meperidine (one-half the potency of morphine), and dextropropoxyphene (one-half the potency of codeine). Methadone, with a potency equal to that of morphine, is used to treat opiate withdrawal symptoms in heroin abusers.

R = H; Morphine
R = CH_3; Codeine

Pentazocine (Talwin)

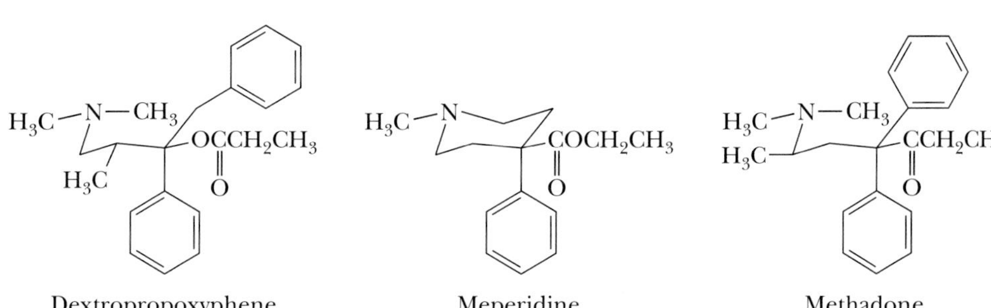

Dextropropoxyphene (Darvon)

Meperidine (Demerol)

Methadone

Opium poppy, the source of morphine. *(© Barbara Woike, PHOTO/NATS)*

(a) List the structural features common to each of these molecules.

(b) The Beckett-Casy rules are a set of empirical rules to predict the structure of molecules that bind to morphine receptors and act as analgesics. According to these rules, to provide an effective morphine-like analgesia, a molecule must have (1) an aromatic ring attached to (2) a quaternary carbon and (3) a nitrogen at a distance equal to two carbon-carbon single bond lengths from the quaternary center. Show that these structural requirements are present in the molecules given in this problem.

Spectroscopy

21.21 Account for the formation of the base peaks in these mass spectra.

(a) Isobutylmethylamine, *m/z* 44 **(b)** Diethylamine, *m/z* 58

21.22 Propose a structural formula for compound (A), molecular formula $C_5H_{13}N$, given its IR and ^{1}H-NMR spectra.

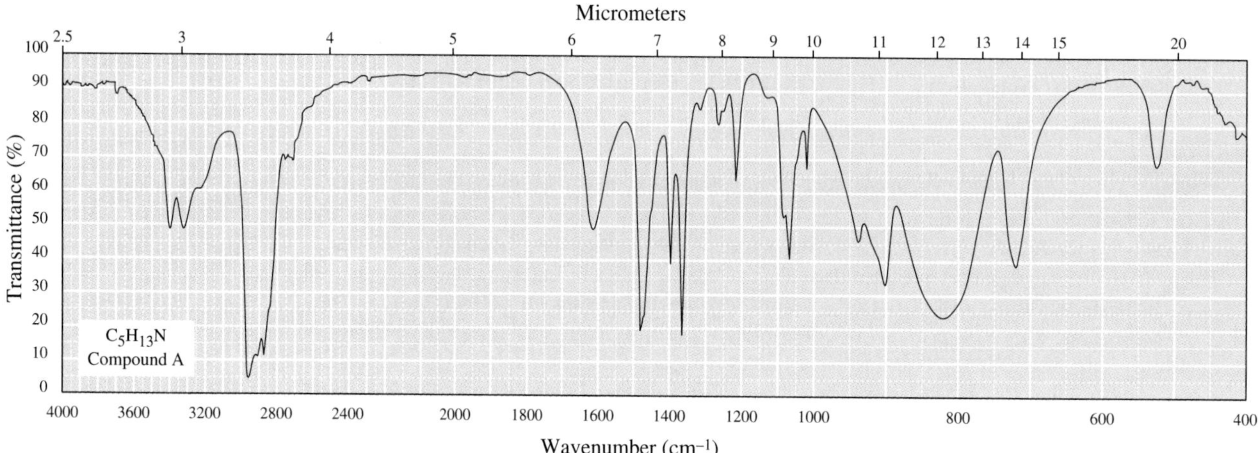

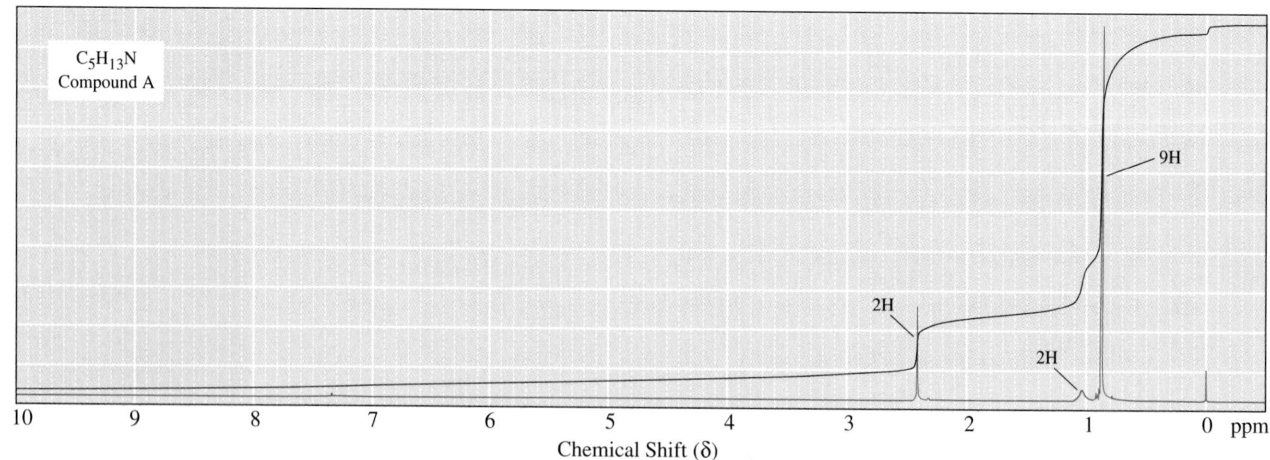

Basicity of Amines

21.23 Select the stronger base from each pair of compounds.

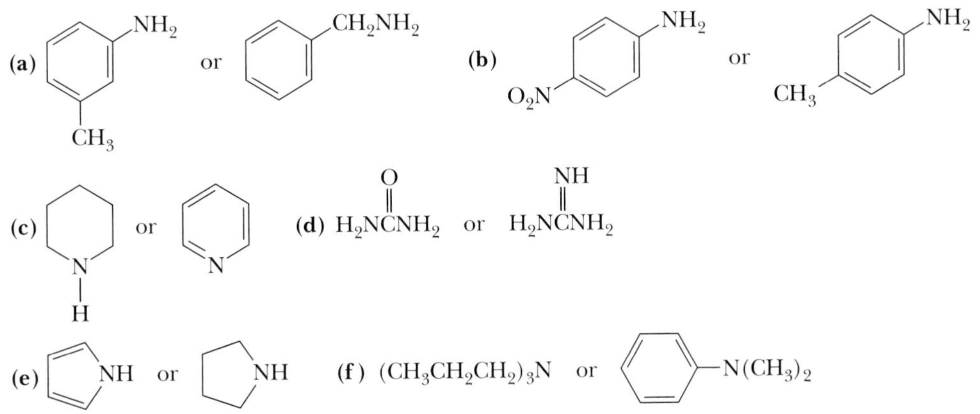

(a) 3-methylaniline (NH_2, CH_3) or benzylamine (CH_2NH_2)

(b) 4-nitroaniline (O_2N, NH_2) or 4-methylaniline (CH_3, NH_2)

(c) piperidine (N–H) or pyridine (N)

(d) $H_2N\overset{O}{\overset{\|}{C}}NH_2$ or $H_2N\overset{NH}{\overset{\|}{C}}NH_2$

(e) pyrrole (NH) or pyrrolidine (NH)

(f) $(CH_3CH_2CH_2)_3N$ or $C_6H_5N(CH_3)_2$

21.24 The pK_a of morpholine is 8.33.

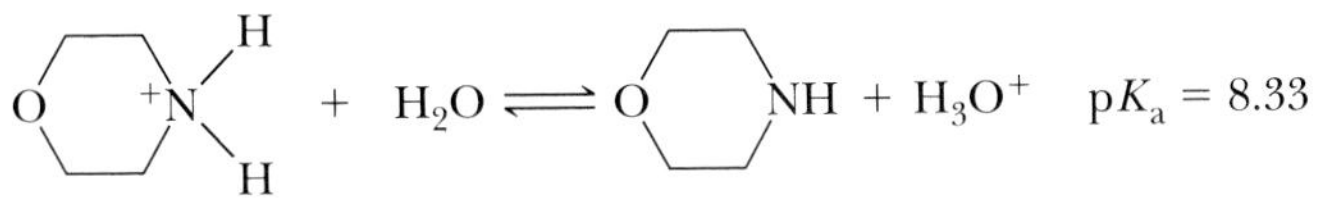

$$\text{Morpholinium ion (O, }^{+}\text{N with H, H)} + H_2O \rightleftharpoons \text{Morpholine (O, NH)} + H_3O^+ \quad pK_a = 8.33$$

Morpholinium ion Morpholine

(a) Calculate the ratio of morpholine to morpholinium ion in aqueous solution at pH 7.0.

(b) At what pH are the concentrations of morpholine and morpholinium ion equal?

21.25 Which of the two nitrogens in pyridoxamine (a form of vitamin B_6) is the stronger base? Explain your reasoning.

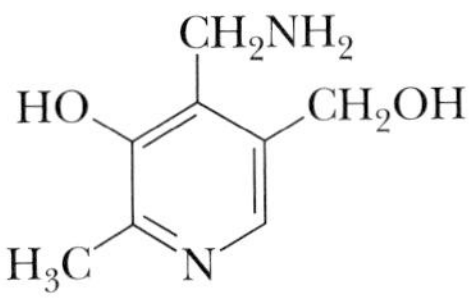

CH_2NH_2, HO, CH_2OH, H_3C, N

Pyridoxamine
(Vitamin B_6)

21.26 Epibatidine, a colorless oil isolated from the skin of the Ecuadorian poison frog *Epipedobates tricolor*, has several times the analgesic potency of morphine. It is the first chlorine-containing, nonopioid (nonopium-like in structure) analgesic ever isolated from a natural source. (See *The Merck Index*, 12th ed., #3647.)

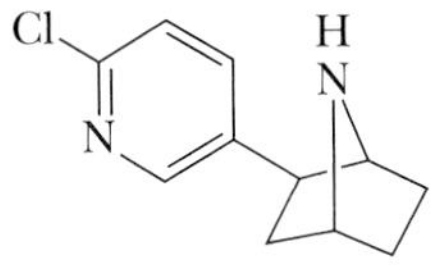

Cl, N, H, N

Epibatidine

(a) Which of the two nitrogen atoms of epibatidine is the more basic?

(b) Mark all stereocenters in this molecule.

(a) Analyze the mechanism of each rearrangement and list their similarities.
(b) Why does the first reaction give ring expansion but not the second?
(c) Suggest a β-aminoalcohol that would give cyclohexanecarbaldehyde as a product?

21.35 Propose a mechanism for this conversion. Your mechanism must account for the fact that there is retention of configuration at the stereocenter. (*S*)-Glutamic acid is one of the 20 amino acid building blocks of polypeptides and proteins (Chapter 27).

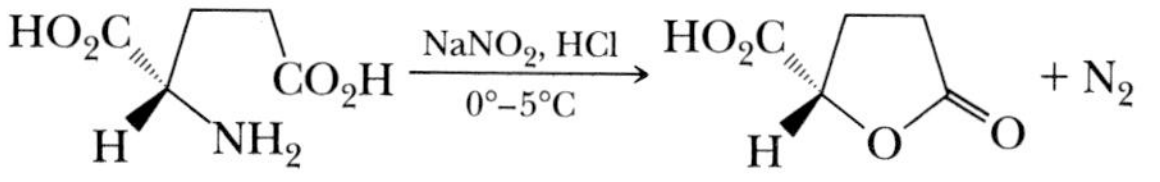

HO_2C ... CO_2H, H, NH_2 $\xrightarrow[0°–5°C]{NaNO_2,\ HCl}$ HO_2C ... H, O, O $+ N_2$

(*S*)-Glutamic acid

21.36 The following sequence of methylation and Hofmann elimination was used in the determination of the structure of this bicyclic amine.

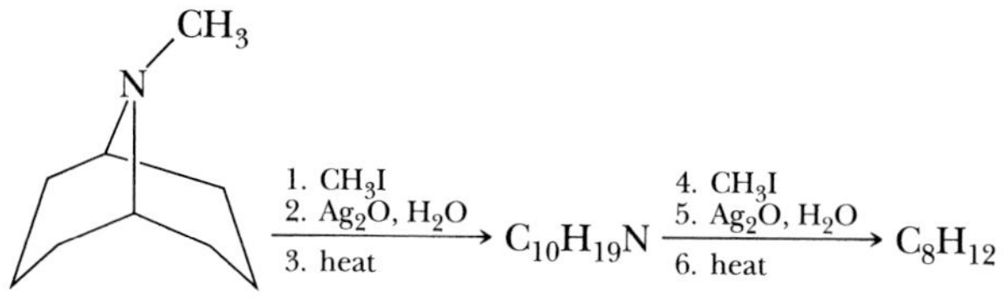

CH_3, N

$(C_9H_{17}N)$ $\xrightarrow[3.\ heat]{1.\ CH_3I \quad 2.\ Ag_2O,\ H_2O}$ $C_{10}H_{19}N$ (A) $\xrightarrow[6.\ heat]{4.\ CH_3I \quad 5.\ Ag_2O,\ H_2O}$ C_8H_{12} (B)

(a) Propose structural formulas for compounds (A) and (B).
(b) Suppose you were given the structural formula of compound (B) but only the molecular formulas for compound (A) and the starting bicyclic amine. Given this information, is it possible, working backward, to arrive at an unambiguous structural formula for compound (A)? For the bicyclic amine?

21.37 Propose a structural formula for compound A, $C_{10}H_{16}$, and account for its formation.

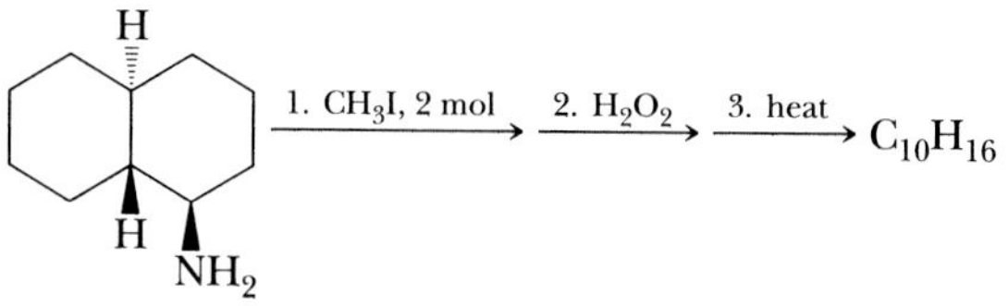

H, H, NH_2 $\xrightarrow{1.\ CH_3I,\ 2\ mol}$ $\xrightarrow{2.\ H_2O_2}$ $\xrightarrow{3.\ heat}$ $C_{10}H_{16}$

21.38 An amine of unknown structure contains one nitrogen and nine carbon atoms. The ^{13}C-NMR spectrum shows only five signals, all between 20 and 60 ppm. Three cycles of Hofmann elimination sequence (1. CH_3I; 2. Ag_2O, H_2O; 3. heat) give trimethylamine and 1,4,8-nonatriene. Propose a structural formula for the amine.

21.39 The Cope elimination of tertiary amine *N*-oxides involves a planar transition state and cyclic redistribution of $(4n + 2)$ electrons. The pyrolysis of acetic esters to give an alkene and acetic acid is also thought to involve a planar transition state and cyclic redistribution of $(4n + 2)$ electrons. Propose a mechanism for pyrolysis of the following ester.

$$CH_3CH_2CH_2CH_2OCCH_3 \text{ (C=O)} \xrightarrow{500°C} CH_3CH_2CH{=}CH_2 + CH_3COH \text{ (C=O)}$$

Butyl acetate — 1-Butene — Acetic acid

21.40 The following transformation is an example of the Carroll reaction, named after the English chemist, M. F. Carroll, who first reported it. Propose a mechanism for this reaction.

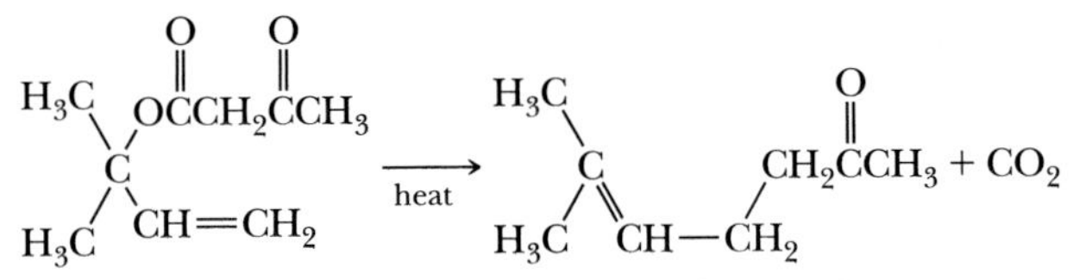

$$\mathrm{(H_3C)_2C(OCCH_2CCH_3)(CH{=}CH_2)} \xrightarrow{\text{heat}} \mathrm{(H_3C)_2C{=}CH{-}CH_2CH_2CCH_3} + CO_2$$

6-Methyl-5-hepten-2-one

Synthesis

21.41 Propose steps for the following conversions using a reaction of a diazonium salt in at least one step of each conversion.

(a) Toluene to 4-methylphenol (*p*-cresol)
(b) Nitrobenzene to 3-bromophenol
(c) Toluene to *p*-cyanobenzoic acid
(d) Phenol to *p*-iodoanisole
(e) Acetanilide to *p*-aminobenzylamine
(f) Toluene to 4-fluorobenzoic acid
(g) 3-Methylaniline (*m*-toluidine) to 2,4,6-tribromobenzoic acid

21.42 Starting materials for the synthesis of the herbicide propranil are benzene and propanoic acid. Show how to bring about this synthesis.

Benzene →(1) chlorobenzene (Cl) →(2) 4-nitrochlorobenzene (Cl, NO_2) →(3) (Cl, Cl, NO_2) →(4) (Cl, Cl, NH_2) →(5) (Cl, Cl, $NHCCH_2CH_3$ with C=O)

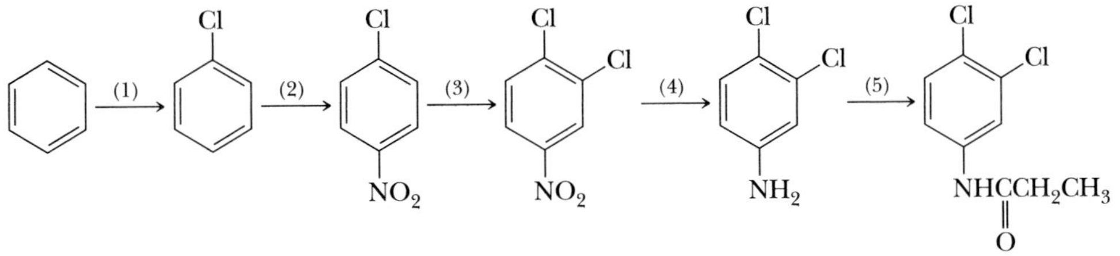

Propranil

21.43 Show how to bring about each step in the following synthesis.

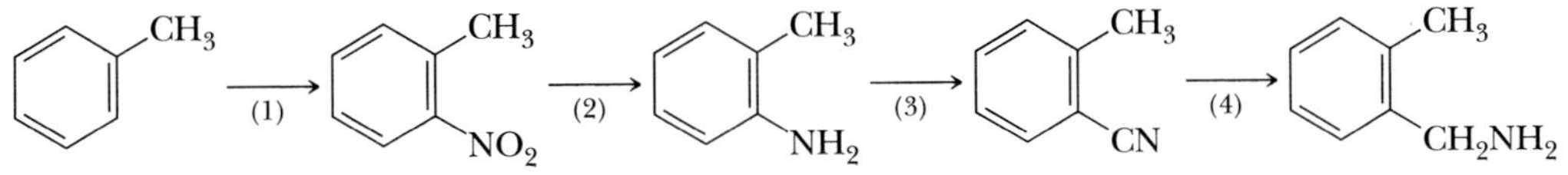

Toluene (CH_3) →(1) (CH_3, NO_2) →(2) (CH_3, NH_2) →(3) (CH_3, CN) →(4) (CH_3, CH_2NH_2)

21.44 Show how to bring about this synthesis.

Toluene (CH_3) → (CH_3, OH, OH)

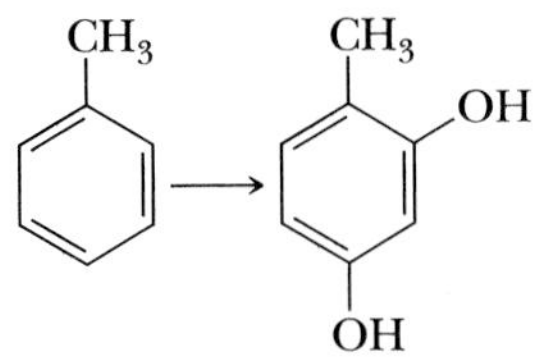

21.45 Following are steps in a conversion of phenol to 4-methoxybenzylamine. Show how to bring about each step in good yield.

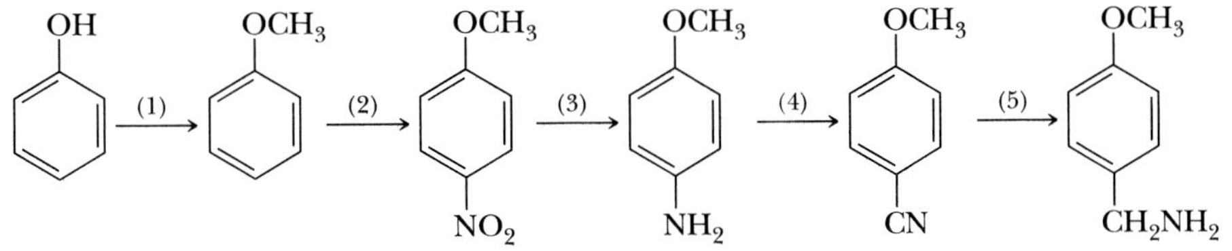

Phenol (OH) →(1) (OCH_3) →(2) (OCH_3, NO_2) →(3) (OCH_3, NH_2) →(4) (OCH_3, CN) →(5) (OCH_3, CH_2NH_2)

21.46 Following is a synthesis of diazepam, a prescription tranquilizer sold under several trade names, the most well known of which is Valium. This drug may well be the most commercially successful of all synthetic drugs.

Show reagents and conditions for the synthesis of diazepam from *p*-chloroaniline.

CONJUGATED DIENES

22

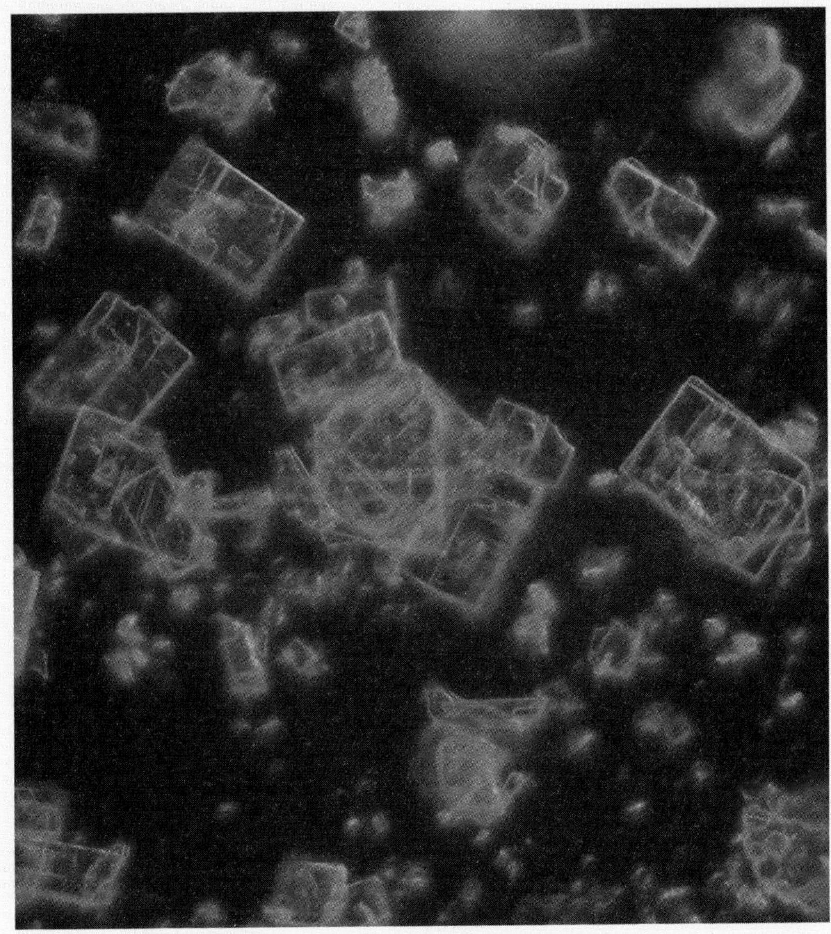

■ Crystals of vitamin D_3 (cholecalciferol, see Problem 22.34) viewed under polarized light. *(© 1997 Herb Charles Ohlmeyer/Fran Heyl Associates)*

In Chapters 5 and 6, we discussed the structure and characteristic reactions of alkenes. We limited this discussion to molecules containing one double bond. In this chapter, we extend our study of alkenes to include molecules that contain two or more conjugated double bonds. As we shall see, although conjugated dienes undergo many of the same reactions characteristic of unconjugated alkenes, they also undergo their own unique set of characteristic reactions.

22.1 Electrophilic Addition to Conjugated Dienes

Conjugated diene A diene in which the double bonds are separated by one single bond.

Conjugated dienes (Section 5.2F) undergo two-step electrophilic addition reactions just as do simple alkenes (Section 6.3). However, certain features are unique to the reactions of conjugated dienes.

A. Conjugate Addition: 1,2-Addition and 1,4-Addition

Addition of one equivalent of HBr to 1,3-butadiene at −78°C gives a mixture of two constitutional isomers: namely, 3-bromo-1-butene and 1-bromo-2-butene.

$$\underset{\text{1,3-Butadiene}}{CH_2{=}CH{-}CH{=}CH_2} + HBr \xrightarrow{-78°C} \underset{\substack{\text{3-Bromo-1-butene}\\(90\%)\\(1,2\text{-addition})}}{CH_2{=}CH{-}\overset{Br}{\overset{|}{C}H}{-}\overset{H}{\overset{|}{C}H_2}} + \underset{\substack{\text{1-Bromo-2-butene}\\(10\%)\\(1,4\text{-addition})}}{\overset{Br}{\overset{|}{C}H_2}{-}CH{=}CH{-}\overset{H}{\overset{|}{C}H_2}}$$

The designations "1,2-" and "1,4-" used here to describe additions to conjugated dienes do not refer to IUPAC nomenclature. Rather, they refer to the four-atom system of two conjugated double bonds and indicate that addition takes place at either carbons 1 and 2 or 1 and 4 of the four-atom system.

The bromobutenes formed by addition of 1 mol of HBr to butadiene can in turn undergo addition of a second mole of HBr to give a mixture of dibromobutanes. Our concern at this point is only with the products of a single addition of HBr.

Addition of one equivalent of Br_2 at −15°C also gives a mixture of 1,2-addition and 1,4-addition products.

$$\underset{\text{1,3-Butadiene}}{CH_2{=}CH{-}CH{=}CH_2} + Br_2 \xrightarrow{-15°C} \underset{\substack{\text{3,4-Dibromo-1-butene}\\(54\%)\\(1,2\text{-addition})}}{\overset{Br}{\overset{|}{C}H_2}{-}\overset{Br}{\overset{|}{C}H}{-}CH{=}CH_2} + \underset{\substack{\text{1,4-Dibromo-2-butene}\\(46\%)\\(1,4\text{-addition})}}{\overset{Br}{\overset{|}{C}H_2}{-}CH{=}CH{-}\overset{Br}{\overset{|}{C}H_2}}$$

We can account for the formation of isomeric products in the addition of HBr in the following way. Electrophilic addition is initiated by reaction of a terminal carbon of one of the double bonds with HBr to form an allylic carbocation intermediate best represented as a resonance hybrid of two contributing structures (Section 8.4E). Formation of this cation is the rate-limiting step. The addition is completed by rapid reaction of the allylic cation with bromide ion. Reaction at one carbon bearing partial positive charge gives the 1,2-addition product; reaction at the other gives the 1,4-addition product.

MECHANISM 1,2- and 1,4-Addition to a Conjugated Diene

$$CH_2{=}CH{-}CH{=}CH_2 + H{-}Br \longrightarrow CH_2{=}CH{-}\overset{+}{C}H{-}CH_2(H) \longleftrightarrow \overset{+}{C}H_2{-}CH{=}CH{-}CH_2(H)$$

A resonance-stabilized allylic carbocation

$\downarrow Br^-$ $\downarrow Br^-$

$$CH_2{=}CH{-}CH(Br){-}CH_2(H) \qquad CH_2(Br){-}CH{=}CH{-}CH_2(H)$$

(1,2-Addition) (1,4-Addition)

EXAMPLE 22.1

Addition of 1 mol of HBr to 2,4-hexadiene gives a mixture of 4-bromo-2-hexene and 2-bromo-3-hexene. No 5-bromo-2-hexene is formed. Account for the formation of the first two bromoalkenes and for the fact that the third bromoalkene is not formed.

Solution

2,4-Hexadiene is a conjugated diene and you can expect products from both 1,2-addition and 1,4-addition. Reaction of the diene with HBr in Step 1, the rate-limiting step, forms a resonance-stabilized 2° allylic carbocation intermediate. Reaction of this intermediate in Step 2 at one of the carbons bearing a partial positive charge gives 4-bromo-2-hexene, a 1,2-addition product; reaction at the other gives 2-bromo-3-hexene, a 1,4-addition product.

$$CH_3CH{=}CH{-}CH{=}CHCH_3 + H{-}Br \longrightarrow CH_3CH{=}CH{-}\overset{+}{C}H{-}CH(H)CH_3 \longleftrightarrow CH_3\overset{+}{C}H{-}CH{=}CH{-}CH(H)CH_3$$

(A resonance-stabilized 2° allylic carbocation)

$\downarrow Br^-$ $\downarrow Br^-$

$$CH_3CH{=}CH{-}CH(Br){-}CH(H)CH_3 \qquad CH_3CH(Br){-}CH{=}CH{-}CH(H)CH_3$$

4-Bromo-2-hexene (1,2-addition) 2-Bromo-3-hexene (1,4-addition)

Formation of 5-bromo-2-hexene requires reaction of the diene with HBr to give a secondary, nonallylic carbocation. The activation energy for formation of this less stable 2° carbocation is considerably greater than that for formation of the resonance-stabilized allylic carbocation, and, therefore, formation of this carbocation and the resulting product, 5-bromo-2-hexene, does not compete effectively with formation of the observed products.

$$CH_3CH{=}CH{-}CH{=}CHCH_3 + H{-}Br \longrightarrow \underset{\text{2° Carbocation, but not allylic}}{CH_3CH{=}CH{-}\overset{H}{\overset{|}{C}H}{-}\overset{+}{C}HCH_3} \xrightarrow{Br^-} \underset{\text{5-Bromo-2-hexene (not formed)}}{CH_3CH{=}CH{-}\overset{H}{\overset{|}{C}H}{-}\overset{Br}{\overset{|}{C}H}CH_3}$$

PROBLEM 22.1

Predict the product(s) formed by addition of 1 mol of Br_2 to 2,4-hexadiene.

B. Kinetic Versus Thermodynamic Control of Electrophilic Addition

We saw in the previous section that electrophilic addition to conjugated dienes gives a mixture of 1,2-addition and 1,4-addition products.

$$\underset{\text{1,3-Butadiene}}{CH_2{=}CH{-}CH{=}CH_2} + HBr \xrightarrow{-78°C} \underset{\substack{\text{3-Bromo-1-butene}\\(90\%)\\(1,2\text{-addition})}}{CH_2{=}CH{-}\overset{Br}{\overset{|}{C}H}{-}\overset{H}{\overset{|}{C}H_2}} + \underset{\substack{\text{1-Bromo-2-butene}\\(10\%)\\(1,4\text{-addition})}}{\overset{Br}{\overset{|}{C}H_2}{-}CH{=}CH{-}\overset{H}{\overset{|}{C}H_2}}$$

$$\underset{\text{1,3-Butadiene}}{CH_2{=}CH{-}CH{=}CH_2} + Br_2 \xrightarrow{-15°C} \underset{\substack{\text{3,4-Dibromo-1-butene}\\(54\%)\\(1,2\text{-addition})}}{\overset{Br}{\overset{|}{C}H_2}{-}\overset{Br}{\overset{|}{C}H}{-}CH{=}CH_2} + \underset{\substack{\text{1,4-Dibromo-2-butene}\\(46\%)\\(1,4\text{-addition})}}{\overset{Br}{\overset{|}{C}H_2}{-}CH{=}CH{-}\overset{Br}{\overset{|}{C}H_2}}$$

Following are some additional experimental observations about the products of electrophilic additions to 1,3-butadiene.

1. For addition of HBr at −78°C and addition of Br_2 at −15°C, the 1,2-addition product predominates over the 1,4-addition product. Generally at lower temperatures, the 1,2-addition products predominate over 1,4-addition products.
2. For addition of HBr and Br_2 at higher temperatures (generally 40° to 60°C), the 1,4-addition products predominate.
3. If the products of low-temperature addition are allowed to remain in solution and warmed to a higher temperature, the composition of the product changes over time and becomes identical to that obtained when the reaction is carried out at higher temperature. The same result can be accomplished at the higher temperature in a far shorter time by adding a Lewis acid catalyst, such as $FeCl_3$ or $ZnCl_2$, to the mixture of low-temperature addition products. Thus, under these higher temperature conditions, an equilibrium is established between 1,2- and 1,4-addition products in which 1,4-addition products predominate.
4. If either the pure 1,2- or pure 1,4-addition product is dissolved in an inert solvent at the higher temperature and a Lewis acid catalyst added, an equilibrium mixture of 1,2- and 1,4-addition product forms. The same equilibrium mixture is obtained regardless of which isomer is used as the starting material.

Chemists interpret these experimental results using the twin concepts of kinetic control and equilibrium control of reactions. We encountered these concepts earlier in Section 18.2A, where we saw that it is often possible to form alternative enolate anions of ketones, depending on whether experimental conditions are chosen for kinetic control or thermodynamic control.

To review briefly, for reactions under **kinetic (rate) control,** the distribution of products is determined by the relative rates of formation of each. We see the operation of kinetic control in the following way. At lower temperatures, the reaction is essentially irreversible and no equilibrium is established between 1,2- and 1,4-addition products. The 1,2-addition product predominates under these conditions because the rate of 1,2-addition is greater than that of 1,4-addition.

Kinetic control For a reaction under kinetic control, the distribution of products is determined by the relative rates of formation of each product.

For reactions under **thermodynamic (equilibrium) control,** the distribution of products is determined by the relative stability of each. We see the operation of equilibrium control in the following way. At higher temperatures, the reaction is reversible and an equilibrium is established between 1,2- and 1,4-addition products. The percentage of each product present at equilibrium is in direct relation to the relative thermodynamic stability of that product. The fact that the 1,4-addition product predominates at equilibrium means that it is thermodynamically more stable than the 1,2-addition product.

Thermodynamic control For a reaction under thermodynamic control, the distribution of products is determined by the relative stabilities of each product.

The structure shown in the potential energy well in the center of Figure 22.1 is the resonance-stabilized allylic cation intermediate formed by proton transfer from HBr to 1,3-butadiene. The dashed lines in this intermediate show the partial double bond character between C2–C3 and C3–C4 in the resonance hybrid. To the left of this intermediate is the activation energy for its reaction with bromide ion to form the less stable 1,2-addition product; to the right is the activation energy for its reaction with bromide ion to form the more stable 1,4-addition product. As shown in this figure, the activation energy for 1,2-addition is less than that for 1,4-addition, and the 1,2-addition product is favored under kinetic control. The 1,4-addition product is the more stable and is favored when the reaction is under thermodynamic control.

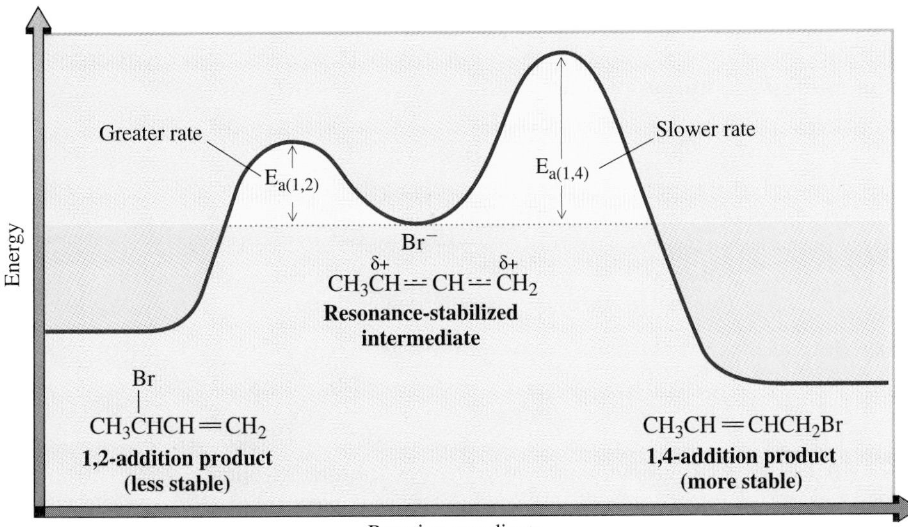

FIGURE 22.1 Kinetic versus thermodynamic control. A plot of potential energy versus reaction coordinate for Step 2 in the electrophilic addition of HBr to 1,3-butadiene. The resonance-stabilized allylic carbocation intermediate reacts with bromide ion by way of the transition state on the left to give the 1,2-addition product. It reacts with bromide ion by way of the alternative transition state on the right to give the 1,4-addition product.

To complete our discussion of electrophilic addition to conjugated dienes and of kinetic versus thermodynamic control, we need to ask the following questions.

1. Why is the 1,2-addition product (the less stable product) formed more rapidly at lower temperatures? First, we need to look at the resonance-stabilized allylic carbocation. We must consider the degree of substitution of both the positive carbon and the carbon-carbon double bond in each contributing structure.

$$CH_2{=}CH{-}\overset{+}{C}H{-}CH_3 \longleftrightarrow \overset{+}{C}H_2{-}CH{=}CH{-}CH_3$$

Less substituted double bond (secondary carbocation)	More substituted double bond (primary carbocation)

A secondary carbocation is more stable than a primary carbocation (Section 6.3), and, if the degree of substitution of the carbon bearing the positive charge were the more important factor, then the Lewis structure on the left would make the greater contribution to the hybrid. A more substituted double bond is more stable than a less substituted double bond (Section 6.6B), and, if the degree of substitution of the carbon-carbon double bond were the more important factor, then the Lewis structure on the right would make the greater contribution to the hybrid.

We know from other experimental evidence that the location of the positive charge in the allylic carbocation is more important than the location of the double bond. Therefore, in the hybrid, the greater fraction of positive charge is on the secondary carbon. Reaction with bromide ion occurs more rapidly at this carbon, giving 1,2-addition, simply because it has a greater density of positive charge.
2. Is the 1,2-addition product also formed more rapidly at higher temperatures, even though it is the 1,4-addition product that predominates under these conditions? The answer is yes. The factors affecting the structure of a resonance-stabilized allylic carbocation intermediate and the reaction of this intermediate with a nucleophile are not greatly affected by changes in temperature.
3. Why is the 1,4-addition product the thermodynamically more stable product? The answer to this question has to do with the relative degree of substitution of double bonds. In general, the greater the degree of substitution of a carbon-carbon double bond, the greater the stability of the compound or ion containing it (Section 6.6B). Following are pairs of 1,2- and 1,4-addition products. In each case, the more stable alkene is the 1,4-addition product.

$$CH_3\overset{Br}{\overset{|}{C}}HCH{=}CH_2 \quad + \quad (H_3C)(H)C{=}C(H)(CH_2Br)$$

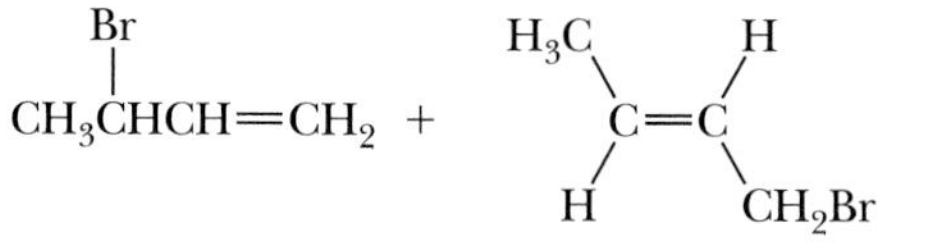

3-Bromo-1-butene (less stable alkene) + (*E*)-1-Bromo-2-butene (more stable alkene)

$$BrCH_2\overset{Br}{\overset{|}{C}}HCH{=}CH_2 \quad + \quad (BrCH_2)(H)C{=}C(H)(CH_2Br)$$

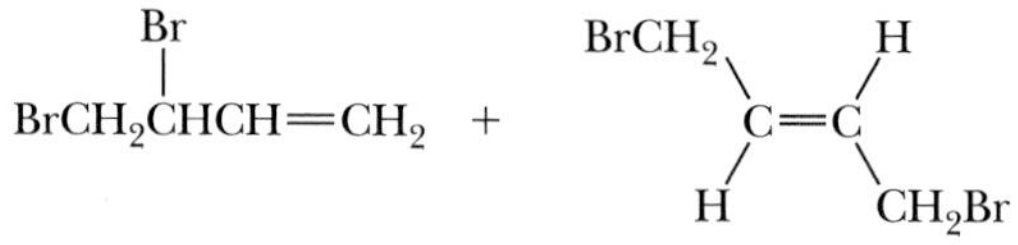

3,4-Dibromo-1-butene (less stable alkene) + (*E*)-1,4-Dibromo-2-butene (more stable alkene)

However, there are cases where the 1,2-addition product is more stable, and would be the product of thermodynamic control. For example, addition of bromine to 1,4-dimethyl-1,3-cyclohexadiene under conditions of thermodynamic control gives 3,4-dibromo-1,4-dimethylcyclohexene, because its trisubstituted double bond is more stable than the disubstituted double bond of the 1,4-addition product.

CH_3 CH_3 $\xrightarrow[\text{high temperature}]{Br_2}$ Br CH_3 Br CH_3 + Br CH_3 Br CH_3

1,4-Addition product (less stable) 1,2-Addition product (more stable)

4. What is the mechanism by which the thermodynamically less stable product is converted to the thermodynamically more stable product at higher temperatures? To answer this question, we must look at the relationships between kinetic energy, potential energy, and activation energy. On collision, a part of the kinetic energy (the energy of motion) is transformed into potential energy, and, if the increase in potential energy is equal to or greater than the activation energy for reaction, then reaction may occur. At the higher temperatures for electrophilic addition of HBr and Br_2 to conjugated dienes, collisions are sufficiently energetic that ionization of the 1,2-addition product occurs to re-form the resonance-stabilized allylic carbocation intermediate. It can then react again with bromide ion to form the thermodynamically more stable 1,4-addition product. At lower temperatures, however, the increase in potential energy on collision is not sufficient to overcome the potential energy barrier to bring about this ionization.

5. Is it a general rule that where two or more products are formed from a common intermediate, the thermodynamically less stable product is formed at a greater rate? The answer is no. Whether the thermodynamically more or less stable product is formed at a greater rate from a common intermediate depends very much on the particular reaction and the reaction conditions.

22.2 The Diels-Alder Reaction

In 1928, Otto Diels and Kurt Alder in Germany discovered another unique reaction of conjugated dienes; they undergo cycloaddition reactions with certain types of carbon-carbon double and triple bonds. For their discovery and subsequent studies of this reaction, Diels and Alder were jointly awarded the Nobel Prize for chemistry in 1950.

The compound with the double or triple bond that reacts with the diene in a Diels-Alder reaction is given the special name of **dienophile** (diene-loving), and the product of a Diels-Alder reaction is given the special name of **Diels-Alder adduct.** The designation **cycloaddition** refers to the fact that two reactants add together to give a cyclic product.

Dienophile A compound containing a double bond (containing one or two C, N, or O atoms) that can react with a conjugated diene to give a Diels-Alder adduct.

Cycloaddition reaction A reaction in which two reactants add together in a single step to form a cyclic product. The best known of these is the Diels-Alder reaction.

Following are two examples of Diels-Alder reactions: one with a compound containing a carbon-carbon double bond, the other containing a carbon-carbon triple bond.

1,3-Butadiene (a diene)	+	3-Buten-2-one (a dienophile)	$\longrightarrow$	4-Cyclohexenyl methyl ketone (a Diels-Alder adduct)
1,3-Butadiene (a diene)	+	Diethyl acetylenedicarboxylate (a dienophile)	$\longrightarrow$	Diethyl 1,4-cyclohexadiene-1,2-dicarboxylate (a Diels-Alder adduct)

Note that the four carbon atoms of the diene and two carbon atoms of the dienophile combine to form a six-membered ring. Note further that there are two more sigma bonds and two fewer pi bonds in the product than in the reactants. This exchange of two (weaker) pi bonds for two (stronger) sigma bonds is a major driving force in Diels-Alder reactions.

We can write a Diels-Alder reaction in the following way, showing only the carbon skeletons of the diene and dienophile. In this representation, curved arrows are used to show that two new sigma bonds are formed, three pi bonds are broken, and one new pi bond is formed. It must be emphasized here that the curved arrows in this diagram are not meant to show a mechanism. Rather they are intended to show which bonds are broken and which new bonds are formed, and how many electrons are involved (six in this case).

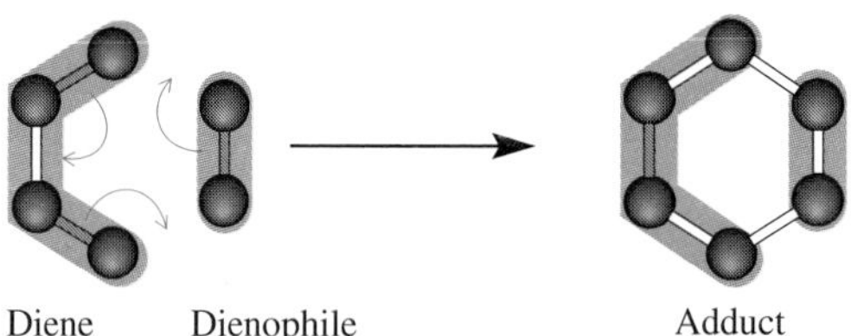

The special values of the reaction discovered by Diels and Alder are that (1) it is one of few reactions that can be used to form six-membered rings, (2) it is one of few reactions that can be used to form two new carbon-carbon bonds at the same time, and, as we will see later in this section, (3) it is stereoselective. For these reasons, the Diels-Alder reaction has proved to be enormously valuable in synthetic organic chemistry.

EXAMPLE 22.2

Draw a structural formula for the Diels-Alder adduct formed by reaction of each diene and dienophile pair.

(a) 1,3-Butadiene and propenal
(b) 2,3-Dimethyl-1,3-butadiene and 3-buten-2-one

Solution

First draw the diene and dienophile so that each molecule is properly aligned to form a six-membered ring. Then complete the reaction to form the six-membered ring Diels-Alder adduct.

(a) 1,3-Butadiene + Propenal (Acrolein) ⟶

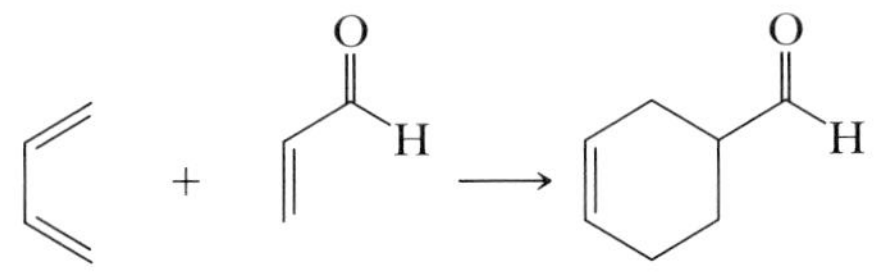

(b) 2,3-Dimethyl-1,3-butadiene + 3-Buten-2-one (Methyl vinyl ketone) ⟶

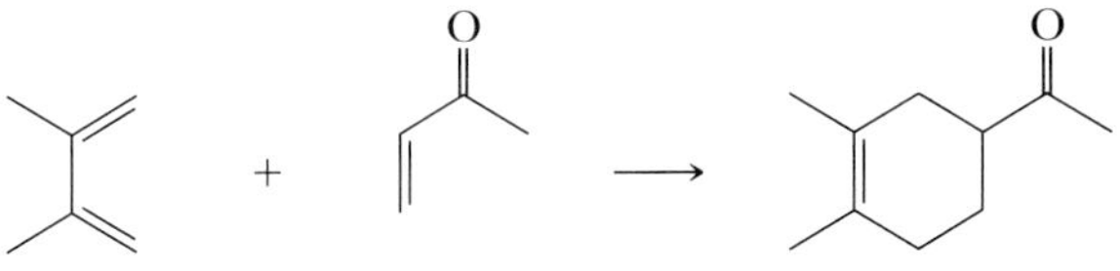

PROBLEM 22.2

What combination of diene and dienophile undergoes Diels-Alder reaction to give each adduct.

(a) $CH{=}CH_2$ (b) CO_2CH_3 CO_2CH_3 (c) CO_2CH_3 CO_2CH_3

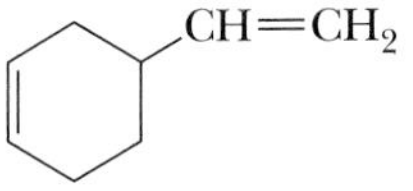

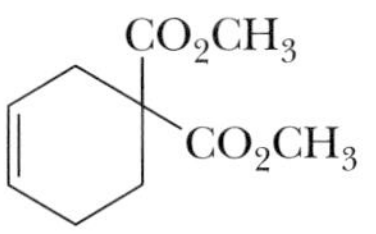

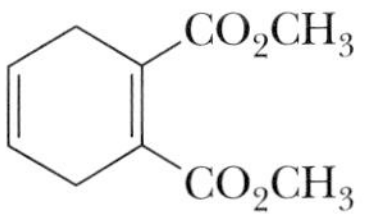

Now let us look more closely at the scope and limitations, stereochemistry, and mechanism of Diels-Alder reactions.

A. The Diene Must Be Able to Assume an s-*cis* Conformation

We can illustrate the significance of conformation of the diene by reference to 1,3-butadiene. For maximum stability of a conjugated diene, overlap of the four unhybridized 2*p* orbitals making up the pi system must be complete, a condition that occurs only when all four carbon atoms of the diene lie in the same plane. It follows then that if the carbon skeleton of 1,3-butadiene is planar, the six atoms bonded to the skeleton of the diene are also contained in the same plane. There are two planar conformations of 1,3-butadiene, referred to as the **s-*trans* conformation** and the **s-*cis* conformation** where the designation **s** refers to the carbon-carbon single bond of the diene. Of these, the s-*trans* conformation is slightly lower in energy and, therefore, slightly more stable.

Although s-*trans*-1,3-butadiene is the more stable conformation, s-*cis*-1,3-butadiene is the reactive conformation in Diels-Alder reactions. In the s-*cis* conformation, carbon atoms 1 and 4 of the conjugated system are close enough to react with the

TABLE 22.1 Electron-Releasing and Electron-Withdrawing Groups

Electron-Releasing Groups	Electron-Withdrawing Groups
$-CH_3$	$-\overset{O}{\overset{\Vert}{C}}H$ (aldehyde)
$-CH_2CH_3$	$-\overset{O}{\overset{\Vert}{C}}R$ (ketone)
$-CH(CH_3)_2$	$-\overset{O}{\overset{\Vert}{C}}OH$ (carboxyl)
$-C(CH_3)_3$	$-\overset{O}{\overset{\Vert}{C}}OR$ (ester)
$-R$ (other alkyl groups)	$-NO_2$ (nitro)
$-OR$ (ether)	$-C{\equiv}N$ (cyano)
$-O\overset{O}{\overset{\Vert}{C}}R$ (ester)	

Placing electron-releasing methyl groups on the diene further facilitates reaction; 2,3-dimethyl-1,3-butadiene and 3-buten-2-one form a Diels-Alder adduct at 30°C.

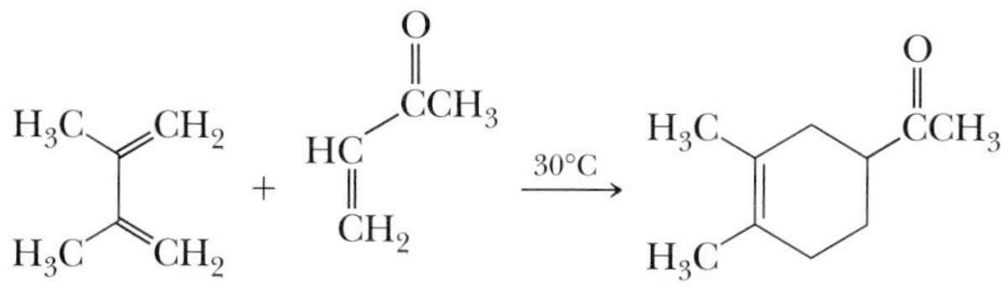

H_3C CH_2 / H_3C CH_2 + $HC(=CH_2)\overset{O}{\overset{\Vert}{C}}CH_3$ $\xrightarrow{30°C}$ H_3C, H_3C ... $\overset{O}{\overset{\Vert}{C}}CH_3$

2,3-Dimethyl-1,3-butadiene 3-Buten-2-one

Several of the electron-releasing and electron-withdrawing groups most commonly encountered in Diels-Alder reactions are given in Table 22.1.

C. Diels-Alder Reactions Can Be Used to Form Bicyclic Systems

Conjugated cyclic dienes, in which the double bonds are of necessity held in an *s-cis* conformation, are highly reactive in Diels-Alder reactions. Two particularly useful dienes for this purpose are cyclopentadiene and 1,3-cyclohexadiene. In fact, cyclopentadiene is reactive both as a diene and as a dienophile, and, on standing at room temperature, it forms a Diels-Alder adduct known by the common name dicyclopentadiene. When dicyclopentadiene is distilled at its normal boiling point of 170°C, a reverse Diels-Alder reaction takes place, and cyclopentadiene is reformed.

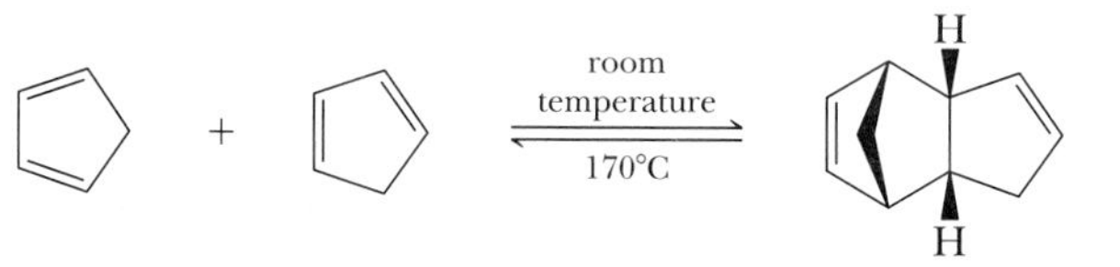

room temperature / 170°C

H H

The diene The dienophile Dicyclopentadiene (endo isomer)

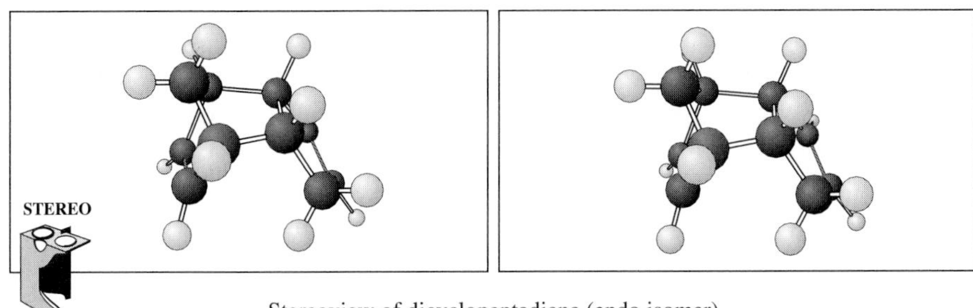

Stereoview of dicyclopentadiene (endo isomer)

The terms "endo" and "exo" are used for bicyclic Diels-Alder products to describe the orientation of substituents of the dienophile in relation to the two-carbon diene-derived bridge. **Exo** (Greek, outside) substituents are on the opposite side from the diene-derived bridge; **endo** (Greek, within) substituents are on the same side.

the double bond derived from the diene

exo (outside) relative to the double bond

endo (inside) relative to the double bond

For Diels-Alder reactions under kinetic control, the endo orientation of the dienophile is favored. Treatment of cyclopentadiene with methyl propenoate (methyl acrylate) gives the endo adduct almost exclusively. The exo adduct is not formed.

Cyclopentadiene + Methyl propenoate → Methyl bicyclo[2.2.1]hept-5-en-endo-2-carboxylate

D. The Configuration of the Dienophile Is Retained

If the dienophile is a *cis* isomer, then the substituents *cis* to each other in the dienophile are *cis* in the Diels-Alder adduct. Conversely, if the dienophile is a *trans* isomer, substituents that are *trans* in the dienophile are *trans* in the adduct.

+ Dimethyl *cis*-2-butenedioate (a *cis*-dienophile) → Dimethyl *cis*-4-cyclohexene-1,2-dicarboxylate

+ Dimethyl *trans*-2-butenedioate (a *trans*-dienophile) → Dimethyl *trans*-4-cyclohexene-1,2-dicarboxylate

E. Mechanism: The Diels-Alder Reaction Is a Pericyclic Reaction

As chemists probed for details of the Diels-Alder reaction, they discovered that there is no evidence for participation of either ionic or radical intermediates. Thus, the Diels-Alder reaction is unlike any reaction we have studied thus far. To account for the stereoselectivity of the Diels-Alder reaction and the lack of evidence for either ionic or radical intermediates, chemists have proposed that reaction takes place in a single step during which there is a cyclic redistribution of electrons. During this cyclic redistribution, bond forming and bond breaking are concerted (simultaneous). To use the terminology of organic chemistry, the Diels-Alder reaction is a **pericyclic reaction,** that is, a reaction that takes place in a single step, without intermediates, and involves a cyclic redistribution of bonding electrons. We can envision a Diels-Alder reaction taking place as shown in Figure 22.2.

Pericyclic reaction A reaction that takes place in a single step, without intermediates, and involves a cyclic redistribution of bonding electrons.

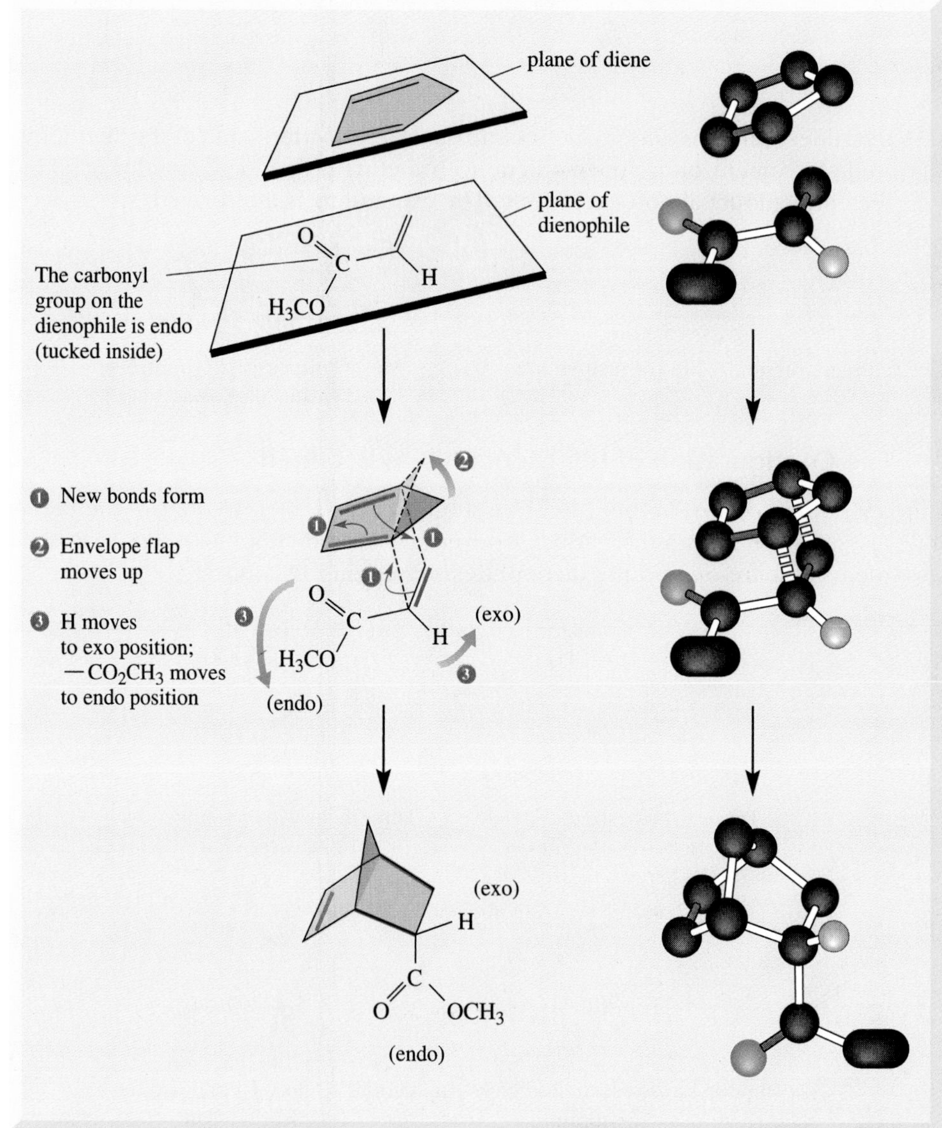

FIGURE 22.2
Mechanism of the Diels-Alder reaction. The diene and dienophile approach each other in parallel planes, one above the other, with the substituents on the dienophile endo to the diene. There is overlap of the pi orbitals of each molecule and syn addition of each molecule to the other. As (1) new sigma bonds form in the transition state, (2) the $—CH_2—$ on the diene rotates upward, (3) the hydrogen atom of the dienophile becomes exo, and (4) the ester group of the dienophile becomes endo.

Example 22.4

Complete the following Diels-Alder reaction, showing the stereochemistry of the product.

Solution

Reaction of cyclopentadiene with this dienophile forms a disubstituted bicyclo[2.2.1]hept-2-ene. The two ester groups are *cis* in the dienophile, and, given the stereoselectivity of the Diels-Alder reaction, they are *cis* and endo in the product.

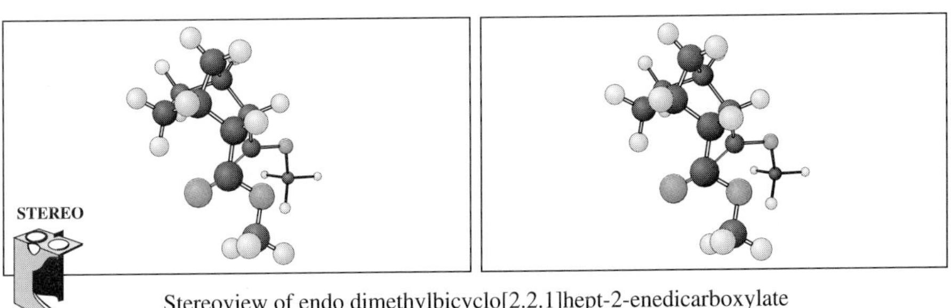

Stereoview of endo dimethylbicyclo[2.2.1]hept-2-enedicarboxylate

Problem 22.4

What diene and dienophile might you use to prepare the following Diels-Alder adduct?

F. A Word of Caution About Electron Pushing

We developed a mechanism of the Diels-Alder reaction and used curved arrows to show the flow of electrons that takes place in the process of bond breaking and bond forming. Diels-Alder reactions involve a four-carbon diene and a two-carbon dienophile and are termed [4 + 2] cycloadditions. We can write similar electron-pushing mechanisms for the dimerization of ethylene by a [2 + 2] cycloaddition to form cyclobutane, and for the dimerization of butadiene by a [4 + 4] cycloaddition to form 1,5-cyclooctadiene.

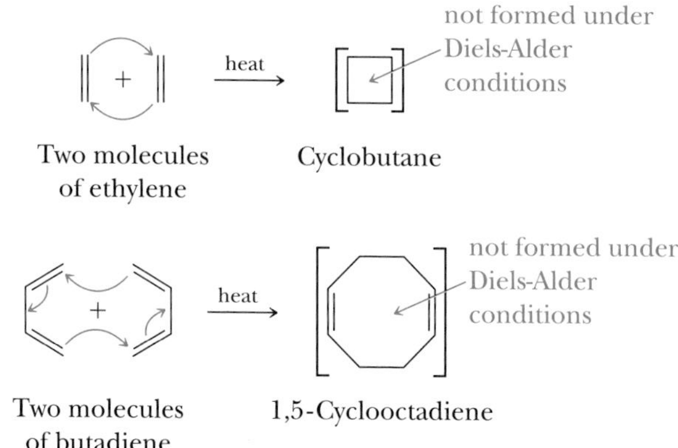

Although [2 + 2] and [4 + 4] cycloadditions bear a formal relationship to the Diels-Alder reaction, neither, in fact, takes place under the thermal conditions required for Diels-Alder reactions. These cycloadditions do occur, but only under experimental conditions different from those required for Diels-Alder reactions and by quite different mechanisms. The point of mentioning them here is to add a note of caution. Although electron pushing is a valuable tool in electron bookkeeping, it is not in itself a full description of a reaction mechanism. Reaction mechanisms are often far more complex than use of curved arrows might suggest.

22.3 Pericyclic Reactions and Transition State Aromaticity

The Diels-Alder reaction is a concerted reaction involving a redistribution of six electrons in a cyclic transition state. The central point here is that there are six electrons and that the transition state is cyclic. We can place the transition state for the Diels-Alder reaction in a larger context referred to as transition state aromaticity. Recall the Hückel criteria for aromaticity (Section 19.2A): the presence of $(4n + 2)$ pi electrons in a ring that is planar and fully conjugated. Just as aromaticity imparts a stability to certain types of molecules and ions, the presence of $(4n + 2)$ electrons in a cyclic transition state imparts a stability to certain types of transition states. Reactions involving 2, 6, 10, 14, and so on electrons in a cyclic transition state have especially low activation energies and take place particularly readily. Transition states involving a cyclic redistribution of $4n$ electrons, on the other hand, are antiaromatic and have especially high activation energies. It is for this reason that the dimerization of ethylene, as shown above, to give cyclobutane (a 4-electron transition state) and of 1,3-butadiene to give 1,5-cyclooctadiene (an 8-electron transition state) do not occur. We now see that, just as the Hückel theory of aromaticity gives us a clearer understanding of the stability of certain types of molecules and ions, it also gives a clearer understanding of certain types of reactions and their transition states.

We have seen three examples of reactions that proceed by cyclic, six-electron transition states:

1. The decarboxylation of β-ketoacids and β-dicarboxylic acids (Section 16.10).
2. The Cope elimination of amine *N*-oxides (Section 21.11)
3. The Diels-Alder reaction (Section 22.2)

We now look at two more examples of reactions that proceed by aromatic transition states.

Chemistry in Action

Catalysts for Diels-Alder Reactions

The Diels-Alder reaction is one of the most valuable reactions for creating new carbon-carbon bonds. Unlike the other reactions studied so far, the Diels-Alder reaction involves a concerted, cyclic redistribution of electrons. Finding a way to catalyze this reaction would seem to be very difficult. Nevertheless, organic chemists have discovered that Lewis acids can catalyze some Diels-Alder reactions. As an example, 1,3-pentadiene and methyl acrylate react to form isomeric Diels-Alder adducts in the ratio 9 : 1.

1,3-Pentadiene + Methyl acrylate ⟶

No catalyst:	90%	10%
$AlCl_3$:	98%	2%

With the Lewis acid catalyst $AlCl_3$, the same reaction is speeded up, and the isomer ratio improves to 49 : 1. Why should $AlCl_3$ have such a dramatic effect?

Of the two partners in this reaction, the Lewis acid interacts preferentially with the dienophile, because of the lone pairs of electrons on the carbonyl oxygen. This causes the dienophile to become polarized. It then interacts in a stepwise fashion, forming the two new carbon-carbon bonds one at a time. Depending on the orientation with which the dienophile approaches the diene, two different allylic cation intermediates result.

a secondary allylic cation intermediate

a primary allylic cation intermediate

In the first orientation, the intermediate is a more stable secondary allylic carbocation and leads to the major product. In the second orientation, the intermediate is a less stable primary allylic carbocation. Thus, using a Lewis acid in a Diels-Alder reaction makes the cycloaddition less concerted and more stepwise in character.

A. The Claisen Rearrangement

The Claisen rearrangement transforms allyl phenyl ethers to *o*-allylphenols. Heating allyl phenyl ether, for example, the simplest member of this class of compounds, at 200–250°C results in a Claisen rearrangement to form *o*-allyphenol. In this rearrangement, an allyl group migrates from a phenolic oxygen to a carbon atom ortho to it. It has been demonstrated by carbon-14 labeling, here shown in color, that during a Claisen rearrangement, carbon 3 of the allyl group becomes bonded to the ring carbon ortho to the phenolic oxygen.

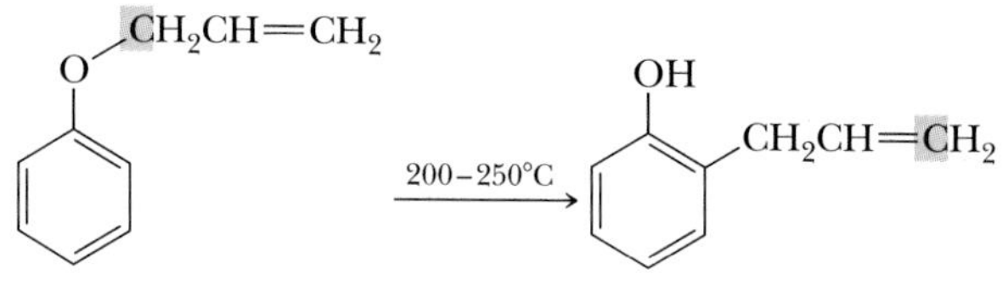

$$\text{Allyl phenyl ether (C}_6\text{H}_5\text{OCH}_2\text{CH=CH}_2) \xrightarrow{200-250°\text{C}} o\text{-Allylphenol (}o\text{-HOC}_6\text{H}_4\text{CH}_2\text{CH=CH}_2)$$

Allyl phenyl ether   *o*-Allylphenol

The mechanism of a Claisen rearrangement involves a concerted redistribution of six electrons in a cyclic transition state. The product of the rearrangement is a substituted cyclohexadienone, which undergoes keto-enol tautomerism to reform the aromatic ring.

MECHANISM The Claisen Rearrangement

Reaction involves a cyclic, six-electron transition state to form a cyclohexadienone intermediate.

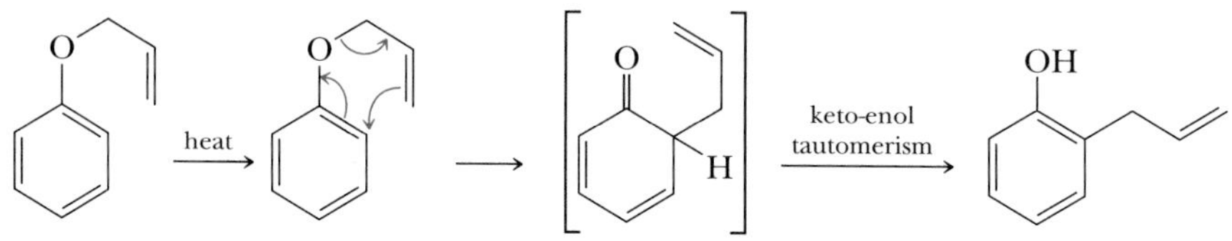

Allyl phenyl ether —heat→ Transition state → [A cyclohexadienone intermediate] —keto-enol tautomerism→ *o*-Allylphenol

Thus we see that the transition state for the Claisen rearrangement bears a close resemblance to that for the Diels-Alder reaction. Both involve a concerted redistribution of six electrons in a cyclic transition state.

EXAMPLE 22.5

Predict the product of Claisen rearrangement of *trans*-2-butenyl phenyl ether.

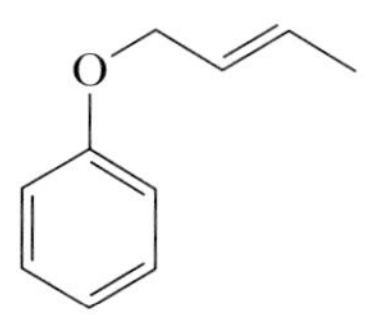

trans-2-Butenyl phenyl ether

Solution

In the six-membered transition state for this rearrangement, it is carbon 3 of the allyl group that becomes bonded to the ortho position of the ring.

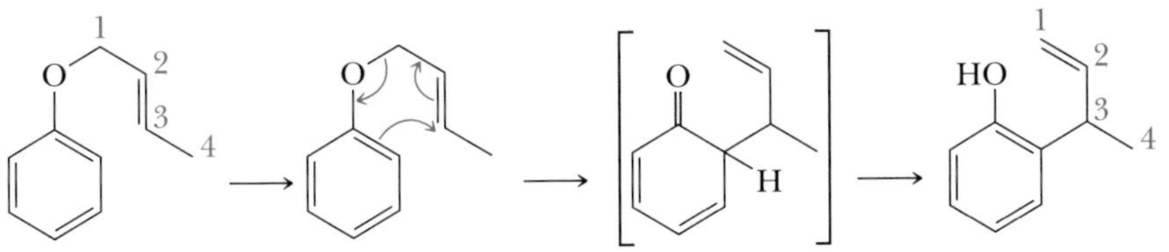

O 1 2 3 4 → O → [O H] → HO 1 2 3 4

The transition state

A cyclohexadienone intermediate

PROBLEM 22.5

Show how to synthesize allyl phenyl ether and 2-butenyl phenyl ether from phenol and appropriate alkenyl halides.

B. The Cope Rearrangement

The Cope rearrangement of 1,5-dienes also takes place via a cyclic, six-electron transition state. In this example, the product is an equilibrium mixture of isomeric dienes. The favored product is the diene on the right, which contains the more highly substituted double bonds.

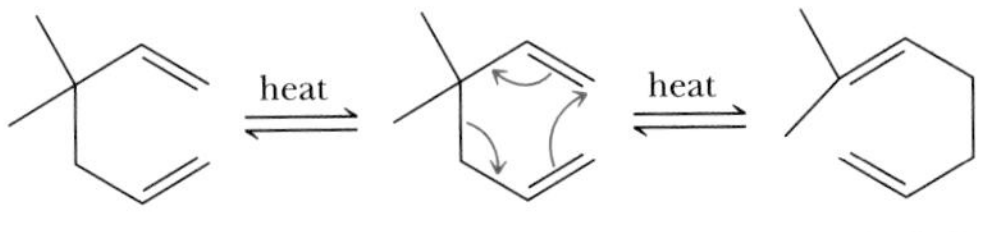

heat ⇌ heat ⇌

3,3-Dimethyl-1,5-hexadiene

Transition state

2-Methyl-2,6-hexadiene

EXAMPLE 22.6

Propose a mechanism for the following Cope rearrangement.

350°C →

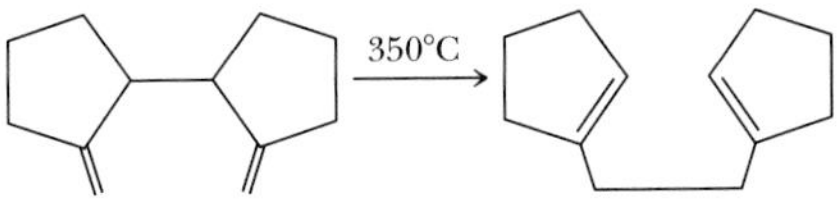

Solution

Redistribution of six electrons in a cyclic transition state gives the observed product.

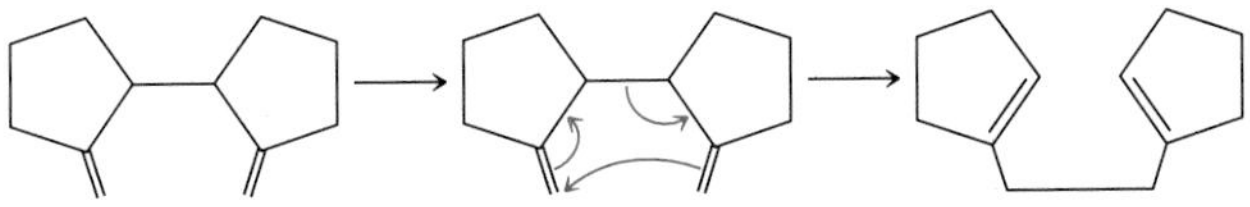

Transition state

PROBLEM 22.6

Propose a mechanism for the following Cope rearrangement.

OH H 320°C → O

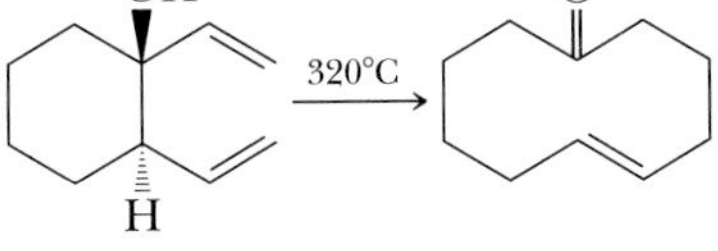

CHEMISTRY IN ACTION

Singlet Oxygen

Molecular oxygen differs from most other stable molecules in having two unpaired electrons. This comes about because, during the "aufbau" of molecular oxygen, there are two highest occupied molecular orbitals (π^*) of the same energy and only two electrons to put in them; thus by Hund's rule, the lowest energy ("ground") state of oxygen has one of these electrons in each orbital and with the same spin. The unpaired electrons cause oxygen to be paramagnetic, that is, attracted by a magnetic field. Most molecules are diamagnetic (repelled by a magnetic field).

The ground state of oxygen is often called a diradical, but in fact oxygen does not really behave as a diradical (it is much less reactive than most radicals). Like protons, electrons have a magnetic moment; two

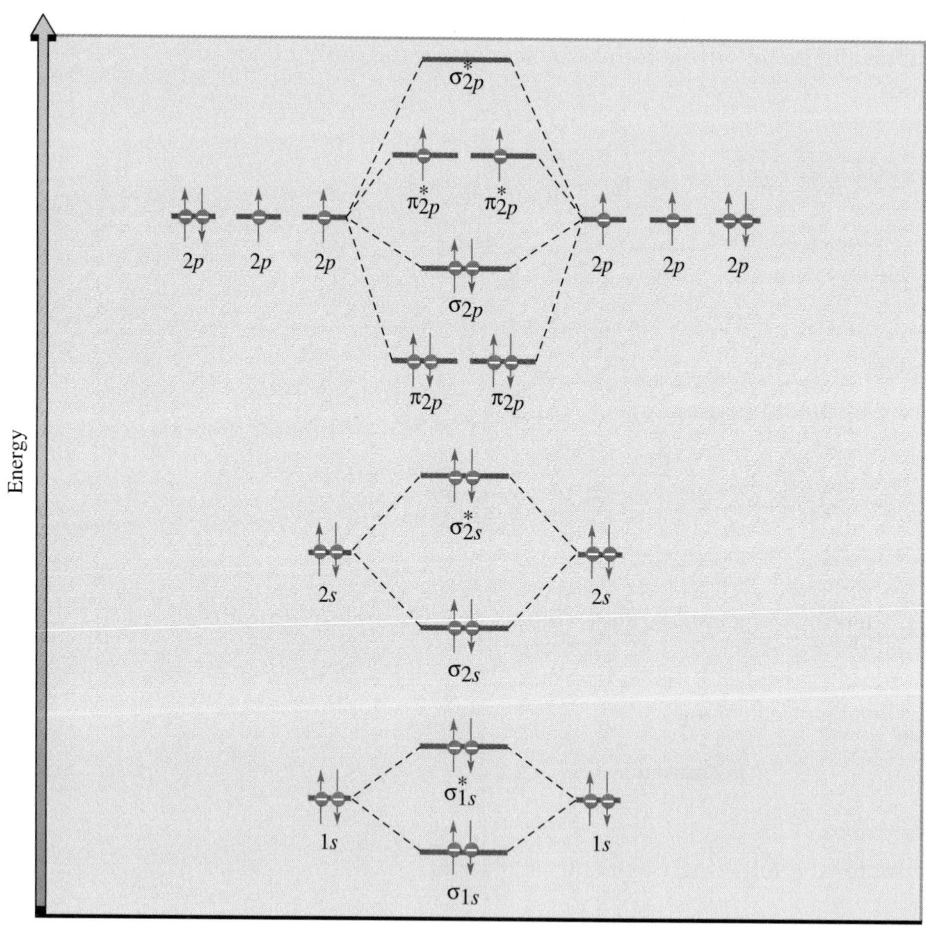

Ground-state molecular orbital energy diagram for O_2.

electrons can take on three orientations in a magnetic field, and the ground state of oxygen is thus called a "triplet" state. Oxygen is one of an extremely small number of compounds that have a triplet ground state.

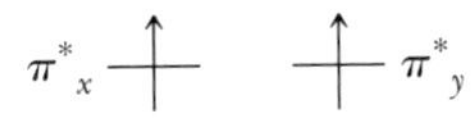

π^*_x ↑ ↑ π^*_y

Highest occupied (antibonding) orbitals in oxygen ground state

However, the triplet state is not the only electronic configuration oxygen can have, only the lowest. There are also two low-energy excited electronic states of oxygen in which the electrons are paired. Only the lower (22 kcal above the ground state) has an appreciable lifetime. Chemists refer to this species as "singlet oxygen"; its lifetime varies with solvent from about four microseconds to almost a tenth of a second. It is symbolized by $^1O_2$ and has a very different reactivity from ground-state oxygen, $^3O_2$.

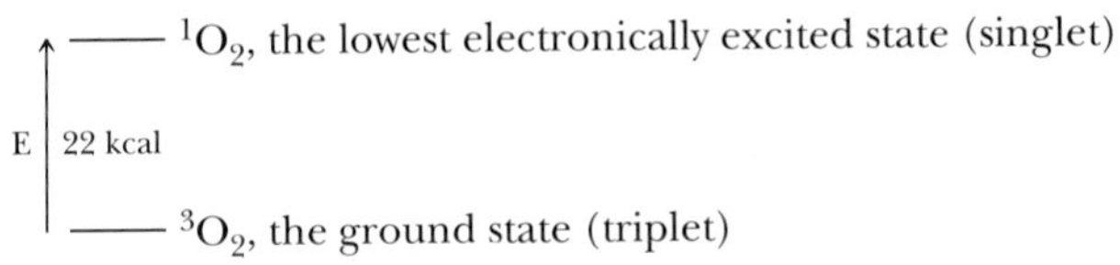

— $^1O_2$, the lowest electronically excited state (singlet)

E | 22 kcal

— $^3O_2$, the ground state (triplet)

Lowest electronic states of oxygen

Singlet oxygen has the Lewis structure shown below, with a double bond between two highly electronegative atoms. As you might expect, it reacts as a dienophile; its reactions resemble those of other Diels-Alder dienophiles. For example, it reacts with cyclohexadiene to give the Diels-Alder adduct peroxide shown below. This is a general and synthetically useful reaction.

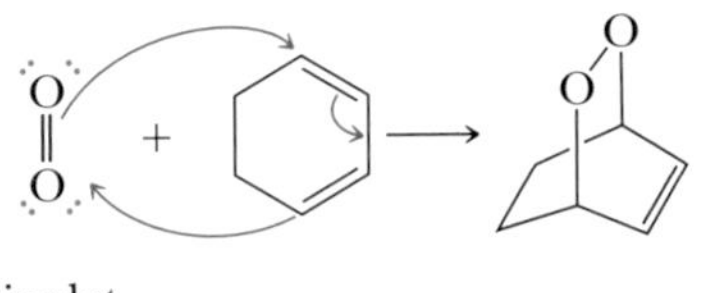

O=O + (cyclohexadiene) ⟶ (bicyclic peroxide)

Singlet oxygen ($^1O_2$)

Singlet oxygen is most often produced photochemically using a dye or other compound called a **photosensitizer** (Sens). Photosensitizers absorb a photon of light ($h\nu$), often in the visible region of the spectrum, and are converted to an electronically excited state (Sens*). This state can transfer its energy to oxygen, producing $^1O_2$. This process can be very efficient. Singlet oxygen can also be formed by a reverse Diels-Alder reaction on warming certain peroxides, as shown below. The resulting singlet oxygen can then react with dienes and other types of reactive acceptors (for example, with 1,4-dimethylnaphthalene at room temperature or below, it produces the endoperoxide). This reaction is called **photosensitized oxidation.**

$$\text{Sens} \xrightarrow[h\nu]{} \text{Sens}^*$$

$$\text{Sens}^* + {}^3O_2 \longrightarrow \text{Sens} + {}^1O_2$$

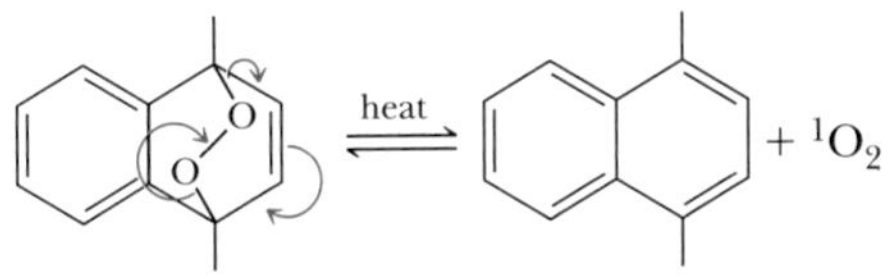

An endoperoxide $\underset{}{\overset{\text{heat}}{\rightleftharpoons}}$ 1,4-Dimethylnaphthalene + $^1O_2$

Singlet oxygen can also react with many biological materials such as unsaturated lipids. This can cause damage to or death of the organism. The biological damage caused by photosensitizers, light, and oxygen is called **photodynamic** damage. These conditions can be used to selectively kill tumor cells; this **photodynamic therapy** has recently been approved by the FDA.

Foote, C. S.; Clennan, E. L. in *Active Oxygen in Chemistry;* Foote, C. S.; Valentine, J. S.; Greenberg, A.; Liebman, J. F.; Eds.; Chapman and Hall, London, 1995; pp. 105–140.

SUMMARY

Conjugated dienes undergo **electrophilic addition** to give **1,2-addition** and **1,4-addition** products (Section 22.1A). The intermediate in electrophilic addition to a conjugated diene is a resonance-stabilized **allylic carbocation.**

For a reaction under **kinetic (rate) control** (Section 22.1B), the distribution of products is determined by the relative rates of formation of each. For a reaction under **thermodynamic (equilibrium) control,** the distribution of products is determined by the relative stabilities of each.

The **Diels-Alder reaction** (Section 22.2) is a **cycloaddition** between a **conjugated diene** and a **dienophile** to give a six-membered ring. Dienophiles contain either a double or triple bond. Diels-Alder reactions are facilitated by electron-withdrawing substituents on one of the reactants (either the diene or dienophile) and electron-releasing substituents on the other reactant. The mechanism is described as a **pericyclic reaction,** that is, one that takes place in a single step, without intermediates, and involves redistribution of bonding electrons in a cyclic transition state. The Diels-Alder reaction is **stereoselective:** (1) the configuration of the dienophile is retained; if substituents on the dienophile are *cis* (or *trans*), they remain *cis* (or *trans*) in the product; (2) formation of the **endo** adduct is favored.

KEY REACTIONS

1. Electrophilic Addition to Conjugated Dienes (Section 22.1)

The ratio of 1,2- to 1,4-addition products depends on whether the reaction is under kinetic control or thermodynamic control.

$$CH_2{=}CHCH{=}CH_2 + HBr \longrightarrow CH_3\underset{}{C}HBrCH{=}CH_2 + CH_3CH{=}CHCH_2Br$$

	$CH_3CH(Br)CH{=}CH_2$	$CH_3CH{=}CHCH_2Br$
Products formed at −78°C (kinetic control):	90%	10%
Products formed at 40°C (thermodynamic control):	15%	85%

2. The Diels-Alder Reaction: A Pericyclic Reaction (Section 22.2)

A Diels-Alder reaction takes place in a single step, without intermediates, and involves redistribution of six pi electrons in a cyclic transition state. The configuration of the dienophile is preserved.

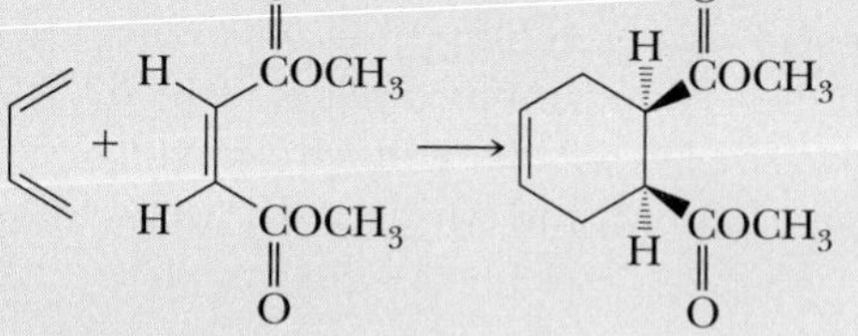

$$\text{butadiene} + \textit{cis}\text{-}HC(COCH_3){=}CH(COCH_3) \longrightarrow \textit{cis}\text{-4,5-diacetylcyclohexene}$$

Formation of the endo adduct is favored.

$$\text{cyclopentadiene} + \textit{cis}\text{-}HC(CO_2CH_3){=}CH(CO_2CH_3) \longrightarrow \text{endo adduct (}CO_2CH_3\text{ groups endo)}$$

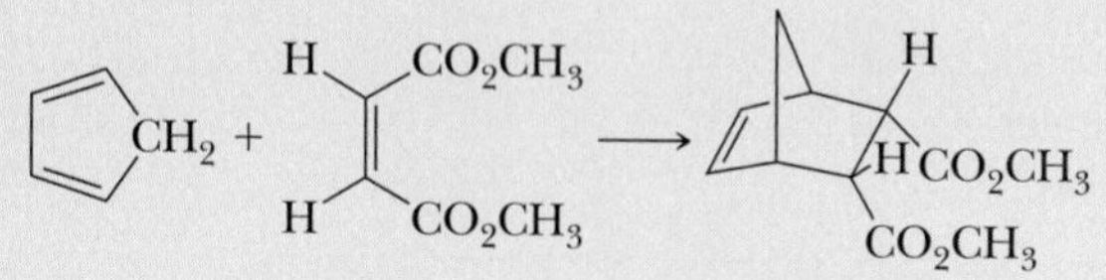

3. The Claisen Rearrangement: A Pericyclic Reaction (Section 22.3A)

Rearrangement of an allyl phenyl ether to an ortho-substituted phenol.

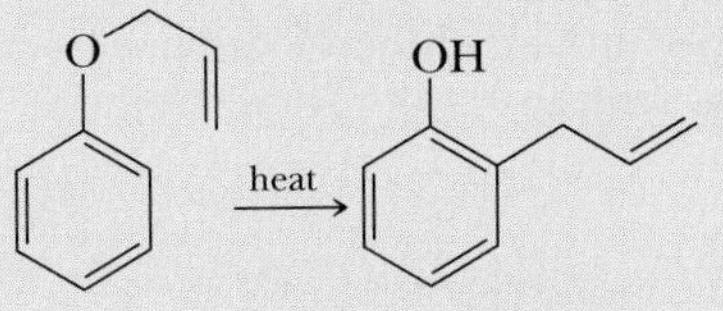

O — heat → OH

4. The Cope Rearrangement: A Pericyclic Reaction (Section 22.3B)

Rearrangement of a 1,5-diene to give an isometric 1,5-diene.

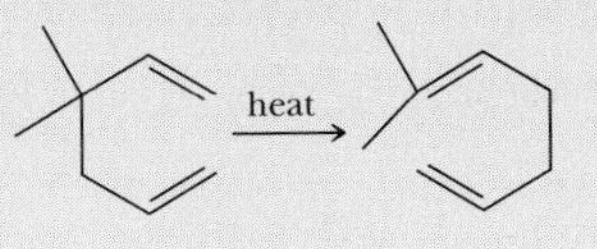

heat →

ADDITIONAL PROBLEMS

Structure and Stability

22.7 If an electron is added to 1,3-butadiene, into which molecular orbital does it go? If an electron is removed from 1,3-butadiene, from which molecular orbital is it taken?

22.8 Draw a potential energy diagram (potential energy versus dihedral angle from 0° to 360°) for rotation about the 2,3 single bond in 1,3-butadiene.

22.9 Draw all important contributing structures for the following allylic carbocations and then rank the structures in order of relative contributions to each resonance hybrid.

(a) 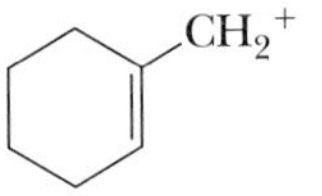 CH_2^+ (b) $CH_2{=}CHCH{=}CHCH_2^+$ (c) $CH_3\overset{+}{C}(CH_3)CH{=}CH_2$

Electrophilic Addition to Conjugated Dienes

22.10 Predict the structure of the major product formed by 1,2-addition of HCl to 2-methyl-1,3-butadiene (isoprene). To arrive at a prediction, first consider proton transfer to carbon 1 of this diene. Second, consider proton transfer to carbon 4 of this diene. Then compare the relative stabilities of the two allylic carbocation intermediates.

22.11 Predict the major product formed by 1,4-addition (conjugate addition) of HCl to isoprene. Follow the reasoning suggested in the previous problem.

22.12 Predict the structure of the major 1,2-addition product formed by reaction of 1 mol of Br_2 with isoprene. Also predict the structure of the major 1,4-addition product formed under these conditions.

22.13 Which of the two molecules shown do you expect to be the major product formed by 1,2-addition of HCl to cyclopentadiene? Explain.

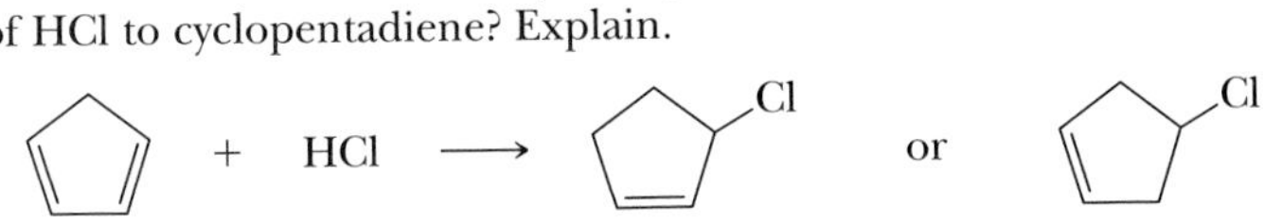

+ HCl ⟶ Cl or Cl

Cyclopentadiene 3-Chlorocyclopentene 4-Chlorocyclopentene

22.14 Predict the major product formed by 1,4-addition of HCl to cyclopentadiene.

22.15 Draw structural formulas for the two constitutional isomers of molecular formula $C_5H_6Br_2$ formed by adding 1 mol of Br_2 to cyclopentadiene.

22.16 What are the expected kinetic and thermodynamic products from addition of one mol of Br_2 to the following dienes?

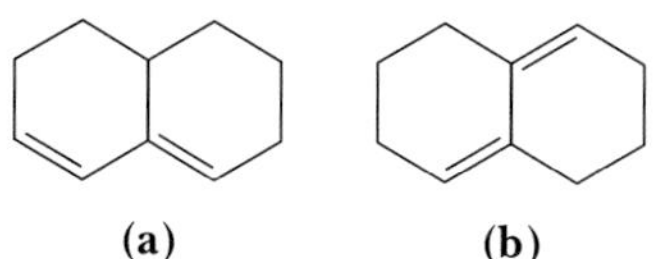

(a) **(b)**

Diels-Alder Reactions

22.17 Draw structural formulas for the products of reaction of cyclopentadiene with each dienophile.

(a) $CH_2{=}CHCl$ **(b)** $CH_2{=}CH\overset{O}{\overset{\|}{C}}OCH_3$

(c) $HC{\equiv}CH$ **(d)** $CH_3O\overset{O}{\overset{\|}{C}}C{\equiv}C\overset{O}{\overset{\|}{C}}OCH_3$

22.18 Propose structural formulas for compounds (A) and (B) and specify the configuration of compound (B).

$$\text{cyclopentene} + CH_2{=}CH_2 \xrightarrow{200°C} \underset{(A)}{C_7H_{10}} \xrightarrow[2.\ (CH_3)_2S]{1.\ O_3} \underset{(B)}{C_7H_{10}O_2}$$

22.19 Under certain conditions, 1,3-butadiene can function both as a diene and a dienophile. Draw a structural formula for the Diels-Alder adduct formed by reaction of 1,3-butadiene with itself.

22.20 1,3-Butadiene is a gas at room temperature and requires a gas-handling apparatus for use in a Diels-Alder reaction. Butadiene sulfone is a convenient substitute for gaseous 1,3-butadiene. This sulfone is a solid at room temperature (mp 66°C) and, when heated above its boiling point of 110°C, decomposes by a reverse Diels-Alder reaction to give s-*cis*-1,3-butadiene and sulfur dioxide. Draw a Lewis structure for butadiene sulfone, and show by curved arrows the path of this reverse Diels-Alder reaction.

$$\text{Butadiene sulfone} \xrightarrow{140°C} \text{1,3-Butadiene} + \underset{\text{Sulfur dioxide}}{SO_2}$$

22.21 The following triene undergoes an intramolecular Diels-Alder reaction to give the product shown. Show how the carbon skeleton of the triene must be coiled to give this product, and show by curved arrows the redistribution of electron pairs that takes place to give the product.

$$CH_2{=}CH\overset{CH_3}{\overset{|}{C}}{=}CHCH_2CH_2CH_2CH_2CH{=}CH_2 \xrightarrow{160°C}$$

22.22 The following trienone undergoes an intramolecular Diels-Alder reaction to give a bicyclic product. Propose a structural formula for the product. Account for the observation that the Diels-Alder reaction given in this problem takes place under milder conditions (at lower temperature) than the analogous Diels-Alder reaction shown in the previous problem?

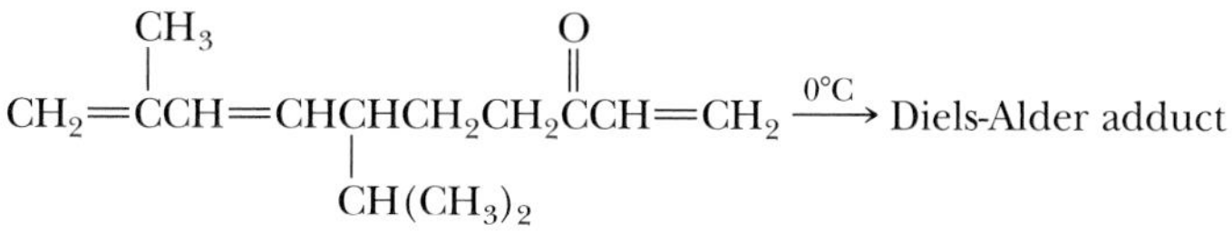

$$CH_2{=}C(CH_3)CH{=}CHCH(CH(CH_3)_2)CH_2CH_2C({=}O)CH{=}CH_2 \xrightarrow{0°C} \text{Diels-Alder adduct}$$

22.23 The following compound undergoes an intramolecular Diels-Alder reaction to give a bicyclic product. Propose a structural formula for the product.

$\xrightarrow{\text{heat}}$ an intramolecular Diels-Alder adduct

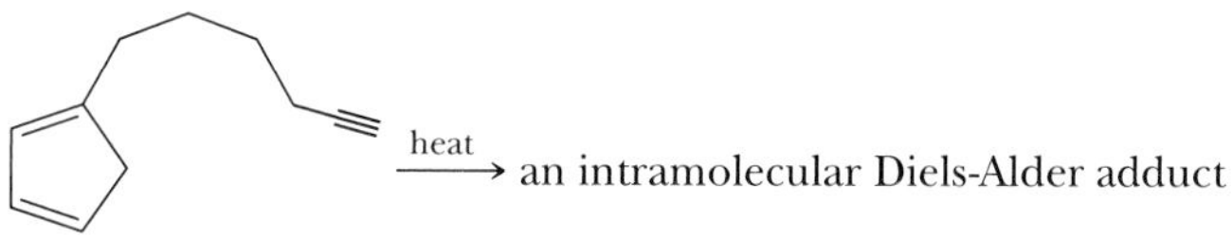

22.24 One of the published syntheses of warburganal (Problem 5.31) begins with the following Diels-Alder reaction. Propose a structure for compound A.

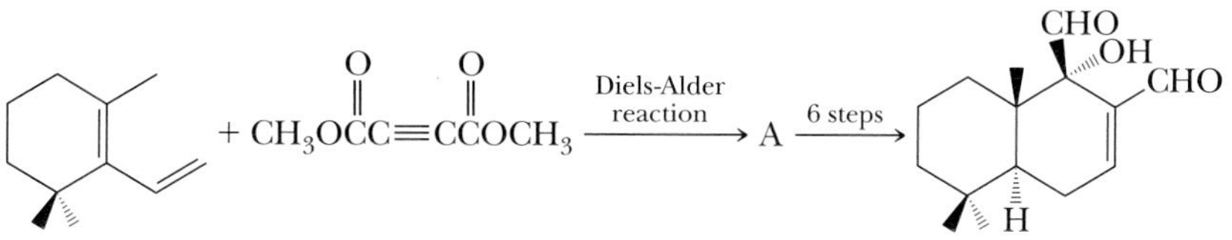

$+ CH_3OC({=}O)C{\equiv}CC({=}O)OCH_3 \xrightarrow{\text{Diels-Alder reaction}} A \xrightarrow{\text{6 steps}}$ Warburganal

CHO, OH, CHO, H

Warburganal

22.25 The Diels-Alder reaction is not limited to making six-membered rings with only carbon atoms. Predict the products of the following reactions that produce heterocycles—rings with atoms other than carbon in them.

(a) $OCCH_3$ (O) diene with $OCCH_3$ (O) + $H-C(=O)-CH_3$ ⟶

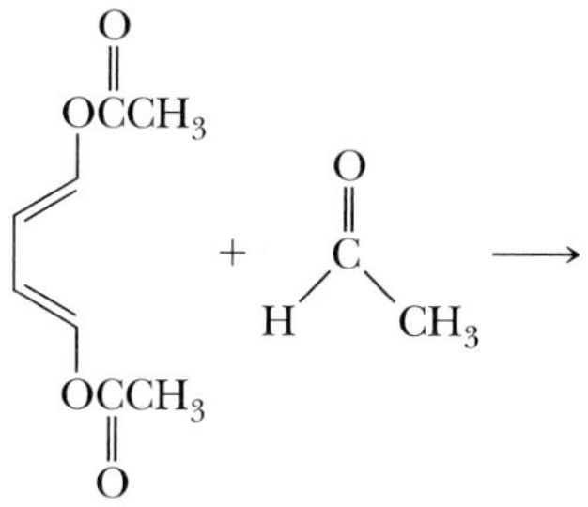

(b) N ring with $C(CH_3)_3$ + O (maleic anhydride) ⟶

(c) + N, N, NCH_3, O, O ⟶

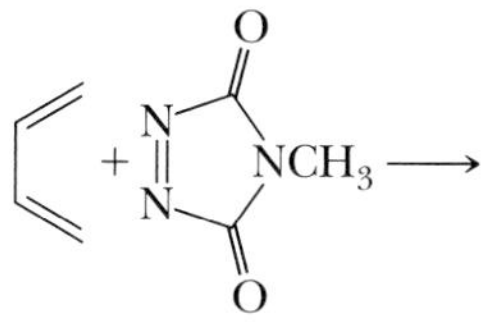

(d) CO_2CH_3, N, N, N, N, CO_2CH_3 + $CH_2{=}CHCH_3$ ⟶

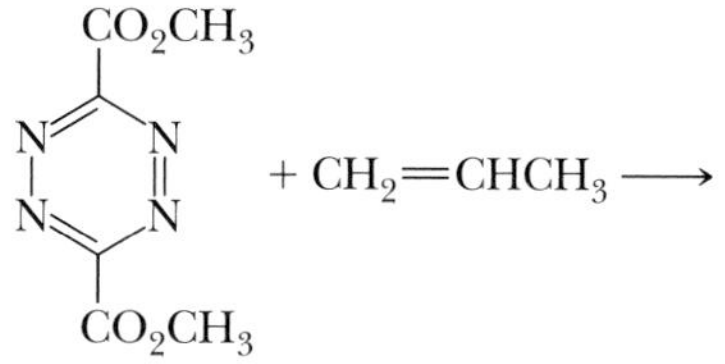

(e) O + $N{\equiv}C$, H, H, $C{\equiv}N$ ⟶

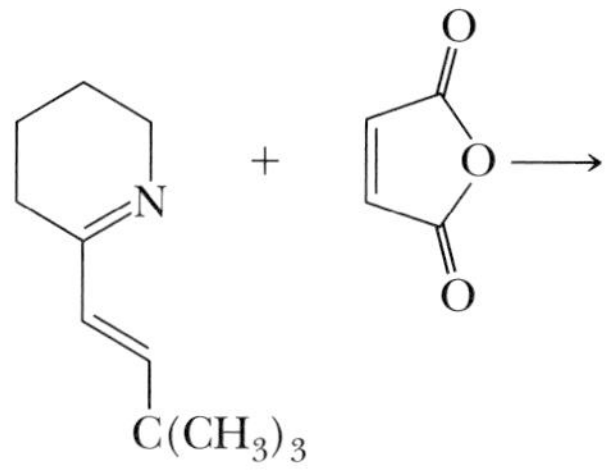

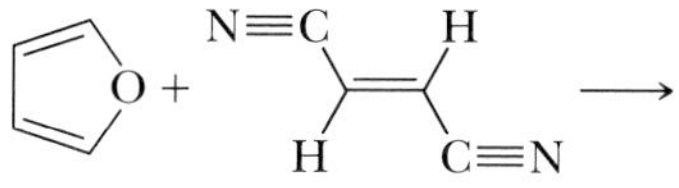

in an s-*cis* conformation. After its formation, it assumes an s-*trans* conformation. Use curved arrows to show the flow of electrons in these photoisomerizations.

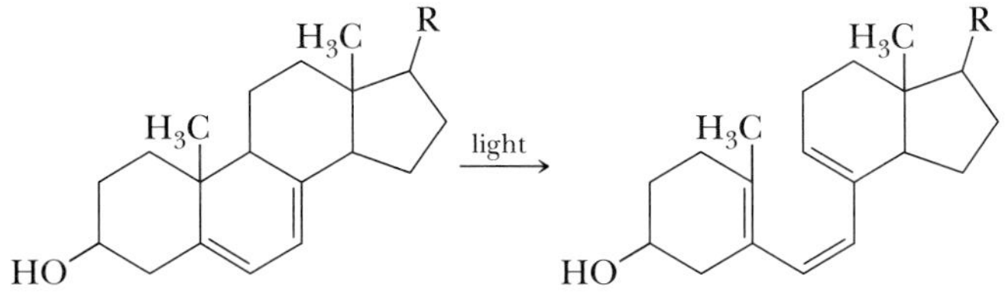

H_3C R H_3C HO light H_3C R H_3C HO

7-Dehydrocholesterol Precalciferol

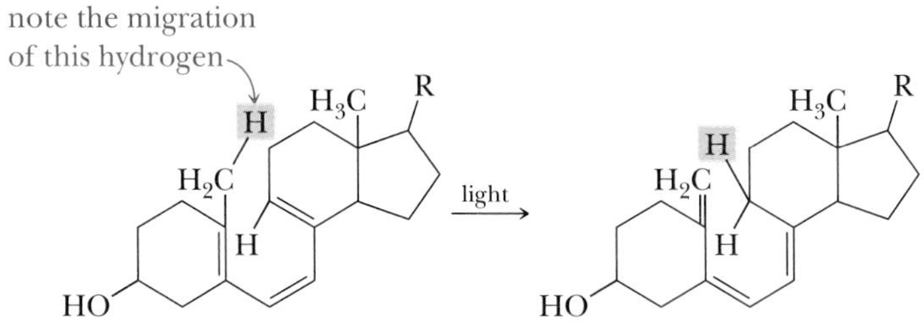

note the migration of this hydrogen

H H_3C R H_2C H HO light H H_3C R H_2C H HO

Precalciferol

Cholecalciferol (Vitamin D_3)
(s-*cis* conformation)

ORGANIC POLYMER CHEMISTRY

23

■ Woven fibers of nylon. *(Harold Rose/Science Photo Library/Photo Researchers)*

The technological advancement of any society is inextricably tied to the materials available to it. Indeed, historians have used the emergence of new materials as a way of establishing a timeline to mark the development of human civilization. As part of the search to discover new materials, scientists have made increasing use of organic chemistry for the preparation of synthetic polymers. The versatility afforded by these polymers allows for the creation and fabrication of materials with ranges of properties unattainable using such materials as wood, metals, and ceramics. Deceptively simple changes in the chemical structure of a given polymer, for example, can change its mechanical properties from those of a sandwich bag to those of a bulletproof vest. Furthermore, structural changes can introduce properties never before imagined in organic polymers. For example, using well-defined organic reactions, one type of polymer can be made into an insulator (for example, the rubber sheath that surrounds electrical cords), or, if treated differently, it can be made into an electrical conductor with a conductivity nearly equal to that of metallic copper!

The years since the 1930s have seen extensive research and development in polymer chemistry, and an almost explosive growth in plastics, coatings, and rubber technology has created a worldwide multibillion-dollar industry. A few basic characteristics account for this phenomenal growth. First, the raw materials for synthetic polymers are derived mainly from petroleum. With the development of petroleum-refining processes, raw materials for the synthesis of polymers became generally cheap and plentiful. Second, within broad limits, scientists have learned how to tailor polymers to the requirements of the end use. Third, many consumer products can be fabricated more cheaply from synthetic polymers than from competing materials, such as wood, ceramics, and metals. For example, polymer technology created the water-based (latex) paints that have revolutionized the coatings industry; plastic films and foams have done the same for the packaging industry. The list could go on and on as we think of the manufactured items that are everywhere around us in our daily lives.

23.1 The Architecture of Polymers

Polymer From the Greek, *poly* + *meros*, many parts. Any long-chain molecule synthesized by linking together many single parts called monomers.

Monomer From the Greek, *mono* + *meros*, single part. The simplest nonredundant unit from which a polymer is synthesized.

Polymers (Greek, *poly* + *meros*, many parts) are long-chain molecules synthesized by linking **monomers** (Greek, *mono* + *meros*, single part) through chemical reactions. The molecular weights of polymers are generally high compared with those of common organic compounds and typically range from 10,000 g/mol to more than 1,000,000 g/mol. The architectures of these macromolecules can also be quite diverse. Types of polymer architecture include linear and branched chains as well as those with comb, ladder, and star structures (Figure 23.1). Additional structural variations can be achieved by introducing covalent crosslinks between individual polymer chains.

Plastic A polymer that can be molded when hot and retains its shape when cooled.

Thermoplastic A polymer that can be melted and molded into a shape that is retained when it is cooled.

Linear and branched polymers are often soluble in solvents, such as chloroform, benzene, toluene, DMSO, and THF. In addition, many linear and branched polymers can be melted to form highly viscous liquids. In polymer chemistry, the term **plastic** refers to any polymer that can be molded when hot and retains its shape when cooled. **Thermoplastics** are polymers that can be melted and become sufficiently fluid that they can be molded into shapes that are retained when they are cooled. **Thermoset plastics** can be molded when they are first prepared, but once they are cooled, they harden irreversibly and cannot be remelted. Because of these very different physical

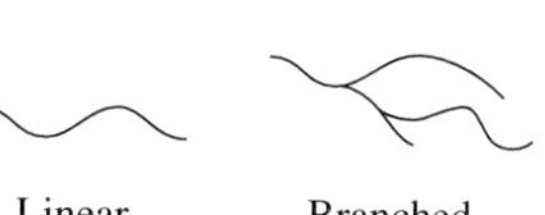

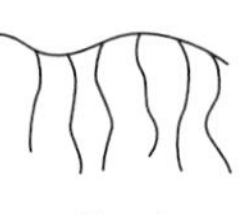

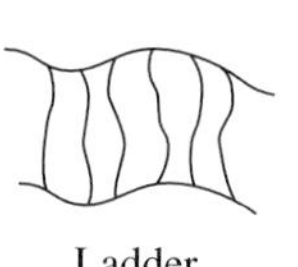

FIGURE 23.1 Various polymer architectures.

characteristics, thermoplastics and thermosets must be processed differently and are used in very different applications.

Thermoset plastic A polymer that can be molded when it is first prepared, but once cooled, hardens irreversibly and cannot be remelted.

The single most important property of polymers at the molecular level is the size and shape of their chains. A good example of the importance of size is a comparison of paraffin wax, a natural polymer, and polyethylene, a synthetic polymer. These two distinct materials have identical repeat units, namely $—CH_2—$, but differ greatly in chain size. Paraffin wax has between 25 and 50 carbon atoms per chain, whereas polyethylene has between 1000 and 3000 carbon atoms per chain. Paraffin wax, such as in birthday candles, is soft and brittle but polyethylene, such as in plastic beverage bottles, is strong, flexible, and tough. These vastly different properties arise directly from the difference in size and molecular architecture of the individual polymer chains.

23.2 Polymer Notation and Nomenclature

The structure of a polymer is shown by placing parentheses around the **repeat unit,** which is the smallest molecular fragment that contains all of the nonredundant structural features of the chain. Thus, the structure of an entire polymer chain can be reproduced by repeating the enclosed structure in both directions. A subscript n, called the **average degree of polymerization,** is placed outside the parentheses to indicate that this unit is repeated n times.

Average degree of polymerization, n A subscript placed outside the parentheses of the simplest nonredundant unit of a polymer to indicate that unit repeats n times in the polymer.

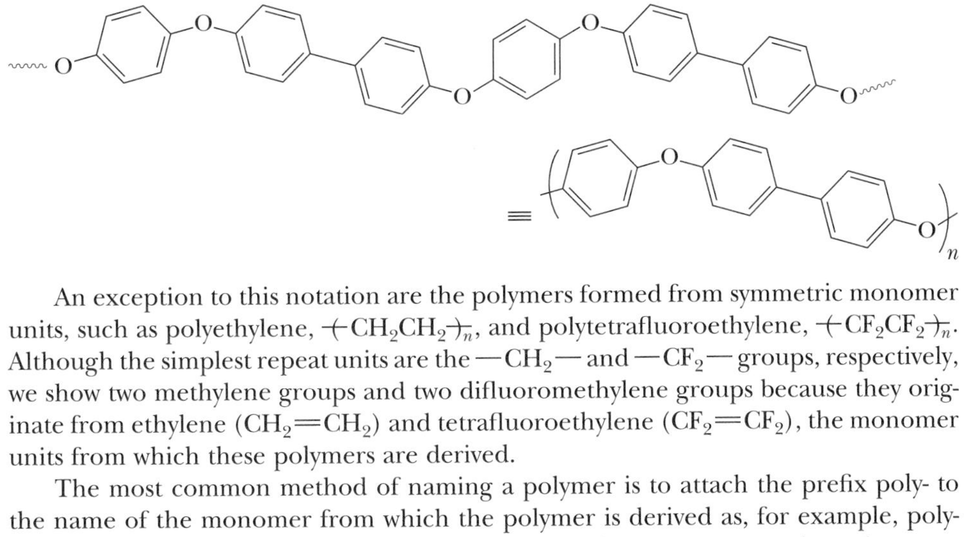

An exception to this notation are the polymers formed from symmetric monomer units, such as polyethylene, $\text{+}CH_2CH_2\text{+}_n$, and polytetrafluoroethylene, $\text{+}CF_2CF_2\text{+}_n$. Although the simplest repeat units are the $—CH_2—$ and $—CF_2—$ groups, respectively, we show two methylene groups and two difluoromethylene groups because they originate from ethylene ($CH_2{=}CH_2$) and tetrafluoroethylene ($CF_2{=}CF_2$), the monomer units from which these polymers are derived.

The most common method of naming a polymer is to attach the prefix poly- to the name of the monomer from which the polymer is derived as, for example, polyethylene and polystyrene. In the case of a more complex monomer or where the name of the monomer is more than one word as, for example, the monomer vinyl chloride, parentheses are used to enclose the name of the monomer.

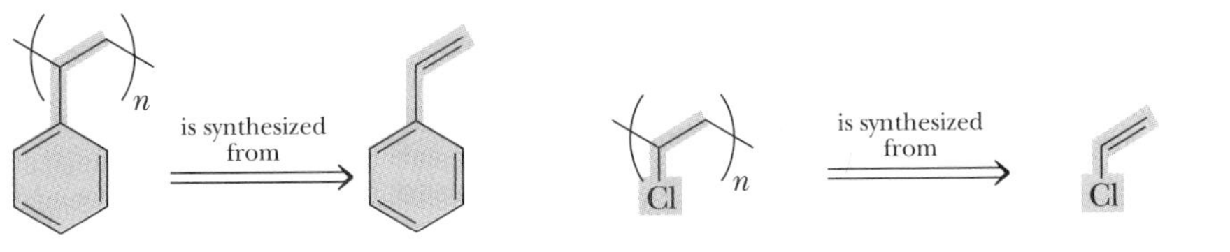

Polystyrene — is synthesized from → Styrene

Poly(vinyl chloride) (PVC) — is synthesized from → Vinyl chloride

EXAMPLE 23.1

Given the following structure, determine the polymer's repeat unit, redraw the structure using the simplified parenthetical notation, and name the polymer.

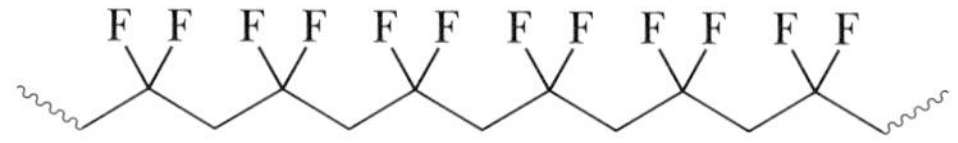

Solution

The repeat unit is $—CH_2CF_2—$ and the polymer is written $\left(CH_2CF_2\right)_n$. The repeat unit is derived from 1,1-difluoroethylene, and therefore, the polymer is named poly(1,1-difluoroethylene). This polymer is used in microphone diaphragms.

PROBLEM 23.1

Given the following structure, determine the polymer's repeat unit, redraw the structure using the simplified parenthetical notation, and name the polymer.

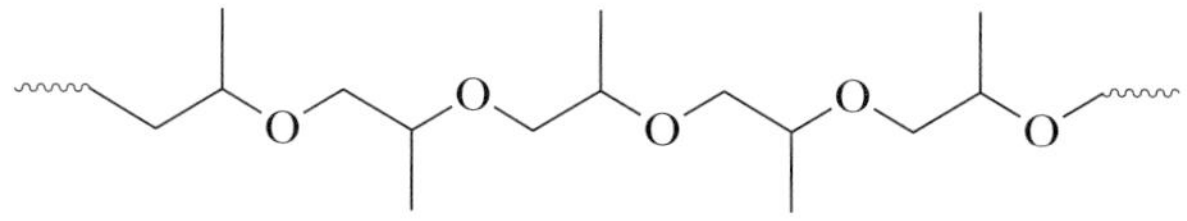

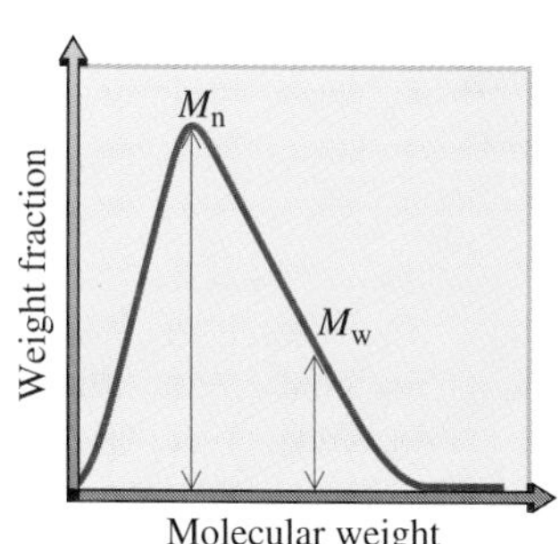

FIGURE 23.2
The distribution of molecular weights in a given polymer sample.

23.3 Molecular Weights of Polymers

All synthetic polymers and most naturally occurring polymers are mixtures of individual polymer molecules of variable molecular weights. When defining molecular weights in polymer chemistry, the two most common definitions used are the number average and weight average molecular weights. The **number average molecular weight, M_n,** is calculated by counting the number of chains of a particular molecular weight, multiplying each number by the molecular weight of its chains, summing these values, and dividing by the total number of polymer chains. The **weight average molecular weight (M_w)** is calculated by recording the total weight of each chain of a particular length, summing these weights, and dividing by the total weight of the sample. Because the larger chains in a sample weigh more than the smaller chains, the weight average molecular weight is skewed to higher values and M_w is always greater than M_n (Figure 23.2).

Both M_n and M_w are useful values and their ratio, M_w/M_n called the **polydispersity index,** provides a measure of the breadth of the molecular weight distribution. When the M_w/M_n ratio is equal to one, all of the polymer molecules in a sample are the same length and the polymer is said to be **monodisperse.** No synthetic polymers are ever monodisperse unless the individual molecules are carefully fractionated using time-consuming, rigorous separation techniques based on molecular size. On the other hand, natural polymers, such as polypeptides and DNA, that are formed using biological processes are monodisperse polymers.

Number average molecular weight, M_n An arithmetic mean calculated by summing the number of chains of a particular molecular weight in a collection of polymer molecules and dividing the sum by the total number of chains.

Weight average molecular weight, M_w The sum of the weights of each chain of a particular length in a collection of polymer molecules divided by the total weight of the sample.

23.4 Polymer Morphology: Crystalline Versus Amorphous Materials

Polymers, like small organic molecules, tend to crystallize upon precipitation or as they are cooled from a melt. Acting to inhibit this tendency are their very large molecules, which tend to slow diffusion, and their sometimes complicated or irregular structures, which prevent efficient packing of the chains. The result is that polymers in the solid state tend to be composed of both ordered **crystalline domains** (crystallites) and disordered **amorphous domains.** The relative amounts of crystalline and amorphous domains differ from polymer to polymer and often depend upon the manner in which the material is processed.

Crystalline domains Ordered crystalline regions in the solid state of a polymer. Also called crystallites.

High degrees of crystallinity are most often found in polymers with regular, compact structures and strong intermolecular forces, such as hydrogen bonding and dipolar interactions. The temperature at which crystallites melt corresponds to the **melt transition (T_m)** of the polymer. As the degree of crystallinity of a polymer increases, its T_m increases and it becomes more opaque due to scattering of the light by the crystalline domains. There is also a corresponding increase in strength and stiffness with increase in crystallinity. For example, poly(6-aminohexanoic acid) has a T_m = 223°C. At and well above room temperature, this polymer is a hard durable material that does not undergo any appreciable change in properties even on a very hot summer afternoon. Its uses range from textile fibers to shoe heels.

Amorphous domains Disordered, noncrystalline regions in the solid state of a polymer.

Melt transitions, T_m The temperature at which crystalline regions of a polymer melt.

Amorphous domains are characterized by the absence of long-range order. Highly amorphous polymers are sometimes referred to as **glassy** polymers. Because they lack crystalline domains that scatter light, amorphous polymers are transparent materials. In addition, they are typically weaker polymers, both in terms of their greater flexibility and their smaller mechanical strength. On being heated, amorphous polymers are transformed from a hard glass to a soft, flexible, rubbery state. The temperature at which this transition occurs is called the **glass transition temperature (T_g).** Amorphous polystyrene, for example, has a T_g = 100°C. At room temperature, it is a rigid solid used for drinking cups, foamed packaging materials, disposable medical wares, tape reels, and so forth. If it is placed in boiling water, it becomes soft and rubbery.

Glass transition temperature The temperature at which a polymer undergoes the transition from a hard glass to a rubbery state.

This relationship between mechanical properties and the degree of crystallinity can be illustrated by poly(ethylene terephthalate) (PET).

Poly(ethylene terephthalate) (PET)

PET can be made with percent crystalline domains ranging from 0% to about 55%. Completely amorphous PET is formed by cooling from the melt quickly. By prolonging the cooling time, more molecular diffusion occurs and crystallites form as the chains become more ordered. The differences in mechanical properties between these forms of PET are great. PET with a low degree of crystallinity is used for plastic beverage bottles, whereas fibers drawn from highly crystalline PET are used for textile fibers and tire cords.

Rubber materials must have low T_g values in order to behave as elastomers. An **elastomer** is a material that, when stretched or otherwise distorted, returns to its original shape when the distorting force is released. If the temperature drops below its

Elastomer A material that, when stretched or otherwise distorted, returns to its original shape when the distorting force is released.

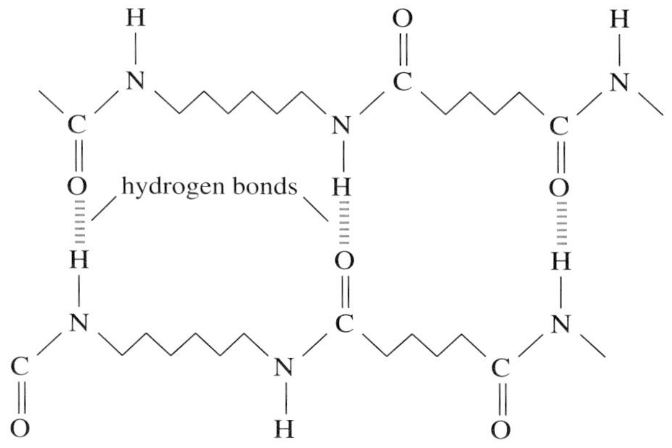

FIGURE 23.3 The structure of cold-drawn nylon 66. Hydrogen bonds between adjacent polymer chains provide additional tensile strength and stiffness to the fibers.

Adipic acid, in turn, is a starting material for the synthesis of hexamethylenediamine. Treatment of adipic acid with ammonia gives an ammonium salt, which, when heated, gives adipamide. Catalytic reduction of adipamide gives hexamethylenediamine. Thus, carbon sources for the production of nylon 66 are derived entirely from petroleum, unfortunately a nonrenewable resource.

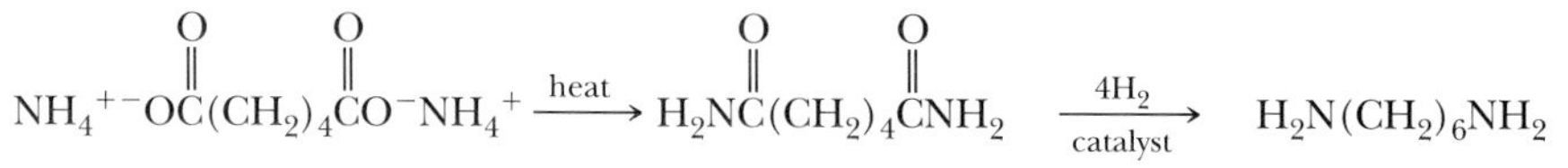

$$NH_4^{+}\,^{-}OC(=O)(CH_2)_4C(=O)O^{-}\,NH_4^{+} \xrightarrow{\text{heat}} H_2NC(=O)(CH_2)_4C(=O)NH_2 \xrightarrow[\text{catalyst}]{4H_2} H_2N(CH_2)_6NH_2$$

Ammonium hexanedioate (Ammonium adipate) — Hexanediamide (Adipamide) — 1,6-Hexanediamine (Hexamethylenediamine)

The nylons are a family of polymers, the members of which have subtly different properties that suit them to one use or another. The two most widely used members of this family are nylon 66 and nylon 6. Nylon 6 is so named because it is synthesized from one six-carbon monomer, namely caprolactam, a seven-membered cyclic amide. In the synthesis of nylon 6, caprolactam is partially hydrolyzed to 6-aminohexanoic acid and then heated to 250°C to bring about polymerization. Nylon 6 is fabricated into fibers, brush bristles, rope, high-impact moldings, and tire cords.

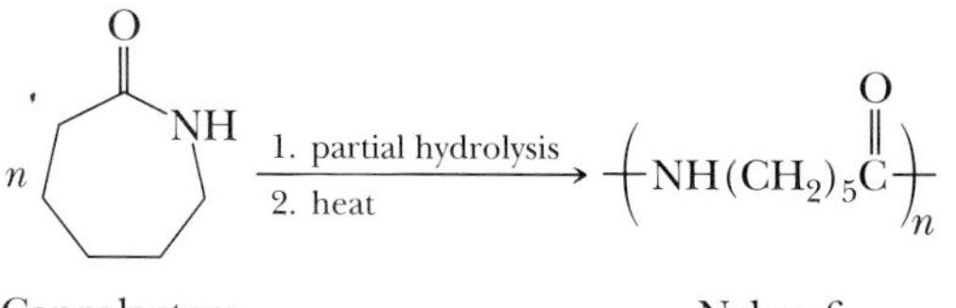

$$n\ \text{Caprolactam} \xrightarrow[\text{2. heat}]{\text{1. partial hydrolysis}} \left[NH(CH_2)_5C(=O) \right]_n$$

Caprolactam — Nylon 6

Aramid A polyaromatic amide; a polymer in which the monomer units are an aromatic diamine and an aromatic dicarboxylic acid.

Based on extensive research into relationships between molecular structure and bulk physical properties, scientists at DuPont reasoned that a polyamide containing aromatic rings would be stiffer and stronger than either nylon 66 or nylon 6, and in early 1960, DuPont introduced Kevlar, a polyaromatic amide **(aramid)** fiber synthesized from terephthalic acid and *p*-phenylenediamine.

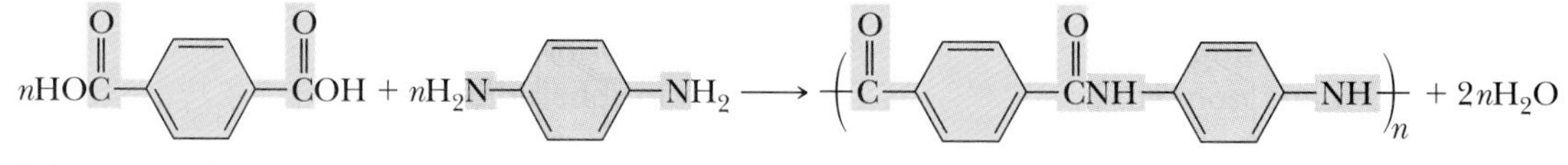

$$nHOC(=O)-C_6H_4-C(=O)OH + nH_2N-C_6H_4-NH_2 \longrightarrow \left[C(=O)-C_6H_4-C(=O)NH-C_6H_4-NH \right]_n + 2nH_2O$$

1,4-Benzenedicarboxylic acid (Terephthalic acid) — 1,4-Benzenediamine (*p*-Phenylenediamine) — Kevlar

One of the remarkable features of Kevlar is its light weight compared with other materials of similar strength. For example, a 3-in. cable woven of Kevlar has a strength equal to that of a similarly woven 3-in. steel cable. Whereas the steel cable weighs about 20 lb/ft, the Kevlar cable weighs only 4 lb/ft. Kevlar now finds use in such articles as anchor cables for offshore drilling rigs and reinforcement fibers for automobile tires. Kevlar is also woven into a fabric that is so tough that it can be used for bulletproof vests, jackets, and raincoats.

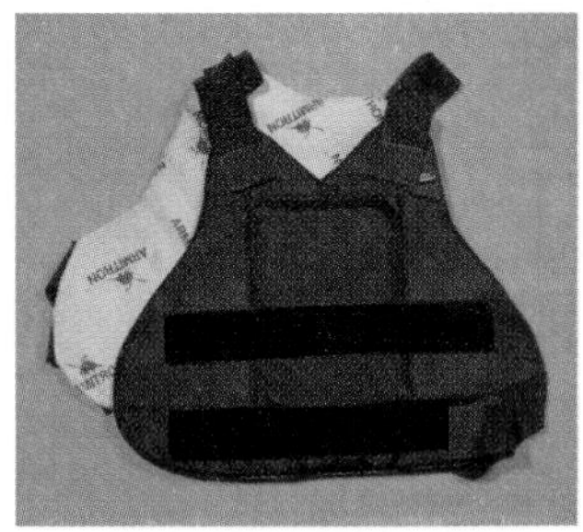

Bulletproof vests have a thick layer of Kevlar. *(Charles D. Winters)*

B. Polyesters

Recall that, in the early 1930s, Carothers and his associates had concluded that polyester fibers from aliphatic dicarboxylic acids and ethylene glycol were not suitable for textile use because their melting points are too low. Winfield and Dickson at the Calico Printers Association in England further investigated polyesters in the 1940s and reasoned that a greater resistance to rotation in the polymer backbone would stiffen the polymer, raise its melting point, and thereby lead to a more acceptable polyester fiber. To create stiffness in the polymer chain, they used terephthalic acid. Polymerization of this aromatic dicarboxylic acid with ethylene glycol gives poly(ethylene terephthalate), abbreviated PET (also PETE).

Polyester A polymer in which each monomer unit is joined to the next by an ester bond, as for example, poly(ethylene terephthalate).

$$n\,HO\overset{O}{\overset{\|}{C}}-C_6H_4-\overset{O}{\overset{\|}{C}}OH + n\,HOCH_2CH_2OH \xrightarrow{\text{heat}} \left[\overset{O}{\overset{\|}{C}}-C_6H_4-\overset{O}{\overset{\|}{C}}OCH_2CH_2O\right]_n + 2n\,H_2O$$

1,4-Benzenedicarboxylic acid (Terephthalic acid) 1,2-Ethanediol (Ethylene glycol) Poly(ethylene terephthalate) (Dacron, Mylar)

The crude polyester can be melted, extruded, and then cold-drawn to form the textile fiber Dacron polyester, outstanding features of which are its stiffness (about four times that of nylon 66), very high strength, and remarkable resistance to creasing and wrinkling. Because the early Dacron polyester fibers were harsh to the touch (due to their stiffness), they were usually blended with cotton or wool to make acceptable textile fibers. Newly developed fabrication techniques now produce less harsh Dacron polyester textile fibers. PET is also fabricated into Mylar films and recyclable plastic beverage containers.

Ethylene glycol for the synthesis of PET is obtained by air oxidation of ethylene to ethylene oxide (Section 11.8A) followed by hydrolysis to the glycol (Section 11.9A).

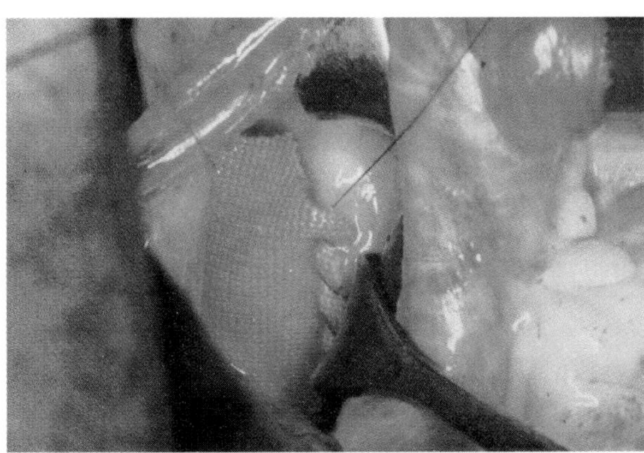

A Dacron patch used in heart surgery. *(Courtesy of Drs. James L. Monro and Gerald Shore and Wolfe Medical Publications, London, England)*

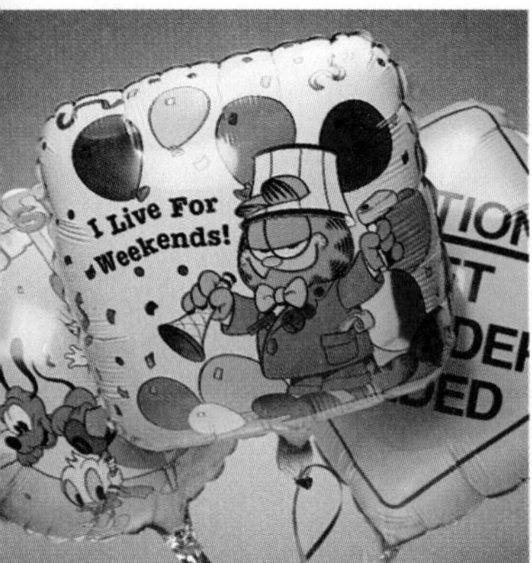

Mylar, a polyester, can be made into extremely strong thin films. Because the film has very tiny pores, it is used for balloons that can be inflated and filled with helium; the helium atoms diffuse through the pores of the film only slowly. *(Charles D. Winters)*

Ethylene is, in turn, derived entirely from cracking either petroleum or ethane derived from natural gas (Section 2.9). Terephthalic acid is obtained by oxidation of *p*-xylene, an aromatic hydrocarbon obtained along with benzene and toluene from catalytic cracking and reforming of naphtha and other petroleum fractions (Section 2.9B).

$$\underset{\text{Ethylene}}{CH_2{=}CH_2} \xrightarrow[\text{catalyst}]{O_2} \underset{\substack{\text{Oxirane}\\\text{(Ethylene oxide)}}}{\overset{O}{CH_2{-}CH_2}} \xrightarrow{H^+,\ H_2O} \underset{\substack{\text{1,2-Ethanediol}\\\text{(Ethylene glycol)}}}{HOCH_2CH_2OH}$$

$$\underset{p\text{-Xylene}}{CH_3{-}C_6H_4{-}CH_3} \xrightarrow[\text{catalyst}]{O_2} \underset{\text{Terephthalic acid}}{HO\overset{O}{\overset{\|}{C}}{-}C_6H_4{-}\overset{O}{\overset{\|}{C}}OH}$$

C. Polycarbonates

Polycarbonate A polyester in which the carboxyl groups are derived from carbonic acid.

Polycarbonates, the most familiar of which is Lexan, are a class of commercially important engineering polyesters. In the production of Lexan, an aqueous solution of the disodium salt of bisphenol A (Problem 20.22) is brought into contact with a solution of phosgene dissolved in methylene chloride. The two solutions are immiscible and no reaction occurs until tetrabutylammonium chloride or other phase-transfer catalyst (Section 8.7) is added. The tetrabutylammonium cation carries the bisphenol A anion into the methylene chloride phase where it reacts smoothly with phosgene to form the polymer. The tetrabutylammonium ion then carries chloride ion back to the aqueous phase.

$$\underset{\substack{\text{Disodium salt}\\\text{of bisphenol A}}}{{}^+Na^-O{-}C_6H_4{-}C(CH_3)_2{-}C_6H_4{-}O^-Na^+} + \underset{\text{Phosgene}}{Cl\overset{O}{\overset{\|}{C}}Cl} \longrightarrow \underset{\substack{\text{Lexan}\\\text{(a polycarbonate)}}}{\left(C_6H_4{-}C(CH_3)_2{-}C_6H_4{-}O\overset{O}{\overset{\|}{C}}O\right)_n} + 2NaCl$$

Lexan is a tough, transparent polymer with high impact and tensile strengths, and retains its properties over a wide temperature range. It has found significant use in sporting equipment, such as bicycle, football, motorcycle, and snowmobile helmets as well as hockey and baseball catchers' face masks. In addition it is used to make light, impact-resistant housings for household appliances and automobile and aircraft equipment, and in the manufacture of safety glass and unbreakable windows (now unfortunately common in schools).

D. Polyurethanes

A urethane, or carbamate, is an ester of carbamic acid, H_2NCO_2H. Carbamates are most commonly prepared by treatment of an isocyanate with an alcohol.

$$\underset{\text{An isocyanate}}{RN{=}C{=}O} + R'OH \longrightarrow \underset{\text{A carbamate}}{RNH\overset{O}{\overset{\|}{C}}OR'}$$

Polyurethanes consist of flexible polyester or polyether units (blocks) alternating with rigid urethane units (blocks). The rigid urethane blocks are derived from a diisocyanate, commonly a mixture of 2,4- and 2,6-toluene diisocyanate. The more flexible blocks are derived from low-molecular-weight (MW 1000 to 4000) polyesters or polyethers with —OH groups at each end of the polymer chain. Polyurethane fibers are fairly soft and elastic and have found use as spandex and Lycra, the "stretch" fabrics used in bathing suits, leotards, and undergarments.

Polyurethane A polymer containing the $-HNCO_2-$ group as a repeating unit.

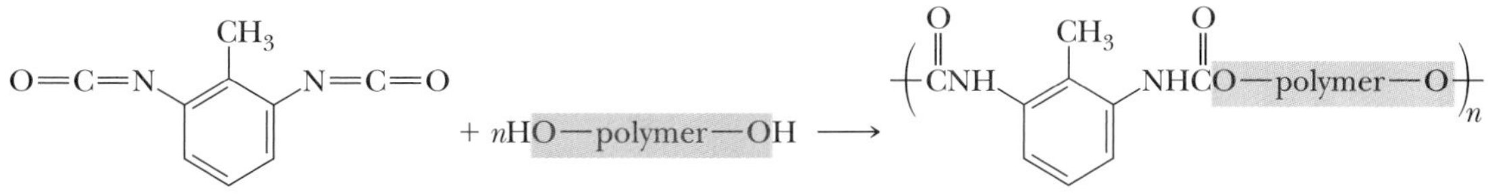

2,6-Toluene diisocyanate + nHO—polymer—OH ⟶ A polyurethane

Low-molecular-weight polyester or polyether

Polyurethane foams for upholstery and insulating materials are made by adding small amounts of water during polymerization. Water reacts with isocyanate groups to produce gaseous carbon dioxide which then acts as the foaming agent.

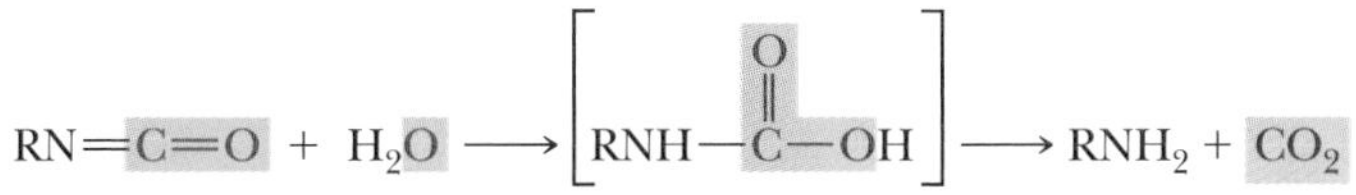

$$RN{=}C{=}O + H_2O \longrightarrow \left[RNH{-}\overset{O}{\overset{\|}{C}}{-}OH\right] \longrightarrow RNH_2 + CO_2$$

An isocyanate — A carbamic acid (unstable)

E. Epoxy Resins

Epoxy resins are materials prepared by a polymerization in which one monomer contains at least two epoxy groups. Within this range, there are a large number of polymeric materials possible, and epoxy resins are produced in forms ranging from low viscosity liquids to high melting solids. The most widely used epoxide monomer is the diepoxide prepared by treatment of 1 mol of bisphenol A (Section 23.5C) with 2 mol of epichlorohydrin (Problem 19.51). To prepare the following epoxy resin, the diepoxide monomer is treated with 1,2-ethanediamine (ethylene diamine). This component is usually called the catalyst in the two-component formulations that can be bought in any hardware store; it is also the component with the acrid smell.

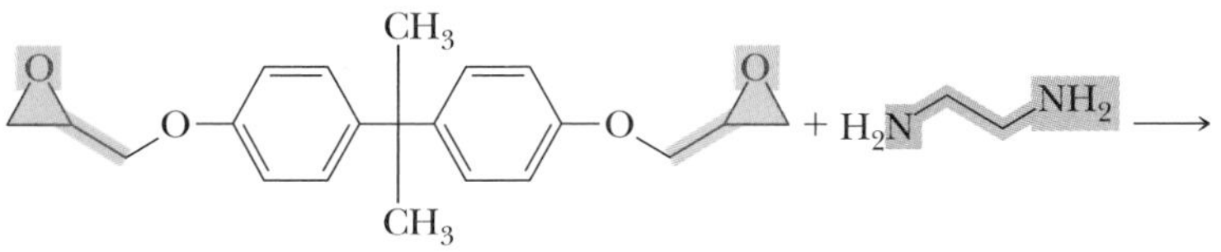

A diepoxide + A diamine ⟶

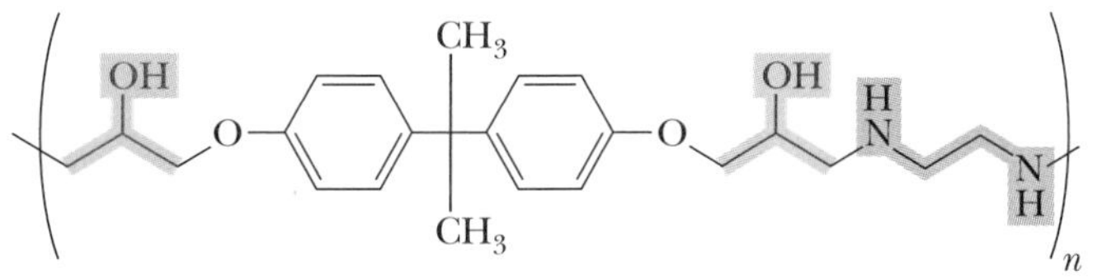

A diepoxide

Epoxy resins are widely used as adhesives and insulating surface coatings. They have good electrical insulating properties which leads to their use for encapsulating electrical components ranging from integrated circuit boards to switch coils and insulators for power transmission systems. They are also used as composites with other materials, such as glass fiber, paper, metal foils, and other synthetic fibers to create structural components for jet aircraft, rocket motor casings, and so on.

EXAMPLE 23.2

Write a mechanism for the acid-catalyzed polymerization of 1,4-benzenediisocyanate and ethylene glycol. To simplify your mechanism, consider only the reaction of one —NCO group with one —OH group.

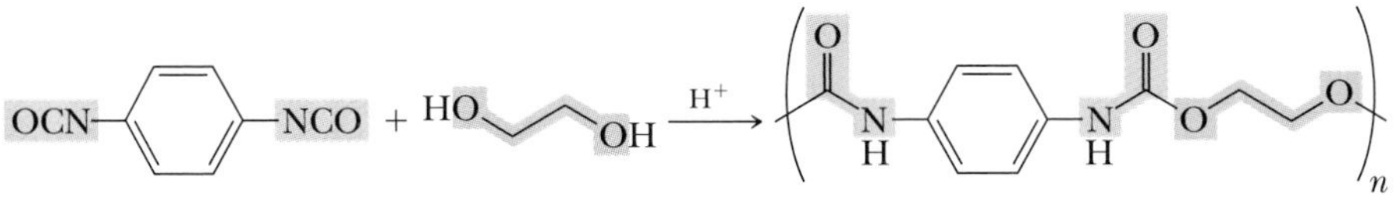

OCN–C₆H₄–NCO + HO–CH₂CH₂–OH $\xrightarrow{H^+}$ (C(=O)–NH–C₆H₄–NH–C(=O)–O–CH₂CH₂–O)$_n$

1,4-Benzenediisocyanate   Ethylene glycol   A polyurethane

Solution

A mechanism is shown in three steps. Proton transfer in Step (1) from the acid, HA, to nitrogen followed by (2) addition of ROH to the carbonyl gives an oxonium ion. Proton transfer in Step (3) from the oxonium ion to A^- gives the carbamate ester.

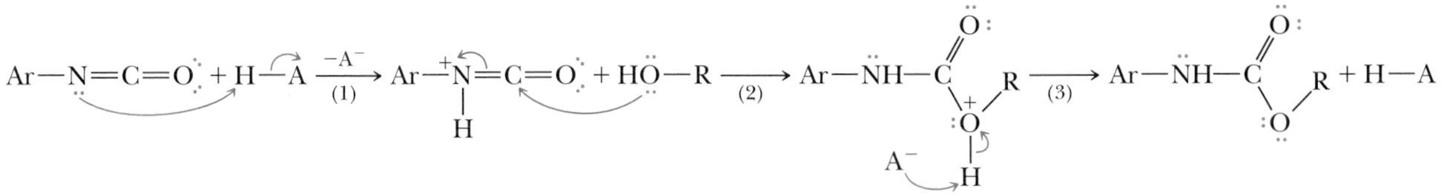

$$\text{Ar—N=C=O} + \text{H—A} \xrightarrow[(1)]{-A^-} \text{Ar—}\overset{+}{\text{N}}\text{H=C=O} + \text{HO—R} \xrightarrow[(2)]{} \text{Ar—NH—C(=O)—}\overset{+}{\text{O}}\text{(H)—R} \xrightarrow[(3)]{A^-} \text{Ar—NH—C(=O)—O—R} + \text{H—A}$$

PROBLEM 23.2

Write the repeating unit of the polymer formed from the following reaction and propose a mechanism for its formation.

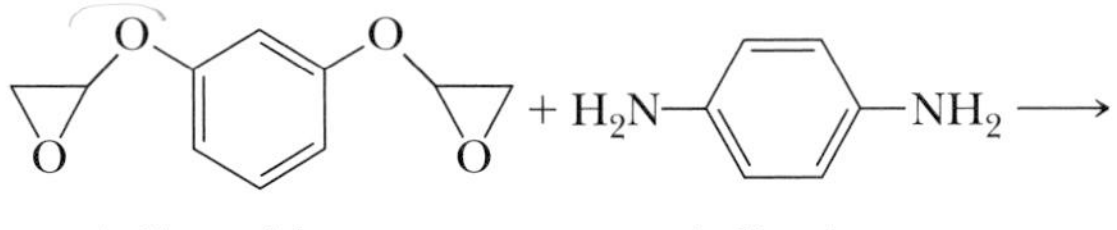

A diepoxide + H_2N–C₆H₄–NH_2 ⟶

A diepoxide   A diamine

23.6 Chain-Growth Polymerizations

Chain-growth polymerization A polymerization that involves sequential addition reactions, either to unsaturated monomers or to monomers possessing other reactive functional groups.

Chain-growth polymerizations involve sequential addition reactions, either to unsaturated monomers or to monomers possessing other reactive functional groups. This mechanism of polymerization differs greatly from the mechanism of step-growth polymerizations. In the latter, all monomers plus the polymer endgroups possess equally reactive functional groups, allowing for all possible combination reactions to

CHEMISTRY IN ACTION

Stitches That Dissolve

Medical science has advanced very rapidly in the last few decades. Some procedures considered routine today, such as organ transplantation and the use of lasers in surgery, were unimaginable 60 years ago. As the technological capabilities of medicine have grown, the demand for synthetic materials that can be used inside the body has increased as well. Polymers have many of the characteristics of an ideal biomaterial: they are lightweight and strong, inert or biodegradable depending on their chemical structure, and have physical properties (softness, rigidity, elasticity) that are easily tailored to match those of natural tissues. Carbon-carbon backbone polymers are degradation resistant and are used widely in permanent organ and tissue replacements.

Whereas most medical uses of polymeric materials require biostability, applications have been developed that use the biodegradable nature of some macromolecules. An example is the use of poly(glycolic acid) and glycolic acid/lactic acid copolymers as absorbable sutures.

$HOCH_2C(=O)OH$ (Glycolic acid) + $HOCH(CH_3)C(=O)OH$ (Lactic acid) $\xrightarrow[-nH_2O]{\text{copolymerization}}$ $-\!\!\left[CH_2C(=O)OCH(CH_3)C(=O)O\right]\!\!-_n$

Glycolic acid

Lactic acid

A copolymer of poly(glycolic acid)-poly(lactic acid)

Traditional suture materials such as catgut must be removed by a health care specialist after they have served their purpose. Stitches of these polyhydroxyacids, however, are hydrolyzed slowly over a period of approximately 2 weeks, and by the time the torn tissues have fully healed, the stitches have fully degraded and no suture removal is necessary. Glycolic and lactic acids formed during hydrolysis of the stitches are metabolized and excreted by existing biochemical pathways.

occur, including monomer with monomer, monomer with dimer, and so forth. In contrast, chain-growth polymerizations involve endgroups possessing reactive intermediates that react with monomer only. The reactive intermediates used in chain-growth polymerizations include radicals, carbanions, carbocations, and organometallic complexes.

The number of monomers that undergo chain-growth polymerizations is large and includes such compounds as alkenes, alkynes, allenes, isocyanates, and cyclic compounds, such as lactones, lactams, ethers, and epoxides. We concentrate on the chain-growth polymerizations of ethylene and substituted ethylenes, and show how these compounds can be polymerized by radical, cation, anion, and organometallic-mediated mechanisms.

$CH_2{=}CHR \longrightarrow -\!\!\left[CH_2CH(R)\right]\!\!-_n$

An alkene

Table 23.1 lists several important polymers derived from ethylene and substituted ethylenes along with their common names and most important uses.

Petri dishes made of Plexiglas.
(© Dan McCoy, Rainbow)

TABLE 23.1 Polymers Derived from Substituted Ethylenes

Monomer Formula	Common Name	Polymer Name(s) and Common Uses
$CH_2{=}CH_2$	ethylene	polyethylene, Polythene: break-resistant containers and packaging materials
$CH_2{=}CHCH_3$	propylene	polypropylene, Herculon: textile and carpet fibers
$CH_2{=}CHCl$	vinyl chloride	poly(vinyl chloride), PVC: construction tubing
$CH_2{=}CCl_2$	1,1-dichloroethylene	poly(1,1-dichloroethylene), Saran: food packaging
$CH_2{=}CHCN$	acrylonitrile	polyacrylonitrile, Orlon: acrylics and acrylates
$CF_2{=}CF_2$	tetrafluoroethylene	poly(tetrafluoroethylene), Teflon: nonstick coatings
$CH_2{=}CHC_6H_5$	styrene	polystyrene, Styrofoam: insulating materials
$CH_2{=}CHCO_2CH_2CH_3$	ethyl acrylate	poly(ethyl acrylate): latex paints
$CH_2{=}C(CH_3)CO_2CH_3$	methyl methacrylate	poly(methyl methacrylate), Lucite, Plexiglas: glass substitutes

A. Radical Chain-Growth Polymerizations

Among the initiators used for radical chain-growth polymerizations are diacyl peroxides, such as dibenzoyl peroxide, which decompose as shown upon heating. In the first step, homolytic cleavage of the weak O—O peroxide bond yields two acyloxy radicals. Each acyloxy radical then decomposes to form an aryl radical and CO_2.

$$C_6H_5C(=O)O{-}OC(=O)C_6H_5 \xrightarrow{\Delta} 2\,C_6H_5C(=O)O\cdot \longrightarrow 2\,C_6H_5\cdot + 2CO_2$$

Dibenzoyl peroxide — A benzoyloxy radical — A phenyl radical

Another common class of initiators used in radical polymerizations are azo compounds, such as azoisobutyronitrile (AIBN), which decompose upon heating or by the absorption of UV light to produce alkyl radicals and nitrogen gas.

$$(CH_3)_2C(C{\equiv}N){-}N{=}N{-}C(C{\equiv}N)(CH_3)_2 \xrightarrow{\Delta \text{ or } h\nu} 2\,(CH_3)_2\dot{C}(C{\equiv}N) + :N{\equiv}N:$$

Azoisobutyronitrile (AIBN) — Alkyl radicals

The chain initiation, propagation, and termination steps for radical polymerization of a substituted ethylene monomer are shown for the monomer $RCH{=}CH_2$.

Dissociation of the initiator produces a radical that reacts with the double bond of a monomer. Once initiated, the chains continue to propagate through successive additions of monomers.

MECHANISM Radical Polymerization of a Substituted Ethylene

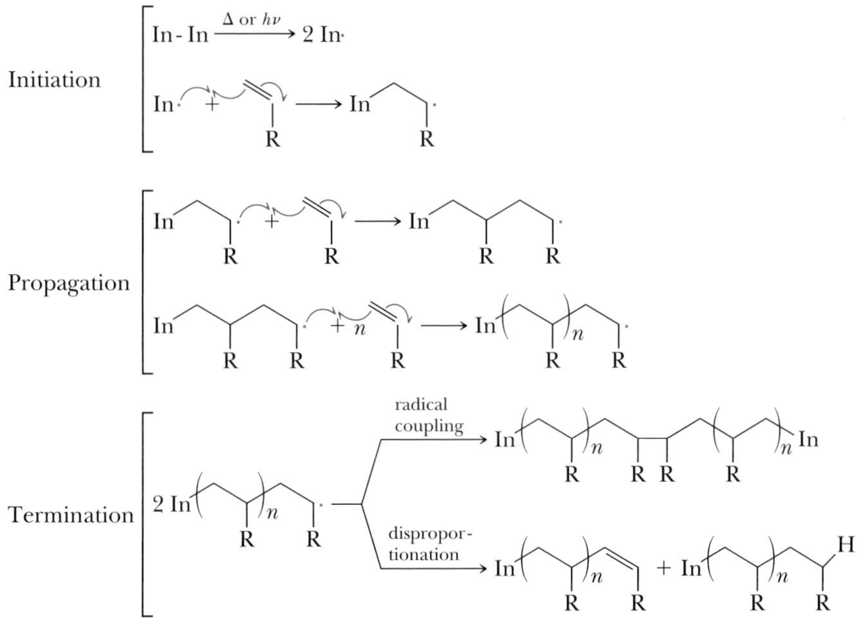

Radical reactions with double bonds almost always give the more stable (more substituted) radical. Because additions are biased in this fashion, the polymerizations of vinyl monomers tend to yield polymers with head-to-tail linkages. Vinyl polymers made by radical processes generally have no more than 1% to 2% head-to-head linkages.

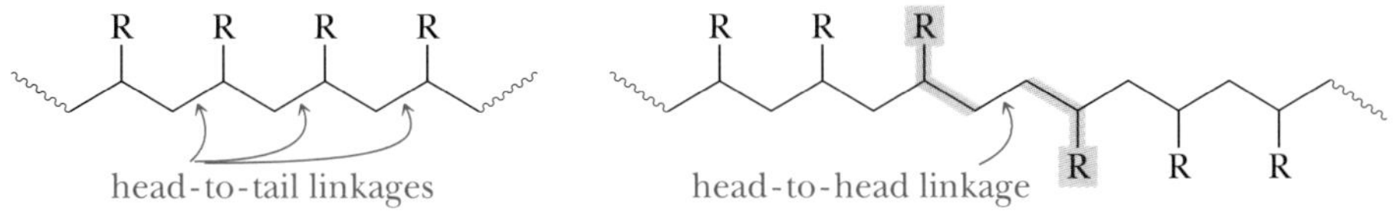

In radical reactions, the chain termination involves combination of radicals to produce a nonradical molecule or molecules. One common termination step is **radical coupling** to form a new carbon-carbon bond linking two growing polymer chains. This type of termination step is a diffusion-controlled process that occurs without an activation energy barrier. Another common termination process is **disproportionation,** which involves the abstraction of a hydrogen atom from the beta position to the prop-

Disproportionation A termination process that involves the abstraction of a hydrogen atom from the beta position to the propagating radical of one chain by the radical endgroup of another chain.

A low-density polyethylene (LDPE) bag. *(Charles D. Winters)*

agating radical of one chain by the radical endgroup of another chain. This process results in two dead chains, one terminated in an alkyl group and the other in an alkenyl group.

Because organic radicals are highly reactive species, it is not surprising that radical polymerizations are often complicated by unwanted side reactions. A frequently observed side reaction is hydrogen abstraction by the radical endgroup from a growing polymer chain, a solvent molecule, or another monomer. These side reactions are called **chain-transfer reactions** because the activity of the endgroup is "transferred" from one chain to another.

Chain-transfer reactions The reactivity of an endgroup is "transferred" from one chain to another during the course of a polymerization.

Chain transfer is illustrated by radical polymerization of ethylene. Polyethylene formed by radical polymerization exhibits a number of butyl branches on the polymer main chain. These four-carbon branches are generated in a "back-biting" chain-transfer reaction in which the radical endgroup abstracts a hydrogen from the fourth carbon back in a favorable six-membered ring transition state. Abstraction of this hydrogen is particularly facile because the transition state associated with the process can adopt a conformation like that of a chair cyclohexane. Continued polymerization of monomer from this new radical center leads to branches four carbons long.

H → H $\xrightarrow{n\,CH_2{=}CH_2}$ n

A six-membered transition state leading to 1,5-hydrogen abstraction

As a result of these various abstraction reactions, polymers synthesized by radical processes can have highly branched structures. The number of butyl branches depends on the relative stability of the propagating-radical endgroup and varies depending on the polymer. Polyethylene chains propagate through highly reactive primary radicals which tend to be susceptible to 1,5-hydrogen abstraction reactions; these polymers typically have 15 to 30 branches per 500 monomer units. In contrast, polystyrene chains propagate through substituted benzyl radicals, which are stabilized by delocalization of the unpaired electron over the aromatic ring. These stabilized radicals are less likely to undergo hydrogen abstraction reactions. Polystyrene typically exhibits only one branch per 4,000 to 10,000 monomer units.

The first commercial process for ethylene polymerization used peroxide catalysts at temperatures of 500°C and pressures of 1000 atm and produced a soft, tough polymer known as low-density polyethylene (LDPE). At the molecular level, chains of LDPE are highly branched due to chain-transfer reactions. Because this extensive chain branching prevents polyethylene chains from packing efficiently, LDPE is largely amorphous and transparent, with only a small amount of crystallites of a size too small to scatter light. LDPE has a density between 0.91 and 0.94 g/cm^3 and a melting point (T_m) of about 108°C. Because its melting point is only slightly above 100°C, it cannot be used for products that will be exposed to boiling water.

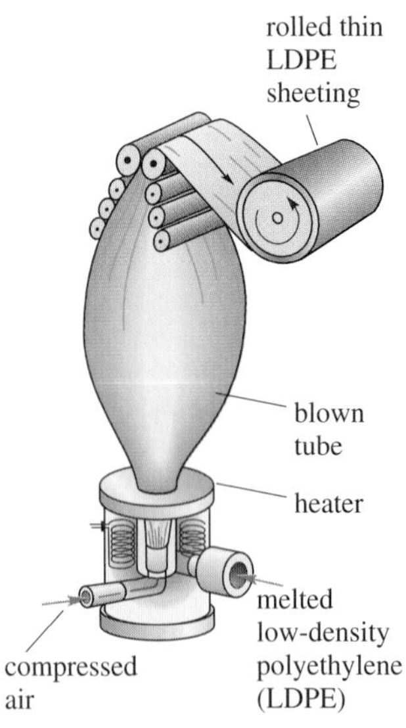

FIGURE 23.4
Fabrication of an LDPE film.

Approximately 65% of all low-density polyethylene is used for the manufacture of films. Fabrication of LDPE films is done by a blow-molding technique illustrated in Figure 23.4. A tube of molten LDPE along with a jet of compressed air is forced through an opening and blown into a giant, thin-walled bubble. The film is then cooled and taken up onto a roller. This double-walled film can be slit down the side

Chemistry in Action

Organic Polymers That Conduct Electricity

The influence of chemical structure on the properties of an organic compound is clearly seen in the electrical conducting properties of certain organic polymers. Most organic polymers are insulators. Polytetrafluoroethylene, for example, with the repeating unit $—CF_2CF_2—$ or polyvinyl chloride with the repeating unit $—CH_2CHCl—$ have conductivities of 10^{-18} S/cm. On the other end of the scale, the conductivity of copper is almost 10^6 S/cm.

Can organic polymers approach the conductivity of copper? Research carried out over the last 20 years shows that the answer is yes. When acetylene is passed through a solution containing certain transition metal catalysts, it can be polymerized to a shiny film of polyacetylene. The approximate structure of this polymer is as follows:

Polyacetylene

By itself, polyacetylene is not a conductor. However, by a process called doping, which involves introducing small amounts of electron-donating or electron-accepting compounds, it is possible to produce a polyacetylene that shows a conductivity of 1.5×10^5 S/cm.

The purpose of the doping agent is either to remove electrons from the pi system (*p*-doping) or add electrons to the pi system (*n*-doping). A *p*-doped polyacetylene can be represented as a conjugated polyalkene chain containing positively charged carbons at several points along the chain:

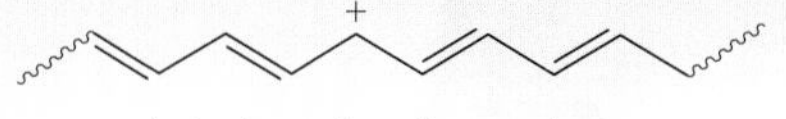

A *p*-doped polyacetylene

We can think of the positive charge as a defect that can move to the left or to the right along the polymer chain, thus giving rise to conductivity.

In crude polyacetylene, the polymer chains are jumbled, pointing in all directions. However, by stretching the film, the chains can be made to line up in a more ordered fashion. The conductivity of doped and oriented polyacetylene chains is greater along the direction of the chain than it is perpendicular to the chain. This result suggests that it is much easier for electrons to travel along a chain than to hop from one chain to the next.

Applications for conducting organic polymers are beginning to be developed. A rechargeable battery with electrodes of *p*-doped and *n*-doped polyacetylene already has been produced. Given the atomic weight of carbon, organic polymer batteries should be lighter than nickel-cadmium or lead-acid batteries. Weight is an important consideration if battery powered electric cars are ever to be made practical. In addition, many metals used in today's batteries (mercury, nickel, lead) are toxic. If research leads to practical organic batteries, waste disposal problems could be considerably lessened.

See R. B. Kaner and A. G. MacDiarmid, *Scientific American*, **258,** Feb., 106, 1988.

to give LDPE film or it can be sealed at points along its length to make LDPE bags. LDPE film is inexpensive, which makes it ideal for packaging such consumer items as baked goods, vegetables and other produce, and for trash bags.

B. Ziegler-Natta Chain-Growth Polymerizations

An alternative method for polymerization of ethylene, which does not involve radicals, was developed by Karl Ziegler of Germany and Giulio Natta of Italy in the 1950s. The early Ziegler-Natta catalysts were highly active, heterogeneous catalysts composed of a

Polyethylene films are produced by extruding the molten plastic through a ring-like gap and inflating the film into a balloon. *(The Stock Market)*

$MgCl_2$ support, a Group IVB transition metal halide, such as $TiCl_4$, and an alkylaluminum compound, such as $Al(CH_2CH_3)_2Cl$. These catalysts brought about polymerization of ethylene and propylene at 1 to 4 atmospheres and at temperatures as low as 60°C. Polymerizations under these conditions do not involve radicals.

$$\underset{\text{Ethylene}}{CH_2{=}CH_2} \xrightarrow[MgCl_2]{TiCl_4/Al(CH_2CH_3)_2Cl} \underset{\text{Polyethylene}}{\left(CH_2CH_2\right)_n}$$

The active catalyst in a Ziegler-Natta polymerization is thought to be an alkyltitanium compound that is formed by alkylation of the titanium halide by $Al(CH_2CH_3)_2Cl$ on the surface of a $MgCl_2/TiCl_4$ particle. Once formed, this species repeatedly inserts ethylene into the titanium-carbon bond to yield polyethylene.

MECHANISM Ziegler-Natta Catalysis of Alkene Polymerization

Step 1: Formation of a titanium-ethyl bond

$MgCl_2/TiCl_4$ particle

$$-Ti-Cl + Al(CH_2CH_3)_2Cl \longrightarrow -Ti-CH_2CH_3 + Al(CH_2CH_3)Cl_2$$

Step 2: Insertion of ethylene into the titanium-carbon bond

$$-Ti-CH_2CH_3 + CH_2{=}CH_2 \longrightarrow -Ti-CH_2CH_2CH_2CH_3$$

Over 60 billion pounds of polyethylene are produced worldwide every year using optimized Ziegler-Natta catalysts, and large-scale reactors can yield up to 60,000 pounds of polyethylene per hour. Production of polymer at this scale is partly due to the mild conditions required for a Ziegler-Natta polymerization and the fact that the polymer obtained has substantially different physical and mechanical properties from that obtained by radical polymerization. Polyethylene from Ziegler-Natta systems, termed **high-density polyethylene (HDPE),** has a higher density ($0.96\ g/cm^3$) and melting point (133°C) than low-density polyethylene, is three to ten times stronger, and is opaque rather than transparent. This added strength and opacity is due to a much lower degree of chain branching and the resulting higher degree of crystallinity of HDPE compared with LDPE.

Approximately 45% of all HDPE used in the United States is blow molded. In blow molding, a short length of HDPE tubing is placed in an open die [Figure 23.5(a)] and the die is closed, sealing the bottom of the tube. Compressed air is then forced into the hot polyethylene/die assembly, and the tubing is literally blown up to take

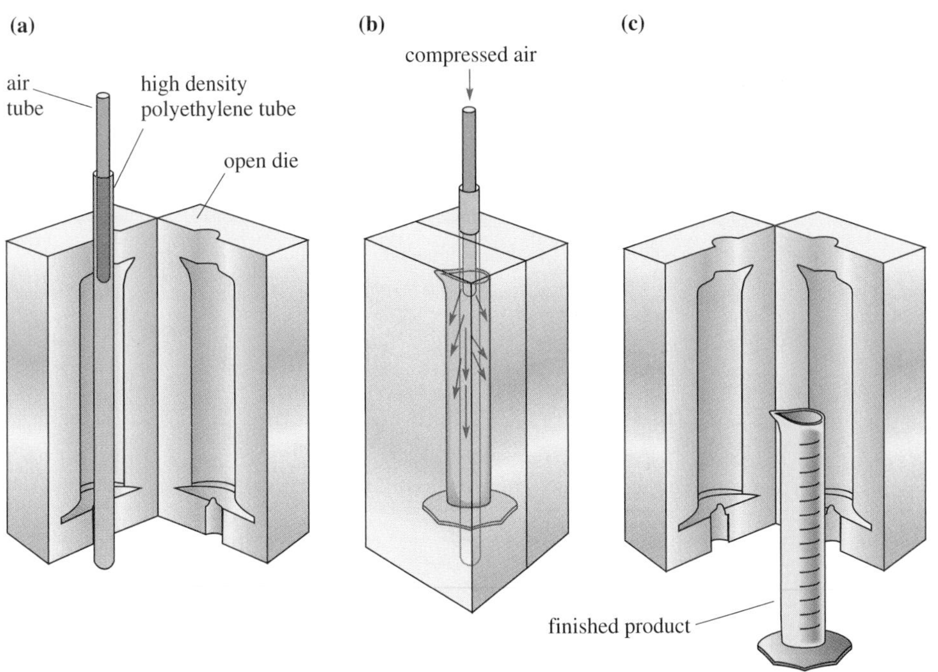

FIGURE 23.5
Blow molding of an HDPE container.

the shape of the mold [Figure 23.5(b)]. After cooling, the die is opened [Figure 23.5(c)] and there is the container!

Even greater improvements in properties of HDPE can be realized through special processing techniques. In the melt state, HDPE chains adopt random coiled conformations similar to those of cooked spaghetti. Engineers have developed special extrusion techniques that force the individual polymer chains of HDPE to uncoil and adopt an extended zig-zag conformation. These extended chains then align with one another to form highly crystalline materials. HDPE processed in this fashion is stiffer than steel and has approximately four times the tensile strength of steel! Because the density of polyethylene ($\approx$1.0 g/cm^3) is considerably less than that of steel (8.0 g/cm^3), these comparisons of strength and stiffness are even more favorable if they are made on a weight basis.

In recent years there have been several important advances made in catalysts used in Ziegler-Natta type polymerizations. One of the most important has been the discovery of soluble complexes that catalyze the polymerization of ethylene and propylene at extraordinary rates. Because these new homogeneous catalysts are substantially different in structure from the early Ziegler-Natta systems, these polymerizations are referred to as **coordination polymerizations.** Catalysts for coordination polymerizations are frequently formed by allowing bis(cyclopentadienyl)dimethylzirconium, [$Cp_2Zr(CH_3)_2$], to react with methaluminoxane (MAO). MAO is a complex mixture of methylaluminum oxide oligomers, [$-(CH_3)AlO-)_n$], formed by allowing trimethylaluminum to react with small amounts of water. It is thought that MAO activates the zirconium by abstracting a methyl anion to form a zirconium cation that is the active polymerization catalyst.

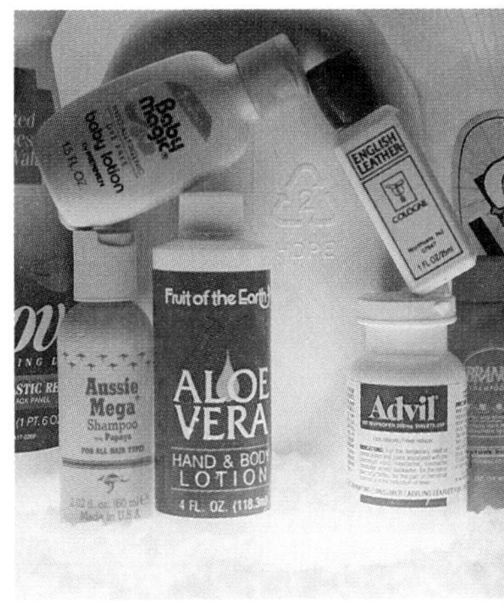

Some common products packaged in high-density polyethylene (HDPE). The symbol for recyclable HDPE materials has a 2 in the center of the triangle of curved arrows and the letters HDPE under the symbol. *(Charles D. Winters)*

MECHANISM Homogeneous Catalysis for Ziegler-Natta Coordination Polymerization

Step 1: Activation of the zirconium catalyst

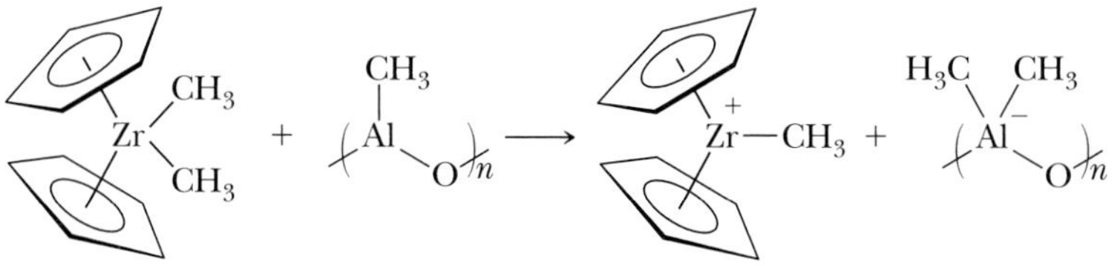

Bis(cyclopentadienyl)-dimethylzirconium + MAO ⟶ A zirconium cation (the active form of the catalyst) + ...

Bis(cyclopentadienyl)-dimethylzirconium — MAO — A zirconium cation (the active form of the catalyst)

Step 2: Insertion of ethylene monomers into the zirconium-carbon bond

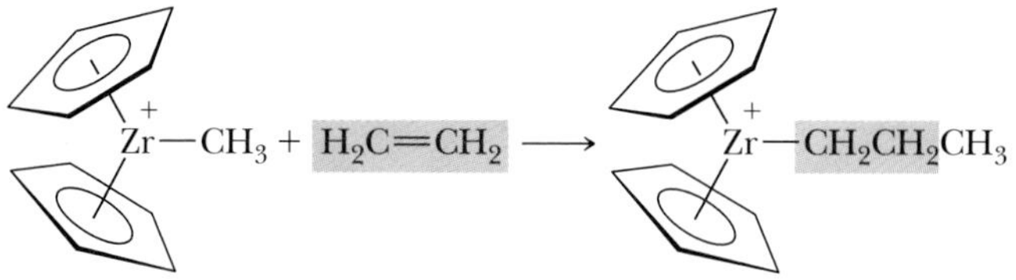

$Cp_2Zr^+{-}CH_3 + H_2C{=}CH_2 \longrightarrow Cp_2Zr^+{-}CH_2CH_2CH_3$

Some of these coordination-polymerization catalysts polymerize up to 20,000 ethylene monomer units per second, a rate reached only by enzyme-catalyzed biological reactions. Another important characteristic of these catalysts is that they show high reactivity toward 1-alkenes, allowing the formation of copolymers, such as that of ethylene and 1-hexene.

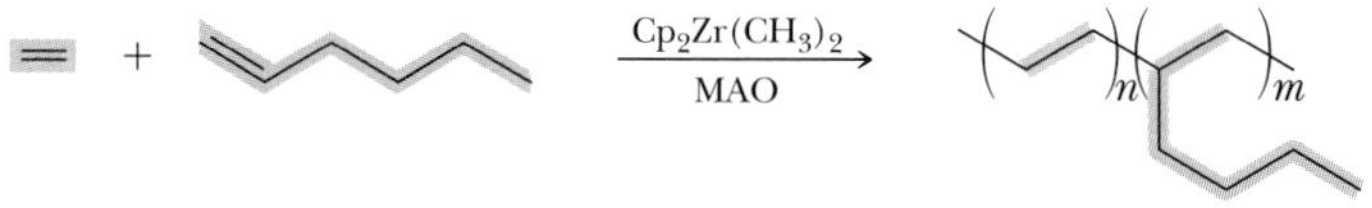

Ethylene + 1-Hexene $\xrightarrow[\text{MAO}]{Cp_2Zr(CH_3)_2}$ Ethylene-(1-hexene) copolymer

Copolymers of this type with these moderate length branches (C_4, C_6, and so on) are called linear low-density polyethylene, or LLDPE. These are useful materials because they have many of the properties of LDPE made from radical reactions but are formed at the substantially milder conditions associated with Ziegler-Natta polymerizations.

C. Stereochemistry and Polymers

Thus far, we have written the formula of a substituted ethylene polymer in the following manner and have not been concerned with the configuration of each stereocenter along the chain.

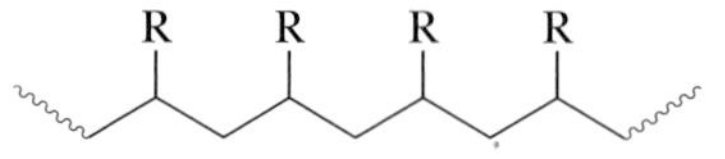

R R R R

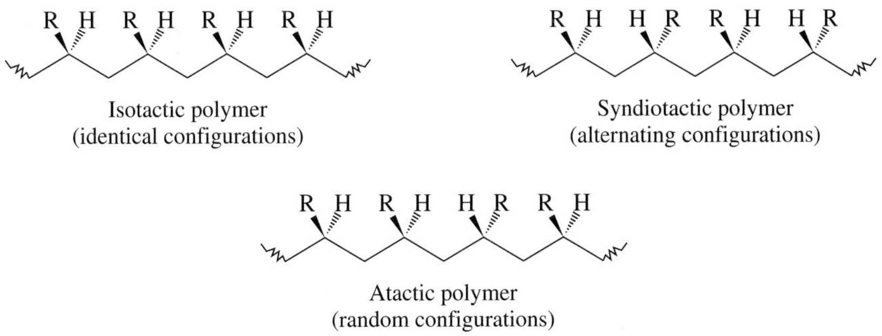

FIGURE 23.6
Relative configurations of stereocenters in polymers with different tacticities.

Nevertheless, the relative configurations of these stereocenters are important in determining the properties of a polymer. Polymers with identical configurations at all stereocenters along the chain are called **isotactic** polymers. Those with alternating configurations are called **syndiotactic** polymers, and those with completely random configurations are called **atactic** polymers (Figure 23.6).

Isotactic polymer A polymer with identical configurations (either all *R* or all *S*) at all stereocenters along its chain, as, for example, isotactic polypropylene.

Syndiotactic polymer A polymer with alternating *R* and *S* configurations at the stereocenters along its chain, as, for example, syndiotactic polypropylene.

Atactic polymer A polymer with completely random configurations at the stereocenters along its chain, as, for example, atactic polypropylene.

In general, the more stereoregular the centers are, that is, the more highly isotactic or highly syndiotactic the polymer is, the more crystalline it is. A random placement of the substituents, such as in atactic materials, results in a polymer that cannot pack well and is usually highly amorphous. Atactic polystyrene, for example, is an amorphous glass, whereas isotactic polystyrene is a crystalline fiber-forming polymer with a high melt transition. The control over the relative configuration, or tacticity, along a polymer backbone is, therefore, an area of important interest in modern polymer synthesis.

D. Ionic Chain-Growth Polymerizations

Chain-growth polymers can also be synthesized using reactions that rely on either anionic or cationic species in the propagation steps. The choice of ionic procedure depends greatly on the electronic nature of the monomers to be polymerized. Vinyl monomers with electron-withdrawing groups, which stabilize carbanions, are used in anionic polymerizations, whereas vinyl monomers with electron-donating groups, which stabilize cations, are used in cationic polymerizations (Table 23.2).

Styrene is conspicuous among the monomers given in Table 23.2 because it can be polymerized using either anionic or cationic techniques as well as radical techniques. This characteristic particular to styrene is due to the fact that the phenyl group can stabilize cationic, anionic, and radical benzylic intermediates.

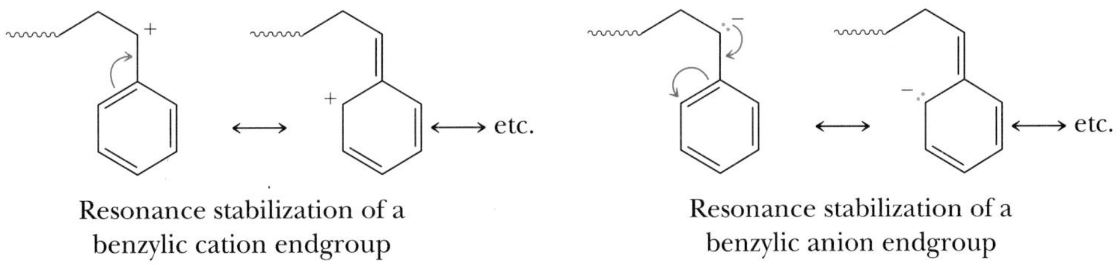

Resonance stabilization of a benzylic cation endgroup

Resonance stabilization of a benzylic anion endgroup

TABLE 23.2 Alkenes Polymerized by Anionic and Cationic Chain-Growth Mechanisms

Anionic polymerizations are most common for monomers subtituted with electron-withdrawing groups.

Styrene	CO_2R Alkyl methacrylates	CO_2R Alkyl acrylates	CN Acrylonitrile	CN, CO_2R Alkyl cyanoacrylates

Cationic polymerizations are most common for monomers substituted with electron-donating groups.

Styrene	Isobutylene	OR Vinyl ethers	SR Vinyl thioethers

Anionic Polymerizations

Anionic polymerizations can be initiated by addition of a nucleophile to an activated alkene. The most common nucleophiles used for this purpose are metal alkyls, such as methyl or *sec*-butyllithium.

The low thermal conductivity of polystyrene makes it a good insulating material. The symbol for recyclable polystyrene materials has 6 in the center of a triangle of curved arrows with the letters PS under the symbol. *(Charles D. Winters)*

MECHANISM Initiation of Anionic Polymerization of Alkenes

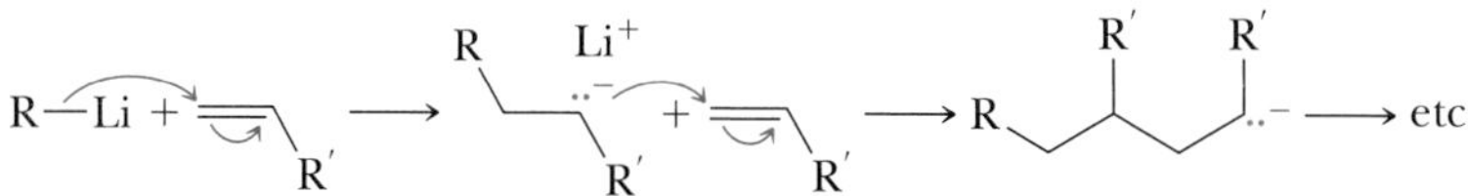

The newly formed carbanion then acts as a nucleophile and adds to another monomer unit, and the propagation continues.

An alternative method for the initiation of anionic polymerizations involves a one-electron reduction of the monomer by lithium or sodium to form a radical anion. The radical anion thus formed is either further reduced to form a dianion, or it dimerizes to form a dimer dianion.

MECHANISM Initiation of Anionic Polymerization of Conjugated Dienes

Butadiene — A radical anion — A dianion

radical coupling to form a dimer

A dimer dianion

In either case, a single initiator can now propagate chains from both ends, by virtue of its two active endgroup carbanions. These reactions are heterogeneous and involve transfer of the electron from the surface of the metal. In order to improve the efficiency of this process, soluble reducing agents such as sodium naphthalide are used. Sodium undergoes electron-transfer reactions with extended aromatic compounds, such as naphthalene, to form soluble radical anions.

Naphthalene $\xrightarrow{\text{THF}}$ Sodium naphthalide (a radical anion)

The naphthalide radical anion is a powerful reducing agent. For example, styrene undergoes a one-electron reduction to form the styryl radical anion, which couples to form a dianion. The latter then propagates polymerization at both ends, growing chains in two directions simultaneously.

Styrene — A styryl radical anion — A distyryl dianion — 1. 2*n* (styrene); 2. H_2O — Polystyrene

The propagation characteristics of anionic polymerizations are similar to those of radical polymerizations, but with the important difference that many of the chain-transfer and termination reactions that plague radical processes are absent. Furthermore, because the propagating chain ends carry the same charge, bimolecular coupling and disproportionation reactions are also averted. An interesting set of circumstances arises when chain-transfer and chain-termination steps are no longer significant. Under these conditions, polymer chains are initiated and continue to grow until either all of the monomer is consumed or some external agent is added to terminate the chains. Polymerizations of this type are called **living polymerizations.**

Living polymer A polymer chain that continues to grow without chain-termination steps until either all of the monomer is consumed or some external agent is added to terminate the chains. The polymer chains will continue to grow if more monomer is added.

The absence of chain-transfer and chain-termination steps in living polymerizations has far-reaching consequences. One of the most visible of these is in the area of molecular weight control. The molecular weight of a polymer originating from living polymerizations is determined directly by the monomer-to-initiator ratio. It is, therefore, relatively easy to obtain polymers of a well-defined size simply by controlling the stoichiometry of the reagents. In contrast, the average sizes of polymer chains formed from nonliving, chain growth processes (radical, Ziegler-Natta, and so on) vary from system to system and are determined by the ratio of the rate of propagation to the rate of termination. In most cases, precise control over the molecular weight of the product obtained in nonliving systems is not possible, because it is very difficult to change one of the rates involved without affecting the other.

After consumption of the monomer under living, anionic conditions, electrophilic terminating agents can be added to functionalize the chain ends. Examples of terminating reagents include CO_2 and ethylene oxide, which, after protonation, form carboxylic acid and alcohol-terminated chains, respectively.

Na^+ CO_2 $CO_2^-Na^+$ $\xrightarrow{H_3O^+}$ CO_2H

$CH_2CH_2O^-Na^+$ $\xrightarrow{H_2O}$ CH_2CH_2OH

In a similar fashion, polymer chains with functional groups at both ends, called **telechelic polymers,** can be prepared by addition of these same reagents (CO_2, ethylene oxide, and so on) to solutions of chains with two active ends initiated by sodium naphthalide.

Telechelic polymer A polymer in which its growing chains are terminated by formation of new functional groups at both ends of its polymer chains. These new functional groups are introduced by adding reagents, such as CO_2 or ethylene oxide, to the growing chains.

EXAMPLE 23.3

Show how to prepare polybutadiene that is terminated at both ends with carboxylate groups.

Solution

Form a growing chain with two active endgroups by treatment of butadiene with 2 mol of lithium metal to form a dianion followed by addition of monomer units and formation of a living polymer. Cap the active endgroups with a carboxylate group by treatment of the living polymer with carbon dioxide.

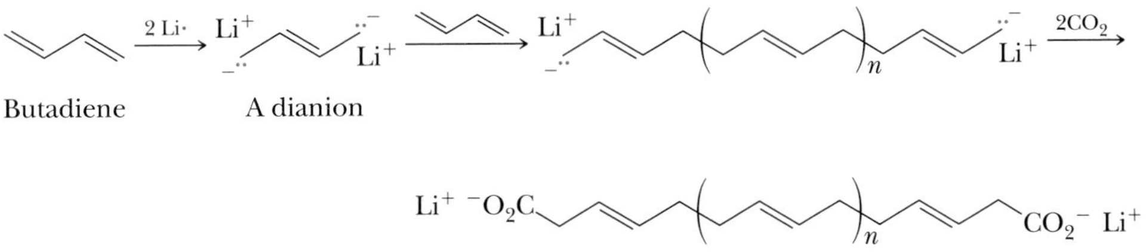

Butadiene A dianion

Li^+ ^-O_2C ... CO_2^- Li^+

PROBLEM 23.3

Show how to prepare polybutadiene that is terminated at both ends with primary alcohol groups.

Cationic Polymerizations

Only alkenes with electron-donating substituents, such as alkyl, aryl, ether, thioether, and amino groups, undergo useful cationic polymerizations. The two most common methods of generating cationic initiators are (1) the reaction of a strong protic acid with an organic monomer and (2) the abstraction of a halide from the organic initiator by a Lewis acid. Cationic chain-growth polymerizations are generally effective only for monomers yielding relatively stable carbocations, that is monomers that form either 3° carbocations or cations stabilized by electron-donating groups, such as ether, thioether, or amino groups.

Initiation by protonation of an alkene requires the use of a strong acid with a nonnucleophilic anion in order to avoid 1,2-addition across the alkene double bond. Suitable acids with nonnucleophilic anions include HF/AsF_5 and HF/BF_3. In the following general equation, initiation is by proton transfer from $H^+BF_4^-$ to the alkene to form a tertiary carbocation, which then continues the cationic chain growth polymerization.

MECHANISM Initiation of Cationic Polymerization of Alkenes by a Proton Acid

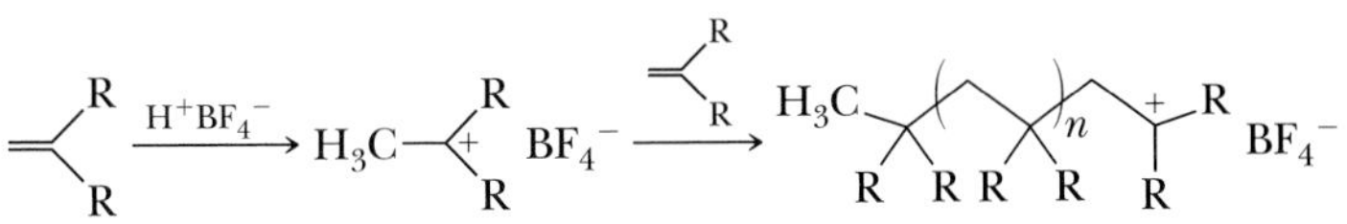

$H^+BF_4^-$ → H_3C (+) BF_4^- → H_3C ... BF_4^-

The second common method for generating carbocations involves the reaction between an alkyl halide and a Lewis acid, such as BF_3, $SnCl_4$, $AlCl_3$, $Al(CH_3)_2Cl$, and

(Text continues on page 938.)

CHEMISTRY IN ACTION

Recycling of Plastics

Polymers, in the form of plastics, are materials upon which our society is incredibly dependent. Durable and lightweight, plastics are probably the most versatile synthetic materials in existence; in fact, their current production in the United States exceeds that of steel. Plastics have come under criticism, however, for their role in the garbage crisis. They comprise 21% of the volume and 8% of the weight of solid wastes, most of which is derived from disposable packaging and wrapping. Of the 53 billion pounds of thermoplastic materials produced in 1993 in America, less than 2% was recycled.

Why aren't more plastics being recycled? The durability and chemical inertness of most plastics make them ideally suited for reuse. The answer to this question has more to do with economics and consumer habits than with technological obstacles. Because curbside pickup and centralized drop-off stations for recyclables are just now becoming common, the amount of used material available for reprocessing has traditionally been small. This limitation, combined with the need for an additional sorting and separation step, rendered the use of recycled plastics in manufacturing expensive compared with virgin materials. Until recently, consumers perceived products made from "used" materials as being inferior to new ones, so the market for recycled products has not been large. However, the growth of the environmental movement over the last few years has resulted in a greater demand for recycled products. As manufacturers adapt to satisfy this new market, plastic recycling will eventually catch up with the recycling of other materials, such as glass and aluminum.

Six types of plastics are commonly used for packaging applications. In 1988, manufacturers adopted recycling code numbers developed by the Society of the Plastics Industry. Because the plastics recycling industry still is not fully developed, only polyethylene terephthalate (PET) and high-density polyethylene (HDPE) are currently being recycled in large quantities, although outlets for the other plastics are being developed (PET soft drink bottles are currently being recycled at a rate of 30%). Low-density polyethylene, which accounts for about 40% of plastic trash, has been slow in finding acceptance with recyclers. Facilities for the reprocessing of polyvinyl chloride (PVC), polypropylene, and polystyrene exist, but are still rare.

These students are wearing jackets made from recycled PET soda bottles. (Charles D. Winters)

The process for the recycling of most plastics is simple, with separation of the desired plastics from other contaminants the most labor-intensive step. PET soft drink bottles, for example, usually have a paper label, adhesive, and an aluminum cap that must be removed before the PET can be reused. The recycling process begins with hand or machine sorting, after which the bottles are shredded into small chips. An air cyclone then removes paper and other lightweight materials. Any remaining labels and adhesives are eliminated with a detergent wash, and the PET chips are then dried and aluminum, the final contaminant, is removed electrostatically. The PET produced by this method is 99.9% free of contaminants and sells for

Recycling Code	Polymer	Common Uses	Uses of Recycled Polymer
1 PET (PETE)	Poly(ethylene terephthalate)	Soft drink bottles, household chemical bottles, films, textile fibers	Soft drink bottles, household chemical bottles, films, textile fibers
2 HDPE	High-density polyethylene	Milk and water jugs, grocery bags, bottles	Bottles, molded containers
3 V	Poly(vinyl chloride) (Vinyl or PVC)	Shampoo bottles, pipes, shower curtains, vinyl siding, wire insulation, floor tiles, credit cards	Plastic floor mats
4 LDPE	Low-density polyethylene	Shrink wrap, trash and grocery bags, sandwich bags, squeeze bottles	Trash and grocery bags
5 PP	Polypropylene	Plastic lids, clothing fibers, bottle caps, toys, diaper linings	Mixed plastic component
6 PS	Polystyrene	Styrofoam cups, egg cartons, disposable utensils, packaging materials, appliances	Molded items such as cafeteria trays, rulers, frisbees, trash cans, videocassettes
7	All other plastics and mixed plastics	Various	Plastic lumber, playground equipment, road reflectors

about half the price of the unused material. Unfortunately, plastics with similar densities cannot be separated with this technology, nor can plastics composed of several polymers be broken down into pure components. However, recycled mixed plastics can be molded into plastic lumber that is strong, durable, and graffiti-resistant.

An alternative to this process, which uses only physical methods of purification, is chemical recycling. Eastman Kodak salvages large amounts of its PET film scrap by a transesterification reaction. The scrap is treated with methanol in the presence of an acid catalyst to give ethylene glycol and dimethyl terephthalate.

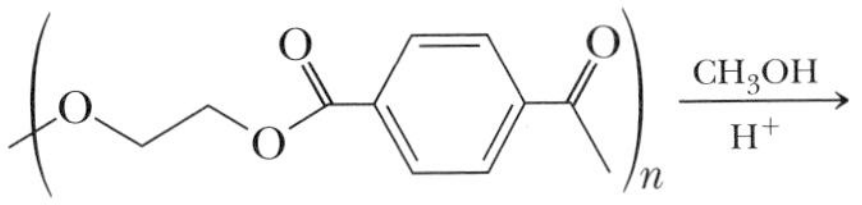

$\xrightarrow[H^+]{CH_3OH}$

Poly(ethylene terephthalate) (PET)

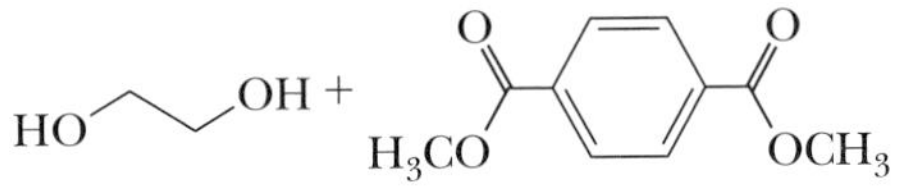

Ethylene glycol + Dimethyl terephthalate

These monomers are purified by distillation or recrystallization and used as feed stocks for the production of more PET film.

$ZnCl_2$. When a trace of water is present, the mechanism of initiation using some Lewis acids is thought to involve protonation of the alkene.

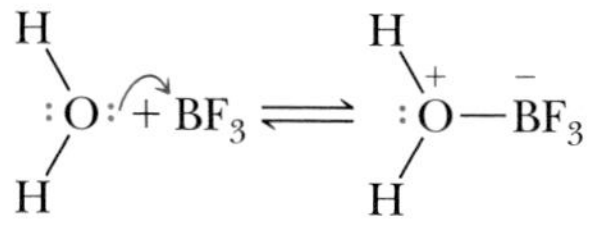

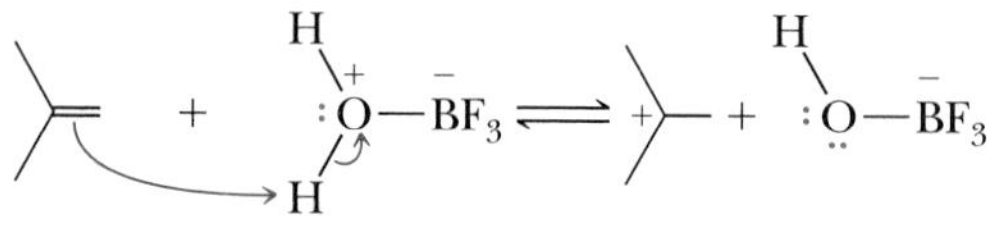

$$H_2\ddot{O}: + BF_3 \rightleftharpoons H_2\overset{+}{O}-\overset{-}{B}F_3$$

$$(CH_3)_2C{=}CH_2 + H_2\overset{+}{O}-\overset{-}{B}F_3 \rightleftharpoons (CH_3)_3C^+ + H\ddot{O}-\overset{-}{B}F_3$$

2-Methylpropene
(Isobutylene)

In the absence of water, the Lewis acid removes a halide ion from the alkyl halide to form the initiating carbocation.

MECHANISM Initiation of Cationic Polymerization of Alkenes by a Lewis Acid

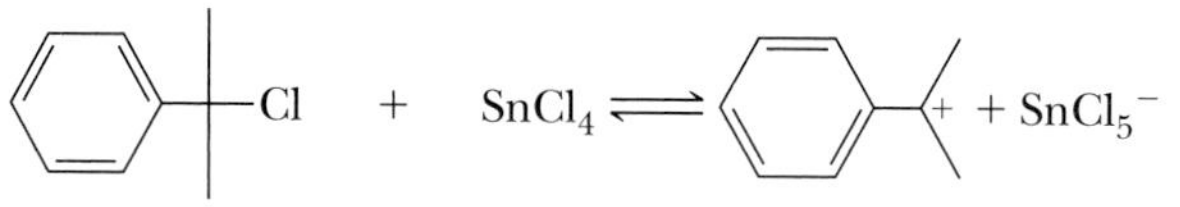

$$C_6H_5C(CH_3)_2Cl + SnCl_4 \rightleftharpoons C_6H_5C^+(CH_3)_2 + SnCl_5^-$$

2-Chloro-2-phenylpropane

The polymerization of alkenes then propagates by the electrophilic attack of the carbocation on the double bond of the alkene monomer. The regiochemistry of the addition is determined by the formation of the more stable (the more highly substituted) carbocation.

EXAMPLE 23.4

Write a mechanism for the polymerization of 2-methylpropene (isobutylene) initiated by treatment of 2-chloro-2-phenylpropane with $SnCl_4$. Label the initiation, propagation, and termination steps.

Solution

Initiation

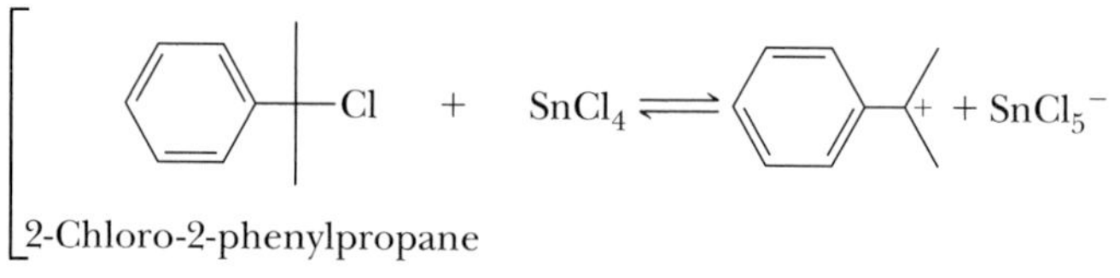

$$C_6H_5C(CH_3)_2Cl + SnCl_4 \rightleftharpoons C_6H_5C^+(CH_3)_2 + SnCl_5^-$$

2-Chloro-2-phenylpropane

Propagation

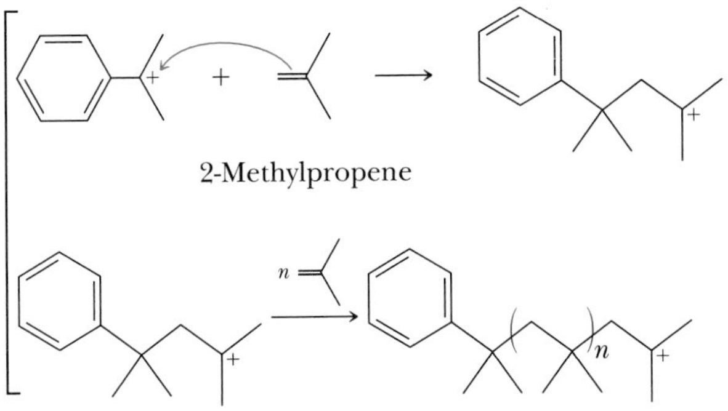

2-Methylpropene

Termination

$\xrightarrow[SnCl_5^-]{H_2O}$ + $H^+SnCl_5^-$

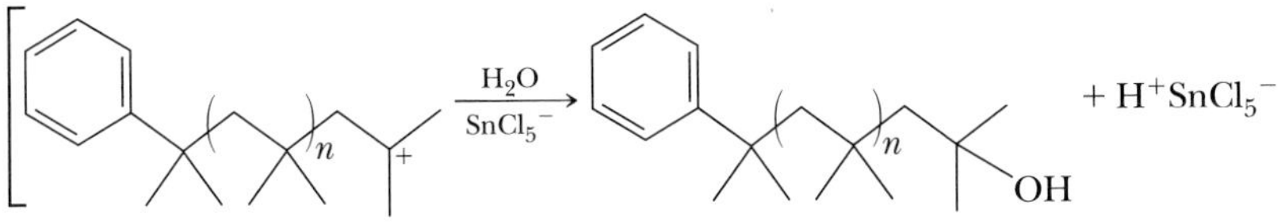

PROBLEM 23.4

Write a mechanism for the polymerization of methyl vinyl ether initiated by 2-chloro-2-phenylpropane and $SnCl_4$. Label the initiation, propagation, and termination steps.

SUMMARY

Polymerization is the process of joining together many small monomer molecules into large, high-molecular-weight **polymers** (Section 23.1). The properties of polymeric materials depend on the structure of the repeat unit, molecular weight (Section 23.3), the chain architecture, the presence or absence of crystalline phases (Section 23.4), tacticity (Section 23.6C), interchain order and packing, and the materials' morphology. **Step-growth polymerizations** involve the stepwise reaction of difunctional monomers (Section 23.5). Further structural variations, such as crosslinks and branches, can be introduced into the resulting polymer by the addition of multifunctional monomers to the reaction mixture. The formation of high-molecular-weight polymers from step-growth processes requires the use of reactions that proceed with very high yields. Important commercial polymers synthesized through step-growth processes include polyamides, polyesters, polycarbonates, polyurethanes, and epoxy resins.

Chain-growth polymerization proceeds by the sequential addition of monomer units to an active chain end (Section 23.6). Important mechanisms for chain-growth polymerizations include radical, anionic, cationic, and transition metal-mediated processes. Chain-growth polymerizations involve initiation, propagation, and termination steps. **Chain-transfer steps** terminate one chain but simultaneously initiate the growth of another. Chain polymerizations that proceed without chain-transfer or chain-termination steps are called **living polymerizations.** Among the transition metal-mediated polymerizations, the **Ziegler-Natta polymerizations** of ethylene and propylene are the most significant (Section 23.6B). These reactions proceed with high specificity to yield polymers that are stereoregular and highly linear. This regularity leads to highly crystalline polymers. When the chains are elongated and oriented through special processing procedures, a polymer with strength and stiffness greater than steel can be obtained.

KEY REACTIONS

1. Step-Growth Polymerization of a Dicarboxylic Acid and a Diamine to Give a Polyamide (Section 23.5A)

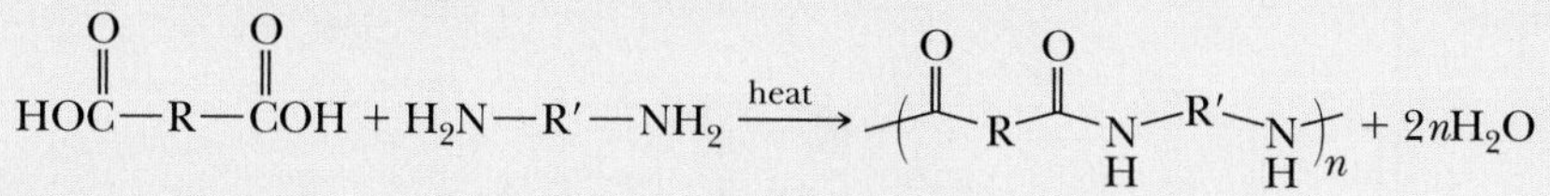

2. Step-Growth Polymerization of a Dicarboxylic Acid and a Diol to Give a Polyester (Section 23.5B)

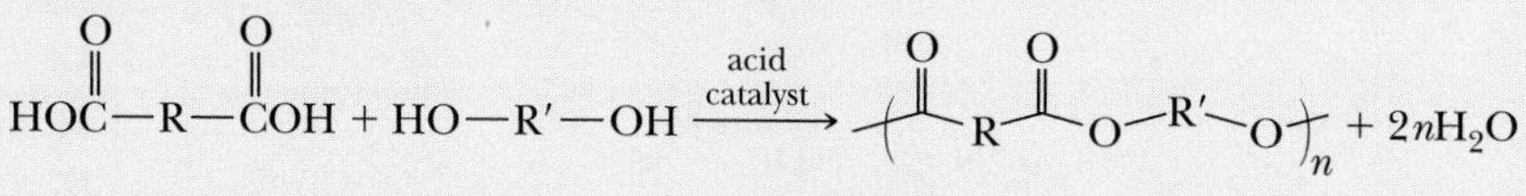

3. Step-Growth Polymerization of a Diacyl Chloride, Such as Phosgene, and a Diol to Give a Polycarbonate (Section 23.5C)

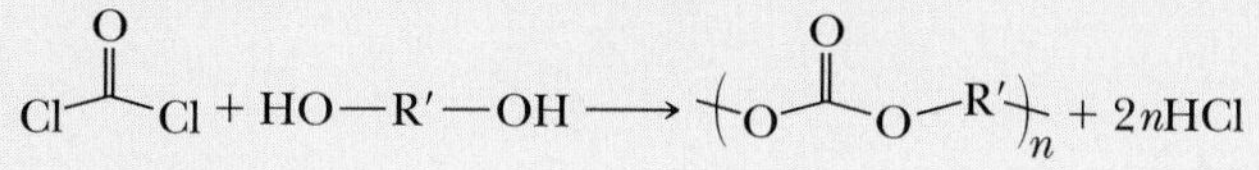

4. Step-Growth Polymerization of a Diisocyanate and a Diol to Give a Polyurethane (Section 23.5D)

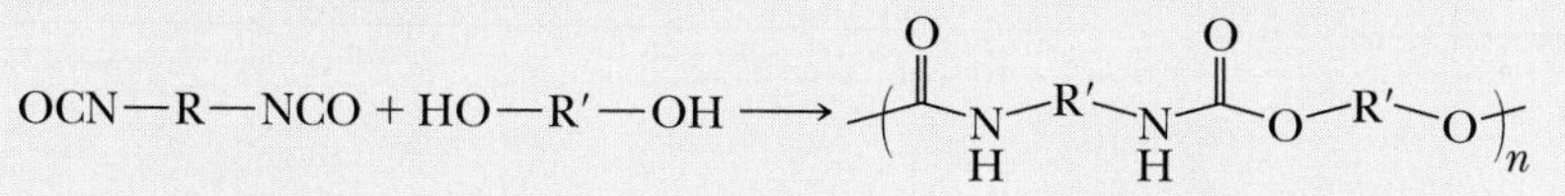

5. Step-Growth Polymerization of a Diepoxide and a Diamine to Give an Epoxy Resin (Section 23.5E)

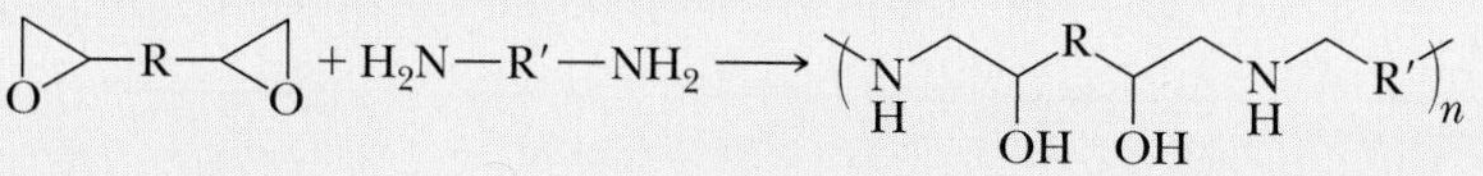

6. Radical Chain-Growth Polymerization of Substituted Ethylenes (Section 23.6A)

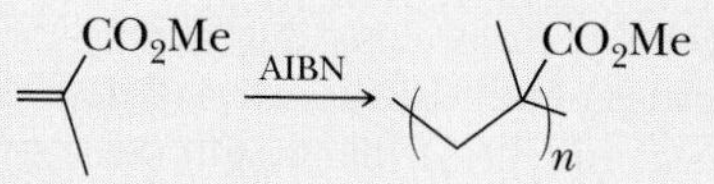

7. Titanium-Mediated (Ziegler-Natta) Chain-Growth Polymerization of Ethylene and Substituted Ethylenes (Section 23.6B)

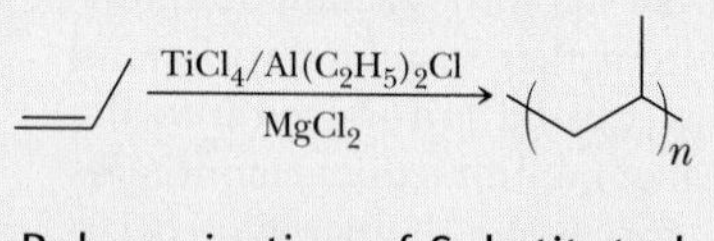

8. Anionic Chain-Growth Polymerization of Substituted Ethylenes (Section 23.6D)

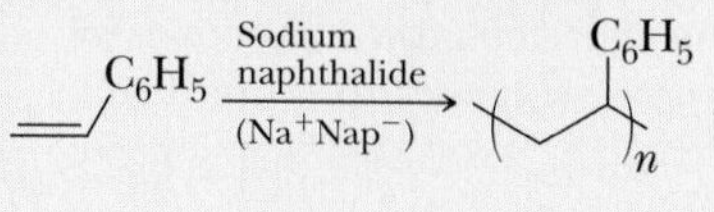

9. Cationic Chain-Growth Polymerization of Substituted Ethylenes (Section 23.6D)

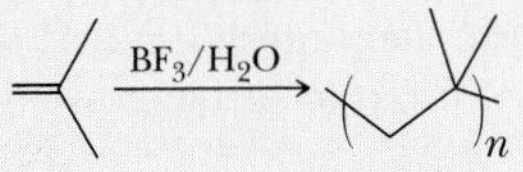

BF_3/H_2O

ADDITIONAL PROBLEMS

Structure and Nomenclature

23.5 Name the following polymers.

(a) 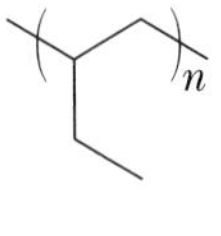 (b) 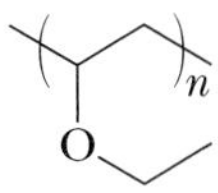 (c) 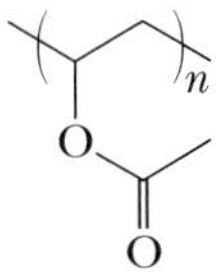 (d) 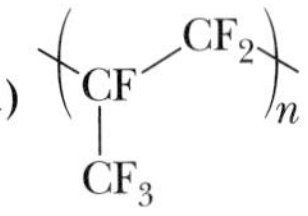 CF, CF_2, CF_3, n

(e) 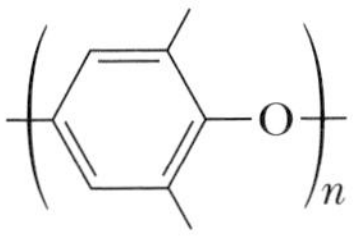 (f) 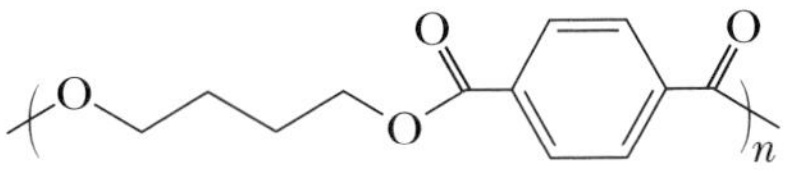 (g) 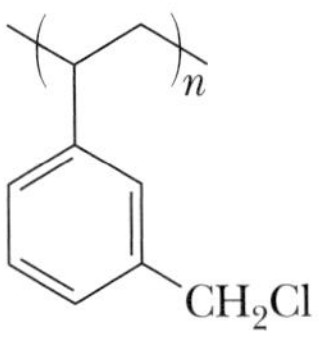 CH_2Cl

(h)

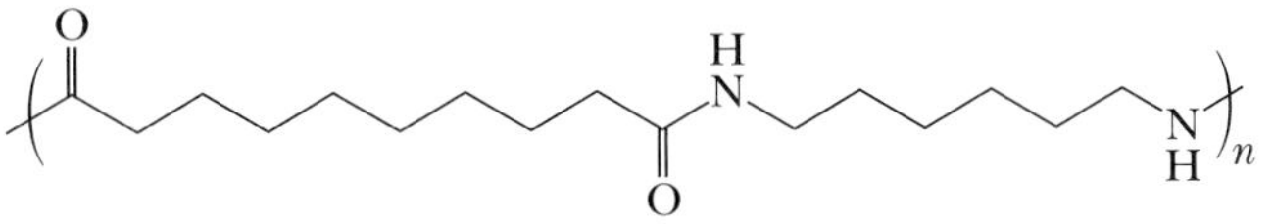

23.6 Draw the structure(s) of the monomer(s) used to make each polymer in Problem 23.5.

Step-Growth Polymerizations

23.7 Draw the structure of the polymer formed in the following reactions.

(a) 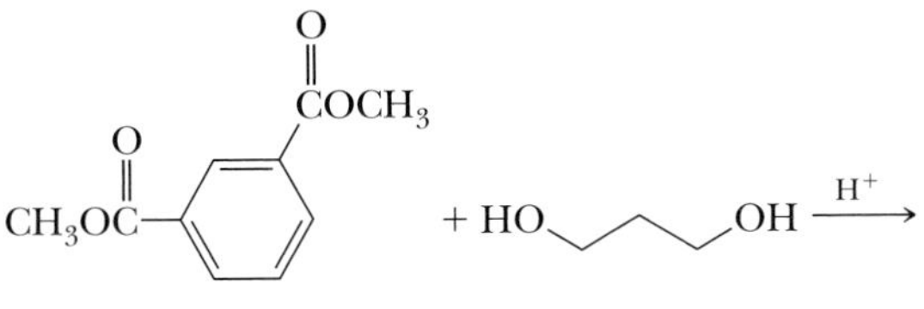 $CH_3OC(=O)$–C_6H_4–$C(=O)OCH_3$ + HO~~OH $\xrightarrow{H^+}$

(b) $CH_3OC(=O)$–C_6H_4–$C(=O)OCH_3$ + HO–CH₂CH(OH)CH₂–OH $\xrightarrow{H^+}$

(c) $\xrightarrow[70°C]{AIBN}$

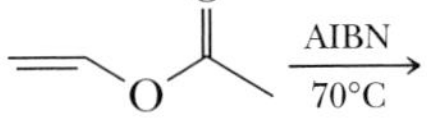

(d) $\xrightarrow{CF_3SO_3H}$

(e) $\xrightarrow{KOH}$

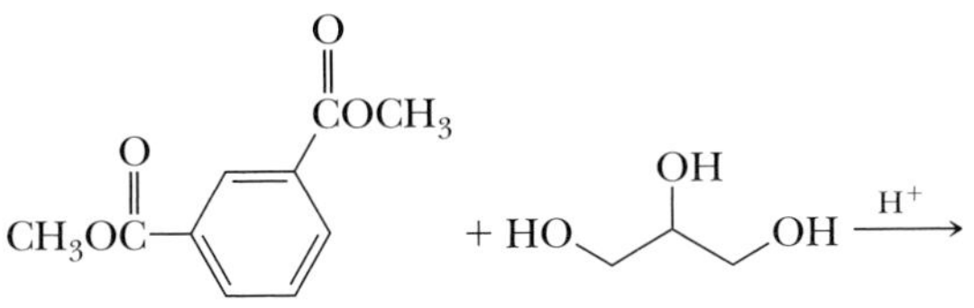

(f) 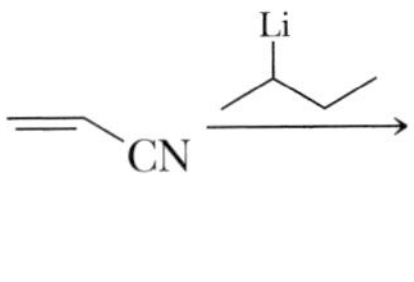 CN $\xrightarrow{Li}$

23.8 At one time, a raw material for the production of hexamethylenediamine was the pentose-based polysaccharides of agricultural wastes, such as oat hulls. Treatment of these wastes with sulfuric acid or hydrochloric acid gives furfural. Decarbonylation of furfural over a zinc-chromium-molybdenum catalyst gives furan. Propose reagents and experimental conditions for the conversion of furan to hexamethylenediamine.

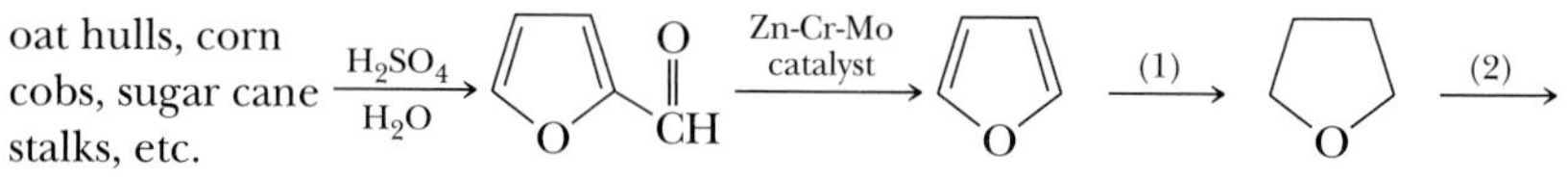

oat hulls, corn cobs, sugar cane stalks, etc. $\xrightarrow[H_2O]{H_2SO_4}$ Furfural $\xrightarrow{\text{Zn-Cr-Mo catalyst}}$ Furan $\xrightarrow{(1)}$ Tetrahydrofuran (THF) $\xrightarrow{(2)}$

$$Cl(CH_2)_4Cl \xrightarrow{(3)} N{\equiv}C(CH_2)_4C{\equiv}N \xrightarrow{(4)} H_2N(CH_2)_6NH_2$$

1,4-Dichlorobutane; Hexanedinitrile (Adiponitrile); 1,6-Hexanediamine (Hexamethylenediamine)

23.9 Another raw material for the production of hexamethylenediamine is butadiene derived from thermal and catalytic cracking of petroleum. Propose reagents and experimental conditions for the conversion of butadiene to hexamethylenediamine.

$$CH_2{=}CHCH{=}CH_2 \xrightarrow{(1)} ClCH_2CH{=}CHCH_2Cl \xrightarrow{(2)} N{\equiv}CCH_2CH{=}CHCH_2C{\equiv}N \xrightarrow{(3)} H_2N(CH_2)_6NH_2$$

Butadiene; 1,4-Dichloro-2-butene; 3-Hexenedinitrile; 1,6-Hexanediamine (Hexamethylenediamine)

23.10 Propose reagents and experimental conditions for the conversion of butadiene to adipic acid. (*Hint:* Review the chemistry of the previous problem.)

$$CH_2{=}CHCH{=}CH_2 \longrightarrow HO_2C(CH_2)_4CO_2H$$

Butadiene; Hexanedioic acid (Adipic acid)

23.11 Polymerization of 2-chloro-1,3-butadiene under Ziegler-Natta conditions gives a synthetic elastomer called neoprene. All carbon-carbon double bonds in the polymer chain have the *trans* configuration. Draw the repeat unit in neoprene.

23.12 Poly(ethylene terephthalate) (PET) can be prepared by this reaction. Propose a mechanism for the step-growth reaction in this polymerization.

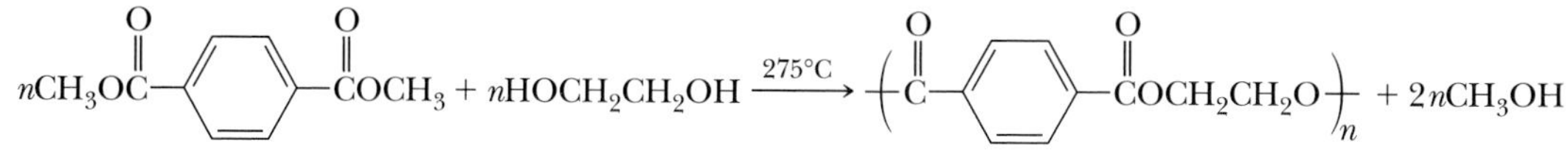

$$n\,CH_3OC(=O){-}C_6H_4{-}C(=O)OCH_3 + n\,HOCH_2CH_2OH \xrightarrow{275°C} {+}C(=O){-}C_6H_4{-}C(=O)OCH_2CH_2O{+}_n + 2n\,CH_3OH$$

Dimethyl terephthalate; Ethylene glycol; Poly(ethylene terephthalate); Methanol

23.13 Identify the monomers required for the synthesis of these step-growth polymers.

(a) $\left[C(=O){-}C_6H_4{-}C(=O){-}O{-}CH_2{-}C_6H_{10}{-}CH_2O\right]_n$

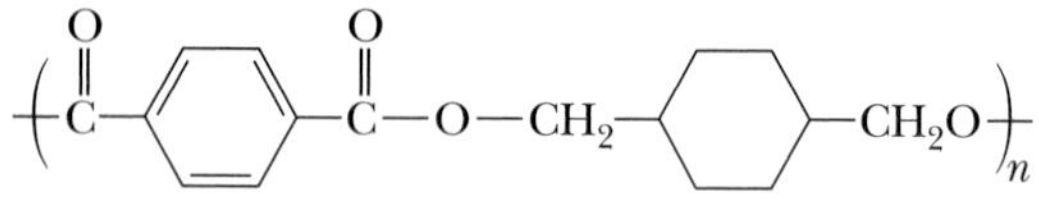

Kodel (a polyester)

(b) $\left[C(=O)(CH_2)_6C(=O){-}NH{-}C_6H_{10}{-}CH_2{-}C_6H_{10}{-}NH\right]_n$

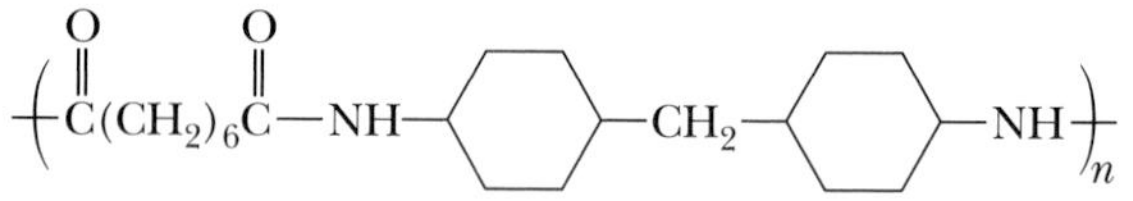

Quiana (a polyamide)

23.14 Nomex, another aromatic polyamide (aramid) is prepared by polymerization of 1,3-benzenediamine and the diacid chloride of 1,3-benzenedicarboxylic acid. The physical properties of the polymer make it suitable for high strength, high temperature applications such as parachute cords and jet aircraft tires. Draw a structural formula for the repeating unit of Nomex.

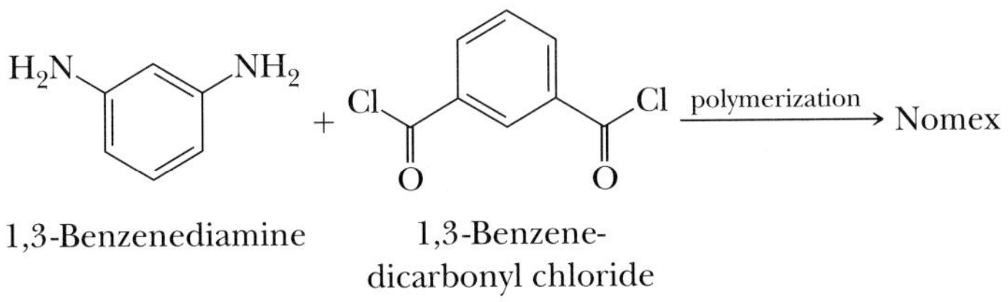

H_2N ... NH_2 + Cl ... Cl $\xrightarrow{\text{polymerization}}$ Nomex

1,3-Benzenediamine   1,3-Benzene-dicarbonyl chloride

23.15 Caprolactam, the monomer from which nylon 6 is synthesized, is prepared from cyclohexanone in two steps. In step 1, cyclohexanone is treated with hydroxylamine to form cyclohexanone oxime. Treatment of the oxime with concentrated sulfuric acid in step 2 gives caprolactam by a reaction called a Beckmann rearrangement. Propose a mechanism for the conversion of cyclohexanone oxime to caprolactam. (*Hint:* Proton transfer to oxygen of the oxime followed by loss of water gives a positively charged, electron deficient nitrogen atom, which then provides the driving force for the skeletal rearrangement.)

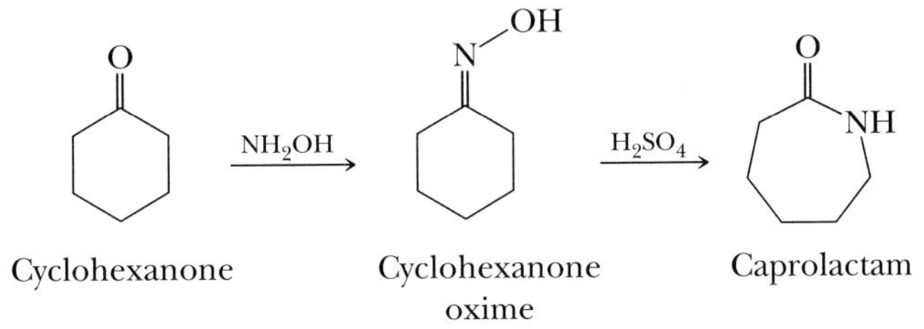

Cyclohexanone $\xrightarrow{NH_2OH}$ Cyclohexanone oxime $\xrightarrow{H_2SO_4}$ Caprolactam

23.16 Nylon 6,10 is prepared by polymerization of a diamine and a diacid chloride. Draw the structural formula of each reactant and for the repeat unit in nylon 6,10.

23.17 Polycarbonates (Section 23.5C) are also formed by using a nucleophilic aromatic substitution route (Section 20.3B) involving aromatic difluoro monomers and carbonate ion. Propose a mechanism for this reaction.

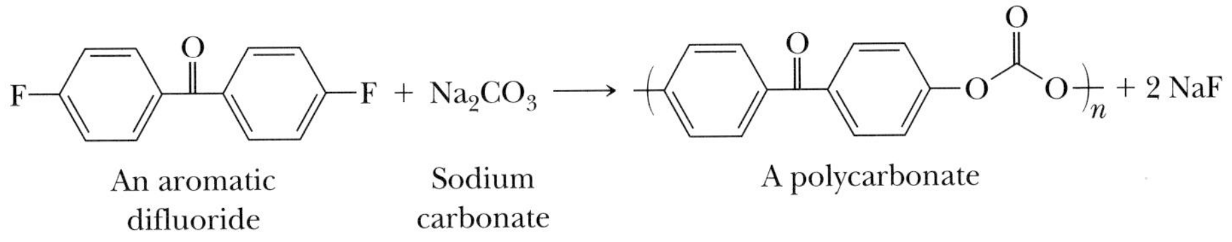

An aromatic difluoride + Na_2CO_3 (Sodium carbonate) $\longrightarrow$ A polycarbonate + 2 NaF

23.18 Propose a mechanism for the formation of this polyphenylurea. To simplify your presentation of the mechanism, consider the reaction of one —NCO group with one $—NH_2$ group.

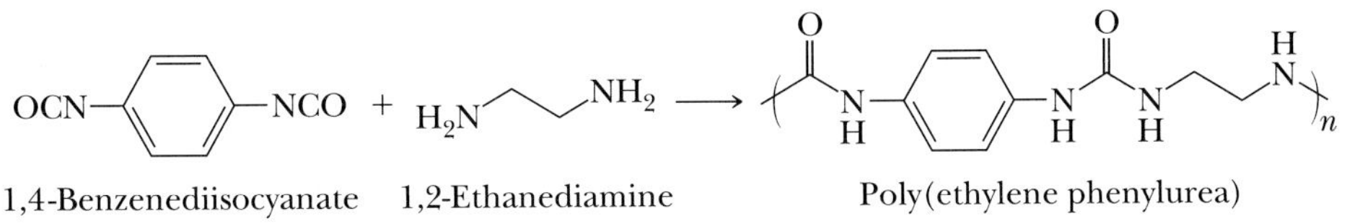

OCN—C_6H_4—NCO + H_2N ... NH_2 $\longrightarrow$ Poly(ethylene phenylurea)

1,4-Benzenediisocyanate   1,2-Ethanediamine   Poly(ethylene phenylurea)

23.19 When equal molar amounts of phthalic anhydride and 1,2,3-propanetriol are heated, they form an amorphous polyester. Under these conditions, polymerization is regioselective for the primary hydroxyl groups of the triol.

Phthalic anhydride + 1,2,3-Propanetriol (Glycerol) $\xrightarrow{\text{heat}}$ a polyester

(a) Draw a structural formula for the repeat unit of this polyester.
(b) Account for the regioselective reaction with the primary hydroxyl groups only.

23.20 The polyester from Problem 23.19 can be mixed with additional phthalic anhydride (0.5 mol of phthalic anhydride for each mol of 1,2,3-propanetriol in the original polyester) to form a liquid resin. When this resin is heated, it forms a hard, insoluble, thermosetting polyester called glyptal.

(a) Propose a structure for the repeat unit in glyptal.
(b) Account for the fact that glyptal is a thermosetting plastic.

23.21 Propose a mechanism for the formation of the following polymer.

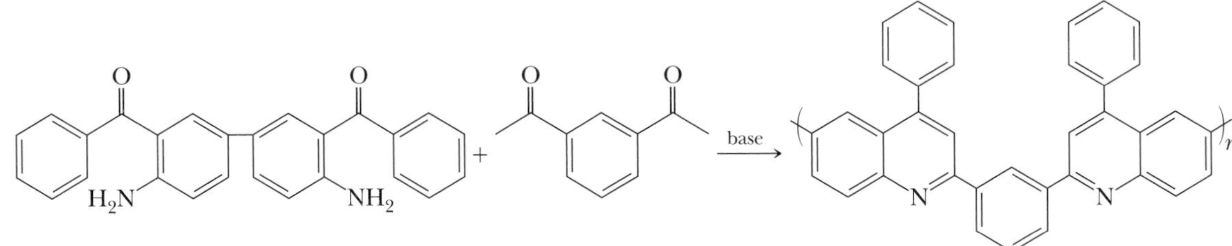

23.22 Draw the structural formula of the polymer resulting from base-catalyzed polymerization of each compound. Would you expect the polymers to be optically active? (*S*)-(+)-lactide is the dilactone formed from two molecules of (*S*)-(+)-lactic acid.

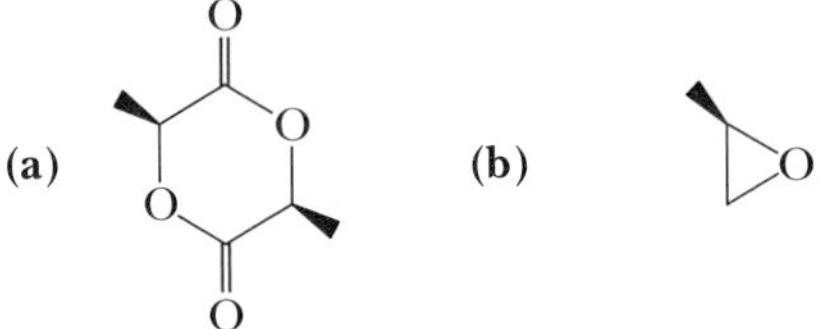

(*S*)-(+)-lactide (*R*)-Propylene oxide

23.23 Poly(3-hydroxybutanoic acid), a biodegradable polyester, is an insoluble, opaque material that is difficult to process into shapes. In contrast, the copolymer of 3-hydroxybutanoic acid and 3-hydroxyoctanoic acid is a transparent polymer that shows good solubility in a number of organic solvents. Explain the difference in properties between these two polymers in terms of their structure.

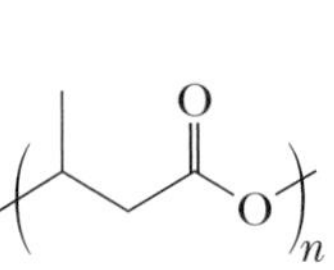

Poly(3-hydroxybutanoic acid)

Poly(3-hydroxybutanoic acid-3-hydroxyoctanoic acid) copolymer

Chain-Growth Polymerizations

23.24 How might you determine experimentally if a particular polymerization is propagating by a step-growth or a chain-growth mechanism?

23.25 Select the monomer in each pair that is more reactive toward cationic polymerization.

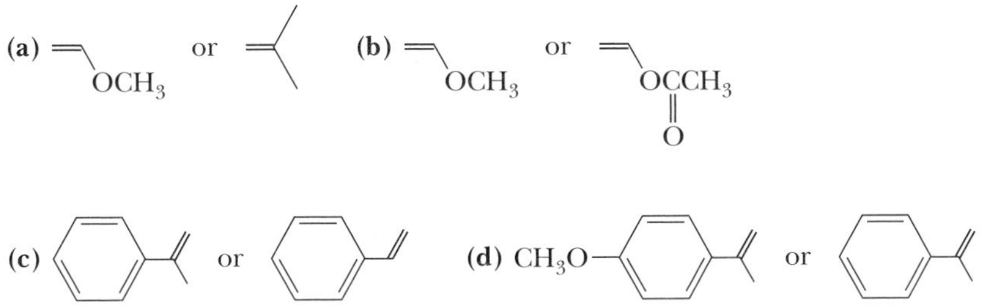

(a) $CH_2{=}CHOCH_3$ or $CH_2{=}C(CH_3)_2$ (b) $CH_2{=}CHOCH_3$ or $CH_2{=}CHOCCH_3$ (with C=O)

(c) $C_6H_5C(CH_3){=}CH_2$ or $C_6H_5CH{=}CH_2$ (d) $CH_3O{-}C_6H_4C(CH_3){=}CH_2$ or $C_6H_5C(CH_3){=}CH_2$

23.26 Polymerization of vinyl acetate gives poly(vinyl acetate). Hydrolysis of this polymer in aqueous sodium hydroxide gives poly(vinyl alcohol). Draw the repeat units of both poly(vinyl acetate) and poly(vinyl alcohol).

23.27 Benzoquinone can be used to inhibit radical polymerizations. This compound reacts with a radical intermediate, R·, to form a less reactive radical that does not participate in chain propagation steps and, thus, breaks the chain.

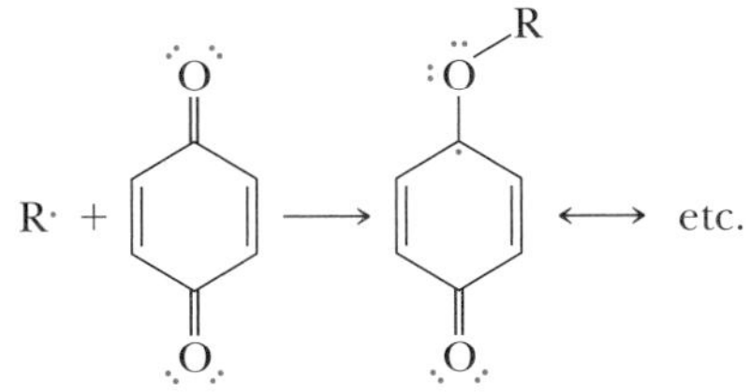

R· + (benzoquinone) ⟶ (R–O– substituted radical) ⟷ etc.

Draw a series of contributing structures for this less reactive radical and account for its stability.

23.28 Following is the structural formula of a section of polypropylene derived from three units of propylene monomer.

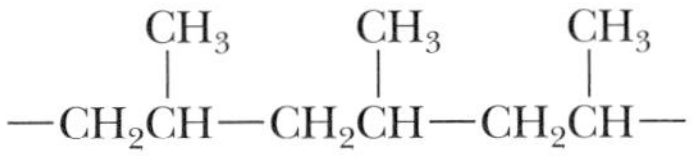

$$-CH_2\underset{CH_3}{CH}-CH_2\underset{CH_3}{CH}-CH_2\underset{CH_3}{CH}-$$

Polypropylene

Draw structural formulas for comparable sections of:

(a) Poly(vinyl chloride) **(b)** Polytetrafluoroethylene
(c) Poly(methyl methacrylate) (Plexiglas) **(d)** Poly(1,1-dichloroethylene)

23.29 Low-density polyethylene (LDPE) has a higher degree of chain branching than high-density polyethylene (HDPE). Explain the relationship between chain branching and density.

23.30 We saw how intramolecular chain transfer in radical polymerization of ethylene creates a four-carbon branch on a polyethylene chain. What branch is created by a comparable intramolecular chain transfer during radical polymerization of styrene?

23.31 Compare the densities of low-density polyethylene (LDPE) and high-density polyethylene (HDPE) with the densities of the liquid alkanes listed in Table 2.4. How might you account for the differences between them?

23.32 Natural rubber is the all *cis* polymer of 2-methyl-1,3-butadiene (isoprene).

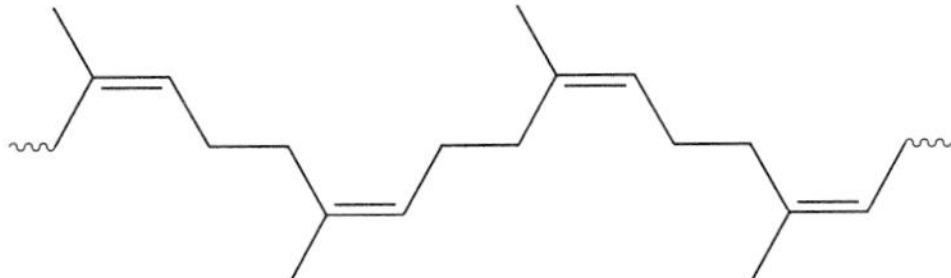

Poly(2-methyl-1,3-butadiene)
(Polyisoprene)

(a) Draw the structural formula for the repeat unit of natural rubber.

(b) Draw the structural formula of the product of oxidation of natural rubber by ozone followed by a workup in the presence of $(CH_3)_2S$. Name each functional group present in this product.

(c) The smog prevalent in many major metropolitan areas contains oxidizing agents, including ozone. Account for the fact that this type of smog attacks natural rubber (automobile tires and so on) but does not attack polyethylene or polyvinyl chloride.

23.33 Radical polymerization of styrene gives a linear polymer. Radical polymerization of a mixture of styrene and 1,4-divinylbenzene gives a crosslinked network polymer of the type shown in Figure 23.1. Show by drawing structural formulas how incorporation of a few percent 1,4-divinylbenzene in the polymerization mixture gives a crosslinked polymer.

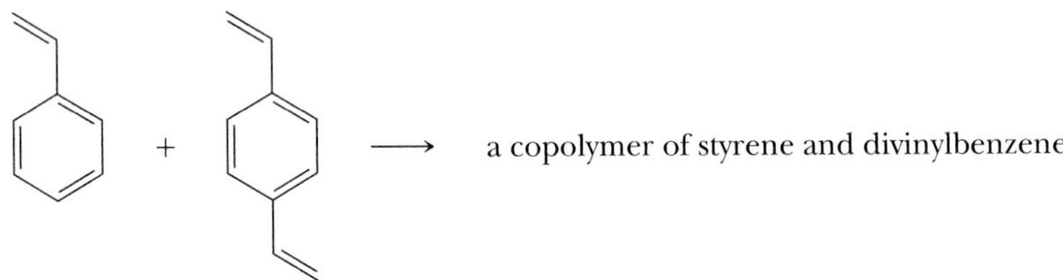

Styrene 1,4-Divinylbenzene

23.34 One common type of cation exchange resin is prepared by polymerization of a mixture containing styrene and 1,4-divinylbenzene (Problem 23.33). The polymer is then treated with concentrated sulfuric acid to sulfonate a majority of the aromatic rings in the polymer.

(a) Show the product of sulfonation of each benzene ring.

(b) Explain how this sulfonated polymer can act as a cation exchange resin.

23.35 The most widely used synthetic rubber is a copolymer of styrene and butadiene called SB rubber. Ratios of styrene to butadiene used in polymerization vary depending on the end use of the polymer. The ratio used most commonly in the preparation of SB rubber for use in automobile tires is 1 mol styrene to 3 mol butadiene. Draw a structural formula of a section of the polymer formed from this ratio of reactants. Assume that all carbon-carbon double bonds in the polymer chain are in the *cis* configuration.

23.36 From what two monomer units is the following polymer made?

$C\equiv N$ $C\equiv N$

CARBOHYDRATES

24

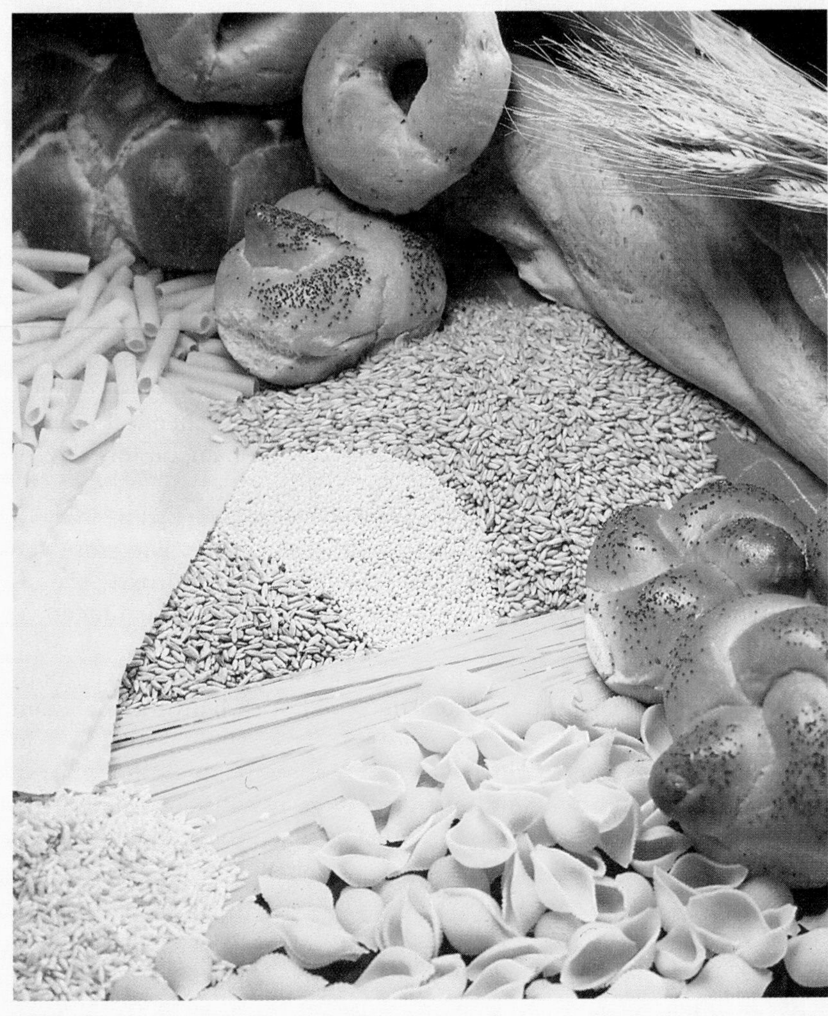

■ Breads, grains, and pasta are sources of carbohydrates. *(Charles D. Winters)*

Carbohydrates are the most abundant organic compounds in the plant world. Among their many vital functions, these compounds are storehouses of chemical energy (glucose, starch, glycogen), components of supportive structures in plants (cellulose) and bacterial cell walls (mucopolysaccharides), and components of nucleic acids (D-ribose and 2-deoxy-D-ribose). Furthermore, bound to plasma membranes of animal cells are large numbers of relatively small carbohydrates that mediate interactions between cells. For example, A, B, and O blood types are determined by specific membrane-bound carbohydrates.

The simpler members of the carbohydrate family are often referred to as saccharides because of the sweet taste of sugars (Latin, *saccharum*, sugar). The name "carbohydrate," or hydrate of carbon, derives from the general formula $C_n(H_2O)_m$ for many members of this class. Two examples of carbohydrates with molecular formulas that can be written alternatively as hydrates of carbon are shown in the table.

Carbohydrate	Molecular Formula	Molecular Formula as a Hydrate of Carbon
glucose (blood sugar)	$C_6H_{12}O_6$	$C_6(H_2O)_6$
sucrose (table sugar)	$C_{12}H_{22}O_{11}$	$C_{12}(H_2O)_{11}$

Not all carbohydrates, however, have this general formula. Some contain too few oxygen atoms to fit this formula, whereas some contain too many. Some also contain nitrogen. But the term carbohydrate has become firmly rooted in chemical nomenclature, and although not completely accurate, it persists as the name of this class of compounds.

Carbohydrate A polyhydroxyaldehyde or polyhydroxyketone, or a substance that gives these compounds on hydrolysis.

At the molecular level, **carbohydrates** are polyhydroxyaldehydes, polyhydroxyketones, or compounds that yield either of these after hydrolysis. Therefore, the chemistry of carbohydrates is essentially the chemistry of hydroxyl groups and carbonyl groups, and of acetal bonds formed between these two functional groups.

However, the notion that carbohydrates have only two types of functional groups belies the complexity of their chemistry. All but the simplest carbohydrates contain multiple stereocenters. For example, glucose, the most abundant carbohydrate in the biological world, contains one aldehyde group, four secondary alcohol groups, one primary alcohol group, and four stereocenters. Working with molecules of this complexity presents enormous challenges to organic chemists and biochemists alike.

24.1 Monosaccharides

A. Structure

Monosaccharide A carbohydrate that cannot be hydrolyzed to a simpler carbohydrate.

Aldose A monosaccharide containing an aldehyde group.

Ketose A monosaccharide containing a ketone group.

Monosaccharides are the monomers from which more complex carbohydrates are constructed. They have the general formula $C_nH_{2n}O_n$, where n varies from 3 to 8. The suffix **-ose** indicates that a molecule is a carbohydrate, and the prefixes tri-, tetr-, pent-, and so forth indicate the number of carbon atoms in the chain. Monosaccharides containing an aldehyde group are classified as **aldoses;** those containing a ketone group are classified as **ketoses.**

Monosaccharides Classified by Number of Carbon Atoms

Name	Formula
triose	$C_3H_6O_3$
tetrose	$C_4H_8O_4$
pentose	$C_5H_{10}O_5$
hexose	$C_6H_{12}O_6$
heptose	$C_7H_{14}O_7$
octose	$C_8H_{16}O_8$

There are only two trioses: glyceraldehyde, which is an aldotriose, and dihydroxyacetone, which is a ketotriose.

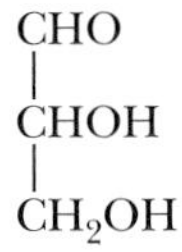

$$\begin{array}{cc} \mathrm{CHO} & \mathrm{CH_2OH} \\ | & | \\ \mathrm{CHOH} & \mathrm{C{=}O} \\ | & | \\ \mathrm{CH_2OH} & \mathrm{CH_2OH} \end{array}$$

Glyceraldehyde (an aldotriose) 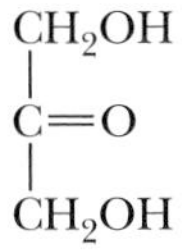 Dihydroxyacetone (a ketotriose)

Often the designations aldo- and keto- are omitted, and these compounds are referred to simply as trioses. Although this designation does not tell the nature of the carbonyl group, at least it indicates that the monosaccharide contains three carbon atoms.

B. Nomenclature

Glyceraldehyde is a common name; the IUPAC name for this monosaccharide is 2,3-dihydroxypropanal. Similarly, dihydroxyacetone is a common name; its IUPAC name is 1,3-dihydroxypropanone. However, the common names for these and other monosaccharides are so firmly rooted in the literature of organic chemistry and biochemistry that they are used almost exclusively whenever these compounds are referred to. Therefore, throughout our discussions of the chemistry of carbohydrates, we use the most common names.

C. Stereoisomerism

Glyceraldehyde contains a stereocenter and exists as a pair of enantiomers. The stereoisomer shown on the left has the *R* configuration and is named (*R*)-glyceraldehyde; its enantiomer, shown on the right, is named (*S*)-glyceraldehyde.

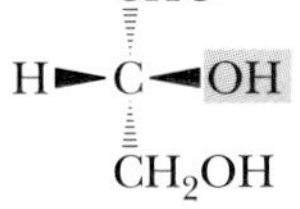

$$\begin{array}{cc} \mathrm{CHO} & \mathrm{CHO} \\ \mathrm{H{-}C{-}OH} & \mathrm{HO{-}C{-}H} \\ \mathrm{CH_2OH} & \mathrm{CH_2OH} \end{array}$$

(*R*)-Glyceraldehyde 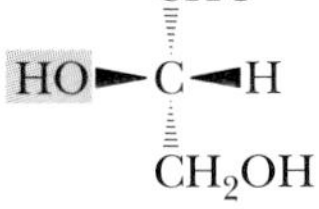 (*S*)-Glyceraldehyde

Fischer projection A two-dimensional representation for showing the configuration of a tetrahedral stereocenter; horizontal lines represent bonds projecting forward and vertical lines represent bonds projecting to the rear.

Structural formulas for monosaccharides can also be drawn as Fischer projections (Section 4.4). In a **Fischer projection** of a monosaccharide, the carbon chain is written vertically with the carbon bearing the carbonyl group toward the top. Horizontal lines represent groups projecting above the plane of the page; vertical lines represent groups projecting behind the plane of the page. Applying these rules gives the following Fischer projections for the enantiomers of glyceraldehyde:

CHO / H►C◄OH / CH_2OH
(*R*)-Glyceraldehyde

convert to a Fischer projection

CHO / HO►C◄H / CH_2OH
(*S*)-Glyceraldehyde

convert to a Fischer projection

CHO / H—+—OH / CH_2OH
(*R*)-Glyceraldehyde

CHO / HO—+—H / CH_2OH
(*S*)-Glyceraldehyde

Fischer projections

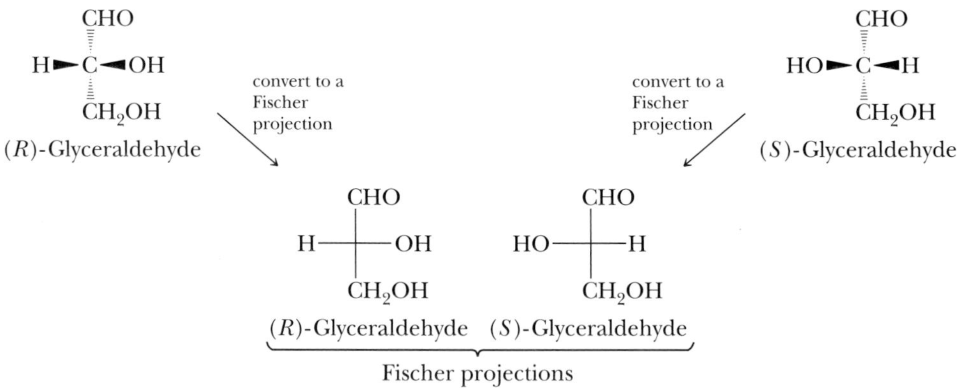

D. D- and L-Monosaccharides

While the *R,S* convention is widely accepted today as a standard for designating configurations, the configurations of carbohydrates are still commonly designated by the D,L convention, proposed by Emil Fischer in 1891. At that time, it was known that one enantiomer of glyceraldehyde has a specific rotation of +13.5°; the other has a specific rotation of −13.5°. Fischer proposed that these enantiomers be designated D and L, respectively. The question, then, was which enantiomer has which specific rotation? Because there was no experimental way to answer the question at that time, Fischer did the only possible thing; he made an arbitrary assignment. He assigned the dextrorotatory enantiomer the configuration shown below, and named it D-glyceraldehyde. He named its enantiomer L-glyceraldehyde. All other carbohydrates were then assigned by reference to this stereocenter, since relative assignments could be determined experimentally. Fischer could have been wrong, but by a stroke of good fortune, he wasn't. In 1952, his assignment of configuration to the enantiomers of glyceraldehyde was proven correct by a special application of x-ray crystallography.

Penultimate carbon The stereocenter of a monosaccharide farthest from the carbonyl group, as, for example, carbon 5 of glucose.

D-Monosaccharide A monosaccharide that, when written as a Fischer projection, has the —OH on its penultimate carbon to the right.

CHO / H►C◄OH / CH_2OH
(D)-Glyceraldehyde
$[\alpha]_D^{25} = +13.5°$

CHO / HO►C◄H / CH_2OH
(L)-Glyceraldehyde
$[\alpha]_D^{25} = -13.5°$

D-glyceraldehyde and L-glyceraldehyde serve as reference points for the assignment of relative configurations to all other aldoses and ketoses. The reference point is the stereocenter farthest from the carbonyl group. Because this stereocenter is the next to the last carbon on the chain, it is called the **penultimate carbon.** A **D-mono-**

TABLE 24.1 Configurational Relationships Among the Isomeric D-Aldotetroses, D-Aldopentoses, and D-Aldohexoses*

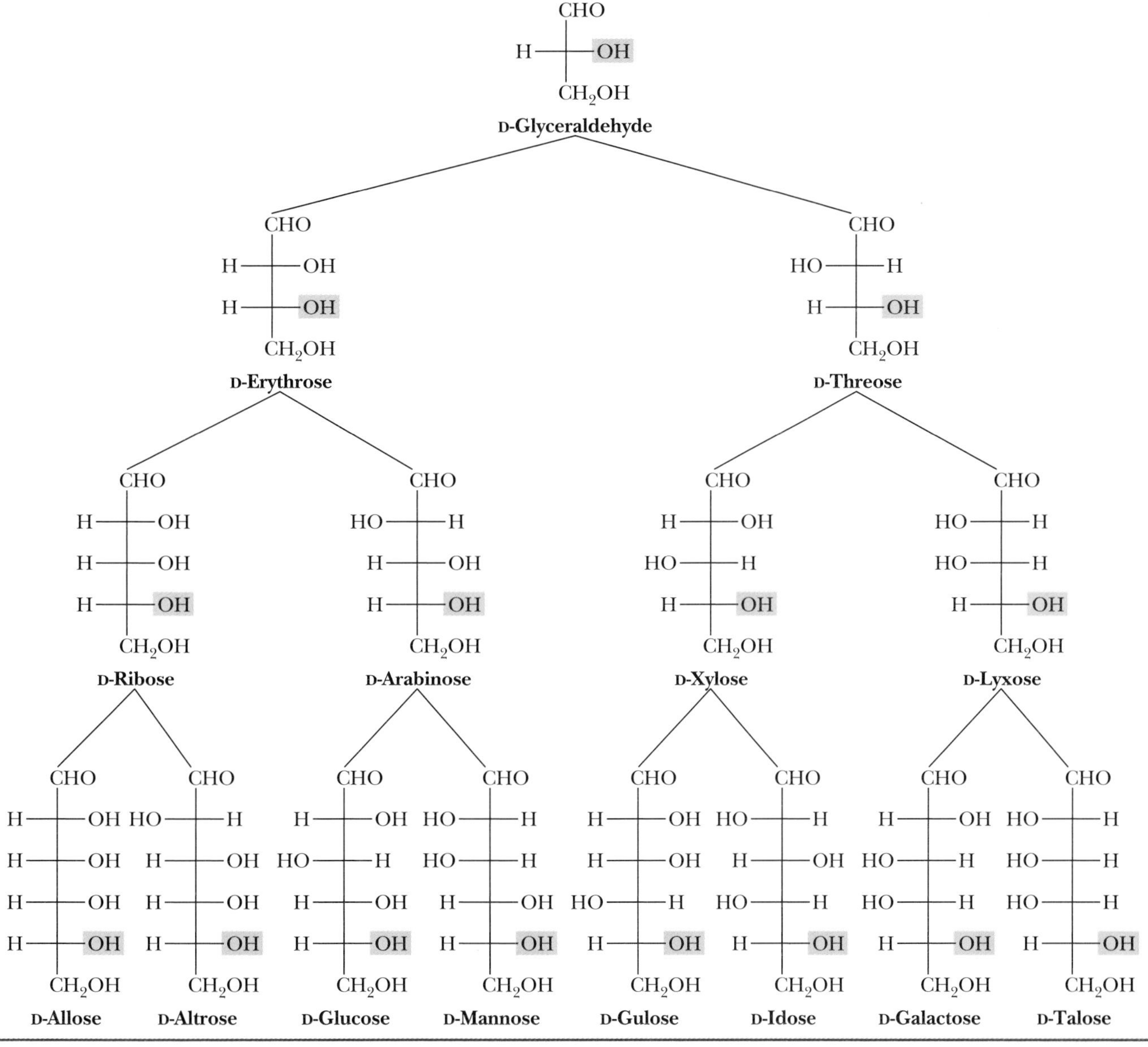

*The configuration of the reference —OH on the penultimate carbon is shown in color.

saccharide is a monosaccharide that has the same configuration at the penultimate carbon as D-glyceraldehyde; an **L-monosaccharide** is a monosaccharide that has the same configuration at the penultimate carbon as L-glyceraldehyde.

L-Monosaccharide A monosaccharide that, when written as a Fischer projection, has the —OH on its penultimate carbon to the left.

Shown in Tables 24.1 and 24.2 are names and Fischer projection formulas for all D-aldo- and D-2- ketotetroses, pentoses, and hexoses. Each name consists of three parts. The letter D specifies the configuration at the stereocenter farthest from the carbonyl

TABLE 24.2 Configurational Relationships Among the D-2-Ketotrioses, D-2-Ketopentoses, and D-2-Ketohexoses*

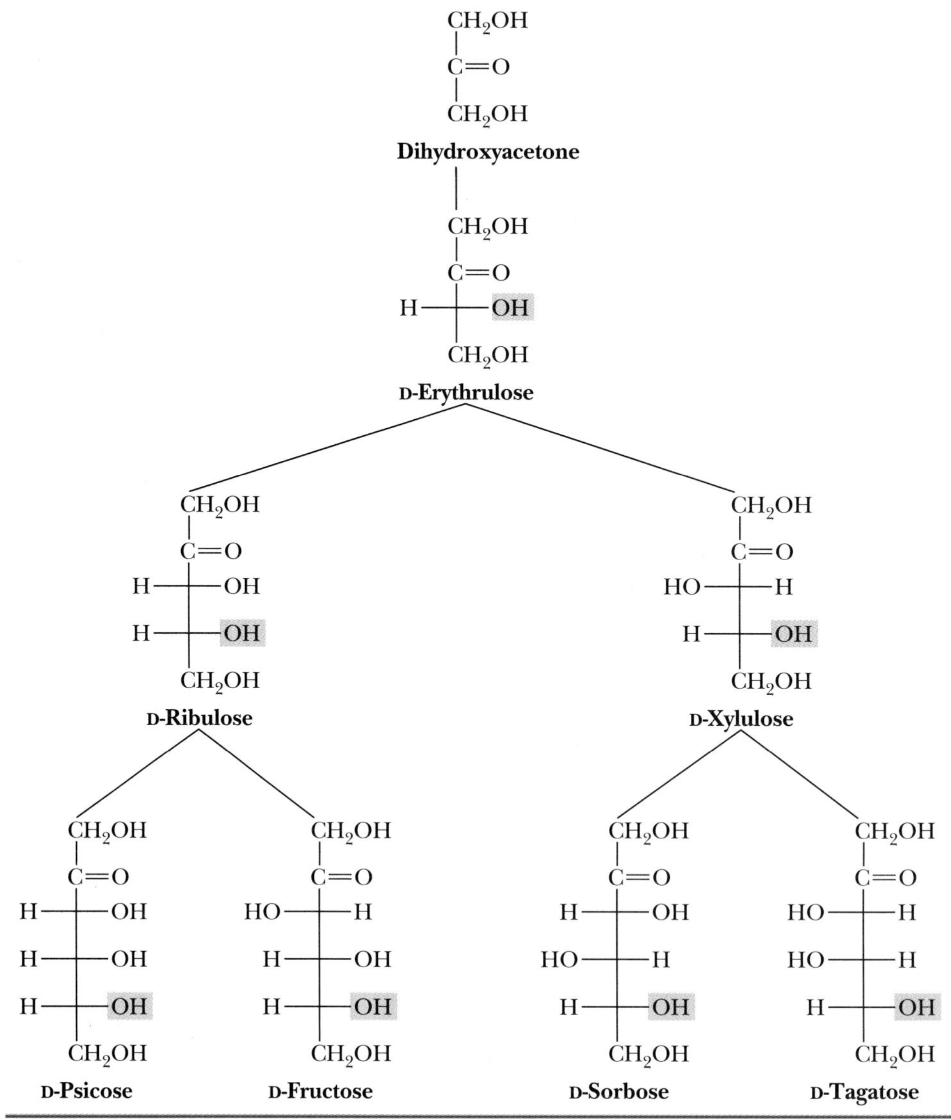

*The configuration of the reference —OH on the penultimate carbon is shown in color.

group. Prefixes, such as rib-, arabin-, and gluc-, are given to these compounds without regard to configuration. The suffix **-ose** shows that the compound is a carbohydrate.

D-Ribose and 2-deoxy-D-ribose, the most abundant pentoses in the biological world, are essential building blocks of nucleic acids (Chapter 28); D-ribose in ribonucleic acids (RNA) and 2-deoxy-D-ribose in deoxyribonucleic acids (DNA).

CHO | H—OH | H—OH | H—OH | CH_2OH

D-Ribose

CHO | H—H | H—OH | H—OH | CH_2OH

2-Deoxy-D-ribose

D-ribose, but without oxygen at carbon 2

The three most abundant hexoses in the biological world are D-glucose, D-galactose, and D-fructose. The first two are D-aldohexoses; the third is a D-ketohexose. Glucose, by far the most common hexose, is also known as dextrose because it is dextrorotatory. Other names for this monosaccharide are grape sugar and blood sugar. Human blood normally contains 65–110 mg of glucose per 100 mL.

Fructose is found combined with glucose in the disaccharide sucrose (table sugar, Section 24.7C). D-Galactose is found combined with glucose in the disaccharide lactose (milk sugar, Section 24.7B).

EXAMPLE 24.1

(a) Draw Fischer projections for the four aldotetroses.

(b) Show which are D-monosaccharides, which are L-monosaccharides, and which are enantiomers.

(c) Refer to Table 24.1 and write names of the aldotetroses you have drawn.

Solution

Following are Fischer projections for the four aldotetroses. D- and L- refer to the configuration of the penultimate carbon, which, in the case of aldotetroses, is carbon 3. In the Fischer projection of a D-aldotetrose, the —OH on carbon 3 is on the right, and, in an L-aldotetrose, it is on the left.

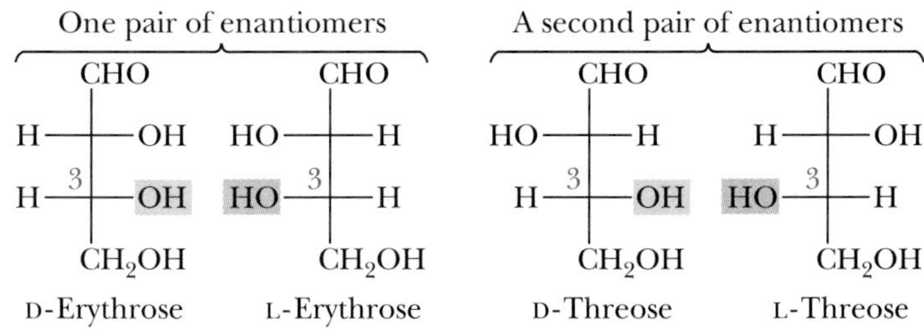

PROBLEM 24.1

(a) Draw Fischer projections of all 2-ketopentoses.

(b) Show which are D-ketopentoses, which are L-ketopentoses, and which are enantiomers.

(c) Refer to Table 24.2, and write names of the ketopentoses you have drawn.

CHEMISTRY IN ACTION

The Genius of Emil Fischer

Emil Fischer was born in a German village near Cologne in 1852. He was educated in German public schools and continued on to the University of Berlin where, against the wishes of his businessman father, he entered a degree program in the sciences. He graduated with a doctorate in chemistry in 1874. Among his first accomplishments were the synthesis of phenylhydrazine, $C_6H_5NHNH_2$, and the discovery that, when treated with this compound, carbohydrates give crystalline derivatives. This discovery made carbohydrates easier to work with. Until that time carbohydrates had been very difficult to characterize because of their tendency to form syrupy mixtures. (Think of honey, which is a concentrated mixture of glucose and fructose.) By using the phenylhydrazine reaction and reaction sequences by which monosaccharide chains can be extended by one carbon at a time, Fischer was able to assign relative configurations to each of the stereocenters in glucose. He published his results in 1891.

Underlying Fischer's work was the theory proposed in 1874 by the Dutch physical chemist Jacobus H. van't Hoff and the French organic chemist Joseph-Achille Le Bel. They suggested that carbon is tetrahedral and that a compound containing a carbon atom bonded to four different groups can exist in enantiomeric forms. This theory was by no means universally accepted. Fischer's establishing the structure of glucose, based as it was on the concept of a tetrahedral carbon atom, offered proof of the theory of van't Hoff and Le Bel and moved it quickly to general acceptance. Thus, in establishing the relative configurations of the stereocenters of glucose, Fischer helped unravel the complexities of carbohydrate structure and at the same time helped establish the validity of a tetrahedral carbon atom. Each accomplishment was as significant as the other. In 1902, Fischer was awarded the Nobel Prize for chemistry for his work on the chemistry of carbohydrates and purines.

Fischer once said that his goal was to be the first to synthesize an "artificial ferment" (what we today call an enzyme) and that he would consider his mission in life complete when that was accomplished. He did not reach that goal nor could he have realized the difficulties and the decades of research it would take before it was accomplished. Fischer died in 1919 of chronic poisoning from his "first chemical love," phenylhydrazine.

E. Amino Sugars

Amino sugars contain an $—NH_2$ group in place of an $—OH$ group. Only three amino sugars are common in nature: D-glucosamine, D-mannosamine, and D-galactosamine.

D-Glucosamine	D-Mannosamine (C-2 stereoisomer of D-glucosamine)	D-Galactosamine (C-4 stereoisomer of D-glucosamine)	*N*-Acetyl-D-glucosamine
CHO	CHO	CHO	CHO
H—C—NH_2	H_2N—C(2)—H	H—C—NH_2	H—C—$NHCCH_3$ (C=O)
HO—C—H	HO—C—H	HO—C—H	HO—C—H
H—C—OH	H—C—OH	HO—C(4)—H	H—C—OH
H—C—OH	H—C—OH	H—C—OH	H—C—OH
CH_2OH	CH_2OH	CH_2OH	CH_2OH

N-Acetyl-D-glucosamine, a derivative of D-glucosamine, is a component of many polysaccharides, including chitin, the hard shell-like exoskeleton of lobsters, crabs, shrimp, and other crustaceans. Several other amino sugars are components of naturally occurring antibiotics. It is the synthesis of these amino sugars that is often the difficult part of antibiotic total synthesis.

24.2 The Cyclic Structure of Monosaccharides

We saw in Section 15.8B that aldehydes and ketones react with alcohols to form **hemiacetals.** We also saw that cyclic hemiacetals form very readily when (1) the hydroxyl and carbonyl groups are parts of the same molecule and (2) their interaction can form a five-membered or six-membered ring. For example, 4-hydroxypentanal forms a five-membered cyclic hemiacetal. Note that 4-hydroxypentanal contains one stereocenter and that a second stereocenter is generated at carbon 1 as a result of hemiacetal formation.

Hemiacetal A molecule containing a carbon bonded to an —OH and an —OR group; the product of adding one molecule of alcohol to the carbonyl group of an aldehyde or ketone.

$CH_3CH(OH)CH_2CH_2CHO$ ≡ (redrawn) ⇌ (cyclic hemiacetal; new stereocenter)

4-Hydroxypentanal (redrawn to show how the cyclic hemiacetal forms) A cyclic hemiacetal

Monosaccharides have hydroxyl and carbonyl groups in the same molecule, and they exist almost exclusively as five- and six-membered cyclic hemiacetals. The new stereocenter resulting from cyclic hemiacetal formation in a carbohydrate is referred to as the **anomeric carbon.** The diastereomers thus formed are given the special name **anomers.**

Anomeric carbon The hemiacetal carbon of the cyclic form of a carbohydrate.

Anomers Carbohydrates that differ in configuration only at their anomeric carbons.

A. Haworth Projections

A common way of representing the cyclic structure of monosaccharides is the **Haworth projection,** named after the English chemist Sir Walter N. Haworth (Nobel laureate, 1937). In Haworth projections, five- and six-membered cyclic hemiacetals are represented as planar pentagons or hexagons, as the case may be, lying perpendicular to the plane of the paper. Groups attached to the carbons of the ring then lie either above or below the plane of the ring. Haworth projections are most commonly written with the anomeric carbon to the right and the hemiacetal oxygen to the back right (Figure 24.1). In Haworth projections, the —OH on the anomeric carbon of the cyclic hemiacetal is either *cis* or *trans* to the terminal $—CH_2OH$, depending on the orientation of the aldehyde group when it reacts with the —OH from carbon 5.

Haworth projection A way to view furanose and pyranose forms of monosaccharides. The ring is drawn flat and viewed through its edge with the anomeric carbon on the right and the oxygen atom of the ring in the rear.

In the terminology of carbohydrate chemistry, the designation β means that the —OH on the anomeric carbon of the cyclic hemiacetal is up, that is, on the same side as the terminal $—CH_2OH$. Conversely, the designation α means that the —OH on the anomeric carbon of the cyclic hemiacetal is down, on the side opposite the terminal $—CH_2OH$.

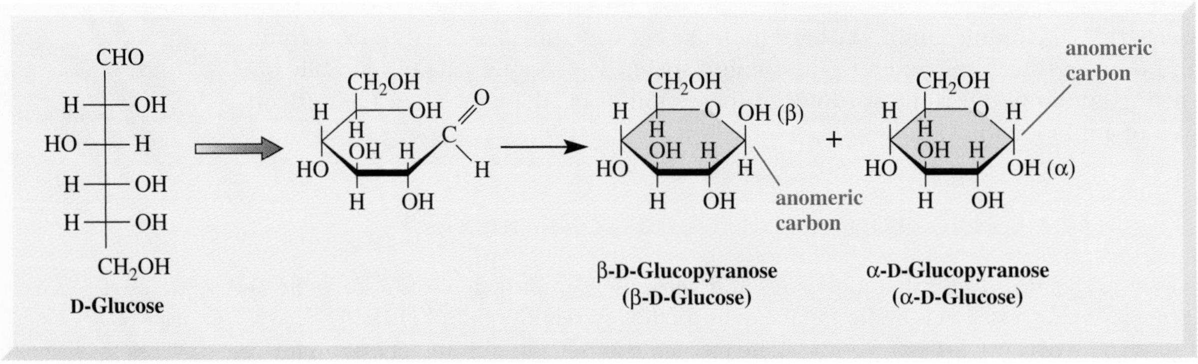

FIGURE 24.1
Haworth projections for α-D-glucopyranose and β-D-glucopyranose.

Furanose A five-membered cyclic form of a monosaccharide.

Pyranose A six-membered cyclic form of a monosaccharide.

Six-membered hemiacetal rings are shown by the infix -pyran- and five-membered hemiacetal rings are shown by the infix -furan-. The terms **furanose** and **pyranose** are used because monosaccharide five- and six-membered rings correspond to the heterocyclic compounds furan and pyran.

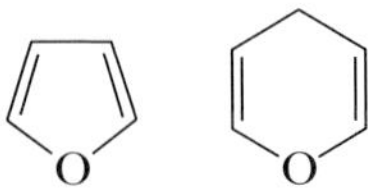

Because the α and β forms of glucose are six-membered cyclic hemiacetals, they are named α-D-glucopyranose and β-D-glucopyranose. However, for convenience, they are often named simply α-D-glucose and β-D-glucose.

You would do well to remember the configuration of groups on the Haworth projection of both α-D-glucopyranose and β-D-glucopyranose as reference structures. By knowing how the open-chain configuration of any other monosaccharide differs from that of D-glucose, you can then construct its Haworth projection by reference to the Haworth projection of D-glucose.

EXAMPLE 24.2

Draw Haworth projections for the α and β anomers of D-galactopyranose.

Solution

One way to arrive at the structures for the α and β anomers of D-galactopyranose is to use the α and β forms of D-glucopyranose as reference and to remember, or discover by looking at Table 24.1, that D-galactose differs from D-glucose only in the configuration of carbon 4. Following are Haworth projections for α-D-galactopyranose and β-D-galactopyranose. Note that in these Haworth projections, all C—H bonds are shown.

Configuration differs from that of D-glucose at C-4

α-D-Galactopyranose
(α-D-Galactose)

β-D-Galactopyranose
(β-D-Galactose)

PROBLEM 24.2

D-Mannose exists in aqueous solution as a mixture of α-D-mannopyranose and β-D-mannopyranose. Draw Haworth projections for these molecules.

The most prevalent forms of D-ribose and other pentoses in the biological world are furanoses. Shown in Figure 24.2 are Haworth projections for α-D-ribofuranose (α-D-ribose) and β-2-deoxy-D-ribofuranose (β-2-deoxy-D-ribose). Units of D-ribose and 2-deoxy-D-ribose in nucleic acids and most other biological molecules are found almost exclusively in the β-configuration.

B. Conformational Representations

A five-membered ring is so close to being planar that Haworth projections are adequate to represent furanoses. For pyranoses, however, the six-membered ring is more accurately represented as a strain-free chair conformation (Section 2.6B). Structural formulas for α-D-glucopyranose and β-D-glucopyranose are drawn in Figure 24.3 as strain-free **chair conformations.** Also shown is the open-chain or free aldehyde form with which the cyclic hemiacetal forms are in equilibrium in aqueous solution. Notice that each of the groups, including the anomeric —OH, on the chair conformation of β-D-glucopyranose is equatorial. Finally, notice that the —OH on the anomeric carbon is axial in α-D-glucopyranose. Because of the equatorial orientation of the —OH

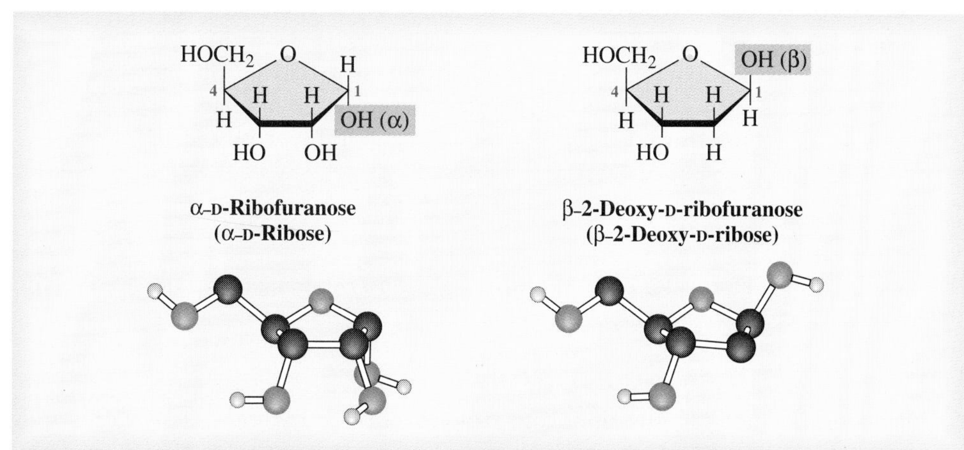

FIGURE 24.2
Haworth projections for the D-ribofuranoses.

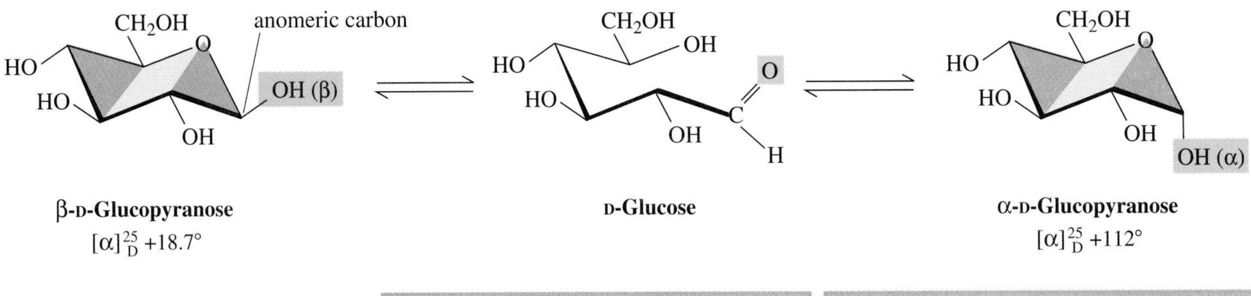

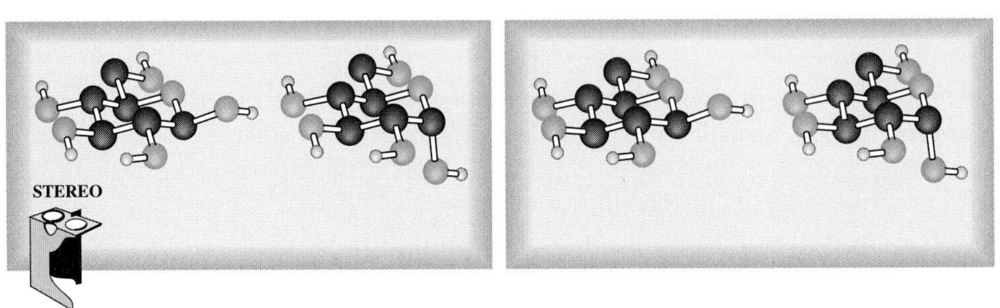

FIGURE 24.3 Chair conformations of α-D-glucopyranose and β-D-glucopyranose.

on its anomeric carbon, β-D-glucopyranose is more stable and predominates in aqueous solution.

At this point, you should compare the relative orientations of groups on the D-glucopyranose ring in the Haworth projection and chair conformation. Notice that the orientations of groups on carbons 1 through 5 in the Haworth projection of β-D-glucopyranose are up, down, up, down, and up, respectively. The same is the case in the chair conformation, which allows all substituents on the chair to be equatorial.

β-D-Glucopyranose (Haworth projection)

β-D-Glucopyranose (chair conformation)

Other monosaccharides also form cyclic hemiacetals. As with the examples we have studied, five- and six-membered cyclic hemiacetals are by far the most common. Shown in Figure 24.4 are the five- and six-membered cyclic hemiacetals of fructose. The β-D-fructofuranose form is found in the disaccharide sucrose (Section 24.7C).

EXAMPLE 24.3

Draw chair conformations for α-D-galactopyranose and β-D-galactopyranose. Label the anomeric carbon in each cyclic hemiacetal.

Solution

D-Galactose differs in configuration from D-glucose only at carbon 4. Therefore, draw the α and β forms of D-glucopyranose and then interchange the positions of the

—OH and —H groups on carbon 4. Shown also are the specific rotations of each anomer.

anomeric carbon

β-D-Galactopyranose (β-D-Galactose) $[\alpha]_D^{25} = +52.8°$ ⇌ D-Galactose ⇌ α-D-Galactopyranose (α-D-Galactose) $[\alpha]_D^{25} = +150.7°$

PROBLEM 24.3

Draw chair conformations for α-D-mannopyranose and β-D-mannopyranose. Label the anomeric carbon atom in each.

C. Mutarotation

Mutarotation is the change in specific rotation that accompanies the interconversion of α- and β-anomers in aqueous solution. As an example, a solution prepared by dissolving crystalline α-D-glucopyranose in water shows an initial rotation of +112° (Table 24.3), which gradually decreases to an equilibrium value of +52.7° as α-D-glucopyranose reaches an equilibrium with β-D-glucopyranose. A solution of β-D-

Mutarotation The change in specific rotation that occurs when an α form or β form of a carbohydrate is converted to a equilibrium mixture of the two forms.

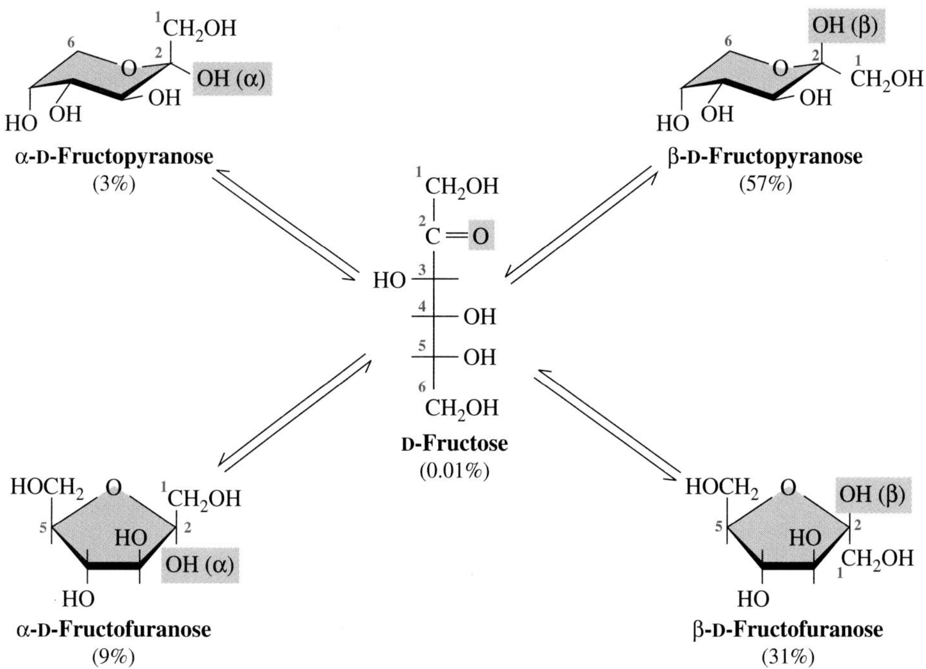

FIGURE 24.4
The major forms of D-fructose at equilibrium in aqueous solution.

TABLE 24.3 Specific Rotations for α- and β-anomers of D-Glucose and D-Galactose Before and After Mutarotation

Monosaccharide	Specific Rotation (degrees)	Specific Rotation After Mutarotation (degrees)	Percent Present at Equilibrium
α-D-glucose	+112.0	+52.7	36
β-D-glucose	+18.7	+52.7	64
α-D-galactose	+150.7	+80.2	28
β-D-galactose	+52.8	+80.2	72

glucopyranose also undergoes mutarotation, during which the specific rotation changes from an initial value of +18.7° (Table 24.3) to the same equilibrium value of +52.7°. Furthermore, it has been determined by modern techniques, most notably ^{13}C-NMR spectroscopy, that only traces of the furanose and open-chain forms are present at equilibrium in aqueous solution. Thus, from the value of the specific rotation after mutarotation, we can calculate that the equilibrium mixture consists of approximately 64% β-D-glucopyranose and 36% α-D-glucopyranose.

Mutarotation is common to all carbohydrates that exist in hemiacetal forms. Also shown in Table 24.3 are specific rotations for freshly prepared solutions of the α and β forms of D-galactopyranose, along with equilibrium values for the specific rotation of each after equilibrium. For this monosaccharide, only traces of the furanose forms and aldehyde are present at equilibrium in aqueous solution.

From analyses of the composition of the equilibrium mixtures of monosaccharides in water, we can make the following generalizations. For most monosaccharides in aqueous solution,

1. Little free aldehyde or ketone is present.
2. Pyranose forms predominate over furanose forms. It is important not to confuse this statement about the form present at equilibrium in aqueous solution with a statement about the form that predominates in a biological system. They may be quite different. For example, although D-ribose and 2-deoxy-D-ribose exist in aqueous solution mainly in the pyranose form, each is found in nucleic acids exclusively in the β-furanose form.

24.3 Physical Properties

Monosaccharides are colorless, crystalline solids. Because hydrogen bonding is possible between their polar —OH groups and water, all monosaccharides are very soluble in water. They are only slightly soluble in alcohol and are insoluble in nonpolar solvents such as diethyl ether, chloroform, and toluene.

Although all monosaccharides are sweet to the taste, some are sweeter than others (Table 24.4). D-Fructose tastes the sweetest, considerably sweeter than sucrose (table sugar). In the production of sucrose, sugar cane or sugar beet is boiled with water, and the resulting solution cooled. Sucrose crystals separate and are collected. Subsequent

TABLE 24.4 Relative Sweetness of Some Carbohydrate Sweetening Agents

Monosaccharide		Disaccharide		Other Carbohydrate Sweetening Agent	
D-fructose	174	sucrose (table sugar)*	100	honey	97
D-glucose	74	lactose (milk sugar)	0.16	molasses	74
D-xylose	0.40			corn syrup	74
D-galactose	0.22				

* Sucrose is taken as a standard for relative sweetness and is assigned a value of 100.

boiling to concentrate the remaining solution followed by cooling yields a dark, thick syrup known as molasses. The sweet taste of honey is due largely to D-fructose, and that of corn syrup, to D-glucose. Lactose has almost no sweetness and is sometimes added to foods as a filler. Some people cannot tolerate lactose well and should avoid these foods.

24.4 Reactions of Monosaccharides

In this section, we discuss the reactions of monosaccharides with alcohols and with methylating, reducing, and oxidizing agents.

A. Formation of Glycosides (Acetals)

Reaction of a monosaccharide hemiacetal with an alcohol forms an acetal as illustrated by the reaction of β-D-glucopyranose (β-D-glucose) with methanol.

β-D-Glucopyranose (β-D-Glucose) + CH_3OH $\xrightarrow[-H_2O]{H^+}$ Methyl β-D-glucopyranoside (Methyl β-D-glucoside) + Methyl α-D-glucopyranoside (Methyl α-D-glucoside)

anomeric carbon

A cyclic acetal derived from a monosaccharide is called a **glycoside,** and the bond from the anomeric carbon to the —OR group is called a **glycoside bond.** Mutarotation is no longer possible in a glycoside because, unlike a hemiacetal, an acetal is not in equilibrium with the open-chain, carbonyl-containing compound in neutral aqueous solution.

Glycosides are named by listing the alkyl or aryl group attached to oxygen followed by the name of the carbohydrate involved in which the ending -e is replaced by -ide. For example, glycosides derived from β-D-glucopyranose are named β-D-glucopyranosides; those derived from β-D-ribofuranose are named β-D-ribofuranosides.

Glycoside A carbohydrate in which the —OH on its anomeric carbon is replaced by —OR.

Glycoside bond The bond from the anomeric carbon of a glycoside to an —OR group.

EXAMPLE 24.4

Draw structural formulas for these glycosides. For each structure, label the anomeric carbon and the glycoside bond.

(a) Methyl β-D-ribofuranoside (methyl β-D-riboside)
(b) Methyl α-D-galactopyranoside (methyl α-D-galactoside)

Solution

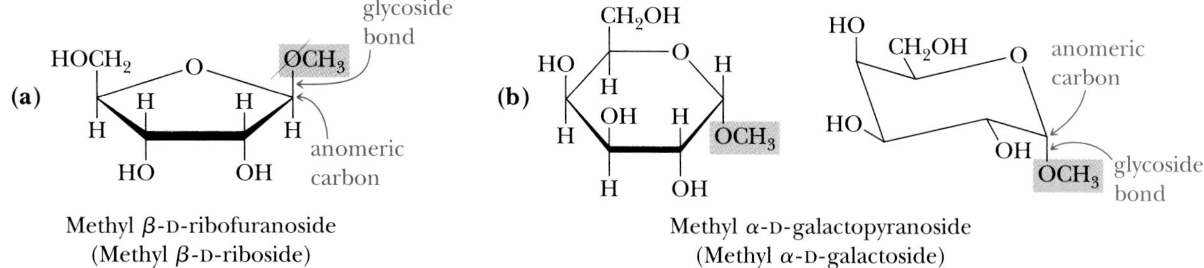

Methyl β-D-ribofuranoside (Methyl β-D-riboside)

Methyl α-D-galactopyranoside (Methyl α-D-galactoside)

PROBLEM 24.4

Draw structural formulas for these glycosides. In each, label the anomeric carbon and the glycoside bond.

(a) Methyl β-D-fructofuranoside (methyl β-D-fructoside)
(b) Methyl α-D-mannopyranoside (methyl α-D-mannoside)

Glycosides are stable in water and aqueous base, but like other acetals (Section 15.8B), they are hydrolyzed in aqueous acid to an alcohol and a monosaccharide. The mechanism of hydrolysis is exactly the reverse of the mechanism shown on the next page.

EXAMPLE 24.5

Propose a mechanism for the reaction of β-D-glucopyranose with methanol in the presence of a mineral acid to give a mixture of methyl α-D-glucopyranoside and methyl β-D-glucopyranoside. In writing this mechanism, show the pyranose ring in a chair conformation.

Solution

Refer to Section 15.8B and the mechanism presented for acid-catalyzed formation of an acetal. Apply the same mechanistic reasoning to this problem. In this solution, the mechanism is divided into two steps.

MECHANISM Acid-Catalyzed Formation of Methyl α- and β-D-Glucopyranosides

Step 1: Proton transfer to the anomeric —OH followed by loss of H_2O gives a resonance-stabilized cation

An oxonium ion

A resonance-stabilized cation

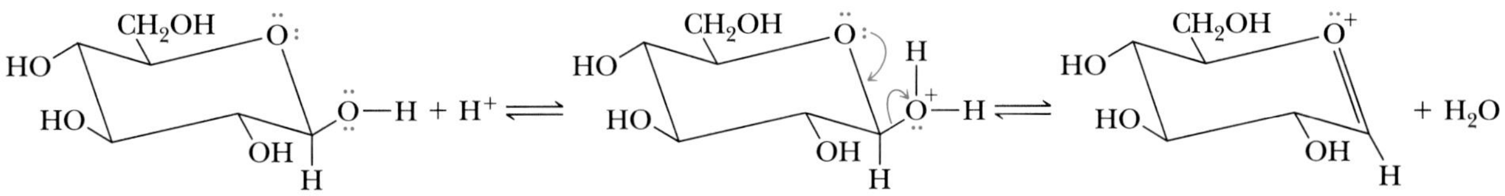

Step 2: Reaction of the resonance-stabilized cation with methanol followed by loss of H^+ gives the α- and β-glucopyranosides

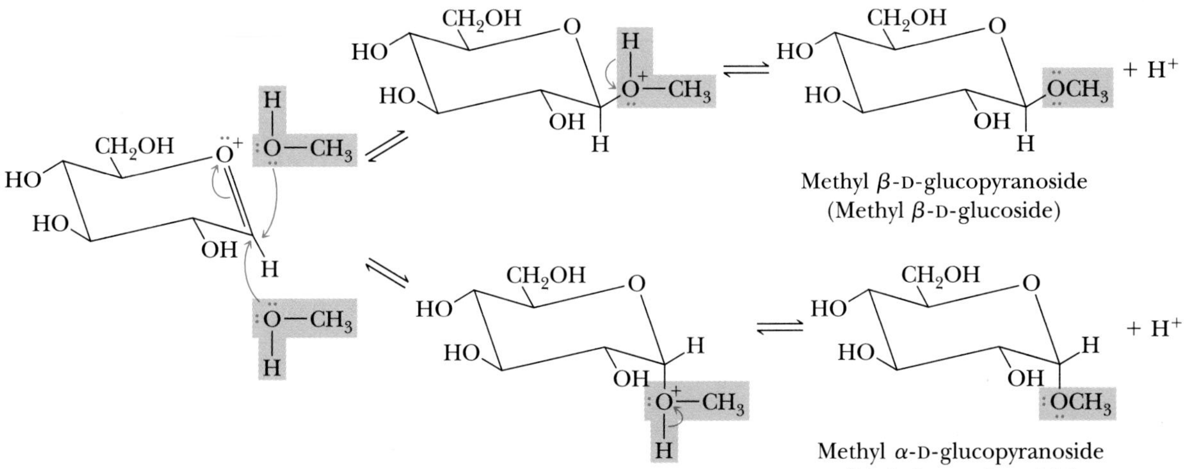

Methyl β-D-glucopyranoside (Methyl β-D-glucoside)

Methyl α-D-glucopyranoside (Methyl α-D-glucoside)

PROBLEM 24.5

Suppose that a β-D-glycopyranose is treated with methanol enriched in oxygen-18. Is the isotopic label found in the resulting methyl glycoside, in the water produced in the reaction, or in both the methyl glycoside and in the water? Explain.

Just as the anomeric carbon of a cyclic hemiacetal undergoes reaction with the —OH group of an alcohol to form a glycoside, it can also undergo reaction with the N—H group of an amine to form an *N*-glycoside. Especially important in the biological world are the *N*-glycosides formed between D-ribose and 2-deoxy-D-ribose, each as a furanose, and the heterocyclic aromatic amines uracil, cytosine, thymine, adenine, and guanine (Figure 24.5). *N*-Glycosides of these pyrimidine and purine bases are structural units of nucleic acids (Chapter 28).

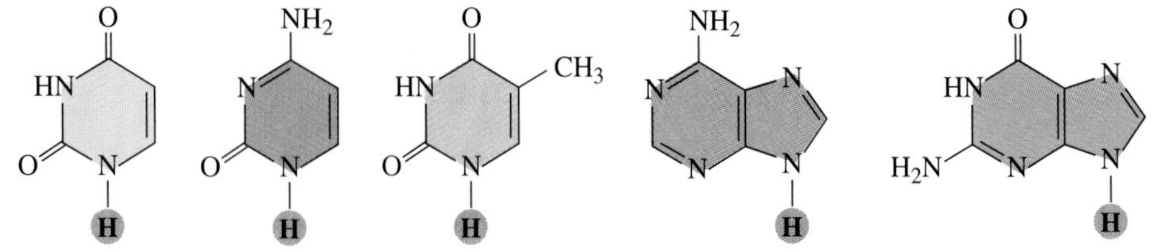

FIGURE 24.5
Structural formulas of the five most important pyrimidine and purine bases found in DNA and RNA. The hydrogen atom shown in color is lost in forming an *N*-glycoside.

Uracil Cytosine Thymine Adenine Guanine

EXAMPLE 24.6

Draw a structural formula for the β-*N*-glycoside formed between D-ribofuranose and cytosine.

Solution

The following are structural formulas for the heterocyclic aromatic amine base, the monosaccharide hemiacetal, and the β-*N*-glycoside.

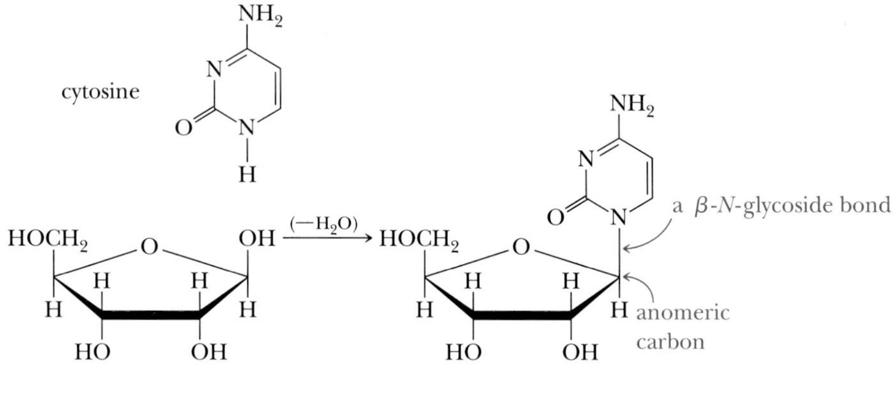

cytosine

β-D-ribofuranose

a β-*N*-glycoside bond

anomeric carbon

PROBLEM 24.6

Draw a structural formula for the β-*N*-glycoside formed between 2-deoxy-D-ribofuranose and adenine.

B. Formation of Methyl Ethers

Treatment of a monosaccharide with methanol or another alcohol in the presence of an acid catalyst converts it into a mixture of α- and β-glycosides. All remaining —OH groups can be converted to methyl ether groups by treatment of the glycoside with dimethyl sulfate in the presence of a strong base such as sodium hydride, NaH (Williamson ether synthesis, Section 11.4A). In this reaction, the function of sodium hydride is to convert each —OH (a poor nucleophile) into an $—O^-$ (a good nucleophile), which then participates in an S_N2 reaction by attacking a carbon of dimethyl sulfate and displacing sulfate ion (a good leaving group).

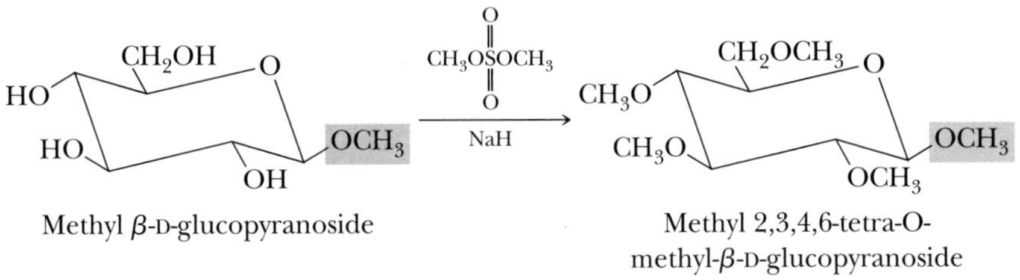

$$\text{Methyl } \beta\text{-D-glucopyranoside} \xrightarrow[\text{NaH}]{CH_3OSO_2OCH_3} \text{Methyl 2,3,4,6-tetra-O-methyl-}\beta\text{-D-glucopyranoside}$$

Methyl β-D-glucopyranoside

Methyl 2,3,4,6-tetra-O-methyl-β-D-glucopyranoside

Because all free —OH groups on the monosaccharide are converted to $—OCH_3$ groups, the process is called permethylation. In naming these compounds, each $—OCH_3$ other than that on the anomeric carbon is named as an O-methyl group. Treatment of a permethylated glycoside with dilute aqueous acid results in hydrolysis of the methyl glycoside bond (acetals are hydrolyzed in dilute aqueous acid; Section 15.8B) but other $—OCH_3$ groups are unaffected (ethers are stable under these conditions; Section 11.6A).

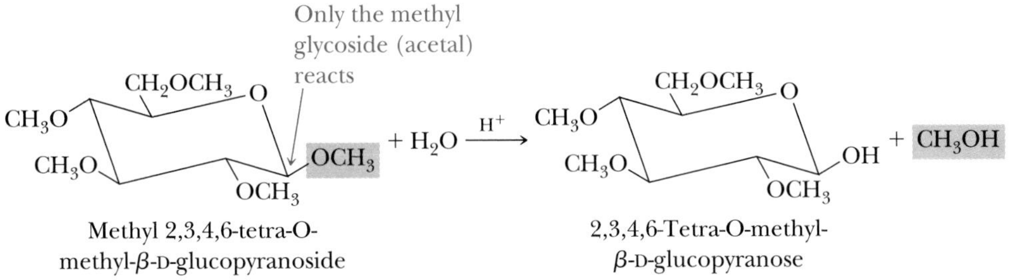

Only the methyl glycoside (acetal) reacts

$$\text{Methyl 2,3,4,6-tetra-O-methyl-}\beta\text{-D-glucopyranoside} + H_2O \xrightarrow{H^+} \text{2,3,4,6-Tetra-O-methyl-}\beta\text{-D-glucopyranose} + CH_3OH$$

Methyl 2,3,4,6-tetra-O-methyl-β-D-glucopyranoside

2,3,4,6-Tetra-O-methyl-β-D-glucopyranose

Permethylation followed by acid-catalyzed hydrolysis of the glycoside bond is an important experimental method for determining the ring size of a monosaccharide glycoside. For example, the fact that the preceding reaction of permethylated methyl β-D-glucoside gives 2,3,4,6-tetra-O-methyl-D-glucose demonstrates that the cyclic structure in the glycoside must have been formed between the carbonyl group at position 1 and the —OH on carbon 5.

Example 24.7

How do you account for the observation that an $—OCH_3$ glycoside bond is hydrolyzed in dilute HCl, but other $—OCH_3$ bonds in a permethylated monosaccharide are not hydrolyzed under these conditions?

Solution

The rate-limiting step in this hydrolysis is formation of a cation. Protonation of the glycoside oxygen followed by loss of methanol gives a resonance-stabilized cation.

protonation here followed by loss of CH_3OH

$$\xrightarrow{H^+} \text{A resonance-stabilized cation} + CH_3OH$$

A resonance-stabilized cation

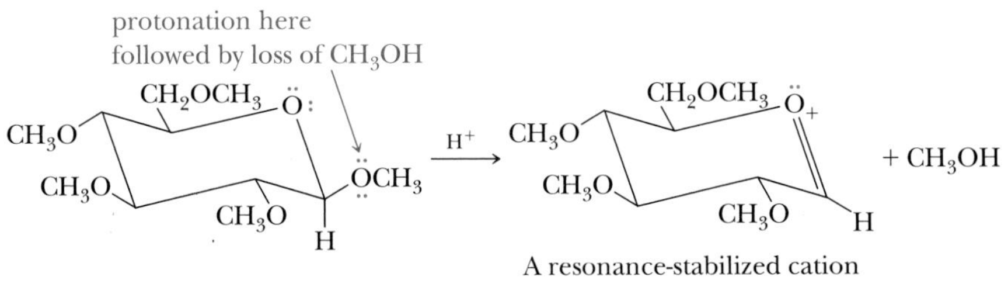

Chemistry in Action

Glycosidase Inhibitors: Natural Defensive Weapons of Plants

Plants and their herbivores coevolve. The continuing adjustments of one to the other reflect the biosynthesis of defensive compounds by plants and the development of detoxification mechanisms by herbivores. Examples of defensive compounds are a group of polyhydroxypiperidines and polyhydroxypyrrolidines recently discovered in several members of the pea family. We discuss one of these compounds, which is a powerful inhibitor of glycosidase activity in insects, mammals, and microorganisms. **Glycosidases** are a class of enzymes that catalyze the hydrolysis of glycoside bonds. These enzymes enable plants and animals to convert di-, tri-, and polysaccharides into monosaccharides, which they then use as sources of energy and sources of carbon for the synthesis of other biomolecules.

Deoxynojirimycin, isolated not only from peas but also from species of bacteria and mulberry trees, is an inhibitor of both α- and β-glycosidase activity in susceptible insects. A possible basis for this inhibition is deoxynojirimycin's similarity in structure to D-glucopyranose, which you can see by drawing deoxynojirimycin in a chair conformation and comparing it with the chair conformation of β-D-glucopyranose. In deoxynojirimycin, the —O— of the cyclic hemiacetal of D-glucopyranose is replaced by —NH—, and the —OH is missing from carbon 2 of the ring (hence the prefix deoxy-). Thus, deoxynojirimycin is a monosaccharide "look-alike." The glycosidase accepts this look-alike molecule as a monosaccharide, upon which the inhibitor blocks the enzyme from catalyzing glycoside hydrolysis.

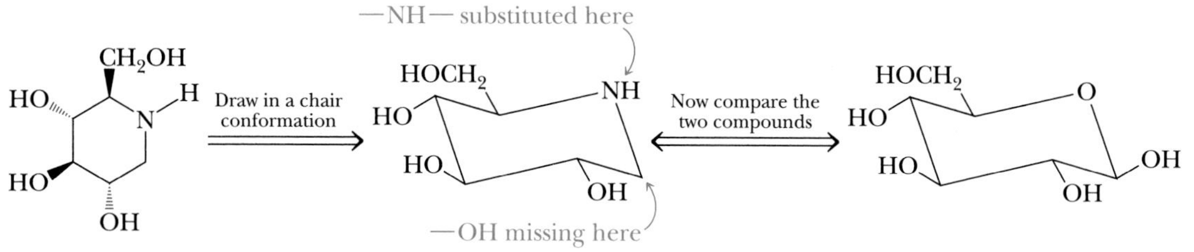

—NH— substituted here

CH_2OH HO H N HO OH — Draw in a chair conformation → $HOCH_2$ HO NH HO OH ⟷ Now compare the two compounds ⟷ $HOCH_2$ HO O HO OH OH

—OH missing here

Deoxynojirimycin

β-D-glucose

Protonation of any of the methyl ether oxygens followed by loss of methanol gives an ordinary primary or secondary carbocation, depending on which carbon loses methanol.

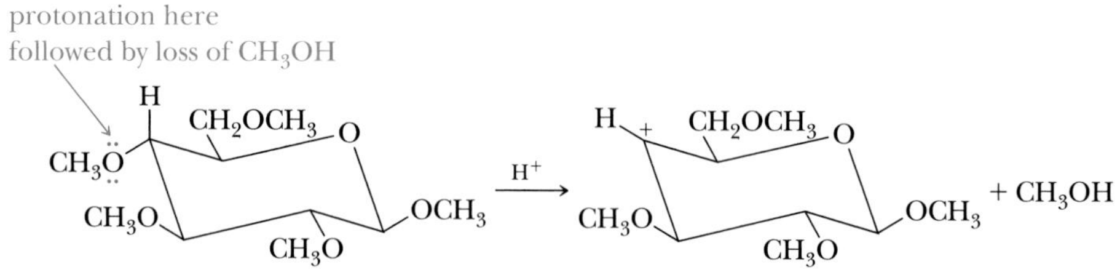

protonation here followed by loss of CH_3OH

H, CH_2OCH_3, O, $CH_3\ddot{O}$, CH_3O, CH_3O, OCH_3 $\xrightarrow{H^+}$ H, +, CH_2OCH_3, O, CH_3O, CH_3O, OCH_3 + CH_3OH

A secondary carbocation

The resonance-stabilized cation formed by loss of the glycoside alcohol is more stable and has a lower activation energy for its formation than any of the alternative cations. Recall from Section 11.6A that hydrolysis of a primary or secondary alkyl ether requires

The fungus infecting sweet peas is carried by the pea weevil, a species of bruchid beetle. Shown here is the bean weevil, a closely related species, which bores into dried beans in storage to create breeding chambers. (Fran Heyl Associates)

How this glycosidase inhibitor functions as a weapon in the chemical defense systems of higher plants is not fully understood. One suggestion is that insect glycosidases are activated when the insect comes in contact with a suitable plant source of food. These enzymes, thus activated, catalyze the release of free sugars in the plant, which then trigger the feeding response. Because deoxynojirimycin inhibits glycosidase activity, its defensive role may be to inhibit the release of free sugars and, thereby, act as a feeding deterrent. Or, it may be a feeding deterrent because it tastes unpleasant to the bruchid beetles, which are serious economic pests to pea crops.

Whatever the biochemical mechanism that deters feeding, many insects have developed a resistance to such inhibitors. Some insects probably have even developed the ability to use them as sources of carbon and nitrogen atoms for their own growth!

See Green, M. B., and Hedin, P. A., eds., "*Natural Resistance of Plants to Pests,*" *ACS Symposium Series*, no. 296, American Chemical Society, Washington, D.C., 1986.

hot concentrated acid and a strong nucleophile, both of which are present in concentrated HBr and HI.

PROBLEM 24.7

Suppose it were possible to convert D-glucose to a permethylated D-glucofuranoside. Draw an open-chain formula for the tetra-O-methyl-D-glucose that would be isolated after this permethylated derivative is treated with dilute aqueous HCl.

C. Reduction to Alditols

The carbonyl group of a monosaccharide can be reduced to a hydroxyl group by a variety of reducing agents, including $NaBH_4$ and hydrogen in the presence of a transition metal catalyst, The reduction products are known as **alditols.** Reduction of D-glucose gives D-glucitol, more commonly known as D-sorbitol. Note that D-glucose is

Alditol The product formed when the C=O group of a monosaccharide is reduced to a CHOH group.

shown here in the open-chain form. Only a small amount of this form is present in solution, but as it is reduced, the equilibrium between cyclic hemiacetal forms and the open-chain form shifts to replace it.

CHO; H—OH; HO—H; H—OH; H—OH; CH_2OH $+ H_2 \xrightarrow{Ni}$ CH_2OH; H—OH; HO—H; H—OH; H—OH; CH_2OH

D-Glucose D-Glucitol (D-Sorbitol)

Many "sugarfree" products contain sugar alcohols, such as D-sorbitol and xylitol. *(Gregory Smolin)*

Sorbitol is found in the plant world in many berries (except grapes) and in cherries, plums, pears, apples, seaweed, and algae. It is about 60% as sweet as sucrose and is used in the manufacture of candies and as a sugar substitute for diabetics. It is also used to keep products moist, because it absorbs water.

Also common in the biological world are erythritol, D-mannitol, and xylitol.

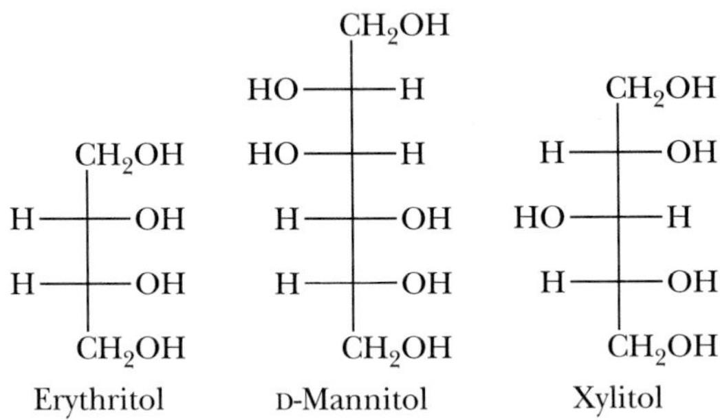

At one time, xylitol was used as a sweetening agent in "sugarless" gum, candy, and sweet cereals. However, it has been removed from the market because tests showed it to be potentially carcinogenic.

EXAMPLE 24.8

D-Glucose is reduced by $NaBH_4$ to D-glucitol. Do you expect the product formed under these conditions to be optically active or optically inactive?

Solution

D-Glucitol is a chiral substance. Given the fact that reduction by $NaBH_4$ does not affect any of the stereocenters nor does the product have a plane of symmetry, the product is optically pure and optically active.

PROBLEM 24.8

D-Erythrose is reduced by $NaBH_4$ to erythritol. Do you expect the alditol formed under these conditions to be optically active or optically inactive?

D. Oxidation to Aldonic and Aldaric Acids

Treatment of an aldose with a buffered solution of Br_2 in water results in oxidation of the aldehyde group and formation of a monocarboxylic acid known as an **aldonic acid.** Oxidation is commonly carried out in the presence of calcium carbonate, whose function is to neutralize the HBr formed during the oxidation. The oxidation product is formed as the calcium salt, which, on treatment with aqueous acid, give the carboxylic acid.

Aldonic acid The product formed when the —CHO group of an aldose is oxidized to a $—CO_2H$ group.

CHO; H—OH; HO—H; H—OH; H—OH; CH_2OH (D-Glucose) $+ Br_2$ $\xrightarrow[2.\ H_3O^+]{1.\ H_2O,\ CaCO_3}$ CO_2H; H—OH; HO—H; H—OH; H—OH; CH_2OH (D-Gluconic acid) $+ CaBr_2 + CO_2$

Monosaccharides (and carbohydrates in general) are classified as reducing sugars or nonreducing sugars according to their behavior toward Ag(I) (Tollens' solution, Section 15.13A) or Cu(II) (Benedict's solution and Fehling's solution). Those that reduce copper(II) ion to Cu_2O or silver(I) to metallic silver are classified as **reducing sugars.** Those that do not reduce these reagents are classified as **nonreducing sugars.** The chemical basis for this classification depends on two features: first, all reducing sugars are hemiacetals in equilibrium with small amounts of open-chain aldose or ketose; second, in dilute aqueous base, the condition of these tests, ketoses are in equilibrium with aldoses via enediol intermediates.

Reducing sugar A carbohydrate that reduces Ag(I) to Ag or Cu(II) to Cu(I).

$$\text{RCH(=O)} + \underset{\text{Tollens' solution}}{Ag(NH_3)_2^+} \xrightarrow{NH_3,\ H_2O} \text{RC(=O)O}^-NH_4^+ + \underset{\text{Precipitates as a silver mirror}}{Ag}$$

Benedict's solution is prepared by dissolving copper(II) sulfate in aqueous sodium citrate. Fehling's solution is prepared by dissolving copper(II) sulfate in aqueous sodium tartrate. The function of the sodium citrate and sodium tartrate is to buffer the pH of the solution and to form a complex ion with copper(II). A positive Benedict's or Fehling's test is indicated by formation of copper(I) oxide, which precipitates as a red solid.

$$\text{RCH(=O)} + Cu^{2+} \xrightarrow{\text{citrate or tartrate buffer}} \text{RC(=O)O}^- + \underset{\text{Precipitates as a brick-red solid}}{Cu_2O}$$

Ketoses also reduce these solutions. Carbon 1 of a ketose is not oxidized directly. Rather, the basic conditions of the test catalyze isomerization of a 2-ketose to an aldose by way of an enediol intermediate. The aldose then gives the positive test with Tollens', Fehling's, or Benedict's solutions.

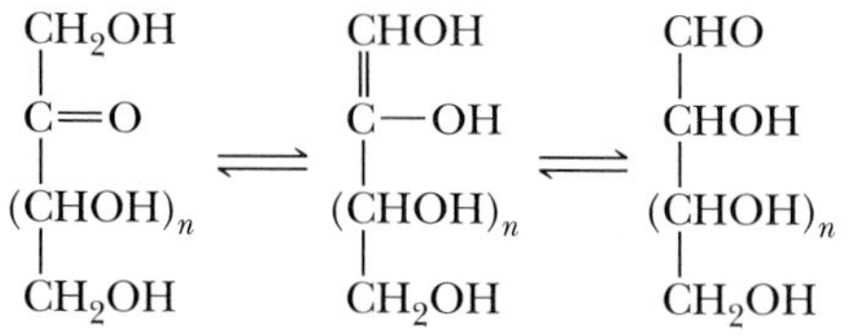

$$\begin{array}{c} CH_2OH \\ | \\ C{=}O \\ | \\ (CHOH)_n \\ | \\ CH_2OH \end{array} \rightleftharpoons \begin{array}{c} CHOH \\ \| \\ C{-}OH \\ | \\ (CHOH)_n \\ | \\ CH_2OH \end{array} \rightleftharpoons \begin{array}{c} CHO \\ | \\ CHOH \\ | \\ (CHOH)_n \\ | \\ CH_2OH \end{array}$$

A 2-ketose — An enediol — An aldose

Warm nitric acid, a more powerful oxidizing agent, oxidizes the —CHO group of an aldose to a —CO_2H group and also oxidizes the terminal —CH_2OH to —CO_2H. A dicarboxylic acid derived from an aldose is called an **aldaric acid.** For example, oxidation of D-glucose by nitric acid yields D-glucaric acid.

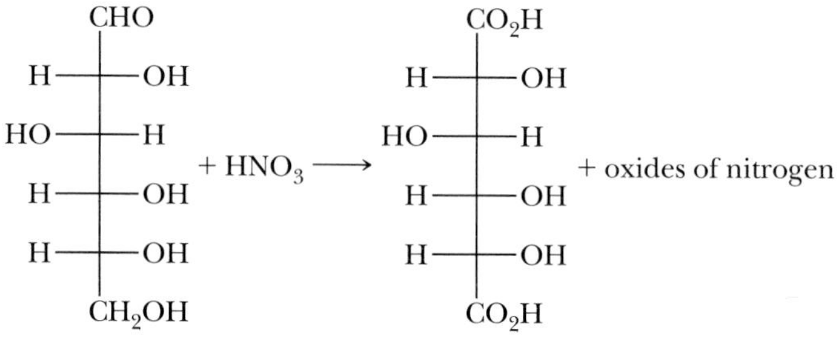

$$\begin{array}{c} CHO \\ H{-}C{-}OH \\ HO{-}C{-}H \\ H{-}C{-}OH \\ H{-}C{-}OH \\ CH_2OH \end{array} + HNO_3 \longrightarrow \begin{array}{c} CO_2H \\ H{-}C{-}OH \\ HO{-}C{-}H \\ H{-}C{-}OH \\ H{-}C{-}OH \\ CO_2H \end{array} + \text{oxides of nitrogen}$$

D-Glucose — D-Glucaric acid

E. Oxidation by Periodic Acid

Oxidation by periodic acid has proven useful in structure determinations of carbohydrates, particularly in determining the size of glycoside rings. Recall from Section 9.10 that periodic acid cleaves the carbon-carbon bond of a glycol in a reaction that proceeds through a cyclic periodic ester. In this reaction, iodine(VII) of periodic acid is reduced to iodine(V) of iodic acid.

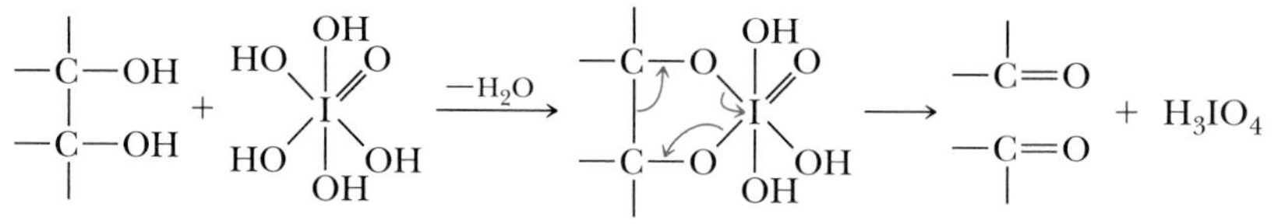

$$\begin{array}{c} -\overset{|}{C}-OH \\ | \\ -\underset{|}{C}-OH \end{array} + (HO)_4I(OH)(=O) \xrightarrow{-H_2O} \text{cyclic periodic ester} \longrightarrow \begin{array}{c} -C{=}O \\ \\ -C{=}O \end{array} + H_3IO_4$$

A 1,2-diol — Periodic acid — A cyclic periodic ester — Iodic acid

Periodic acid also cleaves carbon-carbon bonds of α-hydroxyketone and α-hydroxyaldehyde groups by a similar mechanism. Following are abbreviated structural formulas for these functional groups and the products of their oxidative cleavage by periodic acid. As a way to help you understand how each set of products is formed, each carbonyl in a starting material is shown as a hydrated intermediate that is then oxidized. In this way, each oxidation can be viewed as analogous to oxidation of a glycol.

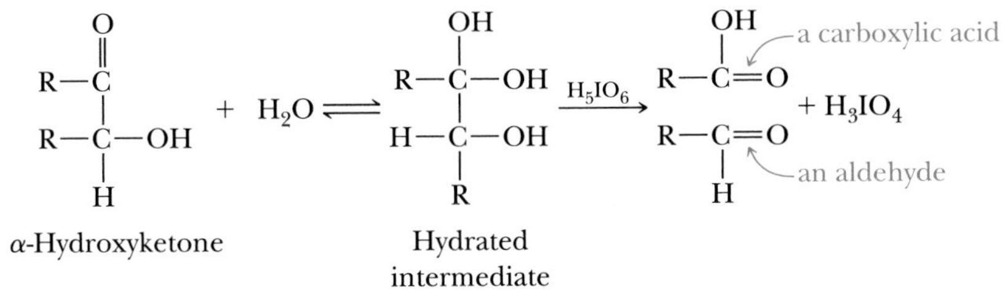

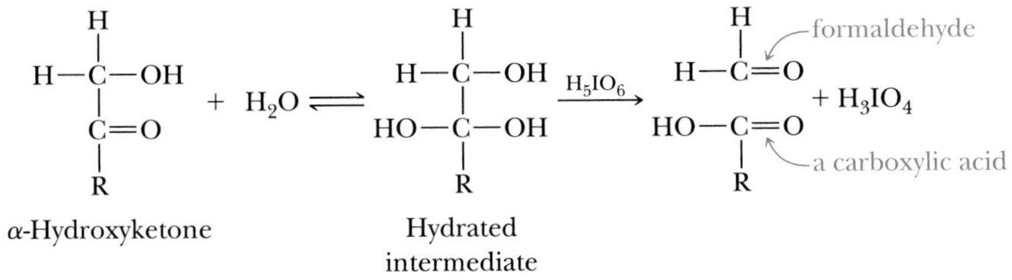

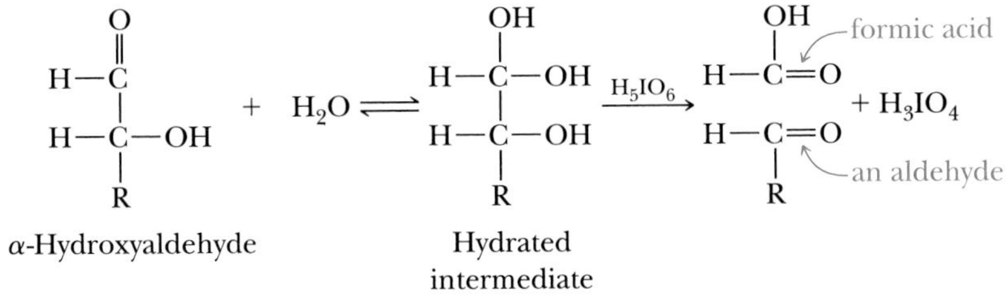

$$R-C(=O)-CH(OH)-R + H_2O \rightleftharpoons R-C(OH)_2-CH(OH)-R \xrightarrow{H_5IO_6} R-C(OH)=O + R-CH=O + H_3IO_4$$

α-Hydroxyketone — Hydrated intermediate — a carboxylic acid — an aldehyde

$$H_2C(OH)-C(=O)-R + H_2O \rightleftharpoons H_2C(OH)-C(OH)_2-R \xrightarrow{H_5IO_6} H_2C=O + HO-C(=O)-R + H_3IO_4$$

α-Hydroxyketone — Hydrated intermediate — formaldehyde — a carboxylic acid

$$H-C(=O)-CH(OH)-R + H_2O \rightleftharpoons H-C(OH)_2-CH(OH)-R \xrightarrow{H_5IO_6} H-C(OH)=O + H-C(=O)-R + H_3IO_4$$

α-Hydroxyaldehyde — Hydrated intermediate — formic acid — an aldehyde

As an example of the usefulness of this reaction in carbohydrate chemistry, oxidation of methyl β-D-glucoside consumes two mol of periodic acid and produces one mol of formic acid. This stoichiometry and the formation of formic acid is possible only if —OH groups are on three adjacent carbon atoms.

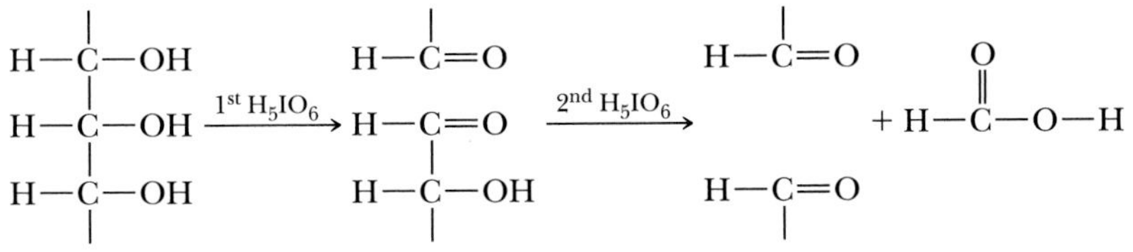

$$-CH(OH)-CH(OH)-CH(OH)- \xrightarrow{1^{st}\ H_5IO_6} -CH=O \;\; H-C(=O)-CH(OH)- \xrightarrow{2^{nd}\ H_5IO_6} -CH=O \;\; H-C(=O)- + H-C(=O)-O-H$$

This is evidence that methyl β-D-glucoside is indeed a pyranoside.

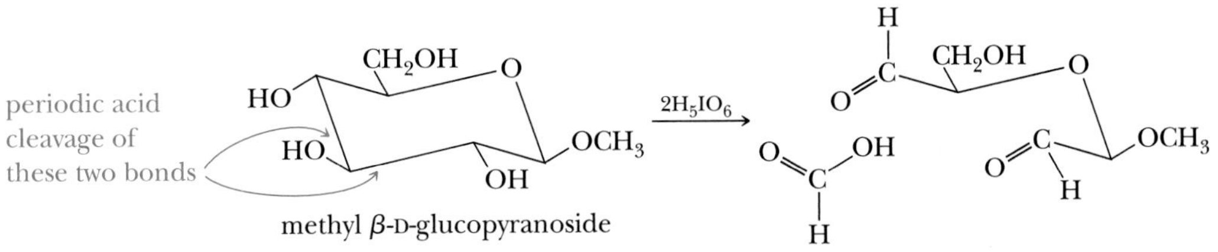

periodic acid cleavage of these two bonds

methyl β-D-glucopyranoside $\xrightarrow{2H_5IO_6}$

Methyl β-D-fructoside consumes only one mol of periodic acid and produces neither formaldehyde nor formic acid. Thus, oxidizable groups exist on adjacent carbons only at one site in the molecule. The fructoside, therefore, must be a five-membered ring (a fructofuranoside).

periodic acid cleaves only this bond

HOCH$_2$ O OCH$_3$ H HO H CH$_2$OH HO H $\xrightarrow{H_5IO_6}$ HOCH$_2$ O OCH$_3$ H H H C=O H C=O CH$_2$OH

Methyl β-D-fructofuranoside

24.5 Glucose Assay: The Search for Specificity

The analytical procedure most often performed in a clinical chemistry laboratory is the determination of glucose in blood, urine, or other biological fluid. The need for a rapid and reliable test for glucose stems from the high incidence of diabetes mellitus. Approximately 2 million known diabetics live in the United States, and it is estimated that another million diabetics are undiagnosed.

Diabetes mellitus is characterized by insufficient blood levels of the polypeptide hormone insulin (Section 27.7C). If the concentration of insulin is insufficient, muscle and liver cells do not absorb glucose, which in turn leads to increased levels of blood glucose (hyperglycemia), impaired metabolism of fats and proteins, ketosis, and possible diabetic coma. Thus, a rapid and reliable procedure for the determination of blood glucose levels is critical for early diagnosis and effective management of this disease.

One of the most successful and widely used glucose assays is based on the fact that aldehydes react with primary aliphatic and aromatic amines to form imines (Section

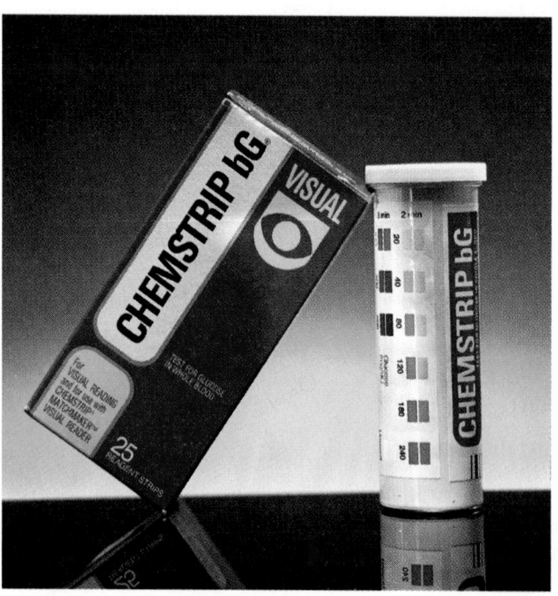

Chemstrip kit used to test for blood glucose. *(Charles D. Winters)*

15.10A). Specifically, when glucose is treated with 2-methylaniline (*o*-toluidene), it forms an imine, which has a blue-green color with an absorption maximum in the visible spectrum at 625 nm. The intensity of the absorption at 625 nm is proportional to glucose concentration.

D-Glucose + 2-Methylaniline (*o*-Toluidine) ⟶ An imine (blue-green) + H_2O

The *o*-toluidine method can be applied directly to serum, plasma, cerebrospinal fluid, and urine and to samples as small as 20 μL (20×10^{-6} L). Although this method can be used to measure glucose concentrations, it has the disadvantage that galactose and mannose, and to a lesser extent lactose and xylose, also react with *o*-toluidine to give colored imines and are, therefore, potential sources of false-positive results. This, however, is generally not a problem because these mono- and disaccharides are normally present in serum and plasma only in very low concentrations.

In recent years, the search for even greater specificity in glucose determinations led to the introduction of enzyme-based assay procedures. What was needed was an enzyme that catalyzes a specific reaction of glucose but not comparable reactions of any other substance normally present in biological fluids. The enzyme **glucose oxidase** meets these requirements. It catalyzes the oxidation of β-D-glucose to D-gluconic acid.

β-D-Glucopyranose + O_2 + H_2O $\xrightarrow{\text{glucose oxidase}}$ D-Gluconic acid + H_2O_2 (Hydrogen peroxide)

Glucose oxidase is specific for β-D-glucose. Therefore, complete oxidation of any sample containing both β-D-glucose and α-D-glucose requires conversion of the α form to the β form. Fortunately, this interconversion is rapid and complete in the short time required for the test. Molecular oxygen, O_2, is the oxidizing agent in this reaction and is reduced to hydrogen peroxide, H_2O_2, the concentration of which can be determined by a number of procedures.

In one procedure, hydrogen peroxide formed in the glucose oxidase-catalyzed reaction is used to oxidize *o*-toluidine to a colored product in a reaction catalyzed by the enzyme peroxidase. The concentration of the colored oxidation product is determined spectrophotometrically and is proportional to the concentration of hydrogen peroxide and, therefore, to the concentration of glucose in the test solution. Several commercially available test kits use the glucose oxidase reaction for qualitative determination of glucose in urine.

$$o\text{-toluidine} + H_2O_2 \xrightarrow{\text{peroxidase}} \text{colored product} + H_2O$$

Any determination of glucose in blood reflects glucose levels during the sampling period only. There is now a simple, convenient laboratory method that can be used to monitor long-term glucose levels. This method depends on the measurement of the relative amounts of hemoglobin and certain hemoglobin derivatives normally present in blood. Hemoglobin A (HbA) is the main type of hemoglobin present in normal red blood cells. In addition, several lesser components are present, including glycosylated hemoglobins (HbA_1). Glycosylated hemoglobins are synthesized within red blood cells in two steps. In Step 1, the free $—NH_2$ group of the β chain of hemoglobin reacts with the carbonyl group of glucose to form an imine. Step 1 is reversible. In a slower, irreversible second step, the imine undergoes a type of keto-enol tautomerism to form a glycosylated hemoglobin.

CHO / —OH / HO— / —OH / —OH / CH_2OH + $H_2N—HbA(\beta)$ ⇌ $CH{=}N—HbA(\beta)$ / —OH / HO— / —OH / —OH / CH_2OH ⟶ $CH_2—NH—HbA(\beta)$ / C=O / HO— / —OH / —OH / CH_2OH

D-Glucose; *N*-Terminal group of the β-chain of hemoglobin A; An imine (unstable); Glycosylated hemoglobin A

Because this slow, irreversible second step occurs continuously throughout the 120-day life span of a typical red blood cell population, levels of glycosylated hemoglobin within this population reflect the average blood glucose levels during that period.

Normal levels of glycosylated hemoglobins fall within the range of 4.5% to 8.5% of total hemoglobin. In cases of uncontrolled or poorly controlled diabetes, the percentage of glycosylated hemoglobins may rise to two or three times these values. Thus, the level of glycosylated hemoglobin can be used to give a picture of the average blood glucose level over the previous 8 to 10 weeks.

24.6 L-Ascorbic Acid (Vitamin C)

A. Structure

The structural formula of L-ascorbic acid (vitamin C) resembles that of a monosaccharide. In fact, this vitamin is synthesized both biochemically by plants and some animals and commercially from D-glucose. Humans do not have the enzyme systems required for the synthesis of L-ascorbic acid and, therefore, for us, it is a vitamin. Ascorbic acid has several roles in the body. It is effective in the treatment and prevention of scurvy, it plays a role in the prevention of anemia, reduces the severity of allergic responses, and, because it stimulates the immune system, it plays a role in the prevention and treatment of infections.

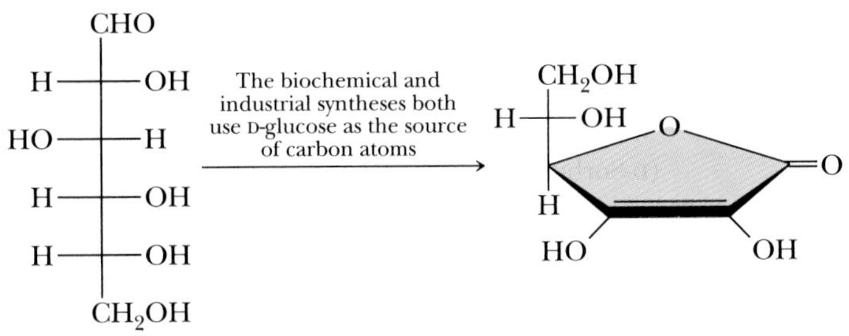

CHO; H—OH; HO—H; H—OH; H—OH; CH_2OH

D-Glucose

The biochemical and industrial syntheses both use D-glucose as the source of carbon atoms →

CH_2OH; H—OH; O; =O; H; HO; OH

L-Ascorbic acid (Vitamin C)

L-Ascorbic acid is very easily oxidized to L-dehydroascorbic acid, a diketone. Both L-ascorbic acid and L-dehydroascorbic acid are physiologically active and are found together in most body fluids.

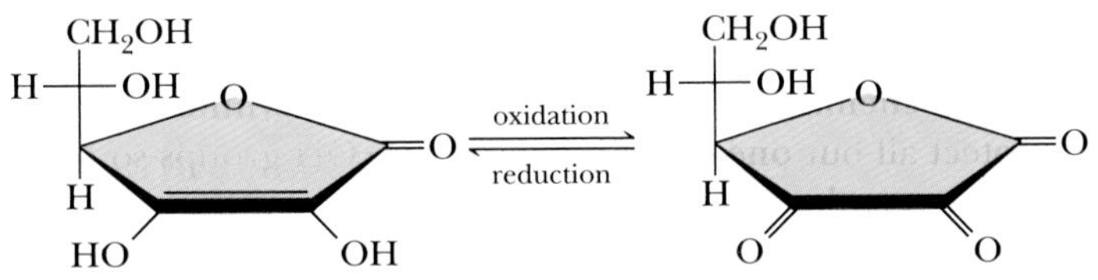

CH_2OH; H—OH; O; =O; H; HO; OH

L-Ascorbic acid (Vitamin C)

oxidation ⇌ reduction

CH_2OH; H—OH; O; =O; H; O; O

L-Dehydroascorbic acid

B. Industrial Synthesis

Approximately 30 million pounds of vitamin C is synthesized every year in the United States, and its synthesis illustrates the use of microbiological fermentation as a synthetic tool where no purely chemical method can carry out a particular step or steps in a regioselective or stereoselective manner. Synthesis of ascorbic acid begins with a readily available and very inexpensive raw material, D-glucose, which is reduced by catalytic hydrogenation to D-sorbitol. Selective oxidation of the secondary alcohol group on carbon 5 of the sorbitol chain is the next step. There is no known way to bring about this selective oxidation economically by purely chemical means. The success of the synthesis depends on the fact that the bacterium *Acetobacter suboxydans* catalyzes this oxidation and converts D-sorbitol to L-sorbose. Note that the product of this oxidation is now of the L series, not because of inversion of configuration at any stereocenter, but because of the rules of the Fischer convention. The carbon chain must be turned so that the more highly oxidized end is toward the top, in which case, the chiral carbon farthest from the carbonyl group has an L configuration. Thus, oxidation of D-sorbitol catalyzed by the enzyme systems of *Acetobacter suboxydans* gives L-sorbose.

fibers. In the production of rayon, cellulose-containing materials are treated with carbon disulfide, CS_2, in aqueous sodium hydroxide. In this reaction, some of the —OH groups on a cellulose fiber are converted to the sodium salt of a xanthate ester, which causes the fibers to dissolve in alkali as a viscous colloidal dispersion.

an —OH group in a cellulose fiber

$$\text{Cellulose—OH} \xrightarrow{\text{NaOH}} \text{Cellulose—O}^-\text{Na}^+ \xrightarrow{\text{S=C=S}} \text{Cellulose—O}\overset{\text{S}}{\overset{\|}{\text{C}}}\text{S}^-\text{Na}^+$$

Cellulose (insoluble in water)

Sodium salt of a xanthate ester (a viscous colloidal suspension)

The solution of cellulose xanthate is separated from the alkali insoluble parts of wood and then forced through a spinneret, a metal disc with many tiny holes, into dilute sulfuric acid to hydrolyze the xanthate ester groups and precipitate regenerated cellulose. Regenerated cellulose extruded as a filament is called viscose rayon thread.

In the industrial synthesis of acetate rayon, cellulose is treated with acetic anhydride (Section 17.6B). Acetylated cellulose is then dissolved in a suitable solvent, precipitated, and drawn into fibers known as acetate rayon. Today, acetate rayon fibers rank fourth in production in the United States, surpassed only by Dacron polyester, nylon, and rayon.

$$\text{(glucose unit)} + 3CH_3\overset{O}{\overset{\|}{C}}O\overset{O}{\overset{\|}{C}}CH_3 \longrightarrow \text{(fully acetylated glucose unit)} + 3CH_3\overset{O}{\overset{\|}{C}}OH$$

A glucose unit in a cellulose fiber

Acetic anhydride

A fully acetylated glucose unit

SUMMARY

Monosaccharides (Section 24.1A) are polyhydroxyaldehydes or polyhydroxyketones. They have the general formula $C_nH_{2n}O_n$ where n varies from 3 to 8. Their names contain the suffix **-ose.** The prefixes tri-, tetr-, pent-, and so on show the number of carbon atoms in the chain. The prefix **aldo-** shows an aldehyde and the prefix **keto-** shows a ketone. In a **Fischer projection,** the carbon chain is written vertically with the most highly oxidized carbon toward the top (Section 24.1C). Horizontal lines show groups projecting above the plane of the page; vertical lines show groups projecting behind the plane of the page.

The **penultimate carbon** is the next to last on the carbon chain of a Fischer projection of a monosaccharide (Section 24.1D). A monosaccharide that has the same configuration at the penultimate carbon as D-glyceraldehyde is called a **D-monosaccharide;** one that has the same configuration at the penultimate carbon as L-glyceraldehyde is called an **L-monosaccharide.**

Monosaccharides exist primarily as cyclic hemiacetals (Section 24.2). The new stereocenter resulting from hemiacetal formation is referred to as the **anomeric carbon.** The stereoisomers thus formed are called **anomers.** A six-membered cyclic hemiacetal is called a **pyranose;** a five-membered cyclic hemiacetal is called a **furanose.** The symbol ***β*-** indicates that the —OH on the anomeric carbon is on the same side of the ring as the terminal $-CH_2OH$. The symbol ***α*-** indicates that —OH on the anomeric carbon is on the opposite side from the terminal $-CH_2OH$. Furanoses and pyranoses can be drawn as **Haworth projections** (Section 24.2A). Pyranoses can also be shown as strain-free **chair conformations** (Section 24.2B). **Mutarotation** (Section 24.2C) is the change in specific rotation that accompanies formation of an equilibrium mixture of *α*- and *β*-anomers in aqueous solution.

A **glycoside** (Section 24.4A) is an acetal derived from a monosaccharide. The name of the glycoside is composed of the name of the alkyl or aryl group bonded to the acetal oxygen atom followed by the name of the monosaccharide in which the terminal -e has been replaced by **-ide.**

An **alditol** (Section 24.4C) is a polyhydroxy compound formed by reduction of the carbonyl group of a monosaccharide to an alcohol group. Reduction of D-glucose, for example, gives D-glucitol. An **aldonic acid** (Section 24.4D) is a carboxylic acid formed by oxidation of the aldehyde group of an aldose. Oxidation of D-glucose, for example, gives D-gluconic acid. **Reducing sugars** (Section 24.4D) reduce Tollens', Fehling's, and Benedict's solutions. **L-Ascorbic acid** (Section 24.6) is synthesized on an industrial scale from D-glucose.

A **disaccharide** (Section 24.7) contains two monosaccharide units joined by a glycoside bond. Terms applied to carbohydrates containing larger numbers of monosaccharides are **trisaccharide, tetrasaccharide, oligosaccharide,** and **polysaccharide. Maltose** is a disaccharide of two molecules of D-glucose joined by an *α*-1,4-glycoside bond. **Lactose** is a disaccharide consisting of D-galactose joined to D-glucose by a *β*-1,4-glycoside bond. **Sucrose** is a disaccharide containing D-glucose joined to D-fructose by a 1,2-glycoside bond.

At the cellular level, the structural basis for the A, B, O blood type classification is a group of relatively small, membrane bound carbohydrates called **blood group substances** or **antigenic determinants** (Section 24.7D). These antigenic determinants have several unusual features including the presence of L-fucose and aminosugars.

Starch (Section 24.8A) can be separated into two fractions given the names amylose and amylopectin. **Amylose** is a linear polymer of up to 4000 units of D-glucopyranose joined by *α*-1,4-glycoside bonds. **Amylopectin** is a highly branched polymer of D-glucopyranose joined by *α*-1,4-glycoside bonds and, at branch points, *α*-1,6-glycoside bonds. **Glycogen** (Section 24.8B), the reserve carbohydrate of animals, is a highly branched polymer of D-glucopyranose joined by *α*-1,4-glycoside bonds and, at branch points, by *α*-1,6-glycoside bonds. **Cellulose** (Section 24.8C), the skeletal polysaccharide of plants, is a linear polymer of D-glucopyranose joined by *β*-1,4-glycoside bonds. **Rayon** (Section 24.8D) is made from chemically modified and regenerated cellulose. **Acetate rayon** is made by acetylation of cellulose.

KEY REACTIONS

1. Formation of Cyclic Hemiacetals (Section 24.2)

A monosaccharide existing as a five-membered ring hemiacetal is a furanose; one existing as a six-membered ring hemiacetal is a pyranose. A pyranose is most commonly drawn as a Haworth projection or as a chair conformation.

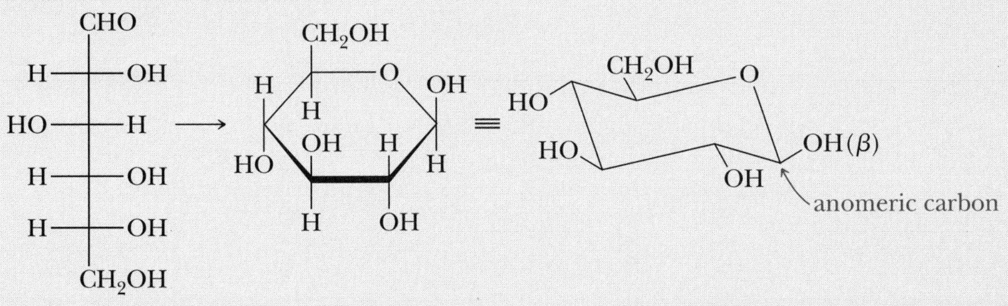

D-Glucose

β-D-Glucopyranose
(*β*-D-Glucose)

2. Mutarotation (Section 24.2C)

Anomeric forms of a reducing monosaccharide are in equilibrium in aqueous solution. Mutarotation is the change in specific rotation that accompanies this equilibration.

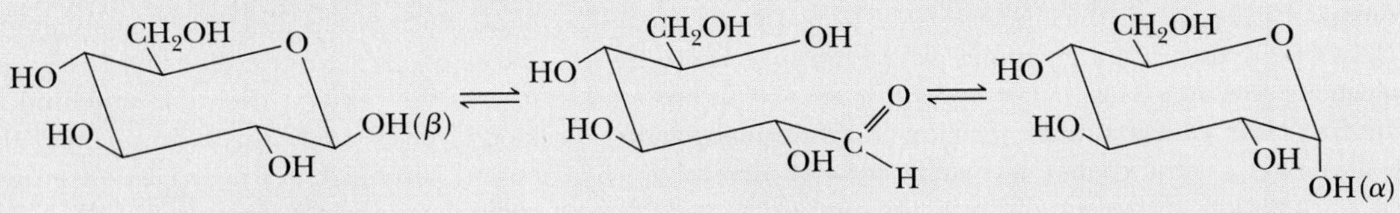

β-D-Glucopyranose $[\alpha]_D^{25}$ + 18.7° ⇌ Open-chain form ⇌ α-D-Glucopyranose $[\alpha]_D^{25}$ + 112°

3. Formation of Glycosides (Section 24.4A)

Treatment of a monosaccharide with an alcohol in the presence of an acid catalyst forms a cyclic acetal called a glycoside. The bond to the new —OR group is called a glycoside bond. Such acetals are not reducing sugars because they are not in equilibrium with the free aldehyde.

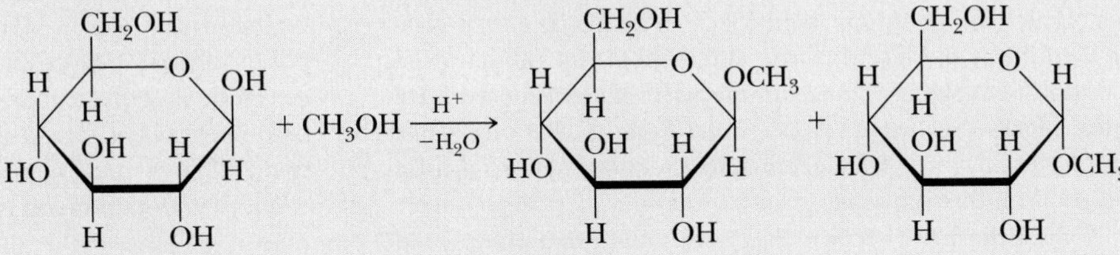

$$+ CH_3OH \xrightarrow[-H_2O]{H^+}$$

4. Formation of *N*-Glycosides (Section 24.4A)

The *N*-glycosides formed between a monosaccharide and a heterocyclic aromatic amine are especially important in the biological world.

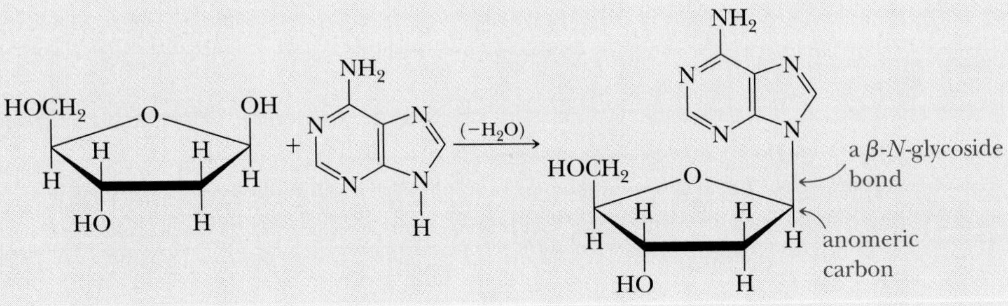

$$\xrightarrow{(-H_2O)}$$

a β-*N*-glycoside bond

anomeric carbon

5. Formation of Methyl Ethers (Section 24.4B)

All —OH groups of a carbohydrate can be converted to methyl ether groups by reaction with dimethyl sulfate in the presence of sodium hydride.

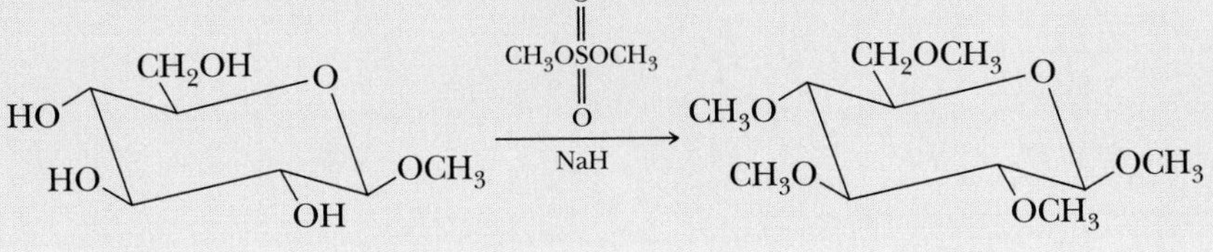

$$\xrightarrow[NaH]{CH_3OSO_2OCH_3}$$

6. Reduction to Alditols (Section 24.4C)

Reduction of the carbonyl group of an aldose or ketose to a hydroxyl group yields a polyhydroxy compound called an alditol.

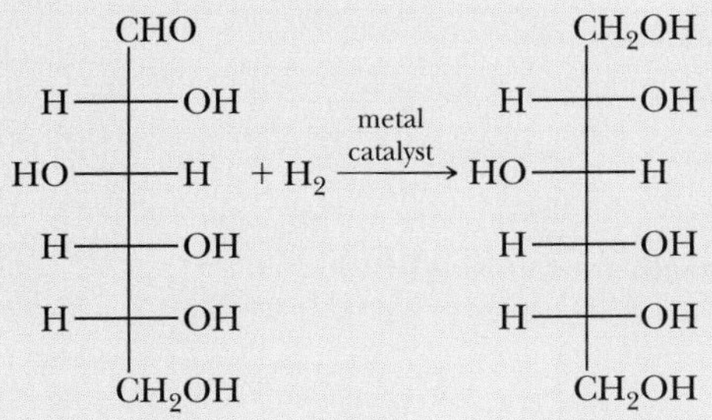

$$
\begin{array}{c}
CHO \\
H-C-OH \\
HO-C-H \\
H-C-OH \\
H-C-OH \\
CH_2OH
\end{array}
+ H_2 \xrightarrow[\text{catalyst}]{\text{metal}}
\begin{array}{c}
CH_2OH \\
H-C-OH \\
HO-C-H \\
H-C-OH \\
H-C-OH \\
CH_2OH
\end{array}
$$

D-Glucose

D-Glucitol
(D-Sorbitol)

7. Oxidation to an Aldonic Acid (Section 24.4D)

Oxidation of the aldehyde group of an aldose to a carboxyl group by a mild oxidizing agent gives a polyhydroxycarboxylic acid called an aldonic acid.

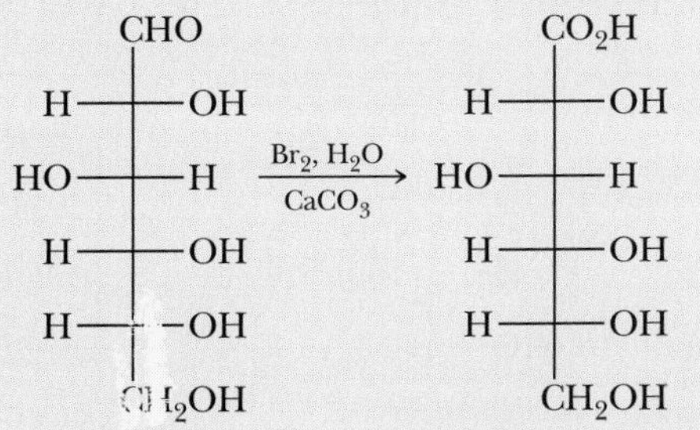

$$
\begin{array}{c}
CHO \\
H-C-OH \\
HO-C-H \\
H-C-OH \\
H-C-OH \\
CH_2OH
\end{array}
\xrightarrow[CaCO_3]{Br_2,\ H_2O}
\begin{array}{c}
CO_2H \\
H-C-OH \\
HO-C-H \\
H-C-OH \\
H-C-OH \\
CH_2OH
\end{array}
$$

D-Glucose

D-Gluconic acid
(an aldonic acid)

8. Oxidation to an Aldaric Acid (Section 24.4D)

Nitric acid oxidation of both the aldehyde and the primary alcohol group of an aldose gives a polyhydroxydicarboxylic acid called an aldaric acid.

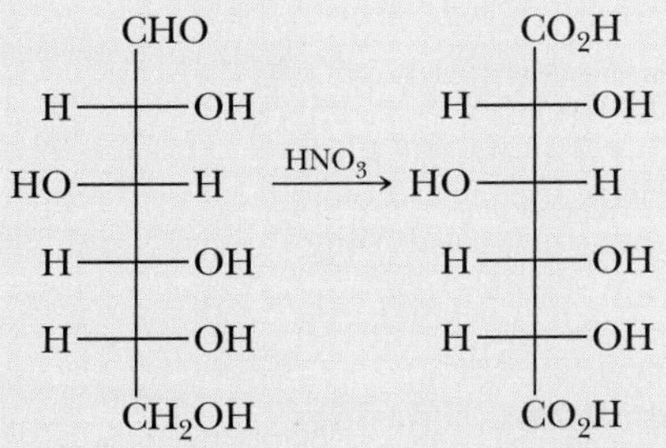

$$
\begin{array}{c}
CHO \\
H-C-OH \\
HO-C-H \\
H-C-OH \\
H-C-OH \\
CH_2OH
\end{array}
\xrightarrow{HNO_3}
\begin{array}{c}
CO_2H \\
H-C-OH \\
HO-C-H \\
H-C-OH \\
H-C-OH \\
CO_2H
\end{array}
$$

D-Glucose

D-Glucaric acid
(an aldaric acid)

9. Oxidation by Periodic Acid (Section 24.4E)

Periodic acid oxidizes and cleaves carbon-carbon bonds of glycol, α-hydroxyketone, and α-hydroxyaldehyde groups.

Periodic acid cleavage at these two bonds

Methyl β-D-glucopyranoside $\xrightarrow{2H_5IO_6}$

ADDITIONAL PROBLEMS

Monosaccharides

24.10 Explain the meaning of the designations D and L as used to specify the configuration of carbohydrates.

24.11 Which compounds are D-monosaccharides and which are L-monosaccharides?

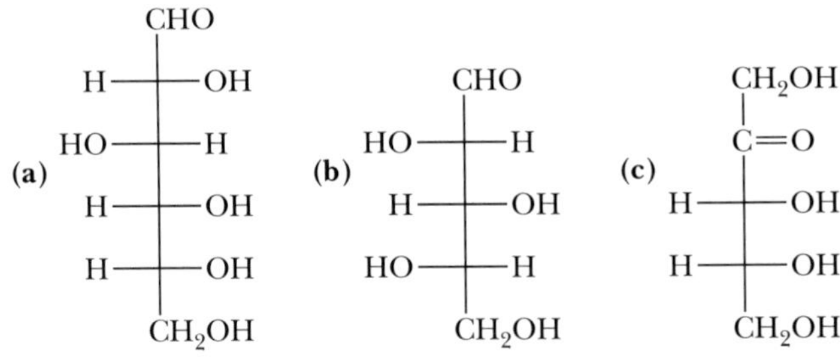

24.12 Classify each monosaccharide in Problem 24.11 using the designations D/L and aldose/ketose and according to the number of carbon atoms it contains. For example, glucose is classified as a D-aldohexose.

24.13 Write Fischer projections for L-ribose and L-arabinose.

24.14 What is the meaning of the prefix deoxy- as it is used in carbohydrate chemistry?

24.15 Give L-fucose a name incorporating the prefix "deoxy-" that shows its relationship to galactose.

The foxglove plant produces the important cardiac medication digitalis. See Problem 24.16. *(© Lois Moulton, Tony Stone Images)*

24.16 2,6-Dideoxy-D-altrose, known alternatively as D-digitoxose, is a monosaccharide obtained on hydrolysis of digitoxin, a natural product extracted from foxglove *(Digitalis purpurea)*. Digitoxin has found wide use in cardiology because it reduces pulse rate, regularizes heart rhythm, and strengthens heartbeat. Draw the structural formula of 2,6-dideoxy-D-altrose.

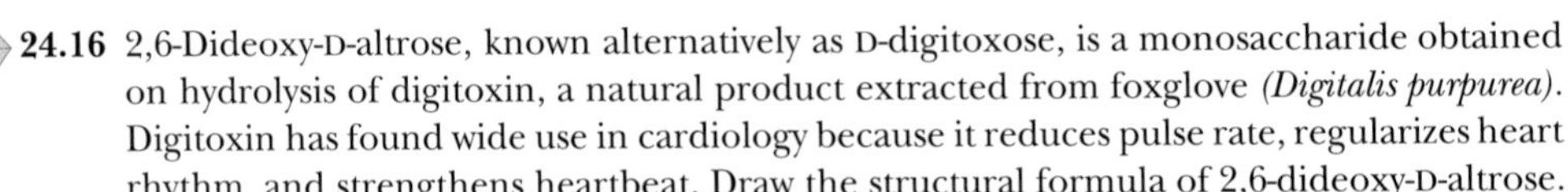

The Cyclic Structure of Monosaccharides

24.17 Build a molecular model of D-glucose and show that its six-membered hemiacetals (α-D-glucopyranose and β-D-glucopyranose) have the configurations shown in Figure 24.1 and not the mirror images of these structures.

24.18 Draw α-D-glucopyranose (α-D-glucose) as a Haworth projection. Now, using only the information given here, draw Haworth projections for these monosaccharides.

(a) α-D-mannopyranose (α-D-mannose). The configuration of D-mannose differs from that of D-glucose only at carbon 2.

(b) α-D-gulopyranose (α-D-gulose). The configuration of D-gulose differs from the configuration of D-glucose at carbons 3 and 4.

24.19 Repeat Problem 24.18, using chair conformations instead of Haworth projections for the monosaccharides.

24.20 Convert each Haworth projection to an open-chain Fischer projection and name the monosaccharide you have drawn.

(a) 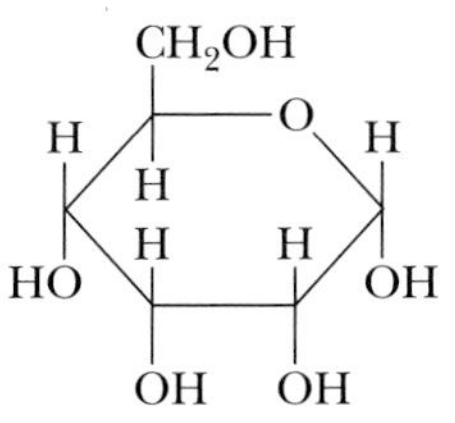 (b)

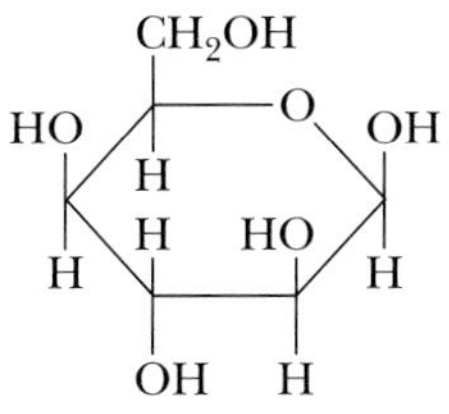

24.21 Convert each chair conformation to an open-chain Fischer projection and name the monosaccharide you have drawn.

(a) 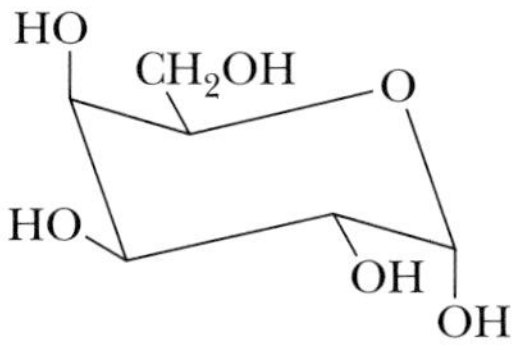 (b)

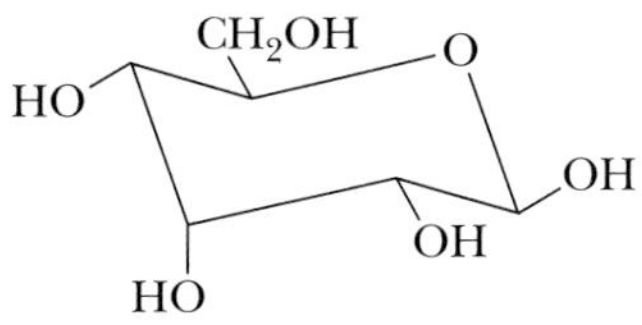

24.22 Explain the phenomenon of mutarotation with reference to carbohydrates. By what means is it detected?

24.23 Following are specific rotations for the anomers of D-mannose and the value after mutarotation. Using these data, calculate the percentage of each anomer present at equilibrium.

$$\alpha\text{-D-mannose } +29.3^\circ \longrightarrow +14.5^\circ$$

$$\beta\text{-D-mannose } -16.3^\circ \longrightarrow +14.5^\circ$$

24.24 It has been proposed that structural characteristics of sweet-tasting compounds are (1) a hydrogen bond donating group, AH; (2) a hydrogen bond-accepting group, B; and (3) that atoms A and B are separated by 2.5 Å to 4.0 Å. Identify the AH and B units in these nonnutritive sweeteners.

(a) 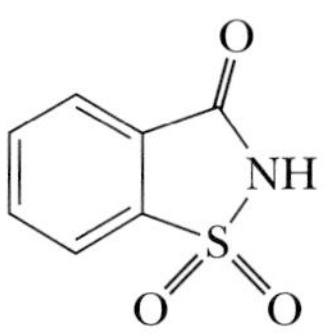 Saccharin (500 × sucrose)

(b) 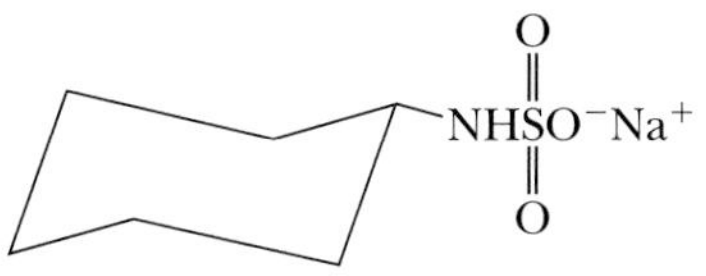 Sodium cyclamate (30 × sucrose)

(c) 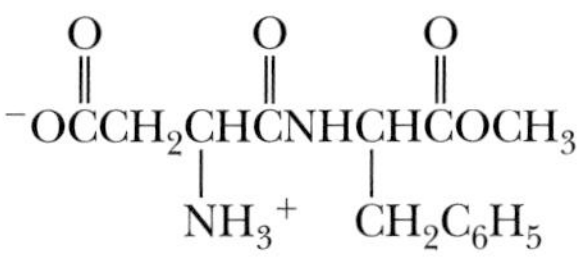 $^{-}OCCH_2CHCNHCHCOCH_3$ (with NH_3^+ and $CH_2C_6H_5$ substituents) Aspartame (160 × sucrose)

24.25 It has been observed that sugars that can form one or more strong intramolecular hydrogen bonds are less sweet than sugars which cannot form such hydrogen bonds. Draw chair conformations of β-D-galactose and β-D-mannose and identify one strong intramolecular hydrogen bond in each molecule.

Reactions of Monosaccharides

24.26 Draw Fischer projections for the product(s) formed by reaction of D-galactose with the following. In addition, state whether each product is optically active or inactive.

(a) $NaBH_4$ in H_2O **(b)** H_2/Pt **(c)** HNO_3, warm

(d) $Br_2/H_2O/CaCO_3$ **(e)** H_5IO_6 **(f)** $C_6H_5—NH_2$

24.27 Repeat Problem 24.26 using D-ribose.

24.28 An important technique for establishing relative configurations among isomeric aldoses and ketoses is to convert both terminal carbon atoms to the same functional group. This can be done either by selective oxidation or reduction. As a specific example, nitric acid oxidation of D-erythrose gives meso-tartaric acid (Table 4.1, Section 4.7). Similar oxidation of D-threose gives (*S,S*)-tartaric acid. Given this information and the fact that D-erythrose and D-threose are diastereomers, draw Fischer projections for D-erythrose and D-threose. Check your answers against Table 24.1.

24.29 There are four D-aldopentoses (Table 24.1). If each is reduced with $NaBH_4$, which yield optically active alditols? Which yield optically inactive alditols?

24.30 Name the two alditols formed by $NaBH_4$ reduction of D-fructose.

24.31 One pathway for the metabolism of glucose-6-phosphate is its enzyme-catalyzed conversion to fructose-6-phosphate. Show that this transformation can be regarded as two enzyme-catalyzed keto-enol tautomerisms.

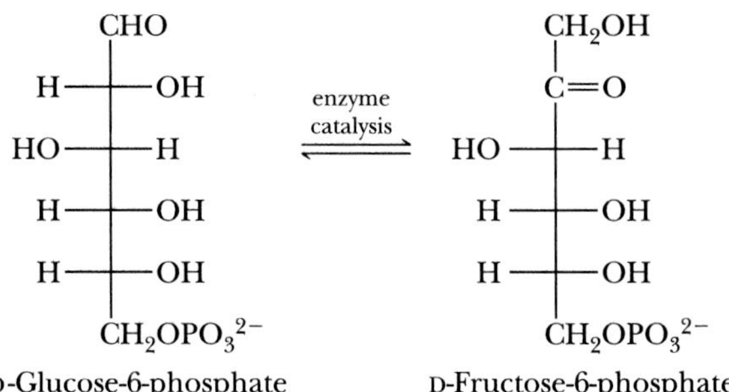

D-Glucose-6-phosphate D-Fructose-6-phosphate

24.32 L-Fucose, one of several monosaccharides commonly found in the surface polysaccharides of animal cells, is synthesized biochemically from D-mannose in the following eight steps:

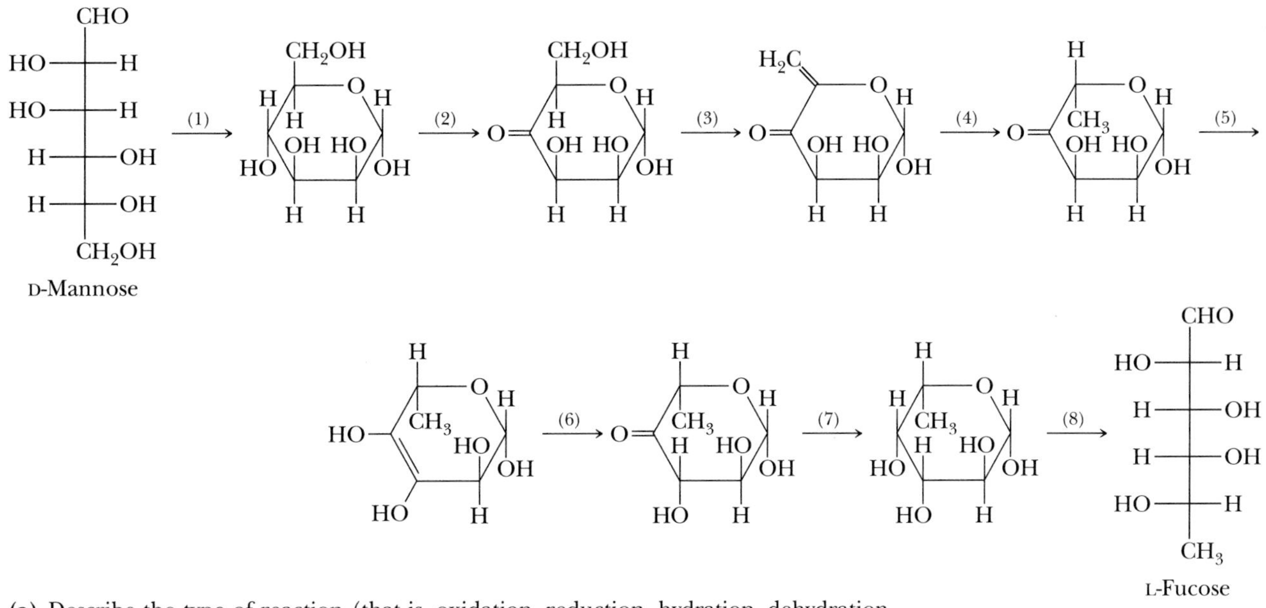

D-Mannose

L-Fucose

(a) Describe the type of reaction (that is, oxidation, reduction, hydration, dehydration, and so on) involved in each step.

(b) Explain why it is that this monosaccharide derived from D-mannose now belongs to the L series.

24.33 Complete the following:

(a) Draw a structural formula for the compound formed when D-ribose is converted to methyl β-D-ribofuranoside and then permethylated with sodium hydride and dimethyl sulfate.

(b) Draw the structural formula for the product of mild acid-catalyzed hydrolysis of the permethylated compound formed in part (a).

(c) With how many moles of periodic acid does the compound in part (b) react?

24.34 Repeat Problem 24.33 for 2-deoxy-D-ribose.

24.35 Account for the fact that, when permethylated monosaccharides are treated with warm aqueous acid, only the methyl glycoside bond is hydrolyzed.

24.36 Treatment of methyl β-D-glucopyranoside with benzaldehyde forms a six-membered cyclic acetal. Draw the most stable conformation of this acetal. Identify each new stereocenter in the acetal? (*Hint:* There are free —OH groups on carbons 2, 3, 4, and 6 of methyl β-D-glucopyranoside. Only two of these —OH groups are properly positioned to give a six-membered cyclic acetal with benzaldehyde.)

24.37 Vanillin, the principal component of vanilla, occurs in vanilla beans and other natural sources as a β-D-glucopyranoside. Draw a structural formula for this glycoside, showing the D-glucose unit as a chair conformation. (See *The Merck Index,* 12th ed., #10069.)

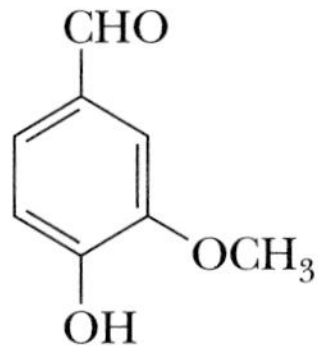

Vanillin

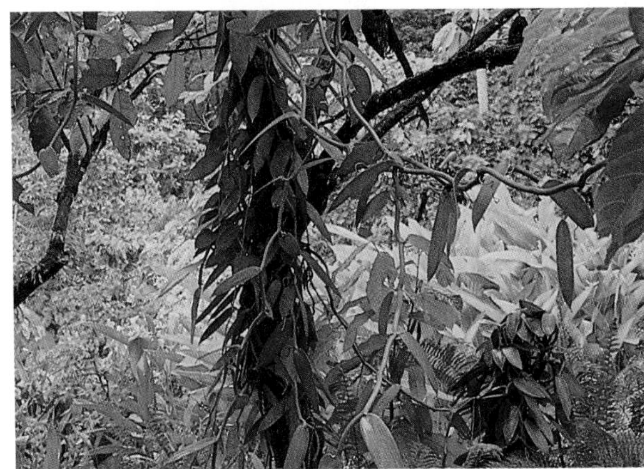

West Indian vanilla plant, *Vanilla pompona.* See Problem 24.37. *(Jane Grushow from Grant Heilman)*

24.38 Hot water extracts of ground willow and poplar bark are an effective pain reliever. Unfortunately, the liquid is so bitter that most persons refuse it. The pain reliever in these infusions is salicin, a β-glycoside of D-glucopyranose and the phenolic —OH group of 2-(hydroxymethyl)phenol. Draw a structural formula for salicin, showing the glucose ring as a chain conformation. (See *The Merck Index,* 12th ed., #8476.)

24.39 Draw structural formulas for the products formed by hydrolysis at pH 7.4 (the pH of blood plasma) of all ester, thioester, amide, anhydride, and glycoside groups in acetyl coenzyme A. Name as many of these compounds as you can.

This is the acetyl group in "acetyl" coenzyme A

$CH_3\overset{O}{\overset{\|}{C}}SCH_2CH_2NH\overset{O}{\overset{\|}{C}}CH_2CH_2NH\overset{O}{\overset{\|}{C}}CH(OH)C(CH_3)_2CH_2O\overset{O}{\overset{\|}{P}}(O^-)O\overset{O}{\overset{\|}{P}}(O^-)OCH_2$–(ribose with adenine, NH_2; 3′-OPO_3^{2-}; OH)

Acetyl coenzyme A (Acetyl-CoA)

Ascorbic Acid

24.40 Assign *R* or *S* configurations to each stereocenter in ascorbic acid.

24.41 Write a balanced half-reaction to show that conversion of ascorbic acid to dehydroascorbic acid is an oxidation. How many electrons are involved in this oxidation?

24.42 Given the fact that ascorbic acid and dehydroascorbic acid are the physiologically active forms of vitamin C, is ascorbic acid a biological oxidizing agent or a biological reducing agent? Explain.

Bark of the willow tree contains the pain reliever salicin. See Problem 24.38. *(Pete Saloutos/TSW/Click/Chicago Ltd.)*

24.43 Ascorbic acid is a diprotic acid with the following acid ionization constants.

$$pK_{a1} = 4.10 \qquad pK_{a2} = 11.79$$

The two acidic hydrogens are those connected with the enediol part of the molecule. Which is the most acidic hydrogen? (*Hint:* Draw separately the anion derived by loss of each acidic hydrogen. Which anion is more stable, that is, which has the greater degree of delocalization of charge?)

Disaccharides and Oligosaccharides

24.44 What is the difference in meaning between the terms glycoside bond and glucoside bond?

24.45 In making candy or sugar syrups, sucrose is boiled in water with a little acid, such as lemon juice. Why does the product mixture taste sweeter than the starting sucrose solution?

24.46 Trehalose, found in young mushrooms and the chief carbohydrate in the blood of certain insects, is a disaccharide consisting of two D-monosaccharide units joined by an α-1,1-glycoside bond.

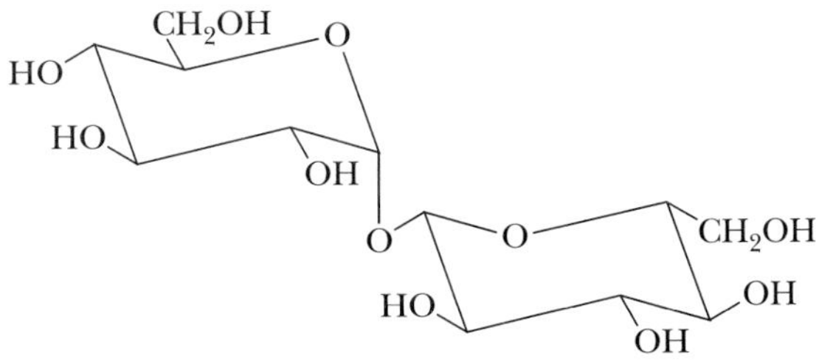

CH_2OH O HO HO OH O O CH_2OH HO OH OH

Trehalose

(a) Is trehalose a reducing sugar?
(b) Does trehalose undergo mutarotation?
(c) With how many moles of periodic acid does trehalose react? How many moles of formaldehyde are formed? How many moles of formic acid are formed?
(d) Draw structural formulas for the two O-methylated monosaccharides formed when trehalose is permethylated with dimethyl sulfate and sodium hydride and then warmed in dilute aqueous acid to hydrolyze the glycoside bond.

24.47 The trisaccharide raffinose occurs principally in cottonseed meal.

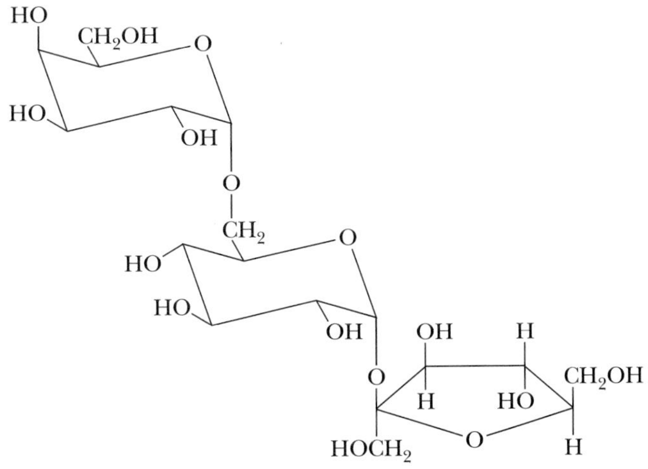

HO CH_2OH O HO OH O CH_2 O HO HO OH OH H O CH_2OH H HO $HOCH_2$ O H

Raffinose

(a) Name the three monosaccharide units in raffinose.
(b) Describe each glycoside bond in this trisaccharide.
(c) Is raffinose a reducing sugar?
(d) With how many moles of periodic acid will raffinose react?

Gentiobiose, a disaccharide, is found in the gentian plant. See Problem 24.48. *(© Larry Ulrich)*

24.48 Gentiobiose is a disaccharide found in a number of natural products including gentian plants *(Gentiana lutea)*. It is a reducing sugar and is hydrolyzed by β-glycosidases (enzymes with catalytic activity limited to hydrolysis of β-glycoside bonds). Reaction of gentiobiose with dimethyl sulfate in the presence of sodium hydride yields an octamethyl derivative, which, when hydrolyzed in warm aqueous acid, gives equimolar amounts of 2,3,4,6-tetra-O-methyl-D-glucose and 2,3,4-tri-O-methyl-D-glucose. Propose a structural formula for gentiobiose.

24.49 Following is the structural formula of laetrile.

Laetrile

(a) Assign an *R* or *S* configuration to the stereocenter bearing the cyano (—CN) group.
(b) Account for the fact that on hydrolysis in warm aqueous acid, laetrile liberates benzaldehyde and HCN.

Polysaccharides

24.50 Following is the Fischer projection for *N*-acetyl-D-glucosamine:

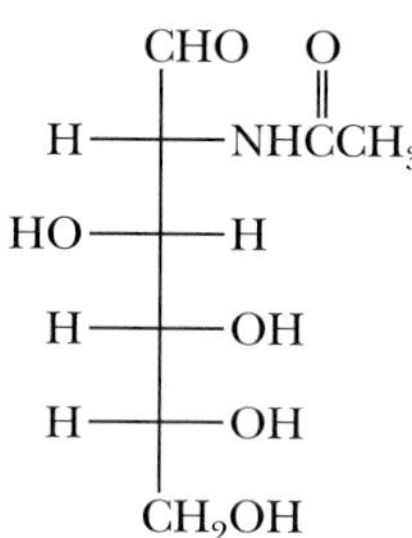

N-Acetyl-D-glucosamine

(a) Draw a chair conformation for the α- and β-pyranose forms of this monosaccharide.
(b) Draw a chair conformation for the disaccharide formed by joining two units of the pyranose form of *N*-acetyl-D-glucosamine by a β-1,4-glycoside bond. If you drew this correctly, you have the structural formula for the repeating dimer of chitin, the structural polysaccharide component of the shell of lobster and other crustaceans.

24.51 Propose structural formulas for the following polysaccharides:

(a) Alginic acid, isolated from seaweed, is used as a thickening agent in ice cream and other foods. Alginic acid is a polymer of D-mannuronic acid in the pyranose form joined by β-1,4-glycoside bonds.
(b) Pectic acid is the main component of pectin, which is responsible for the formation of jellies from fruits and berries. Pectic acid is a polymer of D-galacturonic acid in the pyranose form joined by α-1,4-glycoside bonds.

D-Mannuronic acid	D-Galacturonic acid
CHO	CHO
HO—H	H—OH
HO—H	HO—H
H—OH	HO—H
H—OH	H—OH
CO_2H	CO_2H

D-Mannuronic acid D-Galacturonic acid

24.52 Certain types of streptococci found in the mouth, especially *Streptococcus mutans*, have an enzyme system that uses sucrose as a starting material for the synthesis of high-molecular-weight polysaccharides known as **dextrans.** About 10% of the dry weight of dental plaque is composed of dextran. In one study of the dextran composition of dental plaque, dextran was methylated with methyl iodide in the presence of NaH and then the permethylated polysaccharide was hydrolyzed in dilute aqueous acid. The only monosaccharides obtained were the following four O-methyl derivatives of D-glucose.

Methylated D-glucose	Mole %
2,3,4,6-tetra-O-methyl-D-glucose	14.6
2,4,6-tri-O-methyl-D-glucose	50.5
2,3,4-tri-O-methyl-D-glucose	20.9
2,4-di-O-methyl-D-glucose	14.0

(a) Draw the structural formula of the open-chain form of each of these derivatives of D-glucose.

(b) The isolation of one of these derivatives is evidence that one group of glucose units in this dextran participates in only 1,3-glucoside bonds. Explain. What is the percentage of 1,3-glycoside bonds?

(c) The isolation of a second derivative of D-glucose is evidence that a second group of glucose units participates in only 1,6-glycoside bonds. Explain. What is the percentage of 1,6-glycoside bonds?

(d) The isolation of a third glucose derivative is evidence that a third group of glucose units participates in both 1,3- and 1,6-glycoside bonds and, therefore, serve as branch points in the polysaccharide chain. Explain. What is the percentage of chain branching?

(e) The fourth derivative of D-glucose represents the monosaccharide end of branched chains. Compare the percentage of this terminal monosaccharide unit with the percentage of chain branching you determined in part (d).

(f) From all of this evidence, sketch the polysaccharide of dextran in the same manner as amylopectin is sketched in Figure 24.9.

24.53 Digitalis is a preparation made from the dried seeds and leaves of the purple foxglove, *Digitalis purpurea*, a plant native to southern and central Europe and cultivated in the United States. The preparation is a mixture of several active components, including digitalin. Digitalis is used in medicine to increase the force of myocardial contraction and as a conduction depressant to decrease heart rate (the heart pumps more forcefully but less often).

(a) Describe this glycoside bond
(b) Draw an open-chain Fischer projection of this monosaccharide
(c) Describe this glycoside bond
(d) Name this monosaccharide unit

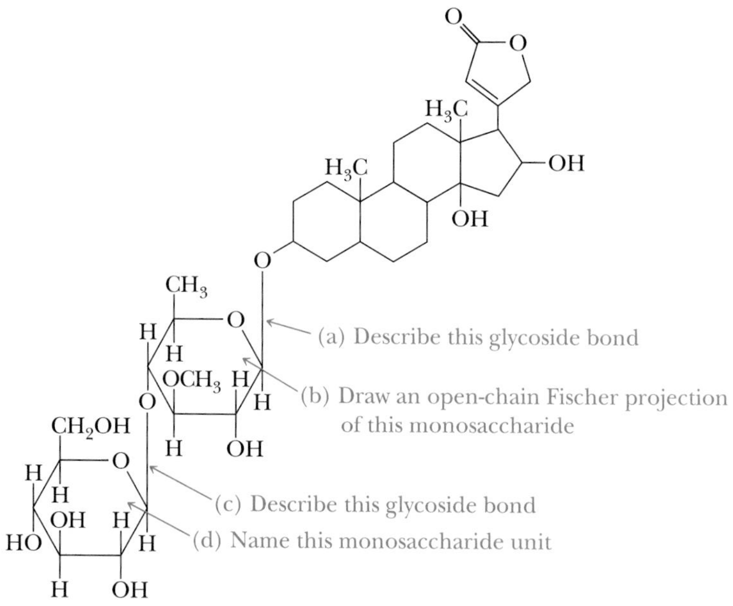

Digitalin

24.54 Following is the structural formula of ganglioside GM_2, a macromolecular glycolipid (meaning that it contains a lipid and monosaccharide units joined by glycoside bonds). In normal cells, this and other gangliosides are synthesized continuously and degraded by lysosomes, which are cell organelles containing digestive enzymes. If pathways for the degradation of gangliosides are inhibited, the gangliosides accumulate in the central nervous system causing all sorts of life-threatening consequences. In inherited diseases of ganglioside metabolism, death usually occurs at an early age. Diseases of ganglioside metabolism include Gaucher's disease, Niemann-Pick disease, and Tay-Sachs disease. Tay-Sachs disease is a hereditary defect that is transmitted as an autosomal recessive gene. The concentration of ganglioside GM_2 is abnormally high in this disease because the enzyme responsible for catalyzing the hydrolysis of glycoside bond (b) is absent.

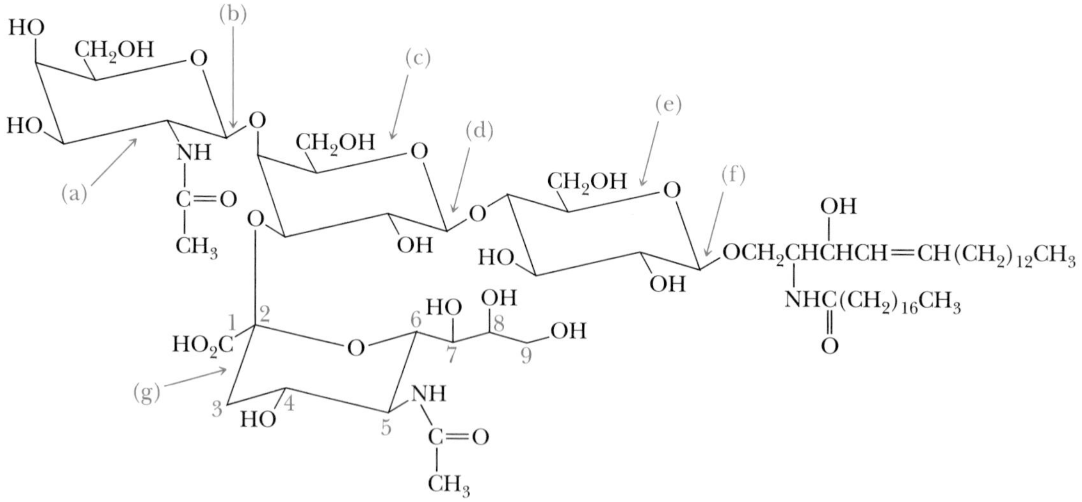

Ganglioside GM_2 or Tay-Sachs ganglioside

(a) Name this monosaccharide unit.
(b) Describe this glycoside bond (α or β, and between which carbons of each unit).
(c) Name this monosaccharide unit.
(d) Describe this glycoside bond.
(e) Name this monosaccharide unit.
(f) Describe this glycoside bond.
(g) This unit is *N*-acetylneuraminic acid, the most abundant member of a family of amino sugars containing nine or more carbons and distributed widely throughout the animal kingdom. Draw the open-chain form of this amino sugar. Do not be concerned with the configuration of the five stereocenters in the open-chain form.

25 LIPIDS

■ Polar bears eat only during a few weeks out of the year and then fast for periods of eight months or more, consuming no food or water during that period. Eating mainly in the winter, the adult polar bear feeds almost exclusively on seal blubber (largely composed of triglycerides) thus building up its own triglyceride reserves. Through the Arctic summer, the polar bear maintains normal physical activity, roaming over long distances, but relies entirely on its body fat for sustenance, burning as much as 1–1.5 kg of fat per day. *(Daniel J. Cox/Tony Stone Images)*

Lipids are a heterogeneous class of naturally occurring organic compounds classified together on the basis of common solubility properties. They are insoluble in water but soluble in aprotic organic solvents, including diethyl ether, methylene chloride, and acetone. Carbohydrates and nucleic acids, as well as amino acids and proteins are largely insoluble in these organic solvents.

Lipids are divided into two main groups. First are those lipids that contain both a relatively large nonpolar hydrophobic region, most commonly aliphatic in nature, and a polar hydrophilic region. Found among this group are the triacylglycerols, phospholipids, prostaglandins, and fat-soluble vitamins. Second are those lipids that contain the tetracyclic ring system called the steroid nucleus. Found among this group are cholesterol, steroid hormones, and bile acids.

25.1 Triglycerides

Animal fats and vegetable oils, the most abundant naturally occurring lipids, are triesters of glycerol and long-chain carboxylic acids. Fats and oils are also referred to as **triglycerides** or **triacylglycerols**. Hydrolysis of a triglyceride in aqueous base followed by acidification gives glycerol and three fatty acids.

Lipids A class of biomolecules isolated from plant or animal sources by extraction with nonpolar organic solvents, such as diethyl ether and methylene chloride.

Triglyceride An ester of glycerol with three fatty acids.

$$\begin{array}{c} CH_2OC(=O)R \\ | \\ R'C(=O)OCH \\ | \\ CH_2OC(=O)R'' \end{array} \xrightarrow[\text{2. HCl, } H_2O]{\text{1. NaOH, } H_2O} \begin{array}{c} CH_2OH \\ | \\ HOCH \\ | \\ CH_2OH \end{array} + \begin{array}{c} RCO_2H \\ R'CO_2H \\ R''CO_2H \end{array}$$

A triacylglycerol (a triglyceride) — 1,2,3-Propanetriol (Glycerol, glycerin) — Fatty acids

A. Fatty Acids

More than 500 different **fatty acids** have been isolated from various cells and tissues. Given in Table 25.1 are common names and structural formulas for the most abundant of these. The number of carbons in a fatty acid and the number of carbon-carbon double bonds in its hydrocarbon chain are shown by two numbers separated by a colon. In this notation, linoleic acid, for example, is designated as an 18:2 fatty acid; its 18-carbon chain contains 2 carbon-carbon double bonds. Following are several characteristics of the most abundant fatty acids in higher plants and animals.

Fatty acid A long, unbranched-chain carboxylic acid, most commonly of 12 to 20 carbons, derived from the hydrolysis of animal fats, vegetable oils, or the phospholipids of biological membranes.

1. Nearly all fatty acids have an even number of carbon atoms, most between 12 and 20, in an unbranched chain.

TABLE 25.1 The Most Abundant Fatty Acids in Fats, Oils, and Biological Membranes

Carbon Atoms/ Double Bonds	Structure	Common Name	Melting Point (°C)
Saturated Fatty Acids			
12:0	$CH_3(CH_2)_{10}CO_2H$	lauric acid	44
14:0	$CH_3(CH_2)_{12}CO_2H$	myristic acid	58
16:0	$CH_3(CH_2)_{14}CO_2H$	palmitic acid	63
18:0	$CH_3(CH_2)_{16}CO_2H$	stearic acid	70
20:0	$CH_3(CH_2)_{18}CO_2H$	arachidic acid	77
Unsaturated Fatty Acids			
16:1	$CH_3(CH_2)_5CH{=}CH(CH_2)_7CO_2H$	palmitoleic acid	32
18:1	$CH_3(CH_2)_7CH{=}CH(CH_2)_7CO_2H$	oleic acid	16
18:2	$CH_3(CH_2)_4(CH{=}CHCH_2)_2(CH_2)_6CO_2H$	linoleic acid	−5
18:3	$CH_3CH_2(CH{=}CHCH_2)_3(CH_2)_6CO_2H$	linolenic acid	−11
20:4	$CH_3(CH_2)_4(CH{=}CHCH_2)_4(CH_2)_2CO_2H$	arachidonic acid	−49

2. The three most abundant fatty acids in nature are palmitic acid (16:0), stearic acid (18:0), and oleic acid (18:1).
3. In most unsaturated fatty acids of fats, oils, and biological membranes, the *cis* isomer predominates; the *trans* isomer is rare.
4. Unsaturated fatty acids have lower melting points than their saturated counterparts. The greater the degree of unsaturation, the lower the melting point. Compare, for example, the melting points of linoleic acid, a **polyunsaturated fatty acid,** and stearic acid, a saturated fatty acid.

Polyunsaturated fatty acid A fatty acid with two or more carbon-carbon double bonds in its hydrocarbon chain.

EXAMPLE 25.1

Draw the structural formula of a triglyceride derived from one molecule each of palmitic acid, oleic acid, and stearic acid, the three most abundant fatty acids in the biological world.

Solution

In this structure, palmitic acid is esterified at carbon 1 of glycerol, oleic acid at carbon 2, and stearic acid at carbon 3.

$$\begin{array}{l} \text{palmitate (16:0)}: \; CH_2OC(=O)(CH_2)_{14}CH_3 \\ \text{oleate (18:1)}: \; CH_3(CH_2)_7CH{=}CH(CH_2)_7C(=O)OCH \\ \text{stearate (18:0)}: \; CH_2OC(=O)(CH_2)_{16}CH_3 \end{array}$$

A triacylglycerol
(a triglyceride)

Beeswax is largely triacontyl palmitate, $CH_3(CH_2)_{14}CO_2(CH_2)_{29}CH_3$, an ester of palmitic acid. *(Charles D. Winters)*

PROBLEM 25.1

(a) How many constitutional isomers are possible for a triglyceride containing one molecule each of palmitic acid, oleic acid, and stearic acid?
(b) Which of these constitutional isomers are chiral?

B. Physical Properties

The physical properties of a triglyceride depend on its fatty acid components. In general, the melting point of a triglyceride increases as the number of carbons in its hydrocarbon chains increases and as the number of carbon-carbon double bonds decreases. Triglycerides rich in oleic acid, linoleic acid, and other unsaturated fatty acids are generally liquid at room temperature and are called **oils,** as, for example, olive oil and corn oil. Triglycerides rich in palmitic, stearic, and other saturated fatty acids are generally semisolids or solids at room temperature and are called **fats,** as, for example butter fat and human fat. Fats of land animals typically contain approximately 40%–50% saturated fatty acids by weight (Table 25.2). Most plant oils, on the other hand, contain 20% or less saturated fatty acids and 80% or more unsaturated fatty acids. The notable exception to this generalization about plant oils are the trop-

Oil A triglyceride that is liquid at room temperature.

Fat A triglyceride that is semisolid or solid at room temperature.

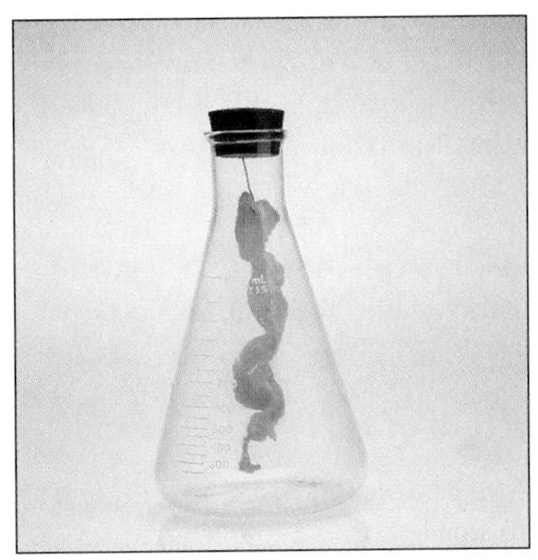

The fat in bacon is partially unsaturated, and like other unsaturated organic compounds, it reacts with Br_2. In these two photos you see the color of the bromine disappear upon reaction. *(Charles D. Winters)*

ical oils (for example, coconut and palm oils), which are considerably richer in the lower molecular-weight saturated fatty acids.

The lower melting points of triglycerides rich in unsaturated fatty acids are related to differences in three-dimensional shape between the hydrocarbon chains of their unsaturated and saturated fatty acid components. Shown in Figure 25.1 is a space-filling model of tripalmitin, a saturated triglyceride. In this model, the hydrocarbon chains lie parallel to each other, giving the molecule an ordered, compact shape. Because of this compact three-dimensional shape and the resulting strength of the dispersion forces between hydrocarbon chains of adjacent molecules, triglycerides rich in saturated fatty acids pack well into crystalline form and have melting points above room temperature.

The three-dimensional shape of an unsaturated fatty acid is quite different from that of a saturated fatty acid. Recall from Section 25.1A that unsaturated fatty acids of

FIGURE 25.1 Tripalmitin, a saturated triglyceride.

TABLE 25.2 Grams of Fatty Acid per 100 g of Triglycerides of Several Fats and Oils*

	Saturated Fatty Acids			Unsaturated Fatty Acids	
Fat or Oil	**Lauric (12:0)**	**Palmitic (16:0)**	**Stearic (18:0)**	**Oleic (18:1)**	**Linoleic (18:2)**
human fat	—	24.0	8.4	46.9	10.2
beef fat	—	27.4	14.1	49.6	2.5
butter fat	2.5	29.0	9.2	26.7	3.6
coconut oil	45.4	10.5	2.3	7.5	trace
corn oil	—	10.2	3.0	49.6	34.3
olive oil	—	6.9	2.3	84.4	4.6
palm oil	—	40.1	5.5	42.7	10.3
peanut oil	—	8.3	3.1	56.0	26.0
soybean oil	0.2	9.8	2.4	28.9	50.7

* Only the most abundant fatty acids are given, other fatty acids are present in lesser amounts.

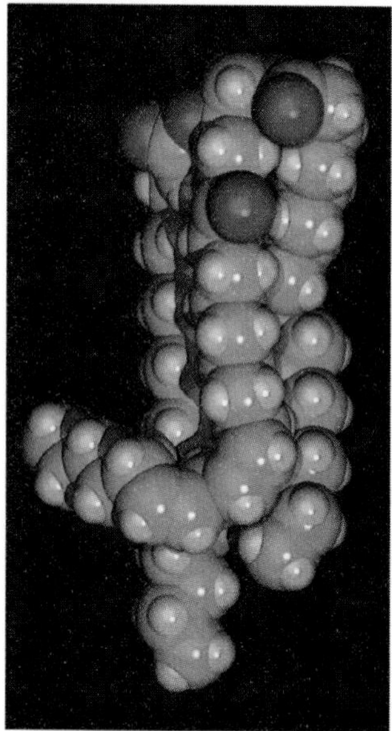

FIGURE 25.2
A polyunsaturated triglyceride.

higher organisms are predominantly of the *cis* configuration. Figure 25.2 shows a space-filling model of a **polyunsaturated triglyceride** derived from one molecule each of stearic acid, oleic acid, and linoleic acid. Each double bond in this polyunsaturated triglyceride has the *cis* configuration.

Polyunsaturated triglycerides have a less ordered structure and do not pack together so closely or so compactly as saturated triglycerides. Intermolecular dispersion forces are weaker, with the result that polyunsaturated triglycerides have lower melting points than their saturated counterparts.

C. Reduction of Fatty Acid Chains

For a variety of reasons, in part convenience and in part dietary preference, conversion of oils to fats has become a major industry. The process is called **hardening** of oils and involves catalytic hydrogenation (Section 6.6A) of some or all of the carbon-carbon double bonds in the polyunsaturated plant oil. In practice, the degree of hardening is carefully controlled to produce fats of a desired consistency. The resulting fats are sold for kitchen use (Crisco, Spry, and others). Margarine and other butter substitutes are prepared by catalytic reduction of corn, soybean, cottonseed, and peanut oils. The resulting product is then churned with milk and artificially colored to give it a flavor, consistency, and color resembling those of butter.

Polyunsaturated triglyceride A triglyceride having several carbon-carbon double bonds in the hydrocarbon chains of its three fatty acids.

Soap A sodium or potassium salt of a fatty acid.

25.2 Soaps and Detergents

A. Structure and Preparation of Soaps

Natural soaps are prepared by boiling lard or other animal fat with sodium hydroxide. The reaction that takes place is called saponification (Latin, *sapo*, soap). At the molecular level, saponification corresponds to base-promoted hydrolysis of ester groups in triglycerides (Section 25.1).

$$\begin{array}{c} CH_2OC(=O)R \\ RC(=O)OCH \\ CH_2OC(=O)R \end{array} + 3NaOH \xrightarrow{\text{saponification}} \begin{array}{c} CH_2OH \\ CHOH \\ CH_2OH \end{array} + 3RC(=O)O^-Na^+$$

A triglyceride (a triester of glycerol) — 1,2,3-Propanetriol (Glycerol; Glycerin) — Sodium soaps

The common triglycerides used today are obtained from beef tallow (from the meat-packing industry) and from coconut and palm oils. After hydrolysis is complete, sodium chloride is added to precipitate the soap as thick curds. The water layer is then drawn off, and glycerol is recovered by vacuum distillation. The crude soap contains sodium chloride, sodium hydroxide, and other impurities. These are removed by boiling the curd in water and reprecipitating with more sodium chloride. After several purifications, the soap can be used without further processing as an inexpensive industrial soap. Other treatments transform the crude soap into pH-controlled cosmetic soaps, medicated soaps, and the like.

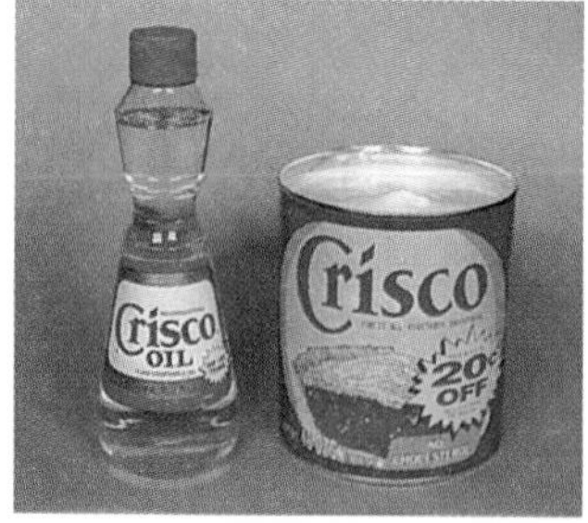

The hydrogenation of an oil to a fat is called hardening because a liquid is converted to a solid. *(Charles D. Winters)*

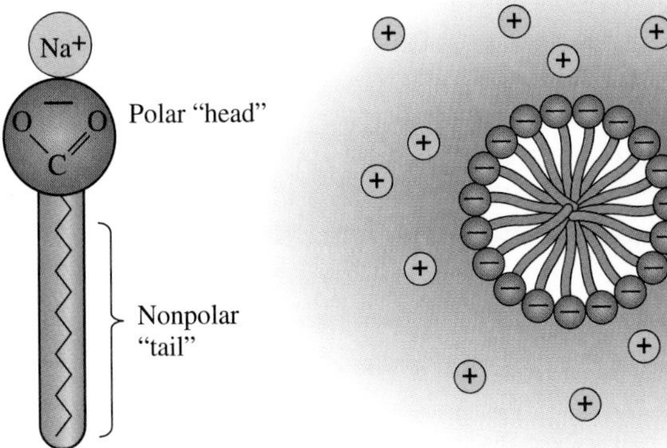

FIGURE 25.3
Soap micelles. (a) A sodium soap molecule. (b) Nonpolar (hydrophobic) hydrocarbon chains are clustered in the interior of the micelle and polar (hydrophilic) carboxylate groups are on the surface of the micelle. Soap micelles repel each other because of their negative surface charges.

B. How Soaps Clean

Soap owes its remarkable cleansing properties to its ability to act as an emulsifying agent. Because the long hydrocarbon chains of natural soaps are insoluble in water, they cluster in such a way as to minimize their contact with surrounding water molecules. The polar carboxylate groups, on the other hand, tend to remain in contact with the surrounding water molecules. Thus, in water, soap molecules spontaneously cluster into **micelles** (Figure 25.3). In a soap micelle, the carboxylate groups form a negatively charged surface and the nonpolar hydrocarbon chains lie buried within the micelle.

Micelle A spherical arrangement of organic molecules in water solution clustered so that their hydrophobic parts are buried inside the sphere and their hydrophilic parts are on the surface of the sphere and in contact with water.

Most of the things we commonly think of as dirt (such as grease, oil, and fat stains) are nonpolar and insoluble in water. When soap and this type of dirt are mixed together, as in a washing machine, the nonpolar hydrocarbon inner parts of the soap micelles "dissolve" the nonpolar dirt molecules. In effect, new soap micelles are formed, this time with nonpolar dirt molecules in the center (Figure 25.4). In this way, nonpolar organic grease, oil, and fat are "dissolved" and washed away in the polar wash water.

Soaps are not without their disadvantages. Foremost among these, they form insoluble salts when used in water containing Ca(II), Mg(II), or Fe(III) ions (hard water).

$$2CH_3(CH_2)_{14}CO_2^-Na^+ + Ca^{2+} \longrightarrow [CH_3(CH_2)_{14}CO_2^-]_2Ca^{2+} + 2Na^+$$

A sodium soap (soluble in water as micelles) — Calcium salt of a fatty acid (insoluble in water)

These calcium, magnesium, and iron salts of fatty acids create problems, including rings around the bathtub, films that spoil the luster of hair, and the grayness and roughness that build up on textiles after repeated washings.

FIGURE 25.4
A soap micelle with a "dissolved" oil or grease droplet.

C. Synthetic Detergents

Once the cleansing action of soaps was understood, a synthetic detergent could be designed. Molecules of a good detergent must have a long hydrocarbon chain, preferably 12 to 20 carbon atoms, and a polar group at one end of the molecule that does not form insoluble salts with Ca(II), Mg(II), or Fe(III) ions present in hard water. Chemists recognized that these essential characteristics of a soap could be produced in a molecule containing a sulfate or sulfonate group instead of a carboxylate group. Calcium, magnesium, and iron salts of monoalkylsulfuric and sulfonic acids are much more soluble in water than comparable salts of fatty acids.

The first synthetic detergent was made from 1-dodecanol (lauryl alcohol) by treating it with sulfuric acid to form a sulfate ester, followed by neutralization with sodium hydroxide to form the detergent commonly known as sodium dodecyl sulfate (SDS).

$$CH_3(CH_2)_{10}CH_2OH \xrightarrow[2.\ NaOH]{1.\ H_2SO_4} CH_3(CH_2)_{10}CH_2O\overset{\overset{O}{\|}}{\underset{\underset{O}{\|}}{S}}O^-Na^+$$

1-Dodecanol (Lauryl alcohol) — Sodium dodecyl sulfate (SDS) (Sodium lauryl sulfate)

The physical resemblance between SDS and natural soap molecules is obvious; both have a long nonpolar (hydrophobic) hydrocarbon chain and a highly polar (hydrophilic) end group. Large-scale commercial production of SDS was not possible because bulk quantities of 1-dodecanol were not available. However, because of its high foaming properties, SDS is used in numerous specialized applications, including shampoos and cosmetics. It is also used in biochemistry to denature proteins and to disrupt biological membranes.

Currently, the most important synthetic detergents are the linear alkylbenzenesulfonates (LAS). One of the most common of these is sodium 4-dodecylbenzenesulfonate. To prepare this type of detergent, a linear alkylbenzene is treated with sulfuric acid to form an alkylbenzenesulfonic acid. The sulfonic acid is then neutralized with NaOH, the product is mixed with builders, and spray-dried to give a smooth flowing powder. The most common builder is sodium silicate.

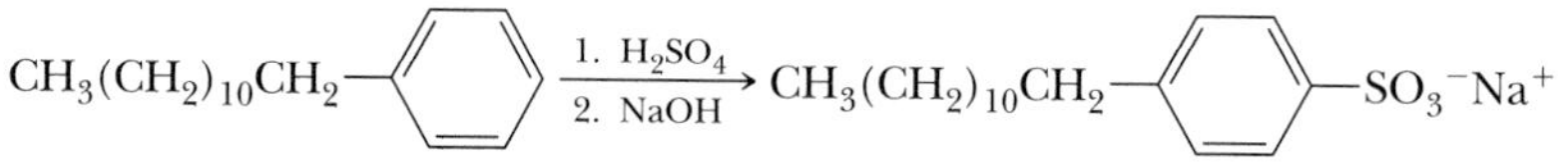

$$CH_3(CH_2)_{10}CH_2-C_6H_5 \xrightarrow[2.\ NaOH]{1.\ H_2SO_4} CH_3(CH_2)_{10}CH_2-C_6H_4-SO_3^-Na^+$$

Dodecylbenzene — Sodium 4-dodecylbenzenesulfonate (an anionic detergent)

Alkylbenzenesulfonate detergents were introduced in the late 1950s, and today they command close to 90% of the market once held by natural soaps.

Among the most common additives to detergent preparations are foam stabilizers, bleaches, and optical brighteners. A common foam stabilizer added to liquid soaps but not laundry detergents (for obvious reasons: think of a top-loading washing machine with foam spewing out the lid!) is the amide prepared from dodecanoic acid (lauric acid) and 2-aminoethanol (ethanolamine). The most common bleach is sodium perborate tetrahydrate, which decomposes at temperatures above 50°C to give hydrogen peroxide, the actual bleaching agent.

$$CH_3(CH_2)_{10}\overset{\overset{O}{\|}}{C}NHCH_2CH_2OH \qquad O{=}B{-}O{-}O^-Na^+\cdot 4H_2O$$

N-(2-Hydroxyethyl)dodecanamide (a foam stabilizer) — Sodium perborate tetrahydrate (a bleach)

Effects of optical bleaches: *(left)* ordinary light; *(right)* black light. *(Charles D. Winters)*

Also added to laundry detergents are optical brighteners, known also as optical bleaches, that are absorbed into fabrics and, after absorbing ambient light, fluoresce with a blue color, offsetting the yellow color due to fabric aging. Quite literally, these optical brighteners produce a "whiter-than-white" appearance. You most certainly have observed the effects of optical brighteners if you have ever been in the presence of black light (UV radiation) and seen the glow of "white" T-shirts, or blouses.

25.3 Prostaglandins

The prostaglandins are a family of compounds all having the 20-carbon skeleton of prostanoic acid.

Prostaglandin A member of the family of compounds having the 20-carbon skeleton of prostanoic acid.

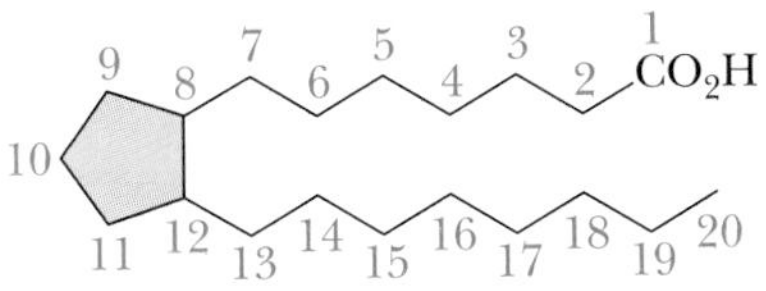

Prostanoic acid

The story of the discovery and structure determination of these remarkable compounds began in 1930 when gynecologists Raphael Kurzrok and Charles Lieb reported that human seminal fluid stimulates contraction of isolated uterine muscle. A few years later in Sweden, Ulf von Euler confirmed this report and noted that human seminal fluid also produces contraction of intestinal smooth muscle and lowers blood pressure when injected into the bloodstream. Von Euler proposed the name *prostaglandin* for the mysterious substance(s) responsible for these diverse effects because it was believed at the time that they were synthesized in the prostate gland. Although we now know that prostaglandin production is by no means limited to the prostate gland, the name nevertheless stuck. By the 1960s, several prostaglandins had been isolated and their structural formulas determined. Within the following decades, chemists have devised strategies for synthesizing not only naturally occurring prostaglandins but also prostaglandin analogs, several of which are now widely used in clinical medicine.

Prostaglandins are not stored as such in target tissues. Rather, they are synthesized in response to specific physiological triggers. Starting materials for the biosynthesis of prostaglandins are polyunsaturated fatty acids of 20 carbon atoms, stored until needed as membrane phospholipid esters. In response to a physiological trigger, the ester is hydrolyzed, the fatty acid released, and the synthesis of prostaglandins initiated. Figure 25.5 outlines the steps in the synthesis of several prostaglandins from the tetraun-

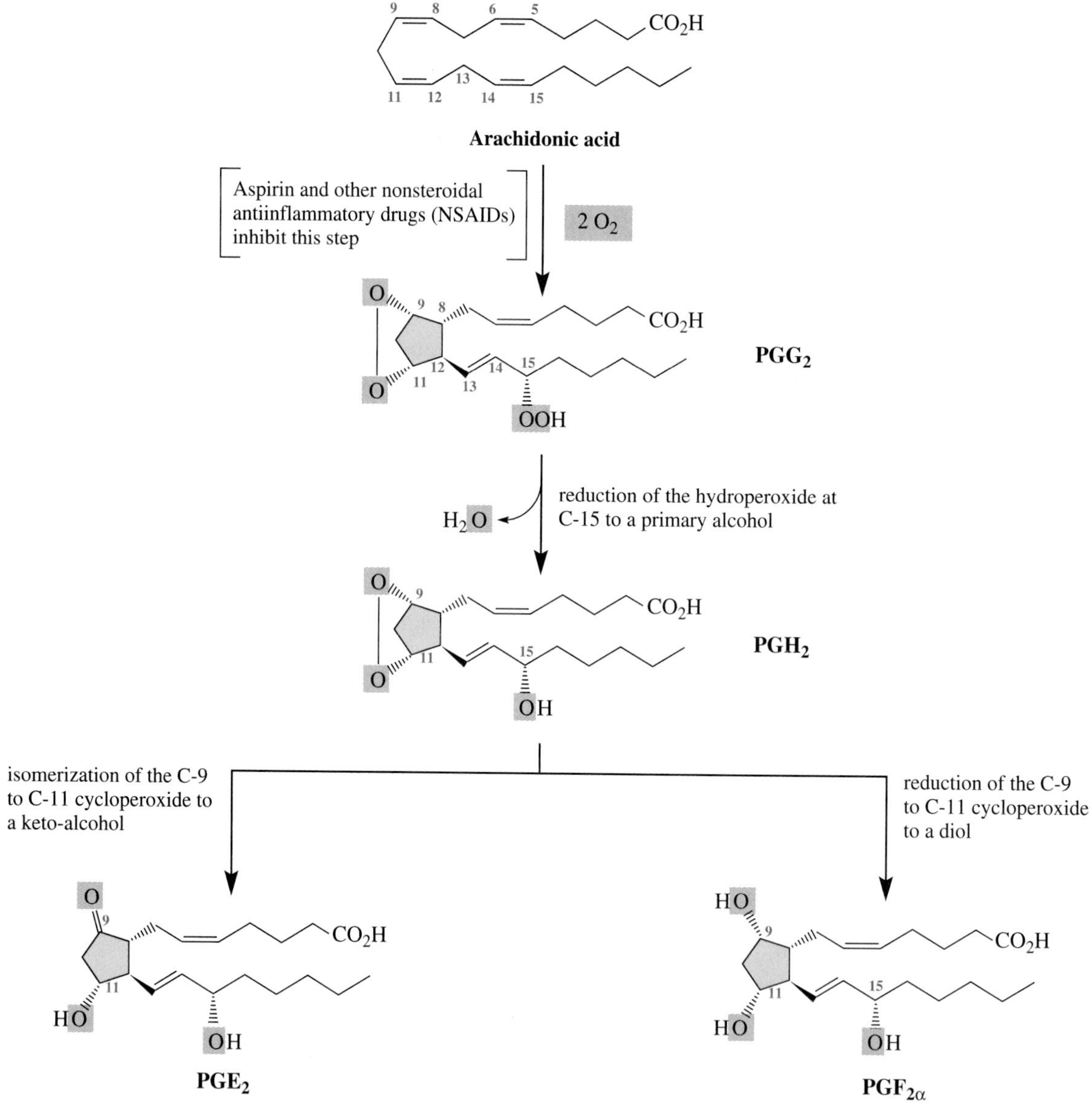

FIGURE 25.5
Key intermediates in the conversion of arachidonic acid to PGE_2 and $PGF_{2\alpha}$.

saturated fatty acid arachidonic acid. A key step in this biosynthesis is the reaction of arachidonic acid with two molecules of O_2 to form prostaglandin G (PGG_2). The antiinflammatory effect of aspirin and certain other nonsteroidal antiinflammatory drugs (NSAIDs) results from their ability to inhibit the action of cyclooxygenase, the first enzyme in the prostaglandin synthesis cascade. This enzyme, with the systematic name prostaglandin H synthase, catalyzes both cyclooxygenation of arachidonic acid and reduction of PGG_2 to PGH_2.

A word about the nomenclature of prostaglandins. They are abbreviated PG with an additional letter and number to indicate the type and series. Those of the G type (for example, PG**G**$_2$) have a cyclic peroxide and a hydroperoxide group; PG**E**s have a β-hydroxyketone group in the five-membered ring; PG**F**s have a 1,3-diol group in the five-membered ring. Those of the **1** series have one double bond in the hydrocarbon side chains; those of the **2** series (for example, PGF$_{\mathbf{2}\alpha}$) have two double bonds in the side chains. The subscript α (for example, in $PGF_{2\alpha}$) indicates that the —OH group at C-9 is below the plane of the five-membered ring and on the same side as the —OH on C-11.

Research on the involvement of prostaglandins in reproductive physiology and the inflammatory process has produced the first clinically useful prostaglandin derivatives. The observations that $PGF_{2\alpha}$ stimulates contractions of uterine smooth muscle led to the suggestion that this substance could be used as a therapeutic abortifacient. One problem with the use of the natural prostaglandins for this purpose is that they are very rapidly degraded within the body. In the search for less rapidly degraded prostaglandins, a number of analogs have been prepared, one of the most effective of which is 15-methyl $PGF_{2\alpha}$. This synthetic prostaglandin is 10 to 20 times more potent than the natural $PGF_{2\alpha}$ and is only slowly degraded. A comparison of these two prostaglandins illustrates how a simple change in the structure of a drug can make a significant change in its biological activity.

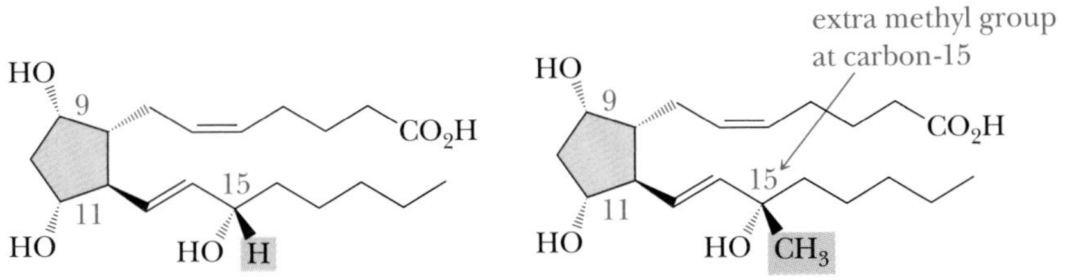

extra methyl group at carbon-15

$PGF_{2\alpha}$ 15-Methyl-$PGF_{2\alpha}$

The PGEs along with several other PGs suppress gastric ulceration and appear to heal gastric ulcers. The PGE_1 analog, misoprostol, is currently used primarily for prevention of ulceration associated with aspirin-like NSAIDs. Misoprostol cannot be administered to pregnant women because of its uterotonic activity.

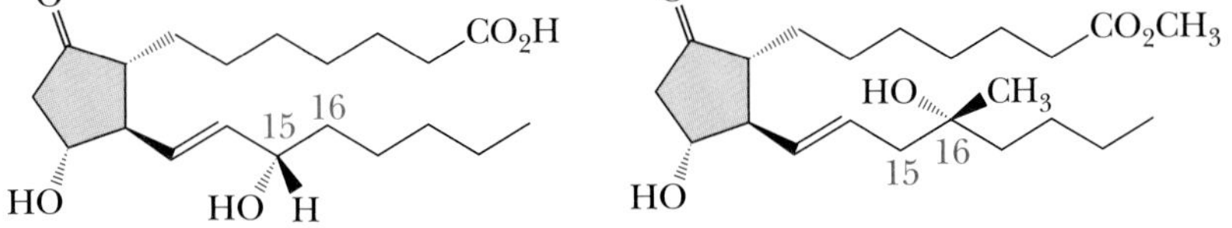

PGE_1 Misoprostol

Prostaglandins are members of an even larger family of compounds called eicosanoids. Eicosanoids contain 20 carbons and are derived from fatty acids. They include not only the prostaglandins but also the leukotrienes, thromboxanes, and prostacyclins. The eicosanoids are extremely widespread, and members of this family of compounds have been isolated from almost every tissue and body fluid.

Leukotriene C_4(LTC_4)
(three conjugated double bonds)

Thromboxane A_2
(a potent vasoconstrictor)

Prostacyclin

Leukotrienes are derived from arachidonic acid and are found primarily in leukocytes (white blood cells). Leukotriene C_4(LTC_4), a typical member of this family, has three conjugated double bonds (hence the suffix -triene) and contains the amino acids L-cysteine, glycine, and L-glutamic acid (Chapter 27). An important physiological action of LTC_4 is constriction of smooth muscles, especially those of the lungs. The synthesis and release of LTC_4 is prompted by allergic reactions. Drugs that inhibit the synthesis of LTC_4 show promise for the treatment of the allergic reactions associated with asthma. Thromboxane A_2 is a very potent vasoconstrictor; its release triggers the irreversible phase of platelet aggregation and constriction of injured blood vessels. It is thought that aspirin and aspirin-like drugs act as mild anticoagulants because they inhibit cyclooxygenase, the enzyme that initiates the synthesis of thromboxane A_2.

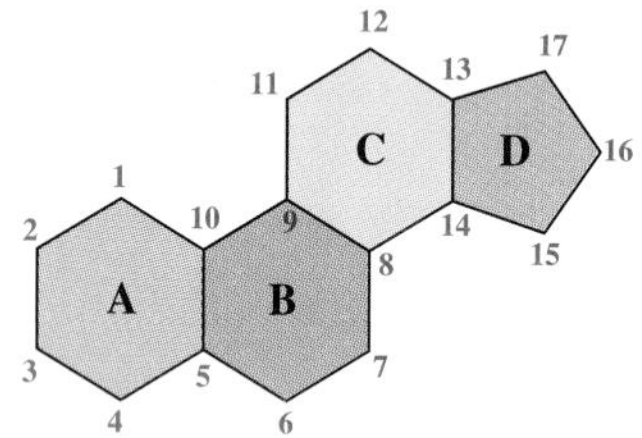

FIGURE 25.6
The tetracyclic ring system characteristic of steroids.

25.4 Steroids

Steroid A plant or animal lipid having the characteristic tetracyclic ring structure of the steroid nucleus, namely three six-membered rings and one five-membered ring.

Steroids are a group of plant and animal lipids that have the tetracylic ring system shown in Figure 25.6. The rings are lettered A, B, C, and D and carbon atoms are numbered 1 through 17. The features common to the tetracyclic ring system of most naturally occurring steroids are illustrated in Figure 25.7.

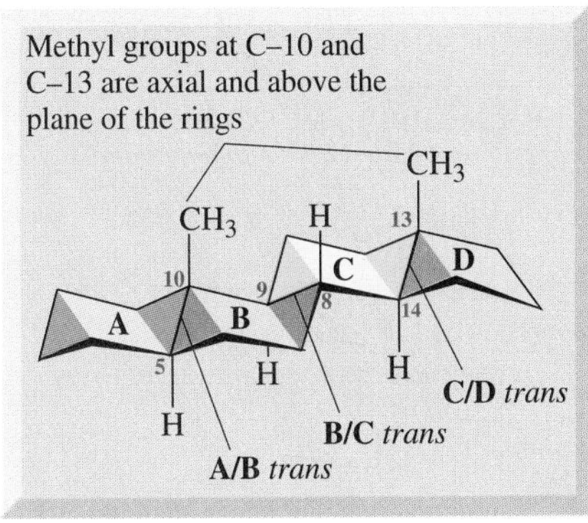

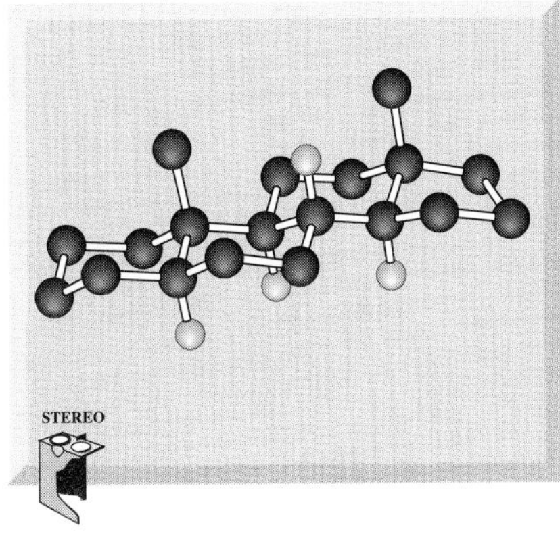

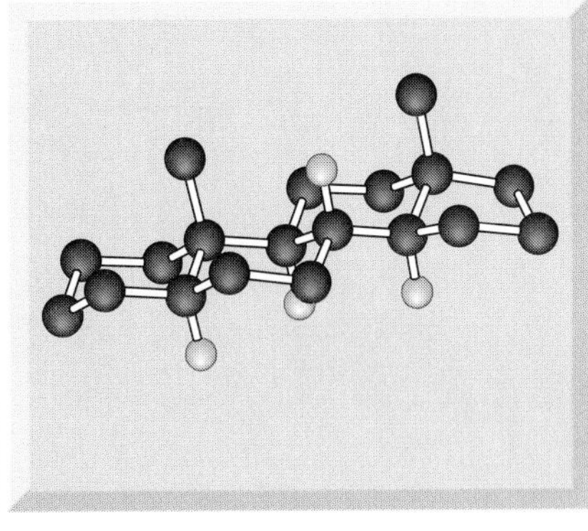

FIGURE 25.7
Features common to the tetracyclic ring of many steroids.

1. The fusion of rings is *trans* and each atom or group at a ring junction is axial. Consider, for example, the orientations of —H at C-5 and —CH_3 at C-10.
2. The pattern of atoms or groups along the points of ring fusion (carbons 5 to 10 to 9 to 8 to 14 to 13) is nearly always *trans*-anti-*trans*-anti-*trans*.
3. Because of the *trans*-anti-*trans*-anti-*trans* arrangement of atoms or groups along the points of ring fusion, the tetracyclic steroid ring system is nearly flat and quite rigid, a feature we return to when we consider the manner in which cholesterol is incorporated into biological membranes.
4. Many steroids have axial methyl groups at C-10 and C-13 of the tetracyclic ring system.

A. Structure of the Major Classes of Steroids

Cholesterol

Cholesterol is a white, water-insoluble, waxy solid found in blood plasma and in all animal tissues. This substance is an integral part of human metabolism in two ways: (1) It is an essential component of biological membranes. The body of a healthy adult contains approximately 140 g of cholesterol, about 120 g of which is present in membranes. Membranes of the central and peripheral nervous systems, for example, contain about 10% cholesterol by weight. (2) It is the compound from which sex hormones, adrenocorticoid hormones, bile acids, and vitamin D are synthesized. Thus, cholesterol is, in a sense, the parent steroid.

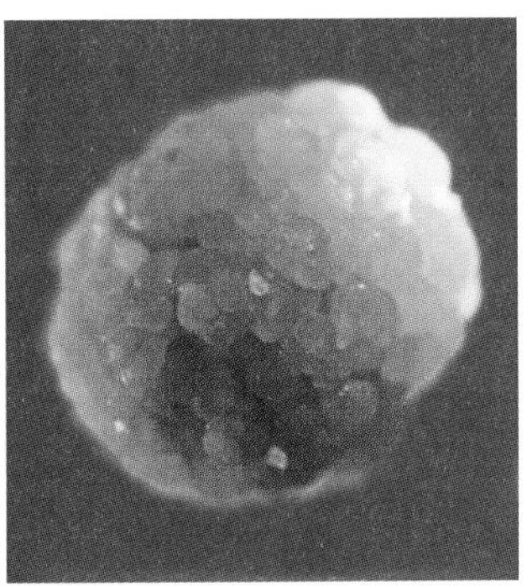

Human gallstones are almost pure cholesterol; this gallstone is about 0.5 cm in diameter. *(© Carolina Biological Supply Company, Phototake NYC)*

Cholesterol has eight stereocenters, and a molecule with this structural feature can exist as 2^8, or 256, stereoisomers (128 pairs of enantiomers). Only one of these stereoisomers is known to exist in nature: the stereoisomer with the configuration shown in Figure 25.8.

Cholesterol is insoluble in blood plasma but can be transported as a complex with a class of proteins called lipoproteins. **Low-density lipoproteins** (LDL) transport cholesterol from the site of its synthesis in the liver to the various tissues and cells of the body where it is to be used. It is primarily cholesterol attached to LDLs that builds up in atherosclerotic deposits in blood vessels. **High-density lipoproteins** (HDL) transport excess and unused cholesterol from cells back to the liver for its degradation to bile acids and eventual excretion in the feces. It is thought that HDLs retard or reduce atherosclerotic deposits.

Low-density lipoprotein (LDL) Plasma particles, density 1.02–1.06 g/mL, consisting of approximately 25% proteins, 50% cholesterol, 21% phospholipids, and 4% triglycerides.

High-density lipoprotein (HDL) Plasma particles, density 1.06–1.21 g/mL, consisting of approximately 33% proteins, 30% cholesterol, 29% phospholipids, and 8% triglycerides.

TABLE 25.3 Selected Steroid Hormones

Structure	Source and Major Effects
Testosterone; Androsterone	Androgens (male sex hormones). Synthesized in the testes; responsible for development of male secondary sex characteristics.
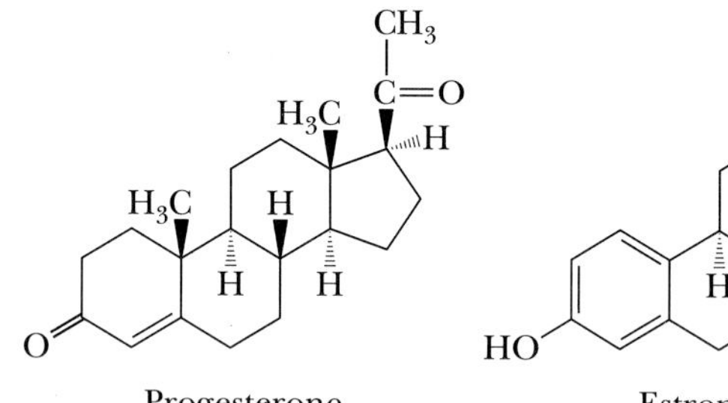 Progesterone; Estrone	Estrogens (female sex hormones). Synthesized in the ovaries; responsible for development of female secondary sex characteristics and control of the menstrual cycle.
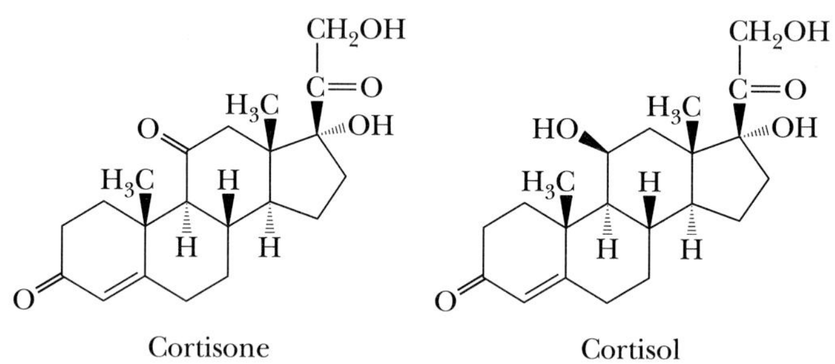 Cortisone; Cortisol	Glucocorticoid hormones. Synthesized in the adrenal cortex; regulate metabolism of carbohydrates, decrease inflammation, and are involved in the reaction to stress.
Aldosterone	A mineralocorticoid hormone. Synthesized in the adrenal cortex; regulates blood pressure and volume by stimulating the kidneys to absorb Na^+, Cl^-, and HCO_3^-.

Steroid Hormones

Given in Table 25.3 are structures of members of each major class of steroid hormones, along with the principal functions of each.

Once the role of progesterone in inhibiting ovulation was understood, its potential as a possible oral contraceptive was realized. Progesterone itself is relatively ineffective when taken orally. As a result of a massive research program in both industrial and academic laboratories, many synthetic progesterone-mimicking steroids became avail-

Estrogen A steroid hormone, such as estrone, which mediates the development of sexual characteristics in females.

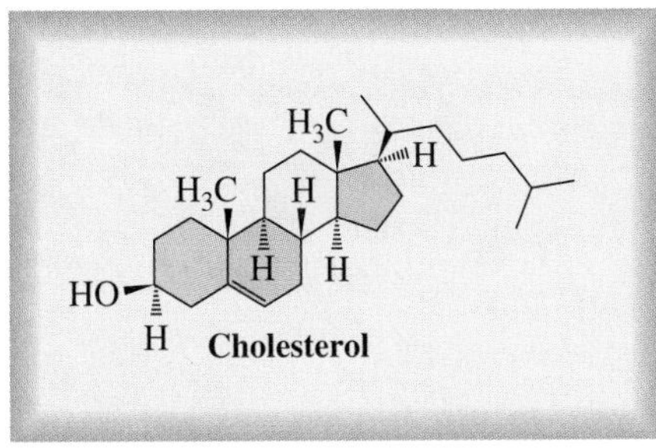

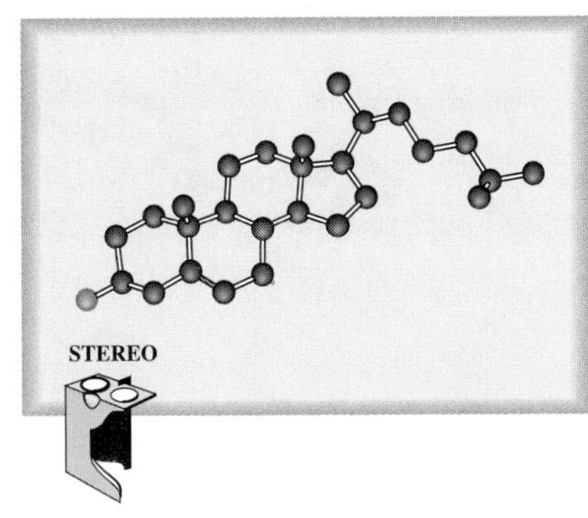

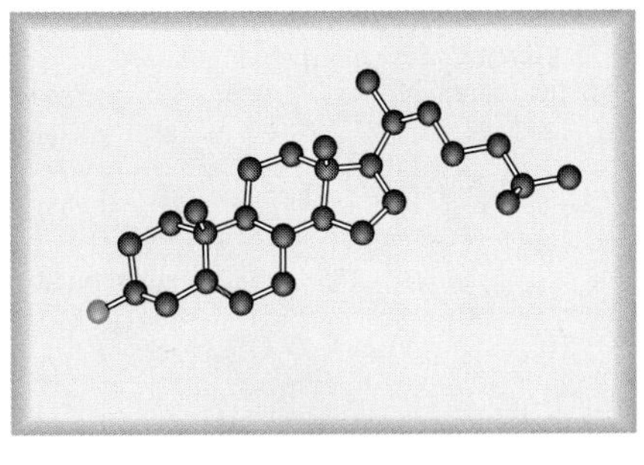

FIGURE 25.8
Cholesterol is found in blood plasma and in all animal tissues.

able in the 1960s. (See "A Conversation with Carl Djerassi," pp. 181–183.) When taken regularly, these drugs prevent ovulation, yet allow women to maintain a normal menstrual cycle. Some of the most effective of these preparations contain a progesterone analog, such as norethindrone, combined with a smaller amount of an estrogen-like material to help prevent irregular menstrual flow during prolonged use of contraceptive pills.

"Nor" refers to the absence of a methyl group here; the methyl group is present in ethindrone

Norethindrone
(a synthetic progesterone analog)

The chief function of testosterone and other androgens is to promote normal growth of male reproductive organs (primary sex characteristics) and development of the characteristic deep voice, pattern of body and facial hair, and musculature (secondary sex characteristics). Although testosterone produces these effects, it is not active when taken orally because it is metabolized in the liver to an inactive steroid. A number of oral anabolic steroids have been developed for use in rehabilitation medicine, particularly when muscle atrophy occurs during recovery from an injury. Among the synthetic anabolic steroids most widely prescribed for this purpose are methandrostenolone and stanozolol. The structural formula of methandrostenolone differs from that of testosterone by introduction of (1) a methyl group at C-17 and (2) an additional carbon-carbon double bond between C-1 and C-2. In stanozolol, ring A is modified by attachment of a pyrazole ring.

Androgen A steroid hormone, such as testosterone, which mediates the development of sexual characteristics of males.

Anabolic steroid A steroid hormone, such as testosterone, which promotes tissue and muscle growth and development.

Methandrostenolone

Stanozolol

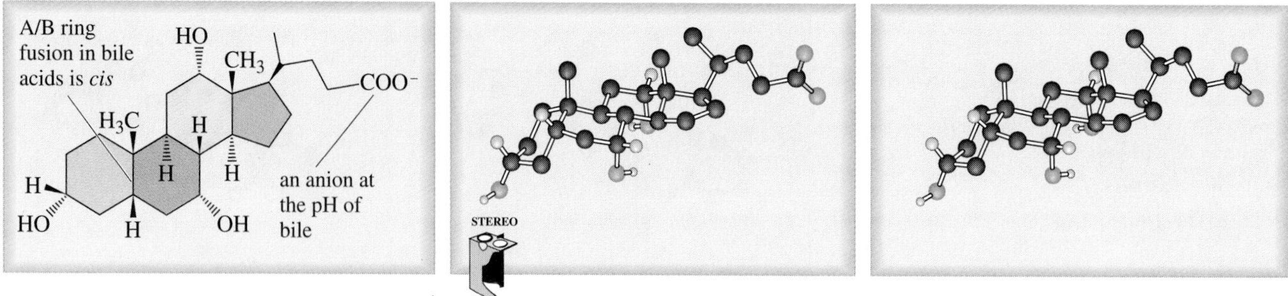

FIGURE 25.9
Cholic acid, an important constituent of human bile.

The misuse of anabolic steroids among certain athletes to build muscle mass and strength, particularly for sports that require explosive action, is common knowledge. The risks associated with abuse of anabolic steroids for this purpose are enormous: heightened aggressiveness, sterility, impotence, and risk of premature death from complications of diabetes, coronary artery disease, and liver cancer.

Bile Acids

Bile acid A cholesterol-derived detergent molecule, such as cholic acid, which is secreted by the gallbladder into the intestine to assist in the absorption of dietary lipids.

Shown in Figure 25.9 is the structural formula of cholic acid, an important constituent of human bile. The molecule is shown as an anion, as it is ionized in bile and intestinal fluids. Bile acids, or more properly, bile salts, are synthesized in the liver, stored in the gallbladder, and secreted into the intestine, where their function is to act as detergents to emulsify dietary fats and thereby aid in their absorption and digestion. Furthermore, bile salts are the end products of the metabolism of cholesterol and, thus, are a principal pathway for the elimination of this substance from the body. A characteristic structural feature of bile salts is a *cis* fusion of rings A/B.

B. Biosynthesis of Steroids

The biosynthesis of cholesterol illustrates a point we first made in our introduction to the structure of terpenes (Section 5.4): in building large molecules, one of the common patterns in the biological world is to begin with one or more smaller subunits, join them together by an iterative process, and then chemically modify the completed carbon skeleton by oxidation, reduction, cross-linking, addition, elimination, or related processes to give a biomolecule with a unique identity. Each step in this process is catalyzed by a specific enzyme.

The building block from which all carbon atoms of steroids are derived is the two-carbon acetyl group of acetyl-CoA. The American biochemist, Konrad Bloch, who shared the 1964 Nobel Prize for medicine and physiology with German biochemist Feodor Lynen for their discoveries concerning the biosynthesis of cholesterol and fatty acids, showed that 15 of the 27 carbon atoms of cholesterol are derived from the methyl group of acetyl-CoA; the remaining 12 carbon atoms are derived from the carbonyl group of acetyl-CoA. Names of key intermediates in this biosynthetic pathway are shown in Figure 25.10, along with the change in size of the carbon skeleton with each step. Doing the carbon-atom bookkeeping, you discover that 18 acetyl groups are required for the synthesis of one molecule of cholesterol. A remarkable feature of this synthetic pathway is that the biosynthesis of cholesterol from acetyl-CoA is completely stereoselective; it is synthesized as only one of 256 possible stereoisomers. We cannot duplicate this exquisite degree of stereoselectivity in the laboratory.

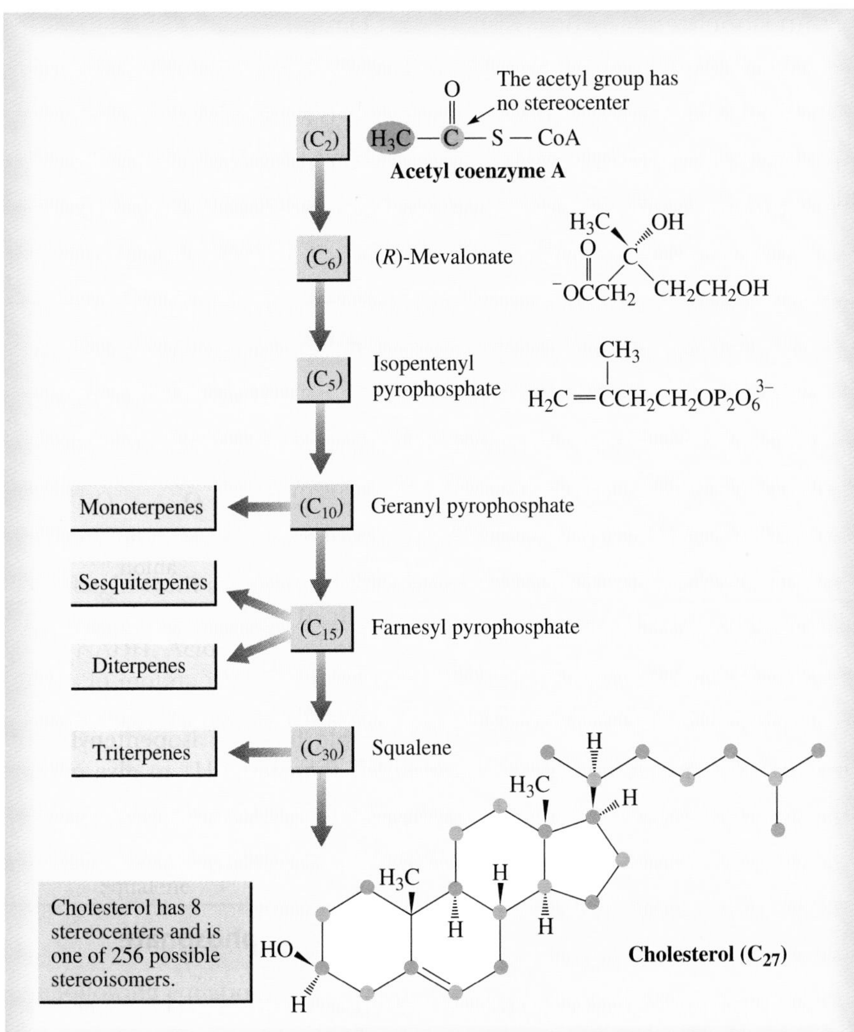

FIGURE 25.10 Several key intermediates in the synthesis of cholesterol from the acetyl groups of acetyl-CoA. Eighteen moles of acetyl-CoA are required for the synthesis of one mole of cholesterol.

Stage 1: From Acetyl-CoA to Isopentenyl Pyrophosphate and Dimethylallyl Pyrophosphate

The first stage in the biosynthesis of cholesterol is the synthesis of isopentenyl pyrophosphate from three molecules of acetyl-CoA (Section 18.4). The enzyme isopentenyl pyrophosphate isomerase catalyzes the interconversion of isopentenyl pyrophosphate and its constitutional isomer, dimethylallyl pyrophosphate, possibly by protonation of carbon 4 to form a 3° carbocation followed by loss of H^+ from carbon 2 to give the isomeric alkene.

$$\underset{\text{Isopentenyl pyrophosphate}}{\overset{4}{C}H_2=\overset{3}{C}(CH_3)\overset{2}{C}H_2\overset{1}{C}H_2OP_2O_6^{3-}} \xrightleftharpoons{\text{Isopentenyl pyrophosphate isomerase}} \underset{\text{Dimethylallyl pyrophosphate}}{CH_3C(CH_3)=CHCH_2OP_2O_6^{3-}}$$

Closure of rings C/D

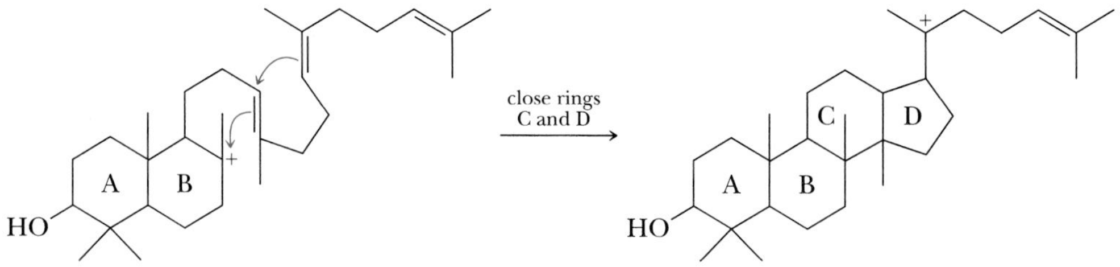

There then follows a series of four concerted 1,2-shifts culminating in loss of H^+ to give the carbon-carbon double bond at the junction of rings B and C.

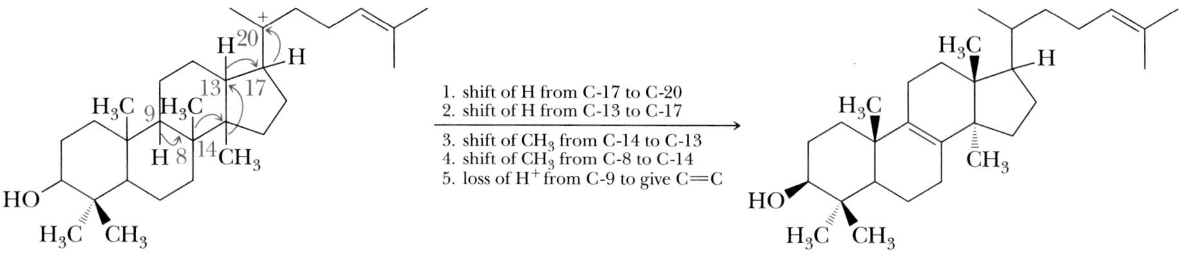

Lanosterol

Stage 4: From Lanosterol to Cholesterol

Conversion of lanosterol to cholesterol involves at least 25 enzyme-catalyzed reactions, the details of which are only poorly understood. This transformation involves loss of three methyl groups, migration of one double bond of lanosterol, and reduction of the other double bond.

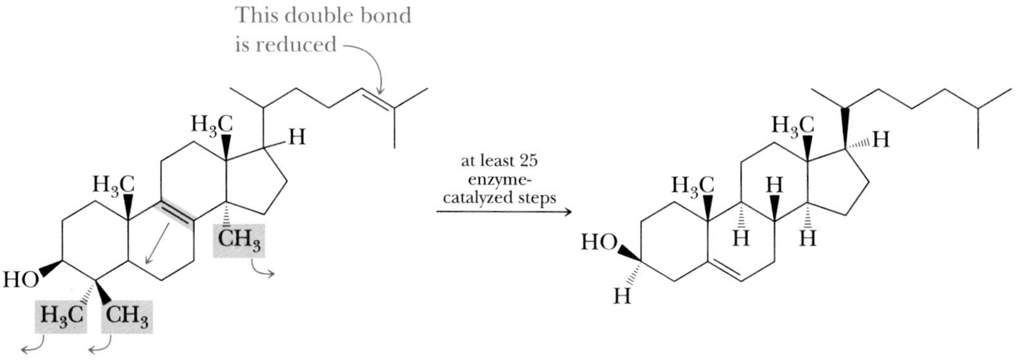

Lanosterol Cholesterol

Cholesterol is then the starting material for the synthesis of bile acids and steroid hormones.

Cholesterol →
- bile acids (e.g., cholic acid)
- sex hormones (e.g., testosterone and estrone)
- mineralocorticoid hormones (e.g., aldosterone)
- glucocorticoid hormones (e.g., cortisone)

CHEMISTRY IN ACTION

Drugs That Lower Plasma Levels of Cholesterol

Coronary artery disease is the leading cause of death in the United States and other Western countries, where about one-half of all deaths can be attributed to atherosclerosis. Atherosclerosis results from buildup and autoxidation of fatty deposits called plaque on the inner walls of arteries. A major component of plaque is cholesterol derived from low-density-lipoproteins (LDL), which circulate in blood plasma. Because more than one-half of total body cholesterol in humans is synthesized in the liver from acetyl-CoA, intensive efforts have been directed toward finding ways to inhibit this synthesis. The rate-limiting step in cholesterol biosynthesis is reduction of 3-hydroxy-3-methylglutaryl CoA (HMG-CoA) to mevalonic acid (Section 18.4). This four-electron reduction is catalyzed by the enzyme HMG-CoA reductase and requires 2 mol of NADPH per mol of HMG-CoA. Beginning in the early 1970s, researchers at the Sankyo Company in Tokyo screened more than 8000 strains of microorganisms and in 1976 announced the isolation of mevastatin, a potent inhibitor of HMG-CoA reductase, from culture broths of the fungus *Penicillium citrinum.* The same compound was isolated by researchers at Beecham Pharmaceuticals in England from cultures of *Penicillium brevicompactum.* Soon thereafter, a second, more active compound called lovastatin was isolated at the Sankyo Company from the fungus *Monascus ruber,* and at Merck Sharpe & Dohme from *Aspergillus terreus.* Both mold metabolites are extremely effective in lowering plasma concentrations of LDL. The active form of each is the 5-hydroxycarboxylic acid formed by hydrolysis of the δ-lactone.

It is thought that these drugs and several synthetic modifications now available inhibit HMG-CoA reductase by forming an enzyme-inhibitor complex that prevents further catalytic action of the enzyme. It is reasoned that the 3,5-dihydroxycarboxylic acid part of each drug binds tightly to the enzyme because it mimics the hemithioacetal intermediate formed after the first two-electron reduction of HMG-CoA.

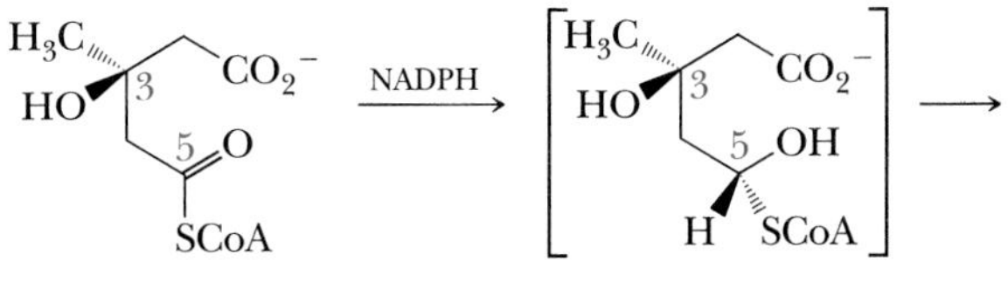

3-Hydroxy-3-methyl glutaryl-CoA (HMG-CoA) —NADPH→ A hemithioacetal intermediate formed by the first two-electron reduction → —NADPH→ Mevalonate

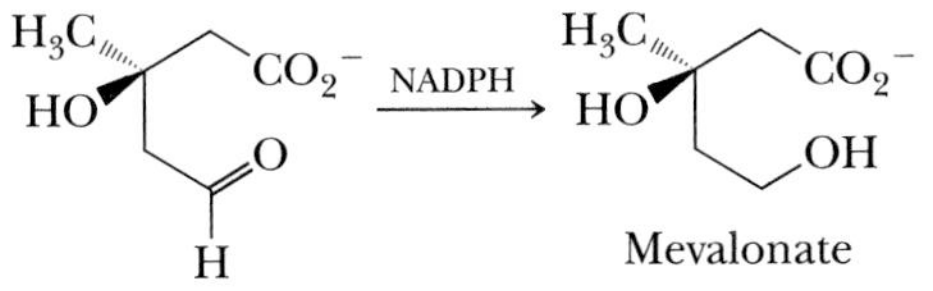

Systematic studies have shown the importance of each part of the drug for effectiveness. It has been found, for example, that the carboxylate anion is essential, and both the 3—OH and 5—OH groups must be free (not masked as ethers). Insertion of a bridging unit other than $-CH_2-CH_2-$ between carbon-5 and the bicyclo[4.4.0] ring system reduces potency as does almost any modification of the bicyclic ring system and its pattern of substitution.

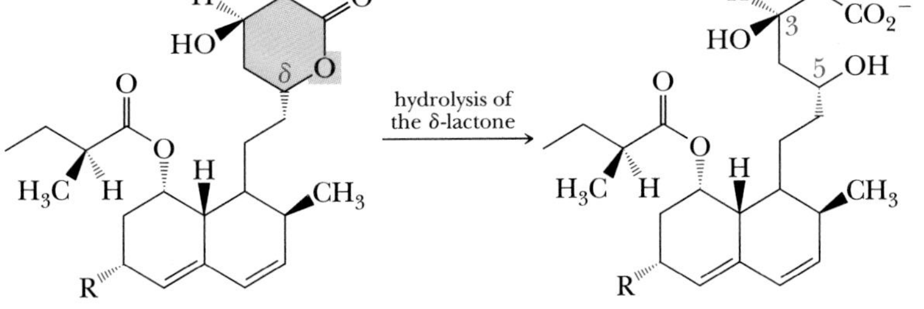

R = H, mevastatin
R = CH_3, lovastatin

—hydrolysis of the δ-lactone→

The active form of each drug

All of these products contain lecithin. *(Charles D. Winters)*

25.5 Phospholipids

A. Structure

Phospholipids, or phosphoacylglycerols as they are more properly named, are the second most abundant group of naturally occurring lipids. They are found almost exclusively in plant and animal cell membranes, which typically consist of about 40%–50% phospholipids and 50%–60% proteins. The most abundant phospholipids are derived from phosphatidic acid (Figure 25.11), a molecule containing glycerol esterified with two molecules of fatty acids and one molecule of phosphoric acid.

The fatty acids most common in phosphatidic acids are palmitic and stearic acids (both fully saturated) and oleic acid (one double bond in the hydrocarbon chain). Further esterification of phosphatidic acid with a low-molecular-weight alcohol gives a phospholipid. Several of the most common alcohols forming phospholipids are given in Table 25.4 (page 1020). All functional groups in this table and in Figure 25.11 are shown as they are ionized at pH 7.4, the approximate pH of blood plasma and of many

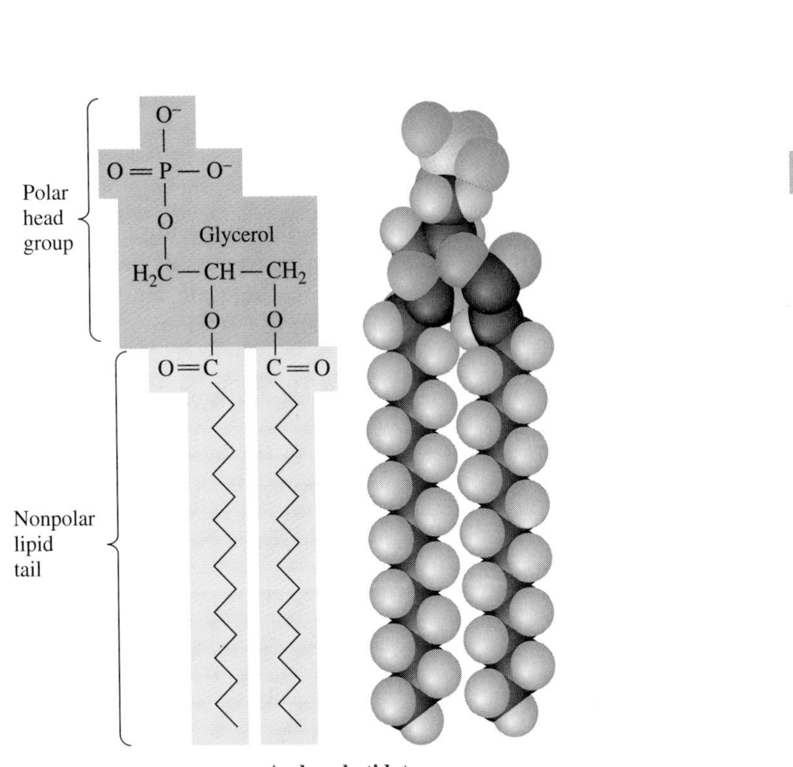

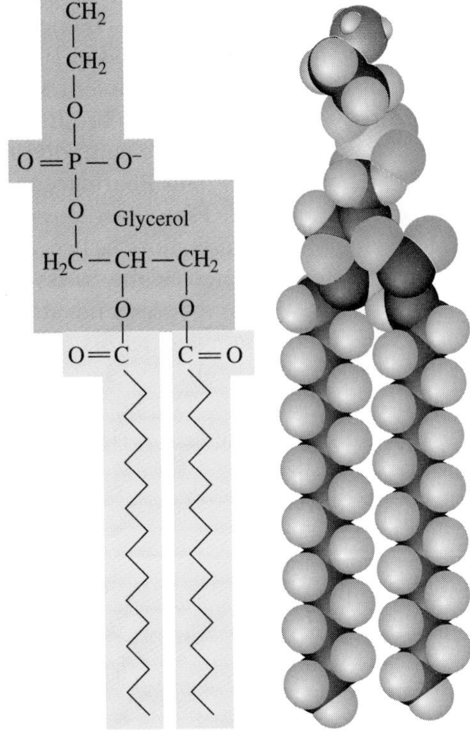

FIGURE 25.11
In a phosphatidic acid, glycerol is esterified with two molecules of fatty acid and one molecule of phosphoric acid. Further esterification of the phosphoric acid group with a low-molecular-weight alcohol gives a phospholipid.

Chemistry in Action

Snake Venom Phospholipases

The venoms of certain snakes contain enzymes called phospholipases. These enzymes catalyze the hydrolysis of carboxylic acid ester bonds of phospholipids. The venom of the eastern diamondback rattlesnake *(Crotalus adamanteus)* and the Indian cobra *(Naja naja)* both contain phospholipase PLA_2, which catalyzes the hydrolysis of esters at carbon 2 of phospholipids. The breakdown product of this hydrolysis, a lysolecithin, acts as a detergent and dissolves the membranes of red blood cells causing them to rupture. Indian cobras kill several thousand people each year.

Milking an Indian cobra for its venom. (Dan McCoy/Rainbow)

```
                 CH2CH2NH3+
                 |
                 O
                 |
               O=P—O−
                 |
                 O
                 |
     CH2—CH—CH2          +      H2O   --PLA2-->
     |    |
     O    O
     |    |  <-- PLA2 catalyzes
   O=C  O=C      hydrolysis
     |    |      of this ester bond
     R1   R2
   A phospholipid

                 CH2CH2NH3+
                 |
                 O
                 |
               O=P—O−
                 |
                 O                              O
                 |                              ‖
     CH2—CH—CH2          +            R2—C—O− + H+
     |    |
     O    OH
     |
   O=C
     |
     R1
   A lysolecithin
```

biological fluids. Under these conditions, each phosphate group bears a negative charge and each amino group bears a positive charge.

B. Shapes of Phospholipids

Phospholipid A lipid containing glycerol esterified with two molecules of fatty acid and one molecule of phosphoric acid.

Figure 25.12 shows a space-filling model of a lecithin (a phosphatidylcholine). It and other phospholipids are important components of biological membranes. These molecules are elongated, almost rodlike structures, with their nonpolar (hydrophobic)

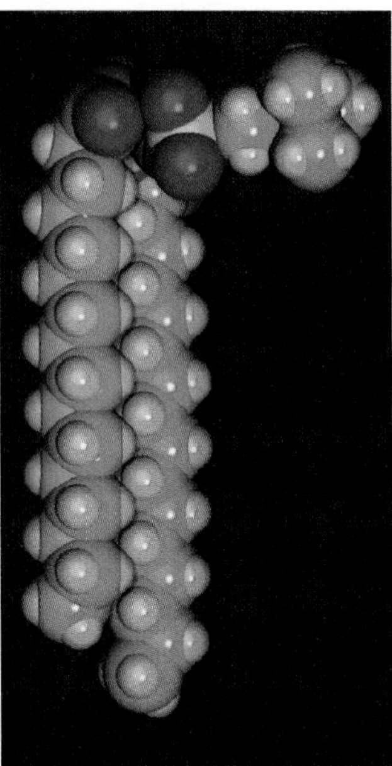

FIGURE 25.12
Space-filling model of a lecithin.

TABLE 25.4 Low-Molecular-Weight Alcohols Most Common to Phospholipids

Alcohols Found in Phospholipids		
Structural Formula	**Name**	**Name of Phospholipid**
$HOCH_2CH_2NH_2$	ethanolamine	phosphatidylethanolamine (cephalin)
$HOCH_2CH_2\overset{+}{N}(CH_3)_3$	choline	phosphatidylcholine (lecithin)
$HOCH_2CH(NH_3^+)CO_2^-$	serine	phosphatidylserine
(inositol ring structure: HO, HO, HO, OH, OH, OH)	inositol	phosphatidylinositol

hydrocarbon chains lying roughly parallel to one another and their polar (hydrophilic) phosphoric ester group pointing in the opposite direction.

C. The Fluid-Mosaic Model of Membrane Structure

Lipid bilayer A back-to-back arrangement of phospholipid monolayers.

Fluid-mosaic model A biological membrane consists of a phospholipid bilayer with proteins, carbohydrates, and other lipids embedded on the surface and in the bilayer.

The most satisfactory current model for the arrangement of phospholipids and proteins in plant and animal membranes is the **fluid-mosaic model** proposed in 1972 by S. J. Singer and G. Nicolson. According to the fluid-mosaic model (Figure 25.13), membrane phospholipids form a lipid bilayer with membrane proteins associated with this bilayer as (1) peripheral proteins both on the inside and outside surfaces of the membrane and as (2) integral proteins penetrating the bilayer. The term "mosaic" signifies that the various components in the membrane coexist side by side, as discrete units, rather than combining to form new molecules or ions. "Fluid" signifies that the protein components of membranes "float" in the bilayer and can move laterally along the plane of the membrane.

25.6 Fat-Soluble Vitamins

Vitamins are divided into two broad classes on the basis of their solubility: those that are fat-soluble (and hence classified as lipids) and those that are water-soluble. The fat-soluble vitamins include A, D, E, and K.

A. Vitamin A

Vitamin A, or retinol, occurs only in the animal world, where the best sources are cod-liver oil and other fish-liver oils, animal liver, and dairy products. Vitamin A in the form of a precursor, or provitamin, is found in the plant world in a group of tetraterpene (C_{40}) pigments called carotenes. The most common of these is β-carotene, abundant in carrots but also found in some other vegetables, particularly yellow ones. β-

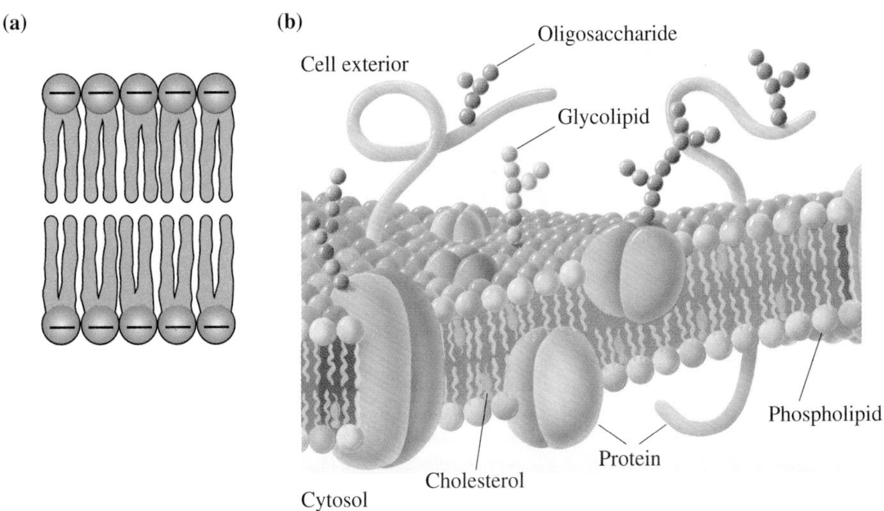

FIGURE 25.13
Fluid mosaic model of a biological membrane, showing the lipid bilayer and membrane proteins oriented on the inner and outer surfaces of the membrane and penetrating the entire thickness of the membrane.

Carotene has no vitamin A activity; however, after ingestion, it is cleaved at the central carbon-carbon double bond to give retinol (vitamin A).

cleavage of this C=C followed by reduction gives vitamin A

β-Carotene

enzyme-catalyzed cleavage in the liver

2 Retinol (Vitamin A)

Probably the best understood role of vitamin A is its participation in the visual cycle in rod cells. In a series of enzyme-catalyzed reactions (Figure 25.14), retinol undergoes a two-electron oxidation to all-*trans*-retinal, isomerization about the C-11 to C-12 double bond to give 11-*cis*-retinal, and formation of an imine with an $-NH_2$ from a lysine unit of the protein, opsin. The product of these reactions is rhodopsin, a highly conjugated pigment that shows intense absorption in the blue-green region of the visual spectrum.

The primary event in vision is absorption of light by rhodopsin in rod cells of the retina of the eye to produce an electronically excited molecule. Within several picoseconds (1 picosec = 10^{-12} sec), the excess electronic energy is converted to vibrational and rotational energy, and the 11-*cis* double bond is isomerized to the more

FIGURE 25.14
The primary chemical reaction of vision in rod cells is absorption of light by rhodopsin followed by isomerization of a carbon-carbon double bond from a *cis* configuration to a *trans* configuration.

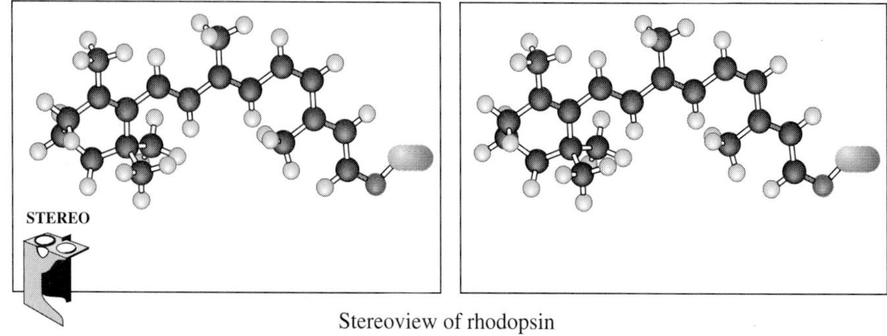

Stereoview of rhodopsin

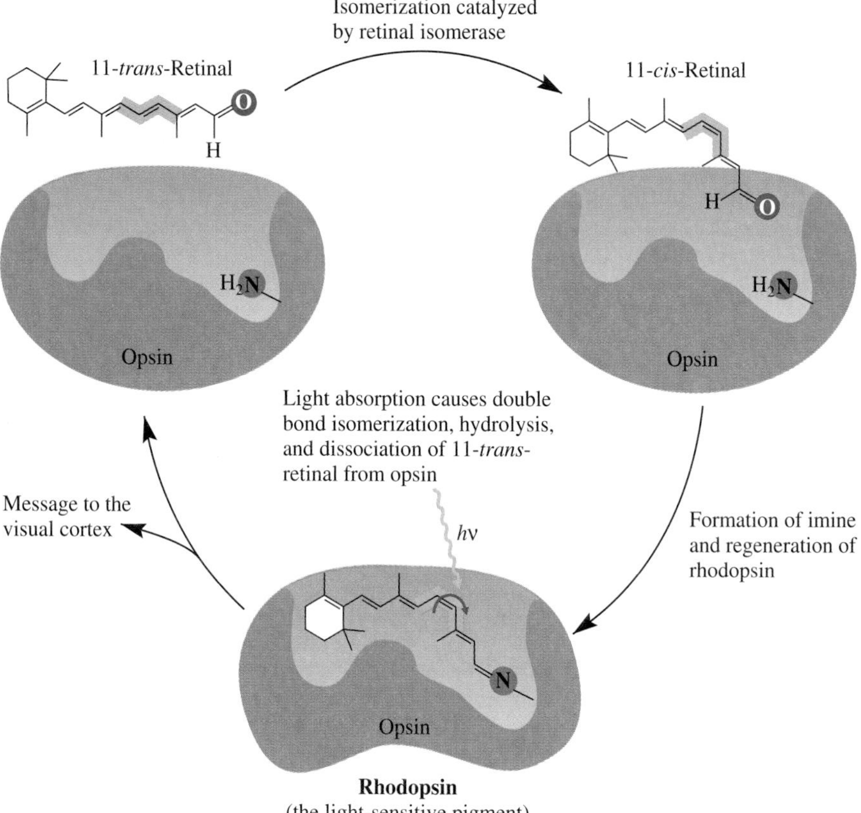

stable 11-*trans* double bond. This isomerization triggers a conformational change in the protein, opsin, that causes firing of neurons in the optic nerve and produces a visual image. Coupled with this light-induced change is hydrolysis of rhodopsin to give 11-*trans*-retinal and free opsin. At this point, the visual pigment is bleached and in a refractory period. Rhodopsin is regenerated by a series of enzyme-catalyzed reactions that converts 11-*trans*-retinal to 11-*cis*-retinal and then to rhodopsin. The visual cycle is shown in abbreviated form in Figure 25.14.

B. Vitamin D

Vitamin D is the name for a group of structurally related compounds that play a major role in the regulation of calcium and phosphorus metabolism. A deficiency of vitamin D in childhood is associated with rickets, a mineral-metabolism disease that leads to bowlegs, knock-knees, and enlarged joints. Vitamin D_3, the most abundant form of the vitamin in the circulatory system, is produced in the skin of mammals by the action of ultraviolet radiation on 7-dehydrocholesterol (cholesterol with a double bond between carbons 7 and 8). In the liver, vitamin D_3 undergoes an enzyme-catalyzed two-electron oxidation at carbon 25 of the side chain to form 25-hydroxyvitamin D_3; the oxidizing agent is molecular oxygen, O_2. 25-Hydroxyvitamin D_3 undergoes further oxidation in the kidneys, also by O_2, to form 1,25-dihydroxyvitamin D_3, the hormonally active form of the vitamin.

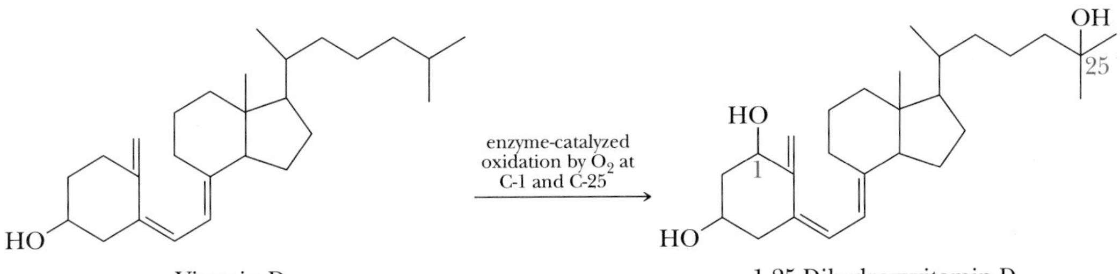

enzyme-catalyzed oxidation by O_2 at C-1 and C-25

Vitamin D_3

1,25-Dihydroxyvitamin D_3

C. Vitamin E

Vitamin E was first recognized in 1922 as a dietary factor essential for normal reproduction in rats, hence its name tocopherol from the Greek, *tocos*, birth and *pherein*, to bring about. Vitamin E is a group of compounds of similar structure, the most active of which is α-tocopherol. This vitamin occurs in fish oil, in other oils such as cottonseed and peanut oil, and in leafy green vegetables. The richest source of vitamin E is wheat germ oil.

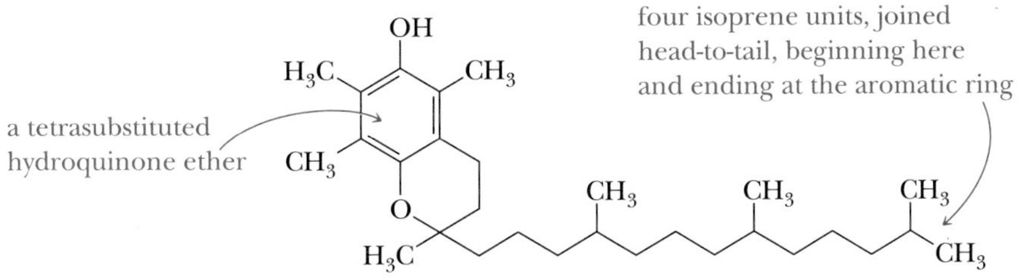

a tetrasubstituted hydroquinone ether

four isoprene units, joined head-to-tail, beginning here and ending at the aromatic ring

Vitamin E (α-Tocopherol)

In the body, vitamin E functions as an antioxidant; it traps peroxy radicals of the type HOO· and ROO· formed as a result of oxidation by molecular oxygen of unsaturated hydrocarbon chains in membrane phospholipids (see the Chemistry in Action box "Autoxidation," pp. 261–262). There is speculation that peroxy radicals play a role in the aging process and that vitamin E and other antioxidants may retard that process. Autoxidation of lipid deposits in arterial plaque is a major contributor to the development of cardiovascular disease, and antioxidants, such as vitamin E, may help inhibit these reactions. Vitamin E is also necessary for the proper development and function of membranes of red blood cells.

CHEMISTRY IN ACTION

Vitamin K, Blood Clotting, and Basicity

Vitamin K is a fat-soluble vitamin, which must be obtained from the diet. A deficiency of vitamin K results in slowed blood clotting, which can be a serious threat to a wounded animal or human. In the process of blood clotting, the natural vitamin, a quinone, is converted to its active hydroquinone form by reduction.

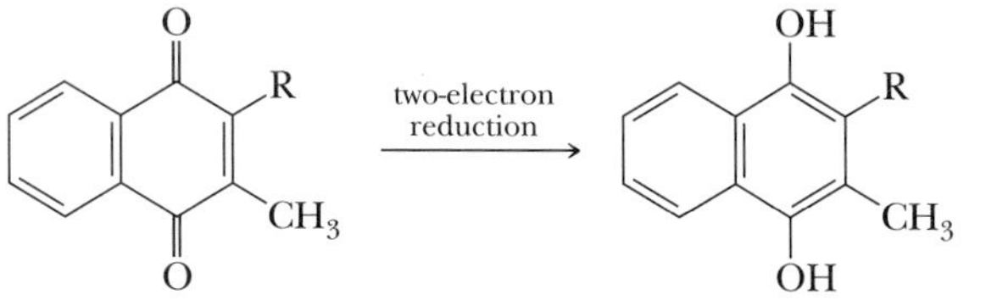

two-electron reduction

Vitamin K_1 quinone → Vitamin K_1 hydroquinone

In the presence of vitamin K hydroquinone, O_2, CO_2, and an enzyme called microsomal carboxylase, side chains of glutamate units in prothrombin, a protein essential for blood-clotting, are modified by addition of carboxyl groups (from CO_2) to form γ-carboxyglutamate units. Note that this reaction is the reverse of the decarboxylation of β-dicarboxylic (malonic) acids seen in Section 16.10B. The two carboxyl groups of the chemically modified glutamate now form a tight bidentate ("two teeth") complex with Ca^{2+} during the blood clotting process. While there is more to be understood about blood clotting, it is at least clear that if prothrombin is not carboxylated, it does not bind calcium and blood does not clot.

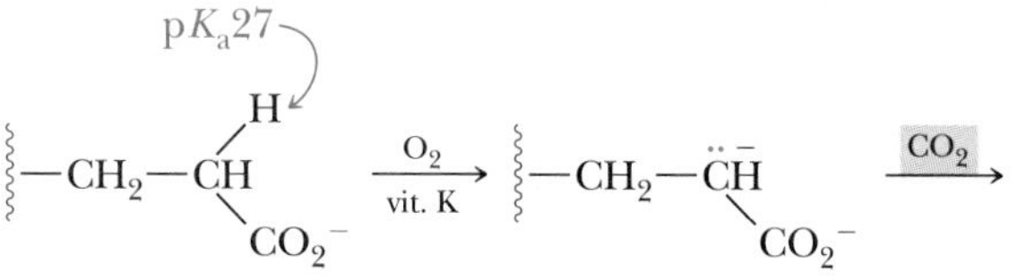

pK_a 27

O_2, vit. K; CO_2

Glutamate side chain of prothrombin

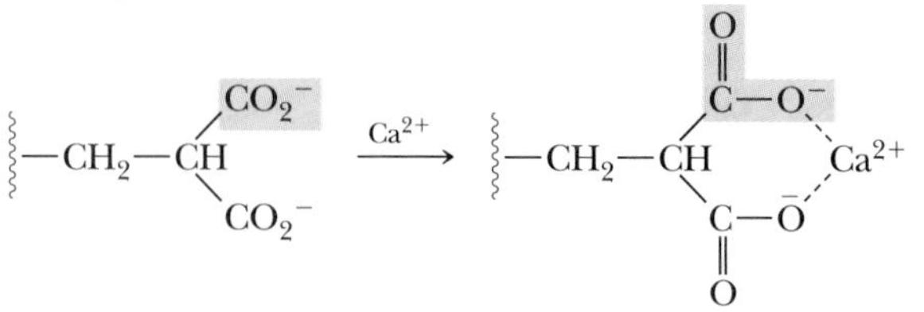

Ca^{2+}

Carboxylated glutamate side chain

Carboxylated glutamate side chain binding calcium ion

These facts have been known for many years. Until quite recently, however, the role of vitamin K_1 in this process remained a mystery. The problem was this. The anion of vitamin K_1 hydroquinone is a weak base (pK_a of approximately 9) derived from a phenol. To remove a proton from a glutamate side chain requires a very strong base derived from a conjugate acid of pK_a approximately 27. How can molecular oxygen increase the base strength of vitamin K_1 hydroquinone by 18 orders of magnitude?

It has recently been discovered that vitamin K_1 hydroquinone anion reacts with oxygen to give the peroxide anion intermediate **1**, which collapses to compound **2**. Notice that compound **2** contains a weak O—O bond in a highly strained four-membered ring. Compound **2** then rearranges to vitamin K_1 base, a strong, sterically hindered alkoxide base.

Reaction of vitamin K with oxygen to form vitamin K alkoxide base

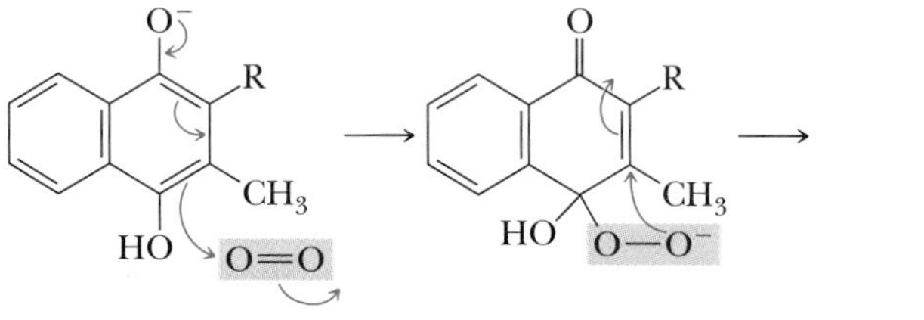

Vitamin K_1 hydroquinone anion (a weak base) → **1** 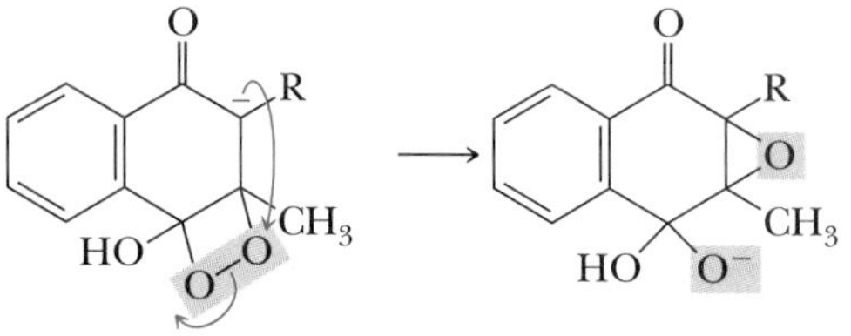 → **2** → Vitamin K_1 base (a very strong base)

The weak O—O bond is, in this way, replaced by a stronger C—O bond in vitamin K_1 base. The extra stability provides the driving force for turning a weak phenoxide base into a strong alkoxide base, which is able to remove a proton from glutamate side chains. This makes possible the addition of CO_2 to form γ-carboxyglutamate side chains which bind calcium ions during the clotting cascade. Thus, it is now understood why O_2, CO_2, and vitamin K_1 are essential for this phase of blood clotting. Synthetic vitamin K analogs are as effective in this process as naturally occurring vitamin K_1.

D. Vitamin K

The name of this vitamin comes from the German word Koagulation, signifying its important role in the blood-clotting process. The natural vitamin has a branched, unsaturated hydrocarbon chain of four isoprene units joined head to tail.

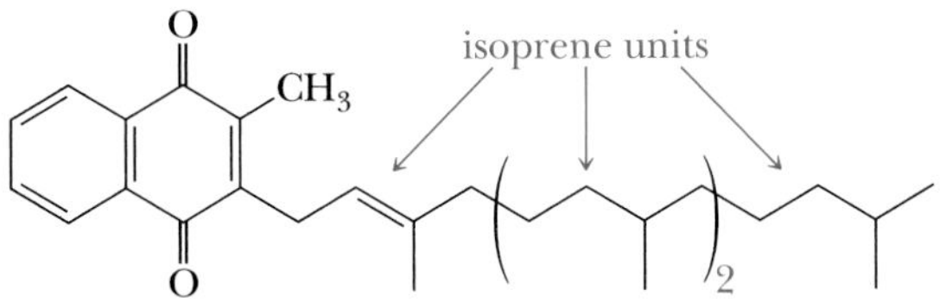

isoprene units

Vitamin K_1

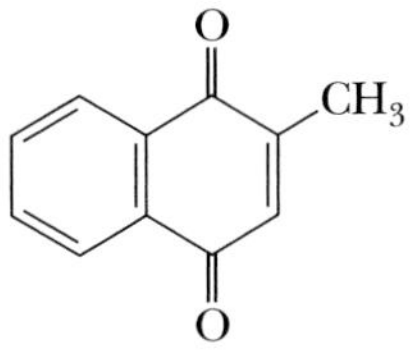

Menadione
(a synthetic vitamin K analog)

Natural vitamins of the K family have for the most part been replaced in vitamin supplements by synthetic preparations. Menadione, one such synthetic material with vitamin-K activity, has only hydrogen in the place of the alkyl chain.

SUMMARY

Lipids are a heterogeneous class of compounds grouped together on the basis of solubility properties; they are insoluble in water and soluble in diethyl ether, acetone, and methylene chloride. Carbohydrates, amino acids, and proteins are largely insoluble in these organic solvents.

Triacylglycerols (triglycerides), the most abundant lipids, are triesters of glycerol and fatty acids (Section 25.1). **Fatty acids** (Section 25.1A) are long-chain carboxylic acids derived from the hydrolysis of fats, oils, and the phospholipids of biological membranes. The melting point of a triacylglycerol increases as (1) the length of its hydrocarbon chains increases and (2) the degree of its saturation increases. Triacylglycerols rich in saturated fatty acids are generally solids at room temperature; those rich in unsaturated fatty acids are generally oils at room temperature (Section 25.1B).

Soaps are sodium or potassium salts of fatty acids (Section 25.2A). In water, soaps form **micelles** (Section 25.2B), which "dissolve" nonpolar organic grease and oil. Natural soaps precipitate as water-insoluble fatty acids in acid solution, and form water-insoluble salts with Mg^{2+}, Ca^{2+}, and Fe^{3+} ions of hard water. The most common and most widely used **synthetic detergents** (Section 25.2C) are linear alkylbenzenesulfonates. Synthetic detergents may also contain one or more of the following: optical brighteners, foam stabilizers, and bleaches.

Prostaglandins are a group of compounds having the 20-carbon skeleton of prostanoic acid (Section 25.3). They are synthesized in response to physiological triggers from phospholipid-bound arachidonic acid and other 20-carbon fatty acids. Also included among the eicosanoids are the **leukotrienes, prostacyclins,** and **thromboxanes.**

Steroids are a group of plant and animal lipids that have a characteristic tetracyclic structure of three six-membered rings and one five-membered ring (Section 25.4). **Cholesterol** is an integral part of animal membranes, and it is the compound from which human sex hormones, adrenocorticoid hormones, bile acids, and vitamin D are synthesized. Cholesterol is transported in the blood by lipoproteins. **Low-density lipoproteins (LDLs)** transport cholesterol from the site of its synthesis in the liver to tissues and cells where it is to be used. **High-density lipoproteins (HDLs)** transport cholesterol from cells back to the liver for its degradation to bile acids and eventual excretion in the feces.

The structural features common to estrogens, androgens, glucocorticoid hormones, and mineralocorticoid hormones are illustrated in Table 25.3. **Oral contraceptive** pills contain a synthetic progestin, for example, norethindrone, which prevents ovulation, yet allows women to maintain an otherwise normal menstrual cycle. A variety of synthetic **anabolic steroids** are available for use in rehabilitation medicine where muscle tissue has weakened or deteriorated due to injury. **Bile acids** differ from most other steroids in that they have a cis configuration at the junction of rings A and B.

The carbon skeleton of cholesterol and those of all biomolecules derived from it originate with the acetyl group (a C_2 unit) of **acetyl-CoA** (Section 25.4B).

Phospholipids (Section 25.5A), the second most abundant group of naturally occurring lipids, are derived from phosphatidic acid, a compound containing glycerol esterified with two molecules of fatty acid and a molecule of phosphoric acid. Further esterification of the phosphoric acid group with a low-molecular-weight alcohol, most commonly ethanolamine, choline, serine, or inositol, gives a phospholipid.

According to the **fluid-mosaic model** (Section 25.5C), membrane phospholipids form a lipid bilayer with membrane proteins associated with the bilayer as both peripheral proteins and as integral proteins.

Vitamin A (Section 25.6A) occurs only in the animal world. The carotenes of the plant world are tetraterpenes (C_{40}) and are cleaved after ingestion into vitamin A. The best-understood role of vitamin A is its participation in the **visual cycle.**

Vitamin D (Section 25.6B) is synthesized in the skin of mammals by the action of ultraviolet radiation on 7-dehydrocholesterol. This vitamin plays a major role in the regulation of calcium and phosphorus metabolism. **Vitamin E** (Section 25.6C) is a group of compounds of similar structure, the most active of which is α-tocopherol. In the body, vitamin E functions as an antioxidant. **Vitamin K** (Section 25.6D) is required for the carboxylation of glutamic acid side chains in the protein prothrombin. Carboxylated prothrombin then forms a tight "bidentate" complex with Ca^{2+}.

ADDITIONAL PROBLEMS

Fatty Acids and Triacylglycerols

25.2 Identify the hydrophobic and hydrophilic region(s) of a triglyceride.

25.3 Explain why the melting points of unsaturated fatty acids are lower than those of saturated fatty acids.

25.4 Which would you expect to have the higher melting point, glyceryl trioleate or glyceryl trilinoleate?

25.5 Explain why olive oil solidifies in the refrigerator, but corn oil does not.

25.6 Draw a structural formula for methyl linoleate. Be certain to show the correct configuration of groups about the carbon-carbon double bonds.

25.7 Explain why coconut oil is a liquid triglyceride, even though most of its fatty acid components are saturated.

25.8 What is meant by the term "hardening" as applied to fats and oils?

25.9 How many moles of H_2 are used in the catalytic hydrogenation of 1 mol of a triglyceride derived from glycerol, stearic acid, linoleic acid, and arachidonic acid?

25.10 Saponification number is defined as the number of milligrams of potassium hydroxide required for saponification of 1.00 g of fat or oil.

(a) Write a balanced equation for the saponification of tristearin.

(b) The molecular weight of tristearin is 890 g/mol. Calculate the saponification number of tristearin.

25.11 The saponification number of butter fat is approximately 230; that of margarine is approximately 195. Calculate the average molecular weight of butter fat and of margarine.

25.12 Characterize the structural features necessary to make a good synthetic detergent.

25.13 Following are structural formulas for a cationic detergent and a neutral detergent. Account for the detergent properties of each. Cationic detergents of this type are also called "quats" because they are quaternary ammonium salts.

$$CH_3(CH_2)_6CH_2\overset{+}{N}(CH_3)_2(CH_2C_6H_5)\ Cl^-$$

Benzyldimethyloctylammonium chloride (a cationic detergent)

$$(HOCH_2)_3CCH_2OC(=O)(CH_2)_{14}CH_3$$

Pentaerythrityl palmitate (a neutral detergent)

25.14 Identify some of the detergents used in shampoos and dish washing solutions. Are they primarily anionic, neutral, or cationic detergents?

25.15 Show how to convert palmitic acid (hexadecanoic acid) into the following:

(a) Ethyl palmitate
(b) Palmitoyl chloride
(c) 1-Hexadecanol (cetyl alcohol)
(d) 1-Hexadecanamine
(e) *N,N*-Dimethylhexadecanamide

25.16 Palmitic acid (hexadecanoic acid) is the source of the hexadecyl (cetyl) group in the following compounds. Each quat (see Problem 25.13) is a mild surface-acting germicide and fungicide and is used as a topical antiseptic and disinfectant.

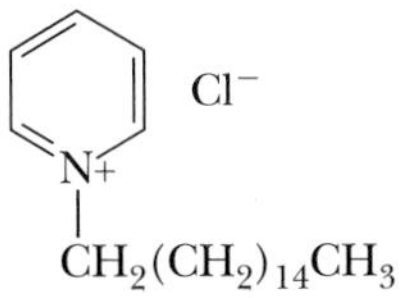

Cl^- N^+ $CH_2(CH_2)_{14}CH_3$

Cetylpyridinium chloride

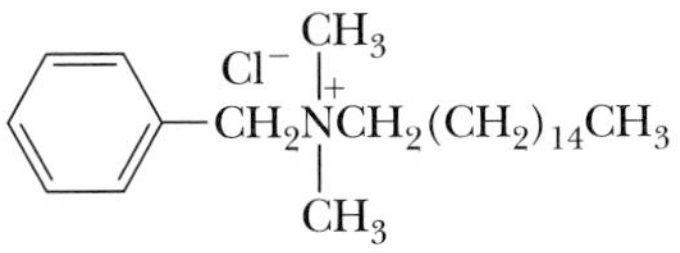

Cl^- CH_3 $-CH_2\overset{+}{N}CH_2(CH_2)_{14}CH_3$ CH_3

Benzylcetyldimethylammonium chloride

(a) Cetylpyridinium chloride is prepared by treating pyridine with 1-chlorohexadecane (cetyl chloride). Show how to convert palmitic acid to cetyl chloride.

(b) Benzylcetyldimethylammonium chloride is prepared by treating benzyl chloride with *N,N*-dimethyl-1-hexadecanamine. Show how this tertiary amine can be prepared from palmitic acid.

25.17 Lipases are enzymes that catalyze the hydrolysis of esters, especially esters of glycerol. Because enzymes are chiral catalysts, they catalyze the hydrolysis of only one enantiomer of a racemic mixture. For example, porcine pancreatic lipase catalyzes the hydrolysis of only one enantiomer of the following racemic epoxyester. Calculate the number of grams of epoxyalcohol that can be obtained from 100 g of racemic epoxy ester by this method.

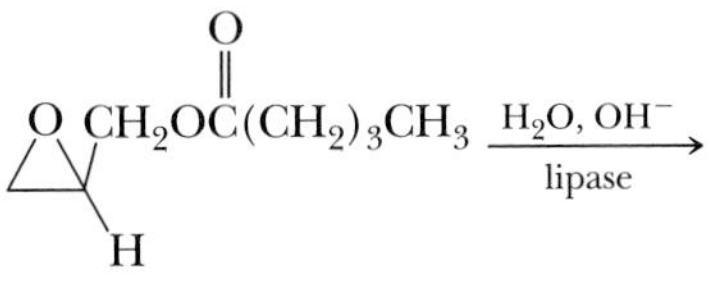

$$CH_2O\overset{O}{\overset{\|}{C}}(CH_2)_3CH_3 \xrightarrow[\text{lipase}]{H_2O,\ OH^-}$$

A racemic mixture

$$CH_2O\overset{O}{\overset{\|}{C}}(CH_2)_3CH_3 + CH_2OH + {}^-O\overset{O}{\overset{\|}{C}}(CH_2)_3CH_3$$

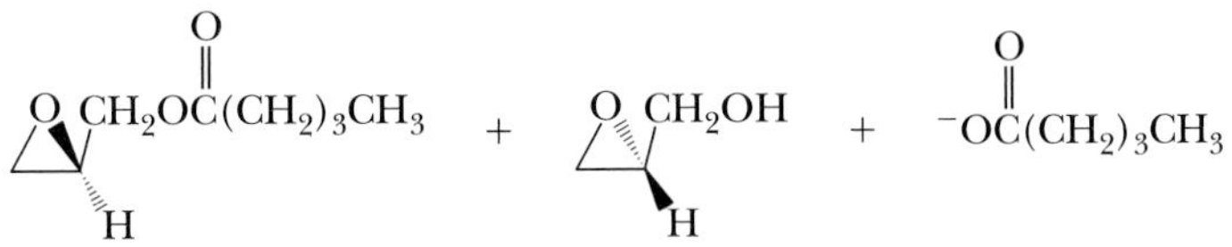

This enantiomer is recovered unhydrolyzed

This epoxyalcohol is obtained in pure form

Prostaglandins

25.18 Examine the structure of $PGF_{2\alpha}$ and

(a) Identify all stereocenters.
(b) Identify all double bonds about which *cis-trans* isomerism can occur.
(c) State the number of stereoisomers possible for a molecule of this structure.

25.19 Doxaprost, an orally active bronchodilator patterned after the natural prostaglandins (Section 25.3), is synthesized in the following series of reactions starting with ethyl 2-oxocyclopentanecarboxylate. Except for the Nef reaction in Step 8, we have seen examples of all other types of reactions involved in this synthesis.

Ethyl 2-oxocyclopentanecarboxylate $\xrightarrow{(1)}$ 2-oxo-1-$(CH_2)_6CO_2CH_3$-cyclopentane-1-$CO_2C_2H_5$ $\xrightarrow{(2)}$ 2-oxo-1-$(CH_2)_6CO_2H$-cyclopentane-1-CO_2H $\xrightarrow{(3)}$

2-$(CH_2)_6CO_2H$-cyclopentanone $\xrightarrow{(4)}$ 2-Br-2-$(CH_2)_6CO_2H$-cyclopentanone $\xrightarrow{(5)}$ 2-$(CH_2)_6CO_2H$-cyclopent-2-enone $\xrightarrow{(6)}$

2-$(CH_2)_6CO_2CH_3$-cyclopent-2-enone $\xrightarrow{(7)}$ 2-$(CH_2)_6CO_2CH_3$-3-CH_2NO_2-cyclopentanone $\xrightarrow[\text{(8)}]{\text{Nef reaction}}$

2-$(CH_2)_6CO_2CH_3$-3-CHO-cyclopentanone $\xrightarrow{(9)}$ 2-$(CH_2)_6CO_2CH_3$-3-(CH=CH–C(=O)–$(CH_2)_4CH_3$)-cyclopentanone $\xrightarrow{(10)}$

2-$(CH_2)_6CO_2H$-3-(CH=CH–C(=O)–$(CH_2)_4CH_3$)-cyclopentanone $\xrightarrow{(11)}$ 2-$(CH_2)_6CO_2H$-3-(CH=CH–C(OH)(CH_3)–$(CH_2)_4CH_3$)-cyclopentanone

Doxaprost (an orally active bronchodilator)

(a) Propose a set of experimental conditions to bring about the alkylation in Step 1. Account for the regioselectivity of the alkylation, that is, that it takes place on the carbon between the two carbonyl groups rather than on the other side of the ketone carbonyl.

(b) Propose experimental conditions to bring about Steps 2 and 3, and propose a mechanism for the loss of carbon dioxide in Step 3.

(c) Propose experimental conditions for bromination of the ring in Step 4 and dehydrobromination in Step 5.

(d) Write equations to show that Step 6 can be brought about using either methanol or diazomethane (CH_2N_2) as a source of the $-CH_3$ in the methyl ester.

(e) Describe experimental conditions to bring about the Michael reaction of Step 7.

(f) The two side chains in the product of Step 7 can be either *cis* or *trans* to each other.

Which of the two do you expect to be the more stable configuration? Account for the fact that the *trans* isomer is formed in this step.

(g) Step 9 is done by a Wittig reaction. Suggest a structural formula for a Wittig reagent that gives the product shown.

(h) Name the type of reaction involved in Step 10.

(i) Step 11 can best be described as a Grignard reaction with methylmagnesium bromide under very carefully controlled conditions. In addition to the observed reaction, what other Grignard reactions might take place in Step 11?

(j) Assuming that the two side chains on the cyclopentanone ring are *trans*, how many stereoisomers are possible from this synthetic sequence?

Steroids

25.20 Examine the structural formulas of testosterone (a male sex hormone) and progesterone (a female sex hormone). What are the similarities in structure between the two? What are the differences?

25.21 Examine the structural formula of cholic acid and account for the ability of this and other bile acids to emulsify fats and oils and thus aid in their digestion.

25.22 Following is a structural formula for cortisol (hydrocortisone). Draw a conformational representation of this molecule.

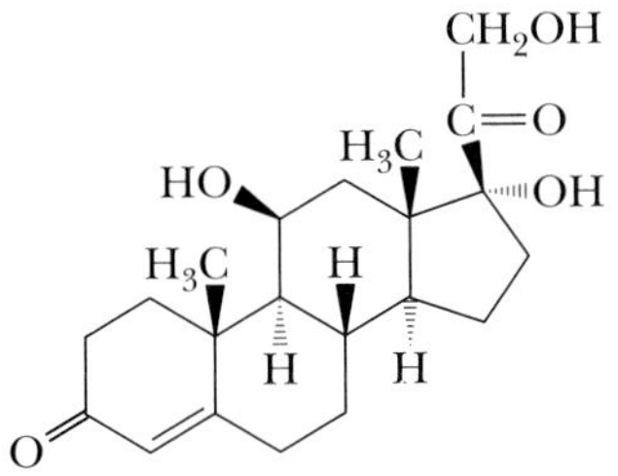

Cortisol
(Hydrocortisone)

25.23 Much of our understanding of conformational analysis has arisen from studies on the reactions of rigid steroid nuclei. For example, the concept of *trans*-diaxial ring opening of epoxycyclohexanes was proposed to explain the stereoselective reactions seen with steroidal epoxides. Predict the product when each of the following steroidal epoxides is treated with $LiAlH_4$.

(a) 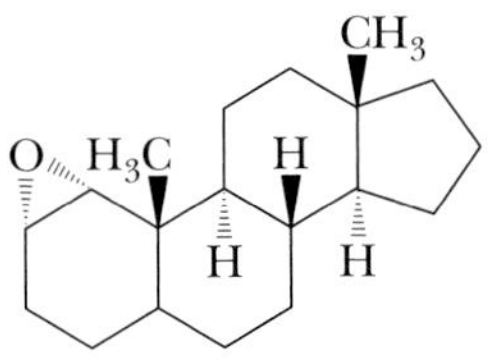 **(b)**

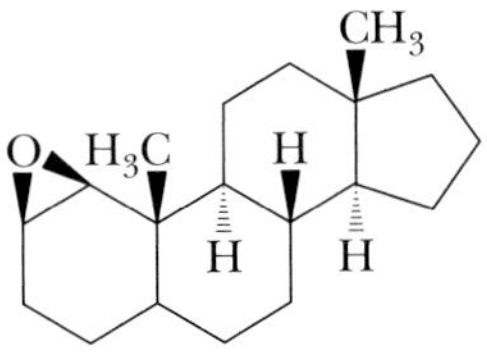

(c) 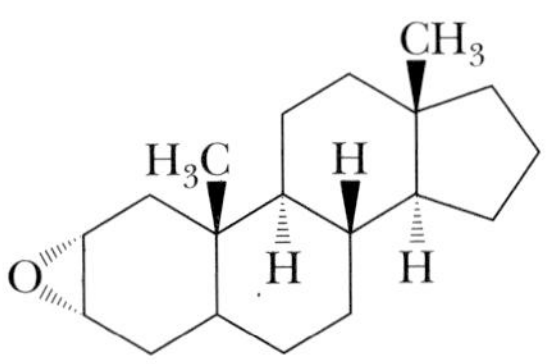 **(d)**

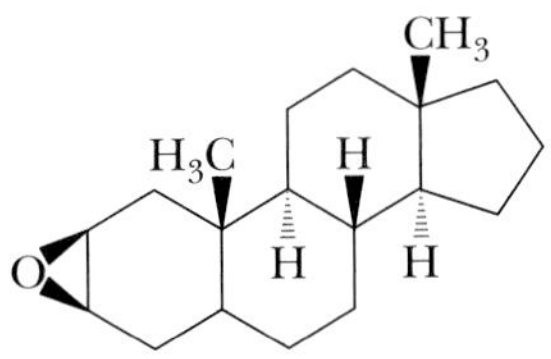

26 The Organic Chemistry of Metabolism

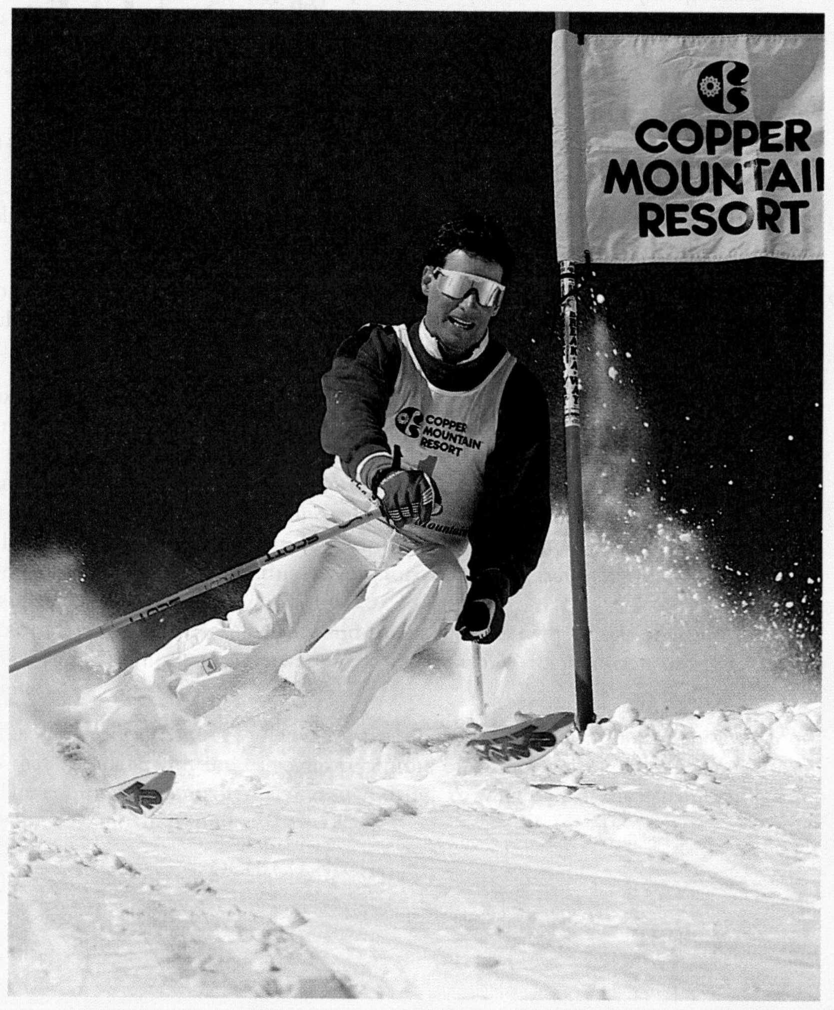

■ Important stages in the production of metabolic energy are glycolysis and β-oxidation. *(© Ben Blankenburg/Stock, Boston)*

We have now studied the typical reactions of the major types of organic functional groups. Further, we have studied the structure and reactions of carbohydrates and lipids, two of the major classes of biomolecules. Now let us see how this background can be applied to the study of the organic chemistry of metabolism. In this chapter, we study two key metabolic pathways, namely β-oxidation of fatty acids and glycolysis. The first of these is a pathway by which the hydrocarbon chains of fatty acids are degraded, two carbons at a time, to acetyl coenzyme A. The second is the pathway by which glucose is converted to pyruvate and then to acetyl coenzyme A.

Those of you who go on to courses in biochemistry will undoubtedly study these metabolic pathways in considerable detail, including their role in energy production and conservation, their regulation, and the diseases associated with errors of particular metabolic steps. Our concern in this chapter is more limited. It is our purpose to show that reactions of these pathways are biochemical equivalents of organic functional group reactions we have already studied in detail. In these pathways, we find examples of keto-enol tautomerism; oxidation of an aldehyde to a carboxylic acid; oxidation of a secondary alcohol to a ketone; a reverse aldol reaction; a reverse Claisen condensation; and formation and hydrolysis of esters, imines, thioesters, and mixed anhydrides. In this chapter, we use the mechanisms we have studied earlier to give us insights into the mechanisms by which these reactions, all of which are enzyme catalyzed, occur.

26.1 Five Key Participants in Glycolysis and β-Oxidation

In order to understand what happens in the β-oxidation of fatty acids and glycolysis, we first need to introduce five of the principal compounds participating in these and a great many other metabolic pathways. Three of these compounds (ATP, ADP, and AMP) are universal carriers of phosphate groups. The other two, NAD^+/NADH and FAD/$FADH_2$, are **coenzymes** involved in the oxidation/reduction of metabolic intermediates.

Coenzyme A low-molecular-weight, nonprotein molecule or ion that binds reversibly to an enzyme, functions as a second substrate for the enzyme, and is regenerated by further reaction.

A. ATP, ADP, and AMP: Agents for Storage and Transfer of Phosphate Groups

Following is the structural formula of adenosine triphosphate (Section 28.1), the most important of the compounds involved in the transfer of phosphate groups. A building block of ATP is adenosine, which consists of a unit of adenine bonded to a unit of D-ribofuranose by a β-*N*-glycoside bond. Bonded to the terminal $—CH_2OH$ of ribose are three units of phosphate: one joined by a phosphoric ester bond and the remaining two joined by phosphoric anhydride bonds.

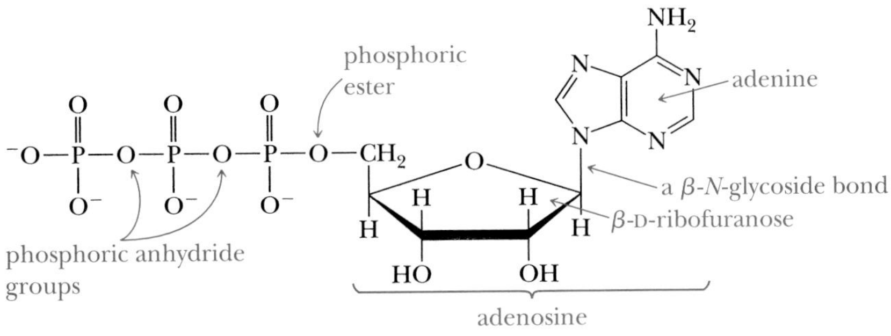

phosphoric ester

adenine

a β-*N*-glycoside bond

β-D-ribofuranose

phosphoric anhydride groups

adenosine

Adenosine triphosphate (ATP)

Hydrolysis of the terminal phosphate group of ATP gives ADP. In the following abbreviated structural formulas, adenosine and its single phosphoric ester group are represented by the symbol AMP.

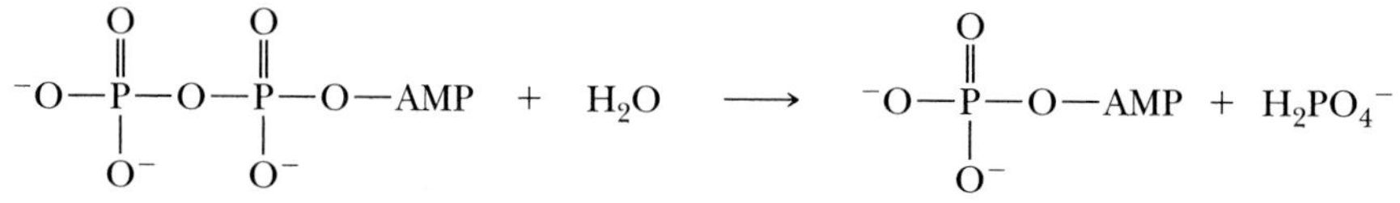

$$^{-}O-\overset{\overset{O}{\|}}{\underset{\underset{O^-}{|}}{P}}-O-\overset{\overset{O}{\|}}{\underset{\underset{O^-}{|}}{P}}-O-AMP + H_2O \longrightarrow {}^{-}O-\overset{\overset{O}{\|}}{\underset{\underset{O^-}{|}}{P}}-O-AMP + H_2PO_4^-$$

Adenosine triphosphate (ATP) — Water (a phosphate acceptor) — Adenosine diphosphate (ADP)

The reaction shown is hydrolysis of a phosphoric anhydride; the phosphate acceptor is water. In the first two reactions of glycolysis, the phosphate acceptors are —OH groups of glucose and fructose, respectively, to form phosphoric esters of these molecules. In glycolysis, there are also two reactions in which the phosphate acceptor is ADP, which is converted to ATP.

B. NAD$^+$/NADH: Agents for Electron Transfer in Biological Oxidation/Reductions

Nicotinamide adenine dinucleotide (NAD$^+$) A biological oxidizing agent. When acting an oxidizing agent, NAD$^+$ is reduced to NADH.

One of the central agents for the transfer of electrons in metabolic oxidations and reductions is the coenzyme **nicotinamide adenine dinucleotide (NAD$^+$)**. NAD$^+$ is constructed of a unit of ADP joined by a phosphoric ester bond to the terminal —CH_2OH of β-D-ribofuranose, which is in turn joined to the pyridine ring of nicotinamide by a β-*N*-glycoside bond.

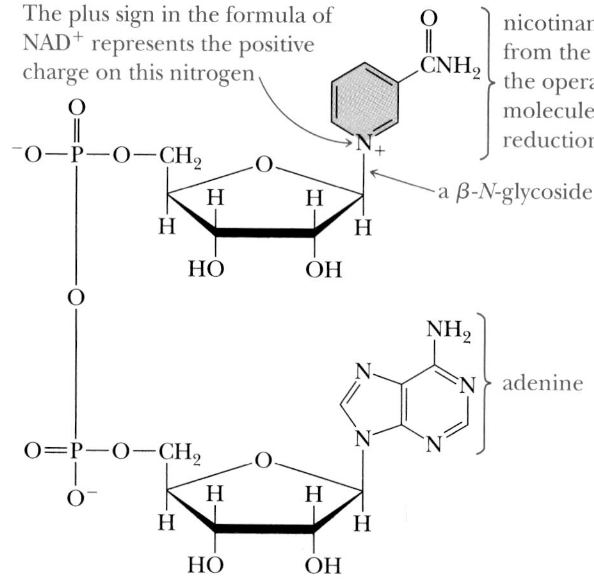

The plus sign in the formula of NAD$^+$ represents the positive charge on this nitrogen

nicotinamide, derived from the vitamin niacin; the operative part of the molecule in oxidation/reduction reactions

a β-*N*-glycoside bond

adenine

Nicotinamide adenine dinucleotide (NAD$^+$)

NAD$^+$ is a two-electron oxidizing agent, as seen in the following balanced half-reaction, and is reduced to NADH. NADH is in turn a two-electron reducing agent and is oxidized to NAD$^+$. In this abbreviated structural formula, the adenine dinucleotide part of the molecule is represented by the symbol Ad.

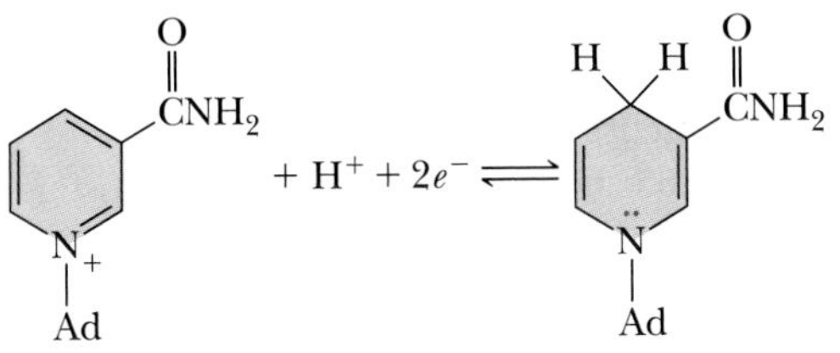

$$\text{NAD}^+ \text{ (oxidized form)} + H^+ + 2e^- \rightleftharpoons \text{NADH (reduced form)}$$

NAD$^+$ is involved in a variety of enzyme-catalyzed oxidation/reduction reactions. The two types of oxidations we deal with in this chapter are oxidation of a secondary alcohol to a ketone and oxidation of an aldehyde to a carboxylic acid. Each is a two-electron oxidation.

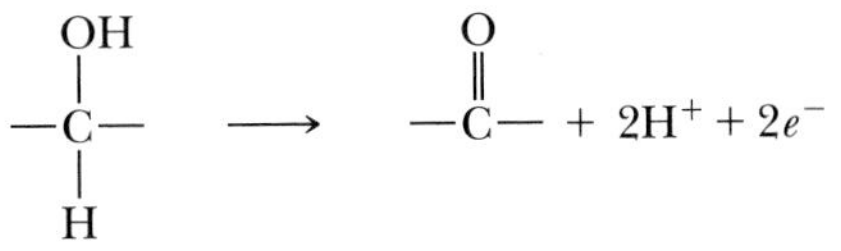

$$\text{A secondary alcohol} \longrightarrow \text{A ketone} + 2H^+ + 2e^-$$

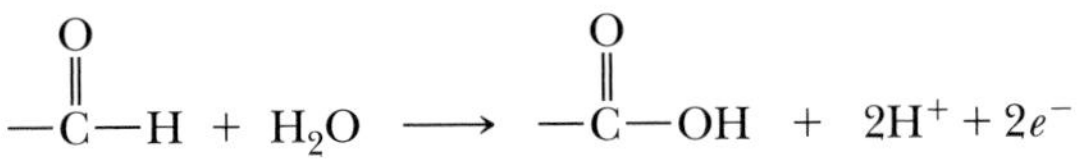

$$\text{An aldehyde} + H_2O \longrightarrow \text{A carboxylic acid} + 2H^+ + 2e^-$$

Oxidation of each functional group involves transfer of a hydride ion to NAD$^+$, here shown by a series of numbered curved arrows: (1) a basic group, B$^-$, on the surface of the enzyme removes H$^+$ from the —OH group; (2) electrons of the H—O bond become the pi electrons of the C=O bond; (3) a hydride ion is transferred from carbon to NAD$^+$ to create a new C—H bond; and (4) electrons within the ring flow to the positively charged nitrogen, which functions as an electron sink.

MECHANISM Oxidation of an Alcohol by NAD$^+$

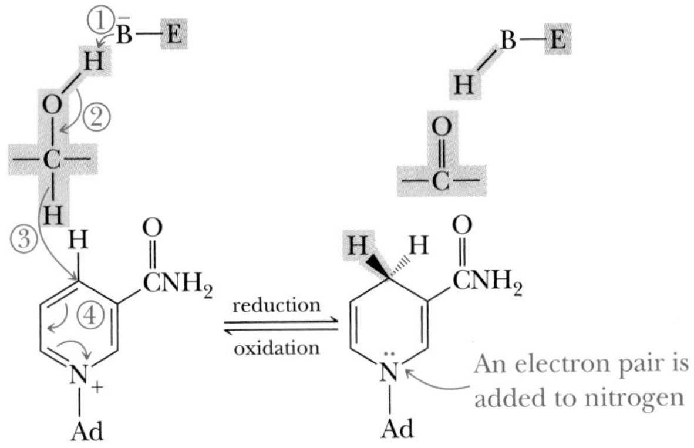

NAD$^+$ ⇌ (reduction / oxidation) NADH

An electron pair is added to nitrogen

The hydride ion, H:$^-$, which is transferred from the secondary alcohol to NAD^+, contains two electrons and, thus, NAD^+ and NADH function exclusively in two-electron oxidations and reductions.

We will see in Section 26.6 how two electrons are transferred from an aldehyde to NAD^+ during the oxidation of an aldehyde to a carboxylic acid.

C. $FAD/FADH_2$: Agents for Electron Transfer in Biological Oxidation/Reductions

Flavin adenine dinucleotide (FAD) A biological oxidizing agent. When acting as an oxidizing agent, FAD is reduced to $FADH_2$.

Like NAD^+/NADH, the coenzyme **flavin adenine dinucleotide (FAD),** and its reduction product $FADH_2$ are also derived from ADP. In FAD, the five-carbon monosaccharide ribitol is covalently bonded to flavin.

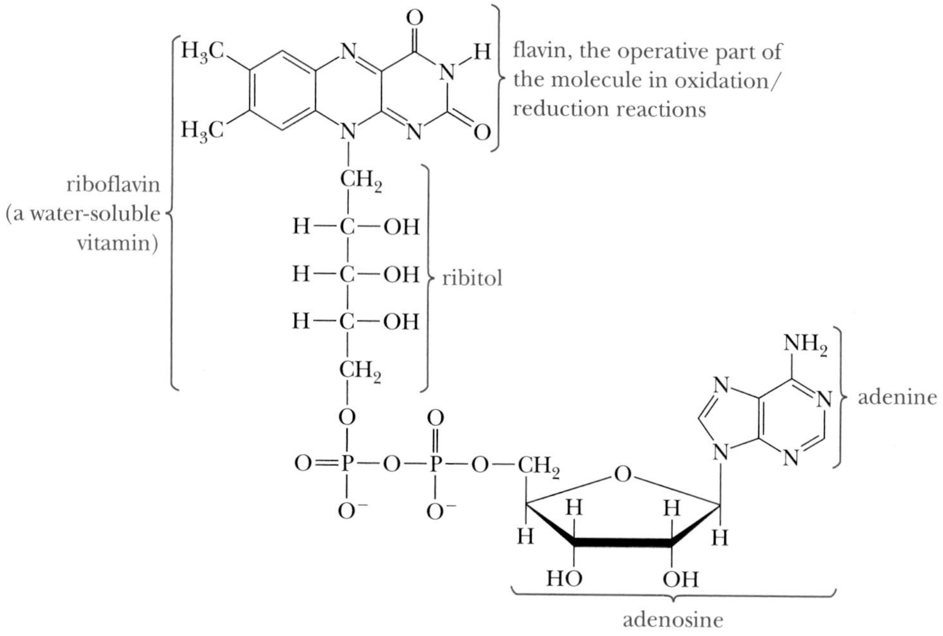

flavin, the operative part of the molecule in oxidation/reduction reactions

riboflavin (a water-soluble vitamin)

ribitol

adenine

adenosine

Flavin adenine dinucleotide, FAD

FAD participates in several types of enzyme-catalyzed oxidation/reduction reactions. Our concern in this chapter is with its role in the oxidation of a carbon-carbon single bond in the hydrocarbon chain of a fatty acid to a carbon-carbon double bond. As seen from balanced half-reactions, the two-electron oxidation of the hydrocarbon chain is coupled with the two-electron reduction of FAD.

Balanced half-reactions:

Oxidation of the hydrocarbon chain: $-CH_2-CH_2- \longrightarrow -CH=CH- + 2H^+ + 2e^-$

Reduction of FAD: $FAD + 2H^+ + 2e^- \longrightarrow FADH_2$

The mechanism by which FAD oxidizes $-CH_2-CH_2-$ to $-CH=CH-$ involves transfer of a hydride ion from the hydrocarbon chain of the fatty acid to FAD, as shown in the following diagram. The individual curved arrows in this figure are numbered 1–6 to help you follow the flow of electrons in this transformation. (1) A basic group on the surface of the enzyme removes a hydrogen from the carbon alpha to the carboxyl group (shown here as R_2), (2) electrons from this C—H bond become the pi electrons of the new carbon-carbon double bond, (3) a hydride ion is transferred from the carbon beta to the carboxyl group to flavin, (4) the pi electrons within flavin are redistributed, (5) electrons of the C=N bond remove a hydrogen from the enzyme, and (6) create a new basic group on the enzyme. You will note that two different groups on the surface of the enzyme are involved in this oxidation: the base B^- removes a proton as shown by arrow 1, and an acid H—B donates a proton as shown by arrow 6.

MECHANISM Oxidation of a Fatty Acid $—CH_2—CH_2—$ to —CH=CH— by FAD

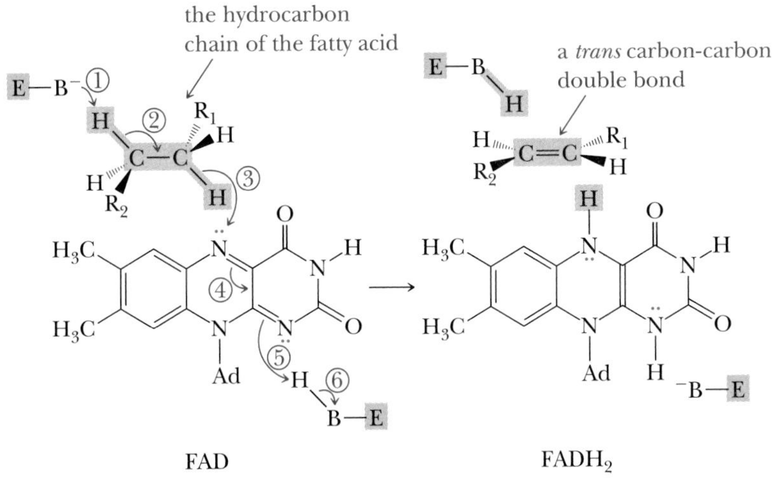

Note that one of the hydrogen atoms added to FAD to produce $FADH_2$ comes as a hydride ion from the hydrocarbon chain; the other comes as a proton from an acidic group on the surface of the enzyme catalyzing this oxidation. Also note that one group on the enzyme functions as a proton acceptor and that another functions as a proton donor.

26.2 Fatty Acids as a Source of Energy

Fatty acids in the form of triglycerides are the principal storage form of energy for most organisms. The principal advantage of storing energy in this form is that the carbon chains of fatty acids, mostly $—CH_2—$ groups, are a more highly reduced form of carbon than the oxygenated chains of carbohydrates and, therefore, the energy yield per gram of fatty acid oxidized is greater than that per gram of carbohydrate.

Complete oxidation of 1 g of palmitic acid yields almost 2.5 times the energy obtained from 1 g of glucose.

	Yield of Energy	
	kcal/mol (kJ/mol)	**kcal/g (kJ/g)**
$C_6H_{12}O_6 + 6O_2 \longrightarrow 6CO_2 + 6H_2O$ Glucose	−686 (−2870)	−3.8 (−15.9)
$CH_3(CH_2)_{14}CO_2H + 23O_2 \longrightarrow 16CO_2 + 16H_2O$ Palmitic acid	−2340 (−9791)	−9.3 (−38.9)

26.3 β-Oxidation of Fatty Acids

β-Oxidation A series of four enzyme-catalyzed reactions that cleaves carbon atoms, two at a time, from the carboxyl end of a fatty acid.

The first phase of the catabolism of fatty acids involves their release from triglycerides, either those stored in adipose tissue or from the diet, by hydrolysis catalyzed by a group of enzymes called lipases. There are two major stages in **β-oxidation of fatty acids:** (A) activation of a free fatty acid in the cytoplasm and its transport across the inner mitochondrial membrane followed by (B) β-oxidation, a repeated sequence of four reactions.

A. Activation of Fatty Acids: Formation of a Thioester with Coenzyme A

Thioester An ester in which one atom of oxygen in the carboxylate group is replaced by an atom of sulfur.

The process of β-oxidation begins in the cytoplasm with formation of a **thioester** between the carboxyl group of a fatty acid and the sulfhydryl group of coenzyme A (Problem 24.39). Formation of this acyl-CoA derivative is coupled with hydrolysis of ATP to AMP and pyrophosphate ion. It is common in writing biochemical reactions to show some reactants and products by a curved arrow set over the main reaction arrow. We use this convention here to show ATP as a reactant, and AMP and pyrophosphate ion as products.

$$\underset{\text{Fatty acid (as anion)}}{R{-}\overset{O}{\overset{\|}{C}}{-}O^-} + \underset{\text{Coenzyme A}}{HS{-}CoA} \xrightarrow{\text{ATP} \quad \text{AMP} + P_2O_7^{4-}} \underset{\text{An acyl-CoA derivative}}{R{-}\overset{O}{\overset{\|}{C}}{-}S{-}CoA} + OH^-$$

The mechanism of this reaction involves attack by the fatty acid carboxylate anion on P=O of a phosphoric anhydride group of ATP to form an intermediate analogous to the tetrahedral carbonyl addition intermediate formed in C=O chemistry; in the intermediate formed in the fatty acid-ATP reaction, the phosphorus attacked by the carboxylate anion becomes bonded to five groups. This intermediate then collapses to give an acyl-AMP, which is a highly reactive mixed anhydride of the carboxyl group of the fatty acid and the phosphate group of AMP.

$$\mathrm{R{-}\overset{O}{\overset{\|}{C}}{-}O^- + Ad{-}O{-}\underset{O^-}{\underset{|}{\overset{O}{\overset{\|}{P}}}}{-}O{-}\underset{O^-}{\underset{|}{\overset{O}{\overset{\|}{P}}}}{-}O{-}\underset{O^-}{\underset{|}{\overset{O}{\overset{\|}{P}}}}{-}O^-} \rightleftharpoons \left[\begin{array}{c}\text{intermediate}\\ \text{with one}\\ \text{phosphorus}\\ \text{bonded to}\\ \text{five groups}\end{array}\right] \longrightarrow \mathrm{R{-}\overset{O}{\overset{\|}{C}}{-}O{-}\underset{O^-}{\underset{|}{\overset{O}{\overset{\|}{P}}}}{-}O{-}Ad + {}^-O{-}\underset{O^-}{\underset{|}{\overset{O}{\overset{\|}{P}}}}{-}O{-}\underset{O^-}{\underset{|}{\overset{O}{\overset{\|}{P}}}}{-}O^-}$$

Fatty acid (as anion) — ATP — An acyl-AMP (a mixed anhydride) — Pyrophosphate

This mixed anhydride then undergoes a carbonyl addition reaction with the sulfhydryl group of coenzyme A to form a tetrahedral carbonyl addition intermediate, which collapses to give AMP and an acyl-CoA (a fatty acid thioester of coenzyme A).

$$\mathrm{CoA{-}SH + R{-}\overset{O}{\overset{\|}{C}}{-}O{-}\underset{O^-}{\underset{|}{\overset{O}{\overset{\|}{P}}}}{-}O{-}Ad} \rightleftharpoons \left[\begin{array}{c}\text{tetrahedral}\\ \text{carbonyl}\\ \text{addition}\\ \text{intermediate}\end{array}\right] \longrightarrow \mathrm{R{-}\overset{O}{\overset{\|}{C}}{-}S{-}CoA + {}^-O{-}\underset{O^-}{\underset{|}{\overset{O}{\overset{\|}{P}}}}{-}O{-}Ad}$$

Coenzyme A — An acyl-AMP — An acyl-CoA — AMP

At this point, the activated fatty acid is transported into the mitochondrion where its carbon chain is degraded by the reactions of β-oxidation.

B. The Four Reactions of β-Oxidation

Reaction 1: Oxidation of a Carbon-Carbon Single Bond to a Carbon-Carbon Double Bond

In the first reaction of β-oxidation, the carbon chain is oxidized, and a double bond is formed between the alpha- and beta-carbons of the fatty acid chain. The oxidizing agent is FAD, which is reduced to $FADH_2$.

$$\mathrm{R{-}\overset{\beta}{CH_2}{-}\overset{\alpha}{CH_2}{-}\overset{O}{\overset{\|}{C}}{-}SCoA + FAD \longrightarrow (H)(R)C{=}C(H)(C({=}O){-}SCoA) + FADH_2}$$

An acyl-CoA — A *trans*-enoyl-CoA

Reaction 2: Hydration of the Carbon-Carbon Double Bond

The second reaction of β-oxidation is enzyme-catalyzed hydration of the carbon-carbon double bond to give an (R)-β-hydroxyacyl-CoA. The reaction is regioselective in that —OH is added to carbon 3 of the chain. It is stereoselective in that only the R enantiomer is formed.

$$\mathrm{(H)(R)C{=}C(H)(C({=}O){-}SCoA) + H_2O \longrightarrow R{-}CH(OH){-}CH_2{-}\overset{O}{\overset{\|}{C}}{-}SCoA}$$

A *trans*-enoyl-CoA — (R)-β-Hydroxyacyl-CoA

Reaction 3: Oxidation of the β-Hydroxy Group to a Carbonyl Group

In the second oxidation step of β-oxidation, the secondary alcohol group is oxidized to a ketone group. The oxidizing agent is NAD^+, which is reduced to NADH.

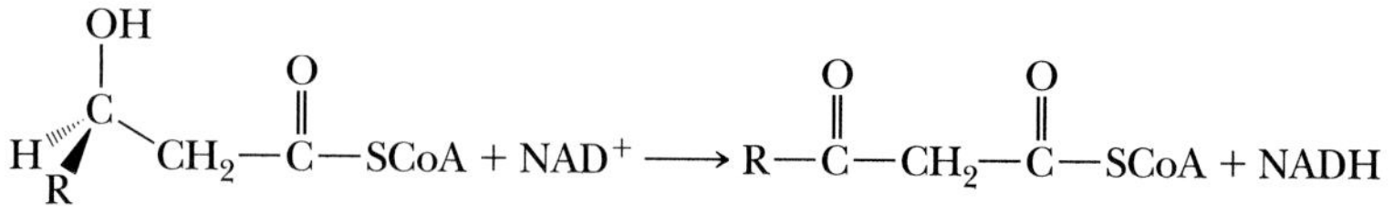

$$\text{(R)-}\beta\text{-Hydroxyacyl-CoA} + NAD^+ \longrightarrow R{-}\overset{O}{\overset{\|}{C}}{-}CH_2{-}\overset{O}{\overset{\|}{C}}{-}SCoA + NADH$$

(*R*)-β-Hydroxyacyl-CoA β-Ketoacyl-CoA

Reaction 4: Cleavage of the Carbon Chain by a Reverse Claisen Condensation

In the final step of β-oxidation, reaction of the β-ketoacyl-CoA with coenzyme A results in cleavage between the alpha- and beta-carbons of the chain to form a molecule of acetyl coenzyme A and a new acyl-CoA, the hydrocarbon chain of which is now shortened by two carbon atoms.

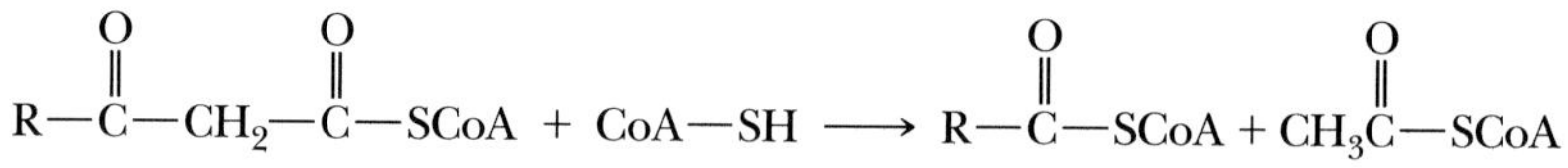

$$R{-}\overset{O}{\overset{\|}{C}}{-}CH_2{-}\overset{O}{\overset{\|}{C}}{-}SCoA + CoA{-}SH \longrightarrow R{-}\overset{O}{\overset{\|}{C}}{-}SCoA + CH_3\overset{O}{\overset{\|}{C}}{-}SCoA$$

β-Ketoacyl-CoA Coenzyme A An acyl-CoA Acetyl-CoA

This reaction begins by attack of a sulfhydryl group of the enzyme thiolase on the ketone to form a tetrahedral carbonyl addition intermediate. This intermediate then collapses to give the anion of acetyl-CoA and the new acyl group bound to the surface of the enzyme by a thioester bond.

MECHANISM A Reverse Claisen Condensation in β-Oxidation of Fatty Acids

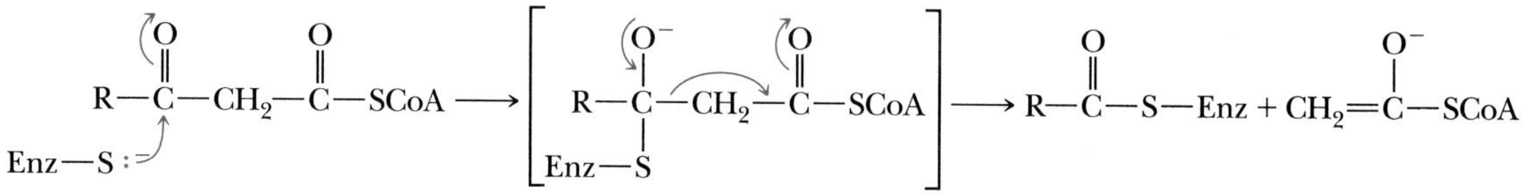

$$R{-}\overset{O}{\overset{\|}{C}}{-}CH_2{-}\overset{O}{\overset{\|}{C}}{-}SCoA \;(+\; Enz{-}S{:}^-) \longrightarrow \left[R{-}\underset{Enz-S}{\overset{O^-}{C}}{-}CH_2{-}\overset{O}{\overset{\|}{C}}{-}SCoA \right] \longrightarrow R{-}\overset{O}{\overset{\|}{C}}{-}S{-}Enz + CH_2{=}\overset{O^-}{\overset{|}{C}}{-}SCoA$$

Tetrahedral carbonyl addition intermediate An enzyme-thioester Enolate anion of acetyl-CoA

If this reaction sequence is read in reverse, it is seen as an example of a **Claisen condensation** (Section 18.3A): the attack of the enolate anion of acetyl-CoA on the carbonyl group of a thioester to form a tetrahedral carbonyl addition intermediate, followed by its collapse to give a β-ketothioester. A similar cleavage can be carried out on a β-ketoester under the conditions of the Claisen condensation (Section 18.3A), because the reaction is reversible. The four steps of β-oxidation are summarized in Figure 26.1.

ADDITIONAL PROBLEMS

β-Oxidation

26.3 Write structural formulas for palmitic, oleic, and stearic acids, the three most abundant fatty acids.

26.4 A fatty acid must be activated before it can be metabolized in cells. Write a balanced equation for the activation of palmitic acid.

26.5 Name three coenzymes necessary for β-oxidation of fatty acids. From what vitamin is each derived?

26.6 We have examined β-oxidation of saturated fatty acids, such as palmitic acid and stearic acid. Oleic acid, an unsaturated fatty acid, is also a common component of dietary fats and oils. This unsaturated fatty acid is degraded by β-oxidation but, at one stage in its degradation, requires an additional enzyme named enoyl-CoA isomerase. Why is this enzyme necessary, and what isomerization does it catalyze? (*Hint:* Consider both the configuration of the carbon-carbon double bond in oleic acid and its position in the carbon chain.)

Glycolysis

26.7 Name two coenzymes required for glycolysis. From what vitamin is each derived?

26.8 Number the carbon atoms of glucose 1 through 6 and show from which carbon atom of glucose the carboxyl group of each molecule of pyruvate are derived.

26.9 How many moles of lactate are produced from three moles of glucose?

26.10 Although glucose is the principal source of carbohydrates for glycolysis, fructose and galactose are also metabolized for energy.

(a) What is the main dietary source of fructose? Of galactose?
(b) Propose a series of reactions by which the carbon skeleton of fructose might enter glycolysis.
(c) Propose a series of reactions by which the carbon skeleton of galactose might enter glycolysis.

26.11 How many moles of ethanol are produced per mole of sucrose through the reactions of glycolysis and alcoholic fermentation? How many moles of CO_2 are produced?

26.12 Glycerol derived from hydrolysis of triglycerides and phospholipids is also metabolized for energy. Propose a series of reactions by which the carbon skeleton of glycerol might enter glycolysis and be oxidized to pyruvate.

26.13 Ethanol is oxidized in the liver to acetate ion by NAD^+.

(a) Write a balanced equation for this oxidation.
(b) Do you expect the pH of blood plasma to increase, decrease, or remain the same as a result of metabolism of a significant amount of ethanol?

26.14 Write a mechanism to show the role of NADH in the reduction of acetaldehyde to ethanol.

26.15 When pyruvate is reduced to lactate by NADH, two hydrogens are added to pyruvate; one to the carbonyl carbon, the other to the carbonyl oxygen. Which of these hydrogens is derived from NADH?

26.16 Review the oxidation reactions of glycolysis and β-oxidation and compare the types of functional groups oxidized by NAD^+ with those oxidized by FAD.

26.17 Why is glycolysis called an anaerobic pathway?

26.18 Which carbons of glucose end up in CO_2 as a result of alcoholic fermentation?

26.19 Which steps in glycolysis require ATP? Which steps produce ATP?

26.20 The respiratory quotient (RQ) is used in studies of energy metabolism and exercise physiology. It is defined as the ratio of the volume of carbon dioxide produced to the volume of oxygen used:

$$RQ = \frac{\text{volume } CO_2}{\text{volume } O_2}$$

(a) Show that RQ for glucose is 1.00. (*Hint:* Look at the balanced equation for complete oxidation of glucose to carbon dioxide and water.)
(b) Calculate RQ for triolein, a triglyceride of molecular formula $C_{57}H_{104}O_6$.
(c) For an individual on a normal diet, RQ is approximately 0.85. Would this value increase or decrease if ethanol were to supply an appreciable portion of caloric needs?

26.21 Acetoacetate, β-hydroxybutyrate, and acetone are commonly known within the health sciences as ketone bodies, in spite of the fact that one of them is not a ketone at all. They are products of human metabolism and are always present in blood plasma. Most tissues, with the notable exception of the brain, have the enzyme systems necessary to use them as energy sources. Synthesis of ketone bodies occurs by the following enzyme-catalyzed reactions. Enzyme names are (1) thiolase, (2) β-hydroxy-β-methylglutaryl-CoA synthase, (3) β-hydroxy-β-methylglutaryl-CoA lyase, and (5) β-hydroxybutyrate dehydrogenase. Reaction (4) is spontaneous and uncatalyzed.

$2CH_3\overset{O}{\overset{\|}{C}}SCoA$ (Acetyl-CoA) —(1), CoA—SH→ $CH_3\overset{O}{\overset{\|}{C}}CH_2\overset{O}{\overset{\|}{C}}SCoA$ (Acetoacetyl-CoA) —(2), $CH_3\overset{O}{\overset{\|}{C}}SCoA$, CoA—SH→

$^-O\overset{O}{\overset{\|}{C}}CH_2C(CH_3)(OH)CH_2\overset{O}{\overset{\|}{C}}SCoA$ ((*S*)-3-Hydroxy-3-methylglutaryl-CoA) —(3), $CH_3\overset{O}{\overset{\|}{C}}SCoA$→ $CH_3\overset{O}{\overset{\|}{C}}CH_2\overset{O}{\overset{\|}{C}}O^-$ (Acetoacetate)

Acetoacetate —(4), H^+, CO_2→ $CH_3\overset{O}{\overset{\|}{C}}CH_3$ (Acetone)

Acetoacetate —(5), NADH, NAD^+→ $CH_3\overset{OH}{\overset{|}{C}}HCH_2\overset{O}{\overset{\|}{C}}O^-$ (β-Hydroxybutyrate)

Describe the type of reaction involved in each step and the type of mechanism by which each occurs.

26.22 Show that (*S*)-3-hydroxy-3-methylglutaryl-CoA is a branch point connecting the synthesis of ketone bodies, terpenes, and cholesterol and the steroid hormones. [*Hint:* Review Section 18.4 (Claisen Condensations in the Biological World) and Section 25.4B (The Biosynthesis of Steroids).]

26.23 A connecting point between anaerobic glycolysis and β-oxidation is formation of acetyl-CoA. Which carbon atoms of glucose appear as methyl groups of acetyl-CoA? Which carbon atoms of palmitic acid appear as methyl groups of acetyl-CoA?

AMINO ACIDS AND PROTEINS

27

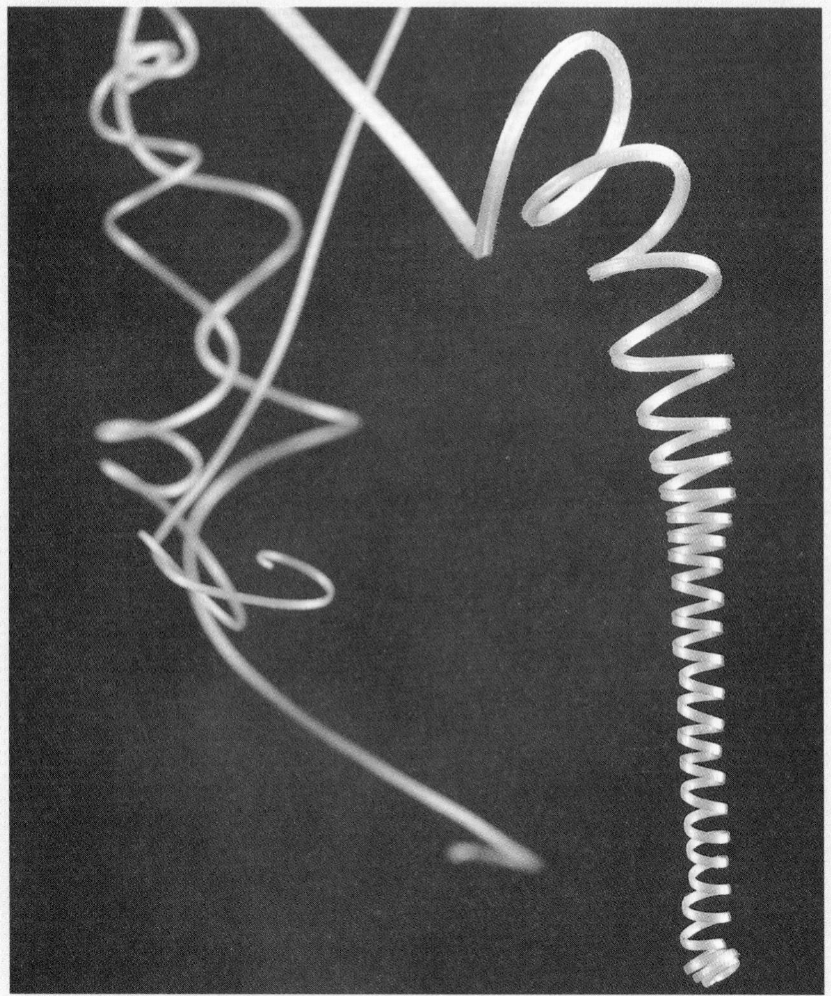

■ The star cucumber *(Sicyos angulatus)* uses left-handed helical tendrils to attach itself to climbing vines. This helical pattern is analogous but the reverse to the right-handed α-helix of polypeptides. *(Runk/Schoenberger/Grant Heilman)*

We begin this chapter with a study of amino acids, compounds whose chemistry is built on amines (Chapter 21) and carboxylic acids (Chapter 16). The fact that molecules of these compounds are difunctional presents a special challenge to chemists because in dealing with a reaction of the carboxyl group of an amino acid, we must also be aware of reactions that the amino group might undergo at the same time, and vice versa. We concentrate in particular on the acid-base properties of amino acids, for it is these properties that are so important in determining many of the properties of proteins, including the catalytic functions of enzymes.

One of the triumphs of organic chemistry has been to synthesize molecules, many of them of almost bewildering complexity. Both this challenge and triumph are well illustrated in our ability to synthesize proteins from their monomer units, namely amino acids. First we examine some of the ways chemists have devised for determining the composition and order of attachment of amino acids in proteins. Then, we examine a few of the many ways for joining amino acids together by amide bonds to synthesize proteins. Finally, in this chapter we look at the three-dimensional structures of several globular proteins and at the noncovalent forces (hydrophobic effects, hydrogen bonding, and salt linkages) that direct specific folding patterns.

27.1 Amino Acids

A. Structure

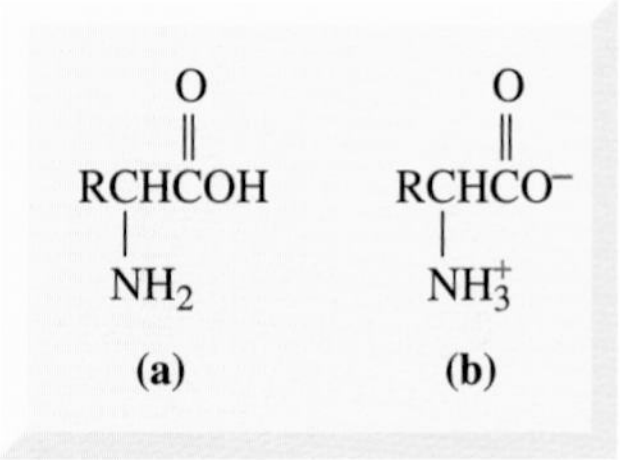

An **amino acid** is a compound that contains both a carboxyl group and an amino group. Although many types of amino acids are known, the **α-amino acids** are the most significant in the biological world because they are the monomers from which proteins are constructed. A general structural formula of an α-amino acid is shown in Figure 27.1.

(a) $RCH(NH_2)C(=O)OH$ (b) $RCH(NH_3^+)C(=O)O^-$

FIGURE 27.1
An α-amino acid. (a) Un-ionized form and (b) dipolar ion (zwitterion).

Although Figure 27.1(a) is a common way of writing structural formulas for amino acids, it is not accurate because it shows an acid ($—CO_2H$) and a base ($—NH_2$) within the same molecule. These acidic and basic groups react with each other to form a dipolar ion or internal salt [Figure 27.1(b)]. A molecule that has both a positive and a negative charge is called a **zwitterion.** Note that a zwitterion has no net charge; it contains one positive charge and one negative charge.

Because they exist as zwitterions, amino acids have many of the properties associated with salts. They are crystalline solids with high melting points and are fairly soluble in water but insoluble in nonpolar organic solvents such as ether and hydrocarbon solvents.

Amino acid A compound that contains both an amino group and a carboxyl group.

α-Amino acid An amino acid in which the amino group is on the carbon adjacent to the carboxyl group.

Zwitterion An internal salt of an amino acid.

B. Chirality of Amino Acids

With the exception of glycine, $H_2NCH_2CO_2H$, all protein-derived amino acids have at least one stereocenter (the α-carbon) and are, therefore, chiral. Figure 27.2 shows Fischer projections for the enantiomers of alanine. Whereas the vast majority of carbohydrates in the biological world are of the D-series, the vast majority of α-amino acids in the biological world are of the L-series.

The alternative *R,S* convention is also used to specify the configurations of amino acids. According to this convention, L-alanine is designated *S*-alanine. Because D- and L- are used more commonly to describe the configurations of amino acids, we use this convention throughout the remainder of the chapter.

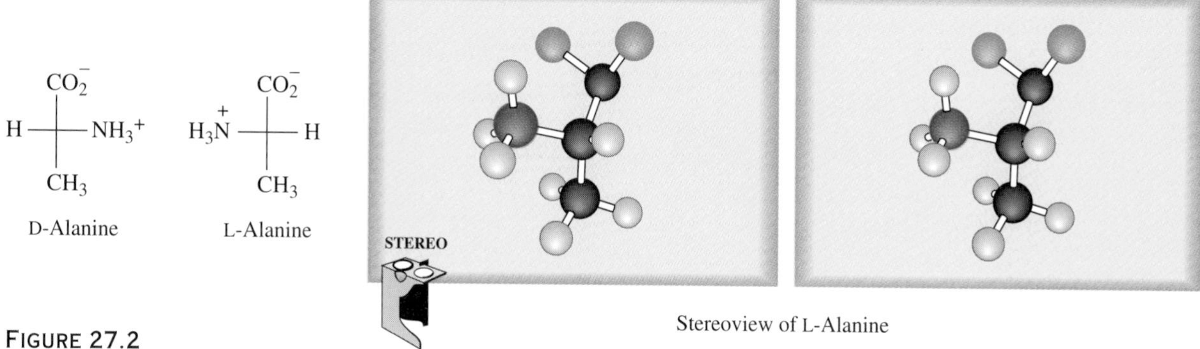

FIGURE 27.2
Alanine. The vast majority of α-amino acids in the biological world have the L-configuration at the alpha carbon.

C. Protein-Derived Amino Acids

Table 27.1 gives names, structural formulas, and standard three-letter and one-letter abbreviations for the 20 common L-amino acids found in proteins. The amino acids in this table could be listed in several ways, for example, alphabetically. However, a more useful classification, and one that is of great value when we discuss the physical properties of amino acids and the three-dimensional shapes of proteins, is to group them according to the polarity of their side chains. For this reason, the amino acids in Table 27.1 are divided into the following four side-chain categories: nonpolar side chains, polar but un-ionized side chains, acidic side chains, and basic side chains. The following structural features of these amino acids should be noted:

1. All 20 of these protein-derived amino acids are α-amino acids, meaning that the amino group is located on the carbon alpha to the carboxyl group.
2. For 19 of the 20 amino acids, the α-amino group is primary. Proline is different; its α-amino group is secondary.
3. With the exception of glycine, the α-carbon of each amino acid is a stereocenter. Although not shown in this table, all 19 chiral amino acids have the same relative configuration at the α-carbon. In the D,L convention, all are L-amino acids. In the *R,S* convention, 18 are *S*-amino acids. Cysteine is an *R*-amino acid, not because it differs in relative configuration from the other 18 but because of the manner in which priority is assigned to groups on the stereocenter.
4. Isoleucine and threonine contain a second stereocenter. Four stereoisomers are possible for each amino acid, but only one is found in proteins.
5. The sulfhydryl group of cysteine and the imidazole group of histidine are partially ionized at pH 7.0, but the ionic form is not the major form present at this pH.

EXAMPLE 27.1

Of the 20 protein-derived amino acids shown in Table 27.1, how many contain (a) aromatic rings, (b) side-chain hydroxyl groups, (c) phenolic —OH groups, and (d) sulfur?

TABLE 27.1 The 20 Common Amino Acids Found in Proteins*

Nonpolar Side Chains

alanine (Ala, A)	$CH_3CH(NH_3^+)CO_2^-$	phenylalanine (Phe, F)	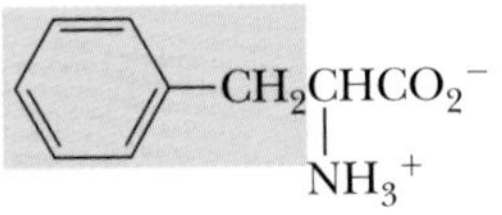 C_6H_5–$CH_2CH(NH_3^+)CO_2^-$
glycine (Gly, G)	$HCH(NH_3^+)CO_2^-$	proline (Pro, P)	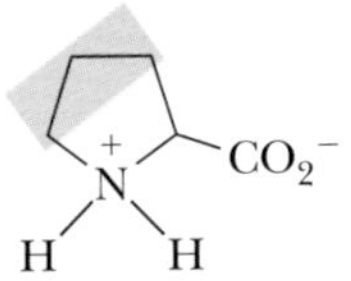 pyrrolidine ring with $\overset{+}{N}H_2$ and CO_2^-
isoleucine (Ile, I)	$CH_3CH_2CH(CH_3)CH(NH_3^+)CO_2^-$	tryptophan (Trp, W)	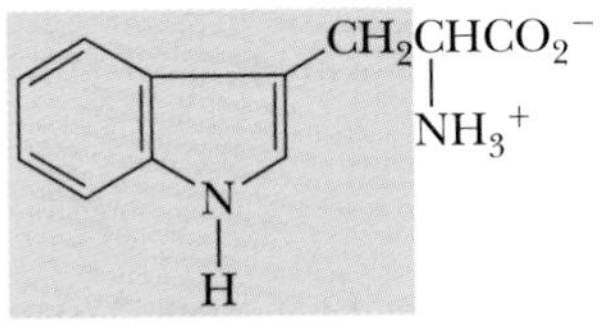 indole (N–H)–$CH_2CH(NH_3^+)CO_2^-$
leucine (Leu, L)	$(CH_3)_2CHCH_2CH(NH_3^+)CO_2^-$		
methionine (Met, M)	$CH_3SCH_2CH_2CH(NH_3^+)CO_2^-$	valine (Val, V)	$(CH_3)_2CHCH(NH_3^+)CO_2^-$

Polar Side Chains

asparagine (Asn, N)	$H_2NC(=O)CH_2CH(NH_3^+)CO_2^-$	serine (Ser, S)	$HOCH_2CH(NH_3^+)CO_2^-$
glutamine (Gln, Q)	$H_2NC(=O)CH_2CH_2CH(NH_3^+)CO_2^-$	threonine (Thr, T)	$CH_3CH(OH)CH(NH_3^+)CO_2^-$

Acidic Side Chains		**Basic Side Chains**	
aspartic acid (Asp, D)	$^-O_2CCH_2CH(NH_3^+)CO_2^-$	arginine (Arg, R)	$H_2NC(=NH_2^+)NHCH_2CH_2CH_2CH(NH_3^+)CO_2^-$
glutamic acid (Glu, E)	$^-O_2CCH_2CH_2CH(NH_3^+)CO_2^-$		
cysteine (Cys, C)	$HSCH_2CH(NH_3^+)CO_2^-$	histidine (His, H)	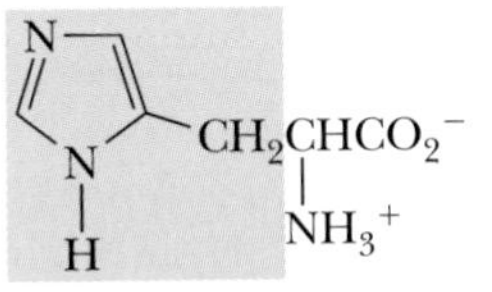 imidazole (N–H)–$CH_2CH(NH_3^+)CO_2^-$
Tyrosine (Tyr, Y)	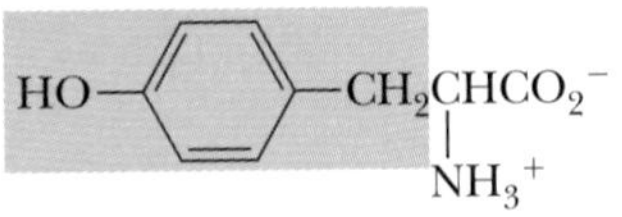 HO–C_6H_4–$CH_2CH(NH_3^+)CO_2^-$	lysine (Lys, K)	$H_3\overset{+}{N}CH_2CH_2CH_2CH_2CH(NH_3^+)CO_2^-$

* Each ionizable functional group is shown in the form present in highest concentration at pH 7.0.

Solution

(a) Phenylalanine, tryptophan, tyrosine, and histidine contain aromatic rings.
(b) Serine and threonine contain side-chain hydroxyl groups.
(c) Tyrosine contains a phenolic —OH group.
(d) Methionine and cysteine contain sulfur.

PROBLEM 27.1

Of the 20 protein-derived amino acids shown in Table 27.1, which contain (a) no stereocenter, (b) two stereocenters?

D. Some Other Common L-Amino Acids

Although the vast majority of plant and animal proteins are constructed from just these 20 α-amino acids, many other amino acids are also found in nature. L-Ornithine and L-citrulline, for example, are found predominantly in the liver and are an integral part of the urea cycle, the metabolic pathway that converts ammonia to urea.

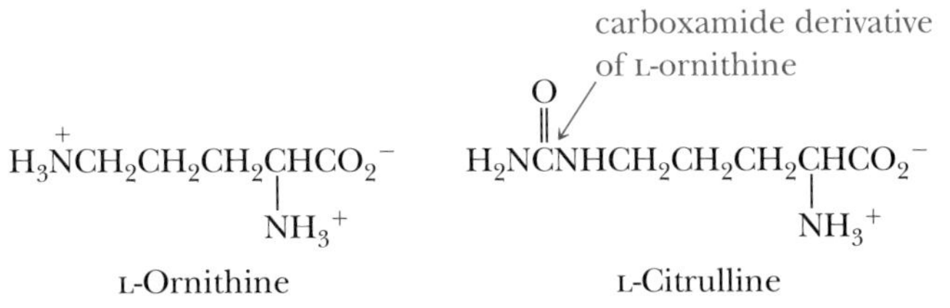

$H_3\overset{+}{N}CH_2CH_2CH_2CH(NH_3^+)CO_2^-$

L-Ornithine

carboxamide derivative of L-ornithine

$H_2NC(=O)NHCH_2CH_2CH_2CH(NH_3^+)CO_2^-$

L-Citrulline

Thyroxine, one of several hormones derived from the amino acid tyrosine, was first isolated from thyroid tissue in 1914. In 1952, triiodothyronine, a compound similar in structure to thyroxine but with only three atoms of iodine, was also discovered in thyroid tissue. Production of these hormones is the principal reason that iodine is required in the human diet; its lack results in the condition called goiter. Sodium iodide is commonly added to table salt to provide the small amounts of iodine required. The principal function of these hormones is to stimulate metabolism in other cells and tissues.

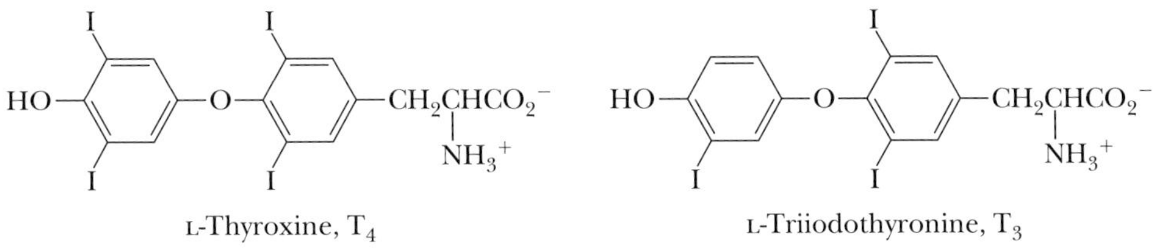

L-Thyroxine, T_4: HO–(3,5-diiodophenyl)–O–(3,5-diiodophenyl)–$CH_2CH(NH_3^+)CO_2^-$

L-Triiodothyronine, T_3: HO–(3-iodophenyl)–O–(3,5-diiodophenyl)–$CH_2CH(NH_3^+)CO_2^-$

4-Aminobutanoic acid (γ-aminobutyric acid, GABA) is a neurotransmitter found in high concentration (0.8 mM) in the brain, but in no significant amounts in any other mammalian tissue. This amino acid, synthesized in neural tissue by decarboxylation of the α-carboxyl group of glutamic acid, is a neurotransmitter in the central nervous system of invertebrates and possibly in humans as well.

$$^{-}O_2CCH_2CH_2\underset{\underset{NH_3^+}{|}}{C}HCO_2^- \xrightarrow[\text{decarboxylation}]{\text{enzyme-catalyzed}} {}^{-}O_2C\overset{\alpha}{C}H_2\overset{\beta}{C}H_2\overset{\gamma}{C}H_2NH_3^+ + CO_2$$

Glutamic acid → 4-Aminobutanoic acid (γ-Aminobutyric acid, GABA)

Only L-amino acids are found in proteins, and only rarely are D-amino acids a part of the metabolism of higher organisms. Several D-amino acids, however, along with their L-enantiomers, are found in lower forms of life. D-Alanine and D-glutamic acid, for example, are structural components of the cell walls of certain bacteria. Several D-amino acids are also found in peptide antibiotics.

27.2 Acid-Base Properties of Amino Acids

A. Acidic and Basic Groups of Amino Acids

Among the most important characteristics of amino acids are their acid-base properties; all are weak polyprotic acids because of their $—CO_2H$ and $—NH_3^+$ groups. Given in Table 27.2 are pK_a values for the ionizable groups of amino acids.

TABLE 27.2 pK_a Values for Ionizable Groups of Amino Acids

Amino Acid	pK_a of α-CO_2H	pK_a of α-NH_3^+	pK_a of Side Chain	Isoelectric Point (pI)
alanine	2.35	9.87	—	6.11
arginine	2.01	9.04	12.48	10.76
asparagine	2.02	8.80	—	5.41
aspartic acid	2.10	9.82	3.86	2.98
cysteine	2.05	10.25	8.00	5.02
glutamic acid	2.10	9.47	4.07	3.08
glutamine	2.17	9.13	—	5.65
glycine	2.35	9.78	—	6.06
histidine	1.77	9.18	6.10	7.64
isoleucine	2.32	9.76	—	6.04
leucine	2.33	9.74	—	6.04
lysine	2.18	8.95	10.53	9.74
methionine	2.28	9.21	—	5.74
phenylalanine	2.58	9.24	—	5.91
proline	2.00	10.60	—	6.30
serine	2.21	9.15	—	5.68
threonine	2.09	9.10	—	5.60
tryptophan	2.38	9.39	—	5.88
tyrosine	2.20	9.11	10.07	5.63
valine	2.29	9.72	—	6.00

Acidity of α-Carboxyl Groups

The average value of pK_a for an α-carboxyl group of a protonated amino acid is 2.19. Thus, the α-carboxyl group of a protonated amino acid is a considerably stronger acid than acetic acid (pK_a 4.76) and other low-molecular-weight aliphatic carboxylic acids. This greater acidity is accounted for by the electron-withdrawing inductive effect of the adjacent $-NH_3^+$ group. Recall that we used similar reasoning in Sections 3.3D and 16.4A to account for the relative acidities of acetic acid and its mono-, di-, and trichloroderivatives.

The ammonium ion has an electron-withdrawing inductive effect

$$\underset{\underset{NH_3^+}{|}}{RCHCO_2H} + H_2O \rightleftharpoons \underset{\underset{NH_3^+}{|}}{RCHCO_2^-} + H_3O^+ \qquad pK_a = 2.19$$

Acidity of Side-Chain Carboxyl Groups

Due to the electron-withdrawing inductive effect of the α-NH_3^+ group, the side-chain carboxyl groups of protonated aspartic acid and glutamic acid are stronger acids than acetic acid (pK_a 4.76). Notice that this acid-strengthening inductive effect decreases with increasing distance of the $-CO_2H$ from the α-NH_3^+; compare the acidities of the α-CO_2H of alanine (pK_a 2.35), the β-CO_2H of aspartic acid (pK_a 3.86), and the γ-CO_2H of glutamic acid (pK_a 4.07).

Acidity of α-Ammonium Groups

The average value of pK_a for an α-ammonium group, α-NH_3^+, is 9.47, compared with a value of 10.76 for primary alkylammonium ions (Section 21.6A). Thus, the α-ammonium group of an amino acid is a slightly stronger acid than a primary alkylammonium ion. Conversely, an α-amino group is a slightly weaker base than a primary alkylamine.

$$\underset{\underset{NH_3^+}{|}}{RCHCO_2^-} + H_2O \rightleftharpoons \underset{\underset{NH_2}{|}}{RCHCO_2^-} + H_3O^+ \qquad pK_a = 9.47$$

$$\underset{\underset{NH_3^+}{|}}{CH_3CHCH_3} + H_2O \rightleftharpoons \underset{\underset{NH_2}{|}}{CH_3CHCH_3} + H_3O^+ \qquad pK_a = 10.60$$

Basicity of the Guanidine Group of Arginine

The side-chain guanidine group of arginine is a considerably stronger base than an aliphatic amine. As we saw in Section 21.6D, guanidine (pK_b 0.4) is the strongest base of any neutral compound. The remarkable basicity of the guanidine group of arginine is attributed to the large resonance stabilization of the protonated form relative to the neutral form.

$$RNH{-}C(NH_2){=}NH_2^+ \longleftrightarrow RNH{-}C({=}NH_2^+)NH_2 \longleftrightarrow RN^+H{=}C(NH_2)NH_2 \overset{H_2O}{\rightleftharpoons} RNH{-}C({=}NH)NH_2 + H_3O^+ \quad pK_a = 12.48$$

The guanidinium ion side chain of arginine is a hybrid of three contributing structures

No resonance stabilization without charge separation

Basicity of the Imidazole Group of Histidine

Because the imidazole group on the side chain of histidine contains six π electrons in a planar, fully conjugated ring, imidazole is classified as a heterocyclic aromatic amine (Section 19.2D). Whereas the unshared pair of electrons on one nitrogen is a part of the aromatic sextet, that on the other nitrogen is not. It is the pair of electrons that is not part of the aromatic sextet that is responsible for the basic properties of the imidazole ring. Protonation of this nitrogen produces a resonance-stabilized cation.

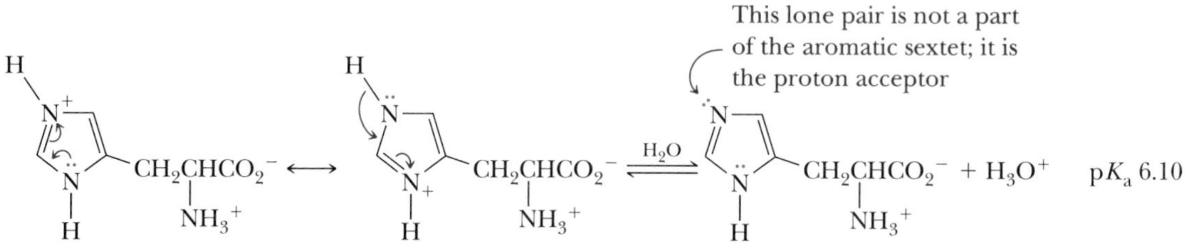

This lone pair is not a part of the aromatic sextet; it is the proton acceptor

$CH_2CHCO_2^-$ (with NH_3^+) $\longleftrightarrow$ $CH_2CHCO_2^-$ (with NH_3^+) $\overset{H_2O}{\rightleftharpoons}$ $CH_2CHCO_2^-$ (with NH_3^+) $+ H_3O^+$ $\quad pK_a$ 6.10

Resonance-stabilized imidazolium cation

B. Ionization of Amino Acids as a Function of pH

In Section 27.2A, we considered each ionizable group of an amino acid separately. Now let us consider how the interaction of ionizable groups within a particular amino acid affects the properties of that amino acid. Given values for the pK_a of each functional group, we can calculate the ratio of each acid to its conjugate base at any given pH. As an example, let us calculate these ratios at pH 7.0. Consider first the ionization of the weak acid, α-CO_2H, to form H_3O^+ and its conjugate base, α-CO_2^-. For the purposes of this calculation and to simplify the mathematics, we use a value of 2.00 for the pK_a of the protonated amino acid.

$$\alpha\text{-}CO_2H + H_2O \rightleftharpoons \alpha\text{-}CO_2^- + H_3O^+ \quad pK_a = 2.00$$

$$K_a = 1.00 \times 10^{-2}$$

The acid ionization constant for this equilibrium is given by the expression:

$$K_a = \frac{[H_3O^+][\alpha\text{-}CO_2^-]}{[\alpha\text{-}CO_2H]}$$

Rearranging this expression gives

$$\frac{[\alpha\text{-}CO_2^-]}{[\alpha\text{-}CO_2H]} = \frac{K_a}{[H_3O^+]}$$

Substituting values of K_a for an α-CO_2H group (1.00×10^{-2}) and the hydrogen ion concentration at pH 7.0 (1.0×10^{-7}) in this equation gives

$$\frac{[\alpha\text{-}CO_2^-]}{[\alpha\text{-}CO_2H]} = \frac{1.00 \times 10^{-2}}{1.00 \times 10^{-7}} = 1.00 \times 10^5$$

Thus, we see that at pH 7.0, the ratio of [α-CO_2^-] to [α-CO_2H] is 10^5 to 1. It is clear that at pH 7.0, an α-carboxyl group is virtually 100% in the ionized or conjugate base form, and it has a charge of -1.

We can also calculate the ratio of acid to conjugate base for an α-ammonium group. For this calculation, let us use a value of 10.0 for the pK_a of an α-ammonium group.

$$\alpha\text{-NH}_3^+ + H_2O \rightleftharpoons \alpha\text{-NH}_2 + H_3O^+ \quad pK_a = 10.00$$

$$K_a = 1.00 \times 10^{-10}$$

The acid ionization expression for this equilibrium can be rearranged as follows.

$$\frac{[\alpha\text{-NH}_2]}{[\alpha\text{-NH}_3^+]} = \frac{K_a}{[H_3O^+]}$$

Substituting values of K_a for an α-NH$_3^+$ group (1.0×10^{-10}) and the hydrogen ion concentration at pH 7.0 (1.0×10^{-7}) gives

$$\frac{[\alpha\text{-NH}_2]}{[\alpha\text{-NH}_3^+]} = \frac{1.00 \times 10^{-10}}{1.00 \times 10^{-7}} = 1.00 \times 10^{-3}$$

Thus, the ratio of α-NH$_2$ to α-NH$_3^+$ at pH 7.0 is approximately 1 to 1000. At this pH, an α-amino group is more than 99.9% in the acid or protonated form and has a charge of +1. As can be seen from these calculations, the dipolar form of an amino acid predominates at pH 7.0.

We have calculated the ratio of conjugate base to weak acid at pH 7.0 for an α-carboxyl group and an α-amino group. We can do the same type of calculation at any other pH. To do this for any weak acid and its conjugate base, it is convenient to transform the acid ionization constant expression in the following way.

$$\text{Weak acid} + H_2O \rightleftharpoons \text{conjugate base} + H_3O^+ \qquad K_a = \frac{[\text{conjugate base}][H_3O^+]}{[\text{weak acid}]}$$

Taking the logarithm of this equation and rearranging gives

This term, by definition, is pH

This term, by definition, is pK_a

$$-\log\,[H_3O^+] = -\log K_a + \log \frac{[\text{conjugate base}]}{[\text{weak acid}]}$$

Thus, in rearranged form, the equation for the ionization of a weak acid to its conjugate base and hydronium ion has the following form, known as the **Henderson-Hasselbalch equation** after the two biochemists who first pointed out its particular usefulness.

Henderson-Hasselbalch equation A mathematical relationship that provides a direct way to calculate the ratio of conjugate base to its weak acid as a function of pH.

Henderson-Hasselbalch equation: $$pH = pK_a + \log \frac{[\text{conjugate base}]}{[\text{weak acid}]}$$

The Henderson-Hasselbalch equation provides a direct way to calculate the ratio of conjugate base to weak acid at any given pH. Figure 27.3 shows the percent of molecules present as un-ionized weak acid at pH values 1 and 2 units greater (more basic), and 1 and 2 units smaller (more acidic) than the value of pK_a. At any pH greater (more basic) than pK_a, the conjugate base predominates. At any pH less (more acidic) than pK_a, the weak acid predominates. When pH = pK_a, the concentrations of weak acid and conjugate base are equal.

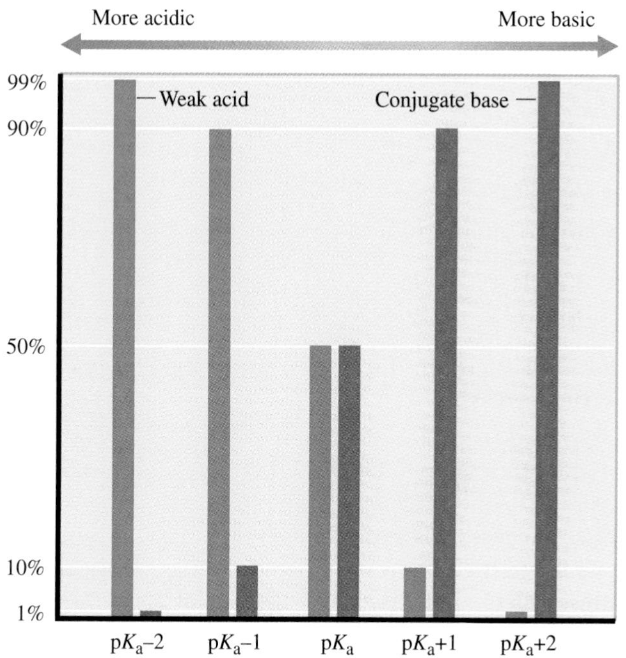

FIGURE 27.3
The dependence of concentration of weak acid and its conjugate base on pH.

EXAMPLE 27.2

Draw a structural formula for L-serine and estimate the net charge on this amino acid at pH values of 3.0, 7.0, and 10.0.

Solution

Start with the Henderson-Hasselbalch equation, substitute values for pH and pK_a, and solve for the ratio of conjugate base to weak acid. The pK_a of the α-carboxyl group of serine is 2.21. At pH of 3.0, which is 0.79 unit greater than its pK_a, the α-carboxyl group is approximately 86% in the ionized (conjugate base) form.

$$\log \frac{[\alpha\text{-CO}_2^-]}{[\alpha\text{-CO}_2\text{H}]} = 3.0 - 2.21 = 0.79$$

Taking the antilog of 0.79 gives the ratio of $[\alpha\text{-CO}_2^-]$ to $[\alpha\text{-CO}_2\text{H}]$ at this pH.

$$\frac{[\alpha\text{-CO}_2^-]}{[\alpha\text{-CO}_2\text{H}]} = \frac{6.17}{1}$$

$$\text{Percent of } \alpha\text{-CO}_2^- \text{ present at pH 3.0} = \frac{[\alpha\text{-CO}_2^-]}{[\alpha\text{-CO}_2\text{H}] + [\alpha\text{-CO}_2^-]} \times 100 = \frac{6.17}{1.00 + 6.17} \times 100 = 86\%$$

The pK_a of the α-amino group of serine is 9.15. Calculation of the ratio of unprotonated to protonated forms for the α-amino group of serine gives the following.

$$\frac{[\alpha\text{-NH}_2]}{[\alpha\text{-NH}_3^+]} = \frac{7.1 \times 10^{-7}}{1.00}$$

Thus, at pH 3.0, which is 6.15 pH units less (more acidic) than its pK_a, the α-amino group is completely in the protonated (positively charged) form.

The same type of calculations can be repeated at pH 7.0 and 10.0. Results are shown on the following structural formulas.

100% → $H_3\overset{+}{N}$–CH(CH$_2$OH)–C(=O)–OH ← 86%	99% → $H_3\overset{+}{N}$–CH(CH$_2$OH)–C(=O)–O^- ← 100%	88% → H_2N–CH(CH$_2$OH)–C(=O)–O^- ← 100%
pH 3.0	pH 7.0	pH 10.0
Net charge +1	Net charge 0	Net charge −1

PROBLEM 27.2

Draw a structural formula for lysine and estimate the net charge on each functional group at values of pH 3.0, 7.0, and 10.0.

C. Titration of Amino Acids

Values of pK_a for the ionizable groups of amino acids are most commonly obtained by acid-base titration and measuring the pH of the solution as a function of added base (or added acid, depending on how the titration is done). To illustrate this experimental procedure, consider a solution containing 1.0 mol of glycine to which has been added enough strong acid so that both the amino and carboxyl groups are fully protonated. Next, this solution is titrated with 1.0 M NaOH; the volume of base added and the pH of the resulting solution are recorded and then plotted as shown in Figure 27.4. The most acidic group and the one to react first with added sodium hydroxide is the carboxyl group. When exactly 0.50 mol of NaOH has been added, the carboxyl group is half neutralized. At this point, the concentration of the positively charged

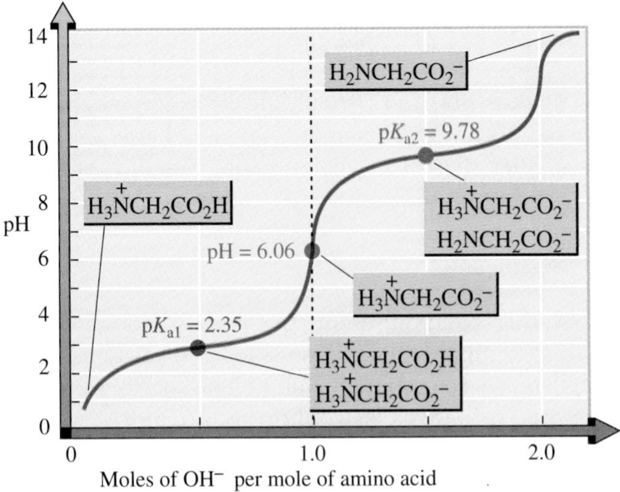

FIGURE 27.4
Titration of glycine with sodium hydroxide.

ion equals that of the dipolar ion, and the pH of 2.35 equals the pK_a of the carboxyl group (pK_{a1}).

$$\text{At pH} = \text{p}K_{a1}\text{:}\quad [\text{H}_3\overset{+}{\text{N}}\text{CH}_2\text{CO}_2\text{H}] = [\text{H}_3\overset{+}{\text{N}}\text{CH}_2\text{CO}_2^-]$$

Positive ion Dipolar ion

The end point of the first part of the titration is reached when 1.0 mol of sodium hydroxide has been added. At this point, the predominant species present is the dipolar ion, and the observed pH of the solution is 6.06.

The next section of the curve represents titration of the $-NH_3^+$ group. When another 0.50 mol of sodium hydroxide has been added (bringing the total to 1.50 mol) half of the $-NH_3^+$ groups are neutralized and converted to $-NH_2$. At this point, the concentrations of the dipolar ion and negatively charged ion are equal, and the observed pH is 9.78, the pK_a of the amino group of glycine (pK_{a2}).

$$\text{At pH} = \text{p}K_{a2}\text{:}\quad [\text{H}_3\overset{+}{\text{N}}\text{CH}_2\text{CO}_2^-] = [\text{H}_2\text{NCH}_2\text{CO}_2^-]$$

Dipolar ion Negative ion

The second end point of the titration is reached when a total of 2.0 mol of sodium hydroxide have been added and glycine is converted entirely to an anion.

D. Isoelectric Point

Isoelectric point (pI) The pH at which an amino acid, polypeptide, or protein has no net charge.

Titration curves such as that for glycine permit us to determine pK_a values for the ionizable groups of an amino acid. They also permit us to determine another important property: isoelectric point. The **isoelectric point, pI,** of an amino acid is the pH at which the majority of molecules in solution have no net charge (they are zwitterions). By examining the titration curve, you can see that the isoelectric point for glycine falls halfway between the pK_a values for the carboxyl and amino groups.

$$\begin{aligned}\text{pI} &= \tfrac{1}{2}\,(\text{p}K_a\ \alpha\text{-CO}_2\text{H} + \text{p}K_a\ \alpha\text{-NH}_3^+)\\ &= \tfrac{1}{2}\,(2.35 + 9.78) = 6.06\end{aligned}$$

At pH 6.06, the predominant form of glycine molecules is the dipolar ion; furthermore, at this pH, the concentration of positively charged glycine molecules equals the concentration of negatively charged glycine molecules.

Given the value for the isoelectric point of an amino acid, it is possible to estimate the charge on that amino acid at any pH. For example, the charge on tyrosine at pH 5.63, its isoelectric point, is zero. A small fraction of tyrosine molecules are positively charged at pH 5.00 (0.63 unit less than its pI), and virtually all are positively charged at pH 3.63 (2.00 units less than its pI). As another example, the net charge on lysine is zero at pH 9.74. At pH values smaller than 9.74 an increasing fraction of lysine molecules are positively charged.

E. Electrophoresis

Electrophoresis The process of separating compounds on the basis of their electric charge.

Electrophoresis, a process of separating compounds on the basis of their electric charges, is used to separate and identify mixtures of amino acids and proteins. Electrophoretic separations can be carried out using paper, starch, agar, certain plastics, and cellulose acetate as solid supports. In paper electrophoresis, a paper strip saturated with an aqueous buffer of predetermined pH serves as a bridge between two electrode vessels (Figure 27.5). Next, a sample of amino acids is applied as a spot on

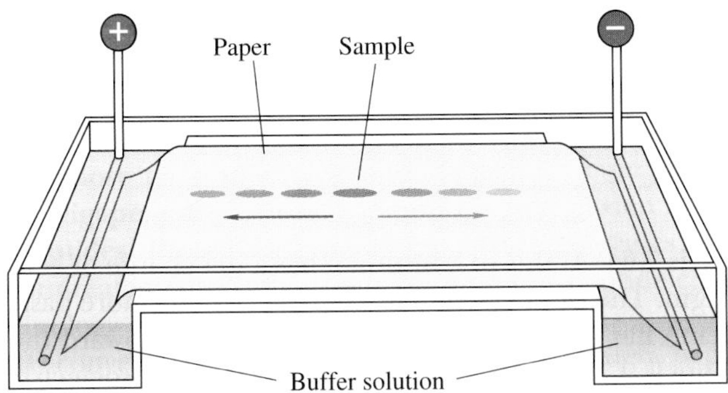

FIGURE 27.5
Electrophoresis of a mixture of amino acids. Those with a negative charge move toward the positive electrode; those with a positive charge move toward the negative electrode. Those with no charge remain at the origin.

the paper strip. When an electrical potential is applied to the electrode vessels, amino acids migrate toward the electrode carrying the charge opposite to their own. Molecules having a high charge density move more rapidly than those with a lower charge density. Any molecule already at its isoelectric point remains at the origin. After separation is complete, the strip is dried and sprayed with a reagent to make the separated components visible.

The reagent most commonly used to detect amino acids is ninhydrin (1,2,3-indanetrione monohydrate). Ninhydrin reacts with α-amino acids to produce an aldehyde, carbon dioxide, and a purple-colored anion with an absorption maximum at 580 nm. This reaction is used very commonly in both qualitative and quantitative analysis of amino acids.

$$\mathrm{RCH(NH_3^+)\overset{O}{\overset{\|}{C}}O^-} + 2\ \text{Ninhydrin} \longrightarrow \text{Purple-colored anion} + \mathrm{R\overset{O}{\overset{\|}{C}}H} + \mathrm{CO_2} + \mathrm{H_3O^+}$$

An α-amino acid — Ninhydrin — Purple-colored anion

Nineteen of the 20 protein-derived α-amino acids have primary amino groups and give the same purple-colored ninhydrin-derived anion. Proline, a secondary amine, gives a different, orange-colored compound.

EXAMPLE 27.3

The isoelectric point of tyrosine is 5.63. Toward which electrode does tyrosine migrate on paper electrophoresis at pH 7.0?

Solution

On paper electrophoresis at pH 7.0 (more basic than the isoelectric point of tyrosine), tyrosine has a net negative charge and migrates toward the positive electrode.

PROBLEM 27.3

The isoelectric point of histidine is 7.64. Toward which electrode does histidine migrate on paper electrophoresis at pH 7.0?

27.4 Primary Structure of Polypeptides and Proteins

Primary structure of proteins The sequence of amino acids in the polypeptide chain, read from the *N*-terminal amino acid to the *C*-terminal amino acid.

The **primary (1°) structure** of a polypeptide or protein refers to the sequence of amino acids in a polypeptide chain. In this sense, primary structure is a complete description of all covalent bonding in a polypeptide or protein.

In 1953, Frederick Sanger of Cambridge University, England, reported the primary structure of the two polypeptide chains of bovine insulin. Not only was this a remarkable achievement in analytical chemistry, but also it clearly established that the molecules of a given protein all have the same amino acid composition and the same amino acid sequence. Today, the amino acid sequences of over 20,000 different proteins are known.

A. Amino Acid Analysis

The first step in determining the primary structure of a polypeptide is hydrolysis and quantitative analysis of its amino acid composition. Recall from Section 17.5D that amide bonds are very resistant to hydrolysis. Typically, samples of protein are hydrolyzed in 6 M HCl in sealed glass vials at 110°C for 24 to 72 hours. This hydrolysis can be done in a microwave oven in a shorter time. Once the polypeptide is hydrolyzed, the resulting mixture of amino acids is analyzed by a technique called ion-exchange chromatography. Amino acids are detected as they emerge from the column by reaction with ninhydrin (Section 27.2). Current procedures for hydrolysis of polypeptides and analysis of amino acid mixtures have been refined to the point where it is possible to obtain amino acid composition from as little as 50 nanomole (50×10^{-9} mol) of polypeptide. Figure 27.7 shows the analysis of a polypeptide hydrolysate by ion-exchange chromatography. Note that during hydrolysis, the side-chain amide groups of asparagine and glutamine are hydrolyzed, and these amino acids are detected as aspartic acid and glutamic acid. For each glutamine or asparagine hydrolyzed, an equivalent amount of ammonium chloride is formed.

B. Sequence Analysis

Once the amino acid composition of a polypeptide has been determined, the next step is to determine the order in which the amino acids are joined in the polypeptide chain. The most common sequencing strategy is to cleave the polypeptide at specific peptide bonds (using, for example, cyanogen bromide or certain proteolytic enzymes), determine the sequence of each fragment (using, for example, the Edman degradation), and then match overlapping fragments to arrive at the sequence of the polypeptide.

Cyanogen Bromide

Cyanogen bromide (BrCN) is specific for cleavage of peptide bonds formed by the carboxyl group of methionine (Figure 27.8). The products of this cleavage are a substituted γ-lactone derived from the *N*-terminal portion of the polypeptide, and a second fragment derived from the *C*-terminal portion of the polypeptide.

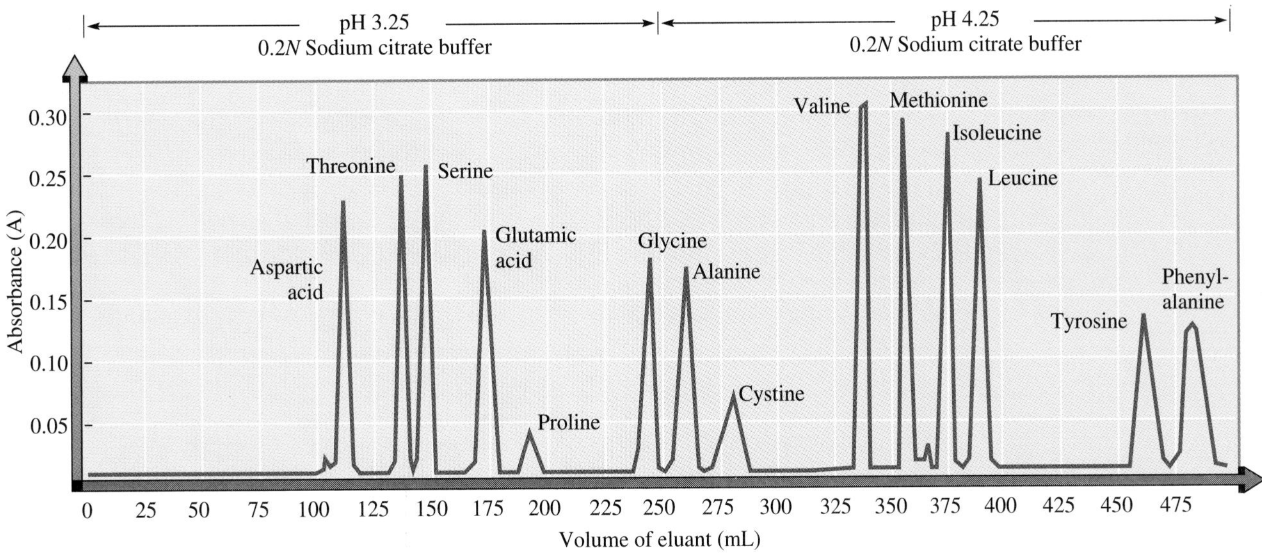

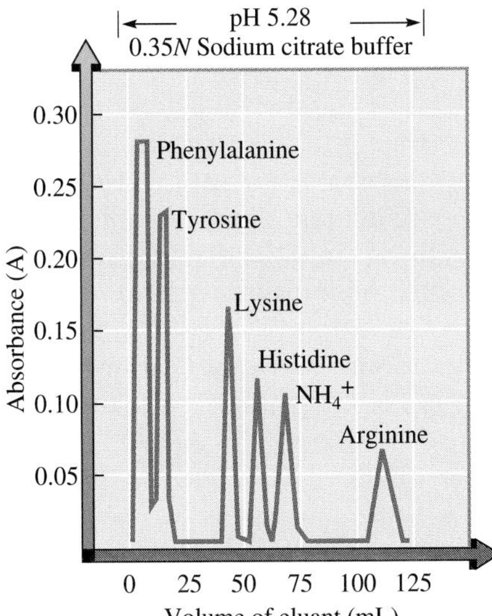

FIGURE 27.7
Analysis of a mixture of amino acids by ion-exchange chromatography using Amberlite IR-120, a sulfonated polystyrene resin. The resin contains phenyl—$SO_3^-Na^+$ groups. The amino acid mixture is applied to the column at low pH (3.25) under which conditions the acidic amino acids (Asp, Glu) are weakly bound to the resin and the basic amino acids (Lys, His, Arg) are tightly bound. Sodium citrate buffers at two different concentrations and three different values of pH are used to elute all amino acids from the column. Cysteine is determined as its disulfide, Cys-S-S-Cys.

This reaction is initiated in Step 1 by nucleophilic attack of the divalent sulfur atom of methionine on the carbon of cyanogen bromide displacing bromide ion. The product of Step 1 is a sulfonium ion. Following in Step 2 is an internal S_N2 reaction in which the oxygen of the methionine carbonyl group attacks the γ-carbon and displaces methyl thiocyanate to form a five-membered ring. Note that the oxygen of a carbonyl group is at best a weak nucleophile. This displacement is facilitated, however, because the sulfonium ion is a very good leaving group and because of the ease with which a five-membered ring is formed.

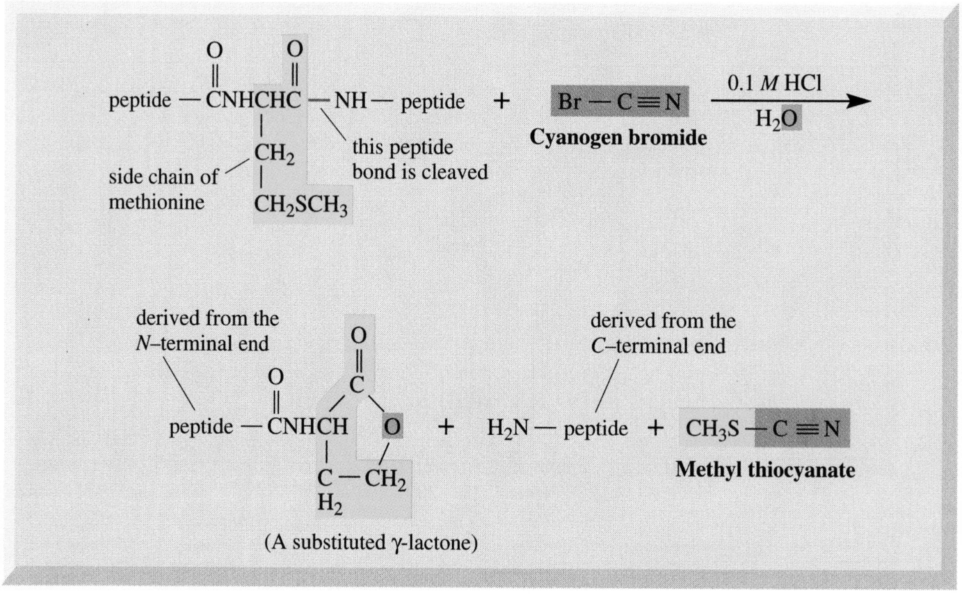

FIGURE 27.8
Cleavage by cyanogen bromide, Br—CN, of a peptide bond formed by the carboxyl group of methionine.

At this point, we should pause to consider the strategy of this reaction. Because CH_3S^- is the anion of a weak acid, it is a very poor leaving group, just as HO^- is a poor leaving group (Section 8.4F). Yet, just as the oxygen atom of an alcohol can be transformed into a better leaving group by converting it into an oxonium ion (by protonation), so too can the sulfur atom of methionine be transformed into a better leaving group by converting it into a sulfonium ion.

Hydrolysis of the imino lactone hydrobromide in Step 3 gives a lactone of the amino acid homoserine and a fragment derived from the C-terminal end of the original polypeptide.

MECHANISM Cleavage of a Peptide Bond at Methionine by Cyanogen Bromide

Step 1: Nucleophilic attack of sulfur on carbon of cyanogen bromide displacing bromine

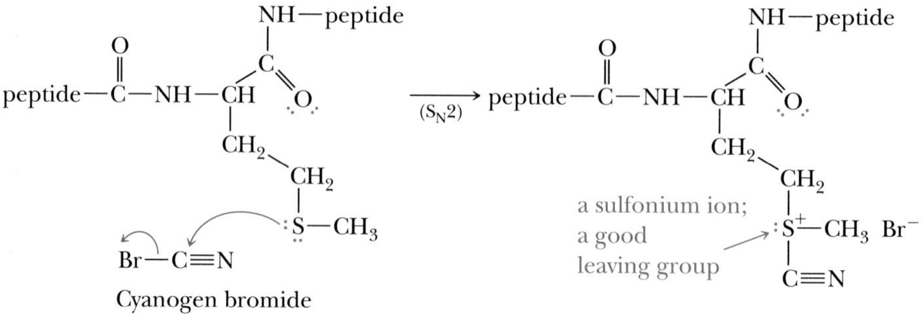

Step 2: Internal S_N2 reaction in which the carbonyl oxygen displaces sulfur

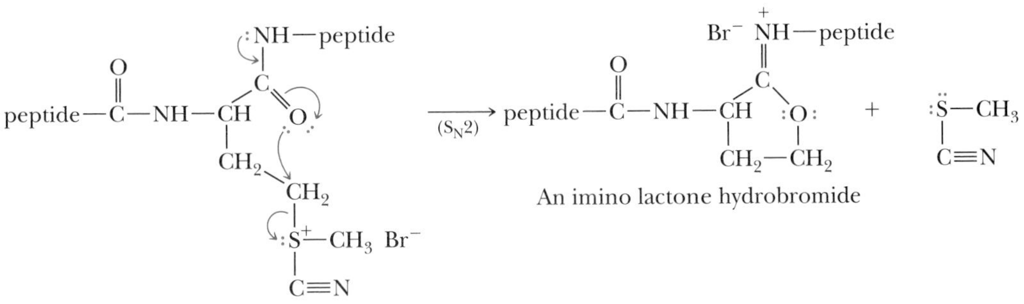

$$\text{peptide}-\overset{O}{\overset{\|}{C}}-NH-CH(C(=O)\text{—NH—peptide})-CH_2-CH_2-\overset{+}{S}(CH_3)(C\equiv N)\ Br^- \xrightarrow{(S_N2)} \text{peptide}-\overset{O}{\overset{\|}{C}}-NH-CH[C(=\overset{+}{N}H\text{—peptide})\,Br^-]\text{—O—}CH_2-CH_2 + :\ddot{S}(CH_3)-C\equiv N$$

An imino lactone hydrobromide

Step 3: Hydrolysis of the imine group to give a γ-lactone and a peptide fragment derived from the *C*-terminal end of the original polypeptide chain

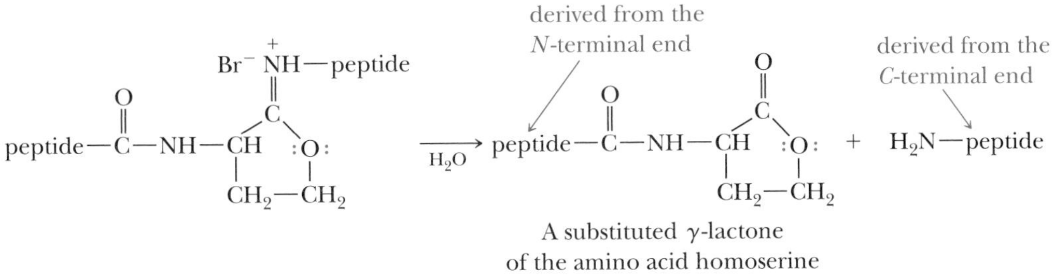

$$\text{peptide}-\overset{O}{\overset{\|}{C}}-NH-CH[C(=\overset{+}{N}H\text{—peptide})\,Br^-]\text{—O—}CH_2-CH_2 \xrightarrow{H_2O} \text{peptide}-\overset{O}{\overset{\|}{C}}-NH-CH[C(=O)]\text{—O—}CH_2-CH_2 + H_2N-\text{peptide}$$

derived from the *N*-terminal end

derived from the *C*-terminal end

A substituted γ-lactone of the amino acid homoserine

Enzyme-Catalyzed Hydrolysis of Peptide Bonds

A group of proteolytic enzymes, among them trypsin and chymotrypsin, can be used to catalyze the hydrolysis of specific peptide bonds. Trypsin catalyzes the hydrolysis of peptide bonds formed by the carboxyl groups of arginine and lysine; chymotrypsin catalyzes the hydrolysis of peptide bonds formed by the carboxyl groups of phenylalanine, tyrosine, and tryptophan (Table 27.3).

EXAMPLE 27.6

Trypsin catalyzes the hydrolysis of peptide bonds formed by the carboxyl groups of arginine and lysine. What structural feature(s) do these side chains have in common?

Solution

Each side chain contains a basic group: lysine contains a primary amino group (pK_a 10.53) and arginine contains the guanidino group (pK_a 12.28).

PROBLEM 27.6

Chymotrypsin catalyzes the hydrolysis of peptide bonds formed by the carboxyl groups of phenylalanine, tyrosine, and tryptophan. What structural feature(s) do these side chains have in common?

TABLE 27.3 Cleavage of Specific Peptide Bonds Catalyzed by Trypsin and Chymotrypsin

Enzyme	Catalyzes Hydrolysis of Peptide Bond Formed by Carboxyl Group of
trypsin	arginine, lysine
chymotrypsin	phenylalanine, tryptophan, tyrosine

EXAMPLE 27.7

Which of these tripeptides is hydrolyzed by trypsin? By chymotrypsin?

(a) Arg-Glu-Ser **(b)** Phe-Gly-Lys

Solution

(a) Trypsin catalyzes hydrolysis of peptide bonds formed by the carboxyl groups of lysine and arginine. Therefore, the peptide bond between arginine and glutamic acid is hydrolyzed in the presence of trypsin.

$$\text{Arg-Glu-Ser} + H_2O \xrightarrow{\text{trypsin}} \text{Arg} + \text{Glu-Ser}$$

Chymotrypsin catalyzes the hydrolysis of peptide bonds formed by the carboxyl groups of phenylalanine, tyrosine, and tryptophan. Because none of these three aromatic amino acids is present, tripeptide (a) is not affected by chymotrypsin.

(b) Tripeptide (b) is not affected by trypsin. Although lysine is present, its carboxyl group is at the *C*-terminal end and not involved in peptide bond formation. Tripeptide (b) is hydrolyzed in the presence of chymotrypsin.

$$\text{Phe-Gly-Lys} + H_2O \xrightarrow{\text{chymotrypsin}} \text{Phe} + \text{Gly-Lys}$$

PROBLEM 27.7

Which tripeptides are hydrolyzed by trypsin? By chymotrypsin?

(a) Tyr-Gln-Val **(b)** Thr-Phe-Ser

Edman Degradation

Edman degradation A method for selectively cleaving and identifying the *N*-terminal amino acid of a polypeptide chain.

Of the various chemical methods developed for determining the amino acid sequence of a polypeptide, the one most widely used is the **Edman degradation,** introduced in 1950 by Pehr Edman of the University of Lund, Sweden. In this procedure, a polypeptide is first treated with phenyl isothiocyanate, $C_6H_5N{=}C{=}S$ and then with acid. The effect of Edman degradation is to selectively remove the *N*-terminal amino acid as a substituted phenylthiohydantoin (Figure 27.9), which is then separated and identified.

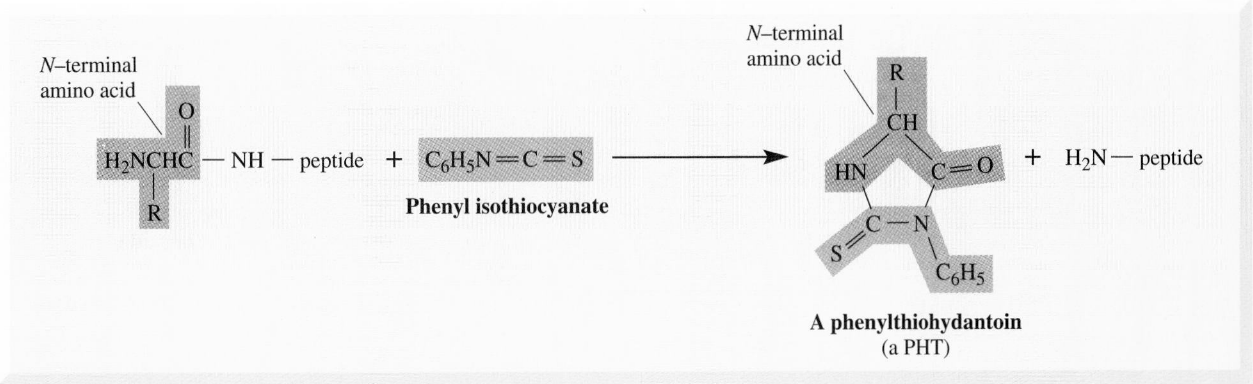

N–terminal amino acid

$$H_2NCH(R)C(=O)—NH—peptide + C_6H_5N{=}C{=}S \longrightarrow$$ A phenylthiohydantoin (a PHT) $+ H_2N—peptide$

Phenyl isothiocyanate

N–terminal amino acid

A phenylthiohydantoin (a PHT)

FIGURE 27.9
Edman degradation. Treatment of a polypeptide with phenyl isothiocyanate followed by acid selectively cleaves the *N*-terminal amino acid as a substituted phenylthiohydantoin.

Edman degradation is initiated in Step 1 by nucleophilic addition of the $—NH_2$ group of the *N*-terminal amino acid to the C═N bond of phenyl isothiocyanate to give a derivative of *N*-phenylthiourea.

In Step 2, the derivatized polypeptide is heated with HCl at 100°C, which brings about cyclization by nucleophilic addition of sulfur to the carbonyl of the adjacent amide group to form a tetrahedral carbonyl addition intermediate. This intermediate then collapses and releases the *N*-terminal amino acid in the form of a thiazolinone derivative. The other product of the reaction is a polypeptide now shortened by one amino acid.

The thiazolinone ring undergoes isomerization in Step 3 and reclosing to give a more stable phenylthiohydantoin, which is then separated and identified by high-resolution chromatography.

MECHANISM Edman Degradation: Cleavage of an *N*-Terminal Amino Acid

Step 1: Nucleophilic addition of the $—NH_2$ group of the *N*-terminal amino acid to the C═N bond of phenyl isothiocyanate to give a derivative of *N*-phenylthiourea

$$H_2N—CH(R)—C(=O)—NH—peptide + C_6H_5—N{=}C{=}S \longrightarrow C_6H_5—NH—C(=S)—NH—CH(R)—C(=O)—NH—peptide$$

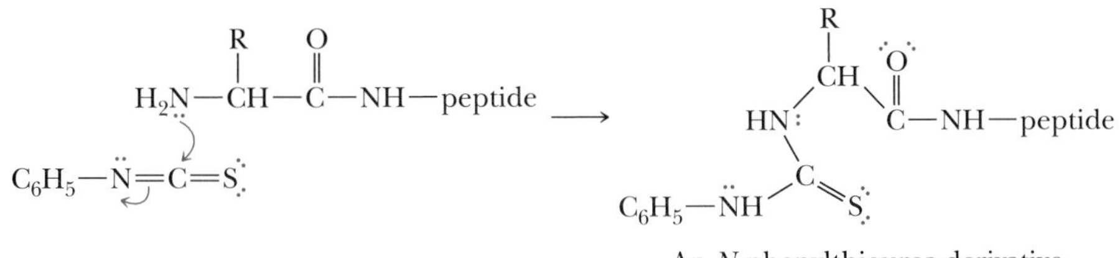

An *N*-phenylthiourea derivative

Step 2: Nucleophilic addition of sulfur to the carbonyl of the adjacent amide group forms a tetrahedral carbonyl addition intermediate, which collapses to give a thiazolinone ring derived from the *N*-terminal amino acid

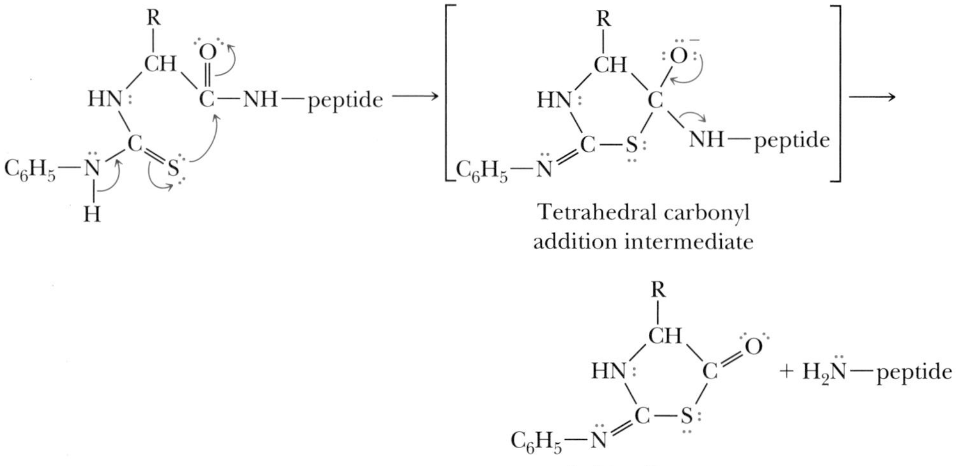

Tetrahedral carbonyl addition intermediate

$+ H_2N$—peptide

A thiazolinone

Step 3: The thiazolinone ring undergoes isomerization by ring opening followed by reclosing to give a more stable phenylthiohydantoin, which is separated and identified by chromatography

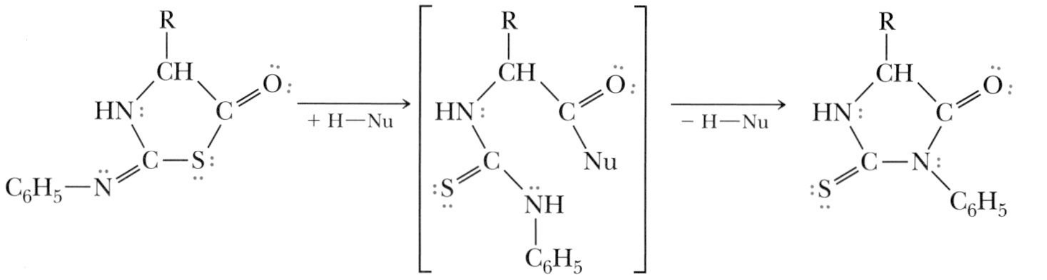

+ H—Nu

− H—Nu

A thiazolinone

A phenylthiohydantoin (a PHT)

The special value of Edman degradation is that it cleaves the *N*-terminal amino acid from a polypeptide without affecting any other bonds in the chain. As a result, Edman degradation can be repeated on the shortened polypeptide, causing the next amino acid in the sequence to be cleaved and identified. In practice, it is now possible to sequence as many as the first 20–30 amino acids in a polypeptide by this method using as little as a few milligrams of material.

Most polypeptides in nature are longer than 20–30 amino acids, the practical limit to the number of amino acids that can be sequenced by repetitive Edman degradation. The special value of cleavage with cyanogen bromide, trypsin, and chymotrypsin is that a long polypeptide chain can be cleaved at specific peptide bonds into smaller polypeptide fragments, and each fragment can then be sequenced separately.

EXAMPLE 27.8

Deduce the amino acid sequence of a pentapeptide from the following experimental results. Note that, under the column "Amino Acid Composition," the amino acids are listed in alphabetical order. In no way does this listing give any information about primary structure.

Experimental Procedure	Amino Acid Composition
pentapeptide	Arg, Glu, His, Phe, Ser
Edman Degradation	Glu
Hydrolysis Catalyzed by Chymotrypsin	
Fragment A	Glu, His, Phe
Fragment B	Arg, Ser
Hydrolysis Catalyzed by Trypsin	
Fragment C	Arg, Glu, His, Phe
Fragment D	Ser

Solution

Edman degradation cleaves glu from the pentapeptide; therefore, glutamic acid must be the *N*-terminal amino acid.

Glu -(Arg, His, Phe, Ser)

Fragment A from chymotrypsin-catalyzed hydrolysis contains phe. Because of the specificity of chymotrypsin, phe must be the *C*-terminal amino acid of fragment A. Fragment A also contains glu, which we already know is the *N*-terminal amino acid. From these observations, conclude that the first three amino acids in the chain must be Glu-His-Phe, and then write the following partial sequence:

Glu-His-Phe -(Arg, Ser)

The fact that trypsin cleaves the pentapeptide means that Arg must be within the pentapeptide chain; it cannot be the *C*-terminal amino acid. Therefore, the complete sequence must be

Glu-His-Phe-Arg-Ser

PROBLEM 27.8

Deduce the amino acid sequence of an undecapeptide (11 amino acids) from the experimental results shown in the accompanying table:

Experimental Procedure	Amino Acid Composition
undecapeptide	Ala, Arg, Glu, Lys_2, Met, Phe, Ser, Thr, Trp, Val
Edman Degradation	Ala
Trypsin-Catalyzed Hydrolysis	
Fragment E	Ala, Glu, Arg
Fragment F	Thr, Phe, Lys
Fragment G	Lys
Fragment H	Met, Ser, Trp, Val
Chymotrypsin-Catalyzed Hydrolysis	
Fragment I	Ala, Arg, Glu, Phe, Thr
Fragment J	Lys_2, Met, Ser, Trp, Val
Reaction with Cyanogen Bromide	
Fragment K	Ala, Arg, Glu, Lys_2, Met, Phe, Thr, Val
Fragment L	Trp, Ser

27.5 Synthesis of Polypeptides

A. The Problem

The problem in peptide synthesis is to join the carboxyl group of amino acid-1 by an amide (peptide) bond to the amino group of amino acid-2.

$$\overset{+}{H_3N}\underset{\substack{|\\ aa_1}}{C}H\overset{\overset{O}{\|}}{C}O^- + \overset{+}{H_3N}\underset{\substack{|\\ aa_2}}{C}H\overset{\overset{O}{\|}}{C}O^- \xrightarrow{?} \overset{+}{H_3N}\underset{\substack{|\\ aa_1}}{C}H\overset{\overset{O}{\|}}{C}NH\underset{\substack{|\\ aa_2}}{C}H\overset{\overset{O}{\|}}{C}O^- + H_2O$$

B. The Strategy

A rational strategy for the synthesis of peptide bonds and polypeptides requires three steps.

1. Protect the α-amino group of amino acid aa_1 to reduce its nucleophilicity so that it does not participate in nucleophilic addition to the carboxyl group of either aa_1 or aa_2.
2. Protect the α-carboxyl group of amino acid aa_2 so that it is not susceptible to nucleophilic attack by the α-amino group of another molecule of aa_2.
3. Activate the α-carboxyl group of amino acid aa_1 so that it is susceptible to nucleophilic attack by the α-amino group of aa_2.

protecting group (Z); activating group (Y); protecting group (X)

$$Z{-}NH\underset{\substack{|\\ aa_1}}{C}H\overset{\overset{O}{\|}}{C}{-}Y + H_2N\underset{\substack{|\\ aa_2}}{C}H\overset{\overset{O}{\|}}{C}{-}X \xrightarrow{\text{form peptide bonds}} Z{-}NH\underset{\substack{|\\ aa_1}}{C}H\overset{\overset{O}{\|}}{C}NH\underset{\substack{|\\ aa_2}}{C}H\overset{\overset{O}{\|}}{C}{-}X + H{-}Y$$

Once dipeptide $aa_1—aa_2$ has been formed, the protecting group Z can be removed and chain growth continued from the *N*-terminal end of the dipeptide. Alternatively, the protecting group X can be removed and chain growth continued from the *C*-terminal end. The range of protecting groups and activating groups is large, and experimental conditions have been found to attach and remove them as desired.

C. Amino-Protecting Groups

The most common strategy for protecting amino groups and reducing their nucleophilicity is to convert them to amides. The reagents most commonly used for this purpose are benzyloxycarbonyl chloride and di-*tert*-butyl dicarbonate. In the terminology adopted by the IUPAC, the benzyloxycarbonyl group is given the symbol Z—, and the *tert*-butoxycarbonyl group is given the symbol BOC—.

$PhCH_2O\overset{O}{\overset{\|}{C}}Cl$	$PhCH_2O\overset{O}{\overset{\|}{C}}—$	$(CH_3)_3CO\overset{O}{\overset{\|}{C}}O\overset{O}{\overset{\|}{C}}OC(CH_3)_3$	$(CH_3)_3CO\overset{O}{\overset{\|}{C}}—$
Benzyloxycarbonyl chloride	Benzyloxycarbonyl (Z—) group	Di-*tert*-butyl dicarbonate	*tert*-Butoxycarbonyl (BOC—) group

Treatment of an amino group with either of these reagents forms a new functional group called a carbamate. A carbamate is an ester of carbamic acid, that is, it is an ester of the monoamide of carbonic acid.

$$\underset{\substack{\text{Benzyloxycarbonyl}\\\text{chloride}\\(Z—Cl)}}{PhCH_2O\overset{O}{\overset{\|}{C}}Cl} + \underset{\text{Alanine}}{H_3\overset{+}{N}\underset{CH_3}{\underset{|}{C}}H\overset{O}{\overset{\|}{C}}O^-} \xrightarrow[\text{2. HCl, } H_2O]{\text{1. NaOH}} \underset{\substack{N\text{-Benzyloxycarbonylalanine}\\(Z—ala)}}{PhCH_2O\overset{O}{\overset{\|}{C}}NH\underset{CH_3}{\underset{|}{C}}H\overset{O}{\overset{\|}{C}}OH}$$

The special advantage of the carbamate group is that it is stable to dilute base but can be removed by treatment with HBr in acetic acid.

$$\underset{\text{A Z-protected peptide}}{PhCH_2O\overset{O}{\overset{\|}{C}}NH—\text{peptide}} \xrightarrow[CH_3CO_2H]{HBr} \underset{\text{Benzyl bromide}}{PhCH_2Br} + CO_2 + \underset{\text{Unprotected peptide}}{H_3\overset{+}{N}—\text{peptide}}$$

A study of the mechanism for removal of this protecting group has shown that the reaction is first order in $[H^+]$ and involves formation of a carbocation and a carbamic acid. A carbamic acid spontaneously loses carbon dioxide to form the free amine. The carbocation may lose a proton to form an alkene or react with an available nucleophile such as halide ion to form an alkyl halide.

MECHANISM Acid-Catalyzed Removal of a Benzyloxycarbonyl Protecting Group

$$PhCH_2{-}O{-}\overset{\overset{\displaystyle :\ddot{O}}{\|}}{C}{-}NH{-}peptide \xrightarrow{H{-}Br} \begin{cases} PhCH_2^+ \xrightarrow{Br^-} PhCH_2Br \\ O{=}\overset{\overset{\displaystyle OH}{|}}{C}{-}NH{-}peptide \longrightarrow CO_2 + H_3\overset{+}{N}{-}peptide \end{cases}$$

A carbamic acid

Note that because acid-catalyzed removal of this protecting group is carried out in nonaqueous media, there is no danger of simultaneous acid-catalyzed hydrolysis of peptide (amide) bonds within the newly synthesized polypeptide. Why? Because water is required for hydrolysis of a peptide bond.

The benzyloxycarbonyl group can also be removed by treatment with H_2 in the presence of a transition metal catalyst (hydrogenolysis, Section 19.6C). In hydrogenolysis of a Z-protecting group, one product is toluene. The other is a carbamic acid, which undergoes spontaneous decarboxylation to give carbon dioxide and the unprotected peptide.

$$PhCH_2O\overset{\overset{\displaystyle O}{\|}}{C}NH{-}peptide + H_2 \xrightarrow{Pd} PhCH_3 + CO_2 + H_2N{-}peptide$$

A Z-protected peptide — Toluene — Unprotected peptide

D. Carboxyl-Protecting Groups

Carboxyl groups are most often protected by conversion to methyl, ethyl, or benzyl esters. Methyl and ethyl esters are prepared by Fischer esterification (Section 16.8A) and are removed by hydrolysis in aqueous base (Section 17.5C) under mild conditions. Benzyl esters are conveniently removed by hydrogenolysis with H_2 over a palladium or platinum catalyst (Section 19.6C). Benzyl groups can also be removed by treatment with HBr in acetic acid.

E. Peptide-Bond Forming Reactions

The reagent most commonly used to bring about peptide bond formation is 1,3-dicyclohexylcarbodiimide (DCC). This reagent is the anhydride of a disubstituted urea, and, when treated with water, it is converted to *N,N′*-dicyclohexylurea (DCU).

$$C_6H_{11}{-}N{=}C{=}N{-}C_6H_{11} + H_2O \longrightarrow C_6H_{11}{-}\underset{\underset{\displaystyle H}{|}}{N}{-}\overset{\overset{\displaystyle O}{\|}}{C}{-}\underset{\underset{\displaystyle H}{|}}{N}{-}C_6H_{11}$$

1,3-Dicyclohexylcarbodiimide (DCC) — *N,N′*-dicyclohexylurea (DCU)

When an amino-protected aa_1 and a carboxyl-protected aa_2 are treated with DCC, this reagent acts as a dehydrating agent; it removes —OH from the carboxyl group and —H from the amino group to form an amide group. More specifically, DCC activates the α-carboxyl group of aa_1 toward nucleophilic acyl substitution by converting its —OH group into a better leaving group.

$$\text{Z—NHCH(R}_1\text{)COOH} + \text{H}_2\text{NCH(R}_2\text{)COOCH}_3 + \text{C}_6\text{H}_{11}\text{—N=C=N—C}_6\text{H}_{11} \xrightarrow{\text{CHCl}_3}$$

Amino-protected aa_1 | Carboxyl-protected aa_2 | 1,3-Dicyclohexylcarbodiimide (DCC)

$$\text{Z—NHCH(R}_1\text{)C(=O)—NHCH(R}_2\text{)C(=O)—OCH}_3 + \text{C}_6\text{H}_{11}\text{—NH—C(=O)—NH—C}_6\text{H}_{11}$$

Amino- and carboxyl-protected dipeptide | *N,N'*-Dicyclohexylurea (DCU)

An abbreviated mechanism for this intermolecular dehydration is shown in Figure 27.10. An acid-base reaction in Step 1 between the carboxyl group of aa_1 and a nitrogen of DCC followed in Step 2 by addition of the carboxylate anion to the C=N double bond results in electrophilic addition to a C=N double bond. The O-acylisourea formed is the nitrogen analog of a mixed anhydride. Nucleophilic addition of the amino group of aa_2 to the carbonyl group of the O-acylisourea in Step 3 generates a tetrahedral carbonyl addition intermediate that collapses in Step 4 to give a dipeptide and DCU.

F. Solid-Phase Synthesis

A major problem associated with polypeptide synthesis is purification of intermediates after each protection, activation, coupling, and deprotection step. If unreacted starting materials are not removed after each step, the final product is contaminated by polypeptides missing one or more amino acids. The required purification steps are not only laborious and time-consuming, but they also inevitably result in some loss of the desired product. These losses become especially severe in the synthesis of larger polypeptides.

A major advance in polypeptide synthesis came in 1962 when R. Bruce Merrifield of Rockefeller University described a **solid-phase synthesis** (alternatively called polymer-supported synthesis) of the tetrapeptide, Leu-Ala-Gly-Ala, by a technique that now bears his name. Merrifield was awarded the 1984 Nobel Prize for chemistry for his work in developing the solid-phase method for peptide synthesis.

The solid support used by Merrifield was a type of polystyrene in which about 5% of the phenyl groups carry a chloromethyl (—CH_2Cl) group in their para positions (Figure 27.11). These chloromethyl groups, like all benzylic halides, are particularly reactive in nucleophilic substitution reactions.

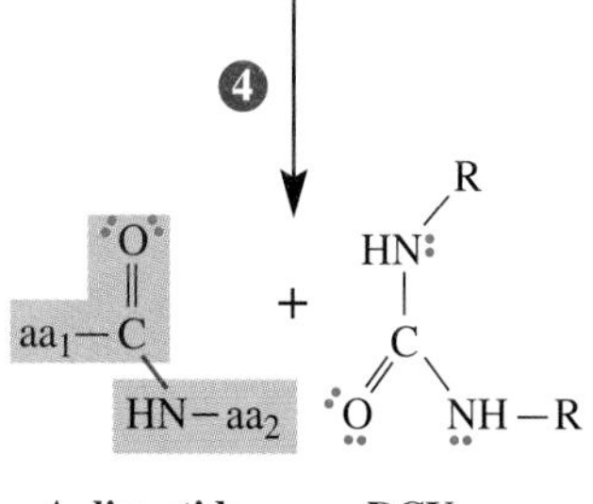

FIGURE 27.10
The role of 1,3-dicyclohexylcarbodiimide (DCC) in the formation of a peptide bond between an amino-protected amino acid (aa_1) and a carboxyl-protected amino acid (aa_2).

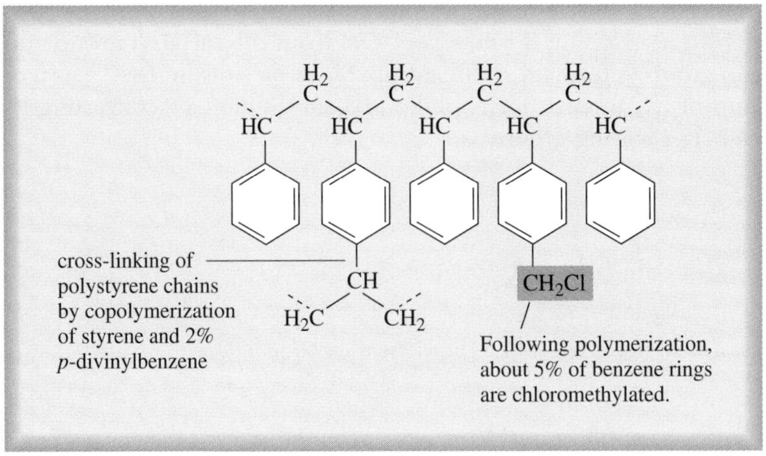

cross-linking of polystyrene chains by copolymerization of styrene and 2% *p*-divinylbenzene

CH_2Cl

Following polymerization, about 5% of benzene rings are chloromethylated.

FIGURE 27.11
The support used for the Merrifield solid phase synthesis is a chloromethylated polystyrene resin.

In the Merrifield method, the *C*-terminal amino acid is joined as a benzyl ester to the solid polymer support, and then the polypeptide chain is extended one amino acid at a time from the *N*-terminal end. The advantage of polypeptide synthesis on a solid support is that the polymer beads with the peptide chains anchored on them are completely insoluble in the solvents used in the synthesis, whereas excess reagents (for example, DCC) and byproducts (for example, DCU) are removed after each step simply by washing the polymer beads. When synthesis is completed, the polypeptide is released from the polymer beads by cleavage of the benzyl ester. The steps in solid-phase synthesis of a polypeptide are summarized in Figure 27.12.

A dramatic illustration of the power of the solid-phase method was the synthesis of the enzyme ribonuclease by Merrifield in 1969. The synthesis involved 369 chemical reactions and 11,931 operations, all of which were performed by an automated machine and without any intermediate isolation stages. Each of the 124 amino acids was added as an *N-tert*-butoxycarbonyl derivative and coupled using DCC. Cleavage from the resin and removal of all protective groups gave a mixture that was purified by ion-exchange chromatography. The specific activity of the synthetic enzyme was 13%–24% of that of the natural enzyme. The fact that the specific activity of the synthetic enzyme was lower than that of the natural enzyme was probably due to the presence of polypeptide byproducts closely related, but not identical, to the natural enzyme. Synthesizing ribonuclease (124 amino acids) requires forming 123 peptide bonds. If each peptide bond is formed in 99% yield, the yield of homogeneous polypeptide is $0.99^{123} = 35\%$. If each peptide bond is formed in 98% yield, the yield is 9%. Thus, even with yields as high as 99% in each peptide bond-forming step, a large portion of the synthetic polypeptides have one or more sequence defects. Many of these, nonetheless, may be fully active or partially active.

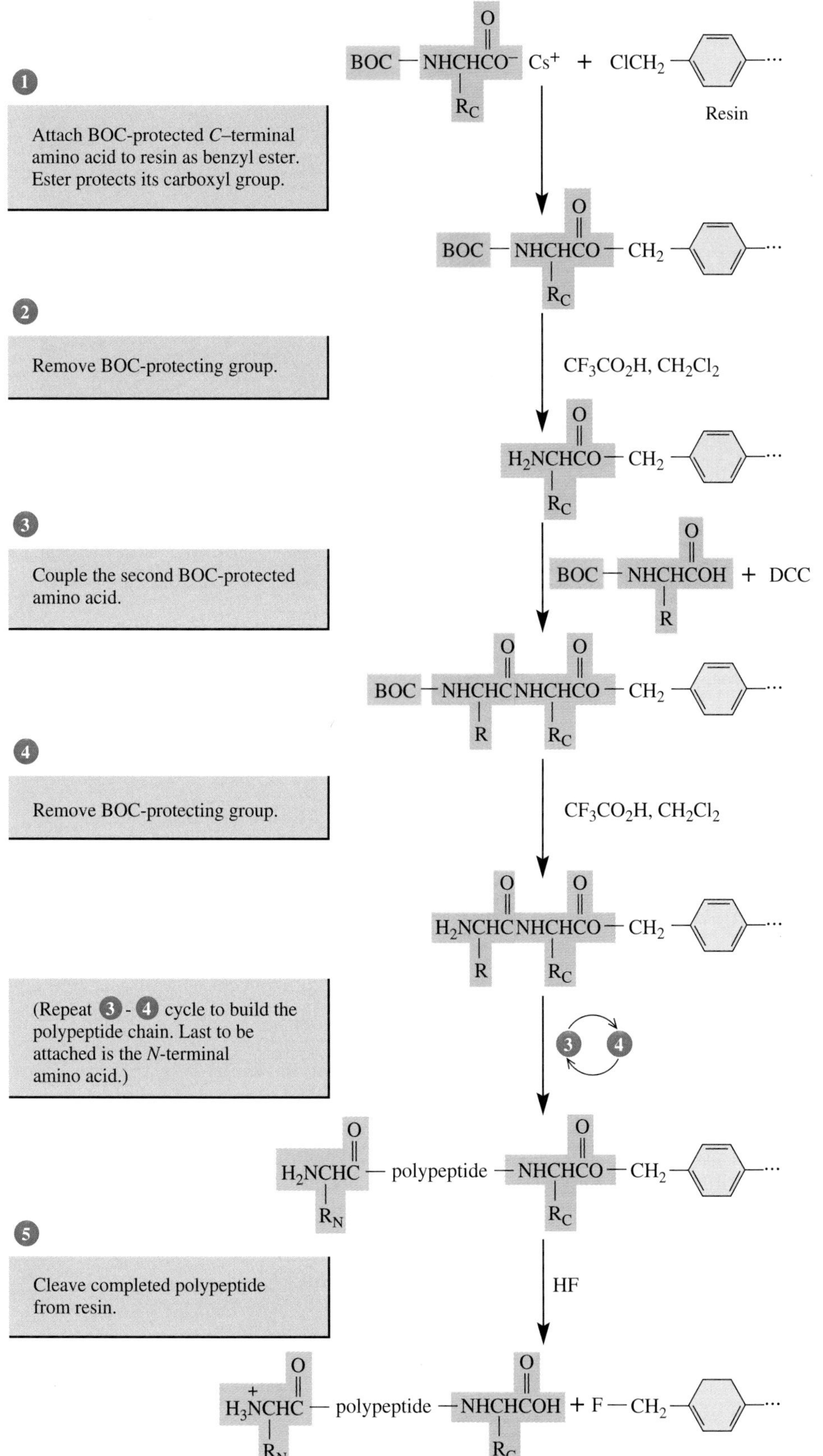

1
Attach BOC-protected C–terminal amino acid to resin as benzyl ester. Ester protects its carboxyl group.
2
Remove BOC-protecting group.
3
Couple the second BOC-protected amino acid.
4
Remove BOC-protecting group.
(Repeat 3 - 4 cycle to build the polypeptide chain. Last to be attached is the N-terminal amino acid.)
5
Cleave completed polypeptide from resin.
Resin
CF3CO2H, CH2Cl2
BOC—NHCHCOH + DCC
HF

FIGURE 27.12
Steps in the Merrifield solid phase polypeptide synthesis.

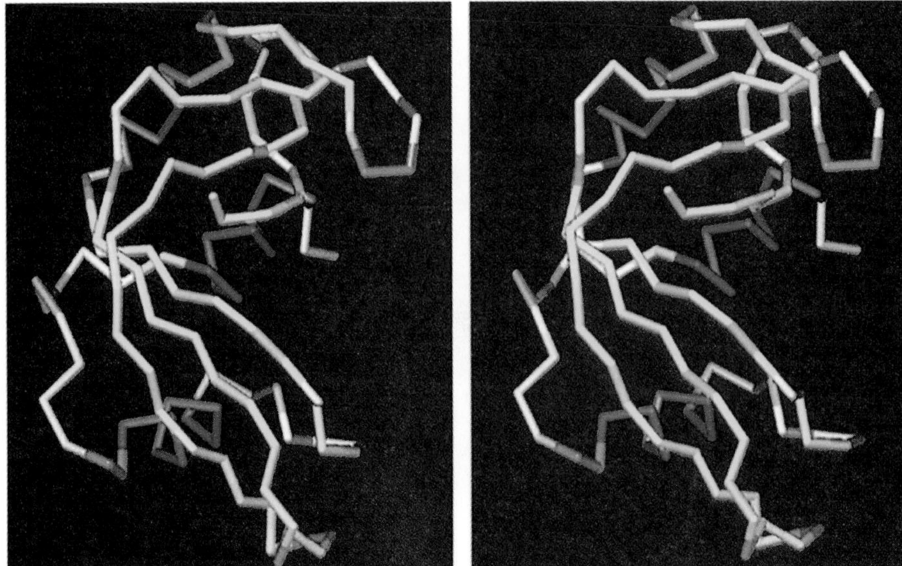

A model of the protein ribonuclease A in stereoview. The red segments are regions of α-helix and the yellow segments are regions of β-pleated sheet, both of which are described in Section 27.6. Other colors represent loop regions.

27.6 Three-Dimensional Shapes of Polypeptides and Proteins

A. Geometry of a Peptide Bond

In the late 1930s, Linus Pauling began a series of studies to determine the geometry of a peptide bond. One of his first discoveries was that a peptide bond is planar. As shown in Figure 27.13, the four atoms of a peptide bond and the two alpha carbons joined to it all lie in the same plane.

Had you been asked in Chapter 1 to describe the geometry of a peptide bond, you probably would have predicted bond angles of 120° about the carbonyl carbon and 109.5° about the amide nitrogen. This prediction agrees with the observed bond angles of approximately 120° about the carbonyl carbon. However, a bond angle of 120° about the amide nitrogen is unexpected. To account for this observed geometry, Pauling proposed that a peptide bond is more accurately represented as a resonance hybrid of these two contributing structures.

(1) ⟷ (2)

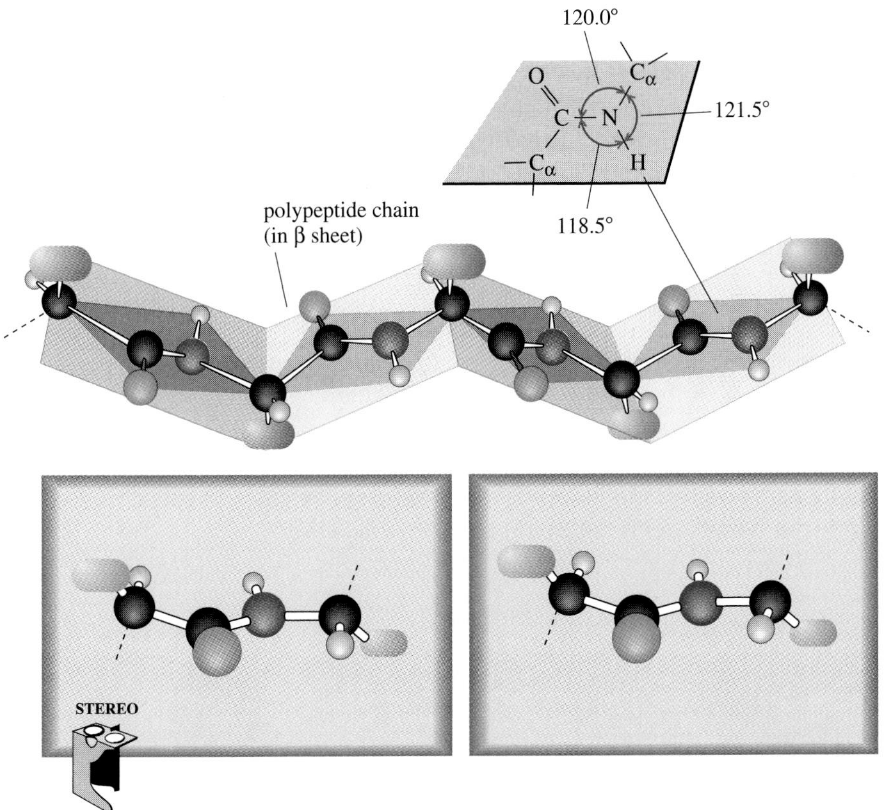

FIGURE 27.13
Planarity of a peptide bond. Bond angles about the carbonyl carbon and the amide nitrogen are approximately 120°.

Contributing structure (1) shows a carbon-oxygen double bond, and structure (2) shows a carbon-nitrogen double bond. The hybrid, of course, is neither of these; in the real structure, the carbon-nitrogen bond has considerable double-bond character. Accordingly, in the hybrid, the six-atom group is planar and rotation about the peptide bond is restricted.

Two conformations are possible for the atoms of a planar peptide bond. In one, the two α-carbons are *cis* to each other; in the other, they are *trans* to each other. The *trans* conformation is more favorable because the alpha carbons with the bulky groups attached to them are farther from each other than they are in the *cis* conformation. Virtually all peptide bonds in naturally occurring proteins studied to date have the s-*trans* conformation.

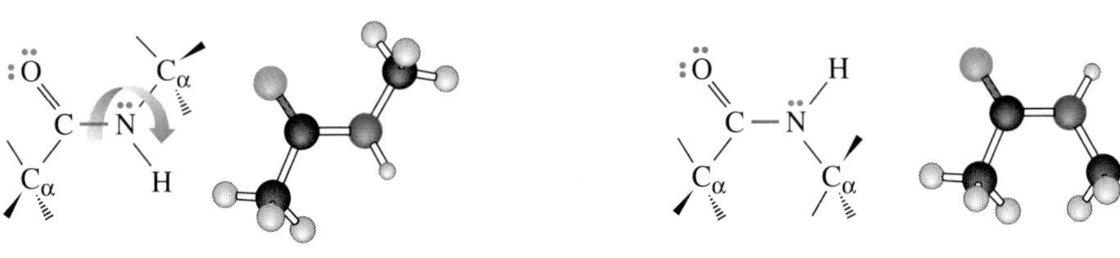

B. Secondary Structure

Secondary structure of proteins The ordered arrangements (conformations) of amino acids in localized regions of a polypeptide or protein.

Secondary (2°) structure refers to ordered arrangements (conformations) of amino acids in localized regions of a polypeptide or protein molecule. The first studies of polypeptide conformations were carried out by Linus Pauling and Robert Corey, beginning in 1939. They assumed that, in conformations of greatest stability, (1) all six atoms in a peptide bond lie in the same plane, and (2) there are hydrogen bonds between the N—H of one peptide bond and the C═O of another as shown in Figure 27.14.

On the basis of model building, Pauling proposed that two folding patterns should be particularly stable: the α-helix and the antiparallel β-pleated sheet. The term "secondary structure" is used to describe α-helix, β-pleated sheet, and other types of periodic conformations in localized regions of polypeptide or protein molecules.

α-Helix A type of secondary structure in which a section of polypeptide chain coils into a spiral, most commonly a right-handed spiral.

The α-Helix

In an **α-helix** pattern shown in Figure 27.15, a polypeptide chain is coiled in a spiral. As you study the α-helix in Figure 27.15, note the following:

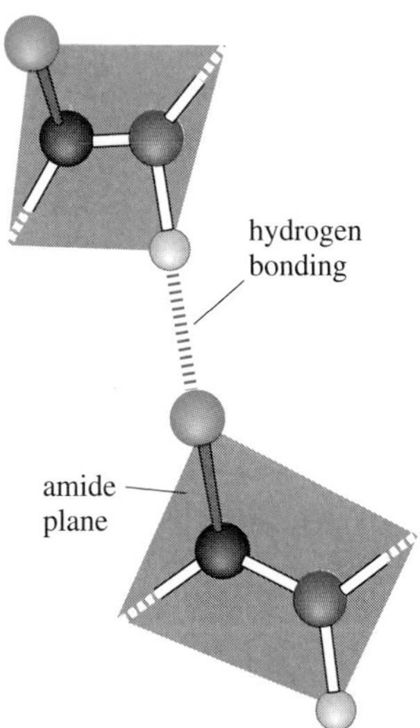

FIGURE 27.14
Hydrogen bonding between amide groups.

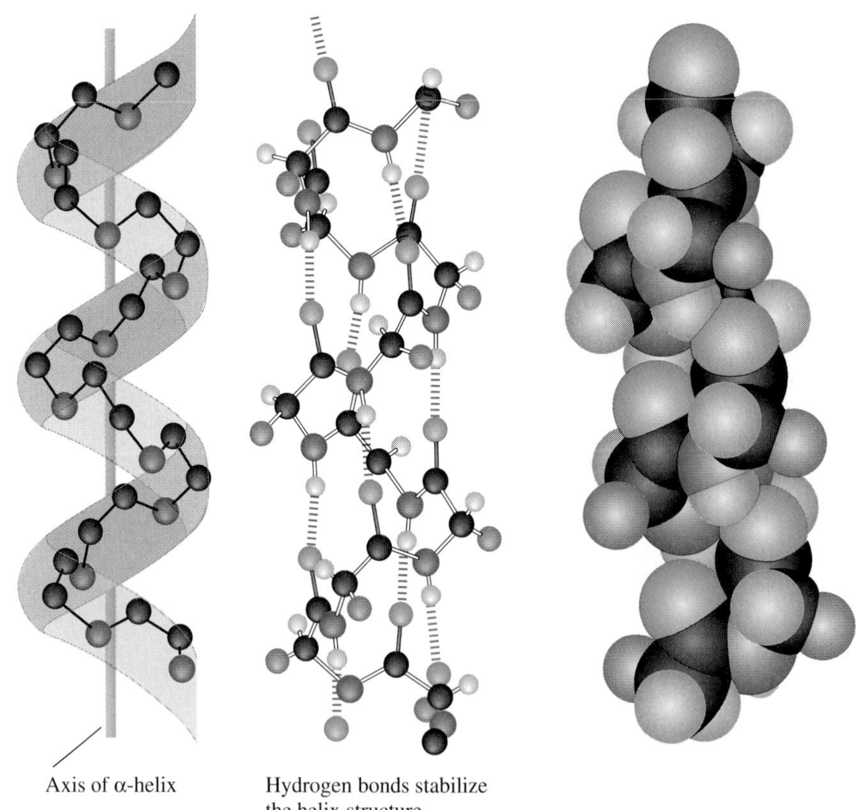

FIGURE 27.15
An α-helix. *(left)* Ball-and-stick model showing carbon-nitrogen backbone. *(middle)* Ball-and-stick model showing intrachain hydrogen bonding along the carbon-nitrogen backbone. *(right)* Space-filling model.

1. The helix is coiled in a clockwise, or right-handed, manner. **Right-handed** means that if you turn the helix clockwise, it twists away from you. In this sense, a right-handed helix is analogous to the right-handed thread of a common wood or machine screw.
2. There are 3.6 amino acids per turn of the helix.
3. Each peptide bond is s-*trans* and planar.
4. The N—H group of each peptide bond points roughly upward, parallel to the axis of the helix, and the C=O of each peptide bond points roughly downward, also parallel to the axis of the helix.
5. The carbonyl group of each peptide bond is hydrogen-bonded to the N—H group of the peptide bond four amino acid units away from it. Hydrogen bonds are shown as dotted lines.
6. All R— groups point outward from the helix.

Almost immediately after Pauling proposed the α-helix conformation, other researchers proved the presence of α-helix conformations in keratin, the protein of hair and wool. It soon became obvious that the α-helix is one of the fundamental folding patterns of polypeptide chains.

The β-Pleated Sheet

An antiparallel **β-pleated sheet** consists of adjacent polypeptide chains running in opposite (antiparallel) directions. In a parallel β-pleated sheet, the polypeptide chains run in the same direction and parallel to each other. Unlike the α-helix arrangement, N—H and C=O groups lie in the plane of the sheet and are roughly perpendicular to the long axis of the sheet. The C=O group of each peptide bond is hydrogen-bonded to the N—H group of a peptide bond of a neighboring chain (Figure 27.16). As you study the section of β-pleated sheet shown in Figure 27.16, note the following:

β-Pleated sheet A type of secondary structure in which sections of polypeptide chains are aligned parallel or antiparallel to one another.

1. The polypeptide chains lie adjacent to each other and run in opposite (antiparallel) directions.
2. Each peptide bond is planar and the alpha carbons are s-*trans* to each other.
3. The C=O and N—H groups of peptide bonds from adjacent chains point toward each other and are in the same plane so that hydrogen bonding is possible between adjacent polypeptide chains.

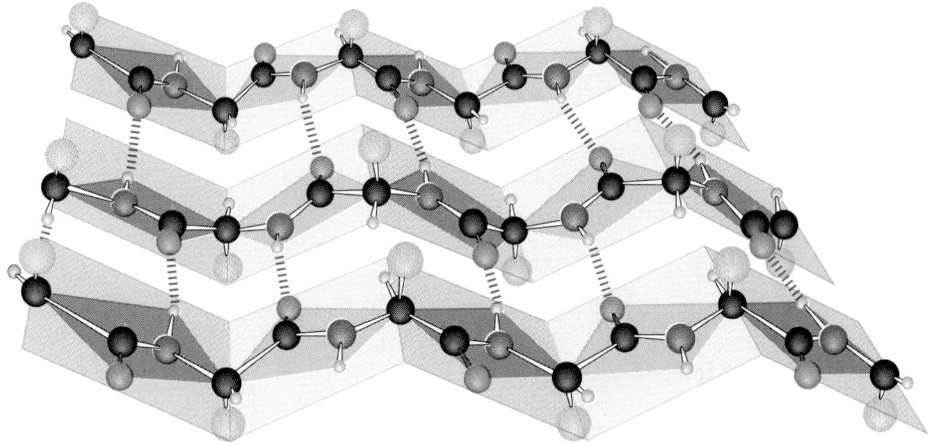

FIGURE 27.16
β-Pleated sheet conformation with three polypeptide chains running in opposite (antiparallel) directions. Hydrogen bonding between chains is indicated by broken lines.

4. The R— groups on any one chain alternate, first above, then below the plane of the sheet, and so on.

The pleated sheet conformation is stabilized by hydrogen bonding between N—H groups of one chain and C=O groups of an adjacent chain. By comparison, the α-helix is stabilized by hydrogen bonding between N—H and C=O groups within the same polypeptide chain.

C. Tertiary Structure

Tertiary structure of proteins The three-dimensional arrangement in space of all atoms in a single polypeptide chain.

Disulfide bond A covalent bond between two sulfur atoms; an —S—S— bond.

Tertiary (3°) structure refers to the overall folding pattern and arrangement in space of all atoms in a single polypeptide chain. No sharp dividing line exists between secondary and tertiary structures. Secondary structure refers to the spatial arrangement of amino acids close to one another on a polypeptide chain, whereas tertiary structure refers to the three-dimensional arrangement of all atoms of a polypeptide chain.

Disulfide bonds play an important role in maintaining tertiary structure. Disulfide bonds are formed between side chains of two cysteine units by oxidation of their thiol groups (—SH) to a disulfide bond (—S—S—) (Section 9.11C). Reduction of a disulfide bond regenerates the thiol groups.

$$\text{—NH—CH(CH}_2\text{SH)—C(=O)—NH—} + \text{—NH—CH(CH}_2\text{SH)—C(=O)—NH—} \underset{\text{reduction}}{\overset{\text{oxidation}}{\rightleftharpoons}} \text{—NH—CH(CH}_2\text{—S—S—CH}_2\text{)CH—}\ldots$$

side chains of cysteine

a disulfide bond

Figure 27.17 shows the amino acid sequence of human insulin. This protein consists of two polypeptide chains: an A chain of 21 amino acids and a B chain of 30 amino acids. The A chain is bonded to the B chain by two interchain disulfide bonds. An intrachain disulfide bond also connects the cysteine units at positions 6 and 11 of the A chain.

As an example of secondary and tertiary structure, let us look at the three-dimensional structure of myoglobin—a protein found in skeletal muscle and particularly abundant in diving mammals, such as seals, whales, and porpoises. Myoglobin and its structural relative, hemoglobin, are the oxygen transport and storage molecules of vertebrates. Hemoglobin binds molecular oxygen in the lungs and transports it to myoglobin in muscles. Myoglobin stores molecular oxygen until it is required for metabolic oxidation.

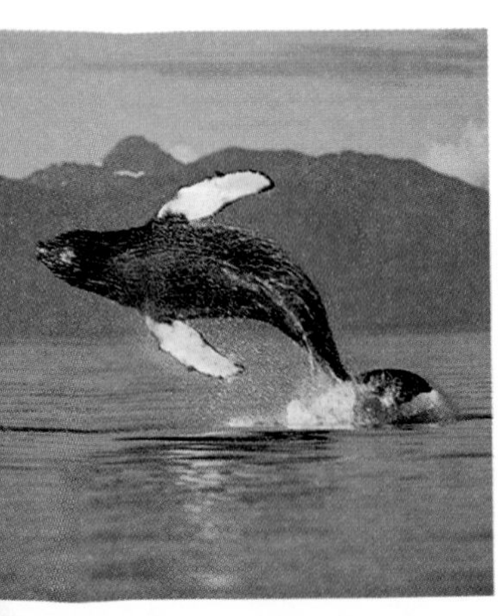

Like all mammals and many other chordates, the humpback whale relies on myoglobin as a storage form of oxygen. *(Stuart Westmoreland/Tony Stone Images)*

Myoglobin consists of a single polypeptide chain of 153 amino acids. Myoglobin also contains a single heme unit. Heme contains one Fe^{2+} ion coordinated in a square planar array with the four nitrogen atoms of a porphyrin molecule (Figure 27.18).

Determination of the three-dimensional structure of myoglobin by x-ray crystallography represented a milestone in the study of molecular architecture. For their

FIGURE 27.17
Human insulin. The A chain of 21 amino acids and B chain of 30 amino acids are connected by interchain disulfide bonds between A7 and B7 and between A20 and B19. In addition, a single intrachain disulfide bond occurs between A6 and A11.

contribution to this research, John C. Kendrew and Max F. Perutz, both of England, shared the 1962 Nobel Prize for chemistry. The secondary and tertiary structure of myoglobin are shown in Figure 27.19. The single polypeptide chain is folded into a complex, almost boxlike shape.

Important structural features of the three-dimensional shape of myoglobin include the following:

1. The backbone consists of eight relatively straight sections of α-helix, each separated by a bend in the polypeptide chain. The longest section of α-helix has 24 amino acids, the shortest has 7. Some 75% of the amino acids are found in these eight regions of α-helix.
2. Hydrophobic side chains of phenylalanine, alanine, valine, leucine, isoleucine, and methionine are clustered in the interior of the molecule where they are shielded

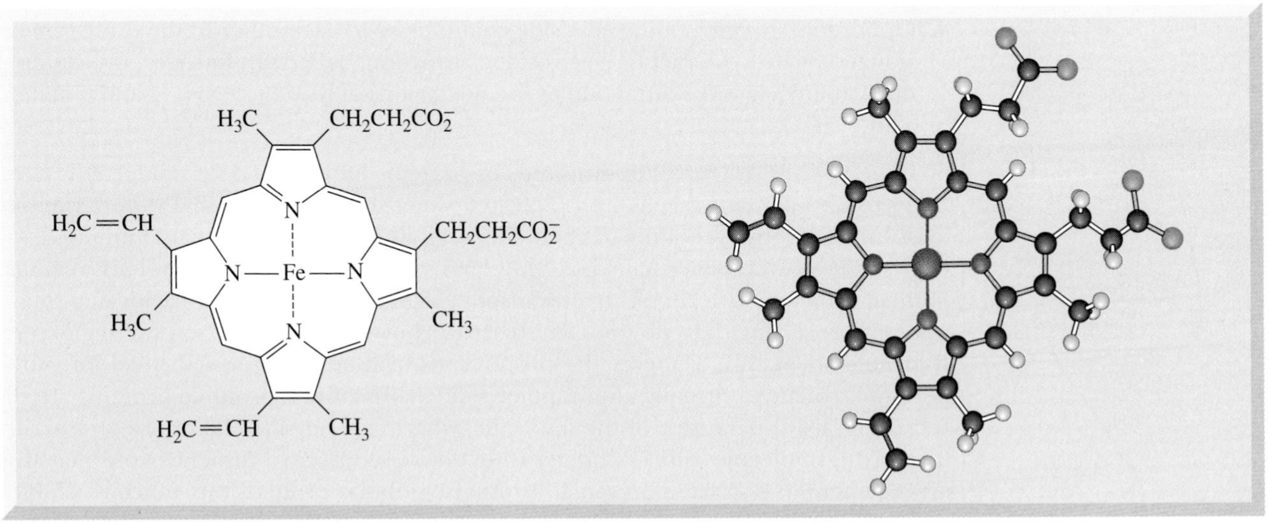

FIGURE 27.18
The structure of heme, found in myoglobin and hemoglobin.

STEREO

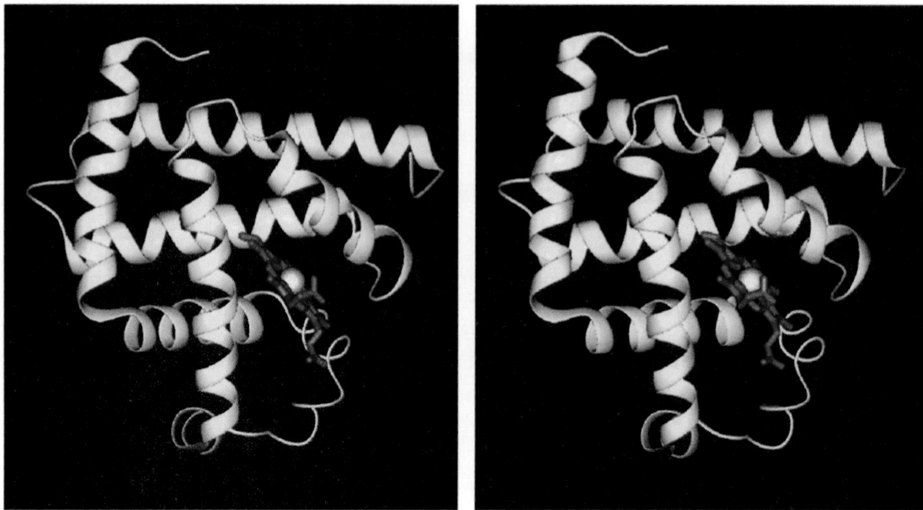

FIGURE 27.19
Ribbon model of myoglobin. The heme group is shown in color. The *N*-terminal amino acid is at the upper left and the *C*-terminal amino acid is at the upper right.

from contact with water. The hydrophobic effect is a major factor in directing the folding of the polypeptide chain of myoglobin into this compact, three-dimensional shape.

3. The outer surface of myoglobin is coated with hydrophilic side chains, such as those of lysine, arginine, serine, glutamic acid, histidine, and glutamine, which interact with the aqueous environment by hydrogen bonding. The only polar side chains that point to the interior of the myoglobin molecule are those of two histidine units, which point inward toward the heme group.
4. Oppositely charged amino acid side chains close to each other in the three-dimensional structure interact by electrostatic attractions called **salt linkages.** An example of a salt linkage is the attraction of the side chains of lysine ($—NH_3^+$) and glutamic acid ($—CO_2^-$).

The tertiary structures of hundreds of proteins have been determined. It is clear that proteins contain α-helix and β-pleated sheet structures, but that wide variations exist in the relative amounts of each. Hen egg white lysozyme with 129 amino acids in a single polypeptide chain has only 25% of its amino acids in α-helix regions. Human cytochrome c, an electron transport protein with 104 amino acids in a single polypeptide chain, has no α-helix structure but does contain several regions of β-pleated sheet. Yet, whatever the proportions of α-helix, β-pleated sheet, or other periodic structure, virtually all nonpolar side chains of water-soluble proteins are directed toward the interior of the molecule, whereas polar side chains are on the surface of the molecule and in contact with the aqueous environment. Note that this arrangement of polar and nonpolar groups in globular proteins very much resembles the arrangement of polar and nonpolar groups of soap molecules in micelles (Figure 25.3). It also resembles the arrangement of phospholipids in lipid bilayers (Figure 25.13).

EXAMPLE 27.9

With which amino acid side chains can the side chain of threonine form hydrogen bonds?

(a) Valine (b) Asparagine (c) Phenylalanine
(d) Histidine (e) Tyrosine (f) Alanine

Solution

The side chain of threonine contains a hydroxyl group that can participate in hydrogen bonding in two ways: its oxygen has a partial negative charge and can function as a hydrogen bond acceptor; its hydrogen has a partial positive charge and can function as a hydrogen bond donor. Therefore, the side chain of threonine can form hydrogen bonds with the side chains of asparagine, histidine, and tyrosine.

PROBLEM 27.9

At pH 7.4, with what amino acid side chains can the side chain of lysine form salt linkages?

D. Quaternary Structure

Most proteins of molecular weight greater than 50,000 g/mol consist of two or more noncovalently linked polypeptide chains. The arrangement of polypeptide chains in an aggregate is known as the **quaternary (4°) structure.** A good example is hemoglobin, a protein that consists of four separate polypeptide chains: two α-chains of 141 amino acids each and two β-chains of 146 amino acids each. The quaternary structure of hemoglobin is shown in Figure 27.20.

Quaternary structure The arrangement of polypeptide monomers into a noncovalently bonded aggregation.

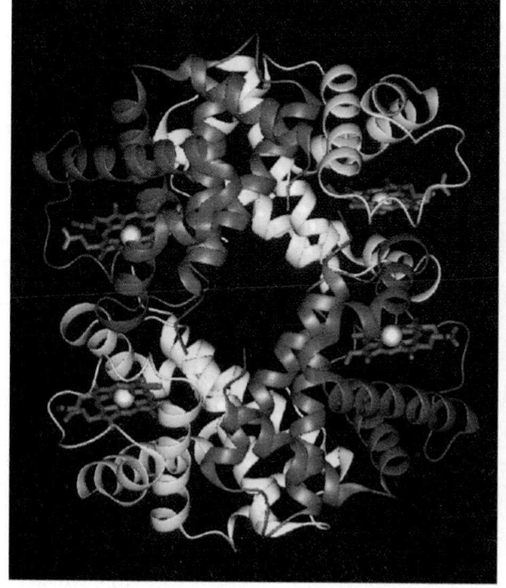

FIGURE 27.20
Ribbon model of deoxyhemoglobin. The α-chains are shown in purple, the β-chains are in yellow, the heme ligands are in red, and the Fe atoms appear as white spheres.

CHEMISTRY IN ACTION

Enzymes Through the Looking Glass

Proteins are polymers of α-amino acids, and because all α-amino acids except glycine are chiral, proteins are also chiral. This chirality results in remarkable stereoselectivity for those proteins that act as enzymes to catalyze organic reactions. For example, if a substrate is chiral, an enzyme normally processes only one of the two substrate enantiomers. An obvious question is: Does an enantiomeric enzyme (one made up of D-amino acids instead of the naturally occuring L-amino acids) catalyze reactions of the enantiomeric substrate? From everything we have learned about chirality during the last 100 years, the answer to this question should be yes. If, experimentally, the answer was found to be no, the change in organic chemistry would be as great as it would be in astronomy if the earth, not the sun, was found to be the center of the solar system.

Until recently, there was no way of obtaining an enzyme built up of D-amino acids. However, with advances in chemical peptide synthesis, it is now possible for chemists and biochemists to complete chemical syntheses of moderately sized enzymes. One such example is HIV-1-protease (a protease is an enzyme that catalyses the hydrolysis of a peptide bond), an enzyme containing 99 amino acids in a single polypeptide chain. This enzyme is essential for the life cycle of the HIV-1 virus, which causes AIDS (acquired immunodeficiency syndrome). Stephen Kent of Scripps Research Institute in La Jolla, California, has synthesized HIV-1-protease from both natural L-amino acids and unnatural D-amino acids. (For technical reasons, two slight modifications were made in each protein's sequence.) Under achiral conditions, the properties of the synthetic D-HIV-1 and L-HIV-1-proteases are identical. For example, their mass spectra and behavior during purification on achiral chromatographic supports are identical. However, the specific rotations of the two polypeptides are equal in magnitude but opposite in sign. Furthermore, an achiral inhibitor of the natural enzyme inhibits both synthetic D- and L-enzymes.

Under chiral conditions, however, the D-HIV-1 and L-HIV-1-proteases are different. Kent observed that the synthetic L-enzyme (as well as the natural L-enzyme isolated from cells infected with the HIV-1 virus) reacts with a synthetic L-polypeptide substrate to catalyze the hydrolysis of a specific peptide bond. The D-enzyme catalyzes the same hydrolysis of a mirror image D-polypeptide substrate. Most importantly, no cross reactivity occurs between the L-enzyme and the D-substrate, or between the D-enzyme and the L-substrate.

Fortunately for organic chemistry, our ideas on stereochemistry are fully confirmed. The mirror-image enzyme reacts only with the mirror-image substrate. This work has biochemical implications as well. For example, it proves that polypeptide chains can fold spontaneously into the complex shapes found in enzymes. In other words, the shape of an enzyme, at least in the case of HIV-1 protease, is encoded in its sequence.

Computer-generated representations of the polypeptide backbones of L-HIV-1-protease dimer and its mirror image, the D-HIV-1-protease dimer, follow. Regions of β-sheets and α-helices are shown above as ribbons. Regions of less regular secondary structure are shown as thin lines.

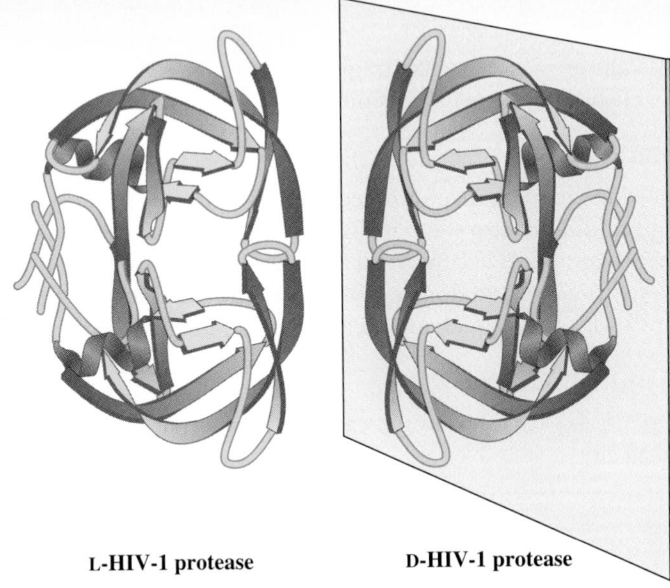

See R. C. deL. Milton, S. C. F. Milton, S. B. H. Kent, *Science,* **256,** 1445, 1992.

TABLE 27.4 Quaternary Structure of Selected Proteins

Protein	Number of Subunits
alcohol dehydrogenase	2
aldolase	4
hemoglobin	4
lactate dehydrogenase	4
insulin	6
glutamine synthetase	12
tobacco mosaic virus protein disc	17

The major factor stabilizing the aggregation of protein subunits is the **hydrophobic effect.** When individual polypeptides fold into compact three-dimensional shapes to expose polar side chains to the aqueous environment and shield nonpolar side chains from water, hydrophobic "patches" may still appear on the surface, in contact with water. These patches can be shielded from water if two or more polypeptides assemble so that their hydrophobic patches are in contact. The number of subunits of several proteins of known quaternary structure are shown in Table 27.4.

Hydrophobic effect The tendency of nonpolar groups to cluster in such a way as to be shielded from contact with an aqueous environment.

SUMMARY

Amino acids are compounds that contain both an amino group and a carboxyl group (Section 27.1A). A **zwitterion** is an internal salt of an amino acid. With the exception of glycine, all protein-derived amino acids are chiral and show enantiomerism (Section 27.1B). In the D,L convention, all are L-amino acids. In the *R,S* convention, 18 are *S*-amino acids. Although cysteine has the same absolute configuration, it is an *R*-amino acid because of the manner in which priorities are assigned about the tetrahedral stereocenter. Isoleucine and threonine contain a second stereocenter. The 20 protein-derived amino acids are commonly divided into four categories (Section 27.1C): nine with nonpolar side chains, four with polar but un-ionized side chains, four with acidic side chains, and three with basic side chains. Acidic side chains are those of aspartic acid (pK_a 3.86), glutamic acid (pK_a 4.07), tyrosine (pK_a 10.07), and cysteine (pK_a 8.00). Basic side chains are those of arginine (pK_a 12.48), histidine (pK_a 6.10), and lysine (pK_a 10.53).

The **Henderson-Hasselbalch equation** (Section 27.2B) provides a direct way of calculating the ratio of conjugate base to its weak acid as a function of pH.

$$pH = pK_a + \log \frac{[\text{conjugate base}]}{[\text{weak acid}]}$$

The **isoelectric point, pI,** of an amino acid, polypeptide, or protein is the pH at which it has no net charge (Section 27.2D). **Electrophoresis** is the process of separating compounds on the basis of their electric charge (Section 27.2E). Compounds having a high charge density move more rapidly than those with a lower charge density. Any amino acid or protein in a solution with a pH that equals the pI of the compound remains at the origin.

A **peptide bond** is the special name given to the amide bond formed between α-amino acids (Section 27.3). A **polypeptide** is a biological macromolecule containing many amino acids, each joined to the next by a peptide bond. By convention, the sequence of amino acids in a polypeptide is written beginning with the ***N*-terminal amino acid** toward the ***C*-terminal amino acid. Primary (1°) structure** of a polypeptide is the sequence of amino acids in the polypeptide chain (Section 27.4).

In **solid-phase synthesis** (Section 27.5F), or polymer-supported synthesis of polypeptides, the *C*-terminal amino acid is joined to a chloromethylated polystyrene resin as a benzyl ester. The polypeptide chain is then extended one amino acid at a time from the *N*-terminal end. When synthesis is completed, the polypeptide chain is released from the solid support by cleavage of the benzyl ester.

A **peptide bond is planar** (Section 27.6A), that is, the four atoms of the amide and the two α-carbons of a peptide bond lie in the same plane. Bond angles about the amide nitrogen and the amide carbonyl carbon are 120°. **Secondary (2°) structure** (Section 27.6B) refers to the ordered arrangement (conformations) of amino acids in localized regions of a polypeptide or protein. Two types of secondary structure are the α-helix and the β-pleated sheet. **Tertiary (3°) structure** (Section 27.6C) refers to the overall folding pattern and arrangement in space of all atoms in a single polypeptide chain. **Quaternary (4°) structure** (Section 27.6D) is the arrangement of polypeptide monomers into a noncovalently bonded aggregate.

KEY REACTIONS

1. Acidity of an α-Carboxyl Group (Section 27.2A)

An α-CO_2H (pK_a approximately 2.19) of a protonated amino acid is a considerably stronger acid than acetic acid (pK_a 4.76) or other low-molecular-weight aliphatic carboxylic acids due to the electron-withdrawing inductive effect of the α-NH_3^+ group.

$$\underset{NH_3^+}{RCHCO_2H} + H_2O \rightleftharpoons \underset{NH_3^+}{RCHCO_2^-} + H_3O^+ \quad pK_a = 2.19$$

2. Acidity of an α-Ammonium Group (Section 27.2A)

An α-NH_3^+ group (pK_a approximately 9.47) of a protonated amino acid is a slightly stronger acid than a primary alkylammonium ion (pK_a 10.76).

$$\underset{NH_3^+}{RCHCO_2^-} + H_2O \rightleftharpoons \underset{NH_2}{RCHCO_2^-} + H_3O^+ \quad pK_a = 9.47$$

3. Reaction of an α-Amino Acid with Ninhydrin (Section 27.2E)

Treatment of an α-amino acid with ninhydrin gives a purple-colored anion with an absorption maximum at 580 nm. Treatment of proline with ninhydrin gives an orange colored solution.

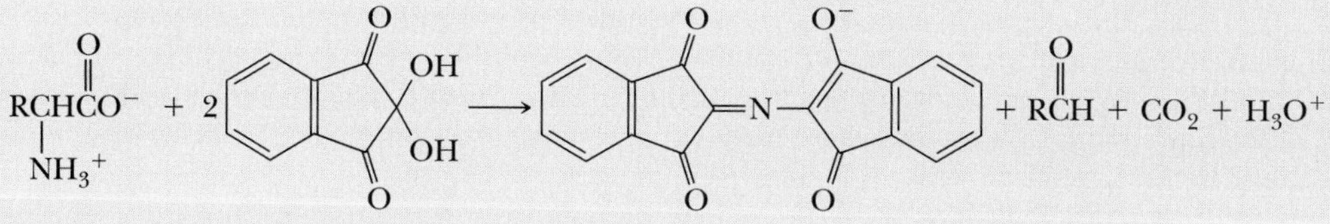

$$\underset{NH_3^+}{RCHCO^-}\ (\text{C=O}) + 2\ \text{Ninhydrin} \longrightarrow \text{Purple-colored anion} + RCH(=O) + CO_2 + H_3O^+$$

An α-amino acid — Ninhydrin — Purple-colored anion

4. Cleavage of a Peptide Bond by Cyanogen Bromide (Section 27.4B)

Cleavage is regioselective for a peptide bond formed by the carboxyl group of methionine.

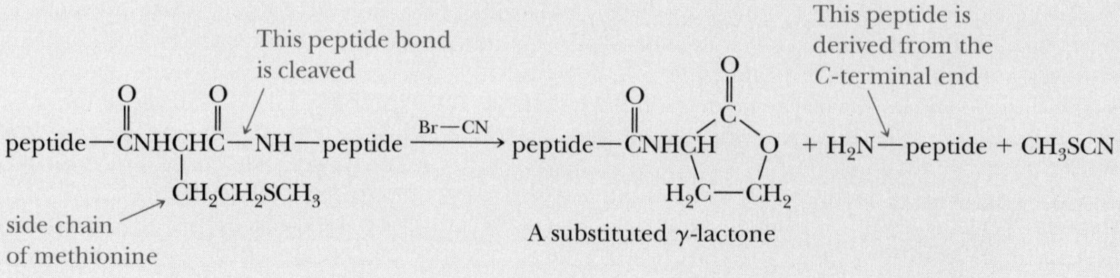

$$\text{peptide}-\overset{O}{\overset{\|}{C}}NH\underset{CH_2CH_2SCH_3}{CH}\overset{O}{\overset{\|}{C}}-NH-\text{peptide} \xrightarrow{Br-CN} \text{peptide}-\overset{O}{\overset{\|}{C}}NHCH\ (\gamma\text{-lactone}) + H_2N-\text{peptide} + CH_3SCN$$

This peptide bond is cleaved

side chain of methionine

A substituted γ-lactone

This peptide is derived from the C-terminal end

5. Edman Degradation (Section 27.4B)

Treatment with phenyl isothiocyanate followed by acid removes the *N*-terminal amino acid as a substituted phenylthiohydantoin, which is then separated and identified. It is possible to sequence as many as 20 to 30 amino acids in a polypeptide chain by repetitive Edman degradation.

$$H_2N\underset{R}{\underset{|}{C}}H\overset{O}{\overset{\|}{C}}NH{-}peptide + Ph{-}N{=}C{=}S \longrightarrow \text{A phenylthiohydantoin} + H_2N{-}peptide$$

Phenyl isothiocyanate — A phenylthiohydantoin (ring: HN–CH(R)–C(=O)–N(Ph)–C(=S)–)

This peptide is derived from the *N*-terminal end

6. The Benzyloxycarbonyl (Z—) Protecting Group (Section 27.5C)

Prepared by treatment of an unprotected α-NH_2 group with benzyloxycarbonyl chloride. Removed by treatment with HBr in acetic acid or by hydrogenolysis.

$$PhCH_2O\overset{O}{\overset{\|}{C}}Cl + H_3\overset{+}{N}\underset{CH_3}{\underset{|}{C}}H\overset{O}{\overset{\|}{C}}O^- \xrightarrow[\text{2. HCl, } H_2O]{\text{1. NaOH}} PhCH_2O\overset{O}{\overset{\|}{C}}NH\underset{CH_3}{\underset{|}{C}}H\overset{O}{\overset{\|}{C}}OH$$

Benzyloxycarbonyl chloride (Z—Cl) — Alanine — *N*-Benzyloxycarbonylalanine (Z—ala)

7. Peptide Bond Formation Using 1,3-Dicyclohexylcarbodiimide (Section 27.5E)

This substituted carbodiimide is a dehydrating agent and is converted to a disubstituted urea. The reaction is efficient and yields are generally very high.

$$Z{-}NH\underset{R_1}{\underset{|}{C}}H\overset{O}{\overset{\|}{C}}OH + H_2N\underset{R_2}{\underset{|}{C}}H\overset{O}{\overset{\|}{C}}OCH_3 + C_6H_{11}{-}N{=}C{=}N{-}C_6H_{11} \xrightarrow{CHCl_3}$$

Amino protected aa_1 — Carboxyl protected aa_2 — 1,3-Dicyclohexylcarbodiimide (DCC)

$$Z{-}NH\underset{R_1}{\underset{|}{C}}H\overset{O}{\overset{\|}{C}}{-}NH\underset{R_2}{\underset{|}{C}}H\overset{O}{\overset{\|}{C}}{-}OCH_3 + C_6H_{11}{-}NH{-}\overset{O}{\overset{\|}{C}}{-}NH{-}C_6H_{11}$$

Amino and carboxyl protected dipeptide — *N,N'*-dicyclohexylurea (DCU)

ADDITIONAL PROBLEMS

Amino Acids

27.10 What amino acids do these abbreviations stand for?

(a) Phe **(b)** Ser **(c)** Asp **(d)** Gln **(e)** His **(f)** Gly **(g)** Tyr

27.11 Why are Glu and Asp often referred to as acidic amino acids?

27.12 Why is Arg often referred to as a basic amino acid? Which two other amino acids are also basic amino acids?

27.13 Referring to Tables 27.1 and Table 27.2, identify the

(a) One achiral amino acid.
(b) Two amino acids that have diastereomers.
(c) Two sulfur-containing amino acids.
(d) Four amino acids with aromatic side chains.
(e) Amino acid with the most basic side chain.
(f) Amino acid with the most acidic side chain.

27.14 As discussed in the Chemistry in Action box "Vitamin K, Blood Clotting, and Basicity" (Chapter 25), vitamin K participates in carboxylation of glutamic acid residues of the blood-clotting protein prothrombin.

(a) Write a structural formula for γ-carboxyglutamic acid.
(b) Account for the fact that the presence of γ-carboxyglutamic acid escaped detection for many years; on routine amino acid analyses, only glutamic acid was detected.

27.15 Isoleucine has two tetrahedral stereocenters, and four stereoisomers are possible. The protein-derived stereoisomer, L-isoleucine, is named (2*S*,3*S*)-(+)-2-amino-3-methylpentanoic acid.

(a) What is the meaning of the designation (+) in this name?
(b) Draw a stereorepresentation showing the configuration of each stereocenter in L-isoleucine.

27.16 The amino acid threonine has two stereocenters. The stereoisomer found in proteins has the configuration 2*S*,3*R*. Draw a Fischer projection of this stereoisomer and also a three-dimensional representation using solid, wedged, and dashed lines.

27.17 Histamine is biosynthesized from one of the 20 protein-derived amino acids. Suggest which amino acid is its biochemical precursor, and the type of organic reaction(s) involved in its biosynthesis (e.g., oxidation, reduction, decarboxylation, nucleophilic substitution).

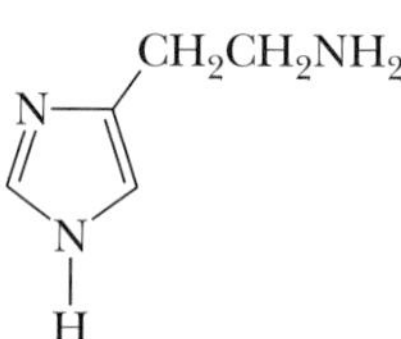

Histamine

27.18 Both norepinephrine and epinephrine are biosynthesized from the same protein-derived amino acid. From which amino acid are they synthesized and what types of reactions are involved in their biosynthesis?

(a) Norepinephrine

(b) Epinephrine (Adrenaline)

27.19 From which amino acid are serotonin and melatonin biosynthesized and what types of reactions are involved in their biosynthesis?

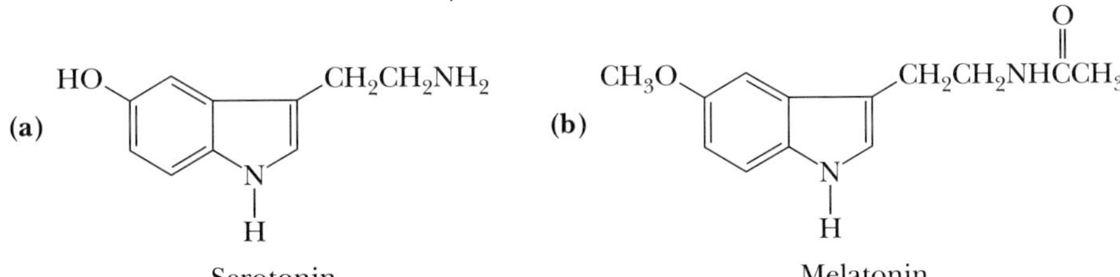

Serotonin

Melatonin

27.20 Following are values of pK_a for *N*-acetylglycine, and for the protonated forms of glycine and glycine methyl ester.

$CH_3\overset{O}{\overset{\parallel}{C}}NHCH_2CO_2H$	$H_3\overset{+}{N}CH_2CO_2H$	$H_3\overset{+}{N}CH_2CO_2CH_3$
N-Acetylglycine	Glycine	Glycine methyl ester
pK_a 3.70	pK_1 2.35, pK_2 9.78	pK_a 7.80

(a) Which is the stronger acid, the carboxyl group of *N*-acetylglycine or the carboxyl group of protonated glycine? How do you account for this difference in acidity?

(b) Which is the stronger acid, the ammonium group of protonated glycine or the ammonium group of protonated glycine methyl ester? How do you account for this difference in acidity?

27.21 For lysine and arginine, the isoelectric point (pI) occurs at a pH where the net charge on the nitrogen-containing groups is +1 and balances the charge of −1 on the α-carboxyl group. Calculate pI for these amino acids.

27.22 For aspartic and glutamic acids, the isoelectric point occurs at a pH where the net charge on the two carboxyl groups is −1 and balances the charge of +1 on the α-amino group. Calculate pI for these amino acids.

27.23 Draw the structural formula for the form of each amino acid most prevalent at pH 1.0.

(a) Threonine **(b)** Arginine **(c)** Methionine **(d)** Tyrosine

27.24 Draw the structural formula for the form of each amino most prevalent at pH 10.0.

(a) Leucine **(b)** Valine **(c)** Proline **(d)** Aspartic acid

27.25 At pH 7.4, the pH of blood plasma, do the majority of protein-derived amino acids bear a net negative charge or a net positive charge?

27.26 Write the zwitterion form of alanine and show its reaction with:

(a) 1 mol NaOH **(b)** 1 mol HCl

27.27 Write the form of lysine most prevalent at pH 1.0 and then show its reaction with the following. Consult Table 27.2 for pK_a values of the ionizable groups in lysine.

(a) 1 mol NaOH **(b)** 2 mol NaOH **(c)** 3 mol NaOH

27.28 Write the form of aspartic acid most prevalent at pH 1.0 and then show its reaction with the following. Consult Table 27.2 for pK_a values of the ionizable groups in aspartic acid.

(a) 1 mol NaOH **(b)** 2 mol NaOH **(c)** 3 mol NaOH

27.29 Account for the fact that the isoelectric point of glutamine (pI 5.65) is higher than the isoelectric point of glutamic acid (pI 3.08).

27.30 Enzyme-catalyzed decarboxylation of glutamic acid gives 4-aminobutanoic acid (Section 27.1D). Estimate the pI of 4-aminobutanoic acid.

27.31 Given pK_a values for ionizable groups in Table 27.2, sketch curves for the titration of (a) glutamic acid with NaOH, and (b) histidine with NaOH.

27.32 Guanidine and the guanidino group present in arginine are two of the strongest organic bases known. Account for this basicity.

27.33 A chemically modified guanidino group is present in cimetidine (Tagamet), a widely prescribed drug for the control of gastric acidity and peptic ulcers. Cimetidine reduces gastric acid secretion by inhibiting the interaction of histamine with gastric H2 receptors. In the development of this drug, a cyano group was added to the substituted guanidino group to significantly alter its basicity. Do you expect this modified guanidino group to be more basic or less basic than the guanidino group of arginine? Explain.

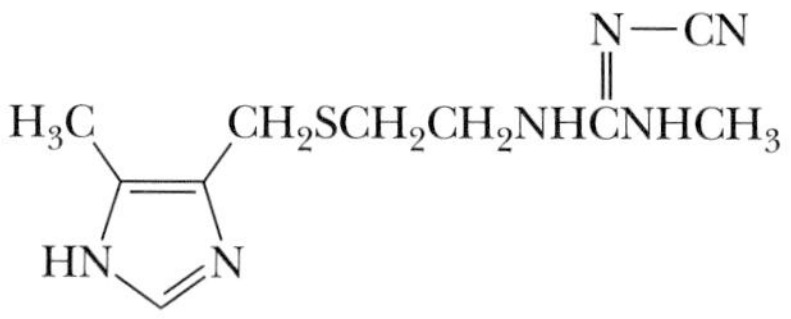

N—CN

H$_3$C CH$_2$SCH$_2$CH$_2$NHCNHCH$_3$

HN N

Cimetidine
(Tagamet)

27.34 Only three amino acids have appreciable absorption in the ultraviolet spectrum. Which three amino acids contribute to the commonly quoted λ_{max} of 280 nm for proteins?

27.35 Draw a structural formula for the product formed when alanine is treated with the following reagents.

(a) Aqueous NaOH **(b)** Aqueous HCl
(c) CH_3CH_2OH, H_2SO_4 **(d)** $(CH_3CO)_2O$, CH_3CO_2Na

(e) 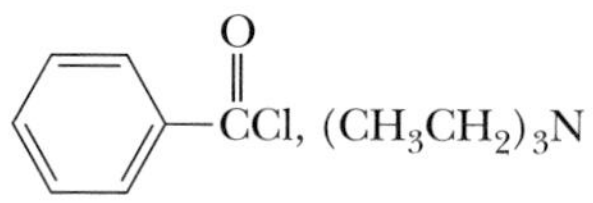 C_6H_5—CCl (O), $(CH_3CH_2)_3N$ **(f)** 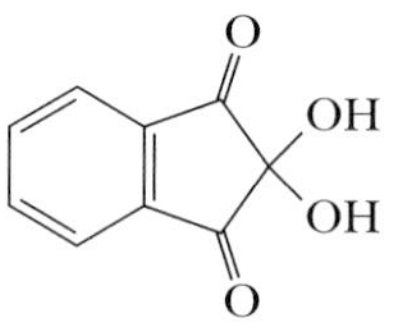 (ninhydrin: O, OH, OH, O)

(g) 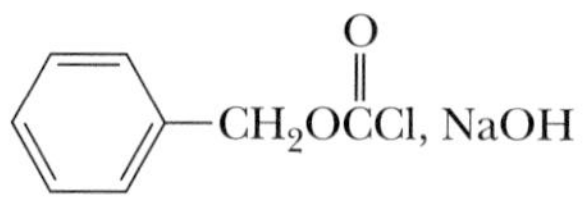 C_6H_5—CH_2OCCl (O), NaOH **(h)** $(CH_3)_3COCOCOC(CH_3)_3$ (O O), NaOH

(i) Product (g) + product (c) + DCC **(j)** Product (h) + product (c) + DCC

27.36 At what pH would you carry out an electrophoresis to separate the amino acids in each mixture of amino acids?

(a) Ala, His, Lys **(b)** Glu, Gln, Asp **(c)** Lys, Leu, Tyr

27.37 Do the following molecules migrate to the cathode or to the anode on electrophoresis at the specified pH?

(a) Histidine at pH 6.8 **(b)** Lysine at pH 6.8

(c) Glutamic acid at pH 4.0 **(d)** Glutamine at pH 4.0
(e) Glu-Ile-Val at pH 6.0 **(f)** Lys-Gln-Tyr at pH 6.0

27.38 Examine the amino acid sequence of human insulin (Figure 27.17). Do you expect human insulin to have an isoelectric point nearer that of the acidic amino acids (pI 2.0–3.0), the neutral amino acids (pI 5.5–6.5), or the basic amino acids (pI 9.5–11.0)?

Primary Structure of Polypeptides and Proteins

27.39 If a protein contains four different SH groups, how many different disulfide bonds are possible if only a single disulfide bond is formed? How many different disulfides are possible if two disulfide bonds are formed?

27.40 How many different tetrapeptides can be made if:

(a) The tetrapeptide contains one unit each of Asp, Glu, Pro, and Phe?
(b) All 20 amino acids can be used, but each only once?

27.41 A decapeptide has the following amino acid composition:

Ala_2, Arg, Cys, Glu, Gly, Leu, Lys, Phe, Val

Partial hydrolysis yields the following tripeptides.

Cys-Glu-Leu + Gly-Arg-Cys + Leu-Ala-Ala + Lys-Val-Phe + Val-Phe-Gly

One round of Edman degradation yields a lysine phenylthiohydantoin. From this information, deduce the primary structure of this decapeptide.

27.42 A tetradecapeptide (14 amino acid residues) gives the following peptide fragments on partial hydrolysis. From this information, deduce the primary structure of this polypeptide. Fragments are grouped according to size.

Pentapeptide Fragments	Tetrapeptide Fragments
Phe-Val-Asn-Gln-His	Gln-His-Leu-Cys
His-Leu-Cys-Gly-Ser	His-Leu-Val-Glu
Gly-Ser-His-Leu-Val	Leu-Val-Glu-Ala

27.43 2,4-Dinitrofluorobenzene, very often known as Sanger's reagent after the English chemist Frederick Sanger who popularized its use, reacts selectively with the *N*-terminal amino group of a polypeptide chain. Sanger was awarded the 1958 Nobel Prize for chemistry for his work in determining the primary structure of bovine insulin. One of the few persons to be awarded two Nobel Prizes, he also shared the 1980 award in chemistry with American chemists, Paul Berg and Walter Gilbert, for the development of chemical and biological analyses of DNAs.

$$O_2N\text{-}C_6H_3(NO_2)\text{-}F + H_2N\underset{R_1}{CH}\overset{O}{\overset{\|}{C}}NH\underset{R_2}{CH}\overset{O}{\overset{\|}{C}}\text{-polypeptide} \longrightarrow$$

polypeptide chain in which the *N*-terminal amino acid is labeled with a 2,4-dinitrophenyl group

2,4-Dinitrofluorobenzene (*N*-Terminal end of a polypeptide chain)

Following reaction with 2,4-dinitrofluorobenzene, all amide bonds of the polypeptide chain are hydrolyzed, and the amino acid labeled with a 2,4-dinitrophenyl group is separated by either paper or column chromatography and identified.

(a) Write the structural formula for the product formed by treatment of the *N*-terminal amino group with Sanger's reagent and propose a mechanism for its formation. (*Hint:* Review nucleophilic aromatic substitution, Section 20.3B).

(b) When bovine insulin is treated with Sanger's reagent followed by hydrolysis of all peptide bonds, two labeled amino acids are detected: glycine and phenylalanine. What conclusions can be drawn from this information about the primary structure of bovine insulin?

(c) Compare and contrast the structural information that can be obtained from use of Sanger's reagent with that from use of the Edman degradation.

27.44 Write structural formulas for the products formed after one cycle of Edman degradation on the tripeptide Ser-Leu-Phe.

27.45 Following is the primary structure of glucagon, a polypeptide hormone of 29 amino acids. Glucagon is produced in the α-cells of the pancreas and helps maintain blood glucose levels in a normal concentration range.

1 5 10 15 20 25 29
His-Ser-Glu-Gly-Thr-Phe-Thr-Ser-Asp-Tyr-Ser-Lys-Tyr-Leu-Asp-Ser-Arg-Arg-Ala-Gln-Asp-Phe-Val-Gln-Trp-Leu-Met-Asn-Thr

Glucagon

Which peptide bonds are hydrolyzed when this polypeptide is treated with the following?

(a) Phenyl isothiocyanate **(b)** Chymotrypsin **(c)** Trypsin **(d)** Br—CN

27.46 Glutathione (G—SH), one of the most common tripeptides in animals, plants, and bacteria, is a scavenger of oxidizing agents. In reacting with oxidizing agents, glutathione is converted to G-S-S-G.

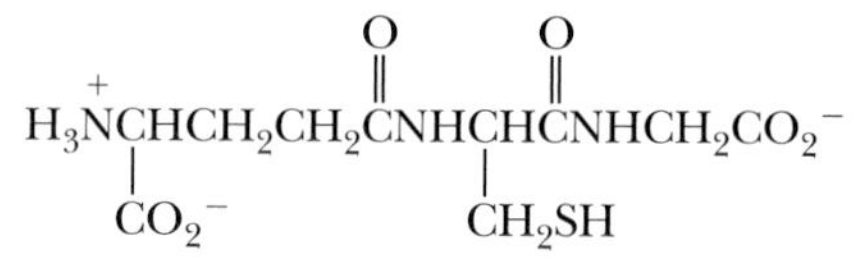

$$H_3\overset{+}{N}\underset{\underset{CO_2^-}{|}}{C}HCH_2CH_2\overset{\overset{O}{\|}}{C}NH\underset{\underset{CH_2SH}{|}}{C}H\overset{\overset{O}{\|}}{C}NHCH_2CO_2^-$$

Glutathione

(a) Name the amino acids in this tripeptide.

(b) What is unusual about the peptide bond formed by the *N*-terminal amino acid?

(c) Write a balanced half-reaction for the reaction of two molecules of glutathione to form a disulfide bond. Is glutathione a biological oxidizing agent or a biological reducing agent?

(d) Write a balanced equation for reaction of glutathione with molecular oxygen, O_2, to form G-S-S-G and H_2O. Is molecular oxygen oxidized or reduced in this process?

Synthesis of Polypeptides

27.47 In a variation of the Merrifield solid-phase peptide synthesis, the amino group is protected by a fluorenylmethoxycarbonyl (FMOC) group. This protecting group is removed by treatment with a weak base such as the secondary amine, piperidine. Write a balanced equation and propose a mechanism for this deprotection.

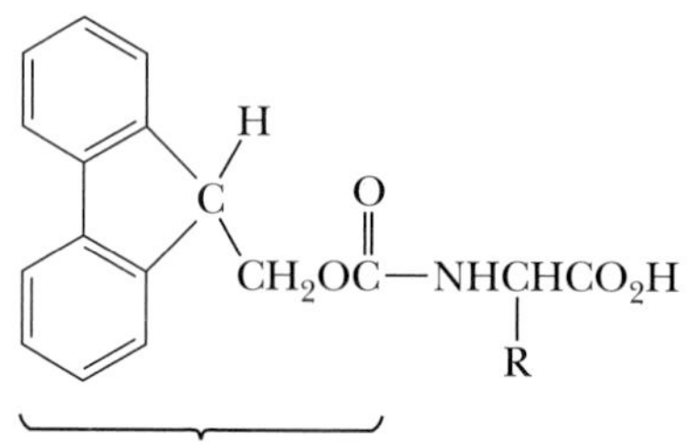

H

C

$CH_2O\overset{\overset{O}{\|}}{C}—NH\underset{\underset{R}{|}}{C}HCO_2H$

Fluorenylmethoxy-carbonyl (FMOC) group

27.48 The BOC protecting group may be added by treatment of an amino acid with di-*tert*-butyl dicarbonate as shown in the following reaction sequence. Propose a mechanism to account for formation of these products.

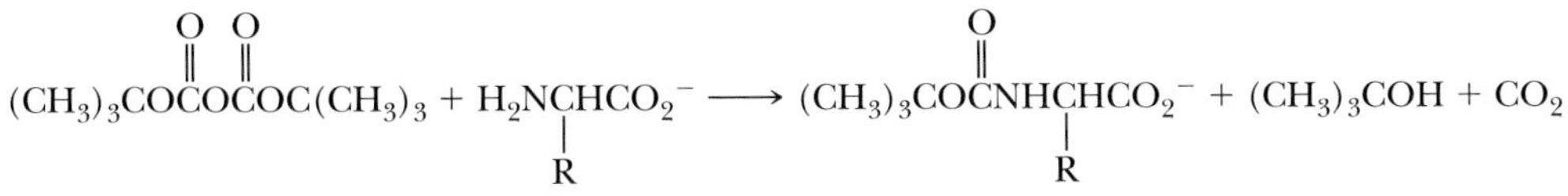

$$\underset{\text{Di-}tert\text{-butyl dicarbonate}}{(CH_3)_3CO\overset{O}{\overset{\|}{C}}O\overset{O}{\overset{\|}{C}}OC(CH_3)_3} + H_2N\underset{R}{\underset{|}{C}}HCO_2^- \longrightarrow \underset{\text{BOC-amino acid}}{(CH_3)_3CO\overset{O}{\overset{\|}{C}}NH\underset{R}{\underset{|}{C}}HCO_2^-} + (CH_3)_3COH + CO_2$$

27.49 In peptide synthesis with BOC-protecting groups, acid is used for deprotection. What is the initial fate of the *tert*-butyl group during acid deprotection? Why is a nucleophile such as anisole often added to the peptide deprotection mixture?

27.50 The side chain carboxyl groups of aspartic acid and glutamic acid are often protected as benzyl esters.

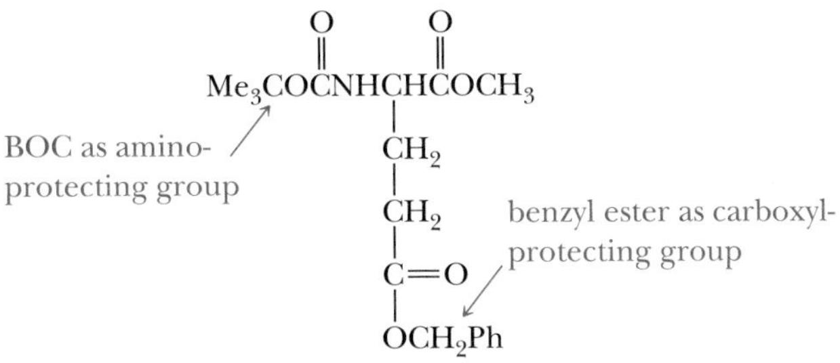

$$Me_3CO\overset{O}{\overset{\|}{C}}NH\underset{\underset{\underset{\underset{OCH_2Ph}{|}}{C=O}}{\underset{|}{CH_2}}}{\underset{|}{CH_2}}{\overset{}{C}}H\overset{O}{\overset{\|}{C}}OCH_3$$

BOC as amino-protecting group

benzyl ester as carboxyl-protecting group

(a) Show how to convert the side-chain carboxyl group to a benzyl ester using benzyl chloride as a source of the benzyl group.

(b) How do you deprotect the side-chain carboxyl under mild conditions without removing the BOC protecting group at the same time?

27.51 In solid-phase peptide synthesis, it is important that the coupling reaction give as close to a 100% yield as possible, otherwise shorter peptides (failure sequences) accumulate. Using a large excess of the BOC-protected amino acid and long reaction times maximize the coupling yield but can also lead to wasteful, expensive, and slow peptide syntheses. Suppose you remove a small amount of the Merrifield resin from the reaction vessel after a coupling reaction. How can you tell, based on the chemistry covered in this chapter, if the coupling reaction has gone to completion?

27.52 Outline a synthesis of the tripeptide Phe-Val-Ala from its constituent amino acids using the Merrifield solid-support synthesis.

27.53 Following is the structural formula of the artificial sweetener aspartame. Each amino acid has the L configuration. L-aspartyl-L-phenylalanine methyl ester has a sweet taste (it is significantly sweeter than sugar) whereas its enantiomer, D-aspartyl-D-phenylalanine methyl ester, has a bitter taste.

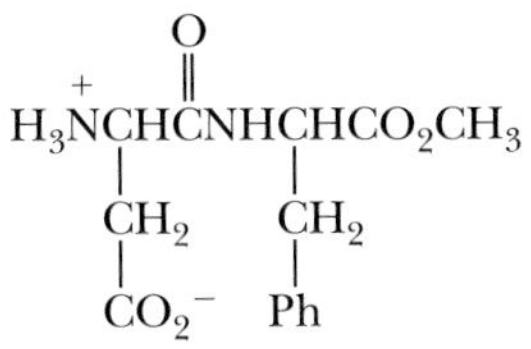

$$\overset{+}{H_3}N\underset{\underset{\underset{CO_2^-}{|}}{CH_2}}{\underset{|}{C}}H\overset{O}{\overset{\|}{C}}NH\underset{\underset{\underset{Ph}{|}}{CH_2}}{\underset{|}{C}}HCO_2CH_3$$

Aspartame

(a) Name the two amino acids in this molecule.

(b) Propose a synthesis of aspartame starting from its constituent amino acids.

Three-Dimensional Shapes of Polypeptides and Proteins

27.54 Following is a diagram of the groups that make up the backbone of a polypeptide chain:

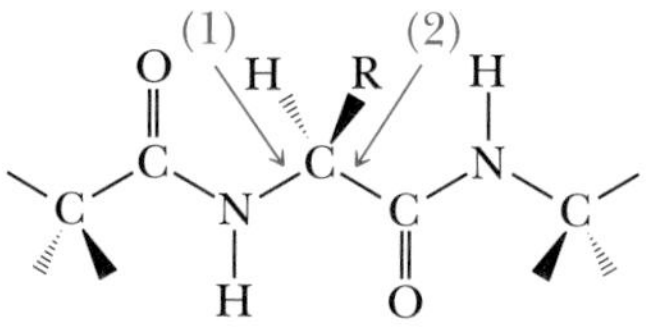

Draw a Newman projection looking down bond (1) with the nitrogen atom toward the front. Also, draw a Newman projection looking down bond (2) with the tetrahedral carbon in the front. What favorable conformations can you identify on the basis of these projections?

27.55 Examine the α-helix comformation. Are amino acid side chains arranged all inside the helix, all outside the helix, or is their arrangement random?

27.56 From the diagram in Figure 27.15, what do you predict to be the direction of the dipole moment of a polypeptide α-helix?

27.57 The term s-*cis* and s-*trans* are not as well defined for a peptide bond involving proline as they are for the peptide bonds involving other naturally occuring amino acids. Draw what you predict to be the more stable and less stable conformations of a peptide bond between proline and another amino acid.

27.58 Denaturation of a protein is a physical change, the most readily observable result of which is loss of biological activity. Denaturation stems from changes in secondary, tertiary, and quaternary structure through disruption of noncovalent interactions including hydrogen bonding and hydrophobic interactions. Two common denaturing agents are sodium dodecyl sulfate (SDS) and urea. What kinds of noncovalent interactions might each reagent disrupt?

NUCLEIC ACIDS

28

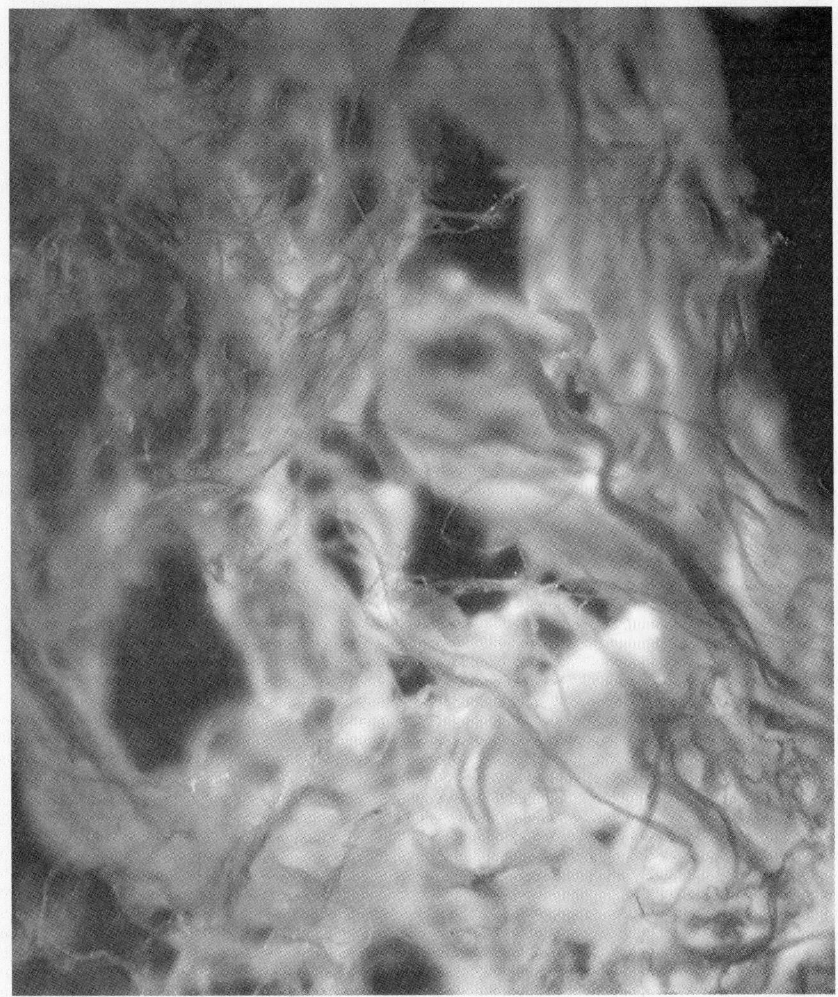

■ DNA fibers. *(Phillip A. Harrington/Fran Heyl Associates)*

The organization, maintenance, and regulation of cellular function requires a tremendous amount of information, all of which must be processed each time a cell is replicated. With very few exceptions, this information, termed **genetic information,** is stored and transmitted from one generation to the next in the form of **deoxyribonucleic acids (DNA).** Genes, the hereditary units of chromosomes, are long

CHEMISTRY IN ACTION

Retroviruses: From RNA to DNA

An important exception to the DNA-RNA-protein expression of genetic information are the retroviruses, simple organisms that store their genetic information in the form of RNA instead of DNA. Viruses are unable to reproduce themselves and must rely, instead, on the biosynthetic machinery of host cells for reproduction. Because host cells do not recognize RNA as a storage form of genetic information, the information encoded in viral RNA must first be transcribed to DNA by a process called reverse transcription. This viral information is then transcribed into forms of RNA recognized by the protein-synthesizing machinery of the host cell. The **human immunodeficiency virus (HIV)** is a retrovirus and produces the condition known as **acquired immune deficiency syndrome (AIDS).**

stretches of double-stranded DNA. If the DNA in a human chromosome in a single cell were uncoiled, it would be approximately 1.8 m in length!

Genetic information is expressed in two stages: transcription from DNA to **ribonucleic acids (RNA),** and then translation for the synthesis of proteins.

$$\text{DNA} \longrightarrow \text{RNA} \longrightarrow \text{protein}$$

Thus, DNA is the repository of genetic information in cells, while RNA serves in the transcription and translation of this information, which is then expressed through the synthesis of proteins.

In this chapter, we examine the structure of nucleosides and nucleotides and the manner in which these monomers are covalently bonded to form nucleic acids. Then, we examine the manner in which genetic information is encoded in molecules of DNA, the function of the three types of ribonucleic acids, and finally how the primary structure of a DNA molecule is determined.

28.1 Nucleosides and Nucleotides

Nucleic acid A biopolymer containing three types of monomer units: heterocyclic aromatic amine bases derived from purine and pyrimidine, the monosaccharides D-ribose or 2-deoxy-D-ribose, and phosphoric acid.

Nucleoside A building block of nucleic acids, consisting of D-ribose or 2-deoxy-D-ribose bonded to a heterocyclic aromatic amine base by a *β-N*-glycoside bond.

Controlled hydrolysis of nucleic acids yields three components: heterocyclic aromatic amine bases, the monosaccharides D-ribose or 2-deoxy-D-ribose (Section 24.1A), and phosphate ions. The five heterocyclic aromatic amine bases most common to nucleic acids are shown in Figure 28.1. Uracil, cytosine, and thymine are referred to as pyrimidine bases after the name of the parent base; adenine and guanine are referred to as purine bases.

A **nucleoside** is a compound containing D-ribose or 2-deoxy-D-ribose bonded to a heterocyclic aromatic amine base by a *β-N*-glycoside bond. The monosaccharide component of DNA is 2-deoxy-D-ribose, while that of RNA is D-ribose. The glycoside bond is between C-1′ (the anomeric carbon) of ribose or deoxyribose and N-1 of a pyrimidine base or N-9 of a purine base. Figure 28.2 shows structural formulas for two

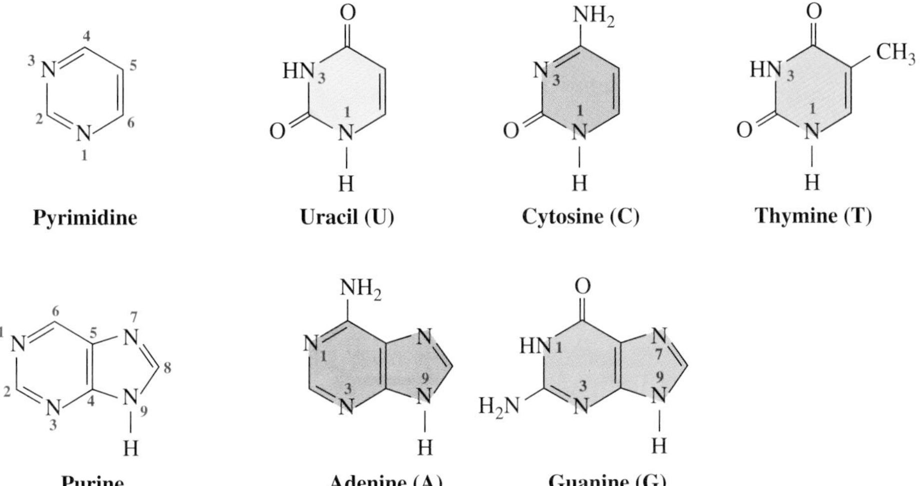

FIGURE 28.1
Names and one-letter abbreviations for the heterocyclic aromatic amine bases most common to DNA and RNA. Bases are numbered according to the patterns of the parent compounds, pyrimidine and purine.

nucleosides: the first derived from ribose and uracil, the second from 2-deoxyribose and adenine.

A **nucleotide** is a nucleoside in which a molecule of phosphoric acid is esterified with a free hydroxyl of the monosaccharide, most commonly either the 3′-hydroxyl or the 5′-hydroxyl (Figure 28.3). A nucleotide is named by giving the name of the parent nucleoside followed by the word "monophosphate." The position of the phosphoric ester is specified by the number of the carbon to which it is attached. Monophosphoric esters are diprotic acids with pK_a values of approximately 1 and 6. Therefore, at pH 7, the two hydrogens of a phosphoric monoester are fully ionized giving the nucleotide a charge of -2.

Nucleotide A nucleoside in which a molecule of phosphoric acid is esterified with an —OH of the monosaccharide, most commonly either the 3′-OH or the 5′-OH.

Nucleoside monophosphates can be further phosphorylated to form nucleoside diphosphates and nucleoside triphosphates. Shown in Figure 28.4 are structural formulas for adenosine 5′-diphosphate (ADP) and adenosine 5′-triphosphate (ATP). Nucleoside diphosphates and triphosphates are also polyprotic acids which are extensively ionized at pH 7.0. Values of pK_{a1} and pK_{a2} for adenosine diphosphate are less than 5.0, and, accordingly, these acidic groups are fully ionized at pH 7.0. The value of pK_{a3} is approximately 6.7 with the result that at pH 7.0, 67% of adenosine diphosphate is present as ADP^{3-} and 33% is present as ADP^{2-}. The pK_a values of the first

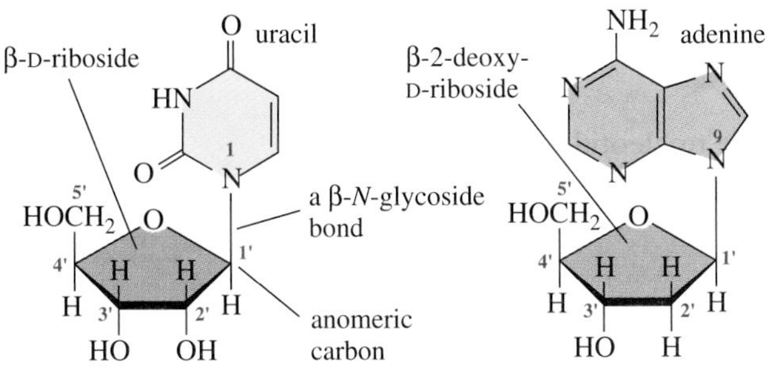

FIGURE 28.2
Structural formulas for two nucleosides. Atom numbers on the monosaccharide rings are primed to distinguish them from atom numbers on the heterocyclic bases.

PROBLEM 28.1

Draw structural formulas for these nucleotides.

(a) 2′-Deoxythymidine 5′-monophosphate
(b) 2′-Deoxythymidine 3′-monophosphate

CHEMISTRY IN ACTION

The Search for Antiviral Drugs

The search for antiviral drugs has been more difficult than the search for antibacterial drugs primarily because viral replication depends on the metabolic processes of the invaded cell. Thus, antiviral drugs are also likely to cause harm to the cells that harbor the virus. The challenge in developing antiviral drugs has been to understand the biochemistry of viruses and to develop drugs that target processes specific to them. Compared with the large number of antibacterial drugs that are available, there are only a handful of antiviral drugs, and they have nowhere near the effectiveness that antibiotics have on bacterial infections.

Acyclovir is one of the first of a new family of drugs for the treatment of infectious diseases caused by DNA viruses called herpesviruses. Herpes infections in humans are of two kinds: herpes simplex type 1, which gives rise to mouth and eye sores, and herpes simplex type 2, which gives rise to serious genital infections. Acyclovir is highly effective against herpesvirus-caused genital infections. The drug is activated *in vivo* by conversion of the primary —OH (which corresponds to the 5′-OH of a riboside or a deoxyriboside) to a triphosphate. Because of its close resemblance to deoxyguanosine triphosphate, an essential precursor for DNA synthesis, acyclovir triphosphate is taken up by viral DNA polymerase to form an enzyme-substrate complex on which no 3′-OH exists for replication to continue. Thus, the enzyme-substrate complex is no longer active (it is a dead-end complex), viral replication is disrupted, and the virus is destroyed.

Perhaps the best known of the new viral antimetabolites is azidothymidine (AZT), an analog of deoxythymidine in which the 3′-OH has been replaced by an azido group. AZT is effective against HIV-1, a retrovirus that is the causative agent of AIDS. It is converted in vivo by cellular enzymes to the 5′-triphosphate, recognized as deoxythymidine 5′-triphosphate by viral RNA-dependent DNA polymerase (reverse transcriptase), and added to a growing DNA chain. There it stops chain elongation because no 3′-OH exists on which to add the next deoxynucleotide. AZT owes its effectiveness to the fact that it binds more strongly to viral reverse transcriptase than it does to human DNA polymerase.

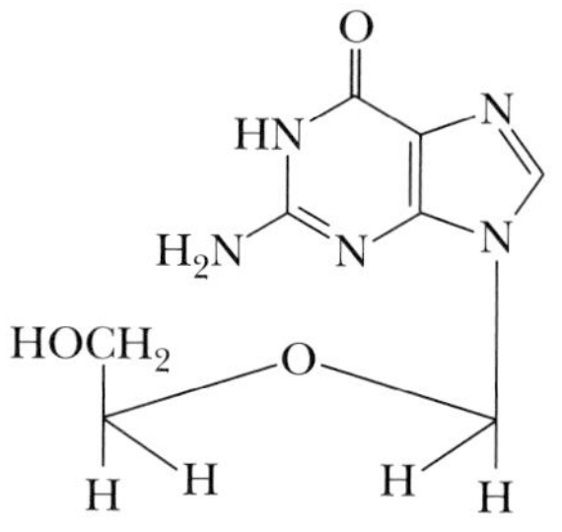

Acyclovir (drawn to show its structural relationship to 2-deoxyguanosine)

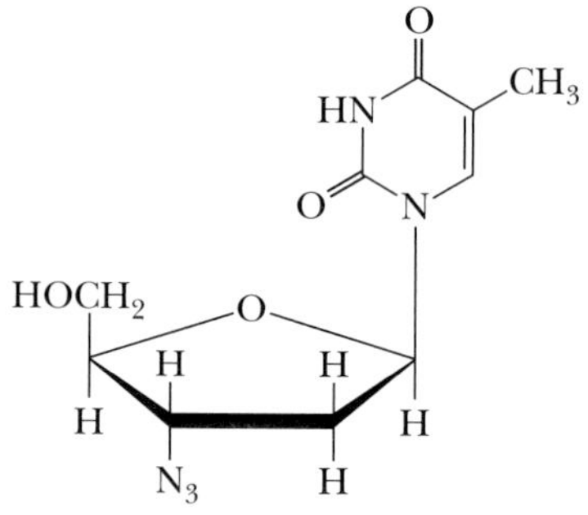

Azidothymidine (AZT)

28.2 The Structure of DNA

In Chapter 27, we saw that the four levels of structural complexity in polypeptides and proteins are primary, secondary, tertiary, and quaternary structure. There are three levels of structural complexity in nucleic acids, and although these levels of structural complexity are somewhat comparable to those in polypeptides and proteins, they also differ in significant ways.

A. Primary Structure: The Covalent Backbone

Deoxyribonucleic acids (DNA) consist of a backbone of alternating units of deoxyribose and phosphate in which the 3′-hydroxyl of one deoxyribose unit is joined by a phosphodiester bond to the 5′-hydroxyl of another deoxyribose unit (Figure 28.5). This pentose-phosphodiester backbone is constant throughout an entire DNA molecule. A heterocyclic aromatic amine base—adenine, guanine, thymine, or cytosine—is bonded to each deoxyribose unit by a *β-N*-glycoside bond. **Primary structure** of DNA refers to the order of heterocyclic bases along the pentose-phosphodiester backbone. The sequence of bases is read from the 5′ end of the chain to the 3′ end.

Primary structure of nucleic acids The sequence of bases along the pentose-phosphodiester backbone of a DNA or RNA molecule read from the 5′ end to the 3′ end.

EXAMPLE 28.2

Draw the structural formula for the DNA dinucleotide TG, which is phosphorylated at the 5′ end only.

Solution

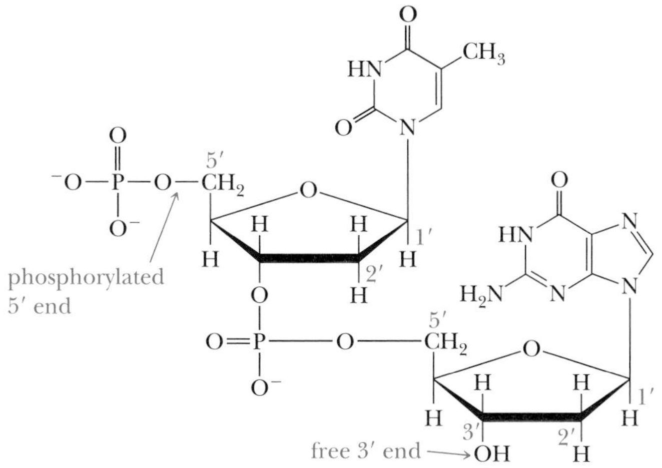

PROBLEM 28.2

Write the structural formula for the section of DNA that contains the base sequence CTG and is phosphorylated on the 3′ end only.

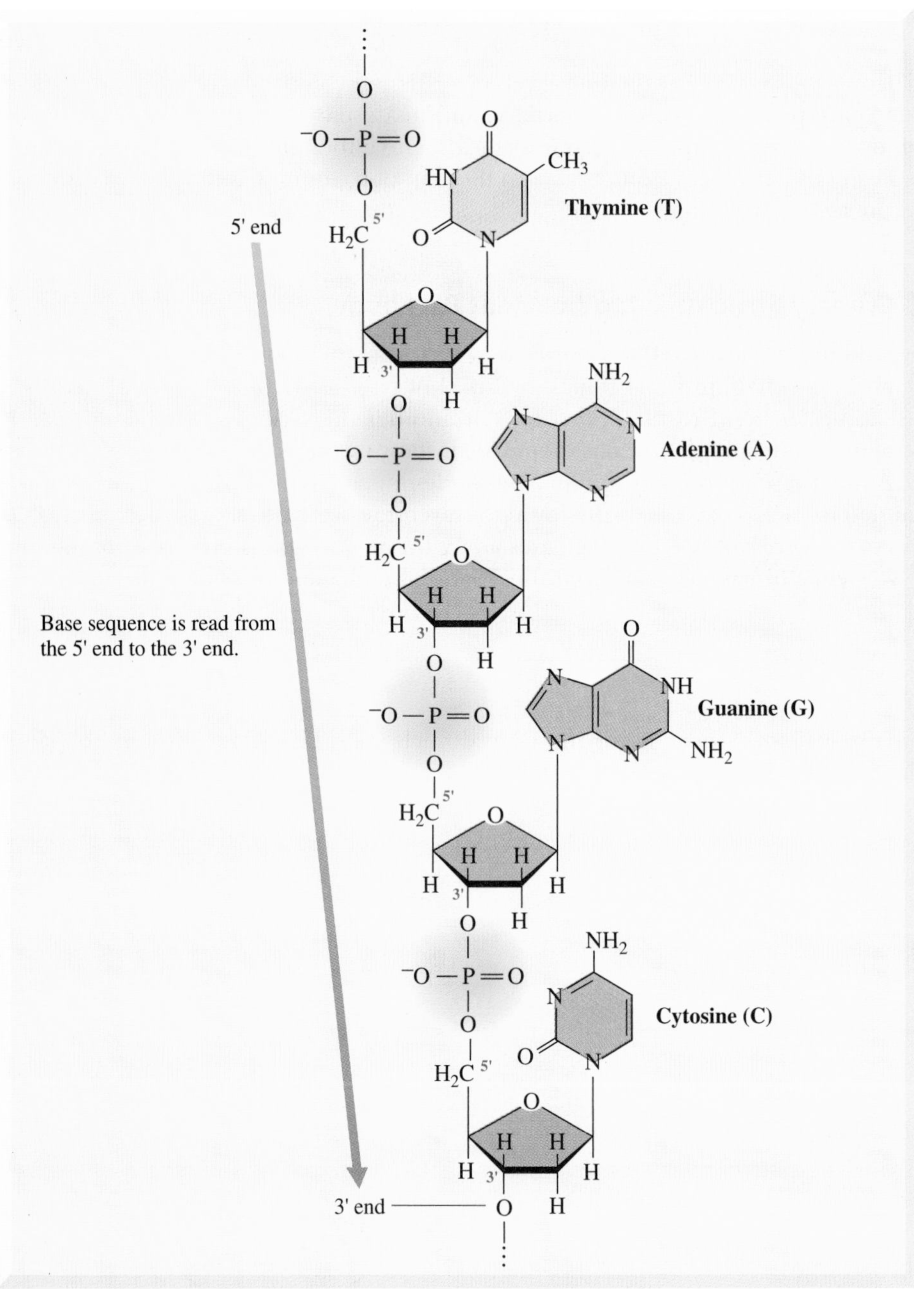

FIGURE 28.5 A tetranucleotide section of a single-stranded DNA.

B. Secondary Structure: The Double Helix

By the early 1950s, it was clear that DNA molecules consist of chains of alternating units of deoxyribose and phosphate joined by 3′,5′-phosphodiester bonds with a base attached to each deoxyribose unit by a *β-N*-glycoside bond. In 1953, the American biologist James D. Watson and the British physicist Francis H. C. Crick proposed a

Watson and Crick with their model of the DNA molecule. *(From M. M. Jones et al.,* Chemistry, Man, and Society, *4th ed. Philadelphia: Saunders College Publishing, 1983, p. 256)*

double-helix model for the **secondary structure** of DNA. Watson, Crick, and Maurice Wilkins shared the 1962 Nobel Prize for physiology and medicine for "their discoveries concerning the molecular structure of nucleic acids, and its significance for information transfer in living material."

The **Watson-Crick model** was based on molecular modeling and two lines of experimental observations: chemical analyses of DNA base compositions and mathematical analyses of x-ray diffraction patterns of crystals of DNA.

Although Rosalind Franklin also took part in this research, her name was omitted from the Nobel list because of her death in 1958 at age 37. The Nobel foundation does not make awards posthumously.

Base Composition

At one time it was thought that the four principal bases occur in the same ratios and perhaps repeat in a regular pattern along the pentose-phosphodiester backbone of DNA for all species. However, more precise determinations of base composition by Erwin Chargaff revealed that bases do not occur in the same ratios in all species (Table 28.1). Researchers drew the following conclusions from this and related data.

Secondary structure of nucleic acids The ordered arrangement of nucleic acid strands.

Watson-Crick model A double-helix model for the secondary structure of a DNA molecule.

TABLE 28.1 Base Composition, in Mole-Percent, of DNA from Several Organisms

	Purines		Pyrimidines				
Organism	**A**	**G**	**C**	**T**	**A/T**	**G/C**	**Purines/ Pyrimidines**
human	30.4	19.9	19.9	30.1	1.01	1.00	1.01
sheep	29.3	21.4	21.0	28.3	1.04	1.02	1.03
yeast	31.7	18.3	17.4	32.6	0.97	1.05	1.00
E. coli	26.0	24.9	25.2	23.9	1.09	0.99	1.04

To within experimental error:

1. The mole-percent base composition in any organism is the same in all cells of the organism and is characteristic of the organism.
2. The mole-percents of adenine (a purine base) and thymine (a pyrimidine base) are equal. The mole-percents of guanine (a purine base) and cytosine (a pyrimidine base) are also equal.
3. The mole-percents of purine bases (A + G) and pyrimidine bases (C + T) are equal.

Analyses of X-Ray Diffraction Patterns

Additional information about the structure of DNA emerged when x-ray diffraction photographs taken by Rosalind Franklin and Maurice Wilkins were analyzed. These diffraction patterns revealed that, even though the base composition of DNA isolated from different organisms varies, DNA molecules themselves are remarkably uniform in thickness. They are long and fairly straight, with an outside diameter of approximately 20 Å and not more than a dozen atoms thick. Furthermore, the crystallographic pattern repeats every 34 Å. Herein lay one of the chief problems to be solved. How could the molecular dimensions of DNA be so regular even though the relative percentages of the various bases differ considerably? With this accumulated information, the stage was set for the development of a hypothesis about DNA structure.

The heart of the Watson-Crick model is the postulate that a molecule of DNA is a complementary **double helix;** it consists of two antiparallel polynucleotide strands coiled in a right-handed manner about the same axis to form a double helix. As illustrated in the ribbon models in Figure 28.6, chirality is associated with a double helix; left-handed and right-handed double helices are related by reflection just as the members of a pair of enantiomers are related by reflection.

Double helix A type of secondary structure of DNA molecules in which two antiparallel polynucleotide strands are coiled in a right-handed manner about the same axis.

To account for the observed base ratios, Watson and Crick postulated that purine and pyrimidine bases project inward toward the axis of the helix and are always paired in a very specific manner. According to scale models, the dimensions of an adenine-thymine base pair are almost identical to the dimensions of a guanine-cytosine base pair, and the length of each pair is consistent with the core thickness of a DNA strand (Figure 28.7). Thus, if the purine base in one strand is adenine, then its complement in the antiparallel strand must be thymine. Similarly, if the purine in one strand is guanine, its complement in the antiparallel strand must be cytosine.

A significant feature of Watson and Crick's model is that no other base pairing is consistent with the observed thickness of a DNA molecule. A pair of pyrimidine bases is too small to account for the observed thickness, whereas a pair of purine bases is too large. Thus, according to the Watson-Crick model, the repeating units in a double-stranded DNA molecule are not single bases of differing dimensions, but rather base pairs of almost identical dimensions.

FIGURE 28.6
A DNA double helix has a chirality associated with it. Right-handed and left-handed double helices are nonsuperposable mirror images.

To account for the periodicity observed from x-ray data, Watson and Crick postulated that base pairs are stacked one on top of the other with a distance of 3.4 Å between base pairs and with ten base pairs in one complete turn of the helix. There is one complete turn of the helix every 34 Å. Shown in Figure 28.8 is a ribbon model of double-stranded **B-DNA,** the predominant form of DNA in dilute aqueous solution and thought to be the most common form in nature. In the double helix, the bases

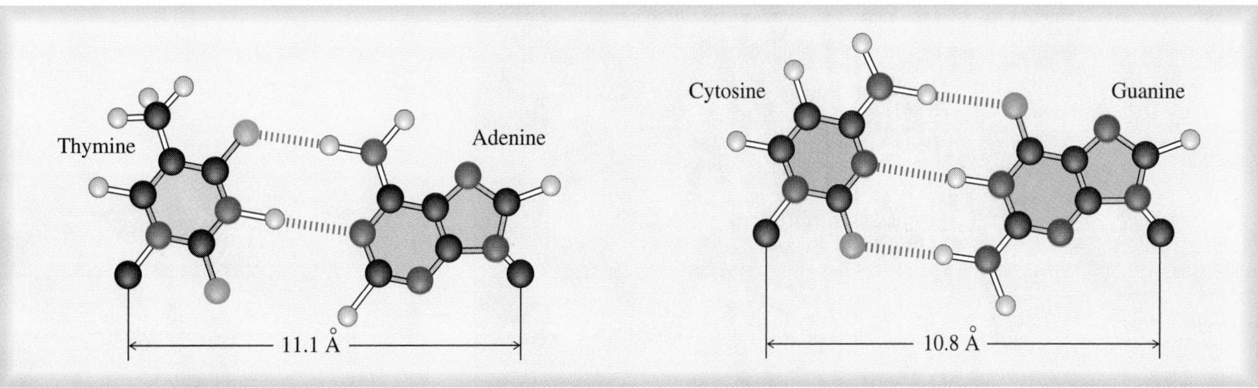

FIGURE 28.7
Base-pairing between adenine and thymine (A-T) and between guanine and cytosine (G-C). An A-T base pair is held by two hydrogen bonds, whereas a G-C base pair is held by three hydrogen bonds. In this representation, base pairs are planar and are viewed as they would be seen looking along the longitudinal axis of a double-stranded DNA molecule.

in each base pair are not directly opposite one another across the diameter of the helix, but rather slightly displaced. This displacement and the relative orientation of the glycoside bond linking each base to the sugar-phosphate backbone leads to two differently sized grooves, a major groove and a minor groove (Figure 28.8). Each groove runs along the length of the cylindrical column of the double helix. The major groove is approximately 22 Å wide; the minor groove is approximately 12 Å wide.

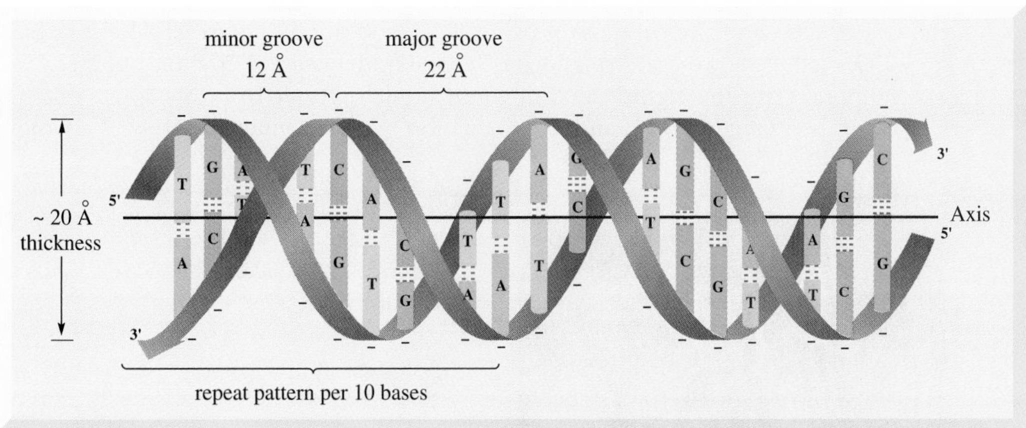

FIGURE 28.8
Ribbon model of double-stranded B-DNA. Each ribbon describes the pentose-phosphodiester backbone of a single-stranded DNA molecule. The strands are antiparallel; one strand runs to the left from the 5′ end to the 3′ end, the other runs to the right from the 5′ end to the 3′ end. Hydrogen bonds are shown by three dotted lines between each G-C base pair and two dotted lines between each A-T base pair.

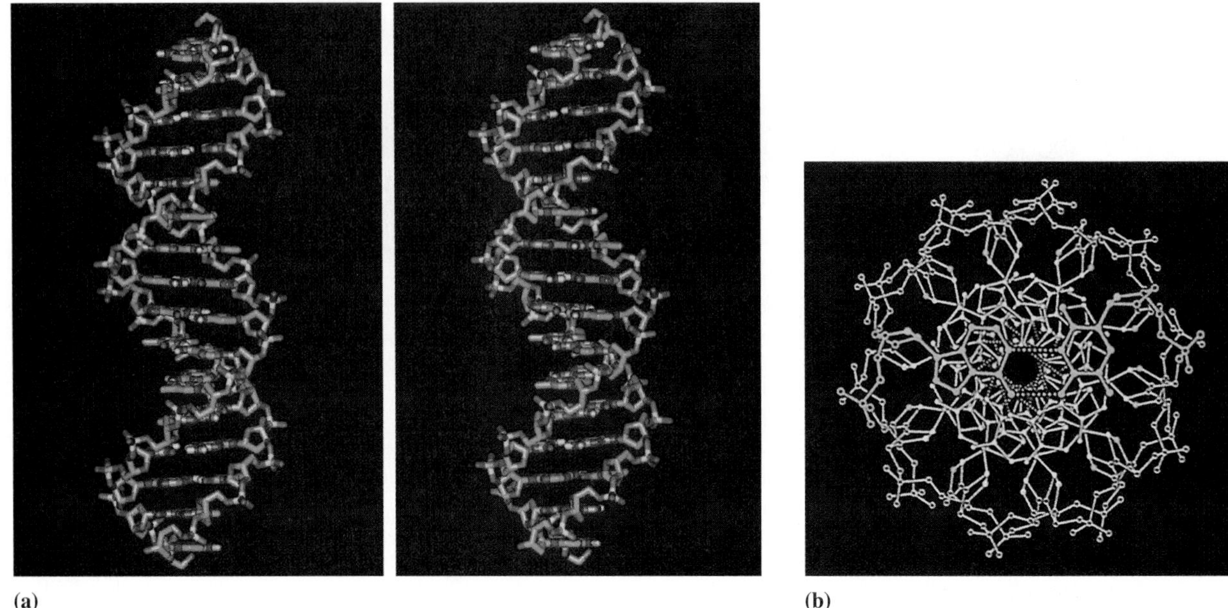

FIGURE 28.9
B-DNA. (a) A stereoview of an idealized model of B-DNA. (b) A view down the axis of the helix. The top-most purine-pyrimidine base pair is shown in green.

Figure 28.9 shows a stereoview illustrating more detail of a B-DNA double helix. The major and minor grooves are clearly recognizable in the stereoview.

Other forms of secondary structure are known that differ in the distance between stacked base pairs and in the number of base pairs per turn of the helix. One of the most common of these, **A-DNA,** also a right-handed helix, is thicker than B-DNA and has a repeat distance of only 29 Å (Figure 28.10). There are 10 base pairs per turn of the helix with a spacing of 2.9 Å between base pairs.

At the structural level, a fundamental difference between B-DNA and A-DNA is the conformation of the furanose form of 2-deoxy-D-ribose. Figure 28.11 shows ball-and-stick models of two conformations of 2′-deoxyadenosine 3′,5′-diphosphate.

To appreciate the significance of these models, recall from Section 2.6B that C—C—C bond angles in a planar conformation of cyclopentane are 108°. This angle

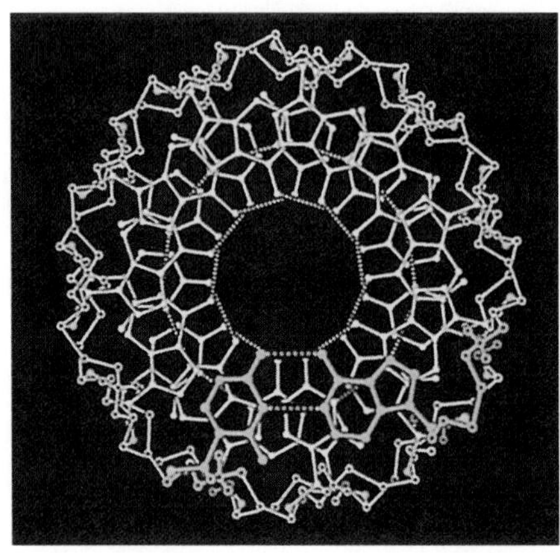

FIGURE 28.10
A-DNA. A view down the axis of the helix. The top-most purine-pyrimidine base pair is shown in green.

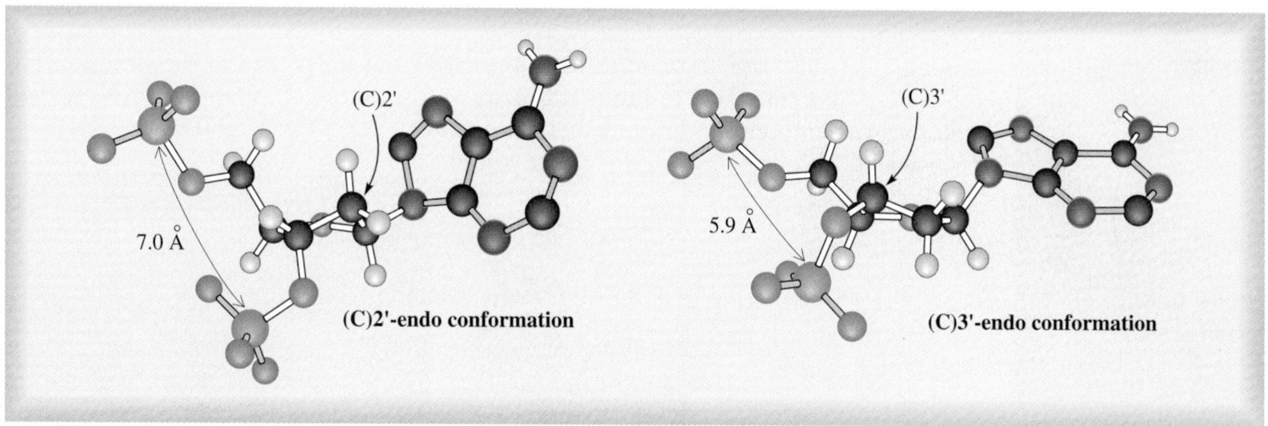

FIGURE 28.11
(C)2′-endo and (C)3′-endo conformations of the furanose ring of 2′-deoxyadenosine 3′,5′-diphosphate. In B-DNA, the phosphate esters are spaced at a distance of 7.0 Å, which leads to a spacing of 3.4 Å between stacked base pairs and 10 base pairs per 34 Å repeat. In A-DNA, the phosphate esters are spaced at a distance of 5.9 Å, which leads to a spacing of 2.9 Å between stacked base pairs, and 10 base pairs per 29 Å repeat.

differs only slightly from the tetrahedral angle of 109.5°, and, hence, there is only negligible angle strain in a planar conformation of cyclopentane. There are, however, ten pairs of fully eclipsed hydrogen interactions. The observed C—C—C bond angles in cyclopentane are approximately 105°, indicating that, in its most stable conformation, the ring is slightly puckered. In forming a puckered or envelope conformation (there are five of them depending on which carbon atom is puckered out of the plane of the other four), C—C—C bond angles are reduced (increasing potential energy), but at the same time eclipsed hydrogen interactions are also reduced (decreasing potential energy).

Although in principle a number of puckered conformations are possible for the furanose ring of a nucleoside, only two are common: the (C)2′-endo conformation and the (C)3′-endo conformation. The designations (C)2′ and (C)3′ indicate which atom of the ring is puckered relative to the plane created by the other atoms of the ring. The designation endo indicates puckering upward, on the same side as the heterocylic aromatic amine base. The alternative to endo is exo, meaning downward, on the side opposite the amine base.

B-DNA exists in the (C)2′-endo conformation, whereas A-DNA exists in the (C)3′-endo conformation. Although the differences in spatial orientation of ring atoms may seem small, they produce significant differences in the spatial orientation of atoms attached to the ring. In particular, compare the relative orientations in space and interatomic distances between the (C)3′-phosphoric ester and (C)5′-phosphoric ester on the furanose ring. The phosphoric esters are significantly closer in the (C)3′-endo conformation than they are in the (C)2′-endo conformation, a difference that is reflected in the shorter repeat distance and reduced stacking distance in the A-DNA helix compared with the B-DNA helix. Both B-DNA and A-DNA are right-handed helices. Another form of DNA, named Z-DNA, has been discovered that has a left-handed double helix.

EXAMPLE 28.3

One strand of a DNA molecule has the base sequence 5′-ACTTGCCA-3′. Write the DNA base sequence for the complementary strand.

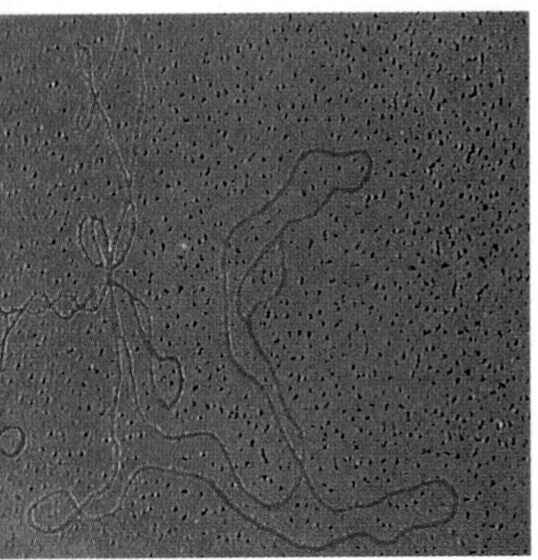

Supercoiled DNA from a mitochondrion. *(Fran Heyl Associates)*

Solution

Remember that base sequence is always written from the 5′ end of the strand to the 3′ end, that A pairs with T, and that C pairs with G. In double-stranded DNA, the two strands run in opposite (antiparallel) directions, so that the 5′ end of one strand is associated with the 3′ end of the other strand.

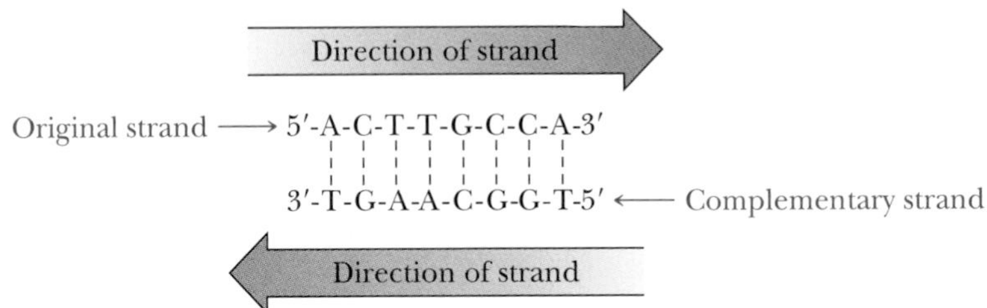

Written from the 5′ end, the complementary strand is 5′-TGGCAAGT-3′.

PROBLEM 28.3

Write the complementary DNA base sequence for 5′-CCGTACGA-3′.

C. Tertiary Structure: Supercoiled DNA

Tertiary structure of nucleic acids The three-dimensional arrangement of all atoms of a nucleic acid, commonly referred to as supercoiling.

The length of a DNA molecule is considerably greater than its diameter, and the extended molecule is quite flexible. A DNA molecule is said to be relaxed if it has no twists other than that imposed by its secondary structure. Said another way, relaxed DNA does not have a clearly defined tertiary structure. We consider two types of tertiary structure, one type induced by perturbations in circular DNA, and a second type introduced by coordination of DNA with nuclear proteins called histones. Tertiary structure, whatever the type, is referred to as **supercoiling.**

Supercoiling of Circular DNA

Circular DNA A type of double-stranded DNA in which the 5′ and 3′ ends of each strand are joined by phosphodiester groups.

Circular DNA is a type of double-stranded DNA in which the two ends of each strand are joined by phosphodiester bonds [Figure 28.12(a)]. This type of DNA, the most prominent form in bacteria and viruses, is also referred to as circular duplex (because it is double-stranded) DNA. One strand of circular DNA may be opened, partially unwound, and then rejoined. The unwound section introduces a strain into the molecule because the nonhelical gap is less stable than hydrogen-bonded, base-paired helical sections. The strain can be localized in the nonhelical gap. Alternatively, it may be spread uniformly over the entire circular DNA by introduction of **superhelical twists,** one twist for each turn of a helix unwound. The circular DNA shown in Figure 28.12(b) has been unwound by four complete turns of the helix. The strain introduced by this unwinding is spread uniformly over the entire molecule by introduction of four superhelical twists [Figure 28.12(c)]. Interconversion of relaxed and supercoiled DNA is catalyzed by groups of enzymes called topoisomerases and gyrases.

Histone A protein, particularly rich in the basic amino acids lysine and arginine, that is found associated with DNA molecules.

Supercoiling of linear DNA in plants and animals takes another form and is driven by interaction between negatively charged DNA molecules and a group of positively charged proteins called **histones.** Histones are particularly rich in lysine and arginine and, therefore, at the pH of most body fluids, have an abundance of positively charged

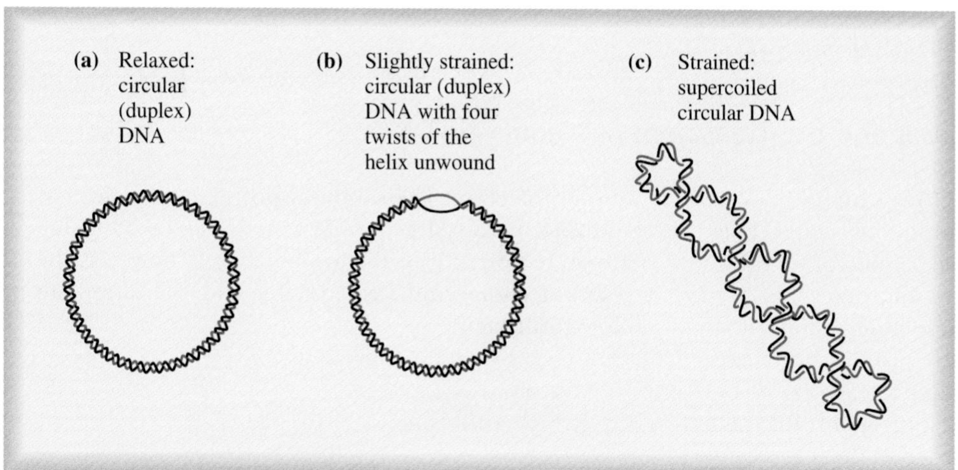

FIGURE 28.12
Relaxed and supercoiled (superhelical) DNA. (a) Circular DNA is relaxed. (b) One strand is broken, unwound by four turns, and the ends then rejoined. The strain of unwinding is localized in the nonhelical gap. (c) Superhelical coiling by four twists distributes the strain of unwinding uniformly over the entire molecule of circular DNA.

sites along their length. The complex between negatively charged DNA and positively charged histones is called chromatin. Figure 28.13 shows a schematic diagram of the structure of chromatin. Histones associate to form core particles about which double-stranded DNA then wraps. Further coiling of DNA produces the chromatin found in cell nuclei.

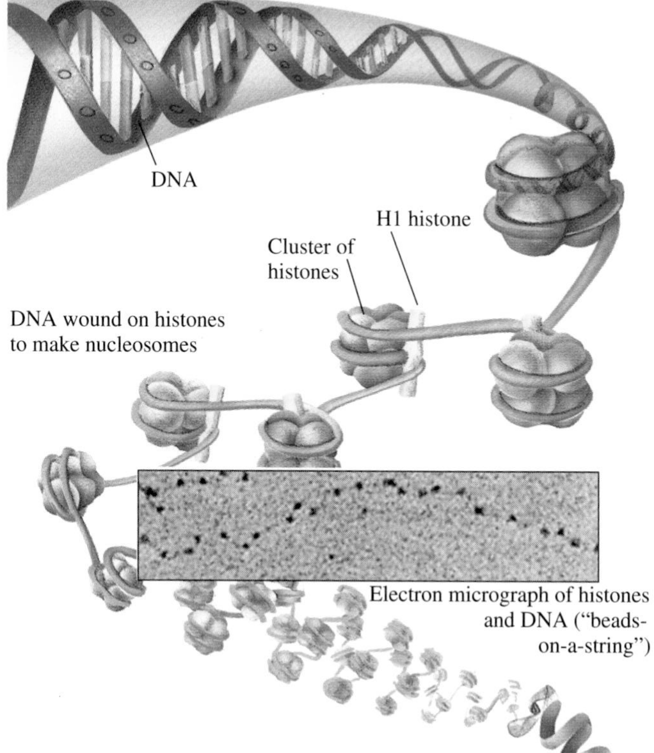

FIGURE 28.13
The structure of chromatin. Chromatin consists of DNA molecules wound around particles of histones in a beadlike structure. Further coiling produces the dense chromatin found in nuclei of plant and animal cells.

CHEMISTRY IN ACTION

Mustard Gases and the Treatment of Neoplastic Diseases

Sulfur mustard (Section 8.5) is a highly toxic gas used in World War I. Autopsies of soldiers killed by sulfur mustard revealed, among other things, very low white blood cell counts and defects in bone marrow development. From these observations, it was realized that sulfur mustards have profound effects on rapidly dividing cells. This became a lead observation in the search for less toxic alkylating agents for use in clinical medicine. Attention turned to the less reactive nitrogen mustards. One of the first compounds tested was mechlorethamine. In reactions of this and other nitrogen mustards, assisted ionization of halogen forms a cyclic aziridinium ion intermediate. This is followed by attack of a nucleophile on a carbon atom of the three-membered ring to give an alkylated product.

$$CH_3{-}\ddot{N}(CH_2{-}CH_2{-}Cl)_2 \xrightarrow[\text{(internal } S_N2)]{-Cl^-}$$

Mechlorethamine

$$CH_3{-}\overset{+}{N}(CH_2CH_2)(CH_2{-}CH_2{-}Cl) \xrightarrow[(S_N2)]{Nu{-}H} CH_3{-}N(CH_2{-}CH_2{-}Nu)(CH_2{-}CH_2{-}Cl)$$

An aziridinium ion intermediate

Mechlorethamine undergoes very rapid reaction with water (hydrolysis) and with other nucleophiles, so much so that within minutes after injection into the body, it has completely reacted. The problem for the chemist, then, was to find a way to decrease the nucleophilicity of nitrogen while maintaining a reasonable water solubility. Substitution of phenyl for methyl reduced the nucleophilicity, but the resulting compound was not sufficiently soluble in water for intravenous injection. The solubility problem was solved by adding a carboxyl group. When the carboxyl group was added directly to the aromatic ring, however, the resulting compound was too stable and, therefore, not biologically active.

$$C_6H_5{-}\ddot{N}(CH_2CH_2Cl)_2$$

The nucleophilicity of nitrogen is acceptable, but the compound is too insoluble in water for intravenous injection

$$HO\overset{O}{\overset{\|}{C}}{-}C_6H_4{-}N(CH_2CH_2Cl)_2$$

The solubility in water is acceptable, but the nucleophilicity of nitrogen is reduced so much that the compound is unreactive

Adding a propyl bridge (chlorambucil) or an aminoethyl bridge (melphalan) between the aromatic ring and the carboxyl group solved both the solubility problem and the reactivity problem. Note that melphalan is a chiral compound. It has been demonstrated that the *R* and *S* enantiomers have approximately equal therapeutic potency.

$$HO\overset{O}{\overset{\|}{C}}CH_2CH_2CH_2{-}C_6H_4{-}\ddot{N}(CH_2CH_2Cl)_2$$

Chlorambucil

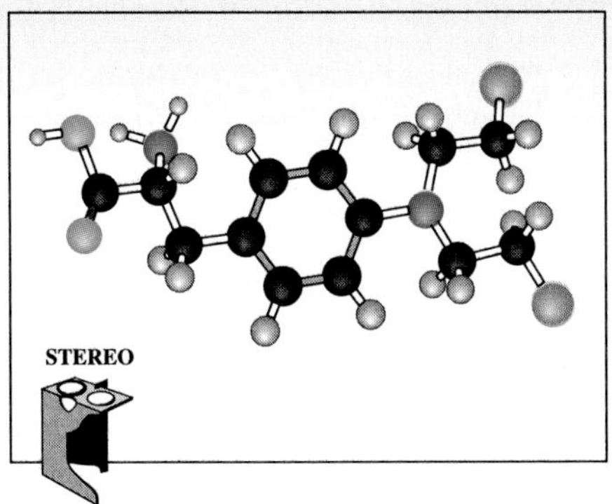

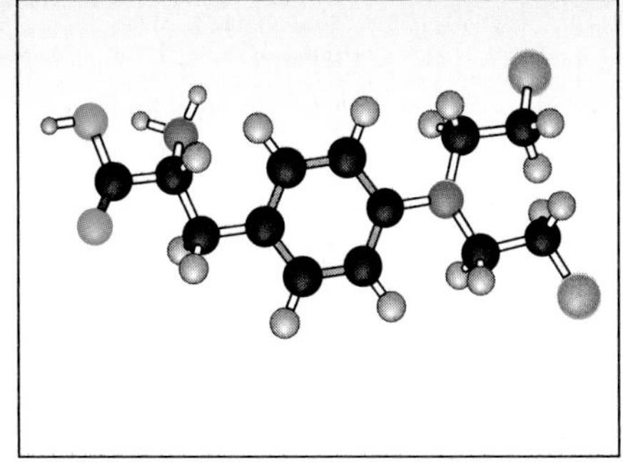

Stereoview of melphalan

$$^{-}\mathrm{O\overset{O}{\overset{\|}{C}}CHCH_2}\text{—}\langle\text{benzene}\rangle\text{—}\mathrm{N(CH_2CH_2Cl)_2}$$
$$\mathrm{NH_3^+}$$

Melphalan

The clinical value of the nitrogen mustards lies in the fact that they undergo reaction with certain nucleophilic sites on the heterocyclic aromatic amine bases in DNA. For DNA, the most reactive nucleophilic site is N-7 of guanine. Next in reactivity is N-3 of adenine, followed by N-3 of cytosine.

keto form

Guanine (G)

The nitrogen mustards are bifunctional alkylating agents; one molecule of nitrogen mustard undergoes reaction with two molecules of nucleophile. When guanine is the target nucleophile, N-7 of each guanine is converted to an ammonium ion. This in turn increases the acidity of guanine and shifts the keto-enol equilibrium from the keto form to the enol form. In the keto form, guanine forms a base pair with cytosine. In the enol form, however, it forms a base pair with thymine which leads to miscoding during DNA replication. The therapeutic value of the nitrogen mustards, then, lies in their ability to form covalent cross-links on DNA strands and to disrupt normal base pairing.

28.3 Ribonucleic Acids

Ribonucleic acids (RNA) are similar to deoxyribonucleic acids (DNA) in that they, too, consist of long, unbranched chains of nucleotides joined by phosphodiester groups between the 3′-hydroxyl of one pentose and the 5′-hydroxyl of the next. There are, however, three major differences in structure between RNA and DNA.

1. The pentose unit in RNA is β-D-ribose rather than β-2-deoxy-D-ribose.
2. The pyrimidine bases in RNA are uracil and cytosine rather than thymine and cytosine.
3. RNA is single-stranded rather than double-stranded.

Following are structural formulas for the furanose form of D-ribose and for uracil.

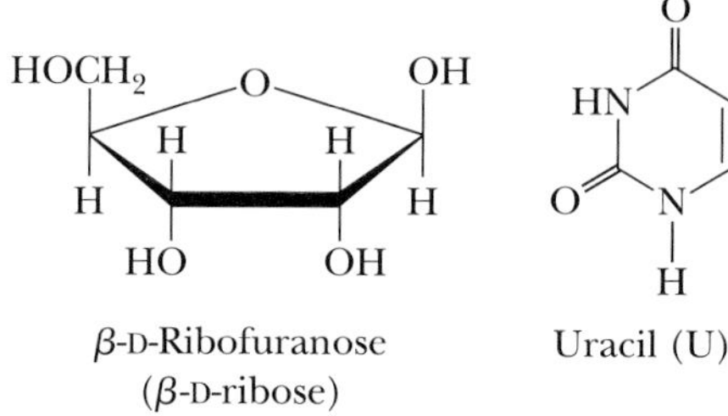

β-D-Ribofuranose (β-D-ribose) Uracil (U)

Cells contain up to eight times as much RNA as DNA, and in contrast to DNA, RNA occurs in different forms and in multiple copies of each form. RNA molecules are classified, according to their structure and function, into three major types: messenger RNA, ribosomal RNA, and transfer RNA. The molecular weight, number of nucleotides, and percent cellular abundance of these types in cells of *E. coli* are summarized in Table 28.2.

A. Ribosomal RNA

Ribosomal RNA (rRNA) A ribonucleic acid found in ribosomes, the sites of protein synthesis.

The bulk of **ribosomal RNA (rRNA)** is found in the cytoplasm in subcellular particles called ribosomes, which contain about 60% RNA and 40% protein. Ribosomes are the sites in cells at which protein synthesis takes place.

B. Transfer RNA

Transfer RNA (tRNA) A ribonucleic acid that carries a specific amino acid to the site of protein synthesis on ribosomes.

Transfer RNA (tRNA) molecules have the lowest molecular weight of all nucleic acids. They consist of 73–94 nucleotides in a single chain. The function of tRNA is to carry

TABLE 28.2 Types of RNA Found in Cells of *E. coli*

Type	Molecular Weight Range (g/mol)	Number of Nucleotides	Percentage of Cell RNA
mRNA	25,000–1,000,000	75–3000	2
tRNA	23,000–30,000	73–94	16
rRNA	35,000–1,100,000	120–2904	82

amino acids to the sites of protein synthesis on the ribosomes. Each amino acid has at least one tRNA dedicated specifically to this purpose. Several amino acids have more than one. In the transfer process, the amino acid is joined to its specific tRNA by an ester bond between the α-carboxyl group of the amino acid and the 3′ hydroxyl group of ribose at the 3′ end of the tRNA.

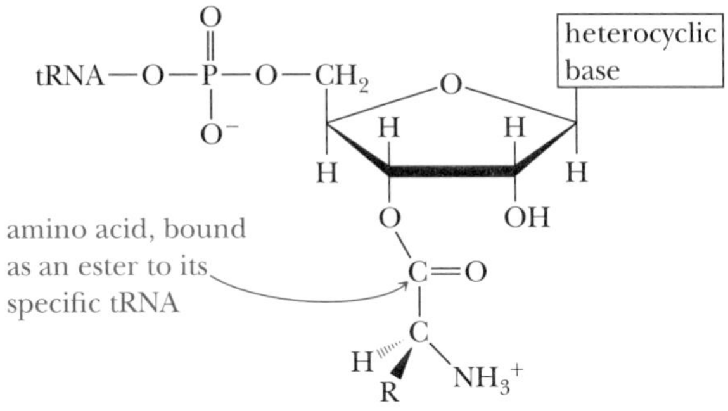

tRNA—O—P(=O)(O⁻)—O—CH₂ (ribose; heterocyclic base at C1′; 3′-O bonded as ester: O—C(=O)—C(H)(R)—NH_3^+; 2′-OH)

amino acid, bound as an ester to its specific tRNA

C. Messenger RNA

Messenger RNAs (mRNA) are present in cells in relatively small amounts and are very short-lived. They are single-stranded, and their synthesis is directed by information encoded on DNA molecules. Double-stranded DNA is unwound, and a complementary strand of mRNA is synthesized along one strand of the DNA template, beginning from the 3′ end. The synthesis of mRNA from a DNA template is called transcription, because genetic information contained in a sequence of bases of DNA is transcribed into a complementary sequence of bases on mRNA. The name "messenger" is derived from the function of this type of RNA, which is to carry coded genetic information from DNA to the ribosomes for the synthesis of new proteins.

Messenger RNA (mRNA) A ribonucleic acid that carries coded genetic information from DNA to the ribosomes for the synthesis of proteins.

EXAMPLE 28.4

Following is a base sequence from a portion of DNA. Write the sequence of bases of the mRNA synthesized using this section of DNA as a template.

3′-AGCCATGTACC-5′

Solution

RNA synthesis begins at the 3′ end of the DNA template and proceeds toward the 5′ end. The complementary mRNA strand is formed using the bases C, G, A, and U. Uracil (U) is the complement of adenine (A) on the DNA template.

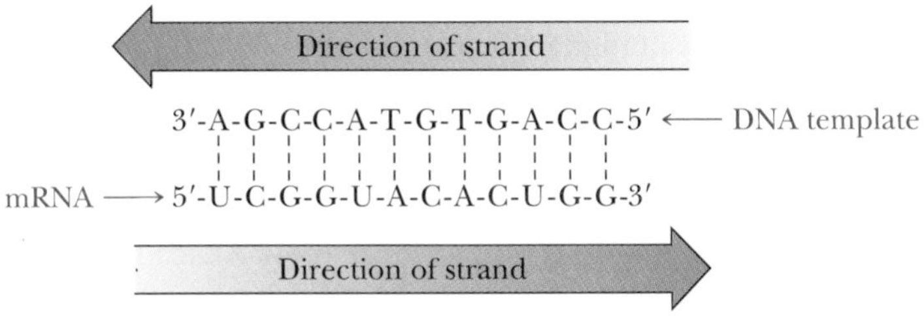

Direction of strand ←

3′-A-G-C-C-A-T-G-T-G-A-C-C-5′ ⟵ DNA template

mRNA ⟶ 5′-U-C-G-G-U-A-C-A-C-U-G-G-3′

Direction of strand →

Reading from the 5′ end, the sequence of mRNA is 5′-UCGGUACACUGG-3′.

PROBLEM 28.4

Here is a portion of the nucleotide sequence in phenylalanine tRNA.

3′-ACCACCUGCUCAGGCCUU-5′

Write the nucleotide sequence of its DNA complement.

28.4 The Genetic Code

A. Triplet Nature of the Code

It was clear by the early 1950s that the sequence of bases in DNA molecules constitutes the store of genetic information and that this sequence directs the synthesis of messenger RNA (mRNA), which, in turn, directs the synthesis of proteins. However, the statement that "the sequence of bases in DNA directs the synthesis of proteins" presents the following problem: How can a molecule containing only four variable units (adenine, cytosine, guanine, and thymine) direct the synthesis of molecules containing up to 20 variable units (the protein-derived amino acids)? How can an alphabet of four letters code for the order of letters in the 20-letter alphabet that occurs in proteins?

An obvious answer is that there is not one base but rather a combination of bases coding for each amino acid. If the code consists of nucleotide pairs, there are $4^2 = 16$ combinations; this is a more extensive code, but it is still not extensive enough to code for 20 amino acids. If the code consists of nucleotides in groups of three, there are $4^3 = 64$ combinations; this is more than enough to code for the primary structure of a protein. This appears to be a very simple solution to a system that must have taken eons of evolutionary trial and error to develop. Yet proof now exists, from comparison of gene (nucleic acid) and protein (amino acid) sequences, that nature does indeed use this simple three-letter or triplet code to store genetic information. A triplet of nucleotides is called a **codon.**

Codon A triplet of nucleotides on mRNA that directs incorporation of a specific amino acid into a polypeptide sequence.

B. Deciphering the Genetic Code

The next question is, Which of the 64 triplets code for which amino acid? In 1961, Marshall Nirenberg provided a simple experimental approach to the problem, based on the observation that synthetic polynucleotides direct polypeptide synthesis in much the same manner as do natural mRNAs. Nirenberg incubated ribosomes, amino acids, tRNAs, and appropriate protein-synthesizing enzymes. With only these components, there was no polypeptide synthesis. However, when he added synthetic polyuridylic acid (poly U), a polypeptide of high molecular weight was synthesized. What was more important, the synthetic polypeptide contained only phenylalanine. With this discovery, the first element of the genetic code was deciphered: the triplet UUU codes for phenylalanine.

Similar experiments were carried out with different synthetic polyribonucleotides. It was found, for example, that polyadenylic acid (poly A) leads to the synthesis of polylysine, and that polycytidylic acid (poly C) leads to the synthesis of polyproline. By 1964, all 64 codons had been deciphered (Table 28.3).

TABLE 28.3 The Genetic Code: mRNA Codons and the Amino Acid Each Codon Directs

First Position (5′ end)	Second Position								Third Position (3′ end)
	U		C		A		G		
U	UUU	Phe	UCU	Ser	UAU	Tyr	UGU	Cys	U
	UUC	Phe	UCC	Ser	UAC	Tyr	UGC	Cys	C
	UUA	Leu	UCA	Ser	UAA	Stop	UGA	Stop	A
	UUG	Leu	UCG	Ser	UAG	Stop	UGG	Trp	G
C	CUU	Leu	CUU	Pro	CAU	His	CGU	Arg	U
	CUC	Leu	CCC	Pro	CAC	His	CGC	Arg	C
	CUA	Leu	CCA	Pro	CAA	Gln	CGA	Arg	A
	CUG	Leu	CCG	Pro	CAG	Gln	CGG	Arg	G
A	AUU	Ile	ACU	Thr	AAU	Asn	AGU	Ser	U
	AUC	Ile	ACC	Thr	AAC	Asn	AGC	Ser	C
	AUA	Ile	ACA	Thr	AAA	Lys	AGA	Arg	A
	AUG*	Met	ACG	Thr	AAG	Lys	AGG	Arg	G
G	GUU	Val	GCU	Ala	GAU	Asp	GGU	Gly	U
	GUC	Val	GCC	Ala	GAC	Asp	GGC	Gly	C
	GUA	Val	GCA	Ala	GAA	Glu	GGA	Gly	A
	GUG	Val	GCG	Ala	GAG	Glu	GGG	Gly	G

* AUG also serves as the principal initiation codon.

C. Properties of the Genetic Code

Several features of the genetic code are evident from a study of Table 28.3.

1. Only 61 triplets code for amino acids. The remaining three (UAA, UAG, and UGA) are signals for chain termination; they signal to the protein-synthesizing machinery of the cell that the primary sequence of the protein is complete. The three chain termination triplets are indicated in Table 28.3 by "Stop."
2. The code is degenerate, which means that several amino acids are coded for by more than one triplet. If you count the number of triplets coding for each amino acid, you will find that only methionine and tryptophan are coded for by just one triplet. Leucine, serine, and arginine are coded for by six triplets, and the remaining amino acids are coded for by two, three, or four triplets.
3. For the 15 amino acids coded for by two, three, or four triplets, it is only the third letter of the code that varies. For example, glycine is coded for by the triplets GGA, GGG, GGC, and GGU.
4. There is no ambiguity in the code, meaning that each triplet codes for one and only one amino acid.

Finally, we must ask one last question about the genetic code: Is the code universal? That is, is it the same for all organisms? Every bit of experimental evidence available today from the study of viruses, bacteria, and higher animals, including humans, indicates that the code is universal. Furthermore, the fact that it is the same for all these organisms means that it has been the same over billions of years of evolution.

At places on the gel where ^{32}P atoms are present, the plate is clouded by emitted gamma rays and a darkened spot appears. Nonlabeled fragments are present on the gel as well, but they do not darken the photographic plate and are not detected.

EXAMPLE 28.7

The following excision fragment is isolated after treatment of the gene coding for bovine rhodopsin with a set of endonucleases and labeling the 5′ end with phosphorus-32.

^{32}P-5′-GCTCTACTACCGTCATCT-3′

Assume conditions of treatment of this excision fragment with dimethyl sulfate followed by piperidine and sodium hydroxide are adjusted so that only single excisions take place. Write the complete set of excision fragments that result.

Solution

The restriction fragment is cleaved into three sets of two fragments.

^{32}P-5′-GCTCT**A**CT**A**CCGTC**A**TCT-3′ ⟶ (cleavage sites (i), (ii), (iii) at the highlighted A's)

(i) ^{32}P-5′-GCTCT + CTACCGTCATCT-3′

(ii) ^{32}P-5′-GCTCTACT + CCGTCATCT-3′

(iii) ^{32}P-5′-GCTCTACTACCGTC + TCT-3′

PROBLEM 28.7

In what order will the excision fragments in Example 28.7 appear on the developed photographic plate? Remember that only the 5′ end of the original restriction fragment is labeled with phosphorus-32.

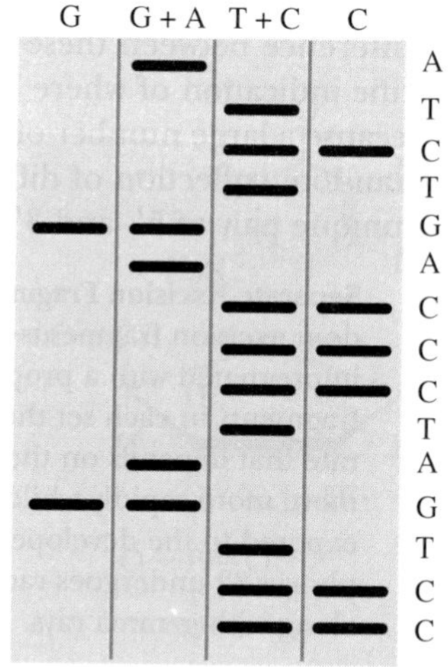

FIGURE 28.14
Gel electrophoresis pattern of a hypothetical restriction fragment labeled in its 5′-end with phosphorus-32. The gel is developed in four lanes: G, G + A, T + C, and C. Reading this pattern from the bottom up gives the nucleotide sequence from the 5′-end.

CHEMISTRY IN ACTION

DNA Fingerprinting

Each human being has a genetic makeup consisting of approximately 3 billion pairs of nucleotides, and, except for identical twins, the base sequence of DNA in one individual is different from that of every other individual. As a result, each person has a unique DNA fingerprint. To determine a DNA fingerprint, a sample of DNA from a trace of blood or skin or other tissue is treated with a set of restriction endonucleases, and the 5′ end of each restriction fragment is labeled with phosphorus-32. The resulting ^{32}P-labeled restriction fragments are then separated by polyacrylamide gel electrophoresis and visualized by placing a photographic plate over the developed gel.

In the DNA fingerprint patterns shown in the figure, lanes 1, 5, and 9 represent internal standards or control lanes. They contain the DNA fingerprint pattern of a standard virus treated with a standard set of restriction endonucleases. Lanes 2, 3, and 4 were used in a paternity suit. The DNA fingerprint of the mother in lane 4 contains five bands, which match with five of the six bands in the DNA fingerprint of the child in lane 3. The DNA fingerprint of the alleged father in lane 2 contains six bands, three of which match with bands in the DNA fingerprint of the child. Because the child inherits only half of its genes from the father, only half of the child's and father's DNA fingerprints are expected to match. In this instance, the paternity suit was won on the basis of the DNA fingerprint matching.

Lanes 6, 7, and 8 contain DNA fingerprint patterns used as evidence in a rape case. Lanes 6 and 7 are DNA fingerprints of semen obtained from the rape victim. Lane 8 is the DNA fingerprint pattern of the alleged rapist. The DNA fingerprint patterns of the semen do not match that of the alleged rapist and excluded the suspect from the case.

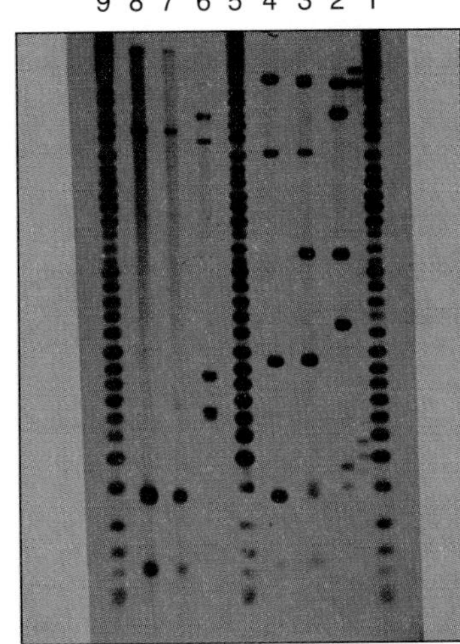

A DNA "fingerprint." (Courtesy of Dr. Lawrence Kobilinsky)

Step 5: **Read the Nucleotide Sequence from the Exposed Photographic Plate.** The darkened spots on the exposed plate appear as a "ladder" from which the sequence of bases in the original nucleotide can be read, starting with the spot farthest from the origin. Figure 28.14 shows the ladder pattern from a polyacrylamide gel electrophoresis of a hypothetical restriction fragment ^{32}P-5′-CCTGATCCCAGTCTA-3′. With current technology, it is possible for an experienced technician to read as many as 400 bases from a gel separation. Automated DNA sequencing machines now available are capable of sequencing up to 10,000 base pairs per day. One of the ultimate challenges now is to sequence the entire human genome with its estimated 3 billion base pairs. That project will require at least a decade. Enormous progress has already been made in this task. Almost every day, a new gene for a human hereditary defect is localized, and many of these genes have been correlated with the proteins they code for. This research may lead to the treatment of many serious human genetic diseases within a short time.

SUMMARY

Nucleic acids are composed of three types of monomer units: heterocyclic aromatic amine bases (more commonly referred to simply as bases) derived from purine and pyrimidine, the monosaccharides D-ribose and 2-deoxy-D-ribose, and phosphoric acid (Section 28.1). A **nucleoside** is a compound containing D-ribose or 2-deoxy-D-ribose bonded to a heterocyclic aromatic amine base by a *β-N*-glycoside bond. A **nucleotide** is a nucleoside in which a molecule of phosphoric acid is esterified with an —OH of the monosaccharide, most commonly either the 3′-OH or the 5′-OH. Nucleoside monophosphate esters are strong diprotic acids and at pH 7.0 have a charge of −2. Nucleoside diphosphates and triphosphates are also strong polyprotic acids and are extensively ionized at pH 7.0. At pH 7.0, adenosine triphosphate is a 50:50 mixture of ATP^{3-} and ATP^{4-}.

The **primary structure** of **deoxyribonucleic acids (DNA)** consists of units of 2-deoxyribose bonded by 3′,5′-phosphodiester bonds (Section 28.2A). A heterocyclic aromatic amine base is attached to each deoxyribose by a *β-N*-glycoside bond. The sequence of bases is read from the 5′ end of the polynucleotide strand to the 3′ end.

The **Watson-Crick model** of the **DNA double helix** was based on analyses of DNA base composition and x-ray diffraction patterns of crystals of DNA. The heart of the Watson-Crick model is the postulate that a molecule of DNA consists of two antiparallel polynucleotide strands coiled in a right-handed manner about the same axis to form a double helix (Section 28.2B). Purine and pyrimidine bases point inward toward the axis of the helix and are always paired G-C and A-T. In **B-DNA,** base pairs are stacked one on top of another with a spacing of 3.4 Å and 10 base pairs per 34 Å helical repeat. In **A-DNA,** bases are stacked with a spacing of 2.9 Å and 10 base pairs per 29 Å helical repeat. In B-DNA, deoxyribose is in the (C)2′-endo conformation whereas in A-DNA it is in the (C)3′-endo conformation.

Tertiary structure of DNA is commonly referred to as **supercoiling** (Section 28.2C). **Circular DNA** is a type of double-stranded DNA in which the ends of each strand are joined by phosphodiester bonds. Opening of one strand followed by partial unwinding and rejoining the ends introduces strain in the nonhelical gap. This strain can be spread over the entire molecule of circular DNA by introduction of **superhelical twists.** Supercoiling in double-stranded DNA in plants and animals is the result of complex formation between negatively charged DNA and positively charged proteins called histones. **Histones** are particularly rich in lysine and arginine and, therefore, have an abundance of positive charges. The association of nuclear DNA and histones produces a pigment called **chromatin.**

There are three important differences between the primary structure of **ribonucleic acids (RNA)** and DNA (Section 28.3). (1) The monosaccharide unit in RNA is D-ribose. (2) Both RNA and DNA contain the purine bases adenine (A) and guanine (G), and the pyrimidine base cytosine (C). As the fourth base, however, RNA contains uracil (U) whereas DNA contains thymine (T). (3) RNA is single-stranded, whereas DNA is double-stranded.

The **genetic code** (Section 28.4) consists of nucleotides in groups of three; that is, it is a triplet code. Only 61 triplets code for amino acids; the remaining three code for termination of polypeptide synthesis.

The five steps in the **Maxam-Gilbert sequence determination** (Section 28.5) are (1) treat double-stranded DNA with restriction endonucleases to cut each chain at very specific sites; (2) label each restriction fragment, usually at its 5′ end, with radioactive phosphorus-32; (3) randomly remove nucleotides from ^{32}P-labeled restriction fragments; (4) separate excision fragments by polyacrylamide gel electrophoresis; and (5) read the nucleotide sequence from an exposed photographic plate.

ADDITIONAL PROBLEMS

Nucleosides and Nucleotides

28.8 Two important drugs in the treatment of acute leukemia are 6-mercaptopurine and 6-thioguanine. In each of these drugs, the oxygen atom at carbon 6 of the parent molecule is replaced by a sulfur. Draw structural formulas for the enethiol forms of 6-mercaptopurine and 6-thioguanine.

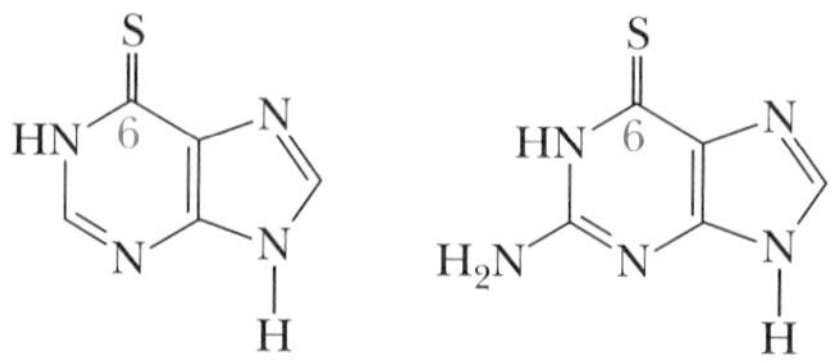

6-Mercaptopurine　　6-Thioguanine

28.9 Following are the structural formulas of cytosine and thymine. Draw two additional tautomeric forms for cytosine and three additional tautomeric forms for thymine.

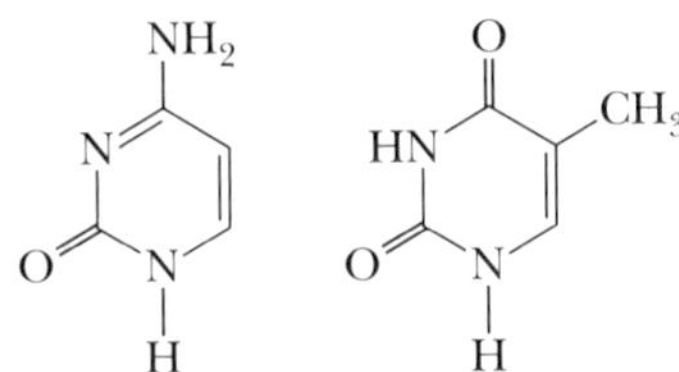

Cytosine (C) Thymine (T)

28.10 Draw structural formulas for a nucleoside composed of:

(a) β-D-Ribose and adenine **(b)** β-D-Deoxyribose and cytosine

28.11 Nucleosides are stable in water and in dilute base. In dilute acid, however, the glycoside bond of a nucleoside undergoes rapid hydrolysis to give a pentose and a heterocyclic base. Propose a mechanism for this acid-catalyzed hydrolysis.

28.12 Estimate the net charge on the following nucleotides at pH 7.4, the pH of blood plasma.

(a) ATP **(b)** GMP **(c)** dGMP

The Structure of DNA

28.13 Why are deoxyribonucleic acids called acids? What are the acidic groups in their structure?

28.14 Human DNA contains approximately 30.4% A. Estimate the percentages of G, C, and T and compare them with the values presented in Table 28.1.

28.15 Draw the structural formula of the DNA tetranucleotide 5′-AGCT-3′. Estimate the net charge on this tetranucleotide at pH 7.0. What is the complementary tetranucleotide to this sequence?

28.16 Write the DNA complement for 5′-ACCGTTAAT-3′. Be certain to label which is the 5′ end and which is the 3′ end of the complement strand.

28.17 Write the DNA complement for 5′-TCAACGAT-3′.

28.18 Write the structural formula of each nucleotide and estimate its net charge at pH 7.4, the pH of blood plasma.

(a) 2′-Deoxyadenosine 5′-triphosphate (dATP)
(b) Guanosine 3′-monophosphate (GMP)
(c) 2′-Deoxyguanosine 5′-diphosphate (dGDP)

28.19 Cyclic-AMP, first isolated in 1959, is involved in many diverse biological processes as a regulator of metabolic and physiological activity. In it, a single phosphoric acid group is esterified with both the 3′- and 5′-hydroxyls of adenosine. Draw the structural formula of cyclic-AMP.

28.20 Discuss the role of the hydrophobic effect in stabilizing:

(a) Double-stranded DNA **(b)** Lipid bilayers **(c)** Soap micelles

28.21 At elevated temperatures, nucleic acids become denatured, that is they unwind into disordered single-stranded DNA. Account for the observation that the higher the G-C content of a nucleic acid, the higher the temperature required for thermal denaturation.

28.22 The Watson-Crick pattern of hydrogen bonding is not the only type of interaction possible for nucleic acids. Draw the structure of an A-T base pair in which the purine uses N-7 instead of N-1 as a hydrogen bond acceptor.

APPENDIX 4

Characteristic ^{1}H-NMR Chemical Shifts

Type of Hydrogen (R = alkyl, Ar = aryl)	Chemical Shift (δ)*	Type of Hydrogen (R = alkyl, Ar = aryl)	Chemical Shift (δ)*
$(CH_3)_4Si$	0 (by definition)	$RC(=O)OCH_3$	3.7–3.9
RCH_3	0.8–1.0	$RC(=O)OCH_2R$	4.1–4.7
RCH_2R	1.2–1.4	RCH_2I	3.1–3.3
R_3CH	1.4–1.7	RCH_2Br	3.4–3.6
$R_2C{=}CRCHR_2$	1.6–2.6	RCH_2Cl	3.6–3.8
$RC{\equiv}CH$	2.0–3.0	RCH_2F	4.4–4.5
$ArCH_3$	2.2–2.5	$ArOH$	4.5–4.7
$ArCH_2R$	2.3–2.8	$R_2C{=}CH_2$	4.6–5.0
ROH	0.5–6.0	$R_2C{=}CHR$	5.0–5.7
RCH_2OH	3.4–4.0	ArH	6.5–8.5
RCH_2OR	3.3–4.0	$RC(=O)H$	9.5–10.1
R_2NH	0.5–5.0	$RC(=O)OH$	10–13
$RC(=O)CH_3$	2.1–2.3		
$RC(=O)CH_2R$	2.2–2.6		

* Values are approximate. Other atoms within the molecule may cause the signal to appear outside these ranges.

APPENDIX 5

Characteristic ^{13}C-NMR Chemical Shifts

Type of Carbon	Chemical Shift (δ)	Type of Carbon	Chemical Shift (δ)
RCH_3	0–40	C_6H_5—R	110–160
RCH_2R	15–55		
R_3CH	20–60		
RCH_2I	0–40	$R\overset{O}{\overset{\|}{C}}OR$	160–180
RCH_2Br	25–65		
RCH_2Cl	35–80	$R\overset{O}{\overset{\|}{C}}NR_2$	165–180
R_3COH	40–80		
R_3COR	40–80	$R\overset{O}{\overset{\|}{C}}OH$	175–185
$RC\equiv CR$	65–85		
$R_2C=CR_2$	100–150	$R\overset{O}{\overset{\|}{C}}H$, $R\overset{O}{\overset{\|}{C}}R$	180–210

H_3PO_4	Phosphoric acid With alcohols; catalyst for regioselective dehydration (9.7) With alcohols; catalyst for intermolecular dehydration to ethers (11.4B) With alkenes and arenes; catalyst for alkylation of arenes (20.1D) With alcohols and arenes; catalyst for alkylation of arenes (20.1D)
K	Potassium With alcohols; formation of potassium alkoxides (9.5)
LDA	Lithium diisopropylamide With aldehydes, ketones, and esters; formation of lithium enolates (18.1) And directed aldol reactions (18.2B)
Li	Lithium; alternatively, lithium metal With alkyl and aryl halides; formation of organolithium reagents (7.7) With alcohols; formation of lithium alkoxides (9.5)
$LiAlH_4$	Lithium aluminum hydride (LAH) With epoxides; regioselective reduction to alcohols (11.10B) With aldehydes and ketones; reduction to alcohols (15.14B) With carboxylic acids; reduction to primary alcohols (16.7A) With esters; reduction to two alcohols (17.11A) With amides; reduction of amines (17.11B) With nitriles; reduction of 1° amines (17.11C)
Mg	Magnesium; alternatively, magnesium metal With alkyl and aryl halides; formation of Grignard reagents (7.7)
N_2H_4	Hydrazine With aldehydes and ketones and KOH; Wolff-Kishner reduction C=O to CH_2 (15.14B)
Na	Sodium; alternatively, sodium metal With alcohols; formation of sodium alkoxides (9.5) With NH_3(l); stereoselective reduction of alkynes to *Z*-alkenes (10.7B)
$NaBH_4$	Sodium borohydride With products of oxymercuration; reduction to alcohols (6.3F) With aldehydes and ketones; reduction to alcohols (15.14B)
$NaNH_2$	Sodium amide With terminal alkynes; formation of alkyne anions (10.4) With alkyltriphenylphosphonium salts; formation of Wittig reagents (15.7) With haloarenes; formation of benzyne intermediates (20.3A)
$NaNO_2$/HCl	Sodium nitrite With 3° aromatic amines; formation of nitrosoarenes (21.9A and B) With 2° amines; formation of *N*-nitrosoamines (21.9C) With 1° alkylamines; formation of unstable alkanediazonium salts (21.9D) With 1° arylamines; formation of arenediazonium salts (21.9E) With β-aminoalcohols; Tiffeneau-Demjanov rearrangement (21.9D)
NBS	*N*-Bromosuccinimide With alkenes; radical allylic bromination (7.6) With alkylarenes; radical benzylic bromination (19.6B)
NH_3	Ammonia With epoxides; regioselective and stereoselective ring opening (11.10B)

	With aldehydes and ketones; formation of imines (15.10A) With acid chlorides; formation of amides (17.7A) With acid anhydrides; formation of amides (17.7B) With esters; formation of amides (17.7C)
O_2	Oxygen; alternatively, molecular oxygen With thiols; oxidation to disulfides (9.11) With ethers; oxidation to hydroperoxides (11.6B) With aldehydes; oxidation to carboxylic acids (15.13A)
O_3	Ozone With alkenes; oxidative cleavage to two carbonyl-containing compounds (6.5C)
OsO_4	Osmium tetroxide With alkenes; syn stereoselective oxidation to a glycol (6.5B)
PBr_3	Phosphorus tribromide With alcohols; formation of alkyl bromides (9.6B)
PCC	Pyridinium chlorochromate With 1° alcohols; oxidation to aldehydes (9.9) With 2° alcohols; oxidation to ketones (9.9)
R_2CuLi	Lithium dialkylcopper(I); alternatively, a Gilman reagent With alkyl, vinylic, and aryl halides; stereoselective cross-coupling (7.7C) With epoxides; regioselective nucleophilic ring opening (11.10B) With acid chlorides; formation of ketones (17.9C) With α,β-unsaturated carbonyl compounds; conjugate addition (18.8B)
RNH_2	A primary amine With epoxides; regioselective and stereoselective ring opening (11.10B) With aldehydes and ketones; formation of imines (15.10A) With acid chlorides; formation of amines (17.7A) With acid anhydrides; formation of amides (17.7B) With esters; formation of amides (17.7C)
RCO_3H	A peroxycarboxylic acid; alternatively, a peracid With alkenes; stereoselective oxidation to epoxides (11.8B) With ketones; Baeyer-Villiger oxidation to esters (15.13B)
RLi	An organolithium reagent With CuI; formation of lithium diorganocopper (Gilman) reagents (7.7C) With aldehydes and ketones; formation of alcohols (15.6A) With carbon dioxide; formation of carboxylic acids (15.6A) With esters; formation of alcohols (17.9A)
RMgX	An organomagnesium reagent; alternatively, a Grignard reagent With aldehydes and ketones; formation of alcohols (15.6A) With carbon dioxide; formation of carboxylic acids (15.6A) With esters; formation of alcohols (17.9A)
R_2NH	A secondary amine With epoxides; regioselective and stereoselective ring opening (11.10B) With aldehydes and ketones; formation of enamines (15.10A) With acid chlorides; formation of amides (17.7A) With acid anhydrides; formation of amides (17.7B) With esters; formation of amides (17.7C)

$(sia)_2BH$	Di(*sec*-isoamyl)borane With terminal alkynes; regioselective and stereoselective hydroboration (10.8)
$SnCl_4$	Tin(IV) chloride; alternatively, stannic chloride Lewis acid catalyst for Friedel-Crafts acylations (20.1C)
$SOCl_2$	Thionyl chloride With alcohols; formation of alkyl chlorides (9.6C) With carboxylic acids; formation of acid chlorides (16.9)
Zn(Hg)/HCl	Amalgamated zinc/hydrochloric acid With aldehydes and ketones; Clemmensen reduction of C=O to CH_2 (15.14C)

Appendix 8

Methods for Forming New Carbon-Carbon Bonds

Acylation of 1,3-dithiane anions	15.9
Acylation of acetoacetic ester anions	18.6
Acylation of enamines	18.5B
Acylation of malonic ester anions	18.7
Addition of HCN to aldehydes or ketones	15.6D
Aldol reactions	18.1, 18.2B
Alkylation of 1,3-dithiane anions	15.9
Alkylation of acetoacetic ester anions	18.6
Alkylation of anions of terminal alkynes	15.6C
Alkylation of cyanide ion	8.1
Alkylation of enamines	18.5A
Alkylation of enolate anions	18.6, 18.7
Alkylation of malonic ester anions	18.7
Alkylation of terminal alkyne anions	10.5
Arenediazonium ions with cyanide ion	21.9E
Claisen condensations	18.3A, 18.3C
Claisen rearrangement	22.3A
Conjugate addition of Gilman reagents	18.8B
Cope rearrangement	22.3B
Dieckmann condensations	18.3B
Diels-Alder reactions	22.5
Friedel-Crafts acylation of arenes	20.1C
Friedel-Crafts alkylation of arenes	20.1C
Gilman reagents with α,β-unsaturated aldehydes and ketones	18.8B
Gilman reagents with acid chlorides	17.9
Gilman reagents with alkyl halides	7.7C
Gilman reagents with epoxides	11.10B
Gilman reagents with vinylic halides	7.7C
Grignard reagents with aldehydes and ketones	15.6
Grignard reagents with carbon dioxide	15.6
Grignard reagents with esters	17.9A
Kolbe carboxylation of phenoxide anions	19.5E
Michael reaction; conjugate addition	18.8A
Organolithium reagents with aldehydes and ketones	15.6B
Polymerization of ethylene and substituted ethylenes	23.6
Wittig reactions	15.7

GLOSSARY

Absorbance (A) (Section 14.3A) A quantitative measure of the extent to which a compound absorbs ultraviolet-visible radiation of a particular wavelength.

Acetal (Section 15.8B) A molecule containing two —OR or —OAr groups bonded to the same carbon.

Aceto group (Section 16.2B) A CH_3CO— group. Also called an acetyl group.

Achiral (Section 4.2) An object that lacks chirality; an object that has no handedness.

Activating group (Section 20.1A) Any substituent on a benzene ring that causes the rate of electrophilic aromatic substitution to be greater than that for benzene.

Activation energy (Section 6.2A) The difference in potential energy between reactants and the transition state.

Acyl group (Section 17.1A) An RCO— or ArCO— group.

Acyl halide (Section 17.1A) A derivative of a carboxylic acid in which the —OH of the carboxyl group is replaced by halogen, most commonly chlorine.

Acylation (Section 18.5) The process of introducing an acyl group, RCO— or ArCO—, onto an organic molecule.

Acylium ion (Section 20.1C) A resonance-stabilized cation with the structure $[RC{=}O]^+$ or $[ArC{=}O]^+$. The positive charge is delocalized over both the carbonyl carbon and the carbonyl oxygen.

Alcohol (Section 1.3A) A compound containing an —OH (hydroxyl) group bonded to an sp^3 hybridized carbon.

Alcoholic fermentation (Section 26.7B) A metabolic pathway that converts glucose to two molecules of ethanol and two molecules of CO_2.

Aldehyde (Section 1.3B) A compound containing a CHO group.

Alditol (Section 24.4C) The product formed when the C=O group of a monosaccharide is reduced to a CHOH group.

Aldonic acid (Section 24.4D) The product formed when the —CHO group of an aldose is oxidized to a —CO_2H group.

Aldose (Section 24.1A) A monosaccharide containing an aldehyde group.

Aliphatic amine (Section 21.1) An amine in which nitrogen is bonded only to alkyl groups.

Aliphatic hydrocarbon (Section 2.1) An alternative word to describe an alkane.

Alkaloid (Section 21.1) A basic nitrogen-containing compound of plant origin, many of which are physiologically active when administered to humans.

Alkane (Section 2.1) A saturated hydrocarbon whose carbon atoms are arranged in an open chain.

Alkene (Introduction, Chapter 5) An unsaturated hydrocarbon that contains a carbon-carbon double bond.

Alkoxy group (Section 11.2) An —OR group, where R is an alkyl group.

Alkyl group (Section 2.3A) A group derived by removing a hydrogen from an alkane.

Alkyl halide (Section 7.1) A compound containing a halogen atom covalently bonded to an sp^3 hybridized carbon atom. Given the symbol R—X.

Alkylation (Section 10.5) Any reaction in which a new carbon-carbon bond to an alkyl group is formed.

Alkyne (Section 10.1) An unsaturated hydrocarbon that contains one or more carbon-carbon triple bonds.

Allene (Section 10.6) The compound $CH_2{=}C{=}CH_2$. Any compound that contains adjacent carbon-carbon double bonds, that is, the arrangement C=C=C.

Allylic (Section 8.4E) Next to a carbon-carbon double bond.

Allylic carbocation (Section 8.4E) Any carbocation in which an allylic carbon bears the positive charge.

Allylic substitution (Section 7.6) Any reaction in which one atom or group of atoms is substituted for another atom or group of atoms at an allylic position.

Amide (Section 17.1D) A compound in which an acyl group is bonded to a trivalent nitrogen atom.

Amino acid (Section 27.1A) A compound that contains both an amino group and a carboxyl group.

α-Amino acid (Section 27.1A) An amino acid in which the amino group is on the carbon adjacent to the carboxyl group.

Amorphous domains (Section 23.4) Disordered, noncrystalline regions in the solid state of a polymer.

Anabolic steroid (Section 25.4A) A steroid hormone, such as testosterone, that promotes tissue and muscle growth and development.

Androgen (Section 25.4A) A steroid hormone, such as testosterone, that mediates the development of sexual characteristics of males.

Angle strain (Section 2.6B) The strain that arises when a bond angle is either compressed or expanded compared to its normal value.

Anion (Section 1.2B) An atom or group of atoms bearing a negative charge.

Annulene (Section 19.2B) A cyclic hydrocarbon with a continuous alternation of single and double bonds.

Anomeric carbon (Section 24.2) The hemiacetal or acetal carbon of the cyclic form of a carbohydrate.

Anomers (Section 24.2) Carbohydrates that differ in configuration only at their anomeric carbons.

Anti stereoselectivity (Section 6.3D) The addition of atoms or groups of atoms on opposite faces of a carbon-carbon double bond.

Antiaromatic compound (Section 19.2A) A monocyclic compound that is planar or nearly so, has one *p* orbital on each atom of the ring, and has $4n$ pi electrons in the cyclic arrangement of overlapping *p* orbitals, where n is an integer. Antiaromatic compounds are especially unstable.

Antibonding molecular orbital (Section 1.8A) A molecular orbital in which electrons have a higher energy than they would in isolated atomic orbitals.

Aprotic solvent (Section 8.2) A solvent that cannot serve as a hydrogen bond donor; nowhere in the molecule is a hydrogen bonded to an atom of high electronegativity.

Ar— (Introduction, Chapter 19) The symbol used for an aryl group, by analogy with R— for an alkyl group.

Aramid (Section 23.5A) A polyaromatic amide; a polymer in which the monomer units are an aromatic diamine and an aromatic dicarboxylic acid.

Arene (Introduction, Chapter 19) An aromatic hydrocarbon.

Aromatic amine (Section 21.1) An amine in which nitrogen is bonded to one or more aryl groups.

Aromatic compound (Introduction, Chapter 19) A term used initially to classify benzene and its derivatives. More accurately, it is used to classify any compound that meets the Hückel criteria for aromaticity (Section 19.2A).

Aryl group (Introduction, Chapter 19) A group derived from an arene by removal of an H; given the symbol Ar—.

Aryl halide (Section 7.1) A compound containing a halogen atom bonded to a benzene ring; given the symbol Ar—X.

Atactic polymer (Section 23.6C) A polymer with completely random configurations at the stereocenters along its chain, as, for example, atactic polypropylene.

Aufbau principle (Section 1.1A) Orbitals fill in order of increasing energy, from lowest to highest.

Average degree of polymerization, n (Section 23.3) A subscript placed outside the parentheses of the simplest nonredundant unit of a polymer to indicate that the unit repeats n times in the polymer.

Axial position (Section 2.6B) A position on a chair conformation of a cyclohexane ring that extends from the ring parallel to the imaginary axis of the ring.

Azeotrope (Section 15.8B) A liquid mixture with constant composition and a boiling point that is different from that of any of its components.

Base peak (Section 12.1) The peak due to the most abundant ion in a mass spectrum; the most intense peak. It is assigned an arbitrary intensity of 100.

Basicity (Section 8.4B) An equilibrium property measured by the position of equilibrium in an acid-base reaction, as, for example, the acid-base reaction between ammonia and water.

Benzyl group $C_6H_5CH_2$— (Section 19.3A) The group derived from toluene by removing a hydrogen from its methyl group.

Benzylic position (Section 15.6) An sp^3 hybridized carbon attached to a benzene ring.

Benzyne intermediate (Section 20.3A) A reactive intermediate formed by β-elimination from adjacent carbon atoms of a benzene ring and having a triple bond in the benzene ring. The second pi bond of the benzyne triple bond is formed by overlap of coplanar sp^2 orbitals on adjacent carbons.

Bicycloalkane (Section 2.4B) An alkane containing two rings that share two carbon atoms.

Bile acid (Section 25.4A) A cholesterol-derived detergent molecule, such as cholic acid, which is secreted by the gallbladder into the intestine to assist in the absorption of dietary lipids.

Bimolecular reaction (Section 8.3) A reaction in which two species are involved in the rate-limiting step.

Boat conformation (Section 2.6B) A puckered conformation of a cyclohexane ring in which carbons 1 and 4 of the ring are bent toward each other.

Bond dipole (μ) (Section 1.2C) A measure of the polarity of a covalent bond. The product of the charge on each atom of a polar bond times the distance between the atoms.

Bond length (Section 1.2C) The distance between atoms in a covalent bond.

Bonding electrons (Section 1.2D) Valence electrons involved in forming a covalent bond, i. e., shared electrons.

Bonding molecular orbital (Section 1.8A) A molecular orbital in which electrons have a lower energy than they would in isolated atomic orbitals.

Brønsted-Lowry acid (Section 3.1) A proton donor.

Brønsted-Lowry base (Section 3.1) A proton acceptor.

Carbanion (Section 16.5A) An anion in which carbon has an unshared pair of electrons and bears a negative charge.

Carbocation (Section 6.3A) A species in which a carbon atom has only six electrons in its valence shell and bears a positive charge.

Carbohydrate (Introduction, Chapter 24) A polyhydroxyaldehyde or polyhydroxyketone, or a substance that gives these compounds on hydrolysis.

α-Carbon (Section 15.11A) A carbon atom adjacent to a functional group.

Carbonyl group (Section 1.3B) A $C{=}O$ group.

Carboxyl group (Section 1.3C) A $-CO_2H$ group.

Carboxylic acid anhydride (Section 17.1B) A functional group in which two acyl groups are bonded to an oxygen.

Cation (Section 1.2B) An atom or group of atoms bearing a positive charge.

Center of symmetry (Section 4.2) A point so situated that identical components of an object are located on opposite sides and equidistant from that point along any axis passing through it.

Chain initiation (Section 7.5B) A step in a chain reaction characterized by the formation of reactive intermediates (radicals, anions, or cations) from nonradical or noncharged molecules.

Chain length (Section 7.5B) The number of times the cycle of chain propagation steps repeats in a chain reaction.

Chain propagation (Section 7.5B) A step in a chain reaction characterized by the reaction of a reactive intermediate and a molecule to give a new reactive intermediate and a new molecule.

Chain termination (Section 7.5B) A step in a chain reaction that involves destruction of reactive intermediates.

Chain-growth polymerization (Section 23.6) A polymerization that involves sequential addition reactions, either to unsaturated monomers or to monomers possessing other reactive functional groups.

Chain-transfer reaction (Section 23.6A) The transfer of the reactivity of an endgroup from one chain to another during the course of a polymerization.

Chair conformation (Section 2.6B) The most stable puckered conformation of a cyclohexane ring; all bond angles are approximately 109.5°, and bonds to all adjacent carbons are staggered.

Chemical shift, δ (Section 13.5) The shift in parts per million (ppm) of an NMR signal from the signal of TMS.

Chiral (Section 4.2) From the Greek, *cheir,* hand; an object that is not superposable on its mirror image; an object that has handedness.

Circular DNA (Section 28.2C) A type of double-stranded DNA in which the 5′- and 3′-ends of each strand are joined by phosphodiester groups.

Cis (Section 2.7A) A prefix meaning on the same side.

***Cis-trans* isomers** (Section 2.7A) Isomers that have the same order of attachment of their atoms but a different arrangement of their atoms in space due to the presence of either a ring or a carbon-carbon double bond.

Codon (Section 28.4A) A triplet of nucleotides on mRNA that directs incorporation of a specific amino acid into a polypeptide sequence.

Coenzyme (Section 26.1) A low-molecular-weight, nonprotein molecule or ion that binds reversibly to an enzyme, functions as a second substrate for the enzyme, and is regenerated by further reaction.

Condensation polymerization (Section 23.5) A polymerization in which chain growth occurs in a stepwise manner between difunctional monomers. Also called step-growth polymerization.

Conformation (Section 2.6A) Any three-dimensional arrangement of atoms in a molecule that results by rotation about a single bond.

Conjugate acid (Section 3.1) The species formed from a base when it accepts a proton from an acid.

Conjugate addition (Section 18.8) Addition of a nucleophile to the β-carbon of an α,β-unsaturated carbonyl compound.

Conjugate base (Section 3.1) The species formed from an acid when it donates a proton to a base.

Conjugated diene (Section 6.6C) A diene in which the double bonds are separated by one single bond.

Constitutional isomers (Section 2.2) Compounds with the same molecular formula but a different order of attachment of their atoms, that is, with a different connectivity.

Contributing structures (Section 1.6A) Representations of a molecule, ion, or radical that differ only in the distribution of valence electrons by resonance.

Coupling (Section 13.12) The magnetic interaction of the nuclear spins of nearby atoms.

Coupling constant (*J*) (Section 13.12) The distance between peaks in a split signal, expressed in hertz. The value of J is a quantitative measure of the magnetic interaction of nuclei whose spins are coupled.

Covalent bond (Section 1.2C) A chemical bond formed between two atoms by sharing one or more pairs of electrons.

Crown ether (Section 11.11) A cyclic polyether derived from ethylene glycol and substituted ethylene glycols.

Crystalline domains (Section 23.4) Ordered crystalline regions in the solid state of a polymer. Also called crystallites.

Curved arrow (Section 1.6B) A symbol used to show the redistribution of valence electrons in resonance contributing structures or reactions.

Cyanohydrin (Section 15.6D) A molecule containing an —OH group and a —CN group bonded to the same carbon.

Cycloaddition reaction (Section 22.2) A reaction in which two reactants add together in a single step to form a cyclic product. The best known of these is the Diels-Alder reaction.

Cycloalkane (Section 2.4A) A saturated hydrocarbon that contains carbon atoms joined to form a ring.

Deactivating group (Section 20.1A) Any substituent on a benzene ring that causes the rate of electrophilic aromatic substitution to be lower than that for benzene.

Decarboxylation (Section 16.10) Loss of CO_2 from a carboxyl group.

Dehydration (Section 9.7) Elimination of water.

Dehydrohalogenation (Section 8.8B) Removal of —H and —X from adjacent carbons; a type of β-elimination.

DEPT-NMR (Section 13.14) Distortionless Enhancement by Polarization Transfer. A spectroscopic technique for distinguishing among ^{13}C signals for CH_3, CH_2, CH, and quaternary carbons in NMR.

Deshielding (Section 13.5) In NMR spectroscopy, an effect produced when electron density is decreased around a nucleus, causing it to absorb toward the left (downfield) on the chart paper.

Dextrorotatory (Section 4.8B) Refers to a substance that rotates the plane of polarized light to the right.

Diamagnetic current (Section 13.6) The circulation of electrons about a nucleus in an applied field. The resulting nuclear shielding is called diamagnetic shielding.

Diastereomers (Section 4.1) Stereoisomers that are not mirror images of each other.

Diaxial interactions (Section 2.6B) Interactions between atoms or groups in axial positions on the same side of a chair conformation of a cyclohexane ring.

Diazonium ion (Section 21.9D) An ArN_2^+ or RN_2^+ ion.

Dielectric constant (Section 8.2) A measure of a solvent's ability to insulate opposite charges from one another.

Dienophile (Section 22.2) A compound containing a double bond (containing one or two C, N, or O atoms) that can react with a conjugated diene to give a Diels-Alder adduct.

Dihedral angle (Section 2.6A) The angle created by two intersecting planes.

Diol (Section 9.2A) A compound containing two hydroxyl groups.

Dipeptide (Section 27.3) A molecule containing two amino acid units joined by a peptide bond.

Dipole moment (Section 1.5) The vector sum of individual bond moments in a molecule. Reported in Debye units (D).

Dipole-dipole interaction (Section 9.3A) The attraction between the positive end of one dipole and the negative end of another.

Disaccharide (Section 24.7) A carbohydrate containing two monosaccharide units joined by a glycoside bond.

Dispersion forces (Section 2.8) Very weak intermolecular coulombic forces of attraction.

Disproportionation (Section 23.6A) A termination process that involves the abstraction of a hydrogen atom from the beta position of the propagating radical of one chain by the radical endgroup of another chain.

Disulfide (Section 11.2) A molecule containing an —S—S— group.

Disulfide bond (Section 27.6C) A covalent bond between two sulfur atoms; an —S—S— bond.

Double bond (Section 5.1) A covalent bond consisting of one sigma (σ) bond and one pi (π) bond.

Double helix (Section 28.2B) A type of secondary structure of DNA molecules in which two antiparallel polynucleotide strands are coiled in a right-handed manner about the same axis.

Double-headed arrow (Section 1.6A) A symbol used to connect resonance contributing structures.

Downfield (Section 13.6) The shift of an NMR signal to the left on the chart paper. A downfield shift results when a nucleus is deshielded and its signal is produced by a weaker applied field.

E (Section 5.2C) From the German, *entgegen,* opposite. Specifies that groups of higher priority on the carbons of a double bond are on opposite sides.

***E,Z* system** (Section 5.2C) A system to specify the configuration of groups about a carbon-carbon double bond.

Eclipsed conformation (Section 2.6A) A conformation about a carbon-carbon single bond in which the atoms or groups on one carbon are as close as possible to the atoms or groups on an adjacent carbon.

Edman degradation (Section 27.4B) A method for selectively cleaving and identifying the *N*-terminal amino acid of a polypeptide chain.

Elastomer (Section 23.4) A material that, when stretched or otherwise distorted, returns to its original shape when the distorting force is released.

Electromagnetic radiation (Section 13.1) Light and other forms of radiant energy.

Electronegativity (Section 1.2C) A measure of the force of an atom's attraction for electrons it shares with another atom in a chemical bond.

Electrophile (Section 6.3A) Any electron deficient species that can accept a pair of electrons from a nucleophile to form a new covalent bond; a Lewis acid.

Electrophilic aromatic substitution (Section 20.1) A reaction in which there is substitution of an electrophile for a hydrogen on an aromatic ring.

Electrophoresis (Section 27.2E) The process of separating compounds on the basis of their electric charge.

β-Elimination (Section 8.8B) A reaction in which a small molecule, such as HCl, HBr, HI, or HOH is split out or eliminated from adjacent carbons.

Enamine (Section 15.10A) An unsaturated compound derived by the reaction of an aldehyde or ketone and a secondary amine followed by loss of H_2O; $CR_2{=}CR{-}NR_2$.

Enantiomeric excess (ee) (Section 4.8D) The difference in the number of moles of each enantiomer in a mixture compared with the total number of moles of both.

Enantiomers (Section 4.1) Stereoisomers that are nonsuperposable mirror images of each other.

Endothermic reaction (Section 6.2A) A reaction in which the energy of the products is higher than the energy of the reactants; a reaction in which heat is absorbed.

Enol (Section 10.8) A compound containing a hydroxyl group attached to a double bonded carbon atom.

Enolate anion (Section 15.11A) An anion derived by loss of a hydrogen from a carbon alpha to a carbonyl group; the anion of an enol.

Epoxide (Section 11.8) A cyclic ether in which oxygen is one atom of a three-membered ring.

Equatorial position (Section 2.6B) A position on a chair conformation of a cyclohexane ring that extends from the ring roughly perpendicular to the imaginary axis of the ring.

Equivalent hydrogens (Section 2.3C) Hydrogens that have the same chemical environment.

Ester (Section 17.1C) A compound in which an acyl group is bonded to —OR or to —OAr.

Estrogen (Section 25.4A) A steroid hormone, such as estrone, that mediates the development of sexual characteristics in females.

Ether (Section 11.1) A compound containing an oxygen atom bonded to two carbon atoms.

Exothermic reaction (Section 6.2A) A reaction in which the energy of the products is lower than the energy of the reactants; a reaction in which heat is liberated.

Fat (Section 25.1B) A mixture of triglycerides that is semisolid or solid at room temperature.

Fatty acid (Section 25.1A) A long, unbranched-chain carboxylic acid, most commonly of 12 to 20 carbons, derived from the hydrolysis of animal fats, vegetable oils, or the phospholipids of biological membranes.

Fingerprint region (Section 14.1D) The portion of the vibrational infrared region that extends from 1000 cm^{-1} to 400 cm^{-1}.

Fischer esterification (Section 16.8A) The process of forming an ester by refluxing a carboxylic acid and an alcohol in the presence of an acid catalyst, commonly H_2SO_4 or HCl.

Fischer projection (Section 4.4) A two-dimensional representation showing the configuration of a stereocenter; horizontal lines represent bonds projecting forward and vertical lines represent bonds projecting to the rear. The only atom in the plane of the paper is the stereocenter.

Fishhook arrow (Section 7.5A) A barbed curved arrow used to show the change in position of a single electron.

Flavin adenine dinucleotide (FAD) (Section 26.1C) A biological oxidizing agent. When acting as an oxidizing agent, FAD is reduced to $FADH_2$.

Fluid-mosaic model (Section 25.5C) A biological membrane consists of a phospholipid bilayer with proteins, carbohydrates, and other lipids embedded in and on the surface of the bilayer.

Formal charge (Section 1.2E) The charge on an atom in an ion or molecule.

Frequency (ν) (Section 13.1) The number of full cycles of a wave that pass a point in a second.

Frost circle (Section 19.2A) A graphic method for determining the relative energies of pi MOs for planar, fully conjugated, monocyclic compounds.

Functional group (Section 1.3) An atom or group of atoms within a molecule that shows a characteristic set of physical and chemical properties.

Furanose (Section 24.2A) A five-membered cyclic form of a monosaccharide.

Geminal (gem) dihalide (Section 10.6) Geminal, from the Latin, *geminatus,* twin. A compound containing two halogens on the same carbon atom.

Gilman reagent (Section 7.7C) A lithium diorganocopper reagent.

Glass transition temperature (Section 23.4) The temperature at which a polymer undergoes the transition from a hard glass to a rubbery state.

Glycol (Section 6.5B) A compound with two hydroxyl (—OH) groups on adjacent carbons.

Glycolysis (Section 26.5) From the Greek, *glyko,* sweet, and *lysis,* splitting. A series of ten enzyme-catalyzed reactions by which glucose is oxidized to two molecules of pyruvate.

Glycoside (Section 24.4A) A carbohydrate in which the —OH on its anomeric carbon is replaced by —OR.

Glycoside bond (Section 24.4A) The bond from the anomeric carbon of a glycoside to an —OR group.

Grignard reagent (Section 7.7A) An organomagnesium compound.

Ground-state electron configuration (Section 1.1A) The electron configuration of lowest energy for an atom, molecule, or ion.

Haloform (Section 7.2B) A compound of the type CHX_3 where X is a halogen.

Hammond's postulate (Section 7.5D) The structure of the transition state for an exothermic step looks more like the reactants. Conversely, the structure of the transition state for an endothermic step looks more like the products.

Haworth projection (Section 24.2A) A way to view furanose and pyranose forms of monosaccharides. The ring is drawn flat and viewed through its edge with the anomeric carbon on the right and the oxygen atom of the ring in the rear.

α-Helix (Section 27.6B) A type of secondary structure in which a section of polypeptide chain coils into a spiral, most commonly a right-handed spiral.

Hemiacetal (Section 15.8B) A molecule containing an —OH and an —OR or —OAr group bonded to the same carbon; the product of adding one molecule of alcohol to the carbonyl group of an aldehyde or ketone.

Henderson-Hasselbalch equation (Section 27.2B) A mathematical relationship that provides a direct way to calculate the ratio of a conjugate base to its weak acid as a function of pH.

Hertz (Hz) (Section 13.1) The unit in which frequency is measured; s^{-1} (read "per second").

Heterocyclic amine (Section 21.1) An amine in which nitrogen is one of the atoms of a ring.

Heterocyclic aromatic amine (Section 21.1) An amine in which nitrogen is one of the atoms of an aromatic ring.

Heterocyclic compound (Section 19.2D) An organic compound that contains one or more atoms other than carbon in its ring.

High-density lipoprotein (HDL) (Section 25.4A) Plasma particles, density 1.06–1.21 g/mL, consisting of approximately 33% proteins, 30% cholesterol, 29% phospholipids, and 8% triglycerides.

High-resolution mass spectrometry (Section 12.2A) Data from a mass spectrometer that is capable of distinguishing between ions that differ in mass by as little as 0.0001 atomic mass unit.

Histone (Section 28.2C) A protein, particularly rich in the basic amino acids lysine and arginine, that is found associated with DNA molecules.

Hofmann's rule (Section 21.10) Any β-elimination that occurs preferentially to give the least substituted double bond as the major product is said to follow Hofmann's rule.

Hückel criteria for aromaticity (Section 19.2A) To be aromatic, a monocyclic compound must have one p orbital on each atom of the ring, be planar or nearly so, and have $4n + 2$ pi electrons in the cyclic arrangement of p orbitals, where n is an integer including zero.

Hund's rule (Section 1.1A) When orbitals of equivalent energy are available but there are not enough electrons to fill all of them completely, then one electron is added to each equivalent orbital before a second electron is added to any one of them.

Hybrid orbital (Section 1.8B) An orbital formed by the combination of two or more atomic orbitals.

Hydration (Section 6.3B) Addition of water.

Hydride ion (Section 15.14B) A hydrogen atom with two electrons in its valence shell; $H:^-$.

Hydrocarbon (Section 2.1) A compound that is composed of only carbon and hydrogen atoms.

α-Hydrogen (Section 15.11A) A hydrogen on an alpha carbon atom.

Hydrogen bonding (Section 9.3A) The attractive interaction between dipoles when the positive end of one of the dipoles is a hydrogen atom bonded to an atom of high electronegativity (most commonly F, O, or N), and the negative end of the other dipole is an atom with a lone pair of electrons (most commonly F, O, or N).

Hydrogenolysis (Section 19.5C) Cleavage of a single bond by H_2, most commonly accomplished by treatment of a compound with H_2 in the presence of a transition metal catalyst.

Hydroperoxide (Section 11.6B) A molecule containing an —OOH group.

Hydrophilic (Section 8.7) From the Greek, meaning water-loving.

Hydrophobic (Section 8.7) From the Greek, meaning water-fearing.

Hydrophobic effect (Section 27.6D) The tendency of nonpolar groups to cluster so as to shield themselves from contact with an aqueous environment.

Hydroxyl group (Section 1.3A) An —OH group.

Hyperconjugation (Section 6.3A) Interaction of electrons in a sigma bonding orbital with the vacant $2p$ orbital of an adjacent positively charged carbon.

Imide (Section 17.1D) A functional group in which two acyl groups, RCO— or ArCO—, are bonded to a nitrogen atom.

Imine (Section 15.10A) A compound containing a carbon-nitrogen double bond; also called a Schiff base.

Inductive effect (Section 3.3D) The polarization of the electron density of a covalent bond due to a nearby atom of high electronegativity.

Infrared spectroscopy (Section 14.1A) A spectroscopic technique in which a compound is irradiated with infrared radiation, absorption of which causes covalent bonds to change from a lower vibrational energy level to a higher one. IR spectroscopy is particularly valuable for determining the kinds of functional groups in an organic molecule.

Intermediate (Section 6.2A) A species, formed between two successive reaction steps, that lies in a potential energy minimum between the two transition states.

Ionic bond (Section 1.2C) A chemical bond resulting from the electrostatic attraction of an anion and a cation.

Ionization potential (IP) (Section 12.1) The minimum energy required to remove an electron from an atom or molecule to a distance where there is no electrostatic interaction between ion and electron.

Isoelectric point (pI) (Section 27.2D) The pH at which an amino acid, polypeptide, or protein has no net charge.

Isomers (Section 4.1) Different compounds with the same molecular formula.

Isotactic polymer (Section 23.6C) A polymer with identical configurations (either all *R* or all *S*) at all stereocenters along its chain, as, for example, isotactic polyethylene.

Keto-enol tautomerism (Section 10.8) A type of isomerism involving keto (from ketone) and enol tautomers.

Ketone (Section 1.3B) A compound containing a carbonyl group bonded to two carbons.

Ketose (Section 24.1A) A monosaccharide containing a ketone group.

Kinetic control (Section 18.2) For a reaction under kinetic control, the distribution of products is determined by the relative rates of formation of each product.

Lactam (Section 17.1D) A cyclic amide.

Lactate fermentation (Section 26.7A) A metabolic pathway that converts glucose to two molecules of lactate.

Lactone (Section 17.1C) A cyclic ester.

Levorotatory (Section 4.8B) Refers to a substance that rotates the plane of polarized light to the left.

Lewis acid (Section 3.5) Any molecule or ion that can form a new covalent bond by accepting a pair of electrons.

Lewis base (Section 3.5) Any molecule or ion that can form a new covalent bond by donating a pair of electrons.

Lewis structure of an atom (Section 1.1B) The symbol of an element surrounded by a number of dots equal to the number of electrons in the valence shell of the atom.

Lindlar catalyst (Section 10.7A) Finely powdered palladium metal deposited on solid calcium carbonate that has been specially modified with lead salts. Its particular use is as a catalyst for the reduction of an alkyne to a *cis*-alkene.

Line-angle drawing (Section 2.4A) An abbreviated way to draw structural formulas in which an angle represents a carbon atom and a line represents a bond.

Lipid (Chapter 25) A biomolecule isolated from plant or animal sources by extraction with nonpolar organic solvents, such as diethyl ether and hexane.

Lipid bilayer (Section 25.5B) A back-to-back arrangement of phospholipid monolayers often forming a closed vesicle or membrane.

Living polymer (Section 23.6D) A polymer chain that continues to grow without chain-termination steps until either all of the monomer is consumed or some external agent is added to terminate the chain. The polymer chains will continue to grow if more monomer is added.

Low-density lipoprotein (LDL) (Section 25.4A) Plasma particles, density 1.02–1.06 g/mL, consisting of approximately 25% proteins, 50% cholesterol, 21% phospholipids, and 4% triglycerides.

Low-resolution mass spectrometry (Section 12.2A) Data obtained from a mass spectrometer that is capable of distinguishing only between ions that differ by one or more atomic mass units.

Markovnikov's rule (Section 6.3A) In the addition of HX, H_2O, or ROH to an alkene, hydrogen adds to the carbon of the double bond having the greater number of hydrogens.

Mass spectrometry (Introduction, Chapter 12) An analytical technique for measuring the mass-to-charge ratio (m/z) of positive ions produced in the gas phase.

Mass spectrum (Section 12.1) A plot of the relative abundance of ions versus their mass-to-charge ratio.

Maxam-Gilbert method (Section 28.5) A method, developed by Allan Maxam and Walter Gilbert, for sequencing DNA molecules.

Melt transition, T_m (Section 23.4) The temperature at which crystalline regions of a polymer melt.

Mercaptan (Section 9.2B) A common name for a compound containing an —SH group; a thiol.

Meso compound (Section 4.5B) An achiral compound possessing two or more stereocenters.

Messenger RNA (mRNA) (Section 28.3C) A ribonucleic acid that carries coded genetic information from DNA to the ribosomes for the synthesis of proteins.

Meta (*m*) (Section 19.3B) Refers to groups occupying 1,3 positions on a benzene ring.

Meta director (Section 20.1A) Any substituent on a benzene ring that directs electrophilic aromatic substitution preferentially to a meta position.

Micelle (Section 25.2B) A spherical arrangement of organic molecules in water solution clustered so that their hydrophobic parts are buried inside the sphere and their hydrophilic parts are on the surface of the sphere and in contact with water.

Mirror image (Section 4.1) The reflection of an object in a mirror.

Molecular ion (M) (Section 12.1) The species formed by removal of a single electron from a molecule.

Molecular spectroscopy (Section 13.2) The study of which frequencies of electromagnetic radiation are absorbed or emitted by substances and the correlation between these frequencies and specific types of molecular structure.

Monomer (Section 23.2) From the Greek, *mono* + *meros,* single part. The simplest nonredundant unit from which a polymer is synthesized.

Monosaccharide (Section 24.1A) A carbohydrate that cannot be hydrolyzed to a simpler carbohydrate.

D-Monosaccharide (Section 24.1D) A monosaccharide that, when written as a Fischer projection, has the —OH on its penultimate carbon to the right.

L-Monosaccharide (Section 24.1D) A monosaccharide that, when written as a Fischer projection, has the —OH on its penultimate carbon to the left.

Mutarotation (Section 24.2C) The change in specific rotation that occurs when an α or β form of a carbohydrate in aqueous solution is converted to an equilibrium mixture of the two forms.

(n + 1) Rule (Section 13.10) The ^{1}H-NMR signal of a hydrogen or set of equivalent hydrogens is split into $(n + 1)$ peaks by a nonequivalent set of n equivalent neighboring hydrogens.

Newman projection (Section 2.6A) A way to view a molecule by looking along a carbon-carbon single bond.

Nicotinamide adenine dinucleotide (NAD^+) (Section 26.1B) A biological oxidizing agent. When acting as an oxidizing agent, NAD^+ is reduced to NADH.

Nitrile (Section 17.1E) A compound containing a —C≡N (cyano) group.

Nitrogen rule (Section 12.3) If a compound has an odd number of nitrogen atoms, its molecular ion will appear at an odd m/z value; if zero or an even number of nitrogen atoms, the molecular ion will be of even mass.

Node (Section 1.7A) Any point in space where the value of a wave function is zero.

Nonbonded interaction strain (Section 2.6A) The strain that arises when atoms not bonded to each other are forced abnormally close to one another.

Nonbonding electrons (Section 1.2D) Valence electrons not involved in forming covalent bonds, i. e., unshared electrons.

Nonpolar covalent bond (Section 1.2C) A covalent bond between atoms whose difference in electronegativity is less than 0.5.

Nuclear magnetic resonance (NMR) spectroscopy (Section 13.2) A spectroscopic technique that gives us information about the number and types of atoms in a molecule, for example, about hydrogens using ^{1}H-NMR spectroscopy, and about carbons using ^{13}C-NMR spectroscopy.

Nucleic acid (Section 28.1) A biopolymer containing three types of monomer units: heterocyclic aromatic amine bases derived from purine and pyrimidine, the monosaccharides D-ribose or 2-deoxy-D-ribose, and phosphoric acid.

Nucleophile (Introduction, Chapter 8) From the Greek, meaning nucleus loving. A molecule or ion that donates a pair of electrons to another atom or ion to form a new covalent bond; a Lewis base.

Nucleophilic acyl substitution (Introduction, Chapter 17) A reaction in which a nucleophile bonded to the carbonyl carbon of an acyl group is replaced by another nucleophile.

Nucleophilic aromatic substitution (Section 20.3B) A reaction in which a nucleophile on an aromatic ring is replaced by another nucleophile.

Nucleophilic substitution (Introduction, Chapter 8) Any reaction in which one nucleophile is substituted for another.

Nucleophilicity (Section 8.4B) A kinetic property measured by the rate at which a nucleophile causes nucleophilic substitution on a reference compound under a standardized set of experimental conditions.

Nucleoside (Section 28.1) A building block of nucleic acids, consisting of D-ribose or 2-deoxy-D-ribose bonded to a heterocyclic aromatic amine base by a β-*N*-glycoside bond.

Nucleotide (Section 28.1) A nucleoside in which a molecule of phosphoric acid is esterified with an —OH of the monosaccharide, most commonly either the 3′-OH or the 5′-OH.

Number average molecular weight, M_n (Section 23.3) An arithmetic mean calculated by summing the number of chains of a particular molecular weight in a collection of polymer molecules and dividing the sum by the total number of chains.

Observed rotation (Section 4.8B) The number of degrees through which a compound rotates the plane of polarized light.

Octane number (Section 2.9B) The percentage of isooctane in a test mixture of isooctane and heptane that has equivalent knock properties to a gasoline being tested.

Octet rule (Section 1.2A) The tendency among atoms of Group IA–VIIA elements to react in ways that achieve an outer shell of eight valence electrons.

Oil (Section 25.1B) In the context of fats and oils, a mixture of triglycerides that is liquid at room temperature.

Oligosaccharide (Section 24.7) A carbohydrate containing from four to ten monosaccharide units, each joined to the next by a glycoside bond.

Optical purity (Section 4.8D) The specific rotation of a mixture of enantiomers divided by the specific rotation of the enantiomerically pure substance.

Optically active (Section 4.8) Refers to a compound that rotates the plane of polarized light.

Orbital (Section 1.1) A region of space where an electron or pair of electrons spends 90%–95% of its time.

Order of Precedence (Section 15.2B) A system for ranking functional groups in order of priority for the purposes of IUPAC nomenclature.

Organometallic compound (Section 7.7) A compound that contains a metal bonded to a carbon atom.

Ortho (*o*) (Section 19.3B) Refers to groups occupying 1,2 positions on a benzene ring.

Ortho-para director (Section 20.2A) Any substituent on a benzene ring that directs electrophilic aromatic substitution preferentially to ortho and para positions.

Oxidation (Section 6.5A) The loss of electrons.

β-Oxidation (Section 26.3B) A series of four enzyme-catalyzed reactions that cleaves carbon atoms, two at a time, from the carboxyl end of a fatty acid by intermediates that are oxidized at the β-position.

Oxonium ion (Section 6.3B) An ion in which oxygen bears a positive charge.

Para (*p*) (Section 19.3B) Refers to groups occupying 1,4 positions on a benzene ring.

Pauli exclusion principle (Section 1.1A) No more than two electrons may be present in an orbital, one with spin $+\frac{1}{2}$, the other with spin $-\frac{1}{2}$.

Peaks (Section 13.10) The units into which an NMR signal is split; two in a doublet, three in a triplet, four in a quartet, and so on.

Penultimate carbon (Section 24.1D) The stereocenter of a monosaccharide farthest from the carbonyl group, as, for example, carbon 5 of glucose.

Peptide bond (Section 27.3) The special name given to the amide bond formed between the α-amino group of one amino acid and the α-carboxyl group of another amino acid.

Pericyclic reaction (Section 22.2D) A reaction that takes place in a single step, without intermediates, and involves a cyclic redistribution of bonding electrons.

Phase-transfer catalyst (Section 8.7) A substance that transfers ions from an aqueous phase into an organic phase and vice versa.

Phenol (Section 19.5A) A compound that contains an —OH bonded to a benzene ring; a benzenol.

Phenyl group C_6H_5— (Section 19.3A) The aryl group derived by removing a hydrogen from benzene.

Phospholipid (Section 25.5A) A lipid containing glycerol esterified with two molecules of fatty acid and one molecule of phosphoric acid.

Phosphoric acid anhydride (Section 17.1B) A functional group in which two phosphoryl groups are bonded to an oxygen atom.

Pi (π) bond (Section 1.8D) A covalent bond formed by the overlap of parallel *p* orbitals.

Pi (π) molecular orbital (Section 1.8D) A molecular orbital formed by the overlap of *p* orbitals on adjacent atoms; its electron density lies above and below the line connecting the atoms.

Plane of symmetry (Section 4.2) An imaginary plane passing through an object dividing it so that one half is the mirror image of the other half.

Plane polarized light (Section 4.8A) Light vibrating in only one plane.

Plastic (Section 23.2) A polymer that can be molded when hot and retains its shape when cooled.

β-Pleated sheet (Section 27.6B) A type of secondary structure in which sections of polypeptide chains are aligned parallel or antiparallel to one another.

Polar covalent bond (Section 1.2C) A covalent bond between atoms whose difference in electronegativity is between 0.5 and 1.9.

Polarimeter (Section 4.8B) An instrument for measuring the ability of a compound to rotate the plane of polarized light.

Polyamide (Section 23.5A) A polymer in which each monomer is joined to the next by an amide bond, as, for example, nylon 66.

Polycarbonate (Section 23.5C) A polyester in which the carboxyl groups are derived from carbonic acid.

Polyester (Section 23.5B) A polymer in which each monomer unit is joined to the next by an ester bond, as, for example, poly(ethylene terephthalate).

Polymer (Section 23.2) From the Greek, *poly* + *meros,* many parts. Any long-chain molecule synthesized by linking together many single parts called monomers.

Polynuclear aromatic hydrocarbon (Section 19.3B) A hydrocarbon containing two or more fused aromatic rings.

Polypeptide (Section 27.3) A macromolecule containing many amino acid units, each joined to the next by a peptide bond.

Polysaccharide (Section 24.8) A carbohydrate containing a large number of monosaccharide units, each joined to the next by one or more glycoside bonds.

Polyunsaturated fatty acid (Section 25.1A) A fatty acid with two or more carbon-carbon double bonds in its hydrocarbon chain.

Polyunsaturated triglyceride (Section 25.1B) A triglyceride having several carbon-carbon double bonds in the hydrocarbon chains of its three fatty acids.

Polyurethane (Section 23.5D) A polymer containing the —$NHCO_2$— group as a repeating unit.

Potential energy (PE) diagram (Section 6.2A) A graph showing the changes in energy that occur during a chemical reaction; potential energy is plotted on the vertical axis and reaction progress is plotted on the horizontal axis.

Primary (1°) amine (Section 21.1) An amine in which one hydrogen of ammonia has been replaced by an alkyl or aryl group.

Primary (1°) carbon (Section 2.3C) A carbon bonded to one other carbon atom.

Primary structure, of nucleic acids (Section 28.2A) The sequence of bases along the pentose-phosphodiester backbone of a DNA or RNA molecule read from the 5′ end to the 3′ end.

Primary structure, of proteins (Section 27.4) The sequence of amino acids in the polypeptide chain, read from the *N*-terminal amino acid to the *C*-terminal amino acid.

Principle of microscopic reversibility (Section 9.7) The sequence of transition states and reactive intermediates in the mechanism of any reversible reaction must be the same, but in reverse order, for the backward reaction as for the forward reaction.

Prostaglandin (Section 25.3) A member of the family of compounds having the 20-carbon skeleton of prostanoic acid.

Protic solvent (Section 8.2) A solvent that is a hydrogen bond donor; the most common protic solvents contain —OH groups.

Pyranose (Section 24.2A) A six-membered cyclic form of a monosaccharide.

Quantum mechanics (Section 1.7A) A branch of science that studies particles and their associated waves.

Quaternary (4°) ammonium ion (Section 21.1) An ion in which nitrogen is bonded to four carbons and bears a positive charge.

Quaternary (4°) carbon (Section 2.3C) A carbon bonded to four other carbon atoms.

Quaternary structure (Section 27.6D) The arrangement of polypeptide monomers into a noncovalently bonded aggregate.

R (Section 4.3) From the Latin, *rectus,* right; used in the *R,S* convention to show that the order of priority of groups on a stereocenter is clockwise.

R— (Section 2.3A) A symbol used to represent an alkyl group.

***R,S* convention** (Section 4.3) A set of rules for specifying configuration about a stereocenter; also called the Cahn-Ingold-Prelog convention.

Racemic mixture (Section 4.8C) A mixture of equal amounts of two enantiomers.

Radical (Section 7.5) Any chemical species that contains one or more unpaired electrons.

Rate-limiting step (Section 6.2A) The step in a multistep reaction sequence that crosses the highest potential energy barrier.

Reaction coordinate (Section 6.2A) A measure of the change in the positions of atoms during a reaction, plotted on the horizontal axis in a potential energy diagram.

Reaction mechanism (Section 6.2A) A step-by-step description of how a chemical reaction occurs.

Rearrangement (Section 6.3C) A change in connectivity of the atoms in a product compared with the connectivity of the same atoms in the starting material.

Reducing sugar (Section 24.4D) A carbohydrate that reduces Ag(I) to Ag or Cu(II) to Cu(I).

Reduction (Section 6.5A) The gain of electrons.

Regioselective reaction (Section 6.3A) A reaction in which one direction of bond forming or bond breaking occurs in preference to all other directions.

Resolution (Section 4.9) Separation of a racemic mixture into its enantiomers. (Section 12.2A) In mass spectrometry, a measure of how well a mass spectrometer separates ions of different mass.

Resonance energy (Section 19.1C) The difference in energy between a resonance hybrid and the most stable of its hypothetical contributing structures in which electrons are localized on particular atoms and in particular bonds.

Resonance hybrid (Section 1.6A) A molecule, ion, or radical described as a composite of a number of contributing structures.

Resonance, in NMR spectroscopy (Section 13.5) The absorption of electromagnetic radiation by a precessing nucleus and the resulting "flip" of its nuclear spin from a lower energy state to a higher energy state.

Restriction endonuclease (Section 28.5) An enzyme that catalyzes hydrolysis of a particular phosphodiester bond within a DNA strand.

Retrosynthesis (Section 10.10) A process of reasoning backwards from a target molecule to a suitable set of starting materials.

Ribosomal RNA (rRNA) (Section 28.3A) A ribonucleic acid found in ribosomes, the sites of protein synthesis.

S (Section 4.3) From the Latin, *sinister,* left; used in the *R,S* convention to show that the order of priority of groups on a stereocenter is counterclockwise.

Saponification (Section 17.5C) Hydrolysis of an ester in aqueous NaOH or KOH to an alcohol and the sodium or potassium salt of a carboxylic acid.

Saturated hydrocarbon (Section 2.1) A hydrocarbon containing only carbon-carbon single bonds.

Secondary (2°) amine (Section 21.1) An amine in which two hydrogens of ammonia have been replaced by alkyl or aryl groups.

Secondary (2°) carbon (Section 2.3C) A carbon bonded to two other carbon atoms.

Secondary structure, of nucleic acids (Section 28.2B) The ordered arrangement of nucleic acid strands.

Secondary structure, of proteins (Section 27.6A) The ordered arrangements (conformations) of amino acids in localized regions of a polypeptide or protein.

Shell (Section 1.1) A region of space around a nucleus in which electrons are found.

Shielding (Section 13.5) In NMR spectroscopy, an effect produced when electron density is increased about a nucleus, causing it to absorb toward the right (upfield) on the chart paper.

Sigma (σ) molecular orbital (Section 1.8A) A molecular orbital in which electron density is concentrated between two nuclei along the axis joining them.

Signal (Section 13.5) A recording in an NMR spectrum of a nuclear magnetic resonance.

Signal splitting (Section 13.10) Splitting of an NMR signal into a set of peaks by the influence of nonequivalent nuclei on the same or adjacent atom(s).

Soap (Section 25.2A) A sodium or potassium salt of a fatty acid.

Solvolysis (Section 8.4G) Any nucleophilic substitution in which the solvent is also the nucleophile.

***sp* Hybrid orbital** (Section 1.8E) A hybrid atomic orbital formed by the combination of one *s* atomic orbital and one *p* atomic orbital.

sp^2 Hybrid orbital (Section 1.8D) A hybrid atomic orbital formed by the combination of one *s* atomic orbital and two *p* atomic orbitals.

sp^3 Hybrid orbital (Section 1.8C) A hybrid atomic orbital formed by the combination of one *s* atomic orbital and three *p* atomic orbitals.

Specific rotation (Section 4.8B) Observed rotation of the plane of polarized light when a sample is placed in a tube 1.0 dm in length and at a concentration of 1 g/mL.

Spiroalkane (Section 2.4B) A cycloalkane in which two rings share only one atom.

Staggered conformation (Section 2.6A) A conformation about a carbon-carbon single bond in which the atoms or groups on one carbon are as far apart as possible from atoms or groups on an adjacent carbon.

Step-growth polymerization (Section 23.5) A polymerization in which chain growth occurs in a stepwise manner between difunctional monomers, as, for example, between adipic acid and hexamethylenediamine to form nylon 66.

Stereocenter (Section 4.2) An atom that has four different atoms or groups of atoms attached to it; also called a stereogenic center.

Stereoisomers (Section 4.1) Isomers that have the same molecular formula and the same connectivity but different orientations of their atoms in space that cannot be interconverted by rotation about a single bond.

Stereoselective reaction (Section 6.3D) A reaction in which one stereoisomer is formed or destroyed in preference to all others that may be formed or destroyed.

Steroid (Section 25.4A) A plant or animal lipid having the characteristic tetracyclic ring structure of the steroid nucleus, namely three six-membered rings and one five-membered ring.

Substitution (Section 7.4) A reaction in which an atom or group of atoms in a compound is replaced by another atom or group of atoms.

Substrate (Section 26.6) A compound upon which an enzyme acts in an enzyme-catalyzed reaction.

Sulfide (Section 11.2) The sulfur analog of an ether; a molecule containing a sulfur atom bonded to two carbon atoms.

Syn stereoselectivity (Section 6.4) The addition of atoms or groups of atoms on the same face of a carbon-carbon double bond.

Syndiotactic polymer (Section 23.6C) A polymer with alternating *R* and *S* configurations at the stereocenters along its chain, as, for example, isotactic polyethylene.

Tautomers (Section 10.8) Constitutional isomers in equilibrium with each other which differ in the location of a hydrogen atom and a double bond relative to a heteroatom, most commonly O, N, or S.

Telechelic polymer (Section 23.6D) A polymer in which its growing chains are terminated by formation of new functional groups at both ends of its chains. These new functional groups are introduced by adding reagents, such as CO_2 or ethylene oxide, to the growing chains.

***C*-Terminal amino acid** (Section 27.4) The amino acid at the end of a polypeptide chain having the free $—CO_2H$ group.

***N*-Terminal amino acid** (Section 27.4) The amino acid at the end of a polypeptide chain having the free $—NH_2$ group.

Terpene (Section 5.4) A compound whose carbon skeleton can be divided into two or more units identical with the carbon skeleton of isoprene.

Tertiary (3°) amine Section (21.1) An amine in which all three hydrogens of ammonia have been replaced by alkyl or aryl groups.

Tertiary (3°) carbon (Section 2.3C) A carbon bonded to three other carbon atoms.

Tertiary structure, of nucleic acids (Section 28.2C) The three-dimensional arrangement of all atoms of a nucleic acid, commonly referred to as supercoiling.

Tertiary structure, of proteins (Section 27.6C) The three-dimensional arrangement in space of all atoms in a single polypeptide chain.

Thermodynamic control (Section 18.2) An experimental condition that permits the establishment of equilibrium between the two or more products of a reaction. The composition of the product mixture is determined by the relative stabilities of the products.

Thermoplastic (Section 23.2) A polymer that can be melted and molded into a shape that is retained when it is cooled.

Thermoset plastic (Section 23.2) A polymer that can be molded when it is first prepared, but once cooled, hardens irreversibly and cannot be remelted.

Thioester (Section 26.3A) An ester in which one atom of oxygen in the carboxylate group is replaced by an atom of sulfur.

Thiol (Section 9.1B) A compound containing an —SH (sulfhydryl) group bonded to an sp^3 hybridized carbon.

Torsional strain (Section 2.6A) The force that opposes the rotation of one part of a molecule about a bond while the other part of the molecule is held fixed.

Trans (Section 2.7A) A prefix meaning across from.

Transesterification (Section 17.6C) Exchange of one —OR or —OAr group of an ester for another —OR or —OAr group.

Transfer RNA (tRNA) (Section 28.3B) A ribonucleic acid that carries a specific amino acid to the site of protein synthesis on ribosomes.

Transition state (Section 6.2A) An unstable species of maximum potential energy formed during the course of a reaction; an energy maximum on a potential energy diagram.

Triglyceride (triacylglycerol) (Section 25.1) An ester of glycerol with three fatty acids.

Tripeptide (Section 27.3) A molecule containing three amino acid units, each joined to the next by a peptide bond.

Ultraviolet-visible spectroscopy (Section 14.3) A spectroscopic technique in which a compound is irradiated with ultraviolet or visible radiation, absorption of which causes electrons to change from a lower energy level to a higher one. Ultraviolet spectroscopy is particularly valuable for determining the extent of conjugation in organic molecules containing pi bonds.

Unimolecular reaction (Section 8.3) A reaction in which only one species is involved in the rate-limiting step.

Upfield (Section 13.6) The shift of an NMR signal to the right on the chart paper. An upfield shift results when a nucleus is shielded and a stronger applied field is required to produce its signal.

Valence electrons (Section 1.1B) Electrons in the valence (outermost) shell of an atom.

Valence shell (Section 1.1B) The outermost electron shell of an atom.

Van der Waals forces (Section 7.3B) A group of intermolecular attractive forces including dipole-dipole, dipole-induced dipole, and induced dipole-induced dipole (dispersion) forces.

Vibrational infrared (Section 14.1A) The portion of the infrared region that extends from 4000 cm^{-1} to 400 cm^{-1}.

Vicinal (vic) dihalide (Section 10.6) Vicinal from the Latin, *vicinalis,* neighbor. A compound containing two halogen atoms on adjacent carbon atoms.

Vinylic carbocation (Section 10.9B) A carbocation in which the positive charge is on one of the carbons of a carbon-carbon double bond.

Vinylic halide (Section 7.1) A compound containing a halogen bonded to one of the carbons of a carbon-carbon double bond.

Watson-Crick model (Section 28.2B) A double-helix model for the secondary structure of a DNA molecule.

Wavenumber ($\bar{\nu}$) (Section 14.1A) A frequency of electromagnetic radiation expressed as the number of waves per centimeter.

Wavelength (λ) (Section 13.1) The distance between consecutive peaks on a wave.

Weight average molecular weight, M_w (Section 23.3) The sum of the weights of chains of a particular length in a collection of polymer molecules divided by the total weight of the sample.

Ylide (Section 15.7) A molecule which, when written in a Lewis structure showing all atoms with complete valence shells, has positive and negative charges on adjacent atoms.

Z (Section 5.2C) From the German, *zusammen,* together. Specifies that groups of higher priority on the carbons of a double bond are on the same side.

Zaitsev's rule (Section 8.8B) The major product of a β-elimination reaction is the most stable, that is, the most highly substituted, alkene.

Zwitterion (Section 27.1A) An internal salt of an amino acid.

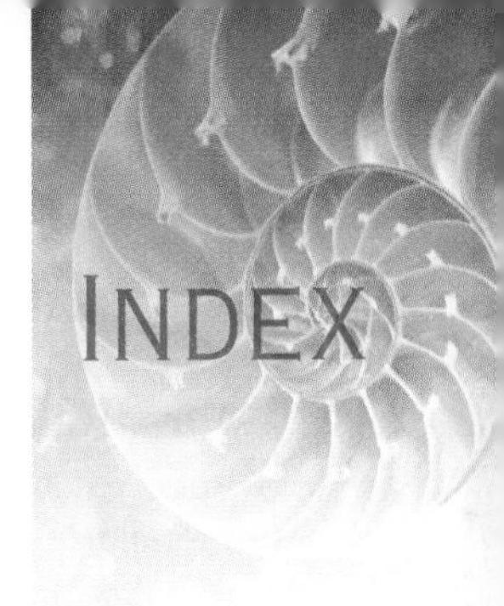

Note: A "B" following a page number indicates a Chemistry in Action box, "F" indicates a figure, "P" indicates a problem, and "T" indicates a table.

A

C

E

N

O

P

S

T

U

The Developing Person
Through the Life Span

Fourth Edition

KATHLEEN STASSEN BERGER

Bronx Community College

City University of New York

With the assistance of

ROSS A. THOMPSON

University of Nebraska—Lincoln

WORTH PUBLISHERS

The Developing Person Through the Life Span, Fourth Edition

Printed in the United States of America

Library of Congress Catalog Card Number: 96-60598

ISBN: 1-57259-106-4

Printing: 2 3 4 5 – 02 01 00 99 98

Executive Editor: Catherine Woods

Developmental Editor: Peter Deane

Design: Malcolm Grear Designers

Art Director: George Touloumes

Production Editor: Timothy Prairie

Production Manager: Barbara Anne Seixas

Layout: Fernando Quiñones

Photo Editor: Elyse Rieder

Graphics Art Manager: Demetrios Zangos

Composition and separations: TSI Graphics, Inc.

Printing and binding: R.R. Donnelley & Sons Company

Cover: *The Clinic*, Francis Livingston, 1987. Oil on panel (detail).

Acknowledgments begin on page IC-1, and constitute an extension of the copyright page.

Worth Publishers

33 Irving Place

New York, NY 10003

Contents

The Developing Person

Through the Life Span

The Clinic, Francis Livingston, 1987. Oil on panel.

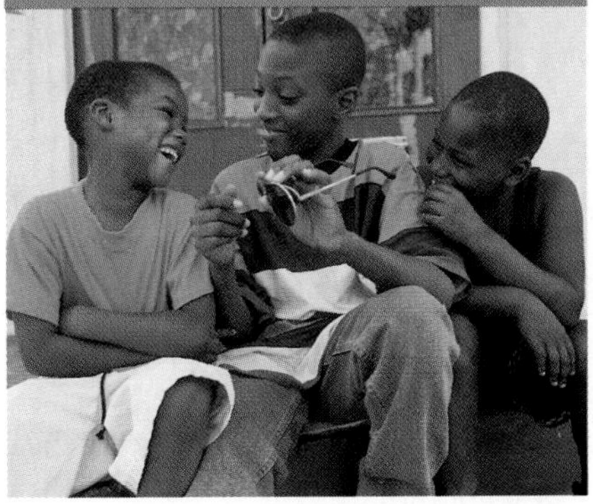

About the Author

Kathleen Stassen Berger received her undergraduate education at Stanford University and Radcliffe College, and earned an M.A.T. from Harvard University and an M.S. and Ph.D. from Yeshiva University. Her broad experience as an educator includes directing a preschool, teaching philosophy and humanities at the United Nations International School, teaching child and adolescent development to graduate students at Fordham University, and teaching social psychology to inmates earning a paralegal degree at Sing Sing Prison.

For the past twenty-five years at Bronx Community College of the City University of New York, Professor Berger has taught introductory psychology, child and adolescent development, adulthood and aging, social psychology, abnormal psychology, and human motivation. Her students—who come from many ethnic, economic, and educational backgrounds and have a wide range of interests—consistently honor her with the highest teaching evaluations. She recently received the Golden Acorn Award for Outstanding Service to Children.

Professor Berger is the author of *The Developing Person Through Childhood and Adolescence,* currently being used at over 460 colleges and universities, and she has contributed articles on developmental topics to the *Wiley Encyclopedia of Psychology* as well as book reviews to various professional journals. Her research interests include adolescent identity, sibling relationships, and employed mothers. She is also active in the community, including serving as president of her local school board. As the mother of four daughters and the daughter of two parents in their 90s, Kathleen Stassen Berger brings to her teaching and writing ample first-hand experience with human development.

Contents in Brief

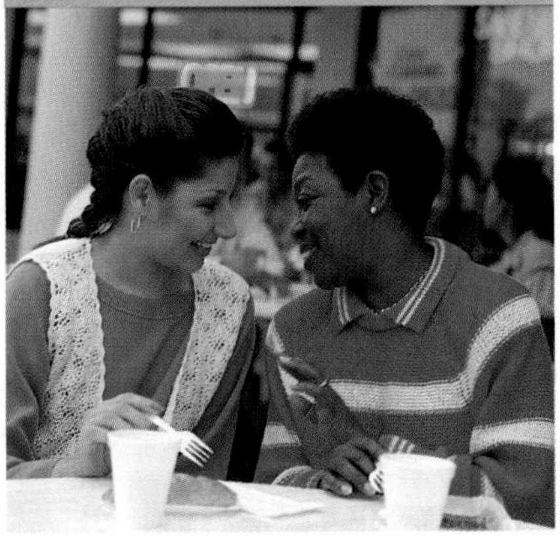

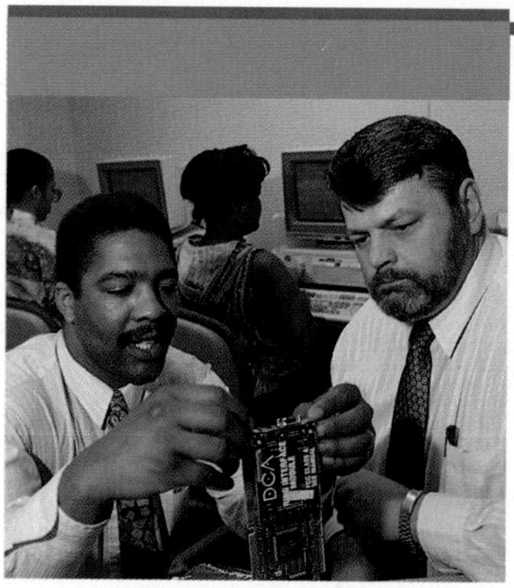

Preface

Teaching and learning are not just my profession and my vocation: they are my passion. I deeply believe that we all live richer, fuller, and more connected lives when we understand ourselves and others and the social contexts that affect us all. Just as deeply, I believe that developmental psychology can be a powerful impetus and guide for moving people toward such understanding.

Consequently, when I began the first edition of this book sixteen years ago, I had a single goal in mind: to reveal the study of life-span development as the intriguing, exciting, and critically important discipline I myself find it to be. Years of teaching and studying had convinced me that a text should respect students' interests and experiences and at the same time reflect the complexity of human development—without being condescending or, alternatively, so overburdened with theoretical and academic details as to be dull and difficult. I wanted to present theory, research, practical examples, and controversial issues in such a way as to inspire critical thinking, insight, and pleasure as well. The response of instructors and students to *The Developing Person Through the Life Span* has been enormously gratifying, encouraging me to believe that the book is fulfilling its goal, and at the same time making me eager to improve as well as update each new edition.

Changes and Highlights of This Edition

Development means change, day-by-day, year-by-year, era-by-era, as well as continuity throughout. Every new life is similar to those that preceded it, but is also affected by new circumstances—scientific, technological, cultural, and so on. Accordingly, this new edition reflects new research, perspectives, and contexts that shape our lives. Revisions are apparent on every page, but four particularly pervasive shifts should be noted.

Social Contexts

Throughout this edition I have given added emphasis both to the ways development is shaped by each individual's specific social context and to the developmental diversity that exists across social contexts. This added emphasis includes particular attention to the impact that income, education, ethnicity, race, sex, and historical period can have on developmental outcomes.

The theme of diversity within universal patterns is evidenced especially in discussions of cultural influences on development. On a wide variety of topics—ranging from infants' sleep schedules, children's perception of spanking, and adolescents' propensity to take risks to adults' selection of career patterns and mates and elders' decisions about whether to retire, where to live, and how to die—I have tried to show how cultural differences in values and customs shape individual development throughout the life span.

Cognitive Development

The cognitive chapters have been reorganized and reoriented to reflect the latest research findings. In infancy, perception and memory are given new prominence, as is theory of mind in preschool development. This current perspective means that Piagetian theory is no longer preeminent, but I have retained many of Piaget's insights that provide illumination into the thought processes of children and adolescents.

Intellectual development over the years of adulthood is portrayed in its multidimensional complexity, influenced by physical and mental health, personal motivation, past and current education, genetic makeup, declines in information-processing capacity, ageist stereotypes, and social opportunities for new learning. Special emphasis is now given to practical intelligence, creativity, expertise, and ongoing education.

Generations and Cohorts

I see each of us as a link in a generational line of ancestors and descendants, with every birth date signifying membership in a particular cohort. Accordingly, intergenerational themes are woven into the text throughout, helping students to more fully appreciate their own location in the developmental path. Families are shown as the multigenerational, extralegal units that they are, defined more by the support and commitment they provide than by genetic overlap or the specifics of household occupancy. I have also given greater attention to differences in historical experience, showing how these differences affect not only development itself but our understanding of it as well.

Public Policy

It is obvious that laws and institutions can enhance or limit development; it is less apparent but equally true that individual decisions—and often the individual voice—can influence the political process. To highlight this interactive view, Public Policy boxes have been included in this edition to focus on some of the most compelling policy issues in human development. Critical issues, both practical and ethical, are presented in their multisided complexity.

New Topics

In addition to the changes and shifts in emphasis noted above, this edition includes a great many new topics that will be of special interest to students.

Here is a sampling of just a few of them:

heritability and what it does and does not mean

implications of the Human Genome Project

teratogenic risk and genetic vulnerability

preventive medicine in infancy

sudden infant death syndrome

the "transient exuberance" of early neural development

infants' surprising long-term memory

effects of early day care

the contextual dimensions of child maltreatment

the effects of spanking

autism and ADHD

IQ testing and multiple intelligences

bullies and their victims

the impact of poverty and homelessness on children

"gateway drugs" and adolescent drug abuse

suicide contagion among adolescents

legal and ethical questions surrounding fertility technology

spouse abuse

adapting to the modern workplace

balancing work and family roles

health differences related to SES, ethnicity, gender, and geography

hormone replacement therapy

the myth of the midlife crisis

caring for disabled adult children

the squaring of the population pyramid

generational equity

new views on intellectual decline in old age

assisted suicide and hastening death

In a number of important ways, the book remains unchanged, including its basic organization. The first part consists of four chapters that deal with, respectively, the definitions and goals of development study, the major theories and methodologies of the field, the interplay of heredity and environment, and prenatal development and birth. The remainder of the book is divided into seven parts that correspond to the seven major periods of life-span development—infancy, early childhood, middle childhood, adolescence, early adulthood, middle adulthood, and late adulthood. Each of these parts consists of a trio of chapters dealing with, respectively, biosocial development, cognitive development, and psychosocial development. This topical organization within a chronological framework fosters students' appreciation of how the various aspects of development are interrelated—of how body, intellect, and personality develop through interaction rather than separately.

The pedagogical aids have also been retained. Thus, at the end of each chapter there is a chapter summary, a list of key terms (with page numbers indicating where the term was introduced), and a series of key questions for reviewing important concepts. At the end of each part there is a full-page chart that provides an overview of the significant biosocial, cognitive, and psychological events covered in that part. A comprehensive glossary at the back of the book lists all the key terms that appear in the text, along with the page number for each term's initial use. One important new pedagogical aid is the addition of an on-page glossary, with definitions of key terms occurring in the margin of the page on which the terms are first used.

Supplementary Materials

As one who has taught many courses in college and graduate school for twenty years, I know that some instructor's aids are not very helpful, and that many of my colleagues ignore them. If this describes you, I urge you to examine the resources available with this book. I think you will be pleasantly surprised by the exceptional quality and usefulness of these supplements.

The *Study Guide* by Richard Straub (University of Michigan, Dearborn) uses the SQ3R format to help students learn more and retain their learning longer. Each chapter includes a review of the key concepts, guided study questions, and section reviews that make students active participants in the learning process. Two practice tests and a challenge test of multiple-choice, true/false, and matching questions help students to determine their degree of mastery of the material. The correct answers to test questions are explained, to ensure understanding.

Each chapter of the *Instructor's Resource Manual* by Richard Straub features a chapter preview and lecture guide, learning objectives, lecture/discussion/debate topics, handouts for group and individual student projects, and supplementary readings from journal articles with introductions and questions. The general resources include course planning suggestions, ideas for term projects, including observational activities, and a guide to commercially available audio-visual and software materials. New to this edition of the Instructor's Resource Manual are internet assignments for instructors to help students access the World Wide Web for better research resources.

A set of acetate *transparencies* of key illustrations, charts, tables, and summary information from the textbook, as well as supplemental material by Richard Straub, is available to adopters.

An extensive *Test Bank*, revised by Carolyn Meyer (Lake Sumter Community College), includes approximately 80 multiple-choice questions and 50 fill-in, true/false, and essay questions for each chapter. Each question is keyed to the textbook topic and page numbers, and its level of difficulty is noted. The *Test Bank* questions are also available with test-generation systems for Windows and Macintosh.

A new supplement, *The Scientific American Frontiers Video Collection for Developmental Psychology*, provides instructors with twelve video segments (approximately 15 minutes each). Use these videos as lecture launchers or to emphasize or clarify material from the text.

The Developing Person Through the Life Span is the textbook that accompanies "Seasons of Life," a telecourse produced by The Annenberg/CPB

Project that first aired on public television in September 1989. A new edition of the telecourse study guide is available from Worth Publishers. Information about the telecourse and its other supplements can be obtained by calling The Annenberg/CPB Project at 1-800-LEARNER.

Finally, students and instructors are encouraged to visit our new World Wide Web site at http://worthpublishers.com/berger for additional learning and teaching resources.

Thanks

This book has benefited from the work of the entire community of scholars involved in human development. Foremost of these is Ross A. Thompson, renowned scholar and professor of psychology at the University of Nebraska, Lincoln, who was my coauthor for the fourth edition of *The Developing Person Through Childhood and Adolescence*. His work and influence are strongly felt throughout this book. I have also learned much from conferences, journals, and conversations with fellow developmentalists. Of course, I am particularly indebted to the many academic reviewers who have read various drafts of this book in each edition, providing suggestions, criticisms, references, and encouragement. Each of them has made the book a better one, and I thank them all. I especially wish to thank those who reviewed this new edition:

Cynthia Berg, University of Utah

Thomas O. Blank, University of Connecticut

Ava Craig, Sacramento City College

Dana H. Davidson, University of Hawaii, Manoa

Linda E. Flickinger, St. Clair County Community College

Deborah Gold, Duke University Medical Center

Sandra Hale, Washington University

Margaret Huyck, Illinois Institute of Technology

Russ Isabella, University of Utah

Leanor Boulin Johnson, Arizona State University

Margaret K. Moroney, Delgado Community College

Anne O'Reilly, West Virginia University

Joseph M. Price, San Diego State University

Tamina Toray, Western Oregon University

Nancy Wargny, Vanier College

Janine A. Watts, University of Minnesota

The editorial, production, and marketing people at Worth Publishers are dedicated to meeting the highest standards of excellence. Their devotion of time, effort, and talent to every aspect of publishing is a model for the industry. When I decided to publish with them, I was told I would have to work twice as hard as I would for any other publisher, and that the result would be many times better. It is true, and I am grateful.

I particularly would like to thank Peter Deane, my editor, who has helped me through every edition of this book, maintaining his perseverance,

brilliance, creativity, and humor despite sometimes compelling reasons not to. Without him, the book would lose much of its elegance and good sense. I also deeply appreciate the efforts of the production staff, and of Timothy Prairie, a meticulous and unflappable production editor, who are responsible for the high quality of the book's appearance. My thanks also go to Barbara Anne Seixas, who bore the burden of trying to keep this book on schedule.

Dedication

The first edition of my first textbook was dedicated to "my students, who have provided inspiration throughout." Later editions were dedicated to various family members and to my editor. With this edition I return to students, but now in a wider perspective and context. *The Developing Person Through the Life Span*, 4th edition, is dedicated to the synergy of all students *and* their teachers, and to the sparks that connect them when learning transforms them.

New York City
May, 1997

Kathleen Stassen Berger

The Developing Person

Through the Life Span

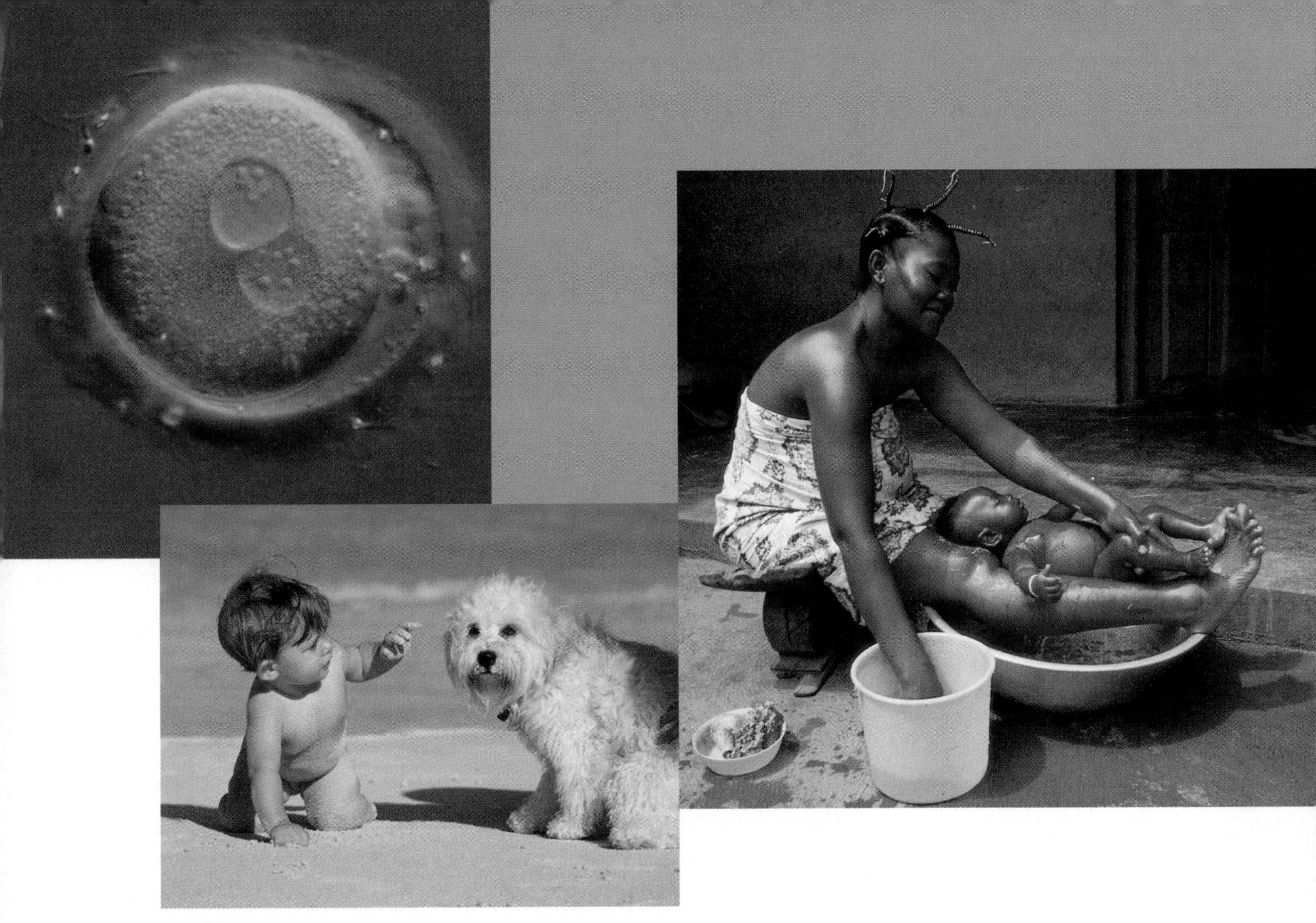

PART I Beginnings

The study of human development has many beginnings, as you will see in the following four chapters. Chapter 1 introduces the scientific study of human development. From it, you will learn what the nature of developmental study is, what kinds of questions developmental scientists ask and try to answer, what the goals of their study are, and how they try to understand human development as it occurs within specific family, social, cultural, and historical contexts. Chapter 2 will introduce you to some of the major theories that guide the study of human development and to the research methods that developmental scientists use to gather data and test their ideas.

A different kind of beginning is described in Chapter 3, which traces the interaction of hereditary and environmental influences. Each human being grows and develops in accordance with chemical guidelines carried on the genes and chromosomes. Interacting with the environment, genes influence everything from the shape of your baby toe to the swiftness of your brain waves to basic aspects of your personality. Thus, understanding the fundamentals of gene-environment interaction is essential to an understanding of human development.

Finally, Chapter 4 details the true beginnings of human life, from the fusing of sperm and ovum to make one new cell to the birth of a new human being, a totally dependent individual who can nevertheless see, hear, and cry, and is ready to engage in social interaction.

Introduction

CHAPTER 1

You are about to begin a fascinating journey through the study of human development. To help prepare you for this journey—which explores development from the moment of conception until the moment of death—this chapter will serve as a kind of roadmap, outlining your route and familiarizing you with the general terrain. More specifically, it will introduce you to the goals and values that underlie the study of human development and suggest some of the practical applications that can arise from developmental science. But before you formally set out on this journey, I would like to speak to you personally for a moment and tell you an unusual developmental story. It involves my brother's son, David.

In many ways, David's childhood and adolescence were typical, marked by a family that cared for him from the moment he was born; schools and teachers that brought out his best, and sometimes his worst; and a social life with peers and the community that gave him both joy and pain. David is now two years out of college, and like many recent college graduates, he looks forward to someday raising a family and, more immediately, to having "a definite career, or at least a decent job." One detail of his young adult years may strike a particularly familiar chord with some of you. As a college sophomore, David fell seriously behind in his classwork, struggled with his writing assignments, and was regularly belittled by one of his professors. As a result, he stopped attending classes, flunked out, and spent more than a year at home doing odd jobs before returning to college. With great effort, David finally earned his B.A. degree, completing his senior year with a 3.7 average. As he expressed it, "College itself was one of my adversities, but I rebounded big time."

The truth is that college was the least of the adversities that David has had to face in his young life. As you will read in more detail later in this chapter, David's development in many ways has been far from typical. He began life with multiple handicaps that seemed to leave him with little chance for survival, let alone for a life approaching normality. His childhood and adolescence were filled with harsh, often heartbreaking obstacles, and he still struggles against unusual odds as his life unfolds.

Most of this book is, of course, about "normal" development—that is, the usual patterns of growth and change that everyone follows to some degree and that no one follows exactly. But later in this chapter we will return to David's unusual story for two reasons.

First, David's struggles and triumphs offer a poignant illustration of the underlying goal of the study of human development: to help each person develop throughout life as fully as possible.

Second, David's example illuminates, with unusual vividness, the basic definitions and central questions that frame the study of human development. Just as suddenly entering an unfamiliar culture can make us more aware of our own daily routines, habits, and assumptions, which tend to go unnoticed precisely because they are so familiar, so, too, can David's story highlight the major factors that influence more typical human development. Let us begin, then, with a brief look at those definitions and questions that form the basis of the study of human development, and then return to David's story.

The Scientific Study of Human Development

Briefly defined, **the scientific study of human development** is the science that seeks to understand how and why people change, and how and why they remain the same, as they grow older. In pursuing this goal, developmental scientists examine all kinds of change—simple growth, radical transformation, improvement, and decline—and all sources of continuity from one day, year, or generation to the next. They consider everything from the genetic codes that lay down the foundations of human development to the countless environmental factors that shape and refine it, from the impact of prenatal life to the influences of the family, school, and peer groups, of health and economic well-being, of career aspirations and opportunities, of marriage, parenthood, and grandparenthood, of friendship, and of religious faith. And they examine all these factors—and untold others—in light of the ever changing social and cultural contexts that give them meaning and force. The scientific study of human development thus involves many academic disciplines, especially biology, education, and psychology, but also history, sociology, anthropology, medicine, and economics.

The Three Domains

To make it easier to undertake this vast interdisciplinary study of developmental change, human development is often separated into three domains, or areas of study (see Figure 1.1). The **biosocial domain** includes brain and body changes and the social influences that guide them. The **cognitive domain** includes thought processes, perceptual abilities, and language mastery, as well as the educational institutions that encourage them. The **psychosocial domain** includes emotions, personality, and interpersonal relationships with family, friends, and the wider community.

All three domains are important at every age. For instance, understanding an infant involves studying his or her health, curiosity, and temperament, as well as dozens of other aspects of biosocial, cognitive, and psychosocial development. Similarly, understanding an adolescent requires studying the physical changes that mark the bodily transition from child to adult, the intellectual development that leads to efforts to think logically about such issues as sexual passion and future goals, and the emerging pat-

the scientific study of human development The science that seeks to understand how and why people change, and how and why they remain the same, as they grow older.

biosocial domain Includes physical growth and development as well as the family, community, and cultural factors that affect that growth and development.

cognitive domain Includes all the mental processes through which the individual thinks, learns, and communicates.

psychosocial domain Includes emotions, personality characteristics, and relationships with other people.

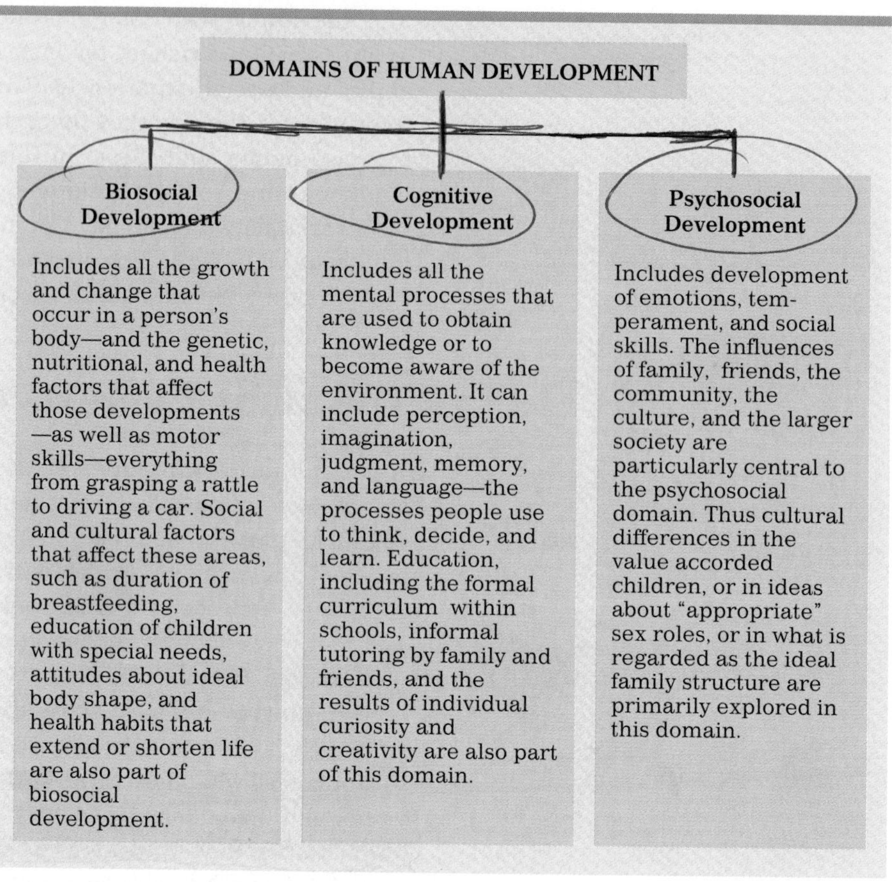

FIGURE 1.1 The division of development into three domains makes it easier to study, but we must remember that very few factors belong exclusively to one domain or another. Development is not piecemeal but holistic: each aspect of development is related to all three domains.

terns of friendship and courtship that prepare the individual for the intimate relationships of adulthood. Understanding an 80-year-old would likewise require examination of all three domains. This would include an assessment, for example, of any impairments that may have developed in strength, stamina, or sensory abilities; of any changes that may have occurred in cognitive abilities and their effect on daily functioning; and of the individual's social milieu—whether, for instance, the person is essentially alone or surrounded by family and friends.

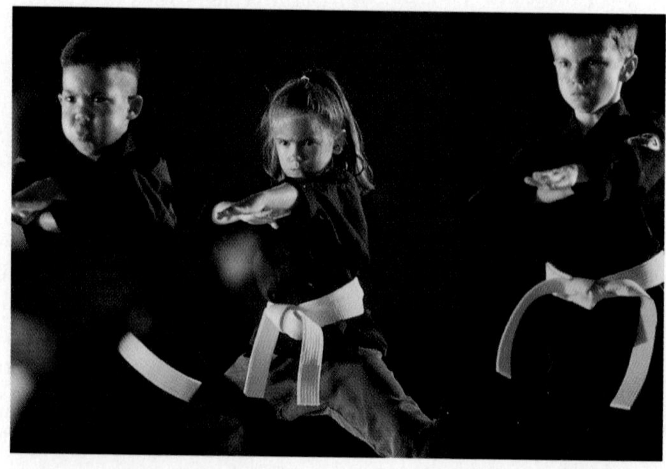

Every aspect of human behavior reflects all three domains. Obviously, biosocial factors—such as nutrition and maturation—are at work here, but so are cognitive and psychosocial ones. For instance, each student's mental concentration or lack of it is critical to karate success, as is the culture's message about who should learn the martial arts. This girl's expression suggests she knows she needs to seem tougher than the boys if she is to succeed.

Inevitably, each domain is affected by the other two: whether or not an infant is well-nourished, for instance, may well affect the baby's learning ability and social experiences. For many adolescents, their perception of their bodies—the way they *think* their bodies look—affects their eating and exercise habits, and these, in turn, affect their physical health and their emotional and social development. And for older adults, the strength of their social network can have an impact on both their health and their mental acuity, while both those factors, in turn, can affect each other as well as the strength of the individual's social network.

The Many Contexts of Development

We often think of development as originating *within* the individual—the result of such internal factors as genetic programming, physical maturation, cognitive growth, and personal choices. However, development is also greatly influenced by forces *outside* the individual, by the physical surroundings and social interactions that provide incentives, opportunities, and pathways for growth. Taken as a whole, these external forces are the *context* of development.

Calling attention to these external influences more than twenty-five years ago, Urie Bronfenbrenner, a leading developmental researcher, began to emphasize what he calls an **ecological approach** to the study of human development (Bronfenbrenner, 1977, 1979, 1986). In essence, this approach regards human development as a "joint function of person and environment" (Bronfenbrenner, 1993). Thus, just as a naturalist studying a flower or a fish needs to examine the organism's supporting ecosystems, Bronfenbrenner argues, developmentalists need to study the ecological systems, or contexts, in which each human being seeks to thrive.

To depict the main ecosystems that support human development, Bronfenbrenner has devised an ecological model that organizes the broad contexts of development in terms of the relative immediacy of their impact on the individual (see Figure 1.2). At the center of this model is the individual. Each immediate social setting that surrounds and shapes that individual is called a *microsystem*. Examples of microsystems include the family, the peer group, the classroom, the workplace, and so on. The *mesosystem* comprises the connections between various microsystems—such as parent-teacher conferences that link home and school. Next comes the *exosystem*, the specific economic, political, educational, and cultural institutions and practices that directly affect the various microsystems and, indirectly but often powerfully, everyone in those microsystems. Surrounding and permeating all these developmental contexts is the *macrosystem*, the overarching traditions, beliefs, and values of the society.

The influences within and between these systems are multidirectional and interactive. For example, research has shown that the quality of life in the family microsystem directly affects a worker's productivity on the job. At the same time, the microsystem of the workplace—specifically the stresses and satisfactions at the office, store, factory, or farm—affects the quality of life at home, including how satisfied a couple is with their marriage and how responsive they are to their children (Barling & Mactwen, 1992; Loscocco & Roschelle, 1991; Zedek, 1992).

ecological approach A perspective on development that takes into account the various physical and social settings in which development occurs.

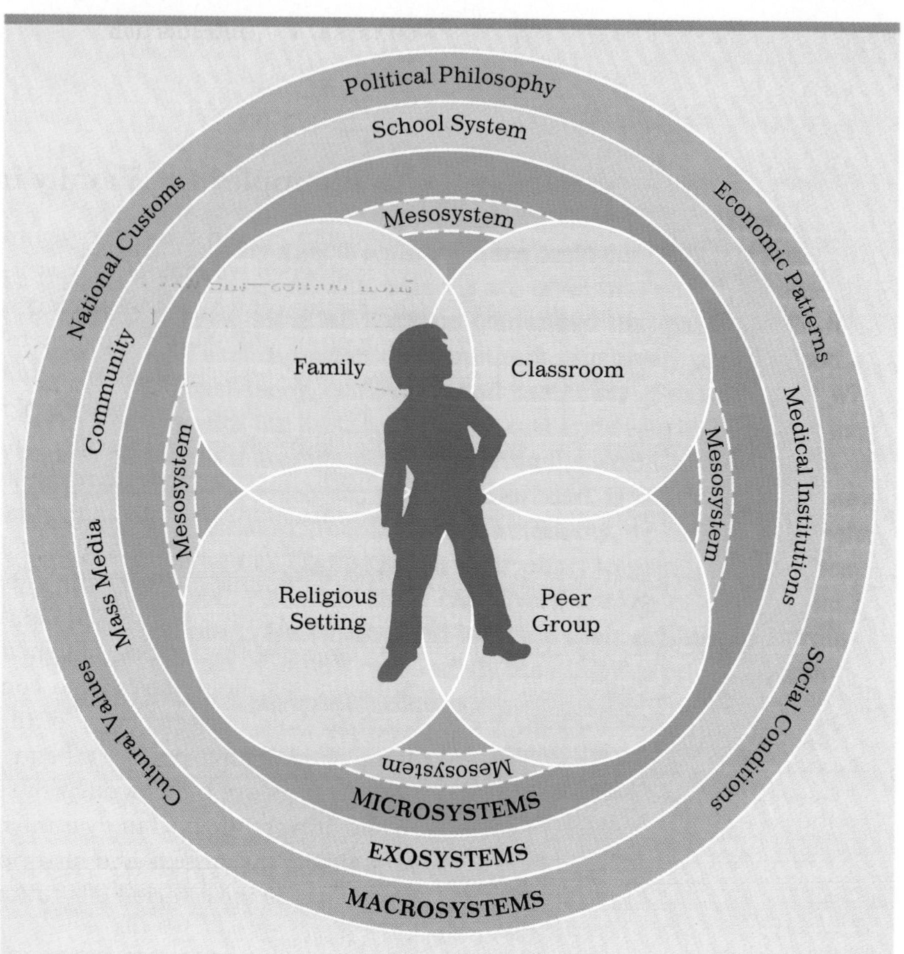

FIGURE 1.2 Each person is significantly affected by interactions between and among a number of overlapping ecosystems. Microsystems are those social systems that intimately and immediately shape human development. The primary microsystems include the family, friendship group, the classroom or workplace, and the neighborhood, and sometimes a church, temple, or mosque as well. The specific interaction between the social systems takes place through the mesosystem, as when parents and teachers coordinate their efforts to educate the child or when employer and employee arrange family-emergency leave. Surrounding the microsystems is the exosystem, which includes all those external networks, such as the community structures and the local educational, medical, employment, and communications systems, that influence the microsystems. And influencing the entire process on a grand scale is the macrosystem, which includes the overarching cultural values, political philosophies, economic patterns, and social conditions that affect all the other systems.

These reciprocal interactions between home and work are also affected by factors in the surrounding exosystem and macrosystem. In North America, for example, hard work and pioneer self-sufficiency are bedrock values, and, for most Americans, being productive and independent is key to their sense of self-worth. Consequently, if the family breadwinner loses a job, that loss may affect family life not only through the financial strain it causes but also through the psychological stress it creates. This link among systems is dramatically reflected during times of economic recession: as unemployment in a community increases, domestic violence and divorce typically increase as well (Dooley & Catalano, 1988).

Bronfenbrenner's ecological approach to development is very useful in highlighting the complex influences, both immediate and distant, on any one person's development. However, Bronfenbrenner's terminology, and his depiction of spheres of influence, nested one inside the other, sometimes lead students to think of these ecosystems as discrete entities, with clear boundaries between them—when, in fact, the various systems are inextricably intermeshed, their effects dynamic, fluid, and overlapping.

To avoid this confusion, the term **social context** will be used in this book to encompass all the ecosystems delineated by Bronfenbrenner. We will thus focus on the many specific contexts of development without trying to locate them specifically within a given ecosystem. Bear in mind, however, that whatever the nomenclature—contexts or systems—the core concept is the same: every individual develops within, is influenced by, and in turn influences the dynamic relationships that exist among many interlocking settings.

social context The entire spectrum of social milieux—including the people, the customs, and the beliefs—that surround each developing person.

An Example: The Family in Social Context

The dynamic and reciprocal complexities of the social context are clearly evident when we look closely at its most basic unit, the family. Universally, the family context is the primary setting for nurturing children to become competent and contributing members of society and for restoring adults after the stresses of daily work. In what manner, and how well, a given family does this depend on a wide array of factors. Some of these factors are rooted directly in the specific family setting—from the number and age of children, parents, and other adults in the family to the emotional climate the interactions of these individuals create. Each family relationship (such as that between husband and wife, parent and child, brother and sister, spouse and in-law) affects all the other family members, and all these relationships are mutually influential, affecting everyone involved.

Some of the factors influencing family relationships are less immediate, such as the values of the community regarding gender roles, or the ways in which the structure of the neighborhood institutions might affect family functioning, or the past and current experiences of the grandparents, who, in turn, are influenced by the customs and core beliefs of earlier times. While universally the family is the basic setting for intimacy and growth, the complexity of contexts and histories that affect each family makes the family one of the most varied institutions on earth (Altergott, 1993).

The social context includes not only the people who inhabit it but also the specific environments that those people create for themselves. The arrangement of rooms within a house, of houses within a neighborhood, and of schools, businesses, hospitals, and so forth within a community all create an ecological niche that in various ways may enhance one individual's development or hinder another's. This deceptively simple backyard setting, for example, may be a good place for a multigenerational family gathering and may provide a secure area for young children to play in. At the same time, its high-fenced privacy, especially for children, may create a sense of restriction and of isolation from those outside the family unit.

A "Difficult" Child

To sharpen our focus a bit, let us look at the contextual interactions that might be affecting a young boy who is "difficult"—disobedient, hostile, demanding, impulsive. A noncontextual analysis of the boy's problem behavior might point to the mother, finding her to be self-absorbed, or cold, or indulgent. Such an explanation would be flawed, however, because it is one-dimensional. As a leading family therapist explains:

> It may be relatively easy to discover that a little boy who misbehaves in school has a mother who doesn't make him behave at home. On closer examination we might see that she doesn't discipline the boy because she is overly involved with him. They're constantly together and interact more like playmates than parent and child. But why is the mother so close to the boy? Why does she need a play-

> mate? Is it because she's emotionally distant from her husband? . . . Perhaps she is deliberately lenient with the boy to counterbalance her husband's overly harsh control. The reason so many family dilemmas defeat us is that we fail to recognize that every family member's behavior is influencing and influenced by the behavior of the rest. [Minuchin, 1993]

Increasingly, developmentalists recognize that not only the nature of the marital relationship but also such factors as the father's involvement in caregiving and the rivalries and tensions of the other siblings are implicated in the child's behavior. And of course, the child, too, is a central player. In the case of the difficult child, the mother's apparent caregiving flaws might be the *result* of the child's intractable behavior more than the *cause* of it. Thus, when considering development within the family setting, the contextual approach attempts to consider the totality of family interactions, with each family member likely to be both "a victim and an architect" of whatever problems the family might have (Patterson, 1982; Patterson & Capaldi, 1991).

A contextual view does not stop there, however, for just as each family member is affected by the interactions of all its members, each family is reciprocally influenced by other social contexts (Bronfenbrenner, 1986). As already noted, the stresses and satisfactions of the workplace can have a significant impact on family interactions. Whether the difficult child's parents feel secure and fulfilled in their work or anxious and frustrated can obviously affect the quality of attention they give their son, as well as their tolerance for certain of his disruptive antics.

Other relevant influences can be found in the contexts of the peer group and the school. The difficult child's friends, for example, may admire and thus encourage his unruly behavior, while the school's demand for obedience and conformity may create tensions that spill out at home (Cairns & Cairns, 1994; Dishion et al., 1995; Patterson et al., 1992). Of course, the influences of these contexts could also be positive: the peer group might provide a setting in which the child learns needed social skills, and the school might provide avenues of success in the classroom or the playground that enhance self-esteem, thus mitigating the child's hostility.

Typically, the contexts of peer group and school affect the family only indirectly, through their influence on the behavior of the child. Sometimes, however, these other contexts have a direct impact on the family. For example, if the parents come to see that their son is actually less rowdy than most of the neighborhood boys, or if they hear from a teacher that he is unusually creative, their perceptions of the boy may change, as they come to appreciate certain of his behaviors that once made them angry. As a result, they may not only perceive him differently but may also treat him differently, and he, in turn, may begin to behave differently.

The same multicontextual, multidirectional analysis could be applied to almost any specific behavior in almost any developing person. In every case, the individual's actions are both cause and consequence of the social context in which development occurs.

As you might imagine, it would be impossible in a single course—indeed, in everyday life—to consider simultaneously all the contextual factors that might bear on any particular aspect of development. However, we will, throughout this book, be examining a great many such factors, exploring the ways in which specific contexts may push a particular avenue of development in one direction or another. More generally, we will also focus atten-

social construction An idea about the way things are or should be that is built more on the shared perceptions of members of a society than on objective reality.

tion on several broad contexts that affect the development of virtually every individual in every phase of development—the historical context, the socioeconomic context, and the cultural–ethnic context. Let us turn now to a brief survey of these major contexts.

The Historical Context

In every era, prevailing assumptions, critical public events, current technologies, and popular trends shape the lives and thoughts of individuals living in that time period. Are your attitudes about hard work and job security, or about the relative importance of money in the bank and independence in one's personal life, quite different from those of your parents or grandparents? One reason they may be is that such attitudes are affected by the economic and social picture that existed when a person first reached adulthood—whether that was during the economically depressed 1930s, the affluent 1950s, or the financially unstable 1990s.

Research into the historical context reveals that as profound economic, political, and technological changes occur over the years, basic concepts about how things "should be" are readily influenced by the era in which such changes arise. Often one or another of our most cherished assumptions is, in fact, a **social construction**, that is, an idea built more on shared perceptions of the members of a society than on objective reality. The obligations of women to be docile housewives and of men to be strong and independent providers are two obvious examples of social construc-

High heels, long dresses, and new appliances were key elements in the social construction of the "happy housewife" of the 1950s, who was expected to devote herself exclusively to the care of the family household. The limitations of this social construction no doubt seem archaic to those who inhabit the current social construction of women "having it all." One statistical indicator of the changed social context: in 1960, there were two college men for every one college woman; since 1980, college women have outnumbered the men.

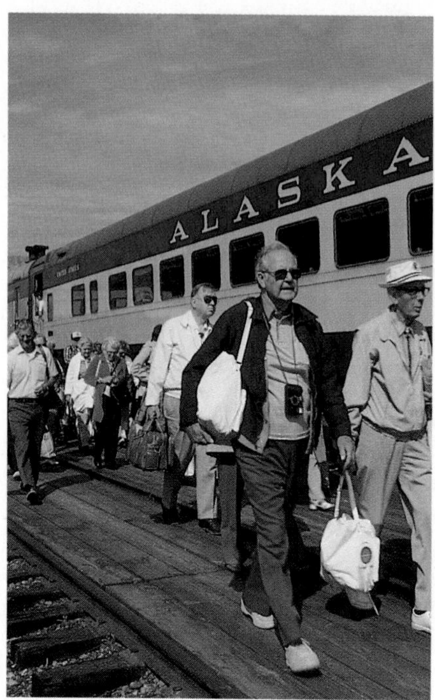

Sunset days or golden years? Once the elderly lived a sheltered life, helping their offspring with babysitting and household work. Now, in the land of the midnight sun, this excursion group of elders prepares to hike past mountains, bears, and glaciers.

tions that, in many cultures and contexts, have lost their consensus in recent times.

Even the most basic ideas about patterns of development change. For example, the very concept of childhood as we know it, that is, as a special and extended stage of life, is a social construction that was virtually nonexistent throughout much of history. In many historical contexts, children were cared for until they could take care of themselves (at about age 7) and then they entered the adult world, working in the fields or at home and spending their leisure time engaged in the activities of grownups. Likewise, the notion of adolescence—as a period between childhood and adulthood when teenagers rebel against authorities, unite with their peers, and seek their own identities—is largely a social construction of the past fifty years, in which mass education in high school and college has kept young people segregated as a group and growing affluence has created a powerful youth market (Boxer et al., 1984).

Old age, of course, has always been recognized as a distinct period of life, but social constructions regarding this period are continually changing, including the idea of when old age actually begins. A hundred years ago, Americans used to think that they were old at age 50, but now many 70-year-olds do not regard themselves as elderly. One particular social construction that has recently taken shape is the idea of *retirement*, not as a forced and regretted consequence of outliving one's usefulness but as a chosen departure from work, a time for indulging in enjoyable leisure activities. Even as recently as fifty years ago, most Americans hoped to—and had to—work until they died (Quadagno & Hardy, 1996). If they became too feeble to do so, they were expected to live out their last days in quiet dependence on their children. Over the years, however, dramatic advances in the quality and availability of health care, changing requirements in the labor force, and growing economic independence acquired through Social Security and pensions have, for many older people, led to a concept of retirement that was unimaginable at the beginning of this century—decades of active retirement including "retirement income," "retirement communities," and "retirement travel."

Cohort Differences

The reality that older Americans today differ in health, aspirations, and experience from their counterparts a few decades ago highlights one of the consequences of the changing historical context: "differences in year of birth expose people to . . . different priorities, constraints, and options" (Elder, 1995). In societies characterized by rapid social change, each new **cohort**—defined as all those persons born within a few years of each other—grows up within a context of priorities, constraints, and options, unlike that of earlier cohorts. Since the social construction of adolescence, for example, the specific behaviors expected of teenagers—from matters of musical taste and hairstyle to more serious concerns about their relationship with the adult world—have changed with every generation, a transformation much more apparent to the young person experiencing it than to the older generations witnessing it. When a 15-year-old rejects advice from a 30-year-old with words such as "You don't understand, everything is different now," there is more than a grain of truth in this retort.

cohort A group of people who, because they were born within a few years of each other, experience many of the same historical and social conditions.

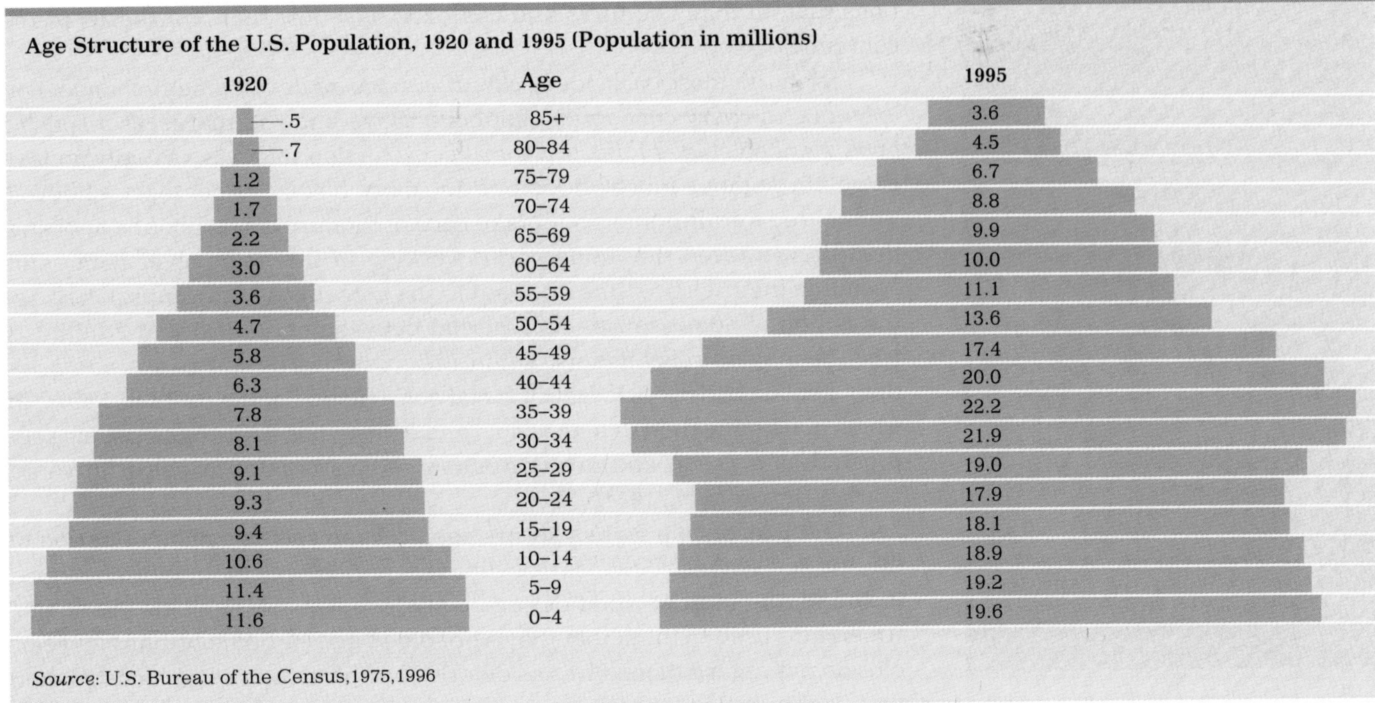

FIGURE 1.3 Unlike earlier times, when each generation was slightly smaller than the one that followed, each cohort today has a unique position that was determined by the reproductive patterns of the preceding generation and by the medical advances that were developed during their own lifetime. As a result of these two factors, the baby boomers, born between 1945 and 1960, represent a huge bulge in the U.S. population. Largely because of the latter factor, the fastest growing group, proportionally, is the oldest old, which has increased sevenfold since 1920. In another three decades, the leading edge of the baby-boom generation, largely intact, will begin moving into this upper age group.

In addition to being influenced by the differing social contexts in which they develop, cohorts can also be affected by differences in their relative sizes (see Figure 1.3). This is dramatically illustrated by the U.S. "baby boom" generation, born in such large numbers between 1946 and 1960 that, when they were teenagers in the 1960s and early 1970s, the entire society tilted toward a youth culture—with much more sexual freedom, political protest, and drug experimentation than the society had ever known before. Now, as those same boomers are beginning to look ahead to retirement, their great numbers make such topics as Social Security, Medicare, the prevention of senility, and assisted suicide urgent social issues. Meanwhile, the current "baby bust" cohort of young adults, born in relatively small numbers from the mid-1960s to the mid-1970s, must cope with an overloaded job market and face the prospect of having their particular social needs overwhelmed by those of their predecessors, who by virtue of their size alone have much more political clout.

The Socioeconomic Context

A second major contextual influence on development is **socioeconomic status**, sometimes called "social class" (as in "middle class" or "underclass"). Socioeconomic status, abbreviated as **SES**, is most accurately measured through a combination of several overlapping variables, including income, education, residence, and occupation. As measured by social scientists, then, the SES of a family consisting of an infant, a full-time mother, and an employed father who earns $10,000 a year could be either lower- or middle-class, depending on whether the household head is an illiterate dishwasher living in an urban slum or a graduate student living on campus and teaching part-time. The point of the distinction between these cases, as should be obvious, is that SES is not just financial: it entails *all* the advantages and disad-

socioeconomic status (SES) An indicator of social class that is based primarily on income, education, and occupation.

vantages, opportunities and limitations, that may be associated with either status. Social class is as much a product of the mind as of the wallet.

Nonetheless, in official government statistics, SES is often measured solely by family income (adjusted for inflation and family size). For example, in 1994 in the contiguous United States, a family of four with an annual income under about $15,000 was considered to have the lowest SES, that is, below the poverty level, which is calculated as the minimum amount needed to pay for basic necessities. (In Alaska and Hawaii, the poverty level is set somewhat higher.)

Looking only at family income is simplistic but nevertheless useful, especially when individuals living in households below the poverty level are concerned. The reason is that inadequate family income both signals and creates a social context of limited opportunities and heightened pressures that make life much more difficult to manage than it is higher up on the socioeconomic ladder (Huston et al., 1994). For example, infant mortality, child neglect, adolescent violence, and adult health problems are each much more common among the poor than among the affluent. Considerable debate among social scientists concerns whether low income alone creates such problems or whether there is a "culture of poverty," a set of values and practices that tends to perpetuate low SES from generation to generation.

Another debate concerning SES reverberates in the larger society, specifically, which generation deserves the greatest portion of public financial support. Four decades ago, the old were the poorest age group in the United States. Now, as you can see from Figure 1.4, the youngest are poorest, with more than one in four children living in poverty. Some argue that this inequity is not only unfair but also foolish, because it will create problems in today's youth that will haunt society in years to come. Others contend that the older generations are not only more deserving of economic support but are more likely to spend it wisely, often helping the younger generations. Indeed, when money transfers within families are examined, it is clear that those over age 65 are more likely to give than to receive (Crystal, 1996).

FIGURE 1.4 The American who sleeps hungry, in a crumbling house, in a crime-ridden community, is more often a young child than an older adult and is almost never middle-aged. Is this fair? As you read this book, you will see that there are many dimensions to this question, and that the question of generational equity is one of growing importance as the U.S. population, proportionally, becomes increasingly old.

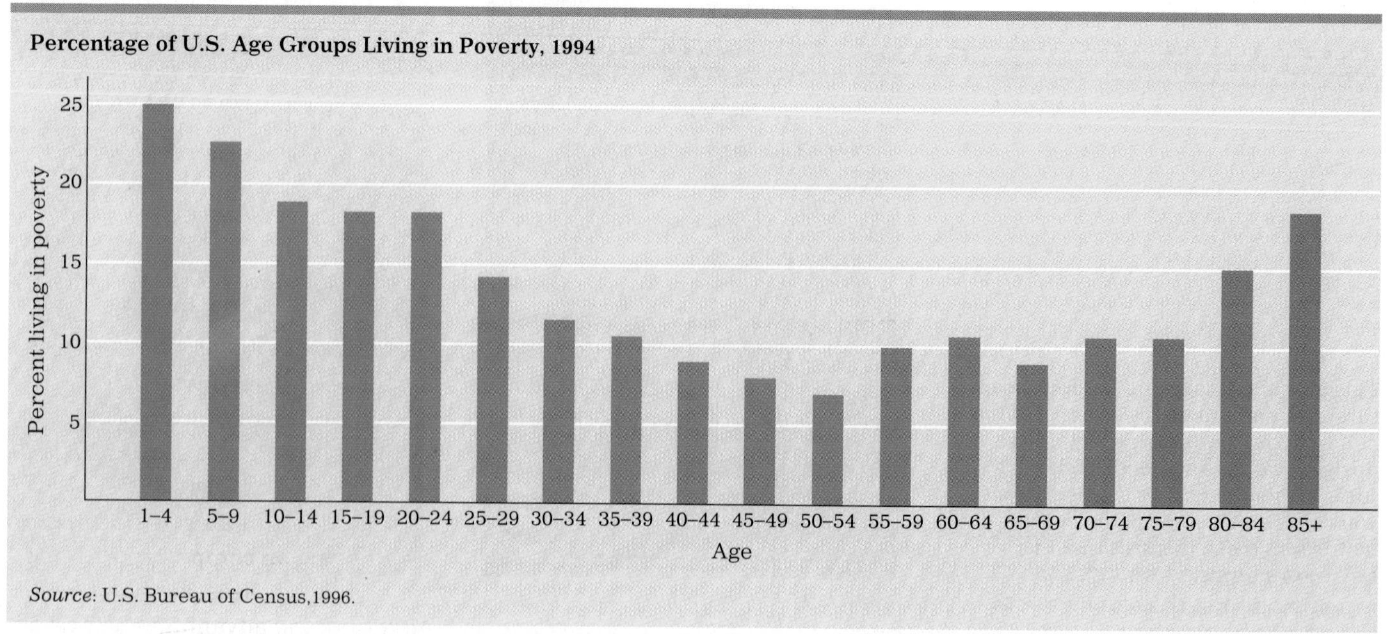

culture The set of shared values, attitudes, customs, and physical objects that are maintained by people in a specific setting as part of a design for living one's daily life.

The Cultural Context

When social scientists use the term **culture**, they refer to the set of values, assumptions, and customs, as well as the physical objects—everything from clothing, dwellings, and cuisine to technologies and works of art—that a group of people have developed over the years *as a design for living* to structure their life together.

When we look closely at various designs for living, it becomes clear that culture guides development in a multitude of interrelated ways. Here is an example. Robert LeVine (1980, 1988, 1989) has noted that in many developing agricultural communities, children are an economic asset because they can contribute to the family's farming and, later, to forming a strong family unit to preserve the land and to care for aging parents. Thus, every infant who survives to childhood and beyond benefits the entire family group. But in many of these communities, nutrition and medical care are poor, leading to high infant-mortality rates. Therefore child-rearing is designed to maximize survival and emphasize family cooperation: its typical features include intensive physical care, feeding on demand, immediate response to crying, close body contact, keeping the child close by at night, and constant care by siblings and other relatives as well as by the mother. All these measures protect the fragile infant from an early death and work to establish such values as the interdependence of family members.

By contrast, according to LeVine, American middle-class parents do not have to be so concerned about infant mortality. Instead, hoping to ensure their children's future success in a technological and urbanized society, they focus their child-rearing efforts on fostering cognitive growth and emotional independence. Accordingly, middle-class American parents typically engage their infants in activities that provide cognitive and social stimulation, talk to their infants more than touch them, put them to sleep by

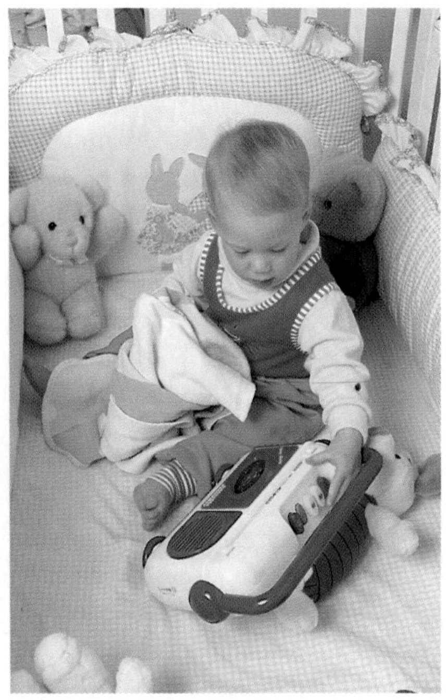

This pair of photographs perfectly captures the point made by Robert LeVine: child-rearing patterns reflect the underlying values and needs of the culture. Without checking the text, see if you can identify some of the cultural features in these pictures that fit LeVine's analysis.

themselves in their own cribs in their own rooms, and often ignore their crying so as not to "spoil" them. Not surprisingly, these contrasting parental strategies produce children with quite different capacities, goals, and expectations, but in both cases, the children become relatively well-prepared for the culture in which they have been raised.

Ethnicity and Culture

An **ethnic group** is a collection of people who share certain attributes, such as ancestry, national origin, religion, and/or language and, as a result, tend to identify with each other and have similar daily encounters with the social world. Racial identity is sometimes an element of ethnicity. However, as social scientists emphatically point out, biological traits (such as hair or skin coloring, facial features, and body type) that distinguish one "race" from another are much less significant to development than the attitudes and experiences that may arise from ethnic or racial consciousness, especially those resulting from minority or majority status. Ethnic identity, then, is more than genetic; it is the product of the social context and the individual's consciousness.

Ethnicity is similar to culture, in that it provides people with shared beliefs, values, and assumptions that can significantly affect their own development as well as how they raise their children. Indeed, sometimes ethnicity and culture overlap. However, people of many ethnic groups can all share one culture, yet maintain their ethnic identities. Within multiethnic cultures, such as those found in most large nations today, ethnic differences are most apparent in matters such as whether children are raised in large extended families or smaller nuclear families; whether they are encouraged toward independence, dependence, or interdependence; whether they view education as all-important or as secondary to family responsibilities; whether they defer to family elders or assert their autonomy; as well as in many other beliefs, values, and behaviors (Harrison et al., 1990).

The Individual and the Social Context

Since each individual develops within many contexts, it is obviously important to understand the special impact that each context has. But it is also important to be cautious about any explanations of personality traits, abilities, or actions that link individual behavior exclusively to any one of these contexts. This is because each of us is often pulled in divergent directions by various contextual influences, the power of which vary from individual to individual, situation by situation, family to family. In short, no one is exactly like the statistically "average" person of his or her generation, socioeconomic status, or culture. Each of us differs in unexpected ways from any stereotypes or generalities that might seem pertinent, and these individual idiosyncrasies demand as much scientific respect and scrutiny as any of the commonalities that link us to any given group

Now let us return to David, a person clearly affected by the social contexts that structured and continue to shape his development, yet who is, just as obviously, unique.

ethnic group A collection of people who share certain background characteristics, such as national origin, religion, upbringing, and language, and who, as a result, tend to have similar beliefs, values, and cultural experiences.

David's Story: Domains and Contexts at Work

David's story begins in 1967, with an event that seems clearly from the biosocial domain. In the spring of that year, in Appalachia, an epidemic of rubella (German measles) struck two more victims—David's mother, who had a rash and a sore throat for a couple of days, and her 4-week-old embryo, who was damaged for life. David was born in November, with a life-threatening heart defect and thick cataracts covering both eyes. Other damage caused by the virus became apparent as time went on, including minor malformations of the thumbs, feet, jaw, and teeth, as well as of the brain.

From a contextual perspective, the larger medical and political contexts had already had a major impact, one determined partly by the particular point in historical time at which David entered the world. Had David been conceived a decade later, the development and widespread use of the rubella vaccine would probably have prevented his mother's contracting the disease, and David would have been born unscathed. On the other hand, had he been born a few years earlier, or in a different part of the world, he would have died, because the medical technology that saved his life would not have been available.

The Early Years: Heartbreaking Handicaps, Slow Progress

As it happened in 1967, heart surgery in the first days after birth saved David's life. However, surgery six months later to open a channel around one of the cataracts failed, completely blinding that eye.

It soon became apparent that David's physical handicaps were contributing to cognitive and psychosocial liabilities as well. Not only did his blindness make it impossible for him to learn by looking at his world, but his parents overprotected him to the point that he spent almost all his early months in their arms or in his crib. An analysis of the family context would have revealed that David's impact on his family, and their effect on him, were harmful in many unintended ways. Like most parents of seriously impaired infants, David's felt guilt, anger, and despair (Featherstone, 1980), and they were initially unable to make constructive plans to foster David's normal development.

Fortunately, however, David's parents came from a socioeconomic background that encouraged them to seek outside help. (David's father is a college professor and his mother is a nurse.) Their first step was to get advice from a teacher at the Kentucky School for the Blind, who told them to stop blaming themselves for David's condition and to stop overprotecting him because of it. If their son was going to learn about his world, he was going to have to explore it. To this end, they were told that, rather than confining David to a crib or playpen, they should provide him with a large rug for a play area. Whenever he crawled off the rug, they were to say "No" and place him back in the middle of it, thus teaching him to use his sense of touch to learn where he could explore safely without bumping into walls or furniture. David's mother dedicated herself to this and other tasks that various other specialists suggested, including exercising his twisted feet and cradling him frequently in her arms as she sang lullabies to provide extra tactile and auditory stimulation.

His father helped, too, taking over much of the housework and care of David's two older brothers, who were 2 and 4 at the time. When he found an opportunity to work in Boston, he took it, partly because the Perkins School for the Blind, located there, had just begun an experimental program for blind toddlers and their mothers. At Perkins, David's mother learned specific methods for developing physical and language skills in children with multiple disabilities, and she, in turn, taught the techniques to David's father and brothers. Every day the family spent hours rolling balls, doing puzzles, and singing with David.

Thus, a smooth collaboration between the family and the educational contexts helped young David develop. However, progress was slow. It became painfully apparent that rubella had damaged much more than his eyes and heart. At age 3, David could not talk, nor chew solid food, nor use the toilet, nor coordinate his fingers well, nor even walk normally. An IQ test showed him to be severely mentally retarded. Fortunately, although most children with rubella syndrome have hearing defects, David's hearing was normal. However, the only intelligible sounds he made mimicked the noises of the buses and trucks that passed by the house.

At age 4, David said his first word, "Dada." Open-heart surgery corrected the last of his heart damage, and an operation brought partial vision to his remaining eye. While sight in that eye was far from perfect, David could now recognize his family by sight as well as by sound and could look at picture books. By age 5, when the family returned to Kentucky, further progress was obvious: he no longer needed diapers or baby food.

David's fifth birthday occurred in 1972, just when the historical view that children with severe disabilities are unteachable was being seriously challenged and many schools were beginning to open their doors to children with problems like David's. David's parents found four schools that would accept him, and in accordance with the family's stress on education, they enrolled him in all four. He attended two schools for children with cerebral palsy: one had morning classes, and the other—forty miles away—afternoon classes. (David ate lunch in the car with his mother on the daily trip.) On Fridays these schools were closed, so he attended a school for the mentally retarded. On Sundays, he spent two hours in church school, his first experience with "mainstreaming"—the then-new idea that children with special needs should be educated with normal children.

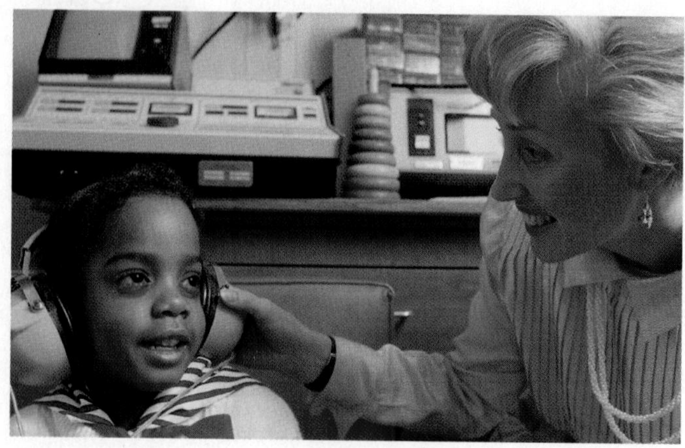

Today's generation of children have many advantages over earlier generations, especially with regard to technological health benefits. The hearing tests this boy is receiving are far more sophisticated than those of even a decade ago, and certain other tests can detect hearing impairments even in infants. When David was a baby, by contrast, hearing tests were such that no one knew how well he could hear until he was about 4 years old.

Childhood and Adolescence: Heartening Progress

By age 7, David's intellectual development had progressed to the point considered adequate for the normal educational system. In some skills, he was advanced; he could multiply and divide in his head. He entered first grade in a public school, one of the first severely disabled children in the United States to be mainstreamed. However, rubella continued to have an obvious impact on his biosocial, cognitive, and psychosocial development. His motor skills were poor (among other things, he had difficulty controlling a pencil); his efforts to learn to read were greatly hampered by the fact that he was legally blind even in his "good" eye; and his social skills were seriously deficient (he pinched people he didn't like and cried and laughed at inappropriate times).

During the next several years, development in the cognitive domain proceeded rapidly. By age 10, David had skipped a year of school and was a fifth-grader. He could read with a magnifying glass—at the eleventh-grade level—and was labeled "intellectually gifted" according to tests of verbal and math skills. At home he began to learn a second language and to play the violin. In both areas, he proved to have extraordinary auditory acuity and memory.

David's greatest problem was in the psychosocial domain. Schools generally ignored the social skills of mainstreamed children, and David's experience was no exception. For instance, David was required to sit on the sidelines during most physical-education classes and to stay inside during most recess periods. Without a chance to experience the normal give-and-take of schoolyard play, David remained more childish than his years. His classmates were not helped to understand his problems, and some of them teased him because he still looked and acted "different."

The efforts of these Special Olympians reflect not only the thrill of competition but also the satisfaction of having one's abilities and interests recognized and accepted. In a highly competitive society like the United States, being forced to the sidelines by social attitudes can be far more devastating psychologically than the limitations imposed by a particular disability.

Because of David's problems with outsiders and classmates, his parents decided to send him to the Kentucky School for the Blind. There, his biosocial, cognitive, and psychosocial development all advanced: David learned to wrestle and swim, mastered algebra with large-print books, and made friends whose vision was worse than his. He mastered not only the regular curriculum but also specialized skills, such as how to travel independently in the city and how to cook and clean for himself. In his senior year he was accepted for admission by a large university in his home state. When David finally graduated from college, it was with a double major in Russian and German, and he has received a master's degree in German.

Looking Back and Looking Forward

Now many of David's worst problems are behind him. In the biosocial domain, doctors have helped to improve the quality of his life: an artificial eye has replaced the blind one; a back brace has helped his posture; and surgery has corrected a misaligned jaw, improving his appearance and his speech. In the cognitive domain, the once severely "retarded" preschooler is looking forward to a career as a translator (an interesting choice for someone who has learned to listen very carefully to what people say because he is unable to read their facial expressions). And in the psychosocial domain, the formerly self-absorbed child is now an outgoing young man, eager for friendship. Although he still lives at home with his parents, he looks forward to "breaking away" soon.

This is not to suggest that David's life is all smooth sailing. In fact, every day presents its struggles, and David, like everyone, has his moments of self-doubt and depression. As he once confided to me when he was in college:

> I sometimes have extremely pejorative thoughts . . . dreams of vivid symbolism. In one, I am playing on a pinball machine that is all broken—glass besmirched, legs tilted and wobbly, the plunger knob loose. I have to really work at it to get a decent score.

Yet David never loses heart, at least not for long. He continues to "really work" on his life, no matter what, and bit by bit, his "score" improves.

In looking at David's life thus far, we can see how the domains and various social contexts interact to affect development, both positively and negatively. We can also see the importance of research and the application of developmental principles. For example, without research that demonstrated the crucial role of sensory stimulation in infant development, David's parents might not have been taught how to keep his young mind actively learning. Nor would David have been educated had not the previous efforts of hundreds of developmental scientists proved that schools could provide effective teaching even for severely handicapped children. David might instead have led an overly sheltered and restricted life, as many children born with his problems once did. Indeed, many children with David's initial level of disability formerly spent their lives in institutions that provided only custodial care.

David's immediate future will likewise be influenced by various social contexts. Changes in the historical context, for instance, have led to increasing sensitivity to the needs of the disabled: the laws of the land now safeguard their right to a normal life in higher education, in housing, and in

employment. And David will continue to draw love and support from the family context. However, as he ventures through adulthood, he will need to find his own social world—a difficult task for most young American adults, especially for those who have an unusual life history and appearance.

At the same time that David's story highlights the influence of developmental domains and social contexts, it also serves to remind us of a universal truth: none of us is simply a product of these influences. Each person is a unique individual who uniquely reacts to, and acts upon, the constellation of contexts that impinges on his or her development. Thus the most important factor in David's past successes may have been David himself, for his determination and stoic courage helped him weather the physical trauma of repeated surgery and the psychological devastation of social rejection. Of all those who should be proud of David's accomplishments—including the scientists, teachers, and family members who directly and indirectly contributed to his growth—the one who should be most proud is David himself. More than anyone else, in the final analysis, David, like each of us, directs his own development.

Two Controversies

As David's case makes abundantly clear, the study of development requires taking into account the interplay of the biosocial, cognitive, and psychosocial domains, within a particular historical time, influenced by familial, socioeconomic, and cultural forces. Not surprisingly, assessing the relative impact of all these factors is no simple matter. In fact, developmentalists often find themselves on one side or another of a number of controversies related to this issue. Two controversies in particular pervade the study of the entire life span.

Nature and Nurture

The central dispute in the study of human development is the nature-nurture controversy. It is the continuing debate over the relative impact of hereditary and environmental influences in shaping various personal traits and characteristics.

Nature refers to the range of traits, capacities, and limitations that each person inherits genetically from his or her parents at the moment of conception. Body type, eye color, and inherited diseases are obvious examples. Nature also includes a host of intellectual and personality characteristics, such as facility with numbers, attraction to novelty, sociability, and tendency to depression, that are powerfully influenced by genes.

Nurture refers to all the environmental influences that come into play after conception, beginning with the mother's health during pregnancy and including all the individual's experiences in the outside world—in the family, school, community, and the culture at large.

The controversy about nature and nurture has taken on many names, among them *heredity versus environment* and *maturation versus learning*. Under whatever name, however, the basic question remains: How much of any given characteristic, behavior, or pattern of development is determined

nature All the genetic influences on development, including those that affect physical characteristics as well as psychological traits, capacities, and limitations.

nurture All the environmental influences on development, from prenatal influences on the embryo to the cultural context at death.

by genetic influences and how much is the result of the myriad experiences that occur after conception? Note that the question is "How much?"—implying that with all characteristics, behaviors, and patterns of development, *both* nature and nurture are influential. All developmentalists agree that, at every point, the *interaction* between nature and nurture is the crucial influence on any particular aspect of development. They note, for example, that intelligence is determined by the interplay of heredity and such aspects of the social and physical environment as schooling and nutrition. Despite their acknowledgment of the interaction between nature and nurture, however, developmentalists can get into heated arguments about the relative importance of each (see Baumrind, 1993; Jackson, 1993; Scarr, 1992).

One of the reasons this controversy is very much alive is that its practical implications are enormous. Consider one example. Although boys and girls in elementary school show similar math aptitude, the mathematical achievement of the typical teenage boy is higher than that of the average teenage girl. Furthermore, high school students who are gifted in math are usually boys, by a 4-to-1 ratio (Benbow & Lubinski, 1993). A closer analysis of the male advantage reveals that, beginning at about age 10, boys are better at spatial skills—the kind required for geometry—and this accounts for much of the difference in math achievement (Johnson & Meade, 1987). In addition, boys are better prepared in math, in that they take more geometry and calculus classes than girls do (Beal, 1994).

It's all in the family. Three of the four musicians in front are brothers—Wynton Marsalis on trumpet, Branford Marsalis on tenor sax, and Jason Marsalis on drums. In the background, on piano, is their father, Ellis Marsalis. Is the extraordinary musical talent of the Marsalis family the result of nurture—a household continually filled with music, music lessons from an early age, a family emphasis on musical traditions—or nature, a genetic gift? The answer, of course, is that both nature and nurture were essential to the development of the Marsalis family's musicianship.

Is nature responsible for these differences? Perhaps some hormonal difference causes early brain differentiation that, at puberty, gives males an advantage (Gaulin, 1993; Jacklin et al., 1988) and, in turn, leads them to take more math. Or is nurture the key factor? Perhaps girls learn that math ability is not considered feminine, and perhaps their parents, teachers, or boyfriends, sharing this view, subtly—or not so subtly—discourage their interest and efforts in math (Eccles & Jacobs, 1986). Support for the latter

hypothesis comes from cross-cultural research on women scientists, which shows that the percentage of physics faculty members who are women ranges from about 1 percent in Japan to 47 percent in Hungary (see Figure 1.5), a diversity that suggests that in this particular example, at least, the cultural context is much more influential than biology (Barinaga, 1994).

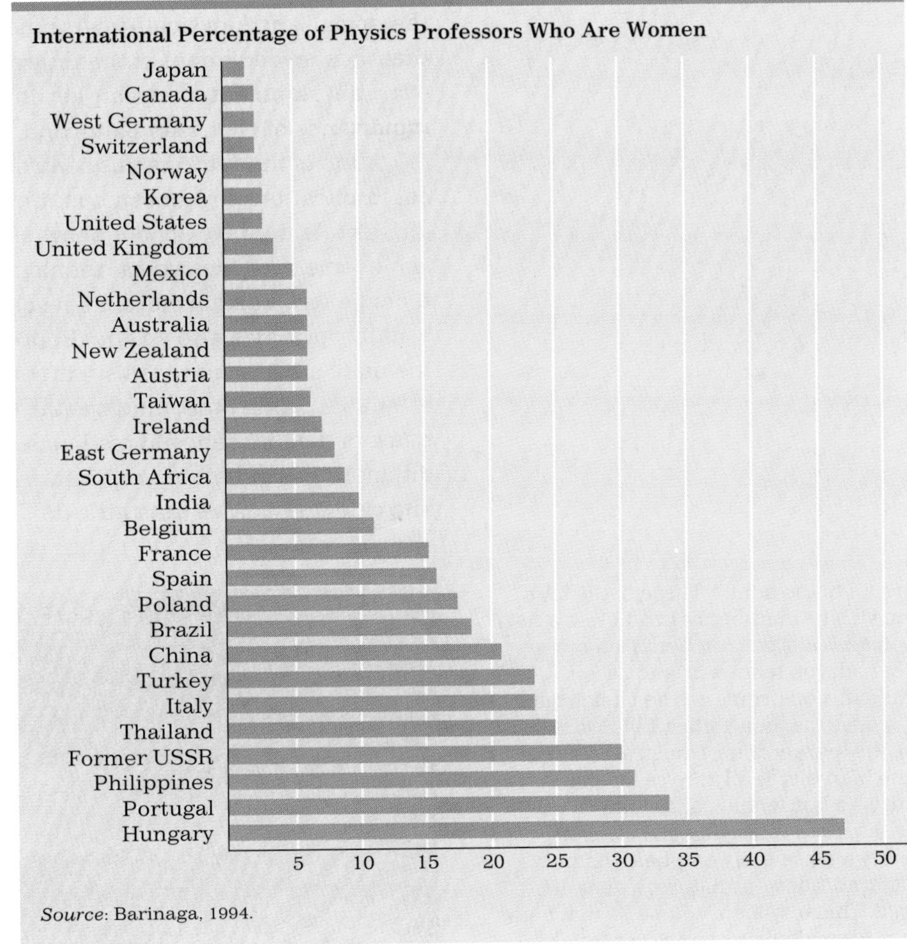

FIGURE 1.5 Nations differ dramatically in how many of their university instructors are women, particularly in the natural sciences and math. The data in this chart, for example, reflects the percentage of female physics professors worldwide. Notice that even within continents, and within ethnic groups, the rates vary by nation. Obviously, nurture—especially the political and economic patterns of each country—is much more at work here than nature.

Whatever the answer, the implications are significant. If boys are naturally better at math than girls, it may be neither wise nor desirable to push girls to study math and related subjects such as physics and engineering. On the other hand, if such gender differences are the result of nurture, we are wasting a major portion of our mathematical potential, as well as limiting the career options for many female workers, by not encouraging girls to develop fully their math abilities.

Another of the many controversial issues that pivots on the nature–nurture debate is the question of sexual orientation. Whereas psychologists once assumed that adult homosexuality resulted from unusual patterns in the mother-father-child relationship, many now believe that homosexuality is at least partly genetic. Included in the growing evidence for this view is research showing that a man is more likely to be gay if his mother's brother or his own brother—especially his identical twin—is homosexual (Hamer et al., 1993; Pool, 1993; Whitam et al., 1993).

But disagreement remains as to whether the evidence for genetic influence on homosexuality is conclusive and, if so, how great a role nature plays (Bailey et al., 1993; Maddox, 1993). Once again, the implications are profound. If homosexuality is primarily the result of nurture, then those who are concerned about the future sexual orientation of the young have reason to examine the influence of school curriculum and television programming regarding this issue. On the other hand, if the primary influences on a person's sexual orientation are genetic, then the issues debated—and perhaps even the need for a debate at all—will change.

Dozens of other examples come to mind. For instance, assumptions about the relative importance of nature and nurture guide society's responses to one of the most pressing developmental problems of the day—the neglect and abuse of millions of children by their parents every year. Policy makers who see the genetic link between biological parent and offspring as the most important factor in children's development tend to believe that maltreated children should stay with their biological parents in all but the most damaging circumstances, and that when the child must be removed from the home, foster care should be temporary and with close relatives. In contrast, those who believe that nurture is the stronger force in children's development tend to believe that seriously maltreated children should be quickly removed from their parents and adopted by competent, loving caregivers. Assumptions about the nature–nurture balance are similarly implicit in all manner of social-welfare policies, as well as in approaches to controversial issues ranging from the effects of alcohol exposure on fetal development to the root causes of senility, with enormous consequences not only for current policies but also for the allocation of scientific attention and research funding.

The First Years of Life: Determining Force or Fading Influence?

A second controversy at the heart of developmental psychology concerns the extent to which the experiences of early life affect later emotional and intellectual development. Are our individual personality characteristics rooted in the events and emotional patterns of our first few years, shaping us for life? Or is personality fluid and malleable, shifting in response to different experiences and perceptions? Is the developmental process like that of a building, with the initial foundation and framework determining the form of the eventual structure? Or is it more like that of a painting, in which the initial sketch may be completely altered as subsequent brushstrokes create the picture?

For much of its history, developmental study endorsed the former view, contending that the first five years of life provide the basic structure for the individual's later personality development. And, in many ways, various theories that emphasize the importance of early experiences have been substantiated by careful research. For example, some years ago Erik Erikson (1963) and John Bowlby (1969), building on the earlier ideas of Sigmund Freud, hypothesized that the nature of an infant's trust in, and attachment to, his or her mother determines whether that person can later sustain other close relationships, such as those with friends or lovers. Recent studies of infant-mother attachment have, in fact, confirmed that the security or

insecurity of this first human bond can have long-term consequences, not only for the child's future relationships, but also for the child's own self-esteem (Egeland et al., 1993; Lamb et al., 1985).

Research in other areas has likewise shown that certain early experiences can have lasting effects. Children who experience devastating poverty or the crushing instabilities and anxieties of homelessness tend to have such extensive early cognitive delays that they are likely to remain intellectually behind for many years to come, even when their circumstances have improved (Rafferty & Shinn, 1991; Ramey & Campbell, 1991). Long-term negative effects, extending well into adulthood, can be even more apparent when young children have undergone acute trauma, such as being severely abused or neglected (Cicchetti, 1990).

For some developmentalists, findings such as these confirm their belief that the early years, good or bad, are far more crucial than the later years for the development of the individual's capabilities and dispositions. In their view, our first experiences in early childhood provide the thrust and direction of our "launching" into life, establishing a trajectory that irrevocably determines the nature of our journey and our eventual destination.

Many other developmentalists hold quite a different view. They believe that our early developmental paths can be changed, even reversed, by our later experiences. Their view is also supported by extensive research, especially by studies showing that children can rebound from difficult early experiences if they have at least minimal amounts of emotional and cognitive support. Even after years of a troubled infancy and childhood, marred by medical problems, extreme poverty, mentally disturbed parents, family disruptions, or abuse, some resilient individuals become quite successful and well-adjusted adults (Elder et al., 1985; Furstenberg et al., 1987; Masten et al., 1990; Rutter, 1989).

For these people, certain protective factors and opportunities—among them finding close friends, attaining higher education, mastering vocational skills, developing a supportive marriage, maintaining spiritual faith—seem to have acted as buffers against devastation. As one study that followed troubled children from birth to age 32 concludes:

> . . . these buffers make a more profound impact on the life course of children who grow up under adverse conditions than do specific risk factors or stressful life events. They appear to transcend ethnic, social class, geographical, and historical boundaries. Most of all, they offer us a more optimistic outlook than the perspective that can be gleaned from the literature on the negative consequences of perinatal trauma, caregiving deficits, and chronic poverty. They provide us with a corrective lens—an awareness of self-righting tendencies that move children toward normal adult development under all but the most persistent adverse circumstances. [Werner & Smith, 1992]

Increasingly, developmental psychologists are coming to agree with the view that life pathways are shaped by later experiences as well as by early ones, and that the first years of life *influence*—but rarely *determine*—later personality and behavior.

Nonetheless, the issue remains controversial because many practical decisions concerning human development hinge on the relative importance of the early or later years. The debate can become especially sharp with respect to public policy, when advocates of "early prevention" find themselves in a tug of war with believers in "targeted remediation." For instance, some

scientific method The sequence and procedures of scientific investigation (formulating questions, collecting data, testing hypotheses, and drawing conclusions) designed to reduce subjective reasoning, biased assumptions, and unfounded conclusions.

child advocates argue that to help prevent juvenile delinquency and other antisocial or violent behavior thought to be associated with inadequate child-rearing, families "at risk" should be provided with help very early on, starting with well-trained nurses who would visit the home frequently, not only to provide medical care for the infant but also to teach the parents how to be responsive and responsible caregivers. Others argue that it is more cost-effective to wait until a child or adolescent actually appears headed for serious trouble, and then to intervene with such measures as personal counseling, intensive job training, and subsidized employment. Still others argue that all intervention is useless, since, in their view, a person's genetic inheritance and prenatal experience completely set the stage for later growth.

In disputes such as this, the evidence, far from settling the issues, generally corroborates many points of view, because, at every age, most observable adult traits can be linked to an intertwining web of immediate and distant events. In most cases, teasing apart the strands of this web, and assessing the relative impact of each factor, is an extremely difficult task that involves subjective interpretation as well as objective judgment—so the controversy continues.

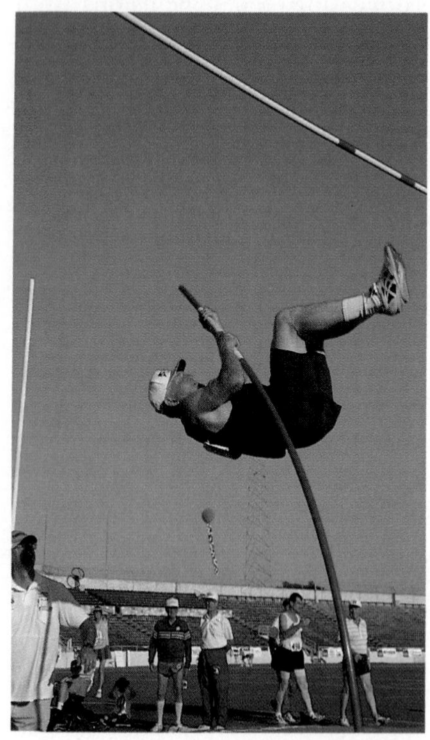

What athletic activity is appropriate for older adults? Scientific answers come not from a seemingly remarkable achievement by one individual but from systematic study of the performance of many individuals in a variety of circumstances. For decades it was assumed that the elderly should avoid all forms of physical exertion. However, research on thousands of older adults has found that a regular regimen of exercise—stretching, running, and especially strength-training—is beneficial to overall health and helps to sustain physical abilities in almost any sports activity.

Developmental Study as a Science

As you can see, this book touches on some of the most intriguing, controversial, and practical issues of life. As you study further, you will encounter many other such topics. Should prospective parents seek genetic counseling? Is it better for the intellectual growth of young children to be in preschool than at home? Which lifestyle choices enhance development and which are destructive? Does cohabitation before marriage harm a couple's later relationship? Are senior workers wiser and more experienced, or slower and less skilled? How do love and friendship change over time? Is it better for an older person to live alone, or with adult children, or in an assisted-living community?

The answers to these and thousands of other developmental questions are vitally important, but by no means obvious. Indeed, part of the excitement of developmental study is to uncover false assumptions, to reveal hidden prejudices, to discover a more satisfying, healthier, or more productive way to develop at any stage of life.

How can such lofty goals be reached? Asking pertinent questions and then systematically gathering and analyzing information that might answer them is the basic process that social scientists use, making every effort to let objective data, and not their wishful or biased ideas, lead them to their conclusions. In short, to answer questions about human development, social scientists follow the **scientific method**, a general procedural model that is designed to promote objectivity.

The scientific method, as it applies to developmental study, involves four basic steps, and sometimes a fifth:

1. *Formulate a research question*. Build on previous research, or on a particular developmental theory, or on personal observation and reflection, and pose a question that has relevance for the study of development.

replicate To repeat, with a different population, the specific design and procedures of a previous scientific study in order to test the validity of that study's conclusions.

2. *Develop a hypothesis.* Reformulate the question into a hypothesis, which is a specific prediction that can be tested.

3. *Test the hypothesis*. Design and conduct a scientific research project that will provide evidence about the truth or falsity of the hypothesis. As you will see in Chapter 2, the research design often includes many specific elements that help make the test of the hypothesis a valid one.

4. *Draw conclusions*. Formulate conclusions directly from the results of the test, avoiding general conclusions that are not substantiated by the test data.

5. *Make the findings available*. Publishing the results of the test is often the fifth step in the scientific method. In this step, the scientist must describe the test procedures and the resulting data in sufficient detail so that other scientists can evaluate the conclusions and, if they wish to, **replicate** the test of the hypothesis—that is, repeat it and obtain the same results—or extend it, using a different but related set of subjects or procedures. Through replication, the conclusions from each test of every hypothesis accumulate, leading to more definitive and extensive conclusions and generalizations.

In actual practice, scientific investigation is less straightforward than these five steps would suggest. The link between theory and hypothesis is sometimes indirect, and the design and execution of research are influenced by human judgment (Bauer, 1992). Throughout the process, human values guide the choice of which topics to examine, which methods to use, and how to interpret the results. However, given all the complexities of studying human development, the scientific approach—curious, conscientious, and creative, rigorous in its investigative methodologies, and open to unexpected, and sometimes unwished-for, findings—is a powerful tool. As you become a student of human development, you will learn to wield this tool with skill and precision, asking questions, examining evidence, and discovering insights that may change the course of not only your own development but that of your children, or your parents, or, perhaps, developing persons everywhere.

SUMMARY

The Scientific Study of Human Development

1. The study of human development explores how and why people change, and how and why they remain the same as they grow older. Developmentalists include researchers from many academic and practical disciplines, especially biology, education, and psychology, who study people of every age and in every social group.

2. Development is often divided into three domains, the biosocial, the cognitive, and the psychosocial. While this division makes it easier to study the intricacies of development, researchers note that development in each domain is influenced by the other two, as body, mind, and emotion always affect one another; indeed, each aspect of development relates to each of the domains.

The Many Contexts of Development

3. An ecological, or contextual, approach focuses on the interactions between the individual and the various settings in which development occurs. Some of those settings involve the physical surroundings—the layout of the neighborhood, the nature of the climate, and so on. Most, however, concern the people who create the social context for the individual's development.

4. Social contexts shift over time, as changing historical events and conditions reshape the circumstances and perspectives surrounding development. It is quite different

being a senior citizen in a era in which only a hardy few reach that status and being one at a time when a sizable cohort are over age 60.

5. Development is also strongly affected by a person's socioeconomic status, cultural values, and ethnicity. The influences of each of these social contexts often overlap, reinforcing, and sometimes contradicting, each other. Ultimately, however, each individual's path is unique, influenced but not determined by the social contexts.

6. The interaction of domains is clearly seen in the example of David, whose problems originating in the biosocial domain quickly affected the other two domains. His example also shows how the individual is affected by, and also affects, the surrounding contexts of family, society, and culture.

Two Controversies

7. All aspects of development are guided by the interaction of hereditary forces and the particular experiences a person has. The relative importance of these factors is a topic of debate often referred to as the nature–nurture controversy.

8. The theory that the early years determine later development is much less influential than it once was, as most developmentalists now recognize that development is shaped and reshaped throughout life. Nevertheless, there are still intense debates over the relative influence of the early years, especially regarding matters of public policy.

Developmental Study as a Science

9. The scientific method is used, in some form, by most developmental researchers. They observe, pose a question, develop a hypothesis, test the hypothesis, and draw conclusions based on the results of the tests.

10. Often the final step of the scientific method is to publish the research in sufficient detail that others can evaluate the conclusions and, if they choose to, replicate the research or extend the findings with research of their own. Science usually begins with individual insight, but ultimately it becomes a very collaborative, communicative activity.

KEY TERMS

the scientific study of human development (4)
biosocial domain (4)
cognitive domain (4)
psychosocial domain (4)
ecological approach (6)
social context (7)
social construction (10)
cohort (11)
socioeconomic status (SES) (12)
culture (14)
ethnic group (15)
nature (20)
nurture (20)
scientific method (25)
replicate (26)

KEY QUESTIONS

1. What is the main focus of the study of human development?

2. What are the three domains into which the study of human development is usually divided?

3. Give examples of the interactions among the various contexts that affect an individual's development.

4. Name and give examples of three broad contextual factors that developmentalists recognize as powerful influences on human development.

5. What is the central dispute in the study of human development? Provide some examples of practical implications of this controversy.

6. What view have many developmentalists come to hold regarding the influence of the early years of life on later development? How is this issue still controversial?

7. What are the advantages of the scientific method?

8. What are the steps of the scientific method?

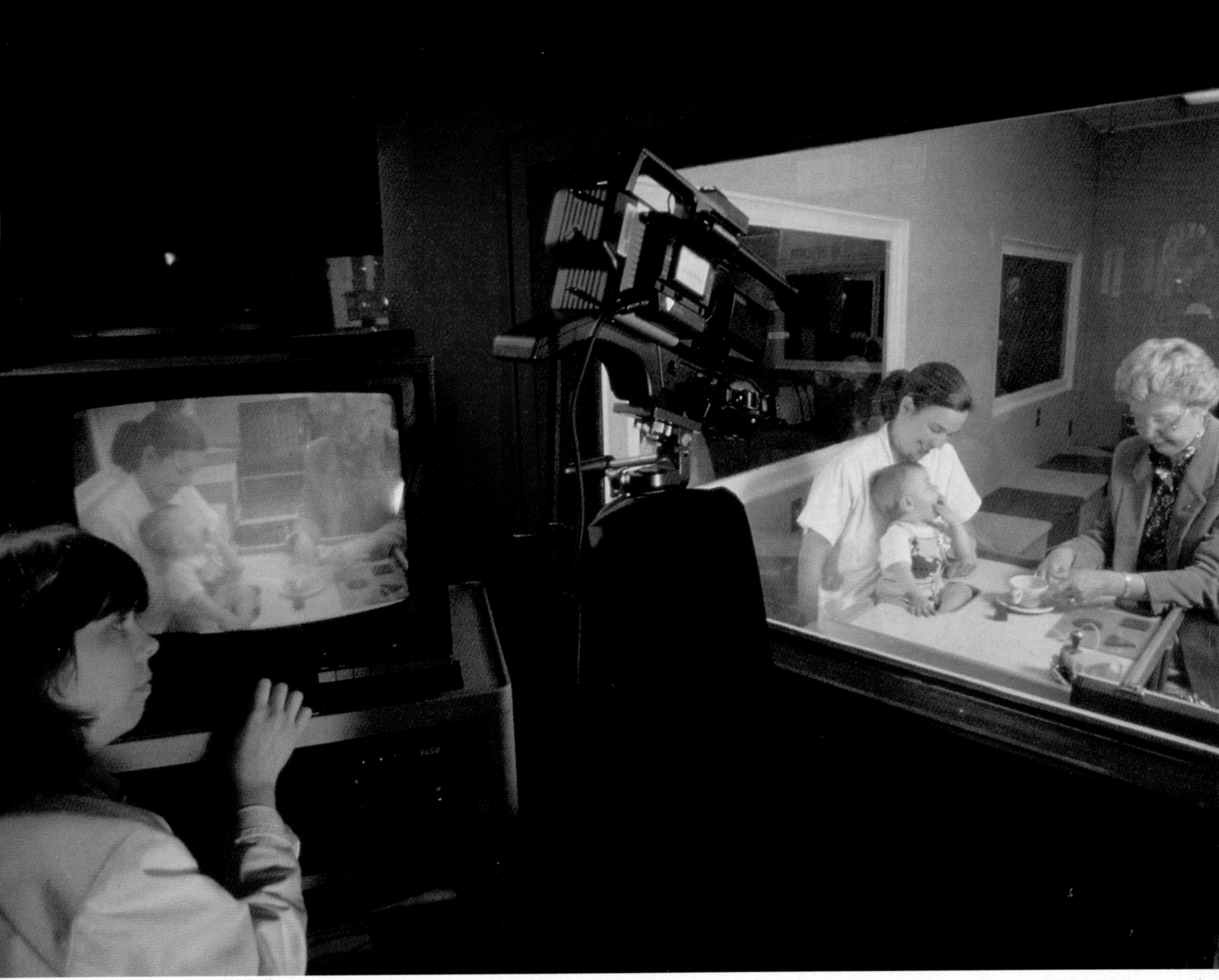

Theories and Methods

CHAPTER 2

As we saw in Chapter 1, the scientific effort to understand human development usually begins with questions. How do we, as individuals, develop into the kind of person we ultimately become? How significant and long-lasting are influences from early childhood? To what degree are we the products of our genetic inheritance, or of our environment? How do we learn to think, reason, create, and understand as we do? What are the unique challenges for personality growth at each stage of the life span?

What Theories Do

To begin to answer these questions and many others, we need some way to select significant facts and organize them in a manner that will take us deeper than our first speculations. In short, we need a theory. A **developmental theory** is a systematic statement of principles that explains behavior and development and guides developmentalists' investigations of new questions. To be more specific, in developmental research, theories have several purposes.

1. Theories provide a broad and coherent view of the complex influences on human development, and thus they offer guidance for practical issues encountered by parents, teachers, psychologists, and others concerned with development. They distinguish certain influences as paramount and others as peripheral, for example, and suggest how to optimize human growth.

2. Theories form the basis for hypotheses—or educated guesses—about behavior and development that can be tested by research studies and either supported or disconfirmed by their results (as you will see later in this chapter).

3. As theories are constantly modified by research findings, they provide a current summary of our knowledge about development. In this respect, developmental study is never complete, because updated theories give rise to new questions and new hypotheses meriting further investigation. Theories thus help us to ask important and relevant questions as well as lead us to useful answers.

developmental theory A systematic set of hypotheses and principles that attempts to explain development and provide a framework for future research.

Theories are central to understanding development. Consider why, for example, most parents devote such time and energy to caring for their children. Is it because parenting is a basic stage of adult development? Or does parental devotion derive from the rewards and reinforcement that children provide? Perhaps parenting impulses are rooted in the intellect, arising from an understanding of, and empathy for, children's needs. Or do cultural expectations, ingrained since childhood, lead us to assume that parenthood is a valued role? Different theories provide different answers to these questions and lead us to view parenting in different ways. Furthermore, the answers provided by developmental theories often have important practical applications. Regarding parenthood, for example, various theories provide insight into such issues as what type of parenting is most beneficial, whether it is better to have children at the beginning of adulthood or closer to middle age, and what can be done to prevent or remediate parental child abuse.

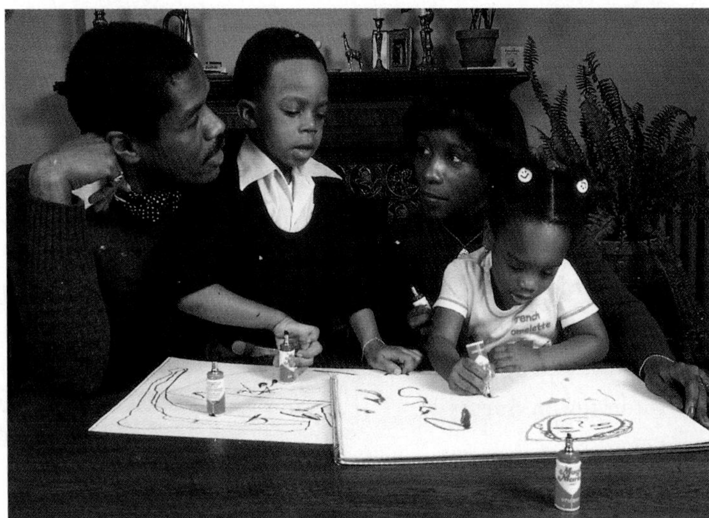

No matter what interaction developmentalists study, they can make their observations from various theoretical perspectives—psychoanalytic, which emphasizes unconscious drives and motives; learning, which emphasizes learned responses to particular situations; cognitive, which emphasizes the individual's understanding of self and others; and sociocultural, which emphasizes the cultural influences on the growth of individual learning and competencies.

Many theories are relevant to the study of development, but in this chapter we will focus on four broad theories that have been most influential and useful to developmental psychology—psychoanalytic theory, learning theory, cognitive theory, and sociocultural theory.

Psychoanalytic Theory

psychoanalytic theory A theory originated by Sigmund Freud that stresses unconscious forces that underlie human behavior.

childhood sexuality Freud's idea that infants and children experience sexual fantasies and erotic pleasures.

psychosexual stages A series of developmental stages proposed by Freud, each originating in sexual interest in, and gratification through, a particular part of the body.

Psychoanalytic theory interprets human development in terms of intrinsic drives and motives, many of which are irrational and unconscious, hidden from our awareness. These basic underlying forces are viewed as influencing every aspect of a person's thinking and behavior, from the smallest details of daily life to the crucial choices of a lifetime. Psychoanalytic theory also sees these drives and motives as providing the foundation for universal stages of development and for specific developmental tasks within those stages, from the formation of human attachments in infancy to the quest for emotional and sexual fulfillment in adulthood.

In addition to being the world's first psychoanalyst, Sigmund Freud was a prolific writer whose many papers and case histories, based largely on his patients' bizarre symptoms and unconscious sexual urges, helped make the psychoanalytic perspective a dominant force for much of the twentieth century.

Freud's Ideas

The psychoanalytic perspective originated with the work of Sigmund Freud (1856–1939), who began to formulate his theories while practicing as a physician in Vienna in the 1890s. Many of Freud's patients suffered from what was then called "hysteria," a disorder that involved various physical symptoms—pain, blindness, convulsions, trembling, or paralysis of specific parts of the body—but had no apparent physical basis. To Freud, it was clear that the source of these symptoms was the mind. To uncover the hidden causes of their problems, Freud would have his patients recline on his office couch and talk about anything and everything that came to mind—daily events, dreams, childhood memories, fears, desires—no matter how seemingly trivial or how unpleasant. From these disclosures and such things as the patients' slips of the tongue and unexpected associations between one idea and another, Freud discerned clues to the deep-seated, and usually unconscious, emotional conflicts that paralyzed one person or terrified another. In most cases, these conflicts seemed to Freud to be related to unconscious and irrational sexual and/or aggressive impulses, some of which appeared to have their origins in the events of infancy.

As a result of his clinical work, Freud proposed a theory of **childhood sexuality**. According to this theory, development in the first six years occurs in three **psychosexual stages** (see Table 2.1, p. 33). Each stage is characterized by the focusing of sexual interest and pleasure on a particular part of the body. In infancy, it is the mouth (the *oral stage*); in early childhood, it is the anus (the *anal stage*); in the preschool years, it is the genitalia (the *phallic stage*).

Freud maintained that in each stage, the sensual satisfaction associated with these body regions is linked to the major developmental needs and challenges that are associated with that particular stage. During the oral stage, for example, the baby not only gains physical nurturance through sucking but also experiences sensual pleasure in the process and becomes emotionally attached to the person who provides these oral gratifications. During the anal stage, pleasures related to control and self-control—initially in connection with defecation and toilet training—are paramount. During the phallic stage, pleasure is derived from genital stimulation, and the young child's interest in physical differences between the sexes leads to the development of gender identity and sexual orientation and to the child's identification with the moral standards of his or her same-sex parent.

The first three of Freud's stages are followed by a five- or six-year period of sexual *latency,* during which sexual forces are dormant. Then, at about age 12, the individual enters a final psychosexual stage, the *genital stage,* which is characterized by mature sexual interests and lasts throughout adulthood.

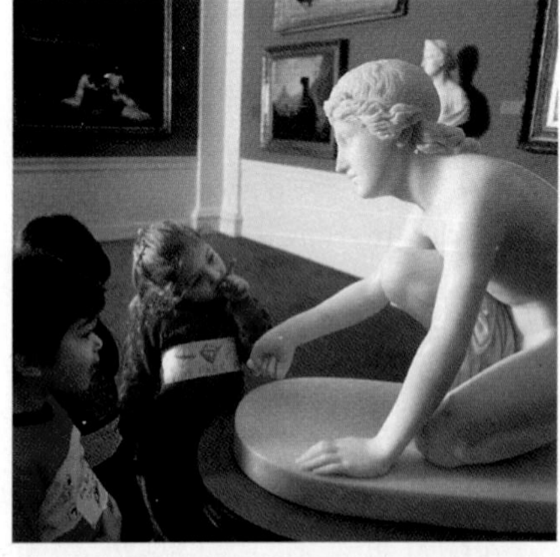

These children's interest in the statue's anatomy may reflect simple curiosity, but Freudian theory would maintain that it is a clear manifestation of the phallic stage of psychosexual development, in which all children become intrigued with their own and others' genitalia and girls are said to feel deprived because they lack a penis.

Perhaps Freud's most influential idea was that each stage has its own potential conflicts between child and parent and that the conflicts arising during the first three stages are all-important in the individual's development. According to Freud, how the child experiences and resolves the conflicts typically occurring in the oral, anal, and phallic stages—especially those related to weaning, toilet training, and childhood sexual curiosity—influences his or her personality growth and determines the individual's lifelong patterns of behavior.

Until his death in 1994, at the age of 92, Erik Erikson continued to write and lecture on psychosocial development. An important feature of his work is its emphasis on psychohistory—the relationship between historical factors and personality development.

Erikson's Ideas

Freud had a number of students who became famous psychoanalytic theorists in their own right. Although they all acknowledged the importance of the unconscious, of irrational urges, and of early childhood, each in his or her own way expanded and modified Freud's ideas. The most notable of these theorists was Erik Erikson (1902–1994), who formulated a comprehensive theory of development.

Erikson spent his childhood in Germany, his adolescence wandering through Italy, his young adulthood in Austria under the tutelage of Freud and Freud's daughter Anna, and his later life in the United States. In America, he studied a wide array of subjects, including students at Harvard, soldiers who suffered emotional breakdowns during World War II, civil rights workers in the South, disturbed and normal children at play, and Native American tribes. Partly as a result of this diversity of experience, Erikson began to think of Freud's stages as too limited and too few. He proposed, instead, eight developmental stages spanning the entire life span. Each stage is characterized by a particular challenge, or developmental **crisis**, that is central to the stage of life in question and must be resolved.

As you can see from Table 2.1, Erikson's first five stages are closely related to Freud's stages. At the same time, they, like the rest of Erikson's stages, differ significantly from Freud's in their emphasis on the person's relationship to the social environment. To highlight this emphasis on social and cultural influences, Erikson called his theory the **psychosocial theory** of human development.

In this psychosocial theory, the resolution of each developmental crisis depends on the interaction of the individual's characteristics and the support provided by the social environment. In the stage of *initiative versus guilt*, for example, children between ages 3 and 6 often want to undertake activities that exceed their abilities and/or the limits set down by their parents. Their efforts to act independently can thus leave them open to feelings of either pride or failure, depending in part on the reactions of their parents and on their culture's expectations regarding children's behavior. Similarly,

It seems quite clear that the toddler here is well into the psychosocial stage Erikson referred to as *autonomy versus shame and doubt*. Whether he emerges from this stage feeling independent or inept depends, in part, on whether his parents encourage his various efforts at self-control or, instead, regularly criticize him for his failures. It seems equally clear that this young girl is trying to negotiate the stage Erikson called *identity versus role confusion*. In this stage, teenagers try out a number of roles (often focusing on appearance) in an effort to discover who they really are.

TABLE 2.1

Comparison of Freud's Psychosexual and Erikson's Psychosocial Stages

Approximate Age	Freud (Psychosexual)	Erikson* (Psychosocial)
Birth to 1 year	*Oral Stage* The mouth, tongue, and gums are the focus of pleasurable sensations in the baby's body, and sucking and feeding are the most stimulating activities.	*Trust vs. Mistrust* Babies learn either to trust that others will care for their basic needs, including nourishment, warmth, cleanliness, and physical contact, or to lack confidence in the care of others.
1–3 years	*Anal Stage* The anus is the focus of pleasurable sensations in the baby's body, and toilet training is the most important activity.	*Autonomy vs. Shame and Doubt* Children learn either to be self-sufficient in many activities, including toileting, feeding, walking, exploring, and talking, or to doubt their own abilities.
3–6 years	*Phallic Stage* The phallus, or penis, is the most important body part, and pleasure is derived from genital stimulation. Boys are proud of their penis, and girls wonder why they don't have one.	*Initiative vs. Guilt* Children want to undertake many adultlike activities, sometimes overstepping the limits set by parents and feeling guilty.
7–11 years	*Latency* Not a stage but an interlude, when sexual needs are quiet and children put psychic energy into conventional activities like schoolwork and sports.	*Industry vs. Inferiority* Children busily learn to be competent and productive in mastering new skills or feel inferior and unable to do anything well.
Adolescence	*Genital Stage* The genitals are the focus of pleasurable sensations, and the young person seeks sexual stimulation and sexual satisfaction in heterosexual relationships.	*Identity vs. Role Confusion* Adolescents try to figure out "Who am I?" They establish sexual, political, and career identities or are confused about what roles to play.
Adulthood	Freud believed that the genital stage lasts throughout adulthood. He also said that the goal of a healthy life is "to love and to work well."	*Intimacy vs. Isolation* Young adults seek companionship and love with another person or become isolated from others by fearing rejection and disappointment. *Generativity vs. Stagnation* Middle-aged adults contribute to the next generation through meaningful work, creative activities, and/or raising a family, or they stagnate. *Integrity vs. Despair* Older adults try to make sense out of their lives, either seeing life as a meaningful whole or despairing at goals never reached.

*Although Erikson described two extreme resolutions to each crisis, he recognized that there is a wide range of outcomes between these extremes and that, for most people, the best resolution to a crisis is neither extreme but a middle course.

all young adults must negotiate the crisis of *intimacy versus isolation*, seeking companionship and love with others. How well they resolve this crisis depends partly on their ability and willingness to share their lives emotionally with another person and partly on external factors, such as the availability of potential partners or socioeconomic conditions (like joblessness and poverty) that might affect an individual's ability to share his or her life.

crisis In psychosocial theory, the central conflict of each developmental stage.

psychosocial theory A theory that stresses the interaction between internal psychological forces and external social influences. Erikson's theory is a psychosocial one.

Evaluation of Psychoanalytic Theory

All developmentalists owe a debt of gratitude both to Freud and to the neo-Freudians who extended and refined his concepts. Many of Freud's ideas are so widely accepted today that they are no longer thought of as his—for example, that unconscious motives affect our behavior, that development occurs in a series of stages, and that the early years are a formative period of personality development. And while much of Freud's thinking has come into question, many have learned from certain of his insights (Emde, 1994).

Moreover, the psychoanalytic approach laid the foundation for current thinking about topics as diverse as mother-infant attachment, the effects of parental discipline, gender identity, moral development, adolescent identity, and a variety of other issues that you will study in subsequent chapters. Although the ideas of Freud and his followers have been modified considerably, they remain suggestive and insightful. Current developmental psychology emphasizes the power of intimate personal relationships, an emphasis that originated, in part, from Freud's stress on patterns of love and hate within the family.

There are, however, many aspects of Freud's psychosexual approach that most contemporary developmentalists find to be inadequate or wrong (Macmillan, 1996). First and foremost, nearly all researchers recognize that Freud, in the words of one critic, "neither thought nor acted like a scientist" (Crews, 1996). His ideas were largely intuited from his clinical work with disturbed patients, and he never subjected any of his theories to scientific testing. Nor has scientific evidence ever been found for many of his major proposals. For instance, Freud's notion that the child's oral and anal experiences during the first two psychosexual stages form the basis for character structure and personality problems in adulthood has found little support in studies of normal children. Similarly, specifics of the phallic stage seem unrelated to a person's future sexual orientation. Most researchers agree that, throughout life, personality characteristics and behavior are affected much more by genetic traits, current life events, and the surrounding social context than by the psychosexual dynamics of early childhood.

Erikson's interpretation of development has fared better than Freud's, perhaps because Erikson's ideas, though arising from Freudian theory, are more comprehensive, contemporary, and apply to a wider range of behavior. Even so, most of the sources of Erikson's theory are, like Freud's, grounded in his own experiences, the recollections of his patients in therapy, and his insights from literature, film, and historical circumstances. In general, psychoanalytic theories do not lend themselves easily to laboratory testing under controlled conditions, a failing that led to the rise of the next theory we will consider.

Learning Theory

Early in the twentieth century, John B. Watson (1878–1958) argued that if psychology was to be a true science, psychologists should study only what they could see and measure. In Watson's words: "Why don't we make what we can *observe* the real field of psychology? Let us limit ourselves to things that can be observed, and formulate laws concerned only with those things. . . . We can observe behavior—what the organism does or says" (Watson, 1930/1967). Many American psychologists agreed, partly because of the difficulty of trying to study unconscious motives and impulses identified in psychoanalytic theory. Actual behavior, by contrast, could be studied far more objectively and scientifically. Thus developed a major theory of American psychology, **behaviorism**. This theory, in turn, gave rise to **learning theory**, which focuses on the ways we learn specific behaviors.

behaviorism A theory that emphasizes the systematic study of observable behavior, especially how it is conditioned.

learning theory A theory that emphasizes the sequences and processes of conditioning that, according to the theory, underlie most of human and animal behavior.

stimulus Anything that elicits a response, such as a reflex or a voluntary action.

response Any behavior (either instinctual or learned) that is elicited by a specific stimulus.

conditioning The process of learning, either through the association of two stimuli or through reinforcement or punishment.

classical conditioning The learning process in which a meaningful stimulus is linked to a neutral one, so that the latter elicits a response similar to that previously elicited by the former.

Laws of Behavior

Learning theorists have formulated laws of behavior that can be applied to any individual at any age, from newborn to octogenarian. These laws provide insights into how mature competencies are fashioned from simple skills and how environmental influences shape individual development. In this view, all development is a process of learning.

The basic laws of learning theory explore the relationship between **stimulus** and **response**, that is, between any experience or event (the stimulus) and the behavioral reaction (the response) with which it is associated. Some responses are automatic, like reflexes. If someone suddenly waves a hand in your face, you will blink; if a hungry dog smells food, it will salivate. But most responses do not occur spontaneously; they are learned. Learning theorists emphasize that life is a continual learning process: new events and experiences evoke new behavior patterns, while old, unproductive responses tend to fade away. One part of this learning process involves **conditioning**, through which a particular response comes to be triggered by a particular stimulus. There are two basic types of conditioning: classical and operant.

Pavlov was a physiologist who received the Nobel Prize in 1904 for his research on digestive processes. It was this line of study that led to his discovery of classical conditioning.

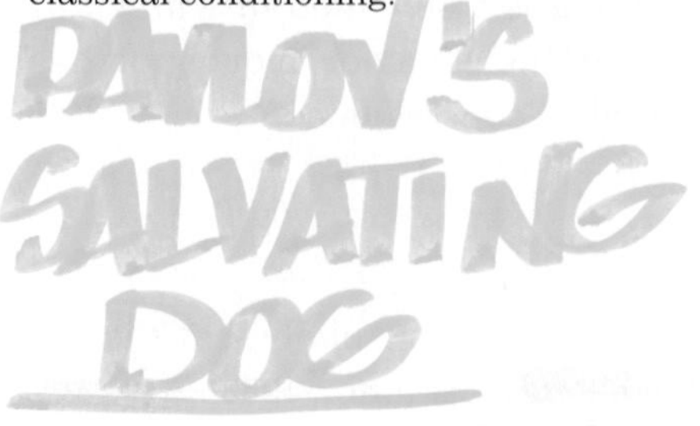

Classical Conditioning

More than ninety years ago, a Russian scientist named Ivan Pavlov (1849–1936) began to study the link between stimulus and response. While doing research on salivation in dogs, Pavlov noted that his experimental dogs began to salivate not only at the sight of food but, eventually, at the sound of the approaching attendants who brought the food. This observation led him to perform his famous experiment in which he taught a dog to salivate at the sound of a bell. Pavlov began by ringing the bell just before feeding the dog. After several repetitions of this association, the dog began salivating at the sound of the bell even when there was no food in sight.

This simple experiment in learning was one of the first scientific demonstrations of **classical conditioning** (also called *respondent conditioning*). In classical conditioning, an organism (any type of living creature) comes to associate a neutral stimulus with a meaningful one, and then *responds* to the former stimulus as if it were the latter. In Pavlov's original experiment, the dog associated the sound of the bell (the neutral stimulus) with food and responded to the sound as though it were the food itself.

Many everyday examples suggest classical conditioning that you yourself have probably experienced: imagining a succulent pizza might make your mouth water; reading a final-exam schedule might make your palms sweat; seeing an erotic photograph might make your heart beat faster. In each instance, the stimulus is connected, or associated, with another stimulus that regularly produced the physiological response in the past. Classical conditioning is also apparent when a child who has been badly frightened by some event—say, an attack by a snarling dog—is returned to the scene of the incident and begins crying, because he or she now associates the location with previous feelings of terror. As Watson (1927) himself noted, emotional responses are especially susceptible to learning through classical conditioning, particularly in childhood.

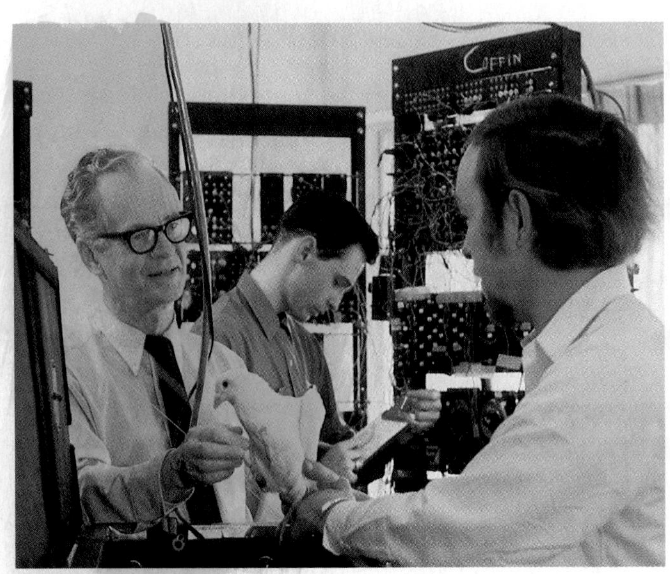

B. F. Skinner is best known for his experiments with rats and pigeons, but he also applied his knowledge to a wide range of human problems. For his daughter, he designed a glass-enclosed crib in which temperature, humidity, and perceptual stimulation could be controlled to make time spent in the crib as enjoyable and educational as possible. He also conceptualized and wrote about an ideal society based on principles of operant conditioning, where, for example, workers at the less desirable jobs earn greater rewards.

Operant Conditioning

The most influential American proponent of learning theory was B. F. Skinner (1904–1990). Skinner agreed with Pavlov that classical conditioning explains some types of behavior. However, Skinner believed that another type of conditioning—**operant conditioning**—plays a much greater role, especially in more complex learning. In operant conditioning, the organism learns that a particular behavior produces a particular consequence. If the consequence is useful or pleasurable, the organism subsequently repeats the behavior to achieve that consequence again. If the consequence is unpleasant, the organism will not repeat the behavior.

In operant conditioning, then, a system of pleasurable consequences (such as rewards) might be used to train an organism to perform a specific behavior that is not in the organism's natural repertoire. (A simple example of this is training a dog to fetch newspapers or jump through a hoop by giving it a treat every time it performs the behavior.) Or a behavior that is already in the organism's repertoire can be operantly conditioned to occur more precisely or quickly. (Early behaviorist experiments typically involved such conditioning efforts as getting rats to run mazes more efficiently for a food reward waiting at the end.) Once the behavior has been learned, the organism will continue to perform the activity even when the pleasurable consequences occur only occasionally rather than consistently. (Sometimes the behavior can become rewarding in itself.) Almost all a person's daily behavior, from socializing with others to earning a paycheck, can be the result of operant conditioning. (Operant conditioning is also called *instrumental conditioning*, bringing attention to the fact that the behavior in question has become an instrument for achieving a particular consequence.)

operant conditioning The learning process in which a person or animal becomes more, or less, likely to perform a certain behavior because of past reinforcement or punishment for similar behavior. (Also called *instrumental conditioning*.)

Types of Reinforcement

In operant conditioning, the process that makes it more likely that the behavior in question will recur is called *reinforcement* (Skinner, 1953). A stimulus that increases the likelihood that a behavior will be repeated is therefore called a *reinforcer*.

positive reinforcer Anything (such as a reward or positive event) that follows a behavior and increases the likelihood that that behavior will recur.

negative reinforcer The removal of an unpleasant stimulus following a particular behavior. This removal increases the likelihood that the behavior will be repeated should the unpleasant stimulus recur.

punishment An unpleasant event that follows from a particular behavior, making it less likely that the behavior will be repeated.

social learning theory The theory that learning occurs through imitation of, and identification with, other people.

Reinforcers may be either positive or negative. A **positive reinforcer** is something pleasant—a good feeling, say, or the satisfaction of a need, or a reward such as a piece of candy or a word of praise. For a grade-conscious student who has studied hard for an exam, getting an "A" would be a positive reinforcer of scholarly effort. A **negative reinforcer** is the removal of an unpleasant stimulus as the result of a particular behavior. When a student's anxiety about test-taking is reduced by extra preparation or, counterproductively, by "getting high," the reduction of anxiety is a negative reinforcer. That is, the anxiety-reduction resulting from the particular behavior increases the probability that the next time the student is worried about a test, he or she will engage in the same behavior. Note that a negative reinforcer differs from a **punishment**, because punishment is an unpleasant event that makes behavior *less* likely to be repeated. For the grade-conscious student, a failing grade on a test might be a punishment that would make the individual subsequently avoid the particular circumstances (like skipping class) that led to the failure.

Reinforcers may also be either extrinsic or intrinsic. *Extrinsic reinforcers* come from the environment in such varied forms as payment for work, a special privilege for behaving a certain way, good grades, and so forth. *Intrinsic reinforcers* come from within the individual, and typically involve feelings of satisfaction for a job well done and perceptions of self-competence. Thus individuals not only obtain reinforcement from others but are also *self-reinforcing* when they act in particular ways.

The beaming faces of these children suggest that they have all been strongly reinforced in their efforts. Both the girls on this championship soccer team and the young recycling champ lying atop part of his collection no doubt experienced the internal reinforcement that comes with doing a task well. In addition, they received the extrinsic reinforcement of public acclaim and, in the girls' case, a trophy.

Social Learning Theory

While learning theorists have traditionally sought to explain behavior primarily in terms of the organism's direct experience, contemporary learning theorists also focus on less direct, though equally potent, forms of learning. They emphasize that people learn new behaviors merely by observing the behavior of others, without directly experiencing any conditioning. These theorists have developed an extension of learning theory called **social learning theory**.

Modeling

An integral part of social learning is **modeling**, in which we observe other people's behavior and then pattern our own after it. This is not simply a case of "monkey see, monkey do." We are more likely to model certain aspects of our behavior, in certain contexts, after that of certain people. Generally, modeling of a particular behavior is most likely to occur in situations in which we are uncertain or inexperienced and the behavior has been previously enacted by someone we consider admirable or powerful or much like ourselves (Bandura, 1977).

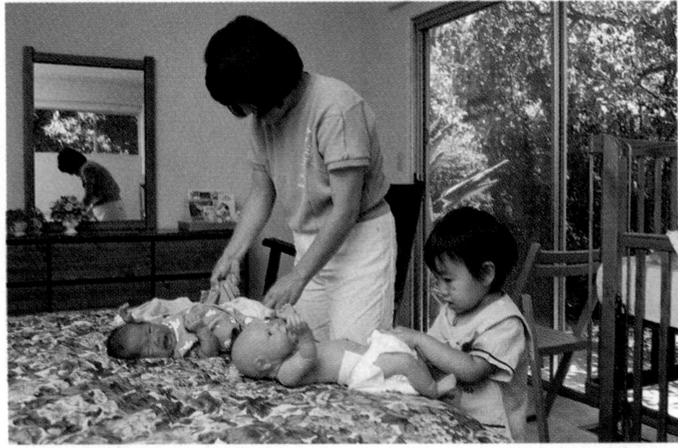

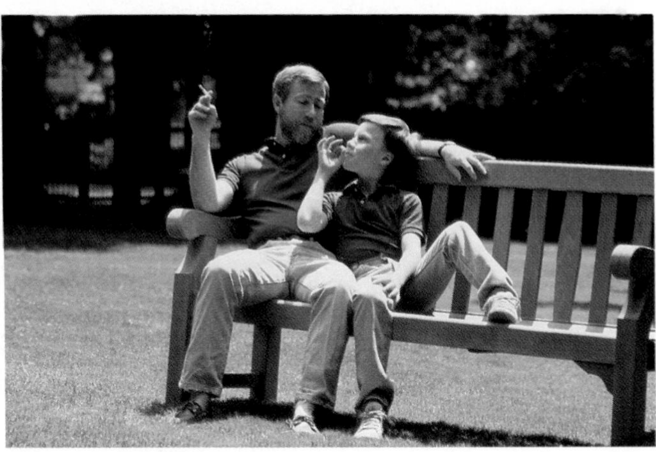

Social learning theory tends to validate the old maxim "Examples speak louder than words." If the moments here are typical for each child, the girl on the left is likely to grow up with a ready sense of the importance of this particular chore of child care. Unfortunately, the boy on the right may become a cigarette smoker like his father—even if his father warns him of the dangers of this habit.

Of course, social learning involves much more than just observing a model and imitating his or her behavior. A person must be motivated to pay attention to the modeled behavior, store information about it in memory (perhaps by mentally rehearsing it), and later retrieve that information when opportunities to use the modeled behavior arise (Bandura, 1977, 1986, 1989). These cognitive and motivational processes help to explain why people's susceptibility to modeling changes as they mature. With increasing age, for example, children become more discriminating observers of other people and are better able to extract general rules of behavior from the specific examples they observe. This is why young children tend to imitate the most obvious behaviors of a wide range of people, whereas adolescents and adults reproduce more subtle behaviors and styles of conduct (such as a "laid-back air" or a more "scholarly" manner) from their observations of selected individuals.

Social learning is also affected by self-understanding, because the standards we set for ourselves, and our confidence in our ability to meet them, influence our motivation to learn from various sources—whether they be peers, mentors, or media stars.

modeling The process in which one person learns from the example of another.

reciprocal determinism The idea that an individual's internal characteristics, environment, and behavior are mutually interactive in determining the person's specific behaviors.

Reciprocal Determinism

Because of these cognitive and motivational influences on social learning, theorists like Albert Bandura (1986, 1989, 1995) regard behavior as an outcome of **reciprocal determinism**—that is, the mutual interaction of the per-

son's internal characteristics, the environment, and behavior itself. One's internal characteristics, such as personal expectations, self-perceptions, and goals, are affected by the social environment, as we have seen, but they also influence that environment. Extroverted individuals, for example, evoke different reactions from others than do withdrawn persons. These reactions, in turn, reinforce the personal qualities in question. Behavior is the result of both personal and environmental factors, but it also influences each. In this concept of reciprocal determinism, social learning theorists seek to include the significant personal and environmental determinants of individual development within a comprehensive theory—a far cry from the early behaviorist focus on dogs salivating or rats running mazes.

Evaluation of Learning Theory

The study of human development has benefited from learning theory in at least two ways. First, the emphasis on the causes and consequences of observed behavior has led researchers to see that many behavior patterns that may seem to be inborn or the result of deeply rooted emotional problems may actually be learned behaviors that can be "unlearned" or learned responses to particular stimuli that can be changed by altering the stimuli. This realization has encouraged many scientists to approach particular problem behaviors, such as temper tantrums, phobic reactions, and harmful addictions, by analyzing and attempting to change the stimulus-response patterns they entail. For example, some smokers automatically light up as a reward to themselves upon completing a task. If such a person wanted to quit smoking, one helpful step might be to get the person to use a substitute—a piece of candy or a stick of gum—as an after-task reward. A similar approach has been adopted by programs that help parents to understand how they unintentionally reinforce or model problem behavior in offspring and to learn more successful child-rearing skills. Teachers, too, have benefited from this insight, developing classroom environments that promote learning and cooperation through reinforcements and modeling influences.

Second, learning theory has contributed considerable scientific rigor to developmental study (Grusec, 1994; Horowitz, 1994). Learning theorists have challenged researchers to define terms precisely, test hypotheses critically, explore alternative explanations for research findings (especially explanations involving environmental influences), and avoid reliance on theoretical concepts (such as unconscious drives or reasoning structures) that cannot be observed and directly tested. This emphasis has made developmental psychology a more scientific—and less speculative and intuitive—field of study.

At the same time, learning theory is often criticized for being inadequate to the task of explaining complex cognitive, emotional, and perceptual dimensions of human development (Grusec, 1992). Critics point out that these developmental processes are influenced not just by the environment, but also by genetic predispositions, biological maturation, internal structures of thought, and the developing person's own efforts to comprehend new experiences. From this perspective, behavioral and social learning theories that focus primarily on learning from the environment provide an important but very incomplete picture of the full range of developmental influences at work throughout life (Cairns, 1994).

Cognitive Theory

All his life Jean Piaget was absorbed with studying the way children think. He called himself a "genetic epistemologist"—one who studies how children gain knowledge about the world as they grow up.

Cognitive theory focuses primarily on the structure and development of the individual's thought processes and the way those processes affect the person's understanding of the world. In turn, cognitive theorists and researchers try to determine how this understanding, and the expectations it creates, affect the individual's attitudes, beliefs, and behavior.

Piaget's Theory

Jean Piaget (1896–1980), a major pioneer of cognitive theory, became interested in thought processes when he was hired to field-test questions that were being considered for a standard intelligence test for children. Piaget was supposed to find the age at which most children could answer each question correctly, but eventually he became more interested in the children's *wrong* answers. What intrigued him was that children who were the same age made similar types of mistakes, suggesting that there is a developmental sequence to intellectual growth. He began to believe that *how* children think is much more important, and more revealing of their mental ability, than *what* they know. Moreover, understanding how people think also reveals how they interpret their experiences and construct their understanding of the world.

Stages of Cognitive Development

Piaget maintained that there are four major stages of cognitive development. Each one is age-related, and as you will see in later chapters, each stage has structural features that permit certain types of knowing and understanding (Piaget, 1952, 1970).

cognitive theory The theory that the way people think and understand shapes their behavior and personality.

sensorimotor stage Piaget's first stage of cognitive development (from birth to about age 2) in which infants use their senses and motor skills to understand their world.

preoperational stage Piaget's second stage of cognitive development (from age 2 to about 6) in which children are unable to grasp logical concepts such as conservation, reversibility, or classification.

concrete operational stage Piaget's third stage of cognitive development (from about age 7 to 11) in which children can reason logically about concrete events and problems but cannot reason about abstract ideas and possibilities.

According to Piaget, infants in the **sensorimotor stage** know the world exclusively through their senses and motor abilities: their understanding of the objects in their world is limited to their sensory experience of them and to the immediate actions they can perform on them. This is a very practical, experience-based kind of early intelligence, but it is limited to the here and now. Once they enter toddlerhood, children become, in Piaget's words, "little scientists," tirelessly experimenting with objects to see how they work and to discover what new uses they can be put to. But as Piaget noted, the methods of the little scientist are limited to trial and error: to discover what happens when an egg is dropped on the floor, for instance, the little scientist has to drop one.

By contrast, preschool children in the **preoperational stage** can begin to think symbolically; that is, they can think about and understand objects using mental processes that are independent of immediate experience. This is reflected in their ability to use language, to think of past and future events, and to pretend. However, they cannot think logically in a consistent way, and thus their reasoning is subjective and intuitive.

School-age children in the **concrete operational stage** can begin to think logically in a consistent way, but only with regard to real and concrete features of their world, not abstract situations. Nevertheless, logical reasoning abilities make the school-age child a more systematic, objective, and scientific kind of thinker.

formal operational stage Piaget's fourth stage of cognitive development (from age 12 through adulthood), characterized by hypothetical, logical, and abstract thought.

In Piaget's final stage, the **formal operational stage**, adolescents and adults, in varying degrees, are able to think hypothetically and abstractly: they can think about thinking and speculate about the possible as well as the real. (As we will see in Chapter 18, some theorists believe that a fifth stage, the *post-formal stage*, emerges in adulthood.) Table 2.2 on page 42 summarizes Piaget's stages.

How Cognitive Development Occurs

Underlying Piaget's stage theory is his basic view of cognitive development as a process that follows universal patterns. This process is guided, according to Piaget, by the need in everyone for cognitive *equilibrium*, that is, a state of mental balance. What Piaget meant is that each person needs to, and continually attempts to, make sense of new experiences by reconciling them with his or her existing understanding.

Equilibrium is experienced when one's present understanding "fits" new experiences, whether this involves a baby's discovery that new objects can be grasped in the same ways as familiar objects or an adult's being able to explain shifting world events in terms of his or her political philosophy. When existing understandings do not seem to fit present experiences, the individual falls into a state of *disequilibrium*, a kind of imbalance that initially produces confusion and then leads to growth, as the person modifies old understandings and constructs new ones to fit the new experience. You may experience disequilibrium, for example, when a friend's argument re-

In Piaget's view, a child's stage of cognitive development influences how the child experiences and understands the world. An infant in the sensorimotor stage "understands" a flower as something that can be looked at, touched, and tasted. A child in the preoperational stage understands a plant as an object that can be named, described, and cared for. A school-age child in the concrete operational stage understands plants as members of a particular kind of life form that can be analyzed through logical reasoning skills, such as classification. The high school student in the formal operational stage can understand plants as part of intertwining ecological systems and can use abstract reasoning to devise horticultural experiments—such as the one shown here, devised to determine if treated sewage sludge is an effective substitute for chemical fertilizer.

TABLE 2.2

Piaget's Stages of Cognitive Development*

Approximate Age	Stage	Characteristics	Major Acquisitions
Birth to 2 years	Sensorimotor	Infant uses senses and motor abilities to understand the world. There is no conceptual or reflective thought; an object is "known" in terms of what an infant can *do* to it.	The infant learns that an object still exists when it is out of sight (*object permanence*) and begins to think through mental actions as well as physical actions.
2–6 years	Preoperational	The child uses *symbolic thinking*, including language, to understand the world. Sometimes the child's thinking is *egocentric*, causing the child to understand the world from only one perspective, his or her own.	The imagination flourishes, and language becomes a significant means of self-expression and of influence from others. Children gradually begin to *decenter*, that is, become less egocentric, and to understand and coordinate multiple points of view.
7–11 years	Concrete operational	The child understands and applies logical operations, or principles, to help interpret experiences objectively and rationally rather than intuitively.	By applying logical abilities, children learn to understand the basic concepts of conservation, number, classification, and many other scientific ideas.
From 12 years on	Formal operational	The adolescent or adult is able to think about abstractions and hypothetical concepts.	Ethics, politics, and social and moral issues become more interesting and involving as the adolescent becomes able to take a broader and more theoretical approach to experience.

*Each of these ways of thinking is described more fully later in this book (see Chapters 6, 9, 12, and 15).

veals inconsistencies in your views, when your favorite chess strategy fails against a skilled opponent, or when your mother does or says something you never expected her to.

Periods of disequilibrium can be disquieting to a child or an adult who suspects that accepted ideas no longer hold true. But they are also exciting periods of mental growth, which is one reason why people of all ages seek new, challenging experiences. By seeking out novel experiences, children are constantly putting their current understanding to the test. Babies poke, pull, and taste everything they get their hands on; preschool children ask thousands of questions; school-age children become avid readers and information collectors; adolescents try out a wide variety of roles and experiences; and adults continually increase their knowledge and expertise in areas that interest them—all because people at every age seek cognitive challenges. Recognition of this active searching for knowledge is the very essence of Piaget's theory of human cognitive development.

Information-Processing Theory

In recent years, another perspective on cognitive growth has influenced a growing number of developmental researchers. Taking its inspiration from modern technology, **information-processing theory** likens many aspects of human thinking to the way computers analyze and process data. In many ways, of course, the human mind is far more sophisticated than the most advanced computer: no computer can match the mind's capacity for reflection, creativity, and intuition. However, information-processing theorists suggest that by focusing on the step-by-step mechanics of human thinking, we might derive a more precise understanding of cognitive development (Klahr, 1989, 1992; Siegler, 1983, 1991). Like computers, for example, humans

information-processing theory A theory of learning that focuses on the steps of thinking—such as sorting, categorizing, storing, and retrieving—that are similar to the functions of a computer.

sensory register A memory system that functions for only a fraction of a second, retaining a fleeting impression of a stimulus on a particular sense organ.

working memory The part of memory that handles current, conscious mental activity. (Also called *short-term memory*.)

knowledge base The part of memory that stores information over a long time, from minutes to decades. (Also called *long-term memory*.)

must store large amounts of information, get access to that information when it is needed, and analyze situations in terms of the particular problem-solving strategies that will yield a correct solution.

One example of how researchers portray the information-processing system can be seen in Figure 2.1. The first step in information processing occurs in the **sensory register**, which stores incoming stimulus information for a split second after it is received to allow it to be selectively processed. Most information that comes into the sensory register is lost or discarded, but what is meaningful is transferred to working memory for further analysis. It is in **working memory** (sometimes called *short-term memory*) that your current, conscious mental activity occurs. This includes, at this moment, your understanding of this paragraph, any previous knowledge you recall that is related to it, and also, perhaps, distracting thoughts about your weekend plans or the interesting person who sat next to you in class today. Working memory is constantly replenished with new information, so thoughts and memories are usually not retained for very long. Some are discarded, while a few are transferred to the knowledge base.

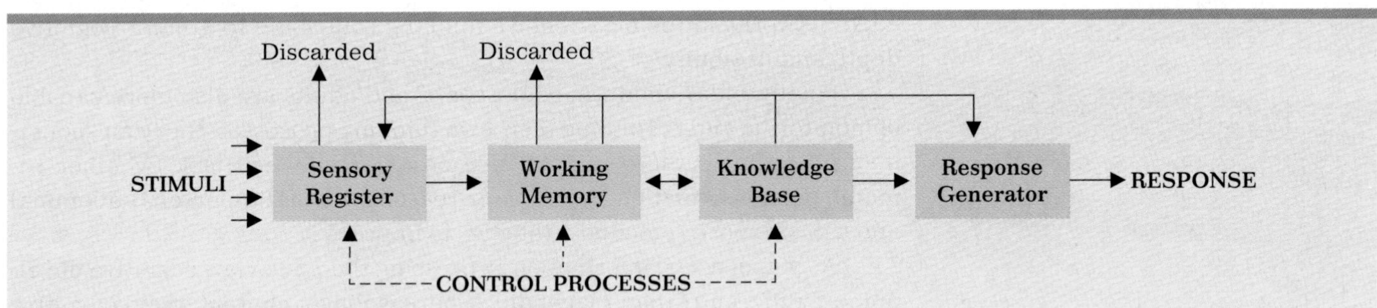

FIGURE 2.1 This is a flow chart of the information-processing system. Solid arrows refer to the transfer of information between system components. Broken arrows refer to influences within the system that affect how information is processed and transferred. (Adapted from Shiffrin and Atkinson, 1969.)

The **knowledge base** (also called *long-term memory*) stores information for days, months, or years and has a virtually limitless capacity. Together with influences from the sensory register and working memory, the knowledge base assists in organizing your reactions to environmental stimuli through the *response generator*, a network of mental processes that organizes behavior.

Developmental Changes

The various components of the information-processing system function differently as individuals mature, which explains in part why learning and memory skills change with development. One obvious change is that the knowledge base expands throughout childhood and adolescence, as children acquire more information about the world. This speeds new learning because it is easier to process new information when it relates to what one already knows.

With development, pivotal changes also occur in the component, called *control processes*, which regulates the analysis and flow of information within the system. When you deliberately use rehearsal or another strategy to remember a phone number, for example, you are using a control process. Control processes are also involved when you try to retrieve someone's name or other specific information from your knowledge base, or lis-

ten for a familiar voice in a crowd, or use a rule-of-thumb to solve a problem. In a sense, control processes assume an executive role in the information-processing system, regulating the analysis and transfer of information within the system.

The developing efficiency of the control processes is most noticeable in young children, as they acquire more sophisticated memory and retrieval strategies, learn to use selective attention, become capable of automatically performing mental activities that formerly required considerable effort (like reading), and develop more effective rules or strategies for problem solving (Kuhn, 1992; Sternberg, 1988).

Other developmental changes contribute to age-related improvements in information-processing skills (Flavell, 1992; Kuhn et al., 1995). As they mature, children develop richer associations among their knowledge networks, making it possible for an idea in one area of their thinking to trigger additional related thoughts and ideas in other knowledge domains. A grade-schooler studying about government, for example, might spontaneously recall things he or she heard about Congress during a TV news show. In contrast with a younger child's tendency to acquire knowledge in a more piecemeal fashion, the integration and association of different knowledge networks in the older child's thinking contribute to greater cognitive depth and flexibility.

Finally, older children, adolescents, and adults are also more capable of monitoring and regulating their own thinking processes: they can spontaneously evaluate their performance on an intellectual task (whether remembering information for a test or reasoning about a personal dilemma) and can often use remedial strategies to improve it.

As you can see, information-processing theorists view cognitive development differently than Piaget did. Whereas Piaget characterized cognitive development as a series of broad stages of mental growth, information-processing theorists tend to think of cognitive development as a more gradual process involving the acquisition of specific strategies, rules, and skills that affect memory, learning, and problem solving.

For the most part, these improvements in information processing are sustained through middle adulthood and are often enhanced by experience. In later adulthood, functioning in every component of the information-processing system slows down, leading to an overall decline in cognitive efficiency. As we will see in Chapter 24, however, researchers have found that the specifics of this decline vary tremendously, from person to person, from ability to ability, and from cohort to cohort, and that declines sufficiently large to interfere with daily functioning are far from inevitable, even at age 80 (Powell, 1994; Schaie, 1996). Moreover, by pinpointing declines in particular intellectual abilities, researchers can provide remediation that is "highly specific to the targeted abilities" and can, in some cases, restore function to what it was many years earlier (Schaie, 1996).

Evaluation of Cognitive Theory

Cognitive theories have revolutionized developmental psychology by focusing attention on active mental processes (Beilin, 1992). The attempt to understand the mental structures and strategies of thought, and to appreciate the internal need for new ones when the old ones become outmoded, has led to a new understanding of certain aspects of human behavior. Thanks to

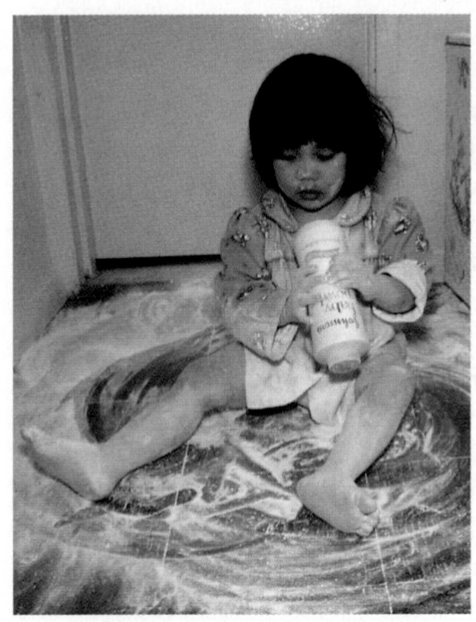

Are children more like scientists or computers in their cognitive growth? Elements of each are apparent in this preschooler's enthusiasm for discovering how baby powder looks, feels, and tastes, and in her attempts to relate these discoveries to prior experiences and events (like tasting and touching snow). Unfortunately, her parents are likely to see her efforts at intellectual discovery as nothing more than making a mess.

the insights provided by cognitive theories, we now have a greater appreciation of the capacities and limitations of the types of thinking that are possible at various ages—and of the ways in which these capacities and limitations can affect behavior.

Cognitive theories also remind us that "intelligence" involves many factors that are not easily summarized in an IQ score. Indeed, true intelligence reflects the remarkably diverse and complex skills and strategies that people evolve through their interactions with the surrounding world.

While Piaget's studies have made a profound contribution to the study of cognitive development, his work has met with criticism. Many people think Piaget was so absorbed by the individual's active search for knowledge that he underestimated the importance of external motivation and instruction and, consequently, underestimated the role of society and home in fostering cognitive development. As we shall see in the next section, sociocultural theorists believe that culture and education can be crucial in providing the proper mix of incentives to spur cognitive growth (Flavell, 1992).

Critics have also found fault with Piaget's depiction of cognitive stages. For example, there are many adults who are very inconsistent in using the skills of abstract thinking that Piaget described as typically developing in adolescence (Klahr et al., 1993). In addition, as we will see in Chapters 6 and 9, new research has shown that infants and preschoolers are far more competent intellectually than Piaget believed (Flavell, 1992). A number of researchers have also pointed out that Piaget's description of cognitive development tends to make it seem comprehensive, as though once a new stage of cognition has been achieved, it will be reflected in all aspects of the individual's thinking. In fact, the cognitive advance may occur in some areas of thinking and not in others, or it may appear on one occasion and not on another. Most cognitive theorists now generally believe that with regard to the pace of cognitive growth in specific areas, each child has "a unique rate of development and possesses his or her own idiosyncrasies" (Thomas, 1993). Information-processing theorists in particular emphasize that the skills and strategies children acquire may be task-specific and do not necessarily generalize to other situations (Case, 1985; Fischer, 1980).

Of course, information-processing theory has its critics as well. Some take issue with the use of the computer metaphor as a model of human thinking. Computers do not have the capacity for reflection, insight, or self-change that people do, and using this metaphor may mislead researchers into neglecting these essential features of human reasoning. These critics find Piaget's portrayal of the child as a little scientist, eagerly generating new understanding by acting on the world, to be a far more appropriate image of children's thinking. Moreover, because the focus of information-processing theorists has been primarily on the development of specific skills and abilities in the context of specific tasks, some critics wonder whether the information-processing approach will lead to useful, general conclusions about the nature of thinking as a whole and its development over time.

In essence, then, Piaget has been faulted for overgeneralizing the nature and progression of cognitive development in four broad stages, while information-processing theorists have been criticized for offering explanations of cognitive change that are too particularized. And both theories have been faulted for not sufficiently recognizing the role of the overall social context in which cognitive development occurs.

sociocultural theory A theory that seeks to explain the growth of individual knowledge and competencies in terms of the guidance, support, and structure provided by the broader cultural context.

guided participation A learning process in which an individual learns through social interaction with a "tutor" (a parent, a teacher, a more skilled peer), who offers assistance with difficult tasks, models problem-solving strategies, and provides explicit instruction when needed.

zone of proximal development The knowledge and understanding that an individual does not yet comprehend on his or her own but could master with guidance.

Sociocultural Theory

Whether they are living in a society that is urban and industrialized, rural and agricultural, or nomadic and pastoral, all people must acquire the skills and knowledge essential to their culture. For children in a small rural Kenyan community, for example, this may mean learning the skills needed for managing farmland and animals, for anticipating and deciphering seasonal rhythms, and for contributing to the well-being of their family and community. In contrast, for children growing up in urban America, acquiring essential cultural knowledge may mean learning the literary, logistical, and mathematical skills required in a technological society and developing the social alertness and self-assertion needed to remain safe while interacting with strangers. Underlying skills in both cultures are specific belief structures that children must also acquire, such as the value of respecting one's elders without question and of putting the needs of one's family and social group ahead of one's own—or, alternatively, the value of routinely questioning the authority of others and of steadfastly securing one's independence and self-interests.

In traditional societies, in which there is typically little social change, the lessons learned in childhood are often good for a lifetime. But in modern technological societies, especially those with strong multicultural influences, the tools, skills, customs, and values through which the culture functions are under continual pressure for change. In such societies, the older generations often find themselves having to learn the innovations of a younger cohort.

Recognizing these crucial aspects of cultural influence, **sociocultural theory** seeks to explain the growth of individual knowledge and competencies in terms of the guidance, support, and structure provided by the broader cultural context. The central thesis of sociocultural theory is that human development is the result of the dynamic interaction between developing persons and their surrounding culture. This view goes beyond mere descriptions of contrasting developmental patterns and looks at the processes through which individuals develop in a cultural context: it recognizes not only the importance of learning and instruction from parents, teachers, peers, colleagues, and mentors in one's immediate environment but also the ways in which these influences are shaped by the beliefs and goals shared by members of the community and the larger society.

The sociocultural theory of development is changing the nature of classroom education today, especially in multiethnic schools such as this one. Increasingly, teachers are recognizing that the most effective methods of instruction take into account the values, practices, and interests of the student's particular culture.

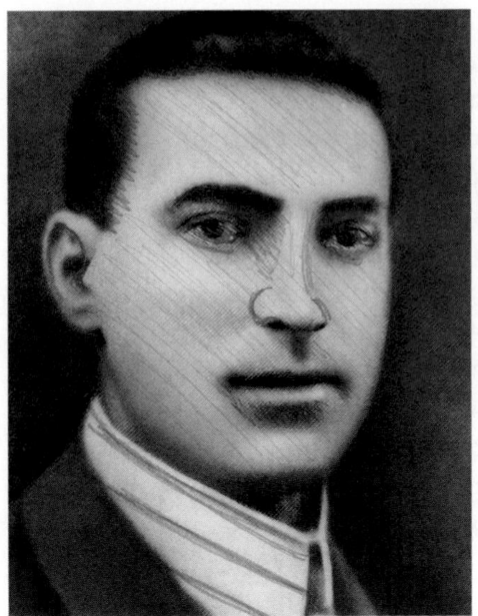

Lev Vygotsky is now recognized as a seminal thinker, whose ideas on the role of culture and history are revolutionizing education as well as expanding the conception of developmental processes. A contemporary of Freud, Skinner, Pavlov, and Piaget, Vygotsky did not attain their eminence in his lifetime, partly because his work, conducted in Stalinist Russia, was largely inaccessible, and partly because he died prematurely at age 38.

Vygotsky's Theory

A major pioneer of the sociocultural perspective was Lev Vygotsky (1896–1934), a psychologist from the former Soviet Union. Although his writings have only recently become available in the West (Vygotsky, 1978, 1987), they have attracted a wide audience, and many current researchers in developmental psychology take a sociocultural view that is deeply influenced by Vygotsky's ideas.

Vygotsky was particularly interested in the development of cognitive competencies. In his view, these competencies result from the interaction between novices and more skilled members of the society, acting as tutors or mentors, in a process called an "apprenticeship in thinking" (Rogoff, 1990). The implicit goal of this apprenticeship is to provide instruction and support for acquiring the knowledge and capabilities that are valued by the culture. In Vygotsky's view, the best way to accomplish this goal is through **guided participation**, in which the tutor engages the learner in joint activities, offering the learner not only instruction but also direct involvement in the learning process.

In every culture, parents and other teachers (not only adults but also more experienced peers) tutor children in practical skills (such as casting a fishing net, or sewing a button on a shirt, or using a TV remote control), social skills (such as shaking hands, or showing deference to elders, or expressing one's wishes in an acceptable manner), and intellectual skills (such as writing in one's native language, or acquiring new knowledge by consulting a village elder or visiting a specific Web site). This apprenticeship may take the form of explicit instruction, or it may occur informally, as children observe older friends or family members carry out the activities of everyday life. Of course, the process also works for adults, who are continually guided by each other, and sometimes by their children, in various skills. The necessity of such learning is demonstrated in modern societies every day, as parents rely on their teenage offspring to tutor them in programming the VCR, or surfing the Net, or interpreting the words on the latest rap CD.

In Bali, elegant, ritualized body movement is a key element in many facets of religious and cultural life. Through guided participation, this young girl is acquiring her culture's highly valued language of gesture as she is being taught a ceremonial dance.

Social interaction not only teaches specific skills; it also provides the context for mastering the culture's tools for further learning, whether they include an alphabet, Roman numerals, an abacus, a telephone, or a computer. Vygotsky believed that, universally, the most important learning tool is the specific *language* of each society, because language provides a powerful means of learning through social interaction. With the mastering of language, our thinking acquires unique potential, enabling us to express our thoughts and ideas to social partners and, in turn, to absorb the ideas of others—and the culture at large—into our own thinking (Vygotsky, 1978, 1987).

In both skill learning and language mastery, the process of social apprenticeship is similar. Typically, a mentor senses the learner's readiness for new challenges and arranges social interactions that help push the learner's skills in new directions. In Vygotsky's terms, the mentor draws the learner into the **zone of proximal development**, which is the range of skills that the learner can exercise with assistance but cannot perform independently. Through sensitive assessment of the learner's abilities and capacity for potential growth, the mentor offers guidance that engages the learner's participation and gradually facilitates the learner's transition from assisted performance to independent performance.

To make this rather abstract-seeming process more concrete, let's take a simple example—a father teaching his 5-year-old daughter to ride a bicycle. He will probably begin by helping his daughter to get the feel of the bicycle, as he slowly rolls her along, firmly supporting her weight and holding her upright while encouraging her to keep her feet on the pedals and to look straight ahead. When she says she feels that she is going to fall, he reassures her that he is right there and suggests that she lean forward a little bit and relax her arms. As she becomes more comfortable and confident, he begins to roll her along more quickly, noting that she is now able to keep her legs pumping in a steady rhythm. Within another lesson or two he is jogging beside her, holding on to just the handlebar, as he feels her control gradually go from dangerously wobbly to slightly shaky. Then comes the moment when he senses that, with a little more momentum, she could maintain her balance by herself. Accordingly, he urges her to pedal faster and slowly loosens his grip on the handlebar until, without her even realizing it, she is riding on her own.

Such excursions in the zone of proximal development are commonplace, not only in childhood but throughout life. Ideally, the learning process follows the same overall pattern in any given instance, with the mentor, sensitively attuned to the learner's continually shifting abilities and motivation, urging the learner on to new levels of competence.

Evaluation of Sociocultural Theory

Sociocultural theory has helped developmentalists deepen their understanding of the diversity in the pathways of growth, leading them to recognize the ways in which the skills, challenges, and opportunities involved in human development vary depending on the values and structures of the society in question. It has also reinforced the idea that in order to understand developmental processes in different cultures, developmental psychologists must understand the values and beliefs of the particular culture, how they affect the culture's members, and how particular competencies fit into the individual's specific cultural context.

In many ways, therefore, sociocultural theory has contributed to a broadening of developmental theory and research, as well as to a deeper appreciation of the cultural specificity of human interactions and growth. Sociocultural theorists have been criticized, however, for overlooking developmental processes that are not primarily social. Vygotsky's theory, in particular, has been viewed as neglecting the role of biological maturation in guiding development, especially with regard to neurological maturation in mental processes (Wertsch, 1985; Wertsch & Tulviste, 1992).

Barbara Rogoff (1990), a leading sociocultural theorist, has noted that Vygotsky did not recognize how much learners affect the context of their own development by, for instance, choosing their own mentors, activities, and settings for learning, or, sometimes, by refusing the guided assistance of others when mastering new skills. No doubt, sociocultural theory provides valuable insights into the social transmission of knowledge, but, as is true of all the other theories, by opening up new vistas on the developmental processes, it may simultaneously obscure others.

The Theories Compared

Each of the theories presented in this chapter has contributed a great deal to the study of human development. Psychoanalytic theory has made us aware of the importance of early childhood experiences and of the impact of the "hidden dramas" that influence our daily lives. Learning theory has shown us the effect that the immediate environment can have on our behavior. Cognitive theory has brought us to a greater understanding of intellectual processes and of how our thinking affects our actions. And sociocultural theory has reminded us of how our development is embedded in a rich and multifaceted cultural context.

Each theory has also been criticized. Psychoanalytic theory has been faulted for being too subjective; learning theory, for being too mechanistic; cognitive theory, for undervaluing the power of direct instruction and overemphasizing rational, logical thought; and sociocultural theory, for neglecting the intrinsic incentives for development.

In reviewing these four theories, consider how they differ in their portrayals of human development. As we have seen, each theory offers its own unique portrait of what the child and/or adult is like—a cauldron of unconscious impulses, a highly trained animal, a little scientist, a computer, an apprentice. Similarly, the theories also vary in how they describe the process of human development. Some theories, like those from the psychoanalytic perspective and Piaget's cognitive theory, view development as a succession of stages of growth, with each stage characterized by its own unique challenges and achievements. For other theories, however, development is a much more gradual and continuous process, and the factors that govern human development (such as learning processes or structures of information-processing) remain more consistent throughout life. The theories also vary in how each addresses the controversies described in Chapter 1, such as the relative importance of nature and nurture or the relative importance of the early years to overall development. Comparing developmental theories in these ways can highlight the unique contributions of each and the ways each theory alerts us to important features of the developmental process (see Table 2.3).

TABLE 2.3

Major Theories and Major Controversies

	Psychoanalytic	Learning	Cognitive	Sociocultural
Basic focus	Psychosexual (Freud) or psychosocial (Erikson) stages	Conditioning through stimulus and response	Thinking, remembering, analyzing	Social context, expressed through people, language, customs
Fundamental depiction of the individual	Battling unconscious impulses and overcoming major crises	Responding to stimuli, reinforcement, and models in the environment	Actively seeking to understand experiences, forming concepts and cognitive strategies	Learning the tools, skills, and values of society through apprenticeships
Emphasis on early experiences?	Yes (especially in Freudian theory)	No (conditioning and reconditioning are lifelong)	No (new concepts and control processes are developed lifelong)	Yes (family and school acculturation are critical)
Relative emphasis on nature and nurture	More nature (biological, sexual impulses are very important, as are parent-child bonds and memories)	More nurture (direct environmental influences produce various behaviors)	More nature (person's own mental activity and motivation are key)	More nurture (specific interaction between mentor and learner, within cultural context)

Each theory by itself is too restricted to encompass the breadth and diversity of human development. As you learned in Chapter 1, the study of human development is multidisciplinary and multicontextual and covers the entire life span, while the four major theories we have just described are rooted primarily in the discipline of psychology and focus primarily on younger persons rather than adults. Further, developmentalists readily acknowledge that all current theories are limited and represent "only the first steps in unraveling the complexities of development" (Parke et al., 1994).

Of necessity, then, the field of human development is moving away from these grand theories toward "minitheories," with various researchers taking useful direction from a variety of theoretical approaches. To cite just a few of the minitheories you will encounter later in this book, for example, humans' biological-social need for each other is sometimes explained in terms of "attachment theory"; the give-and-take among family members is sometimes interpreted through "exchange theory"; older persons' active involvement in life is sometimes described by "activity theory"; the role of instincts in human behavior is sometimes discussed within the framework of "ethological theory." Many developmentalists hope that "a new integration may be emerging in the form of a systems approach that will bring together biological, social, cognitive, and emotional minitheories into a more coherent framework" (Parke et al., 1994). For the meantime, however, most developmentalists take an eclectic perspective, meaning that rather than adopting any of these theories exclusively, they make selective use of many or all of them. For example, when forty-five leaders in the field were asked to describe their approach to developmental studies, "clear theoretical labels were hard to come by," with many describing themselves through some combination of terms, such as "cognitive social learning," "social interactive behaviorist," and even "social evolutionary cognitive behaviorism" (Horowitz, 1994).

In subsequent chapters, as you encounter elaborations and echoes of the four major theories and various minitheories, you will no doubt form and refine your own opinion of the validity of each theory, perhaps developing an eclectic view of your own. To do this in a thoughtful manner, based on facts rather than on personal assumptions, you need to understand more about the methods that scientists use to reduce their own biases and to move from theory to proof to application with as much clarity and honesty as possible.

Research Methods

Between the questions developmental scientists ask and the answers they arrive at lies their methodology, not only the steps of the scientific method outlined in Chapter 1 but also the specific strategies that are used to gather and analyze data. These strategies are a critical intermediary because "the ways that you attempt to clarify phenomena in large measure determine the worth of the solution" (Cairns & Cairns, 1994). To pick a very obvious example, in trying to find out what factors are key to marital happiness, developmental scientists are much more likely to discover information that has wide application if they design their research to include hundreds of people of various backgrounds than if they base their research on the experiences of their close friends. Dozens of less obvious aspects of research design and

observation The unobtrusive watching and recording of the behavior of subjects in certain situations, either in the laboratory or in natural settings.

analysis, sometimes used in creative combinations, can increase the validity of research findings. We will look now at some of the major research tools developmental scientists use to answer the questions posed by various theories, hypotheses, and hunches.

Observation

Scientists can test hypotheses about human development by using **observation**, that is, observing and recording what people do in specific circumstances. Observations can occur either in a laboratory setting that has been especially designed for this purpose or in a naturalistic setting, such as a home, school, playground, neighborhood street, or workplace. Typically, the observing scientist tries to be as unobtrusive as possible, so that the people being studied will act as they normally do.

In *laboratory* observation, scientists (usually situated behind one-way windows, where they can observe unseen) study topics ranging from the rate and duration of eye contact between infant and caregiver in specific contexts, to the play patterns of 3-year-olds in a mixed-sex play group, to the way married couples of various ages negotiate a family issue.

In *naturalistic* observation, scientists observe people in their natural environment. In one study of this type, researchers wanted to test the hypothesis that "maternal responsiveness is affected by cross-cultural differences" (Richman et al., 1992). Accordingly, they arranged for trained observers, familiar with the local language and culture, to compare mothers and their second- and later-born babies in several communities, among them the Gusii in rural Kenya and middle-class whites in suburban Boston. The observations, which spanned several months, were made in the subjects' homes, with the mothers going about their normal household activities and each observer acting like a visiting neighbor, trying to be as casual and unintrusive as possible while recording the mother's and child's behaviors as each responded to the other.

There were, of course, many cross-cultural similarities in maternal responsiveness that were observed. When the infants cried, for example, mothers in both locations were attentive, rarely ignoring their infants' signs of distress and usually responding to them with some form of social interaction—holding, touching, or talking. (As experienced caregivers, they did not

Developmentalists are currently investigating many aspects of children's social behavior. One way they do this is to observe children from behind a one-way window in a laboratory. In this photo, a researcher is observing children in a simulated day-care setting, noting such things as how their play patterns, ability to share, and negotiation strategies are affected by different factors, such as the presence or absence of the teacher.

RESEARCH REPORT

Ways to Make Research More Valid

In scientific investigation, there is always the possibility that the researchers' procedures and/or biases can compromise the validity of their findings. Consequently, scientists often take a number of steps to ensure that their research is as valid as possible. Six of these steps are explained here.

Sample Size

To begin with, in order to make any valid statement about people in general, the scientist must study a group of individuals that is large enough that a few extreme cases will not distort the picture of the group as a whole. Suppose, for instance, that researchers wanted to know the age at which the average American child begins to walk. Since they could not include every American infant in their study, they would work with a large sample group—a *sample population*—determining the age of walking for each member of the sample and then calculating the average for the group.

The importance of an adequate **sample size** can be seen if we assume for the moment that one of the infants in the sample had an undetected disability and did not walk until age 24 months. If the sample size were less than ten infants, that one late walker would, relative to the current standard of 12 months, add more than a month to the age when the "average" child was said to walk. However, if the sample were more than 500 children, one abnormally late walker would not change the results by even one day.

Representative Sample

Since the data collected on one group of individuals might not be valid for other people who are different in significant ways, such as gender, ethnic background, and the like, it is important that the sample population be a **representative sample**, that is, a group of subjects who are typical of the general population the researchers wish to learn about. In a study of when the average American infant begins to walk, the sample population should reflect—in terms of sex ratio, socioeconomic and ethnic background, and so forth—the entire population of American children. Ideally, other factors might be taken into consideration as well. For instance, if there is some evidence that first-born children walk earlier than later- or last-born children, then the sample should include a representative sample of each birth order.

The importance of representative sampling is revealed by its absence in two studies of age of walking (Gesell, 1926; Shirley, 1933) undertaken in the 1920s. Both studies used a relatively small and unrepresentative sample (all the children were white and most were middle-class), and, consequently, both arrived at a norm that is 3 months later than the current one, which was derived from a much more representative sample.

"Blind" Experimenters

A substantial body of evidence suggests that when experimenters have specific expectations about what their research findings will be, those expectations can affect the research results. As much as possible, then, the people who are carrying out the actual testing should be "**blind**," that is, unaware of the purpose of the research. Suppose one hypothesis is that first-born infants walk sooner than later-borns. Ideally, the examiner who tests the infants' walking ability would not know what the hypothesis is and would not even know the age or birth order of the toddlers under study.

Operational Definitions

When planning a study, researchers must establish an *operational definition* of whatever phenomena they will be examining. That is, they must define each variable in terms of specific, observable behavior that can be measured with precision. Even a simple variable such as whether or not a toddler is walking requires an operational definition. For example, does "walking" include steps while holding onto someone or something, or must it occur without support? Is one unsteady step enough to meet the definition, or must the infant be able to move a certain distance without

assume that every cry signaled hunger; they offered a breast or bottle to their crying infants less than 10 percent of the time.) In both locations, mothers also took the baby's developmental stage into account: they were more likely to cradle their crying 4-month-olds than they were to cradle their older babies.

Confirming the researchers' hypothesis, the observers also noted many cultural differences. One of the more intriguing was that American mothers communicated much more with words and much less with physical contact than Kenyan mothers did. This was apparent not only when the babies cried (see Figure 2.2 on page 54) but also when they made other sounds, played with objects, or merely looked at their mothers.

"I'm walking."

Do babies walk (or talk) when they are ready, no matter how little attention their parents provide? Only careful research can provide the answer.

faltering? For a study on age of first walking to be meaningful, the researchers would need to resolve questions like these in a clear and thorough definition.

Understandably, operational definitions become much harder to establish when personality or intellectual variables are being studied, but it is essential that researchers who are investigating, say, "aggression" or "romantic love" or "job satisfaction" define the trait in as precise and measurable terms as possible. Obviously, the more closely operational definitions reflect conceptual definitions, the more objective, valid, and reliable the results of the study will be.

Experimental and Control Groups

In order to test a hypothesis adequately in an experiment, researchers must compare two study groups that are similar in every important way except one: they must compare an **experimental group**, which receives some special experimental treatment, and a **control group**, which does not receive the experimental treatment.

Suppose a researcher hypothesized that infants who are provided with regular exercises that strengthen their legs walk earlier than babies who do not receive such exercise. In order to find out if this is true, the researcher would select two representative groups of children and arrange that one group (the experimental group) receive daily "workouts" devoted to leg-strengthening between their third and twelfth months, while the other group (the control group) would be given no special treatment.

Determining Statistical Significance

Whenever researchers find a difference between two groups, they have to consider the possibility that the difference occurred purely by chance. For instance, in any group of infants, some will walk relatively early and some relatively late. When the researchers in the study divide the sample population into the experimental and control groups, it is possible that, by chance, a preponderance of early walkers ends up in one group or the other.

To determine whether their results are simply the result of chance, researchers apply a test of **statistical significance**. This test takes into account many statistical factors, including the sample size and the average difference between the groups, and yields the *level of significance*, a numerical indication of exactly how likely it is that the particular difference occurred by chance. (Note that the word "significance" here means something quite different from its usual sense; that is, it refers to the validity of a study, not to its value.) Generally, in order to be called statistically significant, the possibility that results occurred by chance has to be less than one in twenty, which is written in decimals as a significance of .05. Often the likelihood of a particular finding's occurring by chance is even rarer, perhaps one chance in a hundred (the .01 level) or one in a thousand (the .001 level).

sample size The number of individuals who are being studied in a research project.

representative sample A select group of research subjects who reflect the relevant characteristics of the larger population that is under study.

blind Refers to researchers who are deliberately kept ignorant of the purpose of the research, or of relevant traits of the research subjects, in order to avoid biasing their data collection.

experimental group Research subjects who experience special conditions or treatments that the control group does not experience.

control group Research subjects who are comparable to the experimental group in every relevant dimension except that they do not experience the special experimental conditions.

statistical significance A mathematical calculation, derived from such factors as sample size and differences between groups, that indicates the likelihood that a particular research result occurred by chance.

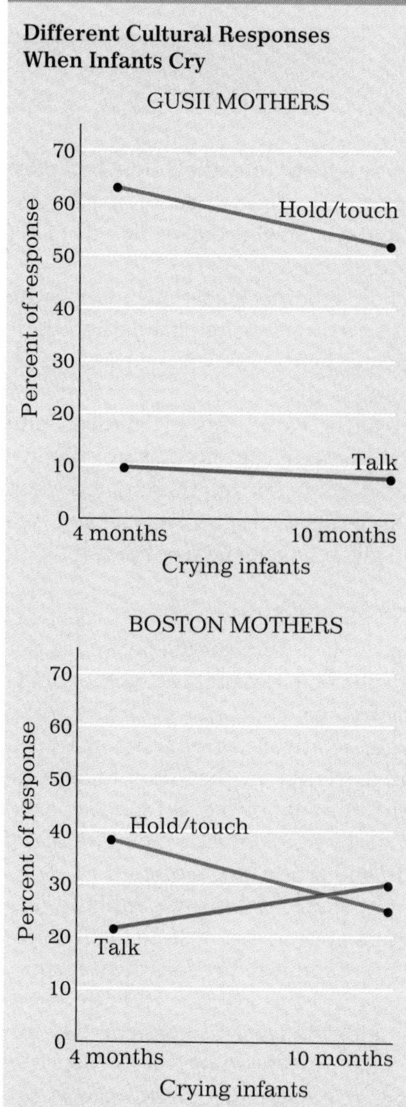

FIGURE 2.2 As is apparent from these data and from other research, African mothers are more physical, and North American mothers more verbal, in raising their children. This does not mean that one group is better than the other. For instance, for 10-month-olds, both touching and talking are quite successful methods of hushing an unhappy baby, and mothers from both groups are equally responsive—albeit in different ways. However, the fact that every child is raised within a culture that encourages some aspects of development more than others is one reason adults have the particular values, abilities, and desires that they do.

variable Any factor or condition that can change or vary from one individual or group or situation to another and thus affect behavior.

The difference is clearly related to cultural views of the mother's role, as the researchers explain:

> Both groups of mothers are responsive to infant signals, but their different behaviors indicate divergent goals and styles. The responsiveness of the Gusii mothers is directed toward soothing and quieting infants rather than arousing them . . . The responsiveness of the Boston mothers, especially as their infants become more communicative later in the first year, is designed to engage the infants in emotionally arousing conversational interaction. Gusii mothers see themselves as protecting their infants, not as playing with or educating them. [Richman et al., 1992]

Limitations of Observation

But what more specific factors might underlie these differing cultural practices? There are several possible explanations, and they reveal the chief drawback of observation—the difficulty of pinpointing the variable that is the direct cause of the behaviors that are observed. A **variable** is any factor or condition that can change or vary from one person or group or situation to another and thus affect behavior. The problem for researchers is that any given human behavior is surrounded by many variables that may or may not be influential. For example, one plausible hypothesis for the above findings points to infant mortality as the key variable. As noted in Chapter 1, in a society in which adequate food and survival are in doubt, as is the case in much of rural Africa, good parenting may place higher value on immediate soothing and physical nurturance. In contrast, in an amply fed group with a relatively low mortality rate, such as the suburban Bostonians, a parental priority is likely to be providing the infant with cognitive stimulation (Le Vine, 1988; Nugent et al., 1989).

Another likely hypothesis, posited by the researchers, is that maternal education may be the key variable, since it is likely to influence attitudes about verbal communication: if the mothers are well-educated and literate, as the Boston mothers were, they may encourage verbal interactions even before the babies can speak a word. Some support for this hypothesis comes from a study of Mexican mothers, who were all from the same low-income neighborhood but who had varying levels of education (Richman et al., 1992). This study found that, indeed, the more education the mothers had, the more verbal they were with their babies. In other words, the study found a *correlation* between maternal education and verbalization with infant offspring.

However, we need to be very careful in interpreting such findings: a **correlation** is a statistic that merely indicates whether two variables are related to each other—specifically, whether changes in one are likely to be accompanied by changes in the other. But in and of itself, *correlation does not indicate causation* (see A Closer Look on the facing page). While it seems logical that being more educated might lead women to be more verbal with their infants, it is possible that some third variable was the underlying cause of the correlation. It could be, for example, that people who are higher in verbal skills tend to spend more years in formal education, and that because of their verbal skills, rather than their education, they tend to talk more to everyone—babies included.

A CLOSER LOOK

Correlation: What It Does, and Does Not, Mean

Correlation is a statistical term that indicates that two variables are somehow related; that is, that one particular variable is likely, or unlikely, to occur when another particular variable occurs. For instance, there is a correlation between height and weight, because, usually, the taller a person is, the more he or she weighs. There is also a correlation between wealth and education, and perhaps even between springtime and falling in love.

Note that the fact that two variables are correlated does not mean that they are related in every instance. Some tall people weigh less than people of average height; some wealthy people never finished high school; some people fall in love in the depths of winter.

Nor does correlation indicate cause. The correlation between education and wealth does not necessarily imply that more education leads to greater wealth. It may be instead that more wealth leads to greater education, since wealthier people can better afford the expense of college. Or there may be a third variable, perhaps intelligence or family background, that accounts for the level of both income and education.

Correlations are one of the most useful tools in psychology and, at the same time, one of the most misused. They are useful because knowing how variables are related helps us understand the world we live in. However, as the respected researcher Sandra Scarr (1985) notes, "the psychological world . . . is a cloud of correlated events to which we as human observers give meaning."

Unless we are cautious in giving that meaning, we are likely to seize on one or another particular correlation as an explanation, without looking for other possible explanations. For example, some research has reported a positive correlation between owning a pet (especially a dog) and having good health. Particularly for the elderly, this research has found that pet ownership correlates with less depression (Garrity et al., 1989) and fewer visits to the doctor (Siegel, 1990). Publicity about such findings has led to the popular notion that pet ownership promotes health and has spawned the development of "pet-facilitated therapy," in which animals are brought to people in hospitals and nursing homes to speed their recovery.

The fact that two variables are strongly correlated does not prove that one caused the other. For example, there is a high correlation between riding motorcycles and wearing black attire, but no one would seriously suggest that a preference for black clothing is what leads people to become bikers. Many correlations found in research—and in everyday life—have no more bearing on cause than this one does, but often that fact is much less obvious.

However, the correlation between pet ownership and good health does not prove that pet ownership *causes* good health. The causal sequence may run in the opposite direction. That is, it may be that people in better health, or with a more positive outlook, are more likely to become pet owners. Or perhaps higher socioeconomic status is the crucial factor, in that those who can afford to care for a pet (particularly a dog) may be better able to afford preventive medical care. The studies cited above tried to take these factors into account, but it is virtually impossible to control for every variable. Perhaps for this reason, many other studies suggest that, over the long term, pet ownership does not cause longer or happier life (Siegel, 1993; Tucker et al., 1995). The lesson here is clear: correlations are often intriguing but, in and of themselves, prove nothing.

correlation A statistical term that indicates a corresponding relation between two variables (when both variables either increase or decrease together, the correlation is positive; when one variable increases as the other decreases, the correlation is negative).

Obviously, then, observation can provide fascinating data and can often generate a number of explanations for the results that it produces. But just as obviously, it cannot definitively link cause and effect, because observational settings, especially naturalistic ones, contain numerous variables that are beyond the researcher's control. To be certain that a particular observation is the result of one variable and not another, and to prove that their speculations are not simply creative hypotheses, scientists must go beyond observation to the experiment.

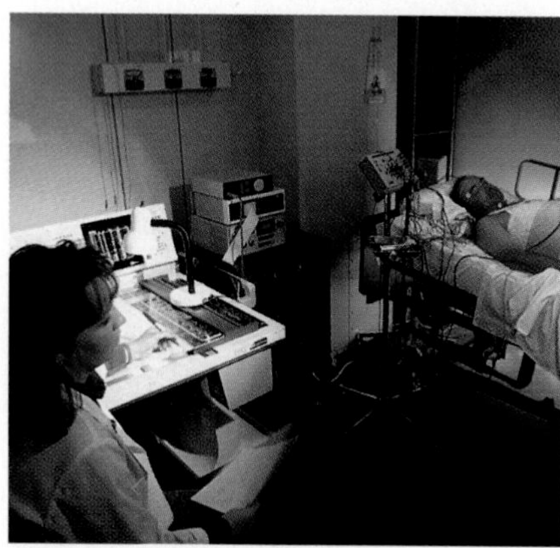

Under highly controlled experimental conditions, such as those in this sleep study, researchers are able to determine precise cause-and-effect relationships. For example, by giving subjects a series of sleep "aids"—ranging from prescription medications to daily exercise to a nightcap of warm milk—and then measuring the subjects' brain activity, organ functioning, and sleep patterns, researchers can determine what factors are likely to promote or inhibit restful sleep. The basic procedure of this particular experiment underlies all experimental research.

The Experiment

Unlike observation, an **experiment** tests a hypothesis in a controlled manner in which the relevant variables are limited and therefore can be manipulated by the experimenter. Typically, the experimenter exposes a group of subjects to the particular variable that is under investigation (for instance, a specific behavior on the part of a caregiver, a new teaching technique, a special diet, a particular social setting, a memory strategy) and then evaluates how they react.

Let's take a simple example. Many older adults complain that they do not sleep well and feel frustrated as a result. One team of researchers (Riedel et al., 1995) hypothesized that these sleep difficulties might occur because older adults do not realize that the elderly need less sleep than younger adults do and that waking up several times during the night is normal for older adults. The researchers also hypothesized that older adults might sleep better, with greater satisfaction, if they practiced "sleep compression," limiting the time spent in bed to the amount of time actually needed for sleep.

To test these hypotheses, the researchers advertised for volunteers over age 65 to participate in a study of sleep patterns, with each participant receiving $100. Of the 234 people who responded to the ad, almost half were eliminated, either because they changed their minds about participating once they learned more about the experiment or because the researchers excluded those with disabilities or medications that might affect sleep. The remaining 125 participants were asked to report data on their sleep for a week, including when they fell asleep and woke up during the night, their total time in bed, and, on a scale of 1 to 10, how satisfied they were with their sleep each night.

At the end of the week-long trial, seventy-five participants identified themselves as insomniacs. They averaged only five hours asleep per night, even though they spent seven and one-half hours in bed, and rated their sleep satisfaction at only 3.5. This was in marked contrast to the average of the fifty noninsomniac elderly, who slept almost seven hours a night, spent less than an hour awake in bed, and rated their sleep satisfaction at 8.

The researchers then divided the insomniacs into three groups of twenty-five each, with each group roughly equivalent in age, sex ratio, and sleep profiles. Two of these groups, the *experimental* groups, received the special experimental treatment, while the third group, the *control* group, received no special treatment. As part of the experimental treatment, the experimental groups were twice shown a fifteen-minute video that explained normal age-related sleep patterns and suggested that insomniacs should compress their total time in bed, going to bed later, getting up earlier, or both. They were also given a brochure that outlined the video's main ideas. After viewing the video the second time, participants were given a quiz about it, to reinforce the message as well as to test comprehension, which was high. Members of one experimental group also received four weekly sessions of sleep counseling, including a prescribed, personalized schedule of sleep compression. At the conclusion of the experimental program, all three groups recorded their sleep patterns, and did so again two months later.

Thus this experiment followed standard experimental practice: first measure the variable of interest, called the *dependent variable* (in this case,

experiment A research method in which the scientist deliberately changes one variable and then observes the results in some other variable.

TABLE 2.4

Remediation of Insomnia in Elders, Experimental Results

	Minutes Total Sleep			Minutes Awake in Bed			Satisfaction		
	Before	*After*	*Increase*	*Before*	*After*	*Decrease*	*Before*	*After*	*Improvement*
Video only	306	350	14%	92	48	48%	3.5	5.7	+2.2
Video and counsel	290	329	13%	68	32	53%	3.6	6.1	+2.5
Controls	314	340	8%	83	64	23%	3.8	4.8	+1.0

Source: Riedel et al., 1995.

sleep patterns), then administer some special treatment, called the *independent variable* (in this case, education and counseling), then measure the dependent variable again to see if any change has occurred. As Table 2.4 illustrates, both experimental groups showed notably more improvement than the control group did. The researchers' hypothesis therefore appears confirmed: understanding the normal sleep patterns of old age and practicing sleep compression can help the elderly reduce insomnia. But before applying that conclusion, we need to consider the limitation of this, and any, experiment.

Limitations of Experiments

As you can see, by comparing groups under controlled conditions, it is possible to make the link between cause and effect quite clear. But the question with experimental results is always: To what degree do the findings from an artificial experimental situation apply in the real world? A major problem with experiments is that the very fact that the process is controlled makes it different in key aspects from normal, everyday life. For instance, the elders who answered the ad for the sleep study may have been unusually ready to do something about their sleep problems, and this very willingness may have been crucial to their improvement. (The fact that the control group improved somewhat despite receiving no treatment suggests that this explanation is plausible.) Further, in eliminating disabled or medicated individuals from their research, the experimenters may have excluded the type of older persons who have the most trouble with sleep. Controlled conditions are essential to establishing a clear link between cause and effect, but they limit the applicability of the results.

In addition, all experiments except those with very young children are hampered by the fact that the participants know they are in an experiment. Simply being in an experiment may make people self-conscious, leading them to behave differently than they normally would. For example, the attention, status, and financial reward these elders received from their participation may have raised their overall levels of daily satisfaction, and this may have spilled over into satisfaction with their sleep.

Further, those who realize that they are part of a scientific study may attempt to help the experimenter get the results they believe the experimenter is looking for. It is quite possible, for example, that some of the elderly in the sleep study—especially those who developed relationships with their counselors—wanted to make both their counselors and themselves look good and therefore reported better results than their case warranted.

Alternatively, experimental subjects sometimes behave unnaturally because they are uncomfortable with the unfamiliar setting or conditions of the experiment. Occasionally, some subjects (usually college students) intentionally alter their behavior to sabotage the experiment. All these possibilities add further layers of potential artificiality to experimental research. For this reason, the experiment is most valuable when it is combined with other forms of research rather than being relied on exclusively.

The Interview or Survey

In an **interview** or **survey**, the researcher asks a series of questions of people and records their answers in order to determine those individuals' knowledge, opinions, or personal characteristics. This seems to be an easy, quick, and direct research method. However, it is more difficult to get valid data through an interview or survey than it seems, because these methods, even more than those of an experiment, are vulnerable to bias on the part of the researcher and the respondents. To begin with, the very phrasing of the questions can influence the answers. A survey on the issue of abortion, for instance, might prompt different responses depending on whether it asked about "terminating an unwanted pregnancy" or "taking the life of an unborn child."

In addition, many people who are interviewed—adults as well as children—give answers that they think the researcher expects, or that they think will make them seem mature or "good." Moreover, adults especially are often sensitive to questions that touch on matters they consider private, including income, politics, religion, and sex. Even when people wish to give completely accurate information, their responses may be flawed because their opinion on a particular question varies from day to day, or because their recollection of events is distorted. Nevertheless, a survey that is well-designed and carefully administered can be extremely valuable, allowing the collection and correlation of data from a large group of people much more quickly, and much less expensively, than other methods can.

The Case Study

An additional research tool is the **case study**, an intensive study of one individual. Typically, the case study is based on interviews with the subject regarding his or her background, present thinking, and actions, and often utilizes interviews of others who know the individual. Observation and standardized tests may furnish additional case-study material.

Case studies can provide a wealth of detail and therefore are rich in possible insights. Particularly for research on the entire life span, the written or oral life histories provided by elders can be provocative and fascinating. Often, they contain material that cannot be plumbed in any other way (Josselson & Lieblich, 1993). However, the interpretation of case-study data depends on the biases as well as the wisdom of the researcher. This is easy to see if you imagine what your view of life in general would be if you were to base it entirely on your knowledge of the two or three people you know best. For the most part, then, the case study is not used to do basic research. It can provide a good starting point, and it is a useful method if the goal is to understand a particular individual very well—as might be done in therapy,

interview A research method in which people are asked specific questions to discover their opinions or experiences.

survey A research method that collects interview information on a large number of people, either through written questionnaires or through personal interviews.

case study A research method that focuses on the life history, attitudes, behavior, and emotions of a single individual.

"You are fair, compassionate, and intelligent, but you are perceived as biased, callous, and dumb."

Reconciling the many possible, sometimes conflicting, views about the subject of a case study requires a talented interpreter, whose views and possible biases must also be taken into account.

for instance. However, no confident conclusions about people in general can be drawn from a study with a sample size of one, or even ten or twenty, no matter how deep and detailed that study is.

Clearly, there are many ways to test hypotheses. Researchers can observe people in naturalistic or laboratory settings, or experimentally alter their reactions under controlled conditions. They can compare one group with another to find significant differences, or correlate one characteristic with another to discover if they are somehow related. They can survey hundreds or even thousands of people about their opinions or knowledge, or interview a smaller number in great depth, or study one life in detail. Because each method has weaknesses, none of these ways of examining a hypothesis is sufficient in itself, but each can bring researchers closer to an understanding of the question being investigated.

Designing Developmental Research

For research to be truly developmental, scientists must discover how and why people change or remain the same *over time*. To learn about the pace and process of change, developmentalists use two basic research designs, cross-sectional and longitudinal.

Cross-Sectional Research

The more convenient, and thus more common, way researchers study development is by doing a **cross-sectional** comparison of people of various ages. In this kind of study, groups of people who are different in age but similar in other important ways (such as their level of education, socioeconomic status, ethnic background, and so forth) are compared on the characteristic under investigation. Any differences on this characteristic that exist between the people of one age and the people of another are, presumably, the result of age-related developmental processes.

One cross-sectional study compared 1,300 young, middle-aged, and older adults on various emotional dimensions, including the propensity to boredom, to see how they changed with age (Lawton et al., 1992). With respect to boredom, the findings suggest that, contrary to commonsense expectations, adults are less subject to boredom as they grow older. For example, the youngest, not the oldest, adults were most likely to agree that, "It is hard to find things that are new and interesting." The youngest were also most likely to seek out stimulation of various kinds, saying it was "very true" that they craved excitement, liked loud music, enjoyed thriller movies, and were inclined to do "something crazy to spice up the week." Older adults tended to say this was only "somewhat true," with many of the oldest adults saying this was "not at all true."

However, with cross-sectional research it is very difficult to ensure that the various comparison groups are similar in every background variable except age, and that age is therefore the explanation for whatever differences are found among the groups. Of course, good scientists try to make the groups as similar as possible. For example, in this study, ethnicity, socioeconomic status, and health were comparable in all three groups, and most of the participants were attending college. However, the young adults were full-time

cross-sectional research In developmental study, research that compares groups of people who are different in age but who are similar in other important ways.

The apparent similarity of these two groups in terms of gender and ethnic composition makes them seem potential candidates for cross-sectional research. However, before we could be sure that any differences between the two groups on any given dimension are the result of age, we would have to be sure the groups are alike in other ways, such as socioeconomic background, religious upbringing, and so forth.

undergraduates, while the older ones were part of an adult-education program called Elderhostel. There is no way to know if this difference affected the results of the study, but it is theoretically possible that those elderly people who become involved in Elderhostel programs are more intellectually curious, motivated, and absorbed than the typical undergraduate and therefore are less in need of exciting external distractions. It is also probable that the Elderhostel subjects are not typical of older adults as a whole, so the findings of this study could not be applied to the general population of elderly adults.

In addition, every cross-sectional study will, to some degree, reflect cohort differences, and in some cases the findings of cross-sectional research may be more a product of historical time than of chronological age. In the present study, for example, adults who grew up before 1950, prior to the electronic era, had a childhood without television, video games, and computers. Perhaps the reason they are less readily bored is that they learned the joys of reading a good book, engaging in deep conversation, writing a thoughtful letter, and so forth, while the younger generation, raised in an era of fast-moving, constantly changing, instant-access diversions that require only a short attention span and can be changed at the touch of a button, has developed a greater need for novelty.

Longitudinal Research

To help discover if age, rather than some other background or historical variable, is the reason for an apparent developmental change, researchers use **longitudinal research**, which involves studying the same individuals over a long period of time. This allows information about people at one age to be compared with information about them at another age, thus enabling researchers to find out how these particular people changed over time.

longitudinal research In developmental study, research that follows the same people over time in order to measure both change and stability with age.

For example, a longitudinal study on the trajectory of boredom and sensation-seeking with age might begin with a group of young adults, perhaps not only asking them survey questions like those in the cross-sectional study just described (such as whether they regularly sought something

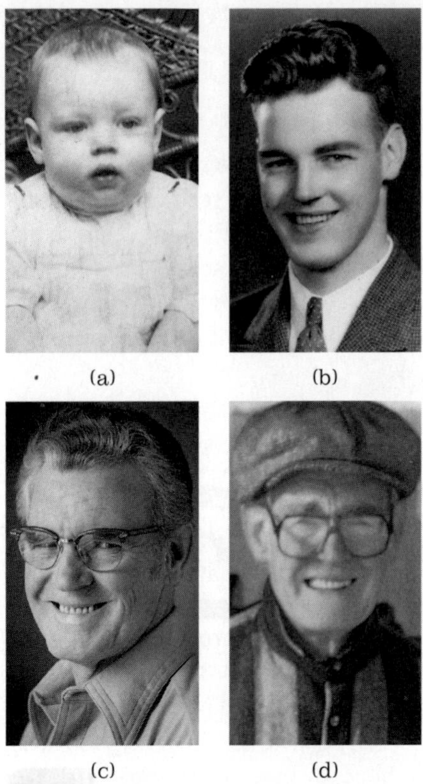

(a) (b) (c) (d)

Longitudinal research has been particularly valuable in studying the consistency of personality over time, providing strong evidence that, in terms of major traits at least, personality is fairly stable, especially across the years of adulthood. Developmentalists familiar with this research would not be surprised by the consistent pattern of hardiness, determination, and caregiving that has marked the life of Walton Hill, who (*a*) was a rare survivor of "blue baby" syndrome in 1914; (*b*) returned to college to earn a teaching degree after having dropped out during the Great Depression to help support his five brothers and sisters; (*c*) earned a doctorate in education at age 54, after his four children had grown and left home; and (*d*), following his retirement from high school teaching, became an administrator at the Pennsylvania Institute of Technology, where he maintains a full schedule at age 83 while continuing the regime of daily care he has provided his wife since she suffered a stroke in 1987.

"crazy to spice up their week") but also interviewing them about their interests, hobbies, daily activities, general frustrations, and so on. Then, every decade or so, the same individuals could be surveyed and questioned again, with their responses compared with those of their younger years. If these subjects were less likely to seek external stimulation as they aged, it would be clear that something about the aging process itself decreased boredom, at least within this group. While such a precise study has not been undertaken, longitudinal studies of overall personality patterns suggest that the propensity to boredom and the need for external stimulation do, in fact, decrease with age (Giambra et al., 1992; Haan et al., 1986).

Longitudinal research is particularly useful in studying developmental trends that occur over a long age span. It has produced valuable and sometimes surprising findings on such questions as children's adjustment to divorce (the negative effects linger, especially for school-age and older boys [Hetherington & Clingempeel, 1992]); the long-term effects of serious birth problems (remarkable resiliency is often apparent [Werner & Smith, 1992]); the role of fathers in child development (even fifty years ago, fathers were far more influential on their children's future happiness than the stereotype of the distant dad implies [Snarney, 1993]); and the consistency of personality in adulthood (personality patterns do not change much from age 30 to 70, but the expression of a person's particular traits can vary a great deal [McCrae & Costa, 1990]).

Clearly, longitudinal research is "a design of choice from the developmental perspective" (Cairns & Cairns, 1994). Nevertheless, it has some serious drawbacks. Over time, some subjects leave the study because they no longer wish to participate, or move far away, or die, and their absence can skew the study's ultimate results. Alternatively, some people who remain in the study may show change of some kind merely by virtue of being in the study ("improving" over a series of tests on some measure, for example, simply because they become increasingly familiar with the tests). The biggest problem of all is that longitudinal investigations are very time-consuming and expensive.

Cross-Sequential Research

As you can see, both cross-sectional and longitudinal research methods allow scientists to look at development over time, and each method has flaws that are, to a large extent, compensated for by the other. Because these two methods are essentially complementary and tend to make up for each other's disadvantages, scientists have devised various ways to use the two methods together. The simplest to understand is **cross-sequential research** (also referred to as *cohort-sequential,* or *time-sequential,* research [Schaie, 1996]). This method begins by studying several groups of people at different ages (a cross-sectional approach) and then follows those people longitudinally for a long time. Using this method, the findings on a group of, say, 50-year-olds can be compared with findings on the same individuals at age 40 and also with findings on groups who were 50-year-olds any number of years before. Cross-sequential research allows scientists to disentangle results related to chronological age from those related to historical period. The need for this method will be especially clear in Chapter 21 when we take up the thorny topic of which intellectual abilities fade in which groups of individuals at what ages.

cross-sequential research In developmental study, research that follows a group of people of different ages over time in order to distinguish differences related to age from differences related to cohort and historical period. (Also called *cohort-sequential research* or *time-sequential research.*)

This chapter has presented you with the basic tools of the trade—theories that provide new hypotheses and explanations, and research methods that can shake out the truth from an assorted grab bag of data. Far from being impractical, these tools are fundamental to the scientific study of human development. Armed with these tools, you are now ready to undertake that study, starting at the very beginning, the moment of conception.

SUMMARY

What Theories Do

1. A theory provides a framework of general principles that can be used to interpret observations. Each developmental theory interprets human development from a somewhat different perspective, but developmental theories attempt to provide a context in which to understand individual experiences and behavior.

Psychoanalytic Theory

2. Psychoanalytic theory emphasizes that our actions are largely ruled by the unconscious—the source of powerful impulses and conflicts that usually lie below the level of our conscious awareness. It also proposes that early experiences can have significant, long-term effects on personality.

3. Freud, the founder of psychoanalytic theory, developed the theory of childhood sexuality to explain how unconscious impulses arise and how they affect behavior during the oral, anal, phallic, and genital stages of psychosexual development of the individual.

4. Erikson proposed a theory of psychosocial development that describes individuals as being shaped by the interaction of personal characteristics and social forces. In this theory, Erikson depicts eight successive stages of psychosocial development, each of which involves a particular developmental crisis.

Learning Theory

5. Learning theorists believe that the focus of psychologists' study should be behavior that can be observed and measured and the environmental bases of that behavior. They are especially interested in the relationship between events and the reactions they are associated with, that is, between stimulus and response.

6. Learning theory emphasizes the importance of various forms of conditioning, a process by which particular stimuli become linked with particular responses. In classical conditioning, a neutral stimulus becomes associated with a meaningful one to produce a particular response. In operant conditioning, reinforcement makes a behavior more likely to occur.

7. Social learning theory recognizes that much of human behavior is modeled after the behavior of others, and that various cognitive and motivational processes influence how we are affected by the behavior of others.

Cognitive Theory

8. Cognitive theorists believe that a person's thought processes—the understanding and analysis of a particular situation—have an important effect on behavior and development.

9. Piaget proposed that intelligence is an active, ongoing process, as people develop general ways of thinking about ideas, objects, and experiences. When a person becomes aware of perceptions or events that cannot be easily comprehended by an existing concept, they experience cognitive disequilibrium. This state propels them to new thought, either a modification of an existing concept or the creation of an entirely new one.

10. Throughout life, according to Piaget, learning is accomplished as the person actively organizes and adapts his or her mental structures. The individual's basic mode of thinking changes over four stages of development: sensorimotor in infancy, preoperational during the preschool years, concrete operational during middle childhood, and, finally, formal operational, beginning in adolescence and continuing lifelong.

11. Information-processing theorists study cognitive development in terms of changes in internal cognitive processes such as working memory and the knowledge base. Growth and refinement of control processes are especially important to cognitive development.

Sociocultural Theory

12. Sociocultural theory explains human development in terms of the guidance, support, and structure provided by the culture. For Vygotsky, learning occurs through the social interactions learners share with more mature members of the society, who provide guided participation in the zone of proximal development.

The Theories Compared

13. Psychoanalytic, learning, cognitive, and sociocultural theories have all contributed to the understanding of human development, yet no one theory is adequate to describe the complexity and diversity of human experience. Most developmentalists, well aware of the criticisms of each of these perspectives, selectively incorporate ideas and generate hypotheses from all of them. In addition, many find various minitheories, related to specific ages or topics, useful for formulating research questions.

Research Methods

14. To check their conclusions and to try to remain as objective as possible, researchers use a variety of methods, among them, adequate sample size, selection of a representative sample population, "blind" experimenters, operational definitions, control groups, and tests of statistical significance.

15. One common method of testing hypotheses is observation, which provides valid information but does not pinpoint cause and effect. The laboratory experiment pinpoints causes but is not necessarily applicable to daily life. Interviews, surveys, and case studies are also useful.

16. In developmental research, ways are needed to detect change over time. Cross-sectional research compares people of different ages; longitudinal research (which is preferable but more difficult to carry out) studies the same individuals over a long time period. Both are valid for the cohorts under examination but not necessarily for other age groups. Scientists use both methods together in cross-sequential research, which may provide more comprehensive findings.

KEY TERMS

developmental theory (29)
psychoanalytic theory (30)
childhood sexuality (31)
psychosexual stages (31)
crisis (32)
psychosocial theory (32)
behaviorism (34)
learning theory (34)
stimulus (35)
response (35)
conditioning (35)
classical conditioning (35)
operant conditioning (36)
positive reinforcer (37)
negative reinforcer (37)
punishment (37)
social learning theory (37)
modeling (38)
reciprocal determinism (38)
cognitive theory (40)
sensorimotor stage (40)
preoperational stage (40)
concrete operational stage (40)
formal operational stage (41)
information-processing theory (42)
sensory register (43)
working memory (43)
knowledge base (43)
sociocultural theory (46)
guided participation (47)
zone of proximal development (47)
observation (51)
sample size (52)
representative sample (52)
blind (52)
experimental group (53)
control group (53)
statistical significance (53)
variable (54)
correlation (54)
experiment (56)
interview (58)
survey (58)
case study (58)
cross-sectional research (59)
longitudinal research (60)
cross-sequential research (61)

KEY QUESTIONS

1. What functions does a good theory perform?

2. What is the major premise of psychoanalytic theory?

3. What is the major difference between Freud's theory and Erikson's theory?

4. What is the major premise of learning theory?

5. According to Piaget, how do periods of disequilibrium lead to mental growth?

6. How do information-processing theorists describe cognitive growth?

7. How does sociocultural theory view development?

8. What are the main differences among the psychoanalytic, learning, cognitive, and sociocultural theories?

9. What are the advantages and disadvantages of testing a hypothesis by observation?

10. What are the advantages and disadvantages of testing a hypothesis by experiment?

11. Compare the advantages of longitudinal research and cross-sectional research and explain how the advantages of each method are combined in cross-sequential research.

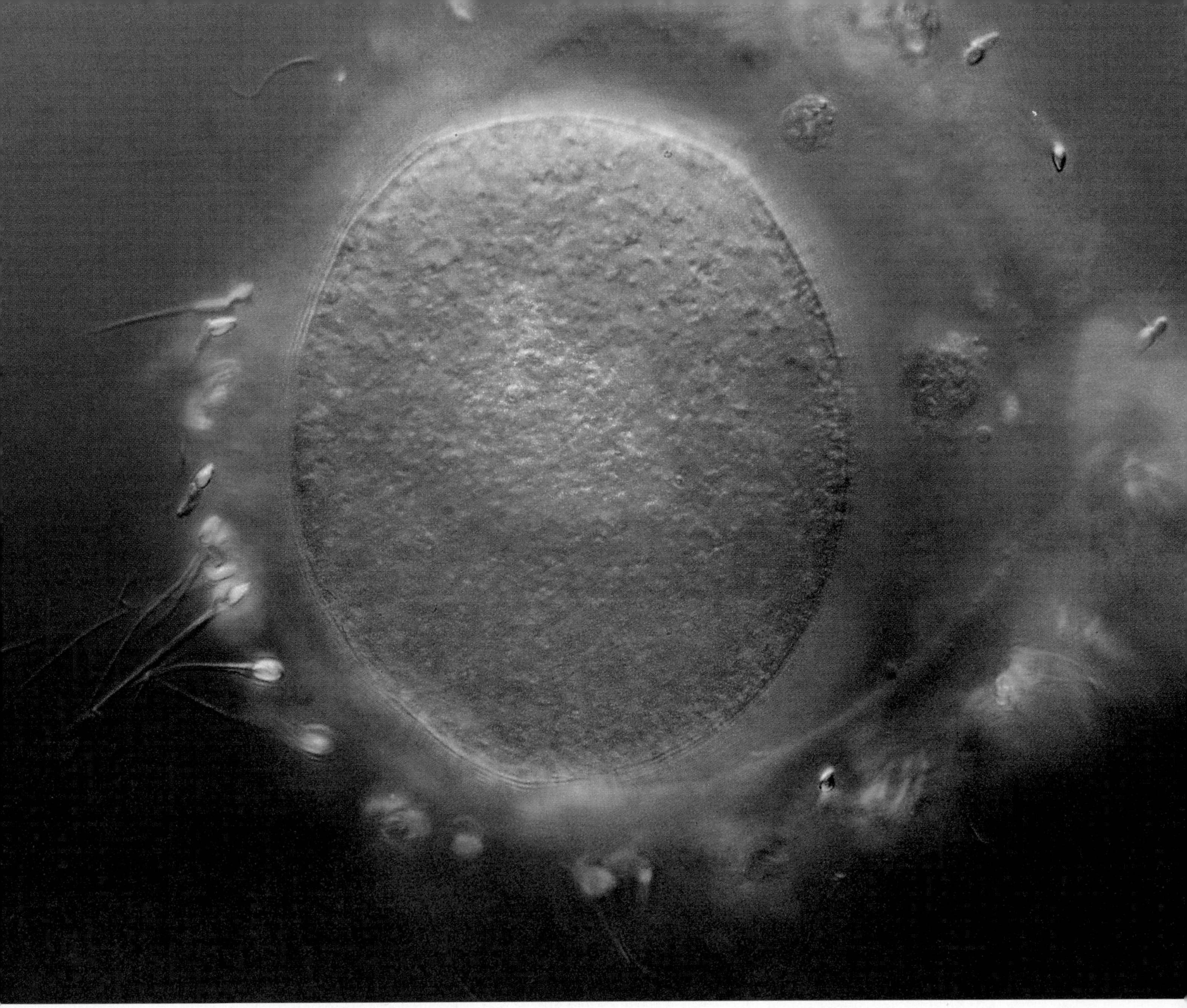

Heredity and Environment

CHAPTER 3

From the very beginning, individual development is driven by the interaction of two prime forces, heredity and environment. At conception, a complex set of genetic instructions takes form to influence every aspect of development, affecting not only obvious characteristics such as sex, coloring, and body shape but also less visible traits, psychological as well as physical—from blood type to bashfulness, from metabolic rate to moodiness, from voice tone to vocational aptitude. Even the timing and pace of certain developmental changes are genetically guided, a fact no less true for the octogenarian than for the embryo.

As we will see throughout this chapter, however, just as no human characteristic is untouched by heredity, no genetic instruction—including those for basic traits such as physical structure and intellectual potential—is unaffected by the environment. Indeed, each person's genetic inheritance and individual experiences are so intertwined that it is virtually impossible to isolate the specific effects of one from those of the other. The interaction between these two factors is lifelong, shaping and changing the individual from the moment of conception until the moment of death.

The Beginning of Development

Human reproductive cells are called **gametes**. Human development is initiated when a male gamete, or **sperm**, penetrates the membrane of a female gamete, or **ovum**. Each of these gametes contains more than a billion chemically coded genetic messages, which, taken together, represent one half of a rough blueprint for human development. When the two reproductive cells subsequently fuse, the two blueprint halves combine, interacting to form a complete set of developmental guidelines.

In the first hour or so after the sperm has penetrated the ovum, the two gametes maintain their separate identities, side by side, enclosed within the ovum's membrane. Then they suddenly merge, their genetic material combines, and a living cell called a **zygote** is formed.

Within hours after its formation, the zygote begins the first stages of development through the processes of *duplication* and *division*. First, all the combined genetic material from both gametes duplicates itself, forming two complete sets of genetic instructions. These two sets move toward opposite sides of the zygote; the cell then divides neatly down the middle, and the zygote thus becomes two cells. In identical fashion, these two cells duplicate

gamete A reproductive cell, that is, a cell that can reproduce a new human being if it combines with a gamete from the other sex. Female gametes are called ova, or eggs; male gametes are called spermatozoa, or sperm.

sperm The reproductive cells of a male, which begin to be produced in a young man's testicles at puberty.

ovum (plural, ova) The reproductive cells of a female, which are present from birth in the ovaries.

zygote The single cell formed from the union of two gametes, a sperm and an ovum.

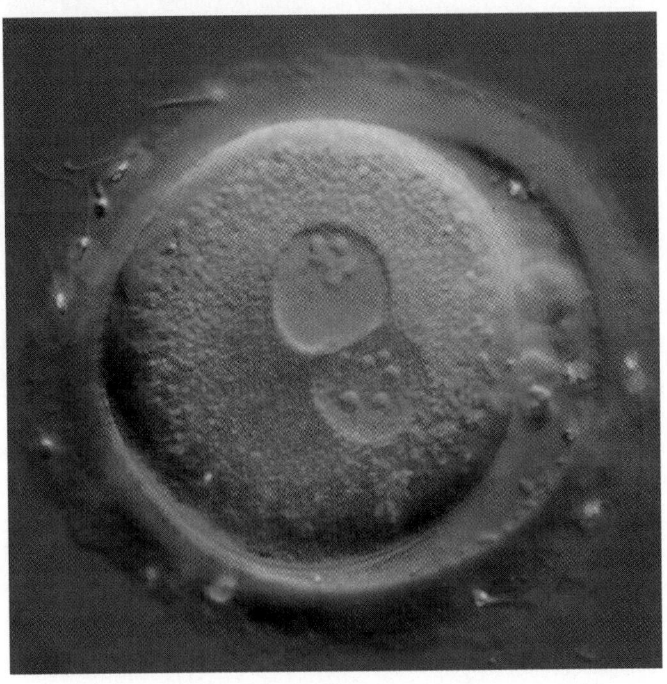

The ovum shown here is about to become a zygote. It has been penetrated by a single sperm, whose nucleus now lies next to the nucleus of the ovum. Shortly, the two nuclei will fuse, bringing together several billion genetic codes that will guide future development.

themselves and divide to become four; these four, in turn, duplicate and divide to become eight; and so on.

Soon, a third process, *differentiation*, is added to the simple duplication and division. Following a genetic timetable, various cells begin to specialize and reproduce at different rates, according to their programmed function. Also following a timetable, some cells will die early in life and others will continually reproduce for decades; still others will remain dormant until puberty, or adulthood, or old age.

At birth, an individual is made up of about 10 trillion cells. By adulthood, the number of cells has increased to between 300 and 500 trillion. But no matter how many cells a person may have, and no matter how much differentiation and specialization has occurred, each body cell carries a copy of the genetic instructions inherited by the one-celled zygote at the moment of conception.

The Genetic Code

gene The basic unit of heredity. Genes, which number about 100,000 in humans, direct the growth and development of every living creature.

chromosome A molecule of DNA that carries genes transmitted from parents to offspring.

DNA (deoxyribonucleic acid) Molecules containing genetic information.

genetic code The sequence of chemical bases in DNA; referred to as a code because it determines the amino-acid sequence in the enzymes and other protein molecules synthesized by the organism.

The basic unit of these genetic instructions is the **gene**. Genes are discrete segments of a **chromosome**, which is a **DNA (deoxyribonucleic acid)** molecule that typically contains thousands of genes as well as other material. Except for gametes, every normal human cell has twenty-three pairs of chromosomes (forty-six chromosomes in all), which collectively carry about 100,000 distinct genes.

The instructions on each gene are "written" in a chemical code, or alphabet, made up of four bases—adenine, guanine, cytosine, and thymine—abbreviated A, G, C, and T. These chemical bases occur in only four pairings, A-T, T-A, G-C, and C-G, which at first would seem to provide a very limited genetic vocabulary. In fact, however, what determines the precise nature of a gene's instructions, called the **genetic code**, is the overall sequence of these base pairs along their segment of the DNA strand (see Figure 3.1). And since there are approximately 3 billion base pairs in the DNA of every human, and thousands of base pairs in every gene, the genetic vocabulary is extremely rich and extensive. Indeed, scientists now engaged in decipher-

FIGURE 3.1 The structure of DNA is like a ladder, with the rungs composed of a pair of bases, some combination of A (adenine), T (thymine), C (cytosine), and G (guanine). Shown here is a DNA molecule in the process of duplicating the code as it creates two new copies. From such a simple but elegant uncoupling and reconnecting, the entire genetic code for a life is transmitted.

ing the complete genetic code (see pp. 92–93) estimate that, when fully transcribed, it will cover at least as many pages as thirteen full sets of the *Encyclopaedia Britannica* (Lee, 1993).

In essence, what this multitude of genetic instructions does is provide the body's cells with directions for the synthesis of hundreds of different kinds of proteins, including enzymes, that are the body's building blocks and regulators. Thus, following these instructions, certain cells become part of the neurons of the brain; others become part of the lens of the eye; others become part of the valves of the heart; and so on—even though every cell contains the entire genetic code of the organism.

In addition to directing the form and location of cells, genes also influence their specific functions. That is, the enzymes created through genetic instructions direct the cells' behavior—influencing, for example, how rapidly the neurons in each particular area of the brain process information and what kind of information that particular area attends to; how quickly and sharply the lens of the eye focuses on diverse forms in various lightings, contrasts, and distances; how readily blood flows through the heart when the body is exerted and when it is at rest. The process doesn't stop there. Through some on-off switching mechanism not yet understood, genes control life itself, instructing cells to grow, to repair damage, to take in nourishment, to multiply, to die. Even certain advances in cognitive development may involve genes that switch on at particular ages, propelling maturation in specific areas of the brain (Gottesman & Goldsmith, 1993; Plomin et al., 1993).

Chromosomes

As already noted, each normal human has forty-six chromosomes, duplicated in every body cell. The chromosomes are arranged in twenty-three distinct pairs, one member of each pair being from the mother and the other being from the father. Each chromosome in a pair serves as the designated site for specific genes, with the genes on one chromosome corresponding in location and function with the genes on its chromosome mate. For the most part, genes are positioned quite precisely on a particular "arm" of a certain chromosome. Thus chromosomes not only carry the genes; they also furnish each gene with a niche opposite a corresponding gene on the matching chromosome, thereby allowing each gene pair to perform its mission in directing development.

The matching of the chromosome pairs occurs at conception. Every human gamete carries twenty-three chromosomes, with each of the twenty-three chromosomes in each sperm corresponding in function to one of the twenty-three chromosomes carried in every ovum. Collectively, the twenty-three chromosomes in the sperm and the twenty-three chromosomes in the

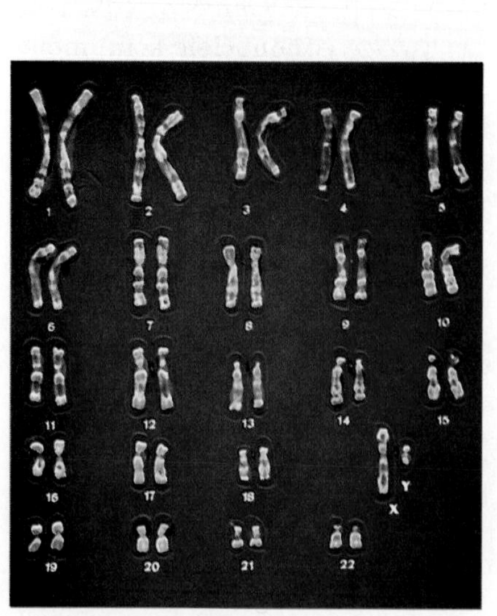

This picture, called a karyotype, shows the forty-six chromosomes from one individual, in this case a normal male. In order to produce a chromosomal portrait such as this one, a cell is removed from the person's body (usually from inside the mouth), processed so that the chromosomes become visible, magnified many times, photographed, and then arranged in pairs according to the length of the upper "arms."

twenty-third pair In humans, the chromosome pair that determines the person's sex.

ovum represent the two halves of the genetic blueprint referred to earlier. When the sperm and ovum unite, their corresponding chromosomes link up in pairs, and the matching genes on each pair member become aligned, providing complete instructions for the development of a new person.

This chromosomal pairing remains lifelong, in every cell—with one important exception. When the human body makes gametes, cell division occurs in such a way that each gamete receives only one member of each chromosome pair. This is why sperm and ova have only twenty-three chromosomes, ensuring that when the chromosomes of a sperm and an ovum combine, the total chromosome number for the new organism will be forty-six.

The Sex Chromosomes

Twenty-two of the twenty-three pairs of human chromosomes are closely matched pairs, each half containing similar genes in almost identical positions and sequence. The **twenty-third pair**, which is the one that determines the individual's sex, is a different case. In the female, the twenty-third pair of chromosomes is composed of two large, X-shaped chromosomes. Accordingly, it is designated *XX*. In the male, the twenty-third pair is composed of one large X-shaped chromosome and one, much smaller, *Y*-shaped chromosome. It is designated *XY*.

Obviously, since a female's twenty-third chromosome pair is XX, every gamete she makes will have either one X or the other. And since a male's twenty-third pair is *XY*, half his sperm will have an X chromosome and half will have a *Y*. Thus the critical factor in the determination of a zygote's sex will be which sperm reaches the ovum first, a *Y* sperm, creating a male (*XY*), or an X sperm, creating a female (XX) (see Figure 3.2).

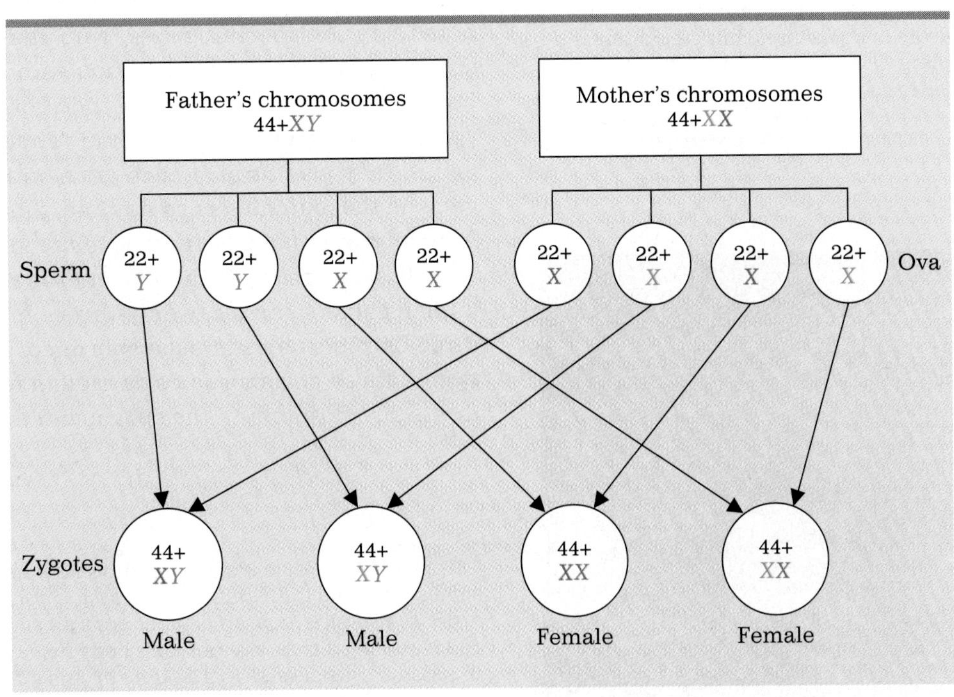

FIGURE 3.2 As you can see, any given couple can produce four possible combinations of sex chromosomes. In terms of the future person's sex, it does not matter at all which of the mother's Xs the zygote inherited. All that matters is whether the *Y* sperm or X sperm fertilized the ovum. However, for X-linked conditions (p. 73), it matters a great deal, since typically one, not both, of the mother's Xs carries the trait.

Human Diversity

Genes accomplish two goals that are essential to the survival of the human race: they ensure both genetic continuity across the species and genetic diversity within it. The vast majority of each person's genes are identical to those of any unrelated person (Plomin, 1994). As a result of the instructions carried by these genes, each new member of the human race shares with every other human common physical structures (such as the pelvic alignment that allows us to walk upright), behavioral tendencies (such as vocalization that allows us to use spoken language), and reproductive potential (allowing each new generation to perpetuate the species by mating with any other human of the other sex). These species-specific characteristics have been fashioned throughout our long evolutionary history, promoting our survival as a species by enabling humans to live successfully on Earth.

The remainder of each person's genes differ in various ways from those of other individuals. The diversity these genes provide over the generations is essential to the ability of our species to adapt to changing environments and needs. The fact that humans differ from one another genetically means that, as a species, we retain the potential to change and evolve. Thus our individual uniqueness fosters the survival of the entire human race.

Given that each sperm or ovum from a particular parent contains only twenty-three chromosomes, one might wonder how every conception can represent the potential for a genetically unique individual. The answer is that when the chromosome pairs divide up during the formation of gametes, which one of each pair will wind up in a particular gamete is a matter of chance, so a vast number of chromosome combinations are possible. According to the laws of probability, there are, in fact, 2^{23}—that is, about 8 million—possible outcomes. In other words, approximately 8 million genetically different ova or sperm can be produced by a single individual.

In addition, just before a chromosome pair divides during the formation of gametes, corresponding segments of the pair are sometimes exchanged, altering the genetic composition of both pair members. Through the recombinations it produces, this *crossing-over* of genes adds greatly to genetic diversity. And finally, when the sperm and ovum unite, the interaction of their chemically coded instructions forms combinations not present in either parent. All things considered, any given mother and father could form over 64 trillion genetically different offspring. Thus it is no exaggeration to say that every conception is, potentially, the beginning of a genetically unique individual.

Twins

Although every zygote is genetically unique, not every newborn is. In about 1 in every 270 pregnancies, the growing cluster of cells splits apart during the first two weeks of development, creating two identical, independent clusters (Bryan, 1992). These cell clusters become **monozygotic twins** (identical twins), so named because they originated from one (mono) zygote. Since they originated from the same zygote, they share identical genetic instructions for physical appearance, psychological traits, vulnerability to certain diseases, and so forth.

monozygotic twins Twins who have identical genes because they were formed from one zygote that split into two identical organisms very early in development.

At age 12, monozygotic twins Kate and Nell are apparently seeking their own separate identities in every way possible, from their choice of hairstyle to their choice of socks. But their genetic identicalness is nevertheless apparent, not only in the misalignment of their teeth and the timing of their puberty, but also in their behavior, as evidenced by their friendly smiles and bitten fingernails. By contrast, Patrick and Colin, like many dizygotic twins, have notably different coloring, features, and behavior. When told to give a "thumbs up," the two 5-year-olds choose different hands, postures, and facial expressions.

Of course, not all twins are monozygotic. **Dizygotic twins** (fraternal twins) begin life as two separate zygotes created by the fertilization of two ova that were ovulated at roughly the same time. Dizygotic conceptions may be as frequent as one in six, but usually only one twin develops past the embryo stage. Dizygotic births occur naturally about once in every sixty births, with considerable variation among different racial and ethnic groups. (Women of African ancestry, for example, are more likely to have dizygotic twins than are women of European ancestry, who, in turn, are more likely to have dizygotic twins than are women of Asian ancestry [Bryan, 1992].) When fertility drugs are used, dizygotic twinning is much more common, occurring about once in every ten births.

Dizygotic twins share no more genes than do any other offspring of the same parents; that is, they share about 50 percent of the genes governing individual differences. They may be of different sexes and be very different in appearance. Or they may look a great deal alike, just as nontwin brothers and sisters sometimes do. Other multiple births, such as triplets and quadruplets, can likewise be monozygotic, dizygotic, trizygotic, quadrazygotic, and so forth (or even some combination of these). Over the past decade, the incidence of multiple births doubled in the United States and in many other nations, again because of the use of fertility drugs (Jewell & Yip, 1995).

dizygotic twins Twins formed when two separate ova are fertilized by separate sperm at roughly the same time. Such twins share about half their genes, just like any other siblings.

polygenic traits Characteristics produced by the interaction of many genes.

multifactorial traits Characteristics produced by the interaction of several genetic and environmental influences.

From Genotype to Phenotype

As we have seen, conception brings together genetic instructions from both parents for every human characteristic. How do these instructions work to influence the specific characteristics a given offspring will inherit? The answer is usually quite complex, because most traits are both **polygenic**—that is, affected by many genes—and **multifactorial**—that is, influenced by many factors, including factors in the environment.

To grasp the complexity of genetic influences we must first distinguish between a person's genetic inheritance—his or her genetic *potential*—and the actual *expression* of that inheritance in the person's physiology, physical appearance, and behavioral tendencies. The sum total of all the genes a per-

son inherits is the person's **genotype**. The sum total of all the genes that are actually expressed in some apparent way—including physical traits, such as bushy eyebrows, and nonphysical traits, such as a hunger for excitement—is the person's **phenotype**. Clearly, we all have many genes in our genotype that are not apparent in our phenotype. In genetic terms, we are **carriers** of these unexpressed genes; that is, they are "carried" in our DNA and can be passed on to our offspring, who will have them in their genotype and may or may not have them expressed in their phenotype.

The phenotype of any given characteristic arises from two levels of interaction: (1) the interaction of the proteins synthesized from the specific genes that affect the characteristic and (2) the ongoing interaction between the genotype and the environment. Let us look first at the types of interaction that can occur among the genes themselves. Then we will consider some of the ways the phenotype is shaped by environmental factors.

genotype A person's entire genetic inheritance, including those characteristics carried by the recessive genes but not expressed in the phenotype.

phenotype An individual's observable characteristics that result from the interaction of the genes with each other and with the environment.

carrier An individual who has in his or her genotype a recessive gene that is not expressed in his or her phenotype. Carriers can pass the gene on to their children, who will express the gene if they receive a similar recessive gene from the other parent.

additive pattern A common pattern of genetic inheritance in which each gene affecting a specific trait makes an active contribution to the final outcome. Skin color and height are additive.

Gene-Gene Interaction

One common pattern of interaction among genes is called **additive** because in this pattern the phenotype reflects the sum of the contributions of all of the genes involved. Many genes affect height and skin color, for instance, and they usually do so in an additive fashion. For example, consider a simplified situation in which a tall man, whose parents and grandparents were all very tall, married a short woman, whose parents and grandparents were all very short. Because of additive interaction of genes affecting height, the couple's children would all be of middling height (assuming that the children's nutrition and physical health were adequate). None of them would be as tall as their father or as short as their mother, because the sum total of all their genes for tallness and all their genes for shortness, when averaged together, would be halfway between the two. Of course, in actuality, many people have both kinds of ancestors, relatively tall ones and relatively short ones, which means that many children are notably taller or shorter than either of their parents, depending on which genes from each parent's varied genotype they happened to inherit.

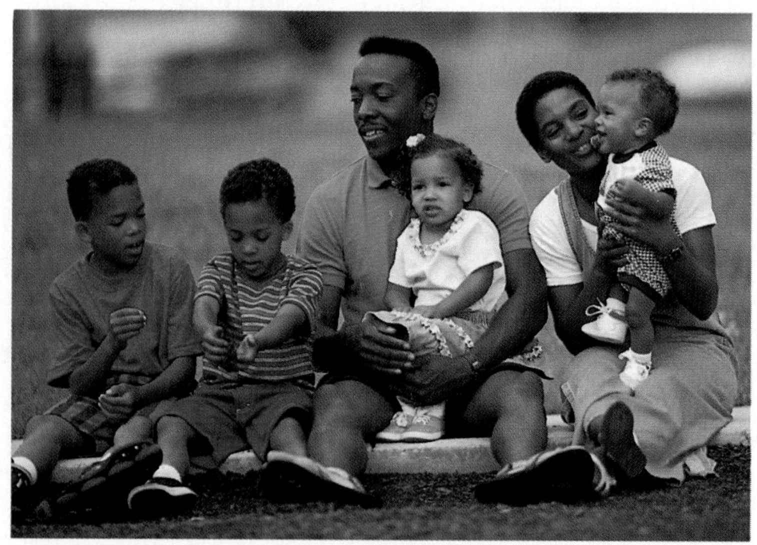

Using terms such as black, white, yellow, red, or brown to denote skin color is misleading, for humans actually exhibit thousands of skin tones, each resulting from the combination of many genes. Depending on which half of their mother's and father's skin color genes children happen to inherit, offspring can be (and in modern nations, often are) paler, ruddier, lighter, darker, more sallow, more olive, or more freckled than either parent. This variation is apparent in many African-American families whose heritage includes ancestors from various parts of both Europe and Africa, and often Asia and pre-Columbian America as well.

nonadditive pattern A pattern of genetic inheritance in which the outcome depends much more on the influence of one gene than of another.

dominant-recessive pattern A pattern of genetic inheritance in which one member of a gene pair (referred to as dominant) acts in a controlling manner, hiding the influence of the other (recessive) gene.

Less often, genes interact in a **nonadditive** fashion: their interaction tends to be an either/or, winner/loser pattern rather than an equal compromise. To be more precise, when a gene pair interacts in a nonadditive pattern, the outcome depends much more on the influence of one gene than on that of the other.

One kind of nonadditive pattern that you may be familiar with is the **dominant-recessive** pattern. In a pair of genes with this pattern, the phenotype reflects the influence of one gene, the *dominant gene*, while the effects of the other gene, the *recessive gene*, are obscured. Indeed, sometimes the dominant gene completely controls the characteristic in question, and the recessive gene is not evident in the phenotype. In other instances, the outcome reflects *incomplete dominance*, with the phenotype influenced primarily, but not exclusively, by the dominant gene.

Hundreds of physical characteristics follow the dominant-recessive pattern. Eye color is one of them. For the sake of illustration, let's simplify greatly and say for the moment that a person inherits only two eye-color genes, one from each parent. We will further simplify and say that each of these genes instructs for either brown or blue eyes and that the gene for brown eyes is dominant and the gene for blue eyes is recessive. (Following traditional practice, we will indicate the dominant gene with an uppercase "B"—for dominant brown—and the recessive gene with a lowercase "b"—for recessive blue.) If both genes are for brown eyes (BB), the person's eyes will be brown. If one gene is for brown eyes and the other for blue (Bb), the person's eyes will also be brown—in this case, because the brown-eye gene is dominant. If both genes are for blue eyes (bb), the person will have blue eyes.

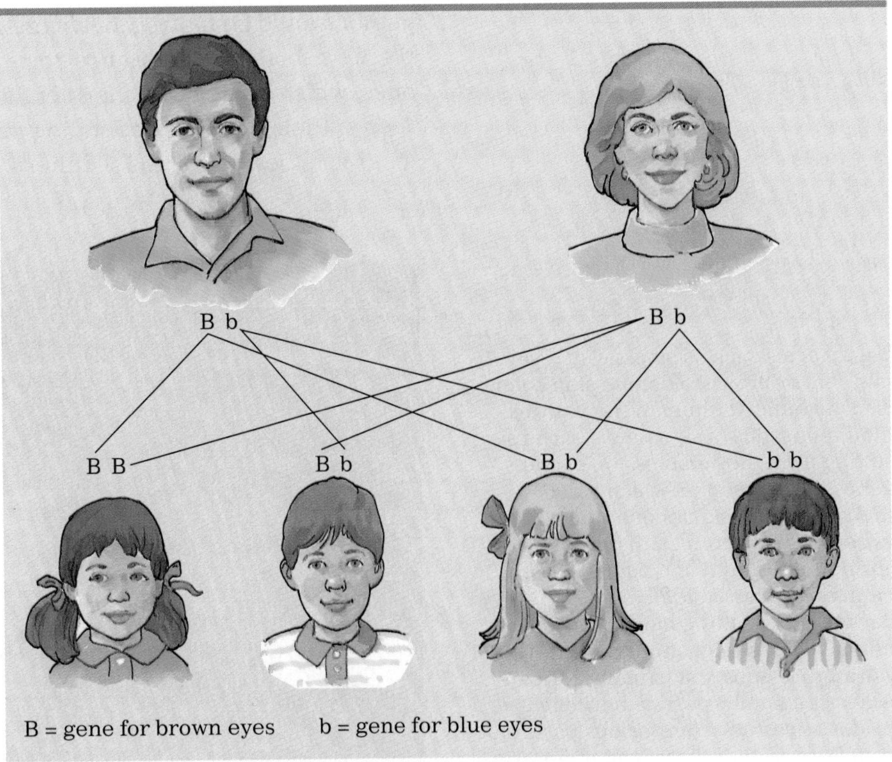

FIGURE 3.3 Two brown-eyed parents who are both carriers for blue eyes can have a blue-eyed child. The odds are 1 in 4 that a child will inherit the brown-eye genes from both parents, and 2 in 4 that a child will inherit one brown-eye gene and one blue-eye gene (one child got the blue gene from the mother and the other from the father, but the outcome is the same). There is 1 chance in 4 that a child will inherit two recessive blue-eye genes and thus have blue eyes. Of course, "chance has no memory," despite the neatness of this illustration. A family of four children born to two brown-eyed carriers for blue eyes might not include any children with blue eyes, or might have two, three, or—in one such family in 256—four blue-eyed offspring.

This last, blue-eye pattern will occur in all of a couple's children if both parents have blue eyes (bb + bb). It has a 50 percent chance of occurring if one parent has blue eyes and the other has brown eyes but carries the blue-eye gene (bb + Bb). A blue-eyed child may also be born to two brown-eyed parents if both parents are carriers of the blue-eye gene. With any such couple (Bb + Bb), the chances are 3 in 4 (BB, Bb, Bb, bb) that any given child will have brown eyes and 1 in 4 that he or she will have blue eyes (see Figure 3.3).

X-Linked Genes

Some genes are called ***X*-linked** because they are located only on the X chromosome. If an X-linked gene is recessive—as are the genes for most forms of color blindness, many allergies, several diseases, and some learning disabilities—the fact that it is on the X chromosome can be critical. Since males have only one X chromosome, the recessive genes they inherit on that chromosome will be expressed in their phenotype because they have no second X chromosome that might carry the counterbalancing gene (see Figure 3.2). This explains why some traits are passed from mother to son (via the X) but not from father to son (since the *Y* does not carry the trait).

An example can help clarify this point. Suppose a man is unable to distinguish red and green. We know that he inherited this particular color blindness from his mother, since the X chromosome that carries the red-green color blindness comes from the mother. If that man has children, none of his sons will inherit his gene for color blindness (because they receive his *Y* chromosome), but all of his daughters will (because they receive his X chromosome). In the unusual situation that the man's wife is a carrier of X-linked color blindness, the daughters have a 50 percent chance of inheriting the disorder because they all will have their color-blind father's X and stand a 50 percent chance of inheriting the maternal X that carries the gene for color blindness. In turn, all the sons of those daughters who do inherit the disorder will themselves be color-blind, while the sons of those who don't inherit it will have a 50 percent chance of being color-blind. This pattern holds true for all X-linked traits: while people of both sexes have them on their genotype, males are far more likely to have them on their phenotype.

As complex as the preceding descriptions of additive, dominant, and recessive gene interaction may seem, they in fact make gene interaction appear much simpler than it actually is. This is because, of practical necessity, we are forced to discuss genes as though they were discretely functioning entities. But as we have seen, what genes actually do is direct the synthesis of hundreds of kinds of proteins that form the body's structures and direct its biochemical functions. In a sense, each body cell is "nothing more than a sea of chemicals" that is continually affected by other proteins and by the enzymes that direct the cell's functioning (Lee, 1993). Thus, no single gene pair directly determines even simple traits, such as eye color or height.

In addition, the patterns of gene interaction are seldom straightforward. Some additive genes contribute more substantially than others, either because they are naturally partially dominant or because their influence is amplified by the presence of certain other genes. When additive genes combine, their final product is not always the simple total of all the contributions.

***X*-linked genes** Genes that are carried on the X chromosome. X-linked genes are more likely to be expressed in the phenotype of males, even though women are more likely to have them in their genotype.

Finally, geneticists have recently discovered that certain genes behave differently depending on whether they are inherited from the mother or the father (Hoffman, 1991). While the full scope and significance of this parental "imprinting" have yet to be determined, it is known that certain of the genes influencing height, insulin production, and several forms of mental retardation affect a child differently—even in opposite ways—depending on which parent they came from.

Such polygenic complexity is particularly apparent in psychological characteristics, including everything from personality traits, such as sociability, assertiveness, moodiness, and fearfulness, to cognitive traits, such as memory for numbers, spatial perception, and fluency of expression. Typically, many pairs of genes, some interacting in the dominant-recessive mode, some additive, and some creating new combinations, affect every behavioral tendency.

Gene-Environment Interaction

Polygenic interaction is only one of the complexities in the relationship between genotype and phenotype. As noted earlier, another key source of complexity is the environment in all its forms.

To understand the wide-ranging impact of the environment on genetic inheritance, you need to know that when social scientists discuss the effects of the **environment**, they are referring to a multitude of variables. As they use the term, "environment" includes everything that can interact with the person's genetic inheritance at every point of life, from the first prenatal moments to the last heartbeat, from the impact of uterine acidity on the first stages of prenatal cell duplication to all the ways the elements in the external world impinge on the individual. These external elements include direct effects, such as those of nutrition, climate, medical care, and family interaction, and indirect effects, such as those of the broad economic, political, and cultural contexts. They also include varying degrees of permanence, from irreversible effects, such as the lifelong toll that severe brain injury has on cognitive ability, to transitory effects, such as the impact that immediate stress has on an individual's mood. In short, the influence of the environment is lifelong, multifaceted, and of varied force—just as the influence of heredity is.

Distinguishing Hereditary and Environmental Influences

Before examining the complex interplay of heredity and the environment, researchers first need to distinguish the developmental impact of these two forces. This is not easy to do, because, with any given trait, both are intertwined at every moment of a person's life. When the trait in question is an obvious physical one, the impact of genes on the phenotype seems fairly easy to identify, as in the case of family resemblances in facial features, eye color, or body type. But when the trait is a psychological one, such as an intellectual ability, artistic talent, or personality trait, the fact that the trait seems to run in families could be explained by nurture just as easily as by nature. How, then, do scientists distinguish genetic from environmental influences on personality characteristics?

environment All the nongenetic factors that can affect the individual's development—everything from the impact of the immediate cell environment on the genes themselves to the effects of nutrition, medical care, socioeconomic status, family dynamics, and the broader economic, political, and cultural contexts.

Studying children in their birth families is not much help, precisely because of the confounding of genetic inheritance and environmental family influences. For instance, if children of highly intelligent parents excel in school, their school performance could, theoretically, be attributed entirely to their genetic inheritance, entirely to the family environment (which is likely to encourage reading, intellectual curiosity, and high academic standards), or to any combination of the two.

One approach to this puzzle has been to study twins. As we have seen, monozygotic twins share all of their genes, while dizygotic twins, like any other two siblings from the same parents, share only half their genes. Thus, if monozygotic twins, on the whole, are found to be much more similar on a particular trait than dizygotic twins are, it seems likely that genes play a significant role in the appearance of that trait. (Of course, this approach assumes, among other things, that each twin growing up in a particular family shares the same environment. In fact, however, twins in the same family sometimes experience quite different environments. Their environments may differ in obvious ways, such as when one twin but not the other has a serious illness or is taught by an extraordinary teacher, and in more subtle, ongoing ways, such as when parents treat the twins differently or when the twins treat each other differently [Reiss et al., 1994].)

Another approach to distinguishing the impact of genes from that of upbringing is to study large numbers of adopted children, comparing their traits with those of both their biological and adoptive parents. Traits that show a strong correlation between adopted children and their biological parents suggest a genetic basis for those characteristics; traits that show a strong correlation between adopted children and their adoptive parents suggest environmental influence. One difficulty with this approach is that adopted children are often placed in families whose socioeconomic, educational, and religious backgrounds are similar to those of their birth families. As a consequence, some of the similarity found between adopted children and their biological parents may be the result of shared culture rather than of shared genes (Plomin, 1990).

The most telling way to try to separate the effects of genes and environment is to combine both approaches, studying identical twins who have been separated at birth and raised in different families. Although it requires painstaking searching to find enough such twin pairs to make statistically significant conclusions, several groups of researchers in the United States, Sweden, England, Denmark, Finland, and Australia have done just that, finding altogether close to a thousand twins reared in separate homes. The results of these researchers' investigations provide dramatic confirmation for the general conclusion reached by more conventional research on thousands of single-born adopted children and on twins raised by their biological parents. And that conclusion is that virtually every psychological characteristic and personal trait is genetically influenced (Bouchard, 1994; Bouchard et al., 1990; Eaves et al., 1989; Pederson et al., 1988; Shaw, 1994). (See A Closer Look, p. 76.) At the same time, these very same studies reinforce another, equally important conclusion: that virtually every psychological characteristic and personal trait is affected, throughout the life span, by one's environment.

A CLOSER LOOK

Twins Reared in Separate Homes

One of the most extensive investigations of twins raised in separate homes is the Minnesota Study of Twins Reared Apart, which over the past twenty years has studied hundreds of twin pairs who were separated early in life (Bouchard, 1994; Bouchard et al., 1990). This study, like others of its kind, has consistently found such striking psychological and behavioral similarities between identical twins that the important role of genes in personality development can no longer be doubted. Typical of the findings in this study are those in the case of Oskar Stohr and Jack Yufe, identical twins born of a Jewish father and Christian mother in Trinidad in the 1930s. Soon after their birth, Oskar was taken to Nazi Germany by his mother to be raised as a Catholic in a household consisting mostly of women. Jack was raised as a Jew by his father, spending his childhood in the Caribbean and some of his adolescence in Israel.

On the face of it, it would be difficult to imagine more disparate cultural backgrounds. In addition, when they were reunited in middle age, the twins certainly had their differences. Oskar was married and a devoted union member. Jack was divorced and owned a clothing store in southern California. But, when the brothers met for the first time in Minnesota,

> similarities started cropping up as soon as Oskar arrived at the airport. Both were wearing wire-rimmed glasses and mustaches, both sported two-pocket shirts with epaulets. They share idiosyncrasies galore: they like spicy foods and sweet liqueurs, are absentminded, have a habit of falling asleep in front of the television, think it's funny to sneeze in a crowd of strangers, flush the toilet before using it, store rubber bands on their wrists, read magazines from back to front, dip buttered toast in their coffee. Oskar is domineering toward women and yells at his wife, which Jack did before he was separated. [Holden, 1980]

Their scores on several psychological tests were very similar, and they struck the investigator as remarkably similar in temperament and tempo.

Other pairs of twins in this study likewise startled the observers by their similarities, not only in appearance and on test scores, but also in mannerisms and dress. One pair of female twins, separated since infancy, arrived in Minnesota, each wearing seven rings (on the same fingers) and three bracelets, a coincidence that might be explained by pure chance, but more likely was partly genetic—that is, genes endowed both women with beautiful hands and, possibly, contributed to an interest in self-adornment.

Case after case in this study has produced similar findings, suggesting that genes affect a much greater number of characteristics than most psychologists, including the leader of the Minnesota study, Thomas Bouchard, originally suspected. Bouchard now finds that genetic variation is significant for "almost every behavioral trait so far investigated from reaction time to religiosity" (Bouchard et al., 1990).

Since Oskar Stohr (left) and Jack Yufe (right) are monozygotic twins, it is no surprise that they are similarly balding, big torsoed, and wrinkling around the neck. And since they were raised on separate continents, with Oskar residing in southern Germany and Jack in southern California, it is not odd that their attire is notably different. However, when the brothers first met in Minnesota, researchers noted hundreds of similarities—including the tint and shape of their glasses and the size of their mustaches—that they had previously thought were strictly a matter of individual choice, influenced by culture, perhaps, but not by genes.

However, the seemingly uncanny similarities between monozygotic twins raised in separate homes should not automatically be attributed directly to genetics. For one thing, when identical twins are reared apart, they are often raised in quite similar homes. Even in the case of Oskar and Jack, says Bouchard, beneath the more dramatic differences in background, their upbringing was basically quite similar. Moreover, personality similarities may foster environmental similarities, as much as vice versa. Large-scale research finds that monozygotic twins tend to evoke similar degrees of warmth and encouragement from the adults who interact with them (Plomin, 1994). Thus a pair of identical twins, making their way in different families, may be similarly influenced by the similar family patterns they themselves help create.

These caveats notwithstanding, most researchers are astonished at the similarities they find in monozygotic twins raised separately (Lykken et al., 1992). Such findings make one wonder anew about the sources of our own individuality. Are our life choices—large and small—mostly an outgrowth of experience and cultural background, or do the roots go much deeper? Could many of the habits, patterns, and values that distinguish each of us be not so much a matter of personal choice as a matter of genetic push? These are intriguing questions to which we may never have definitive answers.

By the way, how many rings do you have on your fingers, and why?

Physical Traits

Environment, as broadly defined on page 74, affects every human characteristic—even physical traits that show a strong genetic influence. Take height, for example. An individual's height potential is genetically directed, yet most adults in developed countries are, on average, taller than their ancestors ever were but often about the same height as their own full-grown children. Why? Because to reach his or her genetically based height potential, a person must have adequate nutrition and good health. In the nineteenth century, these two factors were much less common than they are now, and Americans, for example, were, on average, about 6 inches shorter than they are today (Tanner, 1971). Throughout the twentieth century, however, as nutrition and medical care improved, each generation grew slightly taller than the previous one. Over the past several decades, this trend has stopped, because the prevailing levels of health and nutrition have permitted the vast majority of the population to reach their genetically set height limits. Most children reaching adulthood in the 1990s will, on average, be about as tall as their parents. Of course, in individual cases, environmental factors such as malnutrition, chronic illness, and stress can make a child considerably shorter than the limit set by his or her heredity.

Psychological Traits

The effects of environmental influences such as nutrition on the genetic expression of physical traits are fairly simple to understand. More varied, hidden, and intriguing are the effects of the interactions of environmental influences on psychological traits.

Take shyness, for instance. No doubt shyness is partly inherited: study after study finds that the levels of a personality trait called inhibition (or its opposite, extroversion or sociability) are more similar in monozygotic twins than in dizygotic twins (Bouchard et al., 1990; Eaves et al., 1989; Plomin, 1994; Robinson et al., 1992). In addition, there are biological differences between inhibited and extroverted children (Kagan, 1992; Kagan et al., 1993). Shy children, for example, show quicker startle reactions as infants and show less activity overall as young children.

However, research on adopted children indicates that shyness is affected both by the genetic heritage of the biological parents and by the social atmosphere provided by parents (Loehlin et al., 1982). As this research suggests, a child with a genetic disposition to shyness raised by outgoing parents would have many contacts with other people, and would observe his or her parents socializing freely, even with strangers. Although such a child might initially be hesitant and withdrawn in social encounters, gradually the child might learn to relax in social settings and would become less observably shy. It is not that as life experiences accumulate, genetically based tendencies disappear: a shy person would always feel twinges of inhibition when entering a new school, or when arriving at a party full of strangers. But some shy people are able to use childhood social experiences to learn how to warm up to others and feel more at ease. Alternatively, of course, if this same shy child's parents were also very shy and socially isolated, the child might grow up much more timid socially than he or she would have been with outgoing parents—and considerably more so than most other children.

Culture also plays a role in the expression of shyness. For instance, shyness is more socially acceptable in Chinese culture than in American culture, especially for girls. Thus a shy Chinese female is more likely to still appear shy in adulthood than a shy American male is. Both may feel equally shy, but they may express these feelings quite differently, one always remaining demurely in the background, for example, and the other affecting the air of a detached observer. And, if both act equally shy, the shy Chinese female may be appreciated and protected more than the shy American male (Caspi et al., 1988; Chen et al., 1995).

Thus the expression of a genetically based predisposition toward shyness will ultimately depend on the interactions among parental example, cultural encouragement, the social milieu, cognitive awareness, self-understanding, and individual experience—each of which may exacerbate, diminish, or redirect the impact of the others. The same conclusion applies to other psychological traits that have been found to have strong genetic influences, including intelligence, emotionality, activity level, and even religiosity (Bronfenbrenner & Ceci, 1994; Loehlin, 1992; Plomin, 1994). In each case, various dimensions of the individual's environment can enhance, inhibit, or alter the expression of hereditary predispositions in his or her phenotype.

Psychological Problems

Overall, research reveals that, as in the case of shyness, genes are never the exclusive determinant of any psychological characteristic. This is also true with regard to genetically influenced psychopathologies such as depression, antisocial behavior, phobias, and affective disorders. Here, too, both genes and environment are influential (Gottesman & Goldsmith, 1993; Plomin, 1994; Rutter, 1991).

The most extensive research on this question has been done on schizophrenia, and it shows that relatives of people with schizophrenia have a higher-than-normal risk of developing schizophrenia themselves (Gottesman, 1991) (see Figure 3.4 on the facing page). Most striking is the fact that if one monozygotic twin develops schizophrenia, the chances are about 50 percent that the other will, too—a rate far above the 1 percent incidence in the overall population. Looked at another way, however, the same evidence reveals the importance of the environment: among identical twins, half of those whose twin has schizophrenia are not themselves afflicted. Moreover, most people diagnosed with schizophrenia have no close relatives with the illness (Cromwell, 1993). Obviously, schizophrenia is multifactorial, with environmental elements—possibly a slow-acting virus, head injury, or overall stress—playing a pivotal role.

Alcoholism: A Clear Example of Gene-Environment Interaction

One particularly clear example of the way the environment moderates genetic potential in psychological characteristics can be seen in the case of alcoholism. At various times alcoholism was thought to be a moral weakness, a personality flaw, or a sign of psychopathology. We now know that alcoholism is at least partly genetic, although the specific genes involved, the nature of their interaction, and their precise power have yet to be deter-

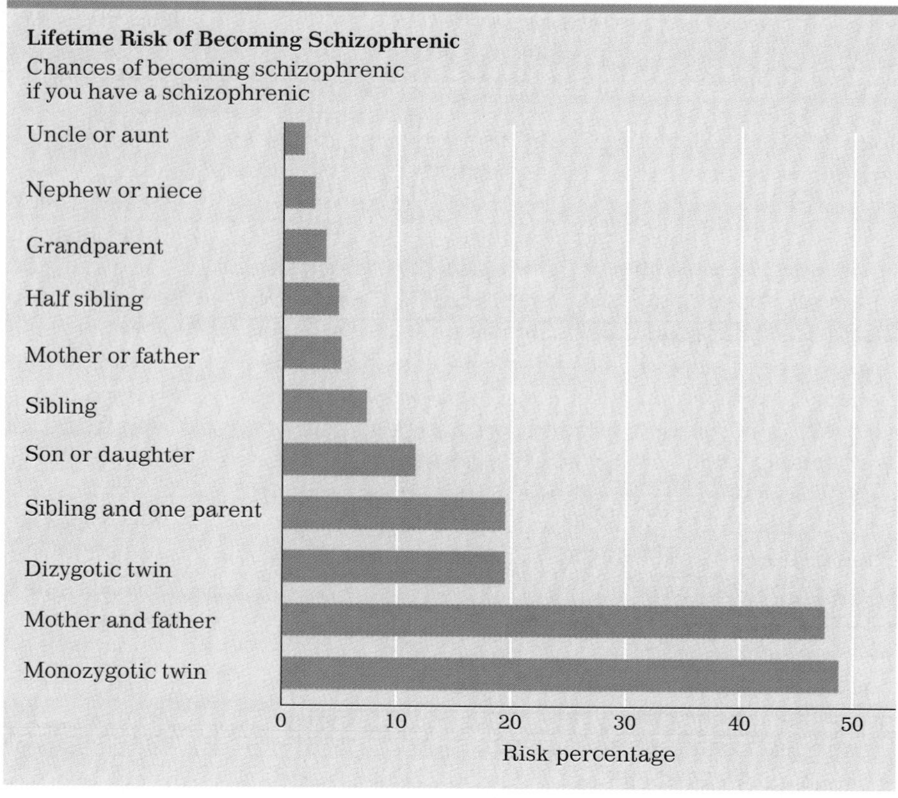

FIGURE 3.4 If a person has a relative who has schizophrenia, that person's lifetime risk of being diagnosed with schizophrenia begins rising from the 1-in-100 chance for the population at large, depending on how genetically close he or she is to the afflicted relative. The highest risk occurs for monozygotic twins: when one twin is diagnosed with schizophrenia, the other has almost a 50 percent chance of eventually being so diagnosed. Note, however, that while this chart shows a clear genetic influence on schizophrenia, the odds also show the effects of environment. For instance, over half the monozygotic siblings whose twin has schizophrenia do not have schizophrenia themselves.

mined (McClearn et al., 1991). We do know that some people's biochemistry makes them highly susceptible to alcoholism; others' biochemistry makes them much less so. Thus, while anyone can abuse alcohol, the addictive pull can be immensely strong or very weak, depending on the person's genetic makeup.

In addition to quite specific metabolic patterns, certain temperamental traits, themselves partly genetic, also correlate with abusive drinking. Among these are a quick temper, a willingness to take risks, and a high level of anxiety. Thus alcoholism is polygenic, with almost every alcoholic inheriting biochemical and temperamental traits that push toward abusive drinking (Bower, 1996).

Obviously, however, environment plays a critical role in the expression of alcoholism (McGue, 1993). If a person with a strong genetic affinity to alcoholism spends a lifetime in an environment where alcohol is unavailable (in a devout family in an Islamic country, for example), the alcoholic tendency in the person's genotype will probably never become manifest in his or her phenotype. On the other hand, if the same person is raised in a culture that promotes the use of alcohol, and he or she is exposed to pressures that lead to its use, that person is much more likely to become an active alcoholic. Even in that situation, however, social influences and individual choices can dramatically alter the outcome. Some alcoholics die of the disease before they are 30; others spend a lifetime going through periods of abuse, controlled drinking, and abstinence; still others recognize the problem by early adulthood, get help, and are sober and productive throughout a long life.

A CLOSER LOOK

Heritability: What It Is, and What It Is Not

In both scientific and popular literature, human traits are sometimes referred to as being "heritable" by a certain percent. For example, among North Americans, height is estimated to be about 90 percent heritable; weight, about 70 percent heritable; and certain personality and intellectual traits, about 50 to 75 percent heritable (Bouchard et al., 1990; Loehlin et al., 1988; Plomin et al., 1990). Such estimates are arrived at by a complex mathematical calculation that uses data from the kinds of twin studies discussed on p. 76. For our present purposes, an understanding of the mathematical basis of heritability estimates is not necessary. What is required, however, is a clear understanding of what heritability is, and what it is not. This is because in popular usage, the idea of heritability is often misunderstood, and sometimes dangerously misapplied.

Heritability refers to the *variation* in a particular trait, in a *particular population*, in a *particular environment*, and it expresses the degree to which that variation can be attributed to genetic differences among the members of the group. Thus heritability is a population statistic that refers to *group differences* in a trait. It does *not* refer to the total genetic component of the trait in question, *nor* does it indicate to what degree genetic inheritance determines that trait in a particular individual. To say that the height of North Americans is 90 percent heritable, for example, means that 90 percent of the variation above and below the average height of all Americans is the result of genes. It does *not* mean that an American who is 70 inches tall owes 63 inches of his or her height to genes and the remaining 7 inches to environmental factors. Similarly, to say that intelligence is between 50 and 75 percent heritable means that within a particular group, between 50 and 75 percent of the variation in intelligence (as measured by IQ tests) is due to genes. Once again, it does *not* mean that between 50 and 75 percent of an individual's intelligence is inherited and the rest due to the environment.

Another important fact is that heritability as calculated for any given trait is not invariant. It varies according to the genetic and environmental similarity or diversity of the group in question. The more genetically diverse the group, and the more uniform the environment, the higher heritability will be. (The heritability of height in North America is as high as it is precisely because of the genetic diversity of the overall population and the relative equality of the nutritional environment.) Correspondingly, the less genetically diverse the group, and the more varied the environment, the lower heritability will be. (If one were in a science-fiction world studying variation in height among a population of clones, the heritability of height would be zero because clones are genetically identical, and any variations in their height would therefore be totally attributable to the differences in the environment.)

Confusion over the meaning of heritability can be particularly harmful when it involves a trait like intelligence. For example, because the heritability of intelligence *within* various ethnic and racial groups in fairly high (again,

The example of alcoholism also illustrates a final factor influencing the interaction between genes and the environment—the individual's age and the particular expectancies the culture holds for persons of that age. Alcoholism may be genetically "present" at birth, but since few cultures allow children to consume alcohol, it is rarely expressed before adolescence. Many other traits also become more apparent as children mature and their capabilities, needs, and interests change, while at the same time, parental restrictions and influence wane (Caspi & Moffitt, 1991; McGue et al., 1993). This is especially the case with adopted children, whose genetic predispositions are sometimes at odds with those of their adoptive parents. When they are very young, adopted children reflect many of their adoptive parents' interests, behaviors, and personality traits. However, with maturity, they often choose friends, hobbies, and habits that express their biological, rather than their familial, heritage. Indeed, as we mature and become increasingly independent and self-determining, we all tend to seek out and construct our own ecological niches, establishing personal environments for ourselves

heritability The variation in a particular trait, in a particular population, and in a particular environment and the degree to which that variation can be attributed to genetic differences among the members of the group.

somewhere between 50 and 75 percent according to most heritability estimates), some people assume that any difference *between* the average IQ scores of different groups reflects genetic differences between the groups. In other words, they assume that groups with lower average IQ scores are genetically inferior in intellectual ability. (This assumption is all too commonly made with regard to the fact that the average IQ scores of black Americans are consistently 10 to 15 points lower than those of white Americans.) The fact is, however, that the degree of within-group heritability of any trait tells us nothing about the genetic determination of between-group differences on that trait. To see why this is so, consider the following example.

Let's say that you plant a random mix of carnation seeds in two window boxes. One box has rich soil and an open southern exposure; the other box has poor soil and a heavily shaded northern exposure. In addition, you water and otherwise tend to the first window box regularly; by contrast, you tend to the second box erratically and inadequately. When the carnations have matured, you will find that those in the first window box vary in height, and that those in the second box do as well. This *within-group* variation will be due mostly to genetic differences, since the environmental conditions for the carnations in the first box were pretty much the same for all the flowers, as were those for the carnations in the second box. At the same time, there will be a difference between the average height of all the carnations in the first box and the average height of all those in the second—but this *between-group* difference will be due to the differences in the environments of the two boxes, that is, the richness of one as opposed to the relative impoverishment of the other. Thus the similar degrees of within-group heritability of height in the two carnation groups has little bearing on the between-group differences in the flowers' height.

The same reasoning applies with regard to between-group differences in average IQ scores. Although the within-group heritability of intelligence may be fairly high, the between-group differences in average IQ scores may be attributable not to genetic differences between the two groups but rather to environmental differences, ranging from socioeconomic and quality-of-life factors (health, safety, opportunities to learn, and so on) to the emphasis and support that each group's culture provides for the kinds of abilities that IQ tests measure.

The danger of similar misinterpretation arises whenever heritability is announced for any characteristic that is particularly desirable or undesirable, whether it be artistic talent or aggressive tendencies, sociability or schizophrenia. Remember, then, whenever you encounter heritability estimates, that they can never be applied to individual cases or used to assess trait differences that exist between groups.

that are compatible with, and reinforcing of, our individual traits and abilities (Scarr, 1994).

In summary, then, it is quite clear that both genes and environment are powerful influences on development, that their interaction is involved in every aspect of development, and that their interaction is complex. On a practical level, this means we should not ignore the fact that there is a genetic component in any given trait—whether it be something wonderful, such as a wacky sense of humor, or something fearful, such as a violent temper, or something quite ordinary, such as the tendency to tire of the same routine. At the same time, we must always recognize that the environment affects every trait in every individual in ways that change as developmental processes unfold. Genes are always part of the tale, influential on every page, but they never determine the plot or the final story. (See A Closer Look for an explanation of the concept of *heritability* and the common misunderstandings associated with it.)

Genetic and Chromosomal Abnormalities

In studying human development, we give particular attention to genetic and chromosomal abnormalities for three reasons. One reason is that, by investigating genetic disruptions of normal development, we can gain a fuller appreciation of the complexities of genetic interaction. A second reason is that an understanding of those who inherit genetic or chromosomal abnormalities is essential to everyone concerned about fostering human development. A lack of such understanding can lead to misinformation and prejudice, which only compound the problems of those affected by such disorders. The third reason is the most practical: the more we know about the origins of genetic and chromosomal abnormalities, the better we understand the risks of their occurring and the better prepared we are to limit their harmful effects.

We will begin by looking at those problems caused directly by the chromosomes. Such genetic abnormalities are, in general, the most serious, but they are also the easiest to detect and prevent.

Chromosomal Abnormalities

Sometimes when gametes are formed, the forty-six chromosomes divide unevenly, producing a sperm or ovum that does not have the normal complement of twenty-three chromosomes. If such a gamete fuses with a normal gamete, the result is a zygote with more or less than forty-six chromosomes. This is not unusual. An estimated half of all zygotes have an odd number of chromosomes. Most of these do not even begin to develop, and most of the rest never come to term, usually because a spontaneous abortion occurs. Once in every 200 births, however, a baby is born with forty-five, forty-seven, or even more chromosomes (Gilbert et al., 1987). In every case, these chromosomal abnormalities lead to a recognizable **syndrome**—a cluster of distinct characteristics that tend to occur together. Individuals who have a particular syndrome do not necessarily have all the distinguishing characteristics, and in any given syndrome, the severity of the symptoms varies from person to person.

In most cases, the presence of an extra chromosome is lethal within the first days or months after birth. There are two major exceptions in which affected individuals often live to adulthood—when the extra chromosome is at the twenty-first pair, where the smallest nonsex chromosome pair is located, or when it occurs at the twenty-third pair.

syndrome A cluster of distinct characteristics that tend to occur together in a given disorder, although the number of characteristics exhibited, and their intensity, vary from individual to individual.

trisomy-21 (Down syndrome) A chromosomal abnormality caused by an extra chromosome at the twenty-first pair. Individuals with this syndrome tend, with much variation, to have round faces and short limbs, and to be slow to develop.

Down Syndrome

The most common of the extra-chromosome syndromes is **trisomy-21**, or **Down syndrome**, in which the individual inherits a third chromosome at the twenty-first pair. Some 300 distinct characteristics can result from the presence of that extra chromosome, but as with all syndromes, no individual with Down syndrome is quite like another, either in the specific symptoms he or she has or in their severity (Cicchetti & Beeghly, 1990; Lott & McCoy, 1992). Despite this variability, almost all people with trisomy-21 have certain facial characteristics—a thick tongue, round face, slanted eyes—as well as

Many of the athletes in this Special Olympics competition have Down syndrome, as one can tell from their distinctive facial features. Other characteristics of Down syndrome show much wide variation, and the intellectual growth and personality development of these individuals will depend on many factors besides the impact of their extra chromosome. Events such as this, for instance, can foster self-esteem and social congeniality.

distinctive hands, feet, and fingerprints. Many also have hearing problems, heart abnormalities, muscle weakness, and short stature.

In terms of psychological development, almost all individuals with Down syndrome experience some mental slowness, but their eventual intellectual attainment varies, from severely retarded to average or even above average. Often—but not always—those who are raised at home and given appropriate cognitive stimulation progress to the point of being able to read and write and care for themselves (and often much more), while those who are institutionalized tend to be, and to remain, much more retarded. In their socioemotional qualities, many children with trisomy-21 are considered unusually sweet-tempered in that they are less likely to cry or complain than are most other children. By middle adulthood, however, individuals with Down syndrome are more likely to develop a form of dementia similar to Alzheimer's disease, severely impairing their limited communication skills and making them much less compliant (Rasmussen & Sobsey, 1994). They are also prone to a host of other problems more commonly found in older persons, including cataracts and certain forms of cancer.

Abnormalities of the Sex Chromosomes

Every newborn infant has at least one *X* chromosome. About 1 in every 500 infants, however, has either a missing sex chromosome, so that the *X* stands alone, or has an *X* chromosome complemented by two or more sex chromosomes. As you can see from Table 3.1, these abnormalities can impair cognitive and psychosocial development, as well as sexual maturation, with each particular syndrome having a specific effect. In many cases, treatment with hormone supplements can alleviate some of the physical problems, and

TABLE 3.1

Common Abnormalities Involving the Sex Chromosomes

Name	Chromosomal Pattern	Physical Appearance*	Psychological Characteristics*	Incidence
Kleinfelter syndrome	*XXY*	Male. Secondary sex characteristics do not develop. For example, the penis does not grow, the voice does not change. Breasts may develop.	Learning-disabled, especially in language skills.	1 in 900 males
(No name)	*XYY*	Male. Prone to acne. Unusually tall.	Tend to be more aggressive than most males. Mildly retarded. especially in language skills.	1 in 1,000 males
Fragile X	Usually *XY*	Male or female. Often, large head, prominent ears. Occasionally, enlarged testicles in males.	Variable. Some individuals apparently normal; others severely retarded, with impaired social skills.	1 in 1,000 males 1 in 2,500 females
(No name)	XXX, XXXX	Female. Normal appearance.	Retarded in almost all intellectual skills.	2 in 1,000 females
Turner syndrome	*XO* (only one sex chromosome)	Female. Short in stature, often "webbed" neck. Secondary sex characteristics (breasts, menstruation) do not develop.	Learning-disabled, especially in abilities related to math and science and in recognition of facial expressions of emotion.	1 in 2,000 females

*There is some variation in the physical appearance of these individuals and considerable variation in their intellectual and temperamental characteristics. With regard to psychological characteristics, much depends on the family environment of the child.

Sources: Borgaonkar, 1994; Dykens et al., 1994; Lee, 1993; McCauley et al., 1987; Rovet et al., 1996.

special education may remedy some of the deficits related to psychological functioning.

Again, however, remember that the specific features of any syndrome vary considerably from one individual to another. In fact, in many cases, the presence of abnormal sex chromosomes goes undetected until a seemingly normal childhood is followed by an abnormally delayed puberty. Many specialists recommend chromosomal analysis as soon as a problem is suspected, followed by carefully individualized counseling and treatment for those with a problem.

The Fragile X

One of the most common problems associated with the sex chromosomes is actually genetic in origin. In some individuals, part of the X chromosome is attached by such a thin string of molecules that it seems about to break off, and thus the problem is called **fragile-*X* syndrome**. This abnormality in the chromosome is caused by the mutation of a single gene, a portion of which contains a DNA sequence (CGG) that is repeated between 200 and 2,000 times (the normal number of repetitions is about 30). Unlike most other known mutations, the mutation involved in fragile X intensifies as it is passed from one generation to the next (Dykens et al., 1994; Hagerman, 1996).

Perhaps because the number of repeats in the faulty gene varies, fragile-X syndrome is highly variable in its effects. Of the females who carry it, most are normal (perhaps because they also carry one normal X chromosome), but a third show some mental deficiency. Among the males who inherit a fragile-X chromosome, there is considerable variation as well: about 20 percent of males are apparently completely normal; about 33 percent are somewhat retarded; and the rest are severely retarded. All told, the cognitive deficits associated with fragile-X syndrome represent the most common form of inherited mental retardation, frequently including inadequate social skills and extreme shyness (Dykens et al., 1994; Hagerman, 1996). While the extreme variability in the effects of this disorder is somewhat unusual, some geneticists believe that the more we learn about other abnormal genes and their interactions, the more diversity we will find in their expression (McKusick, 1994).

Causes of Chromosomal Abnormalities

Chromosomal abnormalities are caused by many factors, some genetic and some environmental (such as viruses contracted by the mother). However, one of the most common known correlates of chromosomal abnormalities is maternal age. According to one detailed estimate, for example, a 20-year-old woman has 1 chance in 500 of having a child with chromosomal abnormalities; a 39-year-old woman has 1 chance in 100; and a 48-year-old woman has 1 chance in 9 (Cefalo & Moos, 1988).

One reason for this may be age-related problems that occur during the final maturation of ova just before ovulation. This theory is supported by a study that analyzed the extra twenty-first chromosome of infants with Down syndrome. The average age of the mothers of these infants was significantly greater than the average age for pregnant women overall, and the study found that in about 90 percent of the cases, the chromosomal error seemed to have originated in the final stages of ovum duplication and division preced-

fragile-*X* syndrome A genetically based condition that results from an abnormality of the X chromosome and that causes mental deficiency in about 30 percent of the women and in about 80 percent of the men who carry it.

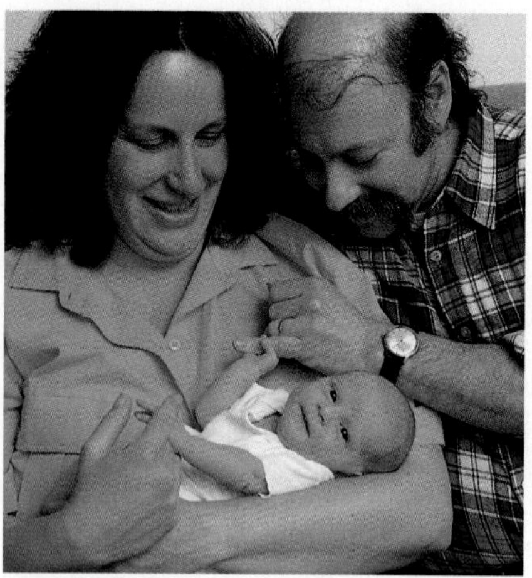

The fact that older parents have a higher risk of conceiving an embryo with chromosomal abnormalities should not obscure another reality. With modern medical care and prenatal testing, pregnancies that occur when the parents are in their 40s can, and almost always do, result in healthy babies.

ing ovulation. In the remaining 10 percent, the cause of the error seemed to be genetic. For this group, the inherited problem was as likely to be found in the sperm as in the ovum, and age was not a factor (Antonarkis et al., 1992).

Harmful Genes

While relatively few people have abnormal chromosomes, everyone has at least twenty genes that could produce serious diseases or handicaps in one's offspring and, in some cases, in oneself. To date, close to 7,000 genetic disorders have been identified, many of them exceedingly rare (McKusick, 1994).

Most of the known genetic disorders are dominant, which means that whenever the gene for such a disorder is inherited, it will be expressed in the person's phenotype. With a few exceptions, most dominant disorders are not seriously disabling, and many others vary tremendously in their manifestation. For example, only about 30 percent of the people with Tourette syndrome exhibit the recurrent uncontrollable tics and explosive verbal outbursts that can make social interaction problematic. The remaining 70 percent experience symptoms ranging from an occasional twitch that is barely noticeable to an impulse to speak inappropriately that can be controlled until a more opportune moment—much the way a person resists the impulse to publicly scratch an itchy private spot.

Although they are fewer in number than dominant genetic disorders, recessive and multifactorial disorders claim many more victims, largely because such disorders pass unchecked from carrier to carrier for many generations and can easily become widespread in a population. Among such common recessive disorders are cystic fibrosis, thalassemia, and sickle-cell anemia, with as many as one in twelve North Americans being a carrier for one or another of these three diseases (see Table 3.2 on the next page).

For the most part, carriers of a specific disorder do not chance to marry other carriers of it. However, about one child in thirty inherits a genetic disorder that is seriously disabling and perhaps life-threatening. Many a prospective parent worries about the possibility of giving birth to a very ill child, especially a child whose suffering is the direct result of the parent's own genetic heritage. Genetic counseling can help relieve such fears as well as make such births less likely. At the same time, it can prepare families and physicians for those births in which a genetic disorder is likely to occur.

TABLE 3.2

Common Genetic Diseases and Conditions

Name	Description	Prognosis	Method of Inheritance	Incidence*	Carrier Detection†	Prenatal Detection?
Alzheimer's disease	Loss of memory and increasing mental impairment.	Eventual death, often after years of dependency.	Some forms are definitely genetic; others are not.	Fewer than 1 in 100 middle-aged adults; 20 percent of all adults over age 80.	No.	No.
Cleft palate, cleft lip	The two sides of the upper lip or palate are not joined.	Correctable by surgery.	Multifactorial. Drugs taken during pregnancy or stress may be involved.	One baby in every 700. More common in Asian-Americans and Native Americans; rare in African-Americans.	No.	Yes, in some cases.
Club foot	The foot and ankle are twisted, making it impossible to walk normally.	Correctable by surgery.	Multifactorial.	One baby in every 200. More common in boys.	No.	Yes.
Cystic fibrosis	Mucous obstructions in body, expecially in lungs and digestive organs due to the lack of an enzyme.	Most live to middle adulthood.	Recessive gene. Also spontaneous mutations.	One white baby in every 2,500. One in 20 white Americans is a carrier.	Usually.	Yes, in some cases.
Diabetes	Abnormal metabolism of sugar because body does not produce enough insulin.	Early onset is fatal unless controlled by insulin. Diabetes in later adulthood increases the risk of other diseases. Controllable by insulin and diet.	Multifactorial. Exact pattern hard to predict because environment is crucial.	About 10 million Americans. Most develop it in late adulthood. One child in 500 is diabetic. More common in Native Americans.	No.	No.
Hemophilia	Absence of clotting factor in blood.	Crippling and death from internal bleeding. Blood transfusions can lessen or even prevent damage.	X-linked recessive. Also spontaneous mutations.	One in 10,000 males. Royal families of England, Russia, and Germany had it.	Yes.	Yes.
Hydrocephalus	Obstruction causes excess water in brain.	Can produce brain damage and death. Surgery can sometimes make survival and normal intelligence possible.	Multifactorial.	One baby in every 100.	No.	Yes.
Muscular dystrophy (13 separate diseases)	Weakening of muscles. Some forms begin in childhood, others in adulthood.	Inability to walk, move; wasting away and sometimes death.	Duchenne's is X-linked; other forms are recessive or multifactorial.	One in every 3,500 males will develop Duchenne's; about 10,000 Americans have some form of MD.	Yes, for some forms.	Yes, for some forms.

*Incidence statistics vary from country to country; those given here are for the United States. All these diseases can occur in any ethnic group. When certain groups have a higher incidence, it is noted here.

†Studying the family tree can help geneticists spot a possible carrier of many genetic diseases or, in some cases, a definite carrier. However, here "Yes" means that a carrier can be detected even without knowledge of family history.

Name	Description	Prognosis	Method of Inheritance	Incidence*	Carrier Detection†	Prenatal Detection?
Neural tube defects (open spine)	Two main forms: anencephaly (parts of the brain and skull are missing) and spina bifida (the lower portion of the spine is not closed).	Often, early death. Anencephalic children are severely retarded; children with spina bifida have trouble with walking and with bowel and bladder control.	Multifactorial; defect occurs in first weeks of pregnancy.	Anencephaly: 1 in 1,000 births; spina bifida: 3 in 1,000. More common in those of Welsh and Scottish descent.	No.	Yes.
Phenylketonuria (PKU)	Abnormal digestion of protein.	Mental retardation, hyperactivity. Preventable by diet.	Recessive gene.	One in 15,000 births. One in 100 European-Americans is a carrier; more common among those of Norwegian and Irish ancestry.	Yes.	Yes.
Pyloric stenosis	Overgrowth of muscle in intestine.	Vomiting, loss of weight, eventual death; correctable by surgery.	Multifactorial.	One male in 200; 1 female in 1,000. Less common in African-Americans.	No.	No.
Sickle-cell anemia	Abnormal blood cells.	Possible painful "crisis"; heart and kidney failure. May now be treatable with drugs.	Recessive gene.	One in 500 African-American babies is affected. One in 10 African-Americans is a carrier, as is 1 in 20 Latinos.	Yes.	Yes.
Tay-Sachs disease	Enzyme disease.	Apparently healthy infant becomes progressively weaker, usually dying by age 5.	Recessive gene.	One in 4,000 births. One in 30 American Jews is a carrier, as is an estimated 1 in 20 French-Canadians and 1 in 200 non-Jewish Americans.	Yes.	Yes.
Thalassemia	Abnormal blood cells.	Paleness and listlessness, low resistance to infection; treatment by blood transfusion.	Recessive gene.	As many as 1 in 10 Greek-, Italian-, Thai-, and Indian-Americans is a carrier.	Yes.	Yes.
Tourette syndrome	Uncontrollable tics, body jerking, verbal obscenities.	Often imperceptible in children; worsens with age. Can be treated with drugs.	Probably dominant gene.	One in 500 births.	Sometimes.	No.

Sources: Bowman & Murray, 1990; Brunn & Brunn, 1994; Caskey, 1992; Connor & Ferguson-Smith, 1991; Lee, 1993; McKusick, 1994; National Academy of Sciences, 1994.

genetic counseling Consultation and testing that enable couples to learn about their genetic heritage and to make decisions about childbearing.

Genetic Counseling

For most of human history, couples at risk for having a child with a chromosomal or genetic problem did not know it. If they had a seriously impaired child, they blamed fate, the heavens, or each other, and went on with the difficult task of raising that child, at the same time risking the birth of another child with the same problem. Today, information obtained through **genetic counseling** enables such couples to learn more about their potential genetic legacy and to make informed decisions about their childbearing and child-rearing future.

Who Should Receive Counseling?

Everyone is a carrier of recessive genes that can cause serious problems in his or her progeny (Motulsky et al., 1994); everyone produces a certain number of gametes that have chromosomal abnormalities; and everyone has, and can pass on, genes that contribute to a wide variety of ailments, from asthma to varicosity. However, this does not mean that everyone who is planning to have children needs to seek genetic counseling. In general, prenatal, preconceptual, or even prenuptial genetic counseling and testing are recommended for the following:

1. individuals who have a parent, sibling, or child with a serious genetic condition;
2. couples who have a history of early spontaneous abortions or stillbirths;
3. couples who are from the same ethnic group or subgroup—especially if the group is a small one with a high rate of intermarriage, and most particularly if the couple are relatives;
4. women over age 34.

When a couple begins counseling, they first help their counselor construct a family history, charting the patterns of health and sickness within

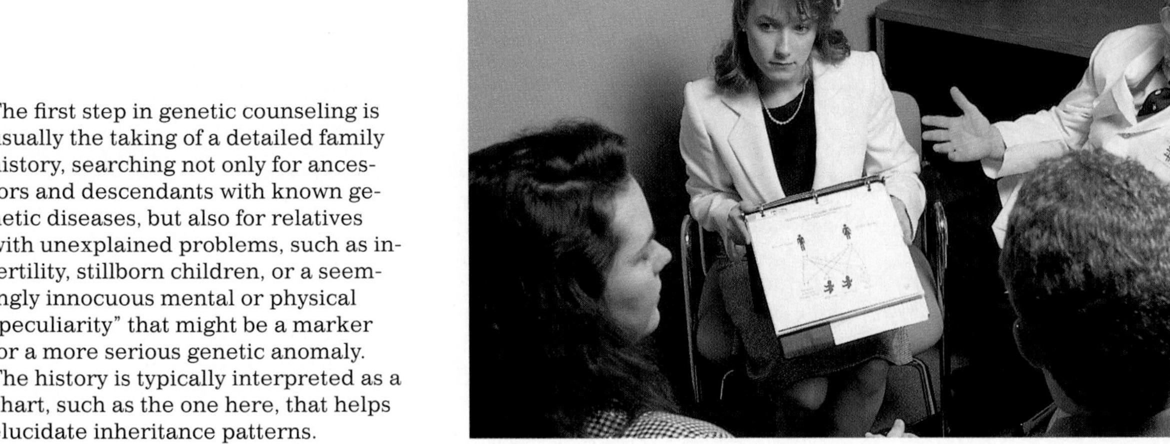

The first step in genetic counseling is usually the taking of a detailed family history, searching not only for ancestors and descendants with known genetic diseases, but also for relatives with unexplained problems, such as infertility, stillborn children, or a seemingly innocuous mental or physical "peculiarity" that might be a marker for a more serious genetic anomaly. The history is typically interpreted as a chart, such as the one here, that helps elucidate inheritance patterns.

their family trees, particularly with regard to early deaths or unexplained symptoms. They then are advised of the conditions for which they might be at particular risk, and they are told what their options would be if testing reveals that the risk is high. They are also informed about the varying reliability of different tests and are cautioned against taking any test results as a guarantee, one way or the other. The couple then decides whether to proceed with testing. Some may prefer not to know what their risks are if the only way to prevent the birth of a child with a serious disorder would be surgical sterility or abortion, choices some couples do not want to make.

Assuming that a couple chooses to be tested, what will be involved?

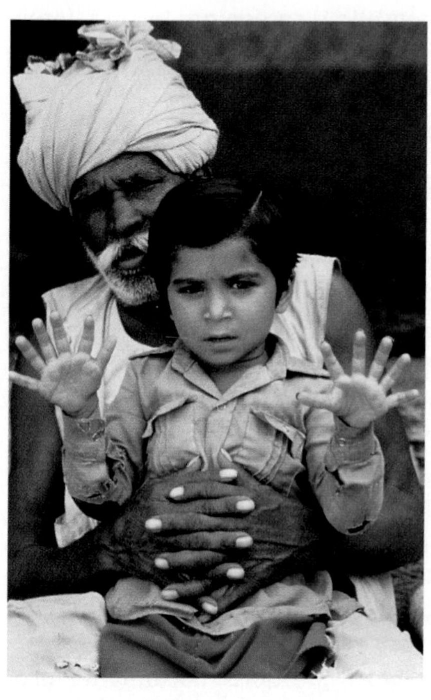

Being born with six fingers is a rare, minor, genetic abnormality, of no consequence if it occurs alone. However, sometimes it is a marker for more serious recessive problems, including dwarfism.

Genetic Testing

The first step in genetic testing is to attempt to find out all the necessary details about the genotypes of the parents-to-be. Sometimes reading the genotype for particular disorders is quite simple. A blood test, for example, detects carriers of the genes for sickle-cell anemia, Tay-Sachs, PKU, hemophilia, and thalassemia, as well as many other less common diseases. Chromosomal analysis of the parents can indicate fragile X, as well as the inherited form of Down syndrome.

Detection of many other disorders is more difficult because the culprit genes have not yet been precisely identified. In many cases, however, research has identified roughly where on a particular chromosome the culprit gene is likely to be located. This may enable scientists to identify certain **markers**, that is, specific gene sequences, harmless in themselves, that typically occur in the same suspect area of the chromosome when the disorder in question is present. In some cases, markers appear in the phenotype as well as in the genotype. For example, an oddly shaped earlobe or finger, or a particular pattern of eye movement, or a particular formation of the toes may signal the presence of a certain problematic gene. Other markers can be detected only through DNA analysis, either of the individual or, for some disorders, of several family members over at least three generations (Lee, 1993).

The information gathered about the genotypes of a particular pair of prospective parents can in many cases lead to a fairly precise calculation of the odds of their having a child with a specific genetic disorder. For example, if one parent has a gene for a dominant disorder, each of the couple's offspring has a 50 percent chance of inheriting that gene and, thus, the disorder. If both parents are carriers of a specific recessive gene that causes a disorder, each of their offspring has a 25 percent chance of inheriting the recessive gene from both parents and therefore of having the disorder. (The principle is the same as in the case [p.72] of two brown-eyed parents who both have the recessive gene for blue eyes and thus have a 1-in-4 chance of having a blue-eyed offspring.) If only one parent is a carrier of a recessive gene, there is no chance that the child will have the disorder.

markers In genetic testing, particular physiological characteristics or gene clusters that suggest that an individual might be a carrier of a harmful gene.

In considering such odds, one must realize that "chance has no memory," which means that the odds in question apply to each child the couple has. If both partners have the recessive gene for sickle-cell anemia, for instance, the couple could have a large number of children, all of whom, some of whom, or none of whom would have the disease.

Of course, even when a couple have been tested for genetic conditions and found to be low-risk, they may still have a child with a genetic disease. One reason for this is that many genetic conditions, including quite common ones such as cystic fibrosis and Tourette syndrome, can occur as a result of a spontaneous mutation (Brunn & Brunn, 1994; Welsh & Smith, 1995). This means that even though an adult is not a carrier for one of these conditions, some of his or her gametes might contain the defective gene for it. Another reason is that many genetic diseases are quite rare, so a particular couple, as well as their genetic counselor, might never suspect that they are carriers until they have a child with the condition. A final reason is that no reliable screening measures have yet been developed for many genetically transmitted diseases.

Fortunately, the pace of genetic counseling, screening, testing, interpretation, and analysis are accelerating as a result of the **Human Genome Project**, a massive international effort to map and sequence the entire set of more than 3 billion base pairs that make up the 100,000 or so genes in the human genotype. Monthly, sometimes daily, researchers are able to detect new genetic vulnerabilities for an increasing number of conditions, not only dominant and recessive ones but also multifactorial ones, including heart disease, diabetes, many forms of cancer, and many types of learning and emotional problems. As defective genes are increasingly identified, scientists are also developing new ways to detect embryos with lethal genes, allowing parents to halt the pregnancy rather than watch their infant die an early, painful death. Scientists also are developing new techniques—including surgery while the fetus is still in the womb—to treat offspring who have serious but not fatal genetic diseases. In addition, an experimental technique with *in vitro* fertilization (the combining of ova and sperm in a laboratory, followed by the transfer of a resulting embryo into the woman's uterus) makes it possible to analyze the genetic traits of the organism during the early stages of cell division, permitting the insertion of only healthy cell masses into the uterus. All these efforts raise hopes for carriers of genetic conditions, but they also raise serious ethical questions, not only for prospective parents but for society at large as well (see Public Policy, pp. 92–93).

Many Alternatives

Many couples who undergo genetic testing have a relatively easy childbearing decision to make, discovering perhaps that only one partner is a carrier of a particular harmful recessive trait, and that therefore none of their children will have the disease. Or perhaps they learn that the odds of their bearing a child with a serious illness are not much higher than those for any other couple, and thus the childbearing decision rests more on psychological or financial factors than on genetic ones.

Even if a couple learns that they both are carriers of a serious condition, or that they are high-risk in other ways, they still have many alternatives, as Figure 3.5 indicates. Some may avoid pregnancy and, perhaps, plan adoption. Some may decide to test the embryo early in pregnancy and, if the results show serious problems, decide to consider abortion or, alternatively, to begin gathering information that will help them deal with the child-care problems that lie ahead. Some may decide to postpone pregnancy until promising treatments—either prenatal or postnatal—are further devel-

Human Genome Project A worldwide effort to construct and decipher a chromosomal map of all 3 billion base pairs of the 100,000 human genes.

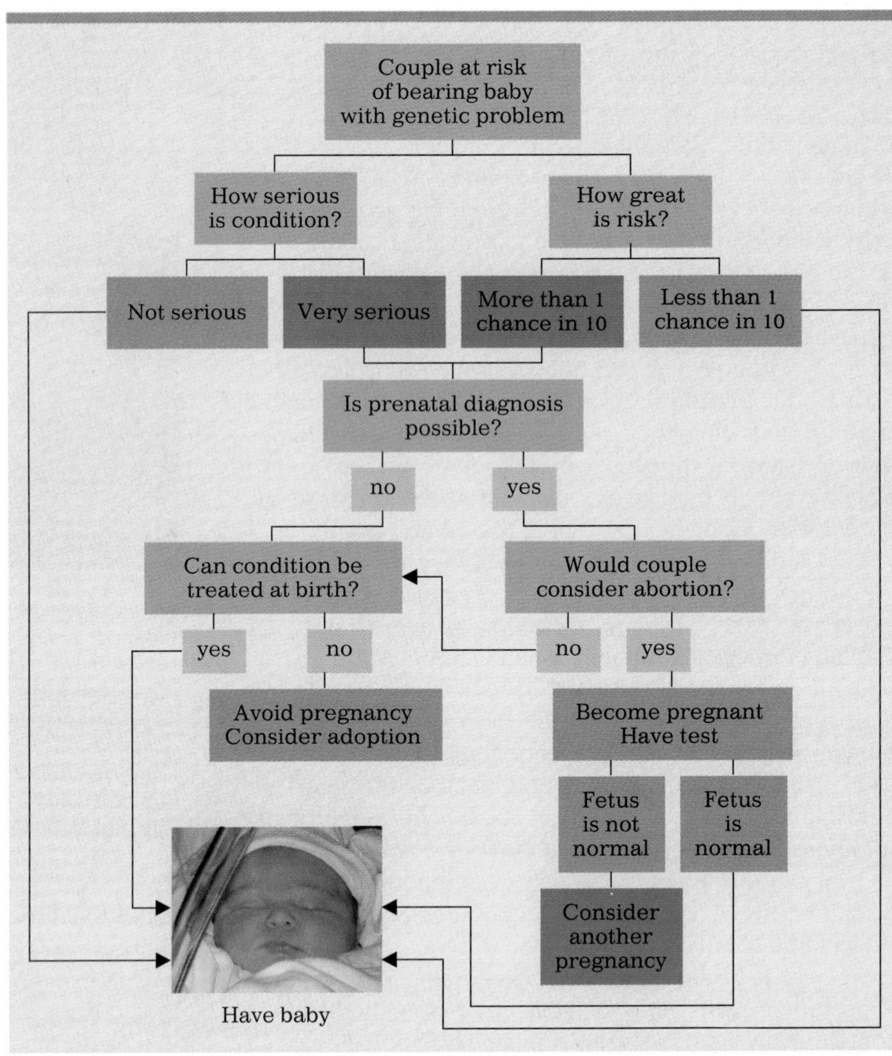

FIGURE 3.5 With the help of a genetic counselor, even couples who know they run a risk of having a baby with a genetic defect might decide to have a child. Although the process of making that decision is more complicated for them than it is for a couple with no family genetic illness and no positive tests for harmful recessive genes, the outcome is usually a healthy baby. In each case, the genetic counselor provides facts and alternatives: every couple must make their own decision. In fact, two couples who have the same potential for producing a child with a genetic defect, and are aware of the same facts regarding the situation, sometimes make opposite decisions because they differ in their attitudes about abortion, in their willingness to raise a child with a genetic abnormality, or in their desire to have their own child rather than an adopted one.

oped. In addition, since many conditions, including such common ones as sickle-cell anemia and cystic fibrosis, vary in severity, some couples may decide on additional testing that might reveal whether a child of theirs would be likely to develop a less serious variant of the disease. Some might choose reproductive alternatives, from artificial insemination with a donor gamete to the experimental type of *in vitro* fertilization mentioned earlier, in which clinicians test the resulting cell mass to determine whether it is genetically normal before inserting it into the uterus (National Academy of Sciences, 1994).

Of course, decisions about conception are not based simply on genetic analysis. Two couples with the identical odds of conceiving zygotes with the identical condition might make quite different decisions, depending on their age, ethnicity, religion, finances, ethics, personal relationship, and the number and health status of any other children they may have (Asch et al., 1996). Thus, genetic counseling begins with quite objective facts and ends with a quite personal decision, influenced by social and cultural factors as well as by statistics and biology, and ultimately dependent on the values of the person who makes the choice.

PUBLIC POLICY

The Human Genome Project

Imagine a time when the mysteries of your heredity could be revealed by analyzing a drop of blood or saliva, or even a snippet of hair. Analyzing a cell from anywhere on your body, a laboratory might be able to provide a complete report of the physiological and psychological characteristics encoded within your DNA.

When going for a medical checkup, you would bring along a computer disk that fully described your hereditary features for the physician to consult while diagnosing or treating your ailments. And when making the most important decisions of your life—whom to marry, where to live, what career to pursue—detailed information about your heredity would shed important light on your options.

Does this sound like a wonderful new world, or a potential nightmare? In either case, certain aspects of this world are fast approaching, partly as a result of the Human Genome Project, a $3 billion, fifteen-year initiative that formally began in 1991. The goal of this worldwide research effort is, quite simply, to map all the 100,000 genes of the human body, indexing the exact chemical instructions of each of the 3 billion base pairs that make up the genes and determining the location of every gene on its carrier chromosome. In this way, scientists hope to crack the genetic code of all our human characteristics, from the hereditary origins of physical structures and physiological systems to the genetic foundations of behavioral tendencies and psychological disorders. Progress on the project has exceeded expectations, with some scientists predicting that it will be completed by 2001, five years ahead of schedule (Marshall, 1995).

The knowledge yielded by this project will have wide-ranging applications, in medicine, criminal justice, public health, education, psychotherapy, and other related fields. Already researchers have located the genetic defects that cause cystic fibrosis, Duchenne's muscular dystrophy, fragile-X syndrome, and many forms of cancer. Scientists believe that with every passing year, the genes associated with other hereditary disorders will be identified, including those for mental illnesses such as depression, dementia, and schizophrenia and for multifactorial disorders such as heart disease, hypertension, and stroke.

In addition, scientists are learning much more about those genes that allow people to live longer and healthier lives. One recent discovery, the gene for a protein called P53, was dubbed "molecule of the year" by *Science* magazine (Culotta & Koshland, 1993). P53 suppresses the growth of tumors and might soon lead to treatment—even prevention and cure—for some fifty-one types of cancer. Other "protective" genes that scientists may soon identify include

The actual work of mapping genes for the Human Genome Project is tedious and time-consuming, as well as expensive. However, the worldwide cooperation in this mapping effort has already paid off, not only in locating some particularly critical genes but also in developing techniques to speed the search. Assisted by the robot in the background, this scientist at Genethon in Paris is cloning thousands of copies of a particular gene colony as part of the process of building a "gene library."

genes that create the vitamin D receptors essential to preventing osteoporosis; genes that enhance the growth of neural axons, slowing the development of senility; genes that reduce a person's unhealthy appetite for fat; genes that foster the production of neurotransmitters that help a person cope with stress efficiently; and a rare gene that protects against the contraction of HIV (Depue et al., 1994; Ducy et al., 1996; Leibowitz & Kim, 1992; Liu et al., 1996; Morrison et al., 1994).

The excitement brought about by these discoveries comes with fear as well as hope, however. Our current worries about safeguarding privacy in a computer-connected world are nothing compared to the concerns that might be unleashed if access to each person's genetic code is not carefully restricted.

The dangers are most apparent with regard to medical insurance. In any nation that does not have universal medical coverage, insurance companies might require a drop of blood upon application for a policy, analyze its DNA messages, and then charge exorbitant rates, or deny coverage altogether, to anyone who appears to have a risk of a serious, chronic illness or who has a harmful recessive gene that might someday be passed on to a dependent child.

Another potential problem relates to employment. Instead of assessing job-related experience or skills, a

prospective employer might ask for a genetic reading to determine a candidate's suitability for a job. Educational opportunities might also depend on DNA "prerequisites." Even the most simple adult privileges, like getting a driver's license or renting an apartment, might require a DNA check and might be denied to those with genetic vulnerability to, say, alcoholism, even if the person in question has never shown any inclination to drink. Obviously, any attempt to discriminate against a particular person because of generalizations based on other people with similar genes should be made illegal. The best way to prevent abuse in this area is to protect such personal information with clear laws and guidelines for its use.

Another area with great potential for healing, and for harm, is "genetic engineering," the altering of a person's faulty genetic instructions through the insertion of a normal gene, thereby enabling ailing cells to be replaced by healthy ones (Anderson, 1995; Lyon & Gorner, 1995). In this technique, normal genes are inserted either indirectly via a blood transfusion or bone marrow transplant or directly into a particular cluster of cells. It is already in experimental use for hemophilia, cystic fibrosis, rheumatoid arthritis, several types of cancer, and dozens of rarer diseases, with some encouraging results. In the most celebrated case, a dying 4-year-old, Ashanti De Silva, received the very first human genetic transfusion, in 1990, to remedy a severe immune deficiency known as SCID. This disease, caused by a miscoded gene on chromosome 20 that fails to instruct for a particular enzyme, had left Ashanti defenseless against any form of infection, requiring her to live in near-isolation. Now, as a result of her treatments, Ashanti not only survives but thrives, experiencing all the scrapes and sniffles of a normal child with no adverse effects.

However, while Ashanti's case may be seen as a sign of hope by the millions of people with inherited and acquired diseases who are hoping for genetic cures, many experts caution against overoptimism: so far, the successes of genetic engineering have been rare and partial (Crystal, 1995). Even in the case of Ashanti, it is difficult to know how much of her recovery is due strictly to genetic therapy, since her treatment has included other effective medical measures as well.

Many researchers, anticipating the eventual success of genetic engineering, raise cautions of a different kind (Anderson, 1995). First, if genetic engineering becomes readily available, some may want to use it to "improve" individuals with nonlethal but "undesirable" traits that do not conform to current social standards. Foreseeing attempts to make a slow child brighter, or a hyperactive child quieter, or a short child taller, those most dedicated to child health emphasize that the goal is "to make sick babies healthy, not normal babies perfect" (March of Dimes, 1992). Many adults with disabilities endorse this goal, stressing that their lives are highly fulfilling, even though others without disabilities might not expect this to be the case.

Second, genetic therapy is extraordinarily expensive. Currently it is financed by private grants and government funding, including $200 million per year within the National Institutes of Health (Welsch & Smith, 1995). If this new area of research leads to successful treatment for, say, the 30,000 people with cystic fibrosis, experts are concerned that only the wealthy would be able to receive it.

The final worry about genetic tinkering goes even deeper. Geneticists stress that wide variation in inherited human characteristics, including traits that a particular parent might prefer not to see in his or her own child, helps keep the entire human community healthy and strong. Indeed, if genetic engineering should ever reach the point of being used to alter gametes, "misguided or malevolent attempts to alter the genetic composition of humans could cause problems for generations" (Anderson, 1995).

For example, our ancestors' genetic diversity in body composition and brain formation helped humans adapt to a wide range of climates, from the arctic to the rain forests, and to overcome social challenges, from the demands of early life on the African savanna to those of modern life in the overcrowded metropolis. Even genes for certain disorders have had a beneficial effect on the species, protecting against various diseases and conditions, including malaria (the gene for sickle-cell anemia) and cholera (the gene for cystic fibrosis) (Gabriel et al., 1994; Weiss, 1993). Shortsighted elimination of some genetic traits today might inadvertently deprive our descendants of characteristics critical for their adaptation and future survival or, even worse, create an unanticipated combination of traits that might trigger the downfall of the entire species.

The Human Genome Project has allotted millions of dollars to exploring the ethical dilemmas inherent in deciphering the genetic code. Scientists hope that—unlike the case of nuclear power, pesticides, and other potentially destructive scientific "miracles" that were blindly embraced initially—the perils as well as the promise of the new genetic technologies will be anticipated, understood, and controlled by citizens and scientists alike.

SUMMARY

The Beginning of Development

1. Conception occurs when a sperm penetrates an ovum, creating a single cell called a zygote. The zygote contains all the genetic material—half from each of the two gametes—needed to create a unique developing person.

The Genetic Code

2. Genes, which contain chemically coded instructions that cells need to specialize and perform specific functions in the body, are arranged on chromosomes. With the exception of gametes, every human cell contains twenty-three pairs of chromosomes, one member of each pair contributed by each parent. Every cell contains a duplicate of the genetic information in the first cell, the zygote.

3. Twenty-two pairs of chromosomes control the development of most of the body. The twenty-third pair determines the individual's sex: zygotes with an *XY* combination will become males; those with an *XX* combination will become females.

4. Genes provide genetic continuity across the human species, ensuring that we all share common physical structures, behavioral tendencies, and reproductive potential. Genes also ensure the genetic diversity that allows our species to continue to evolve through adaptation and natural selection.

5. Each person has a unique combination of genes, with one important exception. Sometimes a zygote separates completely into two or more genetically identical organisms, creating monozygotic (identical) twins, triplets, and so on.

From Genotype to Phenotype

6. The sum total of all the genes a person inherits is the person's genotype. The ways those genes are expressed as physical or nonphysical traits is the person's phenotype. Most human characteristics are polygenic and multifactorial, the result of the interaction of many genetic and environmental influences.

7. The various genes in the genotype interact in many ways to influence the phenotype. Most often genes from both parents contribute to a trait in an additive fashion, but sometimes genes act in a nonadditive pattern. The dominant-recessive pattern is one such case; the phenotype reflects the influence of the dominant gene for some trait while the recessive gene's effects are obscured.

8. Some genes are located only on the X chromosome. Traits controlled by such genes are passed from mother to son but not from father to son because males inherit their only X chromosome from their mother. Females inherit two X chromosomes, one from each parent. For this reason, males are more likely to have recessive traits, such as color blindness, that are X-linked.

9. Genes affect almost every human trait, including intellectual abilities, personality patterns, and mental illness. At the same time, the environment—from the moment of conception throughout life—constantly influences genetic tendencies. Gene-environment interaction is thus ongoing and complex.

Genetic and Chromosomal Abnormalities

10. Chromosomal abnormalities occur when the zygote has too few or too many chromosomes, or when a chromosome has a missing, a nonfunctioning, or an extra piece of genetic material. While most embryos with chromosomal abnormalities are spontaneously aborted early in pregnancy, many of the babies who survive with such defects have extra or missing material on their sex chromosomes.

11. The most common chromosomal abnormality that does not involve the sex chromosomes occurs when an extra chromosome is attached to the twenty-first pair. This causes trisomy-21, or Down syndrome, a varying cluster of problems in physical and intellectual functioning. One of the most common abnormalities associated with the sex chromosomes is fragile-X syndrome, which is often accompanied by some mental deficiency, particularly in males.

12. Every individual carries some genes for genetic handicaps and diseases. However, most carriers of specific disorders do not marry other carriers of it. In addition, many genetically transmitted diseases are polygenic, or multifactorial, so most babies will not inherit a serious genetic defect.

Genetic Counseling

13. Genetic testing and an evaluation of family background can help predict whether a couple will have a child with a genetic problem. If there is a high probability that they will, they can consider several options, such as adoption, remaining childless, or obtaining a prenatal diagnosis and, if the diagnosis confirms a serious problem, considering abortion. In some cases, appropriate postnatal treatment may remedy or alleviate the problem.

14. The Human Genome Project, a massive international effort, expects to map all of the genes in the human body by the year 2001. Though such information raises hopes for cures of genetically transmitted diseases, it also raises ethical questions about the proper use of such information.

KEY TERMS

gamete (65)
sperm (65)
ovum (65)
zygote (65)
gene (66)
chromosome (66)
DNA (deoxyribonucleic acid) (66)
genetic code (66)
twenty-third pair (68)
monozygotic twins (69)
dizygotic twins (70)
polygenic traits (70)
multifactorial traits (70)
genotype (71)
phenotype (71)
carrier (71)
additive pattern (71)
nonadditive pattern (72)
dominant-recessive pattern (72)
***X*-linked genes** (73)
environment (74)
heritability (80)
syndrome (82)
trisomy-21 (Down syndrome) (82)
fragile-*X* syndrome (84)
genetic counseling (88)
markers (89)
Human Genome Project (90)

KEY QUESTIONS

1. How do genes influence one's physical characteristics and behavior?

2. How is each person's genetic uniqueness ensured, and why is this important?

3. What are the differences between additive and nonadditive patterns and dominant and recessive genes?

4. What research strategies are used to determine genetic and environmental influences on psychological characteristics?

5. How does the interaction between heredity and environment occur for physical traits, such as height, and psychological traits, such as shyness?

6. Why is it that some people who have a genetic predisposition for schizophrenia or alcoholism never develop these conditions?

7. What are some of the effects of chromosomal defects?

8. What are some of the factors that determine whether a couple is at risk for having a child with genetic abnormalities?

9. How can genetic counseling help those parents who are at risk for having a child with genetic problems?

10. What is the Human Genome Project, and what are the ethical questions associated with the use of its findings?

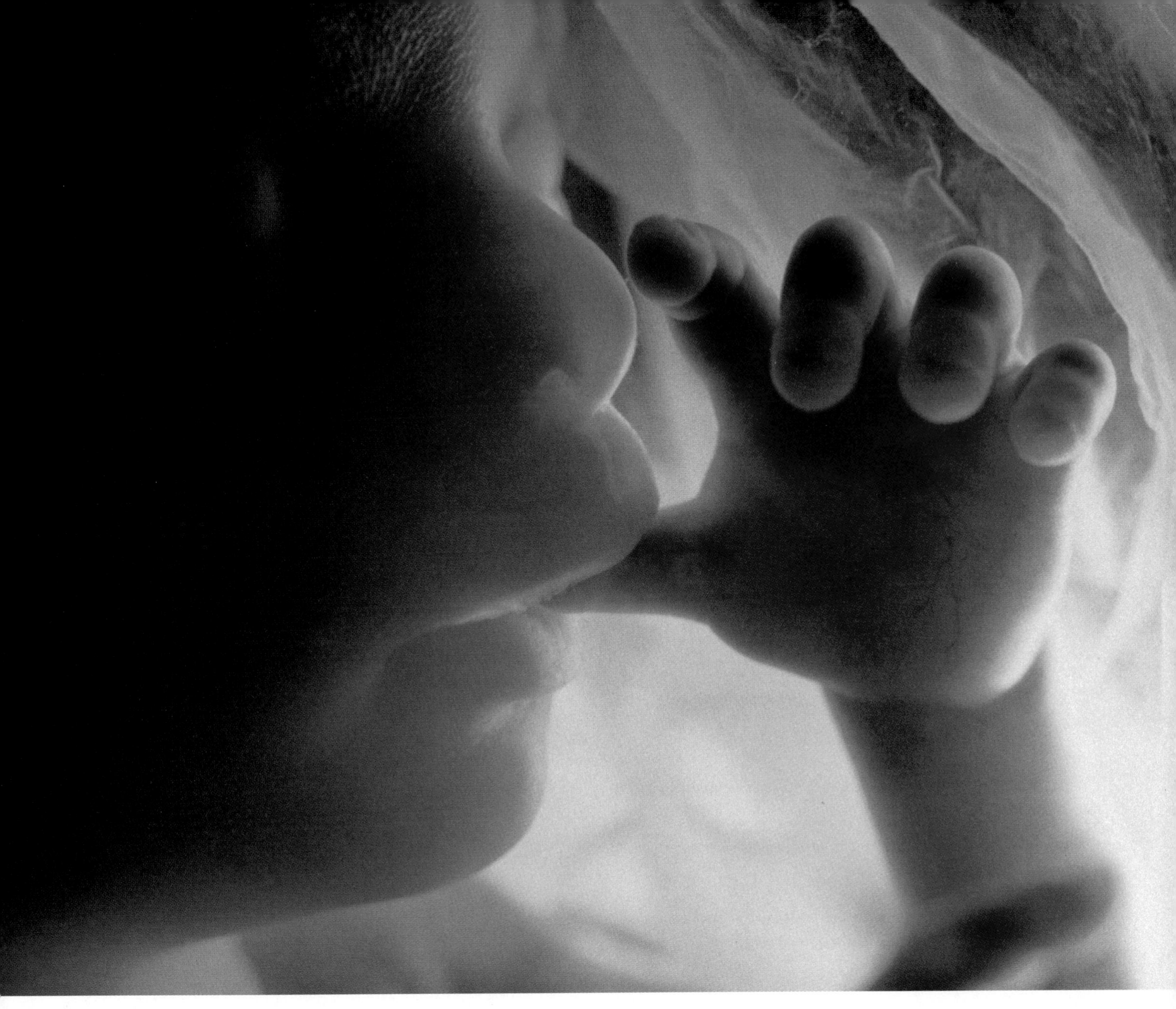

Prenatal Development and Birth

CHAPTER

4

As you learned in Chapter 1, a contextual perspective encourages us to look at any moment of development as a complex interaction among developmental domains within a multifaceted social context. This is no less true for the first 9 months of life than for the 900 or so that follow.

Obviously, the primary focus of a chapter on prenatal development and birth is the astounding physical growth that transforms a single-celled zygote into a fully formed human baby. However, the specifics of this growth, and their impact on later development, are deeply influenced by the various contexts in which they occur. For instance, the mother-to-be's health habits and activities, the community's laws and practices affecting prenatal exposure to toxins and diseases, and the culture's customs regarding birth itself are just some of the many contextual factors that make some newborns much better prepared for a long and healthy life than others. These various factors will be discussed in detail in this chapter, as we examine what is arguably the most important developmental period, and the most significant day, of a human life.

From Zygote to Newborn

The process of human growth from a single-celled zygote into a fully developed baby is generally discussed in terms of three main periods. The first two weeks of development are called the **germinal period**; from the third week through the eighth week is the **period of the embryo**; and from the ninth week until birth is the **period of the fetus**.*

germinal period The first two weeks of development after conception, characterized by rapid cell division and the beginning of cell differentiation.

period of the embryo From approximately the third through the eighth week after conception, when the rudimentary forms of all anatomical structures develop.

period of the fetus From the ninth week after conception until birth, when the organs grow in size and complexity.

The Germinal Period: The First Fourteen Days

Within hours after conception, the one-celled zygote, traveling slowly down the Fallopian tube toward the uterus, begins the process of cell division and growth, first dividing into two cells, which soon become four, then eight, then sixteen, and so on (see photos on the next page). At least through the fourth doubling, each of these cells is identical, and any one of them could

*Technically speaking, the name of the developing human organism changes several times depending on the precise stage of development. While there is no need for you to know all the terms, the curious might be interested to know that the organism that begins as a zygote becomes a morula, a blastocyst, a gastrula, a neurula, an embryo, and a fetus before it finally becomes an infant (Moore, 1988).

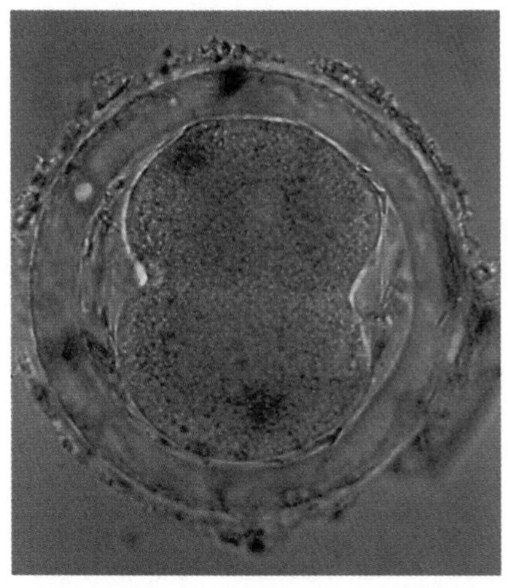

(a)

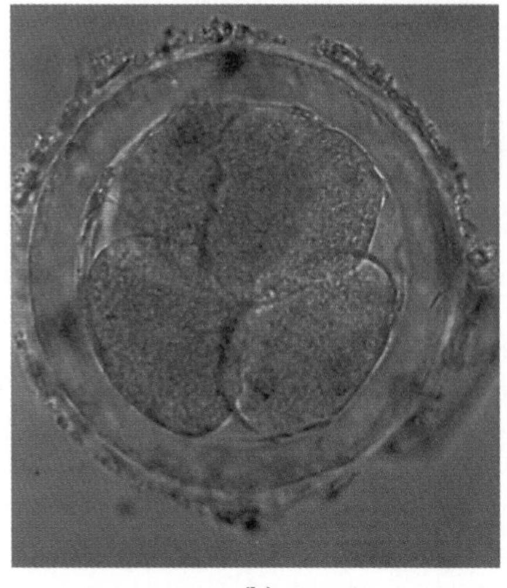

(b)

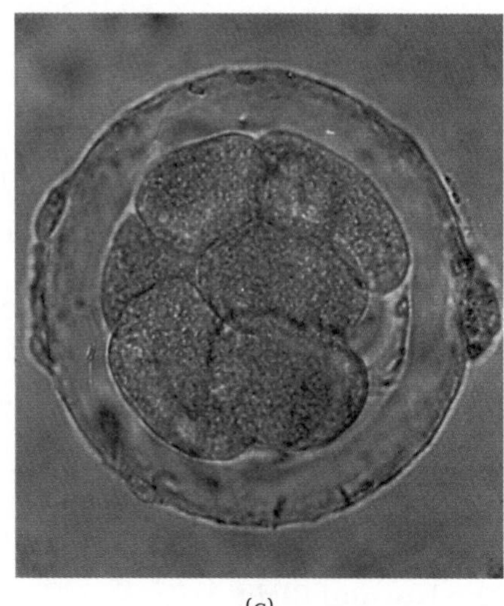

(c)

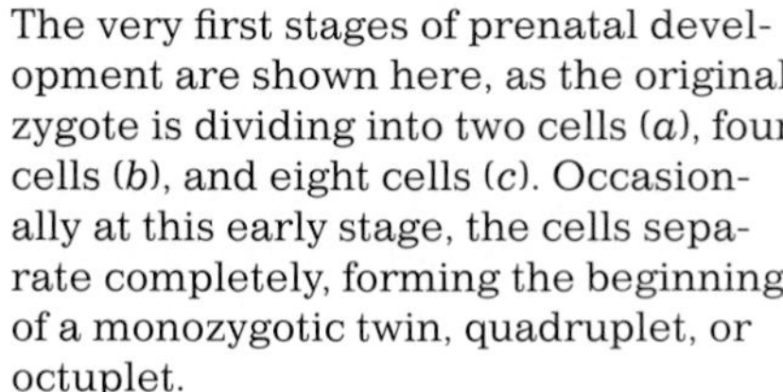

The very first stages of prenatal development are shown here, as the original zygote is dividing into two cells (*a*), four cells (*b*), and eight cells (*c*). Occasionally at this early stage, the cells separate completely, forming the beginning of a monozygotic twin, quadruplet, or octuplet.

become a complete human being. In fact, as explained in Chapter 3, nature sometimes splits the cluster of cells into two or even four distinct segments, and then each segment becomes a monozygotic twin or quadruplet.

Soon, however, the process of **differentiation** occurs, causing clusters of cells to begin to take on distinct traits and gravitate toward particular locations that foreshadow the types of cells they will become. The first clear sign of differentiation occurs about a week after conception, when the multiplying cells (now numbering more than a hundred) separate into two distinct masses, the outer cells forming a protective circle that will become the placenta, and the inner cells forming a nucleus that will become the embryo.

The first task of the outer cells is to achieve **implantation**, that is, to embed themselves in the nurturant environment of the uterus. This is ac-

differentiation The developmental process by which a relatively unspecified cell or tissue undergoes a progressive change to a more specialized cell or tissue.

implantation Beginning about a week after conception, the burrowing of the organism into the lining of the uterus, where it can be nourished and protected during growth.

TABLE 4.1

The Vulnerability of Prenatal Development

The Germinal Period

From the moment of conception until fourteen days later, 58 percent of all developing organisms fail to grow or implant properly, and thus do not survive the germinal period. Most of these organisms were grossly abnormal.

The Period of the Embryo

From fourteen days until fifty-six days after conception, during which time all the major external and internal body structures begin to form, about 20 percent of all embryos are aborted spontaneously, most often because of chromosomal abnormality.

The Period of the Fetus

From the eighth week after conception on, about 5 percent of all fetuses are aborted spontaneously before viability at twenty-two weeks, or are stillborn after 22 weeks.

Birth

Only 31 percent of all conceptions survive prenatal development to become living newborn babies.

Sources: Carlson, 1994; Gilbert et al., 1987.

neural tube The fold of cells that appears in the embryonic disk about three weeks after conception, later developing into the central nervous system.

cephalo-caudal development The sequence of body growth and maturation from head to foot. Human growth, from the embryonic period throughout early childhood, follows this pattern.

proximo-distal development The sequence of body growth and maturation from the spine toward the extremities. Human growth, from the embryonic period through early childhood, follows this pattern.

complished as the cells nestle into the uterine lining, rupturing tiny blood vessels in order to obtain nourishment and to build a connective web of membranes and blood vessels that links the mother and the developing organism, allowing the growth of the next nine months or so. Implantation is far from automatic, however: an estimated 58 percent of all naturally occurring conceptions fail to become properly implanted (Gilbert et al., 1987) (see Table 4.1), thereby ending the new life even before the embryo begins to form or the woman suspects she is pregnant.

The Period of the Embryo: The Third Through the Eighth Week

The start of the third week after conception initiates the *period of the embryo*. At this point the developing organism begins differentiating into three layers, each of which will eventually form key body systems. Soon the first perceptible sign of body formation appears, a fold in the outer layer of cells. This fold will become the **neural tube**, which will later develop into the central nervous system, including the brain and spinal column.

During the embryonic period, growth proceeds in two directions: from the head downward, called **cephalo-caudal development** (literally, "from head to tail"), and from the center (that is, the spine) outward, called **proximo-distal development** (literally, "from near to far"). Thus the most vital organs and body parts form first, before the extremities.

Following this pattern, in the fourth week after conception the head starts to take shape, and a primitive pulsing heart and blood vessels form, making the cardiovascular system the first organ system to show any sign of activity. The head, at first, is simply a featureless protrusion, but within days, eyes, ears, nose, and mouth start to form. By the fifth week, parts of the body more distant from the head and heart develop: buds that will become arms and legs appear, and a tail-like appendage extends from the spine. The embryo is now about ⅕ of an inch long (6 millimeters), about 7,000 times the size of the zygote it was a month before.

At four weeks past conception (*a*), the embryo is only about ⅛ inch long (5 millimeters), but already the head (top right) has taken shape. At five weeks past conception (*b*), the embryo has grown to twice the size it was at four weeks. Its heart, which has been beating for a week now, is visible, as is what appears to be a primitive tail, which will soon be enclosed by skin and protective tissue at the tip of the backbone (the coccyx). By seven weeks (*c*), the organism is about an inch long (2 centimeters). Eyes, nose, the digestive system, and even the first stage of toe formation can be seen. At eight weeks (*d*), the 1½-inch-long (4-centimeter) organism is clearly recognizable as a human fetus.

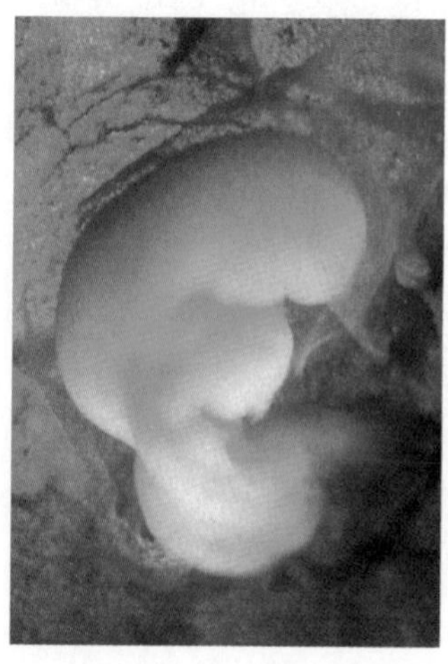
(a)

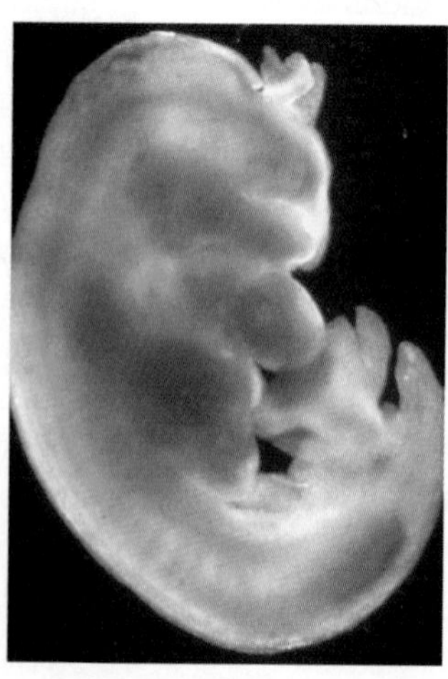
(b)

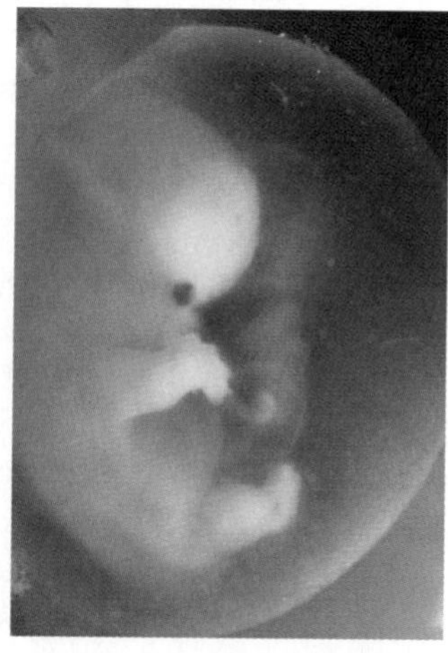
(c)

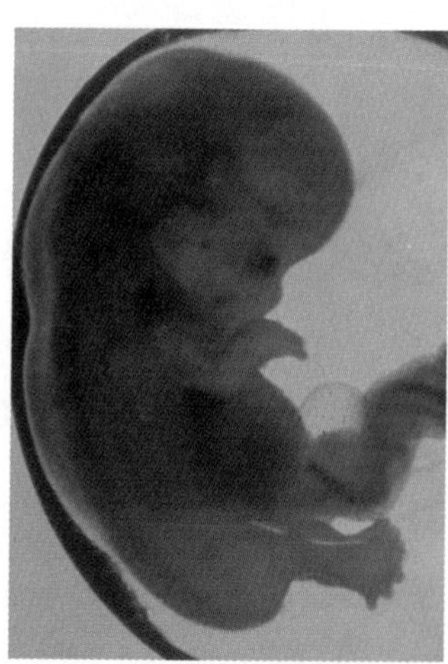
(d)

Continuing in the proximo-distal sequence, the upper arms, then the forearms, the palms, and then webbed fingers appear about five weeks after conception. Legs, feet, and webbed toes, in that order, emerge a few days later, each having the beginning of a skeletal structure (Carlson, 1994).

By eight weeks after conception, the embryo weighs about 1/30 of an ounce (1 gram) and is about 1 inch (2.5 centimeters) long. The head has become more rounded, and the features of the face are fully formed. The embryo has all the basic organs and body parts (except sex organs) of a human being, including elbows and knees and even buds for baby teeth. The fingers and toes have become distinct and separate (at fifty-two and fifty-four days after conception, respectively), and the "tail" is no longer visible, having become incorporated into the lower spine at about fifty-five days after conception. The organism is now ready for another name, the *fetus*, which it is called until the moment of birth.

The Period of the Fetus: The Ninth Week Until Birth

During the third month, the sex organs take discernible shape. The first stage of their development actually occurs at the sixth week, with the appearance of the indifferent gonad, a cluster of cells that can develop into male or female sex organs. If the fetus is male (*XY*), a gene on the *Y* chromosome sends a biochemical signal, sometime around the ninth week, that initiates the development of male sexual organs. If the embryo is female (XX), no such signal is sent, and the indifferent gonad soon begins to develop female sex organs, first the vagina and uterus and then the external structures (Koopman et al., 1991).

By the twelfth week after conception, the external genital organs are fully formed. The fetus now has all its body parts, weighs approximately 3 ounces (87 grams), and is about 3 inches (7.5 centimeters) long. In the next three months, hair, including eyebrows and eyelashes, begins to grow, and fingernails, toenails, and buds for adult teeth form. The heart beat becomes much stronger, and the digestive and excretory systems develop more fully.

The most appreciable development of this middle period involves the brain, which increases about six times in size and, as a result of ongoing neurological maturation, begins to react to stimuli (Carlson, 1994). This process of neurological maturation, which is essential to the regulation of such basic body functions as breathing, sucking, and sleeping, may be the critical factor in the fetus's attaining the **age of viability**, defined as the age at which a preterm fetus has at least some slight chance of survival outside the uterus if expert care is available. The age of viability starts at about twenty-four weeks after conception. Babies born before then rarely survive more than a few hours, because even the most sophisticated respirators and heart regulators cannot maintain life in a fetus whose brain has not yet begun to function. Those few who do survive are virtually always severely brain-damaged (Allen et al., 1993).

age of viability The age (about twenty-four weeks after conception) at which a fetus can possibly survive outside the mother's uterus if specialized medical care is available.

Brain maturation takes a "striking" leap forward at about twenty-eight weeks after conception (Carlson, 1994), when brain-wave patterns shift from a flat pattern to occasional bursts of activity similar to the sleep-wake cycles of a newborn. Largely for this reason, the odds of survival are much better for a fetus who is at least 28 weeks old.

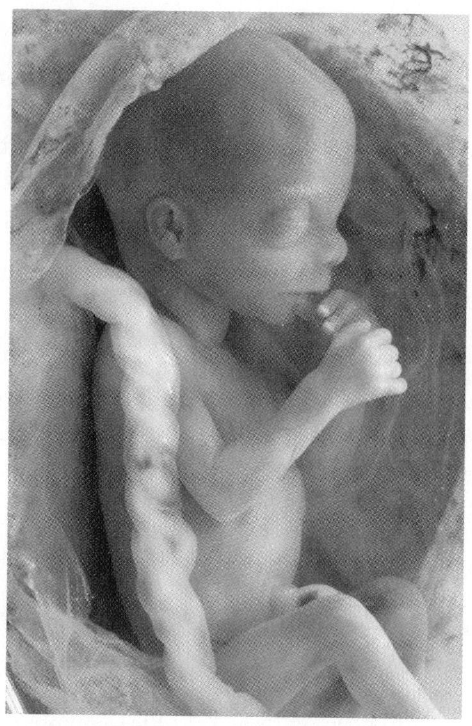

At the end of four months, this fetus, now 6 inches long, looks fully formed, down to the details of eyebrows and fingernails. However, brain development is not yet sufficient to sustain life outside the uterus. For many more weeks, the fetus must depend on the translucent membranes of the placenta and umbilicus (the white cord in the foreground) for survival.

Weight is also crucial to viability: a 24-week-old fetus typically weighs at least 600 grams (about 22 ounces) and, with excellent medical care, has about a 50 percent chance of survival (Allen et al., 1993; Reuss & Gordon, 1995). By twenty-eight weeks, the typical fetus weighs about 1,300 grams (about 3 pounds), and its chances of survival have increased to more than 90 percent.

Of course, attaining the age of viability simply means that life outside the womb is *possible*. Each day of the final three months or so of prenatal growth improves the odds, not only of survival, but of a healthy, happy first few months, for baby and parents as well. A viable infant born at the beginning of this final period is a tiny creature requiring intensive hospital care, dependent on life-support systems for nourishment and for every breath. By contrast, the typical infant born at full term, 266 days after conception, is a vigorous person, ready to thrive at home on mother's milk—with no expert help, oxygenated air, special food, or technical assistance required.

Underlying this distinction between the fragile preemie and the robust newborn is the crucial maturation of the respiratory and cardiovascular systems. In the last months of prenatal life, the lungs begin to expand and contract, exercising the muscles involved in breathing by using the amniotic fluid surrounding the fetus as a substitute for air. At the same time, the valves of the heart go through a final maturation that, at birth, will enable the newborn's circulatory system to function independently. In addition, the fetus usually gains more than 2,000 grams of critical weight in the last three months, increasing, on average, to 7½ pounds (3,400 grams) at birth. An important part of this weight gain is fat, which will provide a protective layer of insulation when the developing person is no longer surrounded by the mother's body warmth. The weight gain of the last weeks also stores nour-

A CLOSER LOOK

Interaction with the Surrounding World

As the foregoing discussion suggests, the fetus is far from passive during development, and its interaction with its mother and the outside world begins well before birth. Biologically, of course, fetal growth is closely linked to the mother-to-be's nutrition and physiology. This means not only that the fetus is dependent on the specifics of the mother-to-be's lifestyle but also that the expectant mother is affected—with nausea, increased urination and digestive upsets, and perhaps gestational diabetes and high blood pressure as well—by the specifics of the fetus's development, with each pregnancy being experienced somewhat differently (Haig, 1995).

Beginning at about nine weeks, the fetus moves its body in response to shifts in the mother's body position. At first, the movement of tiny heels, fists, elbows, and buttocks are imperceptible; then they are felt as faint flutters. Soon, movements are not only easier for the mother to notice but also more predictable, occurring when the mother sits down or especially when she changes position as she sleeps. Sometimes, as the due date approaches, a sudden kick or somersault can occur with such vigor that the mother's delight at feeling life may turn to dismay at a sore rib or interrupted sleep. Such momentary discomfort aside, perception of the fetus's movements usually evokes feelings of wonder. Indeed, many parents-to-be, fathers as well as mothers, delight in rubbing the woman's rippling belly, beginning what may become a lifelong pattern of communication by touch.

Toward the end of prenatal development, the fetal sensory systems begin to function, and again interaction between fetus and mother-to-be is apparent. For example, how much amniotic fluid the fetus swallows depends partly on the taste of that fluid: fetuses swallow sweetened fluid more rapidly than noxious fluid and thus their lungs, digestion, and nutrition are intimately related to the particulars of their mother's diet (Carlson, 1994). At about the twenty-seventh week, the fetus's eyelids open, and the fetus can perceive the faint reddish glow of sunlight or other bright illumination that diffuses through the mother's belly (Kitzinger, 1989).

The most remarkable fetal response to the immediate environment involves hearing. Most mothers-to-be are well aware that their fetus can hear, having felt the developing person quiet down when they sing a lullaby or startle with a kick when a car backfires or a door slams. Further, newborns retain associations with particular sounds they heard prenatally: they typically stop crying, for example, when they are held with an ear close to the mother's heart, presumably because they are comforted by the familiar rhythm they have known for months.

Researchers have confirmed not only that fetuses can hear but that, as newborns, they can remember some of what they heard in the womb. In a series of experiments, pregnant women read a particular children's book aloud every day during the last weeks before birth. Three days after birth, their infants were exposed to tape recordings of the story being read by their own mother and by another baby's mother. Laboratory monitoring of their behavioral reactions indicated that newborns paid greater attention to the recording by their own mother. What's more, they showed greater responsiveness to the recording of their mother reading the familiar story than to a recording of her reading an unfamiliar text (DeCasper & Spence, 1986; Moon & Fifer, 1990). Such results suggest that, at least in some ways, fetuses not only prepare their reflexes and organ systems for physiological functioning at birth; they also begin to accustom themselves to the social world that they soon will join.

ishment and vitamins that will be used in the early days after birth, before the mother's breast milk is fully established.

Preventing Complications

The remarkable nine months that transform a single-celled zygote into a viable human newborn are a time not only of miraculous growth but also of considerable vulnerability, in which normal development can be compromised by many complications and hazards. In this section, we will examine a number of potential prenatal problems, but, as we proceed, keep several facts in mind. First, remember that despite the complexity of prenatal development and the many dangers to the developing organism, the large ma-

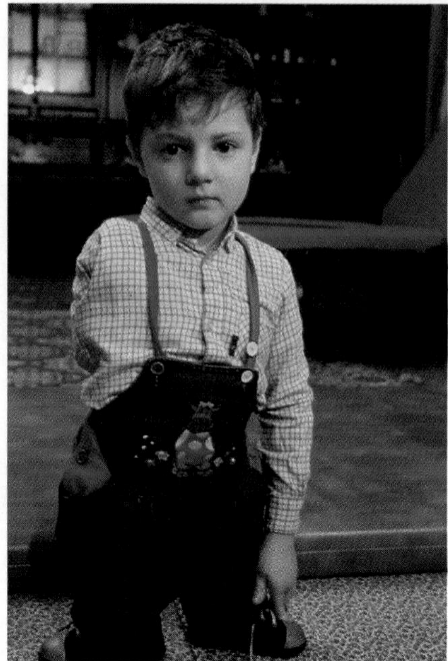

The ability of the environment to disrupt normal prenatal development is made tragically clear by this young boy, one of the many young victims of the 1986 explosion at the Chernobyl nuclear power plant in the Ukraine. As a result of the radioactive contamination spread by the accident, thousands of children were born with major health problems and deformities, ranging from respiratory ailments and malformations of the brain to missing or misshapen limbs.

teratology The scientific study of birth defects caused by genetic or prenatal problems, or by birth complications.

teratogens External agents, such as viruses, drugs, chemicals, and radiation, that can impair prenatal development and lead to abnormalities, disabilities, or even death.

behavioral teratogens Teratogens that tend to damage neural networks in the prenatal brain, affecting the future child's intellectual and emotional functioning.

risk analysis In teratology, the attempt to evaluate all the factors that can increase or decrease the likelihood that a particular teratogen will cause harm.

critical period The period of prenatal development during which a particular organ or body part is most susceptible to teratogenic damage. In many cases the critical period occurs in the first eight weeks of development, when the basic organs and body structures are forming.

jority of babies are born healthy and capable. Further, bear in mind that many potential prenatal hazards can be avoided, or their effects ameliorated, through care taken by an expectant mother and her family and friends, as well as care provided through the community's health services. Despite its complexities, then, prenatal development should be thought of not as a hazardous event but, rather, as a natural process that can be safeguarded and facilitated to ensure good health for the newborn.

Indeed, current successes in preventing or treating prenatal complications make scientists want to improve every newborn's chances for a normal and healthy start in life. This goal is the goal of **teratology,*** which is the study of the factors that can contribute to birth defects.

Scientists now understand a great deal about **teratogens**, the broad range of substances (such as drugs and pollutants) and conditions (such as severe malnutrition and extreme stress) that increase the risk of prenatal abnormalities. These abnormalities include obvious physical problems, such as missing limbs, a cleft palate, or impaired vision, and more subtle impairments, such as brain damage that is not apparent until a particular child begins to show unusual irritability, hyperactivity, or learning difficulties. Some teratogens primarily damage either the body or the brain, and others damage both. Teratogens that harm the brain and therefore impair the child's actions and intellect are called **behavioral teratogens**.

Determining Risk

Although all teratogens increase the risk of prenatal damage, none *always* causes damage. The ultimate effects of any given teratogen depend on the complex interplay of many factors, both destructive and protective.

The science of teratology is thus a science of **risk analysis**, an attempt to weigh all the factors that can increase or decrease the likelihood that a particular teratogen will cause harm. Depending on the specifics, exposure to a teratogen may be low-risk for one fetus and probably do no damage at all, while exposure to the same teratogen might be high-risk for another fetus, almost certainly causing some harm.

Timing

One crucial factor that determines whether a given teratogen will cause harm, and in what way, is the timing of the developing organism's exposure to it. Some teratogens, for example, cause damage only if exposure to them occurs during a specific time span early in pregnancy, when a particular part of the body is undergoing formation. The time span of greatest susceptibility is called the **critical period**. As can be seen in Figure 4.1 on the next page, each body structure has its own critical period: the eyes at about four weeks after conception, the ears and arms at about six weeks, the legs and palate at about seven weeks, and so on. It is essential to note, however, that there is no single critical period with respect to the damage that can be caused by behavioral teratogens: the brain and nervous system can be harmed throughout prenatal development.

*"Teratology" stems from the Greek word *tera*, meaning "monster." Coined over a century and a half ago, the term was born out of ignorance of the true nature of birth defects and is no longer read literally.

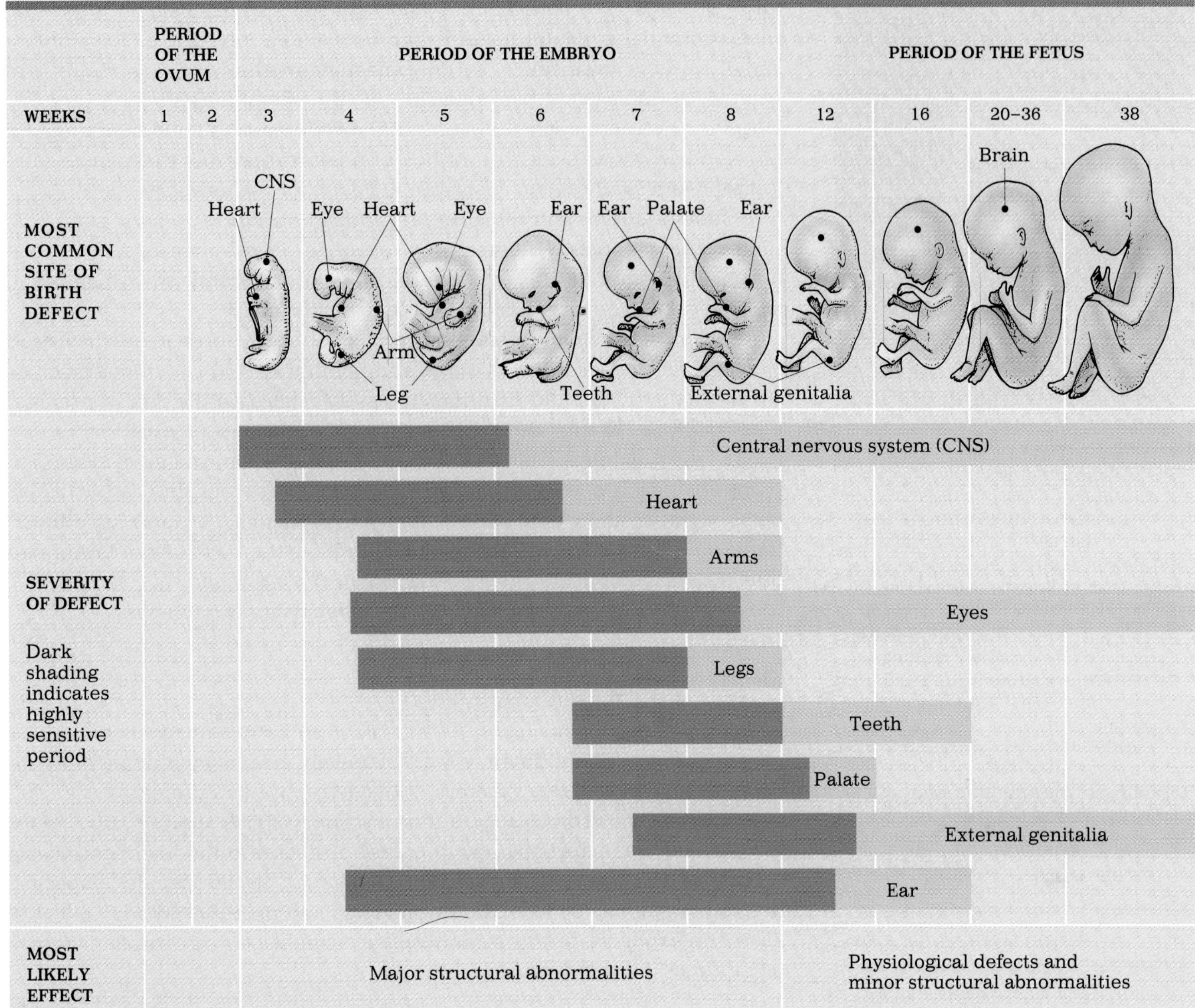

FIGURE 4.1 As this chart shows, the most serious damage from teratogens is likely to occur in the first eight weeks after conception. However, damage to many vital parts of the body, including the brain, eyes, and genitals, can occur during the last months of pregnancy as well.

With those conditions (such as severe malnutrition) or substances (such as heroin) that disrupt and destabilize the overall functioning of the woman's body and therefore alter the normal course of pregnancy, there are two especially critical periods. The first is at the very beginning of pregnancy, when impairment of the woman's body functioning can impede implantation. The second critical period occurs in the final weeks of prenatal development, when the fetus most needs to gain weight and when the risk of placental malfunction and preterm birth are the greatest.

Amount

threshold effect The harmful effect of a substance which is presumably safe until exposure to it reaches a certain level.

A second important factor affecting a specific teratogen's potential for causing damage is the dose and/or frequency of the teratogenic exposure. Indeed, some teratogens have a **threshold effect**; that is, the substance is virtually harmless until exposure to it reaches a certain level, at which point

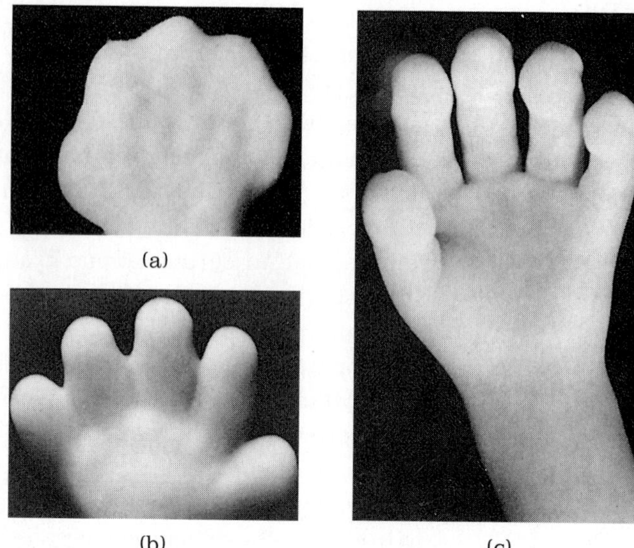

The impact of a potential teratogen partially depends on when the developing organism is exposed to it. This is because there is a critical period in the formation of every body part during which it is especially vulnerable. Shown here is the critical period of hand development: Three stages in finger development: (*a*) notches appear in the hand at day 44; (*b*) fingers are growing but webbed together at day 50; and (*c*) fingers have separated and lengthened at day 52. By day 56, fingers are completely formed, and the critical period for hand development is over. Other parts of the body, including the eyes, heart, and central nervous system, take much longer to complete development, so the critical period when they are vulnerable to teratogens lasts for months rather than days.

it becomes damaging. In fact, a few substances that are actually beneficial in small amounts (such as vitamin A) are teratogenic in large quantities (Sonderegger, 1992).

For most teratogens, experts are reluctant to set the threshold below which the substance is presumably safe, partly because many teratogens have an **interaction effect**; that is, one teratogen intensifies the effects of another. Both marijuana and alcohol, for example, may be threshold drugs, teratogenic at a certain level or over a sustained period of use but not teratogenic in smaller, occasional doses (Waterson & Murray-Lyon, 1990). However, when taken together, their respective thresholds drop, making them harmful in combination at an amount that, for each drug taken separately, would be inconsequential. Similar interaction effects are likely to occur with all other drugs that affect the central nervous system, from nicotine to cocaine (Robins & Mills, 1993).

Genetic Vulnerability

A third factor that determines whether a specific teratogen will be harmful, and to what extent, is the developing organism's genetic vulnerability to the teratogen. When fraternal twins are heavily exposed to alcohol in the mother-to-be's bloodstream, for example, their exposure is equal, yet one will usually be more severely affected teratogenically than the other (Sokol & Abel, 1992). The likely explanation is that dizygotic twins differ in their genetic susceptibility. Researchers are currently working on the theory that this susceptibility involves a genetically related defect in a specific enzyme (alcohol dehydrogenase) that is crucial to the breakdown of alcohol. Similar genetic susceptibilities, in the form of defective enzymes, are suspected in a number of other birth disorders, including cleft palate (when the mother-to-be smokes) and certain growth deformities (when the mother-to-be takes Accutane or an anticonvulsant such as Dilantin). Indeed, one expert believes that "ultimately every teratogen is going to involve a genetic susceptibility" (Holmes, quoted in Kolata, 1995).

interaction effect The intensification of a teratogen's potential for causing harm as a result of its interacting with another teratogen.

Recently, genes have been implicated in the teratogenic effects that sometimes result from a deficiency of folic acid in the mother-to-be's diet. Researchers have known for several years that such a deficiency is often associated with neural-tube defects, either spina bifida, in which the spine does not close properly, or anencephaly, in which part of the brain does not form (see p. 87). They have also known that there sometimes is a genetic component in these defects, since they occur more commonly in certain families and in certain ethnic groups (Irish and English). Recent research suggests that this genetic component may be a defective enzyme that prevents the normal utilization of folic acid, which is essential to the complete development of the neural tube (Mills et al., 1995). Thus, while severe folic-acid deficiency can probably cause a neural-tube defect in any embryo, those pregnancies burdened by the genetic defect in utilizing folic acid are particularly vulnerable.

In some cases, genetic vulnerability to teratogenic hazards is related to the sex of the developing organism (Sonderegger, 1992). Perhaps because the *Y* chromosome carries far fewer genes than the *X* chromosome, leaving males with many genes on their *X* chromosomes that are not paired with a corresponding gene on their other sex chromosome, male embryos, fetuses, and newborns generally are more vulnerable to teratogenic damage. This is evident in the higher rate of males with teratogenic birth defects and the later behavioral problems associated with them, as well as in the higher rate of spontaneous abortions and stillbirths of male embryos and fetuses.

Protective Steps

The most obvious protective step a prospective mother can take is to avoid exposure to teratogens in the days before becoming pregnant and in the months thereafter. If there are teratogens she cannot avoid, she should limit her exposure as much as possible—leaving the room when others smoke cigarettes, keeping her distance from people who might have contagious diseases, and so on.

In addition, two other protective steps seem particularly crucial—maintaining good nutrition and getting prenatal care. Diet before pregnancy as well as during pregnancy is important to defend against specific

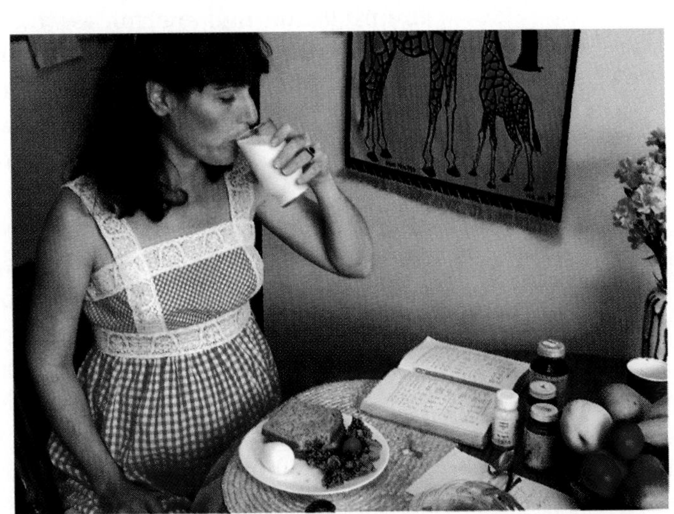

A balanced diet is important throughout pregnancy, but especially toward the end, when among other special requirements, calcium is needed for bones and teeth. Taking calcium pills is usually not recommended: drinking milk or eating other calcium-rich foods—yogurt, cheese, and sardines, for instance—is much better. Other nutritional supplements must be taken cautiously: folic-acid supplements, for example, are definitely recommended; vitamin A supplements are definitely not.

defects and to diminish overall vulnerability (Institute of Medicine, 1990). Ideally, then, a woman should begin pregnancy well-nourished and maintain a well-balanced diet throughout pregnancy. Not only folic acid but many other specific vitamins and minerals, including iron, zinc, calcium, and vitamin A, have been proved to be essential for the normal development of the fetus, and nutritionists are convinced that many other substances, not yet known but present in a varied diet, will someday be proved to be essential.

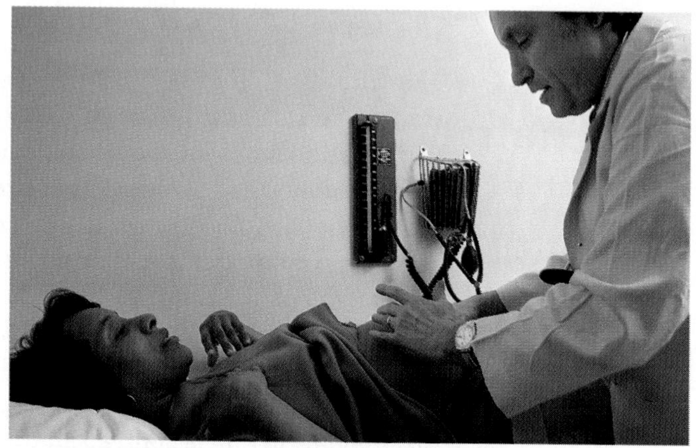

Obtaining good and early prenatal care is an important step in minimizing the risk of complications during prenatal development and birth. As important as the medical monitoring is open communication between the mother-to-be and her physician, especially regarding lifestyle or health habits that might pose a threat to normal prenatal growth.

Early, and competent, preconceptual and prenatal care is also vital. Birth defects of virtually every kind are more common in women who do not visit the doctor until after the third month of pregnancy and occur even more frequently in women who receive no prenatal care at all. One reason for this is that the standard of good prenatal care now includes not only testing and medical treatment but also informing the mother-to-be about lifestyle issues, particularly with regard to dietary needs and the dangers of drug use, and helping her to change whatever bad habits she might have.

Some of the correlation between early medical attention and healthy births is, of course, related to nonmedical factors. That is, women who seek and receive good, early, prenatal care tend to be well educated and financially stable, characteristics that in themselves make complications during pregnancy less likely. Nevertheless, prenatal care, in and of itself, can be key. In some cases, such as early detection of multiple pregnancies or treatment of certain diseases, medical care saves the life and protects the wholeness of the fetus. One clear-cut example is the treatment of maternal syphilis, which can be cured early in pregnancy with no harm to the embryo but will cause blindness, bone damage, and even death if it continues into the second half of pregnancy. More generally, in a study that controlled for the nonmedical factors mentioned above, early prenatal care significantly reduced the rate of low-birthweight infants (Mustard & Roos, 1994). And while no studies of the effects of prenatal care on prenatal drug abuse have been able to successfully control for these confounding factors, one study of births in a drug-infested inner-city community is dramatically suggestive: of those women with no prenatal care at all, 32 percent had used cocaine shortly before birth, compared with 10.5 percent of those who had had inadequate care and only 2.5 percent of those who had had adequate care (McCalla et al., 1995).

Specific Teratogens and Preventive Measures

Because of the many variables involved, risk analysis cannot precisely predict the results of teratogenic exposure in individual cases. However, decades of research have revealed the possible effects of some of the most common, damaging teratogens. More important, much has been learned about the ways individuals, and society, can reduce the risks posed by these teratogens.

Diseases

Many diseases, including a wide array of viruses and virtually all sexually transmitted diseases (Cates, 1995), can harm a fetus, sometimes severely. Here we will focus on two, rubella and HIV, that clearly illustrate the potential for public health measures to prevent birth defects.

Rubella (sometimes called German measles) was one of the first teratogens to be recognized. Long considered a harmless childhood disease, it is now well established that rubella, if contracted by the expectant mother early in pregnancy, is very likely to cause birth handicaps, among them blindness, deafness, heart abnormalities, and brain damage. (Some of these problems and their effects were apparent in David's story in Chapter 1). However, if rubella is contracted during the last three months of pregnancy, typically no discernible damage occurs (Enkin et al., 1989). Thus timing is critically important with respect to the teratogenic potential of this virus.

The seriousness of this teratogen became all too evident in a worldwide rubella epidemic in the mid-1960s. In the United States alone, 20,000 infants had obvious rubella-caused impairments, including hundreds who were born both deaf and blind (Franklin, 1984). Since that epidemic, widespread immunization—either of preschool children (as in the United States) or of all adolescent girls who are not naturally immune and are not pregnant (as in England)—has reduced the rubella threat. As a consequence, between 1990 and 1995, an average of only 15 rubella-syndrome infants were born per year in the United States, and even fewer were born in England and Canada.

No such immunization is available, however, for the most devastating viral teratogen of all: **human immunodeficiency virus (HIV)**. HIV gradually overwhelms the body's natural immune responses, making the individual vulnerable to a host of diseases and infections including, inevitably, the fatal cancers, pneumonias, and other pathologies that together constitute **acquired immune deficiency syndrome (AIDS)**.

In Africa and Asia, AIDS, from its first appearance, has occurred virtually equally among men and women. In the Western world, by contrast, AIDS first appeared predominantly among men, many of whom were homosexual or intravenous drug users. The pattern in Western nations is rapidly changing as the pool of HIV-infected heterosexual men and women grows ever larger. The proportion of American AIDS cases involving heterosexual women, for example, rose from 7 percent in 1985 to 18 percent in 1994 (*MMWR*, February 10, 1995), and heterosexual women of childbearing age are the fastest-growing HIV-positive group (Rosenberg, 1995).

When a woman with HIV becomes pregnant, she risks passing the virus on to her fetus during pregnancy or childbirth. Currently, about one in

rubella A form of measles that, if contracted during pregnancy, can harm the fetus, including causing blindness, deafness, and damage to the central nervous system.

human immunodeficiency virus (HIV) A viral disease agent that gradually overwhelms the body's immune responses, leaving the individual defenseless against a host of pathologies that eventually manifest themselves as AIDS. HIV is carried in the blood and certain other bodily fluids of an infected person and is transmitted chiefly through sexual or direct blood contact.

acquired immune deficiency syndrome (AIDS) The final, terminal stage of HIV degradation of the immune system, which typically appears as serious infections, specific cancers, and the like.

four infants born to an HIV-positive woman contracts the disease. Unless a cure is found or the hopes for new drug therapies are fulfilled, these children will eventually develop AIDS and die—about a third during infancy, another third before kindergarten, and the remaining third by age 20 (Grubman et al., 1995).

The best way to prevent pediatric AIDS, of course, is to prevent adult AIDS. The next best way is to prevent pregnancy in HIV-positive women. Both goals are complicated by the disease's long incubation period—up to ten years or more—during which time a person can transmit the virus without showing any symptoms of the disease and thus without knowing that he or she is infected. Obviously, if a woman does not know that she is carrying HIV, she does not realize her needs to avoid pregnancy and to consider the options—if she does become pregnant—for preventing the birth altogether or for taking measures to reduce the risk of having an HIV-positive infant.

A doctor's early knowledge of the mother-to-be's HIV infection is key to lowering the chances of HIV transmission to the fetus. Such knowledge would allow for a Cesarean delivery, reducing the possibility of exposure during the birth process. In addition, it has recently been discovered that giving HIV-positive pregnant women AZT (zidovudine), a drug that slows the onset of adult AIDS, dramatically reduces prenatal transmission. Indeed, if all newly pregnant women obtained prenatal care and agreed to be tested for the HIV virus, and if those found to be positive were given AZT, the proportion of their infants born with the virus would be reduced from 1 in 4 to 1 in 12 (*MMWR*, April 29, 1994). However, this remarkable result can be achieved only if the outreach to women at risk is much improved.

Medicinal Drugs

Certain widely used medicinal drugs, including tetracycline, anticoagulants, bromides, phenobarbital, retinoic acid (a common treatment for acne, as in Accutane), and most hormones, are teratogenic in some cases. Other newer prescription drugs and possibly nonprescription drugs, including aspirin, antacids, and diet pills, may also be teratogenic. As a result, doctors advise women who are hoping to become pregnant or who already are pregnant to avoid not only these medications but also any medication unless it is essential for their health, and unless the drug is prescribed or recommended by a doctor who is both well-versed in teratology *and* aware the woman is pregnant. For the most part, pregnant women follow this advice, and serious, unexpected defects caused by pharmaceutical drugs are rare today.

Psychoactive Drugs

By contrast, prenatal damage caused by psychoactive drugs—beer and wine, liquor, cigarettes and smokeless tobacco, heroin and methadone, LSD, marijuana, cocaine in any form, and the like—is far too common. All of these drugs slow down the fetus's growth and can contribute to premature labor. And all have the potential to affect the developing brain, producing both short-term and long-term deficits. For a number of days or weeks after birth, infants who were heavily exposed to any of these drugs in the womb

fetal alcohol syndrome (FAS) A cluster of birth defects, including abnormal facial characteristics, slow physical growth, and retarded mental development, that is caused by the mother's drinking excessive quantities of alcohol when pregnant.

fetal alcohol effects (FAE) Subtle impairments of a child's motor and cognitive abilities that are caused by the mother's prenatal consumption of alcohol beyond a certain threshold. Research suggests that this threshold may be anything more than 1 ounce of absolute alcohol per day.

typically sleep fitfully, startle easily, suck voraciously but eat erratically, and show other signs of drug withdrawal. In the years to come, they may exhibit such problems as ongoing learning difficulties, impaired self-control, poor concentration, and overall irritability.

Alcohol

The most prevalent harmful drug in American society is alcohol, and it may also be the most pervasive teratogen. High doses of alcohol (three or more drinks daily, or binge-drinking of five or more drinks on one occasion early in pregnancy) can cause the distinctive symptoms of **fetal alcohol syndrome (FAS)**. These symptoms include overall growth retardation, learning disabilities, behavior problems (such as poor concentration and poor social skills), and abnormal facial characteristics (such as a small head, abnormally wide-spaced eyes, and a flattened nose). Experts believe that FAS is the leading prenatal cause of mental retardation in the industrialized world (Streissguth et al., 1993).

The possible effects of light to moderate drinking during pregnancy are less clear. Some experts believe that any amount of prenatal alcohol exposure can be detrimental, causing subtle impairments in learning ability throughout childhood (Olson et al., 1992). Others contend that drinking that does not exceed one or two servings of beer or wine or one mixed drink a day probably has no measurable effects on prenatal development (Fried et al., 1992; Walpole et al., 1990). However, it is clear that drinking beyond a certain threshold (generally considered to be more than 1 ounce of absolute alcohol) can cause **fetal alcohol effects (FAE)**. While less apparent than the physical anomalies and growth retardation of FAS, FAE is notable in tests of brain functioning. This is clearly shown in an ongoing study of 250 children whose mothers (almost all middle-class and healthy) were not very heavy drinkers during pregnancy. Compared with a control group of 250 children whose mothers had been very light drinkers or abstinent during pregnancy, the children in the first group averaged 5 points lower on IQ tests at age 4 and 7 points lower at age 7 (Streissguth et al., 1990). Another study, this time of French children, also found that, if their mother's daily prenatal consumption exceeded two drinks (usually wine), the children's motor skills and cognitive abilities were negatively affected (Larroque et al., 1995).

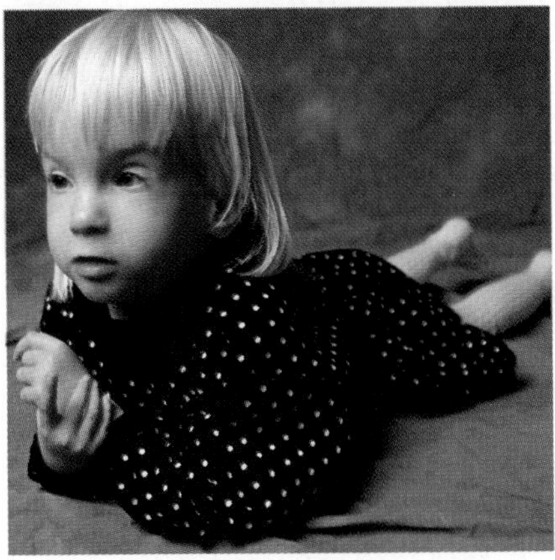

These two children are quite different in age and ethnicity. But this African-American teenager and this Swedish toddler have a great deal in common: both have the distinctive facial features and the retarded mental development of persons with FAS.

Tobacco

Cigarette-smoking is well documented as a prenatal hazard (Martin, 1992). Babies born to regular smokers weigh, on average, approximately 250 grams (about 9 ounces) less than would otherwise be expected, and they are shorter, both at birth and in the years to come. Even the mother-to-be's regular exposure to second-hand smoke has an impact, lowering birthweight by 45 grams (about 2 ounces) on average (Eskenazi et al., 1995). The effect of smoking on birthweight is related to dose and timing, with the most notable damage occurring as the result of heavy smoking in the last trimester of pregnancy (Lieberman et al., 1994).

Smoking also increases the risk of other complications at the end of pregnancy, including stillbirth, premature separation of the placenta, and premature labor. Over the long term, children prenatally exposed to ciga-

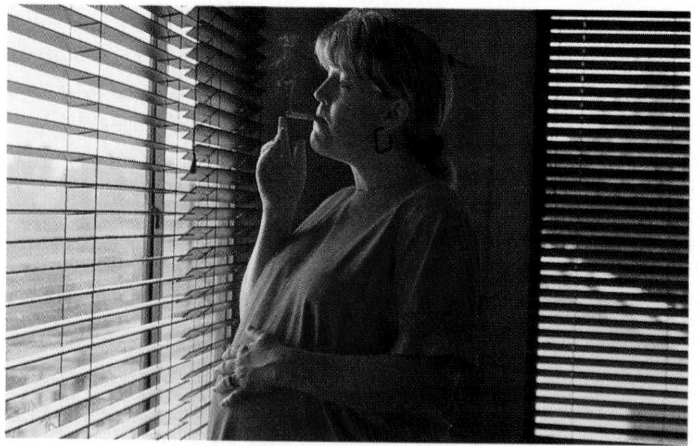

This woman's daydreams about her future baby may not be realized if she smokes heavily throughout her pregnancy. Every puff slows nutrition to the fetus, a cumulative effect that can result in a newborn who weighs a pound or more less than he or she otherwise would have. When a mother-to-be smokes, she always puts her fetus at risk for a number of other prenatal complications, and she may also be compromising her child's later health and intellectual functioning.

rette smoke are more likely to have respiratory problems (Tisi, 1988). (Probably because of subtle changes in the biochemistry of their brains, they are also more likely to become smokers themselves [Kandel et al., 1994].)

Illicit Drugs

For many reasons, the specific effects of each of the various illegal drugs are hard to document. Large-scale, prospective research is difficult, because locating a large representative sample of newly pregnant women who use one, and only one, illicit drug at a steady and measurable dose is virtually impossible. Even if such a group could be found, the researchers would then need to locate a control group of the same number of drug-free women of the same age, health status, and socioeconomic background.

In real life, however, illicit drug users almost always use several drugs and often have many more problems than other pregnant women, including being disproportionately poor, malnourished, and sick, without supportive families or good medical care. One study of infants who were prenatally exposed to cocaine found, for example, that 83 percent were also exposed to tobacco, 43 percent to heavy doses of alcohol, and 7.5 percent to syphilis. Further, 29 percent of the mothers received no prenatal care of any kind (Batemen et al., 1993). Another study found that cocaine-exposed infants were much more likely not only to have been exposed to tobacco but also to have higher blood levels of lead—a known teratogen (Neuspiel et al., 1994). Further, when illicit drug use amounts to addiction, there are almost always additional confounding factors, such as the mother-to-be's erratic sleeping and eating habits, bouts of anxiety, stress, and depression, as well as her increased risk of experiencing accidents, violence, and sexual abuse.

For all these reasons, many of the findings on the long-term effects of prenatal exposure to specific illicit drugs are suggestive but not conclusive. However, it is fairly certain that very heavy marijuana use can affect the fetus's central nervous system, as evidenced by the tendency of affected newborns to emit a high-pitched cry that denotes brain damage (Lester & Dreher, 1989). And it is clear that when a heroin-addicted mother experiences the physiological "highs" and "crashes" of the addiction, such as the reduction of oxygen, irregular heartbeat, and the sweating and chills that occur during withdrawal, the fetus experiences her body destabilization as well. Methadone moderates these effects but is as addictive as heroin. This

means that babies born to heroin-addicted or methadone-treated mothers are born addicted themselves and, in the first days of life, require regulated drug doses to prevent the pain and convulsions associated with withdrawal (Kaltenback & Finnegan, 1992).

It also seems certain that cocaine use during pregnancy causes overall growth retardation in the fetus, increases the risk of problems with the placenta and with labor (Dusick et al., 1993), and often leads to quite specific learning problems in the first months of life (Alessandri et al., 1993; Lester et al., 1991). As cocaine-exposed infants grow older, some research finds that they remain smaller overall, play less happily, and have specific learning problems (Griffith et al., 1994; Metosky & Vondra, 1995). However, attributing these longer-term findings specifically to prenatal cocaine exposure may be invalid, since in many cases such findings are confounded by the effects of the poverty and ongoing addiction of the mother (Hurt et al., 1995; Richardson & Day, 1994).

Despite the ambiguity of much of the research on the teratogenic effects of psychoactive drugs, what solid evidence there is makes it clear that the wisest action for a pregnant woman is to avoid such drugs entirely. Unfortunately, most of the American women in their prime reproductive years who often drink alcohol, smoke cigarettes, or use illicit drugs will continue their drug use in the first weeks of pregnancy before they realize that they are pregnant (Robins & Mills, 1993). Many will stop their drug use by the third month, having realized that they are pregnant, but this is after the early formation of the embryo. To make matters worse, those who are addicts, alcoholics, or heavy users of multiple drugs are likely to continue their drug use even though they know that they are pregnant. Continued drug use in pregnancy is not rare. Random urine testing of every newborn in Rhode Island in 1988 found that 7.5 percent had been exposed to *illicit* drugs in the hours or days before birth, and in Oregon, 5 percent of women giving birth in 1989 admitted to birth attendants that they had recently taken illegal drugs (*MMWR*, April, 1990; Slutsker et al., 1993).

Obviously, the best way to prevent prenatal damage from drugs is to prevent drug addiction and abuse in all quarters of society. Beyond that seemingly impossible task, however, the research points to several quite specific measures. Terotogenic effects of psychoactive drugs accumulate throughout pregnancy, so early prenatal care that includes not only testing for drug use but also effective counseling and treatment for those testing positive would reduce fetal brain damage substantially. Note that since alcohol and tobacco are at least as teratogenic as illegal drugs, they need to be targeted every bit as much as cocaine, heroin, and the like. In addition, since the prenatal effects of psychoactive drugs are dose-related, interactive, and cumulative, each drug that is eliminated, each dose that is reduced, and each day that can be drug-free represents a reduction in the potential damage that can be caused.

Moreover, many experts in child development have noted that babies who are born with cocaine or even heroin in their systems sometimes become quite normal, intelligent children if they receive optimal care (Mayes et al., 1992; Richardson & Day, 1994). Thus measures to protect children from suffering the consequences of their mothers' drug abuse before birth should include various steps (such as parenting education, home visits, or foster care) to ensure sensitive nurturance after birth.

low-birthweight infant A newborn who weighs less than 2,500 grams (5½ pounds) at birth.

preterm Being born three or more weeks before the due date.

Low Birthweight

The final complication we will discuss is **low birthweight**, defined by the World Health Organization as birthweight that is less than 2,500 grams (5½ pounds). Low birthweight is a major, and preventable, health hazard for the fetus, and as such it well illustrates the continuity of development, underscoring the fact that every developmental event is the result of many prior events, and a contributor to many subsequent ones. As you will see, the causes of low birthweight can include a variety of factors that predate birth and even pregnancy, and the consequences of low birthweight extend long past the life-threatening problems that may arise in the early days. Yet, at every point and in many ways, the causes and the consequences of low birthweight can be eliminated or ameliorated if the entire social context surrounding the fetus—mother, family, and community—takes the proper measures.

Causes

One cause of low birthweight is simply that birth occurs too early. Remember that in the last two months of pregnancy, the fetus's body weight doubles, with a typical gain of almost 2 pounds (about 700 grams) occurring in the final three weeks. Thus babies who are **preterm**, defined as being born at least three weeks before the due date, run a high risk of being low-birthweight. (Note, however, that some preterm babies born as much as a month early may be small but not low-birthweight if they were well nourished in their eight or so months of prenatal development.)

Many factors increase the likelihood of early birth. Conditions that disrupt the physiological equilibrium of the mother, such as psychoactive drugs, extreme stress, and chronic exhaustion, can precipitate preterm birth. Infections of various kinds are a major risk factor for early birth because they stimulate the mother's body to produce inflammatory chemicals that can trigger uterine contractions. These infections may also contribute to low birthweight by releasing into the mother's bloodstream toxins that can interfere with fetal development. A minor disease recently linked to early birth is bacterial vaginosis, a common vaginal infection which is easily cured with standard antibiotics. In the United States, this infection alone is responsible for an estimated 8 percent of all preterm births (Meis et al., 1995). Recent research suggests that a surprising but even more common culprit is gum disease, which may lead to a sevenfold increase in the risk of a

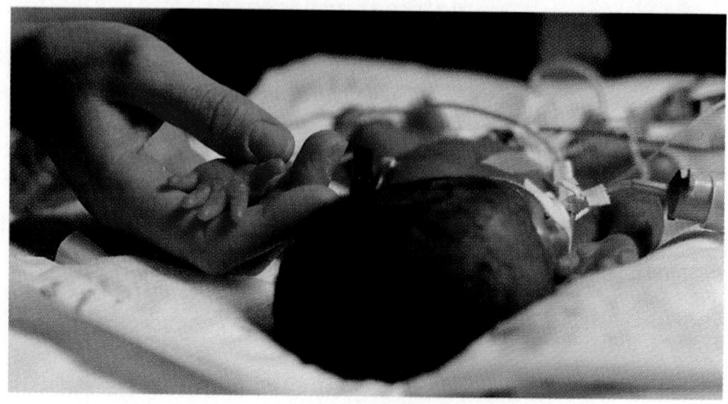

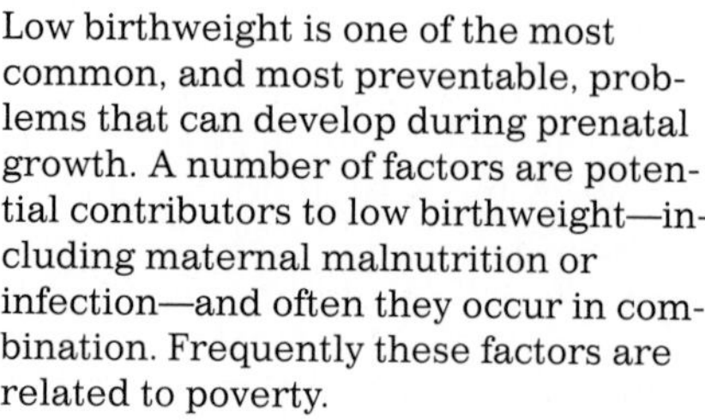
Low birthweight is one of the most common, and most preventable, problems that can develop during prenatal growth. A number of factors are potential contributors to low birthweight—including maternal malnutrition or infection—and often they occur in combination. Frequently these factors are related to poverty.

woman's delivering a preterm low-birthweight infant (Offenbacher, 1996). Physiological factors specific to the pregnancy, such as a placenta that becomes detached from the uterine wall, or a uterus that cannot accommodate further growth, also can trigger early birth. This latter factor helps explain why small women, and women bearing multiple fetuses, are more likely to experience labor weeks before the due date.

Not all low-birthweight babies are preterm. Some simply weigh substantially less than they should, given how much time has passed since conception. They are called small-for-dates, or **small for gestational age (SGA)**.

Why would a fetus grow slowly? One of the most common reasons is maternal cigarette-smoking. As we have seen, every psychoactive drug slows growth, and tobacco is the worst culprit, being implicated in 25 percent of all low-birthweight births in the United States (Chomitz et al., 1995). Another common reason for slow fetal growth is maternal malnutrition. Women who begin pregnancy underweight, who eat poorly during pregnancy, and who do not gain at least 3 pounds per month from the fourth month on run a much higher risk of having a low-birthweight infant. One recent study found, for example, that women who consumed less than 240 micrograms of folic acid a day—slightly more than half the recommended amount—were two to three times more likely to have a low-birthweight infant (Scholl et al., 1996). Indeed, malnutrition is the primary reason young teenagers tend to have small babies: not only do young girls tend to eat sporadically and unhealthily but, because their own bodies are still developing, their typically inadequate diet must support the growth of two.

Rarely is low birthweight the result of a single risk factor. Indeed, most cases of low birthweight are caused by a cascade of multiple risk factors—some slowing growth, others shortening pregnancy. Of particular concern for those interested in child development is that many of these factors are related to poverty (Hughes & Simpson, 1995). Compared with women of higher socioeconomic status, pregnant women at the bottom of the economic ladder are more likely to be ill, malnourished, teenaged, and stressed. They are also more likely to receive late and inadequate prenatal care, to be exposed to higher levels of pollution, to live in overcrowded conditions, and to ingest unhealthy substances, from drugs to spoiled foods. Poverty even plays a role in low birthweight related to malfunction of the placenta or the umbilical cord. Such malfunctions are more likely when pregnancies are closely spaced, and such spacing occurs more often in women of low income, partly because they have less access to family planning services. Thus social-contextual factors are an underlying cause of many of the biological causes of low birthweight. This helps explain the wide national and international variations in the following statistics:

1. Of the more than 20 million low-birthweight infants born worldwide each year, the overwhelming majority are from developing countries (United Nations, 1994).

2. Developing countries in the same general region, with similar ethnic populations, can have markedly different rates of low birthweight. For example, Nicaragua's rate (15 percent) is more than double that of neighboring Costa Rica (6 percent), where per capita income is five times that of Nicaragua (UNICEF, 1995).

3. Overall, ethnic-group differences in low-birthweight rates within nations tend to follow socioeconomic, as well as genetic, patterns

small for gestational age (SGA) A term applied to newborns who weigh substantially less than they should given how much time has passed since conception.

(Kleinman et al., 1991). The most telling example in the United States is the rate of low birthweight for infants of African-American descent, which is 13.3 percent, more than twice the rates for European- or Asian-Americans, who, as a whole, are less likely to be poor. The socioeconomic influences behind these differences are underscored by the fact that, while race of the mother is a relevant factor in these cross-group comparisons, wealthier African-Americans have lower rates of low birthweight than poorer African-Americans (Starfield et al., 1991). Similarly, the rate among Hispanic-Americans of Puerto Rican heritage is 9.2 percent, whereas the rate among those of Cuban descent, who tend to be more affluent, is 6.1 percent (U.S. Bureau of the Census, 1996).

4. Within the United States, the rate of low birthweight in the poorest states (e.g., Louisiana and Mississippi, more than 9 percent) is almost twice that of some richer states (e.g., Alaska and Oregon). Such differences cannot be attributed solely to the greater proportions of African-Americans within the Southern states, for similar state-by-state disparities are seen within ethnic groups. For example, the rate of low birthrate among white newborns in Mississippi is almost twice that among white newborns in Alaska (7 percent compared to 4 percent) (Children's Defense Fund, 1994).

Of course, socioeconomic status is only a rough gauge for other factors that may or may not apply in different cases. This is apparent with another Hispanic group—Americans of Mexican descent, whose low-birthweight rate is only 5.6 percent, which is much better than that of other groups having similar levels of income and education (U.S. Bureau of the Census, 1996). One reason for this difference appears to be that the rate of alcohol, tobacco, and drug use among pregnant Chicanas, especially those who are immigrants, is very low (Singh & Vu, 1996). Another reason may be that many of the very poorest Mexican-Americans live in rural areas and therefore tend to escape many of the risk factors common to crowded, inner-city life. This explanation is supported by one study that found that, in the poorest communities in Chicago, low-birthweight rates among Mexican-Americans were high, almost as high as those of their African-American neighbors (Collins & Shay, 1994).

Consequences of Low Birthweight

The first major challenge facing preterm infants is survival, especially if they are born more than six weeks early and weigh less than 1,500 grams. This is because in several vital ways they are not yet developed enough to easily sustain life outside the womb. For example, these newborns may have difficulty maintaining adequate body heat because they do not have the fat that normally accumulates during the last stages of prenatal growth. Low-birthweight preterm infants are also highly vulnerable to infection. More threatening, they may be unable to take in sufficient oxygen and, especially if they are more than a month preterm, are vulnerable to **anoxia**, a temporary lack of oxygen that can cause brain damage. They may also suffer damage as a result of cerebral hemorrhaging, called "brain bleeds."

anoxia A temporary lack of oxygen. If prolonged, it can cause brain damage or even death.

PUBLIC POLICY

Very-Very-Low-Birthweight Infants

A dramatic improvement has occurred over the past twenty-five years in the survival rate of very tiny infants—those under 1,000 grams (about 2¼ pounds). In 1970, virtually all of them died; now about 50 percent survive. Some of the survivors will live essentially normal lives. Unfortunately, some will not. About one-third will be severely handicapped, and another third will likely have learning difficulties in primary school (Beckwith & Rodning, 1991).

Ironically, the very same medical interventions that save lives sometimes create lifelong handicaps, among them blindness (from the administering of high concentrations of oxygen to aid breathing), cerebral palsy (from brain damage that occurred during the emergency assisted birth), and cognitive deficits (from hemorrhaging during surgery required by heart or respiratory failure). With very-very-low-birthweight infants, a choice must sometimes be made between two risks: the risk of visual and intellectual impairment versus the risk of early death. For all concerned, the choice can be agonizing.

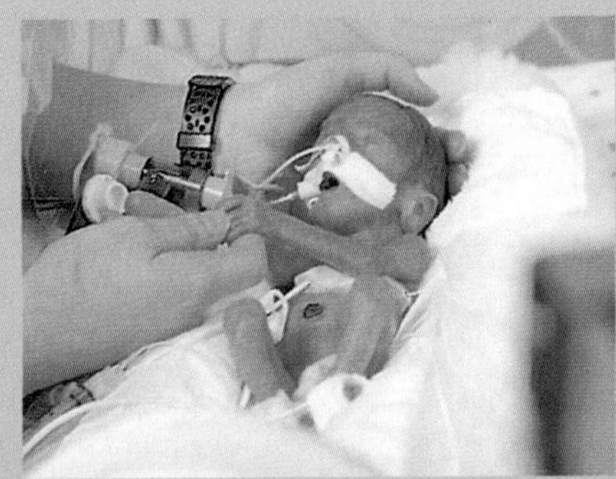

Born weighing only 2 pounds, this infant's heart rate, breathing, temperature, and blood acidity will be monitored continually until he reaches a weight of about 5½ pounds. Although his condition appears to be extremely fragile, current medical technologies give him an excellent chance of survival. However, the medical environment required to meet his most critical physical needs may deprive him of subtle, but important, types of stimulation.

One example makes the point:

> Debbie and Bill['s] . . . daughter Joan was born fifteen weeks prematurely . . . During her four months in [the] Intensive Care Nursery, she suffered the most serious degree of brain hemorrhaging, and her lungs were badly damaged. Joan is home now, but the uncertainty continues. She may have severe brain damage and cerebral palsy. She may never be able to swallow or suck. "When I hold her in my arms, she's my baby and I want her to live," says Debbie. "We appreciate every day we have with her, but sometimes you can't help but wonder whether this is the best for her. We don't know that she'll ever be able to enjoy her life. . . ."
>
> Her husband adds: "There was a time when we were afraid she would die. Now there are times when we're afraid she'll live. Without this technology, she would have died naturally, and we wouldn't have had to ask ourselves

Because of these risks, the immediate care of such infants is dictated by a high level of precaution. Often they are confined to isolettes or are continuously hooked up to one or another piece of medical machinery. At the same time, these infants are subject to a number of other experiences unknown to the normal infant, such as breathing with a respirator, being fed intravenously, and sleeping in the bright lights and noise of an intensive-care unit.

Although these measures are sometimes medically necessary, they often deprive the infant of certain kinds of stimulation, such as the gentle rocking they would have experienced if they still were in the womb, or the regular handling involved in feeding and bathing a newborn. To overcome this deprivation, many hospitals have begun providing low-birthweight infants with regular massage and soothing stimulation, which have been shown to aid weight gain and increase overall alertness (Scafidi et al., 1993). Ideally, hospitals encourage the parents to share in this early caregiving, in recognition of the fact that they too are deprived and stressed. Not only must they cope with the uncertainty surrounding their baby's future well-being but they must often struggle with feelings of inadequacy as parents, and perhaps with sorrow, guilt, and anger as well. For many parents, such

these questions. Maybe that would have been better." [Kantrowitz et al., 1988]

The harsh reality of this example reflects a painful ethical and social dilemma. Parents and doctors alike obviously wish every newborn the fullest chance of survival. But many experts question the practice of using heroic intensive-care procedures on marginally viable infants, whose chances of survival are slim and whose chances of survival without severe lifelong handicaps are slimmer still. They contend that such practice often ignores a number of crucial costs, including the immediate pain and suffering of the infant, the compromised quality of life for both infant and family, and the actual financial cost of care (Tyson, 1995).

Of these three costs, the last is the least difficult to assess. Including the costs of immediate neonatal care, repeated hospitalizations and medications over the first year, and a prorated portion of the expenses for every very-very-low-birthweight infant who dies, the cost of saving a newborn under 1,000 grams is close to half a million dollars (Paneth, 1992). If the same money were spent on the prevention of very-very-low birthweight, there is no doubt that many more lives would be saved and the total hardship and disability would be much less.

Further, the intensive effort to save the life of a tiny baby is seldom matched in the care of the infant once the crisis is over. Few parents are prepared to understand the needs of the very tiny, immature infant, or to cope with the demands of the special infant at home. Medical insurance does not usually cover at-home care, and specialized education for children is sometimes unavailable until they are preschool age—years after such help is required. Many such children spend months, even years, in custodial care in hospitals because no other alternative is available.

Most developmentalists feel that the public-policy choice should not be between ensuring the survival of one tiny infant and providing early prenatal care for a dozen high-risk pregnancies or specialized infant care for a disabled child. Moreover, it is unclear how such choices would be made in a social context of limited public resources. No one would want to suggest that ensuring the survival of any child's life is not worth the cost, but many question the moral myopia of narrowly focusing on high-tech heroics. As one physician put it, "we . . . need to see the folly of ignoring the disparity . . . between unlimited medical life-prolongation in the days and weeks immediately after birth and, later, the total lack of supporting care of vulnerable, often unwanted infants living in unsanitary housing amidst the social chaos of violence and crime in inner city slums" (Silverman, 1990). Perhaps the best approach to this policy dilemma is to recognize that in the earliest stages of life, as at any point in the life span, the best medicine is preventive medicine.

emotions are relieved somewhat if they can cradle and care for their tiny newborn.

For those preterm infants who survive, other problems may lie ahead even after they are home. They are likely to have many more minor medical problems that result in visits to the doctor than are full-term infants, and they are typically behind schedule in many ways, including being slow to smile, to hold a bottle, and to communicate. As the weeks and months go by, short- and long-term difficulties in cognitive development may emerge. Cerebral palsy is a common complication, occurring in about 20 percent of low-birthweight survivors who weighed less than 1,000 grams (2 pounds, 3 ounces) at birth, 15 percent of those who weighed between 1,000 and 1,500 grams, and 7 percent of those who weighed between 1,500 and 2,500 grams (Hack et al., 1995). Infants who were low-birthweight but initially had no obvious impairments sometimes have problems with early cognitive and language development, being more distractible and slower to talk, for example (Byrne et al., 1993; Lukeman & Melvin, 1993).

Long-term learning difficulties, especially distractibility, are likely to occur when low birthweight is accompanied by other medical complications, such as respiratory distress syndrome, anoxia, or cerebral hemorrhaging,

but sometimes they occur even in the absence of these complications (Hack et al., 1995; Szatmari et al., 1993). In cases where low birthweight is related to poverty and other social and family stresses, the potential long-term decrements of low birthweight are likely to be intensified by the continuing effects of these contextual factors. This need not always be so, however. Research with low-birthweight infants whose mothers were mostly young and poor suggests that intensive, long-term intervention—including parent-support programs and high-quality day care beginning at age 1—can result in substantial intellectual gains for the child and notable benefits for the mother-child interaction (Ramey et al., 1992). Such intervention appears particularly likely to aid infants whose birthweight was at least 1,500 grams (Brooks-Gunn et al., 1993).

Overall, unless low birthweight is actually a symptom of some genetic disorder or teratogenic impairment, or is associated with other medical problems soon after birth, the deficits related to it can usually be overcome. As with so much of development, responsive caregiving can make the difference. Only when the infant is very, very tiny—under 1,500 grams—are early problems unlikely to disappear (see Public Policy, p. 116).

The Normal Birth

When the fetus is normal and full-term and the mother is healthy, birth is a fairly simple and quick process. Sometime during the last month of pregnancy, most fetuses change position for the final time, turning upside-down so that their heads are in the mother's pelvic cavity. They are now in position to be born in the usual way, headfirst. At about the 266th day after conception, the fetus's brain signals for the release of specific hormones that pass into the mother's bloodstream, triggering her uterine muscles to begin contracting and relaxing at regular intervals. These contractions gradually push the fetus downward, putting pressure on the cervix to dilate until it is open about 4 inches (10 centimeters), allowing the fetus's head to squeeze through (see Figure 4.2). This process typically lasts eight to twelve hours in first births and four to seven hours in subsequent births, although there is much variation—from a few minutes to a few days. Then the head descends from the uterus and "crowns," as the first bit of scalp becomes visible at the opening of the vagina. The skin surrounding the vagina stretches with each contraction until the head emerges, usually less than an hour after crowning. Within a few seconds of the next contraction, the baby is fully born.

FIGURE 4.2 (*a*) The baby's position as the birth process begins. (*b*) During the first stage of labor, the cervix dilates to allow passage of the baby's head. (*c*) Transition: the baby's head moves through the "birth canal," the vagina. (*d*) The second stage of labor: the baby's head moves through the opening of the vagina and (*e*) emerges completely. The head is turned and the rest of the body emerges.

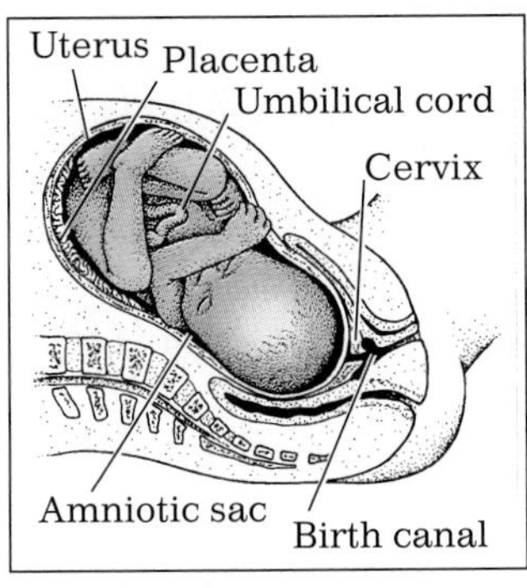

(a)

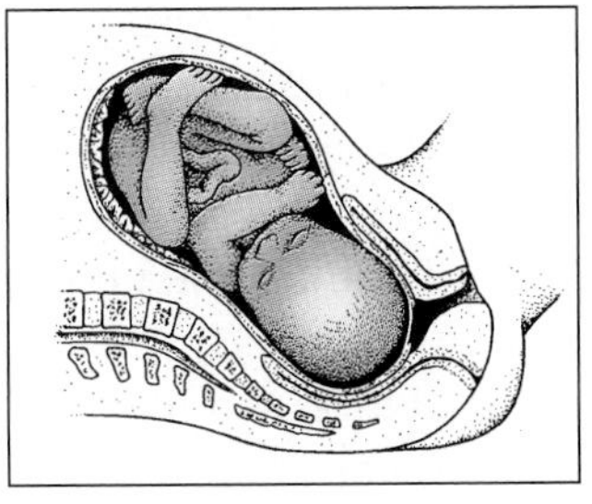

(b)

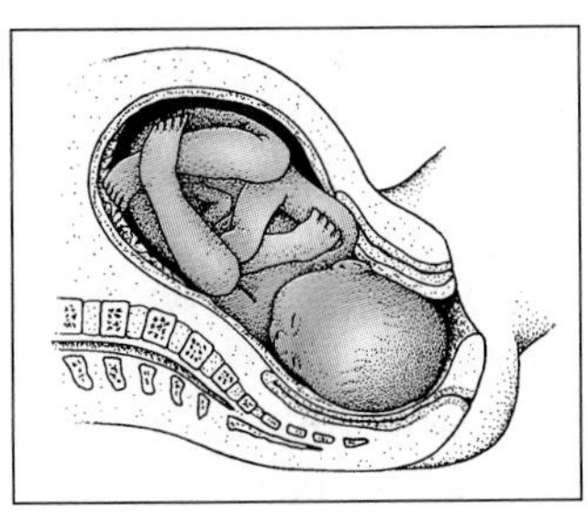

(c)

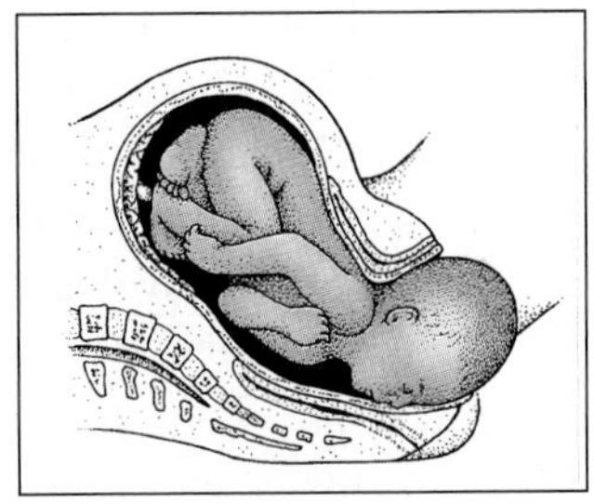

(d)

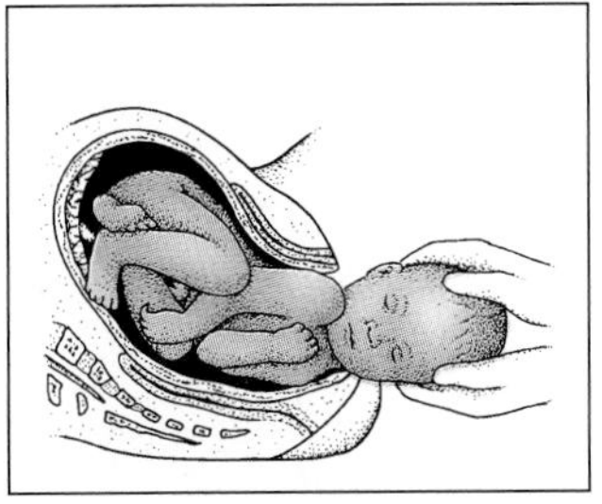

(e)

neonate A newborn baby. Infants are neonates from the moment of birth to the end of the first month of life.

Apgar scale A test devised by Dr. Virginia Apgar to quickly assess the newborn's color, heart rate, reflex irritability, muscle tone, and respiratory effort. This simple method is used one minute and five minutes after birth to determine whether a newborn needs immediate medical care.

The Newborn's First Minutes

People who have never witnessed a birth often picture the newborn being held upside-down and spanked by the attending doctor or midwife to make the baby start breathing. Actually, this is seldom necessary, for newborns, or **neonates**, usually breathe and cry on their own as soon as they are born. In fact, sometimes babies cry as soon as their heads emerge from the birth canal. As the first spontaneous cries occur, the neonate's circulatory system begins to function fully, and soon the infant's color changes from a bluish tinge to pink, as oxygen circulates throughout the system. Nevertheless, there is much for those attending the birth to do. Any mucus that might be in the throat is removed, the umbilical cord is cut, and the infant is wiped dry and wrapped to preserve body heat.

If birth is assisted by a trained health worker (as 98 percent of the births in industrialized nations and 51 percent of the births worldwide are [United Nations, 1994]), the newborn is immediately checked for body functioning. One common method of assessing the newborn's condition is a measure called the **Apgar scale**, which assigns a score of 0, 1, or 2 to the baby's heart rate, breathing, muscle tone, color, and reflexes at precisely one minute after birth and again at five minutes (see Table 4.2). If the five-minute score is 7 or better, the newborn is not in danger; if the score is below 7, the infant needs help establishing normal breathing; if the score is below 4, the baby is in critical condition and needs immediate medical attention to prevent respiratory distress and death. Very few newborns score an immediate perfect 10, but most readily adjust to life outside the womb.

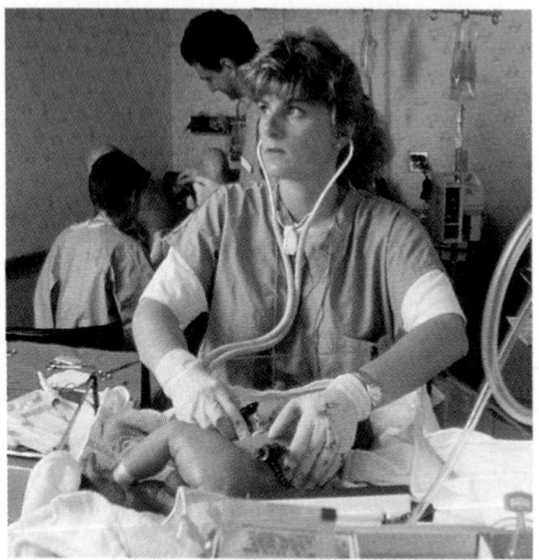

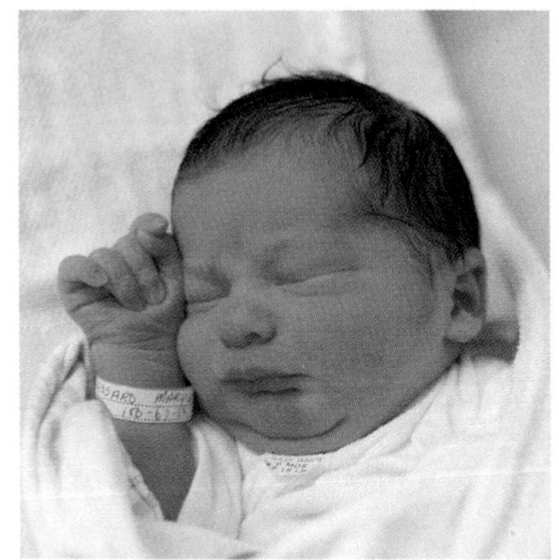

A minute after birth, the newborn on the left is undergoing his first exam, the Apgar scale. From the newborn's ruddy color and obvious muscle tone, it looks as if he will pass with a score of 7 or higher. On the right, 1-day-old James looks bruised, scraped, and squashed—a perfectly normal, beautiful newborn.

TABLE 4.2

Criteria and Scoring of the Apgar Scale

Score	Color	Heartbeat	Reflex Irritability	Muscle Tone	Respiratory Effort
0	blue, pale	absent	no response	flaccid, limp	absent
1	body pink, extremities blue	slow (below 100)	grimace	weak, inactive	irregular, slow
2	entirely pink	rapid (over 100)	coughing, sneezing, crying	strong, active	good, baby is crying

Source: Apgar, 1953.

For many new parents, the moments after birth are very special: the baby is usually quiet and alert, and the parents relieved and joyful. In one father's words:

> Christopher was placed in my wife's arms even before the umbilicus was cut; shortly after it was cut, he was wrapped (still dripping and wonderfully new like a chick out of an egg) and given to me to hold while my wife got her strength back. He was very alert, apparently able to focus his attention on me and on other objects in the room; as I held him he blossomed into pink, the various parts of his body turning from deep purple and almost blue, to pink, to rose. I was fascinated by the colors: time stopped. [quoted in Tanzer & Block, 1976]

Medical Intervention

How closely any given birth approaches the blissful ideal of the foregoing description depends in part on many factors, among them the mother's preparation for birth, gained through prenatal classes, conversations with women friends and relatives, or personal experience; the physical and emotional support provided by birth attendants (both professional and familial); the position of the fetus; and so on. An additional factor affecting the birth experience is the nature and degree of medical intervention it entails. Almost every birth in every developed nation now occurs amid ongoing medical activity. Among many other procedures, this intervention typically includes medication to dull pain or speed contractions, electronic monitoring of both the mother and the fetus, and often surgery, either an episiotomy (a minor incision of the tissue at the opening of the vagina) or, in about 25 percent of births in the United States, a Cesarean section to remove the fetus from the uterus through the mother's abdomen.

All medical intervention surrounding birth is controversial. No doubt, worldwide, the actions of doctors, midwives, and nurses save millions of lives each year, those of mothers as well as of infants (see Table 4.3). Indeed, the lack of such intervention is a major reason that the maternal mortality rates in the least developed nations are vastly greater than those of developed countries. According to a recent report, "an African woman has a one in 21 lifetime risk of dying from birth complications, a woman in Asia has a one in 54 lifetime risk, and a woman in Northern Europe has an almost negligible one in 10,000 lifetime risk" (Nowak, 1995).

But many aspects of this same medical intervention are under serious attack. Essentially, the critics contend that by treating every birth as a med-

TABLE 4.3

Maternal Mortality and Assisted Birth

	Maternal Deaths per 10,000 Births	Percentage of Births with No Trained Health Worker
Industrialized nations	1	2 percent
Developing nations	35	45 percent
Least developed nations	59	72 percent

Source: United Nations, 1994.

ical emergency waiting to happen rather than as a normal, natural event, the emotional development of the new family suffers needlessly.

Let us look at some of the specifics of birth within the United States. About 98 percent of American births occur in a hospital and include most, if not all, of the following particulars. As soon as a woman in labor walks into the hospital, she is placed in a wheelchair and taken to be "prepped" (typically she has her pubic hair shaved), is given an enema, and has her vital signs (blood pressure, temperature, heart rate) checked and recorded. Then she receives a glucose IV inserted in her wrist, has a fetal monitor strapped around her belly, and receives an internal examination to see how far the cervix has dilated—along with instructions and warnings about the need to stay in bed without food or drink while labor progresses. The message of all these initial procedures is clear: once she crosses the hospital threshold, the mother-to-be is not a healthy person undergoing a natural process but a fragile and helpless person, in potential need of emergency intervention and totally under the control of others.

As the hours pass by, an estimated 80 percent of women are given a Pitocin IV drip to intensify and speed up contractions, and 90 percent are given some sort of anesthesia, either through the IV or by means of an epidural, a spinal injection that usually deadens all sensations in the lower body (Davis-Floyd, 1992). As the moment of birth nears, the woman's wishes are increasingly subordinated to the orders of the doctors and nurses, who warn her of the damage she might inflict on her baby if she does not obey. They examine her frequently to check on the dilation of the cervix; they tell her to push or not push as the fetal head begins to move through the birth canal; they respond to their own interpretations of what they see on the fetal monitor rather than to the woman's perceptions of what is occurring within her or to her wishes regarding medication, examination, surgery, and the like. For one American woman in four, these various procedures ultimately result in a Cesarean birth. For the remaining 75 percent, an episiotomy is performed, and after birth, more drugs are administered to the mother to aid delivery of the placenta and repair of the episiotomy.

From several perspectives, there seems little to fault in this medicalized, technological approach to birth, especially since most American women choose to give birth in a hospital rather than at home, want relief of

The Brazilian midwife on the right and the North American obstetrician on the left both use an experienced hand, rather than a fetal monitor, to examine the position of the fetus and the strength of the contractions as the labor begins. Research suggests that the practiced touch of a skilled birth attendant helps the mother to relax, and thus makes contractions more effective and less painful.

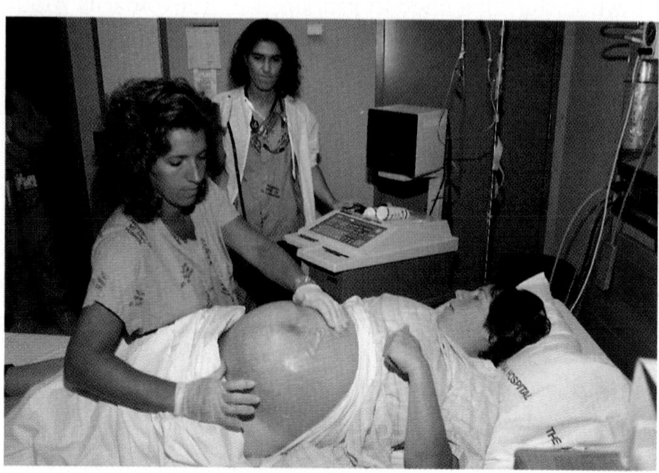

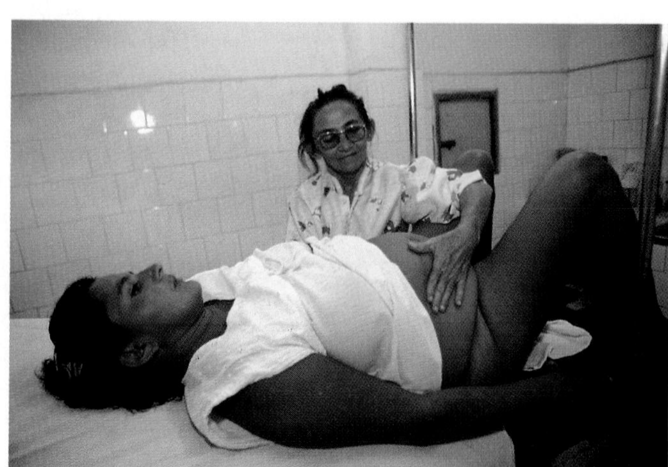

pain instead of a drug-free birth, and expect to be given a certain amount of medical attention rather than to be left alone. Indeed, contrary to the notion that doctors invariably insist on medical procedures despite the wishes of their patients, some women demand anesthesia and even Cesarean sections from medical staff who sometimes are reluctant to give them (Davis-Floyd, 1992).

Further, in many cases, medications make births easier and safer than they would otherwise be, with few long-lasting risks to the mother or newborn. Most important, when a fetus is known to be vulnerable, as a preterm fetus might be, to the stresses of a vaginal birth or to the suppressing effects that anesthesia has on the immature central nervous system, the physician can quickly perform a Cesarean, surgery that is stress-free for the baby.

But one question is not usually asked by the medical personnel directly involved with the birth event: How will the mother, father, and child be affected by hospital procedures in the hours and days after birth? Some negative effects can arise directly from the use of medications, which inevitably remain in the bloodstream of both mother and child and slow down their ability to focus on each other and enjoy their early interactions (Adams, 1989; Brackbill et al., 1988; Murray et al., 1981). Further, specifics such as the pain of the mother's stitches from the episiotomy or from a C-section, and the separations imposed on the new family by medical procedures and hospital protocol, add to the difficulty of the early family relationship.

The most problematic effect of medicalized birth may result, not directly from the medical procedures themselves, but from the psychological impact they may have on the parents-to-be. When the mother and father are knowledgeable and instrumental participants in the birth of their child, they tend to feel powerful as well as "full of love and compassion and support" (Davis-Floyd, 1992). Although such feelings are most likely to arise from a home birth, or one in which medical intervention is minimal, they can also develop in highly technological births if the parents feel that the technologies and medications used are serving them rather than dominating them. On the other hand, if the hospital authorities and procedures make parents feel helpless and ignorant and convince mothers-to-be that they are incapable of giving birth properly and safely without expert intervention, then the postnatal result can be depression and anger that last for days, even weeks, disrupting the family's relationships.

Of course, medical interventions that are disruptive to the family formation can be justified when they are used in response to unavoidable medical emergencies. However, many routine obstetrical procedures are not medically warranted: shaving the pubic hair, for example, is actually more likely to lead to infection; having the mother lie down during labor actually slows the birth process; fetal monitoring and Pitocin are often used routinely, even with no indications of their being needed, and often lead to unnecessary intervention. One recent study found that fetal monitoring, intended to detect complications that could lead to brain damage in the fetus, had a false-alarm rate of 99.8 percent, resulting in many unnecessary Cesarean sections (Nelson et al., 1996). Many other labor practices are likewise rooted in tradition rather than in true medical necessity (Enkin et al., 1989; Tew, 1990).

This is most clearly seen in international comparisons, which reveal the extreme variability of obstetrical practice, a variability that must be caused more by cultural differences than by differences in the rate of emer-

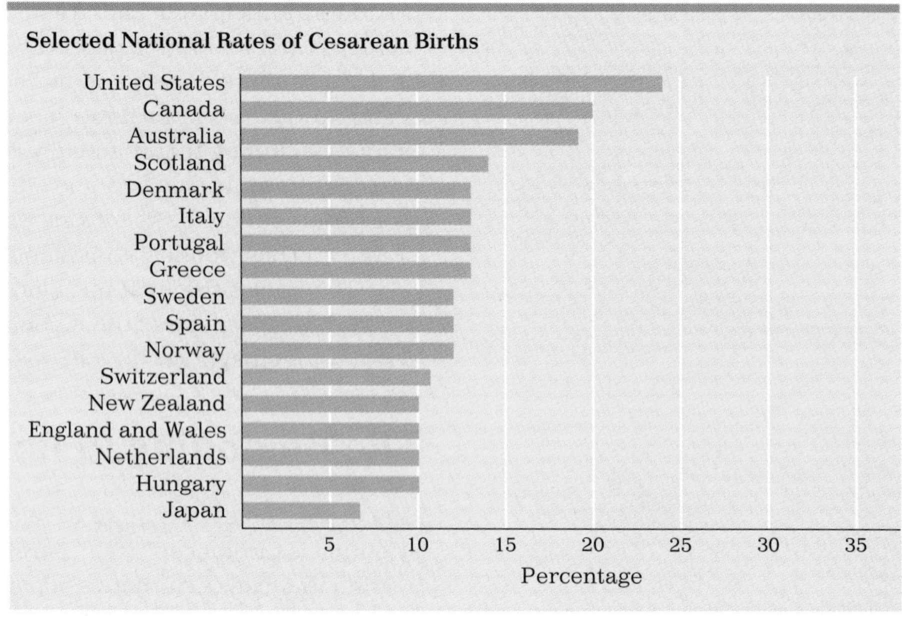

FIGURE 4.3 The United States spends more money on obstetrics than any other nation in the world, largely because of its high-tech, hospital-based approach to birth. This is clearly reflected in its rate of Cesarean births. Given that the United States has higher low-birthweight rates than fourteen other industrialized nations and higher infant mortality rates than twenty-one other nations, many public health experts believe that much of the obstetric expenditure should be directed instead toward improved family planning and prenatal care.

gency conditions. In a study of twenty-three European nations, for example, the rate of obstetrical intervention (e.g., Cesarean sections or forceps deliveries) ranged from 6 to 20 percent. Further, in some countries, women were denied anesthesia that they wanted and that is routinely given elsewhere; in other countries, women were given drugs they did not want. Similar disparities were evident in whether or not the woman could move during labor, whether or not she could deliver her baby in a squatting position, and whether or not she could have her husband, mother, or friends present at birth (Phaff, 1986). Although medical research now affirms that most women can have a vaginal birth in a subsequent pregnancy following a Cesarean section, in the United States and Canada, only 8 percent do so, compared to 40 percent in Norway and Scotland (Goldman et al., 1993).

Within the United States, a careful look at interventions in the birth process, hospital by hospital and doctor by doctor, makes the case even clearer. In the state of Washington, the rate of Cesareans in church and military hospitals is half that in private hospitals, even though private patients are more likely to be in good health and to have had excellent prenatal care. Clearly, the explanation for much of this disparity is that American doctors have a "financial disincentive" to perform Cesareans on poorer patients, because their medical insurance is less likely to cover it (McKenzie & Stephenson, 1993). Even among wealthier patients, the rates vary for reasons that are not medical. A study that compared obstetricians in one Detroit hospital found that their rate of Cesarean sections for private, low-risk patients ranged from 19 to 42 percent. One doctor delivered virtually all his breech babies by C-section, whereas another delivered none of them that way unless other complications occurred (e.g., a premature fetus with anoxia) (Goyent et al., 1989).

Obviously, the costs and benefits of medicalized childbirth vary from birth to birth, and ensuring a long and healthy life for mother and child must outweigh other considerations. However, from a developmental perspective, the cost-benefit equation seems off in its timing: medical emergencies

are often anticipated and treated long before they occur, while normal births are not allowed to proceed without intervention, even though natural births result in less pain and more joy for the new family in the days ahead. A step toward a more balanced equation would be to treat each birth as the unique beginning that it actually is. Then parents-to-be, as well as medical personnel, would analyze and weigh each routine, each medication, and each intervention for its impact—negative and positive—on the prospective newborn. They would act, or simply wait, accordingly.

This raises the final question for this chapter. How important are the psychological aspects of the first hours and days when the parents and child are first together?

The Beginning of Bonding

One of the topics of human development that has captured much popular attention is the concept of the **parent-newborn bond**, the almost instant connection that can occur between parents and their newborn children as they share physical contact in the first hours after birth. Over the past decade, hundreds of newspaper and magazine articles have waxed rhapsodic over the joy and the necessity of forming this special bond. Without it, it has been claimed, the long-term love between parent and child will be diminished, and the child will suffer. Many people have come to believe that bonding is a critically important "magical social glue." As one mother who was deprived of early contact said, "It made me feel like a rotten mother when I didn't get to bond with my first two children. Made me feel they were going to go out and rob a bank" (Eyer, 1992).

The best evidence for a parent-newborn bond comes from studies that reveal the formation of a quite specific and powerful bond between mother and newborn in various species of mammals. Many animal mothers, for instance, nourish and nurture their own young and ignore, reject, or mistreat the young of others.

At least three factors have been identified as contributing to this animal bond: hormones released during and after birth that trigger maternal feelings; the mother's identification of her particular infant by its smell; and the timing of the first physical contact between mother and newborn. The third of these factors can be remarkably precise: in some species, contact must occur within a specific "critical period" in order for bonding to take place. For example, if a baby goat is removed from its mother immediately after birth and returned a few hours later, the mother sometimes rejects it, kicking it and butting it away no matter how pitifully it bleats or how persistently it tries to nurse. However, if the newborn goat remains with the mother who nuzzles and suckles it for the critical first five minutes, and is then separated and later returned, the mother goat welcomes it back (Klopfer, 1971). Sheep and cows react in like fashion, with other species displaying a less pronounced form of the same behavior (Rosenblith, 1992).

parent-newborn bond The strong feelings of attachment between parents and newborns that are said to arise from their initial contact after birth. The long-term importance of this postpartum bond has been greatly, and dangerously, overblown by the popular media.

Does a corresponding sensitive time period exist for bonding in humans? Some early research on a few dozen mothers suggested that it does. In these studies, both initial contact in the moments immediately after birth and opportunities for extended contact over the first several days of life were shown to have a positive effect over the first year. This was especially true for first-time mothers who were very young, poor, or otherwise

stressed, or who had preterm infants who, by traditional standards, would have been deemed too forbiddingly frail, or too dependent on life-support, to be held or played with. The mothers in these studies who had held their infants soon after birth were more attentive and attached to them at age 1 than were the mothers who had barely seen their infants in the early days (Grossman et al., 1981; Klaus & Kennell, 1976; Leifer et al., 1972).

This research is credited with ending several postpartum hospital practices that were once routine, including whisking newborns away to the nursery right after birth, preventing mothers from seeing and holding their newborns for the first twenty-four hours, and barring parents from setting foot in intensive-care units, much less actually touching their preterm babies. All these practices were originally thought to protect mother and child from infection: all are now seen as unnecessary.

Almost no one now questions the wisdom of early contact between mother and child. It can provide a wondrous beginning to the parent-child relationship, as suggested by this mother's account:

> . . . the second he came out, they put him on my skin and I reached down and I felt him and it was something about having that sticky stuff on my fingers . . . it was really important to feel that waxy stuff and he was crying and I made soothing sounds to him . . . And he started calming down and somehow that makes you feel—like he already knows you, he knows who you are—like animals or something, perhaps the smell of each other . . . it was marvelous to hold him and I just touched him for a really long time and then they took him over but something had already happened. Just instant love. [quoted in Davis-Floyd, 1992]

But is this early contact, as has been claimed, essential for the formation of a positive and healthy mother-child bond? Absolutely not. Extensive later research finds that immediate or extended skin-to-skin togetherness makes no specific long-term differences in the mother-child relationship (Lamb, 1982; Myers, 1987).

One social scientist, Diane Eyer, has raised an interesting query: Why was the concept of bonding so quickly accepted, when the research evidence was so sparse? She has concluded that the entire concept of bonding is a social construction, an idea formed as a rallying cry against the medicalization, depersonalization, and patriarchy of the traditional hospital birth. Eyer argues that women and developmental experts were ready to believe that newborns and mothers need to be together from the start, and that therefore it took only a tiny nudge from scientific research for the mystique of early bonding to spring forth into general acceptance. She fears that this zealous acceptance comes at a high price, a standard of instant affection and "active love right after birth . . . that many women find impossible to meet" (Eyer, 1992).

Indeed, some developmentalists argue that too rigidly applying the idea of bonding is hardly better than not promoting it at all. If a medicated mother, exhausted from the birth process, is handed her infant for ten minutes or so while the episiotomy is stitched, and then the baby is removed because "bonding" has supposedly occurred, she may well feel guilty if she has not experienced the surge of emotion that the mystique of early bonding prescribes. Even worse, if an inexperienced mother who accepts the mystique of early bonding is for some reason not allowed to hold her infant in the minutes after birth, all her fears about her own ability to be a good mother may overwhelm her.

Fortunately, the evidence now confirms that the formation of family bonds is flexible. Immediate contact is neither necessary nor sufficient for bonding, as evidenced by the millions of very affectionate and dedicated biological, adoptive, or foster parents who never touched their children when they were newborns.

Does this mean that hospital routines can go back to the old ways, separating mother and newborn? No. As one leading developmentalist states:

> I hope that the weakness of the findings for bonding will not be used as an excuse to keep mothers and their infants separated in the hospital. Although such separation may do no permanent harm for most mother-infant pairs, providing contact in a way that is acceptable to the mother surely does not harm and gives much pleasure to many. It is my belief that anything that may make the postpartum period more pleasurable surely is worthwhile. [Rosenblith, 1992]

Overall, the ebb and flow of bonding research and practice reminds us that there is no "quick fix" or single defining moment in parent-child relationships. Love between a parent and child is affected by their ongoing interactions throughout infancy, childhood, and beyond, as well as by the manifold social contexts in which their relationship flourishes. As the following chapters reveal, while the nature of the parent-infant relationship is critical for healthy development, the specifics of its formation are not.

SUMMARY

From Zygote to Newborn

1. The first two weeks of prenatal growth are the germinal period. During this period, the single-celled zygote grows to an organism more than a hundred cells in size, travels down the Fallopian tube, and implants itself in the uterine lining, where it continues to grow.

2. The cells of the developing organism differentiate into two distinct masses. The outer cells form the membranes that will provide nourishment and protection during the prenatal period. The inner cells will become the embryo.

3. The period from the third through the eighth week after conception is the period of the embryo. The development of the embryo is cephalo-caudal (from the head downward) and proximo-distal (from the inner organs outward). During this period the heart begins to beat and the eyes, ears, nose, and mouth begin to form.

4. At eight weeks after conception, the future baby is only about an inch long. Yet it already has the organs and features of a human baby, with the exception of the sex organs, which take a few more weeks to develop.

5. The fetal period extends from the ninth week after conception until birth. By the twelfth week after conception, the sex organs have taken shape and all the other organs have completed their formation.

6. The fetus attains viability when the brain is sufficiently mature to regulate basic body functions, around the twenty-fourth week after conception. At this point the fetus has about a 50 percent chance of surviving outside the womb with expert care. The average fetus weighs approximately 2 pounds at the beginning of the third trimester and 7½ pounds at the end. The additional pounds, plus maturation of brain, lungs, and heart, ensure survival for more than 99 percent of all full-term babies.

Preventing Complications

7. Many teratogens (substances that can cause birth defects) can harm the embryo and fetus. Diseases, drugs, and pollutants can all cause birth defects. Some cause explicit physical impairment. Others, called behavioral teratogens, harm the brain and therefore impair the child's intellect and actions.

8. In understanding teratology, it is critical to realize that teratogens are risk factors, not inevitable destroyers. Whether a particular teratogen will harm a particular embryo or fetus depends on many factors, including the timing and amount of exposure and the developing organism's genetic vulnerability to the teratogen.

9. There are steps that a prospective mother can take to protect against prenatal complications. She can avoid and limit exposure to teratogens, maintain good nutrition, and seek early and competent prenatal care.

10. As a result of the knowledge derived from teratology, many serious teratogens, including rubella and some prescription drugs, now rarely reach the fetus. However, certain other diseases and psychoactive drugs remain hazards that require prevention on the part of the woman.

HIV is the most deadly of these teratogens; alcohol is the most common.

11. Low birthweight arises from a variety of causes, often occurring in combination. They include the mother's poor health or nutritional status, smoking, drinking, drug use, and age. Many of these factors are associated with poverty.

12. Preterm or small-for-gestational-age babies are more likely than full-term babies to suffer from stress during the birth process and to experience medical difficulties, especially breathing problems, in the days after birth. Some long-term cognitive difficulties may occur as well, depending on the presence of added medical problems or social-contextual challenges. Research has shown that postnatal parent-support programs and high-quality day care beginning at age 1 can result in intellectual gains for the low-birthweight child and enhancement of the infant-mother interaction.

The Normal Birth

13. Birth typically begins with contractions that push the fetus, headfirst, out from the uterus and then through the vagina. The process generally takes between eight and twelve hours in first births and between four and seven hours in subsequent births.

14. The Apgar scale, which rates the neonate's vital signs at one minute after birth and again at five minutes after birth, provides a quick evaluation of the infant's health. If the neonate's five-minute score is 7 or better, all is well.

15. Medical intervention in the birth process can speed contractions, dull pain, and save lives. However, many aspects of the medicalized birth have been faulted as unwarranted and/or as having a negative emotional impact on both infant and parents-to-be.

The Beginning of Bonding

16. Although the idea of early parent-infant bonding has received much popular attention, most developmentalists downplay its importance, stressing that the formation of the parent-infant bond develops continuously over a long period of time. The moments after birth contribute to, but do not determine, the success of the parent-infant relationship.

KEY TERMS

germinal period (97)
period of the embryo (97)
period of the fetus (97)
differentiation (98)
implantation (98)
neural tube (99)
cephalo-caudal development (99)
proximo-distal development (99)
age of viability (100)
teratology (103)
teratogens (103)
behavioral teratogens (103)
risk analysis (103)
critical period (103)
threshold effect (104)
interaction effect (105)
rubella (108)
human immunodeficiency virus (HIV) (108)
acquired immune deficiency syndrome (AIDS) (108)
fetal alcohol syndrome (FAS) (110)
fetal alcohol effects (FAE) (110)
low-birthweight infant (113)
preterm (113)
small-for-gestational-age (SGA) infant (114)
anoxia (115)
neonate (119)
Apgar scale (119)
parent-newborn bond (124)

KEY QUESTIONS

1. What developments occur during the germinal period?

2. What major developments occur during the period of the embryo?

3. What major developments occur during the period of the fetus?

4. In what ways is the fetus capable of responding to the outside world?

5. What factors make a fetus more likely to be harmed by teratogens?

6. What are the three primary protective steps that mothers-to-be can take to promote a healthy birth?

7. Name some of the most common teratogens.

8. How can public health measures provide protection against viral teratogens?

9. What are the effects of maternal drinking on the fetus?

10. How can cigarette-smoking affect prenatal growth?

11. What are some effects of drug abuse on the fetus?

12. What are the causes of low birthweight?

13. What are the most serious problems of low-birthweight infants?

14. What factors can affect the long-term consequences of low birthweight and other birth problems?

15. What vital body signs does the Apgar scale measure? What does the Apgar scale score tell about the health of the newborn?

16. What are the advantages and disadvantages of medication administered during childbirth?

17. How is the formation of the parent-infant bond different in animals than it is in humans?

PART II The First Two Years: Infants and Toddlers

Adults usually don't change much in a year or two. Sometimes their hair gets longer or grows thinner, or they gain or lose a few pounds, or they become a little wiser or more mature. But if you were to be reunited with some friends you hadn't seen for several years, you would no doubt recognize them immediately.

If, on the other hand, you were to care for a newborn twenty-four hours a day for the first month, and then did not see the baby until a year or two later, the chances of your recognizing that child are similar to those of recognizing a best friend who had quadrupled in weight, grown 14 inches, and sprouted a new head of hair. Nor would you find the toddler's way of thinking, talking, or playing familiar. A hungry newborn just cries; a hungry toddler says "more food" or climbs up on the kitchen counter to reach the cookies.

While two years seem short compared to the almost eighty years of the average life span, children in their first two years reach half their adult height, possess cognitive abilities that have surprised even researchers, and express almost any emotion, from jealousy to shame. Two of the most important human abilities, talking and loving, are already apparent. The next three chapters describe these radical and rapid changes.

The First Two Years: Biosocial Development

CHAPTER 5

The forces of biosocial growth and development in the first two years of life are very powerful. Proof of this is visible to any observer, as infants quickly outgrow one set of clothes after another, attempt new behaviors almost daily, and display a rapidly increasing mastery of emerging skills. Evidence is also apparent from laboratory data on brain development, which shows increasing density and complexity of neural networks that are vital to the maturing of physical and mental capacities. All these changes, of course, are biologically rooted, but they are also facilitated by the social context, as parents and others nourish, protect, and encourage the infant's development. In this chapter we will look first at the physical developments in the child's body and brain, and then at the social environment—particularly factors in nutrition—that can either enhance or inhibit those developments.

Physical Growth and Health

Monitoring growth and protecting health are critical throughout childhood, but they are particularly so during infancy, when growth, as well as vulnerability to growth problems and disease, are most pronounced. Throughout childhood, visits to the doctor for a checkup are an annual affair, but in early infancy, visits should occur monthly, not only to keep tabs on the infant's physical progress and to spot the first signs if something is amiss, but also, as you will see, to meet the recommended schedule of immunizations.

Size and Shape

With the exception of prenatal development, infancy is the period of the fastest and most notable increases in size and changes in body proportion. Especially in the first few months of life, babies grow so rapidly that even parents may have difficulty recognizing a photo of their 3-month-old infant as a newborn.

The average North American newborn measures 20 inches (51 centimeters) and weighs a little more than 7 pounds (3.2 kilograms). This means that the average newborn is lighter than a gallon of milk and about as long as the distance from a man's elbow to the tips of his fingers. In the first days of life, as they eliminate more body wastes than they take in nourishment, most newborns lose between 5 and 10 percent of their body weight. Then

they grow rapidly, doubling their birthweight by the fourth month, tripling it by the end of the first year, and growing about an inch (2.5 centimeters) longer each month for the first twelve months. Much of the weight gain in the early months of life is fat, which provides insulation for warmth and a store of nourishment. After eight months or so, weight gain derives more from growth in bone, muscle, and body organs. By age 1, the typical baby weighs about 22 pounds (10 kilograms) and measures almost 30 inches (75 centimeters) (Behrman, 1992).

Growth in the second year proceeds at a slower rate. By 24 months of age most children weigh almost 30 pounds (13 kilograms) and measure between 32 and 36 inches (81 to 91 centimeters), with boys being slightly taller and heavier than girls. In other words, typical 2-year-olds are almost a fifth of their adult weight and half their adult height (see Figure 5.1).

As infants grow, their body proportions change, continuing the same cephalo-caudal patterns that characterized prenatal growth. Most newborns seem top-heavy because their heads are equivalent to about one-fourth of their total length, compared to one-fifth at one year and one-eighth in adulthood. Their legs, in turn, represent only about a quarter of their total body length, whereas an adult's legs account for about half of it. Proportionally, the smallest part of a newborn's body is that part farthest from the head and the most distant from the center, namely, the feet. Over the course of childhood and adolescence, as the body lengthens, the relative size of each part changes. By adulthood, a person's feet, for example, will be about five times as long as they were at birth, while the head will have only doubled in size.

FIGURE 5.1 These figures show the range of height and weight of American children during their first two years. The lines labeled "50th" (the fiftieth percentile) show the average; the lines labeled "90th" (the ninetieth percentile) show the size of children taller and heavier than 90 percent of their contemporaries; and the lines labeled "10th" (the tenth percentile) show the size of children who are taller or heavier than only 10 percent of their peers. Note that girls (red lines) are slightly shorter and lighter, on the average, than boys (blue lines).

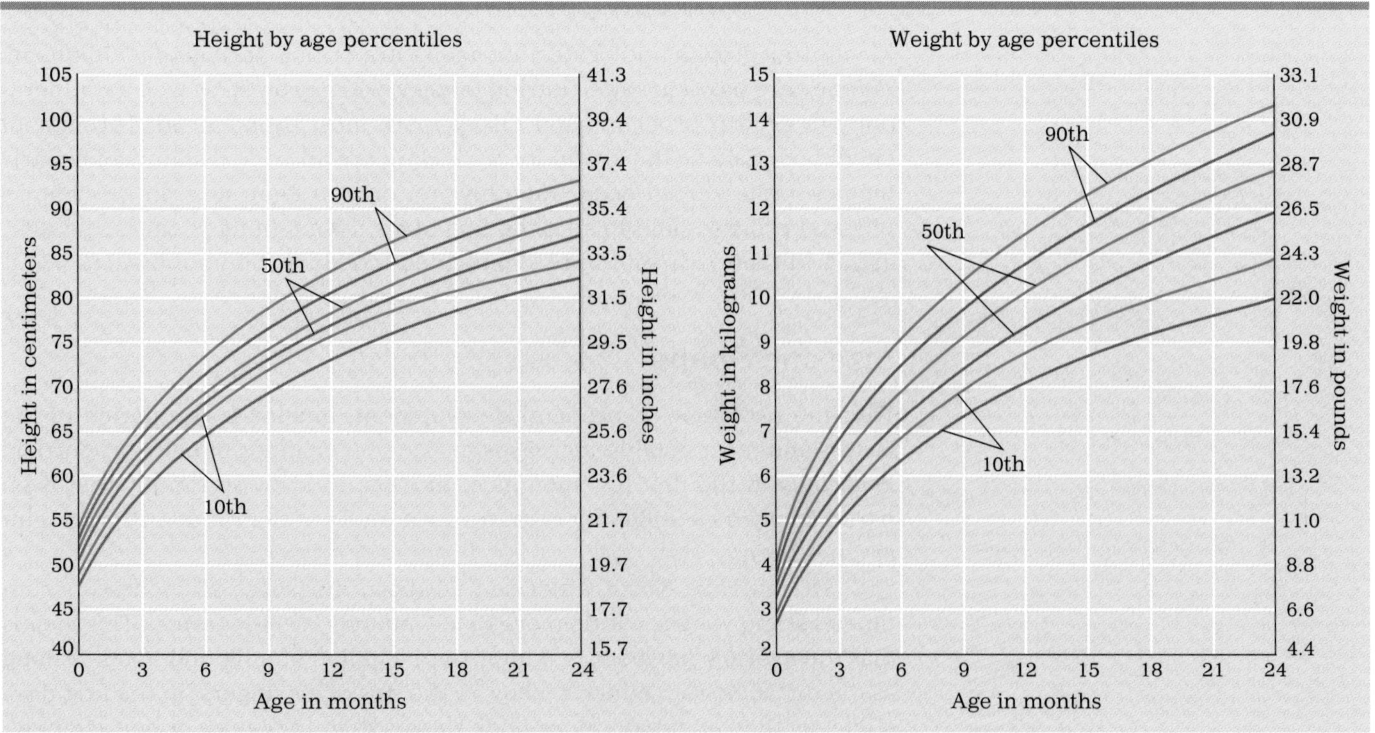

Preventive Medicine

Nowadays, the growth just outlined is typically taken for granted. However, a century ago in developed nations, and just a decade or two ago in less developed ones, not only growth but also basic health and even survival to age 5 were very much in doubt. At any time, a sudden epidemic of an infectious disease—smallpox, whooping cough, polio, diphtheria, or several others—might rapidly spread through young children, putting them at risk of serious complications and death.

Now, deadly childhood diseases are rare: an infant's chance of dying from disease in North America, Western Europe, Japan, or Australia is less than 1 in 100, down from 1 in 20 at midcentury (UNICEF, 1990). Indeed, it is this dramatic reduction in early death, not the extension of life in old age, that is the primary reason for the significant worldwide increase in average life expectancy over the course of this century. In the United States, for instance, life expectancy increased from 55 years in 1920 to 75 years in 1990; in Tanzania, in just the ten years from 1975 to 1985, life expectancy increased from 45 to 51 years.

Many factors have contributed to the reduction of disease-related deaths among young children, ranging from overall improved sanitation to technological breakthroughs for treating high-risk newborns. However, the single most important cause of the dramatic twentieth-century improvement in child survival is immunization. In the less developed nations, for example, the rate of immunization against the leading fatal childhood diseases improved from about 20 percent to 80 percent during the 1980s, reducing deaths from these diseases by three-fourths (UNICEF, 1990). Smallpox, the deadliest disease of all for children, has been completely eradicated worldwide, and polio is close to being wiped out as well. Measles (which can be fatal in the early months when it causes dehydration) is disappearing too. In the United States, only 294 cases of measles were recorded in 1995, the lowest number since incidence tallies began to be kept in 1911 (*MMWR*, February 2, 1996).

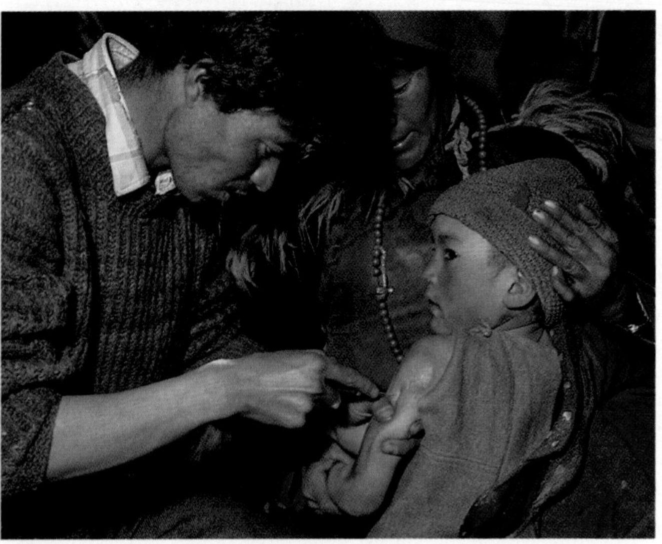

Measles can be a deadly disease for undernourished children and infirm adults in developing nations. This boy's immunization will not only protect him but will also help to protect those in his village in northern India who are too young, too enfeebled, or too frightened to obtain their own shots.

TABLE 5.1

Immunizations Recommended in the First Five Years

	Birth	2 Mo	4 Mo	6 Mo	15 Mo	5 Yr
DPT (diphtheria, pertussis, tetanus)		X	X	X	X	X
HIB (hemophilus influenza type B)		X	X	X	X	
Hepatitis B	X	X			X	
MMR (measles, mumps, rubella)					X	X
Oral polio vaccine		X	X		X	X
BCG* (antituberculosis)	X					

* BCG is not recommended in some developed nations.
Sources: Centers for Disease Control, 1994; UNICEF, 1990.

Worldwide, both the quality and the scope of immunization have improved every decade, such that more than 90 percent of all infants are now immunized against diphtheria, pertussis, tetanus, measles, polio, and mumps (UNICEF, 1995). In developed nations, many infants are immunized against hepatitis B, hemophilus influenza type B (HIB), rubella, and chicken pox as well (see Table 5.1). (Some developed countries also immunize children against tuberculosis, while others, including the United States, test children for exposure and then follow up with further testing and treatment if needed.)

Unfortunately, in the United States, voluntary immunization programs are in danger of becoming victims of their own success. As the once-common diseases of childhood have receded, so has the public's awareness of, and attention to, them. As one health commentator has noted:

> Most parents of infants and preschoolers have no memory of summers when children were kept out of swimming pools for fear of catching polio. . . . Nor do they know firsthand the terrors of breath-robbing whooping cough, the often fatal paralysis of tetanus, or the sometimes fatal throat infection caused by diphtheria . . . [or that] measles can cause life-threatening encephalitis and mental retardation. [Brody, 1993]

Consequently, many physicians fear that parents and policy makers now take immunization too lightly. This is particularly true in the United States, which, in a 1993 survey of 115 nations, ranked 68th in the proportion of preschoolers who were fully immunized (UNICEF, 1995). In 1994, only 60 percent of American 2-year-olds had had all their recommended DPT (diphtheria, pertussis, tetanus), polio, measles, and HIB vaccines, and only 30 percent had received protection against hepatitis. (The children least likely to be fully immunized were those living in the most crowded neighborhoods and therefore those at the highest risk of contagion.) Since one of the prime goals of Health USA 2000, a federally established set of public health objectives, is the full immunization of at least 90 percent of America's preschoolers, experts at the United States Centers for Disease Control and Prevention feel compelled to restate the obvious: "Increased efforts are needed to vaccinate all children" (*MMWR*, May 26, 1995).

Lack of complete immunization obviously puts the child, as well as his or her playmates, at risk, sometimes seriously. In addition to the problems already mentioned, mumps can produce nerve deafness, and HIB is the leading cause of meningitis. Less obviously, lack of immunization jeopardizes the well-being of others in the child's world: infants too young to be immunized may die if they catch a disease from an older child; pregnant women who contract rubella may transmit the virus to their fetuses, causing blindness, deafness, and brain damage; healthy adults who contract mumps or measles may suffer much worse consequences than a child might; and the particularly vulnerable, such as those who are elderly, who have AIDS, or who have cancer, can be killed by any number of the "child" diseases. Chicken pox, for instance, can kill a person whose immune system is depleted by chemotherapy.

The real and potential successes of immunization lead us to a moral that can be drawn again and again in the study of development: a small act of prevention is often much easier, less costly, and less painful than trying to remedy the consequences of indifference and inaction.

Sudden Infant Death Syndrome

Since widespread immunization has now made contagious fatal disease rare in infancy, most infant deaths occur in the first month of life and are related to congenital problems such as heart defects or other inborn abnormalities, to very low birthweight, or to similarly identifiable problems. However, one common cause of infant death is not related to any obvious problem: **sudden infant death syndrome**, or **SIDS**. SIDS typically kills infants who are at least 2 months old and seemingly completely healthy—already gaining weight, learning to shake a rattle, starting to roll over, and smiling at their caregivers. In the United States, SIDS is the second leading cause of infant death (see Table 5.2). Each year more than 4,000 American babies go to sleep and never wake up, victims of a sudden failure to breathe.

The term "sudden infant death" (also called *crib death* or *cot death*) is more a description than a diagnosis, because, despite decades of research, the actual cause of SIDS is still unknown. The diagnosis of sudden infant death is assigned when autopsy suggests that the infant simply stopped

sudden infant death syndrome (SIDS) Death of a seemingly healthy baby who, without apparent cause, stops breathing during sleep.

TABLE 5.2

Infant Deaths in the United States, 1991

Leading Causes	Total
Congenital anomalies	7,685
Sudden infant death syndrome	5,349
Disorders related to short gestation and low birthweight	4,139
Respiratory distress syndrome	2,569
Maternal complications of pregnancy	1,536
Complications of placenta, cord, and membranes	962
Accidents and adverse effects	961
Infections specific to the perinatal period	881
Pneumonia and influenza	607
Intrauterine hypoxia and birth asphyxia	599

Source: National Center for Health Statistics, 1993.

TABLE 5.3

SIDS Risk Factors

	SIDS More Likely	SIDS Less Likely
Characteristics of the Mother		
Age	under 20	over 25
Blood type	O, B, or AB	A
Personal habits	smoker	nonsmoker
Income	poverty-level	middle-class
Education	grade school only	college or higher
Ethnic background	African descent	Asian descent
Characteristics at Birth		
Sex	male	female
Birth order	later-born	first-born
Multiple birth?	yes (twin or triplet)	no (single-born)
Apgar score at 5 min.	7 or lower	8 or higher
Heartbeat	some irregularity	normal
Situation at Death		
Time of year	winter	summer
Age in months	1 to 3	under 1, over 4
Health	has a stuffy nose	no cold, no runny nose
Feeding	bottle-fed	breast-fed
Sleeping Conditions		
Position	sleeps on stomach	sleeps on back
Mattress	soft, natural fibers	firm, synthetic
Blankets, nightclothes	swaddled, tight	allow free movement
Bedroom temperature	heated	cool

Sources: Guilleminault et al., 1982; Haas et al., 1993; Meny et al., 1994; Mitchell et al., 1993; *MMWR*, October 11, 1996; Ponsoby et al., 1993.

breathing, with other possible causes, such as deliberate or accidental suffocation, ruled out. Such determinations need to be done speedily and carefully, since crib deaths sometimes provoke unfounded suspicions from neighbors and police, who assume that the parents must have done something wrong. In recent years, as the diagnosis of SIDS has become more definitive, scientists have intensified their search for a cause—perhaps a subtle neurological or physiological abnormality or some disease agent that is particularly harmful to young infants. A number of potential culprits have been identified, ranging from a bacterium that is occasionally found in raw honey to a brain-stem defect that may impair the infant's arousal response to an excessive intake of carbon dioxide (as when rebreathing exhaled gases trapped under a blanket) (Kinney et al., 1995).

However, as the search for a cause continues, most researchers are increasingly convinced that there is no single cause of SIDS. In all probability, SIDS results from a combination of factors (see Table 5.3), which, as they accumulate, make *certain* infants, partly for genetic reasons, vulnerable. Drawing a profile from the table of risk factors, for example, we can say that a particular (but unidentifiable) 4-month-old boy, who is born in September weighing 5 pounds, who lives with several siblings in a low-income neighborhood, who has a slight case of the sniffles, and whose mother smokes cigarettes and does not breast-feed him, is more likely to die of SIDS than a baby who does not have any of those characteristics. As with all risk analysis, scientists can spot vulnerability but cannot predict actual cases. Indeed, some victims of SIDS have no known risk factors.

However, one important contributing factor has recently been discovered: the infant's sleeping position. All controlled research finds that there

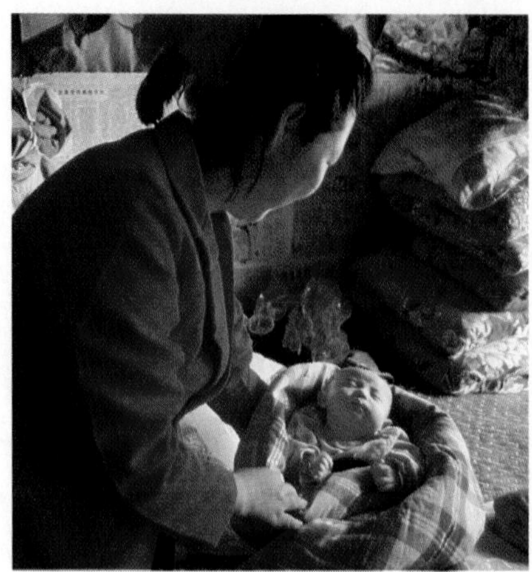

Infants in China are traditionally placed on their backs for sleeping. They also tend to be constantly in the presence of the mother or some other relative, who regularly checks on them as they sleep, adjusting their covers, making them comfortable, and the like. These two factors may help account for the low rate of SIDS in China. The woman in this photo, a seamstress, keeps her baby beside her as she works, a practice, incidentally, that is allowed by many employers in China and is encouraged by the government.

is less of a risk for SIDS when healthy infants sleep on their back than when they sleep on their stomach (Beal & Finch, 1991). Indeed, one comparison study found that an infant's risk of SIDS quadruples if the baby is put to sleep in a prone rather than a supine position (Ponsoby et al., 1993). Additional evidence for the importance of sleeping position comes from China, where infants are almost always put to sleep on their back and where SIDS rates are unusually low, even though babies are generally swaddled as they sleep (a risk factor) (Beal & Porter, 1991). Ironically, putting infants to sleep on their stomach has been recommended by Western pediatricians for decades, on the assumption that, when babies spit up (as almost all sometimes do), they might choke if they are lying on their back (Spock, 1976). While this idea makes sense, and may occasionally be borne out, it is now accepted that putting babies to sleep on their back is the safer course (Willinger et al., 1994), unless they are very young and preterm, in which case the best sleeping position may be on their stomach (Martin et al., 1995).

Another key risk factor in SIDS is ethnic background (see Table 5.3). Generally, within ethnically diverse nations such as the United States, Canada, Great Britain, Australia, and New Zealand, babies of African descent are more likely, and babies of Asian descent less likely, to succumb to SIDS than are babies of European descent. The reasons for this difference may be genetic or they may be related to background variables that correlate with ethnicity but are not caused by it (such as the rate of teenage pregnancy).

An alternative explanation for the correlation between ethnicity and SIDS focuses on quite specific infant-care routines that are widespread in one culture and rare in another. Bangladeshi infants in England, for instance, tend to be low in birthweight and socioeconomic status but nonetheless have lower rates of SIDS than white British infants, despite the fact that the latter are more often of normal birthweight and middle-class. One possible reason may be that Bangladeshi infants tend to be surrounded by many family members in a rich sensory environment, continually hearing noises and feeling the gentle touch of their caregivers, and they therefore do not sleep too deeply for very long. By contrast, their white age-mates tend to sleep in their own private space in an environment of enforced quiet, and "long periods of lone sleep may contribute to the higher rates of SIDS among white infants" (Gantley et al., 1993).

A similar pattern of child care may also account for the low SIDS rate among Chinese-American infants, whose parents not only place their babies to sleep on their back but also tend to them periodically as they sleep, caressing a cheek, repositioning a limb, and so on. Specific child-care practices that vary from one culture to another have only recently been examined as something other than curiosities, but it is easy to imagine that many specifics—perhaps in frequency of feeding, or in sleeping garments, or even in the parents' reaction to thumb-sucking—may likewise have an impact on SIDS (Davies & Gantley, 1994; Farooqi et al., 1993).

As you can see, simple answers to the SIDS tragedy are elusive: risk factors are many, and alternative explanations abound. However, even without a known cause, the prevalence of SIDS can be reduced by limiting exposure to risk factors—reducing low birthweight, encouraging breast-feeding, reducing smoking during pregnancy, putting infants to sleep on their back, and so on. The latter two measures are considered largely responsible for a 20 percent decrease in the rate of SIDS in the United States over the past few years (*MMWR*, October 11, 1996).

neurons Nerve cells of the central nervous system.

axons Nerve fibers that extend from a neuron and transmit impulses from one neuron to the dendrites of another.

dendrites Nerve fibers that extend from a neuron and receive the impulses transmitted from other neurons via their axons.

Brain Growth and Development

As we saw earlier, the newborn's skull is disproportionally large. One reason is that it must accommodate the brain, which at birth has already attained 25 percent of its adult weight. The neonate's body weight, by comparison, is typically less than 5 percent of adult weight, and the rate of brain growth in the first two years is even greater than the rate of overall body growth. By age 2, the brain has already attained about 75 percent of its adult weight, while the 2-year-old's body weight is only about 20 percent of what it will eventually be (Lowrey, 1986).

Weight, of course, provides only a crude index of brain development. More significant are the specific changes that occur in the brain's communication systems and greatly advance the brain's functioning. The brain's communication systems consist primarily of nerve cells, called **neurons**, and intricate networks of nerve fibers, called **axons** and **dendrites**, which interconnect the neurons. At birth, the brain's communication systems contain over 100 billion neurons, far more than the developing person will ever need. By contrast, the networks of axons and dendrites are fairly rudimentary, with relatively few connections established among neurons. During the first months and years, there are major spurts of growth and refinement in the networks of axons and dendrites (see Figure 5.2). These changes are particularly notable in the cortex, the brain's eighth-of-an-inch-deep outer layer (or "gray matter"), which controls perception and thinking (Fischer & Rose, 1994; Greenough, 1993; Greenough et al., 1987).

From birth to age 2, there is, in fact, an estimated fivefold increase in the density of dendrites in the cortex (Diamond, 1990). In some cases, there may be as many as 15,000 new connections established per neuron (Kolb, 1989), a proliferation that enables neurons to become connected to a greatly enlarging variety of other neurons within the brain.

This phenomenal increase in neural connections over the first two years has been called "transient exuberance" (Nowakowski, 1987), a characterization that actually highlights two key aspects of early brain develop-

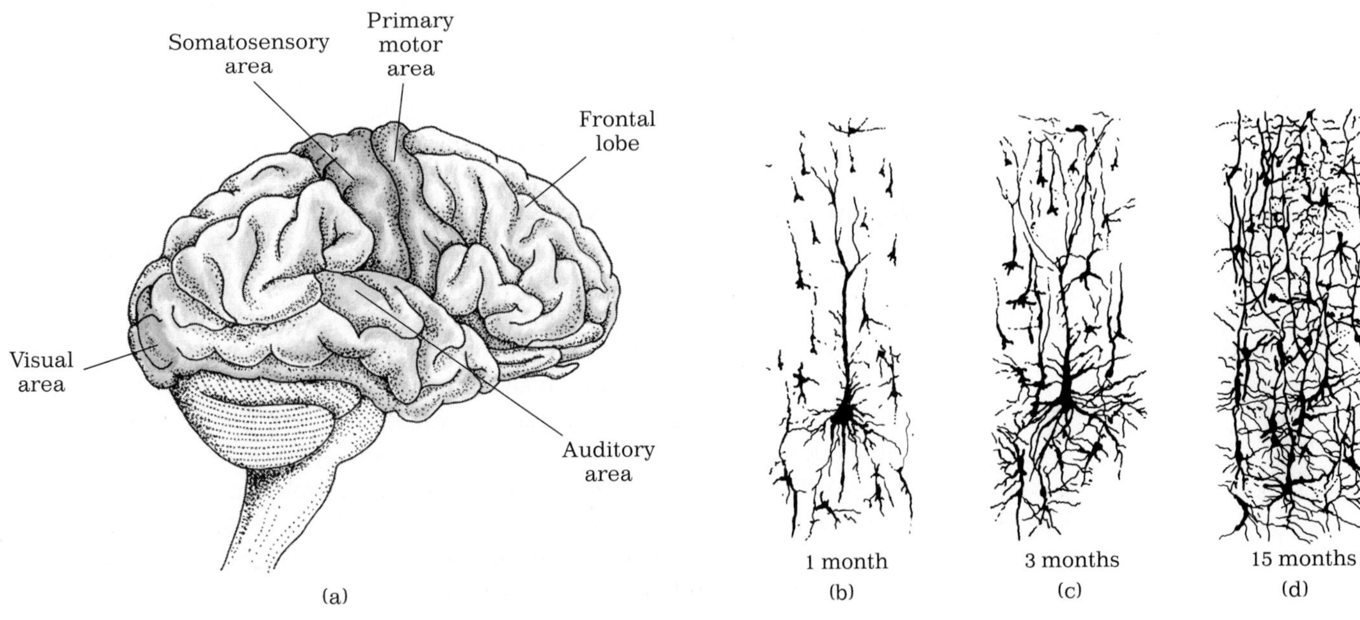

FIGURE 5.2 *(a)* Areas of the brain are specialized for the reception and transmission of different types of information. Research has shown that both experience and maturation play important roles in brain development. For example, myelination of the nerve fibers leading from the visual cortex of the brain will not proceed normally unless the infant has had sufficient visual experience in a lighted environment. The role of maturation is apparent in the growth and development of the neurons that make up the nerve fibers. These cells increase in size and in the number of connections among them as the infant matures, enabling impressive increases in the control and refinement of actions. Drawings *(b)*, *(c)*, and *(d)* illustrate the changes that take place.

ment. "Exuberance," of course, refers to the sheer magnitude of growth in neural connections. "Transient," on the other hand, refers not only to the fact that the rate of growth of neural connections slows as the child grows older but also to the fact that, over time, unused connections atrophy and disappear, in a kind of "pruning" action. Indeed, for many areas of the human brain, a person has more neural connections at age 2 than at any later age. In effect, during infancy, the human brain is prepared to process any type of experience a baby might have. Over the course of the early years, neural pathways that are used become strengthened and develop further connections. Those that are not used die off. Thus, both processes, proliferation and pruning, enhance the efficiency of neural communication while economizing the brain's overall organization (Huttenlocher, 1990; Kolb, 1989; Neville, 1991).

The functioning of the brain's communication networks is also enhanced by a process in which axons become coated with **myelin**, a fatty, insulating substance that speeds the transmission of neural impulses. This process of *myelination*, which proceeds most rapidly from birth to age 4 and continues through adolescence, allows children to gain increasing neurological control over their motor functions and sensory abilities and facilitates their intellectual functioning as well.

The overall maturation of the brain occurs in specific areas at different times and at different rates, affecting various behaviors in the child accordingly. For example, the frontal area of the cortex (located behind the forehead) assists in self-control and self-regulation. This area is immature in the newborn—a young infant cannot stifle a cry of pain or stay awake when drowsiness hits—but as the neurons of the frontal area become myelinated and interconnected during the first year, the baby is better able to regulate everything from reflexive responses to sleep-wake patterns. As development in the frontal area continues, cognitive skills requiring deliberation begin to emerge, along with a basic capacity for emotional self-control (Bell & Fox, 1992; Dawson, 1994; Fox, 1991). As a result, by age 1 the child's emotions are already much more nuanced and predictable, responsive as much to the external world (such as a frightening stranger) as to internal states (hunger and so on). We shall see these developments in more detail in Chapter 7.

The Role of Experience in Brain Development

As indicated above, brain development in the early years is not merely a question of biological maturation. At least a minimal amount of experience is essential, not only in the development of specific abilities but even in the development of the brain structures that make seeing, hearing, and other brain functions possible. This fact is demonstrated clearly by experiments in which animals who are temporarily prevented from using one sensory system or another in infancy become permanently handicapped in that sensory system. For example, if kittens are blindfolded for the first several weeks of life, they never develop normal vision, even though the anatomy of their eyes appears to be normal.

myelin A fatty insulating substance that coats the neurons, facilitating quicker, more efficient transmission of neural impulses.

The reason for this handicap is that, in the absence of visual experience, the neural pathways that transmit signals from the eyes to the visual area of the brain atrophy or fail to develop. If only one eye is temporarily blinded and the other is left normal, the kitten will be able to see but will

never develop the binocular vision that plays a role in depth perception, and thus the adult cat will have great trouble jumping onto a table or leaping from one chair to another. (Surprisingly, the unused visual pathways atrophy at a faster rate if only one eye is blindfolded, apparently because the proliferation of brain connections that occurs for the seeing eye signals to the visual areas of the brain that the neural pathways dedicated to the other eye are unnecessary, speeding their demise [Hubel, 1988].) When similar experiments are performed with older cats, their sight is not affected, because once early experiences have formed and strengthened neurological connections, the connections remain in place.

In very simple terms, these abnormalities occur because the deprivation of certain basic experiences in infancy prevents the development of the normal neural pathways that transmit sensory information, and once the animal's infancy is over, the brain cannot readily build such pathways. As researchers explain the process metaphorically, the "hard-wiring" of the brain—that is, the basic structures that allow the development of specific capacities—is genetically programmed and present at birth. What is required is that these hard-wired structures be "fine-tuned" through the development and integration of the connective neural networks. It is this fine-tuning process that can be affected by the animal's experience or the lack of it.

As best we know, the brain development of humans is similar to that of kittens and other animals in that it becomes fine-tuned through experiences in the first months and years. What are the implications of this fine-tuning process for human development? They are *not* that an infant would benefit from a multimedia extravagance of stimulation during infancy (indeed, an overstimulated baby typically cries and then goes to sleep). Rather, a certain minimal level of stimulation for each of the senses is needed for the brain regions and dendrite connections to develop optimally. In addition, since the areas of the brain dedicated to language and emotion mature rapidly in the first two years, it seems likely that cognitive and emotional experiences during a certain "time window" of neural flourishing foster and shape the brain connections needed for later language learning and emotional expression (Rovee-Collier, 1995). Some of the specifics of these implications—particularly what might happen if an infant were deprived cognitively or socially—may have far-reaching consequences. For the moment, we can say that scientists' understanding of the rapid growth of the human brain suggests that talking to a preverbal infant and showing affection toward a person too immature to love in return may be critical first steps toward developing that small person's human potential.

Regulation of Physiological States

An important function of the brain throughout life is the regulation of **physiological states**, or conditions. Just like an older child or adult, a full-term infant normally exhibits several regularly occurring states, the most distinct being *quiet sleep*, in which breathing is regular and slow (about thirty-six breaths per minute) and muscles seem relaxed; *active sleep*, in which the facial muscles move and breathing is less regular and more rapid (forty-six or

physiological states Refers to various levels of physiological arousal, such as quiet sleep and alert wakefulness.

RESEARCH REPORT

Schedules and the Culture of Infant Care

As the brain matures, the infant's developing ability to regulate alertness and arousal has important implications for the family system. Universally, newborns tend to be notoriously unpredictable: they sleep, wake, and cry with hunger at different hours and varying intervals, day and night. This erratic pattern gradually changes to one of longer and more stable periods of sleep and wakefulness and longer intervals between feedings—a welcome relief to the family. Parents who had run themselves ragged by trying to protect and please their disorganized little person find that they begin to relax and enjoy their baby much more once the child's states become more manageable and predictable, by 3 months or so.

Furthermore, after the child has begun to show regular sleep-wake patterns, parents begin to look forward to a good night's sleep of their own, rather than to mere snatches of rest interrupted by feeding, diapering, and calming the baby. Experienced parents know how important the child's "getting on a schedule" can be: it is one of the first topics they bring up with new parents, and they are quick to offer sympathy and reassurance if the response they get is an exhausted sigh.

But while physiological states become longer and more predictable with maturity in all infants, these universal processes are molded by specific child-rearing practices, which, in turn, reflect the values and assumptions of the family's particular culture. One such value is the importance of schedules *per se*. Most industrialized cultures place great importance on schedules, and on the punctual and predictable meeting of them, while many less developed, more rural cultures tend to be flexible and relaxed about observing timetables. These different approaches to scheduling affect the way parents attempt to "manage" the infant's early months (Triandis, 1994).

For example, the majority of parents in the United States are locked into a rigid workplace schedule and usually make concerted efforts to program their infant's sleeping accordingly, carefully scheduling naps and offering relaxing enticements for sleep, such as gentle rocking, a backrub, or a soft lullaby. Typically, the infant's mealtimes are also scheduled, usually to mesh with sleep and waking, so that soon the infant is having "breakfast," "lunch," "dinner," and "snacks" (even though the menu is actually the same on each occasion) in coordination with evening sleep and daytime nap patterns.

Such concerns about scheduling are much less apparent in nonindustrialized countries. For example, among the Kipsigis, a farming and herding community in Kokwet, Kenya, young infants regularly accompany their mother as she performs her farming, household, and social activities. In the early months, babies are carried around on a front sling, then on the mother's back, and later are allowed to crawl around and play nearby. At night, they sleep beside her. This overall pattern allows Kipsigis infants to sleep and nurse at any hour without disturbing the mother's daily activities or rousing her from bed at night.

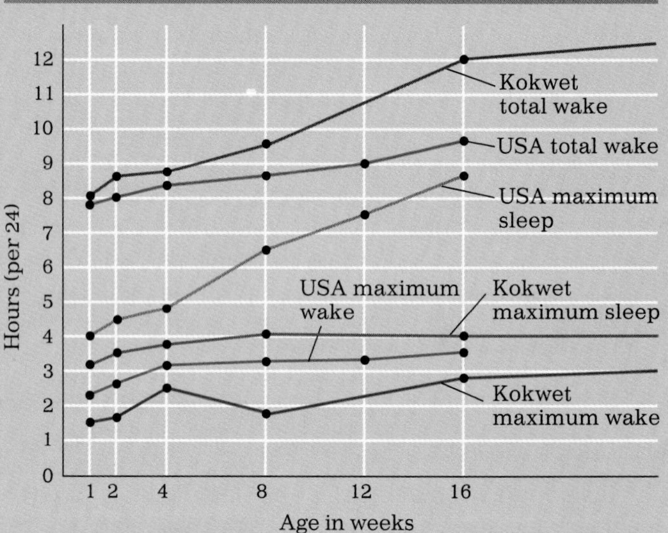

As you can see, the difference in the sleep patterns of Kipsigis infants in Kokwet, Kenya, and infants in the United States is not inborn. For the first month of life, the sleep patterns are almost identical. Then cultural differences in infant-care practices begin to have an impact, as American infants are awake for less total time per day than Kipsigis infants and sleep for longer intervals. By 16 weeks of age, American infants are sleeping notably more per day than the Kipsigis infants, in segments of up to 8 hours, while the Kipsigis babies tend to sleep no more than 4 hours at a stretch.

These diverse cultural attitudes about infant schedules become reflected in the baby's own body rhythms. When Charles Super and Sara Harkness (1982) compared the daily routines of Kipsigis infants with those of babies from middle-class homes in the United States, they found that although newborns from both groups initially exhibited similar patterns of sleep, by the end of the first year, the American babies were, on average, sleeping much more per day than were the Kipsigis (fifteen and a half hours compared to eleven) and for longer stretches at a time (see figure). In addition, by age 1, the American babies slept about seven hours at night, while the unbroken nighttime sleep of the Kipsigis 1-year-olds was only about half that amount. In a sense, middle-class American infants had learned to be "good babies" as defined by their culture, permitting their parents uninterrupted nighttime rest, while a "good" Kipsigis baby could still wake up several times a night.

(continued)

Schedules and the Culture of Infant Care (continued)

Of course, whatever the "scheduling norms" of a particular culture, the infant's own limitations must be taken into account. The most any newborn can sleep at a stretch is four or five hours, with some newborns being physiologically incapable of sleeping more than an hour or two at a time. Some older infants likewise sleep only for short periods. Consequently, an emphasis on sustained nighttime sleeping may overstress the young infant's self-regulatory capacities. As Michael Cole (1992) has noted,

> the rather rigid schedule imposed by modern industrialized lifestyles may be pushing the limits of what the immature human brain can sustain; hence, while the length of a longest sleep period may be a good indicator of physical maturity, pushing those limits may be a source of stress with negative consequences for children who cannot measure up to parental expectations.

If parents expect their infants to sleep soundly through the night, and instead their offspring wake at all hours, the parents are likely to blame both the child and themselves, and to become sleep-deprived as well. This makes them irritable at home, drowsy on the job, and too tired to have a social life. It also is likely to make them obsessed with sleep, talking about it "the way a hungry person talks about food" (Hochschild, 1989). Most American parents, in fact, feel that their infants' sleep patterns are problematic and search for solutions that will make the child sleep longer and awaken less often (Johnson, 1991).

The research we have been examining suggests, however, that at least some of the adaptation should be on the part of the parents—anything from trying to arrange some flextime shift in their work schedule to simply providing a midnight feeding months after the "good" baby is supposed to sleep through the night. Even if such practical adjustments are impossible or inadvisable, changes in attitude and expectations are warranted. An infant's natural rhythms of sleep and arousal may be problematic only in the context of an adult's highly organized, time-stressed, sleep-deprived lifestyle, and when they are problematic, they should not be taken as a sign of either the baby's willfulness or the parent's inadequacy.

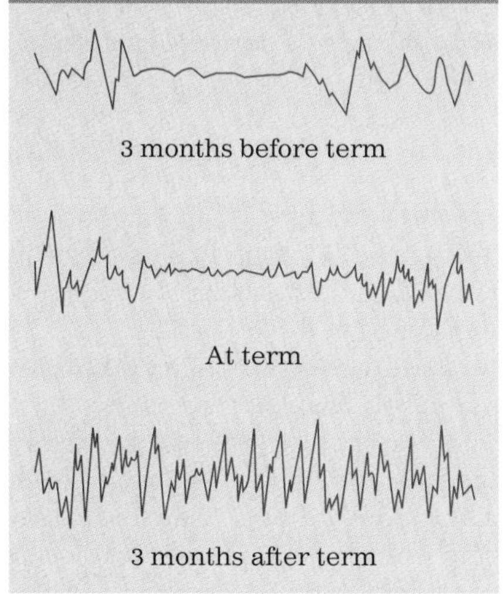

FIGURE 5.3 The more mature pattern of brain-wave activity shows many more bursts of electrical activity and greater overall intensity, as seen in this electroencephalogram of quiet sleep.

electroencephalogram (EEG) A graphic recording of the waves of electrical activity that sweep across the brain's surface.

more breaths per minute); *alert wakefulness*, in which the eyes are bright and breathing is relatively regular and rapid; and *active crying*—which every parent can recognize (Thoman & Whitney, 1990).

Because each state produces a particular pattern of electrical activity in the brain, the patterns can be measured and recorded as an **electroencephalogram (EEG)**, a graphic readout of the electrical impulses, or brain waves, from the neurons. Brain waves change rapidly from about 3 months before term to about 3 months after, reflecting the maturation that is taking place (see Figure 5.3).

As the brain develops, physiological states become more cyclical and distinct. With each passing week, for instance, infants are asleep and awake for longer, more regular periods, because their brain maturation allows deeper sleep, more definite wakefulness, and greater self-regulation of alertness, as noted above. Between birth and age 1, the infant's total daily sleep does not change all that much—from about sixteen hours a day for the newborn to thirteen for the toddler—but the length and timing of sleep episodes more closely match the day-night activities of the family. About a third of all North American 3-month-olds and 80 percent of all 1-year-olds "sleep through the night," defined as sleeping for at least six straight hours during the night. The remainder of each group continue to wake up wanting food and attention (Bamford et al., 1990; Michelsson et al., 1990). However, patterns of infant care can significantly influence the development of sleep-wake cycles in the first year (see the Research Report starting on the previous page). As one might expect, preterm newborns sleep more, but less regularly, throughout the first year.

Motor Skills

We now come to the most visible and dramatic of the physical changes that occur in infancy, those that ultimately allow the child to "stand tall and walk proud." Thanks largely to the changes in body size and proportion and the increasing brain maturation that we have outlined, infants gain dramatically in their ability to move and control their bodies. Consider the transition from the excited, undirected flapping of arms that 2-month-olds exhibit when a toy is dangled in front of them to the typical response of a 6-month-old—a smooth, efficient movement of the arm and shoulder muscles to intercept the toy, together with the finger movements that effectively close around the object. In the course of this four-month transition, infants have learned to (1) scale muscle movement against gravity to hit the target and not overshoot it; (2) compensate for the inertial forces that are transmitted from one muscle group to other muscles (from the shoulder to the arm, for example); (3) anticipate the trajectory of the arm in motion to enable the hand to intercept the target; (4) coordinate moving and braking forces in different muscle groups; and (5) organize these various components into a smooth motor action—while all the time their body is changing in size and strength! Researchers who have studied these "developmental biodynamics" have become convinced not only that this is a remarkable achievement but that its development is a painstaking process of trial-and-error accomplishments that gradually assembles and fine-tunes a sequence of smooth motor actions (Goldfield et al., 1993; Lockman & Thelen, 1993; Thelen et al., 1993). Thus the development of skilled motor behavior—whether it involves learning to walk or to grasp small objects with the fingers—is not simply a matter of waiting for a maturational timetable to unfold, but instead involves the active efforts of the infant to attain competence by mastering and coordinating successive components of each complex skill.

Because of the growing independence they afford the child, motor skills become a "catalyst for developmental change" (Thelen, 1987), as they open new possibilities for the child's discovery of the world. For this reason, especially, it is important to understand the development of these skills—including the usual sequence and timing of their emergence—and the various factors that might cause one child to develop certain skills "behind" or "ahead of" schedule.

Reflexes

The infant's first motor skills are not, technically, skills at all, but **reflexes**, that is, involuntary responses to particular stimuli. The newborn has dozens of reflexes. Some are essential to life itself; others disappear completely in the months after birth; still others provide the foundation for later motor skills. All are important as signs of neurological health and behavioral competence.

Three sets of reflexes are critical for survival and become stronger as the baby matures. One set works to maintain an adequate supply of oxygen. The most obvious reflex in this group is the **breathing reflex**. Normal newborns take their first breath even before the umbilical cord, with its supply of oxygen, is cut. For the first few days, breathing is somewhat irregular and

reflexes Involuntary physical responses to stimuli.

breathing reflex A reflex that ensures an adequate supply of oxygen and the discharge of carbon dioxide by causing the individual to inhale and exhale.

sucking reflex A reflex that causes newborns to suck anything that touches their lips.

rooting reflex A reflex that helps babies find a nipple by causing them to turn their heads toward anything that brushes against their cheek and to attempt to suck on it.

reflexive *hiccups*, *sneezes*, and *spit-ups* are common, as the newborn tries to coordinate breathing, sucking, and swallowing.

Another set of reflexes helps to maintain constant body temperature: when infants are cold, they *cry*, *shiver*, and *tuck in their legs* close to their bodies, thereby helping to keep themselves warm. A third set of reflexes fosters feeding. One of these is the **sucking reflex**: newborns suck anything that touches their lips—fingers, toes, blankets, and rattles, as well as natural and artificial nipples of various shapes. Another is the **rooting reflex**, which helps babies find a nipple by causing them to turn their heads and start to suck when something brushes against their cheek. *Swallowing* is another important reflex that aids feeding, as is *crying* when the stomach is empty.

Other reflexes are not necessary for survival, but they are important signs of normal brain and body functioning. For example, the following five reflexes are present in normal, full-term newborns:

1. When their feet are stroked, their toes fan upward (*Babinski reflex*).
2. When they are held upright with their feet touching a flat surface, they move their legs as if to walk (*stepping reflex*).
3. When they are held horizontally on their stomach, their arms and legs stretch out (*swimming reflex*).
4. When something touches their palms, their hands grip tightly (*Palmar grasping reflex*).
5. When someone bangs on the table they are lying on, newborns usually fling their arms outward and then bring them together on their chests, as if to hold onto something, and they may cry and open their eyes wide (*Moro reflex*).

None of these reflexes remain as involuntary responses after the first few months of life. Why, then, do they exist at all? Some may be vestiges of earlier evolutionary development. The Moro and Palmar grasping reflexes, for example, may have been crucial ways for the young infant to remain close to the mother, especially during startling or unexpected events. Others are the precursors of voluntary movements, or motor skills.

For developmentalists, newborn reflexes are mechanisms for survival, indicators of brain maturation, and vestiges of evolutionary history. For parents, they are mostly delightful and sometimes amazing. This is demonstrated by three star performers: a 2½-week-old infant stepping eagerly forward on legs too tiny to support her body; a 3-day-old infant, still wrinkled from amniotic fluid, contentedly sucking his thumb; and a newborn grasping so tightly that his legs dangle in space.

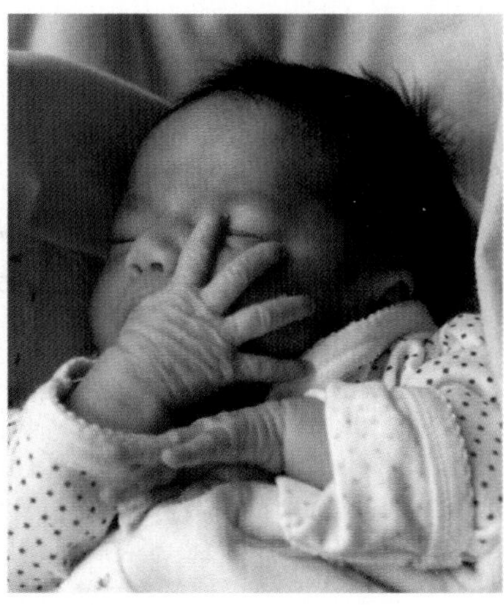

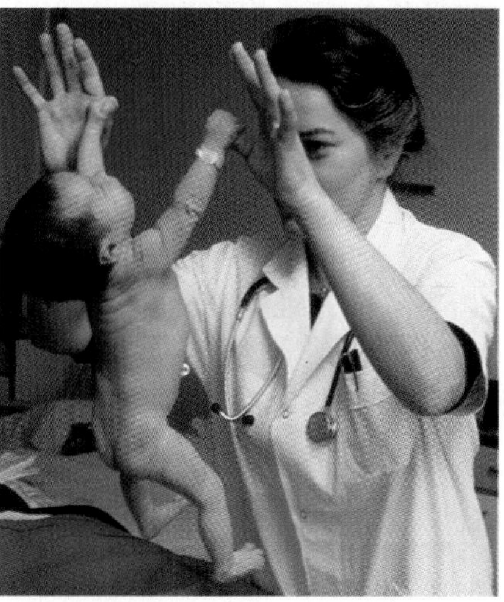

gross motor skills Physical skills involving large body movements such as waving the arms, walking, and jumping.

Gross Motor Skills

Gross motor skills, which involve large body movements, begin to emerge early. Even as newborns, infants placed on their stomachs move their arms and legs swim-fashion and attempt to lift their heads to look around. As they gain muscle strength, they start to wiggle, attempting to move forward by pushing their arms, shoulders, and upper body against the surface they are on. Although these initial efforts usually get them nowhere, or even move them backward, infants persist, and over the next two months or so, they become able to use their arms, and then legs, to inch forward. By 6 months, most infants succeed at this type of locomotion (Chandler, 1990). A few months later, usually between 8 and 10 months after birth, most infants are crawling on "all fours" (sometimes called creeping), coordinating the movement of their hands and knees in a smooth, balanced manner. Within a couple of months, most infants also learn to climb up onto couches and chairs—as well as ledges, window sills, and the like. Some babies do not crawl at all, achieving mobility instead by either scooting along on their buttocks, rolling over and over, doing the "bear walk" (on all four "paws," without letting their knees or elbows touch the ground), or even cruising unsteadily on two feet, moving from place to place by holding onto tables, chairs, or bystanders.

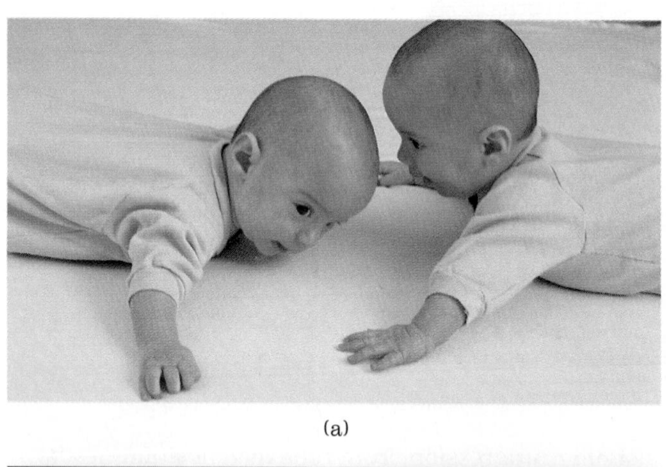

(a)

(b)

(c)

(d)

Nicholas and Daniel are monozygotic twins and consequently reach various stages of motor skills virtually together. The abilities shown here are (*a*) lifting the head and shoulders at 4 months, (*b*) preparing to crawl at 6 months, (*c*) standing with one supporting hand at 8 months, and (*d*) finally walking at 12 months—right on schedule.

As with every new skill, crawling opens new opportunities and challenges. Once infants can locomote on their own, they can propel themselves toward intriguing objects, whether nearby or across the room. They can even leave the room, exploring new areas and gaining a sense of their own independent actions. New hazards are also within reach, from the stairs they might tumble down to the floor polish they might taste. (The prudent parent seals off all dangerous places and substances by 6 months, if not sooner.) Fortunately, with most infants, the advent of crawling coincides with an emerging sense of wariness about the unfamiliar (see Chapter 7), producing a new measure of caution that tempers their curiosity: infants investigate a novel situation tentatively, frequently interrupting their explorations to glance at a parent for signs of encouragement or disapproval. Thus, a combination of motor skills, cognitive awareness, social interaction, and access to new surroundings makes the crawling 9-month-old a quite different baby from the precrawler (Bertenthal & Campos, 1990).

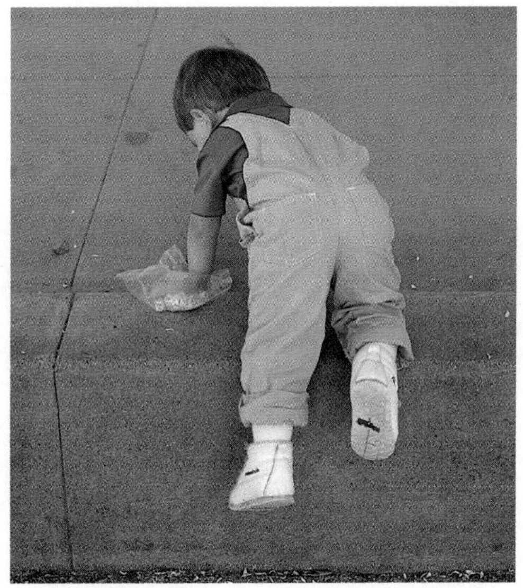

Motor skills develop rapidly during the first two years, partly because infants take advantage of every opportunity to use whatever abilities they have—grasping, crawling, creeping, climbing. In addition, they instinctively know how to proceed. This boy, for instance, naturally moves his left hand and right knee together, rather than trying only one limb at a time or both limbs on the same side of the body.

Walking shows a similar progression, from reflexive, hesitant newborn stepping to a smooth, speedy, coordinated gait (Thelen & Ulrich, 1991). On average, a child can walk while holding a hand at 9 months, can stand alone momentarily at 10 months, and can walk well unassisted at 12 months. In recognition of their accomplishment of walking, infants at this stage are given the additional name **toddler**, for the characteristic way they move their bodies, toddling from side to side. Since their heads and stomachs are relatively heavy and large, they spread out their short little legs for stability, making them seem bowlegged, flatfooted, and unbalanced. Interestingly, once an infant can take steps, walking becomes the preferred mode of movement—except when speed is an issue, and then many new walkers quickly drop to their hands and knees to crawl. Two-year-olds are proficient walkers and almost never crawl except when, with a mocking grin on their face, they pretend to be babies. Within a short time, mastery of walking leads to mastery of running.

toddler A child, usually between the ages of 1 and 2, who has just begun to master the art of walking.

In addition to allowing infants freedom of movement, crawling and walking aid their development in other ways. It is no coincidence that with

infants' increased mobility and independence comes a forward leap in their cognitive awareness (detailed in Chapter 6) and the opening of new dimensions in parent-infant interaction (described in Chapter 7). In addition, in purely practical terms, upright mobility not only raises the child's vistas figuratively but literally gives the child a new perspective on his or her world. It also frees up the child's hands, fostering the development of fine motor skills.

Fine Motor Skills

Fine motor skills, which mostly involve small movements of the arms, hands, and fingers, are more challenging to master because they require the coordination of complex muscle groups. As we have seen, infants are born with a reflexive grasp, but they seem to have no control of it. During their first 2 months, babies will stare and wave their arms at an object dangling within reach, and by 3 months, they can usually touch it. But they cannot yet grab and hold on unless the object is placed in their hands, partly because their eye-hand coordination is so limited. By 4 months, they sometimes grab, but their timing is often off, causing them to close their hand too early or too late, and their grasp tends to be of short duration. Finally, by 6 months, with a concentrated stare and deliberate movements, most babies can reach for, grab, and hold onto almost any object that is the right size, whether it is a bottle, a rattle, or a sister's braids.

Once grabbing is possible, infants explore everything within reach, mastering fine motor skills while they learn about the physical properties of their immediate world. As Eleanor Gibson, a leading researcher in infant perception, describes it, the infant at 6 months has "a wonderful eye-hand-mouth exploratory system," which before age 1 is sufficiently developed that the infant can "hold an object in one hand and finger it with the other, and turn it around while examining it. This is an ideal way to learn about the distinctive features of an object" and, bit by bit, about the tangible world (Gibson, 1988).

Other developing skills contribute to the child's ability to explore. By 4 to 8 months, most infants can transfer objects from one hand to the other. By 8 or 9 months, they can adjust their reach in an effort to catch objects that are tossed toward them, even when the object is thrown fairly fast and from an unusual angle (von Hofsten, 1983).

At the same time, the skill of picking up and manipulating small objects develops. At first, infants use their whole hand, especially the palm and the fourth and fifth fingers, to grasp. Later they use the middle fingers and the center of the palm or the index finger and the side of the palm. Finally, they use thumb and forefinger together, a skill mastered sometime between 9 and 14 months. At this point, infants delight in picking up every tiny object within sight, including bits of fuzz from the carpet and bugs from the lawn.

Development of these fine motor skills is enhanced by the development of gross motor skills, and vice versa. For example, a child who is able to sit steadily becomes more adept at grabbing and manipulating objects (Rochat & Bullinger, 1994; Rochat & Goubet, 1995). At the same time, once a child is able to grab, holding onto chair legs, table tops, and crib rails becomes easier, and this makes standing and even walking more possible. Once walking is possible, toddlers are much more able to poke, pick, and pull at hundreds of tiny things heretofore out of reach.

fine motor skills Physical skills involving small body movements, especially with the hands and fingers, such as picking up a coin and drawing.

Variations in Timing

Although all healthy infants develop the same motor skills in the same sequence, the age at which these skills are acquired varies greatly from infant to infant. Table 5.4 shows the age at which half of all infants in the United States master each major motor skill, and the age at which 90 percent master each skill.

These averages, or **norms**, are based on a large representative sample of infants from a wide range of ethnic groups. Such representativeness is important because norms vary from group to group, as well as from place to place. For example, throughout infancy, African-Americans are more advanced in motor skills than Americans of European ancestry (Rosser & Randolph, 1989). Internationally, the earliest walkers in the world seem to be in Uganda, where, if well nourished and healthy, the typical baby walks at 10 months; some of the latest walkers are in France, where taking one's first unaided steps at 15 months is not unusual.

What factors account for this variation in the acquisition of motor skills? Of primary importance are inherited factors, such as activity level, rate of physical maturation, and body type. The power of this genetic component is suggested by the fact that identical twins are far more likely to sit up, and to walk, on the same day than fraternal twins are. Moreover, there are striking individual differences in the strategies by which infants gradually master and coordinate the various components of motor actions—whether learning to walk or smoothly grasp a toy—that can also affect the timing of these achievements (Thelen et al., 1993). Particular patterns of infant care may also be influential. Indeed, among the Kipsigis of Kenya and other African groups, infants are held next to an adult's body virtually all day long, cradled and rocked as the adult works. This kind of stimulation allows the infant to practice movement while in an upright position and to continually feel the rhythm of an adult's gait, which may well give African

norms The overall usual, or average, standard for a particular behavior. Norms are generally the result of research done on a large sample of a given population.

TABLE 5.4

Age Norms (in Months) for Motor Skills

Skill	When 50% of All Babies Master the Skill	When 90% of All Babies Master the Skill
Lifts head 90° when lying on stomach	2.2	3.2
Rolls over	2.8	4.7
Sits propped up (head steady)	2.9	4.2
Sits without support	5.5	7.8
Stands holding on	5.8	10.0
Walks holding on	9.2	12.7
Stands momentarily	9.8	13.0
Stands alone well	11.5	13.9
Walks well	12.1	14.3
Walks backward	14.3	21.5
Walks up steps (with help)	17.0	22.0
Kicks ball forward	20.0	24.0

Source: The Denver Developmental Screening Test (Frankenburg et al., 1981).

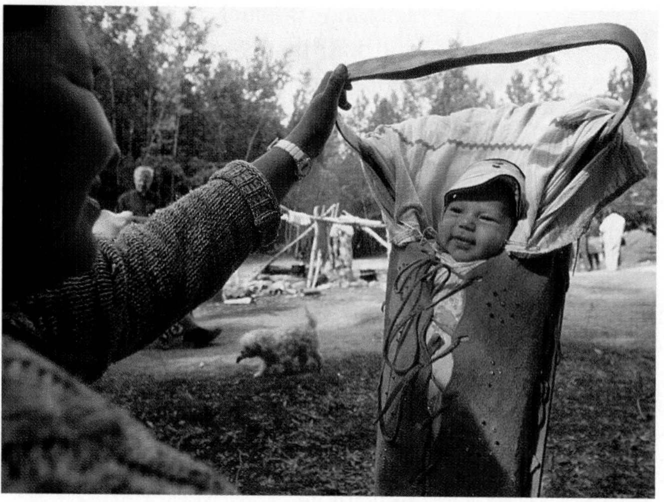

Some practice is essential for development of motor skills, but extensive experience is not necessary, as proven by Algonquin infants from Quebec, Canada, who spend much of their first year in cradle boards but typically sit up, walk, and run within the same age ranges as infants from other cultures.

babies an advantage in gross motor skills over the typical Western infant who spends much of each day in a crib (Bril, 1986).

In short, the age at which a *particular* baby first displays a *particular* skill depends on the interaction between inherited and environmental factors. Each infant has a genetic timetable for maturation, which can be faster or slower than that of other infants from the same ethnic group and even from the same family; and each infant also has a family and culture that provide varying amounts of encouragement, nutrition, and opportunity to practice.

Although variation in the timing of the development of motor skills is normal, a pattern of slow development—that is, development several months behind that of other babies of the same culture and ethnicity—requires attention to make sure that no problem is impeding the child's progress.

Sensory and Perceptual Capacities

Psychologists draw an important distinction between sensation and perception. **Sensation** occurs when a sensory system detects a particular stimulus. **Perception** occurs when the brain tries to make sense out of that stimulus, such that the individual becomes aware of it. This distinction may be clear to you if you have ever done your homework while playing the stereo and realized that you had worked through an entire recording but had actually "heard" only snatches of it. During the gaps in your "hearing," your auditory system was sensing the music—your tympanic membranes, hammer, anvils, stirrups, and the like were vibrating in response to the sound waves coming from the speakers—but you were not perceiving the music; that is, you were not consciously aware of it.

At birth, both sensation and perception are apparent. Newborns see, hear, smell, and taste, and they respond to pressure, motion, temperature, and pain. Most of these sensory abilities are immature and somewhat selective, responding to a fairly narrow range of stimuli. Thus newborns' perceived world is not at all the "great, blooming, buzzing confusion" psychologists once believed it to be (James, 1950; Mehler & Fox, 1985). Further, the stimuli they do

sensation The response of a sensory system to a particular stimulus.

perception The mental processing of sensory information.

respond to—such as visual patterns, the sound of a human voice, and sweet and sour tastes—reveal much about their growing comprehension of the surrounding world.

In this section, we will briefly consider the infant's sensory capacities and basic perceptual abilities. The fuller cognitive dimensions of infant perception will be examined in Chapter 6.

Research on Infant Perception

Over the past twenty years, there has been an explosion of research into infant sensory and perceptual skills. Technological breakthroughs—from brain scans to computer measurement of the eyes' ability to focus—have enabled researchers to measure the capacities of infants' senses and to gain a greater understanding of the relationship between perception and physiology.

The basis of this research is the fact that the perception of an unfamiliar stimulus elicits simple responses, for example, changed heart rate, concentrated gazing, and in the case of infants who have a pacifier in their mouths, intensified sucking. When the new stimulus becomes so familiar that these responses no longer occur, the infant is said to be *habituated* to that stimulus. Employing this phenomenon of **habituation**, researchers have been able to assess infants' ability to perceive by testing their ability to discriminate between very similar stimuli (Bornstein, 1985). Typically, they present the infant with a stimulus—say, a plain circle—until habituation occurs. Then they present another stimulus similar to the first but different in some detail—say, a circle with a dot in the middle. If the infant reacts in some measurable way to the new stimulus (a change of heart rate, a refocusing of gaze), that reaction indicates that the difference in stimulus has been perceived. A somewhat different strategy is to measure the infant's focused attention, as indexed by a fixed gaze, steady heart rate, and so on. Such fixation signifies perception of a stimulus.

Vision

At birth, vision is the least developed of the senses. Newborns focus most readily on objects between 4 and 30 inches (10 and 75 centimeters) away. Their distance vision is about 20/400, which means the baby sees an object 20 feet (6.1 meters) away no better than an adult with 20/20 vision sees the same object 400 feet (122 meters) away. However, distance vision develops rapidly, improving in the first months and reaching 20/40 by 6 months and 20/20 by 12 months (Haith, 1990, 1993). This improvement results more from changes in the brain than from changes in the eye. Distance focusing is not impossible for the newborn (as it would be for an adult with 20/400 vision), but the immaturity of the brain's neural networks makes such focusing slow and difficult (Braddick & Atkinson, 1988). As neurological maturation and myelination allow better coordination of eye movements and more efficient transmission of information between the eyes and the brain, focusing improves. By 6 months, the visual system has matured considerably and more closely approximates adult capacities.

During the same time period, increasing maturation of the visual cortex accounts for improvements in other visual abilities. When 1-month-olds

habituation The process of becoming so familiar with a particular stimulus that it no longer elicits the physiological responses it did when it was originally experienced.

binocular vision The ability to use both eyes together to focus on a single object.

How well can newborns really see? In the first weeks, their vision of Mom is closer to the picture on the left than to the one on the right. The ability to focus improves gradually, and it is not until about the end of the first year that a baby with normal eyesight develops 20/20 vision.

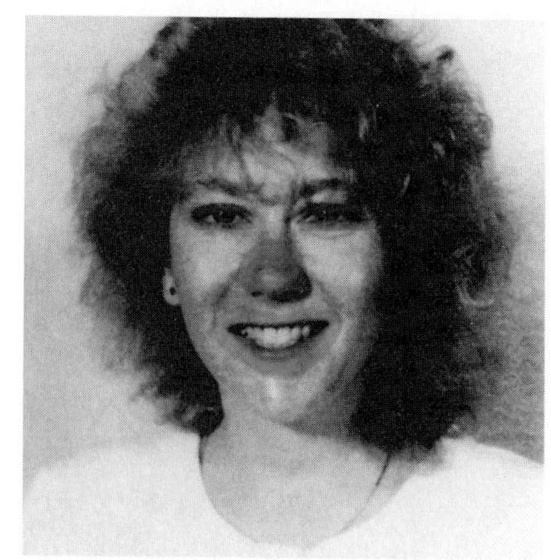

look at something, their gaze often wanders, and their ability to scan the object and attend to the critical areas is quite imperfect. When looking at a face, for example, they look at the peripheral features, such as the hairline. However, by 3 months of age, scanning is more organized, efficient, and centered on important aspects of a stimulus. Thus, when 3-month-olds look at a face, they scan the eyes and mouth regions, which contain more information (Aslin, 1988; Braddick & Atkinson, 1988). **Binocular vision**, that is, the ability to use both eyes together to focus on one object, also develops in the early months, occurring quite suddenly at about 14 weeks, on average (Held, 1993).

As a result of these achievements, depth and motion perception improve dramatically. Evidence of this comes from infants' ability to "track" a moving object, that is, to visually follow its movement (Nelson & Horowitz, 1987). Although some instances of tracking are apparent in the first days of life, this ability is erratic. Most very young babies "lose sight" of an object that moves slowly right in front of their face. One reason for this is that, even with stationary objects, newborns' eyes do not remain focused for long, and they do not focus on edges (Bronson, 1990). Thus continual, smooth tracking of a moving object is virtually impossible. In the months after birth, tracking improves week by week, with large, fast-moving, high-contrast objects being tracked more readily than small, slow-moving, low-contrast objects.

Scientists, aided by Donald Duck, monitor a 7-month-old girl's responses to visual stimuli. As various pictures flash on the screen, the infant's brain activity is recorded by means of the head-band device she is wearing, indicating not only what she sees and how well she sees it, but also which parts of her developing brain are processing visual stimuli.

Color vision is apparent from birth and rapidly becomes refined during the early months. Newborn infants can distinguish among red, green, and white, but are limited in detecting other colors (Adams, 1989; Adams et al., 1986). By 3 to 4 months of age, however, infants can distinguish many more colors and can also differentiate them more acutely, perceiving aqua, for example, as bluish rather than greenish (Bornstein & Lamb, 1992; Haith, 1990).

Infant Visual Preferences

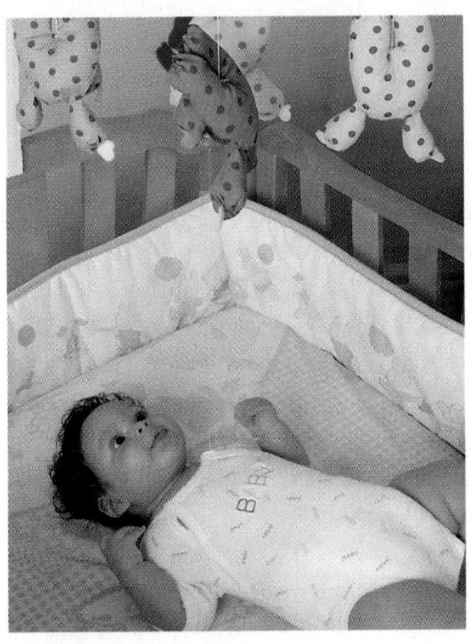

This 3-month-old's concentrated gaze is a sign that her brain is hard at work processing the visual information provided by her mobile. Experiences such as this are not only fascinating to infants but are essential for the normal development of the visual pathways of the cortex.

So far we have described what young infants are *able* to see. But what do they *prefer* to see when given a choice? One clear conclusion from the research on infant visual preferences is that babies seek visual stimulation that offers complexity within their range of perceptual ability. They prefer to look, for example, at novel images rather than at familiar ones, at complex visual patterns rather than at solid colors, and at stimuli with contrast and contour density (like a three-dimensional mask of a face) rather than at something two-dimensional (like a picture of a face). In addition, infants increasingly enjoy visual events that represent incongruity or discrepancy from the usual, such as seeing a familiar crib toy turned upside-down (Haith, 1980, 1990). This preference for visual stimulation may arise from the fact that, as we have seen, visual stimulation is necessary for the full development of the visual system in the early months of life. It may also occur because visual complexity contains more information that provokes the baby's interest—and will stimulate cognitive growth.

These findings have led to a new appreciation of the young infant as a stimulus-seeker who strives to make sense out of his or her surroundings. As Marshall Haith (1990), who has studied infant visual perception for more than thirty years, comments,

> this creature is actively processing whatever lies within its visual province and even looks for more, rather than simply choosing one stimulus or another. It is important for investigators to appreciate the infant as an active processor rather than a selector and to try to figure out what the baby is trying to accomplish rather than how dimensions of the world control its activity.

Hearing

Relative to their vision, newborns' hearing is quite sensitively attuned. Sudden noises startle newborns, making them cry; rhythmic sounds, such as a lullaby or a heartbeat, soothe them and put them to sleep. When they are awake, they turn their heads in an effort to locate the source of a noise (Clarkson et al., 1985), and they are particularly attentive to the sound of conversation. Indeed, as we saw in Chapter 4, newborns can distinguish their mother's voice from the voice of other mothers soon after birth.

By the age of 1 month, infants can also perceive differences between very similar speech sounds. In one experiment, 1-month-old babies activated a recording of the "bah" sound whenever they sucked on a nipple. At first, they sucked diligently, but as they habituated to the sound, their sucking decreased. At this point, the experimenters changed the sound from "bah" to "pah." Immediately the babies sucked harder, indicating by this sign of interest that they had perceived the difference (Eimas et al., 1971). It

(a)

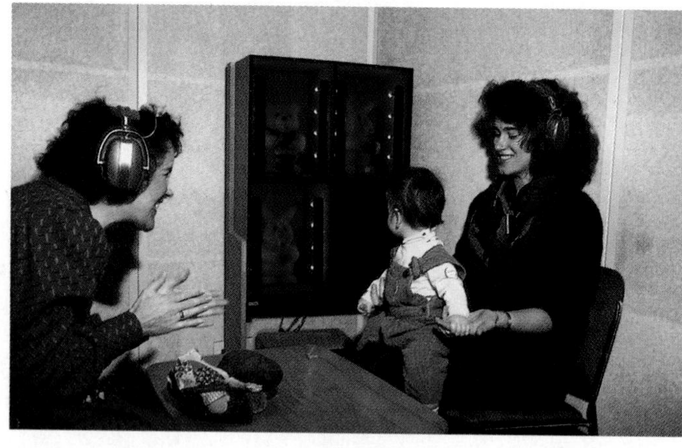

(b)

The procedure pictured here tests an infant's ability to detect changes in speech sounds. While the child is focused on a toy held by the experimenter (*a*), a single speech sound is played repetitively through a loudspeaker. At random intervals the speech sound is changed, and shortly thereafter one of the toys on the infant's right lights up and begins to move (*b*). After this routine is repeated a number of times, the infant learns that a change in speech sounds signals a delightful sideshow across the room. Thereafter, researchers can tell whether the infant discriminates between other speech sounds by whether or not the child looks expectantly over to the showcase after a particular sound is changed. (The experimenter and the child's mother are wearing special headphones that prevent them from hearing the speech sounds, thereby eliminating the possibility of their unwittingly cuing the child to the changes in sound.)

may even be that newborns have some ability to discriminate between vowels (Clarkson & Berg, 1983). More important, young infants can distinguish between speech sounds that are not used in their native language—and that are indistinguishable to adult speakers of their native language. For example, whereas English-speaking adults cannot distinguish between different "t" sounds that are used in Hindi speech, or between various glottal consonants used in some Native American dialects, their infants would be able to differentiate these sounds. This suggests to some researchers that there may be certain innate features to early speech perception (Werker, 1989). Significantly, however, over the course of the first year, and especially with the emergence of early language skills, infants gradually lose this ability (Werker, 1989), and by late childhood, many children simply cannot perceive nuances of pronunciation that are irrelevant to their mother tongue. In a sense, their speech perception becomes developmentally fine-tuned to learning the language sounds of the culture in which they are being raised. Thus sensory experience changes the child's capacity to perceive speech. Similar contextual fine-tuning occurs in many other areas of perception, as we shall see in Chapter 6.

Although very young infants can discriminate among a wide variety of sounds, their hearing is not as good as that of an older child (Trehub et al., 1991). Even at 6 months, when infants can hear high-frequency sounds as well as older children can, their hearing for low-frequency sounds is much less acute (Olsho, 1984; Olsho et al., 1988). Undoubtedly this is one reason most adults use a higher pitch when talking to babies than when talking to other people, as discussed in Chapter 6, and why infants prefer listening to "baby talk" over adult styles of conversation (Cooper & Aslin, 1990; Fernald, 1985; Fernald & Kuhl, 1987). Infant hearing differs in another way from that of older children: infants are less capable of locating sounds in space. Most older children can intuitively determine the location of a sound (say, off to the right or left) based on which ear receives the auditory signal first. But because infants have smaller heads, their ears are closer together, significantly limiting their ability to make this determination. Sound localization ability improves gradually throughout infancy (Morrongiello & Rocca, 1990).

Nutrition

As we have seen, under normal circumstances, infants double their birthweight in the first months, a growth rate that often requires feeding every three or four hours, day and night. The actual feeding "schedule," which can vary considerably from one child to another, is not the crucial factor, however. What matters is the overall quality and quantity of the infant's nutritional intake. Adequate nutrition is essential not only to physical growth but to brain development and skill mastery as well.

The Ideal Diet

At first, infants are unable to eat or digest solid food, but their rooting, sucking, swallowing, and breathing reflexes make them well adapted for consuming the quantities of liquid nourishment that they need. In these early months, breast milk is the ideal infant food (Cunningham et al., 1991). It is always sterile and at body temperature; it contains more iron, vitamin C, and vitamin A than cow's milk; and it also contains antibodies that provide the infant some protection against diseases, such as measles, polio, and specific types of flu, that the mother is immunized against. In addition, the specific fats and sugars in breast milk make it more digestible than any formula, which means that breast-fed babies have fewer allergies and stomach upsets than bottle-fed babies, even when both groups of babies have similar family backgrounds and excellent medical care. Further, breastfeeding decreases the frequency of almost every common infant ailment (Beaudry et al., 1995; Dewey et al., 1995). Finally, research is now discovering many new and possibly crucial substances in breast milk. As one scientist explained, "It's a cocktail of potent hormones and growth factors, most of which we are just beginning to understand" (Frawley, quoted in Angier, 1994). Among the newly discovered ingredients in breast milk are various hormones believed to help regulate growth, encourage attachment, reduce pain, and regulate the brain, liver, intestines, and pancreas. There is also one hormone in breast milk that may affect the timing of sexual maturation.

Given all the benefits of breast milk, doctors worldwide recommend breastfeeding for almost all babies, unless the mother is an active drug addict, is HIV-positive, or is severely malnourished. More precisely, the World Health Organization recommends that

> all infants should be fed exclusively on breast milk for the first four to six months of life. Children should continue to be breast-fed, while receiving adequate complementary foods, through the second year of life and beyond. [UNICEF, 1990]

However, despite the advantages of breast milk, breastfeeding is no longer the most common method of feeding babies. The advent of the rubber nipple, the plastic bottle, canned milk, powdered milk, and premixed infant formulas available in handy six-packs has meant that many infants now survive and thrive without ever tasting breast milk. Fewer than a fifth of all babies born in the United States are breast-fed for six months or more (Brody, 1994).

Although breast milk or formula can be the exclusive food in the first six months, by 6 months or so, "solid" foods should gradually be added to the

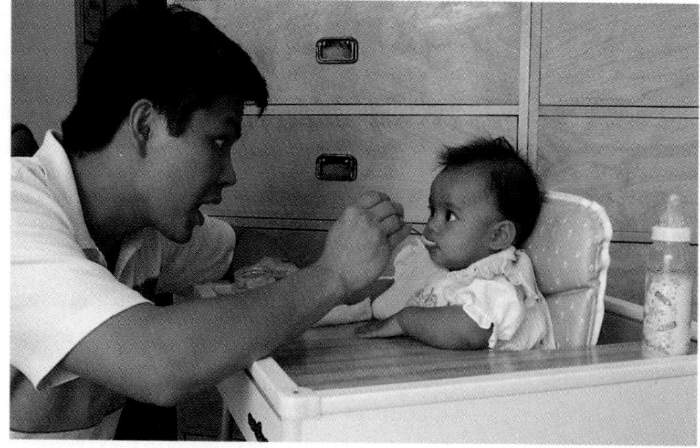

Feeding an infant "solid" foods usually begins in earnest at about 6 months of age, as the baby's digestive system matures and his or her nutritional needs become more complex. The father of this 7-month-old is obviously an experienced feeder, having mastered the "open wide" expression to signal the arrival of the spoon.

diet. Cereals are needed for iron and B vitamins, fruits for vitamins A and C, and when these first solids are well-tolerated, vegetables, meat, and fish can be introduced to provide additional nutrition (Purvis & Bartholmey, 1988). By the time the infant is a year old, the diet should include all the nutritious foods that the rest of the family consumes.

Nutritional Problems

Nutritional problems of every sort, including obesity and life-threatening vitamin deficiencies, occur throughout the life span. In infancy, however, the most common, serious problem is a quite simple one: **protein-calorie malnutrition**, the result of a person's not consuming enough nourishment to thrive. Roughly 7 percent of the world's children are severely malnourished during their early years, with rates running above 50 percent in nations such as Peru, the Sudan, and the Philippines (United Nations, 1994). The consequences for these children can be devastating. In the first year of life, severe protein-calorie malnutrition causes **marasmus**, a condition in which growth stops, body tissues waste away, and the infant eventually dies. During toddlerhood, malnourished children are more likely to suffer **kwashiorkor**, a condition, caused by a deficiency in protein, in which the child's face, legs, and abdomen swell with water, sometimes making the child appear well-fed to anyone who doesn't know the real cause of the bloating. In children with kwashiorkor, the essential organs claim whatever nutrients are available, so other parts of the body become degraded. This includes the children's hair, which usually becomes thin, brittle, and colorless, a telltale sign of systemic malnutrition.

protein-calorie malnutrition A nutritional problem that results when a person does not consume enough nourishment to thrive.

marasmus A disease that afflicts young infants suffering from severe malnutrition. Growth stops, body tissues waste away, and death may eventually occur.

kwashiorkor A disease resulting from a protein deficiency in children. The symptoms include thinning hair and bloating of the stomach, face, and legs.

The primary cause of malnutrition in developing countries is early cessation of breastfeeding. In many of these countries, breastfeeding was usually continued for at least two years, but now it is often stopped much earlier in favor of bottle-feeding, usually with powdered formulas. Under normal circumstances, such formulas are adequate and safe. However,

> for many people in the developing world . . . the hygienic conditions for the proper use of infant formula just do not exist. Their water is unclean, the bottles are dirty, the formula is diluted to make a tin of powdered milk last longer than it should. What happens? The baby is fed a contaminated mixture and soon becomes ill, with diarrhea, which leads to dehydration, malnutrition, and, very often, death. [Relucio-Clavano, quoted in Grant, 1986]

In developed countries, severe malnutrition in infancy is not widespread, even among families with very low income. This is because social programs, though often inadequate in many ways, tend to meet the essential nourishment needs of infants. Even when a particular impoverished family cannot obtain welfare, food stamps, or other governmental assistance, enough help is usually available from neighbors, relatives, and religious groups to prevent the extremes of marasmus or kwashiorkor. However, isolated cases of severe malnutrition during infancy do occur, when emotional and physical stresses on the caregivers (or the devastating effects of drug addiction on parenting) are so overwhelming that the adults ignore the infant's feeding needs or prepare food improperly, and when the malnutrition of such an infant is not noticed by the larger community.

Less apparent, but far more prevalent, than malnutrition is undernutrition. Worldwide, according to United Nations statistics, 188 million children are undernourished, including 56 percent of the children in the least developed countries (UNICEF, 1994). Within developing countries, estimates vary, from about 3 percent to about 15 percent (see Figure 5.4), with many additional children suffering specific vitamin deficiencies, particularly iron-deficiency anemia. In general, whenever an infant gains weight more slowly than would be expected for his or her age, sex, and genetic background, undernutrition should be suspected. Two other warning signs are an infant's being sick often, since illness is both a consequence and a cause of undernutrition, and a caregiver's being inattentive, depressed, or inflexible, since adequate feeding requires frequent, patient, and responsive care (Drotar et al., 1990; Drotar et al., 1994).

Most commonly, undernutrition is caused by a complex interaction of factors, with social and/or family problems being prime underlying factors.

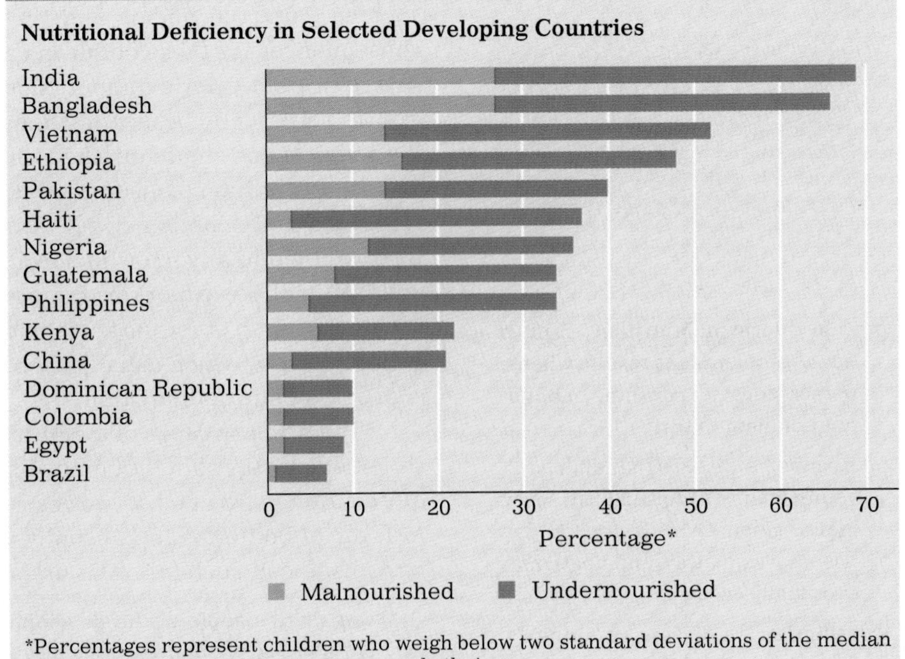

*Percentages represent children who weigh below two standard deviations of the median weight for children of their age, sex, and ethnic group.

Source: UNICEF, 1994.

FIGURE 5.4 In a well-fed community, about 2 percent of all children would weigh below two standard deviations of the norm, simply because they were genetically destined to be small. As you can see from this sample, many children worldwide are underweight. The consequences are lifelong, occurring in every domain. For example, undernourished children are more vulnerable to disease, more likely to drop out of school, and less likely to marry and have children. If they do have children, women who were undernourished when they themselves were young tend to have more difficult pregnancies and a higher rate of low-birthweight infants than women who were well nourished as children—even if women in both groups are well fed as adults. Given the pervasive consequences of undernutrition, public health officials contend that relief efforts should center as readily on undernourished children as on the emaciated, listless, severely malnourished children who easily capture public sympathy.

This is most obvious in regions of the world where everyone is undernourished: typically, the society's socioeconomic policies do not reflect the importance of infant nutrition, and parents may not even realize that a somewhat thin offspring is undernourished.

In developed countries, problems contributing to undernourishment are generally centered in the home. Mothers who are depressed, for example, tend to feed their children erratically and to be highly arbitrary in deciding when an infant has had enough to eat (Drotar et al., 1990). Likewise, families that are inflexible may have trouble adjusting to an infant's feeding pattern if it is irregular (Drotar et al., 1994). Undernutrition also arises from ignorance of the infant's nutritional needs. For example, one common form of undernutrition in the United States is "milk anemia," so named because it arises from parents' giving their toddler a bottle of milk (which has no iron) before every nap and with every meal, inadvertently destroying the child's appetite for other foods that are iron-rich.

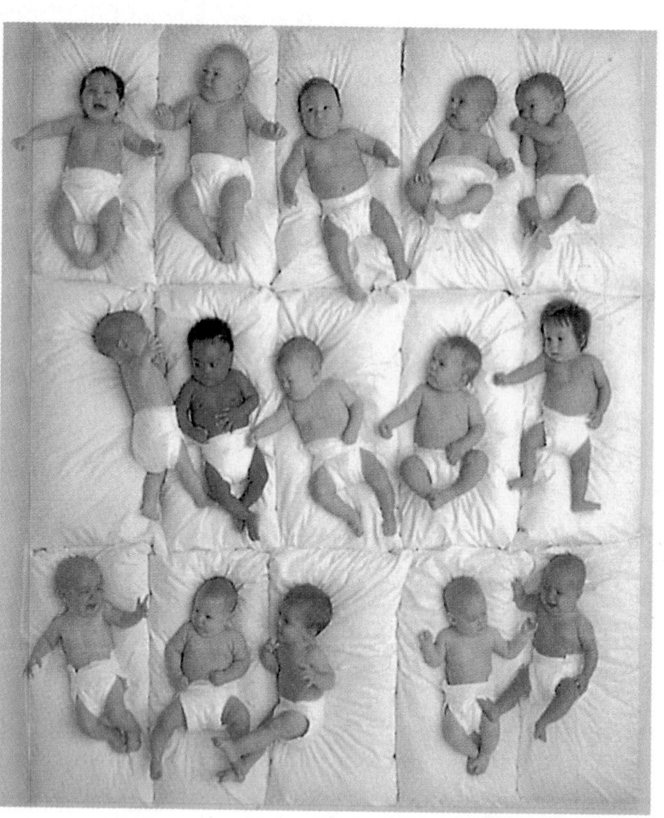

Normal genetic variation is apparent here, in appearance, activity level, and body size. However, if this is a typical sample of American infants, at least one infant is undernourished and one is overnourished. Infants who are undernourished have insufficient energy for normal growth or for expressing curiosity; infants who are overfed are slower to develop motor skills and are more likely to have a variety of health problems—ranging from asthma to heart disease—later in life.

Because of these complexities, programs that see undernutrition as a problem of poor families exclusively, and that seek to solve the problem by providing poor families with free food, are too narrowly focused (Ricciuti, 1991). To be successful in staving off the harm of inadequate nutrition, social policy must consider the entire context, including the need to raise parents' "nutritional consciousness."

The consequences of infant malnutrition and undernutrition extend far beyond the obvious threats to life and health. Since the brain is developing rapidly during these years, an inadequate supply of nutrients can stunt

intellectual development for years to come. Further, malnourished infants and toddlers are less likely to be interested and involved in the sensory, intellectual, and social events that surround them, and thus their brains may not develop as many neural connections as they ordinarily might. For both reasons, it comes as no surprise that longitudinal research on children in Mexico, Kenya, Jamaica, and Barbados, as well as in Europe and North America, reveals that children who are underfed in infancy tend to show impaired learning—especially in their ability to concentrate and in their language skills—throughout childhood and adolescence (Dobbing, 1987; Galler, 1989; Grantham-McGregor et al., 1994). However, these same studies also show that if a severely malnourished infant receives good caregiving as a young child—not only adequate food but also cognitive stimulation and caring social support—many of the deficits caused by the early hunger will disappear (Ricciuti, 1991; Super et al., 1990).

This research also echoes one of the themes of biosocial growth illustrated throughout this chapter: the remarkable unfolding of brain maturation, physical growth, and perceptual development are guided by a genetic plan but require a supportive environment to realize maturational potential. This theme should be kept in mind as we proceed to the next chapter, because the equally remarkable unfolding of cognitive development also reflects the dual influences of biological endowment and social support.

SUMMARY

Physical Growth and Health

1. In their first two years, most babies gain about 20 pounds (9 kilograms) and grow about 15 inches (38 centimeters). The proportions of the body change. The newborn is top-heavy, for the head takes up one-fourth of the body length, partly because the brain, at birth, has attained a high proportion of its adult size in comparison to other parts of the body. In adulthood, the head is about one-eighth of the body length.

2. Improved immunization programs have almost wiped out many of the diseases that once sickened almost every child and killed many. However, the very success of some voluntary programs in the United States may now represent a threat to their effectiveness. No longer faced with reminders of the dangers of childhood diseases, parents may become careless about vaccinating their preschoolers.

3. One cause of infant mortality remains unexplained and unpredictable—sudden infant death syndrome. While the precise cause and ultimate cure of SIDS are yet to be discovered, various measures can be taken to reduce its risk. One of the simplest precautions is putting the infant to sleep on his or her back.

Brain Growth and Development

4. At birth the brain contains over 100 billion nerve cells, or neurons, but the networks of nerve fibers that interconnect them are relatively rudimentary. During the first few years of an infant's life, there are major spurts of growth in these networks, enabling the emergence of new capabilities, including self-regulation and certain cognitive skills.

5. Over the course of the early years, neural pathways in the brain that are used become strengthened and further developed, and those that are not used atrophy.

6. Brain maturation alters the infant's physiological states, bringing deeper sleep, more definitive wakefulness, greater self-regulation of alertness, and increasingly regular sleep patterns. Sleep-wake cycles can, however, be significantly influenced by parental care.

Motor Skills

7. The development of motor abilities during the first two years allows the infant new possibilities in discovering the world. Gross motor skills involve large movements, such as running and jumping; fine motor skills involve small, precise movements, such as picking up a penny.

8. At first, the newborn's motor abilities consist of reflexes. Some reflexes are essential for survival; some provide the foundation for later motor skills; others simply disappear in the first months. However, all reflexes are indexes of brain development.

9. Although the sequence of motor-skill development is the same for all healthy infants, babies—for hereditary, developmental, and environmental reasons—vary in the ages at which they master specific skills.

Sensory and Perceptual Capacities

10. Both sensation and perception are apparent at birth, and both become more developed with time. Newborns are capable of virtually all modes of sensory experience and show early preferences for some types of events over others.

11. At birth, vision is the least developed of the senses, but during the first months of life, distance and binocular vision, focusing skills, and color perception improve considerably. In contrast, hearing is relatively acute at birth.

Nutrition

12. Physical growth, brain development, and the mastery of motor skills all depend on adequate nutrition, and doctors worldwide recommend breast milk as the ideal food for most babies.

13. In developing countries, severe malnutrition can often be attributed to the early cessation of breastfeeding and improper preparation of commercial formulas. Marasmus, which is caused by long-term protein-calorie deficiency, results in a cessation of growth, the wasting away of body tissue, and eventually death. Kwashiorkor, which is caused by long-term protein deficiency, results in bloating and degradation of various parts of the body.

14. Undernutrition is quite common in developing countries and is often apparent in developed countries as well. The consequences vary, depending in part on how long the child was underfed and in part on the child's intellectual stimulation at home and in school.

KEY TERMS

sudden infant death syndrome (SIDS) (135)
neurons (138)
axons (138)
dendrites (138)
myelin (139)
physiological states (140)
electroencephalogram (EEG) (142)
reflexes (143)
breathing reflex (143)
sucking reflex (144)
rooting reflex (144)
gross motor skills (145)
toddler (146)
fine motor skills (147)
norms (148)
sensation (149)
perception (149)
habituation (150)
binocular vision (151)
protein-calorie malnutrition (155)
marasmus (155)
kwashiorkor (155)

KEY QUESTIONS

1. How do the proportions of the infant's body change during the first two years?

2. Why are voluntary immunization programs in the United States now at risk of becoming victims of their own success?

3. What are the leading risk factors for sudden infant death syndrome?

4. What specific changes occur in the brain's communication system during infancy and how do they affect the infant's physical functioning?

5. What role does experience play in the development of the brain's neural pathways?

6. How do the baby's capacities to regulate alertness and arousal change during the first year?

7. Which reflexes are critical to an infant's survival?

8. What is the general sequence of the development of motor skills?

9. What factors account for individual differences in the timing of motor achievements?

10. How do researchers determine whether an infant perceives a difference between two stimuli?

11. What are the sensory capabilities of a newborn infant? How do they change over the first year of life?

12. What kinds of visual experiences do babies typically prefer in early infancy?

13. What are the advantages of breastfeeding?

14. What are some of the consequences of serious, long-term malnutrition?

15. Why does undernutrition occur?

The First Two Years: Cognitive Development

CHAPTER 6

Imagine, for a moment, that you are a newborn infant, and then consider some of the elementary things you have yet to learn about the world around you. Enveloped in a swirl of constantly changing images, sounds, smells, and physical sensations, you must, in the weeks and months ahead, begin determining how these surrounding events fit together, enabling you to develop perceptions of objects, people, and even parts of your own body. You must begin deducing which of these events are enduring features of your everyday experience, and which change from moment to moment, or day to day. You must start figuring out the characteristics of the objects in your world—where they exist relative to you, where and how they move, and whether they are hard or soft, firm or flexible, and so forth. You must begin intuiting the sequences of events, linking causes and the effects they produce, and predicting the consequences of the events you can observe. And these are just the beginning of the incredible variety of cognitive tasks that babies must begin to achieve. No wonder that the grandfather of American psychology, William James, described the young infant's world as a "blooming, buzzing confusion"!

If you are now impressed with the remarkable cognitive growth that must occur early in infancy, consider another thing that has left developmental psychologists in awe: the speed and apparent ease with which this growth occurs. By the end of the first year—and often much sooner—infants have a basic grasp of the fundamental attributes of the objects and people around them (such as their boundaries, their permanence over time and in space, and other properties). They also have a rudimentary concept of number, demonstrate simple problem-solving capacities, and have begun to use language. It is as if the newborn infant is biologically endowed with the necessary intellectual tools for beginning to take in, and understand, the world in all its daunting complexity, and is supremely motivated from birth to do so. By the end of their second year, toddlers are talking in short sentences, thinking through situations before acting on them, and pretending to be someone or something (a mother, an airplane) that they know they are not.

The goal of this chapter is to explore how and why these accomplishments occur during the first two years of life. We will begin where we left off in our discussion of perception in Chapter 5, with the interpretive processes by which infants comprehend sensory experience and make it useful intellectually. To complete our portrayal of the infant's remarkable cognitive accomplishments, we will then consider the growth of memory and other aspects of cognition, as well as the development of sensorimotor intelligence and the emergence of language.

Perception and Cognition

As you learned in Chapter 5, infants possess remarkably acute sensory abilities from their first days: they can see, hear, and otherwise sense with far greater skill than was earlier believed. They also develop early *preferences* for what they experience, showing a hunger for novelty and stimulation, and discriminating easily between familiar and unfamiliar events. But how do infants interpret and make sense of what they experience? Our examination of this side of perception—which is closely related to cognitive growth—will reveal an infant who is an active and eager interpreter of the world.

The first major theorist to realize that infants are active learners and that early learning is based partly on sensory abilities, was Jean Piaget, who was introduced in Chapter 2 and whose depiction of infants' *sensorimotor intelligence* is presented later in this chapter. We begin now, however, with the work of Eleanor and James Gibson, a husband-and-wife team whose understanding of the links between perception and cognition has inspired much of the current research on infants' cognitive growth.

The Gibsons' Contextual View

The Gibsons' central insight regarding perception is that it is far from being an automatic phenomenon that everyone, everywhere, experiences in the same way. Rather, perception is, essentially, an active cognitive process in which each individual selectively interacts with a vast array of perceptual possibilities (Gibson, 1969, 1982; Gibson, 1979).

In the Gibsons' view, all objects have many **affordances**; that is, they "afford," or offer, the perceiver the potential for various activities. What affordances a person actually perceives in any given object depends partly on the individual's developmental level and past experiences, partly on his or her present needs, and partly on the person's cognitive awareness of what the object might be used for. To take a simple example, a lemon affords, among many other possibilities, smelling, tasting, touching, viewing, throwing, and squeezing. Which of these affordances a person perceives depends on the individual and the situation: a lemon might elicit a quite different perceptual response from an artist about to paint a still-life, a thirsty adult in need of a cool, refreshing drink, and a teething baby wanting something to gnaw on.

The idea of affordances thus emphasizes that there is an ecological fit between individual perceptions and the environment, such that affordances do not reside solely in the objective qualities of the object itself but arise in large measure from how the individual subjectively perceives the object (Ruff, 1984). As one psychologist explains:

> If I want to sit down in a sparsely furnished bus station, a floor or a stack of books or a not-too-hot radiator might afford sitting. None of these are chairs, and thus their affordance of "sit-ability" is in relationship to my perception. [Gauvain, 1990]

Affordances are not limited just to objects but also are perceived in the physical characteristics of a setting and the people in it (Reed, 1993).

One of the first affordances that an infant needs to perceive from the environment is *graspability*—whether an object is the right size, shape,

affordances The various opportunities for interaction that an object offers. These opportunities are perceived differently by each person depending on his or her past experiences and present needs.

and texture for grasping, and whether it is within reach. This is vital information for infants, since they learn a great deal about their world by handling various objects (Palmer, 1989; Rochat, 1989). Extensive research has shown that infants perceive graspability long before their manual dexterity enables them to actually grasp successfully. For instance, when 3-month-olds view objects, some graspable and some not, they move their arms excitedly in the direction of those that are the right size and distance for grasping but merely follow ungraspable objects with their eyes (Bower, 1989).

The fact that babies perceive graspability so early helps explain how they explore a face. Once they have some control over their arm and hand movements, and a face comes within their reach, they immediately grab at it. But their grabbing is not haphazard: they do not grab at the eyes or mouth (although they might poke at them), for they already perceive that these objects are embedded, and thus do not afford grasping. A pull at the nose or ears is more likely, because these features do afford grasping. Even better, however, are glasses, or earrings, or a long mustache—all of which are quickly yanked by most babies, who perceive at a glance the graspability these objects afford.

Similarly, from a very early age, infants understand which objects afford suckability, which afford noise-making, which afford movability, and so forth. An impressive feature of this perceptual capacity is the infant's ability to distinguish affordance similarities in dissimilar objects (rattles, flowers, and pacifiers are all graspable) and affordance differences in similar objects (among objects the same color, size, and shape, furry ones are more likely to be patted and rubber ones more likely to be squeezed) (Palmer, 1989).

The affordances infants perceive in the common objects around them evolve as babies gain experience with those objects. A sloping ramp, for example, may afford ascent and descent—but it may also afford falling. Which affordances an infant might perceive in the ramp depends in part on his or her prior experience. In one experiment, Karen Adolph and her colleagues observed two groups of infants as they moved up and down ramps pitched at different inclines: one group consisted of 14-month-olds with plenty of walking experience; the other consisted of 8½-month-olds with crawling (but not walking) experience (Adolph et al., 1993a, 1993b). The researchers had expected that the older infants would respond more cautiously to the incline because they could better perceive the affordance of falling due to their prior experience of walking—and falling—over various surfaces. And they were correct: the 14-month-olds confidently walked down gentle slopes, but when the slopes were made more steep, they negotiated the descent instead by sliding down in a sitting position, often after much hesitation and searching for alternative positions in which to make the descent. By contrast, the 8½-month-olds, regardless of the steepness of the slope, tried to make the descent by crawling rather than sliding, often falling headlong (their mothers were standing by to catch them). Thus, although infants of both ages could perceive the ramp's affordance of descent, only the older infants (with prior walking experience) could perceive its affordance of falling and respond more cautiously as a result.

The Gibsons' ecological view emphasizes, therefore, that early perceptual development involves a growing knowledge of affordances that i

(a)

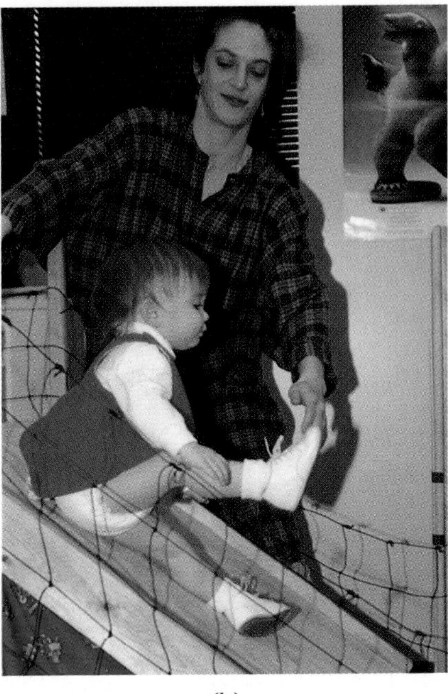

(b)

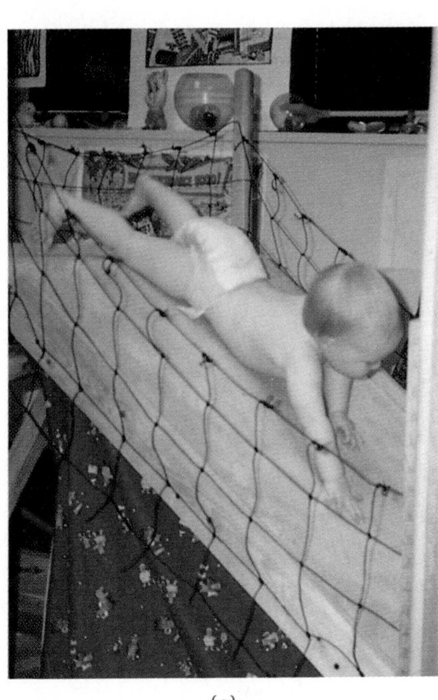

(c)

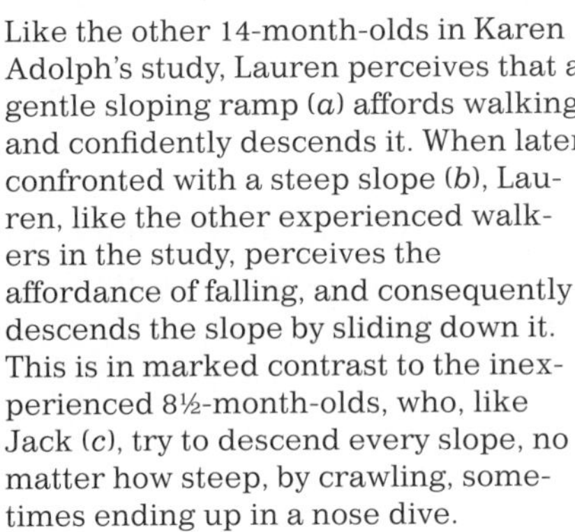

Like the other 14-month-olds in Karen Adolph's study, Lauren perceives that a gentle sloping ramp (*a*) affords walking and confidently descends it. When later confronted with a steep slope (*b*), Lauren, like the other experienced walkers in the study, perceives the affordance of falling, and consequently descends the slope by sliding down it. This is in marked contrast to the inexperienced 8½-month-olds, who, like Jack (*c*), try to descend every slope, no matter how steep, by crawling, sometimes ending up in a nose dive.

acquired through infants' active interactions with the objects, events, and people around them.

Perceiving the Constancy of Objects

A task that is obviously essential to perceiving affordances, and to the infant's overall cognitive growth, is gaining an understanding of the constancy of objects. This is a surprisingly complex accomplishment, as illustrated in this example:

> Consider what we see when we gaze at the family cat stretched out before the fireplace. We see a single, solid, three-dimensional object, located in a particular region of space and separate from both the objects it touches (e.g., the rug on which it lies) and the object it occludes (e.g., the fireplace behind it). We see an object of a particular size, shape, and color, and do so regardless of the distance between us and the cat, the angle of viewing, or the lighting in the room (all factors that change the image in our retina). If an object comes between the cat and us, blocking all except head and tail, we do not perceive the cat as being bisected; rather, our perception is still of a continuous, indivisible object. . . . If the cat stirs itself and strolls from the room we do not expect its [tail] or any other part to remain behind; instead, we realize that the parts of an animate object move together in predictable unison. If the movement takes the cat away from us we correctly perceive that it is receding from us (but not changing size, despite changes in the size of the retinal image); on the other hand, if the cat suddenly makes a run for our lap we perceive not only that it is approaching but that its course guarantees contact within a very short time. (Of course, how we feel about such contact depends more on our attitude toward cats than on perception per se.) [Flavell et al., 1993]

How long does it take a newborn infant to master these complex interpretations of the sensory world? Not long (Flavell et al., 1993; Haith, 1980; Spelke, 1988, 1991). By the age of 3 months, for example, infants are able to distinguish the boundaries of separate three-dimensional objects, and a few months later they can do so even when one object partly overlaps another.

From the angle of her arm and the bend of her hand, it appears that this infant recognizes the constancy of this furry mass, perceiving it as a single entity whether it is standing still, rolling in the sand, or walking along the beach.

This is especially true when the objects are in motion, as they typically are in everyday life. During the first half-year of life, moreover, infants also begin to develop an understanding of **perceptual constancy**, that is, the awareness that the size and shape of an object remain the same despite changes in the object's appearance due to changes in its location. Thus, even though an object (like a cat) looks smaller when seen from a distance, and looks different when perceived from different viewpoints, infants quickly grasp that it is the same object nevertheless.

Dynamic Perception

As we have seen in a number of instances, movement plays a key role in infants' perception of the property of objects and in the development of their perceptual and cognitive skills generally (Bornstein & Lamb, 1992; Flavell et al., 1993). Babies prefer to look at things in motion, whether those objects are a mobile rotating overhead, their own flexing fingers, or their favorite bobbing, talking, human face. They also use movement cues from an early age to discern not only the boundaries of objects but also their rigidity, wholeness, shape, and size. Infants can even form simple expectations of the path that a moving object will follow (Haith et al., 1993; Nelson & Horowitz, 1987). The fact that infants have **dynamic perception**, that is, perception primed to focus on movement and change, works well in a world in which stimuli are constantly moving within the infant's field of vision. Movement captures the baby's attention, highlights certain attributes of an object (like its boundaries), and produces changes in the infant's perception of the object that enable the infant to learn about its other qualities.

perceptual constancy The awareness that the size and shape of an object remain the same despite changes in the object's appearance due to changes in its location.

dynamic perception Perception primed to focus on movement and change.

One indication of dynamic perception is a baby's reaction to the "visual cliff," a surface constructed to look as if there is a sudden drop-off halfway across it. As shown here, when infants who are 8 or more months old are placed on a visual cliff, they hesitate, their heart rate increases, and they typically refuse to crawl over it, even if their mother encourages them to. In contrast, when younger infants with no experience at crawling are placed atop a visual cliff, they show no such signs of trepidation.

Not only objects' movements but also the baby's own movements enhance sensory and perceptual skills (Bertenthal & Campos, 1990). As babies scoot, crawl, creep, walk, and climb around, they are able to perceive and explore many new objects, from many different perspectives, gaining important information about the world around them.

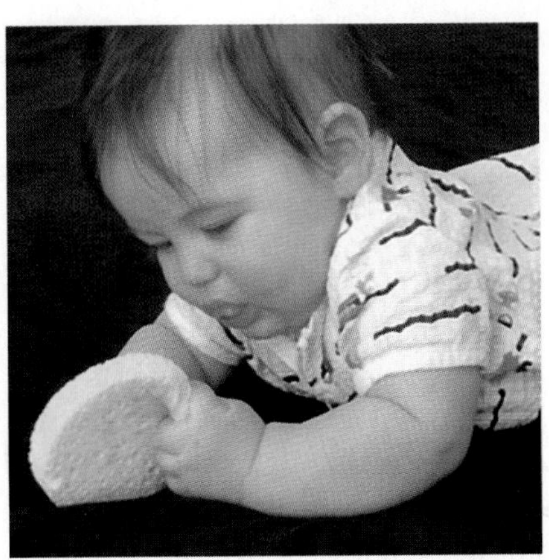

This infant is coordinating an intermodal perception by linking the sight of the sponge with its texture and affordance of squeezability. From the looks of it, the next mode of perception to be coordinated may be taste.

Coordination of the Sensory Systems

Once researchers were alerted to the early development of the infant's perceptual skills, they also began to look closely at the infant's ability to integrate perceptual information from different sensory systems.

One aspect of this is **intermodal perception**, the ability to associate information from one sensory modality (say, vision) with information from another (say, hearing). For example, when we sit near a lighted fireplace, it is through intermodal perception that we realize that the heat, the crackling, the smoky odor, and the flickering light all come from the same source.

Even newborns exhibit some intermodal perception, as when they look to see the source of a sound—though not always in the right direction. By 3 months, however, they not only look in the right direction for the source of what they hear but also have a notion of which sounds are likely to accompany what events. This has been demonstrated in various experiments that test whether infants can "match up" a film they are watching with an appropriate soundtrack, whether the soundtrack is of music, a voice, or simply noises—such as squishing sounds (matched with a film of a sponge being squeezed) or clacking sounds (matched with a film of wooden blocks hitting one another) (Bahrick, 1983). In one experiment, infants about 6 months old simultaneously viewed two films of a person talking, in one film, with a happy expression, in the other, with a sad expression. At the same time, they heard the soundtrack of one of the films, which conveyed either a happy or sad mood. Although the infants at first looked equally at both films, they soon began looking more intently at the one that matched the mood of the soundtrack (Walker, 1982).

Perhaps unsurprisingly, infants make similar discriminations based on the speaker's sex, looking more at the film of a man talking when the soundtrack is of a male voice and more at the film of a woman talking when the soundtrack is of a female voice (Walker-Andrews et al., 1991). Quite startling, however, is the fact that infants look more at speakers whose lip movements match the speech sounds the infants are hearing (Kuhl & Meltzoff, 1988). Basically, the infants are "reading lips" before they understand words!

The fact that infants 6 months old and younger are able to match pictures with sounds in so many different examples suggests that the babies are not simply making a primitive match between visual and auditory rhythms. Rather, they seem to be doing something more complex and more cerebral, turning information from one sensory modality into an expectancy and then matching that expectancy with information from another sensory modality.

Evidence for such cognitive integration of perceptual information also comes from research on a related ability known as **cross-modal perception**, the ability to use information from one sensory modality to imagine something in another—as when you hear the voice of a stranger on the phone and picture the person who is talking, or see a food and imagine how it tastes. In

intermodal perception The ability to associate information from one sensory modality (such as vision) with information from another (such as hearing).

cross-modal perception The ability to use information from one sensory modality to imagine something in another.

infants, of course, cross-modal perception is extremely rudimentary, but it has nevertheless been demonstrated many times (Rose & Ruff, 1987; Spelke, 1987).

The most convincing evidence that very young infants can translate information from one sensory system to another comes from experiments in which infants create a visual expectancy through their sense of touch. Essentially, experimenters allow infants to manipulate an object that is hidden from view and then show them two objects, one of which is the object they have just touched. By analyzing the infants' gaze, researchers can tell whether the infants distinguish the object they manipulated from the one they didn't. In one such experiment, for example, 2½-month-olds touched either a plastic ring or a flat disk and then were shown both objects. The duration of their gazing revealed that most of the infants "recognized" the object they had just manipulated (Streri, 1987).

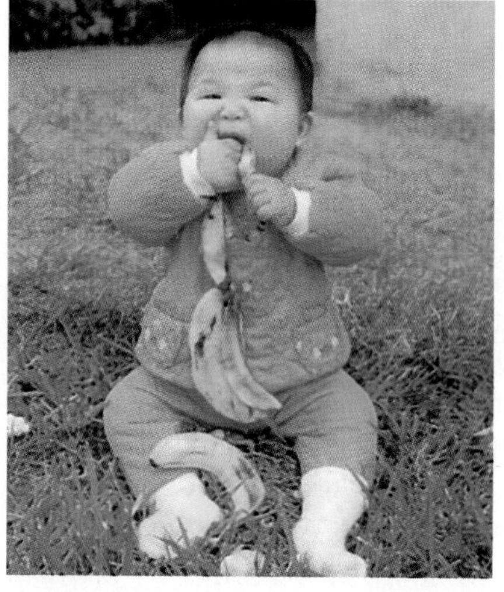

The earliest verifiable instances of cross-modal perception occur when 1-month-olds put something in their mouths and form a concept of how it might look. With experience, infants can reverse the process, imagining how something will taste from the way it looks.

Surprisingly, even 1-month-olds have some cross-modal perceptual abilities. Again, visual expectancy derived from touch is involved, but the touch is by mouth, not by hand. In one experiment, infants sucked for 1 minute on an object, either a rigid one (a lucite cylinder) or a flexible one (a piece of wet sponge). They then observed both objects being manipulated by a pair of black-gloved hands. Analysis of their gazing suggested that they could distinguish the familiar object from the novel one: in other words, just by sucking an object they had gained some understanding of how it might look and move (Gibson & Walker, 1984). This rudimentary cross-modal perceptual skill improves significantly during the first year as infants deduce the qualities of objects more quickly and sensitively with increasing age (Rose & Ruff, 1987).

The remarkable speed and apparent ease with which infants attain all these perceptual accomplishments have led some researchers to conclude either that these basic perceptual skills are innate or that newborns are biologically endowed with powerful capacities and motivation to quickly acquire these skills (Spelke, 1991). However the impressive perceptual development of early infancy is explained, it is quite clear that very young infants do not merely passively absorb sights, sounds, and sensory impressions but also (in a simple way) analyze, interpret, and integrate these perceptions to learn about the world around them. Their early skill in doing so provides the foundation for the equally impressive emergence of early cognitive abilities.

Key Elements of Cognitive Growth

From our discussion so far, it is clear that a considerable portion of infants' knowledge of the world is built upon the development of their perceptual skills. These skills form the basic tools with which infants begin to structure the information they receive through their daily experiences, enabling them to gradually build for themselves a world that is coherent and predictable. Essential to this structuring are several key cognitive elements, including the ability to develop categories, to understand object permanence, to understand cause and effect, and, of course, to remember. Let us now look at each of these elements in the early stages of their development.

Infants have demonstrated a firm concept of the difference between one and two objects, and sometimes even between two and three. This 5-month-old is registering surprise that there are only two dolls in front of her, because seconds before there were three. (The third, as you can see, was surreptitiously removed by the experimenter while the dolls were momentarily hidden behind a screen.) A number of researchers believe that this kind of surprised reaction reflects an innate numerical understanding.

Categories

You have seen that, from a very early age, infants coordinate and organize their perceptions into categories, such as soft, hard, flat, round, rigid, flexible, and so forth. For the preverbal infant, of course, these categories do not have labels such as "soft" and "hard," but they nonetheless represent useful and important ways of conceptualizing the world. Once an object is mentally placed into a category, for example, the infant has a ready set of expectations about it, and can distinguish it from objects that belong to other categories.

How do developmental researchers learn about an infant's categorization abilities when the child cannot verbally label them? Often they measure the infant's habituation, which (as you learned in the preceding chapter) is the gradual subsiding of a baby's initial responses to a novel stimulus as the stimulus becomes more familiar. Young infants usually stare wide-eyed at a new object, for example, but if the object continues to be presented to them, this initial expression of interest wanes, and the baby will, eventually, become uninterested and look away. If, however, a new and different object is presented, the infant will show renewed attention.

As you will recall from Chapter 5, researchers capitalize on this phenomenon to study whether infants can perceptually discriminate between different shapes, colors, sounds, and other sensations. In a similar manner, investigators can explore infant categorization skills by showing a baby several different objects in a particular category (say, circles of different sizes) until habituation occurs. Subsequently, two new objects are presented, one a member of the previous category (that is, another circle) and the other a member of a totally different category (like a square). If the infant shows such signs as an intensified gaze and a change in heart rate, or looks significantly longer at the new object, it can be inferred that the child has discriminated between the objects on the basis of their shape.

This general research format reveals that infants younger than 6 months can categorize objects based on their shape, color, angularity, density, relative size, and number (up to three objects) (Caron & Caron, 1981; Van Loosbroek & Smitsman, 1990; Wynn, 1992). They can similarly categorize different kinds of speech sounds they hear (Quinn & Eimas, 1988).

Taken as a whole, the evidence suggests that young infants are not merely perceiving the difference between shapes like circles and squares, or relative sizes like larger or smaller, or speech sounds like "ba" and "pa"; they are also applying some underlying organizing principles that enable them to develop a concept of what is, or is not, relevant for inclusion in a particular category. Although many researchers believe that a rudimentary understanding of certain categories in the natural world may be biologically based, experience with different kinds of objects and events also plays an important role.

This becomes especially apparent when we observe that as infants get older, they can categorize in increasingly complex ways. Experiments have shown, for example, that by the end of the first year, they can categorize and discriminate faces (based on features like hair length and nose size) (Sherman, 1985), animals (based on tail width or leg length) (Younger, 1990, 1993), and even birds (perceiving parakeets and hawks as similar to each other, but distinct from horses) (Roberts, 1988).

object permanence An infant's realization that objects and people still exist even when they cannot be seen, touched, or heard. The term "object permanence" was coined by Piaget, who believed that this realization does not emerge until 8 months of age.

While the simple categories that infants construct and recognize are nothing like the verbally labeled, and often highly ordered, categories that older children use, they nevertheless form a conceptual foundation for later cognitive accomplishments, and enable young infants to mentally organize their world in increasingly more meaningful ways that are relevant to their day-to-day encounters with objects and people.

Object Permanence

Certainly one of the most important cognitive accomplishments of infancy is the ability to understand that objects (and people) exist independently of one's perception of them. With this understanding, referred to as **object permanence**, infants realize that even when objects like a familiar toy, the family cat, or Mommy cannot be seen or heard, they exist somewhere else in the world; they do not cease to exist simply because they are not immediately apparent. Although this understanding no doubt seems obvious to you, it is not obvious to young infants, whose early awareness of reality is strictly connected to what they can see, hear, and otherwise sense at any given moment. Consequently, the development of an awareness of object permanence has been especially interesting to researchers concerned with cognitive growth in infancy.

One way of testing for awareness of object permanence is to see whether an infant will search for a hidden object. In an experiment devised by Piaget, who was the first to study object permanence, an infant is shown an interesting toy, which is then covered up with a blanket or cloth. If the infant searches under the covering for the toy, he or she realizes the toy still exists, even though it cannot be seen at the moment. Various forms of this experiment have been carried out by many researchers, with fairly consistent results: infants do not search for hidden objects until about 8 months of age. Even then, their search abilities are limited; they cannot easily find an object that has been concealed in one hiding place and then visibly transferred to a second hiding place (they tend to look for the object in the first hiding place). Thus their understanding of object permanence is slow to develop over the first year of life.

One nonexperimental demonstration of object permanence is the delight that 8-month-olds take in simple forms of peek-a-boo. By 11 months of age, the child's firmer grasp of object permanence may lead parents to engage in more sophisticated variations of the game.

But does a failure to search before 8 months of age necessarily reflect the absence of object permanence? After all, many skills are required for an infant to search competently for a hidden object (Harris, 1987; Ruff, 1982). For example, removing a cover to reveal a familiar toy requires the ability to set a goal and to know how to achieve it, and such means-ends understanding does not emerge until sometime in the second half-year of life. In addition, whether or not an infant searches for a hidden toy can depend on the infant's prior experience with searching and on the child's motivation to search. (The infant may have no interest in the toy, or may be tired or distracted by some other event.) The nature of the hiding place can also affect searching, as can the length of the delay between when the object is hidden and when the child is allowed to begin searching. Is it possible, then, that younger babies have an awareness of object permanence that is concealed by other factors when they are given Piaget's test of object permanence?

This intriguing possibility was explored by Renée Baillargeon and her colleagues in a series of experiments using the habituation technique. In one of these, Baillargeon (1987) placed 3½- and 4½-month-old infants directly in front of a large screen that was hinged along its base to the center of a table top. The screen was then repeatedly swung back and forth through a 180-degree arc (see Figure 6.1) until, as habituation occurred, the infants began to look away. At this point, a box was placed directly in the path of the screen's backward descent. Then the screen began to rise again until it reached its full vertical height, concealing the box. Thereafter followed two experimental conditions. In the first, named the "possible event," the screen continued to move until it was intercepted by the box and stopped, as one would expect. In the second experimental condition, named the "impossible event," the screen continued to move through its entire 180-degree arc, as though there were no box to block it. In fact there wasn't: it had been surreptitiously dropped out of the way through a trap door before the screen could hit it. Baillargeon found that 4½-month-old infants stared significantly longer at the "impossible event" than at the "possible event," as if they recognized that this was a novel, unexpected occurrence. She subsequently noted that infants could not have been surprised by this event unless they simultaneously

> (a) believed that the box continued to exist behind the screen, (b) understood that the screen could not rotate through the space occupied by the box, and hence (c) expected the screen to stop in the impossible event and were surprised that it did not. [Baillargeon & DeVos, 1992]

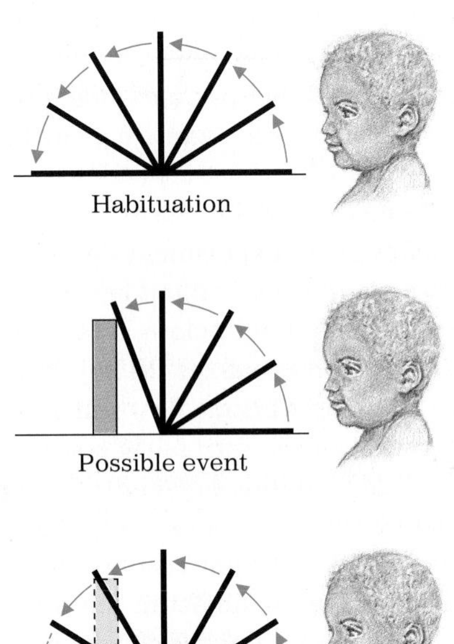

FIGURE 6.1 This illustration shows the basic steps of Renée Baillargeon's test of object permanence—a test that doesn't depend on the infant's searching abilities or motivation to search. First the infant is habituated to the movement of a hinged screen through a 180-degree arc. Next, with the infant observing, a box is placed in the path of the screen. Then the infant witnesses two events: the "possible event," in which the screen's movement through the arc is stopped by the box, and the "impossible event," in which the screen completes its movement through the arc as though the box did not exist. Infants as young as 4½ months old stare longer at the "impossible event," indicating that they are aware that the box does exist even though they cannot see it behind the screen.

Baillargeon and her colleagues have devised a variety of different procedures to substantiate their view that infants possess a basic understanding of object permanence months before they can demonstrate this understanding on a hidden-object task (Baillargeon, 1991; Baillargeon & DeVos, 1992; Baillargeon et al., 1990). In each case, young infants have looked significantly longer at "impossible events" that violated an expectation that objects cannot move through other hidden, solid objects. In addition, these experiments have revealed further aspects of infants' understanding of objects, including an awareness that, although hidden, objects retain their original size, rigidity, and location, and that they can support other, visible, objects. All told, these findings suggest that young infants understand many things about the permanence of objects long before they acquire the motor skills to effectively demonstrate their understanding by searching for a hidden toy.

Memory

Central to the development of any cognitive ability is memory. You have already seen some evidence for the memory skills of infants, ranging from their performance in habituation studies (requiring their ability to remember certain objects as familiar) to their skills in keeping in mind an object that has disappeared from view. But these cases involve memory, sometimes fragmentary, over fairly short spans. The real question is: How good are infants' long-term memory abilities over days, weeks, and months? The consensus of both common sense and research experience is that infants' long-term memory is generally very poor.

However, consistent with the story of much of this chapter, researchers have recently acquired a fuller appreciation of infants' memory abilities. To be sure, young babies have not proven themselves to be reliable recollectors: they have greater difficulty storing new memories, and forget them more easily, than they will by the end of their first year. But research has shown that infants' early memory abilities can be much better than was formerly believed, *if* (1) situations are carefully tailored to the young infant's memory capacity as it might be demonstrated in real life; (2) the infant's motivation to remember is high; and (3) special measures are taken to aid retrieval.

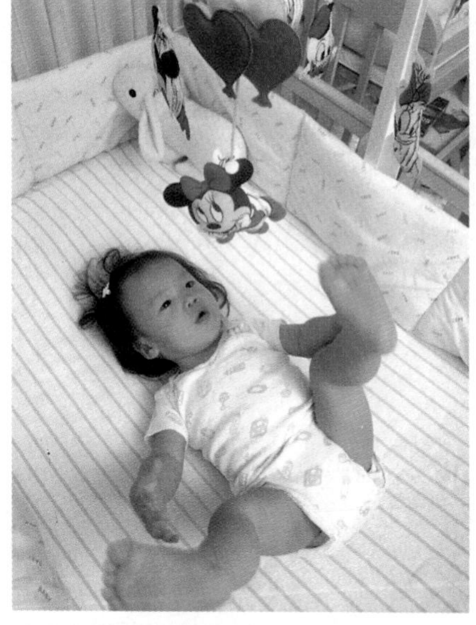

Mobiles, especially moving, musical, colorful ones, are fascinating to infants, because such objects offer exactly the kind of dynamic, intermodal perceptual experience the infant seeks to comprehend. However, this fascination is often likely to turn to boredom within a few weeks, because infants' memories are, in certain respects, better than had once been thought.

Notable among this research has been a series of experiments with 3-month-olds, in which infants learned to make a mobile move by kicking their legs (Rovee-Collier, 1987, 1990; Rovee-Collier & Hayne, 1987). This is a highly reinforcing event for young babies, who delight in controlling colorful, moving objects within easy focusing distance. The infants were tested at home, in their own cribs. A brightly colored mobile was placed overhead, and the infants were connected to the mobile by means of a ribbon tied to one of their feet. In this situation, virtually every infant quickly learned to kick to make the mobile move. Would they later remember this experience?

When the mobile-and-ribbon set-up was reinstalled in their cribs one week later, most infants started to kick, indicating that they remembered the connection between their kicking and the mobile's spinning. But another group of infants who were initially retested *two* weeks later had apparently forgotten the connection. Thus it seemed that, for 3-month-olds, memory fades after only one week.

However, a further experiment demonstrated a remarkable effect: infants *could* remember after two weeks if they were given a brief "reminder" session prior to the retesting (Rovee-Collier & Hayne, 1987). In this reminder session, the infants were not tied to the ribbon and were positioned so that they could not kick. They merely watched as the mobile, activated by a hidden experimenter, moved. The next day, when they were tethered to the mobile and positioned so that they could move their legs, the infants remembered to kick as they had learned to do two weeks before. In a sense, their memory had been "reactivated" by the experience of having watched the mobile move a day earlier.

Similar experiments have been performed with younger infants (as young as 8 weeks) after a longer interval (up to eighteen days) with the same result: a brief "reactivation" of their memory prolonged young infants' retention of an earlier event. Moreover, researchers have also found that early memory improves with extra training or with less time separating training and recall, or with familiar cues (such as a distinctive crib bumper) to help bring to mind the earlier training. The specific timing and context of the re-

RESEARCH REPORT

Long-Term Memory in Infancy?

Most adults cannot remember specific events from their very early years. (The experiences they think they recall are often ones that they were later told about by relatives or friends, or that have been memorialized in family photographs, home movies, and the like.) Many explanations for this "infantile amnesia" emphasize the differences between memory processes in infancy and in later years (Siegler, 1991). In infancy, for example, memories are probably stored and retrieved according to the sensations (smells, sights, sounds) and motor skills associated with specific events or objects, while memories in later childhood and beyond are usually tied to more complex, language-based concepts. Consequently, even if experiences in infancy have been stored in mind, it may be very difficult to gain access to these memories in later years.

Is it possible, however, that infants store long-term memories that can be "reactivated" if meaningful cues are provided? Inspired by their new appreciation of infant memory, researchers have begun exploring whether young children can recall specific experiences that occurred when they were infants. The results have been intriguing.

One study, conducted in a university laboratory, began with 6-month-olds being trained during a single, 20-minute session to reach for a dangling Big Bird toy when it made a noise, first in normal lighting and then in the dark (Perris et al., 1990). Two years later, the children were brought back to the lab and retested on this reaching task, along with a control group of age-mates who had received no training. Prior to the retesting, the trained children were interviewed to see if they had any overt memory of the laboratory setting or the training experience. They did not. Then, 30 minutes before testing, half of all the children who were to participate were randomly selected and given a 3-second exposure to the sound of the toy in the dark—intended as a possible "reactivation" of memory for those in the trained group.

In the retesting that followed, the conditions of the original test were repeated: each child sat in his or her mother's lap and was told the lights would go out. Then, in the dark, the experimenter dangled the noise-making toy in front of the child. Compared with the untrained children, those who had been trained at 6 months were more likely to reach and grab the toy—just as they had been trained to do long ago! In fact, among those who experienced the 3-second "reactivation" session, the trained children reached for the toy almost four times as often as the untrained children. Moreover, their reaction to suddenly being in the dark was "an almost global emotional acceptance," a marked contrast to the discomfort and fussiness exhibited by many children from the control group (and, indeed, by most 2½-year-olds who suddenly find themselves in a dark, unfamiliar room!). Thus, not only the specific behaviors of a single training experience at 6 months, but also its emotional tone, can remain in the memory of a young child for two full years.

activation sessions are critical factors, with some reminders (such as a stationary mobile) having no effect (Borovsky & Rovee-Collier, 1990; Butler & Rovee-Collier, 1989; Hayne & Rovee-Collier, 1995).

Carolyn Rovee-Collier thinks that this kind of "training" occurs quite frequently in an infant's daily experiences, since babies typically experience the same basic events day after day and in familiar circumstances that aid memory renewal. When, for example, a parent regularly shakes a rattle or spins a toy to make it move for the baby, this may help to renew the child's memory for how to create these effects on his or her own. Nevertheless, no matter how specific or frequent the training, young infants recall only for a limited period of time, only under specific conditions, and probably only events in which they themselves have had an active part (Hayne & Rovee-Collier, 1995).

As they mature, infants become capable of retaining information for longer periods of time, and they can do so with less training and less reminding. Thus, toward the end of the first year, a new memory ability is apparent: infants can remember behaviors they have witnessed but have never performed themselves. For example, if 9-month-olds observe someone playing with a toy they themselves have never seen and the next day are presented with the toy, they are likely to play with it in the same manner

It is important to remember that long-term recall was facilitated in many ways in this study—by the children's return to a distinctive testing room, by their reacquaintance with unique procedures and materials, and, ultimately, by a memory reactivation procedure for some of the children. Nevertheless, findings like these, which have been reported by other researchers (Myers et al., 1987), put to rest the idea that all infant memories inevitably disappear completely, and are causing developmentalists to re-examine infant memory to discover exactly how it functions (Lipsitt, 1990). So far, they have come to some of the following conclusions:

Early in life, even under the best of conditions, memory storage and retrieval appear to be fragile and uncertain. Infant recognition and recall of past events is facilitated by reminder events that help to reactivate memory.

The specific learning situation and the infant's motivation to remember are extremely important to later recall. Also, recall is more likely to occur when the task and the situation have some ecological familiarity and personal relevance for the baby.

Very young infants can and do remember, but their recollections probably consist not of words but of sensations and actions—images, smells, movements, and sounds—that are unlikely to be measured by traditional tests of memory.

Improvement in memory ability seems tied to brain maturation and language development, with notable increases in memory capacity and duration occurring at about 8 months of age and again at around 18 months.

Much more needs to be understood, however, including answers to the following questions:

Are there several forms of memory, perhaps a type for basic sensory experiences, another for physical movement, and still another (or others) for more conceptual processes, such as those involving language?

Is the fragility of infant memory primarily a problem of encoding and storage—that is, getting the event into memory—or of retrieval—that is, recalling it after it has been stored?

Do older children and adults seem unable to remember their infancy because these memories are processed primarily through the senses and physical actions, rather than through language and other higher conceptual skills?

Much more research in this area is now under way, promising to yield further revisions in our understandings of what infants retain, and remember, from their earliest experiences.

they saw demonstrated the day before. (Younger infants cannot necessarily do this.) Over the next few months, infants are able to retain more of the details surrounding an observed experience, including the sequence of events, and can duplicate them days later, prompted by only minimal clues associated with the experience—such as a sound or a part of the setting (Bauer & Mandler, 1992). A 1-year-old, for instance, might stare intently as an adult pulls open the refrigerator, takes out an egg carton, and cracks a few eggs into a bowl. The next day (perhaps when the adult has left the refrigerator open while answering the phone), the child might take out the egg carton and crack all the remaining eggs.

By the middle of the second year, toddlers are capable of remembering more complex sequences and can also generalize their memories. In one experiment, 16- and 20-month-olds first watched an experimenter perform various activities, such as putting a doll to bed, making a party hat, and cleaning a table. For each activity, the experimenter used particular props and gave "instructions" for performing the activity. For example, when cleaning the table, the experimenter used a white spray bottle, a paper towel, and a woven, wooden trash basket, and said, "Put on the water. Wipe it. Toss it," as she performed each step. A week later, most toddlers remembered how to carry out the sequence just from hearing the instructions.

They could do this not only when given the same props the experimenter had used but also when given quite different props (for instance, a clear spray bottle, a sponge, and a plastic, lidded garbage can), indicating that they had formed and remembered a general scheme of the event and were not enacting a simple memory triggered by visual props. To make sure that memory, not merely creative play, was involved, the experimenters presented the instructions and the props to toddlers who had not seen the demonstration. These youngsters played with the objects but did not perform the specific actions in sequence (Bauer & Dow, 1994).

Causes and Effects

Another important cognitive accomplishment in infancy is the ability to recognize, and associate, the causes of events with the effects they produce. Infants' ability to understand and anticipate the events they observe (whether shoving one toy into another to send the second toy flying, or clinging to Mommy while she puts on her coat and heads toward the door), as well as to act effectively in the world (whether kicking their legs to make a mobile move or lifting a blanket to find a concealed toy), hinges on the capacity to identify what actions lead to what results. One way of studying an infant's understanding of cause-and-effect relations is to closely observe the child's behavior to see if the child intentionally repeats some action that has produced an interesting result. If, for example, a child squeezes a rubber duck, producing a squeak, and, delighted by the noise, squeezes the duck again, the child has made a cause-and-effect connection between the squeeze and the squeak. But this method of study requires that infants possess the motor skills that many young babies do not yet have. (Even deliberately squeezing a rubber duck demands a degree of hand control that is beyond many young infants.) To overcome this problem in studying young infants' awareness of the connection between causes and effects, researchers have again turned to the familiar habituation procedure.

One commonly used technique presents infants with a **launching event**. In this procedure, the infant sees an object, say, a square, move to the right across a table until it bumps into another object, say, a rectangle: the square stops moving, and the rectangle begins moving to the right. Most adults would view such an event as causal: the square appears to "launch" the rectangle into motion. By contrast, if the rectangle begins to move before it is bumped by the square, or if it doesn't move until a second or two after it is hit, most of us would not regard this as a causal sequence of events. In such cases, the rectangle appears to be moving for other reasons.

Can young babies draw similar inferences? Studies in which infants are first habituated to the initial launching sequence and then shown variations of it (such as the rectangle moving before contact, or after a delay) reveal that 6-month-olds seem to have only a very rudimentary understanding of causal relations but that 10-month-olds can properly interpret the causality of simple launching events like this (Cohen & Oakes, 1993; Leslie, 1984; Leslie & Keeble, 1987; Oakes & Cohen, 1990). As we shall see in our discussion of Piaget's theory, this accomplishment coincides with a period in which infants are very busily engaged in manipulating and interrelating objects in their explorations of their world.

launching event A habituation technique used to determine if a young infant understands the connection between causes and effects.

An understanding of cause-and-effect relations is basic to problem-solving ability, and some research has elicited displays of certain simple de-

ductive skills in infants at about the same time that their causal awareness is blossoming. In one study, Peter Willatts (1989) presented 9-month-olds with a formidable challenge: to obtain an attractive toy that was out of reach on a table in front of them. Fortunately, the toy was resting on a cloth that was within reach; unfortunately, their access to the cloth was blocked by a foam-block barrier. Infants as young as 9 months old skillfully removed the barrier, then deliberately pulled the cloth to obtain the toy. Their strategic behavior contrasts with the actions of a comparison group of infants who were presented with the same situation, except that the out-of-reach toy was *not* resting on the cloth. These infants tended to play with the foam block, but showed no interest in the cloth, since it provided no access to the toy.

The surprising competencies we have surveyed—intermodal and cross-modal integration of perceptions, categorical knowledge, object permanence, memory skill, a capacity for expectations, delayed imitation, an awareness of causal relations, and even simple problem-solving capabilities—can all be considered markers of intelligence in infancy. Although this kind of intelligence is obviously much different from the symbolic, language-based forms of intelligence used by older children and adults, it reflects astonishing abilities in a young being whom earlier researchers had regarded as cognitively very simple.

Active Intelligence: Piaget's Theory

Because it is based largely on laboratory research, the foregoing account of cognitive development may seem to depict the infant as a detached observer, passively deducing an understanding of how the world functions. However, such an image would omit one of the most important characteristics of young babies: their activity and its primary role in learning. This aspect of intelligence is central to Jean Piaget's theory of infants' cognitive development, which Piaget began to develop over sixty years ago, starting with the study of his own three children. Piaget believed that children actively seek to comprehend their world, constructing understandings of it that reflect specific, age-related cognitive stages. He showed that this process begins at birth and accelerates rapidly in the early months of life. To Piaget, infants lack concepts and ideas but are nevertheless intelligent; their intelligence functions exclusively in terms of their senses and motor skills (Gratch & Schatz, 1987). Consequently, Piaget called the first period of cognitive development **sensorimotor intelligence**.

What does it mean to say that infants think exclusively with their senses and motor skills? As Flavell (1985) expresses it, the infant "exhibits a wholly practical, perceiving-and-doing, action-bound kind of intellectual functioning: he does not exhibit the more contemplative, reflective, symbol-manipulating kind we usually think of in connection with cognition." Although Piaget was incorrect in his belief that infants do not have concepts and ideas—as we have seen, infants have a fairly rich conceptual life—and some of his other proposals have been revised by recent findings, his portrayal of the "practical, perceiving-and-doing" side of early intelligence remains valid and provides a vivid encapsulation of the ways in which infants demonstrate their growing cognitive abilities through their actions.

sensorimotor intelligence Piaget's term for the first stage of cognitive development (from birth to about 2 years old). Children in this period primarily use their senses and motor skills to explore and understand their environment.

The Six Stages of Sensorimotor Intelligence

According to Piaget, sensorimotor intelligence develops through six stages, each characterized by a somewhat different way of understanding the world. Following the arrangement laid out in Table 6.1, we will briefly examine these stages in pairs, highlighting the overall progression of infants' cognitive growth—from reflexive and largely self-absorbed cognitive patterns, to increasingly deliberate responses to people and objects, to thinking that suggests the beginning of symbolic understanding.

TABLE 6.1

The Six Stages of Sensorimotor Intelligence

To get an overview of the stages of sensorimotor thought, it helps to group the six stages into pairs.

The first two stages involve the infant's own body.	
Stage One (birth to 1 month)	*Reflexes*—sucking, grasping, staring, listening.
Stage Two (1–4 months)	*The first acquired adaptations*—accommodation and coordination of reflexes—sucking a pacifier differently from a nipple; grabbing a bottle to suck it.
The next two stages involve objects and people.	
Stage Three (4–8 months)	*Procedures for making interesting sights last*—responding to people and objects.
Stage Four (8–12 months)	*New adaptation and anticipation*—becoming more deliberate and purposeful in responding to people and objects.
The last two stages are the most creative, first with action and then with ideas.	
Stage Five (12–18 months)	*New means through active experimentation*—experimentation and creativity in the actions of the "little scientist."
Stage Six (18–24 months)	*New means through mental combinations*—thinking before doing provides the child with new ways of achieving a goal without resorting to trial-and-error experiments.

At the sensorimotor stage of cognitive development, even Father's face is a site for active exploration by all the infant's senses and motor skills.

Stages One and Two

Sensorimotor intelligence begins with newborns' reflexes, such as sucking, grasping, looking, and listening. Through the repeated exercise of these reflexes over the first month, newborns gain important information about the world, information that allows them to begin adapting their reflexes to the specifics of the environment.

Take the sucking reflex, for example. Newborns suck anything that touches their lips. However, at about 1 month, according to Piaget, infants start to adapt their sucking to specific objects, and by 3 months they have organized their world into objects to be sucked for nourishment (breasts or bottle nipples), objects to be sucked for pleasure (fingers or pacifiers), and objects not to be sucked at all (fuzzy blankets and large balls). They also learn that efficient breast-sucking requires squeezing- or suction-sucking, whereas efficient finger- and pacifier-sucking do not. In addition, once infants learn that some objects satisfy hunger and others do not, they will suck contentedly on a pacifier when their stomach is full but will usually spit one out when they are hungry.

Another feature of early sensorimotor intelligence is that infants spend a good deal of time playing with their bodies, seemingly for the pleasure of doing so. They suck their fingers, kick their legs, wave their arms, and stare at their hands again and again. In the process, they gain valuable information. They learn, for instance, that those wiggly little things that regularly come into view and often wind up in their mouths are actually attached to them, and that these things, which they will later know as fingers and toes, are within their control. Information such as this is basic to developing an awareness of body integrity—that is, an awareness of one's body as a whole. The emergence of this awareness is one of the first steps in understanding the world of other people and objects.

Stages Three and Four

In the third stage (4 months to 8 months), infants become more aware of objects and other people, and they begin to recognize some of the specific characteristics of the things in their environment, particularly how objects respond to their actions on them. One way infants show this new awareness is by repeating a specific action that has just elicited a pleasing response from some person or thing. As noted earlier, for example, a baby might accidentally squeeze a rubber duck, hear a squeak, and squeeze the duck again. If the squeak is repeated, the infant will probably laugh and give another squeeze, delighted to be able to control the toy's actions.

During this stage, babies interact diligently with people and objects to produce exciting experiences. Realizing that rattles make noise, for example, babies at this stage shake their arms and laugh when someone puts a rattle in their hands, and just the sight of something that normally delights the infant—a favorite toy, a favorite food, a smiling parent—can trigger an active attempt at interaction. Vocalization of all sorts increases a great deal, for now that babies realize that other people can respond, they love to make a noise, listen for a response, and answer back.

This 7½-month-old knows that a squeal of delight is one way to make the interesting experience of a-tickle-from-Daddy last.

In stage four (8 months to 12 months), babies adapt in new, more deliberate ways. They can better anticipate events that will fulfill their needs and wishes and can set about initiating them. A 10-month-old girl who enjoys playing in the tub might see a bar of soap and bring it to her mother as a sig-

What do you see in this photo? Instead of a mess that needs to be cleaned up, you might discern a demonstration of intellectual ability. Brandon has a goal firmly in mind and is wielding the tools to attain it—an achievement beyond most younger babies. At age 12 months he is about to enter a more elaborate stage of goal-directedness, one in which he might deliberately drop a few peas on the floor, or smash a few noodles on his head, or turn his plate upside-down—all as "experiments in order to see."

nal to start her bath, squealing with delight when she hears the bath water turned on. Similarly, if a 10-month-old boy sees his mother putting on her coat to go out without him, he might begin tugging at it to stop her, or he might signal that he wants her to get his coat too.

Both of these examples reveal anticipation and, even more noteworthy, **goal-directed behavior**—that is, purposeful actions. The baby's greater goal-directedness stems from an enhanced awareness of causes and their effects that develops during this stage, together with the emergence of the motor skills needed to achieve the infant's goals. Thus stage-four babies might see something clear across the room and crawl toward it, ignoring many interesting distractions along the way. Or they might grab a forbidden object—a box of matches, a thumbtack, a cigarette—and cry with rage when it is taken away, even if they are offered a substitute that they normally find fascinating.

As noted earlier, Piaget thought that the concept of object permanence begins to emerge during stage four, because at this point—and usually not before—infants actively search for objects that are no longer in view. Although researchers have since shown that the concept of object permanence actually begins to emerge much earlier, the goal-directed search for toys that have fallen from the crib, rolled under a couch, or disappeared under a blanket does not begin to emerge until the age of about 8 months, just as Piaget had indicated.

Stages Five and Six

Stage five (12 months to 18 months) builds directly on the accomplishments of stage four, as the infant's goal-directed and purposeful activities become more expansive and creative. It is a time of active exploration and experimentation, a time when infants "get into everything," as though trying to discover all the possibilities their world has to offer. Because of the experimentation that characterizes this stage, Piaget referred to the stage-five toddler as a **little scientist** who "experiments in order to see." Having discovered some action or set of actions that is possible with a given object, stage-five infants seem to ask, "What else can I do with this? What happens if I take the nipple off the bottle, or turn over the trash basket, or pour water on the cat?" Their "scientific method" is one of trial and error.

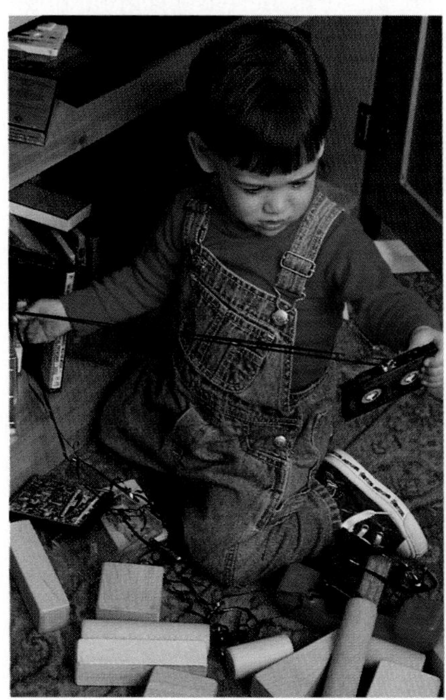

Every "little scientist" seems to be driven by such questions as "What will happen if I pull this?"

In the final stage of sensorimotor intelligence (18 months to 24 months), toddlers begin to anticipate and solve simple problems by using **mental combinations** before they act. That is, they are able to try out various actions mentally without having to actually perform them. Thus the child can invent new ways to achieve a goal without resorting to physical trial-and-error experiments. Consider how Piaget's daughter Jacqueline solved the following problem at the age of 20 months:

> Jacqueline arrives at a closed door with a blade of grass in each hand. She stretches out her right hand toward the knob but sees that she cannot turn it without letting go of the grass. She puts the grass on the floor, opens the door, picks up the grass again and enters. But when she wants to leave the room, things become complicated. She puts the grass on the floor and grasps the doorknob. But then she perceives that in pulling the door toward her she will simultaneously chase away the grass which she placed between the door and the threshold. She therefore picks it up in order to put it outside the door's zone of movement. [Piaget, 1952]

goal-directed behavior Purposeful actions initiated by infants in anticipation of events that will fulfill their needs and wishes.

little scientist Piaget's term for the stage-five toddler who actively experiments to learn about the properties of objects.

mental combinations The mental playing out of a course of action before actually enacting it.

The growth of mental combinations leads to far more than better problem-solving skills. It enables the child to think more flexibly about past and future events, and to anticipate what can occur in a particular situation. Being able to use mental combinations also makes it possible for the child to pretend. A toddler might lie down on the floor and pretend to go to sleep, and then jump up laughing. Or a child might sing to a doll before tucking it into bed. This is in marked contrast to the behavior of the younger child, who treats a doll like any other toy, throwing it, biting it, or banging it on the floor.

These stage-six behaviors all share an important characteristic. They are a step beyond the simple motor responses of sensorimotor thought and a step toward "the more contemplative, reflective, symbol-manipulating activity" (Flavell, 1985) that we usually associate with cognition. As you will see in Chapter 9, these capacities will blossom into the symbolic thought typical of the next period of cognitive development.

Language Development

Mastering the sounds and meanings of one's first language is "doubtless the greatest intellectual feat any one of us is ever required to perform," according to one early developmental scholar (Bloomfield, 1933). Before dismissing this claim as hyperbole, imagine yourself as a tourist in a foreign land, surrounded by natives chattering rapidly in a language quite different from your own. Without extensive experience with that language, you could not even decipher where one word stops and another begins; which nuances of tone and pronunciation are significant and which are merely individual variations; how the string of sounds is put together to make statements or questions; and most important, if the content of the conversation should make you embarrassed, frightened, or delighted. Lost as you might be, at least you would know that spoken sounds have specific meaning, and you could figure out how to begin to learn those meanings. The newborn, of course, does not even know this, yet typically by age 2, "children, bright and dull, pampered or neglected, exposed to Tlingit or to English," all learn language (Wanner & Gleitman, 1982).

Consider, for example, the words, grammar, and conversational skill of one 24-month-old, Sarah, determined to distract her mother, who at the time was intently revising an earlier edition of this textbook.

> Uh, oh. Kitty jumping down.
> What drawing? Numbers? [said as her words were being transcribed]
> Want it, paper.
> Wipe it, pencil.
> What time it is? [said upon seeing a watch]

These sentences show that Sarah has a varied vocabulary and a basic understanding of word order. For example, Sarah said "Kitty jumping down" rather than "Down jumping kitty," or "Jumping kitty down," or "Kitty down jumping." Sarah's speech also shows that she has much to learn, for she incorrectly uses the pronoun "it" and its referent together, omits personal pronouns, and uses reverse word order in asking the time.

But beyond the specifics of English vocabulary and grammar, this excerpt shows that Sarah has learned the universal function of language—to express one's thoughts and wishes to another, and to elicit a response. Despite her mother's preoccupation and nonresponsiveness, Sarah produced seven successive sentences crafted to entice her mother into a dialogue. The final question, "What time it is?" reveals considerable sophistication about the rules of polite conversation: Sarah must have noticed that almost any adult, even a stranger on the street, usually answers that particular question.

Sarah's impressive but imperfect language is quite similar to that of 2-year-olds in many families and cultures. On the basis of detailed studies of hundreds of babies, we know quite a bit about the power of infants' drive to communicate and about the sequence of their emerging verbal skills in the first two years, and Sarah at age 2 is typical.

Steps in Language Development

It seems as if infants are equipped to learn language from birth, partly due to innate readiness and partly because of their auditory experiences during the final prenatal months. Newborns show a preference for hearing speech over other sounds, for hearing "baby talk" over normal speech, and, as we saw in Chapter 4, for hearing their mother's voice over the voices of other adults (Cooper & Aslin, 1990; DeCasper & Fifer, 1980; DeCasper & Spence, 1986). Moreover, young infants can distinguish among many different speech sounds, and can even differentiate between sounds that speakers of their native language cannot differentiate (Werker, 1989). To a young infant, the sound of human speech—whether it comes from Mommy or Daddy, a doting grandparent, or an older sibling—creates special interest and curiosity.

Children the world over follow the same sequence of accomplishments in early language development, although the timing of these accomplishments may vary considerably (see Table 6.2). Prior to their being able to verbalize, of course, infants are very effective at communicating their emotions, preferences, and ideas through grunts, cries, squeals, body movements, gestures, and facial expressions. These early communication skills serve the primary role of *language function*: to understand, and be understood by, others. As you will see, within the first two years of life, this rudimentary ability to communicate evolves into an impressive command of *language structure*, that is, the particular words and rules of the infant's native tongue.

Cries, Coos, and Babbling

Infants are noisy creatures, crying, cooing, and making a variety of other sounds even in the first weeks of life. These noises gradually become more varied over the first months, so that by 5 months, squeals, growls, grunts, croons, and yells, as well as some speechlike sounds, are part of most babies' verbal repertoire. Then, at 6 or 7 months, babies' utterances begin to include the repetition of certain syllables ("ma-ma-ma," "da-da-da," "ba-ba-ba"), a phenomenon referred to as **babbling** because of the way it sounds. In some respects, babbling is universal—all babies do it, and the sounds they

babbling Extended repetition of certain syllables, such as "ba, ba, ba," that begins at about 6 or 7 months of age.

TABLE 6.2

The Development of Spoken Language: The First Two Years*

Newborn	Reflexive communication—cries, movements, facial expressions.
2 months	A range of meaningful noises—cooing, fussing, crying, laughing.
3–6 months	New sounds, including squeals, growls, croons, trills, vowel sounds.
6–10 months	Babbling, including both consonant and vowel sounds repeated in syllables.
10–12 months	Comprehension of simple words; simple intonations; specific vocalizations that have meaning to those who know the infant well. Deaf babies express their first sign; hearing babies use specific gestures (e.g., pointing) to communicate.
13 months	First spoken words that are recognizably part of the native language.
13–18 months	Slow growth of vocabulary, up to fifty words.
18 months	Vocabulary spurt—three or more words learned per week.
21 months	First two-word sentence.
24 months	Multiword sentences. Half of the infant's utterances are two or more words long.

* The ages of accomplishment in this table reflect norms. Many healthy and intelligent children attain these steps in language development earlier or later than indicated here.
Sources: Bloom, 1993; Lenneberg, 1967.

make are similar no matter what language their parents speak. However, over the next few months, babbling begins to incorporate more and more sounds from the native language, perhaps as infants imitate the sounds they hear (Boysson-Bardies et al., 1989; Masataka, 1992). Many cultures assign important meanings to some of these sounds, with "ma-ma-ma," "da-da-da," and the like usually being applied to significant people in the infant's life (see Table 6.3).

TABLE 6.3

First Sounds and First Words: Cross-Linguistic Similarities

	Mother	Father
English	mama, mommy	dada, daddy
Spanish	mama	papa
French	maman, mama	papa
Italian	mamma	babbo, papa
Latvian	mama	tēṭe
Syrian Arabic	mama	baba
Bantu	ba-mama	taata
Swahili	mama	baba
Sanskrit	nana	tata
Hebrew	ema	abba
Korean	oma	apa

Babbling in Deaf Infants

Deaf babies begin to make babbling sounds several months later than other infants do (Oller & Eilers, 1988). However, recent research suggests that deaf infants may actually begin a type of babbling—manually—at about the same time hearing infants begin babbling orally (Pettito & Marentette, 1991). Analysis of videotapes of deaf children whose parents communicate in sign language reveals that before the tenth month, the infants use about a dozen distinct hand gestures—most of which resemble basic elements of the American Sign Language used by their parents—in a rhythmic, repetitive manner analogous to normal babbling. The similar timing of babbling among hearing babies exposed to spoken language and deaf babies exposed to signed language suggests that brain maturation, more than specific maturation of the vocal apparatus, underlies the universal human ability to develop language.

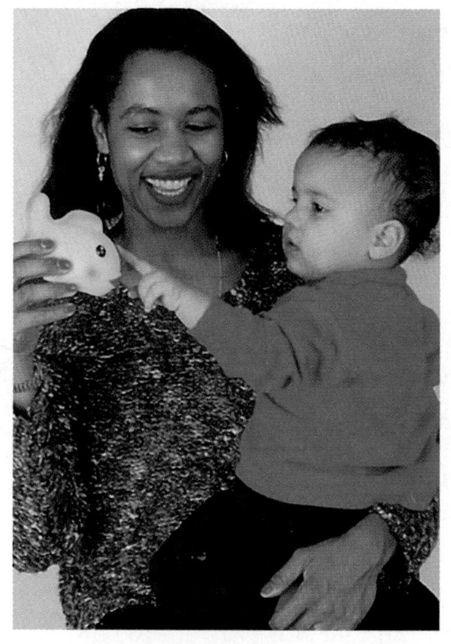

Infants' verbal understanding advances well ahead of their abilities at verbal production. "Fishee" is probably one of dozens of words that this child readily recognizes even though he has yet to say them himself.

Comprehension

At every stage of development, including the preverbal stage, children understand more than they express (Kuczaj, 1986). In fact, according to reports from parents, most children understand more than twenty-five words by the age of 10 months, among them "mommy," "daddy," "no," "hi," "bye," "bath," "book," "car," "kitty," "kiss," and "uh-oh" (Fenson et al., 1994). When asked "Where's Mommy?" for instance, many 10-month-olds will look in her direction, or when asked "Do you want Daddy to pick you up?" will reach out their arms. Of course, context and tone help significantly to supply meaning (Fernald, 1993). For example, when parents see their crawling infant about to touch the electrical outlet, they say "No!" sufficiently sharply to startle and thus halt the infant in his or her tracks. Typically, they then move the child away, pointing to the danger and repeating "No. No." Given the frequency with which the mobile infant's behavior produces this response, it is no wonder that many infants understand "no" months before they can talk.

First Spoken Words

At about 1 year of age, the average baby speaks a few words, not pronounced very clearly or used very precisely. Usually caregivers hear, and understand, the first word before strangers do, which makes it hard to pinpoint exactly what a 12-month-old can say (Bloom, 1993).

Vocabulary increases gradually at first, perhaps by ten words a month. By 16 months of age, the average baby speaks about forty words and comprehends many more (Fensen et al., 1994). Most of these early words are names of specific people and objects in the child's daily world, although some "action" words are included as well (Barrett, 1986; Kuczaj, 1986). At about the fifty-word milestone, vocabulary suddenly begins to build rapidly, sometimes by a hundred or more words a month (Fensen et al., 1994). Toddlers differ in their vocabulary growth: some children (called "referential") primarily learn naming words (such as "dog," "cup," and "ball"), while others (called "expressive") acquire a higher proportion of words that can be used in social interaction (such as "please," "want," and "stop") (Nelson, 1981). Such differences no doubt reflect individual personality, but they may also

reflect cultural emphases. American infants, for example, tend to be more referential than Japanese infants, partly because playing with toys and labeling objects is more central in American families (Fernald & Morikawa, 1993).

At first, infants show marked inaccuracies in the way they connect the few words they know to the people, objects, and events around them. Initially, they tend toward **underextension** of word meanings, applying a word more narrowly than an adult would. "Cat" may be used to name only the family cat, for example, and no other feline. Similarly, once they learn a name for something, they may stubbornly resist alternative labelings, insisting relentlessly, for example, that the little fuzzy, yellow, winged thing they call a "bird" is not a "chick," as Grandpa keeps calling it (Shatz, 1994).

A bit later the opposite tendency may appear, with words being applied far more broadly than their meaning allows. This characteristic, known as **overextension**, or overgeneralization, might lead one child to call anything round "ball," and another to call every four-legged creature "doggie." But once vocabulary begins to expand, toddlers seem to "experiment in order to see" with words just as they do with objects. The "little scientist" becomes the "little linguist," exploring hypotheses about the meaning of specific words. It is not unusual for 18-month-olds to walk down the street pointing to every animal, asking "doggie?" or "horsie?" or "kitty?"—perhaps to confirm their hypotheses about which words go with which specific animals.

As children learn their first words, they usually become adept at expressing intention. Even a single word, amplified by intonation and gestures, can express a whole thought. When a toddler pushes at a closed door and says "bye-bye" in a demanding tone, it is clear that the child wishes to go out. When a toddler holds on to Mother's legs and plaintively says "bye-bye" as soon as the babysitter arrives, it is equally clear that the child is asking Mommy not to leave. A single word that expresses a complete thought in this manner is called a **holophrase**. In the early stages of language development, almost every single-word utterance is a holophrase, making toddlers much more proficient linguists than their limited vocabulary would suggest.

Indeed, it is important to note that vocabulary size is not the only, nor the best, measure of early language learning. Rather, the crux of early language is communication, not vocabulary. If parents are concerned about their 1-year-old son's being nonverbal, they should look at his ability and willingness to make his needs known and to understand what others say. If those skills seem to be normal, and if the child hears enough simple language addressed to him every day (through someone's reading to him, singing to him, talking to him about the sights he sees), he will probably be speaking in sentences before age 2 (Eisenson, 1986).

In fact, one in-depth study of infant language development found that infants who were most adept at expressing their emotions nonverbally (through frowns, smiles, cries, and laughter) were generally slower to talk, but once they began, they progressed to multiword sentences just as rapidly as early talkers did (Bloom, 1993). On the other hand, infants who show signs of language delay (for example, not babbling back when parents babble to them, or not responding to any specific words by age 1) should have their hearing examined as soon as possible. Even a moderate early hearing loss can delay speech acquisition (Butler & Golding, 1986).

underextension The use of a word to refer to a narrower category of objects or events than the term signifies.

overextension The overgeneralization of a word to various objects that share particular—but undefining—characteristics.

holophrase A single word that expresses a complete thought.

Combining Words

Within about six months of speaking his or her first words, a child begins to learn new words more rapidly and, soon after this vocabulary spurt, starts to put words together. As a general rule, the first two-word sentence appears at about 21 months, with some normal infants achieving this milestone at 15 months and others not reaching it until 24 months. Combining words demands considerable linguistic understanding because, in most languages, word order affects the meaning of the sentence. However, even in their first sentences, toddlers demonstrate that they have figured out the basics of subject-predicate order, declaring "Baby cry" or asking "More juice" rather than the reverse. (We will explore other features of language learning in the preschool years in Chapter 9.)

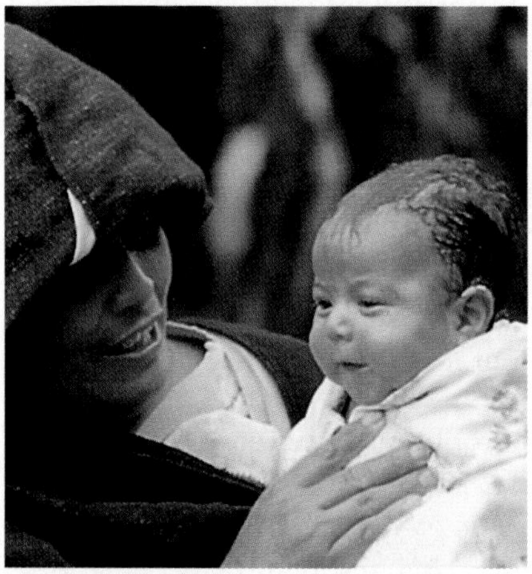

If his infancy is like that of most babies raised in the relatively taciturn Ottavado culture in Ecuador, this 2-month-old will hear significantly less conversation than infants from most other regions of the world, including other localities in South America. According to many learning theorists, such a lack of reinforcement will result in a child who is much less verbal than most children from other cultures. However, each culture tends to encourage the qualities most needed and valued in the culture, and verbal fluency among children is not a priority in this community.

Teamwork: Adults and Babies Teach Each Other to Talk

How do babies learn to talk? Early research on language development tended to take one of two directions, focusing either on the ways parents teach language to their infants or on the emergence of the infant's innate language abilities.

The focus on teaching arose from B. F. Skinner's learning theory, which held that conditioning processes could explain verbal behavior just as well as it could other types of behavior (Skinner, 1957). According to this theory, for example, if babies are reinforced with food and attention when they utter their first babbling sounds, they will soon call "mama," "dada," and "baba" whenever they want their mother, father, or bottle. Similarly, many learning theorists believed that the quantity and quality of parents' talking to their child affect the rate of the child's language development, from the first words through complex sentences.

The focus on innate language ability came from the theories of Noam Chomsky (1968, 1980) and his followers, who believe that language is too complex to be mastered so early and so easily through conditioning. According to Chomsky, the fact that children master the rudiments of grammar so rapidly, and at approximately the same age, implies that the human brain is uniquely equipped with some sort of structure or organization that facilitates language development. Somewhat boldly, Chomsky labeled this theorized facilitator the **language acquisition device**, abbreviated **LAD**. The LAD enables children to quickly and efficiently derive rules of grammar from the speech they hear every day, regardless of whether their native language is English, Chinese, or Urdu. Other theorists have proposed other innate structures to facilitate different features of language learning.

language acquisition device (LAD) Chomsky's term to denote the innate ability to acquire language, including a knowledge of basic aspects of grammar, and a predisposition to attend to and remember the critical, unique aspects of the language.

Research in recent years has suggested that both Skinner's and Chomsky's theories have some validity, but that both miss the mark (Bates & Carnevale, 1994; Bloom, 1991; Golinkoff & Hirsh-Pasek, 1990). One reason is that both theories overlook the social context in which the actual language-learning process occurs, a social context framed by the adult's teaching sensitivity and the child's learning ability. Infants are genetically primed to pick up language, and, on the whole, caregivers are surprisingly skilled at facilitating the infant's language learning.

This process begins very early in life (Locke, 1993). Indeed, on the baby's part, preference for the patterns of the baby's native language begins

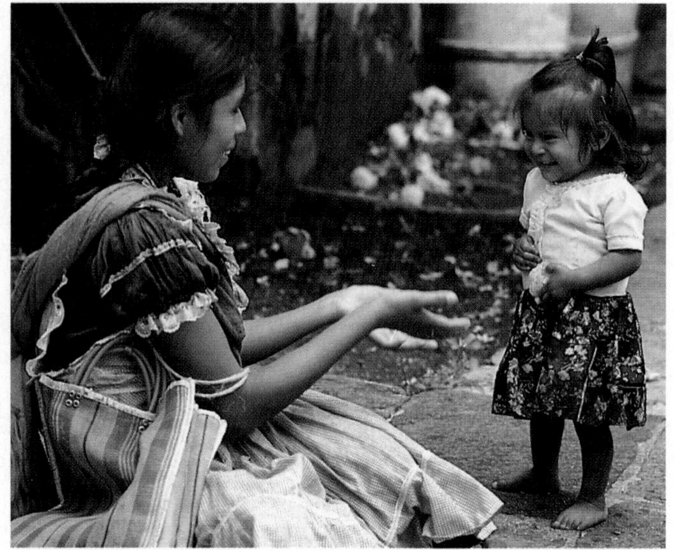

Conversations rich with facial expressions, gestures, and dramatic intonation of a few words are universally found between mothers and toddlers, illustrated here by this winning pair in a Mexican market.

to form even before the baby is born. In one study, newborns whose mothers were monolingual—either English or Spanish—listened to a tape of an unfamiliar female voice speaking one language or the other. The babies' reactions revealed that they preferred hearing whichever language they had grown accustomed to in the womb (Moon et al., 1993). Soon this preference becomes more definite, as infants become attuned to the specific intonations, timing, and phonetic distinctions of the language they hear daily and begin to ignore speech sounds that are not used in their mother tongue (Kuhl et al., 1992). Japanese infants, for example, become less attentive to the difference between "l" and "r" because there is no "l" sound in the Japanese language. Babies raised hearing English, on the other hand, become highly attuned to the distinction between these two letters, noticing the difference long before they can articulate it themselves.

These preferences become incorporated into the neurological expansion and pruning processes that occur within the brain in the early years. While the brain is particularly attuned to speech sounds, word patterns, and linguistic expression in infancy, it usually remains able to perceive and copy the specific pronunciation and cadence of other languages throughout childhood. Generally by adolescence, however, certain sounds become literally impossible for the brain to perceive. Consequently, few people who learn a new language in adulthood ever manage to speak it with no trace of a foreign accent.

While infants learn language in part because of a "deep biological need to interact emotionally with the people that love and care for them" (Locke, 1993), the need of adults to communicate with infants is no less strong. Even strangers on the street feel compelled to smile and talk to a baby, as they never would to an unfamiliar adult. Typically, the type of speech adults use to talk with babies is a special form of language that developmentalists call **baby talk**, or sometimes *Motherese*. Baby talk differs from normal speech in a number of features that are consistent throughout all language communities (Ferguson, 1977): it is distinct in its pitch (higher), intonation (more low-to-high fluctuations), vocabulary (simpler and more concrete), and sentence length (shorter). It also employs more questions, commands, and repeti-

baby talk A term for the special form of language that people typically use when speaking to infants. Nicknamed "Motherese" by developmental psychologists, baby talk is high-pitched, with many low-to-high intonations, is simple in vocabulary, and employs many questions and repetitions.

tions, and fewer past tenses, pronouns, and complex sentences, than adult talk does.

People of all ages, from preschoolers to elders, speak baby talk with infants, and preverbal infants prefer listening to Motherese over normal speech (Cooper, 1993; Fernald, 1985), even if the Motherese is in a language the infant has never heard (Fernald, 1993). Part of the appeal, and the impact, of baby talk may lie in its energy and exaggerated expressiveness. This idea is supported by research showing that the baby talk of depressed mothers is too flat in intonation and too slow in its conversational responses to hold the baby's interest (Bettes, 1988).

The function of baby talk is clearly to facilitate early language learning, for the sounds and words of baby talk are those that infants attend to, and speak, most readily. In addition, difficult sounds are avoided: consonants like "l" and "r" are regularly omitted, and hard-to-say words are given simple forms, often with a "-y" ending. Thus, "father" becomes "daddy," "stomach" becomes "tummy," and "rabbit" becomes "bunny," because if they didn't, infants and parents would have difficulty talking about them. Moreover, the intonations and special emphases of baby talk help infants make connections between specific words and the objects or events to which they refer (Fernald & Mazzie, 1991).

In the earliest stages of baby talk, the conversation is, of course, rather one-sided. Nevertheless, it is important for laying a foundation for the infant's language learning. Babies who are spoken to, sung to, and even read to months before they themselves begin to talk are likely to learn language faster, with a more extensive and elaborate vocabulary, than babies whose caregivers are taciturn or inattentive.

Another aspect of Motherese that promotes early language learning is that the speaker tends to adopt the child's focus of attention. When the child looks at his or her hands, the speaker tends to make a comment using the word "hand"; if the child's focus shifts suddenly to a nearby toy, the speaker is likely to comment on the toy. Talking to a child about whatever the child is focusing attention on at a given moment is an important contributor to vocabulary growth (Akhtar et al., 1991; Bloom, 1993; Tomasello, 1988).

Once the child begins to talk, many conversations between parent and child show the parent interpreting the child's imperfect speech and then responding with short, clear sentences the child can understand, often with special emphasis on important words. Naturalistic observation is the best way to study this interaction, for facial expression and intonation are as much a part of baby talk as the words spoken. However, recorded dialogues like the following one between a mother and her toddler son at bedtime help give the flavor of these exchanges (Halliday, 1979):

Mother: And when you get up in the morning, you'll go for a walk.
Nigel: Tik.
And you'll see some sticks, yes.
Hoo.
And some holes, yes.
Da.
Yes, now it's getting dark.
I wa [repeated thirteen times].
What?

I wa [repeated seven times]. Peaz.
What do you want in bed? Jamie? [his doll]
No!
You want your eiderdown? [quilt]
(grins) Yeah!
Why didn't you say so? Your eiderdown.
Ella [three times]

In most episodes of baby talk, the child is an active participant, responding to the speaker and making his or her needs known. In this one, Nigel asked for his quilt a total of twenty times, persisting until his mother got the point. An analysis of toddlers' speech shows that, especially after the vocabulary spurt begins, early speech is almost never idle conversation. Babies seem intent on communicating their needs and desires, as well as commenting on their own actions.

More generally, the many ways in which adults support early language acquisition—from their nonlinguistic "conversations" with the infant, to their use of baby talk, to the persistent naming of objects and events that capture the child's attention, to their expansion of the child's sounds and words into meaningful communications—are, taken together, part of the structured guidance that theorists like Vygotsky believe to be the basis for cognitive and language growth. As you recall from Chapter 2, Vygotsky and his followers maintain that the child's intellectual competencies emerge through an "apprenticeship in thinking," as skilled mentors provide the child not only active instruction but also guided participation in shared activities that facilitate the development of skills. In many ways, early language development provides an example *par excellence* of how an important feature of intellectual growth—specifically, mastery of one's native tongue—is structured, guided, and nurtured under the sensitive tutelage of the adults in the child's world.

A Social Interaction

As we have seen repeatedly, infants are motivated to understand the world: the same motivation that makes toddlers resemble little scientists makes infants seek to understand the noises, gestures, words, and grammatical systems that describe the world in which they live, as well as to use words to

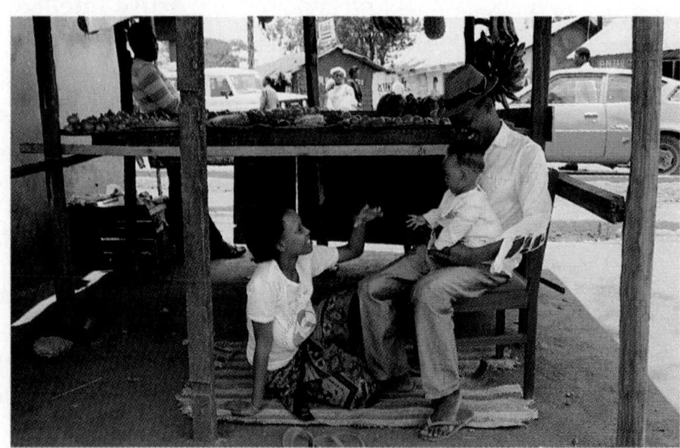

Having completed these first two chapters on infancy, you can probably spot a number of signs that this baby in Nairobi is well nurtured by her family and culture—among them, the delight on both parents' faces, the "breast is best" logo on the mother's shirt, the surrounding objects that stimulate exploration and conceptualization, and the language of gesture that obviously communicates to all three.

engage in social relationships. Central to the achievement of understanding language is verbal interaction. As one researcher writes:

> language . . . could not emerge in any species, and would not develop in any individual, without a special kind of fit between adult behavior and infant behavior. That fit is pre-adapted: It comes to each child as a birthright, both as a result of biological propensities and as a result of social processes learned and transmitted by each new generation. [Kaye, 1982]

The idea that language develops as the outcome of "biological propensities" and "a special kind of fit" highlights the fact that both innate processes and the social context are prerequisites for language development. Humans are biologically destined to communicate, and their brains are primed to develop language. At the same time, verbal interaction between adult and infant is essential, for without a sensitive and responsive conversational partner, a child's language learning will be impeded. Thus parent and baby together accomplish what neither could do alone: teaching a person to talk. The same is true of the other cognitive accomplishments discussed in this chapter. And as we will see in the next chapter, the same parent-infant relationship is at the core of the infant's psychosocial development.

SUMMARY

1. The interaction of the infant's early perceptual, cognitive, and linguistic abilities leads to impressive changes in the baby's capacity to understand and communicate with the world and forms the initial foundation for thinking and learning.

Perception and Cognition

2. Infants quickly grasp the affordances of objects, that is, the activities one can do with them, such as grasping, sucking, squeezing, rolling, shaking, and the like. Affordances change for the infant as his or her experience and repertoire of skills and abilities increase.

3. Infants' perceptual skills contribute to such cognitive understandings as their grasp of the boundaries of objects, of perceptual constancy, and of other physical properties of objects.

4. Both intermodal perception (such as listening to a sound and knowing which object is likely to be the source) and cross-modal perception (such as touching an object and imagining how it might look) are evident in the first months of life.

Key Elements of Cognitive Growth

5. Habituation research has provided important insights into the early growth of categorization skills, as well as into other abilities in infancy. Early categories of shape and sound quickly evolve into categories for faces, animals, and other objects.

6. Young infants have a basic appreciation for object permanence in the early months of life, even though they will not effectively search for a hidden object until later in their first year.

7. Early memory is fragile yet surprisingly capable. Infants easily forget, but their memory can also be "reactivated" to help them remember past events. There is also intriguing evidence for long-term memory capacity in infancy.

8. The ability to understand cause-and-effect relations develops slowly in the first few months of life, but toward the end of the first year, this understanding leads to the beginning of problem-solving abilities.

Active Intelligence: Piaget's Theory

9. From birth to age 2, the period of sensorimotor intelligence in Piaget's theory, infants use their senses and motor skills to understand their environment. They begin by adapting their reflexes, coordinating their actions, and interacting with people and objects. By the end of the first year, they know what they want and have the knowledge and ability to achieve simple goals.

10. In the second year, toddlers find new ways to achieve their goals, first by actively experimenting with physical objects, and then, toward the end of the second year, by manipulating mental images of objects and actions that are not in view.

Language Development

11. Language skills begin to develop as babies communicate with noises and gestures and then practice babbling. Infants say a few words at the end of the first year and thereafter gradually add a few words to their vocabulary each month until about the age of 18 months, when rapid vocabulary acquisition begins.

12. At every age, children understand more words and phrases than they themselves produce. By age 2, if not earlier, toddlers can combine two or three words to make sentences.

13. Children vary in how rapidly they learn vocabulary, as well as in the ways they use words. In the first two years, a child's comprehension of simple words and gestures, and willingness and ability to communicate, are more significant than the size of the child's vocabulary.

14. Language learning is partly the result of the interaction between parent and child. The child is innately primed to learn language, and adults all over the world foster language development by communicating with children using a simplified form of language called baby talk, which suits the child's abilities to understand and use language.

KEY TERMS

affordances (162)
perceptual constancy (165)
dynamic perception (165)
intermodal perception (166)
cross-modal perception (166)
object permanence (169)
launching event (174)
sensorimotor intelligence (175)
goal-directed behavior (178)
little scientist (178)
mental combinations (178)
babbling (180)
underextension (183)
overextension (183)
holophrase (183)
language acquisition device (LAD) (184)
baby talk (185)

KEY QUESTIONS

1. How does an understanding of affordances affect perceptual ability?

2. What evidence shows that infants can coordinate one perception with another?

3. Describe the early growth of categorization skills.

4. Why do developmentalists believe that young infants have an awareness of object permanence even though they cannot search well for a hidden object?

5. How good are the memory abilities of young infants? What factors can strengthen their memory?

6. Explain the growth of an understanding of cause-and-effect relations and how researchers study this accomplishment.

7. What perspective does Piaget's theory bring to the study of cognitive development in infancy?

8. Describe how changes in a baby's actions over time reveal the growth of sensorimotor intelligence.

9. What are the major milestones in the growth of language in infancy? Do all children reach these accomplishments at the same age?

10. How do experiences of social interaction in infancy facilitate language learning?

The First Two Years: Psychosocial Development

CHAPTER 7

Psychosocial development, by definition, involves the interaction between the developing person's *psyche* (emotions, attitudes, and behavior) and his or her *social context* (family, community, and culture). When researchers first began to study the psychosocial development of infants, they assumed that their focus should be primarily on the social-context side of this interaction, specifically, the actions of parents and other caregivers. This assumption stemmed from the firm belief that babies themselves brought little to the interaction, other than a need for food and physical protection.

As you will learn, however, developmentalists now realize that each developing person begins to lead an active emotional and social life early on. In fact, they now see newborns as *innately predisposed to sociability*, capable, in the very first month, of expressing emotions and of responding to the moods, emotions, and actions of others. In this chapter we will examine psychosocial development during infancy quite closely, focusing on the emotional needs and competencies of infants, as well as on the sensitivity that caregivers usually demonstrate in responding to those needs and competencies. In the process, you will see that, just as they do in motor and cognitive development, infants show impressive accomplishments in psychosocial interaction.

Emotional Development

An examination of infants' emotional development provides a valuable window into early psychological growth because it reveals how young infants begin to perceive, understand, and respond to their surroundings. It also shows infants' emotions as important contributors to their social interactions, for a baby's cry, smile, and other expressions are very significant social signals. Indeed, according to the **ethological perspective**, which analyzes how specific behaviors of humans and other animals contribute to the evolutionary survival of each species, infant emotions trigger in caregivers precisely the kind of responses necessary for the newborn's survival. Researchers have discovered, for example, that adults, even if they have never cared for a baby, become physiologically aroused, with focused attention and more rapid heartbeat, upon hearing the sound of a baby's cry (Thompson & Frodi, 1984). Moreover, caregivers respond with greater urgency and more sensitive care the more distressed the baby sounds (Gustafson & Har-

ethological perspective The view that many behaviors and emotions of humans and other animals have an adaptive function that furthers the survival of the species. Ethological studies often shed light on infant emotional development.

ris, 1990; Zeskind & Collins, 1987). The cry thus contributes to the baby's survival by signaling conditions that may be harmful to the infant and by prompting adults to come to the infant's aid. Human infants, otherwise immobile and helpless, display additional emotional signals and behaviors, including reaching, clinging, and grunting, that also summon adults, or keep them nearby, to provide nurturance.

The First Days and Months

The first emotion that can be reliably discerned is distress, most obviously registered in the crying of the infant who is hungry or otherwise uncomfortable. In addition, when newborns a few days or even a few hours old hear a loud noise, or feel a sudden loss of support, or see an object looming toward them, they often cry and look upset (Izard & Malatesta, 1987; Sroufe, 1979). By 4 to 7 months of age, babies exhibit more pronounced distress reactions that are combined increasingly with anger, such as when something attractive is taken away from them or they are prevented from moving (Stenberg & Campos, 1990).

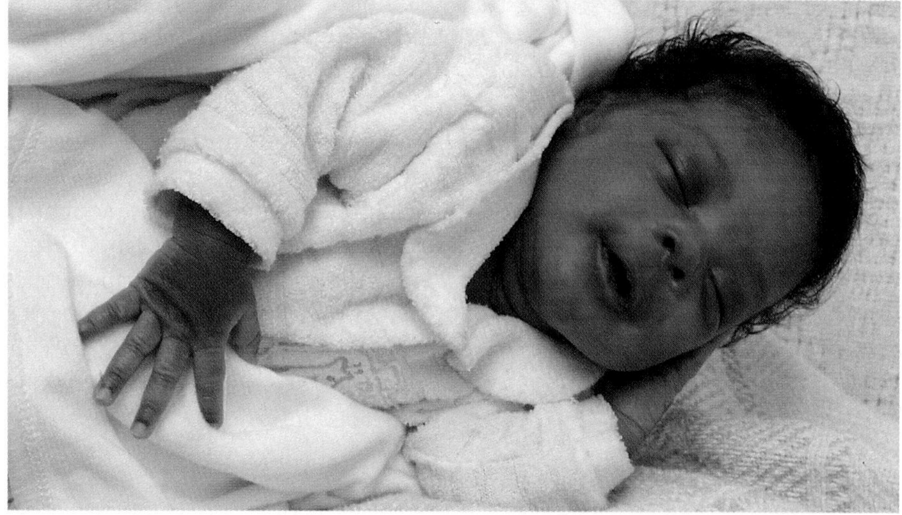

Toni at 1 month of age smiles in her sleep, as many very young babies do in response to the inner satisfaction of a full belly and a comfortable bed. However, not until about 6 weeks of age do babies smile in response to outside stimuli, such as a caregiver's face. Since Toni was born a month preterm, her social smile is likely to occur at 10 weeks of age, for neurological maturation rather than experience outside the womb seems to be the main prerequisite for the first appearance of the social smile.

On the positive side, newborns show wide-eyed looks of interest and surprise when something catches their attention. Smiles also begin early: a half-smile in response to a pleasant noise or a full stomach appears in the first days of life. A **social smile**—a smile in response to a moving face or a human voice—begins to appear at about 6 weeks. By 3 or 4 months, smiles become broader, and babies laugh rather than grin if something is particularly pleasing, especially during social interaction (Malatesta et al., 1989). These patterns are universal, as evident among, say, the hunter-gatherers of the Kalahari as among the upper class of Boston or Paris (Bakeman et al., 1990). Interestingly, some of the most potent elicitors of smiling and laughter in young infants are events (such as shaking a rattle to make a noise) whose occurrence the infant can control (Ellsworth et al., 1993; Lewis et al., 1990). Not surprisingly, some of the earliest anger expressions are observed when changes in circumstances end the infant's control of such an event (such as when a sibling takes a rattle away).

social smile An infant's smile in response to a human face or voice. In full-term infants, this smile first appears at about 6 weeks after birth.

Further Growth in Emotionality

After these early developments, all the basic infant emotions become more differentiated and distinct sometime between 6 and 9 months of age. Infants also show greater range and selectivity in their emotions, owing to their growing cognitive skills and more varied experiences.

This shift is most evident in the various fears and anxieties that the infant experiences. One common reaction, **stranger wariness**, or fear of strangers, is first noticeable at about 6 months of age and is usually full-blown by 10 to 14 months. Perhaps you have observed this fear yourself when offering a 1-year-old a friendly greeting in a supermarket—only to have the child erupt in loud wailing! Contrary to popular belief, however, not all infants experience wariness with every stranger, and those who do vary considerably in the intensity of their reactions. Moreover, many infants respond positively to unfamiliar adults, and some mingle wary and friendly reactions in an unmistakably "coy" demeanor.

How a baby responds to a stranger depends on aspects of the infant (such as temperament and the security of the mother-infant relationship), the stranger (including gender and behavior toward the baby), and the situation (such as the mother's proximity or the infant's current mood) (Thompson & Limber, 1990). A baby may be friendly toward a stranger who keeps at a distance in the mother's presence but react fearfully if the same stranger looms suddenly when the mother is away.

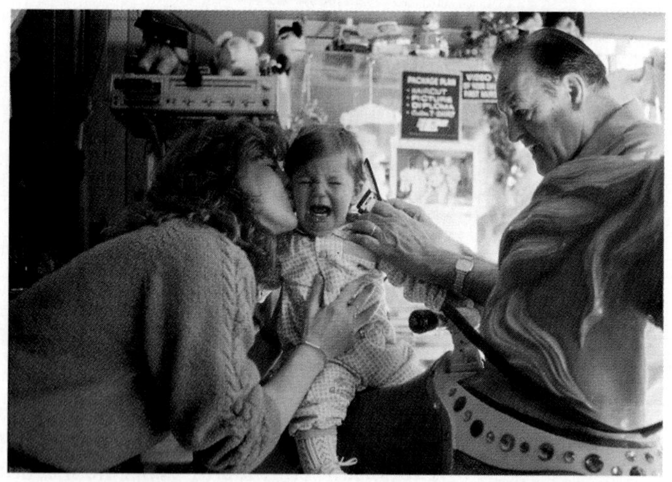

For most toddlers, the approach of a stranger with a buzzing razor triggers a full-blown case of stranger wariness—one that even reassurances and kisses from Mom can't quiet. Had the boy here encountered the stranger in a different context—say, in a friendly conversation with Mom on the street—his reaction might have been a bit different.

A related reaction is **separation anxiety**, the fear of being left by the mother or other caregiver. Separation anxiety emerges at about 8 or 9 months of age, peaks at about 14 months, and then gradually subsides. Whether or not infants will be distressed by separation depends on such factors as the baby's prior experiences with separation and the manner in which the parent departs—leaving abruptly, for example, or in a relaxed fashion with good-byes and reassurance (Thompson & Limber, 1990). Stranger wariness and separation anxiety each reveal how the older infant's emotions are based not on a single event (such as the approach of an unfamiliar person) but on various contextual features of the event in question.

stranger wariness Fear of unfamiliar people, first noticeable in infants at about 6 months of age and usually full-blown by 10 to 14 months.

separation anxiety A child's fear of being left by the caregiver. This emotion emerges at about 8 or 9 months, peaks at about 14 months, then gradually subsides.

As every parent knows, anger also intensifies in toddlerhood. For example, when videotapes of infants being inoculated between the ages of 2

months and 19 months were categorized by "blind" raters who could see only the children's faces, ratings of anger increased dramatically between ages 7 and 19 months. In addition, the duration of anger increased, from a fleeting expression in early infancy to a lengthy demonstration at 19 months (Izard et al., 1987).

Although infants experience anger more intensely as they develop over infancy, they also become better at handling their anger. Instead of simply crying in protest, as a 6-month-old might, they take action (climbing out of a playpen, grabbing for a toy), get help (crying *at* someone, gesturing toward what they want), or comfort themselves (looking away from the source of distress, sucking their fingers) (Gustafson & Green, 1991; Stifter & Braungart, 1995).

Similarly, as infants become older, they smile and laugh more selectively, as well as more quickly and responsively (Lewis & Michalson, 1983; Nwokah et al., 1994). For instance, the sight of almost any human face produces a stare and then a smile in the typical 3-month-old, but the typical 9-month-old may grin immediately at the sight of certain faces—and might remain impassive or burst into tears at the sight of others. At 12 months of age, the infant's immediate smile at seeing a parent's face may be swiftly followed by an explosion of loud squeals if the parent's behavior signals the beginning of a playful interaction.

The overall effect of the many specific changes in the infant's emotional life is that the baby appears to have greater emotional vitality toward the end of the first year (Thompson, 1994). Whether they are expressions of pleasure, fear, distress, or anger, the most striking difference between the emotions of a 6-month-old and those of a 12-month-old is that the older infant's emotions are manifested more quickly, more intensely, and more persistently. As a result, the older infant's emotions become easier for an adult to read and respond to, which allows a more stimulating and mutual emotional exchange.

Social Referencing

At the same time that the infant's emotions are becoming easier for others to read, the infant's capacity to read the emotional expressions of others is increasing, furthering the infant's early emotional development (Feinman, 1985). As early as 5 or 6 months of age, infants associate emotional meaning with different facial expressions, such as happiness and anger, and with different tones of voice, such as encouraging or disapproving ones—and they can do so even in experiments in which the face is a photograph of a stranger and the voice is speaking a language the infant has never heard (Balaban, 1995; Fernald, 1993).

As infants approach the half-year mark, the emotional expressions of others take on new significance, because infants increasingly engage in all kinds of **social referencing**. That is, they increasingly look to trusted adults for emotional cues—a look of calm reassurance, a vocal warning, an expression of alarm or dismay—that tell them how they should react to unfamiliar or ambiguous events. Beginning at about 9 months, as crawling makes them independently mobile, and especially as they begin the active exploration of the "little scientist" (stage five of Piaget's sensorimotor intelligence), infants frequently use social referencing to guide their actions (Derochers et al., 1994). As a result, seemingly self-assured, free-roaming toddlers regularly

social referencing Looking to trusted adults for emotional cues in interpreting a strange or ambiguous event.

glance back at their caregivers, checking for a signal regarding a new toy, a new playmate, or some new activity they are about to engage in.

In many ways, the lessons an infant learns from social referencing become a guide to later activity. This was shown in one experiment that began as a straightforward demonstration of social referencing. In the presence of their mothers, 12-month-olds were presented with a toy robot or a moving, cymbal-clapping monkey, and their mothers were instructed to express either disgust or delight with the toy. As expected, the infants were more likely to avoid or to play with the toy depending on how their mother reacted to it. In the next part of the experiment, the mothers refrained from providing any emotional cues regarding the toy. Nevertheless, the infants once again reacted to the toy in accordance with the guidance they had received in the prior session of social referencing (Hornik et al., 1987).

Over time, the lessons learned from social referencing can have a general shaping effect. For example, if, in his or her explorations and approaches to others, a toddler regularly receives more signals of encouragement and support than of apprehension and prohibition, the child is likely to be more friendly, and less aggressive, than if the opposite messages have been socially referenced (Calkins, 1994). If an infant or toddler sees few signals of any kind, as might happen if the primary caregiver is depressed or overtired, the child tends to become relatively emotionless and passive (Field, 1995).

Self-Awareness

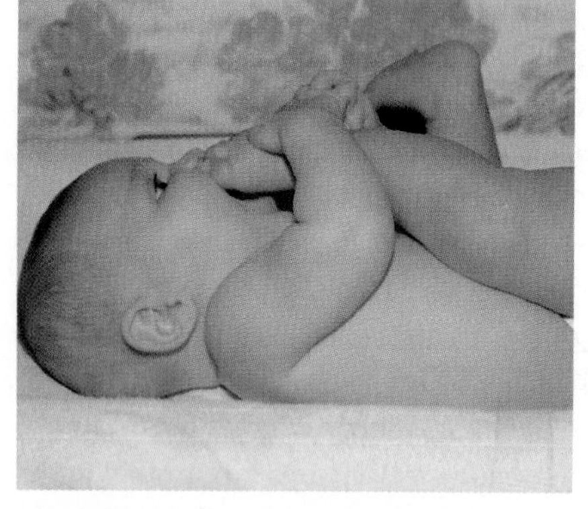

Although Alexandra at 4 months has enviable agility in being able to get her foot to her mouth, she probably does not yet realize that what she is trying to chew on is part of her—a cognitive gap that may result in unexpected discomfort.

A pivotal accomplishment of later infancy is the development of **self-awareness**. The emerging sense of "me and mine" becomes the fertile ground that allows the growth of many self-conscious emotions—from pride and confidence to guilt, shame, and embarrassment. Simultaneously, self-awareness allows a new consciousness of others, which fosters such emotions as defiance and jealously, as well as empathy and feelings of affection that go beyond the pleasure of seeing a familiar face.

The development of this self-awareness is striking when infants of various ages are compared. Young infants have no sense of self: in fact, they do not even have an awareness of their bodies as *theirs* (Lewis, 1990). To them, for example, their hands are interesting objects that appear and disappear: 2-month-olds, in effect, "discover" their hands each time they catch sight of them, become fascinated with their movements, then "lose" them as they slip out of view. Even 8-month-olds often don't seem to know where their bodies end and someone else's body begins, as can be seen when a child grabs a toy in another child's hand and reacts with surprise when the toy "resists." By age 1, however, most infants would be more aware that the other child is a distinct person, and might show this awareness with a smile or a shove if the coveted toy is not immediately forthcoming.

self-awareness A person's sense of himself or herself as a being distinct from others, with particular characteristics.

Evidence of the emerging sense of self was shown in a classic experiment in which babies looked in a mirror after a dot of rouge had been surreptitiously put on their nose (Lewis & Brooks, 1978). If the babies reacted to the mirror images by touching their nose, it was clear that they knew they were seeing their own face. After trying this experiment with ninety-six babies between the ages of 9 and 24 months, the experimenters found a distinct developmental shift. None of the babies under a year reacted to the mark, whereas most of those between 15 and 24 months did. Before their

second birthday, most infants can point to themselves when asked "Where's [child's name]?" and they can use their own name appropriately when pointed at and asked "Who's that?" (Pipp et al., 1987).

The link between the advent of self-awareness and the emergence of certain self-conscious emotions was shown in an extension of the rouge-and-mirror experiment (Lewis et al., 1989). In this study, 15- to 24-month-olds who showed self-recognition in the rouge task described above also looked embarrassed when they were effusively praised by an adult; that is, they smiled and looked away, covered their face with their hands, and so forth. Infants who did not show self-recognition were not embarrassed. These changes from lack of self-recognition to coy smiles seem universal, occurring at about the same time among toddlers from varied backgrounds (Schneider-Rosen & Cicchetti, 1991).

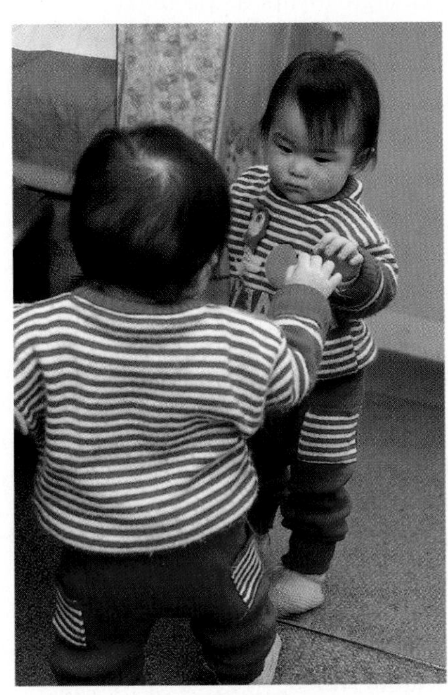

Mirror images make young infants smile and try to touch "the baby." It is not until after they are a year old that children realize that they are looking at themselves.

Self-awareness also changes the intensity and conditions of the toddler's reactions to others, including affection and jealously. Indeed, the infamous toddler temper develops partly because, when children become more aware of themselves, they take frustration and hurt much more personally, and they also realize that they are more able to respond in kind (Dunn & Munn, 1985).

Developing self-awareness also enables toddlers to be self-critical and to have emotional responses such as guilt (Emde et al., 1991). By age 2, for example, most children are aware of the basic "do's" and "don'ts" they should follow, and sometimes show distress or anxiety when they have misbehaved, even when no adult is present. In one experimental demonstration of this, 2-year-olds were "set up" to experience two mild mishaps: they were left alone in a playroom with (1) a doll whose leg was rigged to fall off when the doll was picked up, and (2) a juice drink in a trick cup that dribbled when drunk from. Many of the children responded to their "accidents" with expressions of sadness or tension, accompanied by efforts to repair the damage (Cole et al., 1992). Such reactions are evident at home as well as in the laboratory: mothers report that the toddler's sense of shame and guilt appears for the first time after self-awareness develops (Stipek et al., 1992). With the growth of self-awareness, anger at an injustice (such as another child's getting the first slice of pie), as well as being "sorry" for a misdeed, become part of the child's developing moral sense (Zahn-Waxler et al., 1992).

Self-awareness also permits a child to react in an entirely new way to his or her misdeeds—with pride at going against another's wishes. Shortly before his second birthday, for example, Ricky teased his mother by deliberately pouring a cup of juice onto the rug. That Ricky knew he was being "naughty" was clear from his reaction to his mother's scolding: he was unsurprised and unfazed by it and was quite willing to help his mother clean up the mess. Only when his mother sent him to his room did he protest angrily, apparently not anticipating such punishment. Later that day he told his grandmother, "Juice on a floor." "Juice doesn't go on the floor," she sternly responded. "Yes, juice on a floor, juice on a floor," Ricky laughingly repeated several times, pretending to turn an imaginary cup upside-down. As his grandmother, a noted psychologist, comments:

> The boy's pleasure at watching the juice spill and anger at being sent to his room are emotions that are typical at all periods of infancy, but his obvious pride at his ability to act counter to convention or his mother's wishes is possible only when self-awareness is firmly established. [Shatz, 1994]

The Origins of Personality

Now that we have seen how emotional capacities develop during infancy, the next questions are: How do the infant's emotional and behavioral responses begin to take on the various patterns that form personality? What happens to evoke or create personality traits and social skills during infancy, leading to the emergence of a distinct individual?

Psychological Theory: The Importance of Nurture

In the first half of the twentieth century, the prevailing view among psychologists was that the individual's personality is permanently molded by the actions of his or her parents—most especially the mother—in the early years of childhood. There were two major theoretical versions of how this comes about.

Learning Theory

Those who favored the behaviorist perspective (see Chapter 2) maintained that personality is molded as parents reinforce or punish their child's various spontaneous behaviors. Behaviorists proposed, for example, that if parents smile and pick up their baby at every glimmer of an infant grin, the baby will become a child, and later an adult, with a sunny disposition. Similarly, if parents continually tease the child by, say, removing the nipple as the child is contentedly sucking, or pretending to take away a favorite toy that a toddler is clutching, that child will be likely to develop a suspicious, possessive nature.

The strongest statement of this early view came from John Watson, the leading behaviorist of the time, who cautioned:

> Failure to bring up a happy child, a well-adjusted child—assuming bodily health—falls squarely upon the parents' shoulders. [By the time the child is 3] parents have already determined . . . whether . . . [the child] is to grow into a happy person, wholesome and good-natured, whether he is to be a whining, complaining, neurotic, an anger-driven, vindictive, over-bearing slave driver, or one whose every move in life is definitely controlled by fear. [Watson, 1928]

Later theorists in the behaviorist tradition incorporated the role of social learning, finding that infants tend to imitate personality traits of their parents, even if they are not directly reinforced for doing so. A child might develop a quick temper, for instance, if he or she sees a parent regularly display anger and get respect or obedience from other family members in return. Although these theorists accepted that not all personality traits are directly reinforced in babyhood, "the guiding belief of social learning theorists was that personality is learned" (Miller, 1993).

Psychoanalytic Theory

Beginning with a different set of assumptions about human nature, psychoanalytic theorists (see Chapter 2) also concluded that the individual's personality is formed and permanently fixed in the early years of childhood. Sigmund Freud, who established the framework for their view, felt that the experiences of the first four years of life "play a decisive part in determining

whether and at what point the individual shall fail to master the real problems of life" (Freud, 1918/1963). He also thought that the child's relationship with the mother was "unique, without parallel, established unalterably for a whole lifetime as the first and strongest love-object and as the prototype of all later love relations" (Freud, 1940/1964).

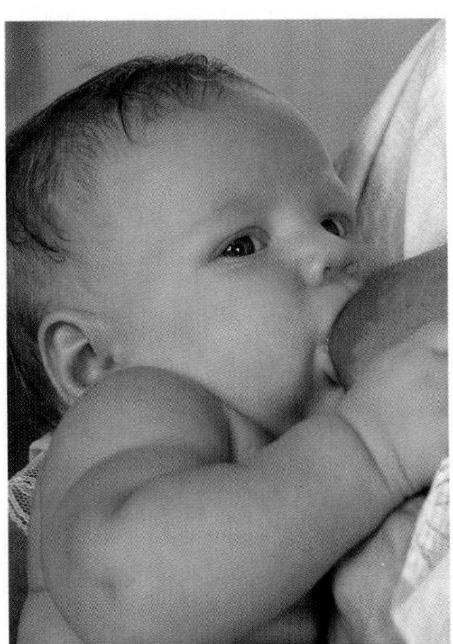

To psychoanalytic theorists, breast-feeding is important not just because it is a source of nourishment but also because the pleasurable, intimate contact it affords strengthens the infant's attachment to the mother and fosters a feeling of "basic trust" in the world.

Freud: Oral and Anal Stages

As we noted in Chapter 2, Freud viewed human development in terms of psychosexual stages that occur at specific ages. According to Freud (1935), psychological development begins with the **oral stage**, so named because in the first year of life the mouth is the infant's prime source of gratification. Not only is the mouth the instrument for attaining nourishment; it is also the main source of pleasure: sucking, especially at the mother's breast, is a joyous, sensual activity for babies.

In the second year, Freud maintained, the infant's prime focus of gratification shifts to the anus, particularly the sensual pleasure taken in stimulation of the bowels and, eventually, the psychological pleasure in controlling them. Accordingly, Freud referred to this period as the **anal stage**. This shift is more than a simple change of focus; it is a shift in the mode of interaction, from the passive, dependent mode of orality to the more active, controlling mode of anality, in which the child has some power. Parents at this time are striving to foster the toddler's self-control in many ways in addition to toileting, and the toddler has the will to comply or resist, a situation that is ripe for a power struggle between adult and child.

Indeed, according to Freud, both the oral and anal stages are fraught with potential conflict that can have long-term consequences. If a mother frustrates her infant's urge to suck—by making nursing a hurried, tense event, or by weaning the infant from the nipple too early, or by continually preventing the child from sucking on fingers, toes, and other objects—the child may be made distressed and anxious. The child may become an adult who is "fixated," or stuck, at the oral stage, excessively eating, drinking, chewing, biting, smoking, or talking in quest of the oral satisfaction denied in infancy. Similarly, if toilet training is overly strict or occurs before the child is mature enough to participate in it (before age 1½ or 2), parent-child interaction may become locked into a conflict over the toddler's resistance or inability to comply, and this conflict, too, may have important consequences for the child's future personality.

Although Freud's ideas concerning orality and anality have been extremely influential, both in psychoanalytic theory and in popular thinking, subsequent research has failed to support the linking of specific conflicts during these stages to later personality traits. Rather, it has been shown that the overall pattern of parental warmth and sensitivity or strict domination is much more important to the child's emotional development than the particulars of either feeding or toilet training. This broader perspective is reflected in the theory of Erik Erikson.

oral stage Freud's term for the first stage of psychosexual development, in which the infant gains pleasure through sucking and biting.

anal stage Freud's second stage of psychosexual development, in which the anus becomes the main source of bodily pleasure, and control of defecation and toilet training are therefore important activities.

Erikson: Trust and Autonomy

As you will remember from Chapter 2, Erik Erikson believed that development occurs through a series of basic crises, or developmental tasks,

trust versus mistrust Erikson's first stage of psychosocial development, in which the infant experiences the world as either secure and comfortable or as unpredictable and uncomfortable.

autonomy versus shame and doubt Erikson's second stage of psychosocial development, in which the toddler struggles between the drive for self-control and shame and doubt about oneself and one's abilities.

throughout the life span (see p. 32). The first crisis of infancy, in Erikson's view, is one of **trust versus mistrust**. In this crisis, the infant learns whether the world is essentially a secure place where basic needs will be readily met or an unpredictable arena where needs are often met only after much crying, and sometimes not at all. Erikson contended that babies begin to develop a sense of security when their mothers sensitively provide food and comfort with "consistency, continuity, and sameness of experience." When the relationship with the mother inspires trust and security, the child experiences confidence in engaging and exploring the world.

The next crisis, which occurs in toddlerhood, is the crisis of **autonomy versus shame and doubt**. Toddlers want autonomy, or self-rule, over their own actions and bodies. If they fail in their effort to gain it, either because they are incapable or because their caregivers are too restrictive and forbidding, they come to feel shame and to doubt their abilities. According to Erikson, the key to the child's successfully meeting this crisis and gaining a sense of autonomy is parental firmness:

> Firmness must protect him [the toddler] against the potential anarchy of his as yet untrained sense of discrimination, his inability to hold on and let go with discretion. As his environment encourages him to "stand on his own feet," it must protect him against meaningless and arbitrary experiences of shame and of early doubt. [Erikson, 1963]

If parents accomplish this, the child is likely to become increasingly self-confident when encountering new challenges as an independent being.

Like Freud, Erikson believed that problems arising in early infancy can last a lifetime. He maintained that the adult who is suspicious and pessimistic, or who always seems burdened by self-doubt, may have been an infant who did not develop sufficient trust, or a toddler who did not achieve sufficient autonomy. However, Erikson also emphasized that experiences later in life can alter or transform the effects of early experiences, and that earlier crises can be reapproached and resolved later in life.

If this girl's explorations result in smashed dinnerware, will her parents react with anger, as though her goal were destruction, or with a firm but understanding caution, as though her goal were discovery? According to Erickson, how parents react to their children's efforts at autonomy can shape how young children resolve the psychosocial crisis of autonomy versus shame and doubt.

temperament Inherent dispositions, such as activity level, intensity of reaction, emotionality, and sociability, that underlie and affect a person's responses to people and things.

Overall, then, traditional psychological theory maintains that personality is primarily shaped by early nurture, particularly the mother's caregiving. This view has been seriously challenged by those who see basic elements of the infant's personality emerge so early that parental influences cannot be credited or blamed for it.

Temperament: The Importance of Nature

As you read in Chapter 3, researchers have determined that each individual has his or her own distinct, genetically based temperament, which permeates virtually every aspect of the person's developing personality. **Temperament** is defined as "relatively consistent, basic dispositions inherent in the person that underlie and modulate the expression of activity, reactivity, emotionality, and sociability" (McCall, in Goldsmith et al., 1987).

Temperament begins in the multitude of genetic codes that guide the development of the brain and is affected by many prenatal experiences, especially those relating to the nutrition and health of the mother. Elements of temperament are evident from birth, and within the first months, temperamental individuality is clearly established. However, although temperament is apparent in the first months of life, as the person develops, the social context and the individual's experiences increasingly influence the nature and expression of temperament.

Dimensions of Temperament

Given the centrality of temperament in determining the kind of individual each person is and how a person interacts with others, many researchers have set out to describe and measure the various dimensions of temperament. Generally, these researchers see the many elements of temperament as clustering into basic features, such as activity, emotionality, and sociability (Buss, 1991), or basic tendencies, such as how people react to external events and how they control their own behavior (Rothbart, 1981, 1991). Many researchers in adult temperament refer to the "Big Five" dimensions of personality: extroversion, agreeableness, conscientiousness, neuroticism, and

One temperamental trait that is evident in the early days and that continues lifelong is a person's typical reaction to new experiences. Some are delighted, some are curious, and some are overcome with distress, whatever the experience might be—from a first bath to a first day at college, from finding oneself on a blanket with nine other babies to finding oneself in a new country surrounded by strangers.

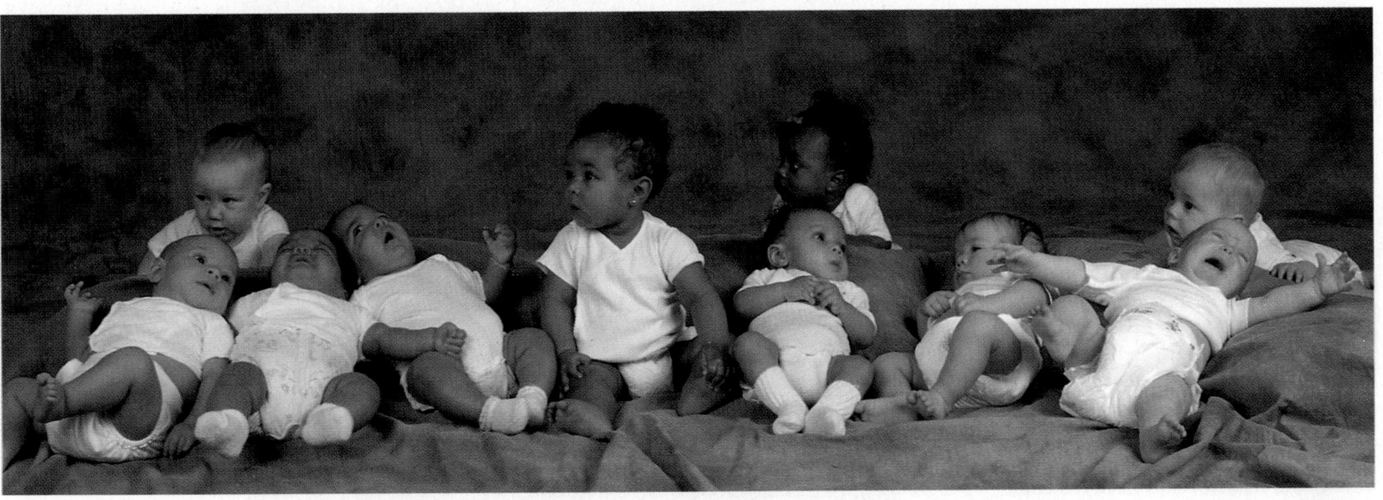

openness (see Chapter 22), and many developmentalists are seeking parallels between adult and child temperament (Kohnstamm, 1996; Marlin et al., 1994).

The most famous, comprehensive, and durable study of children's temperament remains the classic New York Longitudinal Study (NYLS), begun over three decades ago (Thomas & Chess, 1977; Thomas et al., 1963). These researchers interviewed parents of new infants repeatedly and extensively, noting in detail various aspects of infants' behavior and taking steps to reduce the possibility of parental bias:

> For example, if a mother said that her child did not like his first solid food, we asked her to describe his actual behavior. We were satisfied only when she gave a description such as, "When I put the food into his mouth he cried loudly, twisted his head away, and let it drool out."
>
> If we asked what a six-month-old baby did when his father came home in the evening, and his mother said, "He was happy to see him," we pressed for a detailed description: "As soon as he saw his father he smiled and reached out his arms." [Chess et al., 1965]

According to the researchers' initial findings, babies in the first days and months of life differ in nine temperament characteristics:

1. *Activity level.* Some babies are active. They kick a lot in the uterus before they are born, they move around a great deal in their bassinets, and, as toddlers, they are nearly always running. Other babies are much less active.

2. *Rhythmicity.* Some babies have regular cycles of activity. They eat, sleep, and defecate on schedule almost from birth. Other babies are much less predictable.

3. *Approach-withdrawal.* Some babies delight in everything new; others withdraw from every new situation. The first bath makes some babies react in wide-eyed wonder and others tense up and scream; the first play date with a neighbor child makes some crawl toward their new playmate with excitement and makes others try to hide.

4. *Adaptability.* Some babies adjust quickly to change; others are unhappy at every disruption of their normal routine.

5. *Intensity of reaction.* Some babies chortle when they laugh and howl when they cry. Others are much calmer, responding with a smile or a whimper.

6. *Threshold of responsiveness.* Some babies seem to sense every sight, sound, and touch. For instance, they waken at a slight noise or turn away from a distant light. Others seem unaware even of bright lights, loud street noises, or wet diapers.

7. *Quality of mood.* Some babies seem constantly happy, smiling at almost everything. Others seem chronically unhappy: they are ready to protest at any moment.

8. *Distractibility.* All babies fuss when they are hungry, but some will stop if someone gives them a pacifier or sings them a song, while others keep fussing. Similarly, some babies can easily be distracted from their interest in an attractive but dangerous object and diverted to a safer plaything, while others are more single-minded.

RESEARCH REPORT

Early Signs of Shyness

One dimension of temperament, extroversion (or sociability) and its opposite, shyness, has been the focus of extensive research, partly because extroversion/shyness has proven to be one of the most durable and significant traits of human personality. This trait, which relates to such personality dimensions as friendliness, fearfulness, self-confidence, and introspection, has been linked with a variety of life-course events. For example, shy men tend to date, marry, and become fathers later than other men, and shy women are more likely to become full-time homemakers rather than pursue a career. In general, extroverts are more likely to attain higher-paid, stable management positions (Caspi et al., 1988, 1990).

Although every ethnic and racial group has a portion of individuals who are very outgoing and another portion who are unusually shy, the most detailed research concerning extroversion/shyness has been carried out with white American children. Among this group, about 25 percent are "consistently sociable, affectively spontaneous, and minimally fearful," while about 10 percent are "consistently shy, cautious, and emotionally reserved" (Kagan & Snidman, 1991).

The consistency of the extroversion/shyness trait has led some researchers to theorize that the trait might manifest itself even in the first months of life as a general inhibition to unfamiliar objects and experiences. To test this hypothesis, researchers examined ninety-four healthy middle-class infants in a longitudinal study (Kagan & Snidman, 1991). At 4 months the infants were videotaped for 10 minutes as they reacted to several new toys, mobiles, and sounds. Blind observers, watching the tapes, rated the infants as high or low on two behaviors: motor activity (for instance, how much they kicked or waved their arms) and crying (including how quickly they could be soothed if they began to fret). Twenty-three percent of the infants were high in both activities, 37 percent were low in both, and the remaining 40 percent were high in one but not the other.

Then, at 9 months and 14 months, the infants were tested again in several possibly unsettling situations, such as a stranger's inviting the child to play with a metal robot toy or the child's being given unusual things to taste. All the children showed wariness, but some of them were notably more fearful than others—crying and refusing to play with the novel toy, for instance. Again blind observers rated the number of times each infant showed apparent fear. Given the normal variability of infants, and the difficulties in testing them, the results were amazingly clearcut: high motor activity and crying in reaction to the new toys at 4 months correlated strongly with fearfulness in new situations at 14 months. When individual patterns were examined, results were most marked for those who were unusually fearful or fearless. Of the fourteen toddlers who had the lowest fear scores, none had been in the high/high category as infants. Similarly, of the five toddlers who were most fearful, none had been in the low/low category.

Results such as these extend the general finding that extroversion/shyness is an inherited trait (see p. 77); they also provide two further details. First, they suggest that social shyness is only one manifestation of a more general, physiological pattern of inhibition to new stimuli that is apparent in infancy. Second, the fact that the behavior of a specific group of infants was, over time, distinctly more shy or more sociable than that of the average infant suggests that extroversion/shyness may be inherited in more than one way. That is, not only can it be inherited in an additive fashion, producing a simple continuum from very social to very shy with every gradation in between; it may also be that it can be inherited discretely, as blood type is, due to the presence or absence of a certain gene or genes (Kagan, 1989).

9. *Attention span.* Some babies play happily with one toy for a long time. Others quickly drop one activity for another.

The lead NYLS researchers, Alexander Thomas and Stella Chess (1977), believe that "temperamental individuality is well established by the time the infant is two to three months old." In terms of various combinations of personality traits, most young infants can be described as being one of three types: *easy* (about 40 percent), *slow-to-warm-up* (about 15 percent), and *difficult* (about 10 percent). Note, however, that about 35 percent of normal infants do not fit into these well-defined groups.

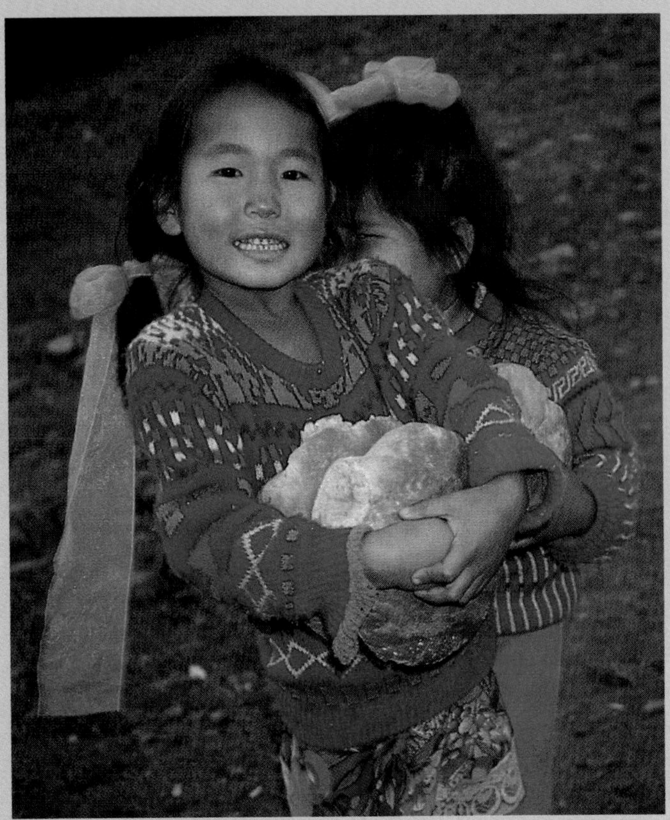

In Mongolia and many other Asian countries, females are expected to display shyness as a sign of respect to elders and strangers. Consequently, if this younger sister is truly as shy as she seems, her parents are less likely to be distressed about her withdrawn behavior than the typical North American parent would be. On the other hand, they may consider the relative boldness of her older sister to be a serious problem.

Of course, as with all inherited tendencies, the reaction of others can modify or exacerbate a person's tendency toward extroversion or shyness. For instance, children who are genetically predisposed to be timid are more likely to become extremely shy if they have a dominating older brother or sister. Parents also make a difference: even if a child becomes an extremely shy toddler, he or she has about a 50/50 chance of no longer behaving with unusual timidness by age 7, with family encouragement of social play with other children being one deciding factor. Even with parental guidance and preschool experience, however, very few shy toddlers become such spontaneous and social 7-year-olds that they would be mistaken for extroverts (Galvin, 1992; Kagan, 1989).

This research has a very practical application. In cultures in which shyness is considered undesirable, the realization that some children may be genetically inhibited should make everyone more accepting and reassuring when a child seems unusually fearful, timid, or quiet.

Interestingly, in Chinese culture, shyness/sensitivity among children is seen as a positive trait, and unlike their Western counterparts, shy Chinese school-age children often are high achievers in almost every aspect of school and also tend to be well-liked by their peers and by adults. However, in early adolescence, some aspects of this pattern shift, with Chinese 12-year-olds tending to most admire those of their classmates who are outgoing and sociable rather than timid. It may be that the hormonal changes of puberty make young people, worldwide, wish to be bold and outgoing, and this universal biological pull makes Chinese adolescents admire their peers whose temperament challenges the cultural value of modesty and self-effacement (Chen et al., 1995).

Stability and Change in Temperament

In a series of NYLS follow-up studies carried into adolescence and adulthood (Carey & McDevitt, 1978; Chess & Thomas, 1990; Thomas et al., 1968), and in other research that looked at the same traits as those in the NYLS (Cowen et al., 1992; Guerin & Gottfried, 1994), temperamental characteristics showed some stability: the easy baby remains a relatively easy child, while the difficult one is more likely to give his or her parents problems. Similarly, the slow-to-warm-up infant who cried on seeing strangers at 8 months may well hide behind Mother's skirt on arriving at nursery school and avoid the crowd in the halls of middle school.

This does not mean that all, or even most, characteristics of temperament remain the same throughout life. Temperament develops over time and can change in many ways. Indeed, some of the NYLS characteristics are not particularly stable. Rhythmicity and quality of mood, for instance, are quite variable, meaning that the infant who has been taking naps on schedule might not do so a few months later, and the baby who seemed consistently happy might become a malcontent if life circumstances change for the worse. Temperamental qualities tend to remain stable over the course of a particular developmental stage but may shift as a new stage brings different challenges. Change itself may follow genetic timetables, and inborn traits may be more apparent during particular developmental periods and under certain conditions (Chess & Thomas, 1990; Plomin et al., 1993).

In addition, there are several ways the environment can affect a child's temperamental characteristics. One way is through the **goodness of fit**, or "match," between the child's temperamental pattern and the demands of the environment. When parents accommodate their child-rearing expectations to their offspring's temperamental style, for example, the result is a more harmonious fit between them, with good outcomes for both child and family. This may involve setting up a child-proof play area in which a high-activity-level child can run off excess energy without safety risk, or it may require allowing extra time for a slow-to-warm-up child to adjust to new situations. By contrast, when the child's temperamental pattern and the caregiving expectations are significantly out of sync, parents and offspring are likely to experience greater conflict, and the child's temperamental style may become more difficult. The same is true of the fit between temperament and other environmental demands. Consider the effects of goodness of fit on one of the original subjects from the NYLS:

> Carl was one of our most extreme cases of difficult temperament from the first months of life through 5 years of age. However, he did not develop a behavior disorder, primarily due to optimal handling by his parents and stability of his environment. His father, who himself had an easy temperament, took delight in his son's "lusty" characteristics, recognized on his own Carl's tendencies to have intense negative reactions to the new, and had the patience to wait for eventual adaptability to occur. He was clear, without any orientation by us, that these characteristics were in no way due to his or his wife's influences. His wife tended to be anxious and self-accusatory over Carl's tempestuous course. However, her husband was supportive and reassuring and this enabled her to take an appropriately objective and patient approach to her son's development.
>
> By middle childhood and early adolescent years, few new situations arose which provoked the difficult temperament responses. The family, school, and social environment was stable and Carl flourished and appeared to be temperamentally easy rather than difficult. . . .
>
> When Carl went off to college, however, he was faced simultaneously with a host of new situations and demands—an unfamiliar locale, a different living arrangement, new academic subjects and expectations, and a totally new peer group. Within a few weeks his temperamentally difficult traits reappeared in full force. He felt negative about the school, his courses, the other students, couldn't motivate himself to study, and was constantly irritable. Carl knew something was wrong, and discussed the situation with his family and us and developed an appropriate strategy to cope with his problem. He limited the new demands by dropping several extracurricular activities, limited his social contact, and policed his studying. Gradually he adapted, his distress disappeared, and he was able to expand his activities and social contacts. . . . [In] the most recent follow-up at age 29 . . . his intensity remains but is now an asset rather than a liability. [Chess & Thomas, 1990]

goodness of fit The quality of the "match" between the child's temperament and the demands of the surrounding environment.

As illustrated in this example, most parents soon learn that how their parenting style affects their child depends on how it "fits" with the child's temperamental style. When trying to teach standards of appropriate behavior to a preschooler, for example, parents may find that if the child is fearful or anxious by temperament, mild criticism of misbehavior is enough to make the child try to avoid the misbehavior in the future. However, if the child is bold and confident, the parents are likely to find that he or she is unaffected by mild criticism and tends to react to more serious punishment by throwing a tantrum. In this case, the most effective approach is probably for the parents to ensure that they have a strong, positive relationship with the child, so that the child tries to please by learning proper behaviors. Research on young children's internalizing of moral standards shows that, particularly for more exuberant children, a secure and warm relationship with the parent is essential (Kochanska, 1995).

Let us now look more closely at the specifics of parent-infant interaction as it changes over the early years.

Parent-Infant Interaction

As the discussion of temperament makes clear, the traditional psychological view of parents as the sole shapers of the child's personality was clearly in error. At the same time, it is also clear that psychological development is not determined solely by the individual's innate characteristics. As developmentalists now emphasize, it is the interaction between the parent and the child that is crucial. This interaction is affected by the personality of the parent and the temperament of the child, as well as by the child's stage of development.

Becoming Social Partners

As you read earlier, even very young infants communicate emotionally, through sounds, movements, and facial expressions. And they are interested in social events virtually from birth: the sound of a human voice, the sight of a human face, and other social stimuli are among the earliest events to capture a young baby's attention, interest, and emotion. But although infants are social from birth, they initially are not ready to play their part in social interactions. First-time parents who look forward to the birth of their child, eagerly anticipating joyous episodes of social exchange after the baby's arrival, are often disappointed to discover that their newborn spends most of the day sleeping, and is often unresponsive even when awake. For most parents, a fixed stare is typically the best they can expect from an attentive newborn.

By the age of 2 to 3 months, a change occurs that parents recognize and rejoice in: the baby begins to respond especially to them. To be sure, other adults can also elicit smiles from the infant, but the appearance of the mother, father, or another familiar caregiver can provoke widened grins, lilting cooing, and other reactions that signify that person's special status to the child. Many parents report a deepening of their own attachment to the baby at this time, as they proceed from the newborn phase, in which they

tended to perceive the child as a delicate guest requiring careful treatment, to a new phase of their relationship, in which they perceive the child as a social partner who can reciprocate their love and attention. This, in turn, provokes a new kind of interaction. Instead of merely gazing intently over the crib rails at the baby, trying to decipher what the infant's needs are, caregivers begin to initiate focused episodes of face-to-face play.

These episodes may occur in a variety of contexts—during a feeding, a diaper change, a bath, or in any other situation. They may be initiated by either the adult or the infant: the caregiver might notice the baby's expression or vocalization and mirror it with his or her own (such as smiling when the baby smiles), or the baby might notice the adult's wide-eyed beaming and break into a grin.

What really distinguishes these episodes of social play from routine caregiving are the moment-by-moment actions and reactions of both partners. To complement the infant's animated but quite limited expressive repertoire, adults use dozens of behaviors that seem to be reserved exclusively for infants. Typically, caregivers open their eyes and mouths wide in exaggerated expressions of mock delight or surprise, make rapid clucking noises or repeated one-syllable sounds ("ba-ba-ba-ba-ba," "di-di-di-di," "bo-bo-bo-bo," etc.), raise and lower the pitch of their voice, change the pace of their movements (gradually speeding up or slowing down), imitate the infant's actions, bring their face close to the baby's and then pull back, tickle, pat, poke, lift, and rock the baby, and do many other simple things. (If you are reading thoughtfully, you probably recognize some of these behaviors as your own natural response when in the presence of a baby—sometimes to your own embarrassment and to the amusement of those around you!) The infant's responses, in turn, typically complement those of the adult: they stare at their partners or look away, vocalize, widen their eyes, smile, move their head forward or back, or turn aside (Stern, 1985).

It appears that episodes of face-to-face play are a universal feature of the early interaction between caregivers and infants, although the frequency and duration of these episodes, as well as the goals of the adults who initiate them, may differ in various cultures. One cross-cultural study of mothers at play with their infants found, for example, that American mothers most often directed the infant's attention to a nearby toy, object, or event, while Japanese mothers focused on establishing mutual intimacy by maintaining eye contact with the infant as well as kissing, hugging, and so on (Bornstein et al., 1992). In another cross-cultural comparison, researchers noted that whereas American mothers employed social overtures that stimulated and excited their babies (such as tickling the baby), mothers from the Gusii community in rural Kenya were more soothing and quieting in their initiatives (Richman et al., 1992). A third study, of mothers from three French-speaking groups in Quebec, Canada—Vietnamese, Haitian, and native-born Quebecois—found notable differences in how the mothers responded to their infants' active curiosity. The Vietnamese mothers tended to be guiding and restrictive across a wide range of their infants' exploratory activities. By contrast, the Haitian mothers encouraged their infants to interact socially with others, while the Quebecois mothers encouraged the exploration of toys—including allowing their infants to mouth objects that the immigrant mothers would have immediately taken away (Sabatier, 1994).

synchrony Carefully coordinated interaction between infant and parent (or any other two people) in which each individual responds to and influences the other's movements and rhythms.

Of course, mothers are not the only ones who engage infants in face-to-face play: fathers are also active partners in play (see A Closer Look, pp. 208–209), and in many non-Western cultures, older siblings and other adults assume an active role in infant care and participate in social play with babies (Tronick et al., 1992; West, 1988).

Developing and Maintaining Synchrony

What accounts for the pleasure that adults and infants both experience from their face-to-face interactions? Many researchers believe that it is the mutual experience of being "in sync"—that is, being socially and emotionally coordinated with the partner. In this respect, therefore, one of the goals of face-to-face play is to develop and maintain **synchrony**, or coordinated interaction between infant and caregiver. Synchrony has been variously described by researchers as the meshing of a finely tuned machine (Snow, 1984), a patterned dance or "dialogue" of exquisite precision (Schaffer, 1984), and an emotional "attunement" of an improvised musical duet (Stern, 1985). It is partly through synchrony that infants learn to express and read emotions (Bremner, 1988) and begin to develop some of the basic skills of social interaction—such as turn-taking—that they will use throughout life.

Even in the early months, synchrony is a partnership. Infants modify their social and emotional expressiveness (smiling, looking, cooing) to match or complement their caregiver's overtures, while adults sensitively modify the timing and pace of their initiatives to accord with their baby's readiness to respond (Cohn & Tronick, 1987). Such coordination, of course, is not necessarily common or constant. In fact, episodes of synchrony occur less than 30 percent of the time in normal mother-infant play. Much of the time, the pair is jointly reestablishing coordinated play following periods of dyssynchrony, caused by the baby's becoming fussy, or by the mother's becoming distracted, or by any number of other factors (Tronick, 1989; Tronick & Cohn, 1989). Thus, infants are learning not only how to socialize during periods of interaction but also, with the caregiver's assistance, how to remedy or "repair" social encounters that are not going well (Gianino & Tronick, 1988).

A moment of perfect synchrony!

A CLOSER LOOK

Fathers and Infants

Traditional views of infant development focused almost exclusively on mothers, partly because the received wisdom in most cultures was that fathers are naturally "remote and authoritarian," too busy with other matters for an intimate relationship with their young children (Poussaint, 1990). And in Western cultures fathers were, historically, removed from most caregiving activities, in response to the cultural expectations as well as to the practical necessities of working long hours away from home while mothers tended house, children, and garden.

Recently, however, as family size has shrunk and mothers have increasingly become employed outside the home, many fathers have taken on a "significant share of the nurturing responsibilities" for their offspring (Poussaint, 1990). This shift is apparent worldwide, including countries such as Ireland and Mexico, where the stereotype was that fathers are above changing diapers or spooning baby food (Bronstein, 1984; Lamb, 1987; Nugent, 1991). Virtually all developmentalists applaud this trend, for fathers who share child-care responsibilities probably enhance the development of their children more than the remote fathers of earlier generations did.

As this change began to occur, it raised some interesting questions about the relationship between fathers and their infants. The first and most urgent was, Could fathers provide adequate care for newborns and young infants? The answer, quick in coming, was a resounding "yes": research found that babies drank just as much formula, emerged from the bath just as clean, and seemed just as content with the caregiving of fathers as with the caregiving of mothers. It was further determined that fathers can provide the necessary emotional and cognitive nurturing as well, coordinating their facial expressions in synchrony, speaking Motherese like a native, and forming secure attachments. In short, "there is perhaps no mystique of motherhood that a man cannot master except for the physical realities of pregnancy, delivery, and breast feeding" (Poussaint, 1990).

Given that fathers *can* master caregiving, the next question was, Why don't more fathers develop this skill? Worldwide, women spend far more time in child care than men do, especially in the early months and particularly if the child is a girl (Lamb, 1987). Even in contemporary marriages, even when both parents work outside the home on weekdays, and even when both agree that child care is a shared responsibility, the reality is that, although fathers do some basic caregiving in the evenings and on weekends, mothers do a great deal more (Bailey, 1994; Pleck, 1985; Thompson & Walker, 1989).

Surprisingly, although the media, and many mothers, tend to blame the fathers for this inequity, mothers may be as responsible for this unequal distribution of labor as fathers. Indeed, many mothers assume the status of the family child-care authority: they serve as a kind of gatekeeper and judge of the father's performance, forbidding or criticizing certain behaviors, permitting and praising others (Kranichfeld, 1987; Pollack & Grossman, 1985). Fathers' limited involvement with babies may be a joint result of maternal and paternal preferences.

In addition, the general social context often works against fathers' being intensely involved in caregiving with infants. The traditional view of child-rearing roles, for example, is reinforced by many cultural pressures. Older relatives may encourage the mother to provide most of the nitty-gritty child care, and the father's friends and colleagues may deride the idea of Daddy's changing diapers. Employers, too, are more likely to recognize, and make allowances for, the woman's role as caregiver than they are the man's. Even when paternity leave is an option, for example, many men do not take advantage of this opportunity because of the stigma—and career sacrifices—they think might occur. Finally, the marriage relationship can affect the father-child relationship: when the couple are happy with their relationship, they are more likely to share child-care duties (Belsky et al., 1991).

As researchers looked more closely at the amount of time mothers and fathers spend with their infants, they discovered another curious difference between the caregiving of mothers and fathers: although fathers provide less basic care, they play more with infants. Moreover, compared with mother's play, father's play is noisier, more boisterous, and idiosyncratic, as fathers make up active and exciting games on the spur of the moment (MacDonald & Parke, 1986).

Even in the first months of the baby's life, fathers are more likely to play by moving the baby's legs and arms in imitation of walking, kicking, or climbing, or by zooming the baby through the air ("airplane"), or by tapping and

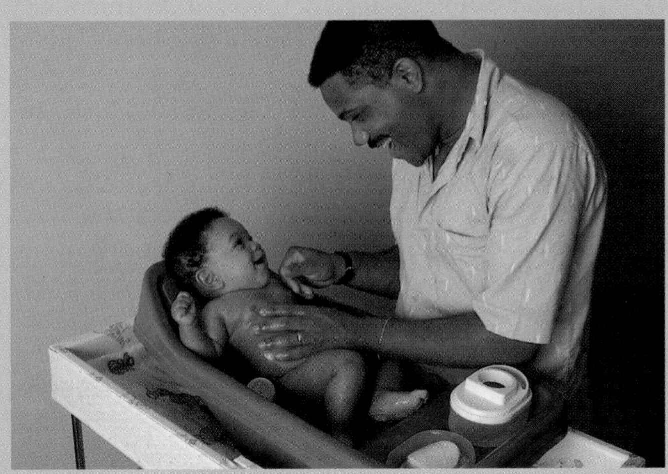

In many modern families, bathing is the caregiving task most often assumed by the father. The reason is illustrated by 4-month-old Christopher and his dad: bathing is not only a necessary, nurturant task, it also offers opportunities for joyous physical play, a father's specialty.

tickling the baby's stomach; mothers, on the other hand, are more likely to talk or sing soothingly, or to combine play with caretaking routines such as diapering and bathing. Generally mothers' play is physically restrained, involving such standard games as peek-a-boo and patty-cake.

These differences between mothers' and fathers' play are not lost on infants. Even young infants typically react with more visible excitement when approached by their fathers than when approached by their mothers. In the first months of life, infants are more likely to laugh—and more likely to cry—in episodes of play with Dad.

As infants grow older, fathers generally increase the time they spend with them, and the fathers' tendency to engage in physical play becomes more pronounced. Fathers are likely to swing their toddlers around, or "wrestle" with them on the floor, or crawl after them in a "chase." They also are apt to tease their children (scaring them with a noise, pretending to take away a favorite toy, saying "I'm the baby now"), often to the mother's disapproval. Fathers also tend to play more creatively. For example, in one study, parents were given a set of objects, including some sponges, Styrofoam chips, plastic containers, and a toy bear, and were asked to play with their 1-year-olds. Mothers tended to be fairly staid in their play, putting the chips into the containers, using the sponge to clean the bear, and so on. Fathers, on the other hand, were more likely to use the objects in unconventional ways, putting chips down the bear's sweater, using the chips as falling snow, playfully tossing sponges at the child, and the like (Pecheux & Labrell, 1994).

What do infants gain from playing with their fathers in addition to having fun? Many things. Most important, playing is a direct avenue to a close relationship. Remember that play with an infant requires mutual engagement and synchronous responsiveness—precisely those qualities that help build a secure attachment. Play with father may also contribute to the growth of unique social skills and tendencies. In one study, 18-month-olds met a stranger while either their father or their mother sat passively nearby. The father's presence made the toddlers much more likely to smile and play with the new person than the mother's presence did, a result especially apparent for the boys. The authors of this study speculated that the child's experience of boisterous, idiosyncratic play with Dad may make the father's presence a cue for playfulness and embolden the child to engage the stranger (Kromelow et al., 1990). Similar speculations have been raised about fathers' teasing, which requires a social response to an unpredictable game—and thereby may increase the infant's excitement and social understanding (Pecheux & Labrell, 1994).

Findings such as these raise anew the question of whether or not gender-specific caregiving with infants may be best. At the moment, researchers have no definitive answers, but some are beginning to shift the emphasis of the question, suggesting that the division of child-rearing labor that may be best for infants is whatever division is best for the parents. Indeed, when their relationship is good, each parent complements, encourages, and enhances the other in "a balanced system of interactive effects between husbands and wives" (Grossman et al., 1988). Even in today's changing world, mothers and fathers together are more likely to meet all their infant's needs—biological, cognitive, and social—than is either one alone.

Generally, repair is not difficult: the signs of dyssynchrony are obvious—the baby's averted eyes, stiffening or abrupt shifting of the body, an unhappy noise—and the alert caregiver can quickly make adjustments, allowing the infant to "recover." Depending on various aspects of their temperament and maturity, of course, some infants take longer than others to recover and to resume synchronous interaction. Since development of the central nervous system improves awareness and timing, 5-month-olds lead the "dance" notably better than do 3-month-olds (Lester et al., 1985).

When initiation and repair of synchrony are difficult, it is usually because the caregiver regularly overstimulates the baby who wants to pause, or ignores the infant's invitation to interact (Isabella & Belsky, 1991). If the infant is repeatedly ignored, he or she may not try as much to respond: offspring of depressed mothers, for example, are less likely to smile and vocalize, not only when interacting with their mothers but also when responding to a nondepressed adult (Field, 1995). Infants with an intrusive, overstimulating caregiver defend themselves more obviously, by turning away or even "shutting down" completely, such as by crying inconsolably. Unfortunately, some caregivers still do not notice the cues, as in this example:

> Whenever a moment of mutual gaze occurred, the mother went immediately into high-gear stimulating behaviors, producing a profusion of fully displayed, high-intensity, facial and vocal . . . social behavior. Jenny invariably broke gaze rapidly. Her mother never interpreted this temporary face and gaze aversion as a cue to lower her level of behavior, nor would she let Jenny self-control the level by gaining distance. Instead she would swing her head around following Jenny's to reestablish the full-face position. Jenny again turned away, pushing her face further into the pillow to try to break all visual contact. Again, instead of holding back, the mother continued to chase Jenny. . . . She also escalated the level of her stimulation more by adding touching and tickling to the unabated flow of vocal and facial behavior. . . Jenny closed her eyes to avoid any mutual visual contact and only reopened them after [she had moved her head to the other side]. All of these behaviors on Jenny's part were performed with a sober face or at times a grimace. [Stern, 1977]

While this example clearly shows the effects of the caregiver's personality, it should be noted that the infant's personality and predispositions also affect the ease of synchrony. For example, some infants are constitutionally more sensitive to stimulation than others; such babies would have particu-

Adults typically use special social behaviors (*a*) with their young infants—leaning in close, opening their eyes and mouths wide in exaggerated expressions of surprise or delight, maintaining eye contact—because they elicit the baby's attention and pleasure. But these behaviors are subdued or absent when the adult is depressed or stressed (*b*), and this makes social interaction much less enjoyable for each partner.

(a)

(b)

lar problems with an intrusive caregiver like Jenny's mother. Fortunately, even with such a mismatch, repair is possible. Sometimes a helpful outsider can teach the caregiver how to more sensitively read the baby's signals, and sometimes the baby and caregiver begin to adjust to each other spontaneously (Stern, 1985). Jenny, for example, eventually became more able to adjust to her mother's sudden overstimulation, and her mother, finding Jenny more responsive, no longer felt the need to bombard her with stimulation as she had earlier. With time, Jenny and her mother established a mutually rewarding relationship.

Attachment

Just as the moment-by-moment harmony between parents and young infants has captured scientific attention, so has the **attachment** between parents and slightly older infants been the subject of extensive research. "Attachment," according to Mary Ainsworth (1973), "may be defined as an affectional tie that one person or animal forms between himself and another specific one—a tie that binds them together in space and endures over time." Not surprisingly, when people are attached to each other, they try to be near one another, and they interact with each other often. Thus infants show attachment through "proximity-seeking" behaviors—such as approaching, following, and climbing into the caregiver's lap—and "contact-maintaining" behaviors—such as clinging and resisting being put down and using social referencing (p. 194) when moving around on their own. Parents show their attachment by keeping a watchful eye on their infant, even when safety does not require it, and by responding affectionately and sensitively to the infant's vocalizations, expressions, and gestures. The attachment bond not only deepens the parent-infant relationship but, over our long evolutionary history, may also have contributed to human survival by keeping infants near their caregivers and keeping caregivers vigilant.

attachment The enduring emotional connection between a person and a particular other that produces a desire for consistent contact as well as feelings of distress during separation.

secure attachment A healthy parent-child connection, signaled by the child's being confident when the parent is present, distressed at the parent's absence, and comforted by the parent's return.

insecure attachment A troubled parent-child connection signaled by the child's overdependence on, or lack of interest in, the parent. Insecurely attached children are not readily comforted by the parent and are less likely to explore their environment than are children who are securely attached.

Measuring Attachment

In studying attachment in England, Uganda, and the United States, Ainsworth discovered that virtually all normal infants develop special attachments to the people who care for them, with some infants much more secure in those attachments than others, a fact confirmed by hundreds of other researchers (Bretherton, 1992; Colin, 1996).

A **secure attachment** is one in which the infant derives comfort and confidence from the caregiver, as evidenced by the infant's attempts to be close to the caregiver and readiness to explore the environment. The caregiver acts as a "secure base," enabling the child to venture forth, perhaps scrambling down from the mother's lap to play with a toy while periodically looking back, vocalizing, or returning for a hug.

By contrast, **insecure attachment** is characterized by the infant's fear, anger, or seeming indifference toward the caregiver. Compared with the securely attached infant, the insecurely attached child has much less confidence, perhaps being unwilling to let go of the caregiver's arms, or perhaps playing aimlessly, with no signs of trying to maintain contact with the caregiver.

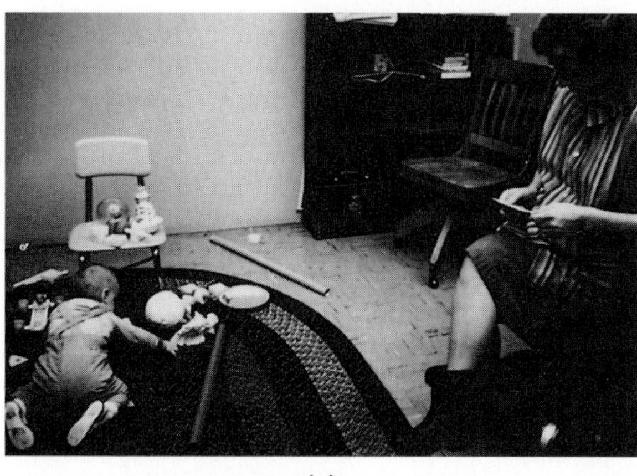
(a)

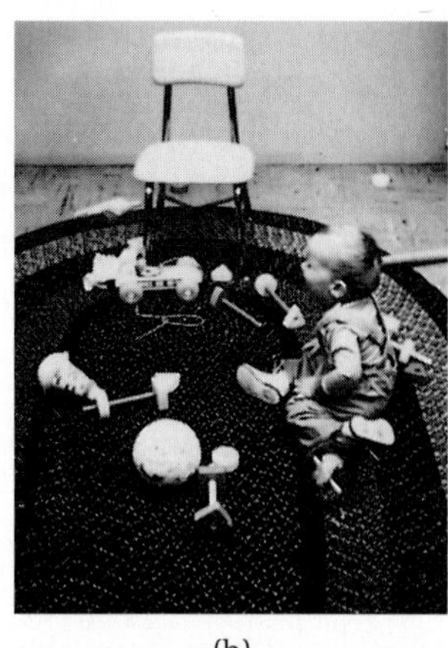
(b)

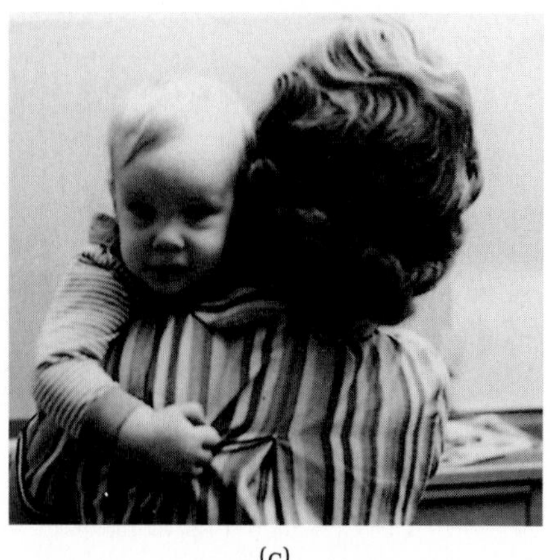
(c)

In this trial of the Strange Situation test, Brian shows every sign of secure attachment. He explores the playroom happily when his mother is present (*a*), cries when she leaves (*b*), and is readily comforted when she returns (*c*).

While there are many ways to measure attachment, Ainsworth developed a classic laboratory procedure, called the **Strange Situation**, which is designed to evoke the infant's reactions to the caregiver—usually the mother—under somewhat stressful conditions. Infants are closely observed in a well-equipped playroom in several successive episodes, with their mother and/or a stranger, and alone. These conditions, which last at most three minutes each, are designed to measure the baby's motivation to be near the caregiver, and to reveal how much the caregiver's presence reestablishes security and confidence.

About two-thirds of all American infants tested in the Strange Situation demonstrate secure attachment. Their mother's presence in the playroom is enough to give them courage to explore the room and investigate the toys; her departure may cause some distress (usually expressed through verbal protest and a pause in playing); and her return is a signal to reestablish positive social contact (with a smile or by climbing into the mother's arms) and then resume playing.

Other infants, however, show one of three types of insecure attachment. Some are anxious and *resistant*: they cling nervously to their mother even before her initial departure and thus are unwilling to explore the playroom; they cry loudly each time she leaves; they refuse to be comforted when she returns, perhaps continuing to sob angrily even when back in her arms. Others are *avoidant*: they engage in little interaction with their mother; they often show no apparent distress when she leaves; and on her return, they tend to avoid reestablishing contact, sometimes even turning their backs. Others are *disoriented*, or disorganized and ambivalent: they show an inconsistent mixture of behavior toward the mother, such as avoiding her just after seeking to be close to her (Main & Solomon, 1986).

Attachment and Context

Strange Situation An experimental condition devised by Mary Ainsworth to assess an infant's attachment. The infant's behavior is observed in an unfamiliar room while a caregiver (usually the mother) and a stranger move in and out of the room.

Ainsworth's procedure for measuring attachment has been used in thousands of studies. From these we have learned that attachment is affected by the quality of care in early infancy (Belsky & Cassidy, 1995; Bretherton & Waters, 1985; Lamb et al., 1985; Thompson, 1997). Among the caregiving features that affect the quality of attachment are (1) general sensitivity to the infant's needs, (2) responsiveness to the infant's specific signals, and (3) talking and playing with the infant in ways that actively encourage the child's growth and development (Ainsworth, 1993; Isabella, 1993). Not surprisingly,

the interactions between mothers and infants who are securely attached have been found to exhibit greater synchrony than those between mothers and infants who are insecurely attached (Isabella & Belsky, 1991). Thus sensitive and responsive caregiving in the early months leads naturally to secure attachment in the later months.

Attachment may also be influenced by the broader family context, including such factors as the extent and quality of the father's involvement in the child's care and the nature of the marital relationship (Easterbrooks & Goldberg, 1984; Goldberg & Easterbrooks, 1984: Pianta et al., 1989). Attachment can also be affected by changes in family circumstances—such as a parent's losing a job—that alter established patterns of family interaction. Temperament is also part of the story. As you read earlier, the goodness of fit between infant temperament and parenting style is a key developmental factor, and it may be the quality of this fit that best predicts whether a secure or insecure attachment will develop (Mangelsdorf et al., 1990).

In essence, then, attachment relationships take shape from the *interaction* of mother and infant within a complex social ecology. As part of the ecology of care, cultural context can also affect the development of attachment, or at least the measurement of it (Sagi & Lewkowicz, 1987; van Ijzendoorn & Kroonenberg, 1988). In cross-cultural comparisons of the Strange Situation, for example, Japanese and Israeli children show a higher rate of anxious/resistant attachment than American infants do, while infants from some Western European countries show higher rates of avoidant attachment. Why do these differences exist? Some researchers believe that particular cultural backgrounds may make the Strange Situation too demanding for certain infants, causing them to exhibit insecure behavior. Japanese mothers, for instance, rarely leave their infants with babysitters, and their offspring are thus less prepared to cope with being with a stranger or alone in the Strange Situation than American infants are (Chen & Miyake, 1986).

Nevertheless, an extensive analysis of cross-cultural data on attachment reveals that, in the Strange Situation, the majority of infants of various nationalities are securely attached (van Ijzendoorn & Kroonenberg, 1988). Most infants worldwide consider their mother's presence a reassuring sign that it is safe to explore the environment, and most infants come back to her for comfort under stress (Sagi et al., 1991). Most infants also show signs of secure attachments to other caregivers—fathers, siblings, day-care providers—although this obviously varies from culture to culture, and also depends on how responsive the caregiver is.

The Importance of Attachment

Why is attachment considered so important? Part of the reason lies in longitudinal research that documents the results of secure and insecure attachment. Studies clearly show that secure attachment at age 1 provides a preview to the child's social and personality development in the years to come. Securely attached infants tend to become children who interact with teachers in friendly and appropriate ways, seeking their help when needed, and who are more competent in a wide array of social and cognitive skills (Belsky & Cassidy, 1995; Turner, 1993). By contrast, insecure infants are more likely to display problems later on. At age 4, for instance, boys who were rated insecurely attached as infants tend to be aggressive, while girls who were rated insecurely attached tend to be overly dependent (Turner, 1991).

PUBLIC POLICY

Attachment Theory and the Infant Day-Care Controversy

When Laura's daughter Heather was 12 months old, her mother began to have serious doubts about Heather's being in a day-care center every weekday. She confided her concerns to a psychologist. "Yesterday was typical," Laura said. "Heather began fussing in the morning when it was time for me to leave, but the teacher said she settled down fine shortly after. But when I came to pick her up at 5, Heather seemed to be more interested in the toys than in me!" Laura needed the income of her job, and she knew that the particular day-care center was safe and that the staff were responsive and stimulating. But she could not help wondering if day care was undermining Heather's attachment to her.

Although researchers disagree about the effects of early day-care experience on attachment, they all concur that infants benefit from high-quality centers with a low adult-child ratio, well-qualified staff, and lots of age-appropriate toys.

Concerns like Laura's have captured the attention of many developmentalists, in many nations. Indeed, until recently, developmentalists in the United States were engaged in a heated debate over the possible detrimental effects of day care, especially during the first year of infancy. On the basis of research suggesting that infants who experienced early, extended day care are more likely to avoid and ignore their mothers in the Strange Situation, one side warned that over twenty hours of nonmaternal care a week in the first year represented a "risk factor" for insecure attachment with parents (Belsky, 1986). The opposing side rejected this warning on several grounds. First, they called attention to a number of weaknesses in the research on which it was based, including the failure to take into account such potentially confounding factors as the nature of the parents' relationship with the child in the home and the quality of child care provided outside the home. Second, they noted that this research found the rate of insecure attachment for infants who received extended nonmaternal care to be no more than 8 to 15 percent higher than for infants who were cared for exclusively by their mothers. Even if confirmed, such a difference could hardly be said to reflect a "risk factor" for insecure attachment. Third, they pointed out that when attachment behavior is measured in the Strange Situation, day-care infants may behave differently—that is, seeming to be indifferent to their mother's comings and goings—because of their regular experience with separation and reunion, not because of insecure attachment (Clarke-Stewart, 1989; Fox & Fein, 1990; Lamb & Sternberg, 1990; Thompson, 1997). Finally, these researchers asserted that with any attachment problems that might be associated with extended day care, the crucial variable is the *quality* of caregiving, not the fact of day care *per se*.

In recent years the debate has calmed considerably, with most researchers—including the initiator of the debate—now agreeing that when infant day care is "of high quality, there should be little reason to anticipate negative developmental outcomes" (Belsky, 1990). In fact, there is strong evidence that, for some children, high-quality day

Does this mean that a secure or insecure attachment in infancy determines whether a child will grow up to be sociable or aggressive, self-directed or dependent, curious or withdrawing? Probably not in itself. Certainly it is true that a sensitive caregiver who fosters a secure attachment in his or her infant is likely to maintain this kind of caregiving as the child matures, encouraging the development of sociability, curiosity, and independence. And it is also true, unfortunately, that the insensitive care that contributes to an insecure attachment is also likely to be maintained, making the child more inclined to be cautious, aggressive, or dependent. But remember that attachment relationships sometimes change. As noted, shifts in family circumstances (a divorce, a new job, a new baby) often establish new patterns of

care is more beneficial to the development of cognitive and social skills than being cared for at home (Clarke-Stewart et al., 1994; Egeland & Hiester, 1995; Field, 1991; National Institute of Child Health and Development, 1996; Roggman et al., 1994). Among the reasons for this is that the caregivers and the setting of high-quality day care are focused expressly on fostering the child's development, with games, toys, and especially designed play spaces. Moreover, high-quality day care offers a wonderful feature that no home provides—a wide assortment of potential playmates and friends.

What, then, are the signs of "high-quality" day care that parents should look for? Researchers have identified four factors that seem essential:

1. *Adequate attention to each infant*. This means both a low caregiver–infant ratio and small group size. An ideal size might be two steady caregivers for a group of five infants.

2. *Encouragement of sensorimotor exploration and incipient language development*. Infants should be provided with a variety of easily manipulated toys and should have a great deal of language exposure, through games, songs, and conversation.

3. *Attention to health and safety*. Cleanliness routines (such as handwashing before meals), accident prevention (such as the absence of small objects that could be swallowed), and safe areas for exploration (such as a clean, carpeted area for crawling and climbing) are all good signs.

4. *Well-trained and professional caregivers*. Ideally, every caregiver should have a degree and certification in early-childhood education and should have worked in this field for several years. Turnover should be low, morale high, enthusiasm evident. Indeed, if the caregivers are knowledgeable and committed, the first three items on this list will follow automatically.

Beyond the specifics that a particular parent might seek in a day-care setting, however, are broader public-policy issues. In many nations, including the United States, employed mothers must find and finance infant care on their own, which typically results in a patchwork caregiving arrangement of babysitters, relatives, and neighbors—some excellent, some dangerous, many mediocre, and almost all untrained and unregulated. Any nation committed to the future development of its youngest citizens must be concerned about the quality of their daily infant care (Clarke-Stewart et al., 1994; Melhuish & Moss, 1991; Peters & Pence, 1992). This is especially true in the case of those children for whom day care might be the only available source of intellectual stimulation and responsive caregiving. As Sandra Scarr (1996), past president of the Society for Research in Child Development, writes:

> For children from middle- and upper-income families—especially stable, two-parent families in nondangerous neighborhoods—day care merely supplements what parents can offer. For children from disturbed and seriously disadvantaged families—especially unstable, one-parent families who live in dangerous neighborhoods—good day care is the most powerful, positive intervention we now have. . . . Most low-income, working families cannot afford to buy decent child care, not to mention good quality care. Federally funded child care assistance is insufficient to meet even present needs, which will expand with welfare reform. Do we in the United States have the political will to provide quality care for poor children?

This final question will become all the more urgent as welfare reforms result in many more children being in need of care outside the home. Since high-quality day care reduces the incidence of many major hazards to the child—from accidental injury to academic failure—it is imperative that the answer to the question be a resounding yes.

family interaction. In addition, as children mature, they face new developmental challenges and experience new social settings, all of which may alter the long-term effects of a secure or insecure attachment in infancy. Thus while an insecure attachment may lead a young child to approach new relationships skeptically and cautiously, later relationships may encourage trust and security that provide the child with new confidence and openness, or vice versa.

This view that early attachment biases, but does not inevitably determine, later social relationships also provides the basis for helpful interventions. In one study with Spanish-speaking immigrant families in the United States, for example, three groups of 1-year-olds were compared: a securely

attached group, an insecurely attached control group, and an insecurely attached experimental group who were visited weekly by an empathic bilingual and bicultural adviser. Within a year, the experimental group of mothers and infants were relating to each other almost as well as the group who were originally securely attached—and far better than the control group on measures of infant anger, maternal responsiveness, and the like (Lieberman et al., 1991). Happy outcomes like these alert us to the fact that although early experiences provide a foundation for later growth, they rarely create developmental pathways that are predetermined and cannot be altered.

The Parents' Side of Attachment

By the time an adult becomes a parent, he or she has a long history of attachment experiences, including relationships with his or her parents, friends, and romantic partners. As you yourself probably know from experience, each new attachment can inspire trust and security or, instead, feelings of insecurity and anxiety—and each relationship further refines the expectations with which one approaches new relationships (Simpson, 1990). Recently, researchers have begin examining the parents' side of attachment to see whether the security or insecurity of parents' own past relationships has any bearing on their attachment with their children.

To explore the parents' side of attachment, Mary Main and her colleagues have devised the Adult Attachment Interview, an hour-long series of questions related to parents' memories of their childhood attachment experiences, their perceptions of trust and security in their early years, and their views of their own parents as well as of their adult relationships (Main & Goldwyn, 1992; Main et al., 1985). Based on what the parents say, and how they say it, they are classified into one of four categories:

Autonomous. Adults who are classified as autonomous value attachment relationships and regard them as influential, but they are also capable of discussing them objectively, whether their own early attachments were positive or negative in quality.

Dismissing. Adults who are dismissing tend to devalue the importance and influence of attachment relationships in their own lives, and tend to idealize their parents without being able to provide specific examples of positive interactions in the past to support their view.

Preoccupied. Adults in this category seem to be very preoccupied with the past and are unable to discuss early attachment experiences ob-

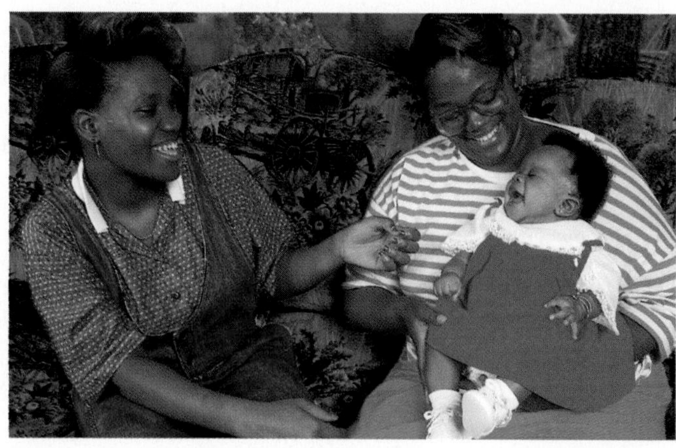

Research suggests that this baby's future attachment to her mother (on left) may depend partly on the kind of attachment her mother had with her own mother many years before. Judging by the grandmother's evident delight at watching her daughter and granddaughter play together, this family appears to be enjoying three generations of secure attachment.

jectively, often showing considerable emotion while talking about their relationships with their parents.

Unresolved. Adults who are unresolved have not yet reconciled their past attachment experiences with the present; these parents are sometimes still coping with parental loss and related experiences.

Researchers have discovered that parents' adult-attachment ratings closely parallel the kind of attachment they form with their children (Crowell & Feldman, 1988, 1991; Fonagy et al., 1991; Zeanah et al., 1993). Autonomous mothers tend to have securely attached infants; dismissing mothers tend to have avoidant babies; and preoccupied mothers tend to have resistant infants. (The parallel is less clear for the unresolved classification, partly because it is often a transitional status for many adults.)

These findings were extended by another study (Benoit & Parker, 1994) that used the Adult Attachment Interview with pregnant women and then with their mothers, and later used the Strange Situation with the first group's 1-year-olds. As you can see from Figure 7.1, this study found that attachment patterns tended to be passed down from one generation to the next, with 64 percent of the families having the exact same status in all three generations.

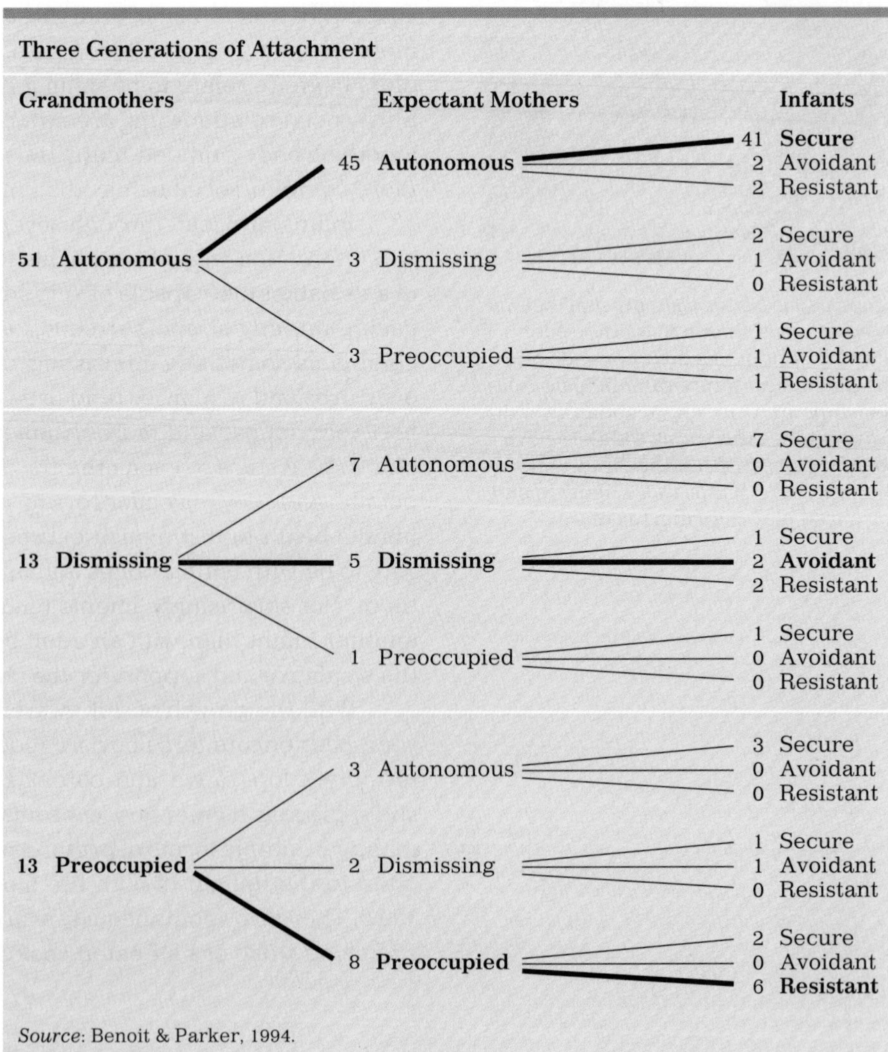

FIGURE 7.1 Using a structured and sensitive interview process, researchers first rated the attachment relationships that pregnant women had had with their mothers as children and then rated the attachment relationships that those mothers had had with their mothers. The ratings comprised three categories, *autonomous* (considered indicative of secure attachment), *dismissing*, and *preoccupied*. More than a year later, blind observers rated the attachment of the formerly pregnant women and their infants using the Strange Situation procedures. The intergenerational transmittal of attachment is clear, especially for the fifty-nine secure infants—fifty-one of whom had mothers who had autonomous relationships with their mothers.

The link between adults' attachment histories and their attachment with their own children could occur for several reasons. It may be that parents who value attachment and can reflect objectively on their own experiences are more sensitive to their offspring and inspire a secure attachment as a result; or it may be that innate temperament predisposes individuals to certain attachment patterns across most of their personal relationships, including those with parents and with children; or it may be that the nature of parents' attachment with their children influences their memories of, and attitudes about, their other attachments. Whatever the reason for the link, it appears that one important contribution to the development of a secure attachment in infants is the parents' attitudes about their own early attachments.

Developing Peer Relations

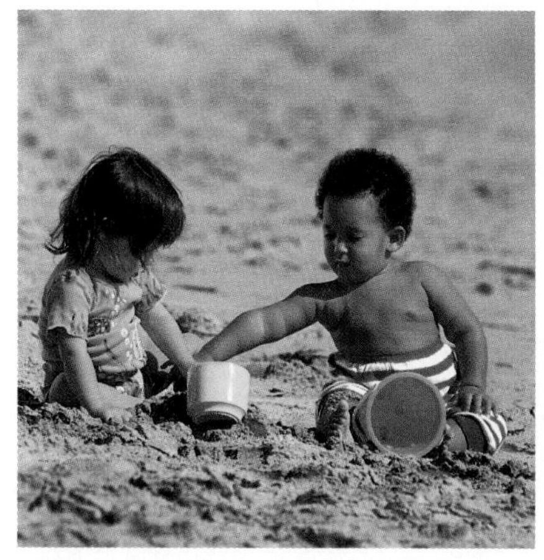

"Your pail is better than my pail" seems to be a typical theme in toddler peer play. What happens next depends on the playmates' temperament, past relationships, and current negotiating skills. Fortunately, even young toddlers have some social sense, and the interaction pictured here seems more likely to end in shared play than in a tug of war.

While parent-child interactions are the main arena for early psychosocial development, interactions with peers also play a vital role. Historically, most toddlers' first experiences in developing peer relations occurred in the home, with slightly older brothers and sisters, or in the backyard or local playground, with neighbor children. Today, smaller family size, greater spacing between births, and, most significant, the fact that the majority of mothers with young children work outside the home, mean that most toddlers begin to relate to peers in a play group or organized day-care setting. Such social relationships are now coming under study, and the results echo a note already sounded many times in this chapter: young children are far more complex social beings than most theorists once realized.

Infants proceed through several stages in their ability to engage in social interaction with other infants (Howes, 1987). Even as young as 6 months of age, babies are capable of very simple social responses to their age-mates: gazing intently at another child's activity, vocalizing animatedly when that child does something interesting, or smiling and touching that child. These overtures and responses tend to be brief and fleeting, but over the rest of the first year, longer and more complex interactions begin to occur. One-year-olds may gesture to each other or exchange words when referring to toys or people, and they may offer toys to, or take toys from, each other. In addition, infants begin to distinguish between familiar and unfamiliar peers, preferring to be with familiar ones and interacting in a more complex fashion with them. Not surprisingly, infants tend to be less skilled when interacting with another infant than with an adult because an infant partner cannot provide the structure and support for the child's social skills that the adult can.

With the growth of self-awareness and emotional vitality in the second year, peer encounters flourish. Toddlers become capable of a broader range of more cooperative and complex social skills, such as turn-taking on a slide, playing elementary coordinated games such as run and chase, and joining in simple forms of pretend play. Toddlers begin to imitate each other, often to the delight of both the model and the imitator (Hanna & Meltzoff, 1993). They also spontaneously share toys, territory (such as a sand box), and even food (such as half-eaten cookies) (Hay et al., 1991).

Of course, toddlers are not always prepared to share, and frequently one toddler seems to covet an object to the degree that another is unwilling to give it up. Often the result is shouts of "No!" and "Mine!" followed by pulling, crying, and hitting. As they play together, however, toddlers develop simple strategies for avoiding such conflicts. Some toddlers become quite skilled tacticians, keeping a favorite toy out of sight, for instance, or distracting a peer by calling attention to some other object.

Increasingly, toddlers recognize each other's individuality, and friendships between two children begin to form even in day-care clusters of six or so children. Signs of obvious affection—hugs, kisses, and big smiles—are sometimes exchanged between 2-year-olds who have not seen each other for a while. Apparently, attachment already begins to extend beyond the home in toddlers who have steady playmates.

At the ripe old age of 2, children still have much to learn in the social sphere. However, taken together, their personality development and expanding emotional range, in combination with their caregiver relationships and peer friendships, make the first two years of life a time for notable psychosocial growth, providing a foundation for the more complex socioemotional achievements to come. The first two years of life comprise a time when a child, who is born with social stirrings, learns to connect those inclinations with specific people in his or her social world. By the end of the second year, the child is very much a person, with individual emotions and family connections, ready for the expanding world of the preschool years.

SUMMARY

Emotional Development

1. The ethological perspective argues that the significance of many behaviors is revealed in light of human evolution. Especially in infancy, many of the infant's behaviors are relevant to the baby's nurturance and survival. For example, infant crying, smiling, and other behaviors help to alert caregivers to the baby's needs and keep them nearby to provide nurturance and protection.

2. In the first weeks and months of life, infants are capable of expressing many emotions, including fear, anger, sadness, happiness, and surprise. Toward the end of the first year, the typical infant expresses emotions more readily, more frequently, and more distinctly. Two common reactions, stranger wariness and separation anxiety, reveal this shift to differentiated emotions. Various contextual events affect the degree of the infant's distress.

3. In the second year, cognitive advances allow infants to become more aware of the causes of events and more conscious of the distinctions between themselves and others. As self-awareness develops, new emotions emerge, such as guilt, pride, and embarrassment. Social referencing begins to occur as one means by which emotions are shaped by social interaction.

The Origins of Personality

4. In the first half of the twentieth century, the prevailing view among psychologists was that the individual's personality is permanently molded by the actions of his or her parents—especially the mother—in the first years of childhood. The early behaviorists, as well as later social learning theorists, believed this occurred as the child experienced or witnessed reinforcing events, day by day, that accumulated to create habits of attitude and action.

5. Freud argued that the child-rearing practices encountered in the oral and the anal psychosexual stages have a lasting impact on the individual's personality and mental health. He believed that the nature of the mother's interaction with the child is central in this process.

6. Erikson built on Freud's psychoanalytic ideas with a broader concept of the first two stages of psychosocial development. According to Erikson, the infant first experiences the crisis of *trust versus mistrust*, discovering whether the immediate world is secure or insecure, and then the crisis of *autonomy versus shame and doubt*, as the infant tries to achieve some measure of independence.

Like Freud, Erikson stressed the lifelong impact of the caregiver's actions during the first two years, but he also emphasized that earlier crises could be resolved later in life.

7. Temperament—a group of basic, early dispositions, some largely influenced by genetics, others more susceptible to environmental influences—is another factor in psychosocial development. Individual temperament tends to be stable over time but can change, sometimes because of changes in the "goodness of fit" between temperament and environmental demands.

Parent-Infant Interaction

8. In addition to the parents' actions and the infant's temperament, developmentalists stress the social partnership between parent and child and its growth during the early months of life. This partnership grows through episodes of face-to-face play.

9. The early parent-child interaction is sometimes characterized by synchrony, a harmony of gesture, expression, and timing that can make early prelanguage play a fascinating interchange.

10. Attachment between parent and child becomes apparent toward the end of the first year. Secure attachment tends to predict curiosity, social competence, and self-assurance later in childhood; insecure attachment tends to correlate with less successful adaptation in these areas. Parents' own attachment histories also tend to correlate with patterns of parent-infant attachment.

Developing Peer Relations

11. In the second year, the toddler's world widens as play groups and day-care settings provide opportunities for children to expand their social understanding and repertoire of social skills.

KEY TERMS

ethological perspective (191)
social smile (192)
stranger wariness (193)
separation anxiety (193)
social referencing (194)
self-awareness (195)
oral stage (198)
anal stage (198)
trust versus mistrust (199)
autonomy versus shame and doubt (199)
temperament (200)
goodness of fit (204)
synchrony (207)
attachment (211)
secure attachment (211)
insecure attachment (211)
Strange Situation (212)

KEY QUESTIONS

1. What does the ethological perspective contribute to our understanding of early development?

2. Which emotions develop in the first year?

3. Which factors influence whether a baby will be afraid of, or friendly toward, a stranger?

4. What are some consequences of the toddler's growing cognitive skills and developing sense of self?

5. According to behaviorists, how do parents affect the formation of personality?

6. What are the similarities between the theories of Freud and Erikson?

7. What are the three most common temperamental patterns in infancy, and how does nurture affect them?

8. Why does temperament sometimes change over time?

9. What are the similarities and differences between mother-infant and father-infant interactions?

10. What do infants learn from parent-infant interaction?

11. What contributes to a secure attachment?

12. Does early attachment determine later psychosocial growth? Why or why not?

13. How do social skills among toddlers develop during the second year of life?

Biosocial Development

Body, Brain, and Nervous System

Over the first two years, the body quadruples in weight and the brain triples in weight. Neurons branch and grow into increasingly dense connective networks between the brain and the rest of the body. As neurons become coated with an insulating layer of myelin, they send messages faster and more efficiently. The infant's experiences are essential in "fine tuning" the brain's responses to stimulation.

Motor Abilities

Brain maturation allows the development of motor skills from reflexes to coordinated voluntary actions, including grasping and walking. At birth, the infant's senses of smell and hearing are quite acute, and although vision at first is sharp only for objects that are about 10 inches away, by age 1, acuity approaches 20/20.

Cognitive Development

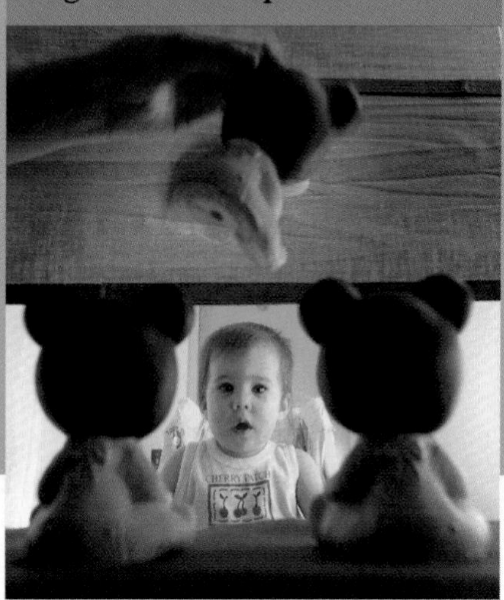

Perceptual Skills

Sensory abilities are linked in both intermodal and cross-modal perception, and the various affordances of objects and people are understood.

Cognitive Skills

The infant's active curiosity and inborn abilities interact with various experiences to develop early categories, such as object size, shape, texture, and even number as well as an understanding of object permanence. Memory capacity, while fragile, grows during the first years, although retrieval is sometimes difficult. The infant progresses from knowing his or her world through immediate sensorimotor experiences to being able to "experiment" on that world mentally, through the use of mental combinations.

Language

Babies' cries are their first communication; they then progress through cooing and babbling. Interaction with adults through "baby talk" teaches them the surface structure of language. By age 1, an infant can usually speak a word or two, and by age 2 is talking in short sentences.

Psychosocial Development

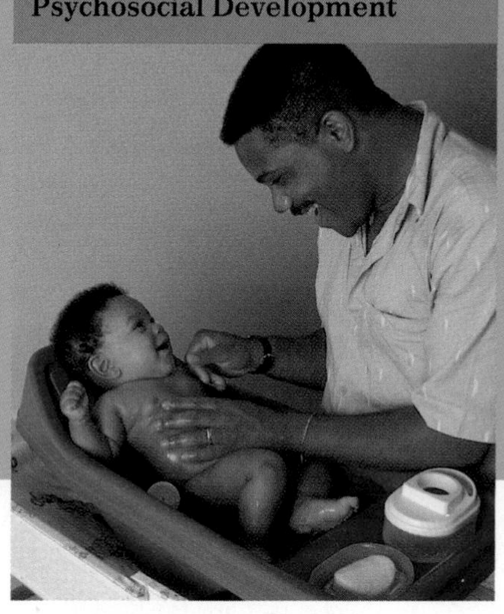

Emotions and Personality Development

Emotions change from quite basic reactions to complex, self-conscious responses. Infants become increasingly independent, a transition explained by Freud in terms of the oral and anal stages, by Erikson in terms of the crises of trust versus mistrust and autonomy versus shame and doubt. While these theories emphasize the parents' role, research finds that much of basic temperament and mood is inborn, and apparent lifelong.

Parent-Infant Interaction

Parents and infants respond to each other first by synchronizing their behavior. Toward the end of the first year, secure attachment between child and parent sets the stage for the child's increasingly independent exploration of the world. Some cultures emphasize an exclusive mother-infant bond; others encourage wider social interaction with the father and other caregivers.

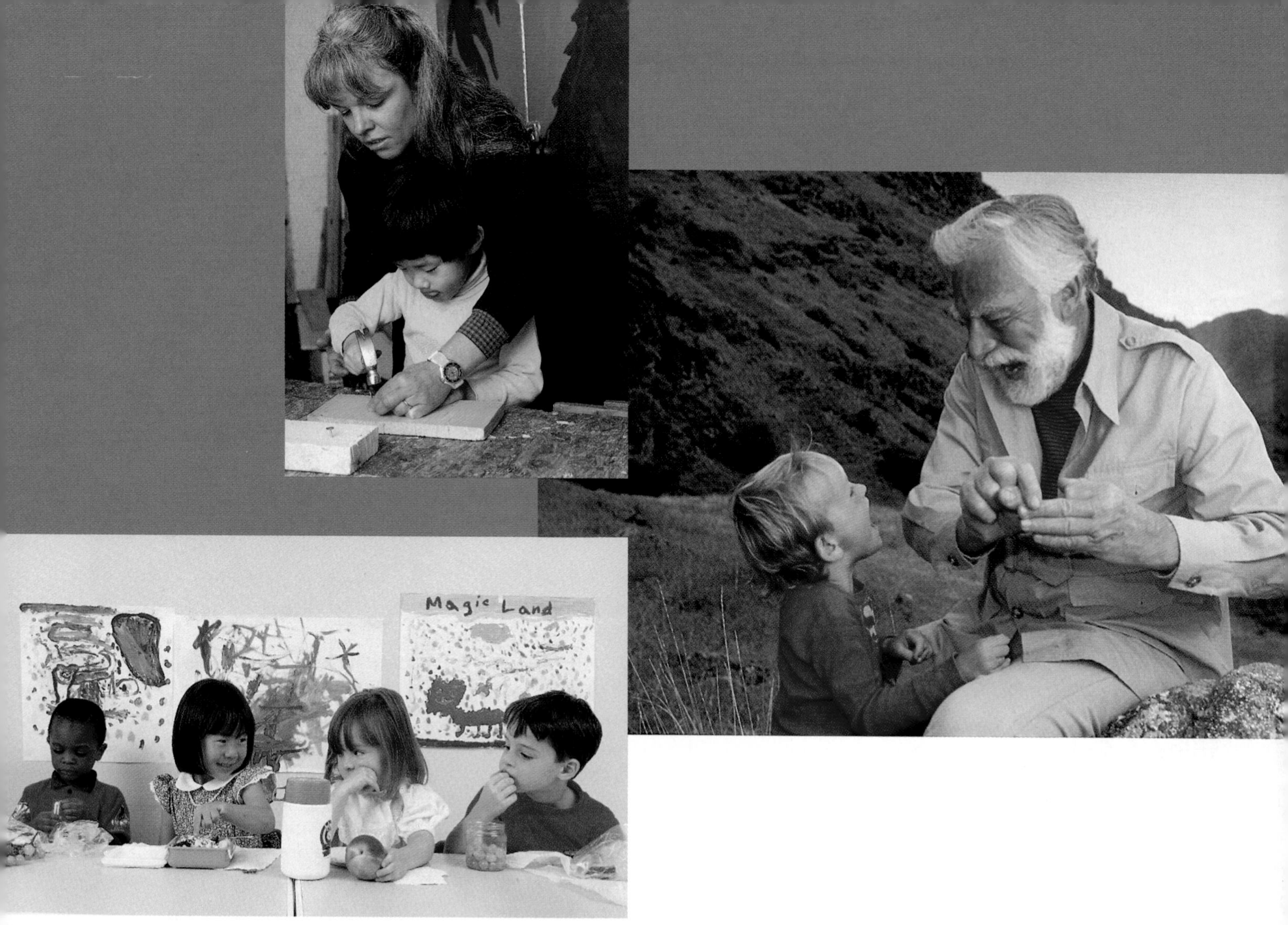

PART III The Play Years

The period from ages 2 to 6 is usually called early childhood, or the preschool period. Here, however, these years are called the play years to underscore the importance of play. Play occurs at every age, of course. But the years of early childhood are the most playful of all, for young children spend most of their waking hours at play, acquiring the skills, ideas, and values that are crucial for growing up. They chase each other and dare themselves to attempt new tasks, developing their bodies; they play with words and ideas, developing their minds; they invent games and dramatize fantasies, learning social skills and moral rules.

The playfulness of young children can cause them to be delightful or exasperating. To them, growing up is a game, and their enthusiasm for it seems unlimited, whether they are quietly tracking a beetle through the grass or riotously turning their play area into a shambles. Their minds seem playful too, for the immaturity of their thinking enables them to explain that "a bald man has a barefoot head," or that "the sun shines so children can go outside to play."

If you expect them to sit quietly, think logically, or act realistically, you are bound to be disappointed. But if you enjoy playfulness, you might enjoy caring for, listening to, and even reading about children between 2 and 6 years old.

The Play Years: Biosocial Development

CHAPTER

8

Between ages 2 and 6, significant biosocial development occurs on several fronts. The most obvious aspects of this development during early childhood, of course, are the striking changes that occur in size and shape, changes that cause many 6-year-olds to find photos of themselves as chubby toddlers unrecognizable. Less obvious but more crucial changes involve the maturation of the brain and central nervous system. This maturation allows the mastery of motor skills that clearly sets the 6-year-old apart from the clumsy toddler and also makes possible the cognitive development that we will discuss in the next chapter. In combination, these changes allow children's exploration and mastery of their world to proceed by leaps and bounds, both literally and figuratively. This growth also, unfortunately, increases children's vulnerability to certain biosocial hazards, including accidental injury and, for some children, abuse.

Let us begin our examination of biosocial development in the preschool years by looking at the way children's body proportions change.

Size and Shape

During the preschool years, children generally become slimmer as the lower body lengthens and some of the fat accumulated during infancy is burned off. The kindergarten child no longer has the protruding stomach, round face, disproportionately short limbs, and relatively large head that are characteristic of the toddler. By age 6, the proportions of a child's body are not very different from those of an adult.

Steady increases in height and weight accompany the changes in body proportions. From age 2 through 6, well-fed children add almost 3 inches (7 centimeters) and gain about 4½ pounds (2 kilograms) per year. By age 6, the average child in a developed nation weighs about 46 pounds (21 kilograms) and measures 46 inches (117 centimeters).

The range of normal development is quite broad. Many children are notably taller or shorter than average, and the spread among age-mates becomes greater with every passing preschool year (see Figure 8.1). Weight is especially variable. For example, by age 6, about 10 percent of American children weigh less than 38 pounds (17 kilograms) and another 10 percent weigh more than 53 pounds (24 kilograms) (Behrman, 1992).

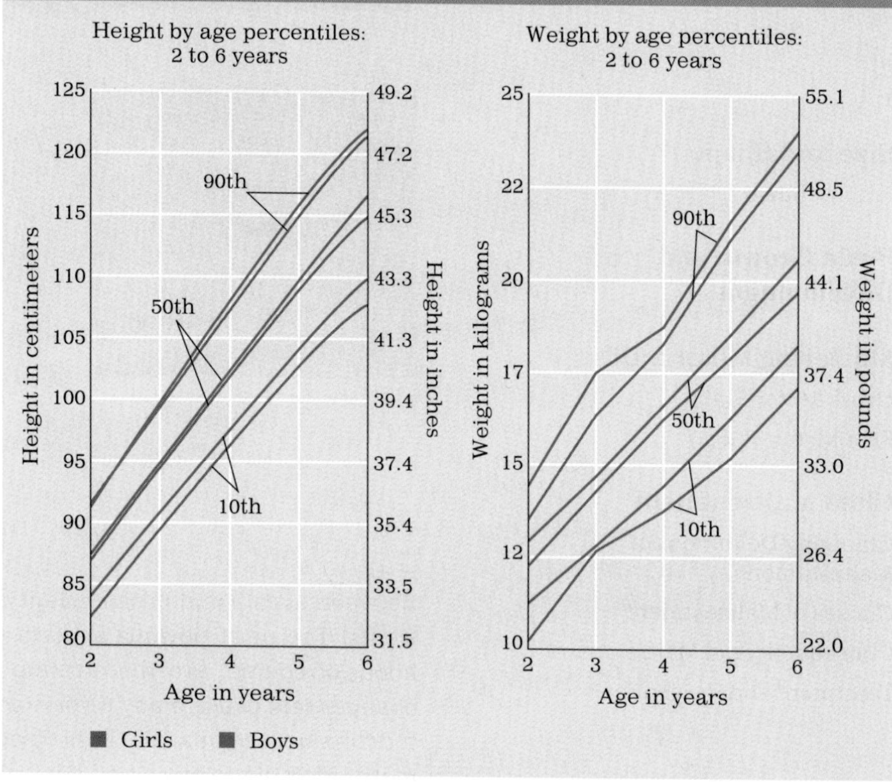

FIGURE 8.1 As these charts show, preschool boys (blue line) and girls (red line) grow more slowly and steadily than they did in the first two years of life. Most children actually lose body fat during these years. The weight that is gained is usually bone and muscle.

Of the many factors that influence growth (see Table 8.1), the three most influential are the child's genetic background, health care, and nutrition. This last factor is largely responsible for the dramatic differences that exist between children in developed and underdeveloped nations: the average 4-year-old in Sweden, for example, is as tall or taller than the average 6-year-old in Bangladesh, where 65 percent of the children experience stunted growth due to poor nutrition (United Nations, 1994). Within developed nations, however, most of the variations that one sees among preschoolers are due to genetic factors.

Generally, boys are more muscular, have less body fat, and are slightly taller and heavier than girls throughout childhood, although this varies depending on the culture and the child's age. For example, even in the early years, boys in several nations of South Asia are markedly taller and heavier than their female counterparts because boys are more highly valued by the society and therefore are more likely to have their nutritional needs taken

TABLE 8.1

Factors Affecting the Height of Preschoolers

Taller Than Average If	Shorter Than Average If
well nourished	malnourished
rarely sick	frequently or chronically sick
African or northern European ancestors	Asian ancestors
mother is nonsmoker	mother smoked during pregnancy
upper SES	lower SES
lives in urban area	lives in rural area
lives at sea level	lives high above sea level
first-born in small family	third- or later-born, large family
male	female

Sources: Eveleth and Tanner, 1976; Lowrey, 1986; Meredith, 1978.

care of first when food is scarce (Poffenberger, 1981). In North America, by contrast, children who are in the heaviest tenth percentile are more likely to be girls than boys, primarily because girls in general have a higher proportion of body fat when they have access to ample food (Lowrey, 1986).

Eating Habits

Whether a child is short or tall, his or her annual height and weight gains are much less from age 2 to 6 than during the first two years of life. In fact, between ages 2 and 3, an average child gains fewer pounds than during any other twelve-month period until age 17 (Rallison, 1986). Since growth is slower during the preschool years, children need fewer calories per pound during this period than they did from birth through toddlerhood, especially if they are among the modern sedentary children who spend much of their time indoors. Consequently, their appetites seem smaller, a fact that causes many parents to worry. In most cases, however, this relative decline in appetite does not represent a medical problem unless the child is unusually thin or is not gaining weight at all. Most parents report a noticeable increase in their children's appetite by age 8 (Achenback & Edelbrock, 1981).

Of course, as at any age, the diet during the preschool years should be a healthful one. The most prevalent specific nutritional deficiency in developed countries during the preschool years is iron-deficiency anemia, a chief symptom of which is chronic fatigue. This problem, which stems from an insufficiency of quality meats, whole grains, and dark-green vegetables, is three times more common among poor families than among others. Although limited financial resources make it harder to purchase high-iron foods, it should also be noted that families of every social class are likely to contribute to the problem by giving their children candy, soda, sweetened cereals, and the like. These items can spoil a small appetite faster than they can a large one, and therefore may keep a child from consuming enough of the foods that contain essential vitamins and minerals.

An additional problem for most American children is that they, like most American adults, eat too few fruits and vegetables and consume too much fat. Whereas preschoolers should obtain no more than 30 percent of their daily calories from fat, the daily diet of six out of seven preschoolers exceeds that limit. Interestingly, both children whose family income is at the poverty level and those whose family income is three times the poverty level are more likely to eat too much fat than those whose family income lies somewhere in between (Thompson & Dennison, 1994).

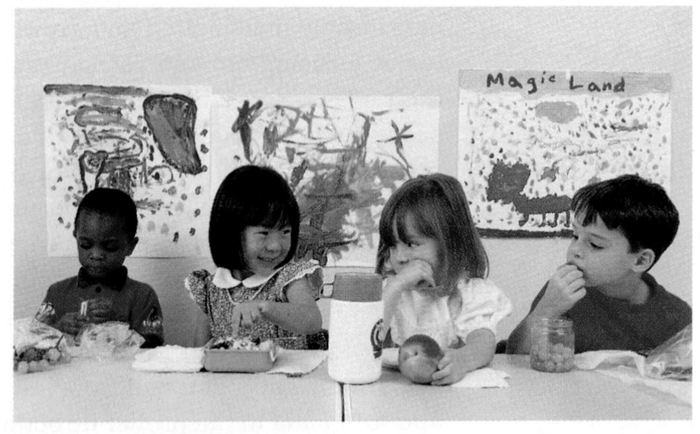

Lifelong food preferences are formed during early childhood, which may be one reason why the two children on the right seem dubious about the contents of the pink lunch box, broccoli and all. Nevertheless, each of these children appears to be a model of healthful eating.

Brain Growth and Development

As explained in Chapter 5, during childhood the brain develops faster than does any other part of the body. By age 2, the brain already weighs 75 percent of its eventual adult weight. By age 5, it has attained 90 percent of its adult weight, while, by comparison, the average 5-year-old's total body weight is about 30 percent that of the average adult's.

Part of this increase in brain size is due to the continued proliferation and creation of communication pathways among the brain's various specialized areas in response to the child's specific experiences (Huttenlocher, 1994). Another aspect of brain growth is ongoing myelination, the insulating process that speeds up the transmission of neural impulses (see p. 139). Finally, several areas of the brain undergo notable expansion, especially those dedicated to control and coordination of the body, the emotions, and the thinking processes.

As a result of all these changes, children during the preschool years develop quicker reactions to stimuli and also are better at controlling those reactions. For instance, compared to a 2-year-old, a 5-year-old would more readily notice when another child begins playing with his or her favorite toy, but would also be less likely to have a stomping-screaming-throwing tantrum over the usurpation. Older preschool-age children are also better at games that require both quick reaction and deliberate action, such as musical chairs, Simon says, steal the bacon, duck-duck-goose, and so on. If older children deign to let them join in any of these games, 2-year-olds are likely to still be thinking about what they should do long after the moment to act has passed.

At about age 5, children also show important gains as a result of growth in the **corpus callosum**, the band of nerve fibers that connects the right and left sides of the brain. This communication link becomes notably thicker due to dendrite growth and myelination and thus becomes more efficient, allowing children to better coordinate functions that involve both sides of the brain and body. A simple example of such coordination is hopping on one foot while using both arms for balance, something few children under age 4 can do. A more complex illustration is older children's increased ability to process information from several areas on opposite sides of the brain at once, connecting and moderating their sensory observations, their emotions, their thoughts, and their controlled reactions.

Another area of important growth involves the visual pathways. Researchers have found that, throughout the preschool years, the areas of the brain associated with the control of eye movements and focusing undergo measurable myelination and growth. As a result, 4-year-olds are much better at focusing on, and recognizing, letters or other figures as they move their eyes across a printed page than are younger children, whose eyes dart around a page much more randomly (Aslin, 1987; Borsting, 1994). Development of the visual pathways, along with improved communication between the left and right sides of the brain, also enhances eye-hand coordination, enabling older preschoolers to copy familiar letters and numbers that they are looking at. By age 5, eye-hand coordination becomes better attuned to left-right distinctions, so children begin to be able to successfully copy a diamond (a task used on psychological tests to reveal the ability to draw diagonals from right to left and from left to right) and to write letters such as "b" and "d" facing in the proper direction (Borsting, 1994).

corpus callosum A network of nerves connecting the left and right hemispheres of the brain.

Researchers have recently discovered that much of this overall brain growth is not linear but rather occurs in spurts and plateaus (Fischer & Rose, 1994). As they have begun to develop precise measures of the varying rates of brain growth, researchers are starting to draw connections between growth spurts and forward leaps in cognitive development. For example, the left hemisphere of the child's brain, where the primary language abilities are usually located, undergoes a growth spurt at around age 2; the right hemisphere, where the primary area for recognition of visual configurations is usually located, undergoes a growth spurt between ages 4 and 5 (Thatcher, 1994), as does the brain overall. Thus, when the corpus callosum expands at around age 5, the child is able to begin forming links between spoken and written language. It is no accident, then, that, worldwide, formal instruction in reading, writing, and arithmetic begins in earnest at about age 6. By that age, children's brains are usually mature enough to begin mastery of these basic literacy skills.

Mastering Motor Skills

As their bodies grow slimmer, stronger, and less top-heavy, and as their brain maturation permits greater control and coordination of their extremities, children between ages 2 and 6 are able to move with greater speed and grace, and become more capable of focusing and refining their activity. The result is an impressive improvement in their various motor skills.

Gross Motor Skills

Gross motor skills, involving large body movements such as running, climbing, jumping, and throwing, improve dramatically during the preschool years (Clark & Phillips, 1985; Du Randt, 1985; Kerr, 1985). The improvement is obvious to anyone who watches a group of children at play. Two-year-olds are quite clumsy, falling down frequently and sometimes bumping into stationary objects. But by age 5, many children are both skilled and graceful. Most North American 5-year-olds can ride a tricycle, climb a ladder, pump a swing, and throw, catch, and kick a ball. Some of them can even ice skate, ski, roller-blade, and ride a bicycle, activities that demand balance as well as coordination. Underlying the development of such skills is a combination of the brain maturation we have just discussed and practice.

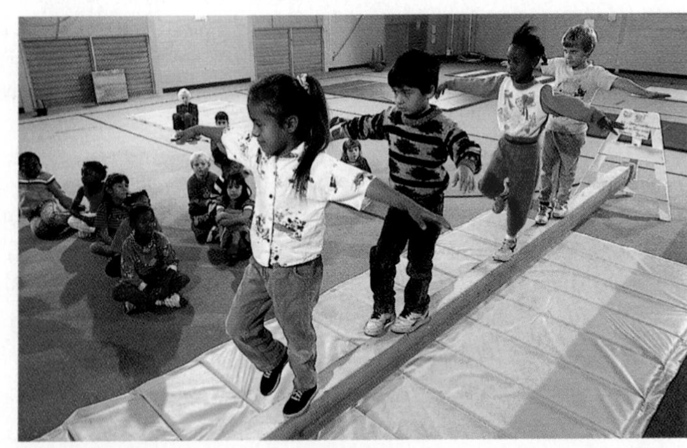

These kindergartners demonstrate an age-related skill. Unlike 3-year-olds, most 5-year olds can walk across a balance beam 4 inches wide, although, as you can see, not without considerable concentration. As you can also see, in this and other skills involving balance, girls tend to be almost a year ahead of the boys. In other gross motor skills—throwing a ball, for example—boys usually surpass the girls.

PUBLIC POLICY

Injury Control Is No Accident

As children gain control of their motor skills, they practice them continually, wherever and however they can. They climb trees and fences; they run along open fields and busy streets; they find ways to play with almost anything they can get their hands on. All of this activity and exploration is healthy in many ways, but it also poses some dangers, exposing children to far greater risks than those that parents usually worry about, such as abduction or leukemia.

In fact, in all but the most disease-ridden or war-torn countries of the world, accidents are, by far, the number-one cause of childhood death. In the United States, a child has about 1 chance in 500 of dying due to an accident before age 15—four times the risk of dying of cancer, the second-leading cause of childhood death. Roughly two-thirds of accidental deaths among children involve non-vehicular causes, such as falling, drowning, choking, poisoning, and so on. These kinds of accidents are particularly prevalent among preschool children, pushing their annual accidental death rate even higher than that of school-age children (16 per 100,000 compared with 9 per 100,000) (U.S. Bureau of the Census, 1996).

Injuries, of course, are even more common. In the United States, a child has more than 1 chance in 4 each year of having an injury that needs medical attention (U.S. Bureau of the Census, 1996). Virtually every child will need stitches or a cast sometime before adolescence, and 44 percent of all serious injuries that require hospitalization, including a disproportionate number of serious brain injuries, occur among children under age 15.

The accident risk for any particular child depends on several factors, including the child's sex, socioeconomic status, and community setting. For instance, no matter where they live, boys, as a group, tend to take more risks and have more injuries and accidental deaths than girls—about one-third more between ages 1 and 5 and twice as many between ages 5 and 14. Sex differences are particularly salient when the neighborhood is filled with attractive hazards, as is the case in crowded cities or rural areas. Farm equipment, especially corn augers, tractors, and gravity boxes, injures 20,000 children and kills 300 each year, most of them boys, with deaths more likely during harvest time, when equipment use is heavy and supervision light (Salmi et al., 1989).

The clearest risk factor of all is socioeconomic status. One study of all childhood deaths in North Carolina found that low-SES children were three times as likely to die an accidental death as other children, with income disparity being particularly pronounced in the fatality rates for preschool children. For example, 1- to 4-year-olds whose families were on welfare were at least four times as likely to be fatally hit by a car, four times as likely to die by choking, and nine times as likely to burn to death, compared with children whose families did not need welfare assistance. Among the reasons cited for these differences were the substandard housing, hazard-filled neighborhoods, and inferior medical care that are typically associated with low socioeconomic status. For reasons that are not clear, when black and white children of the same income were compared, poverty increased the risk of accidental death for white children even more than for black children (Nelson, 1992).

For all children, however, the risk of accidental injury, especially serious injury, could be much lower than it is. With forethought, certain accidents can be avoided completely; and while some scrapes and bruises are inevitable in a normal childhood, proper precautions could significantly reduce the degree of injury. The first step in reducing injury risks, many believe, is to approach the problem in terms of **injury control** instead of "accident prevention." The word "accident," advocates of injury control point out, misleadingly implies that no one was at fault, whereas most serious accidents involve someone's lack of forethought (Christopherson, 1989).

Sometimes forethought means little more than providing adequate adult supervision. But what is "adequate" for controlling injury? There are no absolutes and no guarantees: children sometimes get hurt even when they are being closely watched in their own homes, and sometimes they may be alone in dangerous places all day long without incident. There is also the risk that being overly protected might keep children from developing the independence and self-confidence they need. Nonetheless, there is some consensus on what adequate supervision entails, at least in the case of infants and young children. Pediatricians, child-protection workers, and most parents recognize that a crawling baby cannot be safely left alone anywhere, even for a minute, and that children between the ages of 4 and 8 should never be unsupervised in a neighborhood that has attractive dangers, such as accessible bodies of water (creeks, ponds, or swimming pools) (Peterson et al., 1993). With older children, temperament and past history are important indicators of how long a particular child can safely go unsupervised in areas that pose any kind of risk.

Forethought also involves instituting safety measures in advance to reduce the need for vigilant supervision and to prevent serious injuries when accidents do occur. For instance, compared with adults, children are more likely to drown, choke on a nonfood object, fall off a bicycle, or suffocate in a fire. Advance precautions, such as teaching

children to swim, or removing all swallowable objects from their reach, or requiring them to wear a helmet when bicycling, or installing working smoke alarms in the house, could prevent most such events.

Obviously, the responsibility of injury control is not just the parents' alone. Parents are often unaware of many of the objects and activities that may pose a danger to their children. Not until accurate nationwide data became available did parents realize the extent of the hazards associated with, for example, cribs with widely spaced slats, playgrounds surfaced with concrete, balloons that have popped, or roller-blading without head gear and protective pads. Research is thus needed to identify potential hazards, and information about demonstrated risks needs to be made readily available. In addition, schools, community groups, and legislators can take many educational and legal measures to reduce injuries. The question is, which measures work best?

Overall, broad-based safety education, such as television announcements and poster campaigns, rarely has a direct impact on children's risk-taking, although it may foster a general climate that makes more specific measures likely to work. Similarly, educational programs in schools and preschools may be successful to the extent that they enable children to verbalize safety rules, but they appear to have little effect on children's actual behavior unless the parents also participate in the programs (Garbarino, 1988).

More effective than targeted educational measures in reducing the overall injury rate have been safety laws that include penalties for noncompliance. Among such measures that have led to significant reductions in accidental death rates for children in the United States are laws requiring child-proof safety caps on medicine bottles (resulting in an 80 percent reduction in poisoning deaths of 1- to 4-year-olds); flame retardancy in children's sleepwear (decreasing deaths from burning pajamas and nightgowns 97 percent); fencing around swimming pools (reducing childhood drowning by 51 percent); and the use of car safety seats (helping significantly to promote a 26 percent decline in motor-vehicle deaths of children under age 5 from 1980 to 1991) (National Center for Injury Prevention and Control, 1992, 1993).

Largely as a result of laws like these, the accidental death rate for children between the ages of 1 and 5 in the United States has been cut in half in the past two decades. Nevertheless, this means that 2,600 children in this age group and 3,500 children between the ages of 5 and 14 are still being killed by accidents each year (U.S. Bureau of the Census, 1996).

It is increasingly apparent that the child who escapes serious injury in childhood is not *just lucky*, and that accidents are not *just an accident*. The lessons already learned have reduced accidental deaths dramatically, but good statistics on, and expert analysis of, precisely what causes and prevents accidents are just beginning to be collected (National Center for Injury Prevention and Control, 1992, 1993). As adults become more knowledgeable about the hazards facing children and the ways to reduce their impact, more children will survive their early years intact.

(a)

(b)

In order for parents to safeguard their children from injury, they first need to be aware of safety hazards, and then need to take whatever action is necessary. The parents shown here are to be commended: the parents in (a) not only put a helmet on their child but demonstrate by example the importance of this measure; and as suggested by the smiles in (b), the mother probably has been securing her child in a safety seat from early infancy.

injury control The implementation of educational and legal measures to reduce the risk and impact of childhood injuries.

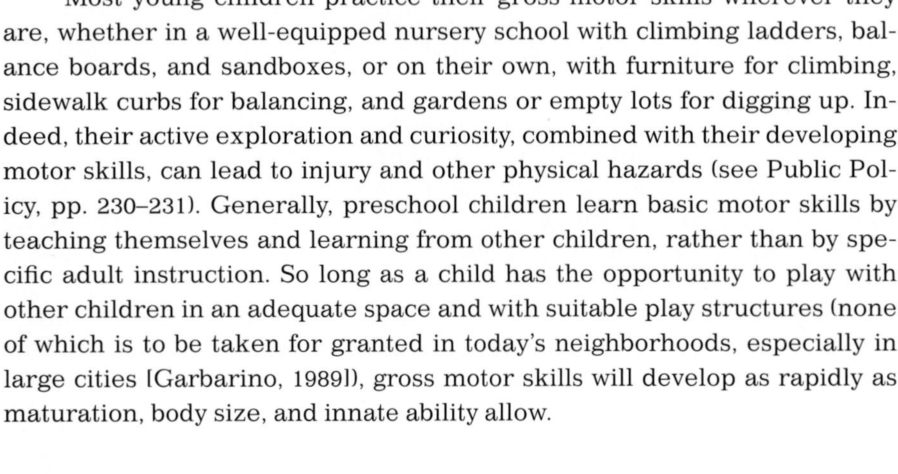

Most young children practice their gross motor skills wherever they are, whether in a well-equipped nursery school with climbing ladders, balance boards, and sandboxes, or on their own, with furniture for climbing, sidewalk curbs for balancing, and gardens or empty lots for digging up. Indeed, their active exploration and curiosity, combined with their developing motor skills, can lead to injury and other physical hazards (see Public Policy, pp. 230–231). Generally, preschool children learn basic motor skills by teaching themselves and learning from other children, rather than by specific adult instruction. So long as a child has the opportunity to play with other children in an adequate space and with suitable play structures (none of which is to be taken for granted in today's neighborhoods, especially in large cities [Garbarino, 1989]), gross motor skills will develop as rapidly as maturation, body size, and innate ability allow.

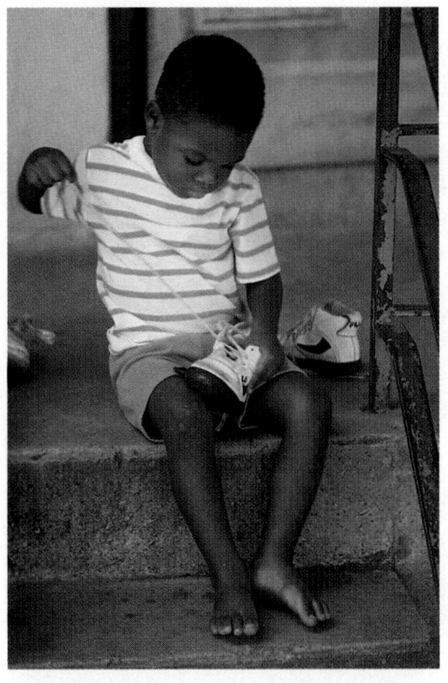

Manipulating shoe laces requires fine motor skills, even for unlacing and knotting and especially for lacing and tying. The latter skill is almost impossible for most 3-year-olds to master. They and their parents have two alternatives: the traditional solution of waiting patiently until hours of practice and months of maturation allow mastery of the skill, or the modern shortcut—sneakers with Velcro closures.

Fine Motor Skills

Fine motor skills, involving small body movements, especially those of the hands and fingers, are much harder for preschoolers to master than gross motor skills. Such things as pouring juice from a pitcher into a glass without spilling, cutting food with a knife and fork, and achieving anything more artful than a scribble with a pencil are difficult even with great concentration and effort. Preschoolers can spend hours trying to tie a bow with their shoelaces, often producing knot upon knot instead. The chief reason many children experience these difficulties is simply that they have not developed the muscular control, patience, and judgment needed for the exercise of fine motor skills, in part because the necessary myelination of the central nervous system is not complete. For many preschoolers, this liability is compounded by their still having short, fat fingers. Unless these limitations are kept in mind when selecting utensils, toys, and clothes for the preschool child, frustration and destruction can result: preschool children may burst into tears when they cannot button their sweaters, or may mash a puzzle piece into place when they are unable to position it correctly.

Fortunately, such frustrations usually fade as the child's persistence at practicing fine motor skills gradually leads to mastery. One fine motor skill that seems particularly linked to later success in school is easy for parents and teachers to encourage—the skill of making meaningful marks on paper.

Children's Art

Drawing is an important form of play. On the simplest level, "the child who first wields a marker is learning in many areas of his young life about tool use" (Gardner, 1980). In addition, in thinking about what to draw, manipulating the pencil, crayon, or brush to execute the thought, and then viewing, and perhaps explaining, the end product, the child is experiencing a sequence of events that not only provides practice with fine motor skills but also enhances his or her sense of accomplishment.

Children's artwork also provides a testing ground for another important skill, self-correction. A developmental study of children's paintings found that whereas 3-year-olds often just plunked their brushes into the paint, pulled them out dripping wet, then pushed them across the paper without much forethought or skill, by age 5 most children took care to get just enough paint on their brushes, planned just where to put each stroke,

and stood back from their work to examine the final result (Allison, 1985). Older children also show an eagerness to practice their skills, drawing essentially the same picture again and again.

Such mastery of drawing skills is related to overall intellectual growth. In general, as children become more skilled and detailed in their drawing, their level of cognitive development rises as well (Bensur & Eliot, 1993; Chappell & Steitz, 1993). While there is no way of knowing to what degree the mastering of drawing skills contributes to cognitive advances—as well as arises from them—it may be that a sketch pad and a box of markers are as much an "educational toy" as traditional alphabet blocks or counting games.

Child Maltreatment

"Forgive us, for we were not there." This pointed epitaph for a 6-year-old girl who was abused and eventually killed by her mother applies to every abused or neglected child. Whenever a child is seriously maltreated, dozens of adults—relatives, neighbors, social workers, and teachers—see signs that something is wrong. Sadly, instead of seeking answers and a solution, many simply raise questions or even worse, ignore the signs completely.

Throughout this chapter and elsewhere in this text, we have assumed that parents naturally want to foster their children's development and protect them from every danger. Yet daily, it seems, the news media report stories of parents who actually cause harm to their offspring. When the harm takes the form of horrific abuse or results in death, the story makes national news, as when six-year-old Elisa's mother

> confessed to killing Elisa by throwing her against a concrete wall. She confessed that she had made Elisa eat her own feces and that she had mopped the floor with her head . . . there was no part of the six-year-old's body that was not cut or bruised. Thirty circular marks that at first appeared to be cigarette burns turned out to be impressions left by the stone in someone's ring. [Van Biema, 1995]

The public is shocked by the brutality and senselessness of cases like Elisa's, and appalled at the pathological perpetrators, indifferent neighbors, and overworked child-welfare workers who often seem to share the blame. Yet sensational cases like Elisa's represent only a small portion of all maltreatment cases. And as experts in child-maltreatment research point out, while the media's focus on the lurid and inexplicably brutal instances of maltreatment triggers justifiable outrage, it also distracts attention from the far more typical, and far too common, cases of maltreatment (Scheper-Hughes & Stein, 1987).

More specifically, most maltreatment does *not* involve serious physical abuse: only about 25 percent of all new cases reported and accepted for protective services involve physical abuse, with a fatality occurring in roughly 1 in 1,000 cases (McCurdy & Daro, 1994). Much more common is a persistent pattern of neglect and psychological abuse that often begins in infancy and, accumulating over the years, harms the child's self-concept, social interactions, and intellectual growth, sometimes permanently. Instead of the irrational rage of Elisa's mother, more frequent are the actions of caregivers who do not understand how to love and guide a child, unaware, for example, that a child needs to be patiently shown how to "do things right." Typical of this kind of maltreatment is the reaction of a mother whose 3-year-old was having trouble carrying an umbrella and was dragging its curved handle on the ground:

> "Carry that umbrella right or I'll slap the [expletive] out of you," she screamed at him. "Carry it right, I said . . . " and then she slapped him in the face, knocking him off balance. [Dash, 1986]

Another example is this father, who has set ideas of how children should behave and believes that his son should be "kept on his toes":

> So his dad teases him a lot . . . [and] plays games with him. If Jon wins, his dad makes fun of him for being an egghead; if he loses, he makes fun of him for being a dummy. It is the same with affection. Jon's dad will call him over for a hug; when Jon responds, his dad pushes him away, telling him not to be a sissy. . . . Jon is tense, sucks his thumb, and is tongue-tied (which his dad teases him about). [Garbarino et al., 1986]

Often child neglect arises partly because of the family's poverty, when inadequate income, dangerous housing, perfunctory medical care, threatening neighborhoods, and an inability to provide the basic necessities combine to put children at multiple risk. These stresses help explain why neglect is more common in single-parent households than in households with two parents. As Leroy Pelton (1994) has noted, "in some cases a mother does not have much choice but to provide her children with inadequate supervision or to deprive them of necessities. A low-income mother with many children cannot easily obtain or pay for a babysitter every time she wants or needs to leave the house. If she leaves her children alone, she is gambling with their safety" and can be arrested for neglect.

The task of social scientists, then, is to get beyond sensationalism and blame and to discover the underlying causes and remedies of all types of child maltreatment. As you will see, compared with a few decades ago, experts today—as well as the general public—are becoming much more aware of the scope, causes, and consequences of maltreatment.

Changing Definitions of Maltreatment

Up until a few decades ago, child maltreatment was thought of mostly in terms of obvious physical assault, which was assumed to occur as a rare outburst of a mentally disturbed person (Zigler & Hall, 1989). However, it is now recognized that maltreatment is neither rare nor sudden, that its perpetrators are usually not deranged, and, as indicated, that severe physical injury is the primary problem in only about a fourth of all cases (Ammerman & Herson, 1989; Cicchetti & Carlson, 1989; McCurdy & Daro, 1994; McGee & Wolf, 1991).

With this recognition has come a broader definition of **child maltreatment** to include all intentional harm to, or avoidable endangerment of, anyone under age 18. Child maltreatment thus includes both **abuse**—deliberate actions that are harmful to a child's well-being—and **neglect**—failure to appropriately meet a child's basic needs. Abuse and neglect are both further subdivided into more specific categories, with abuse including physical, emotional, or sexual abuse (discussed in Chapter 14), and neglect including inattention to the child's basic emotional as well as physical needs. To some extent, each form of maltreatment is distinct, with somewhat different causes, manifestations, and consequences. However, while distinguishing specific forms of maltreatment helps untangle the web of reasons underlying a specific case and helps target the best treatment, the unfortunate reality is that most children who come to the attention of authorities suffer several forms of maltreatment. For these children especially, the severest damage is not to the body but to the spirit, a damage that is difficult to measure or treat (Crittenden et al., 1994; O'Hagen, 1993).

child maltreatment Includes all intentional harm to, or avoidable endangerment of, anyone under age 18.

abuse All actions that are deliberately harmful to an individual's well-being.

neglect A form of child maltreatment in which parents or caregivers fail to meet a child's basic needs.

A CLOSER LOOK

Child Maltreatment in Context

As emphasized throughout this book, every behavior needs to be considered in context. This is especially true when assessing parental nurturance, a topic that involves our deepest and most personal emotions. Developmentalists increasingly believe that "behaviors *per se* can seldom be defined as harmful or beneficial—the immediate, relational, familial, and cultural contexts in which they occur all play a crucial role in determining what effects the behavior may have" (Sternberg & Lamb, 1991).

Understanding child maltreatment in context begins by considering *community standards*. Each community has somewhat different customs and goals regarding child-rearing, which means that sometimes what is maltreatment in one place is not maltreatment in another. For example, while a great majority of American parents sometimes spank, slap, or push their young children and think it justified, only 1 percent of Swedish parents of children under age 6 think physical punishment is ever appropriate (Baumrind, 1995). And, in fact, physical punishment of children of any age is illegal in Sweden. In some Asian, African, and Caribbean countries, by contrast, never to discipline a child by hitting or spanking is tantamount to neglect (Arnold, 1982; Rohner, 1984; Rohner et al., 1991). Physical punishment—administered by mothers as well as fathers—may be especially commonplace and accepted in cultures where men are expected to be aggressive and dominant, where the father's role is that of an "authoritarian tyrant," and where child-rearing styles are highly controlling and nonnurturant. A punitive style of child-rearing in such cultures "may have little connection with the general value accorded to children. Rather, parents may use this style of child-rearing to teach children the code of conduct and behavior they will need to become responsible adults" (Deyoung & Zigler, 1994).

Thus in many cases, before concluding that a particular behavior or practice represents maltreatment, we need to take community standards into account. This is especially important in light of the fact that children everywhere feel loved when their parents raise them not too differently from other children in the same family, neighborhood, and culture, and that parents judge their own child-rearing partly on the basis of the collective wisdom and practice of their peers and partly on how they themselves were raised (Holden & Zambarano, 1992).

Given the diversity of the world's communities, it is "imperative to disentangle the natural from the cultural," distinguishing those practices that hurt any child anywhere from those that are harmful only in a particular place (Woodhead, 1991). Administering a beating is one example: for children in the West Indies, being hit for misbehaving is commonplace, and is taken as a sign of the parent's love and concern. But for West Indian children, as for children everywhere, frequent and severe physical beatings result in feelings of rejection and low self-esteem (Rohner et al., 1991).

Understanding maltreatment in context also requires taking into account the impact of a behavior on the particular child. Since each child is unique, and every child changes over time, a practice that harms one child may not negatively affect another, or may not hurt either of them a few years later. For example, some children need much more supervision than others; some wither more quickly under criticism; some are more likely to be injured, physically or emotionally, by corporal punishment. Thus the seriousness of any specific act of maltreatment depends partly on the age, temperament, and abilities of the child (McGee & Wolfe, 1991).

At what point, then, does imperfect parenting become maltreatment? One measure, of course, is when it can cause serious physical injury. Shaking an infant, for example, is likely to cause permanent brain damage or even death. Beyond that, however, there is no clear line of demarcation. Every case should be judged in terms of its context and developmental history (Cicchetti, 1991; Zigler & Hall, 1989). The question then becomes: Who is judging the behavior, and for what purpose? When the issue is whether a particular adult should be legally labeled as abusive or neglectful, only those cases in which maltreatment seems clearly dangerous, ongoing, and unacceptable to the community merit legal action (Thompson & Jacobs, 1991). On the other hand, when caregivers wonder if they are crossing the line, they need to remember that every parent's caregiving is, indeed, sometimes potentially harmful, no matter what the context, and that whenever doubts about the severity of disciplining arise, it is far better to trust those doubts and err on the side of leniency.

Obviously, historical and cultural norms influence definitions of child maltreatment, with behaviors considered abusive in some eras or cultural settings being regarded as legitimate and acceptable in other times and places (see A Closer Look, p. 235). However, no one can deny that in many nations throughout the world, the rate of child maltreatment has been, and continues to be, alarmingly high. Since 1993, the number of *substantiated* cases of child maltreatment in the United States has exceeded 1 million a year (U.S. Bureau of the Census, 1996), a rate that represents roughly one child in every twenty-two. Disturbing as these figures are, they are well below the actual number of instances of maltreatment, since a great many cases go unreported. The extent of this underreporting is suggested by the results of surveys conducted anonymously with parents over the past decade: between 10 and 12 percent of the parents surveyed acknowledged regularly hitting a child with an object, and 3 percent acknowledged engaging in even more harmful behaviors, including kicking, biting, burning, or beating up (Dodge et al., 1994; Straus, 1994; Straus & Gelles, 1986). These percentages translate into roughly 7 million victims of the first type of abuse and nearly 2 million of the second. Taking into account the fact that many parents would not admit to maltreatment, even anonymously, and that other relatives, neighbors, and even strangers also maltreat children, it is clear that the number of seriously maltreated children is considerably above the official figure.

In a vast majority of the cases substantiated in 1993, the children had endured maltreatment for several years, maltreatment that often was suspected by neighbors and relatives. Sadly, even after their suffering is finally officially recognized, about a third of all U.S. children designated as maltreated receive no special services at all (Daro, 1995). Even with intervention of some sort, most of these children endure new instances of maltreatment. Let us look now at what can be done to stop this problem, first examining underlying causes and then considering treatment possibilities.

Causes of Maltreatment

At first it is hard to imagine any reason why someone would hurt a child entrusted to his or her care. However, research has shown that virtually everything—from the community values to the caregiver's history, from the family culture to the child's temperament—can contribute to the causes of child maltreatment.

The Cultural and Community Context

According to the United Nations, overall concern and protection for the well-being of children varies markedly worldwide. Even countries in the same region of the world, with similar per capita income, differ markedly in measures of children's general health, education, and overall well-being (United Nations, 1994). In addition, day-to-day caregiving is influenced by broad cultural values (Korbin, 1994; Sigler, 1989). Four such values appear to be especially important in protecting children from maltreatment:

1. Children are highly valued, as a psychological joy and fulfillment, as well as an economic asset.

Child maltreatment, a serious social problem in industrialized nations, is rare in Micronesia. One reason, reflected in this photo of a community on Pulap Island, is that family life occurs largely in the open rather than behind closed doors. Neighbors and relatives immediately notice any lapse of care or outbursts of temper, and remedy the problem before it becomes neglect or abuse.

2. Child care is considered the responsibility of the community. If parents are unable to care for their child, other adults willingly take over.

3. Young children are not expected to be responsible for their actions. In some cultures, almost any punishment of children younger than age 3, or even age 7, is considered abusive.

4. Violence in any context—between adults, between children, and between caregiver and child—is disapproved of.

The role of social values in child maltreatment is dramatically highlighted by the different rates of child abuse that existed among the Polynesian people who lived in their traditional home, the Pacific Islands, and those who had emigrated to New Zealand. Among the former, maltreatment was virtually nonexistent (Ritchie & Ritchie, 1981), because their society met the four criteria listed above: children were highly respected, were cared for by many adults, were considered unteachable until they were at least 2 years old, and adults rarely expressed their anger through physical aggression (Reid, 1989; Ritchie & Ritchie, 1981).

However, when Polynesians moved to New Zealand, the rate of child abuse skyrocketed, surpassing the rate of European New Zealanders many times over. The demands of the new lifestyle, designed for nuclear rather than extended families, made it impossible for the parents to continue their relaxed permissiveness, communal authority, and informal, shared child care. Like every immigrant group entering a radically different culture, these Polynesian parents experienced considerable stress until they were able to develop viable new coping strategies, such as learning how to guide children's behavior without resorting to physical punishment, how to replace the freely available caregivers of the past, and how to limit family size so that children did not become an overwhelming financial burden. Until resolved, these contextual stresses often led to a loss of perspective, and abuse resulted.

Poverty often sets the stage for neglect: for some parents, it is hard to give their children the individual love and attention that each needs when they can't even afford to give them a bed of their own. At the same time, many parents are able to be loving and nurturing no matter how difficult their circumstances.

Cultural Factors Affecting Maltreatment in the United States

Comparing the four characteristics of nonmaltreating cultures with the patterns common in the United States explains, to a great extent, why child maltreatment is so prevalent in the United States, and why some U.S. communities have higher rates than others. First, children are often considered to be both a financial and personal burden. Not surprisingly, no matter how maltreatment is defined or counted, it occurs more frequently as family income falls (Pelton, 1994). This is particularly true for neglect and physical abuse, which fall most heavily on children between the ages of 3 and 6 who live in families with an income below the poverty line, an unemployed father, and four or more children. In such families, children obviously add to the financial stress and are more likely to become victims because of it (Wolfner & Gelles, 1993). A 1995 Gallup poll of a representative sample of 1,000 parents found the rate of abuse to be three times as high among families with incomes under $20,000 as among families with incomes of at least $50,000 (Gallup, 1995).

Second, in the United States, social support for parents and young children is scarce. For a variety of reasons, few relatives, neighbors, and friends are willing and able to help with child care. Grandmothers, for instance, once the mainstay of practical help, are now much more likely to live a distance away and to have their lives taken up with careers and friends outside the family. Lacking a supportive network, overburdened parents often take out their problems on their children, with the problems escalating, undetected, until considerable harm has been done. The likelihood of maltreatment is especially high when lack of support results in social isolation, with those who are most isolated being among the most abusive (Corse et al., 1990; Thompson, 1994). Neighborhoods in which there is a strong sense of community and in which there is a network of support through religious groups, schools, and community centers tend to have lower rates of abuse (Garbarino & Kostelny, 1992).

Third, the culture's views concerning the speed with which young children should "grow up" may add to the problem. The emphasis on the infant's and preschooler's ability to learn may cause some parents to forget that young children are also immature and dependent on others. For example, exaggerated expectations can lead some parents to consider irritating but normal behavior to be deliberate and therefore amenable to correction. They may punish their infants for "crying too much" or punish toddlers for being unable to control urination or defecation or punish older children for "immature" behavior, expecting them to get themselves up, dressed, fed, and to school, as well as to avoid "trouble" of all kinds, long before they are sufficiently mature to do so.

Finally, as many have observed, "violence is as American as apple pie." Indeed, by almost any measure, from the prevalence and prominence of aggression on television to the rate of spouse abuse, from the rates of violent crime to the rate of homicide, the United States is one of the most violent nations of the world (Benedek, 1989; Gelles & Straus, 1988; Sigler, 1989).

The Family Context

Each family has its own private culture, including habits, coping styles, and values that affect every family member. Many experts believe that the way these are structured in the individual family system can be pivotal in allowing maltreatment to occur. For example, the daily routine of most families is somewhat flexible, with adults and older children being able to make minor adjustments in their schedules and established roles as the occasion requires. However, the routines of maltreating families are typically at one of two extremes: they are either so rigid in their schedules and role demands that no one can measure up, or they are so chaotic and disorganized that no one can be certain of what is expected, or under what circumstances one can count on receiving appreciation, encouragement, protection—or even food and a clean bed. In such families, hostility and neglect are inevitable (Dickerson & Nadelson, 1989; Panel on Research on Child Abuse and Neglect, 1993).

Similarly, while almost every family experiences crises that disrupt their harmony, most also have ways of coping and readjusting so that the family once again functions well. One crucial element in this restabilization is the family's ability and willingness to avail themselves of social support when needed. Unfortunately, maltreating families tend to shun social support, and problems are especially likely to worsen if a particular family's code includes isolation and distrust of all outsiders, from the neighbor next door to members of the local clergy (Polansky et al., 1985). If family stresses then erupt in violence, children are stuck within

> a family system in which exploitation, loyalty, secrecy and self-sacrifice form the core of the family's value system. In a sense, the victim's survival is dependent on adjusting to a psychotic world where abusive behavior is acceptable but telling the truth about it is sinful. [Carmen, 1989]

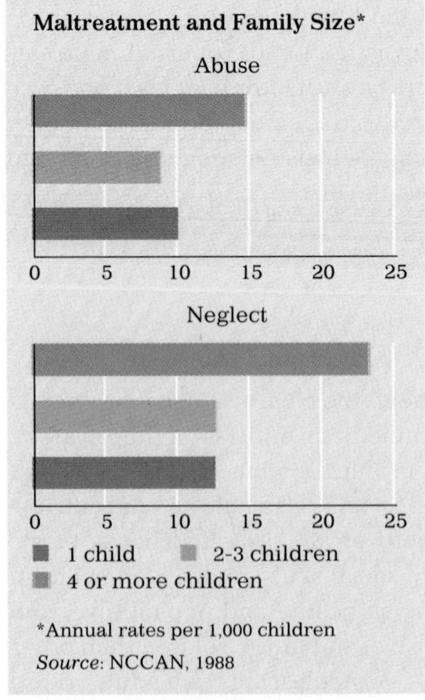

FIGURE 8.2 As suggested by the nursery rhyme about the "old woman who lived in a shoe, she had so many children she didn't know what to do," large families increase the risk of neglecting the children's basic needs.

Another element within the family system that exacerbates maltreatment is abuse or neglect between members other than the perpetrator-victim pair. Child maltreatment is most likely within a home where the relationship between the resident adults—especially that between mother and father or between grandparents and children—is either extremely hostile or emotionally cold or both. Further, that climate of hostility adds significantly to the child's psychological stress, intensifying the effects of whatever direct abuse there may be (Cummings et al., 1994; Fantuzzo et al., 1991).

Destructive family relationships intensify as the child-adult ratio increases (see Figure 8.2), partly because each new child means less money, less space, and less attention for other family members and greater stress for parents. These effects are more pronounced in single-parent families, since it is even harder for one adult to meet all the needs of several children. In the Gallup poll mentioned above, the rate of physical abuse in single-parent families was three times that in two-parent families.

Problems in the Parents

Contrary to popular misconceptions, most maltreating parents are not markedly different from average parents. Like other parents, they love their children and want the best for them. Fewer than 10 percent of them are pathological—so deluded or emotionally and cognitively dysfunctional that they never recognize the basic needs and vulnerabilities of their children.

Overall, however, abusive parents do tend to have personality traits that, in combination with stressful situations in a hostile ecosystem, form a volatile constellation likely to lead to an injured child. Personality tests of abusive parents find that they tend to have lower self-esteem and to be less adaptable than other parents. Adding to this defensive rigidity is immaturity, which makes them more concerned with their own needs and less patient with, or even aware of, the needs of others (Belsky & Vondra, 1989; Christensen et al., 1994).

Maltreating parents are also likely to show cognitive patterns that are somewhat distorted. They tend to view the world in negative ways that affect not so much their general attitudes about child-rearing as their attributions for a child's specific behaviors (Newberger & White, 1989). They tend to see the world as hostile and difficult, which leads them to interpret any signs of their child's discomfort or distress as a personal attack. For instance, they are likely to misinterpret the facial expressions and cries of an infant in distress as displays of anger (Kropp & Haynes, 1987), or to mistake the fearful clinging of a genuinely frightened preschooler as a manipulative demand for attention. This negative attribution, especially when combined with immaturity, makes normal coping with the demands and needs of children very difficult (Heap, 1991).

An additional factor that increases the likelihood of maltreatment is drug dependency. According to official investigation of reported maltreatment, substance abuse is an apparent contributing factor in at least one case in four (McCurdy & Daro, 1994). Undoubtedly, the actual number is higher, given the difficulty any adult has in meeting the insistent, ongoing, and often unpredictable needs of a young child while in the grip of psychoactive drugs.

Consequences of Maltreatment

The more we learn about child maltreatment, the clearer it becomes that its consequences extend far beyond any immediate injuries or deprivation. Compared to well-cared-for children, chronically abused and neglected children tend to be underweight, slower to talk, less able to concentrate, and delayed in academic growth (Cicchetti et al., 1993; Eckenrode et al., 1993). Deficits are particularly apparent in social skills: maltreated children tend to regard other children and adults as hostile and exploitative, and hence they are less friendly, more aggressive, and more isolated than other children (Dodge et al., 1994; Egeland, 1991). As adolescents and adults, those who were severely maltreated in childhood, either physically or emotionally, often engage in self-destructive and/or other harmful behaviors (Crittenden et al., 1994).

The human and financial costs, both to the victim and to society, are virtually impossible to measure. In the United States in the 1980s, the *annual* cost of immediate care (investigation, medical treatment, court costs, emergency shelter) for all reported cases of serious maltreatment was around $500 million, with another $700 million spent on therapeutic services and long-term foster care (Daro, 1988). Additional costs result when victims of maltreatment later require special education for learning disabilities, therapy or institutionalization for emotional problems, and, in some cases, imprisonment for acts of misdirected anger.

In assessing the outcomes of maltreatment, we must neither minimize nor exaggerate. On the one hand, virtually every child who experiences serious, ongoing maltreatment is likely to bear some lifelong scars, including depression, fear of intimacy, difficulty controlling emotions, and low self-esteem (Rutter, 1989). On the other hand, many adults who were victims of childhood abuse or neglect live relatively normal lives, working, marrying, and raising a family.

One potential consequence that must be carefully considered is **intergenerational transmission**, that is, maltreated children growing up to become abusive or neglectful parents themselves. Many people erroneously believe that the transmission of maltreatment from one generation to the next is automatic and unalterable. This assumption is not only false but may be destructive. As one review explains:

> Uncritical acceptance of the intergenerational hypothesis has caused undue anxiety in many victims of abuse, led to biased response by mental health workers, and influenced the outcome of court decisions, even in routine divorce child custody cases. In one such case . . . a judge refused a mother custody rights because it was discovered during the trial that the mother had been abused as a child. Despite the fact that much of the evidence supported the children's placement with their mother, the judge concluded that the mother was an unfit guardian, since everyone "knows" abused children become abusive parents. [Kaufman & Zigler, 1989]

In determining the actual rate of intergenerational transmission, it is critical to study the problem longitudinally rather than retrospectively. Retrospective analyses invariably show high rates of transmission because almost every adult who seriously mistreats his or her child does, in fact, remember a very difficult and sometimes neglectful or abusive childhood. But these analyses, by definition, omit the victims of abuse who do not themselves become abusers. And there are many, many such people.

On the basis of longitudinal studies that begin before the abused individual becomes a parent, experts estimate that between 30 and 40 percent of adults who were abused as children actually become child abusers themselves, a rate many times that of the general population but much less than that generally assumed to be the case (Egeland, 1993; Kaufman & Zigler, 1993). Those parents least likely to perpetuate the abuse they endured as children are those who subsequently had someone who loved and cared for them, such as the other parent or a foster caregiver in childhood or their spouse in adulthood. In addition, those who are able to remember their maltreatment and understand its effects are much better able to avoid abusing their own children. Indeed, those who deny the abuse they suffered as children are actually more likely to maltreat their own offspring (Egeland, 1993).

intergenerational transmission The phenomenon of mistreated children growing up to become abusive or neglectful parents themselves, a consequence that is less common than is generally supposed.

Treatment and Prevention

As the scope and consequences of child maltreatment have become better recognized, efforts at treatment and prevention have greatly expanded, particularly in the area of public awareness. In most countries worldwide, laws now require the reporting of child maltreatment. In addition, in the United States, several public and private national organizations tally reports of abuse and neglect, monitor treatment, and fund research. Attention by the popular press to the problem has increased dramatically, and a professional journal, *Child Abuse and Neglect*, has been in publication for more than two decades. As a result, professionals and the general public have become much more aware of the problem and are more likely to report it.

Indeed, reporting of maltreatment has increased sharply in the United States. In 1975, close to 1 million American children were reported as being mistreated. In 1985, the number was almost 2 million, and in 1993, it rose to nearly 3 million, with, as we have seen, about a third of those cases being substantiated. (Part of the discrepancy between reported and substantiated cases is attributable to false accusations, but most of it arises from a lack of evidence sufficient to prove credible accusations.) Most experts believe that these increases reflect greater community awareness of the problem rather than a three-fold increase in its incidence. One expert argues, in fact, that a decade or more ago, about 5,000 American children died each year as a result of maltreatment, most of it unreported, and that in the 1990s, the death rate has fallen to about 1,100 per year because more cases of serious abuse are reported, allowing children to be saved before they become fatalities (Besharov, 1993). Nevertheless, underreporting remains high, partly because the public still needs to better recognize the signs of maltreatment, and partly because many people are skeptical that alerting the police or social workers will actually help a maltreated child (Finkelhor, 1993).

Treatment

Reporting, investigating, and substantiating maltreatment, and even punishing the perpetrator, do not necessarily stop the harm (Finkelhor, 1992). Even with documented cases, between a third and a half of all victims experience another episode of maltreatment (Daro, 1989). While great strides have been made in recognizing and defining maltreatment, and in understanding its causes and consequences, researchers and practitioners are still struggling with the application of these findings to treatment and prevention.

One of the major challenges is how to tailor treatment to fit the particular family context (Wolfe, 1994). According to one useful analysis, families involved in maltreatment can be subdivided into four categories: vulnerable-to-crisis, restorable, supportable, and inadequate (Crittenden, 1992).

Vulnerable-to-crisis families are experiencing unusual problems and need temporary help to resolve them. For example, a divorce, the loss of a job, the death of a family member, or the birth of a handicapped infant can severely strain some adults' ability to cope with the normal demands and frustrations of child-rearing. Especially if other relatives or friends are unable to relieve the pressure, the relationship between parents and children may deteriorate to the point of abuse or neglect.

vulnerable-to-crisis families Families involved in maltreatment in which the parents are experiencing unusual problems and need temporary help to resolve them.

About a fourth of all maltreating families fall into this vulnerable-to-crisis category. They are relatively easy to help through services such as crisis counseling and parent training that are already available in some parts of the country. In the majority of cases, once the parents learn to cope with their specific difficulty more effectively, a process that usually takes less than a year, they are once again able to provide adequate child-rearing.

Less easily reached are the **restorable families**, who make up about half of all maltreating families. The caregivers in these families seem to have the potential to provide adequate care, and perhaps have done so in the past, but they have many problems, caused by their immediate situation, their past history, and their temperament, that seriously impair their parenting abilities. A single mother, for example, might have untreated medical problems, inadequate housing, and poor job skills, all fraying against a quick temper, which tends to explode when her child is difficult or disobeys her. Or a binge-drinking husband might periodically beat his children, perhaps with the tacit permission of an overly dependent, isolated wife, who herself may have come from an abusive home. Or a teenage couple might be both immature and addicted to drugs, causing them to disregard their infant's basic needs or to seriously overestimate the baby's abilities.

Treatment with restorable families requires a caseworker who has the time and commitment to become a family advocate, mediating and coordinating various services, finding help for every family member who needs it, and providing essential emotional support. In actuality, however, few caseworkers are trained as, or have the time to be, such advocates. Indeed, one study of child-protection workers found that they spent only 11 percent of their time working directly with families—usually in their offices rather than in the family's home—and that half of the approved treatment plans were not implemented, usually because the caseworker or referral agency was too busy (Crittenden, 1992).

Supportable families, who make up about a fifth of all maltreating families, will probably never be able to function adequately and independently until the children are grown. However, with ongoing support, ranging from periodic home visits by a nurse or housekeeper to special residences that include a variety of services—such as free clinics, day-care centers, recreation programs, social workers, and therapists—these families could meet their children's basic needs for physical, educational, and emotional care. Unfortunately, this range of support services is rarely available to, or affordable by, the families that need them most.

Finally, nearly 10 percent of maltreating families are **inadequate families**, so impaired by deep emotional problems or serious cognitive deficiencies that the caregivers can never meet the needs of their children. For children born into these families, long-term foster care is the best solution.

restorable families Families involved in maltreatment in which the caregivers may have the potential to provide adequate care but are experiencing a combination of current stresses and past deficits that seriously impair their parenting abilities.

supportable families Families involved in maltreatment in which the parents can meet their children's needs only through an extensive array of social services that are not normally available.

inadequate families Families involved in maltreatment in which the parents are so impaired by emotional problems and/or cognitive deficiencies that they can never meet their children's needs.

foster care A legally sanctioned, publicly supported arrangement in which children are removed from their original parents and temporarily given to another caregiver.

Foster Care

Historically, many children have been raised by "foster parents"—usually relatives or neighbors—because their biological parents died or were too poor or too ill to care for them. However, in contemporary society, **foster care** generally refers to a legally sanctioned, publicly supported arrangement in which children are officially removed from their original parents and given to another adult to nurture.

Foster care is controversial. Indeed, foster care has long been stereotyped as inadequate care. This stereotype is partly based on the ideology that, no matter what, children are always better off with their biological parents. It is also partly based on the reality that, compared with children overall, foster children tend to do less well in school and have fewer friends, and are more likely to become delinquents and, later, criminals. However, it is not valid to compare foster children with children overall: foster children should be compared instead with children who are left in the care of severely abusive or neglectful biological parents. When this more valid comparison is made, foster children come out ahead. In fact, even though children usually enter foster care with behavioral problems, they are far less likely to be abused by their foster parents than they were by their biological parents. Less than 1 percent of all substantiated abuse occurs in foster care (McCurdy & Daro, 1994).

Research over the past decade has demonstrated that, while it is not good for children to be moved frequently from foster family to foster family, consistent long-term foster placement is preferable to many other arrangements. It is better than allowing a child to remain in a severely abusive family, or permitting a child to be shunted from relative to relative, or, even worse, letting a child be put through a series of foster placements interspersed with stays with an unsupportable or unrestorable biological parent (Fein, 1991; Widom, 1991). Indeed, some foster children do very well, catching up on missed education, learning how to respect themselves and others, and eventually becoming good, nonmaltreating parents.

A positive outcome is especially likely when the foster parents are committed to the child and are provided with the ongoing resources needed to address whatever special emotional, physical, and social needs the child with a history of inadequate care might have (Barth et al., 1994). Those children who fare worst with foster care tend to be those who have endured such severe and extensive maltreatment in their biological family—and hence have such low self-esteem, such impaired social skills, and so much anger—that they would encounter difficulty no matter where they were raised.

Although foster care is sometimes the best solution, it is still the treatment of last resort, to be used when other methods to help a family fail. Unfortunately, decisions made about foster care are too often based on bureaucratic expediency, and sometimes on insensitivity to variations in

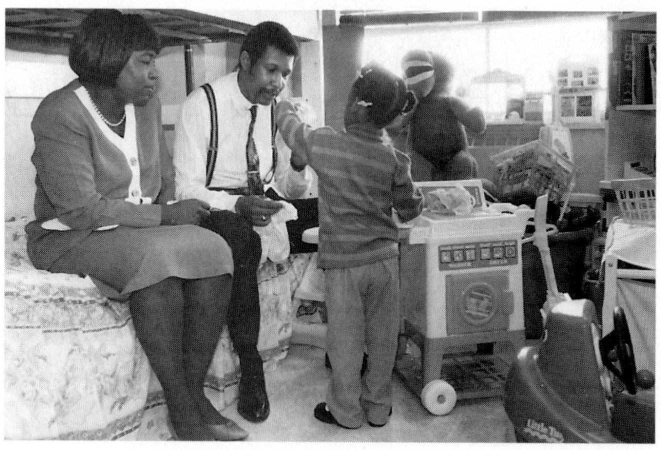

Foster care works well as a response to a crisis of short duration—a few weeks or less—or as a permanent solution to an inadequate home. Unfortunately, many foster families are in limbo between the two. Three-year-old Sharmaine has been with her foster parents, shown here, since she was placed with them as an agitated, inconsolable, crack-exposed newborn. She developed into a happy, spunky daughter until at age 2½, a court ordered her to begin weekend visits with her biological mother and four older siblings in preparation for her eventual return to them. Before each visit, Sharmaine holds on tightly to her father and says, over and over, "Daddy, I don't want to go," and when she returns, her preschool teachers report that she becomes clingy, insecure, and sad. Such children, torn between two opposing principles, number in the hundreds of thousands in the United States.

family and community standards and values (Lindsey, 1991; Pinderhughes, 1991). Many times a child is hastily placed in foster care as a temporary solution, meant to last only until the parents correct whatever problems led to the child's removal. Such temporary placements work against the family's ever improving, because once a child is removed, fewer services—material or psychological—are provided to the biological family than would have been provided if the child were still at home (Lindsey, 1991). The result is that temporary foster care stretches into an uncertain future, and if the child is returned to the biological family, the family typically is neither restored nor sufficiently supported. Thus the child is still vulnerable to another cycle of abuse and neglect.

Prevention

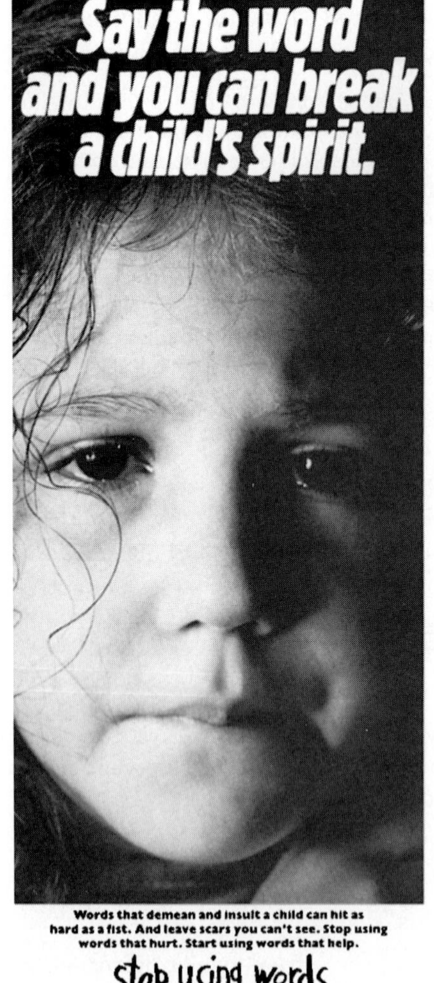

Say the word
and you can break
a child's spirit.
Words that demean and insult a child can hit as hard as a fist. And leave scars you can't see. Stop using words that hurt. Start using words that help.
Stop using words
that hurt.
For helpful information, write National Committee for Prevention of Child Abuse, Box 2866E, Chicago, IL 60690
Ad Council

Public-information campaigns are part of the essential education necessary to begin to prevent abuse and neglect. The more people recognize the many faces of abuse, the fewer the number of children who will suffer.

The most important goal in dealing with child maltreatment is, obviously, to prevent it before it occurs. This is a daunting task that must take into account a wide array of contextual factors, including the cultural values and practices surrounding the family at risk, the various sources of community stress and support, the family's economic status and degree of social isolation, as well as factors within the particular child and the particular parent that increase the child's vulnerability to maltreatment (U.S. Advisory Board on Child Abuse and Neglect, 1993).

A broad variety of programs attempt to do this. As part of their national health care, many developed countries, England and New Zealand among them, provide a network of nurses and social workers who visit all families with young children at home, to encourage good health practices and to screen for potential problems, providing referrals as necessary (Kamerman & Kahn, 1993). One such program, an ambitious state-funded effort in Hawaii titled Healthy Start, has shown highly promising results. Begun as a small demonstration project in 1985, this program now makes home visitation available to the majority of Hawaii's civilian population (Breakey & Pratt, 1991; Hawaii Department of Health, 1992). Parents participate in home visitation voluntarily, and may continue with the program until their child reaches age 5. During regular visits, trained home visitors provide emotional support, model positive parent-infant interaction, and help the mother get in contact with health-care providers and other community agencies who can offer further assistance. The preliminary evaluation of this program has been astonishing: of the 1,204 high-risk families served by Healthy Start between 1987 and 1989, there were only three cases of abuse and six cases of neglect reported to child-protection caseworkers—indicating a less than 1 percent rate of maltreatment in a high-risk segment of the population that usually averages 18 to 20 percent (Hawaii Department of Health, 1992).

Many other nations and U.S. states provide a patchwork of programs, such as parent-newborn bonding sessions, high school classes in parent education, crisis hotlines,* respite care, drop-in centers, special training for teachers and police officers on recognizing abuse and neglect, programs to educate children about sexual abuse, and so on (Willis et al., 1992).

*A national hotline, 1-800-422-4453, or 1-800-4 A CHILD, is open in the United States day and night for questions and problems related to child maltreatment of any kind.

While some early-intervention programs show promise, research has generally revealed only suggestive evidence about what works. Finding out what is truly effective in which specific contexts is crucial, because, as one reviewer of early-intervention programs notes, "the world of public policy is crowded with examples of results that are unexpected and unwanted by-products of well-meaning legislative or administrative initiatives" (Gallagher, 1990). Among these by-products are the dangers of wrongfully stigmatizing certain families as inadequate, of undermining family or cultural patterns that, contrary to conventional wisdom, nurture children, and of creating a sense of helplessness in the family instead of strengthening its self-confidence, skills, and resourcefulness.

Although there is no proven panacea for child maltreatment, every effort must be made to prevent it. Indeed, while good prevention programs are sometimes criticized as being too expensive, their costs pale in comparison with the costs that arise when nothing is done. In evaluating Hawaii's Healthy Start program—which cost an estimated $7,800 for a full five years of service to a single family in 1993—a federal report put it this way:

> The cost of child abuse includes, but may not be limited to, the costs of the immediate consequences of child abuse, such as hospitalization and foster care. A hospital official in Hawaii said that the cost of hospitalizing an abused child for 1 week would range from $3,000 to $15,000. A Hawaii social services official said that providing foster care for 1 year would cost more than $6,000. Adding the costs of the potential long-term consequences of abuse could raise this amount substantially. For example, the Hawaii program estimates the cost of incarcerating a juvenile for 1 year at about $30,000, the cost of providing foster care to an abused child to age 18 at $123,000, and the cost of institutionalizing a brain-damaged child for life at $720,000. [Government Accounting Office, 1992]

At the broadest level, prevention policies need to take into account the entire social context. Since poverty, youth, isolation, and ignorance correlate with unwanted births, inadequate parenting, and mistreated children, public policies that work to raise the lowest family incomes, discourage teenage pregnancies, encourage community involvement, and increase the level of education of prospective parents may provide the most effective prevention of all.

SUMMARY

Size and Shape

1. During early childhood, children grow about 3 inches (7 centimeters) and gain about 4½ pounds (2 kilograms) a year. Normal variation in growth is caused primarily by genes, health care, and nutrition.

Brain Growth and Development

2. Brain maturation, including increased myelination and improved coordination between the two halves of the brain, brings important gains in children's physical abilities and their higher-order cognition. Preschool children develop quicker reactions to stimuli and are better equipped to control those reactions.

3. Better control of eye movements and improved focusing are associated with brain maturation. Development of the visual pathways also enhances eye-hand coordination. These improvements enable children to begin mastery of basic literacy skills at around age 6.

Mastering Motor Skills

4. Gross motor skills improve dramatically during early childhood, making it possible for the average 5-year-old to perform many physical activities with grace and skill. Fine motor skills, such as holding a pencil or tying a shoelace, improve more gradually over this time. Mastery of drawing skills develops steadily in the preschool period.

5. Along with increased motor skills, and the explorations and adventuring that go with them, is an increased risk of accidents and fatalities. Developmentalists emphasize the idea of injury control to contain this danger, stressing that childhood "accidents" are usually the result of a lack of forethought on the part of caregivers.

Child Maltreatment

6. Child maltreatment can take many forms, including abuse or neglect that may be physical and/or psychological. Child abuse and neglect can have diverse consequences for a child's well-being, with many children suffering several forms of maltreatment. About 1 million cases of child maltreatment are reported and confirmed in the United States each year. Experts estimate the actual number to be considerably higher.

7. The causes of child maltreatment are many, including problems in the society (such as negative or exploitative cultural attitudes about violence and about children), in the family (such as social isolation and inadequate means of coping with financial and social stresses or distorted attitudes), and in the parent (such as drug addiction).

8. The consequences of child maltreatment can be far-reaching, impairing the child's learning, self-esteem, social relationships, and emotional management. However, these problems can usually be treated, and it is not inevitable that maltreated children will become maltreating adults.

9. Once maltreatment occurs, careful intervention must support and restore those families that can be helped and provide stable foster care for the minority of families in which the pattern of maltreatment cannot be halted.

10. The most effective strategies for preventing maltreatment emphasize the ecology in which at-risk families live, attempting to enhance community support and address the material as well as emotional needs of troubled families. They also help parents acquire new coping skills and parenting strategies for caring for offspring more competently. One specific method of accomplishing this is home-visitation programs that offer support and assistance to families at risk.

KEY TERMS

corpus callosum (228)
injury control (230)
child maltreatment (234)
abuse (234)
neglect (234)
intergenerational transmission (241)
vulnerable-to-crisis families (242)
restorable families (243)
supportable families (243)
inadequate families (243)
foster care (243)

KEY QUESTIONS

1. How do the size, shape, and proportions of the child's body change during early childhood?

2. What causes variations in height and weight during early childhood?

3. How does growth in the corpus callosum allow older preschoolers to achieve better physical and cognitive coordination?

4. What are some of the major growth spurts in brain development during early childhood and how do these contribute to cognitive advances?

5. How do gross motor skills and fine motor skills compare in their development during early childhood?

6. What measures seem most effective in reducing the rate of accidents in childhood?

7. How is child maltreatment defined? How common is child neglect compared with child abuse?

8. What are some of the factors in the culture, the community, the family, and the parent that can contribute to child maltreatment?

9. What is the estimated probability that a maltreated child will become a maltreated adult? What other long-term consequences of childhood abuse and neglect are likely?

10. What have been found to be the most effective strategies for preventing child maltreatment?

The Play Years: Cognitive Development

CHAPTER 9

One of the delights of observing children in their preschool years is seeing how they construct and express their growing but often fanciful and subjective understanding of the world around them. We are beguiled by the imaginative magical thinking of young children who chatter away with an imaginary playmate, or wonder where the sun sleeps, or comfort a sad parent by offering a lollipop, or confidently claim that they sleep with their eyes open. At the same time, we are often startled by how easily confused preschoolers can be by events with which they have little experience (such as hearing that Mommy is "tied up" at the office), and by how illogically and intuitively they attribute the causes of common occurrences (such as believing that the moon follows them when they are riding in a car at night). Clearly, their approach to the world is often dictated more by their own subjective views than by the world's reality.

For many years, this subjective feature of young children's thinking dominated developmental analyses of preschoolers' cognitive abilities. Researchers maintained that those abilities were sorely limited by preschoolers' **egocentrism**, that is, their prevailing tendency to view the world and others exclusively from their own personal perspective. However, research has highlighted another important side of preschool thought, a side that is suggested by the following episode between a 2-year-old child and his mother, who has been trying to hold his sweet tooth in check:

(Child sees chocolate cake on table.)
Child: Bibby on.
Mother: You don't want your bibby on. You're not eating.
Chocolate cake. Chocolate cake.
You're not having any more chocolate cake either.
Why? [whines] Tired.
You tired? Ooh!
Chocolate cake.
No chance. [from Dunn et al., 1987]

The young child in this episode is definitely *not* being illogical or oblivious to the constraints of the real world, and shows strategic skill in pursuing his goal—from asking for his bib (a noncontroversial request) to eliciting sympathy by feigning fatigue. Indeed, in countless everyday instances, as well as in the findings of numerous research studies, preschoolers reveal themselves to be remarkably perceptive and insightful thinkers, whose grasp of the causes of everyday events, memory of the past, anticipation of

egocentrism Preschoolers' tendency to view the world and others exclusively from their own personal perspective.

the future, intuiting of human psychology, and mastery of language are sometimes astonishing.

These two sides of cognitive growth during the play years are the theme for this chapter. We will begin our examination of them by discussing the growth of basic intellectual skills like number and memory and considering how preschoolers act like theorists in the development of their interpretations of the world. We will then discuss Piaget's depiction of preoperational thinking—which emphasizes young children's intuitive, illogical ways of reasoning—and Vygotsky's depiction of the child as apprentice—which emphasizes the role of parents and others as mentors. Finally, we will consider the remarkable accomplishments of language development and the cognitive benefits of high-quality preschool education.

How Preschoolers Think

As you learned in Chapter 6, researchers long believed that a baby's world was a "blooming, buzzing confusion"—until new research designs allowed them to look closely at infants' capacities for memory, categorization, and thought. In a similar fashion, developmentalists tended to underestimate the cognitive skills of preschoolers—until new research strategies allowed them to go beyond preschoolers' initial intuitive responses to experimental queries and to more closely probe what young children know and understand about the world. Such research has led to a new appreciation of preschoolers' cognitive abilities, many of which begin to emerge earlier than had been thought. At the same time, developmentalists have come to recognize that the emergence of these abilities is partial and fragile—evident in some contexts but not in others. As we will see throughout this chapter, developmentalists have also recognized that how early children attain certain abilities, and how clearly they are able to demonstrate them, can be affected by the intellectual guidance and support children receive from others.

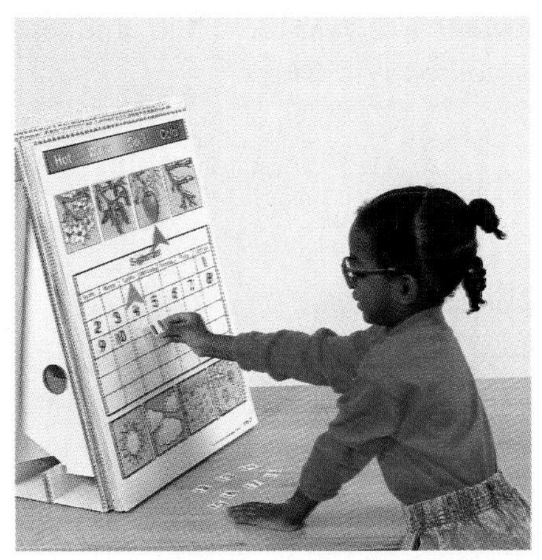

The day, the date, the season, and the weather are all concepts that are part of the curriculum of a good preschool. Young children's ability to grasp these concepts—as well as to develop an understanding of number—is a good deal stronger than researchers or educators once imagined.

Number

Until recently, many developmental researchers believed that the growth of any meaningful numerical understanding was a school-age accomplishment. Indeed, only two decades ago, scientists "denied the preschooler any arithmetic prowess at all" (Gelman & Massey, 1987). In the past few years, however, researchers have come to recognize that preschoolers possess number concepts that are actually fairly sophisticated. As you remember, beginning in early infancy, children have some perceptual awareness of small quantities (such as noticing the difference between two and three objects). At about age 2, children begin to use numbers symbolically, by connecting words with the items they are counting. At first, the connection between numbers and quantity is quite tenuous. Between the ages of 2 and 3, children are likely to omit numbers from their count ("One, two, four, six . . ."), count the same item more than once, and/or leave some items out of their count. By age 4, however, they demonstrate a basic grasp of several key principles, including the *stable-order principle* (when counting, numbers must be said in a fixed order), the *one-to-one principle* (each item being counted gets one number, and only one), and the *cardinal principle* (the last number in a count represents the total).

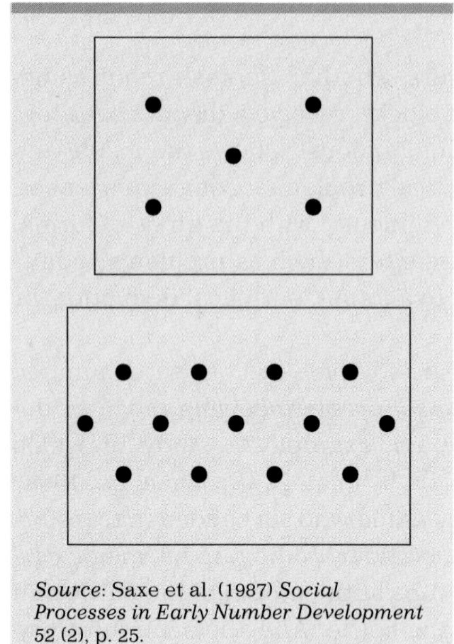

Source: Saxe et al. (1987) *Social Processes in Early Number Development* 52 (2), p. 25.

FIGURE 9.1 In the experimental game of "count the dots," most 2½-year-olds made several errors, even on the five-dot display. But some 4-year-olds did both tasks perfectly. They had already learned the most basic of number rules: count each object once, and only once.

This progression is illustrated by a study (Saxe, 1987) in which 2½- and 4-year-olds were presented with two displays of dots, one of five dots and the other of thirteen, and were asked "to count the dots and touch each dot as you count it" (see Figure 9.1). The younger children seemed to understand the task but typically failed to touch at least two dots in the five-dot display and missed an average of eight dots in the thirteen-dot display. They also miscounted, some saying only "one, two, three" for both the five- and the thirteen-dot display. Most of the 4-year-olds performed perfectly on the five-dot display and made an average of only two mistakes in counting and only two in pointing on the thirteen-dot display. Indeed, many of the 4-year-olds did the thirteen-dot task with no errors.

What contributes to the child's developing understanding of number? One factor is simple maturation, involving not only the underlying brain development that is ongoing throughout childhood but also the emergence of language and the ability to use it to conceptualize and express number. Improved number ability also involves maturation in the sense emphasized by Piaget—that is, the flowering of the child's innate curiosity and exploration of the world of objects, which entails spontaneous organizing and counting during the preschool years in much the same way it involves sucking and fingering during the sensorimotor years.

The overall cultural context, especially the importance the particular culture places on number competence, also plays a role. In societies like the United States, the child's mathematical flowering has a rich environment to sustain it. Preschoolers typically

> accompany their parents on shopping trips; hear numbers used in talk about time, birthdays, and how many presents they will or will not get; and ride elevators in buildings with many floors. . . . They watch the Count on the television program "Sesame Street" talk about his passion for counting different size sets, or Señor Zero looking for nothing so he can count zero things, or even a puppet dressed up in a black leather jacket singing "Born to Add" set to the tune of a popular rock and roll song. [Gelman & Massey, 1987]

In some cultures, another factor that may promote preschoolers' number competence is the structure of the particular language. Indeed, one hypothesis for the overall superiority of East Asian children over European and American youngsters in math is that languages such as Japanese, Korean, and Chinese are much more logical in their labeling of numbers, for instance, making "eleven, twelve, thirteen," and so on, the equivalent of "ten-one, ten-two, ten-three" (Fuson & Kwon, 1992). This linguistic structuring advances young children's intuitive grasp of their number system as soon as they begin to talk.

A final factor in building number competence—a factor we will emphasize in connection with other cognitive abilities as well—is the structure and support provided by parents, other adults, and older children. Parents and offspring frequently use numbers together—counting small quantities, playing number games ("one, two, button my shoe . . ."), pushing television-channel buttons, sorting coins, and measuring small amounts, such as the allotted spoonfuls of cocoa mix to put in a glass of milk. By providing this kind of guided participation in shared activities involving number, parents ensure that the tasks remain challenging to their children but within their capabilities, and help to stimulate the growth of numerical understanding. As we shall see now, this same type of guided participation can help the growth of problem-solving skills.

Problem Solving

Preschoolers are everyday problem solvers, whether the task requires figuring out how to construct a castle out of blocks, resolve a dispute over toys with a friend, or create a picture that simultaneously shows the inside and outside of one's house. Despite their practical problem-solving experiences, however, preschoolers have considerable difficulty with the kinds of formal problem-solving tasks devised by experimenters, such as problems requiring the use of analogy from one situation to another, or logical deduction, or step-by-step planning.

Like most preschoolers, this boy is obviously falling for the old "Look, I broke my finger!" trick. Because of his limited ability to think logically, he may realize that his grandfather's finger is not "really" broken but not be able to figure out how his grandfather takes his finger off and on. It must be magic!

Nevertheless, researchers have learned that, just as with number, some rudimentary skills related to "formal" problem solving begin to develop during the play years. Consider, for example, a study in which preschoolers were asked to solve a problem by analogy (Brown et al., 1986). The children were first told a story about a genie who succeeded in transferring some jewels across a wall and into a bottle by rolling up his magic carpet to make a tube, and then placing the tube at the mouth of the bottle and rolling the jewels through the tube into the bottle. After hearing the story, the children were presented with a problem involving an Easter-bunny puppet who wanted to transport some eggs across a narrow stream and into a basket. The puppet had available several "tools" to solve the problem, including scissors, sticks, and large sheets of heavy paper. On their own, very few 3-year-olds were capable of generalizing what they had heard in the genie story to the rabbit problem, but some 5-year-olds could do so. However, with the assistance of an experimenter who drew the children's attention to the similarities between the story and the problem, and between the genie's magic carpet and the bunny's sheets of paper, children of both ages could solve the problem.

Like the research on number understanding, the results of this experiment suggest that supportive guidance provided by an adult can facilitate the development of cognitive skills in preschoolers. This was further illustrated by a study in which young children were given a sorting task that involved helping a puppet move into his new home by putting doll-house furniture into the proper rooms (Freund, 1990). The children were randomly divided into two groups and given a practice session: one group worked with their mothers on the sorting task; the other group sorted and installed the furniture by themselves, in the company of an experimenter who corrected any errors by properly relocating misplaced furniture while the children watched. When the children later sorted the furniture entirely on their own, the researchers discovered that those who had worked with their mothers sorted better than those who had only received corrective feedback from the experimenter. It was not hard to see why: mothers carefully structured the task for children based on the child's age, as well as the difficulty of the sorting, to keep the demands on the child challenging but manageable. Mothers of 3-year-olds, for example, offered more concrete, specific suggestions to their offspring ("Here is a bed—it goes in the bedroom") than did the mothers of 5-year-olds, while the latter talked more often about planning and the goals of the task ("Let's move in all the furniture that goes in the living room first"). In short, mothers gave their children as much responsibility for problem solving as they thought their youngsters could handle, while providing an overall structure for the children that promoted success.

The importance of parents' providing cognitive structure and guidance is especially clear when we look at memory, another aspect of preschoolers' cognitive capacities that, while marked with notable lapses, has also been underestimated.

Memory

Preschoolers are notorious for having a poor memory, even when compared with children only a few years older. Ask a school-age child what he or she did during the day and you are likely to get a detailed accounting, complete with reflections about why people acted as they did and how their behavior relates to their actions in the past. Ask a preschooler the same question and you are likely to get a quizzical stare, or a noncommittal "Nothing," or a string of seemingly irrelevant details.

However, it is not that preschoolers have deficient memory *per se*. Rather, they have not yet acquired the skills (described in full in Chapter 12) for deliberately storing memories of past events and efficiently retrieving these memories as needed on later occasions. Just as you know how to remember information that you will later need—whether it involves rehearsing an address or a telephone number, organizing your course notes so that one item calls to mind another, or relating a new experience to some past memory to keep it in mind—you also have strategies for retrieving this information when you need it, say, to perform well on a test or to recount the day's events.

Preschoolers, by contrast, are strikingly nonstrategic in their memory: they rarely try deliberately to retain some experience or bit of information in memory, and they seldom know precisely how to recall some hard-to-retrieve bit of past experience (Kail, 1990). As a result, they can sometimes appear strangely incapable of remembering past experiences that older children can recall with ease.

But this does not tell the whole story about memory in young children. In other ways, young children are remarkably capable of storing in mind a representation of past events that they can later use. One way they do so is by retaining **scripts** of familiar, recurrent past experiences. These scripts act as a kind of structure or skeletal outline that facilitates the storage and retrieval of certain memories. By age 3, for example, children can tell you what happens in a restaurant (you order food, eat it, and then pay for it), at a birthday party (you arrive, give presents, play games, have cake and ice cream, and sing "Happy Birthday"), during their bedtime routine (first a bath, then a story, and then lights out), and during other everyday events (Nelson, 1986). Here is one 5-year-old's description of what happens during grocery shopping:

> Um, we get a cart, uh, and we look for some onions and plums and cookies and tomato sauce, onions, and all that kind of stuff, and when we're finished we go to the paying booth, and um, then we, um, then the lady puts all our food in a bag, then we put it in the cart, walk out to our car, put the bags in the trunk, then leave. [quoted in Nelson, 1986]

scripts Skeletal outlines of the usual sequence of certain common recurrent events. Young children use such scripts to facilitate the storage and retrieval of memories related to specific episodes of these events.

This is a typical script in two key respects: it presents a correct sequence of events and it also recognizes the causal flow of events—reflecting an awareness that some events (like putting food in the cart) must precede other events (like going to the "paying booth") (Bauer & Mandler, 1990; Rat-

ner et al., 1990). Preschoolers use scripts not only when recounting familiar routines but also when they engage in pretend play, enacting everyday events such as dinnertime, shopping, or going to work. They may also use scripts in their telling, or retelling, of a story, building the story around their scripts for familiar routines (Nelson & Hudson, 1988).

The fact that preschoolers have scripts for events such as birthdays is especially evident when someone "violates" the script. If this birthday-boy's sister (on his right) had blown out the last candle rather than merely pointing to it, he might have exploded into angry tears.

Scripts are important to early memory development because they provide a framework of general understanding of common events within which memories for specific experiences can be recalled. However, reliance on a routine script to recall a particular event may result in an account that is incomplete or faulty. In such cases, the child's account may reflect knowledge of the relevant script much more than it does an accurate remembrance of the event in question, especially when the event was unusual, complex, or difficult to understand (Farrar & Goodman, 1990; Fivush & Hamond, 1990). A 2½-year-old's recollection of a recent camping trip, for example, might typically begin with a mention of some notable event, such as sleeping in a tent, and then revert to a familiar domestic script, such as "First we eat dinner, then go to bed, and then wake up and eat breakfast" (Fivush & Hamond, 1990). Thus, scripts are an aid to memory, but they can also impair recall of specific experiences.

Another reason young children sometimes appear forgetful is that they often do not attend to the features of an event that an older person would consider pertinent (Bjorklund & Bjorklund, 1992). Every parent has had the experience of taking a preschooler to a memorable event (a circus, a baseball game, a play), only to find that the child's later account of that experience focused on the ticket-taker, the person sitting in the next row, or the refreshments! To young children, the features of an experience that are most important, and memorable, may depend on what momentarily grabs their attention, or on what their mood is. When observing an event entails complexity—such as watching the right players at the right time during a baseball game, or following the sequence of events in a play—preschoolers may be especially inattentive to the things adults expect them to remember.

Research has shown, however, that parents and other caregivers can play a special role in helping children to attend to, and remember, aspects of their experience that may otherwise escape notice. Consider the following conversation between a mother and her 2-year-old daughter after a day at the zoo.

Mother: Brittany, what did we see at the zoo?
Brittany: Elphunts.
That's right! We saw elephants. What else?
(Shrugs and looks at her mother.)
Panda bear? Did we see a panda bear?
(Smiles and nods.)
Can you say "panda bear"?
Panda bear.
Good! Elephants and panda bears. What else?
Elphunts.
That's right, elephants. And also a gorilla.
Gorilla! [quoted in Bjorklund & Bjorklund, 1992]

Conversations like this one are part of the everyday exchanges between parents and offspring that review, reconstruct, and consolidate a young child's memory of the day's events. Many parents use strategies that not only refresh the child's memory but also provide lessons in memory retrieval (Fivush & Hamond, 1990; Hudson, 1990). Parents commonly assist preschoolers' recollections by asking specific questions ("What animals did you see climbing in the tree?") rather than general ones; by reviewing events in their temporal sequence ("First we saw the monkeys, and *then* what did we see?") rather than out of sequence; and by providing children with memory cues ("And then we had lunch at Burger . . ." Child: "King!"). Each of these approaches not only aids young children's memory search but also provides a model of how to better recall past events and experiences.

Parents, of course, vary in how they guide young children's recall of past events. Some parents ("repetitors") tend to ask children specific questions to cue their memory search, and then repeat the child's answer en route to the next inquiry. Other parents ("elaborators") not only do these things but also supplement the child's recall with additional information about the experience, in a sense building additional memories into those the child can already recall (Reese & Fivush, 1993). Research suggests that elaboration is the more effective approach to developing memory retrieval. One study, for example, found that 2-year-olds whose mothers used an elaborative style remembered more and could better answer questions about a prior experience—whether they were asked by their mother or an experimenter (Hudson, 1990). The elaborative style apparently not only aided young children's memory retrieval but also helped to consolidate a more complete account of the shared experience between parent and child.

Memories of Mickey Mouse

Despite the many limitations in young children's memory, developmental researchers have recently discovered that, under certain circumstances, children show surprising evidence for their long-term retention of early

Structured Interview Questions

1. Open-Ended Questions
   I know that you remember a lot about your trip to Disneyworld. I've never gone there before. Can you tell me about Disneyworld?
   What was the very first thing that happened?
   And then what?
2. Directive Questions
   How did you get to Disneyworld?
   Who went with you?
   Where did you stay?
   What did you see at Disneyworld?
   What did you think about that?
   Who did you see there?
   What was that like?
   What rides did you go on at Disneyworld?
   Which one did you like the most?
   What did that feel like?
   What did you like/dislike about it?
   What rides did "X" (other people there) go on?
   What did "X" think of it?
   Did "X" like it?
   What did "X" like about it?
   Did you eat anything there?
   What did you eat?
   Did you buy anything at Disneyworld?
   What did you buy?
   Did anything bad happen at Disneyworld?
   What happened?
   Was there anything you didn't like?
   What didn't you like?
   What was your favorite fun thing at Disneyworld?
   What did you like about it?
   If you got to go to Disneyworld again, what would you want to do the most?

Source: Hamond and Fivush, 1991.

FIGURE 9.2 Preschool children can remember in much more specific detail when asked directive questions than when asked open-ended ones. In fact, after 3- to 6-year-olds had nothing else to say in response to the open-ended questions in this interview about their visit to Disneyworld, they produced, on average, four times more information in response to the directive questions.

experiences. In one particularly revealing study (Hamond & Fivush, 1991), researchers assessed the memories of forty-eight children who had visited Disneyworld at about their third or fourth birthday (specifically, between ages 33 and 42 months or between ages 43 and 54 months). The children were interviewed in their homes, either six months or eighteen months after their Disneyworld visit.

> The experimenter asked each child a structured series of questions about their Disneyworld experience, the first of which was open-ended: "Can you tell me about Disneyworld?" After this question, the experimenter asked a standard series of questions focusing on who went, what rides were ridden, what sights were seen, what was eaten, what presents were bought, and how children felt about their experiences. [The standardized interview is shown in Figure 9.2, at left.] When necessary, these questions were followed with nondirective prompts (e.g., "And what else?"; "And then what?"; "Tell me more about that") designed to elicit further information about the Disneyworld experience.

All the children responded well in their interviews, recalling an impressive amount of information, virtually all of it confirmed as accurate by the parents. Surprisingly, although the older children recalled slightly more information than the younger ones, the age-related differences in the amount of recall were minimal, as were the differences related to the length of time between the Disneyworld visit and the interview. In other words, the children remembered similar amounts of information irrespective of how old they were when they visited Disneyworld, how much time had passed since their visit, and how old they were when they were interviewed!

The most notable finding was that all the children provided much more information in response to directive questions than they did spontaneously. In fact, across both age groups and both retention intervals, nearly 80 percent of all the information elicited from the children came in response to directive questions.

Overall, this study confirms and amplifies other recent research, strongly suggesting that when recollecting personally meaningful material, "even quite young preschoolers can recall a great deal of information if given appropriate cues and prompts" (Hamond & Fivush, 1991). This conclusion is particularly helpful in understanding why preschoolers' memory abilities appear to be so erratic: some of the time, at least, their seemingly vague memories may merely by the result of vague questioning. (The importance of appropriate questioning to young children's ability to provide reliable eye-witness testimony is explored in Public Policy, pp. 257–258.)

Theory of Mind

As we have seen, in a number of areas preschool children have greater competence than developmentalists had previously believed. One additional area in which children sometimes show surprising competence is social understanding. Until fairly recently, most researchers maintained that preschoolers' social understanding is severely limited by their egocentrism, which leads them to see everything from their own point of view and in terms of their own feelings. As a result of recent findings, however, developmentalists now recognize that young children quite frequently are able to transcend this limitation and see things from someone else's perspective.

PUBLIC POLICY

Young Children as Eyewitnesses

Until quite recently, young children in most countries were prohibited from providing courtroom testimony. This was because it was assumed that they are too suggestible and too likely to confuse fact and fiction to provide a truthful and accurate account of stressful events they have experienced or witnessed.

However, as a result of the increased awareness of the prevalence of child maltreatment, including sexual abuse, children in the United States and many other countries are now allowed to testify in court after a judge has determined that they are capable of doing so truthfully and accurately. This judicial acceptance of children's testimony seems essential since many children are the only witnesses to their own devastating abuse. At the same time, critics of this change argue that if children are automatically assumed by the legal system to "provide a truthful and accurate account," they are open to being manipulated by adults for their own ends, such as those of parents in custody battles, or those of overzealous authorities handling alleged abuse cases involving day-care centers (where, in fact, abuse is actually far less likely to occur than it is in homes).

In light of this controversy, what can developmentalists tell the courts about the reliability of children's eyewitness testimony? As the text discussion points out, recent research has shown that, while decidedly uneven in their memory abilities, young children often retain accurate memories of past events and, under the right circumstances, can retrieve them accurately. However, this particular research does not answer several key questions relevant to courtroom testimony: How accurate are the reports provided by children immediately after a stressful event they have witnessed or experienced? Does the accuracy of their reports change over time? How susceptible are young children to leading questions and misleading information that may be part of an investigative interview?

Because researchers cannot ethically subject children to undue stress in order to answer these questions, they have devised methodologies that make use of stressful events that normally occur in a child's life, such as a medical inoculation or a dental exam (Baker-Ward et al., 1993; Goodman et al., 1991; Saywitz et al., 1991). In one series of such studies, children between the ages of 3 and 6 were videotaped while undergoing a medical examination that included, among other things, a DPT inoculation administered by a nurse (Goodman et al., 1990).

> The children's reactions varied widely. Most looked frightened, but some were quite stoic, relatively unfazed, and said "It didn't hurt." Others, however, became nearly hysterical. These children had to be physically restrained, often by two or three people. They cried, screamed, yelled for help, tried to run out of the room, and sobbed afterward while complaining that it hurt. In sum, they reacted as if they were being attacked.

As the only eyewitness to the slaying of a playmate, 4-year-old Jennifer Royal was allowed to testify in open court. Her forthright answers, and the fact that she herself had been wounded, helped convict the accused gunman.

Several days after this event, the children were asked first to tell about the experience, then to answer various questions about it, and finally to look at a photo lineup and identify the nurse who had administered the shot. None of the children offered any false information during the free recall, and, contrary to the concern that emotional arousal might scramble a child's memory, those who showed the most distress during the exam were the ones who provided the most detailed, accurate accounts.

When asked specific questions, all the children were quite good witnesses, particularly about what did, and did not, happen. Notably, none of the children answered "yes" to any of the following four questions: "Did she hit you?" "Did she kiss you?" "Did she put anything in your mouth?" "Did she touch you any place other than your arm?" However, although the children were very clear about what had been done to them, they were less sure about exactly who had done it: on the photo lineup, only half picked the nurse's photograph, while 41 percent picked other photos and 9 percent said they couldn't remember.

The next step in this experiment was to determine the durability of the children's memories of their stressful experience with a stranger. When the children were interviewed again a year later, their overall recall had diminished, but again they reported virtually no significant false memories. For example, none of the children answered that the nurse had hit them, kissed them, or put something in their mouth. The one long-term memory test that most children failed was once again the photo-

(continued on next page)

Young Children as Eyewitnesses *(continued)*

identification task, with only 14 percent correctly identifying the nurse this time around. Most of the rest said they didn't remember what the nurse looked like, and 32 percent picked the wrong photo.

Following up on their findings of impressive long-term recall for a stressful event, these researchers developed several experimental variations that tested whether children's memories can be deliberately distorted. In one, the initial interview after the medical check-up included some questions that were purposely phrased to be misleading ("She touched your bottom, didn't she?" "How many times did she kiss you?" and so on), asked either in a friendly, encouraging manner or in an intimidatingly stern one. The older children (ages 5 to 7) were rarely influenced by these misleading questions, affirming them less than 9 percent of the time whether the interviewer was friendly or stern. The 3- and 4-year-olds, however, were more vulnerable to the adult's tone: although those responding to the friendly questioner affirmed the interviewer's false assumptions only 10 percent of the time, those responding to stern questioning affirmed them 23 percent of the time.

These results accord with other research that finds that, particularly for young children, the social context (including the relationship of the child to the questioner, the age of the questioner, and whether the atmosphere of the interview is intense or relaxed) has a substantial influence on answers offered to memory questions (Baker-Ward et al., 1993; Ceci et al., 1990; Rogoff & Mistry, 1990). At the same time, this research also corroborates other findings showing that, under a variety of circumstances, the great majority of children resist suggestive questioning.

But what happens if a child is not simply misled by questions but is deliberately given false information? This was the topic of another study that employed a medical-exam format, this time with 5-year-olds (Bruck et al., 1995). Immediately after they had been examined and inoculated by a pediatrician, each child met Laurie, who had been a bystander during the inoculation. Laurie, who actually was a research assistant in the study, chatted with the child and then, regardless of what the child's actual reaction to the shot had been, randomly responded to each child in one of three ways: (1) she indicated that the child had been brave; (2) she indicated that the child had been appropriately distressed; or (3) she did not mention the shot at all.

One week later the children were interviewed by another researcher, who asked how they had reacted to their shot. The questioning was straightforward, and the children accurately reported how much they had cried, regardless of what they had been told by Laurie. Then, about a year later, the same children went through a series of three interviews about their medical exam. This time some of the questions they were asked contained misleading information about the details of the exam (such as "When Laurie gave you the shot, was your mom or dad with you?") and about what their reaction to the shot had been. In the final interview, many of the children appeared to have been influenced by the misleading information they had heard in the previous interviews. For example, of the children who had been inaccurately "reminded" that the research assistant had given them the shot, more than one-third reported, on their own, that Laurie had given them the inoculation. Likewise, children who had been told they were "brave" during the inoculation (some of whom had actually been quite distressed) reported less crying than children who weren't given this information. Even more interesting, children who had incorporated misinformation into their account altered other details to make their account consistent. For instance, many of the children who reported that Laurie had administered the inoculation also reported that she had performed other parts of the medical examination as well, such as checking their eyes and nose.

What, then, can we conclude from the varied findings of all this research? As one pair of researchers explain:

> Children are neither as hyper-suggestible and coachable as some pro-defense advocates have alleged, nor as resistant to suggestions about their own bodies as some pro-prosecution advocates have claimed. They can be led, under certain conditions, to incorporate false suggestions into their accounts of even intimate bodily touching, but they can also be amazingly resistant to false suggestions and able to provide highly detailed and accurate reports of events that transpired weeks or months ago. [Ceci & Bruck, 1993]

The research also makes clear that, when children are required to give eyewitness testimony, they should be provided a context that enhances their ability to remember accurately. They should be interviewed by a neutral professional, who encouragingly probes but does not lead in the questioning process, with the interview being videotaped for later use. Or, if testimony must be repeated during the trial, it can occur on closed-circuit television, to spare the child from confronting the accused—a confrontation that can trouble the child for months afterward and does nothing to ensure accuracy (Goodman et al., 1992). Such practices have already been accepted by the United States Supreme Court as fair to defendants, but many states and municipalities do not yet allow these child-friendly processes, and many lawyers still argue that young children are invariably confused and unreliable. The evidence is otherwise: children are not necessarily worse witnesses than adults, as long as their vulnerability to intimidation is fully taken into account.

theory of mind An understanding of mental processes, that is, of one's own or another's emotions, perceptions, and thoughts.

Human emotions, motives, thoughts, and intentions are among the most complicated and thought-provoking phenomena in a young child's world, whether the child is trying to understand a peer's unexpected display of anger, determine whether a sibling will be generous or selfish, or persuade a parent to purchase a desired toy. As a result of their experiences with others, preschoolers develop informal theories about human psychology that attempt to answer basic questions about mental phenomena, such as how a particular person's knowledge and emotions affect that person's actions, and how particular people can differ so markedly in their thoughts, feelings, and intentions, often in response to the same situations. In other words, young children acquire a **theory of mind** that reflects their developing concepts about human mental processes (Frye & Moore, 1991; Wellman, 1990; Whiten, 1991).

Indeed, developmentalists have discovered that young children are not only aware of divergent psychological perspectives but strive hard to understand how, and why, different viewpoints exist. Two-year-olds, for example, have been observed to say things such as "Don't be mad, Mommy," "Mama having a good time?" and even, from one precocious 28-month-old boy, "Maybe Craig would laugh when he saw Beth do that" (Bretherton & Beeghly, 1982). Each of these statements reveals a nonegocentric awareness that other people can have emotions that are not identical to one's own.

By age 3 or 4, young children clearly distinguish between mental phenomena and the physical events to which they refer (for example, you can pet a dog that is in front of you, but not one that is in your thoughts); they appreciate how mental states (like beliefs, expectations, and desires) arise from one's experiences in the real world; they understand that mental phenomena are subjective (others cannot "see" what you are imagining); they recognize that people have differing opinions and preferences (someone might like a game that you dislike); and they realize that beliefs and desires can form the basis for human action (Dad is driving the car fast because he doesn't want to be late for Grandma's dinner) (Flavell et al., 1993; Wellman & Gelman, 1992). They also understand that emotion arises not only from physical events but also from one's goals, expectations, and other mental states: a 4-year-old, for example, might eat lunch by himself at day care to avoid his friends' annoying request to share his dessert (Stein & Levine, 1989).

At age 3, Brittany has already had almost a year of being a big sister to Brian. As a result, her theory of mind—especially with respect to what behaviors please or irritate her baby brother, and which of her behaviors with her brother please or irritate her parents—has advanced considerably.

A preschooler might be told that this is really a person inside a Barney costume, but nonetheless believe that the person becomes Barney once the costume is on. During the play years, the relationship between appearance and reality is a tenuous, tricky one.

An important advance in preschoolers' theory of mind occurs when they realize that mental states may not accurately reflect reality, and that people can be mistaken or fooled. This concept is especially difficult for young preschoolers to grasp when they have themselves been deceived. Consider this example. An adult shows a 3-year-old a candy box and asks the child what is inside. The child says, naturally, that there is candy. But, in fact, the child has been tricked:

> Adult: Let's open it and look inside.
> *Child: Oh . . . holy moly . . . pencils!*
> Now I'm going to put them back and close it up again. [does so]
> Now . . . when you first saw the box, before we opened it, what did you think was inside it?
> *Pencils.*
> Nicky [friend of the child] hasn't seen inside this box. When Nicky comes in and sees it . . . When Nicky sees the box, what will he think is inside it?
> *Pencils.* [adapted from Astington & Gopnik, 1988]

Three-year-olds have considerable difficulty grasping that one's subjective understanding can be different from the way things "really and truly" are—and consequently, when they learn that they have been mistaken, they not only change their mind but also think that they have *always* held the view that they now know to be correct. They even think that others, like Nicky, will intuitively know the "real case."

At age 4, when the brain undergoes the growth spurt described in the preceding chapter, children's theory of mind takes a leap forward (Fischer & Rose, 1994). They begin to grasp the distinction between objective reality and subjective understanding. When they discover that the candy box is filled with pencils, for example, they not only acknowledge that they were earlier mistaken, but they take considerable delight in the prospect of their friends' being similarly fooled.

The growth of children's theory of mind has broader implications for social understanding. As they begin to grasp how a person's thinking can be influenced by past experiences and other people's opinions, as well as how a person's thinking may affect his or her behavior, older preschoolers become far more capable of anticipating and affecting the thoughts, emotions, and intentions of others (Astington, 1993; Flavell et al., 1995). Not surprisingly, this conceptual growth quickly becomes enlisted for various practical purposes, such as persuasion ("If you buy me a TV for my room, Mom, then I won't be fighting with Susie over what to watch anymore!"), sympathy (consoling a sad friend by reminding her of an upcoming birthday party), and teasing (telling an older brother, who is a Shaquille O'Neal fan, that "Shaq would never wear such a stupid-looking T-shirt!"). Preschoolers' understanding of human psychology is rudimentary, of course, but the theory of mind they begin to develop during the play years provides the foundation for the more sophisticated understanding they will develop in the years to come. Simpler though they are, young children's efforts as psychological theorists are sometimes not very much different from those of adults.

We have thus far looked at recent research on preschool thought, particularly in four areas—number, problem solving, memory, and social understanding. In every case, preschoolers show both surprising skill and

notable naiveté. Now we will consider the theories of Jean Piaget and Lev Vygotsky, both of whom provide a set of principles that bring important insights to this cognitive mix.

Piaget's Theory of Preoperational Thought

Piaget describes thinking between about ages 2 and 6 as **preoperational thought**, referring to the fact that preschool children cannot yet perform logical "operations"; that is, they cannot use ideas and symbols to develop logical principles about their experiences. At the simplest level, they cannot regularly apply a general rule, such as "If this, then that" or "If not this, then not that."

The observation that young children are illogical (or, to Piaget, "prelogical") does not mean that they are stupid or ignorant. Rather, it means that their thinking reflects certain characteristics of preoperational thought (Flavell et al., 1993). One such characteristic is **centration**, that is, preschoolers' tendency to focus their analysis on one aspect of a situation to the exclusion of all the others. They may, for example, insist that lions and tigers are not cats because, in their view, "cats" are house pets. Or they may say that their father is a *daddy* only—not a son, or brother, or uncle as well—because they see family members exclusively in the role those individuals play for them. Or upon meeting, say, a 4-year-old and a 5-year-old, the younger of whom is taller, they may assert that the 4-year-old is actually 6, because "bigger is older."

Further, preschoolers are sometimes rather *static* in their reasoning, understanding the world in terms of an either/or framework rather than as a flux of possibilities. As we have seen, for example, 3-year-olds have difficulty appreciating that, at different times, one may have correct or incorrect beliefs about the contents of candy boxes or other aspects of reality. To a child of this age, one's beliefs must be completely consistent with how things "really and truly" are.

Conservation and Logic

These characteristics of preoperational thought were of particular interest to Piaget, and he devised a number of experiments to test and illustrate the ways in which they limit young children's ability to reason logically. His most famous experiments tested children's understanding of **conservation**, the principle that the amount of a substance is unaffected by changes in its shape or placement. Piaget found that this simple principle, taken for granted by older children and adults, is not at all obvious to preschoolers. Rather, preschoolers tend to focus precisely, and exclusively, on any changes in shape or placement that happen to occur and believe that those changes affect the amount. For example, if young children are presented with two identical glasses containing the same amount of liquid, and then are asked to pour the liquid from one of the glasses into a taller, narrower glass and to indicate if one glass has more liquid than the other, they will insist that the taller glass contains more liquid than the remaining original glass does. Similarly, if an experimenter lines up, say, seven pairs of checkers into two rows of equal length and asks a 4-year-old if both rows have the same number of checkers, the child will usually say "Yes." However, if, while

preoperational thought Piaget's term for the second period of cognitive development. This period generally occurs from age 2 to 6, before logical concepts such as conservation, reversibility, or identity are fully understood.

centration The tendency of young children to focus their analysis on one aspect of a situation or object to the exclusion of all others.

conservation The understanding that an amount or quantity is unaffected by changes in its shape or placement.

Tests of Various Types of Conservation

Type of conservation	Initial presentation	Transformation	Question	Preoperational child's answer
Liquids	Two equal glasses of liquid.	Pour one into a taller, narrower glass.	Which glass contains more?	The taller one.
Number	Two equal lines of checkers.	Increase spacing of checkers in one line.	Which line has more checkers?	The longer one.
Matter	Two equal balls of clay.	Squeeze one ball into a long, thin shape.	Which piece has more clay?	The long one.
Length	Two sticks of equal length.	Move one stick.	Which stick is longer?	The one that is farther to the right.

FIGURE 9.3 According to Piaget, until children grasp the basic concept of conservation at about age 6 or 7, they cannot understand that the transformations shown here do not change the total amount of liquid, clay, checkers, and so on. As the text explains, other research shows that achievement of conservation is much more variable in age and other specifics than Piaget envisioned.

the child watches, the experimenter elongates one of the rows by spacing the checkers in it farther apart, and then asks again if the rows have the same number, the child will most likely reply "No," indicating that the longer row has more checkers. In this situation, the child seems to be compelled on the basis of appearance to conclude that the longer row contains a larger amount. Other conservation tasks present young children with similar intellectual challenges and produce similar results (see Figure 9.3).

In all such tests of conservation, Piaget believed, the problem is that preschoolers center on appearances and thus ignore or discount the transformation that has occurred. Instead, they look at the static results of the change and reason intuitively that the longer row, the taller glass, and so on, must contain more. In addition, these tests of conservation also reveal another characteristic of preschoolers' thinking, **irreversibility**. Being irreversible in their thinking means that preschoolers fail to recognize that reversing a transformation brings about the conditions that existed before the transformation process began. In the case of the conservation of liquids, for example, they do not visualize, or suggest, pouring the liquid from the taller glass back into the smaller one to see if the two glasses contain the same amount. (Nor does it occur to them to reverse the actions in any of the other transformations shown in Figure 9.3.)

irreversibility The inability—characteristic of preschoolers' thinking—to recognize that reversing a transformation brings about the same conditions that existed prior to the transformation.

Piaget believed that it is impossible for preoperational children to grasp the idea of conservation and other logical reasoning processes, no matter how carefully they are explained to them. But some later researchers wondered if perhaps the specific nature of Piaget's conservation experiments—including their formality or testlike features—might be af-

fecting children's performance. And, in fact, they found that in playful, gamelike situations (rather than the formal experimental procedures described above), preschoolers often do reveal an accurate grasp of number conservation (Dockrell et al., 1980; McGarrigle & Donaldson, 1974).

In one example, experimenters used a variation of the checkers test in which the elongation of the row of checkers was caused by the action of a "naughty" teddy bear rather than by the deliberate manipulations of an adult (Donaldson, 1978). In this context, preschoolers were more likely to recognize that both rows still contained the same number. The experimenters hypothesized that in the formal experimental situation, young children may assume that if an adult takes the time and trouble to reposition a row of checkers, all the while making sure the child is paying attention, something significant, like the total number of checkers, must be being changed. In the gamelike situation, however, the teddy's "messing up" the display does not lead to this distracting assumption.

Subsequent research has similarly shown that very young children can succeed at other tests of conservation, if the tests are simple and gamelike, and especially if the children are given special training, including explicit verbal instructions with demonstrations. One particularly successful experiment with liquid conservation included both pretend play, in which the children themselves poured liquids from one container to another as part of a game, and specific, step-by-step explanations of why the amount of liquid is the same no matter what the shape of the containers. As a follow-up to this conservation test, the experimenters asked the children who had succeeded very explicit questions to determine the depth of their understanding, such as "How can the water in this one be so much taller and still have the same amount as this one?" The preschoolers were quick and confident in correctly explaining the rationale for conservation (Golumb & McLean, 1984).

In a variety of training experiments, most preschool children are able, for the moment, to follow the examiner's guided instructions and then, apparently, grasp the idea of conservation. Indeed, many 4- and 5-year-olds—though almost no 3-year-olds—still grasp the concept of conservation several weeks after training, demonstrating not only the type of conservation taught to them (such as liquids) but also other types of conservation (such as number or matter) (Field, 1987). Nevertheless, their capacity to exercise these reasoning skills is limited and fragile: when faced with tasks that are more complex or challenging, their tendencies toward centration, egocentrism, and magical thinking reemerge.

In other words, despite the impressive evidence for early competency in reasoning—as well as in number, problem solving, memory, and theory of mind—the thinking and reasoning of the preschooler surely do not approach the systematic, logical, and objective understanding that the typical grade-schooler possesses (Becker, 1989). The "fragile but nonetheless genuine competencies" (Flavell et al., 1993) revealed by recent research highlight how much earlier developmentalists, including Piaget, underestimated the intellectual skills of children during the play years. At the same time, this recent research also confirms how much children have yet to learn.

Now we turn to another major cognitive theorist, Lev Vygotsky, who explains how the gap between the intellectual skills already evident in preschoolers and the knowledge they have yet to learn is bridged.

Vygotsky's Theory of Children as Apprentices

Through shared social activity, adults in every culture guide the development of their children's cognition, values, and skills. Typically, the child's curiosity and interests, rather than the adult's planning for some future need, motivate the process. This seems to be the case as the Guatemalan girl eagerly tries to learn the sewing skills of her mother.

Whether at any given moment a preschooler's thinking shows surprising competencies, confounding ignorance, seemingly far-fetched intuitive imaginings, or a mixture of all of these, it is apparent that young children strive for understanding in a world that fascinates and sometimes confuses them. But they do not strive for understanding in social isolation. As noted on several occasions, their efforts are embedded in a social context, where parents, as well as older children, preschool teachers, and many others, try to guide a young child's cognitive growth by providing challenges for new learning, offering assistance with tasks that may be too difficult, providing instruction, and supporting the child's interest and motivation (Rogoff, 1990; Rogoff et al., 1993). On their part, children ask endless questions and try to include almost anyone of any age in their cognitive quests. In many ways, then, young children are "apprentices in thinking" whose intellectual growth is stimulated and directed by their **guided participation** in social experiences and explorations of their ecological settings (Rogoff, 1990).

The role of the social context in cognitive development is given particular emphasis by the writings of Lev Vygotsky, the Russian psychologist to whom you were first introduced in Chapter 2. As you learned, Vygotsky's ideas have become the basis for several sociocultural theories that emphasize the cultural foundations of growth and development. In contrast to many developmentalists (including Piaget) who tend to regard cognitive growth as a process of *individual* discovery propelled by experience and biological maturation, Vygotsky believed that cognitive growth is driven by cultural processes that shape the experiences, incentives, and goals involved in children's learning. More specifically, Vygotsky saw cognitive growth less as a process of individual discovery than as a *social* activity advanced by the guidance of parents and other teachers who motivate, channel, and structure a child's learning.

To see how Vygotsky's approach works in practical terms, let's look at an example of a young child's guided participation in a challenging activity. Say that a child quits after trying unsuccessfully to assemble a puzzle. Does that mean the task is beyond the child's ability? Not necessarily—that is, if the child can be given guidance that provides motivation, focuses attention, and restructures the task to make its solution more attainable. In this case, an adult or older child might begin such guidance by encouraging the child to look for a likely piece for a particular section ("Does it need a big piece or a little piece?" "Do you see any blue pieces with a line of red?"). Suppose the child finds some pieces the right size, and then some blue pieces with a red line, but again seems stymied. The tutor might then be more directive, selecting out a piece to be tried next, or rotating a piece so that its proper location is more obvious, or actually putting a piece in place with a smile of satisfaction. Throughout, the teacher praises momentary successes, maintains enthusiasm, and helps the child see their joint progress toward the goal of finishing the puzzle.

guided participation A process in which the child learns through social interaction with a "tutor" (a parent, a teacher, a more skilled peer), who offers assistance with difficult tasks, models problem-solving approaches, and provides explicit instruction when needed.

The critical element in guided participation is that the adult and child interact to accomplish the task. Eventually, such guided participation enables the child to succeed independently. Once the child puts the puzzle to-

gether with adult help, chances are he or she will try it again soon, needing less help, or perhaps none at all. At the same time, the tutor gradually encourages the child to do more on his or her own. Each step toward independence is supported by the adult. Assuming that puzzle solving is a valued skill in the culture, the adult might also find a new puzzle for the child to attempt, so that skills mastered in the first instance—such as locating pieces of similar coloring or finding and connecting all the edge pieces first—are transferred to the next instance, and then, eventually, generalized to the world of all possible puzzles.

Such interactive apprenticeships are commonplace: in every culture of the world, adults direct children's attention and provide assistance to teach various skills, and, soon, children who are given such guided practice learn to perform the skills on their own (Rogoff, 1990; Rogoff et al., 1993; Tharp & Gallimore, 1988).

The "Zone of Proximal Development"

Key to the success of these apprenticeship experiences is the tutor's sensitivity to the child's abilities and readiness to learn new skills. According to Vygotsky (1934/1986),

> the only good kind of instruction is that which marches ahead of development and leads it. It must be aimed not so much at the ripe as at the ripening functions. It remains necessary to determine the lowest threshold at which instruction may begin, since a certain ripeness of functions is required. But we must consider the upper threshold as well: instruction must be oriented toward the future, not the past.

As we saw in Chapter 2, Vygotsky believed that, for each developing individual, there is a **zone of proximal development**, that is, a range of *potential* development involving skills that the person can accomplish with assistance but is not yet quite able to perform independently. How and when a person masters these cutting-edge skills depends, in part, on the willingness of tutors to **scaffold**, or sensitively structure, the child's participation in learning encounters (Bruner, 1982; Wood et al., 1976). We have already seen several examples of scaffolding, including a parent's helping a child with a puzzle, or providing a supportive structuring with problem-solving tasks, or guiding a young child's recollections of a shared experience. Other examples, among many, might include a parent's encouraging a child to talk about his or her own feelings in order to deepen the child's emotional understanding, or sensitively rephrasing a child's verbal expressions to foster language learning.

Developmentalists who have observed how parents provide this kind of structure for their child's emergent capabilities have identified a number of steps that contribute to effective scaffolding (Bruner, 1982; Rogoff, 1990; Wood et al., 1976):

1. Recruit the child's interest in the task or activity.
2. Simplify the task by reducing the number of steps required for correct solution (perhaps by helping the child focus only on the best strategy or perhaps by completing certain aspects of the task for the child).
3. Maintain the child's interest in, and enthusiasm for, the task in the face of distraction, waning interest, and/or discouragement.

zone of proximal development Vygotsky's term for the difference between an individual's attained level of development and the person's potential level of development that might be reached with guidance.

scaffold To sensitively structure a child's participation in learning encounters so that the child's learning is facilitated.

The best way to learn almost any practical or intellectual skill is with the help of a "mentor" who guides one's entry into the zone of proximal development, the area between what one can do alone and what one might do with help. Of course, it is essential that the apprentice be eager to enter the zone, as this intent young carpenter certainly seems to be.

4. Anticipate and indicate errors as they occur in the child's performance and provide guidance toward correction.

5. Control frustration, both by encouraging the child's desire to achieve and by reducing his or her unhappiness at making mistakes.

6. Demonstrate or model correct solutions—ideally in a manner that shows how to complete each step along the way.

A person's progress through the zone of proximal development is influenced in other ways besides a parent's scaffolding of cognitive skills. Other family members may also offer incentives for new learning, such as when sibling rivalry provokes a child to figure out what an older brother might be thinking and feeling, and to use this information for teasing, self-defense, or negotiation. Indeed, children who have siblings close to their own age tend to demonstrate a theory of mind earlier than children who do not have near-age siblings (Astington, 1993).

Beyond the home, a child's experiences in day care or preschool provide similar incentives for the growth of social understanding and cognitive competence, especially as the child begins to compare his or her skills with those of other children the same age. More broadly still, culture influences the development of certain cognitive abilities. Every culture values some cognitive skills more than others, so it is not surprising, for example, that children in the Micronesian islands are much better at interpreting weather and navigation signs than are American children, or that American children are more likely than Micronesian children to acquire skills that are well-suited to abstract and scientific reasoning. As Vygotsky reasoned, cognitive growth is facilitated by opportunities and incentives at various levels of a child's social ecology.

Talking to Learn

Vygotsky believed that language is essential to cognitive growth in two crucial ways. The first is through **private speech**, the internal dialogue in which a person talks to himself or herself (Vygotsky, 1987). In adults, private speech is usually silent, but in children, especially preschoolers, it is much more likely to be uttered out loud. With time, this self-talk becomes a whisper, and then becomes inner, private speech.

Researchers studying private speech have found that preschoolers use it to help them think, reviewing what they know, deciding what to do, and explaining events to themselves. Interestingly, many researchers have found that children who have learning difficulties tend to be slower to develop private speech or to use it to guide their behavior (Diaz, 1987). Training in the use of private speech sometimes helps them learn, another sign that language, in this form, aids the learning process.

The second way language advances thinking, according to Vygotsky, is as the *mediator* of the social interaction that is a vital part of learning. Whether it involves explicit instruction or casual conversation, verbal interaction with others helps to refine and extend one's present level of understanding. This function of language is essential to traversing the zone of proximal development, because verbal interaction provides the bridge from the child's current understanding to the almost-understood.

private speech The use of language to form thoughts and analyze ideas, either silently or by talking to oneself.

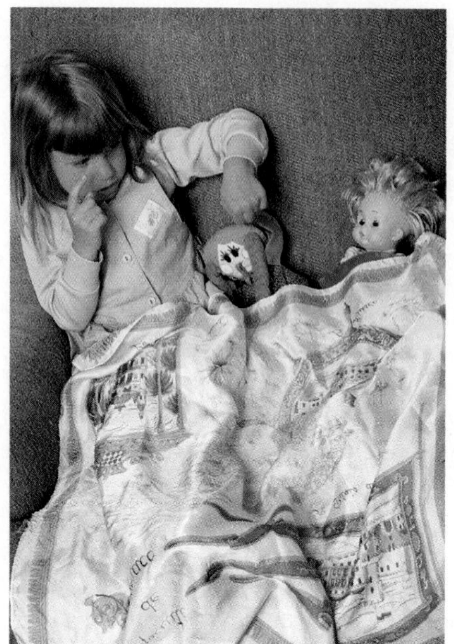

As 3½-year-old Laura explains the rules for taking a nap to her dolls, she is developing her language skills. Most preschoolers talk all the time, to themselves, to their dolls, to the television, and to their parents, even when none of them answer back.

As we will see in the next section, during the preschool years both private and social language erupt in a verbal explosion, as children spend hour upon hour asking question after question, telling stories that seem endless, or just talking and singing to themselves. From Vygotsky's perspective, a child's language development becomes critically important as a cognitive tool, a tool that helps the child master all the other intellectual tools of the culture (van der Veer & Valsiner, 1991). Now let us look explicitly at the course of language development during the preschool years.

Language Development

As noted in Chapter 6, babies normally begin talking at about a year, with language development occurring slowly at first. Toddlers typically add only a few new words to their vocabulary each month, speak in one-word sentences, and sometimes have trouble communicating, frustrating themselves as well as even the most patient caregiver who tries to understand what the 1-year-old wants to say.

During the preschool years, however, as cognitive powers increase, the pace and scope of language learning increase dramatically. Indeed, an "explosion" of language occurs, with vocabulary, grammar, and the practical uses of language showing marked and rapid improvement. As we shall see, the growth of language builds upon, and contributes to, the growth of thinking and reasoning during the preschool years.

Vocabulary

The rapid growth of vocabulary during early childhood is most astonishing, especially when we remember that it takes most infants a year to utter one intelligible word, and several months more before they master a dozen words. Nevertheless, the typical child's vocabulary more than doubles between 18 and 21 months of age, then doubles again over the next three months. In the preschool period, this number increases exponentially. By age 6, children's lexicons contain an average of more than 10,000 words (Anglin, 1993). As one summary describes it:

> Children typically produce their first 30 words at a rate of three to five new words per month. The same children learn their next 30,000 words at a rate of 10 to 20 new words per day. [Jones et al., 1991]

Words are often learned after only one hearing, through a process called **fast mapping**, in which the child immediately assimilates new words by connecting them through their assumed meaning to categories of words he or she has already mastered (see A Closer Look, p. 268).

The learning of new words follows a predictable sequence according to parts of speech. Nouns are generally mastered more readily than verbs, which, in turn, are learned more easily than adjectives, adverbs, conjunctions, or interrogatives. Within parts of speech, the order is predictable as well. For instance, basic general nouns, such as "dog," are learned before specific nouns, such as "collie," or more general categories, such as "animal." The first interrogatives children typically learn are "where?" and "what?" then "who?" followed by "how?" and "why?" (Bloom et al., 1982).

fast mapping A way to grasp the essential meaning of new words by quickly connecting them to words and categories that are already understood.

A CLOSER LOOK

Fast Mapping: Advantages and Disadvantages

Considerable research has attempted to determine how children master vocabulary so quickly. This inquiry begins with the realization that after the first year or so of language acquisition, learning vocabulary is no longer simply an additive process, with each word, in isolation, being added to the child's current stock of words. After age 2, there is an acceleration of vocabulary growth that soon seems like an explosion, with words being added daily in chunks. How does this happen?

One explanation is that the child's mind seems to develop an interconnected set of categories for vocabulary, a kind of mental map that charts the meanings of various words (Golinkoff et al., 1992). Hearing a new word, the child uses the context in which it is being employed to create a quick, partial understanding and then to categorize it, placing it in his or her existing lexicon. Thus children learn new animal names so quickly, for instance, because new names can be mapped close to the old ones ("zebra" is easy to learn if you know "horse," for example); similarly, they learn new color terms by comparing them with those they already know. This process is called "fast mapping," as if, rather than stopping to figure out an exact definition, and waiting until a word has been understood in several contexts, the child simply hears it once or twice and adds it to his or her mental language map (Heibeck & Markman, 1987).

Young children's fast mapping can be aided by the way adults label new things for them (as when a parent points at an animal the child is watching at the zoo and says "See the *lion* resting by the water. It's a *lion*!"). In addition, children make some basic assumptions that enable them to figure out a new word's meaning. They assume, for example, that words refer to whole objects (rather than to their parts) and that each object has only one label (Clark, 1990; de Villiers & de Villiers, 1992; Markman, 1991). While these assumptions sometimes prove misleading (such as when a child retorts, "That's not an animal. It's a *dog*!"), they usually lead to good, provisional definitions.

The quickness of fast mapping is phenomenal: a word can be learned after a single exposure (Dickinson, 1984). Moreover, several new words can be learned over a short time period. In one experiment, 3- and 5-year-olds were first given a multiple-choice vocabulary test on twenty words they were unlikely to know, such as "gramophone," "nurturant," "artisan," "malicious," and "contentment." Their scores were just about what they would be by chance, between five and six correct of the twenty. Then, in two 15-minute sessions, the children viewed cartoons with the twenty words used in context about ten times each. On retesting, the 3-year-olds averaged eight correct answers and the 5-year-olds averaged eleven right. Object words (e.g., "gramophone," which 93 percent got right) were easiest to learn, and emotional-state words (e.g., "contentment," which only 20 percent got right) were hardest. This is not at all surprising, since the language map for objects is well formed during preschool years but the map for emotions is not (Rice, 1990; Rice & Woodsmall, 1988).

Fast mapping has obvious advantages, in that it fosters quick vocabulary acquisition. However, it also means that a child might seem to understand a word because he or she uses it in an appropriate context, when, in fact, the child has no real understanding of the word's meaning or understands it only in a limited way. One very simple, common example is the word "big," a word even 2-year-olds use and seem to understand. In fact, however, young preschoolers often use "big" when they mean "tall," or "old," or "great" ("My love is so big!"), and only gradually use "big" correctly (Sena & Smith, 1990).

If adults realize the difficulty children often have in comprehending exactly what the words they use mean, it becomes easier to understand, and sometimes forgive, the mistakes children make. I can still vividly recall an example of fast mapping that arose when my youngest daughter, then 4, was furious at me.

Sarah had apparently fast-mapped several insulting words into her vocabulary. However, her fast map did not contain precise definitions or reflect the nuances. She first called me a "mean witch," and then a "brat." I smiled at her innocent imprecision. Then she let loose with an X-rated epithet that sent me reeling. Struggling to contain my anger, I tried to convince myself that fast mapping had probably left her with no real idea of what she had just said. "Language like that is never to be used in this house!" I sputtered. My appreciation of the quickness of fast mapping was deepened by her response: "Then how come my big sister called me that this morning?"

The speed with which a child acquires words and relates them to categories and concepts depends partly on the particular conversations the child has with adults, who may or may not stress the linkage between one noun and another (Markman, 1989). When adults do describe categories—not in a formal lesson but simply in the course of normal speech—children

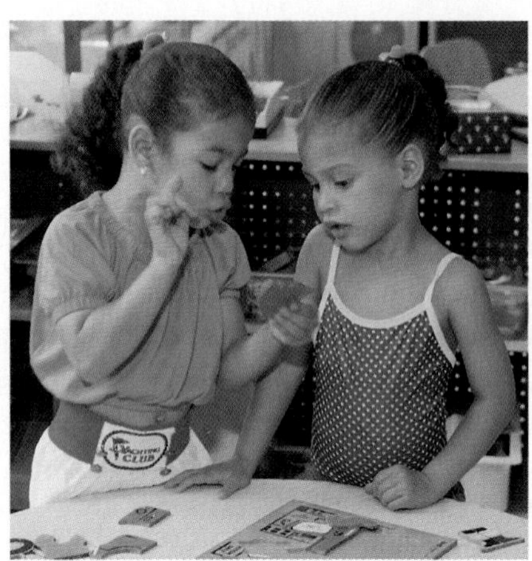

One of the key features of any good preschool program is the frequent opportunity it affords children to hear and express a new vocabulary. While it is of course important for preschool teachers to read to their young students and to regularly engage them in conversation, the best language teachers for children are sometimes other children. With pragmatic skill and personal understanding, they tune into the best topics, and express the most relevant vocabulary, to move language learning along.

are able to map new words more quickly, just as Vygotsky's stress on the pivotal role of social interaction would suggest. For example, when a child meets a dog named Lassie, and repeats "Lassie," a parent might say "Yes, Lassie is a dog, a collie dog," helping the child make the connection between the specific name, the breed, and the kind of animal.

The vocabulary-building process happens so quickly that, by age 5, some children seem to understand and use almost any specific term they hear. In fact, 5-year-olds can learn almost any word or phrase, as long as it is explained to them with specific examples and used in context. One 5-year-old surprised his kindergarten teacher by explaining that he was ambidextrous. When queried, he said, "That means I can use my left or my right hand just the same." In fact, preschoolers are able to soak up language like a sponge, an ability that causes most researchers to regard early childhood as a crucial period for language learning.

The spongelike fast mapping that occurs during these years is so impressive that we need to remind ourselves that young children cannot readily grasp *every* word they hear. Abstract nouns, such as "justice" or "government," are difficult to understand because there is no referent in the child's experience to link them to. Metaphors and analogies are also difficult, because the fast-mapping process is often quite literal, allowing only one meaning per word. When a mother, exasperated by her son's frequent inability to find his belongings, told him that someday he would lose his head, he calmly replied, "I'll never lose my head. If I feel it coming off, I'll find it and pick it up."

Further, although young children can quickly grasp words with objective meanings (such as nouns), they have greater difficulty with words expressing comparisons, such as "tall" and "short," "near" and "far," "high" and "low," "deep" and "shallow" (Reich, 1986). The reason is that children do not understand the *relative* nature of these words. Once they know which end of the swimming pool is the deep one, for instance, preschoolers might obey instructions to stay out of deep puddles by splashing through every puddle they see, insisting that none of them are deep. Words expressing relativities of place and time are difficult as well, such as "here" and "there," and "yesterday" and "tomorrow." More than one pajama-clad child has awakened on Christmas morning and asked "Is it tomorrow yet?"

Grammar

Grammar includes the structures, techniques, and rules that a language uses to communicate meaning. Word order and word form, prefixes and suffixes, intonation and pronunciation, all are part of grammar. Grammar is apparent even in toddlers' two-word sentences, since youngsters always put the subject before the verb.

By age 3, children typically demonstrate extensive grammatical knowledge. They not only put the subject before the verb but also put the verb before the object, explaining "I eat apple" rather than using any of the other possible combinations of those three words. They can form the plural of nouns, the past, present, and future tenses of verbs, the subjective, objective, and possessive forms of pronouns. They can rearrange word order to create questions and can use auxiliary verbs ("I *can* do that"). They are well on their way to mastering the negative, progressing past the simple "no" of

This boy has been having phone conversations with his grandmother since he was 1 year old, although at first he mostly listened, and then cried when the phone was taken away. Now, at almost 3, he chatters away unstoppably, revealing an extensive grasp of vocabulary and grammar. However, he still doesn't necessarily provide all the details that would let his grandmother follow the conversation, sometimes referring to events she has no knowledge of and people she does not know, or telling the end of a story without a beginning, or vice versa.

the 2-year-old ("No sleepy," "I no want it," "I drink juice no") to more complex negatives such as "I want nothing" or "I am not sleepy."

Children's understanding of grammar is revealed when they create original phrases and expressions, like those in Table 9.1. The words in the table show not only children's mastery of grammatical rules but also their ability to apply these rules to create expressions they have never heard before but that convey their thoughts clearly and accurately to others.

How do preschoolers master the basic rules of language so quickly and easily? This impressive accomplishment has inspired many explanations. Recall from Chapter 6 that some developmentalists, following the ideas of Noam Chomsky (1968, 1980), believe that young children are aided in their language learning by a uniquely human brain structure—referred to as a "language acquisition device"—that facilitates their mastery of grammar. This innate mental program provides them with a set of "intuitive" guidelines for quickly deducing the rules of their native language, whether they are learning English, Russian, French, or Mandarin Chinese.

Children's understanding of grammar is also facilitated, of course, by hearing conversations at home that are models of good grammar and by having their parents give them helpful feedback about their language use (Farrar, 1992; Hoff-Ginsberg, 1986, 1990; Tomasello, 1992). Reflecting this fact was one study that followed the language development of two groups of 2-year-olds (Hoff-Ginsberg, 1986). The children in one group had mothers who frequently asked them questions (such as "Where does the duck live?") and then repeated their answers, correctly rephrased (changing the child's reply of "Duck, water," for example, into "Yes, the duck lives on the water"). The mothers in the other group rarely used such strategies. After six months, the researchers found that the children who had received "lessons" in grammar as part of the normal dialogues they had with their mothers were more advanced in their own use of grammar than were the children in the other group.

TABLE 9.1

Children's Knowledge of Grammar in Creating Words

Rule Followed	Word	Context
Add "un" to show reversal.	"unhate"	Child tells mother: "I hate you. And I'll never unhate you."
Use a limiting characteristic as an adjective before a noun to distinguish a particular example.	"plate-egg," "cup-egg" "sliverest seat"	Fried eggs, boiled eggs. A wooden bench.
Add "er" to form comparative.	"salter"	Food needs to be more salty.
Create noun by saying what it does.	"tell-wind"	Child pointing to a weather vane.
Add "er" to mean something or someone who does something.	"lessoner" "shorthander"	A teacher who gives lessons. Someone who writes shorthand.
Add "ed" to make a past verb out of a noun.	"nippled"	"Mommy nippled Anna." Reporting that Mother nursed the baby.
	"needled"	"Is it all needled yet?" Asking if Mother has finished mending the pants.
Add "s" to make a noun out of an adjective.	"plumps"	Buttocks.
Add "ing" to make a participle out of a noun.	"crackering"	Child is putting crumbled crackers into soup, thereby crackering it.

Sources: Examples come from Bowerman, 1982; Clark, 1982; Reich, 1986; and the author's children.

Drawing by Glenn Bernhardt

No, Timmy, not "I sawed the chair." It's "I saw the chair" or "I have seen the chair."

This mother has obviously become accustomed to her son's use of overregularization.

Difficulties with Grammar

Young children learn their grammar "lessons" so well that they often tend to apply the rules of grammar even when they should not. This tendency, called **overregularization**, can create trouble when a child's language is one that has many exceptions to the rules, as English does. For example, one of the first rules of grammar that English-speaking children use is adding "s" to form the plural. Thus many preschoolers, applying this rule, talk about foots, tooths, sheeps, and mouses. They may even put the "s" on adjectives when the adjectives are acting as nouns, as in this dinner-table exchange between a 3-year-old and her father:

Sarah: I want somes.
Father: You want some what?
I want some mores.
Some more what?
I want some more chickens.

Once preschool children learn a rule, they can be surprisingly stubborn in applying it. Jean Berko Gleason reports the following conversation between herself and a 4-year-old:

> She said: "My teacher *holded* the baby rabbits and we *patted* them." I asked: "Did you say your teacher *held* the baby rabbits?" She answered: "Yes." I then asked: "What did you say she did?" She answered again: "She *holded* the baby rabbits and we *patted* them." "Did you say she *held* them tightly?" I asked. "No," she answered, "she *holded* them loosely." [Gleason, 1967]

Although technically wrong, such overregularization is actually a sign of verbal sophistication, since children are, clearly, applying rules of grammar. Indeed, as preschoolers become more conscious of grammatical usages, they may exhibit increasingly sophisticated misapplications of them (de Villiers & de Villiers, 1986). A child who at age 2 says she "broke" a glass may at age 4 say she "braked" one and than at age 5 say she "did broked" another. After children hear the correct form often enough, they spontaneously correct their own speech, so parents can probably best help development of grammar by example rather than explanation or criticism—in this case, for example, by simply responding "You mean you broke it?" While few children will immediately correct their grammar, continual exposure to good grammar speeds up their language mastery (Farrar, 1992).

During the preschool years, children are able to comprehend more complex grammar, and more difficult vocabulary, than they can produce. Thus, while it is a mistake to expect preschoolers to use proper grammar and precise vocabulary, it is also an error to simply mirror the child's speech, "talking down" to his or her level. And while an adultlike understanding of some grammatical forms is beyond many preschoolers, that does not mean that their language-learning abilities are severely limited. Vogotsky's concept of the zone of proximal development is useful here: between the grammar forms that are understood, and those that are, as yet, incomprehensible, lies a zone of potential development that, with adult guidance and the child's natural intellectual curiosity, can be used to expand the child's grammatical comprehension.

Having examined preschoolers' cognitive capabilities, their eagerness to learn, and their ability to learn through guided participation, let us now turn to the final topic of this chapter, preschool education.

overregularization The tendency to apply grammatical rules and structures when they are not called for or when exceptions to them should be used.

Preschool Education

Over the past thirty years, scientists have shown not only that the years before age 6 are a time of rapid learning but also that young children can learn a great deal through preschool education, with noticeable enhancement of their verbal skills and overall intellectual development. The clearest and most extensive evidence for the benefits of preschool education comes from high-quality schools characterized by (1) a low teacher-child ratio, (2) a staff with training and credentials in early childhood education, (3) a curriculum geared toward cognitive development rather than behavioral control, and (4) an organization of space that facilitates creative and constructive play. Research has found that preschools that include cognitive development among their goals but do not have all these costly resources can also foster children's learning, though not as much (Burchinal et al., 1989; Lee et al., 1988).

As a result, then, of the proven benefits of preschool education, as well as the increased need for child day care as mothers enter the labor force in greater and greater numbers, more and more young children are in some sort of education milieu. In the United States in 1970, only 30 percent of married mothers with children under age 6 were in the labor force, and only 20 percent of all 3- and 4-year-olds were in a preschool of some type; in 1994, these figures were 60 percent and 47 percent, respectively (U.S. Bureau of the Census, 1996) (see Figure 9.4). In most other developed countries, the numbers are higher, because their governments sponsor education in early childhood.

In every culture, the goals of preschool education go beyond cognitive preparation for later schooling (Mallory & New, 1994). For example, in Japan, a society that places great emphasis on social consensus and conformity, preschool provides training in the behavior and attitudes appropriate

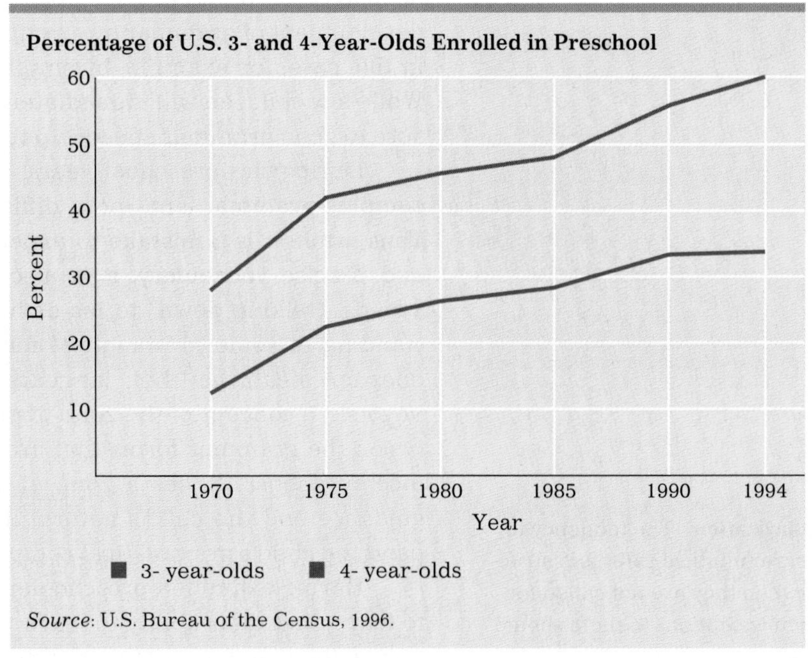

FIGURE 9.4 As you can see, more than half of all 4-year-olds are now in preschool, a marked improvement over the enrollment rates of twenty-five years ago. Other data show that the increases have occurred in every income and ethnic group, with even the least likely to be enrolled—low-income, Spanish-speaking 4-year-olds—now having an enrollment rate of close to 40 percent. Ideally, developmentalists would like to see all preschool-age children attending preschool, whether they are currently cared for at home or in family day care. An educational setting, with trained teachers and stimulating activities, is just as important for 3- and 4-year-olds as for older children.

These photos show happy kindergartners in teacher-directed exercise in two settings, Tokyo and southern California. If you were a stranger to both cultures, with no data other than what you see in the photos, what would you conclude about the values, habits, and attitudes adults hope to foster in these two groups of children?

for group activity: children are encouraged to show concern for others and to contribute cooperatively in group activities. These social skills not only prepare young children for their entry into the formal school system but also socialize attitudes and habits that they will later use in work settings (Peak, 1991). In China, an emphasis on learning how to be part of the group is combined with an emphasis on academic skills as well as creativity in self-expression, both drawn from the culture's Confucian ethic of disciplined study. In the United States, by contrast, preschools are often designed to foster self-confidence and self-reliance—qualities that are valued in our society—and to give children a good academic start through emphasis on language skills (Tobin et al., 1989). In this respect, the goals of preschool education—entailing a mixture of cognitive and social skills—reflect cultural values as well as the needs and capabilities of young children.

An additional goal of at least one form of U.S. preschool—Headstart—is to provide extra assistance and support to children who are disadvantaged by poverty.

Headstart: A Longitudinal Look

Project Headstart, a federally funded program, was inaugurated in 1965 to give children who were thought to be disadvantaged by poverty a "head start" on the skills required in elementary school (Zigler & Berman, 1983; Zigler et al., 1993). Although the specific curriculum varies from school to school, Headstart focuses on activities that will prepare children for formal learning. The program usually requires parental involvement as well.

Various longitudinal studies of Headstart students have found that as they made their way through elementary school, they scored higher on achievement tests and had more positive school report cards than non-Headstart children from the same backgrounds and neighborhoods. By junior high, they were significantly less likely to be placed in special classes or made to repeat a year. In adolescence, Headstart graduates had higher aspirations and a greater sense of achievement than their non-Headstart peers. As they entered adulthood, Headstart graduates were more likely to be in college and less likely to have a criminal record or a dependent child (Haskins, 1989).

Project Headstart A federally funded U.S. preschool educational program designed primarily to give economically disadvantaged children advance preparation (a "head start") for the intellectual and social challenges of elementary school.

Similar findings to these appear in other longitudinal research, begun in the 1960s, on participants in the Perry Preschool Program, a well-financed preschool education project in Ypsilanti, Michigan, that was much like Headstart in design. The latest survey of the Perry subjects, then in their late 20s, indicates that, compared to the study's control group, they have more education, greater earning power, greater family stability, and have required fewer social services (Schweinhart & Weikart, 1993).

Do the same generalities hold for children who are not poor? Yes, to a degree. Longitudinal research on more advantaged children in the United States and elsewhere finds that they also benefit from a quality preschool setting, although the better the home environment, the less pronounced the influence of the preschool is likely to be (Anderson, 1989; Larsen & Robinson, 1989). Such benefits are cumulative: the more months and years a child spends in preschool, the more the cognitive and emotional benefits accrue (Field, 1991).

Considering all we know about cognitive development between the ages of 2 and 6, we should not be surprised at the benefits of a well-run preschool program. Preschool children are surprisingly capable of learning everything from math to grammar to social insights, but their actual learning depends on the kind of guided participation they experience, as well as on the opportunities they have to manipulate objects, learn language, and interact with other children. For children of every background a quality preschool program advances learning, while a poor-quality program—one that provides little more than physical care and supervision—is of little benefit intellectually (Zigler et al., 1993). As you will see in the next chapter, the same can be said of psychosocial development: a great deal depends on the social context surrounding the child.

SUMMARY

How Preschoolers Think

1. Although young children cannot count large amounts and cannot easily add or subtract, their counting reveals an understanding of basic number principles, such as the "stable-order principle," the "one-to-one principle," and the "cardinal principle." Through the many number activities they engage in with their children, parents play an important role in the development of early number skills.

2. Preschoolers sometimes display considerable practical problem-solving skills in their everyday play and in their relations with peers, but they generally do not succeed at formal problem-solving tasks of the kind experimenters devise. However, some research has shown that children's problem-solving abilities can be enhanced when adults provide supportive guidance that is keyed to the child's ability level.

3. Young children are not skilled at deliberately storing or retrieving memories, although they can use scripts of familiar events to "bootstrap" their recollections of particular experiences. Parents and other adults also aid memory by helping children reconstruct their memories of past events, and by modeling the strategies for retrieving memories. Children sometimes display surprising long-term memory ability when adults use directive questions and prompting to help them focus their attention on specific aspects of personally meaningful past events.

4. Children's concepts reveal elementary theories about various features of their life experiences, including human psychology. A preschooler's theory of mind reflects developing concepts about human mental processes, including an understanding of how and why people have different thoughts, feelings, and intentions. Young children show a surprising awareness that people have different viewpoints and knowledge about events, and that subjective states may not be shared. One aspect of their developing theory of mind that is especially challenging to preschoolers is understanding that subjective understanding may not always accurately reflect reality.

Piaget's Theory of Preoperational Thought

5. Piaget described preoperational thought as, essentially, prelogical. This is because preschoolers tend to be irre-

versible in their thought processes, exhibit centration by focusing on one aspect of a situation to the exclusion of others, and reason in a static rather than a dynamic fashion.

6. Although preschoolers do not possess the well-established, systematic logical reasoning skills that they will exhibit during the school-age years, current research reveals that preschoolers are not as illogical as Piaget believed them to be. In particular, young children have been shown to grasp concepts of conservation in certain situations.

Vygotsky's Theory of Children as Apprentices

7. Vygotsky viewed cognitive development as an apprenticeship, in which children acquire cognitive skills through their guided participation in social experiences that stimulate intellectual growth.

8. Vygotsky's view of the zone of proximal development is that there exists, for each child, a range of potential development that is the cutting-edge of new cognitive accomplishments. Social guidance is therefore most helpful in taking the child from what he or she can already do to what he or she is ready to learn next.

9. According to Vygotsky, language fosters cognitive growth by facilitating social interaction that teaches new skills. In addition, children use "private speech" to guide and direct their own actions.

Language Development

10. Language accomplishments during early childhood include learning 10,000 words or more. Children appear to increase their vocabulary so rapidly by connecting unfamiliar words through their context to a mental map of familiar terms.

11. Young children also show marked growth in their understanding of basic grammatical forms. Children of this age, however, have difficulty with abstract words and often misunderstand, or overregularize, grammatical rules.

Preschool Education

12. Whereas once most children used to stay home until they began formal education at about age 6, over the past thirty years insights from developmental psychology and changes in family composition and work patterns have resulted in increases in preschool education throughout the world.

13. The quality of preschool education varies a great deal. Those programs with an educational curriculum led by trained adults have shown a range of long-term benefits, not only in the child's later schooling but also in successful adult development.

KEY TERMS

egocentrism (249)
scripts (253)
theory of mind (259)
preoperational thought (261)
centration (261)
conservation (261)
irreversibility (262)
guided participation (264)
zone of proximal development (265)
scaffold (265)
private speech (266)
fast mapping (267)
overregularization (271)
Project Headstart (273)

KEY QUESTIONS

1. Which number principles are revealed in a young child's counting?

2. How do young children's scripts aid in their recall of specific past experiences?

3. What role do parents play in the development of number skills and memory and in the growth of problem-solving abilities during the preschool years?

4. What advice might a developmental psychologist give a lawyer who was planning to interview a preschool child who had witnessed a crime?

5. Describe the growth of a young child's grasp of mental processes, or theory of mind, through the preschool years.

6. According to Piaget, what are the central features of preoperational thought?

7. In Piaget's view, what are the strengths and weaknesses of preschoolers' reasoning?

8. What key feature of Vygotsky's ideas about cognitive growth sets them apart from those of Piaget and many other developmental psychologists?

9. Give a hypothetical illustration of how a parent fosters a preschool child's new cognitive accomplishmenets in the zone of proximal development.

10. What explanations do developmental psychologists offer for the extremely rapid acquisition of new words during the preschool years?

11. How do young children so quickly and easily master the basic rules of language, or grammar?

12. How do preschool education programs such as Project Headstart affect young children's cognitive development?

The Play Years: Psychosocial Development

CHAPTER 10

Picture a typical 2-year-old and a typical 6-year-old, and consider the psychosocial differences between them. Chances are the 2-year-old still has many moments of clinging, of tantrums, and of stubbornness, vacillating between dependence and self-determination. Further, 2-year-olds cannot be left alone, even for a few moments, in any place where their relentless curiosity might lead them into destructive or dangerous behavior.

Six-year-olds, by contrast, have the confidence and competence to be relatively independent. They can be trusted to do many things by themselves, perhaps getting their own breakfast before school and even helping to feed and dress a younger sibling. They also can show affection with parents and friends without the obvious clinging or exaggerated self-assertion of the younger child. Six-year-olds are able to say goodbye to their parents at the door of the first-grade classroom, where they go about their business, befriending certain classmates and ignoring others, and respecting and learning from their teachers.

It is apparent that in terms of self-confidence, social skills, and social roles, much develops during early childhood. Cognitive growth permits children a greater appreciation of psychological roles, motives, and feelings, deepening their understanding of themselves and of others. At the same time, their social world becomes more diverse, with the introduction of new social partners (in preschool or in the neighborhood) and richer roles for familiar partners (such as parents, siblings, and long-time playmates). Catalysts for psychosocial development thus come from within and around the child, and this chapter examines that development.

The Self and the Social World

Self-concept, self-confidence, and self-understanding, as well as social attitudes, social skills, and social roles, are familiar topics for psychologists who study adults. Increasingly, these same topics are central to researchers studying children, especially those looking at early child development. Between ages 1 and 6, children progress from a dawning awareness that they are independent individuals to a firm understanding of who they are and how their selfhood relates to others. In the course of this progression, children move from the first recognition of themselves in a mirror, to knowing their name, gender, and what belongs to them, to knowing what they need and want from their family and friends and how to get it.

Self-Concept and Social Awareness

The play years are filled with examples of an emerging self-concept, as preschoolers repeatedly explain who they are and who they are not ("I'm a big girl," "I'm not a baby") and assiduously note which possessions are theirs (laying claim to everything from "My teacher" to "My mudpie"). Preschoolers' emerging self-concept is also evident in their relish of many forms of mastery play. As Erik Erikson pointed out (recall Chapter 2), young children's self-concept is largely defined by the expanding range of skills and competencies that demonstrate their independence and initiative, and preschoolers jump at almost any opportunity to show that "I can do it!" An emerging self-concept can also be seen in the initial social interaction between two preschoolers, which typically involves the children's telling each other their names and ages and showing off any interesting toy, garment, or skill they may have.

During the play years, children gradually begin to perceive themselves not just in terms of their physical attributes ("I'm bigger than Natalie!"), or their characteristic behaviors or abilities ("I can run fast!"), but also in terms of their dispositions and traits, seeing themselves, for example, as friendly, shy, happy, or hardworking (Eder, 1989, 1990). By the late preschool years, children possess a self-concept that may include a recognition of certain psychological tendencies, as revealed in this exchange between a 5-year-old and two puppets (manipulated by an experimenter):

Puppet 1: My friends tell me what to do.
Child: *Mine don't.*
Puppet 2: I tell my friends what to do.
Child: *I do too. I like to boss them around.* [Eder, 1990]

Nevertheless, preschoolers' psychological understanding of themselves and others is still very limited (Miller & Aloise, 1989). They do not grasp the complexity of personality or the variability of a person's competencies: they do not appreciate, for example, that a person can be mean to people but kind to animals, or can be good at math but poor in reading. Preschoolers also do not clearly distinguish the different psychological causes of their actions or skills, believing, for example, that ability is self-controlled and can always be changed through effort.

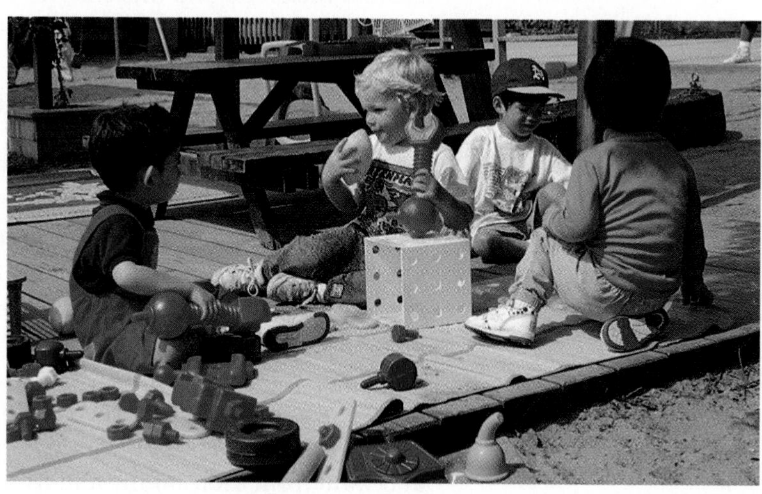

Peer play is an important arena for the growth of self-awareness. In its early stages, self-concept among peers is typically expressed in terms of what belongs to whom. This sets the stage for another important development in peer play—the art of negotiation.

The growth of preschoolers' self-awareness is nowhere more apparent than in their negotiations with others. Just prior to the preschool years, parents typically find themselves dealing with a demanding, stubborn toddler whose primary negotiating skills seem to be insistence and tantrums. But as children's theory of mind expands (see Chapter 9), giving them a better grasp of how they, and other people, think and feel, their negotiations with parents—over what they will wear, what they will eat, when they will go to bed, and so on—evolve from obstinant demands and defiance to bargaining, compromising, and rationalizing (Crockenberg & Litman, 1990; Kuczynski & Kochanska, 1990). As we will see later in this chapter, similar growth is seen in peer relationships.

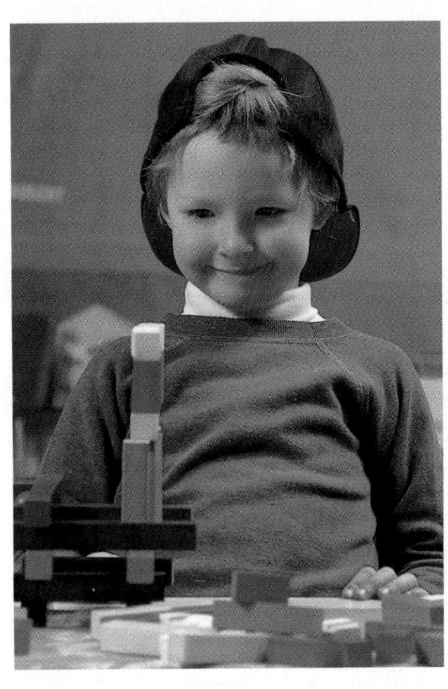

During the play years, pride in the final accomplishment generally overshadows any reasons for self-doubt or self-criticism—such as whether the skyscraper one has just built is recognizable as such to anyone else.

Self-Evaluation

For children of all ages, psychologists emphasize the importance of developing a positive self-concept. Unless their social world makes it impossible, preschoolers usually have no problem in this regard. Typically they form quite general, and quite positive, impressions of themselves. Indeed, much research, as well as anecdotal evidence, shows that preschool children regularly overestimate their own abilities. As every parent knows, the typical 3-year-old believes that he or she can win any race, skip perfectly, count accurately, and make up beautiful songs. In a laboratory test, even when preschoolers had just scored rather low on a game, they confidently predicted that they would do very well the next time (Stipek & Hoffman, 1980). Only when it was specifically pointed out to them how poorly they had done did they revise their estimates downward (Stipek et al., 1984).

As time passes, however, preschoolers become increasingly aware of, and concerned with, how others evaluate their behavior, Gradually, they begin to spontaneously appraise their behavior with the same standards as adults do. In many situations, for example, young children will respond with disappointment or shame when they fail at a task, such as tying their shoes, or when they cause some mishap, such as spilling a cup of juice, even when no adult is present (Cole et al., 1992; Lewis et al., 1992).

At this point, according to Erikson's theory, children have left the stage of autonomy versus shame and doubt (see pp. 32–33) and are in the stage of **initiative versus guilt**. In this stage, which is closely tied to the child's developing sense of self and the awareness of the larger society, preschoolers eagerly take on new tasks and play activities and feel guilty when their efforts result in failure or criticism. Their readiness to take the initiative reflects preschoolers' desire to accomplish things, not simply to assert their autonomy as they did as toddlers. Thus, in a nursery school classroom, the older preschoolers take the initiative to build impressive block towers, whereas younger children in the autonomy stage are more likely to be interested in knocking them down. The enthusiasm of older children to learn and master many things derives, in part, from their growing sense of membership in the larger culture and a desire to acquire the skills of citizen and worker as well as of family member.

According to Erikson, when initiative fails—when eager exploration leads to a broken toy, a crying playmate, or a criticizing adult—the result is guilt, an emotion that is beyond the scope of the infant because it depends on an internalized conscience and a sense of self (Campos et al., 1983).

initiative versus guilt The third of Erikson's eight "crises" of psychosocial development, in which the preschool child eagerly begins new projects and activities—and feels guilt when efforts result in failure or criticism.

Gender Identification

An important feature of self-understanding during the play years is the child's developing understanding of gender roles and personal gender identity. Gender preferences and play patterns emerge early in childhood and typically reflect quite definite ideas about gender. Consider the following account by a leading researcher of gender identity, Sandra Bem (1989), concerning the day her young son Jeremy

> naively decided to wear barrettes to nursery school. Several times that day, another little boy insisted that Jeremy must be a girl because "only girls wear barrettes." After repeatedly asserting that "wearing barrettes doesn't matter; being a boy means having a penis and testicles," Jeremy finally pulled down his pants as a way of making his point more convincingly. The boy was not impressed. He simply said, "Everybody has a penis; only girls wear barrettes."

Although few preschoolers are as well-informed about biology as Jeremy is, all children learn about sex and gender distinctions very early. Most 2-years-olds already know whether they are boys or girls and can identify adult strangers as mommies or daddies. By age 3, children have a rudimentary understanding that they will probably be whatever sex they are for life (although some pretend, hope, or imagine otherwise), and they apply gender labels (Mrs., Mr., lady, man) consistently and are convinced that certain toys (such as dolls and trucks) and certain roles (such as nurses and soldiers) are appropriate for one sex but not the other (Fagot et al., 1992; Levy, 1994; Martin & Little, 1990).

Soon this gender awareness becomes connected to value judgments (Fagot & Leinbach, 1993). For instance, by age 4, children tend to criticize peers who choose toys that are regarded as "gender-inappropriate" (Lobel & Menashri, 1993) and to be proud of themselves when they act in gender-typical ways (Cramer & Skidd, 1992). By age 6, children not only have well-formed ideas (and prejudices) about sex differences but also know which is the better sex (their own) (Huston, 1993). They also insist on dressing in gender-stereotypic ways: shoes for preschoolers are often designed with gender markers such as pink ribbons or blue footballs, and no child would dare wear the shoes meant for the other sex. This insistence on sex differences is evident even in fantasy play, which in every other respect is wide open to possibilities. Even for Halloween, girls typically are beautiful fairy princesses (or mean witches) and boys are superheroes, with rarely a

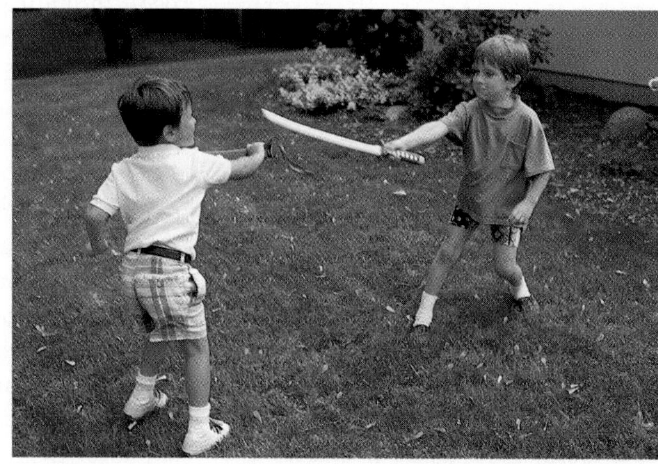

Even in today's world, preschool boys and girls tend to engage in play that is gender-stereotyped, especially when playing with a good friend of the same sex. While some barriers have fallen, they have been more on the girls' side than the boys'. One could imagine these girls sword fighting, for example, but it seems unlikely that the boys would allow themselves to play patty-cake. The question, still unanswered with certainty, is why.

gender-neutral, much less a cross-gender, exception (Ogletree et al., 1993). Stereotypes such as these are often maintained despite the efforts of some parents to break them down and to deemphasize gender distinctions. Consider the following examples cited by Beal (1994):

> A mother who had struggled through medical school, in part so that she would be a strong role model for her daughter, was startled when the little girl told her that she must be a nurse because only men could be doctors. Another little girl said, "Mommy cleans the living room," even though she had only seen her father doing the vacuuming and dusting. When another girl was given a truck to play with, she said, "My Mommy would want me to play with this, but I don't want to." [Bussey & Bandura, 1992]

Children tend not only to maintain but even to strengthen their gender preferences and stereotypes throughout childhood and into adolescence (Etaugh & Liss, 1992; Maccoby, 1990; Powlishta, 1995).

Theories of Gender-Role Development

We have already discussed the nature-nurture issue several times, and you are well aware that developmentalists disagree about what proportion of observed sex differences is biological—perhaps a matter of hormones, of brain structures, or of body size and musculature—and what proportion is environmental (Beal, 1994). However, even for differences that seem most closely related to nurture, theorists hypothesize various reasons for their existence. Specifically, they ask: What is the origin of gender-role differences that children develop during the preschool years? The answers provided by some of the major psychological theories offer very different perspectives on how significant relationships shape this key component of psychosocial growth.

Psychoanalytic Theories

The first major theorist to speculate about gender-role development was Freud (1938). He called the period from about age 3 to 7 the *phallic stage*, because he believed its center of focus is the penis. At about age 3 or 4, said Freud, a boy becomes aware of his penis, begins to masturbate, and develops sexual feelings toward his mother, who has always been an important love object for him. These feelings make him jealous of his father—so jealous, in fact, that, according to Freud, every son secretly wants to replace his father. Freud called this phenomenon the **Oedipus complex**, after Oedipus, son of a king in Greek mythology. Abandoned as an infant and raised in a distant kingdom, Oedipus later returned to his birthplace and, not realizing who they were, killed father and married his mother. When he discovered what he had done, he blinded himself in a spasm of guilt.

According to Freud, little boys feel horribly guilty for having the feelings and thoughts that characterize the Oedipus complex and imagine that their father will inflict terrible punishments on them if he ever finds out about these thoughts. They cope with this guilt and fear by means of **identification**, a defense mechanism through which people imagine themselves to be like a person more powerful than themselves. In a sense, if they cannot replace the father, young boys strive to be *like* the father. As part of their identification with their father, boys copy their father's masculine behavior and adopt his moral standards.

Oedipus complex In the phallic stage of psychosexual development, the sexual desire that boys have for their mother and the related hostility they have toward their father.

identification A defense mechanism that makes a person take on the role and attitudes of someone more powerful than himself or herself.

Freud offered two overlapping descriptions of the phallic stage as it occurs in little girls. One form, the **Electra complex** (also named after a figure in classical mythology), follows the reverse pattern of the Oedipus complex: the little girl wants to get rid of her mother and become intimate with her father. In the other version, the little girl becomes jealous of boys because they have a penis, an emotion called *penis envy*. Somehow the girl decides that her mother is to blame for her "incompleteness," so she becomes angry at her and decides the next best thing to having a penis of her own is to become sexually attractive so that someone with a penis, preferably her father, will love her (Freud, 1933/1965). (See A Closer Look, p. 283.) In both versions, the consequences of this stage are the same for girls as for boys: guilt and fear, which are resolved by the child's adopting gender-appropriate behavior and the moral code of his or her same-sex parent.

Learning Theories

As you remember from Chapter 2, learning theory takes a stance on development that contrasts quite sharply with the psychoanalytic view. This is certainly true regarding the proposed origins of gender-role development during early childhood. Learning theorists believe that virtually all role patterns are learned, rather than inborn, and that parents, teachers, and society are responsible for whatever gender-role ideas and behaviors the child demonstrates.

Preschool children, according to learning theory, are reinforced for behaving in the ways deemed appropriate for their sex and punished for behaving inappropriately. In some ways, research bears this out. Parents, peers, and teachers are all more likely to reward "gender-appropriate" behavior than "gender-inappropriate behavior" (Fagot et al., 1992; Fagot & Hagen, 1991; Huston, 1983). Parents may commend their sons for not crying when hurt, for example, but caution their daughters more about the hazards of rough play.

Interestingly, boys are criticized more than girls for wanting to play with "gender-inappropriate" toys and are rewarded more for playing with toys "for boys." Even between ages 1 and 5, boys are discouraged from wanting to play with dolls (Fagot & Hagen, 1991; Lytton & Romney, 1991). Furthermore, fathers are more likely to expect their girls to be "feminine" and their boys to be "masculine" than mothers are. Thus, gender-role conformity seems to be especially important for males.

Social learning theorists point out that children also learn much about gender behavior by observing and interacting with other people, especially people whom they perceive as nurturing, powerful, and similar to themselves. Parents are crucial models of gender behavior during childhood, and adults in the neighborhood and at school are also influential. In addition, children are strongly influenced by gender-specific behaviors and roles they observe portrayed in the media and enacted in the broader society. They notice, for example, that nursery school teachers are almost always women and that garage mechanics are almost always men. They likewise recognize that, in children's books and on TV, the leaders of their nation are usually men and that people cleaning houses are usually women (Barnett, 1986; Beal, 1994; Crabb & Bielawski, 1994; Huston, 1983).

Electra complex In the phallic stage of psychosexual development, the female version of the Oedipus complex in which girls have sexual feelings for their father and accompanying hostility toward their mother.

A CLOSER LOOK

Professor Berger and Freud

As a woman, and as a mother of four daughters, I have always regarded Freud's theory of female sexual development as ridiculous, not to mention antifemale. I am not alone in this opinion. Psychologists generally agree that Freud's explanation of female sexual and moral development is one of the weaker parts of this theory, reflecting the values of middle-class Victorian society at the turn of the century more than any universal pattern. Many female psychoanalysts (e.g., Horney, 1967; Klein, 1957; Lerner, 1978) are particularly critical of Freud's idea of penis envy. They believe that girls envy, not the male's sexual organ, but the higher status the male is generally accorded. They also suggest that boys may experience a corresponding emotion in the form of womb and breast envy, wishing that they could have babies and suckle them.

However, my own view of Freud's theory as complete nonsense has been modified somewhat by the following experiences with my four daughters when each was in the age range of Freud's phallic stage. The first "Electra episode" occurred in a conversation with my oldest daughter, Bethany, when she was 4 or so.

Bethany: When I grow up, I'm going to marry Daddy.

I: *But Daddy's married to me.*

Bethany: That's all right. When I grow up, you'll probably be dead.

I: *(Determined to stick up for myself) Daddy's older than me, so when I'm dead, he'll probably be dead, too.*

Bethany: That's OK. I'll marry him when he gets born again. [Our family's religious beliefs, incidentally, do not include reincarnation.]

At this point, I couldn't think of a good reply. Bethany must have seen my face fall and taken pity on me.

Bethany: Don't worry, Mommy. After you get born again, you can be our baby.

The second episode was also in a conversation, this time with my daughter Rachel, when she was about 5.

Rachel: When I get married, I'm going to marry Daddy.

I: *Daddy's already married to me.*

Rachel: (With the joy of having discovered a wonderful solution) Then we can have a double wedding!

The third episode was considerably more graphic. It took the form of a "valentine" left on my husband's pillow by my daughter Elissa, who was about 8 at the time. It is reproduced in the next column.

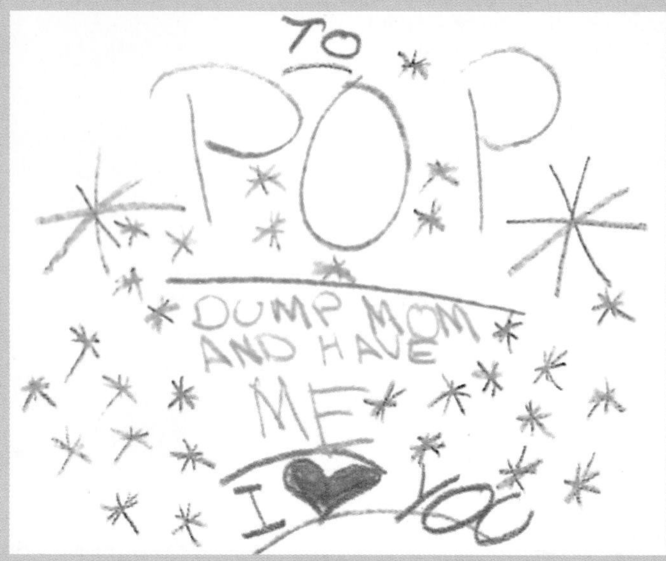

Finally, by the time my youngest daughter, Sarah, turned 5, she also expressed the desire to marry my husband. Her response to my statement that she couldn't marry him because he is already married to me reveals one of the disadvantages of not being able to ban TV in our household: "Oh yes, a man can have two wives. I saw it on television."

I am not the only feminist developmentalist to find Freud's theories on this matter surprisingly perceptive. Nancy Datan (1986) writes about the Oedipal conflict: "I have a son who was once five years old. From that day to this, I have never thought Freud mistaken."

Obviously, these bits of "evidence" do not prove that Freud was correct. But Freud's description of the phallic stage seems not to be as bizarre as it first appears to be.

Theodore Lidz (1976), a respected developmental psychiatrist, offers a plausible explanation of the process evident in my daughters and in many other children. Lidz believes that all children must go through an Oedipal "transition," overcoming "the intense bonds to their mothers that were essential to their satisfactory pre-Oedipal development." As part of this process, children imagine becoming an adult and, quite logically, taking the place of the adult of their own sex whom they know best, their father or mother. This idea must be dispelled before the sexual awakening of early adolescence; otherwise an "incestuous bond" will threaten the nuclear family, prevent the child's extrafamilial socialization, and block his or her emergence as a well-adjusted adult. According to Lidz, the details of the Oedipal transition vary from family to family, but successful desexualization of parent-child love is essential for healthy maturity.

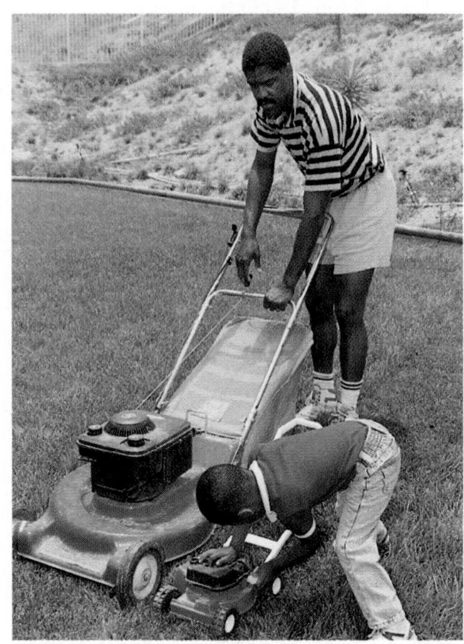

While parents almost all believe that they should, and that they do, treat their boys and girls the same, the fact is that in two-parent households, fathers share more activities with sons, and mothers share more activities with daughters. If the parents divide household tasks in the typical way, with the yard, car, and garbage being the man's responsibility, and cooking, cleaning, and shopping the woman's, the children will most likely follow typical gender roles.

The explanations of gender differences offered by learning theories take on an added dimension when seen through a sociocultural perspective. Traditional cultures, in fact, emphasize gender distinctions even more than modern cultures do, and these distinctions are quickly evidenced in the sex-role patterns adopted by the children. For example, Mexican children idealize the gallant and *macho* man and the graceful and nurturant woman even more than Anglo children do (Stiles et al., 1990). In rural communities throughout the world, girls usually tend the chickens and younger children, while boys tend the larger animals, such as sheep, pigs, and cattle. In many societies where adults have quite distinct gender roles, girls and boys virtually never play together and they attend sex-segregated schools, beginning in kindergarten (Beal, 1994; Whiting & Edwards, 1988).

A Synthesis

The entire topic of early gender differences—a "hot" controversy a decade ago—now finds most contemporary developmentalists in agreement. The learning theory and sociocultural explanations for sex-role development are very persuasive, especially in light of the differing activities that are deemed appropriate for boys and girls in different cultures. Indeed, Freud's emphasis on the deterministic link between sexual organs and gender development seems the product more of Freud's own culture—the repressive Viennese society at the turn of the century—than of a universal process.

However, the idea that *some* gender differences are biologically rather than culturally based has found acceptance on new grounds. Recent research in neurobiology strongly suggests that, in some respects at least, the two sexes are different because of subtle differences in brain development. For example, in females, the corpus callosum tends to be thicker, and overall brain maturation seems to occur more quickly. This could account for the fact that girls tend to be slightly ahead of boys in skills, such as reading and writing (and perhaps overall creativity) (Dudek et al., 1994), that demand simultaneous coordination of various areas of the brain. In males, right-hemisphere activity and dendrite formation tend to be more pronounced

than in females. This could account for the typical superiority of boys in spatial representation, a superiority that eventually leads to an unequal gender ratio among engineers, physicists, and airplane pilots.

These differences are probably not directly genetic but rather are linked to the differing sex hormones that begin to circulate before birth and that continue to influence nuances of brain development throughout childhood (Gaulin, 1993; Hines, 1993). Most of the neurological evidence for this link among normal children is only suggestive, but some confirmation comes from girls with congenital adrenal hyperplasia, which causes their hormone balances to be more similar to a boy's than to a girl's. Even in preschool years, such girls tend to prefer boys' toys and boys' play activities (Berenhaum & Snyder, 1995).

As we learned in Chapter 8, experiences also influence brain development. Thus, if boys' brains are already keyed to be interested in spatial relationships, and if boys have experiences that enhance right-hemisphere development (for example, building structures with Legos, running free in the neighborhood and the like), by middle childhood they may be that much more interested in "masculine" play activities, school subjects, and future careers than are girls who were given dolls and coloring books to play with and were kept at home.

The fact that biological forces may underlie some gender differences in activities and roles might explain why children universally seem to prefer same-sex playmates and certain kinds of toys and games during early and middle childhood. At the same time, the fact that brain development is influenced by children's particular experiences would explain why, in those cultures and families that encourage each child to develop his or her own inclinations, many children grow up to choose roles, express emotions, and develop talents that would be discouraged, even taboo, in a more restrictive culture. Now let us look more broadly at the influence of parents and culture to see how preschoolers are affected by their social context.

Relationships and Psychosocial Growth

The social world of the preschooler is considerably broader than that of the infant. During the play years, even 2-year-olds enter peer networks (in day care, in the neighborhood, or perhaps in a play group organized by parents) that bring them into contact with many new acquaintances and friends. They are also likely to know more adults—their immediate neighbors and others who live nearby, as well as people who play specific roles in their lives, such as the mail carrier, a favorite grocer, and the babysitter. By age 4 or 5, most children interact with dozens of people every day, in preschool and in the community.

Despite their broadening social world, however, psychosocial growth for most children during the play years remains fundamentally guided by the close relationships they share with their parents and siblings, and, secondarily, with peers. These relationships have such an impact because they are based upon a history of personal interaction that creates deep emotional ties and consolidates social expectations, allowing the child to anticipate, say, generosity or rambunctiousness from a close friend, or to rely on a parent for comfort when it is needed. Such long-standing close relation-

relationship perspective Focuses on the diverse ways the quality of children's relationships affects the course of psychosocial development.

ships also foster socioemotional skills (such as how to cheer up a sibling who is sad or how to avoid distracting a parent who is trying to concentrate on a task) and contribute to the growth of self-understanding by providing the child repeated encounters with those who know him or her well.

Developmentalists' awareness that psychosocial growth emerges mainly from experience in close relationships has led to the emergence of the **relationship perspective** (Dunn, 1993; Hartup, 1989; Hinde, 1987; Hinde & Stevenson-Hinde, 1987). With respect to early childhood, this perspective focuses on the ways variations in the quality of children's relationships—such as whether the parent-child relationship inspires security or uncertainty, or whether peer relationships are supportive or undermining—can profoundly affect the course of early psychosocial development. This perspective also highlights the mutual influences of these relationships, pointing out, for instance, that young children who have troubled family relationships are likely to have difficult peer relationships as well.

The Parent-Child Relationship

The parent-child relationship is critical for psychosocial development because of the myriad ways that parents guide the life experience of their offspring. From big decisions—such as what neighborhood the family lives in, whether or not the child attends preschool, and so on—to small ones—such as how to respond to a child's requests for more playtime, more information, or more dessert—parents' child-rearing choices affect the emotional well-being, intellectual growth, and social competence of their children. While it is also true that children profoundly affect the lives of their parents (indeed, simply having a child to care for is the most dramatic change that most adults experience after adolescence), that is probably no more significant an influence on early psychosocial growth than parents' approach to parenting.

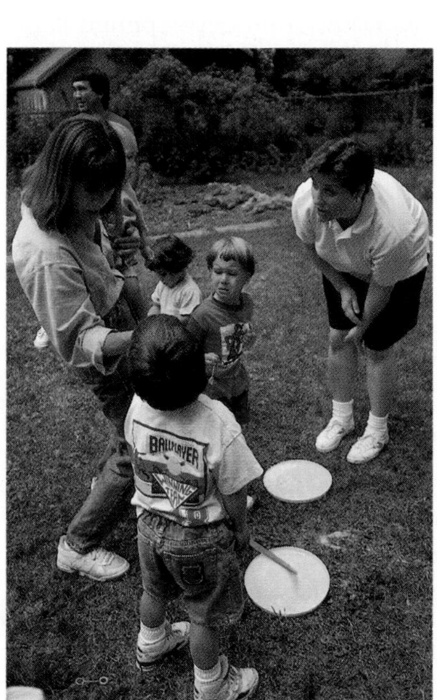

Confronted with a conflict situation such as this, parents might react in a number of ways. Some, for example, might try and punish the initiator of the dispute; others might try to help both parties understand the consequences of their actions; others might leave the combatants to settle things for themselves. Each of these responses is indicative of specific parenting styles that have been identified by researchers.

Styles of Parenting

What kinds of parenting help children to develop a positive sense of themselves, to interact positively with others, and to be competent at school? This question has no simple, universal answer because there is no guaranteed cause-and-effect relationship between how a parent rears a child and how a child turns out. Indeed, parents adopt many effective styles—ranging from quite strict to very permissive, from intensely involved to rather relaxed—and a child reared in one style may not be markedly different from a child reared in another. Conversely, children raised in the same household may differ quite notably in their response to the same parenting style.

However, nearly twenty years of careful research have led to an important insight about the impact of certain parental styles—not that one style is always best, but that some styles are more likely than others to produce confident and competent children. The seminal study in this research was begun in the early 1960s, when Diana Baumrind set out to study 100 middle-class preschool children in California (Baumrind, 1967, 1971). She used many measures of behavior, several of them involving naturalistic observation. First, she observed the children in their nursery school activities and, on the basis of their actions, rated their self-control, independence, self-confidence, and the like. She then interviewed both parents of each child and observed parent-child interaction in two settings, at home and in the

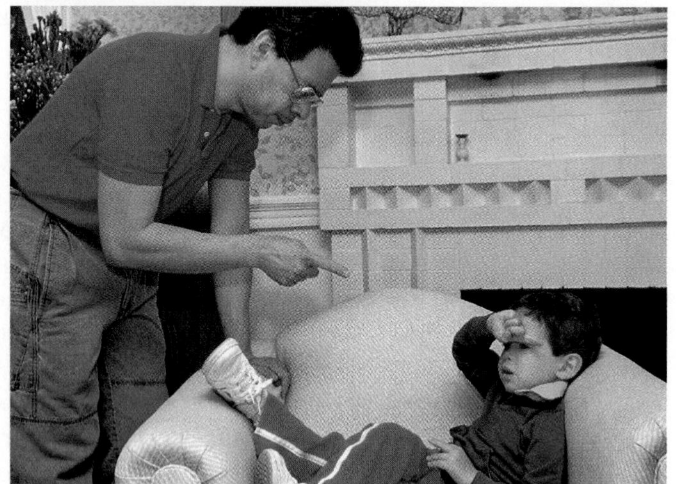

Both authoritarian and authoritative parents might sometimes scold a child. The difference is that the authoritarian parent tolerates no back talk, and is likely to use physical punishment in response to a display of disrespect. The authoritative parent, however, might listen to a child's response, paying close attention not only to the child's words but also to his or her gestures and body language.

laboratory, in order to see if there was any relationship between the parents' behavior with the child and the child's behavior at school.

There were four features of parenting that stood out in Baumrind's observations and interviews. First, parents differed in their warmth, or *nurturance,* toward offspring. Second, they varied also in their strategies to *control* the child's actions through explanation, persuasion, and/or punishment. Third, parents also differed in the quality of *communication* with offspring. Fourth, and finally, they varied in their *maturity demands*—that is, in their expectations for age-appropriate conduct. On the basis of these features, Baumrind delineated three basic patterns of parenting.

authoritarian parenting A style of child-rearing in which standards for proper behavior are high, misconduct is strictly punished, and parent-child communication is low.

permissive parenting A style of child-rearing in which parents rarely punish, guide, or control their children but are nurturant and communicate well with their children.

authoritative parenting A style of child-rearing in which the parents set limits and provide guidance and are willing to listen to the child's ideas and make compromises.

1. **Authoritarian** The parents' word is law, not to be questioned, and misconduct brings strict punishment. Authoritarian parents seem aloof from their children, showing little affection or nurturance. Maturity demands are high, and parent-child communication is rather low.
2. **Permissive** The parents make few demands on their children, hiding any impatience they feel. Discipline is lax. Parents are nurturant, accepting, and communicate well with offspring. They make few maturity demands because they view themselves as available to help their children but not as responsible for shaping how their offspring turn out.
3. **Authoritative** The parents in this category are similar in some ways to authoritarian parents, in that they set limits and enforce rules, but they are also willing to listen receptively to the child's requests and questions. Family rule is more democratic than dictatorial. Parents make high maturity demands on offspring, communicate well with them, and are nurturant.

The characteristics of each of these styles are summarized in Table 10.1.

TABLE 10.1

Characteristics of Three Basic Parenting Styles

	Warmth	Parent-to-Child Communication	Child-to-Parent Communication	Maturity Expectations	Guidelines and Rules	Punishment
Authoritarian	Low	High	Low	High	High	Strict, often physical
Permissive	High	Low	High	Low	Low	Low
Authoritative	High	High	High	High	High	Moderate, withdrawal of privileges

Baumrind and others have continued and extended this research, following the original children as they grew into adulthood and studying thousands of other children of various backgrounds and ages (Baumrind, 1989, 1991; Clark, 1983; Lamborn et al., 1991; Steinberg et al., 1989). The basic conclusions of the original studies have been confirmed: children whose parents are authoritarian are likely to be obedient but not happy; those whose parents are permissive are likely to be even less happy and to lack self-control; those whose parents are authoritative are more likely to be successful, happy with themselves, and generous with others (Darling & Steinberg, 1993; Maccoby, 1992). Follow-up research has also found that the initial advantages of the authoritative approach are likely to become even stronger over time (Steinberg et al., 1994). Authoritative parents, for example, "are remarkably successful in protecting their adolescents from problem drug use and in generating competence" (Baumrind, 1991).

Other Styles

Later research has also found that the original description of only three types of parenting was too limited. While various studies have proposed several new types—more than can be described here—three additional styles merit attention.

Two of these additional styles—*democratic-indulgent* and *rejecting-neglecting*—are subtypes of the permissive pattern. While both of these styles are quite undemanding and uncoercive—that is, the parents rarely control, restrict, or punish the children unless health or safety is obviously jeopardized—**democratic-indulgent** parents are quite warm and responsive. They take a laissez-faire approach in the interests of their children's immediate happiness. By contrast, **rejecting-neglecting** parents are quite cold and unengaged. Although they fall far short of the extreme neglect that characterizes official maltreatment, rejecting-neglecting parents permit the child to do almost anything and seem relatively uninvolved in, and even ignorant about, what the child actually does (Baumrind, 1991; Lamborn et al., 1991; Maccoby & Martin, 1983).

Another distinct type of parenting is called **traditional** (Baumrind, 1989). Parents in this category take somewhat old-fashioned male and female roles, the mother being quite nurturant and permissive, while the father is more authoritarian. Longitudinal research suggests that traditional and democratic-indulgent parenting are midway on the scale of successful parenting—less successful than consistently authoritative parenting but more successful than authoritarian or rejecting-neglecting parenting (Baumrind, 1989).

What can account for differences in parenting style? Many factors play a role, including the parent's specific child-rearing goals as well as his or her beliefs about the nature of children, the proper role of parents, and the best way to raise children (Goodnow & Collins, 1990; Murphey, 1992; Sigel et al., 1992). These combined influences are themselves shaped by factors related to culture, religion, ethnicity, and gender.

A number of other influences also contribute to shaping parenting style. One of these is the parent's personality—his or her quickness to anger, capacity for empathy with offspring, tendency toward optimism, and so

democratic-indulgent parenting A style of parenting that is warm, responsive, and permissive.

rejecting-neglecting parenting A style of parenting that is cold, detached, and permissive.

traditional parenting A style of parenting in which the parents take traditional male and female roles, the mother being primarily nurturant and permissive, while the father is more authoritarian.

forth (Dix, 1991). Another influence on parenting style is the family's economic well-being. Parents who are coping with the stress of poverty and related problems, for example, may not have the psychological energy to be authoritative parents and, instead, may demand obedience, use physical punishment to maintain control, and express less affection (Carter & Midlemiss, 1992; McLoyd, 1990; McLoyd & Flanagan, 1990). Also important is how parents remember their own upbringing (Ainsworth & Eichberg, 1991; Main & Hesse, 1990). Adults are sometimes astonished to find the legacy of their parents' parenting style affecting their interactions with their own offspring. Such a legacy often becomes apparent when particular admonishing phrases, or affectionate gestures, or modes of discipline, which seem to come to the parent naturally, even unconsciously, are suddenly recognized as virtually identical to what the parent experienced as a child.

Finally, children may influence parenting patterns as much as they are influenced by them. Hostile, unruly, or unreliable children, for example, may elicit overcontrolling behavior from adults, while children who are pleasant, self-reliant, and self-controlled may make it easy for parents to be relaxed and flexible in their approach to child-rearing. In fact, with experience, many parents tailor their child-rearing practices to fit their child's unique personality because they learn what works, what is ineffective, and what is overkill. A parent's pointed criticism, for example, may be taken in stride by a child who is assertive and outgoing but may wither one who is temperamentally fearful or inhibited (Kochanski, 1991, 1993).

Discipline

A critical task of parenting is to manage the child's behavior so that the child can grow up safely, competently, and securely. At first glance, the controlling, punishing style of authoritarian parenting might seem the most effective way to meet this challenge. One reason is that, initially, physical punishment seems to be effective: it usually stops the child from misbehaving at the moment and it provides an immediate outlet for a parent's anger or frustration. However, physical punishment does not bring about the kind of long-term compliance and cooperation that most parents want from their children. Indeed, children who regularly experience physical punishment tend themselves to become hostile and aggressive (Weiss et al., 1992). (The effects of spanking are examined in the Research Report on p. 290.)

In the long term, the most effective type of discipline is that typically associated with the authoritative parenting style—not because of the absence of punishment, but rather because of the authoritative parent's commitment to maintaining a reciprocally cooperative interaction with the child (Maccoby, 1994). The affectionate quality of the parent-child relationship, together with the parent's respect for the child as an independent person, lead the parent quite naturally to use praise, encouragement, and other kinds of positive reinforcement for good behavior. The open communication that typifies authoritative families encourages the parent to set clear standards in terms the child can understand. This open communication—in which the child is really listened to—also helps the parent to recognize when the child's failure to comply is for a good reason, such as a misunderstanding or a cognitive or psychological inability to act in an expected way.

RESEARCH REPORT

The Effects of Spanking

More than 90 percent of today's American adults were spanked when they were young, and most consider themselves none the worse for it. Indeed, although the tide of public opinion is changing regarding physical abuse (see Chapter 8), most parents still believe that spanking is acceptable, legitimate, and necessary at times (Holden & Zambarano, 1992; Straus & Gelles, 1986). Nonetheless, history teaches that widespread acceptance of any child-rearing practice does not necessarily prove that it is good for children. Because of the demonstrated link between physical abuse in childhood and later violent aggression (Lewis et al., 1989), many developmentalists now wonder if children who are spanked learn to be more aggressive than children who are not spanked.

To try to answer this question, one research team (Strassberg et al., 1994) studied 273 children and their parents. The parents were from a full range of socioeconomic backgrounds, and roughly a third were single parents. The children were about evenly divided between boys and girls, and about three-fourths were European-American and one-fourth were African-American.

In the spring before their children entered kindergarten, the parents were asked how frequently they had spanked, hit, or beaten their children over the past year. Spanking was defined as involving "an open hand or an object on the child's buttocks in a controlled manner, whereas hitting involves the impulsive or spontaneous use of a fist or closed hand (or object) to strike the child more strongly than one would while spanking." Of the 408 parents surveyed, 9 percent did not use physical punishment at all, 72 percent spanked but did not use more violent punishment, and 19 percent hit or beat, as well as spanked, their preschool child.

Six months later, observers blind to the child's punishment history recorded each child's behavior in kindergarten, taking particular note of acts of aggression. In order to get an accurate snapshot of behavior, the observation was divided into twelve 5-minute segments occurring over several days. Within each segment, the observers recorded how many times the child engaged in one of three types of aggression:

1. *instrumental aggression,* used to obtain or retain a toy or other object;
2. *reactive aggression,* used in angry retaliation against an intentional or accidental act committed by a peer;
3. *bullying aggression,* used in an unprovoked attack on a peer.

Although it is sometimes difficult to differentiate these three types of aggression, the trained observers who independently watched each child agreed 96 percent of the time on the occurrence of aggression and 90 percent of the time on the type of aggression that had occurred.

Analysis of their data revealed that instrumental aggression was not much affected by the type of punishment a child experienced in the home. In other words, a kindergarten child was just as likely to fight over a toy whether he or she was spanked, hit, beaten, or not corporally punished at all. Bullying aggression, as expected, was clearly associated with being violently punished, particularly in the case of "a few extremely aggressive children," mostly boys who were frequently hit or beaten as well as spanked by both parents.

The surprising finding was the clear relationship between spanking and reactive aggression. Compared to children who were not spanked, those children who were spanked were three times as likely to retaliate with an angry shove, punch, kick, or the like to any wrong, real or imagined. In their analysis of the data, the researchers point out that while violent punishment seems to lead a child to be more aggressive under all circumstances, spanking does not seem to model the use of force in general. Rather, it seems to create a quite specific emotional-response pattern in the child—that is, a quick physical reaction to possible attacks. Because the "anger accompanying the spanking is highly salient to the child," the child models "the emotional behavior pattern and not the form of aggression, per se."

Note that this is a correlational study and, as has been pointed out on numerous occasions, correlation does not prove causation. It could be that spanking was more consequence than cause in this study. That is, it may have been that many of the children who were spanked "provoked" this response with angry and aggressive behavior in the home and that their hostile behavior merely carried through into kindergarten. However, several factors—including that girls were almost as likely to be spanked as boys even though they were half as aggressive—suggests that the parents' choice of punishment was related more to their own attitudes and temperament than to the actions of the child. Close analysis of all their data led the researchers to conclude that "in spite of parents' goals, spanking fails to promote prosocial development and, instead, is associated with higher rates of aggression toward peers." This conclusion is in line with other research on the effects of corporal punishment (Straus, 1994) and, increasingly, with the American public's view of discipline practices. In 1968, a survey found that 94 percent of all American adults believed that spanking is sometimes necessary. A repeat of this survey in 1994 found that only 68 percent of American adults approved of spanking (Collins, 1995).

Authoritative parents also provide a positive model for expected behavior, listening and considering before acting. This is in marked contrast to the model of aggressive behavior provided by the authoritarian parent, who spanks first and asks questions later, or not at all ("I don't want to hear any excuses"). Finally, authoritative parents are alert to ways to prevent trouble before it occurs: they may, for example, bring small toys to a restaurant if the child is going to endure a long wait; they dress the child in old clothes for outdoor play; they don't give the child so much responsibility that trouble is inevitable (Holden, 1983; Holden & West, 1989).

In a sense, therefore, a positive parent-child relationship helps to make the process of discipline somewhat easier over the years. As one developmentalist explains, "If parents can do what is necessary early in the child's life to bring about a cooperative, trusting attitude in the child, that parent has earned the opportunity to become a nonauthoritarian parent" (Maccoby, 1984).

Parenting in Context

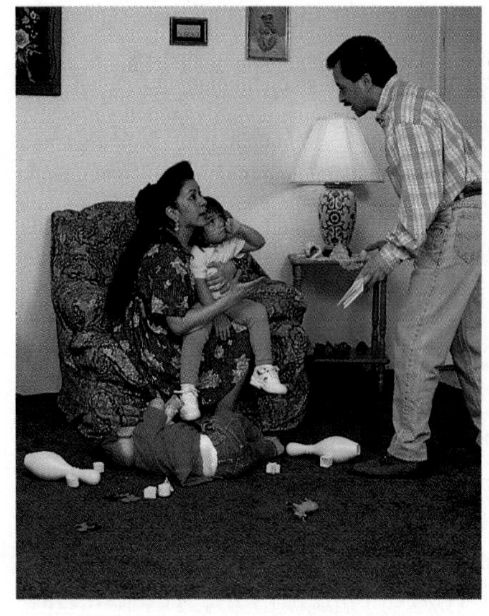

Marital stress can affect children in various ways—by heightening the likelihood that frustrated parents will act punitively with offspring, by increasing children's distress and anger when they hear their parents arguing, and by providing a salient model of conflict within the home. As a consequence it is not surprising that when parents begin fighting, the rest of the family often erupts in conflict and anger as well.

The parent-child relationship obviously does not occur in a psychosocial vacuum. Each of its components is affected by other family members. As we will see in the next section, siblings—their number, age, and gender, as well as the nature of their mutual relationships—strongly influence each parent-child dyad. Each parent-child dyad is also influenced by the relationship between the parents and by the support—or lack of support—they provide each other for good parenting. When the marriage is satisfying and mutually supportive, for example, both parents tend to be authoritative or traditional, together setting high standards and responding with pleasure when their children meet them (Goldberg, 1990).

Moreover, the interactions among all family members are also affected by broader cultural and economic processes that have an impact on each person. The society's attitude toward maternal employment and public day care, the economic stresses on young families, the availability of various forms of social support, and the safety and child-friendliness of the neighborhood are just a few of the factors that influence family dynamics and, in turn, parenting style. Thus, in order to understand how the parent-child relationship affects early psychosocial growth, we need to consider the overall family context and the social ecology in which the family lives.

We also need to bear in mind that these same factors can influence the impact that any particular parenting style might have. For example, although authoritative parenting usually contributes to making children more confident, self-controlled, and successful, in some contexts other styles may be more effective (Darling & Steinberg, 1993; Maccoby, 1992). When adolescents from different ethnic groups are compared, for instance, those with the highest grades come from groups (such as Chinese-Americans) in which parents are more authoritarian, and less authoritative, than parents from other ethnic groups (Steinberg et al., 1991, 1992). Within groups, the impact of a particular pattern may be affected by the ethnic context as well. For instance, a recent study of more than 6,000 adolescents from California found that, as expected, the European-American youths who fared best—both in academic achievement and self-concept—came from authoritative homes, but, unexpectedly, the African-American youths who fared best came from authoritarian homes. If anything, this was even more true for middle-class

African-American youth, particularly those living in predominantly white neighborhoods (Lamborn et al., 1996). One possible explanation for findings such as these points to differing ethnic and cultural perceptions of parenting styles. Several researchers have suggested that for adolescents in some minority groups, a certain degree of strictness is interpreted as supportive and protective, while teenagers from the majority group experience the same parental approach as rejecting (Chao, 1994; Taylor et al., 1993). The importance of the cultural interpretation of the parents' behavior is suggested by another altogether different culture in which permissiveness is the expected pattern—the Aka Pygmy culture of the Central African Republic. In this culture, which is characterized by an emphasis on self-reliance, self-esteem, and shared caregiving, parents raise their children in a notably permissive fashion—and for the most part their children grow up cheerful, independent, and nonaggressive (Hewlett, 1992).

Examples like these—of which there are many—highlight the fact that culture, ethnicity, and community play important roles in parenting patterns, affecting not only parental goals and values but also the appropriateness and effectiveness of various parenting styles. In chaotic and dangerous environments, such as the urban ghettos of the contemporary United States, responsible parenthood in a particular family might well require a high level of parental control, whereas the same family in a more secure and stable setting might find a democratic-indulgent style to be preferable and more effective. Thus, because parenting is embedded within a broader network of practices, beliefs, and supports from outside the family, the effects of parental style must always be regarded within the context of culture and community.

While hidden jealousy and open conflict between siblings are common, so are joint activity and obvious tenderness. Older siblings throughout the world can identify with these two sisters from the Yunnan region of China, who are quite delighted that baby brother accepts their offer of a drink.

Sibling Relationships

All told, sibling relationships are often the most intense, intimate, and long-lasting a person will have, with patterns begun in early childhood sometimes lasting for seventy years or so (Cicirelli, 1995). During the preschool years, having an older sibling is to have an enticing model for behavior, a source of learning, and occasionally a source of comfort and security. Having a younger sibling is an important benchmark of social comparison ("Rosa can't jump rope, but *I* can!") and permits the growth of new social skills of nurturance and dominance.

However, having a sibling, particularly when both children are under age 6, is a mixed blessing. Brothers or sisters who are close in age are more likely to compete with, quarrel with, and physically attack each other—as well as to nurture, cooperate with, and play with each other—than they are with unrelated children the same age (Howe & Ross, 1990). Among the factors that increase the intensity of the relationship, for good or ill, are age and sex. Same-sex children who are close in age are most likely to be rivals as well as buddies.

Another factor affecting the nature of the sibling relationship is the compatibility of the siblings' temperaments. If siblings have contrasting personalities, such as one being strongly emotional and the other being placid, or one being very physically active and the other being sedate, they are more likely to have conflicts (Munn & Dunn, 1989; Stoneman & Brody, 1993).

As you might expect, parental influences are also important in shaping sibling relationships, often from the very first moments that a child becomes

a sibling. One study found that some parents are more likely to discuss a child's new brother or sister as a person, with feelings and desires that can and should be understood, and that taking this step increases the likelihood that the older sibling will be interested in, and nurturant with, the younger one (Dunn, 1988). Parents also influence a sibling relationship through differential treatment of their children. For example, many parents continually bestow more responsibility on the older child. A single parent of a 2-year-old and a 5-year-old might regularly send the older child to the corner store to buy milk (depending, of course, on the safety of the street) and, three years later, when out of milk, might still be sending the older child rather than the younger, new 5-year-old. Especially if such differential treatment also includes numerous instances of having the older keep the younger from harm ("Make sure he doesn't fall while I get something from the bedroom"), the relationship between the two siblings is likely to be dominant-dependent, or perhaps bossy-babyish, long after they are grown (Cicirelli, 1995).

Patterns of differential treatment are not based simply on age. For example, one study of families with school-age children found that in 28 percent of the families, the parents disciplined the younger sibling more; in 17 percent, they disciplined the older sibling more; and in 37 percent, the mother and the father differed in terms of which child they disciplined more. In only 18 percent of the families did both parents say they disciplined their children equally (McHale et al., 1995).

While parents believe that they have good reason to treat their children differently, the children themselves typically do not agree. By the school years, children's complaints of unfairness abound in almost every family, with the older children feeling that the younger ones are spoiled and the younger ones believing that the older ones get special privileges.

The seeds of such sibling rivalry and resentment are usually sown in the preschool years, when two—or more—children are both competing for their mother's attention. A typical example comes from a 2½-year-old, Andy, and Susie, his 14-month-old sister:

> Andy was a rather timid and sensitive child, cautious, unconfident, and compliant. His younger sister, Susie, was a striking contrast—assertive, determined, and a handful for her mother, who was nevertheless delighted by her boisterous daughter. In the course of an observation of Andy and his sister, Susie persistently attempted to grab a forbidden object on a high kitchen counter, despite her mother's prohibitions. Finally, she succeeded, and Andy overheard his mother make a warm affectionate comment on Susie's action: "Susie, you *are* a determined little devil!" Andy, sadly, commented to his mother, "*I'm* not a determined little devil!" His mother replied, laughing. "No! What are you? A poor old boy!" [Dunn, 1992]

Multiplied hundreds of times over during Andy's childhood, such feelings of inadequacy might spur him toward success or failure, bravery or timidity, but they certainly would not have neutral effects. Not all siblings are rivals, but almost all are reciprocally involved in ways that affect the personality and social understanding of each (Dunn & Plomin, 1990).

It is easy to see that, given the attentiveness they bring to their relationship, siblings might contribute more to the development of social skills than anyone else, since they are likely to guide, challenge, and encourage a child's social interactions more frequently and intimately than most others do. Very practical lessons in self-defense, sharing, and negotiation are part of every sibling's childhood.

The Only-Child

This raises questions about the child who is without brothers or sisters, the so-called only-child. How common are such children, and how does their family configuration affect their development? Actually, it is impossible to know for certain how many of today's children are true only-children, since many children spend a good part of their childhood as an only-child and then are joined by a younger sibling or a step-sibling. However, worldwide, the number of both true only-children and temporary only-children is clearly rising.

For example, in the United States the percentage of families with children with only one child under age 18 rose from 33 percent in 1970 to 41 percent in 1994. In fact, spending at least several years with no brothers or sisters at home is now a common experience for young American children: in 1994, about half of all children under age 6 were the only such child in their family (U.S. Bureau of the Census, 1996).

Family composition in many other nations of the world is likewise shifting from many children to only one or two. The example of this trend is most dramatic in China, where the government's strict "one-child policy" to combat overpopulation has resulted in many single-child families. Indeed, in 1991, 95 percent of the student body of most primary schools in Beijing consisted of only-children (Bakken, 1993).

Are only-children handicapped because they have no brothers or sisters? The general public still believes this to be the case. (In China, especially among the older generations, there is widespread prejudice against only-children, especially boys, who are stereotyped as lonely, spoiled, overly dependent on their parents, disrespectful "little emperors," and potential delinquents.) The truth is usually otherwise: only-children are more likely to benefit from increased parental attention than to suffer from lack of siblings (Falbo & Polit, 1986; Falbo & Poston, 1993; Mellor, 1990). Single-child status is particularly beneficial intellectually, with only-children being generally more verbal, more creative, and more likely to attain a college education than children who have one or more siblings. Only-children are particularly advantaged when compared to children from families with four or more children, even when the economic disadvantages of larger families are taken into account. These differences are apparent in China, in Europe, and in North America (Bakken, 1993; Blake, 1989; Yang et al., 1995).

The one area of development in which only-children might be disadvantaged is their social skills, particularly in the development of cooperative play, theory of mind, negotiation strategies, and self-assertion, which usually are enhanced through sibling interactions (Falbo & Poston, 1993). However, as preschool education and public day care become the rule rather than the exception in industrialized nations, most only-children develop social skills that are comparable to those of their contemporaries with siblings.

Peer Relationships and Play

The play years are also a period for developing friendships outside the home. Whereas it was once typical for young children to spend most of their time at home (except for occasional forays with their mother to attend a

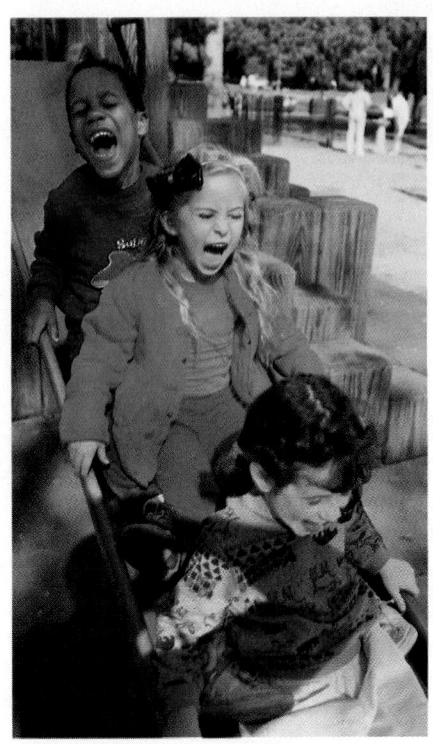

Peer relationships develop during the play years, enabling the cooperation, and the sheer pleasure, evidenced here.

play group or nursery school or to visit another child in the neighborhood), most children today spend a large portion of each weekday in a preschool program or day-care center. As we saw in Chapter 9, high-quality day-care or preschool programs can have benefits for cognitive growth. The same is true of psychosocial development: children in well-run programs acquire a wide range of social skills and become more socially competent as a result of their frequent interactions with other children of the same age (Hayes et al., 1990; Zaslow, 1991; Zigler & Lang, 1990). This can be a mixed blessing, however, for children in group day care not only may learn how to be more helpful and cooperative with peers and become more socially knowledgeable but also may become more assertive and aggressive than children without extensive day-care experience. Peer encounters in day care typically force children to learn to defend their interests, whether keeping their favorite toy or their place in line.

In fact, the play years witness the growth of many social skills (Erwin, 1993; Howes, 1987). This is easy to see when we compare the 2-year-old, whose peer interactions consist mainly of simple cooperative games (like bouncing and catching a ball or hiding and finding a partner), with the more sophisticated interactions of a 5-year-old, who has learned how to gain entry to a play group, can manage conflict through the use of humor, and has structured his or her peer world into friendships and acquaintanceships (Corsaro, 1985). Why peer relationships contribute to such psychosocial growth is fairly obvious. With peers, children are on a much more equal footing than they are with adults, and they must therefore assume greater responsibility for initiating and maintaining harmonious social interaction. In play with peers—whether learning how to share crayons or sand toys, or how to include everybody in the construction of a spaceship, or how to respond to a friend's accusation that "it's not fair!"—children cannot rely on their partners to make all the effort, as they can in encounters with adults. Thus, peer encounters in early childhood afford crucial experiences that teach reciprocity, cooperation, and justice, experiences that would be hard for adults to provide (Eisenberg et al., 1985; Howes, 1987).

In their peer encounters, preschoolers also learn to distinguish among other children in terms of their cooperativeness, friendliness, and "likability" (Denham & Holt, 1993; Denham et al., 1990). As they learn about the characteristics of their peers, children develop friendships based, in part, on their peers' reputations for generosity, aggressiveness, and the like. This, in turn, helps to explain why friendships are remarkably consistent during the play years: young children tend to choose their friends as regular playmates, and their play together, in its complexity, self-disclosure, and reciprocity, becomes distinctively different from play with casual acquaintances (Hinde et al., 1985; Howes, 1983; Park et al., 1993).

Varieties of Play

Most developmentalists believe that play is the work of early childhood. Indeed, during the preschool years, play serves not only as a means of social interaction but also as an avenue for motor development, intellectual growth, and self-discovery. Two types of play, *rough-and-tumble* and *sociodramatic,* are particularly notable for the opportunity they provide for developing social skills.

For many young children, especially boys who know each other well, rough-and-tumble play brings the most pleasure. Many developmentalists believe that this kind of play teaches social skills—such as how to compete without destroying a friendship—that are hard to learn any other way.

Rough-and-Tumble Play

With every passing year from age 2 to 6, children are more likely to engage in a particular form of physical play called **rough-and-tumble play**. The aptness of its name is made clear by the following example:

> Jimmy, a preschooler, stands observing three of his male classmates building a sand castle. After a few moments he climbs on a tricycle and, smiling, makes a beeline for the same area, ravaging the structure in a single sweep. The builders immediately take off in hot pursuit of the hit-and-run phantom, yelling menacing threats of "come back here, you." Soon the tricycle halts and they pounce on him. The four of them tumble in the grass amid shouts of glee, wrestling and punching until a teacher intervenes. The four wander off together toward the swings. [cited in Maccoby, 1980]

One distinguishing characteristic of rough-and-tumble play is its mimicry of aggression, a fact first noted in observations of young monkeys' wrestling, chasing, and pummeling of each other (Jones, 1976). The observers discovered that the key to the true nature of this seemingly hostile behavior was the monkeys' *play face*, that is, a mildly positive facial expression that seemed to suggest that the monkeys were having fun. The play face was an accurate clue, for only rarely, and apparently accidentally, did the monkeys actually hurt each other. (The same behaviors accompanied by a threatening expression usually meant a serious conflict was taking place.)

In human children, too, rough-and-tumble play is quite different from aggression, even though at first glance it may look the same. This distinction is important, for rough-and-tumble play is a significant part of the daily activities of many preschool children. In general, rough-and-tumble play, unlike aggression, is not only fun for children; it is also constructive, developing interactive skills as well as gross motor skills (Pellegrini, 1987). Adults who are unsure whether they are observing a fight that should be broken up or a social activity that should be allowed to continue may be helped by knowing that facial expression is as telltale in children as it is in monkeys: children almost always smile, and often laugh, in rough-and-tumble play, whereas they frown and scowl in real fighting.

Rough-and-tumble play is universal, occurring everywhere children play. It has been observed in Japan, Kenya, and Mexico, as well as in every income and ethnic group in North America, Europe, and Australia (Boulton & Smith, 1989). There are some cultural and situational differences, however. One of the most important is space and supervision: children are much more likely to instigate rough-and-tumble play when they have room to run and chase, and when adults are not directly nearby. In addition, rough-and-tumble play usually occurs among children who have had considerable social experience, often with each other.

rough-and-tumble play Play such as wrestling, chasing, and hitting that mimics aggression but actually occurs purely in fun, with no intent to harm.

Not surprisingly, then, among children in nursery schools, newcomers, younger children, and only-children take longer to join in rough-and-

sociodramatic play Pretend play in which children act out various roles and themes in stories of their own creation.

tumble play than to participate in any other form of play (Garvey, 1976; Shea, 1981). Gender differences are also evident in rough-and-tumble play, and these, too, vary from culture to culture. In some cultures, such as traditional Muslim ones, girls almost never engage in rough-and-tumble play. Among North Americans, girls sometimes engage in such play, but not as often as boys do: one carefully controlled study found that boys spent three times as much time in rough-and-tumble play as girls did (DiPietro, 1981).

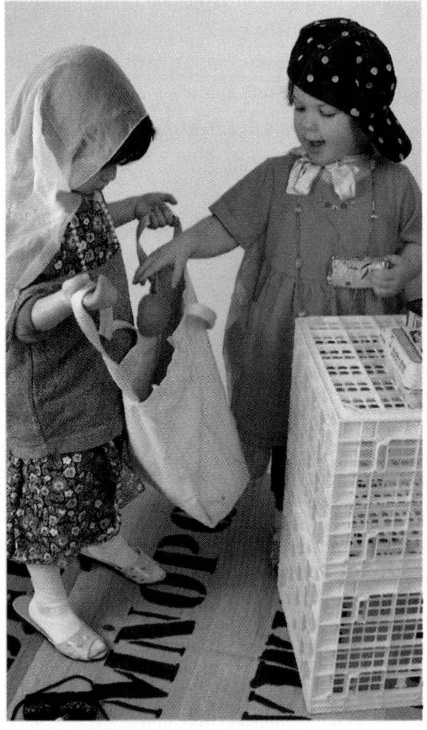

Getting all dressed up to go grocery shopping is a form of sociodramatic play that combines elements of reality and fantasy for 3-year-old Lucy and Rachel. As they grow older, such play will typically become more creative and dramatic, although not necessarily any more realistic.

Sociodramatic Play

In **sociodramatic play**, children act out various roles and themes in stories of their own creation. Besides allowing children to have great fun, sociodramatic play provides a way for children to explore and rehearse the social roles they observe around them and to examine personal concerns in a nonthreatening manner. This function of sociodramatic play is clearly apparent when children enact husband-and-wife stories, or scenarios involving sickness or death, or stories with monsters and superheroes.

The beginnings of sociodramatic play can be seen in a toddler's feeding, cuddling, or punishing a doll or stuffed animal. However, the frequency and complexity of sociodramatic play greatly increase between the ages of 2 and 6 (Howes, 1992; Rubin et al., 1983). One reason is that, as young children are expanding their theory of mind or psychological understanding of other people (see Chapter 9), they practice what they are learning in their play scenarios (Goncu, 1993; Harris & Kavanaugh, 1993; Lillard, 1993a, 1993b, 1994). For instance, young children can use sociodramatic play to try out various means of managing emotions, whether it involves reenacting scary situations in the dark, or offering comfort to a frightened doll, or rehearsing courageous action in a difficult situation (Bretherton, 1989). In a sense, sociodramatic play is a testing ground for early psychological knowledge. Its growth is also related to the development of self-understanding in early childhood. In sociodramatic play, roles are assumed, and then discarded with ease, because of the child's underlying confidence in knowing who he or she is—and is not.

It is appropriate that our final glimpse of the young child is one of the child at play, or, more specifically, a child at play with other children, in an activity that combines the magical, imaginative thinking described in Chapter 9 with the intense social involvement described in this chapter. Between the ages of 2 and 6, the child is a playful creature building various biosocial, cognitive, and psychosocial competencies to form an increasingly independent person. By the end of this period, the child is not only bigger and stronger but also wiser and more socially aware, ready for the challenges of the school years.

SUMMARY

The Self and the Social World

1. An increasing sense of self-understanding helps children to increase their social understanding and to become more skilled in their relationships with others. In turn, social interaction helps young children to learn about who they are and contributes to their self-understanding and self-evaluation.

2. While developmentalists agree that children begin to learn gender roles and gender identity during early childhood, they disagree about how this process occurs. Psychoanalytic theorists stress the fears and fantasies that motivate children to identify with the same-sex parent; learning theorists emphasize the reinforcement and

modeling that children experience at home and elsewhere. Many researchers agree that certain gender differences are biologically based as well.

Relationships and Psychosocial Growth

3. The relationship perspective holds that psychosocial growth emerges mainly from experience in close relationships, which can be mutually influential (such as when young children from troubled families have difficulties in their peer relationships).

4. Parent-child interaction is complex, with no simple answers about the best way to raise a child. However, in general, authoritative parents, who are warm and loving but willing to set and enforce reasonable limits, have children who are happy, self-confident, and competent. Highly authoritarian parents tend to produce aggressive children, while children with very permissive parents often lack self-control.

5. Although parenting style is an important element in shaping a child's development, it is only one of many such elements, and its importance should not be overestimated. Often, children shape their parent's style of child-rearing as well as being shaped by it, chiefly through the effects of their temperament and their changing developmental needs. Parenting is also affected by the nature of the marital relationship and the cultural, ethnic, socioeconomic, and community context of the family.

6. In many respects, children's relationships with their brothers and sisters are uniquely different from relationships with parents and peers. Although siblings may quarrel, they are also likely to show each other more nurturance and cooperation than they show others. One of the most important factors in shaping the relationship between siblings is the relationship each child shares with the parent.

7. Only-children tend to be more verbal and creative than children who have siblings, but they lag behind their peers somewhat in social skills unless they are in a regular play group or preschool.

8. Peer relationships are another important relational influence during the play years. In addition to providing an arena for developing their social skills, peer relationships help preschoolers learn about friendship.

9. Playing with other children requires preschoolers to take responsibility for maintaining social interaction through sharing and reciprocity. These features are evident whether the play is rough-and-tumble or sociodramatic play—the latter also permits children to explore social roles, examine personal concerns, and learn to cooperate.

KEY TERMS

initiative versus guilt (279)
Oedipus complex (281)
identification (281)
Electra complex (282)
relationship perspective (286)
authoritarian parenting (287)
permissive parenting (287)
authoritative parenting (287)
democratic-indulgent parenting (288)
rejecting-neglecting parenting (288)
traditional parenting (288)
rough-and-tumble play (296)
sociodramatic play (297)

KEY QUESTIONS

1. How does self-understanding grow during the play years? What difference does this make for young children's relationships with others?

2. Give some examples that illustrate how young children show gender identification.

3. What are the similarities, and differences, in the views of psychoanalytic and learning theorists about the origins of gender roles in early childhood?

4. What is the relationship perspective? What new insights does it provide for gaining an understanding of early social development?

5. Describe the three basic patterns of parenting, according to Baumrind's research.

6. What are some of the influences that contribute to the complexity of parenting?

7. How do children influence the nature of the parent-child relationship? How is this relationship affected in turn by social factors outside the family?

8. How does having a sibling contribute significantly to the development of social skills?

9. How do peer relationships change during the preschool years? What do peer encounters offer young children that relationships with adults do not?

10. Distinguish between the two primary kinds of play. What are the benefits of sociodramatic play to preschoolers?

The Developing Person So Far: The Play Years, Ages 2 Through 6

Biosocial Development

Brain and Nervous System

The brain continues to develop faster than any other part of the body, attaining 90 percent of its adult weight by the time the child is 5 years old. Myelination proceeds at different rates in various areas of the brain, but the overall coordination between the two halves of the brain, as well as the child's ability to settle down and concentrate when necessary, gradually improves.

Motor Skills

As the child becomes stronger and body proportions become more adultlike, gross motor skills, such as running and jumping, improve dramatically. Fine motor skills, such as writing and drawing, develop more slowly. Gender differences in motor skills become apparent.

Maltreatment

Child abuse and neglect, potential problems at every age, are particularly likely in homes with many children and few personal or community resources. During early childhood, home-visitation programs may be a useful preventive measure.

Cognitive Development

Cognitive Skills

Many cognitive abilities, including some related to number, memory, and problem solving, develop during early childhood. Throughout this period, children begin to develop a theory of mind, as they take into account the ideas and emotions of others. Social interaction, particularly in the form of guided participation, is both a cause and a consequence of this cognitive advancement. At the same time, however, children's thinking can be quite illogical.

Language

Language abilities develop rapidly; by the age of 6, the average child knows 10,000 words and demonstrates extensive grammatical knowledge. Children also learn to adjust their communication to their audience, and use language to help themselves learn. Specific contexts affect the particulars of what, and how much, children say and understand. Preschool education helps children develop language and express themselves.

Psychosocial Development

Emotions and Personality Development

Self-concept emerges, usually with a positive slant. Children boldly initiate new activities, especially if they are praised for their endeavors. As their social and cognitive skills develop, children engage in ever more complex and imaginative types of play, sometimes by themselves and, increasingly, with others.

Parent-Child Interaction

As children become more independent and try to exercise control over their environment, supervising the child's activities becomes more difficult. Some parenting styles are more effective than others in encouraging the child to develop both autonomy and self-control. At the same time, parenting styles are influenced by cultural and community standards, various environmental pressures, and the characteristics of the child.

Gender Roles

Increasingly, children develop stereotypic concepts of sex differences in appearance and gender differences in behavior. The precise role of nature and nurture in this process is unclear.

PART IV The School Years

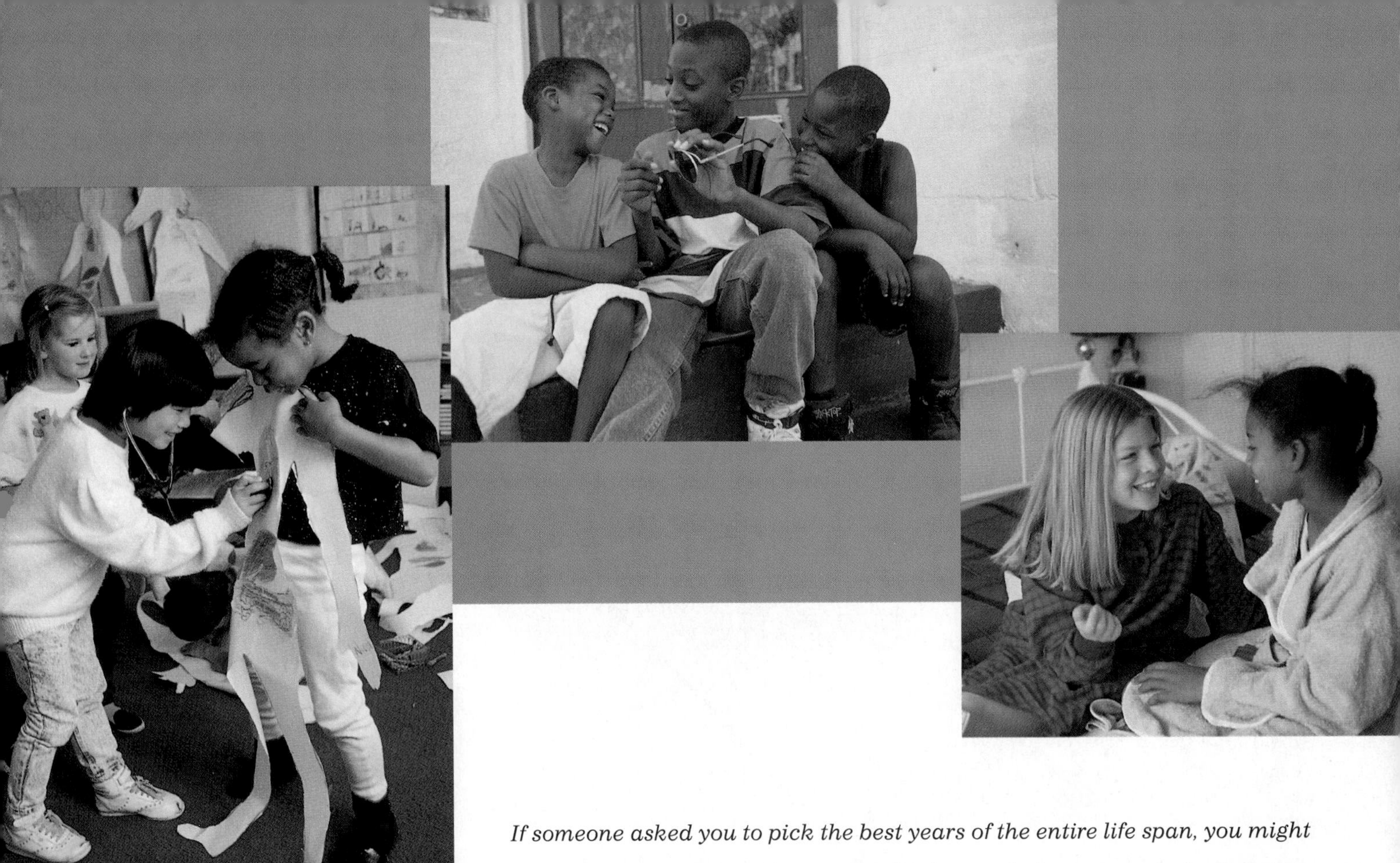

If someone asked you to pick the best years of the entire life span, you might choose the years from 7 to 11 and defend your choice persuasively. To begin with, physical development is usually almost problem-free, making it easy to master dozens of new skills. With regard to cognitive development, most children are able to learn quickly and think logically, providing the topic is not too abstract. Moreover, they are usually eager to learn, mastering new concepts, new vocabulary, and new skills with a combination of enthusiasm, perseverance, and curiosity that makes them a joy to teach.

Finally, the social world of middle childhood seems perfect, for most school-age children think their parents are helpful, their teachers fair, and their friends loyal. In addition, their moral reasoning and behavior have reached that state where right seems clearly distinguished from wrong, with none of the ambiguities that complicate moral issues for adolescents and adults. For most children, then, the future seems filled with promise.

However, school and friendships are so important at this age that two common events can seem crushing: failure in school and rejection by peers. Some lucky children escape these problems; others have sufficient self-confidence or family support to weather them when they arise; and some leave middle childhood with painful memories, feeling inadequate, incompetent, and inferior for the rest of their lives.

The next three chapters celebrate the joys, and commemorate the occasional tragedies, of middle childhood.

The School Years: Biosocial Development

Compared with other periods of the life span, biosocial development in middle childhood seems, on the whole, to be relatively smooth and uneventful. For one thing, disease and death are rarer during these years than during any other period. For another, most children master new physical skills (everything from tree climbing to in-line skating) easily and without much adult instruction, provided their bodies are sufficiently mature and they have an opportunity to practice these skills. In addition, sex differences in physical development and ability are minimal, and sexual urges are quiescent compared to what they will later be. Certainly when bodily development during these years is compared with the rapid and dramatic growth that occurs during infancy and adolescence, middle childhood, overall, seems a period of smooth progress and tranquility. For some children, however, middle childhood can be particularly challenging, because it is a time when certain disabilities first become evident or become more pronounced in their consequences. In this chapter, we will examine the bodily changes and variations that are characteristic of middle childhood, as well as several difficulties that sometimes occur in biosocial development during this period.

Size and Shape

Children grow more slowly during middle childhood than they did earlier or than they will in adolescence. Worldwide, the typical well-nourished child gains about 5 pounds (2¼ kilograms) and grows about 2½ inches (6 centimeters) per year and by age 10 weighs about 70 pounds (32 kilograms) and measures 54 inches (137 centimeters) (Lowrey, 1986). (See Figure 11.1.)

During these years children generally seem slimmer, as they grow taller and their body proportions change. In addition, muscles become stronger, enabling the average 10-year-old, for instance, to throw a ball twice as far as the average 6-year-old can. The capacity of the lungs also increases, so with each passing year children are able to run faster and exercise longer than before. These changes are accentuated in those children who take advantage of their increased strength and endurance and actually exercise.

FIGURE 11.1 As you can see, growth is quite steady throughout middle childhood, except for those girls in the 90th percentile (the heaviest 10 percent). Typically, they begin puberty at about age 10, which accounts for their increasing rate of weight gain at ages 11 and 12.

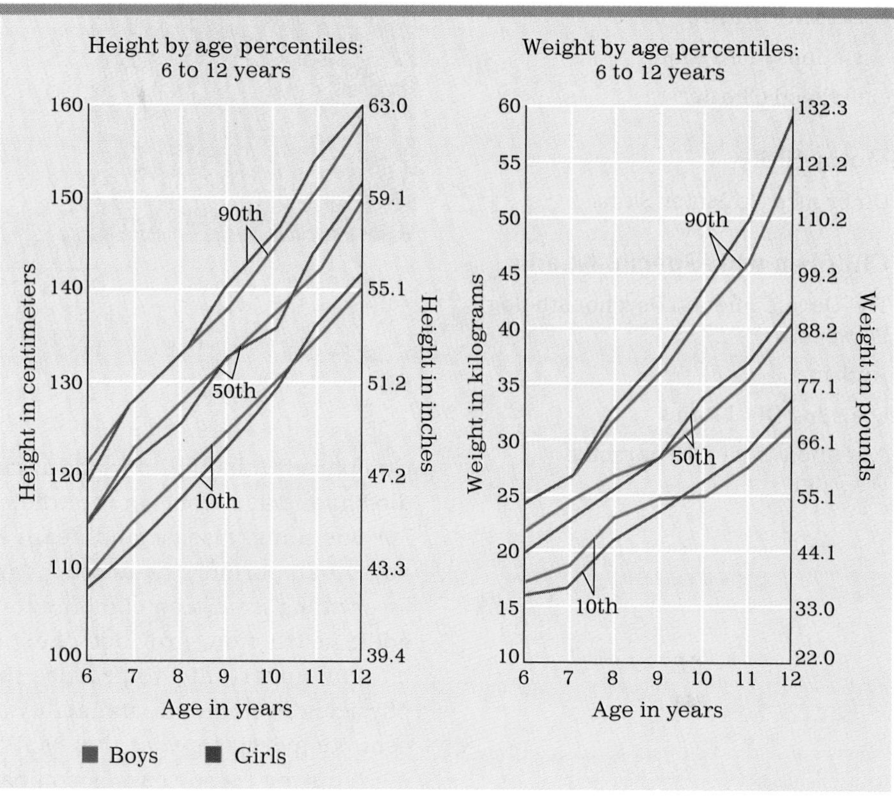

Variations in Physique

In some regions of the world, most of the variation in children's height and weight is caused by malnutrition, with wealthier children being several inches taller than their impoverished contemporaries from the other side of town—whether the town is Hong Kong, Rio de Janeiro, or New Delhi. In developed countries, heredity is the main source of variation, since most children get enough food during middle childhood to grow as tall as their genes allow.

Genetic factors and nutrition affect not only size but rate of maturation as well. This is particularly noticeable at the end of middle childhood, as some 10- and 11-year-olds begin to undergo the changes of puberty and may find that they are ahead of their peers, not only in height, but also in strength and endurance. Among Americans, those of African descent tend to mature somewhat more quickly (as measured by bone growth and loss of baby teeth) and to have longer legs than those of European descent, who, in turn, tend to be maturationally ahead of those with Asian ancestors. Such variations are quite normal and healthy.

While it may be comforting for parents and teachers to know that healthy children come in many shapes and sizes, it is not always comforting to the children themselves. In elementary school, children compare themselves with one another, and those who are "behind" their classmates in areas related to physical maturation may feel deficient. Physical development during this period even affects friendships, which are based partly on physical appearance and physical competence (Hartup, 1983). Consequently, children who look "different" or who are noticeably lacking in physical skills often become lonely and unhappy.

During the school years, variations in children's size and rate of physical maturation are the result of genetic inheritance and nutrition, as well as chronological age.

Childhood Obesity

One difference in size that, from middle childhood on, can seriously affect emotional as well as physical well-being is **obesity**. The precise point at which a particular child is not just chubby but actually obese varies, depending partly on the child's body type, partly on the proportion of fat to muscle, and partly on the culture's standards on this question. However, at least 10 percent of all American children are 20 pounds (9 kilograms) or more above the average weight for their age. By any criterion, they need to slim down. More exact measures of obesity, such as the thickness of fat on the triceps, find that the percentage of American children who are obese is higher than 10 percent and that this number has increased since 1960 (Gortmaker et al., 1987; Wolfe et al., 1994).

Obesity is a physical and medical problem at any stage of life, for the obese person runs a greater risk of serious illness. In children, orthopedic and respiratory problems are especially associated with obesity (Neumann, 1983).

Being overweight is often a psychological problem as well. In middle childhood, fat children are often teased, picked on, and rejected. They know they are overweight, and they are more likely to experience diminished self-esteem, depression, and behavior problems as a result (Israel & Shapiro, 1985; Strauss et al., 1985). Obese children have fewer friends than other children (Strauss et al., 1985), and when they are accepted in a peer

obesity The condition of being significantly and unhealthily overweight.

Tug-of-war is one of the few competitive events in which this fifth-grader is likely to be the first one chosen for the team.

group, it is often at a high price, such as answering to nicknames like "Tubby" or "Blubber" and having to constantly suffer jokes about their shape. A vicious cycle of rejection, isolation, and low self-esteem, leading to inactivity, compensatory overeating, and, in turn, to further rejection by peers, may cause obesity to persist in children. This, in turn, perpetuates psychological problems. Indeed, overweight adults who were obese as children tend to be more distressed and to have more psychophysiological problems than overweight adults who were of normal weight as children (Mills & Adrianopoulos, 1993).

Causes of Obesity

Typically, no one explanation suffices for any particular case of obesity; rather, the problem is generally created through the interaction of a number of influences. Many of these influences usually begin in infancy, continue through childhood, and remain influential in adulthood.

1. *Heredity.* Body type, including the amount and distribution of fat, as well as height and bone structure, is inherited. So are individual differences in metabolic rate and activity level. Therefore, not everyone can be "average" in the ratio of height to weight. Indeed, research on adopted children shows that heredity is at least as strong as environmental factors in predisposing a person toward being overweight (Bray, 1989).

2. *Exercise.* Inactive people burn fewer calories and are more likely to be overweight than are active people. This is even more true in infancy and childhood than during the rest of life. Activity level is influenced not only by heredity but also by the child's willingness to become involved in strenuous play and by the availability of safe places to play.

   Another factor that obviously affects children's activity level is television-watching, large amounts of which correlate with obesity (Dietz & Gortmaker, 1985). While watching TV, children tend to consume many snacks, and of course they burn fewer calories than they would if they were actively playing. In fact, they appear to burn fewer calories when watching TV than when doing *nothing*. One study found that when glued to the tube, children fall into a deeply relaxed state, akin to semiconsciousness, that lowers their metabolism below its normal at-rest rate—on average, 12 percent below normal in children of normal weight and 16 percent below normal in obese children (Klesges, 1993). While watching TV, children also are bombarded with, and swayed by, commercials for junk food. Indeed, 60 percent of the commercials during Saturday morning cartoons on U.S. television are for food products—almost all of them high-fat and high-sugar, shown being consumed by slim children who seem to be having a wonderful time because of what they are eating (Ogletree et al., 1990).

3. *Attitudes toward food.* In some families, parents take satisfaction when their children eat large quantities of food, and they frequently urge them to have another helping. The implied message seems to be that a father's love is measured by how much food he can provide; a mother's love, by how well she can cook; and a child's love, by how much he or she can eat. This is especially true when the parents or grandparents grew up in places where starvation was a real possibility.

Attitudes toward specific foods are important as well. Diets that emphasize fruits, vegetables, and grains are much less likely to lead to excess weight gain than are diets that are high in fat and sugar. Unfortunately, many parents inadvertently discourage healthy attitudes toward food because they give sweets to their children as rewards and buy them commercial snacks—usually the ones the children have seen on TV and insist on having. The popularity of fast-food take-out meals has also added weight to children, just as it has to adults.

4. *Precipitating event.* For many children, the onset of obesity is associated with a critical event or traumatic experience—a hospitalization, a move to a new neighborhood, a parental divorce or death—that creates a sense of loss or diminished self-image and a corresponding need for an alternative source of gratification, in this case, food (Neumann, 1983).

Help for Overweight Children

Clearly, an overweight child needs emotional support for a bruised self-concept, as well as help in losing weight. But reducing is difficult, and psychological encouragement is often scarce, partly because obesity is usually fostered by entrenched family attitudes and habits that promote a fattening diet and, most likely, a sedentary lifestyle. Thus, changes in family patterns, as well as in the child's food intake, are essential to effective treatment. Unfortunately, when parents do try to get their child to eat a healthful, low-fat diet, they often make the mistake of using ultimatums ("You can't go and play until you eat your broccoli") or bribes ("Eat all your spinach and you can have dessert") that boomerang, reinforcing the child's dislike of the food or enhancing the attractiveness of sweets (Birch et al., 1982; Olvera-Ezzell et al., 1990).

Either at their parents' behest or on their own, obese children sometimes try crash diets, which make them irritable, listless, and even sick—adding to their psychological problems without accomplishing much long-term weight loss. One reason is that the body reacts to protect itself during periods of famine. The rate of metabolism becomes slower, enabling the body to maintain its weight with fewer calories, and after a certain amount of initial weight loss, additional pounds become much more difficult to lose (Wing, 1992). To make matters worse, strenuous dieting during childhood can be physically harmful, since cutting down on protein or calcium could hinder important brain and bone growth.

The best way to get children to lose weight is to increase their physical activity. However, exercise is hard for overweight children, especially since they tend to move more slowly and with less coordination than other children (Hills, 1992). Additionally, for these very reasons, obese children are not often chosen to play on teams and are likely to be teased and rebuffed when they try to join in group activities. Parents and teachers can help overweight children to do the kinds of exercise in which their size is not a disadvantage, such as walking to school rather than taking the bus, or bicycling around the neighborhood. Parents can also exercise with their children, not only making activity easier and providing a good model but bolstering the child's self-confidence as well. Children can share responsibility by monitoring their eating, recreation, television-watching, and other activities related to their weight.

The importance of changing the child's eating and exercising patterns is apparent when one realizes that if the childhood weight problem reaches the point that the child is obese, and continues at that level throughout the childhood years, it is likely to last a lifetime. By one estimate, 60 to 80 percent of obese children become obese adults (Lucas, 1991). Treatment of obesity is more successful early in life, before the habits and attitudes contributing to weight gain jeopardize health as well as happiness throughout the life span.

Motor Skills

The fact that children grow more slowly during middle childhood may be part of the reason they become so much more skilled at controlling their bodies during these years. (Compare their physical self-control, for instance, with the clumsiness that typically accompanies sudden changes in body shape and size during toddlerhood or puberty.) School-age children can execute almost any motor skill, as long as it doesn't require very much power or judgment of speed and distance.

Of course, which particular skills a child masters depends, in part, on opportunity and encouragement. The skills of some North American 8- and 9-year-olds may include swinging a hammer well, sawing, using garden tools, sewing, knitting, drawing in good proportion, writing or printing accurately and neatly, cutting fingernails, riding bicycles, scaling fences, swimming, diving, roller-skating, ice-skating, jumping rope, and playing baseball, football, and jacks. Halfway around the world, in Indonesia, children master many of these same skills—though for environmental reasons, they do not learn to ice skate; for cultural reasons, they do not learn to play baseball or football; and in Bali, for religious reasons, they do not learn to swim (water is considered to harbor evil) (Lansing, 1983). At the same time, Indonesian children learn skills not common among North American children, such as cutting wood with sharp knives and weaving intricate baskets.

The impressive gross motor skills of school-age children are well illustrated in these two photos: many adults could not climb a rope or jump one with the skill shown here. What is not clear from these photos is that sex differences in motor skills during the school years are biologically minimal and that whatever notable gender gaps there may be in motor ability usually arise from influences in the social context. Boys, for example, are admired for taking risks, so they practice climbing, not just up ropes high above the gym floor but up trees, cliffs, and most anything else that offers challenge and a bit of danger. Girls are encouraged to be verbal and cooperative, so they develop variations of rope-jumping, hand-clapping, and rhythmic-stepping, while singing "tell me the name of your sweetheart," "Miss Mary had a baby," and so on.

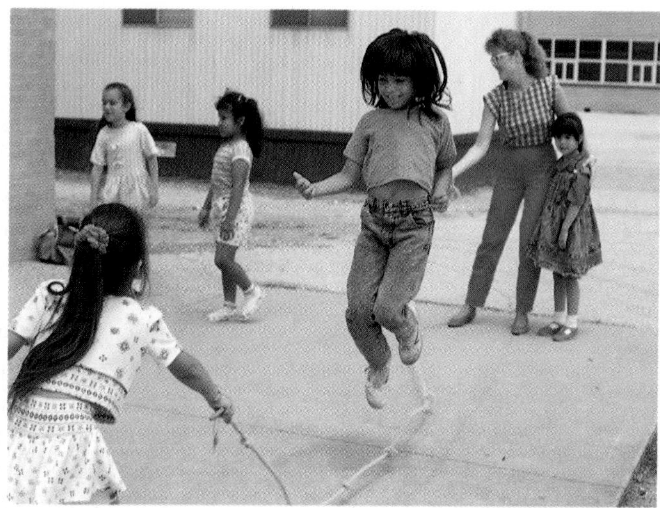

reaction time The time it takes to respond to a particular stimulus.

Differences in Motor Skills

Boys and girls are just about equal in their physical abilities during the elementary-school years, except that boys have greater forearm strength and girls have greater overall flexibility. Consequently, boys have an advantage in sports like baseball, whereas girls have the edge in sports like gymnastics. But for most physical activities during middle childhood, sex is not as important as age and experience: boys can do cartwheels, and girls can hit home runs, if they are given an opportunity to learn and practice these skills.

However, the maxim "Practice makes perfect" does not always hold true. Every motor skill is related to several other abilities, some depending on practice but others relying on body size, brain maturation, or genetically based talent. For example, **reaction time**, the length of time it takes a person to respond to a particular stimulus, is tied to aspects of brain maturation that continue into adolescence. Hand-eye coordination, balance, and judgment of distance are other key abilities that are still developing during the school years.

Unfortunately, the sports that most American adults value, and often push their children into, are not very well-suited for children, because they demand precisely those skills that are hardest for them to master. Even softball is much harder than one might think. Throwing with accuracy and catching both involve more distance judgment and eye-hand coordination than many elementary-school children possess. In addition, catching and batting depend on reaction time. Younger children are therefore apt to drop a ball even if it lands in their mitt, because they are slow to enclose it, and they are similarly likely to strike out by swinging the bat too late. Thus, a large measure of judgment, physical maturity, and experience is required for good ball-playing. As always, of course, underlying hereditary differences among individuals are key. Some children, no matter how hard they try, will never be able to throw or kick a ball with as much strength and accuracy as others, a fact that parents, teachers, and teammates sometimes tend to forget.

Cultural and cohort differences in motor skills are apparent in soccer, a sport that evokes life-and-death passions in many nations but has largely been considered dull in the United States. However, like these Texas schoolchildren, many young Americans now prefer soccer over baseball or football, and they are developing motor skills, such as heading a ball or instep kicking, that few of their parents ever had. Another cultural difference is also apparent: American boys and girls, unlike those in most other nations, often develop the same skills together.

Children with Special Needs

All parents witness the developing accomplishments of their offspring with pride and satisfaction, but for some, these feelings are mingled with worry and uncertainty as their child seems to experience unexplained difficulties in one area of development or another. Often these difficulties are biological in origin, but their worrisome manifestations are more psychological and cognitive than physical. Gradually they deepen to the point that the parent suspects that the child might have a psychological disorder. Consider the experiences of this mother:

> Except for the fact that my daughter has always been a physically-active and strong-willed child, it was not until she entered the first grade that the problems seemed to start. She became very reluctant to attend school, which was very different from her excitement over kindergarten. The stomach aches that she had complained of since age 2 became more frequent and necessitated her going to the nurse's office at school at least once per day if not more often. She could not seem to complete her assignments—sometimes taking an hour to complete just one sentence. She did not seem to willfully refuse to do assignments but seemed preoccupied with daydreams and other, more interesting activities. She was not disruptive in class but did seem to go through a bout of pushing in line and there were some reports of aggressiveness. At home, I could not seem to get her moving in the morning—even to the point where I took her to the babysitter's home in her pajamas because she would not get up and get dressed. Kimberly also began to appear angry and frustrated, having more problems getting along with parents and peers. She had always had best friends, been very sociable, and friendly. She then started to verbalize that people didn't like her and no one wanted to play with her. She would often get very angry at the neighborhood children she played with. As a parent, I found myself constantly yelling and totally frustrated with how to handle my child. [quoted in Thompson, 1995]

In some children, actions like this may be an isolated instance of showing off or of outrageous mischief. In children with ADHD, they are commonplace. When such behavior is accompanied by aggression, the child may be at risk of developing a conduct disorder—possibly becoming the kind of stubborn, disobedient daredevil who is constantly in trouble at home, at school, and in the neighborhood.

What is a parent to do in such a situation? In this case, Kimberly's mother sought the advice of her daughter's teachers and school counselors, who had been equally perplexed that this intellectually above-average child seemed to have such difficulty paying attention in class and cooperating with others. Together, they also contacted developmental experts at a local child-guidance clinic, who, after interviewing her mother, testing Kimberly, and observing her at school, diagnosed Kimberly's problem as attention-deficit hyperactivity disorder (ADHD), which you will read about later in this chapter. With treatment that included medication, parent-child therapy sessions, and structured school activities, Kimberly became a much happier, and more capable, child—although she still had lingering emotional difficulties and attention problems.

Kimberly is one of a great many children who are classified as children with special needs. For these children, the development of new skills, closer friendships, and more mature ways of thinking is impaired by a psychological disorder. Kimberly's problem, ADHD, is just one of a wide variety of such disorders, including, among many others, aggression, anxiety, autism, conduct disorders, depression, learning disabilities, and mental retardation, as well as problems that may arise as a consequence of physical disabilities such as blindness, deafness, or paralysis. As we will see, the study of children with special needs is playing an increasingly important role in the study of development in general.

The Developmental Psychopathology Perspective

In recent years, clinicians who study childhood psychological disorders have allied with developmental psychologists to create the new field of **developmental psychopathology**, which applies the insights from studies of normal development to the study and treatment of childhood disorders (Cicchetti, 1990, 1993; Sroufe & Rutter, 1984). The insights arising from this alliance have been mutually beneficial, for as developmental psychopathologists emphasize, "we can learn more about an organism's normal functioning by studying its pathology and likewise, more about its pathology by studying its normal condition" (Cicchetti, 1990). Indeed, as these developmentalists have discovered, when comparing abnormal and normal children, the distinction between the two often blurs and sometimes even disappears: most normal children, some of the time, act in ways that are decidedly unusual, and most children with psychological disorders are, in other respects, quite normal. In short, children with psychological disorders should be viewed as children first—with the developmental needs that all children share—and only secondarily as children with special challenges.

Further, taking a developmental perspective makes clear that the manifestations of virtually any special problem a child might have will change as the child grows older: a child who seems severely handicapped by a disability at one stage of development may seem much less so at the next, or vice versa. Such shifts are not simply a matter of time passing, but, like all developmental changes, result from the interplay of changes within the individual and forces in the ecological setting (Berkson, 1993). This point is especially important when considering special problems from a life-span perspective, for adulthood affords the person with special needs more possibilities but also fewer protections. For example, many children with seemingly serious disabilities, from blindness to mental retardation, become happy and productive adults if they find a vocational setting where they can perform well. On the other hand, disabilities that make a child unusually aggressive and socially inept may become more serious during adolescence and adulthood, when physical maturity and social demands make control of aggression particularly important (Davidson et al., 1994; Lahey & Loeber, 1994).

Finally, taking a developmental perspective makes diagnosticians much more aware of the social context of a problem. This is apparent in the most recent revision of the *Diagnostic and Statistical Manual of Mental Disorders*, or DSM-IV (1994). This book, which is the official diagnostic guide of the American Psychiatric Association and the result of input from over 2,000 psychiatrists and psychologists throughout the world, now explicitly recognizes that the "nuances of an individual's cultural frame of reference" need to be understood before any disorder can be diagnosed. The manual's descriptions of most of the childhood disorders now include discussions of specific culture, age, and gender features, a direct outgrowth of the developmental psychopathology perspective. For instance, in diagnosing a child's conduct disorder (as evidenced by aggression, destructiveness, and stealing), DSM-IV states that in threatening, impoverished, high-crime, or war-

developmental psychopathology A field of psychology that applies the insights from studies of normal development to the study and treatment of childhood disorders.

ravaged situations, such behaviors may be protective and that the clinician should "consider the social and economic context" before diagnosing such actions as pathological (DSM-IV, 1994).

Because the disorders that developmental psychopathologists study are too great in number to discuss here, we will instead focus on three problems that are biosocially based: autism, learning disabilities, and attention-deficit hyperactivity disorder. Each is particularly instructive about the development of all children.

Autism

One of the most severe disturbances of childhood is called **autism**, from the prefix "auto-," meaning "self." The label "autism" was chosen in 1943 by an American physician, Leo Kanner, to describe children who have an "inability to relate themselves in an ordinary way to people . . . an extreme autistic aloneness that, whenever possible, disregards, ignores, shuts out anything that comes to the child from the outside." Kanner's term was apt: autistic individuals seem unusually restricted by their own perspective and their need for predictable routines, and they seem unable, unwilling, or uninterested in communicating with, or even understanding, others. Classical autism, described by Kanner as including such extreme asocial and uncommunicative behaviors that the person never learns normal speech or forms normal human relationships, is quite rare: it occurs in about 1 of every 2,000 children (DSM-IV, 1994).

In highlighting the similarities between normal and abnormal behavior, however, the developmental psychopathological perspective has helped to show that many more individuals have less severe autistic symptoms. Such individuals are sometimes diagnosed as "high-functioning autistic," or as having "Asperger syndrome," named after a German psychiatrist who described a disorder in 1944 that he also called autism. (Because of World War II, Kanner and Asperger were unaware that they both were reaching similar conclusions and using similar terminology.) Asperger's delineation of autism included some individuals who were quite intelligent and verbal, and his portrayal of autism is, on the whole, less extreme than that of the classical type described by Kanner.

Exactly how impaired must a child be before being diagnosed as autistic? Even today, there is no firm consensus, but experts agree that Kanner's original definition was too narrow and that, when the entire spectrum of autistic disorders is taken into account, as many as 1 child in 100 shows autistic traits (Szatmari, 1992). Both the severe and less severe instances of the disorder are much more common in boys than in girls, generally in a ratio of 2 to 1 or higher.

The Developmental Path of Autism

Autism is truly a developmental disorder, because, although its origin is almost always congenital, its manifestations change markedly with age. As young babies, many autistic children seem quite normal and sometimes unusually "good." Soon, however, severe deficiencies appear in three areas: communication ability, social skills, and imaginative play.

autism A disorder that is chiefly characterized by an inability or unwillingness to communicate with others.

The first two deficiencies are apparent in infancy, as autistic children exhibit a lack of spoken language and a seeming reluctance to respond to others. By the preschool years, these deficits are quite pronounced. Many autistic children continue to be mute, not talking at all, while others engage exclusively in a type of speech called *echolalia*, echoing, word for word, such things as advertising jingles or the questions put to them. Autistic children also avoid eye contact and prefer to play by themselves rather than with others. Moreover, their play patterns are unusual, characterized by repetitive rituals and a decided absence of spontaneous imaginative play. One autistic adult, who has unusual verbal ability, remembers her childhood in the following manner:

> I also liked to sit for hours humming to myself and twirling objects or dribbling sand through my hands at the beach. I remember studying the sand intently as if I was a scientist looking at a specimen under the microscope. I remember minutely observing how the sand flowed, or how long a jar lid would spin when propelled at different speeds. My mind was actively engaged in these activities. I was fixated on them and ignored everything else. [Grandin & Scariano, 1986]

Later Childhood and Beyond

While autistic impairments in language and play patterns remain pronounced from the preschool years on, it is the lack of social understanding that often proves to be the most devastating problem. Autistic children appear to lack a theory of mind, an awareness of the thoughts, feelings, and intentions of other people (Holroyd & Baron-Cohen, 1993; Leslie & Frith, 1988). Consequently, to some autistic children, people seem of no greater interest than objects, because the child is unaware of the internal processes that make people unique and provocative.

As developing persons with autism grow older, the variations among them can be quite marked. On intelligence tests, most young autistic individuals score in the mentally retarded range, but a closer look at their intellectual performance on the various parts of the test often shows marked "scatter," with isolated areas of remarkable skill (such as memory for numbers, or for putting together puzzles). And while some autistic individuals never speak, or have only minimal verbal ability, many who were diagnosed as autistic at age 2 or 3 do learn to express themselves in language by age 6, and some of them demonstrate exceptional mastery of academic skills during the school years. Those in the last group may eventually live self-supporting adult lives, although they always will be less imaginative, more ritualistic, less communicative, and more socially isolated than most people. Indeed, occasionally, high-functioning autistic (or Asperger) individuals may be quite successful in professions in which their attention to routine, concentration on detail, and relative indifference to needless sentiment are an asset. One study of a small sample of individuals with Asperger traits found a dentist, a financial lawyer, a military historian, and a university professor among the group (Gillberg, 1991).

Causes and Treatment of Autism

While the precise cause of autism is not known, twin studies, which show a 50 percent concordance rate among monozygotic twins, make it clear that

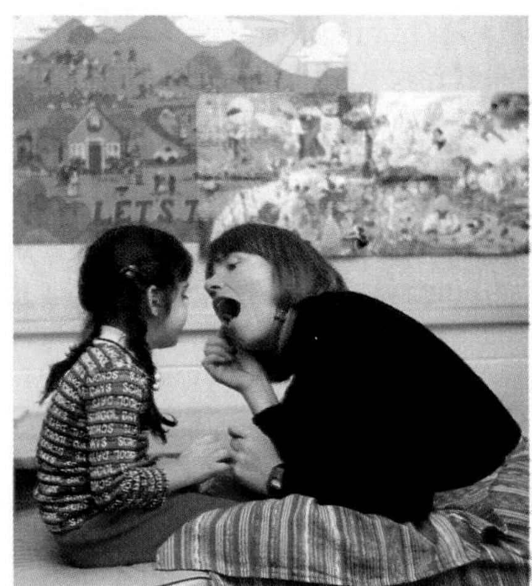

The prime prerequisite in breaking through the language barrier in a nonverbal autistic child, such as this 4-year-old, is to get the child to pay attention to another person's speech. Note that this teacher is sitting in a low chair to facilitate eye contact and is getting the child to focus on mouth movements—a matter of little interest to most children but intriguing to many autistic ones. Sadly, even such efforts were not enough: at age 13 this child was still mute.

genetic factors play a role (Hertzig & Shapiro, 1990; Volkmar, 1991). However, these same concordance rates also make it clear that genes are not the whole story. In all likelihood, a genetic vulnerability in combination with some damage—either prenatal or early postnatal—leads to autism. The role of environmental damage is suggested by the increased prevalence of abnormal neurological patterns, seizures, anoxia, previous exposure to viruses, and hearing abnormalities among children who have been diagnosed as autistic.

Because of the nature and progression of autism, a developmental view of the disorder is essential to its treatment. Since language skills normally develop most rapidly between ages 1 and 4, these are the crucial years for intervention. The most successful treatment methods for autism combine individual attention with behavior techniques that shape particular skills. For example, the child is rewarded immediately for making eye contact, for naming various parts of the face, for using pronouns appropriately, and so on. With such therapy, many autistic children can learn to talk, show appropriate social behaviors, and improve in other ways, with some making sufficient progress to enter normal schools and live normal lives (McEachin et al., 1993). The key seems to be early and very intensive treatment, with the goal of breaking through the communication barrier by age 6.

Now let us look at a much less devastating but much more common problem, one that typically causes the most difficulty in the school years.

Learning Disabilities

Children vary a great deal in how quickly and how well they learn to read, write, add, and so forth. Among those who show notable difficulty in learning, some seem slow in almost every aspect of intellectual development. Their thinking is like that of a normal child several years younger than their present chronological age, and thus they are considered to suffer **mental retardation**, that is, a pervasive delay in cognitive development.

Other children are slow learners in certain areas. They show remarkable "scatter" in their abilities, and they are generally quite competent except in certain skills. These children are said to have a **learning disability**, a failing in a specific cognitive skill that is not attributable to an overall intellectual slowness, to a specific physical handicap such as hearing loss, to severely stressful living conditions, or to a lack of basic education (Silver, 1991).

The most common learning disability is **dyslexia**, a disability in reading. Dyslexic children may seem bright and happy in the early years of school, volunteering answers to some difficult questions, diligently completing their worksheets, sitting quietly and looking at their books. However, as time goes on, it may become clear that they are reading with great difficulty or are not really reading at all. In the severe cases, they are guessing at simple words (occasionally making surprising mistakes) and explaining what they have just "read" by telling about the pictures.

Another common disability is **dyscalcula**, that is, great difficulty in math. This problem usually becomes apparent somewhat later in childhood, at about age 8, when even simple number facts, such as 3 + 3 = 6, are memorized one day and forgotten the next. It soon becomes clear—especially with word problems—that the child is guessing at whether two numbers should

mental retardation A pervasive delay in cognitive development.

learning disability A particular difficulty in mastering one or more basic academic skills, without apparent deficit in intelligence or impairment of sensory functions.

dyslexia A specific learning disability involving unusual difficulty in reading.

dyscalcula A specific learning disability involving unusual difficulty in math.

be added or subtracted, and that almost everything the child knows about math is a matter of rote memory rather than understanding.

Other specific academic subjects that may reveal a learning disability are spelling and handwriting: a child might read at the fifth-grade level but repeatedly make simple spelling mistakes ("kum accros the rode") or take three times as long as any other child to copy something from the chalkboard and still produce a large, illegible scrawl. In addition, although they are not usually labeled as such and given special help, some children are learning-disabled in an underlying skill—such as spatial relations, sequential processing, memory, or attention span—that affects all intellectual areas (Rourke, 1989). Problems in these most basic abilities are especially difficult to spot and treat.

The key criterion for diagnosing a learning disability is a significant discrepancy between measures of overall aptitude and measures of performance in a particular area. For example, if a 9-year-old has an average IQ but reads at the first-grade level, that would indicate a learning disability in the area of reading. The actual process of diagnosis is more complex than this, and various experts use alternative definitions and measures (Aram et al., 1992; Fletcher et al., 1992).

Because of the variations in defining and measuring learning disabilities, it is impossible to determine definitively what percentage of children have such problems. Detailed diagnostic surveys to establish rates of learning disabilities within a large group of seemingly normal children have been carried out only for dyslexia. Such surveys put the incidence of dyslexia at about 7 percent, a number found in large population studies in several nations and, most recently, in a study of all the 6-year-olds in Bergen, Sweden (Gjessing & Karlsen, 1989). Many of these children might not have been recognized as dyslexic in their daily lives if they had not taken a battery of tests for these surveys. Consequently, the rates reflected in these surveys are higher than those arrived at by counting only those children whose learning problems are first noticed in daily life and then are examined, diagnosed, and treated. The estimates of the American Psychiatric Association, for example, put the prevalence of marked learning disabilities at 4 percent for reading and at 1 percent for math, with the prevalence of other learning disabilities being "difficult to establish." In the United States, about 5 percent of all children receive special education for any learning disability (DSM-IV, 1994).

Causes and Remediation of Learning Disabilities

None of these learning problems are caused by a lack of effort on the child's part, although, unfortunately, parents and teachers sometimes treat children who are learning-disabled as though they are not trying hard enough. In fact, the precise causes of learning disabilities are hard to pinpoint (Chalfant, 1989; Gerber, 1993). In many cases, there may be no discrete, identifiable cause. Instead a child might happen to fall at the low end of the wide range of normal abilities in, say, reading or math, just as another child might happen to be at the low end of the normal variation in other traits, such as height, for instance (Shaywitz et al., 1992). In many cases, however, it seems as if some particular part of the learning-disabled child's brain does not function as it does in most people. In the case of dyslexia, for instance, various theories have focused on processes in the visual areas of the brain, such

as those that enable the eye to scan from left to right, or to rapidly process small differences in the shape of letters, or to focus on one word and skip over another (Adams, 1990).

While it is quite logical to imagine that deficiencies in the visual part of the brain are the primary neurological culprit for dyslexia, according to a recent hypothesis, the disability is more likely to originate in certain auditory areas of the brain, such as the one that is dedicated to detecting the differences between sounds spoken very rapidly—such as *p* and *b*, which take less than a twentieth of a second to say—and longer sounds such as *a*, which lasts a full tenth of a second. Poor functioning in this area might affect the development of spoken language and language comprehension. The problem is unlikely to be noticed initially because social context and other cues from the environment would help the child to know, for example, whether a parent had told him to pick up the "pail" or the "mail." But learning to read requires deciphering sounds without many contextual clues. Even when taught the elements of phonics, an affected child might still be lost, because he or she does not hear the difference between the various sounds.

This theory has received support from detailed studies showing that dyslexic individuals have a different pattern of activity in the auditory areas than normal readers do. In addition, their brains contain fewer large cells in an area of the brain that controls the timing of auditory signals (Tallal et al., 1993). Other studies have also found that for many children with dyslexia, the underlying problem is more often auditory than visual (Gerber, 1993; Gjessing & Karlsen, 1989). From this research, a possible clue to treatment emerges: teachers and parents can slow down their speech, and help the child strengthen the ability to hear subtle differences in speech sounds.

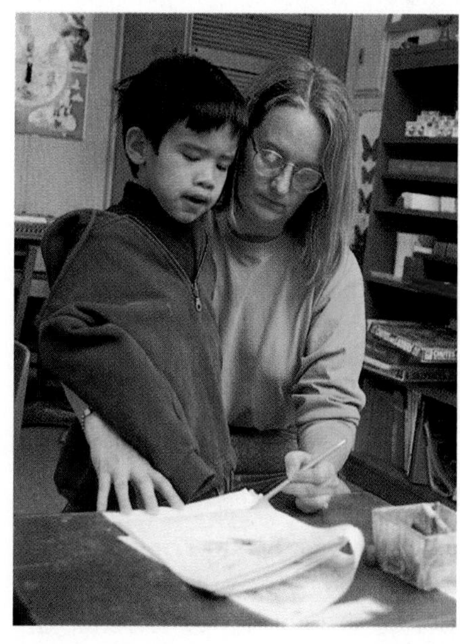

One-on-one contact, literally, is helpful for many learning-disabled children, who benefit also from explicit individualized instruction. In his special class, 7-year-old Alex receives both from his teacher. The one drawback with such special education is the limited opportunity it affords for learning social skills, a failing remedied when children with special needs and their teachers are included within a regular class.

As detailed in Chapters 3 and 4, many prenatal factors can have a detrimental effect on brain functioning. Genetic inheritance is one of them, since learning disabilities tend to run in families (Oliver et al., 1991; Silver & Hagin, 1990). Teratogens, of course, are another factor, with maternal drug use, particularly of cigarettes, alcohol, and cocaine, being among the teratogenic influences that may be associated with later learning disabilities (Needleman & Bellinger, 1994). Prenatal exposure to other toxins, notably mercury and PCBs, is clearly linked to learning disabilities (Jacobson et al., 1992), as is postnatal damage, such as that from convulsions caused by high fever, or from eating or inhaling leaded contaminants (Fergusson et al., 1993).

No matter what the cause of learning disabilities, the way teachers and parents respond to a child who displays difficulties in learning can have an enormous impact on the child's chances of overcoming the problem. If teachers and parents recognize that a child with a learning disability is neither lazy nor stupid, they can help the child become competent through patient, individual tutoring. In addition, attention should be given to the child's social interaction, because social skills and self-esteem are often indirectly affected by a learning disability (Casey et al., 1992; Vaughn et al., 1993). Training the child how to leapfrog, sidestep, or otherwise undercut the disability can minimize its effects, not only aiding the child in making and keeping friends but also obviously helping with academic learning (McIntosh et al., 1991). With such assistance, many children with learning disabilities develop into adults who are virtually indistinguishable from other adults in their educational and occupational achievements (Goodman, 1987).

Recognition that special training might be needed raises the question of how that training might best be provided (Grossman, 1995). The first answer involved special classes, where all the learning-disabled children in a particular school or district were taught together by a special-education teacher. This approach, however, made such children feel singled-out and impaired the development of normal social skills. In response, **mainstreaming** emerged about thirty years ago. In this approach, learning-disabled children were kept with the main group of children, but the regular classroom teacher was asked to be particularly sensitive to their special needs, perhaps using alternative methods to teach them or allowing them extra time to complete assignments and tests.

Unfortunately, mainstreaming too often became a "sink or swim" situation, primarily because many teachers are unable to cope with the special needs of some children, especially in a classroom of thirty or so students. Accordingly, some schools developed a **resource room**, where a child would spend some part of each day with a teacher who had the training and materials to target remediation for whatever disability the child might have. But pulling the child out of the regular class once again undermined social integration.

The most recent solution is called **inclusion**. In this approach, children with learning disabilities are included within the regular class, as in mainstreaming, but they receive targeted help within that setting from a specially trained teacher or paraprofessional who focuses on the special needs of the children all or part of the day. This solution may be the most expensive and may necessitate the most adjustment on the part of the adults, but it also may be the best answer for many children who need both social interaction with their schoolmates and special remediation for their particular learning patterns (Banerji & Dailey, 1995).

mainstreaming An approach to educating children with special needs by putting them in the same "stream"—the general education classroom—as all the other children, rather than segregating them.

resource room A designated room, equipped with special materials and staffed by a trained teacher, where children with special needs spend part of their school day getting help with basic skills.

inclusion An approach to educating children with special needs that includes them in the regular classroom while also providing them with special individualized instruction, typically from a teacher or paraprofessional trained in special education.

attention-deficit hyperactivity disorder (ADHD) A behavior problem characterized by excessive activity, an inability to concentrate, and impulsive, sometimes aggressive, behavior.

Attention-Deficit Hyperactivity Disorder

One of the most puzzling and exasperating of childhood problems is **attention-deficit hyperactivity disorder**, or **ADHD**, in which the child has great difficulty concentrating for more than a few moments at a time and, indeed, is almost constantly in motion (Barkley, 1990; Weiss, 1991). Sitting down to do homework, for instance, an ADHD child might repeatedly look up, ask irrelevant questions, think about playing outside, get up to get a drink of water, sit down, fidget, squirm, tap the table, jiggle his or her legs, and then get up again to get a snack. Often this urge for distraction and diversion is accompanied by excitability and impulsivity.

Many children with ADHD are also prone to aggression, which has led some researchers to propose *ADHDA—attention-deficit hyperactivity disorder with aggression*—as a subtype of this problem. Children who exhibit aggression with ADHD appear to be at increased risk for developing conduct disorders (Dykman & Ackerman, 1993). Attention-deficit disorder can also occur without hyperactivity or aggression. Children with this form of the problem, *ADD*, appear to be prone to anxiety and depression.

The crucial factor underlying ADHD (as well as the other forms of the attention-deficit problem) seems to be a neurological difficulty in screening out irrelevant and distracting stimuli, especially when trying to organize and communicate one's ideas (Riccio et al., 1993; Tannock et al., 1993). This deficit makes "paying attention" difficult, and therefore makes it hard for

PUBLIC POLICY

Lead Poisoning

Taking a developmental view of learning disabilities helps not only in their treatment but, in some cases, in their prevention as well. One such case involves lead poisoning, which can cause specific learning disabilities, hyperactivity, and overall retardation. Although these symptoms are most notable in the school years, their origins are usually to be found in lead exposure that occurred in infancy or early childhood.

Most commonly, children are exposed to lead by breathing or ingesting lead residues, such as those in chips or dust from flaking lead-based paint or those in industrial pollutants. (Lead from automobile fuel, once a major source of childhood exposure, has been eliminated in most nations.) Since lead accumulates in the body, small amounts taken in over a period of time can produce toxicity.

Lead poisoning is most commonly diagnosed through blood analysis. If the lead level is above 70 micrograms per deciliter of blood, the toxic damage may include paralysis, permanent brain damage, and even death. If the level is between 25 and 70 micrograms per deciliter, many less obvious problems may result, including abnormally high activity level, poor concentration, and slow language development.

Whether lower levels, between 10 and 25 micrograms, are toxic is controversial. Some research finds an association between such levels of lead and developmental problems, while other research does not (Boivin & Giordani, 1995; Cooney et al., 1989; Silva et al., 1988). The controversy arises from confounding factors that make it difficult to draw firm conclusions. Looking at the same data, for example, two teams of American scientists both found a correlation between moderate lead levels and behavioral problems, but one team attributed this correlation directly to the effects of the lead (Bellinger & Needleman, 1985; Needleman et al., 1990), while the other team thought the deficits could be better explained by other factors (Ernhart et al., 1989). As the second team points out, for instance, elevated levels of lead are likely to be found in children who live in old houses with peeling paint or who play near the factories that pollute the air—in other words, in children who live in low-income neighborhoods. The proven hazards of poverty (such as troubled families, crowded schools, inadequate nutrition) may be the real culprit, not the hypothetical hazard of low lead levels. The controversy is ongoing. Responding to criticism, the first team conducted further research, controlling for key confounding factors and using bone-lead, rather than blood-lead, levels. (Bone analysis reveals lead accumulated years earlier.) Their study found that 11-year-old boys whose bones showed that they had taken in more than average amounts of lead as preschoolers were more likely to exhibit problem behavior, including aggression and delinquency, and to complain about various ailments such as headaches and stomachaches (Needleman et al., 1996). Other researchers, this time in New Zealand, analyzed children's baby teeth and found a "small but consistent" correlation between bone-lead levels and difficulties in school achievement (Fergusson et al., 1993).

the child to focus on any one thought or experience long enough to process it. Thus a child might impulsively blurt out the wrong answer to a teacher's question or might not have the patience to read and remember a passage in a school textbook. In addition to these difficulties, ADHD (as well as the other attention-deficit disorders) is often accompanied by further learning disabilities (Cantwell & Baker, 1991; Dykman & Ackerman, 1991).

Estimates of the prevalence of ADHD among school-age children vary, from about 1 percent to 5 percent, depending partly on diagnostic criteria, which vary by nation. DSM-IV estimates the prevalence to be between 3 and 5 percent. British doctors are less likely to diagnose children as having ADHD than are American doctors and are more apt to diagnose such children as having a conduct disorder (Epstein et al., 1991; MacArdle et al., 1995). Generally, far more boys than girls are diagnosed with ADHD or conduct disorder, the sex ratio being about four boys for every one girl (Bhatia et al., 1991; Lahey & Loeber, 1994).

Meanwhile, the view of the second team has also been bolstered by other research. One study found that physically abused children are twenty-seven times more likely to have elevated levels of lead in their blood than are nonabused children (Bithoney et al., 1993). This suggests that the underlying problem could be parental maltreatment, leading to neurological and intellectual damage. Not only would such an abused child be less likely to be protected from lead exposure, but such a child would also be more likely to suffer repeated blows to the head, more likely to be severely criticized, and less likely to be listened to. In addition, other research continues to find that socioeconomic factors are powerfully correlated with children's lead levels (Sargent el al., 1995). Thus, while lead ingestion might be a contributor, the root cause of a child's difficulty with learning, friendship, or conduct might be the parent-child relationship or the poverty of the community. Leading experts from the United Kingdom and Australia conclude that "uncertainly remains as to the real impact that lead makes on children's neuropsychological development" (Pocock et al., 1994).

Nevertheless, to be on the safe side, in 1991 the U.S. government lowered the level at which lead is considered toxic to 10 micrograms per deciliter. The Centers for Disease Control and Prevention now recommends that all children under age 6 be tested for lead, and that when levels higher than 10 are found, precautions be taken. Obvious first steps include removal of any deteriorating lead-paint dust in the home, including lead-paint dust in window casements, eliminating lead contaminants from the yard, and keeping children from licking or eating anything that might contain lead (Hilts, 1991).

Fortunately, average blood-lead levels in the United States declined an average of 78 percent from about 1980 to 1990, primarily because all gasoline now sold is lead-free. In addition, less than 1 percent of all the food and soft-drink cans in the United States now use lead solder, compared to 47 percent in 1980. These changes have been particularly beneficial to children under age 6. In 1980, over 53 percent of children between the ages of 1 and 5 had blood-lead levels of 15 micrograms per deciliter or higher; in 1990, only 3 percent did (*MMWR*, August 5, 1994). Nevertheless, there remain an estimated 1.7 million children—mostly low-income, city-dwelling children—whose blood-lead levels are at least 10 micrograms per deciliter. Although not all developmentalists agree that these children are necessarily at risk, most agree that lead testing of all young children and decreasing the pollutants in their environment—not just in their homes but also in the air and water—are wise public health precautions. When certain 3-year-olds consistently run rather than walk and jump up and down and fidget rather than ever being still, it would be reassuring to know that this high activity is age-related normal behavior that will improve in a few years, rather than a toxic reaction that might worsen, leaving the child disadvantaged when he or she is in elementary school and unable to learn as well as his or her classmates.

Causes of Attention-Deficit Hyperactivity Disorder

When confronted with a school-age child who is considerably more active than other children and cannot concentrate very well, it is not easy to explain that child's behavior. However, researchers have identified several factors that, individually or in combination, may contribute to attention-deficit hyperactivity disorder. These include genetic inheritance, possibly related to the rate of activity in specific areas of the brain (Zametkin et al., 1990); prenatal damage, chiefly from teratogens (Needleman & Bellinger, 1994); and postnatal damage, such as from lead poisoning or repeated blows to the head (see Public Policy, pp. 318–319). It should be noted that the once-popular notion that sugar and caffeine are possible contributors to ADHD has been discounted. Indeed, if anything, caffeine seems to slow some hyperactive children down (Ingersoll & Goldstein, 1993).

Whatever the cause of any given case of ADHD, its expression may be intensified or moderated by the child's specific social context. A chaotic and

provocation ecologies Classroom environments that provoke or exacerbate hyperactive behavior by imposing either an unusually rigid classroom structure or none at all.

rarefaction ecologies Classroom environments that ameliorate or diminish hyperactive behavior by providing sufficient classroom structure within a flexible individualized environment.

coercive family setting is particularly likely to worsen the difficulties of a child with ADHD (Buhrmester et al., 1992). Typically, the child is told to behave in ways that are impossible for the child to follow ("Sit still and be quiet") and then is confronted with anger and punishment when he or she fails to comply. Predictably, this unstructured and stressful situation leads to more out-of-control behavior by the child and by the adult.

The school setting can likewise affect the problem. One study showed that some classroom environments, labeled **provocation ecologies**, made the problem worse, while others, called **rarefaction ecologies**, ameliorated the problem. In provocation ecologies, structure was either unusually rigid or completely absent, and noise was either completely forbidden or tolerated to a distracting degree. In rarefaction ecologies, teachers who managed to diminish hyperactivity were flexible in their reactions to minor disruptions (for example, allowing children to ask questions of their neighbors as long as they did so quietly), but they also provided sufficient structure so that children knew what they should be doing and when they should be doing it (Whalen et al., 1979).

Other research has likewise found that teachers who are flexible in their expectations—rather than insisting, for example, that "all third-graders will act like third-graders"—create less tension in children with ADHD and make the expression of hyperactivity less likely. This research also confirms that opposite characteristics need to be avoided in the classroom: too much noise and too much boring quiet (Greene, 1996).

Help for Children with Attention-Deficit Hyperactivity Disorder

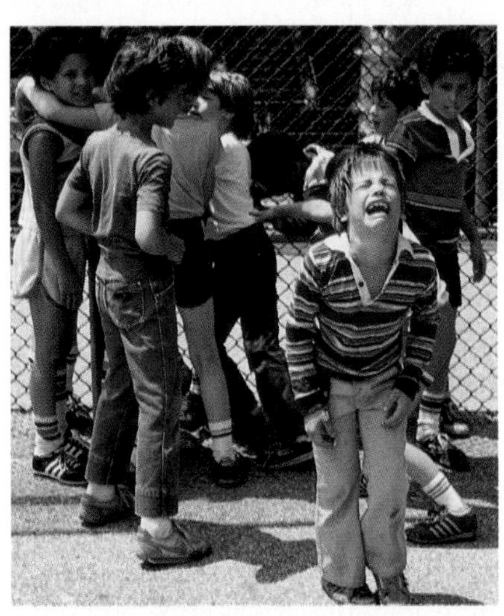

We do not know why this boy is crying. However, the apparent lack of sympathy from his peers suggests social rejection plays a part. Many children have disabilities that impair their interaction with others and would benefit from the same kinds of skills training and practical assistance that children with more obvious academic disabilities often receive.

Not surprisingly, children with attention-deficit hyperactivity disorder are usually annoying to parents and teachers and are rejected by peers (Henker & Whalen, 1989). Many children with the disorder continue to have problems in adolescence, not only with hyperactivity but with academic demands and social skills as well (Barkley et al., 1991; Nussbaum et al., 1990). Many become disruptive and angry. In fact, more than half of all children with attention-deficit hyperactivity disorder have continuing problems as adults in pacing their work, controlling their temper, and developing patience. Some develop psychological disorders (such as antisocial personality) (Weiss, 1991).

As they grow older, however, many people with ADHD learn to cope with these problems, for example, by choosing occupations that suit their skills but that do not emphasize patience and control (Gittelman et al., 1985; Weiss & Hechtman, 1986). In childhood, the most effective forms of help are medication, psychological therapy, and changes in the family and school environments. Ideally, all three are considered.

For reasons not yet determined, certain drugs that stimulate adults, such as amphetamines and methylphenidate (Ritalin), have a reverse effect on many—but not all—children with attention-deficit problems, whether or not they are hyperactive and/or aggressive (Bernaldo deQuiros et al., 1994). For many children, the results are remarkable, allowing them to sit still and concentrate for the first time. While this new ability to pay attention is a welcome relief to many children and parents, it does not necessarily produce gains in intelligence scores or achievement (Swanson et al., 1993).

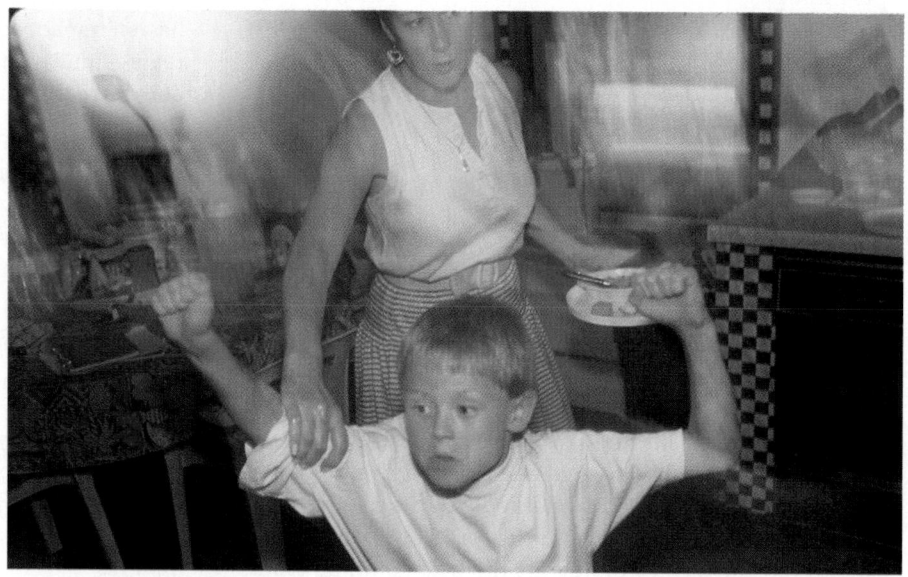

The use of psychoactive drugs to control mental disorders is controversial as well as complicated. However, those who assert that children should never be medicated to quiet their hyperactivity have not met Dusty Nash, age 7, or his mother. Without his daily Benzedrine, Dusty cannot concentrate or even sit quietly for a moment.

Indeed, given the remarkable results that psychoactive drugs provide some children with ADHD, it needs to be emphasized that drug therapy is not a panacea. Unfortunately, drugs are sometimes prescribed for children without proper diagnosis or without follow-up examinations—an abuse that can harm the child. For instance, children are sometimes prescribed an excessively large dosage and become lethargic.

Further, by the time the child has become a candidate for psychoactive drugs, the child's behavior has usually created school, home, and personal problems that drugs alone cannot reverse. Psychoactive drugs should never be given as a one-step solution; instead, they should be part of an ongoing treatment program that involves the child's cognitive and psychosocial worlds (Swanson et al., 1993; Wender, 1987). Usually, the child with attention-deficit hyperactivity disorder needs help overcoming a confused perception of the social world and a bruised ego, while the family needs help with their own management techniques and interaction. Teachers, too, need to be made aware of how their reaction to a child with ADHD can reduce or intensify the child's difficulty.

It should be clear from our discussion of childhood psychopathology that physiological, educational, and social influences can interact to produce problems, and that all such influences must be understood before the impact of these problems can be reduced. Our focus on such problems should highlight the reality that the same interactional approach must characterize attempts to understand and meet the needs of all children. Further, as the developmental psychopathology perspective emphasizes, we must remember that each child has some of the strengths and liabilities typical of children in middle childhood, as well as capabilities and problems that few others share. We will see this truth borne out in the next chapter, on cognitive development, and again in Chapter 13, which includes a discussion of the ability of many children to cope with almost any psychosocial problem they may face.

SUMMARY

Size and Shape

1. Children grow more slowly during middle childhood than at any other time until the end of adolescence. There is much variation in the size and rate of maturation of children as a result of genetic as well as nutritional differences.

2. Overweight children suffer from peer rejection and low self-esteem, which in turn may cause more overeating.

3. Many influences interact to cause obesity. Hereditary factors, inactivity (including excessive TV viewing), family attitudes toward food, and sometimes a critical event are among the chief contributors to this problem. More exercise, rather than severe dieting, is the best solution, along with new attitudes toward food and recreation that are supported by family members.

Motor Skills

4. School-age children can perform almost any motor skill, as long as it doesn't require much strength or refined judgment of distance or speed. The activities that are best for children are ones that demand only those skills that most children of this age can master.

Children with Special Needs

5. The developmental psychopathology perspective applies studies of normal development to an understanding of how children with psychological disorders cope with their particular difficulties. It emphasizes that these youngsters are children first—with the developmental needs that all children share—and, secondarily, are children with special challenges.

6. The developmental psychopathology perspective also stresses that the manifestations of any disorder will change as the child grows older, and that the social context has an impact on the interpretation of a problem. Two contexts—family interactions and school structure—are pivotal in treatment and prognosis.

7. Autism is characterized by a lack of interest in people, delays in language and communication, and a deficiency in imagination that produces an insistence on sameness in the environment. Some researchers believe that autistic individuals lack a theory of mind to account for psychological processes in others. The severity and developmental outcome of autism vary enormously.

8. Children with problems such as a learning disability or attention-deficit hyperactivity disorder need special attention and help to learn to cope with their problems. They particularly need the support of adults who understand their special difficulties and who can provide assistance so that these lifelong conditions do not become lifelong disabilities.

9. Some psychological disorders may originate in genetic or physical problems of some sort, but whether or not the cause is organic, many educational and psychological programs can help children with these disabilities. Psychoactive drugs also help some children, but these should be used carefully and cautiously.

KEY TERMS

obesity (305)
reaction time (309)
developmental psychopathology (311)
autism (312)
mental retardation (314)
learning disability (314)
dyslexia (314)
dyscalcula (314)
mainstreaming (317)
resource room (317)
inclusion (317)
attention-deficit hyperactivity disorder (ADHD) (317)
provocation ecologies (320)
rarefaction ecologies (320)

KEY QUESTIONS

1. What are some of the causes of variation in physical growth in middle childhood?

2. How does obesity affect a child's development?

3. What are the causes of obesity?

4. What are some of the reasons for the notable improvement in children's motor skills during the school years?

5. How does the developmental psychopathology perspective view children with psychological disorders relative to children who develop normally?

6. What are the major characteristics of autism?

7. Why do some researchers believe that autistic individuals lack a theory of mind?

8. What are the symptoms of a learning disability? What are some of the more common types of learning disabilities?

9. What are the advantages and disadvantages of mainstreaming and inclusion for learning-disabled students?

10. What are the possible causes of attention-deficit hyperactivity disorder?

11. What are the arguments for and against the use of psychoactive drugs to control attention-deficit hyperactivity disorder?

12. What other types of treatment are helpful in controlling attention-deficit hyperactivity disorder?

The School Years: Cognitive Development

CHAPTER 12

As we saw in Chapter 9, one of the major challenges of studying cognitive growth in the play years is to avoid underestimating the young child's competencies. Both scientific and casual observation of preschoolers in everyday activities can easily lead to the conclusion that the young child's cognitive skills are quite limited. It is only when preschoolers are observed in the context of carefully designed experiments, involving situations and stimuli that are meaningful and interesting to them, that their impressive cognitive abilities are more fully revealed.

By contrast, we rarely underestimate the cognitive competencies of school-age children. Their everyday skills are, in fact, obvious and quite remarkable, and they expand every year. By age 11, for example, many children can figure out which brand and size of popcorn is the most economical and healthiest, can multiply proper and improper fractions, can memorize a list of fifty new spelling words (or the batting averages of a favorite baseball team), and can use irony appropriately—accomplishments beyond virtually every preschooler. School-age children are also increasingly able to focus their thinking less intuitively and more analytically on the facts and relationships that they perceive in the world. They become astute observers who have acquired "a sense of the game" of thinking. Growing language abilities complement these expanding cognitive skills, so older children can discuss and explain their world and themselves in ways no preschoolers can.

Our goal in this chapter is to understand the cognitive accomplishments of the school years and, more important, to understand *why* these dramatic accomplishments occur. To do this, we will turn again to the insights provided by cognitive researchers regarding how thinking becomes more organized and strategic, as well as to the views of Piaget and his followers regarding how thinking becomes more systematic and logical. We will also discuss further growth in language and consider the impact of schooling on cognitive growth.

The Growth of Thinking, Memory, and Knowledge

Middle childhood witnesses many cognitive changes that make the school-age child a much different kind of thinker than the preschooler. Older children not only know more but are more resourceful in planning and using

their cognitive resources when they must solve a problem, remember a piece of information, or increase their knowledge on a particular topic. In contrast to the fairly fragile cognitive accomplishments of the preschooler, school-age children can apply their thinking and reasoning skills to a variety of cognitive challenges—or can quickly figure out how to do so. In the words of John Flavell (1992), by the middle of childhood, most children have acquired "a sense of the game" of thinking. They know that good thinking entails consideration of evidence, planning, logic, formulation of alternative hypotheses, and consistency, and they try to incorporate these qualities into their own reasoning, as well as in their evaluation of the thinking of others (Flavell et al., 1993). Moreover, during the school years, children become more aware of intellectual strengths and weaknesses, recognizing that one can be "good at" certain things (like math and science) but not as proficient at others.

Information-processing researchers believe that these monumental advances in thinking occur because of basic changes in children's processing and analysis of information (Bjorklund, 1990; Klahr, 1992; Kuhn, 1992). As we will see, these changes are directly related to the growth of selective attention and memory skills, the enhancement of processing speed and capacity, the growth of knowledge, and, finally, the blossoming of "metacognition," or the ability to think about thinking.

Selective Attention

If you were to watch children in a kindergarten and then observe the students in a fifth-grade classroom, many of the differences you would find would be related to the growth of selective attention. In kindergarten, children become easily distracted while listening to a story or printing alphabet letters: they chatter to each other, look around, fidget, and sometimes get up to visit friends or just wander around. By contrast, fifth-graders are likely to be found working independently at their desks or in groups around a table, managing to read, write, discuss, and seek assistance from their teacher without distracting, or being distracted by, other children who are working on different assignments.

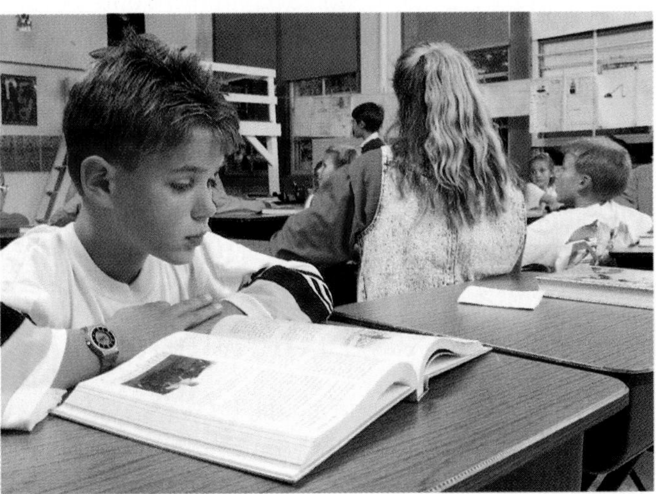

The ability to use selective attention to screen out distractions and focus on the task at hand makes fifth-graders, like this boy, much better students than younger children.

selective attention The ability to concentrate on relevant information and ignore distractions.

storage strategies Procedures for holding information in memory, such as rehearsal (repeating the information to be remembered) and reorganization (regrouping the information to make it more memorable).

This ability to screen out distractions and concentrate on relevant information is called **selective attention**. It is important not only to the task of "concentrating" generally but also to the strengthening of a number of other cognitive abilities. Selective attention plays an important role in memory ability, for example, enabling the individual to focus on the details of an event or task that are most likely to facilitate its recall.

Selective attention is also important for reasoning and problem solving. To solve a challenging problem, a person must first focus on the information that is likely to lead to a solution, and then proceed straightforwardly until a successful outcome is achieved. This is true whether one is figuring out the relative amounts of liquid in different containers or editing another child's writing for spelling and punctuation. Many preschoolers fail at simple problem solving, not because they are ignorant or lazy, but because their selective attention is inconsistent and they become distracted on the way to the solution. Sometimes they even forget the task itself! By contrast, older children are more methodical and strategic: they know when selective attention is called for and know how and where to focus their attention (Flavell et al., 1993; Miller, 1990).

Memory Strategies

Among the most notable cognitive developments over the school years is a marked improvement in memory. In large measure, this improvement is due to children's increasing use of memory strategies, both for putting material into mental storage and for retrieving it on demand at a later time.

The first improvement occurs in **storage strategies**, that is, the procedures an individual uses to put information into memory. From preschool through adolescence, a child's repertoire of storage strategies increases (Kail, 1990). For example, instead of just staring at a group of objects to be remembered, as a preschooler might, 5- or 6-year-olds are likely to repeat the names of the objects over and over (a strategy called rehearsal), and 9- and 10-year-olds are likely to mentally reorganize the objects to make them easier to memorize. Older children are especially likely to use reorganization when memorizing a large number of items. If required to memorize a list of the thirty most populous nations of the world, for instance, an older child might cluster them by region, or sort them alphabetically, or make up

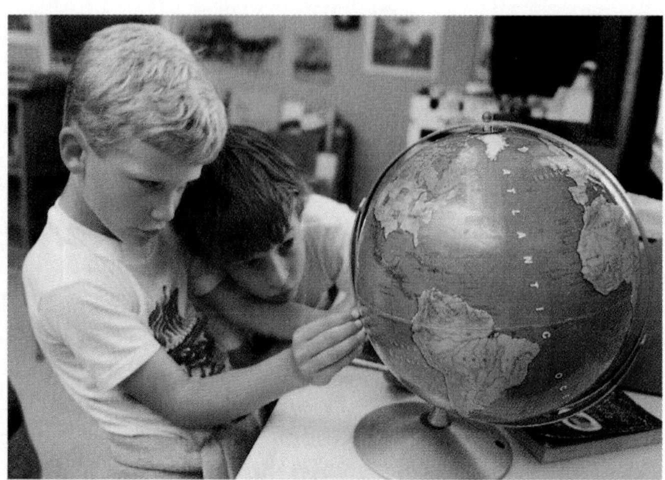

How best to remember names and locations of countries around the world? These boys know that any of a variety of memory strategies would be helpful—including simple rehearsal, grouping countries by region or in alphabetical order, or color-coding them on maps according to their location or other features.

a sentence that uses the first letter of every country. By age 9, if children are given a simple memory task that elicits the spontaneous use of organization strategies, they are then quite capable of incorporating such strategies into their later approaches to more difficult memory problems (Best, 1993).

The next important improvement in memory is the development of **retrieval strategies**, which are particular procedures for accessing previously learned information that doesn't immediately come to mind. The ability to use retrieval strategies begins to emerge in middle childhood and improves steadily thereafter (Kail, 1990). By fifth grade, for example, children usually have a dawning awareness that if something can't initially be recalled, merely "taking a walk down memory lane"—that is, systematically searching one's recollections of other relevant events or information—might prove helpful (Flavell et al., 1993). Somewhat later, they become even more strategic in their retrieval efforts. For example, they try to jog their memories by thinking of clues to stimulate their recall (the first letter of a name or key term), or they attempt to visualize the experience they are trying to remember. By the seventh grade, many students would not panic in a geography test if they could not immediately remember the exact location of Bolivia or Bulgaria. In contrast to a typical fourth- or fifth-grader, who would probably become frustrated by such a memory lapse and give up, a seventh-grader is likely to begin a systematic effort at recall, mentally picturing a map of the world or reconstructing the context of the relevant geography lesson to bring the location to mind.

Reinforcing this memory improvement are two other aspects of cognitive development. First is the selective attention already mentioned: school-age children can actually devote their minds to concentrated study of a list of words or facts in a way that younger children almost never do. Second is peer interaction: school-age children can study in pairs or groups, quizzing and encouraging each other, in ways beyond the ability of the more egocentric preschooler. All told, a marked advance in memory occurs between ages 6 and 12, an advance particularly apparent in older children's ability to remember items over a period of days or longer, with almost no forgetting (Brainerd & Reyna, 1995).

Processing Speed and Capacity

Another cognitive difference between younger and older children is that the latter are quicker thinkers (Hale, 1990; Kail, 1991a, 1991b). In fact, processing speed continues to improve from the preschool years through early adulthood (Kail, 1991b).

Processing speed affects more than just performance time. It also affects processing capacity: being able to think faster makes it possible to think about more things at once. In contrast to a preschooler who becomes easily befuddled when faced with complex tasks or simultaneous demands, the school-age child can mentally coordinate multiple ideas, thoughts, or strategies at the same time. A larger processing capacity means, for example, that a sixth-grader can simultaneously listen to the dinner-table conversation of her parents, respond to the interruptions of her younger siblings, think about her weekend plans, and still remember to ask for her allowance. In school, this increased processing capacity likewise means that she can answer a teacher's question with several relevant points rather than

retrieval strategies Procedures for recollecting previously learned information, such as thinking of related information or trying to create a mental image of the thing to be remembered.

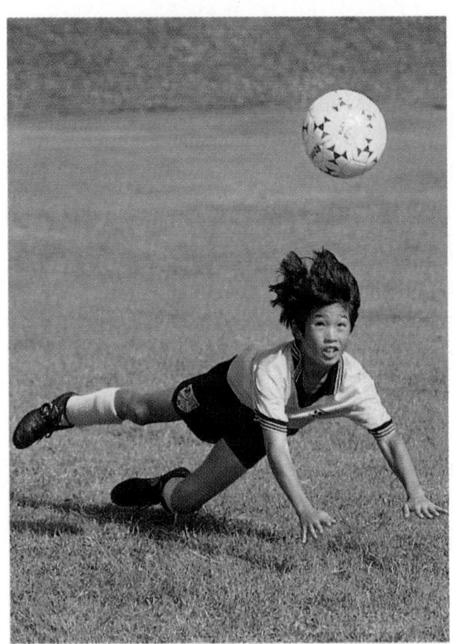

This boy's concentration while heading the ball and simultaneously preparing to fall is a sign that he has practiced the maneuver enough times that he can perform it automatically. Not having to think about what to do on the way down, he can think about what to do when he gets up, such as pursuing the ball or getting back to cover his position.

one and, at the same time, monitor her response for accuracy and clarity and check out her classmates to see if they are paying attention.

Why do processing speed and capacity increase during middle childhood? Certainly neurological maturation helps to account for these changes, especially the ongoing myelination of nerve pathways and maturation of the frontal cortex (Bjorklund & Harnishfeger, 1990; Dempster, 1993; Kail, 1991). Another important reason why processing becomes more efficient is that children learn to use their cognitive resources better (Case, 1985; Flavell et al., 1993). Selective attention is an example: as older children learn to focus their attention on only relevant information in a given task, their thinking becomes less "cluttered" with distractions, thereby enhancing their processing speed and capacity.

Another factor in improved processing is simple experience. Information processing becomes more efficient through **automatization**, the process in which familiar and well-rehearsed mental activities become routine and automatic. Recall when you were young how much concerted effort you needed to read words or to add numbers or to use a foreign vocabulary before you were fluent, or to hit the ball correctly when you first tried to play tennis, or baseball, or pool. As these activities became more familiar, well-practiced, and routine, less mental work was required to carry them out successfully, making it easier to devote your mental energies to other tasks. Consequently, you can now (most likely) comment to yourself mentally about what you are reading—even in a foreign language—or plot strategy while hitting a ball. As children mature and gain experience, more and more mental processes that initially required hard mental labor become automatized, and this increases processing speed and frees up processing capacity, enabling older children to think about many different things at once.

Knowledge Growth

Cognitive growth seems to follow the principle that the more one knows, the more one can learn. Having a **knowledge base**, that is, a body of knowledge or skills in a particular subject area, makes it easier to learn new information in that area because the new information can be integrated with what is already known. Thus, one reason children become better learners in middle childhood is that they have expanded knowledge bases: they already know more about many different domains of knowledge.

Not surprisingly, an expanded knowledge base in a particular area also improves memory ability for that area. What has surprised developmental scientists, however, is *how much* of a difference one's knowledge can have for memory ability—independent of the maturity of thinking, reasoning, and other skills. For example, as you might expect, chess experts are able to remember the board locations of chess pieces better than novices are. But does this remain true even when the experts are children and the novices are adults? This question was explored by Michelene Chi (1978) in a study of young chess experts recruited from a local chess tournament. These children, who were from the third through the eighth grades, were compared on their recall of complex chess positions against a group of adults who were acquainted with the game but were not experts. The children were strikingly more accurate in their recall than were the adults, and their mental organization of the chess pieces into logical, interrelated memory "chunks"

automatization The process by which familiar and well-rehearsed mental activities become routine and automatic.

knowledge base The part of memory that stores information over a long time, from minutes to decades. (Also called *long-term memory.*)

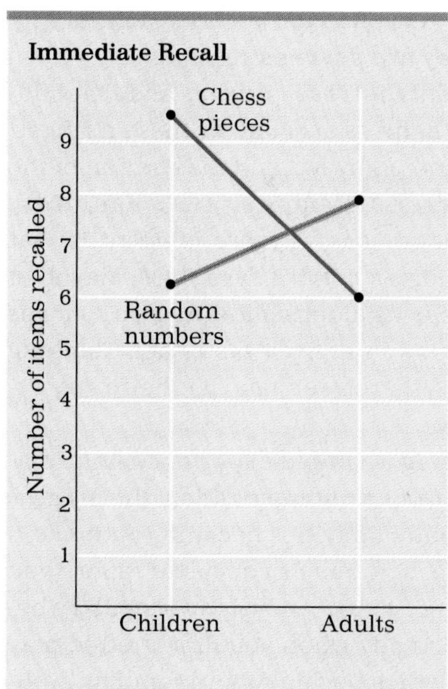

FIGURE 12.1 This graph shows the results of Michelene Chi's (1978) classic test of memory for chess positions. The fact that children who were expert at chess remembered the location of chess pieces better than adults who were novices—even though they did less well than the adults at remembering a series of random numbers—suggests the important role of knowledge in memory ability.

was also more efficient. By contrast, when the same children and adults were compared on a test of number recall—in which the children did not have expertise—the adults were more proficient (see Figure 12.1). The better memory skills of the children were apparent, therefore, only for those domains of thinking in which they had greater knowledge and experience.

Chi's findings are not a surprise to any parent who has been corrected by a 6-year-old dinosaur buff for mistaking a diplodocus for a brontosaurus. But the implications of this study may be surprising, since they suggest that adults are not always more cognitively competent than children. Many differences between schoolchildren's and adult's thinking and reasoning may be due, in fact, to the children's more limited knowledge and experience (Keil, 1984). As the child's storehouse of knowledge increases with age, concepts become more elaborated and interconnected with each other, the mental framework for organizing knowledge becomes more sophisticated, and the young learner can ask better questions to gain new understanding (Chi et al., 1989).

Metacognition

A final reason school-age children advance in learning and reasoning is their developing awareness of cognitive strategies. In essence, during the school years, children become aware that the content of their thinking is partly under their conscious control (Flavell, 1995). As a result of this awareness, and of the example and motivation provided by adults, they realize that learning and problem solving require that they actively manage their thinking, and they increasingly understand what they must do to perform well. What makes it possible for children to manage their thinking is **metacognition**, the ability to evaluate a cognitive task to determine how to best accomplish it—and to monitor and adjust one's performance on that task. In a sense, this "metacognition" means "thinking about thinking," and its development is related to the theory of mind children begin to acquire in the preschool years (see Chapter 9, pp. 256, 259–261).

There are many indicators of developmental growth in metacognition (Flavell et al., 1993; Siegler, 1991). Preschoolers have difficulty judging whether a problem is easy or difficult and thus they devote equal effort to each kind. By contrast, children in the school years know how to identify challenging tasks and they devote greater effort to these challenges, with greater success. During the school years children also acquire a better grasp of which cognitive strategies are well-suited for which cognitive tasks: they might recognize, for example, that rehearsal is a good strategy for memorizing, but that for enhancing understanding, it is less effective than outlining (Lovett & Flavell, 1990).

Moreover, school-age children spontaneously monitor and evaluate their progress in a manner that is rare in preschoolers, so they are usually able to judge when they have adequately learned a set of spelling words or science principles. Older children are also more likely to use external aids—such as making lists or drawing diagrams—to enhance their memorization and problem-solving efforts. In short, older children approach cognitive tasks in a more strategic and planful manner. Their efforts are thus more comprehensive and exhaustive, because they have a heightened metacognitive awareness.

metacognition The ability to evaluate a cognitive task to determine how best to accomplish it and how to monitor one's performance—"thinking about thinking."

Concrete Operational Thought

Another important cognitive characteristic of older children is that they are *logical* thinkers. Whereas preschoolers may invoke intuition and subjective insights to understand the results of a science experiment ("Maybe the caterpillar just felt like becoming a butterfly!"), school-age children seek explanations that are rational, internally consistent, and generalizable ("Does the caterpillar use the air temperature to know when it's time to begin a cocoon?"). This rationalizing process continues throughout middle childhood. Out on the playground, first-graders may argue over the rules of a game by using increasingly loud and assertive protests ("Is!" "Is not!" "*Is*!" "*Is not*!"), whereas fifth-graders temper their arguments with reason and justification ("That *can't* be right, because if it was, we'd have to score points differently!"). In academic and nonacademic contexts, logical thinking is crucial to understanding, to acquiring knowledge, and to communicating clearly with others.

As you know, Piaget was especially interested in the growth of logical reasoning skills during childhood. In his view, the most important cognitive achievement of middle childhood is the attainment of **concrete operational thought**, by which children can reason logically about the things and events they perceive. In Chapter 9 we saw that preschoolers, contrary to Piaget's expectations, sometimes show flashes of concrete operational thinking as demonstrated in their ability to perform successfully on simplified, gamelike tests of conservation. But preschoolers' ability to think logically is fragile at best, and usually limited to tasks on which they have received special training. According to Piaget, between ages 7 and 11, children are firmly in the concrete operational stage. They truly begin to understand logical principles and are able to consistently apply them to concrete—that is, specific, tangible—cases. They therefore become more systematic, objective, scientific—and educable—thinkers.

Logical Principles

To understand the importance of logical principles in the thinking of a concrete operational child, consider the principle of identity. **Identity** is the idea that an object remains the same despite changes in its appearance. Children who understand the principle of identity realize that superficial changes in an object's appearance do not alter its underlying substance or quantity. Mastery of this logical principle is one reason children become able to perform logical operations, such as those required in conservation tasks. In the conservation-of-liquids task (see pp. 261–263), for example, an awareness of identity enables school-age children to realize that pouring the liquid contents of a particular container into a differently sized container does not change the amount of liquid.

Identity is also relevant to mathematical understanding. Once children have a firm grasp of identity, they know that the number 24 is always 24, whether it is arrived at by adding 14 and 10 or 23 and 1. This logical principle also enhances scientific understanding, whether it involves grasping the underlying oneness of the tadpole and frog or understanding the idea that water converted to ice is still H_2O. Identity can be applied in nonacademic ways as well, particularly in everyday social encounters. The principle

concrete operational thought In Piaget's theory, the third period of cognitive development, characterized by the ability to apply logical processes to concrete problems.

identity In Piaget's theory, the logical principle that a given substance remains the same no matter what changes occur in its shape or appearance.

of identity enables school-age children to understand—as most preschoolers cannot—that their mother was once a child and that a baby picture of their mother is, in fact, a picture of their mother (even though she looks quite changed). In all these ways, the underlying logical principle of identity contributes to more acute, and objective, understanding.

This science teacher and student are demonstrating the effects of static electricity. Such demonstrations turn on the logical abilities of concrete-operational children much better than the abstract descriptions of textbooks.

The same is true of two other logical principles, **reversibility** and **reciprocity**. These two principles relate to the fact that a transformation can be restored to its original state by reversing the transformation process or by performing another transformation that compensates for the effects of the original one. Equipped with these two principles in the conservation-of-liquids task, a child can argue that the amount of liquid in the second, differently sized, container is the same as it was in the original container either by pouring the liquid back into the original container to demonstrate equivalence (reversibility) or by noting that the greater (or lesser) height of the liquid in the second container is attributable to that container's narrower (or wider) width (reciprocity). Such logical principles are also essential to understanding different mathematical operations, such as seeing subtraction as the reversing of addition (if $9 + 9 = 18$, then $18 - 9 = 9$) or understanding reciprocity in multiplication (that 4×3 and 3×4 both equal 12). Of course, reversibility and reciprocity also have everyday social relevance, such as when they are applied to social problem solving (as in "Let's start over and be friends again, OK?" or "I'll help you with math, if you teach me to dance").

reversibility The logical principle that something that has been changed can be returned to its original state by reversing the process of change.

reciprocity The logical principle that a change in one dimension of an object can be compensated for by a change in another dimension.

classification The concepts that objects can be organized in terms of categories or classes, as in sorting foods according to whether they are fruits, vegetables, or dairy products. According to Piaget, this concept is mastered during the period of concrete operational thought.

class inclusion The idea that a particular object or person may belong to more than one class. For example, a father can also be someone's brother.

The mastery of logical principles is also crucial to a concept that is basic to formal learning—classification. **Classification** is the concept that objects can be organized in terms of categories, or classes. For example, a child's parents and siblings belong to the class called *family*. *Toys*, *animals*, *people*, and *food* are other everyday classes. A related but more complicated concept is **class inclusion**, the idea that a particular object or person may belong to more than one class. Most preschool children have some understanding of how to apply classification labels, but they do not have a good

grasp of the complex relationships between general and specific categories (Flavell, 1985).

Younger children have particular trouble with the hierarchical relations between categories and subcategories. Consider the following experiment, modeled on a series of experiments conducted by Piaget. An examiner shows children seven toy dogs. Four of them are collies, and the others are a poodle, an Irish setter, and a German shepherd. First, the examiner questions the children to make sure that they know that all the toys are dogs and that they can name each breed. Then comes the crucial question: "Are there more collies or more dogs?" Until the concept of classification is firmly established, most children say "More collies." They cannot simultaneously keep in mind the general category *dog* and the subcategory *collie*, mentally shifting from one to the other. When they can eventually do so, it strengthens their understanding of the world around them. They realize that people, objects, and events can belong to more than one category, and that these categories can be hierarchically organized (a child is simultaneously a "human being," "mammal," and "animal"), overlapping (in many families a child is an "offspring" and a "sibling"), and, at times, diverse (a child is a "family member" and a "class member").

Obviously, a child who can consistently, and thoughtfully, apply logical principles is better equipped to carefully analyze problems, derive correct solutions, and ask follow-up questions that yield further understanding than is a more intuitive, haphazard thinker. The ability to apply logical principles also makes older children more objective thinkers, enabling them to distance themselves from their subjective impressions and personal experiences to derive a more reasoned judgment. Unlike preschoolers, they can arrive at assessments like "Maybe that's the way it seems to be, but really it's. . . . " Moreover, as identity, reversibility, and other logical principles are generalized across different tasks and situations, they provide coherence and consistency to a child's thinking.

Limitations of Piaget's Theory

Piaget's overall depiction of the stages of cognitive development is sometimes taken to suggest that, once begun, a child's movement from one stage to the next generally occurs across the board, in all cognitive domains. However, current researchers maintain that a certain type of reasoning might be apparent in one domain—say, math or science—but not in others, like social understanding.

These researchers believe that cognitive development is considerably more heterogeneous, or inconsistent, than Piaget's descriptions would suggest, particularly in the case of concrete operational thought. According to Flavell (1982), two of the factors that account for this heterogeneity are the hereditary differences among individuals in their abilities and aptitudes and the environmental differences in "cultural, educational, and other task-related experiential background." The sum of these differences, says Flavell, might well produce a great deal of cognitive heterogeneity:

> Imagine, for example, a child or adolescent who is particularly well-endowed with the abilities needed to do computer science, has an all-consuming interest in it, has ample time and opportunity to learn about it, and has an encouraging

> parent who is a computer scientist (whence much of the aptitude, interest, and opportunity, perhaps). The quality and sophistication of the child's thinking in this area might well be higher than that of most adults in any area. It would also likely be much higher in this area than in most other areas of the child's cognitive life. His level of moral reasoning or skill in making inferences about other people might be considerably less developed, for instance. The heterogeneity could be a matter of time constraints as well as a matter of differential aptitudes and interests: that is, time spent at the computer terminal is time not spent interacting with and learning about people.

Thus, while the overall thinking of a school-age child is definitely less intuitive and more logical than it was a few years earlier, the degree to which it is depends on the child, the topic, and the specific context.

Language

As you saw in Chapter 9, the preschool years are the time of a language explosion, in which children's vocabulary, grammar, and pragmatic language skills develop with marked rapidity. Language development between ages 6 and 11 is also remarkable, though much more subtle, as children consciously come to understand more about the many ways language is structured and can be used. This understanding gives them greater control in their comprehension and use of language, and, in turn, enlarges the range of their cognitive powers generally. Their understanding of language is a powerful key to a new understanding of themselves and their world.

Vocabulary and Grammar

During middle childhood, children begin to really enjoy words, as they continue to fast-map new ones into their vocabularies (see p. 268). Indeed, by some estimates, the rate of school-age vocabulary growth exceeds that of the preschool years, with children acquiring as many as twenty words daily to achieve a vocabulary of nearly 40,000 words by the fifth grade (Anglin, 1993). Children's delight in verbal play—clearly demonstrated in the poems they write, the secret languages they create, and the jokes they tell—makes middle childhood a good time to explicitly help children expand their vocabularies, thus providing a foundation for more elaborate self-expression.

One of the most important language developments during middle childhood is a shift in the way children think about words. Gradually they become more analytic and logical in their processing of vocabulary, and less restricted to the actions and perceptual features directly associated with particular words (Holzman, 1983). When a child is asked to say the first word that comes to mind on hearing, say, "apple," the preschooler is likely to be bound to the immediate context of an apple, responding with a word that refers to its appearance ("red," "round") or to an action associated with it ("eat," "cook"). The older child, on the other hand, is likely to respond to "apple" by referring to an appropriate category ("fruit," "snack") or to other objects that logically extend the context ("banana," "pie," "tree"). Moreover, the older child can deduce the meanings of new words that have "apple" as their root (such as "applesauce" and "applecart"), and this ability also helps to account for rapid vocabulary growth (Anglin, 1993).

With a language as irregular as English, it should be no surprise that many children (as well as adults) sometimes generate grammatical errors by applying logic to their language constructions.

Similarly, when they define words, preschoolers tend to use examples, especially examples that are action-bound. For instance, while preschoolers understand that "under," "below," and "above" refer to relative position, they define these words with examples such as "Rover sleeps under the bed," or "Below is to go down under something." Older children tend to define words by analyzing their relationships to other words: they would be more likely to say, for instance, that "under" is the same as "below" or the opposite of "above" (Holzman, 1983).

Older children's more analytic understanding of words is particularly useful as children are increasingly exposed to words that may have no direct referent in their own personal experience. This understanding makes it possible for them to add to their conceptual framework abstract terms such as "mammal" (extracting the commonalities of, say, whales and mice) or foreign terms such as *yen* (relating this unit of currency to the dollar), and to differentiate among similar words such as "big," "huge," and "gigantic," or "jogging," "running," and "sprinting." Thus, the cognitive maturation of middle childhood, coupled with the school experiences that children have, enables children's vocabularies to increase exponentially.

The school-age child's gradual understanding of logical relations also helps in the understanding of complex grammar, such as the correct use of comparatives ("longer," "deeper," "wider"), of the subjunctive ("If you were a millionaire . . ."), and of metaphors (that is, of how a person could be a dirty dog or a rotten egg) (Waggoner & Palermo, 1989). That logical understanding is required for mastering such constructions is clear even with languages in which the particular construction is relatively simple. For instance, the subjunctive form is much less complicated in Russian than in English, but Russian-speaking children master the subjunctive only slightly earlier than do English-speaking children, because the concept *if-things-were-other-than-they-are* must be understood before it can be expressed (de Villiers & de Villiers, 1978, 1992).

School-age children have another decided advantage over younger children when it comes to mastering the more difficult forms of grammar. Whereas preschool children are quite stubborn in clinging to their grammatical mistakes (remember the child in Chapter 9 who "holded" the baby bunnies?), school-age children are more teachable. They no longer judge correctness solely on the basis of their own speech patterns. Assuming that they have had ample opportunity to learn the correct grammar, by the end of middle childhood, children are able to apply the rules of proper grammar when asked to, even if they don't use them in their own everyday speech. Thus, even if they themselves say "Me and Suzy argued," they are able to understand that "Suzy and I argued" is considered correct.

code-switching A pragmatic communication skill that involves a person's switching from one form of language, such as dialect or slang, to another.

Pragmatics

We have already seen that preschoolers have a grasp of some of the pragmatic aspects of language: they change the tone of their voice when talking to a doll, for instance, or when pretending to be a doctor. However, preschoolers are not very skilled at modifying vocabulary, sentence length, semantic context, and nonverbal cues to fit particular situations. The many skills of communicating improve markedly throughout middle childhood.

One of the clearest demonstrations of schoolchildren's improved pragmatic skills is found in their joke-telling, which demands several skills not usually apparent in younger children—the ability to listen carefully; the ability to know what someone else will think is funny; and, hardest of all, the ability to remember the right way to tell a joke. Telling a joke is beyond most preschool children. If asked to do so, they usually just say a word (such as "pooh-pooh") or describe an action ("shooting someone with a water gun") that *they* think is funny.

Further evidence of increased pragmatic skill is shown in children's learning the various forms of polite speech. School-age children realize that a teacher's saying "I would like you to put away your books now" is not a simple statement of preference but a command in polite form (Holzman, 1983). Similarly, compared with 5-year-olds, 7- to 9-year-olds are quicker to realize that when making requests of persons of higher status—particularly persons who seem somewhat unwilling to grant the request—they should use more polite phrases ("Could I please . . . ?") and more indirect requests ("It would be nice if . . . ") than when they are negotiating with their peers (Axia & Baroni, 1985).

At a broader level, children's pragmatic skills are clearly manifest in their ability to engage in **code-switching**, changing from one form of speech to another. Children in middle childhood can engage in many forms of code-switching, from the relatively simple process of censoring profanity when they talk to their parents to switching back and forth from one language to another.

A very obvious example of code-switching is children's use of a formal manner of communicating when they are in the classroom and an informal one when they are with friends outside of school. In general, the formal code

Next time you see a group of school-age children laughing at a friend's joke, you should be impressed with the young person's mastery of pragmatics. Telling a good story requires careful timing, exact vocabulary, and strategic nonverbal cues, as well as a knowledge of one's audience.

is characterized by extensive vocabulary, complex syntax, and lengthy sentences; the informal code, by comparison, has a much more limited use of vocabulary and syntax and relies more on gestures and intonation to convey meaning. The formal code is relatively context-free; the meaning of its statements is explicit. The informal code tends to be context-bound, relying on the shared understandings and experiences of speaker and listener to provide some of the meaning. Switching from one code to another, a dispirited student might tell a teacher, "I am depressed today and I don't feel like doing anything," and later confide to a friend, "I'm down. School stinks." Research has shown that children of all social strata engage in this type of code-switching and that their pronunciation, grammar, and slang all change in the process (Holzman, 1983; Rogers, 1976; Romaine, 1984; Yoon, 1992).

It seems clear that both codes have their place. It is important to be able to explain one's ideas in elaborate and formal terms when appropriate. In fact, two of the basic skills taught during the elementary school years, reading and writing, depend on the comprehension of language in a situation devoid of gestures and intonations. At the same time, it is useful to be able to express oneself informally with one's peers, using more emotive, colloquial, and inventive modes of communication than those of the standard, accepted code. While many adults rightly stress the importance of children's mastery of the elaborated code ("Say precisely what you mean in complete sentences, and no slang"), the code that is used with peers is also evidence of the child's pragmatic skill (Goodwin, 1990).

Bilingualism

Learning a second language is an academic goal in virtually every school system worldwide, with good reason. Few nations are without a sizable minority who speak a language other than that of the majority, and multinational business, travel, and immigration are commonplace today. Thus it would benefit everyone to be at least bilingual. Moreover, learning another language fosters children's overall linguistic and cognitive development, especially if it occurs before puberty (Baker, 1993; Edwards, 1994; Romaine, 1995). Because of their readiness to code-switch, their eagerness to communicate in pragmatic terms, and their ear for nuances of pronunciation, children under 10 years old are the best learners of the spoken form of a foreign language. Thus, from a theoretical viewpoint, early bilingual development is a goal to be sought, not a problem to be solved.

Learning a second language becomes not simply a goal but a necessity when a child's language heritage is different from the language of the larger society. In the United States, an urgent educational and political question is how best to teach the schoolchildren whose first language is not English. In 1990 there were 6.3 million such children between the ages of 5 and 17 (14 percent of the total age group), with 2.2 million of them speaking English "less than very well" (U.S. Bureau of the Census, 1996). Many other countries, including Australia, England, and most of the nations of Europe, have the same problem: a sizable minority of immigrant children who enter school without knowing the usual language of instruction.

In Canada the question is even more complex, since bilingualism is part of a cultural and political struggle that goes to the heart of Canadian

immersion An approach to learning a second language in which the learner is placed in an environment where only the second language is spoken.

identity. Beginning in kindergarten, English-speaking children need to learn French, French-speaking children need to learn English, and children knowing neither language need to learn both, becoming trilingual.

Unfortunately, although this question has been one of intense concern and emotion, no simple answer is apparent. Almost every educational approach has been tried—from total **immersion**, in which the child's instruction occurs entirely in the second language, to "reverse immersion," in which the child is taught in his or her native language until most of childhood is over and the second language can be taught as a "foreign" language. Variations on these approaches present some topics of instruction in one language and other topics in the other language. However, few carefully controlled, longitudinal studies have been done to evaluate the effectiveness of the various approaches.

Of the research that is available, some of the most thorough comes from Canada, where both English and French are official languages that all children are expected to learn (about 40 percent of all Canadian children have English as their first language, about 30 percent have French, and about 30 percent have some other language). Immersion programs seem to work best when children are young. In Canada, immersion has proven successful not only with immigrant children who speak neither English nor French, but also with more than 300,000 English-speaking children who have been taught exclusively in French in their first three years of school and then gradually given more instruction in English. During their period of immersion, these children showed no declines in English skills or in other academic achievement (Edwards, 1994; Lambert et al., 1993).

However, in most such instances of successful immersion, parents voluntarily placed their children in a special program designed to teach the second language. As the Research Report on pages 339–340 points out, variations in attitudes within the family and within the culture are often transmitted to the classroom, in some cases causing the child to cling steadfastly to his or her first language, in others, causing the child to attempt to abandon the mother tongue, and in others, ideally, helping the child to readily learn a second language, perhaps even to become truly bilingual and bicultural.

Bilingual education can be exciting and successful when words are explained using two languages as well as pictures and actions that help students bridge the gap from one language to the other.

RESEARCH REPORT

Bilingual Education

If there were some easy way to accomplish it, nearly everyone could benefit by being fluent in at least two languages. In addition to the practical value of enabling a person to communicate with a larger and more diverse group of people, bilingualism also enhances cognitive flexibility because it makes children aware that ideas can be expressed in many ways (Genesee, 1994; McLaughlin, 1985).

Bilingualism is attained most readily when begun in infancy, with the child hearing and speaking two languages, in the home and/or in day care. While such language mixing sometimes slows a child's early language learning, during the preschool years the typical child's urge to be social and to communicate, coupled with relatively high self-esteem and little concern for social embarrassment, soon fosters rapid learning of every language the child hears, whether it be one, or two, or even three distinct tongues (Goodz, 1994; McLaughlin, 1985).

The same general success can occur in kindergarten, even for those children for whom the second language to be learned is the language of instruction. Most kindergartners begin school already knowing such basics of communication as expressive gestures and turn-taking, as well as social and cognitive strategies for "getting along" in whatever language other children may speak (Saville-Troike et al., 1984). One study of Spanish-speaking children in an English-speaking kindergarten, for example, found eight distinct strategies:

Social Strategies

1. Join a group and act as if you know what is going on, even if you don't.
2. Give the impression, with a few well-chosen words, that you speak the language.
3. Count on your friends to help.

Cognitive Strategies

1. Assume that what people are saying is directly relevant to the situation at hand. GUESS.
2. Use some expression you understand and start talking.
3. Look for recurring parts in the formulas you know.
4. Make the most of what you've got.
5. Work on the big things first; save the details for later.

Obviously, the children need to feel fairly self-confident, and their peers need to be relatively receptive, before these strategies can be used effectively. Unfortunately, such is not always the case for many minority-language children, especially older children.

The success of formal bilingual education, of course, varies greatly from child to child, teacher to teacher, and program to program (Wong Fillmore, 1987). One critical factor is the teacher's ability to create a social milieu that encourages all the children to make friends, to join conversations, and to feel free to guess (Wong Fillmore, 1976, 1987). The presence of this factor alone is the main reason that some young children pick up a second language much more quickly than others.

As children grow older and more self-conscious about making friends, it becomes harder to simply immerse them in a classroom and expect them to learn. One boy recalls his early experiences in a Toronto classroom:

> I did not know what to do when the other students spoke to me because I did not understand them. I was forced to use signs with my hands to communicate with people, just as if I were deaf and dumb. I hated the students who spoke with me. . . . Sometimes there was a joke, and I had to laugh with the others even though I did not know what the joke was, because I was afraid of being laughed at. [quoted in Coelho, 1991]

For school-age children who do not speak the dominant school language, there are three general approaches to teaching them. One is the *second-language* approach (for instance, ESL, or English as a second language), in which a group of children are taught the dominant language in much the same way native-speaking children might be taught a foreign language by an instructor who specifically focuses on language learning. Typically, such a group receives separate instruction in the second language all day, every day, and then soon joins the larger, dominant-language group, perhaps first in classes such as music and gym, then in math and science, and finally in English and social studies. Another approach is *bilingual-bicultural* education, in which the children are educated with others of their culture and are taught in their first language. In this manner, instead of focusing solely on learning the dominant language, they maintain their native language, learning various content areas in it, while studying the new language as an academic subject. Finally, there is *immersion*, in which the child is simply put in with students and teachers who use only the majority language. Each of these approaches has some advantages.

The second-language approach is quite practical, especially if there are children from many language groups in a school. However, in the process of learning the new language, children using this approach may come to feel ashamed of their native language and may even refuse to speak it at home, except under duress. As a result, their relationship with their parents may be seriously strained and their cultural identity may falter (Mills & Mills, 1993). Further, children need to learn "in a context where they can voice their experiences." If they cannot use their native language, and they do not feel comfortable in the new

(continued)

Bilingual Education (continued)

language, their self-esteem, motivation, and education may suffer (Cummins, 1994).

The bilingual-bicultural approach attempts to remedy that problem by providing extensive instruction in both languages and cultures by teachers well-versed in each. This approach sounds ideal, but such programs are difficult to staff and fund, and too often the children feel themselves to be, and are seen as, a separate group. They tend not to socialize with the children from the dominant language group, and, since language is learned best in the context of social communication, this means they are slow to learn the new language and to feel comfortable in the second culture.

By contrast, in immersion programs, social communication among children is almost inevitable, especially if both languages and cultures are relatively high in status and the parents of the children are supportive (Edwards, 1994). Such is the case in Quebec, where thousands of English-speaking children learn French through immersion that starts in kindergarten. By the fifth grade, they are almost as proficient in French as their classmates who are native French speakers. In addition, they do quite well on math, science, and history tests in either language. Similar results are found for Finnish children who are immersed in Swedish, especially if their knowledge of their native language is sound and they enter Swedish immersion in kindergarten or at about age 10. Entering immersion in the middle of elementary school, at about age 7, does not work as well (Paulston, 1992).

However, immersion programs fail if the child feels shy, stupid, or socially isolated, and if the attitude of the school is that the child is deficient because he or she speaks a different language. In such cases, the educational approach might more aptly be called "submersion," because the child is more likely to sink than swim (Edwards, 1994). In the United States, many Latino children instructed only in English become slow learners who repeat a grade or two until they are old enough to drop out of school (McLaughlin, 1985). Typically, their poor performance is blamed on their deficit in English, rather than on the teachers and educational programs that fail to take into account their special language needs. Added to that, limited expectations held by everyone—students, parents, and teachers—are considered a major reason that Spanish-speaking children have the lowest rates of high school graduation—55 percent, compared with 73 percent for African-Americans and 87 percent for whites (U.S. Department of Education, 1993).

Failure for Latinos in immersion programs is, obviously, not inevitable. One notable case in point is that of Richard Rodriguez, a Mexican-American who entered an all-English school in kindergarten, and who eventually mastered the language so well that he studied English literature in graduate school at Berkeley and Columbia and became a well-regarded writer (Lucas et al., 1990). He and his family were determined that he would learn English, no matter what the cost. His teacher advised his parents to stop speaking Spanish to their children, and even though their understanding of English was minimal, they did so. Rodriguez (1983) explains:

> As we children learned more and more English, we shared fewer and fewer words with our parents. Sentences needed to be spoken slowly when a child addressed his mother or father. (Often the parent wouldn't understand.) The child would need to repeat himself. (Still the parent misunderstood.) The young voice, frustrated, would end up saying "Never mind"—the subject was closed. Dinners would be noisy with the clinking of knives and forks against dishes. My mother would smile softly between her remarks; my father at the other end of the table would chew and chew his food, while he stared over the heads of his children.

Despite the drawbacks, in retrospect, Rodriguez approves of his immersion:

> Without question, it would have pleased me to hear my teachers address me in Spanish when I entered the classroom. I would have felt much less afraid. I would have trusted them and responded with ease. But I would have delayed—for how long postponed?—having to learn the language of the public society.

Rodriguez never learned to write Spanish, and as he grew older, he lost his ability to speak it as well. He found himself increasingly distant from his parents and their culture, a not uncommon consequence for immigrant children who succeed in the majority school. The latest wave of immigrants to suffer these consequences are from Southeast Asia. Many of the children in this group eventually become quite successful in school, but they pay a high price—the loss of the ability to communicate with their parents (Wong Fillmore, 1991).

This raises a basic issue: What is the goal of language education? Virtually all programs and evaluations of bilingual education focus on one measure of success: how proficient children are in the new language. Other measures of school success, from math achievement to high school graduation, and measures of personal success, from self-esteem to social understanding, are often ignored (Hakuta & Garcia, 1989; McKeon, 1994). Obviously, school-age children need to learn the language of their society, but as the next chapter explains, they also need to develop their understanding of themselves and of others. Educational practices, including implicit attitudes regarding children's original language, can be instrumental in determining whether "the linguistic, cognitive, and sociocultural resources children bring to school" are used to achieve the larger goal—seeing that "the developing child becomes a fully functioning and valued member of the community" (Genesee, 1994).

Thinking, Learning, and Schooling

The portrait of the school-age child sketched in this chapter is of someone who is thoughtful and eager to learn, able to focus attention, to master logical operations, to remember interrelated facts, and to speak in several linguistic codes. That portrait is universal, describing children aged 6 to 11 the world over.

Agreement on how best to direct and use that learning potential is not universal, however. While schooling of some sort during middle childhood is a feature of every community worldwide, the specifics of who receives instruction, in what, and how, vary enormously. Historically, boys and wealthier children were much more likely to be formally taught than girls or poor children, an inequality still apparent today. In 83 percent of developing countries, fewer girls than boys attend primary school, and although virtually every child in developed countries attends school, less is generally demanded of girls and poor children, particularly in math and science (Minuchin & Shapiro, 1983; UNICEF, 1990). For example, in the United States, more female and poor children drop out of school than do male and nonpoor children, and of those who remain to graduate from high school, far fewer take algebra II, calculus, chemistry, and physics (U.S. Department of Education, 1989).

Another critical variation is in the curriculum offered. Basic literacy is universally sought, but other curriculum elements—science, arts, health, and religion among them—are prominent in some schools and virtually absent in others. Pedagogical techniques vary widely as well, from the strict lecture method, in which students are forbidden to talk, whisper, or even move, to open education, in which students are encouraged to interact and to freely avail themselves of classroom resources, with the teacher acting more as an adviser, guide, and friend than as a knowledge authority and disciplinarian.

Until fairly recently, such variations did not usually trouble the nations involved, because each culture's methods reflected traditions and a national ethos that were taken for granted. As the twenty-first century approaches, however, international economic competition—and the growing need for workers to be literate, skilled, and self-motivated—have intensified public

Educational curricula and methods reflect cultural values and national priorities. The children in this Koranic school in India and in this "classroom" in Somalia are mastering lessons quite different from those that are typically taught in North American, European, and Australian schools. To fairly evaluate any educational system, one must ask whether or not children are learning the skills they will need in their particular social setting.

(a)

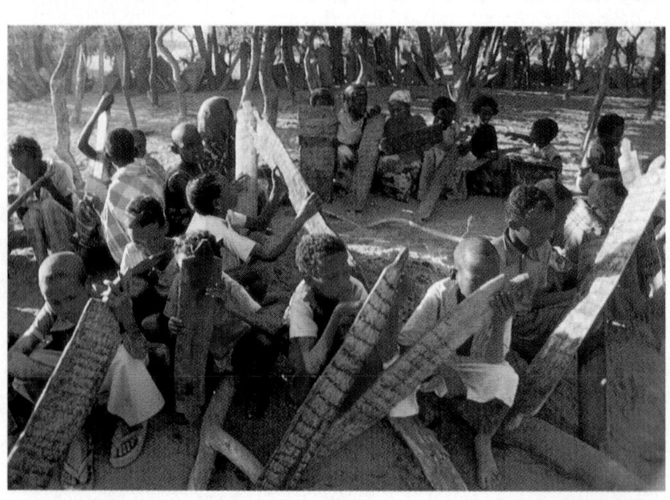

(b)

concern about certain educational processes that now seem to be inadequate.

This concern is particularly apparent in the United States, because achievement scores show American children behind their counterparts in most other industrialized countries, especially in math and science (International Assessment of Educational Progress, 1989; Stevenson & Stigler, 1992). The differences are most marked when children of the United States are compared with those from the Pacific-Rim countries (Japan, Hong Kong, Korea, and the Republic of China). Significantly, while international comparisons generally find that children in countries on the Pacific Rim do especially well in math and science and that, among industrialized nations, children in the United States, Italy, and Spain tend to be near the bottom in those subjects, a closer comparison reveals some interesting differences. For example, high school seniors in both the United States and Japan do better in physics than in biology, while in almost every other country the reverse is true.

More fine-tuned analysis again reveals curious differences: Scottish 13-year-olds, for instance, rank among the lowest internationally in basic arithmetic, but among the highest in understanding statistics and probability, while Irish children are relatively good in arithmetic and poor in probability (National Science Foundation, 1992). Figure 12.2 highlights other international differences in science achievements. In reading ability, especially among gifted children, the advantage of the Pacific-Rim nations over the United States disappears (Stevenson et al., 1993). Such comparisons suggest that international differences in achievement are not primarily the result of children's innate abilities or early nurturance but, rather, are related to specifics within the educational systems and social contexts during the school years (see Research Report, pp. 343–344).

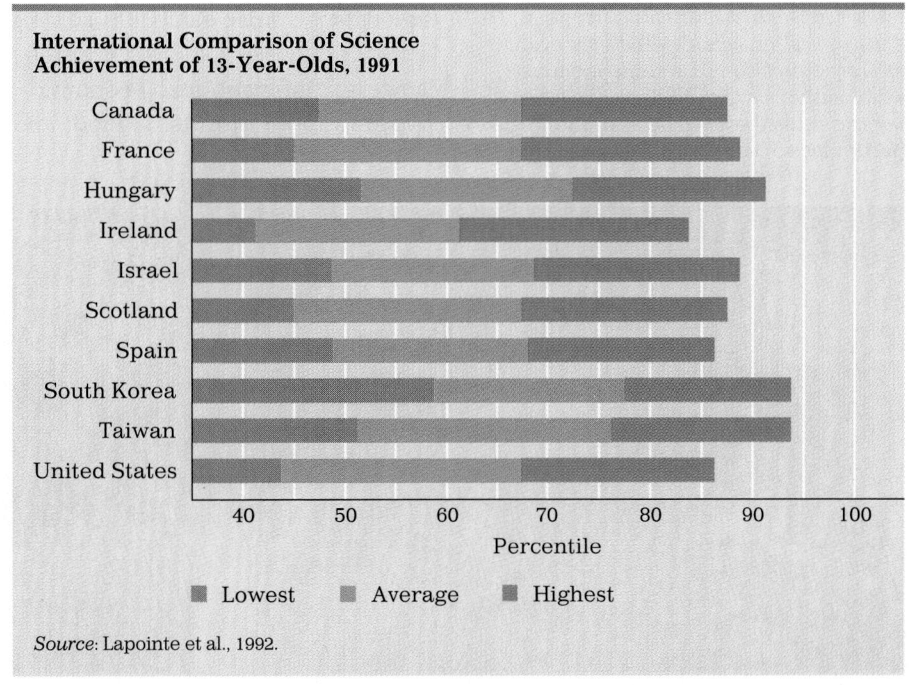

FIGURE 12.2 An international comparison of the science achievement of 13-year-olds shows the gap between the highest- and lowest-scoring countries in 1991 narrowing compared to the scores in the early 1980s, when a child who was in the top 20 percent of U.S. students scored no better than the average South Korean student. However, the gap between the highest and lowest tenth of students reveals marked differences within nations in schooling and preparation. Spain and South Korea are the only countries where the lowest 10th percentile is less than 40 points below the highest 10th percentile.

RESEARCH REPORT

Education in the United States, Japan, and the Republic of China: A Cross-Cultural Comparison

The same international comparisons that find children in the United States close to the bottom in scholastic achievement find children in Japan, Korea, and the Republic of China (Taiwan) at the top (Suter, 1993; United States Department of Education, 1993). To understand these achievement differences, an international team of researchers headed by Harold Stevenson has spent the past twenty years comparing the school achievement, leisure-time use, and academic attitudes of more than 5,000 children and their parents at sixty-four schools in three comparable cities: Minneapolis (USA), Taipei (the Republic of China), and Sendai (Japan) (Stevenson & Stigler, 1992; Stevenson et al., 1993). They found that children in the three cities were similar in aptitude, as measured by intelligence tests when they entered school and by tests of general knowledge when they were in the eleventh grade. However, as other researchers have also found, the Chinese and Japanese students outperformed the Americans in achievement tests, particularly in math and science, at every grade level and across every level of ability. Indeed, the performance of the top 10 percent of American students was only at the level of the average Asian student.

If it was not a matter of ability, why did these differences occur? One reason is the different number of hours the children in the three countries spend in schoolwork. In Japan and the Republic of China, for example, children go to school five and a half days a week, compared to five in the United States; the average school year is one-third longer than in the United States; and the children are in school more hours each day than American children are.

But according to these researchers, even more important than how much time children spend in the classroom is how they spend that time. American schoolchildren in the study devoted far less classroom time to academic activities than Japanese and Chinese students did, more frequently engaging in nonacademic activities (getting out of their seats, chatting with friends, and so on). By fifth grade, American students were spending 64 percent of the schoolday on academic work, whereas Japanese students were academically engaged 87 percent of the time.

One notable result of this difference in the use of classroom time was that American children spent an average of only three hours a week on math, whereas Japanese children spent an average of seven. Also, compared with Japanese and Chinese children, American children were more inclined to work alone than in groups. American teachers, in turn, worked more frequently with individual children or in small groups than with the class as a whole, while teachers in Japan and China strongly emphasized group instruction with every child actively participating. As a consequence, children from the Asian nations had greater overall opportunity to learn from their teachers in class.

Added to the difference in classroom time and work was a notable difference in the time devoted to homework. At grade 5, Americans averaged four hours per week studying at home, compared with thirteen hours for Chinese children and six hours for Japanese (Stevenson & Lee, 1990).

Education also involves the home environment, of course, and the most recent reports from these researchers reveal important differences in parental support and attitudes toward the child's educational progress. According to Stevenson and his colleagues, academic achievement was a much more central concern to Japanese and Chinese parents, who had higher expectations for their offspring and were more involved in fostering their children's success (Stevenson & Stigler, 1992; Stevenson et al., 1993). The Minneapolis mothers, according to these researchers, believed instead that "it is better for children to be bright than to be good students," reflecting the prevailing American belief that high school achievement depends mostly on native ability. Perhaps as a consequence, parents in Japan and the Republic of China were much more involved in the child's education, encouraging and supervising homework (91 percent of Japanese children had their own desk at home), maintaining high standards for academic work, and emphasizing the value and importance of hard work over innate capabilities. In addition, if a child shows signs of falling behind, it is common for Japanese parents to arrange for supplemental courses at special cram schools called *juku* (Rohlen, 1983). Another study, in the United States, revealed that similar family values help to account for the astonishing academic success of the children of immigrant families from Vietnam, Laos, and other Southeast Asian countries (Caplan et al., 1992).

In view of a decade's worth of publicity—often wildly exaggerated—claiming that schools in the United States are failing, that their graduates are ignorant, if not illiterate, and that U.S. industries are unable to compete in the international marketplace because they lack properly educated workers, one might think that American parents would have become at least somewhat concerned about the quality of their children's education. However, a 1990 survey found American parents to be far more satisfied with their children's progress and achievement than were parents in both Japan and the Republic of China, a contrast that, if anything, was even stronger than it was ten years earlier. Perhaps because of their parents' contentment with their achievement, American students spend more time doing nonschool activities than their Asian peers. This is particularly apparent in the eleventh grade,

(continued)

Education in the United States, Japan, and the Republic of China: A Cross-Cultural Comparison (cont.)

Japanese primary schools typically group children by age, not ability, and encourage cooperation among members of each table. The underlying belief is that all children can succeed at mastering quite complex material, as these sixth-graders in Tokyo seem to be doing.

when 74 percent of American students have part-time jobs, compared with 21 percent of Asian students, and 85 percent are dating, compared with 37 percent of their Asian counterparts. Critics of the Asian emphasis on schooling have long maintained that Asian children, worn down by the constant grind and pressure to excel, are highly susceptible to depression and even suicide. In fact, however, American students are the ones more likely to feel stressed, depressed, and aggressive, as well as to be anxious about school (Stevenson et al., 1993).

As this research indicates, educational achievement entails a complex interaction involving a child's aptitude and motivation, the school environment, and support from the home. American parents tend to be most concerned about their child's progress during the preschool years, but then

> they abdicate some of these responsibilities to the teacher once the child enters school. This trend is opposite from that which occurs in Chinese and Japanese families. . . . From the time that the child enters school, life for the Chinese and Japanese child becomes purposeful; the child, the parents, and the teachers begin the serious task of education. [Stevenson & Lee, 1990]

This difference is reflected in the political priorities of each culture as well. For example, in keeping with the culture's regard for education, Japanese teachers are highly trained, greatly esteemed, and well paid (their salary is 2.4 times the average Japanese salary, in contrast with that of American teachers, which is 1.7 times the national average). Similarly, Japanese schools are clean and very well equipped, with almost all having libraries, music rooms, and science laboratories, and 75 percent of them having swimming pools, a luxury enjoyed by only 1 percent of American schools.

It is tempting, of course, to focus on one or two distinctive traits of successful education in another country and to consider them as solutions to domestic educational problems. However, an awareness of the importance of cultural context reminds us that no one national educational system can be transmitted, wholesale, to another. Even the easiest-seeming solutions, such as extending the school day, may contain hidden risks when lifted out of their overall context. As the researchers in the comparison study warn,

> increasing the amount of time spent in academic activities without modifying the content of the curriculum and the manner of instruction might further depress American children's interest in school and increase their dislike of homework. Greater time on task is not the primary basis for the high achievement of Chinese and Japanese children. The answer lies instead in the high quality of experiences that fill this time . . . Chinese and Japanese elementary school classrooms, contrary to common stereotypes, are characterized by frequent interchange between teacher and students, enthusiastic participation by the students, and the frequent use of problems and innovative solutions. [Stevenson & Lee, 1990]

As the text points out, this is precisely the direction suggested by developmental theory, a direction that more and more educators in the United States, Canada, and elsewhere are pursuing. Perhaps combining more time on task and better instructional methods will soon boost American children's achievement. However, such a switch is not likely to occur without a change in the public's attitude, putting more financial resources, better professional training, and higher academic demands on teachers and students. As the researchers write:

> We conclude that the achievement gap is real, that it is persistent, and that it is unlikely to diminish until, among other things, there are marked changes in the attitudes and beliefs of American parents and students about education. American parents appeared to be no more likely in 1990 and 1991 than they were in 1980 to believe that there is an urgent need for educational reform. They did not seem to be incensed by the low levels of performance by American students. Rather they appeared to be pleased with their children's academic achievement, to be satisfied with the job their children's schools were doing, and to believe that children's innate abilities guide their course of progress through school. But the likelihood of improving the nation's competitive position through better education depends, at least in part, on changing such optimistic but ultimately self-defeating views. [Stevenson et al., 1993]

As a result of these international differences, almost every aspect of the schooling process, from teacher selection to public financing, from ethnic diversity to religious instruction, from parental involvement to curriculum balance, has been the target of criticism. While most of the issues in these controversies are more appropriately discussed in a textbook on educational policies or political processes than here, research in cognitive development provides valuable insights into the basic question of how children learn best.

Children encourage and challenge each other to master new knowledge, often better than an adult can. All three children here—doctor, patient, and observer—are likely to remember the use of a stethoscope and the correct location of various organs.

Cognitive Development and Classroom Learning

As we have seen, school-age children are active learners, eager to master logical principles and learning strategies, as well as to develop academic skills and accumulate knowledge. Further, their many skills, from memorizing tricks to concrete operations, are most readily mastered through guided instruction and personal involvement. All this means that passive learning—such as sitting quietly and copying work from the blackboard—and piecemeal learning—such as memorization, by repetition and rote, of the sounds of the alphabet, of the sums of simple numbers, of the names of the continents, and so forth—are not the most appropriate means of instruction overall. Educators influenced by developmental theory, particularly Piaget's, have concluded that the classroom should always be a busy place, in which children's curiosity is met with an array of materials to explore and discuss, such as coins to count, objects to measure, books to read, stories to dramatize.

More recently, the information-processing perspective has led to a reemphasis on explicit instruction, which was sometimes shunted aside in the initial excitement over the Piagetian idea of the child as a self-motivated explorer-scientist. However, the information-processing emphasis on specific skills and a solid knowledge base is quite different from that implicit in the workbooks and rote memorization of old. Student motivation, attention, and mastery of strategies and principles are the key, with the teacher at the ready to provide the necessary knowledge and skills whenever the child is prepared to open a new cognitive door.

An even more recent insight from developmental research has been a recognition of the importance, as highlighted by Vygotsky, of social interaction in the classroom, not only between teacher and student but also among the children themselves, and of the support for academic skills from home. Numerous studies have shown that if their task is structured to encourage cooperation, classmates can draw each other into the zone of proximal development (see pp. 265–266), expanding each other's knowledge as well as, and more often than, any one teacher can (Rogoff, 1990).

This fact was clearly demonstrated in an extensive experiment on the development of reading skills, in which teachers of 3,345 children in Hawaii followed a Vygotskiian model that emphasized social interaction within the zone of proximal development. Instead of the usual practice of having students read silently to themselves, with the teacher asking the entire class some simple comprehension questions, the teachers in this experiment used such strategies as having groups of children read a paragraph aloud together, and then collectively respond to a series of questions designed to elicit discussion as well as confirm understanding. Compared to a control

The School Years: Psychosocial Development

CHAPTER

13

During the school years, emotional and social development occurs in a much more elaborate context than in the closely supervised and circumscribed arenas of the typical younger child. As school-age children explore the wider world of neighborhood, community, and school, independent of parental control, they experience new vulnerability, increasing competencies, ongoing friendships, challenging and sometimes troubling rivalries, deeper social understanding, and conflicting moral values. Personality attributes, coping mechanisms, and future aspirations are all formed by their developing social cognition.

This interplay between increasing competence and independence and an expanding social world is the theme of psychosocial development in middle childhood, and thus the theme of this chapter. First let us look specifically at the emotional growth that characterizes these years, and then at some of the contexts, particularly those associated with family structure, that shape and propel that growth. Finally, we will look at the stresses that many school-age children confront, and at their ways of coping with those stresses.

An Expanding Social World

Throughout the world, school-age children are recognized as markedly more independent and more capable than younger children. As a result, children go to school, or outside to play, or, in many cultures, off to work, out of their parents' view. They meet friends and strangers unknown to their families and experience adventures and challenges that adults often know little about.

A Common Theoretical Thread

This new competence has been recognized by every developmental theorist who has attended to this period. Freud describes middle childhood as the period of **latency**, when children's emotional drives are quieter, their psychosexual needs are repressed, and their unconscious conflicts are submerged, features that make latency "a time for acquiring cognitive skills and assimilating cultural values as the child expands his world to include teacher, neighbors, peers" (Miller, 1983).

latency Freud's term for the period between the phallic stage and the genital stage of psychosexual development. During latency, which lasts from about age 7 to age 11, children's psychosexual drives and unconscious emotional conflicts are relatively quiet, and children direct their attention and energies to the outside social world.

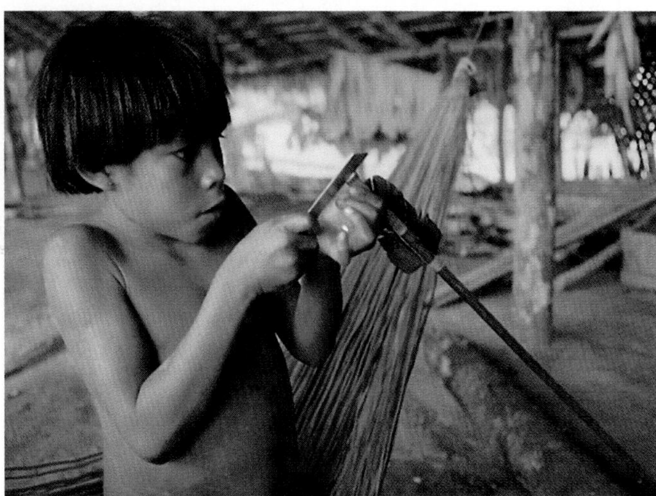

The North American boy at his father's worktable and the Brazilian boy fashioning arrows in the Amazon jungle are both engaged in essentially the same task—attaining competence in their respective cultures, in accordance with the particulars of their social setting.

Erikson (1963) likewise agrees that middle childhood is a quiet period emotionally and that it is productive as well, as the child "becomes ready to apply himself to given skills and tasks." The specific crisis that Erikson describes for this developmental period is **industry versus inferiority**. According to Erikson, as children busily try to master whatever skills are valued in their culture, they develop views of themselves as either competent or incompetent, or, in Erikson's words, as either industrious and productive or inferior and inadequate.

Operating from quite different theoretical bases, developmentalists influenced by behaviorism or social learning theory, or by the cognitive or sociocultural perspectives, are less interested in the school-age child's convoluted emotional life and are more concerned with the step-by-step acquisition of new cognitive abilities, and the steady unfolding of self-understanding, that characterize middle childhood. However, their overview of this period is quite similar to the psychoanalytic depiction: children during middle childhood meet the challenges of the outside world with an openness, insight, and confidence that few younger children possess. Middle childhood is seen as a time when many distinct competencies coalesce. The abilities to learn and to analyze, to express emotions, and to make friends have been in evidence from infancy, but now they come together in a much more focused and consistent manner, helping to form a much stronger, unified, and self-assured personality (Bandura, 1981, 1989; Bryant, 1983; Collins, 1984).

Now let us look at some of the specific manifestations of this developmental period.

industry versus inferiority The fourth of Erickson's eight crises of psychosocial development, in which school-age children attempt to master many skills and develop a sense of themselves as either industrious and competent or incompetent and inferior.

social cognition A person's awareness and understanding of human personality, motives, emotions, intentions, and interactions.

Understanding Others

An integral key to the psychosocial development of school-age children is an advance in **social cognition**, that is, in the understanding of other people and groups. As we saw in Chapter 9, preschoolers first evidence social cognition in a simple theory of mind, when they begin to realize that other people's actions can be motivated by thoughts and emotions that are different

from their own. But such theorizing is prone to error, because young children's grasp of the differences between subjective points of view is limited and fragile. During the school years, children's theory of mind evolves into a complex, multifaceted view of others. Children begin to understand human behavior not just as responses to specific thoughts or desires but as actions that are influenced, simultaneously, by diverse needs and emotions and by complex human relationships and motives (Arsenio & Kramer, 1992; McKeough, 1992).

This developmental progression was shown explicitly in a very simple experiment in which children between the ages of 4 and 10 were shown pictures of various domestic situations and asked how the mother might respond and why (Goldberg-Reitman, 1992). In one scene, for example, a child curses while playing with blocks. As reflected in the following typical responses, in assessing what the mother might do, the 4-year-olds attended only to the immediate behavior, whereas older children recognized the implications and possible consequences of the behavior:

> Age 4: "The mother spanks her because she said a naughty word."
>
> Age 6: "The mother says 'Don't say that again' because it's not nice to say a bad word."
>
> Age 10: "The mother maybe hits her or something because she's trying to teach her . . . because if she grew up like that she'd get into a lot of trouble . . . she might get a bad reputation."

In a variety of similar research studies, as well as in everyday spontaneous examples, younger children are much more likely to focus solely on observable behavior, not on underlying motives, feelings, or social consequences: they know when an adult might protect, nurture, scold, or teach a child but not necessarily why. Older children tend not only to understand the motivational and affective origins of various behaviors but also to analyze the future impact of whatever action they might take. Older children are also aware of the importance of personality traits. They organize their perceptions of a person around the traits they observe in the individual and frequently use those traits as a basis for predicting the person's future behavior and emotional reactions (Gnepp & Chilamkurti, 1988).

During the school years, children's emotional understanding deepens in a number of other ways as well. They begin to appreciate, for example, that emotions have internal causes that can sometimes be personally redirected (such as thinking happy thoughts in a sad situation), and this awareness enables children to better manage their own emotions (Garber & Dodge, 1991; Thompson, 1994). Whereas a young child is likely to become distressed when confronting an older bully, school-age children can sometimes coach themselves into looking—and perhaps even feeling—unafraid. During the school years, moreover, children also realize that someone can feel several emotions simultaneously (and can thus have conflicting or ambivalent feelings), and that people sometimes disguise or mask their emotions to comply with social rules (such as looking delighted after opening a disappointing gift) (Harris, 1989; Harter & Whitesell, 1989; Saarni, 1989).

Children's enhanced emotional understanding also increases their sensitivity to the social purposes of emotional expression and to the possibility that their own emotional expressions—and those of other people—may not reflect what they truly feel. As a consequence, they are harder to

This girl seems hesitant to proceed, perhaps in anticipation of getting a cold shock. However, because of the expanded emotional understanding that is typical of school-age children, she probably realizes that if she stalls much longer, she is bound to get teased. This greater emotional understanding may also help her to control her anxiety long enough to undertake the plunge.

fool: it is the school-age child who shrugs off a parent's sympathy by saying "You're just trying to make me feel better," or dismisses a parent's praise with "You have to say that because I'm your kid!" They also are more aware of the social situations in which it is appropriate to express emotions like anger—and when it is inappropriate to do so (Underwood et al., 1992).

Another outcome of children's increasing social sensitivity is an increased willingness to lie in order to avoid blame or punishment for their own negligence or misbehavior (Graham et al., 1995). One study of 8- and 11-year-olds found that whereas 80 percent of the younger children thought that lying was always bad, and only 40 percent of them said that they ever lied, 72 percent of the older children thought that lying was sometimes acceptable—and all of them admitted to sometimes being untruthful (Peterson et al., 1983).

Understanding Oneself

As we saw in Chapter 10, during the preschool years children have already begun to develop a sense of themselves as individuals with unique characteristics and dispositions. However, their self-understanding is superficial. Over the course of middle childhood, children's thinking about themselves gradually deepens, as their cognitive abilities mature and their social experience widens. In the beginning of the school years, for example, children often explain their actions by referring to the immediate situation; a few years later they more readily relate their actions to their personality traits and feelings (Higgins, 1981). Thus, whereas a 6-year-old might say that she hit him because he hit her first, an 11-year-old might add that she was already upset because she had lost her bookbag and that, besides, he is always hitting people and getting away with it.

Along with their developing self-understanding comes greater self-regulation, as children learn to control their reactions for strategic purposes. They know how to act in various social situations—whether at school, at a concert, or at a ballpark—and they have the self-control to act appropriately. They can use mental distraction to avoid becoming fidgety during a boring concert, for example, and they know how to look attentive in class, even when they are not paying attention.

Moreover, because of their widening social networks, school-age children increasingly perceive themselves in terms of different roles (family member, teammate, student, etc.) and distinct skills (academic, athletic, social, and so forth). As their social networks widen, children also begin to become aware of their belonging to one or another ethnic, religious, or social group, and it is during the school years that children from minority groups begin to take pride in their ethnic identity (Aboud, 1987; Katz, 1987). For many, such pride "bolsters one's self-respect, exalts one's conception of oneself, and inures the individual against the pain incident to low status" (Spencer, 1987).

As children's self-understanding becomes more differentiated, it also becomes more integrated (Harter, 1983), enabling schoolchildren to view themselves in terms of several competencies at once. They might, for example, recognize themselves as weak at playing sports, good at playing a musical instrument, and a whiz at playing Nintendo. Similarly, they might feel that they are basically good at making friends and are considerate of others, but that they have a quick temper that sometimes makes them do things that jeopardize their friendships. In sum, like their understanding of other people, children's self-understanding during the school years becomes psychologically more complex, more discriminating, and more richly textured.

The Rising Tide of Self-Doubt

Children's increasing self-understanding comes at a price. As they begin to see themselves more realistically, with weaknesses as well as strengths, children gradually become more self-critical and their self-esteem dips. One reason is that they more often evaluate themselves through **social comparison**, comparing their skills and achievements with those of others, and often seeing theirs as inferior. Further, as they mature, children are more likely to feel personally to blame for their shortcomings and less likely to believe, as younger children often do, that it is bad luck that makes them do poorly (Powers & Wagner, 1984).

The rising tide of self-doubt is particularly evident at school, where children's perceptions of their intellectual competence decline steadily through the elementary-school years (Eccles et al., 1993; Phillips & Zimmerman, 1990). This is in striking contrast to the self-evaluations of preschoolers, who usually remain buoyantly optimistic and confident of their own abilities, even in the face of negative evaluations by others (Stipek, 1984; Stipek & MacIver, 1989).

Self-Esteem and Poverty

During middle childhood, the development of self-esteem can be severely distorted by poverty. Indeed, many children from low-income homes—especially those in dangerous neighborhoods—come to think of themselves as worthless and their futures as hopeless, and this makes them unmotivated, depressed, and angry (Garbarino et al., 1991).

One reason that poverty has such a debilitating effect on self-esteem during these years is that, being in the stage of concrete operational thinking, school-age children tend to focus their social comparisons on the tangible and thus are highly susceptible to assessing individual and family

social comparison The tendency to assess one's abilities, achievements, social status, and the like by measuring them against those of others, especially those of one's peers.

Aggression

Among the most important peer-group norms that school-age children establish for themselves are those that govern playful and nonplayful aggression—the teasing, insulting, and physical threatening that are at the edge of many episodes of children's social interaction. Indeed, a certain amount of aggression, counteraggression, and reconciliation is present in peer interactions beginning from early childhood, as conflict arises over favorite toys, preferred activities, and popular play partners (Ross & Conant, 1992). The social norms of the school years, however, mean that children often risk social isolation if they do not readily anticipate and defend themselves against sarcastic comments, implied insults, or direct verbal or physical attacks.

The specifics regarding when aggression is appropriate, in what forms, and to what degree depend, of course, on the specifics of the social context (DeRosier et al., 1994; Foster et al., 1996). A study in England, for example, found that children who were most socially accepted were those who "gave as much as they got," sometimes teasing and mocking each other and—girls as well as boys—coming to blows. In fact, reciprocity of aggression was such a part of peer-group acceptance that those who suffered attack without retaliating were rejected as "piss weak" (Davies, 1982). At the same time, arrogance beyond a certain limit was considered out of bounds, for children were quite critical of "getting the snobs, getting the cranks, . . . lying, showing off, getting too full of yourself, posing, . . . wanting everything your way, being spoilt . . ." (Davies, 1982).

A somewhat different distinction between proper and improper aggression was found in a study of first-grade African-American boys: instrumental aggression (e.g., fighting to get one's own way) did not enhance a boy's status, but relatively quick retaliation against an implied threat or insult or a show of force to establish dominance was likely to inspire admiration. As the boys grew older, aggression was less positively viewed, but it still did not undermine acceptance (Coie et al., 1991). Additional research indicates that many boys, especially those from minority groups in the United States, view quick retaliation as a necessary deterrent to future threats (Graham et al., 1992; Herzberger & Hall, 1993; Hudley & Graham, 1993).

As these examples suggest, variation in the norms for aggression can occur by age, by ethnic and socioeconomic group, by neighborhood, and by the specific social situation. In each context, there may be different rules about hitting someone smaller, or boys attacking girls, or girls using physical force, or the appropriate role of friends and bystanders—and different customs about which contacts (e.g., being bumped up against, being shoved, having one's shoe stepped on) or insults (directed at one's relatives, one's physical appearance, or one's intellect) should be ignored, reciprocated, or avenged.

rejected children In childhood peer groups, those children who are actively shunned, teased, or bullied. Some are *aggressive-rejected,* unpopular because of their confrontational behavior, and others are *withdrawn-rejected,* unpopular because of their withdrawn, anxious demeanor.

Lacking Acceptance Among Peers

All children feel left out or unwelcome among their peers on one occasion or another, but a small minority of schoolchildren are unpopular most of the time. Some are merely ignored; others are actively **rejected**. Researchers have found that children in the latter category tend to be either *aggressive-*

Those most likely to reject the oddball are those who feel most vulnerable themselves, a truism for children as well as adults.

rejected—disliked because of their aggressive, confrontational behavior—or *withdrawn-rejected*—disliked because of their withdrawn, anxious demeanor (Bierman et al., 1993; Cillessen et al., 1992; Hymel et al., 1993).

Withdrawn-rejected children are aware of their social isolation, making them lonely and unhappy, and their low self-esteem has a negative impact on their academic achievement and their family relationships. Withdrawn-rejected children are also particularly vulnerable to bullying. Aggressive-rejected children, on the other hand, remain oblivious to their lack of acceptance and tend to overestimate their social competence (Hymel et al., 1993; Parkhurst & Asher, 1992). However, there is no doubt that their peers perceive them as argumentative, disruptive, and uncooperative—a perception that is confirmed by teachers' ratings and direct observations of their behavior with peers (Bierman et al., 1993; Dodge et al., 1990; Patterson et al., 1990).

Several studies have shown that children who are aggressive-rejected are impulsive and immature in their social cognition (Dodge & Feldman, 1990; Perry et al., 1992; Rabiner et al., 1990). Compared with other children, for instance, they tend to misinterpret social situations. They consider a friendly act to be hostile (Dodge et al., 1984; Graham et al., 1992; Hudley & Graham, 1993) or, especially when they feel anxious, interpret accidental harm as intentional (Dodge & Somberg, 1987). For example, they might interpret a compliment as sarcastic, or regard a request for a bite of candy as a demand, or assume that someone's inadvertently stepping on their shoe was intended as an insult. Unfortunately, since the way most children develop their social understanding and skill is from normal give-and-take with peers, aggressive-rejected children are excluded from the very learning situations they need most.

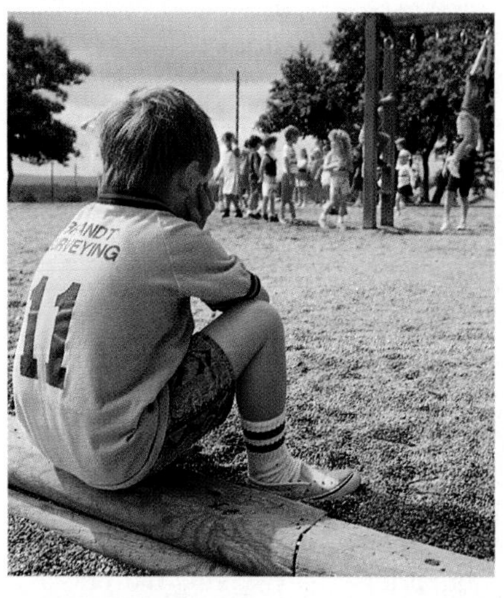

From this angle, it looks as though this boy is either withdrawn-rejected, experiencing problems with peers, or neglected, being overlooked by them. On the other hand, it may be that he simply enjoys being peacefully off by himself.

As rejected children get older, their problems get worse, because peers become more critical of each other as adolescence nears, and withdrawn or aggressive behavior becomes more self-defeating. Aggressive-rejected children are most likely to have adjustment problems—including heightened risk of psychological disorders—in adolescence and adulthood (Coie et al., 1992; Hymel et al., 1990; Parker & Asher, 1987). Often these problems are presaged by chronic bullying (see Research Report, pp. 362–364).

RESEARCH REPORT

Bullies in School

Bullying was once commonly thought to be an unpleasant but normal part of child's play, not to be encouraged, of course, but of little consequence in the long run. However, developmental researchers who have looked closely at the society of children consider bullying to be a very serious problem, one that harms both the victim and the aggressor, sometimes continuing to cause suffering years after the child has grown up (Garrity & Baris, 1996).

One leading researcher in this area is Dan Olweus, who has studied bullying in his native country of Norway and elsewhere for twenty-five years. The cruelty, pain, and suffering that he has documented in that time are typified by the examples of Linda and Henry:

> Linda was systematically isolated by a small group of girls, who pressured the rest of the class, including Linda's only friend, to shun her. Then the ringleader of the group persuaded Linda to give a party, inviting everyone. Everyone accepted; following the ringleader's directions, no one came. Linda was devastated, her self-confidence "completely destroyed."
>
> Henry's experience was worse. Daily, his classmates called him "Worm," broke his pencils, spilled his books on the floor, and mocked him whenever he answered a teacher's questions. Finally, a few boys took him to the bathroom and made him lie, face down, in the urinal drain. After school that day he tried to kill himself. His parents found him unconscious, and only then learned about his torment.

Following the suicidal deaths of three other victims of bullying, the Norwegian government asked Olweus in 1983 to determine the extent and severity of the problem. After concluding a confidential survey of nearly all of Norway's 90,000 school-age children, Olweus reported that the problem was widespread and serious; that teachers and parents were "relatively unaware" of specific incidents of bullying; and that even when adults noticed bullying, they rarely intervened (Olweus, 1992, 1993, 1994). Of all the children Olweus surveyed, 9 percent were bullied "now and then"; 3 percent were victims once a week or more; and 7 percent admitted that they themselves sometimes deliberately hurt other children, verbally or physically.

Several signs indicate this is not an incident of typical bullying: there is only a single aggressor; his victim is actively resisting; and both aggressor and victim seem close to the same size and strength. In advanced bullying, typically a much larger boy or group of children humiliates another child to the point that resistance is either obviously inadequate or completely absent.

As high as these numbers may seem, they are equaled and even exceeded in research done in other countries (Olweus, 1994). For instance, a British study of 8- and 9-year-olds found that 17 percent were victims of regular bullying and that 13 percent were bullies (Boulton & Smith, 1994). A study of middle-class children in a university school in Florida found that 10 percent were "extremely victimized" (Perry et al., 1988). Recently, American researchers have looked particularly at sexual harassment, an aspect of childhood bullying ignored by most adults. Fully a third of all 9- to 15-year-old girls say they have experienced sexual teasing and touching sufficiently troubling that they wanted to avoid school (American Association of University Women Foundation, 1994), and, as puberty approaches, almost every boy who is perceived as homosexual by his peers is bullied, sometimes mercilessly (Slavin-Williams, 1995).

bullying The repeated, systematic efforts to inflict harm on a particular child through physical attack, verbal attack, or social attack.

Researchers define **bullying** as *repeated*, systematic efforts to inflict harm on a particular child through physical attack (such as hitting, punching, pinching, or kicking), verbal attack (such as teasing, taunting, or name-calling),

or social attack (such as deliberate social exclusion or public mocking). Implicit in this definition is the idea of an imbalance of power: victims of bullying are in some way weaker than their harassers and continue to be singled out for attack, in part because they have difficulty defending themselves. In many cases, this difficulty is compounded by the fact that the bullying is being carried out by a group of children. In Olweus's research, at least 60 percent of bullying incidents involved group attacks.

As indicated by the emphasis given to it, the key word in the preceding definition of bullying is "repeated." Most children experience isolated attacks or social slights from other children and come through them unscathed. But when a child must endure such shameful experiences again and again—being forced to hand over lunch money, or to drink milk mixed with detergent, or to lick someone's boots, or to be the butt of insults and practical jokes, with everyone watching and no one coming to the child's defense—the effects can be deep and long-lasting. Not only are bullied children anxious, depressed, and underachieving during the months and years of their torment, but even years later, they have lower self-esteem as well as painful memories.

The picture is somewhat different, but often more ominous, for bullies. Contrary to the public perception that bullies are actually insecure and lonely, at the peak of their bullying they usually have friends who abet, fear, and admire them, and they seem brashly unapologetic about the pain they have inflicted, as they often claim, "all in fun." But their popularity and school success fade over the years, and especially if they are boys, they run a high risk at ending up in prison. In one longitudinal study done by Olweus, by age 24, two-thirds of the boys who had been bullies in the second grade were convicted of at least one felony, and one-third of those who had been bullies in the sixth through the ninth grade were already convicted of three or more crimes, often violent ones (Olweus, 1993). International research likewise finds that children who are allowed to regularly victimize other children are at high risk of becoming violent offenders as adolescents and adults (Junger-Tas et al., 1994).

Unfortunately, bullying during middle childhood seems to be universal: it occurs in every nation that has been studied, is as much of a problem in small rural schools as in large urban ones, and is as prevalent among well-to-do majority children as among poor immigrant children. Also quite common, if not universal, is the "profile" of bullies and their victims. Contrary to popular belief, victims are not distinguished by their external traits: they are no more likely to be fat, skinny, homely, or to speak with an accent than nonvictims are. But they usually are "rejected" children, that is, children who have few friends because they are more anxious and less secure than most children and are unable or unwilling to defend themselves. They also are more often boys than girls and more often are younger children (see figure).

(continued on next page)

The rates of being bullied "now and then" reported by Norwegian schoolchildren have been found to be typical of the rates in many other countries. Although bullying is less common among older children, it is more devastating, because older children depend much more on peers for their self-esteem. Fortunately, when Norwegian adults worked to create a school context that would discourage bullying, the rate was cut in half. No other nation has yet moved in this direction.

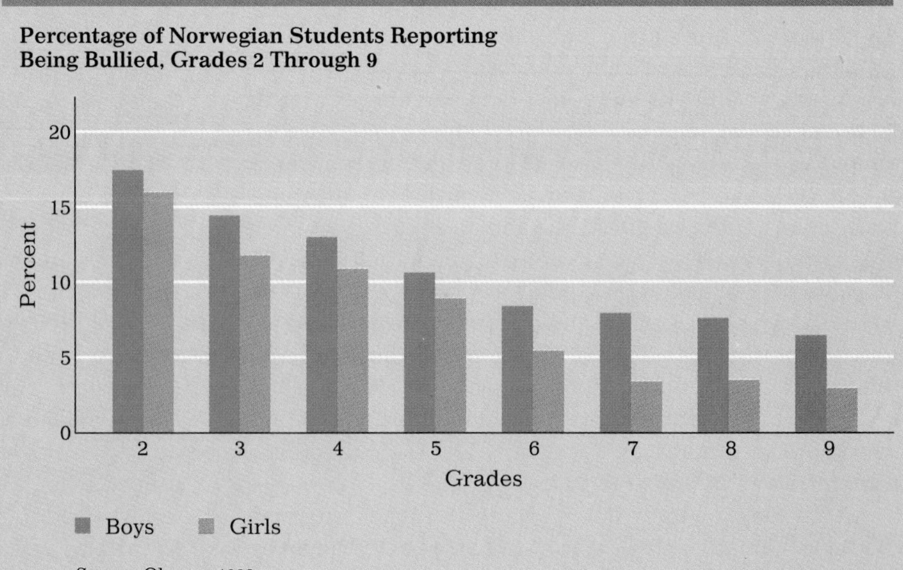

Source: Olweus, 1993.

Bullies in School (continued)

Bullies have traits in common as well, some of which can be traced to their upbringing. The parents of bullies often seem indifferent to what their children do *outside the home* but use "power-assertive" discipline on them *at home*. These children are frequently subjected to physical punishment, verbal criticism, and displays of dominance meant to control and demean them, thereby giving them a vivid model, as well as a compelling reason, to control and demean others (Olweus, 1993). Boys who are bullies are often above average in size, while girls who are bullies are often above average in verbal assertiveness. These differences are reflected in bullying tactics: boys typically use force or the threat of force; girls often mock or ridicule their victims, making fun of their clothes, behavior, or appearance, or revealing their most embarrassing secrets (Lagerspetz & Bjorkquist, 1994).

What can be done to halt these damaging attacks? Many psychologists have attempted to alter the behavior patterns that characterize aggressive or rejected children (Coie & Koeppl, 1990; Crick & Dodge, 1994; Hudley & Graham, 1993). Cognitive interventions seem particularly fruitful: some programs teach social problem-solving skills (such as how to use humor or negotiation to reduce a conflict); others help children reassess their negative assumptions (such as the frequent, fatalistic view of many rejected children that nothing can protect them, or the aggressive child's typical readiness to conclude that accidental slights are deliberate threats); others tutor children in academic skills, hoping to improve confidence and short-circuit the low self-esteem that might be at the root of both victimization and aggression.

These approaches sometimes help individuals. However, because they target one child at a time, they are piecemeal, time-consuming, and costly. Further, they have to work against habits learned at home and patterns reinforced at school, making it hard to change a child's behavior pattern. After all, bullies and their admirers have no reason to learn new social skills if their current attitudes and actions bring them status and pleasure (Patterson, 1994). And "even if rejected children change their behavior, they still face a difficult time recovering accepted positions in the peer group" and gaining friends who will support and defend them (Coie & Cillessen, 1993). The solution to this problem must begin, then, by recognizing that the bullies and victims are not acting in isolation but, rather, are caught up in a mutually destructive interaction within a particular social context (Pierce & Cohen, 1995).

Accordingly, a more effective intervention is to change the social climate within the school, so that bully-victim cycles no longer spiral out of control. That this approach can work was strikingly demonstrated by a government-funded awareness campaign that Olweus initiated for every school in Norway. In the first phase of the campaign, community-wide meetings were held to explain the problem; pamphlets were sent to all parents to alert them to the signs of victimization (such as a child's having bad dreams, no real friends, and coming home from school with damaged clothes, torn books, or unexplained bruises); and videotapes were shown to all students to evoke sympathy for victims.

The second phase of the campaign involved specific actions within the schools. In every classroom, students discussed reasons and ways to mediate peer conflicts, to befriend lonely children, and to stop bullying attacks whenever they saw them occur. Teachers were taught to be proactive, organizing cooperative learning groups so that no single child could be isolated, halting each incident of name-calling or minor assault as soon as they noticed it, and learning how to see through the bully's excuses and to understand the victim's fear of reprisal. Principals were advised that adequate adult supervision during recess, lunch, and bathroom breaks distinguished schools where bullying was rare from those where bullying was common.

If bullying incidents occurred despite such measures, counselors were urged to intervene, talking privately and seriously with bullies and their victims, counseling their parents, and seeking solutions that might include intensive therapy with the bully's parents to restructure patterns of family discipline, reassigning the bully to a different class, grade, or even school, and helping the victim strengthen skills and foster friendships.

Twenty months after this campaign began, Olweus resurveyed the children in forty-two schools. He found that bullying had been reduced overall by more than 50 percent, with dramatic improvement for both boys and girls at every grade level (Olweus, 1992). Developmental researchers are excited because results such as these, in which a relatively simple, cost-effective measure has such a decided impact on a developmental problem, are rare. Olweus (1993) concludes, "It is no longer possible to avoid taking action about bullying problems at school using lack of awareness as an excuse . . . it all boils down to a matter of will and involvement on the part of adults." Unfortunately, at the moment, Norway is the only country to have mounted a nationwide attack to prevent the problem of bullying. Many other school systems, in many other nations, have not even acknowledged the harm caused by this problem, much less shown the "will and involvement" to stop it.

Friendship

While acceptance by the peer group is valued by children, such status is not as important as having friends. Indeed, if they had to choose between (1) having a few close friends but being unpopular with the group, or (2) being popular but friendless, most children would take the friends. Such a choice would be consistent with developmentalists' view of the importance of friendship to children's overall psychosocial development and sense of self-esteem (Hartup, 1996; Parker & Asher, 1993).

Because friendship is so important to them, children spend a good deal of time thinking about its dynamics, and, as a result, their understanding of friendship becomes increasingly abstract and complex during the school years. With their deepened insight into the dimensions of friendship, children learn to balance honesty with protectiveness, mutual dependence with a respect for independence, and competition with cooperation, shared conversation, and shared actions (Berndt, 1989b; Rawlins, 1992). At the same time, older children perceive their friends in psychologically richer ways because of their deeper, more nuanced understanding of themselves (Parker & Gottman, 1989).

These changes are reflected in a study of hundreds of Canadian and Scottish children, from first grade through the eighth, who were asked what made their best friends different from other acquaintances. Children of all ages tended to say that friends did things together and could be counted on for help, but older children were more likely to cite *mutual* help, whereas younger children simply said that their friends helped *them*. Further, the older children considered mutual loyalty, intimacy, and interests, as well as activities, to be part of friendship (Bigelow, 1977; Bigelow & La Gaipa, 1975). Similarly, in a United States study (Berndt, 1981), children were asked "How do you know your best friend?" A typical kindergartner answered:

> I sleep over at his house sometimes. When he's playing ball with his friends he'll let me play. When I sleep over, he lets me get in front of him in 4-squares [a playground game]. He likes me.

By contrast, a typical sixth-grader said:

These girls are having a "sleep over," a common occurrence among school-age children who enjoy the intimacy of staying overnight at each other's homes. Typically, sharing secrets, staying up late, and eating junk food are part of the event, bringing the friends closer together. Notice that all of these activities are much more possible for school-age children than for younger ones, who are more likely to get scared or have a quarrel when they try to sleep over.

However, it is easy to show that this stereotype is false. In the United States, most single mothers are in the labor force, including not only a large majority (77 percent) of those with children aged 6 to 17 but also most (60 percent) of those with younger children (U.S. Bureau of the Census, 1996). Even before the welfare reform enacted in 1996 cut off benefits for many single mothers, less than half of them received government assistance of any kind, including food stamps and medical care. And unmarried mothers from every racial group have fewer children, and are less likely to want more children, than married mothers of the same age, education, and ethnicity. Similar nationwide statistics are unavailable for single fathers, but regarding their employment, all evidence suggests that they are even more likely to be in the labor force than are single mothers (Greif, 1995; Greif et al., 1993). Overall, then, no matter what his or her nationality, ethnicity, or education, a single parent is likely to work hard to fill the role of both major provider and major caregiver, surrendering personal recreation, social life, and sleep to do so.

How do children actually develop in such households? On the whole, just as well as children from two-parent families with similar incomes and adult-child ratios. More specifically, with the exception of children whose parents are recently divorced, children from single-parent families are usually on a par with other children in three crucial areas: school achievement, emotional stability, and protection from serious injury. This generality holds true for preschoolers, school-age children, and adolescents (Dawson, 1991; Entwhisle & Alexander, 1996; Hawkins & Eggebeen, 1991; Smith, 1990).

Thus, the mere fact of growing up in a single-parent family does not necessarily hurt a child's development. And it may be better if the alternative is living with two parents who frequently fight or contending with one parent who is abusive, addicted, or unpredictable (Angel & Angel, 1994; McLanahan & Sandefur, 1994). How well a child does in a single-parent family is influenced by a host of factors, some of which are shown in Table 13.1. Two variables that are particularly critical are family income and social support.

In general, the income of single-parent households is substantially less than that of two-parent households, even when only one of those two parents is employed (McLanahan & Booth, 1991). With or without government assistance, half of all American children in single-parent households live at or below the poverty line (Children's Defense Fund, 1995). As noted repeatedly in this text, low family income can impede child development in many ways, which is why many developmentalists are strong advocates for laws enforcing child-support payments from absent parents and ensuring medical care for poor children, good public schools for everyone, and so on.

The second key variable, social support, is important in any family arrangement, but it is particularly necessary for most single parents, who need not only encouragement but practical assistance from other adults. Often, an important source of such support is the "absent" parent who remains actively involved in the child's daily care. In such cases, the child may be said to be living in a single-parent household but a two-parent family. Another major source of social support is grandparents, who can be vitally important in providing extra income, emotional support, stability, and child care (Thompson et al., 1992). Important forms of support may also be pro-

TABLE 13.1

The Impact of Single Parenthood on Child Development

Likely to Be Harmful if	Likely to Be Beneficial if
Low-income home	Middle-income home
Conflict-filled home	Peaceful home
Parent under age 25	Parent over age 30
Parent not high school graduate	Parent has college education
More than two siblings	Only child, or one sibling
Several changes (e.g., divorce, remarriage, divorce)	Stable family structure
No help from relatives	Grandparents actively helpful
Conflict with other parent	Cordial relations with other parent
Parent has live-in lover	Parent not romantically involved
Parent socially isolated	Parent active with friends, church, etc.
Community hostile to single parents	Community supportive of single parents
If child is under age 5	
More than four caregivers	Two or three caregivers
No steady day care	High-quality preschool
Parent employed 60+ hours a week	Parent has part-time job
If child is over age 5	
Frequent change of school	Child stays in one school
Frequent change of neighborhood	Child stays in one neighborhood
Parent unemployed	Parent employed, flexible hours

Source: Compiled from several sources, among them Angel & Angel, 1994; McLanahan & Sandefur, 1994.

Can single fathers raise children well? Research finds that they can. In fact, some studies show that when the father chooses the custodial role, and when the community is supportive, and the children are aged 6 or older, offspring, particularly boys, thrive under paternal care.

vided by everyone from friends, neighbors, and relatives to employers, church members, storekeepers, and so on. How well children do in single-parent families depends a great deal on the surrounding social context; they can fare very well if they are supported in many ways by their community, or they can be seriously harmed if their parent lives in poverty and isolation.

All the generalities about single parents are as true for fathers as for mothers. Single fathers are increasingly common, more than tripling in the past twenty-five years (from 393,000 in 1970 to 1,314,000 in 1994), and all the research indicates that children fare as well in single-father households as in single-mother households (Thompson et al., 1992). In fact, if anything, the average child growing up in a father-headed household fares better than the average child in a mother-headed household, especially if the child is a boy. Of course, this does not mean that fathers are necessarily better single parents. As you might expect, one critical reason for the single father's overall advantage is economic: generally, father-headed single-parent households are more financially secure than mother-headed single-parent households. Another reason has to do with the particular characteristics of the men themselves: most single-parent fathers were very involved in par-

enting during the marriage and actively sought custody of their children. Hence these particular men are more likely to be good parents than many of the mothers who receive custody of their children more through the customs of the court or the default of their ex-spouses than through their own desire to be the custodial parent (Greif, 1995).

Blended Families

Most divorced parents remarry within a few years, and many unmarried parents eventually marry as well. When remarriage means less loneliness for the custodial parent, improved finances, less conflict with a former spouse, and more stable household organization, it eventually benefits all concerned. However, while the new partners are initially likely to be happy with the remarriage, such is almost never the case for the children, who must suddenly negotiate a new set of family relationships, not only with a stepparent but often also with stepsiblings, stepgrandparents, and so forth, most of whom they would not have chosen on their own (Ganong & Coleman, 1994).

The same factors affecting children in marriage and divorce—parental cooperation, stability, and adequacy of caregiving—affect children in stepfamilies (Keshet, 1988; Kurdek, 1989). Even in the best of circumstances, however, harmony takes time to achieve, as the blended family must develop a new style and culture that all members can live with, each member making certain accommodations to the others. In the process, some members are more likely to benefit, especially younger boys from mother-headed families. Others are more likely to suffer, among them adolescent girls who are particularly likely to resent their new stepfathers, only-children who suddenly lose privacy to new siblings, and children who become indirect targets of the nonremarried parent's jealousy (Allison & Furstenberg, 1989; Giles-Sims & Crosbie-Burnett, 1989; Hetherington & Jodl, 1994; Vuchinich et al., 1991).

Only two members of this newly formed blended family chose each other. The success of their relationship, like that of all parents in blended families, will largely depend on how well the nonchoosing members work out their relationships with stepparents, stepsiblings, and other "acquired" relatives. Generally, the younger the children, and the more years pass, the more likely step-parents and even step-siblings are to be accepted as one's own.

While most remarriages eventually work out for the children as well as for the adults, many do not: the divorce rate is higher for second marriages than first marriages, and this is particularly true if there are young adolescent children involved. Overall, the stress of a second divorce adds disruption to the child's life as well as to the adult's, a fact that itself should caution any single parent hoping to marry primarily in order to give the children a two-parent home.

Other family arrangements, such as those in which foster parents or grandparents are responsible for child-rearing, follow the same general rules. If the family is relatively free of conflict and offers stability as well as ongoing love and guidance, the children in that family are likely to thrive. This fact draws attention to the critical role that the broader social context can play in family functioning: if a particular family type is accepted and supported by ethnic, cultural, or community values, the chances are good that the adults will be able to nurture the children entrusted to them. Indeed, over the past thirty years, as social acceptance of nontraditional families in America has been increasing, the negative effects of alternative families on children have been decreasing (Acock & Demo, 1994).

Coping with Life

As we have seen throughout these three chapters on middle childhood, the expansion of the child's social world sometimes brings with it new and disturbing problems. For example, the beginning of formal education forces any learning disabilities to the surface, making them an obvious handicap; playing with friends beyond the home may result in peer problems, such as rejection and attack that can take a serious toll; leaving the protection of the family can expose the child to social prejudices such as sexism, racism, and classism, in some cases causing shame, self-doubt, and loneliness. Such troublesome problems of middle childhood often piggyback on those chronic stresses that are detrimental at every age, such as living in an impoverished, overcrowded, or violent home, or having a parent who is emotionally disturbed, drug-addicted, or imprisoned. Because of problems such as these, many children fail at school, fight with their friends, fear the future, cry themselves to sleep. Indeed, the entire range of academic and psychiatric difficulties that school-age children sometimes display can be traced, in part, to psychosocial stresses (DeFries et al., 1994; Luthar & Zigler, 1991; Rutter, 1987).

Fortunately, although the potential stresses and hassles are many during middle childhood, so are the coping measures that children develop. As a result, between ages 6 and 11, the overall frequency of various psychological problems decreases while the number of evident competencies—at school, at home, and on the playground—increases (Achenbach et al., 1991). Two factors described in this chapter—the development of social cognition and an expanding social world—seem to combine to buffer school-age children against many of the stresses they encounter. According to some observers, many children seem "stress-resistant," even "invulnerable" and "invincible" (Garmezy, 1985; Werner & Smith, 1992). Let us look more closely at how some children rise above problems that might seem potentially devastating.

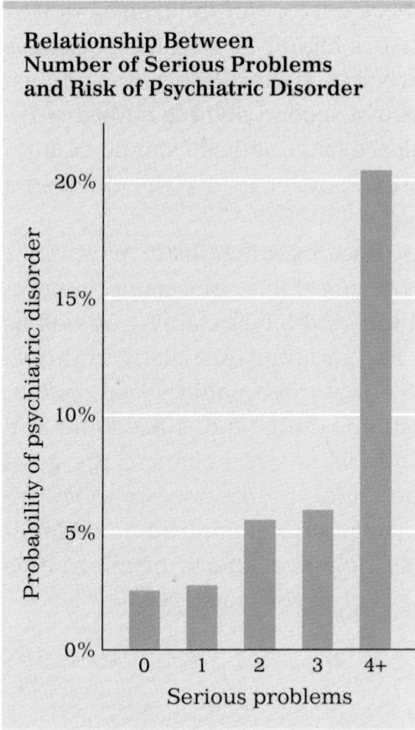

FIGURE 13.2 Rutter found that children who had to cope with one serious problem ran virtually as low a risk of suffering a psychiatric disorder as did children who faced no serious problems. However, when the child had two problems, the chances more than doubled. Four or more problems produced about ten times the likelihood of psychiatric disorders as one. About one child in five who experienced four or more serious stresses actually became emotionally disturbed.

Assessing Stress

The first important point to recognize in studying the effects of various stresses on children is that there is no simple correspondence between a given stress and a given result. Detailed longitudinal studies find that the likelihood that a given stress will produce psychological fallout depends on the number of stresses the child is experiencing concurrently and on their pervasiveness, that is, on the degree to which they affect the overall, long-term patterns of the child's daily life (Fergusson & Lynsky, 1996; Luthar & Zigler, 1991).

Typical of this research is a classic study that found that children coping with one, and only one, serious, ongoing stress (e.g., poverty, large family size, criminal father, emotionally disturbed mother, frequent fighting between the parents) were no more likely to develop serious psychiatric problems than children with none of these liabilities. However, as illustrated in Figure 13.2, when there was more than one risk factor present, "the stresses potentiated each other so that the combination of chronic stresses provided very much more than a summation of the effects of the separate stresses considered singly" (Rutter, 1979). Indeed, the finding of a number of other studies is that, in general, a single chronic problem creates vulnerability in a child without causing obvious harm, but if that vulnerability is subjected to additional stresses—even mild ones that are more often termed "daily hassles" rather than "stressful events"—the child can suffer evident damage (Luthar & Zigler, 1991; Shaw et al., 1994).

The underlying reason is that the impact of stresses depends on how they affect the moment-by-moment tranquillity of daily life. For example, living with an emotionally dysfunctional parent may mean that a child has to prematurely assume many of the responsibilities for his or her own daily care and school attendance, often in the midst of a chaotic household; and/or has to contend with an adult's confused, depressed, or irrational thinking; and/or has to supervise and discipline younger siblings. The net result is a child who never has a moment to play, to relax, to develop his or her own interests and skills.

Sometimes, the child may be partially shielded from these problems by other family adults who give the child stability and nurturance (Garmezy, 1993). However, when the dysfunctional parent is the only adult in the household, the child is particularly likely to suffer. Added problems occur if the dysfunctional parent cannot provide a steady place to live. For school-age children, especially, frequent changes of address are a powerful predictor of low self-esteem, school failure, and parental neglect. The worst situation is when the child has *no* address, except, perhaps, that of a shelter for the homeless (see Public Policy).

Focus on Strengths

Particularly important to a child's ability to cope with problems is that child's competencies, especially well-developed social, academic, or creative skills. Each of these skills can help the child deflect or avoid many of the problems he or she may encounter at home or in the community (Conrad & Hammen, 1993).

PUBLIC POLICY

Homelessness

Between 50,000 and 100,000 American children are homeless each night, about half of them school-age (Jencks, 1994; Masten, 1992). Those literally without a roof over their heads are most often adolescent runaways or "throwaways" whose parents have abused them or disowned them. Homeless children under age 12 usually live with their families in shelters. Although these children have, for the moment, the assurance of a bed and meals, they are troubled in many ways. As one report explains:

> By the time they arrive in a shelter, children may have experienced many chronic adversities and traumatic events. More immediately, children may have gone hungry and lost friends, possessions, and the security of familiar places and people at home, at school, or in the neighborhood. . . . Locations [of shelters] are usually undesirable, particularly with respect to children playing outside. Moreover, necessary shelter rules may strain a child and family life. For example, it is typical for no visitors to be allowed, and for children to be . . . accompanied at all times by a parent. [Masten, 1992]

Despite the anxiety evident in these children, this is a hopeful scene. The woman on the left, formerly homeless, is now living with her children in a stable, community residence and is also studying for her high school degree, with the help of a college student. With secure housing and education or job training, many homeless families can become functional again.

Moreover, a shelter is a temporary solution to homelessness, requiring periodic upheaval as children move to alternate locations in the company of a parent who may be humiliated, depressed, and emotionally exhausted.

Comparing homeless children in middle childhood with their peers of equal SES finds that the homeless children have fewer friends, more fears, more fights, more chronic illnesses, more changes of school, and lower school attendance. They are also about fourteen months behind academically (Masten, 1992; Rafferty & Shinn, 1991). In terms of long-term development, the most chilling result is a loss of faith in life's possibilities: compared even with other impoverished children, they have lower aspirations and less hope for the future or for their fellow humans, expressing doubt that anyone will ever help them. Clinical depression is common, striking almost one homeless child in every three (Bassuk & Rosenberg, 1990). When such attitudes develop in a child, they may take a lifetime to reverse.

Under broader examination, homeless families are distinguished from other impoverished families by their lack of a personal support network to assist them in obtaining food, shelter, and other necessities (Bassuk, 1989; Shinn et al., 1991). It is because of the inability, or unwillingness, of extended family members and friends to provide assistance that homeless families find themselves taking refuge at a city mission, a homeless shelter, or an isolated park. Most homeless families are headed by single mothers struggling to find affordable housing and provide for other family needs, often while trying to cope with the effects of having been physically or emotionally abused.

We have already seen that instability and conflict are the two most devastating home attributes for a child during middle childhood. Obviously, homeless children are overwhelmed by torrents of both. While the public may disagree about the root causes of, and the best solutions to, homelessness, it is apparent from a developmental perspective that something must be done immediately for every homeless child. Every month of education that a homeless child loses, and every blow to self-esteem that he or she suffers, may take years—or perhaps a lifetime—to restore.

There are several reasons why competence can more than compensate for disabling factors. One is self-esteem: if children feel confident in any area of their lives, they are better able to put the rest of their life in perspective. They believe, for example, that despite how others might reject or belittle them, they are not a worthless failure, and that despite the voices of despair within them, life is not hopeless.

More directly, children with better-developed cognitive and social skills are better able to employ various coping strategies, such as changing the conditions that brought about a problem in the first place, or restructuring their own reactions to the problem. Thus, because their coping repertoires "increase and become more differentiated in middle childhood," older children may deal with the stresses of life better than children who are just beginning middle childhood (Aldwin, 1994). For example, when a peer is antagonistic, a 6-year-old is likely to dissolve into tears or to launch a clumsy counterattack, which merely brings further rejection. Older children, on the other hand, are more adept at finding ways to disguise their hurt, or at keeping a bully at bay, or at repairing a broken friendship, or at making new friends to replace the old ones (Compas et al., 1991).

Schools and teachers can obviously play a significant role in the development of competence. Even for children from seriously deprived backgrounds, school achievement can make it possible for them to aspire beyond the limited and constricting horizons that they may encounter in their daily lives.

Social Support

Another important element that helps children deal with problems is the social support they receive (Garmezy, 1993). The companionship and comfort provided by a grandparent, a sibling, or even a family pet can relieve some tension in a child's life (Furman & Buhrmester, 1992; Werner & Smith, 1992). In addition, one of the benefits of the expanding social world of middle childhood is that the child can venture forth to seek out many more potential sources of social support. For example, a child whose parents are fighting bitterly on their way to divorce may spend hours on the phone with a friend whose parents have successfully separated, or may frequently drop in for dinner at a neighbor's house where family harmony still prevails, or may devote himself or herself to helping a teacher, a coach, or a community group.

An additional source of support for many children is religious faith and practice. Especially for children in difficult circumstances—such as the impoverished child in a single-parent family in a dangerous neighborhood—religious faith itself can be psychologically protective. School-age children, almost universally, develop their own theology—influenced by whatever formal religious education they might receive but by no means identical to it—that helps them structure life and deal with worldly problems (Coles, 1990; Hyde, 1990). Their view of a god figure is generally very personal, enabling troubled children to believe that they are being watched over and protected. One example is an 8-year-old African-American girl who, in the 1960s, was one of the first to enter a previously all-white school. She remembers walking past a gauntlet of adults yelling insults:

Although the potential problems that beset children today are legion, including some their parents never knew, the best safeguard is an old-fashioned one—a supportive hand and an attentive ear from a caring adult.

> I was all alone, and those people were screaming, and suddenly I saw God smiling, and I smiled. A woman was standing there, and she shouted at me "Hey you little nigger, what are you smiling at?" I looked right up at her face, and I said "At God." Then she looked up at the sky, and then she looked at me, and she didn't call me any more names. [quoted in Coles, 1990]

In a way, this example illustrates many aspects of children's coping abilities, for it was not only faith but also a measure of self-confidence, social understanding, and skill at deflecting her own emotional reactions that enabled this child to overcome a very real threat.

While adults may wish that all children could have an idyllic childhood, such is almost never the case. Nor is it necessary for healthy development: research on coping in middle childhood clearly suggests that, as they grow older, most children develop ways to deal with all sorts of stress, from minor hassles to major traumatic events.

However, some children are at risk of developing serious psychological disturbances if they are faced with multiple problems that affect their daily routines. To help these children the best strategy may be not only to reduce their stress but also to increase their competencies and to strengthen the social supports surrounding them. If the home situation is problematic, for instance, having access to anyone from a caring teacher to a best friend to a loving grandparent can make a critical difference (Rutter, 1987). Or if a child has a severe reading difficulty, developing the child's talents in other areas, such as math and music, may be as important for the child's overall well-being as specific tutoring to overcome the learning disability. Taking a wider view, measures to change the broader social context, such as making violent neighborhoods safer or improving job opportunities in impoverished communities, may benefit school-age children substantially, if indirectly.

As you will see in the next three chapters, in many ways adolescence is as much a continuation of middle childhood as a radical departure from it. Stresses and strains continue to accumulate during adolescence, and destructive coping mechanisms, such as drug use and dangerous risk-taking, become more accessible. Fortunately, constructive coping methods also increase: personal competencies, family supports, and close friends get most young people through adolescence unharmed.

SUMMARY

An Expanding Social World

1. School-age children develop a multifaceted view of others, becoming increasingly aware of the complex personalities, motives, and emotions that underlie others' behavior. At the same time, they become better able to adjust their own behavior to interact appropriately with others.

2. Children also develop more sophisticated conceptions of themselves and their own behavior. As they become more knowledgeable about their personalities, emotions, abilities, and shortcomings, they evaluate themselves by comparing themselves with others, and this contributes to greater self-criticism and diminished self-esteem.

3. During middle childhood, poverty can be especially detrimental to self-esteem. One reason is that school-age children increasingly compare themselves with others, particularly in terms of visible signs of status, and this leads many children to feel ashamed and hopeless.

The Peer Group

4. Peer relationships provide opportunities for social growth because peers are on an equal footing with each other and must learn to adjust to each other accordingly. During the school years, children create their own subculture, with language, values, and codes of behavior. One such norm establishes rules for conflict and aggression.

5. Friendships become more selective and exclusive as children grow older. However, children may lack acceptance among their peers for various reasons, and this can have long-term consequences for psychosocial growth.

Family Structure and Child Development

6. Family functioning is far more crucial to children's well-being than family structure is. Whether a child lives in a two-parent, one-parent, or blended family is less important than whether the child's home situation is relatively stable, conflict-free, and supportive.

7. While family functioning is key, nonetheless certain family structures—especially conflict-filled, low-income homes and family arrangements that change unpredictably from year to year—tend to be more stressful for children. Transitions between different family structures can be difficult, especially when the detrimental effects are compounded by the stresses of parental conflict and low income. Support from extended kin, friends, and the community, on the other hand, can help ease the difficulty of such transitions.

Coping with Life

8. Almost every child has some difficulties at home, at school, or in the community during middle childhood. Most children cope quite well, as long as the problems are limited in duration and degree.

9. How well particular children cope with the problems in their lives depends on the number and nature of the stresses they experience, the strengths of their various competencies, and the social support they receive.

KEY TERMS

latency (353)
industry versus inferiority (354)
social cognition (354)
social comparison (357)
peer group (358)
society of children (359)
rejected children (360)
bullying (362)
family structure (366)

KEY QUESTIONS

1. How does the child's understanding of other people change during the school years? What difference does this change make for how the child interacts with others?

2. How does a child's self-understanding change from the preschool years through middle childhood?

3. What factors typically cause children to become more self-critical during middle childhood?

4. How does poverty affect development in middle childhood?

5. Why are peer relationships important during the school years? Describe some of the characteristics of the society of children during middle childhood.

6. What are some of the factors that may cause a child to lack acceptance among his or her peers? What can be done to provide assistance to such children?

7. What are the advantages and possible disadvantages of a two-parent home?

8. What are some of the problems that are experienced in single-parent, blended, or other "nontraditional" households? What factors predict whether these families will be stressful, or beneficial, to a child?

9. What factors can help children cope more effectively with stress?

The Developing Person So Far: The School Years, Ages 7 Through 11

Biosocial Development

Growth

During middle childhood, children grow more slowly than they did during infancy and toddlerhood or than they will during adolescence. Increased strength and heart and lung capacity give children the endurance to improve their performance in skills such as swimming and running.

Skills

Slower growth contributes to children's increasing bodily control, and children enjoy exercising their developing skills of coordination and balance. Which specific skills they master depends largely on culture, gender, and inherited ability. Children with special needs typically require intensive education, but in many ways are similar to other children and benefit from interaction with them.

Cognitive Development

Thinking

During middle childhood, children become better able to understand and learn, in part because of growth in their processing capacity, knowledge base, and memory capacity. At the same time, metacognition techniques enable children to organize their learning. Beginning at about age 7 or 8, children also develop the ability to understand logical principles, including the concepts of identity, reciprocity, and reversibility.

Language

Children's increasing ability to understand the structures and possibilities of language enables them to extend the range of their cognitive powers and to become more analytical in their use of vocabulary. Most children develop proficiency in several language codes, and some become bilingual.

Education

Formal schooling begins worldwide, with the specifics of the curriculum depending on economic and societal factors. A child's actual learning also depends on the time allotted to each task and the attitudes of, and specific guided instruction from, teachers and parents.

Psychosocial Development

Emotions and Personality Development

The peer group becomes increasingly important to children as they become less dependent on their parents and more dependent on friends for help, loyalty, and sharing of mutual interests. A child's specific personality patterns can make peer acceptance or rejection an important aspect of life. Children are also increasingly aware of, and involved in, family life, as well as of the world outside the home, and therefore are more likely to feel the effects of family, economic, and political conditions. Whether or not middle childhood will be stressful for a particular child will depend, in part, on the child's temperament, competence, and the social support provided by home and school. Economic factors, especially low SES, also become more influential.

PART V Adolescence

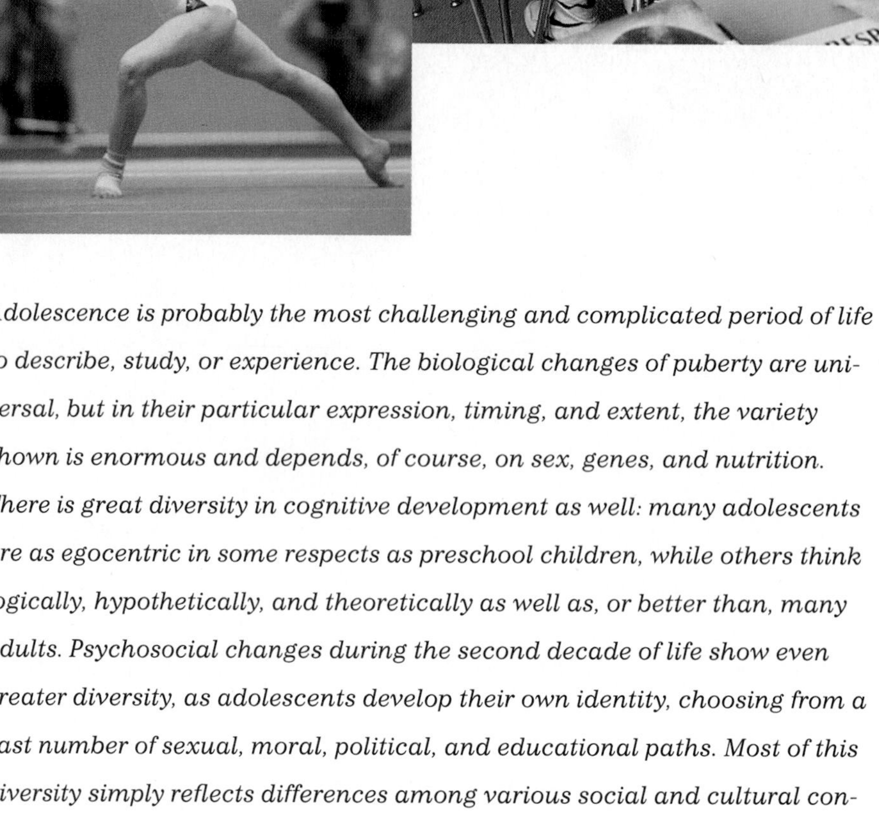

Adolescence is probably the most challenging and complicated period of life to describe, study, or experience. The biological changes of puberty are universal, but in their particular expression, timing, and extent, the variety shown is enormous and depends, of course, on sex, genes, and nutrition. There is great diversity in cognitive development as well: many adolescents are as egocentric in some respects as preschool children, while others think logically, hypothetically, and theoretically as well as, or better than, many adults. Psychosocial changes during the second decade of life show even greater diversity, as adolescents develop their own identity, choosing from a vast number of sexual, moral, political, and educational paths. Most of this diversity simply reflects differences among various social and cultural contexts and the productive variation that typifies the human life course. But for about one adolescent in four, fateful choices are made that handicap, and sometimes destroy, the future.

Yet such differences should not mask the commonality of the adolescent experience, for all adolescents are confronted with the same developmental tasks: they must adjust to their changing body size and shape, to their awakening sexuality, to new ways of thinking, and they must begin to strive for the emotional maturity and economic independence that characterize adulthood. As we will see in the next three chapters, the adolescent's efforts to come to grips with these tasks is often touched with confusion and poignancy.

Adolescence: Biosocial Development

CHAPTER 14

Between the ages of 10 and 20, humans everywhere cross a great divide between childhood and adulthood—biosocially, cognitively, and socioculturally. No one would call this process of becoming an adult simple or easy. In most modern societies, at least, adjusting to the many changes that adolescence entails can be difficult and stressful, turbulent and unpredictable.

Before beginning our discussion of adolescence, however, we should emphasize that while no period of life is problem-free, none—including adolescence—is defined only by its problems. There are moments, it is true, of awkwardness, confusion, anger, and depression in almost every teenager's life. It is also true that many adolescents make serious missteps on the path toward maturity, and some encounter obstacles that halt their progress completely. This chapter and the two that follow will examine some of these problems, putting them in perspective and focusing on causes and prevention.

However, the major theme of these chapters is that the same developmental changes that may be a source of difficulty are also the source of new excitement, challenge, and growth. Seriously troubled adolescents are in the minority. Moreover, many of the so-called problems of adolescence are actually problems more for parents and society than for teenagers themselves: the same music that makes adults shake their heads in disbelief makes young people jump with joy; the "telephone time" that exasperates parents is a social lifeline for teenagers; the sexual awakening that society fears is, for many individuals, the beginning of thrilling intimacy. Thus, any generalization about the nature of adolescence, especially about its turbulence, must be made cautiously and applied with care.

Puberty

The period of rapid physical growth and sexual maturation that ends childhood and brings the young person close to adult size, shape, and sexual potential is called **puberty**. The onset of puberty is triggered by a chain of hormonal effects that bring on a sequence of visible physical changes (see Table 14.1). For girls, these visible changes include, in sequence, the emergence of breast buds, the initial appearance of pubic hair, widening of the hips, peak growth spurt, first menstrual period, the completion of pubic-

puberty The period of early adolescence characterized by rapid physical growth and the sexual changes that make reproduction possible.

TABLE 14.1

Sequence of Puberty

Girls	Approximate Average Age*		Boys
Ovaries increase production of estrogen and progesterone	9	10	Testes increase production of testosterone
Internal sex organs begin to grow larger	9½	11	Testes and scrotum grow larger
Breast "bud" stage	10	12	Pubic hair begins to appear
Pubic hair begins to appear	11	12½	Penis growth begins
Weight spurt begins	11½	13	Spermarche (first ejaculation)
Peak height spurt	12	13	Weight spurt begins
Peak muscle and organ growth (also, hips become noticeably wider)	12½	14	Peak height spurt
Menarche (first menstrual period)	12½	14½	Peak muscle and organ growth (also, shoulders become noticeably broader)
First ovulation	13½	15	Voice lowers
Final pubic-hair pattern	15	16	Readily visible facial hair
Full breast growth	16	18	Final pubic-hair pattern

*Average ages are a rough approximation, with many perfectly normal, healthy adolescents as much as three years ahead or behind these ages. In addition, sequence is somewhat variable. For instance, many girls have some pubic hair before their breasts start to grow, and some boys have visible mustaches before their voices have completely changed.

hair growth, and final breast development. For boys, the visible physical changes of puberty include, in approximate order, growth of the testes, growth of the penis, initial appearance of pubic hair, first ejaculation, peak growth spurt, voice changes, beard development, and completion of pubic-hair growth (Malina, 1990; Rutter, 1980). Generally speaking, the major events of puberty are over in three or four years, although some individuals may gain an additional inch or two of height, and most people gain additional fat and muscle, in early adulthood.

Increases in height, weight, musculature, and body fat are characteristics of all adolescents, but the range of these changes varies considerably, not only between the sexes but also between individuals of the same sex. While all the teenagers in this photo are developing normally, not all of them may feel that way at the moment.

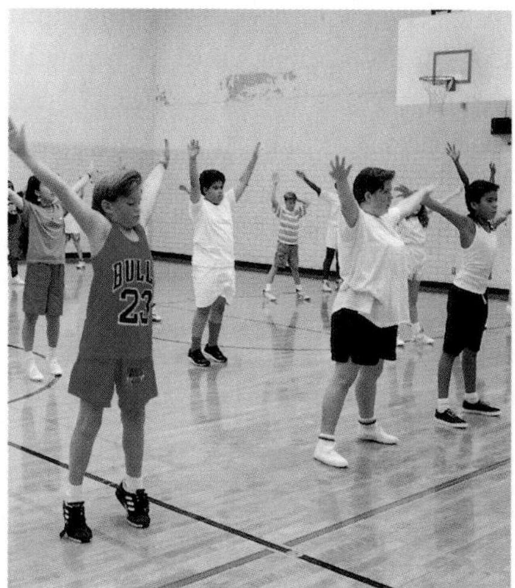

The normal variation in age of puberty is readily apparent in this junior high gym class in Texas.

The Timing of Puberty

The age at which puberty starts is highly variable: normal children begin to notice body changes anywhere between ages 8 and 14. One of the reasons for this variation is sex-based: specific body changes occur a year or two earlier for girls than parallel changes do for boys (Tanner, 1991). A second important factor affecting the age at which puberty begins is genetic inheritance (Brooks-Gunn, 1991a). This is most clearly seen in the case of **menarche**, a girl's first menstrual period. Although most girls reach this milestone between 11 and 14, the age of onset varies from 9 to 18. However, sisters reach menarche, on the average, only 13 months apart, and monozygotic twins differ by a mere 2.8 months.

Perhaps for genetic reasons, the average age of onset of puberty varies somewhat from nation to nation and from ethnic group to ethnic group. Within Europe, for example, onset tends to be relatively late for Belgians and relatively early for Poles (Malina et al., 1988). Within the United States, African-Americans often begin puberty earlier, and Asian-Americans later, than Americans of European ancestry.

Some of these genetic differences may be related to body type. In general, stocky individuals tend to experience puberty earlier than those with taller, thinner builds. Menarche, in particular, seems related to the accumulation of a certain amount of body fat, and thus does not usually occur until a girl weighs about 100 pounds. Females who have little body fat, such as gymnasts, runners, and other athletes, menstruate later than and more irregularly than the average girl, while those who are generally inactive menstruate earlier (Brooks-Gunn & Reiter, 1990; Frisch, 1983). In times of famine, puberty is typically delayed by several years (Golub, 1992).

The effects of nutrition and health are more broadly manifested in one feature of the **secular trend**, the tendency of successive generations to develop in somewhat different ways as the result of improved nutrition and medical advances. The specific feature in question is the tendency of each new generation over the past century or two to experience puberty earlier than their parents did (Tanner, 1991). A century ago, for example, the first visible signs of puberty were typically apparent in males at about age 16 and in females at about age 15. As a result of the secular trend, the current age of puberty more closely approximates the genetic threshold for our species. Indeed, there is considerable evidence that in developed countries, the secular trend has stopped in recent years: in contemporary North America and Europe, the average age of puberty is the same today as it was ten years ago, about age 10 for females and age 11 for boys. Therefore, although most middle-aged American adults reached puberty at an earlier age than their parents and grandparents did, most of today's children will experience puberty at approximately the same age as their parents did. In developing countries, however, the secular trend may be sustained as the quality of the diet and health care continue to improve. Or it may fluctuate, depending on the overall quality of life of the population. This possibility was dramatically illustrated in Japan, where the average age of puberty declined slightly from 1900 to 1935, and then increased from 1935 to 1950, as economic depression, World War II, and the devastations of defeat severely compromised the quality of life. With improving postwar conditions, the secular trend began again and, apparently reaching its genetic threshold, has now leveled off (Tanner, 1991).

menarche A female's first menstrual period, which is the final major change of puberty.

secular trend The tendency of successive generations over the past century or two to experience advanced growth as the result of improved nutrition and medicine.

One intriguing final factor may influence the onset of puberty: the relative degree of warmth or emotional distance in the child's relationship with his or her parents. Many studies have found a correlation between early puberty and parent-adolescent strife. The traditional explanation always was that the combination of the young person's "raging hormones" and emotional immaturity was at the root of such conflict. Surprisingly, longitudinal research suggests that the effects sometimes flow in the other direction: family distance and stress may accelerate onset of puberty (Steinberg, 1988; Surbey, 1990), although the particulars vary a great deal depending on the overall situation (Collins, 1990; Ellis, 1991; Paikoff & Brooks-Gunn, 1991).

Hormones and Puberty

The chain of hormonal effects that initiate puberty begins with a hormonal signal from the hypothalamus, located at the base of the brain. This signal stimulates the pituitary gland (located next to the hypothalamus) to produce hormones that then stimulate the adrenal glands (two small glands near the kidneys at both sides of the torso) and the gonads, or sex glands (the ovaries and testes). One hormone in particular, **GnRH (gonad releasing hormone)**, causes the gonads to dramatically increase production of sex hormones, chiefly estrogen and testosterone. This increase, in turn, loops back to the hypothalamus and pituitary, causing them to increase the production of **GH (growth hormone)** as well as to produce more GnRH, and this, in turn, causes the adrenals and gonads to produce more sex hormones. It should be noted that although **testosterone** is considered the male hormone and **estrogen** the female hormone, both sexes experience a rise in both hormones at puberty. However, the rate of increase is dramatically sex-specific: testosterone skyrockets in boys, up to eighteen times the level in childhood, and estrogen rises up to eight times in girls (Malina & Bouchard, 1991).

GnRH (gonad releasing hormone) A chemical produced at the onset of puberty that triggers heightened activity in the gonads (ovaries and testes), increasing levels of estrogen and testosterone.

GH (growth hormone) A chemical produced by the pituitary gland that stimulates and regulates growth throughout childhood and adolescence. Increases in the amount of GH are one of the signals that puberty is beginning and one of the critical causes of the growth spurt.

testosterone Called the "male hormone," testosterone is, in fact, produced in both sexes throughout the life span, but is produced in much larger amounts in males beginning in puberty. Levels of testosterone correlate with sexual desire and possibly with aggression.

estrogen Called the "female hormone," estrogen (which actually comprises several hormones) is, in fact, present in both sexes but is produced in much larger amounts in females beginning at puberty. Levels of estrogen correlate with sexual interest and with many aspects of the female reproductive cycle, as well as with several aspects of female health.

It is a popular notion that these various hormones are responsible for the emotional changes of puberty as well as the physical ones. To some extent, in fact, they do have an emotional impact (Susman & Dorn, 1991). Rapidly increasing hormone levels, especially of testosterone, cause more rapid arousal of emotions and are also associated with the quicker shifts in the extremes of emotions—from feeling great to suddenly feeling lousy—that seem typical in adolescence. For many girls, the ebb and flow of hormones during the menstrual cycle seem to produce some mood changes, from a relatively positive mood at midcycle to relative sadness or anger a day or two before the next period (Golub, 1992).

However, detailed studies of hormonal fluctuations and adolescent mood suggest that, while hormonal levels may sometimes make a direct contribution to the emotional changes of puberty, it is a small one. A much more potent influence on such emotional changes is the psychological impact of the visible changes of puberty—an impact that, in turn, is powerfully influenced by the values and expectations of the family, the peer group, and the culture (Brooks-Gunn & Reiter, 1991; Nottelman et al., 1990).

Now let us discuss some of the specific physical consequences, both visible and hidden, that result from those hormonal changes. Usually they are subdivided into two aspects: the rapid increases that occur in size during the growth spurt, and the emergence of sexual characteristics.

growth spurt A period of relatively sudden and rapid physical growth of every part of the body such as that which occurs during puberty.

The Growth Spurt

The **growth spurt** is just what the term suggests—a sudden, uneven, and somewhat unpredictable increase in the size of almost every part of the body. The first sign of the growth spurt is increased bone length and density, a process that begins in the ends of the extremities and works toward the center. Thus adolescents' fingers and feet lengthen before their arms and legs do. The torso is the last part of the body to grow, making many adolescents temporarily big-footed, long-legged, and short-waisted.

At the same time that the bones begin to lengthen, the child begins to gain weight much more rapidly than before, because fat begins to accumulate more readily (Malina, 1991). In fact, parents typically notice that their children are emptying their plates, cleaning out the refrigerator, and straining the seams of their clothes even before they notice that their children are growing taller. Toward the end of middle childhood, usually between the ages of 10 and 12, both boys and girls become noticeably heavier, primarily through the accumulation of fat. The specific parts of the body that take on fat and the total amount of fat increase at each site vary a great deal, depending partly on gender (girls accumulate more fat generally, especially on their legs and hips, than boys), partly on heredity, and partly on diet and exercise.

Soon after the onset of the weight increase, a height increase becomes notable, burning up some of the recently accumulated fat and redistributing some of the rest. On the whole, a greater percentage of fat is retained by females, who normally have a higher proportion of body fat than males do. About a year after these weight and height increases take place, a period of muscle increase occurs: consequently, the pudginess and clumsiness exhibited by the typical child in early puberty generally disappear by late pubescence, a few years later. The muscle increase is particularly notable in boys' upper bodies: between ages 13 and 18, arm strength more than doubles (Beunen, 1988).

The typical body proportions of the young adolescent are particularly noticeable in this runner: long legs, long feet, and a relatively short torso.

Overall, the typical girl gains about 38 pounds (17 kilograms) and 9⅝ inches (24 centimeters) between the ages of 10 and 14, while the typical boy gains about 42 pounds (19 kilograms) and about 10 inches (25 centimeters) between the ages of 12 and 16, with girls gaining the most weight in their thirteenth year, and boys, in their fourteenth (Malina & Bouchard, 1991). Note, however, that cross-sectional data, which average out individual growth spurts, are deceptive, because the chronological age for the growth spurt varies considerably from child to child. In any given year between ages 10 and 16, some individuals will not grow much at all because their growth spurt has not begun or is already over, while others will grow very rapidly. Records of individual growth during this period make it obvious why the word "spurt" is used to describe these increases (Tanner, 1991). During the twelve-month period of their greatest growth, many girls gain as much as 20 pounds (9 kilograms) and 3½ inches (9 centimeters), and many boys gain up to 26 pounds (12 kilograms) and 4 inches (10 centimeters).

One of the last parts of the body to grow into final form is the head, which typically reaches adult size and shape several years after final adult shoe size is attained. To the embarrassment of many teenagers, the facial features—especially the markedly larger ears, lips, and nose that differentiate adults from children—increase in size before the head itself takes on the large, more oval shape typical of adults. At least as disturbing to some young people can be the fact that the two halves of the body do not always grow at the same rate: one foot, breast, testicle, or ear can be temporarily larger than the other. None of these anomalies persist very long. Once the growth process starts, every part of the body reaches close to adult size, shape, and proportion in three of four years. Of course, for the adolescent, these few years of waiting for one's body to take on "normal" proportions can seem like an eternity.

Organ Growth

While the torso grows, internal organs also grow. Over the course of adolescence, the lungs increase in size and capacity, actually tripling in weight, allowing the adolescent to breathe more deeply and slowly (a 10-year-old breathes about twenty-two times a minute, while an 18-year-old breathes about eighteen times). The heart doubles in size, and heart rate decreases, slowing from an average of ninety-two beats per minute at age 10 to eighty-two at age 18. In addition, the total volume of blood increases (Malina & Bouchard, 1991).

These changes increase physical endurance in exercise, making it possible for many teenagers to run for miles or dance for hours without stopping for rest. However, the fact that the more visible spurts of weight and height precede the less visible ones of the muscles and organs means that athletic training and weight lifting should match a young person's size of a year or so earlier. Exhaustion and injury might result if the physical demands on a young person's body do not take this lag into account (Thornburg & Aras, 1986).

One organ system, the lymphoid system, including the tonsils and adenoids, actually decreases in size at adolescence, making teenagers less susceptible to respiratory ailments. Mild asthma, for example, generally

primary sex characteristics Those sex organs that are directly involved in reproduction, such as the uterus, the ovaries, the testicles, and the penis.

spermarche A male's first ejaculation of seminal fluid containing sperm.

switches off at puberty. More serious asthma gets better, although only 12 percent of serious asthmatics become totally symptom-free after puberty (Balfour-Lynn, 1986).

The eyes also undergo a change, as the eyeballs elongate, making many adolescents sufficiently nearsighted to require glasses.

Finally, the hormones of puberty cause many relatively minor physical changes that, despite their insignificance in the grand scheme of development, can have substantial psychic impact. For instance, oil, sweat, and odor glands of the skin become much more active during puberty. One result is acne, which occurs to some degree in about 85 percent of all adolescents (Lowrey, 1986). Another result is oilier hair and smellier bodies, which is one reason adolescents spend more money on shampoo and deodorants than any other age group.

Sexual Maturation

While the growth spurt is taking place, another set of changes occurs that transforms boys and girls into men and women. As we have seen, before puberty, the physical differences between boys and girls are relatively minor. At puberty, however, sexual maturation results in many significant body differences. These include changes in both the primary and secondary sex characteristics.

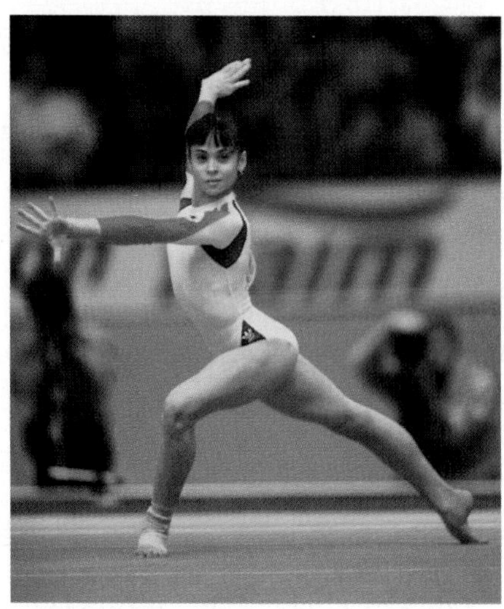

In recent years, adolescent girls have been much more likely to engage in strenuous sports. This has revealed two interesting facts about the adolescent menstrual cycle. First, it is strongly affected by the amount of body fat on the young woman's body; when the athlete has far more muscle than fat, her menstrual period is often irregular or absent altogether. Second, menstrual cramps are real—not just an excuse for laziness or a factor of poor health. Even star athletes are affected by physical changes on or about the first day of their cycle.

Changes in Primary Sex Characteristics

Changes in **primary sex characteristics** involve those sex organs that are directly involved in reproduction. During puberty, all the primary sex organs become much larger. In girls, the uterus begins to grow and the vaginal lining thickens, even before there are visible signs of breast development or pubic hair. In boys, the testes begin to grow and, about a year later, the penis lengthens and the scrotal sac enlarges and becomes pendulous.

By the end of puberty, the young person's sex organs have become sufficiently mature to make reproduction possible. For girls, the specific event that is taken to indicate fertility is the first menstrual period, which, as noted on page 387, is referred to as menarche. For boys, the comparable indicator of reproductive potential is **spermarche**, that is, the first ejaculation of seminal fluid containing sperm. Ejaculation can occur during sleep in a nocturnal emission (a "wet dream"), through masturbation, or through sexual intercourse, with masturbation being the most common cause of the first ejaculation.

Actually, both menarche and spermarche are simply one more step toward full reproductive maturity, which occurs several years later. In fact, a girl's first menstrual cycles are usually *anovulatory*; that is, they occur without ovulation. Even a year after menarche, most young women are still relatively infertile: ovulation is irregular, and if fertilization does occur, the probability of spontaneous abortion is much higher than it will be later, because the uterus is still relatively small (Golub, 1992). In the case of boys, the concentration of sperm usually necessary to fertilize an ovum is not reached until months or even years after the first ejaculation of seminal fluid (Muller et al., 1989). (As many teenagers discover too late, unfortunately, this rela-

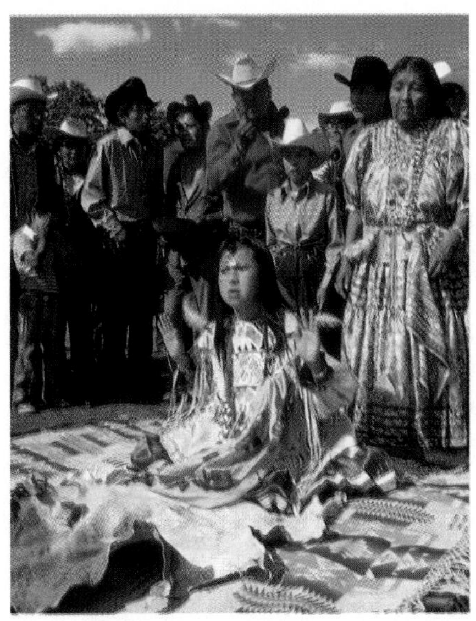

For this Apache girl, puberty brings an elaborate rite of passage, from which she emerges with new status and respect from the community.

tive infertility does not mean that pregnancy is impossible at puberty; it is simply less likely than it will be a few years later.)

In the United States and elsewhere, attitudes toward menarche, menstruation, and spermarche have changed over the past two decades and, for the most part, young people face these events with much less anxiety, embarrassment, or guilt than their parents did. This fact is clearly illustrated by a recent example involving a 13-year-old who, happening to be away from home at menarche, called her mother in tears to announce the event. Her mother, remembering her own experience of menarche and mindful of the shame and misunderstanding that generations of women have experienced regarding menstruation, immediately reassured her about the glory of womanhood, the joy of fertility, the renewal of the monthly cycle, and so on. "I know all that," her daughter protested impatiently. "I'm crying because this means I won't grow much more, and I want to be tall!"

Obviously, how any particular adolescent responds to the changes in his or her body depends on many things, including the teenager's understanding of what is happening, general conversations with parents and peers about puberty, the timing of sexual maturity in relation to others in the peer group, and broader cultural values concerning the meaning of sexual maturation in adolescents. In some cultures, for example, menstruation is heralded, with elaborate rituals, as a young woman's entry into adult status; similar "rites of passage" occur for young men (Brooks-Gunn & Reiter, 1991). In the United States, adolescent girls who were interviewed about how they felt right after menarche mostly reported a combination of positive and negative feelings, including, for some, feelings of fear or upset, mingled with a sense of maturity (Brooks-Gunn & Reiter, 1991). Those who reached menarche earlier than their friends, or who had relatively little information about it, tended to be the ones who were most upset.

No matter what his or her attitude is about the sexual changes of puberty, almost every young person has a strong sense of privacy about them. Virtually no young adolescent discusses these private events with friends or parents of the other sex. Indeed, although most boys are proud to reach spermarche, few of them tell other boys the details of their experiences with masturbation or ejaculation. They are also unlikely to discuss instances of unexpected or unwanted arousal, such as to a photo, to another boy, or to a relative, even though such arousal is quite common (Gaddis & Brooks-Gunn, 1985). For their part, girls, who often promise to tell all their close friends when menarche arrives, are usually more reticent than they had anticipated (Brooks-Gunn et al., 1986). Most girls want their mother to provide practical advice, not generalities about "becoming a woman," and they would prefer that their father not know about the event at all, but if he does, that he not make any comments (Koff & Rierdan, 1995).

Typically, a young woman's first menstrual cycles are irregular and light. However, in a year or two, and particularly in later adolescence, the days before a period can bring moodiness (caused, in part, by increased weight gain and the eruptions of acne), and the first day of a period can be quite painful, necessitating medication, rest, and understanding. Menstrual pain is caused by hormonal changes and often decreases as the body matures, especially if the woman gives birth (Golub, 1992).

Secondary Sex Characteristics

While maturation of the reproductive organs is the most directly sexual development of puberty, changes in many other parts of the young person's body also herald the process of sexual maturation. These changes are called **secondary sex characteristics**, for while they are not directly related to the primary sexual function of reproduction, they are clearly signs of sexual development. Most obviously, the body shapes of males and females, which were almost identical in childhood, become quite distinct in adolescence (Malina, 1990). Males grow taller than females and become wider at the shoulders than at the hips. Females become wider at the hips, an adaptation for childbearing that is apparent even in puberty and becomes increasingly so over the teenage years. Each set of changes forces a psychological recalibration of one's body image (see A Closer Look, pp. 394–395).

Another obvious difference in the shape of the female body, and the one that receives the most attention in Western cultures, is the development of breasts. For most girls, the first sign that puberty is beginning is the "bud" stage of breast development, when a small accumulation of fat causes a slight rise around the nipples. From then on, breasts develop gradually for about four years, with full breast growth not being attained until almost all the other changes of puberty are over (Malina, 1990). Since our culture misguidedly takes breast development to be symbolic of womanhood, girls whose breasts are very small or very large are often distressed: small-breasted girls often feel "cheated," even disfigured; large-breasted girls may become extremely self-conscious as they find themselves the frequent object of unwanted stares and remarks.

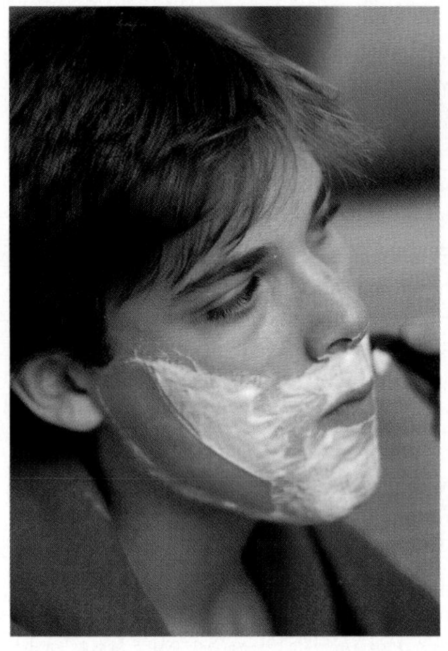

Facial hair is often considered a sign of virility, which explains why many young men shave regularly before it seems warranted. However, since the act of shaving may signal the young male's adult interests and urges, parents should not ridicule it, even when it involves little more than fuzz.

In boys, the diameter of the areola (the dark area around the nipple) increases during puberty. Much to their consternation, about 65 percent of all adolescent boys experience some breast enlargement as well, typically at about age 14 (Smith, 1983). However, their worry is usually short-lived, since this enlargement normally disappears by age 16.

Another secondary sex characteristic that changes markedly is the voice, which becomes lower as the larynx grows. This change, of course, is most noticeable in boys. (Even more noticeable, much to the chagrin of the young male, is an occasional loss of voice control that throws his newly acquired baritone into a high squeak.) Girls also develop somewhat lower voices, a fact reflected in the recognition of a low, throaty female voice as more "womanly."

During puberty, both sexes also experience changes in head and body hair, which usually becomes coarser and darker. In addition, new hair growth occurs under the arms and on the face, as well as in the pubic area. Facial and body hair are generally considered distinct signs of manliness in American society, a notion that is mistaken. The tendency to grow facial and body hair is inherited: how often a man needs to shave, or how hairy his chest is, is determined primarily by his genes rather than by his virility. In addition, girls typically develop some facial hair and more noticeable hair on their arms and legs during puberty—a sign not of masculinity but of sexual maturation, with the specifics of color and density being more genetic than hormonal.

secondary sex characteristics Sexual features not directly involved in reproduction, such as a man's beard or a woman's breasts.

A CLOSER LOOK

Body Image

The physiological changes of puberty necessitate a drastic revision of adolescents' **body image**, that is, their mental conception of, and their attitude toward, their physical appearance. According to many developmentalists, developing a healthy body image is an integral part of becoming an adult (e.g., Erikson, 1968; Simmons & Blythe, 1987). However, few adolescents are satisfied with their physical appearance, most imagining that the bodies appear far less attractive than they actually are.

This self-appraisal can have a major impact on the adolescent's overall sense of self-esteem. Indeed, although self-esteem is obviously influenced by perceived success in athletics, academics, friendship, or other areas that the adolescent considers important, a teenager's assessment of his or her physical appearance is the most important determinant of positive or negative self-esteem. This is especially true of girls, who are also more self-critical than boys of their weight, body type, hair, and even their knees and feet (Duke-Duncan, 1991; Phelps et al., 1993; Rauste-von Wright, 1989). In explaining why it may be that, for teenagers, "self-esteem is only skin deep," one researcher notes that

> the domain of physical appearance . . . is an omnipresent feature of the self, always on display for others and for the self to observe. In contrast, one's adequacy in such domains as scholastic or athletic competence, peer social acceptance, conduct, or morality is not constantly open to evaluation, but rather is more context specific. Moreover, one has more control over whether, when, and how such characteristics will be revealed. [Harter, 1993]

As a result of this intense focus on their physical appearance, many adolescents spend hours examining themselves in front of the mirror—worrying about their complexions, about how their hairstyle affects the appearance of their face, about whether the fit of their clothes makes them look alluring or cool. Some exercise or diet with obsessive intensity (e.g., lifting weights to build specific muscles or weighing food to the gram to better calculate calories). Such self-absorption can seem nearly psychopathological to adults, or even to other adolescents, who tire of hearing each other fretting endlessly about their appearance.

body image A person's mental concept of his or her physical appearance.

Indeed, the quest for the ideal body can become physically as well as psychologically unhealthy. At one time or other, almost every American girl undereats for an extended period, sometimes drastically, in order to be thinner. And roughly 5 percent of male high school seniors use steroids to build up their muscles (Johnston et al., 1995). These young men risk a variety of serious health problems, especially if they obtain the drugs illegally and "stack" one drug with another, as many do. While one misguided motivation for taking steroids may be to excel at sports, one survey found that a third of steroid users did not participate at all in interscholastic athletics, apparently taking steroids solely for the sake of appearance (Johnston et al., 1989). Further, although most adolescents believe that smoking cigarettes puts their health "at risk," more than one in five high school seniors is a daily smoker, partly because smoking reduces appetite. Interestingly, with most drugs, girls are much more cautious than boys and hence less likely to be users, but largely because of the suppressant effect that smoking has on appetite and weight gain, the rate of cigarette-smoking over the past twenty years has been slightly higher among girls than among boys (Johnston et al., 1995).

Before dismissing adolescents' preoccupation with their looks as narcissistic, we should recognize that teenagers' concern for appearance is, in part, a response to the reactions of other people. Parents and siblings, for example, sometimes make memorable and mortifying comments about the growing child's appearance—"You look like a cow," "You're flat as a board," "Your feet look like gunboats"—that they would not dare make to anyone else. Strangers, too, offer commentary (usually unwanted and disconcerting) on adolescents' growing bodies: pubescent girls suddenly hear whistles, catcalls, and lewd suggestions; boys, depending on their level of maturation, often find themselves labeled as studs or wimps. Adolescents may even get subtle messages about their appearance from their teachers: in junior high school, at least, teachers tend to judge new students who are physically attractive as academically more competent than their less attractive classmates (Lerner et al., 1990).

The concern with physical appearance dominates the peer culture as well, with the most obvious impact on self-esteem occurring in early adolescence (Harter, 1993). Teenagers who are unattractive tend to have fewer friends—of either sex—than the average teenager (Rutter, 1980). And, of course, physical attractiveness is an especially important sexual lure during adolescence. In a study that asked close to 200 ninth-graders to rank ten possible qualities of

an ideal man or woman, physical appearance was high on the list for both sexes (Stiles et al., 1987). In fact, for the boys, "good looks" was the most important quality in the ideal woman, followed by being "sexy" and then "fun," with "kind and honest" a distant fourth. For the girls, "kind and honest" and "fun" were tied for first as the most important qualities in the ideal man, followed by "good looks."

Adolescents also receive powerful messages from the broader social ecology regarding the importance of physical appearance. They are daily bombarded with media images of handsome faces and beautiful bodies selling everything from clothes and cosmetics to luncheon meats and auto parts. These images reinforce the cultural ideal that men should be tall and muscular, that women should be thin and shapely, and—at least in the developed Western societies—that both should have features suggesting an Anglo-Saxon heritage.

Obviously, few people fit this so-called ideal. Indeed, many ethnic groups are genetically endowed with shorter stature, or broader hips, or chunkier bodies, and so forth than those called for by this cultural ideal. Often these genetic differences are not very salient in childhood, before hormones evoke the adult body type the person is destined to have. This means that puberty suddenly requires many young adolescents to come to terms with a body form, inherited from their ancestors, that is a far cry from those of the models and movie stars they idolize.

With time, the intense self-preoccupation and concern about body image typical of the pubescent young person lessens, and adolescents gradually become more satisfied with their physical selves (Rauste-von Wright, 1990). By adulthood, most people have learned to accept the discrepancy between the cultural ideal and their own natural appearance, an acceptance that becomes easier once they are assured, over the years, that they and their bodies are loved "just as they are."

Meanwhile, however, teenagers' concern over their body image should not be taken lightly. For most adolescents, thinking that they look terrible makes them feel terrible, even depressed. Instead of ignoring or belittling the adolescent's self-preoccupation, adults should provide whatever practical help seems warranted—such as clothing suggestions, encouragement to exercise, medical treatment for acne, and so forth. Understanding and compliments, instead of criticism and derision, might have far-reaching benefits, not only for the adolescent's body image but also for his or her self-esteem, social acceptance, and overall enjoyment of life.

Many teenagers sacrifice time, money, and health to attain the lean look of the female model or the muscled physique of the macho movie star. Even those teenagers who already have attractive bodies are likely to be highly self-critical, while those who inherit a shape or size that is far from the cultural ideal are likely to be excessively self-conscious and depressed.

Early and Late Maturation

Young people who experience puberty at the same time as their friends tend to view the experience more positively than those who experience it earlier or later (Dubas et al., 1991). Being very early or very late to mature physically can be emotionally stressful. Interestingly, but perhaps predictably, the effects of early or late maturation differ for boys and girls. It is more difficult to be an early-maturing than a late-maturing girl because of the added pressures and expectations faced by girls who are more developed sexually than psychosocially. Conversely, it is more difficult to be a late-maturing than an early-maturing boy because a mature build and physical prowess are key factors in peer status.

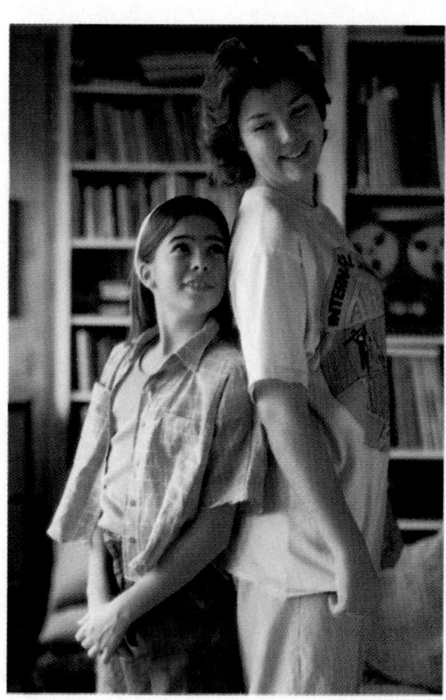

Both 14-year-olds pictured here experienced puberty "off-time." Probably the early-maturer suffered considerable stress at around age 11, when her feet, hips, and breasts provoked unwanted attention. By this point, however, she may already have learned to cope with her body image, and the late-maturing girl may be wishing her body would "hurry up."

Early-maturing girls, who are taller and more developed than their classmates, discover that they have few peers who share their interests or problems. Prepubescent girls call them "boy crazy," and boys tease them about their big feet or developing breasts. Almost every sixth-grade class has an 11-year-old pubescent girl who slouches so she won't look so tall, wears loose shirts so no one will notice her breasts, and buys her shoes a size too small. She is likely to have a poorer body image than other girls and to be at greater risk for developing eating disorders (Graber et al., 1994).

There are additional hazards for the early maturer. If she begins dating, it will probably be with boys who are older, and she may thus begin smoking, drinking, and other "adult activities" at an earlier age (Brooks-Gunn, 1991b). Her self-esteem is likely to fall for a number of reasons: she may feel constantly scrutinized by her parents, criticized by her girlfriends for not spending time with them, and pressured by her dates to be sexually active. Such difficulties for early-maturing girls may be compounded by ongoing family tensions, since some research suggests that family conflict may be a factor in triggering early puberty in girls. More specifically, girls from families high in conflict, or whose parents are divorced, tend to experience menarche earlier than do girls from harmonious homes (Wierson et al., 1993).

Early maturity for girls is not without some advantages, eventually. After a few years of awkwardness and embarrassment, the early-maturing girl is able to advise her less mature girlfriends about topics that they find increasingly important (such as bra sizes, dating behavior, and menstrual cramps), and, perhaps for this reason, by mid-adolescence the early-maturing girl has more close friends than those who have not yet begun to mature (Brooks-Gunn et al., 1986).

In contrast to early-maturing girls, late-maturing boys must watch themselves be outdistanced, first by the girls in their class and then by most of the boys, all the while enduring the patronizing scorn of those who only recently were themselves immature. However, particularly for boys, generalizations based solely on age of puberty are difficult to make. A review of all the research on the timing of puberty in boys finds that, while the effects of early puberty are generally positive, the effects of late puberty are mixed, dependent not only on the particular traits and particular population under study but also on other factors (Downs, 1990). For example, overall, boys who are unusually short, who are not athletic, who appear physically weak or unattractive, and/or who are slow to become sexually involved tend to have lower self-esteem and create more problems for themselves and others dur-

While size and strength obviously play a role in the social status within male peer groups, social skills may be even more salient. The smallest boy's quick wit or athletic ability may secure his position with his friends.

ing adolescence. This occurs no matter what the timing of the boy's development, but the later puberty begins, the more likely each of these liabilities is to occur. Obviously the extent to which a particular boy experiences them will vary—some late developers are quite handsome and athletic—and, further, some peer groups value each of these attributes more than others. Thus, the impact of later maturation varies a great deal, from person to person and from culture to culture.

School, Family, and Cultural Factors

Adolescents are affected not only by how their development meshes with that of others in their class but also by how it compares with that of others in their school. For example, one study showed that tall, early-maturing sixth-grade girls scored higher in measures of social competence when they were enrolled in a middle school where they were among similarly mature seventh- and eighth-graders than when they were the tallest, most mature girls in an elementary school that ended at sixth grade (Nottelmann & Welsh, 1986). The implicit comparisons afforded by a child's peer networks at school may determine whether, and how much, he or she is perceived as "off-time."

The family context is important as well. No matter what their timing, the physical events of puberty and the budding drives for sexual expression and freedom of action typically increase the distance between parent and child—particularly between mother and daughter—adding to the adolescent's adjustment difficulties. If the family has a pattern of dysfunction, or if stressful family events such as divorce coincide with puberty, then the adolescent is especially likely to have difficulty adjusting to the normal changes of adolescence. (The relationship between family patterns and adolescent problems is discussed in Chapter 16.) Even in close families, puberty brings a distancing of child and parents, as the typical eighth-grader spends more time away from parents (often retreating to his or her bedroom) and is less happy when spending time with either parent than when alone or with friends (Larson & Richards, 1994).

In many ways, cultural values can intensify, or lessen, the problems of early or late maturation. The effects of early or late maturation are more apparent, for example, among adolescents of lower socioeconomic status because physique and physical prowess tend to be more highly valued among teenagers of low SES than among middle- or upper-SES teenagers. Correspondingly, alternative sources of status for the early or late maturer, such as academic achievement or vocational aspiration, are less available and less valued among lower-SES adolescents (Rutter, 1980). For those adolescents for whom these alternative sources of positive self-esteem are available and valued, timing of the physical changes of puberty has less significance (Richards et al., 1990).

The impact of school and culture was further revealed in a longitudinal study showing that, unlike their American counterparts, early-maturing girls in Germany do not experience lower self-esteem. The authors of the study speculate that the reason may be that sex education is more extensive in German schools than in American schools, leading to less confusion and embarrassment about sexual maturation, as well as to a reduced likelihood of pregnancy (Silbereisen & Kracke, 1993).

Sexual Abuse

Children are at risk of being sexually abused from the time they are born through adolescence, with the greatest risk occurring between the ages of 7 and 13 (Cappelleri et al., 1993; Finkelhor, 1994). While sexual abuse is destructive at any age, the *meaning* of being sexually victimized may be more profound for older children and adolescents. At a time when teenagers are trying to cope with the changes in body image, self-awareness, and peer relationships that accompany puberty, the experience of sexual victimization can have devastating consequences. Indeed, adolescents may react to maltreatment of all kinds in ways that younger children rarely do, with self-destruction, such as suicide, drug abuse, or running away, or with counterattack, such as vandalism or violence aimed at society or at the perpetrator—including patricide (Ewing, 1990).

Often, adolescents' problems that do not seem directly related to maltreatment are, on closer analysis, revealed to be tied to abuse, especially sexual victimization. For example, researchers studying a representative group of 535 young women in the state of Washington who had become pregnant as teenagers were astonished by the number who had a history of having been sexually victimized: nearly two-thirds reported having been sexually abused before becoming sexually active, with 44 percent reporting having been raped. Compared with women who became pregnant but had not been abused, the teenagers who had been sexually abused reported greater use of drugs and alcohol, had become sexually active at an earlier age, and had higher rates of suspension, expulsion, or dropping out from school. They were also more likely to have experienced other kinds of maltreatment (Boyer & Fine, 1992).

Typical Sexual Abuse

At any age, sexual abuse occurs when someone engages another person in sexual activity without that person's freely given consent. Since children and younger adolescents are vulnerable to the power of adults, and have no or little understanding of the implications of sexual activity, they are incapable of freely consenting to sexual acts (the age of consent is considered to be 16 or 18, depending on state law). Thus, **childhood sexual abuse** includes any erotic activity that arouses an adult and excites, shames, or confuses a young person—whether or not the victim protests and whether or not genital contact is involved. Teasing a child in a sexualized manner, photographing a young person in erotic poses, intrusively questioning a young adolescent about his or her developing body, and invading the privacy of a child's bathing, dressing, or sleeping routines—especially once puberty begins—can all be sexually abusive.

As with other forms of maltreatment (see Chapter 8), the severity of any particular act of sexual abuse depends on many factors, with repeated incidence, distorted adult-child relationships, and impairment of the child's ability to develop normally as key elements. As with physical abuse and neglect, the immediate and obvious impact is less significant, in the long run, than later harm. In particular, childhood sexual abuse has the potential for permanently damaging the ability to establish a warm, trusting, and intimate relationship with another adult.

childhood sexual abuse Any erotic activity that arouses an adult and excites, shames, or confuses a child or young adolescent—whether or not the victim protests and whether or not genital contact is involved.

Sexual victimization in adolescence is often the continuation of less blatant abuse that began during childhood, with the kind of fondling, explicit nudity, and suggestive comments that may confuse a preschooler or young grade-school child. In late childhood or early adolescence, this behavior may escalate into sexual intercourse. Whatever form the sexual abuse takes, overt force is seldom involved, however, because the perpetrator is usually someone who can easily dominate the child, typically the father, a relative, or a trusted family friend. The powerlessness of the child is especially apparent when the victim is a pubescent girl and the perpetrator is her own father. Although the victim may have confused feelings regarding the contact she experiences and any reassurances she receives from the perpetrator, there is no question that this is sexual victimization.

Most sexual abuse is committed by men and by persons known to the child, especially family members (Finkelhor, 1994). While girls are the most common victims, increasingly it is recognized that boys are also often sexually abused (Cappelleri et al., 1993; Finkelhor, 1994). Compared to that of girls, sexual molestation of boys occurs more often outside the home and is committed by someone, most often a male, who is not a family member (Finkelhor, 1994). In addition to the stigma of unwelcome sexual activity, a molested boy is likely to feel shame at the idea of being weak, unable to defend himself, and a participant in homosexual activities, all contrary to the macho image that many boys, especially young adolescents, strive to attain (Bolton et al., 1989). When the sexual abuse of a boy does occur at home, typically by the father or stepfather, the problems of vulnerability and self-esteem are multiplied.

Although mothers and other female relatives seem to be less often perpetrators of obvious sexual abuse, they are sometimes guilty—especially with sons when the father is absent—of sexual teasing and fondling that can evoke feelings of confusion, shame, and victimization (Hunter, 1990). These feelings can later turn outward. Boys who are sexually abused by women are particularly likely to become abusers themselves when they achieve sexual maturity (Biggs et al., 1996; Ryan et al., 1996). With daughters as well as sons, mothers can be part of the problem in a different way: often, they fail to notice, believe, or report that sexual abuse is occurring in the home, or they fail to support and protect the child when the child reports the abuse to authorities. In these circumstances, the victim is doubly victimized—by the perpetrator of the abuse and by the parent who fails to intervene.

As with other types of maltreatment, parents who are in conflict with each other, immature, socially isolated, alcoholic, or drug-abusing are more likely to be sexually abusive or so neglectful that their children are vulnerable to abuse from others. However, unlike other forms of maltreatment, sexual abuse is committed nearly equally by parents of every income and educational level.

Consequences of Sexual Abuse

The psychological effects of sexual abuse depend largely on the extent and duration of the abuse, the age of the child, and the reactions of other people—family as well as authorities—once the abuse is known (Briere & Elliott, 1994). Like younger children, adolescent victims of sexual abuse are likely to become anxious, angry, and fearful after being molested; they may experi-

ence nightmares and problems in school; they are likely to become depressed. Unlike younger children, however, adolescents are also prone to becoming self-destructive through substance abuse or eating disorders, running away from home, risking AIDS through unsafe sex with unknown partners, or even attempting suicide (Briere & Elliott, 1994; Cunningham et al., 1994; Kendall-Tackett et al., 1993). Although these consequences are more apparent in girls, especially those whose sexual abuse is incestuous, boys show similar symptoms (Bolton et al., 1989; Finkelhor, 1990; Hunter, 1990). In addition, adolescent victims of abuse, whether they are male or female, tend to become involved again in abusive relationships, either as the abuser or the abused (Billingham & Sack, 1986; Bolton et al., 1989; Briggs & Hawkins, 1996; Ryan et al., 1996).

These consequences are especially severe when the sexual abuse occurs not once but over a long period, when the child has an especially close relationship with the abuser, and when force is used (Millstein & Litt, 1990). By contrast, if the abuse is a single nonviolent incident and a trusted caregiver believes and reassures the victim, taking steps to make certain the abuse does not occur again, the psychological damage may be short-term. In fact, approximately one-third of sexual-abuse victims never show serious psychological harm, perhaps, researchers believe, because their abuse is not severe and they receive the support of another family member (like the mother) who believes them and offers protection (Finkelhor, 1990). Even with abuse that is more serious, however, children and adolescents can be quite resilient if they are cared for with sensitivity, confidentiality, and respect. While therapy generally improves the chances of recovery, much more research needs to be done before it is known how every victim of childhood sexual abuse can best be helped (Finkelhor & Berliner, 1995).

Health, and Hazards to It

In many ways, adolescence is a healthy time of life. The minor illnesses of childhood—flus, colds, earaches, and high fevers—are much less common, as years of exposure increase immunity to the various viruses that inoculations do not cover. And the main killers of adults, fatal heart disease and terminal cancer, are rare: a 20-year-old's chance of dying from them is only a third that of a 30-year-old and a hundredth that of a 70-year-old.

However, while diseases are uncommon during these years, adolescents are often at risk for health hazards of a different kind, especially as they gain increasing independence and begin to make more of their own decisions. Two such hazards—risky sexual activity and suicidal depression—are discussed in the next two chapters. Here, we will look at two others, poor eating habits and the use of drugs and alcohol.

Nutrition

The rapid body changes of puberty obviously require additional calories, as well as additional vitamins and minerals. In fact, the recommended daily intake of calories is higher for the active adolescent than for a person at any other time during the entire life span, with the greatest calorie requirement occurring at about age 14 for girls and about age 17 for boys (Malina & Bouchard, 1991). And during the growth spurt, the need for calcium, iron,

and zinc for bone and muscle development is about 50 percent greater than that of two years earlier.

In developing nations, where quality food is sufficiently available, most adolescents, most of the time, meet their basic nutritional needs, consuming four or more meals a day, even though they often do not eat breakfast (Leon et al., 1989; Thornburg & Aras, 1986). In fact, skipping breakfast is not usually a nutritional problem at this age, despite what parents and teachers may think. One study that kept an experimental group of high school students on very low-calorie breakfasts, while allowing a control group full, high-calorie breakfasts, found, to the surprise of the experimenters, that there was no difference in the groups' performance on a battery of cognitive and psychological tests (Cromer et al., 1990).

However, while adolescents generally eat enough, they do not always eat the right foods. Only 14 percent of American teenagers consume the recommended five servings of fruits and vegetables a day, and 66 percent eat too many fried or fatty foods (*MMWR*, March 4, 1994). The one specific nutrient that many seemingly well-nourished adolescents are deficient in is iron: fewer than half consume the recommended daily dose of 15 milligrams. Because each menstrual period depletes the body of some iron, females between the ages of 15 and 17 are more likely to suffer from iron-deficiency anemia than any other subgroup of the population (Baynes & Bothwell, 1990). This means, for one thing, that if a teenage girl seems apathetic and lazy, she should have her hemoglobin checked before it is assumed that she suffers from a poor attitude or other psychological difficulties. It also means that if a teenager is pregnant, she will need iron supplements as well as folic acid, under the guidance of a doctor, nurse, or midwife, to protect her fetus from nutritional deficiencies.

While it is true that most adolescents in developed countries are well nourished most of the time, it is also true that most experience periods of overeating, undereating, or nutritional imbalance. For many reasons, the typical adolescent is especially vulnerable to food fads and strange diets. These can be harmful at every age, but they are particularly so during the rapid growth of early adolescence, when the body must have sufficient nourishment to reach full growth potential. Teenagers who systematically deprive themselves of basic nourishment throughout the growth spurt will become shorter and less well proportioned adults (Tanner, 1978).

In their efforts to see that their teenage children are well-nourished, many parents feel they are engaged in a lost cause: they urge fruits, vegetables, and lean meat, while their offspring stoke up on sugary, greasy, salty, fast foods. However, given the high calorie needs of the growing young person, fast foods are not unequivocally bad. For instance, milk shakes are rich in calcium, French fries, in vitamin C, and pizza, in iron, and all provide ample calories necessary to sustain rapid growth.

Indeed, a sizable minority of adolescents have problems that interfere with normal, healthy eating. Childhood obesity often worsens in adolescence if the young person experiences increased social rejection and a further lowering of self-esteem. Addictive drug use also affects eating patterns, altering the appetite and digestive processes and depriving the young person of growth and energy.

Another problem, particularly for girls, is a preoccupation with being thin, which can lead to serious undernourishment, halting both growth and sexual maturation. In many modern industrialized nations, slimness is associated with beauty, health, and wealth, leading many people to be overly focused on their weight. It is not surprising, then, that a majority of adolescent girls wish they were thinner. In fact, one large longitudinal study of younger adolescents found that even in sixth grade many girls were unhappy with their weight. By ninth grade, not only were most girls dissatisfied with their weight but they "cared very much" about it. No other problem with appearance was of such concern (Simmons & Blyth, 1987).

Another study found that older girls (aged 14 to 18) typically wanted to be about 12 pounds lighter (Brooks-Gunn et al., 1989). Amazingly, this held true across the board—regardless of the girls' maturation status or their exercise levels (see Table 14.2). For example, girls who matured late, and thus had relatively thin, girlish bodies, wanted to lose almost as many pounds as those more womanly shaped girls who had matured on time. Similarly, competitive swimmers, whose bodies were quite muscular and who needed some fat to help them with buoyancy and endurance, wanted to lose weight just as much as the nonathletes, who had somewhat more fat on their bodies and less reason to need it. Even the thinnest group, late-maturing girls who practiced daily in professional dance schools, wished they weighed 10 pounds less. As we will see in Chapter 17, these attitudes are sometimes precursors of serious, even potentially fatal, eating disorders.

TABLE 14.2

Adolescent Girls' Actual and Desired Weight*

	Matured on Time		Matured Late	
	Actual	Desired	Actual	Desired
Dancers	116	102	108	98
Swimmers	130	117	128	114
Nonathletes	125	114	121	111

*For all six groups, average height was between 5'4" and 5'5½".
Source: Brooks-Gunn et al., 1989.

Drug Use

Abuse of drugs, including alcohol and medications, always harms physical as well as psychological development. *Use* of drugs, however, is another matter, sometimes harmful, sometimes not, depending in part on the maturation of the drug-user as well as on the reason for the use (Gerstein & Green, 1993). In this chapter, we will focus primarily on adolescent drug use, which may or may not lead to long-term abuse but quite frequently poses a threat to the adolescent's health and immediate well-being. (Our major discussion of serious abuse and addiction occurs in Chapter 17.)

Disquieting Trends

In recent decades, the use of alcohol and other drugs has become a part of many young people's lives in every industrialized nation in the world, as chronicled by a large number of studies. One of the most notable of these studies is an annual, detailed, confidential survey of more than 15,000 American high school seniors from a representative cross-section of over 100 high schools, along with a comparable sample of eighth- and tenth-graders. Since its inception in 1975, this survey has consistently shown that nine out of ten seniors had at least sipped alcohol, that two out of three had smoked at least one cigarette, and that, at a minimum, over 40 percent had tried at least one illegal drug (Johnston et al., 1995).

More troubling than these particular rates, which no doubt include instances of one-time-only experimentation, are recent findings on use "within the past thirty days," a statistic that reflects more regular drug use. After a long period of decline in the regular use of most drugs (except cocaine) during the 1980s, the number of students admitting to drug use within the past thirty days began to increase in the early 1990s and is continuing to climb (see Figure 14.1). From 1991 to 1995, for example, the regular use of marijuana rose significantly for all age groups, reaching 21 percent for twelfth-graders. There were also increases in the regular use of other illegal

FIGURE 14.1 As this graph shows, a decade-long decline in regular drug use among U.S. high school seniors came to an end in 1992. Also apparent is the fact that the use of all drugs follows the same general pattern. Thus, although teenagers use legal drugs more than they use illegal drugs, drinking alcohol and smoking cigarettes are affected by the same cohort conditions as swallowing pills or smoking marijuana (or snorting cocaine or shooting heroin, not shown here because of their relatively low use rate). The cohort factor that seems most related to the recent upsurge in adolescent drug use is a decline in teenagers' belief that drug use is harmful.

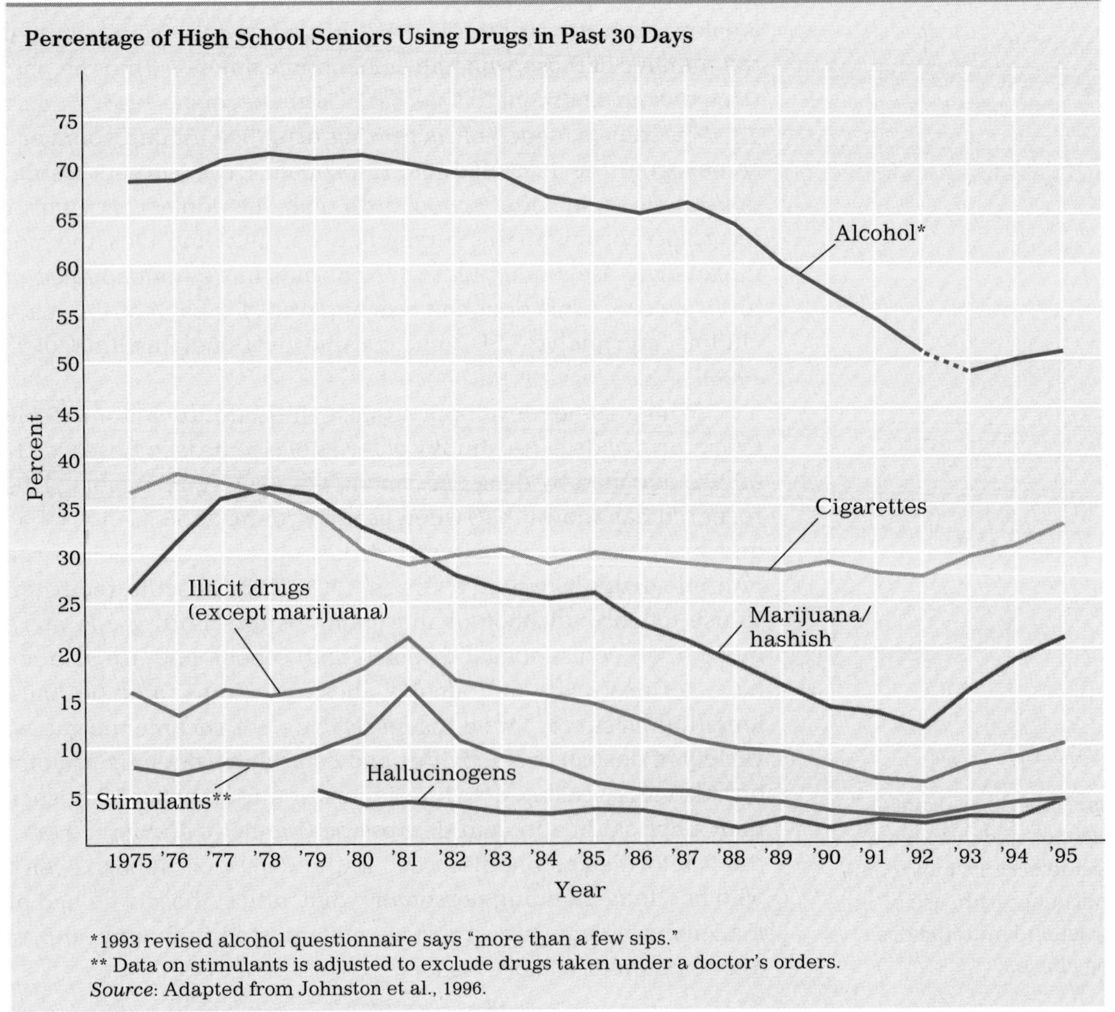

*1993 revised alcohol questionnaire says "more than a few sips."
** Data on stimulants is adjusted to exclude drugs taken under a doctor's orders.
Source: Adapted from Johnston et al., 1996.

drugs, with 10 percent having used hallucinogens (including LSD and PCP), nonprescription stimulants, cocaine, heroin, or the like within the past month.

Repeated use of legal drugs is also cause for concern. According to the high school survey, after a decade-long dip in the 1980s, smoking increased again in the 1990s, reaching a fifteen-year high in 1995, with 16 percent of tenth-graders and 22 percent of seniors reporting *daily* tobacco use (Johnston et al., 1995). Regular use of alcohol has risen only slightly since 1991 but nevertheless is at a high level, with over half of all seniors reporting use within the past thirty days. The one "bright spot" in the survey concerns alcohol abuse, as indicated by high school seniors who had had five or more drinks on one occasion during the two weeks preceding the survey. The percentage of seniors who acknowledged such abuse averaged 39 percent in the 1970s, 37 percent in the 1980s, and hit an all-time low in 1993 at 27.5 percent. From there it has inched upward, to 30 percent in 1995.

A particularly disquieting feature of the 1995 school survey is the early onset of drug use it reveals. About half of all eighth-graders have already had at least one drink (54 percent) and smoked at least one cigarette (46 percent), and about 20 percent have tried marijuana, twice the number who had in the 1991 survey. The increase in the repeated use of these drugs at age 13 or so is especially troubling. Between 1991 and 1995, the number of eighth-graders who had had five or more drinks in a single sitting within the past two weeks rose from 13 to 15 percent; the number of those who had smoked cigarettes within the past thirty days rose from 14 to 19 percent; and the number of those who had smoked marijuana within the past thirty days tripled from 3 percent to 9 percent (Johnston et al., 1995).

Further evidence of increasing drug use among younger adolescents comes from the 1995 National Household Drug Survey. (Being a survey of households, it includes school drop-outs, the adolescents most likely to use drugs.) According to this survey, between 1992 and 1995, the monthly rate of illegal drug use among 12- to 17-year-olds more than doubled, rising from 5 to 11 percent. This doubling was apparent for each of the major drugs, including marijuana, LSD, and cocaine (National Institute of Drug Abuse, 1996).

While some developmentalists hesitate to automatically condemn every form of drug use among older adolescents (see Research Report), they all are alarmed by drug use among very young teenagers. This is because research has shown that, when used by young adolescents, tobacco, alcohol, and marijuana act as **gateway drugs**, opening the door to regular use of multiple drugs (Gerstein & Green, 1993). To be specific, teenagers who begin to use tobacco, alcohol, or marijuana before ninth grade are significantly more likely to use illegal drugs in high school, and are also more likely to have serious drug- and alcohol-abuse problems later on. Indeed, a large longitudinal survey found that early drug use became the gateway to a wide variety of destructive activities and conditions, not only use of illegal drugs but also risky sexual activities, alienation from school, antisocial behavior, poor physical health, and depression (Kandel & Davies, 1996). Similarly, in the California sample discussed in the Research Report, even for children who had long-standing personality difficulties, those who had already tried marijuana by age 14 were more maladjusted, unhappy, and rebellious by age 18 (Shedler & Block, 1990).

gateway drugs The three drugs—tobacco, alcohol, and marijuana—that young adolescents most commonly experiment with, use of which may lead to multiple drug use or drug abuse.

RESEARCH REPORT

Distinctions in Adolescent Drug Use

In their efforts to gain a comprehensive understanding of adolescent drug use, psychologists have learned an important lesson: too simplistic a view of adolescents' drug use makes the problem more difficult to deal with. Some adults seem not only alarmed but frightened at adolescent drug use. They act as if drug addiction occurs the moment a teenager takes one drag, swallow, or snort, and as if drug-using teenagers are necessarily a threat to themselves and society. In contrast, many adolescents take the view that they are just experimenting and that they are immune to the addictive pull, or destructive impact, of any substance.

The reality is usually more complex, as was suggested by the findings of a fifteen-year longitudinal study that followed children from the age of 3. By age 18, most had tried marijuana as well as alcohol and tobacco. This led the researchers to wonder if such experimentation was normative in "psychologically healthy, sociable, and reasonably inquisitive individuals" (Shedler & Block, 1990).

To explore this question, the researchers divided their sample into three groups: *abstainers*—those who had never tried any illegal drug; *experimenters*—those whose marijuana use was no more frequent than once a month and who had tried no more than one other illegal drug; and *frequent users*—those who used marijuana once a week or more and who had used at least one other illegal drug. (Those who did not fit into any of these groups were omitted from the analysis.)

To no one's surprise, the researchers found the typical frequent user to be a "troubled adolescent, an adolescent who is interpersonally alienated, emotionally withdrawn, and manifestly unhappy, and who expresses his or her maladjustment through undercontrolled, overtly antisocial behavior." Somewhat surprisingly, however, the typical abstainer was found to be not much better off—a "relatively tense, overcontrolled, emotionally constricted individual who is somewhat socially isolated and lacking in interpersonal skills." The experimenters, by contrast, were the most outgoing, straightforward, cheerful, charming, and poised of the three groups. Compared to the other two, they were least likely to distrust others or to keep them at a distance.

The longitudinal data in this study reveal that these patterns were not caused by the drug use but reflected preexisting personality characteristics. Even as young children, the frequent drug users and the abstainers tended to be more tense, distressed, and insecure, while the experimenters were more curious, open, happy, warm, and responsive. Further, the parents of the future frequent users and the future abstainers had much in common: they tended to be cold and unresponsive, pressuring their children to achieve but not encouraging them for what they did.

This study by no means suggests that drug use during adolescence should be looked on benignly. The authors emphasize that frequent drug use, especially in early adolescence, is not only a sign of preexisting problems but most likely makes them worse. The authors also stress that, for adolescents who are emotionally vulnerable, abstinence is the best choice, because in their case especially, drug experimentation may lead to drug addiction.

However, the authors do take strong exception to the present thrust of drug prevention in the schools and media. They suggest that telling teenagers that a single indulgence will lead to addiction and that they should "just say no" is likely to foster not abstinence but either undue anxiety or the notion that adults are hopelessly misguided. In all likelihood, this approach is ignored by most adolescents, since it contrasts starkly with their own views of drug use, which emphasize wise judgment and personal discretion rather than a blanket moral condemnation (Nucci et al., 1991). Moreover, the authors contend that this approach focuses on the symptom rather than the problem:

> Current efforts at drug "education" seem flawed on two counts. First, they are alarmist, pathologizing normative adolescent experimentation and limit-testing, and perhaps frightening parents and educators unnecessarily. Second, and of far greater concern, they trivialize the factors underlying drug abuse, implicitly denying their depth and pervasiveness. [Shedler & Block, 1990]

It seems, then, that the problem of drug abuse during adolescence is really two problems (Dryfoos, 1990; Muisener, 1994). One applies to all adolescents, whose poor judgment about when and how to experiment with drugs, and whose behavior when "under the influence," might lead to fatal accidents or other serious consequences. The second applies to those adolescents who use drugs as an attempt to solve or forget long-standing problems. For them, drugs may bring temporary relief but, as time goes on, only add to their difficulties with growing up. Many of these teenagers have other problems as well—with school, with sexual relationships, with the law—that are made worse by drug abuse. They need much more help than a lecture on the evils of addiction.

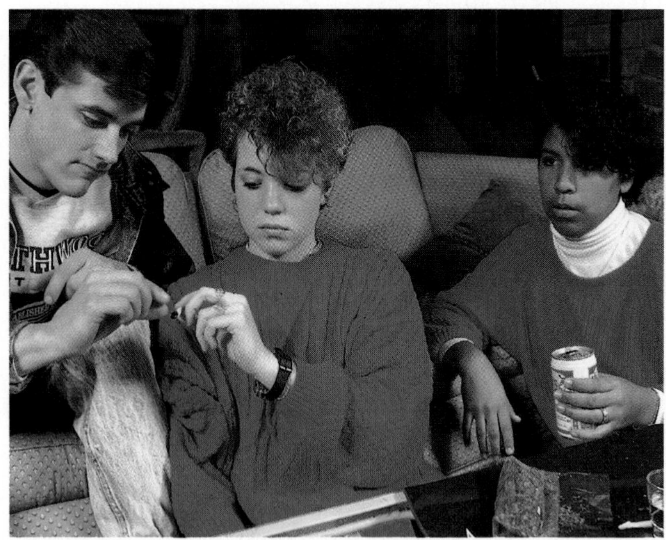

While many parents rightly fear the physical addiction and thoughtless risk-taking that drug abuse can lead their children into, this pot-smoking, alcohol-drinking trio illustrates another danger. Psychoactive drugs suspend the need to engage in meaningful dialogue, while distorting and anesthetizing emotions. Regular use of psychoactive drugs brings some adolescents into adulthood with diminished self-understanding and immature social responsiveness.

Of course, while early use of gateway drugs makes later abuse more likely, this outcome is not inevitable. Some early users become early quitters or never become heavy users. However, even if drug use does not progress to more serious harm, each of the three gateway drugs can have quite a specific impact on health and well-being, with the younger adolescent particularly vulnerable to the drugs' effects.

Tobacco use decreases food consumption and interferes with the absorption of nutrients, both of which can limit the growth spurt, causing young steady smokers to be significantly shorter and smaller than they would have been if they hadn't started smoking. In addition, nicotine, the toxic chemical found in tobacco, is probably the most physically addictive drug of all (Schilit & Gomberg, 1991), and the great majority of adolescents who become regular smokers are quickly hooked. Proof of this comes from self-reports of young people, aged 10 to 18, who had smoked at least one cigarette a day for the previous thirty days. Most of them (74 percent) said that they smoked partly because "it's really hard to quit" (*MMWR*, October 21, 1994). The fact that they are already addicted in adolescence is, obviously, a serious health problem, since continued smoking is a risk factor for a wide variety of diseases in adulthood and is the leading preventable cause of death. This fact, however, does not seem obvious to adolescents, especially eighth-graders, with 50 percent of them not seeing any "great risk" in smoking (Johnston et al., 1996). Sadly, but not surprisingly, these attitudes are reflected in cigarette use (see Figure 14.2).

Alcohol consumption, even in moderation, can be especially destructive in adolescence. The primary reason is that, even in small doses, alcohol loosens inhibitions and impairs judgment—a potentially dangerous effect in a person who may already be psychologically off-balance because of the physical, sexual, and emotional changes he or she is experiencing. For instance, a survey of 46,000 high school students, mostly "middle American" (middle-class, Midwestern), found that, compared to those who never or rarely drank, teenagers who had drunk six or more times in the previous month were more than twice as likely to be sexually active (70 percent compared with 32 percent), more than twice as likely to engage in antisocial behaviors such as stealing, fighting in groups, and vandalizing property (49

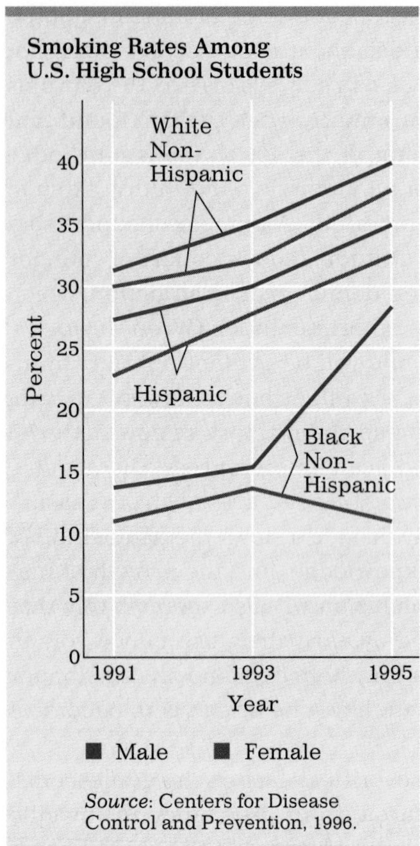

FIGURE 14.2 The increase in smoking among teenagers from 1991 to 1995 is attributed to advertising campaigns that expressly target youth, an attribution tobacco manufacturers reject. Proof one way or the other will be whether the trend reverses itself once Joe Camel and his ilk disappear from the public scene.

percent compared with 19 percent), almost four times as likely to be excessively absent from school (23 percent compared with 6 percent), and more likely to ride in a car with a driver who has been drinking (86 percent compared with 56 percent) (Benson, 1993).

Finally, marijuana seriously slows down key thinking processes, particularly those related to memory and abstract reasoning. Further, these effects may continue to be evident days after actual use. While such impairments would be a problem at any age, they are especially damaging in early adolescence, when academic learning requires greater memory ability and a higher level of abstract thinking. In addition, over time, the temporary relaxation of repeated marijuana highs may translate into a general lack of motivation and an indifference toward the future, creating apathy at the very time the young person should be focusing his or her energies on meeting the challenges of adolescence (Schilit & Gomberg, 1991).

What can be done about drug use among adolescents, particularly among young adolescents? As long as drugs are available, the majority of adolescents will try them, and many will abuse them. The goal, then, as most developmentalists see it, is to delay drug experimentation as long as possible, increasing the adolescent's chances of becoming realistically informed about the risks of drug use and of developing the reasoning ability to limit, if not avoid altogether, the use of various drugs in various contexts. To this end, a wide array of measures—including improving health education classes, raising the cost of, and restricting access to, alcohol and cigarettes, enforcing drunk-driving laws, and teaching parents how to communicate with their teenagers—have been tried and been shown to be at least partially effective in postponing or decreasing drug use and in preventing or diminishing serious consequences when experimentation occurs (Dielman, 1994; Gerstein & Green, 1993; Grossman et al., 1994; Wagenaar & Perry, 1994).

Of course, a variety of factors in the social context work in the opposite direction to encourage drug use, especially for younger adolescents, who are wide open to influences from their peers, the media, and the larger culture. One striking example comes from the world of advertising. As a result of ads and promotions that featured "Joe Cool," a cartoonlike figure that caught the attention of children, the percentage of 10- to 18-year-old smokers who smoked Camels leaped from 8 percent in 1989 to more than 13 percent in 1993, a 64 percent increase. During the same period, the percentage of adults who smoked Camels held almost steady (*MMWR*, August 19, 1994). Specially targeted advertising is also seen as the reason for the dramatic increase in smoking among young black males, who experienced the greatest rise in smoking overall (Feder, 1996) (see Figure 14.2).

Beyond the specifics of advertising, young people's likelihood of using drugs directly follows their peers' attitudes about the acceptability and riskiness of using various drugs, especially if their parents do not discuss the issues with them (Mounts & Steinberg, 1995). The school survey reported above shows this relationship for marijuana use over the past twenty years, with trends in actual use following, by about a year, overall shifts in the way young people perceive the dangers of the drug (Johnston et al., 1995). Unfortunately, antidrug attitudes have softened markedly among young Americans over the past several years, perhaps because each cohort experiences "generational forgetting," rendering them oblivious to the harmful effects of

drug use as witnessed by the cohort that preceded them (Johnston, quoted in Goldberg, 1996). More accepting attitudes about drug use are clear not only in the annual high school survey but also in surveys of 9- to 18-year-olds sponsored by Partnership for a Drug-Free America (1996), which found that compared with teenagers at the beginning of the 1990s, teenagers today "are less likely to consider drug use harmful and risky [and] more likely to believe that drug use is widespread and tolerated." A number of experts believe that, because of this overall shift in attitude, today's cohort of younger adolescents are going to have "their own epidemic" of drug addiction, which is already beginning with increased early experimentation (Wren, 1996).

The consequences of such an epidemic will be evident not only in increased addiction but in reduced learning as well. As one review put it, drug use undermines "the attainment of skills and the mastery of new material [that are] so important to self-esteem and academic progress, the acquisition of new material, and the type of learning that requires abstract reasoning" (Schilit & Gomberg, 1991). Thus, early drug use slows the acquisition of knowledge, the ability to reflect on that knowledge, and the growth of mature judgment that leads to reasoned conclusions. It also seems likely that these same cognitive impairments caused by early drug use play a role in the later development of drug abuse. This may help explain why "the longer a person delays initiating drug use, the less likely he or she is to become a chronic user" (Schilit & Gomberg, 1991).

Indeed, maturation and reason are critical factors in adolescents' being able to avoid many potential hazards, from pregnancy to juvenile delinquency, from crash diets to overwhelming despair. In the next chapter we will look at cognitive maturation, including logic, planfulness, objective analysis, and at education. Together these can foster wise choices instead of destructive ones.

SUMMARY

Puberty

1. While the sequence of pubertal events is similar for most young people of both sexes and in every culture, the timing of puberty shows considerable variation. Normal young people experience their first body changes any time between the ages of 8 and 14.

2. The individual's sex, genes, body type, nutrition, and physical and emotional health all affect the age at which puberty begins. The secular trend over the last 100 to 200 years to earlier puberty is a result of improved nutrition and medicine. Girls typically begin puberty ahead of boys, and children with more body fat begin earlier than those who are lean. Consequently, some young women are, essentially, full-grown by age 13, while some young men still are growing at age 18.

3. Puberty is initiated by the production of hormones in the brain. Four of the most important hormones are gonad releasing hormone, growth hormone, testosterone, and estrogen. While performing their biological functions, these hormones can make adolescents' moods more volatile. However, their impact on emotions is powerfully mitigated or potentiated by the adolescent's perception of the obvious changes occurring in his or her body and the particular circumstances in his or her life.

4. The growth spurt—first in weight, then in height—provides the first obvious evidence of puberty, although some hormonal changes precede it. During the year of fastest growth, an average girl grows about 3½ inches (9 centimeters), and an average boy, about 4 inches (10 centimeters).

5. The growth spurt usually affects the extremities first and then proceeds toward the torso. By the end of puberty, the lungs and heart also change in size and capacity.

6. During puberty, all the sex organs grow larger as the young person becomes sexually mature. Menarche in girls and spermarche in boys are the events usually taken to indicate reproductive potential, although full fertility is reached years after these initial signs of maturation. Young people's attitudes toward their bodies' physical changes vary depending on social, cultural, and family factors.

7. Most secondary sex characteristics—including changes in the breasts and voice and the development of pubic, facial, and body hair—appear in both sexes, although there are obvious differences in the typical development of males and females.

8. As the body changes, so must the individual's body image. For many adolescents, this is problematic, because their actual new shape and appearance are not what they expected or what the cultural ideal promotes.

9. While most teenagers do not have an especially difficult adolescence, early-maturing girls and late-maturing boys are more likely to experience stress because of their off-time physical development. For both sexes, the ecological context—including family interactions, the transition from elementary to junior high school, and cultural values—can ameliorate or exacerbate the problem.

10. Coming at a time when the child is confronted with the physical changes of puberty and their psychological impact, sexual abuse in adolescence can be particularly devastating. The effects of sexual abuse largely depend on the nature of the abuse, its duration, and the emotional support the adolescent receives.

Health, and Hazards to It

11. To fuel the growth of puberty, adolescents experience increasing nutritional demands for vitamins and minerals as well as for calories—more than at any other period of life.

12. Cultural pressures make many girls particularly concerned and unhappy about their weight, setting the stage for potentially dangerous dieting and possible eating disorders.

13. Drug use of some kind occurs among most adolescents, and after a long period of decline, drug use of all kinds has increased in the 1990s. This rise has been accompanied by a gradual liberalizing of teenagers' attitudes toward drug use—particularly the gateway drug marijuana—that may foreshadow further increases.

KEY TERMS

puberty (385)
menarche (387)
secular trend (387)
GnRH (gonad releasing hormone) (388)
GH (growth hormone) (388)
testosterone (388)
estrogen (388)
growth spurt (389)
primary sex characteristics (391)
spermarche (391)
secondary sex characteristics (393)
body image (394)
childhood sexual abuse (398)
gateway drugs (404)

KEY QUESTIONS

1. What is the usual sequence of biological changes during puberty?

2. What is the "secular trend" and why does it occur?

3. What factors make puberty occur early, late, or on time?

4. What are the main changes that characterize the growth spurt?

5. How is the sexual maturation of males and females similar, and how is it different?

6. How do adolescents react to the changes in their body shape and size?

7. How is body image related to development of self-esteem?

8. How does the age at which puberty occurs interact with psychosocial development?

9. What are the potential consequences of sexual abuse in adolescence, and what factors affect their severity?

10. What are the nutritional needs and possible nutritional problems of adolescence?

11. What have been the recent trends with respect to drug use in adolescence?

12. How might drug use be harmful even if the drugs are legal and used only occasionally?

Adolescence: Cognitive Development

CHAPTER 15

Having a conversation with a 15-year-old about international politics, or the latest rage in music, or the meaning of life is obviously quite different from having a conversation on the same topic with an 8-year-old. Because of advances in their cognitive abilities, adolescents are increasingly aware of both world concerns and personal needs—others' as well as their own—and they are also more adultlike in their use of analysis, logic, and reason.

In acquiring these new intellectual perspectives, however, young people not only move closer to adulthood but also become open to new vulnerabilities, as their thoughts take a more comprehensive, less practical sweep. Many adolescents appear tough-minded: at least their frequent displays of sarcasm, cynicism, and arrogance give this impression. But the opposite is more likely to be true. Adolescents often are self-absorbed, idealistic, troubled by their own introspections, and hypersensitive to criticism, real or imagined.

Recognizing this peculiar mixture of intellectual bravado and fragile self-centeredness is especially important for parents and teachers as they try to guide and advance adolescents' thinking abilities and social problem-solving skills. As we will see later in this chapter, when the school structure, the academic curriculum, or parental advice does not recognize the adolescent mindset, education is likely to falter, and good personal guidance is likely to be ignored or resisted. Recognizing the nature of their mindset is also crucial for adolescents themselves, lest they be misled by their bravado to make risky choices that may compromise their future.

Adolescent Thought

Every basic skill of thinking, learning, and remembering that advances during the school-age years (see Chapter 12) continues to progress during adolescence (Keating, 1990). Selective attention becomes more skillfully deployed, for instance, enabling adolescents to do their homework surrounded by peers or blaring music (or both), if they are motivated to do so. Expanded memory skills and a rapidly growing knowledge base advance adolescents' ability to relate new ideas and concepts to old ones, enhancing their understanding of everything from calculus and chemistry to fads and friendship. Language mastery continues, vocabulary grows at an accelerating rate, and many adolescents begin to establish a personal style in their

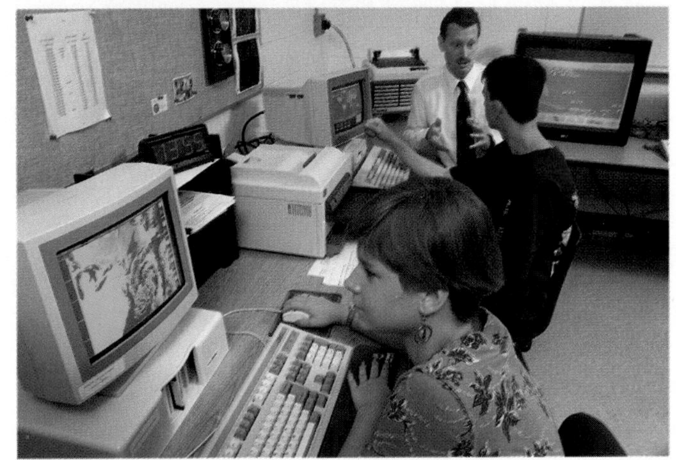

Thanks to a satellite feed from high-tech computers and monitors, these students are learning the art and science of weather forecasting. Such an educational event is well-suited to the adolescent mind, which in its expansive knowledge of relevant factors, ability to remember several interrelated variables, and capacity to imagine possibilities is notably different from that of the younger schoolchild.

speech and writing: poets, diarists, and debaters emerge in every classroom. In addition, advances in metacognition—the ability to think about thinking—deepens adolescents' ability to play at "the game of thinking" (Flavell, 1992) (see p. 326).

However, the fact that adolescence includes an ongoing maturation of earlier cognitive skills does not capture the essence of adolescent thought. Something new emerges during puberty and beyond, a different type of thinking recognized by all theorists, researchers, and practitioners who specialize in the study of adolescent cognition.

New Intellectual Powers

Piaget described the reasoning that characterizes adolescence as **formal operational thought**—in his theory, the fourth and final stage of cognitive development, which arises from a combination of maturation and experience. Information-processing researchers likewise see in adolescents' thinking a new and higher level of cognition attained by the accumulated improvement in specific skills. Sociocultural theorists similarly point to mental advances resulting from the transition from primary school to secondary school that occurs worldwide in early to middle adolescence. But no matter what the explanation, almost everyone agrees that adolescent thought is qualitatively different from children's thought (Andrich & Styles, 1994; Byrnes, 1988; Kitchener & Fischer, 1990; Overton, 1990; Inhelder & Piaget, 1958; Siegler, 1991).

For many developmentalists, the single most distinguishing feature of adolescent thought is the capacity to think in terms of *possibility* rather than only in terms of reality. As Flavell and his colleagues (1993) explain:

> The elementary school child's characteristic approach to many conceptual problems is . . . an earthbound, concrete, practical-minded sort of problem-solving approach, one that persistently fixates on the perceptible and inferable reality right there in front of him. . . . The child usually begins with reality and moves reluctantly, if at all, to possibility; in contrast, the adolescent or adult is more apt to begin with possibility and only subsequently proceed to reality. . . . [R]eality is seen as that particular portion of the much wider world of possibility that happens to exist or hold true in a given problem situation.

This ability to think in terms of possibility allows adolescents to fantasize, speculate, and hypothesize much more readily and on a much grander scale than children, who are still tied to the tangible reality of the here and now. Adolescents can, and do, break free from the earthbound, traditional reasoning of the schoolchild, soaring into contrary notions and ethereal dreams quite apart from conventional wisdom.

formal operational thought Piaget's term for the fourth period of cognitive development, characterized by hypothetical, logical, and abstract thought.

hypothetical thought Thought that involves propositions and possibilities that may or may not reflect reality.

Hypothetical Thinking

It may seem paradoxical, but such liberation of thought actually leads to a more logical cast of mind. Consider the following two propositions.

> If elephants are bigger than dogs
> And dogs are bigger than mice
> Then elephants are bigger than mice.
>
> If mice are bigger than dogs
> And dogs are bigger than elephants
> Then mice are bigger than elephants. [adapted from Moshman & Franks, 1986]

Evaluating these two propositions, a school-age child is likely to conclude that the first is logical because each part of it is true, and it leads to a true conclusion. The same child is also likely to reject the second proposition because it conflicts with reality. By contrast, an adolescent is more willing to play with possibilities by imagining a world in which enormous mice tower over minuscule elephants—and conclude that the second proposition, while inconsistent with the actual world, is nevertheless perfectly logical, because the word "if" opens the way to untold possibilities (Moshman, 1990). In reasoning this way, the adolescent is demonstrating a capacity for **hypothetical thought**, that is, thought that involves reasoning about propositions that may or may not reflect reality. For the younger child, imagined possibilities (such as in pretend play) are always tied to the everyday world as he or she knows it or wishes it to be. For the adolescent, possibility takes on a life of its own in which the here and now is only one among many alternative possibilities.

The adolescent's ability to ignore the real and think about the possible is clear in this example from Flavell and his colleagues (1992). If an impoverished college student is offered $50 to argue in favor of the view that government should *never* give or lend money to impoverished college students, chances are that he or she could earn the money by providing a convincing (if insincere) argument. By contrast, school-age children have great difficulty arguing against their personal beliefs and self-interest.

The ability to divorce oneself from what one believes and argue as if one believed something different or contrary makes adolescents much more interesting and adept as participants in intellectual bull sessions or as partners in debate. It also leads them—sometimes unhelpfully—to turn every assumption on its head and critically examine the logic of every belief.

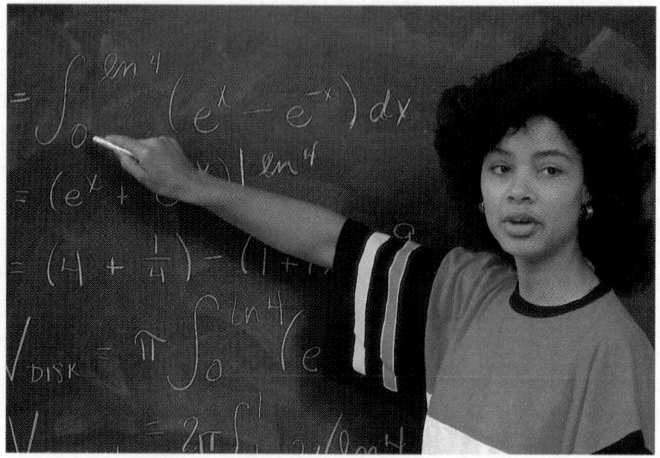

This high school student explains a calculus problem, a behavior that requires a level of hypothetical and abstract thought beyond that of any concrete operational child—and beyond that of most adults.

deductive reasoning Reasoning that draws logical inferences and conclusions from general premises.

This newfound hypothetical thinking, which on any given issue might lead a person to simultaneously perceive that all ideas are possible, justifiable, and at the same time questionable, can make reflection about any serious issue a complicated and wrenching process. Such complications are illustrated on a personal level by one high school student who wanted to keep her friend from making a life-threatening decision but did not want to judge her, because

> to . . . judge [someone] means that whatever you are saying is right and you know what's right. You know it's right for them and you know it's right in every situation. [But] you can't know if you are right. Maybe you are right. But then, right in what way? [quoted in Gilligan et al., 1990]

The capacity to think hypothetically also means that adolescents can analyze current realities within the framework of abstract values, such as those concerning love, justice, the role of the divine, and the meaning of human life. Consequently, they are much less likely to accept current conditions just because "that's the way things are": instead, they are willing to criticize "the way things are" in light of hypotheses about how things would be in a world in which justice was realized, people were always sincere, or the meaning of human life was truly recognized. This is one reason that adolescence is often a time of agonized reflection about the world and one's place in it, confronting the adolescent with thoughts and feelings that are novel, provocative, and sometimes frightening.

Deductive Reasoning

As we saw in Chapter 12, during the school years, children make great strides in *inductive reasoning*, becoming increasingly able to use their accumulated knowledge of concrete facts, as well as their personal experience, to reach solid conclusions about their practical experiences with objects, people, natural events, and school subjects. In essence, their reasoning is of the order "If it looks like a duck and quacks like a duck, then it must be a duck."

During adolescence, as young people develop their capacity to think hypothetically, they soon become more capable of **deductive reasoning** (Byrnes, 1988). That is, they can begin with a general premise or theory and draw logical inferences from it, and then test the validity of those inferences.

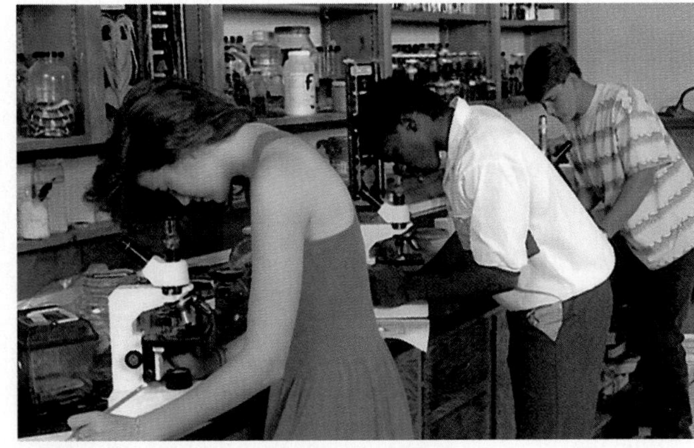

One common high school chemistry-lab assignment is to analyze an unknown substance by testing its reaction to various other substances and conditions. The kind of deductive experimental reasoning this involves is a hallmark of formal operational thought. (Without such lab experiments, high school science becomes a matter of memorizing symbols rather than understanding science.)

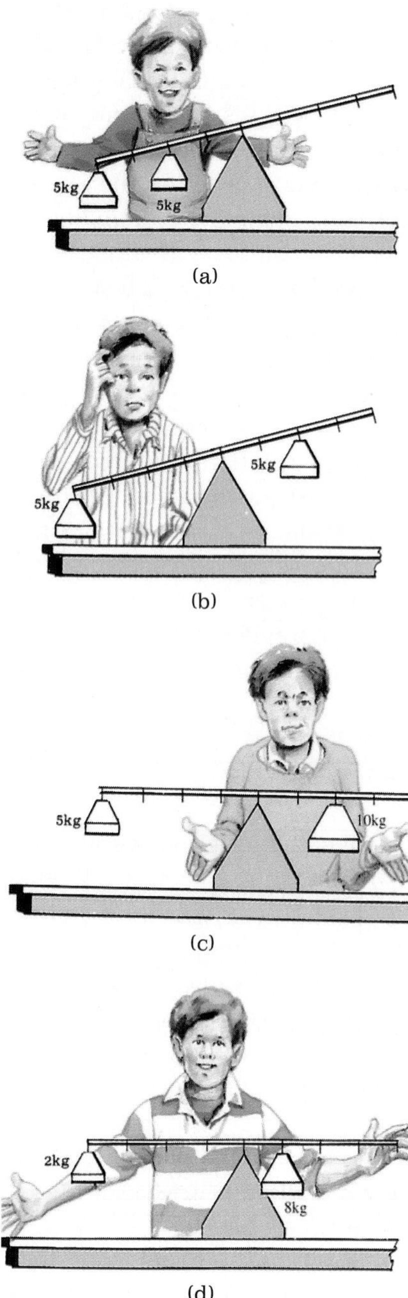

FIGURE 15.1 These drawings illustrate Piaget's balance-scale test of formal reasoning, as it is described in the text and is attempted by a 4-year-old (*a*), a 7-year-old (*b*), a 10-year-old (*c*), and a 14-year-old (*d*).

To see the developmental path of inductive and deductive reasoning more clearly, let's look at an example of how children of various ages might try to figure out what species dolphins belong to. Preschoolers are quick to believe that dolphins are fish—because dolphins, like other fish, live in the water—and they then become resistant to any contrary suggestions regarding these obviously fishlike creatures. School-age children might have heard that dolphins are mammals and might have noticed that dolphins breathe air. Accordingly, with those two bits of evidence, they might deduce that dolphins are mammals rather than fish and lord it over any younger child so foolish as to disagree.

In contrast, the adolescent is likely to consider many alternative hypotheses, recognizing that until the evidence is in, dolphins could be some form of fish, or mammal, or reptile, or even a bizarre type of bird (perhaps a distant relative of the penguin). Proceeding logically and deductively on the premise that dolphins are mammals, the adolescent looks for the general principle that distinguishes mammals from other categories and reasons that "If dolphins are mammals, then they nurse their young." This would lead the adolescent to seek evidence that mother dolphins indeed suckle their pups and to ignore irrelevant data, such as that dolphins swim like fish or have lateral fins that look like vestigial wings. The route of deductive reasoning is from the general principle, which may or may not agree with received wisdom or with everyday observation, to the specific application, rather than vice versa.

Deductive thinking is one of the hallmarks of formal operational thought. Piaget, like theorists from the other cognitive perspectives, believed that the person who thinks at this stage is able to follow the formal rules of logic, understanding such propositions as "If mice are bigger than dogs" and then applying both deductive and inductive reasoning to address various questions related to those propositions.

Piaget devised a number of famous tasks involving principles of chemistry, physics, and the like to study how children of various ages reasoned hypothetically and deductively. In one experiment, for example, children were asked to balance a scale with weights that could be hooked onto the scale's arms (see Figure 15.1) (Inhelder & Piaget, 1958). Mastering this problem requires realizing that the heaviness of the weights and their distance from the center interact together to affect balance. This understanding is, of course, completely beyond the ability of preschoolers, who, in Piaget's research, randomly hung different weights on different hooks.

By age 7, however, children realized that the scale could be balanced by putting the same amount of weight on both arms, but they didn't realize that the distance of the weights from the center of the scale is also important. By age 10 (near the end of the concrete operational stage), children were often able to realize the importance of the weights' location on the arms, but their efforts to coordinate weight and distance-from-the-center to balance the scale involved trial-and-error experimentation, not logical deduction.

Finally, by about age 13 or 14, some children hypothesized the general principle that there is an inverse relationship between a weight's distance from the center of the scale and the force that it exerts. By systematically testing this hypothesis, they correctly concluded that a mathematical relation exists between weight and distance-from-the-center and were able to

solve the problem accurately and efficiently. Piaget attributed these changes in problem solving to the growth of logical reasoning skills pertinent to each stage of cognitive growth.

These hypothetical, and deductive, reasoning skills help to make the adolescent a more flexible, resourceful thinker. At least this is true for some adolescents, some of the time. Recent research has shown that the growth of formal reasoning abilities is far slower and less complete than many developmentalists—most particularly Piaget—believed it to be (Keating, 1990). Many older adolescents, as well as adults, do poorly on standard tests of deductive reasoning skills, such as the balance-scale task, which suggests that these cognitive gains are not always accomplished during adolescence nor are they necessarily acquired by all people. Moreover, a teenager who is quite adept at using deductive reasoning to figure out a math problem may have considerably greater difficulty deducing the solution to a problem in biology, or assessing the ethical dimensions of various approaches to national health insurance, or determining the most effective way to deal with a complex and ambiguous social situation. In other words, adolescents seem to apply formal logic to some situations but not to others, with their reasoning depending much more on their idiosyncratic intellectual endowments, experiences, talents, and interests than on the application of formal reasoning skills alone.

Thinking About Oneself

Advancing to the realm of the possible, the hypothetical, and the abstract has important personal consequences for adolescents, permitting them a new degree of self-scrutiny. Teenagers ruminate at length about how they are regarded by others; they try to sort out conflicting feelings and motives about parents, school, and close friends; and they think deeply about their future possibilities. Analyzing one's thoughts and feelings, forecasting one's future, and reflecting on one's experiences underlie the greater reflection and self-awareness—and enhanced capacity for self-centeredness—that distinguish adolescence.

These new ventures in introspection are an essential part of the adolescent's expanding self-awareness, but they are often distorted by **adolescent egocentrism**, a self-view in which adolescents tend to regard themselves as much more central and significant on the social stage than they actually are (Elkind, 1967, 1984). In addition, they tend to hypothesize what others might be thinking (especially about them) and then take their hypothesis to be fact. According to David Elkind (1978), who first described this trait, adolescent egocentrism occurs because adolescents fail

> to differentiate between the unique and the universal. A young woman who falls in love for the first time is enraptured with the experience, which is entirely new and thrilling. But she fails to differentiate between what is new and thrilling to herself and what is new and thrilling to humankind. It is not surprising, therefore, that this young lady says to her mother, "But, Mother, you don't know how it feels to be in love."

There are several possible outgrowths of adolescent egocentrism. One is the **invincibility fable**, in which some young people feel that they will never fall victim, as others do, to the consequences of dangerous or illegal behavior. Consequently, they take all kinds of risks, falsely secure in their

adolescent egocentrism A characteristic of adolescent thinking that sometimes leads young people to focus on themselves to the exclusion of others, believing, for example, that their thoughts, feelings, or experiences are unique.

invincibility fable The fiction, fostered by adolescent egocentrism, that one is immune to common dangers, such as those associated with unprotected sex, drug abuse, or high-speed driving.

personal fable The egocentric idea, held by many adolescents, that one is destined for fame and fortune and/or great accomplishments.

imaginary audience Refers to the egocentric idea that one is constantly being scrutinized by others. This belief often leads adolescents to fantasize about people's reactions to their appearance and behavior.

The invincibility fable leads some teenagers to try drugs older people would not, which is one reason each new generation seems to have a new "drug of choice." Even when the harmful effects of a particular drug are well known, some adolescents continue to use it, convinced that they can "handle it."

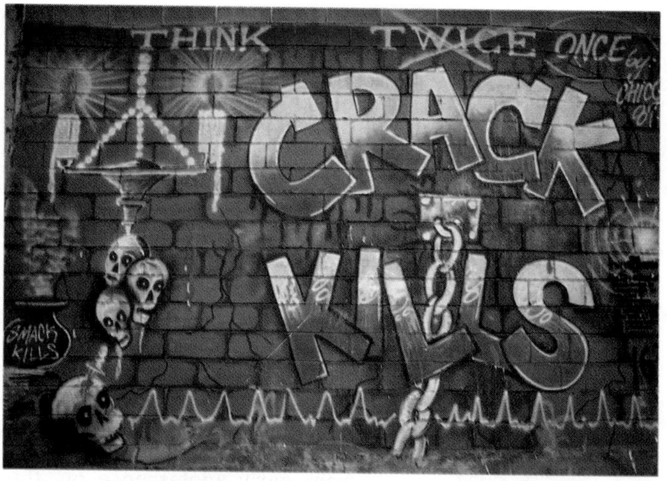

belief in their own immunity. Many adults find evidence for the invincibility fable in, among other things, teenagers' high rates of smoking (despite most adolescents' awareness of the health risks of smoking), unsafe sexual behavior (despite risks of pregnancy and of sexually transmitted diseases), and dangerous driving.

Another example of adolescent egocentrism is the **personal fable**, through which adolescents imagine their own lives as unique, heroic, or even mythical. They perceive themselves as different from others, distinguished by unique experiences, perspectives, and values. Sometimes adolescents see themselves as destined for honor and glory, discovering a cure for cancer, authoring a masterpiece, influencing the social order. As one high school student expresses it, her goal is to

> affect . . . as many people as possible. . . . I see myself in a big way saying big things. I see myself going to school for a long time and learning a lot. I want to write a book that is very, very solid and hard to say No, you are wrong. So that I can give it to the president and say, Look, Mr. President, you are wrong and you are going to hurt all these people and you are going to hurt yourself. I want to say something big. I want to change the world. [Gilligan et al., 1990]

Other adolescents see themselves destined for fame and fortune, becoming a rock or movie star, a sports hero, a business tycoon, or whatever else will make millions (sometimes having already decided that a high school education is a waste of time).

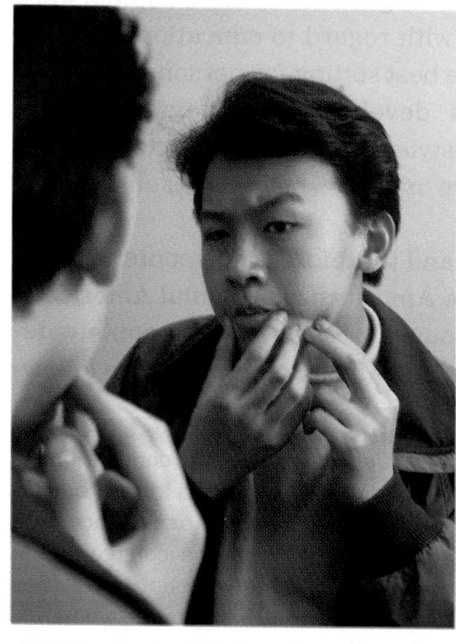

Although it is generally girls who are considered to be overly aware of minor flaws in their complexion or attire, the truth is that boys also pay exaggerated attention to their appearance. The cognitive capacity to think about oneself in egocentric terms makes many young people of both sexes spend hours combing their hair, adjusting their clothing, and searching for blemishes.

A third dimension of adolescent egocentrism is called the **imaginary audience**, created as adolescents fantasize about how others react to their appearance and behavior. The imaginary audience arises from adolescents' assumption that other people are as intently interested in them as they themselves are. At times, this can cause teenagers to enter a crowded room with the air of regarding themselves as the most attractive human being alive. More often, it makes them view their having something as trivial as a slight facial blemish or a spot on their shirt as an unbearable embarrassment that everyone will take note of and judge. The acute self-consciousness resulting from the imaginary audience reveals that young people are often not at ease with the broader social world, which is part of the reason many of them devote so much attention to their hair, clothing, and other aspects of their physical appearance before going out in public.

ego-involvement learning An educational strategy that bases academic grades on individual test performance, with students competing against each other.

task-involvement learning An educational strategy that bases academic grades on the mastery of certain competencies and knowledge, with students being encouraged to learn cooperatively.

procedures. Particularly antithetical to the facilitation of a smooth person-environment fit may be the bureaucratic structure of the educational setting: many secondary schools attempt to educate more than a thousand students, each of whom travels from room to room to learn in 40-minute segments from numerous teachers—some of whom do not even know the names, much less the personality traits, intellectual interests, and personal aspirations, of their students (Carnegie Council, 1989; Eccles et al., 1993).

In addition, the emphasis of most American schools is on what some educators refer to as **ego-involvement learning**, that is, learning in which academic grades are based solely on individual test performance, and students are ranked against each other. In this setting, students who try to succeed and fail to do so experience embarrassment as well as a low grade, while exceptionally good students risk being ostracized as "brainiacs," "geeks," or "nerds." In such competitive conditions, many students—especially girls and students from minority backgrounds—find it easier, and psychologically safer, not to try, therefore avoiding the potential pains of both success and failure. Particularly when young people first enter secondary school, academic self-confidence typically dips, with many students feeling less able, less conscientious, and less motivated than they did in elementary school (Eccles, 1993; Hickson et al., 1994; Stipek, 1992).

One possible antidote to this environment is **task-involvement learning**, in which grades are based on acquiring certain competencies and knowledge that everyone, with enough time and effort, is expected to attain. Unlike ego-involvement learning, task-involvement learning, which typically utilizes team research projects, in-class discussion groups, and after-school study groups, allows all students to improve if they cooperate, and one person's success can foster another's. In this situation, when the task is assisting, rather than surpassing, one's peers, the social interaction that teenagers cherish is actually constructive for education.

Of course, such an approach needs to be skillfully coordinated within the classroom and within the school, so that each individual is challenged and motivated by the group (Johnson & Johnson, 1994). Indeed, an exclusive reliance on cooperative learning may hinder individual learning, especially

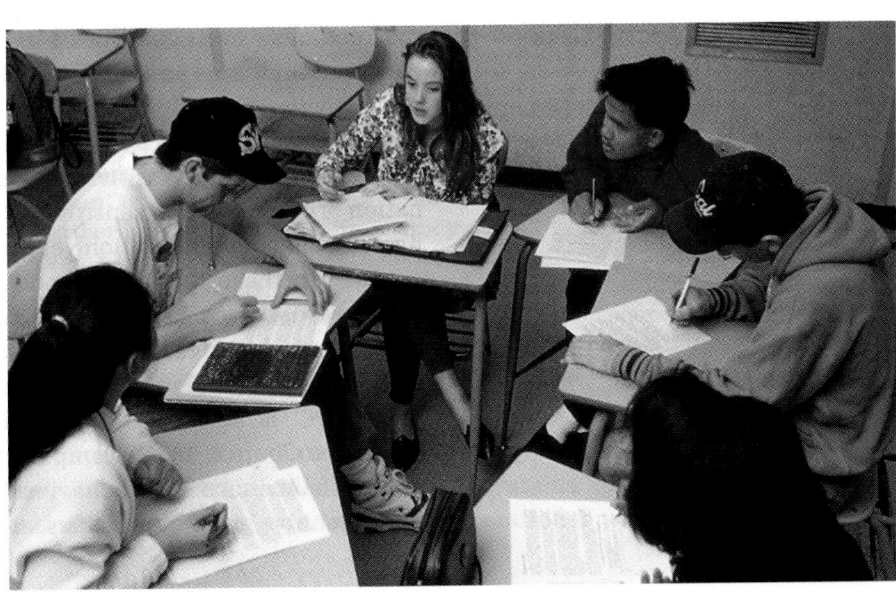

These high school seniors are discussing a group writing project, for which they will all get credit. Such cooperative learning is highly suited to the social motivations of many young people, who learn more, study harder, and comprehend more deeply when they work together.

Recipe for a Well-Structured Classroom

1 Heaping cup of cooperative learning
2 Tablespoons of competition
2 Tablespoons of individual learning

Blend ingredients gently, with thoughtfulness and care.
Amount of each ingredient may be altered to suit individual taste.

Source: Johnson & Johnson, 1994.

FIGURE 15.2 Like every good cook, every good teacher adjusts classroom rules and routines to suit the particular needs of the moment, always keeping in mind that adolescents learn far more from each other than from a distant authority.

for students from families and cultures that value competition and autonomy much more than cooperation and interdependence. However, when students are socially motivated and personally sensitive, as many adolescents are, the best recipe for academic growth includes task-involvement learning as its main ingredient (see Figure 15.2).

Cooperative learning offers an added benefit for students whose secondary school provides their first extended exposure to people of differing backgrounds—economic, ethnic, religious, and racial—and thus to ideas, assumptions, and perspectives that may be quite different from their own. In a cooperative setting, these differences can expand learning opportunities, bringing to the classroom a variety of perspectives that is exciting and enriching rather than threatening. In a competitive setting, however, these differences can lead to rivalry, social distance, and open hostility, as insecure adolescents protect their own self-concept by exaggerating group differences and rejecting anyone who represents a challenge to their sense of identity. When we look at the demographics of the future, with more and more migration from one nation to another, it certainly benefits young people to deepen their cooperative understanding of other groups.

Enhancing Education

At a broader level, schools must be organized so that each student's cognitive needs are likely to be met. How can this be accomplished? First, it is important to realize that schools can make a decided difference—in academic achievement, in self-image, and in future success. Too often, poor student achievement has been blamed on everything but the schools—on the students' native intelligence, on low SES, on minority background, or on family disorganization. All these factors are relevant, of course, but many studies over the past fifteen years have shown that, when these factors are controlled, some schools simply educate much more effectively than others. And virtually every study finds that those schools that are more effective share a central characteristic—educational goals that are high, clear, and attainable, and that are supported by the entire staff, from the principal on down (Mortimore, 1995; Reynolds & Cuttance, 1992; Rutter et al., 1979; Wilson & Corcoran, 1988).

Initially, it may sound contradictory for goals to be both high *and* within reach of everyone. However, many studies find that this combination is not only possible; it is pivotal, particularly for students from whom teachers have traditionally demanded little, such as those from minority ethnic groups or impoverished families (Carnegie Council, 1989; Page & Valli, 1990; Wilson & Corcoran, 1988). A variety of practices—including serious attendance and homework demands, teacher involvement in curriculum decisions, manageable class size, after-school tutoring, and sports and club activities that foster student involvement—make it more likely that teachers will be able to expect and get high achievement from all their students (Liston & Zeichner, 1991; Maughan et al., 1990; Oakes & Lipton, 1990). Perhaps most important, the goals and relevancy of the curriculum must be made explicit. Many young people do not realize the connection between specific course work and later accomplishment: advanced math, foreign language, and laboratory science too often fall by the wayside as irrelevant, especially when they compete with the more compelling needs of adolescents to look good and feel accepted.

adolescents the critical judgment skills that will lead them to reduce their risks or to constrain their options (passing curfew laws, strictly enforcing regulations about underage smoking and drinking, requiring parental consent for contraceptive or abortion services) until their judgment improves (e.g., Baron, 1989; Mann et al., 1989).

Supporting the other side of the debate is research finding that, in many cases, the judgment of adolescents is as good as that of adults. For example, developmentalists have found in careful research that when adolescents are compared with adults on hypothetical situations pertaining to pregnancy, medical treatment, or drug use, they show comparable ability to obtain pertinent information, to evaluate and compare risks and benefits, to anticipate consequences, and to provide reasonable justification for their choices (Beyth-Marom et al., 1993; Jacobs & Ganzel, 1995; Mann et al., 1989). Indeed, when it comes to specific information, adults sometimes are more ignorant about specifics than adolescents, such as knowing about the addictive potential of marijuana or the efficacy of the contraceptive sponge. From this perspective, it is hypocritical to restrict adolescents' freedom to decide while allowing adults to freely make the same choices (Melton & Russo, 1987; Moshman, 1993).

Assuming for the moment that adolescents have the capacity to make certain types of personal decisions as well as adults do, four cautions are in order before it is decided that adolescents should be given carte blanche. First, being able to evaluate a hypothetical situation maturely is a long way from actually thinking through the long-term implications of one's own choices and then acting wisely. Many adolescents take dangerous risks even when they know the potential consequences of their actions (Bauman et al., 1984; Gilbert et al., 1986; Paikoff, 1992). Second, the same research that finds that, *on balance,* adolescents display mature decision-making analysis and good choices also finds that some adolescents, especially those who are under age 16 and who have less education and fewer adults to talk with, show no such maturity (Dryfoos, 1990). Third, just because adolescents sometimes make the same decisions as adults do on certain personal issues does not necessarily make such decisions wise. It may be that adults need education on these issues just as much as adolescents do. Finally, some adolescents' weighing of the risks and benefits of dangerous behavior may entail assumptions about dangerous behavior that others—even peers—do not share (see Public Policy, pp. 428–429).

Taken together, these cautions make clear that, on many issues, adolescents need guidance with their decision making, especially given the many complexities they must face and the fact that puberty is beginning earlier than ever before. This is particularly clear in adolescents' decision making on sexual issues.

Sexual Decision Making: Unanticipated Outcomes

Sexual interest during adolescence is a normal (even essential) part of development, with the particular expression of that interest affected by a host of factors, including biology and culture, family and friends, as well as cognition. Study after study confirms that beliefs, values, and reasoning processes affect what kind of sexual activity adolescents engage in, when, and with whom. Unfortunately, for many adolescents—especially younger ones—their minds are not as ready for sex as their bodies are.

sexually transmitted diseases (STDs) Diseases spread by sexual contact. Such diseases include syphilis, gonorrhea, herpes, chlamydia, and AIDS.

Worldwide, more teenagers are becoming sexually active, and at younger ages, than ever before. The United States seems to be leading this trend. A little over two decades ago, 30 percent of all American 17-year-olds, and 60 percent of all 19-year-olds, had had sexual intercourse (National Center for Health Statistics, 1982). Currently, half of all American 16-year-olds have had sexual intercourse, as have almost 80 percent of 19-year-olds (Benson, 1993; Guttmacher Institute, 1994). Data on younger adolescents are scarce, but one study of all public school students in a poor New England city found that 19 percent of the 12-year-olds had already experienced sexual intercourse (Barone et al., 1996). Sexual activity at a young age raises many issues, including two potential problems that are clearly detrimental to development, sexually transmitted disease and unwanted pregnancy.

Sexually active teenagers have higher rates of gonorrhea, herpes, syphilis, and chlamydia—the most common **sexually transmitted diseases (STDs)**—than any other age group (Centers for Disease Control, 1993). Most cases of STDs are not serious if promptly treated, but some untreated STDs can cause lifelong sterility and life-threatening complications. Sexually active adolescents also risk exposure to HIV, a risk that is increased by three factors that are common among teenagers: infection with other STDs, having more than one partner, and not using condoms during sex. Recent data reveal that, by their senior year of high school, 27 percent of American teenagers have already had four or more sexual partners and that only half used a condom at last intercourse (*MMWR*, March 24, 1995).

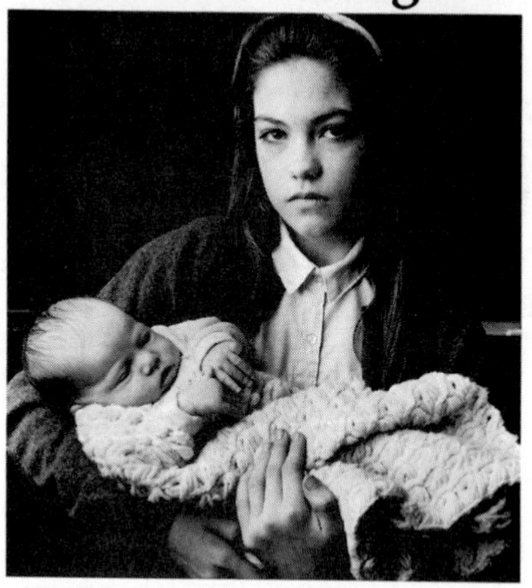

The one on the left will finish high school before the one on the right.
Adolescent pregnancy isn't just a problem in America, it's a crisis. To learn more about a social issue that concerns all of us, write: Children's Defense Fund, 122 C Street, N.W., Washington, D.C. 20001.
The Children's Defense Fund.

Unfortunately, the fact that teenage mothers are less likely to finish high school, find a job, or enter a satisfying marriage is sometimes lost on the young sexually active woman. Unless she is guided to think of the long-term consequences of her decisions, she is more likely to worry about immediate problems — such as what her partner will think if she insists that he wear a condom or what she will say to her parents if she gets pregnant.

The second risk for adolescents is unwanted pregnancy, the negative consequences of which are largely defined by the social context. In the United States in 1960, for example, the teenage birth rate was half again as high as it is today (90 per 1,000 compared to 60 per 1,000 in 1993). Few considered this a major problem, however, because about 80 percent of American teen mothers in 1960 were married, and most of them expected to be housewives all their lives. Today the reverse is true: 80 percent of teenage mothers are unmarried, and almost all—married or not—expect to enter the labor force.

Therefore, in the social conditions of today, teenage pregnancy is a detriment to development, and is an acute problem in the United States, where unmarried teenagers become pregnant more often than in any other nation. Each year, a million American teenagers, about 90 percent of them unwed, conceive. This means that before reaching age 20, one American girl out of three will become pregnant. Very few—perhaps 10 percent—of these pregnancies are intended by both partners (Kost & Forrest, 1995; Marsiglio, 1993). About 35 percent are aborted by choice, another 15 percent are aborted spontaneously (through miscarriage), and the remaining 50 percent are carried to term, with both mother and child likely to experience a wealth of problems (Bachrach et al., 1993).

For the mothers, the most powerful long-term consequence of teenage parenthood is interference with their educational and vocational achievement and a narrowing of their social and personal growth. This is true no matter what the mother's level of family support, income, or intellectual capacity. Having a baby before age 20 reduces the mother's eventual academic achievement by three years, on average (Klepinger et al., 1995), reduces the likelihood that the mother will ever marry, and, if she gets married because she is having a baby, increases her chances of being abused, abandoned, or eventually divorced (Waite & Lillard, 1991).

PUBLIC POLICY

Adolescent Risk-Taking

As we have noted, adolescents often take serious risks with their health and well-being, even though they are aware of these risks. Whether their behavior involves substance abuse, unprotected sex, reckless driving, smoking, or delinquency, adolescents are generally aware of the personal consequences, and potential costs, of their actions (Bauman et al., 1984; Levinson et al., 1995; Gilbert et al., 1986; Paikoff, 1992). This has led many developmentalists to question the maturity and rationality of adolescents' reasoning and judgment processes. Some maintain, for example, that teenagers are influenced by the "invincibility fable" that often accompanies adolescent egocentrism; others assert that adolescent judgment suffers from excessive peer pressure, uncontrolled emotional impulses, or an irrational need to display masculine bravado (since much adolescent risk-taking is by young men). Against this backdrop of professional wisdom, parents and educators merely ask "What's *wrong* with these kids?"

One researcher argues, however, that risky behavior by many adolescents may be entirely rational. Using the principles of rational choice theory—which hypothesizes that people make decisions on the basis of their own personal cost-benefit analyses—William Gardner (1991; Gardner et al., 1991) proposes that many teenagers believe that the benefits of their dangerous behavior outweigh the potential costs.

Gardner takes the problem of AIDS prevention as an example. Since virtually all adolescents know that the decision to have sex without using a condom incurs the risk of contracting the deadly HIV virus, their risk-benefit calculus involves comparing the immediate benefits of unprotected sex, such as ensuring maximum pleasure, against its most obvious cost, the threat to life expectancy. For most of us, the outcome of this analysis is obvious: abstinence or condom use seems like a small price to pay compared to the threat of contracting a fatal disease. But Gardner argues that many adolescents may not view the calculus that way:

> Individuals who perceive that they are likely to die from a cause other than AIDS [may] discount the risks of AIDS, because they are risking a shorter life expectancy. The reward for abstaining from a preferred act of unprotected sex is increased life expectancy. But a high probability that one might avoid HIV infection tonight, only to die tomorrow of another cause, diminishes this reward. . . . The life expectancy of an American adolescent varies significantly as a function of gender, ethnic group, and economic status. For example, black and other minority males are far more likely to die from homicide than are their white peers. . . . Life can be startlingly brief in many American neighborhoods. . . . Rational choice theory predicts that young people facing significant competing death risks in these ecologies will be disposed to accept health risks such as AIDS. [Gardner et al., 1991]

In short, if your friends and peers are dying all around you from violence, and you imagine that you might soon die this way too, you may perceive little reason to postpone or diminish present pleasures in order to ward off some distant threat.

Moreover, Gardner argues that for most adolescents generally, the future they put at risk is a murky one at best. Compared with adults, who see their futures outlined by

The likely consequences for the child at every stage of development are even more troublesome. In research that controlled for the mother's education, income, and race, investigators found that babies of teenagers are at higher risk of prenatal and birth complications, including low birthweight and brain damage, than are infants whose mothers are older when they first give birth. As they move through childhood, those born to young mothers are mistreated more often and become high achievers less often. In adolescence, they are more likely to become drug abusers, delinquents, drop-outs, and—against their mother's advice—parents themselves (Furstenberg et al., 1987; Hoffman et al., 1993).

Why are the rates of STDs and unwanted pregnancy so much higher among adolescents than adults, even though adults are more active sexually? A logical explanation would be that adolescents are uninformed about sex, or that contraception is unavailable to them. But this explanation does

These Brazilian boys are "surfistas," who ride on top of high-speed train cars in Rio de Janeiro. The risk of falling or hitting high-voltage wires intensifies the thrill, adds to one's status—and kills some teenagers each year. Nearly every culture has high-risk activities that certain adolescents engage in even though they are fully aware of the dangers. In some cases, it may be that risk-takers feel that, given their future prospects, they have little to lose.

the educational, marital, child-rearing, and professional commitments they have already made, adolescents look at their future with considerable uncertainty about what their lives will be like. For most middle- or upper-income youth, at least, there is reason to be optimistic, even in the face of indeterminacy: their prospects, through largely undefined, seem promising. But for many lower-income youth, life's prospects may seem not only indeterminate but also not very hopeful. In such cases, the perceived cost of risking the future for immediate and dangerous pleasures may seem slight.

Gardner argues that if appropriate public policies are instituted, this unfortunate risk-benefit calculus might change. Some of the policies he urges are fairly conventional: greater AIDS education emphasizing abstinence and safe sex, and intensive counseling to adolescents who are at greatest risk of becoming infected. He also supports a program of free, frequent HIV testing as one means to reassure high-risk adolescents that they are uninfected because "abolition of uncertainty would increase a person's estimate of his or her own life expectancy, which would increase the cost of additional unprotected sexual contact." But Gardner also argues that wise public policy must look beyond the immediate concern with AIDS education and counseling and address how adolescents—especially those from disadvantaged groups—assess their own future life prospects:

> If we do, we may notice that we are creating a future for many of our young that is uncertain, impoverished, and dangerous. This future engenders indifference to the health consequences of behavioral choices—how much should one sacrifice to protect a life that might be nasty, brutish, and short? And, as rational choice theory indicates, the longer we defer action to prevent AIDS among adolescents, the more difficult prevention becomes. [Gardner et al., 1991]

not fit the reality of the 1990s. Today, some form of sex education is nearly universal in American high schools. In addition, the media are quite open about sexuality—particularly the problems it can cause—and condoms are available in drug stores, supermarkets, and even in some schools.

Clearly, then, information about, and protection from, STDs and pregnancy are within reach of today's teenager. However, various studies have found that merely understanding the facts of sexuality and knowing how to obtain contraception do not necessarily correlate with more responsible and cautious sexual behavior (Hanson et al., 1987; Howard & McCabe, 1990). Nor do adolescents always learn from their mistakes. After being treated for an STD, they are two to three times more likely than adults to become reinfected (Bates, 1995). Similarly, programs to help adolescent mothers usually succeed in improving their parenting skills and furthering their education, but they are less successful in reducing repeat pregnancies (Polit, 1989).

Reasoning About Sex

To understand why knowing sexual facts does not always affect adolescent sexual behavior, we need to consider the limitations that typify adolescents' judgment, particularly that of younger adolescents (Gordon, 1990). Young adolescents, who are only beginning to acquire the reasoning skills needed for mature decision making, may have difficulty envisioning alternatives, and then evaluating each one, to choose the best option among many. Instead, they may focus on immediate considerations rather than on future ones. Consequently, they may focus on the concrete difficulty and inconvenience of using contraception, failing to measure them against the hypothetical difficulty and inconvenience of pregnancy.

Similarly, when imagining possible parenthood, they may fail to measure the status and love that they think a baby might bring against the consuming responsibility that caring for an infant entails. This failure to think through possible consequences helps explain an interesting finding. Although few teenage girls actually want a baby, most who do not practice contraception think that having a child would not necessarily be so bad (Hanson et al., 1987). This view is captured in the following response of an eighth-grader who was asked what she thought her reaction would be if she were to become a mother in the near future:

> I think having a baby would really make me a little happier because it would make me have something of him, . . . I know I'm young and everything, but I know I could take care of it because my mother had me when she was young. I am just about the same age when she got pregnant. And she did okay, I'm here. [quoted in Taylor et al., 1995]

In other words, not having carefully considered the future implications of an unwanted pregnancy, many teenagers do not regard the possibility of one as something to be assiduously avoided. By eleventh grade, the girl quoted above had had a child and had dropped out of school. And contrary to her stated confidence that she could take care of a child if she were to become a teenage mother, she was forced to temporarily give up custody of her baby because of her inability to care for the child properly.

Another problem with adolescents' thinking about sex is that their sense of personal invincibility is compounded by difficulty in grasping the logical probability of risk assessment. This combination makes clear thinking about risks much more difficult. Many, particularly those under age 16, seriously underestimate the chances of pregnancy or of contracting a disease, reasoning that either is unlikely to occur from just one episode of unprotected intercourse (Voydanoff & Donnelly, 1990). Then, if no misfortune occurs, the adolescent does the same thing again, reinforced by the belief that past experience has already shown that he or she will not succumb as others do. Similarly, adolescents generally believe that other teenagers are more likely to contract the AIDS virus than they themselves are, with females being especially likely to underestimate their own risk (Moore & Rosenthal, 1991). This is particularly unfortunate since recent findings suggest that females are, for a number of reasons, twice as vulnerable to HIV infection as males (Zierler, 1994).

Because of their faulty risk assessment, many teenagers are incredulous when a pregnancy occurs. Consequently, they take longer than adults to confirm the pregnancy, to seek advice, to obtain prenatal care, or to have an abortion. Each of the delays increases the actual risk of problems.

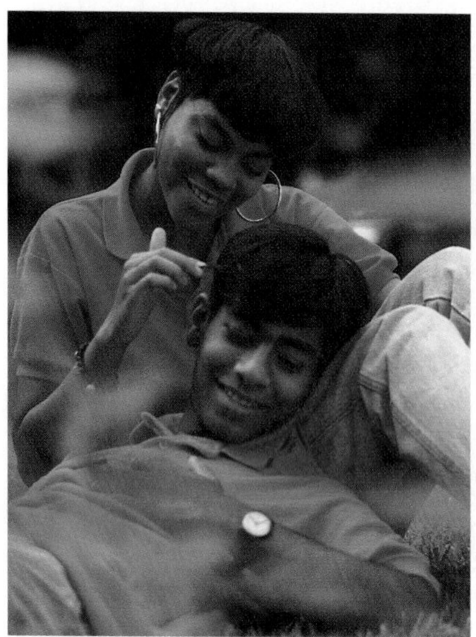

As adults try to warn adolescents of the dangers of early sex, they must also recognize the joys of young love. If they don't, their advice is likely to be rejected as either hypocritical or ignorant.

Promoting Better Judgment

How can adolescents be helped to make more rational decisions about their sexual activity? A first step, according to many experts, is for adults to be more rational in *their* thinking. To begin with, they need to realize that, at some point, almost every adolescent will have some sort of sexual experience. Next, adults have to recognize that adolescents need more than just the facts about sex to guide them: they also need an opportunity to develop the attitudes and social skills that can help them refrain from sexual activity until they are mature enough to understand, and take responsibility for, the consequences of their actions.

Ideally, providing such an opportunity should be the task of parents. Obviously, parents do have some influence on their child's sexual behavior: their overall relationship with the child, and with each other, and their overall values and religious convictions regarding sex, health, and the child's future all have an effect (Hanson et al., 1987). However, few parents do well as sex educators. For one thing, many parents do not begin discussing sexual issues with their child until long after the child has been informed, or misinformed, by friends, intuition, or personal experience. Further, when they do discuss sexual questions, parents often close their eyes to the realities and try to lay down the law rather than help the child think about options. One study summarized the parents' typical message to their daughters as "Don't, and if you do, we don't want to hear about it" (Treboux & Busch-Rossnagel, 1990). Finally, many parents are themselves uninformed, especially if their teenage years occurred before the threat of AIDS, before the new forms of contraception were available, and before "good" girls were sexually interested or "real" boys were sexually careful.

Sex Education

Traditional sex education may be similarly unhelpful, especially when the teacher tries to fit a lot of factual information, mostly about biology, into a ten-hour course (the usual time allotted). Adolescents are often put off by sex-education courses that are merely "organ recitals" or that seem to have little bearing on their actual sexual dilemmas and pressures.

In recognition of these realities, a new form of sex education is emerging that attempts to take into account adolescent thinking patterns, particularly the difficulty many adolescents have in making mature personal choices, in drawing connections between the immediate lure and excitation of sexual behavior and the possible long-term consequences of it, and in placing decisions about sexuality into a broader view of human values and relationships (Katchadourian, 1990). For example, to make the consequences of sexual activity more immediate, the new approach to sex education often uses first-hand accounts in place of textbook examples. Typically, someone with AIDS talks to a class about safe sex; a teenage parent discusses the responsibilities of parenthood; older teenagers explain the kind of social and individual pressures that might push them into becoming sexually active. Intimate topics, such as the proper use of condoms, are discussed candidly and in specific detail.

While concrete examples are stressed, they are not presented in isolation. Formal thinking is also fostered through discussions and specific exercises that help adolescents to weigh alternatives, recognize the rights and

Classes in sex education need to be carefully constructed to be effective. Both classes shown here seem to be efforts in the right direction, but in the one on the left, the humorous emphasis on abstinence may strike some of the students as a naive scare tactic. On the right, a woman explains some of the realities of pregnancy, a personal approach that may be more effective. Note, however, that the woman appears to be much older and more middle-class than the students and perhaps is of a different ethnic background. Also note that there are no boys in the class. Educational leaders are learning to tailor programs to the specific social context, a nuanced approach that is proving far more successful than the "one size fits all" measures sometimes mandated by state legislators or local boards of education.

needs of others, and analyze risks. This new approach also makes extensive use of role-playing, allowing adolescents to act out the various kinds of sexual pressures and problems that they might face in the future and to develop the social skills for responding to them. For example, one of the main concerns of young adolescents is how to say "no" without hurting the other person's feelings or being emotionally rejected. Acting out scenarios that reveal how hard it is for many individuals to be assertive or, alternatively, to accept rejection, helps everyone see the importance of mutual agreement. Students also have the opportunity for private conversations with the teacher or a medical practitioner, so that those who are already experiencing specific problems, such as sexual abuse or a sexually transmitted disease, can get help. Teens in these programs are also strongly encouraged to have explicit discussions with their parents, in part because research makes it clear that many adolescents do not know how their parents really feel about teenage sexual activity and pregnancy (Freeman & Rickels, 1993). Taken as a whole, these various strategies reflect the belief that, as with other aspects of cognition, allowing teenagers the time and opportunity to bring the many aspects of adolescent reasoning to bear on a given issue allows a more mature understanding to develop.

Evidence that such an intensive approach works is accumulating. For example, one British program that provided thirty hours of social-skills training, as well as factual information, reduced the percentage of high school students who became sexually active in the two years after the program from 53 to 42 percent (Mellanby et al., 1995). A review of four similar programs in the United States that were designed to promote abstinence—or, in the absence of abstinence, contraception—found that three of the four were successful in convincing some sexually inexperienced young people to postpone sexual activity and in convincing some who were sexually active to use contraception (Frost & Forrest, 1995). Another study that involved 1,880 teenage boys likewise found that sex education that combined information with social-skills training succeeded in reducing the incidence of intercourse, the number of partners the boys had, and the practice of unsafe sex (Ku et al., 1992). The author of this study concluded:

> In many communities, concerned parents or community members have feared that education about sex or AIDS may increase sexual activity by condoning contraception; this analysis does not indicate such an association. Education about resisting sexual activity can and does coexist with education about contraception and condom use. . . . Abstinence and safe sex can be taught together, just as safe driving and using seat belts can be taught together. The point is that people can be taught how to avoid dangerous situations and, at the same time, how to prevent harm in case they are caught in one.

Overall, the research on these new forms of sex education suggests that they are most effective with adolescents who have not yet become sexually active. For these young people, accurate information and social-skills training not only help them postpone becoming sexually active but also promotes use of contraception and monogamy once they do become sexually active. On the other hand, this research finds that classroom education is unlikely to change the behavior of those teenagers who have already established risky sexual patterns (Kirby et al., 1991). Obviously, then, the timing and specifics of such programs need to be tailored to fit the specific social context. For example, surveys find considerable variation, from school to school and from state to state, in age of first sexual activity (*MMWR*, March 24, 1995). Accordingly, explicit instruction might begin at age 11 in some neighborhoods in Delaware and South Carolina, where sexual activity begins early, and at age 14 in some communities in Utah and Nebraska, where sexual activity begins later. Ideally, sex education begins before students become sexually active, but not so long before that the instruction seems irrelevant. Similarly, these programs need to take into account students' religious values and enlist the parents' involvement, since both factors are key to fostering healthy sexual behavior.

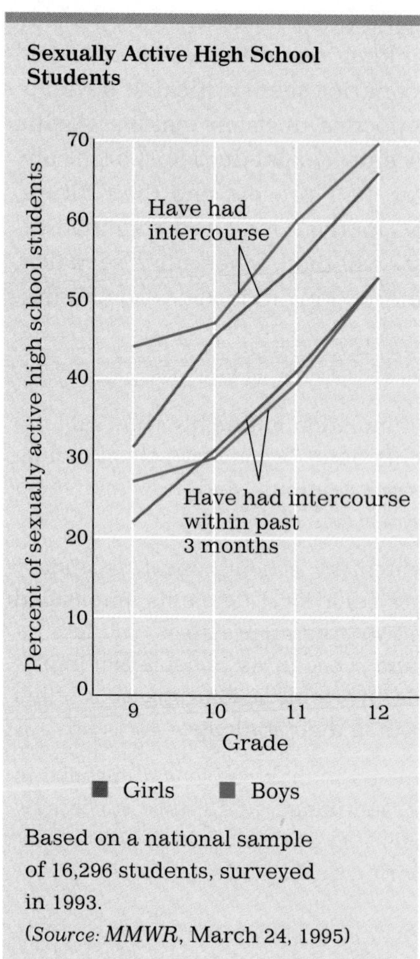

FIGURE 15.3 The narrow sex gap in sexual experience suggests two positive trends. The first is that the double standard—which dictated that "good girls don't do it," but "real boys score"—is disappearing. The second is that, especially as adolescents grow older, it becomes more likely that those who are sexually active will develop ongoing monogamous relationships rather than engaging in casual sex.

Another variable in the social context is the gender of the participants. Traditionally, sex education was focused toward girls, because they were regarded as the moral gatekeepers who bore the responsibility of saying "no" to boys' sexual advances, or at least of ensuring that pregnancy did not occur. Among today's young people, however, sexual decision making is more likely to be shared by both parties, and sexual activity is more apt to be part of a mutual relationship. This is evidenced, in part, by data showing that, unlike in their parents' generation, girls are as sexually active as boys are (see Figure 15.3) (*MMWR*, March 24, 1995). This means that boys should learn the same cautionary messages that girls do. Indeed, some studies find that effective sex education influences boys more than girls, at least if success is measured by postponement of intercourse (Eisen & Zellman, 1992; Frost & Forrest, 1995; Howard & McCabe, 1990).

More and more school systems are revising their sex-education programs to make them more practical and more focused on social interaction. And these efforts are having an impact. In the United States, the pregnancy rate among sexually experienced teenagers reached a high in 1980 (247 per 1,000) and has declined ever since, down 16 percent by 1990 (to 207 per 1,000). However, because the total percentage of sexually active teenagers increased, and the abortion rate fell by about 5 percent, the number of babies born to teen mothers actually rose by 12 percent over the 1980s. Recently, the teenage birth rate itself has been falling, from 62 per 1,000 in 1991 to 59 per 1,000 in 1994. Since the abortion rate has also been falling, this indicates that fewer teenagers overall are becoming pregnant (National Center for Health Statistics, 1996). While such shifts have multiple causes, evidence

suggests that sex education deserves part of the credit. For one thing, young women who have had no sex education, either because they are too young or because they are school drop-outs, are now more likely to become pregnant than those who have had sex education. Another, more telling bit of evidence regards condom use. For both sexes, the proportion of sexually active, unmarried teenagers using condoms has increased dramatically over the past ten years, with rates far surpassing those of teenagers in previous generations and even surpassing those of sexually active, unmarried adults. Further, although only about half the condom users of any age use them completely correctly, some research finds that teenagers are more likely to do so than adults, a further indication that the newer forms of sex education have had an impact (Guttmacher Institute, 1994).

The example of sexual decision making can be seen as applicable to all forms of adolescent decision making, outside the schools as well as inside them. Given the nature of adolescent thinking, open discussion of logical possibilities and the use of hypothetical scenarios seem critical to developing the cognitive maturity required for effective decision making. Again, cognition is not the whole story for adolescents' sexual decisions or for any of the choices teenagers must make about their present and their future. But programs that address the emotional, motivational, and cognitive dimensions of adolescent decision making—whether concerning sexuality, drinking, drug use, or delinquency—are more likely to foster wise and mature judgment.

SUMMARY

Adolescent Thought

1. During adolescence, young people become better able to speculate, hypothesize, and deduce from theories, emphasizing possibility as much as reality. Unlike the younger child, whose thought is tied to tangible reality and "the way things are," adolescents can build formal systems and general theories that transcend (and sometimes ignore) practical experience. Their reasoning can be formal and abstract, rather than empirical and concrete. This ability also enables them to consider abstract principles, like those concerning love, justice, and the meaning of human life.

2. Adolescent thought is marked, and sometimes marred, by adolescent egocentrism. This egocentrism is exemplified by the invincibility and personal fables, both of which may cause teenagers to engage in risky behavior. This egocentrism also helps to account for the self-consciousness of this phase of life.

Schools, Learning, and the Adolescent Mind

3. Adolescence is typically a period of mixed openness and self-consciousness, in which adolescents find themselves eager for intellectual stimulation but highly vulnerable to self-doubt. Many students enter secondary school feeling less competent, less conscientious, and less motivated than they did in elementary school.

4. Compared with elementary schools, most secondary schools have more rigid behavioral demands, intensified competition, more punitive grading practices, and less individualized attention and procedures. Schools that implement cooperative, task-involvement learning show high levels of academic growth in their students.

5. Schools can be organized to make a clear difference in academic achievement, self-image, and future success by setting educational goals that are high, clear, and attainable, and that are supported by the entire staff.

Moral Development

6. Moral reasoning becomes more complex during adolescence. Kohlberg proposed that this development occurs through six stages of increasing complexity, from the elemental "might makes right" to the recognition of universal ethical principles. Despite some criticisms leveled against it, Kohlberg's theory seems to be generally valid, and although males and females may analyze moral problems somewhat differently, neither sex is generally more competent at resolving moral dilemmas.

Adolescent Decision Making

7. Adolescents seem to have mixed abilities with regard to good decision making. On the one hand, adolescence witnesses the growth of many cognitive skills that are essential to good judgment. On the other hand, judgment skills alone do not lead to good decisions, and self-image, peer pressure, and heightened emotion may produce poor decision making.

8. Cognitive and motivational factors can make it difficult for teenagers to make good judgments about their sexual activity, as reflected in the high rates of sexually transmitted diseases and unwanted pregnancy during adolescence. Adolescents who believe that they are not susceptible to AIDS or pregnancy, or who focus exclusively on their own interests and immediate needs, may not take appropriate precautions.

9. The best way to help adolescents avoid problems with sexual behavior may be to help them to develop the social skills to postpone sexual exploration until they are able to reason more maturely. Role models provided by peers seem especially influential, and education that encourages thinking, role-playing, and discussion about sexuality appears to be more effective than traditional sex-education programs.

KEY TERMS

formal operational thought (412)
hypothetical thought (413)
deductive reasoning (414)
adolescent egocentrism (416)
invincibility fable (416)
personal fable (416)
imaginary audience (416)
person-environment fit (418)
ego-involvement learning (420)
task-involvement learning (420)
preconventional moral reasoning (423)
conventional moral reasoning (423)
postconventional moral reasoning (423)
sexually transmitted diseases (STDs) (427)

KEY QUESTIONS

1. What new thinking abilities tend to emerge during adolescence?

2. What is deductive reasoning, and how does it compare with the inductive reasoning of the younger child?

3. What advances occur in the theoretical reasoning of the adolescent? What are the shortcomings that remain in theoretical thinking at this age?

4. What are some of the characteristics of adolescent egocentrism?

5. What are some of the reasons for a decline in academic achievement in high school?

6. How should schools be organized to foster better academic success and supportive social interactions among teens?

7. What are the stages of moral reasoning proposed by Kohlberg, and how do they differ in terms of the reasoning that typifies them?

8. Do adolescents make good decisions, and what factors besides cognitive ability may affect their decision making?

9. What are some of the explanations for the high rates of sexually transmitted diseases and pregnancy in adolescence?

10. What are some of the ways of helping adolescents avoid the hazards of early sexual experience?

Adolescence: Psychosocial Development

CHAPTER 16

The physical changes of puberty initiate adolescence by transforming a childish body into an adult one; and the cognitive developments described in the preceding chapter enable the young person to begin to transcend the realm of concrete thought and to think abstractly and hypothetically. However, it is psychosocial growth—relating to parents with new independence, to friends with new intimacy, to society with new commitment, and to oneself with new understanding—that eventually brings the young person adult status and maturity. Taken as a whole, psychosocial development during adolescence can best be understood as a quest for self-understanding and for identity—that is, for answers to a question that seldom arises in younger years: "Who am I?"

The Self and Identity

The momentous changes that occur during the teenage years—the biological changes of puberty, the transition to larger and more impersonal schools, the negotiation of new sexual and peer relationships, and the beginning of long-term decisions concerning the future—each challenge the adolescent to define who he or she really is (Larson & Ham, 1993). The first step in this process of self-definition is usually an attempt to establish the integrity of one's personality, that is, to see one's emotions, thinking, and behavior as consistent from one situation or relationship to another. Later, the adolescent's search for identity broadens, as the young person attempts to integrate his or her sense of self with the many roles that are associated with becoming a young adult.

Multiple Selves

In the process of trying to find their "true self," many adolescents experience the emergence of "possible selves"—that is, diverse perceptions of who they really are, who they are in different groups or settings, who they would like to become, and who they are afraid of becoming (Markus & Nurius, 1986; Markus et al., 1990). Many teenagers, for example, realize that they play different roles with different people in different circumstances and that, accordingly, their behavior may range from reserved to rowdy, from cooperative to antagonistic, from loving to manipulative. Aware of the in-

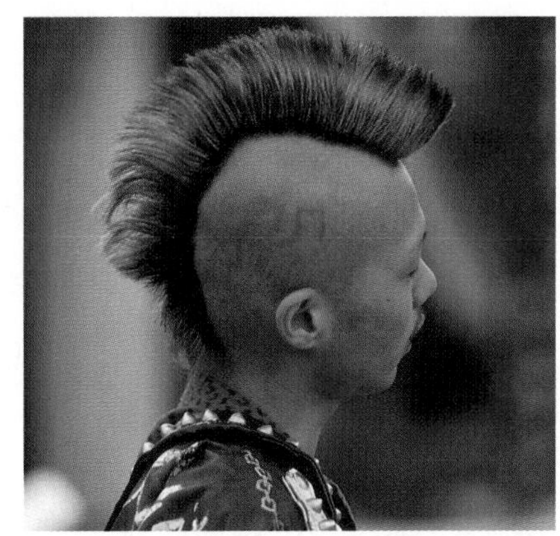

Defining oneself as distinct from adult authority and tradition is one of the tasks of adolescence, although not every adolescent does it in such visible ways as these two young people. Note that the girl is literally closing her eyes to, turning her back on, and keeping her distance from her mother—all figures of speech that denote stubborn independence. The boy's "doin' my thing" is even more obvious in its statement of independence, especially when we realize that the young man lives not in Chicago, London, Toronto, or some other English-speaking city, but in Tokyo.

evitable inconsistencies among these multiple selves, many adolescents ask which one—if any—is the "real me." As one teenager put it, "I'd *like* to be friendly and tolerant all of the time. That's the kind of person I *want* to be, and I'm disappointed when I'm not" (Harter, 1990).

Along with trying to sort through their multiple selves, adolescents frequently take on a **false self**, behaving in ways that they feel are contrary to their core being—even if they are not sure what that core being is. According to one group of researchers (Harter et al., 1996), such false behavior manifests itself in three distinct types. The first type arises from the adolescent's perception that his or her real self is rejected by parents and peers, a perception that is often coupled with the adolescent's own self-dislike. Adolescents who adopt a false self for this reason tend to feel worthless, depressed, and hopeless and engage in high levels of false behavior to hide their true nature and gain the acceptance of others. They also report low levels of real self-understanding. Such adolescents warrant concern.

The second type of false self arises from a wish to impress or please others. This type of false self is quite common among adolescents, and those who behave falsely to gain favor appear to be less debilitated psychologically, and to have greater self-understanding, than those who behave falsely in an effort to achieve validation of self from others. The third type of false behavior is experimentation: the adolescent tries out different behaviors "just to see how it feels." Compared with adolescents who engage in the first two types of false behavior, adolescents who experiment with false selves report the highest levels of self-esteem and self-knowledge. These adolescents are the ones most ready to begin the quest for fuller identity.

Identity

As they try to sort through their multiple selves, adolescents know that they are moving toward the assumption of adult roles and responsibilities and making choices that can have long-term implications for their lives. As a consequence, they begin to think of themselves in far more complex ways than they did a few years earlier. Whereas younger children, for example, may describe themselves primarily in terms of their skills in school, with friends, and perhaps on the athletic field, adolescents distinguish their scholastic competence from their job skills, athletic ability, romantic appeal, moral conduct, peer acceptance, and other aspects of who they are (Harter, 1993b). They also begin to think about their career options, political identifi-

false self A constellation of behaviors that is contrary to one's core being.

cations, religious commitments, and sexual values, and how these fit together with their expectations for the future and beliefs acquired in the past.

In attempting to reconcile these increasingly diverse and complex facets of selfhood, adolescents tackle the psychosocial challenge referred to by Erik Erikson as **identity versus role confusion**. The specific task of this challenge is to integrate the various aspects of one's self-understanding into a coherent **identity**, that is, a definition of oneself that is unique and internally consistent. For developmentalists like Erikson, the search for identity represents a basic human need that begins to be felt in adolescence. Indeed, according to Erikson, the search for identity is the primary crisis of adolescence, a crisis in which the young person struggles to reconcile a quest for "a conscious sense of individual uniqueness" with "an unconscious striving for a continuity of experience . . . and a solidarity with a group's ideals" (Erikson, 1968).

In other words, the young person seeks to establish himself or herself as a separate individual while at the same time maintaining some connection with the meaningful elements of the past and accepting the values of a group. In the process of "finding themselves," adolescents attempt to develop sexual, moral, political, and religious identities that are relatively stable, consistent, and mature. This identity ushers in adulthood, as it bridges the gap between the experiences of childhood and the personal goals, values, and decisions that permit each young person to take his or her place in society (Erikson, 1975).

identity versus role confusion Erikson's term for the psychosocial crisis of adolescence, in which adolescents must determine who they are, combining their self-understanding and social roles into a coherent identity.

identity As used by Erikson, a person's self-definition as a separate individual in terms of roles, attitudes, beliefs, and aspirations.

identity achievement Erikson's term for a person's knowing who he or she is as a unique individual. Identity achievement includes a self-definition that encompasses sexual, moral, political, and vocational identity.

foreclosure Erikson's term for premature identity formation, in which a young person accepts parental values and goals without exploring alternative roles.

negative identity Erikson's term for a chosen identity that is in opposition to the identity one's parents or society expects one to adopt.

Identity Statuses

The ultimate goal, called **identity achievement**, is achieved when adolescents attain their new identity through "selective repudiation and mutual assimilation of childhood identifications" (Erikson, 1968). That is, the adolescent, ideally, establishes his or her own goals and values by abandoning some of those set by parents and society and accepting others.

Many young people, however, short-circuit this quest, never reconsidering parental values and childhood identifications. The result often is **foreclosure**, in which the adolescent accepts earlier roles and parental values wholesale, rather than exploring alternatives and truly forging a unique personal identity. A typical example might be the young man who from childhood has wanted to, or perhaps was pressured into wanting to, follow in his father's footsteps, as, say, a doctor. He might diligently study chemistry and biology in high school, take premed courses in college, and then perhaps discover in his third year of medical school (or at age 40, when his success as a surgeon seems unfulfilling) that what he really wanted to be was an archaeologist or a poet.

Other adolescents may find that the roles their parents and society expect them to fill are unattainable or unappealing, yet they may be unable to find alternative roles that are truly their own. Adolescents in this position often take on a **negative identity**, that is, an identity that is the opposite of the one they are expected to adopt. The child of a teacher, for instance, might drop out of high school, despite having the capacity to do college-level work. The child of devoutly religious parents might defy his or her upbringing by stealing, taking drugs, and the like.

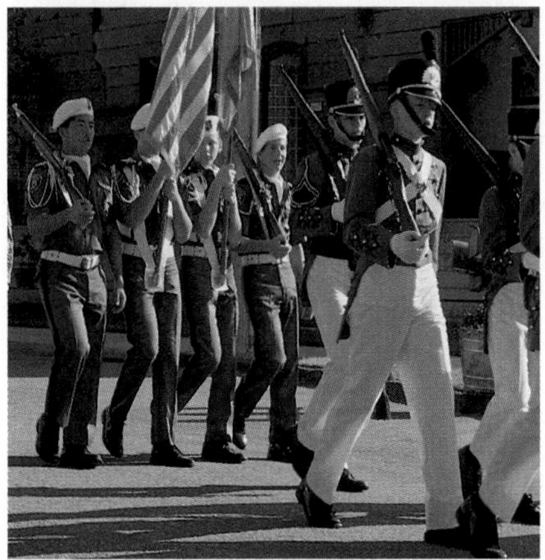

Despite the obvious differences, both photos depict the same phenomenon: teenage boys in uniform, parading their postponement of personal identity achievement.

Other young people experience **identity diffusion**: they typically have few commitments to goals or values—whether those of parents, peers, or the larger society—and are often apathetic about trying to find an identity. These young people may have difficulty meeting the usual demands of adolescence, such as completing school assignments, finding a job, and thinking about the future.

Finally, in the process of finding a mature identity, many young people seem to declare a **moratorium**, a kind of time-out during which they experiment with alternative identities without trying to settle on any one. In some cases, a society may provide formal moratoriums for adolescents through various institutions. In the United States, the most obvious example of an institutional moratorium is college, which usually requires young people to sample a variety of academic areas before concentrating in any one. Being a full-time student also forestalls pressure from parents and peers to choose a career and mate. Other institutions that permit a moratorium are the military, the Peace Corps, and various internships, many of which enable older adolescents to travel, acquire valuable skills, and test themselves while delaying lifetime commitments.

Research on Identity Status

Following Erikson's lead, many other developmentalists have found the concept of identity a useful one in understanding adolescence. Foremost among these is James Marcia, who has defined the four major identity statuses (achievement, foreclosure, diffusion, and moratorium) in sufficiently precise terms that he and other investigators can interview an adolescent and determine his or her overall identity status (Marcia, 1966, 1980).

Dozens of studies have compared adolescents' identity statuses with various measures of their cognitive or psychological development and have found that each identity status is typified by a number of distinct characteristics (see Table 16.1). For example, each of the four identity statuses correlates with a somewhat different attitude toward parents: the diffused adolescent is withdrawn, perhaps deliberately avoiding parental contact by sleeping or listening to music on headphones when the rest of the family is together; the moratorium adolescent is not withdrawn as much as independent, busy with his or her own interests; both the forecloser and the achiever are loving, but the forecloser evidences more respect and deference, while the achiever treats parents with more concern, behaving toward them as an equal or even as a caregiver rather than only as a care-receiver.

Table 16.1 shows some revealing combinations of statuses and traits. Note, for instance, that both adolescents who have achieved identity and those who have foreclosed their search for self-definition have a strong sense of ethnic identification, seeing themselves as proud to be Irish, Italian, Latino, or whatever. However, those who have foreclosed are relatively high in prejudice, while the identity achievers are relatively low, presumably because they are sufficiently secure in their ethnic background that they do not need to denigrate that of others.

Extensive research, much of it longitudinal, confirms that many adolescents go through a period of foreclosure or diffusion, and then a moratorium, before they finally commit themselves to a mature identity. The process can take ten years or more, with many college students still not

identity diffusion Erikson's term for uncertainty and confusion regarding what path to take toward identity formation. Identity diffusion typically leads adolescents to become erratic, apathetic, and disoriented.

moratorium Erikson's term for a pause in identity formation that allows young people to explore alternatives without making final choices. College or military service can provide a moratorium.

TABLE 16.1

Attitudes, Relationships, and Emotions Typical of the Various Identity Statuses

	Foreclosure	Diffusion	Moratorium	Achievement
Anxiety	repression of anxiety	moderate	high	moderate
Attitude toward parents	loving and respectful	withdrawn	trying to distance self	loving and caring
Self-esteem	low (easily affected by others)	low	high	high
Ethnic identity	strong	medium	medium	strong
Prejudice	high	medium	medium	low
Moral stage	preconventional or conventional	preconventional or conventional	postconventional	postconventional
Dependence	very dependent	dependent	self-directed	self-directed
Cognitive processes	simplifies complex issues; refers to others and to social norms for opinions and decisions	complicates simple issues; refers to others in both personal and ideological choices	thoughtful; procrastinates, especially in decisions; avoids referring to others' opinions or to social norms	thoughtful; makes decisions by both seeking new information and considering others' opinions
College	very satisfied	variable	most dissatisfied (likely to change major)	high grades
Relations with others	stereotyped	stereotyped or isolated	intimate	intimate

Source: Adapted from research reviewed by Berzonsky, 1989; Kroger, 1993; Marcia, 1980; Streitmatter, 1989.

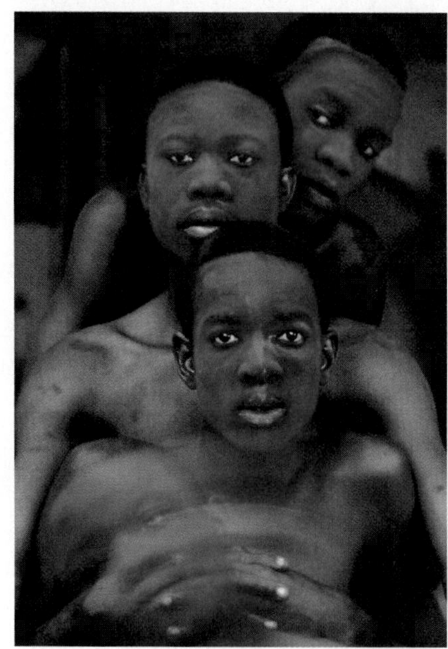

These boys are participating in a puberty ritual in the Congo. The blue dye on their faces indicates that they are temporarily dead, to be reborn as men once the ritual is over. Such rites of passage, which are based on strong cultural cohesion regarding social roles and responsibility, make adolescence a quick and distinct transition for all concerned.

clear about who they are or what they want to do (Marcia, 1980; Waterman, 1985). Further, research finds that within the four categories of identity formation, subcategories are evident. For example, some individuals in diffusion are apathetic, seeming not to care about anything, while others are alienated, rebelling against everything. Similarly, in foreclosure, some prematurely take on the identities their parents urge on them, while others choose totalitarian groups—such as a religious cult or a doctrinaire political organization—that take over all independent decision making (Archer & Waterman, 1990). Also, some adolescents in foreclosure have permanently closed the door to other options; some have closed it only temporarily and are likely to reopen it in a few years (Kroger, 1995).

Culture and Identity

There is also no doubt that the ease or difficulty of finding an identity is very much affected by forces outside the individual. One of the most influential of these is the surrounding society, which can aid identity formation primarily in two ways: by providing values that have stood the test of time and that continue to serve their function, and by providing social structures and customs that ease the transition from childhood to adulthood. Whether these factors are present and actually do make the search for identity easy depends primarily on the degree to which the members of the society are agreed on basic values, and the degree to which the individual is exposed to social change.

A CLOSER LOOK

Identity for Minority Adolescents

In most contemporary societies, finding an identity is extremely complex: many alternative paths are open and a multitude of choices must be made. For members of minority ethnic groups in democratic societies, identity achievement may be particularly complicated, and often painful (Phinney, 1993; Phinney et al., 1990; Spencer & Markstrom-Adams, 1990). On the one hand, democratic ideology espouses a color-blind, open society, in which background is irrelevant to achievement and all citizens develop their potential according to their individual merits, particular characteristics, and personal goals. On the other hand, most minority ethnic groups place major emphasis on ethnicity and expect their teenagers to honor their roots and take pride in their heritage. Thus, identity formation requires finding the right balance between transcending one's background and immersing oneself in it. In Erikson's words, during adolescence "each new generation links the actuality of a living past with that of a promising future" (Erikson, 1968).

Making this link is particularly difficult for many children from minority groups. Often, their expectations of living by the democratic ideal are thwarted by social prejudice and institutionalized racism. As a result, "many ethnic minority youth . . . may have to deny large parts of themselves to survive, may internalize negative images of their group, and may fail to adopt an ethnic cultural identity" (Hill et al., 1994). In some cases, the inability to resolve issues of identity successfully may lead minority youth to adopt a negative identity—rejecting wholesale the traditional values of both their ethnic group and the majority culture. Or it may lead them to foreclose on identity prematurely, choosing the values of one culture exclusively (Phinney et al., 1990). More often the latter is the case. In one study, African-American, Native American, Mexican-American, and Asian-American adolescents were all more likely to have foreclosed on identity questions than were European-Americans (Streitmatter, 1988). Thus, they had stopped searching for their own values and selfhood, perhaps because the process was too painful or confusing.

The primary effect of such foreclosure is to prevent mature reconciliation of both wellsprings of identity, the minority and the majority culture. For example, in a study of African-American adolescents (Miller, 1989) bused into high schools that were 95 percent white, one student remarked:

> I don't consider myself to be a minority because my [White] friends, they don't even consider or even look at it as me being a different color, just being regular, being just like them. They [other bused students] prefer to be Black, they want to just hang around with the Blacks, they don't want nothing to do with the Whites . . . I'm not like that . . . I attended the ski club and I asked if anyone else wanted to get into it, and you should have seen their faces, it was hysterical. What is this kid talking about, the ski club? It's a bunch of "honkies" gonna be there.

By contrast, another of the bused students, who socialized almost entirely with other minority students and thought both the school rules and the principal were unfair, said:

> I think this school is prejudiced. I didn't want to come out here. . . . And this school does not do things that Black people can get into. Like at our prom, we wanted to have a D.J. that could play White music and Black music. But they [White students] didn't want that. They wanted a band, which we can't comprehend.

Relationships with parents and other relatives are often particularly stressful for minority adolescents. Many minority groups revere family closeness—respecting elders and accepting self-sacrifice for the sake of the family—more than most members of the majority culture do (Harrison et al., 1990). Yet in many Western societies, this ideal clashes with the majority culture's emphasis on adolescent freedom and self-determination, exacerbating the normal conflict that arises when a teenager's need for independence and friendship is incompatible with parental

In a culture where virtually everyone holds the same moral, political, religious, and sexual values, and social change is slow, identity is easy to achieve. Unless some central feature of their personality—their temperament, sexual interests, aspirations, or the like—is antithetical to the social roles and values endorsed by the culture, most young people in a traditional society simply accept the social roles and values they grew up with, the only roles and values they have ever known. In modern industrial and postindustrial societies, by contrast, rapid social change, a broad diversity of values and goals, and an ever expanding array of identity choices can make iden-

demands and cultural expectations. Some minority adolescents (mostly girls) give in to parental control (perhaps docilely living at home until marriage), while others (mostly boys) rebel completely (perhaps leaving home in a mad fury). In both cases, the normal search for identity is sacrificed. Finding a balance between these two extremes has its own price. As one Chinese-American girl lamented:

> I don't know who I am. Am I the good Chinese daughter? Am I an American teenager? I always feel I am letting my parents down when I am with my friends because I act so American, but I also feel that I will never really be an American. I never really feel comfortable with myself anymore. [quoted in Olson, 1988]

While adolescents from every minority group encounter additional complications in finding their own identity, the particulars vary depending on the subculture. Asians feel the stress of becoming an "achieving Asian," with academic pressure that precludes much of a social life and may, in fact, undermine school success (Natagata, 1989). African-Americans and Latino-Americans who are academically talented may feel the opposite pressure, hiding their ability in order not to be perceived as a "brainiac" by their peers (Fordham & Ogbu, 1986; Matute-Bianchi, 1986). In addition, members of minority groups often experience intense criticism from peers if they make an effort to join the white majority culture: typically they are branded as an "oreo," a "banana," an "apple"—of color on the outside but white inside. One productive solution is to achieve two identities rather than one. Adolescents who can identify with both their own ethnic culture and the mainstream culture become more successful academically as well as healthier psychologically (LaFromboise et al., 1993).

Of course, all these generalities are themselves stereotypic, overlooking both individual and group differences. Depending on the particular national origin, socioeconomic status, and acculturation of the family, pressures differ. The Cuban-American is quite distinct from the Mexican-American; the impoverished inner-city African-American differs from the middle-class, suburban one; the newly arrived Cambodian is very different from the third-generation Japanese, just as the Aleut from Northern Canada is different from the Navajo in New Mexico.

Even within each subgroup, further distinctions arise. As one ninth-grade Mexican immigrant complained:

> There is so much discrimination and hate. Even from other kids from Mexico who have been here longer. They don't treat us like brothers. They hate even more. It makes them feel like natives. They want to be American. They don't want to speak Spanish to us, they already know English and how to act. If they're with us, other people will treat them more like wetbacks, so they try to avoid us. [quoted in Olson, 1988]

In addition, historical changes can present each generation with a somewhat different array of possibilities and problems. The past forty years in America, for example, have seen swings from legal and de facto segregation and discrimination to enforced integration and affirmative action, and, more recently, from the ideal of the racial and ethnic "melting pot" to a new emphasis on racial and ethnic heritage and a retreat from affirmative action. Because of such historical shifts, each generation and each individual must wage a new and different struggle to achieve an ethnic identity. Ethnic identity, like identity of all sorts, takes time to find, with individuals usually going to extremes of assimilation and separation before reaching a mature self-affirmation (Cross, 1991). For some minority adolescents, it may take years, even decades, before they can sort through the divergent historical roots, gender roles, vocational aspirations, religious beliefs, and political values of their surrounding cultures and can forge for themselves a sense of identity that they feel to be truly their own (Staples & Johnson, 1993).

tity formation a formidable task. As A Closer Look, which starts on the facing page, points out, some aspects of identity formation can be particularly difficult for minority-group members, who may feel pulled in opposite directions by the values and customs of their group and those of the majority culture.

Now let us look at two other forces—the immediate family and the peer group—that are instrumental not only in the young person's quest for self-understanding and identity but in all other aspects of navigating the passage to adulthood.

generational stake The need of each generation of family members to view their own intergenerational interactions from their own perspective. As a typical result, parent-adolescent conflicts are viewed quite differently by each party.

Family and Friends

Sailing the changing seas of development is never done alone. At every turn, a voyager's family, friends, and community each provide provisions and directions, ballast for stability, and an anchor when it is time to rest or, through example and pressure, reasons to move full speed ahead. In adolescence, when the winds of change blow particularly strong, parents and peers become especially powerful influences, for good or ill.

Parents

Often adolescence is characterized as a time of waning adult influence, when the values and behaviors of young people are said to become increasingly distant and detached from those of their parents and other adults. According to all reports, however, the *generation gap,* as the differences between the younger generation and the older one have been called, is not very wide. Younger and older generations, in fact, have very similar values and aspirations. This is especially true when adolescents are compared, not with adults in general, but with their own parents (Steinberg, 1990).

Numerous studies have shown substantial agreement between parents and adolescents on political, religious, educational, and vocational opinions and values (Coleman & Hendry, 1990; Youniss, 1989). Most young people, for instance, favor the same candidate for president or prime minister and follow the same religion as their parents do. Other similarities between parents and adolescents are also apparent. For example, regardless of academic potential, adolescents who do relatively well in high school and college tend to be the offspring of parents who value education and did well in high school and college themselves. Similarly, whether or not an adolescent experiments with drugs is highly correlated with his or her parents' attitudes and behavior regarding drugs (Denton & Kampfe, 1994). Indeed, virtually every aspect of adolescent behavior is directly affected by the family.

The fact that the generation gap tends to be small by objective measures does not mean that parents and their children perceive it that way. In fact, members of each generation in the parent-adolescent relationship view that relationship somewhat differently because of their generation's views and needs. In effect, each group has its own **generational stake** in the family (Bengston, 1975). Because of their different developmental stages,

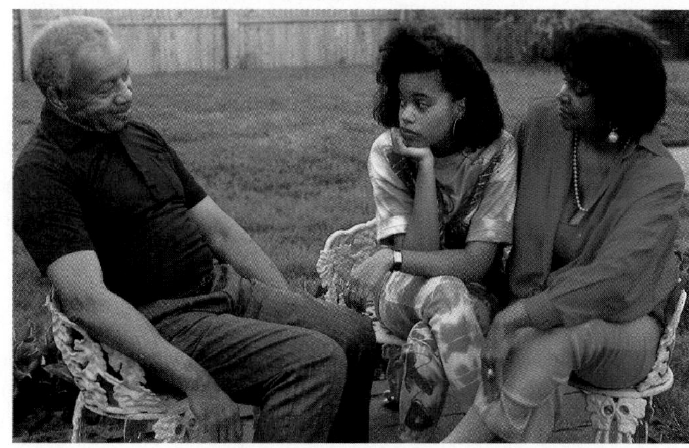

Parents and teenagers are generally inclined to view their interaction somewhat differently. Parents tend to believe that their own experience makes them well suited to advise their offspring during any "phase" they might be going through. Adolescents, on the other hand, often consider their parents misguided and "out of touch." Thus, for obvious reasons, parents are more likely to be proud of their children than vice versa.

each generation has a natural tendency to see the family in a certain way. Parents are concerned about continuity of their own values, so they tend to minimize the import of whatever differences occur, blaming them on hormones or peer influences rather than on anything long-lasting. Adolescents, on the other hand, are concerned with shedding many parental restraints and forging their own independent identity, so they are likely to exaggerate problems. Thus a conflict about a curfew may be seen by the teenager as evidence of the parents' outmoded values, or lack of trust, whereas the parents may see it merely as a problem of management, the latest version of trying to get the child to bed on time.

Harmony and Conflict

A certain amount of conflict between parents and adolescents, therefore, is likely to occur in most families throughout the world, as the young person's drive for independence clashes with the parents' traditional control (Schlegal & Barry, 1991). The extent of the conflict, and how much it dominates the parent-child relationship, depend on many factors. Perhaps the most important of these is the parenting style used in the home.

As you recall from Chapter 10, researchers have distinguished three broad types of parenting: the *authoritarian* style, in which the parent expects to be obeyed and shows little affection or nurturance; the *permissive* style, in which the parent is affectionate but nondemanding; and the *authoritative* style, in which the parent sets limits but is also nurturant and communicates well with offspring. As at earlier stages of development, authoritative parenting is the style most likely to foster the adolescent's achievement and self-esteem and to promote a positive parent-adolescent relationship (Baumrind, 1991; Coleman & Hendry, 1990; Steinberg, 1990). On the whole, adolescents who have authoritative parents are, for example, more self-reliant, perform better at school, and show fewer problem behaviors (like delinquency or drug and alcohol abuse) compared with other teenagers (Lamborn et al., 1991; Steinberg et al., 1994). Especially harmful are parents who are permissive because they do not seem to care what their child does (*rejecting-neglecting*): their teenagers are likely to lack confidence and to be depressed, low-achieving, and delinquent (Baumrind, 1991; Lamborn et al., 1991; Steinberg et al., 1994).

Authoritative parents contribute to better psychosocial development in adolescence because they are capable of relaxing their control—to accommodate the adolescent's growing capacity and desire for independent decision making—while continuing to provide firm guidance, warmth, and acceptance. By contrast with a teenager's need to establish emotional detachment from parents who are overcontrolling or uninterested, the offspring of authoritative parents can achieve independence in the context of a more supportive family climate (Furman & Holmbeck, 1995; Lamborn & Steinberg, 1993; Ryan & Lynch, 1989).

While these generalizations tend to hold across most ethnic groups, some ethnic differences in parent-adolescent interaction are apparent. A number of studies suggest that, compared with European-American parents, ethnic-minority parents are more inclined toward an authoritarian-type parenting style—in which they exercise strict control over the daily lives of their adolescent offspring—and that their offspring are less nega-

tively affected by this style than European-American teenagers are (Baldwin et al., 1990; Chao, 1994; Chiu, 1987; Dornbusch et al., 1987). One reason for this difference may be that cultural values endorsing family cohesiveness and parental authority cause many ethnic-minority teens to perceive strict parenting as a sign of caring rather than as a sign of heedless domineering. Another reason may be that ethnic-minority families are more likely to live in neighborhoods where drugs, crime, and violence are prevalent, and in such circumstances, strict parenting may be highly protective and is likely to be perceived as such by offspring. This latter analysis has been used in particular to explain the fact that, contrary to the general pattern, inner-city African-American adolescents tend to fare better, academically and psychologically, with an authoritarian-type parenting style than with an authoritative one. In a related manner, the protection explanation may also apply to the recent surprising finding that middle-class black teens living in predominantly white neighborhoods may also fare better with authoritarian parenting (Lamborn et al., 1996). As the authors of this finding note:

> As members of a visibly salient ethnic group that has experienced a long history of discrimination in the United States, African-American youth may be especially vulnerable to risk in predominantly white communities. . . . African-American families may feel the need to protect youth in affluent as well as disadvantaged communities—albeit for potentially different reasons.

Another factor that influences the frequency of parent-adolescent conflict is the teenager's developmental status. Conflict is most likely to emerge in early adolescence and, especially for girls, is more likely to involve early-maturing offspring than late-maturing offspring (Caspi et al., 1993; Montemayor, 1986; Steinberg, 1990). Such conflict is also more likely to involve mothers than fathers. The reason these are the typical combatants is not hard to see. Teen-parent conflict usually involves bickering over habits of daily life—hair, neatness, and clothing issues—that traditionally fall under the mother's supervision. And it is the relatively young adolescent who feels compelled to make a statement—with green hair or blaring music—that establishes in unmistakable terms that a new stage has arrived. While the emergence of parent-child conflict in early adolescence is the general pattern, there is some evidence that for Chinese-, Korean-, and Mexican-American teens, stormy relations with parents may not surface until later in adolescence, perhaps because their respective cultures' encouragement of dependency in children and emphasis on family closeness delay the quest for autonomy (Greenberger & Chen, 1996; Molina & Chassin, 1996).

Family conflict and parent-child distancing also arise, as all family members recalibrate their expectations for one another in light of the child's growing physical maturity (Collins, 1990; Collins & Russell, 1991; Steinberg, 1988). In general, adolescents believe that they should be granted the privileges of adult status much earlier, and more extensively, than parents do (Holmbeck & O'Donnell, 1991). Further, adolescents believe that many of the issues of controversy between themselves and their parents are matters of personal privacy and freedom and therefore ought not to be interfered with by parents. Parents, on the other hand, consider such issues as sleeping late on weekends, engaging in long phone conversations, wearing tight or torn clothing, and leaving one's room a mess to be within their authority—at least for critical commentary, if not for outright control (Smetana & Asquith, 1994). While recurrent bickering and squabbling over mundane

If parents try to lay down the law about obedience, hard work, and responsibility every moment of the adolescent's life, the adolescent—especially if a boy—is likely to rebel, either with open defiance or passive resistance. Many a parent has ended yet another lecture wondering "I don't understand what his problem is. I've told him again and again."

matters may seem petty, the underlying issue—adolescents' freedom to make their own decisions—is not. This persistent conflict can lower the quality and harmony of family life. When parents and children are asked to cite specific conflicts, most families can describe about three or four recurrent disputes, each arising about once a week (Smetana et al., 1991).

A further reason for conflict is teenagers' changing perceptions of their parents' wisdom. With a growing capacity to perceive themselves from multiple viewpoints—in different roles, in idealized images, and as the "real me"—adolescents also develop a new view of their parents as "real" people, with faults and limitations like everyone else (Steinberg, 1990). In other words, in the same way that teenagers experience a gap between their ideal self and their real self, they may experience a gap between their images of an "ideal parent" and how they perceive their parents truly to be (Collins, 1990; Feldman & Gehring, 1988). This can make them highly critical of their parents' beliefs, expectations, and authority as they recognize the fallability and compromises in the life commitments their parents have made—without having yet made such commitments of their own. A teenage son or daughter may be harshly critical, for example, of the apparent hypocrisy of a parent's moral beliefs, without realizing how difficult it can be to apply moral principles to the complex realities of adult life.

A final factor in parent-adolescent conflict is birth order. First-borns have more conflicts with parents than later-borns do, although later-borns may actually be more rebellious (Small et al., 1988; Sulloway, 1996). As one older adolescent explained to a younger sibling, "You should be grateful to me. I had to break them in, so you have it easy."

In most families, fortunately, conflict and disagreement gradually subside with time. Parents' first reaction to the emergence of assertiveness in their adolescent is usually to be more assertive themselves—reiterating general rules the child seems to have forgotten, exaggerating the dangers of independence, and insisting on the obedience and respect that the young person suddenly seems disinclined to give. But eventually parents tend to be more yielding, in part because they recognize that their child is growing up, in part because some of the adolescent's arguments start to make sense, and in part because the young person begins to act in more mature ways. Usually by late adolescence, parent-child relations are more harmonious, a trend found among children of both sexes, adopted as well as nonadopted children, in all family types, and in many nations (Hurrelmann & Engel, 1989; Larson et al., 1996; Maughan & Pickles, 1990; Montemayor, 1986).

All told, then, conflict typically occurs in early adolescence and tends to center on issues of self-discipline and self-control, with disagreements being more likely to involve the adolescent's musical tastes, domestic neatness, and sleeping habits than his or her views on world politics or deep moral concerns (Barber, 1994). Indeed, the large majority of adolescents report feeling loved and accepted by their parents and perceive them as role models and sources of guidance and support (Steinberg, 1990). Those who do not follow this generally benign pattern—the roughly 20 percent of all families who find that conflict and hostility appear and reappear throughout adolescence (Montemayor, 1986; Offer & Offer, 1975)—are those who are the most vulnerable to the special problems discussed later in this chapter (Dryfoos, 1990).

Peers and Friends

The socializing role of peers, which becomes prominent during the latter part of middle childhood, becomes even more influential during early adolescence (Berndt, 1989). From "hanging out" with a large group at school or the mall to whispered phone conversations with a trusted confidant, relations with peers and with close friends are a vital part of the transition from childhood to adulthood.

Adolescent peers help each other negotiate the tasks and trials of adolescence in many ways. As B. Bradford Brown (1990) explains, "teenagers construct a peer system that reflects their growing psychological, biological, and social-cognitive maturity and helps them adapt to the social ecology of adolescence." Among the special roles performed by peer relationships and close friendships, Brown finds the following four most noteworthy:

1. As the typical physical changes of adolescence confront the young person with new feelings, experiences, and challenges to self-esteem, the peer group functions both as a source of information on these matters and as a self-help group of contemporaries who are going through the same sorts of struggles. As we shall shortly see, peers can be especially helpful in negotiating new relationships with members of the other sex, which can be daunting and scary without the support of same-sex friends.

Peers provide each other welcome guidance regarding what to wear, as shown by these two groups of friends. Although there may have been some early-morning phone consultation regarding who was wearing what, the overall fashion mode in both cases—"black-with-chains" and "casual prep"—was a foregone conclusion.

2. The peer group also provides support in adjusting to changes in the social ecology of adolescence, especially the movement to larger, more impersonal middle and junior high schools, with their more heterogeneous student populations and less nurturant adult supervision. As Brown observes, "most major changes in peer groups can be seen as efforts to cope with the new school structure thrust on youngsters at adolescence. The depersonalized and complex routine of secondary school increases the young teenager's need for sources of social support and informal exchanges."

3. In the search for self-understanding and a stable sense of identity, the peer group functions as a kind of mirror in which adolescents check their reflection, uniting with friends who share many of their own dispositions, interests, and capabilities. Peers also help adolescents to define who they are by helping them define who they are not. As adolescents associate themselves with this or that subgroup (the jocks, the brains, or the druggies, for instance), they are rejecting other subgroups—and the particular self-definitions that would go with them.

4. Finally, the peer group serves as a sounding board for exploring and defining one's values and aspirations. By experimenting with different viewpoints, philosophies, and attitudes toward themselves and the world, adolescents can begin to crystallize, in the context of others who are doing the same, the values that are truest to them.

Because of the nature and importance of these peer roles, the characteristics of peer friendships that are most important to adolescents are loyalty and intimacy (Berndt & Savin-Williams, 1992; Savin-Williams & Berndt, 1990). Friends are obligated to stand up for each other and never to speak behind a friend's back. Most important, they are expected to listen and share personal thoughts and feelings without ridicule or betrayal of trust. As one eighth-grader expressed it:

> I can tell Karen things and she helps me talk. If we have problems at school, we work them out together. And she doesn't laugh at me if I do something weird—she accepts me for who I am. [from Berndt & Perry, 1990]

Peer Pressure

The largely constructive role of peers that we have just outlined runs counter to the notion of **peer pressure**, the idea that the norms of the peer group force adolescents to act in ways that they otherwise would not. In fact, the negative power of peers during adolescence is often exaggerated. For one thing, social pressure to conform is very strong only for a short period, rising dramatically in early adolescence, until about age 14, and then declining (Coleman & Hendry, 1990). In addition, peer-group pressure may be functional in some cases, helping to ease the transition for the young person who is trying to abandon childish modes of behavior, including dependence on parents, but who is not ready for full autonomy.

peer pressure Social pressure to conform to one's friends or contemporaries in behavior, dress, and attitude. Peer pressure can be positive or negative in its effects.

It is also important to realize that peer pressure is not necessarily negative. For example, one study of 373 Wisconsin junior and senior high school students found that peer pressure to study hard and get good grades was as

apparent as pressure to dress appropriately, and that peers were more likely to discourage cigarette smoking than to encourage it (Brown, 1990; Brown et al., 1986).

The reality that peer pressure can be positive does not negate another reality: young people sometimes lead each other into trouble. When no adults are present, the excitement of being together and the desire to defy adult restrictions can result in risky, forbidden, and destructive behavior (Baumrind, 1987; Csikszentmihalyi & Larson, 1984; Dishion et al., 1995). However, although parents sometimes blame other teenagers for leading their "innocent, self-controlled, law-abiding" child astray, the truth tends to be otherwise. A young person usually chooses to associate with friends whose values and interests he or she shares, and collectively they become involved in escapades that none of them would engage in alone. Fortunately, most peer-inspired misbehavior is a short-lived experiment rather than a foreshadowing of long-term delinquency.

Overall, the teenager who argues that he or she must engage in a particular activity, dress a certain way, or hang out in certain parts of town because "everyone else does it" is trying to lighten the burden of responsibility for some style, demeanor, or philosophy that she or he is trying out. In a way, therefore, peers act as a buffer between the relatively dependent world of childhood and the relatively independent world of young adulthood.

When teenagers first start forming couples, they often spend time with other couples, to ease the uncertainty about what to say, when to touch, and so on. Interestingly, it is not unusual for a breakup in one couple to trigger a breakup in the other, often with a new alliance between the same individuals.

Boys and Girls Together

One of the most important ways adolescents help each other is by easing the transition into relationships with the other sex. As you remember, voluntary sex segregation is the common practice among children during most of early and middle childhood. Neither sex pays much attention to the other. Then, as puberty begins, boys and girls begin to notice one another in a new way.

Usually, the first sign of heterosexual attraction is not an overt positive interest but a seeming dislike. A study of adolescents in New Zealand (Kroger, 1989), for example, found that a typical feeling among girls at age 11 was that "boys are a sort of disease"; at age 13, that "boys are stupid although important to us"; at age 15, that "boys are strange—they hate you if you're ugly and brainy but love you if you are pretty but dumb"; and at age 16, that "boys are a pleasant change from the girls." Likewise, a typical feeling among boys at age 11 was that "girls are a pin prick in the side"; at age 13, that "girls are great enemies"; at age 15, that "girls are the main objective"; and at age 16, that "girls have their good and bad points—fortunately, the good outnumber the bad." Similar patterns of "warming up" to the other sex have been found in many nations, although the pace of change depends on various factors. The hormonal changes of puberty obviously affect when this warming up occurs, although they are probably not as powerful in this regard as the influence of the culture and peers, and the availability of someone to warm up to (Coe et al., 1988).

Typically, the peer group plays an important part in the "warming-up" process. Friendships within the group become a launching pad for romantic involvements, providing peers with security and role models while sparing them the embarrassment of being alone for an extended period of time with a member of the other sex without knowing what to do or say. The peer group also provides witnesses and companions of the same sex who will

help the young person evaluate whether so-and-so (male) is really nice or a nerd, whether so-and-so (female) is cute or stuck up, and, especially important, whether a particular attraction is mutual or not. The role of peers in the "warming-up" process is clearly reflected in this account of two young adolescent boys working as a team to test the waters of romance:

> We started calling girls we liked on the phone, one at a time. We'd each call the girl the other one liked and ask if she wanted to go with the other one . . . Usually they wouldn't say too much. So sometimes we would call her best friend to see if she could tell us anything. Then they would call each other and call us back. If we got the feeling after a few calls that she really was serious about No, then we might go on to our next choice, if we had one. [quoted in Adler et al., 1992]

Even if a boy and girl manage to set up a date just for the two of them, they are unlikely to be alone for long. Their friends might find out every detail of their plans and then sit behind them at the movie, hang out near them in the school yard, or walk on the other side of the street as they go from one place to another. As soon as the date is over, the daters are obliged to call their friends, bragging about or bemoaning every detail.

The typical adolescent friendship circle is quite large and fluid. For example, one study found that the average tenth-grader had ten friends of the same sex and seven friends of the other sex, and that between the beginning and end of the school year, more than half of the old friends had been replaced by new ones (Fischer et al., 1986). The overall shifts in heterosexual friendship patterns within the peer group were documented in another study of students in a large, multiethnic public school outside Chicago (Csikszentmihalyi & Larson, 1984). Not only did these adolescents gradually spend more time with the other sex, but they enjoyed it more: freshmen were happiest when they were with companions of their own sex or in mixed-sex groups, while juniors and seniors were happiest when with members of the "opposite" sex.

As romantic relationships develop, a social support network of friends continues to be important. For many adolescents, the first experiences with intimacy are fraught with problems, especially the likelihood of rejection. Indeed, the fear of rejection keeps many adolescents from trying to form intimate relationships, and when rejection does occur, adolescents are likely to experience it as devastating (Mitchell, 1986). Having friends to cushion this pain by offering reassurance and solace and by validating one's feelings and sense of self-worth is essential.

For those adolescents who are gay or lesbian, the added complications of finding both romantic partners and friends as confidants usually slow down the entire process. Generally, feeling at ease with one's sexual identity takes longer if one's orientation is toward one's own sex, in part because finding an accepting peer network is more difficult. Especially in homophobic cultures, many young men with homosexual feelings deny these feelings altogether, or they try to change or conceal them by becoming heterosexually involved (Savin-Williams, 1995). Similarly, many young women who will later identify themselves as lesbian spend their teenage years relatively oblivious to, or in denial of, their specific sexual urges. However, because female friendships in general tend to be close and intimate, lesbian adolescents are more often able to establish strong friendships with same-sex heterosexual peers than homosexual teenage boys are (Bell et al., 1981; D'Augelli & Hershberger, 1993; Savin-Williams & Berndt, 1990).

Overall, then, research suggests that peers are a positive aid in every major task of adolescence, from adjusting to the physical changes of puberty, to searching for identity, to forming romantic attachments. These findings help put the much overemphasized problems of "peer pressure" in perspective. Many parents worry that their children might be transformed during adolescence by the pressure of their friends, becoming, perhaps, sexually promiscuous, or drug-addicted, or delinquent. In some cases, of course, parents are right to worry and may even have to intervene. However, while certainly some young people do things with their friends that they would not do alone or with a different peer group (the first drag on a cigarette or the first swig of beer is almost always related to the urging of friends), in general, peers are more likely to complement the influence of parents during adolescence than to pull in the opposite direction (Brown, 1990). In some ways, this is hardly surprising. The values and beliefs that teenagers tend to share with their parents also affect their choices of friends and activities: adolescents who have been raised to value academic achievement, for instance, are likely to choose friends who have the same goals. Moreover, parents affect peer relationships indirectly, depending on how they monitor the adolescent's activities and on how flexible they are in accommodating the teen's changing needs and aspirations (Brown et al., 1993). Parents who are unable to relax their authority to adjust to a teenager's desire for greater independence, for example, are likely to find that teenager becoming increasingly peer-oriented, even to the extent of sacrificing schoolwork or personal standards in order to maintain popularity with friends (Fuligni & Eccles, 1993). On the other hand, when parents recognize their teenager's growing maturity and accordingly become more democratic in family decision making, their offspring are likely to regard home and peer experiences as mutually supportive, rather than in opposition to each other (Jacobs et al., 1993).

Special Problems

As we have seen, for the majority of adolescents, the voyage to young adulthood is fairly smooth sailing, and they are able to ride out whatever rough seas may arise. For a minority of teenagers, however, the rough seas can become a boiling tempest that threatens development, or even life itself. When serious problems do occur, they usually are a consequence of a wide range of factors—including long-standing problems in the home, inherited vulnerabilities in temperament, and exacerbating cultural or historical conditions, as well as the special strains of adolescence and the particular peer and neighborhood context (Robins & Rutter, 1990; Smith & Rutter, 1995).

Two areas of particular concern are juvenile delinquency and adolescent suicide.

Juvenile Delinquency

Data on arrests, worldwide, show that a person is more likely to be arrested in the second decade of life than in any other time period. For instance, international statistics show that overall arrest rates are highest at about age 16 and then decline slowly with every passing year (Smith, 1995). In the

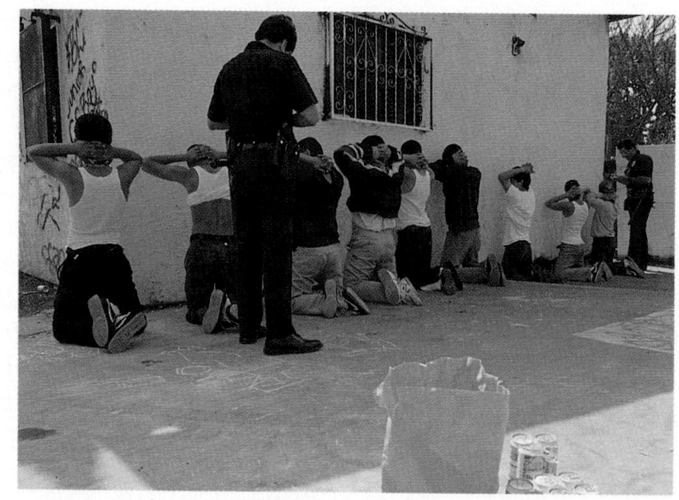

What happens if a police officer spots a teenager carrying a brown paper bag? In this case, the teenager was a member of a Los Angeles street gang, so he and his fellow gang members were stopped and ordered to kneel against the wall while being searched. Scenes like this cause many researchers to question whether custom and bias distort the statistics on patterns of delinquency. A white teenager, especially a girl, in a suburban neighborhood, carrying the same bag, would probably never have been stopped, much less made to kneel in order to be searched.

United States, 30 percent of all arrests, and 42 percent of all arrests for serious crimes, are of persons between the ages of 10 and 20 (U.S. Department of Justice, 1995).

However, simply looking at arrest statistics is not sufficient to reveal the prevalence of delinquency, that is, how widespread law-breaking might be among adolescents. If, as some contend, a small minority of repeat offenders is committing a highly disproportionate share of juvenile offenses, then the prevalence of law-breaking among juveniles in general might be much lower than the high arrest statistics would seem to suggest (Farrington, 1990).

On the other hand, many juvenile crimes never come to the attention of the police, and the police do not arrest some young people whom they apprehend. Indeed, one researcher found that even among self-admitted repeat offenders, only 20 percent were ever arrested (Henggeler, 1989). For this reason, law-enforcement data on gender and ethnic differences among youth offenders are particularly questionable. Overall, official statistics show that males are more than three times as likely to be arrested as females, that African-Americans are three times as likely to be arrested as European-Americans, and that European-Americans, in turn, are three times as likely to be arrested as Asian-Americans (U.S. Department of Justice, 1995).

But on the basis of longitudinal studies that ask teenagers, confidentially, about their own misbehavior, law-breaking that could lead to arrest seems part of normal adolescence, and sex and ethnic differences seem much less dramatic than indicated by the official arrest data. When all illegal acts—including such minor infractions as underage drinking, disorderly conduct, breaking a curfew, playing hooky, buying cigarettes, and the like—are included, virtually every adolescent is a repeat offender. In addition, at least in North America, Great Britain, and Australia, most self-report studies reveal that somewhat more serious crimes—such as property damage, stealing, causing bodily harm, and buying or selling illegal drugs—are also common, being committed at least once by about 80 percent of all adolescents (Binder et al., 1988; Farrington et al., 1990; Moffit, 1997).

This is not a new phenomenon. Lyle Shannon (1988) found similar patterns in a study of three cohorts of young people born in 1942, 1949, and 1955

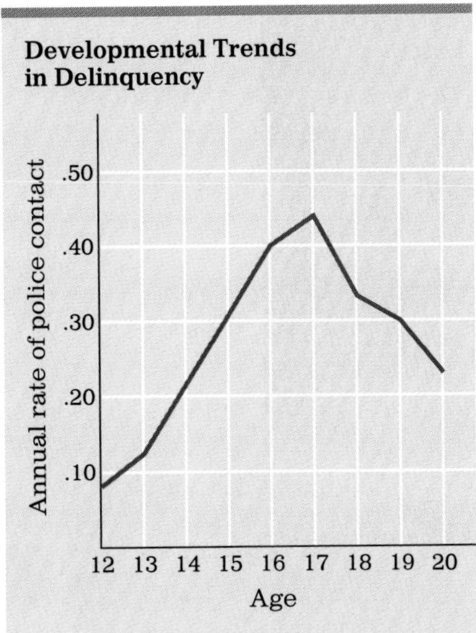

FIGURE 16.1 While these data come only from one group—boys in a Wisconsin town—two characteristics are equally true, no matter where the information is collected. First, the rate of police contact rises and then falls during the years from age 10 to 20. Second, almost every American boy has almost three police interactions by age 20. Of course, not every transgression is caught: the actual average number of delinquent acts is much higher.

in one Wisconsin town. Using longitudinal research that included self-reports, interviews, and police statistics, he determined that well over 90 percent of the boys and 65 to 70 percent of the girls had engaged in illegal acts.

The period of greatest increase in illegal activity appears to be age 15 or 16, with a rate about three times that of age 12 (Shannon, 1988). Thereafter it levels off and then drops, so that by age 20 it is about half the rate at age 17 (see Figure 16.1). Interestingly, being caught and being punished do not seem to affect the frequency of law-breaking or the time span it includes: those whose law-breaking ends as adulthood approaches usually cite maturity, rather than any contact with the criminal justice system, as their reason for stopping their criminal behavior (Quay, 1987; Rutherford, 1986; Shannon, 1988).

Obviously, however, this does not mean that adolescent law-breaking should be allowed to run its course. For one thing, some adolescent crime is very serious. In the United States, of all those arrested in 1993 for serious crimes—including murder, rape, assault, and robbery—6 percent were under age 15 and another 20 percent were aged 15 to 18 (U.S. Department of Justice, 1995). Similar rates are found in other nations: arrests for sudden, impulsive, destructive crimes tend to occur most often at adolescence. Further, although many who engage in repeated delinquency at age 15 cease their criminal activity before adulthood, a significant minority do not, going on to become career criminals. That minority are recognizable once they are well along the path of criminality. They are among the earliest of their cohort to drink alcohol and smoke cigarettes, among the least involved in school activities and most involved in "hanging out" with older, law-breaking youth, and, finally, are arrested many times for increasingly serious offenses throughout their teen years.

Dealing with Delinquency

Unfortunately, attempts to discover the best prevention of juvenile delinquency are fraught with problems. Some headway has been made toward identifying a child at risk for serious delinquency. According to a number of studies, the children most at risk are those who by age 10 have learning difficulties in school, especially in reading, *and* who face significant stresses at home (such as very low income, a large family, or an alcoholic-abusing parent) that weaken family bonds. They are especially at risk of delinquency if, as early as age 6, they have signs of attention-deficit disorder and antisocial behavior. By age 10, in fact, most delinquency-prone children are already showing aggression (Kupersmidt & Coie, 1990; Moffitt, 1993; Sampson & Lamb, 1993).

However, while it is quite possible to distinguish those children who are at high risk of delinquency (those who are vulnerable in all three ways—biological, familial, and social), it is virtually impossible to differentiate those who will actually become serious adult criminals (Arbuthnot et al., 1987; Trasler, 1987). One reason is that so many factors bear on the outcome—from prenatal brain damage to the availability of guns, on the negative side, and from being a first-born child to attending a well-structured junior high school, on the positive side—that it is very difficult to identify and weigh all the factors that might be relevant for a particular person.

Thus, if an adolescent is statistically at risk of committing a serious crime someday, he or she may never actually do so, for reasons that never enter the calculation, and an adolescent who has already broken the law and been caught may not necessarily be one of the ones who most need special treatment. Finally, the array of intervention options is very broad—ranging from vocational help to incarceration in adult prisons, from reading lessons to family therapy—yet none of them has been shown to be consistently effective for every delinquent, and some, such as frightening delinquents into going straight by taking them on prison tours, actually backfire, making prison look like a tough guy's rite of passage (Lundman, 1993).

Nevertheless, from a developmental and scientific point of view, we have gained some important insights. Most experts agree that those adolescents whose crimes include repeated antisocial acts—the lone mugger as opposed to the adolescent who joins friends for a joy ride in a stolen car—need more intensive intervention. With this subgroup, carefully structured rehabilitation—which may involve removal from the community to a place where the adolescent can learn academic and interpersonal skills he or she has thus far failed to acquire—can help (Garrett, 1985; Guerra et al., 1994; Morris & Braukmann, 1987).

For most delinquents, however, such drastic programs are not needed and may even prove harmful, segregating them with a peer group that prizes possessions more than people and that survives by means of deceit and self-centeredness rather than trust. In addition, since weak social bonds are a factor that contributes to criminal behavior, completely breaking delinquents' connection to home, school, and neighborhood by removing them from the community may only make the problem worse (Sampson & Lamb, 1993).

While no response to adolescent crime is always successful, the best outcomes result when the teenager is encouraged to understand the social consequences of his or her offense—as when the punishment is restitution or the performance of unpaid community service (Schneider, 1990). Research has shown that another powerful deterrent is **parental monitoring**, that is, parental vigilance regarding where one's child is and what he or she is doing and with whom (Fletcher et al., 1995; Patterson et al., 1989; Sampson & Lamb, 1993). Most effective in the long run would be an ecological approach that simultaneously aids families, schools, and the community, as well as the adolescent. As two leading researchers write in the conclusion of a 500-page book devoted entirely to the problem of youth and aggression:

> Rather than waiting until violence has been learned and practiced, and then devoting increased resources to hiring policemen, building more prisons, and sentencing three-time offenders to life imprisonment, it would be more effective to redirect the resources to early violence prevention programs, particularly for young children and early adolescents. [Slaby & Eron, 1994]

What should such prevention programs include? Key elements that most developmentalists would endorse include the following (Bensen, 1993; Dryfoos, 1990; Henggler, 1989; Pepler & Slaby, 1994):

1. early intervention to help parents establish strong attachment bonds with their children and discipline them in an effective manner;

2. strengthening of schools so that fewer young people have academic difficulties and more have high aspirations;

parental monitoring Parental watchfulness and awareness regarding where offspring are and what they are doing.

3. support of community institutions (religious groups, Scouts, sports leagues, and so on) so that they can provide constructive challenges for young people.

As daunting as the implementation of these measures would seem to be, it represents a far better way to try to prevent serious delinquency than attempting to reform every offender or to answer the frequent outcry for a cop on every block and a cell for every law-breaker. True prevention of juvenile delinquency begins early, from birth and even conception, and continues throughout childhood.

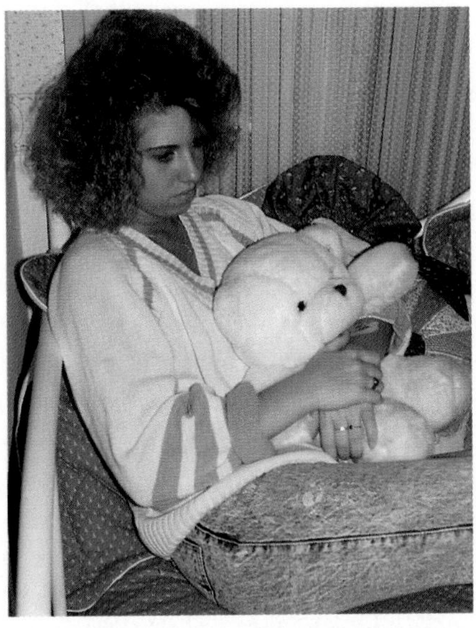

Depression can be an occasional feature of normal adolescence. When it is sustained, or when it is accompanied by a loss of interest in social relationships, a decline in school achievement, or an evident wish for the dependency of the young child, it may be a sign that an adolescent is at risk for suicide. Those around the troubled teen should regard depression seriously and take preventive action.

Adolescent Suicide

The most perplexing tragedy that can occur in adolescence is suicide. From a life-span perspective, teenagers are just beginning to explore life's possibilities. Even if they experience some troubling event—failing a class, ending a romance, fighting with a parent—surely they must realize that better days lie ahead. This perspective, however, is not shared by suicidal adolescents, who are so overwhelmed with depression or anger that, for a few perilous hours or days, death seems the only solution.

Each year millions of young people around the world contemplate, plan, and sometimes attempt suicide. A review of studies from many nations finds that **suicidal ideation**, that is, thinking about committing suicide, is so common among high school students that it might be considered normal (Diekstra, 1995). In these studies, the prevalence rate of suicidal ideation ranged from 15 to 53 percent. Whether the prevalence rate in a given study was at the high or low end of this range depended largely on the specifics of the research, such as whether adolescents were asked if they had "ever" seriously thought about suicide or if they had done so "in the past year." For example, a survey of 1,600 students from twenty-three American states, asking only about the past year, found that 24 percent had thought seriously about suicide, 13 percent had planned how to carry it out, and 9 percent had made at least one suicide attempt in 1993. A third of those attempts were sufficiently serious to require medical attention (*MMWR*, March 24, 1995).

As an index of despair, such numbers are distressingly high. Fortunately, most suicide attempts in adolescence do not result in death. The reasons for this are seldom clear, however. Even on clinical examination, it is often impossible to determine whether what appears to have been a failed suicide attempt was in fact a token gesture at suicide, an attempt to manipulate others, or a potentially lethal attempt to kill oneself that, for one reason or another, did not reach completion. In acknowledgment of this ambiguity, many experts in the field use the term **parasuicide** to denote any deliberate act of self-destruction that does not result in death (Diekstra, 1995). Experts prefer this term over "attempted suicide" precisely because it does not assume the intention of the self-destructive act. This is an important consideration because, particularly in adolescence, the "true" intent of most self-destructive acts—carried out in a state of extreme emotional agitation and confusion—is not clear even to the individuals who have performed them. Many who make what appears to be a "real" attempt to kill themselves are not only relieved to realize that they did not die but later wonder what they could possibly have been thinking when they acted.

suicidal ideation Obsessive or extensive thinking about committing suicide.

parasuicide A deliberate act of self-destruction that does not end in death. Parasuicide can include everything from an overdose of aspirin (rarely deadly) to a single-occupant car crash (often fatal).

Whether or not a suicidal ideation eventually leads to a parasuicide, or to death, depends on a multitude of factors. Four of the most influential variables are the following:

1. the availability of lethal methods;
2. the extent of parental supervision;
3. the use of alcohol and other drugs;
4. the attitudes about suicide that are held by the adolescent's family, friends, and culture.

Two other factors that influence the outcome of self-destructive impulses are gender and culture.

Suicide, Gender, and Culture

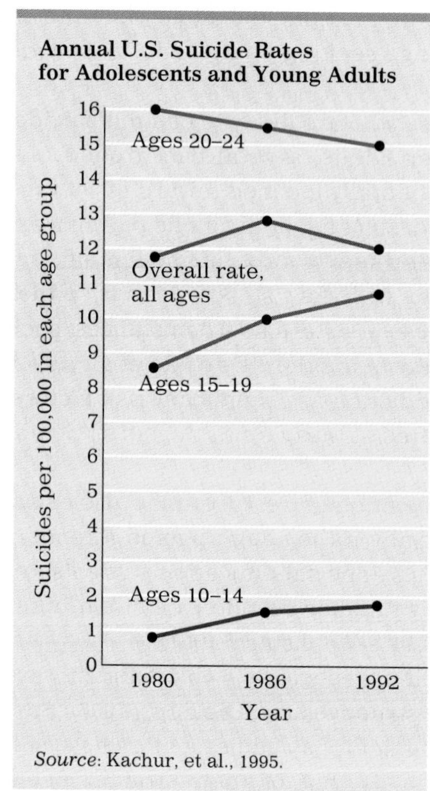

FIGURE 16.2 Charting suicide rates in the United States reveals that much of the media hype about the "epidemic" of teenage suicide is misplaced. The rates have not changed dramatically over the past fifteen years, and teenagers are less likely to kill themselves than adults are—particularly those who are just entering adulthood or those (not shown here) in their 70s and 80s. Obviously, this fact does not diminish the tragedy of adolescent suicide, and the fact that teenage suicide is edging up even as suicide overall is edging down is cause for increased concern and greater efforts at prevention.

At every age, but particularly in adolescence, males are more likely to kill themselves than females are, while the rate of parasuicide is notably higher for females than for males. For example, in the study of 1,600 American high school students cited above, only 5 percent of the boys had made at least one suicide attempt in the past year, as compared with 13 percent of the girls. However, during that same year in the United States, five times as many male 15- to 19-year-olds killed themselves as did females in the same age group. These differences are explained in part by the four key variables listed above: boys are more likely to have guns, are less supervised by adults, and are more likely to abuse alcohol and drugs. In addition, the idea of violent death is interwoven with the American culture's idea of manliness, a topic we will discuss in more detail in the next chapter.

The same factors are at least a partial explanation for cross-cultural variation in the suicide rates among adolescent boys worldwide. Currently, boys in Sri Lanka, Finland, and Hungary are at greatest risk (40 suicides per year per 100,000 youths), while those in Japan, Greece, and Italy are at least risk (10 suicides per year per 100,000 youths). In the middle of these extremes are France, Australia, the United States, and Canada (20 suicides per year per 100,000 youths) (WHO, 1993). Within the United States, the risk factors cited above may also explain ethnic differences in the suicide rates. Native American youths have the highest rates, followed by European-Americans, then African-Americans (whose rate is rising most rapidly), Hispanic-Americans, and finally, Asian-Americans and Pacific Islander–Americans. (Interestingly, these rankings are quite different for older groups. Among people over age 60, Native Americans have the lowest rates of suicide and Asian-Americans, the second highest.)

Overall, the rate of youth suicide in North America has risen over the past few decades, in part because rising rates of divorce, unwed motherhood, and dual-earner families have meant less parental supervision and because adolescents have greater access to alcohol, drugs, and guns. In 1992, more than 3,000 adolescents aged 15 to 19 killed themselves, triple the rate of forty years earlier. Only about 500 younger adolescents killed themselves in that year, but this number represented a doubling of the rate since 1980 (see Figure 16.2) (Kachur et al., 1995; Leenaars & Lester, 1995). For both older and younger adolescents, death by gunshot between 1980 and 1992 almost doubled, while suicides through poisoning, hanging, and cutting held steady. Curiously, during the same period, the much higher suicide rate among 20- to 24-year-olds fell slightly.

Causes of Suicide

What can lead an adolescent to take his or her own life? Although suicide is commonly thought of as a response to a specific and immediate psychological blow, in fact, self-destruction at any point in the life span is usually the final outcome of long-standing problems in the individual, as well as in the person's family and social environment (Brent et al., 1994; Faberow, 1991; Maris, 1994; Shneidman et al., 1994).

Particularly for adolescents, suicide often arises from a family context that is characterized by "anger, ambivalence, rejection, and/or communication difficulties" (Curran, 1987). Against this backdrop of long-standing family tension and conflict, any number of negative events—from a sudden decrease in parental attention (as might occur with a parent's divorce or remarriage) to some occasion that causes the adolescent to feel shame, loathing, or rage—can act as the final straw, plunging the adolescent into an emotional tailspin of despair or revenge-seeking that leads to suicide (Adams et al., 1994; DeMan et al., 1993).

There is no specific "profile" of those who are likely to commit suicide. However, for adolescents, as well as for adults, several background elements increase the risk: being temperamentally inclined toward fits of rage or bouts of depression; having depressed, suicidal, or alcoholic parents; experiencing the early loss of a loved one through divorce, abandonment, imprisonment, or death; and growing up with few steady friends, either because of one's personal traits or because of external circumstances, such as moving frequently from one community to another (DeWilde et al., 1994). At every age, alcohol abuse and drug addiction are significant risk factors, with the suicide rate of alcoholics estimated at thirty times the norm (Schilit & Gomberg, 1991).

During high school and college, educational pressures may also cause depression and suicidal ideation, especially if the young person attends a large and impersonal school. Three groups—special-education students, recent drop-outs, and high-achievers—have higher suicide rates than other students (Davis & Sandoval, 1991; Leenaars & Wenkstern, 1991).

For some adolescents, puberty itself may become problematic in ways that lead to suicidal ideation and self-destructive acts (Diekstra, 1995). Sexual awakening can be particularly troubling for gay adolescents, whose suicide rate is about three times as high as that of their straight peers (D'Augelli & Dark, 1994). Sexually abused youth, male or female, are also at high risk of suicide, especially when they are blamed by their abuser and feel that no one will believe them if they ask for help (Finkelhor, 1994).

Finally, adolescents are susceptible to **cluster suicide**, in which one suicide—especially of a famous person or of a well-known peer—leads others to attempt, and sometime commit, suicide. Often it is not the initial suicide itself that precipitates the cluster but the response of peers, schools, and the media who glorify the life and death of the victim. Large public memorial services, with outpourings of praise and grief, or public statements that somehow justify the death or blame it on others, make vulnerable peers imagine that suicide is an attractive, reasonable, and common option. Because of this contagion effect, some experts suggest that suicides not be reported in the media. However, the consensus is that reports of suicide are necessary to stop gossip and speculation that excite interest, but that the

cluster suicide A series of suicides or suicide attempts that is precipitated by one initial suicide, usually that of a famous person or of a well-known peer.

Hypothetical News Reports with High and Low Potential for Promoting Suicide Contagion

High Potential for Promoting Suicide Contagion

Hundreds turned out Monday for the funeral of John Doe, Jr., 15, who shot himself in the head late Friday with his father's hunting rifle. Town Moderator Brown, along with State Senator Smith and Selectman's Chairman Miller, were among the many well-known persons who offered their condolences to the City High School sophomore's grieving parents, Mary and John Doe, Sr.

Although no one could say for sure why Doe killed himself, his classmates, who did not want to be quoted, said Doe and his girlfriend, Jane, also a sophomore at the high school, had been having difficulty. Doe was also known to have been a zealous player of fantasy video games.

School closed at noon Monday, and buses were on hand to transport students who wished to attend Doe's funeral. School officials said almost all the student body of 1,200 attended. Flags in town were flown at half staff in his honor. Members of the School Committee and the Board of Selectmen are planning to erect a memorial flag pole in front of the high school. Also, a group of Doe's friends intend to plant a memorial tree in City Park during a ceremony this coming Sunday at 2:00 p.m.

Doe was born in Otherville and moved to this town 10 years ago with his parents and sister, Ann. He was an avid member of the high school swim team last spring, and he enjoyed collecting comic books. He had been active in local youth organizations, although he had not attended meetings in several months.

Low Potential for Promoting Suicide Contagion

John Doe, Jr., 15, of Maplewood Drive, died Friday from a self-inflicted gunshot wound. John, the son of Mary and John Doe, Sr., was a sophomore at City High School.

John had lived in Anytown since moving here 10 years ago from Otherville, where he was born. His funeral was held Sunday. School counselors are available for any students who wish to talk about his death.

In addition to his parents, John is survived by his sister, Ann.

Source: MMWR, April 22, 1994.

accounts should be strictly factual with no emotional overtones (*MMWR*, April 22, 1994). (A hypothetical example of the kind of news report that tends to glorify the suicide victim and promote suicide contagion is shown above, along with an example of the kind of report that is unlikely to promote such contagion.)

Responding to the Cry for Help

Many, many adolescents, at one time or another, fall into at least one of the risk categories outlined above, and certainly all merit special attention to help them ride out whatever stresses they are experiencing. But how can we recognize those who are in danger of attempting suicide, and what can we do to prevent them from making a fatal mistake?

One psychologist, Edwin Shneidman, who has devoted his entire career to the study of suicide, believes that every suicide is preceded by clues—verbal, behavioral, and situational—that are "not too difficult to recognize" (Shneidman & Mandelkorn, 1994). Beyond the background factors described above, the following are warning signs that must be taken seriously:

1. *A sudden decline in school attendance and achievement, especially in students of better-than-average ability.* While about a third of the young people who attempted suicide had recently failed or dropped out of school, only 11 percent were in serious academic difficulty before their precipitous decline.

2. *Talk about suicide.* About 80 percent of all those who commit suicide talk about it before they do it, often portraying suicide in idealized, romantic terms.

3. *Withdrawal from social relationships.* The adolescent who decides that suicide is the solution sometimes seems less depressed than previously and may cheerfully say something to the effect of "It's been nice knowing you." A joking or serious "goodbye" accompanied by a sudden desire to be alone is a serious sign. Another potentially serious sign is an adolescent's responding to a failed romance with statements such as "There will never be anyone else for me."

4. *Running away.* The adolescent who literally runs away, or who retreats from normal life through drinking or drugging to the point of oblivion, is at high risk.

5. *Attempted suicide.* An attempted suicide, however weak it might seem, is an effort to communicate serious distress and therefore must be taken seriously. If nothing changes in the adolescent's social world, a parasuicide may turn out to have been a trial run for the real thing. Many adolescent suicides follow failed attempts.

In fact, an attempted suicide is a "late clue," usually preceded by clues that should have signaled to others the need to take action (Faberow, 1994). As Shneidman explains, reading these clues is everyone's "moral responsibility, something akin to omnipresent fire prevention" (Shneidman & Mandelkorn, 1994). What should be done when the alarm sounds? Shneidman stated it quite clearly: "the way to save a person's life is also to . . . put your knowledge of the person's plan to commit suicide into a social network—to let others know about it, to break the secret, to talk to the person, to talk to others, to offer help, to put action around the person, to show response and concern, and, if possible, to offer love" (Shneidman, 1978).

Conclusion

As this trio of chapters draws to a close, let us look again at the years from age 10 to 20. Except perhaps for the very first months of life, no other period is characterized by changes so multifaceted and inexorable. Nor is the developing person likely to experience a sequence of changes more fascinating, or more potentially confusing, than those that adolescents typically undergo. Their developmental tasks—to grow to adult size and sexuality, to adjust to different educational expectations and intellectual patterns, to develop autonomy from parents and intimacy with friends, to achieve a sense of identity and purpose—are too complex to be accomplished without some unanticipated surprises. No wonder every young person, in every family and culture, experiences some disruption (Schlegal & Barry, 1991).

As we have seen, most adolescents, most families, and most cultures survive this transition fairly well. Parents and children bicker and fight, but they still respect and love each other. In America, many teenagers skip school, eat unwisely, drink too much, experiment with drugs, break laws, feel depressed, rush sexual activity, conform to peer pressure, disregard their parents' wishes—but all these behaviors typically stay within limits.

They do not occur too often or last too long; they do not lead to lifelong or life-threatening harm. For most young people, the teenage years overall are happy ones, as they escape serious problems and discover the rewards of maturity.

Unfortunately, while all adolescents have some difficulties, those with one serious problem seem to have several others as well (Cairns & Cairns, 1994; Dryfoos, 1990). For instance, girls who become teenage mothers also tend to be those from troubled families, likely to leave school and experiment with hard drugs before age 16. Boys who become repeat delinquents also tend to be alienated from their families, failing in school, drug-abusing, and lacking in close friends. Suicidal adolescents typically are heartbreakingly lonely, with inadequate social support from family, friends, and school.

In almost every case, these clusters of problems stem from earlier developmental events, beginning with genetic vulnerability and prenatal insults and continuing with family disruptions and discord in early childhood and with learning disabilities and aggressive behavior in elementary school—all within a community that does not provide adequate intervention. With the inevitable stresses of puberty, such early handicaps become worse and more obvious, as well as more resistant to change. If these chronic patterns are not somehow altered, they are not "grown out of"; they persist into adulthood and begin to disrupt the development of the next generation, when those who should become responsible workers and nurturing parents are unable to do so.

Fortunately, an encouraging theme is apparent in all three adolescent chapters. No developmental trajectory is set in stone by previous events, and adolescents are, by nature, innovators, idealists, and risk-takers, open to new patterns, goals, and lifestyles. Research on effective schools, on teenage drug programs, on the positive role of friendship, and on identity achievement shows that some individuals take a path distinct from the limitations and burdens of their past. For some of those who find adolescence sorrowful, humiliating, and problem-filled, the next stage of life—early adulthood—can be a time for growth, achievement, and fulfillment.

SUMMARY

The Self and Identity

1. The growth of self-understanding takes on a new dimension in adolescence, as teenagers begin to recognize and sort through their "multiple selves," which may reflect different, and sometimes contradictory, aspects of their personality in different settings and different circumstances. As they search for their real self among these diverse multiple selves, adolescents often intentionally act out a false self, sometimes to gain basic acceptance, sometimes to impress others or win approval, and sometimes to test a new role.

2. According to Erikson, the psychosocial crisis of adolescence is *identity versus role confusion*. Ideally, adolescents resolve this crisis by developing a sense of both their own uniqueness and their relationship to the larger society, establishing a sexual, political, moral, and vocational identity in the process. The difficulty of this process depends partly on the society: if its basic values are consistent and widely accepted, the adolescent's task is fairly easy.

3. Sometimes the pressure to resolve the identity crisis is too great, and instead of exploring alternative roles, young people foreclose their options. Others may take on a negative identity, defying the expectations of family and community, while some teenagers experience identity diffusion, making few commitments to goals, values, or a particular self-definition. Many young people declare a moratorium from settling on a mature identity.

4. In industrial and postindustrial societies, social change is rapid, and identity possibilities are endless. Conse-

quently, identity achievement can be more difficult, especially for those—such as members of minority groups—who are caught between diverse cultural patterns.

Family and Friends

5. Parents are an important influence on adolescents: the generation gap within families is usually not very large, especially with regard to basic values. Children tend not to stray too far from parental values and ideals, and parents have a personal stake in minimizing whatever conflicts there are.

6. Conflict can emerge in parent-adolescent relationships, however, depending on the adult's parenting style and the teenager's stage of development. In particular, the onset of puberty forces a recalibration of mutual expectations that can lead parents and adolescents to see their relationship much differently. As at earlier stages of development, authoritative parenting is generally most apt to foster self-esteem and a positive parent-child relationship. Among some ethnic groups, and in certain conditions, a parental style that is more authoritarian but still nurturing may be beneficial.

7. The peer group is an important source of information and encouragement for adolescents. The adolescent subculture provides a buffer between the world of children and the world of adults, allowing, for example, a social context for the beginning of heterosexual relationships.

8. Thus parents and peers are both important social influences on the adolescent, filling complementary rather than conflicting roles. However, especially as they grow older, adolescents spend much more time with peers (who usually listen to them) than with parents (who often tell them what to do). While most close friendships in early adolescence are with the same sex, by late adolescence friendships typically include members of the other sex, as romantic relationships begin to develop.

Special Problems

9. Misbehavior that is illegal is more common in adolescence than in any other period of the life span, but serious delinquency involving assault, theft, or rape is not. Progress has been made toward identifying a child at risk for serious delinquency: those children who by age 10 have learning difficulties in school, act aggressively toward other children, and have significant stresses at home are more likely to become delinquents.

10. Ideally, prevention of juvenile delinquency should include programs to (1) provide early intervention to help parents to discipline their children effectively, (2) support schools' efforts to remedy academic difficulties and foster academic aspirations, and (3) develop neighborhood support networks.

11. Suicidal ideation is very common among high school students, with a small minority engaging in deliberate acts of self-destruction (parasuicide). Wide variation in teenage suicide rates is evident among ethnic and national groups, but worldwide, girls are more likely to attempt suicide and boys are more likely to complete it.

12. Most adolescent suicides are preceded by a long sequence of negative events, including family problems and breakdowns in family communication, drug and/or alcohol abuse, and sometimes a critical event that triggers a suicide attempt. Suicide prevention requires heeding the preliminary warning signs and responding quickly to cries for help.

KEY TERMS

false self (438)
identity versus role confusion (439)
identity (439)
identity achievement (439)
foreclosure (439)
negative identity (439)
identity diffusion (440)
moratorium (440)
generational stake (444)
peer pressure (449)
parental monitoring (455)
suicidal ideation (456)
parasuicide (456)
cluster suicide (458)

KEY QUESTIONS

1. Why is the gulf between "real self" and "false self" behavior so concerning to adolescents?

2. What are some of the difficulties that adolescents might experience on the way toward identity formation?

3. How might identity formation be difficult for minority-group adolescents?

4. Which parenting characteristics are most helpful to adolescents? Why?

5. What are some of the factors that lead to conflict in parent-adolescent relationships?

6. What functions does the peer group assume for teenagers?

7. What social and cultural characteristics tend to be associated with delinquency?

8. What are some of the psychosocial and cognitive patterns associated with adolescent suicide?

Biosocial Development

Physical Growth

Between the ages of 8 and 14, puberty begins with increases in various specific hormones that trigger a host of changes. Within a year of the hormonal increases, the first perceptible physical changes appear—enlargement of the girl's breasts and the boy's testes. About a year later, the growth spurt begins. During adolescence, boys and girls gain in height, weight, and musculature. The growth that occurs during these years usually proceeds from the extremities to the torso and may be uneven.

Changes in Sex Organs and Secondary Sex Characteristics

Toward the end of puberty, menarche in girls and ejaculation in boys signals reproductive potential. On the whole, males become taller than females and develop deeper voices and characteristic patterns of facial and body hair. Females become wider at the hips; breast development continues for several years. Puberty that is early or late can be stressful, although the specifics depend on gender, personality, and culture.

Cognitive Development

Adolescent Thinking

Adolescent thought emphasizes the possible more than the actual, with a newly emerging ability to think hypothetically, to reason deductively, and to explain theoretically. At the same time, adolescent egocentrism, along with feelings of uniqueness and invincibility, can cloud teenagers' judgment, as well as make them extraordinarily self-absorbed.

Education

The specific intellectual advancement of each teenager depends greatly on education. Each culture and each school emphasizes different subjects, values, and modes of thinking, a variation which makes some adolescents much more sophisticated in their thoughts and behavior than others. The interplay between education and egocentrism helps explain why some teenagers are at greater risk for STDs, AIDS, and pregnancy than others.

Psychosocial Development

Identity

One of the major goals of adolescence is self-understanding and self-determination, culminating in identity achievement. Achieving identity can be affected by personal factors—including relationships with family and peers—the nature of the society, and the economic and political circumstances of the times. Identity achievement can be especially problematic for members of a minority group in a multiethnic society.

Peers and Parents

The peer group becomes increasingly important in fostering independence and interaction, particularly with members of the other sex. Parents and young adolescents are often at odds over issues centering on the child's increased assertiveness or lack of self-discipline. These difficulties usually diminish as teenagers become more mature and parents allow more autonomy. While most adolescents try drugs, break the law, and sometimes get depressed, the minority who have serious problems in these areas often come from a troubled family and a debilitating social context. More supportive communities can moderate these problems.

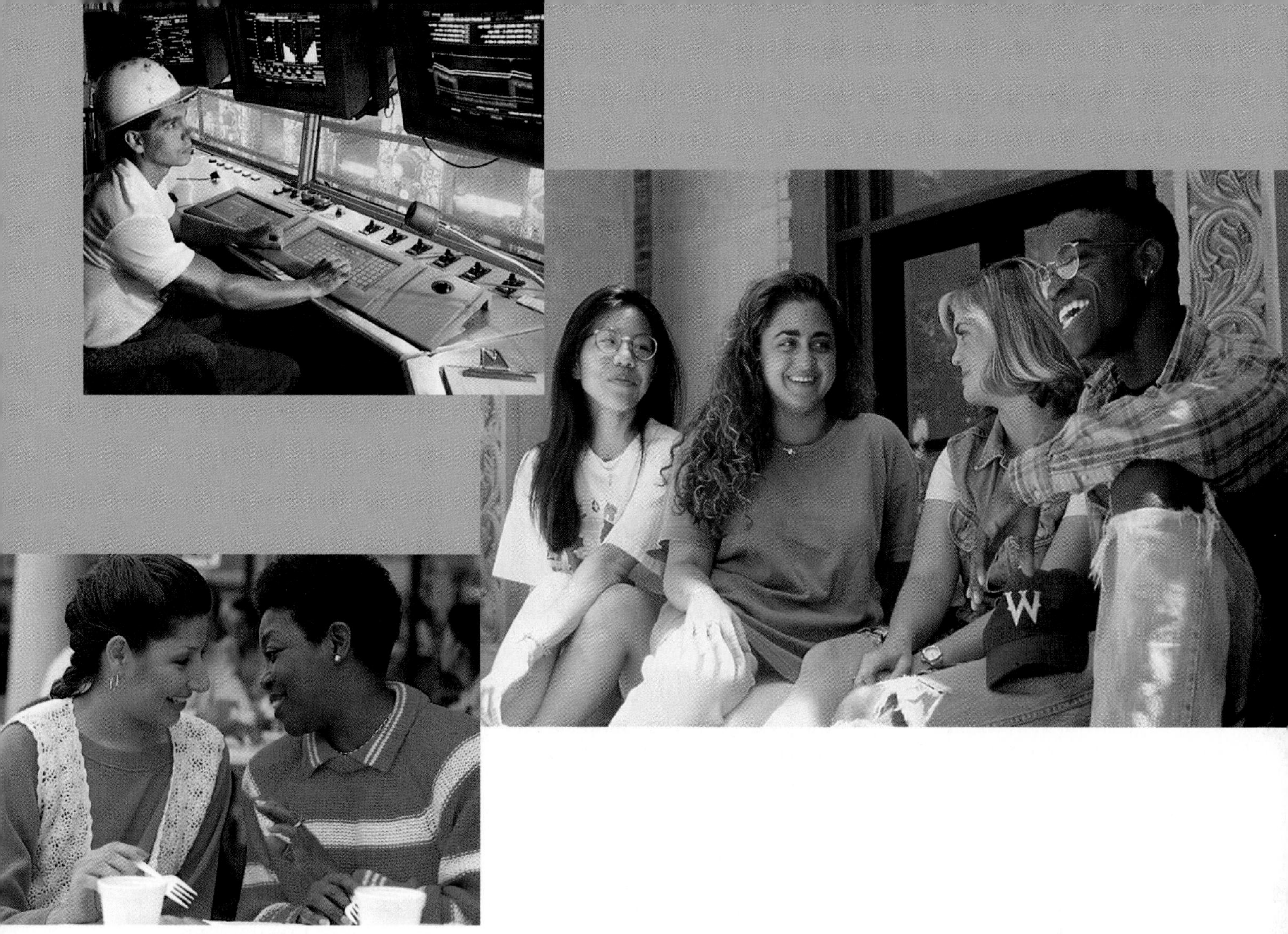

PART VI Early Adulthood

As young children, we look forward to the day when we will be "all grown up," imagining that when we attain adult size, we will automatically master the roles, privileges, and responsibilities of adulthood. As young teenagers, we likewise impatiently await our high school graduation or 18th or 21st birthday, anticipating that independence, and the competence to cope with it, will be bestowed when we arrive at these "official" milestones.

But young adults, who must make their own decisions about career goals, intimate relationships, social commitments, and moral conduct, usually find these aspects of independence, though exciting, far from easy to deal with. This is especially true today because the array of lifestyle choices seems so vast and varied. No matter which of the roles of adulthood they choose to take on, or how thoughtfully and eagerly they strive to play them, they are bound to be confronted with stresses, set-backs, and second thoughts. Yet for most young adults, it is problems faced and usually solved, and limitations accepted or overcome, that make the decades from ages 20 to 40 an exhilarating period when people often feel they are living to the fullest. The next three chapters describe how many young adults cope with the engrossing, multidimensional realities of early adulthood.

Early Adulthood: Biosocial Development

CHAPTER

17

In terms of physiological development, early adulthood, from age 20 to roughly age 40, can be considered the prime of life. Our bodies are stronger, taller, and healthier than during any other period. The first years of young adulthood (the early 20s) are the best ones for hard physical work, for problem-free reproduction, and for peak athletic performance. As we will see, although the advancing years of early adulthood are accompanied by some declines throughout the body, whatever difficulties young adults experience in biosocial development are usually related to factors other than aging per se.

Before examining the biosocial development of early adulthood in detail, let us clarify the connection between chronological age and the aging process. In childhood these two variables are highly correlated. As we saw in earlier biosocial chapters, the normal annual increases in height and weight over the course of childhood occur in a predictable progression, and even the changes of puberty, while variable in age of onset, are predictable in sequence, direction, and scope, once final maturation begins.

Sometime between ages 15 and 30, overall growth stops and **senescence**, defined as the gradual decline in physical functioning related to age, begins. That much is true for everyone: no one at the end of early adulthood, at age 40, has a body that functions as well in every aspect as it did at the beginning of adulthood, at age 20. However, throughout adulthood, the connection between senescence and chronological age is tenuous, with marked variation from person to person and organ to organ, due to the influences of genes, the environment, and personal lifestyle. Even within a given individual, notable deviations from the norm may occur, including increases in capacity when the usual pattern is for decline or sudden drops in capacity when steady maintenance is the norm. Keep this variability in mind, especially when attempting to apply the generalities in these chapters to those individuals—including yourself—whom you know best.

Growth, Strength, and Health

senescence The state of physical decline, in which the body gradually becomes less strong and efficient with age.

For most people, the genetic and hormonal forces propelling growth have stilled by age 20 or so. For example, noticeable increases in height have stopped by the beginning of early adulthood. In fact, most girls reach their maximum height by age 16, and most boys reach theirs by age 18, the chief exception being late-maturing boys who may undergo final skeletal growth

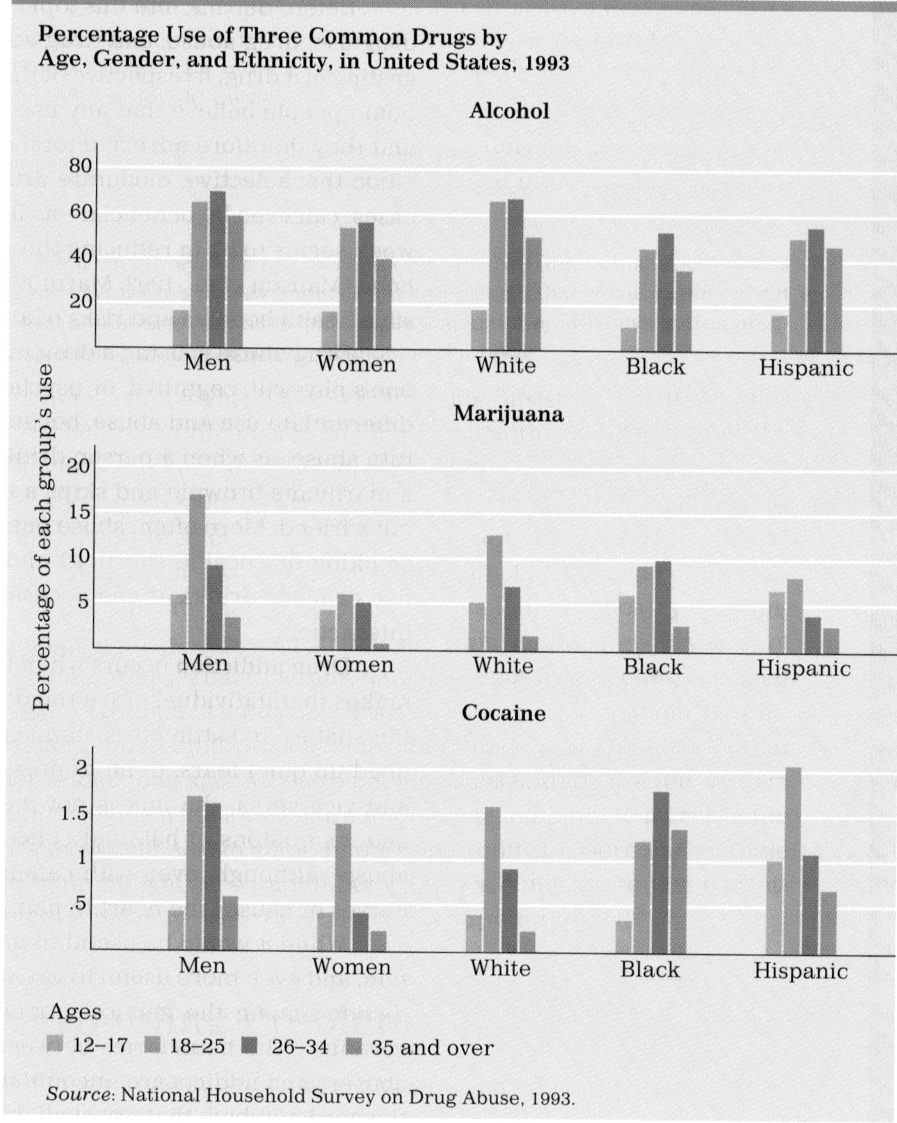

FIGURE 17.2 The data depicted in these graphs clearly show gender and ethnic differences in drug use at various ages. Overriding these differences is the clear fact that young adults are the group most likely to use and abuse drugs.

tween ages 19 and 23, with the late teens and early twenties being the time of heaviest alcohol and marijuana use, and about age 23 being the time of heaviest use of other drugs, including cocaine. Often drug abuse ceased before age 30 (69 percent of the marijuana smokers and 67 percent of the cocaine users had quit by that age, as had 11 percent of the drinkers).

Another survey, conducted at 104 colleges of all sizes in every region of the United States, found that undergraduates are particularly vulnerable to drug abuse (Presley, 1995). Within the two weeks preceding the survey, 48 percent of the men and 30 percent of the women had consumed five or more drinks at one occasion. Within the month preceding the survey, 10 percent of the undergraduates had used marijuana and 1 percent had used cocaine (in the year preceding the survey, the use rates for these two drugs were 25 percent and 6 percent, respectively).

The harmful consequences of this use and abuse of drugs were many. In answer to a question regarding harm "because of their drug or alcohol use" within the past year, many students admitted that they had been hung-

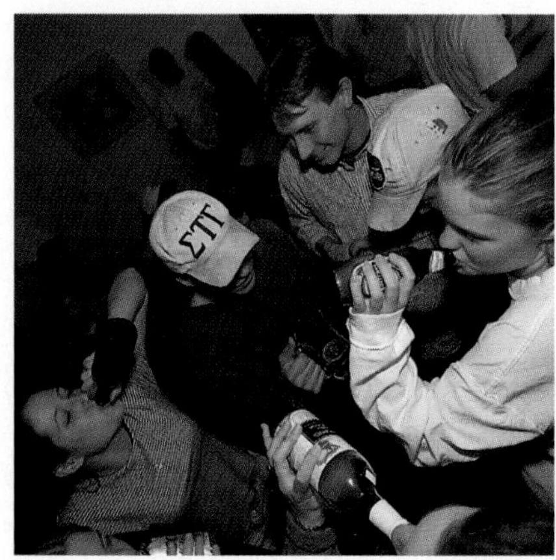

For some young adults, social camaraderie demands drinking to the point of being "blasted," "wasted," "plastered," and sometimes, quite literally, "smashed." This form of socializing was once considered a "male thing" but in recent years it has been taken up by young women at an alarming rate, particularly on college campuses.

over (60 percent) and nauseous (48 percent), had done things that they later regretted (37 percent), had driven a car under the influence (33 percent, with only 2 percent arrested for driving while intoxicated), had gotten into a fight (30 percent), had missed class (27 percent), had suffered a memory loss (25 percent), had performed poorly on a test (20 percent), had been physically hurt (15 percent), had been taken advantage of sexually (8 percent of the men, 10 percent of the women), had taken sexual advantage of someone else (11 percent of the men, 2 percent of the women), had thought of suicide (5 percent), and had attempted it (1 percent). Further, 11 percent thought that they might have an alcohol or drug problem, and 8 percent of the men and 4 percent of the women had tried to stop their own abuse and had failed, an ominous sign of addiction.

Why are the rates of drug use, and presumably of abuse, highest in the early years of young adulthood? Remember from Chapter 3 that a user becomes an abuser or addict not only because of a genetic predisposition but also because of personality and social factors. The temperament of those most likely to misuse drugs includes attraction to excitement, intolerance of frustration, and vulnerability to depression (Brook et al., 1992; Kaplan & Johnson, 1992). How powerfully those three characteristics—sensation-seeking, quick temper, and depressive mood—affect a particular person is influenced partly by genes and partly by early family experiences. But the force of such traits is also affected by the person's age; and almost everyone is more vulnerable to these tendencies in adolescence and early adulthood than at other developmental periods.

In addition, for many young adults, the social context produces a convergence of factors that encourage drug use and abuse. For one thing, availability of drugs and the opportunity to use them increase as many young adults are free to make their own lifestyle choices, no longer supervised or even observed by their parents. At the same time, young adults experience a number of pressures that drugs might temporarily relieve, not only the pressures to complete an education, establish a career, and find a mate but also the more immediate need to feel sophisticated and socially at ease. Also encouraging drug use are certain group activities—large parties, rock concerts, spectator sports—at which excessive drug use is tolerated and even expected, sometimes by one's own companions. Indeed, a careful longitudinal study found that the single most important correlate of drug use among young adults—even more important than life stress, temperament, and personal attitudes—was having friends who used drugs (Jessor et al., 1991). Finally, young adults are least likely to be regularly exposed to a powerful deterrent to serious drug and alcohol abuse—religious faith and practice (Brunswick et al., 1992). Americans in their 20s join religious groups, attend services, and pray less often than do those in any other age group (Gallup, 1993).

As a consequence of this accumulation of factors, the rate of both abuse and addiction increases from about age 18 to 26 (Glantz & Pickens, 1992), and the harm is measurable: compared to others their age, young adult drug users are more likely to avoid, fail, or drop out of college, to be employed below their potential and then lose or quit those jobs, to be involved in transitory, uncommitted sexual relationships, to die violently (a topic we will discuss shortly), and to suffer from serious eating disorders, a subject we will now examine.

Even if drug abuse in young adulthood does not lead to addiction, it nonetheless can take a serious toll on development. The conclusion for all young adults is that their age puts them at risk, and that the sooner they recognize an alcohol or other drug problem in themselves or their friends, the sooner they will be able to treat the problem and get on with their lives.

Dieting as a Disease

For many young women, dressing in the morning is an obsessive ritual of checking for the slightest sign of weight gain from the day before. Developmentalists decry the fact that vast numbers of American women are held hostage to a cultural "ideal" of thinness that is both abnormal and unhealthy.

The processes of homeostasis work to maintain body weight just as they do to maintain sufficient oxygen in the blood or normal body temperature. Obvious mechanisms such as pangs of hunger and the feeling of fullness, and less obvious ones such as fluctuations in hormonal levels and neurotransmitter activity, regulate the urge to eat. Healthy and active adults and children tend to "automatically" consume sufficient calories to maintain their required energy level. In addition, many scientists believe that each person has a certain "set point," that is, a particular weight that the individual's homeostatic processes strive to maintain (Schlundt & Johnson, 1990). A person's set point is powerfully influenced by genes, which strongly affect metabolism as well as the storage and distribution of body fat (Bouchard, 1989; Skunkard et al., 1990). Nurture, particularly factors that establish childhood eating habits, affects the set point as well. The set point is not rigidly fixed, however. Throughout adulthood, it can change somewhat through the effects of diet, age, illness, hormones, and exercise, a potential for a shift that makes some experts think the term "settling point" is more accurate (Bouchard & Bray, 1996).

Recognizing that everyone has his or her own natural weight, medical experts now maintain that people can vary by 20 pounds or so from others who are the same height and still be in the healthy range. To assess whether a person is too thin or too fat, clinicians calculate his or her **body-mass index (BMI)**, the ratio of a person's weight in kilograms divided by his or her height in meters squared (see Table 17.3). A healthy BMI is somewhere between 19 and 25.

Cultural norms and specifics of the social context can undermine the body's natural tendency toward maintaining a healthy weight, causing a person to eat substantially more or less than his or her homeostatic controls call for. A prime example is the cultural notion that an ideal body is "fat-free," a notion that causes many young adult women to strive for an elusive and unhealthy thinness, with a BMI far below the healthy range. Such striving is common among young women in nearly all developed nations, most of which are gripped by a cultural obsession with female thinness that is fueled at every turn by the popular media. White American women seem particularly vulnerable to this obsession (Cash & Henry, 1995).

The result is that many women needlessly feel dissatisfied with their bodies and attempt to lose weight. A national survey of 15,000 American women, for example, found that 44 percent of those between the ages of 17 and 60 were currently dieting. On average, they hoped to lose 30 pounds (14 kilograms)—a goal that, if attained, would have made most of them underweight (Williamson et al., 1992). Another survey of dieting American women found that almost half of them had a body-mass index under 25—and thus were not at all overweight—and that those who were young, well-educated, and employed were the ones most likely to diet and the least likely to need to (Biener & Heaton, 1995).

body-mass index (BMI) The ratio of a person's weight in kilograms divided by his or her height in meters squared.

TABLE 17.3

How to Calculate Body-Mass Index (BMI)

Many current studies of the relationship between body size and overall health focus on the ratio of body weight to height, referred to as the body-mass index, or BMI. The formula for calculating body-mass index is:

$$\text{BMI} = \frac{w}{h^2}$$

w standing for weight in kilograms (pounds divided by 2.2)

h standing for height in meters (inches divided by 39.4)

Thus, if a person measures 1.65 meters (5 feet, 5 inches), his or her height squared is 2.72. At various weights, that person's BMI is:

45 kilograms [99 pounds] ÷ 2.72 = 16.6 [dangerously underweight, anorexic]

50 kilograms [110 pounds] ÷ 2.72 = 18.4 [underweight]

55 kilograms [121 pounds] ÷ 2.72 = 20.2 [normal weight]

60 kilograms [132 pounds] ÷ 2.72 = 22.1 [normal weight]

65 kilograms [143 pounds] ÷ 2.72 = 23.9 [normal weight]

70 kilograms [154 pounds] ÷ 2.72 = 25.7 [overweight]

75 kilograms [165 pounds] ÷ 2.72 = 27.6 [obese—new United States standard]

80 kilograms [176 pounds] ÷ 2.72 = 29.4 [obese]

85 kilograms [187 pounds] ÷ 2.72 = 31.3 [severely obese—world standard]

Overall, the BMI for both men and women should be between 19 and 25, with more muscular people on the higher end of that range and less muscular people on the lower end (since muscle weighs more than fat). Below 18 is considered anorexic; above 28 is usually considered obese. The World Health Organization has set a BMI of 30 as severely obese.

Of course, from a developmental point of view, it is important to avoid becoming overweight. As we will see in Chapter 20, excess weight, especially in combination with a lack of exercise, leads to serious health risks. But when a person's self-selected target weight is substantially lower than the natural, healthy weight sought by the body's homeostatic mechanisms, dieting is not only unnecessary from a health standpoint but is almost always bound to lead to failure and frustration. In addition, the disturbances in natural equilibrium that this kind of dieting creates seem to make overweight more likely later on. Millions of young women share a self-defeating set of circumstances—feeling fat, reading diet books, dieting, doing spot exercises, losing weight, and then, as their bodies attempt to compensate, gaining weight, which leads to unhappiness and a new obsessive-compulsive cycle. As one expert explains:

> It is as if women wear blinders to shut out half the truth about diets. On the one hand, women know how to lose weight—and they do so. In any room filled with chronic dieters, you will find them all knowledgeable about calories and nutrition. Dieters are not short on information on "proper eating." They know everything possible about all kinds of diets, from the liquid protein to the "sensible," moderate balanced diets. Most women have been on numerous and varied diets and lost weight on all of them. This is the part they remember. The second half of the experience—gaining back more than they lost—is routinely forgotten. [Gutwill, 1994]

The fact that many diets fail may actually be a good thing for biological health, because if women were able to lose all the weight they wished, and keep it off, they might impair their health. Especially for women, being too thin is disruptive of normal development: insufficient fat halts the natural hormonal rhythms that are the hallmark of healthy womanhood, making menstruation irregular and reproduction difficult or impossible (Hsu, 1990).

Not only is strict leanness uncalled for in terms of good health, but repeated dieting can be very unhealthy as well as counterproductive. This is especially true when it involves crash diets, which nearly all result in nutritional imbalance, energy loss, and vulnerability to disease. For many dieters, the health consequences are even more deleterious when, frustrated, guilty, or depressed at having their best efforts "sabotaged" by their body's natural chemistry, they become dependent on diet drugs (Zerbe, 1993). Women are particularly likely to abuse illegal or prescribed stimulants that work for a few weeks until tolerance builds, requiring the person to quit or become hooked, or to rely on over-the-counter appetite suppressants that can produce such side effects as insomnia, tenseness, anxiety, and, in megadoses, psychosis.

Eating Disorders

For some dieters, especially those who are young and well-educated, their obsession with food and weight control turns into a serious eating disorder, creating an addiction no less powerful or shameful to the addict than alcoholism or heroin. One such problem is **anorexia nervosa**, an affliction characterized by self-starvation, sometimes to the point of death. Typically, a high-achieving female who is in early adulthood (or sometimes in early puberty) restricts her eating so severely that her BMI goes below 18. She may weigh a bony 80 pounds or less but still be exercising and complaining about being fat. Approximately 1 percent of young adult and adolescent females are anorexic (DSM-IV, 1994).

About three times as common as anorexia, especially among young female adults, is the other major eating disorder of our time, **bulimia nervosa**, which involves compulsive binge eating followed by purging through vomiting or taking massive doses of laxatives. Such behaviors are performed on occasion by many young adult women, with some studies finding that half of all college women have binged and purged at least once (Fairburn & Wilson, 1993). To warrant a clinical diagnosis of bulimia, such behaviors must occur at least once a week for three months, and the person must experience uncontrollable urges to overeat and show a distorted self-judgment based largely on body size. Between 1 and 3 percent of American women are clinically bulimic during early adulthood (DSM-IV, 1994). While people who suffer from bulimia are usually close to normal in weight and therefore unlikely to starve to death, they can experience a wide range of serious health problems, including severe damage to the gastrointestinal system and cardiac arrest from the strain of electrolyte imbalance (Hsu, 1990).

College women are at particular risk for eating disorders, and college athletes—who, in theory at least, should be most concerned about health and fitness—are even more vulnerable than others (Cohn & Adler, 1992; Guthrie, 1991; Thompson & Sherman, 1993). Men athletes—especially wrestlers, rowers, and swimmers—are vulnerable as well.

Intertwined with the physical consequences of excessive dieting are the psychological ones, including low self-esteem and depression, that can act as a stimulus for an eating disorder and then as a reason to continue this destructive pattern. Fasting, binging, and purging "have powerful effects as immediate reinforcers—that is, in relieving states of emotional distress and tension" (Gordon, 1990). The result is that the person becomes en-

anorexia nervosa A serious eating disorder in which a person restricts eating to the point of emaciation and possible starvation. Most victims are high-achieving females in early puberty or early adulthood.

bulimia nervosa An eating disorder in which the person, usually female, engages repeatedly in episodes of binge eating followed by purging through induced vomiting or use of laxatives.

meshed in an increasingly destructive and addictive cycle, usually requiring outside intervention—individual psychotherapy or group therapy (such as Overeaters Anonymous)—to break the chains of depression and dieting, binging and purging. Without help, the problem can get very serious, not only causing the physical damage outlined above but also creating a despair so severe that it can lead to suicide (Marx, 1993).

Why do women torture themselves so with eating habits that are contrary to health and happiness? The most obvious explanation is, as outlined above, the cultural pressure to meet the current, and arbitrary, "slim and trim" female ideal, a pressure exerted particularly on unmarried young women seeking autonomy from their parents' nurturance. Other possible explanations range from the psychoanalytic theory that women with eating disorders have a conflict with their parents, who provided their first nourishment, to the more sociological explanation that, as women enter the workplace, they try to project a strong, self-controlled, masculine image. Recently it has been learned that a genetic component may be present as well. The concordance rate for anorexia among female monozygotic twins is 75 percent, compared with a 27 percent rate among dizygotic female twins (Treasure & Holland, 1993). Further, individuals with eating disorders tend to have a family history of depression and to exhibit a higher-than-normal rate of depression before the onset of the disorder. They also tend to have a family history of alcoholism, or drug addiction, or both (Marx, 1993; Miller, 1993).

No matter what the explanation, it seems clear that for many of today's young women, dieting is a disease and that during the young adult years, when women are supposed to be at their peak, many jeopardize their health with distorted ideas of how their bodies should appear.

Violent Death

Just as contemporary notions about "feminine" appearance may promote the excessive dieting and eating disorders that afflict many young women, stereotypes about "manly" behavior may lead to a problem that afflicts mostly young men—**violent death**, that is, death from accident, homicide, or suicide.

Indeed, relative to all other age groups, young adult males are at increased risk for virtually every kind of violent death, from car crashes to gang shoot-outs, from jumping off roofs to overdosing on drugs. More specifically, between his 15th and 35th birthday, one American male in every forty-four dies violently (U.S. Bureau of the Census, 1996). Other nations of the American continents also have high rates, with those of Canada, Mexico, and Chile being as high as the U.S. rate, although the specific mixture of suicide/homicide/accident varies from country to country, with Canada, for example, having far more suicides than homicides (Leenaars & Lester, 1995). Most European and Asian countries have somewhat lower rates of violent death, and age-specific data are unavailable for most Middle Eastern or African nations. However, worldwide, every nation or city that tallies such statistics finds that young men in their early 20s are much more likely to die from violence than from disease or famine, and are more likely to die violently than women of their own age and national group, with a sex ratio at least 3:1 and sometimes as much as 10:1 (World Health Organization, 1994).

violent death Death from accident, homicide, or suicide. It afflicts mostly young men and is often brought about by stereotypically "manly" attitudes and behavior.

RESEARCH REPORT

Male Violence and the Social Context

Although young men worldwide are more likely to engage in unnecessary bravado and irrational risk-taking than young women or older adults are, the chances of a particular young man dying a violent death, and the specific type of death he risks, depend on many factors within him and within his family, neighborhood, and culture. Maleness is much more hazardous to some men in some social settings than in others.

This variability is apparent in international comparisons of homicide rates (see figure). One reason for these differences is the variation in cultural ideals of what constitutes a "real man," variations that can, for example, encourage young men to master the art of gentlemanly compromise or propel them to exhibit an uncompromising macho facade. Family discipline techniques, school curricula, television heroes, religious values, and alcohol availability also vary markedly from nation to nation and from subculture to subculture, and these differences have obvious effects on how predisposed young men are to various expressions of violence.

An additional contextual factor frequently cited for the variation in homicide rates, and especially for the extremely high rates of homicide among American young men, is the availability of firearms. In 1993, 42 percent of all American adults admitted that there was a firearm of sorts in their household, with most such households having more than one gun (U.S. Bureau of Justice Statistics, 1995). In theory, such weapons are equally available to all adults in the family, but young adult males are the most likely to use them and to be injured by them. Overall, the firearm-death rate in the United States is highest for both sexes at ages 20 to 24, with the rate among young adult men almost ten times higher than that among women (U.S. Bureau of the Census, 1996).

Variation in International Homicide Rates (Males 15–24 Years of Age)

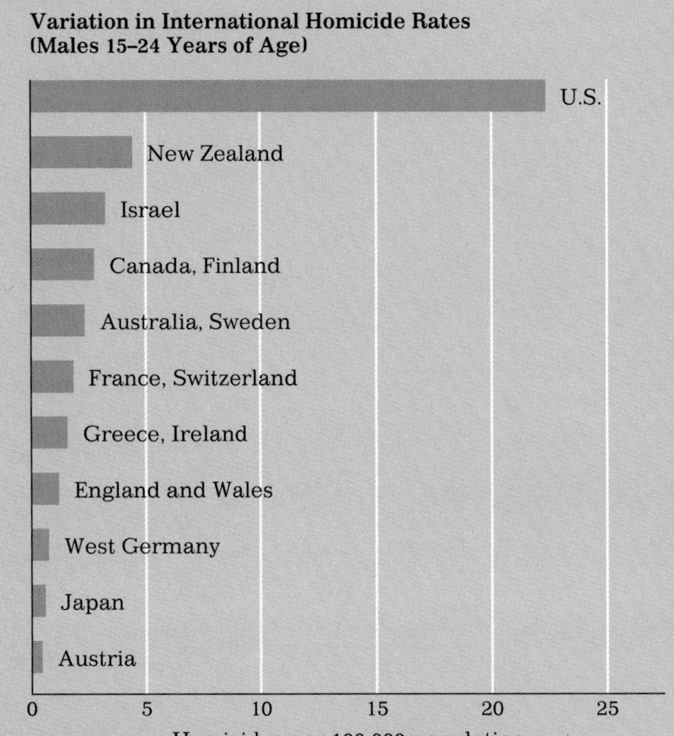

Source: National Center for Health Statistics, World Health Organization, and country reports for 1986–1987.

Values and Violence

According to social scientist Miriam Miedzian (1991), the tendency of young men to perpetrate and to suffer harm is due to a cascade of biosocial factors. Among these are factors that influence every young man, such as higher testosterone levels, and less common ones that nevertheless affect males disproportionately, such as dyslexia, attention-deficit disorder with hyperactivity, and certain genetic abnormalities. In addition, cultural and familial factors, including child maltreatment, divorce, movie and television violence, the glorification of war, and the lure of drug abuse all seem to promote destructive behavior more in males than in females.

In the final analysis, social values are at the root of the problem. A society that turns positive masculine tendencies such as courage, indepen-

While some argue that "guns don't kill people, people kill people," public-health and law-enforcement experts emphasize that the presence of a gun often transforms nonlethal aggressive impulses into deadly ones (Centers for Disease Control, 1992). Research tends to verify this. Consider the violent-death statistics in the mid-1980s from two demographically, economically, and geographically similar cities in the Pacific Northwest—Seattle, Washington, and Vancouver, British Columbia. The data reveal that the residents of Vancouver were no less aggressive than those of Seattle: both cities had similar rates of violent assault and similar rates of arrest, conviction, and punishment. However, murder in Seattle was almost twice as common as in Vancouver (Sloane et al., 1988). The most likely explanation for this disparity lies in the relative prevalence and legality of guns: Seattle had no restrictions on gun ownership or use—except the need for a permit to carry a loaded, concealed weapon on public property—while in Vancouver, gun possession was severely curtailed.

Gun availability also seemed to affect suicide rates. Overall suicide rates were similar in the two cities, but fewer young men killed themselves in Vancouver than in Seattle. The researchers suggest that, for the youthful Canadians, not having a gun at hand allowed time for self-destructive impulses to subside before it was too late, while their armed, suicide-prone peers 60 miles southward had no such buffer (Sloane et al., 1990).

Within nations, violent death varies by subgroup, largely because of cohort, ethnic, and socioeconomic forces. For example, overall, African-Americans are less likely to kill themselves than European-Americans are, a disparity that increases with age to the point that, among 80-year-olds, the whites' rate is five times that of blacks (Kachur et al., 1995). In general, the suicide rate of young Native Americans is well above the national rate and that of Hispanic-Americans is well below it, although marked differences occur within specific tribes and Latino subgroups (Centers for Disease Control, 1992). Among the reasons for these differences are that, compared to European-Americans, African-Americans tend to have more extensive family and friendship networks, which helps protect against the sense of isolation that is typically a precondition to suicide. Further, the suicide rates for Hispanic-Americans and Native Americans are partly explained by religious influences: the Catholic tradition of most Hispanic-Americans holds suicide to be a mortal sin, whereas in some Native American traditions, suicide can be viewed as an honorable act.

Another notable ethnic variation in violent death is that homicide is the leading cause of death for young African-American and Latino men, while among European-American men, accidents are the number-one cause. The reason is primarily economic: white young men are more likely to be middle-class and living in the suburbs, and hence are more likely to own and drive a car—a potentially dangerous tool in the hands of a young man. By contrast, daily life for many young black men includes survival in the inner city—overcrowded, crime-torn, drug-infested, and job-poor—where violent impulses often intensify to the breaking point, turning even minor disagreements between friends into deadly confrontations. Indeed, a black newborn male has one chance in twenty of being murdered before reaching age 75 (U.S. Bureau of the Census, 1996).

Whatever the economic, cultural, or other explanations for national and ethnic differences in violent death, one thing is clear: it is not maleness per se that puts a young man at risk. Instead, it is a lethal combination of biological factors and cultural values, encouraging young men to act, or not act, in ways that carry the risk of death (Wilson & Daly, 1993).

dence, and competitiveness (all of which might have an evolutionary biological base) into such negative male traits as recklessness, callousness, and "an egocentric and often obsessive need to be dominant and to win" is bound to suffer violent consequences (Miedzian, 1991). Young men taught to avoid being a wimp or a sissy at all costs will eventually pay a price.

Living up to a tough-guy image, for example, makes it hard for a young man to back down from a confrontation, to back away from a dangerous challenge, or to admit that he needs help—especially emotional help—even if doing so would remove him from a life-threatening situation. Of course, cultural and familial influences make it much harder for some men to seek help than for others (see Research Report, beginning on the facing page).

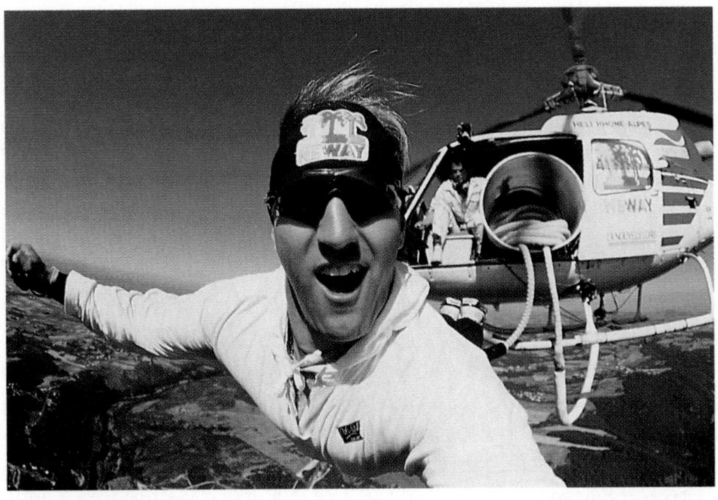

Would you pay $200 to jump out of a helicopter with a bungee cord around your middle? If your immediate answer is yes, chances are that you are male, in your 20s, and already testing the limits of life-threatening entertainment.

Social values can also lead to protective measures. For example, in 1980 in the United States, a 16-year-old boy had more than 1 chance in 150 of dying in a motor-vehicle accident before he reached age 25, with death most likely caused by his own, or a friend's, drunk driving. Between then and 1993, public anger and sorrow led to a series of steps to reduce drunk driving, among them intensified driver education, public service announcements, activism by private groups such as MADD and SADD (Mothers/Students Against Drunk Driving), peer pressure for appointing designated drivers, and the enforcement of various pertinent laws (which, among other benefits, doubled the national DWI arrest rate). Partly as a result, the automobile crash rate involving young drunk men fell by half between 1980 and 1993, as did the motor-vehicle death rate of young men (U.S. Bureau of the Census, 1995).

Although such statistics are encouraging, noteworthy here is that young men still drink too much and still take foolhardy and sometime fatal risks while under the influence, drowning because they drank and swam alone at night, killing each other because a minor disagreement escalated into a drunken fight, and so on (Cotten et al., 1994; Howland et al., 1995; Wilson & Daly, 1993). However, various social pressures and policies have translated into far less danger when young men take the wheel, and literally thousands of young Americans who would have died now live to reach a more cautious age, such as 25 or 30.

It should be noted that gender is not necessarily protective for any of these three troubling problems: many young women die of violence, many young men develop abnormal eating habits, and many young adults of both sexes abuse drugs. However, it certainly seems true that all three hazards spring partly from cultural pressures to fit a particular stereotype and that these pressures affect men and women in divergent ways. It also seems true that changes in social values and specific practices can reduce the harm.

Given the grimness of the three troubling problems we have been considering, we should end this chapter with a reminder that all the self-destructive behaviors just discussed are evident in only an unfortunate minority of young adults. Most young people, no matter what their ethnic group, economic status, or gender, manage the transition from adolescence

to adulthood well and pass through their early adulthood healthy and robust. As we will see in the next two chapters, although early adulthood is not easy, most young people become increasingly capable of understanding and coping with their lives during their prime years.

SUMMARY

Growth, Strength, and Health

1. While young adults do not grow significantly taller in their 20s, they typically grow stronger and fuller as their bodies reach adult size. In terms of overall health, as well as peak physical condition, early adulthood is the prime of life.

2. With each year from 20 to 40, all the body systems gradually become less efficient (though at different rates) and homeostasis takes increasingly longer to reach.

3. However, because of organ reserve, none of these changes is particularly troublesome or even noticeable for most people most of the time. Even athletic performance, while slowed somewhat, can remain at a high level.

The Sexual-Reproductive System

4. As middle age approaches, the speed of sexual responses slows down in men, but not in women. These modest declines usually have no negative effect on the man's sexual experiences, and, in some cases, they may enhance the woman's experience.

5. About 15 percent of all couples have fertility problems. One common reason is that the man's sperm are insufficient in quantity or motility. Another common problem is that the woman's ova do not reach the uterus because the Fallopian tubes are blocked or because ovulation itself does not occur.

6. The normal aging process is rarely the primary cause of infertility in the two decades of early adulthood, but age can be a contributing factor. While most couples can conceive a child even in their early 40s, those who have fertility difficulties should get medical assistance early so that the age-related declines in the reproductive system do not make existing problems worse. A number of procedures, ranging from surgery to remove obstructions to conception to IVF (*in vitro* fertilization) and AID (artificial insemination by donor), offer potential answers to infertility.

Three Troubling Problems

7. Young adults are more likely to abuse alcohol and illicit drugs than are people of any other age, often doing themselves or others serious harm.

8. Eating disorders are also more common during young adulthood than at other ages, as some young women feel a compulsion to be thinner than their bodies naturally tend to be.

9. Suicide, homicide, and fatal accidents are a serious problem for young adults, especially for young men in American society. The reasons are at least as much cultural as biological, as revealed by ethnic differences in the rates of these three causes of violent death.

KEY TERMS

senescence (467)
homeostasis (469)
organ reserve (470)
infertility (474)
pelvic inflammatory disease (PID) (475)
endometriosis (475)
in vitro fertilization (476)
drug use (479)
drug abuse (479)
drug addiction (479)
body-mass index (BMI) (482)
anorexia nervosa (484)
bulimia nervosa (484)
violent death (485)

KEY QUESTIONS

1. In what specific ways is early adulthood the prime of life?

2. How is the physical performance of a 20-year-old athlete likely to be different from that of a 40-year-old?

3. As a person ages, what are the changes that occur in organ reserve? How do these changes affect a person's activities?

4. What are some of the factors that tend to diminish fertility toward the end of early adulthood?

5. What can be done to prevent and remedy the main causes of infertility?

6. Why are young adults particularly susceptible to drug use and abuse?

7. How can concern about being fat become a health hazard?

8. What are the sex differences in the rate of violent deaths, and how do you explain them?

Early Adulthood: Cognitive Development

CHAPTER 18

Over the course of adulthood, many changes occur in our thinking processes. There are changes in our store of knowledge and experience, in how fast we think, in what we think about, in how efficiently we process new information, in how deeply or reflectively we relate new experiences to previous ones, and in how we use our intellectual skills. Unlike the relatively "straightforward" cognitive growth of childhood and adolescence, these changes are *multidirectional*: some abilities increase, others wane, and some remain stable throughout this period (Uttal & Perlmutter, 1989). Understanding adult cognitive development thus involves appreciating how thinking and reasoning reflect an interplay of growth and decline in intellectual abilities.

Developmental theorists have used at least three different approaches to explain the cognitive changes that occur in adulthood. The *postformal approach* picks up where Piaget left off, emphasizing the possible emergence of a new stage of thinking and reasoning in adulthood that builds on the skills of formal operational thinking. The *psychometric approach*, which analyzes components of intelligence such as those measured by IQ tests, examines whether these components improve or decline during adulthood. The *information-processing approach*, which studies the encoding, storage, and retrieval of information throughout life, considers whether the efficiency of these processes changes as the individual grows older. In a sense, the three approaches can be viewed as studying changes in *thinking*, *knowing*, and *processing* during adulthood (Rybash et al., 1986).

All three approaches provide valuable insights into cognitive development across adulthood, but to examine each one in each of the cognitive chapters for early adulthood, middle adulthood, and late adulthood would be repetitive and, potentially, confusing. Therefore, we will concentrate on each approach separately. In this chapter, our primary focus will be on the postformal approach, the psychometric approach will be emphasized in our discussion of middle adulthood cognition in Chapter 21, and the information-processing approach to cognition in late adulthood will be considered in Chapter 24. To begin the present chapter, however, we will consider in broader terms the nature of the adult cognitive changes that each approach seeks to explain.

Adult Thinking

Adult thinking seems different from adolescent thinking in many ways. While adolescents often try to distill universal truths from their personal experiences and tend to think about resolving the world's problems in terms of rational absolutes, adult thinking is more personal, practical, and integrative. Similarly, adults are less inclined toward the "game of thinking" (see Chapter 15), as their intellectual skills become enlisted in the occupational and interpersonal demands that shape adult life, and thus become more specialized and experiential. Broader experience also leads most adults to accept, and adapt to, the contradictions and inconsistencies of everyday experience, rather than decrying them or trying to resolve them definitively. Indeed, one hallmark of mature adult thinking is the realization that most of life's answers are provisional rather than necessarily enduring. As Gisela Labouvie-Vief (1992) explains, adult thinking

> is less and less considered a purely objective, impersonal, and rational activity. Instead, it embraces dimensions that are subjective, interpersonal, and nonrational. By establishing a dialogue with those dimensions, thinking becomes rebalanced . . .

Taken together, these characteristics of adult thinking are referred to by many developmentalists as postformal thought.

Postformal Thought

Postformal thought involves reasoning that is adapted to the subjective, real-life contexts to which it is applied. Labouvie-Vief (1985, 1986) points out that the traditional models of advanced thought stressed objective, logical thinking and devalued the importance of subjective feelings and personal experience. This kind of objective thinking, she maintains, is adaptive for the schoolchild, the adolescent, and the "novice adult," because it permits them to "categorize experience in a stable and reliable way." However, purely objective, logical thinking may be maladaptive in trying to understand, and deal with, the complexities and commitments of the adult world. For the adult, subjective feelings and personal experiences must be taken into account, or the result will be reasoning that is "limited, closed, rigidified in relation to the complex human dimensions of everyday experience." In this view, truly mature thought involves the interaction between abstract, objective forms of processing and expressive, subjective forms that arise from sensitivity to context.

An additional dimension of postformal thought is an understanding that one's own perspective is only one of many potentially valid views and that knowledge is not necessarily absolute or fixed (Sinnott, 1989). A contextual awareness emerges that recognizes that life entails inconsistencies, including the inconsistencies between intellectual analysis and emotional realities, and often requires a relativistic, flexible perspective.

postformal thought A type of adult thinking that is suited to solving real-world problems. Postformal thought is less abstract and absolute than formal thought, more adaptive to life's inconsistencies, and more dialectical—capable of combining contradictory elements into a comprehensive whole.

To demonstrate the development of this form of thought, Labouvie-Vief and her colleagues presented subjects between the ages of 10 and 40 with brief narratives that tested problem-solving logic. Because the researchers were more interested in their subjects' problem-solving approach than in their specific solutions, the tests were designed to be superficially simple

and logical but to allow for deeper interpretations outside the straightforward propositions of the text. One such story went as follows:

> John is known to be a heavy drinker, especially when he goes to parties. Mary, John's wife, warns him that if he comes home drunk one more time, she will leave him and take the children. Tonight John is out late at an office party. John comes home drunk.—Does Mary leave John?

In arriving at their answers, all the young adolescents and many of the older ones reasoned strictly according to the basic premise of the story: in the case of the drunken husband, it was evident to them that Mary would leave John because that is what she said she would do. Older respondents, of course, recognized the explicit logic of the story, but they resisted the limitedness of the narrative's logical premise and explored the real-life possibilities and contextual circumstances that might apply—whether, for example, Mary's warning was a plea rather than a final ultimatum, whether John was apologetic or abusive upon his return home, whether Mary had somewhere to go, what the history of the marriage relationship might be, and so forth. At the most advanced level, adults tried to "engage in an active dialogue" with the text, forming multiple perspectives as a result (Adams & Labouvie-Vief, 1986). This appreciation and reconciliation of both objective and subjective approaches to real-life problems are the hallmarks of adult adaptive thought.

Within this overall pattern of cognitive progression, other research finds that the difference between the reasoning maturity of adolescents and that of young adults is particularly apparent when the problems to be solved are emotionally charged. In such cases, younger thinkers exhibit lower levels of reasoning, even when they have demonstrated reasoning ability equal to that of young adults on problems that are low in emotional content. In one study (Blanchard-Fields, 1986), for example, adolescents, young adults, and middle-aged adults were presented with three reasoning tasks that varied in emotional saliency. Each task involved accounts of a fictional event presented from two conflicting points of view. The first pair of accounts told of the war between "North Livia" and "South Livia" as reported by a partisan historian for each of the opposing sides. The second pair of accounts related a teenage boy's unwilling visit to his grandparents, with one account presented by the boy's parents (who said that they had reasoned with their son and convinced him to go) and the other side presented by the boy himself (who said that his parents had lectured him on his duty and had forced him to go). The third pair of accounts came from a couple who were faced with an unplanned pregnancy, with one account presenting the woman's proabortion stance and the other presenting the man's antiabortion stance.

After the subjects read these accounts, they were asked questions regarding what the conflict was about, who was to blame, how the conflict was resolved, and who emerged the winner. The subjects were also asked to explain how the discrepancies in the conflicting accounts arose and how to decide who was more credible. (A typical discrepancy in the war task was the assertion by the North Livian historian that a particular battle had turned the tide "heavily in favor of the North," while the historian for South Livia described the same battle as a "minor" setback for the South. Among the discrepancies in the visit task was the boy's assertion that, although he was as polite as possible at his grandparents', he was bored and felt forced into

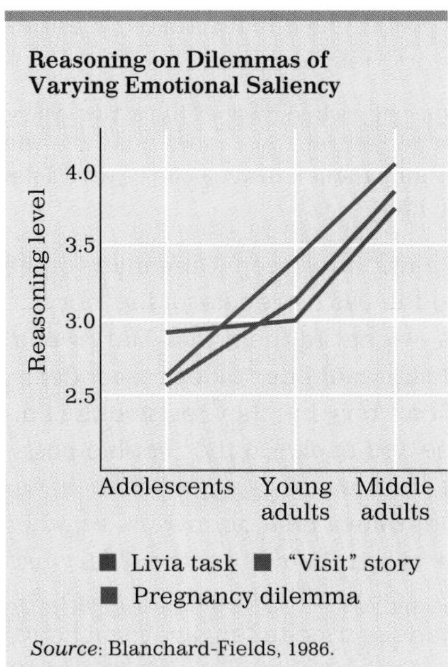

FIGURE 18.1 In the Blanchard-Fields study of age-related reasoning on social dilemmas, older, more mature thinkers scored higher because they were better able to take into account the interpretive biases of each party's version of events. As you can see, this was especially true when the social dilemmas were emotionally charged.

everything he did while there—versus his parents' perception that he had a good time at his grandparents and enjoyed "the family closeness.")

The subjects' responses were scored in accordance with six levels of reasoning, ranging from an absolutist approach in which only one perspective was recognized as correct (level 1), to a recognition of interpretive discrepancies and a weakening insistence on external truth (level 3), to a multiple-perspectives stance that recognized the need to "weigh discrepant sources of information in order to arrive at the best answer for the particular situation" (level 6). Despite considerable variation in reasoning performance, the overall results of this study support the cognitive progression outlined above. More specifically, only 16 percent of adolescents scored above level 3, compared with 36 percent of young adults and 61 percent of middle-aged adults. What was particularly striking, however, was the difference in adolescents' and young adults' reasoning on the emotionally charged problems (see Figure 18.1). Not only did the young adults reason more maturely than the younger subjects on the visit and pregnancy tasks (both considered highly emotional), but the adolescents actually reasoned less well on these two tasks than they did on the war task (considered to be low in emotional content), even though their reasoning on this task was virtually as good as the young adults'. According to the author of this study,

> The more emotionally salient context appeared to be more disruptive for the younger than for older thinkers, thus affecting their performance, particularly their ability to differentiate an event from its interpretation. . . . [D]evelopmentally mature adults are able to account for subjective factors in their thinking. This is in contrast with youthful thinkers, who assume that thinking is based on an objective structure of reality that is juxtaposed to subjectivity. Instead of resulting in more "objective" conclusions and resolutions of reasoning dilemmas, this form of thinking tends to maximize the chances of subjective errors. Therefore, maturity in thinking is not just a return to more subjective modes of thinking. Instead, by accounting for subjectivity, the mature adult becomes a more objective and powerful thinker. [Blanchard-Fields, 1989]

dialectical thought Thought that is characterized by understanding the pros and cons, advantages and disadvantages, and possibilities and limitations inherent in every idea and course of action. In daily life, dialectical thinking involves the constant integration of one's beliefs and experiences with all the contradictions and inconsistencies one encounters.

thesis A proposition or statement of belief; the first stage of the dialectical process.

antithesis A proposition or statement of belief that opposes the thesis; the second stage of the dialectical process.

synthesis The reconciliation of thesis and antithesis into a new and more comprehensive level of truth; the third stage of the dialectical process.

Dialectical Thought

Some theorists consider **dialectical thought** the most advanced form of cognition (Basseches, 1984, 1989; Leadbeater, 1986; Riegel, 1975). The word "dialectical" refers to the philosophical concept that every idea, every truth, bears within itself the suggestion of the opposite idea or truth. In terms used by philosophers, each new idea, or **thesis**, implies an opposing idea, or **antithesis**. Dialectical thinking involves considering both of these poles of an idea simultaneously and then forging them into a **synthesis**, that is, a new idea that integrates both the original idea and its opposite. The idea of the dialectical process also emphasizes that, because ideas are always initiating new syntheses, constant change is inevitable, and the dialectical process is continual. Moreover, because each new synthesis is a deepening and refinement of the idea that initiated it, dialectical change results in developmental growth.

For our purposes here, we may say that in daily life, dialectical thinking involves the constant integration of one's beliefs and experiences with all the contradictions and inconsistencies one encounters. The result of dialectical thinking is a continuously evolving view of oneself and the world, a view that recognizes that few, if any, of life's most important questions have single, unchangeable, correct answers.

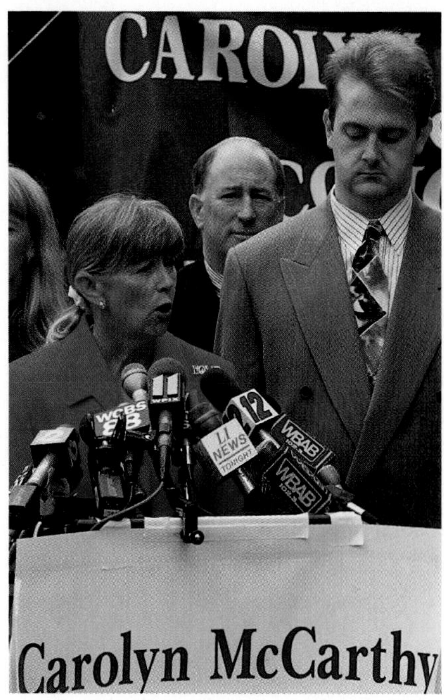

In dialectical thinking, an individual develops new thoughts that seem opposed to his or her original thinking. Eventually, the individual arrives at a new cognitive pattern that incorporates both the original idea and the opposing one. In 1994, Carolyn McCarthy thought of herself primarily as a wife, mother, and nurse (thesis)—until her husband was senselessly murdered, and her son seriously wounded, by a gunman who went on a shooting rampage with an assault rifle on a Long Island commuter train. She then began questioning many of the basic assumptions of her life and of the surrounding social order (antithesis). In particular, she opposed her congressman—whom she, as a lifelong Republican, had previously supported—because he was against gun control. This led to a synthesis in which she herself ran for Congress, as a Democrat, winning the seat to become a caregiver and public advocate for a much wider community.

This does not mean that dialectical thinkers adopt the idea that "everything is relative" and stop there, unable to commit themselves to broader values. On the contrary, a dialectic view explicitly recognizes the limitations of extremely relativistic positions such as "If you think it is true, then it's true for you" (Leadbeater, 1986). Truly dialectical thinkers, in fact, acknowledge both the subjective nature of reality *and* the need to make firm commitments to values that they realize are likely to change over time. They recognize that while many viewpoints may be potentially valid, some can be better justified or defended than others and thus provide a better basis for thoughtful decisions.

Let us see how the dialectical process might work in a simple example. Take the aphorism "Honesty is the best policy," which many people accept uncritically. A dialectical thinker, too, might begin by agreeing with this thesis but would then consider the opposite idea: that honesty can cause hurt feelings, or foolish behavior, or destructive emotions, and so it is sometimes *not* the best policy. From these conflicting conclusions, a dialectical thinker might synthesize a new idea: that honesty is a desirable goal in human relationships because it fosters trust and intimacy, but honesty should not conflict with respect for the other person. The dialectical process does not stop here, however, for this new synthesis is itself constantly refined by new real-life situations. Does "respect for the other person" mean complimenting someone for a poor achievement in which he or she feels pride and has invested hard work, for example? The answer to this question might vary depending on such factors as whether the person in question is a child or an adult, has achieved the maximum he or she is capable of or can improve, is spurred on by constructive criticism or is discouraged by it, and so on. In each new case, the dialectical thinker attempts to ascertain how, why, and in what form of expression honesty is best, recognizing all the while that whatever choices he or she makes may have to be subsequently reconsidered in light of new information or changing circumstances. This is different from an extreme relativism, which asserts that honesty may be best for some people and not for others. You can see that dialectical thought is a complicated process, seeking answers that are integrative rather than simple or fixed.

Now let us look at another example of dialectical thought as it applies to an experience familiar to many: the fading of a love affair (Basseches, 1984). A nondialectical thinker is likely to see each of the partners in a relationship as having stable, independent traits and to define their relationship in terms of the enduring interaction of these traits in compatible ways. Faced with a troubled romance, then, the nondialectical thinker is likely to conclude that one partner or the other is at fault, or that the relationship was a mistake from the beginning because the two partners are basically incompatible. By contrast, the dialectical thinker sees relationships as constantly changing rather than as stable, and understands that the partners are changed by their relationship as much as they create it. Thus the dialectical thinker realizes that the personalities, needs, and circumstances of any relationship change over time, making alterations in the relationship necessary and inevitable. A troubled romance may occur, therefore, not because the partners are fundamentally incompatible but because both have changed without adapting their relationship accordingly. Rather than concluding that one partner is at fault, therefore, the dialectical thinker can

pose a new alternative: both partners can accommodate their relationship to the changes that have already occurred in it and in themselves, creating, in effect, a new relationship between them.

A recognition of the continually evolving quality of human relationships gives the dialectical thinker a broader and more flexible perspective on many aspects of personal and social interaction. This perspective, in turn, makes the individual better able to adapt to the flux of life and to perceive the disequilibrium of new demands, roles, and responsibilities as new opportunities for growth and synthesis rather than as sources of stress and dysfunction.

Postformal Thinking: A Fifth Cognitive Stage?

These characteristics of postformal thought provide a rich portrayal of new forms of thinking that emerge in early adulthood. They suggest—optimistically—that new cognitive capacities emerge in young adults that enable them to confront life experiences more adaptively, realistically, and competently than they could as adolescents.

But this formulation of adult cognition also has its critics. Although there is research evidence that dialectical and relativistic thought increase during the adult years (e.g., Kramer & Melchior, 1990; Kramer & Woodruff, 1986), in general, research on postformal thought has not kept pace with theories about it. Critics note that we have especially little knowledge about the maintenance of postformal thought into late adulthood, when the narrowing of certain cognitive competencies may alter styles of reasoning and thought (see Chapter 24).

On the question of whether postformal thought represents a distinct fifth stage of cognitive development, theorists are also divided. On the one hand, some claim that the unique features of postformal reasoning justify claims that a Piagetian stage beyond formal operational thought has been discovered, a stage that includes more creativity and social understanding (Commons & Richards, 1984; Sinnott, 1993). On the other hand, critics contend that the characteristics of postformal thought are not universal and do not necessarily build on the prior accomplishments of formal operations, and thus do not resemble the characteristics of Piaget's earlier stages (Kramer, 1983; Rybash et al., 1986). On balance, it is probably wisest to regard postformal thinking as a constellation of several styles of thought that are based on life experience, education, and other factors associated with adult maturity. Indeed, the picture of adaptive, dialectical thinking presented by postformal theory should probably also be viewed as ideal rather than normative. In describing the best accomplishments of young adult cognitive development, theorists have provided a conception of postformal reasoning that some adults use regularly, that others never use, and that many use irregularly or in specific areas.

Adult Moral Reasoning

According to many researchers, the responsibilities, experiences, and concerns of adulthood can also affect moral reasoning, propelling a person from a lower moral stage to a higher one. Indeed, according to James Rest,

Regardless of whether one is pro-life or pro-choice, the American debate over abortion reveals that complex moral issues are not easily resolved through logic or moral principles alone, but must include the confrontation of ethics with life experience.

dramatic and extensive changes occur in young adulthood (the 20s and 30s) in the basic problem-solving strategies used to deal with ethical issues. . . . These changes are linked to fundamental reconceptualizations in how the person understands society and his or her stake in it. [Rest, 1993]

According to Rest's research (1993), one catalyst for such shifts in moral reasoning is the college experience itself, especially if the student becomes immersed in course work that includes discussion of moral issues or is preparing for a profession (such as law or medicine) in which questions of ethics are prominent.

In the view of some, however, academic debate of ethical issues can, at best, be only a beginning of new growth in moral reasoning. Lawrence Kohlberg maintains that in order to be capable of "truly ethical" reasoning, a person must have "the experience of sustained responsibility for the welfare of others and the experiences of irreversible moral choice which are the marks of adult personal experience" (Kohlberg, 1973). The development of faith follows a similar path, for the same reasons (see A Closer Look on pp. 498–499).

The challenges and dilemmas of adult responsibilities, in tandem with the emerging relativistic and dialectical features of adult thought, can lead to new and different qualities of moral reasoning. Carol Gilligan has looked particularly at the relationship between adult life experiences and a broader understanding of moral issues. As we saw in Chapter 15, Gilligan believes that in matters of moral reasoning, males tend to be more concerned with the question of rights and justice, whereas females are more concerned with personal relationships, tending to put human needs above justice principles. According to Gilligan, however, as all people's experience of life expands, and especially as they become committed to, and responsible for, the needs of others, they begin to realize that moral reasoning based chiefly on justice principles or on individual human needs is inadequate to resolve real-life moral dilemmas (Gilligan, 1981, 1982). Consequently, they begin to construct principles that are relative and changeable, seeking a synthesis of ethical principles with life experience (Gilligan & Murphy, 1979; Murphy & Gilligan, 1980).

One young man whom Gilligan studied illustrates this shift very well. In late adolescence he was able to reason abstractly, and at an advanced level, about the moral dilemma involving Heinz and his dying wife (see p. 424), citing the principle that life is to be valued more than money. But seven years later, in early adulthood, he had become much more aware of the personal implications of his answer:

This is a very crisp little dilemma and you can latch onto that principle pretty fast and say that life is more important than money. But then, when you reflect back on how you really act in your own life, you don't use that principle, or I haven't yet used that principle to operate on. And none of the people who answer that dilemma that way use that principle to operate on because they were blowing $7,000 a year for their education at Harvard instead of giving it to the Children's Fund to give porridge to the kids in Botswana and to that extent answering the dilemma with that principle is not hypocritical, it's just that you don't recognize it. I hadn't recognized it at the time, and I am sure they didn't recognize it either. [Gilligan & Murphy, 1979]

Similarly, a woman stated that she once thought there were no absolute principles of right and wrong: "I went through a time when I thought

A CLOSER LOOK

The Development of Faith

Thinking about religious matters is another aspect of adult cognitive development that has interested some researchers. Like morality, faith obviously is not only a cognitive process: it involves practice as well as preaching, and it arises from religious experience as well as religious education. Nonetheless, one view of faith is as a developmental process; as a person has more experience trying to reconcile religion with daily life, his or her faith may reach higher levels.

The most detailed description of the development of faith comes from James Fowler (1981, 1986), who delineates six stages of faith. It should be noted that when Fowler describes "faith," he does not necessarily mean religious faith. He agrees with Paul Tillich (1958) that all humans need to have faith in something, whether that something is a god figure, philosophical principles, country, or simply oneself. Faith gives humans a reason for living their daily lives, a way of understanding the past, a hope for the future. It is whatever each person really cares about, his or her "ultimate concern," in Tillich's words.

Stage One: Intuitive-Projective Faith

Stage-one faith is magical, illogical, imaginative, and filled with fantasy, especially about the power of God and the mysteries of birth and death. It is typical of children ages 3 to 7.

Stage Two: Mythic-Literal Faith

At this stage, the individual takes the myths and stories of religion literally and believes simplistically in the power of symbols. In a religious context, this stage usually involves reciprocity: God sees to it that those who follow his laws are rewarded and that those who do not are punished. Stage two is typical of middle childhood, but it also occurs in adulthood. For example, Fowler cites the case of a woman who says extra prayers at every chance, in order to put them "in the bank." Whenever she needs divine help, she thinks she can withdraw some of her accumulated credit.

Stage Three: Synthetic-Conventional Faith

A nonintellectual, tacit acceptance of cultural or religious values in the context of interpersonal relationships is typical of stage three. Unlike stage-two faith, stage-three faith serves to coordinate the individual's involvements in a complex social world, providing a sense of identity and adding significance to the rituals and symbols of daily life. For example, Fowler describes one man who puts his faith in his relationship with his family, a man whose personal rules include "being truthful with my family. Not trying to cheat them out of anything . . . I'm not saying that God or anybody else set my rules. I really don't know. It's what I feel is right." Because of his commitment to his family, he has learned to accept the "rat race" of his daily work. These responses are typical of the conformist stage of faith, which is conventional, concerned about other people, and values "what feels right" more than what makes intellectual sense.

Stage Four: Individual-Reflective Faith

By contrast, stage-four faith is characterized by intellectual detachment from the values of the culture and from the approval of significant other people. The experience of college can be a springboard to stage four as the young person learns to question the authority of parents, teachers, and other powerful figures and to rely, instead, on his or her own understanding of the world. An unexpected experience in adulthood, such as divorce, the loss of a job, or the death of a child, can also lead to stage four. The adult's understanding of faith ceases to be a matter of acceptance of the usual order of things and becomes, instead, an active commitment to a life goal and lifestyle that differs from that of many other people.

Fowler's example of someone at the fourth stage of faith is Jack, whose time in the army allowed him to know people from other backgrounds and gradually to develop a personal philosophy. Jack explains:

> I began to see that the prejudice against blacks that I had been taught, and that everybody in the projects where I grew up believed in, was wrong. I began to see that us poor whites being pitted against poor blacks worked only to the advantage of the wealthy and powerful. For the first time I began to think politically. I began to have a kind of philosophy.

Jack's ability to articulate his own values, distinct from those of family, friends, and culture, makes his faith an individual-reflective faith.

Stage Five: Conjunctive Faith

This type of faith incorporates both powerful unconscious ideas (such as the power of prayer and the love of God) and rational, conscious values (such as the worth of life compared with that of property) and is characterized by a willingness to accept contradictions. It involves a synthesis of

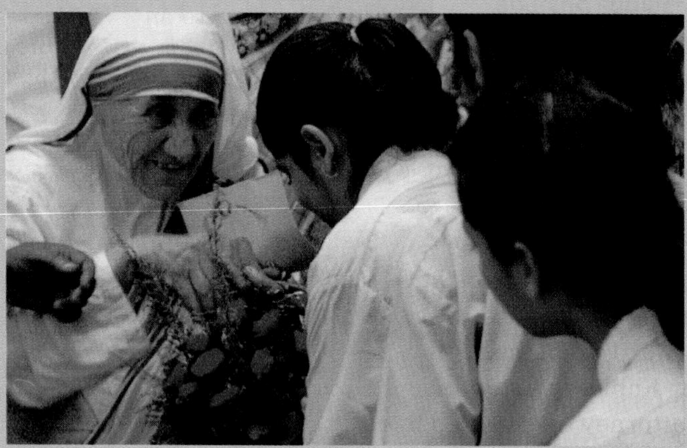

An understanding of the stages of the development of faith begins with the realization that levels of spirtual reasoning are not measured in terms of religious doctrine. Mother Theresa is probably at Fowler's highest stage, but not all Roman Catholics share her convictions. Similarly, in any group of worshippers, be they in a mosque, temple, or church, some will be at the first stages of faith and some will be in the final one, depending on their experiences and maturation, not on their devotion to particular items of creed or ritual.

the magical understanding of symbols and myths that characterized stage two and the conceptual clarity of stage four. Fowler cites one woman at this stage who believes strongly in God, but adds, "I don't think it matters a bit what you call it. I think some people are so fed up with the word God that you can't talk to them about God." Her recognition that the word "God" may be distracting and misleading is typical of the ability of the stage-five thinker to articulate paradoxes and contradictions in faith. Also typical of this stage is an openness to new truths; this woman explains her beliefs by referring to Jesus, George Fox, Krishna Murti, and Carl Jung. Fowler says this cosmic perspective rarely comes before middle age.

Stage Six: Universalizing Faith

People at stage six have a powerful vision of universal compassion, justice, and love that compels them to live their lives in a way that, to most people, seems either saintly or foolish. They put their own personal welfare aside, and sometimes even sacrifice their lives, in an effort to enunciate universal values. Often, a transforming experience converts an adult to stage six, as happened to Moses when he saw the burning bush, and to Mohammed, Buddha, and Paul. Fowler mentions some twentieth-century people who have reached this level, among them Mohandas Gandhi, Martin Luther King, Jr., and Mother Theresa, each of whom radically redefined their lives after a particular experience produced a new understanding of human community. Clearly, a person reaching stage six of faith is an exceedingly rare individual.

Indeed, the scarcity of people at the upper stages of Fowler's hierarchy might make one wonder how useful it is. Moreover, it may be galling to read that there are "higher" stages of faith than most adults are likely to reach—especially when some of the "lower" levels of thinking can be seen as no less valid than the "higher" levels. Describing levels of faith seems to imply values about the nature and object of belief. Further, a more intuitive personalized form of spiritual belief may be more reflective of deep religious experience than are the more abstract later stages of faith outlined by Fowler (Reich, 1993). In Fowler's defense it should be noted that he never explicitly says that the higher stages are better. In fact, he explains:

> Each stage has its proper time of ascendancy. For persons in a given stage at the right time *for their lives*, the task is the full realization and integration of the strengths and graces of that stage rather than rushing on to the next stage. Each stage has the potential for wholeness, grace and integrity, and for strengths sufficient for either life's blows or blessings. [Fowler, 1981]

If Fowler is correct, faith, like other aspects of cognition, may progress from a quite simple, self-centered, one-sided perspective to a more complex, altruistic, and multisided view.

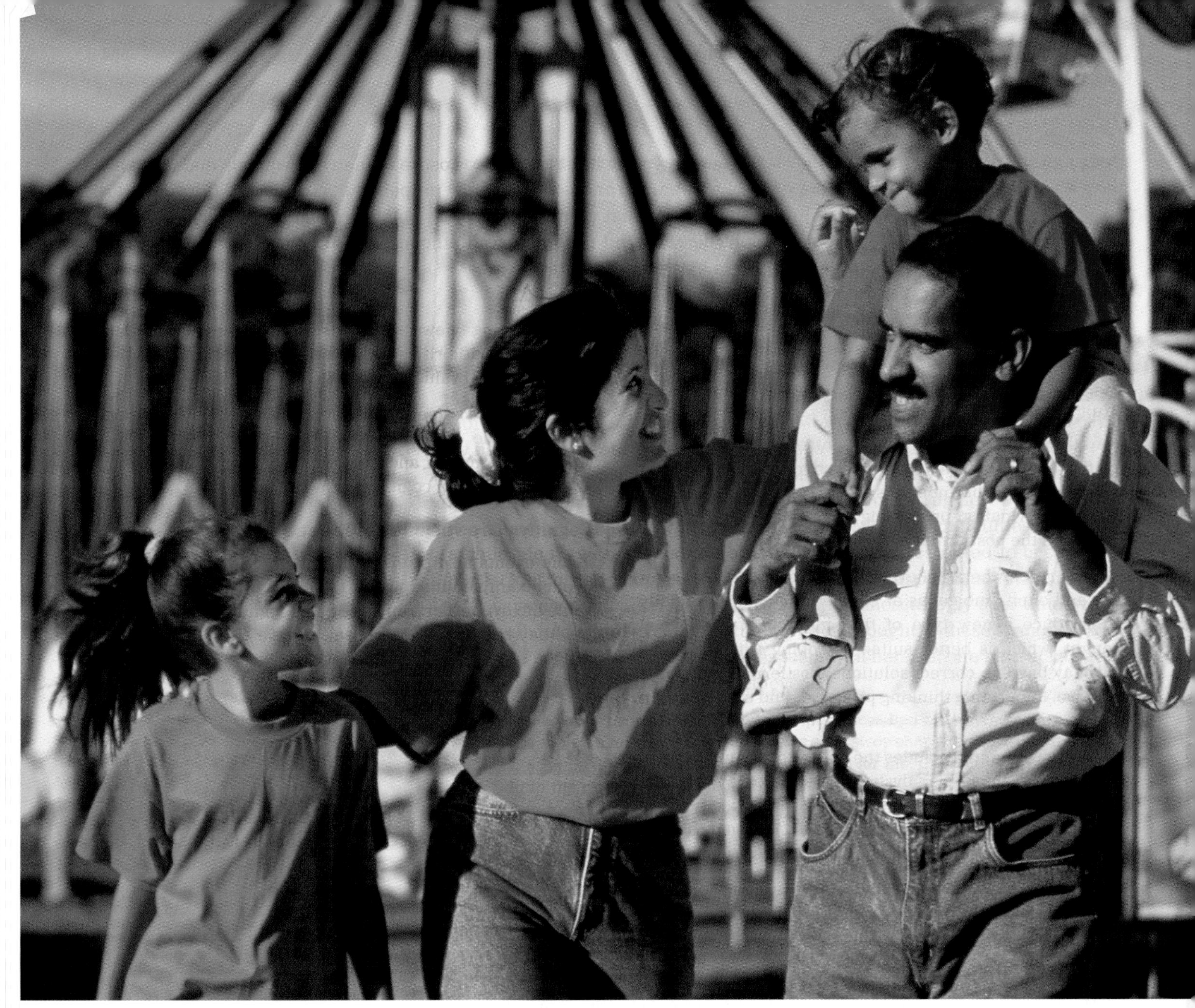

Early Adulthood: Psychosocial Development

CHAPTER 19

In terms of psychosocial development, the hallmark of contemporary adult life is diversity. No longer limited by the pace of biological maturation or bound by parental restrictions, adults are much freer to choose their own developmental paths—and the array of paths regarding career, marriage, parenthood, lifestyle, and friendship in the final years of the twentieth century is mind-boggling.

To sort out the various paths through adulthood, let us begin with several themes that underlie and organize the complexity and diversity of human life.

The Tasks of Adulthood

Two universal psychosocial needs drive adult development. Various traditional theorists describe these needs in somewhat different terms. Abraham Maslow (1968), a humanist theorist, cites the need for *love and belonging*, which, if met, is followed by the need for *success and esteem*. Other theorists have described these two needs in terms of *affiliation and achievement*, or *affection and instrumentality*, or *communion and agency* (Bakan, 1966). Freud (1935) put the same duality even more simply, explaining that a healthy adult was one who could *love and work*. Although these various formulations differ from each other in subtle ways, they all highlight the fact that human adults have both a need to connect with other people in mutually nurturing relationships and a need for independent accomplishment.

The clearest statement of these two needs comes from Erik Erikson. After resolving the identity crisis, Erikson maintains, young adults next experience the crisis of **intimacy versus isolation**, which arises from a powerful drive to share one's personal life with someone else, a drive that, if unfulfilled, carries the risk of profound aloneness. As Erikson (1963) explains:

> The young adult, emerging from the search for and the insistence on identity, is eager and willing to fuse his identity with others. He is ready for intimacy, that is, the capacity to commit himself to concrete affiliations and partnerships and to develop the ethical strength to abide by such commitments, even though they call for significant sacrifices and compromises.

intimacy versus isolation The sixth of Erikson's eight stages of development. Adults seek to find someone with whom to share their lives, in an enduring and self-sacrificing commitment. Without such commitment, they risk profound aloneness, isolated from their fellow humans.

In Eriksonian theory, the crisis of intimacy versus isolation is followed by the crisis of **generativity versus stagnation**, which involves the need to be productive in some meaningful way, usually through work or parenthood. Without a sense of generativity, says Erikson, adults experience life as empty and purposeless and are filled with "a pervading sense of stagnation and personal impoverishment."

Stages of Adulthood: Levinson (1978)

Early Adult Transition
Ages 17 to 22
Leave adolescence, make preliminary choices for adult life.

Entering the Adult World
Ages 22 to 28
Initial choices in love, occupation, friendship, values, lifestyle.

Age 30 Transition
Ages 28 to 33
Changes in life structure. Either a moderate change, or, more often, a severe and stressful crisis.

Settling Down
Ages 33 to 40
Establish a niche in society, progress on a timetable, both in family and career accomplishments.

Midlife Transition
Ages 40 to 45
Life structure comes into question, usually a time of crisis in the meaning, direction, and value of each person's life. Neglected parts of the self (talents, desires, aspirations) seek expression.

Entering Middle Adulthood
Ages 45 to 50
Choices must be made, a new life structure formed. Person must commit to new tasks.

Ages and Stages

Fifty years ago or so, the pursuit of intimacy and generativity, as well as adult development overall, seemed to follow a progression of stages in much the same way that childhood development does. The typical middle-class man, for example, finished his education by his early 20s, chose his lifetime occupation, married, bought a house, and had children by age 30, and then achieved generativity by devoting himself to climbing the career ladder in order to be "a good provider." The typical middle-class woman was married in her early 20s, had several children by age 30, then achieved generativity through her role as wife, mother, and homemaker, with the specific steps of her life dependent primarily on the age of her offspring. These stages of development seemed so nearly inevitable that many social scientists confidently assigned them labels and constructed their view of adult development around them (see chart at left).

Such patterns were, in fact, followed by many adults at midcentury. However, any attempt today to match one particular age with one particular stage of adult development would be quite limited. Erikson, for example, saw intimacy as a need met primarily during the very beginning of adulthood, and generativity as a need met between roughly ages 25 and 65, but contemporary theorists and researchers see both needs evident throughout adulthood (Wrightsman, 1994). Most social scientists who study adulthood also emphasize that whatever "stages" there might be can be experienced in almost any order and can be experienced more than once—as in the case of parents who divorce and then begin new families in their 40s, or workers who retire in their 50s and start new careers in their 60s.

The Social Clock

Although most developmentalists no longer take a strict stage view of adulthood, they do recognize that, to varying degrees, adult development is influenced by the **social clock**, the culturally set timetable that establishes when various events and endeavors in life are appropriate. Each culture, each subculture, and every historical period has a somewhat different social clock, with variations in the "best" age to become independent of one's parents, to finish schooling, to establish a career, to have children, and so on (Keith, 1990; Settersten & Hagestad, 1996).

Internationally, social-clock norms vary in scope and stringency. Societies in developed regions tend to be quite age-stratified, in that certain privileges, responsibilities, and expectations are connected to a certain age. There is a legal age for driving, drinking, voting, getting married, and even for signing a mortgage; and there is an expected age for marriage, childbearing, completion of childbearing, grandparenthood, and so on. Societies in undeveloped regions tend to be much less age-stratified. In these soci-

generativity versus stagnation Erikson's seventh stage of development, in which adults seek to be productive through vocation, avocation, or child-rearing. Without such productive work, adults stop developing and growing.

social clock Refers to the idea that the stages of life, and the behaviors "appropriate" to them, are set by social standards rather than by biological maturation. For instance, "middle age" begins when the culture believes it does, rather than at a particular age in all societies.

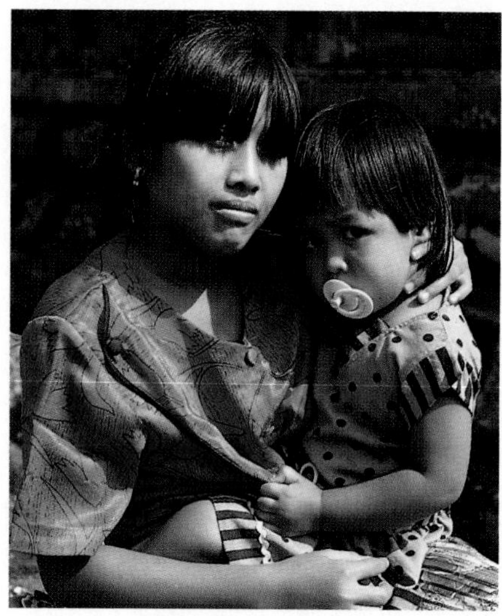

Social-clock settings in developed nations tend to be notably different from those in developing nations. In developed nations, for example, the social clock now permits grandmothers to be, for example, college graduates and discourages teenagers from becoming mothers. This is in marked contrast to developing nations, such as Indonesia, where grandmothers never go to college and young teenagers, like this Javanese girl, often become mothers.

eties, status differences related to age may be recognized (people are more respected if they have accumulated years or grandchildren, for instance), but the idea that 30- to 40-year-olds have something in common that 20-year-olds do not, including certain rights, is incomprehensible (Keith et al., 1994).

A prime influence on this cultural clock-setting—worldwide—is socioeconomic status: the lower a person's SES, the younger the age at which he or she is expected to reach life's major milestones, from becoming independent to becoming "old." The influence of socioeconomic status is particularly apparent with regard to the age at which women become wives and mothers. In many developed nations, women from low-SES backgrounds are pressured to be married by age 18 and to stop childbearing by age 30, while women from high-SES backgrounds may not feel pressure to marry until age 30 or to stop childbearing until age 40.

Internationally, the social-clock settings for marriage and childbearing vary even more. In most South American nations, for instance, marriage is legal at age 12 for females and 14 for males, and more than half of all brides in Brazil, Ecuador, Paraguay, and Venezuela are under age 22. By contrast, men and women in Germany cannot legally marry until they are at least 18. Most wait much longer to wed, with the median age of marriage being 27 for women and 29 for men (United Nations, 1995).

Currently, in the United States, the social clock allows for much greater diversity in all areas of life than it once did. Nonetheless, the prescribed ages for many adult roles still cluster in early adulthood, and in whatever sequence a particular individual takes on specifics tasks and roles in adulthood, the themes of intimacy and generativity are universally apparent.

Intimacy

To meet the need for intimacy, or affiliation, or communion, an adult may take on many roles—friend, lover, and spouse among them. Each intimate relationship involves a progression, from initial attraction to close connection and then to ongoing commitment. Each role demands some personal

sacrifice, a giving of oneself to others. In the process of becoming more open and more vulnerable to others, the individual gains deeper self-understanding and avoids the isolation caused by too much self-protection. As Erikson explains, the young adult must

> face the fear of ego loss in situations which call for self-abandon: in the solidarity of close affiliations . . . sexual unions, in close friendship and in physical combat, in experiences of inspiration by teachers and of intuition from the recesses of the self. The avoidance of such experiences . . . may lead to a deep sense of isolation and consequent self-absorption. [Erikson, 1963]

The two main sources of intimacy in early adulthood are close friendship and sexual partnership. (A third source, ongoing family ties across the generations, is discussed in Chapter 22.) First let us look at friendship, a bond that is almost universal in early adulthood, and then discuss the more variable bond of sexual intimacy.

Friendship

Friends are very important at every stage of development, a reality increasingly recognized and studied by researchers. Even more than family members, friends are a buffer against stress and a source of positive feelings (Antonucci, 1990). One probable reason for this is that friends choose each other, often for the very qualities (understanding, tolerance, loyalty, affection, humor) that make them good companions, trustworthy confidants, and reliable sources of emotional support. In addition, the fact that friendship ties are voluntary, in contrast to the obligatory basis of family ties, makes close friendship a validation of one's personal worthiness and thus is an invaluable source of self-esteem (Allan, 1989; Fehr, 1996).

Because at the start of adulthood most individuals are relatively free of overriding commitments (such as marriage, dependent children, or filial obligation to aging parents), they find it relatively easy to form extensive and varied social networks. Typically, they explore a broad range of goals and interests in many settings—at college, work, or social gatherings, and among political, cultural, athletic, or religious groups. Within this wide array of social contexts, they are usually able to make many acquaintances who can provide companionship, information, advice, and sympathy to ease the numerous challenges that early adulthood entails.

But what factors help move acquaintanceship forward to become close friendship? The entry into friendship generally involves four factors that act as "gateways to attraction": (1) physical attractiveness (even in platonic same-sex relationships), (2) apparent availability as a friend, (3) absence of "exclusion criteria" (that is, absence of specific traits that a person does *not* want in a friend), and (4) frequent exposure (Fehr, 1996). (The last factor is surprisingly powerful. The lifelong friends people make in college, for example, are more likely, given the other factors, to be those who chanced to sit next to them in class rather than fifteen seats away or who lived on their floor of a dorm rather than on the floor above.)

In part because they are open to new acquaintances and are exposed to many other people about the same age, young adults make friends readily. In fact, almost never do young adults feel bereft of friendship. One study found, for example, that less than 2 percent of young adults said they were without close friends (Jessor et al., 1991). Those young adults who live alone

are particularly likely to develop an elaborate friendship circle and to spend as much leisure time with their friends as they do by themselves (Alwin et al., 1985). Not surprisingly, during early adulthood, both men and women tend to be more satisfied with the size and functioning of their friendship networks than with almost any other part of their lives.

Gender Differences in Friendship

As might be expected, in friendships with members of their own sex, men and women tend to differ in many ways—in what they do together, in what they say to each other, and in how they feel about each other (Fehr, 1996). In general, men's friendships are based on shared activities and interests. In contrast, friendships between women tend to be more intimate and emotional, based on shared confidences and practical assistance in times of crisis.

One sign of this gender difference is in conversations. Female friends tend to talk about personal matters—secrets from their past, problems with their bodies, difficulties with their romantic or family relationships. These self-disclosing conversations are the heart of their social interaction. In contrast, male friends tend to talk about external matters—sports, work, politics, and vehicles—and their conversation tends to be peripheral to their social interaction:

> Men's adult friendships . . . are often geared toward accomplishing things and having something to show for their time spent together—practical problems solved, the house painted or the deck completed, wildlife netted, cars washed or tuned, tennis, basketball, poker, or music played, and so on. Shared talk may occur during these pursuits, but it is not the principle focus. [Rawlins, 1992]

As a result of these basic gender differences, men and women tend to have different expectations of their friendships. Women tend to expect that they can freely reveal their weaknesses and problems to friends and be met with an attentive and sympathetic ear, and if necessary, a shoulder to cry on. Men, on the other hand, not only are less likely to talk about their weaknesses and problems, but, when they do discuss them, they expect practical solutions rather than sympathy (Tannen, 1990). Indeed, men's conversations, much more than women's, are meant to showcase their strengths and expertise, whether it involves their ability to do their job, their knowledge of

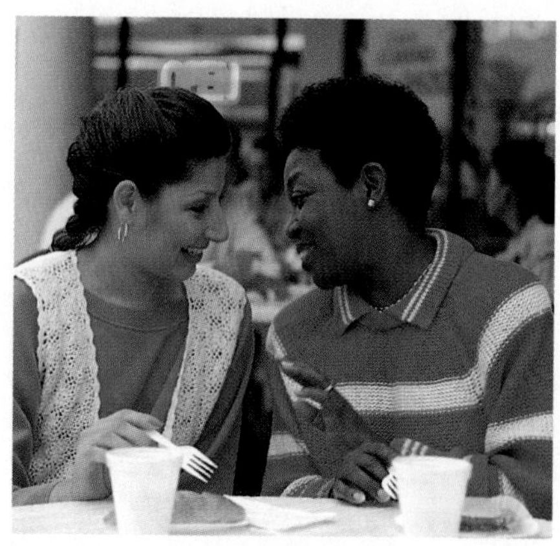

Friendship patterns vary from person to person, of course, and gender stereotypes regarding these patterns are often wide of the mark. Nonetheless, on the whole, friendships between men tend to take a different direction than those between women. Men typically do things together—with outdoor activities frequently preferred, especially if they lend themselves to showing off and friendly bragging. Women, on the other hand, tend to spend more time in intimate conversation, perhaps commiserating about their problems rather than calling attention to their accomplishments.

the latest advances in computer technology, or their insight into why this or that team won't make it into the playoffs. Thus, whereas friendship for women may be important as a way of coping with problems via shared fears, sorrows, and disappointments, for men, friendship serves primarily as a way of maintaining a favorable self-concept.

Closely related to this gender difference in self-revelation is the fact that men's friendships are more clearly tinged with open competition, as reflected in the teasing and needling that often characterize them and in male friends' willingness to embarrass each other as a way of indicating the solidity of the friendship (Sharkey, 1993). Men learn to handle this rivalry with some care. In the words of one: "I'm sure each of us wants to gain greater status in the eyes of the other . . . but I can't think of any competition where one comes out a winner and the other a loser" (Rawlins, 1992).

Many social scientists have asked why men's friendships seem so much less intimate than women's (e.g., Allan, 1989; Fox et al., 1985; Rawlins, 1992). One reason is that intimacy is grounded in mutual vulnerability, a characteristic that is discouraged by the cultural pressure on men to be strong and hide their weaknesses and fears. Another reason is that, from childhood, boys seem inclined to be more active and girls more verbal, and this early difference may lay the groundwork for interaction patterns in adulthood (Fehr, 1996; Tannen, 1990). A third reason is homophobia: many men avoid any expression of affection toward other men because they fear its association with homosexuality. Ironically, the most open expressions of affection between men tend to occur in situations where they are banded together in the name of aggression, such as intensely competitive team sports or military combat, perhaps because in such situations, few people would think to question a man's masculinity.

Male-Female Friendship

Overall, the basic differences in male and female same-sex friendships pose important opportunities and problems for cross-sex friendships. Cross-sex friendships offer men and women an opportunity to learn more about their commonalities, as well as to gain practical skills that may have traditionally been "reserved" for the other sex—from cooking to car repair, from child care to money management. Further, to the extent that men and women do indeed have separate experiences and perceptions, cross-sex friendships expand each partner's perspective.

At the same time, cross-sex friendships also have special problems, as each sex tends to have its own expectations for the friendship process. For example, a woman can be genuinely upset at what her male friend considers good-natured teasing (the kind he applies to his male friends), while a man is likely to wonder why his female friend continues to talk about her problems rather than taking *his* advice on how to solve them (Tannen, 1990). An additional complicating factor is that cross-sex friendships are both "enriched and plagued by fluctuating and unclear sexual boundaries" (Swain, 1992). Men are more often inclined to try to sexualize a platonic friendship, while women are more likely to be offended when a friend tries to cross the sexual boundary. Sometimes there is ambiguity about sexual attraction on

both sides, which can lead to confusion and tension. Even when sexual attraction is not an issue for either friend, cross-sex friends may have difficulty explaining to others the exact nature of their relationship, particularly if one of them is romantically involved with someone else or if they both work together and one is subordinate to the other (O'Meara, 1989, 1994). Few people, for instance, ever believe that a male boss and his female secretary are "just close friends."

All told, the typical female friendship pattern may be better in terms of meeting intimacy needs. Even though men are sometimes critical of women for gossiping, sharing secrets, or simply spending too much time on the phone with their friends (Allan, 1989), these patterns help prevent the loneliness and self-absorption that Erikson describes as a danger of the stage of intimacy versus isolation. If there is a down side to the female pattern, it is that, being more intimate and committed in their friendships, women tend to be more distressed than men when a friendship ends.

However, the male pattern of keeping one's distance while sharing useful information, productive activities, and reciprocal assistance may be more effective and efficient in at least one area of life—the workplace. In fact, in the work setting, where "the shifting sands of promotion and demotion render some friends expendable," a woman's tendency to seek mutual loyalty among confidantes who know each other's secrets can undermine her job performance (Rawlins, 1992). Women may be handicapped vocationally if they cannot treat their work colleagues with some distance and dispatch, just as men may be handicapped psychologically if they cannot share their problems with their personal friends.

Friendship and Marriage

For most young adults, the friendship circle remains robust until they marry, and then it typically shrinks. The main reason is a practical one: in the busy years of simultaneously building a marriage, establishing a home, pursuing a career, and raising children, there is only limited time and energy for friendship (Jessor et al., 1991; Rawlins, 1992).

Marriage brings other complications to the friendship network. In some cases, friends are seen as a potential threat to the marriage, particularly in the case of cross-sex friendships. A man's major concern in this regard is that his mate's relationship with a close male friend may involve sexual attraction, while women become more upset if their partner seems emotionally involved with another woman (Buss, 1994; Rawlins, 1992).

Even close friends of the same sex are sometimes considered rivals by newly married spouses, partly because neither sex necessarily understands the typical friendship patterns of the other. Many husbands feel excluded and even subverted by their wife's eagerness to discuss intimate personal information, including marital matters, with her friends, while many wives feel neglected because of their husband's desire to be off with his buddies (Allan, 1989). Typically, over a period of years, both partners become less jealous of outsiders, as well as better able to develop "couple friends," sharing activities and confidences with another married pair.

The Development of Love and Marriage

Although having close friends is one important way to satisfy one's need for affiliation, for most adults, having a close relationship with a mate is an even more important goal. Humans try to find one partner, one "significant other," one person with whom to bond throughout life.

Living Together

Whereas traditional signs of anticipated partnership involved engagement announcements and wedding bells, in contemporary times, many couples take their first steps toward commitment with an informal sharing of domestic life. They might first deepen their intimacy by spending occasional nights and weekends together—learning what it is like to be in each other's company in a domestic setting around the clock. If the partners enjoy the experience but are not prepared—financially, legally, or emotionally—to marry, this intimacy often leads to their living together, or **cohabitation**.

Cohabitation, of course, is not just for the young. In some cultures—Sweden, Jamaica, and Puerto Rico among them—cohabitation is the norm throughout adulthood, equivalent in many ways to marriage. Our focus here, however, will be on cohabitation among young adults as a prelude to, or a try-out for, marriage.

Official statistics indicate that 40 percent of all young adult North Americans acknowledge cohabiting before their first marriage; the actual numbers are undoubtedly higher (London, 1991). One large survey found that half of all cohabiting couples have definite plans to marry their current partner, a third are thinking about it, and only one out of five plans not to marry their current roommate, mostly because they plan never to marry anyone (Bumpass et al., 1991). Not only do many couples consider cohabitation a first step toward marriage; they also believe that cohabitation improves the chances of marital success. In the aforementioned survey, cohabiting adults were asked to check the most important reasons to live together: half cited "ensuring compatibility before marriage."

Despite such youthful hopes, cohabitation does not appear to strengthen marriage. In fact, the opposite seems true. In many studies in North America as well as in Western Europe, marriages that are preceded by cohabitation typically are less happy and less durable (Booth & Johnson, 1988; Nock, 1995). Of course, correlation does not prove causation, and many factors confound any clear link between cohabitation and marital outcome. For instance, couples who decide to cohabit are already at higher risk of divorce than couples who do not, since they tend to be less conventional, less religious, and lower in SES than couples who do not live together without benefit of marriage (DeMaris & Rao, 1992).

However, even when such factors are taken into account, studies still find that cohabiters who later marry are overrepresented in the ranks of the divorced. One possible explanation is that cohabitation may decrease the sense of commitment the act of marriage traditionally entails, because, for a couple who have been living together, uniting formally is a much less dramatic turning point in the relationship than it is for couples who share a household for the first time after the wedding (Thompson & Colella, 1992).

cohabitation Literally, "cohabitation" means living together. It is used primarily to refer to unrelated adults of the opposite sex who share the same house or apartment, presumably sharing the same bed as well.

One indication of this is the fact that wives who cohabited before marriage are three times as likely to have an extramarital affair as are those who did not live with their husbands before the wedding (Forste & Tanfer, 1996).

Although relatively little research has been done on the outcome of cohabitation among gays and lesbians, it is clear that long-term homosexual relationships, once rare, or at least more hidden, are now more common and open. An estimated 2 to 5 percent of all American adults spend some part of their lives in gay or lesbian partnerships (Laumann et al., 1994), choosing such commitments either exclusively or in a sequence that includes heterosexual relationships (Gonsiorek & Weinrich, 1991; Kurdek, 1992).

Whatever one's sexual orientation, establishing a committed relationship of sexual bonding and intimacy is a highly complex undertaking (see A Closer Look on pp. 518–519). Now let us look specifically at marriage.

Marriage

In much of the world, marriage is not what it once was—that is, a legal, and usually religious, arrangement sought at the onset of adulthood as the sole avenue for sexual expression, as the only legitimate context for childbearing, and as a lifelong source of intimacy and support. Among the statistics that make this point are the following: in the United States today, the proportion of adults who are unmarried is higher than at any other time in this century; only 10 percent of brides are virgins; 30 percent of all births are to unmarried mothers; at least another 10 percent of first births are conceived before marriage; the divorce rate is 49 percent of the marriage rate; and the rate of first marriages in young adulthood is the lowest in fifty years (Kahn & London, 1991; U.S. Bureau of the Census, 1996).

Indeed, between ages 20 and 30, the majority of Americans are unmarried, with 60 percent of the men in this age group and 46 percent of the women never having wed and an additional 3 percent of men and 5 percent of women already divorced and not yet remarried. Including the roughly 15 percent who have not yet married by age 40, most American adults now

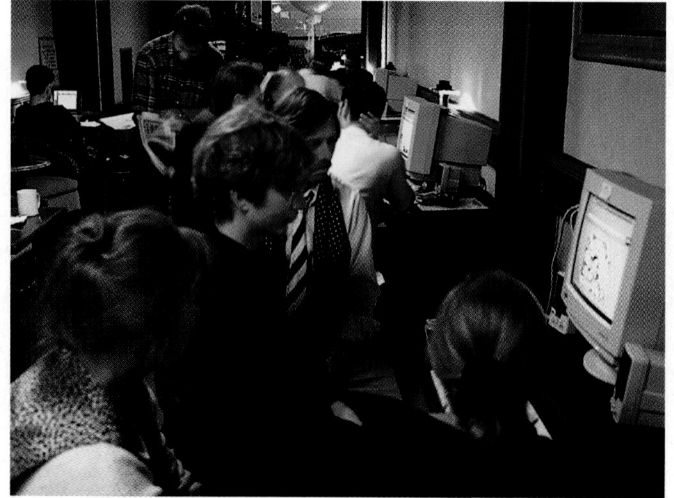

Once, almost every adult married before age 30, pressured by parents, religious institutions, and the social clock to do so. Now some young adults begin a traditional married life, others live together anticipating such a life, and still others explore interpersonal partnership in unconventional ways. For all three groups, psychologists observe an underlying urge for intimacy and commitment, even in this "Internet cafe," where the World Wide Web has replaced the church social or the college mixer.

A CLOSER LOOK

The Dimensions of Love

People sometimes talk and act as though romantic love were a simple, universally understood experience—as though the lyric "All you need is love" says it all. In fact, over the life span, love takes many forms, affected not only by personal preferences and the mutual interactions between the two participants in an ongoing relationship but also by developmental stages, gender differences, socioeconomic forces, and historical and cultural context—all of which makes love complex and often confusing (Fehr, 1983; Sternberg & Barnes, 1988).

To begin with, love styles are quite individualistic, affected by the person's temperament and the history of his or her childhood attachments. Thus some people seem susceptible to passionate and irrational love at first sight; others prefer a loving affection that develops slowly over time; others seek a playful, uncommitted love (Lee, 1988). Still others seek some combination of these, or change from one form to another as they mature.

In addition, partners in a love relationship almost never have precisely the same needs, and thus both almost always experience differing, and varying, degrees of lust and loyalty, jealousy and trust, hot romance and placid companionship—a complex, shifting mix that often leads to misunderstanding and the accusation that the other partner does "not know the meaning of love" (Dion & Dion, 1988; Hatfield, 1988).

One theory that might decrease such misunderstandings comes from Robert Sternberg, who says that we should recognize that love does not have one simple form but instead has three distinct components: (1) passion, (2) intimacy, and (3) commitment. Sternberg believes that the relative presence or absence of these three components typifies seven different forms of love (see table). Further, he finds that the emergence and prominence of each of these components tends to follow a usual developmental pattern as a relationship matures.

Seven Forms of Love

	Passion	Intimacy	Commitment
Liking		•	
Infatuation	•		
Empty love			•
Romantic love	•	•	
Fatuous love	•		•
Companionate love		•	•
Consummate love	•	•	•

As many researchers have found, *passion* generally is highest early in a relationship. This is the period of "falling in love," an intense physical, cognitive, and emotional onslaught characterized by excitement, ecstasy, and euphoria. Such moonstruck joy is often a bittersweet business, however, beset with uncertainties about intimacy and commitment and "fueled by a sprinkling of hope and a large dollop of loneliness, mourning, jealousy and terror" (Hatfield, 1988).

The truth is that, early in a relationship, while physical intimacy and feelings of closeness may be strong, true emotional intimacy is not high. Indeed, research finds a negative correlation between passionate love and several characteristics of intimacy, including openness, honesty, and trust (Aron & Westbay, 1996). The probable reason for this lack of emotional intimacy is that the partners have not shared enough experiences and emotions to be able to understand each other very well. Nor have they gained the firm sense of commitment that comes when various obstacles have been overcome together.

If the relationship is to grow, personal *intimacy* must intensify: a good, lasting partnership depends on a high level of trust, openness, and acceptance of the other's limi-

spend, on average, about half of their twenty years between ages 20 and 40 single (U.S. Bureau of the Census, 1996). Similar trends are apparent in virtually every industrialized nation (Burns & Scott, 1994).

Nevertheless, marriage remains the most enduring evidence of couple commitment, celebrated in every culture of the world by the wedding—complete with special words, clothes, blessings, food, and drink, and, usually, many guests and ostentatious expense. The hoped-for outcome, of course, is a love that deepens over the years, cemented by events such as bearing and raising children, weathering economic and emotional ups and downs, surviving serious illnesses or other setbacks, and sharing a social life

tations that enables "honest, graceful, complete, and patient communication" without anger or guilt (Hatfield, 1988). Attaining such a high level of intimacy is no simple matter. Every time one partner allows the other to share in a secret, to witness a vulnerability, or understand a hidden shame, he or she takes the risk of misunderstanding, rejection, and pain. For the same reasons, true sexual intimacy, in which each partner expresses his or her intensely personal sexual wishes and finds that the other understands and is accepting, may take months, even years, to reach, no matter how intense the passion or how frequent the sexual activity in the early phases of the relationship (Wallerstein & Blakeslee, 1995).

As intimacy continues through time, the third aspect of love, *commitment*, is gradually established, expressed, and strengthened. Commitment grows through a series of day-to-day decisions to spend time together, care for each other, share possessions, and overcome problems even when that involves some personal sacrifice. Devotion and mutual dependence are among the dominant traits of commitment (Aron & Westbay, 1996). Signs of commitment range from formal acknowledgments such as engagement rings, weddings, and childbearing to the routine aspects of daily life—from shared morning meals to shared checking accounts, from working out compromises about leisure activities to working out a division of household work.

As you can see from Sternberg's chart, in the Western ideal, when commitment is added to passion and intimacy, the result is consummate love. For developmental reasons, however, this ideal is difficult to achieve. In large measure, passion is fueled by unfamiliarity, unexpectedness, uncertainty, and risk. Consequently, the growing familiarity and security that contribute to intimacy may dampen passion. In the beginning of passion, a reciprocated touch of the hand, a certain smile, or a mere glance at the lover's body will produce sexual excitement. As lovers get used to their physical relationship, maintaining the same level of arousal would require increasing degrees of physical, psychological, and fantasy stimulation.

In the same way, early in a relationship, the simplest shared confidence can trigger a rush of feelings of trust, but once a certain level of intimacy is taken for granted, deeper sharing is required to promote similarly palpable feelings of togetherness. In other words, with time, passion tends to fade and intimacy tends to grow and then stabilize, even as commitment develops (Sternberg, 1988). This developmental pattern is true over the years for all types of couples, married, unmarried, and remarried, heterosexual and homosexual, young and old (Ganong & Coleman, 1994; Kurdek, 1992).

Recognizing the dynamic nature of love relationships—their continually changing boundaries, demands, and satisfactions—is key to making them endure. As Sternberg (1988) notes:

> "Living happily every after" need not be a myth, but if it is to be a reality, the happiness must be based upon different configurations of mutual feelings at various times in a relationship. Couples who expect their passion to last forever, or their intimacy to remain unchallenged, are in for disappointment. The theory suggests that we must constantly work at understanding, building, and rebuilding our love relationships. Relationships are constructions, and they decay over time if they are not maintained and improved. We cannot expect a relationship simply to take care of itself, any more than we can expect that of a building. Rather, we must take responsibility for making our relationships the best they can be.

and financial commitments. Ideally, over the long run, the marriage is mutually beneficial, with each spouse taking distinct and complementary roles, both strengthened by their relationship.

Given the high rate at which these expectations are dashed, however, many experts and lay people have tried to figure out what makes a marriage work. One important developmental factor is clear: the age and maturity of the partners. In general, the younger marriage partners are when they first wed, the less likely their marriage is to succeed (Greenstein, 1995).

One explanation for the failure rate of marriages between younger people is that, as Erikson points out, intimacy is hard to establish until iden-

tity is secure. Because many older adolescents and young adults are still figuring out their values and roles, a young couple might initially see themselves as compatible only to find their values and roles diverging as they become more mature. Further, the compromise and interdependence that are part of establishing intimacy are not only hard to achieve until one has a clear notion of self and has experienced independence; they may also be hard to recognize as essential. In a series of seven studies of college students, those who were less advanced on a measure based on Erikson's identity and intimacy stages tended to define love in terms of passion rather than intimacy or commitment—butterflies and excitement rather than openness, trust, and loyalty (Aron & Westbay, 1996).

Another factor influencing marital success is the degree to which a couple is homogamous or heterogamous. When studying various cultures around the world, anthropologists draw a distinction between the custom of **homogamy**, that is, marriage within the same tribe or ethnic group, and **heterogamy**, or marriage outside the group. In industrialized nations, homogamy and heterogamy are more a matter of the degree to which the partners are similar in interests, attitudes, and goals, as well as in background variables such as cohort, religion, SES, ethnicity, and local origin. In general, the more homogamous a marriage is, the more likely it is to succeed, partly because the couple's being "on the same page" in many dimensions reduces the potential for tension and disagreement.

One study of 168 young couples found that **social homogamy**, defined as similarity in leisure interests and role preferences, is particularly important to marital success (Houts et al., 1996). For instance, if both spouses had a similar level of interest in, say, picnicking, dancing, swimming, going to the movies, listening to music, eating out, and entertaining friends, the partners tended to be more "in love" and more committed to the relationship. Similarly, if both agreed on who should make meals, pay bills, buy groceries, and so on, then ambivalence and conflict were reduced.

However, the authors of this study criticize those who promote the idea that "finding a mate compatible on many dimensions is an achievable goal." In reality, the authors say, "individuals who are seeking a compatible mate must make many compromises if they are to marry at all," because a high level of marital homogamy is extremely rare. Indeed, according to their research, the odds of finding someone who shares a similar interest in at least three of one's favorite leisure activities and holds a similar view regarding at least three of one's most important role preferences is less than 1 percent. Most successful couples share a few crucial values or interests and otherwise learn to compromise, adjust, or agree to disagree.

A third factor affecting the fate of a marriage is **marital equity**, the extent to which the two partners perceive a rough equality in the partnership. According to one theory, called **exchange theory**, marriage is an arrangement in which each person contributes something useful to the other, something the other would find difficult to attain alone (Edwards, 1969). The marriage becomes a stable and happy one when both partners consider the exchange fair. Historically, the two sexes traded quite gender-specific commodities: men provided social status and financial security, while women provided homemaking, sex, and children.

In many modern marriages, however, the equity that is sought involves shared contributions of a similar kind: instead of husbands earning all the

homogamy As used by developmentalists, the term refers to marriage between individuals who tend to be similar with respect to such variables as attitudes, interests, goals, SES, religion, ethnic background, and local origin.

heterogamy As used by developmentalists, the term refers to marriage between individuals who tend to be dissimilar with respect to such variables as attitudes, interests, goals, SES, religion, ethnic background, and local origin.

social homogamy The similarity with which a couple regard leisure interests and role preferences.

marital equity Refers to the marriage partners' perception of the relative equality of their respective contributions to the marriage.

exchange theory The theory that marriage is an arrangement in which each person contributes something useful to the other, something the other would find difficult to attain alone.

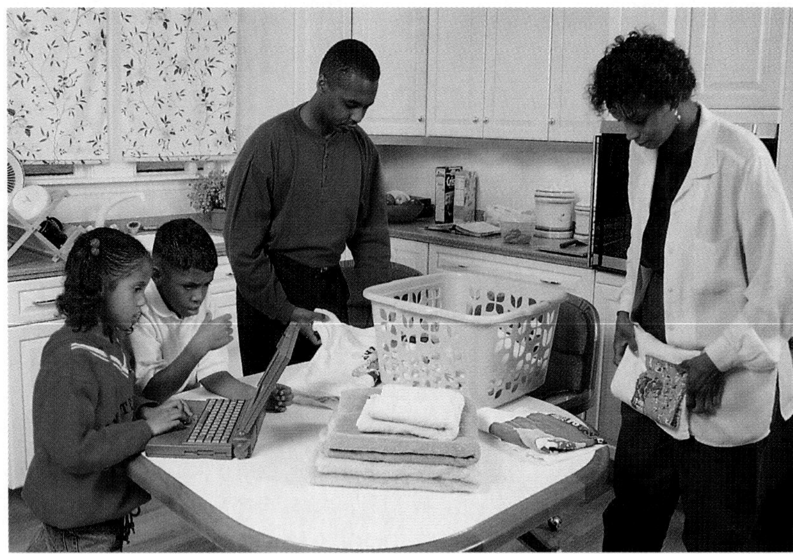

The kind of gender equity shown here presages fewer wars between the sexes in future years.

money and wives doing all the domestic work, both are now expected to do both. Similarly, both partners expect sensitivity to their needs and equity regarding dependence, sexual desire, shared confidences, and so on. Obviously, this combination is not easy to achieve, as divorce statistics reveal.

Divorce

Throughout this book, adopting a contextual view of development has enabled us to recognize that many events that seem isolated, personal, and transitory are actually interconnected, socially mediated, and have enduring implications. Divorce is a prime example. The ending of a marriage does not occur in a social vacuum but, rather, is influenced by factors in the overall social context as well as in the immediate family context, affecting the lives of many people for years to come (White, 1990).

One indication of the impact that the social context has on divorce is the wide variation in divorce rates from nation to nation. The United States has the highest rate of any major country: almost one out of every two marriages ends in divorce. Many other industrialized countries (including Canada, Sweden, Great Britain, and Australia) have a divorce rate of about one in three, while others (including Japan, Italy, Israel, and Spain) have markedly lower rates, with less than one in five marriages ending in divorce (U.S. Bureau of the Census, 1996).

Historical variations are as marked as national ones. Worldwide, divorce rates increased over most of the past fifty years but have stabilized recently. In England, for example, the divorce rate more than doubled between 1970 and 1980 but has since increased by less than 10 percent. In the Netherlands, the rate of divorce tripled between 1970 and 1992 (though it is still less than half the U.S. rate) and then leveled off (U.S. Bureau of the Census, 1996). This worldwide trend is particularly apparent in the United States (see Figure 19.1 on p. 522), where a rapid increase that began in the 1970s reached a high of about 50 percent in 1980 and has hovered near or at that level ever since.

Certainly some of those cultural and historical variations occur because of differences and fluctuations in laws and social norms (White, 1990).

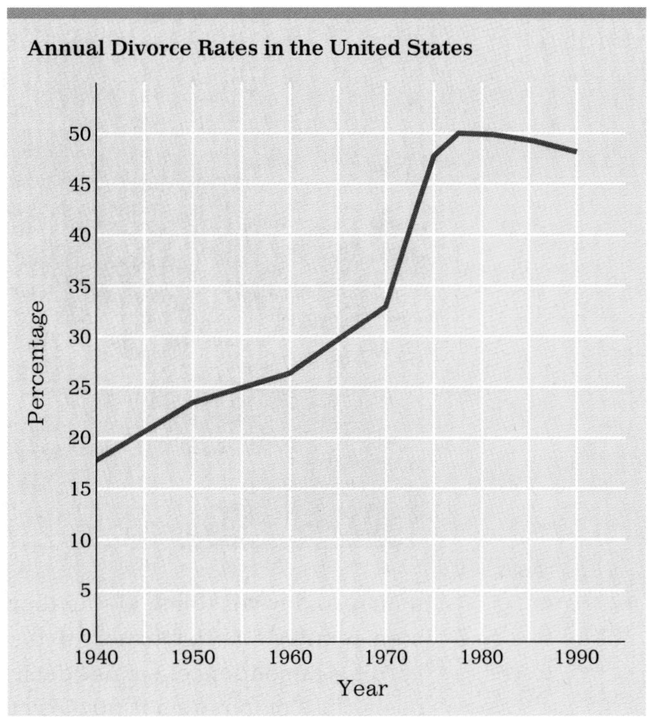

FIGURE 19.1 That the divorce rate has leveled off in recent years in the United States can be seen as an encouraging sign—except for the fact that the leveling-off has occurred at about one divorce for every two marriages, a rate that is higher than that of any other nation in the world.

For example, in the United States, laws that had "preserved" a vast number of troubled marriages by permitting divorce only on proof of serious spousal misconduct gave way in the 1970s to "no fault" laws that granted divorce simply on the assertion of incompatibility. One result was a sudden increase in the number of divorces, as many marriages that had limped along for years were finally dissolved. Similar booms occurred later in other countries when their divorce laws also eased.

Although this boom is now over, the divorce rate in the United States will probably continue to remain high because of another historical trend: married couples today are less likely than their counterparts in earlier cohorts to consider their marriages very happy (Glenn, 1991).

The Role of Expectations

Many developmentalists believe that an underlying explanation for the increasing divorce rate is a cognitive shift that has led most spouses today to expect a great deal more from each other than spouses in the past did. In earlier decades marital equity was judged on the basis of firm gender roles. As long as both partners did their jobs, the marriage usually survived. As one woman, married in 1909, advised newlyweds on her seventy-first wedding anniversary:

> Don't stop on the little things. Be satisfied whatever happens. Ben didn't commit adultery, he's not a gambler, not a liar . . . So what's there to complain about? [Elevenstar, 1980]

Husbands and wives in the past usually did not expect to really understand each other: they generally assumed that masculinity and femininity

are opposites and that the sexes therefore are naturally a mystery to each other. Today, marriage partners have a much more flexible view of marriage roles and responsibilities and are likely to expect each other to be a friend, lover, and confidant as well as a wage-earner and caregiver.

Ironically, while couples expect more from a relationship than couples once did, they may at the same time devote less of themselves to a marriage. As one social scientist reviewing the research on declining happiness in marriage notes, it would seem that

> when the probability of marital success is as low as it is in the United States today, to make a strong, unqualified commitment to a marriage—and to make the investment of time, energy, and foregone opportunities that entails—is so hazardous that no totally rational person would do it. . . . If this reasoning is sound, the current low probability of marital success in the United States will tend to be self-perpetuating, and disseminating information on the low probability may have a negative effect on marital success. Indeed, the institution of marriage may be as healthy as it is only because of the unrealistic optimism of many persons who marry. [Glenn, 1991]

In a later analysis, the same social scientist suggests that the solution to this dilemma is not to raise expectations *for* oneself but to raise expectations *of* oneself, replacing the "unfettered pursuit of self-interest" with a willingness to "commit fully to the marriage and make the sacrifices and investments needed to make it succeed" (Glenn, 1996).

Uncoupling

What impact does divorce have on development, in the short term and over time? Initially, the consequences of divorce are usually worse than either partner anticipated in almost every dimension—health, happiness, self-esteem, financial stability, social interaction, and child-rearing (Kitson & Morgan, 1990). The longer a couple has been together, the more intimate they once were, and the more commitments they shared—such as joint property, mutual friends, and, most important, children—the more stress a breakup brings.

There are two reasons for the unanticipated problems. First, before the breakup, unhappy partners are often so focused on what is missing in their relationship that they are "hardly aware of needs currently being well served" (Glenn, 1991). Thus when they do separate, they find that they have suddenly lost benefits they had not noticed they had. Second, even in troubled relationships, emotional dependence almost inevitably deepens over time, so feuding as well as friendly ex-partners are often surprised by the currents of emotion that remain after the breakup. Hostile ex-partners often have to face rejection that is now unrestrained, while amicable ex-partners find that their attempts to start a new life are undermined by feelings of regret and doubt (Tschann et al., 1989).

Of course, sometimes relationships are so destructive that a breakup—the more radical the better—is a welcome relief. (See A Closer Look, p. 524, on the topic of spouse abuse.) And it is also true that sometimes every aspect of love—passion, intimacy, and commitment—dies long before a formal breakup occurs, minimizing the pain at separation. Nonetheless, ending a long-term marital relationship is almost always difficult.

A CLOSER LOOK

Spouse Abuse

Violence in intimate relationships is very common. Surveys in the United States and Canada find that each year about 12 percent of all spouses push, grab, shove, or slap the other, and that between 1 and 3 percent use more extreme measures, hitting, kicking, beating up, or making threats with a knife or a gun (Dutton, 1992; Straus & Gelles, 1990). Abuse is also common among unmarried couples living together, whether heterosexual, gay, or lesbian.

What leads to such harmful behavior between people who supposedly love each other? Many contributing factors have been identified, including social pressures that create stress, cultural values that condone violence, personality pathologies (such as poor impulse control), and drug and alcohol addiction (Gelles, 1993; McKenry, 1995; O'Leary, 1993; Straus & Yodanis, 1996; Yllo, 1993). From a developmental perspective, one critical factor is a history of child maltreatment. The child who is physically punished, often and harshly, who is sexually abused, or who witnesses regular spousal assault is at increased risk of becoming an abuser or a victim (Straus & Yodanis, 1996).

A more detailed examination of spouse abuse reveals that it occurs in two forms. The first form of spouse abuse is called **common couple violence**. This form of abuse entails outbursts of yelling, insulting, and physical attack, but it is not part of a systematic campaign of dominance. The perpetrators in common couple violence are as likely to be women as men, with both partners sometimes becoming involved in a violent argument. Indeed, in many cultural groups, some interspousal violence is acceptable, with 25 percent of all Americans agreeing that it is sometimes "okay" to slap a spouse. Thus common couple violence is less likely to be pathological (the act of a mentally ill person who is incapable of a normal love relationship) than to result from a combination of personality factors (including a tendency to express anger physically, low self-esteem, and inadequate social skills) and a cultural context that allows, or even condones, occasional violent outbursts between partners.

Common couple violence can sometimes evolve into worse abuse, but more often the couple gradually learns more constructive ways to resolve conflicts, on their own or through marriage counseling. Thus, in common couple violence, there is hope for the relationship.

There is almost no hope, however, when couples are locked in the second type of spousal abuse: patriarchal terrorism. **Patriarchal terrorism** occurs when one partner, almost always the man, uses a range of methods to isolate, degrade, and punish the other. Patriarchal terrorism leads to the *battered-wife syndrome*, which includes the woman's being not only periodically beaten but also psychologically and socially broken, living in perpetual fear and self-loathing, without friends or family to turn to, increasingly vulnerable to permanent injury and death. In nearly all cases, patriarchal terrorism becomes more extreme the longer the relationship endures, because the cycle of violence and submission feeds on itself. Each act that renders the wife helpless adds to the man's feeling of control and the woman's feeling that she cannot, must not, fight back.

Many people find it difficult to understand why a woman would stay in such a relationship. There are two prime reasons: she has been conditioned, step-by-step, to accept the abuse, and she has been systematically isolated from those who might encourage her to leave. In addition, if the couple have children, the husband typically uses them as hostages by threatening to kill them if the woman leaves. In many cases, such threats eventually backfire, as mothers who endure abuse themselves finally become brave enough to leave when the damage to the children is glaringly obvious.

Since a battered wife, by definition, cannot break the cycle of abuse on her own, her escape from it requires outside assistance. Such intervention has become increasingly available over the past twenty years, although the woman must still reach out for help. Recognition of patriarchal terrorism in the United States has led to a much tougher approach by law-enforcement agencies, with police more likely to arrest perpetrators of domestic violence and judges more likely to issue and enforce orders of protection. It also has led to a network of shelters for battered women and their children. These shelters provide not only a haven from violence but also intensive counseling and assistance to help repair the woman's emotional damage and to protect her against further abuse.

Partly because of these interventions, the overall incidence of wife-battering has decreased, from 4 percent to 3 percent of all couples according to an American survey (Straus & Gelles, 1990). The number of marital homicides is also down nationally, although precise nationwide statistics are hard to come by. City-by-city statistics are sometimes available, with Cleveland, for instance, recording twenty-four spousal murders in 1980 and only three in 1994 (Butterfield, 1996). The problem is a long way from being solved, however, and increased vigilance is necessary to stop patriarchal terrorism when it occurs and to prevent further bouts even after a woman leaves or a divorce is finalized. Crucial here is to break through the isolation. The battered spouse must ask for help, and then others must provide legal and emotional protection as long as it takes for the spouse to become able to function independently.

The Developmental Pattern of Divorce

In the first year of divorce, many ex-spouses become even angrier and more bitter toward each other than they were in the last months of the marriage. Often this is not entirely their fault: the legal system fosters contention over alimony, the division of property, and/or child custody. Then, too, the need to preserve a sense of self-esteem in the face of a failed marriage almost requires that each spouse direct some fury and blame at the other.

Another adjustment problem is that the ex-spouses' social circle almost always shrinks in the first year after divorce: former friends and in-laws find it difficult to remain on good terms with both halves of a severed couple; neighborhood friends are lost when one or both spouses move away; and casual friends and work colleagues tend to distance themselves when the emotional neediness of the newly divorced person suddenly escalates. The cumulative result is the loss not only of friendships but of vital support at a time of high vulnerability.

Given all this, it is not surprising that newly divorced people are more prone to loneliness, disequilibrium, promiscuous sexual behavior, and erratic patterns of eating, sleeping, working, and drug and alcohol use. In most cases, such effects generally dissipate within a few years (Larson et al., 1995). However, for many people, especially those who do not remarry, certain effects linger over the years (Kitson & Morgan, 1990). National surveys find that single divorced adults of every age are least likely to be "very happy" with their lives, not only when they are compared to married people but also when they are compared to never-married or widowed adults (Lee et al., 1991; Myers, 1993). One reason for this long-term effect is that divorced individuals who do not remarry generally have less income (especially true for women with children) and a smaller social circle (especially true for men) than their never-divorced peers. Even taking finances and friendship into account, however, those who are divorced, especially if they have been divorced more than once, have higher rates of depression and poor health (Kurdek, 1991; Larson et al., 1995).

Extensive research confirms that the presence of children is a key factor that makes adjustment to divorce more difficult. Especially in the first year or two after the divorce, children often create additional direct stresses as they become more demanding, disrespectful, or depressed. The presence of children also adds financial pressures, forces the ex-spouses to maintain contact with each other, visibly reminds them of what might have been (or of what actually was), and makes remarriage less likely (Maccoby et al., 1990).

The financial burden of child-rearing usually falls heaviest on custodial mothers, because whatever child-support payments they receive and salary they earn do not offset their increased child-care expenses and the loss of the father's direct contribution to total household expenses. Some of this disparity is caused by fathers who provide no child support at all—that is, about 40 percent of all divorced fathers. Recent legal measures against "deadbeat dads"—garnishing of wages, withholding of a driver's license, interstate enforcement cooperation, and even imprisonment—are closing this gap (Beller & Graham, 1993), but currently only about half of divorced fathers pay the full court-ordered amount over the years. In fairness, those who are delinquent are more likely to have little money themselves. One

common couple violence A form of abuse in which one or both partners of a couple engage in outbursts of verbal and physical attack.

patriarchal terrorism The form of spouse abuse in which the husband uses violent methods of accelerating intensity to isolate, degrade, and punish the wife.

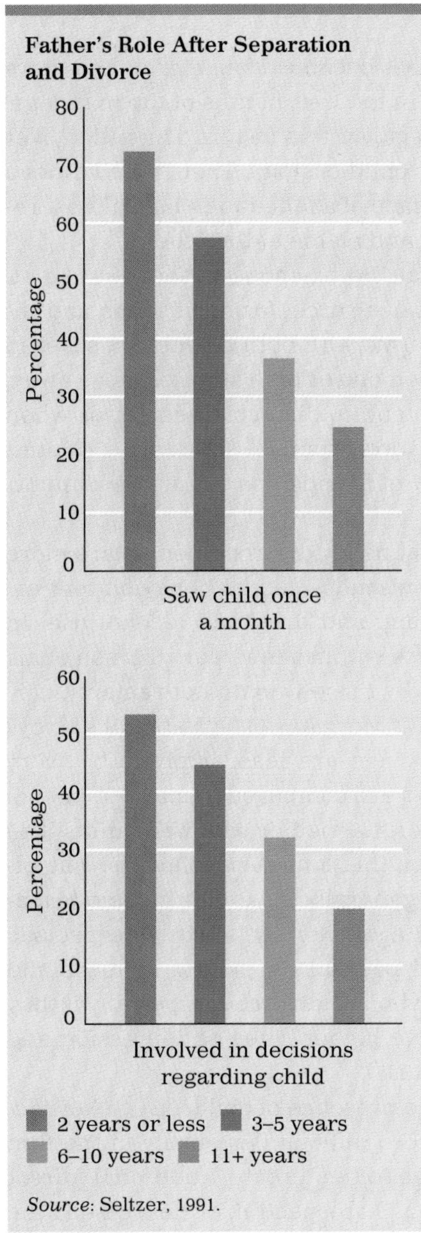

FIGURE 19.2 As graphically shown here, time tends to weaken the bonds between divorced fathers and their growing children, even when both generations wish it were otherwise.

detailed study found that two-thirds of the fathers who earned more than $20,000 a year paid full child support, but only 22 percent of those who earned less than $10,000 did (Meyer & Bartfield, 1996).

Even if both parents contribute as much time and money to child-rearing as they did before the divorce, one or both of them lose in the process, as two dwellings and added child care are unavoidable. Custodial parents, of course, are more likely to experience a severe loss of income and free time, but noncustodial parents not only must pay for themselves and for part of their children's care but, because they live elsewhere, they cannot maintain the intimate bonds that come from daily and nightly interactions.

This physical distance, leading to a psychic distance, is particularly distressing for many of today's fathers who had been involved caregivers from infancy. Only a minority of them maintain intimate ongoing relationships with their children (Arendell, 1995). Compounding the problem is the fact that, as children develop, their interests and emotional needs change, and parental anticipation of, and adjustments to, these changes are difficult without frequent, ongoing interaction with the child. Further, many custodial mothers express their anger at their reduced financial circumstances by limiting the father's physical and emotional access to the child—either directly by forbidding contact without support payments, or indirectly by creating practical obstacles to visitation or otherwise trying to undermine the father's role. Such measures, while understandable, are destructive in many ways, including economic. In general, the less a father sees his children, the less he contributes financially (Peters et al., 1993).

The net result for divorced fathers is that most of them become peripheral to their children's lives. One national study (Seltzer, 1991) found fathers less involved with their children with every passing year (see Figure 19.2). As time goes on, these trends continue, with many aging divorced fathers having virtually no contact with their grown children, risking a lonely late adulthood. This is not inevitable, of course. Some divorced fathers remain very active in their children's lives, and some remarry, becoming involved with stepchildren or new biological children. (The likely benefits and hazards of remarriage are discussed in Chapter 22.) Many, however, miss out on one important way to meet the other basic need of adulthood: generativity.

Generativity

The motivation to achieve is one of the strongest, and most frequently studied, of human motives. The observable expression of achievement motivation, of course, varies a great deal from person to person and culture to culture. For example, some people and some cultures are much more competitive or cooperative than others; some seek tangible signs of success, while others strive for less materialistic attainments; some emphasize college degrees or official titles, while others strive for neighborhood respect or family attainments. But in one way or another, for our self-esteem as adults, we all need to feel successful at something that makes our lives seem productive and meaningful. Adults meet their need for achievement, confronting what Erikson describes as the crisis of generativity versus stagnation, primarily through work and parenthood.

The Importance of Work

From a developmental perspective, a paycheck is only one of the many possible benefits derived from employment. Work also provides a structure for daily life, a setting for human interaction, a source of status and fulfillment. Even more crucial, work can satisfy generativity needs, allowing the individual to develop and use personal skills and talents, to express his or her unique creative energy, to aid and advise coworkers, and to contribute to the larger community by providing a needed product or service.

The pleasure of "a job well done" is universal. Research from many cultures confirms that job satisfaction correlates more with challenge, creativity, and productivity than with high pay and easy work (Myers, 1993; Wicker & August, 1995). In fact, those who work primarily for the money are also those least likely to enjoy their work. Also contributing to job satisfaction and the fulfillment of generativity needs are organizational structures that give workers respect and some degree of control over their work and their work environment. Management policies, such as team-building, on-the-job training, flexible scheduling, and so on, all encourage the sense of efficacy that makes workers feel productive (Cascio, 1995).

Of course, generativity needs can be met through other activities such as artistic creation, doing volunteer work, and fulfilling family responsibilities. But in most Western cultures today, the crisis of generativity versus stagnation is confronted directly within the workplace, with the outcome affecting the worker's family life and mental health. Generally, adults who feel fulfilled and productive at work are more likely to be satisfied with their marriages and happily involved with their children (Barnett et al., 1995; Ferber & O'Farrell, 1991; Larson & Richards, 1994; Zedeck, 1992). Correspondingly, dissatisfaction, a sense of powerlessness, stress, frustration, anger, and other negative feelings arising in the workplace can spill over into the home, with detrimental effects on all. A satisfactory job, then, is often more than an economic necessity; it is frequently a developmental imperative.

The Historical Context

As recently as the 1950s in the industrialized world, finding the right job seemed a fairly straightforward developmental task that could be accomplished by following a culturally prescribed pattern. As described by the then-leading theorist of life-span vocational development, there were four stages of the career cycle: exploration, establishment, maintenance, and decline (retirement) (Super, 1957). In early adulthood, a person was expected to explore the vocational possibilities, decide on a lifelong occupation, get the necessary training, and then, by age 25, establish a "permanent position" (Super & Thompson, 1981). Maintaining the same vocation, usually with the same employer, the individual was expected to climb steadily up the career ladder—achieving predictable advancements gained at predictable times—or to make linear progress along one vocational path, achieving increasing seniority, responsibility, and salary and ending up with "forty years and a gold watch" at age 65 (Vondracek & Kawasaki, 1995).

This pattern fit well with the stable and not very demanding low-tech job market that was typical at midcentury. Most men at that time worked in manufacturing or sales in large companies or family businesses. Most jobs,

particularly in manufacturing, "required low levels of expertise, so that employees became interchangeable," easily trained or replaced (Caruso, 1992). Consequently, young men tried to obtain a basic education and then find an employer or a union that would guarantee security, income, and eventually a pension. For the vast majority of young women at midcentury, of course, a "career" meant marriage, children, and housekeeping, so the task for them in early adulthood was to find a husband who was "a good provider," enabling them to be model homemakers.

Obviously, not everyone followed this job-cycle pattern. Many workers found this lock-step approach to the career path unsuitable to their needs, and most minority workers simply found such careers unavailable (Leong, 1995). But the structure of the job market endorsed this pattern as the natural one, with guidelines for promotion, health benefits, and retirement all drawn up with the stable, long-term employee in mind.

New Patterns

As the twenty-first century approaches, every aspect of the employment scene—the work, the workers, the employers, and the typical career sequence—is changing worldwide. As a result, the developmental pattern of work is no longer the unbroken line of advancement it once was.

Much of this change begins with alterations in the world economy. In the poorest regions of the world, the shift is from agriculture to industry, as multinational corporations move in to take advantage of cheap labor. Developed nations meanwhile are shifting from industry-based economies to information and service economies. Indicative of this restructuring is the fact that by 2005 the manufacturing and mining components of the U.S. economy are expected to shrink by a third from their already reduced levels, while rapid expansion is expected in jobs that require the worker to provide information, a service, or treatment rather than a particular product (U.S. Bureau of the Census, 1996). For example, among the fastest-growing occupations are physical therapist, human service worker, computer engineer, home health caregiver, systems analyst, medical assistant, and special-education teacher.

As a result of all this change, certain job categories are appearing or disappearing, seemingly overnight; educational requirements for work are shifting every few years; and such corporate strategies as downsizing and outsourcing are becoming standard practices. Consequently, the work path for individuals is much less predictable and secure than it once was, with a widening employment and income gap between those with knowledge and expertise and those without (Reich, 1992).

Diversity in the Workplace

Another major change in the contemporary employment scene is that the workplace is more diverse in every way than it was a few decades ago. The most marked change is in the sex ratio. Once very few women were employed; now in developed nations almost half the civilian labor force is female: 40 percent in Japan, 45 percent in Canada and England, 46 percent in the United States, and 49 percent in Sweden (U.S. Bureau of Labor Statistics, 1996). Increasing numbers of working women are apparent in every occupa-

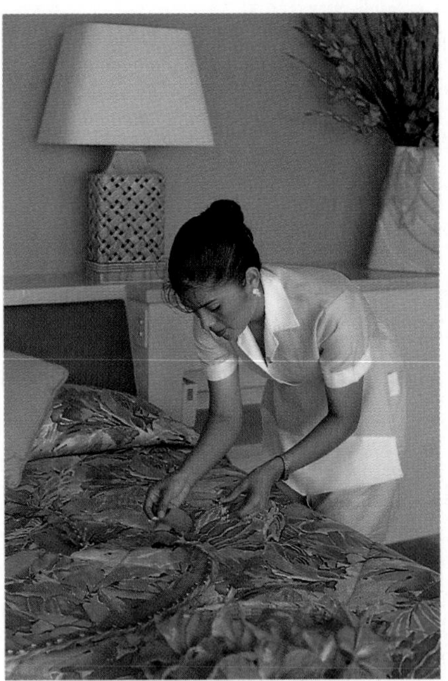

In the modern workplace, sex segregation is more likely at the lower end of the pay scale than nearer the top. While there are many more women lawyers and male teachers than there were even twenty years ago, few women today are truck drivers and almost no men are hotel housekeepers.

tion. In the United States in 1995, for example, a fourth of all the lawyers, physicians, computer scientists, and chemists were female. Some occupations continue to be "sex-segregated," of course, with 93 percent of the registered nurses being female and 97 percent of the firefighters and airplane pilots being male. Even in these occupations, however, the sex barriers are decreasing: twice as many of the other sex are nurses, pilots, or firefighters as were fifteen years ago (U.S. Bureau of the Censes, 1996).

In recent years, ethnic diversity is increasing as well. In every developed nation except France, the total immigrant segment of the labor force is increasing, with rates currently at 25 percent in Australia, 19 percent in Canada, 9 percent in the United States, Austria, and Germany, 3 percent in England, and 1 percent in two nations that formerly had virtually no immigrant workers, Italy and Japan (U.S. Bureau of the Census, 1996).

A national breakdown of the employed immigrants in the United States is not available, but from overall statistics it is obvious that ethnic diversity in the American work force is increasing. Between 1989 and 1994, more than 6 million immigrants, most of them of working age, entered the United States, about the same number over that five-year period as immigrated over the thirty years from 1950 to 1980. During that five-year period, fifty-four nations each sent at least 10,000 immigrants to the United States, with more emigrating from Asia than from any other continent. In the 1990s, the top six nations sending immigrants to the United States were China, the Philippines, Vietnam, India, Mexico, and the Dominican Republic (U.S. Bureau of the Census, 1996). The influx from the last two of these countries, as well as the increasing numbers of young adults among the native-born Latino population, are the reasons that, between 1980 and 1995, the number of Hispanics in the U.S. work force doubled from 6 to 12 million. For a variety of reasons, including increased educational levels and affirmative-action policies, racial diversity in the work force increased as well: between 1980 to 1995, the number of African-Americans in the work force rose from 11 to 15 million (U.S. Bureau of Labor Statistics, 1992, 1996).

Typical in many ways of today's labor force, these two workers in a food-processing plant check quality rather than perform manual labor, as their counterparts fifty years ago did. The other major difference is that their historical counterparts were almost always married white men. In large companies today, almost half of the employees are female and/or nonwhite, except at top management, where a glass ceiling keeps boardrooms almost exclusively male and white.

Implications for Development

What do all these changes mean for today's workers, particularly for young adults just starting out in the work world? Two implications stand out: the first is that, on the whole, it is a mistake to plan on climbing one specific career ladder, rung by rung, for forty years. Instead of training for one job in one career that will last a lifetime, young adults are urged to develop basic skills, particularly in communication, logical thought, and human relations, that will enable them to be flexible, learning new skills, mastering new tasks, working with new people, and changing jobs often (Reich, 1992).

Indeed, given the fluidity of the job picture, every adult should expect at some point to be retrained and transferred within a restructuring company, to quit or be fired by a downsizing employer, to start his or her own small business, or to be hired by an employer in an industry that did not exist a decade ago. Instead of a smooth career path, almost every young adult will experience several periods of unemployment, self-employment, temporary work, or further education, and most middle-aged workers will have at least one such episode as well. Indeed, if current rates continue, the typical American worker, employed from age 20 to 65, will have ten employers and seven distinct occupational titles before retirement (U.S. Bureau of the Census, 1996).

The second implication, arising from the growing diversity of the workplace, is the need for greater sensitivity to cultural differences. Increasingly, potential tension between the culture of the employee and the culture of the organization must be recognized and reconciled (Leong & Brown, 1995). For example, an organization might expect adherence to a hierarchical chain of command and formal channels of communication, or it might expect free-wheeling teamwork and open cooperation across job titles. Depending on their particular background, certain workers might find one or the other of these approaches alien to their own culture's normative style of interaction. Similarly, the worker's culturally influenced attitudes regarding such matters as punctuality, appropriate attire, cross-sex interaction, and the like, may differ from those endorsed by the organization. Ideally, whatever cultural values and assumptions workers bring to the job need first to be recognized and appreciated and then incorporated when they are beneficial to the job and respectfully revised when they are not.

Unfortunately, another implication of the rapidly changing job picture is that women and members of minorities may continue to experience difficulty in breaking through the **glass ceiling**, the invisible barrier to career advancement that has little to do with the individual's skills and abilities and a great deal to do with cultural assumptions and biases. The glass ceiling is easy to demonstrate in the overall work scene. For example, of the 700,000 managers of marketing, advertising, and public relations in the United States today, only 2 percent are African-American, only 3 percent are Hispanic-American, and only 36 percent are women (U.S. Bureau of the Census, 1996). In addition to whatever other biases may account for such numbers, research suggests that women and minority-group members are likely to be held to higher-than-normal performance standards with regard to promotion (Biernat & Kobrynowicz, 1997).

glass ceiling An invisible barrier experienced by many women in male-dominated occupations—and by many minority workers in majority-dominated occupations—that halts promotion and undercuts their power at a certain managerial level.

As obvious as the glass ceiling is looked at in terms of statistics and research, it is often hard for an individual to know if—and to prove that—he or

mentor In career development, a mentor is a more experienced coworker or supervisor who provides advice, instruction, and support. In many professions, the mentor system operates informally, with great impact. Finding a good mentor may make the difference between success or failure in a new career.

she has hit it when denied promotion into higher management on the grounds of personality, experience, commitment, and the like. As a result of the increasing flux of the job market, the diminishment of job security, the decreasing support for affirmative action, and the likely need for employees to switch jobs and/or career tracks fairly frequently, it will be even easier for those in power to keep the glass ceiling in place.

Particularly for minority and female workers, one way to obtain help on the job is to have a **mentor**, an experienced worker who acts as teacher, sponsor, protector, counselor, role model, and friend (Wunsch, 1994). Many companies and universities assign mentoring pairs, specifying how mentor and mentee should move from initial contact to final independence (Galbraith & Cohen, 1995; Wunsch, 1994). While such programs are useful, it is rare for such a formal pairing to function as intended. Instead, new employees more often seek out advisors as needed, using a designated mentor as a safety net, a person required to give help if no better source is readily available. Generally, new employees of whatever background or gender are likely to succeed to the extent that they reach out to not just one but to several experienced coworkers for advice and support (Feij et al., 1995; Hesketh, 1995; Wood & Bandura, 1996).

Finally, preparing for the coming job market requires the same human relations skills we have already seen as being needed in friendship and marriage. According to surveys of employers, only 9 percent of all dismissed new employees lose their jobs because they could not learn the work. Instead attitude, absenteeism, and inability to adapt to the work environment are the most common problems leading to dismissal (Cascio, 1995). For their part, workers need to choose a work climate—not just a vocation—that encourages all forms of ongoing education, with networking, scaffolding, on-the-job training, summer seminars, and so on. Such workplaces will allow the individual to achieve what work can be at its best, not just a job with a paycheck but work that is generative.

In the early 1940s, the pressures of the wartime economy led to the mass hiring of women. This young adult was happily employed as a welder in the Bethlehem shipyards in 1943. Three years later, she, like most other women who had been wartime workers, was out of the labor force and considered suited only for the role of wife and mother.

Work and Family

Gender roles in work and family have shifted dramatically over the twentieth century. In the first decades of the century, custom and sometimes law dictated that married women were to be full-time housewives and mothers. In some professions, including teaching, female employees were required to quit as soon as they married; in others, being married was an automatic bar to employment. Husbands, by contrast, were designated by society as "the family breadwinner," happily avoiding baby care and housework, bragging that they had never changed a diaper, scrubbed an oven, or even boiled an egg.

A temporary relaxation of this gender divide occurred during World War II, when the drafting of millions of men into the armed services required their replacement in the labor force by millions of women. These women worked at all manner of jobs, including those in the factories that produced the machinery of war. As exemplified by the popular song "Rosie the Riveter," their efforts were seen as vital to the nation's survival. But as soon as the men came home after the war, the gender divide was restored. Women workers quit or were fired. Postwar legislation granted ex-soldiers free college education, employment, and low-cost mortgages, which they

used to buy homes where their wives raised a baby-boom generation. By 1960 in the United States, 82 percent of the married mothers of young children were not in the labor force but 98 percent of their husbands were.

Compare that with the picture in the United States in 1995, when most married mothers were employed, including 77 percent of those whose youngest child was in school and 60 percent who had an infant (U.S. Bureau of the Census, 1996). Most (95 percent) of their husbands were in the labor force as well, but most also shared domestic tasks with their wives, sometimes providing major child care.

Benefits and Problems

Dual-earner, dual-caregiving couples face many potential problems related to the coordination of work and family roles and responsibilities. Before discussing some of these, however, we need to make clear that two difficulties once attributed to such families—neglected children and overloaded mothers—are usually not problems at all.

First, contemporary children generally do not suffer when both parents work outside as well as inside the home, a research conclusion that has been evident for twenty years regarding school-age children and now is found to be true for younger children as well (Gottfried & Gottfried, 1994; Hoffman, 1989). In fact, if anything, children typically gain a number of benefits from both parents being employed and caregivers, most notably higher family income and more active relationships with their fathers. They also witness more flexible role models—seeing their mothers as competent workers and their fathers as competent caregivers.

There is no evidence that children in dual-income families suffer from neglect. In fact, one study of more than 2,000 single- and dual-income families found that, as the number of hours a mother spent in the work world increased, the amount of time the children spent with their parents *increased*. This was partly because fathers were more involved with the children when the mother worked and partly because the employed mothers tended to include the children when doing such chores as preparing meals, washing dishes, and so on, instead of doing these chores alone, as unemployed mothers were more likely to do (Bryant & Zick, 1996).

Second, women who are simultaneously wife, mother, and employee do not necessarily experience **role overload**, as the stress of multiple obligations is called. Of course, adults of both sexes can be overloaded with multiple demands—as might happen when one spouse wants the other to provide home-cooked meals, a picture-perfect house, and romantic evenings during the same week that the other spouse's job requires extensive overtime and several small children are insisting on undivided attention.

However, **role buffering** in dual-earner families is more prevalent than role overload. For both sexes, each role (spouse, parent, and employee) provides a buffer that reduces disappointments in other roles. In addition, the fact that two people share obligations cushions the inevitable troubles, such as a seriously sick child or a period of unemployment, that virtually every parent and worker encounters. Generally those adults who, over the years, balance marital, parental, and vocational roles are healthier, happier, and more successful than those who function in only one or two of them (Barnett & Baruch, 1987; Larson & Richards, 1994; Orbuch & Custer, 1995; Snarey,

role overload The stress of multiple obligations that may occur for a parent in a dual-earner family.

role buffering The common situation in dual-earner families in which one role that a parent plays reduces the disappointments that may occur in other roles.

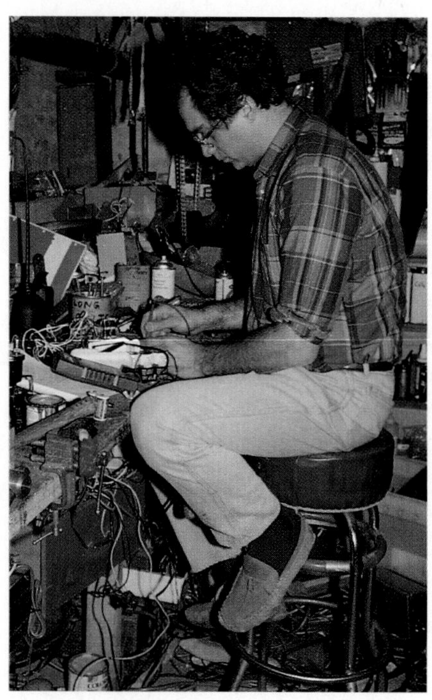

Contemporary couples often feel that their domestic labor is underappreciated by their spouse, a perception that leads to more marital discord than the traditional problems of sexual incompatibility or alcohol abuse. Whoever vacuums, dusts, folds laundry, or empties waste baskets typically believes that his or her partner has no idea how time-consuming and enervating such tasks can be. Meanwhile, the creative work—decorating, gardening, trying new recipes, fixing the car, or puttering in a basement workshop—is often resented rather than valued as the family-supportive effort that it is.

1993). If an adult cannot perform all three roles, two are usually better than one. Single parents fare better if they are employed; adults without children are happier if they are in a committed relationship and if they have a career that is important to them.

Those reassurances aside, however, the dual-income couple with children does face potential problems that, if not resolved, can negatively affect marital happiness and family development (Parcel & Menaghan, 1995). One major problem involves the perception of marital equity, both in household responsibilities and in income production. When either spouse feels a marked inequality in these areas (whether or not an outside observer would agree), resentment builds and the marriage suffers.

Distress about unequal child-care responsibilities is especially likely when the children are small. Almost every dual-earner couple is stressed by the requirements of the "second shift" of domestic work that awaits their return home (Hochschild, 1989). Almost always, employed wives do much more housework and child care than employed husbands, even when they work as many or more hours outside the home (Brayfield, 1992). In addition, their tasks are those repetitive ones that must be redone daily—meal preparation, cleaning, child care—while the husband's tasks tend to be more flexible in timing, such as taking care of the car or of household repairs. Typically, both partners say they believe in gender equity more than they actually practice it, with wives resenting that husbands do much less than an equal share, while husbands feel that the fact that they do so much more domestic work than their own fathers or their unmarried friends goes unappreciated. For example, in one two-career marriage, the husband thought he was doing his fair share because he took care of all the tasks related to the first floor of the house—the floor that had the car, his workshop, and their dog—while the wife seethed that "her floor" had the kitchen, all the living and sleeping areas, and, not incidentally, their 3-year-old son (Hochschild, 1989).

There are some signs that this aspect of marital equity may be achieving a better balance, at least among today's younger couples. In one recent study of young two-income families, the wives logged an average of twenty-six hours a week on domestic work and forty-two hours on the job, while the husbands worked twenty-one hours at home (including car and yard care) and forty-eight and a half hours on the job (Barnett & Rivers, 1996). Overall, the particular male-female breakdown of domestic work depends on many factors that vary from couple to couple. In general, wives perform a disproportionate share of domestic labor in a first marriage with several children, and work is more fairly distributed when the spouses are college-educated or African-American (Orbuch & Custer, 1995; Pittman & Blanchard, 1996).

Another kind of marital equity that can be especially troubling for men is when the wife earns more, and is more committed to her career, than the husband, an inequality that is increasingly common, and particularly devastating, among men who are blue-collar workers (Orbuch & Custer, 1995; Staines et al., 1986).

Many couples resolve this economic disparity by trying to equalize their employment over the long term. Research in ten European nations as well as in the United States and Canada finds that wives tend to have jobs that are similar to or slightly lower in status than their husband's work (Philliber & Vannoy-Hiller, 1990; Smits et al., 1996). This means that, when a

family logistics The task of coordinating the many overlapping elements of family life and work schedules, a task that is particularly urgent for dual-earner families.

woman could earn substantially more than her mate, her husband's status tends to become an artificial ceiling to her own vocational rise. On the other hand, if the husband's status is substantially higher than hers, it may facilitate her vocational rise (perhaps allowing her to finance more education or to find work through his contacts) or it may lead her to depart from the job market completely. In many marriages, one spouse supports the other's education or career climb, and then, years later, the situation is reversed, leading to equity over the long term rather than at any particular stage.

The second problem for dual-earner families is more immediate, even urgent. **Family logistics**—coordinating births, job changes, further training, relocations, housework, child care, work schedules, medical leave, vacations, and such specifics as who will dress Mary, mop up Johnny's spilled milk, and feed the baby just as the phone is ringing and the school bus is honking—require a level of mutual agreement, coordination, and planning that was unnecessary in earlier generations. Logistical complications are increasing. As one researcher explains:

> Given the current economic climate, everyone's use of time and the ways of organizing work and family relationships are changing. The rapid expansion of the service sector and the growing reliance on contingent workers mean that more and more parents will move away from the conventional 8-hour, 5-day per week job. This structural transformation complicates the mutual negotiation of work and family roles, and it implies that strategies for juggling multiple roles will have to become more creative. [Brayfield, 1995]

In the family context, creativity means finding the best solution for the particular family in question. Some succeed by following traditional gender roles; others practice equity in all aspects; and still others switch traditional gender roles, with the wife acting as the primary wage-earner and the husband becoming a stay-at-home dad. In all cases, both spouses need to be able to take on the other's tasks when necessary and to recognize that no particular pattern is inevitably beneficial or harmful. Over the decades of any family, new babies arrive and older children grow up, job opportunities emerge or disappear, financial burdens increase or decrease. If, at each stage along the way, both spouses are able to appreciate each other's contributions and to readjust their roles as needed instead of clinging to a rigid pattern or resenting an unequal partnership, the family is likely to thrive more readily.

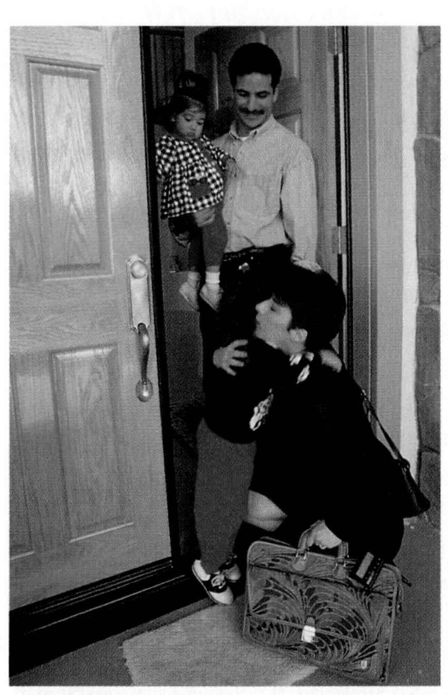

Seeing the breadwinner off to work was photographic cliché fifty years ago—except that Mom and Dad played opposite roles. Today an estimated one in ten two-parent families, like this one, include a father who works inside the home and a mother who works outside it. Usually the children, and the marriage, have as good or better a chance at success with this new pattern as with the traditional one.

Looking to families in the future, the larger social context is crucial, as many institutions follow policies based on the typical family of 1960 and must now adjust to current patterns. Large corporations and employers already differ markedly in how "family friendly" they are, with some having flextime, subsidized day care, personal leave to discharge family responsibilities, voluntary overtime, health benefits that include preventive care and coverage for all family members, and so on.

Nations vary on these factors as well. Some provide every resident free child care, universal health benefits, and paid maternity leave, sometimes for a year; others mandate that employers provide such benefits; others, such as the United States, furnish only minimal support, such as the first "family leave" law, not passed until 1993, which legalize maternity leave—but for only twelve weeks, without pay, and only if the employer has at least fifty full-time workers. National variations obviously have deep cultural and political roots, with implicit values in every law and custom that affect family life, and with potential for adjustment as work patterns change. Many de-

velopmentalists would agree with John Snarey, author of a four-generation study of fatherhood and generativity, who hopes for a more supportive social context in the future. Snarney (1993) concludes:

> . . . as more childrearing fathers and mothers launch their children from the nest and begin to take on positions of broader responsibility, they will draw on their generative values to promote positive institutional changes . . . enlightened workplace policies which acknowledge the priority of family life and promote the common good of children, parents, and society as a whole.

Parenthood

When one stranger asks another "What do you do?" the answer is rarely "Raise children." Yet adults are as likely to be parents as to be employed, and many of them consider the successful rearing of their children to be their most important achievement. As Erikson points out, while generativity can take many forms, its chief form is "establishing and guiding the next generation," usually through parenthood. Caring for children fulfills an important adult need, according to Erikson (1963), for "the fashionable insistence of dramatizing the dependence of children on adults often blinds us to the dependence of the older generation on the younger one." The mature adult "needs to be needed," and the interdependence of parents and children is a lifelong process, which begins at conception and continues throughout late adulthood.

Since the parental role has been thoroughly explored in the first half of this book, here we need only quickly summarize what has been apparent throughout. Bearing one's own biological offspring, nurturing them, and then finally launching them into the adult world has a major impact on each parent's own development, bringing different psychosocial challenges and satisfactions at each stage of the children's growth. During a child's infancy, for example, parents experience the joy of getting to know a new human being who is just beginning to take shape as an individual—and they also experience the overwhelming task of providing that infant with constant care. During the preschool years, the primary challenge for parents involves authority, over the child and within the marriage. The preschool years are often the period of greatest direct conflict between husband and wife and between parent and child, due to the pressure of increasing financial burdens, multiplying household tasks, and shifting roles, as well as the child's growing need to assert his or her independence. Similarly, adolescence presents a whole new set of parenting issues, as the young person begins to demand the privileges of adulthood and in the process often calls the parent's values into question, sometimes undermining the parent's self-esteem (Cowan & Cowan, 1992; Hooker et al., 1996; Silverberg & Steinberg, 1990).

In short, every parent is tested and transformed by the experience of raising a child, and just when adults think that they have mastered the role of parenting, their child's advance into the next stage of development requires some major rethinking on that score.

Alternative Forms of Parenthood

For the roughly one-third of all North American adults who will become stepparents, adoptive parents, or foster parents at some point in their lives,

the challenges and satisfactions of parenthood are much the same as they are for biological parents. At the same time, however, these alternative forms of parenthood are open to special problems.

The core problem facing many nonbiological parents involves the strength of the parent-child bond. Strong bonds between parent and child are particularly hard to create when a child is old enough—a year or more—to have formed definite attachments to other caregivers who are still available to the child. This is usually the case with stepchildren, since the average age of a new dependent stepchild is about 9 years. Stepmothers often enter the marriage with visions of healing a broken family through love and understanding, while stepfathers typically believe that their new children will welcome a benevolent disciplinarian who can bring some order to their lives. Neither of these expectations usually develops. If all goes well, the stepparent usually becomes a close friend, an "intimate outsider," who nonetheless remains much more distant from the stepchild's personal life than he or she initially imagined would be the case (Hetherington & Clingempeel, 1992; Papernow, 1988).

Indeed, many stepchildren are fiercely loyal to the absent parent, sabotaging any newcomer's effort to fill the traditional parent-role, perhaps by directly challenging his or her authority ("You're not my *real* father, so don't tell me what to do"), or continually intruding on the couple's privacy, or evoking guilt by getting hurt, sick, lost, drunk, arrested, and so on (Pasley & Ilhinger-Tallman, 1987). Such childish reactions, often unconscious, may cause the stepparent, or both parents, to overreact in ways that further alienate the child. Even if the stepparent is patient and understanding, it often takes years before children adjust to the changes in the family dynamics that any new spouse creates. It might take even longer before the children express the appreciation and affection that a caring stepparent deserves. The dilemmas of stepparenting are reflected in one stepfather's advice for taking on stepchildren:

> Don't ever expect to replace the natural father in their eyes; win their respect; treat them as your own; love them and discipline them as your own; let nothing come between you and your woman—especially the kids. [quoted in Giles-Sims & Crosbie-Burnett, 1989]

This father's final caution is particularly poignant because disobedient and difficult stepchildren, especially those aged 10 and older, are often a precipitating factor in a second divorce (White & Booth, 1985).

Some foster and older adopted children are likewise attached to their birthparents, an attachment that can be especially volatile because of the destructive treatment many of them have experienced. Other foster children present the opposite problem: they were never attached to anyone and thus rebuff the foster and adoptive parents' attempts to win them over. The attachment between foster parents and children is further handicapped by the fact that the bond can be suddenly severed for reasons that have nothing to do with the quality of foster care, ranging from the biological mother's completing a drug-treatment program and thus qualifying for another attempt at motherhood to a policy shift at the child welfare agency that requires placing the child somewhere else.

Stepparents face different but equally significant threats to their relationship with their children (Papernow, 1993). One potential problem is divorce, which occurs in about half of all marriages involving stepchildren.

Usually such divorces halt the children's relationship with their ex-stepparents, who have no legal visitation rights regarding their ex-stepchildren, no matter how strong or long-lasting their emotional bond with them. Even if the marriage continues, at some point the child's other biological parent may take over custody, formally or informally, legally or not. These various realities make both stepparents and foster parents less likely to invest themselves completely in the parent-child relationship.

Adoptive families have an advantage here: they are legally connected for life. Nevertheless, during adolescence, their emotional bonds may abruptly stretch and loosen, for many adoptive children become intensely rebellious and rejecting of family control, even as they insist on information about, or reunification with, their birthparents. The reasons—whether to test their parents' devotion, or to follow the lead of their native temperament, or to discover their roots, or to establish an identity independent from their adoptive family—are understandable. But the result is often a painful demonstration that the parent-child relationship is more fragile than the law pretends (Rosenberg, 1992).

One sign of the difficulties with parent-child attachments in all three situations is that stepchildren, foster children, and adoptive children tend to leave home—running away, marrying, joining the military, being sent away to school, or moving out on their own—earlier than adolescents living with one or both biological parents (Aquilino, 1991). Early home-leaving is particularly likely for adopted children, as they seek their own identity distinct from parental expectations.

All these potential complications certainly make nonbiological parenthood riskier than it is generally pictured to be. However, we must not exaggerate the difficulties here. Most adoptive and foster parents cherish their parenting experiences so much that they try for more of the same, typically seeking a second child within a few years after the arrival of the first. Similarly, once stepparents realize that they cannot fill the shoes of the absent biological parent, they usually find satisfaction in the role they do play. On their part, the children usually reciprocate, if not immediately, then later on when they have a clearer understanding of the voluntary sacrifices their nonbiological parents have made (Keshet, 1988; Rosenberg, 1992).

Indeed, for some stepparents, foster parents, and adoptive parents, the rewards of their work go beyond the immediate household. This is exemplified in the reflections of one American mother of an adopted Korean child, who writes of her deepened understanding of the "global family" and of the bonds that connect one human being with another:

> We [adoptive parents], like these children whom we claim so adamantly as our kids, have deeper roots than we knew, an enlarged sense of family, another place in the heart, and a rich and varied history of facing life issues we would never have encountered without them . . . I hear news about the mudslide [in Brazil, that buried the shacks of 50 families], or the orphans or the starving children, with a refrain at the end: These could be my kids. [Register, 1991]

Perhaps even more than biological parenthood, alternative routes to parenthood tend to make adults more humble, less self-absorbed, and more aware of the problems facing children everywhere. When this occurs, adults become true exemplars of generativity as Erikson and others (1986) described it, characterized by the virtue that is, perhaps, the most important of all—caring for others.

SUMMARY

The Tasks of Adulthood

1. Adult development is remarkably diverse, yet it appears to be characterized by two basic needs. The first need is for intimacy, achieved through friendships and love relationships. The second is for generativity, usually achieved through satisfying work and/or parenthood.

2. Traditional patterns of development that followed specific age-related stages have been replaced to a large extent by more varied and flexible patterns. The culturally set social clock still influences behavior but less profoundly than it used to.

Intimacy

3. During early adulthood, a primary source of intimacy is friendship, with notably different needs being served by men's and women's same-sex friendships. Cross-sex friendships present unique challenges and benefits.

4. For most people, the deepest source of intimacy is found through sexual bonding with a mate, a bonding that frequently involves cohabitation and/or marriage.

5. Of the many factors that can affect the success or failure of a marriage, three are particularly notable: the age of the partners at marriage, the similarity of their background, values, and interests, and the couple's perception of the balance of equity in the marriage.

6. Physical abuse is not rare in married and cohabitating couples. Common couple violence generally resolves itself peacefully over time, whereas patriarchal terrorism leads to a devastating cycle of abuse.

7. The divorce rate has risen dramatically over the past fifty years, internationally and especially in the United States. The primary factors contributing to this increase include changing divorce laws and changing expectations regarding both the partners' respective marital roles and the permanency of marriage.

8. Divorce is emotionally draining on both partners and is particularly difficult for those who have children. While some divorced parents manage to maintain good relationships with each other and with their children, more typically mothers have more work and less money than before, and fathers become estranged from their children.

Generativity

9. For most adults, work is an important source of satisfaction and esteem, as the pleasure of a job well done helps meet the need to be generative.

10. Significant new patterns are emerging in today's world of work. A shift from a manufacturing- to a service-based economy in the United States means workers must acquire flexible skills that will enable them to perform a variety of different jobs over their careers. Increased diversity in the workplace also creates a need for cultural sensitivity between employer and employee.

11. Changing gender roles have brought about significant shifts in the demographics of working women. Most parents today are part of dual-earner, dual-caregiving couples, presented with challenges and benefits that traditional families did not face.

12. Parenthood is the other common expression of generativity. The specific challenges and satisfactions parents experience depend in part on the child's stage of development. Nonbiological parents experience the same challenges and satisfactions as biological parents but they also are likely to be faced with special problems.

KEY TERMS

intimacy versus isolation (509)
generativity versus stagnation (510)
social clock (510)
cohabitation (516)
homogamy (520)
heterogamy (520)
social homogamy (520)
marital equity (520)
exchange theory (520)
common couple violence (524)
patriarchal terrorism (524)
glass ceiling (530)
mentor (531)
role overload (532)
role buffering (532)
family logistics (534)

KEY QUESTIONS

1. How does achieving intimacy and generativity fulfill essential needs of young adults?

2. How does the social clock affect life choices?

3. What are some typical gender differences in friendship patterns?

4. What are some of the major factors that are likely to affect marital outcome?

5. Why has the divorce rate risen dramatically over the past fifty years?

6. How do ex-spouses typically fare after divorce?

7. Why is work important beyond supplying income?

8 What major changes are occurring in today's workplace? What factors enhance job satisfaction?

9. What work and family issues must be confronted in a dual-income family?

10. What special problems are often associated with alternative parenthood?

PART VI The Developing Person So Far: Early Adulthood, Ages 20 Through 40

Biosocial Development

Growth, Strength and Health

Noticeable increases in height have stopped by age 20, but increases in muscle strength continue until about age 30. All body systems and senses function at optimal levels as the individual enters adulthood, and declines in organ reserve and sensory acuity are so gradual that the onset of senescence is rarely noticed.

Sex and Gender Differences

For both sexes, sexual responsiveness remains high in early adulthood: the only notable changes are that men experience some slowing of their responses with age and woman become more likely to experience orgasm. In both sexes, problems with fertility increase with age. While disease is rare, the years of early adulthood are peak times for hazards that are chosen by individuals and encouraged by the culture, specifically drug abuse, violent death (particularly for men), and destructive dieting (particularly for women).

Cognitive Development

Adult Thinking

As an individual takes on the responsibilities and commitments of adult life, thinking may become more adaptive, practical, and dialectical to take into account the inconsistencies and complexities encountered in daily experiences. Partly as a result, moral thinking becomes deeper and religious faith becomes more reflective, with more appreciation of diverse viewpoints and also more commitment to one's own convictions. Years of education is the variable that has been shown to have one of the most powerful effects on the depth and complexity of adult thinking. Significant life events can also precipitate cognitive development.

Psychosocial Development

Intimacy

The need for affiliation is fulfilled by friends and, often, by a romantic commitment to a partner. Friendships are important throughout adulthood but are particularly so for individuals who are single. The developmental course of marriage depends on several factors, including the presence and age of children and whether the interests and needs of the partners converge or diverge over time. Divorce, if it is to occur, is powerfully affected by cultural pressures.

Generativity

The need for achievement can be met both by finding satisfying work and by parenthood, including several types of nonbiological parenthood. The labor market is changing radically and individuals should expect to experience several job changes and an increasing need for knowledge and flexibility within a diverse group of coworkers.

PART VII Middle Adulthood

Max Roach has been a leading jazz drummer for over fifty years. His approach to his work at age 73 clearly reflects the idea of selective optimization with compensation: "I joined a health club . . . I thought I'd tune up, you know, tone up. Playing my instrument is a lot of exercise. All four limbs going. . . . I don't play the way I did back in the 52nd Street days. We were playing long, hard hours in all that smoke. It would kill me now if I played like I did then. Now I play concerts, and the show goes on for just an hour. But I'll tell you something: I'm going to play until the sticks get too heavy for me to hold up."

pensation is that individuals set their own goals, assess their own abilities, and then figure out how to accomplish what they want to achieve despite the limitations and declines of life (Baltes & Baltes, 1990).

One example of this process comes from the late Artur Rubinstein, a world-famous concert pianist who continued to perform in his 80s. When asked how he managed to sustain his level of performance, Rubinstein explained that he limited his repertoire somewhat to pieces he knew he could perform well (selection) and that he practiced before a concert more than he might have when he was younger (optimization). And because he could no longer play fast passages at the brilliant tempos he was once capable of, he deliberately played slow passages slower than he formerly would have, thereby making his playing of the fast passages seem faster than it was (compensation) (Schroots, 1996).

More common examples are provided by the many elders who become much more forceful in structuring their lives so that they can do what they want and do it well. In a study of some of the strategies used by people over age 80, for example, Johnson and Baer (1993) report a woman who does her food shopping at a distant store because it is near the end of the bus line, ensuring that there will be seats available for her and her groceries on the return trip. Similarly, a man who continues to drive a car "plots out the streets and plans his exact route in order to avoid getting lost, while another drives around the block rather than making a left turn at a busy intersection." According to the authors, such routinization of daily activities is a common strategy used by the elderly to gain greater control over their environment.

Integrity Versus Despair

The most comprehensive self-theory comes from Erik Erikson, who even in his 90s was still writing about the vitality of life. Erikson called the final crisis in his theory of development the crisis of **integrity versus despair**. His depiction of the diversity of late life is framed in terms of what he saw as the universal attempt of older adults to integrate and unify their unique personal experiences with their vision of the future of their community. Some develop a sense of pride and contentment with their past and present lives, as well as a "shared sense of 'we' within a communal mutuality" (Erikson et al., 1986). Others experience despair, "feeling that the time is now short, too short for the attempt to start another life and to try out alternate roads to recovery" (Erikson, 1963).

As at each of Erikson's eight stages, tension between the two opposing aspects of the developmental crisis helps move the person forward. This is particularly apparent in this eighth stage, when

> life brings many, quite realistic reasons for experiencing despair; aspects of the present that cause unremitting pain; aspects of a future that are uncertain and frightening. And, of course, there remains inescapable death, that one aspect of the future which is both wholly certain and wholly unknowable. Thus, some despair must be acknowledged and integrated as a component of old age. [Erikson et al., 1986]

Ideally, coming to grips with the idea of one's own death leads to a new view of survival, through children, grandchildren, and the human community as a whole. This vision of the integrity of the generational life cycle allows a "life-affirming involvement" in the present.

integrity versus despair The last of Erikson's eight stages of development. During late adulthood, according to Erikson, people feel either that their lives have had meaning and they look back on their past experiences with a sense of integrity and wholeness, or they despair at their past and therefore dread the future.

social stratification theories Theories emphasizing that social forces, particularly those related to a person's social stratum or social category, limit individual choices and affect the ability to function.

In order to reach integrity, an individual must take stock of the personal choices and events that have shaped his or her life and integrate them into a meaningful whole. For the most part, even those who earlier in adulthood

> experienced periods of profound unhappiness and restlessness, which they attributed to misguided decisions concerning spouse, career, child-rearing . . . look back now, quite satisfied with how they have chosen to live their lives—with the people they married, with the ways they raised their children, with the kinds of work they did. [Erikson et al., 1986]

Some, however, are mired in bitterness and blame, fearful of death, unable to accept either the past or the future.

What kind of vision enables a person to reach integrity rather than despair? As we have seen throughout this book, there is no one true path, no one lifestyle, no one cultural route toward personal or community wholeness, but each person must believe in the direction his or her own life has taken within the context of the individual's particular culture. As Erikson (1963) explains, people of every background and income can reach integrity, "each aware of the relativity of all the various lifestyles which have given meaning to human striving . . . [yet knowing] that for him all human integrity stands or falls with the one style of integrity of which he partakes."

In other words, instead of comparing themselves with others, those who achieve integrity become self-affirming and self-actualizing, able to judge their life by their culture's standards and to find it good.

This model shows the homes of the 25,000 residents of Sun City West, Arizona, one of hundreds of retirement communities throughout the country, including the original Sun City, a larger community only 2 miles away. Although voluntary age segregation could be considered disengagement from the mixed-age communities where most people dwell, it does not necessarily lead to passivity. Most Sun City residents maintain active schedules of work and play as well as shopping, exercising, worshiping, and socializing, all within the security of the compound.

Social Stratification Theories

A quite different perspective on late adulthood is proposed by those who endorse **social stratification theories**. These theories maintain that social forces limit individual choice and direct life at every stage, particularly in late adulthood, when a person's ability to function depends largely on which stratum of society he or she is located in.

Stratification by Age

One form of social stratification theory focuses on *age stratification*. According to this theory, industrialized nations segregate the oldest generations, giving them limited roles and circumscribed opportunities in the society in order to make way for upcoming generations—as is clearly exemplified in mandatory-retirement practices. The most controversial version of this theory is *disengagement theory* (Cumming & Henry, 1961), which holds that with the approach of old age, a person's social sphere becomes increasingly narrow, as traditional roles become less available or less important and the person's social circle shrinks because friends die or move away. People anticipate and adjust to this narrowing of the social sphere by disengaging—relinquishing many of the roles they have played, withdrawing from society and developing a more passive style of interaction.

Disengagement theory provoked a storm of protest, particularly from gerontologists who insisted that older people need to and want to substitute new involvements and new friends for the ones they lose with retirement, relocation, and so forth. If the elderly do disengage and withdraw, these critics countered, they do so unwillingly (Hochschild, 1975; Rosow, 1985). In

fact, the critics of disengagement theory have been so cogent that the theory "no longer has any serious defenders" (Uhlenberg, 1996). The dominant view today is that the more active the elderly are and the more roles they play, the greater their life satisfaction and the longer their life (Harlow & Cantor, 1996).

A more developmental version of age stratification has been offered by Matilda Riley (1971, 1987). Riley explains that a great deal of an elderly person's experience depends not only on chronological age but also on the specifics of the individual's cohort, including the historical and technological contexts in which the cohort finds itself. Particularly influential is cohort size. Small cohorts are more likely to be marginalized by being encouraged or forced to abandon their long-standing social roles and to disengage from society. By contrast, larger cohorts, because of the weight of their numbers and votes, have more power and hence can create social structures and laws that allow them more roles, higher income, and better health. Many other factors—"racial and gender composition, educational background, marital and family history, work history, religious experiences, savings and pension accumulation, war experiences, political behavior, and so forth"—vary from one cohort to another, and each of these factors affects the way the surrounding social forces shape and structure the lives of a particular generation of elders (Uhlenberg & Miner, 1996). To the extent that a particular cohort reaching age 60 differs from the cohorts preceding and following it, it will find itself occupying a distinct age-based niche.

Stratification by Gender and Race

All social stratification theories focus on the ways in which people organize themselves—and are organized by society—according to their particular characteristics and circumstances. Social stratification theories focus particularly on how such organization can limit people's individuality and life choices. As we have just seen, cohort is one such organizational grid. Many argue that, especially in late adulthood, sex and race are also powerful and limiting stratification categories.

Feminist theory, for example, draws attention to the values underlying the gender divisions promoted by society. Feminists are particularly concerned about late adulthood because "the study of aging, by sheer force of demography, is necessarily a woman's issue" (Ray, 1996). Currently in the United States, women make up nearly two-thirds of the population over age 65 and nearly three-fourths of the elderly poor. Feminist theorists point out that, since most social structures and economic policies have been established by men, women's perspectives and needs are devalued by the larger society. This is a particular problem in later years because economic policies—such as pension plans that are fixed to a lifetime of continuous employment or medical insurance that pays more for hospitalization for acute illness than for repeated clinic visits to treat chronic illness—are largely responsible for pushing many women into poverty in their old age. Feminist theorists likewise note that because women are given the lifelong role of caregivers, many elderly women are expected to provide full-time care for frail relatives (husbands, siblings, or handicapped children), even if it strains their own health. Indeed, in many ways, social policies and values converge in such a way as to make later life particularly burdensome for women (Barusch, 1994).

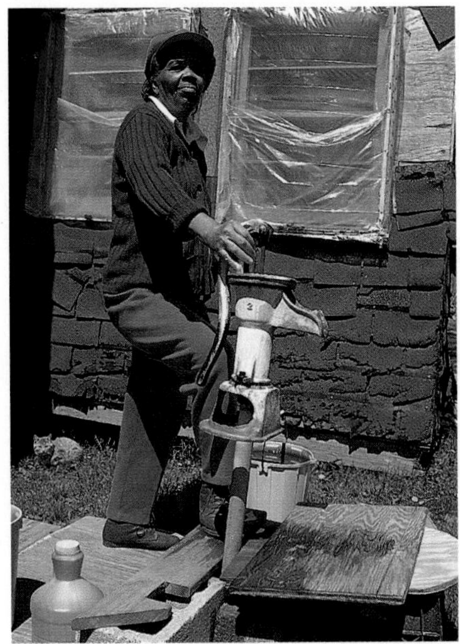

A glance at this woman at her outdoor pump might evoke sympathy, as her home's lack of plumbing suggests that she is experiencing late adulthood in poverty, in a rural community that probably offers few social services. As social stratification theories point out, factors of race and gender put her at additional risk of problems as she ages. However, a deeper understanding of her life might reveal many strengths—from deep religious faith to strong family ties and gritty survival skills—that will sustain her despite the lack of privileges and safeguards that others might have.

A similar theory is *critical race theory*, which sees race not as something inborn but rather as "a social construct whose practical utility is determined by a particular society or social system" (King & Williams, 1995). According to this theory, long-standing racism and racial discrimination shape the experiences and attitudes of both racial minorities and racial majorities in ways that structure life events for both groups—often without their conscious awareness (Bell, 1992).

In old age, this structuring includes poverty and frailty for those minority-race members who were excluded from the economic mainstream all their lives and who were, and continue to be, without ready access to needed health care. In old age, these same minority members are also less likely to have access to social services and amenities, including senior-citizen centers, nursing homes, and the like (Skinner, 1995). The consequences of racial structuring in old age take more subtle forms as well. For example, in America many of the elderly are homeowners who can choose to sell their home for profit, live in it rent-free, or obtain a "reverse mortgage," receiving a monthly payment in exchange for title to the house when they die. However, decades of housing discrimination mean that African-Americans are much less likely to be homeowners than are European-Americans, and even when they do own a home, it is likely to be worth less than homes owned by others of the same income level (Myers & Chung, 1996).

The central view of these social stratification theories—that membership in a certain social structure can place older individuals at risk of experiencing a number of dangers, including disability, poverty, and isolation—certainly seems valid when it comes to poverty or poor health. In every nation, an elderly person's economic and health status seems closely connected to minority or majority status as well as to gender. However, many contemporary analysts argue that this view unfairly stigmatizes members of minority groups, who often have strengths in late adulthood (e.g., intergenerational supports and religious faith) that other groups are less likely to have (Blakemore & Boneham, 1994). Similarly, the fact that women are more likely to be caregivers and kinkeepers means that they are less likely to be lonely and depressed in old age (Barusch, 1994).

With all the obvious inequities of stratification based on race and gender, it is noteworthy that one measure of despair—suicide—is less common among minorities and among women. In fact, of all the age, sex, and racial groups in the United States, the lowest suicide rate of all is that of black elderly women, an infinitesimal 2 per 100,000, compared with 94 per 100,000 for elderly white males (National Center for Health Statistics, 1994).

dynamic life-course theory The theory that each person's life is an active, ever changing, largely self-propelled process, occurring within specific social contexts that themselves are constantly changing. For this reason, developmental predictions based on past history may be useful, but they are never completely accurate, since each person constantly makes choices that affect the next step of development.

Dynamic Life-Course Theory

All the foregoing theories provide avenues of insight into development in late adulthood, and each contributes an important perspective to the theory that has the broadest applicability to late adulthood, **dynamic life-course theory**. According to this theory, each person's life is an active, ever changing, largely self-propelled process, occurring within specific social contexts which themselves are constantly changing.

Underlying this dynamic flux are the major factors emphasized throughout this text—behavioral genetics, intrafamily and intergenerational

Popular conceptions of middle adulthood are riddled with clichés like "midlife crisis," "middle-aged spread," and "autumn years" that conjure up a sense of dullness, resignation, and perhaps a touch of despair. Yet the tone of these clichés is far from reflecting the truth of the development that can and often does occur between the ages of 40 and 60. Many adults feel healthier, smarter, more pleased with themselves and their lives during these two decades than they ever did.

Of course, such a rosy picture does not apply to everyone. Some middle-aged adults are burdened by health problems, or a decline in intellectual powers, or unexpected responsibilities for aged parents or adult children. Some feel trapped by choices made in early adulthood. But the underlying theme of the next three chapters is that in middle age, much of the quality of one's life is directly related to how one views it and to decisions, sometimes new ones, about how to live it. There are still many turning points ahead where new directions can be set, new doors opened, and a healthier and happier life story written.

Middle Adulthood: Biosocial Development

CHAPTER 20

In Chapter 17 we saw that the first two decades of adulthood can be considered the prime of life as far as biosocial development is concerned, with the effects of aging being, for many people, barely perceptible. Between the ages of 40 and 60, aging continues at the same steady rate, but the levels of change that are now reached are more difficult to ignore. As we will see, however, adults in middle age can do a great deal to safeguard their vitality and to remedy, or compensate for, many of the physiological declines they experience, discovering in the process that though aging is inevitable, it is not inevitably bad.

Normal Changes in Middle Adulthood

As people advance in age past 40, physiological signs of aging are apparent. Hair usually turns noticeably gray and thins appreciably; skin becomes drier and more wrinkled; and body shape changes, as "middle-age spread" develops across the abdomen and pockets of fat settle on various other parts of the body—the upper arms, the buttocks, even the eyelids.

The size of the body may change as well. As back muscles, connecting tissues, and bones lose strength, the vertebrae collapse somewhat, causing some individuals to lose nearly an inch in height by age 60 (Whitbourne, 1985). In addition, overeating and underexercising cause many people to become noticeably overweight, a condition that is more prevalent during middle age than during any other period of life. In the United States, this problem has grown worse in recent decades. In 1961, 29 percent of all 50-year-olds were overweight, with a BMI of 27 or higher (see Chapter 17 for a discussion of body-mass index). In 1991, 40 percent were overweight to that degree (National Center for Health Statistics, 1995).

With the exception of weight gain that is excessive, midlife changes in appearance have no significant health consequences. The consequences for self-image, however, can be substantial, particularly for women, because for them, youth and physical attractiveness are much more closely linked to social status and self-esteem than they are for men (Katchadourian, 1987). How important this link is for a particular woman depends partly on how much her culture emphasizes youth and beauty relative to other attributes.

The message of these ads is clear: old is ugly. American women are bombarded daily with similar messages from all the media, instilling in them the idea that they have two choices, to fight age or be unattractive—a lose-lose proposition.

In Western cultures, of course, youthful allure is promoted as a prime value in women and its fading can be very distressful. The French writer Simone de Beauvoir probably spoke for many middle-aged women of her cohort when she confessed in her 50s:

> I loathe my appearance now: the eyebrows slipped down toward the eyes, the bags underneath, and the air of sadness around the mouth that wrinkles always bring. Perhaps the people I pass in the street see merely a woman . . . who simply looks her age, no more, no less. But when I look I see my face as it was, attacked by the pox of time for which there is no cure. [de Beauvoir, 1964]

Most developmentalists would find these feelings both understandable and unfortunate. As we will see throughout the rest of this book, the overall impact that aging has on the individual in middle age, and in late adulthood, depends in large measure on the individual's attitude toward growing old. For those who develop a constructive, adaptive attitude, the difficulties of the aging process can be greatly diminished and the pleasures significantly enhanced.

Regarding the "pox of time," many developmentalists might wish that both men and women held a view closer to the one expressed by another writer, Germaine Greer, who became middle-aged several decades after de Beauvoir did. An advocate of the self-affirming ideas of the women's movement, Greer declared of middle age:

> Now, at last, we can escape from the consciousness of glamour; we can really listen to what people are saying, without worrying whether we look pretty doing it. . . . We ought to be turning ourselves loose, freeing ourselves from inauthentic ideas of beauty, from discomfort borne in order to be beautiful. [Greer, 1986]

This is not to suggest that developmentalists are in favor of people just "letting themselves go" in middle age. On the contrary, they emphasize that, throughout adulthood, the benefits of staying in shape are far more than cosmetic. What they urge is a more balanced view of the changes of aging, one that is, literally and figuratively, more than skin deep. Whether such a balanced view will be increasingly common as the next century dawns remains to be seen, but as the bulging population of baby boomers moves into middle age, it is likely that the views of the wider culture will reflect more of the realities, and less of the prejudices, concerning aging.

Now let us look at some of the physical changes of middle age that are more than just cosmetic and that, to varying degrees, affect physical functioning.

Declines in the Sensory Systems

Sometime between ages 40 and 60, virtually all adults notice that the functioning of their sense organs is less acute than it once was. Although all the sensory systems decline, age-related deficits are most obvious in the two most crucial systems: hearing and vision (Kline & Scialfa, 1996).

Hearing

Pete Townshend is a prime example of the dangers posed by prolonged exposure to loud noise. Night after night of high-decibel rocking with The Who eventually brought him to a state of almost complete deafness.

The hearing losses that adults experience involve three biological factors: sex, genes, and age. Audiology tests reveal that women, on average, begin to show hearing deficits at around age 50. Men, by contrast, begin to show some deficits by age 30, and they lose hearing twice as fast as women (Pearson et al., 1995). The hearing loss most closely associated with age involves a reduction of auditory acuity in the inner ear. This type of loss appears to be genetically influenced and varies considerably in age of onset and intensity. As a result of this normal hearing loss, some individuals find that by age 50 they are frequently having to ask others to repeat themselves or are having to turn the TV volume louder than others might have set it. By age 70, almost everyone realizes that his or her hearing is less sharp.

Another factor that contributes to hearing loss over time is prolonged exposure to noise. The damage caused by such exposure is cumulative, and its effects are imperceptible at first. Consequently, many young adults perform very noisy work for long periods without wearing protective headphones or habitually listen to music at ear-splitting levels, unaware of the hearing deficits they may be creating for themselves in the future.

With normal age-related hearing changes, the ability to distinguish pure tones declines faster than the ability to understand conversation: often the first sign of a hearing loss is a 40-year-old's difficulty hearing a doorbell or a telephone ringing in the distance, or a 50-year-old's tendency not to be awakened by the sound of an alarm clock. Speech sounds are usually understood until late adulthood (see Table 20.1). When speech-related hearing losses do become evident, they usually involve high-frequency noises, as might occur in rapid conversation with a woman or small child (Meisami, 1994).

Fortunately, most hearing losses in middle adulthood are easy to remedy. Usually only a minor accommodation is needed, such as asking others to speak up, adjusting the ring of the telephone, and the like. In the case of more serious loss, today's tiny digitally programmed hearing aids can usually correct the problem much more efficiently than the hearing aids of even a few years ago, and a range of special telecommunication devices for the

TABLE 20.1

Hearing Loss at Age 50

	Men	Women
Can understand even a whisper	65%	75%
Can understand soft conversation but cannot understand a whisper	28%	22%
Can understand loud conversation but cannot understand soft conversation	5%	2%
Cannot understand even loud conversation	2%	1%

glaucoma An eye disease, increasingly common after age 40, that begins without apparent symptoms and often causes eventual blindness. Early detection and treatment can prevent vision impairment from glaucoma.

deaf (TDDs) are now available. Thanks to such technological advances, many middle-aged people who notice their hearing is fading and fear deafness in old age can now anticipate decades of nearly normal hearing in most circumstances (Tyler & Schum, 1995).

Vision

The standard measure of visual acuity—the ability to focus on objects at various distances—shows more variation from person to person across adulthood than do measures of auditory ability. The main reason for this is that, after puberty, hereditary factors affect focusing ability much more than age does.

As we saw in Chapter 17, older adults are more likely to need corrective lenses than young adults are, and changes in the shape of the cornea affect the kind of lenses they need. People who require glasses before age 20 tend to be nearsighted; that is, they have difficulty in distance viewing. By contrast, older adults tend to be farsighted and to require glasses for reading. In addition, older adults are likely to have *astigmatism*, which involves loss of elasticity in the lens and a consequent difficulty in changing focus from near to far distance. As a result, at some point in middle age, many people find that they need bifocals, or two different pairs of glasses, for close and far viewing.

Several other aspects of vision, among them depth perception, eye-muscle resilience, color sensitivity, and adaptation to darkness, decline steadily with age and are generally notable by age 50 (Kline & Scialfa, 1996; Meisami, 1994). Each of these changes can affect daily life. Decreasing depth perception makes people more likely to misstep going up and down stairs; eye-muscle weakness makes it harder to focus on small print for several hours; decreased color sensitivity means that clothes no longer match as well and multicolored signs are harder to read; slower adaptation to darkness increases the time it takes to begin to find one's way in a dark room after coming in from the bright light, or, more ominously, to see the road at night after experiencing the momentary blindness caused by oncoming headlights. These changes are particularly likely to become apparent after age 50. It is noteworthy, however, that middle-aged adults seem to adjust to these changes without major difficulty. Serious accidents, either in a fall or while driving a car, are much more common in late adolescence or late adulthood than in middle adulthood.

Having always had 20/20 vision, Kirby Puckett, a ten-time all-star outfielder, never felt the need to have an eye exam. At the age of 34, he awoke one morning to discover a permanent black spot in the vision of his right eye. It was his first manifestation of glaucoma, and the end of his baseball career. Mr. Puckett is now an activist for increased awareness of glaucoma, a disease that is most prevalent among African-Americans and that can be easily detected in its earliest stages by a simple test.

Although most age-related vision problems in middle age are minor, and relatively easy to correct or compensate for, one—glaucoma—can be very serious. **Glaucoma** is an eye disease characterized by a hardening of the eyeball due to an increase of fluid within it. This hardening puts pressure on the optic nerve, which over time can destroy the nerve. Glaucoma becomes increasingly common after age 40, and in the United States, Canada, and Great Britain, it is the leading cause of blindness by age 70. The incidence of glaucoma is especially high among African-Americans, who are also more likely to suffer serious impairment when it does occur (Wilson, 1989). Luckily, the serious consequences of glaucoma can usually be prevented by early treatment; unfortunately, the disease has no obvious early warning signs. There is, however, a simple optometric test (a puff of air to detect increasing pressure within the eyeball) that spots glaucoma in the early stages, and it should be part of routine health care for every middle-aged person.

immune system The complex system of antibodies, cells, and tissues that protect the body against illnesses and infections.

autoimmune diseases Illnesses that occur when the body attacks its own healthy cells as though they are foreign bodies.

Changes in Vital Body Systems

Systemic declines, as outlined in our discussion of early adulthood, continue to reduce the efficiency and the underlying organ reserve of the lungs, the heart, the digestive system, and so forth, making people more vulnerable to disease (see Table 20.2). Declines are also evident in the **immune system**, which defends the body against foreign invaders such as bacteria, parasites, and viruses, as well as against such internal threats as cancer and infections. The decline of the immune system actually begins in adolescence but is not readily apparent until middle age, when recovery from everything from chicken pox to major surgery takes longer. With age, the immune system is also more likely to mistake certain of the person's own body cells as foreign invaders and attack them. The result can be one of a number of **autoimmune diseases**, such as rheumatoid arthritis or lupus, which become more common in middle age (Miller, 1996; Sternberg, 1994). (One additional systemic decline occurs in the sexual-reproductive system, which will be discussed at the end of this chapter.)

For most middle-aged people in developed nations, none of these changes in the body's vital systems is critical. Indeed, in a nationwide U.S. survey of people between the ages of 45 and 64, more than half rated their health as excellent or very good, and only one-fifth rated their health as merely fair or poor (Benson & Marano, 1994). Reflecting this self-reporting of good health is the fact that, thanks to better health habits and disease prevention, death rates among middle-aged Americans have declined dramat-

TABLE 20.2

The Increments of Chronic Disease

Age	Stage	Atherosclerosis (Hardening of Arteries)	Cancer	Arthritis	Diabetes	Emphysema	Cirrhosis
20	Start	Elevated cholesterol	Carcinogen exposure	Abnormal cartilage staining	Obesity, genetic susceptibility	Smoker	Drinker
30	Discernible	Small lesions on arteriogram	Cellular metaplasia*	Slight joint space narrowing	Abnormal glucose tolerance	Mild airway obstruction	Fatty liver on biopsy
40	Subclinical	Larger lesions on arteriogram	Increasing metaplasia	Bone spurs	Elevated blood glucose	Decrease in surface area and elasticity of lung tissue	Enlarged liver
50	Threshold	Leg pain on exercise	Carcinoma *in situ*	Mild articular pain	Sugar in urine	Shortness of breath	Upper GI hemorrage
60	Severe	Angina pectoris	Clinical cancer	Moderate articular pain	Drugs required to lower blood glucose	Recurrent hospitalization	Fluid in the abdomen
70	End	Stroke, heart attack	Cancer spreads from site of origin	Disabled	Blindness; nerve and kidney damage	Intractable oxygen debt	Jaundice; hepatic coma
Prevention or Postponement		No cigarettes; no obesity; exercise	No cigarettes; limit pollution; diet; early detection	No obesity; exercise; minimize stress on any one joint	No obesity; exercise; diet	No cigarettes; exercise; limit pollution	No heavy drinking; diet

*Abnormal replacement of one type of cell by another.
Source: Adapted from Fries & Crapo, 1980.

"Structurally, you're sound. It's your facade that's crumbling."

In addition to maintaining better health habits, such as cutting down on fatty foods and not smoking, many middle-aged people are coming to realize the benefits of active conditioning. A moderate diet and a program of regular exercise can produce wonders of restoration, even for a facade like this.

ically over the past fifty years, especially those relating to the two leading killers of this age group, heart disease and cancer. Despite the advent of AIDS (which killed 12,000 Americans in their 40s in 1994), the death rate of people between ages 40 and 60 in 1994 was only half what it was in 1940. Current estimates are that only 3 out of every 100 American 40-year-olds will die before age 50 and that only 8 out of every 100 50-year-olds will die before age 60 (U.S. Bureau of the Census, 1996).

Health Habits over the Years

As signs from their bodies, advice from their doctors, and the celebration of a fortieth or fiftieth birthday drive home the reality of aging, middle-aged adults become more likely to improve some of their health habits (Katchadourian, 1987). For example, in the United States, while 30 percent of middle-aged men and 24 percent of middle-aged woman currently are smokers, another 30 percent of men and 20 percent of women of this age group are former smokers, most of whom quit between the ages of 30 and 50 (National Center for Health Statistics, 1995). Many middle-aged adults have also moderated their consumption of fatty foods and alcohol.

These improvements may be partly developmental, reflecting a greater wisdom, or at least a greater inclination toward moderation, attendant to aging. However, they also reflect a historical shift that now places greater emphasis on disease prevention. For example, since 1970, the overall rate of smoking at any age in the United States has declined from about 40 percent to about 25 percent, and the average quantity of hard liquor (excluding beer and wine) consumed per adult has decreased by about 35 percent. Over the same period, the American diet overall has become healthier, with consumption of red meat and whole milk down about 10 percent and consumptions of fruits and vegetables, and of flour and cereal, up by 25 and 50 percent, respectively (Chung et al., 1992; National Center for Health Statistics, 1995; U.S. Bureau of the Census, 1996).

Such positive signs can be misleading, however. As the next section makes clear, much of the U.S. population continues to be at risk of premature death because of poor health habits.

Lifestyle Factors and Disease

Over the past twenty-five years, several large longitudinal studies have followed the lives of thousands of healthy adults, noting the relationship between their lifestyles—particularly their health habits—and the later incidence of disease and death. This research has found that, especially in middle age, more than half of all instances of death and disease are the direct result of lifestyle factors, not age (Deeg et al., 1996). If every 40-year-old ate better, exercised more, and were a nonsmoker, the great majority of this age group would live to at least age 70, and in many cases, to age 80 or 90 or even 100. Further highlighting the degree to which we hold our health in our own hands, a recent study by the Harvard School of Public Health determined that 65 percent of all cancer deaths were attributable to lifestyle factors, including smoking (30 percent), poor diet and obesity (30 percent), and lack of exercise (5 percent) (Willett & Trichopoulos, 1996).

RESEARCH REPORT

Smoking and Health

As noted, in the United States, about 30 percent of all middle-aged men, and 24 percent of all middle-aged women, still smoke cigarettes, at significant peril to their health. The chief danger, of course, is lung cancer, which is by far the leading cause of cancer deaths. Lung-cancer death rates reflect smoking patterns of several decades earlier. This explains why the rate for men, which rose from 35 per 100,000 in 1960 to a plateau of around 63 per 100,000 in the 1980s, has now begun to fall slightly, dropping to 60 per 100,000 in 1993. At the same time, the rate for women, which was 5 per 100,000 in 1960, has risen steadily, reaching 28 per 100,000 in 1993 (National Center for Health Statistics, 1994). In fact, the lung-cancer death rate for women now exceeds the death rate for breast, uterine, and ovarian cancers combined. This is a marked contrast to fifty years ago, when twice as many women died from breast cancer as from lung cancer (U.S. Bureau of the Census, 1996).

Smoking is a known risk factor for most other serious diseases that beset adults, including cancer of the bladder, kidney, mouth, and stomach, as well as heart disease, stroke, and emphysema. Marijuana and low-nicotine cigarettes increase the risk of the same diseases, although researchers are uncertain whether they are equally, more, or less harmful. Second-hand smoke is also deleterious to health—so much so that it is now considered "the third leading preventable cause of death after active smoking and alcohol" (Cahalan, 1991). One clear example of this is the fact that nonsmokers have a 30 percent higher risk of lung cancer if they are married to smokers than if they are married to nonsmokers (Brownson et al., 1992).

For all these health reasons, cigarette smoking in most developed countries has been declining in recent years. Nevertheless, the smoking rates in most European nations still surpass the U.S. rate, reaching about 50 percent among men and 30 percent among women in several countries, including Germany, Denmark, Poland, Holland, Switzerland, and Spain (United Nations, 1991). In most developing nations, unfortunately, the news is worse, with the respective rates for males in China, Argentina, and Indonesia, for example, at 55, 58, and 75 percent and rising. The rates for women in developing countries are much lower (3, 18, and 5 percent, respectively, in the countries just cited), but they too are on the rise. Thus, while smoking-related deaths are likely to decline in developed nations over the next twenty years, they are likely to increase in the rest of the world.

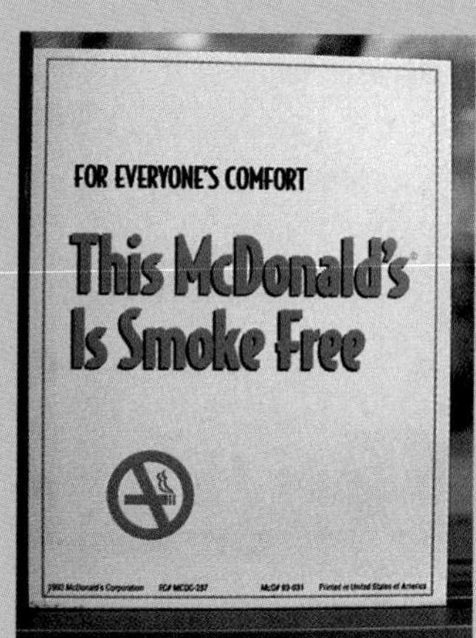

Health, as much as any aspect of life, is affected by social norms. At mid-century, smoking was the "thing to do," and the dancing Old Gold cigarette packs shown here were typical of cigarette advertising that dominated every medium, including television, in the 1950s. An immediate result of such promotion was the highest rate of cigarette smoking in American history, followed thirty years later by the highest rates of lung cancer. Today, "smoke-free" is in, and the rate of smoking among adults has dropped by half.

All smoking diseases are dose- and duration-sensitive: a middle-aged man who smoked two packs a day for forty years is three times as likely to die of lung cancer as a contemporary who smoked one pack a day for thirty years. At any age, a person who quits smoking reduces his or her health risk substantially. However, there is one surprise in the statistics. Although African-American smokers tend to begin smoking later and smoke fewer cigarettes than European-Americans, they die of smoking-related causes at higher rates. While genetic vulnerability might initially seem a reasonable explanation for this, research suggests another likely reason: 90 percent of black smokers (compared with 30 percent of white smokers) smoke menthol cigarettes, which deliver more nicotine and presumably more of all the other carcinogens in tobacco than do regular cigarettes (McCarthy et al., 1995). In addition, early diagnosis and treatment may, for various reasons, be less available to African-Americans, as discussed on page 557.

Alcohol

Contrary to the traditional assumptions of temperance advocates, adults who drink wine, beer, spirits, or other alcohol in moderation—no more than two servings a day—tend to live longer than those who never drink. The major benefit of moderate alcohol consumption is a reduction in coronary heart disease. One possible reason for this effect is that alcohol increases the blood's supply of HDL (high-density lipoprotein), often referred to as "good cholesterol." HDL is instrumental in reducing LDL (low-density lipoprotein), the "bad" cholesterol that causes clogged arteries and increases the chance of blood clots (Hein et al., 1996; Rimm et al., 1996). Small amounts of alcohol, taken with food, may also reduce tension and aid digestion. Moderate drinkers tend to experience less stress and less depression than either abstainers or heavy drinkers (Lipton, 1994).

Whatever the potential health benefits associated with alcohol, however, they come with notable risks. The clear fact is that many people do not and cannot drink in moderation, and middle-aged people are among the most vulnerable to chronic alcohol abuse, with alcohol dependence being most common at about age 40 (Cahalan, 1991). Heavy daily drinking over the years is the main cause of cirrhosis of the liver, which kills 15,000 American adults under age 65 each year. It also puts a stress on the heart and stomach, destroys brain cells, hastens the calcium loss that causes osteoporosis, and seems to make many forms of cancer, including breast cancer, more likely. Further, even moderate alcohol consumption may pose a health risk, in that it tends to increase cigarette-smoking and overeating. More immediately, alcohol is implicated in about half of all accidents, suicides, and homicides. All told, alcohol consumption, in one form or another, is responsible for about 108,000 deaths in the United States each year, which is 5 percent of total mortality (Archer et al., 1995). Thus the fact that moderate drinking has health benefits should not delude anyone: for many adults, not only alcoholics, alcohol is a major health risk.

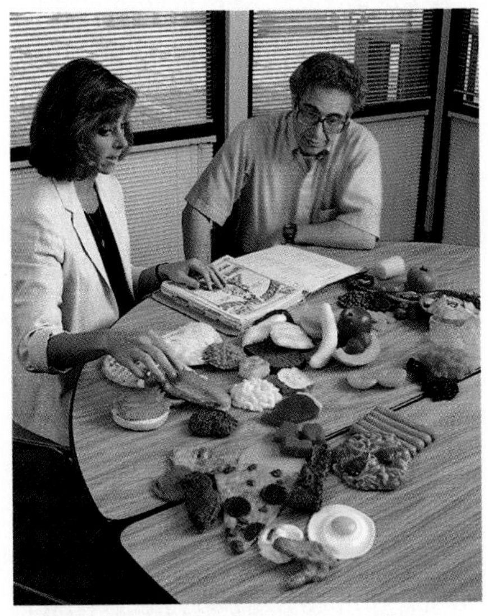

This session between a dietician and a heart-attack patient illustrates two factors that powerfully correlate with a healthy diet. The first is one-on-one training, with a knowledgeable professional able to individualize and specify nutritional advice. The second, unfortunately, is that it often takes a brush with death to make people examine and alter their lifestyle.

Nutrition

As we have seen, nutrition plays a central role in development throughout the life span. During middle adulthood, dietary factors are particularly important in the onset and progress of both of the major killers of the middle-aged, heart disease and cancer. Adults in industrialized countries typically consume about 40 percent of their daily calories as fat, much of it animal fat (in whole milk, cheese, butter, beef, pork, and eggs), which is high in cholesterol. This fat consumption is a major contributor to coronary heart disease, particularly in middle age, and it also makes several types of cancer more likely. Further, American adults currently consume only about 20 grams a day of fiber (which is found primarily in fruits, vegetables, and grains), making several forms of cancer, particularly colon cancer, more likely. Consequently, it is recommended that adults reduce the fat content of their daily diet to below 30 percent and that they increase their consumption of fiber to more than 30 grams per day, including at least five daily servings of fruits and vegetables (Healthy People 2000, 1990; National Cancer Institute, 1992).

TABLE 20.3

Healthy weight ranges

Height	Weight
4′10″	91–119
4′11″	94–124
5′0″	97–128
5′1″	101–132
5′2″	104–137
5′3″	107–141
5′4″	111–146
5′5″	114–150
5′6″	118–155
5′7″	121–160
5′8″	125–164
5′9″	129–169
5′10″	132–174
5′11″	136–179
6′0″	140–184
6′1″	144–189
6′2″	148–195
6′3″	152–200
6′4″	156–205
6′5″	160–211
6′6″	164–216

Source: U.C. Berkeley Wellness Letter, March 1996.

Weight

At least 40 percent of all middle-aged Americans are obese, defined as having a BMI of 27 or higher (see p. 483). This is a significantly greater middle-aged obesity rate than that of thirty years ago (29 percent), and for the currently middle-aged, it represents a marked increase in obesity since they were young adults (17 percent) (National Center for Health Statistics, 1995). Obesity is a definite risk factor for heart disease, diabetes, and stroke, and it is a contributing factor for arthritis, the most common disability for older adults.

While these dangers of obesity have long been known, it was generally assumed that little harm could come from being "somewhat" overweight. However, experts now believe that even those who are only slightly overweight may be increasing their risk of virtually every cause of disease, disability, and death (Lee et al., 1993; Manson et al., 1995). Many of these experts, in fact, contend that every excess pound increases these risks, and their view is strongly supported by research on lower animals such as mice and rats, whose life span has been doubled by reducing their normal calorie intake by roughly a third (Snyder, 1989). As a result of this new view, recommended-weight tables no longer allow for a few extra pounds with each decade (see Table 20.3).

If it is true that even a little extra fat is hazardous to health, then as middle age approaches, people of normal weight need to reduce their calorie intake below that of their younger years. The reason is that between ages 20 and 50 a person's metabolism normally slows down by about a third, which means that merely eating at the same level as at younger ages would cause ballooning weight gain (Ausman & Russell, 1990).

Exercise

Activity, even moderate activity, reduces the risk of serious illness and death. Simply walking briskly for twenty minutes three times a week—something only a fourth of all middle-aged adults do—has marked health benefits (Siegel et al., 1995). Even such modest exercise as washing the car, pushing a stroller, raking leaves, and so on, benefits the body in many ways. Unfortunately, according to a 1996 U.S. surgeon general's report, less than half of all Americans exercise regularly, and more Americans are overweight today than in any previous decade. Both inactivity and excess fat are increasingly common as people grow older, beginning at age 20 through age 60.

Even better than moderate or "incidental" exercise is working out three or more times a week for at least thirty minutes at a level that is sufficiently strenuous to raise the pulse to about 75 percent of its maximum capacity. Such exercise, which includes swimming, jogging, fast walking, bicycling, and the like, increases heart and lung capacity, lowers blood pressure, increases HDL in the blood, and, even if weight remains the same, reduces the ratio of body fat to body weight. Indeed, such exercise is also the best method of weight reduction: it is not only burns calories but also decreases the appetite and increases metabolism, so the person continues to benefit for several hours after a workout is over (Going et al., 1994). An addi-

Regular aerobic exercise, complete with stretches, is likely to be even more beneficial to the health of middle-aged men than to the health of the young women who usually frequent such classes.

tional advantage from exercising, especially in middle-aged and older people, is enhanced cognitive functioning, probably because of improved blood circulation to the brain (Stones & Kozma, 1996). Perhaps for this reason, exercise decreases depression and hostility, making a person psychologically healthier as well as physically more fit. Overall, regular exercise prolongs life and improves vitality, by increasing "stamina, strength, suppleness, balance, bone mass . . . and well being" (Coni et al., 1992). Indeed, since physical activity makes people look good, feel good, and stay healthy, the puzzle is why most adults—especially as they grow older and presumably wiser—would rather park themselves hour after hour in front of the TV than go for a swim, a run, or even a walk.

Variations in Health

Overall statistics about health in midlife are generalities that veil many variations. For example, worldwide, individuals who are relatively well-educated, financially secure, and living in or near cities tend to live longer, with fewer chronic illnesses or disabilities, and to feel healthier than those who are less well educated, with less money, and living in rural areas. Further, within every nation, certain regions seem healthier than others: in the United States, middle-aged people living in the West and Midwest are healthier than those in the South and Middle Atlantic region; in Canada, those in Ontario are healthiest; in Great Britain, health among the middle-aged tends to improve as one moves from north to south (Cruickshank & Beevers, 1989).

The reasons for such differences range from the quality of the environment and of available health care to more personal factors relating to the individuals who happen to live in a given region. For example, genetic, dietary, religious, socioeconomic, medical, and cultural patterns of their respective populations may explain why fatal heart attacks are twice as common in Mississippi and West Virginia as in Utah and Colorado (Smith, 1987; U.S. Bureau of the Census, 1996). Indeed, many specifics of ethnicity, income, and sex affect health dramatically in middle age. Before describing these, let us clarify what "health" means.

Four Measures of Health

There are at least four distinct measures of health: mortality, morbidity, disability, and vitality.

In one sense, the most solid indicator of the health of a given age group is its **mortality**, or death, rate, as measured by the number of deaths each year per thousand individuals. Mortality statistics are based on legally required death certificates, which indicate age and sex of the deceased as well as the immediate cause of death. This measure allows international and historical comparisons of the health of any age or gender group.

However, while such mortality statistics are obviously useful for developmental comparisons, a much more comprehensive measure of health is **morbidity**, defined as disease of all kinds, "the numerous acute and chronic problems that course through an individual's life, that remit or repeat, worsen or stand still, simply accumulate in number or interact synergistically" (Verbrugge, 1989). Morbidity can be *acute,* that is, sudden and severe, ending in either death or recovery, or it can be *chronic,* extending over a long time period.

To truly portray a person's overall health, the picture must be broadened to include two additional measures, disability and vitality. **Disability** refers to a person's inability to act in "necessary, expected, and personally desired ways" (Verbrugge, 1994). A victim of heart disease, for example, might be unable to walk more than a block without stopping to rest. Morbidity does not always result in disability, however: by taking appropriate measures, some people with heart disease not only walk but run, longer and faster than they did before becoming ill. From the perspective of society as a whole, disability is more costly than mortality or morbidity, because when a person is unable to perform the tasks of daily life, society must intervene through the provision of individual caregivers, institutional care, special equipment, and medical devices (Jette, 1996; Nusselder et al., 1996). The cost of nursing-home care, which is only a fraction of the total cost of disability care, was $50 billion in the United States in 1989 (Cutler & Sheiner, 1994).

The final measure of health, **vitality**, refers to how healthy and energetic—physically, intellectually, and socially—an individual actually feels. Vitality is, of course, a subjective measure—some people say their health is very good even when they have several chronic diseases and obvious disabilities—but for that very reason, vitality is probably more important to quality of life than any other measure (Stewart & King, 1994). Although there is no consensus about how to define quality of life, most experts, as well as the general public, now agree that the goal of medicine should be extending and improving vitality rather than simply postponing mortality, preventing morbidity, or remediating disability. Indeed, the motto of those who study aging is to "add life to years" not just "years to life" (Timiras, 1994).

mortality A measure of health that refers to the number of deaths each year per thousand members of a given population.

morbidity A measure of health that refers to the rate of diseases of all kinds in a given population.

disability A measure of health that refers to the inability to perform activities that most others can.

vitality A measure of health that refers to how healthy and energetic—physically, intellectually, and socially—an individual actually feels.

Ethnicity and Health

Ethnicity, with its attendant genetic and cultural factors, is a powerful influence on all four measures of health in middle age (James et al., 1992; Jones, 1989; Markides & Black, 1996). American mortality data on five ethnic groups make the point clearly: between the ages of 45 and 55, African-

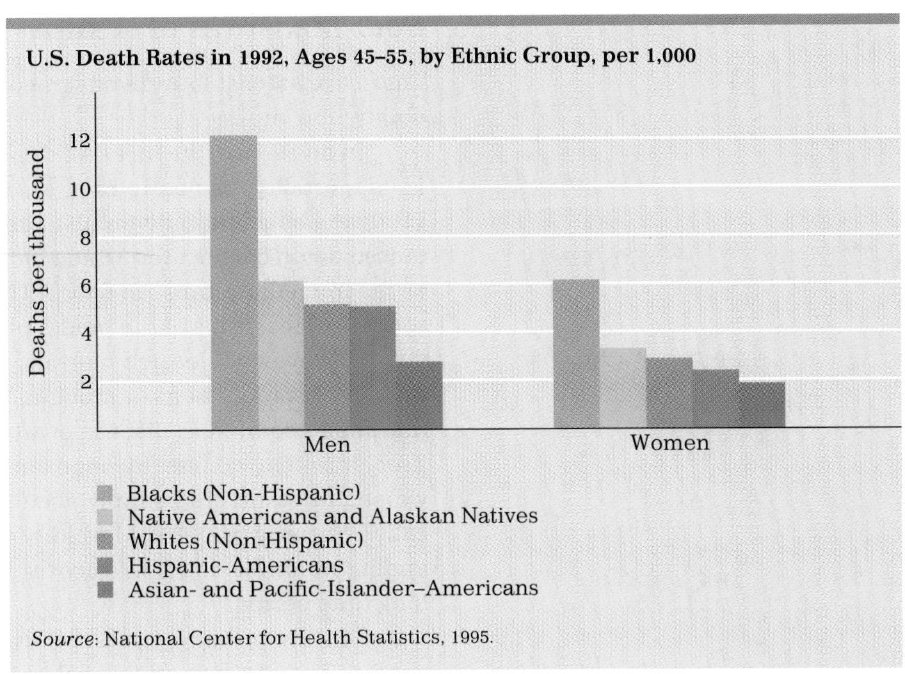

FIGURE 20.1 Racial differences in death rates among Americans are probably caused more by environmental factors than by heredity, according to many developmentalists. Evidence for this is shown here in the contrast between the rates of death for Native Americans and Asian-Americans, who are quite similar genetically. Sex differences in death rates are even more marked than racial differences (women die at half the rate that men do), but the reasons are more controversial. Some believe that a biological explanation (the second X chromosome or female hormones) underlies women's lower death rates. Others assert social and psychological factors (such as that women are more attuned to their bodies and are more likely than men to seek medical help when it is warranted).

Americans die at twice the rate of European-Americans, who themselves die at twice the rate of Asian-Americans. In between these groups, Native Americans have about a 20 percent higher chance of dying than European-Americans do, and Hispanic-Americans have a 20 percent lower chance (see Figure 20.1).

In general, morbidity and disability follow the same ethnic patterns in middle age as does mortality. These patterns are further reflected in self-reports of health status. For example, close to 60 percent of all middle-aged Americans of European or Hispanic backgrounds say that their health is good or excellent, compared with only 45 percent of African-Americans (U.S. Department of Health and Human Services, 1990).

The overall health of specific groups within each of these five broad ethnic groupings differs as well. Among middle-aged Hispanic-Americans, for instance, the death rate of Cuban-Americans is quite low, while that of Puerto Ricans is relatively high. Among Asian-Americans, Japanese-Americans tend to live longer than Indian-Americans. One particularly interesting intragroup difference is the fact that in all minority groups, the illness and death rates among recent immigrants are lower than among long-time U.S. residents. For example, among Chicanos, the group for which the most reliable data are available, those adults born in the United States have higher mortality rates than those born in Mexico (Markides & Black, 1996). Several factors may explain this difference. One is self-selection: on the whole, only the more hardy individuals emigrate. Another factor is health habits, which tend to be healthier in those less assimilated, particularly with regard to alcohol use, exercise, and diet. Additional factors may be optimism and family support, which reduce stress and which likewise tend to be stronger among those who are less assimilated.

Genes and cultural habits undoubtedly play a role in mortality and morbidity, especially in the occurrence of particular diseases. For example, the prevalence of hypertension and a preference for fried foods and fatty

For every American ethnic group, health is affected by two cultures. For example, overall, the death rate of Japanese-Americans is lower than that of most other Americans but it is higher than that of Japanese who have not emigrated. Every ethnic group also has certain illnesses to which it is particularly or rarely prone. For the Japanese, the rate of heart disease is quite low but the rate of stomach cancer is quite high.

meats make stroke-related morbidity and mortality three times higher for African-Americans than for any other ethnic group (Johnson et al., 1990; Wilson, 1989). Likewise, the incidence of skin cancer is much higher among European-Americans than among any other American group, undoubtedly because genetically they have less protective melanin in their skin and because they spend the most time in the sun trying to get a tan.

Additional Factors Related to Ethnic Health Differences

Genes and culture are not the only explanations for ethnic differences in health, however. Such factors as education, socioeconomic status (SES), and the pressures and opportunities provided by the larger society are also key elements in these differing rates of health. Among American white men ages 25 to 44, the annual death rate for those who have not graduated from high school is more than twice as high as for those with at least one year of college (6.79 per 1,000 compared with 3.01 per 1,000) (National Center for Health Statistics, 1995). Similarly, in the Netherlands, where there is relative genetic and cultural homogeneity, rates of chronic lung disease, heart disease, and diabetes are three times as high among the least educated as among those with a college education (Mackenbach et al., 1996).

In both these cases, higher levels of education most likely result in a greater awareness of health issues and requirements. Much more significant, however, are the specific life choices, stresses, opportunities, and access to health care that tend to vary with SES, as both indexed and enhanced by education. It is therefore no coincidence that among Hispanic- and Asian-American groups, the healthiest subgroups—Cuban- and Japanese-Americans—tend to be the most educated and to have the highest SES.

Now, let us consider the gap between the health of black and white (non-Hispanic) Americans. This gap exists at almost every age and on virtually every measure, but it widens or narrows depending on age. For example, the racial disparity in the death rate is greater in early and middle adulthood than it is between ages 1 and 15 or between ages 65 and 85. By age 85, the gap disappears. If anything, at the very oldest ages, surviving blacks

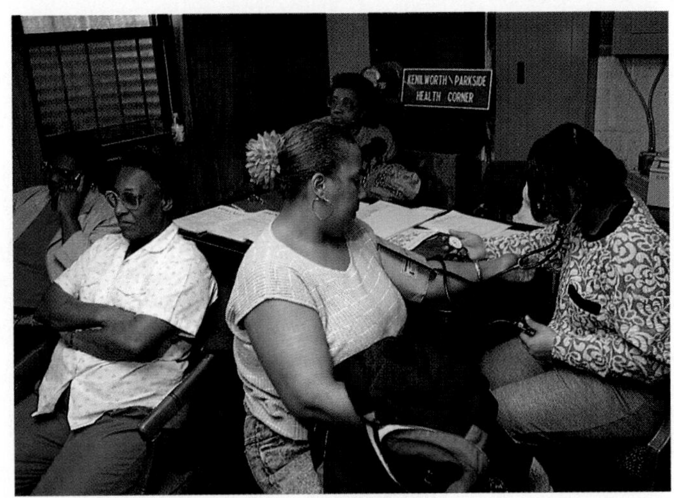

The health of African-Americans is adversely affected by three factors suggested here. First, African-American women are more likely than women of other groups to be overweight, and second, they are more likely to have high blood pressure. The third factor is reflected by the absence in this clinic. Men of all races are less likely to obtain preventive care and health screening than women, but this sex difference is particularly notable among African-Americans.

have a lower death rate than surviving whites, irrespective of SES (Sorlie et al., 1995; Wilkinson, 1992).

There are many possible explanations for this pattern, but it is unlikely that genetic predispositions or African-American culture affect middle-age blacks much differently than they affect children or the elderly (Markides & Black, 1996). A more plausible explanation is that extrinsic factors, particularly racism and poverty, are particularly harmful between adolescence and age 65. Because they experience less job discrimination and more family nurturance, younger children and the elderly may not suffer the full brunt of discrimination. Support for the idea that prejudice affects health comes from research showing that exposure to racial discrimination can substantially affect the blood pressure of young adult African-Americans. This again suggests that social and psychological factors, more than genetic ones, are related to certain ethnic differences in health (Krieger & Sidney, 1996).

Further evidence for the environmental links to the higher disease rates of blacks is provided by two recent studies. In one, researchers measured blood pressure in adults of West African ancestry now living in seven communities: three in Africa, three in the Caribbean, and one in the United States. Rates of hypertension—a serious risk factor for stroke and heart disease—rose the farther the adult was from rural Africa, with 33 percent of the African-Americans in a Chicago neighborhood having a blood pressure reading of 140/90 or higher, compared with only 15 percent of those in rural Cameroon (Cooper et al., 1997). The other study found that the risk for high blood pressure, heart disease, and stroke among blacks varied by place of birth, and that these differences in risk rates were greater than the overall difference between the risk rates of blacks and whites (Alderman, 1996). Analyzing the death certificates and census data of people born in the New York area, researchers found that the death rate from cardiovascular disease was nearly equal for blacks and whites. The death rate for New York–area blacks born in the South, however, was significantly higher, while the rate for New York–area blacks born in the Caribbean was significantly lower. (Among men between the ages of 25 and 44, blacks born in the South had a 30 percent higher death rate from heart disease than blacks born in the New York area and a 400 percent higher death rate than Caribbean-born blacks.) Again, genes seem to be less influential than environmental factors, such as diet, health habits, and perhaps different experiences with, and reactions to, racial discrimination.

Finally, the incidence of cancer in middle-aged American black women is actually slightly less than in white women, but their death rate is

25 percent higher, a racial gap that has actually increased over the past forty years (National Center for Health Statistics, 1994). Again, factors related to education, income, and racism, such as awareness of early signs of cancer, involvement in routine screening, and access to health care once cancer is found, seem more plausible explanations of this discrepancy than any possible genetic differences.

Overall, whenever groups in a given society are differentiated by SES, race, language, or other such factors, then that society will "structure the life experiences of their members so that advantages and disadvantages tend to cluster cross-sectionally and accumulate longitudinally" (Blane, 1995). This structuring will cause people from certain groups to experience multiple health risks, and some individuals will sicken and die at higher rates than others, not because of their genes and personal choices but because of the social niche they occupy.

Sex Differences in Health

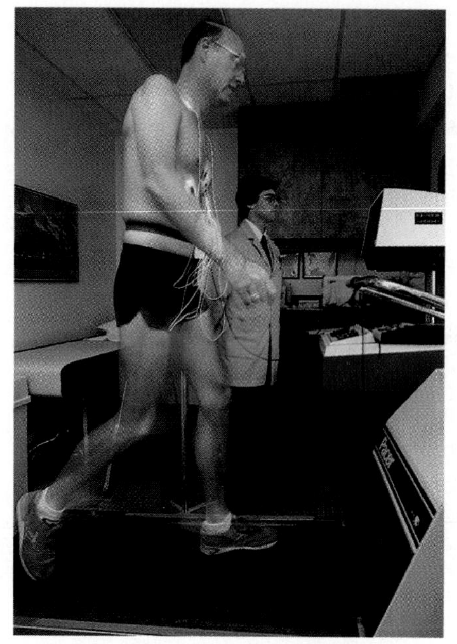

Until recently, heart disease was regarded by most doctors as a man's disease, and diagnostic procedures such as the stress test shown here were performed almost exclusively on men. As a result, heart disease tended not to be recognized in women until it had reached an advanced stage.

As you can see from Figure 20.1, in middle age, mortality rates continue to favor females, with men twice as likely to die of any cause and three times as likely to die of heart disease. Not until age 85 are the rates equivalent. By contrast, beginning in middle age and beyond, women have higher morbidity and disability rates than men, a difference particularly apparent after menopause.

The fact that middle-aged men are more likely to experience sudden, possible fatal disease and that women of the same age are more likely to experience chronic, disabling illnesses is reflected in an unfortunate gender difference: "even though women make up the primary patient load for health care professionals . . . men have been the primary subjects of medical research" (Parrott & Condit, 1996). This gender bias has not been due to deliberate sexism. Rather, it has arisen because, traditionally, the focus in the medical community has been on acute illnesses rather than chronic conditions, on preventing death rather than avoiding disability. Thus relatively little research has been devoted to such diseases as arthritis, osteoporosis, lupus, or migraine headaches—each of which is a common chronic condition that affects far more women than men, but none of which typically produce sudden death.

Even with diseases that can be fatal, there is often a gender bias in both research and treatment, in part because of an age bias that favors younger adults. For example, over the entire course of adulthood, heart disease is the leading cause of mortality in both sexes, killing women at the same rate (one death in every three) as men (National Center for Health Statistics, 1995). However, for women, heart disease is more often chronic, and death rates from heart attack do not begin to rise until after menopause, while men are more vulnerable to fatal heart attacks throughout adulthood. As a result, in middle age, men die from heart disease three times as often as women, a fact that has traditionally caused heart failure to be seen as a man's problem. Consequently, key longitudinal and large-scale studies of heart disease have excluded women; women with heart disease have been less likely to receive specific diagnostic and therapeutic procedures; and once heart disease was diagnosed, women were more likely to die (D'Hoore et al., 1994; Steingart et al., 1991).

Contributing to these negative outcomes is the fact that the medical approach to heart disease has been standardized on a male model, and the medical community has only recently begun to recognize that diagnostic procedures, heart surgery, and protective factors may produce different effects in women than in men (Altman, 1991). Because women's responses to exercise are not identical to men's, for example, standard treadmill tests often give false results for women. More telling, the mortality and morbidity rates for both bypass surgery and angioplasty are notably higher for women than for men. In part, this may be because women have smaller arteries, but it also may be that women's heart problems are often not correctly diagnosed until they have resulted in a medical emergency (Kahn et al., 1990).

Fortunately, this pattern of gender bias in health research and care is changing across the board, as research increasingly focuses on gender and genetic variations in mortality, morbidity, and disability. In 1993, the National Institutes of Health launched a fifteen-year, $625 million longitudinal study of 160,000 women between the ages of 50 and 79, specifically designed to overcome the neglect that women's health issues have suffered. As the director of the institutes noted, "For far too long . . . men were the normative standard for medical research and treatment. The corollary for this, of course, is that men's hormones set the standard for us all" (*New York Times*, March 31, 1993). This study has deliberately sampled women of various educational levels and ethnic backgrounds and should provide data that will help the medical profession tailor their preventive, diagnostic, and therapeutic measures to the particular patient, not to the standard John Doe, a 70-kilogram white American male.

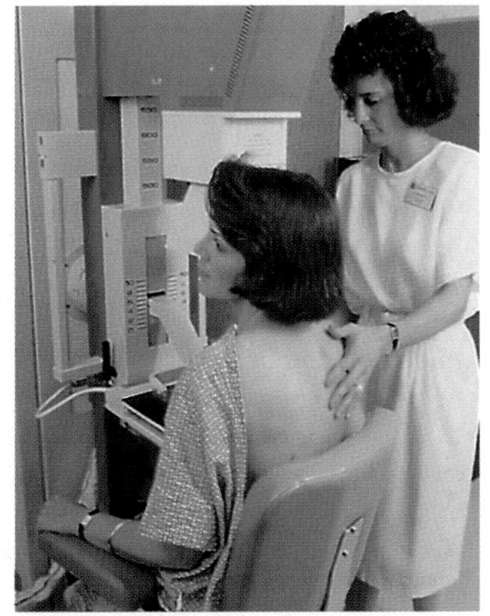

While menopause brings relief from possible complications of birth control, pregnancy, and menstruation, it also brings increased risk for heart disease, osteoporosis, and breast cancer. Fortunately, preventive medicine can halt most problems before they do serious damage. Unfortunately, many women do not get the care they need until symptoms emerge. Less than half of all American women over age 50 get their annual mammograms, even though doing so would save thousands of lives.

The Sexual-Reproductive System

As the sexual-reproductive system continues to age during middle adulthood, sexual responses gradually become slower and less distinct. We will soon discuss the consequences of these ongoing changes, but first let us look at one change that is definitive: the cessation of reproductive potential in women.

The Climacteric

Sometime between ages 42 and 58 (the average age is 51), women usually reach **menopause**, as ovulation and menstruation stop and the production of several hormones, especially estrogen, progesterone, and testosterone, drops considerably. Strictly speaking, menopause is dated one year after a woman's last menstrual period (Carlson et al., 1996).

The term "menopause" is also sometimes loosely used to refer to the **climacteric**, a phase preceding actual menopause and lasting about six years, during which the woman's body adjusts to much lower levels of estrogen. This adjustment is marked by a variety of biological and psychological symptoms.

The first symptom of the climacteric in women is typically shorter menstrual cycles. Over the course of her reproductive years, a woman's menstrual cycle changes, on average, from about every thirty days in her 20s and early 30s to twenty-two to twenty-eight days in her late 30s and early 40s. Then, toward the end of her 40s, a woman's periods become erratic in

their timing, with some being missed entirely, and may be unusually heavy in some cycles and unusually light in others. The timing of ovulation typically varies as well. Instead of occurring midcycle as in earlier years, ovulation sometimes occurs quite early or late in a cycle, or it may not occur at all. This explains why fertility declines markedly as midlife approaches. It also explains why some women who thought they knew their body's rhythm well enough to avoid pregnancy without using contraception find themselves with a "change of life" baby (Carlson et al., 1996).

The most obvious symptoms of the climacteric are hot flashes (suddenly feeling hot), hot flushes (suddenly looking hot), and cold sweats (feeling cold and clammy). These symptoms are all caused by **vasomotor instability**, that is, a temporary disruption in the body mechanisms that constrict or dilate the blood vessels to maintain body temperature. Lower estrogen levels produce many other changes in the female body, including drier skin, less vaginal lubrication during sexual arousal, and loss of some breast tissue. Two other changes caused by reduced estrogen levels pose serious health risks. One is loss of bone calcium, which can eventually lead to **osteoporosis**, a porosity and brittleness of the bones that makes them break easily. The other is an increase of fat deposits in the arteries, setting the stage for coronary heart disease. Osteoporosis and coronary heart disease are both experienced by about half of all woman who survive to late adulthood.

Many women also find that, in the climacteric, their moods change inexplicably from day to day. Usually this moodiness is blamed directly on the changes occurring in hormone levels, but, in fact, it may be caused primarily by the tiredness that occurs if hot flashes interrupt sleep night after night (Carlson et al., 1996).

All these symptoms are variable, with the sudden onset of menopause caused by a hysterectomy being most likely to produce marked symptoms, for both biological and psychosocial reasons.

How troubling natural menopause is depends, in part, on factors in the social context, from the cultural values placed on a woman's reproductive capacity to the prevailing medical views of the menopause experience. For much of this century and earlier, the Western medical profession tended to apply a disease model to menopause, describing menopausal women as "diseased, sexless, irritable, and depressed" (Golub, 1992). The result of such negative views of menopause created a kind of self-fulfilling prophecy, leading women to focus unduly on the symptoms of menopause, fearing the worst in their least discomfort or slightest mood swing, and having their fears confirmed by their physicians.

Today, menopause is no longer regarded as a disease, nor are its symptoms regarded as onerous, although, obviously, some women are more troubled by them than are others. In the United States, only about 10 percent of all women going through natural menopause, as well as about 90 percent going through surgically induced menopause, experience symptoms sufficiently difficult that they require medical treatment, specifically, estrogen replacement. In most cases, this treatment takes the form of **hormone replacement therapy (HRT)**, which includes progesterone as well as estrogen. This is because estrogen replacement alone increases the risk of uterine cancer, and the addition of progesterone negates that risk (Bellantoni & Blackman, 1996).

In recent years, research has shown that continued use of HRT beyond menopause can have a number of health benefits. Accordingly, many doc-

menopause The time in middle age, usually around age 50, when a woman's menstrual periods cease completely and the production of estrogen drops considerably. Strictly speaking, menopause is dated one year after a woman's last menstrual period.

climacteric Refers to the various biological and psychological changes that accompany menopause.

vasomotor instability The temporary disruption of the homeostatic mechanisms of the vascular system, which usually constrict or dilate the blood vessels to maintain body temperature. Vasomotor instability causes moments of feeling suddenly hot or cold, a typical experience during menopause.

osteoporosis A loss of calcium within the bone that makes the bone more porous and fragile. It occurs somewhat in everyone with aging, but serious osteoporosis is more common in elderly women than men. Osteoporosis is the main reason the elderly suffer broken hip bones much more often than the young.

hormone replacement therapy (HRT) Treatment to compensate for hormone reduction at menopause or following surgical removal of the ovaries. Such treatment, which usually involves estrogen and progesterone, minimizes menopausal symptoms and diminishes the risk of heart disease and osteoporosis in later adulthood.

tors today recommend HRT not primarily to treat the symptoms of menopause but to treat the physiological consequences of living forty years or so with diminished hormone levels. Long-term use of HRT reduces by half the risk of coronary artery disease (the leading cause of death for women over age 55) and reduces by at least half the incidence of hip fractures (now experienced by a third of all women who reach age 90) (Carlson et al., 1996; Grodstein et al., 1996). A recent longitudinal study of 8,877 elderly women finds another possible benefit from long-term HRT: reduced risk of developing Alzheimer's disease, with those taking estrogen for fifteen years or more particularly likely to benefit. Other evidence finds that estrogen supplements can prevent tooth loss (Grodstein et al., 1996).

All this is not to suggest that HRT is entirely risk-free. Hormone supplements might increase the risk of some cancers, and evidence regarding HRT use for twenty years or more is not yet available, making many 50-year-olds wonder if some unforeseen disaster will occur at age 80 or 100 if they take synthetic hormones all those years. Certainly every woman, whether or not she takes estrogen, should lower her risks of heart disease and osteoporosis by a lifetime of healthy habits, such as eating a low-fat, high-calcium diet, avoiding cigarettes, and engaging in regular exercise (with aerobics especially important for the heart and weight-bearing exercise important for the bones). Decisions regarding the use of HRT need to be made on a case-by-case basis that includes specific risk analysis. Osteoporosis, for example, is most prevalent in small-framed women of European descent, and heart disease is most prevalent in larger women of African descent, especially those with high blood pressure. So women from these groups particularly need to check their bone density, heart functioning, and family health history because HRT may reduce their risk of disease substantially. Indeed, many physicians now recommended HRT for all menopausal women, unless there are counterindications. Medical opinion is "definitely turning in favor" of HRT from menopause throughout late adulthood (Carlson et al., 1996).

Reproductive Changes in Context

With or without HRT, menopause signals the loss of reproductive potential. Is this, in itself, a stressful and depressing experience for women? In this day and age, no. Traditionally, in virtually all cultures, childbearing was particularly important for women, since most women attained social status directly from their role as mother: indeed, the more children a couple had, the more fortunate they were considered to be. In these circumstances, the psychological impact of declining fertility and menopause may have been substantial. Especially if a couple had only one or two children, menopause may have been greeted with considerable sorrow as the final "closing of the gates" of reproduction, as psychoanalyst Helene Deutsch (1945) once described it.

However, historical changes have meant that the end of childbearing is now determined less by age than by personal factors, such as the number of children a couple already has or the couple's financial situation. In fact, for the most part, the end of childbearing occurs through a conscious decision that is usually made when a woman is in her 30s, long before reproduction becomes biologically impossible. Nearly half of all American couples in

which the woman is between the ages of 35 and 44 have chosen to become surgically sterile (U.S. Bureau of the Census, 1996). For about a third of these couples, the man has had a vasectomy; in the rest, it is the woman who has been sterilized (Marquette et al., 1996). Most other couples, as well as sexually active unpartnered women, avoid pregnancy by other means: the annual birth rate for women in their early 40s is less than 1 per 100, compared with a rate of more than 1 in 10 for women in their 20s (U.S. Bureau of the Census, 1996). Menopause, then, as the time when menstruation ceases and sexual activity is no longer accompanied by fear of pregnancy and the inconveniences and risks of contraception, is more often welcomed than regretted.

Male Menopause?

For men in middle age, there is no sudden downward shift in reproductive ability or hormonal levels, as there is with women. Thus, physiologically, men experience nothing like the female climacteric. Most men continue to produce sperm indefinitely, and although there are important age-related declines in the number and motility of sperm (see Chapter 17), men are theoretically (and in some cases, actually) able to father a child in late adulthood. Similarly, the average levels of testosterone and other hormones decline gradually, not suddenly, with age (Mobbs, 1996).

Strictly speaking, then, there is no "male menopause." However, this phrase may have been coined to refer to another phenomenon: testosterone can dip markedly if a man suddenly becomes sexually inactive or unusually worried, as might happen if he were faced with unemployment, marital problems, serious illness, or unwanted retirement. Levels of testosterone correlate with levels of sexual desire and speed of sexual responses, so a man with low testosterone might find himself unable to have an erection when he wanted to. Thus, the effects of this dip, especially when added to whatever age-related declines have already occurred, may make a man highly anxious about his sexual virility, which, in turn, may reduce his testosterone level even more. By the same token, a middle-aged man who lands a new, ego-enhancing job or adds some novelty to his sex life may experience a rise in self-esteem, testosterone, and desire. Underlying such situational peaks and valleys, however, is a steady, gradual decline.

Age-Related Changes in Sexual Expression

One usual way to measure sexual activity is in terms of the frequency of intercourse and orgasm. By this measure, the gradual declines in sexual activity that occur over the course of early adulthood continue during middle age, though with wide individual differences, including some people who stop having intercourse altogether and others who continue to have intercourse on a regular basis (Edwards & Booth, 1994).

Even for the sexually active, however, the specifics of their activity change with age, especially for men. Sexual stimulation takes longer, and needs to be more direct, than at earlier ages. Further, as Herant Katchadourian (1987), a physician who studies sexuality, writes about men, "orgasmic reactions become less intense with age . . . contractions are fewer, ejaculation is less vigorous, and the volume of the ejaculate is smaller."

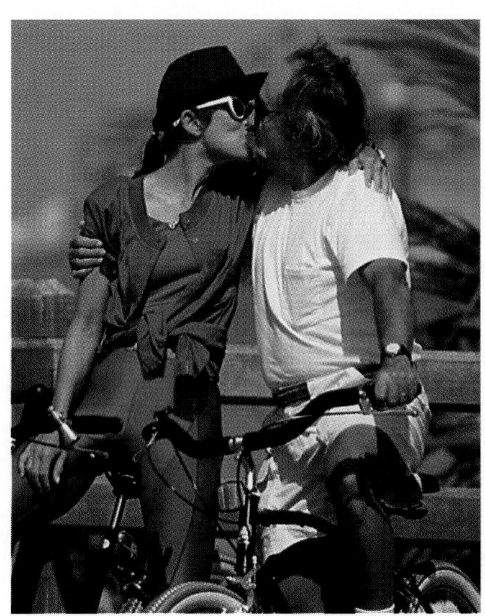

Throughout adulthood, continued, pleasurable sexual experiences depend much less on the partners' age than their attitudes toward each other and toward sex itself. As many experts have noted, the most important human sexual organ is the brain.

To a 20-year-old male, such prospects may seem dismaying, but to the middle-aged male actually experiencing these differences, the reality is much less troubling. A study of middle-aged men found that although their levels of desire and their frequency of ejaculation decreased with age, almost all were satisfied with their sex life. It wasn't until age 60 or so that more men agreed rather than disagreed with the statement that "men's sexual interest declines with age." Even then, most 60-year-olds were satisfied with their sex life, although the minority who were very dissatisfied with their sex life rose from about 5 percent throughout middle age to about 10 percent after age 60 (McKinlay & Feldman, 1994).

Changes in women's orgasmic ability are harder to measure, but many researchers think a woman's eroticism is at least as strong in middle age as in early adulthood. As one group reports:

> The woman's capacity for orgasm is not impaired in any way by aging as long as there is no other health problem complicating the picture. In fact, many women report being more easily orgasmic in their post-menopausal years than they were previously, although this effect may relate more to psychosocial components of sexual responsivity (e.g., no worries about becoming pregnant) than to biological factors. [Masters et al., 1994]

After menopause, signs of arousal, including vaginal lubrication, may be less apparent, but none of these changes need impair a sexual relationship. As Katchadourian explains regarding women, "while the intensity of the physiological responses clearly diminish over time, the subjective experience of orgasm continues to be highly satisfying, though not as explosive as in previous years." Thus sexual activity usually continues and brings pleasure for most middle-aged people.

One final fact seems to be suggested by the research: middle-aged couples do not move from active, happy sex lives to troubled or nonexistent ones unless their relationship is plagued by other problems that may be reflected, but not caused, in the bedroom. Quite simply, people who have active sex lives in young adulthood are most likely to have active sex lives in middle and late adulthood, and couples who were never comfortable with their sexual relations are likely to end them, perhaps with relief, in later years. For middle-aged and older adults, both overall enjoyment of sex and preference for particular sexual activities are more closely related to past sexual interests and desires, and perhaps current health problems, than to more global variables such as current income, education, age, or life satisfaction (Edwards & Booth, 1994; Masters et al., 1994). Throughout life, it seems that sexual activity itself helps promote sexual interest and excitement; correspondingly, absence of sexual activity results in lower levels of sex hormones and a loss of sexual interest.

Again, as we saw in early adulthood, it seems that, as adults grow older, biosocial development is much less indexed by chronological age than it was in earlier periods of development. Personal choices become increasingly important in affecting the course of development. In the next chapter, we will see that choice can influence the development of our intellectual skills as well. At least in some abilities, to some extent, we can choose to be smarter, wiser, or more expert, and then take action to become so.

SUMMARY

Normal Changes in Middle Adulthood

1. A person's appearance undergoes gradual but notable changes as middle age progresses, including more wrinkles, less hair, and new fat, particularly on the abdomen. With the exception of excessive weight gain, changes in appearance have little impact on health.

2. Hearing gradually becomes less acute, with noticeable losses being more likely for high-frequency sounds, particularly in men. Vision also becomes less sharp with age. Two particular difficulties for many middle-aged people are reading small print and adjusting to glare at night.

3. During middle age, declines in all the body's systems become apparent, but generally they are not sufficient to impair normal functioning. Overall, health is generally quite good, with the death rate for today's middle-aged adults being significantly lower than for earlier cohorts.

4. Although there is evidence that middle-aged Americans today are more conscious of good health habits, many people still put their health at risk by smoking cigarettes, drinking alcohol excessively, eating poorly, gaining weight, and maintaining a sedentary lifestyle.

Variations in Health

5. Variations in health—which can be measured in terms of mortality, morbidity, disability, and vitality—arise from a combination of many factors, chief among them race, ethnicity, socioeconomic status, and gender.

6. Both genetic and cultural factors affect the overall health of various ethnic groups to a large extent, but social and psychological factors may be even more influential, with members of certain ethnic groups in certain settings much more prone to health risks.

7. Beginning in middle age, women have higher morbidity and disability rates than men. One reason for this difference may be a gender bias in health research that has favored the study of heart disease in men. More research studies today, though, are examining patterns of illness among women.

The Sexual-Reproductive System

8. At menopause, as a woman's menstrual cycle stops, ovulation ceases and levels of estrogen are markedly reduced. This hormonal change produces various symptoms and possible problems, although most women find the experience of menopause much less troubling than they had expected it to be. Hormone replacement therapy has been shown to have many health benefits for postmenopausal women.

9. Men do not have sudden age-related drops in hormone levels or in fertility. In this sense, there is no "male menopause."

10. As a man's sexual responses slow down with age, many couples find that they engage in intercourse less often. However, active sexual relationships can, and often do, continue throughout adulthood, to the satisfaction of both sexes.

KEY TERMS

glaucoma (546)
immune system (547)
autoimmune diseases (547)
mortality (553)
morbidity (553)
disability (553)
vitality (553)
menopause (558)
climacteric (558)
vasomotor instability (559)
osteoporosis (559)
hormone replacement therapy (HRT) (559)

KEY QUESTIONS

1. What changes in appearance typically occur during middle age, and what is their impact?

2. What are the reasons one person might have a greater hearing loss than another in middle adulthood?

3. What are the likely changes in a person's vision during middle adulthood?

4. How do changes in the immune system affect people as they grow older?

5. How do lifestyle factors affect middle-aged adults' risks for developing serious health problems?

6. How does ethnicity affect health? Explain variations of health between and within ethnic groups.

7. What are the major differences in health between middle-aged men and women and what explains these differences?

8. How does age affect a couple's sex life?

Middle Adulthood: Cognitive Development

CHAPTER 21

Overall, would you say that adults become more intelligent, or less, as they grow older? On one hand, the growth of new, postformal styles of thought (such as those described in Chapter 18) and broader experience might yield intellectual progress. On the other hand, the biological and perceptual changes that accompany aging might lead to intellectual decline. No matter which way you answer, you are in good scientific company. Investigators who have spent their professional lives gathering and reviewing research on this question have reached opposite conclusions. Researchers on one side assert that intellectual decline during adulthood is inevitable, with undeniable decreases occurring in the flexibility of thinking and the speed and efficiency of problem solving. Researchers with a different viewpoint contend that, throughout life, intelligence is quite plastic, molded by health, education, life experiences, and intellectual activity, and that depending on the impact of these factors, various aspects of intelligence can either increase or decrease.

How could researchers looking at the same issue, and evaluating much of the same evidence, come to such different conclusions? Which view is more accurate? Is there any synthesis that combines these opposing points of view? This chapter attempts to answer these questions.

Decline in Adult Intelligence?

For most of the twentieth century, psychologists were convinced that intelligence reaches a peak in adolescence, and then gradually declines during adulthood (Woodruff-Pak, 1989). This belief was based on what seemed to be solid evidence. For instance, all literate American draftees in World War I were given an intelligence test, called Alpha, that tested a variety of cognitive skills. When the scores of men of various ages were compared, one conclusion seemed obvious: the average American male reached an intellectual peak at about age 18, stayed at that level until his mid-20s, and then began to show a decline (Yerkes, 1923).

Similar results came from a classic study of 1,191 subjects between the ages of 10 and 60, chosen from nineteen carefully selected, insular New England villages. (The purpose of the sampling procedure was to achieve an ethnically homogeneous group of adults who had had fairly similar life experiences. Thus age would be the only significant difference among the test-takers.) IQ tests from this group showed intellectual ability peaking be-

tween ages 18 and 21, and then slowly and steadily declining to the point that the average 55-year-old scored the same as the average 14-year-old (Jones & Conrad, 1933). The case for age-related decline in intelligence was considered proven beyond a reasonable doubt.

Contrary Evidence

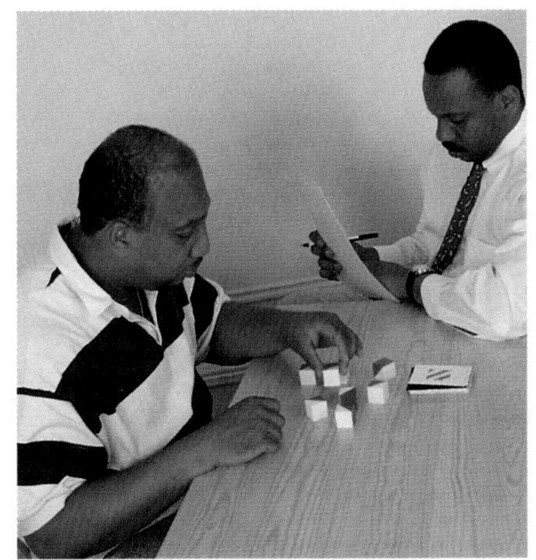

The Wechsler Adult Intelligence Test is a timed, one-on-one exam that involves ten separate subtests, including the spatial-design item shown here. In general, middle-aged subjects do as well or better on this test than they did as teenagers, because their careful and deliberate attention compensates for their slowdown in response rate. Note that the tester keeps track of the time it takes to solve the problem, giving extra points for quick solutions. By late adulthood, slowed reaction time and the anxiety it begins to cause in test-taking reduce IQ scores.

The first evidence to contradict the assumption that intelligence declines with age was uncovered by Nancy Bayley and Melita Oden (1955). They were analyzing the adult development of the children originally selected by Lewis Terman in 1921 for his study of child geniuses, a group that has been studied by a succession of researchers over the past seventy years. As Bayley later explained, she fully expected to find a decline in these subjects' cognitive development because in previous cross-sectional studies "the invariable findings had indicated that most intellectual functions decrease after about 21 years of age" (Bayley, 1966). But on several tests of concept mastery, including questions that involved use of synonyms, antonyms, and analogies, the scores of these gifted individuals increased between ages 20 and 50.

Bayley decided to follow this clue by retesting a less gifted (or more typical) group of adults who had also been tested as children. (These subjects, as members of the Berkeley Growth Study, had been selected in infancy to be representative of the infant population of Berkeley, California.) Bayley's results again showed a general increase in intellectual functioning from childhood through young adulthood. Instead of reaching a plateau at age 21, the typical person at age 36 was still improving on the most important subtests of the **Wechsler Adult Intelligence Scale (WAIS)**, the most commonly used adult IQ test. Specifically, subjects continued to improve on tests of vocabulary, comprehension, and information and declined on only two of the ten subtests—arithmetic, which measures speed and accuracy of mathematical ability, and picture completion, which requires the person to spot an element that is missing from a picture of a common object (such as an ankle from a foot, or a pair of legs from a bee). Bayley (1966) concluded that "intellectual potential for continued learning is unimpaired through 36 years."

Cross-Sectional, Longitudinal, and Cross-Sequential Studies

Why did Bayley find such a different pattern of intellectual aging? Recall that her study used a longitudinal design (studying the same people repeatedly as they grew older), whereas earlier studies were cross-sectional (studying groups of people that differed only in age).

As we saw in Chapter 2, developmentalists now recognize that cross-sectional research can sometimes yield a misleading picture of adult development, not only because it is impossible to select adults who are similar to each other in every important aspect except age, but also because of the cohort effects created by each group's own unique history of life experiences. Cohort effects, in particular, complicate the interpretation of adult differences in intellectual performance. Adults who grew up during the Great Depression, for example, or during World War II, might have acquired different cognitive skills than younger cohorts who grew up during the 1950s, say, or the 1980s. Among other influences, the quality of public education, the variety of cultural opportunities, and the dissemination of information

Wechsler Adult Intelligence Scale (WAIS) The most commonly used assessment of adult IQ.

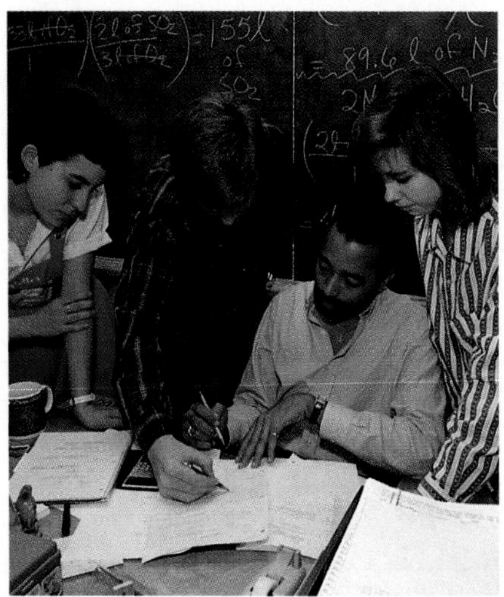

Use it or lose it—pertinent advice for any adult who wants to maintain intellectual ability. Reading, studying, and—best of all—explaining what you know to someone else increases measured ability as well as achievement.

in the popular media have provided advantages to later-born cohorts. Even more significant, elderly adults who grew up at a time when most 16-year-olds quit school to begin working are likely to differ intellectually from those who grew up later, when a high school education was more normative.

K. Warner Schaie was one of the earliest researchers to recognize the potentially distorting impact that cohort effects can have on cross-sectional research. In 1956, as part of his doctoral research, Schaie tested a cross-sectional sample of 500 adults aged 20 to 70 on five "primary mental abilities:" Verbal Meaning (vocabulary comprehension), Spatial Orientation, Inductive Reasoning, Number Ability, and Word Fluency. His initial results showed some gradual decline in ability with age, but Schaie wondered whether cohort effects accounted for these cross-sectional findings. Therefore, he retested his original subjects at seven-year intervals, in 1963, 1970, 1977, 1984, and 1991 (Schaie, 1996). The results of Schaie's research, known as the **Seattle Longitudinal Study**, confirmed and extended what Bayley had found many years earlier: most people improve in primary mental abilities during most of adulthood. Cross-sectional research had, apparently, given misleading testimony regarding adult intelligence.

But longitudinal findings might be misleading also (Cunningham & Tomer, 1990; Horn & Hofer, 1992). People who are retested several times on similar tests might improve their performance as they become practiced at taking the tests. Moreover, it is hard to retain adults in a longitudinal sample over long periods of time: often 50 percent or more of Schaie's adults could not be found for the next testing period. Adults move away, become ill or die, or may refuse to return for another round of testing. Like those who show up at class reunions, the subjects who return for long-term longitudinal studies may be the most stable, healthy, well-functioning adults who are happy to be retested. Those who are ill or who are troubled by their declining abilities are more likely to drop out. Thus, while cross-sectional studies may *overestimate* adult intellectual decline, longitudinal designs may *underestimate* it.

To correct for some of these problems, Schaie went one step further, developing a new research design called *cross-sequential*, or *cohort-sequential*, research, which is a combination of cross-sectional and longitudinal approaches (see Chapter 2). Each time he retested his original subjects he also tested a new group of adults at each age interval, and then he followed this new group longitudinally as well.

The cross-age comparisons made possible by this accumulation of longitudinal and cross-sectional data over many years allow analysis of possible effects of retesting, cohort differences, and other influences on adult changes in intelligence. As shown in Figure 21.1 (p. 568), from age 20 until the late 50s, cognitive abilities are more likely to increase than decrease, with the exception of arithmetic skills (how quickly and accurately a person can do math), which begin to shift slightly downward by age 40. After age 60

> 7-year decrements are statistically significant throughout. These data suggest that average decline in psychological competence may begin for some as early as the mid-50s, but that it is typically of small magnitude until the 70s are reached. [Schaie, 1990]

In his most recent follow-up of these individuals, Schaie adds that "it is not until the 80s are reached that the average older adult will fall below the middle range of performance for young adults" (Schaie, 1996).

Seattle Longitudinal Study The study led by K. Warner Schaie in which adults are tested and retested every seven years to reveal patterns of change in intellectual abilities.

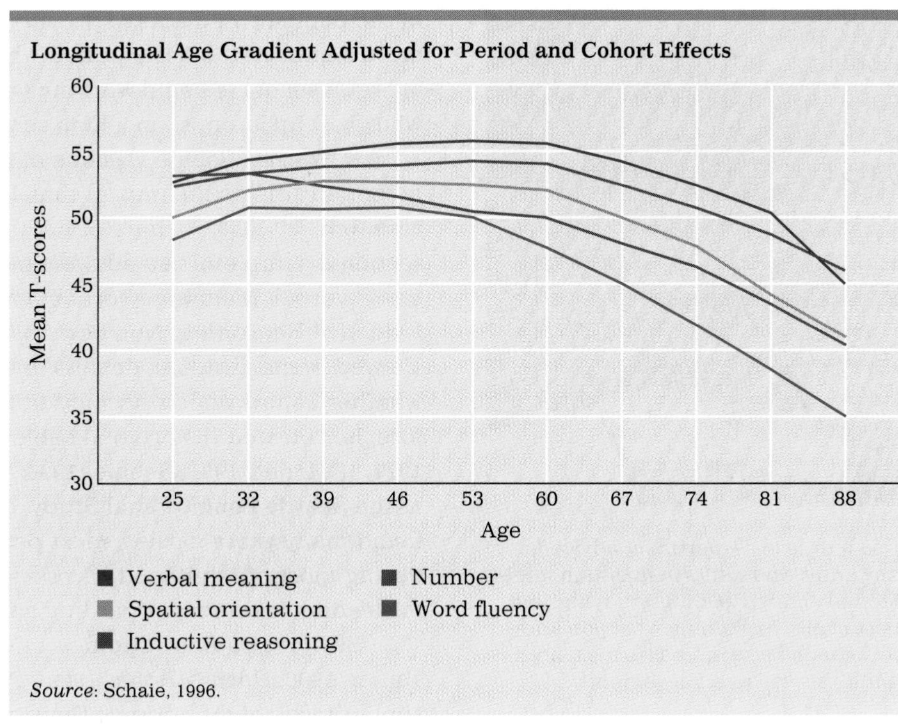

FIGURE 21.1 Cross-sectional data on intellectual abilities at various ages would show much steeper declines than those shown here; longitudinal research, on the other hand, would show more notable rises. Because Schaie's research is cross-sequential, the trajectories it depicts are more revealing. In part, this is because the methodology takes into account the cohort and historical effects and thus holds educational experiences constant. As you can see in this more accurate reflection of intellectual development, the age-related differences from age 20 to 60 are very small.

Fluid and Crystallized Intelligence

Although tracking *average overall* performance throughout adulthood is useful in measuring general intellectual functioning, it masks the differential changes that occur in distinct aspects of intelligence, called fluid intelligence and crystallized intelligence. As its name implies, **fluid intelligence** is flexible reasoning used to draw inferences and understand relations between concepts. It is made up of those basic mental abilities—inductive reasoning, abstract thinking, speed of processing, and the like—required for understanding any subject matter, particularly material that is unfamiliar. Someone high in fluid intelligence is quick and creative with words, numbers, and intellectual puzzles. Among the questions that might be used to test fluid intelligence are the following (Horn, 1985):

> What comes next in these series?*
> 4 5 6 3 4 5 2 3 4 5 6
> B D A C Z B Y A

fluid intelligence Those types of basic intelligence that make learning of all sorts quick and thorough. Underlying abilities such as short-term memory, abstract thought, and speed of thinking are all usually considered part of fluid intelligence.

crystallized intelligence Those types of intellectual ability that reflect accumulated learning. Vocabulary and general information are examples. Some developmental psychologists think crystallized intelligence increases with age, while fluid intelligence declines.

Another standard type of item that is used to measure fluid intelligence is the timed assembly of puzzles, with credit given for completion within 2 minutes, bonus points for completion within 1 minute, and no credit after 2 minutes.

Crystallized intelligence is the accumulation of facts, information, and knowledge that comes with education and experience within a particular culture. The size of vocabulary, the knowledge of chemical formulas, and long-term memory for dates in history are all indications of crystallized intelligence (Horn, 1982). Test items designed to measure crystallized intelligence might include questions such as these:

*The correct answers to these two items are: "1" and "X."

What is the meaning of the word "temerity"?†
What do you do with a mango?
What word is associated with bathtubs, prizefighting, and weddings?
In what year was the Magna Carta enacted?

Originally, psychologists thought that fluid intelligence is primarily genetic and that crystallized intelligence is primarily learned (Cattell, 1963). Now most experts think that this nature-nurture distinction is invalid, in part because the acquisition of crystallized intelligence is affected by the quality of fluid intelligence (Horn, 1985). For instance, the strength of your present vocabulary is partly the result of your reading speed and of your ability to make logical associations among words—both of which are related to fluid intelligence.

Over the years of adulthood, fluid intelligence declines markedly. So also do related abilities like processing speed and short-term memory as measured on psychological tests (Horn & Hofer, 1992). This decline is temporarily disguised by an increase in crystallized intelligence, which continues to expand throughout most of adulthood. Take verbal ability as an example. Once you have acquired, via fluid intelligence, a working knowledge of your native language, you are likely to remember it all your life, as long as you continue to use it. This ability to speak, read, and write your native language makes it easier to enlarge your crystallized understanding of it, because you can relate new words to, or define them with, words already familiar to you. However, if you had to learn to write and speak a new phrase in a language quite different from any you knew, it would take you longer the older you were. The reason is that your crystallized knowledge of your native language would be of little use to you, so you could not compensate for the decline in your fluid ability to learn new material.

The marked contrast between the decline in fluid intelligence and the maintenance or even increase in crystallized intelligence is clearly demonstrated on the WAIS. In the Wechsler test, total IQ is a weighted average of two types of intelligence, *Verbal* (vocabulary, information, and the like) and *Performance* (puzzles, visual perception, and so forth). As you can see from the cross-sectional data in Figure 21.2 (p. 570), Verbal IQ remains in the average range throughout adulthood, with small fluctuations and only a minimal decline from the early 20s to the early 70s. Since cross-sectional data can give an exaggerated picture of decline, even these small declines might be reversed if subjects in the younger age groups were followed longitudinally. In contrast, Performance IQ falls markedly over the course of adulthood, dropping by an average of 25 points, from around 101 to 76 (that is, to nearly a level of mental retardation). Although longitudinal data might show a slower decline, it is obvious that Performance IQ declines notably (Kaufman, 1990).

What causes the declines in fluid intelligence during adulthood? The primary cause is the gradual accumulation of irreversible damage to brain structures that results from disease, injury, and biological changes associated with age (Horn & Hofer, 1992). Because these structures guide the attentional and processing capabilities that are necessary for quick, flexible reasoning, their impairment results in a progressive decline in the speed and efficiency of thinking.

†The correct answers to these test items are "Boldness," "Eat It," "Ring," and "1215."

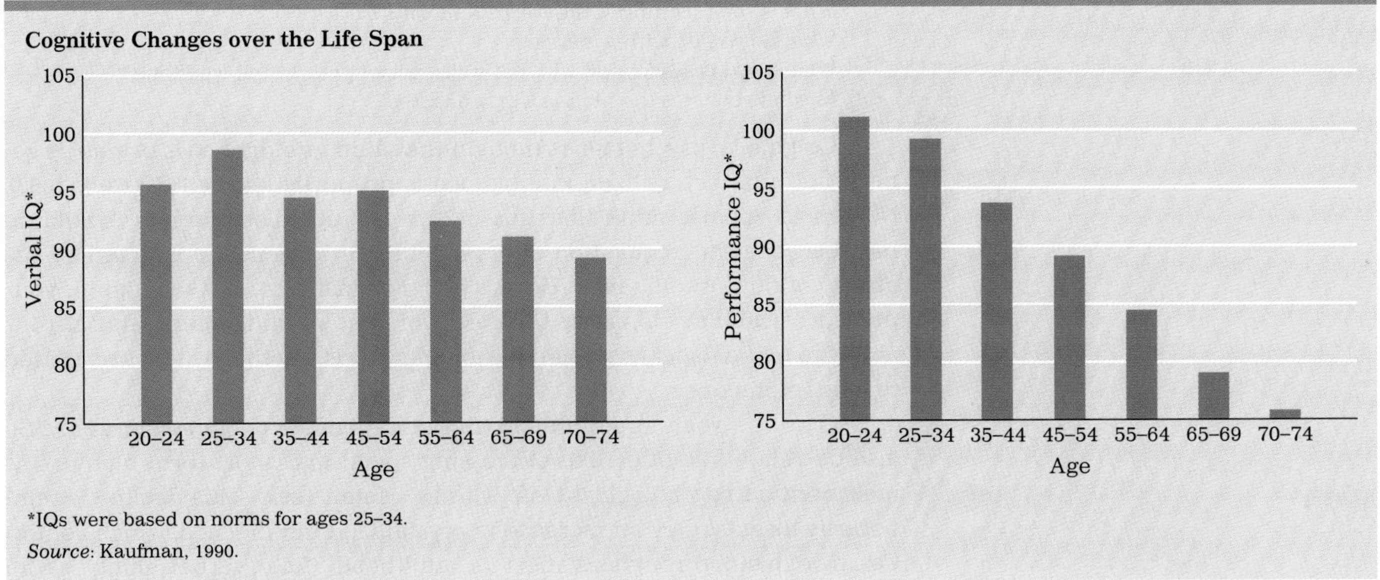

FIGURE 21.2 Whether intellectual abilities increase or decline over the years of adulthood depends on which ability is measured—as is apparent to every older adult whose knowledge of history far overshadows that of their teenage grandchildren, who nonetheless can beat them four times over at a video game. Even clusters of abilities, such as those of crystallized versus fluid intelligence, or the verbal items of the Wechsler (left) versus the Performance items (right), follow distinct developmental patterns.

Crystallized intelligence, too, is affected by these changes, but the knowledge structures underlying crystallized intelligence are overlapping and interconnected, such that losses can be more easily compensated. If an adult happens to forget a name or date, that information can be easily retrieved in other ways (such as by looking it up, asking someone, or using mnemonic devices to aid remembering). However, because the maintenance of crystallized intelligence depends partly on how it is used, Horn and Hofer (1992) have also found that individual differences in crystallized intelligence increase with age. In other words, the consequences of remaining involved with stimulating people and events—or of being socially isolated and withdrawn—become increasingly important to the functioning of crystallized intelligence in adulthood. Other research has likewise shown that variability in experiences and lifestyle can have a substantial impact on the trajectory of overall adult IQ. In short, a 50-year-old who lives alone and spends much of his or her time watching television reruns is more likely to demonstrate declines in IQ than a 50-year-old who has stimulating social relationships, a challenging job, and an active intellectual life.

An Assessment

Many other researchers have studied adult cognitive development through a variety of methods. Overall, the results of cross-sectional, longitudinal, and cross-sequential research, and of experimental as well as observational study, have led to four general conclusions about intellectual development from age 20 or so to beyond age 90 (Baltes & Baltes, 1990; Ceci, 1990; Horn & Hofer, 1992; Powell, 1994; Salthouse, 1985, 1991, 1992; Schaie, 1996):

1. *Stability.* In general, intellectual abilities increase slightly from the beginning of adulthood until midlife and then remain stable throughout middle adulthood. This overall stability begins to give way to selective declines in later adulthood.

2. *Decline.* Age-related declines in intellectual functioning are first noted in aspects of cognition related to quick thinking and speedy reaction time. These declines are sometimes apparent as early as age 30 (for instance, when an adult competes against a teenager in a video game) and affect fluid-intelligence scores by age 40. However, they do not usually begin to affect overall intelligence scores until about age 60. They become part of a pervasive slowing and decline in every component of information processing at about age 80 (speed of processing is covered in detail in Chapter 24).

3. *Education.* Intellectual functioning as measured by IQ tests is powerfully influenced by years of schooling in childhood, adolescence, and early adulthood. Part of the correlation between education in childhood and intelligence in adulthood is obviously the result of genes: that is, intelligent adults were intelligent children and therefore were more likely to stay in school. But some of a person's level of intellectual functioning throughout adulthood is a direct result of education received years earlier: that is, given comparable genetic abilities in any particular group, those with more education are more likely to maintain a higher level of their intellectual functioning into late adulthood.

4. *Variability.* Especially during adulthood, intelligence follows many patterns of development. Its course is *multidimensional, multidirectional, and plastic*—all concepts we will now discuss.

Multidimensional Intelligence: Not One Intelligence, but Many

Historically, psychologists as well as laypeople have thought of intelligence as a single entity, a single ability that people possess in greater or lesser amounts. In this vein, a leading theoretician, Charles Spearman (1927), argued for the concept of *general intelligence*, which he called *g*. Although it cannot be measured directly, he contended, *g* can be inferred from various abilities that can be tested, such as vocabulary, memory, and reasoning. Just the fact that psychologists give IQ tests with various subtests, and then calculate an overall IQ for the person tested, implies that intelligence can be thought of as an integrated whole.

The idea that what is called "intelligence" is a single entity continues to influence thinking on this subject. However, an increasing number of researchers question whether the concept of *g* is valid, especially as it refers to a general quality that can be tested and measured in such a way that two people with the same score can be considered intellectual equals. Virtually every psychologist who looks at patterns of intellectual development in adulthood finds it more useful to examine several distinct intellectual abilities, all important facets of cognition. In fact, over seventy different intellectual abilities can be measured by currently available tests (Carroll, 1993). Two psychologists, Howard Gardner and Robert Sternberg, have studied the many facets of intelligence and have developed persuasive theories.

Multiple Intelligences

As noted in Chapter 12, Howard Gardner (1983) believes there are seven distinct intelligences: linguistic, logical-mathematical, musical, spatial, body-kinesthetic, social-understanding, and self-understanding. According to Gardner, each of these intelligences has its own neurological network in a particular part of the brain, which explains why brain-damaged people can be unusually inept or amazingly skilled in any one of them, while showing subnormal abilities in other areas. Every normal person has the capacity to achieve at least minimal proficiency in all seven abilities, but for genetic reasons will be more gifted in some abilities than in others. In addition, each culture values some intelligences more than others, which leads parents to encourage, children to develop, schools to emphasize, and adults to maintain whichever abilities are culturally valued. Meanwhile, the less valued abilities, in most individuals, remain relatively underdeveloped and fade faster in adulthood.

Although many researchers are studying Gardner's multiple intelligences in schoolchildren, no longitudinal or cross-sequential research on these seven intelligences in adults has yet been published. However, notice that two, linguistic intelligence and logical-mathematical intelligence (which are similar to Schaie's Verbal Meaning and Inductive Reasoning), are particularly valued in contemporary American culture. If Gardner is correct that people develop the abilities that their social context encourages, then that would help to explain why these two abilities tend to increase among Americans through early and middle adulthood. Gardner's theory would also help explain why dimensions of intelligence not particularly valued by middle-aged Americans, such as musical or kinesthetic ability, fade more quickly.

Robert Sternberg proposes that intelligence comprises three fundamental aspects: the analytic, the creative, and the practical (Sternberg, 1988). The *analytic*, or academic, aspect consists of mental processes that foster efficient learning, remembering, and thinking. These include planning, strategy selection, attention, effective information processing, as well as verbal and logical skills. Such skills and talents are particularly valued at

Both these adults at work—a scientist investigating the genetics of breast cancer and a Turkese helmsman navigating the South Pacific without instruments—demonstrate highly specialized intellectual abilities. Imagine the difficulty of trying to create a single IQ test that would allow both individuals to demonstrate their intelligence equally. Similar though less obvious disparities exist among those whose minds are unusually creative, analytic, or practical and likewise pose a problem in adequately assessing IQ.

CHAPTER 21

Overall, would you say that adults become more intelligent, or less, as they grow older? On one hand, the growth of new, postformal styles of thought (such as those described in Chapter 18) and broader experience might yield intellectual progress. On the other hand, the biological and perceptual changes that accompany aging might lead to intellectual decline. No matter which way you answer, you are in good scientific company. Investigators who have spent their professional lives gathering and reviewing research on this question have reached opposite conclusions. Researchers on one side assert that intellectual decline during adulthood is inevitable, with undeniable decreases occurring in the flexibility of thinking and the speed and efficiency of problem solving. Researchers with a different viewpoint contend that, throughout life, intelligence is quite plastic, molded by health, education, life experiences, and intellectual activity, and that depending on the impact of these factors, various aspects of intelligence can either increase or decrease.

How could researchers looking at the same issue, and evaluating much of the same evidence, come to such different conclusions? Which view is more accurate? Is there any synthesis that combines these opposing points of view? This chapter attempts to answer these questions.

Decline in Adult Intelligence?

For most of the twentieth century, psychologists were convinced that intelligence reaches a peak in adolescence, and then gradually declines during adulthood (Woodruff-Pak, 1989). This belief was based on what seemed to be solid evidence. For instance, all literate American draftees in World War I were given an intelligence test, called Alpha, that tested a variety of cognitive skills. When the scores of men of various ages were compared, one conclusion seemed obvious: the average American male reached an intellectual peak at about age 18, stayed at that level until his mid-20s, and then began to show a decline (Yerkes, 1923).

Similar results came from a classic study of 1,191 subjects between the ages of 10 and 60, chosen from nineteen carefully selected, insular New England villages. (The purpose of the sampling procedure was to achieve an ethnically homogeneous group of adults who had had fairly similar life experiences. Thus age would be the only significant difference among the test-takers.) IQ tests from this group showed intellectual ability peaking be-

Multiple Intelligences

As noted in Chapter 12, Howard Gardner (1983) believes there are seven distinct intelligences: linguistic, logical-mathematical, musical, spatial, body-kinesthetic, social-understanding, and self-understanding. According to Gardner, each of these intelligences has its own neurological network in a particular part of the brain, which explains why brain-damaged people can be unusually inept or amazingly skilled in any one of them, while showing subnormal abilities in other areas. Every normal person has the capacity to achieve at least minimal proficiency in all seven abilities, but for genetic reasons will be more gifted in some abilities than in others. In addition, each culture values some intelligences more than others, which leads parents to encourage, children to develop, schools to emphasize, and adults to maintain whichever abilities are culturally valued. Meanwhile, the less valued abilities, in most individuals, remain relatively underdeveloped and fade faster in adulthood.

Although many researchers are studying Gardner's multiple intelligences in schoolchildren, no longitudinal or cross-sequential research on these seven intelligences in adults has yet been published. However, notice that two, linguistic intelligence and logical-mathematical intelligence (which are similar to Schaie's Verbal Meaning and Inductive Reasoning), are particularly valued in contemporary American culture. If Gardner is correct that people develop the abilities that their social context encourages, then that would help to explain why these two abilities tend to increase among Americans through early and middle adulthood. Gardner's theory would also help explain why dimensions of intelligence not particularly valued by middle-aged Americans, such as musical or kinesthetic ability, fade more quickly.

Robert Sternberg proposes that intelligence comprises three fundamental aspects: the analytic, the creative, and the practical (Sternberg, 1988). The *analytic*, or academic, aspect consists of mental processes that foster efficient learning, remembering, and thinking. These include planning, strategy selection, attention, effective information processing, as well as verbal and logical skills. Such skills and talents are particularly valued at

Both these adults at work—a scientist investigating the genetics of breast cancer and a Turkese helmsman navigating the South Pacific without instruments—demonstrate highly specialized intellectual abilities. Imagine the difficulty of trying to create a single IQ test that would allow both individuals to demonstrate their intelligence equally. Similar though less obvious disparities exist among those whose minds are unusually creative, analytic, or practical and likewise pose a problem in adequately assessing IQ.

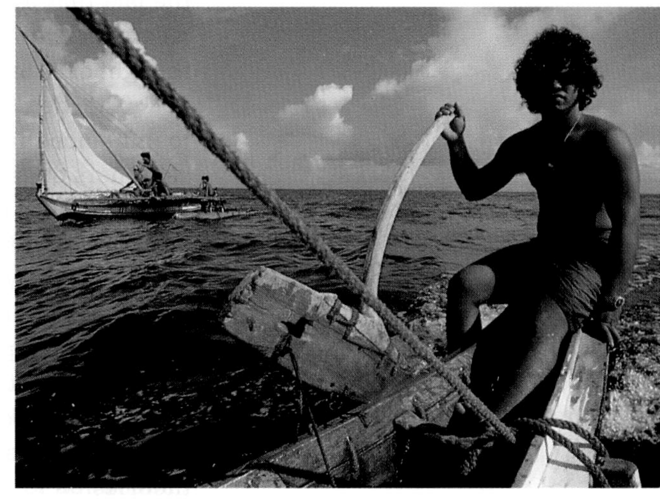

the beginning of adulthood, in college and graduate school or in learning a new job. The *creative* aspect involves the capacity to be intellectually flexible and innovative when dealing with new situations. This aspect is prized whenever life circumstances or challenges change, and, over the long run, creativity is a better predictor of accomplishment than traditional measures of IQ (Csikszentmihalyi, 1996). Finally, the *practical* aspect involves the capacity to adapt one's behavior to the contextual demands of a given situation, including the expectations and needs of the particular people in question and the particular skills that are called for.

In colloquial terms, the first of these three aspects of intelligence is stressed in formal education and the last is stressed in the "school of hard knocks." To put it another way, the first is "book smarts," the latter is "street smarts." As you might imagine, it is the last of these three that is particularly valued by adults in middle adulthood and that now has become the leading topic of research in adult cognition (Berg & Sternberg, 1985; Sternberg, 1988; Sternberg et al., 1995).

Practical Intelligence

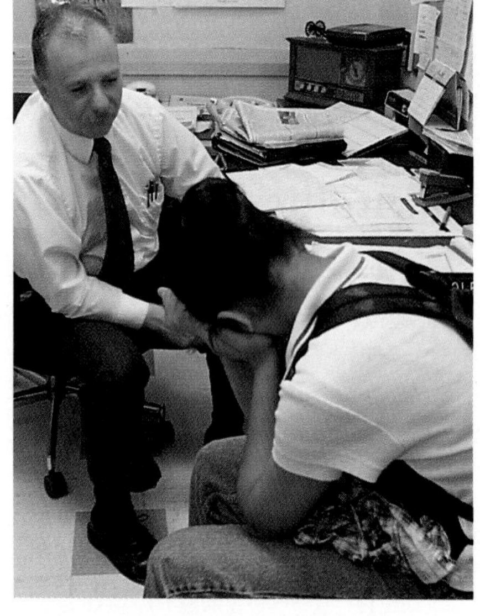

What kinds of abilities are needed to listen to, understand, and advise the diverse adolescent population of a large urban school? Whatever the abilities might be, they spring from long hours of practical experience, not primarily from abstract learning.

Many developmentalists who study adult intelligence believe that traditional assessment measures rely too heavily on little-used knowledge and skills, and that decline in these abilities with age reflects their waning due to lack of use (Denney, 1982, 1989; Labouvie-Vief, 1985). Certainly most adults do not occupy their time trying to define obscure words or deduce the next element in a number sequence but instead try to solve the real-world challenges of managing a home, advancing a career, balancing family finances, analyzing information from the newspaper or television, and addressing the needs of family members, neighbors, and colleagues. According to contextual theorists of adult development like Roger Dixon (1992; Dixon & Baltes, 1986) and Stephen Ceci (1990), assessments of adult intelligence should encompass the skills and knowledge that are relevant to practical functioning. This has led to a new approach to adult intelligence—the study of **practical intelligence**, or the intellectual skills used in everyday problem solving (Marsiske & Willis, 1995; Poon et al., 1989; Sternberg et al., 1995).

In contrast to the examples of measures of fluid and crystallized intelligence on pages 568 and 569, consider, for example, these items from a measure of practical intelligence that focuses on real-life "domestic" problem solving (Denney, 1989, 1990):

> Let's say that one evening you go to the refrigerator to get something cold to drink. When you open the refrigerator, you notice that it is not cold inside, but rather, is warm. What would you do?
>
> Let's say that a young man who is living in an apartment building finds that the heater in his apartment is not working. He asks his landlord to send someone out to fix it and the landlord agrees. But, after a week of cold weather and several calls to the landlord, the heater is still not fixed. What should the young man do?
>
> Let's say that a 60-year-old man who lives alone in a large city needs to go across town for a doctor's appointment. He cannot drive because he doesn't have a car and he doesn't have any relatives who live nearby who could drive him. What should he do?

practical intelligence The intellectual skills used in everyday problem solving.

multidirectional A way of describing variations in specific cognitive abilities over time.

interindividual variation Differences among individuals that are the result of the uniqueness of each person's genetic makeup and particular environment.

Other measures of practical intelligence concern career decisions, consumer behavior, conflict resolution, finding one's way in complex physical spaces, and managing an office. Problems in areas like these not only concern real-life situations but also involve the kinds of intellectual abilities and knowledge that most adults consider to be increasingly important as one matures.

Not only do most adults value such practical abilities more and more as they grow older, but most also think that they steadily improve in this aspect of intelligence even into old age. One study found that most (76 percent) believed that their thinking abilities improved with age, and very few (4 percent) thought that they declined. When confronted with evidence for overall decline in IQ, the subjects explained that when they answered that they thought their thinking abilities had improved, they had in mind their intelligence in practical matters (Williams et al., 1983). Thus many adults believe that when they were young they were foolish with love or money, or naive in human relations, or ignorant about career management, but that gradually their practical intelligence, at least as it relates to matters of understanding themselves and others, improved. This belief is substantiated by research (see Research Report). As one review concludes, "practical knowledge and problem-solving skills exhibit progressive changes, advancing at least from early adulthood to middle age, and perhaps beyond" (Cornelius et al., 1989).

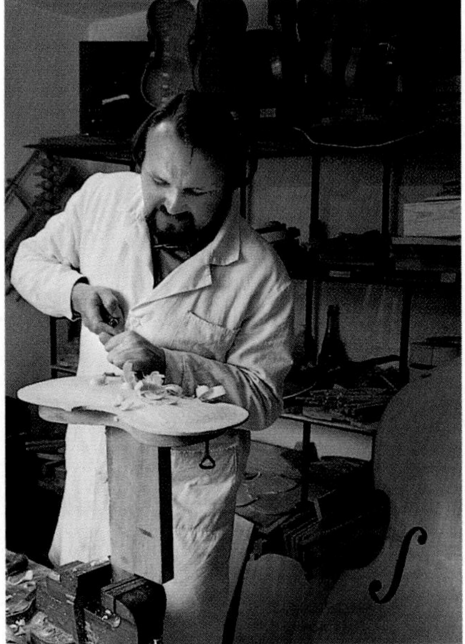

Over the life span, each individual's cognitive abilities vary in both direction and slope from those of other individuals the same age. Even more than the universal passage of time, the interaction of hereditary factors and vocational specialization makes our mental powers flow in unique ways. For example, few of us have ever crafted a double bass, but if we had done this for years, our cognitive patterns and strengths would be quite different than they are.

The Multidirectionality of Intelligence

No matter which particular list of intelligences one considers, it is clear that the abilities involved are **multidirectional** in that they can follow different trajectories with age. Thus short-term memory generally falls quite steadily, while vocabulary generally rises. Other abilities—such as mathematical reasoning—might rise, fall, and rise again, depending on how much they are used in daily life. Still others might hold steady until a sudden drop occurs because of such factors as illness and depression. Since virtually every pattern is possible, it is misleading to ask whether intelligence, in general, either increases or decreases. An either/or question is too simplistic.

Interindividual Variation

One reason many patterns exist, of course, is that each individual is genetically unique and has unique experiences, and both of these factors affect the pattern of a particular person's intellectual development (see A Closer Look, p. 576). The result of this **interindividual variation**, as shown by longitudinal research, is that some individuals decline in some or all mental abilities by age 40, while others are just as capable at age 70 as they were at earlier ages.

Often interindividual variations are related to changes in family and career responsibilities. The middle-aged housewife who finds that an "empty nest" brings empty days, and who suffers from, say, hypertension, might begin to show noticeable declines in certain kinds of intellectual functioning. On the other hand, someone who replaces a stimulating career as a parent or a worker with some equally stimulating form of retirement is likely, given continued good health, to maintain fairly steady levels of cogni-

RESEARCH REPORT

Adaptive Competence in Adulthood

Evidence for a slowing down of behavior and thinking in later adult years does not tell the whole story of cognitive change, because adults often find ways to compensate for their loss of speed and efficiency. Indeed, many researchers believe that a hallmark of "successful aging" is the ability to strategically use one's intellectual strengths to compensate for declining capacities associated with age (Charness, 1989; Dixon, 1992; Salthouse, 1987). Paul Baltes and his colleagues call this **selective optimization with compensation**, and they believe that it accounts for the ability of many older adults to maintain the levels of performance of their younger years (Baltes, 1987; Baltes & Baltes, 1990). This kind of adaptive competence can be found not just in research laboratories but in everyday workplaces and recreational settings as well.

Take waiting on tables in a restaurant, for example, a job that demands a wide range of cognitive skills, including knowledge of menu items, memory for ordering and delivery procedures, simultaneous management of several tables (each at a different stage of the meal), the ability to combine, order, and prioritize different tasks, and monitoring social relations with customers and coworkers—as well as physical stamina! Adolescents and young adults in these roles have an advantage over much older adults in their physical dexterity and endurance and in their cognitive speed and flexibility, and this may account for why restaurant managers sometimes prefer younger employees. But are older employees less efficient, or do they have ways of compensating for the declines they experience?

Marion Perlmutter and her colleagues sought an answer to this question by analyzing the skills required for successful performance in restaurant work and then assessing these skills in a sample of sixty-four restaurant and cafeteria employees who varied in age and prior work experience (Perlmutter et al., 1990). These workers were assessed on tests of memory ability, physical strength and dexterity, knowledge of the technical and organizational requirements of the job, and social capacities. They were also observed during different periods of the work day, such as during "rush" and "nonrush" hours, to determine their effectiveness. Perlmutter and her colleagues wanted to know if younger and older employees would differ in their overall job performance—and if so, whether this difference was due to differences in their physical and cognitive skills, their work experience, or both.

They were surprised to discover that, independent of age, the amount of prior work experience had little impact on the employees' work performance or on their physical or cognitive skills. Apparently, after one has learned the basic requirements of the job, additional experience does not necessarily yield better performance (Ceci & Cornelius, 1990). However, the employee's age (independent of prior experience) made a significant difference. Younger employees, as expected, had better physical skills, better memory abilities, and greater efficiency in computation (such as when calculating the check). Nevertheless, when the two groups were compared on their work performance, older employees outperformed their younger counterparts in the number of customers served during rush and nonrush periods of the day.

Perlmutter found that this was consistent with the reports of some of the restaurant managers she interviewed. For example,

> it was consistently reported that older workers chunk tasks to save steps by combining orders for several customers at several tables and/or by employing time management strategies such as preparing checks while waiting for food delivery. . . . Although younger experienced food servers may have the knowledge and skills necessary for such organization and chunking, they do not seem to use the skills as often, perhaps because they do not believe they need to.

Thus older employees devised cognitive strategies to compensate for the narrowing of some of their other job-related skills. These researchers concluded "that this evidence of adaptive competence in adulthood represents functional improvements that probably are common, particularly in the workplace." And indeed, this appears to be so. Salthouse (1984) found, for example, that older skilled typists could perform at speeds comparable to younger typists by using cognitive strategies (such as reading ahead) to compensate for age-related declines in their perceptual motor skills. Through such selective optimization with compensation, older adults like these learned to maintain their performance levels, but through strategies that were well-suited to the changing complexion of their cognitive capacities.

selective optimization with compensation The individual's devising of alternative strategies to compensate for age-related declines in ability.

tive functioning well into late adulthood. Educational level, income, and marital status are additional contributors to individual differences in intellectual aging: to be well-educated, financially secure, and happily married to a stimulating spouse has intellectual as well as emotional benefits (Schaie, 1990a).

A CLOSER LOOK

Individual Profiles of Adult Intellectual Change

The portrayal of adult intelligence as multidimensional, multidirectional, and individually variable can seem terribly abstract when it is based on group averages. But one of the advantages of longitudinal studies is that individual profiles of intellectual change can be separated out from group trends, allowing the reasons for growth, decline, and stable functioning to be examined on a case-by-case basis. Case studies like these reveal that the reasons for intellectual growth or decline are complex and multifaceted, based on the unique experiences of adult life.

Using data from his Seattle Longitudinal Study, K. Warner Schaie (1989) has provided the following portrayals of changes in word recognition (a measure of crystallized intelligence) in two pairs of adults of comparable age. More important, information about each person's occupation, health, marital status, and significant life events was used to assess why patterns of growth, decline, or stability emerged.

> The first two profiles [figure (a)] represent two young-old women who throughout life functioned at very different levels. Subject 155510 is a high school graduate who has been a homemaker all of her adult life and whose husband is still alive and well-functioning. She started our testing program at a rather low level, but her performance has had a clear upward trend. The comparison participant subject (154503) had been professionally active as a teacher. Her performance remained fairly level and above the population average until her early sixties. Since that time she has been divorced and retired from her teaching job; her performance in 1984 dropped to an extremely low level, which may reflect her experiential losses but could also be a function of increasing health problems.
>
> The second pair of profiles [figure (b)] shows the 28-year performance of two old-old men now in their eighties. Subject 153003, who started out somewhat below the population average, completed only grade school and worked as a purchasing agent prior to his retirement. He showed virtually stable performance until the late sixties; his performance actually increased after he retired, but he is beginning to experience health problems and has recently become a widower, and his latest assessment was below the earlier stable level. By contrast, subject 153013, a high school graduate who held mostly clerical types of jobs, showed gain until the early sixties and stability over the next assessment interval. By age 76, however, he showed substantial decrement that continued through the last assessment, which occurred less than a year prior to his death.

Looking at the relative scores of each pair of subjects at their first testing, it is clear that no one could have predicted their later-life intellectual performance solely on the basis of the changes in group averages that occur with increasing age. For each pair, the influences of education, occupation, and idiosyncratic events like a divorce, problems with physical health, and death of a spouse contributed to unique profiles of intellectual growth. For one, in fact, intellectual *growth* rather than decline was the rule in later life. The lesson: intellectual changes are complexly interwoven into the variety of changing life circumstances that a given individual experiences during the adult years, resulting in variations in intellectual development that defy precise prediction.

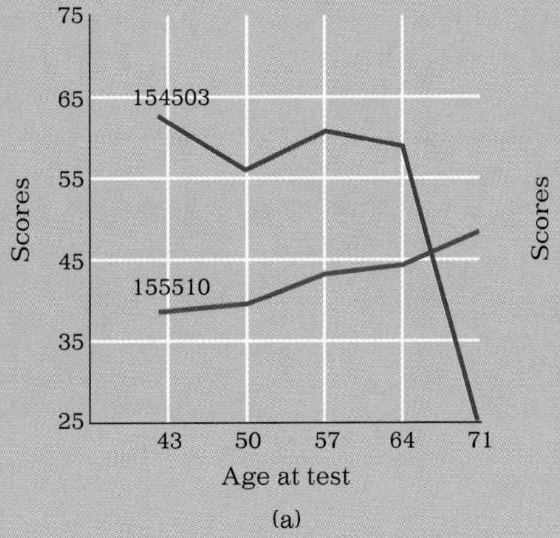

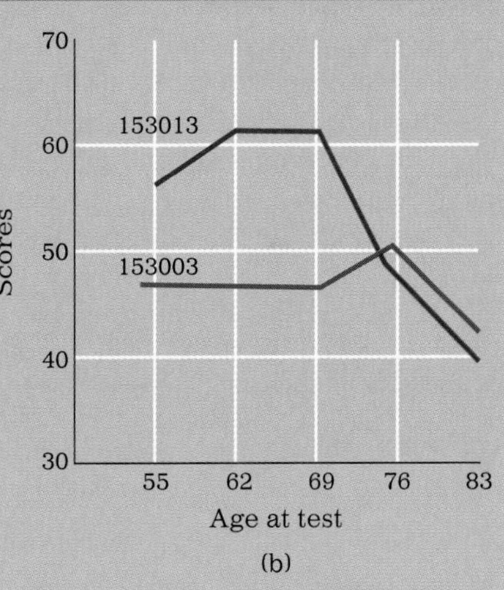

These figures index changes in word recognition scores (which are used as a measure of crystallized intelligence) for two pairs of comparable adults over time. Notice how distinctly different the profiles of individual change for each person are—even though each is the same age and part of the same birth cohort. These differences underscore how much intellectual change in adulthood is affected by occupational, marital, health, and other experiences that vary from one person to another.
Source: Schaie, 1989.

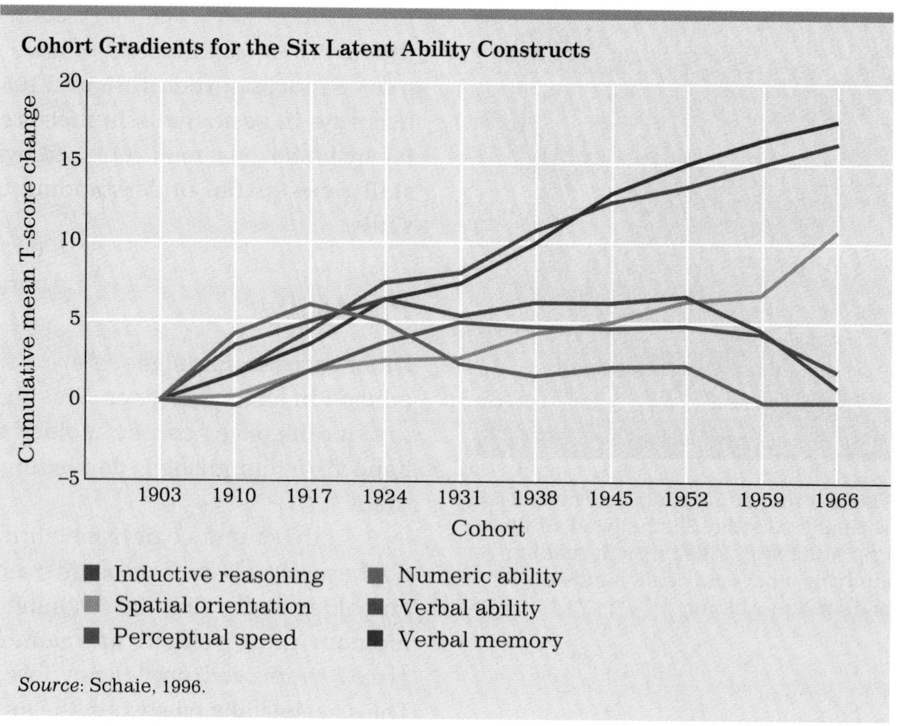

FIGURE 21.3 Schaie's multicohort, multiperiod research enables him to compare adult intellectual abilities among people born in 1903 and every seven years after that, including those who are young adults in the 1990s. The results show that most intellectual abilities have actually improved in recent decades, the product of an educational system that stresses reasoning and verbal communication more than quick answers or memorization.

Cohort Differences

As noted earlier, the particular gains and losses that adults demonstrate in their cognitive development are partially affected by factors specific to their cohort. The clearest evidence for cohort differences in the patterns of cognitive growth has come from the Seattle Longitudinal Study of Americans born over a seventy-seven-year period. As you can see in Figure 21.3, each successive cohort scored higher in the two abilities currently most prized among educators, Verbal Meaning and Inductive Reasoning. The most probable explanation for these results is that the later cohorts had more years of schooling and were more likely to have had teachers who encouraged them as students to think for themselves and express their ideas rather than to memorize facts and others' opinions, as teachers of earlier cohorts tended to do. The improvements for recent cohorts are reflected in scores on IQ tests as well, which have shown a steady upward drift over most of the twentieth century (Neisser et al., 1996).

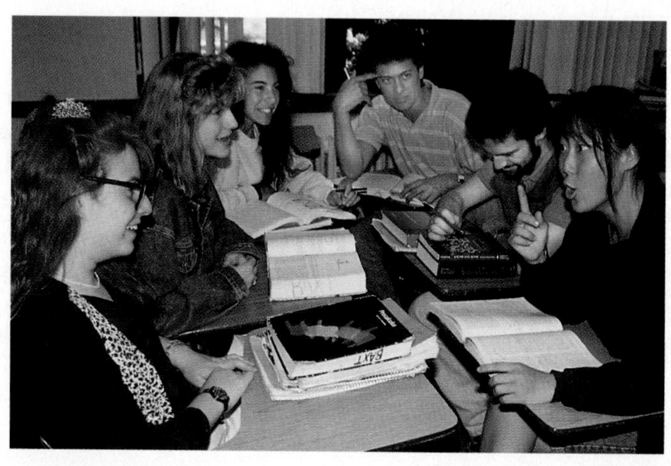

All the signs of good education by modern standards are apparent—an impassioned and articulate student, an attentive small group, and various reference books and papers. Such classroom activities prepare students well for the modern workplace, and would have been totally out of place in the schools of seventy years ago, when the goal was obedient assembly-line employees. As a result, students today score higher on analytic and communicative abilities and lower on memory and arithmetic measures of intelligence.

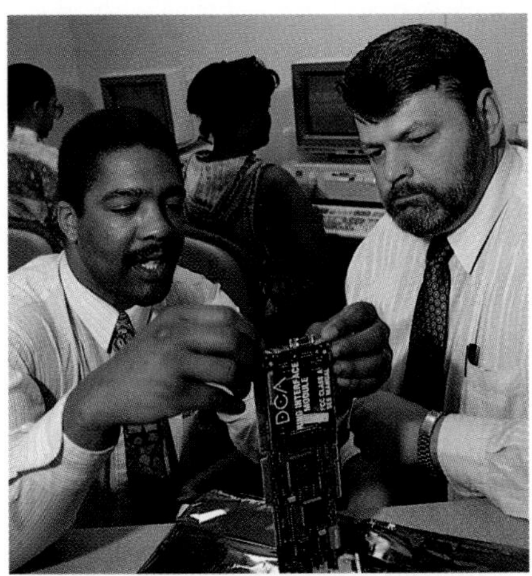

The fact that adult minds are plastic is verified every day. Virtually all adults upgrade their skills through reading, studying, and, best of all, hands-on, one-on-one training.

Changes in teaching strategy are also a likely explanation for the fact that arithmetic skills tend to have peaked in the cohorts who attended primary school between 1915 and 1935, with later generations not as quick or accurate in basic math. In more recent years, drill in multiplication tables, for instance, has been replaced by "new math"—to the benefit of reasoning skills and spatial understanding but to the detriment of basic arithmetic skills.

Plasticity

Differences in intelligence are not inflexible: they can be altered through education and experience. This characteristic is called **plasticity**, to suggest that intelligences can be molded in many ways. Abilities can become enhanced or diminished, depending on how, when, and why a person uses them.

The fact that cognitive abilities are plastic means that they can also be deliberately altered through training. Numerous studies of middle-aged and older adults find that training in a specific area—such as techniques of memory improvement or mathematical problem solving—can result in greatly improved proficiency levels (Schaie, 1996; Verhaegen et al., 1992). This is especially true when the area of training is relevant to the older person's actual life. To be sure, intellectual plasticity declines to some extent with increasing age (Baltes & Lindenberger, 1988). But a capacity for change and improvement is always there. (Some of the specifics of the research on this topic are reviewed in Chapter 24.)

Expertise

Recognizing the plasticity of intelligence, some developmentalists believe that as we age, we each develop **expertise** at whatever is important to us—in our work, in our leisure activities, in our relationships with others. That is, we tend to become selective "experts," developing specialized competencies in activities that are personally meaningful, whether fixing a car, preparing gourmet meals, diagnosing an illness, or mastering fly fishing (Dixon et al., 1985). This helps to explain the multidimensional and individualized patterns of adult intellectual skills, as adults acquire expertise in some areas while remaining novices in others.

plasticity In developmental psychology, a term used to indicate that a particular characteristic is shaped by many environmental influences. Many mental abilities, once thought to be firmly fixed before adulthood, are now known to have much more plasticity than was once believed.

expertise The acquisition of knowledge in a specific area. As individuals grow older, they concentrate their learning in certain areas that are of the most importance to them, becoming experts in these areas while remaining relative novices in others.

What Makes an Expert?

When developmentalists use the term "expert," they do not mean someone who is extraordinarily gifted at a particular task. They simply mean someone who is significantly better at a task than people who have not put time and effort into doing that task. The difference between experts and novices cannot be reduced merely to differences in the amount of experience and knowledge, however. Research suggests that there are several distinctions to be made between those who are experts and those who are less skilled (Charness, 1986, 1989; Chi et al., 1988; Rybash et al., 1986; Salthouse, 1985, 1987; Walsh & Hershey, 1993).

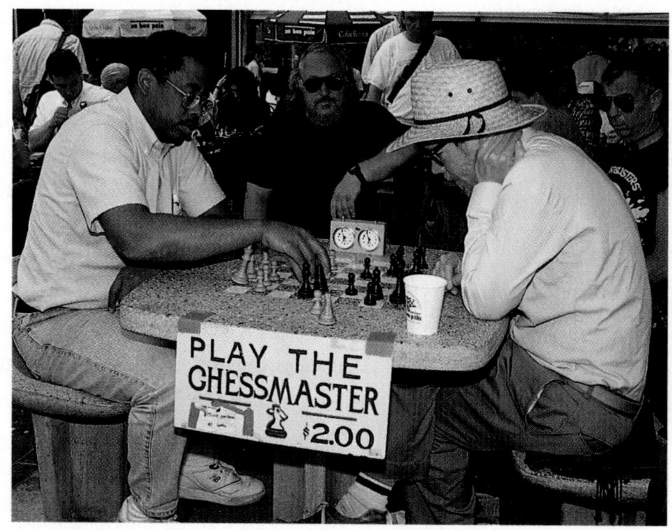

Learning the basic moves in chess is relatively simple: an adult can do it in a few hours. Becoming a master, however, takes many years and is directly related to time spent playing. Expertise—in chess, in music, in sports, and probably in almost any other human endeavor—is strongly correlated to hours of practice, the result of both opportunity and motivation. Future chess masters spend time every day playing with rivals, playing alone, and reading up on, discussing, and even dreaming about, chess.

Expert vs. Novice

First of all, novices tend to rely on formal procedures and rules to guide them. Experts, on the other hand, rely more on their accumulated experience and on the circumstances of the immediate context, and they are therefore more intuitive and less stereotyped in their performance. For example, when they look at X-rays, expert physicians interpret them more accurately than do young doctors, though often they cannot verbalize exactly how they arrived at their diagnosis. As one pair of researchers explain:

> The expert physician, with many years of experience, has so "compiled" his knowledge that a long chain of inference is likely to be reduced to a single association. This feature can make it difficult for an expert to verbalize information that he actually uses in solving a problem. Faced with a difficult problem, the apprentice fails to solve it at all, the journeyman solves it after long effort, and the master sees the answer immediately. [Rybash et al., 1986]

In much the same way, the expert artist, or musician, or scientist is not simply a practiced technician but is an intuitive creator as well (Charness, 1986; Csikszentmihalyi, 1996).

Second, many elements of expert performance are automatic; that is, the complex action and thought they involve have become routine, making it appear that most aspects of the work in question are performed "instinctively." In fact, experts process incoming information more quickly and analyze it more efficiently than nonexperts, and then they act in well-rehearsed ways that make their efforts appear almost nonconscious.

Third, the expert has better strategies, and more of them, for accomplishing a particular task. In fact, this may be the crucial difference between a skilled and an unskilled person. The superior strategies of the expert also permit the selective optimization with compensation examined in the Research Report on p. 575. As noted there, many developmentalists regard the capacity to accommodate to changes in ability over time as an essential element in "successful aging" (Charness, 1989; Charness & Bosman, 1990).

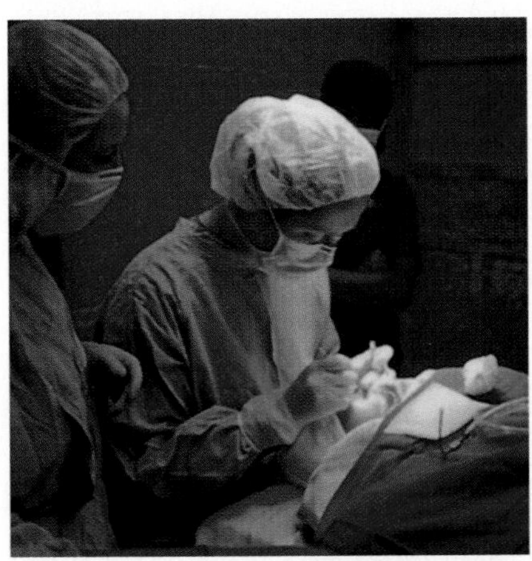

The importance of experience is obvious for a surgeon. Almost anyone would agree that if a child needs to have a cleft palate fixed (as shown here), the more practice the surgeon has had with this particular operation, the better. Less obvious, but equally valuable, is expertise in other areas. For example, for the auto mechanic, the airplane pilot, the cook, and the parent, years of practice produce a combination of intuition, creativity, and wisdom that makes the job easier and the results better.

Finally, perhaps because of all these differences, experts are more flexible. The expert artist, musician, or scientist, for example, is more creative and curious in his or her work, deliberately experimenting and enjoying the surprise when things do not go according to plan (Csikszentmihalyi, 1996; John-Steiner, 1986). Another example of the flexibility arising from expertise comes from surgeons (Salthouse, 1985). Obviously, since no two patients are exactly alike, every type of operation has the potential for sudden, unexpected complications. Not only will the expert surgeon be more likely than the novice to notice little telltale signs (an unexpected lesion, an oddly shaped organ, a rise or drop in a vital sign) that may signal a possible problem, but the expert will be better able to deviate from standard textbook procedure to devise strategies to overcome the problem. Similarly, experts in all walks of life seem better able to adapt to individual cases and exceptions to the rule—somewhat like an expert chef who adjusts ingredients, temperature, technique, and timing as things develop, and virtually never follows a recipe exactly.

How, then, can we answer the question with which we began this chapter? Perhaps the best short answer is that adults both gain and lose mental powers in a lifelong process (Schultz & Heckhausen, 1996). As the longitudinal evidence reviewed in the first half of the chapter indicates, certainly on some basic abilities, many adults improve over most of adulthood, and show no decline by age 60. Indeed, as you will see again in Chapter 24, some people never show evident decline and continue to master new areas of knowledge even in late adulthood. The suggestion that we are most likely to improve and develop expertise in those areas that capture our attention and intelligences is also heartening.

Nevertheless, there is strong evidence for certain age-related cognitive declines. Adults do not learn new material as quickly in middle age as they did at age 20, which means that, when learning a new skill, older adults need to be more selective, to be more motivated, and to allow more time for practice if they hope to achieve as well as the younger trainee. On the whole, however, it does seem clear that individual differences are much more critical in determining the course of cognitive development than is chronological age alone. In other words, you would be better able to predict an adult's intellectual competencies by sampling his or her background, interests, and motivation than by knowing his or her age. And, as we will see in Chapter 24, even in old age, many cognitive declines seem more closely related to particular personal circumstances, such as health and social context, than to age.

SUMMARY

Decline in Adult Intelligence?

1. On the basis of many large cross-sectional studies, psychologists once believed that intelligence inevitably declined in adulthood. Within the past forty years, longitudinal research has led to the opposite conclusion, that intelligence may improve during adulthood.

2. Cross-sequential research, which attempts to distinguish the general aging process from the specific experiences of each generation, finds that cohort effects as well as chronological age have considerable impact on measurements of intelligence during middle adulthood.

3. Despite evidence for cognitive growth during middle adulthood, the impact of age on intelligence is still controversial. Some psychologists believe that while crystallized intelligence, which is based on accumulated knowledge, increases with time, one's fluid, flexible reasoning skills inevitably decline with age.

4. For the most part, researchers agree that, overall, intellectual abilities increase slightly until midlife and remain relatively stable through the rest of middle adulthood. However, decrements involving speed of thinking may begin to appear as early as age 30. Researchers also agree that there is considerable variation in the specific patterns of intellectual development and that a key factor in this variation is education.

Multidimensional Intelligence: Not One Intelligence, but Many

5. Most researchers now think that, rather than there being a single entity of general intelligence, there are several distinct intelligences.

6. The study of practical intelligence is based on the view that typical measures of adult intelligence do not include the skills employed in everyday problem solving. Most adults believe that their practical intelligence improves as they grow older, and research supports this belief.

The Multidirectionality of Intelligence

7. Adult intellectual competence is multidirectional, characterized by interindividual variation and plasticity. Its demonstration can be affected by cohort differences related to past education.

Expertise

8. As people grow older, they may become more expert in whatever types of intelligence or skills they choose to develop. Meanwhile, abilities that are not exercised may fall into decline.

9. In addition to being more experienced, experts are better thinkers than novices in many ways. Experts are more intuitive and flexible, and they use better strategies to perform whatever task is required. Their cognitive processes are more specialized, as well as automatic, often seeming to require little conscious thought.

KEY TERMS

Wechsler Adult Intelligence Scale (WAIS) (566)
Seattle Longitudinal Study (567)
fluid intelligence (568)
crystallized intelligence (568)
practical intelligence (573)
multidirectional (574)
interindividual variation (574)
selective optimization with compensation (575)
plasticity (578)
expertise (578)

KEY QUESTIONS

1. What evidence suggests that intelligence declines during adulthood?

2. What evidence suggests that intelligence increases during adulthood?

3. What are the advantages of longitudinal research on adult intelligence?

4. What are the advantages of cross-sequential research on adult intelligence?

5. How is fluid intelligence different from crystallized intelligence? How does each change in adulthood?

6. What are the four conclusions that researchers generally agree on regarding cognitive development in middle adulthood?

7. What differences would you expect to find in your own intelligence ten years from now? Why would they occur?

8. What is practical intelligence? Why is it important to a complete picture of adult intelligence?

9. What are the differences between an expert and a novice?

Middle Adulthood: Psychosocial Development

CHAPTER

22

Although no dramatic biological shifts occur to signal it, our society recognizes a point called **midlife** at about age 40, when the average adult has about as many years of life ahead as have already passed. Midlife ushers in **middle age**, which lasts until about age 60.

Middle age is often said to represent a time of personal crisis, or, at the least, to be a troubling transition between the prime of life and the decline of old age, as the individual measures his or her goals, accomplishments, and commitments in terms of "time left" in life. The popular image of midlife upheaval—standardly referred to as "the midlife crisis"—depicts last-ditch efforts to recapture youth and last-chance efforts to start a new life. However, as you will see, in midlife, continuities of love relationships, family commitments, work involvements, and personality patterns often seem more salient than any changes that occur.

Changes During Middle Age

That middle age might be a time of crisis is certainly a plausible idea, for a number of potentially troubling personal changes often cluster in the 40s. The most obvious is simply that the social clock is signaling that one is beginning to grow old. Beginning with the "Big Four-O," birthdays tend to be seen in a new perspective—a measure less of time lived than of time remaining. This perceptual shift is often highlighted by the death or serious illness either of a close relative from the next older generation—perhaps a parent or a favorite aunt or uncle—or of a friend or colleague, possibly someone only a few years older than oneself. Such events bring not only feelings of personal loss but also thoughts about one's own mortality (Katchadourian, 1987).

For many middle-aged parents, an additional source of upheaval is the need to make important, and not always easy, adjustments in their parental roles. Typically, at just about the time that parents enter midlife, children become adolescents, demanding greater independence and putting their parents' authority, and sometimes their values, to the test. No sooner have parents adjusted to these changes than their children set out on their own as adults, perhaps distancing themselves from their parents initially and then calling forth a different form of nurturance and closeness when they make their parents become parents-in-law and grandparents.

midlife The point, at about age 40, when the adult has approximately as many years of life ahead as have already passed. Midlife ushers in middle age, which lasts until about age 60.

middle age The years from age 40 to age 60.

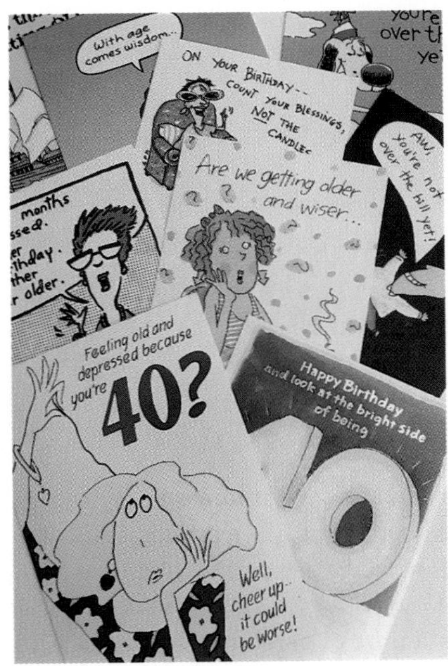

As these birthday greetings suggest, the advent of middle age is assumed to be a source of high anxiety and turmoil. For most people, however, the actual experience of middle age is not strikingly different from that of early adulthood and seldom produces anything like the proverbial "midlife crisis."

Further, for many adults, midlife is a time to reexamine whatever earlier choices they may have made regarding intimacy and generativity. Many who thought that time was on their side in the quest for certain life goals suddenly see the door to the future closing. Some reassess the balance between work and family: those who have been single-mindedly pursuing a career often become concerned about the loved ones they have neglected; those who have devoted themselves to child-rearing often worry about what they will do when their children are gone. Marriages may also be examined, perhaps with the thought that middle age is the last chance for extramarital affairs, divorce, or remarriage.

All these midlife shifts are recognized by most developmentalists who study adulthood. However, few developmentalists today believe that such changes commonly produce a midlife crisis involving radical reexamination and sudden transformation (Hunter & Sundel, 1989; Wrightsman, 1994). Instead, a wealth of research reveals a great deal of developmental variability among people the same age and a great deal of developmental continuity within each individual. In short, how people react to the challenges and changes in middle adulthood has much more to do with their overall developmental history than it has to do with calendar milestones. As two experts in the psychology of aging comment, chronological age, as a predictor of development, is like "an initially appealing false lover who tells you everything and nothing" (Birren & Schroots, 1996).

Personality Throughout Adulthood

Throughout adulthood, the major source of developmental continuity is the stability of personality. Paradoxically, the stability of personality is also a major reason for developmental discontinuity when it occurs. Our personality traits lead us to seek, interpret, and then react to life events in ways that are distinctly our own. Longitudinal research over the decades of adulthood has found that basic personality traits endure despite the external vicissitudes of life. As two leading researchers explain:

> [P]eople undoubtedly do change across the life span. Marriages end in divorce, professional careers are started in mid-life, fashions and attitudes change with the times. Yet often the same traits can be seen in new guises: Intellectual curiosity merely shifts from one field to another, avid gardening replaces avid tennis, one abusive relationship is followed by another. Many of these changes are best regarded as variations on the "uniform tune" played by individuals' enduring dispositions. [McCrae & Costa, 1994]

To better understand this, we need to draw a distinction between the basic personality traits that seem stable throughout life and the variable expression of them, an expression that is affected by maturation and experience as well as culture.

Stable Traits: The Big Five

Big Five The five basic clusters of personality traits that remain quite stable throughout adulthood: extroversion, agreeableness, conscientiousness, neuroticism, and openness.

Extensive longitudinal and cross-sectional research among men and women of many nations and ethnicities finds five basic clusters of personality traits—referred to as the **Big Five**—that remain quite stable throughout adulthood (Digman, 1990; Eaves et al., 1989; Loehlin, 1992; Paunonen et al., 1992).

The Marine sergeant in the top photo and the park ranger below have much in common. Both are middle-aged women, serving their country in uniform. However, if their expressions here are at all indicative, they have quite different personality traits, and each has picked a vocational niche that is compatible with those traits and reinforces them—a niche in which the other would probably feel quite out of place.

Although various experts use somewhat different terms to describe each of these clusters, there is a consensus that they include the following traits:

1. *extroversion*—the tendency to be outgoing, assertive, and active;
2. *agreeableness*—the tendency to be kind, helpful, and easygoing;
3. *conscientiousness*—the tendency to be organized, deliberate, and conforming;
4. *neuroticism*—the tendency to be anxious, moody, and self-punishing;
5. *openness*—the tendency to be imaginative, curious, and artistic, welcoming new experiences.

Whether a given individual ranks high or low in each of the Big Five is determined by the interacting influences of genes, culture, early child-rearing, and the experiences and choices made during late adolescence and early adulthood. In childhood, adolescence, and the first years of adulthood, there is likely to be some fluctuation in the strength of these traits, as the social context and personal choices can evoke new personality patterns that were not apparent in early childhood. However, by about age 30, the Big Five usually become quite stable, and remain so throughout the life span.

Certainly one reason for this stability is genetic, with some personality characteristics already apparent in infancy. Other predispositions may not be triggered until puberty, perhaps, but they still can reflect the unfolding of genetic traits. By adulthood, the interaction of genetic predispositions and environmental influences has produced a personality that is relatively stable.

An equally influential reason for this stability is that, by age 30, most people have created and settled into an **ecological niche**—a particular lifestyle and social context, including vocation, mate, neighborhood, and daily routines—that evokes and reinforces their particular personality needs and interests. The fact that people choose their surroundings to suit their temperament leads two personality researchers to quip: "Ask not how life's experiences change personality; ask instead how personality shapes lives" (McCrae & Costa, 1990).

Examples of Niche-Picking

Using some simplified examples, let us illustrate how an individual's ecological niche is shaped by, and might interact with, his or her particular traits. For instance, by age 30, those high in extroversion would likely have found mates who share, or at least appreciate, their outgoingness, and they would have established a busy social life with a wide circle of friends and acquaintances. Their jobs would allow them to interact with many people, perhaps working as a salesperson, a politician, a teacher, a personnel director. Similarly, the details of their lifestyle would foster social contact: they would probably have phones in every room that would ring often (with call-waiting, call-transfer, and answering machines), as well as e-mail. They would probably live in a busy neighborhood, join an amateur sports league or social club, volunteer at their local community center, and engage in occasional adventures such as group mountain-hiking or bicycle touring.

ecological niche The particular lifestyle and social context adults settle into that are compatible with their individual personality needs and interests.

While both of these explanations of gender convergence may seem intriguing, longitudinal survey research suggests that a cohort explanation may be the most salient. It is true that many middle-aged and older adults are less tightly bound by gender roles than they were when they were younger. However, over the twentieth century, every decade has witnessed a loosening of gender restrictions (Snarey, 1993). Thus some of the self-reported gender crossover may actually be a trend experienced by adults of all ages but felt more strongly by older adults who remember the more rigid roles they played when younger. The current cohort of middle-aged adults seems less marked in their convergence of sex roles because male and female roles are already less sharply defined than they once were (Helson et al., 1995; Moen, 1996).

Family Dynamics in Middle Adulthood

Family ties across the generations are particularly important for those in the middle-aged generation, precisely because they are the "generation in the middle," between aging parents and adult children. Often they have grandchildren and sometimes grandparents as well.

It is easy to underestimate the critical family role of the middle generation in modern Western cultures. This is because, in comparison with the traditional extended family found in many other cultures—in which several generations live under the same roof and are intensely involved in each others' lives—the various generations of the modern family can seem geographically and emotionally remote from each other. Superficially, statistics back up this notion: by age 50, for example, most American adults today live only with a spouse or a friend from the same generation rather than with their aging parents or adult children.

However, the fact that the generations do not sleep under the same roof does not mean that family links are weak. Researchers in many developed countries find that relatives typically have frequent contact with each other. They stay in touch by telephone and travel (much easier and less expensive than in earlier times) and provide each other with substantial help, ranging from advice and emotional support to gifts, loans, babysitting, home

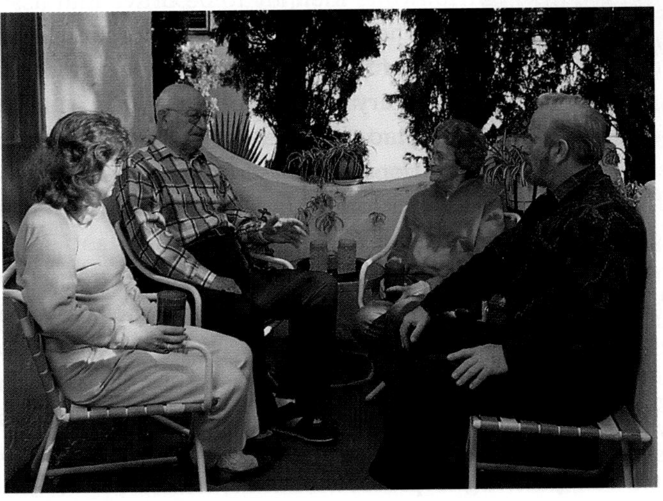

Years after the nest has emptied, the links between parents and their adult children generally remain strong, with frequent communication between the generations and regular visits. In many cases, the relative independence of the two generations works to make their relations smoother than they were when everyone was sharing the same household.

repair, and health care (Barresi & Menon, 1990; Crimmims & Ingegneri, 1990; Farkas & Hogan, 1995). Indeed, while it is tempting to idealize the notion of households in which relatives of various ages live together, the truth is that many contemporary family members are more supportive of each other, with less tension and trouble, when they live apart than when they live together (Coward et al., 1992; Fry, 1995; Umberson, 1992).

Middle-Aged Adults and Their Aging Parents

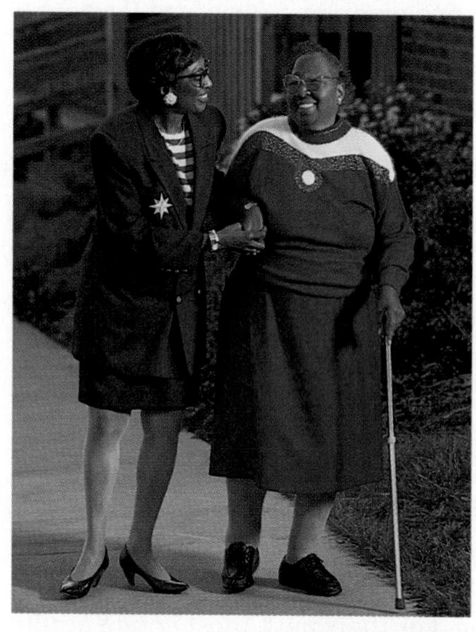

Many adults find their understanding and love of their parents increasing as they themselves become older. Mother-daughter and father-son relationships are particularly likely to become closer than they were when the child was an adolescent.

The relationship between most middle-aged adults and their parents improves with time. One reason is that, as adult children mature, they develop a more balanced view of the relationship as a whole, especially with regard to their years of growing up. This change is particularly apparent in men who had a tendency as adolescents and young adults to blame their fathers for not having been sufficiently helpful and understanding. In middle age, they are likely to reevaluate the "old man" and become more appreciative of his good qualities and more accepting of his limitations (Farrell & Rosenberg, 1981). Women, too, become more appreciative of the older generation, as they better understand their own role in safeguarding family traditions and maintaining the links between the generations (Helson & Moane, 1987). Moreover, the perspective of time can lead to a measure of forgiveness, as both generations acknowledge past mistakes in their relationships with each other. The middle-aged adult may be aided in this by having gained a firsthand understanding of the pressures of being a parent.

The improvement in relationships between middle-aged children and their parents is particularly likely to occur in current times, because most of the elderly are healthy, active, and independent. They typically prefer not to live with their adult children, and, thanks to a changed economic picture in the past several decades, most of them can afford to live on their own. This provides a measure of freedom and privacy that enhances the relationship between the two generations.

Positive relationships between the middle and older generations are also apparent in subcultures in which intergenerational dependence is encouraged and coresidence is still the custom. For example, Hispanic- and Asian-Americans are much more likely to live in three-generation families, for the most part harmoniously. When conflicts do occur, the source is usually either a very dependent and demanding elderly person or a rebellious teenager, not the generation in the middle (Johnson, 1995; Mindel et al., 1988).

As alluded to above, the middle generation typically assumes the role of maintaining the links between the generations. In effect, the middle generation becomes the **kinkeepers**, maintaining communication from one relative to another, planning and hosting family get-togethers, providing help if someone is sick or in some other crisis. Traditionally, this role has been taken on by the middle generation, in part because they are past the labor-intensive stage of raising young children and in part because they have not yet experienced any of the diminishing of strength or resources that may occur in old age. In the past, the role of kinkeeper was much more likely to have been filled by a woman than by a man. Although this continues to be true, in today's world, middle-aged men are also likely to participate in kinkeeping.

kinkeepers The people who celebrate family achievements, gather the family together, and keep in touch with family members who no longer live nearby.

Middle-Aged Adults and Their Adult Children

Parents generally maintain close relationships with their children, even when the latter are fully grown, independent, and living away from the family home. A seven-nation survey of thousands of middle-aged adults found that the majority (75 percent) contacted their adult children several times a week and only a tiny minority (less than 5 percent) communicated with them only once a year or less. Most adults of both sexes and both generations said that they would turn to each other for help, in everything from handling emotional or financial problems to arranging furniture (Farkas & Hogan, 1995).

Typically, financial help and a variety of other support services (from home repair to washing the laundry) are more likely to flow from the middle-aged parents to the young adults than in the other direction. However, middle-aged parents benefit in many other ways. They usually take great pride in their adult children's accomplishments, and their self-esteem is enhanced by the fact that their offspring are successful adults (although the parents are sometimes less than enthusiastic if their grown children are markedly more successful, confident, and happy than they themselves were at the same age) (Ryff et al., 1994).

Middle-aged parents also benefit from the younger generation's serving as a "cohort bridge," a source of information and advice about new developments in the culture (Green & Boxer, 1986). Many a middle-aged woman has gone to college because her adult children urged her to; many a middle-aged man has begun to take better care of himself because his children provided specific health information. For their part, most young adults benefit not only from the material aid and advice their parents provide but also from the self-confidence that results from realizing that their parents treat them as adults.

Middle-Aged Adults and Their Partners

Having an intimate relationship over the years of middle adulthood is a source of happiness, comfort, and self-respect. For the majority of middle-aged adults, their most intimate relationship is with their spouse, who is also their closest friend. For a growing minority of divorced or never-married middle-aged adults, intimacy is achieved through cohabitation with a partner. In the 1990s, as many American 40- to 60-year-olds are living with opposite-sex mates as 20- to 40-year-olds are, and the number of gay and lesbian middle-aged couples living together also seems to be increasing (Chevan, 1996; Huyck, 1995).

Many of the remaining single, middle-aged adults find intimacy in a close romantic relationship that does not include formally sharing a household. For a number of reasons, research on the exact developmental impact of these nonmarital relationships is sparse, so it is difficult to draw conclusions about them. However, it is clear that adults who choose to be in close relationships are better off, financially and emotionally, throughout the life span than adults who tend to remain unattached (Antonucci & Akiyama, 1995; Ross, 1995). Now let us look at the close relationship that has been intensely studied, marriage in middle age.

Marriage

Throughout adulthood, in every nation, marriage is the family relationship that seems most closely linked to personal happiness, health, and companionship (Hu & Goldman, 1990; Myers, 1993). This does not mean that single people are necessarily unhappy, or that married people are necessarily happy. Indeed, in recent decades, the "happiness gap" between married and single adults has narrowed: while about 35 percent of the married adults over age 40 claim they are very happy, so do 20 percent of their never-married peers, double the rate of over twenty years ago (Lee et al., 1991).

Generally, the marital relationship is a potent source of self-esteem, with a person's overall happiness strongly correlated with his or her marital happiness (Myers, 1993). The power of a marriage in this respect can be measured by the decline that occurs in overall happiness when a marriage becomes seriously troubled. Individuals in a long-term marriage are particularly vulnerable when deep marital rifts occur, experiencing greater loneliness, lower self-esteem, and deeper depression than couples having equivalent troubles in a young marriage or a cohabiting relationship. This is probably so because the longer one has been married, the more one's self is defined by the relationship and the more threatening marital troubles are to one's sense of self-integrity (Helson, 1992; Keith & Schafer, 1991). In fact, being in an unhappy relationship—married or not—is even more likely to lead to feelings of sadness and despair than being unattached (Ross, 1995).

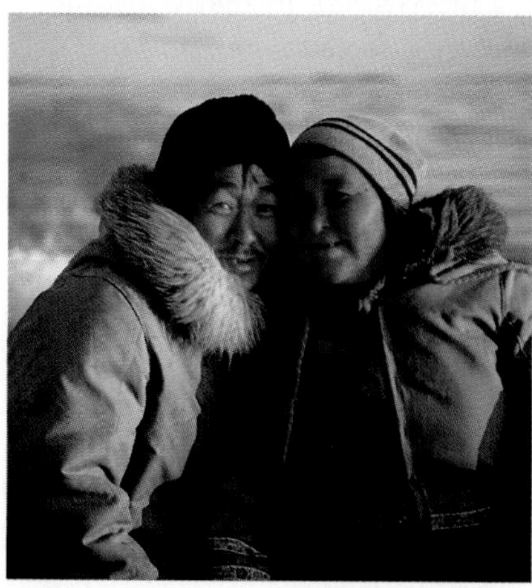

During the empty-nest period, many middle-aged parents regain their freedom to frolic, invigorating their marriage with renewed closeness and the sharing of activities that the earlier demands of work and family life may have curtailed.

Most of the research indicates that, after the first ten years or so, marital happiness is more likely to gradually increase than to decrease (Glenn, 1991). There are many possible reasons for marriages to be better as the decades pass by (Berry & Williams, 1987; Pina & Bengston, 1993; Ward, 1993). One is financial. Overall, stress and conflicts concerning money are a major source of marital tension, but by the empty-nest stage, financial pressures usually have lessened. At this point husbands have often reached a more predictable job pattern and employed wives can devote more time and energy to their work, thus increasing their pay. Meanwhile, there are fewer expenses, because the children are, at least partly, self-supporting. Another reason for marital improvement over the course of middle age is certain shifts in household patterns. Once the overwhelming demands of raising young children have subsided, or at least become routinized, fights over equity in domestic work are less pointed. Reduction of tension also results from the children's growing independence, which requires the parents to make fewer judgment calls over which they might disagree and provides more opportunities for a sense of parental accomplishment.

In addition, in typical marriages, doing things together—everything from fixing up the home to taking a vacation—contributes to marital satisfaction. If the children have left the nest and an employer no longer mandates extra hours at work, there is much more time for such activities. As a result, married couples can recapture some of the close companionship and marital intimacy that were difficult to sustain during the hectic child-rearing, career-building years.

As a qualification to all this good news, it should be noted that one other reason research on long-term marriages generally finds higher levels of satisfaction is that many unhappy marriages end before becoming "long-term." Thus any survey of long-term marriages involves a selective sample.

Divorce and Remarriage

Overall, the reasons for divorce in middle adulthood, and the problems it causes, are similar to those associated with divorce in early adulthood (see pp. 521–526). However, divorce in middle adulthood is generally more difficult because a divorce at this time is more likely to occur after years of marriage or after a second marriage—which means that the loss of self-esteem is usually greater. In addition, the chances for remarriage are slimmer.

Overall, most divorced people remarry—on average, within five years of being divorced. In fact, in the United States, divorced people are more likely to marry than single people who are the same age, and almost half of all marriages (46 percent) are remarriages for at least one of the spouses (U.S. Bureau of the Census, 1996). Remarriage is more likely if the divorced person is relatively young, in part because there are more potential partners still available, with the typical remarrying bride being age 34 and the groom age 37 (see A Closer Look). Thus a sizable percentage of middle-aged adults—more men than women—are in the early years of their second marriage.

Remarriage often brings initial happiness and other benefits: divorced women typically become financially more secure, and divorced men typically become healthier and more social once they have a new partner. In terms of parenthood, remarriage for men, in particular, often improves relationships with offspring, creates new bonds with custodial stepchildren, or occasions the birth of a baby—an event that typically strengthens the second marriage while loosening the holdover emotional bonds of the first.

Popular wisdom to the contrary, however, there is no guarantee for either sex that love is better the second time around: remarried people generally report lower average rates of happiness than people in first marriages, and their divorce rate by age 50 is 20 percent higher (Booth & Edwards, 1992; Glenn, 1991; U.S. Bureau of the Census, 1996). One reason is that some lonely divorced people marry too quickly, "on the rebound" (sometimes to the first person who seems interested). In many instances, however, an important factor is the disruptive effects of stepchildren. For remarriages involving stepchildren, one study found the divorce rate within three years after marriage to be 17 percent, compared with 10 percent for childless remarriages and 6 percent for first-time marriages (White & Booth, 1985). Other research confirms that stepchildren are a source of conflict between husband and wife but suggests that such conflict is lower in the early years of the marriage and is less likely to occur if the couple has a biological child of their own (MacDonald & Demaris, 1995).

From one perspective, the high divorce rate of the remarried may mean that these individuals have learned from their first marital experience and are quicker to recognize problems and thus are quicker to either resolve the conflict or sever the tie. That this may be the case is suggested by the fact that, compared with people in first marriages, remarried people are more likely to describe their marriages as either very happy or quite unhappy, with less middle ground (Ganong & Coleman, 1994).

It may also be that remarriages break up more often than first marriages because of the temperament or values of the people who get divorced and married more than once. That is, some individuals may be temperamentally prone to divorce (perhaps by virtue of being unusually impatient,

A CLOSER LOOK

Supply and Demand in the Middle-Age Marriage Market

In terms of finding a partner, middle-aged women are disadvantaged in the remarriage market, partly because middle-aged men tend to marry younger women and partly because men die at younger ages. For example, for every divorced man between the ages of 40 and 44, there are two unmarried (divorced, single, or widowed) women plus six such women between the ages of 30 and 39. By contrast, for every divorced women in the 40-to-44 age group, there is only one unmarried man of similar age and only two between ages 45 and 54. Because a middle-aged woman must compete for these men with all the other unmarried women of her age and younger, her chances of finding a marriage partner diminish with every passing year.

This unevenness of the sex ratio of available partners varies from subgroup to subgroup. Partly because single men are more likely to immigrate than single women, Hispanic women already in the United States can be more selective: 54 percent of all unmarried Spanish-surname adults living in the United States are men. Of course, this puts women living in Mexico, Puerto Rico, Cuba, and Latin America at an additional disadvantage.

By contrast, for African-American women, the marriage odds are stacked against them in three ways. First, only 46 percent of African-American adults are men. Added to that, 8 percent of all black grooms marry nonblack women, while only 4 percent of black brides marry outside their race. Finally, for every group, marriages are more likely between couples of equal educational status, or, if unequal, involve a better-educated man. This bias works against all educated women but is particularly pronounced among African-American women, since they are significantly more likely to attain a degree than African-American men are. The lower marriage rate of black women helps explain the fact that middle-aged black women tend to have higher friendship bonds and stronger cross-generational ties that other middle-aged persons (Sterk-Elifson, 1994). When Savannah, one of the four African-American women in the novel *Waiting to Exhale*, complains, "None of us has a man," Gloria responds, "Men ain't everything." Savannah continues:

> If I had a man and it was your birthday . . . and Robin and Bernie called me up to come over here to help you celebrate, I'd still be here girl. So don't ever think a man would have that much power over me that I'd stop caring about my friends. [McMillan, 1992]

Many men complained that the movie Waiting to Exhale *portrayed them unfairly. However, as in* The First Wives Club *and* Thelma and Louise, *a Hollywood distortion can capture an emotional truth: when men are scarce, insensitive, or abusive, women depend on each other.*

dissatisfied, or adventurous). This explanation is supported by the fact that the more often a person has been married, the more likely his or her current marriage is to end in divorce. For such people, divorce may be much less troublesome than having to accept a mate as he or she is.

Grandparenthood

Grandparenthood most often begins in middle age, although its timing and prevalence obviously vary somewhat from culture to culture and cohort to cohort. In the United States in 1995, almost two out of every three Americans become a grandparent between ages 40 and 60. Another smaller group become grandparents "off-time," typically before age 40 (about 10 percent overall) or occasionally after age 60. The rest never become grandparents, usually because they themselves are childless.

When the first grandchild arrives, almost every grandparent reacts to his or her new status with pride and wonderment. As one grandfather exulted, "Now I'm immortal." However, while *becoming* a grandparent is an affirmation, the actual experience of *being* a grandparent is much more variable, ranging from fulfilling to frustrating, from pivotal to peripheral. Generally, ongoing grandparent-grandchild relationships take one of three forms: *remote*, *involved*, or *companionate* (Cherlin & Furstenberg, 1986; Gratton & Haber, 1996).

The role of **remote grandparent** was the one typically adopted by American grandparents a century ago, and it is still prevalent in some traditional cultures. Remote grandparents are distant but esteemed elders who are honored, respected, and obeyed by children, grandchildren, and great-grandchildren. Often, remote grandparents control the family land or other wealth, and even when they do not, they may see themselves as the patriarch or matriarch of the family heritage, maintaining traditional values.

The **involved grandparent** is active in the day-to-day life of the grandchildren. Involved grandparents live in or near the grandchildren's household and see their grandchildren daily, filling up their "empty nest" with a new generation of fledglings or overcoming the loneliness of widowhood with close grandchild-grandparent bonds. This pattern, which was especially prevalent among grandmothers for most of the twentieth century, is relatively rare today, at least among grandparents who were born in the United States. However, comparative statistics confirm what many observers have noted, namely, that many immigrant grandparents take an involved role with their grandchildren, especially when all the family members speak a common language (Perez, 1994). Of all American households, only 4 percent of those with a native-born head of household include grandchildren. By contrast, 25 percent of those households with a Mexico-born head, 18 percent of those headed by an Asian refugee, and 15 percent of those headed by someone born in the Caribbean include three or four generations (Glick et al., 1997).

The **companionate grandparent** is independent and autonomous regarding the parent generation, including maintaining a separate household. The generational interaction is constructed to be mutually enjoyable but optional. Grandparents are free to choose when and how this interaction occurs, to select which grandchildren get the most attention, to schedule visits by invitation, to refuse requests for babysitting. They play with and even "spoil" their grandchildren, but they are not free to discipline them—especially in ways, or for reasons, that the parents would not.

Most grandparents today seek the companionate role in that they "strive for love and friendship rather than demand respect and obedience" (Gratton & Haber, 1996). Several historical changes help to explain this preference. First, the increased geographical mobility of both adult generations often places grandparents at a significant distance from their grandchildren, making visiting less frequent and the relationship less intimate. In addition, as adults live longer and healthier lives, more are actively involved in work, friendship, and community activities. Middle-aged women, who typically were the doting, involved grandmothers of the past, are now usually employed (75 percent are in the labor market, compared with only 35 percent in 1960). As a result, grandparents have far less energy, time, or need to

remote grandparents Grandparents who are distant but who are honored, respected, and obeyed by the younger generations in their families.

involved grandparents Grandparents who remain active in the everyday activities of their grandchildren, seeing them daily.

companionate grandparents Grandparents whose relationships with their children and grandchildren are characterized by independence and friendship, with visits occurring by the grandparents' choice.

make their grandchildren central to their lives. Further, as the relationship between parents and children has become more egalitarian, the deference once automatically accorded to the older generation has diminished. Consequently, because their advice about child-rearing is likely to go unheeded and their attempts to command respect are likely to backfire, grandparents follow the primary rule of generational harmony, "noninterference," and hence stay silent on matters of child-rearing. Often the parent generation facilitates this approach by deliberately hiding any child-rearing problems from them (Cherlin & Furstenberg, 1986).

Most grandparents are quite comfortable with a companionate role: they like to boast about their grandchildren's achievements, display their photographs in their living rooms, and provide their grandchildren with fun, treats, and laughter—and leave the parenting responsibilities to the parents (Erikson et al., 1986). As one grandparent confessed, "Glad to see them come and glad to see them go."

In fact, companionate grandparents today generally enjoy grandparenting more than remote grandparents, who are likely to be frustrated if they cannot maintain family rules and traditions, and more than involved grandparents, who can feel overburdened by the demands of their involvement (Jendrek, 1994; Shore & Hayslip, 1994; Thomas, 1986).

These generalities should not obscure the tremendous diversity of grandparent-grandchild relationships. Some of this diversity results from differences in personality and ethnic traditions. This is highlighted by the case of one 18-month-old who had grandparents from very different ethnic backgrounds, one pair Latin, the other Nordic:

> Her Latin grandparents tickled, frolicked with, and cajoled her. Her Nordic grandparents (who loved her no less) let her "be." Her Latin mother thought her in-laws were "cold and hard" while her Nordic father thought his in-laws were "driving her crazy." The youngster was perfectly content with both sets of grandparents. [Kornhaber, 1986]

Another factor contributing to diversity in grandparenting is the developmental stage of the grandchild: all things being equal, grandparents are closer to younger grandchildren than to adolescent ones. The developmental stage of the oldest generation is a factor as well. Middle-aged grandparents,

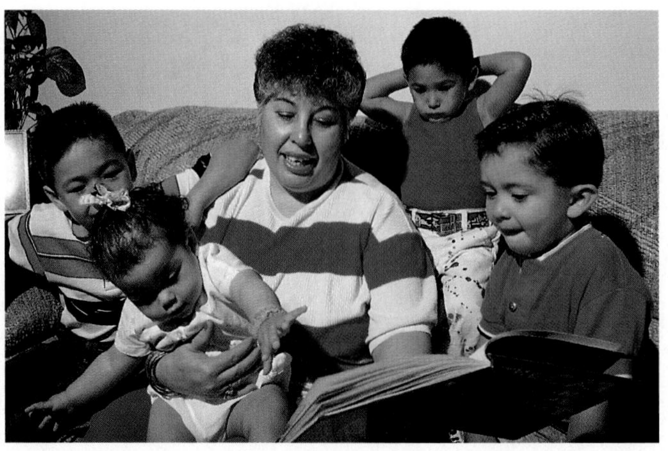

The Hispanic and Korean cultures from which these children's parents come traditionally endorse the involved-grandparent role—and the children's grandmother, who provides the children's daily care, seems to thrive in filling it.

for all the reasons listed above, have less time for active grandparenting unless they are retired. If the oldest generation comes to live with their descendants, it is usually when the grandparents are quite old and the youngest grandchildren are teenagers. Hence remote grandparenting is disrespected, involved grandparenting is unneeded, and companionate grandparenting falls short because of the lack of mutually shared interests. More often, it is the child and the grandchild generations who become caregivers of the grandparents, rather than vice versa, as it was in former times.

Many developmentalists are saddened at the trend toward relatively uninvolved grandparenting in middle age. While this trend certainly provides more independence for each generation and protects everyone from unwelcome closeness (Gratton & Haber, 1996; Thompson et al., 1989), every generation can suffer when distance diminishes the sense of generational continuity and interdependence. This is particularly true in immigrant and minority groups, in which grandparents traditionally are the "keepers of the community," transmitting the stories, traditions, and values that allow each new generation to understand its cultural heritage (Barusch & Steen, 1996; Cherlin & Furstenberg, 1986). Many immigrant grandparents today are rebuffed in this role by grandchildren who neither listen nor obey, in part because they believe that their grandparents' emphasis on the ways of the "old country" undercuts their own attempt to adjust to the new country. This causes distress on all sides, particularly for the involved grandparent whose identity is tied to the new generation. As one 60-year-old Cambodian woman explained:

> I'm afraid they might not be what I want them to be because in this country the children are very unpredictable. . . . I don't like to talk too much, because the more you talk the less respect they have toward you. [quoted in Detzner, 1996]

Many immigrant grandparents are troubled that their grandchildren are allowed too much freedom. They complain, for example, that parents "aren't allowed to punish their kid with sticks," or they insist that the grandchildren "must be respectful to the elders and polite to the guests. They must bow their heads . . ." (quoted in Detzner, 1996).

Grandparents as Surrogate Parents

While grandparents today are usually peripheral, and this usually is what all generations want, a range of social problems can sometimes catapult grandparents into an extremely involved grandparenting role. When the parents are poor, young, unemployed, drug- or alcohol-addicted, or newly divorced—a group of risk factors that includes many families—grandparents are sometimes called upon to become **surrogate parents**, taking over the work, the cost, and the worry of raising the children. This is particularly likely to occur when the children are in need of more intensive involvement. Drug-affected infants or rebellious school-age boys, for example, are more likely to live with grandparents than normal preschool girls are.

No doubt many middle-aged and older grandparents provide excellent surrogate care, furnishing the stability, guidelines, and patience that the distressed parent lacks (Solomon & Marx, 1995). But this caregiving can impair their own health and well-being, with an increased risk of illness as

surrogate parents The role some grandparents play for their grandchildren due to their children's extreme social problems.

well as marital problems (Jendrek, 1994; Minkler & Roe, 1996; Shore & Hayslip, 1994). Surrogate parenting takes a particular toll if it lasts for years or even for generations, especially among families already burdened by poverty. As one 53-year-old great-grandmother lamented:

> I have been taking care of all these kids for a mighty long time. Sandy [her daughter] needs so much help with her children. LaShawn [her granddaughter] I raised from a baby. Now she got two kids and I'm doing it again. I bathe, feed them, and everything. Three generations I raised. Lord Almighty! I'm tired, tired, tired. Sick too . . . [quoted in Burton, 1995]

Grandparenting After a Divorce

Following a divorce involving children, the custodial parent's parents are likely to see more of their grandchildren in the short term—often providing emergency housing, financial support, and surrogate parenting. But as time goes on, it is not unusual for the relationship to become more distant, sometimes suddenly so if the custodial parent remarries and moves away. Unwelcome distance from grandchildren is even more common for the parents of the noncustodial parent.

Some grandparents are so troubled by being cut off from their grandchildren that they sue for visitation rights. Since 1970, every state in the United States has enacted laws that require continuation of a close relationship already formed if a child has lived with the grandparent. Some states go much further, mandating grandparent visitation rights even if the grandchild has never had a relationship with the grandparent, even if the parents were never married, and even if both parents are opposed to it. In other states, the courts are much more restrictive, ruling that grandparents must prove that the child will benefit sufficiently to justify undermining the primacy of the parent-child relationship (Hartfield, 1996).

Developmentalists are on both sides of this dilemma. Intergenerational nurturance is valued as an important mainstay for many grandparents and grandchildren, but, on the other hand, simply being connected by biology does not prove that a relationship is beneficial to the child.

The Sandwich Generation

As just suggested in our discussion of grandparenthood, under some circumstances family ties can become quite burdensome. Because of their position in the generational hierarchy, the middle-aged can be called on to help both older and younger generations of their family. Some feel squeezed from both sides, and thus the middle-aged have been called the **sandwich generation**.

sandwich generation The generation "in between," having both grown children and elderly parents. Many middle-aged people feel pressured by the needs and demands of their adult children on the one hand, and of their elderly (and perhaps ailing or widowed) parents on the other.

Middle-aged adults are often particularly surprised by the need to care for their young adult children, long after they thought the nest would be empty. In fact, not infrequently, instead of flying away, young adults are compelled by financial circumstances to stay on, or return, prompting one observer to describe "the swollen nest" phase of life, a period after the child-rearing phase (Ginn & Arber, 1994).

At any given moment, almost half of all middle-aged adults who have grown children find that at least one of them is still living with them. That this arrangement results from the need of the younger, not the older, generation is confirmed by two findings. First, middle-aged adults in good health are twice as likely to have adult children living with them as are those in poor health. Second, in such households, parents continue to pay most household bills and to do most of the household work, with the nesting children seldom contributing substantially to either (Ward & Spitze, 1996).

Neither generation is particularly happy living in a swollen nest. Parents, especially, regard independent living as the natural order of things, and both generations feel the loss of privacy and the increase in conflict that sharing a household entails (Alwin, 1996; White & Rogers, 1997). But the circumstances of today's cohort of young adults—including longer education, lower salaries, higher unemployment, fewer marriages, and more single parenthood—often make it difficult for the young to leave the nest and necessitate return sojourns after venturing forth (Goldscheider & Goldscheider, 1994; Ward et al., 1992).

As might be expected, an added stress occurs when the adult children living with their parents have serious disabilities (Roberto, 1993). This is an increasingly frequent occurrence for two reasons. The first is that medical advances have saved the lives of many such children who, thirty years ago, would have died in infancy. The second is that, as society has increasingly recognized the individuality of those with severe disabilities, large institutions—which often simply warehoused such persons—have been closed.

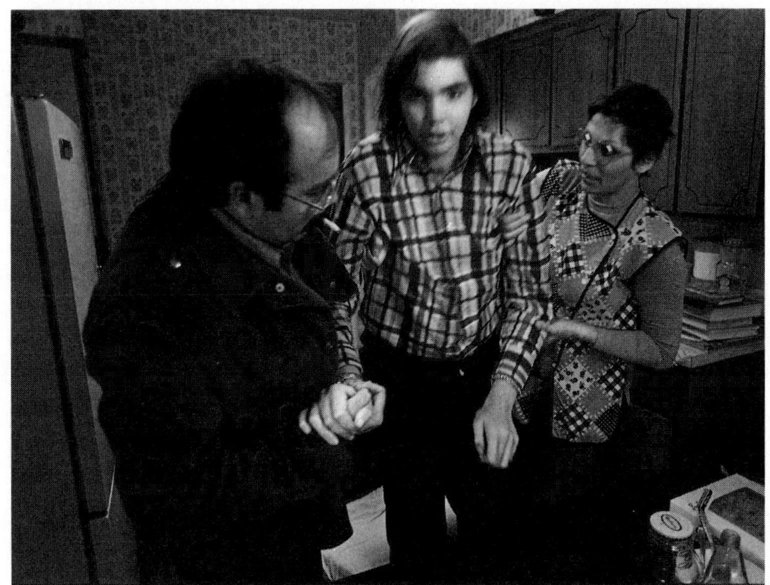

At the age of 16, Alan Green developed a heart condition that put him in a coma for several weeks and left him with severe multiple disabilities. At the age of 26, he is living with his parents, dependent on them for nearly all his needs. For long periods, Alan was depressed and angered by his condition, adding greatly to the stress his care created for his family. Particularly frustrating for both Alan and his parents has been the lack of publicly funded treatment-and-education programs for those over the age of 21. Recently Alan managed to get into a daily life-skills workshop sponsored by the Association for Children with Learning Disabilities, and, for the first time since his illness, is showing signs of happiness. Says his mother: "He's actively participating in life. That's so much more than he had before."

As a result of both these improvements, many parents must care for a disabled child long after the age when children usually become independent. Typically, healthy siblings and other relatives provide understanding, but it is rare for them to take in a disabled sibling, even if the parents can no longer provide care (Seltzer et al., 1991).

The other side of the sandwich involves the middle generation's being called on to care for elderly relatives. This is nothing new, particularly for middle-aged women. Typically, one member of the middle generation, usually a daughter, takes on the task of kinkeeping and caregiving when required. Today, however, this role is less often voluntary and more often disruptive than it was in earlier times. This is because many middle-aged women are employed and family size is smaller, making it less likely that a person with both the time and temperament for this role will naturally emerge. Thus, when an elderly person needs ongoing care from someone in the middle generation, often a daughter feels she must quit her job, or a daughter-in-law finds she is caring for someone who never raised her, or a son finds that he is the only adult child available to be the nurse. When this occurs, it seems less a part of the natural rhythm of life and more an interruption, and hence a burden. (Specifics of elderly care are discussed in more detail in Chapter 25.)

As you can imagine, any of the above problems might make middle-aged adults feel squeezed. Daughters-in-law are particularly likely to feel unfairly burdened with elder care (Ingersol-Dayton, 1996). However, most middle-aged adults feel no such squeeze, for many reasons (Rosenthal et al., 1996; Soldo, 1996). Young adult children living at home may not contribute much to the household, but their independence means that the ongoing burden they represent is minimal. Disabled adult children take their toll, but usually by the time they have reached adulthood, their care is somewhat less difficult than earlier and their parents have adjusted to the task: the problem for most such parents is not so much the daily caregiving as the fear that they will be unable to provide well when they are old and the uncertainty over what will happen to their children when they themselves die.

Further, as the older generation lives longer and healthier lives, major caregiving of the oldest-old is often borne by other older adults—spouses, siblings, friends, or elderly children. Middle-aged adults do provide some care to generations on both sides, but they are less "caught in a web of generational obligation" than buoyed by the rewards of each role (Soldo, 1996; Stephens et al., 1994). Thus, only a minority of the middle generation is unfairly, unexpectedly, involuntarily sandwiched. Most experience pleasure in their maturing children, take comfort in their closer relationships with their active and healthy parents, and find satisfaction with their family relationships overall.

SUMMARY

Changes During Middle Age

1. Although middle age often brings personal changes, these changes usually do not lead to a "midlife crisis."

2. After about age 30, several personality traits, referred to as the Big Five, tend to remain quite stable, and these strongly influence the course of development.

3. The stability of personality is partly genetic, partly the result of early life experiences, and partly the result of the individual's creation of an ecological niche. However, personality changes occur as well. Two changes are notable. People generally shift toward self-improvement and greater generativity, and both sexes typically experience a gender crossover in personality traits.

Family Dynamics in Middle Adulthood

4. In contemporary Western culture, the middle generation plays a critical role within the family, providing emotional and material support to older and younger family members and serving as the link that connects one relative to another. Middle-aged adults generally find that their relationships with their own parents and with their young adult children improve.

5. Most middle-aged adults are involved in a satisfying intimate relationship, usually through marriage. Adults whose marriages have survived to middle age generally are happily wed, deriving a strong sense of self-esteem from their marriage. Decreasing family and work responsibilities may allow a couple to devote more time to each other.

6. When a marriage ends through divorce in middle age, the experience may be more difficult than at an earlier phase of life because middle-aged people are less likely to find new partners.

7. Middle adulthood is the time people are likely to become first-time grandparents. Grandparents commonly take the role of remote, involved, or companionate figures in their grandchildren's lives, with grandparents in today's society typically less involved with the younger generation than they were in the past. One exception to this trend occurs when, due to problems in their offsprings' lives, grandparents are called upon to act as surrogate parents to their grandchildren.

8. Family ties can be particularly burdensome for some middle-aged adults, and they may feel sandwiched between the financial and emotional needs of their parents and their adult children. As people live longer and healthier lives, though, the burden of caring for the oldest-old is increasingly being borne by the younger elderly.

KEY TERMS

midlife (583)
middle age (583)
Big Five (584)
ecological niche (585)
kinkeepers (589)
remote grandparent (594)
involved grandparent (594)
companionate grandparent (594)
surrogate parents (596)
sandwich generation (597)

KEY QUESTIONS

1. What are some of the major shifts that occur in midlife?

2. What are the five trait clusters in personality that tend to be quite stable after about age 30?

3. How does one's ecological niche contribute to stability of personality?

4. What age-related shifts in personality and gender roles commonly occur in middle adulthood?

5. What are some of the reasons for the improved relations that typically occur between middle-aged adults and both their parents and children?

6. How do divorce and remarriage affect adults in midlife?

7. What are the different forms of grandparent-grandchild relationships?

8. What are some of the stresses that may make middle-aged adults feel that they are part of the sandwich generation?

The Developing Person So Far: Middle Adulthood, Ages 40 Through 60

Biosocial Development

Normal Changes

Changes in the appearance of the skin, hair, and body shape are benign, but can be disconcerting. Losses of acuity in hearing and vision are usually gradual, and individuals usually learn to compensate quite easily. Overall wellness is influenced by variables such as sex, ethnicity, SES, and long-term health habits. Genetics clearly play a role as well, but social context and individual choice are more powerful influences on vitality and morbidity than heredity in middle age.

The Sexual Reproductive System

During their late 40s and early 50s women experience the climacteric, when their body adjusts to changing hormonal levels. Menopause signals the end of a woman's reproductive potential, as well as the beginning of reduced natural hormones. Men experience no comparable dramatic decline in hormones or reproductive ability, but the gradual diminution of their sexual responses continues. Both sexes adjust to changes in their sexual interaction.

Cognitive Development

Adult Intelligence

Some intellectual abilities improve with age, while others decline. Typically, fluid intelligence decreases, and crystallized intelligence increases. Reaction time and speed of thinking slow down; practical intelligence deepens. Overall, cohort differences and individual variations are more important influences on the development of adult intelligence than age alone.

Expertise

Adult intelligence tends to flourish in areas of the individual's particular interests, because motivation leads to years of practice and involvement. The result is the development of expertise, characterized by cognitive processes that are intuitive, automatic, and flexible.

Psychosocial Development

Changes During Middle Age

Middle age is characterized by more stability than change in personality, as the Big Five personality traits combine with each person's construction of an ecological niche to engender continuity. Personality changes that do occur, such as a narrowing of the gap between masculine and feminine personality traits, result from historical shifts and personal efforts at self-improvement.

Family Dynamics in Middle Age

Middle-aged adults usually have rewarding relationships with their adult children and grandchildren, without the stress that responsibility for child-rearing creates. Marriages tend to become less conflicted. Usually the oldest generation does not require extensive caregiving. However, some middle-aged adults are squeezed to meet the caregiving and financial demands of both the older and younger generations who demand care.

PART VIII Late Adulthood

What emotions do you anticipate experiencing as you read about development in late adulthood? Given the myths that abound regarding old age, you may well expect to feel discomfort, depression, resignation, and sorrow. Certainly there are instances in the next three chapters when such emotions would be appropriate. However, your most frequent emotion in learning about late adulthood is likely to be surprise. For example, you will learn in Chapter 23 that most centenarians are active, alert, and happy; in Chapter 24, that marked intellectual decline is the fate of only a minority of the elderly, who are sometimes victims of conditions that can be prevented; in Chapter 25, that relationships between the older and younger generations are neither as close as some sentimentalists idealize them to be nor as distant as some critics claim. Overall, late adulthood is much more a continuation of earlier patterns than a break from them, and, instead of falling into a period of lonely isolation, most older adults become more social and independent than ever.

Nevertheless, this period of life, more than any other, seems to be a magnet for misinformation and prejudice. Why is this so? Think about this question when the facts, theories, and research of the next three chapters are not what you expected them to be.

Late Adulthood: Biosocial Development

CHAPTER 23

Imagine that you are a scientist about to study the usual course of life from age 60 on. For practical reasons, you decide to begin your research in the United States and to exclude people currently in hospitals and nursing homes. With those two limitations, you select a representative cross-section of 1,600 older adults from every region, race, religion, and SES—from those in their early 60s to those pushing 100, some employed full-time, some retired, and some bedridden—and ask them to answer dozens of questions relating to their well-being and happiness.

Like every honest scientist, you recognize that you are beginning your research with certain expectations and hypotheses. Do you think, for example, that you will find that most of the subjects in your sample are handicapped by disabilities, suffering from illness, lonely, or depressed? Or do you expect that most of them will turn out to be able, active, healthy, and happy? And how do you predict they will answer a question such as, "Overall, how satisfied are you with your life?" Before reading further, take a minute and fill in your predictions below.

	Your Prediction
Completely satisfied	________ percent
Very satisfied	________ percent
Somewhat satisfied	________ percent
Not very satisfied	________ percent
Not at all satisfied	________ percent

Now compare your expectations and predictions with the data from an actual study of people much like those in your hypothetical survey (Herzog, 1991). Overall, the findings were heartening. Even though 62 percent of the sample had two or more chronic conditions (such as arthritis and high blood pressure), most subjects said that their health does not limit their activities at all, and only 15 percent said their health limits them a great deal. Few were lonely; most visited with friends at least once a week and talked with them on the phone every day. Most telling, the overall results on the question about life satisfaction were more positive than those of some samples of college students. Specifically, twelve times as many older adults said they were "completely" or "very" satisfied with their lives (32 percent and 42 percent, respectively) as said they were "not very" or "not at all" satisfied (5 percent and 1 percent, respectively). About a fifth of the subjects (21 percent) were in between, answering that they were "somewhat satisfied."

If your predictions were far from the mark, you might think that this study was a fluke. It wasn't. Other research finds very similar results (Costa et al., 1987; Deeg, 1995; Havik, 1991; Myers, 1993). If you are wondering how much the exclusion of nursing-home residents might have skewed the results, the answer is, "Very little." Although the general public estimates that about one older adult in three is institutionalized, the actual number in the United States is only about one in twenty. (The proportion of institutionalized elderly is slightly higher in some other developed countries, such as Canada and Australia, and lower in others, such as Japan and the United Kingdom, but no nation in the world has more than 10 percent of its older population in institutions such as hospitals or nursing homes [Cutler & Sheiner, 1994].) Further, as we will see in Chapter 25, most elderly Americans who enter nursing homes stay there less than a year and then return to the community, and many of those who remain as long-term residents are, contrary to the common view, quite satisfied with their lives. Nonetheless, as two leading scientists who study old age explain:

> Common beliefs about the aging process result in negative stereotypes—oversimplified and biased views of what old people are like. The "typical" old person is often viewed as uninterested in (and incapable of) sex, on the road to (if not arrived at) senility, conservative and rigid. The stereotype would have us believe that old people are tired and cranky, passive, without energy, weak, and dependent on others. [Schaie & Willis, 1996]

Ageism

Why are people's perceptions of late adulthood often much more negative than the reality? The reason is that many of us are prejudiced about older people. **Ageism**, or categorizing and judging people solely on the basis of their chronological age, is similar in many respects to racism and sexism and, like those biases, "is a way of pigeon-holing people and not allowing them to be individuals with unique ways of living their lives" (Butler et al., 1991).

Ageism can target people of any age. Teenagers, for example, are frequently branded as being irresponsible and trouble-prone. One result is that 146 of the 200 largest American cities have curfew laws that apply only to teenagers, a manifestation of ageism that restricts law-abiding adolescents. However, ageism clearly does most damage when it targets the old. It fosters stereotypes that make it difficult to see the elderly as they actually are, and it permits policies and attitudes that discourage the elderly from participating in work and social activities, thus isolating many of them from the rest of us. Worst of all, ageism makes many of the elderly ashamed of their age.

Fortunately, this picture is changing rapidly, in part because of the work being done in **gerontology**, the scientific study of old age. New theoretical perspectives have pointed to ways in which aging is a social construction, and in part because extensive longitudinal research has shown that, beneath the surface of this social construction, there is great developmental variety (Riley et al., 1994). To point up this variety and allow for a more comprehensive view of old age, gerontologists now try to distinguish aging in terms of quality of aging. For example, some draw a distinction between the **young-old** and the **old-old**, a distinction based not exclusively on age but on

ageism A term that refers to prejudice against the aged. Like racism and sexism, ageism works to prevent elderly people from being as happy and productive as they could be.

gerontology The study of old age. This is one of the fastest-growing special fields in the social sciences.

young-old Healthy, vigorous, financially secure older adults who are well-integrated into the lives of their families and their communities.

old-old Older adults who suffer from severe physical, mental, or social deficits and thus require supportive services such as nursing homes and hospital stays.

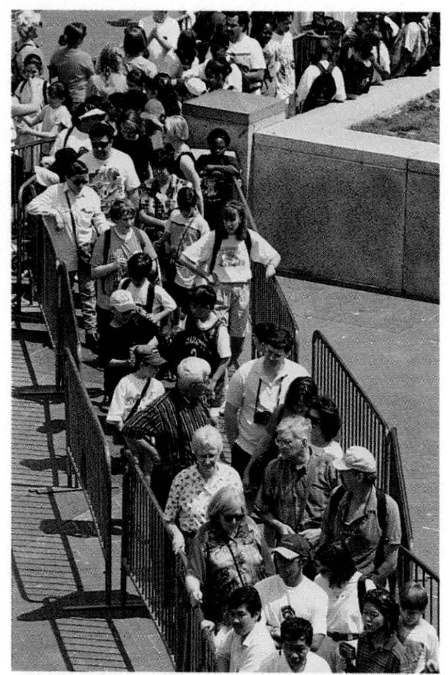

A cross-section of the U.S. population, like these people about to visit the Statue of Liberty, always includes some old, some young, and some in between. In the future, however, when the baby-boomers reach old age, a representative sample of the population will, proportionally, include a great many more of the elderly than are present here. It seems safe to predict that as the elderly come to represent an ever increasing proportion of the general population, the culture's extreme emphasis on youth, and its attendant ageist biases, will fade considerably.

characteristics related to health and social well-being. The young-old, who make up the large majority of the elderly, are, for the most part, "healthy and vigorous, relatively well-off financially, well integrated into the lives of their families and communities, and politically active" (Neugarten & Neugarten, 1986). The old-old are those who suffer "major physical, mental, or social losses" and who are likely to require supportive services or to spend their days in nursing homes or hospitals. Another set of terms commonly used to draw similar distinctions is *optimal aging, usual aging,* and *impaired aging* (Powell, 1994).

As a result of their efforts, cultural bias that associates youth with health and vigor and age with disease and fragility is weakening. In addition, as the proportion of older adults in the population increases, with the baby boomers the next cohort to reach late adulthood, the sheer weight of numbers will force politicians and the general public to reconsider any ageist assumptions they may still have (see Public Policy on pp. 608–609). Now let us look beyond stereotypes, at what the aging process actually entails.

The Aging Process

The first step in understanding the aging process is to distinguish between **primary aging**, which comprises the irreversible changes that occur with time, and **secondary aging**, which comprises the changes that are caused by particular conditions or illnesses. We all, even the youngest of us, experience primary aging—inexorably, minute by minute, day by day. However, secondary aging entails changes that, though they correlate with age, are not inevitable: they can be prevented, and when they do occur, they can sometimes be reversed or remediated. We will explore the relationship between normal aging and the pathology of aging later in this chapter. But first, let us look at some of the specifics of the overall aging process.

primary aging The universal and irreversible physical changes that occur to living creatures as they grow older.

secondary aging The specific physical illnesses or conditions that are more common in aging but are caused by health habits, genes, and other influences that vary from person to person.

Appearance

Looking "old" is something most adults dread and try to avoid, by many means, no matter what age they are. However, there comes a time for everyone when the superficial changes of aging are beyond concealment. One of the most obvious of these changes occurs in the skin, which becomes dryer, thinner, and less elastic, producing marked wrinkling and making blood vessels and pockets of fat much more apparent (Kligman et al., 1985). In addition, the dark patches of skin known as "age spots" are visible in about 25 percent of adults by age 60, about 70 percent by age 80, and in almost everyone by age 100. Hair also undergoes obvious changes, continuing to become thinner and grayer and, in many people, eventually becoming white. Interestingly, loss of hair pigment is the physical change that correlates most closely with chronological age, a clearer index to age than skin changes or hearing acuity and a much stronger one than blood pressure, heart size, or vital capacity (Balin, 1994).

Squaring the Pyramid

One reason that ageism in the United States used to go relatively unchallenged was that, in the past, the number of elderly was relatively small, less than 5 percent of the total population—too few, with too little political or economic clout, to prevent stereotyping by younger and much larger generations. Demeaning clichés like "second childhood," "dirty old man," and "doddering and senile," or patronizing compliments like "spry, with all his [her] marbles," suggest that those few who survived to age 80 or so could easily be dismissed, overlooked as individuals with their own personal combination of specific strengths and weaknesses.

In recent years, **demography**—the study of populations—has revealed that this numerical weakness has been rapidly diminishing worldwide. Traditionally, demographers statistically sort and stack populations according to age, with the youngest at the bottom and the oldest at the top. In the past, the resulting picture was a **demographic pyramid**, as seen on page 12.

There were two reasons for this "wedding-cake" shape. First, each generation of young adults in the past gave birth to more than enough children to replace themselves, typically four or more children per couple, thereby creating larger cohorts at the bottom of the pyramid. Second, before immunization, antibiotics, or even modern sanitation, a sizable number of each cohort died before advancing to the next level of the pyramid, with the death rate of adults particularly high after age 45. This explains why, in the U.S. pyramid for 1920, each five-year cohort from middle age on is about 20 percent smaller than the next-younger age group.

Today, however, because of falling birth rates and increased longevity, the age-sorted depiction of the U.S. population is becoming closer to a **demographic square**. As you can see in the figure, currently in the United States the size of half the cohorts is within 5 percent of the neighboring cohort. The two deviant age clusters are explainable by unusual events: (1) the Great Depression and World War II, which resulted in a dramatic drop in the number of babies born between 1930 and 1945, and (2) postwar prosperity, including GI mortages, which gave rise to a baby boom after the war, particularly between 1950 and 1965. Unless there is a dramatic increase in birth rates in the coming years or an enormous wave of immigration (neither of which seems likely), demographers will soon chart not a pyramid but a square or rectangle, with each generation replacing themselves fully and then living to about age 90.

Whereas in 1900, people over age 60 (the six oldest cohorts) numbered about 5 million, or 6 percent of the total population, in 1995 the number of people over age 60 was 44 million, or 17 percent of the population. Further, the oldest-old (the three cohorts over age 75) are now the fastest-growing group, increasing from slightly more than 1 million people in 1900 to 15 million (5.6 percent of the total population) in 1995 (U.S. Bureau of the Census, 1996). The proportions of those over age 60 in other nations varies, with the rate in developed countries like Japan, Britain, Spain, and France at 20 percent and the rate in many developing nations like Angola, Ethiopia, and Syria at less than 4 percent. But if the trends of the past twenty years continue, the proportion of those over 60 will double worldwide, from the current 8 percent to 16 percent by 2030. This increase will occur not only because the elderly will live longer but also because, as more babies survive, the birth rate will fall.

What are the public policy implications of these demographic shifts? As birth rates fall, young adults postpone the age at which they complete their education, select mates, and begin raising a family. For most people in the nineteenth century, all three of these milestones were passed by age 18. Today they are reached at age 21, 25, 30, or even later. At the other end of the life span, the retirement age is falling in North America, having already dropped, on average, from age 65 to 62. In some European nations the retirement age is even lower (Guillemard & Rein, 1993; Quadagno & Hardy, 1996). The net result of all these trends is that by the year 2030, the population in developed nations will be divided roughly into thirds: one-third below age 30, one-third aged 30 to 59, and one-third aged 60 and older, with only the middle third fully productive in the labor force.

This raises the specter of a highly unbalanced **dependency ratio**, that is, the ratio of self-sufficient, productive adults to dependents (children and elderly). Traditionally, the dependency ratio is calculated by dividing the number of the population who are aged 15 to 65 by the number who are under or over that age. Because of the declining birth rate since 1970 and the small size of the Depression-era cohort now just beginning late adulthood, the current dependency ratio is actually higher in most industrialized countries than it has been for a century, with about two adults for every one dependent. That is a much more manageable ratio than that in many developing nations, where the prevailing pattern of couples having numerous children makes the current ratio about 1:1 (see figure). But unless there is a dramatic increase in the death rate of those who are now middle-aged, every developed nation will experience a steep rise in its dependency ratio. In the United States the rise will be quite sharp beginning in the year 2010, when the first of the baby boomers become elderly (Easterline, 1996).

What will be the result of this shift in the dependency ratio when the pyramid becomes a square? Some experts warn about "catastrophic consequences" of increased costs for the medical care of the older generations and of decreased public funds for the needs of the younger generation (Howe, 1995). If the young continue to enter the labor market later and the old continue to exit earlier, the entire tax and caregiving burden will fall on the shoulders of the

shrinking middle cohorts. Some predict that this heavy load will cause a revolt, with those in the middle cohorts refusing to fund public support for either the old or young. Cloaking themselves in the banner of "family values," it is said, employed adults will provide minimal care for their own parents and children and no one else. Doomsayers contend this is already occurring in the United States, as taxpayers refuse to fund public education or senior health care, even at the levels of a decade ago.

Others argue that such rebellion is not inevitable, citing countries where increasing numbers of the elderly have inspired social policies that benefit all the generations (Pampel, 1994). Among these policies are publicly funded heath programs, continuing education, and housing within the community, as well as private and church-sponsored programs that use the elderly as volunteers to help people of all ages (Caro et al., 1993; Laslett, 1991). The result has been fewer elderly who are disabled and dependent, and more who are productive participants in society.

Indeed, the entire notion of counting everyone over age 65 as a "dependent" in calculating the ratio is questionable, especially since most of this age group are independent, caring for their own health and economic needs, and more likely to contribute to society by voting, obeying the law, and participating in community and religious groups than are those in any other generation (Posner, 1995).

Overall, most developmentalists believe that the oldest cohorts can work cooperatively to benefit people of every age, as they already do in many cultures. There is one prerequisite: in order for the entire social structure to meet the challenges brought about by a growing older population, society must understand what the effects of aging actually are, separating ageist myths from reality. Then we can see "beyond the gray horizon" (Seltzer, 1995), anticipating whatever dependent burdens may arise and deploying whatever strengths the oldest cohorts may have.

The number of adults capable of supporting "dependents" relative to the number of old and young people is called the dependency ratio. Although much concern has been raised about the growing proportion of dependent elderly in developed nations, as you can see those nations have a favorable dependency ratio because they have relatively few children. The biggest dependency burden is felt by adults in Nigeria, Iraq, and other developing nations, where the dependency ratio is about 1:1.

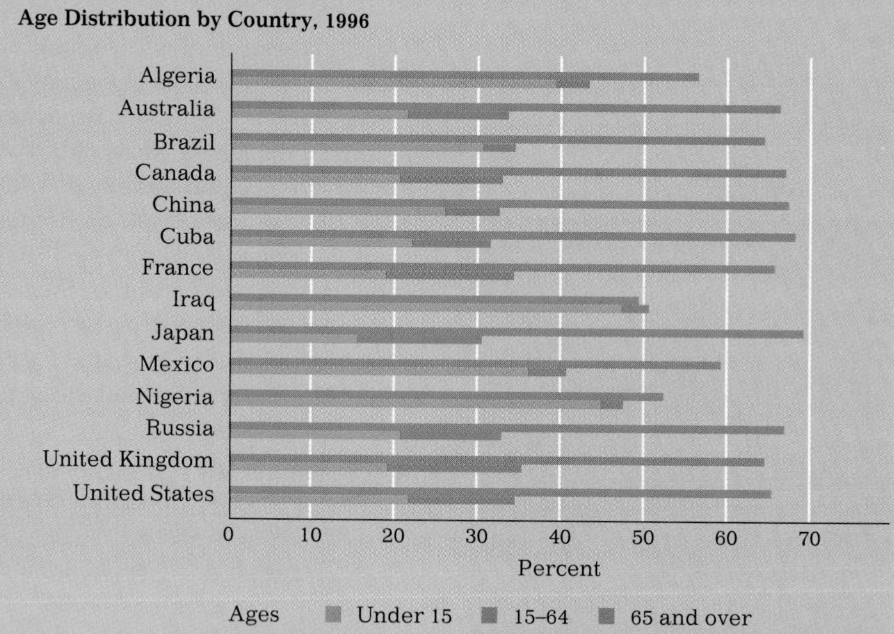

Source: U.S. Bureau of the Census, 1996.

demography The study of populations and social statistics associated with these populations.

demographic pyramid The shape that traditionally resulted when populations were graphed by numbers of individuals in each age group; the largest population group was the youngest and the smallest group was the oldest, giving the graph the shape of a pyramid.

demographic square The shape that social scientists project will develop, in the near future, when populations are graphed by numbers of individuals in each age group. The traditional pyramid shape is becoming more square, with equal numbers of older and younger persons in the population.

dependency ratio The ratio of self-sufficient, productive adults to dependents—children and elderly.

Other obvious physical changes in late adulthood include alteration of overall body height, shape, and weight (Whitbourne, 1985). Most older people are more than an inch shorter than they were in early adulthood, because, as noted in Chapter 20, the vertebrae begin settling closer together in middle age. Further, the muscles that hold the vertebrae have become less flexible, making it harder to stand as straight as in earlier years. Body shape is affected by redistribution of fat, which collects less in the arms, legs, and upper face and more in the torso (especially the abdomen) and the lower face (especially the jowls and chin).

Body weight is often reduced in late adulthood, partly because muscle tissue becomes reduced, a difference particularly notable in men, who have relatively more muscle and less body fat than women. For both sexes, the reduction in muscle strength is especially apparent in the legs, necessitating slower walking and sometimes use of a cane or walker. Another reason for lower body weight is osteoporosis, the loss of bone calcium that causes bones to become more porous and fragile. Bone loss usually occurs at a rate of approximately 1 to 2 percent a year after menopause in women and after age 55 in men (Nordin & Need, 1990). Osteoporosis is the main reason some older people walk with a marked stoop, and it is also the main reason the elderly are much more likely to break a bone, particularly the hip bone, when they fall.

All these changes in appearance have serious social and psychological implications: in an ageist society, those who look old are treated as old, in a stereotypic way. This fact is poignantly underscored by the reactions of the elderly themselves. Most older people consider their personality, values, and attitudes quite stable and, except for acknowledging that they may have slowed down a bit, do not feel that they have changed all that much from their younger days (Troll & Skaff, 1997). Therefore, when older people see a recent photograph of themselves, or catch an unguarded glimpse of themselves in the mirror, or merely notice how others treat them, they are often taken with surprise and regret, even in late-late adulthood. As one 92-year-old woman described this experience:

> There's this feeling of being out of one's skin. The feeling that you are not in your own body. . . . Whenever I'm walking downtown, and I see my reflection in a store window, I'm shocked at how old it is. I never think of myself that way. [quoted in Kaufman, 1986]

Similar feelings were expressed by a man, also 92, who needed a cane to get around:

> I look like a cripple. I'm not a cripple mentally. I don't feel that way. But I am physically. I hate it. . . . You know, when I hear people, particularly gals and ladies, their heels hitting the pavement . . . I feel so lacking in assurance—why can't I walk that way? . . . I have the same attitude now, toward life and living, as I did 30 years ago. That's why this idea of not being able to walk along with other people—it hurts my ego. Because inside, that's not really me. [quoted in Kaufman, 1986]

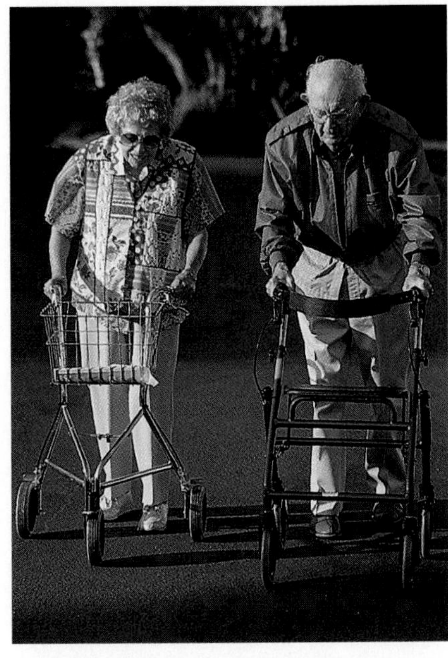

Just because this couple can no longer take carefree strolls, hand in hand, does not mean that romance no longer beats in their hearts. Assumptions based on appearance alone are at the root of ageism, just as they are with racism.

When elderly people associate appearance and identity, or depend on the reactions of others to validate their self-concept (as we all do sometimes), the realization that they look like, or are treated like, an old person may make them act and think like the stereotype of the elderly—with harmful consequences.

Sense Organs

For many of the healthy elderly, the most troubling part of aging is not how well they appear to others but how well they connect with others. Such social connection depends on their ability to use their senses. Unfortunately, all the senses become less sharp with each decade, and notable deficits are universal by age 70 (Meisami, 1994). Up until a century ago, these sensory losses were often devastating: many people who survived to old age were rendered isolated and vulnerable by their inability to see and hear well, if by nothing else. Today, however, most of the visual and auditory losses of the aged can be corrected or at least remedied. While this does not mean that the old will be able to see or hear as well as they did in earlier years, it does mean that few need to lose vision or hearing completely.

Vision

Although only 10 percent of the aged do not need corrective lenses, another 80 percent can see quite well with glasses. The remaining 10 percent have serious vision problems even with corrective lenses. This group, which includes only about 5 percent of those younger than age 80 but more than a third of those older than 80, is most likely to have one of the three major eye diseases of the elderly: cataracts, glaucoma, and senile macular degeneration. Together, these diseases in the elderly account for over half the cases of legal blindness in the United States.

Cataracts involve a thickening of the lens that causes vision to become cloudy, opaque, and distorted. Even by age 50, about 10 percent of adults have some lens clouding, with 3 percent experiencing a partial loss of vision. By age 70, 30 percent have some visual loss because of cataracts. These losses are initially treatable with glasses, and then, as they get progressively worse, by a simple surgical procedure that is almost always successful (Meisami, 1994).

Glaucoma, a problem for 1 percent of those in their 70s and 10 percent of those in their 90s, involves damage to the optic nerve due to a build-up of fluid within the eye. This condition can be relieved with special eye drops, but if left untreated, it can eventually destroy vision (see p. 546).

Senile macular degeneration involves deterioration of the retina. This problem, which affects one in twenty-five people between the ages of 66 and 74, and one in six of those who are older, is hard to treat medically and is therefore the leading cause of legal blindness among the elderly. However, people with senile macular degeneration, like those with severe cataracts, are rarely completely blind. They can be helped in a variety of ways to function well with the little vision they have. Their ability to get around on their own can be enhanced, for example, by mobility training, which teaches them how to guide themselves by using auditory and tactile cues (such as particular street noises, the contours of the sidewalk, the location of furniture in their rooms). Sometimes their daily activities are improved just by installing strong lighting throughout the house and a powerful magnifying glass by the reading chair. Those who are entirely unable to read can benefit from a *Kurzweil reader*, a computer that scans printed text and "speaks" the words. In these and other ways, the population of elderly whose vision is extremely impaired can be helped to maintain independence.

cataracts A common eye disease among the elderly involving a thickening of the lens; it can cause distorted vision if left untreated.

glaucoma A disease of the eye that can destroy vision if left untreated. It involves hardening of the eyeball due to a fluid build-up within the eye.

senile macular degeneration A disease of the eye involving deterioration of the retina.

Hearing

Hearing problems affect a third of the elderly overall, including one in four of those between ages 65 and 80 and almost half of those over age 80 (Havik, 1986). Unfortunately, most elderly people wait five years or more from the time they are aware of a serious hearing loss to the time they seek help. Moreover, if the problem is not an easily remedied one (such as impacted ear wax), they are usually reluctant to use technological aids such as a hearing aid. Indeed, although most people using the new types of hearing aids are well satisfied with them, and although nearly a third of the elderly could benefit from them, less than 10 percent use them (Mulrow, 1990).

One reason for their reluctance is that the hearing aid is regarded as a symbol of agedness, and many people would rather miss the sounds of daily life than risk being considered old. Ironically, old is usually what their poor hearing causes them to be thought of as anyway, often with unanticipated consequences. Individuals who frequently mishear and misunderstand conversation or who often ask "What did you say?" are likely to be thought of as a bit dottering and to be excluded from much social give-and-take, thereby being deprived of important cognitive stimulation. In addition, they themselves are likely to withdraw socially because of others' annoyance at not being heard, and they are also likely to become suspicious of inaudible conversation that they perceive to be about them (Busse, 1985). The net effect is that, much more than the visually impaired, "hard of hearing individuals are often mistakenly thought to be retarded or mentally ill . . . [and] are more subject to depression, demoralization, and even at times psychotic symptomology" (Butler et al., 1991).

In addition to problems hearing conversation, the elderly have other quite specific hearing losses: in detecting where sound is coming from, in deciphering electronically transmitted speech (especially by the telephone), and in noticing high-frequency sounds, a problem that first appears with pure tones, such as that of a doorbell, but then emerges with speech, particularly with consonants and with women's and children's speech. These problems can be ameliorated, even without resorting to a hearing aid. For example, the hearing-impaired elderly person can learn to focus on a speaker's face to pick up cues from the speaker's lip movements and facial expressions. For their part, those speaking to the elderly need to speak more slowly and distinctly, in a lower register, and to be aware of the difficulties that background noise poses for the hard-of-hearing.

These three friends are using key strategies to enhance their hearing: sit close together and pay attention to facial expressions and lip movement.

Another hearing problem that sometimes develops is *tinnitis*, a buzzing or rhythmic ringing in the ears that is experienced by 10 percent of the elderly (Coni et al., 1992). Since the problem originates in the ear and brain, the only treatment at the moment is surgery, and that is not always successful. In many cases, those with tinnitis become so used to the ringing that they are barely aware of it, just as one eventually no longer notices the familiar sounds of a ticking clock or of street traffic.

Other Body Systems

We have discussed sight and hearing extensively not only because these are the main channels for individuals' interaction with their world but also because the age-related changes they undergo mirror those that occur in the other body systems. As we have seen, every body system becomes less efficient with age, with a gradual reduction in capacity and organ reserve that begins relatively early in adulthood and becomes readily apparent, in most people, by late adulthood. As you remember from Chapter 20, with age, the heart beats more slowly, the arteries harden, the digestive organs become less efficient, the lungs lose capacity, sexual responses become slower, and so on. While these changes occur over several decades, the pace of decline speeds up in later life (Kanungo, 1994).

For optimal functioning, all these changes require adjustment, not merely passive acceptance. Adjustment is an active process that involves finding the right balance between maintaining one's normal activities and modifying one's routines to fit diminished capacities. For the healthy young-old, the changes may be minor, such as eating smaller, more frequent meals and devoting more time to stretching and warm-ups before heavy exercise. For some of the oldest-old, energy and effort may need to be conserved, requiring that, beyond the basic routines of life, each day be limited to only one or two activities—having lunch with a friend *or* working in the garden *or* visiting a grandchild.

As with adjustment to hearing losses, the critical factor is the older person's recognition of the need to make a change and then acting on that recognition in a timely fashion. But the older person is not the only one who must do the adjusting: family and friends need to adjust as well. For instance, adult children need to realize that while a visit from a grandchild may be an invigorating experience for an elderly person, a visit from several active youngsters at once may be exhausting. Similarly, others need to realize that, for an elderly person, having a conversation on an ordinary telephone or in a noisy restaurant may be a frustrating or even impossible experience instead of the relaxing exchange it was meant to be.

Aging and Disease

As the distinction between primary and secondary aging makes clear, aging and disease are not synonymous. Indeed, most aged people, most of the time, consider their health to be good or excellent and, on physical examination, they, in fact, are found to be quite well. Whether a particular elderly person is seriously ill, somewhat ailing, or in fine health depends not on the person's age but on his or her genetic makeup and past and current life-

style (including eating and exercise habits). Psychosocial factors such as how much social support the person has and whether or not the person feels a sense of control over his or her daily life can also be crucial.

Nevertheless, it is undeniable that the incidence of *chronic diseases* (long-standing illnesses or conditions that are generally irreversible) increases significantly with age, as does the risk of various *acute diseases* (sudden illnesses that can be deadly). One reason for this increased incidence of disease is that the older a person is, the more likely he or she is to have accumulated several risk factors for a variety of illnesses. For example, decades of smoking, heavy drinking, and inactivity greatly increase the risks of osteoporosis and heart disease in many elderly of both sexes.

But the relationship between aging and disease is not just a matter of the cumulative risks arising from a history of bad lifestyle habits. Because of the reduced efficiency of the body's systems due to primary aging, older persons are more susceptible to disease, take longer to recover from illness, and are more likely to die of every disease they might have (see Figure 23.1). If a younger person contracts pneumonia, for example, he or she almost always is fine again in a few weeks, but if a person already seriously weakened by very old age comes down with pneumonia, it is often the immediate cause of death. Even the flu can kill an older person, which is why flu shots are strongly recommended for everyone over age 65 but not for younger people unless they have some chronic condition—such as heart or kidney disease—that makes them particularly vulnerable.

In short, no one reaches age 100 disease-free. Although many of the very old are still able to take care of their basic needs within their social world, all have at least a few chronic illnesses, and many have to have daily medication, pacemakers, artificial hips, and the like to enable them to continue to function (Adler, 1995).

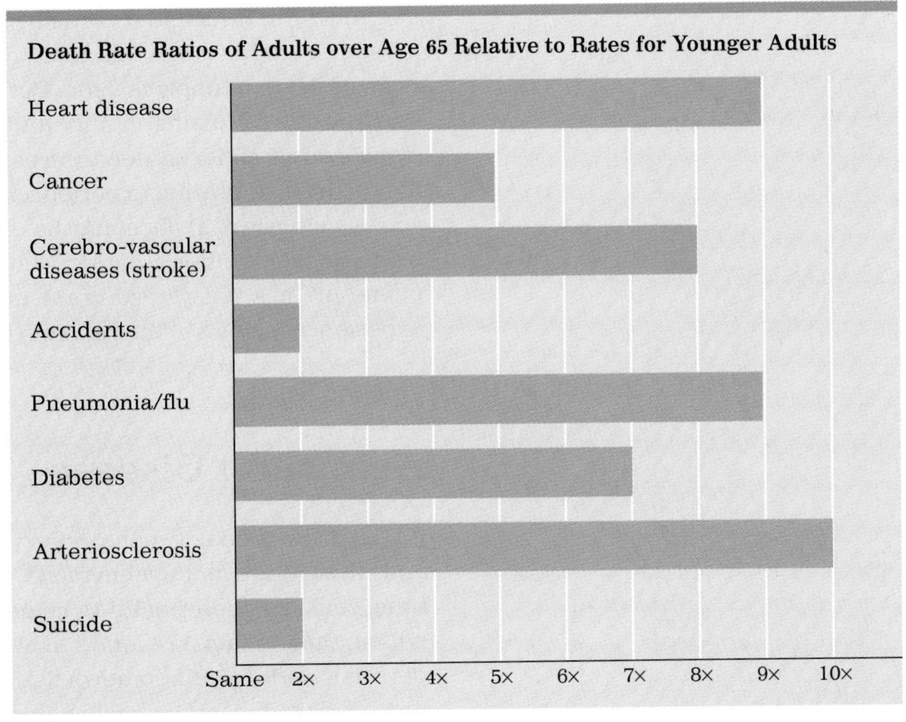

FIGURE 23.1 The death rate from the eight leading causes of death is significantly higher for elderly people than for younger people. This chart shows approximate ratios between the death rates for Americans aged 65 and older and those under 65. A finer analysis reveals some interesting age differences. For example, elderly pedestrians are much more likely to be killed in auto accidents than other adult pedestrians are, whereas younger adults are more likely to be killed in auto accidents when they are drivers or passengers.

Variations in the Incidence of Disease

A closer examination of the relationship between aging and specific diseases reveals variations in the incidence of disease that are not completely understood. For example, while the rate of cancer worldwide increases dramatically with age, the international rates of cancer in general, as well as the rates of specific cancers, are surprisingly variable. Breast cancer occurs in Canada at four times the rate in Japan, but the stomach-cancer rate in Japan is five times that in Canada. In Australia, men are almost twice as likely to die of cancer as are women, whereas in New Zealand—with a similar population in roughly the same part of the world—the cancer death rate is almost equal for both sexes (Higginson et al., 1992). Similarly, a 20-year-old woman's chance of eventually contracting breast cancer is powerfully influenced by where she spends the next forty or so years of her life. If, for example, a young British woman moves to Canada, her breast-cancer risk will be higher than if she had moved to Australia, but her lower risk in Australia would nonetheless be higher than if she had lived her adult life in Japan (Kliewer & Smith, 1995).

In the United States, cancer death rates among older European-Americans are lower than those among older African-Americans but are higher than those among older Hispanic-Americans, Asian-Americans, and Native Americans (Jones, 1989). Similar variation is evident in disability rates, which are higher, and begin at younger ages, for African-Americans and Hispanic-Americans than for European- and Asian-Americans (Skinner, 1995). Although it seems likely that some combination of genetic, environmental, and economic factors underlies these variations, with health habits clearly playing a significant role, at present that is about all that can be said by way of explanation.

Certain sex differences are also of unclear origin. As already noted in Chapter 20, middle-aged men have higher death rates, but middle-aged women have higher rates of chronic disease, a trend that continues into late adulthood (Ory & Warner, 1990). Of twenty-four common chronic conditions, men are markedly higher in only three: hearing impairments, heart conditions, and visual losses. Women are higher in eleven: arthritis, bronchitis, cataracts, constipation, corns, eczema, indigestion, ingrown nails, migraines, ulcers, and varicose veins (U.S. Bureau of the Census, 1996). This pattern continues through very old age: even at age 80, men die at a higher rate than women, but women are more likely to be disabled.

Let us now look more closely at the complex relationship between primary aging and secondary aging with two examples, heart disease and cancer, the leading causes of death in late adulthood.

Heart Disease and Cancer

Normal aging reduces the functioning of the heart—especially in times of exercise and stress—partly because it reduces the strength of the heart muscle and lengthens the time the heart needs to relax between contractions. Normal aging also reduces the elasticity of the cardiovascular system. But aging in itself does not cause heart disease. Most aged people have quite healthy, although aging, hearts, capable of sustaining life for many more years (Harris, 1986). However, many older people also show a number of risk factors related to heart disease, including elevated blood pressure, a high cholesterol level, obesity, lack of exercise, and a history of smoking.

Over time, the interaction of these accumulating risk factors with the general weakening of the heart and relevant genetic weaknesses makes the elderly increasingly vulnerable to heart disease (Lau, 1991). Thus it is not surprising that heart disease causes about 40 percent of all deaths over age 65, with an increasing proportion of such deaths being the result of gradual heart failure rather than sudden massive heart attack.

A similar relationship exists between aging and cancer, the cause of about almost 25 percent of all deaths in the aged. The predisposing factors that may lead to cancer—such as genetic vulnerability and environmental insults (such as exposure to asbestos or tobacco smoke)—typically predate old age by half a century or more, and the cancer process probably begins decades before the disease is evident. As people age, however, the latent cancer is more likely to grow, in part because of declines in the immune system. Further, because of the aging of the body overall, cure is more difficult. For both reasons, a person older than 85 is more than a hundred times as likely to die of cancer within a year as a person aged 30. In fact, among those over age 85 in the United States, the rate of cancer death between 1970 and 1993 increased by 63 percent for men and 29 percent for women (U.S. Bureau of the Census, 1996). This is largely because the cohort that currently makes up the old-old was the first group to enthusiastically take up cigarette smoking as the "modern," sophisticated thing to do (Thun et al., 1995). Compared with earlier cohorts, they also ate more animal fat and were exposed to more pollution.

There is another reason that cancer rates are rising among the oldest-old. Simply put, far fewer of the young-old are dying from other diseases, particularly heart disease and stroke (for 35- to 75-year-olds, the rate for the former is now half what it was in 1970, while the rate for the latter is down about a third). Thus more people are reaching the age when cancer finally overcomes them.

Postponement of Disease

As you can see, although both heart disease and cancer are examples of secondary aging, they become more likely to occur with every passing year because of primary aging. While some aged hearts are quite strong, none are as strong as they were when the person was young. And while most people are not diagnosed with cancer, autopsies of old people "usually reveal incipient or well-developed tumors, although they have not been the causes of death" (Holliday, 1995). The most striking example of this is prostate cancer, which upon autopsy is evident in virtually every man who dies at age 80 or older but is the cause of death in less than 1 in 1,000 instances. Thus, primary aging, which cannot be stopped, is eventually the cause of many of the cancers and other diseases of aging.

Although it is impossible to prevent the diseases that constitute secondary aging, it is possible in many cases to postpone their onset. Ideally, say gerontologists, this possibility will lead to the **compression of morbidity**, that is, the shortening of the number of days and months that a person is disabled or in pain because of various chronic illnesses and, correspondingly, the lengthening of the period of vital, successful aging.

compression of morbidity A limiting of the time a person spends ill or infirm.

James Fries (1994) illustrates the idea of the compression of morbidity with an example of identical twin brothers (see Figure 23.2). Both have the

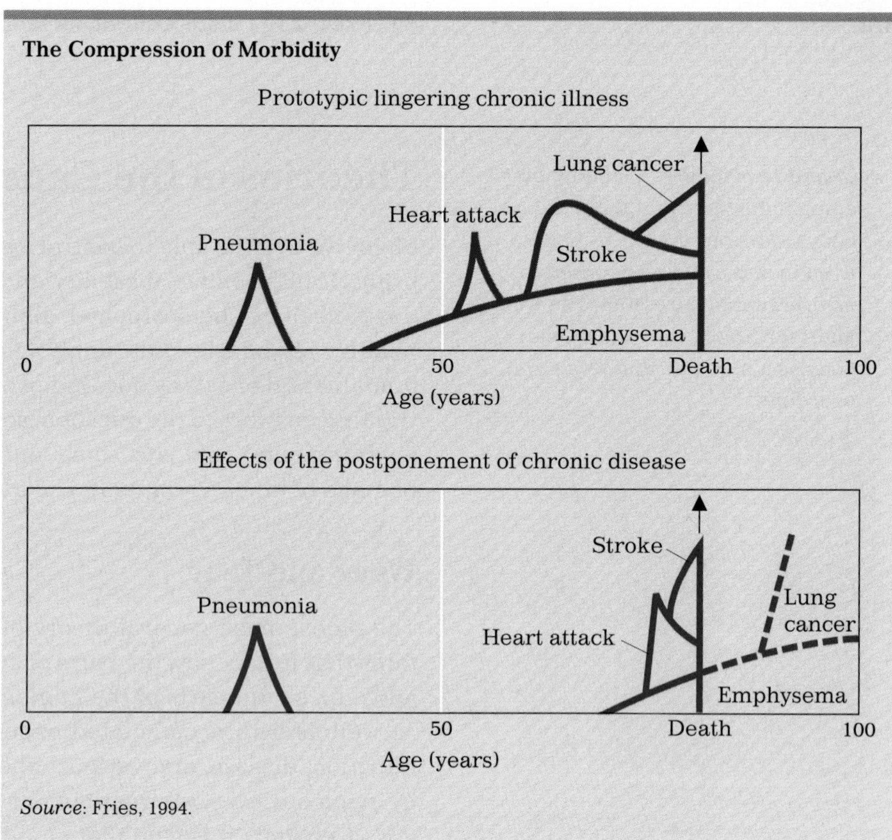

FIGURE 23.2 The interplay of primary and secondary aging is shown in this schematic of the illness and death of a pair of monozygotic twins. Both are equally subject to certain illnesses—so both experience a bout of pneumonia at about age 25. Both also carry the same genetic clock, so they die at age 80. However, genetic vulnerabilities to circulatory, heart, and lung problems affect each quite differently. The nonexercising smoker suffers from an extended period of morbidity, as his various illnesses become manifest when his organ reserve is depleted, beginning at about age 45, and worsening over the next thirty years. By contrast, his twin's healthy lifestyle keeps disability and disease at bay until primary aging is well advanced. Indeed, he dies years before the emergence of lung cancer—which had been developing throughout late adulthood but was slowed by the strength of his organ reserve and immune system.

same genetic vulnerabilities to disease and both are exposed to the same pathogens, but one "smokes like a chimney, is fat, doesn't exercise, and has a poor diet," while the other has "fairly good health habits." When both get pneumonia at about age 30, they recover quickly because their organ reserves and immune systems have only just begun to age. However, the other illnesses that both are genetically predisposed to—emphysema, heart attack, stroke, and lung cancer—occur much later in the health-conscious twin. The other twin is sick from middle age on, with chronic disease. By his 70s, he has several serious illnesses. Meanwhile, because of his genes, the health-conscious twin has the same illnesses, but due to his lifestyle, he experiences them to a lesser extent. Indeed, by the time he dies, his genetic vulnerability to cancer has not yet manifested itself.

Notice that the time the healthier twin spent seriously ill was much shorter than that of the sick twin. This compression of morbidity has other consequences in addition to purely biological ones. When an elderly person is healthier, he or she is also more likely to be intellectually alert and socially active—in other words, to be among those experiencing a successful aging as a young-old person.

The compression of morbidity is the most powerful accomplishment of medical science over the past century. Good prevention, detection, and, most important, treatment measures have meant that a typical old person alive today has gained several times over, living with less pain, more mobility, better vision, more and stronger teeth, sharper hearing, clearer thinking, and so on (Bunker et al., 1995). Dramatic reductions in premature death related to heart disease and strokes have been caused primarily by a better lifestyle, as have the reductions in cancer death rates up until about age 70. Such improvements for the young-old raise important questions regarding the aging process itself, questions we will now turn to.

wear-and-tear theory A theory of aging that states that the human body wears out merely by being lived in and by being exposed to environmental stressors. The wear-and-tear theory sees the human body as a machine that wears out over time.

Theories of the Causes of Aging

Might we ever be able to control aging sufficiently that late adulthood never brings frailty, senility, disability, and pain? Can aging be slowed down so that death itself can be postponed, allowing the average person to live 90 or 100 healthy years instead of simply 75 or 85? Underlying these questions is the fundamental one: Why does aging occur? As you will see, there are many intriguing answers to the question, some implicating our interactions with our environmental contexts, some, our genetic makeup, and some, the simple passage of time (Arkin, 1991; Holliday, 1995; Kanungo, 1994).

Wear and Tear

The oldest, most general theory of aging, the **wear-and-tear theory**, maintains that just as, say, the parts of an automobile begin giving out as mileage adds up, so the parts of the human body deteriorate with each year of use, as well as with accumulated exposure to pollution, radiation, inadequate nutrition, disease, and various other stresses. According to this theory, just by living our lives, we wear out our bodies to the point of dying.

Certainly it is true that an athlete who puts repeated stress on his or her shoulders or knees is likely to have damaged joints by middle adulthood; or that someone who regularly works outdoors in strong sunlight without protection is likely to have damaged skin; or that an industrial worker who inhales asbestos and smokes cigarettes over many years will eventually have damaged lungs. By late adulthood, everyone's body has parts that are showing signs of wear. Indeed, one of the major advances of modern medical technology is in the replacement of worn-out parts, from vital organs to artificial knees and hips (now almost routine for the 4 percent of the population who require them) to dentures and tooth implants (which aid eating, speaking, and appearance in an estimated 75 percent of those over age 65 [Bunker et al., 1995]).

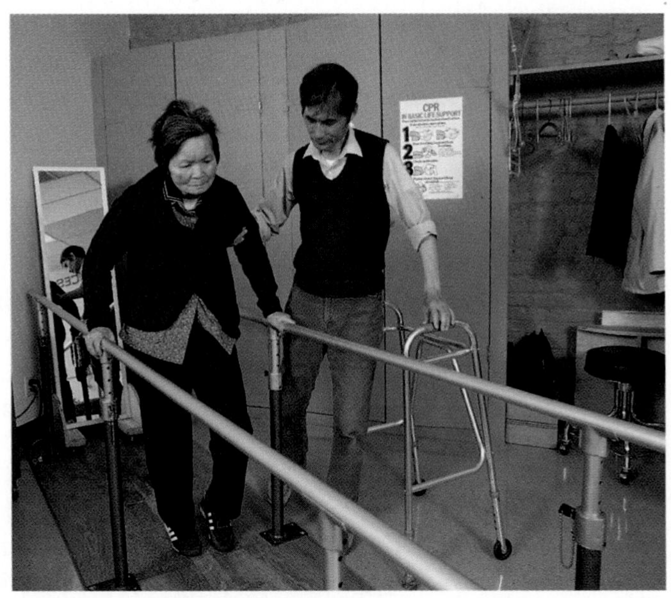

Although wear-and-tear theory might predict otherwise, the single most critical failure of body functions that accelerates aging is loss of mobility. We now know that after a stroke or other mobility-restricting event, the best therapy is to start walking again.

Overall, however, the analogy to a machine's wearing out simply does not hold up. In many respects, the human body is its own repair shop, replacing or mending many of its damaged parts. In addition, unlike most machines, many parts of the human body benefit from use. As we have seen, the heart functions better if the person regularly makes it work faster than normal; the respiratory system benefits from routine exertion; the sexual arousal system is more likely to function in old age if the person has been sexually active throughout adulthood; the digestive system benefits from raw fruits and vegetables that require vigorous digestive activity. It seems clear, then, that the notion of wear and tear applies to some diseases and problems in some organs and body parts, but it is not very helpful in explaining the aging process overall.

More promising theories of aging begin at the cellular level, suggesting that some occurrence in the cells themselves causes aging.

Cellular Accidents

One cellular theory of aging proposes that senescence is the result of the accumulation of accidents that occur during cell reproduction. With the exception of certain types of cells, notably the nerve cells of the ears and eyes and neurons of the brain, the cells of the human body continue to reproduce throughout life. An obvious example is the outer cells of the skin. Under normal conditions, these cells are entirely replaced every few years; the process occurs much more rapidly when a cut or scrape is healing. Thanks to precise functioning of DNA and RNA, each replacement cell is the exact copy of an old cell, or ought to be.

However, "an ever growing number of chemical agents discharged in our environment, [and] . . . an increased possibility for the interaction of different toxicants," as well as radiation from the sun and other sources, all cause mutations in the DNA structure of more and more cells as a person ages (Cooper et al., 1991). Mutations also occur in the normal process of DNA repair. As a consequence, the instructions for creating new cells become imperfect, so the new cells are not quite exact copies of the old. Over time, such changes may result in aging of the skin, or benign skin changes, or possibly cancer. Throughout the body, cellular imperfections and the body's declining ability to detect and correct them can result in harmless changes, small declines in function, or, sometimes, potentially fatal damage.

A related aspect of this theory of aging arose from the observation that some electrons of certain molecules in the body sometimes separate from their atoms, resulting in atoms with unpaired electrons. These atoms, called **free radicals**, are highly unstable and capable of reacting violently with other molecules in the cell, sometimes splitting them or tearing them apart. The most critical damage caused by free radicals is that done to DNA molecules by free radicals of oxygen. The actions of these oxygen free radicals can produce errors in cell maintenance and repair that, over time, may eventually cause such diseases as cancer, diabetes, and arteriosclerosis.

free radicals Atoms that, as a result of metabolic processes, have an unpaired electron. Free radicals are believed to damage cells, affect organs, accelerate diseases, and decrease the ability of DNA to maintain and repair the body.

In addition, free radicals formed from oxygen molecules appear to aggravate several illnesses, including cancers caused by other factors. Consequently, many doctors are selective in giving pure oxygen to patients. Furthermore, although oxygen free radicals are reproduced naturally, as

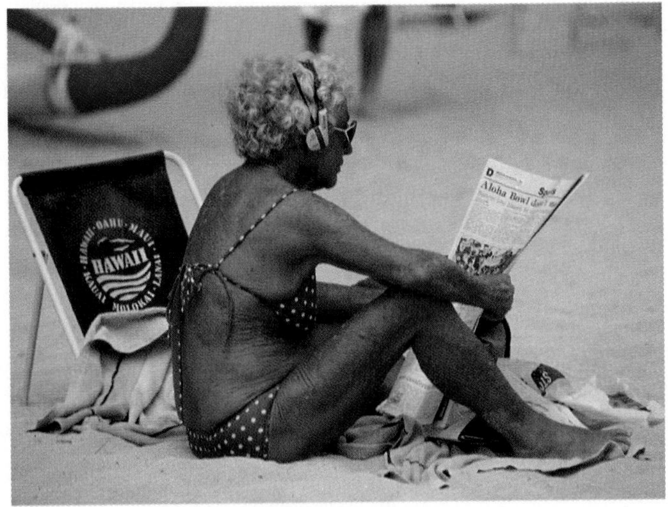

The "beauty benefits" of devoted sunbathing are transient; the damage is cumulative. Age spots, wrinkles, and leathery skin texture associated with aging become exaggerated from lengthy exposure to the sun. Worse, prolonged sun exposure greatly increases the risk of skin cancer.

part of the normal growth and maintenance of the body, they are also produced at a faster rate in reaction to infections and inflammation in the intestinal tract. The production of free radicals is also accelerated by exposure to ultraviolet radiation (Holliday, 1995), which is one of the reasons scientists are concerned about the loss of the ozone layer protecting the earth's atmosphere. It seems, then, that since free radicals damage cells, affect organs, and accelerate diseases, and since the number of free radicals in the body increases over time, the gradual accumulation of free-radical damage may be one of the causes of the aging process (Harman, 1992).

Error Catastrophe

When the systems of the body, especially the immune system, are in shape, the effects of cellular damage tend to be held in check by other cells whose function it is to destroy seriously damaged cells and take over the work that imperfect cells no longer perform. For example, three major enzymes are involved in the removal of oxygen free radicals. In addition, a variety of **antioxidants** nullify the effects of oxygen free radicals by forming a bond with their unattached oxygen electron. These antioxidants include vitamins C and E and beta-carotene, which is an "active sink for radicals" (Holliday, 1995).

In these and many other ways, the body makes use of self-healing processes. Given a healthy lifestyle, cell errors accumulate slowly, with little overall harm caused. However, according to some theories, as the immune system declines and the processes of repair become less efficient, the constellation of errors can become so extensive or affect such critically important cells that the body can no longer control or isolate the errors, leading to what has been called an **error catastrophe**. At this point the normal, healthy aging process gives way, disease overtakes the person, and death occurs.

This theory of accumulating errors in cell reproduction is one possible explanation for the fact that cancer, a disease in which normal cell reproduction somehow goes awry and cells reproduce so rapidly that malignant tumors are formed, is much more common in humans, and indeed all mammals, as they reach old age. It may also explain why the older a person is, the more likely he or she is to die of a cascade of illnesses, rather than sim-

antioxidants Compounds that nullify the effects of oxygen free radicals by forming a bond with their unattached oxygen electron.

error catastrophe A key idea in a theory of aging that holds that, while the body can isolate and repair a certain number of errors in cell duplication, at some point, accumulating errors can no longer be controlled and fatally impair the body's ability to function.

ply one. As noted by the chief mortality statistician of the Centers for Disease Control:

> Death among the oldest old appears to have a somewhat opportunistic character, with many chronic conditions competing to be the precipitating cause. The particular cause of death is less the result of a clearly defined etiological path than the random result of a more generalized deterioration of the capacity for life. [Rosenberg et al., 1991]

The Decline of the Immune System

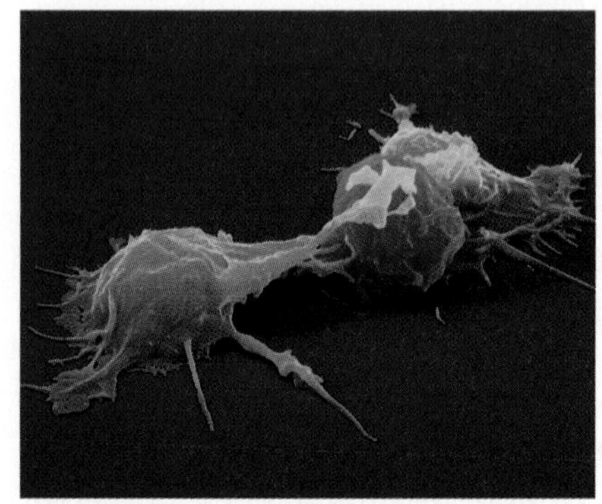

The immune system is always at war, attacking invading bacteria, viruses, and other destructive agents. Here two "natural killer" cells are overwhelming a leukemia cell. How healthy we are and how long we live are directly related to the strength and efficiency of our immune system.

As noted in Chapter 20, the immune system works in part by recognizing foreign or abnormal substances in the circulatory system, and then isolating and destroying them. It does this mainly with two types of "attack" cells. The first are called **B-cells** because they are manufactured in the bone marrow. B-cells create antibodies that attack specific invading bacteria and viruses. Since these antibodies remain in the system, we do not get measles, mumps, or specific strains of flu more than once. The second type of attack cells are called **T-cells** because they are manufactured by the thymus gland. T-cells produce specific substances that attack infected cells of the body. T- cells also help the B-cells produce more efficient antibodies and strengthen other aspects of the immune system as well.

B-cells and T-cells are further specialized, with some responding to antigens to which the body has been previously exposed and others primed to defend against new diseases and tumors. Further, these two types of cells compose only part of the "very complex defense mechanism of the immune response," necessitated by the fact that humans are "slow breeding complex organisms continuously exposed to infection by rapidly-growing pathogens or parasites" (Holliday, 1995).

Many, but not all, aspects of the immune system become less efficient with age (Miller, 1996). The first notable change in the immune system involves the thymus gland. A key part of the immune system in children, the thymus begins to shrink during adolescence and by age 50 is only 15 percent of the size it was at puberty (Holliday, 1995). Partly for this reason, over the course of adulthood there is also a reduction in the production and power of T- and B-cells, as well as in the efficiency of the mechanisms that regulate them. These depletions are another reason why most forms of cancer become much more common with age, and why various other illnesses—from chicken pox to food poisoning to the latest strain of influenza—are much more serious in an adult than in a child. By late adulthood, the "flu" can be fatal.

Additional support for the idea that impairments of the immune system are closely involved in aging as well as disease comes from research on AIDS, the course of which seems directly linked to cellular measures of the declining immune system. Some specific cancers and signs of dementia that occur in younger persons with AIDS are also much more common in later life.

In short, the immune system's gradual loss of the ability to repair damage to the cells and to protect them against invasion is a key dimension of the aging process. Even in the absence of illness, individuals with stronger immune systems outlive their contemporaries whose immune systems are less strong, suggesting to some researchers that the decline of the immune system may be *the* cause of aging (Miller, 1996).

B-cells Cells manufactured in the bone marrow that create antibodies for isolating and destroying invading bacteria and viruses.

T-cells Cells created in the thymus that produce substances that attack infected cells in the body.

The immune system is also a candidate to explain the sex differences in morbidity and mortality as people age. Throughout life, females tend to have stronger immune systems than males: their thymus is larger, and laboratory tests reveal that their immune responses are more efficient. This advantage may be a mixed blessing, however, because women are more vulnerable to autoimmune diseases such as rheumatoid arthritis and lupus, in which an overactive immune system attacks the person's own body (Carlson et al., 1996; Verbrugge, 1990).

The Genetics of Aging

When she died in 1997, at age 122, Jeanne Calment of southern France was the oldest living person in the world, and one of the very few documented instances of someone's living out the maximum human life span.

Errors in cellular duplication seem to be one part of the explanation for primary aging in that they are cumulative, universal, and pervasive, occurring in all the body systems. Similarly, a slowdown in the body's repair system is a partial explanation for the increased rate of disease with aging and could account for the compression of morbidity in that, at a certain point, the immune system becomes too depleted to protect the body's various systems.

However, while the two preceding theories appear to be partly true, they do not seem to explain the many changes of aging that begin to occur in early adulthood, nor do they account for the predictable acceleration of death that universally occurs with age. If cellular error and immune-system decline were the only factors involved in primary aging, we would expect that some individuals would be so lucky (experiencing no error catastrophe) or so protected (having a strong immune system) that they would show no signs at all of aging until age 100 or more, while other unlucky ones would suddenly age at 30. This is not the case.

Noting the incompleteness of these explanations, some theorists propose that aging is incorporated into the genetic plans of all species. In other words, aging is not a mistake. It is part of the normal, natural development within each species, part of the genetic plan. Evidence for this includes the finding that every living species seems to have a genetically inbred **maximum life span**. That is, for each species there is an upper limit to the length of time the members of that species can live, even if they have maximum protection against all the "extrinsic factors, such as nutrition, temperature, radiation, pollution, and . . . stress as well as intrinsic factors such as hormones and free radicals" (Kanungo, 1994). For instance, under *ideal* circumstances, hamsters live, at most, 4 years; rabbits, 13; sheep, 20; house cats, 28; brown bears, 37; chimpanzees, 44; Indian elephants, 70; finback whales, 80; and humans, approximately 120.

Maximum life span, of course, is quite different from **average life expectancy**, which is the number of years the average newborn of a particular population of a given species is likely to live. In humans, life expectancy varies according to historical, cultural, and socioeconomic factors that affect frequency of death in childhood, adolescence, or middle age. In the United States in 1993, average life expectancy at birth was about 72 years for men and 79 years for women (U.S. Bureau of the Census, 1996). Americans who already are 60 years old, and thus no longer at any risk of an early death, are expected to live to 79 if male and to 83 if female. Current average life expectancy is about four times what it was for people of ancient times, and twenty-eight years more than it was at the turn of the century. The reason for this improvement is not that the maximum life span for the human

maximum life span The oldest age to which members of a species can live, under ideal circumstances. For humans, that age is approximately 120 years.

average life expectancy The number of years the average newborn of a particular species is likely to live. In humans, this age has tended to increase over time; in the United States in 1993, the average life expectancy at birth was 72 years for men and 79 for women.

species has increased but that people are less likely to die in infancy or childhood, thereby increasing the average length of life.

As best we can tell from historical records, the maximum human life span a millennium or two ago was a few years past 100, just as it is today. Far fewer people reached old age then as do now, but if childhood diseases, accidents, infections, warfare, and famine had not killed them, many more would have. At that point, however, primary aging would have taken over and those who survived into old age would have died of aging-related causes, just as the oldest-old do today. Thus, in the same way that we are genetically programmed to reach various levels of biological maturation at fixed times, we may be genetically programmed to die after a fixed number of years.

The Genetic Clock

This 16-year-old South African boy, embraced by his 81-year-old grandmother, has progeria, a genetic disorder that produces accelerated aging, including baldness, wrinkled skin, arthritis, and heart and lung difficulties, and early death.

According to one version of the genetics theory of aging, the DNA that directs the activity of every cell in the body also regulates the aging process, not as a mutation over time but as a proper function. Our genetic makeup acts, in effect, as a **genetic clock**, triggering hormonal changes in the brain (similar to the hormonal changes that produce puberty, for instance) and regulating the cellular reproduction and repair process. As the genetic clock gradually "switches off" the genes that promote growth, the theory speculates, it also switches on the genes that promote aging. Aging processes continue to accumulate until one or more body systems can no longer function, and a natural death occurs.

Genetic regulation of aging is suggested by several genetic diseases that include premature signs of aging and early death as part of their symptoms. Down syndrome is the most common: people with Down syndrome who survive childhood almost always die by middle adulthood, with symptoms of heart disease and Alzheimer's disease, a type of dementia that occurs most frequently in old age. Children born with a rare genetic disease called **progeria** have a normal infancy but by age 5 stop growing and begin to look like old people, with wrinkled skin and balding heads. They develop many other signs of premature aging during middle childhood and die by their teens, seemingly of heart diseases typically found in the elderly (Brown et al., 1990).

Evidence for a genetic clock that limits the life span also comes from laboratory research, particularly from the work of Leonard Hayflick (Hayflick, 1979; Hayflick & Moorhead, 1961). In this research, cells cultured from human embryos were allowed to "age under glass" by providing them with all the necessary nutrients for cell growth and replication and protecting them from external stress or contamination. In such ideal conditions, it was believed, the cells would multiply forever. Instead, the cells stopped replicating after about fifty divisions. Cells similarly cultured from children showed a smaller number of doublings before they stopped dividing, and cells from adults divided even fewer times. The total number of cell divisions was shown to be roughly related to the age of the donor.

This research has been repeated by hundreds of scientists, using many techniques and various types of cells from people and animals of various ages. The result is always that the cells stop replicating at a certain point,

genetic clock According to one theory of aging, a regulatory mechanism in the DNA of cells regulates the aging process.

progeria A rare genetic disease that causes young children to age prematurely and to die by their teens.

Hayflick limit The number of times a human cell is capable of dividing into two new cells. Leonard Hayflick determined that the limit for most human cells is approximately fifty divisions, suggesting that the life span is limited by our genetic program, which does not allow cells to reproduce themselves indefinitely.

referred to as the **Hayflick limit**. Even in ideal conditions, the replication of cells of living creatures is roughly equal to that occurring in the maximum life span of their particular species. Cells from people with progeria, Down syndrome, and other genetic conditions characterized by accelerated aging show fewer numbers of doublings than would be expected given the age of the donors (Tice & Setlow, 1985).

In all cases, at the point where cell division stops, the cells are different from young cells in many ways. This provides new support for the idea that DNA and RNA are responsible for cell death, not only because of random errors but, more important, because of programmed senescence.

Conclusion

While most scientists now believe that aging is genetic, and that Hayflick's idea of a genetic clock that governs the life of each cell is intriguing, very few researchers believe that, except in rare cases such as progeria, aging is directly controlled by one or several specific genes. Instead, they maintain that aging, like every aspect of life, results from the interaction of many genes, with each other and with external forces. There is widespread agreement that the causes of primary aging itself, as well as the specific diseases of secondary aging, are the result of "multiple cellular pathways," with no one factor acting in isolation (Cristofalo, 1996).

Given the complexities that underlie primary aging, it seems likely that any significant extension of the maximum human life span lies in the distant future. However, the diseases related to secondary aging can be prevented or slowed, health can be increased, and average life expectancy can be extended. An analysis of the death ratio in Sweden, where health care has been excellent and universal for decades, and where accurate death records have been available for a century, shows that average life expectancy has been increasing and is continuing to (Vaupel & Lundstrom, 1994). Not only are fewer Swedish babies and children dying (decreased early death is the primary reason for increases in average life expectancy worldwide over the past century) but older people are living longer. At age 95, the number of remaining years for Swedes was about two years in 1945 and is about three now. The authors of this study conclude:

> The available evidence, taken together, suggests that, if historical rates of progress in reducing mortality rates continue to prevail in the future, newborn children today can expect to live about 90 years on average. If, as health and biomedical knowledge develops, progress accelerates so that age-specific mortality rates come down at an average rate of about 2 percent per year, then the typical newborn today in developed countries will live to celebrate his or her 100th birthday.
>
> Whether progress in reducing mortality rates will continue at historical levels or even accelerate is, of course, an open question. Even more uncertainty envelops an equally important question: if our children survive to become centenarians, what will their health be like during their extra life span? Will the added years be active, healthy years or years of decrepitude, disability, and misery? [Vaupel & Lundstrom, 1994]

This last question lingers. The theory that morbidity can be compressed is a hopeful one, but many people now shudder at the idea of late adulthood lasting fifty years or more, or at the concept of a society in which the elderly outnumber the young by two to one. To relieve some of this anxiety, let us look at some people who have lived to be very old.

(a) (b) (c) (d) (e) (f)

Three remote regions of the world are renowned for the longevity of their people. In Vilcabamba, Ecuador, (*a*) 87-year-old Jose Maria Roa stands on the mud from which he will make adobe for a new house, and (*d*) 102-year-old Micaela Quezada spins wool. In Abkhasia in the Republic of Georgia, companionship is an important part of late life, as shown by (*b*) Selakh Butka, 113, posing with his wife Marusya, 101, and (*c*) Ougula Lodara talking with two "younger" friends. Finally, Shah Bibi (*e*) at 98, and Galum Mohammad Shad (*f*), at 100, from the Hunza area of Pakistan, spin wool and build houses. Alexander Leaf, the physician who studied these people, believes that the high social status and continued sense of usefulness of the very old in these cultures may be just as important in their longevity as the diet and exercise imposed by the geographical conditions in each region.

May You Live So Long

Three remote areas of the world—one in Georgia in the former Soviet Union, one in Pakistan, and one in Peru—have become famous for having large numbers of people who enjoy unusual longevity. In these places, late adulthood is not only long but is also usually quite vigorous.

One researcher described the Abkhasia people in Georgia as follows:

> Most of the aged [those about age ninety] work regularly. Almost all perform light tasks around the homestead, and quite a few work in the orchards and gardens, and care for domestic animals. Some even continue to chop wood and haul water. Close to 40 percent of the aged men and 30 percent of the aged women report good vision; that is, that they do not need glasses for any sort of work, including reading or threading a needle. Between 40 and 50 percent have reasonably good hearing. Most have their own teeth. Their posture is unusually erect, even into advanced age. Many take walks of more than two miles a day and swim in mountain streams. [Benet, 1974]

Among the people described in this report are a woman said to be over 130 who drinks a little vodka before breakfast and smokes a pack of cigarettes a day; a man who sired a child when he was 100; and another man who was a village storyteller with an excellent memory at a reported age of 148.

A more comprehensive study (Pitskhelauri, 1982) finds that all the regions famous for long-lived people share four characteristics:

1. Diet is moderate, consisting mostly of fresh vegetables and herbs with little consumption of meat and fat. A prevailing belief is that it is better to leave the dining table a little bit hungry than too full.

A CLOSER LOOK

Making an Effort to Stay Vital

Several times in the past few pages, as well as in Chapter 20, nutrition and exercise have been suggested as important factors in the aging process. Not only are the most common diseases of old age, coronary heart disease and cancer, related to nutrition but many others are as well. For instance, for Type II diabetes, both the incidence and control of the disease are strongly related to diet. More generally, antioxidants in many foods defend against free radicals; specific proteins and enzymes promote optimal functioning of the entire body; and exercise not only improves circulation but promotes health on every level. However, maintaining a good diet and healthy activity is not easy in late adulthood.

Nutrition

With aging, ensuring good nutrition becomes more complex, primarily because two countervailing forces are at play. On the one hand, the need for vitamins and minerals increases with age, as the body's ability to break down food and utilize its nutrients decreases, due largely to the growing inefficiency of the digestive system. On the other hand, daily caloric requirements decrease by about 100 calories per decade after age 45, so the average 75-year-old has to consume at least 10 percent fewer calories per day than at middle age to maintain weight and energy (Blumberg, 1996).

Because more nutrients need to be packed into fewer calories, a varied and healthful diet, emphasizing fresh fruits and vegetables, lean meats and fish, and complex carbohydrates (cereals and grains), is even more essential in late adulthood than earlier in life. Taking in adequate amounts of water is particularly important for the elderly, because aging cells hold water less efficiently and aging digestive systems need more water to function well.

Further complicating the picture is the fact that the senses of smell and taste diminish with age, making food less appealing. A number of external factors may also affect the nutrition of certain segments of the elderly: (1) poverty (high-quality nutrients are more expensive); (2) living alone (those who eat alone tend to eat quick, irregular meals); (3) dental problems (missing teeth and gum disease make people eat softer food and less of it).

Furthermore, many of the elderly take drugs that affect nutritional requirements. For example, aspirin (taken daily by many who have arthritis) increases the need for vitamin C; antibiotics reduce the absorption of iron, calcium, and vitamin K; antacids can reduce absorption of protein; oil-based laxatives deplete vitamins A and D; and so on (Rollins & Thompson, 1993). Alcohol, especially in large amounts, is very detrimental to good nutrition, depleting B vitamins, calcium, magnesium, and vitamin C in particular (Iber, 1990). Alcohol and caffeine, as well as many prescription drugs, reduce the water in the body, making the elderly especially vulnerable to dehydration (Reiff, 1989).

What are the hazards of inadequate nourishment? One may be faster aging. Antioxidants, including vitamins C and E and beta-carotene (most commonly found in carrots), as well as some enzymes and minerals, notably selenium, might slow down the disease and aging processes. Deficiencies, on the other hand, might speed them up (Mertz, 1989).

Another hazard may be impaired memory. Some research suggests that those elderly with clear deficits in B vitamins, particularly B_{12} and folic acid, are more likely to have memory deficiencies than those whose intake was adequate (Wahlin et al., 1996).

A third hazard is reduced potency, or, in some cases, excessive potency, of various drugs. Proper medication often requires taking the drugs with food in order to mix the chemicals, reduce stomach upsets, and allow proper elimination. Failure to do so commonly produces adverse reactions (Bressler, 1993).

A variety of studies confirm that nutritional deficits are destructive at any age but are particularly so in late adulthood (Blumberg, 1996). However, taking vitamins and minerals in pill form needs to be done cautiously, if at all. Since the digestive system of the elderly both absorbs food less efficiently and excretes wastes less quickly, the dangers of vitamin imbalance and overuse are more serious in old age than earlier. Some vitamins, such as vitamin A, are toxic in large doses, and large doses of vitamins, in

2. Work continues throughout life. In these rural areas, even very elderly adults help with farm work and household tasks, including child care.

3. Family and community are important. All the long-lived are well integrated into families of several generations and interact frequently with friends and neighbors.

4. Exercise and relaxation are part of the daily routine. Most of the long-lived take a stroll in the morning and another in the evening

general, may well harm an individual by upsetting the natural nutritional balance in the system, creating vitamin needs and dependencies where there were none before.

Given this state of affairs, it seems reasonable to conclude that the elderly should consume a healthful diet and, beyond that, proceed cautiously. Diagnosis of particular individual vitamin deficiencies should be the only basis for undertaking any expensive or unusual vitamin regimen.

Exercise

Many of the elderly are reluctant to exercise as much as they might. In fact, the frequency of any exercise—from formal competitive sports to taking a stroll—decreases with every decade beginning at age 20.

Even for the very old, however, physical activity is beneficial, not only for the cardiovascular system but also for the respiratory, digestive, and virtually every other body system. In late adulthood, of course, the pace of exercise must be carefully adjusted to match the declines that have occurred in heart and lung functioning. For some, this means that jogging replaces running; for others, that brisk walking replaces jogging; for others, that strolling replaces brisk walking. Nonetheless, regular exercise—three or more times a week for at least half an hour—is even more important in late adulthood than earlier to help maintain the strength of the heart muscle and the lungs. Activities that involve continuous rhythmic movements for an extended period are more beneficial than those that require sudden, strenuous effort (Berg & Cassells, 1990).

Surprisingly, an exercise regimen for the elderly does not preclude weight-lifting, if done carefully. In fact, although the elderly lose muscle at an increasingly rapid rate, they can retain or gain muscle rapidly as well if they undertake strength training, including working out with weight machines (Felsenthal et al., 1994; Fiatarone et al., 1990; Raloff, 1996). The result is usually not bulging muscles but something much more important—greater mobility, greater leg strength, and thus fewer falls (and less damage if a fall occurs) and longer life. In fact, lower body movement is one of the best predictors of vitality in old age.

While most members of health clubs are adults in their 20s and 30s, the ones who benefit most are in their 70s and 80s. Rapid muscle gain, improved mobility, intellectual quickness, and longer life are all benefits from targeted strengthening in late adulthood.

With muscle strength, as with many other aspects of aging with vitality, it is almost never too late to improve. Reporting on a ten-week program of muscle training with frail nursing-home residents between the ages of 72 and 98, Raloff (1996) notes: "Individuals more than doubled the strength of trained muscles and increased their stair-climbing power by 28 percent when they exercised their legs with resistance training three times a week." Perhaps the day will come when "Let's get pumped" is as much a motto for the old as it is for young hard-bodies.

(often up and down mountains), most take a midday nap, and most spend several hours socializing in the evening, telling stories and discussing the day's events.

Beyond these four aspects of the individual's life, geography and tradition may be influential as well. All three places famous for long-lived people are in rural, mountainous regions that are at least 3,000 feet above sea level. This minimizes pollution and maximizes lung and heart fitness, for even walking in these regions can be considered aerobic exercise. Furthermore,

in all three, the aged are respected, and strong traditions ensure the elderly an important social role.

Some researchers suggest that another factor may account for these cases of unusual longevity—stretching the truth. None of the communities of the long-lived have birth or marriage records from the nineteenth century that are verifiable, at least to the satisfaction of critics. In fact, beginning at about age 70, many people in these areas systematically exaggerate their age, although how greatly is debatable (Leaf, 1982; Thorson, 1995). It may be that persons who claim to be 100 years old are only in their 80s, and that those who are supposedly long past 100 are only a little bit past it.

This does not render the earlier reports useless, for no one doubts that an unusual number of very old healthy people thrive in these isolated areas of the world. While their genetic clocks almost certainly do not allow them to live to age 148, their habits and culture do allow a surprising number of them to reach 100.

Research on those who reach late-late adulthood in countries where records are accurate show lifestyle patterns similar in many ways to those of the long-lived of isolated areas: a lifetime of moderate diet, hard work, optimistic attitude, intellectual curiosity, and social involvement is typical of the oldest-old in developed countries as well (Adler, 1991; Beard, 1991; Poon, 1992).

Taken together, these and many other studies of the very old lead to a ready conclusion: if people reach late adulthood in good health, their attitudes and activities may be even more important in determining the length and quality of their remaining years than purely physiological factors. It also leads those of us who are much younger to push our own horizons, diminishing our self-defeating ageism as we look toward our future.

SUMMARY

Ageism

1. Prejudices about the elderly are common and destructive, for they result in the old living lives that are more limited and isolated than they need to be. Contrary to the stereotype, most of the aged are happy, healthy, and active. Fortunately, ageism is weakening as gerontologists work to provide a more comprehensive picture of old age and as the sheer number of the aged in the general population increases.

The Aging Process

2. Primary aging occurs throughout the life span, even from birth. Secondary aging involves changes caused by particular conditions or illnesses, which may correlate with age but are not the inevitable result of age.

3. The many apparent changes in skin, hair, and body shape that began earlier in adulthood continue. In addition, most older people are somewhat shorter and weigh less than they did, and they walk more stiffly. Such changes in appearance can affect the self-concept of the older person.

4. Vision is almost always impaired by late adulthood: nine out of ten of the elderly need glasses. Those over age 80 are likely to experience at least one of the three major eye diseases of the elderly—cataracts, glaucoma, or senile macular degeneration. Most vision problems can be corrected, to some extent. Problems with hearing affect about a third of the elderly, often causing social isolation as well as feelings of rejection.

5. The age-related declines of the major body systems and organ reserve eventually reach a point—different for everyone—at which some of the routines of daily life need adjusting. For example, although exercise is just as important during late adulthood as earlier, its pace needs to be slower.

Aging and Disease

6. The aging process is not synonymous with the disease process. We should not assume that illness is an expected, and thus an accepted, companion during the later years. Unfortunately, many of the elderly—and some of their physicians—attempt to overlook problems that need medical attention because they believe their symptoms are just part of growing old. It is true, however, that aging makes people more susceptible to most chronic and critical diseases and makes recovery slower.

7. Although there is a correlation between aging and disease, there is much variation in the incidence of every specific disease. Some diseases affect men more than women, for instance, and populations of some countries have higher rates of certain cancers and lower rates of others, while the rates may be the reverse in other nations. Almost every elderly person has some chronic ailments, but the severity and specifics depend on the individual's health habits, environments, and genes.

8. Many gerontologists believe postponement of the onset of various diseases associated with aging will lead to the compression of morbidity, reducing the period of pain and disability before death. The result will be the extension not of the maximum human life span but of the period of vital, healthy aging.

Theories of the Causes of Aging

9. There are many theories that address the environmental and genetic causes of aging. One theory is that as we use our bodies we wear them out, just as a machine wears out with extended use. This wear-and-tear theory does not, however, explain much of what research finds—that activity promotes longer life and healthier aging.

10. Cellular theories of aging seem more plausible. Perhaps the DNA duplication and repair processes are affected by radiation and other factors, leading to an accumulation of errors when new cells are made. These errors may eventually produce an error catastrophe, in which the body suddenly reaches a point at which it is much more vulnerable to disease and death.

11. The decline in the immune system is thought by some to be the cause of aging, as it contributes to the elderly's increasing vulnerability to disease. As the thymus shrinks and production of both B- and T-cells decreases, the body becomes less able to fight against diseases.

12. The maximum human life span may well be fixed by a kind of genetic clock that switches the aging process on at some point. The theory that genes may be responsible for aging is buttressed by evidence that several conditions that are accompanied by premature aging, such as Down syndrome and progeria, are caused by genetic abnormalities.

13. Further evidence for programmed senescence is found in the Hayflick limit. Even in ideal conditions, cells in the laboratory stop reproducing themselves after a certain number of divisions. This number decreases as the age of the cell donor increases.

May You Live So Long

14. In three regions of the world, parts of the former Soviet Union, Pakistan, and Peru, large numbers of people seem to live to be very old. Moderate diet, high altitude, hard work, and traditional respect for the aged characterize all three places.

KEY TERMS

ageism (606)
gerontology (606)
young-old (606)
old-old (606)
primary aging (607)
secondary aging (607)
demography (608)
demographic pyramid (608)
demographic square (608)
dependency ratio (608)
cataracts (611)
glaucoma (611)
senile macular degeneration (611)
compression of morbidity (616)
wear-and-tear theory (618)
free radicals (619)
antioxidants (620)
error catastrophe (620)
B-cells (621)
T-cells (621)
maximum life span (622)
average life expectancy (622)
genetic clock (623)
progeria (623)
Hayflick limit (624)

KEY QUESTIONS

1. How is ageism comparable to racism or sexism?

2. Are the effects of secondary aging inevitable? Why or why not?

3. What changes occur in the sense organs in old age and how can they be ameliorated?

4. What is the relationship between aging and disease?

5. Why is the wear-and-tear theory not very helpful in explaining the aging process?

6. In what ways do the cellular theories of aging seem plausible?

7. How is the immune system affected by aging? Why is the decline of the immune system a plausible theory for the cause of aging?

8. What does Leonard Hayflick's research on cells from individuals of various ages tend to show?

9. What are some of the characteristics of people who live to a very old age?

10. What type of exercise is appropriate for the elderly and why?

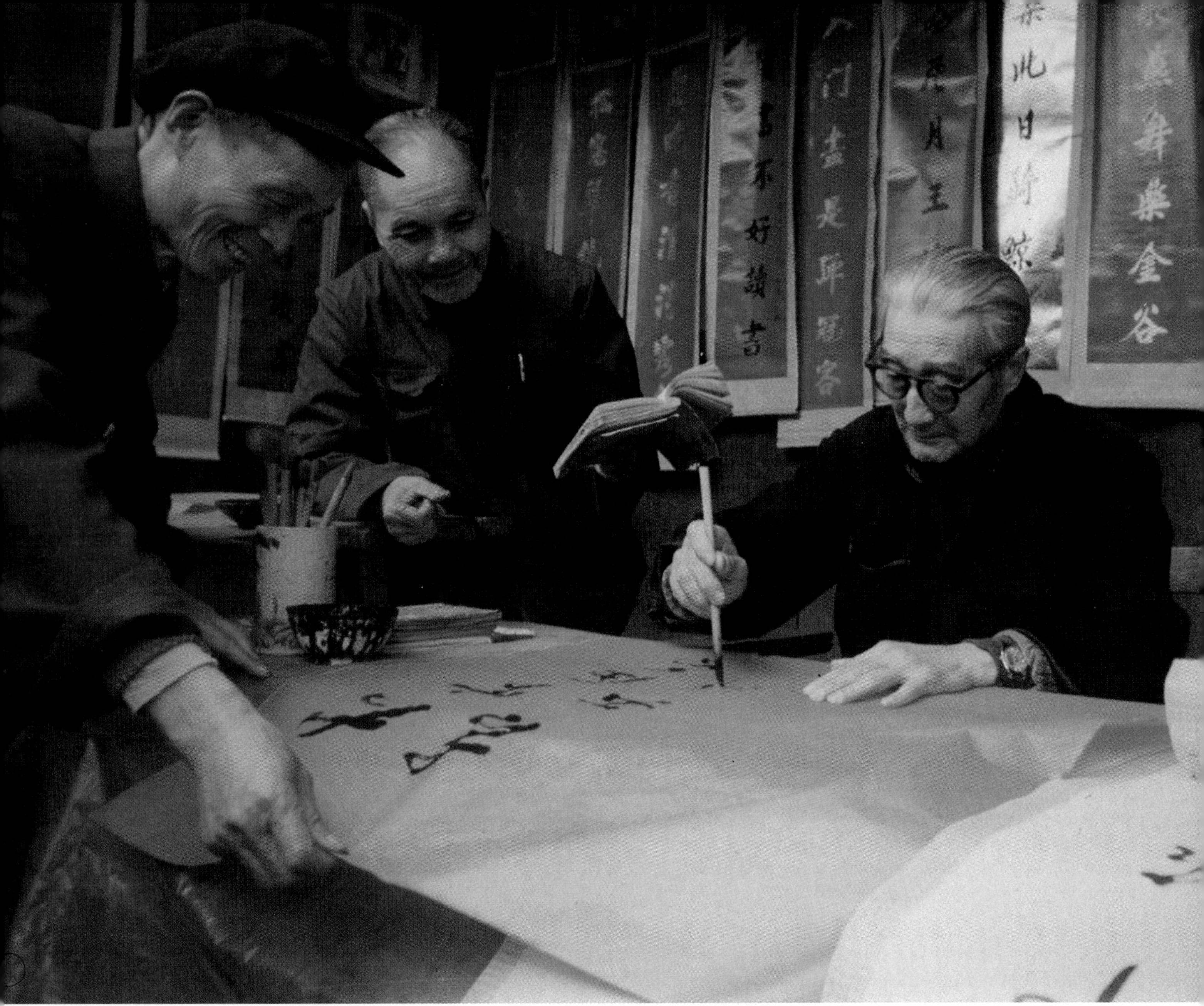

Late Adulthood: Cognitive Development

CHAPTER

24

As we saw in Chapter 21, over the course of adulthood some cognitive abilities increase, others wane, and some remain stable. By late adulthood, the fact that cognitive development is multidirectional becomes even more apparent, as the cumulative effects of increasing experience and declining biological capacities exert opposing forces. On one hand, older adults have a broader history of personal experience and knowledge to draw on when facing the intellectual challenges of later life, and this period may witness a deepening wisdom. On the other hand, physical impairments, perceptual declines, decreases in energy, and slowed reaction time exact an increasing toll on cognitive competence, causing many aspects of intelligence to show obvious declines during late adulthood. In this chapter, we complete the story of cognitive development in adulthood by examining the factors that are particularly influential in shaping cognition during old age.

Intellectual Changes in Information Processing

Although most intellectual abilities for most people increase or remain stable throughout early and middle adulthood, beyond age 60 everyone begins to experience some cognitive decrements. In Schaie's Seattle Longitudinal Study (described in Chapter 21), the average older adult began to show significant declines on all five "primary mental abilities" (Verbal Meaning, Spatial Orientation, Inductive Reasoning, Number Ability, and Word Fluency), with particularly notable declines in the underlying abilities of processing speed and numeric ability (Schaie, 1996).

However, while age-related declines are experienced by everyone, not everyone experiences them the same way. Variability in intellectual ability from person to person, apparent throughout childhood and adulthood, is even greater in later life, as some individuals reach new peaks of creative production, many others seem relatively unchanged, and some sink into mindless confusion (Powell, 1994). Further, each particular cognitive ability tends to show a somewhat different rate of age-related decline, partly because the various parts of the brain are affected differently by aging.

Why do cognitive decrements occur and exactly how are they apparent? Many researchers believe that answers can be found by examining age-

sensory register A memory system that functions for only a fraction of a second during sensory processing, retaining a fleeting impression of the stimulus that has just impinged on a particular sense organ (e.g., the eyes). If a person looks at an object, for example, and then closes his or her eyes, the visual image of the object is briefly maintained. Significant information is transferred to working memory. (Also called *sensory store*.)

related changes in the various processes by which the human mind records, stores, and retrieves information. More specifically, researchers have sought to understand intellectual changes in late adulthood by studying changes in the components of information processing—the sensory register, working memory, knowledge base, and control processes.

Sensory Register

As introduced in Chapter 2, the **sensory register** (also called the *sensory store*) holds incoming sensory information for a split second after it is received, allowing it to be selectively processed by other components of the information-processing system. (The workings of the sensory register are revealed in the momentary afterimage, visual or auditory, that remains in the brain after you perceive something. If, for example, you hear someone say several words and then realize that the person was speaking to you, you can usually remember what the person said, even though you weren't paying attention at the time.) Research suggests that the effects of aging may create small decrements in the sensitivity of the sensory register. As a result, it takes longer to register sensory information, and the information fades more quickly once it has been registered (Fozard, 1990). In general, this age-related slowdown is relatively slight and can easily be overcome (such as by asking others to speak slowly, or by looking longer and more intently at crucial sights) (Albert & Moss, 1996; Poon, 1985).

However, in order for sensory information to become registered, it must, obviously, cross the sensory threshold. That is, the sensory systems must be able to detect the relevant stimuli. Due to the kinds of sensory-system declines outlined in Chapter 23, some older people cannot register certain information—like the details of a dimly lit room or a soft conversation spoken against a noisy background—because they cannot detect the sensory stimuli in question. Since such deficits are usually progressive, eventually even loud conversation or the major features of a poorly lighted room may be missed by a substantial proportion of the oldest-old.

The effect of this sensory loss on cognition can be powerful. For example, in a study that compared the cognitive abilities of individuals ranging in age from 70 to 100, it was found that the differences in the subjects' visual

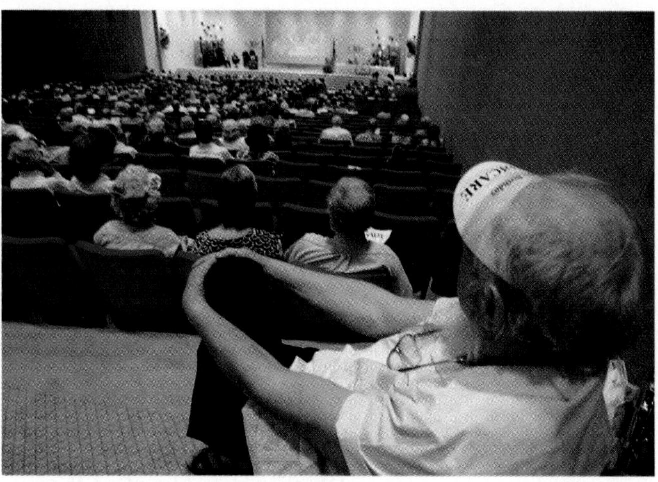

Everyone attending this Medicare meeting no doubt has a less efficient sensory register than they did in their younger years. One effect of this decrement is that spoken words that are only partially heard are likely to fade from consciousness before the listener can decipher them on the basis of their context.

and auditory acuity accounted for half of the variance in their cognitive scores (Lindenberger & Baltes, 1994). In other words, if one person scored 20 points higher on a given cognitive test than another, typically about 10 of those points could be traced to the better sight or hearing of the "smarter" person. Thus, while the sensory register itself declines only a small amount, the declines in sensory acuity can be large, and when they are, cognition clearly suffers.

Working Memory

Once information is perceived, it must be placed in working memory in order to be utilized. As discussed in Chapters 2 and 12, **working memory** is the processing component through which current, conscious mental activity occurs. Like a personal computer, working memory has two interrelated functions (Salthouse, 1990). The first function is to temporarily store information so that it can be consciously used, somewhat like the information displayed on a PC monitor. As with a PC, the working-memory component is constantly being replenished with new information (from the immediate sensory register or from the knowledge base) for current use. Also as with a PC, the working-memory component discards or transfers to the knowledge base old information when it is no longer relevant to current tasks or when there is no more room in working memory.

The second function of working memory is to process information. Using the information in current storage, working memory enables integrative reasoning, mental calculations, the drawing of inferences, and other cognitive processes. In a sense, working memory functions both as a temporary information repository and as an analytical processor of information.

In terms of both storage and processing functions, older adults have smaller working-memory capacity than younger adults do. Older individuals are particularly likely to experience difficulty simultaneously holding several items of new information in mind while analyzing them in complex ways. They also experience difficulty holding new material in mind when it is mixed in with distracting material. For example, if an experimenter gives adults something to remember, such as a list of objects or a passage of poetry, and then distracts them by having them count backward by 3s from 150 to 0, older adults have far worse recall of the initial material than do younger adults (Parkin, 1993).

Similarly, it is harder for older adults to perform a task when they must simultaneously ignore distracting information embedded in the task. This was demonstrated by an experiment in which a group of 62- to 75-year-olds and a group of 17- to 22-years-olds were asked to read aloud only the italicized words in a passage that began:

> *Bertha McKee was working as a* cold Van Dyck *volunteer at the* gallery Van Dyck *information booth* college in a carton *at the museum.* Gallery in a carton *She brushed off the* Van Dyck College *snow which had fallen . . .* [Carlson et al., 1995]

In order to successfully perform this task, the older adults had to read the passage three times as slowly as the young adults.

The same declines in working-memory processes are apparent with more meaningful material, such as remembering and analyzing the content

working memory That part of the information-processing system that handles current, conscious mental activity. Working memory is constantly receiving new information, so thoughts and memories are either discarded or transferred to the more permanent knowledge base.

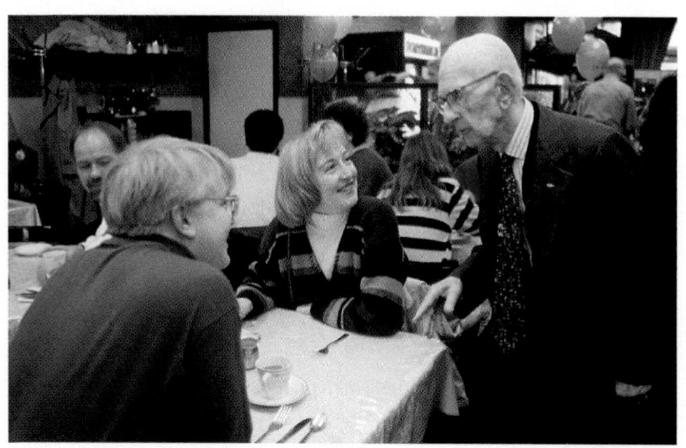

At the age of 104, Sidney Amber took up a career as a maitre d' at Sears Fine Food Restaurant in San Francisco, California, where he worked until his death six years later. Because his job involved social skills and established routines more than processing new information, whatever deficits he may have experienced in working memory over his many years were less crucial to his performance than if he had been, say, a waiter or a cashier.

of a paragraph or an essay. In one study (Light & Capps, 1986), younger and older adults (averaging ages 24 and 71, respectively) were asked to listen to short sentences and designate the probable antecedent for ambiguous pronouns. In each case, the context provided clear clues to the likely answer. For example, in "*Henry spoke at a meeting while John drove to the beach. He brought along a surfboard,*" it is likely that "He" refers to John, whereas in "*Henry spoke at a meeting while John drove to the beach. He lectured on the administration,*" it is likely that "He" refers to Henry. On tasks such as these, older and younger adults performed equally. However, when a sentence or two separated the pronoun and its antecedent—as in "*Henry spoke at a meeting while John drove to the beach. It was a nice day, and there was the sound of activity in the streets. He brought along a surfboard*"—older adults were significantly less adept at identifying who "He" referred to, because the intervening sentence made it more difficult for them to integrate information from the first and the third sentences while keeping the entire paragraph in mind. As the authors of this study concluded, "under conditions of high memory load . . . memory problems may masquerade as comprehension problems."

All these examples show that increased processing demands, whether caused by the complexity of the task or by distractions, strain the working memory of older adults much more than they do the working memory of younger adults. However, if processing demands are not influential and speed of performance is not a consideration, older adults can understand and remember the meanings of text passages and other material almost as well as younger adults.

Knowledge Base

The **knowledge base** is the storehouse of all the information that was ever put into memory, including information learned a minute or two earlier (short-term memory) and information learned days or decades ago (long-term memory). The knowledge base, is, of course, accessed through memory. Correspondingly, its functioning is assessed by determining which facts, observations, and concepts the person can retrieve from memory under which circumstances.

knowledge base That part of the information-processing system that stores long-term information and has a virtually limitless capacity.

When groups of older and younger adults are asked to deliberately remember anything that might be in the knowledge base—from long-stored information, such as a famous date in world history, to something they re-

cently learned, such as the date and time for an upcoming doctor's appointment—the average score of the younger adults is almost always better. Indeed, a major review of the accumulated research on memory came to the conclusion that both short-term and long-term memory are diminished in older adults, a conclusion that would come as no surprise to the aged themselves (LaVoie & Light, 1994; Light, 1991).

However, this conclusion must be qualified. As we will now see, accessing material from the knowledge base is not a simple process; some kinds of knowledge, under particular circumstances, are much easier for older adults to retrieve than other kinds.

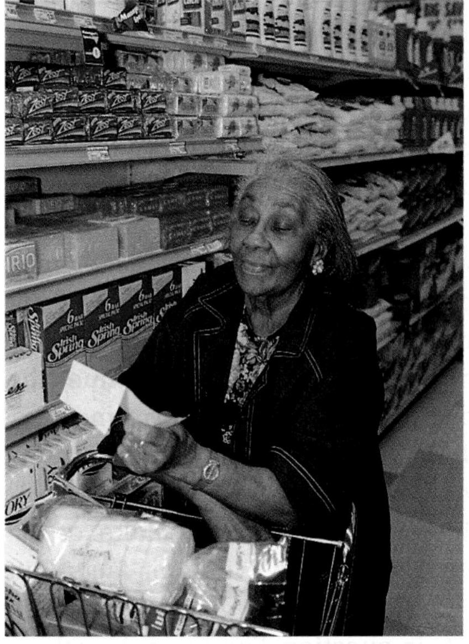

As memory becomes less efficient with age, older adults usually need a shopping list when buying the week's groceries. However, this generality overlooks the distinction between explicit and implicit memory. If this woman were asked to memorize and recite her list without looking at it (explicit memory), she might well forget a number of the items. But if she were to go through the store's aisles without consulting the list, carefully scrutinizing the shelves, her implicit memory would most likely prompt her to select all the items she had written down.

Implicit and Explicit Memory

Memory, both short- and long-term, takes two distinct forms, each originating in a different area of the brain (Mayford et al., 1996; Schacter & Tulving, 1994). **Implicit memory** is a kind of "unconscious" or "automatic" memory involving habits, emotional responses, routine procedures, and the senses. For the most part, the contents of implicit memory were not deliberately memorized for later recall, and thus they may be difficult to retrieve in response to questioning, even though they are retrievable through other avenues. For example, if you were asked to describe the distinctive facial features of your best friend in third grade, you might find the task nearly impossible—but you would no doubt be able to immediately recognize that friend in a class photo. By contrast, **explicit memory** involves words, data, concepts, and the like. Most of what is in explicit memory was consciously learned, often by linking it with verbal information already in store and often for the purposes of later recall. Partly for this reason, the contents of explicit memory are easier to retrieve in response to questioning.

Researchers agree that implicit memory is much less vulnerable to age-related deficits than explicit memory is (Java, 1996). Indeed, on tests of implicit memory involving such tasks as picture recognition, older adults who are intellectually sharp overall show no evidence of decline (Cherry & Stadler, 1995; Fastenau et al., 1996; Mitchell, 1993).

The following experiment illustrates the age-related differences in explicit and implicit memory competency. In a series of trials, subjects were asked to look for an asterisk to appear in one of four boxes on a video screen in front of them and then to push a button under that box as quickly as possible. In each series, the asterisks appeared in a particular sequence, such that the same pattern was repeated ten times. The subjects were not told that the asterisks would appear in a particular sequence. Nonetheless, the older and younger adults recognized the patterns equally quickly, as measured by how much faster they pushed the buttons toward the end of the trials (when they could anticipate which box the asterisk would appear in next) than at the beginning, before implicit memory had allowed them to perceive the patterns. However, the age equality evident in implicit memory was not evident in explicit memory. When subjects were asked "What was the pattern?" older adults were less likely to correctly describe the pattern sequence than were younger adults (Howard & Howard, 1992). Thus, the memory for the pattern was in the older adults' knowledge base, as evidenced by their performance on the trials, even though it was not readily available to recall in response to explicit questions.

implicit memory Unconscious or automatic memory involving habits, emotional responses, routine procedures, and the senses.

explicit memory Memory that involves consciously learned words, data, and concepts.

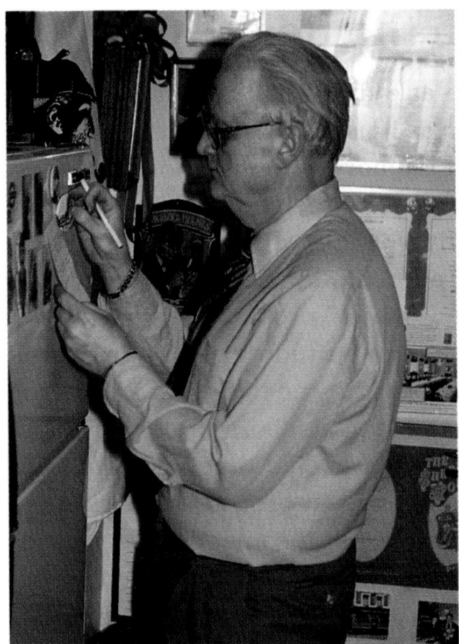

Many of the deficits that older adults exhibit in laboratory tests of memory ability are offset in daily life by their use of memory strategies—from putting up lists of things to do in a conspicuous place to arranging visual reminders, such as always leaving their medication dispenser and a water glass next to the kitchen sink.

Why the dramatic difference? Younger adults, it seems, were likely to put excessive trust in their memories ("I have an internal alarm that always goes off at the right time") and therefore were less likely to use mnemonic devices. Older adults, with a heightened awareness of the unreliableness of memory, used reminders, such as a note on the telephone or a shoe near the door, and thus almost always called on time.

The experimenters then attempted to increase the rate of forgetting in the older adults by asking for only one call per week, at a time designated by the researcher, and by making the subjects promise to avoid using any visible reminders of the appointment. This time the results were, indeed, much different: about half the elderly and an equal proportion of the young failed to call at the appointed time. And it seems likely that the number of old people who forgot the appointment might have been greater, since some of the older subjects continued to use some sort of memory-priming measure, such as carrying the phone number in a visible place in their wallet, for instance, despite instructions to keep it out of sight. One of the researchers concluded:

> With more effort, we are sure we can bring old people's memory to its knees . . . but that hardly seems to be the point of this research. The main lesson of this venture into the dangerous real world is that old people have learned from experience what we have so consistently shown in the laboratory—that their memory is getting somewhat poorer—and they have structured their environment to compensate. [Moscovitch, 1982]

In another experiment that highlighted the limitations of laboratory research, adults of various ages were taught a novel memory technique. (In this technique, called the "method of loci," the person creates a mental picture of bizarre locations in which to "place" the items to be remembered.) The researchers found that many of the older adults quietly resisted using the new method, despite the fact that the laboratory experiment required it and that, within the narrow confines of the research conditions, the strategy ensured better recall. Instead, the older adults, to the detriment of their memory scores, used their own memory strategies or tried to combine theirs with the new technique. Half of the gap between the memory scores of the older adults and the younger ones could be traced directly to "noncompliance" rather than to age-related decline (Verhaeghen & Marcoen, 1996).

Resistance to learning new methods certainly would impair older adults in laboratory experiments, but it would not necessarily be detrimental in daily life. In many everyday situations, familiar recall strategies are often the most efficient.

The same can be said for other aspects of cognition in everyday life. For example, the fact that older adults tend to use top-down problem-solving strategies does not necessarily mean that their solution to a problem is inferior to that of younger adults. A lifetime of experience has given many older adults a wealth of problem-solving strategies and a depth to their knowledge base that few younger adults might have. This combination can enable older adults to arrive at answers and decisions that are just as good as those of younger adults, even though they may seek less information and consider fewer variables in their problem solving than the younger adults do (Meyer et al., 1995; Walsh & Hershey, 1993).

dementia Irreversible loss of intellectual functioning caused by organic brain damage or disease. Dementia becomes more common with age, but even in the very old, dementia is abnormal and pathological. Sometimes dementia is misdiagnosed, since reversible conditions such as depression and drug overdose can cause the symptoms of dementia.

The fact that older adults are slower to abandon old techniques to try new ones, or that they are less likely to consider peripheral information, or that they take longer to access information from their knowledge base is usually irrelevant in daily life, especially in situations that are not highly novel. As a noted scientist explains, the decisions that people make in everyday life are usually complex enough that

> decision time is controlled more by "appropriate programming" that uses our brains efficiently than by raw speed of information processing. . . . In most cases involving everyday activity, the young-old contrast should not be thought of as a contrast between a fast and a slow computer, but as a contrast between a fast computer with a limited library of programs and a slow computer with a large library. [Hunt, 1993]

As we saw in Chapter 21, a "hallmark of successful aging" is the capacity to strategically compensate for intellectual declines associated with aging, what Paul Baltes and his colleagues call "selective optimization with compensation" (Baltes & Baltes, 1990). Using mnemonic devices and written reminders, allowing additional time for problem solving and repeating instructions that might be confusing, focusing on cognitive tasks that are meaningful and ignoring those that are irrelevant—all are compensatory methods that most older adults typically use to optimize their strengths. As a result, the cognitive demands of daily life are well within the intellectual capacity of most older adults. (As the Research Report on page 646 indicates, this general principle can apply even in the difficult circumstances of a nursing home.)

This optimistic conclusion applies to "most," not all, of the elderly. A minority find that even the simplest cognitive tasks of everyday life become too confusing, too stressful, too overwhelming. This minority suffer from dementia, which we will now discuss.

Dementia

In ordinary conversation, loss of intellectual ability in elderly people is often referred to as "senility," with "senile dementia" used as a synonym for Alzheimer's disease. However, these terms are ageist in that they inaccurately emphasize the factor of age and ignore the many other factors that can cause failing cognition in the old. A better and more precise term for pathological loss of brain functioning is **dementia**—literally, severely impaired judgment, memory, or problem-solving ability (Edwards, 1993). Traditionally, when dementia occurred before age 60, it was called *presenile dementia*; when it occurred after age 60, it was referred to as *senile dementia* or *senile psychosis*. However, these terms are also ageist because the symptoms and consequences of dementia are similar at any age.

To be specific, dementia can be caused by more than fifty diseases and circumstances. In all cases, the general symptoms are similar: severe memory loss, rambling conversation and language lapses, confusion about place and time, inability to function socially or professionally, and changes in personality. The sequence and severity of these symptoms vary, as does the incidence, with age being one factor that increases the risk of onset (see Table 24.1). Although some forms of dementia are reversible, all are chronic and most are degenerative, becoming steadily worse as the years go by.

TABLE 24.1

Estimated Prevalence of Dementia

Age	Prevalence
60–64	1%
65–69	2%
70–74	3%
75–79	5%
80–84	11%
85–89	21%
90–95	33–40%

Based on studies in fifteen nations. Although there is some cross-cultural variance in both the definition and prevalence of dementia, all nations find that dementia is rare at age 60—about 1 person in 100 at age 60—and common after age 90—occurring in more than 1 person in 3.
Sources: Jorm et al., 1987; O'Connor, 1989; Preston, 1986; Ritchie et al., 1992.

RESEARCH REPORT

Practical Competence in a Nursing Home

The picture that research presents of cognitive functioning in late adulthood is fairly optimistic: despite age-related declines in memory and abstract reasoning, older adults can acquire strategies for adapting to these changes so that they can completely manage the demands of everyday life. But remember that this picture is based largely on studies of healthy, well-educated older adults living independently in the community. These adults are chosen as research subjects because they tend to be cooperative, are free of serious health problems that might confound the results, and are quite capable of giving informed consent. But they may also provide an overly rosy view of cognitive functioning in old age.

Consider another group, the elderly residents of a nursing home. Quite clearly, their living conditions often do not foster the kinds of practical competencies that are experienced by older adults living independently. Indeed, many nursing homes reinforce passive, dependent, and predictable behavior, and discourage behavior that is active, independent, or innovative. For example, residents who do not manage their own personal care or hygiene—who, say, just sit staring at their food when it is placed before them—are likely to receive help and attention from the staff; those who manage for themselves are likely to be ignored. Similarly, those who stick to the nursing home's schedules and routines are much more likely to be praised than those who, against the rules, attempt to get a midnight snack, or want to go shopping on a day not designated as shopping day, or try to keep a pet in their room. When older patients ask for an explanation of some medicine or therapy or, worse, refuse to cooperate with some aspect of their medical treatment, they are likely to be labeled as mentally impaired and disruptive, and to be treated accordingly. (Similar behavior in younger persons is much more likely to be regarded as a sign of mental alertness.) One review sums up, with frightening clarity, the conditions that prevail in many nursing homes:

> . . . the individual . . . gives up control over the most mundane daily activities, when to sleep, wake, visit, perform toileting activities, bathe, and shop. The patient is exposed to infantilization and numbing bureaucratic and health routines that are of obscure purpose due to the invariably poor communication and misinformation given to placate the patient. Information is withheld or distorted under the assumption that it will not be understood or well-tolerated by the patient. [White & Janson, 1986]

All of these contextual factors lead directly to a dearth of intellectual stimulation and, consequently, to intellectual decline.

Declines in cognitive competence are intensified if, in the name of protection from self-harm, the patient is subjected to physical restraints. As one report explains:

> Far from protecting patients from harm, restraints inflict it. Physical risks include bed sores, infections, reduced circulation, muscle weakness, pneumonia, loss of appetite and incontinence caused by immobility . . . Psychosocial risks, more difficult to quantify, include humiliation, fear of abandonment, impairment to self-image, agitation, panic, and disorientation. [Collopy et al., 1991]

Obviously, all these effects can have a devastating impact on cognitive functioning and competency.

However, research has shown that when nursing-home patients are encouraged to manage on their own as much as possible, many learn to take more control of their activities, developing their own schedules and social lives as well as becoming more responsible for their daily care. Such an approach is called "therapeutic risk-taking," allowing patients the freedom to make mistakes in order to facilitate their physical and mental health (Aronson, 1994).

Other research shows that nursing-home residents can learn to improve their perspective-taking skills. One commonly cited characteristic of nursing-home residents is their egocentrism: on Piagetian tasks, they often show themselves to be limited to their own perspective; in daily life, they typically seem greatly absorbed in their own problems and needs, with little awareness of those of their fellow residents. The goal of one research project was to reduce this egocentrism (Zaks & Labouvie-Vief, 1980). The researchers tested the ability of nursing-home residents to understand other points of view and then divided the residents into three groups that were equal in terms of age, health, and performance on the tests. One group was given special training in social understanding. The members participated in discussions and role-playing, centering on problems that might occur in the home (such as what could be done if one roommate liked to watch television late at night and the other liked to sleep). A second group discussed such problems but did not role-play. The third group, the control group, had no special training at all.

On retesting after the training period, the residents in the first group who had the most active social learning training were markedly less egocentric than the other two groups, and they improved in their ability to communicate with each other. As the authors of the study concluded, a substantial part of the egocentrism of older institutionalized adults may be caused by their lack of social interaction rather than by their cognitive inability to see other points of view.

Overall, when cognitive deficits appear in older people, trying to find a way to remedy them is a better approach than sympathetically accepting the deficits as inevitable. This is as true for nursing-home residents as for older people who live independently, and the social environment of the nursing home can play an important role in enhancing—or blunting—practical competence in nursing-home residents.

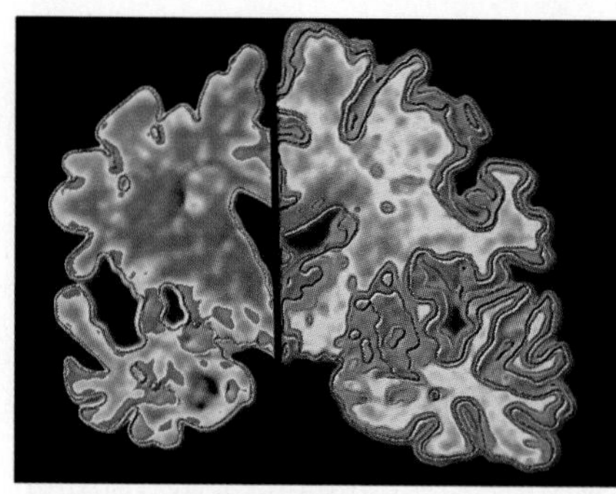

This computer graphic shows a vertical slice through a brain ravaged by Alzheimer's disease (*left*) compared with a similar slice of a normal brain (*right*). The diseased brain is shrunken as the result of the degeneration of neurons. Not viewable in this cross-section are tangles of protein filaments within the nerve cells as well as plaques that contain decaying dendrites and axons.

Alzheimer's Disease

The most common form of dementia in every industrialized nation except Japan is **Alzheimer's disease**, a disorder characterized by the proliferation of certain abnormalities in the cerebral cortex, called plaques and tangles, that destroy normal brain functioning. Alzheimer's disease runs usually through a progressive course of identifiable stages, beginning with general forgetfulness and ending in total mindlessness.

Until recently, the term "Alzheimer's disease" (AD) was reserved for symptoms of dementia in people under age 60, while the same symptoms in people over age 60 were termed SDAT, or senile dementia of the Alzheimer type. However, new techniques for analyzing brain tissue on autopsy (the only sure way to diagnose AD) show that, physiologically, there are no age-related distinctions to be made with Alzheimer's disease and that the amount of plaques and tangles correlates not with the victim's age but with the degree of intellectual impairment before death.

There are some age-related characteristics associated with Alzheimer's disease, however. When the disease appears in middle age, it usually progresses more quickly, reaching the last phase within three to five years, while in late adulthood it can take ten years or more to run its course. Alzheimer's disease in middle age is relatively rare and is probably caused by a genetic abnormality. (Many adults with trisomy-21 [Down syndrome] develop Alzheimer's disease in middle age.) Alzheimer's disease in late adulthood is increasingly common and probably is not genetic. According to a compilation of thirteen studies from several nations (Ritchie et al., 1992), the incidence begins to rise at about age 65, going from 1 in 100 to about 1 in 5 in those over age 85.

Stages of Alzheimer's Disease

As noted, Alzheimer's disease begins with a general forgetfulness concerning recent events or recently acquired information, particularly the names of people and places. Typically, a person in the first stage of the disease will put something away and shortly thereafter be unable to remember where it is. The person may also be unable to recall people's names after being introduced to them.

In this early stage, most people recognize that they have a memory problem and try to cope with it, writing down names, addresses, appointments, shopping lists, and other items much more than they once did. This first stage is often indistinguishable from "benign senescent forgetfulness," the normal explicit memory loss described earlier (Powell, 1994). Many people with Alzheimer's disease never progress beyond this first stage of dementia, remaining somewhat forgetful for the rest of their lives. On autopsy, almost all of the elderly have some plaques and tangles apparent in their cortexes, but those who suffer from Alzheimer's disease have far more such abnormalities than those who do not.

Alzheimer's disease The most common form of dementia, characterized by gradual deterioration of memory and personality, and marked by plaques and tangles in the brain. The underlying cause is unknown, as is the cure. Alzheimer's disease is not part of the normal aging process.

In the second stage of AD, the individual's confusion is more generalized and there are noticeable deficits in the person's concentration and short-term memory. The speech of people at this stage is often aimless and repetitious, their vocabulary is much more limited, and they frequently mix up words, using "tunnel" when they mean "bridge," for instance. They are

likely to read a newspaper article and forget it completely the next moment, or to put down their keys or glasses and within seconds have no idea where they could be. If they are suspicious by nature, they may accuse others of having stolen what they themselves have mislaid and forgotten. Then, "in the firm conviction of having been robbed, the patient starts hiding everything, but promptly forgets the hiding place. This reinforces the belief that thieves are at work" (Wirth, 1993).

In general, personality changes that occur in the second stage of AD tend to be—like the increased suspicion just cited—long-standing traits, which become more pronounced the less they are controlled by rational thought. A person given to tidiness may become compulsively neat; a person with a quick temper may begin to display explosive rages; a person who is asocial may become even more remote. Unlike the memory lapses in the first stage, memory loss in the second stage is sufficiently severe that many people forget they have a memory problem. Typical is the case of a man who, in stage one, began to run into financial problems because of his fading memory. In stage two, he was forced to turn over all his financial decisions to others, having no responsibility beyond putting his signature on documents. When asked if he was depressed, he replied that he didn't have any reason to be. He knew that he had had problems in the past, but now, he said, "I sign the papers. I'm in charge" (Foley, 1992).

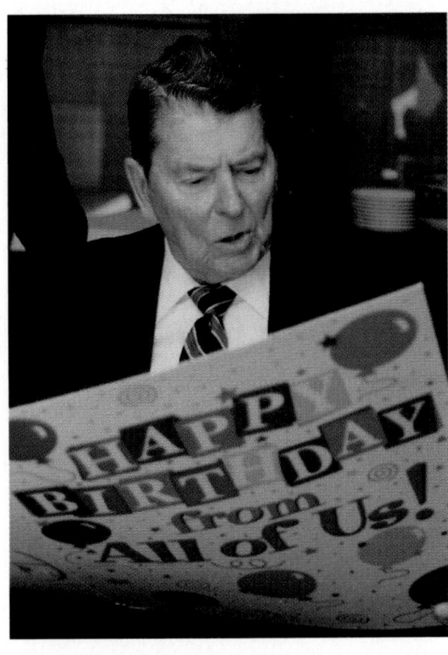

Shown here in 1997 on his 86th birthday, former President Ronald Reagan has helped to bring the problem of Alzheimer's disease to national attention. In an advanced stage of the disease, President Reagan can no longer even talk on the phone or recognize his former colleagues. In many ways, the greater suffering in most cases of Alzheimer's disease is experienced by those who knew and cared for the individual before the onset of the disease. When asked if President Reagan still recognized him, a close friend replied, "No. But much worse, I no longer recognize him."

The third stage of AD begins when memory loss becomes truly dangerous as well as debilitating. Individuals at this point can no longer manage their basic daily needs. They may take to eating a single food, such as bread, exclusively, or they may forget to eat entirely. Often they fail to dress properly, or at all, going out barefoot in winter or walking about the neighborhood naked. They are likely to turn away from a lighted stove or a hot iron and completely forget about it for the rest of the day, creating a clear fire hazard. They might go out on some errand and then lose track not only of the errand but also of the way back home. And they would not be able to ask neighbors for help because they wouldn't recognize them.

In the fourth stage of AD, patients need full-time care. They not only cannot take care of themselves; they also do not respond normally to others, sometimes becoming irrationally angry or paranoid. At the end, they can no longer put even a few words together to communicate. Saddest of all, they cannot recognize even their closest loved ones.

As people increasingly survive to old age, more and more will develop Alzheimer's disease, eventually living in a state of total dependency. Thus the victims of Alzheimer's disease include not only the patient but the patient's family, who typically are the main source of care until the last stages of the disease. Often compounding the demands of caregiving for family members is the difficulty in being understanding and patient with a person who seems in good health but is unable to behave or think normally and who sometimes becomes angry at efforts to help (Gatz et al., 1990). Almost always, the time comes when the family must seek assistance in caregiving, but paying for home-care aides or for nursing-home care is usually problematic because few insurance plans cover the long-term care of those with Alzheimer's disease. (Long-term care for the elderly who can no longer care for themselves is discussed in more detail in the next chapter.)

It is estimated that currently in the United States, 4 million elders have Alzheimer's disease and that, unless a cure is found, the number will double

multi-infarct dementia (MID) The form of dementia characterized by sporadic, and progressive, loss of intellectual functioning. The cause is repeated infarcts, or temporary obstructions of blood vessels, preventing sufficient blood from reaching the brain. Each infarct destroys some brain tissue. The underlying cause is an impaired circulatory system.

subcortical dementias Dementias, such as Parkinson's disease, Huntington's disease, and multiple sclerosis, that originate in the subcortex. These diseases begin with impairments in motor ability, and produce cognitive impairment in later stages.

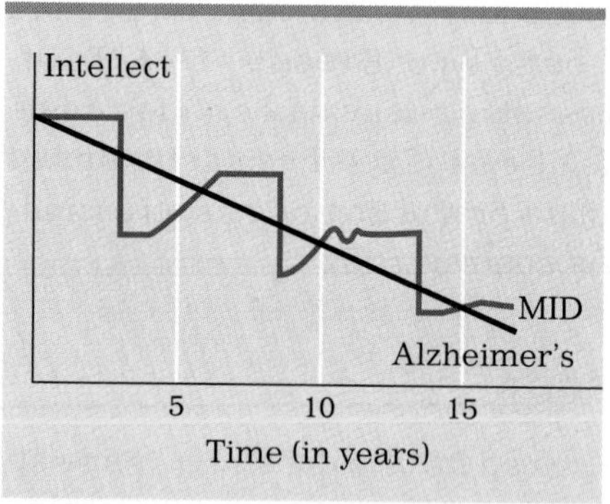

FIGURE 24.2 As shown on this chart, cognitive decline is apparent in both Alzheimer's disease and multi-infarct dementia. However, the pattern of decline for each disease is different. Victims of Alzheimer's disease show steady, gradual decline while those who suffer from MID get suddenly much worse, improve somewhat, and then experience another serious loss.

by the year 2025 (Binstock et al., 1992). Fortunately, although no cure seems forthcoming, scientists are learning many ways to postpone or slow the course of Alzheimer's disease. As we have already seen, estrogen replacement seems to delay AD in women. Other research shows that, when taken in the early stages, drugs to stop the loss of certain brain chemicals can improve cognition for AD patients (Giacobini, 1995; Hagino et al., 1995). Overall, drug treatment for Alzheimer's disease is in the very beginning stages, but it is a promising start (Marx, 1996).

Multi-Infarct Dementia

The second most common type of dementia is **multi-infarct dementia (MID)**, which in the United States is, by itself, responsible for about 15 percent of all cases of dementia and, in combination with Alzheimer's disease, accounts for another 25 percent. MID occurs because an infarct—a temporary obstruction of the blood vessels—prevents a sufficient supply of blood from reaching a particular area of the brain. This causes destruction of brain tissue, in what is commonly called a silent stroke, or ministroke. The immediate symptoms can include blurred vision, shaky or paralyzed limbs, slurred speech, or obvious mental confusion, but these manifestations typically disappear in hours or even minutes and are often so slight that the person is unaware that anything has happened. Nevertheless, brain damage has occurred (Edwards, 1993).

The underlying cause of the blood-vessel obstructions that lead to MID is systemic arteriosclerosis (hardening of the arteries). People who have problems with their circulatory systems, including those with heart disease, hypertension, numbness or tingling in their extremities, and diabetes, are at risk for arteriosclerosis and MID. Therefore, measures to improve circulation, such as exercise, or to control hypertension and diabetes, such as diet and drugs, help to prevent MID and to slow or halt the progression of the disease if it occurs.

The progression of MID is quite different from that of Alzheimer's disease (see Figure 24.2). Sometimes the person with MID shows a sudden drop in intellectual functioning following an infarct. Then, as other neurons take over some of the functions of the damaged area, the person becomes better.

However, as the name of the disease denotes, *multiple* infarcts typically occur, making it harder and harder for the remaining parts of the brain to compensate. If heart disease, major stroke, or other illnesses do not kill the MID victim, and ministrokes continue to occur, the person's behavior eventually becomes indistinguishable from that of someone suffering from Alzheimer's disease. On autopsy, however, it is clear that parts of the brain have been destroyed while other parts seem normal; the widespread proliferation of plaques and tangles characteristic of Alzheimer's disease is not present.

Subcortical Dementias

Many other dementias originate in the subcortex, which does not directly involve thinking and memory. **Subcortical dementias** cause a progressive loss of motor control but, initially at least, typically leave the thinking processes intact. Among these dementias are Parkinson's disease, Hunting-

Parkinson's disease A chronic, progressive disease that is characterized by muscle tremors and rigidity, and sometimes dementia, caused by a reduction of dopamine production in the brain.

ton's disease, and multiple sclerosis. All of these diseases can and often do eventually lead to dementia, but all begin with a person in full possession of his or her mental faculties and with clear indications that a serious, chronic illness has taken hold. In the later stages of subcortical dementia, when mental functioning has become affected, there can be great variability in a given person's mental impairment, depending on such factors as the time of day, the degree of stress or activity the person is experiencing, and the particular medical treatment being used. Further, in subcortical dementias, short-term memory and the ability to learn new material are usually much better than long-term memory, which is the opposite of the case for people with cortical degeneration (Derix, 1994).

The best-known of the subcortical dementias is **Parkinson's disease**, which initially causes rigidity and/or tremor of the muscles and often eventually leads to dementia (Edwards, 1993). Parkinson's disease is most common in the aged but it is not exclusively an old person's disease: an estimated 8 percent of the victims are under age 40 when the disease is first diagnosed. Parkinson's disease is related to the degeneration of neurons in an area of the brain that produces dopamine, a neurotransmitter essential to normal brain functioning. The reason for this degeneration is not known. Among the factors implicated as contributors to Parkinson's disease are genetic vulnerability and certain viruses. An interesting finding is that dementia in Parkinson's disease is not usually evident until the destruction of brain cells has reached a certain threshold. It is likely that mental impairment occurs when, and only when, the normal ability of the brain to compensate for neuron loss is overloaded. Since cognitive reserve declines with age, it is not surprising that Parkinson's disease that begins after age 60 is more likely to lead to dementia than the early-onset type (Edwards, 1993). Treatment for Parkinson's disease now includes taking daily doses of dopamine medication, which slows the course of the disease. Experimental treatment using cells from aborted fetuses to stimulate production of neurotransmitters in the brain has also helped but is very controversial.

Other Organic Causes of Dementia

As you have read, many other diseases also can involve dementia. Some affect the cortex first, impairing the person's mental processes while leaving motor skills intact. One of these is Pick's disease, which involves atrophy of the frontal and temporal lobes of the brain (Edwards, 1993). The initial symptoms of Pick's disease are personality changes, including loss of social skills and motivation, followed by loss of language and memory. Eventually the individual lapses into a completely vegetative state. Pick's disease is always fatal, running its course in two to fifteen years. Its cause is unknown, but genetic factors appear to be implicated.

Various toxins and infectious agents can also affect the cortex. Chronic alcoholism, for example, can lead to "wet-brain," or Korsakoff's syndrome, which is marked by severely impaired short-term memory. Almost half of all AIDS patients develop a brain infection that produces dementia, as do many people in the last stages of syphilis. Consumption of meat from "mad-cows" infected with bovine encephaly is believed to produce dementia and death, either because of a slow-acting virus or a prion, a protein particle that acts as a disease agent (Collinge et al., 1996).

Reversible Dementia

It is not uncommon for the elderly to be assumed to be suffering from one form of dementia or another when, in fact, their "symptomatic" behaviors are being caused by some other factor, such as medication, alcohol abuse, mental illness, or depression, all of which can be treated once the problem is determined.

Polypharmacy

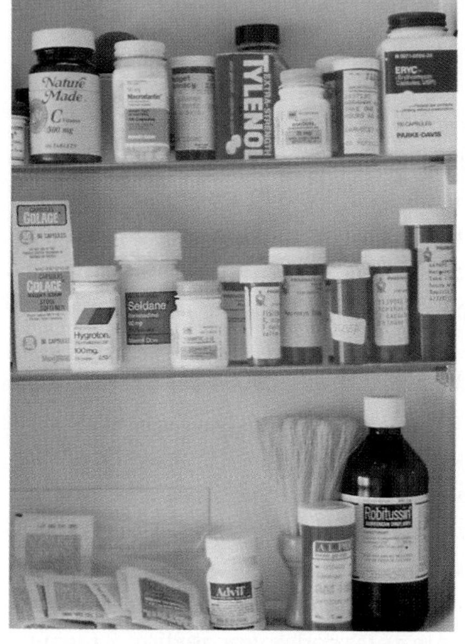

The vast majority of the elderly have medicine cabinets that look very much like this one, crammed with prescribed and over-the-counter medications for a variety of chronic conditions. In some cases, overdoses or intermixings of these medications can produce symptoms that resemble those of dementia.

Noncompliance with the prescribed use of medications is a serious problem for many of the elderly (Higbee, 1994), a problem that can lead to drug-related loss of intellectual functioning. Adults over 65, who represent about 12 percent of the American population, use 30 percent of all the drugs prescribed in the United States and 50 percent of all drugs sold over the counter (Beizer, 1994). In most cases, they are taking several different drugs for a variety of conditions. Such **polypharmacy** can result in drug interactions that produce symptoms of dementia, from confusion to psychotic behavior. An added problem is that the appropriate dose of most prescription drugs is usually determined by tests on younger adults, yet the dosage that is appropriate for 30-year-olds may be an overdose for the elderly, whose physiological ability to get rid of excess drugs is impaired (Beizer, 1994). Not surprisingly, adverse reactions to any single drug are between three and five times as common in adults over age 60 as they are in younger adults (Beizer, 1994; Butler et al., 1991).

Further, many of the drugs commonly taken by the elderly (such as most of those to reduce high blood pressure, or combat Parkinson's disease, or mitigate pain) can, by themselves, slow down mental processes, especially if the drugs are taken on an empty stomach or if, inadvertently, a double dose is taken. When this happens, the solution is simple—moderation or elimination of the problem prescription—but this solution obviously requires that the specific problem first be recognized.

Another problem that can impair mental functioning is malnutrition. Deficiencies of vitamin B_{12}, thiamin, niacin, and folic acid can cause confusion and loss of memory, and can further exaggerate the effects of various drugs. Malnutrition, which is almost as likely to occur among the well-off elderly as among the poor, is most common in the elderly overall because many older people lose their appetite, have tooth and gum problems, or have digestive difficulties that undercut adequate nutrition.

Alcohol

It is estimated that between 10 and 15 percent of older Americans abuse alcohol and that a third of this group become alcoholics in old age (Butler, 1991). Alcoholism among the elderly is more likely to take the form of steady, measured drinking rather than heavy drinking at a single sitting (La Rue et al., 1985). This type of "maintenance" drinking may be difficult for others to notice, since they might assume that some unsteady movement or slightly slurred speech or lapse of memory is attributable to age rather than to alcohol. It is also likely to go unnoticed because many elderly live alone and have no steady work obligations.

polypharmacy The use of multiple medications for various diseases and conditions.

Ounce for ounce, alcohol abuse causes greater cognitive impairment in old age than in earlier stages of adulthood. At the same blood alcohol levels, older adults show greater impairment of reaction time, memory, and decision making than younger adults. When these functions are already affected by age, the further declines produced by alcohol may reach the point where serious errors of judgment are likely to occur. And when alcohol abuse is chronic, disruptions in the functioning of the central nervous system impair learning, reasoning, perception, and other mental processes, producing alcohol dementia (Thompson et al., 1987). Over the long term, alcohol abuse can lead to Korsakoff's syndrome, with severe short-term memory loss caused by lesions in the brain.

Psychological Illness

In general, psychological illnesses are less common in the elderly than in younger adults (Gatz et al., 1996). Nonetheless, about 10 percent of the elderly who are diagnosed as demented are actually experiencing psychological, rather than physiological, illness. In some cases, the person is merely unusually anxious, which, as anyone who has taken a final exam under pressure knows, can make even a bright and healthy person forget important information. For many older people, the anxiety that occurs on arrival at a hospital or nursing home is sufficient to cause substantial disorientation and loss of memory. If the anxious new patient is tested immediately, a misdiagnosis of organic brain damage is possible. And if psychotropic medicine is overprescribed or overused, the result can be a person who seems continually demented (Sherman, 1994).

Depression in the elderly is also often misread as dementia. Although major depression is less common in late adulthood than earlier, at some time in their later years, many adults experience symptoms of minor depression that are sufficiently debilitating to resemble those of dementia (Blazer et al., 1991). In clinical assessments of the elderly, even mild depression, like anxiety, can diminish overall cognitive performance, even though it does not reduce underlying ability (Powell, 1994; Rabbitt et al., 1995). Of course, some individuals might become anxious or depressed because they notice signs of mental deterioration, or at least they think they do. On the whole, however, it seems that depression causes intellectual decline more

Depression in the elderly often goes unrecognized and untreated because it is mistaken as a natural component of aging, or as a sign of dementia. This is particularly unfortunate because depression is now fairly easy to treat, especially in its milder forms, which are more common among the elderly.

often than declines in intellectual ability affect mood. In fact, one of the symptoms of depression in late adulthood is exaggerated attention to small memory losses, which is quite the opposite reaction from that of people who truly suffer from dementia, who are often blithely unaware of their serious problems.

Depression in late adulthood, as at younger ages, is one of the most treatable mental illnesses. Psychotherapy and carefully used pharmacotherapy usually bring about notable improvement in a few weeks. However, even more than at younger ages, most depressed older people are not treated because no one recognizes their depression as a curable disease. Instead, many caregivers consider depression a natural consequence of aging, or they confuse the symptoms with those of brain disease. If the depressed person has recently lost a loved one, the symptoms of depression may mistakenly be attributed to bereavement. It is normal, of course, for the elderly who are in mourning to be sad and to have difficulty eating and sleeping, but the symptoms of bereavement do not normally include strong feelings of guilt and self-deprecation, nor do they last longer than a few months.

Sadly, one consequence of untreated depression among the elderly is that the suicide rate is higher for those over age 60 than for any other age group, with rates in the United States particularly high among white and Hispanic men and Asians of both sexes (U.S. Department of Health and Human Services, 1995). In many other nations—Canada, Chile, Hungary, France, and Japan, among them—the suicide rate for elderly men is higher than for any other age group.

In most cases, the precipitating event for the suicide is a social loss, with retirement or widowhood being the most common precipitator, especially for men who live alone (Canetto, 1992). A related cause of suicide is illness, particularly cancer or illnesses that affect the brain, or the fear that normal symptoms of aging are the first sign of such an illness. As at every age, confusion brought on by alcohol or drugs increases the risk of suicide.

Dementia can also be caused by brain injuries, brain tumors, and head injuries that result in an excess of fluid pressing on the brain. In these cases, surgery can often remedy the problem and restore normal cognitive functioning.

New Cognitive Development in Later Life

So far in this chapter we have mainly considered possible declines in the intellectual functioning of older adults. What about positive changes? Can older adults develop new interests, new patterns of thought, a deeper wisdom? Many of the major theorists on human development believe that they can. For example, Erik Erikson finds that the older generation are more interested in the arts, in children, and in the whole of human experience. They are the "social witnesses" to life and thus are more aware of the interdependence between one generation and another (Erikson et al., 1986). Abraham Maslow maintains that older adults are much more likely than younger people to reach what he considers the highest stage of development—*self-actualization*—which includes heightened aesthetic, creative, philosophical, and spiritual understanding (Maslow, 1970). Let us look, then, at some of these areas of development during late adulthood.

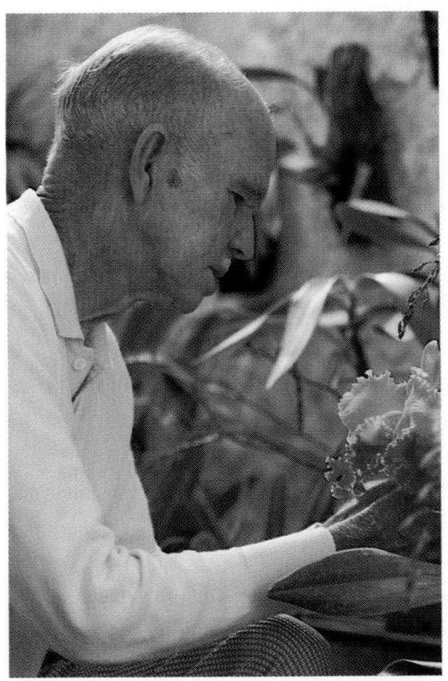

Gerontologists have found that, in old age, people often develop a heightened aesthetic sense and a deeper appreciation of nature. Perhaps this is because older adults focus more on the fundamental aspects of life, or perhaps it is simply that they have more time to savor delights that were overlooked in the hustle and bustle of earlier years.

Aesthetic Sense

Many people seem to appreciate nature and aesthetic experiences in a deeper way as they get older. As three leading gerontologists explain: "The elemental things of life—children, friendship, nature, human touching (physical and emotional), color, shape—assume greater significance as people sort out the more important from the less important. Old age can be a time of emotional and sensory awareness and enjoyment" (Butler et al., 1991).

For many older people, this heightened appreciation leads to active expression. They may begin gardening, bird-watching, pottery, painting, or playing a musical instrument—and not simply because they have nothing better to do. The importance that creativity can have for some in old age is wonderfully expressed by a 79-year-old man, unfamous, little educated, yet joyful at his workbench:

> This is the happiest time of my life. . . . I wish there was twenty-four hours in a day. Wuk hours, awake hours. Yew can keep y' sleep; plenty of time for that later on . . . That's what I want all this here time for now—to make things. I draw and I paint too . . . I don't copy anything. I make what I remember. I tarn wood. I paint the fields. As I say, I've niver bin so happy in my whole life and I only hope I last out. [quoted in Blythe, 1979]

For this man, and for many people, the impulse to create did not suddenly arise in late adulthood; it was present, although infrequently expressed, in earlier years. What does seem to occur in late adulthood is a deepening need to express and develop that impulse, perhaps because, as the years left to live become fewer, those people who bore a dream of creative expression decide to defer that dream no longer.

One of the most famous examples of late creative development is found in Anna Moses, a farm wife and mother of ten. For most of her life, she expressed her artistic sensitivity by stitching quilts and doing embroidery during the long farm winters. At age 75, arthritis made needlework impossible, so she took to "dabbling in oil" instead. Four years later, three of her oil paintings, displayed in a local drugstore, caught the eye of a New York City art dealer who happened to be passing by. He bought them, drove to Anna Moses' house to buy fifteen more, and began to exhibit them in the city. One year later, at age 80, "Grandma Moses" had her first one-woman show in New York, receiving international recognition for her unique "primitive" style. She continued to paint, "incredibly gaining in assurance and artistic discretion," into her 90s (Yglesias, 1980).

For those who have been creative all their lives, old age is often a time of continuing productivity, even of renewed inspiration. Famous examples of this include Michelangelo's painting the amazing frescoes in the Pauline chapel at age 75, Verdi's composing *Falstaff* when he was 80, and Frank Lloyd Wright's completing the Guggenheim Museum, an innovative architectural masterpiece, when he was 91. In a recent study of extraordinarily creative people, it was found that almost none of the subjects felt that their ability, their goals, or the quality of their work was much impaired with age. What had changed was their sense of urgency about their work, which was sharpened by their realization that fewer years lay ahead and that their energy and physical strength were diminishing (Csikszentmihalyi, 1996). As the researcher observed of these individuals, "In their seventies, eighties, and nineties, they may lack the fiery ambition of earlier years, but they are just as focused, efficient, and committed as before . . . perhaps more so."

The Life Review

In old age, many older people become more reflective and philosophical than they once were. In most cases, this reflectivity is personally centered, as the individual attempts to put his or her life in perspective, assessing accomplishments and failures in terms of what the person perceives to be the overall scheme of life.

One form of this attempt to put one's life into perspective is called the **life review**, in which an older person recalls and recounts various aspects of his or her life, remembering the highs and lows, and comparing the past with the present. In general, the life-review process connects one's own life with the future as one tells one's story to younger generations. At the same time, it renews links with past generations as one remembers what one's parents, grandparents, and even great-grandparents did and thought. One's relationship to humanity, to nature, to the whole of life also becomes a topic of reflection, as various memories are revived, reinterpreted, and finally reintegrated to achieve a better understanding of the entire life course (Kotre, 1995).

Sometimes the life review takes the simple form of nostalgia, reminiscence, or storytelling, which may be quite helpful to the older person, although not always easy for others to listen to. Yet it may be crucial to the older person's self-worth that others recognize the significance of these reminiscences. As Butler and his colleagues explain:

> We have been taught that this nostalgia represents living in the past and a preoccupation with self and that it is generally boring, meaningless, and time-consuming. Yet as a natural healing process it represents one of the underlying human capacities on which all psychotherapy depends. The life review should be recognized as a necessary and healthy process in daily life as well as a useful tool in the mental health care of older people. [Butler et al., 1991]

In some cases, the reflectivity of old age may lead to, or intensify, attempts to put broader historical, social, and cultural contexts of life into perspective. It is interesting to note that when Wayne Dennis (1966) studied the production of professionals in sixteen different fields, he found that in two of them—history and philosophy—production peaked in the 60s and 70s. Certainly one of the most famous examples is Will and Ariel Durant's *The Story of Civilization*—a monumental, ten-volume history of civilization written mostly in late adulthood—followed by *The Lessons of History* and *Interpretations of Life*, published when the Durants were in their 80s.

Wisdom

Wisdom is one of the most positive attributes commonly associated with older people. Indeed, the idea that wisdom may be one of the benefits of old age has become a "hoped-for antidote to views that have cast the process of aging in terms of intellectual deficit and regression" (Labouvie-Vief, 1990).

The question is, What is wisdom, really, and is it a common feature of old age? Wisdom is clearly an elusive concept, and any definition of it is bound to be at least partly subjective. In addition, whether any given individual is perceived as wise depends very much on the immediate social context in which that person's thoughts or actions are being judged. Given these obstacles to definitional precision, one of the more comprehensive, "all-purpose" definitions of wisdom is that offered by Paul Baltes, a develop-

life review The examination of one's own past life that many elderly people engage in. According to Butler, the life review is therapeutic, for it helps the older person to come to grips with aging and death.

Some cultures recognize and appreciate, far more than others, the contributions of wisdom drawn from years of experience and the "pragmatics of life."

mentalist specializing in cognitive gerontology. Baltes defines wisdom as "expert knowledge in the fundamental pragmatics of life permitting exceptional insight and judgment involving complex and uncertain matters of the human condition" (Baltes et al., 1992). Baltes and his colleagues argue that five features distinguish wisdom from other forms of human understanding (Dittmann-Kohli & Baltes, 1990):

1. rich factual knowledge that concerns the broad topic of human experience;
2. knowledge of the "pragmatics of life"—that is, practical and procedural knowledge about the conditions of life and its variations;
3. a contexual approach to understanding life that takes into account its broader ecological, social, and historical dimensions;
4. acceptance of the uncertainty in defining and solving life's problems and of the unpredictability of one's own future life course;
5. recognition of individual differences in values, goals, and priorities, leading to flexibility and relativism in tackling the contradictions of life experience.

Wisdom thus involves both the elements of dialectical thinking that emerge in early adulthood and the refinement of thinking that comes with years of personal experience. But is wisdom a typical characteristic of older adults' thinking?

In one effort to study wisdom, Smith and Baltes (1990) asked sixty adults of various ages to assess the lives of four fictitious persons who each faced a difficult decision regarding the future. Here is an example of one story concerning a young adult.

> Elizabeth, 33 years old and a successful professional for 8 years, was recently offered a major promotion. Her new responsibilities would require an increased time commitment. She and her husband would also like to have children before it is too late. Elizabeth is considering the following options: She could plan to accept the promotion, or she could plan to start a family.

The other three stories concerned dilemmas over parental responsibilities at home, accepting early retirement, and intergenerational commitments. After hearing these stories, subjects were asked to formulate a course of action for each fictitious person and to think aloud as they did so, indicating when they thought additional information was needed about certain issues. Their responses were subsequently transcribed and rated by a panel of human-service professionals according to whether they exhibited the characteristics of wisdom described above.

Not unexpectedly, wisdom appeared to be in fairly short supply. Of the 240 responses to their hypothetical stories, Smith and Baltes found that only 5 percent were judged as truly wise. Somewhat more surprisingly, the distribution of responses judged to be wise was fairly even across young, middle-aged, and old adults in the sample. That is, wisdom was not reserved for later life but could be found at any phase of adulthood, depending, presumably, on one's life experiences and reflective insight about them.

On balance, then, it seems fair to conclude that the mental processes in late adulthood can be adaptive and creative, not necessarily as efficient as thinking at younger ages but more appropriate to the final period of life. An illustrative and exemplary case in point is the following poem, written by Henry Wadsworth Longfellow at age 80.

> But why, you ask me, should this tale be told
> Of men grown old, or who are growing old?
> Ah, Nothing is too late
> Till the tired heart shall cease to palpitate;
> Cato learned Greek at eighty; Sophocles
> Wrote his grand *Oedipus*, and Simonides
> Bore off the prize of verse from his compeers,
> When each had numbered more than four score years,
> And Theophrastus, at four score and ten,
> Had just begun his *Characters of Men*.
> Chaucer, at Woodstock with the nightingales,
> At sixty wrote the *Canterbury Tales*;
> Goethe at Weimar, toiling to the last,
> Completed *Faust* when eighty years were past.
> These are indeed exceptions, but they show
> How far the gulf-stream of our youth may flow
> Into the arctic regions of our lives
> When little else than life itself survives.
> Shall we then sit us idly down and say
> The night hath come; it is no longer day?
> The night hath not yet come; we are not quite
> Cut off from labor by the failing light;
> Some work remains for us to do and dare;
> Even the oldest tree some fruit may bear;
> And as the evening twilight fades away
> The sky is filled with stars, invisible by day.

SUMMARY

Intellectual Changes in Information Processing

1. Although thinking processes become slower and less sharp once a person reaches late adulthood, there is much individual variation in this decrement, and each particular cognitive ability shows a different rate of age-related decline.

2. The sensory register declines relatively little in late adulthood. However, working memory shows notable declines, especially when one must simultaneously store and process information in complex ways. One reason for this loss is that processing takes longer with age.

3. With increasing age, adults experience greater difficulty accessing information from both short- and long-term memory. However, knowledge stored in implicit memory is more easily retrieved than are the facts and concepts stored in explicit memory. Past knowledge is more accessible the more it was initially "overlearned" and subsequently used.

4. Control processes also are less effective with age, particularly when measured in laboratory tests of intellectual functioning. The two possible reasons for this deficit are either that the aging brain is less capable of strategizing the best use of mental ability or that older persons do not know how to organize, memorize, and analyze information as well as they might.

5. Another reason older adults, on average, do not perform as well as younger adults on tests of cognitive functioning is that more of the older group have negative self-perceptions of their mental skills that undermine their motivation to succeed. Studies have also shown that older adults' cognitive performance can be negatively affected by exposure to ageist stereotypes.

Reasons for Age-Related Changes

6. With age, the brain's communication processes slow down, as measured by a notable decrease in reaction time in the elderly. When older adults are given more time to remember, analyze, and answer an intellectual problem, their performance improves markedly. Researchers have also found that both physical and intellectual activity can halt or reverse some of the cognitive slowdown that occurs in old age.

Cognitive Functioning in the Real World

7. In daily life, most of the elderly are not seriously handicapped by cognitive difficulties. Usually, once they recognize problems in their memory or other intellectual abilities, they learn to compensate with selective optimization; that is, they learn to build on strengths and shore up weaknesses. Even in nursing homes, recognition of ways in which older adults can manage their intellectual efforts is likely to produce positive results.

Dementia

8. Dementia, whether it occurs in late adulthood or earlier, is characterized by memory loss—at first minor lapses, then more serious forgetfulness, and finally such extreme losses that recognition of closest family members fades.

9. The most common cause of dementia is Alzheimer's disease, an incurable ailment that becomes more prevalent with age. While some cases of Alzheimer's disease are genetic, for the most part the cause is unknown. Drug therapy is beginning to offer some promise for the prevention and treatment of Alzheimer's disease.

10. Multi-infarct dementia is caused by a series of ministrokes that occur when impairment of blood circulation destroys portions of brain tissue. Measures to improve circulation and to control hypertension can prevent or slow the course of this form of dementia.

11. In addition to Alzheimer's disease and multi-infarct dementia, subcortical abnormalities, such as that leading to Parkinson's disease, are a leading cause of dementia. Other diseases that may lead to dementia are Pick's disease, alcoholism, and AIDS.

New Cognitive Development in Later Life

12. Many people become more responsive to nature, more interested in creative endeavors, and more philosophical as they grow older. The life review is a personal reflection that many older people undertake, remembering earlier experiences and putting their entire lives into perspective.

13. Wisdom is commonly thought to increase in life as a result of experience. This idea is yet to be confirmed by psychological research.

KEY TERMS

sensory register (632)
working memory (633)
knowledge base (634)
implicit memory (635)
explicit memory (635)
control processes (637)
dementia (645)
Alzheimer's disease (647)
multi-infarct dementia (MID) (649)
subcortical dementias (649)
Parkinson's disease (650)
polypharmacy (651)
life review (655)

KEY QUESTIONS

1. How is each part of the information-processing system—sensory register, working memory, knowledge base, and control processes—affected by age?

2. Compare age differences in short- and long-term memory and in explicit and implicit memory.

3. What are the physiological reasons for age-related declines in cognition and in what ways can these declines be moderated?

4. What are the psychological reasons for age-related declines in cognition and how can these declines be moderated?

5. How valid is it for people to refer to all older people as "senile" and for everyone to expect to develop dementia if they live long enough?

6. What are the similarities and differences between Alzheimer's disease and MID?

7. What can be done to prevent or reverse the various causes of dementia?

8. What is the purpose and the result of the life review?

9. What evidence is there that creativity, aesthetic appreciation, and wisdom increase with age?

Late Adulthood: Psychosocial Development

CHAPTER 25

Viewing development from age 60 on, one is struck by the vast array of possibilities and outcomes. As we saw in Chapter 23, some people in their 70s or even 80s are physically fit enough to run marathons, while others are completely incapacitated. In Chapter 24, we saw similar diversity in cognitive functioning. Such developmental variety will be even more apparent in this chapter. Research shows that, in psychosocial development, old age acts less as a leveler of individual differences than as a magnifier of them, expanding the diversity of human experience (Crystal & Waehrer, 1996; Ferraro & Farmer, 1996; Light et al., 1996; Schroots, 1993).

Theories of Psychosocial Development in Late Adulthood

One way to understand and organize this diversity is to begin with pertinent theories of psychosocial development in late adulthood. To simplify, these theories can be considered in three clusters: self-theories, social stratification theories, and life-course theories.

Self-Theories

Self-theories begin with the premise that adults make choices, confront problems, and interpret reality in such a way as to define, become, and express themselves as fully as possible. As Abraham Maslow (1968) described it, they attempt to *self-actualize*, to achieve their full potential. Such theories emphasize "human intentionality and the active part played by the individual in developing selfhood" (Marshall, 1996).

self-theories Theories of late adulthood holding that individuals seek to achieve their full potential, to become self-actualized.

continuity theory The theory that each person experiences late adulthood and behaves toward others in much the same way as at earlier periods of life.

Continuity Theory

One well-respected self-theory is **continuity theory**, which focuses on the ways the integrity of self is maintained as individuals experience the events and changes of old age. According to this theory, each person copes with late adulthood in much the same way that he or she coped with earlier periods of life, ideally with "a healthy capacity to see inner change as connected to the individual's past" (Atchley, 1991). Reinforced by the ecological niches that individuals have carved out for themselves, the so-called Big Five per-

Older adults who do volunteer work usually do not decide to do so just because they have free time on their hands. Their volunteerism is more likely to be a continuation of a lifelong pattern of being helpful to others and of having a sense of responsibility to the community at large.

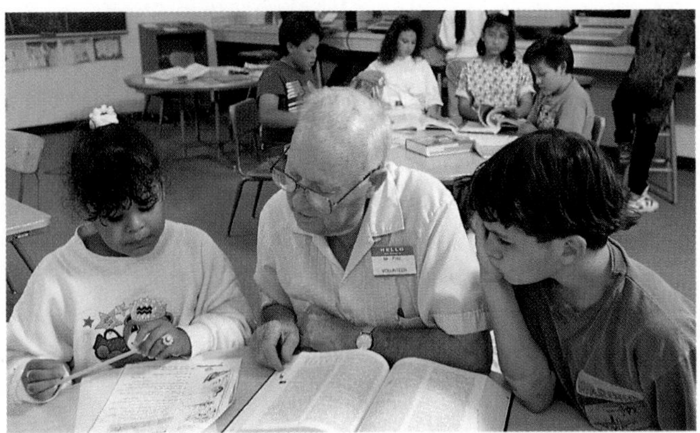

sonality traits (described in Chapter 22) are maintained throughout old age as they were in younger years. The outgoing person, for example, maintains a circle of friends and makes new acquaintances, even at age 90 in a nursing home; the socially withdrawn person continues to shy away from others. Correspondingly, when the elderly confront a new problem, such as high blood pressure, how they react will depend a great deal on their individual temperament. Someone who strongly tends toward neuroticism, for example, might spend hours morose and alone, imagining an impending stroke lurking behind every stress. The hypertensive person who is highly extroverted, by contrast, is likely to seek out advice and suggestions, organizing a support group or joining an exercise class.

In addition, habits and values developed in childhood and early adulthood generally tend to be maintained in late adulthood. For example, for those elderly who, over the years, gained inner strength and guidance from religious faith, such faith is likely to continue to be an important factor. This is so not because they see the end of life drawing nearer but because they have found, and continue to find, faith to be a reliable source of support (Idler, 1994). Similarly, attitudes toward everything from drugs, sexuality, and money to neatness, privacy, and the role of government are likely to reflect lifelong continuity (Binstock & Day, 1996).

In recent years, continuity theory has received substantial confirmation from behavioral genetics. Longitudinal studies of monozygotic and dizygotic twins find that, contrary to the logical idea that genetic influences weaken as life experiences accumulate, some traits seem even more apparent in late adulthood than earlier. Various life events—from how early a person retires to how often a person marries—seem to be at least as much affected by genetics as by life circumstances (Saudino et al., 1997). The explanation for this, some developmentalists hypothesize, is that, as older adults emerge from the harness of family and work obligations, their temperament receives freer rein and they can become more truly themselves. As one 103-year-old woman observed: "My core has stayed the same. Everything else has changed" (Troll & Skaff, 1997).

Selective Optimization

selective optimization with compensation The individual's devising of alternative strategies to compensate for age-related declines in ability.

Paul and Margaret Baltes emphasize that people can choose to cope with the undeniable physical and cognitive losses of late adulthood through **selective optimization with compensation**, a concept central to self-theories. As we saw in Chapter 21, the basic idea of selective optimization with com-

Max Roach has been a leading jazz drummer for over fifty years. His approach to his work at age 73 clearly reflects the idea of selective optimization with compensation: "I joined a health club . . . I thought I'd tune up, you know, tone up. Playing my instrument is a lot of exercise. All four limbs going. . . . I don't play the way I did back in the 52nd Street days. We were playing long, hard hours in all that smoke. It would kill me now if I played like I did then. Now I play concerts, and the show goes on for just an hour. But I'll tell you something: I'm going to play until the sticks get too heavy for me to hold up."

pensation is that individuals set their own goals, assess their own abilities, and then figure out how to accomplish what they want to achieve despite the limitations and declines of life (Baltes & Baltes, 1990).

One example of this process comes from the late Artur Rubinstein, a world-famous concert pianist who continued to perform in his 80s. When asked how he managed to sustain his level of performance, Rubinstein explained that he limited his repertoire somewhat to pieces he knew he could perform well (selection) and that he practiced before a concert more than he might have when he was younger (optimization). And because he could no longer play fast passages at the brilliant tempos he was once capable of, he deliberately played slow passages slower than he formerly would have, thereby making his playing of the fast passages seem faster than it was (compensation) (Schroots, 1996).

More common examples are provided by the many elders who become much more forceful in structuring their lives so that they can do what they want and do it well. In a study of some of the strategies used by people over age 80, for example, Johnson and Baer (1993) report a woman who does her food shopping at a distant store because it is near the end of the bus line, ensuring that there will be seats available for her and her groceries on the return trip. Similarly, a man who continues to drive a car "plots out the streets and plans his exact route in order to avoid getting lost, while another drives around the block rather than making a left turn at a busy intersection." According to the authors, such routinization of daily activities is a common strategy used by the elderly to gain greater control over their environment.

Integrity Versus Despair

The most comprehensive self-theory comes from Erik Erikson, who even in his 90s was still writing about the vitality of life. Erikson called the final crisis in his theory of development the crisis of **integrity versus despair**. His depiction of the diversity of late life is framed in terms of what he saw as the universal attempt of older adults to integrate and unify their unique personal experiences with their vision of the future of their community. Some develop a sense of pride and contentment with their past and present lives, as well as a "shared sense of 'we' within a communal mutuality" (Erikson et al., 1986). Others experience despair, "feeling that the time is now short, too short for the attempt to start another life and to try out alternate roads to recovery" (Erikson, 1963).

As at each of Erikson's eight stages, tension between the two opposing aspects of the developmental crisis helps move the person forward. This is particularly apparent in this eighth stage, when

> life brings many, quite realistic reasons for experiencing despair; aspects of the present that cause unremitting pain; aspects of a future that are uncertain and frightening. And, of course, there remains inescapable death, that one aspect of the future which is both wholly certain and wholly unknowable. Thus, some despair must be acknowledged and integrated as a component of old age. [Erikson et al., 1986]

Ideally, coming to grips with the idea of one's own death leads to a new view of survival, through children, grandchildren, and the human community as a whole. This vision of the integrity of the generational life cycle allows a "life-affirming involvement" in the present.

integrity versus despair The last of Erikson's eight stages of development. During late adulthood, according to Erikson, people feel either that their lives have had meaning and they look back on their past experiences with a sense of integrity and wholeness, or they despair at their past and therefore dread the future.

A glance at this woman at her outdoor pump might evoke sympathy, as her home's lack of plumbing suggests that she is experiencing late adulthood in poverty, in a rural community that probably offers few social services. As social stratification theories point out, factors of race and gender put her at additional risk of problems as she ages. However, a deeper understanding of her life might reveal many strengths—from deep religious faith to strong family ties and gritty survival skills—that will sustain her despite the lack of privileges and safeguards that others might have.

A similar theory is *critical race theory*, which sees race not as something inborn but rather as "a social construct whose practical utility is determined by a particular society or social system" (King & Williams, 1995). According to this theory, long-standing racism and racial discrimination shape the experiences and attitudes of both racial minorities and racial majorities in ways that structure life events for both groups—often without their conscious awareness (Bell, 1992).

In old age, this structuring includes poverty and frailty for those minority-race members who were excluded from the economic mainstream all their lives and who were, and continue to be, without ready access to needed health care. In old age, these same minority members are also less likely to have access to social services and amenities, including senior-citizen centers, nursing homes, and the like (Skinner, 1995). The consequences of racial structuring in old age take more subtle forms as well. For example, in America many of the elderly are homeowners who can choose to sell their home for profit, live in it rent-free, or obtain a "reverse mortgage," receiving a monthly payment in exchange for title to the house when they die. However, decades of housing discrimination mean that African-Americans are much less likely to be homeowners than are European-Americans, and even when they do own a home, it is likely to be worth less than homes owned by others of the same income level (Myers & Chung, 1996).

The central view of these social stratification theories—that membership in a certain social structure can place older individuals at risk of experiencing a number of dangers, including disability, poverty, and isolation—certainly seems valid when it comes to poverty or poor health. In every nation, an elderly person's economic and health status seems closely connected to minority or majority status as well as to gender. However, many contemporary analysts argue that this view unfairly stigmatizes members of minority groups, who often have strengths in late adulthood (e.g., intergenerational supports and religious faith) that other groups are less likely to have (Blakemore & Boneham, 1994). Similarly, the fact that women are more likely to be caregivers and kinkeepers means that they are less likely to be lonely and depressed in old age (Barusch, 1994).

With all the obvious inequities of stratification based on race and gender, it is noteworthy that one measure of despair—suicide—is less common among minorities and among women. In fact, of all the age, sex, and racial groups in the United States, the lowest suicide rate of all is that of black elderly women, an infinitesimal 2 per 100,000, compared with 94 per 100,000 for elderly white males (National Center for Health Statistics, 1994).

dynamic life-course theory The theory that each person's life is an active, ever changing, largely self-propelled process, occurring within specific social contexts that themselves are constantly changing. For this reason, developmental predictions based on past history may be useful, but they are never completely accurate, since each person constantly makes choices that affect the next step of development.

Dynamic Life-Course Theory

All the foregoing theories provide avenues of insight into development in late adulthood, and each contributes an important perspective to the theory that has the broadest applicability to late adulthood, **dynamic life-course theory**. According to this theory, each person's life is an active, ever changing, largely self-propelled process, occurring within specific social contexts which themselves are constantly changing.

Underlying this dynamic flux are the major factors emphasized throughout this text—behavioral genetics, intrafamily and intergenerational

interactions, educational experiences, work patterns, cultural expectations, and so on. In late adulthood, the interaction of these factors has led to an incredible diversity of developmental pathways. This diversity adds enormous complexity to the study of late adulthood, because any research finding, or social trend, or gerontological theory that applies to one individual or group may be entirely inappropriate for another. Indeed, as one gerontologist expresses it, "current and future generations of the elderly are part of a quiet revolution—a revolution of older individuals representing the broadest range of ethnic, racial, cultural, regional, religious, political, and socioeconomic diversity ever witnessed in American society" (Burton et al., 1991). As you read about the specific events of psychosocial development in the rest of this chapter, keep in mind the multifaceted diversity of older persons as well as the universal impact of the aging processes.

Generativity in Late Adulthood

The theme of generativity—seeking to be both productive and caring—was thought by Erikson to be the focus of middle adulthood (Erikson, 1963). This placement of generativity needs in one period of life fit well with the concept, held for most of the past fifty years or so, that the life span is a straightforward progression through three basic periods. First comes a period of education, followed by a period of full-time employment and/or parenthood, and then, finally, a period of leisure—sometimes glorified as the golden years, sometimes dreaded as the emptiness before death. Cultural norms for age stratification reinforced these divisions (Hagestad & Neugarten, 1984), and most people followed the norms, finishing school by early adulthood, working intensely until age 65 or so, and then retiring, often pushed to do so because of mandatory-retirement laws. In short, several possible avenues of generativity were closed to the elderly, who mostly had to look to their grandchildren to express their generativity needs.

Today, however, these age-based divisions and norms are increasingly seen as "vestigial remains of an earlier era when most people had died before their work was finished or their last child had left home" (Riley & Riley, 1994). While about two-thirds of the U.S. population still agree on target ages for when a person "should" leave school (age 26), settle into a job (29), and retire (60), these norms are now relatively loose and flexible (Settersten & Hagestad, 1996). As you will see, society is currently moving away from a linear view of life and toward one that recognizes a "circling back, with multiple strands" (Schuller, 1995). Age segregation is being replaced by an age-integrated approach that allows more varied generativity (Riley & Riley, 1994).

Employment and Retirement

In the United States, mandatory retirement is now illegal. Although some predicted its demise would result in people staying in the labor force well beyond age 65, in fact, American adults now retire at younger and younger ages, some as early as age 55, most before age 65, and almost all by age 70. In the United States in 1995, for instance, only 13 percent of adults over age 65 (both men and women) were in the labor force, compared with 58 percent of

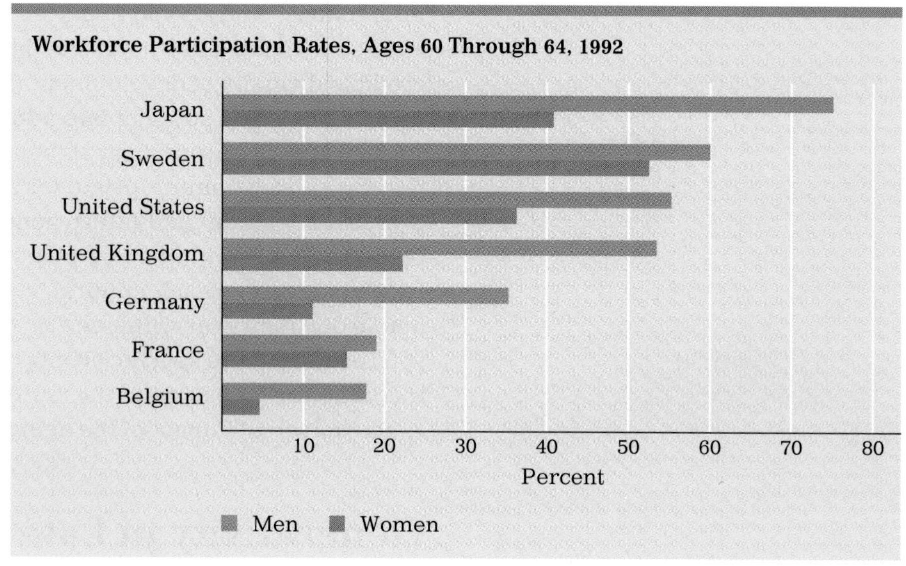

FIGURE 25.1 A few years ago, every able-bodied worker remained in the labor force until at least age 65. Now older workers opt for buy-outs, employers hire younger and cheaper labor, and governments provide extensive benefits—in some European nations resulting in widespread retirement at age 55.

those aged 55 to 65 and 84 percent of those aged 25 to 55 (U.S. Bureau of the Census, 1996). Similar trends are apparent in every developing nation (Ilmarinen, 1995) (see Figure 25.1). Even these statistics tend to exaggerate the labor-force participation of those aged 55 and older because labor-force participation for this group often includes temporary, part-time, or self-employed jobs that serve as a bridge to retirement. By age 70, even those who are still employed are likely to have scaled back considerably: more than half are self-employed, more than half are working part-time, and almost no one is working full-time for someone else (Quinn & Kozy, 1996).

Most older adults leave the work force because they want to, and most of those who continue to work do so part-time and prefer it that way (Hobbs & Damon, 1996). Research accumulated over the past thirty years finds that, contrary to some expectations, most elders who cut back on work or retire completely not only adjust well to their change in employment status but even improve in health and happiness (Betancourt, 1991; Herzog et al., 1991; Reitzes et al., 1996). This is in dramatic contrast to the situation in the first half of the twentieth century, when the elderly depended on their jobs for survival and hung on to them for dear life, rightly fearing poverty and depression if they quit. Poverty in old age is much less common today, primarily because pensions are almost universal. For example, in 1950, only one out of every three older Americans received any pension income, including Social Security; by 1992, 93 percent were receiving Social Security and about half had private pensions as well (Hobbs & Damon, 1996).

Obviously, this bright, or at least benign, picture does not hold for all elderly Americans. For many of the elderly, the picture is blighted by the consequences of race or gender stratification. Blacks are particularly unlikely to retire in order to live a life of leisure. Rather, they retire mostly because they lost their job through downsizing or restructuring and could not find new work or because disability kept them from work (Bound et al., 1996; Gibson, 1993; Hayward et al., 1996).

Many women suffer because of gender-biased policies. Emotionally, women actually tend to adjust to sudden retirement better than men because they are likely to have social ties to many people not connected with work. However, because full pension benefits are tied to sustained full-time

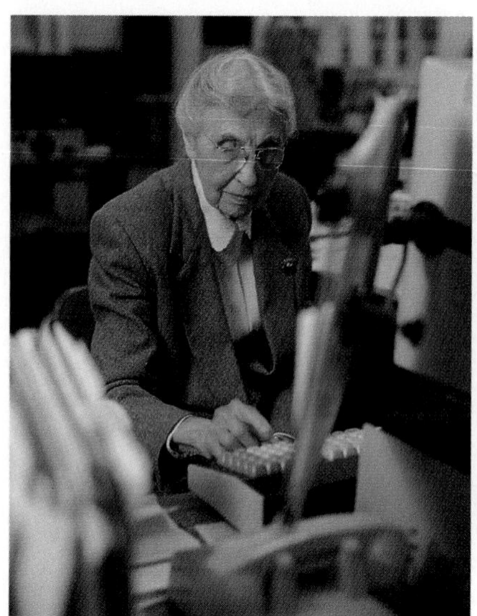

Unlike most of the older adults who continue to work during the traditional "retirement years," Audrey Stubbart puts in a full 40-hour week at her job. At age 100, she is a proofreader and columnist for *The Examiner* in Independence, Missouri.

employment, many mothers and wives who have worked part-time or on an interrupted schedule experience a drastic drop in income when they retire—unless they have a husband with a good pension (Moen, 1996). For both sexes and all races, early retirement, especially when induced by a generous severance payout, often brings pleasure at age 60 but poverty at age 80 (Quadagno & Hardy, 1996). Thus retirement is not always a blessing.

Fortunately, more and more of the elderly find their own way to retire gradually and supplement their income if need be. After leaving their career job, some become self-employed. Others take on a different job with shorter hours and less pay, a shift that is sometimes facilitated by such employment practices as flextime and job-sharing. The financial situation for widows and divorcees is also improving, as many pension policies today more adequately reflect their needs.

Whether they continue working or not, whether they have generous pensions or marginal incomes, most elders remain productive. Their activities may not be financially rewarding but, as we will see, they are worthwhile in many other ways—learning, helping, and creating.

Continuing Education

For many of the elderly, retirement offers an opportunity for pursuing educational interests. At any given time, about one out of twenty adults aged 60 and older is enrolled in classes of some sort, ranging from courses in the practical arts to those leading to advanced college degrees. Given that career advancement is seldom their motive for renewed schooling, the elderly are more likely to study for the joy of learning, or to master some specific skill, than younger adults are.

The eagerness of the elderly to learn is best exemplified by the rapid growth of *Elderhostel*, a program in which people aged 55 and older live on college campuses and take special classes, usually during college vacation periods. Begun in New England in 1975 with 220 students, Elderhostel now operates at 1,800 sites throughout the United States and Canada, with an annual enrollment of more than 250,000 paid students (Elderhostel, 1997). Evaluation from the participants has been very positive, which accounts for the expansion of the program. Except for the age requirement, admission is open to almost anyone with the interest and tuition. Thousands of other learning programs in the United States and Canada are filled with retirees. And in Europe, *Universities of the Third Age*, dedicated to the older learner, have opened in a least a dozen nations. A survey in Norway found that 10 percent of those over age 67 had attended a course in the past three years (Ingebretsen & Endestad, 1995).

While some elderly students eagerly enroll in programs designed specifically for them, many hesitate to enter academic classes populated by mostly younger students. This is unfortunate, because many of the elderly who do enter the standard classroom outperform younger students and enjoy the experience once they start, and many younger students and professors benefit from the involvement and example of older students. One man surprised himself by taking drawing, painting, and Spanish classes at a community college, explaining:

> When I first retired, I couldn't wait to pack up and go to a warm climate and just goof off. But now, retirement is an enormous challenge. Once you start learning about yourself, you get the feeling that anything is possible. [Goldman, 1991]

Not all Elderhostel courses take place in the classroom. This group of U.S. students is on a three-week expedition studying the wildlife of Kangaroo Island, Australia.

Although younger adults may imagine that the aged generally take courses to further their self-fulfillment or to explore the meaning of life, many elders enroll for quite practical reasons, such as to better understand their grandchildren or their finances or to learn skills needed for a hobby or a job (Jeanneret, 1995).

Unfortunately, say some critics, the United States has done much too little to make elder education available to all those who want it:

> If continuing education for adults has long been a stepchild of higher education, then older adult education must be considered an orphan . . . Since older-adult education does not bring in grants for universities or offer career training paths, it tends not to be taken seriously. [Moody, 1993]

Given what is known about the potential for older adults to be actively and intelligently involved in society, and about education as a catalyst for personal growth, formal education is a step in the right direction. However, more needs to be done to make it easier for the elderly to get such education, especially those whose access to education in the past has been limited (Glendenning, 1995).

Volunteer Work

Many older adults feel a strong commitment to their community and, even more than younger people, believe that older people should be of service to others. For example, when a cross-section of close to 3,000 Americans were asked whether older adults have an obligation to help others and serve the community, about twice as many older adults as younger adults strongly agreed. Strong disagreement with this idea was expressed by only 6 percent of those over age 60, whereas 12 percent of those under 60 strongly disagreed (Herzog & House, 1991). Perhaps because of their perspective on life, or because of their patience and experience, older adults are particularly likely to do volunteer work assisting the very young, the very old, or the sick.

Much of this volunteer work is informal. In stable neighborhoods that have a sizable number of elderly, the more capable residents often run errands, fix meals, repair broken appliances, and perform other services that generally make it possible for the disabled elderly to continue to live at home. For example, in one four-block residential area, routine sharing and socializing allowed three widows in their late 80s, one of them blind, one ob-

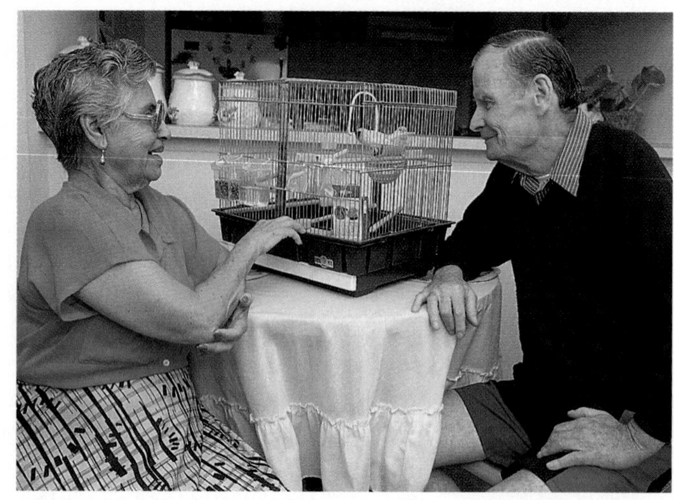

Some volunteer programs are trying a new approach known as "timedollars." Volunteer caregivers earn credits for the time they volunteer—visiting, shopping, doing housekeeping, providing transportation, and the like—and are then able to redeem those credits when they themselves are in need of services. Felina Mendoza, shown here visiting a man who is just home from the hospital, is a member of Friend-to-Friend, a timedollar program in Miami, Florida. Its 800 participants earn an average of 8,000 hours of service credits each month.

viously failing cognitively, and one physically disabled, to remain in the community. Their neighbors, most of whom were elderly themselves, shopped for them, cooked for them, and simply checked in on them often enough to make them feel part of the neighborhood, while still allowing them some privacy and independence (Rosel, 1986). The elderly also provide a good deal of care for relatives. A recent large-scale survey found that about 30 percent of the elderly provide regular personal care to someone, most often to an elderly relative, and that 15 percent provide child care, again usually for relatives (Kincade et al., 1996).

This same survey found that about a fifth of the elderly were involved with more structured volunteering, often through churches, hospitals, or schools. This number is substantial, but it is less than volunteerism rates for younger adults with similar work and family obligations. And, surprisingly, the volunteerism rates for the elderly who work part-time are higher than those for the elderly who are fully retired. Many gerontologists believe that more of the elderly would do volunteer work if they recognized their potential to be of service and if more social agencies likewise recognized this potential, not only recruiting the elderly specifically but also making an effort to match each person's talents with the needs of the community (Herzog & Morgan, 1993). That the elderly often do not recognize the service they can provide is reflected in the reaction one older woman had when she and her husband were asked to work with stroke victims:

> Of what possible use could we be, a seventyish couple who had none of the seemingly necessary skills, no knowledge of speech or physical therapy, no special social skills either?

Several months after this couple overcame their self-doubt and began the volunteer work, the woman wrote:

> The real question was [not] "What can we do?" [but] "What can we be?" Can we be warm, caring unstroke-damaged human beings, to meet . . . with stroke-damaged ones, exchanging concerns, playing word games, encouraging them to feel at ease, to talk and tease and laugh with us and with each other . . . I wish everyone who feels so useless and lonely could have such a completely satisfying experience as Tom and I are having. [quoted in Vickery, 1978]

That volunteer organizations can benefit by appealing directly to retired seniors is made clear by the success of several organizations that now

do so. One such organization is SCORE (Service Corps of Retired Executives), whose volunteers provide advice to owners of small businesses. In 1995, more than 12,000 volunteers helped more than 250,000 entrepreneurs (Goodno, 1996). Another is Foster-Grandparents, which is designed to help an older person "adopt" a child who needs the help a grandparent could provide. A third program, particularly suited for the elderly who are unable to, or are afraid to, leave their home is "Phone Pal." Volunteers in this program daily call children who are home alone after school, checking to see if everything is all right and helping the children focus on doing homework if necessary. Religious institutions already provide volunteer opportunities for many of the elderly, particularly within the African-American community. Strengthening those institutions and encouraging other religious organizations to follow their example would tap valuable resources and improve life satisfaction as well as health for the elderly (Coke & Twaite, 1995). The critical factor in all these programs is cultural attitude: once people appreciate that the elderly are more often caregivers than care receivers, many more of the elderly will become involved in volunteer programs (Caro & Bass, 1995; Uhlenberg, 1996).

Political Involvement

By many measures, the elderly are more politically active than any other age group (see Figure 25.2). On the whole, they know more about national

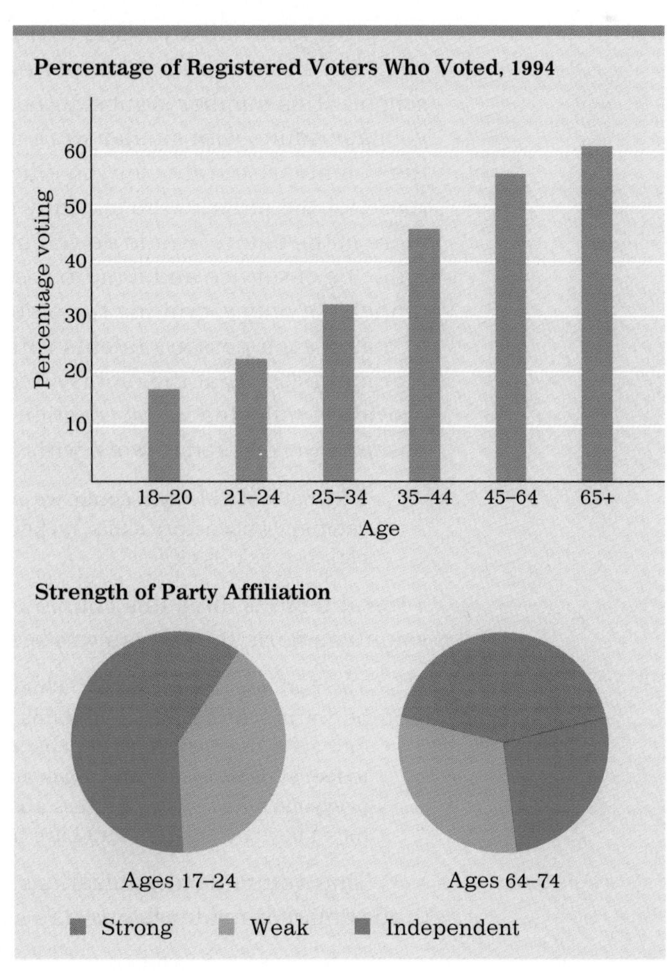

FIGURE 25.2 Compared with the youngest members of the electorate, older Americans are three times as likely to vote and twice as likely to have strong party loyalties—a combination of factors that makes their political clout far stronger than their numbers in the population would predict. Every elected official pays attention to this group, whether they are advocating in self-interest, as when fighting against cuts in Social Security or Medicaid, or in the interests of the future of the community, as when calling for prudent land use and better public education.

and local issues, write more letters to their elected representatives, vote in more elections, feel stronger about party loyalty, and belong to more groups that lobby for their interests (Torres-Gil, 1992). The only measurement of political participation on which they fall short is in door-to-door campaigning, although even here, many are active pavement pounders.

All this political participation translates into considerable power, especially when the elderly organize themselves into political action groups. The major U.S. organization representing the elderly is the American Association of Retired Persons (AARP), the largest organized interest group in America. In 1995, the AARP had a membership of more than 35 million (members must be over 50 but need not be retired), employed nearly 50 congressional lobbyists, and involved over 400,000 volunteers in various projects related to advancing the interests of the elderly or to benefiting the overall community. The political influence of this organization is one reason Social Security is called "the third rail" of domestic politics, not to be touched by the budget-cutting that affects almost every other domestic subsidy, from public assistance to highway repair.

While the AARP is the major senior organization, it certainly is not the only one. The National Committee to Preserve Social Security and Medicare and the National Council for Senior Citizens have 5 million members each, and the Gray Panthers—a more radical group that is likely to demonstrate in the streets—claims 400,000 members. Dozens of other smaller special-interest groups center on issues of elder concern (Binstock & Day, 1996).

All this political activism makes many younger adults suspect that senior citizens wield unprecedented power and use it unfairly and narrowly to advance their own economic interests. However, the idea that the elderly are narrowly focused on their self-interest is inaccurate. True, elders in many nations tend to support financial assistance for the aged (Myles & Quadagno, 1991), but many elders also understand wider social concerns and are willing to vote against their own age group if a greater good is at stake. When a tax on the Social Security income of wealthier elders was proposed in 1993, for example, the idea received more support from the old than from the young (Kosterlitz, 1993). Of course, questions of **generational equity**, that is, whether each generation contributes and receives their fair share of society's wealth, can be asked of every age group (see Public Policy, pp. 674–675).

The Social Convoy

The phrase **social convoy** highlights the truism that we travel our life course in the company of others (Antonucci, 1985). At various points, other people join and leave our convoy, but just like members of a wagon train headed west, we could never make the journey successfully by ourselves. Furthermore, the bonds formed as we journey together help us in good times and bad. It is more pleasant to share triumphs with those who know how important the victory is; it may be critical for our survival in times of defeat and sorrow to have familiar confederates whom we have helped in the past.

For older adults particularly, the social network's continuity over time is an important affirmation of who they are and what they have been. Friends who "knew them when" are particularly valuable, as are family members who share a lifetime of experiences.

generational equity Equal contributions from, and fair benefits for, each generation.

social convoy Collectively, the family members, friends, and acquaintances who move through life with an individual.

PUBLIC POLICY

Poverty, Age, and Equity

In the second half of the twentieth century, the economic circumstances of the elderly have changed dramatically. More than thirty-five years ago, Michael Harrington's *The Other America* drew a chilling picture of the extent of poverty among adults who were then above age 65.

> Fifty percent of the elderly exist below minimal standards of decency, and this is a figure much higher than that for any other age group. . . . We have given them bare survival, but not the means of living honorable and satisfying lives. [Harrington, 1962]

Harrington was not far from wrong. In actuality, one out of every three elderly Americans was then living below the poverty line, and a substantial additional number were "near poor," living at the very edges of poverty.

Since that time, various economic, demographic, and political changes have raised both the personal income and the living standards of many of the elderly throughout the world. In the United States, for instance, the federal government's "war on poverty" in the 1960s extended Social Security and provided a range of medical and social benefits to the aged, reducing the proportion of the elderly below the poverty line to about one in nine in 1995. Money spent for the aged represented more than a third of the U.S. 1994 federal budget; money spent for defense represented only 18 percent (U.S. Congress, 1994).

Private firms have followed the federal lead with subsidies ranging from pensions that afford many older adults a comfortable retirement to symbolic measures such as weekday discounts at movie theaters and fast-food restaurants.

Ironically, during the same years that the American elderly were growing richer, thanks, in part, to increases in their government benefits, children were growing poorer, due, in part, to cuts in government benefits to young families, including low-cost mortgages, public child care, and income supplements. More than one American child in five (21 percent) now lives below the poverty line (see figure), compared with one in seven (15 percent) in the 1970 (U.S. Bureau of the Census, 1996). This economic disparity between young and old has been exacerbated by periods of inflation that have undercut every public benefit except Social Security, which automatically rises with the cost of living.

A backlash against this reversal of financial fortunes has led to calls for generational equity, defined as equal contributions from, and fair benefits for, each generation. Some demand that the government limit public money that flows to elders at the expense of the children, pointing out that the current distribution of benefits is particularly imbalanced for nonwhite minorities, who tend to be overrepresented among the young and underrepresented among the old. For example, only 26 percent of non-Hispanic whites in the United States are under age 20, compared with 39 percent of Hispanic-Americans, 38 percent of Native Americans, 36 percent of African-Americans, and 32 percent of Asian-Americans. And while 15 percent of non-Hispanic whites are over the age of 65, only about 7 percent of the four minority groups just mentioned are (U.S. Bureau of the Census, 1996).

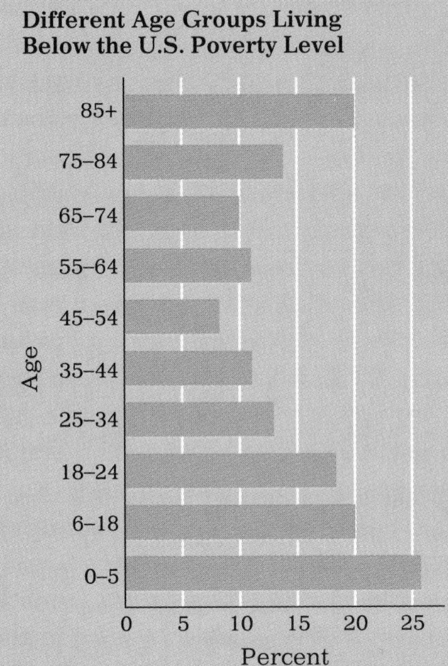

Source: Children's Defense Fund, 1995; Moon & Mulvey, 1996; U.S. Bureau of the Census, 1996.

If any age group is unfairly burdened by poverty, it is the youngest, whose ill health, miseducation, and undernutrition can affect them for years to come.

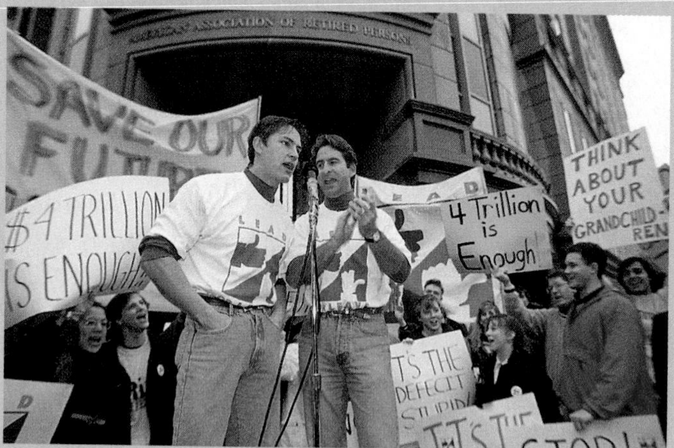

Social Security is the only major public assistance that rises automatically with inflation—a fact not lost on many young adults who see their Social Security taxes rise and their education benefits fall.

One frequently mentioned candidate for generational equity is health care: costly public outlay for doctors and hospitalization of the aged results in grossly inadequate funding of preventive medicine in childhood and adolescence (Callahan, 1990). At their fiercest, critics argue that the young are soaked to enrich the old, that greedy elders are living off the regressive Social Security taxes that deplete the wages of younger workers, and that "age wars" will soon ensue unless we "cut the graybeards a smaller slice of the pie" (Becker, 1994).

Gerontologists believe this framing of the problem is an "ideological smokescreen" that obscures the real needs of the poor, whatever age they are (Adams & Dominick, 1995). In fact, income inequality in the United States increases with age, as some of the elderly are among the richest, and others are among the poorest, in the nation (O'Rand, 1996). Most are in the middle, but even for them, the specter of poverty seems frighteningly close. The average older person now has enough income for food, shelter, and other basic expenses, but few are wealthy. Despite government and private help, those elderly who need costly medications or out-patient care must pay sizable portions of their income and savings or "spend down," that is, get rid of all their assets until they are so poor that they qualify for full government subsidy.

Thus for most of the nonpoor elderly, whatever extra savings they might have are their bulwark against two dreaded circumstances, lingering helplessness and abject poverty. Often this defense fails: many of the most vulnerable older individuals, especially the widowed, the divorced, the nonwhite, and the physically frail, still suffer cruel effects of poverty, living in dangerous and dilapidated housing with insufficient food or heat. This is particularly likely to be the fate of the oldest-old, whose poverty rates are similar to those of the youngest-young. Compared with those just starting their lives, however, the older poor suffer an additional burden: they have little hope of marked financial improvement.

However, even this broader perspective on the distribution of wealth and poverty misses a critical contribution that is implicit in a life-span view of human development—that is, that every age and cohort has its own particular and legitimate economic needs that other generations might fail to appreciate (Deutscher, 1988; Torres-Gil, 1992). For instance, whether one gives greater legitimacy to federally funded day-care programs, or to low-interest government loans for college tuition, or to protecting entitlements for the elderly depends largely on the immediate interests of one's own age group—yet in the long term, each of these subsidies works to the betterment of all.

Taking a developmental view, then, instead of a battle among age groups, each generation must try to balance its own needs with those of the others, because *interdependence* is at the heart of intergenerational relationships. It is thus both unfair and counterproductive to pit one generation against another on the basis of a "flawed and dangerous premise" of exclusionary self-interest, such as "that America's younger generations are suffering *because* of the elderly population" (Minkler, 1991). Instead, the fact that fewer of the elderly are poor today benefits not only the older generations but their descendants and the entire nation. Correspondingly, the fact that more children are poor today imperils not only the children themselves but also the whole community. The crucial needs of every generation need to be seen in full perspective and addressed accordingly: the solution to impoverished youth cannot be found in scapegoating the old.

We will begin our examination of the social convoy by looking at the closest relationship of all, the long-term marriage.

Long-Term Marriages

Research suggests that long-term marriages improve with age, and that many long-married older adults are happier with their partners than they were as newlyweds. This blissful couple, married over forty years, would no doubt concur.

For many older people, a spouse is the best buffer against many of the potential problems of old age. Most (about 60 percent) of all Americans currently aged 60 and older are married, and they tend to be healthier, wealthier, and happier than those who never married, or who are divorced or widowed (Myers, 1993).

The great majority of elderly married couples are in long-term marriages, which raises an important question: Do marriage relationships change as people grow old, and if so, how? According to longitudinal as well as cross-sectional research, both continuity and discontinuity are apparent in these relationships. The single best predictor of the nature of a marriage in its later stages is its nature early on: while the absolute levels of conflict, sexual activity, or emotional intensity drop over time, over time couples who were high or low in these dimensions early on tend to remain so relative to their position on each of the other dimensions.

If there is change in a long-term marriage, usually it is positive. Cross-sectional research comparing long-term marriages in middle age with long-term marriages in old age finds that the latter have far less conflict and that the partners experience more pleasure together than the younger couples do (Levenson et al., 1993). Even making "plans for the future" is more positively viewed by the older couples. These cross-sectional findings are supported by other research finding that most older married couples believe that their marriage improved over the years (Erikson et al., 1986; Glenn, 1991). Older husbands and wives tend to be happier with each other, and with their marriages, than they have been since they were newlyweds, a continuation of an upward trend that begins about ten years after the wedding (see Figure 25.3).

One reason for this improvement may be traced to the effects of the couple's children, who are the prime source of conflict for middle-aged couples and the prime source of shared pleasure for elderly couples (Levenson et al., 1993). Another reason for improvement may be that the accumulation of shared life experiences makes husbands and wives become more compatible. That is, all the shared contextual factors—living in the same community, raising the same children, and dealing with the same financial and spiritual circumstances—tend to change both partners in similar ways, bringing long-married couples closer together in personality, perspectives, and values (Caspi et al., 1992). Indeed, in many aspects of shared life, from dividing the housework to deciding where to go on vacation, the longer a couple is married, the more likely both are to believe that their relationship is a fair and equitable one (Keith & Schafer, 1991; Suitor, 1992).

The perception of fairness among older married couples is particularly striking when one of them becomes seriously disabled, as occurs eventually in about half of all marriages that last fifty years or more. As we will see later in this chapter, the enormous burden of caring for a seriously ill person over a long period of time cannot be minimized, especially if financial wor-

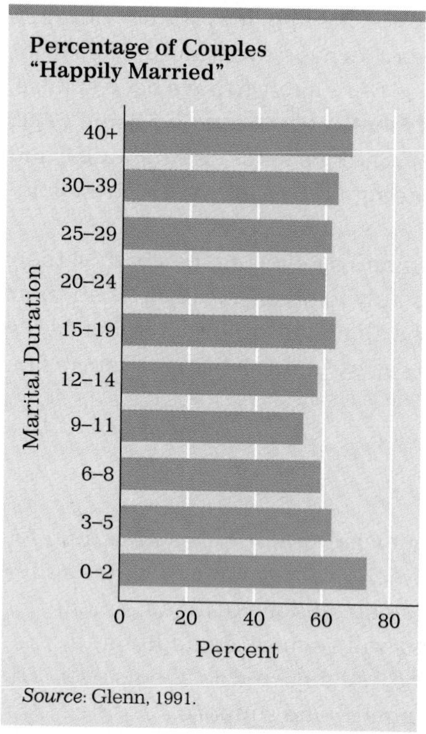

FIGURE 25.3 As you can see, duration of marriage has some effect on likelihood of wedded bliss, with the best years often being the early and late ones.

ries are part of the burden, as they often are (Booth & Johnson, 1994). However, if the demands on the healthier spouse are not overwhelming, most older couples adjust fairly well. Indeed, one study of seventy-six older couples in which one spouse or the other was ill found that although sickness did change the marriage relationship somewhat, it did not affect overall marital satisfaction. As the researcher learned from her interviews:

> With the illness of one spouse, when additional demands are placed on the marriage, the interdependence that had developed over the years appears to provide the means to meet these needs, usually without reservations. . . . [S]ome conflict was observed in these interviews but . . . disagreements are handled with good-humored joking, sarcastic remarks, or teasing, rather than overt arguments. [Johnson, 1985]

Generally, older spouses accept their mutual frailties, tending to each other's physical and psychological needs as best they can, usually with feelings of affection rather than of simple obligation. The affection that accompanies such caregiving was shown in another study (Seltzer & Li, 1996), which found that wives who cared for their dependent husbands usually felt closer to their spouses and less burdened by the experience in the later stages of caregiving than at the start. The data showed that this positive result was a direct consequence of being a spousal caregiver—rather than of simply being a female caregiver, or of simply getting used to the caregiving task—because the same study also found that daughters providing care for aging parents felt more burden and distance in the later stages of caregiving.

Our focus on mutual caregiving should not distort our understanding of late-life marriages. Intimacy, companionship, and even passionate love are part of the marital relationship for many older couples (Melton et al., 1995). One example makes the point. When one happily married elderly couple were asked about their sex life, the husband responded:

> We have sex less frequently now, but it's satisfying to me. Now that we are both home, we could spend all our time in bed. But it's still more amorous when we go away. When we travel, it's like a second honeymoon.

His wife added:

> Sex has been important in our marriage, but not the most important. The most important thing has been our personal relationship, our fondness, respect, and friendship. [Wallerstein & Blakeslee, 1995]

Widowhood

Half of all married older adults will, obviously, experience the death of a spouse, one of the most serious stresses a person can undergo. Most surviving spouses are widows. Because the average adult woman in the United States lives six years longer than the average man, and the average husband is three years older than his wife, the average American wife will eventually spend nine years of her life as a widow. Of the 33 million Americans who were 65 or older in 1995, almost 9 million were widows while less than 2 million were widowers (U.S. Bureau of the Census, 1996).

For both widows and widowers, the first months alone are generally hardest, for obvious reasons. The death of a mate usually means not only the

loss of a close friend and lover but also lower income, less status, a broken social circle, and disrupted daily routines. Perhaps even more crucial for today's older widows—whose primary roles were likely to have been spouse, caregiver, and homemaker—is the loss of identity (DeGarmo & Kitson, 1996). In addition, routines of meal preparation, visiting friends, taking walks, and even regular bedtimes typically falter, adding up to substantial physiological stress.

It is not surprising, then, that in the months following the death of their spouse, widows and widowers are more likely to be physically ill than their married contemporaries are. Widowers particularly are also at a markedly increased risk of death, either by suicide or by natural causes (Hemström, 1996; Osgood, 1992).

Sex Differences in Adjustment

In general, living without a spouse is somewhat easier for widows than for widowers. One reason is that, most likely, elderly women expect to outlive their husbands and, to some degree, anticipate and make arrangements for some of the adjustments that widowhood will require. In addition, the recently widowed usually have friends and neighbors who themselves are widows and who are ready to provide sympathy and support.

Another gender difference also makes adjustment more difficult for widowers. Many of the men who now are elderly grew up with restrictive notions of masculine behavior, and tended to depend on their wives to perform the basic tasks of daily living (such as cooking and cleaning) and to be their main source of emotional support and social interaction. When their wives die, they often find it hard to reveal their feelings of weakness and sorrow to another person, to ask for help, or even to invite someone over to chat (Wilson, 1995).

Although women have an easier time coping with the emotions of losing a spouse, they have a harder time with the financial consequences. They tend to have smaller pensions than their husband did (they are eligible for only half of their husband's Social Security benefits) and are likely to have less knowledge about savings and investments. In many cases, widowhood precipitates poverty (Moen, 1996).

Widows must also come to grips with a hard truth: they are probably never again going to find a man to be their close friend and lover. Men, however, once they have gotten over their initial depression, discover that they are a much-sought-after item—the unmarried, mature male. There are many women who would happily fix their meals, clean their houses, and marry them, if possible. Statistics on remarriage bear this out: widowers are far more likely to remarry (often to women considerably younger than themselves) than widows are.

The fact that men have far more potential partners to choose from is not the only reason for this imbalanced remarriage pattern, however. Older widows often do not want to marry. After having been without the traditional wifely responsibilities for a time, many are reluctant to take them up again and enjoy their new independence (Wilson, 1995). In fact, after an initial period of grief, many widows cope quite successfully, adding new mem-

What is in store for the elderly woman whose husband dies? Probably not a new man: unattached older males are "scarce as hen's teeth." Instead, a string of new female friends, such as these women sunning themselves in Miami Beach, is usually eager to welcome one more to their social circle.

bers to their social network and feeling the pleasure of accomplishment because they have created a new sense of self (Lieberman, 1992).

The Never-Married and the Divorced

Of those currently over age 65 in the United States, only 4 percent (1.3 million) have never married, making this the most married cohort in history (U.S. Bureau of the Census, 1996). Even though the never-married elderly chose a different path through life, they tend to be particularly content with their independent state of affairs. Since they have spent a lifetime without a spouse, they usually have developed long-standing, alternative social patterns and activity preferences that keep them active and happy in late adulthood as long as their health is reasonably good. A large portion of the never-married are gays and lesbians, many of whom have long-time companions as well as extensive social networks.

Never-married older persons, particularly women, tend to be much involved with their relatives, caring for an aged parent, living with a sibling, or actively helping nieces and nephews (O'Brien, 1991). They also value their independence, and, even though they usually live alone, they are not lonely. In fact, one study of loneliness found that never-married women compared favorably with married ones (see Table 25.1) (Essex & Nam, 1987).

The situation is usually much different for those who divorce during late adulthood. Late-life divorce can be devastating, disrupting long-

TABLE 25.1

Loneliness in Elderly Women

How often lonely	Never	Hardly ever	Sometimes	Often
Married women	38%	33%	23%	6%
Never married	29%	44%	24%	3%
Formerly married	15%	33%	36%	16%

planned retirement communities Housing units designed exclusively for older adults.

naturally occurring retirement communities (NORCs) Neighborhoods or apartment complexes where more than half the residents happen to be elderly.

standing lifestyle patterns, habits, and self-esteem. In fact, for women who divorce in old age, the loss of identity is usually as great or greater over the long run than the loss caused by widowhood (DeGarmo & Kitson, 1996). However, divorce is rare in late life, except with marriages that have recently been entered into. Thus most of the nearly 2 million divorced older adults in the United States (about 6 percent of the elderly) have been single for decades and adjusted to the major disruptions of the break-up long ago.

Particularly for those divorced older women who suffered from social isolation, role overload, or financial stress when they were younger, life is likely to improve with age. Especially if they have successfully raised children as single mothers, or have succeeded in a career despite all odds, they feel quite proud of past accomplishments. Further, they are less lonely than they were as young divorcees, when married couples tended to exclude them and when the demands of survival and child-rearing undercut their social lives. Now their friendship circle grows as more and more of their contemporaries become widows. Adult children of divorced mothers typically keep in contact, often with grandchildren in tow.

As a group, older divorced men do not fare so well, however. Because women are usually the kinkeepers and social secretaries of a family, and because father-child contact typically diminishes over time when mothers have custody, many former husbands find themselves isolated from children, grandchildren, and old friends (Keith, 1986). Formerly married men who have not remarried have a higher rate of physical and psychological problems than any other category of seniors (Kurdek, 1991).

Friends

By late adulthood, many members of the social network have been part of a person's convoy for decades. This helps explain a surprising finding: older people's satisfaction with life bears relatively little relationship to the quantity or quality of their contact with the younger members of their own family, but it shows substantial correlation with the quantity and quality of their contact with friends (Antonucci, 1985; Sabin, 1993; Ulbrich & Bradsher, 1993).

Particularly important is the quality of friendship. Having at least one close friend in whom to confide acts as a buffer against the loss of status and roles that comes with such common experiences as retirement and widowhood. Given the importance of friendship, it is comforting to know that most elderly people have at least one close friend.

These women have been close friends for over sixty years. Their shared history and decades of mutual support are, most likely, a treasure that increases with age. As one time-honored prescription expresses it: "Make new friends and keep the old; the first is silver and the other gold."

As at younger ages, women seem to do more befriending than men. If an older adult is a married man, typically his closest friends are all relatives. Married women also are usually close to several family members but, in addition, are likely to have a close female friend who is not related. Not surprisingly, unmarried older adults have more close friends who are not relatives than older married adults do. For men, such friends are equally likely to be male or female; for unmarried women, the friend is almost always female (Akiyama et al., 1996).

Also as at every age, the strength of late-life friendship correlates with feelings of well-being and self-esteem (Crohan & Antonucci, 1989). In this respect, each year that a particular friendship lasts increases the shared

Aging in Place

The social convoy—particularly with regard to the need for friends—is one factor underlying a notable characteristic of most older adults: they like to "age in place" (Pynoos & Golant, 1996). Indeed, a nationwide survey found that 86 percent of the American elderly want to "stay in their present home and never move" (AARP, 1990). This preference is clearly reflected in United States statistics: the older an adult is, the less likely he or she is to move, with the relocation rate of households headed by young adults being seven times that of households headed by older adults. When older Americans do move, they more often move nearby, with only one out of four crossing state lines, usually relocating near old friends or family members (Hobbs & Damon, 1996). The much-publicized "snowbirds," who move from the North and Northeast to the Sunbelt states, are an affluent minority.

The need to be near familiar people may be particularly important for the current cohort of older Americans, about a fourth of whom are immigrants or first-generation Americans. For them, having neighbors, grocery stores, places of worship, social clubs, and other cultural sources reflecting their own ethnic and linguistic background often makes the difference between social interaction and lonely isolation. Elders from ethnic minorities are particularly reluctant to move if they anticipate being exposed to racial or religious prejudice from their new neighbors and if moving means leaving siblings, cousins, and so on behind.

Retirement Communities

The fact that the elderly want to stay put has come as something of a surprise to many government bureaucrats, private builders, and gerontologists, who, during the 1970s, championed the construction of tens of thousands of government-subsidized housing units designed for the growing aged population (Dobkin, 1992). **Planned retirement communities**, where all the residents must be over a certain age, are now widely available, not only in the temperate climate of the Sunbelt but also in every state of the United States and every province of Canada.

Planned retirement communities usually are a bargain, not only because they are publicly subsidized but also because they do not need expensive amenities such as large private yards, large kitchens, and easy access to schools and places of employment. Other advantages include security, quiet, and nearby medical care. In addition, many retirement communities offer services, from housekeeping to communal dining, for those who want them. Most important of all, age-segregated housing provides a natural setting for making friends, since all the residents are about the same age and often share similar hopes, memories, and problems.

Given the advantages of planned retirement communities, many gerontologists are trying to figure out why relatively few of the elderly—only about 8 percent—move into them (Pynoos & Golant, 1996). Ironically, those who are most likely to benefit from retirement communities—those who are over 75, or widowed, or relatively poor—are least likely to move (AARP, 1990). This is particularly true for the urban-dwelling African-American and Puerto Rican elderly—despite the fact that fear of crime and dangerous traffic make many of them "vulnerable to a hostile environment . . . not only aging in place [but] stuck in place, prisoners in their own homes" (Skinner, 1992).

Percent of U.S. Population Who Changed Residences Within a Year

Age group (years)	Moved	Moved to another state
1–4	22	3
5–9	17	3
10–14	13	2
15–19	17	3
20–24	34	5
25–29	30	5
30–44	17	3
45–64	9	2
65–74	6	1
75–84	5	1
85 and older	6	1

Source: U.S. Bureau of the Census, 1996.

While most of the elderly are reluctant to join planned retirement communities, an estimated 27 percent of elderly Americans live in **NORCs**, or **naturally occurring retirement communities**, defined as neighborhoods or apartment complexes where more than half the residents are elderly. Many NORCs originated in the 1950s, in the new housing developments designed for couples with young children. As their baby-boom offspring grew up and moved out, the parents often stayed put, consulting with their neighbors on the problems of aging just as they had once consulted with them on techniques of toilet-training their toddlers or on curfews for their teenagers (Hunt & Ross, 1990).

Once a NORC develops, it often attracts other older people, who form an active social network, with card parties, carpools to the mall, low-impact aerobics classes, and quilting circles to replace the babysitting co-ops and PTA meetings of the past. NORC residents tend to be pleased but somewhat surprised at this turn of events. As one man notes:

> We were younger when we moved in but we kept staying, and we all got older. If I were stuck in a retirement home, I'd say "Heavens get me away from all these old people" but I have had quadruple-bypass surgery, and it's good to be around people who understand that. [Lewin, 1991]

Once again, note that the crucial variable here is other people who provide support. Many of the elderly, even when they become frail and their homes become decrepit, prefer to stay put rather than move away from the neighborhood convoy (Pynoos & Golant, 1996).

memories, intimacies, and gratitude for past support. Thus, while older people sometimes enjoy relationships with much younger people, they particularly cherish their long-term friendships with those they knew in their youth (Rawlins, 1995). Unfortunately, this preference can eventually lead to sorrow, as some treasured friends die, and others become disabled, making the simple act of getting together much more arduous. In the latter case, many of the elderly turn to letters, cards, and phone calls to sustain their friendships, making the post office and the telephone as vital to social life in old age as it was in early adolescence.

The oldest-old adjust to unwelcome changes in the social convoy in other ways as well. A study of community-dwelling elderly between the ages of 70 and 104 found that, with age, the size of their social network decreased overall but that the number of people they deemed so close "it would be hard to imagine life without them" remained the same. Thus new friends were added to the inner circle when death or distance intervened. Probably for the same reasons, after age 85, the closest friends were somewhat less likely to be members of the nuclear family and more likely to be nonrelatives or more distant younger relatives. Friends were particularly helpful in buffering the oldest-old who were widowed.

Overall, these social-convoy patterns demonstrated "proactive selection, compensation, and optimization to enhance old age" (Lang & Carstensen, 1994). Thus successful aging requires that people keep themselves from becoming socially isolated, a task that most of the elderly manage to accomplish.

Siblings

One of the understudied and vital aspects of late adulthood is the relationship between siblings, who often are quite close in late adulthood (Kendig et al., 1988; Moyer, 1992). In one study of 300 older adults, 77 percent considered at least one of their siblings to be "a close friend" (Connidis, 1994).

Of course, closeness is not automatic: sibling rivalries set in childhood often continue throughout life (Greer, 1992). In some cases, the death of the parents, rather than freeing siblings from a lifelong rivalry, actually increases it. Even the death of a sibling does not necessarily release the survivor from jealousy (Cicirelli, 1995).

More often, however, sibling conflicts fade and closeness increases in late adulthood. One researcher described the usual pattern between siblings as an "hourglass effect," close during childhood, increasingly distant until middle adulthood, then gradually closer again as late adulthood progresses (Bedford, 1995). Siblings who respected and liked each other throughout adulthood but who never spent much time together often become close confidants and even share residences once again when a spouse dies. Relationships between sisters are likely to be particularly close.

The specifics of sibling relationships in late adulthood are strongly influenced by family values set in childhood. If the family encouraged the idea that one should "protect your sister" or "fight for your brother," this value is likely to reemerge in old age, especially if frailty or loneliness threatens (Cicirelli, 1995).

Other Generations

The typical older adult has many family members. In fact, because more people today are living longer, more older people are part of multigenerational families than at any time in history (Uhlenberg, 1996). Sometimes the family spans five generations, usually in what is called the **beanpole family**, consisting of more generations than in the past but with only a few members in each generation. More often, older people are members of three- or four-generation families.

As we have seen, many of the elderly live alone, a fact that has made some bemoan the demise of the extended family living under one roof. However, interviews and studies of the elderly themselves find that the majority of those living alone prefer it that way. Nevertheless, the generations of each family tend to see each other and help each other frequently. In one multinational study, most family members said they talk to each other regularly, and 90 percent said they would solicit help or advice from another generation within their family if a problem arose (Farkas & Hogan, 1995). Thanks to modern-day communications and travel, aging parents and middle-aged children typically stay very much in touch with each other even if they live great distances from each other (Bengston et al., 1996).

While intergenerational relationships are clearly of positive value, they also are likely to include tension and conflict. Almost all elders who have children and grandchildren devote time and attention to them, sometimes to the frustration of all parties. Fewer older adults stop parenting simply because their children are full-grown, independent, or even married or parents themselves. As one 82-year-old woman succinctly put it: "No matter how old a mother is, she watches her middle-aged children for signs of improvement" (Scott-Maxwell, 1968).

Generally, the mother-daughter relationship is simultaneously very close and vulnerable to tension. For example, in one study of forty-eight mother-daughter pairs whose average ages were 76 and 44, respectively, 75 percent of the mothers and almost 60 percent of the daughters said that the other was one of the three most important persons in their life. At the same time, 83 percent of the mothers and 100 percent of the daughters readily acknowledged having recently been "irritated, hurt, or annoyed" by the other. The mothers were more likely to blame someone else for the irritation ("Her husband kept on turning up the radio every time I turned it down"), and the daughters were more likely to blame their mother for intruding on their lives ("She tells me how to discipline my kids") (Fingerman, 1996).

For the adult children, an added source of tension springs from their sense of **filial obligation**, that is, the sense of duty and need to protect and care for their aging parents (Blieszner & Hamon, 1992). Often the children and their parents do not share the same view about how much care is appropriate. Many elders, quite rightly, value their independence and resist their children's every attempt to help them—from offering to provide transportation to taking over financial affairs. Many other elders expect and demand more care than their children are willing or able to provide. Such differences obviously can lead to frustration, friction, and feelings of guilt.

All these overlapping family concerns highlight the complexity of the ongoing relationships among the generations of a family. Most families evi-

beanpole family A multigenerational family having only a few members in each generation.

filial obligation The sense, experienced by most adult children, of a duty and need to protect and care for their aging parents.

Family members from different generations still rely on each other for many forms of support as well as enjoy each other's company, as suggested by this family reunion in Wisconsin. Note, however, that each generation tends to have closer contact with those of about the same age. When widowed elders seek a close family confidant, they are more likely to choose a sibling or cousin than a child or grandchild.

dence three signs of intergenerational closeness—mutual assistance, frequent contact, and shared affection, each one signifying a somewhat different aspect of the intergenerational bond (Mangen et al., 1988). In general, assistance arises both from the need for it and from the desire and ability to provide it; in-person contact is more dependent on geographical proximity than on any other factor; and affection is strongly influenced by a family's past history of mutual love and respect. This means that a family can be high on the first two signs of closeness—assistance and contact—without necessarily being high on warmth and affection.

Contrary to popular perception, since most of the elderly are quite capable of caring for themselves, assistance typically flows from the elder generation to their children instead of vice versa. In fact, most elders are pleased to be able to buy things for their children and help out occasionally with the grandchildren, and they do not expect large gifts in return, even if someday their financial circumstances change (Hamon & Blieszner, 1990). The older generation also typically enjoys social contact with the younger generations, but on their own turf and terms. Most would rather the children visit them at their invitation for a few hours rather than come uninvited, or stay too long, or expect the elders to do the traveling.

As you might imagine, the most complex aspect of the relationship among the generations involves the exchange of advice and respect. It is not unusual for there to be a "generation gap" between the elderly and their children and grandchildren (Hagestad, 1984). When the generations differ in their views on such areas as politics, sex, child-rearing, and religion, family members, in the name of "good relations," may confine their conversation to "demilitarized zones" involving topics that are unlikely to provoke anger or hurt feelings. Frequently this results in superficial relations, with the different generations hesitating to offer constructive criticism or advice, even when it might be needed.

Thus the price of intergenerational harmony may be intergenerational distance. This was clearly the case for the following couple, who felt that they were viewed as "outmoded and irrelevant" rather than "wise or expert." They explained:

frail elderly Older people who are physically infirm, very ill, or cognitively impaired.

activities of daily life (ADLs) Actions that are important to independent living, typically comprising five tasks: eating, bathing, toileting, walking, and dressing. The inability to perform these tasks is a sign of frailty.

instrumental activities of daily life (IADLs) Actions that are important to independent living and that require some intellectual competence and forethought. These are even more critical to self-sufficiency than ADLs.

> Our grandson just got married. They both have fancy taste and fancy plans, and they mean to have it all. When we asked about a baby, they said they wouldn't even think about having a child until they could afford a full-time nanny to raise it. Imagine planning to have children so that you won't have to raise them! Whose children are they anyway? Why have them? We love him so much, and we don't want to hurt him, so we didn't say anything. As long as we keep rather quiet, he thinks we're sweet and lovable—and rather silly . . . [Erikson et al., 1986]

The Frail Elderly

So far in these chapters we have emphasized the majority of the elderly—either those aging in the "normal" way, alert and active, financially secure, supported by friendship and family ties, or those who are so successful that they can retire with new financial and social freedom. These two groups are quite different from the group we focus on now—the **frail elderly**, a group that includes the physically infirm, the very ill, and the cognitively impaired.

No single demarcation differentiates the frail from their hardy and successful contemporaries. However, beyond simple vulnerability and fragility, the crucial sign of frailty is probably an inability to perform, safely and adequately, the various tasks of self-care. Gerontologists often refer to the **activities of daily life**, abbreviated as **ADLs**, typically comprising five tasks: eating, bathing, toileting, walking, and dressing. If a person needs assistance with even one of these, he or she may be considered frail, although for some purposes (such as medical insurance compensation or government research on dependency), frailty is not considered to begin until a person is unable to perform three or more.

Equally important, if not more important, to independent living are the **instrumental activities of daily life**, or **IADLs**, actions that require some intellectual competence and forethought (Willis, 1996). As one might expect, specific IADLs vary somewhat from culture to culture. For most of the elderly in developed nations, IADLs include shopping for groceries, paying bills, making phone calls, taking appropriate medications, and keeping ap-

Severe arthritis puts this woman in the category of the frail, meaning that she needs assistance with her ADLs. She is not frail of spirit, however, and with the help of an aide, she is able to live a robust and relatively independent life.

TABLE 25.2

Instrumental Activities of Daily Living

Domain	Exemplar task
Managing medications	Determining how many doses of cough medicine can be taken in 24-hour period Completing a patient medical history form
Shopping for necessities	Ordering merchandise from a catalog Comparison of brands of a product
Managing one's finances	Comparison of Medigap Insurance Plans Completing tax return for income tax form
Using transportation	Computing taxi rates Interpreting driver's right-of-way laws
Using the telephone	Determining amount to pay from phone bill Determining emergency phone information
Maintaining one's household	Following instructions for operating a household appliance Comprehending appliance warrantee
Meal preparation and nutrition	Evaluating nutritional information on food label Following recipe directions

Source: Willis, 1996.

pointments (Table 25.2). For some of the elderly in other nations, tending the family livestock, cultivating the family garden, mending clothes, and baking bread might be among the culture's list of IADLs. Everywhere, however, there is a normative decline in the ability to perform such tasks, a decline that sometimes reaches sufficient impairment that a person becomes cognitively frail, dependent on others for help with meal preparation, medications, shopping, housekeeping, and so forth.

Increasing Prevalence of Frailty

Worldwide, the frail elderly are a minority, even in those nations where health care prolongs life for the most vulnerable individuals. At any given moment, no more than 15 to 25 percent of the world's senior citizens overall are frail by any measure. In terms of total populations, the frail elderly represent about 4 percent of the population in the most developed nations and less than 1 percent in the least developed. However, in every nation, the number of frail elderly and the degree of their frailty are increasing, for three reasons.

As we have seen, the first reason is that as average life expectancy increases and fewer individuals die young, more people reach old-old age. In the United States, for instance, the number of Americans aged 85 and older doubled in the twenty years from 1975 to 1995 (a time period when the number of teenagers actually fell 7 percent), making the oldest age group the fastest-growing segment of the population (U.S. Bureau of the Census, 1996). As more people reach old age, the absolute number of frail individuals will increase.

The increasing frailty with age was detailed by one extensive study of functional capacity in the year before death (Lentzner et al., 1992). As you

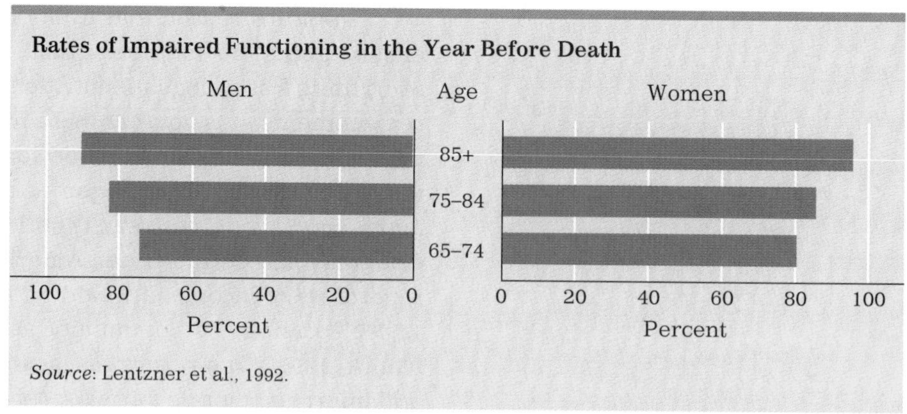

FIGURE 25.4 While most elderly people, most of the time, function quite well, the older a person is and the closer to death, the more likely it is that he or she will be unable to perform all the normal activities of daily life. As you can see, after age 85 less than one person in ten is functioning at full capacity in the last year of life.

can see from Figure 25.4, the great majority of elderly experience some degree of frailty before they die, and after age 85, less than 10 percent are fully functioning in their last year of life. Note, however, that the figure includes all those who needed *some* assistance, which means that a portion of the frail old remain independent right up to death. Nonetheless, the older a person is, the greater the chances that he or she will need major assistance before dying. Among the oldest-old in their final year, half require assistance with all five ADLs and more than a third exhibit cognitive impairment as well.

These general findings are supported by a British study that used random samples to compare functioning in the last year of life in 1969 and 1987. The data showed that most who died in 1987 were more likely to be over age 75 (54 percent compared with 40 percent in 1969) and were likely to be physically frail, in and out of hospitals, in the last year of life (Seale & Cartwright, 1994).

The second reason for the rise in frailty is that, as we have seen, the medical establishment is still geared more toward death prevention than toward life enhancement, focusing on dramatic, life-saving intervention for acute illness rather than on the prevention and treatment of chronic illness. Since chronic problems—everything from Alzheimer's disease and arthritis to ulcers and varicose veins—are the ones that most commonly sap the elderly's strength, pride, and independence, the result of these medical priorities is the increasing prevalence of morbidity even as mortality rates fall. Just using the knowledge already available, frailty could be decreased substantially if medical efforts were redirected to ameliorating or preventing nonfatal diseases. For instance, current therapies, if fully utilized, could cut in half the rate of urinary incontinence, which is a problem for up to a fourth of all the elderly. Focusing the attention of general practitioners on the available solutions to this problem would thus improve the quality of life for about 10 million older Americans (Coni et al., 1992).

However, when the balance in medical care is weighted toward adding "years to life" rather than "life to years," the compression of morbidity discussed in Chapter 23 is unlikely to occur. Rather, the push to extend life for its own sake means that, in many cases, "the time between the onset of chronic morbidity and death is long, measured in years and decades. Discomfort and limitations become everyday matters—perpetual for some people, episodic in others" (Verbrugge, 1994).

The third reason that frailty is increasing is that many nations fail to ensure access to measures that could prevent or reduce impairment—everything from adequate nutrition to safe housing, from hearing aids to hip replacements. It is not that these measures are not available. Indeed, many nations spend substantial money on services for the elderly. However, while the specifics vary from nation to nation, the central problem is that adequate services do not reach the most frail. For example, in 1965 the U.S. Congress passed the **Older Americans Act**, providing every older person, regardless of income, a wealth of benefits—from subsidized meals at over 14,000 locations to community services offered by over 20,000 agencies (Quirk, 1991). One result was, as intended, better health, less dependence, and improved morale for many seniors.

However, these benefits are neither comprehensive nor free, with housing and health benefits particularly likely to require copayment. In addition, obtaining these benefits requires some mobility, planning, and initiative. This combination of factors tends to exclude precisely the poor, uneducated, or isolated who are most likely to become frail (Estes et al., 1996). For example, social services of every kind—from senior-citizen centers to visiting nurses—are relatively scarce in rural areas, where a disproportionate share of the elderly reside (McLaughlin & Jensen, 1993). This is particularly the case for "elders of color," such as the African-American elderly in the Deep South and the Native American and Chicano elderly in the Southwest. Because of their age, race, SES, and location, these older people in rural communities can be said to live in "quadruple jeopardy" for frailty (Agency for Health, Policy and Research, 1994; Bane, 1991). Such risk often tips into reality, with a higher proportion of such individuals being both physically and mentally frail (Skinner, 1995).

Ironically, for those elderly who have access to, but less need for, public assistance, unquestioned acceptance of these benefits sometimes leads toward frailty rather than away from it. As critics note, although the Older Americans Act was intended to "make elderly persons as a whole economically solvent and independent managers of their own affairs," in many cases, its "welfare-oriented articulation further transformed them into a state-dependent class" (Lynott & Lynott, 1996).

Age and Self-Efficacy

Because we are focusing on impairments and on the social structures that foster them, we need to stress again that long life does not inevitably include years of frailty. Nor necessarily does being female or a minority member or poor, or all three. True, membership in any of these categories puts a person at increased risk of frailty. But it is sometimes unclear whether a particular impoverished frail person is an example of the poor becoming sick or the sick becoming poor. In addition, although it is apparent that elders in the bottom fifth of income rankings—half of whom are minority women—report being in poor health three times more often than those in the top fifth (Crystal, 1996), some poor minority women are nonetheless quite independent, not frail.

To better understand how a person avoids frailty, we need to look not only at demographics but also at self-theory. An active drive for autonomy, control, and independence is one of the best defenses against dependence.

Older Americans Act Legislation passed by the U.S. Congress in 1965 that provides every older person, regardless of income, a wealth of benefits—from subsidized meals to community services.

A CLOSER LOOK

Between Old, Fragile, and Frail—Protective Buffers

Frailty is not automatically defined by either age or illness. Both advanced years and specific infirmities may make someone more fragile, but neither necessarily makes that person frail, because the health and independence of the elderly depend not only on intrinsic impairment but also on extrinsic resources (Davies, 1991). Many elderly persons never become helpless because four protective factors—their attitude, their social network, their physical setting, and their financial resources—act as a buffer, preventing or postponing the progression from fragility to frailty.

Consider the hypothetical example of two 80-year-old childless widows who have the same failing eyesight and advanced osteoporosis. One widow might live alone in an old, rundown house in an isolated neighborhood. Among the particulars of her residence and daily life are uneven hardwood floors covered with braided scatter rugs, a flight of steep stairs separating the bedroom and the kitchen, dimly lit rooms and hallways, and rumors of a recent robbery two blocks away. After falling and fracturing her wrist on the way to the toilet one night, she is now apprehensive about walking around anywhere without help, and she refuses to go downstairs to prepare meals. She, of course, never ventures outside anymore and is afraid to answer the door or the phone. Further, she no longer tries to wash or dress herself, or even feed herself as much as she should, citing some lingering pain in her fingers and the fact that "no one cares anyway."

Obviously, this widow is very frail, requiring ongoing care. At present, a home attendant comes every morning to bathe her and prepare the day's food, but the attendant is worried about the woman's depression. This is a valid concern, since suicide is more prevalent among the elderly than among any other age group and is particularly common among those over 75 who live alone. This widow is on the waiting list for a nursing home, where she will probably become even more frail, since nursing homes often discourage independent functioning (Van Nostrand, 1991).

The other widow, by contrast, might have had the financial resources and foresight to have purchased, with two old friends, a large co-op apartment near a small shopping center. As all three are aging, they have reduced their vulnerability by outfitting their home with precautionary amenities such as bright lighting, sturdy furniture strategically placed to aid mobility, secure grab rails in the bathroom to ease bathing and toileting, wall-to-wall carpeting nailed to the floor, a telephone programmed to dial numbers at the push of one button, a stove that automatically shuts off after a certain time, a front door that buzzes until it is properly locked with the key, and so forth.

In addition, all three women gladly compensate for each other's impairments: the one who sees best reads the fine print on all the medicine bottles, legal papers, and cooking directions; the one who is the sturdiest sweeps, mops, and vacuums; and our poorly sighted, osteoporotic widow, who has excellent hearing, responds to the phone, the doorbell, the alarm clock, the oven timer. All three regularly eat, converse, and laugh together—a practice that is good for the digestion as well as the spirit.

Unlike the first widow, who will soon be institutionalized, the second widow with the same physical problems is safe and happy in her apartment, caring for herself, socializing with friends, shopping in the community, and so forth. For her, protective buffers will continue to defend against many factors that could otherwise be disabling. For example, she will be motivated, encouraged, and financially able to obtain good medical care and enabling accessories, such as corrective eye-drops and special glasses, or calcium supplements and a hip replacement, or even, if both major disabilities worsen, home delivery of books-for-the-blind and the purchase of a small, motorized wheelchair.

The lesson here is that a certain degree of fragility and vulnerability does not necessarily translate into an equivalent degree of frailty. Just as a fine crystal goblet—admired, lovingly handled, and carefully stored in soft cloth—is unlikely to break despite its fragility, so an older person, surrounded by crucial buffering, is less likely to become frail.

Correspondingly, loss of control invites further weaknesses in many domains (Rodin & Timko, 1992). As one team of researchers note:

> To a sizable number of chronically ill older persons, their disability is of less salience than might be expected, because they can shift their priorities to other behavioral options. Perceived control and perceived self-efficacy are associated with positive health practices on the one hand, and with the absence of chronic disease and with good functional ability and self-perceived health on the other hand. [Deeg et al., 1996]

A CLOSER LOOK

Elder Maltreatment

For many reasons, the frail elderly are particularly vulnerable to abuse. Not only do they depend on others for their physical care but they also are often confused about much in their lives, from the names of those around them to the state of their finances. Further, many who are abused or neglected are ashamed to admit it, and when they do, the accusation is sometimes dismissed as paranoid. Given all this, it seems certain that the rate of maltreatment for the frail elderly is much higher than that for elders overall, which is estimated to be 2 to 4 percent in the United States and 1 to 4 percent in Canada (Hornick et al., 1992).

Although some of the substantiated maltreatment of the elderly is perpetrated by professional caregivers, con-artists, mean-spirited strangers, and the like, **elder maltreatment** is primarily a family affair. For example, one detailed study of all confirmed cases of elder abuse in Illinois found that 87 percent of the perpetrators were family members, most often a middle-aged child (39 percent), sometimes another relative (24 percent), and least often a spouse (14 percent) (Hwalek et al., 1996).

In many ways, elder abuse and neglect parallel child and spouse maltreatment, both in kind and in cause, ranging from direct physical attack to ongoing emotional neglect and involving such underlying factors as the social isolation and powerlessness of the victim, mental impairment or drug addiction of the perpetrator, and inadequate education and poverty within the household (Pillemer & Finkelhor, 1988).

There are differences as well. The typical case of elder maltreatment begins benignly, as an outgrowth of a mutual caregiving relationship within the family (Steinmetz, 1988). For example, an elder may begin to financially assist someone of the younger generation, who then gradually takes control of and misuses more and more of the elder's assets; or a younger family member may assume care of an increasingly frail relative, only to become so overwhelmed by the task that gross neglect and abuse seem inevitable.

Occasionally, an elder who becomes mentally impaired also becomes the perpetrator of abuse. For example, one Chinese-American woman, age 73, was admitted to the hospital with bruises and a broken wrist. On careful questioning, she admitted that her husband, suffering from Alzheimer's disease, had battered her when she tried to care for him (Elder Abuse Project, 1991).

These examples make it clear that elder abuse within the family must be diagnosed, treated, and prevented case by case. As one noted researcher explains, "elder abuse is not a monolithic problem that can be solved with blanket programs, such as those resulting from mandatory reporting" (Pillemer, 1991). Typically, abuse is "only one component of a larger set of complicated problems," part of a complex "social, psychological, and economic equilibrium that has taken a lifetime to develop" (Hornick et al., 1992). The best solutions would be the provision of extensive public and personal safety nets of support for those elderly who are frail or powerless, so that no one—caregiver or care-receiver—gets to the point of abuse. If abuse does occur, a social worker, not simply a prosecutor, needs to confront all the problems revealed by the abuse.

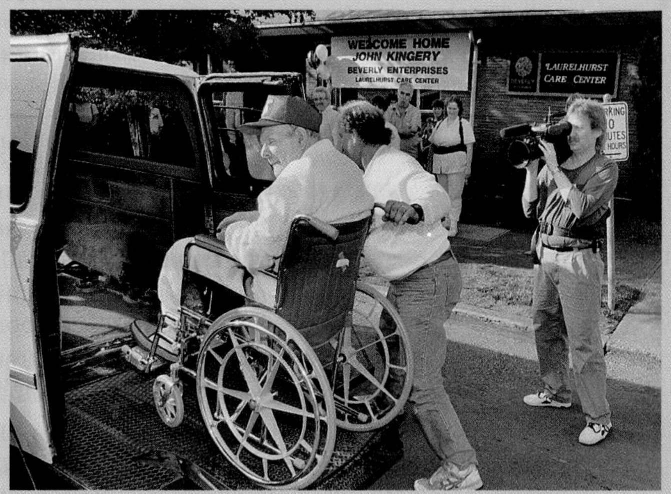

This elderly man was found alone, amnesiac and incontinent, tied in his wheelchair at an Idaho dog track. It was later discovered that his own 40-year-old daughter had abandoned him there, hundreds of miles from her home in Oregon, hoping to rid herself of the stress of having to take care of him. The daughter was put on trial on a variety of charges, but her defense describes the kind of tangled family dynamics that often surround elder abuse: the father has late-stage Alzheimer's disease, requiring extensive care and boundless patience, and the daughter says that she was sexually and physically abused by both parents when she was young. Until a year before this incident, she had not seen her father for twenty years. Then one day he arrived at her doorstep, needing help.

elder maltreatment Elder abuse and neglect, ranging from direct physical attack to ongoing emotional neglect and involving such underlying factors as the social isolation and powerlessness of the victim, mental impairment or drug addiction of the perpetrator, and inadequate education and poverty within the household.

feel relief rather than an obligation to do their share. Second, care-receivers and caregivers, as noted earlier, often disagree about the nature and extent of care that is needed and whether the caregiver has the right to set the daily schedule, regulate menus, arrange doctor's visits, and so forth. Such disagreements are bound to cause strain, not only in the caregiver who feels frustrated but also in the care-receiver who wants to be self-determining. Finally, in an effort to contain costs, social agencies rarely offer services unless they are urgently and obviously needed. Most difficult to obtain are those services, such as respite care or support groups, that are designed primarily for the

caregiver. Such supports are especially needed if the care-receiver is cognitively impaired, a burden that is never easy to bear. As one review of research on caregiving to people with dementia concludes, "there is nothing—social support included—that can make everything all right" (Pearlin et al., 1996).

As a result of these circumstances, many caregivers feel that they are "left to care alone," resentfully resigning themselves to years of sacrifice (Cicirelli, 1992; Kaye & Applegate, 1990). Indeed, many studies show that the amount of stress, resentment, and ill health experienced by caregivers correlates more with their subjective interpretation of the support they experience than with any other variable—including how impaired the aged person actually is (Biegel et al., 1991; Kaye & Applegate, 1990; Uchino et al., 1992). Sadly, as A Closer Look on page 692 explains, when caring for a frail elder does create feelings of resentment and entails social isolation, the consequence sometimes is elder maltreatment.

Nursing Homes

Many older Americans and their relatives feel that nursing homes should be avoided at all costs, usually because they believe that all nursing homes are horrible (Pyke & Bergston, 1996). Some nursing homes are indeed horrible. The worst tend to be those profit-making ventures where most patients are subsidized entirely by Medicare. The only way for these institutions to turn a profit is to cut down on expenses. Consequently, they are staffed by overworked, poorly trained aides who provide minimal, often dehumanizing care.

Overall, however, the kinds of abuses that occurred in the 1950s and 1960s, when the sudden, unregulated expansion of the nursing-home industry triggered a rash of shoddy care and maltreatment, have been greatly reduced by increased professionalism and government oversight. Indeed, some nursing homes are excellent. Consider the case of Laura Hunter (Hunter & Memhard, 1981), for example. She had reached the age of 80 "bedridden, arthritic, and crotchety . . . relying on drugs, and incapacitated by fears of impending change, illness, and death, [she] clung to the radio for company and turned away even her closest family and friends." A period of hospitalization convinced her children that something had to be done. After much soul-searching, they decided to place her in a nursing home, one of the best in the country. Residents there had their own rooms and were able to carry on private lives, having friends in to visit, for instance, or going to

Good nursing homes encourage residents to participate in regular physical exercise, and the best provide physical therapy for those who need it. Good nursing homes also encourage contact with the outside community, including visits from volunteers to allow friendship between the generations. This type of social contact is especially important, since a disproportionate number of nursing-home residents have no living relatives.

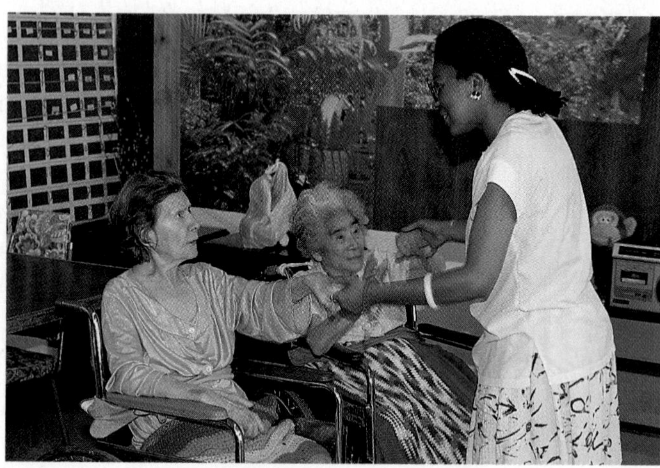

sleep when they wanted to. The staff encouraged the residents to stay active, and the residents had their own council, which helped determine residence policies.

Initially, Laura was despondent and uncommunicative, but gradually she came out of her shell and joined the community life around her. Among other things, she made friends, joined a book-review club and an exercise class, won election to the residence council, worked as a reporter for the residence newspaper, and developed a romance. She also kept a journal of all those activities. Reading it, one gets the impression of a spunky, good-humored lady with a love of life. She needed good nursing-home care, surrounded by people who could become her friends, to help her express that love.

Such an ideal example, while true, should not blind us to the problems that are more typical.

Ongoing Problems

One problem with many nursing homes is that they concentrate almost exclusively on the physical maintenance of their residents and give insufficient attention to the residents' psychological needs, such as social interaction and a feeling of social control. Too often the staff pay more attention to those patients who docilely wait for their needs to be met, thereby reinforcing the dependence of many patients who could learn to fill some of these needs themselves. Correspondingly, the more independent patients suffer from inattention and, even worse, they learn that the best way to cope with life is by becoming passive and relinquishing control. An immediate consequence is that they become less active; in the longer term, they become less healthy, lose self-esteem, and die earlier than they would have (Baltes & Wahl, 1992).

This attempt of the nursing-home staff to increase the dependency of residents is called **infantilization**, because, in effect, the staff treat the old as if they were infants. They determine the residents' daily activities—including when they go to bed and when they get up—choose their daily menus, and structure their ADLs, including bathing and dressing. They are even likely to use baby talk with them. For the most part, infantilization occurs in the name of efficiency. As one critic observes, "the natural tendency of caregivers is to err on the side of dependence, as the institution runs more smoothly if staff rather than residents organize daily schedules . . . for example, it is more efficient for staff to [diaper and] change a patient than to rely on the patient's schedule of voiding" (Whitbourne & Wills, 1994).

Such efficiency comes at the cost of the elderly's dignity and independence. At the same time, those residents who resist this infantilization, who attempt to retain control of their lives by ignoring rigid rules and trying to set their own schedules, may pay a different price:

> It is likely that they will be labeled "troublemakers" and have even more of the autonomy they seek withheld from them. A vicious cycle is set into motion, in which the more these individuals protest, the more they stand to lose and the more likely it is that their resolve will give way to compliance at best or isolation at worst. Unlike their counterparts who entered the institution with a more compliant attitude, these defiant individuals will find great difficulty in accepting institutional control and do so only at a great emotional cost. [Whitbourne & Wills, 1994]

infantilization The treatment of elderly people in institutions as if they were infants.

Increasingly, professionals are becoming involved in developing good nursing-home care, where the goal is to help each patient gain as much independence, control, and self-respect as possible (Goldsmith, 1994; Gordon & Stryker, 1996). Thus it is possible to find good care, if one knows what to look for, and can afford it. For those who need nursing-home care, the quality and suitability of that care can make the difference between the final years being full and satisfying or a desolation. In fact, good care can even lead to a longer life.

Let us end with the story of a 98-year-old woman whose great-grandson, Rob, related that his great-grandmother "began to fail. We had no idea why and thought, well, maybe she is growing old." The family reluctantly decided it was time to move her from her own suburban home where she had lived for decades into a nursing home. Fortunately, the nursing home was one that encouraged independence and did not assume that declines in functioning are always signs of "final failing." Indeed, the doctors there discovered the woman's pacemaker was not working properly. As Rob explains, "We were very concerned to have her undergo surgery at her age, but we finally agreed. . . . Soon she was back to being herself, a strong, spirited, energetic, independent woman. It was the pacemaker that was wearing out, not great-grandmother" (Adler, 1995).

This story contains a lesson for us all, one that underlies many of the lessons learned throughout this book. Whenever an older person seems to be failing, or a preschooler is unusually aggressive, or a teenager is depressed, or an adult is overloaded with work and family, the tendency is to think that such problems "go with the territory" of being at a particular age. And there is some truth in that, for all these possible problems are more common at the stages mentioned. However, the overall theme of the life-span perspective is that, at every age, people can be "strong, spirited, and energetic," able to live their life to the fullest no matter how young or old they are.

SUMMARY

Theories of Psychosocial Development in Late Adulthood

1. Several self-theories hold that adults make personal choices in ways that allow them to become fully themselves. For instance, the continuity theory emphasizes that the stability of personality traits and behavior patterns makes the changes that occur with age much less disruptive than they might appear to be. According to Erikson's theory, individuals in the final stage of development seek to integrate the various earlier stages and to understand their contribution to the future of humankind.

2. Social stratification theories maintain that social forces limit personal choices. The individual's ability to function well in old age depends in large measure on what social stratum the individual is in. Some social scientists argue that lifelong stratification by gender or race limits an elder's ability to function well. While social stratification theories have some merit, they do not necessarily consider the complexity and diversity of late adulthood.

3. Dynamic life-course theory sees human development as an ever changing process, influenced by social contexts, which themselves are constantly changing, as well as by genetic and historical factors that are unique to each person.

Generativity in Late Adulthood

4. More and more people are retiring at earlier ages than ever before, partly because the financial incentives to do so have increased and partly because the jobs available have decreased. Many retired people continue their education or perform volunteer work in their communities. Both of these activities enhance the health and well-being of the elderly and benefit the larger society.

5. The elderly are politically active and influential, which is one reason for their success in protecting their economic benefits. Fortunately, most older adults agree on the need for generational equity.

The Social Convoy

6. Marriage provides important social support in old age. Older adults in long-standing marriages tend to be quite satisfied with their relationships and safeguard each other's health. As a result, married elders tend to live longer, happier, and healthier lives than unmarried elders.

7. The loss of a spouse is one of the most serious stresses the elderly can experience. Widowers are more likely to experience health problems but are more likely to remarry. Widows are more likely to have financial difficulties but also find comfort through an expanded network of friends.

8. Friendship continues to be important in late adulthood, as a source of happiness and as a buffer against trouble. Particularly among the never-married or the no-longer-married, long-term friendships are particularly valued, which is one reason siblings often draw closer in late adulthood. Partly because friends and neighbors are vital parts of the social convoy, most older adults prefer to remain in their neighborhoods rather than move to a planned retirement community or move in with a grown child.

9. Many older people are part of multigenerational families, sometimes with two generations over age 60, each maintaining dependence as well as mutual support. Typically the young-old are more likely to give advice and assistance to younger generations rather than to be the beneficiaries of filial obligation.

The Frail Elderly

10. Most older people will eventually become frail, unable to care for their daily needs. Those who are poor, female, and/or over age 85 are particularly likely to experience an extended period of frailty. As more people reach very old adulthood, the number needing help from family and society will increase.

11. The frail are usually cared for by a close relative—typically their spouse, daughter, or daughter-in-law. Despite the personal sacrifices this care entails, most relatives consider such care an expression of family commitment. For a minority of adults, however, caring for a dependent and needy older person leads to frustration, anger, and maltreatment.

12. For the elderly who eventually need more care than their family can readily provide, and therefore must enter a nursing home, the quality of their final years of life can vary enormously, depending on the quality of the home. The best homes recognize the individuality of the elderly and encourage their independence.

KEY TERMS

self-theories (661)
continuity theory (661)
selective optimization with compensation (662)
integrity versus despair (663)
social stratification theories (664)
dynamic life-course theory (666)
generational equity (673)
social convoy (673)
planned retirement communities (681)
naturally occurring retirement communities (NORCs) (681)
beanpole family (683)
filial obligation (683)
frail elderly (685)
activities of daily life (ADLs) (685)
instrumental activities of daily life (IADLs) (685)
Older Americans Act (688)
elder maltreatment (692)
infantilization (694)

KEY QUESTIONS

1. Describe the key elements of self-theories of late adulthood.

2. Give examples of three different social stratification theories.

3. How are older people typically involved with their community?

4. What changes typically occur in long-term marriages in late adulthood?

5. What specific needs of the elderly do friends and neighbors fill?

6. What are the advantages and disadvantages of various living arrangements in late adulthood?

7. How do the elderly relate to relatives other than their husband or wife?

8. What accounts for the increasing prevalence of the frail elderly?

9. What factors are critical in providing good care for the frail elderly?

PART VIII The Developing Person So Far: Late Adulthood, Age 60 Onward

Biosocial Development

The Aging Process

Because of declines in organ reserve, the immune system, and overall muscle strength, older adults are at greater risk of chronic and acute diseases, heart disease, and cancer. However, risk is also related to long-standing health habits and quality of health care. Postponement of many of the illnesses linked with age is possible, allowing a "compression of morbidity." Research on the causes of aging indicates that genes play a prime role. Specific theories of aging related to the immune system, a "genetic clock," damage from free radicals, and cellular error are plausible but not definitive.

Cognitive Development

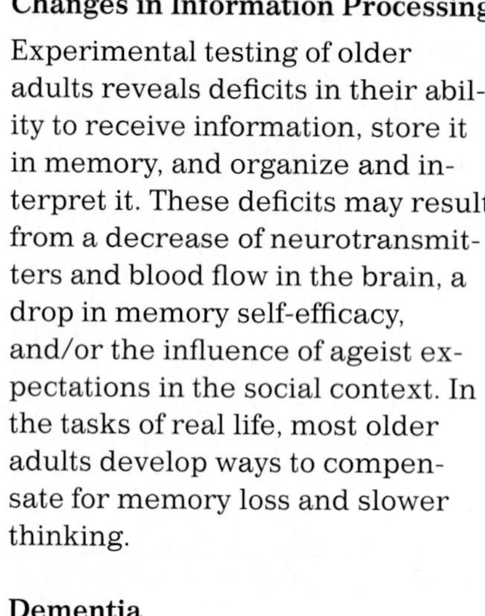

Changes in Information Processing

Experimental testing of older adults reveals deficits in their ability to receive information, store it in memory, and organize and interpret it. These deficits may result from a decrease of neurotransmitters and blood flow in the brain, a drop in memory self-efficacy, and/or the influence of ageist expectations in the social context. In the tasks of real life, most older adults develop ways to compensate for memory loss and slower thinking.

Dementia

Dementia, with its progressive impairment of cognitive functioning, is not inevitable in old age but it does become more common, especially in the very old. Symptoms of dementia may be caused by Alzheimer's disease, problems in the circulatory system, other diseases, depression, or drugs.

New Cognitive Development

Many older individuals develop or intensify their aesthetic and philosophical interests and values in later life.

Psychosocial Development

Generativity

Elders usually remain active whether working or retired. Most find ways to expand their horizons after retirement, with education, volunteering, political involvement, and socializing.

The Social Convoy

Older adults' satisfaction with life depends in large part on continuing contact with friends and family. Generally, marital satisfaction continues to improve. The greatest source of social support is likely to be other elders, either relatives or friends, particularly those of long-standing importance.

The Frail Elderly

As people increasingly live to a very old age, the number needing assistance in the activity of daily life grows. Ideally, this assistance encourages elders to be as active and independent as possible.

Death and Dying

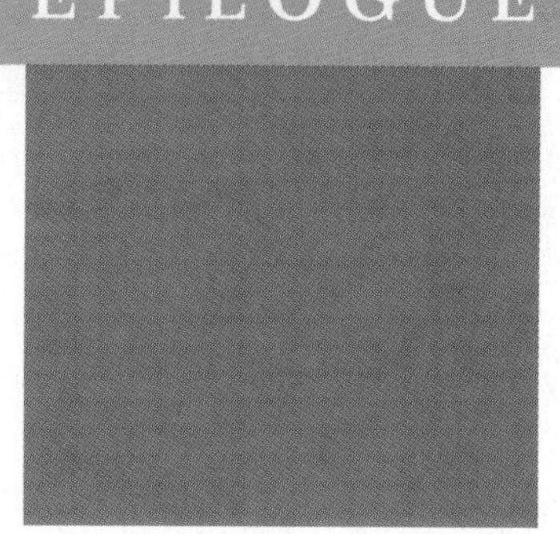

EPILOGUE

One goal of the study of human development, as outlined in Chapter 1, is to help each person realize his or her full potential. According to many developmental theorists, achieving an understanding of death and dying is essential to the complete realization of self. By shedding light on this final stage in human development, **thanatology**, the study of death, can thus be "an important means of affirming the value and love of life" (Moller, 1996).

The Social Context of Dying

One thing thanatology teaches us is that although death comes to everyone, our perceptions of death are highly variable. Throughout many cultures, past and present, death has been seen as a quite social moment, witnessed and shared by the community as a whole. In most African traditions, for example, elders take on an important new status through death, joining the ancestors who watch over not only their own descendants but over the entire village as well. Therefore everyone in the village participates in a funeral, preparing the body and providing food and money for the deceased's journey to the ancestral realm. Mourning the death of the individual becomes an occasion for the affirmation of the entire community, as members jointly celebrate their connection with each other and with their collective past (Opoku, 1989). This sense of connectedness is aptly captured in an African proverb that states "When an old person dies, a library burns."

In many Muslim nations, death affirms not so much faith in the group as faith in Allah. In Islam, religious teachings emphasize that the achievements, problems, and pleasures of this life are transitory and ephemeral, and that everyone should be mindful of, and ready for, death at any time. Therefore, for believers, caring for the dying and the dead is a holy reminder of their own mortality. Specific rituals—including reciting prayers, carrying the coffin, and attending the funeral—are performed by devout strangers as well as by relatives and friends (Knappert, 1989; Muwahidi, 1989).

Among Buddists, disease and death are among the inevitable sufferings of life, suffering that may bring enlightenment. Therefore the task of the individual who is dying is to gain insight from the experience, and the task of mourners is to recognize that death is part of the journey to rebirth.

thanatology A field of research that studies death.

This colorful Balinese funeral procession on its way to a Buddhist cremation is a marked contrast to the somber memorial service that is more common in the West. No matter what form it takes, community involvement in death and dying seems to benefit the living.

Among Hindus and Sikhs, helping the dying to relinquish their ties to this world and prepare for the next is a particularly important obligation for the immediate family. A holy death is one that is welcomed by the dying person, who should be placed on the ground at the very last moment, chanting prayers and surrounded by family members also reciting sacred texts. Such a holy death is believed to ease entry into the next life. Therefore two practices by no means common in Europe or North America are critically important: (1) knowing in advance that one is going to die so proper preparations can be made, and (2) having all one's family "present when the soul leaves" (Firth, 1993).

Preparations for death are not emphasized in the Jewish tradition because hope for life should never be extinguished. For the same reason, the dying person should never be left alone. When a person dies, he or she should be buried within a day, and expressions of mourning are appropriate within certain temporal guidelines. The close family is expected to mourn at home for a week, and then to reduce pleasurable activities for a year (Katz, 1993).

In the Christian tradition, death can be seen as the entrance to Heaven or Hell, and is thus welcomed or feared as the case may be. Particular customs, such as preserving the body as much as possible for bodily resurrection, or celebrating the "passing" with food and drink, vary from place to place, and denomination to denomination (Power, 1993).

Historical Differences

Throughout most of history, death was an accepted, familiar event that usually occurred at home (Ariès, 1981). Family members of all ages had intimate contact with deaths that resulted from childbirth, from disease and infection, from accidents, and from the consequences of old age. In general, family members were the ones who tended to the dying person and then to the corpse: in most cases they built the coffin, dug the grave, and buried the body themselves.

In mid-twentieth-century Western Europe and particularly in North America, with the ascendancy of modern medicine, death came to be withdrawn from everyday life, as though in an effort to deny it. More and more, people died alone in hospitals rather than at home among family. The disposition of the deceased passed into the hands of professionals, who sanitized and euphemized death in an effort to disguise its reality. They embalmed and made up the corpse to give it a normal and "healthy" look; they coined terms like "slumber room" (referring to the room in which the corpse was displayed) that gave no hint of death; they supervised the burial and formalized the grieving.

This denial of death likewise came to permeate the medical profession. Doctors and nurses routinely resisted telling terminal patients the truth about their condition, and, in fact, avoided the dying as much as possible. Before the end of biological life, the dying, in effect, experienced a "social death," a kind of institutionalized isolation in which they found themselves shunned by their medical caregivers and constricted in their intimacy with family and friends by hospital procedures and protocol (Kastenbaum, 1992).

At the close of the twentieth century, although hospital deaths are even more common than they were twenty years ago, sheer obdurate denial—the

conviction that death should not, would not, could not occur—is less common than it was (Walter, 1993). A major factor leading to this change in attitude about death was the pioneering work of Elisabeth Kübler-Ross, a physician who was asked in 1965 by four seminary students to help them do research on the feelings and needs of people close to death. When Kübler-Ross approached her professional colleagues for permission to interview the dying, they responded with anger and shock, and even denied that any of their patients were terminally ill. "It suddenly seemed that there were no dying patients in this huge hospital" (Kübler-Ross, 1969). Finally, the first interview was obtained, and thereafter Kübler-Ross found many dying people who were eager to talk about their feelings. What she learned surprised many who thought they knew how dying persons might react to hearing the truth about their condition.

The Dying Person's Emotions

Death can have many personal meanings. It can be seen as "a biological event, a rite of passage, an inevitability, a natural occurrence, a punishment, extinction, the enforcement of God's will, the absurd, separation, reunion . . . a reasonable cause for anger, depression, denial, repression, guilt, frustration, relief . . . " (Kalish, 1985). Such a diverse range of perspectives is seen not only among various individuals and cultures but also often within a single dying person. The role of doctors, nurses, and other witnesses can be crucial here.

One of the first things Kübler-Ross learned was how important informing the dying of their condition can be. She discovered that doctors sometimes told a patient's immediate family of the patient's terminal illness and then explicitly instructed them to keep "the facts from the patient in order to avoid an emotional outburst." In many cases, the patients eventually guessed their fate but were unable to talk about their feelings because family and staff continued to pretend that all would be well. The result of this "conspiracy of silence" was increased isolation and sorrow for both the patients and their families (Moller, 1996). In other instances, patient and family were told of the probability of death in such an abrupt and insistent manner that all hope was destroyed. And sometimes the truth was hidden from everyone until the last moment, allowing the dying no time to put their affairs in order or to share their final expressions of love with their family.

Kübler-Ross's research on death and dying (1969, 1975) led her to propose that when they are told the truth by an empathetic listener, the dying go through five emotional stages in confronting their impending death. The first is *denial*, in which they refuse to believe that their condition is terminal. Typically, they convince themselves that their laboratory tests were inaccurate, or that the disease will have an unexpected remission. The second stage is *anger*—at everyone else for not caring enough, or for caring too much, or simply for being alive and well. The third stage is *bargaining*, in which a person tries to negotiate away the death, promising God or fate to, say, pray more or live a better life. When bargaining appears to have failed, *depression* sets in, causing the dying person to mourn his or her own impending death and to be unwilling to make any plans or to take an interest in medical treatment. Finally, *acceptance* can occur. Death is understood as the last stage of this life and, perhaps, the beginning of the next—a transition, not a trauma. Kübler-Ross (1969) writes:

> Acceptance should not be mistaken for a happy stage. It is almost void of feelings. It is as if the pain had gone, the struggle is over, and there comes a time for "the final rest before the long journey" as one patient phrased it. This is also the time during which the family usually needs more help, understanding, and support than the patient . . .

Kübler-Ross's findings have been investigated by many other researchers, and most have not found the same five stages occurring in sequence. More typically, denial, anger, and depression appear and reappear during the dying process (Kastenbaum, 1992), depending largely on the specific context of the death. For example, denial often occurs when the illness is one of the forms of cancer that has periods of remission. Anger often predominates when the dying feel that others are responsible for their condition, or are not sympathetic to it, as many victims of AIDS feel. Some studies show that depression increases rather than decreases as death nears, although what appears to be depression related to dying may, in many cases, be a side effect of pain killers or other drugs.

The age of the dying person also affects the way he or she feels (Stillion, 1995; Wass, 1995). Young children, not understanding the concept of death, are usually upset by the thought of dying because it involves separation from those they love. A dying child therefore needs constant companionship and reassurance. The developing cognitive competencies of the school-age child often lead the very ill young person to become absorbed with learning the facts about his or her illness and treatment and about the "mechanics" of dying.

Adolescents tend to think not about the distant future but about the quality of present life. Thus, to the dying or seriously ill adolescent, the effect of their condition on their appearance and social relationships may be of primary importance. For the young adult, coping with dying often produces great rage and depression at the idea that, just as life is about to begin in earnest, it must end. For the middle-aged adult, death is an interruption of important obligations and responsibilities, so most middle-aged people who know they are dying try to make sure that others will take over their obligations. An older adult's feelings about dying depend a great deal on the particular situation. If one's spouse has already died, and if the terminal illness brings pain and infirmity, acceptance of death is comparatively easy.

From all this it is clear that Kübler-Ross's five stages make feelings about death seem much more predictable and universal than they actually are (Kastenbaum, 1992). It is also clear that there is no one approach to death that is universally "best" and that any specific prescription for the proper approach to death is limited, partly by the cultural context and partly by the particular needs and circumstances of the individuals who are directly involved.

The need to understand a dying person's unique emotions and values has been reinforced as more and more thanatologists have described exceptions to Kübler-Ross's stages. As one critic of the "stages of death" notion has observed:

> Just as no one would listen to music because it is "right" but because it is beautiful, so no one should be urged to accept a specific metaphysical notion because it is "right" . . . but only because it enriches someone's life and makes a positive difference in how he or she . . . endures pain or deprivation. The criterion, it is suggested, for either adopting or rejecting a certain concept of death is not "correctness" but social, spiritual and emotional relevance. [Klatt, 1991]

As another explains, "every life is different from any that has gone before it, and so is every death" (Nuland, 1994). Death is the final expression of each developing person's individuality, as well as their final connection with their family, friends, and community. The surprise is in how varied the emotions of the dying person can be, changing from day to day. Negative emotions such as terror and fury are only part of the picture; laughter and even joy can appear as well.

Deciding How to Die

Not everyone has the desire, or even the opportunity, to decide how to die. Most healthy young people, as we saw in Chapter 15, believe they are invincible, immortal, so the whole concept of deciding how to die is alien to them, unless they are suicidal—in which case preparations for death are a sign that they need psychological help. Other children, adolescents, and young adults—far too many—die a sudden, violent death from an unexpected accident or a tragic homicide. Young people sometimes die of illness, with some foreknowledge that death might occur, but such deaths are rare, outnumbered 3:1 by violent deaths. Among people under the age of 30 in the United States, death from illness occurs at an annual rate of less than once in every 4,000 people.

As the years accumulate, however, serious diseases begin to appear in people. Each year after the age of 65, a person has about one chance in twenty of dying after a period of illness. Even without illness, most older adults prepare in some way for death. Especially if they do suffer an illness, though—particularly one for which the prognosis involves the possibility of death—older people often draw up a will, reconcile with friends, and end each family visit with loving good-byes. Such preparations are not only normal, but also psychologically healthy.

As part of their preparation for death, many adults express the hope that they will die swiftly, with little pain and great dignity. All too often, their hope is denied. Increasingly, modern medicine can sustain life beyond its time, holding off death with all manner of technological interventions—able even to maintain organ functioning after brain death has occurred. Obviously, when there is a reasonable chance for recovery, extraordinary and "heroic" medical procedures make good sense; even people of advanced age who have life-threatening illnesses sometimes get well when given appropriate surgery, chemotherapy, resuscitation, or the like. However, when death is inevitable and near, the same procedures may rob the patient of "a good death."

Many people who die in hospitals or other medical institutions (where 85 percent of North Americans and 70 percent of Britons now die) are likely to depart this life after an extended period of confused semiconsciousness, attached to an assortment of machines, tubes, and intravenous drips. Often they die in pain, largely because analgesic medications tend to be underprescribed (because of fear of addiction!) (Solomon et al., 1993). Even if pain relief is adequate, other discomforts (nausea, bed sores, constipation, shortness of breath), psychic terrors (nightmares and hallucinations brought on by the pain medication), and fears that dignity will be lost and "medicine will take over their death" are almost inevitable (Cassel, 1996).

A CLOSER LOOK

The Hospice

One effort to help ensure that more of the terminally ill die a good death began in London during the 1950s, when a dedicated woman named Cecily Saunders opened the first modern *hospice*, a place where the terminally ill could come to die in peace. Conceived in response to the dehumanization of the typical hospital death, hospices provide the dying with skilled medical care—which includes pain-killing medication but shuns artificial life-support systems—and a setting where their dignity as human beings, and that of their family members, is respected (Saunders, 1978).

In a hospice, both the dying person and the family are considered "the unit of care" (Lattanzi-Licht & Connor, 1995). Consequently, family and friends are encouraged to visit at any time. Often, one family member or close friend, called a *lay primary caregiver*, is present much of the time and is responsible for some of the routine care. This arrangement makes the dying person feel less alone and helps the caregiver to be involved rather than excluded, as he or she would be in most hospital settings. When death comes, the staff continues to tend to the family's psychological and other needs.

In some cases, a dying person's home can become a hospice, allowing the individual the emotional comfort of being in familiar surroundings. In addition, having the home as the hospice can actually prolong life, since it is free of the infections and contagious diseases that the sick often contract, and die of, in hospitals. When hospice care does occur in the home, doctors and nurses visit regularly, to give comfort as well as medication and therapy and to instruct family members in how to provide daily care.

Obviously, the hospice has much to recommend it. However, the hospice concept is not accepted uncritically. First of all, to be accepted by a hospice, patients must be diagnosed as terminally ill; that is, they must have no reasonable chance of recovery and death must be anticipated within six months or so. Such a diagnosis can be made for only a minority of the dying. Second, the patients and their families must accept this diagnosis, agreeing that longer life or a cure is virtually impossible, and that a good death is the only remaining choice. Understandably, even for some who are extremely ill, hope is so crucial that they would rather have one last operation, with all odds against success, than wait for death in a hospice.

Further, hospice care does not always reach its goal of meeting all the needs of the dying and their families. One reason is cost: while the well-functioning hospice uses less high-technology equipment and fewer surgical procedures than a comparable hospital would to treat the same patients, good hospice care is labor-intensive. All the providers of hospice care—doctors, nurses, psychologists, social workers, clergy, and volunteers—must be well-trained and must be available to provide continuity of individualized care until the patient dies. Meeting such staffing demands is costly, and most insurance plans are less likely to cover hospice care than hospital care. In addition, even with careful staff development, burnout is a common problem, and replacements are not easy to find (Lafer, 1991).

For a variety of reasons, hospices are much better prepared to meet the needs of a relatively young patient with cancer than those of a relatively old patient with, say, a combination of illnesses including heart disease, emphysema, and diabetes, all of which shorten life but none of which necessarily brings death. More and more hospices are accepting the latter type of patient—22 percent of all hospice patients have a diagnosis other than cancer—but problems arise if persons with such chronic illnesses do not die "on schedule" but continue to receive supportive hospice care indefinitely. For one thing, such a situation may undermine the emotional preparation for death. Sometimes a patient is actually discharged only to deteriorate, changing from a person who is chronically ill but not really dying to a person who quickly sickens and dies (Lattanzi-Licht & Connor, 1995).

Thus, hospice care benefits some, but by no means all, dying individuals. Overall, the most important benefit derived from the hospice concept may be its bringing to acceptance three general principles that were once overlooked by medical personnel: (1) death is a family affair; (2) a dedicated interdisciplinary team can provide the best care for any sick person; and (3) **palliative care**, that which relieves suffering and safeguards dignity, is a worthy medical goal. Thanks to the work of hospices, more doctors and nurses now see part of their mission with the dying to be helping them prepare for death, keeping the patient and the family informed and comfortable.

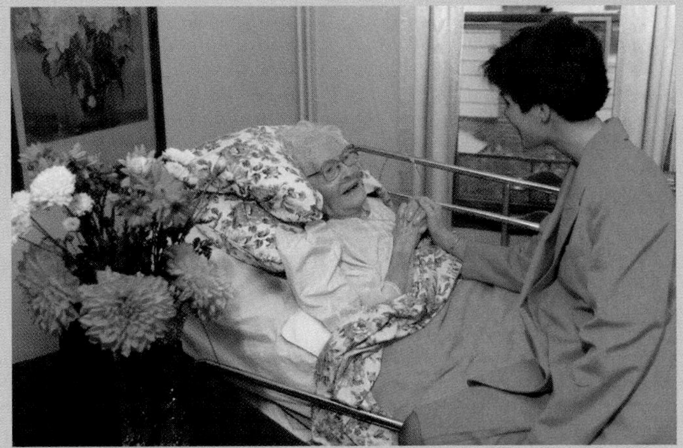

A particularly appealing aspect of hospices is the attention they pay to the "extras" that make their patients' remaining days more pleasant, such as colorful bedclothes, freshcut flowers, and frequent visits from volunteers. Here, a volunteer hospice worker holds the hand of a 92-year-old patient with terminal congestive heart failure.

The decision whether or not to prolong life encompasses many issues. One relatively simple issue is whether to give a dying patient enough morphine to relieve his or her pain, a decision that may also slow respiration and hasten death. A much more complex dilemma occurs when strongly held opinions conflict. The medical mandate to fight death, the religious conviction that death comes when God chooses, the legal precedent that individual rights must be preserved, and the social-policy concern that extraordinary measures be granted only to those who are likely to live many productive years all come into play when there is a question of prolonging the life of the dying. In the midst of such turmoil, sometimes a patient wants to die but the family wants "no expense spared" to keep the patient alive (Jecker & Schneiderman, 1996).

There is a growing consensus, both in law and in hospital practice, that the ultimate authority regarding what measures are to be used in terminal cases should be made by the person most directly concerned, the patient. To this end, many people, long before death is imminent, make a living will, a document that indicates what kind of medical intervention they want should they become terminally ill and incapable of expressing their wishes. The problem, of course, is that those who write a living will rarely know the specifics of their dying. If a particular treatment—itself painful or debilitating—has only one chance in ten of prolonging life, are those odds high enough to merit using it? What about one chance in 100? What if the treatment is only discomforting? And if the patient survives, what quality of life is acceptable? Only full mental and physical capacity, or something much less? Such quantitative and qualitative issues are at the heart of the decision to fight death or accept it, and opinions vary from person to person (Jecker & Schneiderman, 1996). Unfortunately, dying patients themselves are often in no condition to be able to address them.

Helping Others Die

Deciding to accept death is one thing, but acting on that decision on behalf of another person is quite another. Almost everyone—healers, judges, theologians, and the general public—agrees that when vital-organ failure occurs in a terminally ill person who has already experienced severe pain, fearful confusion, and loss of consciousness, medical personnel are not obligated to restore breathing, restart the heart, and so forth. Such **passive euthanasia**—defined as mercifully allowing a person to die by not doing something that might extend his or her life—is permitted when, at the patient's or family's request, the orders *DNR (do not resuscitate)* have been placed on a person's hospital chart. Similarly, it is becoming more common to provide the dying with ample morphine and other medication to reduce pain, even if thus improving the quality of their last days of life might possibly hasten death.

palliative care Care that relieves suffering and safeguards dignity.

passive euthanasia The practice whereby a person is allowed to die by withholding some procedure or drug that would have allowed life to continue a bit longer. Passive euthanasia is practiced in many hospitals and hospices, when the extension of life seems only to prolong misery.

However, doctors and nurses often disagree regarding the specifics of passive euthanasia. For instance, many medical experts and professional societies make no practical or ethical distinction between ordinary and extraordinary measures to delay death, or between not starting life support when it seems pointless and stopping it when it is no longer warranted. The closer doctors and nurses are to the actual care of terminal patients, the more troubling this lack of distinctions becomes. For example, most doctors

bereavement overload An emotional state that may result when a person experiences the death of several loved ones over a relatively short period of time and is unable to reach acceptance of the first death before having to mourn the second.

their emotions without being overcome by them. When the mourning period was over, people were helped to pick up the pieces of their lives again, neither forgetting nor dwelling on their loss.

In recent times, mourning became a more private, and less emotional, affair. The bereft are often encouraged to "bear up"; friends and relatives do less consoling than advising—to keep busy, to remarry, to look on the bright side; the large funeral has generally given way to a small memorial service; the deceased is less likely to be buried in a commemorative family plot and more likely to be cremated without ceremony. If current trends continue, according to one observer, we may eventually reach the point where death becomes little more than a minor annoyance, to be handled as efficiently and unemotionally as possible (Kastenbaum, 1992).

What are the results of these trends? Certainly they do not abolish grief. They merely privatize and mask its expression, which itself can harm the bereaved. According to a review of the research, one result is an increasing tendency toward social isolation for those who have just lost a loved one—exactly the opposite of a healthy reaction. Another result is physical illness. Many of the bereaved find themselves feeling sick more often in the days and months after a death. This sickness sometimes propels them to a doctor, not only for medication but also for sympathy and attention—exactly what a proper mourning period would have brought. Grief that lingers can even precipitate death, particularly from heart disease, cirrhosis, and, especially for men, suicide (Moller, 1996).

Unexpressed mourning also harms the larger community, particularly children. As one thanatologist writes:

> [T]he funeral provides the setting in which both private sorrow and public loss can be both expressed and shared. . . . As a social ceremony, it serves to bring together the community. As such, it serves as an important vehicle of cultural transmission. The contemporary impulse to preclude funerals from society or to exclude children from funerals can also have unintended consequences. In addition to cutting children off from direct expressions of love, concern, and support at this time of family crisis, it may deprive them of the opportunity to learn about life's most basic fact—death. The social meaning and intrinsic value of human life itself, moreover, may be implicitly denied by the failure to acknowledge our mortality. [Fulton, 1995]

Compiling a scrapbook of memorabilia, building a memorial garden, or designing and sewing a square on a quilt are among the many creative actions that people sometimes take to fight the numbness after a loved one dies. The results are intended to express to the world that "A special person lived and died." When others acknowledge that statement, as occurred here at the Washington Mall at the displaying of the AIDS Memorial Quilt, the survivors no longer feel that someone's existence passed unnoticed.

By contrast, public expression of grief can bind a community together. For example, the gay community has become much more visible, political, and united in the face of the AIDS epidemic, as each public memorial service and each new square created for the AIDS Memorial Quilt has contributed to a sense of shared loss (Fulton, 1995).

The crucial problem for many mourners in contemporary Western cultures is that outsiders sometimes do not understand and sympathize with their grief. For example, it might seem that an elderly person should accept the death of a very elderly mother or father, or that a prospective parent should not be unduly distressed over an unsuccessful pregnancy. However, every death—especially a death that changes the generational line—is potentially grievous. The older person who now becomes the oldest generation of the family, or the parent whose hopes for the next generation are unfulfilled, can be, temporarily at least, devastated by his or her loss.

Because the elderly are likely to experience the death of a number of close friends and relatives in a fairly short span of time, they are particularly vulnerable to **bereavement overload**, as each new death starts the mourning process up again before the earlier ones have been completed.

Among the most difficult deaths to accept, both by the dying and the living, are senseless, unexpected deaths long before old age. Here lies a young man shot and killed in gang retribution. Despite bringing flowers, his friends probably find it hard to let him "rest in peace."

When this happens, other people need to be particularly sympathetic. Further, elderly widows and widowers often have a more difficult time adjusting than their younger counterparts do, in part because age has already diminished the size of their social circle, as well as their ability to fashion a new life for themselves (Sable, 1991).

At the same time, deaths that are expected, especially if they occur in old age after a long illness, sometimes bring less grief than outsiders or even the mourner expected. Just as each death is unique, each mourner experiences bereavement somewhat differently (Moller, 1996).

The sudden death of someone who is not "supposed to" die is the most difficult to bear. The clearest example is the death of a child, especially one who has lived long enough to have a distinct personality and position in the family. If the death is a violent and sudden one, as most young people's deaths are, the loss is particularly devastating. Parents and siblings are often racked by powerful and personal emotions of guilt, denial, and anger, as well as sorrow.

Healing is particularly difficult in the event of sudden violent death, and sometimes coroners, police, and hospital personnel make matters worse. In one case, emergency-room staff prevented the mother of a 17-year-old boy who was killed in an auto crash from seeing her son's body until she, first, had answered many questions and, second, had promised not to "do anything silly." She writes,

> I desperately needed to hold him, to look at him, to find out where he was hurting. These instincts don't die immediately with the child. The instinct to comfort and cuddle, to examine and inspect the wounds, to try to understand, most of all, to hold. But my lovely boy was draped on an altar, covered with a purple robe, and all expressions of love and care were denied to me. And I don't know when that wound will heal. [Awoonor-Renner, 1993]

Recovery

What can others do, then, to help the bereaved person? The first step is simply to be aware that powerful and complicated emotions are likely: a friend should listen, sympathize, and not ignore the mourner's pain. Seeing the

dead person, visiting the grave, lighting a candle, cherishing a memento—all might, or might not, be desired. The second step is to understand that bereavement is often a lengthy process, demanding sympathy, honesty, and social support for months or even years. As time passes, the bereaved person should become involved in other activities but should not be expected to forget the person to whom they were attached: memories of the deceased and sorrow usually continue.

It is also important to recognize that each culture imparts to its people distinct customs and traditional wisdom for dealing with death. Those who grow up in a culture where one is expected to "bear up" stoically in grief might be doubly distressed if at some point they are advised to cry and cannot; those whose cultures endorse wailing loudly at death, on the other hand, might become confused and angry if they are told to be quiet—as sometimes happens in hospitals governed by different cultural practices (Firth, 1993). It also must be recognized that while culture influences patterns of grieving, it does not determine them, and the handling of grief can show much variation within a particular ethnic group (see Table 1).

TABLE 1

Ethnic Background and Spousal Grieving

Question: What do you feel is the minimum number of visits one should make to a spouse's grave during the first year, not counting the burial service?

Response	African-Americans	Japanese-Americans	Mexican-Americans	Anglo-Americans
Unimportant	39	7	11	35
1–2 times	32	18	19	11
3–5 times	16	18	12	18
6+ times	13	58	59	35
(Don't know, etc.)	11	6	3	19

Source: Kalish & Reynolds, 1981.

However the bereaved work through their emotions of grief, the experience may give them a deeper appreciation of themselves as well as of the value of human relationships. In fact, a theme frequently sounded by those who work with the bereaved is that there are lessons in the processes of dying and mourning that we all could learn. The most central of these is the value of intimate, caring relationships. As one counselor expresses it:

> I often have heard phrases such as 'I wish I had told him I loved him' or 'I wish we could have resolved our differences earlier.' There may be things we need to say, appreciations that need to be expressed, distances to bridge. . . . Loving and being loved is not just something that happens to us. It is a creative art that must be worked in a variety of ways. [Sanders, 1989]

It is fitting to end this book with just such a reminder of the creative work of loving. As first described in Chapter 1, the study of the process of human development is a science—a topic to be researched, understood, and explained in order to enhance human lives. But the process of actually living one's own life is an art as well as a science, with strands of love and sorrow and recovery that are woven into each person's unique tapestry. Death, when accepted, grief, when allowed expression, and bereavement, when it leads to recovery, give added meaning to birth, growth, development, and all human relationships.

Glossary

abuse All actions that are deliberately harmful to an individual's well-being. (234)

achievement tests Tests designed to measure how much a person has learned in a specific subject area. (346)

acquired immune deficiency syndrome (AIDS) The final, terminal stage of HIV degradation of the immune system, which typically appears as serious infections, specific cancers, and the like. (108)

active euthanasia The practice whereby someone intentionally acts to terminate the life of a suffering person. (E–8)

activities of daily life (ADLs) Actions that are important to independent living, typically comprising five tasks: eating, bathing, toileting, walking, and dressing. The inability to perform these tasks is a sign of frailty. (685)

additive pattern A common pattern of genetic inheritance in which each gene affecting a specific trait makes an active contribution to the final outcome. Skin color and height are additive. (71)

adolescent egocentrism A characteristic of adolescent thinking that sometimes leads young people to focus on themselves to the exclusion of others, believing, for example, that their thoughts, feelings, or experiences are unique. (416)

affordances The various opportunities for interaction that an object offers. These opportunities are perceived differently by each person depending on his or her past experiences and present needs. (162)

ageism A term that refers to prejudice against the aged. Like racism and sexism, ageism works to prevent elderly people from being as happy and productive as they could be. (606)

age of viability The age (about twenty-four weeks after conception) at which a fetus can possibly survive outside the mother's uterus if specialized medical care is available. (100)

Alzheimer's disease The most common form of dementia, characterized by gradual deterioration of memory and personality, and marked by plaques and tangles in the brain. The underlying cause is unknown, as is the cure. Alzheimer's disease is not part of the normal aging process. (647)

anal stage Freud's second stage of psychosexual development, in which the anus becomes the main source of bodily pleasure, and control of defecation and toilet training are therefore important activities. (198)

anorexia nervosa A serious eating disorder in which a person restricts eating to the point of emaciation and possible starvation. Most victims are high-achieving females in early puberty or early adulthood. (484)

anoxia A temporary lack of oxygen. If prolonged, it can cause brain damage or even death. (115)

antioxidants Compounds that nullify the effects of oxygen free radicals by forming a bond with their unattached oxygen electron. (620)

antithesis A proposition or statement of belief that opposes the thesis; the second stage of the dialectical process. (494)

Apgar scale A test devised by Dr. Virginia Apgar to quickly assess the newborn's color, heart rate, reflex irritability, muscle tone, and respiratory effort. This simple method is used one minute and five minutes after birth to determine whether a newborn needs immediate medical care. (119)

aptitude tests Tests designed to measure potential, rather than actual, accomplishment. (347)

assisted suicide Suicide in which the means of death is provided by a second party. (E–8)

attachment The enduring emotional connection between a person and a particular other that produces a desire for consistent contact as well as feelings of distress during separation. (211)

attention-deficit hyperactivity disorder (ADHD) A behavior problem characterized by excessive activity, an inability to concentrate, and impulsive, sometimes aggressive, behavior. (317)

authoritarian parenting A style of child-rearing in which standards for proper behavior are high, misconduct is strictly punished, and parent-child communication is low. (287)

authoritative parenting A style of child-rearing in which the parents set limits and provide guidance and are willing to listen to the child's ideas and make compromises. (287)

autism A disorder that is chiefly characterized by an inability or unwillingness to communicate with others. (312)

autoimmune diseases Illnesses that occur when the body attacks its own healthy cells as though they are foreign bodies. (547)

automatization The process by which familiar and well-rehearsed mental activities become routine and automatic. (329)

autonomy versus shame and doubt Erikson's second stage of psychosocial development, in which the toddler struggles between the drive for self-control and shame and doubt about oneself and one's abilities. (199)

average life expectancy The number of years the average newborn of a particular species is likely to live. In humans, this age has tended to increase over time; in the United States in 1993, the average life expectancy at birth was 72 years for men and 79 for women. (622)

axons Nerve fibers that extend from a neuron and transmit impulses from one neuron to the dendrites of another. (138)

babbling Extended repetition of certain syllables, such as "ba, ba, ba," that begins at about 6 or 7 months of age. (180)

baby talk A term for the special form of language that people typically use when speaking to infants. Nicknamed "Motherese" by developmental psychologists, baby talk is high-pitched, with many low-to-high intonations, is simple in vocabulary, and employs many questions and repetitions. (185)

B-cells Cells manufactured in the bone marrow that create antibodies for isolating and destroying invading bacteria and viruses. (621)

beanpole family A multigenerational family having only a few members in each generation. (683)

behavioral teratogens Teratogens that tend to damage neural networks in the prenatal brain, affecting the future child's intellectual and emotional functioning. (103)

behaviorism A theory that emphasizes the systematic study of observable behavior, especially how it is conditioned. (34)

bereavement overload An emotional state that may result when a person experiences the death of several loved ones over a relatively short period of time and is unable to reach acceptance of the first death before having to mourn the second. (E–10)

Big Five The five basic clusters of personality traits that remain quite stable throughout adulthood: extroversion, agreeableness, conscientiousness, neuroticism, and openness. (584)

binocular vision The ability to use both eyes together to focus on a single object. (151)

biosocial domain Includes physical growth and development as well as the family, community, and cultural factors that affect that growth and development. (4)

blind Refers to researchers who are deliberately kept ignorant of the purpose of the research, or of relevant traits of the research subjects, in order to avoid biasing their data collection. (52)

body image A person's mental concept of his or her physical appearance. (394)

body-mass index (BMI) The ratio of a person's weight in kilograms divided by his or her height in meters squared. (482)

breathing reflex A reflex that ensures an adequate supply of oxygen and the discharge of carbon dioxide by causing the individual to inhale and exhale. (143)

bulimia nervosa An eating disorder in which the person, usually female, engages repeatedly in episodes of binge eating followed by purging through induced vomiting or use of laxatives. (484)

bullying The repeated, systematic efforts to inflict harm on a particular child through physical attack, verbal attack, or social attack. (362)

carrier An individual who has in his or her genotype a recessive gene that is not expressed in his or her phenotype. Carriers can pass the gene on to their children, who will express the gene if they receive a similar recessive gene from the other parent. (71)

case study A research method that focuses on the life history, attitudes, behavior, and emotions of a single individual. (58)

cataracts A common eye disease among the elderly involving a thickening of the lens; it can cause distorted vision if left untreated. (611)

centration The tendency of young children to focus their analysis on one aspect of a situation or object to the exclusion of all others. (261)

cephalo-caudal development The sequence of body growth and maturation from head to foot. Human growth, from the embryonic period throughout early childhood, follows this pattern. (99)

child maltreatment Includes all intentional harm to, or avoidable endangerment of, anyone under age 18. (234)

childhood sexual abuse Any erotic activity that arouses an adult and excites, shames, or confuses a child or young adolescent—whether or not the victim protests and whether or not genital contact is involved. (398)

childhood sexuality Freud's idea that infants and children experience sexual fantasies and erotic pleasures. (31)

chromosome A molecule of DNA that carries the genes transmitted from parents to offspring. (66)

classical conditioning The learning process in which a meaningful stimulus is linked to a neutral one, so that the latter elicits a response similar to that previously elicited by the former. (35)

class inclusion The idea that a particular object or person may belong to more than one class. For example, a father can also be someone's brother. (332)

classification The concepts that objects can be organized in terms of categories or classes, as in sorting foods according to whether they are fruits, vegetables, or dairy products. According to Piaget, this concept is mastered during the period of concrete operational thought. (332)

climacteric Refers to the various biological and psychological changes that accompany menopause. (558)

cluster suicide A series of suicides or suicide attempts that is precipitated by one initial suicide, usually that of a famous person or of a well-known peer. (458)

code-switching A pragmatic communication skill that involves a person's switching from one form of language, such as dialect or slang, to another. (336)

cognitive domain Includes all the mental processes through which the individual thinks, learns, and communicates. (4)

cognitive theory The theory that the way people think and understand shapes their behavior and personality. (40)

cohabitation Literally, "cohabitation" means living together. It is used primarily to refer to unrelated adults of the opposite sex who share the same house or apartment, presumably sharing the same bed as well. (516)

cohort A group of people who, because they were born within a few years of each other, experience many of the same historical and social conditions. (11)

common couple violence A form of abuse in which one or both partners of a couple engage in outbursts of verbal and physical attack. (524)

companionate grandparents Grandparents whose relationship with their children and grandchildren are characterized by independence and friendship, with visits occurring by the grandparents' choice. (594)

compression of morbidity A limiting of the time a person spends ill or infirm. (616)

concrete operational thought In Piaget's theory, the third period of cognitive development, characterized by the ability to apply logical processes to concrete problems. (40, 331)

conditioning The process of learning, either through the association of two stimuli or through reinforcement or punishment. (35)

conservation The understanding that an amount or quantity is unaffected by changes in its shape or placement. (261)

continuity theory The theory that each person experiences late adulthood and behaves toward others in much the same way as at earlier periods of life. (661)

control group Research subjects who are comparable to the experimental group in every relevant dimension except that they do not experience the special experimental conditions. (53)

control processes That part of the information-processing system that regulates the analysis and flow of information within the system, such as using memory and retrieval strategies, selective attention, and rules or strategies for problem solving. Control processes become increasingly efficient with development but usually show some specific individual declines in late adulthood. (Also called *executive function.*) (637)

conventional moral reasoning Kohlberg's term for the middle two stages of moral thinking, in which the individual considers social standards and laws to be the primary moral values. (423)

corpus callosum A network of nerves connecting the left and right hemispheres of the brain. (228)

correlation A statistical term that indicates a corresponding relation between two variables (when both variables either increase or decrease together, the correlation is positive; when one variable increases as the other decreases, the correlation is negative). (54)

crisis In Erikson's psychosocial theory, the central conflict of each developmental stage. (32)

critical period The period of prenatal development during which a particular organ or body part is most susceptible to teratogenic damage. In many cases the critical period occurs in the first eight weeks of development, when the basic organs and body structures are forming. (103)

cross-modal perception The ability to use information from one sensory modality to imagine something in another. (166)

cross-sectional research In developmental study, research that compares groups of people who are different in age but who are similar in other important ways. (59)

cross-sequential research In developmental study, research that follows a group of people of different ages over time in order to distinguish differences related to age from differences related to cohort and historical period. (Also called *cohort-sequential research* or *time-sequential research.*) (61)

crystallized intelligence Those types of intellectual ability that reflect accumulated learning. Vocabulary and general information are examples. Some developmental psychologists think crystallized intelligence increases with age, while fluid intelligence declines. (568)

culture The set of shared values, attitudes, customs, and physical objects that are maintained by people in a specific setting as part of a design for living one's daily life. (14)

deductive reasoning Reasoning that draws logical inferences and conclusions from general premises. (414)

dementia Irreversible loss of intellectual functioning caused by organic brain damage or disease. Dementia becomes more common with age, but even in the very old, dementia is abnormal and pathological. Sometimes dementia is misdiagnosed, since reversible conditions such as depression and drug overdose can cause the symptoms of dementia. (645)

democratic-indulgent parenting A style of parenting that is warm, responsive, and permissive. (288)

demography The study of populations and social statistics associated with these populations. (608)

demographic pyramid The shape that traditionally resulted when populations were graphed by numbers of individuals in each age group; the largest population group was the youngest and the smallest group was the oldest, giving the graph the shape of a pyramid. (608)

demographic square The shape that social scientists project will develop, in the near future, when populations are graphed by numbers of individuals in each age group. The traditional pyramid shape is becoming more square with equal numbers of older and younger persons in the population. (608)

dendrites Nerve fibers that extend from a neuron and receive the impulses transmitted from other neurons via their axons. (138)

dependency ratio The ratio of self-sufficient, productive adults to dependents—children and elderly. (608)

developmental psychopathology A field of psychology that applies the insights from studies of normal development to the study and treatment of childhood disorders. (311)

developmental theory A systematic set of hypotheses and principles that attempts to explain development and provide a framework for future research. (29)

dialectical thought Thought that is characterized by understanding the pros and cons, advantages and disadvantages, and possibilities and limitations inherent in every idea and course of action. In daily life, dialectical thinking involves the constant integration of one's beliefs and experiences with all the contradictions and inconsistencies one encounters. (494)

differentiation The developmental process by which a relatively unspecified cell or tissue undergoes a progressive change to a more specialized cell or tissue. (98)

disability A measure of health that refers to the inability to perform activities that most others can. (553)

dizygotic twins Twins formed when two separate ova are fertilized by separate sperm at roughly the same time. Such twins share about half of their genes, just like any other siblings. (70)

DNA (deoxyribonucleic acid) Molecules containing genetic information. (66)

dominant-recessive pattern A pattern of genetic inheritance in which one member of a gene pair (referred to as dominant) acts in a controlling manner, hiding the influence of the other (recessive) gene. (72)

drug abuse The use of a drug to the extent that it impairs one's physical, cognitive, or social well-being. (479)

drug addiction A condition of drug dependence, such that the absence of the given drug in the individual's system produces a drive—physiological, psychological, or both—to ingest more of the drug. (479)

drug use Any ingestion of a drug irrespective of the amount or frequency of use or of the drug's legality or effect. (479)

dynamic life-course theory The theory that each person's life is an active, ever changing, largely self-propelled process, occurring within specific social contexts that themselves are constantly changing. For this reason, developmental predictions based on past history may be useful, but they are never completely accurate, since each person constantly makes choices that affect the next step of development. (666)

dynamic perception Perception primed to focus on movement and change. (165)

dyscalcula A specific learning disability involving unusual difficulty in math. (314)

dyslexia A specific learning disability involving unusual difficulty in reading. (314)

ecological approach A perspective on development that takes into account the various physical and social settings in which development occurs. (6)

ecological niche The particular lifestyle and social context adults settle into that are compatible with their individual personality needs and interests. (585)

egocentrism Preschoolers' tendency to view the world and others exclusively from their own personal perspective. (249)

ego-involvement learning An educational strategy that bases academic grades on individual test performance, with students competing against each other. (420)

elder maltreatment Elder abuse and neglect, ranging from direct physical attack to ongoing emotional neglect and involving such underlying factors as the social isolation and powerlessness of the victim, mental impairment or drug addiction of the perpetrator, and poverty and inadequate education within the household. (692)

Electra complex In the phallic stage of psychosexual development, the female version of the Oedipus complex in which girls have sexual feelings for their father and accompanying hostility toward their mother. (282)

electroencephalogram (EEG) A graphic recording of the waves of electrical activity that sweep across the brain's surface. (142)

endometriosis A condition in which fragments of the uterine lining become implanted and grow on the surface of the ovaries or the Fallopian tubes, blocking the reproductive tract and leaving many women with fertility problems. (475)

environment All the nongenetic factors that can affect the individual's development—everything from the impact of the immediate cell environment on the genes themselves to the effects of nutrition, medical care, socioeconomic status, family dynamics, and the broader economic, political, and cultural contexts. (74)

error catastrophe A key idea in a theory of aging that holds that, while the body can isolate and repair a certain number of errors in cell duplication, at some point, accumulating errors can no longer be controlled and fatally impair the body's ability to function. (620)

estrogen Called the "female hormone," estrogen (which actually comprises several hormones) is, in fact, present in both sexes but is produced in much larger amounts in females beginning at puberty. Levels of estrogen correlate with sexual interest and with many aspects of the female reproductive cycle, as well as with several aspects of female health. (388)

ethnic group A collection of people who share certain background characteristics, such as national origin, religion, upbringing, and language, and who, as a result, tend to have similar beliefs, values, and cultural experiences. (15)

ethological perspective The view that many behaviors and emotions of humans and other animals have an adaptive function that furthers the survivial of the species. Ethological studies often shed light on infant emotional development. (191)

exchange theory The theory that marriage is an arrangement in which each person contributes something useful to the other, something the other would find difficult to attain alone. (520)

experiment A research method in which the scientist deliberately changes one variable and then observes the results in some other variable. (56)

experimental group Research subjects who experience special conditions or treatments that the control group does not experience. (53)

expertise The acquisition of knowledge in a specific area. As individuals grow older, they concentrate their learning in certain areas that are of the most importance to them, becoming experts in these areas while remaining relative novices in others. (578)

explicit memory Memory that involves consciously learned words, data, and concepts. (635)

false self A constellation of behaviors that is contrary to one's core being. (438)

family logistics The task of coordinating the many overlapping elements of family life and work schedules, a task that is particularly urgent for dual-earner families. (534)

family structure The legal and biological connections between members of a particular family. (366)

fast mapping A way to grasp the essential meaning of new words by quickly connecting them to words and categories that are already understood. (267)

fetal alcohol effects (FAE) Subtle impairments of a child's motor and cognitive abilities that are caused by the mother's prenatal consumption of alcohol beyond a certain threshold. Research suggests that this threshold may be anything more than 1 ounce of absolute alcohol per day. (110)

fetal alcohol syndrome (FAS) A cluster of birth defects, including abnormal facial characteristics, slow physical growth, and retarded mental development, that is caused by the mother's drinking excessive quantities of alcohol when pregnant. (110)

filial obligation The sense, experienced by most adult children, of a duty and need to protect and care for their aging parents. (683)

fine motor skills Physical skills involving small body movements, especially with the hands and fingers, such as picking up a coin and drawing. (147)

fluid intelligence Those types of basic intelligence that make learning of all sorts quick and thorough. Underlying abilities such as short-term memory, abstract thought, and speed of thinking are all usually considered part of fluid intelligence. (568)

foreclosure Erikson's term for premature identity formation, in which the young person accepts parental values and goals without exploring alternative roles. (439)

formal operational thought Piaget's term for the fourth period of cognitive development, characterized by hypothetical, logical, and abstract thought. (41, 412)

foster care A legally sanctioned, publicly supported arrangement in which children are removed from their original parents and temporarily given to another caregiver. (243)

fragile-X syndrome A genetically based condition that results from an abnormality of the X chromosome and that causes mental deficiency in about 30 percent of the women and in about 80 percent of the men who carry it. (84)

frail elderly Older people who are physically infirm, very ill, or cognitively impaired. (685)

free radicals Atoms that, as a result of metabolic processes, have an unpaired electron. Free radicals are believed to damage cells, affect organs, accelerate diseases, and decrease the ability of DNA to maintain and repair the body. (619)

gamete A reproductive cell, that is, a cell that can reproduce a new human being if it combines with a gamete from the other sex. Female gametes are called ova, or eggs; male gametes are called spermatozoa, or sperm. (65)

gateway drugs The three drugs—tobacco, alcohol, and marijuana—that young adolescents most commonly experiment with, use of which many lead to multiple drug use or drug abuse. (404)

gene The basic unit of heredity. Genes, which number about 100,000 in humans, direct the growth and development of every living creature. (66)

generational equity Equal contributions from, and fair benefits for, each generation. (673)

generational stake The need of each generation of family members to view their own intergenerational interactions from their own perspective. As a typical result, parent-adolescent conflicts are viewed quite differently by each party. (444)

generativity versus stagnation Erikson's seventh stage of development, in which adults seek to be productive through vocation, avocation, or child-rearing. Without such productive work, adults stop developing and growing. (510)

genetic clock According to one theory of aging, a regulatory mechanism in the DNA of cells regulates the aging process. (623)

genetic code The sequence of chemical bases in DNA; referred to as a code because it determines the amino acid sequence in the enzymes and other protein molecules synthesized by the organism. (66)

genetic counseling Consultation and testing that enable couples to learn about their genetic heritage and to make decisions about childbearing. (88)

genotype A person's entire genetic inheritance, including those characteristics carried by the recessive genes but not expressed in the phenotype. (71)

germinal period The first two weeks of development after conception, characterized by rapid cell division and the beginning of cell differentiation. (97)

gerontology The study of old age. This is one of the fastest-growing special fields in the social sciences. (606)

GH (growth hormone) A chemical produced by the pituitary gland that stimulates and regulates growth throughout childhood and adolescence. Increases in the amount of GH are one of the signals that puberty is beginning and one of the critical causes of growth spurt. (388)

glass ceiling An invisible barrier experienced by many women in male-dominated occupations—and by many minority workers in majority-dominated occupations—that halts promotion and undercuts their power at a certain managerial level. (530)

glaucoma A disease of the eye that can destroy vision if left untreated. It involves hardening of the eyeball due to a fluid build-up within the eye. (546, 611)

GnRH (gonad releasing hormone) A chemical produced at the onset of puberty that triggers heightened activity in the gonads (ovaries and testes), increasing levels of estrogen and testosterone. (388)

goal-directed behavior Purposeful actions initiated by infants in anticipation of events that will fulfill their needs and wishes. (178)

goodness of fit The quality of the "match" between the child's temperament and the demands of the surrounding environment. (204)

gross motor skills Physical skills involving large body movements such as waving the arms, walking, and jumping. (145)

growth spurt A period of relatively sudden and rapid physical growth of every part of the body such as that which occurs during puberty. (389)

guided participation A learning process in which the child learns through social interaction with a "tutor" (a parent, a teacher, a more skilled peer), who offers assistance with difficult tasks, models problem-solving strategies, and provides explicit instruction when needed. (47, 264)

habituation The process of becoming so familiar with a particular stimulus that it no longer elicits the physiological responses it did when it was originally experienced. (150)

Hayflick limit The number of times a human cell is capable of dividing into two new cells. Leonard Hayflick determined that the limit for most human cells is approximately fifty divisions, suggesting that the life span is limited by our genetic program, which does not allow cells to reproduce themselves indefinitely. (624)

heritability The variation in a particular trait, in a particular population, and in a particular environment and the degree to which that variation can be attributed to genetic differences among the members of the group. (80)

heterogamy As used by developmentalists, the term refers to marriage between individuals who tend to be dissimilar with respect to such variables as attitudes, interests, goals, SES, religion, ethnic background, and local origin. (520)

holophrase A single word that expresses a complete thought. (183)

homeostasis The adjustment of the body's systems to keep physiological functions in a state of equilibrium. As the body ages, it takes longer for these homeostatic adjustments to occur, making it harder for older bodies to adapt to stresses. (469)

homogamy As used by developmentalists, the term refers to marriage between individuals who tend to be similar with respect to such variables as attitudes, interests, goals, SES, religion, ethnic background, and local origin. (520)

hormone replacement therapy (HRT) Treatment to compensate for hormone reduction at menopause or following surgical removal of the ovaries. Such treatment, which usually involves estrogen and progesterone, minimizes menopausal symptoms and diminishes the risk of heart disease and osteoporosis in later adulthood. (559)

Human Genome Project A worldwide effort to construct and decipher a chromosomal map of all 3 billion base pairs of the 100,000 human genes. (90)

human immunodeficiency virus (HIV) A viral disease agent that gradually overwhelms the body's immune responses, leaving the individual defenseless against a host of pathologies that eventually manifest themselves as AIDS. HIV is carried in the blood and certain other bodily fluids of an infected person and is transmitted chiefly through sexual or direct blood contact. (108)

hypothetical thought Thought that involves propositions and possibilities that may or may not reflect reality. (413)

identification A defense mechanism that makes a person take on the role and attitudes of someone more powerful than himself or herself. (281)

identity In Piaget's theory, the logical principle that a given substance remains the same no matter what changes occur in its shape or appearance. (331)

identity As used by Erikson, a person's self-definition as a separate individual in terms of roles, attitudes, beliefs, and aspirations. (439)

identity achievement Erikson's term for a person's knowing who he or she is as a unique individual. Identity achievement includes a self-definition that encompasses sexual, moral, political, and vocational identity. (439)

identity diffusion Erikson's term for uncertainty and confusion regarding what path to take toward identity formation. Identity diffusion typically leads adolescents to become erratic, apathetic, and disoriented. (440)

identity versus role confusion Erikson's term for the psychosocial crisis of adolescence, in which adolescents must determine who they are, combining their self-understanding and social roles into a coherent identity. (439)

imaginary audience Refers to the egocentric idea that one is constantly being scrutinized by others. This belief often leads adolescents to fantasize about people's reactions to their appearance and behavior. (416)

immersion An approach to learning a second language in which the learner is placed in an environment where only the second language is spoken. (338)

immune system The complex system of antibodies, cells, and tissues that protect the body against illnesses and infections. (547)

implantation Beginning about a week after conception, the burrowing of the organism into the lining of the uterus, where it can be nourished and protected during growth. (98)

implicit memory Unconscious or automatic memory involving habits, emotional responses, routine procedures, and the senses. (635)

in vitro fertilization (IVF) A technique in which ova (egg cells) are surgically removed from a woman and fertilized with sperm in a laboratory dish. After the original fertilized cells (the zygotes) have divided several times, they are inserted into a woman's uterus (usually, but not necessarily, the ova's provider) for implantation or are frozen for later use. This method of reproduction is used to bypass problems that cause infertility, such as blocked Fallopian tubes. (476)

inadequate families Families involved in maltreatment in which the parents are so impaired by emotional problems and/or cognitive deficiencies that they can never meet their children's needs. (243)

inclusion An approach to educating children with special needs that includes them in the regular classroom while also providing them special individualized instruction, typically from a teacher or paraprofessional trained in special education. (317)

industry versus inferiority The fourth of Erikson's eight crises of psychosocial development, in which school-age children attempt to master many skills and develop a sense of themselves as either industrious and competent or incompetent and inferior. (354)

infantilization The treatment of elderly people in institutions as if they were infants. (694)

infertility The inability to conceive a child after one year of regular intercourse without contraception. (474)

information-processing theory A theory of learning that focuses on the steps of thinking—such as sorting, categorizing, storing, and retrieving—that are similar to the functions of a computer. (42)

initiative versus guilt The third of Erikson's eight "crises" of psychosocial development, in which the preschool child eagerly begins new projects and activities—and feels guilt when efforts result in failure or criticism. (279)

injury control The implementation of educational and legal measures to reduce the risk and impact of childhood injuries. (230)

insecure attachment A troubled parent-child connection signaled by the child's overdependence on, or lack of interest in, the parent. Insecurely attached children are not readily comforted by the parent and are less likely to explore their environment than are children who are securely attached. (211)

instrumental activities of daily life (IADLs) Actions that are important to independent living and that require some intellectual competence and forethought. These are even more critical to self-sufficiency than ADLs. (685)

interaction effect The intensification of a teratogen's potential for causing harm as a result of its interacting with another teratogen. (105)

intergenerational transmission The phenomenon of mistreated children growing up to become abusive or neglectful parents themselves, a consequence that is less common than is generally supposed. (241)

integrity versus despair The last of Erikson's eight stages of development. During late adulthood, according to Erikson, people feel either that their lives have had meaning and they look back on their past experiences with a sense of integrity and wholeness, or they despair at their past and therefore dread the future. (663)

interindividual variation Differences among individuals that are the result of the uniqueness of each person's genetic make-up and particular environment. (574)

intermodal perception The ability to associate information from one sensory modality (such as vision) with information from another (such as hearing). (166)

interview A research method in which people are asked specific questions to discover their opinions or experiences. (58)

intimacy versus isolation The sixth of Erikson's eight stages of development. Adults seek to find someone with whom to share their lives, in an enduring and self-sacrificing commitment. Without such commitment, they risk profound aloneness, isolated from their fellow humans. (509)

invincibility fable The fiction, fostered by adolescent egocentrism, that one is immune to common dangers, such as those associated with unprotected sex, drug abuse, or high-speed driving. (416)

involved grandparents Grandparents who remain active in the everyday activities of their grandchildren, seeing them daily. (594)

IQ tests Aptitude tests designed to measure a person's intelligence, defined as mental age divided by chronological age (hence, intelligence quotient). (347)

irreversibility The inability—characteristic of preschoolers' thinking—to recognize that reversing a transformation brings about the same conditions that existed prior to the transformation. (262)

kinkeepers The people who celebrate family achievements, gather the family together, and keep in touch with family members who no longer live nearby. (589)

knowledge base That part of the information-processing system that stores long-term information and has virtually limitless capacity. (Also called *long-term memory*.) (43, 329, 634)

kwashiorkor A disease resulting from protein-calorie deficiency in children. The symptoms include thinning hair and bloating of the stomach, face, and legs. (155)

language acquisition device (LAD) Chomsky's term to denote the innate ability to acquire language, including a knowledge of basic aspects of grammar and a predisposition to attend to and remember the critical, unique aspects of the language. (184)

latency Freud's term for the period between the phallic stage and the genital stage of psychosexual development. During latency, which lasts from about age 7 to age 11, children's psychosexual drives and unconscious emotional conflicts are relatively quiet, and children direct their attention and energies to the outside social world. (353)

launching event A habituation technique used to determine if a young infant understands the connection between causes and effects. (174)

learning disability A particular difficulty in mastering one or more basic academic skills, without apparent deficit in intelligence or impairment of sensory functions. (314)

learning theory A theory that emphasizes the sequences and processes of conditioning that, according to the theory, underlie most of human and animal behavior. (34)

life review The examination of one's own past life that many elderly people engage in. According to Butler, the life review is therapeutic, for it helps the older person to come to grips with aging and death. (655)

little scientist Piaget's term for the stage-five toddler who actively experiments to learn about the properties of objects. (178)

longitudinal research In developmental study, research that follows the same people over time in order to measure both change and stability with age. (60)

low-birthweight infant A newborn who weighs less than 2,500 grams (5 1/2 pounds) at birth. (113)

mainstreaming An approach to educating children with special needs by putting them in the same "stream"—the general education classroom—as all the other children, rather than segregating them. (317)

marasmus A disease that afflicts young infants suffering from severe malnutrition. Growth stops, body tissues waste away, and death may eventually occur. (155)

marital equity Refers to the marriage partners' perception of the relative equality of their respective contributions to the marriage. (520)

markers In genetic testing, particular physiological characteristics or gene clusters that suggest that an individual might be a carrier of a harmful gene. (89)

maximum life span The oldest age to which members of a species can live, under ideal circumstances. For humans, that age is approximately 120 years. (622)

menarche A female's first menstrual period, which is the final major change of puberty. (387)

menopause The time in middle age, usually around age 50, when a woman's menstrual periods cease completely and the production of estrogen drops considerably. Strictly speaking, menopause is dated one year after a woman's last menstrual period. (558)

mental combinations The mental playing out of a course of action before actually enacting it. (178)

mental retardation A pervasive delay in cognitive development. (314)

mentor In career development, a mentor is a more experienced coworker or supervisor who provides advice, instruction, and support. In many professions, the mentor system operates informally, with great impact. Finding a good mentor may make the difference between success or failure in a new career. (531)

metacognition The ability to evaluate a cognitive task to determine how best to accomplish it and how to monitor one's performance—"thinking about thinking." (330)

middle age The years from age 40 to age 60. (583)

midlife The point, at about age 40, when the adult has approximately as many years of life ahead as have already passed. Midlife ushers in middle age, which lasts until about age 60. (583)

modeling The process in which one person learns from the example of another. (38)

monozygotic twins Twins who have identical genes because they were formed from one zygote that split into two identical organisms very early in development. (69)

moratorium Erikson's term for a pause in identity formation that allows young people to explore alternatives without making final choices. College or military service can provide a moratorium. (440)

morbidity A measure of health that refers to the rate of diseases of all kinds in a given population. (553)

mortality A measure of health that refers to the number of deaths each year per thousand members of a given population. (553)

multidirectional A way of describing variations in specific cognitive abilities over time. (574)

multifactorial traits Characteristics produced by the interaction of several genetic and environmental influences. (70)

multi-infarct dementia (MID) The form of dementia characterized by sporadic, and progressive, loss of intellectual functioning. The cause is repeated infarcts, or temporary obstructions of blood vessels, preventing sufficient blood from reaching the brain. Each infarct destroys some brain tissue. The underlying cause is an impaired circulatory system. (649)

myelin A fatty insulating substance that coats the neurons, facilitating quicker, more efficient transmission of neural impulses. (139)

naturally occurring retirement communities (NORCs) Neighborhoods or apartment complexes where more than half the residents happen to be elderly. (681)

nature All the genetic influences on development, including those that affect physical characteristics as well as psychological traits, capacities, and limitations. (20)

negative identity Erikson's term for a chosen identity that is in opposition to the identity one's parents or society expects one to adopt. (439)

negative reinforcer The removal of an unpleasant stimulus following a particular behavior. This removal increases the likelihood that the behavior will be repeated should the unpleasant stimulus recur. (37)

neglect A form of child maltreatment in which parents or caregivers fail to meet a child's basic needs. (234)

neonate A newborn baby. Infants are neonates from the moment of birth to the end of the first month of life. (119)

neural tube The fold of cells that appears in the embryonic disk about three weeks after conception, later developing into the central nervous system. (99)

neurons Nerve cells of the central nervous system. (138)

nonadditive pattern A pattern of genetic inheritance in which the outcome depends much more on the influence of one gene than of another. (72)

norms The overall usual, or average, standard for a particular behavior. Norms are generally the result of research done on a large sample of a given population. (148)

nurture All the environmental influences on development, from prenatal influences on the embryo to the cultural context at death. (20)

obesity The condition of being significantly and unhealthily overweight. (305)

object permanence An infant's realization that objects and people still exist even when they cannot be seen, touched, or heard. The term "object permanence" was coined by Piaget, who believed that this realization does not emerge until 8 months of age. (169)

observation The unobtrusive watching and recording of the behavior of subjects in certain situations, either in the laboratory or in natural settings. (51)

Oedipus complex In the phallic stage of psychosexual development, the sexual desire that boys have for their mother and the related hostility they have toward their father. (281)

old-old Older adults who suffer from severe physical, mental, or social deficits and thus require supportive services such as nursing homes and hospital stays. (606)

Older Americans Act Legislation passsed by the U. S. Congress in 1965 that provides every older person, regardless of income, a wealth of benefits—from subsidized meals to community services. (688)

operant conditioning The learning process in which a person or animal becomes more, or less, likely to perform a certain behavior because of past reinforcement or punishment for similar behavior. (Also called *instrumental conditioning*.) (36)

oral stage Freud's term for the first stage of psychosexual development, in which the infant gains pleasure through sucking and biting. (198)

organ reserve The extra capacity of the heart, lungs, and other organs that makes it possible for the body to withstand moments of intense or prolonged stress. With age, organ reserve is gradually depleted, but the rate of depletion depends on the individual's general state of health. (470)

osteoporosis A loss of calcium within the bone that makes the bone more porous and fragile. It occurs somewhat in everyone with aging, but serious osteoporosis is more common in elderly women than men. Osteoporosis is the main reason the elderly suffer broken hip bones much more often than the young. (559)

overextension The overgeneralization of a word to various objects that share particular—but undefining—characteristics. (183)

overregularization The tendency to apply grammatical rules and structures when they are not called for, or when exceptions to them should be used. (271)

ovum (plural, ova) The reproductive cells of a female, which are present from birth in the ovaries. (65)

palliative care Care that relieves suffering and safeguards dignity. (E–7)

parasuicide A deliberate act of self-destruction that does not end in death. Parasuicide can include everything from an

overdose of aspirin (rarely deadly) to a single-occupant car crash (often fatal). (456)

parent-newborn bond The strong feelings of attachment between parents and newborns that are said to arise from their initial contact after birth. The long-term importance of this postpartem bond has been greatly, and dangerously, overblown by the popular media. (124)

parental monitoring Parental watchfulness and awareness regarding where offspring are and what they are doing. (455)

Parkinson's disease A chronic, progressive disease that is characterized by muscle tremors and rigidity, and sometimes dementia, caused by a reduction of dopamine production in the brain. (650)

passive euthanasia The practice whereby a person is allowed to die by withholding some procedure or drug that would have allowed life to continue a bit longer. Passive euthanasia is practiced in many hospitals and hospices, when the extension of life seems only to prolong misery. (E–7)

patriarchal terrorism The form of spouse abuse in which the husband uses violent methods of accelerating intensity to isolate, degrade, and punish the wife. (524)

peer group A group of individuals of roughly the same age and social status who play, work, or learn together. (358)

peer pressure Social pressure to conform to one's friends or contemporaries in behavior, dress, and attitude. Peer pressure can be positive or negative in its effects. (449)

pelvic inflammatory disease (PID) A common result of recurring pelvic infections in women. Pelvic inflammatory disease often leads to blocked Fallopian tubes, which, in turn, can lead to infertility. (475)

perception The mental processing of sensory information. (149)

perceptual constancy The awareness that the size and shape of an object remain the same despite changes in the object's appearance due to changes in its location. (165)

period of the embryo From approximately the third through the eighth week after conception, when the rudimentary forms of all anatomical structures develop. (97)

period of the fetus From the ninth week after conception until birth, when the organs grow in size and complexity. (97)

permissive parenting A style of child-rearing in which parents rarely punish, guide, or control their children but are nurturant and communicate well with their children. (287)

person-environment fit Refers to the degree to which a particular environment is conducive to the growth of a particular individual. (418)

personal fable The egocentric idea, held by many adolescents, that one is destined for fame and fortune, and/or great accomplishments. (417)

phenotype An individual's observable characteristics that result from the interaction of the genes with each other and with the environment. (71)

physiological states Refers to various levels of physiological arousal, such as quiet sleep and alert wakefulness. (140)

planned retirement communities Housing units designed exclusively for older adults. (680)

plasticity In developmental psychology, a term used to indicate that a particular characteristic is shaped by many environmental influences. Many mental abilities, once thought to be firmly fixed before adulthood, are now known to have much more plasticity than was once believed. (578)

polygenic traits Characteristics produced by the interaction of many genes. (70)

polypharmacy The use of multiple medications for various diseases and conditions. (651)

positive reinforcer Anything (such as a reward or positive event) that follows a behavior and increases the likelihood that that behavior will recur. (37)

postconventional moral reasoning Kohlberg's term for the two highest stages of moral thinking, in which the individual follows moral principles that may supersede the standards of society or the wishes of the individual. (423)

postformal thought A type of adult thinking that is suited to solving real-world problems. Postformal thought is less abstract and absolute than formal thought, more adaptive to life's inconsistencies, and more dialectical—capable of combining contradictory elements into a comprehensive whole. (492)

practical intelligence The intellectual skills used in everyday problem solving. (573)

preconventional moral reasoning Kohlberg's term for the first two stages of moral thinking, in which the individual reasons in terms of his or her own welfare. (423)

preoperational thought Piaget's term for the second period of cognitive development. This period generally occurs from age 2 to 6, before logical concepts such as conservation, reversibility, or identity are fully understood. (40, 261)

preterm Being born three or more weeks before the due date. (113)

primary aging The universal and irreversible physical changes that occur to living creatures as they grow older. (607)

primary sex characteristics Those sex organs that are directly involved in reproduction, such as the uterus, the ovaries, the testicles, and the penis. (391)

private speech The use of language to form thoughts and analyze ideas, either silently or by talking to oneself. (266)

progeria A rare genetic disease that causes young children to age prematurely and to die by their teens. (623)

Project Headstart A federally funded U. S. preschool educational program designed primarily to give economically disadvantaged children advance preparation (a "head start") for the intellectual and social challenges of elementary school. (273)

protein-calorie malnutrition A nutritional problem that results when a person does not consume enough nourishment to thrive. (155)

provocation ecologies Classroom environments that provoke or exacerbate hyperactive behavior by imposing either an unusually rigid classroom structure or none at all. (320)

proximo-distal development The sequence of body growth and maturation from the spine toward the extremities. Human growth, from the embryonic period through early childhood, follows this pattern. (99)

psychoanalytic theory A theory originated by Sigmund Freud that stresses unconscious forces that underlie human behavior. (30)

psychosexual stages A series of developmental stages, proposed by Freud, each originating in sexual interest in, and gratification through, a particular part of the body. (31)

psychosocial domain Includes emotions, personality characteristics, and relationships with other people. (4)

psychosocial theory A theory that stresses the interaction between internal psychological forces and external social influences. Erikson's theory is a psychosocial one. (32)

puberty The period of early adolescence characterized by rapid physical growth and the sexual changes that make reproduction possible. (385)

punishment An unpleasant event that follows from a particular behavior, making it less likely that the behavior will be repeated. (37)

rarefaction ecologies Classroom environments that ameliorate or diminish hyperactive behavior by providing sufficient classroom structure within a flexible individualized environment. (320)

reaction time The time it takes to respond to a particular stimulus. (309)

reciprocal determinism The idea that an individual's internal characteristics, environment, and behavior are mutually interactive in determining the person's specific behaviors. (38)

reciprocity The logical principle that a change in one dimension of an object can be compensated for by a change in another dimension. (332)

reflexes Involuntary physical responses to stimuli. (143)

rejected children In childhood peer groups, those children who are actively shunned, teased, or bullied. Some are aggressive-rejected, unpopular because of their confrontational behavior, and others are withdrawn-rejected, unpopular because of their withdrawn, anxious demeanor. (360)

rejecting-neglecting parenting A style of parenting that is cold, detached, and permissive. (288)

relationship perspective Focuses on the diverse ways the quality of children's relationships affects the course of psychosocial development. (286)

remote grandparents Grandparents who are distant but who are honored, respected, and obeyed by the younger generations in their families. (594)

replicate To repeat, with a different population, the specific design and procedures of a previous scientific study in order to test the validity of that study's conclusions. (26)

representative sample A select group of research subjects who reflect the relevant characteristics of the larger population that is under study. (52)

resource room A designated room, equipped with special materials and staffed by a trained teacher, where children with special needs spend part of their school day getting help with basic skills. (317)

response Any behavior (either instinctual or learned) that is elicited by a specific stimulus. (35)

restorable families Families involved in maltreatment in which the caregivers may have the potential to provide adequate care but are experiencing a combination of current stresses and past deficits that seriously impair their parenting abilites. (243)

retrieval strategies Procedures for recollecting previously learned information, such as thinking of related information or trying to create a mental image of the thing to be remembered. (328)

reversibility The logical principle that something that has been changed can be returned to its original state by reversing the process of change. (332)

risk analysis In teratology, the attempt to evaluate all the factors that can increase or decrease the likelihood that a particular teratogen will cause harm. (103)

role buffering The common situation in dual-earner families in which one role that a parent plays reduces the disappointments that may occur in other roles. (532)

role overload The stress of multiple obligations that may occur for a parent in a dual-earner family. (532)

rooting reflex A reflex that helps babies find a nipple by causing them to turn their heads toward anything that brushes against their cheek and to attempt to suck on it. (144)

rough-and-tumble play Play such as wrestling, chasing, and hitting that mimics aggression but actually occurs purely in fun, with no intent to harm. (296)

rubella A form of measles that, if contracted during pregnancy, can harm the fetus, including causing blindness, deafness, and damage to the central nervous system. (108)

sample size The number of individuals who are being studied in a research project. (52)

sandwich generation The generation "in between," having both grown children and elderly parents. Many middle-aged people feel pressured by the needs and demands of their adult children, on the one hand, and of their elderly (and perhaps ailing or widowed) parents on the other. (597)

scaffold To sensitively structure a child's participation in learning encounters so that the child's learning is facilitated. (265)

scientific method The sequence and procedures of scientific investigation (formulating questions, collecting data, testing hypotheses, and drawing conclusions) designed to reduce subjective reasoning, biased assumptions, and unfounded conclusions. (25)

the scientific study of human development The science that seeks to understand how and why people change, and how and why they remain the same, as they grow older. (4)

scripts Skeletal outlines of the usual sequence of certain common recurrent events. Young children use such scripts to facilitate the storage and retrieval of memories related to specific episodes of these events. (253)

Seattle Longitudinal Study The study led by K. Warner Schaie in which adults are tested and retested every seven years to reveal patterns of change in intellectual abilities. (567)

secondary aging The specific physical illnesses or conditions that are more common in aging but are caused by health habits, genes, and other influences that vary from person to person. (607)

secondary sex characteristics Sexual features not directly involved in reproduction, such as a man's beard or a woman's breasts. (293)

secular trend The tendency of successive generations over the past century or two to experience advanced growth as the result of improved nutrition and medicine. (387)

secure attachment A healthy parent-child connection, signaled by the child's being confident when the parent is present, distressed at the parent's absence, and comforted by the parent's return. (211)

selective attention The ability to concentrate on relevant information and ignore distractions. (327)

selective optimization with compensation The individual's devising of alternative strategies to compensate for age-related declines in ability. (575)

self-awareness A person's sense of himself or herself as a being distinct from others, with particular characteristics. (195)

self-theories Theories of late adulthood holding that individuals seek to achieve their full potential, to become self-actualized. (661)

senescence The state of physical decline, in which the body gradually becomes less strong and efficient with age. (467)

sensorimotor intelligence Piaget's term for the first stage of cognitive development (from birth to about 2 years old). Children in this period primarily use their senses and motor skills to explore and understand their environment. (175)

senile macular degeneration A disease of the eye involving deterioration of the retina. (611)

sensation The response of a sensory system to a particular stimulus. (149)

sensorimotor stage Piaget's first stage of cognitive development (from birth to about age 2) in which infants use their senses and motor skills to understand their world. (40)

sensory register A memory system that functions for only a fraction of a second during sensory processing, retaining a fleeting impression of the stimulus that has just impinged on a particular sense organ (e.g., the eyes). If a person looks at an object, for example, and then closes his or her eyes, the visual image of the object is breifly maintained. Significant information is transferred to working memory. (Also called *sensory store.*) (43, 632)

separation anxiety A child's fear of being left by the caregiver. This emotion emerges at about 8 or 9 months, peaks at about 14 months, then gradually subsides. (193)

sexually transmitted diseases (STDs) Diseases spread by sexual contact. Such diseases include syphilis, gonorrhea, herpes, chlamydia, and AIDS. (427)

small for gestational age (SGA) A term applied to newborns who weigh substantially less than they should given how much time has passed since conception. (114)

social clock Refers to the idea that the stages of life, and the behaviors "appropriate" to them, are set by social standards rather than by biological maturation. For instance, "middle age" begins when the culture believes it does, rather than at a particular age in all societies. (510)

social cognition A person's awareness and understanding of human personality, motives, emotions, intentions, and interactions. (354)

social comparison The tendency to assess one's abilities, achievements, social status, and the like by measuring them against those of others, especially those of one's peers. (357)

social construction An idea about the way things are or should be that is built more on the shared perceptions of members of a society than on objective reality. (10)

social context The entire spectrum of social milieux—including the people, the customs, and the beliefs—that surround each developing person. (7)

social convoy Collectively, the family members, friends, and acquaintances who move through life with an individual. (673)

social homogamy The similarity with which a couple regard leisure interests and role preferences. (520)

social learning theory The theory that learning occurs through imitation of, and identification with, other people. (37)

social referencing Looking to trusted adults for emotional cues in interpreting a strange or ambiguous event. (194)

social smile An infant's smile in response to a human face or voice. In full-term infants, this smile first appears at about 6 weeks after birth. (192)

social stratification theories Theories emphasizing that social forces, particularly those related to a person's social stratum or social category, limit individual choices and affect the ability to function. (664)

society of children The social culture of children, including the games, vocabulary, dress codes, and rules of behavior that characterize their interaction. (359)

sociocultural theory A theory that seeks to explain the growth of individual knowledge and competencies in terms of the guidance, support, and structure provided by the broader cultural context. (46)

sociodramatic play Pretend play in which children act out various roles and themes in stories of their own creation. (297)

socioeconomic status (SES) An indicator of social class that is based primarily on income, education, and occupation. (12)

sperm The reproductive cells of a male, which begin to be produced in a young man's testicles at puberty. (65)

spermarche A male's first ejaculation of seminal fluid containing sperm. (391)

statistical significance A mathematical calculation, derived from such factors as sample size and differences between groups, that indicates the likelihood that a particular research result occurred by chance. (53)

stimulus Anything that elicits a response, such as a reflex or a voluntary action. (35)

storage strategies Procedures for holding information in memory, such as rehearsal (repeating the information to be remembered) and reorganization (regrouping the information to make it more memorable). (327)

Strange Situation An experimental condition devised by Mary Ainsworth to assess an infant's attachment. The infant's behavior is observed in an unfamiliar room while a caregiver (usually the mother) and a stranger move in and out of the room. (212)

stranger wariness Fear of unfamiliar people, first noticeable in infants at about 6 months of age and usually full-blown by 10 to 14 months. (193)

subcortical dementias Dementias, such as Parkinson's disease, Huntington's disease, and multiple sclerosis, that originate in the subcortex. These diseases begin with impariments in motor ability, and produce cognitive impariment in later stages. (649)

sucking reflex A reflex that causes newborns to suck anything that touches their lips. (144)

sudden infant death syndrome (SIDS) Death of a seemingly healthy baby who, without apparent cause, stops breathing during sleep. (135)

suicidal ideation Obsessive or extensive thinking about committing suicide. (456)

supportable families Families involved in maltreatment in which the parents can meet their children's needs only through an extensive array of social services that are not normally available. (243)

surrogate parents The role some grandparents may play for their grandchildren due to their children's extreme social problems. (596)

survey A research method that collects interview information on a large number of people, either through written questionnaires or through personal interviews. (58)

synchrony Carefully coordinated interaction between infant and parent (or any other two people) in which each individual responds to and influences the other's movements and rhythms. (207)

syndrome A cluster of distinct characteristics that tend to occur together in a given disorder, although the number of characteristics exhibited, and their intensity, vary from individual to individual. (82)

synthesis The reconciliation of thesis and antithesis into a new and more comprehensive level of truth; the third stage of the dialectical process. (494)

task-involvement learning An educational strategy that bases academic grades on the mastery of certain competencies and knowledge, with students being encouraged to learn cooperatively. (420)

T-cells Cells created in the thymus that produce substances that attack infected cells in the body. (621)

temperament Inherent dispositions, such as activity level, intensity of reaction, emotionality, and sociability, that underlie and affect a person's responses to people and things. (200)

teratogens External agents, such as viruses, drugs, chemicals, and radiation, that can impair prenatal development and lead to abnormalities, disabilities, or even death. (103)

teratology The scientific study of birth defects caused by genetic or prenatal problems, or by birth complications. (103)

testosterone Called the "male hormone," testosterone is, in fact, produced in both sexes throughout the life span, but is produced in much larger amounts in males beginning in puberty. Levels of testosterone correlate with sexual desire and possibly with aggression. (388)

thanatology A field of reserach that studies death (E–1)

theory of mind An understanding of mental processes, that is, of one's own or another's emotions, perceptions, and thoughts. (259)

thesis A proposition or statement of belief; the first stage of the dialectical process. (494)

threshold effect The harmful effect of a substance which is presumably safe until exposure to it reaches a certain level. (104)

toddler A child, usually between the ages of 1 and 2, who has just begun to master the art of walking. (146)

traditional parenting A style of parenting in which the parents take traditional male and female roles, the mother being primarily nurturant and permissive, while the father is more authoritarian. (288)

trisomy-21 (Down syndrome) A chromosomal abnormality caused by an extra chromosome at the twenty-first pair. Individuals with this syndrome tend, with much variation, to have round faces and short limbs, and to be slow to develop. (82)

trust versus mistrust Erikson's first stage of psychosocial development, in which the infant experiences the world as either secure and comfortable or as unpredictable and uncomfortable. (199)

twenty-third pair In humans, the chromosome pair that determines the person's sex. (68)

underextension The use of a word to refer to a narrower category of objects or events than the term signifies. (183)

variable Any factor or condition that can change or vary from one individual or group or situation to another and thus affect behavior. (54)

vasomotor instability The temporary disruption of the homeostatic mechanisms of the vascular system, which usually constrict or dilate the blood vessels to maintain body temperature. Vasomotor instability causes moments of feeling suddenly hot or cold, a typical experience during menopause. (559)

violent death Death from accident, homicide, or suicide. It afflicts mostly young men and is often brought about by stereotypically "manly" attitudes and behavior. (485)

vitality A measure of health that refers to how healthy and energetic—physically, intellectually, and socially—an individual actually feels. (553)

vulnerable-to-crisis families Families involved in maltreatment in which the parents are experiencing unusual problems and need temporary help to resolve them. (242)

wear-and-tear theory A theory of aging that states that the human body wears out merely by being lived in and by being exposed to environmental stressors. The wear-and-tear theory sees the human body as a machine that wears out over time. (618)

Wechsler Adult Intelligence Scale (WAIS) The most commonly used assessment of adult IQ. (566)

working memory The part of the information-processing system that handles current, conscious mental activity. Working memory is constantly receiving new information, so thoughts and memories are either discarded or transferred to the more permanent knowledge base. (43, 633)

X-linked genes Genes that are carried on the X chromosome. X-linked genes are more likely to be expressed in the phenotype of males, even though women are more likely to have them in their genotype. (73)

young-old Healthy, vigorous, financially secure older adults who are well-integrated into the lives of their families and their communities. (606)

zone of proximal development Vygotsky's term for the difference between an individual's attained level of development and the person's potential level of development that might be reached with guidance. (47, 265)

zygote The single cell formed from the union of two gametes, a sperm and an ovum. (65)

Bibliography

AARP. (1990). *Understanding senior housing for the 1990s: An American Association of Retired Persons survey of consumer preferences, concerns and needs.*Washington, DC: American Association of Retired Persons.

Abbey, Antonia, Andrews, Grank M., & Halman, L. Jill. (1992). Infertity and subjective well-being. The mediating roles of self-esteem, internal control, and interpersonal conflict. *Journal of Marriage and the Family, 54,* 408–417.

Aboud, F. (1987). The development of ethnic self-identification and attitudes. In J.S. Phinney & M.J. Rotheram (Eds.), *Children's ethnic socialization: Pluralism and development.* Newbury Park, CA: Sage.

Achenbach, Thomas M., & Edelbrock, Craig S. (1981). Behavioral problems and competencies reported by parents of normal and disturbed children aged four through sixteen. *Monographs of the Society for Research in Child Development, 46* (Serial No. 188).

Achenbach, Thomas M., Howell, Catherine T., Quay, Herbert C., & Conners, C. Keith. (1991). National survey of problems and competencies among four- to sixteen-year-olds. *Monographs of the Society for Research in Child Development, 56,* (Serial No. 225) 3.

Acock, Alan C., & Demo, David H. (1994). *Family diversity and well-being.* Thousand Oaks, CA: Sage.

Adams, Cynthia, & Labouvie-Vief, Gisela. (1986, November 20). *Modes of knowing and language processing. Symposium on developmental dimensions of adult adaptation: Perspectives on mind, self, and emotion.* Paper presented at the meeting of the Gerontological Association of America, Chicago.

Adams, Dalia M., Overholser, James C., & Lehnert, Kim L. (1994). Perceived family functioning and adolescent suicidal behavior. *Journal of the American Academy of Child and Adolescent Psychiatry, 33,* 498–507.

Adams, Marilyn Jager. (1990). *Beginning to read: Thinking and learning about print.* Cambridge, MA: MIT Press.

Adams, Paul, & Dominick, Gary L. (1995). The old, the young, and the welfare state. *Generations, 14,* 38–42.

Adams, Paul L., Milner, Judith R., & Schrept, N.A. (1984). *Fatherless children.* New York: Wiley-Interscience.

Adams, Russell J. (1989). Obstetrical medication and the human newborn: The influence of alphaprodine hydrochloride on visual behavior. *Developmental Medicine and Child Neurology, 31,* 650–656.

Adams, Russell J., Maurer, D., & Davis, M. (1986). Newborns' discrimination of chromatic from achromatic stimuli. *Journal of Experimental Child Psychology, 41,* 267–281.

Adler, Lynn Peters. (1995). *Centenarians: The bonus years.* Santa Fe, NM: Health Press.

Admiraal, Pieter. (1996). Euthanasia and assisted suicide. In David C. Thomasma & Thomasine Kushner (Eds.), *Birth to death.* Cambridge, England: Cambridge University Press.

Adolph, K.E., Eppler, M.A., & Gibson, E.J. (1993). Development of perception of affordances. In C. Rovee-Collier & L.P. Lipsitt (Eds.), *Advances in infancy research* (Vol. 8). Norwood, NJ: Ablex.

Adolph, K.E., Eppler, M.A., & Gibson, E.J. (1993). Crawling versus walking infants' perception of affordances for locomotion over sloping surfaces. *Child Development, 64,* 1158–1174.

Agency for Health Care Policy and Research. (1994). *Health status and access to care of rural and urban populations* (No. 94-0031). Rockville, MD: Author.

Ainsworth, Mary D. Salter. (1973). The development of infant-mother attachment. In Bettye M. Caldwell & Henry N. Ricciuti (Eds.), *Review of child development research* (Vol. 3). Chicago: University of Chicago Press.

Ainsworth, Mary D. Salter. (1993). Attachment as related to mother-infant interaction. In C. Rovee-Collier & L.P. Lipsitt (Eds.), *Advances in infancy research* (Vol. 8). Norwood, NJ: Ablex.

Ainsworth, Mary D. Salter, & Eichberg, C. (1991). Effects on infant-mother attachment of mother's unresolved loss of an attachment figure, or other traumatic experience. In C. Murray Parkes, J. Stevenson-Hinde, & P. Marris (Eds.), *Attachment across the life cycle.* London: Routledge.

Akhtar, N., Dunham, F., & Dunham, P. (1991). Directive interactions and early vocabulary development: The role of joint attentional focus. *Journal of Child Language, 18,* 41–49.

Akiyama, Hiroko, Elliot, Kathryn, & Antonucci, Toni C. (1996). Same-sex and cross-sex relationships. *Journal of Gerontology, 51B,* 374–382.

Albert, Marilyn S., & Moss, Mark B. (1996). Neuropsychology of aging: Findings in humans and monkeys. In Edward L. Schneider & John W. Rowe (Eds.), *Handbook of the biology of aging.* San Diego, CA: Academic Press.

Aldwin, Carolyn M. (1994). *Stress, coping, and development.* New York: Guilford.

Alessandri, Steven, Sullivan, Margaret Wolan, Imaizumi, Sonia, & Lewis, Michael. (1993). Learning and emotional responsivity in cocaine-exposed infants. *Developmental Psychology, 29,* 989–997.

Allan, Graham. (1989). *Friendship: Developing a sociological perspective.* Boulder, CO: Westview.

Allen, E.M., Mitchell, E.H., Stewart, A.W., & Ford, R.P.K. (1993). Ethnic differences in mortality from sudden infant death syndrome in New Zealand. *British Medical Journal, 306,* 13–16.

Allison, Clara. (1985). Development direction of action programs: Repetitive action to correction loops. In Jane E. Clark & James H. Humphrey (Eds.), *Motor development: Current selected research.* Princeton, NJ: Princeton Book Company.

Allison, Paul D., & Furstenberg, Frank F. (1989). How marital dissolution affects children: Variations by age and sex. *Developmental Psychology, 25,* 540–549.

Altergott, Karen. (1993). *One world, many families.* Minneapolis, MN: National Council on Family Relations.

Altman, Lawrence K. (1991, August 6). Men, women, and heart disease: More than a question of sexism. *New York Times*, pp. C1, C8.

Alwin, Duane F. (1996). Coresidence beliefs in American society—1973 to 1991. *Journal of Marriage and the Family, 58,* 393–403.

Alwin, Duane F., Converse, Philip E., & Martin, Steven S. (1985). Living arrangements and social integration. *Journal of Marriage and the Family, 47,* 319–334.

Amato, Paul R. (1993). Children's adjustment to divorce: Theories, hypotheses, empirical support. *Journal of Marriage and the Family, 55,* 23–38.

Amato, Paul R., & Booth, Alan. (1996). A prospective study of divorce and parent-child relationships. *Journal of Marriage and the Family, 58,* 356–365.

Amato, Paul R., & Keith, Bruce. (1991). Parental divorce and adult well-being: A meta-analysis. *Journal of Marriage and the Family, 53,* 43–58.

Amato, Paul R., & Rezac, Sandra J. (1994). Contact with nonresident parents, interparental conflict, and children's behavior. *Journal of Family Issues, 15,* 191–207.

American Psychiatric Association. (1994). *Diagnostic and Statistical Manual of Mental Disorders—DSM-IV.* Washington, DC: Author.

Ammerman, Robert T., & Hersen, Michel (Eds.). (1989). *Treatment of family violence.* New York: Wiley.

Anderson, B. (1989). Effects of public day care: A longitudinal study. *Child Development, 60,* 857–866.

Anderson, W. French. (1995). Gene therapy. *Scientific American, 273,* 124–128.

Andrich, David, & Styles, Irene. (1994). Psychometric evidence of intellectual growth spurts in early adolescence. *Journal of Early Adolescence, 14,* 328–344.

Angel, Ronald J., & Angel, Jacqueline L. (1994). *Painful inheritance: Health and the new generation of fatherless families.* Madison: University of Wisconsin Press.

Angier, Natalie. (1994, May 24). Mother's milk found to be potent cocktail of hormones. *New York Times*, p. C1.

Anglin, Jeremy M. (1993). Vocabulary development: A morphological analysis. *Monographs of the Society for Research in Child Development, 58*(10, Serial No. 238).

Antonarakis, S.E., Petersen, M.B., McInnis, M.G., Adelsberger, P.A., Schinzel, A.A., Binkert, F., Pangalos, C., Raoul, O., Slaugenhaupt, S.A., Hafez, M., Cohen, M.M., Roulsston, D., Schwartz, S., Mikkelsen, M., Tranebjaerg, L., Greenberg, F., Hoar, D.I., Rudd, N.L., Warren, A.C., Metaxotou, C., Bartsocas, C., and Chakravarti, A. (1992). The meiotic stage of nondisjunction in trisomy 21: Determination using DNA polymorphisms. *American Journal of Human Genetics, 50,* 544–550.

Antonucci, Toni C. (1985). Personal characteristics, social support, and social behavior. In R.H. Binstock & E. Shanas (Eds.), *Handbook of aging and the social sciences* (2nd ed.). New York: Van Nostrand Reinhold.

Antonucci, Toni C. (1990). Attachment, social support, and coping with negative life events. In E.M. Cummings, A.L. Greene, & K.H. Karraker (Eds.), *Life-span developmental psychology: Vol. 11. Stress and coping across the life-span.* Hillsdale, NJ: Erlbaum.

Antonucci, Toni C., & Akiyama, Hiroko. (1995). Convoys of social relations: Family and friendships within a life span context. In Rosemary Blieszner & Victoria Hilkevitch Bedford (Eds.), *Handbook of aging and the family.* Westport, CT: Greenwood Press.

Apgar, Virginia. (1953). A proposal for a new method of evaluation in the newborn infant. *Current Research in Anesthesia and Analgesia, 32,* 260.

Aquilino, William S. (1991). Family structure and home-leaving: A further specification of the relationship. *Journal of Marriage and the Family, 53,* 999–1010.

Aram, Dorothy M., Morris, Robin, & Hall, Nancy E. (1992). The validity of discrepancy criteria for identifying children with developmental language disorders. *Journal of Learning Disabilities, 25,* 549–554.

Arbuthnot, Jack, Gordon, Donald A., & Jurkovic, Gregory J. (1987). Personality. In Herbert C. Quay (Ed.), *Handbook of juvenile delinquency.* New York: Wiley.

Archer, Loran, Grant, Bridget F., & Dawson, Deborah A. (1995). What if Americans drank less? The potential effect on the prevalence of alcohol abuse and dependence. *American Journal of Public Health, 85,* 61–66.

Archer, Sally L., & Waterman, Alan S. (1990). Varieties of identity diffusions and foreclosures: An exploration of the subcategories of the identity statuses. *Journal of Adolescent Research, 5,* 96–111.

Arendell, Terry. (1995). *Fathers and divorce.* Newbury Park, CA: Sage.

Ariès, Philippe. (1981). *The hour of our death.* New York: Knopf.

Arking, Robert. (1991). *Biology of aging: Observations and principles.* Englewood Cliffs, NJ: Prentice Hall.

Arlin, Patricia K. (1984). Adolescent and adult thought: A structural interpretation. In Michael L. Commons, Francis A. Richards, & Cheryl Armon (Eds.), *Beyond formal operations: Late adolescent and adult cognitive development.* New York: Praeger.

Armistead, Lisa, Kempton, Tracy, Lynch, Sean, Forehand, Rex, Nousiainen, Sarah, Neighbors, Bryan, & Tannenbaum, Lynne. (1992). Early retention: Are there long-term beneficial effects? *Psychology in the Schools, 29,* 342–347.

Arnold, Elaine. (1982). The use of corporal punishment in child-rearing in the West Indies. *Child Abuse and Neglect, 6,* 141–145.

Aron, Arthur, & Westbay, Lori. (1996). Dimensions of the prototype of love. *Journal of Personality and Social Relationships, 70,* 535–551.

Aronson, Miriam K. (1994). *Reshaping dementia care.* Thousand Oaks, CA: Sage.

Arsenio, W.F., & Kramer, R. (1992). Victimizers and their victims: Children's conceptions of the mixed emotional consequences of moral transgressions. *Child Development, 63,* 915–927.

Asch, David A., Hershey, John C., Pauly, Mark V., Patton, James P., Jedriziewski, Kathryn M., & Mennuti, Micheal T. (1996). Genetic screening for reproductive planning: Methodological and conceptual issues in policy analysis. *American Journal of Public Health, 86,* 684–690.

Aslin, Richard N. (1987). Visual and auditory development in infancy. In Joy Doniger Osofsky (Ed.), *Handbook of infant development* (2nd ed.). New York: Wiley.

Aslin, Richard N. (1988). Visual perception in early infancy. In Albert Yonas (Ed.), *Perceptual development in infancy*. Hillsdale, NJ: Erlbaum.

Astington, Janet Wilde. (1993). *The child's discovery of the mind*. Cambridge, MA: Harvard University Press.

Astington, Janet Wilde, & Gopnik, A. (1988). Knowing you've changed your mind: Children's understanding of representational change. In J.W. Astington, P.L. Harris, & D.R. Olson (Eds.), *Developing theories of mind*. Cambridge, England: Cambridge University Press.

Atchley, R.C. (1991). Family, friends, and social support. In Robert C. Atchley (Ed.), *Social forces and aging*. Belmont, CA: Wadsworth.

Attridge, M., Berscheid, E., & Simpson, J.A. (1995). Predicting relationship stability from both partners versus one. *Journal of Personality and Social Relationships, 69*, 254–268.

Ausman, L.M., & Russell, R.M. (1990). Nutrition and aging. In E.L. Schneider & J.W. Rowe (Eds.), *Handbook of the biology of aging* (3rd ed.). San Diego, CA: Academic Press.

Awooner-Renner, Sheila. (1993). I desperately needed to see my son. In Donna Dickenson & Malcolm Johnson (Eds.), *Death, dying & bereavement*. London: Sage.

Axia, Giovanna, & Baroni, Rosa. (1985). Linguistic politeness at different age levels. *Child Development, 56*, 918–927.

Bachrach, C.A., Clogg, C.C., & Carver, K. (1993). Outcomes of early childbearing: Summary of a conference. *Journal of Research on Adolescence, 3*, 337–348.

Bahrick, L.E. (1983). Infants' perception of substance and temporal synchrony in multimodal events. *Infant Behavior and Development, 6*, 429–451.

Bahrick, H.P. (1984). Semantic memory content in permastore: Fifty years of memory for Spanish learned in school. *Journal of Experimental Psychology: General, 113*, 1–35.

Bailey, J. Michael, Pillard, Richard C., & Knight, Robert. (1993). At issue: Is sexual orientation biologically determined? *CQ Researcher, 3*, 209.

Bailey, William T. (1994). Fathers' involvement and responding to infants: "More" may not be "better". *Psychological Reports, 74*, 92–94.

Baillargeon, R. (1987). Object permanence in 3.5- and 4.5-month-old infants. *Developmental Psychology, 23*, 655–664.

Baillargeon, R. (1991). Reasoning about the height and location of a hidden object in 4.5- and 6.5-month-old infants. *Cognition, 38*, 13–42.

Baillargeon, R., & DeVos, J. (1992). Object permanence in young infants: Further evidence. *Child Development, 62*, 1227–1246.

Baillargeon, R., Graber, M., Decops, J., & Black, J. (1990). Why do young infants fail to search for hidden objects? *Cognition, 36*, 255–284.

Bakan, D. (1966). *The duality of human existence: Isolation and communion in Western man*. Boston: Beacon.

Bakeman, Roger, Adamson, Lauren B., Konner, Melvin, & Barr, Ronald G. (1990).!Kung infancy: The social context of object exploration. *Child Development, 61*, 794–809.

Baker, Colin. (1993). *Foundation of bilingual education and bilingualism*. Clevedon, England: Multilingual Matters.

Baker-Ward, Lynne, Gordon, Betty N., Ornstein, Peter A., Larus, Deanna M., & Clubb, Patricia A. (1993). Young children's long-term retention of a pediatric examination. *Child Development, 64*, 1519–1533.

Bakken, B. (1993). Prejudice and danger: The only child in China. *Childhood, 1*, 46–61.

Balaban, Marie T. (1995). Affective influences on startle in five-month-old infants: Reactions to facial expressions of emotion. *Child Development, 66*, 28–36.

Balfour-Lynn, Lionel. (1986). Growth and childhood asthma. *Archives of Diseases of Childhood, 60*, 231–235.

Balin, Arthur K. (Ed.). (1994). *Practical handbook of human biologic age determination*. Boca Raton, FL: CRC Press.

Baltes, Margret M., & Wahl, Hans-Werner. (1992). The behavior system of dependency in the elderly: Interaction with the social environment. In Marcia G. Ory, Ronald P. Abeles, & Paul Darby Lipman (Eds.), *Aging, health, and behavior*. Newbury Park, CA: Sage.

Baltes, Paul B. (1987). Theoretical propositions of life-span developmental psychology: On the dynamics between growth and decline. *Developmental Psychology, 23*, 611–626.

Baltes, Paul B., & Baltes, Margret M. (1990). Psychological perspectives on successful aging: The model of selective optimization with compensation. In Paul B. Baltes & Margret M. Baltes (Eds.), *Successful aging: Perspectives from the behavioral sciences*. Cambridge, England: Cambridge University Press.

Baltes, Paul B., & Lindenberger, Ulman. (1988). On the range of cognitive plasticity in old age as a function of experience: 15 years of intervention research. *Behavior Therapy, 19*, 283–300.

Baltes, Paul B., & Smith, Jacqui. (1990). Toward a psychology of wisdom and its ontogenesis. In Robert J. Sternberg (Ed.), *Wisdom: Its nature, origins, and development*. Cambridge, England: Cambridge University Press.

Baltes, Paul B., Smith, Jacqui, & Staudinger, Ursula. (1992). Wisdom and successful aging. In T. Sonderegger (Ed.), *Psychology and aging: Nebraska Symposium on Motivation* (Vol. 39). Lincoln: University of Nebraska Press.

Bamford, F.N., Bannister, R., Benjamin, C.M., Hillier, V.F., Ward, B.S., & Moore, W.M.O. (1990). Sleep in the first year of life. *Developmental and Child Neurology, 32*, 718–734.

Bandura, Albert. (1977). *Social learning theory*. Englewood Cliffs, NJ: Prentice-Hall.

Bandura, Albert. (1981). Self-referent thought: A developmental analysis of self-efficacy. In John H. Flavell & Lee Ross (Eds.), *Social cognitive development: Frontiers and possible futures*. Cambridge, England: Cambridge University Press.

Bandura, Albert. (1986). *Social foundations of thought and action: A social cognitive theory*. Englewood Cliffs, NJ: Prentice-Hall.

Bandura, Albert. (1989). Social cognitive theory. In R. Vasta (Ed.), *Annals of child development* (Vol. 6). Greenwich, CT: JAI Press.

Bandura, Albert. (Ed.). (1995). *Self-efficacy in changing societies*. New York: Cambridge University Press.

Bane, Share DeCroix. (1991). Rural minority populations. *Generations, 27*(3), 63–65.

Banerji, Madhabi & Dailey, Ronald A. (1995). A study of the effects of the inclusion model on students with specific learning disabilities. *Journal of Learning Disabilities, 28*, 511–522.

Barber, B.K. (1994). Cultural, family, and personal contexts of parent-adolescent conflict. *Journal of Marriage and the Family, 56*, 375–386.

Barinaga, Marcia. (1994). Surprises across the cultural divide. *Science, 263*, 1468–1472.

Barkley, R.A. (1990). Attention deficit disorders: History, definition, and diagnosis. In M. Lewis & S.M. Miller (Eds.), *Handbook of developmental psychopathology*. New York: Plenum Press.

Barkley, R.A., Anastopoulos, A.D., Guevremont, D.D., & Fletcher, K.E. (1991). Adolescents with ADHD. Patterns of behavioral adjustment, academic functioning, and treatment utilization. *Journal of the American Academy of Child Adolescent Psychiatry, 30*, 752–761.

Barkow, Jerome H., Cosmides, Leda, & Tooby, John. (Eds.). (1992). *The adapted mind: Evolutionary psychology and the generation of culture*. New York: Oxford University Press.

Barling, Julian, & McEwan, Karyl. (1992). Linking work experiences to facets of marital functioning. *Journal of Organizational Behavior, 13*, 573–583.

Barnett, Mark A. (1986). Sex bias in the helping behavior presented in children's picture books. *Journal of Genetic Psychology, 147*, 343–351.

Barnett, Ronald. (1994). *The limits of competence: Knowledge, higher education, and society*. Buckingham, England: Society for Research into Higher Education.

Barnett, Rosalind C., & Baruch, Grace K. (1987). Determinants of fathers' participation in family work. *Journal of Marriage and the Family, 49*, 29–40.

Barnett, Rosalind C., Raudenbush, Stephen, Brennan, Robert T., Pleck, Joseph H., & Marshall, Nancy L. (1995). Change in job and marital experiences and change in psychological distress: A longitudinal study of duel earner couples. *Journal of Personality and Social Psychology, 69*, 839–850.

Barnett, Rosalind C., & Rivers, Caryl. (1996). The new dad works the "second shift," too. *Radcliff Quarterly, 82*, 9.

Baron, J. (1989). *Teaching decision-making to adolescents*. Hillsdale, NJ: Erlbaum.

Barone, Charles, Ickovics, Jeannette R., Ayers, Tim S., Katz, Sharon M., Voyce, Charlene K., & Weissberg, Roger P. (1996). High risk sexual behavior among young urban students. *Family Planning Perspectives, 28*, 69–74.

Barresi, Charles M., & Menon, Geeta. (1990). Diversity in black family caregiving. In Zev Harel, Edward A. McKinney, & Mischel Williams (Eds.), *Black aged*. Newbury Park, CA: Sage.

Barrett, Martyn D. (1986). Early semantic representations and early word-usage. In Stan A. Kuczaj & Martyn D. Barrett (Eds.), *The development of word meaning: Progress in cognitive developmental research*. New York: Springer-Verlag.

Barth, Richard. (1994). *From child abuse to permanency planning*. New York: Aldine de Gruyter.

Barusch, Amanda Smith. (1994). *Older women in poverty: Private lives and public policies*. New York: Springer.

Barusch, Amanda Smith, & Steen, Peter. (1996). No place without a home: Southeast Asian grandparents in refugee families. *Generations, 20*, 45–48.

Basseches, Michael. (1984). *Dialectical thinking and adult development*. Norwood, NJ: Ablex.

Basseches, Michael. (1989). Dialectical thinking as an organized whole: Comments on Irwin and Kramer. In Michael L. Commons, Jan D. Sinnott, Francis A. Richards, & Cheryl Armon (Eds.), *Adult development: Vol. 1. Comparisons and applications of developmental models*. New York: Praeger.

Bassuk, E.L. (1989). Homelessness: A growing American tragedy. *Division of Child, Youth, and Family Services Newsletter, 12*, 1–13.

Bassuk, E.L., & Rosenberg, L. (1990). Psychosocial characteristics of homeless children and children with homes. *Pediatrics, 85*, 257–261.

Bateman, David A., Ng, Stephen K.C., Hansen, Catherine A., & Heagarty, Margaret C. (1993). The effects of intrauterine cocaine exposure in newborns. *American Journal of Public Health, 83*, 190–193.

Bates, Betsy. (1995). STD reinfection greatest in teens. *Pediatric News, 29*(12), 8.

Bates, Elizabeth, & Carnevale, George F. (1994). Developmental psychology in the 1990s: Research on language development. *Developmental Review, 13*, 436–470.

Bauer, H.H. (1992). *Scientific literacy and the myth of the scientific method*. Urbana: University of Illinois Press.

Bauer, Patricia J., & Dow, Gina. (1994). Episodic memory in 16- and 20-month-old children. Specifics are generalized but not forgotten. *Developmental Psychology, 30*, 403–417.

Bauer, Patricia J., & Mandler, J.M. (1990). Remembering what happened next: Very young children's recall of event sequences. In R. Fivush & J.A. Hudson (Eds.), *Knowing and remembering in young children*. Cambridge, England: Cambridge University Press.

Bauer, Patricia J., & Mandler, J.M. (1992). Putting the horse before the cart: The use of temporal order in recall of events by one-year-old children. *Developmental Psychology, 28*, 441–452.

Bauman, K.E., Fisher, L.A., Bryan, E.S., & Chenoweth, R.L. (1984). Antecedents, subjective expected utility, and behavior: A panel study of adolescent cigarette smoking. *Addictive Behaviors, 9*, 121–136.

Baumrind, Diana. (1967). Child-care practices anteceding three patterns of preschool behavior. *Genetic Psychology Monographs, 75*, 43–88.

Baumrind, Diana. (1971). Current patterns of parental authority. *Developmental Psychology, 4*(Monograph 1), 1–103.

Baumrind, Diana. (1987). A developmental perspective on adolescent risk-taking behavior. In C.E. Irwin (Ed.), *Adolescent social behavior and health*. San Francisco: Jossey-Bass.

Baumrind, Diana. (1989). Rearing competent children. In William Damon (Ed.), *New directions for child development: Adolescent health and human behavior*. San Francisco: Jossey-Bass.

Baumrind, Diana. (1991). The influence of parenting style on adolescent competence and substance use. *Journal of Early Adolescence, 11*, 56–95.

Baumrind, Diana. (1993). The average expectable environment is not good enough: A response to Scarr. *Child Development, 64*, 1299–1317.

Baumrind, Diana. (1995). *Child maltreatment and optimal caregiving in social contexts*. New York: Garland.

Bayley, Nancy. (1966). Learning in adulthood: The role of intelligence. In Herbert J. Klausmeier & Chester W. Harris (Eds.), *Analysis of concept learning*. New York: Academic Press.

Bayley, Nancy, & Oden, Melita. (1955). The maintenance of intellectual ability in gifted adults. *Journal of Gerontology, 10,* Section B (1), 91–107.

Baynes, R.D., & Bothwell, T.H. (1990). Iron deficiency. *Annual Review of Nutrition, 10,* 133.

Beal, Carole R. (1994). *Boys and girls: The development of gender roles.* New York: McGraw-Hill.

Beal, S.M., & Finch, C.F. (1991). An overview of retrospective case control slides investigating the relationship between prone sleep positions and SIDS. *Journal of Paediatrics and Child Health, 27,* 334–339.

Beal, S.M., & Porter, C. (1991). Sudden infant death syndrome related to climate. *Acta Paediatrica Scandinavica, 80,* 278–287.

Beard, Belle Boone. (1991). *Centenarians: The new generation.* Westport, CT: Greenwood Press.

Beaudry, Micheline, Dufour, R., & Marcoux, Sylvie. (1995). Relation between infant feeding and infections during the first six months of life. *Journal of Pediatrics, 126,* 191–197.

Beck, Ulrich, & Beck-Gernsheim, Elizabeth. (1995). *The normal chaos of love.* Cambridge, England: Polity Press.

Becker, G.S. (1994, March 28). Cut the graybeards a smaller slice of the pie. *Business Week,* p. 20.

Becker, Joseph. (1989). Preschoolers' use of number words to denote one-to-one correspondence. *Child Development, 60,* 1147–1157.

Beckwith, Leila, & Rodning, Carol. (1991). Intellectual functioning in children born preterm: Recent research. In Lynn Okagaki & Robert J. Sternberg (Eds.), *Directors of development: Influences on the development of children's thinking.* Hillsdale, NJ: Erlbaum.

Bedford, Victoria Hilkevitch. (1995). Sibling relationships in middle and old age. In Rosemary Blieszner & Victoria Hilkevitch Bedford (Eds.), *Handbook of aging and the family.* Westport, CT: Greenwood Press.

Beecham, Clayton T. (1989). Natural childbirth: A step backward? *Female Patient, 14,* 56–60.

Behrman, Richard E. (1992). *Nelson textbook of pediatrics.* Philadelphia: Saunders.

Beilin, H. (1992). Piaget's enduring contribution to developmental psychology. *Developmental Psychology, 28,* 191–204.

Beizer, Judith L. (1994). Medications and the aging body: Alteration as a function of age. *Generations, 18,* 13–18.

Bell, A.P., Weinberg, M.S., & Mammersmith, S. (1981). *Sexual preference: Its development in men and women.* Bloomington: University of Indiana Press.

Bell, Derrick. (1992). *Faces at the bottom of the well: The permanence of racism.* New York: Basic Books.

Bell, M.A., & Fox, N.A. (1992). The relations between frontal brain electrical activity and cognitive development during infancy. *Child Development, 63,* 1142–1163.

Bellantoni, Michele F., & Blackman, Marc R. (1996). Menopause and its consequences. In Edward L. Schneider & John W. Rowe (Eds.), *Handbook of the biology of aging.* San Diego, CA: Academic Press.

Beller, A.H., & Graham, J.W. (1993). *Small change: The economics of child support.* New Haven, CT: Yale University Press.

Bellinger, David D., & Needleman, Herbert L. (1985). Prenatal and early postnatal exposure to lead: Developmental effects, correlations, and implications. *International Journal of Mental Health, 14,* 78–111.

Belsky, Jay. (1986). Infant day care: A cause for concern? *Zero to Three, 6,* 1–7.

Belsky, Jay. (1990). Infant day care, child development, and family policy. *Society, 27*(5), 10–12.

Belsky, Jay, & Cassidy, J. (1995). Attachment theory and evidence. In M. Rutter, D. Hay, & S. Baron-Cohen (Eds.), *Developmental principles and clinical issues in psychology and psychiatry.* Oxford, England: Blackwell.

Belsky, Jay, Steinberg, Laurence, & Draper, Patricia. (1991). Childhood experience, interpersonal development, and reproductive strategy: An evolutionary theory of socialization. *Child Development, 62,* 647–670.

Belsky, Jay, & Vondra, Joan. (1989). Lessons from child abuse: The determinants of parenting. In Dante Cicchetti & Vicki Carlson (Eds.), *Child maltreatment: Theory and research on the causes and consequences of child abuse and neglect.* Cambridge, England: Cambridge University Press.

Bem, Sandra L. (1989). Genital knowledge and gender constancy in preschool children. *Child Development, 60,* 649–662.

Benedek, Elissa. (1989). Baseball, apple pie, and violence: Is it American? In Leah J. Dickerson & Carol Nadelson (Eds.), *Family violence: Emerging issues of national crisis.* Washington, DC: American Psychiatric Press.

Benet, Sula. (1974). *Abkhasians: The long-lived people of the Caucasus.* New York: Holt, Rinehart & Winston.

Bengston, Vern L. (1975). Generation and family effects in value socialization. *American Sociological Review, 40,* 358–371.

Bengston, Vern L., Reedy, Margaret N., & Gordon, Chad. (1985). Aging and self-conceptions: Personality processes and social contexts. In James E. Birren & K. Warner Schaie (Eds.), *Handbook of the psychology of aging* (2nd ed.). New York: Van Nostrand.

Bengston, Vern, Rosenthal, Carolyn, & Burton, Linda. (1996). Paradoxes of families and aging. In Robert H. Binstock & Linda K. George (Eds.), *Handbook of aging and the social sciences.* San Diego, CA: Academic Press.

Benoit, Diane, & Parker, Kevin C. (1994). Stability and transmission of attachment across three generations. *Child Development, 65,* 1444–1456.

Benson, Peter L. (1993). *The troubled journey: A portrait of 6th–12th grade youth.* Minneapolis, MN: Search Institute.

Benson, V., & Marano, M. (1994). *Current estimates from the National Health Interview Survey 1993,* Vital Health Statistics (Series 10, No. 190). Washington, DC: National Center for Health Statistics.

Bensur, Barbara, & Eliot, John. (1993). Case's developmental model and children's drawings. *Perceptual and Motor Skills, 76,* 371–375.

Berenbaum, Sheri, & Snyder, Elizabeth. (1995). Early hormonal influences on childhood sex-typed activity and playmate preferences: Implications for the development of sexual orientation. *Developmental Psychology, 31,* 31–42.

Berg, Cynthia A., & Sternberg, Robert J. (1985). A triarchic theory of intellectual development during adulthood. *Developmental Review, 5,* 334–370.

Berg, Robert L., & Cassells, Joseph E. (1990). *The second fifty years: Promoting health and preventing disability*. Washington, DC: National Academy Press.

Berkson, Gershon. (1993). *Children with handicaps: A review of behavioral research*. Hillsdale, NJ: Erlbaum.

Bernaldo de Quiros, Guillermo, Kinsbourne, Marcel, Palmer, Roland L., & Rufo, Dolores Tocci. (1994). Attention deficit disorder in children: Three clinical variants. *Journal of Developmental and Behavioral Pediatrics, 15*, 311–319.

Berndt, Thomas J. (1981). Relations between social cognition, nonsocial cognition, and social behavior. In John H. Flavell & Lee Ross (Eds.), *Social cognitive development: Frontiers and possible futures*. Cambridge, England: Cambridge University Press.

Berndt, Thomas J. (1989). Friendships in childhood and adolescence. In William Damon (Ed.), *Child development today and tomorrow*. San Francisco: Jossey-Bass.

Berndt, Thomas J., & Perry, T.B. (1990). Distinguishing features of early adolescent friendship. In Raymond Montemeyer, Gerald R. Adams & Thomas P. Gullota (Eds.) *From childhood to adolescence: A transitional period?* Newbury Park, CA: Sage.

Berndt, Thomas J., & Savin-Williams, R.C. (1992). Peer relations and friendships. In P.H. Tolan & B.J. Kohler (Eds.), *Handbook of clinical research and practice with adolescents*. New York: Wiley.

Berry, Ruth E., & Williams, Flora L. (1987). Assessing the relationship between quality of life and marital and income satisfaction: A path analytic approach. *Journal of Marriage and the Family, 49*, 107–116.

Bertenthal, Bennett I., & Campos, Joseph J. (1990). A systems approach to the organizing effect of self-produced locomotion during infancy. In Carolyn Rovee-Collier & Lewis P. Lipsitt (Eds.), *Advances in infancy research* (Vol. 6). Norwood, NJ: Ablex.

Berzonsky, Michael D. (1989). Identity style: Conceptualization and measurement. *Journal of Adolescent Research, 4*, 268–282.

Besharov, Douglas J. (1993). Overreporting and underreporting are twin problems. In Richard J. Gelles & Donileen R. Loseke (Eds.), *Current controversies in family violence*. Newbury Park, CA: Sage.

Best, Deborah L. (1993). Inducing children to generate mnemonic organizational strategies: An examination of long term retention and materials. *Developmental Psychology, 29*, 324–336.

Betancourt, Raoul Louis. (1991). *Retirement and men's physical and social health*. New York: Garland.

Bettes, Barbara A. (1988). Maternal depression and motherese: Temporal and intonational features. *Child Development, 59*, 1089–1096.

Beunen, G.P., Malina, R.M., Van't Hof, M.A., Simons, J., Ostyn, M., Renson, R., & Van Gerven, D. (1988). *Adolescent growth and motor performance: A longitudinal study of Belgian boys*. Champaign, IL: Human Kinetics Books.

Beyth-Marom, Ruth, Austin, Laurel, Fischhoff, Baruch, & Palmgren, Claire. (1993). Perceived consequences of risky behaviors: Adults and adolescents. *Developmental Psychology, 29*, 539–548.

Bhatia, M.S., Nigam, V.R., Bohra, N., & Malik, S.C. (1991). Attention deficit disorder with hyperactivity among paediatric outpatients. *Journal of Child Psychology and Psychiatry and Allied Disciplines, 32*, 297–306.

Bialystok, E. (1988). Levels of bilingualism and levels of linguistic awareness. *Developmental Psychology, 24*, 560–567.

Biegel, D.E., Sales, E., & Schultz, R. (1991). *Family caregiving in chronic illness*. London: Sage.

Biener, Lois, & Heaton, Alan. (1995). Women dieters of normal weight: Their motives, goals, and risks. *American Journal of Public Health, 85*, 714–717.

Bierman, Karen Lynn, Smoot, D.L., & Aumiller, K. (1993). Characteristics of aggressive-rejected, aggressive (nonrejected), and rejected (nonaggressive) boys. *Child Development, 64*, 139–151.

Biernat, Monica, & Kobrynowicz, Diane. (1997). Gender- and race-based standards of competence: Lower minimum standards but higher ability standards for devalued groups. *Journal of Personality and Social Psychology, 72*, 544–557.

Bigelow, B.J. (1977). Children's friendship expectations: A cognitive developmental study. *Child Development, 48*, 246–253.

Bigelow, B.J., & La Gaipa, J.J. (1975). Children's written descriptions of friendship: A multidimensional analysis. *Developmental Psychology, 11*, 857–858.

Billingham, Robert E., & Sack, Alan R. (1986). Courtship and violence: The interactive status of the relationship. *Journal of Adolescent Research, 1*, 315–326.

Binder, Arnold, Geis, Gilbert, & Bruce, Dickson. (1988). *Juvenile delinquency: Historical, cultural and legal perspectives*. New York: Macmillan.

Binstock, Robert H., & Day, Christine L. (1996). Aging and politics. In Robert H. Binstock & Linda K. George (Eds.), *Handbook of aging and the social sciences*. San Diego, CA: Academic Press.

Binstock, Robert H., Post, Stephen G., & Whitehouse, Peter J. (1992). The challenges of dementia. In Robert H. Binstock, Stephen G. Post, & Peter Whitehouse (Eds.), *Dementia and aging: Ethics, values, and policy choices*. Baltimore: Johns Hopkins University Press.

Birch, Leann L., Birch, D., Marlin, D.W., & Kramer, L. (1982). Effects of instrumental consumption on children's food preference. *Appetite, 3*, 125–134.

Birren, James E., & Schroots, Johannes, J.F. (1996). History, concepts, and methods in psychology of aging. In James E. Birren & K. Warner Schaie (Eds.), *Handbook of the psychology of aging*. San Diego, CA: Academic Press.

Bithoney, William G., Vandeven, Andrew M., & Ryan, Amy. (1993). Elevated lead levels in reportedly abused children. *Pediatrics, 122*, 719–720.

Bjorklund, D.F. (Ed.). (1990). *Children's strategies: Contemporary views of cognitive development*. Hillsdale, NJ: Erlbaum.

Bjorklund, D.F., & Bjorklund, B.R. (1992, August). "I forget." *Parents*, pp. 62–68.

Bjorklund, D.F., & Harnishfeger, K.K. (1990). The resources construct in cognitive development: Diverse sources of evidence and a theory of inefficient inhibition. *Developmental Review, 10*, 48–71.

Blake, Judith. (1989). *Family size and achievement*. Berkeley: University of California Press.

Blakemore, Ken, & Boneham, Margaret. (1994). *Age, race and ethnicity: A comparative approach*. Buckingham, England: Open University Press.

Blanchard-Fields, Fredda. (1986). Reasoning on social dilemmas varying the emotional saliency: An adult developmental perspective. *Psychology and Aging, 1*, 325–333.

Blanchard-Fields, Fredda, & Abeles, Ronald P. (1996). Social cognition and aging. In James E. Birren & K. Warner Schaie (Eds.), *Handbook of the psychology of aging*. San Diego: Academic Press.

Blanchard-Fields, Fredda, Jahnke, Hather Casper, & Camp, Cameron. (1995). Age differences in problem-solving style: The role of emotional salience. *Psychology and Aging, 10*, 173–180.

Blane, David. (1995). Editorial: Social determinants of health—socioeconomic status, social class, and ethnicity. *American Journal of Public Health, 85*, 903–904.

Blankenhorn, David. (1995). *Fatherless America: Confronting our most urgent social problem.* New York: Basic Books.

Blazer, D., Burchett, B., Service, C., & George, L.K. (1991). The association of age and depression among the elderly: An epidemiological exploration. *Journal of Gerontology: Medical Sciences, 46*, M210–M215.

Blieszner, Rosemary, & Hamon, Raeann R. (1992). Filial responsibility: Attitude, motivators, and behaviors. In Jeffrey W. Dwyer & Raymond R. Coward (Eds.), *Gender, families and elder care.* Newbury Park, CA: Sage.

Bloom, L. (1991). *Language development from two to three.* New York: Cambridge University Press.

Bloom, L. (1993). *The transition from infancy to language: Acquiring the power of expression.* New York: Cambridge University Press.

Bloom, L., Merkin, S., & Wootten, Janet. (1982). Wh– questions: Linguistic factors that contribute to the sequence of acquisition. *Child Development, 53*, 1084–1092.

Bloomfield, L. (1933). *Language.* New York: Henry Holt.

Blumberg, Jeffrey B. (1996). Status and functional impact of nutrition in older adults. In Edward L. Schneider & John W. Rowe (Eds.), *Handbook of the biology of aging*, 4th ed. San Diego, CA: Academic Press.

Blythe, Ronald. (1979). *The view in winter: Reflections on old age.* New York: Penguin.

Boivin, Michael J. & Giordani, Bruno. (1995). A risk evaluation of the neuropsychological effects of childhood lead toxicity. *Developmental Neuropsychology, 11*, 157–180.

Bolton, Frank G., Morris, Larry A., & MacEacheron, Ann E. (1989). *Males at risk: The other side of child sexual abuse.* Newbury Park, CA: Sage.

Booth, Alan, & Edwards, John N. (1992). Starting over: Why remarriages are more unstable. *Journal of Family Issues, 13*, 179–194.

Booth, Alan, & Johnson, E. (1988). Premarital cohabitation and marital success. *Journal of Family Issues, 9*, 387–394.

Borgaonkar, Digamber S. (1994). *Chromosomal variation in man.* New York: Wiley.

Bornstein, Marc H. (1985). Habituation of attention as a measure of visual information processing in human infants: Summary, systematization, and synthesis. In Gilbert Gottlieb & Norman A. Krasnegor (Eds.), *Measurement of audition and vision in the first year of postnatal life: A methodological overview.* Norwood, NJ: Ablex.

Bornstein, Marc H., & Lamb, M.E. (1992). *Development in infancy* (3rd ed.). New York: McGraw-Hill.

Bornstein, Marc H., Tamis-LeMonda, C.S., Tal, J., Ludemann, P., Toda, S., Rahn, C.W., Pecheux, M.-G., Azuma, H., & Vardi, D. (1992). Maternal responsiveness to infants in three societies: The United States, France, and Japan. *Child Development, 63*, 808–821.

Borovsky, D., & Rovee-Collier, Carolyn K. (1990). Contextual constraints on memory retrieval at six months. *Child Development, 61*, 1569–1583.

Borsting, Eric. (1994). Overview of vision and visual processing development. In Mitchell Scheiman & Michael Rouse (Eds.), *Optimetric management of learning-related vision problems.* St. Louis, MO: Mosby.

Bouchard, C., & Bray, G.A. (Eds.). (1996). *Regulations of body weight: Biological and behavioral mechanisms.* New York: Wiley.

Bouchard, C., Trembley, A., Nadeau, A. Despres, J.P., Theriault, G., Boulay, M.R., Lortie, G., Leblanc, C., & Fournier, G. (1989). Genetic effect in resting and exercise metabolic rate. *Metabolism, Clinical and Experimental, 38*, 364–370.

Bouchard, Thomas J. (1994). Genes, environment, and personality. *Science*, 264, 1700–1701.

Bouchard, Thomas J., Lykken, David T., McGue, Matthew, Segal, Nancy L., & Tellegen, Auke. (1990). Sources of human psychological differences: The Minnesota Study of Twins Reared Apart. *Science, 250*, 223–228.

Boulton, Michael, & Smith, Peter K. (1989). Issues in the study of children's rough-and-tumble play. In Marianne N. Bloch & Anthony D. Pellegrini (Eds.), *The ecological context of children's play.* Norwood, NJ: Ablex.

Bound, John, Schoenbau, Michael, & Waidmann, Timothy. (1996). Race differences in labor force attachment and disability status. *Gerontologist, 36*, 311–321.

Bower, Bruce. (1996). Alcohol-loving mice spur gene search. *Science News, 149*, 340.

Bower, T.G.R. (1989). *The rational infant: Learning in infancy.* New York: Freeman.

Bowerman, Melissa. (1982). Reorganizational processes in lexical and syntactic development. In Eric Wanner & Lila R. Gleitman (Eds.), *Language acquisition: The state of the art.* Cambridge, England: Cambridge University Press.

Bowlby, John. (1969). *Attachment* (Vol. 1). New York: Basic Books.

Bowman, James E., and **Murray, Robert F.** (1990). *Genetic variation and disorders in peoples of African origin.* Baltimore, MD: Johns Hopkins University Press.

Boxer, Andrew M., Gershenson, Harold P., & Offer, Daniel. (1984). Historical time and social change in adolescent experience. *New Directions for Mental Health Services, 22*, 83–95.

Boyer, Debra, & Fine, David. (1992). Sexual abuse as a factor in adolescent pregnancy and child maltreatment. *Family Planning Perspectives, 24*, 4–11, 19.

Boysson-Bardies, B., Halle, P., Sagart, L., & Durand, C. (1989). A crosslinguistic investigation of vowel formants in babbling. *Journal of Child Language, 16*, 1–17.

Brackbill, Yvonne, McManus, Karen, & Woodward, Lynn. (1988). *Medication in maternity: Infant exposure and maternal information.* Ann Arbor: University of Michigan Press.

Braddick, Oliver, & Atkinson, Janette. (1988). Sensory selectivity, attentional control, and cross-channel integration in early visual development. In Albert Yonas (Ed.), *Perceptual development in infancy.* Hillsdale, NJ: Erlbaum.

Brainerd, C. J., & Reyna, V.F. (1995). Learning rate, learning opportunities, and the development of forgetting. *Developmental Psychology, 31,* 251–262.

Bray, G.A. (1989). Obesity: Basic considerations and clinical approaches. *Disease-a-Month, 35,* 449–537.

Brayfield, April A. (1992). Employment resources and housework in Canada. *Journal of Marriage and the Family, 54,* 19–30.

Brayfield, April A. (1995). Juggling jobs and kids: The impact of employment schedules on fathers' caring for children. *Journal of Marriage and the Family, 57,* 321–332.

Breakey, G., & Pratt, B. (1991). Healthy growth for Hawaii's "Healthy Start": Toward a systematic statewide approach to the prevention of child abuse and neglect. *Zero to Three* (Bulletin of the National Center for Clinical Infant Programs), *11,* 16–22.

Bremner, J. Gavin. (1988). *Infancy.* Oxford, England: Blackwell.

Brent, David A., Johnson, Barbara A., Perper, Joshua, Connolly, John, Bridge, Jeff, Bartle, Sylvia, & Rather, Chris. (1994). Personality disorder, personality traits, impulsive violence, and complete suicide in adolescents. *Journal of the American Academy of Child and Adolescent Psychiatry, 33,* 1080–1086.

Bressler, Robin. (1993). Adverse drug reactions. In Robin Bressler & Michael D. Katz (Eds.), *Geriatric pharmacology.* New York: McGraw-Hill.

Bretherton, Inge. (1989). Pretense: The form and function of make-believe play. *Developmental Review, 9,* 383–401.

Bretherton, Inge. (1992). The origins of attachment theory: John Bowlby and Mary Ainsworth. *Developmental Psychology, 28,* 759–775.

Bretherton, Inge, & Beeghly, M. (1982). Talking about internal states: The acquisition of an explicit theory of mind. *Developmental Psychology, 18,* 906–921.

Bretherton, Inge, & Waters, Everett. (1985). Growing points of attachment theory and research. *Monographs of the Society for Research in Child Development, 50*(1–2, Serial No. 209).

Briere, J.M., & Elliott, D.M. (1994). Immediate and long-term impacts of child sexual abuse. *The Future of Children, 4,* 54–69.

Briggs, Freda, & Hawkins, Russell M.F. (1996). A comparison of the childhood experiences of convicted male child molesters and men who were sexually abused in childhood and claimed to be nonoffenders. *Child Abuse & Neglect, 20,* 221–233.

Bril, B. (1986). Motor development and cultural attitudes. In H.T.A. Whiting & M.G. Wade (Eds.), *Themes in motor development.* Dordrecht, The Netherlands: Martinus Nijhoff.

Brody, Jane E. (1993, August 11). Personal health: Skipping vaccinations puts children at risk. *The New York Times,* p. C11.

Brody, Jane E. (1994, April 6). The value of breast milk. *The New York Times,* p. C11.

Bronfenbrenner, Urie. (1977). Toward an experimental ecology of human development. *American Psychologist, 32,* 513–531.

Bronfenbrenner, Urie. (1979). *The ecology of human development: Experiments by nature and design.* Cambridge, MA: Harvard University Press.

Bronfenbrenner, Urie. (1986). Ecology of the family as a context for human development research perspectives. *Developmental Psychology,* 22, 723–742.

Bronfenbrenner, Urie. (1993). The ecology of cognitive development: Research models and fugitive findings. In Robert H. Wozniak & Kurt W. Fischer (Eds.), *Development in context: Acting and thinking in specific environments.* Hillsdale, NJ: Erlbaum.

Bronfenbrenner, Urie, & Ceci, Stephen J. (1994). Nature-nurture reconceptualized in developmental perspective: A bioecological model. *Psychological Review, 10,* 568–586.

Bronson, Gordon W. (1990). Changes in infants' visual scanning across the 2- to 14-week age period. *Journal of Experimental Child Psychology, 49,* 101–125.

Bronstein, Phyllis. (1984). Differences in mothers' and fathers' behaviors toward children: A cross-cultural comparison. *Developmental Psychology, 20,* 995–1003.

Brook, Judith S., Cohen, Patricia, Whiteman, Martin, & Gordon, Ann S. (1992). Psychosocial risk factors in the transition from moderate to heavy use or abuse of drugs. In Meyer Glantz & Roy Pickens (Eds.), *Vulnerability to drug abuse.* Washington, DC: American Psychological Association.

Brooks, George A., & Fahey, Thomas D. (1984). *Exercise physiology: Human bioenergetics and its application,* New York: Wiley.

Brooks-Gunn, Jeanne. (1991). Maturational timing variations in adolescent girls. In Richard M. Lerner, Ann C. Petersen, & Jeanne Brooks-Gunn (Eds.), *Encyclopedia of adolescence* (Vol. 2). New York: Garland.

Brooks-Gunn, Jeanne, Attie, I., Burrow, C., Rosso, J.T., & Warren, M.P. (1989). The impact of puberty on body and eating concerns in athletic and nonathletic contexts. *Journal of Early Adolescence, 9,* 269–290.

Brooks-Gunn, Jeanne, Klebanov, P.K., Liaw, F., & Spiker, D. (1993). Enhancing the development of low-birthweight, premature infants: Changes in cognition and behavior over the first three years. *Child Development, 64,* 736–753.

Brooks-Gunn, Jeanne, & Reiter, Edward O. (1990). The role of pubertal processes. In Shirley S. Feldman & Glenn R. Elliott (Eds.), *At the threshold: The developing adolescent.* Cambridge, MA: Harvard University Press.

Brooks-Gunn, Jeanne, Warren, M.P., Samelson, M., & Fox, R. (1986). Physical similarity of and disclosure of menarchal status to friends: Effects of grade and pubertal status. *Journal of Early Adolescence, 6,* 3–14.

Brown, A.L., Kane, M.J., & Echols, K. (1986). Young children's mental models determine analogical transfer across problems with a common goal structure. *Cognitive Development, 1,* 103–122.

Brown, B.B. (1990). Peer groups and peer cultures. In S.S. Feldman & G.R. Elliott (Eds.), *At the threshold: The developing adolescent.* Cambridge, MA: Harvard University Press.

Brown, B.B., Lohr, Mary Jane, & McClenahan, Eben L. (1986). Early adolescents' perception of peer pressure. *Journal of Early Adolescence, 6,* 139–154.

Brown, B.B., Mounts, N., Lamborn, S.D., & Steinberg, L. (1993). Parenting practices and peer group affiliation in adolescence. *Child Development, 64,* 467–482.

Brown, Steven M. (1993). Motivational effects on test scores of elementary students. *Journal of Educational Research, 86,* 133–136.

Brown, W. Ted, Jenkins, E.C., Gross, A.C., Chan, C.B., Krawczun, M.S., Duncan, C.J., Sklower, S.L., & Fisch, G.S. (1987). Further evidence for genetic heterogeneity in the fragile X syndrome. *Human Genetics, 75,* 311–321.

Brown, W. Ted, Zebrower, Michael, & Kieras, Fred J. (1990). Proferia: A genetic disease model of premature aging. In David E. Harrison (Ed.), *Genetic effects on aging II*. Caldwell, NJ: Telford.

Brownson, Ross C., Alavanja, Michael C.R., Hock, Edward T., & Loy, Timothy S. (1992). Passive smoking and lung cancer in nonsmoking women. *American Journal of Public Health, 82,* 1525–1530.

Bruck, Maggie, Ceci, Stephen J., Francoeur, Emmett, & Barr, Ronald. (1995). "I hardly cried when I got my shot!": Influencing children's reports about a visit to their pediatrician. *Child Development, 66,* 193–208.

Bruner, Jerome S. (1982). The organization of action and the nature of adult-infant transaction. In M. von Cranach & R. Harre (Eds.), *The analysis of action*. Cambridge, England: Cambridge University Press.

Brunswick, Ann F., Messerie, Peta A., & Titus, Stephen P. (1992). Predictive factors in adult substance abuse: A prospective study of African-American adolescents. In Meyer Glantz & Roy Pickens (Eds.), *Vulnerability to drug abuse*. Washington, DC: American Psychological Association.

Bruun, Ruth Dowling, & Brunn, Bertel. (1994). *A mind of its own: Tourette's syndrome—a story and a guide*. New York: Oxford University Press.

Bryan, E. (1992). *Twins, triplets, and more*. New York: St. Martin's Press.

Bryan, Janet, & Luszcz, Mary A. (1996). Speed of information processing as a mediator between age and free-recall performance. *Psychology and Aging, 11,* 3–9.

Bryant, B.K. (1985). The neighborhood walk: Sources of support in middle childhood. *Monographs of the Society for Research in Child Development, 50* (3, Serial No. 210).

Bryant, W. Keith, & Zick, Cathleen D. (1996). An examination of parent-child shared time. *Journal of Marriage and the Family, 58,* 227–238.

Buhrmester, D., Camparo, L., Christensen, A., Gonzalez, L.S., & Hinshaw, S.P. (1992). Mothers and fathers interacting in dyads and triads with normal and hyperactive sons. *Developmental Psychology, 28,* 500–509.

Bumpass, Larry L., Sweet, James A., & Cherlin, Andrew. (1991). The role of cohabitation in declining rates of marriage. *Journal of Marriage and the Family, 53,* 913–927.

Bunker, John P., Frazer, Howard S., & Mosteller, Frederick. (1995). The role of medical care in determining health: Creating an inventory of benefits. In Benjamin C. Amick, III, Sol Levine, Alvin R. Tarlov, & Diana Chapman Walsh (Eds.), *Society and health*. New York: Oxford University Press.

Burchinal, Margaret, Lee, Marvin, & Ramey, Craig. (1989). Type of day care and preschool intellectual development in disadvantaged children. *Child Development, 60,* 128–137.

Burelson, Brant R. (1982). The development of comforting communication skills in childhood and adolescence. *Child Development, 53,* 1578–1588.

Burke, D.M., MacKay, D.G., Worthley, J.S., & Wade, E. (1991). On the tip of the tongue: What causes word finding failures in young and older adults? *Journal of Memory and Language, 30,* 542–579.

Burns, Ailsa. (1992). Mother-headed families: An international perspective and the case of Australia. *Society for Research in Child Development: Social Policy Report, 6,* 1–22.

Burns, Ailsa, & Scott, Cath. (1994). *Mother-headed families and why they have increased*. Hillsdale, NJ: Erlbaum.

Burton, Linda M. (1995). Intergenerational patterns of providing care in African-American families with teenage childbearers: Emergent patterns in an ethnographic study. In Vern L. Bengtson, K. Warner Schaie, & Linda M. Burton (Eds.), *Adult intergenerational relations: Effects of societal change*. New York: Springer.

Burton, Linda M., Dilworth-Anderson, Peggye, & Bengston, Vern L. (1991). Creating culturally relevant ways of thinking about diversity and aging. *Generations, 15*(4), 67–72.

Buss, A.H. (1991). The EAS theory of temperament. In J. Strelau & A. Angleitner (Eds.), *Explorations of temperament*. New York: Plenum Press.

Buss, David M. (1994). *The evolution of desire: Strategies of human mating*. New York: Basic Books.

Busse, Ewald W. (1985). Normal aging: The Duke longitudinal studies. In M. Bergener, Marco Ermini, & H.B. Stahelin (Eds.), *Thresholds in aging*. London: Academic Press.

Bussey, K., & Bandura, A. (1992). Self-regulatory mechanisms governing gender development. *Child Development, 63,* 1236–1250.

Butler, J., & Rovee-Collier, Carolyn K. (1989). Contextual gating of memory retrieval. *Developmental Psychobiology, 22,* 533–552.

Butler, Robert N., & Golding, J. (1986). *From birth to five: A study of the health and behaviour of Britain's 5-year-olds*. Oxford, England: Pergamon Press.

Butler, Robert N., Lewis, Myrna, & Sunderland, Trey. (1991). *Aging and mental health: Positive psychosocial and biomedical holdings* (4th ed.). New York: Merrill.

Butterfield, Fox. (1996, August 19). Barrooms' decline underlies a drop in adult killings. *New York Times*, p. A1.

Byne, William. (1994, May). The biological evidence challenged. *Scientific American, 270,* 50–55.

Byrne, Joseph, Ellsworth, Christine, Bowering, Elizabeth, & Vincer, Michael. (1993). Language development in low birth weight infants: The first two years. *Developmental and Behavioral Pediatrics, 14,* 21–27.

Byrnes, J.P. (1988). Formal operations: A systematic reformulation. *Developmental Review, 8,* 66–87.

Cahalan, Don. (1991). *An ounce of prevention: Strategies for solving tobacco, alcohol, and drug problems*. San Francisco: Jossey-Bass.

Cairns, Robert B. (1994). The making of a developmental science: The contributions and intellectual heritage of James Mark Baldwin. In Ross D. Parke, Peter A. Ornstein, John J. Rieser, & Carolyn Zahn-Waxler (Eds.), *A century of developmental psychology*. Washington, DC: American Psychological Association.

Cairns, Robert B., & Cairns, Beverly D. (1994). *Lifelines and risks: Pathways of youth in our time*. Cambridge, England: Cambridge University Press.

Calkins, Susan D. (1994). Origins and outcomes of individual differences in emotional regulation. *Monographs of the Society for Research in Child Development, 59*(2-3, Serial No. 240), 53–72.

Call, Vaughn, Sprecher, Susan, & Schwartz, Pepper. (1995). Incidence and frequency of marital sex in a national sample. *Journal of Marriage and the Family, 57,* 639–652.

Callahan, Daniel. (1990). Afterword. In Paul Homer & Martha Holstein (Eds.), *A good old age*? New York: Simon & Schuster.

Campbell, Ruth, & Sais, Efisia. (1995). Accelerated metalinguistic (phonological) awareness in bilingual children. *British Journal of Developmental Psychology, 13*, 61–68.

Campos, Joseph J., Barrett, Karen C., Lamb, Michael L., Goldsmith, H. Hill, & Stenberg, Craig. (1983). Socioemotional development. In Paul H. Mussen (Ed.), *Handbook of child psychology: Vol. 2. Infancy and developmental psychobiology*. New York: Wiley.

Canetto, Silvia Sara. (1992). Gender and suicide in the elderly. *Suicide and Life-Threatening Behavior, 22*, 80–96.

Cantwell, Dennis P., & Baker, Lorian. (1991). Association between attention deficit-hyperactivy disorder and learning disorders. *Journal of Learning Disabilities, 24*, 88–95.

Caplan, N., Choy, M.H., & Whitmore, J.K. (1992). Indochinese refugee families and academic achievement. *Scientific American, 267*, 36–42.

Cappelleri, J.C., Eckenrode, J., & Powers, J.L. (1993). The epidemiology of child abuse: Findings from the Second National Incidence and Prevalence Study of Child Abuse and Neglect. *American Journal of Public Health, 83*, 1622–1624.

Carey, William B., & McDevitt, Sean C. (1978). Stability and change in individual temperament diagnoses from infancy to early childhood. *Journal of the American Academy of Child Psychiatry, 17*, 331–337.

Carlson, Bruce M. (1994). *Human embryology and developmental biology*. St. Louis, MO: Mosby.

Carlson, Karen J., Eisenstat, Stephanie A., & Ziporyn, Terra. (1996). *The Harvard guide to women's health*. Cambridge, MA: Harvard University Press.

Carlson, Michelle C., Hasher, Lynn, Connelly, S. Lisa, & Zacks, Rose T. (1995). Aging, distraction, and the benefits of predictable location. *Psychology and Aging 10*, 427–436.

Carmen, Elaine. (1989). Family violence and the victim-to-patient process. In Leah J. Dickstein & Carol C. Nadelson (Eds.), *Family violence: Emerging issues of national crisis*. Washington, DC: American Psychiatric Press.

Carnegie Council on Adolescent Development. (1989). *Turning points: Preparing American youth for the 21st century*. New York: Carnegie Corporation.

Caro, Francis G., & Bass, Scott A. (1995). Increasing volunteering among older people. In S.A. Bass (Ed.), *Older and active: How Americans over age 55 are contributing to society*. New Haven, CT: Yale University Press.

Caro, Francis G., Bass, Scott A., & Chen, Yung-Ping. (1993). Introduction: Achieving a productive aging society. In Scott A. Bass, Francis G. Caro, & Yung-Ping Chen (Eds.), *Achieving a productive aging society*. Westport, CT: Auburn House.

Caron, Albert J., & Caron, Rose F. (1981). Processing of relational information as an index of infant risk. In S.L. Friedman & M. Sigman (Eds.), *Preterm birth and psychological development*. New York: Academic Press.

Carraher, T.N., Carraher, D.W., & Schliemann, A.D. (1985). Mathematics in the streets and in schools. *British Journal of Developmental Psychology, 3*, 21–29.

Carraher, T.N., Schliemann, A.D., & Carraher, D.W. (1988). Mathematical concepts in everyday life. In G.B. Saxe & M. Gearhart (Eds.), *New directions for child development: Vol. 41. Children's mathematics*. San Francisco: Jossey-Bass.

Carroll, J.B. (1993). *Human cognitive abilities: A survey of factor-analytic studies*. Cambridge, England: Cambridge University Press.

Carter, D.B., & Middlemiss, W.A. (1992). The socialization of instrumental competence in families in the United States. In J.L. Roopnarine & D.B. Carter (Eds.), *Annual advances in applied developmental psychology: Vol. 5. Parent-child socialization in diverse cultures*. Norwood, NJ: Ablex.

Caruso, Richard E. (1992). *Mentoring and the business environment: Asset or liability*? Newcastle-upon-Tyne: Athenaeum Press.

Cascio, Wayne F. (1995). Whither industrial and organizational psychology in a changing world of work? *American Psychologist, 50*, 928–939.

Case, Robbie. (1985). *Intellectual development: Birth to adulthood*. Orlando, FL: Academic Press.

Casey, Rosemary, Levy, Susan E., Brown, Kimberly, & Brooks-Gunn, J. (1992). Impaired emotional health in children with mild reading disability. *Journal of Developmental and Behavioral Pediatrics, 13*, 256–260.

Cash, Thomas F., & Henry, Patricia. (1995). Women's body images: The results of a national survey in the U.S.A. *Sex Roles, 33*, 19–28.

Caskey, C.T. (1992). DNA-based medicine: Prevention and therapy. In D.J. Kevles & L. Hood (Eds.), *The code of codes: Scientific and social issues in the Human Genome Project*. Cambridge, MA: Harvard University Press.

Caspi, Avshalom, Elder, Glen H., & Bem, Daryl J. (1988). Moving away from the world: Life-course patterns of shy children. *Developmental Psychology, 24*, 824–831.

Caspi, Avshalom, Elder, Glen H., & Herbener, Ellen S. (1990). Childhood personality and the prediction of life-course patterns. In Lee N. Robins & Michael Rutter (Eds.), *Straight and devious pathways from childhood to adulthood*. Cambridge, England: Cambridge University Press.

Caspi, Avshalom, Herbener, Ellen S., & Ozer, Daniel J. (1992). Shared experiences and the similarities of personalities: A longitudinal study of married couples. *Journal of Personality and Social Psychology, 62*, 281–291.

Caspi, Avshalom, & Moffitt, Terrie. (1991). Individual differences are accentuated during periods of social change: The sample case of girls at puberty. *Journal of Personality and Social Psychology, 61*, 157–168.

Caspi, Avshalom, Lynam, Donald, Moffit, Terrie, & Silva, Phil A. (1993). Unraveling girls' delinquency: Biological, dispositional, and contextual contributions to adolescent misbehavior. *Developmental Psychology, 29*, 19–30.

Cassel, Christine K. (1996). Physician-assisted suicide: Progress or peril? In David C. Thomasma & Thomasine Kushner (Eds.), *Birth to death*. Cambridge, England: Cambridge University Press.

Cates, Willard, Jr. (1995). Sexually transmitted diseases. In B.P. Sachs, R. Beard, & E. Papiernik (Eds.), *Reproduction health care for women and babies*. Oxford, England: Oxford University Press.

Cattell, R.B. (1963). Theory of fluid and crystalized intelligence: A critical experiment. *Journal of Educational Psychology 54*, 1–22.

Ceci, Stephen J. (1990). *On intelligence . . more or less: A bioecological treatise on intellectual development*. Englewood Cliffs, NJ: Prentice Hall.

Ceci, Stephen J. (1991). How much does schooling influence cognitive and intellectual development? *Developmental Psychology, 27*, 703–722.

Ceci, Stephen J., & Bruck, M. (1993). The suggestibility of the child witness. *Psychological Bulletin, 113*, 403–439.

Ceci, Stephen J., & Cornelius, Steven W. (1990). Commentary. *Human Development, 33*, 198–201.

Ceci, Stephen J., Toglia, Michael P., & Ross, David F. (1990). The suggestibility of preschoolers' recollections: Historical perspectives on current problems. In Robyn Fivush & Judith A. Hudson (Eds.), *Knowing and remembering in young children*. Cambridge, England: Cambridge University Press.

Cefalo, Robert C., & Moos, Merry-K. (1988). *Preconceptual health promotion*. Rockville, MD: Aspen.

Centers for Disease Control. (1992). *Setting the national agenda for injury control in the 1990s*. Washington, DC: Department of Health and Human Services, Public Health Service.

Centers for Disease Control. (1993). *Annual report: 1992*. Division of STD/HIV prevention. Atlanta, GA: Author.

Centers for Disease Control and Prevention. (1990, April). Statewide prevalence of illicit drug use by pregnant women. *Morbidity and Mortality Weekly Report, 39*.

Centers for Disease Control and Prevention. (1993, December 31). Recommendations of the International Task Force for Disease Eradication. *Morbidity and Mortality Weekly Report, 42*, 17.

Centers for Disease Control and Prevention. (1994, January 28). General recommendations on immunization: Recommendations of the Advisory Committee on Immunization Practices. *Morbidity and Mortality Weekly Report, 43*.

Centers for Disease Control and Prevention. (1994, March 4). Health risk behaviors among adolescents who do and do not attend school—United States, 1992. *Morbidity and Mortality Weekly Report, 43*, 129–132.

Centers for Disease Control and Prevention. (1994, April 22). Suicide contagion and the reporting of suicide. *Morbidity and Mortality Weekly Report, 43* (No. RR–6), 13–18.

Centers for Disease Control and Prevention. (1994, April 29). Zidovudine for the prevention of HIV transmission from mother to infant. *Morbidity and Mortality Weekly Report, 43*, 285–287.

Centers for Disease Control and Prevention. (1994, August 19). Changes in the cigarette brand preferences of adolescent smokers—United States, 1989–1993. *Morbidity and Mortality Weekly Report, 43*, 577–581.

Centers for Disease Control and Prevention. (1994, October 21). Reasons for tobacco use and symptoms of nicotine withdrawal among adolescent and young adult tobacco users—United States, 1993. *Morbidity and Mortality Weekly Report, 43*, 745–750.

Centers for Disease Control and Prevention. (1995). *Health United States 1994*. Hyattsville, MD: U.S. Department of Health and Human Services.

Centers for Disease Control and Prevention. (1995). Measles—United States, 1994. *Morbidity and Mortality Weekly Report, 44*, 486–487, 493–494.

Centers for Disease Control and Prevention. (1995). Recommended childhood immunization schedule—United States, 1995. *Morbidity and Mortality Weekly Report, 44*, 2.

Centers for Disease Control and Prevention. (1995, February 10). Update: AIDS among women—United States, 1994. *Morbidity and Mortality Weekly Report, 44*.

Centers for Disease Control and Prevention. (1995, March 24). Youth risk behavior surveillance—United States, 1993. *Morbidity and Mortality Weekly Report, 44* (No. SS-1), 1–56.

Centers for Disease Control and Prevention. (1996, October 11). Sudden Infant Death Syndrome—1983–1994. *Morbidity and Mortality Weekly Report, 45*, 859–865.

Chalfant, J.C. (1989). Learning disabilities: Policy issues and promising approaches. *American Psychologist, 44*, 392–398.

Chandler, Lynette A. (1990). Neuromotor assessment. In Elizabeth D. Gibbs & Douglas M. Teti (Eds.), *Interdisciplinary assessment of infants*. Baltimore: Brookes.

Chao, Ruth K. (1994). Beyond parental control and authoritarian parenting style: Understanding Chinese parenting through the cultural notion of training. *Child Development, 65*, 1111–1119.

Chappell, Patricia A., & Steitz, Jean A. (1993). Young children's human figure drawings and cognitive development. *Perceptual and Motor Skills, 76*, 611–617.

Charlish, Anne. (1991). *Birth-tech: Tests and technology in pregnancy and birth*. New York: Facts on File.

Charness, Neil. (1986). Expertise in chess, music, and physics: A cognitive perspective. In L.K. Obler & D.A. Fein (Eds.), *The neuropsychology of talent and special abilities*. New York: Guilford.

Charness, Neil. (1989). Age and expertise: Responding to Talland's challenge. In Leonard W. Poon, David C. Rubin, & Barbara A. Wilson (Eds.). *Everyday cognition in adulthood and later life*. Cambridge, England: Cambridge University Press.

Charness, Neil, & Bosman, Elizabeth A. (1990). Expertise and aging: Life in the lab. In Thomas M. Hess (Ed.), *Aging and cognition: Knowledge, organization and utilization*. Amsterdam: North-Holland/Elsevier.

Chen, Kevin, & Kandel, Denise. (1995). The natural history of drug use from adolescence to the mid-thirties in a general population sample. *American Journal of Public Health, 85*, 41–47.

Chen, S.-J., & Miyake, K. (1986). Japanese studies of infant development. In H. Stevenson, H. Azuma, & K. Hakuta (Eds.), *Child development and education in Japan*. New York: Freeman.

Chen, Xinyin, Rubin, Kenneth H., & Zhen-yun, Li. (1995). Social functioning and adjustment in Chinese children: A longitudinal study. *Developmental Psychology, 31*, 531–539.

Cherlin, Andrew, & Furstenberg, Frank F., Jr. (1986). *The new American grandparent: A place in the family, a life apart*. New York: Basic Books.

Cherry, Katie E., & Stadler, Michael A. (1995). Implicit learning of a nonverbal sequence in younger and older adults. *Psychology and Aging, 10*, 379–394.

Chess, Stella, & Thomas, Alexander. (1990). Continuities and discontinuities in development. In Lee N. Robins & Michael Rutter (Eds.), *Straight and devious pathways from childhood to adulthood*. New York: Cambridge University Press.

Chess, Stella, Thomas, Alexander, & Birch, Herbert. (1965). *If your child is a person*. New York: Viking Press.

Chevan, Albert. (1996). As cheaply as one: Cohabitation in the older population. *Journal of Marriage and the Family, 58,* 656–667.

Chi, Micheline T.H. (1978). Knowledge structures and memory development. In R.S. Siegler (Ed.), *Children's thinking: What develops?* Hillsdale, NJ: Erlbaum.

Chi, Micheline T.H., Glaser, R., & Farr, M.J. (1988). *The nature of expertise.* Hillsdale, NJ: Hove & London.

Chi, Micheline T.H., Hutchinson, J.E., & Robin, A.F. (1989). How inferences about novel domain-related concepts can be constrained by structured knowledge. *Merrill-Palmer Quarterly, 35,* 27–62.

Chickering, Arthur W. (1981). Conclusion. In Arthur W. Chickering (Ed.), *The modern American college: Responding to the new realities of diverse students and a changing society.* San Francisco: Jossey-Bass.

Children's Defense Fund. (1994). *The state of America's children yearbook 1994.* Washington, DC: Publications Department.

Chiriboga, D.A. & Catron, L.A. (1992). *Divorce: Crisis, challenge or relief?* New York: New York University Press.

Chiriboga, David. (1992). Paradise lost: Stress in the modern age. In May L. Wykle, Eva Kahana, & Jerome Kowal (Eds.), *Stress and health among the elderly.* New York: Springer.

Chomitz, Virginia Rall, Cheung, Lilian W.Y., & Lieberman, Ellice. (1995). The role of lifestyle in preventing low birth weight. *The Future of Children: Low Birth Weight, 5,* 121–138.

Chomsky, Noam. (1968). *Language and mind.* New York: Harcourt, Brace, World.

Chomsky, Noam. (1980). *Rules and representations.* New York: Columbia University Press.

Christensen, Myra J., Brayden, Robert M., Dietrich, Mary S., McLaughlin, F. Joseph, Sherrod, Kathryn B., & Altemeier, William A. (1994). The prospective assessment of self-concept in neglectful and physically abusive low income mothers. *Child Abuse and Neglect, 18,* 225–232.

Christophersen, Edward R. (1989). Injury control. *American Psychologist, 44,* 237–241.

Chung, Chin S., Villafuerte, Arnold, Wood, William, & Lew, Ruth. (1992). Trends in prevalence of behavioral risk factors: Recent Hawaiian experience. *American Journal of Public Health, 82,* 1544–1546.

Cicchetti, Dante. (1990). The organization and coherence of socioemotional, cognitive, and representational development: Illustrations through a developmental psychopathology perspective on Down syndrome and child maltreatment. In R.A. Thompson (Ed.), *Nebraska Symposium on Motivation: Vol. 36. Socioemotional development.* Lincoln: University of Nebraska Press.

Cicchetti, Dante. (1991). Defining psychological maltreatment: Reflections and future directions (Editorial). *Development and Psychopathology, 3,* 1–2.

Cicchetti, Dante. (1993). Developmental psychopathology: Reactions, reflections, projections. *Developmental Review, 13,* 471–502.

Cicchetti, Dante, & Beeghly, Marjorie. (1990). *Children with Down syndrome: A developmental perspective.* Cambridge, England: Cambridge University Press.

Cicchetti, Dante, & Carlson, Vicki. (Eds.). (1989). *Child maltreatment: Theory and research on the causes and consequences of child abuse and neglect.* Cambridge, England: Cambridge University Press.

Cicchetti, Dante, Toth, S.L., & Hennessy, K. (1993). Child maltreatment and school adaptation: Problems and promises. In Dante Cicchetti & S.L. Toth (Eds.), *Advances in applied developmental psychology series: Vol. 8. Child abuse, child development, and social policy.* Norwood, NJ: Ablex.

Cicirelli, Victor G. (1995). *Sibling relationships across the life span.* New York: Plenum Press.

Cillessen, A.H.N., van Ijzendoorn, H.W., van Lieshout, C.F.M., & Hartup, W.W. (1992). Heterogeneity among peer-rejected boys: Subtypes and stabilities. *Child Development, 63,* 893–905.

Clark, Eve V. (1982). The young word maker: A case study of innovation in the child's lexicon. In Eric Wanner & Lila R. Gleitman (Eds.), *Language acquisition: The state of the art.* Cambridge, England: Cambridge University Press.

Clark, Eve V. (1990). On the pragmatics of contrast. *Journal of Child Language, 17,* 417–431.

Clark, Jane E., & Phillips, Sally J. (1985). A developmental sequence of the standing long jump. In Jane E. Clark & James H. Humphrey (Eds.), *Motor development: Current selected research.* Princeton, NJ: Princeton Book Company.

Clark, Robert D. (1983). *Family life and school achievement: Why poor black children succeed or fail.* Chicago: University of Chicago Press.

Clarke-Stewart, K. Alison. (1989). Infant day care: Maligned or malignant? *American Psychologist, 44,* 266–273.

Clarke-Stewart, K. Alison, Gruber, Christian P., & Fitzgerald, Linda May. (1994). *Children at home and in day care.* Hillsdale, NJ: Erlbaum.

Clarkson, Marsha G., & Berg, W. Keith. (1983). Cardiac orienting and vowel discrimination in newborns: Crucial stimulus parameters. *Child Development, 54,* 162–171.

Clarkson, Marsha G., Clifton, Rachel K., & Morrongiello, Barbara A. (1985). The effects of sound duration on newborns' head orientation. *Journal of Experimental Child Psychology, 39,* 20–36.

Clemens, Andra W., & Axelson, Leland J. (1985). The not-so-empty nest: The return of the fledgling adult. *Family Relations, 34,* 259–264.

Clinchy, Blythe McVicker. (1993). Ways of knowing and ways of being: Epistemological and moral development in undergraduate women. In Andrew Garrod (Ed.), *Approaches to moral development: New research and emerging themes.* New York: Teachers College Press.

Clinchy, Blythe McVicker, & Zimmerman, Claire. (1982). Epistemology and agency in the development of undergraduate women. In Pamela J. Perun (Ed.), *The undergraduate woman: Issues in educational equity.* Lexington, MA: Lexington Books.

Coe, Christopher, Kayashi, Kevin T., & Levine, Seymour. (1988). Hormones and behavior at puberty: Activation or concatenation? In Megan R. Gunnar & W. Andrew Collins (Eds.), *Development during the transition to adolescence.* Hillsdale, NJ: Erlbaum.

Coelho, Elizabeth. (1991). Social integration of immigrant and refugee children. In J. Porter (Ed.), *New Canadian voices.* Toronto: Wall & Emerson.

Coffey, C., Wilkinson, W., Paraskos, I., Soady, S., Sullivan, R., Patterson, L., Figiel, W., Webb, M., Spritzer, C., & Djang, W. (1992). Quantitative cerebral anatomy of the aging human brain: A cross-sectional study using magnetic resonance imaging. *Neurology, 42,* 527–536.

Cohen, L.B., & Oakes, L.M. (1993). How infants perceive a simple causal event. *Developmental Psychology, 29,* 421–433.

Cohen, S., Tyrrell, D.A.J., & Smith, A.P. (1993). Life events, perceived stress, negative affect and susceptibility to the common cold. *Journal of Personality and Social Psychology, 64,* 131–140.

Cohn, Jeffrey F., & Tronick, Edward Z. (1987). Mother-infant face to face interaction: The sequence of dyadic states at 3, 6, and 9 months. *Developmental Psychology, 23,* 68–77.

Cohn, Lawrence D.S., & Adler, Nancy E. (1992). Female and male perception of ideal body shapes. *Psychology of Women Quarterly, 16,* 69–79.

Coie, John D., & Cillessen, A.H.N. (1993). Peer rejection: Origins and effects on children's development. *Current Directions in Psychological Science, 2,* 89–92.

Coie, John D., Dodge, Kenneth A., Terry, Robert, & Wright, Virginia. (1991). The role of aggression in peer relations: An analysis of aggression episodes in boys' play groups. *Child Development, 62,* 812–826.

Coie, John D., & Koeppl, G.K. (1990). Adapting intervention to the problems of aggressive and disruptive rejected children. In S.R. Asher & J.D. Coie (Eds.), *Peer rejection in childhood.* Cambridge, England: Cambridge University Press.

Coie, John D., Lochman, J.E., Terry, R., & Hyman, C. (1992). Predicting early adolescent disorder from childhood aggression and peer rejection. *Journal of Consulting and Clinical Psychology, 60,* 783–792.

Coke, Marguerite M., & Twaite, James A. (1995). *The black elderly: Satisfaction and quality of later life.* New York: Haworth Press.

Colby, Anne, Kohlberg, Lawrence, Gibbs, John, & Lieberman, Marcus. (1983). A longitudinal study of moral development. *Monographs of the Society for Research in Child Development, 48*(1–2, Serial No. 200).

Cole, Michael. (1992). Culture in development. In M.H. Bornstein & M.E. Lamb (Eds.), *Developmental psychology: An advanced textbook* (3rd ed.). Hillsdale, NJ: Erlbaum.

Cole, Pamela M., Barrett, Karen C., & Zahn-Waxler, Carolyn. (1992). Emotion displays in two-year-olds during mishaps. *Child Development, 63,* 314–324.

Coleman, J.C., & Hendry, L. (1990). *The nature of adolescence* (2nd ed.). London: Routledge.

Coleman, M., & Ganong, L.H. (1991). Remarriage and stepfamily research in the 1980s: Increased interest in an old family form. In A. Booth (Ed.), *Contemporary families: Looking forward, looking back.* Minneapolis, MN: National Council on Family Relations.

Coles, Robert. (1990). *The spiritual life of children.* Boston: Houghton Mifflin.

Colin, Virginia L. (1996). *Human attachment.* Philadelphia: Temple University Press.

Collinge, John, Sidle, Katie C., Meads, Julie, Ironside, James, & Hill, Andrew F. (1996). Molecular analysis of prion strain variation and the aetiology of "new variant" CJD. *Nature (London) 383,* 685–690.

Collins, Claire. (1995, May 11). Spanking is becoming the new don't. *New York Times,* p. C8.

Collins, J.W., & Shay, D.K. (1994). Prevalence of low birth weight among Hispanic infants with United States-born and foreign-born mothers: The effect of urban poverty. *American Journal of Epidemiology, 139,* 184–192.

Collins, W. Andrew (Ed.). (1984). *Development during middle childhood: The years from 6 to 12.* Washington, DC: National Academy Press.

Collins, W. Andrew. (1990). Parent-child relationships in the transition to adolescence: Continuity and change in interaction, affect, and cognition. In R. Montemayor, G. Adams, & T. Gullotta (Eds.), *From childhood to adolescence: A transitional period? Advances in adolescent development: Vol. 2. The transition from childhood to adolescence.* Beverly Hills, CA: Sage.

Collins, W. Andrew, & Russell, G. (1991). Mother-child and father-child relationships in middle childhood and adolescence: A developmental analysis. *Developmental Psychology, 11,* 99–136.

Collopy, Bart, Boyle, Philip, & Jennings, Bruce. (1991, March–April). New directions in nursing home ethics. *Hastings Center Report, 21,* 1–15 (Special Supplement).

Commons, Michael L., & Richards, Francis A. (1984). A general model of stage theory. In Michael L. Commons, Francis A. Richards, & Cheryl Armon (Eds.), *Beyond formal operations.* New York: Praeger.

Compas, Bruce E., Banez, Gerard A., Malcarne, Vanessa, & Worsham, Nancy. (1991). Perceived control and coping with stress: A developmental perspective. *Journal of Social Issues, 47,* 23–34.

Comstock, George W. (1994). Tuberculosis: Is the past once again prologue? *American Journal of Public Health, 84,* 1729–1731.

Coni, Nicholas K. (1991). Ethical dilemmas faced in dealing with the aged sick. In Frederick C. Ludwig (Ed.), *Life span extension.* New York: Springer.

Coni, Nicholas K., Davison, William, & Webster, Stephen. (1992). *Aging: The facts.* Oxford, England: Oxford University Press.

Connidis, I.A. (1994). Sibling support in older age. *Journal of Gerontology: Social Sciences, 49,* S309–S317.

Connor, J.M., & Ferguson-Smith, M.A. (1991). *Essential medical genetics.* Oxford, England: Blackwell Scientific Publications.

Conrad, Marilyn, & Hammen, Constance. (1993). Protective and resource factors in high- and low-risk children: A comparison of children with unipolar, bipolar, medically ill, and normal mothers. *Development and Psychopathology, 5,* 593–607.

Cooney, George H., Bell, A., McBride, W., & Carter, C. (1989). Low-level exposures to lead: The Sydney Lead Study. *Developmental Medicine and Child Neurology, 31,* 640–649.

Coontz, Stephanie. (1992). *The way we never were: American families and the nostalgia trap.* New York: Basic Books.

Cooper, Richard, Rotime, Charles, Ataman, Susan, McGee, Daniel, Osotimehin, Babatunde, Kadiri, Soloman, Muna, Walinjom, Kingue, Samuel, Fraser, Henry, Forrester, Terrence, Bennett, Franklyn, & Wilks, Rainford. (1997). The prevalence of hypertension in seven populations of West African Origin. *American Journal of Public Health, 87,* 160–168.

Cooper, Ralph L., Goldman, Jerome M., & Harbin, Thomas J. (1991). *Aging and environmental toxicology.* Baltimore: Johns Hopkins University Press.

Cooper, R.O. (1993). The effect of prosody on young infants' speech perception. In C. Rovee-Collier & L.P. Lipsitt (Eds.), *Advances in infancy research* (Vol. 8). Norwood, NJ: Ablex.

Cooper, R.P., & Aslin, R.N. (1990). Preference for infant-directed speech in the first month after birth. *Child Development, 61,* 1584–1595.

Copeland, L., Wolraich, M., Lindgren, S., Milich, R., & Woolson, R. (1987). Pediatricians' reported practices in the assessment and treatment of attention deficit disorders. *Developmental and Behavioral Pediatrics, 8,* 191–197.

Cornelius, Steven W., Kenny, Sheryl, & Caspi, Avshalom. (1989). Academic and everyday intelligence in adulthood: Conceptions of self and ability tests. In Jan D. Sinnott (Ed.), *Everyday problem solving: Theory and applications.* New York: Praeger.

Corsaro, W.A. (1985). *Friendship and peer culture in the early years.* Norwood, NJ: Ablex.

Corse, S.J., Schmid, K., & Trickett, P.K. (1990). Social network characteristics of mothers in abusing and nonabusing families and their relationships to parenting beliefs. *Journal of Community Psychology, 18,* 44–59.

Costa, Paul T., & McCrae, Robert R. (1989). Personality continuity and the changes of adult life. In Martha Storandt & Gary R. Vandenbos (Eds.), *The adult years: Continuity and change.* Washington, DC: American Psychological Association.

Costa, Paul T., Zonderman, A.B., McCrae, R.R., Coroni-Huntley, J., Locke, B.Z., & Barbano, H.E. (1987). Longitudinal analyses of psychological well-being in a national sample: Stability of mean levels. *Journal of Gerontology, 42,* 50–55.

Cotman, Carl W., & Neeper, Shawne. (1996). Activity-dependent plasticity and the aging brain. In Edward L. Schneider & John W. Rowe (Eds.), *Handbook of the biology of aging* (4th ed.). San Diego, CA: Academic Press.

Cotten, N.U., Resnick, J., Browne, D.C., Martin, S.L., McCarraher, D.R., & Woods, J. (1994). Aggression and fighting behavior among African-American adolescents: Individual and family factors. *American Journal of Public Health, 84,* 618–622.

Cowan, C.P., & Cowan, P.A. (1992). *When partners become parents.* New York: Basic Books.

Coward, Raymond T., Horne, Claydell, & Dwyer, Jeffrey W. (1992). Demographic perspectives on gender and family caregiving. In Jeffrey W. Dwyer & Raymond T. Coward (Eds.), *Gender, families and elder care.* Newbury Park, CA: Sage.

Cowen, Emory I., Wyman, Peter A., & Work, William C. (1992). The relationship between retrospective reports of early child temperament and adjustment at ages 10–12. *Journal of Abnormal Child Psychology, 20,* 39–50.

Crabb, Peter B., & Bielawski, Dawn. (1994). The social representation of material culture and gender in children's books. *Sex Roles, 30,* 69–70.

Cramer, Phebe, & Skidd, Jody E. (1992). Correlates of self-worth in preschoolers: The role of gender-stereotyped styles of behavior. *Sex Roles, 26,* 369–390.

Crews, Frederick. (1996). The verdict on Freud. *Psychological Science, 7,* 63–68.

Crimmins, E.M., & Ingegneri, D.G. (1990). Interaction and living arrangements of older parents and their children: Past trends, present determinants, future implications. *Research on Aging, 2,* 3–35.

Cristofalo, Vincent J. (1996). Ten years later: What have we learned about human aging from studies of cell cultures? *Gerontologist, 36,* 737–741.

Crittenden, Patricia M. (1992). The social ecology of treatment: Case study of a service system for maltreated children. *American Journal of Orthopsychiatry, 62,* 22–34.

Crittenden, Patricia M., Claussen, Angelika H., & Sugarman, David B. (1994). Physical and psychological maltreatment in middle childhood and adolescence. *Development and Psychopathology, 6,* 145–164.

Crnic, Keith A., & Booth, Carolyn L. (1991). Mothers' and fathers' perceptions of daily hassles of parenting across early childhood. *Journal of Marriage and the Family, 53,* 1042–1050.

Crockenberg, S., & Litman, C. (1990). Autonomy as competence in 2-year-olds: Maternal correlates of child defiance, compliance, and self-assertion. *Developmental Psychology, 26,* 961–971.

Crockin, Susan L. (1993). The legal response to the new reproductive technologies. In Machelle M. Seibel, Ann A. Kiessling, Judith Bernstein, & Susan R. Levin (Eds.), *Technology and infertility: Clinical, psychosocial, legal and ethical aspects.* New York: Springer-Verlag.

Crohan, Susan E., & Antonucci, Toni C. (1989). Friends as a source of social support in old age. In Rebecca G. Adams & Rosemary Bleiszner (Eds.), *Older adult friendship.* Newbury Park, CA: Sage.

Cromer, B.A., Tarnowski, K.J., Stein, A.M., Harton, P., & Thornton, D.J. (1990). The school breakfast program and cognition in adolescents. *Journal of Developmental and Behavioral Pediatrics, 11,* 295–300.

Cromwell, Rue L. (1993). Searching for the origins of schizophrenia. *Psychological Science, 4,* 276–279.

Cross, W.W., Jr. (1991). *Shades of black: Diversity in African-American identity.* Philadelphia: Temple University Press.

Crowell, J.A., & Feldman, S.S. (1988). Mothers' internal models of relationships and children's behavioral and developmental status: A study of mother-child interaction. *Child Development, 59,* 1273–1283.

Crowell, J.A., & Feldman, S.S. (1991). Mothers' working models of attachment relationships and mother and child behavior during separation and reunion. *Developmental Psychology, 27,* 597–605.

Cruickshank, J.K., & Beevers, D.G. (1989). *Ethnic factors in health and disease.* London: Wright.

Crystal, Ronald. (1995). Transfer of genes to humans: Early lessons and obstacles to success. *Science, 270,* 404–409.

Crystal, Stephen. (1996). Economic status of the elderly. In Robert H. Binstock & Linda K. George (Eds.), *Handbook of aging and the social sciences.* San Diego, CA: Academic Press.

Crystal, Stephen, & Waehrer, Keith. (1996). Later life economic inequality in longitudinal perspective. *Journal of Gerontology, 51B,* S307–S318.

Csikszentmihalyi, Mihaly. (1996). *Creativity.* New York: HarperCollins.

Csikszentmihalyi, Mihaly, & Larson, R. (1984). *Being adolescent.* New York: Basic Books.

Culotta, Elizabeth, & Koshland, Daniel E. (1993). P53 sweeps through cancer research. *Science, 262,* 1958–1961.

Cummings, E., & Henry, W. (1961). *Growing old: The process of disengagement.* New York: Basic Books.

Cummings, E.M., & Davies, P. (1994). *Children and marital conflict: The impact of family dispute and resolution.* New York: Guilford.

Cummings, E.M., Hennessy, K.D., Rabideau, G.J., & Cicchetti, D. (1994). Responses of physically abused boys to interadult anger involving their mothers. *Development and Psychopathology, 6,* 31–41.

Cummins, Jim. (1991). Interdependence of first- and second-language proficiency. In Ellen Bialystok (Ed.), *Language processing in bilingual children.* Cambridge, England: Cambridge University Press.

Cummins, Jim. (1994). Knowledge, power, and identity in teaching English as a second language. In Fred Genesee (Ed.), *Educating second-language children: The whole child, the whole curriculum, the whole community.* Cambridge, England: Cambridge University Press.

Cunningham, Renee M., Stiffman, Arlene Rubin, & Dore, Peter. (1994). The association of physical and sexual abuse with HIV risk behaviors in adolescence and young adulthood: Implications for public health. *Child Abuse and Neglect, 18,* 233–245.

Cunningham, Walter R., & Tomer, Adrian. (1990). Intellectual abilities and age: Concepts, theories and analyses. In Eugene A. Lovelace (Ed.), *Aging and cognition: Mental processes, self-awareness and interventions.* Amsterdam: North-Holland/Elsevier.

Curran, David K. (1987). *Adolescent suicidal behavior.* Washington, DC: Hemisphere.

Cutler, David M., & Sheiner, Louise M. (1994). Policy options for long-term care. In David A. Wise (Ed.), *Studies in the economics of aging.* Chicago: University of Chicago Press.

Damon, William. (1984). Self-understanding and moral development from childhood to adolescence. In William M. Kurtines & Jacob L. Gewirtz (Eds.), *Morality, moral behavior, and moral development.* New York: Wiley.

Damon, William, & Hart, Daniel. (1992). Social understanding, self-understanding, and morality. In Marc Bornstein & Michael Lamb (Eds.), *Developmental psychology: An advanced textbook.* Hillsdale, NJ: Erlbaum.

Darling, N., & Steinberg, L. (1993). Parenting style as context: An integrative model. *Psychological Bulletin, 113,* 487–496.

Daro, Deborah. (1988). *Confronting child abuse.* New York: Free Press.

Daro, Deborah. (1995). *Current trends in child abuse reporting and fatalities: Results of the 1994 annual fifty state survey.* Chicago: National Committee to Prevent Child Abuse.

Dash, L. (1986, January 26). Children's children: The crisis up close. *The Washington Post,* pp. A1, A12.

Datan, Nancy. (1986). Oedipal conflict, platonic love: Centrifugal forces in intergenerational relations. In Nancy Datan, Anita L. Greene, & Hayne W. Reese (Eds.), *Life-span developmental psychology: Intergenerational relations.* Hillsdale, NJ: Erlbaum.

D'Augelli, Anthony R., & Dark, Lawrence J. (1994). Lesbian, gay, and bisexual youths. In Leonard D. Eron, Jacquelyn H. Gentry, & Peggy Schlegel (Eds.), *Reason to hope: A psychosocial perspective on violence and youth.* Washington, DC: American Psychological Association.

D'Augelli, A.R., & Hershberger, S.L. (1993). Lesbian, gay and bisexual youth in community settings: Personal challenges and mental health problems. *American Journal of Community Psychology, 21,* 421–448.

Davajan, Val, & Israel, Robert. (1991). Diagnosis and medical treatment of infertility. In Annette L. Stanton & Christine Dunkel-Schetter (Eds.), *Infertility.* New York: Plenum Press.

Davidson, Philip W., Cain, Nancy N., Sloane-Reeves, Jean E., & Van Speybroech, Alec. (1994). Characteristics of community-based individuals with mental retardation and aggressive behavioral disorders. *American Journal of Mental Retardation, 98,* 704–716.

Davies, A. Michael. (1991). Function in old age: Measurement, comparability, and service planning. *Proceedings of the 1988 International Symposium on Aging* (DHHS Publication No. 91-1482) (Series 5, No. 6). Hyattsville, MD: Department of Health and Human Services.

Davies, Bronwyn. (1982). *Life in the classroom and playground.* London: Routledge & Kegan Paul.

Davies, D.P., & Gantley, M. (1994). Ethnicity and the aetiology of sudden infant death syndrome. *Archives of Disease in Childhood, 70,* 349–353.

Davis, Janet M., & Rovee-Collier, Carolyn K. (1983). Alleviated forgetting of a learned contingency in 8-week-old infants. *Developmental Psychology, 19,* 353–365.

Davis, John M., & Sandoval, Jonathan. (1991). *Suicidal youth.* San Francisco: Jossey-Bass.

Davis-Floyd, Robbie E. (1992). *Birth as an American rite of passage.* Berkeley: University of California Press.

Dawson, Deborah A. (1991). Family structure and children's health and well-being: Data from the 1988 national health interview study on child health. *Journal of Marriage and the Family, 53,* 573–584.

Dawson, Geraldine. (1994). Frontal electroencephalographic correlates of individual differences in emotional expression in infants. *Monographs of the Society for Research in Child Development, 59* (Serial No. 240).

de Beauvoir, Simone. (1964). *Force of circumstances* (Richard Howard, Trans.). New York: Putnam.

DeCasper, Anthony J., & Fifer, William P. (1980). Of human bonding: Newborns prefer their mothers' voices. *Science, 208,* 1174–1175.

DeCasper, Anthony J., & Spence, M.J. (1986). Prenatal maternal speech influences newborns' perception of speech sounds. *Infant Behavior and Development, 9,* 133–150.

Deeg, Dorly J.H. (1995). Research and the promotion of quality of life in older persons in the Netherlands. In Eino Heikkinen, Jorma Kussinen, & Isto Ruoppila (Eds.), *Preparation for aging.* New York: Plenum Press.

Deeg, Dorly J.H., Kardaun, Jan W.P.F., & Fozard, James L. (1996). Health, behavior, and aging. In James E. Birren & K. Warner Schaie (Eds.), *Handbook of the psychology of aging.* San Diego, CA: Academic Press.

DeFries, John C., Plomin, Robert, & Fulker, David W. (Eds.). (1994). *Nature and nurture during middle childhood.* Cambridge, MA: Blackwell.

DeFriese, Gordon H., & Woomert, Alison. (1992). Informal and formal health care. In Marcia G. Ory, Ronald P. Ables, & Paula Darby Lipman (Eds.), *Aging, health, and behavior.* Newbury Park, CA: Sage.

DeGarmo, David S., & Kitson, Gay C. (1996). Identity relevance and disruption as predictors of psychological distress for widowed and divorced women. *Journal of Marriage and the Family, 58*, 983–997.

DeMan, A.F., Labreche-Cauthier, L., & Leduc, C.P. (1993). Parent-child relationships and suicidal ideation in French-Canadian adolescents. *Journal of Genetic Psychology, 154*, 17–23.

DeMaris, Alfred, & Rao, K. Vaninadha. (1992). Premarital cohabitation and subsequent marital stability in the United States: A reassessment. *Journal of Marriage and the Family, 54*, 178–190.

Demo, David H. (1992). Parent-child relations: Assessing recent changes. *Journal of Marriage and the Family, 54*, 104–117.

Demo, David H., & Acock, A.C. (1991). The impact of divorce on children. In A. Booth (Ed.), *Contemporary families: Looking forward, looking back*. Minneapolis, MN: National Council on Family Relations.

Dempster, Frank N. (1993). Resistance to interference: Developmental changes in a basic processing mechanism. In M. L. Howe and R. Pasnak (Eds.), *Emerging themes in cognitive development: Vol I. Foundations*. New York: Springer-Verlag.

Denham, S.A., & Holt, R.W. (1993). Preschoolers' likability as cause or consequence of their social behavior. *Developmental Psychology, 29*, 271–275.

Denham, S.A., McKinley, M., Couchoud, E.A., & Holt, R. (1990). Emotional and behavioral predictors of preschool peer ratings. *Child Development, 61*, 1145–1152.

Denney, Nancy Wadsworth. (1982). Aging and cognitive changes. In Benjamin B. Wolman & G. Sticker (Eds.), *Handbook of developmental psychology*. Englewood Cliffs, NJ: Prentice-Hall.

Denney, Nancy Wadsworth. (1989). Everyday problem solving: Methodological issues, research findings, and a model. In Leonard W. Poon, David C. Rubin, & Barbara A. Wilson (Eds.), *Everyday cognition in adulthood and late life*. Cambridge, England: Cambridge University Press.

Denney, Nancy Wadsworth. (1990). Adult age differences in traditional and practical problem solving. In Eugene A. Lovelace (Ed.), *Aging and cognition: Mental processes, self-awareness and interventions*. Amsterdam: North-Holland/Elsevier.

Dennis, Wayne. (1966). Creative productivity between the ages of 20 and 80 years. *Journal of Gerontology, 21*, 1–8.

Denton, Rhonda, & Kampfe, Charlene M. (1994). The relationship between family variables and adolescent substance abuse: A literature review. *Adolescence, 29*, 475–495.

Depue, Richard A., Luciana, Monica, Arbisi, Paul, Collins, Paul, & Leon, Arthur. (1994). Dopamine and the structure of personality: Relation of agonist-induced dopamine activity to positive emotionality. *Journal of Personality and Social Psychology, 67*, 485–498.

Derix, Mayke. (1994). *Neuropsychological differentiation of dementia syndromes*. Berwyn, PAQ: Lisse.

Derochers, Stephen, Ricard, Marcelle, Dexarie, Therese Gouin, & Allard, Louise. (1994). Developmental syncronicity between social referencing and Piagetian sensorimotor causality. *Infant Behavior and Development, 17*, 303–309.

DeRosier, Melissa E., Cillessen, Antonius H.N., Coie, John D., & Dodge, Kenneth. (1994). Group social context and children's aggressive behavior. *Child Development, 65*, 1068–1079.

Detzner, Daniel F. (1996). No place without a home: Southeast Asian grandparents in refugee families. *Generations, 20*, 45–48.

Deutsch, Helene. (1944–1945). *The psychology of women: A psychoanalytic interpretation* (Vol. 2). New York: Grune & Stratton.

Deutscher, Irwin. (1988). Misers and wastrels: Perceptions of the Depression and yuppie generations. In Suzanne K. Steinmetz (Ed.), *Family and support systems across the life span*. New York: Plenum Press.

de Villiers, Jill G., & de Villiers, Peter A. (1978). *Language acquisition*. Cambridge, MA: Harvard University Press.

de Villiers, Jill G., & de Villiers, Peter A. (1986). *The acquisition of English*. Hillsdale, NJ: Erlbaum.

de Villiers, Peter A., & de Villiers, Jill G. (1992). Language development. In M.H. Bornstein & M.E. Lamb (Eds.), *Developmental psychology: An advanced textbook* (3rd ed.). Hillsdale, NJ: Erlbaum.

Dewey, Kathryn G., Heinig, M. Jane, & Nommsen-Rivers, Laurie A. (1995). Differences in morbidity between breast-fed and formula-fed infants. *Journal of Pediatrics, 126*, 696–702.

DeWilde, E.J., Keinhorst, C.W.M., Diekstra, R.F.W., & Wolders, W.H.G. (1994). Social support, life events, and behavioral characteristics of psychologically distressed adolescents at high risk for attempting suicide. *Adolescence, 29*, 49–60.

Deyoung, Yolanda, & Zigler, Edward F. (1994). Machismo in two cultures: Relation to punitive child-rearing practices. *American Journal of Orthopsychiatry, 64*, 386–395.

D'Hoore, William D., Sicotte, Claude, & Tilquin, Charles. (1994). Sex bias in the management of coronary artery disease in Quebec. *American Journal of Public Health, 84*, 1013–1015.

Diamond, A. (1990). Neuropsychological insights into the meaning of object concept development. In S. Carey & R. Gelman (Eds.), *The epigenesis of mind: Essays on biology and cognition*. Hillsdale, NJ: Erlbaum.

Diamond, Marion Cleeves. (1988). *Enriching heredity*. New York: Free Press.

Diaz, Rafael M. (1987). The private speech of young children at risk: A test of three deficit hypotheses. *Early Childhood Research Quarterly, 2*, 181–197.

Diaz, Rafael M., & Klinger, Cynthia. (1991). Toward an explanatory model of the interaction between bilingualism and cognitive development. In Ellen Bialystok (Ed.), *Language processing in bilingual children*. Cambridge, England: Cambridge University Press.

Dickerson, Leah J., & Nadelson, Carol (Eds.). (1989). *Family violence: Emerging issues of national crisis*. Washington, DC: American Psychiatric Press.

Dickinson, David K. (1984). First impressions: Children's knowledge of words gained from a single exposure. *Applied Psycholinguistics, 5*, 359–374.

Dickstein, S., & Parke, R.D. (1988). Social referencing in infancy: A glance at fathers and marriage. *Child Development, 59*, 506–511.

DiClemente, Ralph J. (1990). The emergence of adolescents as a risk group for human immunodeficiency virus infection. *Journal of Adolescent Research, 5*, 7–17.

Diehl, Manfred, Coyle, Nathan, & Labouvie-Vief, Gisela. (1996). Age and sex differences in strategies of coping and defense across the life span. *Psychology and Aging, 11*, 127–139.

Diekstra, Rene. (1995). Depression and suicidal behaviors in adolescence: Sociocultural and time trends. The positive effects of schooling. In Michael Rutter (Ed.), *Psychosocial disturbances in young people: Challenges for prevention*. Cambridge, England: Cambridge University Press.

Dielman, T.E. (1994). School-based research on the prevention of adolescent alcohol use and misuse: Methodological issues and advances. *Journal of Research on Adolescence, 4*, 271–294.

Dietz, William H., Jr., & Gortmaker, Steven L. (1985). Do we fatten our children at the television set? Obesity and television viewing in children and adolescents. *Pediatrics, 75*, 807–812.

Digman, J.M. (1990). Personality structure: Emergence of the five-factor model. *Annual Review of Psychology, 41*, 417–440.

Dion, Kenneth L., & Dion, Karen K. (1988). Romantic love: Individual and cultural perspectives. In Robert J. Sternberg & Michael L. Barnes (Eds.), *The psychology of love*. New Haven, CT: Yale University Press.

DiPietro, Janet Ann. (1981). Rough and tumble play: A function of gender. *Developmental Psychology, 17*, 50–58.

Dishion, Thomas J., Andrews, David W., & Crosby, Lynn. (1995). Antisocial boys and their friends in early adolescence: Relationship characteristics, quality, and interactional processes. *Child Development, 66*, 139–151.

Dittmann-Kohli, Freya, & Baltes, Paul B. (1990). Toward a neofunctionalist conception of adult intellectual development: Wisdom as a prototypical case of intellectual growth. In Charles N. Alexander & Ellen J. Langer (Eds.), *Higher stages of human development*. New York: Oxford University Press.

Dix, T. (1991). The affective organization of parenting: Adaptive and maladaptive processes. *Psychological Bulletin, 110*, 3–25.

Dixon, Roger A. (1992). Contextual approaches to adult intellectual development. In Robert J. Sternberg & Cynthia A. Berg (Eds.), *Intellectual development*. New York: Cambridge University Press.

Dixon, Roger A., & Baltes, Paul B. (1986). Toward life-span research on the functions and pragmatics of intelligence. In Robert J. Sternberg & Richard K. Wagner (Eds.), *Practical intelligence*. Cambridge, England: Cambridge University Press.

Dixon, Roger A., Kramer, Dierdre A., & Baltes, Paul B. (1985). Intelligence: A life-span developmental perspective. In Benjamin B. Wolman (Ed.), *Handbook of intelligence: Theories, measurements, and applications*. New York: Wiley.

Dobbing, John. (Ed.). (1987). *Early nutrition and later achievement*. London: Academic Press.

Dobkin, Leah. (1992). If you build it, they may not come. *Generations, 16*(2), 31–32.

Dockrell, J., Campbell, R., & Neilson, I. (1980). Conservation accidents revisited. *International Journal of Behavioral Development, 3*, 423–439.

Dodge, Kenneth A., & Feldman, E. (1990). Issues in social cognition and sociometric status. In S.R. Asher & J.D. Coie (Eds.), *Peer rejection in childhood*. Cambridge, England: Cambridge University Press.

Dodge, Kenneth A., & Somberg, Daniel R. (1987). Hostile attributional biases among aggressive boys are exacerbated under conditions of threats to self. *Child Development, 58*, 213–224.

Dodge, Kenneth A., Murphy, Roberta R., & Buchsbaum, Kathy. (1984). The assessment of intention-cue detection skills in children: Implications for developmental psychopathology. *Child Development, 55*, 163–173.

Dodge, Kenneth A., Pettit, Gregory S., McClaskey, C.L., & Brown, M.M. (1986). Social competence in children. *Monographs of the Society for Research in Child Development, 51*(2, Serial No. 213).

Dodge, Kenneth A., Coie, J.D., Pettit, G.S., & Price, J.M. (1990). Peer status and aggression in boys' groups: Developmental and contextual analyses. *Child Development, 61*, 1289–1309.

Dodge, Kenneth A., Pettit, Gregory S., & Bates, John E. (1994). Effects of physical maltreatment on the development of peer relations. *Development and Psychopathology, 6*, 43–55.

Donaldson, Margaret. (1978). *Children's minds*. New York: Norton.

Dooley, David, & Catalano, R. (1988). Psychological effects of unemployment. *Journal of Social Issues, 44*, 1–191.

Dornbusch, S.M., Ritter, P.L., Leiderman, P.H., Roberts, D.F., & Fraleigh, M.J. (1987). The relation of parenting style to adolescent school performance. *Child Development, 58*, 1244–1257.

Downs, A. Chris. (1990). The social biological constraints of social competency. In Thomas P. Gullotta, Gerald R. Adams, & Raymond R. Montemayor (Eds.), *Developing social competency in adolescence*. Newbury Park, CA: Sage.

Drigotas, Stephen M., & Rusbult, Caryl E. (1992). Should I stay or should I go? A dependence model of relationships. *Journal of Personality and Social Psychology, 62*, 62–87.

Drotar, Dennis, Eckerle, Debby, Satola, Jackie, Pallotta, John, & Wyatt, Betsy. (1990). Maternal interactional behavior with nonorganic failure-to-thrive infants: A case comparison study. *Child Abuse and Neglect, 14*, 41–51.

Drotar, Dennis, Pallotta, John, & Eckerle, Debbie. (1994). A prospective study of family environments of children hospitalized for nonorganic failure-to-thrive. *Developmental and Behavioral Pediatrics, 15*, 78–85.

Dryfoos, Joy. (1990). *Adolescents at risk: Prevalence and prevention*. New York: Oxford University Press.

Dubas, Judith Semon, Graber, Julia A., & Petersen, Anne C. (1991). A longitudinal investigation of adolescents' changing perceptions of pubertal timing. *Developmental Psychology, 27*, 580–586.

Dubler, Nancy Neveloff. (1993). Commentary: Balancing life and death—proceed with caution. *American Journal of Public Health, 83*, 23–25.

Ducy, P., Desbois, C., Boyce, B., Pinero, G., Story, B., Dunstan, C., Smith, E., Bonadio, J., Goldstein, S., Gundberg, C., Bradley, A., & Karsenty, G. (1996). Increased bond formation in osteocalcin-deficient mice. *Nature, 382*, 448–451.

Dudek, Stephanie Z., Strobel, M.G., & Runco, Mark A. (1994). Cumulative and proximal influences on the social environment and children's creative potential. *Journal of Genetic Psychology, 154*, 487–499.

Duggar, Celia W. (1991, March 9). Neighbors ask, how could parents let that baby starve? *The New York Times*, pp. 25–26.

Duke-Duncan, Paula. (1991). Body image. In Richard M. Lerner, Ann C. Petersen, & Jeanne Brooks-Gunn (Eds.), *Encyclopedia of adolescence* (Vol. 1). New York: Garland.

Dunn, Judy. (1988). *The beginnings of social understanding.* Cambridge, MA: Harvard University Press.

Dunn, Judy. (1993). *Young children's close relationships: Beyond attachment.* Newbury Park, CA: Sage.

Dunn, Judy, & Munn, Penny. (1985). Becoming a family member: Family conflict and the development of social understanding in the second year. *Child Development, 56,* 480–492.

Dunn, Judy, & Plomin, Robert. (1990). *Separate lives: Why siblings are so different.* New York: Basic Books.

Dunn, Judy, Bretherton, I., & Munn, P. (1987). Conversations about feeling states between mothers and their young children. *Developmental Psychology, 23,* 132–139.

Du Randt, Ross. (1985). Ball-catching proficiency among 4-, 6-, and 8-year-old girls. In Jane E. Clark & James H. Humphrey (Eds.), *Motor development: Current selected research.* Princeton, NJ: Princeton Book Company.

Dusick, Anna M., Covert, Robert F., Schreiber, Michael D., Yee, Gloria T., Browne, Susan P., Moore, Christine M., & Tebbert, Ian R. (1993). Risk of intracranial hemorrhage and other adverse outcomes after cocaine exposure in a cohort of 323 very low birth weight infants. *Journal of Pediatrics, 122,* 438–445.

Dutton, Donald G. (1992). Theoretical and empirical perspectives on the etiology and prevention of wife assault. In Ray D. Peters, Robert J. McMahon & Vernon L. Quinsey (Eds.) *Aggression and violence throughout the lifespan.* Newbury Park: Sage.

Dykens, Elisabeth M., Hodapp, Robert M., & Leckman, James F. (1994). *Behavior and development in fragile X syndrome.* Thousands Oaks, CA: Sage.

Dykman, Roscoe, & Ackerman, Peggy T. (1991). Attention deficit disorder and specific reading disability: Separate but often overlapping disorders. *Journal of Learning Disabilities, 24,* 96–103.

Dykman, Roscoe, & Ackerman, Peggy T. (1993). Behavioral subtypes of attention deficit disorder. *Exceptional Children, 60,* 132–141.

Easterbrooks, M. Ann, & Goldberg, W.A. (1984). Toddler development in the family: Impact of father involvement and parenting characteristics. *Child Development, 55,* 740–752.

Easterlin, Richard A. (1996). Economic and social implications of demographic patterns. In Robert H. Binstock & Linda K. George (Eds.), *Handbook of aging and the social sciences.* San Diego, CA: Academic Press.

Eaves, L.J., Eysenck, H.J., & Martin, N.G. (1989). *Genes, culture, and personality.* London: Academic Press.

Eccles, J.S. (1993). School and family effects on the ontogeny of children's interests, self-perceptions, and activity choices. In J.E. Jacobs (Ed.), *Nebraska Symposium on Motivation: Vol. 40. Developmental perspectives on motivation.* Lincoln: University of Nebraska Press.

Eccles, J.S. & Jacobs, J.E. (1986). Social forces shape math attitudes and performance. *Signs, 11,* 367–389.

Eccles, J.S., Midgley, C., Wigfield, A., Buchanan, C.M., Reuman, D., Flanagan, C., & MacIver, D. (1993). Development during adolescence: The impact of stage-environment fit on young adolescents' experiences in schools and in families. *American Psychologist, 48,* 90–101.

Eccles, J.S., Wigfield, A., Harold, R.D., & Blumenfeld, P. (1993). Age and gender differences in children's self- and task perceptions during elementary school. *Child Development, 64,* 830–847.

Eckenrode, J., Laird, M., & Doris, J. (1993). School performance and disciplinary problems among abused and neglected children. *Developmental Psychology, 29,* 53–62.

Eder, R.A. (1989). The emergent personologist: The structure and content of 3.5-, 5.5-, and 7.5-year-olds' concepts of themselves and other persons. *Child Development, 60,* 1218–1228.

Eder, R.A. (1990). Uncovering young children's psychological selves: Individual and developmental differences. *Child Development, 61,* 849–863.

Edwards, Allen Jack. (1993). *Dementia.* New York: Plenum Press.

Edwards, John N. (1969). Familiar behavior as social exchange. *Journal of Marriage and the Family, 31,* 518–526.

Edwards, John N., & Booth, Alan. (1994). Sexuality, marriage, and well-being: The middle years. In Alice S. Rossi (Ed.), *Sexuality across the life course.* Chicago: University of Chicago Press.

Edwards, John R. (1994). *Multilingualism.* London: Routledge.

Egeland, Byron. (1991). A longitudinal study of high risk families: Issues and findings. In Raymond H. Starr & Davia A. Wolfe (Eds.), *The effects of child abuse and neglect.* New York: Guilford.

Egeland, Byron. (1993). A history of abuse is a major risk factor for abusing the next generation. In Richard J. Gelles & Donileen R. Loseke (Eds.), *Current controversies in family violence.* Newbury Park, CA: Sage.

Egeland, Byron, & Hiester, Marnie. (1995). The long-term consequences of infant daycare and mother-infant attachment. *Child Development, 66,* 474–485.

Egeland, Byron, Carlson, Elizabeth, & Sroufe, L. Alan. (1993). Resiliance as process. *Development and Psychopathology, 5,* 517–528.

Eichorn, Dorothy H., Hunt, Jane V., & Honzik, Marjorie P. (1981). Experience, personality, and IQ: Adolescence to middle age. In Dorothy Eichorn, John A. Clausen, Marjorie P. Honzik, & Paul H. Mussen (Eds.), *Present and past in middle life.* New York: Academic Press.

Eimas, Peter D., Sigueland, Einar R., Jusczyk, Peter, & Vigorito, James. (1971). Speech perception in infants. *Science, 171,* 303–306.

Eisen, M., & Zellman, G. L. (1992). A health beliefs field experiment: Teen talk. In B.C. Miller et al. (Eds.), *Preventing adolescent pregnancy.* Newbury Park, CA: Sage.

Eisenberg, N., Lunch, T., Shell, R., & Roth, K. (1985). Children's justifications for their adult and peer-direction compliant (prosocial and nonprosocial) behaviors. *Developmental Psychology, 21,* 325–331.

Eisenson, Jon. (1986). *Language and speech disorders in children.* New York: Pergamon Press.

Elder, Glen H., Jr., Nguyen, Tri Van, & Caspi, Avshalom. (1985). Linking family hardship to children's lives. *Child Development, 56,* 361–375.

Elder, Glen H., Jr., Rudkin, Laura, & Conger, Rand D. (1995). Intergenerational continuity and change in rural America. In Vern L. Bengtson, K. Warner Schaie, & Linda M. Burton (Eds.), *Adult intergenerational relations: Effects of societal change.* New York: Springer.

Elder Abuse Project. (1991, October/November). Aging today. *American Society on Aging*, pp. 14–15.

Elderhostel: United States catalog. (1997, Summer). Boston: Elderhostel Inc.

Elevenstar, D. (1980, January 8). Hapy couple a tribute to old-fashioned virtues. *The Los Angeles Times*, p. 2.

Elkind, David. (1967). Egocentrism in adolescence. *Child Development, 38*, 1025–1034.

Elkind, David. (1978). *The child's reality: Three developmental themes*. Hillsdale, NJ: Erlbaum.

Elkind, David. (1984). *All grown up and no place to go*. Reading, MA: Addison-Wesley.

Elkind, David. (1996). Inhelder and Piaget on adolescence: A postmodern appraisal. *Psychological Science, 7*, 216–220.

Ellis, Nancy Borel. (1991). An extension of the Steinberg accelerating hypothesis. *Journal of Early Adolescence, 11*, 221–235.

Ellsworth, C.P., Muir, D.W., & Hains, S.M.J. (1993). Social competence and person-object differentiation: An analysis of the still-face effect. *Developmental Psychology, 29*, 63–73.

Emde, Robert N. (1994). Individual meaning and increasing complexity: Contributions of Sigmund Freud and Rene Spitz to developmental psychology. In Ross D. Parke, Peter A. Ornstein, John J. Rieser, & Carolyn Zahn-Waxler (Eds.), *A century of developmental psychology*. Washington, DC: American Psychological Association.

Emde, Robert N., Biringen, Z., Clyman, R.B., & Oppenheim, D. (1991). The moral self of infancy. *Developmental Review, 11*, 251–270.

Emery, Robert E. (1988). *Marriage, divorce, and children's adjustment*. Newbury Park, CA: Sage.

Enkin, Murray, Keirse, Marc J.N.C., & Chalmers, Iain. (1989). *Effective care in pregnancy and childbirth*. Oxford, England: Oxford University Press.

Entwhisle, Doris R. & Alexander, Karl L. (1995). A parent's economic shadow: Family structure versus family resources as influences on early school achievement. *Journal of Marriage and the Family, 57*, 399–409.

Epstein, M.A., Shaywitz, S.E., Shaywitz, B.A., & Woolston, J.L. (1991). The boundaries of attention deficit disorder. *Journal of Learning Disabilities, 2*, 78–86.

Ericsson, K. Anders. (1990). Peak performance and age: An examination of peak performance in sports. In Paul B. Baltes & Margret M. Baltes (Eds.), *Successful aging: Perspectives from the behavioral sciences*. Cambridge, England: Cambridge University Press.

Erikson, Erik H. (1963). *Childhood and society* (2nd ed.). New York: Norton.

Erikson, Erik H. (1968). *Identity, youth, and crisis*. New York: Norton.

Erikson, Erik H. (1975). *Life history and the historical moment*. New York: Norton.

Erikson, Erik H., Erikson, Joan M., & Kivnick, Helen Q. (1986). *Vital involvement in old age*. New York: Norton.

Ernhart, Claire B., Sokol, Robert J., Ager, Joel W., Morrow-Tlucak, Mary, & Martier, Susan. (1989). Alcohol-related birth defects: Assessing the risk. In Donald Hutchings (Ed.), *Prenatal abuse of licit and illicit drugs*. New York: New York Academy of Sciences.

Erwin, Phil. (1993). *Friendship and peer relations in children*. New York: Wiley.

Eskenazi, Brenda, Prehn, Angela W., & Christianson, Roberta E. (1995). Passive and active maternal smoking as measured by serum cotinine: The effect on birthweight. *American Journal of Public Health, 85*, 395–398.

Essex, Marilyn J., & Nam, Sunghee. (1987). Marital status and loneliness among older women. *Journal of Marriage and the Family, 49*, 93–106.

Estes, Carroll L., Linkins, Karen W., & Binney, Elizabeth A. (1996). The political economy of aging. In Robert H. Binstock & Linda K. George (Eds.), *Handbook of aging and the social sciences*. San Diego, CA: Academic Press.

Etaugh, Claire, & Liss, Marsha B. (1992). Home, school, and playroom: Training ground for adult gender roles. *Sex Roles, 26*, 129–147.

Eveleth, Phillis B., & Tanner, James M. (1976). *Worldwide variation in human growth*. Cambridge, England: Cambridge University Press.

Ewing, Charles Patrick. (1990). *Kids who kill*. Lexington, MA: Lexington Books.

Eyer, D. (1992). *Maternal-infant bondings: A scientific fiction*. New Haven, CT: Yale University Press.

Faberow, Norman L. (1994). Preparatory and prior suicidal behavior factors. In Edwin S. Schneidman, Norman L. Faberow, & Robert E. Litman (Eds.), *The psychology of suicide* (rev. ed.). Northwale, NJ: Aronson.

Fagot, Beverly I., & Hagan, R. (1991). Observations of parental reactions to sex-stereotyped behaviors: Age and sex effects. *Child Development, 62*, 617–628.

Fagot, Beverly I., & Leinbach, Mary D. (1993). Gender-role development in young children: From discriminating to labeling. *Developmental Review, 13*, 205–224.

Fagot, Beverly I., Leinbach, Mary D., & O'Boyle, C. (1992). Gender labeling, gender stereotyping, and parenting behaviors. *Developmental Psychology, 28*, 225–230.

Fairburn, Christopher G., & Wilson, G. Terence. (Eds). (1993). *Binge eating: Nature, assessment and treatment*. New York: Guilford.

Falbo, T., & Polit, D.F. (1986). Quantitative review of the only-child literature: Research evidence and theory development. *Psychology Bulletin, 100*, 176–189.

Falbo, T., & Poston, D.L. (1993). The academic, personality, and physical outcomes of only children in China. *Child Development, 64*, 18–35.

Fang, Jing, Madhavean, Shanta, & Alderman, Micheal H. (1996). The association between birthplace and mortality from cardiovascular causes among black and white residents. *New England Journal of Medicine, 335*, 1545–1551.

Fantuzzo, J., DePaola, L., Lambert, L., Martino, T., Anderson, G., & Sutton, S. (1991). Effects of inter-parental violence on the psychological adjustment and competencies of young children. *Journal of Consulting and Clinical Psychology, 59*, 258–265.

Farkas, Janice I., & Hogan, Dennis P. (1995). The demography of changing intergenerational relationships. In Vern L. Bengtson, K. Warner Schaie, & Linda M. Burton (Eds.), *Adult intergenerational relations: Effects of societal change*. New York: Springer.

Farooqi, S., Perry, I.J., & Beevers, D.G. (1993). Ethnic differences in infants. *Pediatric and Prenatal Epidemiology, 7,* 245–252.

Farrar, M.J. (1992). Negative evidence and grammatical morpheme acquisition. *Developmental Psychology, 28,* 90–98.

Farrar, M.J., & Goodman, G.S. (1990). Developmental differences in the relation between scripts and episodic memory: Do they exist? In R. Fivush & J.A. Hudson (Eds.), *Knowing and remembering in young children.* Cambridge, England: Cambridge University Press.

Farrell, Michael P., & Rosenberg, Stanley D. (1981). *Men at midlife.* Boston: Auburn House.

Farrington, David P., Loeber, Rolf, Elliot, Delbert S., Hawkins, J. David, Kendell, Denise B., Klein, Malcolm W., McCord, Joan, Rowe, David C., & Tremblay, Richard E. (1990). Advancing knowledge about the onset of delinquency and crime. In *Advances in clinical child psychology* (Vol. 13). New York: Plenum Press.

Fastenau, Philip S., Denburg, Natalie L., & Abeles, Norman. (1996). Age differences in retrieval: Further support for the resource-reduction hypothesis. *Psychology and Aging, 11,* 140–146.

Fawzy, F.I., Fawzy, N.W., Hyun, C., Elashoff, R., Guthrie, D., Fahey, J.L., & Morton, D.L. (1993). Malignant melanoma: Effects on an early structured psychiatric intervention, coping, and affective state on recurrence and survival six years later. *Archives of General Psychiatry, 50,* 681–689.

Featherstone, Helen. (1980). *A difference in the family.* New York: Basic Books.

Fehr, Beverley. (1993). How do I love thee . . . let me consult my prototype. In Steve Duck (Ed.), *Individuals in relationships* (Vol. 1). Newbury Park, CA: Sage.

Fehr, Beverley. (1996). *Friendship processes.* Thousand Oaks, CA: Sage.

Feij, Jan A., Whitely, William T., Diero, Jose M., & Taris, Tom W. (1995). The development of cancer enhancing strategies and content innovation: A longitudinal study of new workers. *Journal of Vocational Behavior, 46,* 231–256.

Fein, Edith. (1991). Issues in foster family care: Where do we stand? *American Journal of Orthopsychiatry, 61,* 578–583.

Feinman, S. (1985). Emotional expression, social referencing, and preparedness for learning in infancy—Mother knows best, but sometimes I know better. In G. Ziven (Ed.), *The development of expressive behavior.* Orlando, FL: Academic Press.

Feiring, Candice, & Lewis, Michael. (1989). The social network of girls and boys from early through middle childhood. In Deborah Belle (Ed.), *Children's social networks and social supports.* New York: Wiley.

Feldman, S. Shirley, & Gehring, T.M. (1988). Changing perceptions of family cohesion and power across adolescence. *Child Development, 59,* 1034–1045.

Felmlee, D., Sprecher, S., & Bassin, E. (1990). The dissolution of intimate relationships: A hazard mode. *Social Psychology Quarterly, 53,* 13–30.

Felner, R.D., & Terre, L. (1987). Child custody dispositions and children's adaptation following divorce. In L.A. Weithorn (Ed.), *Psychology and child custody determinations.* Lincoln: University of Nebraska Press.

Felsenthal, G., Garrison, S.J., & Steinberg, F.U. (Eds.). (1994). *Rehabilitation of the aging patient.* Baltimore: Williams & Wilkins.

Fenson, Larry, Dale, Philip S., Resnick, J. Steven, Bates, Elizabeth, Thal, Donna J., & Petchick, Stephen J. (1994). Variability in early communicative development. *Monographs of the Society for Research in Child Development, 59* (Serial No. 242).

Ferber, Marianne A., & O'Farrell, Brigid. (1991). *Work and family: Policies for a changing work force.* Washington, DC: National Academy Press.

Ferguson, Charles A. (1977). Baby talk as a simplified register. In Catherine E. Snow & Charles A. Ferguson (Eds.), *Talking to children: Language input and requisition.* Cambridge, England: Cambridge University Press.

Fergusson, David M., & Lynskey, Michael T. (1996). Adolescent resiliency to family adversity. *Journal of Child Psychology and Psychiatry and Allied Disciplines, 37,* 281–292.

Fergusson, David M., Horwood, L. John, & Lynskey, Michael T. (1993). Early dentine lead levels and subsequent cognitive and behavioral development. *Journal of Child Psychology and Psychiatry and Applied Disciplines, 34,* 215–227.

Fernald, Anne. (1985). Four-month-old infants prefer to listen to motherese. *Infant Behavior and Development, 8,* 181–195.

Fernald, Anne. (1993). Approval and disapproval: Infant responsiveness to vocal affect in familiar and unfamiliar languages. *Child Development, 64,* 657–674.

Fernald, Anne, & Kuhl, P. (1987). Acoustic determinants of infant preference for motherese speech. *Infant Behavior and Development, 10,* 279–293.

Fernald, Anne, & Mazzie, Claudia. (1991). Prosody and focus in speech to infants and adults. *Developmental Psychology, 27,* 209–221.

Ferraro, Kenneth F., & Farmer, Melissa M. (1996). Double jeopardy, aging as leveler, or persistent health inequality? A longitudinal analysis of white and black Americans. *Journal of Gerontology, 51B,* S319–S328.

Fiatarone, M.A., Marks, E.C., Ryan, N.D., Meredith, C.N., Lipsitz, L.A., & Evans, W.J. (1990). High-intensity strength training in nonagenarians: Effects on skeletal muscle. *JAMA, Journal of the American Medical Association, 263,* 3029–3034.

Field, D. (1987). A review of preschool conservation training: An analysis of analysis. *Developmental Review, 7,* 210–251.

Field, Tiffany M. (1987). Affective and interactive disturbances in infants. In Joy Doniger Osofsky (Ed.), *Handbook of infant development* (2nd ed.). New York: Wiley.

Field, Tiffany M. (1991). Quality infant day-care and grade school behavior and performance. *Child Development, 62,* 863–870.

Field, Tiffany M. (1995). Infants of depressed mothers. *Infant Behavior and Development, 18,* 1–13.

Fingerman, Karen L. (1996). Sources of tension in the aging mother and adult daughter relationship. *Psychology and Aging, 11,* 591–606.

Finkelhor, David. (1992). New myths about the child welfare system. *Child, Youth, and Family Service Quarterly, 15*(1), 3–5.

Finkelhor, David. (1993). The main problem is still underreporting, not overreporting. In Richard J. Gelles & Donileen R. Loseke (Eds.), *Current controversies in family violence.* Newbury Park, CA: Sage.

Finkelhor, David. (1994). Current information on the scope and nature of child sexual abuse. *The Future of Children, 4,* 31–53.

Finkelhor, David, & Berliner, Lucy. (1995). Research on the treatment of sexually abused children: A review and recommendations. *Journal of the American Academy of Child & Adolescent Psychiatry, 34*, 1408–1423.

Firth, Shirley. (1993). Approaches to death in Hindu and Sikh communities in Britain. In Donna Dickenson & Malcolm Johnson (Eds.), *Death, Dying and Bereavement*. London: Sage.

Firth, Shirley. (1993). Cross-cultural perspectives on bereavement. In Donna Dickenson & Malcolm Johnson (Eds.), *Death, Dying and Bereavement*. London: Sage.

Firth, Uta. (1989). A new look at language and communication in autism. *British Journal of Disorders of Communication, 24*, 123–150.

Fischer, Judith L., Sollie, Donna L., & Morrow, K. Brent. (1986). Social networks in male and female adolescents. *Journal of Adolescent Research, 1*, 1–14.

Fischer, Kurt W. (1980). A theory of cognitive development: The control of hierarchies of skill. *Psychological Review, 87*, 477–531.

Fischer, Kurt W., & Rose, Samuel P. (1994). Dynamic development of coordination of components in brain and behavior: A framework for theory and research. In G. Dawson & Kurt W. Fischer (Eds.), *Human behavior and the developing brain*. New York: Guilford.

Fisk, John E., & Warr, Peter. (1996). Age and working memory: The role of perceptual speed, the central executive, and the phonological loop. *Psychology and Aging, 11*, 316–323.

Fivush, R., & Hamond, N.R. (1990). Autobiographical memory across the preschool years: Toward reconceptualizing childhood amnesia. In R. Fivush & J.A. Hudson (Eds.), *Knowing and remembering in young children*. Cambridge, England: Cambridge University Press.

Flavell, John H. (1982). Structures, stages, and sequences in cognitive development. In W. Andrew Collins (Ed.), *Minnesota Symposia on Child Psychology: Vol. 15. The concept of development*. Hillsdale, NJ: Erlbaum.

Flavell, John H. (1985). *Cognitive development* (2nd ed.). Englewood Cliffs, NJ: Prentice Hall.

Flavell, John H. (1992). Cognitive development: Past, present, and future. *Developmental Psychology, 28*, 998–1005.

Flavell, John H., Miller, P.H., & Miller, S.A. (1993). *Cognitive development* (3rd ed.). Englewood Cliffs, NJ: Prentice Hall.

Flavell, John H., Green, Frances L., & Flavell, Eleanor R. (1995). Young children's knowledge about thinking. *Monographs of the Society for Research in Child Development, 60* (Serial no. 243).

Fletcher, Anne C., Darling, Nancy, & Steinberg, Laurence. (1995). Parental monitoring and peer influences on adolescent substance use. In Joan McCord (Ed.) *Coercion and punishment in long-term perspectives*. New York: Cambridge University Press.

Fletcher, Jack M., Francis, David J., Rourke, Byron P., Shaywitz, Sally E., et al. (1992). The validity of discrepancy-based definitions of reading disabilities. *Journal of Learning Disabilities, 25*, 555–561.

Foley, Joseph M. (1992). The experience of being demented. In Robert H. Binstock, Stephen G. Post, & Peter J. Whitehouse (Eds.), *Dementia and aging: Ethics, values, and policy choices*. Baltimore: Johns Hopkins University Press.

Fonagy, P., Steele, H., & Steele, M. (1991). Maternal representations of attachment during pregnancy predict the organization of infant-mother attachment at one year of age. *Child Development, 62*, 891–905.

Fordham, S., & Ogbu, J.U. (1986). Black students' school success coping with the burden of acting white. *Urban Review, 18*, 176–206.

Forste, Renata, & Tanfer, Koray. (1996). Sexual exclusivity among dating, cohabiting, and married women. *Journal of Marriage and the Family, 58*, 33–47.

Foster, Sharon, Martinez, Charles, & Kulberg, Andrea. (1996). Race, ethnicity, and children's peer relations. In Thomas H. Ollendick & Ronald J. Prinz (Eds.), *Advances in clinical child psychology* (Vol. 18). New York: Plenum Press.

Fowler, James W. (1981). *Stages of faith: The psychology of human development and the quest for meaning*. New York: Harper & Row.

Fowler, James W. (1986). Faith and the structuring of meaning. In Craig Dykstra & Sharon Parks (Eds.), *Faith development and Fowler*. Birmingham, AL: Religious Education Press.

Fox, Margery, Gibbs, Margaret, & Auerbach, Doris. (1985). Age and gender dimensions of friendship. *Psychology of Women Quarterly, 9*, 489–502.

Fox, Nathan A. (1991). If it's not left, it's right: Electroencephalograph asymmetry and the development of emotion. *American Psychologist, 46*, 863–872.

Fox, Nathan A., & Davidson, R.J. (1984). Hemispheric substrates of affect: A developmental model. In N.A. Fox & R.J. Davidson (Eds.), *The psychobiology of affective development*. Hillsdale, NJ: Erlbaum.

Fox, Nathan A., & Fein, Greta G. (1990). *Infant day care: The current debate*. Norwood, NJ: Ablex.

Fozard, James L. (1990). Vision and hearing in aging. In James E. Birren & K. Warner Schaie (Eds.), *Handbook of the psychology of aging* (3rd ed.). San Diego, CA: Academic Press

Frankenburg, W.K., Frandel, A., Sciarillo, W., & Burgess, D. (1981). The newly abbreviated and revised Denver Developmental Screening Test. *Journal of Pediatrics, 99*, 995–999.

Franklin, Deborah. (1984). Rubella threatens unborn in vaccine gap. *Science News, 125*, 186.

Fraser, M.W., Pecora, P.J., & Haapala, D.A. (Eds.). (1991). *Families in crisis: The impact of intensive family preservation services*. New York: Aldine de Gruyter.

Freeman, Ellen W., & Rickels, Karl. (1993). *Early childbearing: Perspectives on black adolescents and pregnancy*. Newbury Park, CA: Sage.

Freud, Sigmund. (1935). *A general introduction to psychoanalysis* (Joan Riviare, Trans.). New York: Modern Library.

Freud, Sigmund. (1938). *The basic writings of Sigmund Freud* (A.A. Brill, Ed. and Trans.). New York: Modern Library.

Freud, Sigmund. (1963). *Three case histories*. New York: Collier. (Original work published 1918)

Freud, Sigmund. (1964). *An outline of psychoanalysis: Vol. 23. The standard edition of the complete psychological works of Sigmund Freud* (James Strachey, Ed. and Trans.). London: Hogarth Press. (Original work published 1940)

Freud, Sigmund. (1965). *New introductory lectures on psychoanalysis* (James Strachey, Ed. and Trans.). New York: Norton. (Original work published 1933)

Freund, L.S. (1990). Maternal regulation of children's problem-solving behavior and its impact on children's performance. *Child Development, 61*, 113–126.

Fried, Peter A., O'Connell, Colleen, & Watkinson, Barbara. (1992). 60- and 72-month follow-up of children prenatally exposed to marijuana, cigarettes, and alcohol: Cognitive and language assessment. *Developmental and Behavioral Pediatrics, 13*, 383–391.

Fries, James F. (1994). *Living well: Taking care of your health in the middle and later years.* Reading, MA: Addison-Wesley.

Fries, James F., & Crapo, Lawrence M. (1981). *Vitality and aging.* San Francisco: Freeman.

Frisch, Rose E. (1983). Fatness, puberty, and fertility: The effects of nutrition and physical training on menarche and ovulation. In Jeanne Brooks-Gunn & Anne C. Petersen (Eds.), *Girls at puberty: Biological and psychosocial aspects.* New York: Plenum Press.

Frost, Jennifer J., & Forrest, Jacqueline Darroch. (1995). Understanding the impact of effective teenage pregnancy prevention programs. *Family Planning Perspectives, 27*, 188–195.

Fry, Christine L. (1995). Kinship and individuation: Cross-cultural perspectives on intergenerational relations. In Vern L. Bengtson, K. Warner Schaie, & Linda M. Burton (Eds.), *Adult intergenerational relations: Effects of societal change.* New York: Springer.

Frye, Douglas, & Moore, C. (1991). *Children's theories of mind.* Hillsdale, NJ: Erlbaum.

Frye, Douglas, Braisby, Nicholas, Lowe, John, Marouda, Cline, & Nicholls, Jon. (1989). Young children's understanding of counting and cardinality. *Child Development, 60*, 1158–1171.

Fuhrman, Teresa, & Holmbeck, Grayson N. (1995). A contextual-moderator analysis of emotional autonomy and adjustment in adolescence. *Child Development, 66*, 763–811.

Fuligni, A.J., & Eccles, J.S. (1993). Perceived parent-child relationships and early adolescents' orientation toward peers. *Developmental Psychology, 29*, 622–632.

Fulton, Robert. (1995). The contemporary funeral: Functional or dysfunctional? In Hannelore Wass & Robert A. Neimeyer (Eds.), *Dying: Facing the facts.* Washington, DC: Taylor & Francis.

Furman, Wyndol, & Buhrmester, D. (1992). Age and sex differences in perceptions of networks of personal relationships. *Child Development, 63*, 103–115.

Furstenberg, Frank F., Jr., & Cherlin, Andrew J. (1991). *Divided families: What happens to children when parents part.* Cambridge, MA: Harvard University Press.

Furstenberg, Frank F., Jr., & Nord, Christine Winquist. (1985). Parenting apart: Patterns of childbearing after marital disruption. *Journal of Marriage and the Family, 47*, 893–912.

Furstenberg, Frank F., Jr., Brooks-Gunn, Jeanne, & Morgan, Philip S. (1987). *Adolescent mothers in later life.* New York: Cambridge University Press.

Fuson, Karen C. (1988). *Children's counting and concepts of number.* New York: Springer-Verlag.

Fuson, Karen C., & Kwon, Youngshim. (1992). Korean children's understanding of multidigit addition and subtraction. *Child Development, 63*, 491–506.

Gaddis, Alan, & Brooks-Gunn, Jeanne. (1985). The male experience of pubertal change. *Journal of Youth and Adolescence, 14*, 61.

Galambos, N.L. (1992). Parent-adolescent relations. *Current Directions in Psychological Science, 1*, 146–149.

Galbraith, Michael W., & Cohen, Norman H. (1995). *Mentoring: New strategies and challenges.* San Francisco: Jossey-Bass.

Galinsky, Ellen. (1981). *Between generations: The six stages of parenthood.* New York: Berkley.

Gallagher, James J. (1990). The family as a focus for intervention. In Samuel J. Meisels & Jack P. Shonkoff (Eds.), *Handbook of early childhood intervention.* Cambridge, England: Cambridge University Press.

Galler, Janina. (1989). A follow-up study of the influence of early malnutrition on development: Behavior at home and at school. *Journal of the American Academy of Child and Adolescent Psychiatry, 28*, 254–261.

Gallup, George, Jr. (1996). *The Gallup Poll public opinion 1995.* Wilmington, DE: Scholarly Resources. Copyright 1996 by the Gallup Organization.

Galvin, Ruth Mehrtens. (1992). The nature of shyness. *Harvard Magazine, 94*(4), 40–45.

Ganong, Lawrence H., & Coleman, Marilyn. (1994). *Remarried family relationships.* Thousand Oaks, CA: Sage

Gantley, M., Davies, D.P., & Murcett, A. (1993). Sudden infant death syndrome: Links with infant care practices. *British Medical Journal, 306*, 16–20.

Garbarino, James. (1988). Preventing childhood injury: Developmental and mental health issues. *American Journal of Orthopsychiatry, 58*, 25–45.

Garbarino, James. (1989). An ecological perspective on the role of play in child development. In Marianne N. Bloch & Anthony D. Pellegrini (Eds.), *The ecological context of children's play.* Norwood, NJ: Ablex.

Garbarino, James, & Kostelny, Kathleen. (1992). Child maltreatment as a community problem. *Child Abuse and Neglect, 16*, 144–164.

Garbarino, James, Guttmann, Edna, & Seeley, James Wilson. (1986). *The psychologically battered child.* San Francisco: Jossey-Bass.

Garbarino, James, Kostelny, Kathleen, & Dubrow, Nancy. (1991). *No place to be a child: Growing up in a war zone.* New York: Lexington Books.

Garbarino, James, Dubrow, N., Kostelny, K., & Pardo, C. (1992). *Children in danger: Coping with the consequences of community violence.* San Francisco: Jossey-Bass.

Garber, J., & Dodge, K.A. (Eds.). (1991). *The development of emotional regulation and dysregulation.* Cambridge, England: Cambridge University Press.

Gardner, Howard. (1980). *Artful scribbles: The significance of children's drawings.* New York: Basic Books.

Gardner, Howard. (1983). *Frames of mind: The theory of multiple intelligences.* New York: Basic Books.

Gardner, William P. (1991). A theory of adolescent risk taking. In N. Bell (Ed.), *Adolescent and adult risk taking. The eighth Texas Tech symposium on interfaces in psychology.* Lubbock: Texas Tech University Press.

Gardner, William, Millstein, Susan G., & Wilcox, Brian L. (Eds). (1991). Adolescents in the AIDS epidemic. *New directions for child development, 50.* San Francisco: Jossey-Bass.

Garmezy, Norman. (1985). Stress-resistant children: The search for protective factors. In J. E. Stevenson (Ed.), *Recent research in developmental psychopathology.* Oxford, England: Pergamon Press.

Garmezy, Norman. (1993). Vulnerability and resiliance. In David C. Funder, Ross D. Parke, Carol Tomlinson-Keasy, & Keith Widaman (Eds.), *Studying lives through time.* Washington, DC: American Psychological Association.

Garrett, C.J. (1985). Effects of residential treatment on adjudicated delinquents. *Journal of Research on Crime and Delinquency, 22,* 287–308.

Garrity, Carla & Baris, Mitchell A. (1996). Bullies and victims. *Contemporary Pediatrics, 13,* 90–114.

Garrity, R.F., Stallones, L., Marx, M.B., & Johnson, T.P. (1989). Pet ownership and attachment as supportive factors in the health of the elderly. *Anthrozoos, 3,* 35–44.

Garvey, Catherine. (1976). Some properties of social play. In Jerome S. Bruner, Alison Jolly, & Kathy Sylva (Eds.), *Play.* New York: Basic Books.

Gatz, Margaret, Bengston, Vern L., & Blum, Mindy J. (1990). Caregiving families. In James E. Birren & K. Warner Schaie (Eds.), *Handbook of the psychology of aging* (3rd ed.). San Diego, CA: Academic Press.

Gatz, Margaret, Kasl-Godley, Julia E., & Karel, Michele J. (1996). Aging and mental disorders. In James E. Birren & K. Warner Schaie (Eds.), *Handbook of the psychology of aging.* San Diego, CA: Academic Press.

Gaulin, S.J.C. (1993). How and why sex differences evolve, with spatial ability as a paradigm example. In Marc Haug, Richard Whalen, Claude Aron, & Kathie Olsen (Eds.), *The development of sex differences and similarities in behavior.* Boston: Kluwer.

Gauvain, Mary. (1990). Review of Kathleen Berger, *The developing person through childhood and adolescence* (3rd ed.). New York: Worth.

Ge, Xiaojia, Lorenz, Frederick O., Conger, Rand D., Elder, Glen H., Jr., & Simons, Ronald L. (1994). Trajectories of stressful life events and depressive symptoms during adolescence. *Developmental Psychology, 30,* 467–483.

Gelles, Richard J. (1993). Through a sociological lens: Social structure and family violence. In Richard J. Gelles & Donileen R. Loseke (Eds.), *Current controversies on family violence.* Thousand Oaks, CA: Sage.

Gelles, Richard J., & Straus, M.A. (1988). *Intimate violence.* New York: Simon & Schuster.

Gelman, Rochel, & Massey, Christine M. (1987). Commentary. *Monographs of the Society for Research in Child Development,* 52 (Serial No. 216).

Genesee, Fred. (1994). *Educating second-language children: The whole child, the whole curriculum, the whole community.* Cambridge, England: Cambridge University Press.

Gerber, Adele. (1993). *Language related learning disabilities.* Baltimore: Brookes.

Gerstein, Dean R., & Green, Lawrence W. (Eds.). (1993). *Preventing drug abuse.* Washington, DC: National Academy of Science.

Gesell, Arnold. (1926). *The mental growth of the pre-school child: A psychological outline of normal development from birth to the sixth year including a system of developmental diagnosis.* New York: Macmillan.

Giacobini, Ezio. (1995). Alzheimer's disease: Major neurotransmitter deficits. Can they be corrected? In Israel Hanin, Mitsuo Yoshida, & Abraham Fisher (Eds.), *Alzheimer's and Parkinson's diseases: Recent developments.* New York: Plenum Press.

Giambra, Leonard M., Camp, Cameron J., & Grodsky, Alicia. (1992). Curiosity and stimulation-seeking across the adult life span: Cross-sectional and 6- to 8-year longitudinal findings. *Psychology and Aging, 7,* 150–157.

Gianino, A., & Tronick, Edward. (1988). The Mutual Regulation Model: The infant's self and interactive regulation and coping and defensive capacities. In T. Field, P. McCabe, & N. Schneiderman (Eds.), *Stress and coping* (Vol. 2). Hillsdale, NJ: Erlbaum.

Gibson, Eleanor. (1969). *Principles of perceptual learning and development.* New York: Appleton-Century-Crofts.

Gibson, Eleanor. (1982). The concept of affordances in development: The renascence of functionalism. In W. Andrew Collins (Ed.), *Minnesota Symposia on Child Psychology: Vol. 15. The concept of development.* Hillsdale, NJ: Erlbaum.

Gibson, Eleanor. (1988). Levels of description and constraints on perceptual development. In Albert Yonas (Ed.), *Perceptual development in infancy.* Hillsdale, NJ: Erlbaum.

Gibson, Eleanor, & Walker, Arlene S. (1984). Development of knowledge of visual-tactile affordances of substance. *Child Development, 55,* 453–460.

Gibson, James J. (1979). *The ecological approach to visual perception.* Boston: Houghton Mifflin.

Gibson, Rose C. (1993). The black American retirement experience. In James Jackson, Linda Chatters, & Robert Joseph Taylor (Eds.), *Aging in black America.* Newbury Park, CA: Sage.

Giele, J.Z. (1982). Women's work and family roles. In J.Z. Giele (Ed.), *Women in the middle years.* New York: Free Press.

Gilbert, Enid F., Arya, Sunita, Loxova, Renata, & Opitz, John M. (1987). Pathology of chromosome abnormalities in the fetus: Pathological markers. In Enid F. Gilbert & John M. Opitz (Eds.), *Genetic aspects of developmental pathology.* New York: Liss.

Gilbert, M.A., Bauman, K.E., & Udry, J.R. (1986). A panel study of subjective expected utility for adolescent sexual behavior. *Journal of Applied Social Psychology, 16,* 745–756.

Giles-Sims, Jean, & Crosbie-Burnett, Margaret. (1989). Adolescent power in stepparent families: A test of normative resource theory. *Journal of Marriage and the Family, 51,* 1065–1078.

Gillberg, Christopher. (1991). Clinical and neurobiological aspects of Asperger syndrome in six family studies. In Uta Frith (Ed.), *Autism and Asperger syndrome.* Cambridge, England: Cambridge University Press.

Gilligan, Carol. (1981). Moral development. In Arthur W. Chickering (Ed.), *The modern American college: Responding to the new realities of diverse students and a changing society.* San Francisco: Jossey-Bass.

Gilligan, Carol. (1982). *In a different voice: Psychological theory and women's development.* Cambridge, MA: Harvard University Press.

Gilligan, Carol, & Murphy, John M. (1979). Development from adolescence to adulthood: The philosopher and the dilemma of the fact. In William Damon (Ed.), *New directions for child development* (Vol. 5). San Francisco: Jossey-Bass.

Gilligan, Carol, Murphy, John M., & Tappan, Mark B. (1990). Moral development beyond adolescence. In Charles N. Alexander & Ellen J. Langer (Eds.), *Higher stages of human development*. New York: Oxford University Press.

Ginn, Jay, & Arber, Sara. (1994). Midlife women's employment and pension entitlement in relation to coresident adult children in Great Britain. *Journal of Marriage and the Family, 4*, 813–819.

Gittelman, R., Mannuzza, S., Shenker, R., & Bonagura, N. (1985). Hyperactive boys almost grown up: Psychiatric status. *Archives of General Psychiatry, 42*, 937–947.

Gjessing, Hans-Jorgen, & Karlsen, Bjorn. (1989). *A longitudinal study of dyslexia*. New York: Springer-Verlag.

Glantz, Meyer, & Pickens, Roy. (Eds.). (1992). *Vulnerability to drug abuse*. Washington, DC: American Psychological Association.

Gleason, Jean Berko. (1967). Do children imitate? *Proceedings of the International Conference on Oral Education of the Deaf*, 2, 1441–1448.

Glendenning, Frank. (1995). Education for older adults: Lifelong learning, empowerment, and social change. In Jon F. Nussbaum & Justine Coupland (Eds.), *Handbook of communication and aging research*. Mahwah, NJ: Erlbaum.

Glenn, Norval D. (1991). The recent trend in marital success in the United States. *Journal of Marriage and the Family, 53* 261–270.

Glenn, Norval D. (1996). Values, attitudes, and the state of American marriage. In David Popenoe, Jean Bethke Elshtain, & David Blankenhorn (Eds.), *Promises to keep: Decline and renewal of marriage in America*. Lanham, MD: Rowman & Littlefield.

Glick, Jennifer E., Bean, Frank D., & Van Hook, Jennifer V.W. (1997). Immigration and changing patterns of extended family household structure in the United States. *Journal of Marriage and the Family, 59*, 177–191.

Gnepp, Jackie, & Chilamkurti, Chinni. (1988). Children's use of personality attributions to predict other people's emotional and behavioral reactions. *Child Development, 59*, 743–754.

Goedegebuure, Leo, Kaiser, Frans, Maassen, Peter, Meek, Lynn, Van Vught, Frans, & de Weert, Egbert. (1994). *Higher education policy: An international comparative perspective*. Oxford, England: Pergamon Press.

Going, S.B., Williams, D.P., Lohman, T.G., & Hewitt, M.J. (1994). Age, body composition, and physical activity: A review. *Journal of Aging and Physical Activity, 2*, 38–66.

Goldberg, Carey. (1996, August 21). Survey reports more drug use by teen-agers. *New York Times*, A12.

Goldberg, Wendy A. (1990). Marital quality, parental personality, and spousal agreement about perceptions and expectations for children. *Merrill-Palmer Quarterly, 36*, 531–556.

Goldberg, Wendy A., & Easterbrooks, M.A. (1984). Role of marital quality in toddler development. *Developmental Psychology, 20*, 504–514.

Goldberg-Reitman, Jill. (1992). Young girls' conception of their mother's role: A neo-structural analysis. In Robbie Case (Ed.), *The mind's staircase: Exploring the conceptual underpinning of children's thought and knowledge*. Hillsdale, NJ: Erlbaum.

Goldfield, E.G., Kay, B.A., & Warren, W.H., Jr. (1993). Infant bouncing: The assembly and tuning of action systems. *Child Development, 64*, 1128–1142.

Goldman, Connie. (1991). Late bloomers: Growing older or still growing? *Generations, 15*, 41–44.

Goldman, Gail, Pineault, Raynald, Potvin, Louise, Blais, Regis, & Bilodeau, Henriette. (1993). Factors influencing the practice of vaginal birth after cesarean section. *American Journal of Public Health, 83*, 1104–1108.

Goldscheider, Francis K., & Goldscheider, Calvin. (1994). Leaving and returning home in 20th century America. *Population Bulletin, 48*, (no. 4) 1–35.

Goldsmith, H. Hill, & Rothbart, M.K. (1991). Contemporary instruments for assessing early temperament by questionnaire and in the laboratory. In J. Strelau & A. Angleitner (Eds.), *Explorations in temperament*. New York: Plenum Press.

Goldsmith, H. Hill, Buss, A.H., Plomin, R., Rothbart, M. Klevjord, Thomas, A., Chess, S., Hinde, R.A., & McCall, R.B. (1987). Roundtable: What is temperament? Four approaches. *Child Development, 58*, 505–529.

Goldsmith, Seth B. (1994). *Essentials of long-term care administration*. Gaithersburg, MD: Aspen.

Golinkoff, Roberta Michnick, & Hirsh-Pasek, Kathy. (1990). Let the mute speak: What infants can tell us about language acquisition. *Merrill-Palmer Quarterly, 36*, 67–91.

Golinkoff, Roberta Michnick, Hirsh-Pasek, Kathy, Bailey, Leslie M., & Wenger, Neill R. (1992). Young children and adults use lexical principles to learn new nouns. *Developmental Psychology, 28*, 99–108.

Golub, S. (1992). *Periods: From menarche to menopause*. Newbury Park, CA: Sage.

Golumb, C., & McLean, L. (1984). Assessing cognitive skills in pre-school children of middle and low income families. *Perceptual and Motor Skills, 58*, 119–125.

Gomberg, Edith S. Lisansky, & Nirenberg, Ted D. (Eds.). (1993). *Women and substance abuse*. Norwood, NJ: Ablex.

Goncu, A. (1993). Development of intersubjectivity in social pretend play. *Human Development, 36*, 185–198.

Gonsiorek, John C., & Weinrich, James D. (1991). The definition and scope of sexual orientation. In John C. Gonsiorek & James D. Weinrich (Eds.), *Homosexuality: Research implications for public policy*. Newbury Park, CA: Sage.

Goodman, G.S., Rudy, L., Bottoms, B.L., & Aman, C. (1990). Children's concerns and memory: Issues of ecological validity in the study of children's eyewitness testimony. In Robyn Fivush & Judith A. Hudson (Eds.), *Knowing and remembering in young children*. Cambridge, England: Cambridge University Press.

Goodman, G.S., Hirschman, J.E., Hepps, D., & Rudy, L. (1991). Children's memory for stressful events. *Merrill-Palmer Quarterly, 37*, 109–158.

Goodman, G.S., Taub, E.P., Jones, D.P.H., England, P., Port, L.K., Rudy, L., & Prado, L. (1992). Testifying in criminal court: Emotional effects on child sexual assault victims. *Monographs of the Society for Research in Child Development, 57*(Serial No. 229).

Goodman, N.C. (1987). Girls with learning disabilities and their sisters: How are they faring in adulthood? *Journal of Clinical Child Psychology, 16*, 290–300.

Goodno, Christine. (1996). *Profile—The SCORE Association.* Washington, DC: National SCORE Office.

Goodnow, Jacqueline J., & Collins, W. Andrew. (1990). *Development according to parents: The nature, sources, and consequences of parents' ideas.* Hillsdale, NJ: Erlbaum.

Goodwin, M.H. (1990). *He-said-she-said: Talk as social organization among black children.* Bloomington: Indiana University Press.

Goodz, Naomi S. (1994). Interactions between parents and children in bilingual families. In Fred Genesee (Ed.), *Educating second-language children: The whole child, the whole curriculum, the whole community.* Cambridge, England: Cambridge University Press.

Gordon, Debra Ellen. (1990). Formal operational thinking: The role of cognitive-developmental processes in adolescent decision-making about pregnancy and contraception. *American Journal of Orthopsychiatry, 60,* 346–356.

Gordon, George Kenneth, & Stryker, Ruth. (1994). *Creative long-term care administration.* Springfield, IL: Thomas.

Gortmaker, S.L., Dietz, W.H., Sobol, A.M., & Wehler, C.A. (1987). Increasing pediatric obesity in the United States. *American Journal of Diseases of Children, 141,* 535–540.

Gottesman, Irving I. (1991). *Schizophrenia genesis.* New York: Freeman.

Gottesman, Irving I., & Goldsmith, H.H. (1993). Developmental psychopathology of antisocial behavior: Inserting genes into its ontogenesis and epigenesis. In C.A. Nelson (Ed.), *Minnesota Symposia on Child Psychology: Vol. 27. Threats to optimal development: Integrating biological, psychological, and social risk factors.* Hillsdale, NJ: Erlbaum.

Gottfried, Adele Eskeles, & Gottfried, Allen W. (1994). Impact of redefined families on children's development: Conclusions, conceptual perspectives, and social implications. In Adele Eskeles Gottfried & Allen W. Gottfried (Eds.), *Redefining families: Implications for children's development.* New York: Plenum Press.

Government Accounting Office. (1992). *Child abuse: Prevention programs need greater emphasis.* Washington, DC: U.S. Government Printing Office.

Goyert, Gregory L., Bottoms, Sidney F., Treadwell, Marjorie C., & Nehra, Paul C. (1989). The physician factor in cesarean birth rates. *New England Journal of Medicine, 320,* 706–709.

Graber, Julia A., Brooks-Gunn, Jeanne, Paikoff, Roberta L., & Warren, Michelle P. (1994). Prediction of eating problems: An 8-year study of adolescent girls. *Developmental Psychology, 30,* 823–834.

Graham, Sandra, Hudley, C., & Williams, E. (1992). Attributional and emotional determinants of aggression among African-American and Latino young adolescents. *Developmental Psychology, 28,* 731–740.

Graham, Sandra, Weiner, Bernard, & Benesh-Weiner, Marijana. (1995). An attributional analysis of the development of excuse giving in aggressive and nonaggressive African-American boys. *Developmental Psychology, 31,* 274–284.

Grandin, Temple, & Scariano, Margaret M. (1986). *Emergence labeled autistic.* Norato, CA: Arena.

Grant, James P. (1986). *The state of the world's children.* Oxford, England: Oxford University Press.

Grantham-McGregor, Sally, Powell, Christine, Walker, Susan, & Chang, Susan. (1994). The long term follow up of severely malnourished children who participated in an intervention program. *Child Development, 65,* 428–439.

Gratch, Gerald, & Schatz, Joseph. (1987). Cognitive development: The relevance of Piaget's infancy books. In Joy Doniger Osofsky (Ed.), *Handbook of infant development* (2nd ed.). New York: Wiley.

Gratton, Brian, & Haber, Carole. (1996). Three phases in the history of American grandparents: Authority, burden, companion. *Generations, 20,* 7–12.

Green, A.L., & Boxer, Andres M. (1986). Daughters and sons as young adults. In Nancy Datan, Anita Greene, & Hayne W. Reese (Eds.), *Life-span developmental psychology: Intergenerational relations.* Hillsdale, NJ: Erlbaum.

Green, A.L., Cummings, E.M., & Karraker, K.H. (Eds.). (1991). *Life-span developmental psychology: Perspective on stress and coping.* Hillsdale, NJ: Erlbaum.

Greenberger, Ellen, & Chen, Chuansheng. (1996). Perceived family relationships and depressed mood in early and late adolescence: A comparison of European and Asian Americans. *Developmental Psychology, 32,* 707–716.

Greene, Ross W. (1996). Students with attention deficit hyperactivity disorder and their teachers: Implications of a goodness of fit perspective. In Thomas H. Ollendick & Ronald J. Prinz (Eds.), *Advances in clinical child psychology* (Vol. 18). New York: Plenum Press.

Greenough, W.T. (1993). Brain adaptation to experience: An update. In M.H. Johnson (Ed.), *Brain development and cognition.* Oxford, England: Blackwell.

Greenough, W.T., Black, J.E., & Wallace, C.S. (1987). Experience and brain development. *Child Development, 58,* 539–559.

Greenstein, Theodore N. (1995). Gender ideology, marital disruption, and the employment of married women. *Journal of Marriage and the Family, 57,* 31–42.

Greer, Germaine. (1986, May). Letting go. *Vogue, 176,* 141–143.

Greer, Jane. (1992). *Adult sibling rivalry.* New York: Crown.

Greif, Geoffrey L., DeMaris, Alfred, & Hood, Jane C. (1993). Balancing work and single fatherhood. In Jane C. Hood (Ed.), *Men, work, and family.* Newbury Park, CA: Sage.

Griffith, Dan R., Azuma, Scott D., & Chasnoff, Ira J. (1994). Three-year outcome of children exposed prenatally to drugs. *Journal of the American Academy of Child and Adolescent Psychiatry, 33,* 20–27.

Grodstein, Francine, Goldman, Marlene B., & Cramer, Daniel W. (1994). Infertility in women and moderate alcohol use. *American Journal of Public Health, 84,* 1429–1432.

Grodstein, Francine, Colditz, G.A., & Stampfer, M.J. (1996). Postmenopausal hormone use and tooth loss: A prospective study. *JAMA, Journal of the American Medical Association, 127,* 370–377.

Grossman, Frances K., Pollack, William S., & Golding, Ellen. (1988). Fathers and children: Predicting the quality and quantity of fathering. *Developmental Psychology, 24,* 82–91.

Grossman, John H. (1986). Congenital syphillis. In John L. Sever & Robert L. Brent (Eds.), *Teratogen update: Environmentally induced birth defect risks.* New York: Liss.

Grossman, Herbert. (1995). *Special education in a diverse society.* Needham, MA: Allyn and Bacon.

Grossman, K., Thane, K., & Grossman, K.E. (1981). Maternal tactile contact of the newborn after various postpartum conditions of mother-infant contact. *Developmental Psychology, 17,* 159–169.

Grossman, Michael, Chaloupka, Frank J., Saffer, Henry, & Laixuthai, Adit. (1994). Effects of alcohol price policy on youth: A summary of economic research. *Journal of Research on Adolescence, 4,* 347–364.

Grubman, Samuel, Gross, Elaine, Lerner-Weiss, Nancy, & Hernadez, Miriam. (1995). Older children and adolescents living with perinatally acquired human immuno-deficiency virus infection. *Pediatrics, 95,* 657–663.

Grusec, Joan E. (1992). Social learning theory and developmental psychology: The legacies of Robert Sears and Albert Bandura. *Developmental Psychology, 28,* 776–786.

Grusec, Joan E. (1994). Social learning theory and developmental psychology: The legacies of Robert R. Sears and Albert Bandura. In Ross D. Parke, Peter A. Ornstein, John J. Rieser, & Carolyn, Zahn-Waxler (Eds.), *A century of developmental psychology.* Washington, DC: American Psychological Association.

Guerin, Diana Wright, & Gottfried, Allen W. (1994). Temperamental consequences of infant difficultness. *Infant Behavior and Development, 17,* 413–421.

Guerra, Nancy G., Tolan, Patrick H., & Hammond, W. Rodney. (1994). Prevention and treatment of adolescent violence. In Leonard D. Eron, Jacquelyn H. Gentry, & Peggy Schlegel (Eds.), *Reason to hope: A psychosocial perspective on violence and youth.* Washington, DC: American Psychological Association.

Guillemard, A.M., & Rein, M. (1993). Comparative patterns of retirement: Recent trends in developed societies. *Annual Review of Sociology, 19,* 469–503.

Guilleminault, C., Boeddiker, Margaret Owen, & Schwab, Deborah. (1982). Detection of risk factors for "near miss SIDS" events in full-term infants. *Neuropediatrics, 13,* 29–35.

Gustafson, G.E., & Green, J.A. (1991). Developmental coordination of cry sounds with visual regard and gestures. *Infant Behavior and Development, 14,* 51–57.

Gustafson, G.E., & Harris, K.L. (1990). Women's responses to young infants' cries. *Developmental Psychology, 26,* 144–152.

Guthrie, Sharon R. (1991). Prevalence of eating disorders among intercollegiate athletes: Contributing factors and preventative measures. In David R. Black (Ed.), *Eating disorders among athletes.* Reston, VA: American Alliance for Health, Physical Education, Recreation and Dance.

Gutmann, David. (1987). *Reclaimed powers: Toward a new psychology of men and women in later life.* New York: Basic Books.

Gutwill, Susan. (1994). The diet: Personal experience, social condition, and industrial empire. In Women's Therapy Centre Institute (Eds.), *Eating problems.* New York: Basic Books.

Haan, Norma. (1985). Common personality dimensions or common organizations across the life span. In Joep M.A. Munnichs, Paul H. Mussen, Erhard Olbrich, & Peter G. Coleman (Eds.), *Life span and change in a gerontological perspective.* Orlando, FL: Academic Press.

Haan, Norma, Millsap, R., & Hartka, E. (1986). As time goes by: Change and stability in personality over fifty years. *Psychology and Aging, 1,* 220–232.

Haas, Joel E., Taylor, James S., Bergman, Abraham B., & van Belle, Gerald. (1993). Relationship between epidemiologic risk factors and clinicopathologic findings in sudden infant death syndrome. *Pediatrics, 91,* 106–112.

Hack, Maureen, Klein, Nancy, & Taylor, H. Gerry. (1995). Long-term developmental outcomes of low birth weight infants. *The Future of Children: Low Birth Weight, 5,* 176–196.

Hacker, Douglas J. (1994). An existential view of adolescence. *Journal of Early Adolescence, 14,* 300–327.

Hagberg, J.M. (1987). Effects of training on the decline of VO_2 max with aging. *Federation Proceedings, 46,* 1830–1833.

Hagerman, Randi J. (1996). Biomedical advances in developmental psychology: The case of Fragile X syndrome. *Developmental Psychology, 32,* 416–424.

Hagestad, G.O. (1986). The aging society as a context for family life. *Daedalus, 115,* 119–139.

Hagestad, G.O., & Neugarten, B.L. (1984). Age and the life course. In R. Binstock & E. Shanas (Eds.), *Handbook of aging and the social sciences* (2nd ed.). New York: Van Nostrand Reinhold.

Hagino, Nobuyoshi, Ohkura, Takeyoshi, Isse, Kunihiro, Akasuwa, Kenji, & Hamamoto, Makoto. (1995). Estrogen in clinical trials for dementia of Alzheimer type. In Israel Hanin, Mitsuo Yoshida, & Abraham Fisher (Eds.), *Alzheimer's and Parkinson's diseases: Recent developments.* New York: Plenum Press.

Haig, David. (1995). Prenatal power plays. *Natural History, 104,* 39.

Haith, Marshall M. (1980). *Rules that babies look by.* Hillsdale, NJ: Erlbaum.

Haith, Marshall M. (1990). Perceptual and sensory processes in early infancy. *Merrill-Palmer Quarterly, 36,* 1–26.

Haith, Marshall M. (1993). Preparing for the 21st century: Some goals and challenges for studies of infant sensory and perceptual development. *Developmental Review, 13,* 354–371.

Haith, Marshall M., Wentworth, N., & Canfield, R.L. (1993). The formation of expectations in early infancy. In C. Rovee-Collier & L.P. Lipsitt (Eds.), *Advances in infancy research* (Vol. 8). Norwood, NJ: Ablex.

Hakuta, K., & Garcia, E. (1989). Bilingualism and education. *American Psychologist, 44,* 374–379.

Hale, S. (1990). A global developmental trend in cognitive processing speed. *Child Development, 61,* 653–663.

Halliday, M.A.K. (1979). One child's protolanguage. In Margaret Bullowa (Ed.), *Before speech: The beginning of interpersonal communication.* Cambridge, England: Cambridge University Press.

Hamer, Dean H., Hu, Stella, Magnuson, Victoria L., Hu, Nan, & Pattatucci, Angela M.L. (1993). A linkage between DNA markers on the X chromosome and male sexual orientation. *Science, 261,* 321–327.

Hamon, Raeann R., & Blieszner, Rosemary. (1990). Filial responsibility expectations among adult child-older parent pairs. *Journal of Gerontology: Psychological Sciences, 45,* 110–112.

Hamond, Nina R., & Fivush, Robyn. (1991). Memories of Mickey Mouse: Young children recount their trip to Disneyworld. *Cognitive Development, 6,* 433–448.

Hanna, E., & Meltzoff, A.N. (1993). Peer imitation by toddlers in laboratory, home, and day-care contexts: Implications for social learning and memory. *Developmental Psychology, 29,* 701–710.

Hanninen, T., Reinikainen, K.J., Helkala, E., Kkoivisto, K., Mykkanen, L., Laakso, M., Pyorala, K., & Riekkinen, P.J. (1994). Subjective memory complaints and personality traits in normal elderly subjects. *Journal of the American Geriatrics Society, 42,* 1–4.

Hanson, Sandra L., Myers, David E., & Ginsberg, Alan L. (1987). The role of responsibility and knowledge in reducing teenage out-of-wedlock childbearing. *Journal of Marriage and the Family, 49,* 241–256.

Harlow, Robert E., & Cantor, Nancy. (1996). Still participating after all these years: A study of life task participation in later life. *Journal of Personality and Social Psychology, 71,* 1235–1249.

Harman, D. (1992). Free radical theory of aging. *Mutation Research, 275,* 257–266.

Harrington, Michael. (1962). *The other America: Poverty in the United States.* New York: Macmillan.

Harris, P.L. (1987). The development of search. In Philip Salapatek & Leslie Cohen (Eds.), *Handbook of infant perception: Vol. 2. From perception to cognition.* Orlando, FL: Academic Press.

Harris, P.L. (1989). *Children and emotion: The development of psychological understanding.* Oxford, England: Blackwell.

Harris, P.L., & Kavanaugh, R.D. (1993). Young children's understanding of pretense. *Monographs of the Society for Research in Child Development, 58*(Serial No. 231).

Harris, Raymond. (1986). *Clinical geriatric cardiology.* Philadelphia: Lippincott.

Harrison, Algea O., Wilson, Melvin N., Pine, Charles J., Chan, Samuel Q., & Buriel, Raymond. (1990). Family ecologies of ethnic minority children. *Child Development, 61,* 347–362.

Hart, Daniel, Yates, Miranda, Fegley, Suzanne, & Wilson, Gerry. (1995). Moral committment in inner city adolescents. In Melanie Killen & Daniel Hart (Eds.), *Morality in everyday life: Developmental perspectives.* Cambridge, England: Cambridge University Press.

Hart, Stuart N., & Brassard, Marla R. (1991). Psychological maltreatment: Progress achieved. *Development and Psychopathology, 3,* 61–70.

Harter, Susan. (1983). Developmental perspectives on the self-system. In Paul H. Mussen (Ed.), *Handbook of child psychology: Vol. 4. Socialization, personality and social development.* New York: Wiley.

Harter, Susan. (1990). Processes underlying adolescent self-concept formation. In Raymond Montemayor, Gerald R. Adams, & Thomas P. Gullotta (Eds.), *From childhood to adolescence: A transitional period?* Newbury Park, CA: Sage.

Harter, Susan. (1993). Visions of self: Beyond the me in the mirror. In J.E. Jacobs (Ed.), *Nebraska Symposium on Motivation: Vol. 40. Developmental perspectives on motivation.* Lincoln: University of Nebraska Press.

Harter, Susan, & Whitesell, N.R. (1989). Developmental changes in children's understanding of single, multiple, and blended emotion concepts. In C. Saarni & P.L. Harris (Eds.), *Children's understanding of emotion.* Cambridge, England: Cambridge University Press.

Harter, Susan, Marold, Donna B., Whitesell, Nancy R., & Cobbs, Gabrielle. (1996). A model of the effects of perceived parent and peer support on adolescent false self behavior. *Child Development, 67,* 360–374.

Hartfield, Bernadette W. (1996). Legal recognition of the value of intergenerational nurturance: Grandparent visitation statutes in the nineties. *Generations, 20,* 53–56.

Hartup, Willard W. (1983). Peer relations. In Paul H. Mussen (Ed.), *Handbook of child psychology: Vol. 4. Socialization, personality and social development.* New York: Wiley.

Hartup, Willard W. (1989). Social relationships and their developmental significance. *American Psychologist, 44,* 120–126.

Hartup, Willard W. (1996). The company they keep: Friendships and their developmental significance. *Child Development, 67,* 1–13.

Haskins, Ron. (1989). Beyond metaphor: The efficacy of early childhood education. *American Psychologist, 44,* 274–282.

Hatfield, Elaine. (1988). Theories of romantic love. In Robert J. Sternberg & Michael L. Barnes (Eds.), *The psychology of love.* New Haven, CT: Yale University Press.

Havik, Richard J. (1986, September 19). *Aging in the eighties: Impaired senses for sound and light in persons age 65 years and over.* (No. 125, DHHS Publication No. 86–1250). Hyattsville, MD: National Center for Health Statistics, Public Health Service.

Havik, Richard J. (1991). Physical, social, and mental vitality. *Proceedings of the 1988 International Symposium on Aging* (Ser. 5, No. 6, DHHS Publication No. 91–1482). Hyattsville, MD: U.S. Department of Health and Human Services.

Hawaii Department of Health. (1992). *Healthy Start: Hawaii's system of family support services.* Honolulu: Hawaii Department of Health.

Hawkins, Alan J., & Eggebeen, David J. (1991). Are fathers fungible? *Journal of Marriage and the Family, 53,* 958–972.

Hay, D.F., Caplan, M., Castle, J., & Stimson, C.A. (1991). Does sharing become increasingly "rational" in the second year of life? *Developmental Psychology, 27,* 987–993.

Hayes, C.D., Palmer, J.L., & Zaslow, M.J. (Eds.). (1990). *Child care choices.* Washington, DC: National Academy Press.

Hayflick, Leonard. (1979). Cell aging. In Arthur Cherkin (Ed.), *Physiology and cell biology of aging.* New York: Raven Press.

Hayflick, Leonard, & Moorhead, Paul S. (1961). The serial cultivation of human diploid cell strains. *Experimental Cell Research, 25,* 585.

Hayne, Harlene, & Rovee-Collier, Carolyn K. (1995). The organization of reactivated memory in infancy. *Child Development, 66,* 893–906.

Hayward, Mark D. (1996). Race inequities in men's retirement. *Journals of Gerontology, 51B,* S1–S10.

Heap, Kari Killen. (1991). A predictive and follow-up study of abusive and neglectful families by case analysis. *Child Abuse and Neglect, 15,* 261–273.

Heaton, Tim, & Albrecht, Stan L. (1991). Stable unhappy marriages. *Journal of Marriage and the Family, 53,* 747–758.

Heibeck, Tracy H., & Markman, Ellen M. (1987). Word learning in children: An examination of fast mapping. *Child Development, 58,* 1021–1034.

Hein, H.O., Suadicani, P., & Gyntelberg, F. (1996). Alcohol consumption, serum low density lipoprotein cholesterol concentration, and risk of ischaemic heart disease: Six year followup in the Copenhagen male study. *British Medical Journal, 312,* 736–741.

Held, Richard. (1985). Binocular vision—Behavioral and neuronal development. In Jacques Mehler & Robin Fox (Eds.), *Neonate cognition: Beyond the blooming buzzing confusion.* Hillsdale, NJ: Erlbaum.

Helson, Ravenna. (1992). Women's difficult times and the rewriting of the life story. *Psychology of Women Quarterly, 16,* 331–347.

Helson, Ravenna, & Moane, Geraldine. (1987). Personality change in women from college to midlife. *Journal of Personality and Social Psychology, 52,* 1176–1186.

Helson, Ravenna, & Wink, Paul. (1992). Personality change in women from the early 40s to the early 50s. *Psychology and Aging, 7,* 46–55.

Helson, Ravenna, Stewart, Abigail J., & Ostrove, Joan. (1995). Identity in three cohorts of midlife women. *Journal of Personality and Social Psychology, 69,* 554–557.

Hemstrom, Orjan. (1996). Is marriage dissolution linked to differences in mortality risks for men and women? *Journal of Marriage and the Family, 58,* 366–378.

Henggeler, S.W. (1989). *Delinquency in adolescence.* Newbury Park, CA: Sage.

Henker, B., & Whalen, C.K. (1989). Hyperactivity and attention deficits. *American Psychologist, 44,* 216–223.

Hertzig, M.E., & Shapiro, T. (1990). Autism and pervasive developmental disorders. In M. Lewis & S.M. Miller (Eds.), *Handbook of developmental psychopathology.* New York: Plenum Press.

Herzberger, S.D., & Hall, J.A. (1993). Consequences of retaliatory aggression against siblings and peers: Urban minority children's expectations. *Child Development, 64,* 1773–1785.

Herzog, Regula A. (1991). Measurement of vitality in the American's Changing Lives study. *Proceedings of the 1988 International Symposium on Aging.* (Series 5, No. 6, DHHS Publication No. 91-1482). Hyattsville, MD: U.S. Department of Health and Human Services.

Herzog, Regula A., & House, James S. (1991). Productive activities and aging well. *Generations, 15*(1), 49–54.

Herzog, Regula A., & Morgan, James N. (1992). Age and gender differences in the value of productive activities. *Research on Aging, 14,* 169–198.

Herzog, Regula A., House, James S., & Morgan, James N. (1991). Relation of work and retirement to health and well-being in older age. *Psychology and Aging, 6,* 202–211.

Hesketh, Beth. (1995). Personality and adjustment styles: A theory of work adjustment approach to career enhancing strategies. *Journal of Vocational Behavior, 46,* 274–282.

Hetherington, E. Mavis. (1989). Coping with family transitions: Winners, losers, and survivors. *Child Development, 60,* 1–14.

Hetherington, E. Mavis, & Clingempeel, W. Glenn. (1992). Coping with marital transitions. *Monographs of the Society for Research in Child Development, 57*(2–3, Serial No. 227).

Hewlett, B.S. (1992). The parent-infant relationship and social-emotional development among Aka Pygmies. In J.L. Roopnarine & D.B. Carter (Eds.), *Annual advances in applied developmental psychology: Vol. 5. Parent-child socialization in diverse cultures.* Norwood, NJ: Ablex.

Hickson, Joyce, Land, Arthur J., & Aikman, Grace. (1994). Learning style differences in middle school pupils from four ethnic backgrounds. *School Psychology International, 15,* 349–359.

Higbee, Martin D. (1994). Consumer guidelines for using medications wisely. *Generations, 18,* 43–48.

Higgins, E. Tory. (1981). Role taking and social judgment: Alternative developmental perspectives and processes. In John H. Flavell & Lee Ross (Eds.), *Social cognitive development: Frontiers and possible futures.* Cambridge, England: Cambridge University Press.

Higginson, John, Muir, Calum S., & Munoz, Nubia. (1992). *Human cancer: Epidemiology and environmental causes.* Cambridge, England: Cambridge University Press.

Hill, Hope M., Soriano, Fernando I., Chen, S. Andrew, & LaFromboise, Teresa D. (1994). Sociocultural factors in the etiology and prevention of violence among ethnic minority youth. In Leonard D. Eron, Jacquelyn H. Gentry, & Peggy Schlegel (Eds.), *Reason to hope: A psychosocial perspective on violence and youth.* Washington, DC: American Psychological Association.

Hills, Andrew P. (1992). Locomotor characteristics of obese children. *Child: Care, Health and Development, 18,* 29–34.

Hilts, Philip J. (1991, October 8). Lower lead limits are made official. *New York Times,* p. C3.

Hinde, R.A. (1987). *Individuals, relationships, and culture.* Cambridge, England: Cambridge University Press.

Hinde, R.A., & Stevenson-Hinde, J. (1987). Interpersonal relationships and child development. *Developmental Review, 7,* 1–21.

Hinde, R.A., Titmus, G., Easton, D., & Tamplin, A. (1985). Incidence of "friendship" and behavior toward strong associates versus nonassociates in preschoolers. *Child Development, 56,* 234–245.

Hines, Marc. (1993). Hormonal and neural correlates of sex-typed behavioral development in human beings. In Marc Haug, Richard Whalen, Claude Aron, & Kathie Olsen (Eds.), *The development of sex differences and similarities in behavior.* Boston: Kluwer.

Hobbs, Frank B., & Damon, Bonnie L. (1996). *65+ in the United States.* Washington, DC: U.S. Government Printing Office.

Hochschild, A. (1975). Disengagement theory: A critique and proposal. *American Sociological Review, 40,* 553–569.

Hochschild, Arlie. (1989). *The second shift: Working parents and the revolution at home.* New York: Viking Press.

Hodgson, J.L., & Buskirk, E.R. (1981). The role of exercise in aging. In D. Danon, N.W. Schock, & M. Marios (Eds.), *Aging: A challenge to science and society.* London: Oxford University Press.

Hoff-Ginsberg, E. (1986). Function and structure in maternal speech: Their relation to the child's development of syntax. *Developmental Psychology, 22,* 155–163.

Hoff-Ginsberg, E. (1990). Maternal speech and the child's development of syntax: A further look. *Journal of Child Language, 17,* 85–99.

Hoffman, Lois Wladis. (1989). Effects of maternal employment in the two-parent family. *American Psychologist, 44,* 283–292.

Hoffman, Lois Wladis. (1991). The influence of the family environment on personality: Accounting for sibling differences. *Psychological Bulletin, 110,* 187–203.

Hoffman, S.D., Foster, E.M., & Furstenberg, F.F. (1993). Reevaluating the costs of teenage childbearing. *Demography, 30,* 1–13.

Hogan, Robert, Hogan, Joyce, & Roberts, Brent W. (1996). Personality measurement and employment decisions: Questions and answers. *American Psychologist, 51*, 469–477.

Hokado, R., Saito, T.R., Wakafuji, Y., Takahashi, K.W., & Imanichi, T. (1993). The change with age of the copulatory behavior of the male rats age 67 and 104 weeks. *Experimental Animal, 42*, 75.

Holden, Constance. (1980). Identical twins reared apart. *Science, 207*, 1323–1328.

Holden, G.W. (1983). Avoiding conflict: Mothers as tacticians in the supermarket. *Child Development, 54*, 233–240.

Holden, G.W., & West, M.J. (1989). Proximate regulation by mothers: A demonstration of how differing styles affect young children's behavior. *Child Development, 60*, 64–69.

Holden, G.W., & Zambarano, R.J. (1992). The origins of parenting: Transmission of beliefs about physical punishment. In I.E. Sigel, A.V. McGillicuddy, & J.J. Goodnow (Eds.), *Parental belief systems: The psychological consequences for children* (2nd ed.). Hillsdale, NJ: Erlbaum.

Holliday, Robin. (1995). *Understanding aging*. Cambridge, England: Cambridge University press.

Holmbeck, G.N., & O'Donnell, K. (1991). Discrepancies between perceptions of decision making and behavioral autonomy. In R.L. Paikoff (Ed.), *New directions for child development: No. 51. Shared views in the family during adolescence*. San Francisco: Jossey-Bass.

Holmes, Ellen Rhoads, & Holmes, Lowell D. (1995). *Other cultures, elder years*. Thousand Oaks, CA.: Sage.

Holroyd, Sarah, & Baron-Cohen, Simon. (1993). Brief report: How far can people with autism go in developing a theory of mind? *Journal of Autism and Developmental Disorders, 23*, 379–385.

Holzman, Mathilda. (1983). *The language of children: Development in home and in school*. Englewood Cliffs, NJ: Prentice Hall.

Hooker, Karen, Fiese, Barbara H., Jenkins, Lisa, Morfei, Milene Z., & Schwagler, Janet. (1996). Possible selves among parents of infants and preschoolers. *Developmental Psychology, 32*, 387–389.

Horn, John L. (1982). The aging of human abilities. In Benjamin B. Wolman (Ed.), *Handbook of developmental psychology*. Englewood Cliffs, NJ: Prentice Hall.

Horn, John L. (1985). Remodeling old models of intelligence. In Benjamin B. Wolman (Ed.), *Handbook of intelligence: Theories, measurements, and applications*. New York: Wiley.

Horn, John L., & Hofer, Scott M. (1992). Major abilities and development in the adult period. In Robert J. Sternberg & Cynthia A. Berg (Eds.), *Intellectual development*. New York: Cambridge University Press.

Horney, Karen. (1967). *Feminine psychology*. Harold Kelman (Ed.). New York: Norton.

Hornick, Joseph P., McDonald, Lynn, & Robertson, Gerald B. (1992). Elder abuse in Canada and the United States: Prevalence, legal and service issues. In Ray De B. Peters, Robert J. McMahon, & Vernon L. Quinsey (Eds.), *Aggression and violence throughout the life span*. Newbury Park, CA: Sage.

Hornick, R., Risenhoover, N., & Gunnar, M. (1987). The effects of maternal positive, neutral, and negative affective communications on infant responses to new toys. *Child Development, 58*, 937–944.

Horowitz, Frances Degen. (1994). John B. Watson's legacy: Learning and environment. In Ross D. Parke, Peter A. Ornstein, John J. Rieser, & Carolyn Zahn-Waxler (Eds.), *A century of developmental psychology*. Washington, DC: American Psychological Association.

Houts, Renate M., Robins, Elliot, & Huston, Ted L. (1996). Compatibility and the development of premarital relationships. *Journal of Marriage and the Family, 58*, 7–20.

Howard, Anne. (1992). Work and family crossroads spanning the career. In Sheldon Zedeck (Ed.), *Work, families and organizations*. San Francisco: Jossey-Bass.

Howard, Darlene V., & Howard, James H., Jr. (1992). Adult age differences in the rate of learning serial patterns: Evidence from direct and indirect tests. *Psychology and Aging, 7*, 232–241.

Howard, Marion, & McCabe, Judith Blamey. (1990). Helping teenagers postpone sexual involvement. *Family Planning Perspectives, 22*, 21–26.

Howe, Neil. (1995). Why the graying of the welfare state threatens to flatten the American Dream—or worse: Age-based benefits as our downfall. *Generations: Quarterly Journal of the American Society on Aging, 19*, 15–19.

Howe, Neil, & Ross, H.S. (1990). Socialization, perspective-taking, and the sibling relationship. *Developmental Psychology, 26*, 160–165.

Howes, Carollee. (1983). Patterns of friendship. *Child Development, 54*, 1041–1053.

Howes, Carollee. (1987). Social competence with peers in young children: Developmental sequences. *Developmental Review, 7*, 252–272.

Howes, Carollee. (1992). *The collaborative construction of pretend*. Albany: State University of New York Press.

Howland, Jonathan, Hingson, Ralph, Mangione, Thomas W., Bell, Nicole, & Bak, Sharon. (1996). Why are most drowning victims men? Sex differences in aquatic skills and behaviors. *American Journal of Public Health, 86*, 93–96.

Hsu, L.K. George. (1990). *Eating disorders*. New York: Guilford.

Hu, Tuanreng, & Goldman, Noreen. (1990). Mortality differentials by marital status: An international comparison. *Demography, 27*, 233–250.

Hubel, David H. (1988). *Eye, brain, and vision*. New York: Scientific American Library.

Hudley, C., & Graham, S. (1993). An attributional intervention to reduce peer-directed aggression among African-American boys. *Child Development, 64*, 124–138.

Hudson, J.A. (1990). The emergence of autobiographical memory in mother-child conversation. In R. Fivush & J.A. Hudson (Eds.), *Knowing and remembering in young children*. Cambridge, England: Cambridge University Press.

Hughes, Dana, & Simpson, Lisa. (1995). The role of social change in preventing low birth weight. *The Future of Children: Low Birth Weight, 5*, 87–102.

Hunt, Earl. (1993). What do we need to know about aging? In John Cerella, John Rybash, William Hoyer, & Michael L. Commons (Eds.), *Adult information processing: Limits on loss*. San Diego, CA: Academic Press.

Hunt, M.E., & Ross, L. (1990). Naturally-occurring retirement communities: A multiattribute examination of desirability factors. *Gerontologist, 30*, 667–674.

Hunter, Laura Russell, & Membard, Polly Hunter. (1981). *The rest of my life*. Stamford, CT: Growing Pains Press.

Hunter, M. (1990). *Abused boys: The neglected victims of sexual abuse*. Lexington, MA: Lexington Books.

Hunter, Ski, & Sundel, Martin. (Eds.). (1989). *Midlife myths: Issues, findings and practice implications*. Newbury Park, CA: Sage.

Hurrelmann, K., & Engel, W. (1989). *The social world of adolescents: International perspectives*. Berlin: de Gruyter.

Hurt, Hallam, Brodsky, Nancy L., Betancourt, Laura, Braitman, Leonard E., Malmud, Elsa, & Gianetta, Joan. (1995). Cocaine-exposed children: Follow up through 30 months. *Journal of Developmental and Behavioral Pediatrics, 16*, 29–35.

Huston, Aletha C. (1983). Sex-typing. In P.H. Mussen (Ed.), *Handbook of child psychology: Vol. 4. Socialization, personality and social development*. New York: Wiley.

Huston, Aletha C., McLoyd, Vonnie C., & Coll, Cynthia Garcia. (1994). Children and poverty: Issues in contemporary research. *Child Development, 65*, 275–282.

Huttenlocher, Peter R. (1990). Morphometric study of human cerebral cortex development. *Neuropsychologia, 28*, 517–527.

Huttenlocher, Peter R. (1994). Synaptogenesis in human cerebral cortex. In Geraldine Dawson & Kurt W. Fischer (Eds.), *Human behavior and the developing brain*. New York: Guilford.

Huyck, Margaret Hellie. (1995). Marriage and close relationships of the marital kind. In Rosemary Blieszner & Victoria Hilkevitch Bedford (Eds.), *Handbook of aging and the family*. Westport, CT: Greenwood Press.

Hwalek, Melanie A., Neale, Anne Victoria, Goodrich, Carolyn Stahl, & Quinn, Kathleen. (1996). The association of elder abuse and substance abuse in the Illinois elder abuse system. *Gerontologist, 36*, 694–700.

Hyde, Kenneth E. (1990). *Religion in childhood and adolescence: A comprehensive review of the research*. Birmingham, AL: Religious Education Press.

Hymel, S., Rubin, K.H., Rowden, L., & LeMare, L. (1990). Children's peer relationships: Longitudinal prediction of internalizing and externalizing problems from middle to late childhood. *Child Development, 61*, 2004–2021.

Hymel, S., Bowker, A., & Woody, E. (1993). Aggressive versus withdrawn unpopular children: Variations in peer and self-perceptions in multiple domains. *Child Development, 64*, 879–896.

Iber, Frank L. (1990). Alcoholism and associated malnutrition in the elderly. In Derek M. Prinsley & Harold H. Sandstread (Eds.), *Nutrition and aging*. New York: Liss.

Idler, Ellen L. (1994). *Cohesiveness and coherance: Religion and the health of the elderly*. New York: Garland.

Ilmarinen, Juhani. (1995). A new concept for productive aging at work. In Eino Heikkinen, Jorma Kuusinen, & Isto Ruoppila (Eds.), *Preparation for aging*. New York: Plenum Press.

Ingebretsen, Reidun, & Endestad, Tor. (1995). Lifelong learning experiences from Norway. In Eino Heikkinen, Jorma Kuusinen, & Isto Ruoppila (Eds.), *Preparation for aging*. New York, Plenum Press.

Ingersol-Dayton, Berig, & Starrels, Marjorie E. (1996). Caregiving for parents and parents in law: Is gender important? *Gerontologist, 36*, 438–491.

Ingersoll, Barbara D., & Goldstein, Sam. (1993). *Attention deficit disorder and learning disabilities: Realities, myths and controversial treatments*. New York: Doubleday.

Inhelder, Bärbel, & Piaget, Jean. (1958). *The growth of logical thinking from childhood to adolescence*. New York: Basic Books.

Institute of Medicine. (1990). *Nutrition during pregnancy*. Washington, DC: National Academy of Sciences.

International Assessment of Educational Progress. (1989). *A world of differences: An international assessment of math and science*.

Isabella, R.A. (1993). Origins of attachment: Maternal interactive behavior across the first year. *Child Development, 64*, 605–621.

Isabella, R.A., & Belsky, J. (1991). Interactional synchrony and the origins of infant-mother attachment: A replication study. *Child Development, 62*, 373–384.

Israel, A.C., & Shapiro, L.S. (1985). Behavior problems of obese children enrolling in a weight reduction program. *Journal of Pediatric Psychology, 10*, 449–460.

Izard, Carroll E., & Malatesta, C.Z. (1987). Perspectives on emotional development I: Differential emotions theory of early emotional development. In Joy Doniger Osofsky (Ed.), *Handbook of infant development* (2nd ed.). New York: Wiley.

Izard, Carroll E., Hembree, E.A., & Huebner, R.R. (1987). Infants' emotional expressions to acute pain: Developmental change and stability of individual differences. *Developmental Psychology, 23*, 105–113.

Jacklin, Carol Nagy, Wilcox, K.T., & Maccoby, Eleanor E. (1988). Neonatal sex-steroid hormone and intellectual abilities of six-year-old boys and girls. *Developmental Psychobiology, 21*, 567–574.

Jackson, Jacquelyne Faye. (1993). Human behavioral genetics, Scarr's theory, and her views on interventions: A critical review and commentary on their implications for African American children. *Child Development, 64*, 1318–1332.

Jacobs, J.E., & Ganzel, A.K. (1995). Decision-making in adolescence: Are we asking the wrong question? In P.R. Pintrich & M.L. Maehr (Eds.), *Advances in achievement and motivation: Vol. 8. Motivation in adolescence*. Greenwich, CT: JAI Press.

Jacobs, J.E., Bennett, M.A., & Flanagan, C. (1993). Decision-making in one-parent and two-parent families: Influence and information selection. *Journal of Early Adolescence, 13*, 245–266.

Jacobs, P.A. (1991). The fragile X syndrome [Editorial]. *Journal of Human Genetics, 28*, 809–810.

Jacobs, Selby C., Kosten, Thomas R., Kasl, Stanislav V., Ostfeld, Adrian M., & Berkman, Lisa. (1987–1988). Attachment theory and multiple expressions of grief. *Omega, 18*, 41–52.

Jacobson, Joseph L., Jacobson, Sandra W., Padgett, Robert J., Brummitt, Gail A., & Billings, Robin L. (1992). Effects of prenatal PCB exposure on cognitive processing efficiency and sustained attention. *Developmental Psychology, 28*, 297–306.

James, Sherman, Keenan, Nora L., & Browning, Steve. (1992). Socioeconomic status, health behaviors, and health status among blacks. In K. Warner Schaie, Dan Blazer, & James S. House (Eds.), *Aging, health behaviors, and health outcomes*. Hillsdale, NJ: Erlbaum.

James, William. (1950). *The principles of psychology* (Vol. 1). New York: Dover. (Original work published 1890)

Java, Rosalind I. (1996). Effects of age on state of awareness following implicit and explicit word-association tasks. *Psychology and Aging, 11,* 108–111.

Jeanneret, Rene. (1995). The role of preparation for retirement in the improvement of the quality of life for elderly people. In Eino Heikkinen, Jorma Kuusinen, & Isto Ruoppila (Eds.), *Preparation for aging.* New York: Plenum Press.

Jecker, Nancy S., & Schneiderman, Lawrence J. (1996). Stopping futile medical treatment: Ethical issues. In David C. Thomasma & Thomasine Kushner (Eds.). *Birth to death.* Cambridge, England: Cambridge University Press.

Jencks, Christopher. (1994). *The homeless.* Cambridge, MA: Harvard University Press.

Jendrek, Margaret Platt. (1994). Grandparents who parent their grandchildren: Effects on lifestyle. *Journal of Marriage and the Family, 55,* 609–622.

Jessor, Richard. (1992). Risk behavior in adolescence: A psychosocial framework for understanding and action. *Developmental Review, 12,* 374–390.

Jessor, Richard, Donovan, John E., & Costa, Frances M. (1991). *Beyond adolescence: Problem behavior and young adult development.* Cambridge, England: Cambridge University Press.

Jette, Alan M. (1996). Disability trends and transitions. In Robert H. Binstock & Linda K. George (Eds.), *Handbook of aging and the social sciences.* San Diego, CA: Academic Press.

Jewell, S.E., & Yip, R. (1995). Increasing trends in plural births in the United States. *Obstetrics & Gynecology, 85,* 229–232.

Jockin, Victor, McGue, Matt, & Lykken, David T. (1996). Personality and divorce: A genetic analysis. *Journal of Personality and Social Psychology, 71,* 288–299.

Johnson, C.I., & Baer, B.M. (1993). Coping and a sense of control among the oldest old. *Journal of Aging Studies, 7,* 67–80.

Johnson, Charles D. (1992). Projecting unmet need for prenatal care. In Jonathan B. Kotch, Craig H. Blakely, Sarah S. Brown, & Frank Y. Wong (Eds.), *A pound of prevention: The case for universal maternity care in the U.S.* Washington, DC: American Public Health Association.

Johnson, Clifford Merle. (1991). Infant and toddler sleep: A telephone survey of parents in the community. *Journal of Developmental and Behavioral Pediatrics, 12,* 108–114.

Johnson, Clifford Merle, Mirands, Leticia, Sherman, Arloc, & Weill, James D. (1991). *Child poverty in America.* Washington, DC: Children's Defense Fund.

Johnson, Colleen L. (1995). Cultural diversity in the late-life family. In Rosemary Blieszner & Victoria Hilkevitch Bedford (Eds.), *Handbook of aging and the family.* Westport, CT: Greenwood Press.

Johnson, David W., & Johnson, Roger T. (1994). *Learning together and alone: Cooperative, competitive, and individualistic learning* (4th ed.). Boston: Allyn & Bacon.

Johnson, Edward S., & Meade, Ann C. (1987). Developmental patterns of spatial ability: An early sex difference. *Child Development, 58,* 725–740.

Johnson, Harold R., Gibson, Rose C., & Luckey, Irene. (1990). Health and social characteristics. In Zev Brown, Edward A. McKinney, & Michael Williams (Eds.), *Black aged.* Newbury Park, CA: Sage.

John-Steiner, Vera. (1986). *Notebooks of the mind: Explorations of thinking.* Albuquerque: University of New Mexico Press.

Johnston, Janet R. (1994). High-conflict divorce. *The Future of Children.*

Johnston, Janet R., Kline, Marsha, & Tschann, Jeanne. (1989). Ongoing post-divorce conflict in families contesting custody: Do joint custody and frequent access help? *American Journal of Orthopsychiatry, 59,* 576–592.

Johnston, Lloyd D., O'Malley, Patrick M. & Bachman, Jerald G. (1995). *National survey results on drug use from the Monitoring the Future study, 1975–1994: Volume I. Secondary School students.* (NIH Pub. No. 96–4027). Rockville, MD: National Institute on Drug Abuse.

Johnston, Lloyd D., O'Malley, Patrick M. & Bachman, Jerald G. (1996). *National survey results on drug use from the Monitoring the Future study, 1975–1995: Volume I. Secondary School students.* (NIH Pub. No. 97–4139). Rockville, MD: National Institute on Drug Abuse.

Jones, C.J., & Meredith, W. (1996). Patterns of personality change across the life span. *Psychology of Aging, 11,* 57–65.

Jones, Harold E., & Conrad, Herbert S. (1933). The growth and decline of intelligence: A study of a homogeneous group between the ages of ten and sixty. *Genetic Psychology Monographs, 13,* 223–298.

Jones, Lovell A. (1989). *Minorities and cancer.* New York: Springer-Verlag.

Jones, N. Burton. (1976). Rough-and-tumble play among nursery school children. In Jerome S. Bruner, Alison Jolly, & Kathy Sylva (Eds.), *Play.* New York: Basic Books.

Jones, Susan S., Smith, Linda B., & Landau, Barbara. (1991). Object properties and knowledge in early lexical learning. *Child Development, 62,* 499–516.

Jorm, A.F., Christensen, H., Henderson, A.S., Korten, A.E., MacKinnon, A.J., & Scott, R. (1994). Complaint of cognitive decline in the elderly: A comparison of reports by subjects and informants in a community survey. *Psychological Medicine, 24,* 365–374.

Jorm, A.F., Korten, A.E., & Henderson, A.S. (1987). The prevalence of dementia: A quantitative integration of the literature. *Acta Psychiatrica Scandinavica, 76,* 465–479.

Josselson, Ruthellen, & Lieblich, Amiz. (1993). *The narrative study of lives.* Newbury Park, CA: Sage.

Juel-Nielsen, Neils. (1980). *Individual and environment: Monozygotic twins reared apart.* New York: International Universities Press.

Jung, C.G. (1961). *Memories, dreams, recollections.* New York: Vintage. (Original work published 1933)

Junger-Tas, J., Terlouw, G. J. & Klein, M. W. (1994). *Delinquent behavior among young people in the Western World.* New York: Kugler.

Kach, Nick, Mazurek, Kas, Patterson, Robert S., & DeFaveri, Ivan. (1991). *Essays on Canadian education.* Calgary, Alberta: Detselig.

Kachur, S. Patrick, Potter, Lloyd B., James, Stephen P., & Powell, Kenneth E. (1995). *Suicide in the United States: 1980–1992* (Violence Surveillance Summary Series No. 1). Atlanta, GA: National Center for Injury Prevention and Control.

Kagan, Jerome. (1989). Temperamental contributions to social behavior. *American Psychologist, 44*, 668–674.

Kagan, Jerome. (1992). Yesterday's premises, tomorrow's promises. *Developmental Psychology, 28*, 990–997.

Kagan, Jerome, & Snidman, Nancy. (1991). Infant predictors of inhibited and uninhibited profiles. *Psychological Science, 2*, 40–44.

Kagan, Jerome, Arcus, Doreen, & Snidman, Nancy. (1993). The idea of temperament: Where do we go from here? In Robert Plomin & Gerald E. McClearn (Eds.), *Nature, nurture, and psychology*. Washington, DC: American Psychological Association.

Kahn, Joan R., & London, Kathryn A. (1991). Premarital sex and the risk of divorce. *Journal of Marriage and the Family, 53*, 845–855.

Kahn, P.H., Jr. (1992). Children's obligatory and discretionary moral judgments. *Child Development, 63*, 416–430.

Kahn, S.S., Nessim, S., Gray, R., Czer, L.S., Chaud, A., & Matloff, J. (1990). Increased mortality of women in coronary artery bypass surgery: Evidence for referral bias. *Annals of Internal Medicine, 112*, 561–567.

Kail, R. (1990). *The development of memory in children* (3rd ed.). New York: Freeman.

Kail, R. (1991). Developmental changes in speed of processing during childhood and adolescence. *Psychological Bulletin, 109*, 490–501.

Kail, R. (1991). Processing time declines exponentially during childhood and adolescence. *Developmental Psychology, 27*, 259–266.

Kalish, Richard A. (1985). The social context of death and dying. In Robert H. Binstock & Ethel Shanas (Eds.), *Handbook of aging and the social sciences*. New York: Van Nostrand Reinhold.

Kalish, Richard A., & Reynolds, David. (1981). *Death and ethnicity: A psychocultural study*. New York: Baywood Press.

Kaltenback, K., & Finnegan, L.P. (1992). Methadone maintenance during pregnancy: Implications for perinatal and developmental outcome. In T.B. Sonderegger (Ed.), *Perinatal substance abuse: Research findings and clinical implications*. Baltimore: Johns Hopkins University Press.

Kamerman, S.B., & Kahn, A.J. (1993). Home health visiting in Europe. *The Future of Children, 3*, 39–52.

Kandel, Denise B., & Davies, Mark. (1996). High school students who use crack and other drugs. *Archives of General Psychiatry, 53*, 71–80.

Kandel, Denise B., Wu, Ping, & Davies, Mark. (1994). Maternal smoking during pregnancy and smoking by adolescent daughters. *American Journal of Public Health, 84*, 1407–1413.

Kanner, Leo. (1943). Autistic disturbances of affective contact. *Nervous Child, 2*, 217–250.

Kantrowitz, Barbara, Wingert, Pat, & Hager, Mary. (1988, May 16). Premies. *Newsweek*.

Kaplan, Cynthia, Heneghan, Randi J., Trunca, C., & Rochelson, B. (1987). Femoral cylinder index in the diagnosis of the Ullrich-Turner syndrome. In Enid F. Gilbert & John M. Opitz (Eds.), *Genetic aspects of developmental pathology*. New York: Liss.

Kaplan, Howard B., & Johnson, Robert J. (1992). Relationship between circumstances surrounding initial illicit drug use and escalation of drug use: Moderating effect of gender and early adolescent experiences. In Meyer Glantz & Roy Pickens (Eds.), *Vulnerability to drug abuse*. Washington, DC: American Psychological Association.

Kasl, Stanislav V. (1992). Stress and health among the elderly: Overview of the issues. In May L. Wykle, Eva Kahana, & Jerome Kowal (Eds.), *Stress and health among the elderly*. New York: Springer.

Kastenbaum, Robert J. (1986). *Death, society, and the human experience*. Columbus, OH: Merrill.

Kastenbaum, Robert J. (1992). *The psychology of death*. New York: Springer-Verlag.

Katchadourian, Herant A. (1987). *Fifty: Midlife in perspective*. New York: Freeman.

Katchadourian, Herant A. (1990). Sexuality. In S.S. Feldman & G.R. Elliott (Eds.), *At the threshold: The developing adolescent*. Cambridge, MA: Harvard University Press.

Katz, Jeanne Samson. (1993). Jewish perspectives on death, dying and bereavement. In Donna Dickenson & Malcolm Johnson (Eds.), *Death, dying & bereavement*. London: Sage.

Katz, Joseph, & Sanford, Nevitt. (1979). Curriculum and personality. In Nevitt Sanford (Ed.), *College and character*. Berkeley, CA: Montaigne.

Katz, P.A. (1987). Developmental and social processes in ethnic attitudes and self-identification. In J.S. Phinney & M.J. Rotheram (Eds.), *Children's ethnic socialization: Pluralism and development*. Newbury Park, CA: Sage.

Kaufman, A.S. (1990). *Assessing adolescent and adult intelligence*. Boston: Allyn & Bacon.

Kaufman, Joan, & Zigler, Edward. (1989). The intergenerational transmission of child abuse. In Dante Cicchetti & Vicki Carlson (Eds.), *Child maltreatment: Theory and research on the causes and consequences of child abuse and neglect*. Cambridge, England: Cambridge University Press.

Kaufman, Joan, & Zigler, Edward. (1993). The intergenerational hypothesis is overstated. In Richard J. Gelles & Donileen R. Loseke (Eds.), *Current controversies in family violence*. Newbury Park, CA: Sage.

Kaufman, K.L., Johnson, C.F., Cohn, D., & McCleery, J. (1992). Child maltreatment prevention in the health care and social service system. In D.J. Willis, E.W. Holden, & M. Rosenberg (Eds.), *Prevention of child maltreatment: Developmental and ecological perspectives*. New York: Wiley.

Kaufman, Sharon R. (1986). *The ageless self*. Madison: University of Wisconsin Press.

Kaye, Kenneth. (1982). *The mental and social life of babies: How parents create persons*. Chicago: University of Chicago Press.

Kaye, Leonard, & Applegate, Jeffrey S. (1990). *Men as caregivers to the elderly*. Lexington, MA: Lexington Books.

Keating, D.P. (1990). Adolescent thinking. In S.S. Feldman & G.R. Elliott (Eds.), *At the threshold: The developing adolescent*. Cambridge, MA: Harvard University Press.

Keil, F. (1984). Mechanisms of cognitive development and the structure of knowledge. In Robert J. Sternberg (Ed.), *Mechanisms of cognitive development*. New York: Freeman.

Keith, Jennie. (1990). Age in social and cultural context: Anthropological perspectives. In Robert H. Binstock & Linda K. George (Eds.), *Handbook of aging and the social sciences* (3rd ed.). San Diego, CA: Academic Press.

Keith, Jennie, Fry, Christine L., Glascock, Anthony P., Ikels, Charlotte, Dickerson-Putman, Jeannette, Harpending, Henry C., & Draper, Patricia. (1994). *The aging experience: Diversity and commonality across cultures.* Thousand Oaks, CA: Sage.

Keith, Pat M. (1986). Isolation of the unmarried in later life. *International Journal of Aging and Human Development, 23,* 81–96.

Keith, Pat M., & Schafer, Robert B. (1991). *Relationships and well-being over the life stages.* New York: Praeger.

Kendig, Hal L., Coles, R., Pittelkow, Y., & Wilson, S. (1988). Confidants and family structure in old age. *Journal of Gerontology, 43,* 31–40.

Keniston, Kenneth, & The Carnegie Council on Children. (1977). *All our children: The American family under pressure.* New York: Harcourt Brace Jovanovich.

Kerr, Robert. (1985). Fitts' law and motor control in children. In Jane E. Clark & James H. Humphrey (Eds.), *Motor development: Current selected research.* Princeton, NJ: Princeton Book Company.

Keshet, Jamie. (1988). The remarried couple: Stresses and successes. In William R. Beer (Ed.), *Relative strangers.* Totowa, NJ: Rowman & Littlefield.

Kincade, Jean E., Rabiner, Donna J., Bernard, Shulamit L., Woomert, Alison, Konrad, Thomas R., DeFriese, Gordon H., & Ory, Marcia G. (1996). Older adults as a community resource: Results from the national survey of self-care and aging. *Gerontologist, 36,* 474–482.

King, Gary, & Williams, David R. (1995). Race and health: A multidimensional approach to African-American health. In Benjamin C. Amick III, Sol Levine, Alvin R. Tarlov, & Diana Chapman Walsh (Eds.), *Society and health.* New York: Oxford University Press.

King, P.M., Kitchner, K.S., Davison, M.L., Parker, C.A., & Wood, P.K. (1983). The justification of beliefs in young adults: A longitudinal study. *Human Development, 26,* 106–116.

King, Valerie. (1994). Non-resident father involvement and child well-being: Can dads make a difference? *Journal of Family Issues, 15,* 78–96.

Kinney, Hannah C., Filiano, James J., Sleeper, Lynn A., Mandell, Frederick, Valdes-Dapena, Marie, & White, W. Frost. (1995). Decreased muscarinic receptor binding in the arcuate nucleus in sudden infant death syndrome. *Science, 269,* 1446–1450.

Kirby, Douglas, Barth, Richard P., Leland, Nancy, & Fetro, Joyce V. (1991). Reducing the risk: Impact of a new curriculum on sexual risk-taking. *Family Planning Perspectives, 21,* 253–263.

Kitchener, Karen S., & Fischer, Kurt S. (1990). A skill approach to the development of reflective thinking. In D. Kuhn (Ed.), *Developmental aspects of teaching and learning thinking skills: Vol. 21. Contributions to human development.* Basel: Karger.

Kitchener, Karen S., & King, Patricia M. (1990). The Reflective Judgment Model: Ten years of research. In Michael L. Commons, Cheryl Armon, Lawrence Kohlberg, Francis A. Richards, Tina A. Grotzer, & Jan D. Sinnott (Eds.), *Adult development: Vol. 2. Models and methods in the study of adolescent and adult thought.* New York: Praeger.

Kitson, Gay C., & Morgan, Leslie A. (1990). The multiple consequences of divorce: A decade review. *Journal of Marriage and the Family, 52,* 913–924.

Kitzinger, Sheila. (1989). *The complete book of pregnancy and childbirth.* New York: Knopf.

Klahr, David. (1992). Information-processing approaches to cognitive development. In M.H. Bornstein & M.E. Lamb (Eds.), *Developmental psychology: An advanced textbook* (3rd ed.). Hillsdale, NJ: Erlbaum.

Klahr, David, Fay, A.L., & Dunbar, K. (1993). Heuristics for scientific experimentation: A developmental study. *Cognitive Psychology, 25,* 111–146.

Klatt, Heinz-Jahchim. (1991). In search of a mature concept of death. *Death Studies, 15,* 177–187.

Klaus, Marshall H., & Kennell, John H. (1976). *Maternal-infant bonding: The impact of early separation or loss on family development.* St. Louis, MO: Mosby.

Klein, Melanie. (1957). *Envy and gratitude.* New York: Basic Books.

Klein, T.W. (1988). *Program evaluation of the Kaemhameha Elementary Education Program's reading curriculum in Hawaii public schools: The cohort analysis 1978–1986.* Honolulu: Center for Development of Early Education.

Kleinman, J.C., Fingerhut, L.A., & Prager, K. (1991). Differences in infant mortality by race, nativity status, and other maternal characteristics. *American Journal of the Diseases of Children, 145,* 194–199.

Klepinger, Daniel H., Lundberg, Shelly, & Plotnick, Robert D. (1995). Adolescent fertility and the educational attainment of young women. *Family Practice Perspectives, 27,* 23–28.

Klesges, Robert. (1993). Effects of television on metabolic rate: Potential implications for childhood obesity. *Pediatrics, 91,* 281–286.

Kliewer, E.V., & Smith, K.R. (1995). Breast cancer mortality among immigrants in Australia and Canada. *Journal of the National Cancer Institute, 87,* 1154–1161.

Kligman, Albert M., Grove, Gary L., & Balin, Arthur K. (1985). Aging of the human skin. In Caleb E. Finch & Edward L. Schneider (Eds.), *Handbook of the biology of aging* (2nd ed.). New York: Van Nostrand.

Kline, Donald W., & Scialfa, Charles T. (1996). Visual and auditory aging. In James E. Birren & K. Warner Schaie (Eds.), *Handbook of the psychology of aging.* San Diego, CA: Academic Press.

Kline, Marsha, Tschann, Jeanne M., Johnston, Janet R., & Wallerstein, Judith S. (1989). Children's adjustment in joint and sole physical custody families. *Developmental Psychology, 25,* 430–438.

Klopfer, P. (1971). Mother love: What turns it on? *American Scientist, 49,* 404–407.

Knappert, Jan. (1989). The concept of death and afterlife in Islam. In Arthur Berger, Paul Badham, Austin H. Kutscher, Joyce Berger, Ven. Michael Petty, & John Beloff (Eds.), *Perspectives on death and dying: Cross-cultural and multidisciplinary views.* Philadelphia: Charles Press.

Kochanska, Grazyna. (1991). Socialization and temperament in the development of guilt and conscience. *Child Development, 62,* 1379–1392.

Kochanska, Grazyna. (1993). Toward a synthesis of parental socializations and child temperament in early development of conscience. *Child Development, 64,* 325–347.

Kochanska, Grazyna. (1995). Children's temperament, mother's discipline, and security of attachment: Multiple pathways to emerging socialization. *Child Development, 66,* 597–615.

Kohlberg, Lawrence. (1963). Development of children's orientation towards a moral order (Part I). Sequencing in the development of moral thought. *Vita Humana, 6*, 11–36.

Kohlberg, Lawrence. (1973). Continuities in childhood and adult moral development revisited. In Paul B. Baltes & K. Warner Schaie (Eds.), *Life-span developmental psychology: Personality and socialization*. New York: Academic Press.

Kohlberg, Lawrence. (1981). *Essays on moral development* (Vol. 1). New York: Harper & Row.

Kohlberg, Lawrence. (1981). *The philosophy of moral development*. New York: Harper & Row.

Kohnstamm, Geldolph A., Halverson, Charles F., Havil, Valerie L., & Mervielde, Ivan. (1996). Parents' free descriptions of child characteristics: A cross cultural search for the developmental antecedents of the big five. In Sara Harkness & Charles M. Super (Eds.), *Parents' cultural belief systems: The origins, expressions, and consequences*. New York: Guilford.

Kolata, Gina. (1995, May 23). Molecular tools may offer clues to reducing risks of birth defects. *The New York Times*.

Kolb, B. (1989). Brain development, plasticity, and behavior. *American Psychologist, 44*, 1203–1212.

Kondratas, A. (1991). Ending homelessness: Policy changes. *American Psychologist, 46*, 1226–1231.

Koopman, Peter, Gubbay, John, Vivian, Nigel, Goodfellow, Peter, & Lovell-Badge, Robin. (1991). Male development of chromosomally female mice transgenic for Sry. *Nature (London), 351*, 117–122.

Korbin, J.E. (1994). Sociocultural factors in child maltreatment. In G.B. Melton & F. Barry (Eds.), *Safe neighborhoods: Foundations for a new national strategy on child abuse and neglect*. New York: Guilford.

Kornhaber, Arthur. (1986). Grandparenting: Normal and pathological—A preliminary communication from the grandparent study. *Journal of Geriatric Psychiatry, 19*, 19–37.

Kornhaber, Arthur, & Woodward, Kenneth L. (1981). *Grandparents/grandchildren: The vital connection*. Garden City, NY: Anchor.

Kost, Kathryn, & Forrest, Jacqueline Darroch. (1995). Intention status of United States births in 1988: Differences by mother's socioeconomic and demographic characteristics. *Family Planning Perspectives, 27*, 11–17.

Kosterlitz, J. (1993). Golden silence? *National Journal*, pp. 800–804.

Kotlowitz, Alex. (1991). *There are no children here*. New York: Doubleday.

Kotre, John. (1984). *Outliving the self: Generativity and the interpretation of lives*. Baltimore: Johns Hopkins University Press.

Kotre, John. (1995). *White gloves: How we create ourselves through memory*. New York: Free Press.

Kozol, Jonathan. (1991). *Savage inequalities*. New York: Crown.

Kramer, Deirdre A. (1983). Post-formal operations? A need for further conceptualization. *Human Development, 26*, 91–105.

Kramer, Deirdre A., & Melchior, Jacqueline. (1990). Gender, role conflict, and the development of relativistic and dialectical thinking. *Sex Roles, 23*, 553–575.

Kramer, Deirdre A., & Woodruff, Diana S. (1986). Relativistic and dialectical thought in three adult age groups. *Human Development, 29*, 280–290.

Kranichfeld, Marion L. (1987). Rethinking family power. *Journal of Family Issues, 8*, 42–56.

Krebs, D.L., Vermueulen, S.C.A., Carpendale, J.I., & Denton, K. (1991). Structural and situational influences on moral judgement: The interaction between stage and dilemma. In W.M. Kurtines & J.L. Gewirtz (Eds.), *Handbook of moral behavior and development* (Vol. 2). Hillsdale, NJ: Erlbaum.

Krieger, Nancy, & Sidney, Stephen. (1996). Racial discrimination and blood pressure: The cardia study of young black and white adults. *American Journal of Public Health, 86*, 1370–1378.

Kroger, Jane. (1989). *Identity in adolescence: The balance between self and other*. London: Routledge.

Kroger, Jane. (1993). Ego identity: An overview. In J. Kroger (Ed.), *Discussions on ego identity*. Hillsdale, NJ: Erlbaum.

Kroger, Jane. (1995). The differentiation of "firm" and "developmental" foreclosure identity statuses: A longitudinal study. *Journal of Adolescent Research, 10*, 317–337.

Kromelow, Susan, Harding, Carol, & Touris, Margot. (1990). The role of the father in the development of stranger sociability during the second year. *American Journal of Orthopsychiatry, 6*, 521–530.

Kropp, Joseph P., & Haynes, O. Maurice. (1987). Abusive and nonabusive mothers' ability to identify general and specific emotion signals of infants. *Child Development, 58*, 187–190.

Ku, Leighton C., Sonenstein, Freya L., & Pleck, Joseph H. (1992). The association of AIDS education and sex education with sexual behavior and condom use among teenage men. *Family Planning Perspectives, 24*, 100–106.

Kübler-Ross, Elisabeth. (1969). *On death and dying*. New York: Macmillan.

Kübler-Ross, Elisabeth. (1975). *Death: The final stage of growth*. Englewood Cliffs, NJ: Prentice Hall.

Kuczaj, Stan A. (1986). Thoughts on the intentional basis of early object word extension: Evidence from comprehension and production. In Stan A. Kuczaj & Martyn D. Barrett (Eds.), *The development of word meaning: Progress in cognitive developmental research*. New York: Springer-Verlag.

Kuczynski, L., & Kochanska, G. (1990). Development of children's noncompliance strategies from toddlerhood to age 5. *Developmental Psychology, 26*, 398–408.

Kuhl, P.K., & Meltzoff, A.N. (1988). Speech as an intermodal object of perception. In A. Yonas (Ed.), *Minnesota Symposia on Child Psychology: Vol. 20. Perceptual development in infancy*. Hillsdale, NJ: Erlbaum.

Kuhl, P.K., Williams, K.A., Lacerda, F., Stevens, K.N., & Lindblom, B. (1992). Linguistic experience alters phonetic perception in infants by 6 months of age. *Science, 255*, 606–608.

Kuhn, Deanna. (1992). Cognitive development. In M.H. Bornstein & M.E. Lamb (Eds.), *Developmental psychology: An advanced textbook* (3rd ed.). Hillsdale, NJ: Erlbaum.

Kuhn, Deanna, Garcia-Mita, Merce, Zohar, Arat, & Anderson, Christopher. (1995). Strategies of knowledge acquisition. *Monographs of the Society for Research in Child Development, 60* (Serial No. 245).

Kuhse, Helga. (1996). Voluntary euthanasia and other medical end-of-life decisions: Doctors should be permitted to give death a helping hand. In David C. Thomasma & Thomasine Kushner (Eds.), *Birth to death*. Cambridge, England: Cambridge University Press.

Kupersmidt, J.B., Coie, J.D. (1990). Preadolescent peer status, aggression, and school adjustment as predictors of externalizing problems in adolescence. *Child Development, 61,* 1350–1362.

Kurdek, Lawrence A. (1989). Relationship quality for newly married husbands and wives: Marital history, stepchildren, and individual difference predictors. *Journal of Marriage and the Family,* 52, 1053–1064.

Kurdek, Lawrence A. (1991). The relations between reported well-being and divorce history, availability of a proximate adult, and gender. *Journal of Marriage and the Family, 53,* 71–78.

Kurdek, Lawrence A. (1992). Relationship status and relationship satisfaction in cohabiting gay and lesbian couples. *Journal of Social and Personal Relationships, 9,* 125–142.

Labouvie-Vief, Gisela. (1985). Intelligence and cognition. In James E. Birren & K. Warner Schaie (Eds.), *Handbook of the psychology of aging* (2nd ed.). New York: Van Nostrand Reinhold.

Labouvie-Vief, Gisela. (1986, November 20). *Mind and self in life-span development. Symposium on developmental dimensions of adult adaptation: Perspectives on mind, self, and emotion.* Paper presented at the meeting of the Gerontological Association of America, Chicago.

Labouvie-Vief, Gisela. (1990). Wisdom as integrated thought: Historical and developmental perspectives. In Robert J. Sternberg (Ed.), *Wisdom: Its nature, origins, and development.* Cambridge, England: Cambridge University Press.

Labouvie-Vief, Gisela. (1992). A neo-Piagetian perspective on adult cognitive development. In Robert J. Sternberg & Cynthia A. Berg (Eds.), *Intellectual development.* New York: Cambridge University Press.

LaFramboise, T.D., Coleman, H.L.K., & Gerton, J. (1993). Psychological impact of biculturalism: Evidence and theory. *Psychological Bulletin, 114,* 395–412.

Lagerspetz, Kirsti & Bjorkquist, Kaj. (1994). In L. Rowell Huesmann (Ed.) *Aggressive Behavior.* New York: Plenum.

Lagrand, Louis E. (1981). Loss reactions of college students: A descriptive analysis. *Death Education,* 5, 235–248.

Lahey, Benjamin B., & Loeber, Rolf. (1994). Framework for a developmental model of oppositional defiant disorder and conduct disorder. In Donald K. Routh (Ed.), *Disruptive behavior disorders in childhood.* New York: Plenum Press.

Lamb, Michael E. (1982). Maternal employment and child development: A review. In Michael E. Lamb (Ed.), *Nontraditional families: Parenting and child development.* Hillsdale, NJ: Erlbaum.

Lamb, Michael E. (1987). *The father's role: Cross-cultural perspectives.* Hillsdale, NJ: Erlbaum.

Lamb, Michael E., & Sternberg, Kathleen J. (1990). Do we really know how day care affects children? *Journal of Applied Developmental Psychology, 11,* 351–379.

Lamb, Michael E., Thompson, R.A., Gardner, W.P., & Charnov, E.L. (1985). *Infant-mother attachment.* Hillsdale, NJ: Erlbaum.

Lambert, Wallace E., Genesee, Fred, Holobow, Naomi & Chartrand, Louise. (1993). Bilingual education for majority English-speaking children. *European Journal of Psychology of Education, 8,* 3–22.

Lamborn, Susie D., & Steinberg, Laurence. (1993). Emotional autonomy redux: Revisiting Ryan and Lynch. *Child Development, 64,* 483–499.

Lamborn, Susie D., Mounts, Nina S., Steinberg, Laurence, & Dornbusch, Sanford M. (1991). Patterns of competence and adjustment among adolescents from authoritarian, authoritative, indulgent, and neglectful families. *Child Development, 62,* 1049–1065.

Lamborn, Susie D., Dornbusch, Sanford M., & Steinberg, Laurence. (1996). Ethnicity and community context as moderators of the relations between family decision making and adolescent adjustment. *Child Development, 67,* 283–301.

Lamm, S.S., & Fisch, M.L. (1982). *Learning disabilities explained.* Garden City, NY: Doubleday.

Landale, Nancy S., & Fennelly, Katherine. (1992). Informal unions among mainland Puerto Ricans: Cohabitation or an alternative to legal marriage. *Journal of Marriage and the Family, 54,* 269–280.

Lang, Frieder R., & Carstensen, Laura L. (1994). Close emotional relationships in late life: Further support for proactive aging in the social domain. *Psychology and Aging, 9,* 315–324.

Langford, Peter E., & Claydon, Leslie R. (1989). A non-Kohlbergian approach to the development of justifications for moral judgements. *Educational Studies, 15,* 261–279.

Lansing, L. Stephen. (1983). *The three worlds of Bali.* New York: Praeger.

LaPointe, A.E., Mead, N.A., & Askew, J.M. (1992). *Learning mathematics.* Princeton, NJ: Educational Testing Service.

Larroque, Beatrice, Kaminski, Monique, Dehaene, Philippe, Subtil, Damien, Delfosse, Marie-Jo, & Querleu, Denis. (1995). Moderate prenatal alcohol exposure and psychomotor development at preschool age. *American Journal of Public Health, 85,* 1654–1661.

Larsen, Jean M., & Robinson, Clyde C. (1989). Later effects of preschool on low-risk children. *Early Childhood Research Quarterly, 4,* 133–144.

Larson, David, Swyers, James, & Larson, Susan. (1995). *The costly consequences of divorce: Assessing the clinical, economic, and public health impact of marital disruption in the United States.* Rockville, MD: National Institute for Healthcare Research.

Larson, Reed W., & Ham, Mark. (1993). Stress and "storm and stress" in early adolescence: The relationship of negative events with dysphoric affect. *Developmental Psychology, 29,* 130–140.

Larson, Reed, & Richards, Maryse H. (1994). *Divergent realities: The emotional lives of mothers, fathers, and adolescents.* New York: Basic Books.

La Rue, Asenath, Dessonwille, Connie, & Jarvik, Lissy F. (1985). Aging and mental disorders. In James E. Birren & K. Warner Schaie (Eds.), *Handbook of the psychology of aging,* New York: Van Nostrand Reinhold.

Laslett, Peter. (1991). *A fresh map of life: The emergence of the third age.* Cambridge, MA: Harvard University Press.

Lattanzi-Licht, Marcia, & Connor, Stephen. (1995). Care of the dying: The hospice approach. In Hannelore Wass & Robert A. Neimeyer (Eds.), *Dying: Facing the facts.* Washington, DC: Taylor & Francis.

Lau, Christopher. (1991). The impact of aging on cardiovascular function and reactivity. In Ralph L. Cooper, Jerome M. Goldman, & Thomas J. Harbin (Eds.), *Aging and environmental toxicology.* Baltimore: Johns Hopkins University Press.

Laumann, Edward O., Gagnon, John H., Michael, Robert T., & Michaels, Stuart. (1994). *The social organization of sexuality: Sexual practices in the United States.* Chicago: University of Chicago Press.

LaVoie, Donna, & Light, Leah L. (1994). Adult age differences in repetition priming: A meta-analysis. *Psychology and Aging, 9,* 539–553.

Lawton, M. Powell, Klebank, Morton H., Rajagopal, Doris, & Dean, Jennifer. (1992). Dimensions of affective experience in three age groups. *Psychology and Aging, 7,* 171–184.

Leadbeater, B. (1986). The resolution of relativism in adult thinking: Subjective, objective, or conceptual. *Human Development, 29,* 291–300.

Leaf, Alexander. (1982). Long-lived populations: Extreme old age. *Journal of the American Geriatric Society, 30,* 485–487.

Lee, Gary R., Seccombe, Karen, & Shehan, Constance L. (1991). Marital status and personal happiness: An analysis of trend data. *Journal of Marriage and the Family, 53,* 839–844.

Lee, I-Min, Manson, JoAnn E., Hennekens, Charles H., & Paffenbarger, Ralph S. (1993). Body weight and mortality: A 27-year follow-up of middle-aged men. *JAMA, Journal of the American Medical Association,* 270, 2823–2828.

Lee, John Alan. (1988). Love-styles. In Robert J. Sternberg & Michael L. Barnes (Eds.), *The psychology of love.* New Haven, CT: Yale University Press.

Lee, Thomas F. (1993). *Gene future: The promise and perils of the new biology.* New York: Plenum Press.

Lee, Valerie E., Brooks-Gunn, Jeanne, & Schnur, Elizabeth. (1988). Does Head Start work? A 1 year follow-up comparison of disadvantaged children attending Head Start, no preschool, and other preschool programs. *Developmental Psychology, 24,* 210–222.

Leenaars, Antoon A., & Lester, David. (1995). The changing suicide pattern in Canadian adolescents and youth, compared to their American counterparts. *Adolescence, 30,* 539–547.

Leibowitz, Sarah F., & Kim, Taewon. (1992). Impact of a galanin antagonist on exogenic galanin and natural patterns of fat ingestion. *Brain Research, 599,* 148.

Leifer, A.D., Leiderman, P.H., Barnett, C.R., & Williams, J.A. (1972). Effects of mother-infant separation on maternal attachment behavior. *Child Development, 43,* 1203–1218.

Lenneberg, Eric H. (1967). *Biological foundations of language.* New York: Wiley.

Lentzner, Harold R., Pamuk, Elsie R., Rhodenhiser, Richard R., & Powell-Griner, Eve. (1992). The quality of life in the year before death. *American Journal of Public Health, 82,* 1093–1098.

Leon, Gloria R., Perry, Cheryl L., Mangelsdorf, Carolyn, & Tell, Grethe J. (1989). Adolescent nutritional and psychological patterns and risk for the development of an eating disorder. *Journal of Youth and Adolescence, 18,* 273-282.

Leong, Frederick T.L. (Ed.). (1995). *Career development and vocational behavior of racial and ethnic minorities.* Mahwah, NJ: Erlbaum.

Leong, Frederick T.L., & Brown, Michael T. (1995). Theoretical issues in cross-cultural career development: Cultural validity and cultural specificity. In W. Bruce Walsh & Samuel H. Osipow (Eds.), *Handbook of vocational psychology: Theory, research and practice* (2nd ed.). Mahwah, NJ: Erlbaum.

Lerner, H.E. (1978). Adaptive and pathogenic aspects of sex-role stereotypes: Implications for parenting and psychotherapy. *American Journal of Psychiatry, 135,* 48–52.

Lerner, Richard A., Delaney, Mary, Hess, Laura E., Jovanovic, Jasna, & von Eye, Alexander. (1990). Adolescent physical attractiveness and academic competence. *Journal of Early Adolescence, 10,* 4–20.

Leslie, A.M. (1984). Spatiotemporal continuity and the perception of causality in infants. *Perception, 13,* 297–305.

Leslie, A.M., & Frith, U. (1988). Autistic children's understanding of seeing, knowing, and believing. *British Journal of Developmental Psychology, 6,* 315–324.

Leslie, A.M., & Keeble, S. (1987). Do six-month-olds perceive causality? *Cognition, 25,* 265–288.

Lester, Barry M., & Dreher, Melanie. (1989). Effects of marijuana use during pregnancy on newborn cry. *Child Development, 60,* 765–771.

Lester, Barry M., Hoffman, Joel, & Brazelton, T. Berry. (1985). The rhythmic structure of mother-infant interaction in term and preterm infants. *Child Development, 56,* 15–27.

Lester, Barry M., Corwin, Michael J., Sepkoski, Carol, Seifer, Ronald, Peucker, Mark, McLaughlin, Sarah, & Golub, Howard L. (1991). Neurobehavioral syndromes in cocaine-exposed newborn infants. *Child Development, 62,* 694–705.

Levenson, Robert W., Carstensen, Laura R., & Gottman, John M. (1993). Long-term marriage: Age, gender, and satisfaction. *Psychology and Aging, 8,* 301–313.

LeVine, Robert A. (1980). A cross-cultural perspective on parenting. In M.D. Fantini & R. Cardenas (Eds.), *Parenting in a multicultural society.* New York: Longman.

LeVine, Robert A. (1988). Human parental care: Universal goals, cultural strategies, individual behavior. In R.A. LeVine, P.M. Miller, & M.M. West (Eds.), *Parental behavior in diverse societies.* San Francisco: Jossey-Bass.

LeVine, Robert A. (1989). Cultural influences in child development. In William Damon (Ed.), *Child development today and tomorrow.* San Francisco: Jossey-Bass.

Levinson, Daniel J. (1978). *The seasons of a man's life.* New York: Knopf.

Levinson, Ruth Andrea, Jaccard, James, & Beamer, LuAnn. (1995). Older adolescents' engagement in casual sex: Impact of risk perception and psychosocial motivations. *Journal of Youth and Adolescence, 24,* 349–364.

Levy, Becca. (1996). Improving memory in old age through implicit self-stereotyping. *Journal of Personality and Social Psychology, 71,* 1092–1106.

Levy, Gary D. (1994). Aspects of preschoolers' comprehension of indoor and outdoor gender-typed toys. *Sex Roles, 30,* 391–405.

Levy, Gary D., & Carter, D.B. (1989). Gender schema, gender constancy, and gender-role knowledge: The roles of cognitive factors in preschoolers' gender-role stereotype attributions. *Developmental Psychology, 25,* 444–449.

Lewin, Tamar. (1991, July 21). Communities and their residents age gracefully. *The New York Times,* pp. A1, A16.

Lewin, Tamar. (1994). Births to young teenagers decline, agency says. *The New York Times,* p. A18.

Lewis, M. (1990). Social knowledge and social development. *Merrill-Palmer Quarterly, 36,* 93–116.

Lewis, M., & Brooks, J. (1978). Self-knowledge and emotional development. In M. Lewis & L.A. Rosenblum (Eds.), *The development of affect*. New York: Plenum Press.

Lewis, M., & Michalson, L. (1983). *Children's emotions and moods*. New York: Plenum Press.

Lewis, M., Sullivan, M.W., Stanger, C., & Weiss, M. (1989). Self development and self-conscious emotions. *Child Development, 60*, 146–156.

Lewis, M., Alessandri, S.M., & Sullivan, M.W. (1990). Violation of expectancy, loss of control, and anger expressions in young infants. *Developmental Psychology, 26*, 745–751.

Lewis, M., Alessandri, S.M., & Sullivan, M.W. (1992). Differences in shame and pride as a function of children's gender and task difficulty. *Child Development, 63*, 630–638.

Lidz, Theodore. (1976). *The person: His and her development throughout the life cycle* (rev. ed.). New York: Basic Books.

Lieberman, A.F. (1993). *The emotional life of the toddler*. New York: Free Press.

Lieberman, Alicia F., Weston, Donna R., & Pawl, Jeree H. (1991). Preventive intervention and outcome with anxiously attached dyads. *Child Development, 62*, 199–209.

Lieberman, Ellice, Gremy, Isabelle, Lang, Janet, & Cohen, Amy. (1994). Low birthweight at term and the timing of fetal exposure to maternal smoking. *American Journal of Public Health, 84*, 1127–1131.

Lieberman, Morton A. (1992). Limitations of the psychological stress model. In May L. Wykle, Eva Kahana, & Jerome Kowal (Eds.), *Stress and health among the elderly*. New York: Springer.

Light, John M., Grigsby, Jill S., & Bligh, Michelle C. (1996). Aging and heteogeneity: Genetics, social structure, and personality. *Gerontologist, 36*, 165–173

Light, Leah L. (1991). Memory and aging: Four hypotheses in search of data. *Annual Review of Psychology, 42*, 333–376.

Light, Leah L., & Capps, Janet L. (1986). Comprehension of pronouns in young and older adults. *Developmental Psychology, 22*, 580–585.

Lillard, A.S. (1993). Pretend play skills and the child's theory of mind. *Child Development, 64*, 348–371.

Lillard, A.S. (1993). Young children's conceptualization of pretense: Action or mental representational state? *Child Development, 64*, 372–386.

Lillard, A.S. (1994). Making sense of pretense. In C. Lewis & P. Mitchell (Eds.), *Children's early understanding of mind*. Hillsdale, NJ: Erlbaum.

Lindenberger, Ulman, & Baltes, Paul B. (1994). Sensory functioning and intelligence in old age: A strong connection. *Psychology and Aging, 9*, 339–355.

Lindsey, Duncan. (1991). Factors affecting the foster care placement decision: An analysis of national survey data. *American Journal of Orthopsychiatry, 61*, 272–281.

Lipsitt, Lewis P. (1990). Learning and memory in infants. *Merrill-Palmer Quarterly, 36*, 53–66.

Lipton, Robert I. (1994). The effect of moderate alcohol use on the relationship between stress and depression. *American Journal of Public Health, 84*, 1913–1917.

Liston, Daniel P., & Zeichner, Kenneth M. (1991). *Teacher education and the social conditions of schooling*. New York: Routledge.

Liu, R., Paxton, W.A., Choe, S., Ceradini, D., Martin, S.R., Horuk, R., MacDonald, M.E., Stuhlmann, H., Koup, R.A., & Landau, N.R. (1996). Homozygous defect in HIV-1 coreceptor accounts for resistance of some multiply-exposed individuals to HIV-1 infection. *Cell, 86*, 367–378.

Lobel, T.E., & Menashri, J. (1993). Relations of conceptions of gender-role transgressions and gender constancy to gender-typed toy preferences. *Developmental Psychology, 29*, 150–155.

Locke, J.L. (1993). *The child's path to spoken language*. Cambridge, MA: Harvard University Press.

Lockman, J.J., & Thelen, E. (1993). Developmental biodynamics: Brain, body, behavior connections. *Child Development, 64*, 953–959.

Loehlin, John C. (1992). *Genes and environment in personality development*. Newbury Park, CA: Sage.

Loehlin, John C., Willerman, Lee, & Horn, Joseph M. (1982). Personality resemblances between unwed mothers and their adopted-away offspring. *Journal of Personality and Social Psychology, 42*, 1089–1099.

Loehlin, John C., Willerman, Lee, & Horn, Joseph M. (1988). Human behavior genetics. *Annual Review of Psychology, 39*, 101–133.

London, K. (1991). Cohabitation, marriage, marital dissolution, and remarriage: United States 1988. *Advance data, 194*. Washington, DC: U.S. Government Printing Office.

Loomis, Laura Spencer, & Landale, Nancy S. (1994). Nonmarital cohabitation and childbearing among black and white American women. *Journal of Marriage and the Family, 56*, 949–962.

Loscocco, Karyn, & Roschlee, Anne R. (1991). Influences on the quality of work and nonwork life: Two decades in review. *Journal of Vocational Behavior, 39*, 182–225.

Lott, I.T., & McCoy, E.E. (1992). *Down syndrome: Advances in medical care*. New York: Wiley-Liss.

Lovelace, Eugene A. (1990). Aging and metacognitions concerning memory function. In Eugene A. Lovelace (Ed.), *Aging and cognition: Mental processes, self-awareness, and interventions*. Amsterdam: North-Holland/Elsevier.

Lovett, S.B., & Flavell, J.H. (1990). Understanding and remembering: Children's knowledge about the differential effects of strategy and task variables on comprehension and memorization. *Child Development, 61*, 1842–1858.

Lowrey, George H. (1986). *Growth and development of children* (8th ed.). Chicago: Year Book Medical Publishers.

Lucas, A.R. (1991). Eating disorders. In M. Lewis (Ed.), *Child and adolescent psychiatry: A comprehensive textbook*. Baltimore: Williams & Wilkins.

Lucas, Tamara, Hense, Rosemary, & Donato, Ruben. (1990). Promoting the success of Latino language-minority students: An exploratory study of six high schools. *Harvard Educational Review, 60*, 315–340.

Lukeman, Diane, & Melvin, Diane. (1993). Annotation: The preterm infant: Psychological issues in childhood. *Journal of Child Psychology and Psychiatry, 34*, 837–849.

Lund, Dale. (1984–1988). Longitudinal study on caregivers. Salt Lake City: University of Utah, Gerontology Center.

Lundman, Richard J. (1993). *Prevention and control of juvenile delinquency* (2nd ed.) New York: Oxford University Press.

Luthar, Suniya S., & Zigler, Edward. (1991). Vulnerability and competence: A review of research on resilience in childhood. *American Journal of Orthopsychiatry, 61*, 6–22.

Lykken, D.T., McGue, M., Tellegen, A., & Bouchard, T.J., Jr. (1992). Emergenesis: Genetic traits that may not run in families. *American Psychologist, 47*, 1565–1577.

Lynott, Robert J., & Lynott, Patricia Passuth. (1996). Tracing the course of theoretical development in the sociology of aging. *Gerontologist, 36*, 749–760.

Lyon, Jeff, & Gorner, Peter. (1995). *Altered fates: Gene therapy and the retooling of human life.* New York: Norton.

Lytton, Hugh, & Romney, D.M. (1991). Parents' differential socialization of boys and girls: A meta-analysis. *Psychological Bulletin, 109*, 267–296.

MacArdle, Paul, O'Brien, Gregory, & Kolvin, Israel. (1995). Hyperactivity: Prevalence and relationship with conduct disorder. *Journal of Child Psychology and Psychiatry and Allied Disciplines, 36*, 279–303.

Maccoby, Eleanor Emmons. (1980). *Social development: Psychological growth and the parent-child relationship.* New York: Harcourt Brace Jovanovich.

Maccoby, Eleanor Emmons. (1984). Socialization and developmental change. *Child Development, 55*, 317–328.

Maccoby, Eleanor Emmons. (1990). Gender and relationships: A developmental account. *American Psychologist, 45*, 513–520.

Maccoby, Eleanor Emmons. (1992). The role of parents in the socialization of children: An historical overview. *Developmental Psychology, 28*, 1006–1017.

Maccoby, Eleanor Emmons. (1994). The role of the parents in the socialization of children: An historical overview. In Ross D. Parke, Peter A. Ornstein, John J. Rieser, & Carolyn Zahn-Waxler (Eds.), *A century of developmental psychology.* Washington, DC: American Psychological Association.

Maccoby, Eleanor Emmons, & Martin, John A. (1983). Socialization in the context of the family: Parent-child interaction. In H. Mussen (Ed.), *Handbook of child psychology: Vol. 4. Socialization, personality and social development.* New York: Wiley.

Maccoby, Eleanor Emmons, & Mnookin, R.H. (1992). *Dividing the child: Social and legal dilemmas of custody.* Cambridge, MA: Harvard University Press.

Maccoby, Eleanor Emmons, Depner, Charlene E., & Mnookin, Robert H. (1990). Coparenting in the second year after divorce. *Journal of Marriage and the Family, 52*, 141–155.

MacDonald, Kevin, & Parke, Ross D. (1986). Parent-child physical play: The effect of sex and age of children and parents. *Sex Roles, 15*, 367–378.

MacDonald, William L., & DeMaris, Alfred. (1995). Remarriage, stepchildren, and marital conflict: Challenges to the incomplete institutionalization hypothesis. *Journal of Marriage and the Family, 57*, 387–398.

Mackenbach, Johan P., Looman, Caspar W.N., & van der Meer, Joost B.W. (1996). Differences in the misreporting of chronic conditions by level of education: The effect on inequalities in prevalence rates. *American Journal of Public Health, 86*, 706–711.

Macmillan, Malcolm. (1996). *Freud evaluated: The completed arc.* Cambridge, MA: MIT.

Maddox, J. (1993). Wilful public misunderstanding of genetics. *Nature (London), 364*, 281.

Main, Mary, & Goldwyn, R. (1992). Interview-based adult attachment classifications: Related to infant-mother and infant-father attachment. (Unpublished.)

Main, Mary, & Hesse, E. (1990). Parents' unresolved traumatic experiences are related to infant disorganized attachment status: Is frightened and/or frightening parental behavior the linking mechanism? In M.T. Greenberg, D. Cicchetti, & E.M. Cummings (Eds.), *Attachment in the preschool years.* Chicago: University of Chicago Press.

Main, Mary, & Solomon, J. (1986). Discovery of an insecure-disorganized/disoriented attachment pattern. In T.B. Brazelton & M.W. Yogman (Eds.), *Affective development in infancy.* Norwood, NJ: Ablex.

Main, Mary, Kaplan, N., & Cassidy, J. (1985). Security in infancy, childhood, and adulthood: A move to the level of representation. *Monographs of the Society for Research in Child Development, 50*(Serial No. 209).

Malatesta, C.Z., Culver, C., Tesman, J.R., & Shepard, B. (1989). The development of emotional expression during the first two years of life. *Monographs of the Society for Research in Child Development, 54*(1–2, Serial No. 219).

Malina, Robert M. (1990). Physical growth and performance during the transitional years (9–16). In Raymond Montemayor, Gerald R. Adams, & Thomas P. Gullotta (Eds.), *From childhood to adolescence: A transitional period?* Newbury Park, CA: Sage.

Malina, Robert M. (1991). Growth spurt, adolescent. In Richard M. Lerner, Ann C. Petersen, & Jeanne Brooks-Gunn (Eds.), *Encyclopedia of adolescence* (Vol. 1). New York: Garland.

Malina, Robert M., & Bouchard, Claude. (1991). *Growth, maturation, and physical activity.* Champaign, IL: Human Kinetics Books.

Malina, Robert M., Bouchard, Claude, & Beunen, G. (1988). Human growth: Selected aspects of current research on well-nourished children. *Annual Review of Anthropology, 17*, 187–219.

Mallory, B.L., & New, R.S. (1994). *Diversity and developmentally appropriate practice: Challenges for early childhood education.* New York: Teachers College Press.

Mangelsdorf, S., Gunnar, M., Kestenbaum, R., Lang, S., & Andreas, D. (1990). Infant proneness-to-distress temperament, maternal personality, and mother-infant attachment: Associations and goodness of fit. *Child Development, 61*, 820–831.

Mangen, David J., Bengston, Vern L., & Landry, Pierre H. (1988). *Measurement of intergenerational relations.* Newbury Park, CA: Sage.

Mann, L., Harmoni, R., & Power, C. (1989). Adolescent decision-making: The development of competence. *Journal of Adolescence, 12*, 265–278.

Manning, Wendy D. (1993). Marriage and cohabitation following premarital conception. *Journal of Marriage and the Family, 55*, 839–850.

Manning, Wendy D., & Landale, Nancy S. (1996). Racial and ethnic differences in the role of cohabitation in premarital childbearing. *Journal of Marriage and the Family, 58*, 63–77.

Manson, JoAnn E., Tosteson, H., & Saherfield S. (1992). The primary prevention of myocardial infarction. *New England Journal of Medicine, 326*, 1406–1416.

Manson, JoAnn E., Willett, Walter C., Stampfer, Meir J., Colditz, Graham A., Hunter, David J., Hankinson, Susan E., Hennekens, Charles H., & Speizer, Frank E. (1995). Body weight and mortality among women. *New England Journal of Medicine, 333*, 677–685.

March of Dimes Birth Defects Foundation. (1992, September). *Genetic testing and gene therapy: National survey findings.* New York: Author.

Marcia, James E. (1966). Development and validation of ego identity status. *Journal of Personality and Social Psychology, 3*, 551–558.

Marcia, James E. (1980). Identity in adolescence. In J. Adelson (Ed.), *Handbook of adolescent psychology*. New York: Wiley.

Maris, Ronald W. (1991). The developmental perspective of suicide. In Antoon Leenaars (Ed.), *Lifespan perspectives on suicide.* New York: Plenum Press.

Markides, Kyriakos S., & Black, Sandra A. (1996). Race, ethnicity, and aging: The impact of inequality. In Robert H. Binstock & Linda K. George (Eds.), *Handbook of aging and the social sciences.* San Diego, CA: Academic Press.

Markman, E.M. (1989). *Categorization and naming in children: Problems of induction.* Cambridge, MA: MIT Press.

Markman, E.M. (1991). The whole object, taxonomic, and mutual exclusivity assumptions as initial constraints on word meanings. In J.P. Byrnes & S.A. Gelman (Eds.), *Perspectives on language and cognition.* Cambridge, England: Cambridge University Press.

Markus, H., & Nurius, P. (1986). Possible selves. *American Psychologist, 41*, 954–969.

Marmot, M.G., Rose, G., Shipley, M.J., & Thomas, B.J. (1981). Alcohol and mortality: A U-shaped curve. *Lancet, 1*, 580–583.

Marquette, Catherine M., Koonin, Lisa M., Antarch, Libby, Garguillo, Paul M., & Smith, Jack C. (1995). Vasectomy in the United States, 1991. *American Journal of Public Health, 85*, 644–649.

Marshall, Eliot. (1995). Human Genome Project: A strategy for sequencing the genome 5 years early. *Science, 267*, 783–784.

Marshall, Victor W. (1996). The state of theory in aging and the social sciences. In Robert H. Binstock & Linda K. George (Eds.), *Handbook of aging and the social sciences.* San Diego, CA: Academic Press.

Marsiske, Michael, & Willis, Sherry L. (1995). Dimensions of everyday problem-solving in older adults. *Psychology and Aging, 10*, 269–283.

Martin, C.L., & Little, J.K. (1990). The relation of gender understanding to children's sex-typed preferences and gender stereotypes. *Child Development, 61*, 1427–1439.

Martin, J.C. (1992). The effects of maternal use of tobacco products or amphetamines on offspring. In T.B. Sonderegger (Ed.), *Perinatal substance abuse: Research findings and clinical implications.* Baltimore: Johns Hopkins University Press.

Martin, Roy P., Wisenbaker, Joseph, & Huttenen, Matti. (1994). Review of factor analytic studies of temperament measures based on the Thomas-Chess structural model: Implications for the big five. In Charles F. Halverson, Geldolph Kohnstramm, & Roy P. Martin (Eds.), *The developing structures of temperament and personality from infancy to adulthood.* Hillsdale, NJ: Erlbaum.

Marx, Jean L. (1991). Zeroing in on individual cancer risk. *Science, 252*, 612–626.

Marx, Jean L. (1996). Searching for drugs that combat Alzheimer's. *Science, 273*, 50–53.

Masataka, N. (1992). Early ontogeny of vocal behavior of Japanese infants in response to maternal speech. *Child Development, 63*, 1177–1185.

Maslow, Abraham H. (1968). *Toward a psychology of being* (2nd ed.). Princeton, NJ: Van Nostrand.

Maslow, Abraham H. (1970). *Motivation and personality* (2nd ed.). New York: Harper & Row.

Masten, Ann S. (1992). Homeless children in the United States: Mark of a nation at risk. *Current Directions in Psychological Science, 1*, 41–43.

Masten, Ann S., Best, K.M., & Garmezy, Norman. (1990). Resilience and development: Contributions from children who overcome adversity. *Development and Psychopathology, 2*, 425–444.

Masters, William H., & Johnson, Virginia E. (1981). Sex and the aging process. *Journal of the American Geriatrics Society, 29*, 385–390.

Masters, William H., Johnson, Virginia E., & Kolodny, Robert C. (1994). *Heterosexuality.* New York: HarperCollins.

Matute-Bianchi, M.E. (1986). Ethnic identities and patterns of school success and failure among Mexican-descent and Japanese-American students in a California high school: An ethnographic analysis. *American Journal of Education, 91*, 233–255.

Maughan, Barbara, & Pickles, Andres. (1990). Adopted and illegitimate children growing up. In Lee N. Robins & Michael Rutter (Eds.), *Straight and devious pathways from childhood to adulthood.* Cambridge, England: Cambridge University Press.

Maughan, Barbara, Pickles, Andres, Rutter, Michael, & Ouston, J. (1990). Can schools change? I. Outcomes at six London schools. *School Effectiveness and School Improvement, 1*, 188–210.

Mayes, L.C., Granger, R.H., Bornstein, M.H., & Zuckerman, B. (1992). The problem of prenatal cocaine exposure: A rush to judgment. *JAMA, Journal of the American Medical Association, 267*, 406–408.

Mayford, Mark, Bach, Mary Elizabeth, Huand, Yan-You, Wang, Lei, Hawkins, Robert D., & Kandel, Eric R. (1996). Control of memory formation through regulated expression of a CaMKII transgene. *Science, 274*, 1678–1683.

McAdams, Dan P., de St. Aubin, Ed, & Logan, Regina L. (1993). Generativity among young, midlife, and older adults. *Psychology and Aging, 8*, 221–230.

McCabe, P.M., Schneiderman, N., Field, T.M., & Sklyler, J.S. (Eds.). (1991). *Stress, coping and disease.* Hillsdale, NJ: Erlbaum.

McCalla, Sandra, Feldman, Joseph, Webbeh, Hassan, Ahmadi, Ramin, & Minkoff, Howard L. (1995). Changes in perinatal cocaine use in an inner-city hospital, 1988 to 1992. *American Journal of Public Health, 85*, 1695–1697.

McCarthy, William J., Caskey, Nicholas H., Jarvik, Murray E., Gross, Todd M., Rosenblatt, Martin R., & Carpenter, Catherine. (1995). Menthol vs nonmenthol cigarettes: Effects on smoking behavior. *American Journal of Public Health, 85*, 67–72.

McCauley, Elizabeth, Kay, Thomas, Ito, Joanne, & Treder, Robert. (1987). The Turner syndrome: Cognitive deficits, affective discrimination, and behavior problems. *Child Development, 58*, 464–473.

McClearn, Gerald E., Plomin, Robert, Gora-Maslak, Grazyna, & Crabbe, John C. (1991). The gene chase in behavioral science. *Psychological Science, 2*, 222–229.

McCrae, Robert R., & Costa, Paul T., Jr. (1990). *Personality in adulthood*. New York: Guilford.

McCrae, Robert R., & Costa, Paul T., Jr. (1994). The stability of personality: Observations and evaluations. *Current Directions in Psychological Science, 3*, 173–175.

McCurdy, Karen, & Daro, Deborah. (1994). *Current trends in child abuse reporting and fatalities: The results of the 1993 annual fifty-state survey*. Chicago: National Committee to Prevent Child Abuse.

McEachin, John J., Smith, Tristram, & Lovaas, O. Ivar. (1993). Long-term outcome for children with autism who received early intensive behavioral treatment. *American Journal on Mental Retardation, 97*, 359–372.

McFadden, Robert D. (1990, June 19). Tragic end to adoption of crack baby. *The New York Times*, pp. B1, B4.

McGarrigle, J., & Donaldson, Margaret. (1974). Conservation "accidents." *Cognition, 3*, 341–350.

McGee, Robin A., & Wolfe, David A. (1991). Psychological maltreatment: Toward an operational definition. *Development and Psychopathology, 3*, 3–18.

McGue, Matthew. (1993). From proteins to cognitions: The behavioral genetics of alcoholism. In Robert Plomin & Gerald E. McClearin (Eds.), *Nature, nurture, and psychology*. Washington, DC: American Psychological Association.

McGue, Matthew, Bouchard, Thomas J., Jr., Iacono, William G., & Lykken, David T. (1993). Behavioral genetics of cognitive ability: A life-span perspective. In Robert Plomin & Gerald E. McClearin (Eds.), *Nature, nurture, and psychology*. Washington, DC: American Psychological Association.

McHale, Susan M., Crouter, Ann C., McGuire, Shirley A., & Updegraff, Kimberly A. (1995). Congruence between mothers' and fathers' differential treatment of siblings: Links with family relations and children's well-being. *Child Development, 66*, 116–128.

Mcintosh, Ruth, Vaughn, Sharon, & Zaragoza, Nina. (1991). A review of social interventions for students with learning disabilities. *Journal of Learning Disabilities, 24*, 451–458.

McKenry, Patrick, Julian, Teresa W., & Gavazzi, Stephen M. (1995). Toward a biosocial model of domestic violence. *Journal of Marriage and the Family, 57*, 307–320.

McKenzie, Lisa, & Stephenson, Patricia A. (1993). Variation in cesarean section rates among hospitals in Washington state. *American Journal of Public Health, 83*, 1109–1112.

McKeon, Denise. (1994). Language, culture, and schooling. In Fred Genesee (Ed.), *Educating second-language children: The whole child, the whole curriculum, the whole community*. Cambridge, England: Cambridge University Press.

McKeough, Anne. (1992). A neo-structural analysis of children's narrative and its development. In Robbie Case (Ed.), *The mind's staircase: Exploring the conceptual underpinning of children's thought and knowledge*. Hillsdale, NJ: Erlbaum.

McKinlay, John B., & Feldman, Henry A. (1994). Age-related variation and interest in normal men: Results from the Massachusetts male aging study. In Alice S. Rossi (Ed.), *Sexuality across the life course*. Chicago: University of Chicago Press.

McKusick, Victor A. (1994). *Mendelian inheritance in humans* (10th ed.). Baltimore: Johns Hopkins University Press.

McLanahan, Sara S., & Booth, K. (1991). Mother-only families: Problems, prospects, and politics. In A. Booth (Ed.), *Contemporary families: Looking forward, looking back*. Minneapolis, MN: National Council on Family Relations.

McLanahan, Sara S., & Sandefur, G. (1994). *Growing up with a single parent: What hurts, what helps*. Cambridge, MA: Harvard University Press.

McLaughlin, Barry. (1984). *Second language acquisition in childhood: Vol. 1. Preschool children* (2nd ed.). Hillsdale, NJ: Erlbaum.

McLaughlin, Barry. (1985). *Second language acquisition in childhood: Vol. 2. School-age children* (2nd ed.). Hillsdale, NJ: Erlbaum.

McLaughlin, D.K., & Jensen, L. (1993). Poverty among older Americans: The plight of nonmetropolitan elders. *Journals of Gerontology: Social Sciences, 48*, S44–S54.

McLoyd, V.C. (1990). The impact of economic hardship on black families and children: Psychological distress, parenting, and socioemotional development. *Child Development, 61*, 311–346.

McLoyd, V.C., & Flanagan, C. (Eds.). (1990). *New directions for child development: No. 46. Economic stress: Effects on family life and child development*. San Francisco: Jossey-Bass.

McMillan, Terry. (1992). *Waiting to Exhale*. New York: Viking Penguin.

Mehler, Jacques, & Fox, Robin. (Eds.). (1985). *Neonate cognition: Beyond the blooming buzzing confusion*. Hillsdale, NJ: Erlbaum.

Meis, Paul J., Goldenberg, Brian, Mercer, Moawad, Atef, Das, Anita, McNellis, Donald, Johnson, Francee, Iams, Jay D., Thom, Elizabeth, & Andrews, William W. (1995). The preterm prediction study: Significance of vaginal infections. *American Journal of Obstetrics and Gynecology, 173*, 1231–1235.

Meisami, Esmail. (1994). Aging of the sensory system. In Paola S. Timiras (Ed.), *Physiological basis of aging and geriatrics* (2nd ed.). Boca Raton, FL: CRC Press.

Melhuish, Edward, & Moss, Peter. (1991). *Day care for young children: International perspectives*. London: Routledge.

Mellanby, Alex R., Phelps, Fran A., Chrichton, Nicola J., & Tripp, John H. (1995). School sex education: An experimental programme with educational and medical benefit. *British Medical Journal, 311*, 414–417.

Mellor, Steven. (1990). How do only children differ from other children. *Journal of Genetic Psychology, 151*, 221–230.

Melton, G.B., & Russo, N. (1987). Adolescent abortion: Psychological perspectives on public policy. *American Psychologist, 42*, 69–72.

Melton, Michael A., Hersen, Michel, Van Sickle, Timothy D., & Van Hasselt, Vincent. (1995). Parameters of marriage in older adults: A review of the literature. *Clinical Psychology Review, 15*, 891–904.

Menken, Jane, Trussell, James, & Larsen, Ulla. (1986). Age and infertility. *Science, 233*, 1389–1394.

Meny, Robert G., Carroll, John L., Carbone, Mary Terese, & Kelly, Dorothy H. (1994). Cardiorespiratory recordings from infants dying suddenly and unexpectedly at home. *Pediatrics, 93*, 44–49.

Meredith, Howard V. (1978). Research between 1960 and 1970 on the standing height of young children in different parts of the world. In Hayne W. Reese & Lewis P. Lipsitt (Eds.), *Advances in child development and behavior* (Vol. 12). New York: Academic Press.

Mertz, W. (1989). Minerals. In A. Horwitz, D.M. Macfadyen, H. Munro, N.S. Scrimshaw, B. Steen, & T.F. Williams (Eds.), *Nutrition in the elderly*. Oxford, England: Oxford University Press.

Metosky, Patti, & Vondra, Joan. (1995). Prenatal drug exposure and play and coping in toddlers: A comparison. *Infant Behavior and Development, 18*, 15–25.

Meyer, Daniel R., & Bartfeld, Judi. (1996). Compliance with child support orders in divorce cases. *Journal of Marriage and the Family, 58*, 201–212.

Meyer, Bonnie J.F., Russo, Connie, & Talbot, Andrew. (1995). Discourse comprehension and problem solving: Decisions about the treatment of breast cancer by women across the life span. *Psychology and Aging, 10*, 84–103.

Meyer, D.R., & Garasky, S. (1993). Custodial fathers: Myths, realities, and child support policy. *Journal of Marriage and the Family, 55*, 73–89.

Michelsson, Katarina, Rinne, Arto, & Paajanen, Sonja. (1990). Crying, feeding and sleeping patterns in 1- to 12-month-old infants. *Child: Care, Health, and Development, 116*, 99–111.

Miedzian, Miriam. (1991). *Boys will be boys: Breaking the link between masculinity and violence*. New York: Doubleday.

Miller, Patricia H. (1990). The development of strategies of selective attention. In D.F. Bjorklund (Ed.), *Children's strategies: Contemporary views of cognitive development*. Hillsdale, NJ: Erlbaum.

Miller, Patricia H. (1993). *Theories of developmental psychology*. New York: Freeman.

Miller, Patricia H., & Aloise, P.A. (1989). Young children's understanding of the psychological causes of behavior: A review. *Child Development, 60*, 257–285.

Miller, Randi L. (1989). Desegregation experiences of minority students: Adolescent coping strategies in five Connecticut high schools. *Journal of Adolescent Research, 4*, 173–189.

Miller, Richard A. (1996). Aging and the immune response. In Edward L. Schneider & John W. Rowe (Eds.), *Handbook of the biology of aging*. San Diego, CA: Academic Press.

Mills, James L., McPartlin, Joseph M., Kirke, Peadar N., & Lee, Young J. (1995). Homocysteine metabolism in preg-nancies complicated by neural-tube defects. *Lancet, 345*, 149–151.

Mills, Jon K., & Andrianopoulos, Georgia D. (1993). The relationship between childhood onset obesity and psychopathology in adulthood. *Journal of Psychology, 127*, 547–551.

Mills, Richard W., & Mills, Jean. (1993). *Bilingualism in the primary school*. London: Routledge.

Millstein, S.G., & Litt, I.F. (1990). Adolescent health. In S.S. Feldman & G.R. Elliott (Eds.), *At the threshold: The developing adolescent*. Cambridge, MA: Harvard University Press.

Milne, Ann M., Myers, David E., Rosenthal, Alvin S., & Ginsburg, Alan. (1986). Single parents, working mothers, and the educational achievement of school children. *Sociology of Education, 59*, 125–139.

Milunsky, Aubrey. (1989). *Choices, not chances*. Boston: Little, Brown.

Mindel, C.J., Haberstein, R.W., & Roosevelt, W., Jr. (Eds.). (1988). *Ethnic families in America* (3rd ed.). New York: Elsevier.

Minkler, Meredith. (1991). Generational equity or interdependence? *Generations, 15*(4), 36, 40–42.

Minkler, Meredith, & Roe, Kathleen M. (1996). *Generations, 20*, 34–38.

Minuchin, Patricia, & Shapiro, Edna K. (1983). The school as a context for social development. In Paul H. Mussen (Ed.), *Handbook of child psychology: Vol. 4. Socialization, personality and social development*. New York: Wiley.

Minuchin, Salvador, & Nichols, Michael P. (1993). *Family healing: Tales of hope and renewal from family therapy*. New York: Free Press.

Mitchell, D.B. (1993). Implicit and explicit memory for pictures: Multiple views across the lifespan. In P. Graf & M.E.J. Masson (Eds.), *Implicit memory: New directions in cognition, development and neuropsychology*. Hillsdale, NJ: Erlbaum.

Mitchell, E.A., Ford, R.P.K., Steward, A.W., & Taylor, B.J. (1993). Smoking and the sudden infant death syndrome. *Pediatrics, 91*, 893–896.

Mitchell, John J. (1986). *The nature of adolescence*. Calgary, Alberta, Canada: Detselig.

MMWR *(Morbidity and Mortality Weekly Report)*: See **Centers for Disease Control and Prevention.**

Mobbs, Charles V. (1996). Neuroendocrinology of aging. In Edward L. Schneider & John W. Rowe (Eds.), *Handbook of the biology of aging*. San Diego, CA: Academic Press.

Moen, Phyllis. (1996). Gender, age, and the life course. In Robert H. Binsock & Linda K. George (Eds.), *Handbook of aging and the social sciences* (4th ed.). San Diego, CA: Academic Press.

Moffitt, Terrie E. (1993). The neuropsychology of conduct disorder. *Development and Psychopathology, 5*, 135–151.

Moffitt, Terrie E. (1997). Adolescence—Limited and life-course-persistent offending: A complementary pair of developmental theories. In Terence P. Thornberry (Ed.), *Development theories of crime and delinquency*. New Brunswick, NJ: Transaction.

Molina, Brooke S.G., & Chassin, Laurie. (1996). The parent-adolescent relationship at puberty: Hispanic ethnicity and parent alcoholism as moderators. *Developmental Psychology, 32*, 675–686.

Moller, David Wendell. (1996). *Confronting death: Values, institutions, and human mortality*. New York: Oxford University Press.

Montemayor, Raymond. (1986). Family variation in parent-adolescent storm and stress. *Journal of Adolescent Research, 1*, 15–31.

Moody, Harry R. (1993). A strategy for productive aging: Education in later life. In Scott A. Bass, Francis G. Caro, & Yung-Ping Chen (Eds.), *Achieving a productive aging society*. Westport, CT: Auburn House.

Moody, Harry R. (1995). Meaning and late life learning. In Eino Heikkinen, Jorma Kuusinen, & Isto Ruoppila (Eds.), *Preparation for aging*. New York: Plenum Press.

Moon, Christine, & Fifer, William P. (1990). Syllables as signals for 2-day-old infants. *Infant Behavior and Development, 13*, 377–390.

Moon, Christine, Cooper, Robin Panneton, & Fifer, William P. (1993). Two-day olds prefer their native language. *Infant Behavior and Development, 16*, 495–500.

Moore, Keith L. (1988). *The developing human: Clinically oriented embryology* (4th ed.). Philadelphia: Saunders.

Moore, Keith L. (1989). *Before we are born: Basic embryology and birth defects* (3rd ed.). Philadelphia: Saunders.

Moore, Susan, & Rosenthal, Doreen. (1991). Adolescent invulnerability and perceptions of AIDS risk. *Journal of Adolescent Research, 6*, 164–180.

Morris, Edward K., & Braukmann, Curtis J. (Eds.). (1987). *Behavioral approaches to crime and delinquency: A handbook of applications, research, and concepts.* New York: Plenum.

Morrison, N.A., Qi, J.C., Tokita, A., Kelly, P.J., Crofts, L., Niguyen, T.V., Sambrook, P.N., & Eisman, J.A. (1994). Prediction of bone density from vitamin D receptor alleles. *Nature (London), 367,* 284–287.

Morrongiello, B.A., & Rocca, P.T. (1990). Infants' localization of sounds within hemifields: Estimates of minimum audible angle. *Child Development, 61,* 1258–1270.

Mortenson, Thomas G. (1997). In the newsletter Postsecondary Education Opportunity.

Mortimer, David. (1994). *Practical laboratory andrology.* New York: Oxford University Press.

Mortimore, Peter. (1995). The positive effects of schooling. In Michael Rutter (Ed.), *Psychosocial disturbances in young people: Challenges for prevention.* Cambridge, England: Cambridge University Press.

Moscovitch, Morris. (1982). Neuropsychology of perception and memory in the elderly. In Fergus I.M. Craik & Sandra Trehub (Eds.), *Aging and cognitive processes.* New York: Plenum Press.

Moshman, D. (1990). The development of metalogical understanding. In W.F. Overton (Ed.), *Reasoning, necessity, and logic: Developmental perspectives.* Hillsdale, NJ: Erlbaum.

Moshman, D. (1993). Adolescent reasoning and adolescent rights. *Human Development, 36,* 27–40.

Moshman, D., & Franks, B.A. (1986). Development of the concept of inferential validity. *Child Development, 57,* 153–165.

Motulsky, Arno. (1994). *Assessing genetic risks: Implications for health and social policy.* Washington, DC: National Academy Press.

Mounts, Nina S., & Steinberg, Laurence. (1995). An ecological analysis of peer influence on adolescent grade point average and drug use. *Developmental Psychology, 31,* 915–922.

Moyer, Marth Sebastian. (1992). Sibling relationships among older adults. *Generations, 27*(3), 55–60.

Muisener, Philip P. (1994). *Understanding and treating adolescent substance abuse.* Thousand Oaks, CA: Sage.

Muller, J., Nielson, C. Thoger, & Skakkebaek, N.E. (1989). Testicular maturation and pubertal growth in normal boys. In I.M. Tanner & M.A. Preece (Eds.), *The physiology of human growth.* Cambridge, England: Cambridge University Press.

Mulrow, C.D., Aguilar, C., Endicott, J.E., Tuley, M.R., Velez, R., Charlip, W.S., Rhodes, M.C., Hill, J.A., & DeNino, L.A. (1990). Quality-of-life changes and hearing impairment: A randomized trial. *Annals of Internal Medicine, 113,* 188–194.

Munn, D., & Dunn, J. (1989). Temperament and the developing relationship between siblings. *International Journal of Behavioral Development, 12,* 433–451.

Murphey, D.A. (1992). Constructing the child: Relations between parents' beliefs and child outcomes. *Developmental Review, 12,* 199–232.

Murphy, John M., & Gilligan, Carol. (1980). Moral development in late adolescence and adulthood: A critique and reconstruction of Kohlberg's theory. *Human Development, 23,* 77–104.

Murphy, Lois Barclay, & Moriarty, Alice E. (1976). *Vulnerability, coping, and growth: From infancy to adolescence.* New Haven, CT: Yale University Press.

Murphy, Patricia Ann. (1986–1987). Parental death in childhood and loneliness in young adults. *Omega, 17,* 219–228.

Murray, Ann D., Dolby, Robyn M., Nation, Roger L., & Thomas, David B. (1981). Effects of epidural anesthesia on newborns and their mothers. *Child Development, 52,* 71–82.

Mustard, Cameron A., & Roos, Noralou P. (1994). The relationship of prenatal care and pregnancy complications to birthweight in Winnipeg, Canada. *American Journal of Public Health, 84,* 1450–1457.

Muwahidi, Ahmad Anisuzzaman. (1989). Islamic perspective on death and dying. In Arthur Berger, Paul Badham, Austin H. Kutscher, Joyce Berger, Ven. Michael Petty, & John Beloff (Eds.), *Perspectives on death and dying: Cross-cultural and multi-disciplinary views.* Philadelphia: Charles Press.

Mydans, Seth. (1997, February 2). Legal euthanasia: Australia faces a grim reality. *The New York Times, International,* p. 3.

Myers, B.J. (1987). Mother-infant bonding as a critical period. In M.H. Bornstein (Ed.), *Sensitive periods in development: Interdisciplinary perspectives.* Hillsdale, NJ: Erlbaum.

Myers, David G. (1993). *The pursuit of happiness.* New York: Avon Books.

Myers, N.A., Clifton, R.K., & Clarkson, M.H. (1987). When they were very young: Almost-threes remember two years ago. *Infant Behavior and Development, 10,* 123–132.

Myers, Samuel L., & Chung, Chanjin. (1996). Racial differences in home ownership and home equity among preretirement-aged households. *Gerontologist, 36,* 350–360.

Myles, J., & Quadagno, J. (Eds.). (1991). *States, labor markets, and the future of old-age policy.* Philadelphia: Temple University Press.

Natagata, D. (1989). Japanese-American children and adolescents. In G.T. Gibbs & L.N. Huang (Eds.), *Children of color.* San Francisco: Jossey-Bass.

National Academy of the Sciences. (1994). *Assessing genetic risks: Implications for health and social policy.* Washington DC: National Academy Press.

National Cancer Institute. (1992). *5-A-DAY for better health* (RFA No. CA-92-17). Bethesda, MD: Author.

National Center for Education Statistics. (1993). *The condition of education, 1993.* Washington, DC: U.S. Department of Education.

National Center for Health Statistics. (1991, June). Family structure and children's health: United States, 1988. *Vital Health Statistics* (Series 10, No. 178). Washington, DC: U.S. Department of Health and Human Services.

National Center for Health Statistics. (1993). *Vital statistics of the United States: Vol. 2. Mortality* (Part A). Washington, DC: U.S. Government Printing Office.

National Center for Health Statistics. (1995). *Health, United States, 1994.* Hyattsville, MD: Public Health Service.

National Center for Injury Prevention and Control. (1992). *Position papers from the third National Injury Control Conference.* Atlanta, GA: Centers for Disease Control.

National Center for Injury Prevention and Control. (1993). *Injury control in the 1990s: A national plan for action.* Atlanta, GA: Centers for Disease Control and Prevention, U.S. Department of Health and Human Services.

National Household Survey on Drug Abuse. (1993). United States Substance Abuse and Mental Health Services Administration.

Needleman, Herbert L., & Bellinger, David. (1994). *Prenatal exposure to toxicants: Developmental consequences.* Baltimore: Johns Hopkins University Press.

Needleman, Herbert L., Schell, Alan, Bellinger, David, Leviton, Alan, & Allred, Elizabeth N. (1990). The long-term effects of exposure to low doses of lead in childhood: An 11-year follow-up report. *New England Journal of Medicine, 322,* 83–89.

Needleman, Herbert L., Riess, Julie A., Tobin, Michael J., Biesecker, Joel B., & Greenhouse, Joel B. (1996). Bone lead levels and delinquent behavior. *JAMA, Journal of the American Medical Association, 275,* 403–404.

Neisser, Ulric, Boodoo, Gwyneth, Bouchard, Thomas J., Boykin, A. Wade, Brody, Nathan, Ceci, Stephen J., Halpern, Diane F., Loehlin, John C., Perloff, Robert, Sternberg, Robert J., & Urbina, Susana. (1996). Intelligence: Knowns and unknowns. *American Psychologist, 51,* 77–101.

Nelson, Charles A., & Horowitz, Frances Degen. (1987). Visual motion perception in infancy: A review and synthesis. In Philip Salapatek & Leslie Cohen (Eds.), *Handbook of infant perception: Vol. 2. From perception to cognition.* New York: Academic Press.

Nelson, Karin B., Dambrosia, James M., Ting, Tricia Y., & Grether, Judith K. (1996). Uncertain value of electronic fetal monitoring in predicting cerebral palsy. *New England Journal of Medicine, 334,* 613–618.

Nelson, Katherine. (1981). Individual differences in language development: Implications for development and language. *Developmental Psychology, 17,* 171–187.

Nelson, Katherine. (Ed.). (1986). *Event knowledge: Structure and function in development.* Hillsdale, NJ: Erlbaum.

Nelson, Katherine, & Hudson, J. (1988). Scripts and memory: Functional relationships in development. In F.E. Weinert & M. Perlmutter (Eds.), *Memory development: Universal changes and individual differences.* Hillsdale, NJ: Erlbaum.

Nelson, Melvin D. (1992). Socioeconomic status and childhood mortality in North Carolina. *American Journal of Public Health, 82,* 1131–1133.

Neugarten, Bernice L., & Neugarten, Dail A. (1986). Changing meanings of age in the aging society. In Alan Pifer & Lynda Bronte (Eds.), *Our aging society: Paradox and promise.* New York: Norton.

Neumann, C.G. (1983). Obesity in childhood. In M.D. Levine, W.B. Carey, A.C. Crocker, & R.T. Gross (Eds.), *Developmental-behavioral pediatrics.* Philadelphia: Saunders.

Neuspiel, Daniel R., Markowitz, Morri, & Drucker, Ernest. (1994). Intrauterine cocaine, lead and nicotine exposure and fetal growth. *American Journal of Public Health, 84,* 1492–1495.

Neville, Helen J. (1991). Neurobiology of cognitive and language processing: Effects of early experience. In K.R. Gibson & A.C. Petersen (Eds.), *Brain maturation and cognitive development: Comparative and cross-cultural perspectives.* New York: Aldine de Gruyter.

Newberger, Carolyn Moore, & White, Kathleen M. (1989). Cognitive foundations for parental care. In Dante Cicchetti & Vicki Carlson (Eds.), *Child maltreatment: Theory and research on the causes and consequences of child abuse and neglect.* Cambridge, England: Cambridge University Press.

NICHD Early Child Care Research Network. (1996, April 20). Infant child care and attachment security: Results of the NICHD study of early child care. *Proceedings of the Symposium, International Conference on Infant Studies.* Bethesda, Maryland: National Institute of Child Health and Development.

Nordin, B.E.C., & Need, A.G. (1990). Prediction and prevention of osteoporosis. In M. Bergener, M. Ermini, & H.B. Stahelin (Eds.), *Challenges of aging.* San Diego, CA: Academic Press.

Nottelmann, Edith D., & Welsh, C. Jean. (1986). The long and the short of physical stature in early adolescence. *Journal of Early Adolescence, 6,* 15–27.

Nottelmann, Edith D., Inoff-Germain, Gale, Susman, Elizabeth J., & Chrousos, George P. (1990). Hormones and behavior at puberty. In John Bancroft & June Machover Reinisch (Eds.), *Adolescence in puberty.* New York: Oxford University Press.

Nowak, Rachel. (1995). New push to reduce maternal mortality in poor countries. *Science, 269,* 780–782.

Nowakowski, R.S. (1987). Basic concepts of CNS development. *Child Development, 58,* 598–595.

Nucci, L., Guerra, N., & Lee, J. (1991). Adolescent judgments of the personal, prudential, and normative aspects of drug usage. *Developmental Psychology, 27,* 841–848.

Nugent, J. Kevin. (1991). Cultural and psychological influences on the father's role in infant development. *Journal of Marriage and the Family, 53,* 475–485.

Nugent, J. Kevin, Lester, B.M., & Brazelton, T.B. (1989). *The cultural context of infancy* (Vol. 1). Norwood, NJ: Ablex.

Nuland, Sherwin B. (1994). *How we die.* New York: Knopf.

Nussbaum, N.L., Grant, M.L., Roman, M.J., Poole, J.H., & Bigler, E.D. (1990). Attention deficit disorder and the mediating effect of age on academic and behavioral variables. *Developmental and Behavioral Pediatrics, 11,* 22–26.

Nusselder, Wilma J., van der Veldon, Koos, van Sonsbeek, Jan L. A., Lenior, Maria E., & van den Bos, Geertrudis A.M. (1996). The elimination of selected chronic disease in a population: The compression and expansion of morbidity. *American Journal of Public Health, 86,* 187–194.

Nwokah, Evangeline E., Hsu, Hui-Chin, Dobrowolka, Olga, & Fogel, Alan. (1994). The development of laughter in mother-infant communication: Timing parameters and temporal sequences. *Infant Behavior and Development, 17,* 23–35.

Oakes, Jeanne, & Lipton, Martin. (1990). *Making the best of schools: A handbook for parents, teachers, and policymakers.* New Haven, CT: Yale University Press.

Oakes, L.M., & Cohen, L.B. (1990). Infant perception of a causal event. *Cognitive Development, 5,* 193–207.

Oberlé, I., Rousseau, F., Heitz, D., Kretz, C., Devys, D., Hanauer, A., Boule, J., Bertheas, M.F., & Mandel, J.L. (1991). Instability of a 550-base pair DNA segment and abnormal methylation in fragile X syndrome. *Science, 252,* 1097–1102.

O'Brien, Mary. (1991). Never married older women: The life experience. *Social Indicators Research, 24,* 301–315.

O'Connor, D. W. (1989). The prevalence of dementia as measured by the Cambridge Mental Disorders of the Elderly examination. *Acta Psychiatrica Scandinavica, 79,* 190–198.

Offenbacher, S., Katz, V., Fertik, G., Connins, J., Boyd, D., Maynor, G., McKaig, R., & Beck, J. (1996). Periodontal infection as a possible risk factor for preterm low birth weight. *Journal of Periodontology, 67,* 1103–1113.

Offer, Daniel, & Offer, Judith. (1975). *From teenage to young manhood.* New York: Basic Books.

Ogletree, Shirley Matile, Denton, Larry, & Williams, Sue Winkle. (1993). Age and gender differences in children's Halloween costumes. *Journal of Psychology, 127,* 633–637.

O'Hagan, Kieran. (1993). *Emotional and psychological abuse of children.* Toronto: University of Toronto Press.

O'Leary, K. Daniel. (1993). Through a psychological lens: Personality traits, personality disorders, and levels of violence. In Richard J. Gelles & Donileen R. Loseke (Eds.), *Current controversies on family violence.* Thousand Oaks, CA: Sage.

Oliver, J.M., Cole, Nancy Hodge, & Hollingsworth, Holly. (1991). Learning disabilities as functions of familial learning problems and developmental problems. *Exceptional Children, 57,* 427–440.

Oller, D. Kimbrough, & Eilers, Rebecca. (1988). The role of audition in infant babbling. *Child Development, 59,* 441–449.

Olsho, Lynn Werner. (1984). Infant frequency discrimination. *Infant Behavior and Development, 7,* 27–35.

Olsho, Lynn Werner, Koch, E.G., Carter, E.A., Halpin, C.F., & Spetner, N.B. (1988). Pure tone sensitivity in human infants. *Journal of the Acoustical Society of America, 84,* 1316–1324.

Olson, Heather Carmichael, Sampson, Paul D., Barr, Helen, Streissguth, Ann P., & Bookstein, Fred L. (1992). Prenatal exposure to alcohol and school problems in late childhood: A longitudinal prospective study. *Development and Psychopathology, 4,* 341–359.

Olson, L. (1988). *Crossing the schoolhouse border: Immigrant students and the California public schools.* San Francisco: California Tomorrow.

Olvera-Ezzell, Norma, Power, Thomas G., & Cousins, Jennifer H. (1990). Maternal socialization of children's eating habits: Strategies used by obese Mexican-American mothers. *Child Development, 61,* 395–400.

Olweus, Dan. (1992). Bullying among schoolchildren: Intervention and prevention. In Peters, R.D., McMahon, R.J. & Quincy, V.L. (Eds.). *Aggression and violence throughout the life span.* Newbury Park, CA: Sage.

Olweus, Dan. (1993). Victimization by peers: Antecedents and long-term outcomes. In K.H. Rubin & J.B. Asendorf (Eds.), *Social withdrawal, inhibition, and shyness in childhood.* Hillsdale, N.J.: Erlbaum.

Olweus, Dan. (1993). *Bullying at school: What we know and what we can do.* Oxford, England; Blackwell.

Olweus, Dan. (1994). Bullying at school: Basic facts and effects of a school based intervention program. *Journal of Child Psychology and Psychiatry, 35,* 1171–1190.

O'Meara, J.J. (1989). Cross-sex friendship: Four basic challenges of an ignored relationship. *Sex Roles, 21,* 525–543.

O'Meara, J.J. (1994). Cross sex friendship's opportunity challenge: Uncharted terrain for exploration. *Personal Relationship Issues, 2,* 4–7.

Opoku, Kofi Asare. (1989). African perspectives on death and dying. In Arthur Berger, Paul Badham, Austin H. Kutscher, Joyce Berger, Ven. Michael Petty, & John Beloff (Eds.), *Perspectives on death and dying: Cross-cultural and multidisciplinary views.* Philadelphia: Charles Press.

O'Rand, Angela M. (1996). The cumulative stratification of the life course. In Robert H. Binstock & Linda K. George (Eds.), *Handbook of aging and the social sciences.* San Diego, CA: Academic Press.

Orbuch, Terri L., & Custer, Lindsay. (1995). The social context of married women's work and its impact on black husbands and white husbands. *Journal of Marriage and the Family, 57,* 333–345.

Ory, Marcia G., & Warner, Huber R. (1990). *Gender, health, and longevity.* New York: Springer.

Osgood, Nancy J. (1992). *Suicide in later life.* Lexington, MA: Lexington Books.

Ouslander, J.G. (1989). Urinary incontinence: Out of the closet. *JAMA, Journal of the American Medical Association, 261,* 2695–2696.

Overton, William F. (1990). *Reasoning, necessity, and logic: Developmental perspectives.* Hillsdale, NJ: Erlbaum.

Page, Reba, & Valli, Linda. (Eds.). (1990). *Curriculum differentiation: Interpretive studies in U.S. secondary schools.* Albany: State University of New York Press.

Paikoff, Roberta L. (1990). Attitudes toward consequences of pregnancy in young women attending a family planning clinic. *Journal of Adolescent Research, 5,* 467–468.

Paikoff, Roberta L., & Brooks-Gunn, Jeanne. (1991). Do parent-child relationships change during puberty? *Psychological Bulletin, 110,* 47–66.

Palmer, Carolyn F. (1989). The discriminating nature of infants' exploratory actions. *Developmental Psychology, 25,* 885–893.

Pampel, F.C. (1994). Population aging, class context, and age inequality in public spending. *American Journal of Sociology, 100,* 153–195.

Panel on Research on Child Abuse and Neglect, National Research Council. (1993). *Understanding child abuse and neglect.* Washington, DC: National Academy Press.

Paneth, Nigel. (1992). The role of neonatal intensive care in lowering infant mortality. In Jonathan B. Kotch, Craig H. Blakely, Sarah S. Brown, & Frank Y. Wong (Eds.), *A pound of prevention: The case for universal maternity care in the U.S.* Washington, DC: American Public Health Association.

Papernow, Patricia L. (1988). Stepparent role development: From outsider to intimate. In William R. Beer (Ed.), *Relative strangers.* Totowa, NJ: Rowman & Littlefield.

Papernow, Patricia L. (1993). *Becoming a stepfamily: Patterns of development in remarried families.* San Francisco: Jossey-Bass.

Parcel, Toby L., & Menaghan, Elizabeth G. (1995). *Parents' jobs and children's lives.* New York: Aldine de Gruyter.

Park, K.A., Lay, K. L., & Ramsay, L. (1993). Individual differences and developmental changes in preschoolers' friendships. *Developmental Psychology, 29,* 264–270.

Parke, Ross D., Ornstein, Peter A., Rieser, John J., & Zahn-Waxler, Carolyn. (1994). The past as prologue: An overview of a century of developmental psychology. In Ross D. Parke, Peter A. Ornstein, John J. Rieser, & Carolyn Zahn-Waxler (Eds.), *A century of developmental psychology.* Washington, DC: American Psychological Association.

Parker, Jeffrey G., & Asher, Steven R. (1987). Peer relations and later personal adjustment: Are low-accepted children at risk? *Psychological Bulletin, 102,* 357–389.

Parker, Jeffrey G., & Asher, Steven R. (1993). Friendship and friendship quality in middle childhood: Links with peer group acceptance and feelings of loneliness and social dissatisfaction. *Developmental Psychology, 29,* 611–621.

Parker, Jeffrey G., & Gottman, J.M. (1989). Social and emotional development in a relational context. In T.J. Berndt & G.W. Ladd (Eds.), *Peer relationships in child development.* New York: Wiley.

Parkhurst, J.T., & Asher, S.R. (1992). Peer rejection in middle school: Subgroup differences in behavior, loneliness, and interpersonal concerns. *Developmental Psychology, 28,* 231–241.

Parkin, Alan J. (1993). *Memory: Phenomena, experiment and theory.* Oxford: Blackwell.

Parrott, Roxanne Louiselle, & Condit, Celeste Michelle. (1996). *Evaluating women's health messages: A resourcebook.* Thousand Oaks, CA: Sage.

Partnership for a Drug-Free America. (1996). Survey.

Pascarella, Ernest T., & Terenzini, Patrick T. (1991). *How college affects students: Findings and insights from twenty years of research.* San Francisco: Jossey-Bass.

Pasley, Kay, & Ihinger-Tallman, Marilyn. (1987). *Remarriage and stepparenting.* New York: Guilford.

Patterson, C.J., Kupersmidt, J.B., & Griesler, P.C. (1990). Children's perceptions of self and of relationships with others as a function of sociometric status. *Child Development, 61,* 1335–1349.

Patterson, Gerald R. (1982). *Coercive family processes.* Eugene, OR: Castalia Press.

Patterson, Gerald R., & Capaldi, D. (1991). Antisocial parents: Unskilled and vulnerable. In Paul E. Cowan & Mavis Hetherington (Eds.), *Family transitions.* Hillsdale, NJ: Erlbaum.

Patterson, Gerald R., DeBaryshe, Barbara D., & Ramsey, Elizabeth. (1989). A developmental perspective on antisocial behavior. *American Psychologist, 44,* 329–335.

Patterson, Gerald R., Reid, J.B., & Dishion, T.J. (1992). *Antisocial boys.* Eugene, OR: Castalia Press.

Paulston, Christina Bratt. (1992). *Sociolinguistic perspectives on bilingual education.* Cleveden, England: Multilingual Matters.

Paunonen, Sampo V., Jackson, Douglas N., Trzebinski, Jerzy, & Forsterling, Friedrich. (1992). Personality structure across cultures: A multimodal evaluation. *Journal of Personality and Social Science, 62,* 447–456.

Peak, L. (1991). *Learning to go to school in Japan: The transition from home to preschool life.* Berkeley: University of California Press.

Pearlin, Leonard I., Aneshensel, Carol S., Mullan, Joseph T., & Whitlatch, Carol J. (1996). Caregiving and its social support. In Robert H. Binstock & Linda K. George (Eds.), *Handbook of aging and the social sciences.* San Diego, CA: Academic Press.

Pearson, J.D., Morell, C.H., Gordon-Salant, S., Brant, L.J., Metter, E.J., Klein, L., & Fozard, J.L. (1995). Gender differences in a longitudinal study of age-associated hearing loss. *Journal of the Acoustical Society of America, 97,* 1196–1205.

Pearson, Jane L., Hunter, Andrea G., Ensminger, Margaret E., & Kellam, Sheppard G. (1990). Black grandmothers in multigenerational households: Diversity in family structure and parenting involvement in the Woodlawn community. *Child Development, 61,* 434–442.

Pecheux, Marie Germaine, & Labrell, Florence. (1994). Parent-infant interactions and early cognitive development. In Andre Vyt, Henriette Bloch, & Marc H. Bornstein (Eds.), *Early child development in the French tradition: Contributions from current research.* Hillsdale, NJ: Erlbaum.

Pedersen, Frank A. (1981). Father influences viewed in a family context. In Michael E. Lamb (Ed.), *The role of the father in child development* (2nd ed.). New York: Wiley.

Pederson, Nancy L., Plomin, Robert, McClearn, Gerald E., & Friberg, L. (1988). Neuroticism, extraversion, and related traits in adult twins reared apart and reared together. *Journal of Personality and Social Psychology, 55,* 950–957.

Pellegrini, Anthony D. (1987). Rough and tumble play: Developmental and educational significance. *Educational Psychologist,* pp. 23–44.

Pelton, L.H. (1994). The role of material factors in child abuse and neglect. In G.B. Melton & F. Barry (Eds.), *Safe neighborhoods: Foundations for a new national strategy on child abuse and neglect.* New York: Guilford.

Pepler, Debra J., & Slaby, Ronald J. (1994). Theoretical and developmental perspectives on youth and violence. In Leonard D. Eron, Jacquelyn H. Gentry, & Peggy Schlegel (Eds.), *Reason to hope: A psychosocial perspective on violence and youth.* Washington, DC: American Psychological Association.

Perez, L. (1994). The household structure of second-generation children: An exploratory study of extended family arrangements. *International Migration Review, 28,* 736–747.

Perlmutter, Marion, Kaplan, Michael, & Nyquist, Linda. (1990). Development of adaptive competence in adulthood. *Human Development, 33,* 185–197.

Perris, Eve Emmanuel, Myers, Nancy Angrist, & Clifton, Rachel Kern. (1990). Long-term memory for a single infancy experience. *Child Development, 61,* 1796–1807.

Perry, Constance M., & McIntire, Walter G. (1995). Modes of moral judgment among early adolescents. *Adolescence, 30,* 707–715.

Perry, D.G., Perry, L.C., & Kennedy, E. (1992). In C.U. Shantz & W.W. Hartup (Eds.), *Conflict in child and adolescent development.* Cambridge, England: Cambridge University Press.

Perry, David G., Kusel, Sara J., & Perry, Louise C. (1988). Victims of peer aggression. *Developmental Psychology, 24,* 807–814.

Perry, William G., Jr. (1981). Cognitive and ethical growth: The making of meaning. In Arthur W. Chickering (Ed.), *The modern American college: Responding to the new realities of diverse students and a changing society.* San Francisco: Jossey-Bass.

Peters, D.L., & Pence, A.R. (1992). *Family day care: Current research for informed public policy.* New York: Teachers College Press.

Peters, H.E., Arrgys, L.M., Maccoby, E.E., & Mnookin, R.H. (1993). Enforcing divorce settlements: Evidence from child support compliance and award modifications. *Demography, 30,* 719–735.

Peterson, Bill E., & Stewart, Abigail J. (1996). Antecedents and contexts of generativity motivation at midlife. *Psychology and Aging, 11,* 21–33.

Peterson, C., Peterson, J., & Seeto, D. (1983). Developmental changes in ideas about lying. *Child Development, 54,* 1529–1535.

Peterson, James L., & Zill, Nicholas. (1986). Marital disruption, parent-child relationships, and behavior problems in children. *Journal of Marriage and the Family, 48,* 295–307.

Peterson, Lizette, Ewigman, Bernard, & Kivlahan, Coleen. (1993). Judgments regarding appropriate child supervision to prevent injury: The role of environmental risk and child age. *Child Development, 64,* 934–950.

Petitto, Anne, & Marentette, Paula F. (1991). Babbling in the manual mode: Evidence for the ontogeny of language. *Science, 251,* 1493–1496.

Pfeiffer, Eric. (1977). Sexual behavior in old age. In Ewald W. Busse & Eric Pfeiffer (Eds.), *Behavior and adaptation in late life* (2nd ed.). Boston: Little, Brown.

Pfeiffer, Eric, & Davis, Glenn C. (1972). Determinants of sexual behavior. *Journal of American Geriatrics Society, 20,* 151–158.

Phaff, J.M.L. (Ed.). (1986). *Perinatal health services in Europe: Searching for better childbirth.* London: Croom Helm.

Phelps, LeAdelle, Johnston, Lisa Swift, Jimenez, Dayana P., Wilczenski, Felicia L., Andrea, Ronald K., & Healy, Robert W. (1993). Figure preference, body dissatisfaction, and body distortion in adolescence. *Journal of Adolescent Research, 8,* 297–310.

Phillips, D.A., & Zimmerman, M. (1990). The developmental course of perceived competence and incompetence among competent children. In R.J. Sternberg & J. Kolligian (Eds.), *Competence considered.* New Haven, CT: Yale University Press.

Phinney, Jean S. (1993). Multiple group identities: Differentiation, conflict, and integration. In J. Kroger (Ed.), *Discussions on ego identity.* Hillsdale, NJ: Erlbaum.

Phinney, Jean S., Lochner, Bruce T., & Murphy, Rodolfo. (1990). Ethnic identity development and psychological adjustment in adolescence. In Arlene Rubin Stiffman & Larry E. Davis (Eds.), *Ethnic issues in adolescent mental health.* Newbury Park, CA: Sage.

Piaget, Jean. (1952). *The origins of intelligence in children* (M. Cook, Trans.). New York: International Universities Press.

Piaget, Jean. (1970). *The child's conception of movement and speed* (G.E.T. Holloway & M.J. Mackenzie, Trans.). New York: Basic Books

Pianta, Robert, Egeland, Byron, & Ericson, Martha Farrell. (1989). The antecedents of maltreatment: Results of the mother-child interaction project. In Dante Cicchetti & Vicki Carlson (Eds.), *Child maltreatment: Theory and research on the causes and consequences of child abuse and neglect.* Cambridge, England: Cambridge University Press.

Pierce, Karen A. & Cohen, Robert. (1995). Aggressors and their victims: Toward a contextual framework for understanding children's aggressor-victim relationships. *Developmental Review, 15,* 292–310.

Pillemer, Karl A., & Finkelhor, David. (1988). The prevalence of elder abuse: A random sample survey. *Gerontologist, 29,* 51–57.

Pillemer, Karl, & Suitor, J. Jill. (1991). Relationship with children adds distress in the elderly. In Karl Pillemer & Kathleen McCartney (Eds.), *Parent-child relations throughout life.* Hillsdale, NJ: Erlbaum.

Pina, Darlene L., & Bengston, Vern. (1993). The division of household labor and wives' happiness: Ideology, employment, and perceptions of support. *Journal of Marriage and the Family, 55,* 901–912.

Pinderhughes, Ellen E. (1991). The delivery of child welfare services to African-American clients. *American Journal of Orthopsychiatry, 61,* 599–605.

Pinker, S. (1984). *Language learnability and language development.* Cambridge, MA: Harvard University Press.

Pinker, S. (1987). The bootstrapping problem in language acquisition. In B. MacWhinney (Ed.), *Mechanisms of language acquisition.* Hillsdale, NJ: Erlbaum.

Pipp, S., Fischer, K.W., & Jennings, S. (1987). Acquisition of self- and mother knowledge in infancy. *Developmental Psychology, 23,* 86–96.

Pitskhelauri, G.Z. (1982). *The long-living of Soviet Georgia* (Gari Lesnoff-Caravaglia, Trans.). New York: Human Sciences Press.

Pittman, Joe F., & Blanchard, David. (1996). The effects of work history and timing of marriage on the division of household labor: A life-course perspective. *Journal of Marriage and the Family, 58,* 78–90.

Pleck, J.H. (1985). *Working wives/working husbands.* Beverly Hills, CA: Sage.

Plomin, Robert. (1989). Environment and genes: Determinants of behavior. *American Psychologist, 44,* 105–111.

Plomin, Robert. (1990). *Nature and nurture: An introduction to human behavioral genetics.* Pacific Grove, CA: Brooks/Cole.

Plomin, Robert, Lichtenstein, Paul, Pederson, Nancy L., McClearn, Gerald, & Nesselroade, John R. (1990). Genetic influence on life events during the last half of the life span. *Psychology and Aging, 5,* 25–30.

Plomin, Robert, Emde, R.N., Braungart, J.M., Campos, J., Corley, R., Fulker, D.W., Kagan, J., Reznick, J.S., Robinson, J., Zahn-Waxler, C., & DeFries, J.C. (1993). Genetic change and continuity from fourteen to twenty months: The MacArthur Longitudinal Twin Study. *Child Development, 64,* 1354–1376.

Plomin, Robert, Chipuer, Heather M., & Neiderhiser, Jenae M. (1994). Behavioral genetic evidence for the importance of nonshared environment. In E. Mavis Hetherington, David Reiss, & Robert Plomin (Eds.), *Separate social worlds of siblings: The impact of nonshared environments on development.* Hillsdale, NJ: Erlbaum.

Pocock, Stuart J., Smith, Marjorie, & Baghurst, Peter. (1994). Environmental lead and children's intelligence: a systematic review of the epidemiological evidence. *The British Medical Journal, 309,* 1189–1196.

Poffenberger, Thomas. (1981). Child rearing and social structure in rural India: Toward a cross-cultural definition of child abuse and neglect. In Jill E. Korbin (Ed.), *Child abuse and neglect: Cross-cultural perspectives.* Berkeley: University of California Press.

Polansky, N.A., Gaudin, J.M., Ammons, P.W., & David, K.B. (1985). The psychological ecology of the neglectful mother. *Child Abuse and Neglect, 9,* 265–275.

Polit, Denise F. (1984). The only child in single-parent families. In Toni Falbo (Ed.), *The single-child family.* New York: Guilford.

Polit, Denise F. (1989). Effects of a comprehensive program for teenage parents: Five years after project redirection. *Family Planning Perspectives, 21,* 164–169.

Pollack, William S., & Grossman, Frances K. (1985). Parent-child interaction. In L. L'Abate (Ed.), *The handbook of family psychology and therapy* (Vol. 1). Homewood, IL: Dorsey Press.

Ponsoby, Anne-Louise, Dwyer, Terence, Gibbins, Laura E., Cochrane, Jennifer A., & Wang, Yon-Gan. (1993). Factors potentiating the risk of sudden infant death syndrome associated with the prone position. *New England Journal of Medicine, 329,* 377–382.

Pool, Robert. (1993). Evidence for the homosexuality gene. *Science, 261,* 291–292.

Poon, Leonard W. (1985). Differences in human memory with aging: Nature, causes, and clinical implications. In James E. Birren & K. Warner Schaie (Eds.), *Handbook of the psychology of aging.* New York: Van Nostrand Reinhold.

Poon, Leonard W. (1992). *The Georgia Centenarian Study.* Amityville, NY: Baywood.

Poon, Leonard W., Rubin, David C., & Wilson, Barbara A. (Eds.). (1989). *Everyday cognition in adulthood and late life.* Cambridge, England; Cambridge University Press.

Posner, Richard A. (1995). *Aging and old age.* Chicago: University of Chicago Press.

Poussaint, Alvin F. (1990). Introduction. In Bill Cosby, *Fatherhood.* New York: Berkley Books.

Powell, Douglas H. (1994). *Profiles in cognitive aging.* Cambridge, MA: Harvard University Press.

Power, Rosemary. (1993). Death in Ireland: Death, wakes and funerals in contemporary Irish society. In Donna Dickenson & Malcolm Johnson (Eds.), *Death, dying & bereavement.* London: Sage.

Powers, Stephen, & Wagner, Michael J. (1984). Attributions for school achievement of middle school students. *Journal of Early Adolescence, 4,* 215–222.

Powlishta, Kimberly K. (1995). Gender bias in children's perceptions of personality traits. *Sex Roles, 32,* 17–28.

Presley, Cheryl A., Meilman, Philip, & Lyerla, Rob. (1995). *Alcohol and drugs on American college campuses.* Carbondale: Southern Illinois University Press.

Preston, George A.N. (1986). Dementia in elderly adults: Prevalence and institutionalization. *Journal of Gerontology, 41,* 261–267.

Price, Sharon J., & McHenry, Patrick C. (1988). *Divorce.* Newbury Park, CA: Sage.

Purvis, George A., & Bartholmey, Sandra J. (1988). Infant feeding practices: Commercially prepared baby foods. In Reginald C. Tsang & Buford L. Nicholos (Eds.), *Nutrition during infancy.* Philadelphia: Hanley & Belfus.

Pyke, Karen D., & Bengtson, Vern L. (1996). Caring more or less: Individualistic and collectivist systems of family eldercare. *Journal of Marriage and the Family, 58,* 379–392.

Pynoos, Jon, & Golant, Stephen. (1996). Housing and living arrangements for the elderly. In Robert H. Binstock & Linda K. George (Eds.), *Handbook of aging and the social sciences.* San Diego, CA: Academic Press.

Quadagno, Jill, & Hardy, Melissa. (1996). Work and retirement. In Robert H. Binstock & Linda K. George (Eds.), *Handbook of aging and the social sciences.* San Diego, CA: Academic Press.

Quay, Herbert C. (1987). Institutional treatment. In Herbert C. Quay (Ed.), *Handbook of juvenile delinquency.* New York: Wiley.

Quinn, Joseph F., & Kozy, Michael. (1996). The role of bridge jobs in the retirement transition: Gender, race, and ethnicity. *Gerontologist, 36,* 363–372.

Quinn, P.C., & Eimas, P.D. (1988). On categorization in early infancy. *Merrill–Palmer Quarterly, 32,* 331–363.

Quirk, Daniel A. (1991). The aging network: An agenda for the nineties and beyond. *Generations, 15*(3), 23–26.

Rabbitt, Patrick, Donlan, Christopher, McInnes, Lynn, Watson, Peter, & Bent, Nuala. (1995). Unique and interactive effects of depression, age, socioeconomic advantage, and gender on cognitive performance of normal healthy older people. *Psychology and Aging,* 10, 307–313.

Rabiner, D.L., Lenhart, L., & Lochman, J.E. (1990). Automatic versus reflective social problem solving in relation to children's sociometric status. *Developmental Psychology, 26,* 1010–1016.

Rafferty, Yvonne, & Shinn, Marybeth. (1991). The impact of homelessness on children. *American Psychologist, 46,* 1170–1179.

Rallison, Marvin L. (1986). *Growth disorders in infants, children, and adolescents.* New York: Wiley.

Raloff, Janet. (1996). Vanishing flesh: Muscle loss in the elderly finally gets some respect. *Science News, 150,* 90–91.

Ramey, C.T., & Campbell, F.A. (1991). Poverty, early childhood education, and academic competence: The Abecedarian experiment. In A.C. Huston (Ed.), *Children in poverty: Child development and public policy.* Cambridge, England: Cambridge University Press.

Ramey, C.T., Bryant, D.B., Wasik, B.H., Sparling, J.J., Fendt, K.H., & Lavange, L.M. (1992). The Infant Health and Development Program for low birth weight, premature infants: Program elements, family participation, and child intelligence. *Pediatrics, 89,* 454–465.

Ramsey, P. (1976, August). The enforcement of morals: Nontherapeutic research on children. *Hastings Center Report,* pp. 21–30.

Randall, Vernellia R. (1994). Ethnic Americans, long-term health care providers, and the patient self-determination act. In Marshall B. Kapp (Ed.), *Patient self-determination in long-term care: Implementing the PSDA in medical decisions.* New York: Springer.

Rasmussen, Dianne E., & Sobsey, Dick. (1994). Age, adaptive behavior, and Alzheimer disease in Down syndrome: Cross-sectional and longitudinal analyses. *American Journal on Mental Retardation, 99,* 151–165.

Ratner, H.H., Smith, B.S., & Padgett, R.J. (1990). Children's organization of events and event memories. In R. Fivush & J.A. Hudson (Eds.), *Knowing and remembering in young children.* Cambridge, England: Cambridge University Press.

Rauste-von Wright, Maijaliisa. (1989). Body image satisfaction in adolescent girls and boys: A longitudinal study. *Journal of Youth and Adolescence, 18,* 71–83.

Rawlins, William K. (1992). *Friendship matters.* Hawthorne, NY: Aldine de Gruyter.

Rawlins, William K. (1992). Friendships in later life. In Jon F. Nussbaum & Justine Coupland (Eds.), *Handbook of communication and aging research.* Mahwah, NJ: Erlbaum.

Ray, Ruth E. (1996). A postmodern perspective on feminist gerontology. *The Gerontologist. 36.*

Rebach, Howard. (1992). Alcohol and drug use among American minorities. In Joseph E. Trimble, Catherine S. Bolek, & Steve J. Niemcryyk (Eds.), *Ethnic and multicultural drug abuse.* Binghamton, NY: Haworth Press.

Reed, Edward S. (1993). The intention to use a specific affordance: A conceptual framework for psychology. In Robert H. Wozniak & Kurt W. Fischer (Eds.), *Development in context: Acting and thinking in specific environments.* Hillsdale, NJ: Erlbaum.

Reese, E., & Fivush, R. (1993). Parental styles of talking about the past. *Developmental Psychology, 29,* 596–606.

Reese, Hayne W., & Rodeheaver, Dean. (1985). Problem solving and complex decision making. In James E. Birren & K. Warner Schaie (Eds.), *Handbook of the psychology of aging* (2nd ed.). New York: Van Nostrand Reinhold.

Register, Cheri. (1991). *Are those kids yours?* New York: Free Press.

Reich, Peter A. (1986). *Language development.* Englewood Cliffs, NJ: Prentice Hall.

Reich, K.H. (1993). Cognitive developmental approaches to religiousness: Which version for which purpose? *International Journal for the Psychology of Religion, 3,* 145–171.

Reich, Robert B. (1992). *The work of nations: Preparing ourselves for 21st century capitalism.* New York: Random House.

Reid, Barbara Van Steenburgh. (1989). Socialization for moral reasoning: Maternal strategies of Samoans and Europeans in New Zealand. In Jaan Valsiner (Ed.), *Child development in cultural context.* Toronto: Hogrefe & Huber.

Reiff, T.R. (1989). Body composition with special reference to water. In A. Horwitz, D.M. Macfadyen, H. Munro, N.S. Scrimshaw, B. Steen, & T.F. Williams (Eds.), *Nutrition in the elderly.* Oxford, England: Oxford University Press.

Reisberg, Barry. (1981). *Brain failure. An introduction to current concepts of senility.* New York: Free Press.

Reiss, David, Plomin, Robert, Hetherington, E. Mavis, Howe, George W., Rovine, Michael, Tryon, Adeline, & Hagen, Margaret Stanley. (1994). The separate worlds of teenage siblings: An introduction to the study of nonshared environments and adolescent development. In E. Mavis Hetherington, David Reiss, & Robert Plomin (Eds.), *Separate social worlds of siblings: The impact of nonshared environments on development.* Hillsdale, NJ: Erlbaum.

Reitzes, Donald C., Mutran, Elizabeth J., & Fernandez, Maria E. (1996). Does retirement hurt well being? Factors influencing self-esteem and depression among retirees and workers. *Gerontologist, 26,* 649–656.

Rest, James R. (1986). *Moral development: Advances in research and theory.* New York: Praeger.

Rest, James R. (1993). Research on moral judgment in college students. In Andrew Garrod (Ed.), *Approaches to moral development: New research and emerging themes.* New York: Teachers College Press.

Rest, James R., & Thoma, Stephen J. (1985). Relation of moral judgment development to formal education.*Developmental Psychology, 21,* 709–714.

Reuss, M. Lynne, & Gordon, Howard R. (1995). Obstetrical judgments of viability and perinatal survival of extremely low birthweight infants. *American Journal of Public Health, 85,* 362–366.

Revkin, Andrew C. (1989, September). Crack in the cradle. *Discover, 10*(9), 62–69.

Reynolds, D., & Cuttance, P. (Eds.). (1992). *School effectiveness: Research, policy, and practice.* London: Cassell.

Riccio, Cynthia A., Hynd, George W., Cohen, Morris J., & Gonzalez, Jose J. (1993). Neurological basis of attention deficit hyperactivity disorder. *Exceptional Children, 60,* 118–124.

Ricciuti, Henry N. (1991). Malnutrition and cognitive development: Research policy linkages and current research directions. In Lynn Okagaki & Robert J. Sternberg (Eds.), *Directors of development.* Hillsdale, NJ: Erlbaum.

Rice, Mabel L. (1990). Preschoolers QUIL: Quick incidental learning of words. In G. Conti Ramsden & C. Snow (Eds.), *Children's language* (Vol. 7). Hillsdale, NJ: Erlbaum.

Rice, Mabel L., & Woodsmall, L. (1988). Lessons from television: Children's word learning when viewing. *Child Development, 59,* 420–429.

Richards, Maryse H., Boxer, Andrew M., Petersen, Anne C., & Albrecht, Rachel. (1990). Relation of weight to body image in pubertal girls and boys from two communities. *Developmental Psychology, 26,* 313–321.

Richardson, Gale A., & Day, Nancy L. (1994). Detrimental effects of prenatal cocaine exposure: Illusion or reality? *Journal of the American Academy of Child and Adolescent Psychiatry, 33,* 28–34.

Richman, Amy L., Miller, Patrice M., & LeVine, Robert A. (1992). Cultural and educational variations in maternal responsiveness. *Developmental Psychology, 28,* 614–621.

Riedel, Brant W., Lichstein, Kenneth L., & Dwyer, William O. (1995). Sleep compression and sleep education for older insomniacs: Self-help versus therapist guidance. *Psychology and Aging, 10,* 54–63.

Riegel, Klaus F. (1975). Toward a dialectical theory of development. *Human Development, 18,* 50–64.

Riley, Matilda White. (1971). Social gerontology and the age stratification of society. *Gerontologist, 11,* 79–87.

Riley, Matilda White. (1987). On the significance of age in sociology. *American Sociological Review, 52,* 1–14.

Riley, Matilda White, & Riley, John W., Jr. (1994). Age integration and the lives of older people. *Gerontologist, 34,* 110–115.

Rimm, E.B., Stampfer, Meir J., Klatsky, Arthur, & Grobbee, Diederick. (1996). Review of moderate alcohol consumption and reduced risk of coronary heart disease: Is the effect due to beer, wine or spirits? *British Medical Journal, 312,* 731–736.

Risman, B.J. (1987). Intimate relationships from a microstructural perspective: Men who mother. *Gender and Society, 1,* 6–32.

Ritchie, Jane, & Ritchie, James. (1981). Child rearing and child abuse: The Polynesian context. In J.E. Korbin (Ed.), *Child abuse and neglect: Cross-cultural perspectives.* Berkeley: University of California Press.

Ritchie, K., Kildea, D., & Robine, J.M. (1992). The relationship between age and the prevalence of senile dementia: A meta-analysis of recent data. *International Journal of Epidemiology, 21,* 763–769.

Roberto, K.A. (1993). Family caregivers of aging adults with disabilities: A review of the caregiving literature. In K.A. Roberto (Ed.), *The elderly caregiver: Caring for adults with developmental disabilities.* Newbury Park, CA: Sage.

Roberts, K. (1988). Retrieval of a basic-level category in prelinguistic infants. *Developmental Psychology, 24,* 21–27.

Robins, Lee N. (1995). Editorial: The natural history of substance abuse as a guide to setting drug policy. *American Journal of Public Health, 85,* 12–13.

Robins, Lee N., & Mills, James L. (1993). Effects of in utero exposure to street drugs. *American Journal of Public Health, 83* (Suppl.), 1–32.

Robins, Lee N., & Rutter, Michael. (1990). *Straight and devious pathways from childhood to adulthood.* Cambridge, England: Cambridge University Press.

Robinson, Ira, Ziss, Ken, Ganza, Bill, Katz, Stuart, & Robinson, Edward. (1991). Twenty years of the sexual revolution, 1965–1985: An update. *Journal of Marriage and the Family, 53,* 216–220.

Robinson, J.L., Kagan, J., Reznick, J.S., & Corley, R. (1992). The heritability of inhibited and uninhibited behavior: A twin study. *Developmental Psychology, 28,* 1030–1037.

Rochat, Philippe. (1989). Object manipulation and exploration in 2- to 5-month-old infants. *Developmental Psychology, 25,* 871–884.

Rochat, Philippe, & Bullinger, Andre. (1994). Posture and functional action in infancy. In Andre Vyt, Henriette Bloch, & Marc H. Bornstein (Eds.), *Early child development in the French tradition: Contributions from current research.* Hillsdale, NJ: Erlbaum.

Rochat, Philippe, & Goubet, Nathalie. (1995). Development of sitting and reaching in 5- to 6-month-old infants. *Infant Behavior and Development, 18,* 53–68.

Rodin, Judith, & Timko, Christine. (1992). Sense of control, aging, and health. In Marcia G. Ory, Ronald P. Abeles, & Paula Darby Lipman (Eds.), *Aging, health, and behavior.* Newbury Park, CA: Sage.

Rodriguez, Richard. (1983). *Hunger of memory.* New York: Bantam.

Rogers, Sinclair. (1976). The language of children and adolescents and the language of schooling. In Sinclair Rogers (Ed.), *They don't speak our language.* London: Edward Arnold.

Roggman, Lori A., Langlois, Judith H., Hubbs-Tait, Laura, & Rieser-Danner, Loretta A. (1994). Infant day-care, attachment, and the "filedrawer problem". *Child Development, 65,* 1429–1443.

Rogoff, Barbara. (1990). *Apprenticeship in thinking: Cognitive development in social context.* New York: Oxford UniversityPress.

Rogoff, Barbara, & Mistry, Jayanthi. (1990). The social and functional context of children's remembering. In Robyn Fivush & Judith A. Hudson (Eds.), *Knowing and remembering in young children.* Cambridge, England: Cambridge University Press.

Rogoff, Barbara, & Morelli, Gilda. (1989). Perspectives on children's development from cultural psychology. *American Psychologist, 44,* 343–348.

Rogoff, Barbara, Mistry, Jayanthi, Goncu, Artin, & Mosier, Christine. (1993). Guided participation in cultural activity by toddlers and caregivers. *Monographs of the Society for Research in Child Development, 58*(Serial No. 236).

Rohlen, Thomas P. (1983). *Japan's high schools.* Berkeley: University of California Press.

Rohner, Ronald P. (1984). Toward a conception of culture for cross-cultural psychology. *Journal of Cross-Cultural Psychology, 15,* 111–138.

Rohner, Ronald P., Kean, Kevin J., & Cournoyer, David E. (1991). Effects of corporal punishment, perceived caretaker warmth, and cultural beliefs on the psychological adjustment of children in St. Kitts, West Indies. *Journal of Marriage and the Family, 53,* 681–693.

Roll, J. (1989). *Lone parent families in the European community.* London: Family Policy Studies Center.

Rollins, Carol J., & Thompson, Cynthia. (1993). Nutrition. In Robin Bressler & Michael D. Katz (Eds.), *Geriatric pharmacology.* New York: McGraw-Hill.

Romaine, Suzanne. (1984). *The language of children and adolescents: The acquisition of communication competence.* Oxford, England: Blackwell.

Rose, Susan A., & Ruff, Holly A. (1987). Cross-modal abilities in infants. In J. Doniger Osofsky (Ed.), *Handbook of infant development* (2nd ed.). New York: Wiley.

Rosel, Natalie. (1986). Growing old together: Neighborhood, communality among the elderly. In Thomas R. Cole & Sally A. Gadow (Eds.), *What does it mean to grow old?* Durham, NC: Duke University Press.

Rosenberg, Elinor B. (1992). *The adoption life cycle.* Lexington, MA: Lexington Books.

Rosenberg, Harry M., Chevarley, Frances, Powell-Griner, Eve, Kochankek, Kenneth, & Feinleib, Manning. (1991). Causes of death among the elderly: Information from the death certificate. *Proceedings of the 1988 International Symposium on Aging* (Series 5, No. 6, DHHS Publication No. 91–1482). Hyattsville, MD: U.S. Department of Health and Human Services.

Rosenberg, M. (1986). Self-concept from middle childhood through adolescence. In J. Suls & A.G. Greenwald (Eds.), *Psychological perspective on the self.* Hillsdale, NJ: Erlbaum.

Rosenberg, Philip S. (1995). Scope of the AIDS epidemic in the United States. *Science, 270,* 1372–1375.

Rosenblith, Judy F. (1992). *In the beginning: Development from conception to age two* (2nd ed.). Newbury Park, CA: Sage.

Rosow, Irving. (1985). Status and role change through the life cycle. In Robert H. Binstock & Ethel Shanas (Eds.), *Handbook of aging and the social sciences* (2nd ed.). New York: Van Nostrand.

Ross, Catherine R. (1995). Reconceptualizing marital status as a continuum of social attachment. *Journal of Marriage and the Family, 57,* 129–140.

Ross, H.S., & Conant, C.L. (1992). The social structure of early conflict: Interaction, relationships, and alliances. In C.U. Shantz & W.W. Hartup (Eds.), *Conflict in child and adolescent development.* Cambridge, England: Cambridge University Press.

Rosser, Pearl L., & Randolph, Suzanne M. (1989). Black American infants: The Howard University normative study. In J. Kevin Nuegent, Barry M. Lester, & T. Berry Brazelton (Eds.), *The cultural context of infancy: Vol 1. Biology, culture, and infant development.* Norwood, NJ: Ablex.

Rossi, Alice S. (1994). *Eros and caritas: A biopsychosocial approach to human sexuality and reproduction.* Chicago: University of Chicago Press.

Rossi, Alice S. (Ed.). (1994). *Sexuality across the life course.* Chicago: University of Chicago Press.

Rotenberg, Ken J., & Mann, Luanne. (1986). The development of the norm of reciprocity of self-disclosure and its function in children's attraction to peers. *Child Development, 57,* 1349–1357.

Rotenberg, Ken J., & Sliz, Dave. (1988). Children's restrictive disclosure to friends. *Merrill-Palmer Quarterly, 34,* 203–215.

Rothbart, M.K. (1981). Measurement of temperament in infancy. *Child Development, 52,* 569–578.

Rothbart, M.K. (1991). Temperament: A developmental framework. In J. Strelau & A. Angleitner (Eds.), *Explorations in temperament.* New York: Plenum Press.

Rourke, B.P. (1989). *Non-verbal learning disabilities: The syndrome and the model.* New York: Guilford.

Rovee-Collier, Carolyn K. (1987). Learning and memory in infancy. In J. Doniger Osofsky (Ed.), *Handbook of infant development* (2nd ed.). New York: Wiley.

Rovee-Collier, Carolyn K. (1990). The "memory system" of prelinguistic infants. In A. Diamond (Ed.), *The development and neural bases of higher cognitive functions*. New York: New York Academy of Sciences.

Rovee-Collier, Carolyn K. (1995). Time windows in cognitive development. *Developmental Psychology, 31,* 147–169.

Rovee-Collier, Carolyn K., & Hayne, H. (1987). Reactivation of infant memory: Implications for cognitive development. In H.W. Reese (Ed.), *Advances in child development and behavior* (Vol. 20). New York: Academic Press.

Rovet, Joanne, Netley, Charles, Keenan, Maureen, Bailey, Jon, & Steward, Donald. (1996). The psychoeducational profile of boys with Klinefelter syndrome. *Journal of Learning Disabilities, 29,* 180–196.

Rubin, Kenneth H., Fein, Greata G., & Vandenberg, Brian. (1983). Play. In Paul H. Mussen (Ed.), *Handbook of child psychology: Vol. 4. Socialization, personality and social development*. New York: Wiley.

Ruff, Holly A. (1982). The development of object perception in infancy. In Tiffany M. Field, Aletha Huston, Herbert C. Quay, Lillian Troll, & Gordon E. Finley (Eds.), *Review of human development*. New York: Wiley.

Ruff, Holly A. (1984). An ecological approach to infant memory. In Morris Moscovitch (Ed.), *Infant memory*. New York: Plenum Press.

Rutherford, Andrew. (1986). *Growing out of crime*. Middlesex, England: Penguin.

Rutter, Michael. (1979). Protective factors in children's responses to stress and disadvantage. In Martha Whalen Kent & Jon E. Rolf (Eds.), *Primary prevention of psychopathology: Vol. 3. Social competence in children*. Hanover, NH: University Press of New England.

Rutter, Michael. (1980). *Changing youth in a changing society: Patterns of development and disorder*. Cambridge, MA: Harvard University Press.

Rutter, Michael. (1987). Psychosocial resilience and protective mechanisms. *American Journal of Orthopsychiatry, 57,* 316–331.

Rutter, Michael. (1989). Intergenerational continuities and discontinuities. In Dante Cicchetti & Vicki Carlson (Eds.), *Child maltreatment: Theory and research on the causes and consequences of child abuse and neglect*. Cambridge, England: Cambridge University Press.

Rutter, Michael. (1991). Nature, nurture, and psychopathology: A new look at an old topic. *Development and Psychopathology, 3,* 125–136.

Rutter, Michael, & Rutter, Marjorie. (1993). *Developing minds: Challenge and continuity across the life span*. New York: Basic Books.

Ryan, Gail, Miyoshi, Thomas J., Metzner, Jeffrey L., Krugman, Richard D., & Fryer, George E. (1996). Trends in a national sample of sexually abusive youths. *Journal of the American Academy of Child and Adolescent Psychiatry, 35,* 17–25.

Ryan, R.M., & Lynch, J.H. (1989). Emotional autonomy versus detachment: Revisiting the vicissitudes of adolescence and young adulthood. *Child Development, 60,* 340–356.

Rybash, John M., Hoyer, William J., & Roodin, Paul A. (1986). *Adult cognition and aging: Developmental changes in processing, knowing, and thinking*. New York: Pergamon Press.

Ryff, Carol D., Lee, Young Hyun, Essex, Marilyn J., & Pamela, A. (1994). My children and me: Midlife evaluations of grown children and of self. *Psychology and Aging, 9,* 195–205.

Saarni, C. (1989). Children's understanding of strategic control of emotional expression in social transactions. In C. Saarni & P.L. Harris (Eds.), *Children's understanding of emotion*. Cambridge, England: Cambridge University Press.

Sabatier, Colette. (1994). Parental conceptions of early development and developmental stimulation. In Andre Vyt, Henriette Bloch, & Marc H. Bornstein (Eds.), *Early child development in the French tradition: Contributions from current research*. Hillsdale, NJ: Erlbaum.

Sabin, E.P. (1993). Social relationships and mortality among the elderly. *Journal of Applied Gerontology,* 44–60.

Sable, Pat. (1991). Attachment, loss of spouse, and grief in elderly adults. *Omega, 23,* 129–142.

Sagi, Abraham, & Lewkowicz, K.S. (1987). A cross-cultural evaluation of attachment research. In L.W.C. Tavecchio & M.H. van Ijzendoorn (Eds.), *Attachment in social networks*. Amsterdam: Elsevier.

Sagi, Abraham, van Ijzendoorn, Marinus H., & Koren-Karie, Nina. (1991). Primary appraisal of the Strange Situation: A cross-cultural analysis of preseparation episodes. *Developmental Psychology, 27,* 587–596.

Salmi, L.R., Weiss, H.B., Peterson, P.L., Spengler, R.F., Sattin, R.W., & Anderson, H.A. (1989). Fatal farm injuries among young children. *Pediatrics, 83,* 267–271.

Salthouse, Timothy A. (1984). Effects of age and skill in typing. *Journal of Experimental Psychology: General, 113,* 345–371.

Salthouse, Timothy A. (1985). Speed of behavior and its implications for cognition. In James E. Birren & K. Warner Schaie (Eds.), *Handbook of the psychology of aging* (2nd ed.). New York: Van Nostrand Reinhold.

Salthouse, Timothy A. (1987). The role of experience in cognitive aging. In K. Warner Schaie (Ed.), *Annual review of gerontology and geriatrics* (Vol. 7). New York: Springer.

Salthouse, Timothy A. (1990). Working memory as a processing resource in cognitive aging. *Developmental Review, 10,* 101–124.

Salthouse, Timothy A. (1991). *Theoretical perspectives on cognitive aging*. Hillsdale, NJ: Erlbaum.

Salthouse, Timothy A. (1992). *Mechanisms of age-cognition relations in adulthood*. Hillsdale, NJ: Erlbaum.

Salthouse, Timothy A. (1993). Speed mediation of adult age differences in cognition. *Developmental Psychology, 29,* 722–738.

Salthouse, Timothy A. (1996). General and specific speed mediation of adult age differences in memory. *Journals of Gerontology, 51B,* 30–42.

Sampson, Robert J., & Laub, John N. (1993). *Crime in the making: Pathways and turning points through life*. Cambridge, MA: Harvard University Press.

Sandelowski, Margarete. (1993). *With child in mind: Studies of the personal encounter with infertility*. Philadelphia: University of Pennsylvania Press.

Sanders, Catherine M. (1989). *Grief: The mourning after*. New York: Wiley.

Santrock, John W., Warshak, Richard A., & Elliott, Gary L. (1982). Social development and parent-child interaction in father-custody and stepmother families. In Michael E. Lamb (Ed.), *Non-traditional families: Parenting and child development*. Hillsdale, NJ: Erlbaum.

Sargent, James D., Brown, Mary Jean, Freeman, Jean L., Baiely, Adrian, Goodman, David, & Freeman, Daniel H. (1995). Childhood lead poisoning in Massachusetts communities: Its association with sociodemographic and housing characteristics. *American Journal of Public Health, 85*, 528–534.

Saudino, Kimberly J., McClearn, G.E., Pedersen, Nancy L., Lichtenstein, Paul, & Plomin, Robert. (1997). Can personality explain genetic influences on life events? *Journal of Personality and Social Psychology, 72*, 196–206.

Saunders, C.M. (1978). *The management of terminal disease*. London: Edward Arnold.

Saville-Troike, Muriel, McClure, Erica, & Fritz, Mary. (1984). Communicative tactics in children's second language acquisition. In Fred R. Eckman, Lawrence H. Bell, & Diane Nelson (Eds.), *Universals of second language acquisition*. Rowley, MA: Newbury House.

Savin-Williams, Ritch C. (1995). An exploratory study of pubertal maturation timing and self-esteem among gay and bisexual male youths. *Developmental Psychology, 31*, 56–64.

Savin-Williams, Ritch C., & Berndt, T.J. (1990). Friendship and peer relations. In S.S. Feldman & G.R. Elliott (Eds.), *At the threshold: The developing adolescent*. Cambridge, MA: Harvard University Press.

Saxe, Geoffrey, Guberman, Steven R., & Gearhart, Maryl. (1987). Social processes in early number development. *Monographs of the Society for Research in Child Development*, 52(Serial No. 216).

Saywitz, K.J., Goodman, G.S., Nicholas, E., & Moan, S.F. (1991). Children's memories of a physical examination involving genital touch: Implications for reports of sexual abuse. *Journal of Consulting and Clinical Psychology, 59*, 682–689.

Scafidi, Frank A., Field, Tiffany, & Schanberg, Saul M. (1993). Factors that predict which preterm infants benefit most from massage therapy. *Journal of Developmental and Behavioral Pediatrics, 14*, 176–180.

Scarr, Sandra. (1985). Constructing psychology: Making facts and fables for our times. *American Psychologist, 40*, 499–512.

Scarr, Sandra. (1992). Developmental theories for the 1990s: Development and individual differences. *Child Development, 63*, 1–19.

Scarr, Sandra. (1996). Families and day care: Both matter for children. *Contemporary Psychology, 41*, 330–331.

Schacter, D.L., & Tulving, E. (Eds.). (1994). *Memory systems*. Cambridge, MA: MIT Press.

Schaffer, H. Rudolf. (1984). *The child's entry into a social world*. New York: Academic Press.

Schaie, K. Warner. (1983). The Seattle Longitudinal Study. A twenty-one year investigation of psychometric intelligence. In K. Warner Schaie (Ed.), *Longitudinal studies of adult psychological development*. New York: Guilford.

Schaie, K. Warner. (1989). Individual differences in rate of cognitive change in adulthood. In Vern L. Bengston & K. Warner Schaie (Eds.), *The course of later life*. New York: Springer.

Schaie, K. Warner. (1989). Perceptual speed in adulthood: Cross-sectional and longitudinal studies. *Psychology and Aging, 4*, 443–453.

Schaie, K. Warner. (1990). Intellectual development in adulthood. In James E. Birren & K. Warner Schaie (Eds.), *Handbook of the psychology of aging*. San Diego, CA: Academic Press.

Schaie, K. Warner. (1996). *Intellectual development in adulthood: The Seattle Longitudinal Study*. Cambridge, England: Cambridge University Press.

Schaie, K. Warner, & Willis, Sherry L. (1996). *Adult development and aging*. New York: HarperCollins.

Scheibel, Arnold B. (1996). Structural and functional changes in the aging brain. In James E. Birren & K. Warner Schaie (Eds.) *Handbook of the Psychology of Aging*. San Diego: Academic Press.

Scheper-Hughes, Nancy, & Stein, H. (1987). Child abuse and the unconscious in American popular culture. In N. Scheper-Hughes (Ed.), *Child survival*. Dordrecht, The Netherlands: Reidel.

Schilit, Rebecca, & Gomberg, Edith S. Lisansky. (1991). *Drugs and behavior: A sourcebook for the helping professions*. Newbury Park, CA: Sage.

Schlegal, Alice, & Barry, Herbert. (1991). *Adolescence: An anthropological inquiry*. New York: Free Press.

Schlundt, David G., & Johnson, William G. (1990). *Eating disorders: Assessment and treatment*. Boston: Allyn & Bacon.

Schneider, Anne L. (1990). *Deterrence and juvenile crime*. New York: Springer-Verlag.

Schneider-Rosen, Karen, & Cicchetti, Dante. (1991). Early self-knowledge and emotional development: Visual self-recognition and affect reactions to mirror self-images in maltreated and non-maltreated toddlers. *Developmental Psychology, 27*, 471–478.

Scholl, Theresa O., Hediger, Mary L., Schall, Joan I., Khoo, Chor-San, & Fischer, Richard L. (1996). Dietary and serum folate: Their influence on the outcome of pregnancy. *American Journal of Clinical Nutrition, 63*, 520–525.

Schroots, Johannes J.F. (1993). *Aging, health, and competence: The next generation of longitudinal research*. Amsterdam: Elsevier.

Schroots, Johannes J.F. (1996). Theoretical developments in the psychology of aging. *Gerontologist, 36*, 741–748.

Schuller, Tom. (1995). Life after work: Lines, boundaries, and spaces. In Eino Heikkinen, Jorma Kuusinen, & Isto Ruoppila (Eds.), *Preparation for aging*. New York: Plenum Press.

Schulz, Richard, Musa, Donald, Staszewski, James, & Siegler, Robert S. (1994). The relationship between age and major league baseball performance: Implications for development. *Psychology and Aging, 9*, 274–286.

Schweinhart, L.J., & Weikart, David. (Eds.). (1993). *Significant benefits: High/Scope Perry preschool study through age 27*. Ypsilanti, MI: High/Scope Press.

Scott-Maxwell, Florida. (1968). *The measure of my days*. New York: Knopf.

Seale, Clive, & Cartwright, Ann. (1994). *The year before death*. Aldershot, England: Avebury.

Segal, Bernard. (1992). Ethnicity and drug-taking behavior. In Joseph E. Trimble, Catherine S. Bolek, & Steve J. Niemcryyk (Eds.), *Ethnic and multicultural drug abuse*. Binghamton, NY: Haworth Press.

Seibel, Machelle M. (1993). Medical evaluation and treatment of the infertile couple. In Machelle M. Seibel, Ann A. Kiessling, Judith Bernstein, & Susan R. Levin (Eds.), *Technology and infertility: Clinical, psychosocial, legal and ethical aspects*. New York: Springer-Verlag.

Seibel, Machelle M. (moderator). (1993). Panel discussion: Ethical issues in fertility. In Machelle M. Seibel, Ann A. Kiessling, Judith Bernstein, and Susan R. Levin (Eds.), *Technology and infertility: Clinical, psychosocial, legal and ethical aspects*. New York: Springer-Verlag.

Seifer, R., & Sameroff, Arnold J. (1987). Multiple determinants of risk and invulnerability. In E.J. Anthony & B.J. Cohler (Eds.), *The invulnerable child*. New York: Guilford.

Seltzer, G.B., Begun, A., Seltzer, M.M., & Krauss, M.W. (1991). Adults with mental retardation and their aging mothers: Impacts on siblings. *Family Relations, 40,* 310–317.

Seltzer, Judith A. (1991). Legal custody arrangements and children's economic welfare. *American Journal of Sociology, 96,* 895–929.

Seltzer, Judith A. (1991). Relationships between fathers and children who live apart: The father's role after separation. *Journal of Marriage and the Family, 53,* 79–102.

Seltzer, Judith A., & Bianchi, S.M. (1988). Children's contact with absent parents. *Journal of Marriage and the Family, 50,* 663–677.

Seltzer, Marsha M., & Li, Lydia Wailing. (1996). The transitions of caregiving: Subjective and objective definitions. *The Gerontologist, 36.*

Seltzer, Mildred M. (Ed.). (1995). *The impact of increased life expectancy: Beyond the gray horizon*. New York: Springer.

Sena, Rhonda, & Smith, Linda B. (1990). New evidence on the development of the word Big. *Child Development, 61,* 1034–1052.

Settersten, Richard A., & Hagestad, Gunhild. (1996). What's the latest? Cultural deadlines for educational and work transitions. *Gerontologist, 36,* 602–613.

Shannon, Lyle W. (1988). *Criminal career continuity: Its social context*. New York: Human Sciences Press.

Sharkey, William F. (1993). Who embarrasses whom? Relational and sex differences in the use of intentional embarrassment. In Pamela J. Kalbfleisch (Ed.), *Interpersonal communication: Evolving interpersonal relationships*. Hillsdale, NJ: Erlbaum.

Shatz, Marilyn. (1994). *A toddler's life*. New York: Oxford University Press.

Shaw, Daniel S., Vondra, Joan I., Hommerding, Katherine Dowdell, Keenan, Kate & Dunn, Marija. (1994). Chronic family adversity and early child behavior problems. A longitudinal study of low income families. *Journal of Child Psychology and Psychiatry, 35,* 1109–1122.

Shaywitz, Sally E., Excobar, M.D., Shaywitz, Bennett A., Fletcher, J.M., & Makuch, R. (1992). Evidence that dyslexia may represent the lower tail of a normal distribution of reading disability. *New England Journal of Medicine, 326,* 145–151.

Shea, John D.C. (1981). Changes in interpersonal distances and categories of play behavior in the early weeks of preschool. *Developmental Psychology, 17,* 417–425.

Shedler, Jonathan, & Block, Jack. (1990). Adolescent drug use and psychological health: A longitudinal inquiry. *American Psychologist, 45,* 612–630.

Sherman, David S. (1994). Geriatric psychopharmacotherapy: Issues and concerns. *Generations, 18,* 34–39.

Sherman, T. (1985). Categorization skills in infants. *Child Development, 53,* 183–188.

Shiffrin, R.M., & Atkinson, R.C. (1969). Storage and retrieval processes in long-term memory. *Psychological Review, 76,* 179–193.

Shinn, M., Knickman, J.R., & Weitzman, B.C. (1991). Social relationships and vulnerability to becoming homeless among poor families. *American Psychologist, 46,* 1180–1187.

Shirley, Mary M. (1933). *The first two years: A study of twenty-five babies* (Institute of Child Welfare Monograph No. 8). Minneapolis: University of Minnesota Press.

Shneidman, Edwin S. (1978). Suicide. In G. Lindzey, Calvin S. Hall, & R.F. Thompson (Eds.), *Psychology* (2nd ed.). New York: Worth.

Shneidman, Edwin S., & Mandelkorn, Philip. (1994). Some facts and fables of suicide. In Edwin S. Shneidman, Norman L. Faberow, & Robert E. Litman (Eds.), *The psychology of suicide* (rev. ed.). Northwale, NJ: Aronson.

Shneidman, Edwin S., Faberow, Norman L., & Litman, Robert E. (1994). *The psychology of suicide* (rev. ed.). Northwale, NJ: Aronson.

Shore, R. Jerald, & Hayslip, Bert, Jr., (1994). Custodial grandparenting: Implications for children's development. In Adele Eskeles Gottfried & Allen W. Gottfried (Eds.), *Redefining families: Implications for children's development*. New York: Plenum Press.

Siegel, J.M. (1993). Companion animals: In sickness and in health. *Journal of Social Issues, 49,* 157–167.

Siegel, Paul. Z., Brackbill, Robert J., & Health, Gregory W. (1995). The epidemiology of walking for exercise: Implications for promoting activity among sedentary groups. *American Journal of Public Health, 85,* 706–710.

Siegler, Robert. (1983). Information processing approaches to development. In Paul H. Mussen (Ed.), *Handbook of child psychology: Vol. 1. History, theory, and methods*. New York: Wiley.

Siegler, Robert. (1991). *Children's thinking* (2nd ed.). Englewood Cliffs, NJ: Prentice Hall.

Sigel, I.E., McGillicuddy-DeLisi, A.V., & Goodnow, J.J. (Eds.). (1992). *Parent belief systems* (2nd ed.). Hillsdale, NJ: Erlbaum.

Sigler, Robert T. (1989). *Domestic violence: An assessment of community attitudes*. Lexington, MA: Lexington Books.

Silbereisen, Rainer K., & Kracke, Barbel. (1990). Variation in maturational timing and adjustment in adolescence. In Sandy Jackson & Hector Rodriguez-Tome (Eds.), *Adolescence and its social worlds*. Hove, England: Erlbaum.

Silva, Phil A., Hughes, Pauline, Williams, Sheila, & Faed, James M. (1988). Blood lead, intelligence, reading attainment and behavior in eleven-year-old children in Dunedin, New Zealand. *Journal of Child Psychology and Psychiatry, 29,* 43–52.

Silver, A.A., & Hagin, R.A. (1990). *Disorders of learning in childhood*. New York: Wiley.

Silver, L.B. (1991). Developmental learning disorders. In M. Lewis (Ed.), *Child and adolescent psychiatry: A comprehensive textbook*. Baltimore: Williams & Wilkins.

Silverberg, S.B., & Steinberg, Lawrence. (1990). Psychological well-being of parents with early adolescent children. *Developmental Psychology, 26,* 658–666.

Silverman, W.A. (1990). Setting a limit in the treatment of neonates: All or none? In G. Duc, A. Huch, & R. Huch (Eds.), *The very low birthweight infant*. New York: George Thieme.

Simmons, R.G., & Blyth, Dale A. (1987). *Moving into adolescence: The impact of pubertal change and school context*. New York: Aldine de Gruyter.

Simpson, J.A. (1990). Influence of attachment styles on romantic relationships. *Journal of Personality and Social Psychology, 59*, 971–980.

Sinclair, David. (1989). *Human growth after birth*. New York: Oxford University Press.

Singh, Gopal K., & Yu, Stella M. (1994). Birthweight differentials among Asian Americans. *American Journal of Public Health, 84*, 1444–1449.

Sinnott, Jan D. (Ed.). (1989). *Everyday problem solving*. New York: Praeger.

Sinnott, Jan D. (1993). Creativity and postformal thought: Why the last stage is the creative stage. In C. Adams-Price (Ed.), *Creativity and aging: Theoretical and empirical approaches*. New York: Springer.

Skinner, B.F. (1953). *Science and human behavior*. New York: Macmillan.

Skinner, B.F. (1957). *Verbal behavior*. New York: Appleton-Century-Crofts.

Skinner, John H. (1992). Aging in place: The experience of African American and other minority elders. *Generations, 16*(2), 49–52.

Skinner, John H. (1995). Ethnic racial diversity in long term care use and service. In Zev Havel & Ruth E. Dunkle (Eds.), *Matching people with services in long-term care*. New York: Springer.

Skolnick, A. (1991). *Embattled paradise: The American family in an age of uncertainty*. New York: Basic Books.

Skunkard, A.J., Harris, J.R., Pedersen, N.I., & McClearn G.E. (1990). A separated twin study of the body mass index. *New England Journal of Medicine, 322*, 1483–1487.

Slaby, Ronald J., & Eron, Leonard D. (1994). Afterword. In Leonard D. Eron, Jacquelyn H. Gentry, & Peggy Schlegel (Eds.), *Reason to hope: A psychosocial perspective on violence and youth*. Washington, D.C: American Psychological Association.

Sloane, John Henry, Kellerman, Arthur L., Reay, Donald T., Ferris, James A., Koepsell, T., & Rivara, Frederick P. (1988). Handgun regulation, crime, assault and homicide: A tale of two cities. *New England Journal of Medicine, 319*, 1256–1262.

Sloane, John Henry, Rivara, Frederick P., Reay, Donald T., Ferris, James A., Path, M.R.C., & Kellerman, Arthur L. (1990). Firearm regulations and rates of suicide: A comparison of two metropolitan areas. *New England Journal of Medicine, 322*, 369–373.

Slutsker, Laurence, Smith, Richard, Higginson, Grant, & Fleming, David. (1993). Recognizing illicit drug use by pregnant women: Reports from Oregon birth attendants. *American Journal of Public Health, 83*, 61–64.

Small, Stephen A., Eastman, Gay, & Cornelius, Steven. (1988). Adolescent automomy and parental stress. *Journal of Youth and Adolescence, 17*, 377–391.

Smetana, Judith G., & Asquith, P. (1994). Adolescents' and parents' conceptions of parental authority and adolescent autonomy. *Child Development, 65*, 1147–1162.

Smetana, Judith G., Killen, M., & Turiel, E. (1991). Children's reasoning about interpersonal and moral conflicts. *Child Development, 62*, 629–644.

Smetana, Judith G., Yau, Jenny, Restrepo, Angela, & Braeges, Judith L. (1991). Adolescent-parent conflict in married and divorced families. *Developmental Psychology, 27*, 1000–1010.

Smith, David J. (1995). Youth crime and conduct disorders: Trends, patterns and causal explanations. In Michael Rutter & David J. Smith (Eds.), *Psychosocial disorders in young people: Time trends and their causes*. New York: Wiley.

Smith, Glenn E., Petersen, Ronald C., Ivnik, Robert J., Malec, James F., & Tangalos, Eric G. (1996). Subjective memory complaints, psychological distress, and longitudinal change in objective memory performance. *Psychology and Aging, 11*, 272–279.

Smith, Jacqui, & Baltes, Paul B. (1990). Wisdom-related knowledge: Age/cohort differences in response to life-planning problems. *Developmental Psychology, 26*, 494–505.

Smith, M. Brewster. (1983). Hope and despair: Keys to the socio-psychodynamics of youth. *American Journal of Orthopsychiatry, 53*, 388–399.

Smith, Thomas Ewin. (1990). Parental separation and the academic self-concepts of adolescents: An effort to solve the puzzle of separation effects. *Journal of Marriage and the Family, 52*, 107–118.

Smith, Wrynn. (1987). *Cancer: A profile of health and diseases in America*. New York: Facts on File.

Smits, Jeroen, Ultee, Wout, & Lammers, Jan. (1996). Effects of occupational status differences between spouses on the wife's labor force participation and occupational achievement: Findings from 12 European countries. *Journal of Marriage and the Family, 58*, 101–116.

Snarey, John R. (1993). *How fathers care for the next generation: A four-decade study*. Cambridge, MA: Harvard University Press.

Snarey, John R., Reimber, Joseph, & Kohlberg, Lawrence. (1985). Development of social-moral reasoning among kibbutz adolescents: A longitudinal cross-cultural study. *Developmental Psychology, 21*, 3–17.

Snow, Catherine E. (1984). Parent-child interaction and the development of communicative ability. In Richard L. Schiefelbusch & Joanne Pickar (Eds.), *The acquisition of communicative competence*. Baltimore: University Park Press.

Snyder, D.L. (1989). *Dietary restriction and aging*. New York: Liss.

Sokol, R.J., & Abel, E.L. (1992). Risk factors for alcohol-related birth defects. Threshold, susceptibility, and prevention. In T.B. Sonderegger (Ed.), *Perinatal substance abuse: Research findings and clinical implications*. Baltimore: Johns Hopkins University Press.

Soldo, Beth J. (1996). Cross pressures on middle-aged adults: A broader view. *Journals of Gerontology, 51B*, 271–279.

Solomon, J.C., & Marx, J. (1995). *The psychology of grandparenthood: An international perspective*. London: Routledge.

Solomon, Mildred Z., O'Donnell, Lydia, Jennings, Bruce, Guifoy, Vivian, Wolff, Susan M., Nolan, Kathleen, Jackson, Rebecca, Koch-Weser, Dieter, & Donnelley, Strachan. (1993). Decisions near the end of life: Professional views on life-sustaining treatments. *American Journal of Public Health, 83*, 14–23.

Sonderegger, Theo B. (Ed.). (1992). *Perinatal substance abuse: Research findings and clinical implications.* Baltimore: John Hopkins University Press.

Sorlie, Paul D., Backlund, Eric, & Keller, Jacob B. (1995). US mortality by economic, demographic and social characteristics: The national longitudinal mortality study. *American Journal of Public Health, 85,* 949–956.

Spanier, Graham, & Thompson, Linda. (1984). *Parting: The aftermath of separation and divorce.* Beverly Hills, CA: Sage.

Spearman, Charles. (1927). *The abilities of man.* New York: Macmillan.

Spelke, Elizabeth. (1987). The development of intermodal perception. In Philip Salapatek & Leslie Cohen (Eds.), *Handbook of infant perception: Vol. 2. From perception to cognition.* Orlando, FL: Academic Press.

Spelke, Elizabeth. (1988). Where perceiving ends and thinking begins: The apprehension of objects in infancy. In A. Yonas (Ed.), *Minnesota Symposia on Child Psychology: Vol. 20. Perceptual development in infancy.* Hillsdale, NJ: Erlbaum.

Spelke, Elizabeth. (1991). Physical knowledge in infancy: Reflections of Piaget's theory. In S. Carey & R. Gelman (Eds.), *The epigenesis of mind: Essays on biology and cognition.* Hillsdale, NJ: Erlbaum.

Spencer, M.B. (1987). Black children's ethnic identity formation: Risk and resilience of castelike minorities. In J.S. Phinney & M.J. Rotheram (Eds.), *Children's ethnic socialization: Pluralism and development.* Newbury Park, CA: Sage.

Spencer, M.B., & Markstrom-Adams, C. (1990). Identity processes among racial and ethnic minority children in America. *Child Development, 61,* 290–310.

Spock, Benjamin. (1976). *Baby and child care.* New York: Pocket Books.

Sroufe, L. Alan. (1979). Socioemotional development. In Joy Doniger Osofsky (Ed.), *Handbook of infant development.* New York: Wiley.

Sroufe, L. Alan, & Rutter, M. (1984). The domain of developmental psychopathology. *Child Development, 55,* 17–29.

Staines, Graham L., Pottick, Kathleen J., & Fudge, Deborah A. (1986). Wives' employment and husbands' attitude toward work and life. *Journal of Applied Psychology, 71,* 118–128.

Stanton, Annette L., & Dunkel-Schetter, Christine. (Eds.). (1991). *Infertility.* New York: Plenum Press.

Staples, Robert, & Johnson, Leanor B. (1993). *Black families at the crossroads.* San Francisco: Jossey-Bass.

Starfield, B., Shapiro, S., Weiss, J., Liang, K.Y., Ra, K, Paige, D., & Wang, X.B. (1991). Race, family income and low birthweight. *American Journal of Epidemiology, 134,* 1167–1174.

Stein, N.L., & Levine, L.J. (1989). The causal organization of emotional knowledge: A developmental study. *Cognition and Emotion, 3,* 343–378.

Steinberg, Lawrence. (1988). Reciprocal relation between parent-child distance and pubertal maturation. *Developmental Psychology, 24,* 122–128.

Steinberg, Lawrence. (1990). Interdependency in the family: Autonomy, conflict, and harmony in the parent-adolescent relationship. In Shirley S. Feldman & G.R. Elliot (Eds.), *At the threshold: The developing adolescent.* Cambridge, MA: Harvard University Press.

Steinberg, Lawrence, Elmen, J.D., & Mounts, N.S. (1989). Authoritative parenting, psychosocial maturity and academic success among adolescents. *Child Development, 60,* 1424–1436.

Steinberg, Lawrence, Mounts, Nina S., Lamborn, Susan D., & Dornbusch, Sanford M. (1991). Authoritative parenting and adolescent adjustment across various ecological niches. *Journal of Research on Adolescents, 1,* 19–36.

Steinberg, Lawrence, Dornbusch, Sanford M., & Brown, B.B. (1992). Ethnic differences in adolescent achievement: An ecological perspective. *American Psychologist, 47,* 723–729.

Steinberg, Lawrence, Lamborn, Susie D., Darling, Nancy, Mounts, Nina A., & Dornbusch, Sanford M. (1994). Over-time changes in adjustment and competence among adolescents from authoritative, authoritarian, indulgent, and neglectful families. *Child Development, 65,* 754–770.

Steinfels, Peter. (1993, February 14). Help for the helping hands in death. *New York Times,* Sect. 4, pp. 1,6.

Steingart, R.M., Packer, M., Hamm, P., Coglianese, M.E., Gersh, B., Geltman, E.M., Sollano, J., Katz, S., Moye, L., & Basta, L.L. (1991). Sex differences in the management of coronary artery disease. *New England Journal of Medicine, 325,* 226–230.

Steinmetz, Suzanne. (1988). *Duty bound: Elderly abuse and family care.* Beverly Hills, CA: Sage.

Stenberg, C.R., & Campos, J.J. (1990). The development of anger expressions in infancy. In N.L. Stein, B. Leventhal, & T. Trabasso (Eds.), *Psychological and biological approaches to emotion.* Hillsdale, NJ: Erlbaum.

Stephens, Mary Ann Parris, Franks, Melissa M., & Townsend, Aloen L. (1994). Stress and rewards in women's multiple roles: The case of women in the middle. *Psychology and Aging, 9,* 45–52.

Sterk-Elifson, Claire. (1994). Sexuality among African-American women. In Alice S. Rossi (Ed.), *Sexuality across the life course.* Chicago: University of Chicago Press.

Stern, Daniel N. (1977). *The first relationship: Mother and infant.* Cambridge, MA: Harvard University Press.

Stern, Daniel N. (1985). *The interpersonal world of the infant.* New York: Basic Books.

Sternberg, Hal. (1994). Aging of the immune system. In Paola S. Timiras (Ed.), *Physiological basis of aging and geriatrics* (2nd ed.). Boca Raton, FL: CRC Press.

Sternberg, Kathleen J., & Lamb, Michael E. (1991). Can we ignore context in the definition of child maltreatment? *Development and Psychopathology, 3,* 87–92.

Sternberg, Robert J. (1988). Intellectual development: Psychometric and information-processing approaches. In M.H. Bornstein & M.E. Lamb (Eds.), *Developmental psychology: An advanced textbook* (2nd ed.). Hillsdale, NJ: Erlbaum.

Sternberg, Robert J. (1988). *The triarchic mind: A new theory of human intelligence.* New York: Viking Press.

Sternberg, Robert J., & Barnes, Michael L. (Eds.). (1988). *The psychology of love.* New Haven, CT: Yale University Press.

Sternberg, Robert J., Wagner, Richard K., Williams, Wendy M., & Horvath, Joseph A. (1995). Testing common sense. *American Psychologist, 50,* 912–927.

Stevenson, Harold W., & Lee, Shin-Ying. (1990). Contexts of achievement: A study of American, Chinese, and Japanese children. *Monographs of the Society for Research in Child Development, 55*(1–2, Serial No. 221).

Stevenson, Harold W., & Stigler, Robert W. (1992). *The learning gap: Why our schools are failing and what we can learn from Japanese and Chinese education.* New York: Summit Books.

Stevenson, Harold W., Chen, Chuansheng, & Lee, Shin-Ying. (1993). Mathematics achievement of Chinese, Japanese, and American children: Ten years later. *Science, 259,* 53–58.

Stewart, Robert B. (1990). *The second child: Family transitions and adjustment.* Newbury Park, CA: Sage.

Stifter, Cynthia, & Braungart, Julia M. (1995). The regulation of negative reactivity in infancy: Function and development. *Developmental Psychology, 31,* 448–455.

Stiles, Deborah A., Gibbons, Judith L., Hardardottir, Sara, & Schnellmann, Jo. (1987). The ideal man or woman as described by young adolescents in Iceland and the United States. *Sex Roles, 17,* 313–320.

Stiles, Deborah A., Gibbons, Judith L., & Schnellmann, Jo De La Garza. (1990). Opposite-sex ideal in the U.S.A. and Mexico as perceived by young adolescents. *Journal of Cross-Cultural Psychology, 21,* 180–199.

Stillion, Judith M. (1995). Death in the lives of adults: Responding to the tolling of the bell. In Hannelore Was & Robert A. Neimeyer (Eds.), *Dying: Facing the facts.* Washington, DC: Taylor & Francis.

Stipek, Deborah J. (1984). Young children's performance expectations: Logical analysis or wishful thinking? In J. Nicholls (Ed.), *The development of achievement motivation.* Greenwich, CT: JAI Press.

Stipek, Deborah J. (1992). The child at school. In M.H. Bornstein & M.E. Lamb (Eds.), *Developmental psychology: An advanced textbook* (3rd ed.). Hillsdale, NJ: Erlbaum.

Stipek, Deborah J., & Hoffman, J. (1980). Development of children's performance-related judgments. *Child Development, 51,* 912–914.

Stipek, Deborah J., & MacIver, D. (1989). Developmental change in children's assessment of intellectual competence. *Child Development, 60,* 521–538.

Stipek, Deborah J., Roberts, Theresa A., & Sanborn, Mary E. (1984). Preschool-age children's performance expectations for themselves and another child as a function of the incentive value of success and the salience of past performance. *Child Development, 55,* 1983–1989.

Stipek, Deborah J., Recchia, Susan, & McClintic, Susan. (1992). Self-evaluation in young children. *Monographs of the Society for Research in Child Development, 57*(Serial No. 226), 1–79.

Stoller, Eleanor Palo. (1992). Gender differences in the experiences of caregiving spouses. In Jeffrey W. Dwyer & Raymond T. Coward (Eds.), *Gender, families and elder care.* Newbury Park, CA: Sage.

Stoneman, Z., & Brody, G.H. (1993). Sibling temperaments, conflict, warmth, and role asymmetry. *Child Development, 64,* 1786–1800.

Stones, Michael J. & Kozma, Albert. (1996). Activity, exercise and behavior. In James E. Birren and K. Warner Schaie (Eds.) *Handbook of the psychology of aging.* San Diego, CA: Academic Press.

Strassberg, Zvi, Dodge, Kenneth A., Pettit, Gregory S., & Bates, John E. (1994). Spanking in the home and children's subsequent aggression toward kindergarten peers. *Development and Psychopathology, 6,* 445–462.

Straus, Murray A. (1994). *Beating the devil out of them: Corporal punishment in American families.* Lexington, MA: Lexington Books.

Straus, Murray A., & Gelles, Richard J. (1986). Societal change and change in family violence from 1975 to 1985 as revealed by two national surveys. *Journal of Marriage and the Family, 48,* 465–479.

Straus, Murray A., & Gelles, Richard J. (1990). *Physical violence in American families: Risk factors and adaptation to violence in 8,415 families.* New Brunswick, NJ: Transaction Books.

Straus, Murray A., & Yodanis, Carrie L. (1996). Corporal punishment in adolescence and physical assaults on spouses in later life: What accounts for the link? *Journal of Marriage and the Family, 58,* 825–841.

Strauss, C.C., Smith, K., Frame, C., & Forehand, R. (1985). Personal and interpersonal characteristics associated with childhood obesity. *Journal of Pediatric Psychology, 10,* 337–343.

Streissguth, Ann P., Barr, Helen M., & Sampson, Paul D. (1990). Moderate prenatal alcohol exposure: Effects on child IQ and learning problems at age 7½ years. *Alcohol Clinical Experimental Research, 14,* 662–669.

Streissguth, Ann P., Bookstein, Fred L., Sampson, Paul D., & Barr, Helen M. (1993). *The enduring effects of prenatal alcohol exposure on child development: Birth through seven years, a partial least squares solution.* Ann Arbor: University of Michigan Press.

Streitmatter, Janice L. (1988). Ethnicity as a mediating variable of early adolescent identity development. *Journal of Adolescence, 11,* 335–346.

Streitmatter, Janice L. (1989). Identity status development and cognitive prejudice in early adolescents. *Journal of Early Adolescence, 9,* 142–152.

Streri, A. (1987). Tactile discrimination of shape and intermodal transfer in 2- to 3-month-old infants. *British Journal of Developmental Psychology, 5,* 213–220.

Suitor, J. Jill. (1991). Marital quality and satisfaction with the division of household labor across the family life cycle. *Journal of Marriage and the Family, 53,* 221–230.

Sulloway, Frank. (1996). *Born to rebel: Radical thinking in science and social thought.* New York: McKay.

Super, C.M., & Harkness, S. (1982). The infant's niche in rural Kenya and metropolitan America. In L.L. Adler (Ed.), *Cross-cultural research at issue.* New York: Academic Press.

Super, C.M., Herrera, M.G., & Mora, J.O. (1990). Long-term effects of food supplementation and psychosocial intervention on the physical growth of Colombian infants at risk of malnutrition. *Child Development, 61,* 29–49.

Super, Donald E. (1957). *The psychology of careers.* New York: Harper and Row.

Super, Donald E., & Thompson, A. S. (1981). *The adult career concerns inventory.* New York: Teachers College.

Surbey, M. (1990). Family composition, stress, and human menarche. In F. Bercovitch & T. Zeigler (Eds.), *The socioendocrinology of primate reproduction.* New York: Liss.

Susman, Elizabeth J., & Dorn, Lorah D. (1991). Hormones and behavior in adolescence. In Richard M. Lerner, Ann C. Petersen, & Jeanne Brooks-Gunn (Eds.), *Encyclopedia of adolescence* (Vol.2). New York: Garland.

Suter, Larry. (Ed.). (1993). *Indicators of science and mathematics education, 1992*. Washington, DC: National Science Foundation.

Swain, S.O. (1992). Men's friendships with women: Intimacy, sexual boundaries, and the informant role. In P.M. Nardi (Ed.), *Gender in intimate relationships*. Belmont, CA: Wadsworth.

Swann, William B., Stein-Seroussi, Alan, & Giesler, R. Brian. (1992). Why people self-verify. *Journal of Personality and Social Psychology, 62*, 392–401.

Swanson, James M., McBurnett, Keith, Wigal, Tim, & Pfiffner, Linda J. (1993). Effect of stimulant medication on children with attention deficit disorder: "A review of reviews." *Exceptional Children, 60*, 154–161.

Swanson, Jane L. (1995). The process and outcome of career counseling. In W. Bruce Walsh & Samuel H. Osipow (Eds.), *Handbook of vocational psychology: Theory, research, and practice* (2nd ed). Hillsdale, NJ: Erlbaum.

Szatmari, Peter. (1992). The validity of autistic spectrum disorders: A literature review. *Journal of Autism and Developmental Disorders, 22*, 583–600.

Szatmari, Peter, Saigal, Saroj, Rosenbaum, Peter, & Campbell, Dugal. (1993). Psychopathology and adaptive functioning among extremely low birthweight children at eight years of age. *Development and Psychopathology, 5*, 345–357.

Tallal, Paula, Miller, S., & Fitch, R.H. (1993). Neurological basis of speech: A case for the preeminence of temporal processing. In Paula Tallal (Ed.), *Annals of the New York Academy of Sciences: Vol. 682. Temporal information processing in the nervous system: Special reference to dyslexia and aphasia*. New York: New York Academy of Sciences.

Tannen, Deborah. (1990). *You just don't understand*. New York: Morrow.

Tanner, James M. (1971). Sequence, tempo, and individual variation in the growth and development of boys and girls aged twelve to sixteen. *Daedalus, 100*, 907–930.

Tanner, James M. (1978). *Fetus into man: Physical growth from conception to maturity*. Cambridge, MA: Harvard University Press.

Tanner, James M. (1991). Growth spurt, adolescent. In Richard M. Lerner, Ann C. Petersen, & Jeanne Brooks-Gunn (Eds.), *Encyclopedia of adolescence* (Vol. 1). New York: Garland.

Tanner, James M. (1991). Menarche, secular trend in age of. In Richard M. Lerner, Ann C. Petersen, & Jeanne Brooks-Gunn (Eds.), *Encyclopedia of adolescence* (Vol. 2). New York: Garland.

Tannock, Rosemary, Purvis, Karen L., & Schachar, Russell J. (1993). Narrative abilities in children with attention deficit hyperactivity disorder and normal peers. *Journal of Abnormal Child Psychology, 21*, 103–117.

Tanzer, Deborah, & Block, Jean Libman. (1976). *Why natural childbirth? A psychologist's report on the benefits to mothers, fathers and babies*. New York: Schocken.

Taylor, J.L., Miller, T.P., & Tinklenberg, J.R. (1992). Correlates of memory decline: A 4-year longitudinal study of older adults with memory complaints. *Psychology and Aging, 7*, 185–193.

Taylor, Jill McLean, Gilligan, Carol, & Sullivan, Amy M. (1995). *Between voice and silence: Women and girls, race and relationship*. Cambridge, MA: Harvard University Press.

Taylor, Ronald D., Casten, Robin, & Flickinger, Susan M. (1993). Influence of kinship social support on the parenting experiences and psychosocial adjustment of African-American adolescents. *Developmental Psychology, 29*, 382–388.

Taylor, R.J., Chatters, L.M., Tucker, M.B., & Lewis, E. (1991). Developments in research on black families: A decade review. In A. Booth (Ed.), *Contemporary families: Looking forward, looking back*. Minneapolis, MN: National Council on Family Relations.

Tew, Marjorie. (1990). *Safer childbirth: A critical history of maternity care*. New York: Routledge, Chapman & Hall.

Tharp, Roland G., & Gallimore, Ronald. (1988). *Rousing minds to life: Teaching, learning, and schooling in social context*. Cambridge, England: Cambridge University Press.

Thatcher, Robert W. (1994). Cyclic cortical reorganization: Origins of human cognitive development. In Geraldine Dawson & Kurt W. Fischer (Eds.), *Human behavior and the developing brain*. New York: Guilford.

Thelen, Esther. (1987). The role of motor development in developmental psychology: A view of the past and an agenda for the future. In Nancy Eisenberg (Ed.), *Contemporary topics in developmental psychology*. New York: Wiley.

Thelen, Esther, & Ulrich, Beverly D. (1991). Hidden skills. *Monographs of the Society for Research in Child Development, 56* (Serial No. 223).

Thelen, Esther, Corbetta, D., Kamm, K., Spencer, J.P., Schneider, K., & Zernicke, R.F. (1993). The transition to reaching: Mapping intention and intrinsic dynamics. *Child Development, 64*, 1058–1098.

Thoman, E.B., & Whitney, M.P. (1990). Behavioral states in infants: Individual differences and individual analyses. In J. Colombo & J. Fagen (Eds.), *Individual differences in infancy*. Hillsdale, NJ: Erlbaum.

Thomas, Alexander, & Chess, Stella. (1977). *Temperament and development*. New York: Brunner/Mazel.

Thomas, Alexander, Chess, Stella, & Birch, Herbert G. (1963). *Behavioral individuality in early childhood*. New York: New York University Press.

Thomas, Alexander, Chess, Stella, & Birch, Herbert G. (1968). *Temperament and behavior disorders in children*. New York: New York University Press.

Thomas, Hoben. (1993). Individual differences in children, studies, and statistics: Application of an Empirical Bayes methodology. In Mark L. Howe & Robert Pasnak (Eds.), *Emerging themes in cognitive development: Vol. 1. Foundations*. New York: Springer-Verlag.

Thomas, Jeanne L. (1986). Age and sex differences in perceptions of grandparents. *Journal of Gerontology, 41*, 417–423.

Thompson, Elizabeth, & Colella, Ugo. (1992). Cohabitation and marital stability: Quality or commitment? *Journal of Marriage and the Family, 54*, 259–267.

Thompson, Frances E., & Dennison, Barbara. (1994). Dietary sources of fats and cholesterol in U.S. children aged 2 through 5 years. *American Journal of Public Health, 84*, 799–806.

Thompson, Larry W. (1994). *Correcting the code*. New York: Simon & Schuster.

Thompson, Larry W., Gong, Vincent, Haskins, Edmund, & Gallagher, Dolores. (1987). Assessment of depression and dementia during the late years. In K. Warner Schaie (Ed.), *Annual review of gerontology and geriatrics* (Vol. 7). New York: Springer.

Thompson, Linda, & Walker, Alexis J. (1989). Gender in families: Women and men in marriage, work, and parenthood. *Journal of Marriage and the Family, 5,* 845–871.

Thompson, Ron A., & Sherman, Roberta. (1993). *Helping athletes with eating disorders.* Bloomington, IN: Human Kinetics Books.

Thompson, Ross A. (1990). Emotion and self-regulation. In R.A. Thompson (Ed.), *Nebraska Symposium on Motivation: Vol. 36. Socioemotional development.* Lincoln: University of Nebraska Press.

Thompson, Ross A. (1990). *Social support and the prevention of child maltreatment.* Paper commissioned by the U.S. Advisory Board on Child Abuse and Neglect, Washington, DC.

Thompson, Ross A. (1990). Vulnerability in research: A developmental perspective on research risk. *Child Development, 61,* 1–16.

Thompson, Ross A. (1991). Infant day care: Concerns, controversies, choices. In Jacqueline V. Lerner & Nancy L. Galambos (Eds.), *Employed mothers and their children.* New York: Garland.

Thompson, Ross A. (1992). Developmental changes in research risk and benefit: A changing calculus of concerns. In B. Stanley & J.E. Sieber (Eds.), *Social research on children and adolescents: Ethical issues.* Newbury Park, CA: Sage.

Thompson, Ross A. (1994). Fatherhood and divorce. *The Future of Children, 4.*

Thompson, Ross A. (1994). Emotional regulation: A theme in search of definition. In Nathan A. Fox (Ed.) The development of emotional regulation: Biological and behavioral considerations. *Monographs of the Society for Research in Child Development.* (Serial no. 240.)

Thompson, Ross A. (1995). Personal communication from one of Professor Thompson's students.

Thompson, Ross A. (1997). Early sociopersonality development. In William Damon (Ed.), *Handbook of child psychology* (5th ed., Vol. 3). New York: Wiley.

Thompson, Ross A., & Frodi, A.M. (1984). The sociophysiology of infants and their caregivers. In W.M. Waid (Ed.), *Sociophysiology.* New York: Springer-Verlag.

Thompson, Ross A., & Jacobs, Janis E. (1991). Defining psychological maltreatment: Research and policy perspectives. *Development and Psychopathology, 3,* 93–102.

Thompson, Ross A., & Limber, S.P. (1990). "Social anxiety" in infancy: Stranger and separation anxiety. In H. Leitenberg (Ed.), *Handbook of social anxiety.* New York: Plenum Press.

Thompson, Ross A., Tinsley, Barbara R., Scalora, Mario J., & Parke, Ross D. (1989). Grandparents' visitation rights. *American Psychologist, 44,* 1217–1222.

Thompson, Ross A., Scalora, M.J., Castrianno, L., & Limber, S.P. (1992). Grandparent visitation rights: Emergent psychological and psycholegal issues. In D.K. Kagehiro & W.S. Laufer (Eds.), *Handbook of psychology and law.* New York: Springer-Verlag.

Thornburg, Herschel D., & Aras, Ziya. (1986). Physical characteristics of developing adolescents. *Journal of Adolescent Research, 1,* 47–78.

Thorson, J.A. (1995). *Aging in a changing society.* Belmont, CA: Wadsworth.

Thun, Michael J., Day-Lally, Cathy A., Calle, Eugenia E., Flanders, W. Dana, & Heath, Clark W. (1995). Excess mortality among cigarette smokers: Changes in a 20-year interval. *American Journal of Public Health, 85,* 1223–1230.

Tice, Raymond R., & Setlow, Richard B. (1985). DNA repair and replication in aging organisms and cells. In Caleb E. Finch & Edward L. Schneider (Eds.), *Handbook of the biology of aging* (2nd ed.). New York: Van Nostrand.

Tiefer, Leonore. (1995). *Sex is not a natural act and other essays.* Boulder, CO: Westview Press.

Tillich, Paul. (1958). *Dynamics of faith.* New York: Harper & Row.

Timiras, Paola. (1994). *Physiological basis of aging and geriatrics* (2nd ed.). Boca Raton, FL: CRC Press.

Tinetti, M.E., & Powell, L. (1993). Fear of falling and low self-efficacy: A cause of dependence in elderly persons [Special issue]. *Journal of Gerontology, 48,* 35–38.

Tisi, Gennaro M. (1988). Pulmonary problems: Smoking, obstructive lung disease, and other lung disorders. In Dorothy Reycroft Hollingsworth & Robert Resnik (Eds.), *Medical counseling before pregnancy.* New York: Churchill-Livingstone.

Tobin, J.D., Wu, D.Y.H., & Davidson, D. (1989). *Preschool in three cultures.* New Haven, CT: Yale University Press.

Tomasello, M. (1988). The role of joint attentional processes in early language development. *Language Sciences, 10,* 69–88.

Tomasello, M. (1992). The social bases of language acquisition. *Social Development, 1,* 67–87.

Torres-Gil, Fernanda M. (1992). *The new aging.* New York: Auburn House.

Trasler, Gordon. (1987). Biogenetic factors. In Herbert C. Quay (Ed.), *Handbook of juvenile delinquency.* New York: Wiley.

Treasure, J.L., & Holland, A.J. (1993). What discordant twins tell us about the etiology of anorexia nervosa. In E. Ferrari, F. Branbilla, & S.B. Solerte (Eds.), *Primary and secondary eating disorders.* Oxford, England: Pergamon Press.

Treboux, Dominique, & Busch-Rossnagel, Nancy. (1990). Social network influences on adolescent sexual attitudes and behaviors. *Journal of Adolescent Research, 5,* 175–189.

Trehub, S.E., Schneider, B.A., Thorpe, L.A., & Judge, P. (1991). Observational measures of auditory sensitivity in early infancy. *Developmental Psychology, 27,* 40–49.

Triandis, Harry C. (1994). *Culture and social behavior.* New York: McGraw-Hill.

Troll, Lillian E., & Skaff, Marilyn McKean. (1997). Perceived continuity of self in very old age. *Psychology and Aging, 12,* 162–169.

Tronick, Edward Z. (1989). Emotions and emotional communication in infants. *American Psychologist, 44,* 112–119.

Tronick, Edward Z., & Cohn, Jeffrey F. (1989). Infant-mother face-to-face interaction: Age and gender differences in coordination and the occurrence of miscoordination. *Child Development, 60,* 85–92.

Tronick, Edward Z., Morelli, G.A., & Ivey, P.K. (1992). The Efe forager infant and toddler's pattern of social relationships: Multiple and simultaneous. *Developmental Psychology, 28,* 568–577.

Tschann, Jeanne M., Johnston, Janet R., & Wallerstein, Judith S. (1989). Resources, stressors, and attachment as predictors of adult adjustment after divorce: A longitudinal study. *Journal of Marriage and the Family, 51,* 1033–1047.

Tucker, Joan S., Friedman, Howard S., Tsai, Catherine M., & Martin, Leslie R. (1995). Playing with pets and longevity among older people. *Psychology and Aging, 10,* 3–7.

Turner, B.F., & Adams, C.G. (1988). Reported change in preferred sexual activity over the adult years. *Journal of Sex Research, 25,* 289–303.

Turner, Patricia J. (1991). Relations between attachment, gender, and behavior with peers in preschool. *Child Development, 62,* 1475–1488.

Turner, Patricia J. (1993). Attachment to mother and behavior with adults in preschool. *British Journal of Developmental Psychology, 11,* 75–89.

Tyler, Richard S., & Schum, Donald J. (Eds.). (1995). *Assistive devices for persons with hearing impairment.* Boston: Allyn & Bacon.

Tyson, Jon. (1995). Evidence-based ethics and the care of premature infants. *The Future of Children: Low Birth Rate, 5,* 197–213.

Tzeng, Meei-Shenn. (1992). The effects of socioeconomic heterogamy and changes on marital dissolution for first marriages. *Journal of Marriage and the Family, 54,* 609–619.

Uchino, Bert N., Kiecolt-Glaser, Janice K., & Cacioppo, John T. (1992). Age-related changes in cardiovascular response as a function of chronic stressor and social support. *Journal of Personality and Social Psychology, 63,* 839–846.

Udry, J. Richard. (1981). Marital alternatives and marital disruption. *Journal of Marriage and the Family, 43,* 889–897.

Uhlenberg, Peter. (1996). The burden of aging: A theoretical framework for understanding the shifting balance of caregiving and care receiving as cohorts age. *Gerontologist, 36,* 761–767.

Uhlenberg, Peter, & Miner, Sonia. (1996). Life course and aging: A cohort perspective. In George H. Binstock & Linda K. George (Eds.), *Handbook of aging and the social sciences.* San Diego, CA: Academic Press.

Ulbrich, P.M., & Bradsher, J.E. (1993). Perceived support, help seeking, and adaptation to stress among older black and white women living alone. *Journal of Aging and Health, 5,* 265–286.

Umberson, Debra. (1992). Relationship between adult children and their parents: Psychological consequences for both generations. *Journal of Marriage and the Family, 54,* 664–674.

Underwood, M.K., Coie, J.D., & Herbsman, C.R. (1992). Display rules for anger and aggression in school-age children. *Child Development, 63,* 366–380.

UNICEF. (1990). *Children and development in the 1990s: A UNICEF sourcebook.* New York: United Nations.

UNICEF. (1994). *The state of the world's children, 1994.* New York: Oxford University Press.

UNICEF. (1995). *The state of the world's children, 1995.* New York: United Nations.

United Nations. (1991). *Human development report, 1991.* New York: Oxford University Press.

United Nations. (1994). *The state of the world's children, 1994.* New York: Oxford University Press.

United Nations. (1995). *Demographic yearbook: 1993.* New York: Department for Economic and Social Information and Policy Analysis.

U.S. Advisory Board on Child Abuse and Neglect. (1993). *Neighbors helping neighbors: A new national strategy for the protection of children.* Washington, DC: U.S. Government Printing Office.

U.S. Bureau of the Census. (1976). *Historical statistics of the United States: Colonial times to 1970.* Washington, DC: U.S. Department of Commerce.

U.S. Bureau of the Census. (1991). *Statistical abstract of the United States, 1991* (111th ed.). Washington, DC: United States Department of Commerce.

U.S. Bureau of the Census. (1992). *Statistical abstract of the United States, 1992* (112th ed.). Washington, DC: United States Department of Commerce.

U.S. Bureau of the Census. (1994). *Statistical abstract of the United States, 1993* (114th ed.). Washington, DC: United States Department of Commerce.

U.S. Bureau of the Census. (1995). *Statistical abstract of the United States, 1995* (115th ed.). Washington, DC: U.S. Department of Commerce.

U.S. Bureau of the Census. (1996). *Statistical abstract of the United States, 1996* (116th ed.). Washington, DC: U.S. Department of Commerce.

U.S. Bureau of Labor Statistics. (1996). *Comparative labor force statistics for ten countries, 1959–1995.* Washington, DC: Author.

U.S. Congress, Congressional Budget Office. (1994). *The economic and budget outlook: Fiscal years 1995–1999.* Washington, DC: U.S. Government Printing Office.

U.S. Department of Education. (1989). *High school and beyond: 1987 transcript study.* Washington, DC: National Center for Educational Statistics.

U.S. Department of Education. (1991). *The condition of education, 1991: Vol 1. Elementary and secondary education.* Washington, DC: National Center for Educational Statistics.

U.S. Department of Health and Human Services. (1990). *Healthy people 2000: National health promotion and disease prevention objectives* (DHHS Publication No. PHS–50212). Washington, DC: Author.

U.S. Department of Health and Human Services. (1990). *Long-term care for the functionally dependent elderly* (Series 13: Data from the National Health Survey, No. 104). Hyattsville, MD: Author.

U.S. Department of Justice. (1995). *Justice sourcebook, 1995.* Washington DC: Bureau of Justice Statistics.

Uttal, David H., & Perlmutter, Marion. (1989). Toward a broader conceptualization of development: The role of gains and losses across the life span. *Developmental Review, 9,* 101–132.

Vaillant, George E. (1977). *Adaptation to life.* Boston: Little, Brown.

Vaillant, George E. (1993). *The wisdom of the ego.* Cambridge, MA: Harvard University Press.

Van Biema, David. (1995, December 11). A shameful death. *Time,* 33–36.

Vandell, Deborah Lowe. (1987). Baby sister/baby brother. Reactions to the birth of a sibling and patterns of early sibling relations. *Journal of Children in Contemporary Society, 19*(3/4), 13–37.

Vandell, Deborah Lowe, & Hembree, Sheri E. (1994). Peer social status and friendship: Independent contributors to children's social and academic adjustment. *Merrill Palmer Quarterly, 40,* 461–477.

Vandenberg, Steven G. (1987). Sex differences in mental retardation and their implications for sex differences in ability. In June Machover Reinisch, Leonard A. Rosenblum, & Stephanie A. Sanders (Eds.), *Masculinity/femininity: Basic perspectives.* New York: Oxford University Press.

Vander Linde, Eleanor, Morrongiello, Barbara A., & Rovee-Collier, Carolyn K. (1985). Determinants of retention in 8-week-old infants. *Developmental Psychology, 21,* 601–613.

van der Veer, Rene, & Valsiner, Jaan. (1991). *Understanding Vygotsky: A quest for synthesis.* Oxford, England: Blackwell.

van der Voort, T.H.A., & Valkenburg, P.M. (1994). Television's impact on fantasy play: A review of research. *Developmental Review, 14,* 27–51.

van Ijzendoorn, M.H., & Kroonenberg, P.M. (1988). Cross-cultural patterns of attachment: A meta-analysis of the Strange Situation. *Child Development, 59,* 147–156.

van Loosbroek, E., & Smitsman, A.W. (1990). Visual perception of numerosity in infancy. *Developmental Psychology, 26,* 916–922.

Van Nostrand, Joan F. (1991). Long-term care in the United States: Issues in measuring nursing home outcomes. *Proceedings of the 1988 International Symposium on Aging.* (Series 5, No. 6, DHHS Publication No. 91–1482). Hyattsville, MD: U.S. Department of Health and Human Services.

Vaughn, Brian E. (1987). Maternal characteristics measured prenatally are predictive of ratings of temperamental "difficulty" on the Carey Infant Temperament Questionnaire. *Developmental Psychology, 23,* 152–161.

Vaughn, Sharon, Zaragoza, Nina, Hogan, Anne, & Walker, Judy. (1993). A four-year longitudinal investigation of the social skills and behavior problems of students with learning disabilities. *Journal of Learning Disabilities, 26,* 404–406.

Vaupel, James W., & Lundstrom, Hans. (1994). Longer life expectancy? Evidence from Sweden of reductions in mortality rates at advanced ages. In David A. Wise (Ed.), *Studies in the economics of aging.* Chicago: University of Chicago Press.

Veatch, Robert M. (1995). The definition of death: Problems for public policy. In Hannelore Was & Robert A. Neimeyer (Eds.), *Dying: Facing the facts.* Washington, DC: Taylor & Francis.

Vega, W.A., Kolody, B., Hwang, J., & Noble, A. (1993). Prevalence and magnitude of perinatal substance exposures in California. *New England Journal of Medicine, 329,* 850–854.

Verbrugge, Lois M. (1989). The dynamics of population aging and health. In Stephen J. Lewis (Ed.), *Aging and health: Linking research and public policy.* Chelsea, MI: Lewis Publishers.

Verbrugge, Lois M. (1990). The twain meet: Empirical explanations of sex differences in health and mortality. In Marcia G. Ory & Huber R. Warnen (Eds.), *Gender, health, and longevity.* New York: Springer.

Verbrugge, Lois M. (1994). Disability in late life. In Ronald P. Abeles, Helen C. Gift, & Marcia G. Ory (Eds.), *Aging and quality of life.* New York: Springer.

Verhaeghen, Paul, & Marcoen, Alfons. (1996). On the mechanisms of plasticity in young and older adults after instruction in the methods of loci: Evidence for an amplification model. *Psychology and Aging, 11,* 164–178.

Verhaeghen, Paul, Marcoen, Alfons, & Goossens, L. (1992). Improving memory performance in the aged through mnemonic training: A meta-analytic study. *Psychology and Aging, 7,* 242–251.

Vickery, Florence E. (1978). *Old age and growing.* Springfield, IL: Thomas.

Volkmar, F.R. (1991). Autism and the pervasive developmental disorders. In M. Lewis (Ed.), *Child and adolescent psychiatry: A comprehensive textbook.* Baltimore: Williams & Wilkins.

Vondracek, Fred W., & Kawasaki, Tomotsugu. (1995). Toward a comprehensive framework for adult career development theory and intervention. *Handbook of vocational psychology: Theory, research, and practice.* Hillsdale, NJ: Erlbaum.

von Hofsten, Claes. (1983). Catching skills in infancy. *Journal of Experimental Psychology: Human Perception and Performance, 9,* 75–85.

Voydanoff, Patricia, & Donnelly, Brenda Wixson. (1990). *Adolescent sexuality and pregnancy.* Newbury Park, CA: Sage.

Vuchinich, S., Hetherington, E.M., Vuchinich, R.A., & Clingempeel, W.G. (1991). Parent-child interaction and gender differences in early adolescents' adaptation to stepfamilies. *Developmental Psychology, 27,* 618–626.

Vygotsky, Lev S. (1978). *Mind in society: The development of higher psychological processes.* Cambridge, MA: Harvard University Press.

Vygotsky, Lev S. (1986). *Thought and language.* Cambridge, MA: MIT Press. (Original work published 1934)

Vygotsky, Lev S. (1987). *Thinking and speech* (N. Minick, Trans.). New York: Plenum Press.

Wagenaar, Alexander, & Perry, Cheryl L. (1994). Community strategies for the reduction of youth drinking: Theory and application. *Journal of Research on Adolescence, 4,* 319–346.

Waggoner, J.E., & Palermo, D.S. (1989). Betty is a bouncing bubble: Children's comprehension of emotion-descriptive metaphors. *Developmental Psychology, 25,* 152–163.

Wahlin, Ake, Winblad, Bengt, Hill, Robert D., & Backman, Lars. (1996). Effects of serum vitamin B12 and folate status on episodic memory performance in very old age: A population-based study. *Psychology and Aging, 11,* 487–496.

Wainryb, Cecilia, & Turiel, Elliot. (1995). Diversity in social development: Between or within cultures? In Melanie Killen & Daniel Hart (Eds.), *Morality in everyday life: Developmental perspectives.* Cambridge, England: Cambridge University Press.

Waite, L. J., & Lillard, L. A. (1991). Children and marital disruption. *American Journal of Sociology, 96,* 930–953.

Walker, Arlene S. (1982). Intermodal perception of expressive behaviors by human infants. *Journal of Experimental Child Psychology, 33,* 514–535.

Walker, Lawrence J. (1988). The development of moral reasoning. *Annals of Child Development, 55,* 677–691.

Walker, Lawrence J., de Vries, Brian, & Trevethan, Shelley D. (1987). Moral stages and moral orientations in real-life and hypothetical dilemmas. *Child Development, 58,* 842–858.

Walker, Lawrence J., Pitts, Russell C., Hennig, Karl H., & Matsuba, M. Kyle. (1995). Reasoning about morality and real-life moral problems. In Melanie Killen & Daniel Hart (Eds.), *Morality in everyday life: Developmental perspectives.* Cambridge, England: Cambridge University Press.

Walker-Andrews, A.S., Bahrick, L.E., Raglioni, S.S., & Diaz, I. (1991). Infants' bimodal perception of gender. *Ecological Psychology, 3,* 55–75.

Wallerstein, Judith S., & Blakeslee, Sandra. (1990). *Second chances: Men, women, and children a decade after divorce.* New York: Ticknor & Fields.

Wallerstein, Judith S., & Blakeslee, Sandra. (1995). *The good marriage.* Boston: Houghton Mifflin.

Walpole, I., Zubrick, S., & Pontre, J. (1990). Is there a fetal effect with low to moderate alcohol use before or during pregnancy? *Journal of Epidemiology and Community Health, 44,* 297–301.

Walsh, David A., & Hershey, Douglas A. (1993). Mental models and the maintenance of complex problem solving-skills in old age. In John Cerella, John Rybash, William Hoyer, & Michael L. Commons (Eds.), *Adult information processing: Limits on loss.* San Diego, CA: Academic Press.

Walter, Tony. (1993). Modern death: Taboo or not taboo? In Donna Dickenson & Malcolm Johnson (Eds.), *Death, dying & bereavement.* London: Sage.

Wanner, Eric, & Gleitman, Lila R. (Eds.). (1982). *Language acquisition: The state of the art.* Cambridge, England: Cambridge University Press.

Ward, Russell A. (1993). Marital happiness and household equity in later life. *Journal of Marriage and the Family, 55,* 427–438.

Ward, Russell A., & Spitze, Glenna. (1996). Gender differences in parent-child coresidence experiences. *Journal of Marriage and the Family, 58,* 718–725.

Ward, Russell A., Logan, John, & Spitze, Glenna. (1992). The influence of parent and child needs on coresidence in middle and later life. *Journal of Marriage and the Family, 54,* 209–221.

Wasik, B.H., Bryant, D.M., & Lyons, C.M. (1990). *Home visiting.* Newbury Park, CA: Sage.

Wass, Hannelore. (1995). Death in the lives of children and adolescents. In Hannelore Wass & Robert A. Neimeyer (Eds.), *Dying: Facing the facts.* Washington, DC: Taylor & Francis.

Waterman, Alan S. (1985). Identity in the context of adolescent psychology. In Alan S. Waterman (Ed.), *Identity in adolescence: Processes and contents: Vol. 30. New directions in child development.* San Francisco: Jossey-Bass.

Waterson, E.J., & Murray-Lyon, Iain M. (1990). Preventing alcohol related birth damage: A review. *Social Science and Medicine, 30,* 349–364.

Watson, John B. (1927, March). What to do when your child is afraid (interview with Beatrice Black). *Children,* pp. 25–27.

Watson, John B. (1928). *Psychological care of the infant and child.* New York: Norton.

Watson, John B. (1967). *Behaviorism* (rev. ed.). Chicago: University of Chicago Press. (Original work published 1930)

Webster, Harold, Freedman, Mervin B., & Heist, Paul. (1979). Personality change in students. In Nevitt Sanford & Joseph Axelrod (Eds.), *College and character.* Berkeley, CA: Montaigne.

Weiss, B., Dodge, K.A., Bates, J.E., & Pettit, G.S. (1992). Some consequences of early harsh discipline: Child aggression and a maladaptive social information processing style. *Child Development, 63,* 1321–1335.

Weiss, Gabrielle. (1991). Attention deficit hyperactivity disorder. In M. Lewis (Ed.), *Child and adolescent psychiatry: A comprehensive textbook.* Baltimore: Williams & Wilkins.

Weiss, Gabrielle, & Hechtman, Lily Trokenberg. (1986). *Hyperactive children grow up: Empirical findings and theoretical considerations.* New York: Guilford.

Weiss, Kenneth A. (1993). *Genetic variation and human disease: Principles and evolutionary approaches.* Cambridge, England: Cambridge University Press.

Weitzman, Lenore J. (1985). *The divorce revolution: The unexpected social and economic consequences for women and children in America.* New York: Free Press.

Welford, Alan T. (1980). On the nature of higher-order skills. *Journal of Occupational Psychology, 53,* 107–110.

Wellman, H.M. (1990). *The child's theory of mind.* Cambridge, MA: MIT Press.

Wellman, H.M., & Gelman, S.A. (1992). Cognitive development: Foundational theories of core domains. *Annual Review of Psychology, 43,* 337–375.

Welsh, Michael, & Smith, Alan E. (1995). Cystic fibrosis. *Scientific American, 273,* 52–59.

Wender, Paul H. (1987). *The hyperactive child, adolescent and adult: Attention deficit disorder through the lifespan.* New York: Oxford University Press.

Werker, J.F. (1989). Becoming a native listener. *American Scientist, 77,* 54–59.

Werner, Emmy E. (1993). Risk resilience and recovery: Perspectives from Kauai longitudinal study. *Development and Psychopathology, 5,* 503–515.

Werner, Emmy E., & Smith, Ruth S. (1982). *Vulnerable but invincible: A study of resilient children.* New York: McGraw-Hill.

Werner, Emmy E., & Smith, Ruth S. (1992). *Overcoming the odds: High risk children from birth to adulthood.* Ithaca, NY: Cornell University Press.

Wertsch, J.V. (1985). *Vygotsky and the social formation of mind.* Cambridge, MA: Harvard University Press.

Wertsch, J.V., & Tulviste, P. (1992). L.S. Vygotsky and contemporary developmental psychology. *Developmental Psychology, 28,* 548–557.

West, M.M. (1988). Parental values and behavior in the outer Fiji islands. In R.A. LeVine, P.M. Miller, & M.M. West (Eds.), *New directions for child development: No. 40. Parental behavior in diverse societies.* San Francisco: Jossey-Bass.

Whalen, C.K., Henker, B., Collins, B.E., Finck, D., & Dotemoto, S. (1979). A social ecology of hyperactive boys: Medication effects in systematically structured classroom environments. *Journal of Applied Behavioral Analysis, 12,* 65–81.

Whitam, Frederick L., Diamond, Milton, & Martin, James. (1993). Homosexual orientation in twins: A report on 61 pairs and three triplet sets. *Archives of Sexual Behavior, 22,* 187–206.

Whitbourne, Susan Krauss. (1985). *The aging body.* New York: Springer-Verlag.

Whitbourne, Susan Krauss, & Wills, Karen-Jo. (1994). Psychological issues in institutional care of the aged. In Seth B. Goldsmith (Ed.), *Essentials of long-term care administration.* Gaithersburg, MD: Aspen.

White, Charles B., & Janson, Philip. (1986). Helplessness in institutional settings: Adaptation or inotropic disease. In Margaret M. Baltes & Paul B. Baltes (Eds.), *The psychology of control and aging.* Hillsdale, NJ: Erlbaum.

White, Lynn K. (1990). Determinants of divorce: A review of research in the eighties. *Journal of Marriage and the Family, 52,* 904–912.

White, Lynn K., & Booth, Alan. (1985). The quality and stability of remarriages: The role of stepchildren. *American Sociological Review, 50,* 689–698.

White, Lynn K. & Rogers, Stacy J. (1997). Strong support but uneasy relationships: Coresidence and adult children's relationships with their parents. *Journal of Marriage and the Family, 59,* 62–76.

Whiten, A. (Ed.). (1991). *Natural theories of mind.* Oxford, England: Blackwell.

Whiting, Beatrice Blyth, & Edwards, Carolyn Pope. (1988). *Children of different worlds: The formation of social behavior.* Cambridge, MA: Harvard University Press.

Wickelgren, Ingrid. (1996). For the cortex, neuron loss may be less than thought. *Science, 273,* 48–50.

Wicker, Allan W., & August, Rachel A. (1995). How far should we generalize? The case of a workload model. *Psychological Science, 6,* 39–44.

Widom, Cathy Spatz. (1991). The role of placement experience in mediating the criminal consequences of early childhood victimization. *American Journal of Orthopsychiatry, 61,* 195–209.

Wilkinson, Richard G. (1992). National mortality rates: The impact of inequality. *American Journal of Public Health, 82,* 1082–1084.

Willatts, P. (1989). Development of problem-solving in infancy. In A. Slater & G. Bremner (Eds.), *Infant development.* Hove, England: Erlbaum.

Willett, W.C., & Trichopoulos, D. (1996). Nutrition and cancer: A summary of the evidence. *Cancer Causes Control, 7,* 178–180.

Williams, Sharon, Denney, Nancy Wadsworth, & Schadler, Margaret. (1983). Elderly adults' perception of their own cognitive development during the adult years. *International Journal of Aging and Human Development, 16,* 47–158.

Williamson, David F., Serdula, Mary K., Anda, Robert F., Levy, Alan, & Byers, Tim. (1992). Weight loss attempts in adults: Goals, duration, and rate of weight loss. *American Journal of Public Health, 82,* 1251–1257.

Willinger, M., Hoffman, H.J., & Hartford, R.B. (1994). Infant sleep position and risk for sudden infant death syndrome. *Pediatrics, 93,* 814–819.

Willis, D.J., Holden, E.W., & Rosenberg, M. (Eds.). (1992). *Prevention of child maltreatment: Developmental and ecological perspectives.* New York: Wiley.

Willis, Sherry L. (1996). Everyday cognitive competence in elderly persons: Conceptual issues and empirical findings. *Gerontologist, 36,* 595–601.

Wilson, B.L., & Corcoran, T.B. (1988). *Successful secondary schools.* New York: Falmer Press.

Wilson, C.L., & Sindelar, P.T. (1990). Direct instruction in math word problems: Students with learning disabilities. *Exceptional Children, 57,* 512–519.

Wilson, Gail. (1995). "I'm the eyes and she's the arms": Changes in gender roles in advanced old age. In Sara Arber & Jay Ginn (Eds.), *Connecting gender and aging.* Buckingham, England: Open University Press.

Wilson, Geraldine S. (1989). Clinical studies of infants and children exposed prenatally to heroin. In Donald Hutchings (Ed.), *Prenatal abuse of licit and illicit drugs.* New York: New York Academy of Sciences.

Wilson, Geraldine S. (1992). Heroin use during pregnancy: Clinical studies of long-term effects. In T.B. Sonderegger (Ed.), *Perinatal substance abuse: Research findings and clinical implications.* Baltimore: Johns Hopkins University Press.

Wilson, James Q. (1983). Raising kids. *Atlantic Monthly, 252*(4), 45–56.

Wilson, James Q., & Herrnstein, Richard J. (1985). *Crime and human nature.* New York: Simon & Schuster.

Wilson, Jerome. (1989). Cancer incidence and mortality differences of black and white Americans. In Lovell A. Jones (Ed.), *Minorities and cancer.* New York: Springer-Verlag.

Wilson, Margo, & Daly, Martin. (1993). Lethal confrontational violence among young men. In Nancy J. Bell & Robert W. Bell (Eds.), *Adolescent risk taking.* Newbury Park, CA: Sage.

Wilson, Melvin N. (1989). Child development in the context of the black extended family. *American Psychologist, 44,* 380–385.

Wilson, M. Roy. (1989). Glaucoma in blacks: Where do we go from here? *JAMA, Journal of the American Medical Association, 261,* 281–282.

Wing, R.R. (1992). Weight cycling in humans: A review of the literature. *Annals of Behavioral Medicine, 14,* 113–119.

Wirth, H.P. (1993). Caring for a chronically demented patient within the family. In W. Meier-Ruge (Ed.), *Dementing brain disease in geriatric medicine.* Switzerland: Karger.

Wolf, Susan M. (1996). Gender, feminism, and death: Physician assisted suicide and euthanasia. In Susan M. Wolf (Ed.), *Feminism and bioethics: Beyond reproduction.* New York: Oxford University Press.

Wolfe, D.A. (1994). The role of intervention and treatment services in the prevention of child abuse and neglect. In G.B. Melton & F. Barry (Eds.), *Safe neighborhoods: Foundations for a new national strategy on child abuse and neglect.* New York: Guilford.

Wolfe, Wendy S., Campbell, Cathy C., Fongillo, Edward A., Haas, Jere D., & Melnick, Thomas A. (1994). Overweight schoolchildren in New York State: Prevalence and characteristics. *American Journal of Public Health, 84,* 807–813.

Wolfner, Glenn D., & Gelles, Richard J. (1993). A profile of violence toward children: A national study. *Child Abuse and Neglect, 17,* 197–212.

Wong Fillmore, Lily. (1976). *The second time around: Cognitive and social strategies in second language acquisition.* Doctoral dissertation, Stanford University (cited in McLaughlin, 1984).

Wong Fillmore, Lily. (1987, April 25). *Becoming bilingual: Social processes in second language learning.* Paper presented at the Society for Research in Child Development, Baltimore.

Wong Fillmore, Lily. (1991). Second-language learning in children: A model of language learning in social context. In E. Bialystok (Ed.), *Language processing in bilingual children.* Cambridge, England: Cambridge University Press.

Wood, D., Bruner, Jerome S., & Ross, G. (1976). The role of tutoring in problem solving. *Journal of Child Psychology and Psychiatry, 17,* 89–100.

Wood, Robert, & Bandura, Albert. (1996). Social cognitive theory and organizational management. In Richard M. Steers, Lyman W. Porter, & Gregory A. Bigley (Eds.), *Motivation and leadership at work.* New York: McGraw-Hill.

Woodhead, Martin. (1991). Psychology and the cultural construction of "children's needs." In Martin Woodhead, Paul Light, & Ronnie Carr (Eds.), *Child development in social context: Vol. 3. Growing up in a changing society*. London: Routledge.

Woodruff-Pak, Diana S. (1989). Aging and intelligence: Changing perspectives in the twentieth century. *Journal of Aging Studies, 3,* 91–118.

World Health Organization (WHO). (1993). *World Health Statistics Quarterly, 46,* No.1.

World Health Organization (WHO). (1994). *World Health Statistics Quarterly, 47,* No. 1.

Wren, Christopher S. (1996, February 20). Marijuana use by youths continues to rise. *The New York Times,* p. A11.

Wright, Paul H. (1982). Men's friendships, women's friendships, and the alleged inferiority of the latter. *Sex Roles, 8,* 1–20.

Wrightsman, Lawrence S. (1994). *Adult personality development* (Vols. 1 and 2). Thousands Oaks, CA: Sage.

Wunsch, Marie A. (1994). *Mentoring revisited: Making an impact on individuals and institutions.* San Francisco: Jossey-Bass.

Wykle, May L., Kahana, Eva, & Kowal, Jerome. (Eds.). (1992). *Stress and health among the elderly.* New York: Springer.

Wynn, K. (1992). Addition and subtraction by human infants. *Nature (London), 358,* 749–750.m

Yang, Bin, Ollendick, Thomas, Dong, Qi, Xia, Yong, & Lin, Lei. (1995). Only children and children with siblings in the People's Republic of China: Levels of fear, anxiety, and depression. *Child Development, 66,* 1301–1311.

Yerkes, R.M. (1923). Testing and the human mind. *Atlantic Monthly, 131,* 358–370.

Yglesias, Helen. (1980). Moses, Anna Mary Robertson (Grandma). In Barbara Sicherman & Carol Hurd Green (Eds.). *Notable American women: The modern period.* Cambridge, MA: Belknap Press.

Yoon, Keumsil Kim. (1992). New perspective on intrasentential code-switching: A study of Korean-English switching. *Applied Psycholinguistics, 13,* 433–449.

Younger, B.A. (1990). Infant categorization: Memory for category-level and specific item information. *Journal of Experimental Child Psychology, 50,* 131–155.

Younger, B.A. (1993). Understanding category members as "the same sort of thing": Explicit categorization in ten-month-old infants. *Child Development, 64,* 309–320.

Youniss, James. (1989). Parent-adolescent relationships. In William Damon (Ed.), *Child development today and tomorrow.* San Francisco: Jossey-Bass.

Zahn-Waxler, C., Radke-Yarrow, M., Wagner, E., & Chapman, M. (1992). Development of concern for others. *Child Development, 28,* 126–136.

Zaks, Peggy M., & Labouvie-Vief, Gisela. (1980). Spatial perspective taking and referential communication skills in the elderly: A training study. *Journal of Gerontology, 35,* 217–224.

Zametkin, A.J., Nordahl, T.E., Gross, M., King, A.C., Semple, W.E., Rumsey, J., Hamburger, S., & Cohen, R.M. (1990). Cerebral glucose metabolism in adults with hyperactivity of childhood onset. *New England Journal of Medicine, 323,* 1361–1366.

Zarbatany, L., Hartmann, D.P., & Rankin, D.B. (1990). The psychological functions of preadolescent peer activities. *Child Development, 61,* 1067–1080.

Zaslow, Martha J. (1991). Variation in child care quality and its implications for children. *Journal of Social Issues, 47,* 125–138.

Zaslow, Martha J., Pederson, Frank A., Cain, Richard L., Suwalksy, Joan T.D., & Kramer, Eva L. (1985). Depressed mood in new fathers: Associations with parent-infant interaction. *Genetic, Social, and General Psychology Monographs, 111,* 133-150.

Zeanah, C.H., Benoit, D., Barton, M. Regan, C., Hirshberg, L.M., & Lipsitt, L.P. (1993). Representations of attachment in mothers and their one-year-old infants. *Journal of the American Academy of Child and Adolescent Psychiatry, 32,* 278–286.

Zedeck, Sheldon. (Ed.). (1992). *Work, families, and organizations.* San Francisco: Jossey-Bass.

Zeskind, P.S., & Collins, V. (1987). Pitch of infant crying and caregiver responses in a natural setting. *Infant Behavior and Development, 10,* 501–504.

Zierler, Sally. (1994). Women, sex, and HIV. *Epidemiology, 5,* 565–567.

Zigler, Edward, & Berman, Winnie. (1983). Discerning the future of early childhood intervention. *American Psychologist, 38,* 894–906.

Zigler, Edward, & Hall, Nancy W. (1989). Physical child abuse in America: Past, present, and future. In Dante Cicchetti & Vicki Carlson (Eds.), *Child maltreatment: Theory and research on the causes and consequences of child abuse and neglect.* Cambridge, England: Cambridge University Press.

Zigler, Edward, & Lang, M.E. (1990). *Child care choices.* New York: Free Press.

Zigler, Edward, Styfco, Sally, & Gilman, Elizabeth. (1993). National Head Start program for disadvantaged preschoolers. In E. Zigler & S. Styfco (Eds.), *Head Start and beyond: A national plan for extended childhood intervention.* New Haven, CT: Yale University Press.

Zill, Nicholas. (1988). Behavior, achievement, and health problems among children in stepfamilies: Findings from a national survey of child health. In E. Mavis Hetherington & Josephine D. Aresteh (Eds.), *Impact of divorce, single parenting, and stepparenting on children.* Hillsdale, NJ: Erlbaum.

Zinsmeister, K. (1990, June). Growing up scared. *Atlantic Monthly,* pp. 49–66.

Illustration Credits

PHOTOGRAPHS

Author Photo

p. vi Siva Bonatti.

Part Openers

(left to right) **Part I p. 1** Lennart Nilsson/Bonnier Fakta/Stockholm; Ulli Seer/Image Bank; Lisl Dennis/Image Bank; Frank Siteman/Tony Stone Images; Lori Adamski Peek/Tony Stone Images; Bob Daemmrich/Stock, Boston. **Part II pp. 128-129** Petit Format/Photo Researchers; Michael Newman/Photo Edit; Design Conceptions; Laura Dwight; Steve Starr/Stock, Boston; Hazel Hankin/Stock, Boston. **Part II summary p. 221** Elizabeth Crews; Joseph McNally/Sygma; Laura Dwight. **Part III pp. 222-223** Laura Dwight; Laura Dwight; Michael Nichols/Magnum; John Eastcott/Image Works; Bob Daemmrich/Stock, Boston; Erika Stone. **Part III summary p. 299** Bob Daemmrich/Stock, Boston; Dave Bartruff/Stock, Boston; Lawrence Migdale. **Part IV pp. 300-301** Paul Barton/The Stock Market; Paul Conklin/Monkmeyer Press; Ken Lax/Photo Researchers; Russel D. Curtis/Photo Researchers; Joel Gordon; David Young-Wolff/Photo Edit. **Part IV summary p. 381** Kaz Mori/Image Bank; Tony Freeman/PhotoEdit; James Wilson/Woodfin Camp. **Part V pp. 382-383** Sybil Shackman/Monkmeyer Press; Will McIntyre/Photo Researchers; H. Marais-Barrit/Gamma-Liaison; Butch Martin/Image Bank; Bob Martin/Allsport; James Schnepf/Gamma-Liaison. **Part V summary p. 463** Stephen Wilkes/Image Bank; L. Pickerell/Image Works; Richard Hutchings/Photo Researchers. **Part VI pp. 464-465** *(top)* Tom Carroll/International Stock; *(bottom)* Tom McCarthy/Folio, Inc.; Gary Conner/Photo Edit; Dave Bartruff/Stock, Boston; Paul Conklin/Monkmeyer Press; Hank Morgan/Photo Researchers. **Part VI summary p. 539** Carl Schneider/Gamma-Liaison; Paul Conklin/Monkmeyer Press; Andy Levin/Photo Researchers. **Part VII pp. 540-541** *(top)* Tom McCarthy/Picture Cube; *(bottom)* Tony Freeman/Photo Edit; Paul Conklin/Photo Edit; Myrleen Ferguson/Photo Edit; Mike Yamashita/Woodfin Camp; Paula Lerner/Picture Cube. **Part VII summary p. 599** Steve Starr/Stock, Boston; Cindy Charles/Photo Edit; Charles Gupton/Stock, Boston. **Part VIII pp. 600-601** *(top)* Grantpix/Monkmeyer Press; *(bottom)* Paul Conklin/Monkmeyer Press; Jim Harrison/Courtesty of Elderhostel; Jim Cartier/Photo Researchers; Michael Newman/Photo Edit; Susan Greenwood/Gamma-Liaison. **Part VIII summary p. 697** Grantpix/Photo Researchers; Stirling Dickenson/Woodfin Camp; Kindra Clineff/Picture Cube.

Chapter 1

Chapter Opener p. 2 Bob Daemmrich/Stock, Boston; **p. 5** Zigy Kaluzny/Tony Stone Images; **p. 8** Ida Wyman/Monkmeyer Press; **p. 10** New York Public Library Picture Collection; **p. 11** Porterfield/Chickering/Photo Researchers; **p. 14** *(left)* Gianni Tortoli/Photo Researchers; *(right)* Cindy Charles/Photo Edit; **p. 17** Michael McGovern/Picture Cube; **p. 18** John Sohm/Chromosohm/Stock Market; **p. 21** Jack Vartoogian; **p. 25** Bob Daemmrich/Stock, Boston.

Chapter 2

Chapter Opener p. 28 Richard T. Nowitz/Photo Researchers; **p. 30** *(left)* Erika Stone; *(right)* Camilla Smith/Rainbow; **p. 31** *(top)* Archiv/Photo Researchers; *(bottom)* Fredrik D. Bodin/Stock, Boston; **p. 32** *(top)* UPI/Corbis-Bettmann; *(bottom)* Evan Johnson/Jeroboam; *(bottom)* Richard Hutchings/Photo Edit; **p. 35** Sovfoto; **p. 36** Sam Falk/Monkmeyer Press; **p. 37** *(left)* Frank Siteman/Tony Stone Images; *(right)* James D. Wilson/Gamma-Liaison; **p. 38** *(left)* Elizabeth Crews; *(right)* Richard Hutchings/Photo Edit; **p. 40** Yves Debraine/Black Star; **p. 41** *(top)* Sally Cassidy/Picture Cube; *(top)* David Austen/Stock, Boston; *(bottom)* Lew Merrim/Monkmeyer Press; *(bottom)* Tony Freeman/Photo Edit; **p. 45** Anthony Jaladoni/Monkmeyer Press; **p. 46** James Wilson/Woodfin Camp; **p. 47** *(top)* Courtesy of Dr. Michael Cole, Laboratory of Comparative Human Cognition; *(bottom)* Owen Franken/Stock, Boston; **p. 51** Jeff Greenberg/Picture Cube; **p. 55** Renato Rotolo/Gamma-Liaison; **p. 56** Will and Deni McIntyre/Photo Researchers; **p. 60** *(left)* Bob Daemmrich/Image Works; *(right)* Ulrich Tutsch; **p. 61** *(all)* Terry Wild Studio.

Chapter 3

Chapter Opener p. 64 and p. 66 Lennart Nilsson/Bonnier Fakta/Stockholm; **p. 67** Photo Researchers; **p. 70** *(left)* Laura Dwight; *(right)* Sheila Brady/Monkmeyer Press; **p. 71** Lori Adamski Peek/Tony Stone Images; **p. 76** Robert Burroughs; **p. 83** Joel Gordon; **p. 85** Laura Dwight; **p. 88** Will and Deni McIntyre/Photo Researchers; **p. 89** DPA/Image Works; **p. 92** Philippe Plailly/Eurelios/Science Photo Library/Photo Researchers.

Chapter 4

Chapter Opener p. 96 From Lennart Nilsson, *A Child is Born*, 2nd ed., Delacorte Press, © 1990. Photograph coutesy of Lennart Nilsson/Bonnier Fakta/Stockholm; **p. 98** *(all)* Petit Format/Science Source/Photo Researchers; **p. 99** *(left, center*

left, and center right) Petit Format/Science Source/Photo Researchers; (*right*) National Medical Slide/Custom Medical Stock Photo; **p. 101** S.J. Allen/International Stock; **p. 103** RIA/Gamma-Liaison; **p. 105** (*all*) Carnegie Institute of Washington; **p. 106** Andrew Brilliant/Picture Cube; **p. 107** William Hubbell/Woodfin Camp; **p. 110** (*both*) George Steinmetz; **p. 111** Amy C. Etra/Photo Edit; **p. 113** Dan McCoy/Rainbow; **p. 116** John Ficara/Woodfin Camp; **p. 119** (*left*) Henry Schleichkorn/Custom Medical Stock Photo; (*right*) Laura Dwight; **p. 121** (*left*) Doug Goodman/Monkmeyer Press; (*right*) Viviane Moos.

Chapter 5

Chapter Opener p. 130 Lisl Dennis/Image Bank; **p. 133** Simon Fraser/Science Photo Library/Photo Researchers; **p. 137** L.K. Tai/Woodfin Camp; **p. 144** (*left and center*) Elizabeth Crews; (*right*) Petit Format/Photo Researchers; **p. 145** (*all*) Laura Dwight; **p. 146** Sybil Shackman/Monkmeyer Press; **p. 149** Michael Greenlar/Image Works; **p. 151** (*top*) Courtesy of A. Slater; (*bottom*) Peter Menzell; **p. 152** Laura Dwight; **p. 153** (*both*) Peter Mcleod; **p. 155** Laura Dwight; **p. 157** Michael Tchereukoff/Image Bank.

Chapter 6

Chapter Opener p. 160 Superstock; **p. 164** (*all*) Courtesy of Karen Adolph; **p. 165** (*top*) Ulli Seer/Image Bank; (*bottom*) Courtesy of J. Campos, B. Bertenthal & R. Kermoinanan; **p. 166** Laura Dwight; **p. 167** Elizabeth Crews; **p. 168** Joseph McNally/Sygma; **p. 169** (*both*) Elizabeth Crews/Image Works; **p. 171** Michael Newman/Photo Edit; **p. 176** Design Conceptions; **p. 177** Laura Dwight; **p. 178** (*both*) Laura Dwight; **p. 182** Laura Dwight; **p. 184** Elliott Varner Smith; **p. 185** Hazel Hankin/Stock, Boston; **p. 187** Betty Press/Woodfin Camp.

Chapter 7

Chapter Opener p. 190 Kindra Clineff/Picture Cube; **p. 192** Laura Dwight; **p. 193** Susan Lapides/Design Conceptions; **p. 195** Courtesy of Bill Davis; **p. 196** Sybil Shackman/ Monkmeyer Press; **p. 198** Tom McCarthy/Picture Cube; **p. 199** Nancy Sheehan/Picture Cube; **p. 200** Kory Addis/Picture Cube; **p. 203** Leong Ka Tai/Material World; **p. 207** Betts Anderson/Unicorn Stock Photos; **p. 209** Laura Dwight; **p. 210** (*left*) Elizabeth Crews; (*right*) Bruce Plotkin/Image Works; **p. 212** Courtesy of Mary Ainsworth; **p. 214** Lawrence Migdale; **p. 216** Hazel Hankin; **p. 218** Steve Starr/Stock, Boston.

Chapter 8

Chapter Opener p. 224 Lawrence Migdale; **p. 227** Laura Dwight; **p. 229** Bob Daemmrich/Stock, Boston; **p. 231** (*top*) Renate Hiller/Monkmeyer Press; (*bottom*) Carol Palmer/ Picture Cube; **p. 232** Robert L. W. Ginn/Unicorn Stock Photos; **p. 233** Adam Nadel/AP/Wide World Photos; **p. 237** Wolfgang Kaehler; **p. 238** Porter Gifford/Gamma-Liaison; **p. 244** José R. Lopez/New York Times.

Chapter 9

Chapter Opener p. 248 Barbara Campbell/Gamma-Liaison; **p. 252** Michael Nichols/Magnum; **p. 254** Elizabeth Crews; **p. 257** James Kamp/Life Magazine © Time Inc.; **p. 259** Laura Dwight; **p. 260** Photo Works/Monkmeyer Press; **p. 264** Dave Bartruff/Stock, Boston; **p. 266** Laura Dwight; **p. 267, p. 269** Laura Dwight; **p. 270** Elizabeth Crews; **p. 273** (*left*) Tom Prettyman/Photo Edit; (*right*) Fujifotos/Image Works.

Chapter 10

Chapter Opener p. 276 Bob Daemmrich/Stock, Boston; **p. 278** Lawrence Migdale; **p. 279** Mel Digiacomo/Image Bank; **p. 280** (*left*) Erika Stone; (*right*) John Coletti/Stock, Boston; **p. 284** (*left*) Joel Gordon; (*right*) Spencer Grant/Photo Researchers, Inc.; **p. 286** Joel Gordon; **p. 287** Robert Brenner/Photo Edit; **p. 292** John Eastcott/Image Works; **p. 295** David M. Grossman; **p. 296** Myrleen Ferguson/Photo Edit; **p. 297** Laura Dwight.

Chapter 11

Chapter Opener p. 302 Elaine Rebman/Photo Researchers; **p. 305** (*top*) Jeff Greenberg/Photo Researchers; (*bottom*) Richard Orton/Picture Cube; **p. 308** (*both*) Bob Daemmrich/ Stock, Boston; **p. 309** Bob Daemmrich/Stock, Boston; **p. 310** Nancy Acevedo/Monkmeyer Press; **p. 314** Alan Carey/Image Works; **p. 316** Elizabeth Crews; **p. 320** Hazel Hankin/Stock, Boston; **p. 321** José Azel/Aurora.

Chapter 12

Chapter Opener p. 324 Tony Freeman/Photo Edit; **p. 326** HMS Images/Image Bank; **p. 327** Elizabeth Crews; **p. 329** Kaz Mori/Image Bank; **p. 332** Leif Skoogfors/Woodfin Camp; **p. 336** Joel Gordon; **p. 338** Paul Conklin/Monkmeyer Press; **p. 341** (*left*) John Elk/Stock, Boston; (*right*) Mike Yamashita/Woodfin Camp; **p. 344** Elizabeth Crews; **p. 345** Russel D. Curtis/Photo Researchers; **p. 347** Bob Daemmrich/Stock, Boston; **p. 350** George Ancona/International Stock.

Chapter 13

Chapter Opener p. 352 David Young-Wolff/Photo Edit; **p. 354** (*left*) Arnold John Kaplan/Picture Cube; (*right*) Victor Englebert/Photo Researchers; **p. 356** Ellis Herwig/Stock, Boston; **p. 359** Joel Gordon; **p. 361** Bob Daemmrich/Stock, Boston; **p. 362** George White Location Photography; **p. 365** David Young-Wolff/Photo Edit; **p. 367** Bob Daemmrich/Stock, Boston; **p. 373** James Wilson/Woodfin Camp; **p. 374** Michael Newman/Photo Edit; **p. 377** Katherine McGlynn/Image Works; **p. 379** Peter Chapman/Stock, Boston.

Chapter 14

Chapter Opener p. 384 Leif Skoogfors/Woodfin Camp; **p. 386** David Young-Wolff/Photo Edit; **p. 387** Bob Daemmrich/Stock, Boston; **p. 389** Stephen Wilkes/Image Bank; **p. 391** Bob Martin/Allsport; **p. 392** Bill Gillette/Stock, Boston; **p. 393** Janeart/Image Bank; **p. 395** (*top*) Fran Heyl & Associates; (*bottom*) Michael Newman/Photo Edit; **p. 396** Joel Gordon; **p. 397** Sybil Shackman/Monkmeyer Press; **p. 401** Billy E. Barnes/ Photo Edit; **p. 406** Mark M. Walker/Picture Cube.

Chapter 15

Chapter Opener p. 410 Scott Thode/Material World; **p. 412** L. Pickerell/Image Works; **p. 413** Will McIntyre/Photo Researchers; **p. 414** Bill Bachmann/Image Works; **p. 417** (*top*) Joel Gordon; (*bottom*) Sybil Shackman/Monkmeyer Press; **p. 419** H. Marais-Barrit/Gamma-Liaison; **p. 420** Elizabeth Crews; **p. 422** (*left*) Charles Harbutt/Actuality; (*right*) Bob Daemmrich/Image Works; **p. 427** Courtesy of the Children's Defense Fund; **p. 429** Miguel Fairbanks; **p. 431** Richard Hutchings/Photo Researchers; **p. 432** James Schnepf/Gamma-Liaison.

Chapter 16

Chapter Opener p. 436 Lawrence Migdale; *(right)* Catherine Kornow/Woodfin Camp; *(left)* Robert Brenner/Photo Edit; **p. 440** *(left)* Renato Rotolo/Gamma-Liaison; *(right)* Elizabeth Crews; **p. 441** Daniel Laine/Actuel; **p. 444** Rhoda Sidney/Monkmeyer Press; **p. 447** Tony Freeman/Photo Edit; **p. 448** *(left)* Joel Gordon; *(right)* Dan Walsh/Picture Cube; **p. 450** Butch Martin/Image Bank; **p. 453** Alon Reininger/Woodfin Camp; **p. 456** R. J. Mathews/Unicorn Stock Photos.

Chapter 17

Chapter Opener p. 466 Carl Schneider/Gamma-Liaison; **p. 470** Roger Tully/Tony Stone Images; **p. 471** Tom Carroll/International Stock; **p. 472** Jonathan Daniel/Allsport; **p. 473** Paul Conklin/Monkmeyer Press; **p. 477** Hank Morgan/Photo Researchers; **p. 481** Paula Scully/Gamma-Liaison; **p. 482** Rick Kopstein/Monkmeyer Press; **p. 488** Greg Riffi/Gamma-Liaison.

Chapter 18

Chapter Opener p. 490 Terry Wild Studio; **p. 495** Steve Berman/Gamma-Liaison; **p. 497** Paul Conklin/Monkmeyer Press; **p. 499** *(top)* Bikas Das/AP/Wide World Photos; *(bottom)* Margot Granitsas/Photo Researchers; **p. 501** Howard Dratch/Image Works; **p. 504** Gary Conner/Photo Edit; **p. 505** *(left)* Dave Bartruff/Stock, Boston; *(right)* Pat Carter/AP/Wide World Photos.

Chapter 19

Chapter Opener p. 508 Scott Barrow/International Stock; **p. 513** *(left)* Bob Daemmrich/Stock, Boston; *(right)* Tom McCarthy/Folio, Inc.; **p. 517** Jonathan Elderfield/Gamma-Liaison; **p. 521** Blair Seltz/Photo Researchers; **p. 529** *(left)* Terry Wild Studio; *(right)* Cliff Hollenbeck/International Stock; **p. 530** Andy Levin/Photo Researchers; **p. 531** Corbis-Bettmann; **p. 533** T. Michaels/Image Works; **p. 534** Spencer Grant/Photo Researchers.

Chapter 20

Chapter Opener p. 542 George Goodwin/Monkmeyer Press; **p. 544** Robert Ullmann/Monkmeyer Press; **p. 545** Steve Granitz/Retna; **p. 546** Ann Heisenfelt/AP/Wide World Photos; **p. 549** *(left)* Springer/Corbis-Bettmann; *(right)* Francene Keery/Stock, Boston; **p. 550** Nathan Benn/Stock, Boston; **p. 552** Tom McCarthy/Picture Cube; **p. 555** Mike Yamashita/Woodfin Camp; **p. 556** Cynthia Johnson/Gamma-Liaison; **p. 557** Will McIntyre/Photo Researchers; **p. 558** K. Nomachi/Photo Researchers; **p. 562** Steve Starr/Stock, Boston.

Chapter 21

Chapter Opener p. 564 Michael Newman/Photo Edit; **p. 566** Laura Dwight/Photo Edit; **p. 567** Paul Conklin/Photo Edit; **p. 572** *(left)* Cindy Charles/Photo Edit; *(right)* Nicholas Devore III/Photographers/Aspen, Inc.; **p. 573** Peter Byron/Monkmeyer Press; **p. 574** A. Ramey/Photo Edit; **p. 577** Jeff Greenberg/Picture Cube; **p. 578** Steven Rubin/Image Works; **p. 579** Paula Lerner/Picture Cube.

Chapter 22

Chapter Opener p. 582 Jim Pickerell/Folio, Inc.; **p. 584** Robert Ullmann/Monkmeyer Press; **p. 585** *(top)* Jim Daniels/Picture Cube; **p. 585** *(bottom)* Ulrike Welsch/Photo Edit; **p. 587** Myrleen Ferguson/Photo Edit; **p. 588** Mary Kate Denny/Photo Edit; **p. 589** Charles Gupton/Stock, Boston; **p. 591** *(top)* Mark Anderman/Terry Wild Studio; *(bottom)* David Hiser/Photographers/Aspen, Inc.; **p. 593** Randee St. Nicholas/Archive Photos; **p. 595** Tony Freeman/Photo Edit; **p. 598** Michael Weisbrot.

Chapter 23

Chapter Opener p. 604 Frederic Reglain/Gamma-Liaison; **p. 607** Paul Gish/Monkmeyer Press; **p. 610** Grantpix/Monkmeyer Press; **p. 612** Cary Wolinsky/Stock, Boston; **p. 618** Lawrence Migdale/Photo Researchers; **p. 620** Catherine Karnow/Woodfin Camp; **p. 621** Meckes/Ottawa/Photo Researchers; **p. 622** Robert Ricci/Gamma-Liaison; **p. 623** David Barrit/Gamma-Liaison; **p. 625** *(all)* John Launois/Black Star; **p. 627** Jerry Wachter/Photo Researchers.

Chapter 24

Chapter Opener p. 630 Xinhua-Chine Nouvelle/Gamma-Liaison; **p. 632** Paul Howell/Gamma-Liaison; **p. 634** Kermani/Gamma-Liaison; **p. 635** Michael Newman/Photo Edit; **p. 636** *(left)* Corbis-Bettmann; **p. 636** *(center)* David Redfern/Retna; **p. 636** *(right)* Charles Peterson/Retna; **p. 640** Grantpix/Photo Researchers; **p. 644** Bernard Wolf/Monkmeyer Press; **p. 647** Alfred Pasieka/Science Photo Library/Photo Researchers; **p. 648** Mike Guastella/Star File; **p. 651** Barbara Alper/Stock, Boston; **p. 652** Richard Sobol/Stock, Boston; **p. 654** Stirling Dickenson/Woodfin Camp; **p. 656** Jim Cartier/Photo Researchers.

Chapter 25

Chapter Opener p. 660 D. Young-Wolff/Photo Edit; **p. 662** Cont/Reninger/Woodfin Camp; **p. 663** Christian Him/Retna; **p. 664** Giboux/Gamma-Liaison; **p. 666** Sonda Dawes/Image Works; **p. 669** Tim Pott Photography; **p. 670** Jim Harrison/Courtesty of Elderhostel; **p. 671** Susan Greenwood/Gamma-Liaison; **p. 675** Cynthia Johnson/Gamma-Liaison; **p. 676** Bill Weems/Woodfin Camp; **p. 679** Paul Conklin/Monkmeyer Press; **p. 680** Kindra Clineff/Picture Cube; **p. 684** Bob Daemmrich/Image Works; **p. 685** James Schnepf/Gamma-Liaison; **p. 690** Eastcott/Momatiuk/Woodfin Camp; **p. 692** Oreganian/Gamma-Liaison; **p. 693** Charles Gupton/Stock, Boston; **p. 693** Lester Sloan/Woodfin Camp; **p. 698** Brenda Tharp/Photo Researchers.

Epilogue

Chapter Opener p. E-2 Steven M. Stone/Picture Cube; **p. E-6** C. Ampanie/Image Works; **p. E-9** Phyllis Picardi/International Stock; **p. E-10** Vanessa Vick/Photo Researchers; **p. E-11** A. Ramey/Unicorn Stock Photos.

FIGURES AND TABLES

Chapter 2

Fig. 2.1 Shiffrin, R.M., & Atkinson, R.C. (1969). Storage and retrieval processes in long-term memory. *Psychological Review, 76,* 179-193. Copyright © 1969 American Psychological Association. Reprinted by permission. **Fig. 2.2** Richman, A.L., Miller, P.M., & Le Vine, R.A. (1992). Cultural and educational variations in maternal responsiveness. *Developmental Psychology, 28,* 614-621. Copyright © 1992 American Psychological Association. Reprinted by permission. **Table 2.4** Riedel,

B.W., et al., (1995). Sleep compression and sleep education for older insomniacs: Self-help versus therapist guidance. *Psychology and Aging, 10,* 54-63. Copyright © 1995 American Psychological Association. Reprinted with permission.

Chapter 3

Fig. 3.4 Data from *Schizophrenia genesis: The origins of madness* by I.I. Gottesman. Copyright © 1991 by W.H. Freeman and Company. Used with permission.

Chapter 4

Fig. 4.1 From Moore, K.L., & Persand, T.V.N.: *The Developing Human: Clinically Oriented Embryology,* 1993. © W.B. Saunders Co., Philadelphia. Reprinted with permission.

Chapter 5

p. 141 Super, C.M., & Harkness, S. (1982). The infant's niche in rural Kenya and metropolitan America. In L.L. Adler (Ed.), *Cross-cultural research at issue.* New York: Academic Press. **Table 5.4** Frankenburg, W.K., et al. (1981). The newly abbreviated and revised Denver Developmental Screening Test. *Journal of Pediatrics, 99,* 995-999.

Chapter 7

p. 204 Excerpt from Chess, S., & Thomas, A. (1990). Continuities and discontinuities in development. In Lee N. Robbins & Michael Rutter (Eds.), *Straight and devious pathways from childhood to adulthood.* New York: Cambridge University Press. Reprinted with permission of Cambridge University Press. **Fig. 7.1** Benoit, D., & Parker, K.C. (1994). Stability and transmission of attachment across three generations. *Child Development, 65,* 1444-1456. Copyright © 1994 Society for Research in Child Development.

Chapter 9

Fig. 9.1 Saxe, G., Guberman, S.R., & Gearhart, M. (1987). Social processes in early number development. *Monographs of the Society for Research in Child Development,* 52(Serial No. 216). Copyright © 1987 Society for Research in Child Development. **Fig. 9.2** Hamond, N.R., & Fivush, R. (1991). Memories of Mickey Mouse: Young children recount their trip to Disneyworld. *Cognitive Development, 6,* 433-448.

Chapter 12

Fig. 12.1 Chi, M.T.H. (1978). Knowledge structures and memory development. In R.S. Siegler (Ed.), *Children's thinking: What develops?* Hillsdale, NJ: Erlbaum. **Fig. 12.2** LaPointe et al., (1992). *Learning mathematics.* Princeton, NJ: Educational Testing Service.

Chapter 15

Fig. 15.2 Johnson, D.W., & Johnson, R.T. (1994). *Learning together and alone: Cooperative, competitive, and individualistic learning (4th ed.).* Boston: Allyn & Bacon. Copyright © 1994 by Allyn & Bacon. Reprinted with permission.

Chapter 16

Fig. 16.1 Adapted from Lyle W. Shannon (1988). *Criminal career continuity: Its social context.* New York: Human Sciences Press. Reprinted with permission of Plenum Publishing Corp.

Chapter 17

Fig. 17.1 Reprinted with permission from Meisami, Esmail. (1994). Aging of the sensory system. In Paola S. Timiras (Ed.), *Physiological basis of aging and geriatrics (2nd ed.).* Copyright CRC Press, Boca Raton, Florida, © 1994.

Chapter 18

Fig. 18.1 Blanchard-Fields, F. (1986). Reasoning on social dilemmas varying in emotional saliency: An adult developmental perspective. *Psychology and Aging, 1,* 325-333. Copyright © 1986 American Psychological Association. Reprinted with permission. **Table 18.1** Perry, W.G., Jr. (1981). Cognitive and ethical growth: The making of meaning. In Arthur W. Chickering (Ed.), *The modern American college: Responding to the new realities of diverse students and a changing society.* Copyright 1981 Jossey-Bass Inc., Publishers.

Chapter 19

Fig. 19.2 Adapted from Seltzer, Judith A. (1991). Legal custody arrangements and children's economic welfare. *American Journal of Sociology, 96,* 895-929.

Chapter 21

Figs. 21.1 and 21.3 Schaie, K.W. (1996). *Intellectual development in adulthood: The Seattle longitudinal study.* Cambridge, England: Cambridge University Press. Copyright © 1996. Reprinted with permission of Cambridge University Press. **p. 576** Schaie, K.W. (1989). Perceptual speed in adulthood: Cross-sectional and longitudinal studies. *Psychology and Aging, 4,* 443-453. Copyright © 1989 American Psychological Association. Reprinted with permission.

Chapter 23

Fig. 23.2 Fries, James F. (1994). *Living well: Taking care of your health in the middle and later years.* Reading, MA: Addison-Wesley Longman.

Chapter 24

Fig. 24.1 Smith et al. (1996). Subjective memory complaints, psychological distress, and longitudinal change in objective memory performance. *Psychology and Aging, 11,* 272-279. Copyright © 1996 American Psychological Association. Adapted with permission. **Fig. 24.2** Adapted with permission of The Free Press, a division of Simon & Schuster from *Brain Failure* by Barry Reisberg. Coyright © 1981 Barry Reisberg.

Chapter 25

Fig. 25.3 Glenn, N.D. (1991) The recent trend in marital success in the United States. *Journal of Marriage and the Family, 53,* 261-270. **Table 25.1** Essex, M.J. & Nam, S. (1987). Marital status and loneliness among older women. *Journal of Marriage and the Family, 49,* 93-106. **Fig. 25.4** Lentzner et al. (1992). The quality of life in the year before death. *American Journal of Public Health, 82,* 1093-1098.

Name Index

Subject Index

Periodic Table of the Elements

	IA	IIA	IIIB	IVB	VB	VIB	VIIB	VIIIB	VIIIB	VIIIB	IB	IIB	IIIA	IVA	VA	VIA	VIIA	Noble gases
1								Hydrogen 1 H 1.0079										Helium 2 He 4.0026
2	Lithium 3 Li 6.941	Beryllium 4 Be 9.0122											Boron 5 B 10.811	Carbon 6 C 12.011	Nitrogen 7 N 14.0067	Oxygen 8 O 15.9994	Fluorine 9 F 18.9984	Neon 10 Ne 20.1797
3	Sodium 11 Na 22.9898	Magnesium 12 Mg 24.3050											Aluminum 13 Al 26.9815	Silicon 14 Si 28.0855	Phosphorus 15 P 30.9738	Sulfur 16 S 32.066	Chlorine 17 Cl 35.4527	Argon 18 Ar 39.948
4	Potassium 19 K 39.0983	Calcium 20 Ca 40.078	Scandium 21 Sc 44.9559	Titanium 22 Ti 47.867	Vanadium 23 V 50.9415	Chromium 24 Cr 51.9961	Manganese 25 Mn 54.9380	Iron 26 Fe 55.845	Cobalt 27 Co 58.9332	Nickel 28 Ni 58.6934	Gopper 29 Cu 63.546	Zinc 30 Zn 65.39	Gallium 31 Ga 69.723	Germanium 32 Ge 72.61	Arsenic 33 As 74.9216	Selenium 34 Se 78.96	Bromine 35 Br 79.904	Krypton 36 Kr 83.80
5	Rubidium 37 Rb 85.4678	Strontium 38 Sr 87.62	Yttrium 39 Y 88.9059	Zirconium 40 Zr 91.224	Niobium 41 Nb 92.9064	Molybdenum 42 Mo 95.94	Technetium 43 Tc (97.907)	Ruthenium 44 Ru 101.07	Rhodium 45 Rh 102.9055	Palladium 46 Pd 106.42	Silver 47 Ag 107.8682	Cadmium 48 Cd 112.411	Indium 49 In 114.818	Tin 50 Sn 118.710	Antimony 51 Sb 121.760	Tellurium 52 Te 127.60	Iodine 53 I 126.9045	Xenon 54 Xe 131.29
6	Cesium 55 Cs 132.9054	Barium 56 Ba 137.327	Lanthanum 57 La 138.9055	Hafnium 72 Hf 178.49	Tantalum 73 Ta 180.9479	Tungsten 74 W 183.84	Rhenium 75 Re 186.207	Osmium 76 Os 190.2	Iridium 77 Ir 192.22	Platinum 78 Pt 195.08	Gold 79 Au 196.9665	Mercury 80 Hg 200.59	Thallium 81 Tl 204.3833	Lead 82 Pb 207.2	Bismuth 83 Bi 208.9804	Polonium 84 Po (208.98)	Astatine 85 At (209.99)	Radon 86 Rn (222.02)
7	Francium 87 Fr (223.02)	Radium 88 Ra (226.0254)	Actinium 89 Ac (227.0278)	Rutherfordium 104 Rf (261.11)	Dubnium 105 Db (262.11)	Seaborgium 106 Sg (263.12)	Bohrium 107 Bh (262.12)	Hassium 108 Hs (265)	Meitnerium 109 Mt (266)	Ununnilium 110 Uun (269)	Unununium 111 Uuu (272)	Ununbium 112 Uub (277)						

Metals

Metalloids

Nonmetals

Cerium 58 Ce 140.115	Praseodymium 59 Pr 140.9076	Neodymium 60 Nd 144.24	Promethium 61 Pm (144.91)	Samarium 62 Sm 150.36	Europium 63 Eu 151.965	Gadolinium 64 Gd 157.25	Terbium 65 Tb 158.9253	Dysprosium 66 Dy 162.50	Holmium 67 Ho 164.9303	Erbium 68 Er 167.26	Thulium 69 Tm 168.9342	Ytterbium 70 Yb 173.04	Lutetium 71 Lu 174.967
Thorium 90 Th 232.0381	Protactinium 91 Pa 231.0388	Uranium 92 U 238.0289	Neptunium 93 Np (237.0482)	Plutonium 94 Pu (244.664)	Americium 95 Am (243.061)	Curium 96 Cm (247.07)	Berkelium 97 Bk (247.07)	Californium 98 Cf (251.08)	Einsteinium 99 Es (252.08)	Fermium 100 Fm (257.10)	Mendelevium 101 Md (258.10)	Noblelium 102 No (259.10)	Lawrencium 103 Lr (262.11)

Note: Atomic masses are 1993 IUPAC values (up to four decimal places).